Marine Technology Reference Book

Marine Technology Reference Book

Edited by
Nina Morgan, BSc, PhD

Butterworths
London · Boston · Singapore · Sydney
Toronto · Wellington

PART OF REED INTERNATIONAL P.L.C.

All rights reserved. No part of this publication may be reproduced in any material form (including photocopying or storing it in any medium by electronic means and whether or not transiently or incidentally to some other use of this publication) without the written permission of the copyright owner except in accordance with the provisions of the Copyright, Designs and Patents Act 1988 or under the terms of a licence issued by the Copyright Licensing Agency Ltd, 33–34 Alfred Place, London, England WC1E 7DP. Applications for the copyright owner's written permission to reproduce any part of this publication should be addressed to the Publishers.

Warning: The doing of an unauthorised act in relation to a copyright work may result in both a civil claim for damages and criminal prosecution.

This book is sold subject to the Standard Conditions of Sale of Net Books and may not be re-sold in the UK below the net price given by the Publishers in their current price list.

First published 1990

© Butterworth & Co. (Publishers) Ltd. 1990

The cover illustration is reproduced by courtesy of The Motor Ship

British Library Cataloguing in Publication Data
Marine technology reference book.
  1. Marine engineering
  I. Morgan, Nina
  623.87

ISBN 0–408–02784–3

Library of Congress Cataloging-in-Publication Data
Marine technology reference book / edited by Nina Morgan.
    p.   cm.
  Includes bibliographical references.
  ISBN 0–408–02784–3:
  1. Ocean engineering.   I. Morgan, Nina.
  TC1645.M27   1990
  620.4′162– –dc20

Photoset by TecSet Ltd, Wallington, Surrey
Printed and bound in Great Britain by Courier International Ltd., Tiptree, Essex

Preface

Marine Technology covers such a broad range of scientific, engineering, technical and legal topics, that even the experts can become confused! Marine technologists working in one area often find it difficult to understand the work of colleagues working elsewhere and non-scientists have been known to recoil in horror at the complexity of it all. I hope the Marine Technology Reference Book will change all that.

The Marine Technology Reference Book is designed to serve as a first point of reference for those wishing to know the basics of this very diverse topic, and to provide accessible references for those who wish to read about particular subjects in more detail. The areas covered have been chosen to provide a comprehensive overview of marine technology and the chapters have been written by specialists with the aim of providing a learned, but lucid, introduction to their own areas of interest.

The book is aimed at the marine technology community in its widest sense, especially people expert in one aspect of marine technology who would like to know what goes on in other branches of the subject. However, I hope it will also serve as a beacon to illuminate the complexities of the topic for advanced science and engineering students who have an interest in marine matters and anyone involved with marine financial and legal work, including bankers, marine insurers and those concerned with marine policy.

Many people have worked together to make this book possible. The specialist authors gave generously of their time and expertise and worked to tight deadlines without protest (Never? Well, hardly ever!). Discussions with many marine technologists helped me to formulate the initial outline for this book. Among them Gareth Owen of Heriot-Watt University in Edinburgh, Scotland, James Paffit in Chichester, UK, Ian Cullen of South Shields College, UK, Andrew Sinclair of Vine and Able in London, UK and Joseph Morgan of the East–West Center at the University of Hawaii in the United States, deserve special mention. And finally, a book such as this could not be produced without the help of the in-house editors and designers in the book production department at Butterworths and the assistance of Dr Sheila Shepherd who worked long and hard in the final stages of production.

The book has been written by specialists for non-specialists. We hope it will provide an accessible introduction and a useful reference to this wide-ranging and fascinating discipline.

Nina Morgan
Shalford near Guildford,
Surrey, UK
February, 1990

Contents

Preface

List of Contributors

1 Ocean Environments
Introduction. Oceanic provinces. Sea water. Oceanic circulation. The atmosphere. Biological oceanography. Waves. Regional oceanography. Conclusions. References

2 Offshore Structures
Introduction. Fixed offshore structures. Environmental loading. Structural response of fixed structures. Floating offshore structures. Analysis with multiple degrees of freedom. References and further reading

3 Ships and Advanced Marine Vehicles
Basic naval architecture and ship design. Introduction. Buoyancy and stability. Resistance and propulsion. Manoeuvring. Ship motions and seakeeping. Ship structures. The ship design process. References. **High speed and multihull marine vehicles.** Introduction. Planing craft. Hydrofoil craft. Hovercraft. Multihull vessels. Propulsion. Comparative performances. References

4 Diving and Underwater Vehicles
Introduction. Diving. Atmospheric diving suits and systems. Underwater vehicles – introduction. Selection of underwater intervention methods for offshore oil and gas operations. Further reading

5 Marine Risers and Pipelines
Introduction. Marine risers. Pipelines. References

6 Marine Engines and Auxiliary Machinery
Machinery arrangements. Propulsion diesel engines. Steam plant. Gas turbines. Nuclear power. Propeller shaft drives. Electrical power generation. Steam generation. Ancilliary machinery. Propulsion. Further reading

7 Marine Control Systems
The requirement for control. Control methodology. The ship as a dynamic system. Ship controllability. The role of the classification societies. Heading control – autopilots. Track-keeping and position control. Roll control – ship stabilizers. Machinery control. Integrated ship control. Assessing controllability – ship trials. References

8 Mooring and Dynamic Positioning
Introduction. Mooring. Dynamic positioning. References

9 Marine Salvage
Introduction. Contractual salvage. Salvage related organizations. Types of salvage. Salvage operations. Salvage methods and risks. Towing. Pollution control. References

10 Corrosion and Defect Evaluation
Introduction. The chemistry of corrosion. The mechanisms of corrosion. Corrosion resistance of marine alloys. Corrosion protection. Defects and non-destructive testing. Codes, test methods and standards. References. Codes and standards

11 Marine Safety
Introduction. Onboard safety information. Operational deck safety. Safety in cargo operations. Tanker work safety. Safety in offshore working practice. Marine emergencies. Marine survival. Search and rescue operations. Acknowledgements. References. Further reading. Merchant shipping notices

12 Electronic Navigation and Radar
Navigation systems. Introduction. The model of the Earth. Hydrography. Location of observer. The beginnings of electronic navigation. Navigation in the modern world. **Acoustic sensing and positioning systems.** Introduction. Acoustic theory. Acoustic system components. Sensor systems. Acoustic navigation and positioning systems. Other acoustic systems. Conclusion. **Earth-based electronic navigation systems.** Introduction. Propagation of low and very low frequency waves. Racal-Decca navigator. Loran C. Omega. Conclusions. **Earth-based electronic precise**

positioning systems. Introduction. Categorization of systems. Laser systems. Microwave systems. UHF systems. HF systems. LF systems. **Satellite navigation.** Introduction. Satellite orbits. The Navy Navigation Satellite System. Navstar Global Positioning System. Starfix. Geostar. Navsat. **Radar.** The history of radar. Parts of radar equipment. Cathode ray tube. Antenna. Time synchronized components. General aspects of radar performance. Collision avoidance interpretation of radar data. Computer aided systems. Radar in vessel traffic systems. Electronic chart display systems. References

13 Maritime Law
Introduction. Maritime zones. Offshore oil and gas exploration. Mining of manganese nodules in the international seabed area. Shipping. Fishing. Military uses of the sea. Conclusions. References and further reading

Index

List of Contributors

V J Abbott
Senior Lecturer, Institute of Marine Studies, Polytechnic of the South West, Plymouth

R L Allwood, BSc, PhD
Senior Lecturer, Cranfield Institute of Technology, Cranfield

Charles V Betts, MA, MPhil, CEng, FRINA, RCNC
Professor of Naval Architecture, University College, London

Capt. D J Bray, MNI
Senior Lecturer, Department of Maritime Studies, Lowestoft College, Lowestoft

Capt. C Brown, BSc, MSc, FNI
Senior Lecturer, Department of Marine Science and Technology, Institute of Marine Studies, Plymouth

Robin R Churchill, LLM, PhD
Lecturer in Law, University of Wales
Institute of Science and Technology, Cardiff

Brian R Clayton, BSc (Eng), PhD, CEng, FIMechE, FRINA
Reader in Mechanical Engineering, University College, London

S Grimes, MISTC
Paega Technical Writing, Paignton

David J House, Master Mariner
Marine Author and Lecturer in Nautical Studies, Whitehouse, Greenhalgh Lane, Greenhalgh, Preston

J Kearon, Master Mariner, BSc, MRINA
Nautical Surveyor, The Salvage Association, London

Geoffrey J Lyons, BSc, PhD, CEng, MRINA, AMIMechE
Lecturer, Department of Mechanical Engineering, University College, London

I McCallum, MA, PhD, CEng, MIEE, MRINA, MSNAME
Maritime Dynamics Ltd, Llantrisant

K MacCallum, Master Mariner, BSc, MSc, ARICS
Marine and Hydrographic Consultant
Formerly Principal Lecturer, Plymouth Polytechnic, Plymouth

Nina Morgan, BSc, PhD
Freelance Journalist and Editor, 88 Station Road, Shalford, Guildford

Minoo H Patel, BSc (Eng), PhD, MRINA, MIMechE
Head, Department of Mechanical Engineering, University College, London

R L Reuben, BSc, PhD, CEng, MIM
Department of Offshore Engineering, Heriot Watt University, Edinburgh

S Waddingham, BSc
Marine Consultant, Galatron (Marine Services) Ltd, Bexley

B Whiting, BSc, PhD, MRIN, FRAF
Senior Lecturer, Polytechnic of East London, Dagenham

C T Wilbur, CEng, MIMarE
Freelance Editor, 30 Ryecroft, Haywards Heath

1 Ocean Environments

Nina Morgan

Contents

1.1 Introduction
1.2 Oceanic provinces
    1.2.1 Major oceans
    1.2.2 Marginal ocean basins
    1.2.3 The depth of the oceans
    1.2.4 Topography
    1.2.5 Continental margins
    1.2.6 Marine sediments
    1.2.7 Sea bed stability
1.3 Sea water
    1.3.1 Chemistry of sea water
    1.3.2 Salinity
    1.3.3 Dissolved gases
    1.3.4 The pH of sea water
    1.3.5 Mineralization
    1.3.6 Temperature distribution
    1.3.7 Density distribution
    1.3.8 Light transmission
    1.3.9 Underwater acoustics
1.4 Oceanic circulation
    1.4.1 Surface currents
    1.4.2 Deep ocean circulation
    1.4.3 Tides and tidal currents
1.5 The atmosphere
    1.5.1 Characteristics of the atmosphere
    1.5.2 Winds
    1.5.3 Weather systems
    1.5.4 The ionosphere
1.6 Biological oceanography
    1.6.1 Characteristics of the marine lifestyle
    1.6.2 Types of marine life
    1.6.3 Distribution of organisms
    1.6.4 Effect of organisms on marine technology
1.7 Waves
    1.7.1 Characteristics of waves
    1.7.2 Types of waves
1.8 Regional oceanography
    1.8.1 Marginal ocean basins
    1.8.2 The deep ocean
    1.8.3 The coastal ocean
    1.8.4 Estuaries
    1.8.5 The surf zone
    1.8.6 Equatorial and tropical oceans
    1.8.7 Polar oceans
    1.8.8 Temperate oceans
1.9 Conclusions
References and further reading

1.1 Introduction

Seventy-one percent of the Earth's surface is covered by water, yet it is often said that we know more about the surface of the Moon than the bottom of the ocean basins. The Southern Hemisphere is dominated by the oceans, with 81% of its surface covered by water. Although there is a higher proportion of land to water in the Northern Hemisphere, roughly 60% is under water. In spite of the fact that the oceans dominate the Earth, it is important to keep them in perspective. In relation to the radius of the Earth, the heights reached on land, 8.8 km (5.50 miles) on Mt Everest and the depths reached in the ocean basins, 11 km (6.88 miles) in the Marianas Trench are very insignificant. When viewed from space, the oceans appear as a thin film of water on a nearly smooth globe, interrupted occasionally by continents.

The oceans greatly influence the weather systems and ecology of the Earth. The oceans have the potential to provide many sources of renewable energy, including that from the waves or by the exploitation of thermal gradients between the surface and the depths. The oceans also contain significant amounts of minerals, some of which are in short supply on land. Sea bed mining of mineral nodules is one possibility, but there is also the potential for extracting various metallic compounds from the low levels that are dissolved in sea water.

The oceans are also a valuable source of food. However, future development will require increasingly careful conservation of fisheries and the further development of mariculture or fish farming.

Efficient exploitation and conservation of the oceans poses great technical challenges for scientists and engineers who must develop materials, structures and equipment for use in the harsh environment of the oceans.

1.2 Oceanic provinces

1.2.1 Major oceans

The continental land masses divide the oceans into three major bodies which are linked by the Southern Ocean: the Pacific, the Atlantic and the Indian Oceans (*Figure 1.1*). In addition there are a number of marginal ocean basins.

The Pacific is the deepest and largest ocean basin. It covers more than one-third of the Earth's surface and

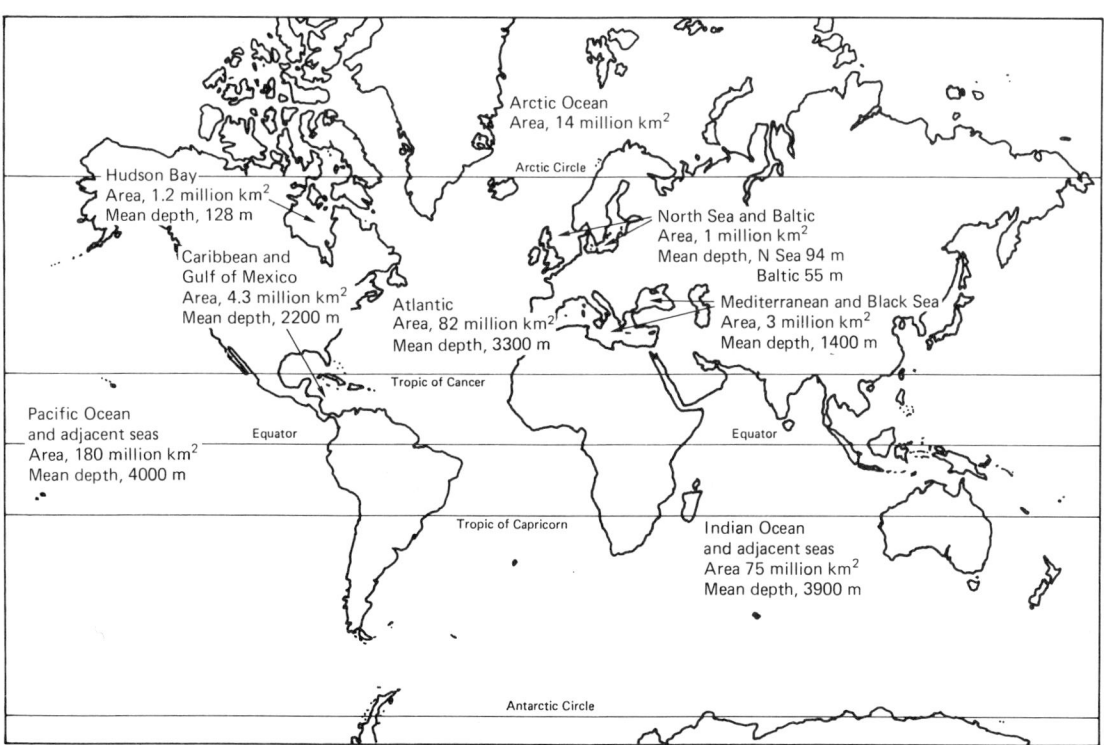

Figure 1.1 The major oceans (from Tait, 1981, 3rd Edn, London, Butterworth; based on a map of the world by courtesy of G Philip and Son Ltd reproduced with permission)

contains more than one-half of the Earth's free water. Islands are abundant in the Pacific, especially in the southern and western parts. The smallest are low lying sand islands and many of these are associated with coral reefs. The majority of the Pacific islands are volcanic and many of the volcanoes are still active.

Most of the volcanic activity is related to plate tectonics. With the exception of the Pacific Antarctic Ridge in the south, the Pacific is ringed by plate boundaries, either subduction zones, where oceanic crust is destroyed, or transform faults, where two plates slide past each other. As a result, the Pacific continental margins are generally narrow and the Pacific is much less affected by runoff from the surrounding continents than are other major ocean basins.

The flow of fresh water and sediments from the eastern margin of the Pacific is largely blocked by mountain ranges. Therefore, most of the sediments and fresh water flowing into the Pacific come from Asia.

The Atlantic is a narrow S-shaped basin that connects the North and South Polar Oceans. It is relatively shallow and contains few islands. In contrast to the Pacific, the Atlantic is not surrounded by plate boundaries and is bordered by wide continental margins. As a result, the Atlantic receives large amounts of fresh water and sediments from rivers.

Much of the Indian Ocean is located in the Southern Hemisphere, although parts of it extend to about 24° north. It is the smallest of the major ocean basins. Continental shelves surrounding the Indian Ocean are relatively narrow, but it receives large amounts of sediments and fresh water into its northern part from three large rivers: the Ganges, the Brahmaputra and the Indus. Because of this input of sediment, it is intermediate in depth between the Atlantic and Pacific.

The Indian Ocean contains relatively few islands. There are a few volcanic islands and groups of carbonate islands and atolls (islands grouped around a shallow lagoon, usually located on top of a submerged volcano). The largest island, Madagascar, is a fragment of the African continent.

1.2.2 Marginal ocean basins

Marginal ocean basins are large oceanic depressions located near continents. They are typically more than 2 km (1.25 miles) deep, which means that their bottom waters are at least partially isolated from the main ocean basins. They are usually floored by oceanic crust covered by a thick layer of sediments and are separated from the open ocean by submarine ridges or sills, islands or parts of continents.

There are three main types of marginal seas: basins associated with island arcs and submarine volcanic ridges, such as many small basins around Indonesia; basins located between continents, for example the Mediterranean; and long narrow basins located within a continent, such as the Red Sea and the Gulf of California.

1.2.3 The depth of the oceans

The ocean depths are generally in the range of 2.5–6.0 km (1.56–3.75 miles) with an average of 3.8 km (2.38 miles) (Neuman and Pierson, 1966). This range is exceeded in the oceanic trenches which generally reach depths of 7.5–11 km (4.69–6.88 miles) (the Marianas trench); the continental plateaux range from + 1 km to − 0.2 km (0.63 to 0.13 miles) in height (Neumann and Pierson, 1966).

1.2.4 Topography

In the early days of ocean exploration depths were measured by individual soundings. This involved lowering a weighted line and determining the length of line needed to reach the bottom. This method was time consuming and inaccurate because the effects of ocean currents and wind meant that the lines were not always vertical.

The development of the echo sounder to measure ocean depths in the early 1900s was a major breakthrough. These operate by accurately recording the transmission time of a sound pulse at the sea surface and the arrival time of its echo reflected off the sea bottom. The velocity of sound in sea water, which is well known, is then used to calculate the depth.

Systematic mapping of the ocean basins using echo sounders began in the late 1940s and led to greatly improved maps of the sea floor. Now accurate bathymetric data can be obtained by sounding systems which are mounted on ships or towed behind them. One example of this type of system is Sea Beam, which combines a hull mounted narrow beam echo sounder with an echo processor. The system produces an on line high resolution contour map of a swath of sea floor which is roughly three-quarters of the water depth wide for each traverse of the ship (Alexandrou and De Moustier, 1988).

The ability to map the sea floor more accurately has greatly increased knowledge of its structure. In the mid-1800s, only the large ocean bottom features, such as the Mid Atlantic Ridge, which occur on well travelled routes, were known. Now, intermediate scale features, such as hills, valleys, channels and banks, as well as large scale features, can be studied in detail using wide swath sonar techniques, such as GLORIA (Geophysical Long Range Inclined Asdic; Hill and McGregor, 1988). In addition, the development of underwater photography has made it possible to study the microrelief of the sea floor, including features such as ripple marks and grooves. Some of the major features of the sea floor are illustrated in *Figure 1.2*.

1.2.4.1 Mid ocean ridges

The mid ocean ridges form a continuous feature running the length of the Atlantic and into the Indian and Pacific Oceans, and cover 23% of the Earth's surface.

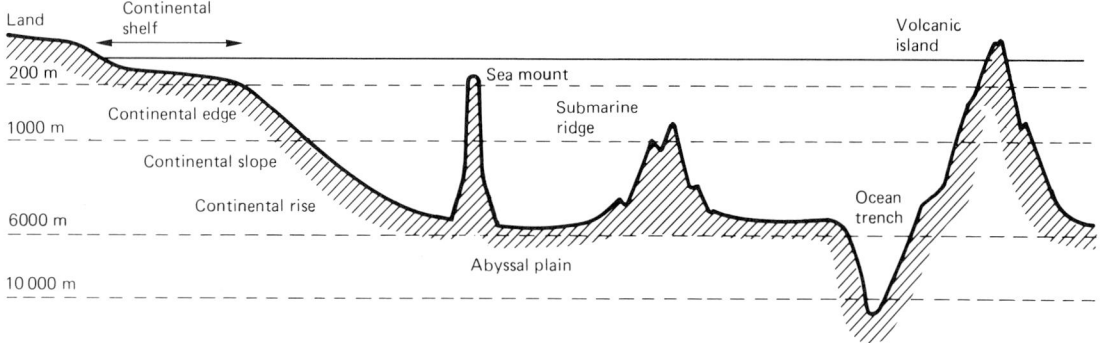

Figure 1.2 Terms applied to parts of the sea floor (from Tait, 1981, Butterworth; reproduced with permission)

They are the sites where new oceanic crust is created by the extrusion of magma. Ridges are regions of high heat flow and are characterized by zones of frequent, shallow focus earthquakes.

The Mid Atlantic Ridge rises 1–3 km (0.63–1.88 miles) above the ocean floor. It is 1500–2000 km (937–1250 miles) wide and has a steep sided central valley, 25–50 km (15.63–31.25 miles) wide and 1–2 km (0.63–1.25 miles) deep. The highest peaks on the Mid Atlantic Ridge come within 2 km (1.25 miles) of the ocean surface. Volcanoes and volcanic islands, such as Iceland, are associated features.

The East Pacific Rise is much less rugged and forms a large low bulge on the sea floor. In contrast to the Mid Atlantic Ridge it has no central valley. It rises 2–4 km (1.25–2.50 miles) above the sea floor.

The Mid Indian Ocean Ridge is rugged and of high relief, although it is not as high or narrow as the Mid Atlantic Ridge.

1.2.4.2 Fracture zones

The ocean floor is cut by hundreds of large scale fracture zones. These are narrow zones of irregular topography, up to 4000 km (2500 miles) long. They are associated with extension of the oceanic crust and form where parts of the crust move past each other. They often displace sections of mid ocean ridges. Some geologists believe certain fractures can be traced on to the continents (Ludman and Coch, 1982).

1.2.4.3 Trenches

Trenches are the deepest parts of the ocean floor. They are narrow features, tens of kilometres wide, which can extend 3–4 km (1.88–2.50 miles) below the ocean floor. The deepest part of the ocean, the Marianas Trench at a depth of 11 km (6.88 miles), is an example. Trenches are associated with subduction zones or areas where one plate moves beneath another and is destroyed. Subduction zones are generally located near ocean basin margins (*Figure 1.3*). They are most common in the Pacific, where they are associated with island arcs.

1.2.4.4 Abyssal plains

Abyssal plains are huge areas of extremely flat ocean bottom which occur near continents. They have a thick cover of sediments and are common at the seaward margins of deep sea fans. Abyssal plains are generally found in marginal seas such as the Gulf of Mexico and the Caribbean and in small ocean basins around the Pacific Ocean margin.

1.2.4.5 Marine canyons

Mid ocean canyons are steep walled but flat floored depressions, up to several kilometres wide, which occur on abyssal plains.

1.2.4.6 Other features

Guyots are flat topped sea mounts, rising from the ocean floor. They occur in all the ocean basins, but are most common in the Pacific.

Volcanoes and volcanic islands are numerous in some ocean basins such as the Pacific. They can rise 1 km (0.63 miles) or more above the ocean floor. Volcanism is commonly associated with mid ocean ridges and island arcs. However, active volcanoes also occur in the middle of ocean basins over hot spots. Not all volcanic activity forms volcanoes. Extensive lava flows can form archipelagic aprons around volcanic island groups. These aprons cover the previous topography and may extend over large areas.

1.2.4.7 Microrelief

The small scale topographic features, or microrelief, of the sea floor can be observed in underwater photographs.

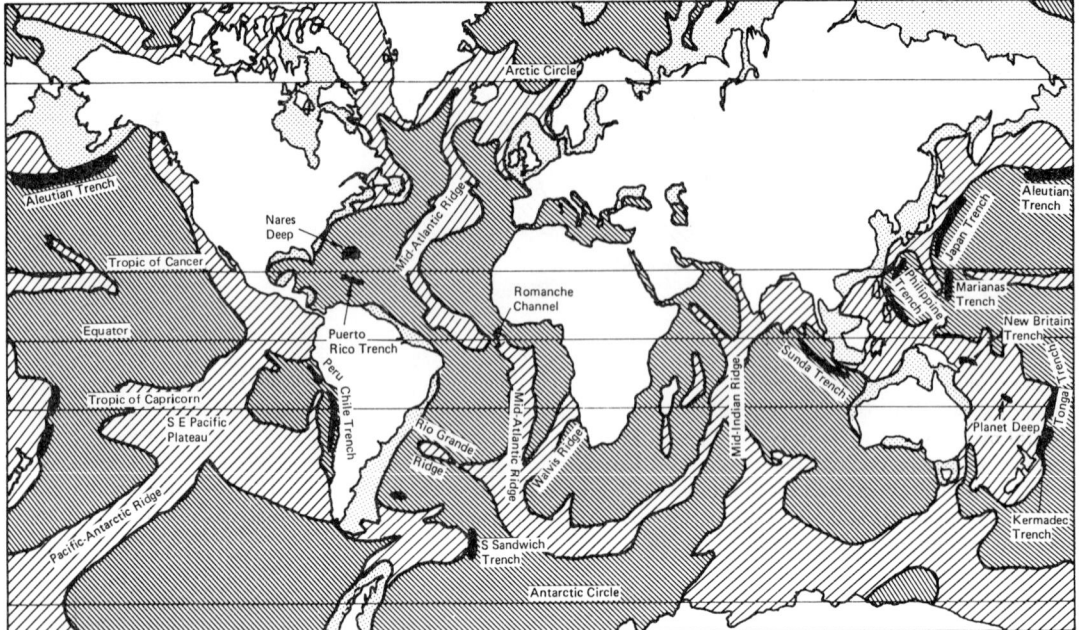

Figure 1.3 Main areas of continental shelf, mid ocean ridges and trenches or subduction zones (from Tait 1981; Butterworth; based on a map of the world by courtesy of G Philip and Son Ltd; reproduced with permission)

Microrelief is usually the result of physical, chemical and biological processes at the interface between the water and the sea bed. Important influences include temperature, pressure, the type of sediment which forms the sea bottom and the effects of wave and current action.

Deep ocean bottom currents are sometimes strong enough to create ripples in sea floor sediments. Chemical processes can lead to the precipitation of phosphate and iron and the production of manganese nodules. In both shallow and deep water, the activities of marine organisms can leave their mark.

1.2.5 Continental margins

Continental margins are made up of a continental shelf, a continental slope and a continental rise.

A continental shelf is a shallow submarine terrace bordering the continents which represents the submerged edge of the continental landmass. A coastal plain is an emergent area of continental shelf. Continental shelves generally extend from the sea level to depths of 150–200 m (488–650 ft), although some, such as those around Antarctica, are deeper. The average width of a continental shelf is approximately 70 km (43.75 miles) (Gross, 1972) but they may be much wider in some areas, such as around the Arctic Sea (Gross, 1972; Neumann and Pierson, 1966). Wide continental shelves occur in areas on continental margins where mountain building has not occurred for many millions of years. Continental shelves slope very gently, typically at 0.1° (Neumann and Pierson, 1966) towards the continental shelf break. There the slope steepens abruptly on to the continental slope. The continental shelf break occurs on average at depths of around 130 m (423 ft).

Continental slopes are the edges of the continents and extend to depths of 220 and 2400 m (715 ft–1.53 miles). Typical slope angles are approximately 4°. The slope becomes less steep near the base of the continental slope where it meets the continental rise.

Submarine canyons are elongated depressions on the continental shelf and slope. Some are associated with old river beds. Others, most commonly on the continental slope, form winding V-shaped valleys with many tributaries and may extend with meandering patterns across the continental rise. Continental rises are formed by the deposition of thick accumulations of sediments eroded from the continents. They cover the transition between the edges of the continents and the ocean basins.

1.2.6 Marine sediments

1.2.6.1 Types of sediments

Marine sediments are derived from several sources. Terrigenous sediments originate by erosion of the continents. They are carried into the marine environment by rivers, waves, wind or, in the case of glacio–marine sediments, are released from floating icebergs. The raw materials for biogenic sediments in the marine environment are the skeletal remains, such as bones, shells and teeth, of marine organisms. Volcanic sediments originate from submarine volcanoes but can also contain ash from land or sea eruptions. Extraterrestrial sediments include microscopic particles from meteors. Hydrogenic deposits are formed by chemical reactions in sea water which result in the precipitation of minerals. These will be discussed in Section 2.1.

1.2.6.2 Distribution of sediments

Figure 1.4 illustrates the distribution of sediment types on the ocean floor. The thickest sediments are found near continental margins and areas of old oceanic crust. A few areas of the ocean bottom have no sediment cover. This may be because they are too new to have accumulated sediment, such as mid ocean ridges, or because they have been swept clean of sediment by strong bottom currents.

Proximity to the source of sediment controls the distribution of nearshore sediments. For example, thick land derived sediments are restricted to areas near continental margins. Large amounts of terrigenous sediments also occur on adjacent deep sea plains. Their transportation is discussed in Section 1.2.6.3. Glacio–marine sediments are restricted to high latitudes where icebergs are common.

Interaction with the oceanic environment also plays an important part in the distribution of deep sea sediments. Carbonate sediments are deposited in areas where there are minimal terrigenous sediments and where water temperatures are warm enough to promote the abundant growth of carbonate secreting organisms.

Carbonate sediments generally occur in low to middle latitudes. The distribution of carbonate sediments is also controlled by the carbonate compensation depth. This is the depth at which the rate of dissolution of calcium carbonate by carbon dioxide in deep sea waters exceeds its rate of deposition. The carbonate compensation depth is 4–5 km (2.50–3.13 miles) in the Pacific and slightly shallower in the Atlantic and Indian Oceans. Carbonate oozes, which are extremely fine grained sediments composed of the skeletons of calcareous organisms, accumulate at a rate of 1–4 cm (0.40–1.60 inches) per thousand years.

Siliceous ooze, derived from the skeletons of silica producing organisms, occurs in areas of high biological productivity which are generally related to the upwelling of nutrient rich water. Siliceous sediments are not dissolved by CO_2 and can thus be deposited below the carbonate compensation depth. Siliceous oozes accumulate at a rate of 0.2–0.8 cm (0.08–0.32 inches) per thousand years.

In areas of low biological productivity below the carbonate compensation depth, reddish pelagic clays are deposited. The chief constituents are very fine

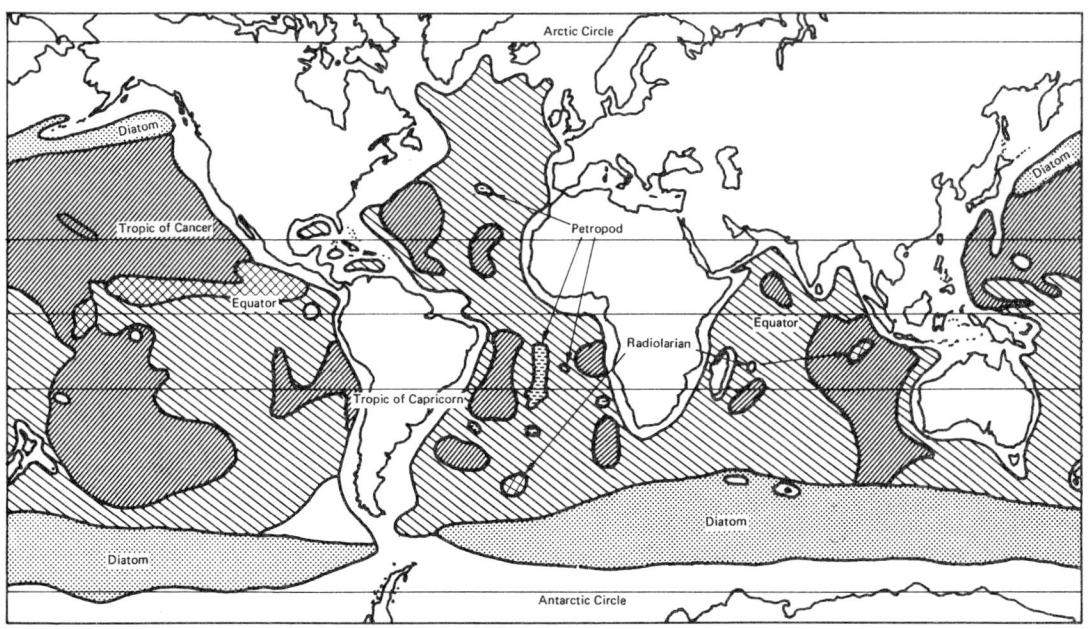

Figure 1.4 Distribution of sediment types on the ocean floor (from Tait, 1981, Butterworth; based on a map of the world by courtesy of G Philip & Son Ltd; reproduced with permission)

particles of clay minerals, quartz and mica which are derived mostly from terrigenous sources and transported by the wind. The sediment is blown on to the surface of the ocean and sinks very slowly. The red coloration is due to oxidation during deposition. Pelagic clays accumulate at a rate of 0.1–1.5 cm (0.04–0.60 inches) per thousand years.

1.2.6.3 Transportation of sediments

The most rapid accumulation of sediments occurs near the continents and most of the sediment reaches the ocean via rivers.

When the sediment reaches the ocean its transport is controlled mainly by physical processes. Particle size and current speed are the most important controlling forces. In general, large grains settle rapidly and finer grains more slowly. Fine clays can take up to 50 years to settle. The settling time of sediments is greatly affected by the action of waves and currents and only 10% of the sediment reaching the shelf remains in suspension long enough to reach the deep sea.

Currents and waves distribute sediments along the coast. Coarse sediments are deposited within 5–6 km (3.13–3.75 miles) of the shore. However, relict coarse sediments, deposited when the shoreline migrated seawards during times of lowered sea level, are common farther offshore.

Nearshore carbonate sediments accumulate on shallow continental shelves where land derived sediment is minimal and the sea is warm enough to promote the abundant growth of carbonate secreting organisms. Coarse carbonate sediments accumulate in areas of turbulence and strong currents, for example around reefs. Fine grained carbonate sediments are deposited in lower energy environments, such as black reef lagoons.

Sediments deposited by turbidity currents and those present in deep sea fans are exceptions to the general rule that land derived sediments are confined to continental shelves. Deep sea fans occur at the base of large canyons cut in the continental slope, generally aligned with the mouths of large rivers. The deep sea fan sediments were deposited at times of lowered sea level when the rivers extended across the continental shelf.

Turbidity currents are gravity driven currents consisting of dilute mixtures of sediment and water which have a density greater than the surrounding water. They are set off by earthquakes, floods, slumping of unconsolidated deposits at river mouths and major storms. They reach velocities of at least 30 km (18.75 miles) per hour (Skinner and Porter, 1987) and are capable of moving thick bodies of coarse sediment to depths of 4–5 km (250–3.13 miles). Deposits from turbidity currents show a characteristic fining upward of grain size due to the continuous loss of energy of the transporting current.

1.2.7 Sea bed stability

Knowledge of the composition and stability of the offshore sediments is critical in many parts of the marine technology industry, including the siting of offshore structures and the laying of pipes.

General information about the nature of the sea bed nearshore can be obtained from published Admiralty charts, geological and government topographic survey maps and records and from records and charts from port authorities and research laboratories concerned with coastal processes.

For more specific information about a particular site, it may be necessary to carry out a survey. These typically seek to obtain detailed information concerning water depth, sea bed topography and subsea bed information to depths of up to 2 km (1.25 miles).

Echo sounders, which can operate in depths between 15 cm and 4.5 km (6 inches–2.81 miles) are used to map the bathometry in detail (Terrett, 1989). For detailed topographic information, side scan sonar, which works by measuring the travel time of pulses emitted as a fan-like acoustic beam spreading across the sea bed below and to the side of the ship, can be used to identify objects which stand above the sea bed. On good records, it is also possible to discriminate between mud, sand, gravel and hard rock on the sea bed.

To obtain details of the sediment below the surface, cores can be taken. However, this can be very expensive and time consuming. A high resolution seismic survey can provide much information at a relatively low cost. With high resolution boomer equipment it is possible to recognize the base of marine mud, base of alluvium, surface of weathered rock and surface of sound rock. With sparker equipment, which provides more energy but less resolution, sediment layers and geological features down to a depth of 100–150 m (325–488 ft) can be detected. (Terret, 1989; Gardline Surveys, 1988).

Correct interpretation of the data obtained from any type of site survey is essential. Data interpretation is a very specialized skill. A number of companies specialize in carrying out and interpreting site surveys.

1.3 Sea water

1.3.1 Chemistry of sea water

Sea water consists of roughly 96.5% water and 3.5% salts, some dissolved organic matter and gases and a few particles. The structure of the water molecule itself has a very important effect on the behaviour and physical properties of the oceans.

Water (H_2O) is a polar molecule with an asymmetrical structure. The hydrogen bonds in water are important in the determination of its physical and chemical properties. Each oxygen atom has three electron pairs. One of these pairs is shared with the two hydrogen atoms to form a covalent bond, formed when each atom retains its electrons but shares them with adjacent atoms.

The characteristic polar nature of the water molecule arises because the oxygen side is negatively charged due to the two unshared electron pairs and the hydrogen side is positively charged.

Water consists of two different types of molecular aggregates. In one type, the hydrogen atoms form weak hydrogen bonds with the oxygens of adjacent water molecules. These clusters of water molecules form and reform rapidly and are less dense than free water molecules. At depths below 10 km (6.25 miles), at pressures above 1000 atmospheres, the clustered molecules disappear. They are also broken up when water evaporates.

In contrast, the unstructured portion of water consists of closely packed molecules, which accounts for its relatively high density. These free water molecules can move and rotate without restriction and their interactions with adjacent water molecules are relatively weak. The free molecules surround the structured portions of water.

The nature of hydrogen bonding in the water molecules means that the properties of water are very different from the hydrogen compounds of other molecules. The unusual structure of the water molecule accounts for many of the physical properties of water. For example, the free molecules, which surround the structured clusters of water molecules assist flow, leading to the fluidity of water. The molecular structure of water also affects its behaviour during heating and cooling.

The continuous breaking and forming of the weak hydrogen bonds means that water has a high heat capacity, or ability to store large amounts of heat energy with relatively small temperature changes. Much of the heat absorbed by water is used to change its internal structure. The large amount of water on the surface of the Earth prevents wide variations in surface temperatures and so prevents rapid climatic change.

Heat is taken up in water in two ways. One form is sensible heat, energy which results in a rise in temperature. The temperature change results from the increased vibration of the water molecules. The other form of heat is latent heat, the energy necessary to break water bonds.

When water is cooled, its density is influenced by the reduced molecular vibration, which tends to increase density, and the formation of bulky molecular clusters, which also tends to reduce density. Initially the density of water increases as the temperature is reduced. However, at about 4°C, as the number of clusters increases, the density of fresh water reaches a maximum. Below this temperature, the density decreases.

Water also has high values of latent heat of fusion and evaporation, surface tension, heat conduction, dielectric constant and transparency. It also has great ability to dissolve impurities. It is often referred to as the universal solvent.

1.3.2 Salinity

Salinity is expressed in parts per thousand and defined as the number of grams of salt dissolved in 1 kg (2.2 lbs) of sea water. On average the salinity of sea water is 35 parts per thousand. Six constituents, chloride, sodium, sulphate, magnesium, calcium and potassium, make up 99% of sea salts. Their relative weight percentages are shown in *Table 1.1*.

Table 1.1 Major constituents of sea water (data from Tait, R.V. (1981), Butterworth reproduced with permission)

Constituent	g/kg
Sodium	10.770
Magnesium	1.300
Calcium	0.412
Potassium	0.399
Strontium	0.008
Chloride	19.340
Sulphate as SO_4	2.710
Bromide	0.067
Carbon, present as bicarbonate, carbonate and molecular carbon dioxide	0.023 at pH 8.4 to 0.027 at pH 7.8

The constituents dissolved in sea water occur as ions, that is, electrically charged particles. The ions in the salts attract and become surrounded by the polar water molecules. This causes the salt crystals to dissolve and this is why water is such a good solvent. The movement of ions through the water controls the electrical conductivity of sea water; the more dissolved ions, the greater the conductivity. Thus, it is possible to measure salinity very accurately by measuring conductivity.

Measurements of salinity can be used in a number of ways. For example, salinity may be related to settling velocities and flocculation – deflocculation processes in clay minerals or to define the limits of an estuary. Salinity measurements are also used with temperature measurements to derive ocean density and to identify the source and extent of water masses.

The major constituents in sea salts behave conservatively in that their relative proportions and concentrations do not change except as a result of physical processes at the ocean surface which alter the amount of fresh water present in the oceans. Examples include evaporation and freezing of ice, which removes fresh water, or precipitation and river discharge which adds fresh water. After a volume of water leaves the ocean surface, its temperature and salinity cannot be changed except through mixing with surrounding water which has different temperature and salinity characteristics. The total heat and salinity are, however, conserved. In contrast, constituents whose relative abundances are changed as they are taken up or released by biological processes are said to be non-conservative processes.

Weathering of rocks on land is a major source of dissolved salts. Volcanic activity on the sea floor is another. In addition, some constituents dissolved in sea water are derived from chlorine and sulphur dioxide gases released to the atmosphere from volcanic eruptions on land.

The residence time of an element in sea water is related to its chemical behaviour. Elements, such as aluminium, which reacts readily with clay particles, have a residence time ranging from a few hundred to a few thousand years. Less reactive elements, such as chloride, have residence times of millions of years (Gross, 1972).

Dissolved constituents can be removed from sea water by various processes. For example, sodium and chloride are removed as a result of the formation of evaporite deposits. Calcium is removed through biological processes, especially the formation of calcareous shells. Potassium and aluminium react with clay and are thus removed from sea water. Magnesium and sulphate are removed as a result of their reaction with newly formed oceanic crust (Gross, 1972).

Evidence from the fossil record indicates that the types and abundances of marine organisms have not varied greatly during the past 1.5 billion years. This suggests that the salinity and composition of sea water have remained similarly stable.

In general, salinity is higher at the surface and there is a halocline, or marked decrease in salinity with depth at 100 m–1.5 km (325 ft–0.94 miles). However, it is density, which is influenced more by temperature than salinity, which ultimately determines the vertical stability of the water column. Therefore, the general relationship of a decrease in salinity with depth does not always hold. For example, in the tropics a salinity maximum often occurs below the upper layers of water due to the intrusion of a thin layer of highly saline and slightly denser water from subtropical areas.

Salinity differences at the surface are caused by regional variations in evaporation and precipitation. The highest salinity surface waters occur in areas of excess evaporation. This is common, for example, in central regions of ocean basins in the subtropics or in partially landlocked seas in arid regions, such as the Red Sea. The lowest salinities occur in areas where precipitation exceeds evaporation, for example in coastal areas which receive large amounts of river discharges and in equatorial regions, where there is a large amount of precipitation.

1.3.3 Dissolved gases

Sea water is saturated with most atmospheric gases, including oxygen, nitrogen, carbon dioxide and rare gases such as helium, argon, krypton and xenon. These atmospheric gases dissolve into sea water at the air–water interface. The solubility of gases in sea water is a function of temperature, salinity and pressure. In general, the gases become less soluble in sea water as temperature and salinity increase.

Many of the gases, such as nitrogen and the rare gases, do not react with the sea water and are not involved in any biological processes. They are found at near saturation throughout the oceans. However, the distribution of gases such as oxygen and carbon dioxide is greatly affected by biological activity. When oxygen consumption exceeds the rate of supply, reducing conditions occur. Below the photic zone, the presence of oxygen depends mainly on advection from surface waters.

Carbon dioxide enters the sea from the atmosphere and is also produced by the respiration of plants and animals. Its abundance is controlled by the physical properties of the sea water, by photosynthesis and respiration and by the formation and destruction of carbonate shells. Carbonate precipitation in plant and animal tissues causes a net loss of CO_2. However, because hydrogen ions in sea water are freely accepted and given up by carbonate and bicarbonate, they form a buffer against sharp changes in acidity.

1.3.4 The pH of sea water

The pH of sea water is related to its salinity. There are more dissolved cations than anions and as a consequence sea water is weakly alkaline (pH 8.1–8.2) (Neumann and Pierson, 1966).

1.3.5 Mineralization

Temperature, gas content of sea water and pressure are important determinants of the chemical reactions which lead to the precipitation of phosphate, iron and manganese oxide nodules on the sea floor. These nodules are precipitated under oxidizing conditions. Many elements in sea water exist partly as multi-ion complexes or as ion pairs. This affects their ability to participate in chemical reactions. The availability of electrons in water determine whether oxidation or reduction dominates.

Many elements, such as iron, manganese and cobalt, are able to participate in electron transfer reactions and can exhibit a variable valency. Higher valency types are formed by oxidation. They are generally less soluble and are therefore precipitated. For example, the manganese in managanese nodules is the higher valency species MN^{4+}, whereas soluble manganese is MN^{2+}.

1.3.6 Temperature distribution

Zones of equal ocean surface temperatures are generally orientated east–west. Temperatures are highest around the Equator because of the large amount of incoming solar radiation absorbed in that region. Temperatures become gradually lower towards the poles. This pattern is modified by the effect of ocean currents, such as the Gulf Stream, which carry warmer waters into northerly latitudes.

This general trend is modified by seasonal and daily variations. Ocean temperatures change seasonally due to variations in solar radiation. The largest seasonal differences in temperature occur in the mid latitudes. This is because there is only a relatively small seasonal variation in solar radiation in equatorial regions and in polar areas the year round presence of ice acts as a buffer against large temperature changes.

There are important vertical variations in temperature throughout the oceans. In the surface zone, which ranges from about 100 to 500 m (325–1625 ft) deep (Gross, 1972), the temperature throughout is similar to the temperature at the surface. This zone is in continuous contact with the atmosphere at the surface where the effect of winds, waves and cooling of surface waters causes mixing to extend down through the layer.

The thermocline, or zone of rapidly decreasing temperature underlying this upper layer, generally occurs at depths of 500 m–1.0 km (1625 ft–0.63 miles) although it can be as shallow as 100 m (325 ft). Thermoclines are especially prominent in open ocean areas where the surface layers are strongly heated for much of the year. They are very stable because density gradients inhibit vertical mixing and so separate the surface waters from the deep zone.

The deep zone contains about 80% of the ocean water and generally occurs below depths of 1 km (0.63 miles) or more. However, at high latitudes the thermocline is absent and deep zone waters have direct contact with the atmosphere. The temperatures in the deep zone average 3.5°C (Gross, 1972).

1.3.6.1 Ice

Ice forms in the oceans when surface waters are chilled below their freezing point. Sea ice is always present around the Arctic and the Antarctic. During the winter the area covered by ice expands.

The salinity of sea water lowers its freezing point below 0°C and changes the properties so that maximum water density occurs at the freezing point. Water with a salinity of 35 parts per thousand freezes at −1.91°C. When sea water begins to freeze, the surface takes on a greasy appearance due to the presence of flat ice crystals. As freezing continues, individual plates of ice develop and eventually aggregate to form sheet ice. The rate of ice formation increases with low salinity and in shallow water. Ice also forms faster when the surface of the sea is calm, for example when there is minimal air or water movement, or when old ice is present.

As sea water begins to freeze, crystals of pure ice form. This increases the salinity of the remaining liquid. As the ice continues to form, some of the brine becomes trapped in the open structure of the ice crystals. The more slowly ice forms, the easier it is for the brine to escape and the resulting ice has a lower salinity. At very low temperatures, ice forms rapidly and the ice has a higher salinity. The average salinity of newly formed ice is around 7 parts per thousand (Gross, 1972). However, the salinities of sea ice are always lower than those of the surrounding water.

Icebergs, which are a major danger to shipping in the North Atlantic and the Antarctic, originate from the fresh water glacial ice that covers Greenland and Antarctica. Icebergs are moved by currents and many migrate out of the polar regions. Large icebergs, with an average height of about 30 m (98 ft) (Clayton and Bishop, 1982) are carried as far south as the Grand Banks off Newfoundland. Icebergs from Antarctica are found between 50 and 40° south. Icebergs generally survive for four years in the open ocean (Gross, 1972).

1.3.7 Density distribution

The density of sea water depends on pressure, temperature and salinity. The zone of minimum density of the surface waters occurs at the Equator. This corresponds to the region of maximum surface water temperatures. Cooling, evaporation and ice formation all tend to increase the density of sea water.

Density, like temperature, varies vertically in the oceans. In the surface zone the density remains nearly constant at about 1.0–1.025 g/cm^3 (0.04 16/in^3). Density increases markedly with increasing depth in the pycnocline, a zone where water density changes significantly with increasing depth. The pycnocline is usually synonymous with the thermocline. The former is a much more stable layer than is the surface zone. It serves as an efficient vertical barrier to the passage of water, salts and gases.

Spatial variation of density below the surface can occur in the presence of currents when cold high density surface water at high latitudes sinks gradually along isopycnals, or surfaces of constant density, which are inclined at a small angle to the air–water interface.

1.3.7.1 Pressure

Pressure depends on the density and on the height of the fluid column. As a general rule, pressure increases with depth in water at a rate of about 10^5 Pa per 10 m. Thus at a depth of 100 m (325 ft) the pressure is about 1.1 MPa. At the bottom of the Challenger Deep, about 11 km (6.88 miles) the pressure is roughly 110 MPa (Clayton and Bishop, 1982).

1.3.8 Light transmission

The optical properties of sea water influence the colour, underwater visibility and biological activity of the oceans. When sunlight reaches the surface of the water, some is reflected and some passes through the air–water interface where it is either scattered or absorbed.

When the sea surface is calm, 2–40% of the total direct light is reflected as the sun moves from overhead to low on the horizon. About 8% of the diffuse light is reflected throughout the day.

When light is absorbed, the radiant energy is converted to internal energy and there is a rise in water temperature. When light passes through sea water, its intensity is reduced. The rate at which light is absorbed in water depends on the wavelength of the light. The reduction in intensity varies with depth and turbidity.

Fifty percent of the solar radiation which enters the oceans is absorbed within 1 cm (0.4 in) of the surface. Ninety percent is absorbed in the upper 40 m (130 ft). Light in the blue part of the spectrum is the least absorbed and thus penetrates to greater depths, which is why clear oceans are predominantly blue. Near the coasts, the presence of suspended sediment, organic particles or marine organisms, inhibits the transmission of light in the blue and violet ranges. In these areas the water colour is blue–green or even yellowish.

Even in the clearest oceans most of the solar radiation is absorbed at depths less than 100 m (325 ft). In turbid coastal waters, light may not penetrate below 10 m (32.5 ft). Therefore, divers may be restricted by limited visibility in the oceans, even if powerful artificial light sources are available (Clayton and Bishop, 1982).

1.3.9 Underwater acoustics

Water is a very efficient medium for the transmission of sound. The velocity of sound in water, typically about 1.5 km/s (0.94 miles/s), is about 4.5 times greater than in air because water is less compressible than air. As a result, only a small amount of sound energy originating in air will penetrate the air–water interface. This is why it is not possible for a diver and the crew of the support ship to communicate directly. However, submerged transducers can be used to both receive and transmit sound. This is the principle behind the use of echo sounders.

The velocity of sound is related to the bulk modulus of elasticity and the density of water. These properties vary with temperature, pressure and salinity and the velocity of sound increases as they increase. The effects of temperature and pressure are the most significant. The velocity of sound increases by about 200 m/s (650 ft/s) between the surface and the deepest part of the ocean.

The velocity of sound waves varies in both the vertical and horizontal directions. For example, because the relative velocity of sound decreases with decreasing temperature, sound waves will be refracted downwards as they travel from the warmer surface layers into deeper colder layers. However, at depths of around km (0.63 miles) the decrease of temperature with depth is relatively small and the effects of pressure on velocity become more important. As pressure increases, so does the sound velocity. The combined effects of downward refraction due to velocity decreasing above 1 km (0.63 miles) and the upward refraction due to the velocity increases below that depth cause the sound energy to be trapped within the water mass and to travel over very long ranges in a horizontal sound channel. The SOFAR (sonar fixing and ranging) technique makes use of this effect.

An understanding of the acoustic properties of sea water is important in the development and use of various echo sounding systems, including the Sea Beam bathymetric survey system (Alexandrou and De Moustier, 1988) and in sonar systems such as GLORIA, which are being used on a large scale for mapping of the sea bed (Hill and McGregor, 1988). Use of acoustic properties of sea water is also made in naval operations, particularly in submarine and anti-submarine warfare, where underwater sound is used extensively by both submarines and surface vessels. For example, acoustic arrays are used to infer the change of water masses passing by (acoustic tomography) and upward pointing beams are used to measure changes in surface levels. In addition, sound receivers are used to monitor sound in the ocean and to detect rainfall.

1.4 Oceanic circulation

The flow of water in the oceans is highly variable both in time and space. Eddies, with scales ranging from millimetres to 100 km (63 miles), often dominate the motion. However, patterns of circulation, of currents, can be recognized throughout the oceans. There are three main types of currents: surface, deep ocean and tidal currents. Surface currents are driven primarily by winds, which act at the air–sea interface. In contrast, deep ocean circulation is driven primarily by temperature and density differences between waters at different latitudes. Tidal currents are caused by the gravitational forces of the Sun and Moon acting on the thin layer of water which covers much of the Earth.

Increasing concern about pollution of the oceans has led to an increased interest in models of oceanic currents. Pollutants enter the oceans in various ways, for example from rivers, the atmosphere or from spills or dumping. Oceanic currents can transport pollutants over long distances and if they end up in areas of stagnant water, their residence time can be long. A greater understanding of current systems is necessary in order to combat effectively the effects of pollution.

Because of high thermal inertia and slow mixing, ocean waters preserve their original temperature and salinity characteristics. Therefore, the basic features of the currents can be obtained indirectly by mapping such features as the surface temperature distribution. In addition, many current systems result in a subtle topography on the ocean surface. Remote sensing of the oceans using satellite technology has made it possible to investigate ocean characteristics on a global scale.

Estimates of the sea surface temperature can be made using high resolution thermal infrared imagery. In this technique an infrared radiometer is used to measure the intensity of radiation emitted from the sea in the infrared band in a broad swath below the satellite. In order to interpret the images correctly, they must be correlated with direct measurements of sea surface temperatures (Tolmazin, 1985).

The surface topography of the oceans can be measured remotely by means of satellite altimetry. In this technique a pencil beam microwave radar is used to measure distances between the spacecraft and the surface of the ocean. The accuracy is within a few centimetres (Tolmazin, 1985). When the geoid (the shape assumed by the oceans when they are at rest) is subtracted from the total height, the resultant topography highlights some of the features of oceanic circulation; for example, the Gulf Stream can be recognized.

1.4.1 Surface currents

Surface currents are caused by the interaction of wind and water at the air–sea interface. The main surface currents are shown in *Figure 1.5*. Current patterns are similar in all of the major ocean basins. They closely resemble the patterns of surface winds if the influences of the land masses and the Coriolis effect of the Earth's rotation are taken into account.

Most of the equatorial ocean is dominated by the North and South Equatorial Currents which flow westwards, driven mainly by the Trade Winds. These are separated by the eastward flowing Equatorial Countercurrent, which is related to the light and variable winds of the doldrums.

East–west elongated gyres, or closed circulatory current systems, are associated with the equatorial currents. Each current gyre includes a major east–west trending current at higher latitudes which flows in the opposite direction to the Equatorial Current, the Equatorial Countercurrent. These higher latitudes currents are relatively slow moving.

Each of the major ocean basins north or south of the Equatorial Countercurrent systems contains a major gyre. Within each gyre, boundary currents flow parallel to the continental margins, usually in a north–south direction. The boundary currents on the western sides of the ocean basins, such as the Gulf Stream in the Atlantic and the Kuroshio in the Pacific, are the strongest currents in the ocean. They flow in a poleward direction, northwards in the northern hemisphere and southwards in the southern and extend to depths of around 1 km (0.63 miles), and are thus usually too deep to flow over onto the continental shelves. Their speeds range approximately 0.7–1.4 m/s (1.4–2.8 knots) (Clayton and Bishop, 1966) and they move fastest in a 50–75 km (31–47 miles) wide band at the surface. Boundary currents are separated from adjacent slower moving water by marked changes in water colour, temperature and salinity. The waters in the western boundary currents are derived from Trade Wind belts and tend to be depleted in nutrients. There is little or no upwelling.

Western boundary currents are strong because the rotation of the Earth, which tends to compress the gyres against the continents, causes the surface of the sea to slope more steeply on the western sides of oceans. This effect is intensified by the Trade Winds, which generally blow in a westerly direction around the Equator and tend to pile up surface waters on the western sides of basins.

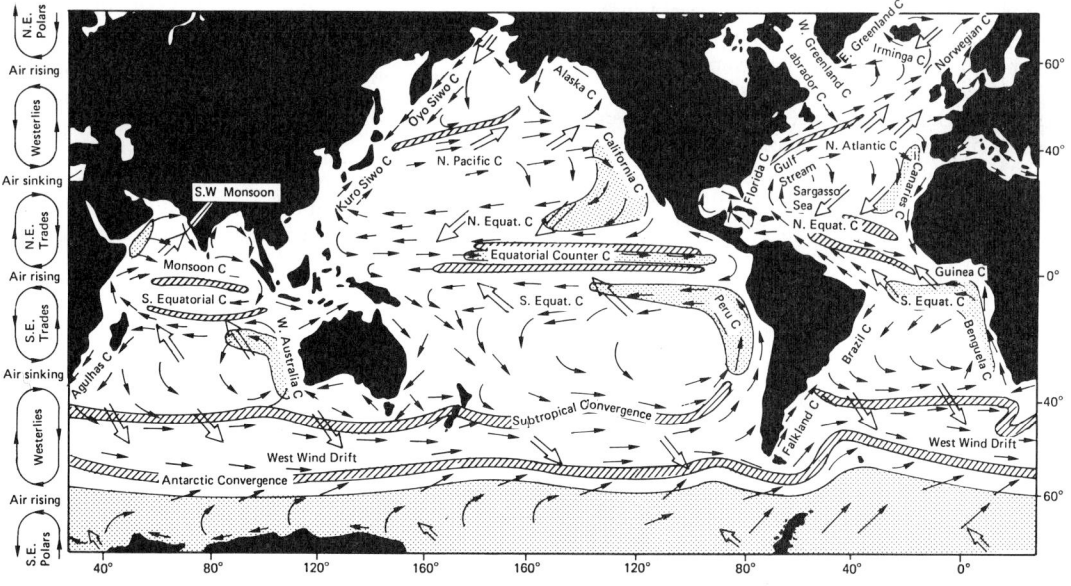

Figure 1.5 The major surface currents (from Tait, 1981, Butterworth; reproduced with permission)

Boundary currents on the eastern sides of ocean basins are much weaker and slower and are relatively shallow. They extend to depths of less than 500 m (1625 ft) and can therefore flow over continental margins. The waters in eastern boundary currents are derived from the mid latitudes and the boundaries separating them from other currents are diffuse. Coastal upwelling, caused by local wind conditions, is common.

Smaller current gyres, which circulate in a direction opposite to the subtropical gyres, are present in the subpolar and polar regions of all the ocean basins. The subpolar gyres in the Northern Hemisphere flow in an anticlockwise direction. They are very well developed because they are surrounded by land masses. In contrast, currents flow uninterrupted around Antarctica in the Antarctic Circumpolar Current, because there are no land masses in the subpolar regions of the Southern Hemisphere.

1.4.1.1 Coriolis effect

The Coriolis effect describes the force imparted to moving fluids due to the rotation of the Earth. In the Northern Hemisphere currents are deflected to the right and in the Southern they are deflected to the left. There is little or no deflection for currents moving east or west within about 5° either side of the Equator.

The deflection due to the Coriolis effect increases as the speed of the wind or current increases. Therefore, faster currents are deflected to a greater degree than slow currents.

1.4.1.2 Geostrophic currents

Geostrophic currents are formed by a balance of wind, Coriolis and gravitational forces. For example, the Gulf Stream, which flows at a speed of approximately 1 m/s (3.25 ft/s), is subject to a Coriolis force which balances a pressure gradient. This results in a change in the level of about 1 m (3.25 ft) across the Gulf Stream and allows it to flow without continually diverting to the right. Flow along boundaries is also generally balanced by pressure gradients across the flow, which results in tilted density surfaces. Flows in geostrophic balance are called geostrophic currents.

Convergence (zones where currents flow towards each other) and divergence (areas where currents separate) result in a subtle topography on the surface of the ocean. It also results in a balance between the wind stress which is driving the currents, the pressure gradient and the Coriolis forces on the moving water. Zones of convergence cause water to pile up and thicken the surface layer. Water responds to this raised oceanic topography by tending to move away from the higher surfaces in response to gravitational forces. However, the water is also deflected by the Coriolis effect. The balance results in a circulation about the raised topography, or geostrophic currents. These will be strongest on the steepest surface slopes and weakest on the more gentle slopes.

1.4.1.3 The Ekman spiral

The Ekman spiral describes the movement of near surface water due to the effects of wind. When wind blows across water, it exerts a force on the surface caused by frictional shear stress and the surface pressure gradient in the atmosphere. This sets a thin layer of water in motion, estimated in idealized models to be at 45° to the wind direction (Clayton and Bishop, 1982; Gross, 1972). These surface currents move at about 2% of the speed of the wind. The process continues downwards and momentum is transferred to successively deeper layers.

The transfer of momentum downwards is inefficient and energy is lost in the process. This means that current speed decreases with increasing depth. In addition, the Coriolis effect causes each successive deeper layer to be deflected further to the right in the Northern Hemisphere or further to the left in the Southern Hemisphere. This results in a spiralling current direction as depth increases.

Eventually, a depth is reached where the velocity of the layers is reduced to a very low value in a direction opposite to the surface current. This depth of frictional resistance is about 180 m (585 ft) for a wind of 7 m/s (23 ft/s) at 5° latitude, and 60 m (195 ft) for the same wind at a latitude of 50° because of the greater Coriolis force.

Ekman's model is based on an infinitely deep and uniform ocean. In shallow waters, wind generated currents are not deflected as much as predicted. In addition, an Ekman spiral requires one or two days of steady wind to develop fully. As a result, the deflection of the layers is usually found to be less than predicted by the model.

1.4.1.4 Ocean eddies

As a consequence of instability, interaction with topography or variable wind forcing, ocean currents are highly variable. Eddies are large scale, usually greater than 100 km (63 miles), fluctuating currents which develop rotating patterns. They can be generated in various ways. For example, numerous eddies are generated in the friction zone between the boundary currents and the continents. They can also be caused by phenomena such as wind forcing.

1.4.2 Deep ocean circulation

Surface currents involve a surface layer which is about 1 km (0.63 miles) thick. Below this level the currents are slower, but may extend to the bottom of the ocean. Circulation in the deep ocean is a convective circulation, which is due mainly to the effects of temperature and salinity on the density of sea water. These density driven currents are also called thermohaline circulation.

The depth of a water mass is determined by its density relative to the densities of the water masses around it. When a water mass reaches an equilibrium level in terms of density, it tends to spread out to form a thin layer. This is a reflection of the fact that less energy is needed to move water along a surface of equal density than is needed to move water across it. The surfaces of constant density are essentially horizontal. Thus subsurface currents, which move along surfaces of constant density, are more or less horizontal.

The horizontal currents in the deep ocean are driven by large scale sinking of dense water masses formed primarily in high latitudes. Large volumes of cold dense bottom waters form in partially isolated polar areas where the already highly saline water is subjected to cooling or freezing. These cold high salinity water masses are very dense and therefore sink. As the dense water masses sink, they flow slowly along the continental shelf, down the continental slope and on to the bottom of the ocean basins. After they sink, they are isolated from the atmosphere and solar radiation for up to 200 years (Clayton and Bishop, 1982).

These dense near bottom currents, with speeds of 1–2 cm/s (0.4–0.8 in/s), move much more slowly than surface currents. Dense water formed in the Antarctic flows northwards into the deep basins of the Pacific, Indian and South Atlantic Oceans. Dense waters formed in the Arctic flow southwards. Arctic water is slightly less dense than Antarctic, so when the two masses meet, the former flows on top of the latter. Large scale mixing occurs in the South Atlantic off the east coast of South America.

1.4.2.1 Stratification in the ocean

The oceans are strongly stratified and consist of a number of horizontal layers of water which have uniform properties of temperature, salinity and density. These characteristics may change abruptly from one layer to the next, but within the layers they are constant to within the accuracy of measurement.

This stratification leads to the formation of a stable vertical water column which increases in density with depth. Density, in turn, is dependent on temperature and salinity.

The ocean waters may be divided into three vertical zones:

1. The surface zone, which is generally saturated with oxygen, warm, well lighted, less dense and more saline.

2. The pycnocline, where water density changes markedly with increasing depth. This zone forms a more stable layer than the overlying surface zone. It partially isolates the deep ocean from surface processes because low density surface waters cannot readily move through it. However, deep ocean waters may move slowly upward through the pycnocline as part of the worldwide deep ocean circulation.

3. The deep zone, which contains about 80% of the ocean water. It is dark, cold and generally isolated from short term changes occurring at the surface. Dense cold masses of water formed at high latitudes sink to join the deep zone. These masses supply dissolved oxygen (*Figure 1.6*).

1.4.3 Tides and tidal currents

The effects of tides are felt most strongly in coastal areas. The normal tidal movement relative to still water level is usually between 0.5 and 2 m (1.13 and 6.50 ft) (Clayton and Bishop, 1982), but local geographical features can often produce larger variations.

In narrow gulfs or channels, the tidal range can be very great. For example the tidal range in the Bristol Channel in the UK is up to 15 m (49 ft) and the range in

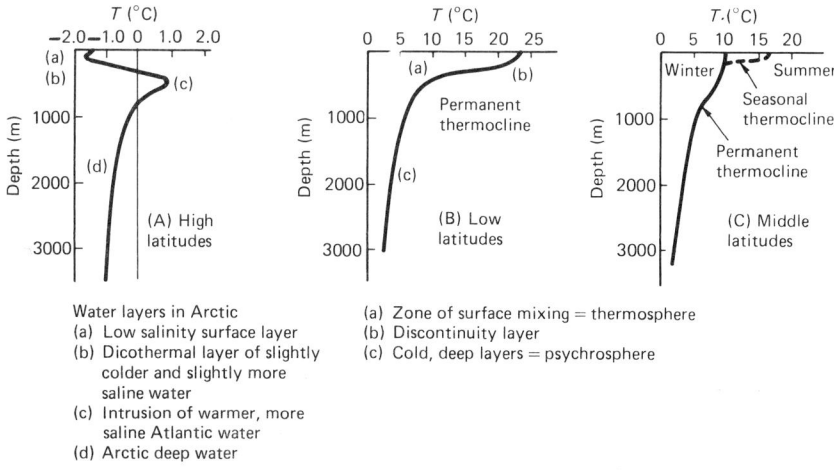

Figure 1.6 Temperature profiles in the deep ocean (from Tait, 1981, Butterworth; reproduced with permission)

the Bay of Fundy in Canada is up to 21 m (68 ft) (Clayton and Bishop, 1982). In contrast, almost land locked seas, such as the Mediterranean, have little or no tidal movement.

All bodies in the Universe exert forces on the water covering the Earth, but only the Moon and Sun are of major importance. The influence of the Sun is due to its great mass. However, the Moon, although smaller than the Sun, is much closer to it and the effect of its influence is roughly twice that of the Sun (Clayton and Bishop, 1982).

The gravitational attraction of the Moon is greatest on the side of the Earth nearest the Moon. However, centrifugal forces, which act in the opposite direction to the force of gravitational attraction and which keep the Earth and the Moon apart, are equal over the surface of the Earth. On the side of the Earth nearest the Moon, the gravitational attraction of the latter exceeds the centrifugal force. Therefore, the water is pulled towards the Moon and forms a bulge on the surface of the Earth. On the opposite side of the Earth, the centrifugal forces exceed the gravitational forces of the Moon. These centrifugal forces cause a second bulge of water to form on this side. These bulges change geographical position as the Earth rotates and as the position of the Moon changes. They create unbalanced forces on the surface of the Earth and these forces generate the tides.

This model provides a very simplified explanation for why the tides act as they do. For a more accurate understanding of the tides, the tidal bulges can be considered as long period waves with the two water bulges representing the crests. Their wavelength is half of the Earth's circumference or approximately 20 000 km (12 500 miles). If the tides were to keep up with the Moon in the Southern Ocean, the only zone in which waves could propagate fully around the Earth's axes, they would have to move at a velocity of 1600 km/h (1000 ft/s) at the Equator. In order for a wave to propagate at this speed, the ocean would need to be 22 km (13.75 miles) deep. Because it is much shallower, an average 4 km (2.5 miles), the tidal bulges move as forced waves whose speed is determined by the movements of the Moon. In practice, the position of the tidal bulges is determined by a balance between the attraction of the Sun and the Moon and the frictional effects of the ocean bottom and by the shape and topography of the ocean basins.

The tidal day is 24 h 50 min long. This time corresponds to the time between successive passes of the Moon over any point on the Earth. Because the tidal day is slightly longer than the solar day (the time it takes for the Earth to rotate about its own axis) high and low tides do not occur at the same time each day.

In some areas, for example parts of the Gulf of Mexico, tides are diurnal, that is there is only one high tide and one low tide each day. Many areas have semidiurnal tides, or two high and two low tides that are approximately equal per day. A number of places exhibit mixed tides. They have two high tides and two low tides per day, but the heights of the two high and the two low tides are different. The higher of the two high tides is called higher high water (HHW) and the lower of the high tides is referred to as lower high water (LHW). Similarly, the low tides can be classified as higher low water (HLW) and lower low water (LLW).

The times of diurnal, semidiurnal and mixed tides are relatively easy to predict because the high tides tend to occur at a known time after the Moon passes. Tidal predictions for ports are routinely available.

The tidal range, or difference between the highest and lowest tide levels, is also variable. Spring tides, which have a higher tidal range than average, occur every two weeks, near the times of the full and new Moons. This is when the positions of the Sun and Moon are such that the effect of each on the Earth's water reinforces each other. Neap tides, which have a lower tidal range than average, occur during the first- and third-quarters of the Moon, when the Sun and Moon are positioned so that their effects oppose each other. The highest spring tides occur when the Moon is nearest to the Earth.

1.4.3.1 Tides in ocean basins

Every basin has its own natural period of standing waves. Because of their dimensions, and hence their natural periods, each ocean basin responds more readily to some constituents of the tide generating forces than others. The tide in a bay, harbour or partially enclosed sea is greatly influenced by the magnitude of the ocean tide at its mouth as well as by the natural period of its basin and by the cross section of the opening through which the tide wave must pass. Where the natural period of a basin is near the tidal period, it is possible to set up large standing waves which give rise to exceptional tidal ranges. For example, in the Bay of Fundy, the natural period is about 12 h. Spring tides there have ranges of 15 m (49 ft) because a standing wave is combined with a narrowing valley which funnels the tide and increases its range.

Tides in coastal areas affect estuaries and their tributaries. The tide enters an estuary as a progressive wave and travels upstream. In many estuaries, the wave is gradually reduced in height by friction of the channel and the opposing river flow. However, in some estuaries, the wave reaches the end of the estuary and is reflected. River discharges may modify tidal currents in estuaries and tidal currents can be affected by winds. The times and heights of tides in bays and coastal areas are commonly altered by storms. For these reasons, predicted tides based solely on astronomical factors may be inaccurate. The characteristics of coastal oceans, estuaries and fjords are discussed more fully in Section 1.8.

1.4.3.2 Tidal currents

Tidal currents are horizontal water movements associated with the rise and fall of the sea surface due to tidal forces. Not every coast has tidal currents and some areas have these but no tidal displacement of the surface.

Most of the water motion associated with tides is horizontal. The flood current occurs when the water flows in with the tide. The ebb current corresponds to a change in direction of the current which causes the water to flow out. Each time the current changes direction there is a period of no current, or slack water.

Tidal currents are altered by winds or river run off. In addition, different tides will have different tidal currents. The strength of a tidal current depends on the volume of water that must flow through an opening and the size of the opening.

In general, tidal currents are strongest in the coastal regions and are much weaker in the open ocean. In the deep oceans tidal currents typically move at about 5 cm/s (2 in/s). In shallow water, these currents may have speeds of several metres per second in tidal races. Tidal currents are free to change direction continually in the open ocean because they are not restricted by the coastline. Slack water periods in the open ocean are times when the current is at a minimum

1.5 The atmosphere

The oceans and the atmosphere, the Earth's two fluid outer layers, are closely linked. Not only are they both affected by forces such as gravity, heat flow and the rotation of the Earth, but there are also extensive exchanges of internal and mechanical energy between them. The oceans and the atmosphere are so closely related and affect each other to such a great extent that no description of the ocean environment would be complete unless atmospheric processes are also taken into account.

1.5.1 Characteristics of the atmosphere

The atmosphere is a mixture of gases. The relative abundance by volume of gases in a dry atmosphere is nitrogen, 78%, oxygen, 21%, argon, 0.9% and carbon dioxide, methane and inert gases, 0.1% (Gross, 1972). The atmosphere is very well mixed. Except for variable amounts of water vapour, ozone (O_3, a three atom molecule of oxygen) and dust, its composition is generally constant.

Water vapour plays a major role in atmospheric circulation. It is lighter than nitrogen gas, the most abundant constituent of the atmosphere. As a result, the addition of water vapour to the atmosphere decreases the density of the air and causes it to rise. In contrast, removing water vapour from the air makes it denser and causes it to sink.

Air also becomes less dense when it is warmed and warm air can hold more water vapour than cold air. Water vapour, itself, increases in density as it cools and the release of heat by condensing water vapour warms the surrounding air. This is an important source of energy for atmospheric processes (Gross, 1972).

The atmosphere is density stratified. The densest air is closest to the surface of the Earth. This is because gases are very compressible and the greater pressure near the surface of the Earth means that the gases in the atmosphere are more compressed and denser. Atmospheric pressure decreases by a factor of 0.5 for every 6 km (3.75 miles) of altitude.

The atmosphere is divided vertically into four layers on the basis of temperature. The bottom layer, which is closest to the surface of the Earth is the troposphere. The average top of the troposphere is 12 km (7.5 miles). However, the troposphere extends up to 16 km (10 miles) in the tropics but is only 9 km (5.63 miles) high or less in the polar regions. The upper boundary of the troposphere is called the tropopause.

Within the troposphere, air temperature drops with increased elevation because the atmosphere is warmed near the surface of the Earth and cooled near the tropopause. Most of the weather on the Earth occurs in the troposphere. Vertical movement is a characteristic of the troposphere and thunderstorms are a dramatic manifestation of this motion.

The stratosphere is the layer above the troposphere. It ranges from approximately 12 to 50 km (7.5 to 31.25 miles). There is no large scale mixing or vertical movement in the stratosphere. Temperature remains constant in the lower part of the stratosphere, up to about 20 km (12.5 miles). Above that level there is a sharp increase in temperature up to the stratopause, which marks the top of the stratosphere. The ozone layer, which absorbs ultraviolet radiation from the Sun occurs in the stratosphere (Lutgens and Tarbuck, 1982).

The mesophere is the layer above the stratopause. Here temperature decreases with height. At the mesopause, which occurs at about 80 km (50 miles) above the surface of the Earth and marks the top of the mesosphere, the temperature is around $-100°C$. Above the mesopause, in the thermosphere, the temperature increases with height as a result of short wave solar energy (Lutgens and Tarbuck, 1982).

1.5.1.1 Heat budget

Most of the heat on the Earth is derived from the Sun. On average, about 340 W/m^2 reaches the top of the atmosphere. Because the average surface temperatures of the Earth are approximately constant, it is assumed that a similar amount of energy is radiated back into space each year.

About 33% of the solar radiation striking the top of the atmosphere is reflected back into space by clouds, which typically cover about 50% of the Earth's surface.

About half of the Sun's radiation which strikes the top of the atmosphere reaches the surface of the Earth. About 4% of the Earth's heat is lost by radiation from the surface. Most of this back radiation is absorbed by clouds. This absorption moderates the effect of the heat loss and prevents large temperature variations on the Earth.

The lower few thousand metres of the atmosphere are heated primarily by the latent heat from water evaporated from the oceans which is released to the atmosphere when water vapour condenses.

The maximum amount of incoming solar energy, averaged over the year, is received in low latitudes. However, energy is radiated back into space at a nearly constant rate. As a result, most of the heat in the Earth is received between 40° North and 40° South. There is a net heat loss of 40–90°C in both hemispheres.

1.5.2 Winds

The winds result from the heating of the Earth in the low latitudes and cooling near the poles. Air warmed near the equator rises. As it cools, water vapour condenses and falls as rain. The drier air then flows towards the polar regions where it is cooled. The dry cool air sinks and flows along the Earth's surface to the tropics, where the process is repeated. This north–south circulation provides the energy to induce winds (Clayton and Bishop, 1982). Easterlies and westerlies result from the deflection of the basic motion by the Coriolis effect. This effect increases towards the poles, but is zero at the Equator. As a result, wind patterns are deflected to the right in the Northern Hemisphere and to the left in the Southern. A general model of atmospheric circulation is shown in *Figure 1.7* (Clayton and Bishop, 1982).

There are three main circulation cells in each hemisphere. In both hemispheres there are easterly winds at 5–25° latitude, westerlies at 35–55° latitude and polar easterlies at 60–85° latitude (Clayton and Bishop, 1982).

Air tends to accumulate or converge at around 30° North and 30° South and form a zone of high pressure, with calms or variable winds and clear skies (the Horse Latitudes). As this air sinks, it spreads along the surface of the Earth and part of it moves towards the Equator in the form of trade winds. The trade winds blow from the south east in the Southern Hemisphere and from the north east in the Northern Hemisphere. They converge on the equatorial zone at the intertropical convergence zone, the Doldrums.

The belt of low pressure around the equator in the region of 5° North to 5° South causes air streams to converge and rise. This results in light and variable surface winds and a cloudy climate with heavy rainfall.

The remaining air from the Horse Latitudes flows toward the poles in the form of westerlies. At about 50° of latitude it converges with colder denser air from the polar regions which is moving south towards the

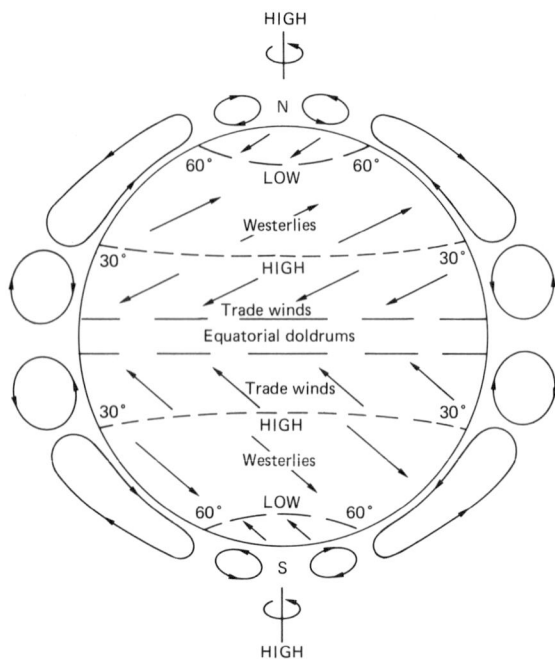

Figure 1.7 General model of atmospheric circulation (redrawn from Clayton and Bishop, 1982, E & F.N. Spon Ltd, London; reproduced with permission)

Equator. The zone of convergence is called the polar front and it marks a sharp boundary between air masses. The latitudes affected by the polar front experience highly variable weather.

Wind patterns are also altered by the presence of continents. This is particularly true in the Northern Hemisphere where there are larger amounts of land. In the summer, the land becomes warmer than the ocean. The air over the land is warmed and forms a low pressure area. The air over the ocean is relatively cooler and forms a high pressure cell over the oceans. This deflects the wind patterns. The situation is reversed in the winter.

The effects of differential heating of the land and ocean surfaces are also responsible for the monsoons, which are large scale seasonal changes in winds which occur in some areas. On a smaller scale, these effects affect the winds along the coasts daily. During the day the land warms more quickly than the ocean surface. This results in sea breezes or winds blowing from the sea towards the land, during the afternoon. At night the land cools more quickly than the ocean and winds blowing from the land towards the sea or land breezes are generated.

1.5.3 Weather systems

Climate is determined by the location of areas of rising or sinking air masses. In the mid latitudes, air masses commonly sink over the ocean. There the climate is characterized by low rainfall and high evaporation and high atmospheric pressure. Near the Equator, where air masses commonly rise due to warming and high water vapour content, the climate is characterized by high rainfall and cloudiness and low atmospheric pressure.

Cyclones and anticyclones are large rotating masses of air. Cyclones can form at the polar front where cold dense air moving towards the Equator meets warmer moist air moving towards the poles. Circulation develops around an area of low pressure. The cold air circulates around the low pressure zone faster than the warm air and eventually overtakes the warm air front and cuts it off. This causes the low pressure cell to intensify and winds to grow stronger. Wind speeds greater than 120 km/h (75 mph) can occur. These types of storms occur fairly frequently in winter when the temperature contrasts across the polar front are greatest.

Tropical cyclones form in a similar way along the intertropical convergence zone. Because the temperature contrasts are not so great, individual tropical cyclones do not generally affect a very large area. However, extremely intense hurricanes, or typhoons, often develop in tropical areas in the summer and autumn.

The energy of a hurricane increases with the abundance of moisture in the air. At the centre, or eye, of the storm is an area of very low pressure. The storm follows a very erratic path which is difficult to predict.

Tornadoes are the most violent storms in the mid latitudes. They are tight cyclones, a few hundred metres in diameter at ground level. They have zones of extremely low pressure at their centres and very high wind speeds (90–150 m/s (325–536 km/h), 29–49 ft/s (203–335 mph)). They form in the warm sector of cyclonic systems just ahead of the cold front. They usually travel parallel to the cold front. Over the sea, tornadoes form water spouts in areas where cold continental air pushes over warm water, for example off the east coast of the USA, China and Japan (Clayton and Bishop, 1982).

1.5.3.1 Wind speeds

Wind speeds are usually measured at a standard altitude of 10 m (32 ft) above the local land or sea surface. For marine work, wind speeds are generally quoted in terms of the Beaufort Scale. This is shown in *Table 1.2* (Clayton and Bishop, 1982).

1.5.4 The ionosphere

The ionosphere is an electrically charged layer which is located in the altitude range 80–400 km (50–250 miles). Within the ionosphere, molecules of nitrogen and atoms of oxygen are ionized as they absorb high energy, short wavelength solar energy. These elements are also ionized at lower and higher altitudes but are most concentrated in the ionosphere.

The ionosphere contains three layers of varying ion density. These are, from top to bottom, the D, E, and F layers. The production of ions is dependent upon direct solar radiation. Therefore the concentration of charged particles varies from day to night. This is especially true in the lower D and E zones, which weaken and disappear at night then reappear during the day. The uppermost F layer is always present.

The presence of the ionosphere affects the reception of radio signals. The F layer is largely responsible for reflecting radio signals back to Earth. However, during the day the D layer is present and absorbs most of the radio signals. At night, when the D layer disappears, radio waves reach the F layer easily and are reflecting back to the surface of the Earth. As a result, signals from long distances can be received more easily at night than during the day.

1.5.4.1 Effects on marine communication and navigation systems

Radio waves, which may be very strong, can be caused by electrical discharges in the atmosphere. They can be reflected by the ionosphere and interfere with reception over a wide area.

Occasionally precipitation particles may become electrically charged. When these particles make contact with the receiving antenna they discharge to Earth via the antenna and the receiver. This causes interference (Sonnenburg, 1988).

1.6 Biological oceanography

The marine ecosystem is complex. The distribution of marine organisms is related to the availability of suitable chemical and physical conditions in and around the ocean. However, the life processes of marine organisms cause them to interact with and alter their environment. For example, biochemical processes change the composition of sea water when organisms remove certain constituents from sea water and incorporate them in living tissue. One example of this process is organisms removing calcium carbonate from sea water to make shells. After organisms die, they form particles which sink through the water column and some decompose as they sink, others decompose on the ocean bottom. In this way, they are responsible for the vertical transport of many chemical constituents.

1.6.1 Characteristics of the marine lifestyle

The density of sea water is much greater than the density of air. Thus, marine organisms do not require heavy skeletons for support. Instead, they can rely on the surrounding water. As a result, many marine organisms have no skeletons and consist largely of

Table 1.2 The Beaufort scale (data from Clayton and Bishop, 1982 E. & F.N. Spon Ltd London; reproduced with permission)

Number	Wind speed at 10 m (km/h)	Wind description	Mean wave height (m)	Noticeable effect of wind on land	At sea
0	< 2	calm	none	smoke vertical; flags still	surface mirror smooth
1	2–5	light air	< 0.5	smoke drifts; vanes static	ripples; no foam crests
2	5–11	light breeze	0.5	wind felt on face; leaves, flags rustle, vanes move	waves are short and more pronounced
3	11–19	gentle breeze	1	leaves and twigs in motion; light flags extended	crests begin to break
4	19–30	moderate breeze	1.5	raises dust; moves small branches	longer waves; many white caps
5	30–40	fresh breeze	2	small trees sway	pronounced waves; white caps everywhere
6	40–50	strong breeze	3	large branches move; telephone wires 'sing'	large waves; extensive foam crests
7	50–60	moderate gale	4	whole trees in motion	sea heaps up; foam blows
8	60–75	fresh gale	5	twigs break off; progress impeded	waves increase; dense foam streaks
9	75–85	strong gale	6	chimney pots removed	higher waves; foam clouds
10	85–100	whole gale	7.5	trees uprooted; structural damage	long hanging crests; great foam patches
11	100–120	storm	8.5	widespread damage	ships hidden in troughs; air filled with spray; sea covered with foam
12	120	hurricane	> 8.5	countryside devastated	ships swamped; sea a maelstrom

water. One problem they do face is that of sinking. If an organism is more dense than the surrounding water it will sink unless it swims or is able to maintain buoyancy by changing its density. Fish accomplish this by means of a gas-filled bladder. Zooplankton change their buoyancy by storing fat. In addition, irregularly shaped organisms sink more slowly than streamlined ones. Organisms which live in cold waters sink more slowly, because cold water is more viscous than warm water.

The body fluids of most marine organisms are similar in salt content to sea water. Therefore organisms which live in the open ocean do not need to regulate the composition of their body fluids. However, organisms living in regions where the salinity is variable, such as coastal waters or in estuaries, face varying salinity and must be able to regulate the salt content of their body fluids.

Marine organisms adopt several types of strategies to avoid being eaten by predators. Some are very small in size. Others are almost transparent and therefore hard to see. Some organisms migrate vertically. They sink into deeper dark waters during the day and swim to the surface at night to feed. Schooling, or travelling in large groups of high density, is another strategy to avoid predation.

16.2 Types of marine life

Marine organisms can be conveniently classified according to their lifestyles and the habitats in which they live. There are five main habitats: the intertidal, the neritic or coastal ocean, the oceanic or upper regions of the open ocean, the mesopelagic, or middle depths of the open ocean, and the abyssal or deep ocean. Within these main habitats different organisms adapt to various lifestyles. The relationship between marine habitats and marine lifestyles is illustrated in *Figure 1.8*.

The main marine lifestyles are nektonic, planktonic and benthonic. The nekton are animals capable of swimming sufficiently strongly to give them independence from ocean currents. Many nektonic animals are large. They have skeletons to which muscles are attached and against which the body shape is contorted to achieve a net propulsion force. Many phyla are represented in the nekton including fish, birds and marine mammals, as well as some of the cephalopod molluscs such as octopuses and squids.

The plankton include animals and plants which are freely drifting. Although most have some primitive means of locomotion such as flagella, plankton are at the mercy of and are carried by the currents.

The benthonic organisms are a diverse group of organisms which live on the sea bottom. Some are mobile, and usually move by moving along a substrate rather than swimming. Examples include arthropods, such as crabs and lobsters, and echinoderms such as starfish, which move along on tiny tube feet. Others, like worms, burrow through the substrate. Many others, such as sea anemones, barnacles and mussels are sessile. That is, they live attached to a substrate.

Many benthonic organisms have a planktonic larval stage.

1.6.3 Distribution of organisms

The distribution of organisms is controlled mainly by water movements and the availability of light and nutrients. The physical properties of the sea water are also important controls.

1.6.3.1 Physical properties of sea water

Most marine organisms are sensitive to small changes in ambient conditions of salinity and temperature. In many open ocean species, the range of tolerance to changes in salinity and temperature is small. In contrast, organisms which inhabit coastal areas are adapted to a more variable environment and have a greater tolerance for change. Organisms in estuarine environments exhibit extremes of tolerance.

1.6.3.2 Water depth

Two important factors, pressure and light, are controlled by water depth. In addition, temperature varies with depth.

Light generally penetrates less than 100 m (325 ft) in the open ocean (see Section 1.3.8). The zones where light penetrates are known as the photic zones, the regions below the photic zone, aphotic. The distribution of light is a major controlling factor of the primary production of nutrients in the ocean. In turn, the primary production of organic matter by plants is one of the major controlling factors on the distribution of marine organisms.

Light is essential for photosynthesis, the process by which plants store energy from sunlight to produce carbohydrates which can be used as energy by other organisms. Animals can live below the photic zone because they can feed on organic matter which sinks from above. Plants, on the other hand, must have access to light.

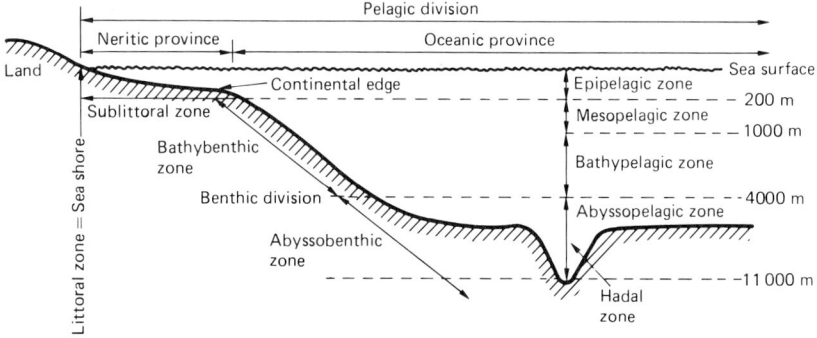

Figure 1.8 Marine habitats and lifestyles (from Tait, 1981, Butterworth; reproduced with permission)

Pressure undergoes wide variation within the marine environment. It ranges from atmospheric pressure at the sea surface to more than 1000 atmospheres in deep sea trenches (Neshyba, 1987). Most marine animals function over a wide pressure range, within certain limits. For example, most abyssal animals do not function well at near surface pressures.

There is some evidence from net tows in the open ocean to suggest that the range of a given species is determined more by temperature than by pressure. These observations have demonstrated that many types or organisms which live near the surface at high latitudes live at greater depths at lower latitudes, where surface temperatures are higher (Neshyba, 1987).

1.6.3.3 Water masses

Faunal assemblages are correlated with distinct water masses. The major surface currents (see Section 1.4.1) outline the biological provinces in the open ocean.

Upwelling or vertical mixing returns nutrients which have sunk below the photic zone back to the surface. Because of the abundance of nutrients, areas of upwelling have the greatest production of organic matter and support the largest amounts of marine life. Areas of upwelling are associated with areas of divergence. For example, in the Indian Ocean, divergence and upwelling are caused by monsoon winds. In the open ocean, upwelling is associated with divergences along the Equator and at the border between the Antarctic and mid oceanic current gyres. Coastal upwelling is associated with the eastern margins of ocean basins.

Each current gyre supports a distinct assemblage of organisms. The subpolar gyres are characterized by variable light intensity and high levels of nutrients. They contain relatively few species, but have a large biomass. Upwelling occurs at divergences within the gyres.

The subtropical gyres have a relatively constant water temperature and light intensity and are low in nutrients. They contain a large number of species, but the biomass is low. Nutrients are slowly replenished by inward transport of the nutrient rich waters from areas of upwelling at the margins of the gyres.

The abundance of organisms in the equatorial zone is variable. The zone contains large numbers of species and an intermediate to high biomass. This is partly due to upwelling along the Equator.

The phytoplankton, which are the primary producers, tend to be widespread throughout the oceans. They appear to be spread by currents from one gyre to the next and seem to be able to survive unfavourable conditions. The most important limit to their distribution appears to be the decrease in light with depth.

On the other hand, zooplankton, the minute marine animals, are much more restricted in their distributions. Many open ocean zooplankton species occur only in a single gyre. A few species are restricted to particular current systems. Mixing of zooplankton from different current systems occurs in the western boundary currents. These areas contain the highest diversity of species.

Biological distribution data can be a valuable aid in the interpretation of circulation. There is a strong correlation among species found in samples taken from a single water mass, regardless of the latitude at which that mass is sampled. The main patterns of species distribution in the open ocean are repeated in all taxons and at all trophic levels.

Zooplankton are some of the most useful indicator species. This is because they have a relatively long lifespan, are fairly easy to capture and have a wide vertical distribution and limited motility.

1.6.4 Effect of organisms on marine technology

Organisms affect marine technological systems in many ways. For example, marine invertebrates may bore through man made structures and lead to fouling, corrosion and fatigue failures. Marine vertebrates have been known to bite and sever underwater cables and mooring ropes. Even the noises made by animals, such as the snapping of shrimps and the clicks and whistles of dolphins may affect the performance of communications systems.

1.6.4.1 Fouling

Fouling occurs when marine organisms attach themselves to marine structures. The pattern of fouling is variable. It is dependent on water depth, geographical location, the species of animals and plants present, salinity, temperature, season and the amount of light reaching the surface of the marine structure. Where organisms find favourable conditions, fouling can be a severe problem. Some types of fouling are seasonal, but many types occur throughout the year.

In general fouling begins when slime composed of bacteria and diatoms accumulates on a structure. Weeds grow from spores and become trapped in the slime. The commonest weeds are the green algae *Enteromorpha sp.* and the brown algae *Ectocarpus sp.* (British Ship Research Association, 1972). Finally, animals such as barnacles, tube worms and hydroids become attached.

Fouling can occasionally serve as a protection against corrosion; for example, it can help to prevent corrosion of sheet steel piles (Chandler, 1985). However, more often fouling leads to increased corrosion and may cause pitting in passive–active alloys such as stainless steel.

Fouling also causes many other types of problems in the marine industry. Fouling on ships' hulls may lead to increased drag and hence to increased fuel consumption in order to maintain speed. The weight of marine organisms may affect the operation of offshore structures. Some types of organisms settle inside pipes and tubes and cause problems by restricting the flow of sea water.

Because fouling is such an important threat to the success of many marine industries, some attempts have been made to map its occurrence. In addition, much research has gone into the development of antifouling paints, to discourage the settlement of organisms on the surfaces of marine structures (British Ship Research Association, 1972).

1.6.4.2 Biological corrosion

Biological processes can affect corrosion in several ways. They can influence the nature or rate of corrosion by their effect on anodic or cathodic process. Their metabolic processes can create acids which can increase corrosion. Organisms can also produce deposits or films which affect surface reactions.

Bacteria, particularly sulphate reducing bacteria, have an important influence on marine structures because they are able to cause corrosion of steel in the absence of oxygen.

However, other organisms, including algae, iron oxidizing bacteria and slime forming bacteria are also important. The effect of organisms on corrosion is discussed further in Section 2.3.10 of Chapter 10.

1.7 Waves

Waves are disturbances of the water surface. They are an important feature at the interface between the ocean and the atmosphere. They affect the transfer of momentum, energy, heat, moisture, particles and gases between the two media and promote mixing, bubbles and spray. They also induce mixing, which results in evaporation and increased gas solubility. Waves also provide a means of energy transfer to the surface layers of the oceans.

Waves affect most aspects of marine technology. An understanding of waves is essential for the design and construction of coastal structures, such as jetties and docks, offshore structures and marine vehicles. Knowledge of wave behaviour and its effects is also critical for salvage and rescue operations. Waves are discussed further in Chapter 2.

1.7.1 Characteristics of waves

In order to understand the motion of waves, an idealized wave is often described. Although these simple uniform waves are rare in the ocean, they provide a useful way of analysing the more complicated waves found in nature.

The main features of an idealized wave are shown in *Figure 1.9*. The highest point of the wave is called the crest and the lowest part the trough. The vertical distance between the crest and the succeeding trough is the wave height (H). The horizontal distance between

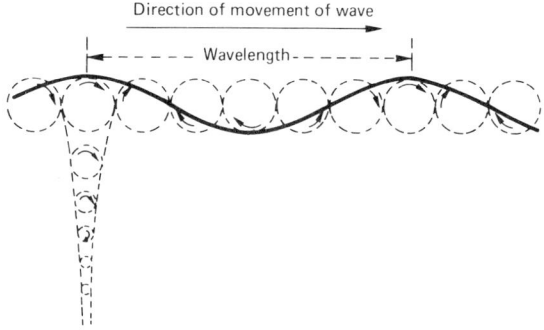

Figure 1.9 An idealized wave (from Tait, 1981, Butterworth; reproduced with permission)

successive troughs or successive crests is the wavelength (L). The time taken for successive crests or troughs to pass a fixed point is the wave period (T). The wave frequency ($1/T$) is the number of wave troughs or wave crests which pass a fixed point in a given length of time. The speed of the wave (C) is equal to L/T.

When the wave height is low relative to the wavelength, the crests and troughs tend to be rounded. They may be approximated mathematically by a sine wave. As the wave height increases, the crests become steeper and are more similar to trochoid curves.

Wave stability is measured by the wave steepness (wave height/wavelength (H/L)). When wave steepness is greater than about 1:7, waves become unstable and break.

In deep water, as waves pass, the water mass moves in almost closed vertical circular orbits. The water is moved forward as the crest passes, then downward and finally backward with the trough. There is a very small amount of net forward displacement of the water with each wave.

The water at the surface moves in nearly stationary circular orbits whose diameter is approximately equal to the wave height. As the water depth shallows, the diameter of the circular orbits at the surface decreases. At depths below half the wavelength, there is little water movement due to waves.

Deep water waves, which form in water which is deeper than half the wavelength, are scarcely affected by the sea bed. However, shallow water waves, especially where water depths are less than 1/20 of the wavelength, are strongly affected. In shallow water waves, horizontal back and forth motion occurs near the bottom, and water movements at shallower depths are in the form of elliptical orbits, rather than the circular orbits characteristic of deep water waves.

Waves can also be characterized by the types of forces which cause them. Forced waves are waves where the disturbing force is applied continuously. The tides are examples of very long wavelength forces waves. In contrast, free waves move independently of the force which caused them. An impact wave, generated by a volcanic explosion or a submarine landslide is one example. Winds are the most common disturbing force which generates waves. Wind generated waves have characteristics of both free and forced waves.

Waves may also be classified by the forces which act to restore them to an equilibrium, or still water, condition. Waves with wavelengths less than 1.7 cm (0.68 in) and periods less than 0.1 s, are known as capillary waves. For these waves, surface tension at the ocean surface is the main restoring force. Gravity is the dominant restoring force for waves with periods between 1 s and 5 min. Gravity waves are the most common waves seen in the ocean.

Very long waves with periods greater than 5 min, such as tides and seismic sea waves (tsunamis), act like shallow water waves because of their long wavelength in relation to the depth of the ocean. These very long waves are affected both by gravity and the Coriolis effect.

1.7.2 Types of waves

1.7.2.1 Ocean waves

The energy of the wind forms waves on the ocean surface. Gentle breezes form ripples on the surface of the sea. These provide a rough surface which allows the wind to move the water. If the wind dies, ripples disappear very quickly. However, if the wind continues to blow the ripples are gradually transformed into larger waves. These newly formed waves are generally short and choppy. As the wind continues to blow, the waves gather energy through the pushing and dragging effect of the wind on the sea surface. Choppy seas provide a roughened surface, which carries more momentum from the wind than do smooth swells. The roughness of the sea, the specific wave form and the relative speed of the wind and waves control how much energy will be gained by the waves. When waves are first formed in the sea, they are steep, chaotic and sharp crested.

The size of the waves which are formed depends on the wind speed, the time that the wind blows in a constant direction and the fetch, or the distance over which the wind blows in a constant direction. The largest waves are formed by strong steady winds blowing over long periods of time in the same direction over large bodies of water.

Ocean waves reach a maximum size when the energy supplied by the wind is equal to the energy lost by the breaking waves. This is known as a fully developed sea. The waves break when they reach their theoretical limit of stability which occurs when the ratio of wave height/wavelength reaches 1:7. These oversteepened waves break in order to dissipate excess energy provided by the wind. They can also break when their tops are blown off by the wind.

As waves develop, their speed eventually exceeds the wind speed. When this happens, the steepness of the wave decreases as the wavelength increases. When waves travel out of the generating area, or the wind dies, they are transformed into swells, smoother waves with longer crests and periods. Long period swells can travel for long distances because they do not lose very much energy due to viscous forces. For example, swells which originated in Antarctic storms have been detected as far away as the Alaskan coast.

1.7.2.2 Shallow water waves

As waves approach the coast where water depths are shallower, they begin to be affected by the sea bottom. At this point they change from deep water to shallow water waves. The wavelength and speed decrease, but the wave period remains constant. After an initial slight decrease, wave height increases rapidly as water depths decrease and wave crests move closer together.

As the waves enter the shallower water, the part of the wave still in deeper water moves faster than the part which has entered the shallower water. As a result, the crests of the waves in shallow water rotate so that they are more parallel to the bathymetric contours, a process known as wave refraction. This is why breakers are nearly parallel to the coastline when they reach the beach, even though the waves may have originated from many different directions.

When waves enter shallow water their wavelength decreases but the wave height and consequently the wave steepness increases. When the wave height reaches about 0.8 of the water depths, the wave breaks (*Figure 1.10*). Surf is a belt of nearly continuous breaking waves along a shore or over a submerged bank or bar.

The breakers in a surface zone are of several different types. This is because they result from the interaction of different waves types with an irregular bottom topography. Spilling breakers are oversteepened waves where the unstable top spills in front of the wave. They are concave on both sides of the crest and when the crest breaks, the wave form continues. Spilling breakers are formed over a gently sloping sea bottom when the wave form advances but the wave energy

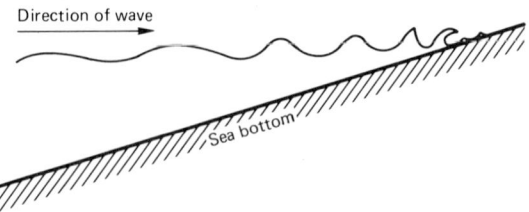

Figure 1.10 Changes of waveform on entering shallow water. At a depth of about half the wavelength, the waves become closer and higher. With decreasing water depth, the fronts of the waves become steeper until they are unstable and their crests topple forward (from Tait, 1981, Butterworth; reproduced with permission)

gradually decreases. In plunging breakers the wave crest curls over and forms a large air pocket. The wave form is convex behind the crest but concave in front of it. Plunging breakers form when small swells move over a gently sloping bottom with irregular topography. Surging breakers, like plunging breakers, are convex behind the crest and concave in front of it. They form over a steeply sloping topography. A wall of water moves back as the wave retreats.

The height of the surf is related to the height and steepness of the waves offshore and to the offshore bottom topography. Surf can be very low in protected areas but can reach heights of up to 20 m (65 ft) at exposed points (Gross, 1972).

Waves generate currents and turbulence along beaches and are responsible for the seasonal changes seen on beaches. The waves stir up sediment which is transported parallel to the coast via longshore currents generated by the waves and tides. During periods of low, long period swells, which are more typical of summer, sand is accreted onto beaches which then increase in height and width. During winter, when high choppy waves are more common, beaches are eroded.

Waves also cause erosion of cliffs. Waves breaking at the base of cliffs often carry gravel and sand from the sea floor. Even the hardest cliff faces are eventually eroded by this continuous abrasion. Soft rocks and sediments are eroded very quickly by waves. The amount of erosion of these types of materials is probably more dependent on the capacity of inshore currents to remove the products than on the rate of erosion.

1.7.2.3 Seiches

Seiches are stationary waves which play an important role in tidal phenomena. In a standing wave, part of the water surface does not move vertically and acts as a hinge about which the rest of the water surface tilts. The node is the part of the wave which remains stationary and the parts of the water which undergo the greatest vertical movement are the antinodes.

In seiches, the water flows in one direction for a certain period, then reverses and flows in the opposite direction. The wave form alternately appears and disappears and the water moves back and forth rather than in the nearly circular orbit charcteristic of progressive waves. Seiches occur in steepsided basins such as the Red Sea.

1.7.2.4 Internal waves

Internal waves are, in principle, similar to surface waves, but they occur within the ocean rather than at the ocean surface. Internal waves may be formed at any region in which the density varies with depths. They are often most clearly apparent at an interface between water layers of different densities, such as the pycnocline. They may cause mixing below the surface.

Internal waves can cause bending of surface waves such as breakers by producing currents at the sea surface.

The surfaces where internal waves form have only small differences in density. As a result, because the energy they contain is proportional to the density gradient, internal waves can have much greater wave heights than surface waves, which occur at a boundary which marks much greater density differences. They may reach 100 m (325 ft) in height and produce currents of 10–50 cm/s (0.4–20 in/s). Internal waves generally travel much more slowly than surface waves, typically at speeds of 0.3–2 m/s (0.98–6.5 ft/s), and usually have less energy (Gross, 1972).

1.7.2.5 Tsunamis

Tsunamis are caused by sudden movements of the ocean bottom due to earthquakes or volcanic eruptions. They have very long periods and behave like shallow water waves, even when they pass through the deep ocean. They travel at high speeds and in deep water generally exhibit very low wave crests. In the open seas, they are often not detected by shipping. However, when they encounter a coast they sometimes rapidly steepen to form a rapidly moving wall of water which may be very destructive. At other times the principal effect of a tsunami may be general coastal flooding.

1.7.2.6 Storm surges

Storm surges are long period surface waves caused by strong winds. In a storm surge the winds drive water along a coast and cause the sea level to rise. This causes flooding of low lying coasts.

The approximate height of a storm surge can be predicted on the basis of the wind speed and direction, fetch, water depth and shape of the ocean basins. Currents, tides and seiches set up by storms complicate the calculations and make accurate predictions difficult.

Some storm surges take the form of a large wave which moves with the storm which caused it. Initially the sea level at a coastline in front of the storm surge may fall slightly. However, as the centre of the storm passes, there is a sharp rise in water level (the surge) which lasts 2.5–5 h. After the storm passes, sea level continues to rise and fall as oscillations set up by the storm pass. These oscillations are similar to free surface waves. Because they occur after the storm has subsided they are often difficult to predict and can cause a good deal of damage on low lying coasts.

1.8 Regional oceanography

Oceanographic principles and processes operate in all regions of the oceans, but the effects differ from place to place. This is primarily due to obvious differences in the character of the various oceanic zones. The deep oceans differ considerably from coastal or shallow marine regions. Tropical oceans respond to various phenomena differently from polar regions. There are also some semi-enclosed seas which have distinct characteristics of their own.

Because the oceans affect and are affected by climatic conditions on land, marine regions closer to land generally respond more rapidly to changes in various parameters, such as temperature and salinity, than do ocean areas farther from coasts.

1.8.1 Marginal ocean basins

Marginal ocean basins are transitional between continents and deep ocean basins. They are nearly all located in areas of active crustal deformation. They receive the discharge from the rivers on the adjacent continents which contain large amounts of sediment. As a result, marginal seas contain about one-sixth of all known ocean sediments. Because they are adjacent to large land masses, marginal seas are subjected to greater extremes of climate than is the open ocean (see Section 1.2.2).

1.8.2 The deep ocean

Deep ocean regions are characterized by uniformity of conditions over great distances. There are well recognized water masses that cover thousands of square miles of surface area and extend to considerable depths. Ocean currents in the deep ocean travel quite slowly and, as a result, mixing processes are generally ineffective in producing rapid changes in conditions such as sea surface temperature and surface salinity. Salinity changes in the range of hundredths of parts per thousand are the rule in most ocean areas at considerable distances from land. Tidal effects are of little consequence in deep ocean regions. This further reduces mixing processes.

The time scale of changes in deep oceans is generally of the order of a thousand or more years, although there are some recognizable seasonal changes in temperature zones. However, these changes are limited to the surface zone and rarely extend as deep as the lower part of the thermocline. Below the thermocline, there are hardly any seasonal changes in the deep water.

Deep oceans, on the other hand, are generally lacking in nutrients and have sometimes been referred to as ocean deserts. Unlike coastal waters, the deep ocean is not well oxygenated throughout its depth. The beautiful deep blue colour of deep oceans indicates low biological production.

1.8.3 The coastal ocean

It is difficult to establish a seaward boundary to the coastal ocean, but as a general rule waters above the continental shelf might be considered coastal. Because the distance from the land to the outer edge of the continental shelf differs considerably throughout the world, no specific distance from land can be used to delineate the boundary between coastal and deep ocean areas.

In some regions, such as the North Atlantic Ocean, other oceanographic characteristics provide a recognizable boundary. The characteristics of Gulf Stream waters differ greatly from inshore waters. Gulf Stream waters are warmer and more saline. They have a characteristic deep blue colour which clearly distinguishes them from coastal waters. To a certain extent, the same differences are found along the east coast of Japan, where the Kuroshio Current marks the beginning of the deep ocean. The waters landward of the Kuroshio are coastal.

Processes in coastal oceans, particularly those in temperate climatic zones, operate at far shorter time scales than they do in the deep ocean. Tidal cycles are short, only 6 h for semidiurnal tides, and mixing processes are rapid. Conditions near shore can change dramatically with a change in the tide. Rising flood tides bring oceanic water into bays and shallow coastal areas. Falling ebb tides spread less saline waters, which have been brought into the area by river flow, farther out to sea.

Longer cycles, such as those that result from seasonal changes in temperature zones, are also effective in changing oceanic characteristics. Summer air temperatures warm coastal waters fairly rapidly, and melting snow in the mountains feeds streams, resulting in an abundance of fresh water flowing into the coastal ocean. This results in a decrease in both salinity and sea surface temperature.

In some coastal regions the seasonal outflow of river water results in well defined coastal currents, which are distinctly seasonal. The Davidson Current, off the coast of California, is an example. This current flows inshore of the California Current in the late spring and summer, but is absent in the winter months. It flows northwards, in contrast to the more continuous California Current, which flows southwards.

Because coastal waters are shallow, mixing processes frequently extend from the surface to the bottom. Hence, the thermocline may disappear in coastal waters under some circumstances. This is particularly true in the winter months when storms are more frequent and mixing from wave action is more effective. During the summer, the surface waters become warmer, mixing becomes less effective and the thermoclines re-establish themselves.

Both the rise and fall of water due to tidal effects and the horizontal movement of water in the tidal currents have considerable effect on coastal oceans but are virtually absent in the deep ocean. Coastal current velocities can be very great, much more rapid than the movements of the deep ocean currents. Where tidal currents are funnelled through narrow straits, speeds of

18 km/h (11.25 mph) or more have been measured. On the other hand, typical ocean current speeds in the deep ocean may be measured in tenths of knots, although places in the Gulf Stream may flow at 5–7 km/h (3.13–4.38 mph).

Salinities in coastal oceans may vary from close to zero at the mouths of large rivers, to 40 or more parts per thousand in regions where evaporation greatly exceeds precipitation. In areas such as the Red Sea and the Persian Gulf, coastal waters are characterized by very high salinities and sea surface temperatures. In some coastal regions in the Arctic, low salinity water is characteristic of the summer months when sea ice melts and river flows become more active.

Biological productivity is generally high in coastal oceans as compared to deep oceans. There is usually an ample supply of nutrients provided by run off from the land. The waters are frequently well oxygenated from the surface to the bottom. The characteristic turbid appearance of waters close to land and the generally green colour indicates the presence of large quantities of phytoplankton and zooplankton.

The most productive commercial fisheries are found close to land, in coastal waters above continental shelves or offshore banks. These areas contain both pelagic species, which are sometimes of commercial importance, and species which live at or near the sea bottom. Except for tuna, which inhabit deep ocean near surface waters, commercial fisheries do not exist in the deep ocean.

1.8.4 Estuaries

Estuaries are particular features of some coastal oceans. They are defined as semi-enclosed bodies of water having a free connection with the open sea and within which sea water is diluted measurably with fresh water run off. Coastal plain estuaries frequently exist at the mouths of large rivers. Coastal oceans are affected by fresh water outflow, which can sometimes be detected for many miles seaward. Similarly, the mouths of rivers frequently contain measurable quantities of salt water of oceanic origin.

Circulation in estuaries is often complex. In the simplest type, the salt wedge estuary, fresh water of river origin flows seaward at the surface and oceanic water flows landward below the outgoing river water. The salt water moves as a wedge. Although this model implies that there is no mixing between surface fresh water and the wedge of salt water, this is rarely the case.

This is because the velocity of the outflowing river will cause some turbulence. In addition, where the tidal range is great, additional mixing takes place. In some cases, the mixing is complete enough to break down the vertical stratification. In the Northern Hemisphere, this often results in an inflow of sea water on the right side of the estuary and an outflow of the river water on the left side, rather than a surface outflow and a deep inflow. This pattern is reversed in the Southern Hemisphere. The salinity is then constant from top to bottom, but the horizontal gradient differs dramatically.

1.8.4.1 Fjords

Fjords are particular types of estuaries produced from glacial scouring which are found in high latitudes. In terms of ocean circulation, the chief characteristic of a fjord is the presence of a shallow sill at the seaward end. Fresh water outflow is limited to depths above the sill, while incoming salt water flows over the sill and fills up the basin behind the sill. The salt water of oceanic origin frequently becomes stagnant because its circulation is limited by the shallow sill depth. In many cases, the deep water in the fjord is very deficient in oxygen. When it is brought to the surface by an occasional storm, it has a distinctly rank odour.

1.8.4.2 Circulation in estuaries

Circulation in estuaries, including fjords, is controlled by the velocity of the river flow, the degree of tidal mixing and the width and depth of the river. Because the width, depth and flow of the river vary seasonally, the characteristics of the estuary may change considerably. For instance, it has been noted that in the pronounced drought in the midwestern USA in 1988, the salinity of the Mississippi River increased notably as the water flow, width and depth of the river decreased. This led to an increase in salinity in the estuary and there were consequent biological effects.

1.8.5 The surf zone

The narrow area within which waves are translated into surf is an important subregion of the coastal ocean. When waves enter water shallow enough to change the oscillatory motion of water particles so that the mass movement of the water is towards land, a surf zone is created.

The surf has important effects on land, particularly on sandy beaches. Beaches are normally subject to erosion during the winter months when waves are created by local storms. These short wavelength, steep waves produce powerful surf which erodes beaches. Consequently, in temperate latitudes beaches are commonly narrower in the winter months and occasionally they disappear completely. On the other hand, in the summer, the waves which arrive have a long period and a low swell. The breaking swell tends to transport sand from offshore areas onto the beach. This widens the beach and results in the summer beach profile.

Although waves are subjected to refraction as they enter shallow water, the wave crests are rarely lined up exactly parallel to the land when they break. When the direction of the swell or surf is at an acute angle to the direction of the coast, longshore currents are set up. These currents move sand which has been eroded from the beach down current, where it is eventually deposited on another beach. Thus up-current beaches erode while down-current beaches accrete. The effect of longshore currents is more or less continuous erosion of beaches. This creates problems in beach areas where structures have been built.

The surf zone is also subjected to the effects of storm waves and tsunamis. When tsunamis or storm surges arrive at high tide, the damage to coastal structures can be considerable because areas not exposed to the sea are affected.

1.8.6 Equatorial and tropical oceans

Ocean waters located close to the Equator are characterized by great uniformity of sea surface temperatures because the air temperatures in these regions are uniformly warm and vary little from winter to summer. On the other hand, salinities in some regions where evaporation greatly exceeds precipitation can be considerably above the norm for open oceans. For example, in some parts of the Red Sea and the Persian Gulf, salinities greater than 40 parts per thousand are common. In other tropical seas, salinities may be somewhat lower than average, due to large amounts of rain. Thus, in terms of surface water salinities, tropical oceans can be divided into the wet tropics and the dry tropics.

Surface waters in the tropical and equatorial regions are characterized by high sea surface temperatures and a shallow thermocline. Surface waters are generally not well mixed because winter weather does not act to cool and sink the surface water. Unless there are violent storms, the warm surface water forms a shallow zone above the rapidly cooling water in the thermocline.

There are some exceptions. For example, where Trade Winds with their generally high wind velocities prevail, the surface layers are much better mixed and the thermocline is deeper.

Equatorial and tropical regions frequently have monsoon climates which are characterized by a rainy season and a dry season. In nearshore areas, the effect of the rainy season lowers surface salinities considerably but these effects are barely noticeable in the deep tropical and equatorial oceans.

The winter and summer monsoons are also characterized by different wind directions. Because ocean currents are frequently wind driven, particularly in the tropics, current patterns reverse from winter to summer. *Figures 1.11* and *1.12* illustrate the effects of both winter and summer currents and generally uniform sea surface temperatures in south-east Asian seas.

1.8.7 Polar oceans

Oceanic regions in polar climates are characterized by uniformly low temperature water and the presence of sea ice during certain seasons of the year. Although sea surface temperatures are always low, seasonal conditions are readily noticeable. Continuous sunlight during the summer melts the sea ice and warms the surface water. This has dramatic effects on the local ecosystems. Biological productivity in the summer can be very high, but it is virtually absent in the winter. The increased availability of light for photosynthesis, coupled with high nutrient levels in some areas results in a pronounced spring phytoplankton bloom.

The variable extent of sea ice in winter and summer and the effects of freezing and thawing affect salinity. When sea water freezes, high salinity brine is trapped in pockets in the sea ice, which is actually frozen pure water. Sea ice typically has a salinity, including the entrapped brine pockets, of 7 parts per thousand *versus* a normal average salinity of 35 parts per thousand. Thus, the freezing of sea water causes the surrounding water to have a greater than average salinity. This surrounding water also has a low temperature. As a result it is denser than the surrounding water and sinks. When melting occurs, the low salinity melt water is less dense than the surrounding water and floats as a thin, almost pure water layer on the surface.

These general characteristics of polar oceans: uniformly low temperatures, the presence of sea ice and density differences due to freezing and thawing, are common to both the Arctic and Antarctic oceans. However, the two oceans differ considerably in a number of ways.

1.8.8 Temperate oceans

The North and South Atlantic, North and South Pacific and the central and southern parts of the Indian Ocean are characterized by distinct seasonal changes in some oceanographic parameters. The seasonal effects of cold winters and warm or hot summers affect sea surface temperatures. As a consequence, in summer the combination of warm air temperatures and the lack of strong winds and storms creates a sharp, shallow thermocline. The sea surface temperature is relatively high and the surface layer does not extend to any great depth because mixing effects are slight. The reverse is true in winter, when surface water is cooled and storms are more abundant. This combination deepens the thermocline and there is a tendency for the surface water to cool to the same temperature as the deeper water below. Under extreme winter conditions in shallow areas, the thermocline virtually disappears. Conditions in the spring and autumn are somewhere in between the extremes of winter and summer.

Salinities vary little from winter to summer in deep temperate oceans far from land. However, they may show some slight effect due to the excess of evaporation over precipitation in the dry season and the reverse during the rainy season.

Water movements are generally sluggish at the centres of temperate oceans and the relatively constant atmospheric high pressure systems influence the water characteristics.

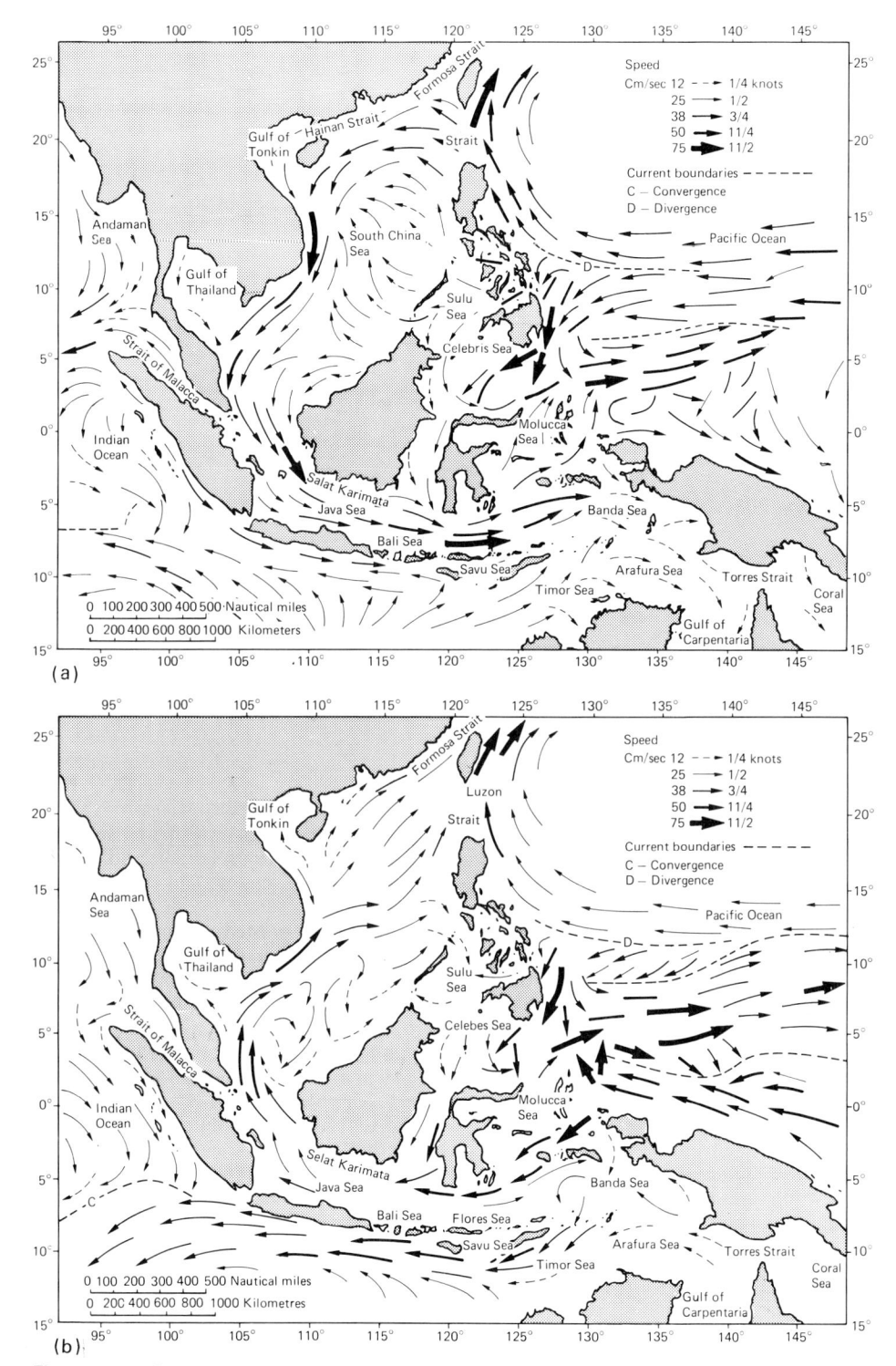

Figure 1.11 Surface currents in south east Asia in (a) February and (b) August (redrawn from Morgan, J.R. and Valencia, M.J., 1983, University of California Press: Berkeley California; reproduced with permission)

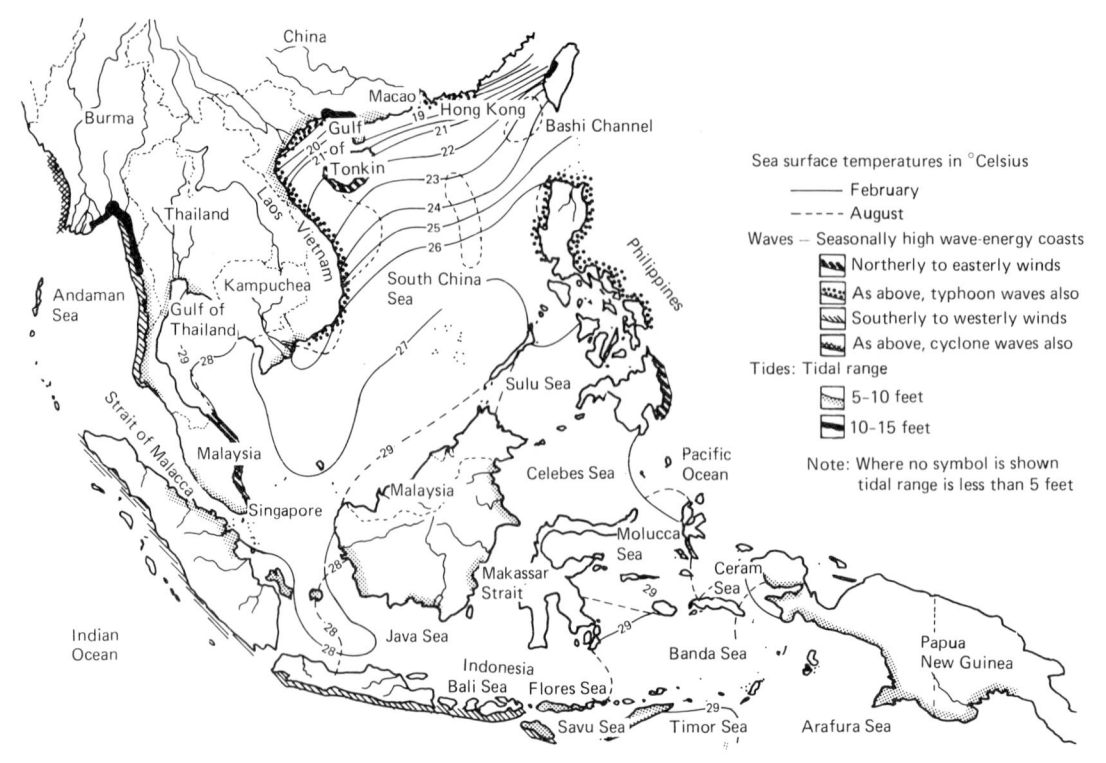

Figure 1.12 Sea surface temperatures and significant waves and tides in south east Asia (redrawn from Morgan, J.R. and Valencia, M.J., 1983, University of California Press: Berkeley, California; reproduced with permission)

Although the Mediterranean and the Black Seas are both parts of the North Atlantic Ocean, their physical oceanography differs greatly from that of the North Atlantic as a whole. They are discussed below.

1.8.8.1 The Mediterranean

The Mediterranean is a shallow inland sea of the North Atlantic, which occupies a depression in the Earth's surface between Europe and Africa. It is virtually cut off from the North Atlantic by the narrow and shallow Strait of Gibraltar. Its only other outlets are through the Turkish Straits to the north-east and Suez Canal to the south-east. Water moves generally from the Black Sea into the Mediterranean at the surface and into the Black Sea at depth. The flow into and out of the Red Sea through the Suez Canal is negligible.

Although the Mediterranean is complicated in detail, its general characteristics are relatively simple. It is basically divided by the Strait of Sicily, a shallow ocean ridge south of Sicily, into Eastern and Western basins. The Eastern Basin is deeper with depths of over 3 km (1.88 miles) in places and is divided by the Mediterranean Ridge System, which separates it into a number of smaller basins and intervening troughs. The Western Basin is more uniform in depth. It consists of the Balearic and Tyrrenian Basins which are separated from each other by the islands of Sardinia and Corsica. The Balearic is the more shallow and regular of the two.

The climate in the eastern Mediterranean is typically hot and dry. There is a substantial excess of evaporation over precipitation which results in high salinites. The increased salinity increases the density of the water in the eastern Mediterranean causing it to sink to the bottom. It fills the Eastern Basin and eventually overflows the sill at the Strait of Sicily. The high salinity water then moves into the Western Basin and fills it up to the depth of the sill at Gibraltar. The deep saline water then overflows the sill and enters the Atlantic as a tongue of highly saline warm water. Its characteristic temperature and salinity give it a density intermediate between the North Atlantic surface water and the colder water below. The Mediterranean water can be traced for hundreds of miles into the Atlantic as an intermediate water mass.

The ratio of evaporation to precipitation in the eastern Mediterranean is approximately 4:1. Hence, dense water is being formed virtually continuously and at a rapid rate. The outflow of deep Mediterranean water is compensated for by an inflow of less saline water at the surface from the Atlantic. Thus there is a rapid flow of Atlantic water into the Mediterranean through the Strait of Gibraltar and an equally swift outflow of Mediterranean water. The flushing of the Mediterranean is estimated to occur roughly every 75 years.

Although this is a remarkably rapid rate of flushing for such a large body of water, the Mediterranean is in danger of becoming excessively polluted and valuable fishery resources are endangered. The seeming inconsistency between a rapid exchange of water and excessive pollution, which is usually associated with insufficient water movement, can be explained by the fact that great quantities of industrial pollution and sewage are being dumped into the sea by the inhabitants of the surrounding land masses. In addition, the 75 year flushing rate refers to the entire sea. The complicated nature of the land–sea boundaries produces bays, inlets and other bodies of water which may act as cul de sacs. The northern parts of the Aegean and Adriatic Seas, for instance, are not as likely to benefit from the high exchange rate of Mediterranean and Atlantic water as the Ionian Sea and the Strait of Sicily which are part of the main Mediterranean. The small tidal range, about 30 cm (1 ft) also contributes to the pollution problem because it prevents flushing of small bays and inlets.

1.8.8.2 The Black Sea

North east of the Mediterranean is a smaller body of water with completely different oceanographic conditions. The Black Sea is bounded by Turkey to the south, Bulgaria to the west and the southern part of the Soviet Union to the north and east. Its only outlet is through the Bosporus to the Sea of Marma, then through the Dardanelles to the Aegean Sea between Greece and Turkey. A broad continental shelf underlies the northern Black Sea and a 600–900 m (0.38–0.56 miles) deep basin is found in its southern part.

Because precipitation greatly exceeds evaporation in the Black Sea there is no production of deep saline water. Instead, a layer of lighter, less saline water is produced. This overlies the deeper water, which originates in the Mediterranean. The deep water is stagnant. Occasionally a storm will create an overturn of the deep and surface water. When the deep water is brought back to the surface it can have a noticeable odour, indicative of its lack of dissolved oxygen. Sea life in the Black Sea is generally limited to the better oxygenated surface water layer.

1.9 Conclusions

This chapter provides a broad outline of the nature and physical properties of the oceans. The oceans encompass a wide range of environments and conditions which provide challenges to the ingenuity and skill of marine technologists of all disciplines. The chapters which follow discuss in more detail how these challenges are being met.

References and further reading

Alexandrou, D. and De Moustier, C. (1988) Adaptive noise cancelling applied to sea beam sidelobe interference rejection, *Ocean Engineering*, **13**, 70–76

British Ship Research Association (1972) *Recommended Practice for the Protection and Painting of Ships*, British Ship Research Association, Wallsend, UK

Chandler, K.A. (1985) *Marine and Offshore Corrosion*, Butterworth, London

Clayton, B.R. and Bishop, R.E.D. (1982) *Mechanics of Marine Vehicles* E & F.N. Spon Ltd, London

Gardline Surveys (1988) *Company Profile*, Gardline Surveys, Great Yarmouth, UK

Gross, M.G. (1972) *Oceanography, A View of the Earth*, Prentice-Hall, Englewood Cliffs, New Jersey

Hill, G.W. and McGregor, B.A. (1988) Small-scale mapping of the Exclusive Economic Zone using wide-swath side-scan sonar, *Marine Geodesy*, **12**, 41–53

Ludman, A. and Coch, N.K. (1982) *Physical Geology*, McGraw-Hill, New York

Lutgens, F.K. and Tarbuck E.J. (1982) *The Atmosphere*, 2nd edn, Prentice-Hall, Englewood Clifs, New Jersey

Neshyba, S. (1987) *Oceanography: Perspectives on a Fluid Earth*, Wiley, New York

Morgan, J.R. and Valencia, M.J. (eds) (1983) *Atlas for Marine Policy in Southeast Asian Seas*, University of California Press, Berkeley

Neumann, G. and Pierson, W.J. Jr (1966) *Principles of Physical Oceanography*, Prentice-Hall, Englewood Cliffs, New Jersey

Skinner, B.J. and Porter, S.C. (1987) *Physical Geology*, Wiley, New York

Sonnenburg, G.J. (1988) *Radar and Electronic Navigation*, 6th edn, Butterworth, London

Tait, R.V. (1981) *Elements of Marine Ecology*, 3rd edn, Butterworth, London

Terrett, F.L. (1989) Coastal and maritime engineering. In Blake, L.S. (1987) *Civil Engineer's Reference Book*, 4th edn, Butterworth, London, pp. 31/1–31/35

Tolmazin, D. (1985) *Elements of Dynamic Oceanography*, Allen and Unwin, London

Williams, J. (1962) *Oceanography: An Introduction to Marine Sciences*, Little, Brown, Boston

2 Offshore Structures

Minoo H Patel

Contents

2.1 Introduction
2.2 Fixed offshore structures
    2.2.1 Types of rigs
    2.2.2 Drilling depth capability
    2.2.3 Types of structures
    2.2.4 Selection of platform type
    2.2.5 Cost considerations
    2.2.6 New design concepts
    2.2.7 Other equipment
    2.2.8 Subsea engineering
    2.2.9 Subsea maintenance
2.3 Environmental loading
    2.3.1 Introduction
    2.3.2 Concepts of fluid mechanics
    2.3.3 Viscous forces and the Reynolds number
    2.3.4 Wind and current loads
    2.3.5 Wave loads on offshore structures
    2.3.6 Vortex shedding induced loads
    2.3.7 Seabed proximity and slamming
    2.3.8 Wave loads on large bodies – diffraction theory
2.4 Structural response of fixed structures
    2.4.1 Quasi-static loading and response
    2.4.2 Dynamic response
    2.4.3 Structural failure and fatigue life calculations
    2.4.4 Design and certification requirements
2.5 Floating offshore structures
    2.5.1 General characteristics
    2.5.2 Hydrostatic analysis and stability
    2.5.3 Dynamic response analysis
    2.5.4 Effects of moorings
2.6 Analysis with multiple degrees of freedom
    2.6.1 Basic equations
    2.6.2 Coefficient matrices and waveforces

References and further reading

2.1 Introduction

Over the last five decades, traditional uses of the oceans, such as transportation and fishing, have expanded to include the exploitation of hydrocarbons below the sea bed and the potential for large scale mineral gathering and energy extraction. In the long term man's utilization of the oceans as a resource will stimulate engineering developments in other areas.

The design of offshore structures used for oil and gas production offers some indication of the technical problems which will have to be faced. For example, these structures encounter largely steady environmental forces from wind and current flows and their own weight. They are also subjected to high levels of cyclic load due to gravity waves. The steel with which they are fabricated will have to survive sea water corrosion for long periods because large parts of the structures will be difficult to access and maintain. The associated equipment on the sea bed must be capable of functioning in the presence of very high pressures at depth.

These problems are compounded by the uncertainties of predicting the most extreme environment likely to be encountered by the structure as well as the uncertainty of categorizing the geology and recoverable resources of the hydrocarbon reservoir that the steel structure is designed to develop. All of these interacting problems offer unique challenges for advanced scientific analysis and engineering design.

It is difficult to predict exactly how the resources of the oceans will be exploited in the future, but the design, construction, installation, operation and maintenance of structures to carry out activities in the oceans will continue to provide challenging engineering problems. Such developments will need to be supported by a better physical understanding of the mechanics of the environment, aerodynamic and hydrodynamic loadings on structures, their consequent static and dynamic response, the material behaviour of the structure and a multitude of other aspects. This chapter gives an introduction to some of the technologies that will need to be developed further, in order to design effective structures for working in the oceans.

2.2 Fixed offshore structures

2.2.1 Types of rigs

2.2.1.1 Jack up rigs

The technical evolution of the modern offshore industry can be measured by the depth at which it has been able to carry out exploration drilling and by the structures or vessels that have made such drilling possible. Initially exploration drilling was carried out from shallow water fixed platforms which were initially made of timber and were piled to the sea bed. This was followed by drilling from barges or drilling tenders mounted in sheltered locations.

The water depth capability of drilling equipment has gradually increased first by the use of jack up rigs. A typical modern jack up rig is shown in *Figure 2.1*. Jack up rigs are generally of triangular construction containing steel framed legs at each corner which can be jacked up or down by electric or hydraulic machinery. The jack up hull is transported to the drilling site either by loading on to a barge or by towing the self floating hull. The legs are jacked down at the drill site, and the platform hull is raised a sufficient distance above mean water level to prevent waves hitting the underside of the platform. When the drilling programme has been completed, the hull is lowered down to the sea surface or on to a transportation barge and the rig is moved to the next drilling location. Some jack ups have also been used for oil production in shallow water with modern jack ups able to operate in severe weather at water depths of up to 107 m (350 ft).

2.2.1.2 Semisubmersible rigs

Exploration drilling in waters of up to 18 m (60 ft) depth was also carried out from floating vessels with one or more pontoons or caissons supporting a deck with vertical columns. Such platforms were floated on to the drilling site and then ballasted down on to the sea bed to carry out a drilling programme. During the flotation phase of deployment, it was noticed that such column stabilized platforms had exceptionally low motion response to waves. This feature was utilized to develop so called semisubmersible floating platforms which did not have to be ballasted down to the sea bed in order to carry out drilling.

Figure 2.2 shows a perspective view of one such modern vessel with two parallel submerged pontoons at keel level supporting a large deck area using eight vertical surface piercing columns. The columns, pontoons and deck are braced to form a space frame structure which can be moored on location by a spread of catenary mooring ropes or chains. The deck contains a central opening called the moonpool through which drilling activities take place. Most semisubmersible vessels are equipped with accommodation, helicopter pads for crew transport, propulsion engines, pipe handling equipment, cranes and a variety of marine services to support crew and drilling operations.

Drilling into the sea bed is carried out through a vertical pipe called the marine riser which connects the area under the moonpool to a subsea well head. A drill bit on the end of the drill pipe is passed through the riser to penetrate the sea bed and carry out the actual drilling. A lubricating fluid called drilling mud is pumped to the drill face at high pressure through the hollow drill pipe and returned through the annulus between the drill pipe and riser. The drilling mud pressure is used to drive rotating cones on the drill bit with the drill bit and pipe itself rotating at up to 250 r.p.m. The drilling mud serves several purposes. It lubricates the drill bit, drives the drill bit cones, sweeps

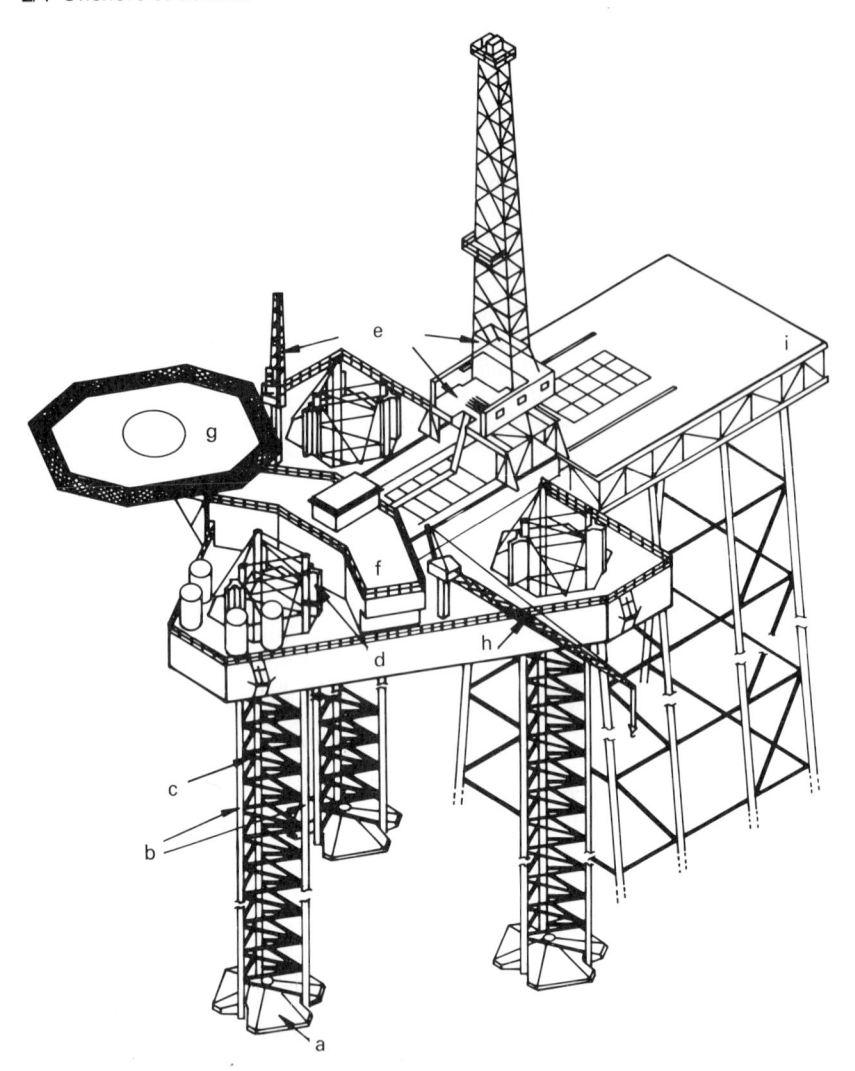

Figure 2.1 Typical jack up rig; a – spud cans;
b – elevating racks; c – legs; d – gear units; e – drilling
derrick and equipment; f – accommodation;
g – helicopter pad; h – cranes; i – nearby jacket platform

out drill cuttings to the surface and provides a hydrostatic pressure head to contain high pressure oil or gas that may be encountered. The small but significant heave motions of the floating drilling rig due to ocean waves are compensated for by a slip joint at the base of the drilling derrick. A tensioner also provides an upward force on the riser to keep it in tension and prevent buckling of its slender structure. The moorings of a drilling semisubmersible have to maintain station to less than 7% of the water depth, in order to continue drilling.

The maximum water depth and weather conditions that a semisubmersible vessel can drill in is governed by its mooring system, by the amount of drill pipe and riser pipe that it can carry and by heave compensator limits imposed by the vessel's motion response and prevailing weather. A catenary line mooring system is impractical above water depths of about 457 m (1500 ft), and in such cases a dynamically positioned vessel is required. Dynamic positioning requires a vessel, whether a ship or a semisubmersible, to be installed with rotating thruster pods and propulsion machinery which is able to work against and balance the environmental disturbing forces due to wind, current and wave drift. Dynamic positioning is also discussed in Chapter 8.

Dynamic positioning has been most commonly employed on drill ships which utilize position signals from sensors, such as a taut vertical wire to the sea bed or from acoustic beacons on the sea bed around the

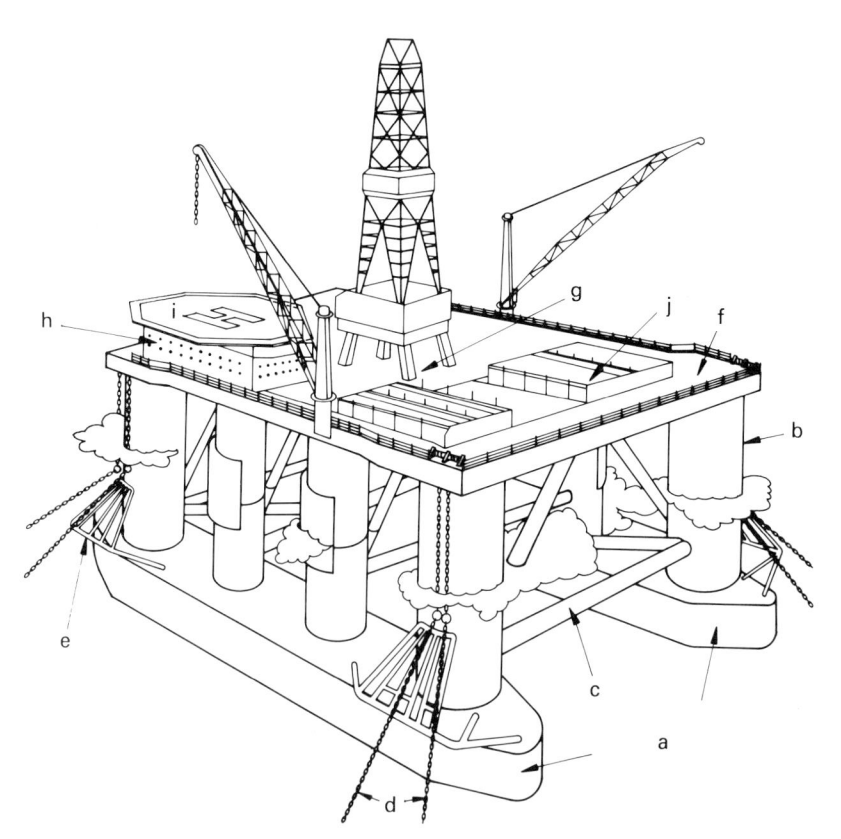

Figure 2.2 Typical semi-submersible vessel:
a – submerged pontoons; b – surface piercing deck support columns; c – bracing members; d – mooring lines; e – anchor racks; f – deck structure; g – moonpool; h – accommodation; i – helicopter pad; j – drill pipe racks

well head, to obtain an error signal representing horizontal offset from above a subsea well head. The vessel is fitted with thrusters controlled from an automated computer based algorithm which is programmed to reduce this offset and maintain the vessel as close to directly above the well head as is possible. *Figure 2.3* shows a schematic view of this arrangement with the drill ship connected by a riser to the blow out preventor (BOP) stack, prior to that being connected to the well head. Acoustic beacons provide a position reference to the drill ship and the data is then used by a computer to control the main propulsion system and lateral thrusters to resist wind, current and wave drift forces on the vessel. It is necesary to maintain position within a watch circle that has a radius of less than about 7% of the water depth to continue drilling.

Several modern semisubmersibles have also been equipped with dynamic positioning systems. The maximum water depth to which a dynamically positioned drilling vessel can work is theoretically unlimited although in practice limitations are imposed by the length of marine riser that can be carried on the vessel and by

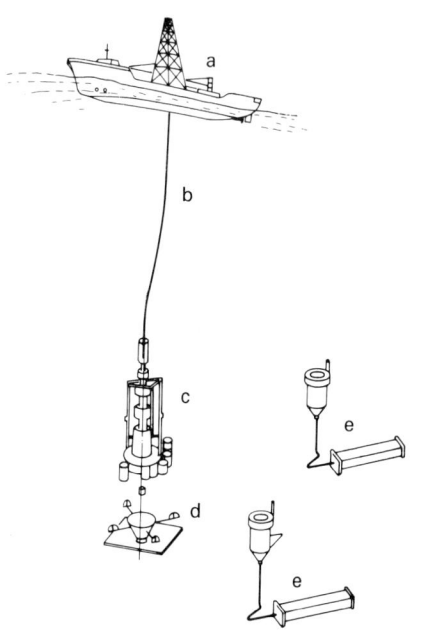

Figure 2.3 Dynamically positioned drill ship: a – drill ship; b – riser; c – blow out preventer; d – well head on sea bed; e – acoustic positioning beacons

tensioner and heave compensator performance. Dynamically positioned vessels, nevertheless, have to disconnect the riser in extreme weather conditions, to ride out storms and then to reconnect in calmer weather.

2.2.2 Drilling depth capability

The expansion of the worldwide drilling fleet has been matched by a rapid increase in the water depths at which drilling can be carried out. This is illustrated most vividly by *Figure 2.4* which presents water depth against year with two lines showing the maximum drilling depth capability, as well as the actual water depth records that were achieved in the years between 1952 and 1985. It can be seen that in the period 1970–1985, the water depth at which drilling was carried out increased from around 1366 m (1200 ft) up to 2128 m (6981 ft). Some drilling in very high water depths has been carried out for geological research into the structure of the land mass below the oceans.

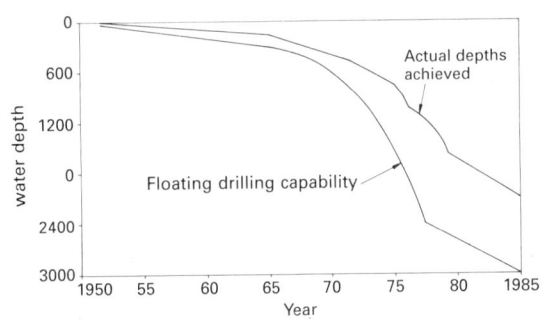

Figure 2.4 Deep water drilling capability and achievements

The water depths at which exploration drilling is carried out is an indicator of future requirements for oil production. Production platforms have to be capable of carrying large amounts of development drilling and oil process equipment. The process equipment has to separate water, gas and sand from crude oil emerging from the reservoir, to prepare the crude oil for transportation and to provide a transportation pumping station. At the same time, the hydrocarbon reservoir depletion procedures may require gas lift, which involves the use of gas to assist in raising the oil if the reservoir pressure is insufficient. Water injection may be required to increase reservoir pressure and therefore permit more oil to be extracted. It is common for one platform to be used to deplete a large plan area of the reservoir by using directional drilling or employing step out satellite wells.

All of these and many other requirements mean that the complexity of equipment on board a production platform can resemble a chemical plant or refinery. At any one point in time, therefore, the depth capability of production platforms tends to be considerably lower than the drilling depth capability. Furthermore, the engineering equipment demands of development drilling, oil production, enhanced recovery and oil export are such that the water depth capability with time for production platforms has not increased as steeply as it has for drilling vessels.

This is illustrated by *Figure 2.5* which shows the evolution of timber or steel framed bottom standing oil production structures, usually called jackets. The figure illustrates jacket size and water depth capability starting with 1947 through to a large deepwater jacket in 1978.

2.2.3 Types of structures

2.2.3.1 Jacket structure

It is instructive to examine the structure of a typical jacket for deployment at moderate water depth in the North Sea as illustrated by *Figure 2.6*. The jacket consists of a steel framed tubular structure, connected to the sea bed by piles which are driven through pile guides on the outer members of the jacket. The jacket topsides consist of a series of modules which house drilling equipment, production equipment, living quarters, gas flare stack and revolving cranes. The topsides also have facilities for living quarters, survival craft, hotel and catering facilities, and a helicopter pad for transfer of crews from and to a shore base. The topside modules are supported on a module support frame, which is mounted on the water surface piercing jacket structure. The drilling and production tubulars are brought up from the sea bed to the topsides through conductor guides located within the jacket framing. Crude oil and gas are brought up to the surface from the reservoir for processing and then pumped back down through an export pipeline, either to a tanker loading buoy or by subsea pipeline to a shore terminal.

The detailed design of the framing for a jacket structure can vary considerably, depending on requirements of strength, fatigue and launch procedure. The main structural members and bracing can be connected by X or K shaped joints with sizes that can range from an extensive space frame of small diameter X braced or K braced members to an alternative sparser framing with larger diameter members. *Figure 2.5* gives some indication of the variety of design options that are employed.

A jacket designer has to cope with an extensive list of constraints imposed on a jacket structure through its life. The life cycle consists of design, construction, load out, launch, installation, piling and hook up phases of the platform before it comes into oil production service. This is followed by a 10–25 year operational life through which the platform has to be maintained followed by the ecologically desirable requirement to remove and dispose of the platform after the reservoir has been depleted.

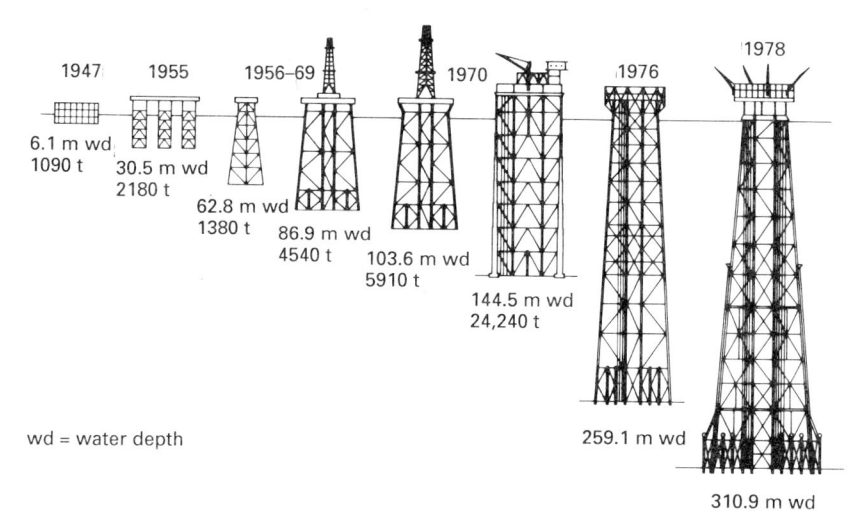

Figure 2.5 Evolution of deep water production capability (from Lee 1980)

Labels on figure:
- 1947, 6.1 m wd, 1090 t
- 1955, 30.5 m wd, 2180 t
- 1956–69, 62.8 m wd, 1380 t
- 86.9 m wd, 4540 t
- 1970, 103.6 m wd, 5910 t
- 144.5 m wd, 24,240 t
- 1976, 259.1 m wd
- 1978, 310.9 m wd

wd = water depth

The complexity of any one of these phases is illustrated by considering the procedure for launching and installing a jacket. *Figure 2.7* shows two ways in which this has been done in the past. *Figure 2.7(a)* illustrates a jacket which has been loaded out from a fabrication yard onto a barge and then towed out to a launching site during which the jacket is subjected to large dynamic loads. The jacket must then be up-ended into an upright position, either by using a floating crane, as in step 5, or by differential ballasting of jacket structural members and then sunk down on the sea bed, in preparation for piling.

Figure 2.7(b) illustrates another approach to the installation problem by utilizing a self floating jacket with main members which are large enough that their buoyancy can support the jacket weight, therefore dispensing with a launch barge. Such a jacket is towed out on to the site and then up-ended using differential buoyancy in structural members, as shown by steps 2–5 in *Figure 2.7(b)*. The jacket can then be ballasted down in preparation for piling. There are several pros and cons for a designer to consider in choosing between self floating and barge launched jackets. Most large jackets constructed for the North Sea in the last few years have been of the barge launch type. They often have to be installed with removable auxiliary buoyancy chambers in order to facilitate launch and up-ending. The only self floating jacket that has been installed recently in the North Sea is for the Magnus Field.

Since barge launched jackets require less buoyancy, they utilize less steel and are in general easier to fabricate. However, such jackets require an appropriate transport barge and are subjected to large dynamic loads in the launch process. On the other hand, self floaters do not require a launch barge, but need to use additional steel to provide extra buoyancy. The larger diameter of the buoyancy members also induces additional unnecessary wave forces during the whole life of the structure.

A compromise between self floaters and barge launch jackets can be achieved by using a structure which has removable buoyancy chambers. These chambers need to be installed for the load out, launch and up-ending phases only and can be removed after the installation phase is complete, thus largely eliminating the disadvantage of higher wave forces during operation. A typical barge launched jacket structure, such as the jacket on the Cerveza Field, weighs 24 000 tonnes. The capabilities of an available fabrication yard for load out or launch of a self floater as well as the availability of launch barges influences this design decision.

Assessment of jacket foundations and pile design offer another technical challenge. Prior to commencing jacket design, the sea bed soil has to be surveyed and its characteristics evaluated. The designers then determine the number of piles, the required depth of penetration and whether the piles should be drilled or driven by pile hammers. Drilled piling implies that holes for the piles are drilled before the piles are installed and grouted *in situ*. However, recent developments in underwater pile hammers have been so rapid that driven piles are now almost invariably used. Continuing improvements in pile hammer capability have also influenced jacket design and piling installation.

Jacket topside weights in the central and northern North Sea vary between 17 000 tonnes for the North Cormorant jacket up to 35 000 tonnes for the Statfjord and Brae platforms. This topside weight has to be broken down into a number of modules which can be installed offshore on to the module support frame on top of the jacket by crane barges. The design of the

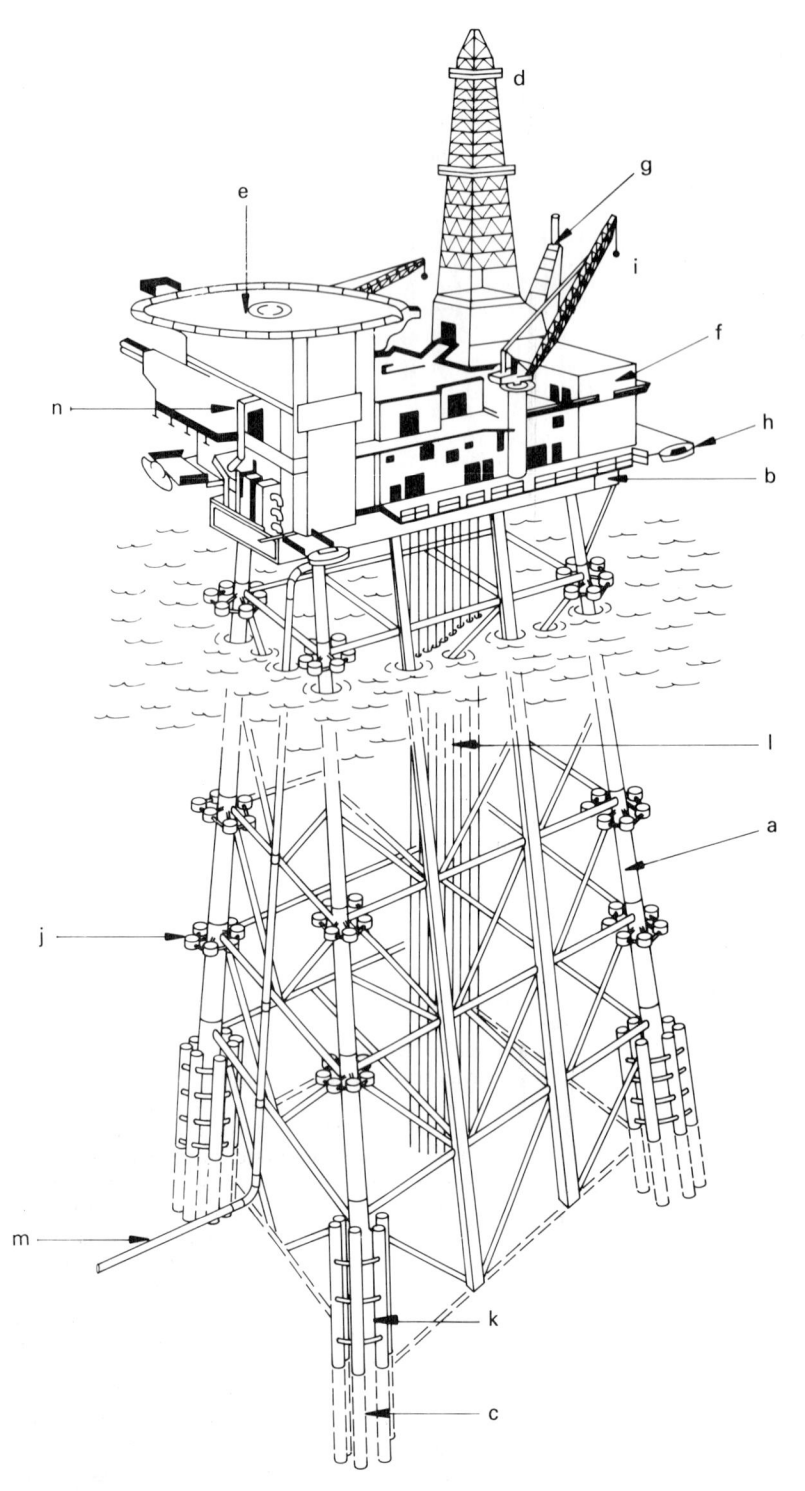

Figure 2.6 Typical offshore drilling and production platform: a – jacket; b – module support frame; c – piles; d – drilling derrick; e – helicopter pad; f – drilling and production equipment; g – flare stack; h – survival craft; i – revolving cranes; j – pile guides; k – pile sleeves; l – drilling and production risers; m – export pipeline; n – accommodation

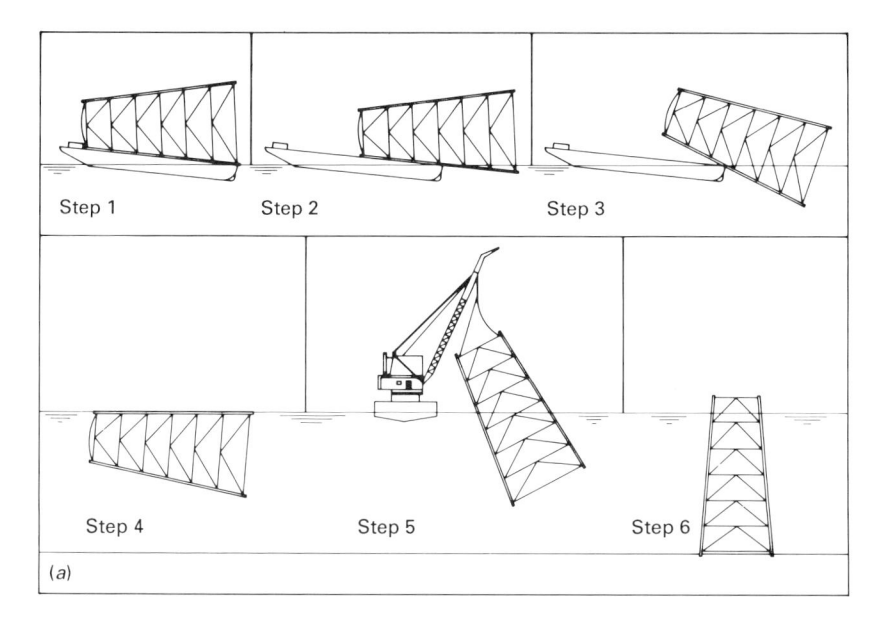

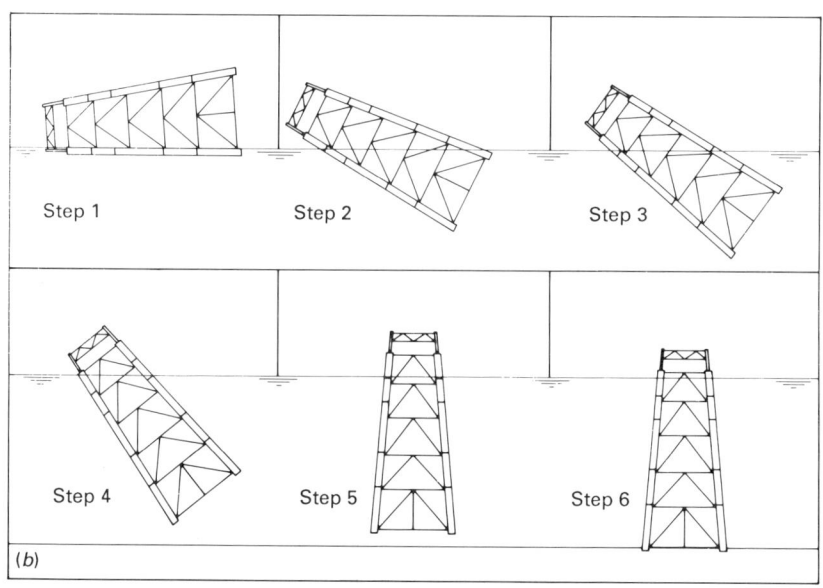

Figure 2.7 Installation of barge launched and self floating jackets: a – barge launched jacket; b – self-floating jacket

jacket and module arrangement is greatly influenced by the availability of crane barges and by their maximum lift capacities.

Recent construction of several semisubmersible crane vessels with large lift load capacities is beginning to have a significant influence on offshore installation practice. The semisubmersible configuration offers a low down time to weather and the very large size of the vessel means that lift loads of up to 12 000 tonnes can currently be lifted. It is, therefore, feasible to construct and lift fewer modules each of larger tonnage and, therefore, reduce the complexity and cost of module installation and hook up operations. The availability of larger module sizes also reduces hook up times and enables earlier oil or gas production to be achieved.

The use of steel frame jacket structures in deeper waters and harsher environments requires the designer to ensure that inspection and maintenance of underwater parts of the platform are planned for during the design process. Such inspection and consequent repair can be a major and costly activity since it has to be carried out under water by divers or with remotely operated equipment. The emphasis in engineering design and economic evaluation of modern jacket structures is shifting away from the historical approach of considering initial capital expenditure only and towards much more attention being paid to designing for economic inspection and maintenance of the structure during service.

2.2.3.2 Concrete platforms

Offshore oil developments in the North Sea have also pioneered a completely different design of fixed platform based on the use of concrete to create a large heavy structure which can rest on the sea bed and remain stable under its own weight. The first such platform was installed in the North Sea on the Ekofisk Field in 1973 and since then over 17 platforms have been installed up to 1982. *Table 2.1* gives a summary of concrete platforms that have been installed worldwide.

Although concrete gravity structures are considered relatively expensive compared to steel frame jackets, the structures do offer an attractive alternative to jackets in hostile waters like the North Sea and in situations where the uncertainty of a tanker export system demands a certain amount of oil storage on the platform. Other advantages of concrete gravity platforms over jackets are that the structures can be constructed onshore or in sheltered waters, with all the

Table 2.1: Summary of concrete platforms (from Furnes and Loset 1981)

Type of design and location	Main function	Design wave height (m)	Depth (m)	Approx. concrete volume (m^3)	Base diameter (m)	Storage capacity (million)	Installation year
1 Doris Ekofisk 1 (N)	Storage	24.0	70	90 000	92	1.000	1973
2 Condeep Beryl A (UK)	Drilling, production, storage	29.5	120	55 000	100	0.930	1975
3 Condeep Brent B (UK)	Drilling, production, storage	30.5	142	65 000	100	1.000	1975
4 Doris Frigg CDP 1 (UK)	Drilling, compression, production	29.0	96	60 000	101	–	1975
5 Sea Tank Brent C (UK)	Drilling, production, storage	30.5	142	105 000	100	0.650	1978
6 Sea Tank Frigg TP1 (UK)	Production	29.0	104	70 000	72	–	1976
7 Sea Tank Cormorant A (UK)	Drilling, production, storage	30.5	152	115 000	100	1.000	1978
8 Condeep Brent D (UK)	Drilling, production, storage	30.5	142	65 000	100	1.000	1976
9 Andoc Dunlin A (UK)	Drilling, production, storage	30.5	152	89 000	104	0.850	1977
10 Condeep Statfjord A (N)	Drilling, production, storage	30.5	149	88 000	110	1.300	1977
11 Condeep Frigg TCP2 (N)	Treatment, compression, production	29.0	104	50 000	100	–	1977
12 Doris Frigg MP2 (UK)	Compression station	29.0	94	60 000	101	–	1976
13 Doris Ninian (UK)	Drilling and production	31.2	139	142 000	140	–	1978
14 Pub 3 Petrobras	Drilling, production, storage	11.0	15	15 000	52	0.125	1977
15 Pub 2 Petrobras	Drilling, production, storage	11.0	15	15 000	52	0.125	1989
16 Pag 3 Petrobras	Drilling, production, storage	11.0	15	15 000	52	0.125	1978
17 Condeep Statfjord B (N)	Drilling, production, storage	30.5	149	130 000	169	1.500	1981

topsides installed, hooked up and tested prior to floating out and towing the structure to its offshore location. Installation then only requires ballasting the platform down on to the sea bed and consolidating the foundation below the platform by pumping grout into the spaces between the platform base and foundation.

The elimination of steel piling and of having a concrete structure tolerant to overloading and to degradation due to exposure to sea water offers other advantages over the use of steel. It has been demonstrated that concrete used in coastal installations in the 1930s and 1940s has survived, essentially unaffected by exposure to sea water up to the present day, whereas conventional steel structures would have been susceptible to sea water corrosion and require substantial levels of maintenance and protection. These advantages have to be set against the fact that concrete gravity structures are relatively expensive. They often actually employ a bigger mass of steel in their re-inforcing members than would be required by an equivalent steel frame jacket structure. Concrete gravity structures are likely to suffer from foundation settlement during their working lives which can reduce the air gap between the mean water level and the underside of the structure. Another disadvantage with concrete gravity structures is that no feasible means of removing the structure has been defined at present.

Figure 2.8 shows a typical design for a Condeep concrete gravity structure which consists of a cellular concrete caisson or base with three columns supporting a topside structure. The base of the platform contains steel skirts, which penetrate a short distance into the foundation during installation of the platform. The concrete caisson at the platform base is used for oil

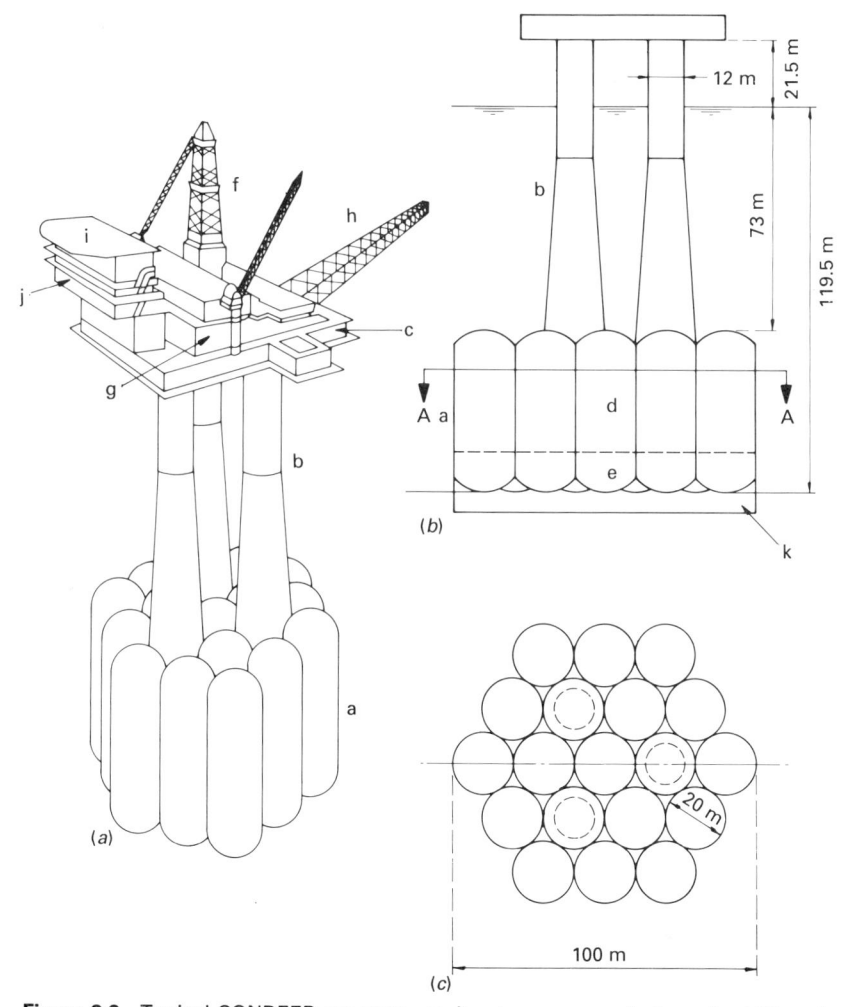

Figure 2.8 Typical CONDEEP concrete gravity structure: (a) perspective view; (b) side view; (c) section AA: a – sea bed caisson; b – surface piercing columns; c – module support structure; d – oil storage; e – sand ballast; f – drilling derrick; g – drilling and production equipment; h – flare stack; i – helicopter pad; j – accommodation; k – steel skirt (from Furnes and Loset 1981)

storage. *Figure 2.8* shows a Condeep design for the Beryl A platform which was installed in June 1975. This platform consists of a base caisson of 100 m (305 ft) maximum width with circular cellular oil storage tanks of 20 m (65.6 ft) diameter and water surface piercing support towers that are 94.5 m (310 ft) high. These towers are used for drilling and oil production conductors (pipe lines) and for oil transfer piping. Some concrete production platforms are designed with transverse support frames between their vertical columns to support exposed conductor pipes passing through the base caisson into the sea bed.

Concrete gravity platforms are constructed by a technique known as slip forming, which uses a structure surrounding the cross-section of the platform in plan view to contain concrete as it is poured and while it sets. As the concrete sets and more is poured, the slip forming structure is raised and the platform is built or slip formed from the ground up. Steel re-inforcing members are set into the concrete and prestressed to give appropriate overall structural properties. Once the entire platform has been constructed, it is towed out to a deep water site, ballasted down and the preprepared complete topsides are mated to the top. The topsides can be pretested and hooked up in the module fabrication yard, thus ensuring that very little offshore hook up or testing is required. Once the topsides have been installed, the platform is deballasted up and towed out to the installation site where it is ballasted down on to the sea bed. Skirt piles of approximately 4–6 m (13–20 ft) height at the base of the platform cut into the foundation under the weight of the platform. The seal between the foundation and the platform is consolidated by pumping grout into the interface region.

2.2.4 Selection of platform type

There are many technical arguments for and against the selection of steel frame jackets or concrete gravity platforms for a particular location and oil field. Although concrete gravity platforms are expensive to build, they do offer advantages of low maintenance and high deck payload growth capability during their lifetime. Nevertheless, the construction of new concrete gravity platforms has slowed down since the early 1980s and compared to steel jacket structures very few are being considered or are likely to be used for future developments.

The total cost for developing an offshore oil or gas field clearly pays a major part in the selection of the production platform. In the early growth years of the offshore industry when very large oil deposits were being found, the economics of each development permitted large expensive structures to be built. However, as more of the larger reservoirs were discovered and developed, the remaining reservoirs have a higher probability of containing smaller oil deposits and being in deeper waters and harsher environments. At the same time, the cost of fixed steel frame jackets or gravity structures tends to increase in an approximately exponential fashion with water depth, primarily due to the larger platform structure and the greater volume of material required for structural integrity in deeper water.

Figure 2.9 illustrates a typical variation of jacket weight with water depth for two locations: the Gulf of Mexico and California as well as the North Sea. The jacket weight on the vertical axis can be considered to be representative of the platform cost whereas the lines denoting the North Sea and Gulf of Mexico and California are illustrative of the influence of environmental conditions on jacket weight. The full lines denote current developments and the dashed lines are an estimate of future projections using the same kind of technology.

Figure 2.9 shows that the effect of more extreme environmental conditions for the North Sea in the 100–200 m (328–656 ft) water depth range increases jacket weight and, therefore, cost by a factor between 3 and 6. The figure also illustrates that a jacket weighing 30 000 tonnes in the Gulf of Mexico or California could be used in up to 320 m (1050 ft) water depth whereas that jacket weight would only suffice for approximately 175 m (574 ft) water depth in the North Sea.

2.2.5 Cost considerations

The approximately exponential increase of jacket weight and therefore cost with increasing water depth has set the scene for the development of new technology in oil production to try to reduce the rate of increase of cost with water depth. This need for reducing costs has prompted feasibility studies on a wide variety of new types of production platforms based on compliant structures which are fixed to the sea bed but respond with some degree of flexibility to the ocean environment. The consequent reduction in the volume of steel or concrete material that connects the surface facility to the sea bed can lead to a potential reduction in costs. This is illustrated by *Figure 2.10* which presents the cost of design, fabrication and installation for different platform types plotted against water depth. The cost curve of a typical North Sea jacket is also plotted on the same axis for comparison with the shaded areas around each line denoting the cost range that might be obtained with different design and installation options.

It can be seen that new platform concepts such as the tensioned buoyant platform and semisubmersible or tanker based floating production systems, all of which are described later, offer significant reductions in the rate of increase of cost with water depth. It is now becoming increasingly obvious in the industry that oil production in deep water up to the end of this century is likely to be carried out by compliant or floating production platforms in all cases except for the very largest of fields or for fields in shallow water. In deep waters above 457 m (1500 ft), a compliant or floating produc-

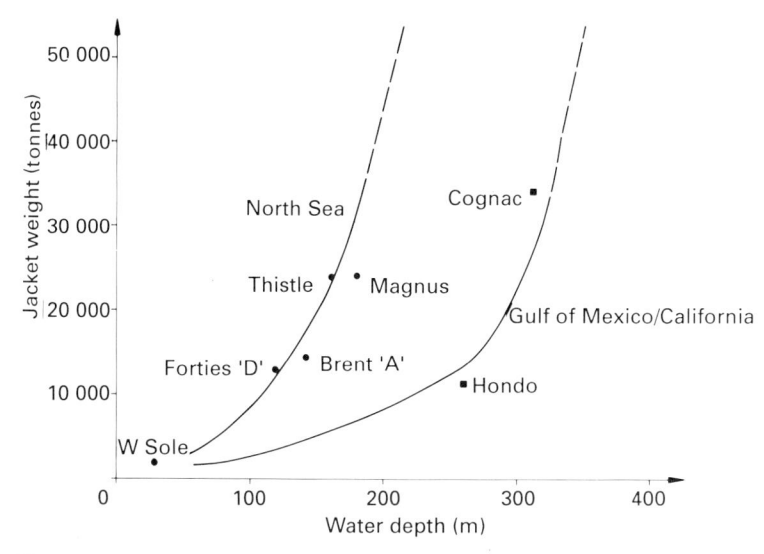

Figure 2.9 Variation of platform jacket weight with water depth (from Thornton 1979)

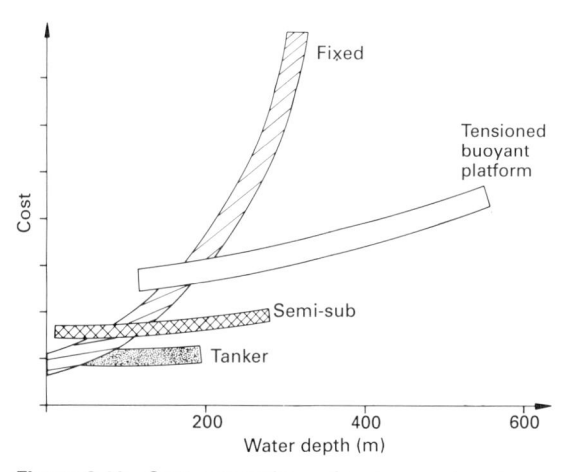

Figure 2.10 Cost comparisons for deep water structures (from Hamilton and Perrett, 1986)

tion system may in fact be the only means of producing a field. There are a number of existing and proposed alternatives for cheaper compliant or floating production systems. *Table 2.2* gives a summary of developments in this area since the mid-1970s.

One of the most popular floating production platforms is shown in *Figure 2.11* and is based around a converted or new semisubmersible vessel which is catenary moored to the sea bed and is connected to a subsea manifold or well head by either rigid tensioned vertical multi-tube risers or by flexible risers. Such systems are technically suitable for depth ranges of 70–250 m (230–820 ft), approximately. They are, however, limited by a number of factors and technical risks. Semisubmersibles only have a relatively small deck payload carrying capacity and, therefore, the range of available oil process options is restricted compared to what is possible on fixed platforms. The compliancy of the platform and the catenary mooring system poses a risk of the vessel losing station and damaging its risers during severe environmental conditions or after damage to a mooring line. The platform cannot directly offer oil storage facilities and the heave, roll and pitch motions of the platform to waves can limit or degrade processing operationg. Nevertheless, catenary moored semisubmersibles offer an economic means of exploiting small reservoirs which have sufficiently good hydrocarbon field characteristics so as to require only a small amount of processing on the surface facility.

A technically more advanced form of compliant structure is offered by the tension or tethered buoyant platform, shown in *Figure 2.12*. The surface platform configuration resembles that of a semisubmersible but is tethered to the sea bed by vertical legs which are kept in tension by excess buoyancy in the platform. The excess buoyancy is of the order of 15–25% of the platform displacement and tensions the tethers to such an extent that heave, roll and pitch motions of the platform to ocean waves are virtually eliminated. The platform does, however, experience surge, sway and yaw due to wave action but the vertical tethers are a reliable means of keeping the platform permanently on station above the subsea wells. These wells are connected to the surface facility by an array of risers.

Oil processing is carried out on the platform and the hydrocarbon products are pumped back down to the sea bed to an export pipe line. In 1984 Conoco Ltd installed the world's first tethered buoyant platform in 148 m (485 ft) of water on the Hutton Field in the

Table 2.2: Floating production installations

Field	Location	Water depth (m)	Startup	date	Operator	Equipment
Argyll	UK	76	June	1975	Hamilton Bros	Semi
Enchova	Brazil	122	August	1977	Petrobras	Semi
Castellon	Spain	119	August	1978	Shell	SALS
Dorado	Spain	94	May	1978	Eniepsa	Semi
Garoupa	Brazil	120	February	1979	Petrobras	Tank
Enchova	Brazil	189	June	1979	Petrobras	Semi
Casablanca	Spain	134	December	1979	Chevron	Semi
Nilde	Sicily	96	April	1980	Agip	SALS
Buchan	UK	117	May	1981	BP	Semi
Hutton	UK	148	September	1984	Conoco	TBP
Duncan	UK	79	December	1984	Hamilton Bros	Semi
Balmoral	UK	128	June	1986	Sun Oil	Semi
Cyprus	UK	122	March	1988	BP	SWOPS

Installations Semi: Semisubmersible surface platform
SALS: Single anchor leg system with tanker
SWOPS: Single well oil production system
TBP: Tensioned buoyant platform

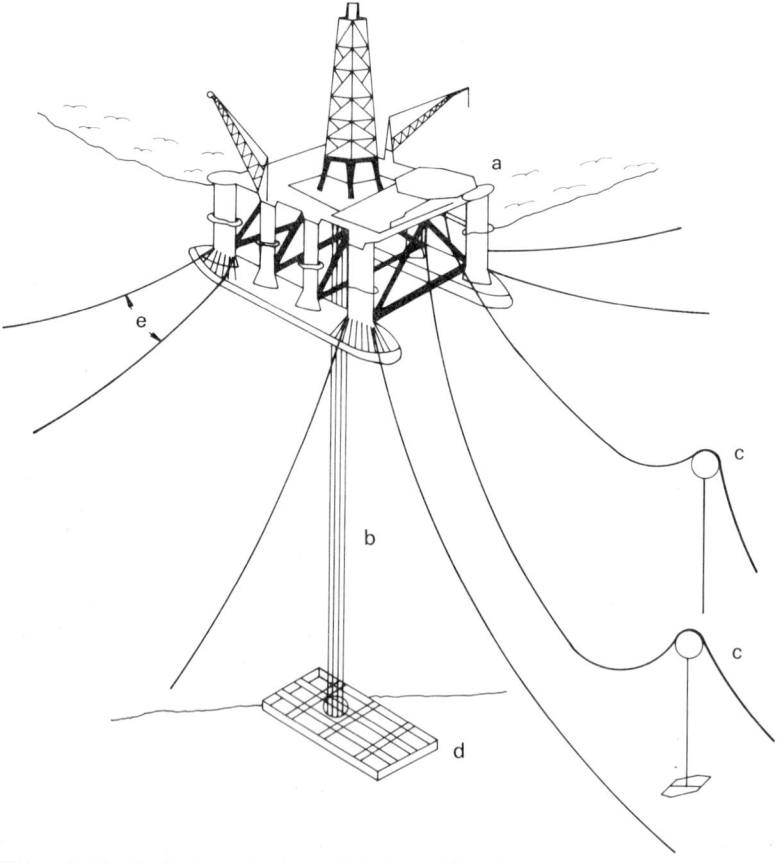

Figure 2.11 Typical semi-submersible based floating production system: a – surface platform; b – multi-tube vertical drilling and production riser; c – flexible production risers; d – sea bed template; e – catenary moorings

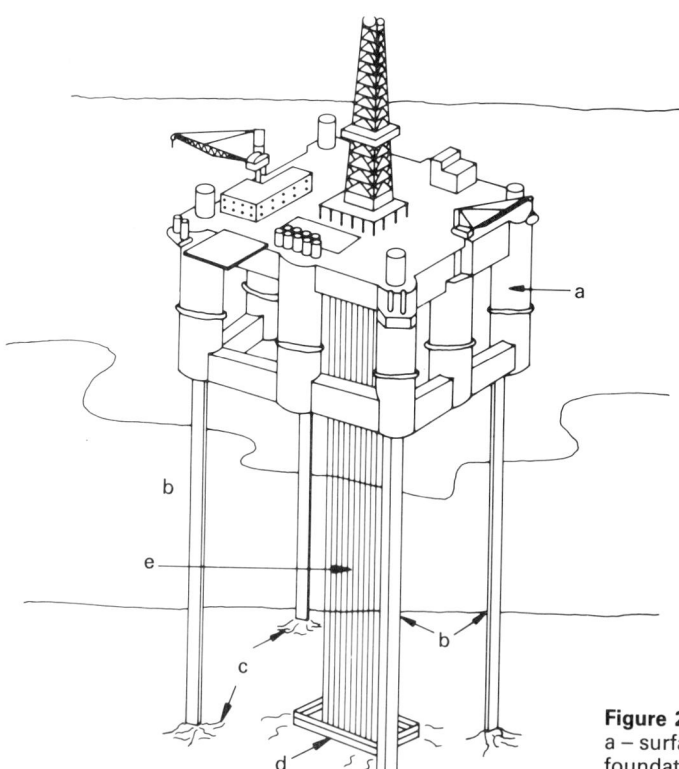

Figure 2.12 Typical tensioned buoyant platform: a – surface platform; b – tensioned tethers; c – tether foundations; d – template on sea bed; e – marine risers

North Sea. Due to the success of this development, it is likely that tethered buoyant platforms will be considered for a number of future field developments. The probable depth range of tensioned buoyant platforms is from approximately 120 m (394 ft) up to as high as 1500 m (4921 ft).

It is also possible to develop small isolated fields in water depths of up to 170 m (558 ft) in calm weather areas by the use of articulated column or single anchor leg storage and tanker systems. *Figures 2.13* and *2.14* illustrate two such systems, the first of which is an articulated oil production and loading column. Crude oil is extracted from the reservoir and transferred up the articulated tower to the tethered tanker for processing and storage. The processed hydrocarbons can then be transported by a shuttle tanker mooring up to the production tanker shown in the figure. An alernative scenario would entail the process equipment being mounted on a larger articulated column and the processed crude pumped to the tanker which shuttles to and from a shore loading terminal. *Figure 2.14* shows a single anchor leg system which uses a yoke structure, a buoyancy tank and a tensioned riser connection to the sea bed to moor a tanker. The tanker houses crude oil processing and storage facilities, with oil export again taking place by shuttle tanker or oil export pipeline.

Table 2.2 shows that two systems of the single anchor leg type are in operation at present – in the Castellon Field off Spain and the Nilde Field off Sicily.

Another form of economically viable compliant structure is offered by the guyed tower. The guyed tower is a slender truss steel structure, as shown in *Figure 2.15*, which is supported on the sea bed at its base by a spud can and foundation and held upright by multiple guy lines made up of wire ropes or chains. The guy lines are connected to anchor piles with heavy clump weights attached to the lines between the anchor piles and the tower to modify the horizontal stiffness of the tower/guy line system. In normal, or operating weather conditions, the guy lines restrain the platform and keep it upright without lifting the clump weights off the sea bed. In more extreme weather, the guy lines are designed such that the clump weights are lifted off the sea bed and a larger restoring force is available to the tower to resist increased wave forces.

The guyed tower offers all of the advantages of a fixed jacket in that there is a continuous structure between the sea bed and surface facility but at a significantly lower cost. One guyed tower was installed in 1983 in the Gulf of Mexico on the Lena Field. This tower has a steel weight of 24 100 tonnes, a topside weight of 19 640 tonnes and is installed in 305 m

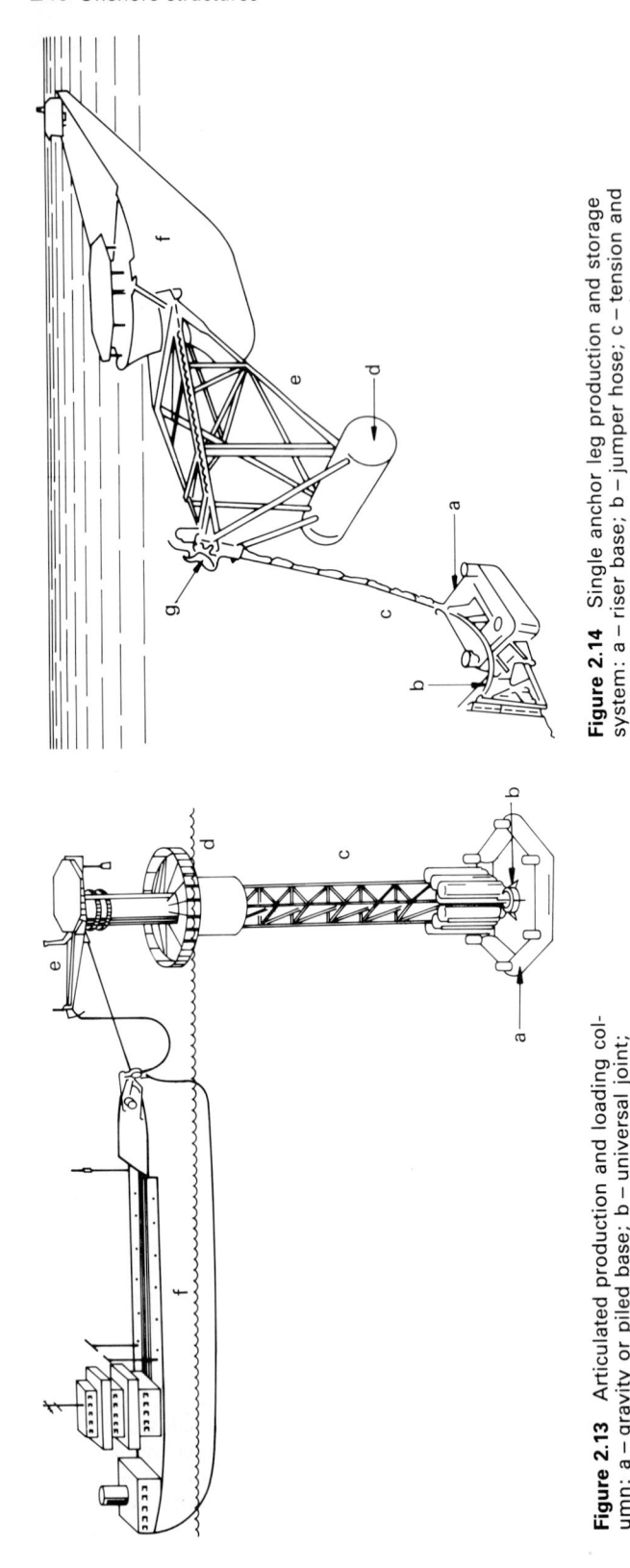

Figure 2.13 Articulated production and loading column: a – gravity or piled base; b – universal joint; c – tower; d – buoyancy tanks; e – loading and mooring arm; f – tanker with production equipment and storage

Figure 2.14 Single anchor leg production and storage system: a – riser base; b – jumper hose; c – tension and production risers; d – buoyancy tank; e – yoke; f – tanker with production equipment and storage; g – universal joint

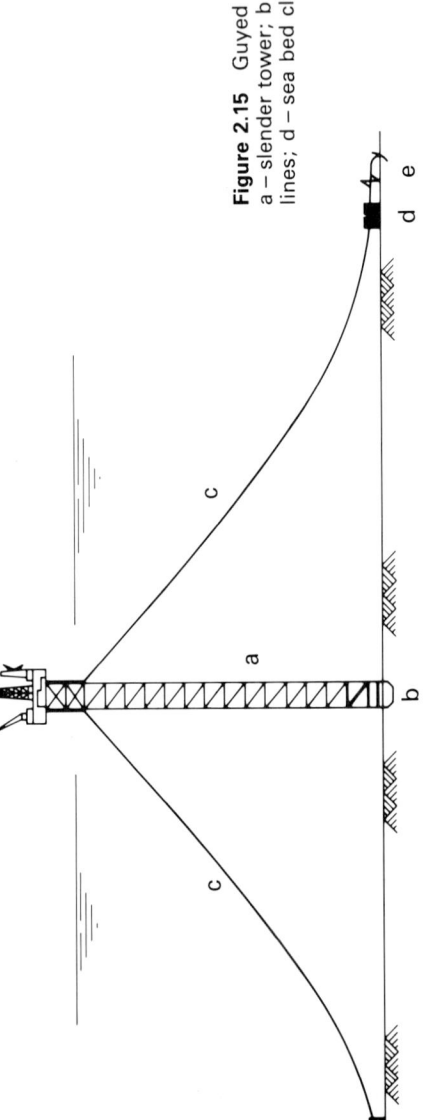

Figure 2.15 Guyed tower production platform: a – slender tower; b – spud can foundation; c – mooring lines; d – sea bed clump weights; e – anchors

(1000 ft) water depth. It utilizes galvanized steel spiral wound wire ropes as guy lines and 179 tonne anchor weights with 20 guy lines laid symmetrically around the plarform.

2.2.6 New design concepts

Apart from floating production platforms types that have already been installed, there are a large number of design concepts which have been evaluated to varying degrees of sophistication.

As an example, *Figure 2.16* shows one such concept – a concrete tripod shaped gravity platform in the form of a single water surface piercing tower with a tripod substructure. This platform uses the tripod support structure to provide the necessary resistance to overturning without the cost of a massive gravity structure.

Other options that have been proposed are shown in *Figure 2.17* and *2.18*. *Figure 2.17* shows a concrete floating production platform which uses an array of catenary anchor lines and clump weights for station keeping. *Figure 2.18* shows the design of a concrete articulated tower, which consists of a cellular concrete base piled to the sea bed. An articulated tower is sited above the concrete base with a universal joint at the base of the tower. The tower is kept upright by excess buoyancy in the structure, with the tower being able to undergo small angular motions in a compliant manner due to wave action. It is probable that such an articulated tower can be used in a depth range 100–250 m (328–820 ft). There are a large number of other novel production platform alternatives that have been proposed in recent years and considered in various research papers.

2.2.7 Other equipment

The development of an offshore oil field also requires a variety of other floating vessels and engineering equipment. An example of such vessels are the exposed location single buoy mooring, the ELSBM, shown in *Figure 2.19*. This buoy enables shuttle tankers to moor up to it and to load crude oil, which is transported to the ELSBM from a nearby oil production platform. Exploration and development of an offshore oil field also requires the services of a range of ship shape and semisubmersible vessels. These include survey vessels for the initial geophysical mapping, drilling vessels to carry out exploration drilling, supply boats to transport equipment and sometimes ferry crew on and off drilling vessels. Once a production platform has been installed, or just prior to its installation, diver support vessels are required to carry out subsurface and sea bed work. During the operating life time of a platform or a group of platforms, maintenance and fire-fighting vessels are required.

The installation phase of a conventional production platform requires crane vessels to install topsides modules. Typical ship shape and semisubmersible crane vessels are shown in *Figures 2.20* and *2.21*. The design and operation of these specialized vessels raises a number of unusual engineering analysis requirements. Offshore oil fields also utilize a network of pipelines and control cables on the sea bed, either joining wells within a field or joining a field to an onshore crude oil reception facility. All of these pipelines and control cables have to be installed, maintained and repaired. A fleet of semisubmersible and ship shape pipe laying vessels, survey vessels and associated pipe laying methods and repair techniques are part of the technology of offshore field development.

2.2.8 Subsea engineering

The design, installation and maintenance of subsea well heads, controls, manifolds and protection frames are carried out within a separate subject area called subsea engineering. Some subsea equipment presents difficult design problems since it operates in a harsh high pressure environment and is extremely difficult to access and repair. High levels of redundancy and well thought out maintenance and repair procedures have to be used with such equipment.

Marine riser pipes that connect compliant platform surface facilities to subsea well heads offer another field of interesting analysis and physical problems. Such riser pipes carry crude oil between the well head and surface platform and yet have to accommodate the movement of the compliant platform relative to the sea bed. Under extreme environmental conditions, the riser may have to be disconnected to prevent its failure. The design of such marine risers requires application of the physics of fluid–structure interaction, which is itself not fully understood. Risers are discussed further in Chapter 5.

2.2.9 Subsea maintenance

When equipment installed on the sea bed does fail or needs to be maintained, manned or unmanned intervention is necessary, in order to replace components or carry out repairs. The offshore industry has stimulated development of a range of manned and unmanned vehicles for subsea intervention, survey and repair work. Subsea vehicles and diving are discussed in Chapter 4.

2.3 Environmental loading

2.3.1 Introduction

Offshore structures are subjected to both steady and time dependent forces due to the action of winds, currents and waves. Winds exert predominantly steady forces on the exposed parts of offshore structures

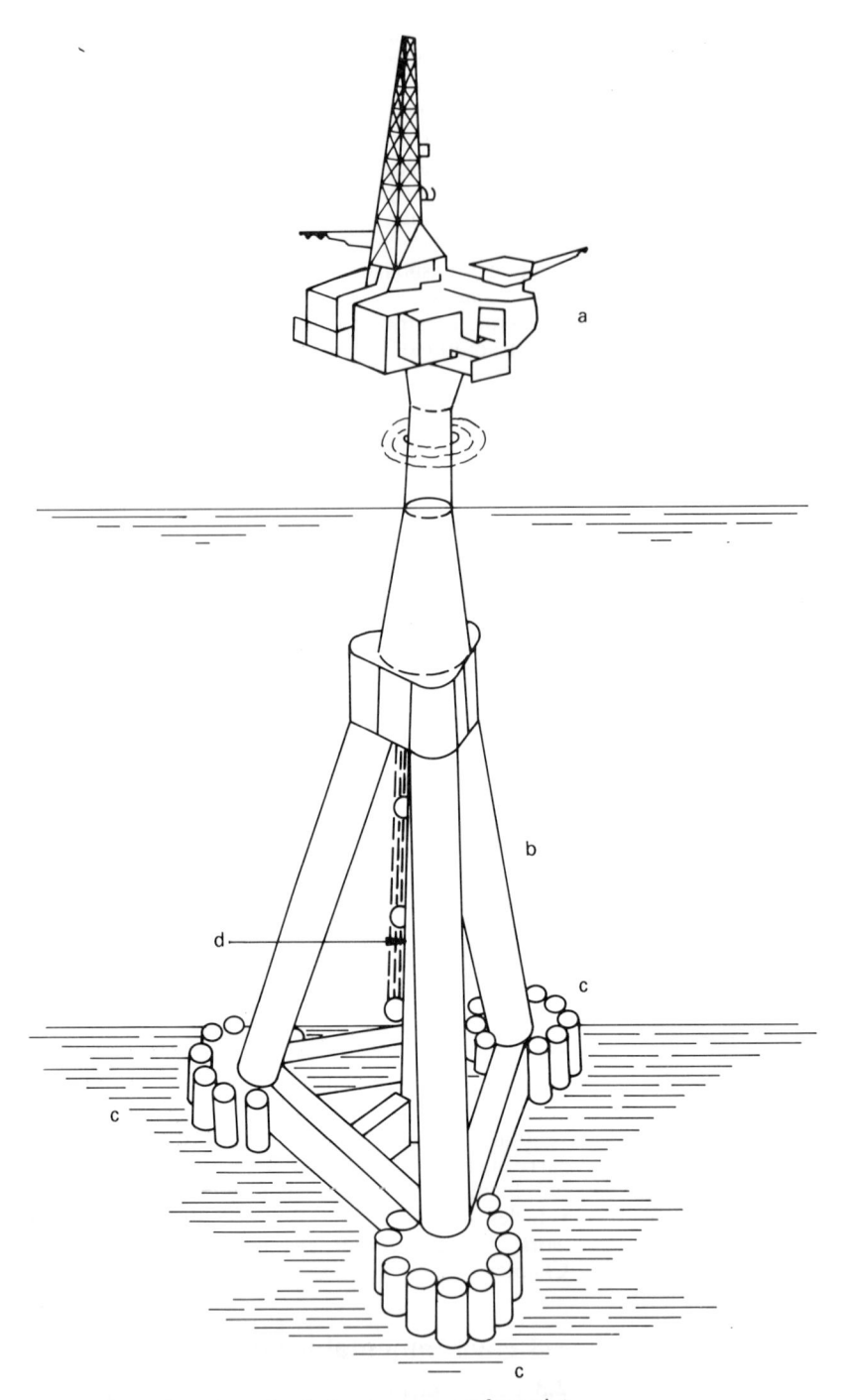

Figure 2.16 Concrete tripod structure: a – surface platform; b – concrete tripod structure; c – gravity bases; d – drilling and production risers

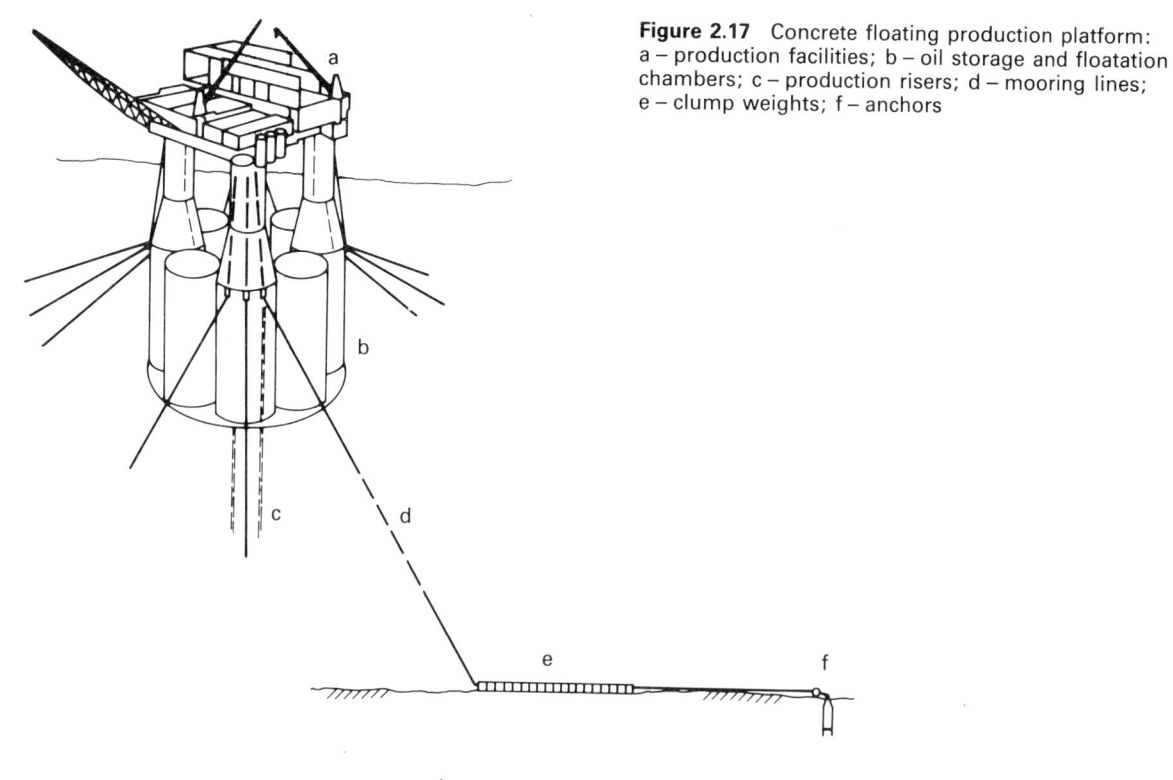

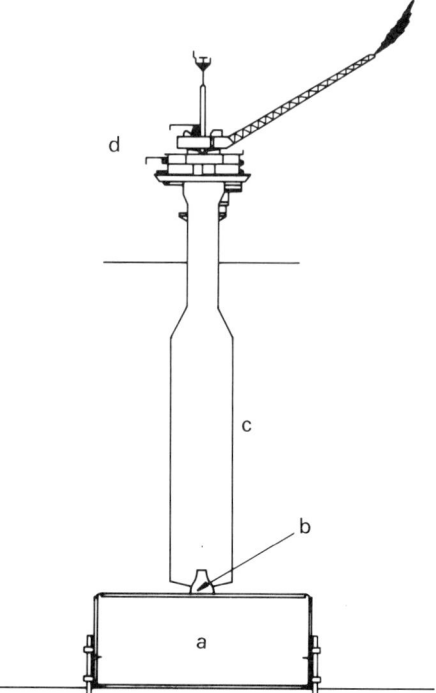

Figure 2.17 Concrete floating production platform: a – production facilities; b – oil storage and floatation chambers; c – production risers; d – mooring lines; e – clump weights; f – anchors

Figure 2.18 Concrete articulated tower: a – piled oil storage base; b – universal joint; c – tower; d – production facilities

although there are significant gust or turbulence components in winds which induce high unsteady local forces on structural components as well as a low frequency total force on the whole structure.

Ocean currents also exert predominantly steady forces on submerged structures together with the localized effects of vortex shedding which can induce substantial unsteady forces on structural members. However, gravity waves induce by far the largest force on most offshore structures. The applied force is periodic in nature although non-linear wave properties in the wave loading mechanism can also induce superharmonic force components. Both of these secondary forces can be significant if they excite resonance in a compliant structure.

In general, an air or water flow incident on an offshore structure will exert a force that arises from two primary mechanisms. A steady or unsteady flow will directly exert a correspondingly steady or unsteady flow with a line of action that is parallel to the incident flow direction. Such forces are sometimes called in line forces. However, the localized interaction of steady or unsteady flow with a structural member will also cause vortices to be shed in the flow and will induce unsteady transverse forces with lines of action that are perpendicular to the incident flow direction. Both of these force mechanisms are considered in this chapter.

The design of offshore structures requires that methods are available to translate a definition of environmental conditions into the resultant steady and time

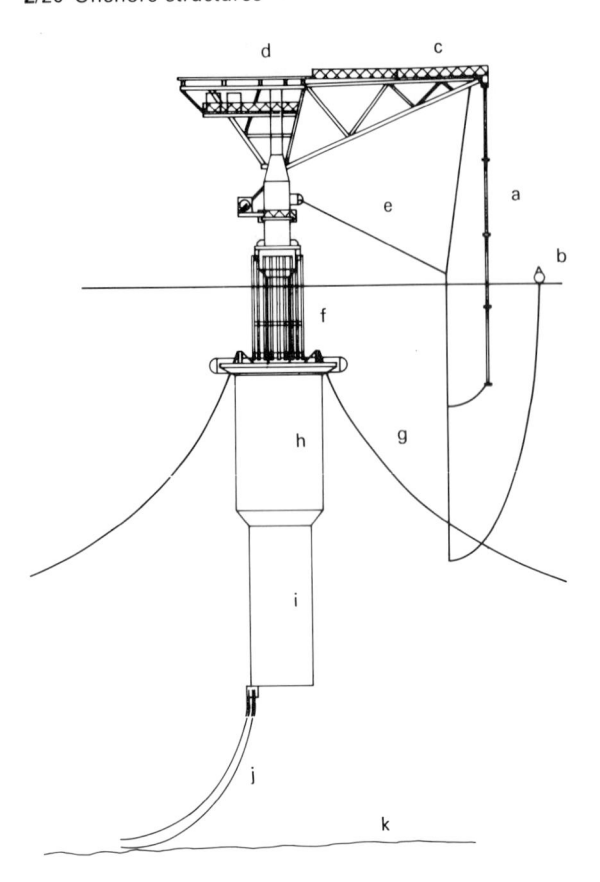

Figure 2.19 Typical exposed location single buoy mooring (ELSBM): a – oil loading hose; b – mooring hawser and loading hose pick up buoy; c – revolving loading arm; d – helicopter pad; e – mooring hawser; f – fenders; g – buoy mooring chains; h – buoyancy compartments; i – ballast compartments; j – oil transfer hose

dependent forces on the structure. This section is concerned with describing and developing methods of calculating these forces for fixed offshore structures. A general treatment of fluid flow around streamlined and bluff bodies is followed by a description of specific methods to calculate resultant body forces. However, offshore structures for deeper waters tend to be more flexible structures with lower natural frequencies of response. In such cases, it is necessary to calculate the loading resulting from wave action on members which themselves may have motions of significant magnitude compared to water particle motions. This would also apply to compliant or fully floating offshore platforms and is considered further in Section 2.5.

Consider an arbitrary body that is stationary in a uniform steady fluid flow. The presence of the body will modify the velocity distribution of the fluid as it passes the body and the resultant pressure distribution can be integrated over the body to yield a total force. The component of this force perpendicular to the undisturbed flow direction is defined as the lift or transverse force while the component in the undisturbed flow direction is defined as the drag or in line force. *Figures 2.22(a)* and *(b)* show two typical bodies – a streamlined aerofoil section and a circular cylindrical bluff body. For the aerofoil section, the lift to drag ratio can be as high as 40 whereas for the bluff body, the drag force is high and the mean lift force is zero although vortex shedding induces significant periodic lift forces.

The flow induced pressures, p, lift L and drag, D on such streamlined and bluff bodies can be non-dimensionalized to yield pressure, lift and drag coefficients which are defined as:

$$C_p = \frac{p - p_a}{\frac{1}{2}\rho U^2}$$

$$C_L = \frac{L}{\frac{1}{2}\rho U^2 S} \quad (2.1)$$

$$C_D = \frac{D}{\frac{1}{2}\rho U^2 S}$$

respectively, where ρ is fluid density, U is fluid velocity and S is normally the frontal area projected in the flow direction although for thin streamlined bodies such as aerofoils, S is defined as the plan area of the body.

Now, the general turbulent flow around an arbitrary bluff body cannot be calculated in a mathematically rigorous manner at present because of the non-linear convective acceleration terms on the left-hand sides of the Navier Stokes Equations. However, a number of empirical techniques have been developed to enable forces due to flow around bluff bodies to be quantified for engineering design purposes. The fundamental concepts of fluid mechanics that underpin these techniques are described further in this section.

2.3.2 Concepts of fluid mechanics

The concept of similarity in fluid mechanics relates to the behaviour of fluid flow in systems of different physical magnitudes. For example, a question may be posed as: 'How is the flow of a fluid of velocity U around a cylinder of diameter $2a$ related to the flow of a fluid of velocity $2U$ around a cylinder of diameter a?'

There are three preconditions that must apply before considering the similarity of two systems of fluid flow. These are that the systems are geometrically similar; the systems have similar internal constitutions, that is both fluids may be Newtonian although of different physical properties; and that they have similar conditions at the system boundaries.

2.3.3 Viscous forces and the Reynolds number

Consider a flow of velocity U incident on a body of characteristic length L where the above three conditions are satisfied. The fluid can be assumed to be

Environmental loading 2/21

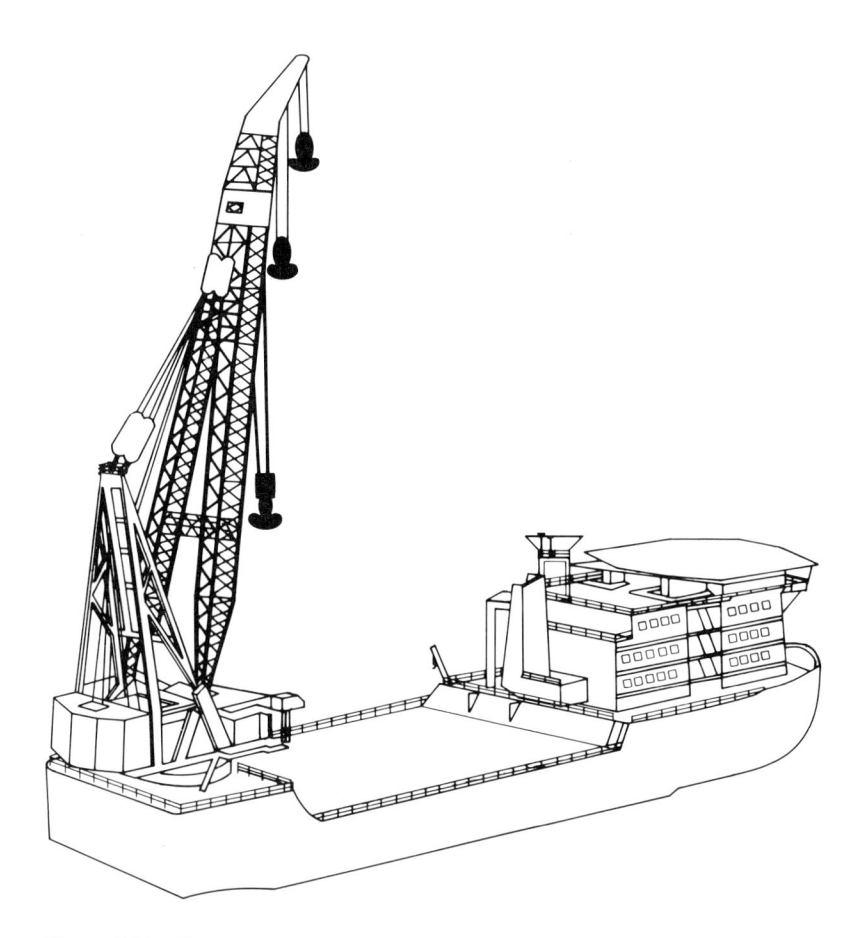

Figure 2.20 Ship shape crane vessel

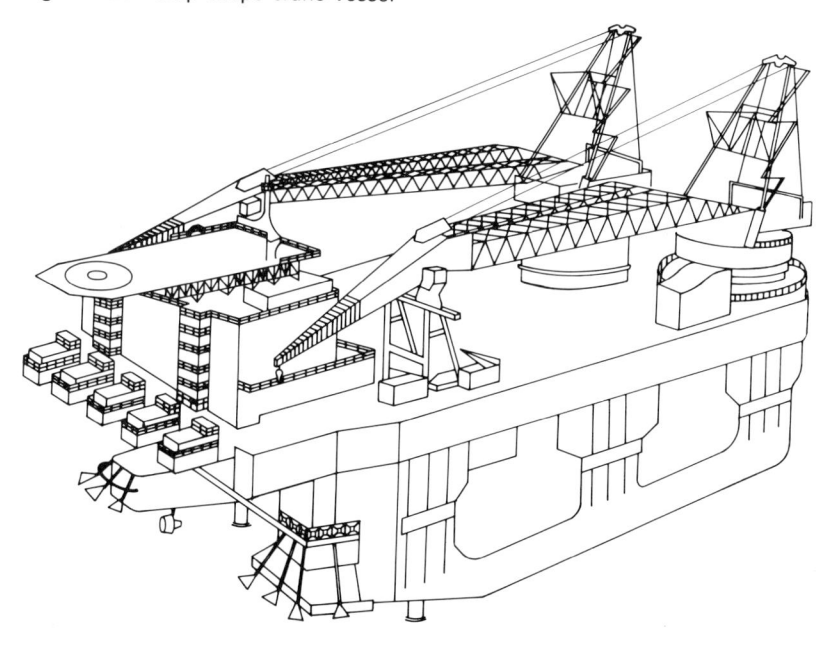

Figure 2.21 Semi-submersible crane vessel

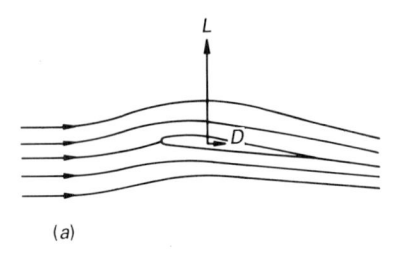

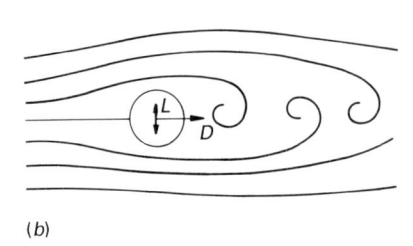

Figure 2.22 Streamlines around (a) aerofoil section and (b) circular cylinder

homogeneous and incompressible with a constant dynamic viscosity, and density, ρ. The motion of the fluid around the body will be governed by forces due to accelerations (inertia forces) and also due to viscosity (viscous forces). A ratio of these two forces will yield a numerical characteristic of the relative magnitudes of these two mechanisms in the flow. Now the volume of a representative element of fluid in the flow pattern will be proportional to ρL^3. Corresponding components of acceleration at corresponding elements will be proportional to U^2/L, U being the velocity and L being proportional to the radius of curvature of the fluid flow paths. Then the inertia force being proportional to mass times acceleration, can be written as:

$$\rho L^3 \cdot (U^2/L) = \rho U^2 L^2 \quad (2.2)$$

For viscous forces, the velocity gradient in the fluid element being considered will be proportional to U/L; so the shear stress will be proportional to U/L and the viscous force which is given by shear stress times area of element will be proportional to:

$$\frac{\mu U}{L} \times L^2 = \mu U L$$

Thus the ratio of inertia to viscous forces is:

$$\frac{\rho U^2 L^2}{\mu U L} = \frac{UL}{(\mu/\rho)} = \frac{UL}{\nu} \quad (2.3)$$

This is defined as the Reynolds number. Two flow systems with the same Reynolds number have similar flows and is thus a condition of similarity in flows where only inertia and viscous forces play a part. This answers the question posed earlier in that the two flow systems are similar (or equivalent) since the ratios of internal inertia to viscous forces are equal. Thus, given a solution for one system, the solution for the other system is simply obtained by a suitable scaling of parameters. For most flows of practical interest, viscous forces are small, giving values of Reynolds numbers of up to 5×10^6.

The Reynolds number and the definition of force coefficients in Equations 2.1 can also be derived using dimensional analysis. Now the drag force, D, on an arbitrary body in a fluid flow will be expected to be a function of fluid density, ρ, velocity U, the dynamic viscosity, μ, and the characteristic length of the body, L. Then:

$$D = f(\rho, U, L, \mu) \quad (2.4)$$

and the function can be expected to have a form:

$$D = k \, \rho^a \, U^b \, L^c \, \mu^d \quad (2.5)$$

where a, b, c, d and k are constants to be determined. Dimensionally, this equation can be written in terms of mass, length and time (M', L' and T') as:

$$\frac{M'L'}{T'^2} = \left[\frac{M'}{L'^3}\right]^a \left[\frac{L'}{T'}\right]^b L'^c \left[\frac{M'}{L'T'}\right]^d \quad (2.6)$$

Equating powers of M, L and T yields:

$$1 = a + d$$
$$1 = -3a + b + c - d$$
$$-2 = -b - d \quad (2.7)$$

which can be simplified to:

$$a = 1 - d, \, b = 2 - d, \, c = 2 - d$$

Equation 2.5, therefore, becomes:

$$\frac{D}{\rho U^2 L^2} = \text{constant} \cdot \left[\frac{UL}{\nu}\right]^{-d}$$

which can then be written as:

$$C_D = \frac{D}{\frac{1}{2}\rho U^2 S} = f\left[\frac{UL}{\nu}\right] \quad (2.8)$$

where the characteristic length squared is conveniently written as an area, S, and a factor of half is introduced. The drag coefficient is thus a function of the Reynolds number only. The most important implication of flow similarity is that the force coefficients on bodies of similar shape in uniform motion in an infinite incompressible fluid can be simply represented as a function of the Reynolds number for all such bodies. The form of the function depends solely on the shape of the bodies and the presence of an unconfined fluid. The concept of Reynolds number also applies to unidirectional phases within an oscillating flow.

For a viscous incompressible uniform fluid with a free surface in a uniform gravitational (body force) field, there are two numerical conditions for the similarity of two systems of fluid flow. The first is equality of the Reynolds number, UL/ν. The second, however, is a ratio of inertia forces to body (gravitational) forces acting on the fluid. This can be written as the non-dimensional number:

$$\frac{\text{inertia force}}{\text{gravitational force}} = \frac{\rho L^3 \times U^2/L}{\rho L^3 \times g} = \frac{U^2}{gL} \quad (2.9)$$

and the square root of this ratio is called the Froude number. It refers essentially to the wave making properties of a body at the fluid surface; waves being an oscillation obtained by an interplay of fluid inertia and restoring gravity forces.

Thus for any incompressible fluid flow the pressure, lift and drag coefficients are functions of the non-dimensional Reynolds and Froude numbers. The Reynolds number is always significant whereas the Froude number is of interest for flows with free surfaces. The similarity of non-dimensional numbers is also a key issue in representing prototype situations in model form for wind tunnel and wave tank testing.

2.3.3.1 Boundary layers

A relative velocity between a flow field and a solid body is governed by a boundary condition that the fluid layer immediately adjacent to the body does not move relative to the body. This is often called the no slip boundary condition. For flows around streamlined bodies or upstream segments of flow around bluff bodies, the no slip boundary condition gives rise to a thin layer of fluid adjacent to the surface where the flow velocity relative to the surface increases rapidly from zero at the surface to the local stream velocity at the outer edge of the layer. Such a thin sheared layer is appropriately called the boundary layer.

For many practical situations (e.g. flow around ships or aircraft) the boundary layer thickness is very small compared with distance from the forward end of the vehicle over which the flow has developed. Hence, the velocity gradients within a boundary layer in a direction perpendicular to the surface are very large in comparison to velocity gradients parallel to the surface. These former velocity gradients induce large shear stresses from the action of viscosity within the boundary layer fluid whereas, outside this layer, the velocity gradients and, therefore, the shear stresses are sufficiently small that the fluid can be regarded as being inviscid.

Thus, many problems concerned with flow of a fluid of small viscosity around a streamlined body can be solved by considering the fluid as inviscid everywhere except within a thin boundary layer region adjacent to the body and in the wake that forms behind it.

In the boundary layer and wake, however, the rates of shear strain are high so that the effects of viscosity and the associated shear stresses must be accounted for. The value of this shear stress at the body surface contributes to a frictional drag force which is of considerable engineering significance for streamlined bodies.

Now as long as the boundary layer is thin, the bulk of the flow around the body is very close to that predicted by inviscid flow theory with the body surface as a boundary so that matching that flow to the boundary layer flow does not present a serious problem. There are, however, cases where the boundary layer is not sufficiently thin or it separates from the surface to form a thick wake. In such situations, the boundary layer has a substantial effect on the bulk of the flow around the body and on the force experienced by the body. Boundary layer separation and the formation of a thick wake are characteristic features of flow around the bluff bodies typically used as members of offshore structures.

Figuire 2.33(a) presents typical development of a boundary layer flow around a streamlined shape. The boundary layer grows downstream from the leading edge of the shape.

During early development of the boundary layer, the flow within it is smooth with streamlines approximately parallel to the surface. This is defined as a laminar boundary layer and has a profile characterized by the shape given in *Figure 2.23(b)*. As the laminar boundary

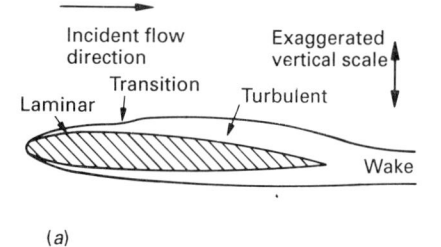

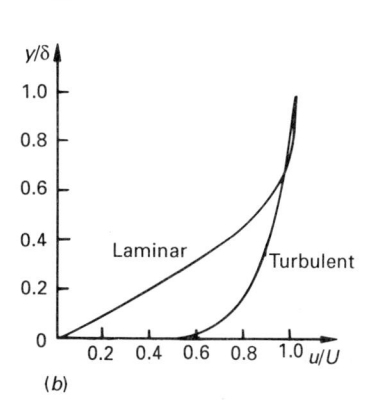

Figure 2.23 Boundary layer development: (a) boundary layer structure; (b) typical laminar and turbulent boundary layer velocity profiles

layer develops along the surface, at some stage, transition occurs to a boundary layer flow characterized by significant random fluctuations in direction and velocity superimposed on a mean flow. This mean flow still remains parallel to the surface but the turbulent fluctuations in the flow can be over 10% of the magnitude of the mean velocity. Such a sheared flow is called a turbulent boundary layer. It has a fuller mean velocity profile as shown in *Figure 2.23(b)*.

The laminar boundary layer is very sensitive to disturbances in the presence of a positive (i.e. opposing) pressure gradient and will readily separate or become turbulent. Its sensitivity increases with the increase of the Reynolds number. For Reynolds numbers less than 10^3, the laminar boundary layer tends to separate more readily from the surface than it becomes turbulent, and complete breakaway of flow may result, but in some cases the boundary layer may re-attach itself in the turbulent state. The precise sequence of events depends on the pressure distribution over the surface, the state of the surface (i.e. whether rough or wavy), and turbulence in the incident flow. Laminar separation followed by re-attachment can also occur at a higher Reynolds number with an example of offshore interest being flow around a circular cylinder as shown later.

2.3.3.2 Boundary layer separation

The phenomenon of boundary layer separation also plays a very important part in flows around typical members of offshore structures. Consider the boundary layer of *Figure 2.24* moving into a region of rising pressure (as over the rear of a circular cylinder or aircraft wing). As the flow progresses towards the right, it tends to lose velocity, as it would in an inviscid flow, but because of the shear stresses associated with viscosity there is some interchange of momentum between adjacent fluid layers with momentum passing from the faster to the slower layers. Consequently, the velocity decrease in the slower moving streamlines nearer the surface is not as great for a given increase of pressure as Bernoulli's equation would predict. Nevertheless, the flow slows down and may eventually reverse in direction near the surface. The boundary layer is then said to separate. The boundary layer at separation appears to move away from the surface with a large eddy forming between it and the surface. Such eddies are unstable and tend to move downstream from the surface with new eddies forming to replace them. The wake behind the body is then filled with a stream of eddies some of which can be as large as the body itself (see *Figure 2.22(b)*).

These eddies and wakes involve large mechanical energy losses and result in a substantial drag force on the body, mainly manifested by high suctions over the rear of the body. In contrast, the streamlined shape of *Figure 2.22(a)* has a small wake and a small drag force if the angle of incidence is such that no boundary layer separation occurs. In this case, the drag force is largely due to tangential viscous stresses.

2.3.3.3 Bluff body flows and Reynolds number

Now, the state of knowledge in boundary layer theory is such that calculation for the behaviour of such layers on aircraft and ship hulls can be made with sufficient accuracy for engineering purposes. The calculations are, however, restricted to streamlined bodies whereas the members of offshore structures are invariably bluff bodied to the extent that flow about these bodies is in a complex coupled inviscid and viscous regime. Nevertheless, a qualitative understanding of such bluff body flows is provided by considering the flow around a circular cylinder over a range of Reynolds numbers. *Figures 2.25(a)* to *(f)* illustrate these flow patterns around a circular cylinder whereas *Figure 2.26* presents the resultant variation of drag coefficient with Reynolds number. The flow regime corresponding to Reynolds number Re, of less than 5, is a Stokean creeping motion which is of little practical significance since it corresponds to very small cylinder diameters or very slow flow velocities. An analytic solution for this flow does, however, exist. Over the Reynolds number range 5–40, separation of the laminar boundary layer occurs at the point marked L and two stable closed eddies are formed at the rear of the cylinder with the downstream wake remaining laminar.

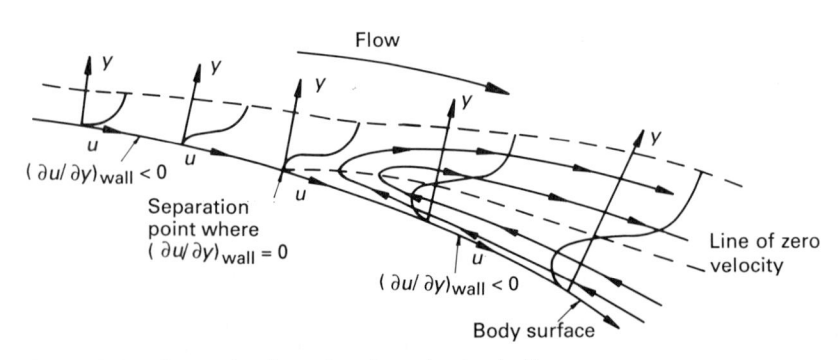

Figure 2.24 Separating boundary layer in steady flow

(a) Re < 5

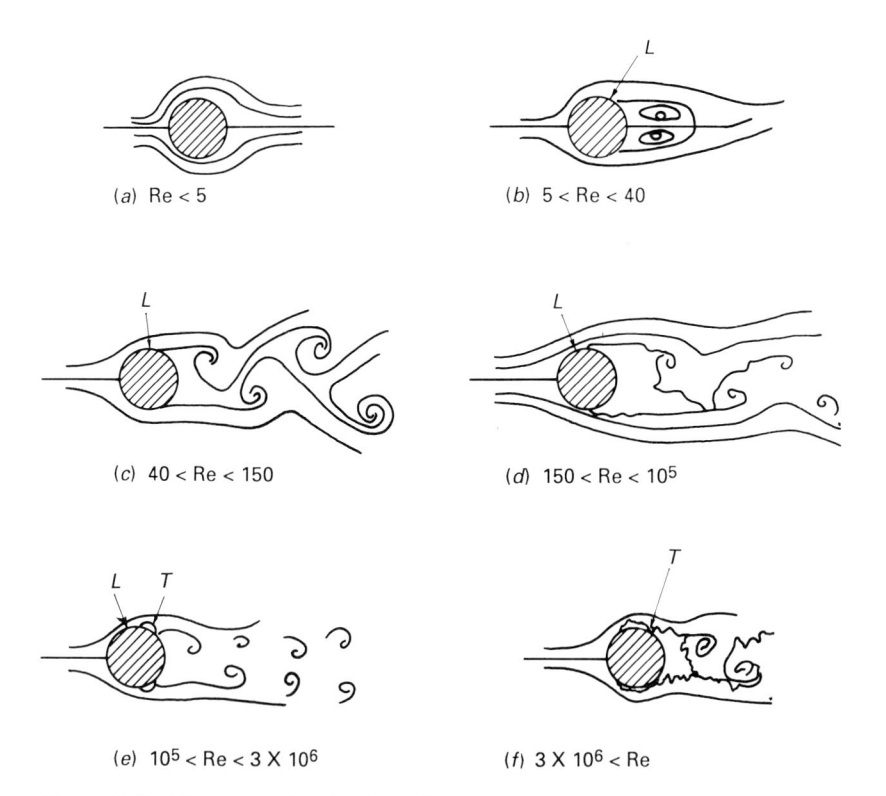

(b) 5 < Re < 40

(c) 40 < Re < 150

(d) 150 < Re < 10^5

(e) 10^5 < Re < 3 × 10^6

(f) 3 × 10^6 < Re

Figure 2.25 Flow around a circular cylinder for a range of Reynolds numbers. L and T denote laminar and turbulent boundary layer separation (from ESDU 1986)

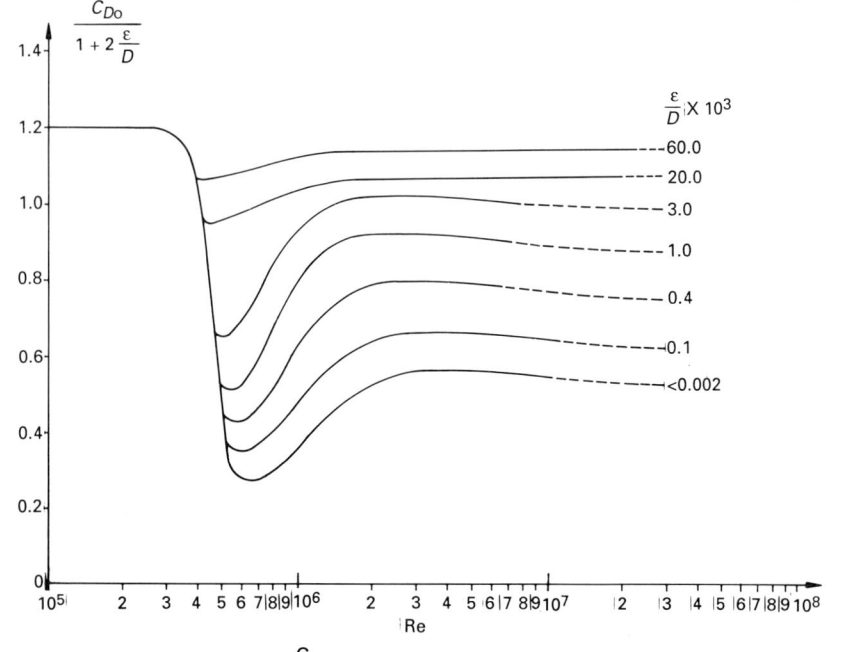

$\dfrac{\varepsilon}{D} \times 10^3$

---60.0
----20.0
----3.0
----1.0
----0.4
----0.1
----<0.002

Figure 2.26 Variation of $\dfrac{C_{Do}}{1 + 2\varepsilon/D}$ with Re (From ESDU 1986)

In the Reynolds number range 40–150, the disturbances downstream of the cylinder increase in intensity and well developed vortices appear. The eddies downstream of the cylinder begin to be shed and re-appear periodically to form a series of vortices known as the Karman vortex street. Boundary layer separation still remains laminar. As Re increases in the range 150–10^5, the flow pattern remains as above but the wake becomes turbulent. The non-symmetric vortex formation for $Re > 40$ gives rise to oscillating lift forces with a frequency defined by a Strouhal number (see Section 2.3.6).

For the Reynolds number range of 10^5–3×10^6, the flow in the wake is entirely turbulent. On the cylinder, a laminar separation is followed by a re-attachment and turbulent separation. The turbulent boundary layer, because of its fuller velocity profile and higher turbulence, has a higher kinetic energy and, therefore, remains attached for longer distances round the cylinder before separating. The width of the wake therefore decreases and thus the pressure drag decreases giving the net effect of a sharp drop in the drag coefficient. For Reynolds numbers greater than 3×10^6, the flow, including the cylinder surface boundary layer, is fully turbulent with vortex shedding occurring at a frequency defined by the Strouhal number.

2.3.4 Wind and current loads

The fluid loading mechanisms of wind and current on the exposed or submerged members of offshore structures are identical apart from the different natures of the incident flow fields and the physical properties of the media.

2.3.4.1 Wind forces

Wind forces on offshore structures account for approximately 15% of the total forces from wind, current and waves acting on the structure. In deep water, wind induced forces can have a larger relative effect on the base overturning moment of bottom emplaced structures because of the large moment arm involved.

Since the presence of high winds is always accompanied by high seas, the loadings from wind and waves are always applied simultaneously in the design process. The general equation for the wind force on a slender member of an open space framed offshore structure exposed to a wind is:

$$F = \tfrac{1}{2} \rho_a U_n^2 S C_D \qquad (2.10)$$

where U_n is the flow velocity incident on the member due to the wind, ρ_a is the air density, S is the frontal area (facing the wind) and C_D is a drag coefficient obtained from experiment and experience. A value of C_D in the range 1.1–1.3 is applied for most bluff bodies although a C_D value of 1.5 is often used for design purposes. Two kinds of wind speeds are considered in the design of offshore structures, these being the sustained and gust wind speeds.

The sustained wind speed is defined as the average wind speed over a time interval of 1 min measured at an elevation of 10 m (32 ft) above still water level. It is usual to define a sustained wind speed value that is statistically likely to occur once in 50 or 100 years (the return period) for design purposes. The wind velocity above still water level (SWL) varies significantly with height due to the boundary layer induced by viscosity. The speed at a height z above SWL can be related to the speed at 10 m height by the relationship:

$$U_z = U_{10} \left(\frac{z}{10}\right)^{0.113} \qquad (2.11)$$

where U_{10} is the 1 min mean sustained wind speed 10 m above still water level.

The gust wind speed is defined as the average wind speed over a time interval of 3 s measured at an elevation of 10 m (32 ft) above still water level. In the same sense as above, the 50 or 100 year return period gust wind speed can be stipulated for design purposes. Adjustments for elevation above still water level are given by the equation:

$$(U_z)_{\text{GUST}} = (U_{10})_{\text{GUST}} \left(\frac{z}{10}\right)^{0.100} \qquad (2.12)$$

These power law equations are semi-empirical and are continually being updated by new data. Values of sustained and gust wind speeds are obtained from meteorological records or from recommendations made by certifying authorities in the offshore area of interest. Thom (1973) presents some analysis techniques for obtaining 50 and 100 year return period design wind speeds from meteorological data for a shorter interval. Thom also presents typical distributions of extreme winds over open oceans in the presence of storms. In general, design will be more accurate if the 100 year return period wind speed criteria are based on local meteorological data sources.

As explained before, actual values of drag coefficient, C_D, used in Equation (2.10) are approximately 1.1–1.3 but higher values can be used in design to incorporate a safety factor. Generally, for long, slender members with length to diameter ratios greater than 5, C_D for sharp edged sections is taken in the range 1.5–2.0 whereas for cylindrical sections of diameters less than 0.3 m (0.975 ft), $C_D = 1.2$ and for diameters greater than 0.3 m (0.975 ft), $C_D = 0.7$. Also for shorter members:

$$C_{DS} = C_D \left[0.5 + \frac{0.1\,L}{D} \right] \qquad (2.13)$$

where C_D and C_{DS} are drag coefficients for long and short members, respectively, L is member length and D is member diameter.

If members are located behind each other in the direction of the wind, a shielding effect can be taken into account by reducing the effective area of members down wind according to the distance between the members, provided that this distance is less than seven times the upwind member diameter. More detailed guidance on these is provided by ESDU (1986) for circular cylinders.

Trusses or columns which are partly open can have their appropriate drag coefficients reduced. For complex structures and in cases where wind loads are vitally important (e.g. the stability of towed or semisubmersible structures), wind tunnel experiments are often necessary to predict design loads. Miller and Davies (1982) and Hoerner (1965) give a comprehensive survey of fluid dynamic drag coefficients for offshore structures and for a wide variety of other shapes.

Finally, for some streamlined structures, typically like a hangar, substantial lift loads can be induced. The design value for such forces can be given by the equation:

$$L = C_L \tfrac{1}{2} \rho_a U^2 S' \qquad (2.14)$$

where S' is the plan area of the structure and C_L is an appropriate lift coefficient obtained from wind tunnel tests or aerodynamic calculations. Although the sustained and gust wind speeds offer some quantification of the statistically random nature of wind turbulence, this feature can be defined further by writing a power spectral density of the in line (parallel to the mean wind direction) component of wind velocity fluctuations. This in line component is generally much larger than the two components perpendicular to the wind direction and its spectral density is given by Davenport (1961) to be of the form:

$$S_{vv}(f) = \frac{4 k \bar{f}}{[1 + \bar{f}^2]^{4/3}} \cdot \frac{\bar{U}_{10}^2}{f} \qquad (2.15)$$

where f is frequency in Hz, $\bar{f}$ is a normalized frequency ($\bar{f} = fL/U_{10}$), k is the sea surface drag coefficient (taken to be 0.005 for open sea) and $\bar{U}_{10}$ is the hourly mean wind velocity at 10 m (32 ft) height. L denotes a representative length scale taken to be 1200 m (3900 ft) here. *Figure 2.27* presents a typical in line gust velocity spectrum for $\bar{U}_{10}$ = 30 m/s (98 ft/s) and L = 1200 m (3937 ft).

This in line wind gust velocity spectrum can be readily applied to a member of an open space frame structure through Equation (2.10), since the member diameter will be small compared to turbulent fluctuation length scales. The transformation from velocity spectrum to wind force will require a linearization of the velocity dependency in Equation (2.10) or the conversion of the frequency domain spectrum into a velocity time history. For very large or monolithic structures, however, calculation of gust forces must take account of the fact that a gust fluctuation length scale which is smaller or of the same dimension as the structure will tend to apply very much smaller loads. A calculation approach using aerodynamic admittance

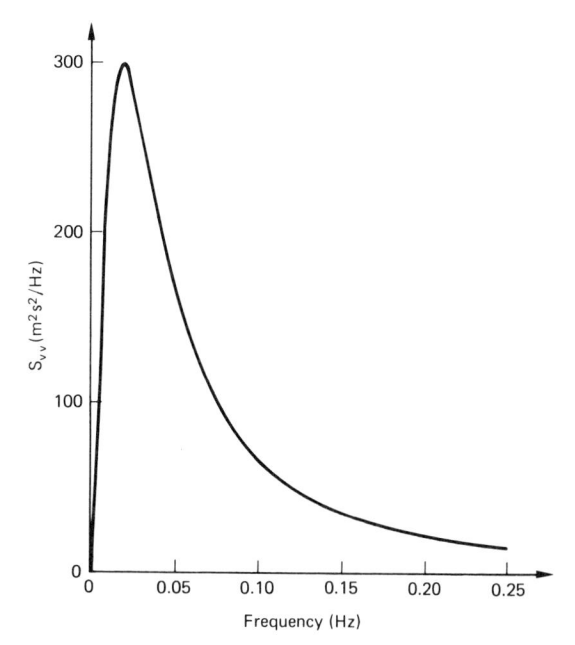

Figure 2.27 Typical wind gust spectrum with L = 1200m, V_{10} = 30m/s and k = 0.005

functions is necessary here. This technique is described by Davenport (1967) but is only really necessary for calculating wind forces on very large bulky structures.

2.3.4.2 Current forces

Current forces on offshore structures are calculated using the same methods as for wind forces alone apart from the fact that currents do not exhibit similar fluctuations to those of wind gusts. The current velocities used are much smaller (<2.5 m/s or 8 ft/s) than those for wind force calculations although, of course, the water density is higher. The magnitude of lift and drag forces due to current may be increased substantially if currents can also affect gravity waves by refraction which results in a change in the apparent wavelength and direction of incident waves.

Drag force coefficient data for currents is obtained in the same way as for forces due to winds. In critical applications, however, these data might need to be confirmed by model tests. Furthermore, for most large structures, a substantial current can induce frictional and pressure drag as well as a residual drag which is due primarily to wave making at the surface. This residual drag must also be accounted for in calculations.

2.3.5 Wave loads on offshore structures

2.3.5.1 Gravity waves

Gravity waves form part of a large family of surface water waves ranging from wind generated waves in the oceans, to tidal action, water surface oscillations in harbours, flood waves in rivers, tidal bores, hydraulic jumps, waves generated by siesmic disturbances and so on. This subsection is concerned primarily with gravity waves generated by wind action. The unsteady fluid velocities, accelerations and pressures induced by such waves are responsible for a predominant proportion of the environmental forces acting on offshore structures, subsea pipes and associated submerged structures.

Gravity waves can exist at the interface between any two fluids of different densities, e.g. air and water, and are driven by an interaction between inertia of the fluid at the interface and its potential energy under gravitational action. The physical mechanisms underlying gravity wave formation by wind action and decay by internal losses are complex. Winds blowing over a water surface generate waves with the length over which this occurs being called a fetch and the resultant forced waves classified as wind waves or a sea. Gravity waves are progressive (or travelling) in nature and can, therefore, move out of their area of formation and when running free of the winds that generated them are called a swell. Such swell waves lose energy very gradually due to air resistance, internal friction and by friction with the sea bed in shallow water. Thus swell waves can travel substantial distances across the Earth's surface with little reduction in intensity. Gravity waves are, however, dispersive in nature; their speed (wave or phase speed) is a function of wave period or wave length so that they spread out (disperse) along their direction of wave propagation.

In nature, very many components of periodic waves with different wave heights, periods and directions of travel occur at the same time in a given area. The superposition of all of these wave components coupled with their dispersive behaviour leads to a randomly varying sea surface elevation which can be treated by stastistical methods. There are a number of wave theories which are used to mathematically represent long crested periodic waves with varying degrees of accuracy. The simplest of these theories is a linearized mathematical model of gravity waves called Linear Wave Theory, that permits the determination of irregular ocean waves by superposition of regular wave motions at different periods and from different directions.

2.3.5.2 Linear wave theory

The main difficulty in the study of ocean waves is that one of the boundaries, the free surface, on which boundary conditions have to be evaluated is also one of the unknowns. Mathematically, this difficulty arises from the non-linear convective, inertia terms in the Navier Stokes equations which occur as products of velocities and their spatial gradients. However, since these products of velocities tend to be small for waves of small height compared to other terms in the equation of motion, a convenient idealization is obtained by assuming small wave height and neglecting convective inertia terms. This is the basis of linear wave theory described here. Other more exact mathematical solutions do not use linearization and account for the convective terms by methods based on power series.

Initially, some basic parameters describing the characteristics of a progressive wave are defined. The wave number, k, is defined as:

$$k = \frac{2\pi}{\lambda} \tag{2.16}$$

where λ is the wave length. Wave frequency can be defined as f in Hz, ω in radians per second or period T in seconds. These parameters are related by the equations:

$$\omega = 2\pi f, \quad T = \frac{2\pi}{\omega}, \quad T = \frac{1}{f} \tag{2.17}$$

The wave speed, c (that is, its phase velocity) of any progressive wave is related to the wave length, λ, and frequency, f, by the equation:

$$c = \lambda f \tag{2.18}$$

Linear wave theory, originally developed by G B Airy (1845), offers a powerful analytical solution for gravity waves. The following assumptions have to be invoked:

1. The water is assumed to be of uniform density with constant water depth.
2. Viscosity and surface tension effects are neglected.
3. The square of particle velocities are considered to be negligible but vertical accelerations are not neglected. The wave heights are accordingly assumed to be very low, hence the theory is often called small amplitude wave theory.
4. The wave motion is assumed to be irrotational.

The two dimensional linear gravity wave has to satisfy the equation of continuity which for irrotational flow can be written in terms of velocity potential, ϕ, as:

$$\frac{\partial^2 \phi}{\partial x^2} + \frac{\partial^2 \phi}{\partial y^2} = 0 \tag{2.19}$$

and the dynamic equation of motion (or Bernoulli's Equation) with irrotational flow, also, is:

$$\frac{p}{\rho} = -gy - \frac{\partial \phi}{\partial t} - \frac{1}{2}\left[\left(\frac{\partial \phi}{\partial x}\right)^2 + \left(\frac{\partial \phi}{\partial y}\right)^2\right] \tag{2.20}$$

with respect to the reference axes with the origin at the still water surface, the y axis pointing vertically upwards and the x axis pointing to the right. Variable P denotes pressure, ρ is water density, t is time and g is the acceleration due to gravity. The Bernoulli Equation incorporates the assumptions of inviscid and irrotational flow.

The boundary condition of zero flow through the sea bed at $y = -d$ gives:

$$\left(\frac{\partial \phi}{\partial y}\right)_{y=-d} = 0 \tag{2.21}$$

The boundary condition at the free air/water surface poses difficulties however. The unknown water surface shape, defined by a function $\eta = f(x, y, t)$, can be regarded as a flow boundary to which the local velocity vector must be tangential. Thus if n_η is a local normal to the free surface, then:

$$\left(\frac{\partial \phi}{\partial n_\eta}\right)_s = 0 \tag{2.22}$$

This is referred to as a kinetic free surface boundary condition.

A further dynamic free surface boundary condition can also be specified as constant pressure on the free surface shape $\eta = f(x, y, t)$.

The non-linearity of the above system of equations arises from the fact that the free surface boundary condition is applied on an unknown surface which will be obtained only after the solution of the governing equations and boundary-conditions.

Now the water pressure at the sea surface can be obtained from Equation 2.20 as:

$$\frac{p_s}{\rho} + g\eta + \left(\frac{\partial \phi}{\partial t}\right)_s + \frac{1}{2}\left[\left(\frac{\partial \phi}{\partial x}\right)^2 + \left(\frac{\partial \phi}{\partial y}\right)^2\right]_s = 0 \tag{2.23}$$

with subscript s denoting quantities defined at the water surface.

Differentiating Equation 2.23 with respect to time and neglecting velocity squared terms gives:

$$\frac{1}{\rho}\frac{\partial p_s}{\partial t} + g\left(\frac{D\eta}{Dt}\right)_s + \left(\frac{\partial^2 \phi}{\partial t^2}\right)_s = 0 \tag{2.24}$$

where the derivative of η has strictly to be the substantive derivative:

$$\frac{D\eta}{Dt} = \frac{\partial \eta}{\partial t} + u\frac{\partial \eta}{\partial x} + v\frac{\partial \eta}{\partial y} \tag{2.25}$$

Ignoring surface tension gives:

$$p_s = p_a \text{ and } \frac{\partial p_s}{\partial t} = \frac{\partial p_a}{\partial t} = 0 \tag{2.26}$$

for constant atmospheric pressure with subscript, a, denoting quantities just above the water surface. Furthermore, for waves of small (mathematically infinitesimal) height:

$$\frac{D\eta}{Dt} = \frac{\partial \eta}{\partial t} = \left(\frac{\partial \phi}{\partial t}\right)_{y=0} \tag{2.27}$$

Substituting Equations 2.26 and 2.27 into Equation 2.24 yields the equation:

$$g\left(\frac{\partial \phi}{\partial y}\right)_s + \left(\frac{\partial^2 \phi}{\partial t^2}\right)_s = 0 \tag{2.28}$$

This is called the linearized free surface boundary condition.

The removal of velocity squared terms limits the applicability of linear theory to infinitesimal waves in mathematical terms. Note that the linearization process not only makes the equations linear but also fixes the free surface boundary. For engineering purposes, however, the theory can be used for small to moderate wave heights.

The Linear Wave Theory problem can thus be restated as:
Governing equation:

$$\frac{\partial^2 \phi}{\partial x^2} + \frac{\partial^2 \phi}{\partial y^2} = 0 \tag{2.29}$$

Boundary conditions
Sea bed:

$$\frac{\partial \phi}{\partial y} = 0 \text{ at } y = -d$$

Surface: $\tag{2.30}$

$$g\left(\frac{\partial \phi}{\partial y}\right)_s + \left(\frac{\partial^2 \phi}{\partial t^2}\right)_s = 0 \text{ at } y = 0$$

The above is an elliptic partial differential equation with the solution domain limited only by horizontal planes $y = 0$ and $-d$ and with no limits in the x direction. The solution for ϕ is, therefore, likely to be a simple harmonic function of x and will be of the form:

$$\phi = F(y) \, e^{-i(kx-\omega t)} \tag{2.31}$$

Substituting this into Equation 2.29 and dividing by the exponential term yields the ordinary differential equation:

$$\frac{d^2 F}{dy^2} - k^2 F = 0 \tag{2.32}$$

This has the solution:

$$F = Ae^{ky} + Be^{-ky} \tag{2.33}$$

Applying the boundary condition:

$$\left(\frac{\partial \phi}{\partial y}\right)_{y=-d} = \left(\frac{dF}{dy}\right)_{y=-d} = 0 \tag{2.34}$$

yields:

$$Ake^{-kd} - Bke^{kd} = 0$$

$$\therefore Ake^{-kd} = Bke^{kd} = \tfrac{1}{2} C \text{ say}$$

and $A = \dfrac{C}{2k} e^{kd}$, $B = \dfrac{C}{2k} e^{-kd}$ (2.35)

Substituting Equations 2.33 and 2.35 into Equation 2.31 yields:

$$\phi = \dfrac{C}{k} \cosh k(y + d) \, e^{-i(kx - \omega t)} \quad (2.36)$$

where the constant C still remains to be evaluated. This can be done by considering oscillatory variations due to the wave in Equation 2.23. Since the pressure in the medium above the wave surface is taken as uniform and velocity squared terms are neglected, the oscillatory wave surface elevation η is given by the equation:

$$\eta = -\dfrac{1}{g} \left(\dfrac{\partial \phi}{\partial t}\right)_{y=0} \quad (2.37)$$

Substituting Equation 2.36 into the above equation for $y = 0$ gives:

$$\eta = -\dfrac{\omega C}{gk} \cosh kd \cdot ie^{-i(kx - \omega t)}$$

Combining the terms from the above equation into constant a yields:

$$\eta = -\dfrac{\omega C}{gk} \cosh kd \quad (2.38)$$

thus giving:

$$\eta = a \cdot ie^{-i(kx - \omega t)} \quad (2.39)$$

where a can be regarded as a wave amplitude. Substituting Equation 2.38 into 2.36 then gives:

$$\phi = -\dfrac{ag}{\omega} \dfrac{\cosh k(y + d)}{\cosh kd} e^{-i(kx - \omega t)} \quad (2.40)$$

Substituting Equations 2.36 or 2.40 into the free surface boundary condition given by 2.30 also yields the equation:

$$\omega^2 = gk \tanh kd \quad (2.41)$$

Taking real parts of the complex terms in Equations 2.39 and 2.40 then gives a useful Linear Wave Theory solution as:

$$\phi = -\dfrac{ag}{\omega} \dfrac{\cosh k(y + d)}{\cosh kd} \cos(kx - \omega t) \quad (2.42)$$

with

$$\eta = a \sin(kx - \omega t) \quad (2.43)$$

Now using the progressive wave relationship:

$$c = f\lambda = \dfrac{\omega}{k} \quad (2.44)$$

and Equation 2.41, the velocity potential can also be written as:

$$\phi = -ac \dfrac{\cosh k(y + d)}{\sinh kd} \cos(kx - \omega t) \quad (2.45)$$

Furthermore, for deep water ($d \to +\infty$), the hyperbolic terms reduce to exponentials and Equations 2.42, 2.45 and 2.41 become:

$$\phi = -\dfrac{ag}{\omega} e^{ky} \cos(kx - \omega t)$$

or

$$\phi = -ac \, e^{ky} \cos(kx - \omega t) \quad (2.46)$$

and

$$\omega^2 = gk$$

respectively. The above equations are now used to explore some of the properties of deep water waves.

Simple relationships between wave period, T, wave length, λ and speed, c, can readily be defined for deep water from Equations 2.41 and 2.46. These can be written as:

$$c = \sqrt{\dfrac{g\lambda}{2\pi}}$$

$$T = \sqrt{\dfrac{2\pi\lambda}{g}} \quad (2.47)$$

$$c = \dfrac{gT}{2\pi}$$

For example, a deep water wave of 10 s period will have a wave length of 156.1 m (507 ft) and a speed of 15.61 m/s (51.2 ft/s) whereas a wave of 6 s period will have a wave length of 56.2 m (184 ft) and a speed of 9.37 m/s (30.7 ft/s). Thus long waves travel faster than short ones so as to overtake them and produce a continually varying water surface. This dispersive nature of gravity waves leads to a number of interesting properties.

Table 2.3 summarizes the velocity potentials, stream function and derived properties for linear deep water gravity waves. Equivalent expressions for shallow water are also listed. The expressions for shallow water theory differ from the deep water equations in having hyperbolic variations of vertical and horizontal particle displacements, velocities and accelerations with depth. The resulting difference in wave fluid behaviour is illustrated in *Figure 2.28* which shows wave particle orbits with depth for deep and shallow water. In shallow water, particle motions describe ellipses instead of circles. The ellipses become shallower with decreasing depth and at the bottom the motion is purely horizontal. In terms of numerical values, the shallow water equations tend to their deep water simplifications for water depth $D > \lambda/2$.

Table 2.3: Summary of water wave properties

Quantity	Deep water	Shallow water ($d/\lambda < 1/2$)
Surface profile	$\eta = a \sin(kx - \omega t)$	$\eta = a \sin(kx - \omega t)$
Velocity potential	$\phi = -ace^{ky}\cos(kx - \omega t)$	$\phi = -ac \dfrac{\cosh k(y+d)}{\sinh kd} \cos(kx - \omega t)$
Stream function	$\psi = +ace^{ky}\sin(kx - \omega t)$	$\psi = +ac \dfrac{\sinh k(y+d)}{\sinh kd} \cos(kx - \omega t)$
Horizontal water velocity	$u = \partial\phi/\partial x$ $= +kcae^{ky}\sin(kx - \omega t)$	$u = +ack \dfrac{\cosh k(y+d)}{\sinh kd} \sin(kx - \omega t)$
Vertical water velocity	$v = \partial\phi/\partial y$ $= -kcae^{ky}\cos(kx - \omega t)$	$v = -ack \dfrac{\sinh k(y+d)}{\sinh kd} \cos(kx - \omega t)$
Horizontal acceleration	$= -k^2c^2ae^{ky}\cos(kx - \omega t)$	$= -k^2c^2a \dfrac{\cosh k(y+d)}{\sinh kd} \cos(kx - \omega t)$
Vertical acceleration	$= -k^2c^2ae^{ky}\sin(kx - \omega t)$	$= -k^2c^2a \dfrac{\sinh k(y+d)}{\sinh kd} \cos(kx - \omega t)$
Wave speed	$c = \dfrac{\lambda}{T} = \dfrac{g}{\omega} = \left(\dfrac{g\lambda}{2\pi}\right)^{1/2} = \dfrac{gT}{2\pi}$	$c^2 = \dfrac{g\lambda}{2\pi} -\textit{anh } kd$ *
Wave length	$\lambda = \dfrac{2\pi c^2}{g} = \dfrac{2\pi g}{\omega^2} = \dfrac{gT^2}{2\pi}$	
Wave number	$k = \dfrac{2\pi}{\lambda} = \dfrac{\omega^2}{g} = \dfrac{g}{c^2} = \dfrac{gT^2}{gT^2}$	
Pressure ** $P' - P_a$	$= +\rho g a e^{ky} \sin(kx - \omega t)$	$= +\rho g a \dfrac{\cosh k(y+d)}{\cosh kd} \sin(kx - \omega t)$
Horizontal pressure gradient** $\partial p/\partial x$	$= +\rho g a k e^{ky} \cos(kx - \omega t)$	$= +\rho g a k \dfrac{\cosh k(y+d)}{\cosh kd} \cos(kx - \omega t)$
Vertical pressure gradient** $\partial p/\partial y$	$= +\rho g a k e^{ky} \sin(kx - \omega t)$	$= \rho g a k \dfrac{\sinh k(y+d)}{\cosh kd} \sin(kx - \omega t)$
Maximum wave slope	$\theta = ka = \dfrac{2\pi a}{\lambda}$	

* For very shallow water ($d \leq \lambda/20$), this expression becomes
$c^2 = gd$
with $c^2 = 0.968\, gd$ at $d = \lambda/20$.
For deeper water ($d > \lambda/2$), the expression becomes
$c^2 = \dfrac{g\lambda}{2\pi}$
with $c^2 = 0.996\, g\lambda/(2\pi)$ for $d = \lambda/2$.
** These terms exclude hydrostatic pressure

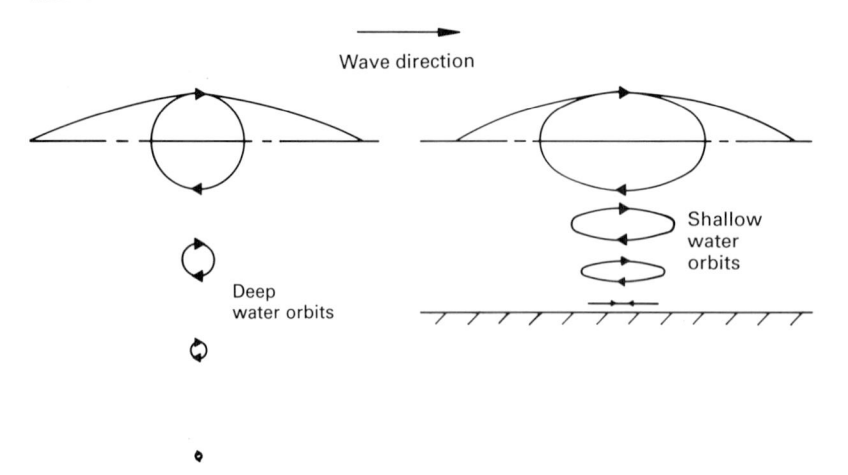

Figure 2.28 Wave particle orbits

2.3.5.3 Role of wave theories in design

Wave theories allow information on waves (height, period and water depth) to be translated into numerical values of fluid velocity, acceleration and pressure at any point under the wave. This presupposes an adequate knowledge of the wave climate at an offshore site. Such wave data has been obtained in the past by visual observations, wave recording instruments or from deductions based on meteorological data. Draper (1963) and Hogben and Lumb (1967) give typical design data and the procedures for deriving such data from ocean wave statistics. Wave data which is used as an input to design computations can be presented in two ways.

1. The design wave approach: the design wave approach is applied by defining a wave of large height, H, and a corresponding wave period range which has a sufficiently low probability of occurrence for it to be used as the maximum wave that the structure will encounter. The low probability of occurrence is achieved by statistically defining the highest wave that will be encountered in a large number (50 or 100) of years. Thus the structure can be designed to withstand this rare occurrence which is likely to be the worst encountered condition over the 20 year or so lifetime of the structure. Since the joint probability distribution of wind, current and wave action over such long time scales is unknown, it is conventional to apply the maximum 50 or 100 year return period values of these parameters simultaneously and acting from the same direction. This approach is realistic from the viewpoint of designing against structural failure due to large waves but does not permit fatigue damage to be considered within the design. There are well known statistical techniques that may be used to estimate 50 or 100 year return period wave data from oceanographic observations made over a limited 1–3 year time scale.

The design wave approach is used for both fixed and floating offshore structures. On fixed structures in shallow water, the design wave is applied as a quasi-static loading with the wave stepped past the structure through several points in its cycle. Table 2.4 presents examples of design wave and wind conditions that have been applied for four offshore operating areas of the world.

2. Design to a statistical wave description: in this approach, the occurrence of waves incident on the offshore structure is expressed in terms of a probability distribution which describes the probability of occurrence of waves with specified height, period and direction. Such probability data can also be expressed as the average number of waves encountered per year for each range of wave height, period and direction. The relative energy in simultaneously occurring period components is described by unidirectional or directional wave spectra. Such a wave description still contains extreme design waves as part of the statistical data.

Use of a statistical wave description in design does permit fatigue inducing loads to be adequately characterized. The biggest drawback of this approach, however, is that it relies on the linearity of superposition of wave components and is, therefore, likely to be in some error for high waves. The requirement for linear superposition is more likely to be valid for calculating loads due to small waves on large structures where linear wave theory applies and linear inertia loading predominates over quadratic drag forces. For structures with members of small diameter, non-linear drag forces are significant and the extension of the statistical wave description to loading has to be done with caution. Current design practice for offshore structures tends to use both of the above approaches and takes advantages of the merits of each.

Table 2.4: Typical design wave and wind parameters for 100 year return period conditions

Parameter	Offshore area					
	Malaysia	Persian Gulf	Gulf of Mexico	Southern North Sea	Central North Sea	Northern North Sea
Maximum wave height (m)	10.1	10.7	11.9	15.2	22.9	30.5
Maximum 1 minute mean wind speed (m/s)	31.9	27.3	38.1	51.5	51.5	51.5

2.3.5.4 Flow regimes for wave loads

The procedure for calculating wave forces on off shore structures can be split up into fundamentally different approaches depending on the size of the structural member and the height and wavelength of incident waves. These parameters can be written in the form of two ratios: structural member diameter to wave length (D/λ) and wave height to structural member diameter (H/D). Table 2.5 uses approximate values for the ratio K (which is a measure of H/D) to separate the wave regimes into its constituent parts.

For small structural members where $D/\lambda < 0.2$, Morison's equation is used to estimate forces due to wave action with the implicit assumption that the diameter of the member is small enough in relation to the wave length so as not to alter incident wave characteristics to any significant extent. Morison's Equation, which was first put forward by Morison *et al.* (1950), is based on the assumption that wave forces can be expressed as the sum of a drag force due to wave fluid velocity and an inertia force due to wave accelera-

Table 2.5: Wave loading regimes

↓K D/λ →	< 0.20	> 0.20
$K > 25$	Drag dominated flow regime. Morison's Equation with C_M and C_D values required for computing wave forces. Drag coefficient is function of Reynolds number. For Re $> 1.5 \times 10^6$, $C_M = 1.8$, $C_D = 0.62$; For $10^5 <$ Re $< 1.5 \times 10^2$, $C_M = 1.8$, C_D varies from 1.0 to 0.6	
$5 < K > 25$	Intermediate regime between drag and inertia domination. Morison's Equation applicable but published C_M and C_D values exhibit wide scatter. Flow behaviour and consequent loading complex and uncertain. For Re $> 1.5 \times 10^6$, $C_M = 1.8$, $C_D = 0.62$	
$K < 5$	Inertia dominated regime. Morison's Equation or diffraction theory for computing wave forces. $C_M = 2.0$ Effect of C_D is negligible	Morison's Equation unsuitable for computing wave forces. Diffraction theory required

C_M inertia coefficient, C_D drag coefficient

$$\text{Re} = \frac{U_m D}{\nu} \quad K = \frac{U_m T}{D}$$

U_m peak velocity, T wave period, ν kinematic viscosity, Re Reynolds number, K Keulegan Carpenter number

tion. On the other hand, for larger structural members (with $D/\lambda > 0.2$), diffraction theories are necessary to account for the reflection and radiation of waves from the structural member. These potential flow methods cannot account for viscous drag forces.

The second parameter of interest is the ratio H/D. Its importance is based on the fact that drag forces on structures, say a pipe of circular cross section, in an oscillatory wave flow are dominated by the separation of flow behind the cylinder and the formation of large vortices. For a small H/D ratio (< 1.5), the wave height (and thus orbital diameter) are not unidirectional long enough for the flow to initiate separation and develop or shed vortices. In this case, drag forces are very small, acceleration dependent inertia forces dominate and potential flow diffraction theory can be used to predict wave forces with confidence.

At the other extreme for $H/D > 8$, approximately, the wave flow will have been unidirectional long enough for a substantial vortex flow to develop. Drag forces will then be large and a Morison formulation, which accounts for these, must be used. An intermediate region, where $1.5 < H/D < 8$ also exists where the flow regime can be highly complicated and wave forces are difficult to compute. The two flow regimes corresponding to H/D of < 1.5 and > 8 are illustrated in *Figure 2.29*. A non-dimensional parameter called the Keulegan–Carpenter number is a more rational measure of the ratio of water particle motion double amplitude to cylinder diameter than H/D which is only valid close to the water surface. The Keulegan–Carpenter number is defined as:

$$K = \frac{U_m T}{D} \quad (2.48)$$

where U_m is the maximum normal velocity in the oscillatory flow of period T about the cylinder of diameter, D.

The approximate range of numerical values encountered are 0–30 m (0–97 ft) for H and 60–620 m (195–2015 ft) for λ. Thus a subsea pipe or the space frame of a tower jacket structure (with typical D, 0.4–3 m (1.3–9.7 ft)) would give large H/D and small D/λ, indicating a drag dominated wave loading with use of Morison's equation being appropriate for computing the loading. Alternatively, a large monolithic gravity structure (with typical $D = 80$ m (260 ft)) would yield small H/D and large D/λ; thus pointing to an inertia dominated loading regime and requiring potential flow diffraction theory for computing wave loads. The combination of both H/D and D/λ being small can occur but the other extreme of both H/D and D/λ being large cannot because of the height limitation of stable waves which constrains the ratio H/λ to a maximum value of approximately 1/7.

Table 2.5 illustrates the wave loading regimes described above. The Keulegan–Carpenter number K is used as a measure of ratio H/D.

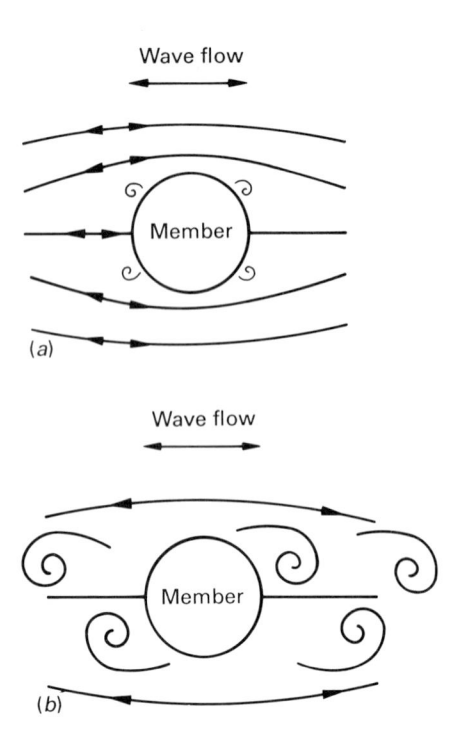

Figure 2.29 Inertia and drag dominated flows

2.3.5.5 The Morison Equation

The mechanics for wave loading on slender tubular subsea pipe lines or members of offshore jacket structures is described here. Such structures (with $D/\lambda < 0.2$) will be subjected to so called Morison wave loading which is based on the assumption that wave properties are not affected by the presence of the structure and that the total wave force can be expressed as a sum of inertia forces (due to wave fluid acceleration) and drag forces (due to wave fluid velocity).

An equation to represent this loading can be written as:

$$dF = C_M \rho\, dV\, \dot{U}_n + C_D \tfrac{1}{2} \rho\, dS |U_n| U_n \quad (2.49)$$

$$\text{inertia force} \qquad \text{drag force}$$

where dF is the total wave force on a pipeline element of volume dV, and frontal area, dS, U_n, $\dot{U}_n$ are instantaneous wave fluid velocities and accelerations normal to the pipe axis, ρ is the fluid density and C_M and C_D are inertia and drag coefficients. The inertia coefficient is conventionally taken as the added mass coefficient plus one to account for the Froude–Krylov or undisturbed pressure force due to wave fluid acceleration. The drag force term has a modulus sign to ensure that the drag force remains in the same direction as the wave velocity.

Although the Morison Equation was originally developed for vertical piles in shallow water, for engineering purposes, it is assumed to be valid for cylinders of arbitrary orientation in deep and shallow water with the proviso that the coefficients C_M and C_D are chosen experimentally for the appropriate condition. Application of the Morison Equation can be illustrated by developing expressions for the wave force on vertical and horizontal pipe segments.

Consider an element dy, of a vertical circular cylinder of radius r, as shown in *Figure 2.30*. The total force dF, acting on this element in the direction of wave propagation is the sum of inertia and drag force components. This force can be integrated to yield the total force F and moment about the sea bed as:

$$F = C_M \rho \pi r^2 \int_{-d}^{0} \dot{u}\, dy + C_D \rho r \int_{-d}^{0} |u|u\, dy$$

and (2.50)

$$F = C_M \rho \pi r^2 \int_{-d}^{0} (d+y)\dot{u}\, dy + C_D \rho r \int_{-d}^{0} (d+y)|u|u\, dy$$

where u and $\dot{u}$ are horizontal components of wave velocity and accelerations. Using expressions for u and $\dot{u}$ for Linear Wave Theory with $x = 0$ and $\theta = \omega t$, Equations (2.50) can be readily integrated to yield:

$$F = -\frac{2\pi \rho r H^2 \lambda}{T^2} \{K_2\, C_M \cos\theta + K_1\, C_D |\sin\theta|\sin\theta\}$$

and

$$M = -\frac{2\rho r H^2 \lambda^2}{T^2} \{K_4\, C_M \cos\theta + K_3\, C_D |\sin\theta|\sin\theta\}$$

where

$$K_1 = \frac{1}{16\sinh^2 kd}[2kd + \sinh 2kd] \qquad (2.51)$$

$$K_2 = \frac{\pi r}{2H}$$

$$K_3 = -\frac{1}{64\sinh^2 kd}\{2k^2d^2 + 2kd\sinh 2kd + 1 - \cosh 2kd\}$$

$$K_4 = \frac{\pi r}{4H\sinh kd}\{1 + kd\sinh kd - \cosh kd\}$$

Consider also wave forces on horizontal submerged cylinders as shown in *Figure 2.31*. The horizontal and vertical wave force components on the cylinder of *Figure 2.31(a)* with axis parallel to wave crest lines are simply written as

$$F_H = C_m \rho \pi r^2 L\, \dot{u} + C_D \rho L |u|u$$

and (2.52)

$$F_v = C_m \rho \pi r^2 L\, \dot{v} + C_D \rho L |v|v$$

where L is the cylinder length and the expressions for u, v, $\dot{u}$, $\dot{v}$ are obtained from *Table 2.3*.

If on the other hand, the horizontal cylinder is lying with its axis in the direction of wave propagation, as in *Figure 2.31(b)*, then the above expression for horizontal force is no longer applicable. The expression for vertical force per unit length can still be used but the calculation does not have the benefit of the simplification due to $x = 0$. The equation for vertical force on the cylinder can then be written as:

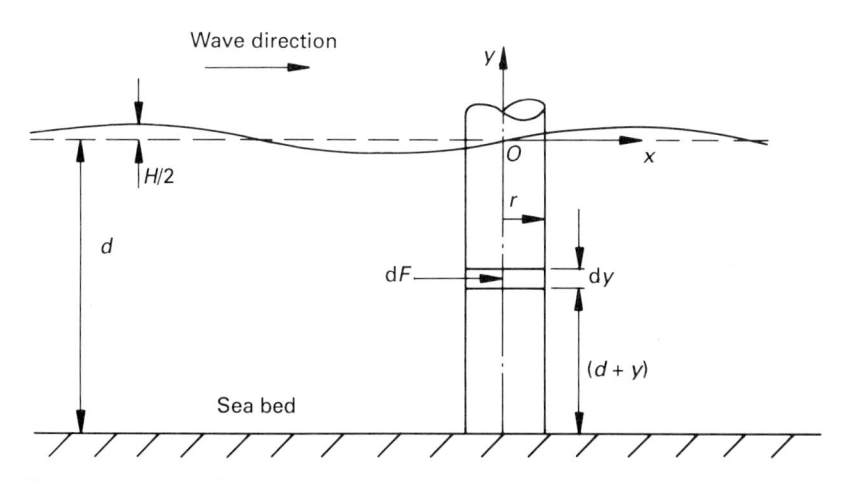

Figure 2.30 Wave force on vertical cylinder

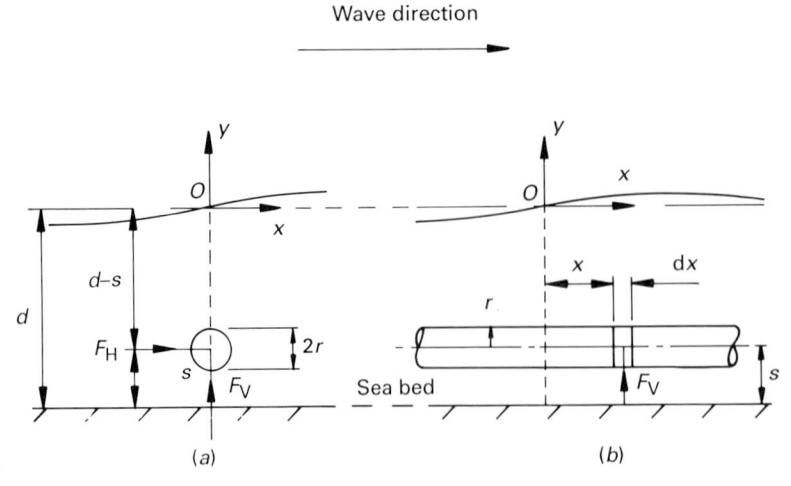

Figure 2.31 Wave force on horizontal cylinders

$$F_v = \int_{x_1}^{x_2} \{\rho\, C_D\, r|v|v\, dx + \rho\, C_m\, \pi\, r^2\, \dot{v}\, dx\} \quad (2.53)$$

for a cylinder segment from $x = x_1$ to $x = x_2$. Evaluation of this equation requires that expressions for v and $\dot{v}$ be integrated with respect to x. Integration of the modulus sign is carried out by splitting up the integration limits into positive and negative integrand regions and evaluating each integrand separately.

The Morison Equation can also be applied to arbitrarily inclined circular cylinders although the original formulation by Morison et al. (1950) was only intended for vertical piles. The extension of the Morison Equation to inclined cylinders can be defined ambiguously depending on whether wave velocity components incident on the cylinder are first resolved into a total normal velocity component before computing forces, called the normal component approach or the wave velocity components are combined with cylinder projected areas to compute force components in the three orthogonal directions, called the projected area method. Morison's Equation may be written in vector form for the normal component approach with wave force component normal to the cylinder axis on an element of length, ds, given by:

$$dF = C_M\, \rho\, \pi\, r^2\, \dot{q}_n\, ds + C_d\, \rho\, r|q_n|q_n\, ds \quad (2.54)$$

using the definition diagram of *Figure 2.32* with a wave velocity vector q_n normal to the cylinder axis at the element of length ds. An equivalent equation for the projected area method is:

$$dF = C_M\, \rho\, \pi\, r^2\, \dot{q}_n\, ds + C_D\, \rho \begin{bmatrix} |u|u\, dA_x \\ |v|v\, dA_y \\ |w|w\, dA_z \end{bmatrix} \quad (2.55)$$

where $q = ui + vj + wk$ is the instantaneous wave velocity vector and dA_x, dA_y and dA_z are the projections of the element surface area in the x, y and z directions, respectively. Note that the inertia term of Equation 2.55 is still evaluated using the normal component approach and that the drag terms from the two equations will yield different results. The wave forces obtained will, however, depend on the added inertia and drag coefficients selected and it may be argued that the coefficients that should be selected for Equation 2.54 must also be derived by decomposing experimental data through the formulation of Equation 2.54. The same may be said for Equation 2.55. However, in terms of the physics of wave loading and the ease of applicability of the resultant equation, use of the normal component approach and Equation 2.54 is recommended and is nearly always used in design calculations. However, many fundamental questions on the mechanics of wave loading on inclined cylinders still remain open and are the subject of continuing research.

A further point to note is that the formulations of Equations 2.54 and 2.55 do not give the total force on an isolated cylindrical member. There are additional

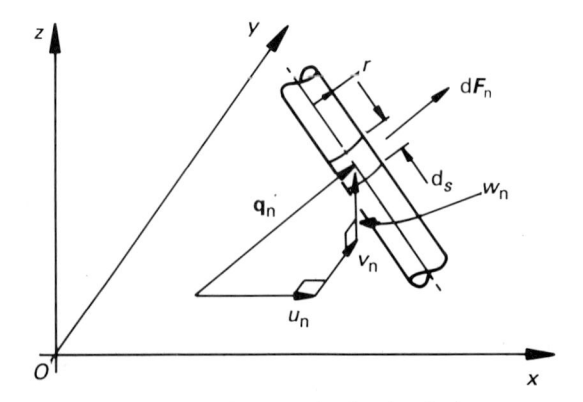

Figure 2.32 Wave force on inclined cylinder

axial forces due to the wave pressure added mass and drag from the flow at the cylinder ends. In particular, the axial wave pressure force at each end is obtained as the product of the local wave pressure and the cylinder cross-sectional area.

Figure 2.33 shows a typical variation of inertia force, drag force and the wave elevation within a wave cycle for a vertical cylinder. The non-linear behaviour of the drag force is shown up by the shape of the curve close to the horizontal axis. The expression of Equation 2.51 for wave force can be rewritten as:

$$F = A|\sin\theta|\sin\theta + B\cos\theta$$

where $\theta = \omega t$ and A, B are appropriate constants. The maximum value of F is given by:

$$\frac{dF}{d\theta} = 2A\sin\theta\cos\theta - B\sin\theta = 0 \quad (2.56)$$

for $0 < \theta < \pi$ with turning points given by:

$$\sin\theta = 0 \quad \cos\theta = \frac{B}{2A}$$

These points give the location within the wave cycle and the magnitude of the instantaneous maximum values of wave force.

2.3.5.6 Drag and inertia coefficients

The value of drag coefficients, C_D to be inserted in Morison's Equation can only be obtained experimentally. In theory, the value of the inertia coefficient C_M, can be calculated (it is, for example, 2.0 for a smooth cylinder in an ideal fluid). However, measured values are used in practice, particularly when drag is the dominant force. For irregular shapes, the inertia coefficient can be calculated from diffraction theory.

Many experiments have been designed to measure values of drag and inertia coefficients in steady or planar oscillatory flows but their results can only be used for wave force prediction with caution. The most useful experimental measurements have been made in circumstances which model full scale conditions within the ocean environment. Measurements made offshore on large test structures in real sea ways are especially valuable.

One problem facing the user of Morison's Equation is the large scatter in values of the inertia and drag coefficients. However, there is a useful degree of correlation between the coefficients and two flow parameters, Keulegan–Carpenter nunber K, and Reynolds number Re. Nevertheless, the scatter and hence some uncertainty remain. *Table 2.5* gives a summary of generally accepted added inertia and drag coefficients for flow around circular cylinders in unconfined flow remote from solid boundaries such as the sea bed for pipelines. The effect of such sea bed proximity on coefficient values is examined in more detail later.

Sarpkaya (1976) has carried out many systematic studies of the variations of inertia and drag forces for circular cylinders in planar oscillatory flow. *Figures 2.34* and *2.35* present plots of C_M and C_D against Reynolds number and Keulegan–Carpenter number for smooth cylinders from data prepared by Sarpkaya and Isaacson (1981). *Figures 2.36* and *2.37* also present this data plotted as a function of the Keulegan–Carpenter number with parameter β being the ratio of the Reynolds number to Keulegan–Carpenter number. It can be seen that the C_M and C_D data show consistent and physically meaningful correlations with the Keulegan–Carpenter number.

Marine growth in the ocean environment causes the surfaces of offshore structures, particularly near the splash zone, to become highly roughened. This has the effect of increasing the effective diameter of individual

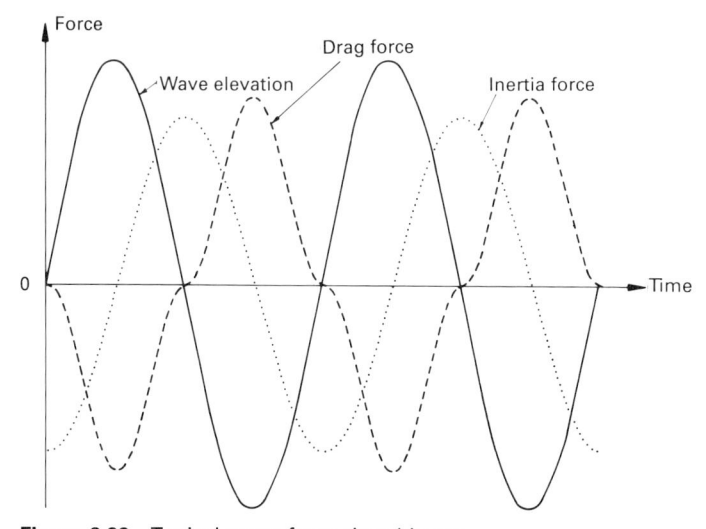

Figure 2.33 Typical wave force time history

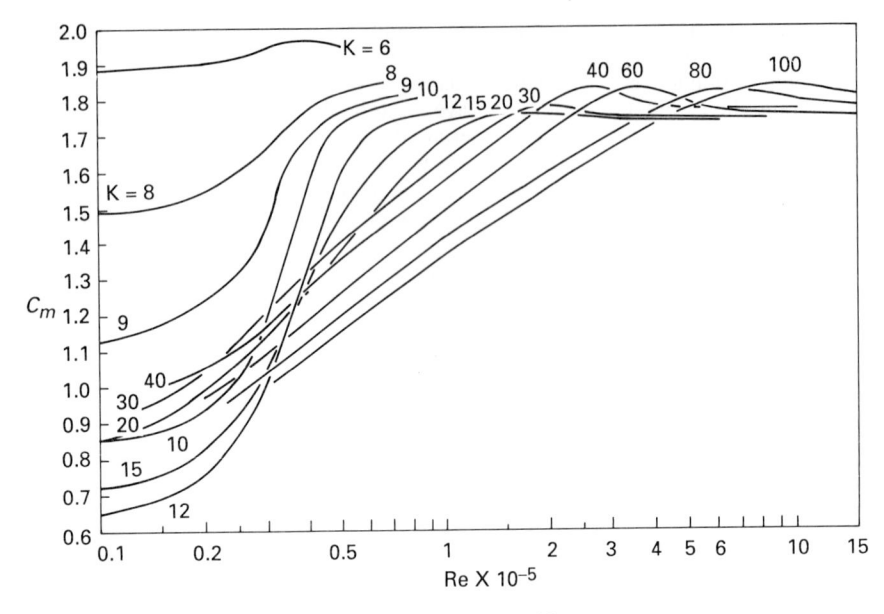

Figure 2.34 Added inertia coefficient against Reynolds number for various values of Keulegan–Carpenter number (from Sarpkaya and Isaacson 1981)

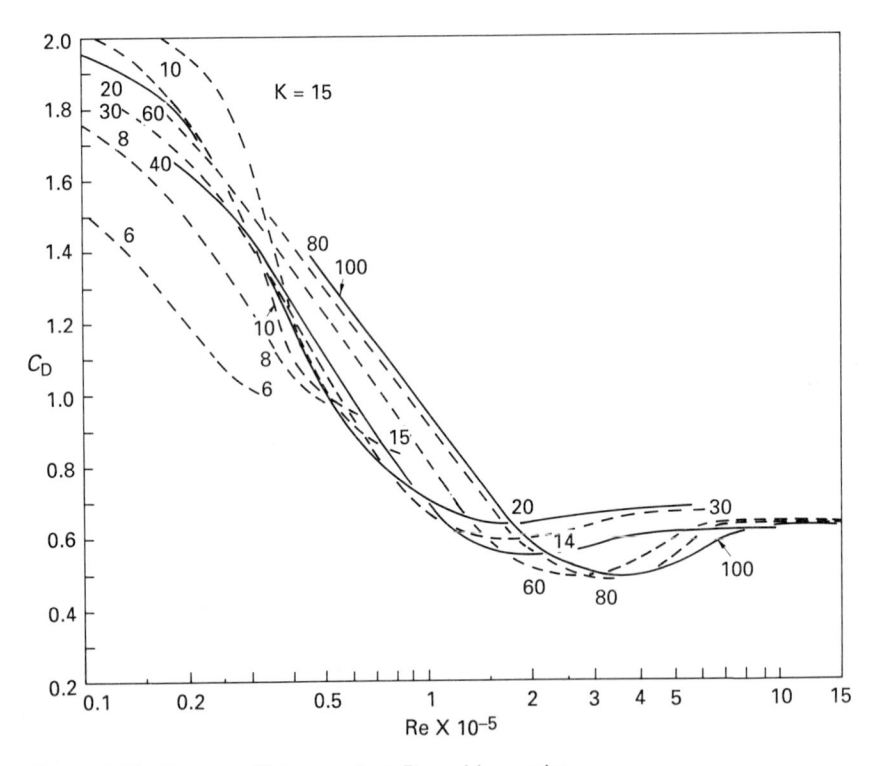

Figure 2.35 Drag coefficient against Reynolds number for various values of Keulegan–Carpenter number (from Sarpkaya and Isaacson 1981)

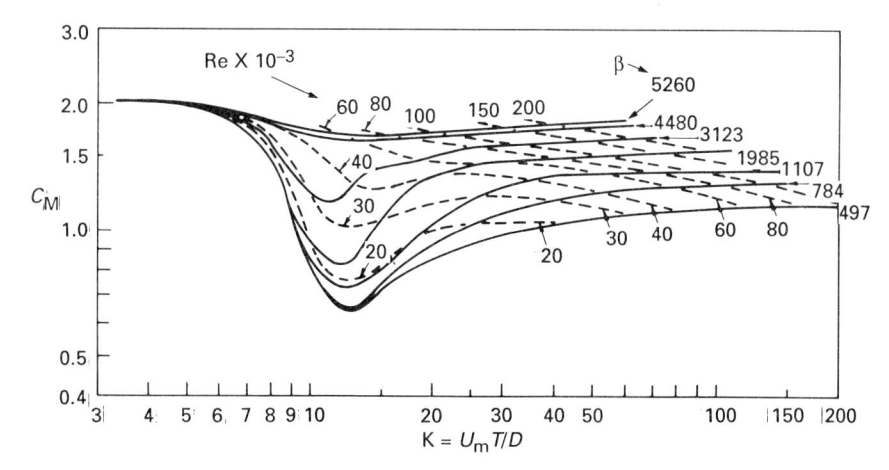

Figure 2.36 Added inertia coefficient plotted against Keulegan–Carpenter number for various values of Reynolds number (Re) and frequency parameter, β (from Sarpkaya and Isaacson 1981)

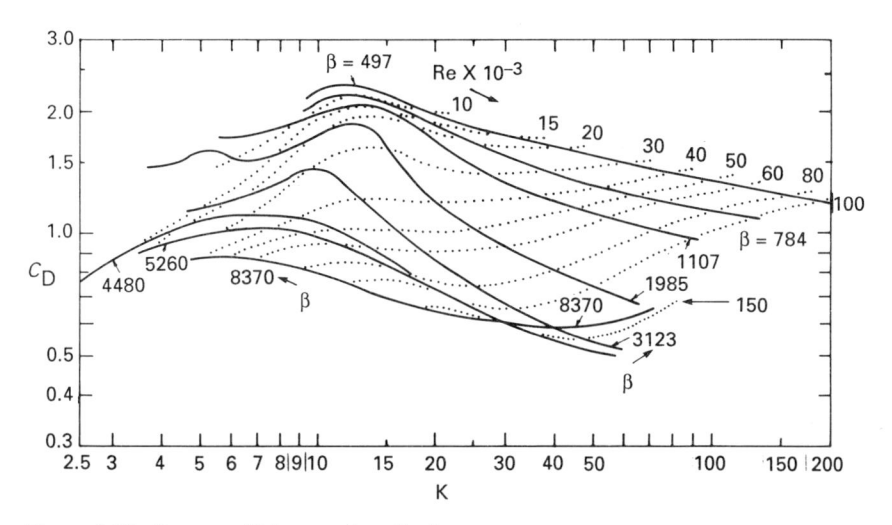

Figure 2.37 Drag coefficient against Keulegan–Carpenter number for various values of Reynolds (Re) and frequency parameter, β (from Sarpkaya and Isaacson 1981)

members and to modify the inertia and drag coefficients due to the changed interaction between the cylinder boundary layer and the roughness elements.

Roughness due to marine growth will normally increase the skin friction due to a turbulent boundary layer but there exists a critical height below which the roughness has no effect on friction. This critical roughness height is also determined by the height for which the Reynolds number of the roughness corresponds to the development of an eddying wake. As the roughness height increases it causes a progressive increase of drag, because of the energy loss in the eddying flow behind the protuberances. In a sense, the extra drag due to the roughnesses is predominantly a form drag, derived from their pressure distributions rather than a frictional drag, although over a surface embracing a number of roughnesses the net effect is that of an increase of skin friction. When the height of the roughnesses is large enough, the extra drag due to them becomes very much greater than the initial drag of the smooth surface. It is characteristic of the drag coefficient of a bluff body, due largely to the shedding of eddies, that it is practically independent of Reynolds number over a wide range. This also applies for large enough roughnesses where the drag coefficient becomes independent of the Reynolds number. Fully developed roughness flow is then said to be established.

Figure 2.26 shows the effects of roughness on the steady flow drag coefficient. For planar oscillatory flow, Sarpkaya and Isaacson (1981) present recommended data for inertia and drag coefficients as a function of roughness Reynolds number and for a Keulegan–Carpenter range 20–100.

There are three other physical features that influence wave loading on cylindrical members of offshore structures. These are summarized below.

There is very little design information available at present on the effects of flow interference in waves between groups of closely spaced cylinders although the physics of such phenomena are being researched at present. Recent work has shown that planar oscillatory wave flow about circular cylinders induces motion of vortex pairs that can induce substantial velocity magnifications in the flow around the cylinder. The presence of other smaller cylinders around a larger diameter cylinder can interact with these velocity magnifications to give rise to significant interference effects. Grass *et al.* (1987) presents research work in this area.

The presence of a current superimposed on a wave seems to influence C_D but not C_M probably because the current sweeps away the vortices shed in the flow by the obstruction. With a current running parallel to the wave direction, the current velocity should first be added vectorially to wave particle velocity before the calculation of wave forces by the Morison Equation. It is generally felt that use of coefficient values for zero current leads to conservative results for design.

2.3.6 Vortex shedding induced loads

Vortex shedding from a circular cylinder assumes importance here because of the to and fro circular motion of fluid particles during wave motion. This motion initially produces vortices in the lee of the cylinder which are then forced back on to the body when the direction of flow reverses. The effect of this motion can produce very large variations in an oscillating lift force as well as increasing the mean drag force.

The formation of vortices from a circular cylinder in uniform steady flow has been studied extensively. An empirical expression for f_s, the frequency of vortex shedding can be written as:

$$f_s = S\left[1 + \frac{19.7}{Re}\right]\frac{U}{D} \quad (2.57)$$

where S is a Strouhal number, U is the steady velocity, D is the diameter of the structural member and the equation is approximately valid over a Reynolds number Re, range 60–2×10^5. Except for very small values of Re, the Reynolds number based variation can be neglected. Then:

$$S = \frac{f_s D}{U} \quad (2.58)$$

For circular cylinders, $S = 0.2$ for $10^3 < Re < 2 \times 10^5$. Above $Re = 2 \times 10^5$, there is considerable variation in S.

Consider the horizontal component of wave particle velocity at the water surface. For deep water, the equation can be written as:

$$u = \frac{\pi H}{T}\cos\theta \quad (2.59)$$

If U is taken as the average value of u above, we have:

$$U = \frac{\pi H}{2T} \quad (2.60)$$

Since unidirectional flow persists for a time $T/2$ in a wave cycle of period T and the time interval between successive vortices being shed is $1/(2/f_s)$ (there are two vortices per cycle of shedding), then the condition for at least one vortex to have time to form is:

$$\frac{T}{2} > \frac{1}{2f_s}$$

But

$$\frac{1}{2f_s} = \frac{D}{US} = \frac{2TD}{\pi H S}$$

Taking $S = 0.2$ (2.61)

$$\frac{T}{2} > \frac{1}{2}\frac{10}{\pi}\frac{TD}{H}$$

$$\therefore \frac{H}{D} > \frac{10}{\pi}$$

H/D can also be written as $U_m T/D$ where U_m is the maximum velocity in the cycle to yield the Keulegan–Carpenter number of Equation 2.48. This parameter was first introduced by Keulegan and Carpenter (1958) for correlating measured inertia and drag coefficients. The role of the Keulegan–Carpenter number in quantifying the number of vortices shed per cycle of wave oscillation leads to a physically more meaningful correlation between inertia and drag coefficients with the Keulegan–Carpenter number than is apparent with Reynolds number. These arguments also suggest that it is beneficial to keep the Keulegan–Carpenter number constant between full scale and model scale testing conditions.

The formation of vortices as described above leads to an oscillating transverse lift force on a segment of circular cylinder in a flow. This lift force dL, for a segment of length dS, on a cylinder of radius r, can be written as:

$$dL = C_L \tfrac{1}{2} \rho\,(2r)\,U_m^2\,ds \quad (2.62)$$

where U_m is the peak flow velocity in a cycle, ρ is fluid density and C_L is the lift coefficient. Once again experimental data in planar oscillatory flow provides the necessary information for values of C_L. *Figure 2.38* presents values of C_L against the Reynolds number at a range of Keulegan–Carpenter numbers for smooth cylinders from Sarpkaya and Isaacson (1981). The force oscillation can be taken to occur at the Strouhal frequency given by taking $S = 0.2$ in Equation 2.58. For smooth cylinders with Reynolds numbers greater than 1.5×10^5, a value of $C_L = 0.20$ may be used. *Figure 2.39* (also from Sarpkaya and Isaacson 1981) gives equivalent data for roughened cylinders with C_L taken to be 0.25 for large K and R together with a Strouhal number of 0.22.

The oscillatory nature of the lift force can induce dynamic excitation of the loaded member. It is known that as the vortex shedding period of the cylinder approaches its natural period, this lift force increases considerably due to cylinder motion and a dynamic instability can result. In many practical situations, structural failure has been known to occur in this fashion. Blevins (1977) gives further details.

2.3.7 Sea bed proximity and slamming

For pipelines laying on the sea bed or spanning some small distance above the sea bed, wave and current induced flow around the line is substantially distorted by the presence of the sea bed boundary. Experimental work for such flows shows that the Morison Equation can still be used together with Equation 2.62 for lift forces provided that the added inertia, drag and lift coefficients are recognized as being strong functions of pipe distance from the sea bed. The above comments also apply for pipes that are partially buried or embedded in the sea bed. The presence of large flow velocities close to the sea bed due to a partially suspended pipe can cause sea bed scour and change local sea bed contours.

Applicable added inertia, drag and lift coefficients for pipes in close proximity to the sea bed are generally obtained from experimental data, much of which is proprietary to companies within the industry. However, the inertia coefficient for a pipe resting on the sea bed can be derived from diffraction theory to be 3.29 Yamamoto *et al.* 1974). This work shows that $C_M = 3.29$ for a cylinder touching a plane wall but that this reduces rapidly to the potential flow value of 2.00 for a gap length equal to one cylinder diameter on coefficient changes due to sea bed proximity.

The best practical guide, however, is supplied by Det Norske Veritas (1982). *Figure 2.40* to *2.43* present significant increases in inertia, drag and lift coefficients with sea bed proximity. The increases occur because the presence of the sea bed restricts fluid flow largely to one side of the body and, therefore, presents a greater restriction to the flow. These increased forces must be accounted for in calculations for pipeline stability.

Just as structural members close to the sea bed require special attention, cylindrical members that are close to the sea surface need to be considered further. In particular, horizontal structural members in the splash zone are susceptible to vertical forces due to buoyancy changes and wave slam. These loads are impulsive in nature and can generate appreciable dynamic magnification.

Slamming loads occur when a moving body strikes and enters a water surface or conversely when a moving water surface strikes and submerges a body. If a

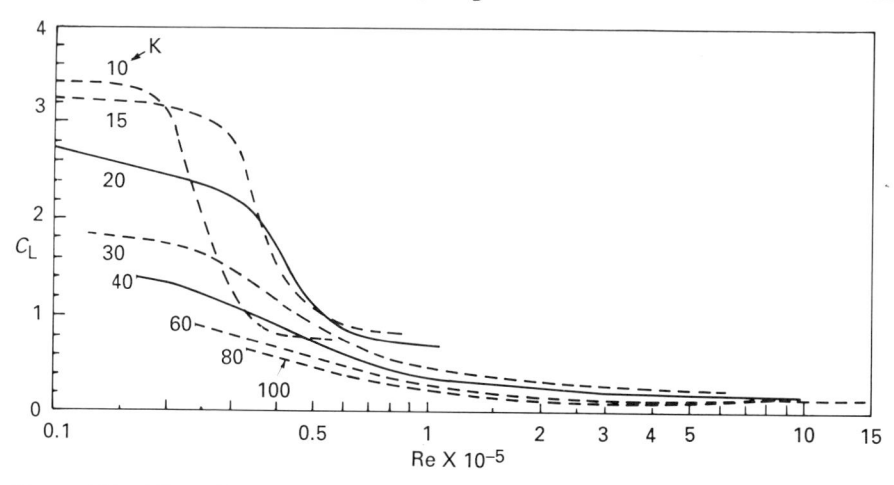

Figure 2.38 Lift coefficient against Reynolds number for various values of Keulegan–Carpenter number, K (from Sarpkaya and Isaacson 1981)

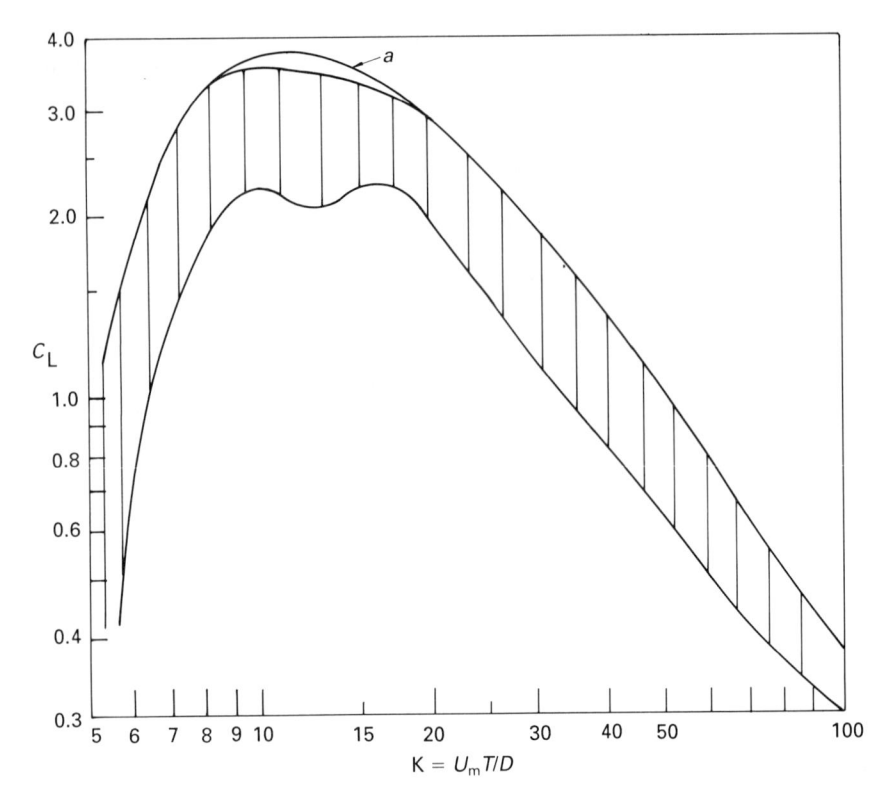

Figure 2.39 Lift coefficient for rough cylinders as a function of Keulegan–Carpenter number for various roughness heights. Line a denotes smooth cylinder data for β from 1000 to 2000. Cross-hatched region denotes occurrence of C_L data for rough cylinders with $L/D = 1/200$ (from Sarpkaya and Isaacson 1981)

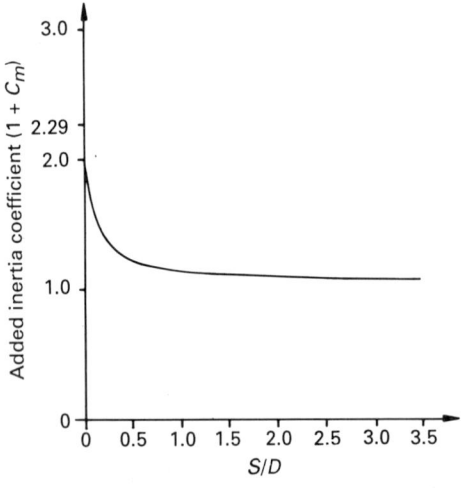

Figure 2.40 Variation of added inertia coefficient with wall proximity for a circular cylinder. S is distance of cylinder bottom above seabed and D is cylinder diameter (from Det Norske Veritas 1982)

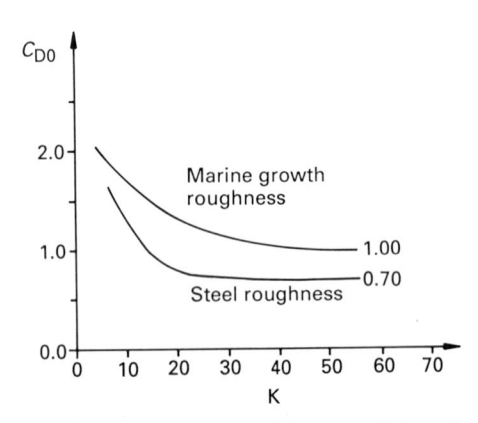

Figure 2.41 Variations of drag coefficient C_{Do} with Keulegan–Carpenter number, K_c for a circular cylinder in unconfined flow (from Det Norske Veritas 1982)

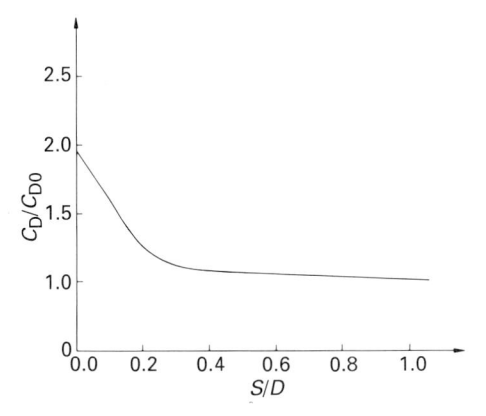

Figure 2.42 Variation of drag coefficient C_D with wall proximity ratio S/D for a circular cylinder in oscillatory supercritical flow with $K_c > 20$ and Reynolds number range of from 10^5 to 2×10^6. Notation as in *Figure 2.41* (from Det Norske Veritas 1982)

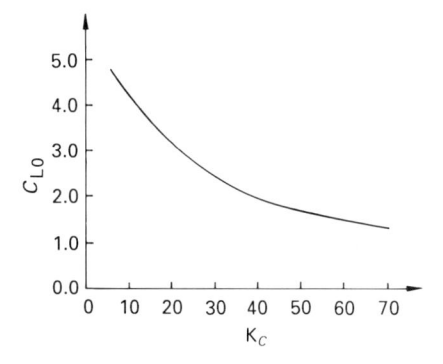

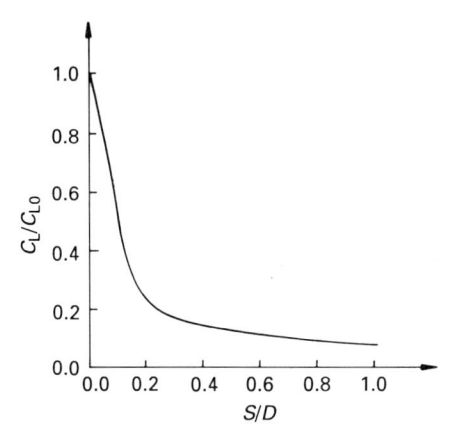

Figure 2.43 Variation of lift coefficients for a circular cylinder with Keulegan–Carpenter number and wall proximity, C_{L0} is lift coefficient for unconfined flow and C_L is lift coefficient for flow close to a wall. Remaining notation as in *Figure 2.41* (from Det Norske Veritas 1982)

structural member presents a large face parallel to, or nearly parallel to, the wave surface and if the relative velocity of impact is high, slamming forces are both impulsive and large, usually much larger than the other forces (due to buoyancy, drag and inertia) experienced in fully immersed flow of the same velocity.

Peak slamming forces on the member occur due to the change of fluid momentum caused by the impact. This change is proportional to the square of the velocity of impact and the slamming force F_s can be written as:

$$F_s = C_s \tfrac{1}{2} \rho U^2 (2r) \qquad (2.63)$$

per unit length for a cylinder of radius r. Hence C_s is the slamming coefficient and U is the velocity of impact between the water surface and structural member. There is a large amount of scatter in the values of C_s obtained from experimental and theoretical work on this problem. Von Karman and Wattendorf (1929) and Kaplan and Silbert (1976) describe two theoretical approaches to the problem of predicting C_s which yield a value of π. Experimental determination of C_s values is particularly difficult because of the problem of separating the dynamic response of the test rig from the rigid structure values calculated by the theory and also because the dynamic response of the test rig strongly influences the measured peak load. The measured value of C_s is also very sensitive to even very small angles between the wave crest and cylinder axis. As a result, model test results for slamming coefficients show large variations in a range 2–5.2. However, work by Miller (1980) gives support to using an average C_s value of 3.5 for design.

2.3.8 Wave loads on large bodies – diffraction theory

The calculation of wave forces on bodies with dimensions that span a significant proportion of the average length ($> 0.2\lambda$) must account for the fact that the wave flow will be influenced by the presence of the body. The solution of the Laplace governing equation with the usual sea bed and free surface boundary condition together with the additional condition of no flow through the body surface will give rise to additional component waves in the solution which are called scattered or diffracted waves. The wave force on the body is then due to the incident wave as well as the scattered wave field arising from the presence of the body. Furthermore, the Froude–Krylov force in such a flow field with a large body can no longer be written as ρVU because the wave induced pressure gradient around the body is no longer constant. At the same time, the conditions that require the diffraction theory to be used ($D/\lambda > 0.2$) also imply that the ratio of wave height to body dimension H/D is less than unity since $H/\lambda < 1/7$. For such small values of H/D, drag forces will be small since flow amplitudes will be less than the body dimension and flow separation will not occur.

These fortuitous circumstances mean that a potential flow solution of the wave diffraction problem will represent a physically realistic situation.

A statement of the full wave diffraction problem in three dimensions can be written using an axes system with Oxy in the still water surface and Oz pointing vertically upwards. The governing equation is:

$$\frac{\partial^2 \phi}{\partial x^2} + \frac{\partial^2 \phi}{\partial y^2} + \frac{\partial^2 \phi}{\partial z^2} = 0 \qquad (2.64)$$

with the linearized free surface boundary condition at $z = 0$ of:

$$\frac{\partial^2 \phi}{\partial t^2} + g \frac{\partial \phi}{\partial z} = 0 \qquad (2.65)$$

and the sea bed boundary condition at $z = -d$ of:

$$\frac{\partial \phi}{\partial z} = 0 \qquad (2.66)$$

together with the no flow through the immersed body boundary condition of:

$$\frac{\partial \phi}{\partial n} = 0 \qquad (2.67)$$

at the body surface where n denotes a direction normal to this surface. All of the assumptions inherent in the governing equations and the linearized free surface boundary condition, therefore, apply to this linear diffraction problem, that is, irrotational and inviscid flow of small wave amplitude.

The solution to this wave diffraction problem is obtained by writing the velocity potential as the sum of incident and scattered potentials, ϕ_i and ϕ_s respectively. Thus:

$$\phi = \phi_i + \phi_s \qquad (2.68)$$

Furthermore, the scattered waves generated by the presence of the body boundary have to be restricted in the mathematical formulation to out going waves only. This requires that the scattered wave potential satisfies the condition:

$$\frac{\partial \phi_s}{\partial r} + \frac{1}{c} \frac{\partial \phi_s}{\partial t} = 0 \qquad (2.69)$$

where r is radial distance from a point on the body surface and c is wave speed. This can be written (Sommerfield 1949 and Stoker 1957) as:

$$\lim_{r \to \infty} r^{1/2} \left[\frac{\partial \phi_s}{\partial r} - ik \phi_s \right] = 0 \qquad (2.70)$$

where the factor $r^{1/2}$ takes account of the directional spreading of waves.

Since the incident wave potential ϕ_i is known, the boundary condition at the body surface can also be written as:

$$\frac{\partial \phi_s}{\partial n} = - \frac{\partial \phi_i}{\partial n} \qquad (2.71)$$

Once ϕ_s is obtained as a solution to this problem, wave induced pressures can be obtained from the linearlized Bernoulli Equation and integrated to obtain forces and moments.

This section describes three solutions to the linear wave diffraction problem; the first of this is an analytic solution for vertical circular cylinders in arbitrary water depth followed by two numerical computational approaches based on boundary integral and boundary element methods.

2.3.8.1 Analytic solution for vertical cylinder

This solution was developed initially by Havelock (1940) for deep water and developed subsequently for arbitrary water depths by MacCamy and Fuchs (1954). The latter solution is presented here because it gives wave forces on the cylinder in a readily usable form.

Using the definition diagram of *Figure 2.44*, the problem can be re-expressed in (r, ϕ, z) polar coordinates with the Laplace Equation becoming:

$$\frac{\partial^2 \phi}{\partial r^2} + \frac{1}{r} \frac{\partial \phi}{\partial \theta} + \frac{1}{r^2} \frac{\partial^2 \phi}{\partial \phi^2} + \frac{\partial^2 \phi}{\partial z^2} = 0 \qquad (2.72)$$

and the body surface boundary condition being:

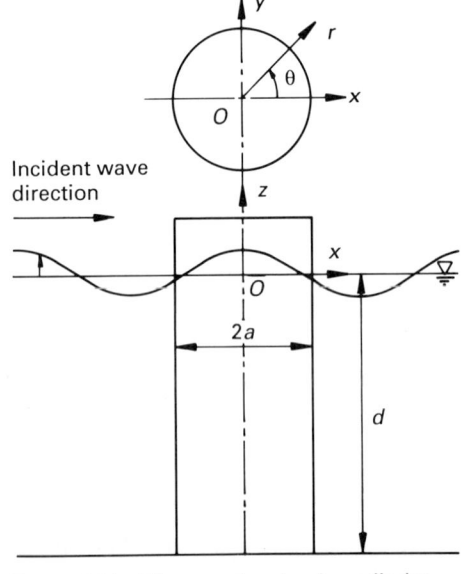

Figure 2.44 Wave past a circular cylinder

$$\frac{\partial \phi}{\partial r} = \frac{\partial \phi_i}{\partial r} + \frac{\partial \phi_s}{\partial r} = 0 \text{ at } r = a$$

The incident wave potential is given by:

$$\phi_i = A \frac{\cosh k(z+d)}{\cosh kd} \left[\sum_{m=0}^{\infty} \beta_m J_m(kr) \cos(m\theta) \right] e^{-i\omega t} \quad (2.73)$$

where $A = -ag/\omega$, $\beta_m = 1$ for $m = 0$, $\beta_m = 2i^m$ for $m \geq 1$, and $J_m(kr)$ is the Bessel function of the first kind of order m and argument kr. Equation 2.73 is derived from the velocity potential of Equation 2.40 by writing $x = r \cos \theta$ and expanding the exponential as an infinite series of Bessel functions. An expression for the scattered velocity potential which already satisfies the Laplace equation and the sea bed and free surface boundary condition is written as:

$$\phi_s = A \frac{\cosh k(z+d)}{-\cosh kd} \left[\sum_{m=0}^{\infty} \beta_m B_m H_m^{(1)}(kr) \cos(m\theta) \right]$$
$$e^{-i\omega t} \quad (2.74)$$

where B_m are unknown coefficients and $H_m^{(1)}(kr)$ is a Hankel function of the first kind. It can be shown that ϕ_s satisfies the far field radiation condition through the asymptotic value of $H_m^{(1)}(kr)$ for large r. Substituting Equations 2.73 and 2.74 into the boundary condition at the cylinder surface in Equation 2.72 then gives the coefficients:

$$B_m = -\frac{J'_m(ka)}{H_m^{(1)'}(ka)} \quad (2.75)$$

where ' denotes differentiation with respect to the argument (ka). This yields the full expression for velocity potential as

$$\phi = A \frac{\cosh k(z+d)}{\cosh kd} \left[\sum_{m=0}^{\infty} \beta_m \left\{ J_m(kr) - \frac{J'_m(ka)}{H_m^{(1)}(ka)} \right. \right.$$
$$\left. \left. H_m^{(1)}(kr) \right\} \cos(m\theta) \right] e^{-i\omega t} \quad (2.76)$$

Evaluation of pressures and integration over the body surface yields forces and base overturning moment expressions of the form:

$$\frac{F}{\rho g Had} = 2 \frac{G(ka)}{ka} \frac{\tanh(kd)}{kd} \cos(\omega t - \delta) \quad (2.77)$$

and

$$\frac{M}{\rho g Had^2} = 2 \frac{G(ka)}{ka} \left[\frac{kd \sinh(kd) + 1 - \cosh(kd)}{(kd)^2 \cosh(kd)} \right]$$
$$\cos(\omega t - \delta) \quad (2.78)$$

where $G(ka) = [J_1'^2(ka) + Y_1'^2(ka)]^{-1/2}$

and $\delta = -\tan^{-1}\left[\frac{Y_1'(ka)}{J_1'(ka)}\right]$

Figure 2.45 presents a curve of calculated maximum horizontal force against non-dimensional wave number (ka). The above analytic results are compared with data from a boundary integral numerical computation described in the next section and also against model test data.

2.3.8.2 Boundary integral numerical technique

For offshore structures of arbitrary geometry, it becomes necessary to develop general numerical solutions for design purposes. One such solution that is extensively used in the offshore industry is obtained by modelling the scattered or diffracted wave flow by oscillating point sources placed on facets covering the body surface. If the total incident and scattered wave potential is written as:

$$\phi = \text{Real} \{\phi'(x, y, z) \, e^{-i\omega t}\} \quad (2.79)$$

the spatial variation of potential ϕ' can be split up into its incident and diffracted components:

$$\phi' = \phi'_i + \phi'_s \quad (2.80)$$

Then Lamb (1975) shows that the scattered wave potential may always be represented as a continuous distribution of point sources over the immersed body surface through a summation integral of the form:

$$\phi'_s(x, y, z) = \frac{1}{4\pi} \iint_S f(\xi, \eta, \zeta) \, G(x, y, z; \xi, \eta, \zeta) \, dS \quad (2.81)$$

where (x, y, z) denotes a point in the fluid with (ξ, η, ζ) denoting a point on the body surface, $f(\xi, \eta, \zeta)$ is the source strength distribution on this surface and dS is a differential area on the body surface. The Green's function $G(x, y, z; \xi, \eta, \zeta)$ must be such that it satisfies the Laplace condition, the radiation condition as well

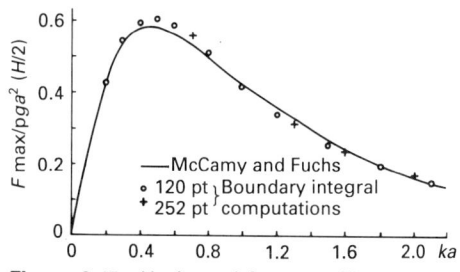

Figure 2.45 Horizontal force coefficient for circular cylinder $d/a = 4.0$, Full lines denote the exact solution by McCamy and Fuchs (1954). Circles and crosses denote boundary element calculations with 120 and 252 facets respectively.

as the sea bed and free surface boundary conditions. Such a condition is given by Wehausen and Laitone (1960) and can be written in an infinite series form as:

$$G(x, y, z; \xi, \eta, \zeta) = \frac{2\pi(v^2 - k^2)}{k^2 d - v^2 d + v} \cosh[k(z + d)].$$

$$\cosh[k(\eta + d)] [Y_o(kr) - i J_o(kr)]$$

$$+ 4 \sum_{k=1}^{\infty} \frac{k^2 + v^2}{k^2 d + v^2 d - v} \cos[\mu_k(z + d)]$$

$$\cos[\mu_k(\eta + d)] K_o(\mu_k r) \quad (2.82)$$

where μ_k are positive roots of equation:

$$\mu_k \tan \mu_k d + v = 0$$
$$r = \sqrt{(x - \xi)^2 + (y - \eta)^2},$$
$$v = k \tanh kd$$

and

$$k = 2\pi/\lambda$$

J_o and Y_o denote Bessel functions of the first and second kind respectively, and of zero order and K_o denotes the modified Bessel function of the second kind of zero order. The scattered velocity potential ϕ'_s is determined by imposing a no flow boundary condition on the body surface. This condition can be implemented by imposing the condition that at an element of the immersed body surface, the flow due to a local source plus the flow due to all other sources on the body surface must be equal and opposite to the flow at the element due to the undisturbed wave. This can be written mathematically as:

$$\frac{1}{2} f(x, y, z) - \frac{1}{4\pi} \iint_S f(\xi, \eta, \zeta) \cdot \frac{\partial G}{\partial n}$$

$$(x, y, z; \xi, \eta, \zeta) \, dS = -W(x, y, z) \quad (2.83)$$

where the undisturbed incident wave flow is given by:

$$W(x, y, z) = \left[n_z \frac{\sinh k(z + d)}{\cosh kd} + i n_x \frac{\cosh k(z + d)}{\cosh kd} \right] e^{ikx}$$

$$(2.84)$$

In Equation 2.83, the factor of half in the first term is due to the fact that half the surface source fluid will be streaming into the body's interior volume. A negative sign on the left hand side of this equation defines fluid flow into the surface from the exterior.

Now the unknown source strengths $f(x, y, z)$ are found by satisfying Equation 2.83 at all points on the body surface. However, since the derivation of $\partial G/\partial n$ is complicated for an arbitrary body, a closed form solution is not possible and a discretized numerical method is adopted instead (Garrison and Chow 1972). The submerged surface of the body is divided into a lattice of facets and Equation 2.83 is satisfied at nodal points on each facet to reduce the integral equation to a finite number of simultaneous linear equations. Taking f_i as the source strength at the centroid (x_i, y_i, z_i) of the ith facet of area δS_i, Equation 2.83 can be written as:

$$-f_i + \alpha_{ij} f_j = 2W_i \quad (2.85)$$

where

$$\alpha_{ij} = \frac{1}{2\pi} \sum_{j=1}^{N} \frac{\partial G}{\partial n} (x_i, y_i, z_i; \xi_i, \eta_i, \zeta_i) \quad (2.86)$$

for a total of N facets with the summation excluded for $i = j$. The derivative of G is evaluated using the equation:

$$\frac{\partial G}{\partial n} = \frac{\partial G}{\partial x} n_x + \frac{\partial G}{\partial y} n_y + \frac{\partial G}{\partial z} n_z \quad (2.87)$$

where n_x, n_y and n_z are components of the unit normal vector to the facet surface. Once the coefficient matrix α_{ij} is obtained from Equation 2.86, Equation 2.85 is solved for all f_i by matrix inversion and the scattered potentials obtained through the equation:

$$\phi'_{si} = \frac{1}{4\pi} f_j \sum_{j=1}^{N} G(x_i, y_i, z_i; \xi_i, \eta_i, \zeta_i) \, \delta S_i \quad (2.88)$$

Then taking the hydrodynamic pressure as:

$$p(x, y, z, t) = -\rho \frac{\partial \phi}{\partial t} \quad (2.89)$$

$$= \text{Re} \{i\rho\omega [\phi'_i(x, y, z) + \phi'_s(x, y, z)] e^{-i\omega t}$$

the total hydrodynamic forces and moments are given by the integrals:

$$F(t) = -\iint_S p(x, y, z, t) \, n \, dS$$

and $\quad (2.90)$

$$M(t) = -\iint_S p(x, y, z, t) \, (r \times n) \, dS$$

where n denotes the unit normal to the surface and r denotes the moment arm vector. Numerical values are computed by discretizing these equations and using the scattered wave potential from Equation 2.88.

The numerical integrations defined by Equations 2.86 and 2.88 pose some difficulties due to singularities in the Green's function form. However, the integrals are readily evaluated by using numerical integration schemes designed to cope with singularities. Furthermore, for some so called irregular wave frequencies, the matrix to be inverted for solving Equation 2.85 becomes non-positive definite and no unique solution of the boundary integral problem is possible. This feature of the solution is not due to the numerical discretization employed but arises inherently from the source distribution representation of the scattered wave potential (John 1950 and Murphy 1978). Irregular wave frequencies generally correspond to wave lengths which are smaller than the size of the body and are, therefore, usually at frequencies that are too high to be of concern in most wave loading calculations. Boundary integral techniques also suffer from numerical problems when modelling re-entrant structure geometries or structures with small holes or sharp corners; these problems are trigered by numerical problems due to the close proximity of adjacent surface panel sources.

Apart from these specific problem areas, however, boundary integral methods offer a robust and numerically reliable means for calculating wave forces on offshore structures. This has been recognized in the offshore industry through widespread use of diffraction analysis for calculating wave forces on both large and small bodies.

Figure 2.94 shows a typical facet geometry for a floating vessel hull. *Figure 2.45* presents a comparison of wave force amplitude for an upright vertical cylinder calculated from the exact solution of MacCamy and Fuchs (1954) and also using the boundary integral method for 120 and 252 facets. The agreement between the two is seen to be reasonably good although recent improvements in discretization techniques and numerical integration accuracy have enabled virtually exact agreement to be attained with use of fewer facets (Eatock–Taylor and Waite 1978).

2.3.8.3 Boundary element numerical solution

This approach seeks to obtain a solution for the scattered potential ϕ'_s, by discretizing the fluid around the submerged body into finite elements in an analogous manner to the use of this technique in elastic problems.

The problem is formulated in numerical terms by defining a functional which may be considered to be related to the total energy of the fluid domain and the work done on it across its boundaries. From the calculus of variations, it can be shown that the solution to this boundary value problem is obtained by minimizing the value of this functional which can be written as:

$$\Pi = \int_V \frac{1}{2}\left[\left[\frac{\partial \phi'_s}{\partial x}\right]^2 + \left[\frac{\partial \phi'_s}{\partial y}\right]^2 + \left[\frac{\partial \phi'_s}{\partial x}\right]^2\right] dV$$
$$+ \int_S \left[\frac{1}{2}\alpha \phi'^2_s + \beta \phi'_s\right] dS \qquad (2.91)$$

where $\alpha = -\omega^2/g$ on the free surface, $\alpha = -i\omega/c$ on the radiation boundary and zero elsewhere with $\beta = \partial \phi'_s / \partial n$ on the body surface and zero on all other boundaries. Here dV is an element of the fluid volume and dS is an element of the entire fluid boundary including free surface, sea bed, body surface and far fluid boundary. The finite element solution is implemented by dividing the fluid volume into smaller volumes and specifying unknown potentials ϕ_e at nodal points on these element volumes. A typical element geometry is shown in *Figure 2.46*. The scattered wave potential ϕ_s' within the element can then be written as:

$$\phi_s' = (N)^T \phi_e \qquad (2.92)$$

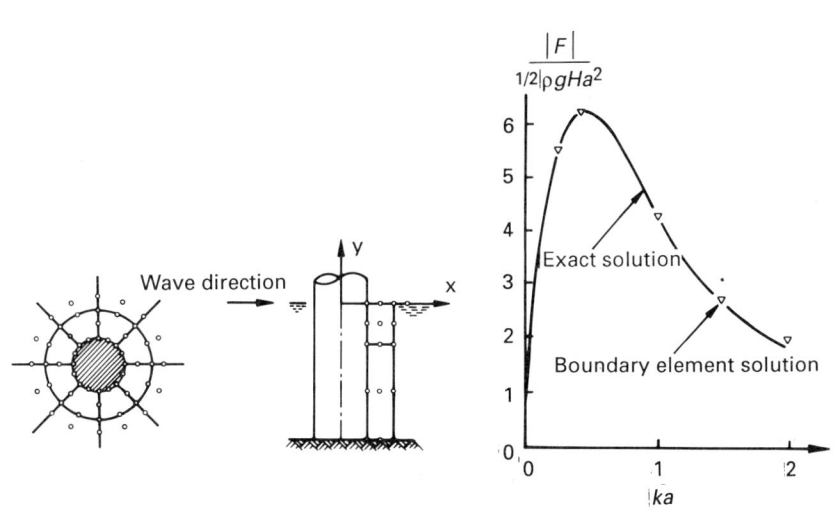

Figure 2.46 Comparison of boundary element numerical techniques and exact solution for horizontal forces on an upright cylinder for $d/a = 5$ (from Zienkiewicz *et al.* 1978)

where N is an interpolation function column vector. Substituting Equation 2.92 into Equation 2.91 and minimizing the functional Π leads to the condition:

$$\frac{\partial \Pi}{\partial (\phi_e)} = \int_{V_e} \left\{ \left[\frac{\partial N}{\partial x}\right]\left[\frac{\partial N}{\partial x}\right]^T + \left[\frac{\partial N}{\partial y}\right]\left[\frac{\partial N}{\partial y}\right]^T \right.$$

$$\left. + \left[\frac{\partial N}{\partial z}\right]\left[\frac{\partial N}{\partial z}\right]^T \right\} \phi'_e \, dV + \int_{S_e} \left[a(N)(N)^T \phi_e \right.$$

$$\left. + q(N) \right] dS = 0 \qquad (2.93)$$

where subscript e refers to integration over a single element only. Equation 2.93 is then converted to a discretized form for each element and all the element equations are assembled into a global matrix equation of the form:

$$A \, \phi_e = B \qquad (2.94)$$

where A and B can be written in integral form as:

$$A = -\frac{\omega^2}{g} \int_{S_s} (N)(N)^T \, dS - \frac{i\omega}{c} \int_{S_r} (N)(N)^T \, dS$$

$$+ \int_V \left\{ \left[\frac{\partial N}{\partial x}\right]\left[\frac{\partial N}{\partial x}\right]^T + \left[\frac{\partial N}{\partial y}\right]\left[\frac{\partial N}{\partial y}\right]^T \right.$$

$$\left. + \left[\frac{\partial N}{\partial z}\right]\left[\frac{\partial N}{\partial z}\right]^T \right\} dV \qquad (2.95)$$

and

$$B = -\int_{S_b} \frac{\partial \phi'_i}{\partial n} (N) \, dS$$

where subscripts b, r and s refer to the body, the radiation surface and the free surface boundaries respectively. Equation 2.94 can be solved by matrix inversion, the scattered potentials recovered by Equation 2.92 and forces and moments calculated through the equations outlined earlier. There is some freedom in applying the technique as to the discretization density and element shapes to be used as well as the type of interpolation functions; these matters are reviewed by Zienkiewicz et al. (1978).

One feature of the finite element solution that has to be given special attention is concerned with satisfying the far field radiation condition in theory at an infinite distance from the body. This poses a difficulty in the numerical technique which has been circumvented by a number of approaches, the simplest of which is to assume that the radiation condition at infinity is applied directly at a radiation boundary at a finite distance from the body. Alternatively, the flows inside and outside this radiation boundary can be matched by using an analytic series solution or a boundary integral solution for the exterior region. A third technique involves the use of infinite elements where the radiation condition can be applied directly.

Zienkiewicz et al. (1978) review these methods and present a comparison of results. *Figure 2.46* presents a typical dicretization and comparison of results between the boundary element and exact solution for a vertical circular cylinder. It can be seen that the techniques are in good general agreement.

The boundary element numerical technique for solution of the diffraction problem offers some potential advantages for practical design problems. The technique can take account of variations in water depth around a structure by appropriate choice of boundary elements. The method can also accommodate a wider range of complicated structure geometries and has the potential to incorporate non-linear wave behaviour and coupling between structural elasticity and the dynamics of the surrounding fluid. These gains are, however, obtained at the cost of greater computational cost and ambiguities on the optimum calculation method to be used; the latter is primarily concerned with choice of the far field solution to be used and its matching with the near field boundary elements.

2.4 Structural response of fixed structures

This subsection describes the final stages of the design process for an offshore structure where static and dynamic environmental forces calculated from the methods outlined earlier are converted to structural stresses and stress load cycles. These are then used to ensure sufficient structural strength and to estimate fatigue lives. Static and dynamic structural analyses of offshore structures are complicated by non-linear effects due predominantly to soil structure interaction in the former case and non-linearities arising from gravity wave mechanics and drag force loading in the latter. These methods of analysis and the effects of non-linear behaviour are considered in detail in this section.

2.4.1 Quasi-static loading and response

2.4.1.1 Methods of analysis

Shallow water steel framed jackets can be analysed for structural stresses with acceptable accuracy by using a quasi-static environmental loading coupled with an elastic structural analysis modified for non-linear behaviour in the soil foundation. In this approach, wind and current are assumed to apply static loads with wave action applying a dynamic loading which is translated into dynamic structural stresses through a quasi-static stress analysis. The technique assumes that resonant frequencies of structural vibrations are sufficiently separated from wave frequencies so that dynamic magnification has a negligible effect on calculated stresses.

In the early days (before the 1960s), structural analysis of jackets was carried out using hand calculations with simple frame theory. However, the advent of computers and the development of finite element structural analysis methods has revolutionized structural calculations in the offshore industry. Finite element analysis program packages such as NASTRAN and STRUDL are used in conjunction with wave loading and foundation analysis programs to carry out complex computer aided design calculations for three dimentional steel frame structures as well as for monolithic concrete structures.

The basic theory underlying one such finite element analysis called the direct stiffness method is described here. The basis of the method is that the structure to be analyzed is discretized into a number of small elements with a framed jacket structure represented by an assembly of beam elements whereas a concrete gravity structure could also be described by beam elements or alternatively by cylindrical shell elements. The displacements of nodal points on the elements are taken as unknowns and the finite element method is implemented using the following steps:

1. Initially physical data for the structure configuration, member properties and support constraints are identified. The structure is then discretized into finite elements and the unknown nodal displacements are identified and numbered using a systematic indexing procedure.
2. This is followed by evaluation of member stiffness matrices using local member axes. The member stiffness matrices are then transformed into global structure axes and assembled into a global stiffness matrix.
3. The loads acting on the members and nodes are also assembled into a global load vector. This requires that forces acting on each member other than at the nodes are transformed into equivalent joint loads before being added on to the known loads at the nodes such that the structure is considered as being loaded at the nodes only.
4. It is then necessary to re-arrange the equation relating the nodal displacement vector and stiffness matrix to the load vector in order to substructure out degrees of freedom which are constrained (that is their nodal displacements are known or are zero) in order that a matrix equation for the unknown displacements only can be formed.
5. The penultimate step in the procedure is to use the stiffness matrix and load vector to arrive at a solution for the displacement at each of the nodes.
6. In the final step, the displacements are used in combination with the equivalent joint loads to compute member forces and resultant internal stresses.

The mathematical formulation of the above steps is described below for a three dimensional (space frame) beam element idealization of a typical jacket structure shown in *Figure 2.47*. The jacket structure consists of up to 600 beam elements to represent each of the tubular members making up the frame. A global axes system $Oxyz$ is first defined with respect to the whole structure and all of the structural nodes are numbered in a systematic manner around the structure. A typical numbering system could, as shown in *Figure 2.47*, start from a horizontal level at the origin with increasing node numbering in a prescribed direction around the nodes at that level followed by a similar procedure at the next level down and so on. Once this node definition is complete, structure data on node coordinates relative to global axes, member lengths, areas of cross-section and second moments of area can be systematically identified in terms of the node numbering. This is followed by definition of six degrees of freedom for each node – three in translation and three in rotation parallel to and about the global axes directions Ox, Oy and Oz. Thus the number of degrees of freedom will be six times the number of nodes although a small proportion of these freedoms (e.g. at the base of the jacket of *Figure 2.47* will be constrained to be zero or prescribed in other ways to account for the effect of foundations.

The next step in the analysis is to define the stiffnesses of each member in the frame work, this being done in terms of principal member axes $O_m x_m y_m z_m$ as defined in *Figure 2.48*. The single beam element in this figure has 12 degrees of freedom, three translations and rotations along and about member axes directions at each end of the members. These are shown in *Figure 2.48* with the rotational degrees of freedom denoted by double-headed arrows showing the direction that the right-hand screw rule axis would point to generate the rotation. These degrees of freedom are systematically numbered by indices 1–12 with, for example, index 3 denoting displacement of the member at end j parallel to the $O_m z_m$ axis and index 6 denoting clockwise rotation at end j about the positive $O_m z_m$ axis.

Now a stiffness matrix S_{mi} for the ith frame member can be defined to be:

$$F_{mi} = S_{mi} D_{mi} \qquad (2.96)$$

where D_{mi} is the vector of 12 displacements (1–12 in *Figure 2.48*), F_{mi} is a 12 × 1 vector of forces along these index directions and S_{mi} (corresponding to the 12 possible member end displacements) are obtained by defining the actions (or forces) required to obtain unit translation or rotation for each degree of freedom with all other degrees of freedom restrained to zero displacement or rotation. This is illustrated diagrammatically in *Figure 2.49*. *Figure 2.49(a)* demonstrates how a unit displacement along degree of freedom, 1, requires a force of EA_x/L along 1 and will also result in a force of $-EA_x/L$ along degree of freedom 7 where E, A_x and L are the member Young's Modulus, area of cross-section and length, respectively. Similarly, unit rotation about the $O_m y_m$ axes only (index 5) leads to forces

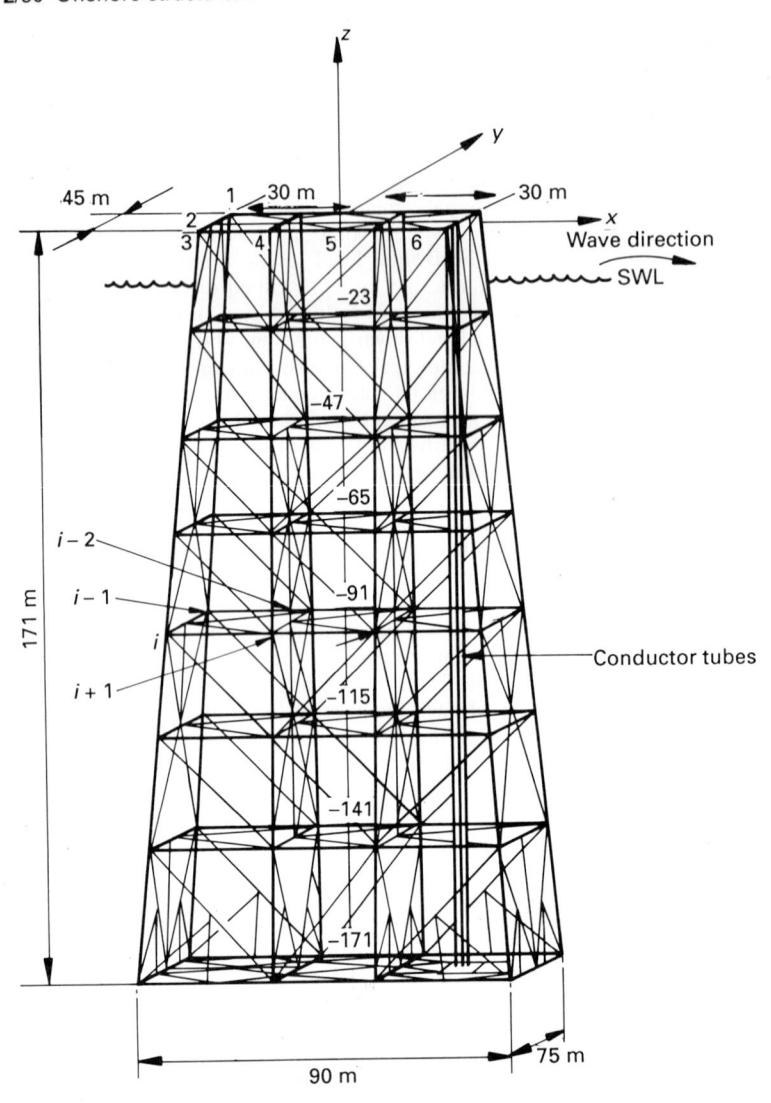

Figure 2.47 Typical idealization of a jacket structure

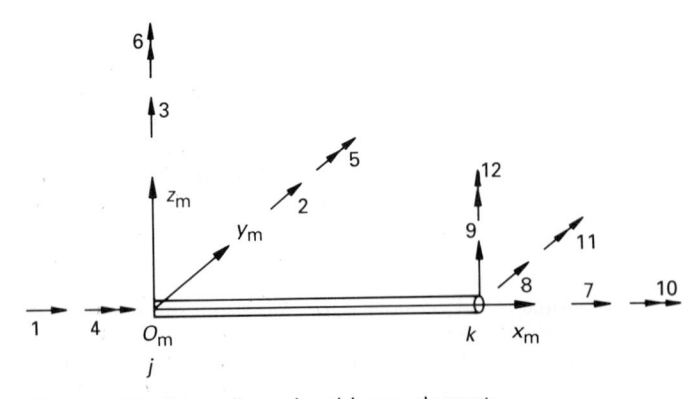

Figure 2.48 Three dimensional beam element

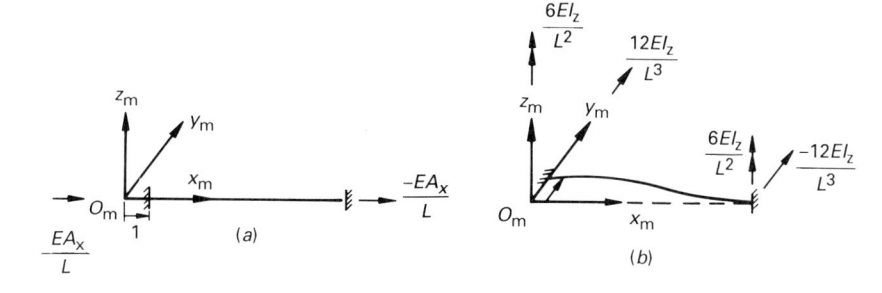

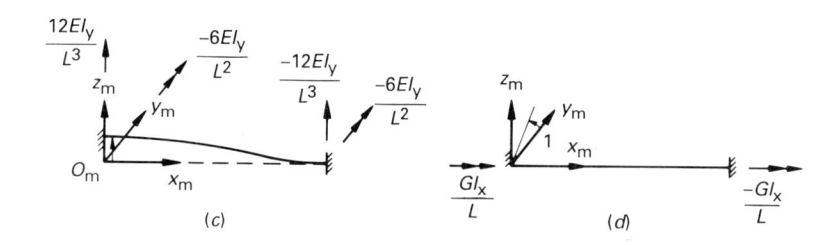

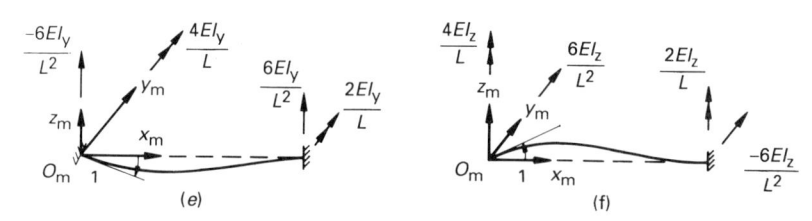

Figure 2.49 Forces and moments required for unit displacement indices 1 to 6 on beam element. L – beam length; A_x – area of cross-section; I_x, I_y, I_z – second moments of area of cross-section about $O_m x_m$, $O_m y_m$ and $O_m z_m$ axes through centroid of cross-section; E – Youngs modulus of beam material

$-6EI_y/L^2$ and $+6EI_y/L^2$ along indices 3 and 9 as well as moments $4EI_y/L$ and $2EI_y/L$ along indices 5 and 11, respectively, as shown in *Figure 2.49(e)*. Similar diagrams can be obtained for unit displacements in the degrees of freedom corresponding to indices 7–12 in *Figure 2.48*.

These effects can be summarized into the 12 × 12 member stiffness matrix S_{mi} given in *Table 2.6*. The matrix rows and columns are numbered to enable a particular stiffness term to be identified with respect to the index numbering of displacements and forces. The matrix is symmetrical and can be partitioned as shown to separate the effects of each set of six displacements on themselves and on each other.

In order to use the stiffness matrix of all frame members in a global structural analysis, the stiffness matrix for each member needs to be transformed from member to global axes. The transformation is illustrated by *Figure 2.50* with generalized member axes $O_G x_G y_G z_G$ for member i with end nodes j and k and global axes $Oxyz$. It is possible that the $O_G x_G y_G z_G$ axes may not be principal axes of the member cross-section and that a rotation of angle θ (clockwise about the positive $O_G x_G$ axis) may need to be applied to transform generalized member axes to principal axes $O_m x_m y_m z_m$. *Figure 2.50* also illustrates how axes $O_m x_m y_m z_m$ can be transformed to being parallel to axes $Oxyz$ by two rotations, through angle ϕ to make axis $O_m x_m$ parallel to the Oxy plane and then through angle ψ to make $O_m x_m$ parallel to Ox.

Now the rotation of each axis through a prescribed angle can be readily defined by vector algebra. Thus, the rotation to transform generalized member axes to principal axes through rotation θ can be written as:

Table 2.6: Member stiffness matrix

$$S_{Mi} = \begin{bmatrix}
 & 1 & 2 & 3 & 4 & 5 & 6 & 7 & 8 & 9 & 10 & 11 & 12 \\
1 & \dfrac{EA_x}{L} & 0 & 0 & 0 & 0 & 0 & -\dfrac{EA_x}{L} & 0 & 0 & 0 & 0 & 0 \\
2 & 0 & \dfrac{12EI_z}{L^3} & 0 & 0 & 0 & \dfrac{6EI_z}{L^2} & 0 & -\dfrac{12EI_z}{L^3} & 0 & 0 & 0 & \dfrac{6EI_z}{L^2} \\
3 & 0 & 0 & \dfrac{12EI_y}{L^3} & 0 & -\dfrac{6EI_y}{L^2} & 0 & 0 & 0 & -\dfrac{12EI_y}{L^3} & 0 & -\dfrac{6EI_y}{L^2} & 0 \\
4 & 0 & 0 & 0 & \dfrac{GI_x}{L} & 0 & 0 & 0 & 0 & 0 & -\dfrac{GI_x}{L} & 0 & 0 \\
5 & 0 & 0 & -\dfrac{6EI_y}{L^2} & 0 & \dfrac{4EI_y}{L} & 0 & 0 & 0 & \dfrac{6EI_y}{L^2} & 0 & \dfrac{2EI_y}{L} & 0 \\
6 & 0 & \dfrac{6EI_z}{L^2} & 0 & 0 & 0 & \dfrac{4EI_z}{L} & 0 & -\dfrac{6EI_z}{L^2} & 0 & 0 & 0 & \dfrac{2EI_z}{L} \\
7 & -\dfrac{EA_x}{L} & 0 & 0 & 0 & 0 & 0 & \dfrac{EA_x}{L} & 0 & 0 & 0 & 0 & 0 \\
8 & 0 & -\dfrac{12EI_z}{L^3} & 0 & 0 & 0 & -\dfrac{6EI_z}{L^2} & 0 & \dfrac{12EI_z}{L^3} & 0 & 0 & 0 & -\dfrac{6EI_z}{L^2} \\
9 & 0 & 0 & -\dfrac{12EI_y}{L^3} & 0 & \dfrac{6EI_y}{L^2} & 0 & 0 & 0 & \dfrac{12EI_y}{L^3} & 0 & \dfrac{6EI_y}{L^2} & 0 \\
10 & 0 & 0 & 0 & -\dfrac{GI_x}{L} & 0 & 0 & 0 & 0 & 0 & \dfrac{GI_x}{L} & 0 & 0 \\
11 & 0 & 0 & -\dfrac{6EI_y}{L^2} & 0 & \dfrac{2EI_y}{L} & 0 & 0 & 0 & \dfrac{6EI_y}{L^2} & 0 & \dfrac{4EI_y}{L} & 0 \\
12 & 0 & \dfrac{6EI_z}{L^2} & 0 & 0 & 0 & \dfrac{2EI_z}{L} & 0 & -\dfrac{6EI_z}{L^2} & 0 & 0 & 0 & \dfrac{4EI_z}{L}
\end{bmatrix}$$

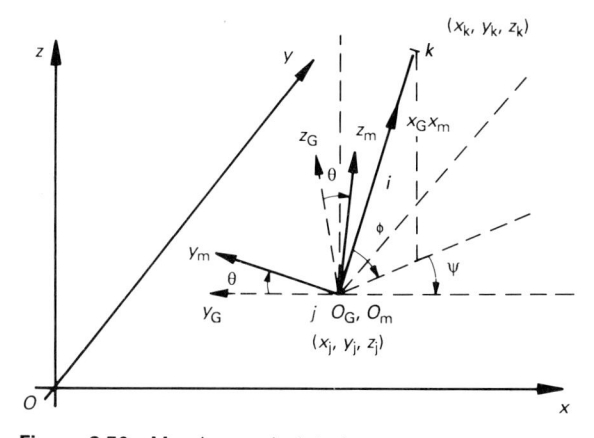

Figure 2.50 Member and global axes

$$R_\theta = \begin{bmatrix} 1 & 0 & 0 \\ 0 & \cos\theta & \sin\theta \\ 0 & -\sin\theta & \cos\theta \end{bmatrix} \quad (2.97)$$

and the transformations required to rotate the $O_m x_m$ axes first parallel to the global Oxy plane and then parallel to the global Ox axis can be shown to be:

$$R_\phi = \begin{bmatrix} C_{xz} & C_y & 0 \\ -C_y & C_{xz} & 0 \\ 0 & 0 & 1 \end{bmatrix} \quad (2.98)$$

and

$$R_\psi = \begin{bmatrix} \dfrac{C_x}{C_{xz}} & 0 & \dfrac{C_z}{C_{xz}} \\ 0 & 1 & 0 \\ -\dfrac{C_z}{C_{xz}} & 0 & \dfrac{C_x}{C_{xz}} \end{bmatrix} \quad (2.99)$$

respectively, where:

$$C_x = \frac{x_k - x_j}{L}, \quad C_y = \frac{y_k - y_j}{L}, \quad C_z = \frac{z_k - z_j}{L}$$

$$C_{xz} = \sqrt{C_x^2 + C_z^2} \quad (2.100)$$

and

$$L = \sqrt{(x_k - x_j)^2 + (y_k - y_j)^2 + (z_k - z_j)^2}$$

The three successive rotations required for general transformation of axes can then be written as:

$$R = R_\theta R_\phi R_\psi$$

$$= \begin{bmatrix} C_x & C_y & C_z \\ \dfrac{C_x C_y \cos\alpha - C_z \sin\alpha}{C_{xz}} & C_{xz}\cos\alpha & \dfrac{C_y C_z \cos\alpha + C_x \sin\alpha}{C_{xz}} \\ \dfrac{C_x C_y \sin\alpha - C_z \cos\alpha}{C_{xz}} & -C_{xz}\sin\alpha & \dfrac{C_y C_z \sin\alpha + C_x \cos\alpha}{C_{xz}} \end{bmatrix}$$

and applied to the four sets of three degrees of freedom for each member (see Figure 2.47) by the total rotation matrix:

$$R_T = \begin{bmatrix} R & 0 & 0 & 0 \\ 0 & R & 0 & 0 \\ 0 & 0 & R & 0 \\ 0 & 0 & 0 & R \end{bmatrix} \quad (2.102)$$

The member stiffness matrix S_{mi} for member i can be transformed to the stiffness for global structure axis S_{msi} by the matrix equation:

$$S_{msi} = R_{Ti}^T S_{mi} R_{Ti} \quad (2.103)$$

It is usual during or just after this transformation to assemble the full global stiffness matrix from its constituent member matrices which are now in relation to global axes. This assembly process can be illustrated using the matrix equation:

$$\begin{bmatrix} f_1 \\ f_2 \\ --- \\ f_i \\ --- \\ f_m \end{bmatrix} \begin{bmatrix} & & & & & 0 \\ & & & & & \\ --- & & & & & \\ & & & & & \\ --- & & & & & \\ & 0 & & & & \end{bmatrix} \begin{bmatrix} d_1 \\ d_2 \\ --- \\ d_i \\ --- \\ d_m \end{bmatrix}$$

where f_i and d_i are the vectors of forces and displacements along the m degrees of freedom of the total structure with f_{6j-5} to f_{6j} containing the forces and moments at the jth node and d_{6j-5} to d_{6j} containing the resultant displacements and rotations. The stiffness matrix of each member is then assembled within the appropriate row and column range corresponding to its degrees of freedom with overlapping terms summed to yield the assembled joint stiffness matrix. In a space frame analysis, the upper and lower 6 × 6 partition of each 12 × 12 member stiffness matrix will overlap and be added to the matrices of other adjacent members connected to the frame at that node.

In the direct stiffness method, two fundamental equations govern structure reaction forces and member internal and applied forces. These equations can be obtained by considering the response of the structure to variations in the support and free joint displacements. The column vector of support reaction forces A_R can be written as:

$$A_R = -A_{RC} + S_{RF} D_F \qquad (2.105)$$

where A_{RC} is the column vector of actual or equivalent joint loads applied directly to the supports and S_{RF} is the global stiffness matrix relating the forces at restrained degrees of freedom to displacements D_F at the free degrees of freedom.

The column vector of internal forces for member i, A_{MI} can also be written as:

$$A_{mi} = A_{mLi} + S_{mi} D_{mi} \qquad (2.106)$$

with all terms applying to member axes for each member i. Here A_{mLi} is the column vector of fixed end actions, S_{mi} and D_{mi} are row i of the stiffness matrix and the column of member end displacements, respectively.

It now becomes necessary to define loads acting on the structure in order to define an applied force vector which is made up of forces on the member (between its ends) and forces acting at the ends or nodes. Matrix methods of structural analysis require that the forces acting on the members be converted first to forces at the fixed ends and then to equivalent joint loads acting on the member ends which are used as nodes in the finite element analysis method.

These fixed end forces and equivalent joint loads are first calculated in member axes before being transformed to structure axes as described above.

A typical beam element can be subjected to concentrated loads as well as uniformly or non-uniformly distributed loads along its length. These can readily be translated into forces at the beam fixed ends by using equilibrium considerations. *Table 2.7* presents fixed end forces or actions for typical loadings. The fixed end actions due to a linearly distributed force $w(x)$, given in the last row of *Table 2.7*, is often used in jacket structural analysis.

The fixed end actions in member axes for the ith member are written in a matrix A_{mL} of order $12 \times m$ in which the ith column gives the vector of 12 forces and moments at the ends of the ith member, m being the total number of degrees of freedom of the structure. The elements of matrix A_{mL} are transformed to the equivalent joint load vector A_E by first transforming A_{mL} into global axes through the rotational transformation:

$$A_{msi} = R_{Ti}^T A_{mLi} \qquad (2.107)$$

and then incorporating the negative of the vector A_{msi} into an assembly of the equivalent joint load vector A_E. The negative sign arises because the fixed end actions and equivalent joint loads at the member ends are in equilibrium so that the equivalent joint loads can simply be taken as the negative of the fixed end actions.

This assembly operation is similar to that shown for the stiffness matrix in Equation 2.104 with the square member stiffness matrices replaced by column vectors of member forces. Finally, the column vector A_J, of forces applied directly at each of the joints is then summed with A_E to obtain the total combined load vector:

$$A_C = A_E + A_J \qquad (2.108)$$

Table 2.7: Fixed end forces and moments on beams for typical loading conditions

Loading	R_A	R_B	M_A	M_B
Concentrated lateral force P downwards at A from left hand end and B from right hand end	$\dfrac{PB^2(3A+B)}{L^3}$	$\dfrac{PA^2(A+3B)}{L^3}$	$\dfrac{PAB^2}{L^2}$	$-\dfrac{PA^2B}{L^2}$
Concentrated moment M with axis at right angles to beam at A from left hand end and B from right hand end	$\dfrac{6MAB}{L^3}$	$-\dfrac{6MAB}{L^3}$	$\dfrac{MB}{L^2}(2A-B)$	$\dfrac{MA}{L^2}(2B-A)$
Uniformly distributed lateral force w over a length A of the left hand end	$\dfrac{wA}{2L^3}(2L^3 - 2A^2L + A^3)$	$\dfrac{wA^3}{2L^3}(2L-A)$	$\dfrac{wA^2}{12L^2}(6L^2 - 8AL + 3A^2)$	$\dfrac{wA^3}{12L^2}(4L - 3A)$
Distributed lateral force of intensity $w_o x / L$	$\dfrac{3w_o L}{20}$	$\dfrac{7w_o L}{20}$	$\dfrac{w_o L^2}{30}$	$-\dfrac{w_o L^2}{20}$
Linearly distributed lateral force of intensity $w(x) = w_1 + x(w_2 - w_1)/L$	$\dfrac{L(7w_1 + 3w_2)}{20}$	$\dfrac{L(3w_1 + 7w_2)}{20}$	$\dfrac{L^2(3w_1 + 2w_2)}{60}$	$-\dfrac{L^2(2w_1 + 3w_2)}{60}$

The stiffness matrix S_m and load vector A_C are now known in the matrix equation:

$$S_M D_M = A_C \qquad (2.108)$$

Prior to solution of this equation, however, the matrix equation corresponding to the free and restrained degrees of freedom needs to be separated out. If the free and restrained degrees of freedom are denoted by suffixes F and R, respectively, Equation (2.109) becomes:

$$\begin{bmatrix} S_{FF} & S_{FR} \\ S_{RF} & S_{RR} \end{bmatrix} \begin{bmatrix} D_F \\ D_R \end{bmatrix} = \begin{bmatrix} A_F \\ A_R \end{bmatrix} \qquad (2.110)$$

or

$$S_{FF} D_F + S_{FR} D_R = A_F \qquad (2.111)$$
$$S_{RF} D_F + S_{RR} D_R = A_R \qquad (2.112)$$

Here only the free joint displacements D_F and the support reaction A_R are unknowns whereas the restrained support displacements will be known or will be zero.

In computer based implementation of this technique, equations for the free joint displacements are separated out from those for the restrained displacements by re-arranging the rows and columns of the stiffness matrices and the order of elements in the desplacement and load column vectors in a consistent manner such that Equations 2.111 and 2.112 can be separated out. Equation 2.111 can then be solved for D_F using the matrix inversion:

$$D_F = S_{FF}^{-1}(A_F - S_{FR} D_R). \qquad (2.113)$$

The unknown support reactions can also be calculated by resubstituting D_F into the equation:

$$A_R = -A_{RC} + S_{RF} D_F + S_{RR} D_R \qquad (2.114)$$

where A_{RC} is the column vector of actual and equivalent joint loads applied directly to the supports and the restrained displacements D_R are known and usually zero.

Internal member forces (or member end actions) A_m are then recovered by using the combined vector D_m in the equation:

$$A_{mi} = A_{mLi} + S_{mi} R_{Ti} D_{mi} \qquad (2.115)$$

for each member i. The end forces and bending moments in A_{mi} are then readily converted to a distribution of stresses within the member.

It is useful at this stage to give an indication of the way in which the above equations are used to permit easier programming of the direct stiffness method for solution by a digital computer. Consider a space frame made up of n_j joints with each joint being analysed for six degrees of freedom, three translations and three rotations. If n_r degrees of freedom of the structure are restrained, the number of free degrees of freedom to consider are:

$$n = 6n_j - n_r \qquad (2.116)$$

For the jth joint, the degrees of freedom corresponding to translation along the x, y and z directions can be systematically numbered as $6j - 5$, $6j - 4$ and $6j - 3$, respectively, with rotations about the x, y and z axis numbered as $6j - 2$, $6j - 1$ and $6j$, respectively.

Thus a member i with joint numbers j and k at its ends will have end displacements $j1-j6$ and $k1-k6$ as shown in *Figure 2.51* and indexed as follows:

$$\begin{array}{lll} j1 = 6j-5 & j2 = 6j-4 & j3 = 6j-3 \\ j4 = 6j-2 & j5 = 6j-1 & j6 = 6j \\ k1 = 6k-5 & k2 = 6k-4 & k3 = 6k-3 \\ k4 = 6k-2 & k5 = 6k-1 & k6 = 6k \end{array} \qquad (2.117)$$

The above index system is used with arrays within computer programs to simplify the extensive matrix manipulation that is required to solve for displacements. A typical example is given by the manner in which the first column of the stiffness matrix of member i in global axes contributes to the global joint stiffness matrix. This contribution is described by the equations:

$$(S_J)j_n j_1 = \Sigma S_{ms} + (S_{ms_{n,1}})i$$
$$(S_J)k_n j_1 = (S_{m_{n+6;1}})i \qquad (2.118)$$

for $n = 1, 2, 3, 4, 5$ and 6

The quasi-static structural finite element analysis described above is implemented in the design through the following steps for a steel framed jacket structure:

1. The geometry of the structure is first defined with respect to a three dimensional coordinate axis system with the beam elements and their nodes and nodal displacements numbered in a consistent manner. Initial values of member diameter and wall thickness are defined usually from experience or from a preliminary design study to examine the properties of different structural configurations. If

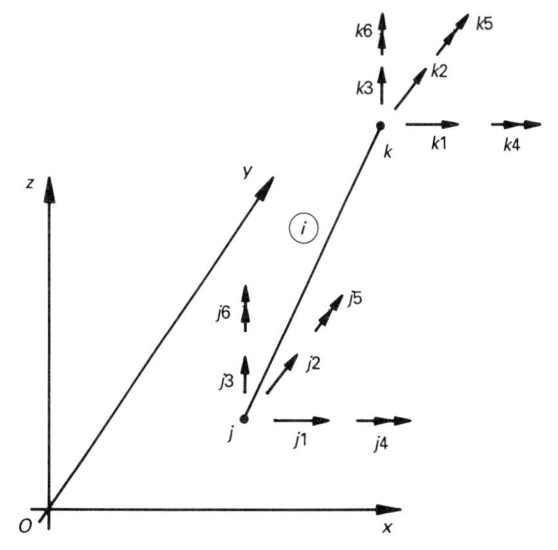

Figure 2.51 Index notation for computer inplementation of direct stiffness method

the cross-sectional properties of a structural member vary significantly between its two ends, then it is modelled by using more than one beam element. Local joint reinforcements are normally disregarded although more detailed local analysis and inclusion of eccentric connection detail is worthwhile at joints between large diameter stiffened members.

2. The loads experienced by the jacket structure are then defined. These consist of the various vertical live loads and equipment dead loads from the platform top sides and environmental loads from wind, current and waves. Advanced versions of finite element analysis packages for offshore applications are capable of calling subprograms which can compute the loading due to current and waves on individual cylindrical members; the latter is most usually done using Stoke's fifth order wave theory. Wind force values are usually input as data. The analysis programs can automatically compute member self weights and the buoyancy forces due to flooded or non-flooded members. Current forces are calculated by summing wave and current velocities vectorially before applying the general Morison Equation force formulation defined by Equations 2.54 or 2.55.

3. The soil structure interaction has a strong effect on the structural response of a piled jacket. This interaction is modelled within the finite element analysis by replacing each pile or closely spaced pile group by a substructure consisting of lateral stiffnesses in two directions, an axial stiffness and a moment stiffness. These stiffnesses can be deduced by linearizing the response from the non-linear pile and soil interaction analysis that is carried out using data obtained from soil investigation. More details of this procedure are given later.

4. The finite element analysis is then implemented by stepping the design wave through the structure with small increments of wave phase (or time). Calculations are carried out for several different wave periods and for simultaneous wind and current in the wave direction as well as for different permutations of top sides loads and equivalent soil stiffness properties. For each loading condition, the analysis will yield maximum shear force and overturning moment at the base of the structure together with maximum end force moments, joint displacements and rotations for each member. Some programs will also give external support reactions and carry out force and moment equilibrium checks at each joint.

5. The member forces are then used to compute member axial, bending, combined axial and bending, tension and shear stresses. It is common practice to use the analysis program to carry out an automatic check of these stress levels against allowable values recommended by a specified certifying authority.

The design procedure outlined above is then run repeatedly to optimize the jacket design for the number of piles, member wall thicknesses and detailed member configuration.

Application of the above technique to gravity structures differs only in two ways. The structural analysis can be carried out through a simple beam element model using a typical idealization of the type shown in *Figure 2.52* or alternatively more representative cylindrical shell elements can be used to model the caisson cellular structure and vertical walls. In both cases, it is necessary to include some idealization of the foundation elasticity either by equivalent springs as in *Figure 2.52* or by continuing the finite element idealization into the foundation using solid elements as described by Penzien and Tseng (1978).

Furthermore, wave forces are calculated using diffraction theory only or a combined Morison/diffraction approach. It is usual to calculate the forces separately and to define them as a distribution of equivalent nodal loads for input to the analysis. Since the structural deflections of these relatively rigid structures are very small compared to water particle displacements, the drag force non-linearity in the Morison Equation (Equation 2.54 or 2.55) can be explicitly computed and does not present analysis problems.

2.4.1.2 Foundation effects

For piled jackets, the structural calculations described above have to be supplemented by a lateral pile loading analysis which quantifies the coupled response of an embedded pile and non-linear material behaviour of the surrounding soil. When a single pile is loaded by a horizontal force due to environmental loading of the jacket structure of which it is a part, the pile will deflect against the surrounding soil and take up a typical deflection and bending moment profile shown in *Figure 2.53*. The magnitude of deflection, rotation and bending moment at the end of the pipe will depend on a balance of forces between the pile and surrounding soil. Determination of this behaviour requires data on the lateral soil reaction on the pile as a function of depth together with a means of structural analysis of the pile itself due to varying soil loading. Taking the former first, the soil reaction is a function of soil type, pile properties, its loading, deflection and the soil depth and may be expressed as:

$$p = -E(x, y)\, y \qquad (2.119)$$

where p is the distributed soil reaction force, y is the lateral pile deflection and E is the modulus of elasticity of the soil which will vary with soil deflection y, and depth x. From basic beam bending theory, the structural behaviour of the pile can be written as:

$$E_p I_p \frac{d^4 y}{dx^4} = w \qquad (2.120)$$

where $E_p I_p$ is the elastic modulus of the pile and w is the distributed lateral load on the pile. Combining Equa-

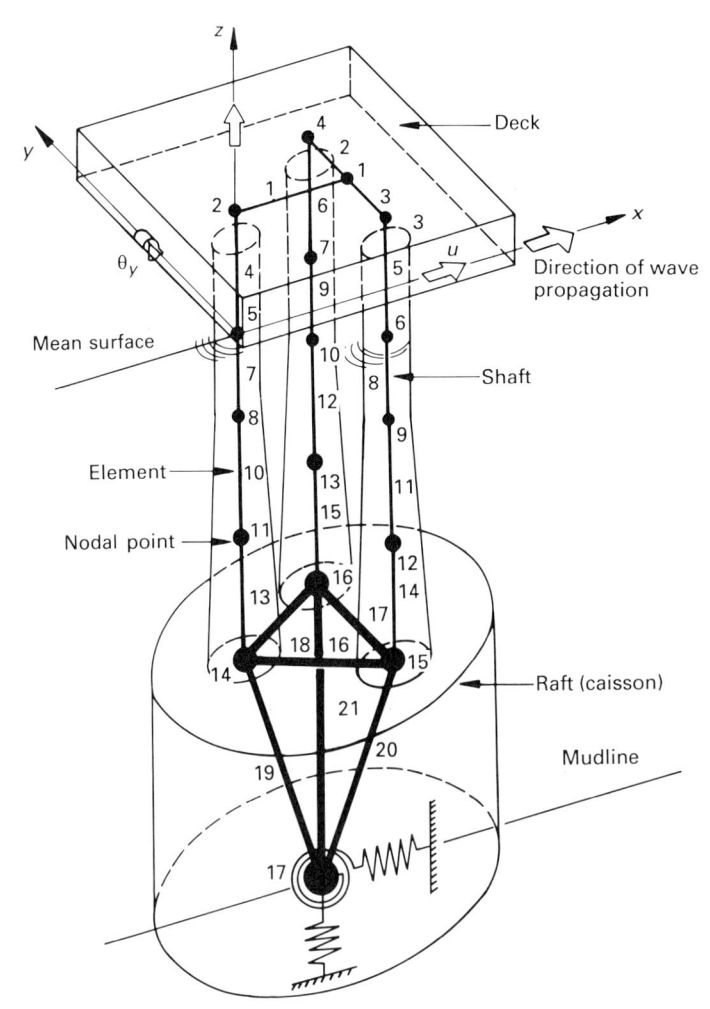

Figure 2.52 Typical simple idealization of gravity structure (from Bell *et al.* 1976)

tions 2.119 and 2.120 yields the basic differential equation for a laterally loaded pile as:

$$E_p I_p \frac{d^4 y}{dx^4} + E(x, y) y = 0 \qquad (2.121)$$

One of the greatest difficulties in solving this equation arises from the fact that the soil modulus E varies both with pile deflection y, and depth x. For many soil types, E increases with depth and can be modelled by the equation:

$$E = k x \qquad (2.122)$$

where k is a constant. A simple but very approximate method for calculating pile behavior is obtained by assuming that the pile is completely fixed against rotation at some point below the mud line (such as A in *Figure 2.53*). The position of this point is selected from experience and the forces exerted on the pile by the soil are neglected above this point. Despite the desirable simplicity of this method, it does not reflect the physical behavior of the pile and soil and is not recommended for use.

An alternative method for computing pile/soil interaction is to solve Equation 2.121 by a finite difference numerical scheme although an analytical solution can be obtained for the special case of constant E. In a finite difference numerical scheme, Equation 2.121 is written as:

$$E_p I_p \left[\frac{y_{i-2} - 4y_{i-1} + 6y_i - 4y_{i+1} + y_{i+2}}{\delta^4} \right] + E_i y_i = 0 \qquad (2.123)$$

where the pile length shown in *Figure 2.54* is descretized into n elements of length δ each. Equation 2.123 is rewritten as:

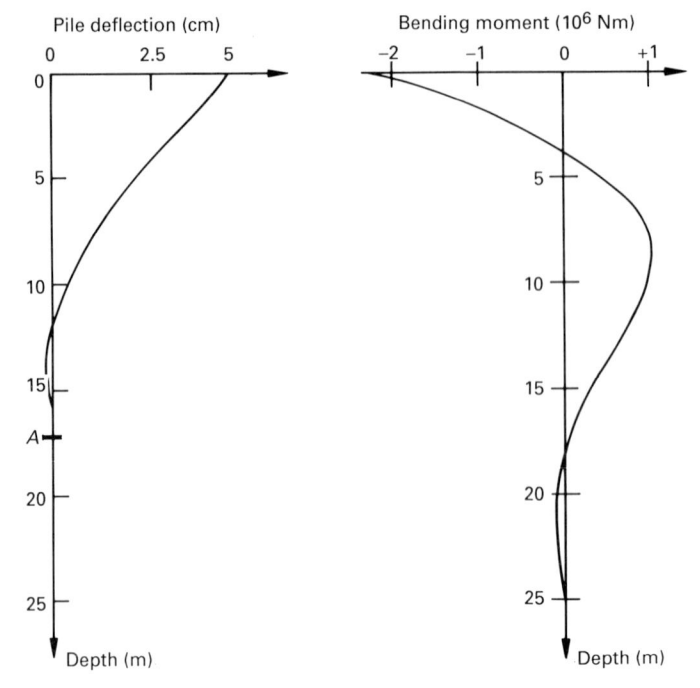

Figure 2.53 Deflection and moment of an embedded pile

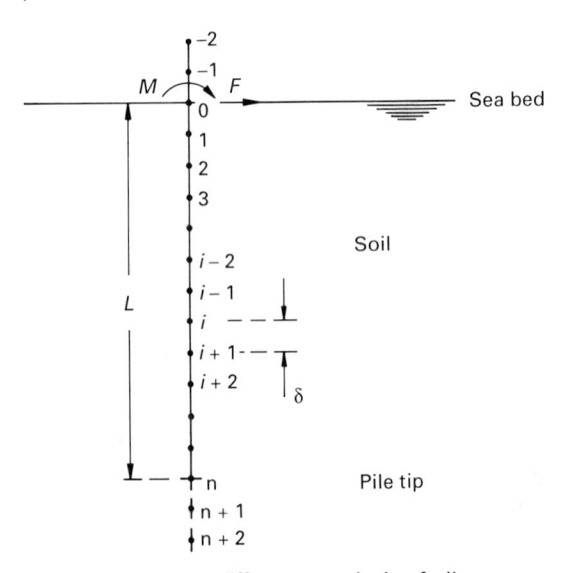

Figure 2.54 Finite difference analysis of pile

$$Y_{i-2} - 4y_{i-1} + \left[6 + \frac{E_i L^4}{E_p I_p n^4}\right] y_i + 4y_{i+1} + y_{i+2} = 0 \quad (2.124)$$

using $\delta = L/n$ from the discretization of *Figure 2.54*. The recurrence relation of Equation 2.124 represents $i-1$ equations when applied at the discrete points $2-n$ along the pile length.

Boundary conditions at the pile top and bottom ends will yield four additional equations. The shear force F, and bending moment M, at the top of the pile yield the equations:

$$F = E_p I_p \frac{d^3 y}{dx^3} \text{ or } -y_{-2} + 2y_{-1} - 2y_2 + y_3 = \frac{FL^3}{E_p I_p n^3}$$

and $\quad (2.125)$

$$M = E_p I_p \frac{d^2 y}{dx^2} \text{ or } y_2 - 2y_1 + y_{-1} = \frac{ML^2}{E_p I_p n^2}$$

At the pile bottom, a free pile tip gives zero shear force and bending moment to yield:

$$E_p I_p \frac{d^3 y}{dx^3} = 0 \text{ or } -y_{n-1} + 2y_n - 2y_{n+2} + y_{n+3} = 0$$

$\quad (2.126)$

$$E_p I_p \frac{d^2 y}{dx^2} = 0 \text{ or } y_n - 2y_{n+1} + y_{n+2} = 0$$

Equilibrium of horizontal shear forces and bending moments on the pile give two additional equations to bring the total number of equations to $n + 5$. Solution of these equations by a finite difference scheme yields $(n + 5)$ values of pile lateral deflection y of which the values at -2, -1, $n +2$ and $n + 3$ are only used to ensure numerical accuracy of the derivatives in the formulation and do not have any physical significance.

These deflections and its derivatives together with the resultant values of F and M provide the necessary information on pile – soil interaction.

However, the results of the analysis are dependent on the soil modulus of elasticity function $E(x, y)$. One form of E that is commonly used is:

$$E = E_L \left[\frac{x}{L}\right]^N \quad (2.127)$$

where E_L is the value of the modulus at the pile bottom and N is an empirical index equal to or greater than zero. $N = 0$ corresponds to a constant modulus whereas $N = 1$ gives a linear variation as in Equation 2.122. In general, N is taken between 0 and 0.15 for clay soils and $N = 1$ for granular soils. Analytic solutions for Equation 2.121 are available for $N = 0$.

For real soils, however, the relationship between soil reaction and deflection is non-linear with the reaction reaching a limiting value when the deflection is sufficiently large. *Figure 2.55* presents typical soil reaction against deflection curves for a consolidated clay soil and demonstrates the flattening out of the reaction/deflection curves at large deflections. These curves are conventionally referred to as $p-y$ curves.

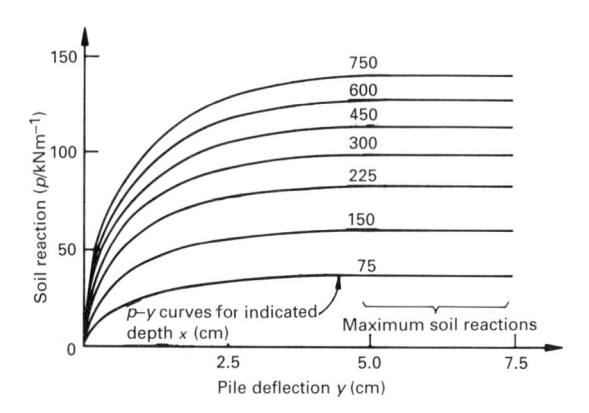

Figure 2.55 Typical $p-y$ curves for a consolidated clay

The non-linear soil reaction curves are accommodated in analysis by rewriting Equation 2.121 as:

$$\frac{d^2M}{dx^2} + S(x)\frac{d^2y}{dx^2} + E(x, y) y = 0 \quad (2.128)$$

where M is the pile moment at depth x and the additional effect of a variation of axial force S with depth x is accounted for (see Reese 1977 for further details). This equation can be solved by a finite difference technique similar to that described above. The solution requires inputs of $p-y$ curves at various depths (see *Figure 2.55*) although the method implicitly assumes that the soil behaviour at a particular depth is independent of that in adjacent layers. Equation 2.128 and the $p-y$ curves are solved using an iterative finite difference technique. The resultant shear forces and bending moments at the pile surface are derived as functions of pile top displacement and rotation, respectively, and combined with a jacket finite element analysis as in *Figure 2.56* to complete definition of the pile and structure interaction.

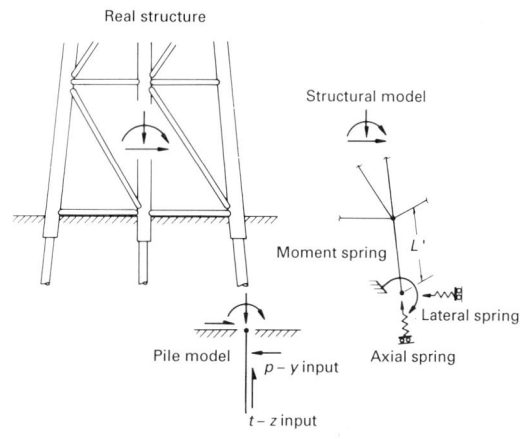

Figure 2.56 Spring model to represent pile

The $p-y$ data necessary for the analysis can be obtained in three different ways: full size or model scale lateral loading tests on instrumented piles are one of the most reliable methods although practical considerations often restrict $p-y$ data sources to laboratory tests or empirical correlations based on soil survey data.

Other pile and foundation analysis also use an elastic continuous mathematical model of the surrounding soil together with a finite beam element model of the pile to quantify pile deflections and soil behaviour; Poulos and Davis (1980) give further details. Analysis of piles for jacket platforms often need to account for pile groups and their interaction with the soil; Focht and Kock (1973) and O'Neill et al. (1977) describe extensions of the techniques described above for pile groups.

2.4.2 Dynamic response

Dynamic analysis of fixed offshore structures becomes increasingly necessary as their installed water depth increases and the structural configuration becomes more slender due to the combined effects of water depth and economic pressure on costs. A dynamic analysis is also able to quantify the dynamic magnification effects of structural resonances and to provide more accurate load cycle data for fatigue calculations.

The finite element static analysis described in Section 2.4.1.1 can be readily extended to dynamic analysis by using Lagrange's Equations to derive an equation of motion in matrix form for a multidegree of freedom dynamic system. Lagrange's Equation is written as:

$$\frac{\partial}{\partial t}\left[\frac{\partial T}{\partial \dot{q}_i}\right] - \frac{\partial T}{\partial q_i} + \frac{\partial V}{\partial q_i} = Q_i \qquad (2.129)$$

where T, V and Q_i are kinetic energy, potential energy and generalized force respectively acting on a system whose motion is described by generalized coordinates q_i.

Now a discretized multidegree of freedom linear system can have its kinetic and potential energies written as:

$$T = \frac{1}{2}\sum_{j=1}^{n}\sum_{i=1}^{n} m_{ij}\,\dot{q}_i\dot{q}_j \qquad (2.130)$$

and

$$V = \frac{1}{2}\sum_{j=1}^{n}\sum_{i=1}^{n} k_{ij}\,q_iq_j \qquad (2.131)$$

where n is the number of degrees of freedom of the system. These energies can also be written in matrix form as:

$$T = \frac{1}{2}\dot{q}^T M \dot{q} \qquad (2.132)$$

and

$$V = \frac{1}{2}q^T K q \qquad (2.133)$$

Substituting Equations (2.132) and (2.133) into Equation (2.129) leads to the matrix equation:

$$M\ddot{q} + Kq = Q \qquad (2.134)$$

since the differential $\partial T/\partial q_i$ is zero. M and K are matrices of order $(n \times n)$ with q and $\ddot{q}$ being column vectors of order $(n \times 1)$. The generalized force Q contains all non-conservative forces and external forces and with:

$$Q = F - C\dot{q} \qquad (2.135)$$

the general equation of motion becomes:

$$M\ddot{q} + C\dot{q} + Kq = F \qquad (2.136)$$

where C is a velocity dependent $(n \times n)$ damping coefficient matrix and F is an $(n \times 1)$ vector of external forces.

Now all the elements of an offshore structure that may need to be modelled by Equation 2.136 have continuous mass and stiffness properties that have to be represented by descretized matrices. A continuous infinite degree of freedom system can be reduced to a finite number of coordinates by using the concept of dimensionless shape functions which provides a mathematical relationship between a continuously varying function and a finite set of coordinates describing the function's behaviour.

Consider a beam element of an offshore structure that has an arbitrary but continuous lateral deflection given by the function $u(x, t)$ where x is distance along the member and t is time. $u(x, t)$ can be represented in terms of finite coordinates $q_1(t)$, $q_2(t)$,...,$q_i(t)$,...,$q_n(t)$ through the equation:

$$u(x, t) = q_1(t)\psi_1(x) + q_2(t)\psi_2(x) + .. \\ + q_i(t)\psi_i(x) + .. + q_n(t)\psi_n(x) \qquad (2.137)$$

where $\psi_i(x)$ are dimensionless shape functions which satisfy the geometric boundary conditions for the member.

Now the potential energy term V in Equation 2.133 denotes the flexural strain energy which for a beam element of length L can be written as:

$$V = \frac{1}{2}\int_0^L EI(x)\left[\frac{\partial^2 v(x, t)}{\partial x^2}\right]^2 dx \qquad (2.138)$$

Substituting Equation 2.137 into Equation 2.138 gives:

$$V = \frac{1}{2}\sum_{j=1}^{n}\sum_{i=1}^{n} k_{ij}\,q_i q_j \qquad (2.139)$$

where

$$k_{ij} = \int_0^L EI(x)\frac{\partial^2 \psi_i(x)}{\partial x^2} \cdot \frac{\partial^2 \psi_j(x)}{\partial x^2} dx \qquad (2.140)$$

For a beam element of a space frame structure with 12 degrees of freedom as shown in *Figure 2.48*, the shape functions ψ_1 to ψ_{12} and their first and second derivatives with respect to x are presented in *Table 2.8*. Use of the shape functions of *Table 2.8* in Equation 2.140 leads to the identical stiffness matrix as given by *Table 2.6* and is an alternative consistent method for deriving it.

A consistent mass matrix can also be obtained by writing the kinetic energy in terms of the shape functions as:

$$T = \frac{1}{2}\int_0^L m(x)\,\dot{u}(x, t)^2\, dx \qquad (2.141)$$

Substituting Equation 2.137 into Equation 2.141 gives:

$$T = \frac{1}{2}\sum_{j=1}^{n}\sum_{i=1}^{n} m_{ij}\,\dot{q}_i\,\dot{q}_j \qquad (2.142)$$

where

$$m_{ij} = \int_0^L m(x)\psi_i(x)\psi_j(x)dx \qquad (2.143)$$

Use of the shape functions of *Table 2.8* in Equation (2.143) yields a consistent mass matrix for a space frame beam element (see *Figure 2.48*) which is given in *Table 2.9*.

The mass matrix can also be modelled by taking the mass of each beam element and distributing it as lumped masses at the element nodes. This lumped mass

Table 2.8: Shape function for a three dimensional beam element

$\psi_i(x)$	$\dfrac{\partial \psi_i}{\partial x_m}$	$\dfrac{\partial^2 \psi_i}{\partial x_m^2}$
$\psi_1(x_m) = (x_m/L)$	$1/L$	
$\psi_2(x) = 1 - 3\left[\dfrac{x_m}{L}\right]^2 + 2\left[\dfrac{x_m}{L}\right]^3$	$-\dfrac{6x_m}{L^2} + \dfrac{6x_m^2}{L^3}$	$-\dfrac{6}{L^2} + \dfrac{12x_m}{L^3}$
$\psi_3(x) = 1 - 3\left[\dfrac{x_m}{L}\right]^2 + 2\left[\dfrac{x_m}{L}\right]^3$	$-\dfrac{6x_m}{L^2} + \dfrac{6x_m^2}{L^3}$	$-\dfrac{6}{L^2} + \dfrac{12x_m}{L^3}$
$\psi_4(x) = (x_m/L)$	$1/L$	0
$\psi_5(x) = -x_m\left[1 - \dfrac{x_m}{L}\right]^2$	$-1 + \dfrac{4x_m}{L} - \dfrac{3x_m^2}{L^2}$	$\dfrac{4}{L} - \dfrac{6x_m}{L^2}$
$\psi_6(x) = -x_m\left[1 - \dfrac{x_m}{L}\right]^2$	$-1 + \dfrac{4x_m}{L} - \dfrac{3x_m^2}{L^2}$	$-\dfrac{4}{L} + \dfrac{6x_m}{L^2}$
$\psi_7(x) = \left[1 - \dfrac{x_m}{L}\right]$	$-1/L$	0
$\psi_8(x) = 3\left[\dfrac{x_m}{L}\right]^2 - 2\left[\dfrac{x_m}{L}\right]^3$	$-\dfrac{6x_m}{L^2} + \dfrac{6x_m^2}{L^3}$	$\dfrac{6}{L^2} - \dfrac{12x_m}{L^3}$
$\psi_9(x) = 3\left[\dfrac{x_m}{L}\right]^2 - 2\left[\dfrac{x_m}{L}\right]^3$	$\dfrac{6x_m}{L^2} + \dfrac{6x_m^2}{L^3}$	$-\dfrac{6}{L^2} + \dfrac{12x_m}{L^3}$
$\psi_{10}(x) = \left[1 - \dfrac{x_m}{L}\right]$	$-1/L$	0
$\psi_{11}(x) = \left[\dfrac{x_m^2}{L}\right]\left[1 - \dfrac{x_m}{L}\right]$	$\dfrac{2x_m}{L} - \dfrac{3x_m^2}{L^2}$	$\dfrac{2}{L} - \dfrac{6x_m}{L^2}$
$\psi_{12}(x) = \left[\dfrac{x_m^2}{L}\right]\left[\dfrac{x_m}{L} - 1\right]$	$-\dfrac{2x_m}{L} - \dfrac{3x_m^2}{L^2}$	$-\dfrac{2}{L} + \dfrac{6x_m}{L^2}$

formulation gives a diagonal mass matrix where each diagonal element is made of all elements influenced by the acceleration of that degree of freedom. In the lumped mass formulation, a non-zero mass is only present for translational degrees of freedom with rotational degrees of freedom having zero inertia because of the concentration of the masses at the nodal points. The lumped mass matrix has less theoretical justification than the consistent mass formulation but is easier to apply.

Whether the mass matrix for a beam element is of the lumped or consistant mass formulation, its terms must include the physical mass of the structure, its added mass due to lateral accelerations, the masses of piles within tubulars, the mass of any attachments such as pile guides, conductor mounts and so on. The definition of the mass matrix for elements piercing the free surface also needs special care to ensure that the added mass of the submerged element segment is correctly distributed between the nodes.

The damping matrix C is used to quantify the internal structural and foundation damping of the structure together with the linear (due to wave radiation) or non-linear (due to drag) damping induced by vibration

Table 2.9: Consistent mass matrix for a beam element: ρ = density of beam material, A = area of cross-section, L = beam length, I_p = polar second moment of area

$$m = \frac{\rho A L}{420}\begin{bmatrix}
140 & & & & & & 70 & & & & & \\
 & 156 & & & & 22L & & 54 & & & & -13L \\
 & & 156 & & -22L & & & & 54 & & 13L & \\
 & & & \frac{140 I_p}{A} & & & & & & \frac{70 I_p}{A} & & \\
 & & -22L & & 4L^2 & & & & -13L & & -3L^2 & \\
 & 22L & & & & 4L^2 & & 13L & & & & -3L^2 \\
70 & & & & & & 140 & & & & & \\
 & 54 & & & & 13L & & 156 & & & & -22L \\
 & & 54 & & -13L & & & & 156 & & 22L & \\
 & & & \frac{70 I_p}{A} & & & & & & \frac{140 I_p}{A} & & \\
 & & 13L & & -3L^2 & & & & 22L & & 4L^2 & \\
 & -13L & & & & -3L^2 & & -22L & & & & 4L^2
\end{bmatrix}\begin{matrix}1\\2\\3\\4\\5\\6\\7\\8\\9\\10\\11\\12\end{matrix}$$

$$\ \ \ 1\ \ \ 2\ \ \ 3\ \ \ 4\ \ \ 5\ \ \ 6\ \ \ 7\ \ \ 8\ \ \ 9\ \ 10\ \ 11\ \ 12$$

of the structure with fluid surrounding it.

The equation of motion, Equation 2.134, derived from the Lagrange Equations can be extended to include damping and exciting force separately for a beam element by writing the beam moment due to internal energy losses as:

$$M_e(x, t) = c_e\, I(x)\, \frac{\partial^3 u(x, t)}{\partial^2 x\, \partial t} \qquad (2.144)$$

where $I(x)$ is the second moment of area of the beam section, c_e is a damping coefficient and $\partial^3 u/(\partial^2 x\, \partial t)$ is time rate of change of beam curvature. Therefore, the total work performed by these internal non-conservative forces is:

$$W_{nc_{int}} = -\int_0^L c_e\, I(x)\, \frac{\partial^3 u(x, t)}{\partial^2 x\, \partial t} \cdot \delta\!\left[\frac{\partial^2 u(x, t)}{\partial x^2}\right] dx \qquad (2.145)$$

and also by a distributed transverse non-conservative loading $f(x, t)$ is:

$$W_{nc_{ext}} = \int_0^L f(x, t)\, \delta[u(x, t)]\, dx \qquad (2.146)$$

where δ is an operator denoting variation of the quantity following it.

Substituting Equation 2.137 into Equations 2.145 and 2.146 and adding gives:

$$\delta W_{nc_{tot}} = \sum_{i=1}^{n}\left[f_i - \sum_{j=1}^{n} C_{ij}\, \dot{q}_j\right]\delta q_i \qquad (2.147)$$

where
$$f_i = \int_0^L w(x, t)\, \psi_i(x)\, dx \qquad (2.148)$$

and

$$C_{ij} = \int_0^L c_s\, I(x)\, \frac{\partial^2 \psi_i(x)}{\partial x^2}\, \frac{\partial^2 \psi_j(x)}{\partial x^2}\, dx \qquad (2.149)$$

Comparing Equation 2.147 with Equation 2.129 leads to:

$$Q_i = f_i - \sum_{i=1}^{n} C_{ij}\, \dot{q}_j \qquad (2.150)$$

which can be written in matrix form as:

$$Q = F - C\dot{q} \qquad (2.151)$$

and leads to the full equation of motion, Equation 2.136.

The precise form of the structural damping matrix C is very difficult to quantify and it may not only be a function of velocities in real structures. A commonly adopted approach is to represent the structural damping as a linear combination of the mass and stiffness matrices so that:

$$C = \mu_1 M + \mu_2 K \qquad (2.152)$$

with μ_1, μ_2 obtained from assuming certain small percentages of critical damping in the first two modes of vibration.

The force vector F of Equation 2.136 is obtained by first calculating the wave force on a small segment of a member element of the framed structure. This segment force df can be written as:

$$df = \rho \, dV \, C_M \, \dot{q}_n + \tfrac{1}{2} \rho \, C_D \, dA |q_n - \dot{x}_n|(q_n - \dot{x}_n) \tag{2.153}$$

where dV and dA are the volume and frontal area of the segment, $\dot{x}_n$ is the velocity of the segment normal to the member axis and q_n, $\dot{q}_n$ are wave fluid velocities and acceleration components normal to the member axis. C_M and C_D are added inertia and drag coefficients. The wave force on the segment is written as a distributed force and the resultant fixed end actions and equivalent joint loads are calculated using numerical integration based on equations in the last row of *Table 2.7*. The equivalent joint loads are transformed into global axes and assembled into a force vector F using the techniques given in Section 2.4.1.1.

Equation 2.153 can be linearized by writing $\sigma_{\dot{v}_r}$ as the root mean square of the relative velocity $|\dot{q}_n - \dot{x}_n|$, see Malhotra and Penzien (1970), so that:

$$df = \rho \, dV \, C_M \, \dot{q}_n + \tfrac{1}{2} \rho \, C_D \, dA \, \sqrt{(8/\pi)} \sigma_{\dot{v}_r} (\dot{q}_n - \dot{x}_n) \tag{2.154}$$

The matrix equation of motion, Equation 2.136, for dynamic response can be solved in either the frequency or time domain. For a solution in the frequency domain, it is necessary to use linear wave theory to obtain wave properties and to linearize the drag force term. Once these simplifications are made, a frequency domain solution offers readily usable transfer functions of structure displacements, internal forces and stresses which can be applied to deriving statistical response results.

The frequency domain solution can be obtained in two ways. A direct substitution of the form of solution:

$$q = q_o \, e^{-i\omega t} \tag{2.155}$$

in Equation 2.136 will yield the unknown displacements through matrix inversion of the term in square brackets in the equation:

$$[k - \omega^2 M - i\omega C] q_o = F_o \tag{2.156}$$

where

$$F = F_o \, e^{-i\omega t} \tag{2.157}$$

is the linearized form of the wave exciting force. Solution of Equation 2.156 requires the inversion of a large matrix with complex number elements. This is a numerically cumbersome operation and has prompted the development of computationally more efficient solution techniques.

One such method uses modal superposition. This technique relies on the fact that despite the large number of degrees of freedom of a typical offshore structure, its dynamic response is limited to only a few modes of vibration. The modal superposition technique uses this feature to reduce the solution computation time. The method initially requires determination of the undamped natural frequencies and vibration modes of the structure. The equation for free undamped motion is:

$$M \ddot{q} + K q = 0 \tag{2.158}$$

Assuming a solution of the form:

$$q = q_o \, e^{-i\omega t} \tag{2.159}$$

and substituting into Equation 2.158, yields:

$$(K - M \omega^2) q_o = 0 \tag{2.160}$$

Apart from the trivial solution of $q_o = 0$, Equation 2.160 has solutions given by:

$$\text{Det} |K - M \omega^2| = 0 \tag{2.161}$$

which leads to a polynomial of degree n where n is the number of degrees of freedom in the governing Equation 2.136. Solution of the determinant yields n values of ω which are the natural frequencies of the system. Each natural frequency ω_r, has a corresponding set of values of q_{or} given by:

$$(K - M \omega_r^2) q_{or} = 0 \tag{2.162}$$

although the elements of vector q_{or} are only defined relative to each other rather than in absolute terms. For systems with many degrees of freedom, the determination of the eigen values ω_r and the eigen vectors q_{or} requires considerable computational effort although modern computer based algorithms make the task simpler than it would be otherwise (see Bathe and Wilson (1973) for a survey of available techniques).

It can also be shown that the eigen vectors for any two vibration modes satisfy an orthogonality condition. Consider vibration modes r and s. From Equation 2.162 these can be written as:

$$K \, q_{or} = M \, \omega_r^2 \, q_{or} \tag{2.163}$$

and

$$K \, q_{os} = M \, \omega_s^2 \, q_{os} \tag{2.164}$$

Taking the transpose of Equation 2.163 gives:

$$(K \, q_{or})^T = \omega_r^2 \, (M \, q_{or})^T$$

or $\tag{2.165}$

$$q_{or}^T \, K = \omega_r^2 \, q_{or}^T \, M$$

since K and M are symmetric and $K = K^T$ and $M = M^T$. If Equation 2.165 is post-multiplied by q_{os} and Equation 2.164 is premultiplied by q_{or}^T then:

$$q_{or}^T \, K \, q_{os} = \omega_r^2 \, q_{or}^T \, M \, q_{os} \tag{2.166}$$

and

$$q_{or}^T \, K \, q_{os} = \omega_s^2 \, q_{or}^T \, M \, q_{os} \tag{2.167}$$

Subtraction of Equation 2.167 from Equation 2.166 yields:

$$(\omega_r^2 - \omega_s^2) \, q_{or}^T \, M \, q_{os} \tag{2.168}$$

so that for $\omega_r \neq \omega_s$, the orthogonality condition for the modes of vibration are obtained as:

$$q_{or}^T M q_{os} = 0 \quad (2.169)$$

and

$$q_{or}^T K q_{os} = 0 \quad (2.170)$$

The eigen vectors are often normalized by using a version of Equation 2.169 with $r = s$ so that for the rth normalized eigen vector ϕ_r, the condition:

$$\phi_r^T M \phi_r = 1 \quad (2.171)$$

holds. This can be achieved by calculating the scale factor:

$$S_r = q_{or}^T M q_{or} \quad (2.172)$$

and getting the normalized eigen vector as:

$$\phi_r = \frac{q_{or}}{\sqrt{(S_r)}} \quad (2.173)$$

With this normalization procedure, if all the eigen vectors ϕ_r are collected in an $m \times m$ square matrix Φ, then, the condition:

$$\Phi^T M \Phi = I \quad (2.174)$$

where I is an $m \times m$ matrix with unit diagonal terms and zero elsewhere.

The scene is now set for further development of the modal superposition technique. The eigen vectors or mode shapes can be said to be building blocks of the dynamic response of a general N degree of freedom system to the extent that it is efficient to describe the response in terms of these mode shapes. Thus the solution q of an N degree of freedom dynamic system described by the matrix equation:

$$M \ddot{q} + C \dot{q} + K q = F(t) \quad (2.175)$$

can be written in terms of eigen vectors (or mode shapes) through the matrix equation:

$$q = \Phi Y \quad (2.176)$$

where Φ is the $N \times N$ matrix of normalized eigen vectors and Y is the $(N \times 1)$ column vector of modal amplitudes. Thus Equation 2.176 transforms from geometric variables q to generalized coordinates Y which describe the amplitudes of excited modes during dynamic response. These generalized coordinates are called normal coordinates.

Equation 2.175 written in terms of the normal coordinates Y by substitution of Equation 2.176 and pre-multiplying by the transpose of the ith eigen column vector ϕ_i^T, to give:

$$\phi_i^T M \phi \ddot{Y} + \phi_i^T C \phi \dot{Y} + \phi_i^T K \phi Y = \phi_i^T F(t) \quad (2.177)$$

Now, as a consequence of the orthogonality condition, all components except for the ith column vector term vanish in the mass and stiffness terms. The orthogonality does not apply to the damping term but the modal superposition technique assumes that terms other than in the ith mode also vanish in the terms of Equation 2.177 that are derived from the damping matrix. Thus the ith mode is entirely decoupled and satisfies the equation:

$$M_i \ddot{Y}_i + C_i \dot{Y}_i + K_i Y_i = F_i(t)$$

where $M_i = \phi_i^T M \phi_i$

$$k_i = \phi_i^T K \phi_i \quad (2.178)$$

$$C_i = \phi_i^T C \phi_i$$

and $F_i(t) = \phi_i^T F(t)$

Equation 2.178 can be solved for all the normalized coordinates Y_i and the geometric coordinates q can be recovered from Equation 2.176. The major advantage of this modal superposition technique is that the structure dynamic response is usually made up of the superposition of only a few of all the modes that are likely to be excited. The number of times Equation 2.178 needs to be evaluated can, therefore, be reduced for computational efficiency by evaluating it only for significant modes. On the other hand, the technique requires that eigen frequencies and vectors be evaluated prior to obtaining a solution.

It can be shown that a damping matrix of the form of Equation 2.152 will satisfy the orthogonality condition, since the mass and stiffness matrices will each satisfy this condition.

Thus, if the damping matrix is approximated to the form of Equation 2.152, the modal superposition technique will uncouple the damping term without resorting to the approximation of neglecting off-diagonal terms in Equation 2.177.

Note that the uncoupled equations of motion in normal coordinates given by Equation 2.178 can be solved by any method in the frequency or time domain.

Solution of the equation of motion by frequency domain and modal superposition techniques as described above are only applicable to linear systems with harmonic excitation whereas the equations of motion for drag dominated structures can have a substantial non-linear character as illustrated by Equation 2.153 for wave forces. This non-linear nature can be preserved by solving the equations of motion through a time step procedure. This also permits the wave kinetics to be more accurately represented by Stoke's fifth order theory for regular waves at least. The disadvantage of this approach is its complexity and the fact that output results are available only as time histories. Time domain solutions are also computationally expensive.

Two methods for time step integration are described here. The first, called the central difference method, is an explicit algorithm which is based on a Taylor series expansion.

The equation of motion in matrix form is taken as:

$$M \ddot{x} + C \dot{x} + K x = F \quad (2.179)$$

in the usual notation. x_i and x_{i+1} are taken to be the values of vector x at times i and $i + 1$ such that

$\Delta t = t_{i+1} - t_i$. Then expanding x_{i+1} and x_i as a Taylor series gives:

$$x_{i+1} = x_i + (\Delta t) \dot{x}_i + \tfrac{1}{2} (\Delta t)^2 \ddot{x}_i + \ldots$$
$$x_{i-1} = x_i - (\Delta t) \dot{x}_i + \tfrac{1}{2} (\Delta t)^2 \ddot{x}_i - \ldots \quad (2.180)$$

By subtracting and adding Equations 2.180 and neglecting terms of power $(\Delta t)^3$ or higher, we get:

$$\dot{x}_i = (x_{i+1} - x_{i-1})/(2\Delta t)$$
and $\quad (2.181)$
$$\ddot{x}_i = \{x_{i+1} - 2x_i + x_{i-1}\}/(\Delta t)^2$$

Direct substitution in Equation 2.179 with $x = x_i$ gives:

$$\left[M + \frac{1}{2}(\Delta t) C\right] x_{i+1} = (\Delta t)^2 F_i + \left[2M - (\Delta t)^2 K\right] x_i$$
$$+ \left[\frac{1}{2}(\Delta t) C - M\right] x_{i-1} \quad (2.182)$$

Equation 2.182 is used for successive time steps to derive the displacement of time $i + 1$ from displacements at time i and $i - 1$. The central difference method requires the condition:

$$\frac{\Delta t}{T_n} \leq 0.318 \quad (2.183)$$

to be satisfied for numerical stability where T_n is the period of the highest vibration mode of the system.

The second time step integration method considered here is the Newmark–β technique which assumes that the displacement and velocity at the end of a time interval can be related to the displacement, velocity and acceleration at the beginning of the time interval by the equation:

$$\dot{x}_{i+1} = \dot{x}_i + \tfrac{1}{2}(\Delta t) [\ddot{x}_i + \ddot{x}_{i+1}]$$
and $\quad (2.184)$
$$x_{i+1} = x_i + (\Delta t) \dot{x}_i + (\tfrac{1}{2} - \beta)(\Delta t)^2 \ddot{x}_i + \beta (\Delta t)^2 \ddot{x}_{i+1}$$

The variable β does have physical significance, in that $\beta = 1/4$ in Equation 2.184 corresponds to a constant acceleration variation from i to $i + 1$ whereas $\beta = 1/6$ converts the equation to apply for linearly varying acceleration.

Writing Equation 2.179 for time periods $i - 1$, i and $i + 1$ gives:

$$M \ddot{x}_{i-1} + C \dot{x}_{i-1} + K x_{i-1} = F_{i-1} \quad (2.185)$$
$$M \ddot{x}_i + C \dot{x}_i + K x_i = F_i \quad (2.186)$$
$$M \ddot{x}_{i+1} + C \dot{x}_{i+1} + K x_{i+1} = F_{i+1} \quad (2.187)$$

Multiplying Equations 2.185 and 2.187 by $\beta(\Delta t)^2$, Equation 2.186 by $(1-2\beta)(\Delta t)^2$ and adding gives:

$(\Delta t)^2 M [\{\beta \ddot{x}_{i+1} + (\tfrac{1}{2} - \beta) \ddot{x}_i\} - \{\beta \ddot{x}_i + (\tfrac{1}{2} - \beta) \ddot{x}_{i-1}\}$
$+ \{\tfrac{1}{2} \ddot{x}_i + \tfrac{1}{2} \ddot{x}_{i-1}\}] + (\Delta t)^2 C [\{\tfrac{1}{2} \dot{x}_i\} + \{\tfrac{1}{2} \dot{x}_{i-1}\}$
$+ \{\beta(\dot{x}_{i+1} - \dot{x}_i)\} + \{(\tfrac{1}{2} - \beta)(\dot{x}_i - \dot{x}_{i-1})\}]$
$+ (\Delta t)^2 K [\beta x_{i+1} + (1 - 2\beta) x_i + \beta x_{i-1}]$
$= (\Delta t)^2 [\beta F_{i+1} + (1 - 2\beta) F_i + \beta F_{i-1}] \quad (2.188)$

Substituting from Equation 2.184, simplifying and re-arranging gives:

$[M + \tfrac{1}{2}(\Delta t) C + \beta (\Delta t)^2 K] x_{i+1}$
$= (\Delta t)^2 [\beta F_{i+1} + (1 - 2\beta) F_i + \beta F_{i-1}]$
$+ [2M - (\Delta t)^2 (1 - 2\beta) K] x_i$
$- [M - \tfrac{1}{2}(\Delta t) C + \beta (\Delta t)^2 K] x_{i-1} \quad (2.189)$

Thus the displacements at time $i + 1$ can be obtained from the displacement at times i and $i - 1$. Displacement x_1 at time Δt can be obtained by a special case of Equation 2.189 obtained from Equations 2.186, 2.187 and 2.184 as:

$[M \tfrac{1}{2}(\Delta t) C + \beta (\Delta t)^2 K] x_1$
$= (\Delta t)^2 \beta F_1 + (\Delta t)^2 [(\tfrac{1}{2} - \beta)I + (\tfrac{1}{4} - \beta) \Delta t \, C \, M^{-1}] F_0 \quad (2.190)$

with $x_0 = \dot{x}_0 = 0$ at $t = 0$

Comparison of Equations (2.181) and (2.189) shows that the central difference and Newmark β methods are equivalent for $\beta = 0$. It can be shown from stability analysis that the Newmark β method is unconditionally stable for $\beta > 1.4$. For $\beta > 1/4$, the stability conditions are given by the equations

$$\frac{\Delta t}{T_n} = \begin{cases} 0.318 \text{ for } \beta = 0 \\ 0.450 \text{ for } \beta = \dfrac{1}{8} \\ 0.551 \text{ for } \beta = \dfrac{1}{6} \end{cases} \quad (2.191)$$

Once the condition for stability is met for either method, the accuracy of the solution must be investigated to ensure that a sufficiently small value of Δt is used for accuracy but is yet not too small that computation time is prohibitive.

Felippa and Park (1978) present more information on these and other time step integration techniques whereas Godeau et al. (1977) presents some results for nonlinear behaviour of a fixed offshore structure in irregular waves.

Despite the more exact solution of structural dynamics offered by time step integration, research has continued to examine ways in which frequency domain solutions can be extended to predict the effects of the drag force nonlinearity by using an iterative procedure based on minimization of mean square error (see Penzien and Tseng 1978 and Taudin 1978). Eatock–Taylor and Rajgopalan (1982) present a perturbation technique to examine the higher harmonics generated due to the drag force non-linearity.

2.4.3 Structural failure and fatigue life calculations

The safe structural design of a jacket structure ultimately depends on the detailed design of individual tubular joints. The analysis of Section 2.4.1 is readily able to ensure that members between joints do not fail from buckling or overstressing in tension or bending. There are, however, several failure mechanisms that have to be designed against in the definition of a tubular joint. A tubular joint brace loading the main member in compression can initiate failure by local buckling or by punching shear of the main member wall. Tensioned braces can cause failure due to crack growth or lamellar tearing of the main member wall.

However, a large amount of structural design studies, finite element analyses and model tests have ensured that these failure mechanisms can be almost completely avoided. The majority of useful recommendations made by such work have been incorporated into codes of practice of which those issued by the American Petroleum Institute (1987) and the British Standards Institute (1982) are good examples.

At present, the largest source of structural failures in operation is due to fatigue damage compounded by the effects of corrosion. A metal will fail under repeated cycles of relatively low stress levels due to the growth of internal cracks that can be initiated from local fabrication defects or at stress concentration sites. Such metal fatigue failures are quantified from experimental data by the use of S–N curves which are plots of stress range S, against number of cycles to failure N, with both axes plotted in logarithmic form. The fatigue behaviour of a variety of steels can, therefore, be characterized by S–N curves of the type shown in *Figure 2.57* obtained from British Standard 6235 (1982). These curves show that steel specimens in air do exhibit a low stress range level (called the fatigue limit) below which the metal would have a theoretically infinite life although this property disappears if the effects of combined stress range cycles and corrosion are considered.

The Palmgren–Miner cumulative fatigue damage rule is used to quantify fatigue damage caused by a large number of cycles at low stress ranges. The rules give the resultant cumulative damage ratio D, as:

$$D = \sum_{i=1}^{N} \frac{n_i}{N_i} \leq 1 \qquad (2.192)$$

where N denotes the total number of stress range intervals being considered, n_i is the number of cycles experienced by the structure within the ith stress range and N_i is the number of cycles to failure given by the S–N curve for this ith stress range. Failure will occur when parameter D reaches unity. If the number of stress range cycles used in the calculation correspond to one year's operation then the fatigue life of the structure in years can be given by the reciprocal of the cumulative damage.

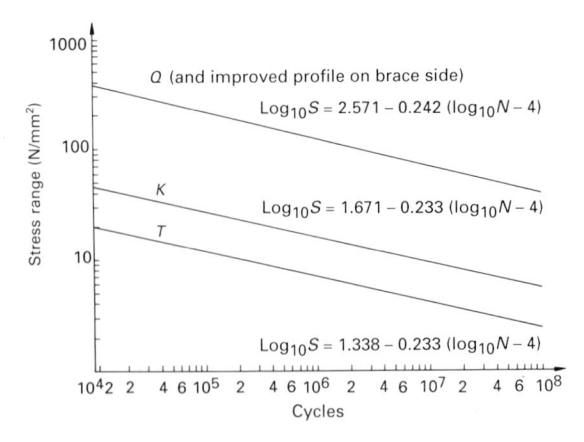

Figure 2.57 Typical S–N lines for welded tubular joints. Q line used for general fatigue life calculations with K and T lines used for an appropriate punching shear method (from British Standards Institution 1982)

Lines shown:
- Q (and improved profile on brace side): $\log_{10}S = 2.571 - 0.242(\log_{10}N - 4)$
- K: $\log_{10}S = 1.671 - 0.233(\log_{10}N - 4)$
- T: $\log_{10}S = 1.338 - 0.233(\log_{10}N - 4)$

In design practice, the fatigue life is calculated using a deterministic approach illustrated by the left hand column of the flow diagram of *Figure 2.58*. The starting point for such an analysis is a significant wave height against average zero crossing period scatter diagram wave climate description typically averaged for one year's weather conditions. The number of occurrences (expressed in parts per thousand) in each row of the scatter diagram is converted to number of waves in one year by using the appropriate average zero crossing period and added to build up a variation of maximum wave height (converted from the significant height in the scatter diagram) against number of waves that exceeded this height as in box number 2 of *Figure 2.58*. Each wave height range in this figure has an associated period range that is also presented with the figure. The wave heights and period ranges are used through the wave theory, loading and response analyses described earlier to yield an equivalent curve of stress range against number of loading cycles that will be applied at this range in box 3. It is usual practice in the calculation to select a wave period for any wave height that will produce the largest stress although the choice of an average between the range of wave periods will yield a less conservative result. A curve such as in box 3 of *Figure 2.58* is obtained for each of the fatigue prone zones on the structure. A stress concentration factor is then applied to the data of box 3 and combined with the Palmgren–Miner damage rule and a material S–N curve to yield the annual cumulative damage, D, and fatigue life of $1/D$ years.

A probabilistic approach can also be applied to estimating the stress range exceedence of box 8 by using a parameterized spectral density to represent each significant wave height-zero crossing period range in the scatter diagram of box 1. A linear structural response analysis is then carried out to yield the stress spectrum (as in box 6) at all points of interest. A

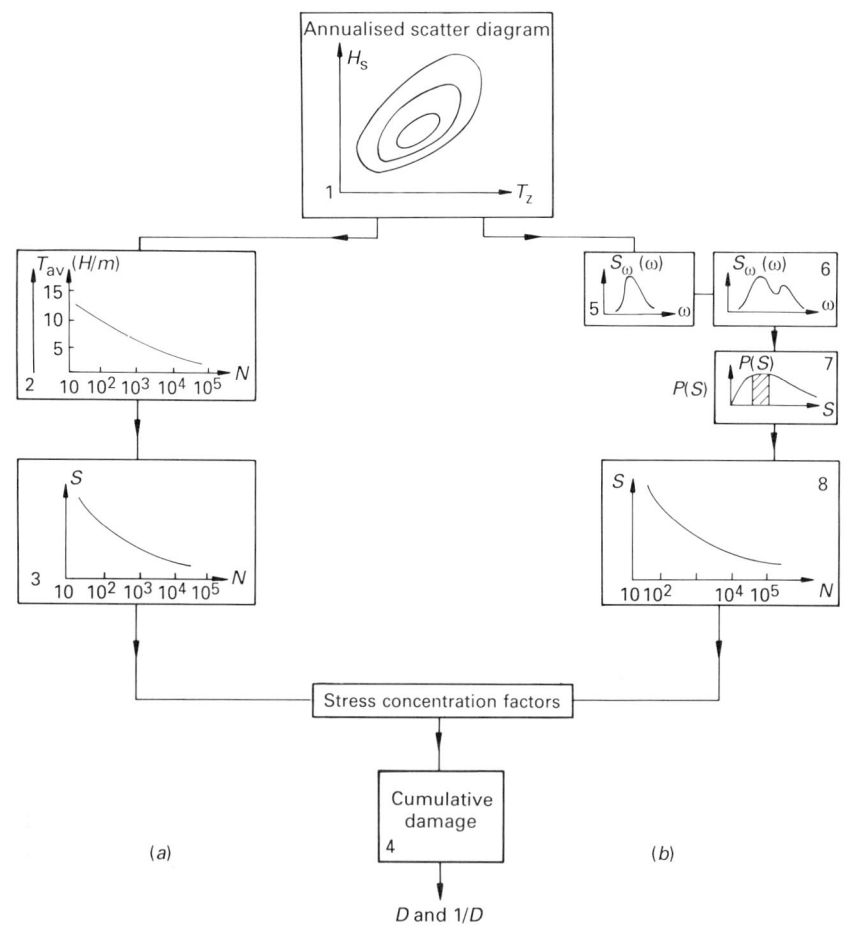

Figure 2.58 Fatigue life calculation procedures: (a) deterministic; (b) stochastic

Rayleigh probability distribution of stress peaks (box 7) is then assumed to estimate the probability of occurrence of each stress range from which the figure of box 8 can be derived and the fatigue analysis completed. It should be noted that the average zero crossing period of the stress spectrum in box 6 will be different from that of the wave spectrum of box 5 and will need to be calculated in order to determine the numbers of stress cycles applied in a year.

The effect of sea water corrosion and the value of stress concentration factors to be used are two of the most important sources of uncertainty in fatigue analysis of an offshore structure. The American Petroleum Institute (1987) and the British Standards Institution (1982) give some guidance on these matters.

Much recent progress has been made on quantifying fatigue crack growth in tubular joints by the application of the theory of fracture mechanics coupled with more accurate in service crack detection techniques (see Dover and Connolly (1986) and Dover and Wilson (1986)). Their combined use is leading to more reliable monitoring and rectification of fatigue crack growth during the life of an offshore structure.

2.4.4 Design and certification requirements

It is important to appreciate that the design procedures for jacket structures outlined previously are only a small part of the total design process. In order to illustrate this point, *Figure 2.59* presents a flow chart showing the design procedures that need to be followed from the initial specification through to commencing operation of a typical offshore structure. The jacket has to have sufficient strength as it is assembled during the fabrication stage and loaded out of the yard. It has also to meet the naval architectural and structural requirements of tow out, up ending and installation as well as surviving a 20–40 year life of in service conditions. Some of the supplementary design tasks not covered in this chapter includes the response of the structure to earthquakes, the provision of corrosion protection and in service structural monitoring. The design procedure

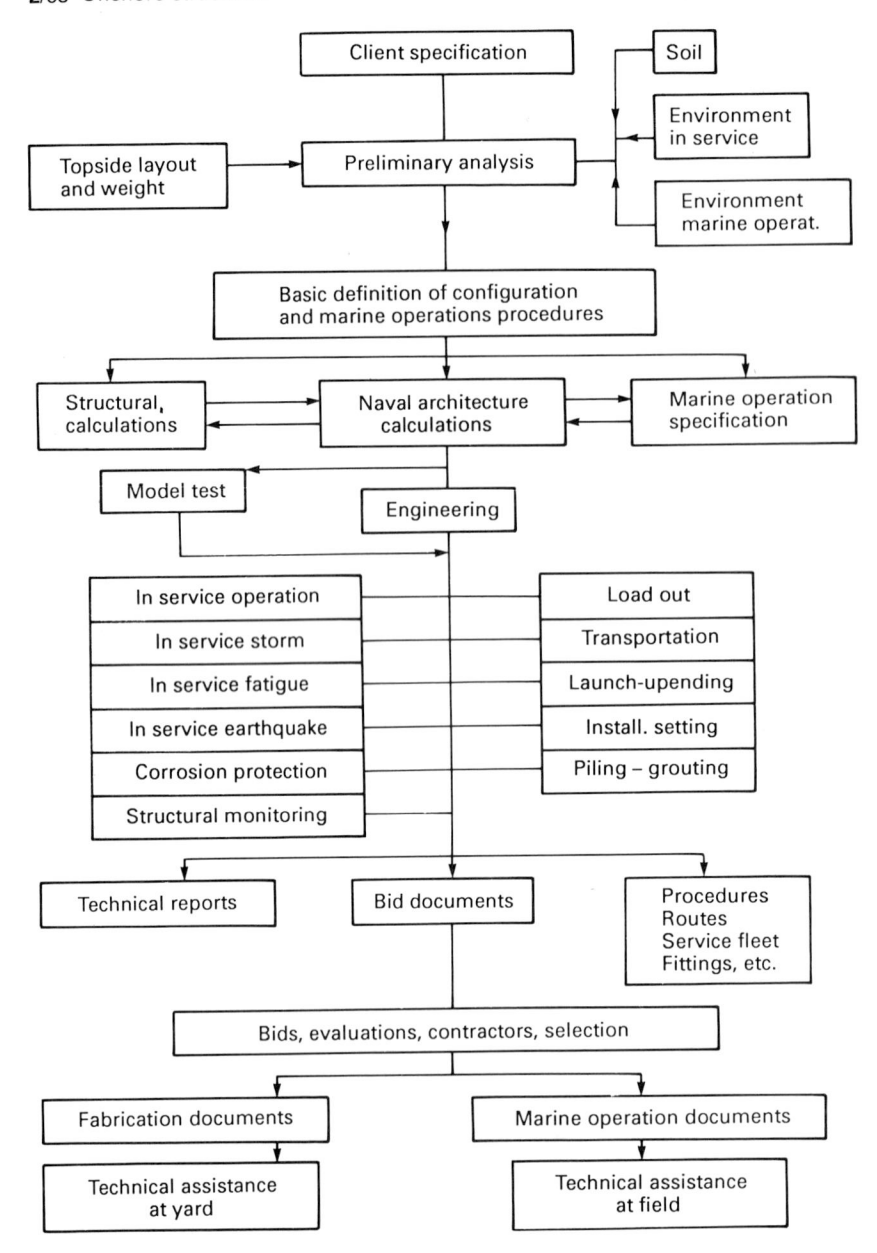

Figure 2.59 Design procedure for jacket structure

for large jackets invariably contains a model test phase for critical operations such as up-ending during installation. The documentation of the material, structural and welding details of the design during its certification, fabrication and service life pose an engineering management problem.

Certifying authorities play a key role in the design procedure for an offshore structure. The major certifying authorities in the UK, Norway and the USA have built up extensive codes of practice which reflect research work, in service experience and the results of failure investigations over many years of operation. Certifying authorities also provide an independent check of many of the calculations and decisions that need to be made during a typical design. There tends to be close technical collaboration between research establishments, designers and the operators of offshore structures.

2.5 Floating offshore structures

2.5.1 General characteristics

Floating offshore structures such as semisubmersibles, ships and loading buoys are playing an increasing role in offshore operations as the escalating cost of bottom placed structures becomes uneconomic for small fields or greater water depths. Compliant structures such as tensioned buoyant platforms, articulated towers and single anchor leg moored vessels are also examples of floating structures with a greater level of mooring restraint than a freely or catenary moored floating vessel. The wave induced motions of these vessels and consequent structural loads are crucial to their operating effectiveness. In particular, wave induced motions contribute to down time in severe weather when excessive vessel motions may prevent connection of the vessel to a drilling or production work site on the sea bed. Thus the design of a floating vessel with low wave induced motions, a high level of station keeping in response to steady environmental forces, large equipment payload and low cost offers the best combination of characteristics for improving the effectiveness of offshore operations.

It is shown in Section 2.5.3 that the fluid structure interactions of a floating vessel can be reduced to the classical equations of motion for a linear multidegree of freedom system consisting of mass, linear or non-linear damping and stiffness terms. It is, however, instructive first to examine the hydrostatics of floating offshore structures which plays a dominant role in the design of most such structures.

2.5.2 Hydrostatic analysis and stability

2.5.2.1 Basic concepts

Hydrostatic pressures within a fluid at rest can exert very large forces on the submerged parts of offshore structures, particularly at large water depths. Hydrostatic pressures also impart rather more subtle properties which affect the stability of floating bodies. Both of these features have a profound influence on the design of floating and fixed offshore structures and are considered here in more detail.

A fluid may be defined as a substance that deforms due to the effect of a shear stress, however small. It follows from this statement that no shearing stresses can exist in a fluid at rest with the reactions between adjacent layers of fluid being confined to normal stresses only. These normal stresses are called pressures and defined as the normal force per unit area on an infinitesimal plane surface at any orientation in the fluid. By resolving pressure induced forces on an infinitesimal tetrahedral fluid element with three mutually perpendicular faces, it can readily be shown that the pressure at a point in a fluid in equilibrium is the same in all directions. This is also true for a fluid in bulk motion in which there are no shearing stresses.

Although the above properties apply both to liquids and gases, the hydrostatics of offshore structures is principally concerned with liquids of large density (compared to gases) which induces substantial pressure gradients in the earth's gravitational field. A qualitative characteristic that distinguishes liquids from gases is the fact that a gas will expand in volume indefinitely to fully fill its containing volume whereas a liquid will be of an essentially constant volume with no definite shape.

Sea water at rest in the oceans, relative to the earth, is subjected primarily to a uniform gravitational force field acting towards the earth's centre. Thus, there are no external forces to consider apart from the weight of sea water and the resultant hydrostatic pressure p, at a point distance z below the water surface can be proved to be:

$$p = p_a + \rho g z \tag{2.193}$$

by considering force equilibrium of a vertical column of fluid of constant cross section from the water surface down to depth z. In Equation 2.193, p_a is atmospheric pressure, ρ is sea water density and g is the acceleration due to gravity. It demonstrates that there is a pressure increase of 1 atmosphere for every 10.08 m (32 ft) increase of depth for sea water of density 1025 kg/m^3.

Taking horizontal force equilibrium on any horizontal cylinder of fluid can also show that in a fluid at rest under gravity, the pressure is the same at all points in a horizontal plane.

Now the pressure field defined by Equation 2.193 will exert forces on structures immersed in the fluid. For an arbitrary plane surface, the pressure field will induce a system of parallel forces of differing magnitude acting normal to the surface. The point in the plane where the resultant of these pressure forces acts is defined as the centre of pressure. For Cartesian axes Oxy in an arbitrary plane surface, the coordinates (ξ, η) of the centre of pressure can be obtained by the following integrations:

$$\xi = \frac{\iint p\, x\, dx\, dy}{\iint p\, dx\, dy}$$

$$\eta = \frac{\iint p\, y\, dx\, dy}{\iint p\, dx\, dy} \tag{2.194}$$

where the function $p = f(x, y)$ defines the linear pressure gradient of Equation 2.193 in terms of axes Oxy. Also the total pressure induced force T, on the plane surface is given by:

$$T = \iint p\, dx\, dy \tag{2.195}$$

The total thrust and centre of pressure on a curved surface are similarly found by integration of the pressure field of Equation 2.193 over the surface using an appropriate coordinate system. Both analytic and numerical integration techniques may be utilized depending on the complexity of the problem.

The resultant of hydrostatic pressure induced forces on the surfaces of a fully or partly submerged body immersed in a fluid at rest is called the buoyancy force. The magnitude of this force can be calculated using the summation of pressure induced forces defined above but a simpler and more elegant approach is offered by the principle of Archimedes. This states that the buoyancy force on a wholly or partially submerged free body in a fluid at rest is equal and opposite to the weight of fluid displaced by the body and acts in a vertical line through the centre of gravity of the displaced fluid. For a fluid of constant density, this centre of gravity will be at the centre of volume of the displaced fluid.

This result can be readily proved. Consider an arbitrary boundary n, drawn in B, a fluid at rest, in *Figures 2.60(a)* and *(b)*. Since the fluid is at rest, the pressure gradient within it will be supporting the weight of fluid contained inside boundary A such that the total force F_B across the boundary will be vertical and given by:

$$F_B = \rho g V \qquad (2.196)$$

where V is the fluid volume contained in boundary A and ρ is the density of fluid B. F_B will also act through the centre of gravity of the fluid in volume V. Now, if the fluid volume inside the boundary was replaced by a solid body of identical shape, the above pressure induced force, the buoyancy force, will still act. The point through which it acts is also called the centre of buoyancy of the submerged volume.

It is important to appreciate that Archimedes' principle only applies to freely floating bodies such as in *Figure 2.60(a)* and *(b)*. There are, however, many structures used in offshore engineering that are submerged but not freely floating. *Figures 2.60(c)* and *(d)* gives two examples: the former being a vertical riser pipe connection between the atmosphere and a drill hole or hydrocarbon reservoir below the sea bed whereas the latter could be a subsea one atmosphere habitat. Since a perfectly vertical riser pipe is only subjected to horizontal pressure induced forces, the net vertical buoyancy force on it must be zero. On the other hand, the subsea habitat could be said to have a negative (downward acting) buoyancy force. It is clear, therefore, that the buoyancy force for bodies which are not freely floating must be calculated from first principles by integration of pressure induced forces on their surface. This exercise is often necessary in the design of offshore structures.

2.5.2.2 Hydrostatic stability

A freely floating body, such as in *Figure 2.61*, will be in vertical equilibrium if its weight W, acting through its centre of gravity G, is equal to the buoyancy force $\rho g V$ acting through the centre of buoyancy (or volume) B, where V is the body's submerged volume. However, the relative positions of points G and B will govern the body's stability in roll and pitch. The body will be in stable, neutral or unstable equilibrium if the centre of buoyancy B, is above, coincides with or is below the centre of gravity G. A body in stable equilibrium will, if given a small displacement and then released, return to its original position. If the equilibrium is unstable, the body will not return to its stable position but will move further from it. In neutral equilibrium, the body will neither return to its original position nor increase its displacement following the initial disturbance but will simply adopt a new position. The condition for stable equilibrium requires a restoring couple to return the vessel to its original position following angular perturbation.

This restoring couple can be written as:

$$M = \rho g V (KB - KG) \sin\theta = W(KB - KG) \sin\theta \qquad (2.197)$$

where θ is the rotation angle of the body from its equilibrium position.

Now, for a fully submerged rigid body without free surfaces in internal tanks, the positions of points B and G relative to the body will remain fixed and the stability will be fully defined by the above considerations. However, for a freely floating body at the free surface, the shape of the submerged volume and hence the

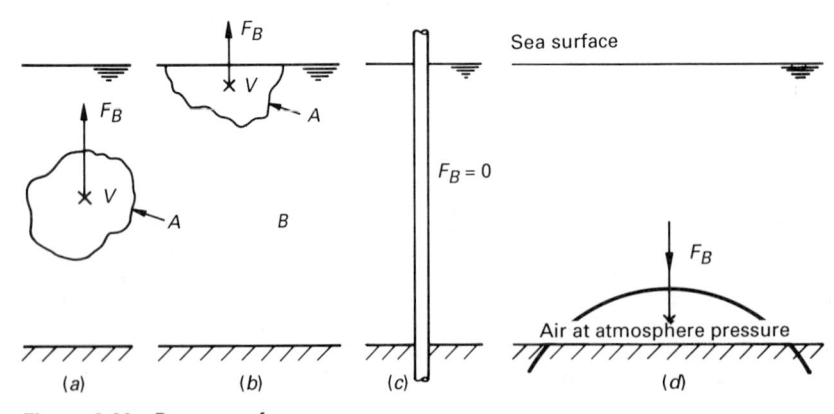

Figure 2.60 Buoyancy forces

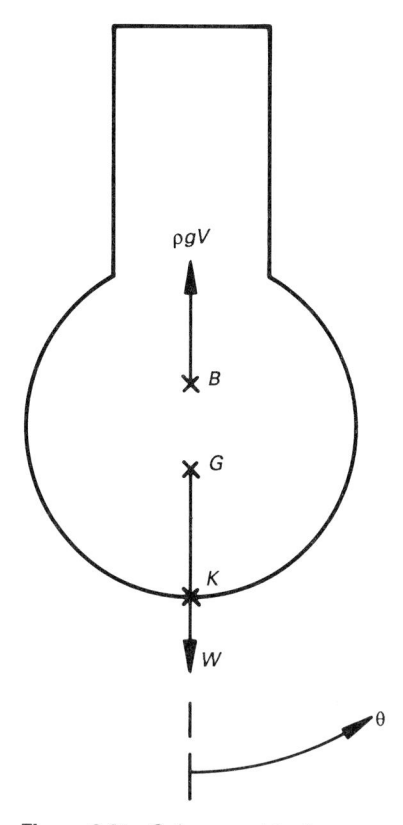

Figure 2.61 Submerged body

position of the centre of volume (or buoyancy) will shift with inclination of the body.

This aspect complicates consideration of the stability of a floating body at the free surface and requires the problem to be examined from first principles. Consider the arbitrary body in *Figure 2.62* floating at the free surface with submerged volume V.

The shape of the surface boundary of this volume is described with respect to Cartesian axes $Oxyz$ that are fixed in the body with Oxy initially in the water plane. At equilibrium, the constant body weight W will be equal to the buoyancy force $\rho g V$, with the latter acting through the centre of buoyancy B. Now if the body is given a small inclination β, about the Oy axis, the condition that the submerged volume must remain constant before and after inclination is:

$$V = \int_A S \, dA = \int_A (S - x\beta) \, dA \quad (2.198)$$

where the function $S = f(x, y)$ describes the z coordinate of the submerged body surface relative to axes $Oxyz$. Simplifying Equation 2.198 and carrying out a similar procedure for rotation about the x axis leads to the conditions that:

$$\int_A x \, dA = 0 \, , \quad \int_A y \, dA = 0 \quad (2.199)$$

which proves that rotation of the body at constant submerged volume, and, therefore with equilibrium maintained, can only take place about axes through the centroid of the water plane area. This point is defined as the centre of flotation.

The coordinates of the centre of buoyancy before inclination $(\bar{x}, \bar{y}, \bar{z})$ can be defined by the equations:

$$\bar{x} = \frac{1}{V} \int_A S x \, dA$$

$$\bar{y} = \frac{1}{V} \int_A S y \, dA \quad (2.200)$$

$$\bar{z} = \frac{1}{V} \int_A \frac{1}{2} S^2 \, dA$$

After an inclination of β about the Oy axis, the centre of buoyancy shifts to coordinates $(\bar{x}', \bar{y}', \bar{z}')$:

$$\bar{x}' = \frac{1}{V} \int_A (S - x\beta) x \, dA$$

$$\bar{y}' = \frac{1}{V} \int_A (S - x\beta) y \, dA \quad (2.201)$$

$$\bar{z}' = \frac{1}{V} \int_A \frac{1}{2}(S - x\beta)(S + x\beta) \, dA$$

with the moments being taken about the initial coordinate axes. The movement of the centre of buoyancy parallel to the y axis can then be obtained as:

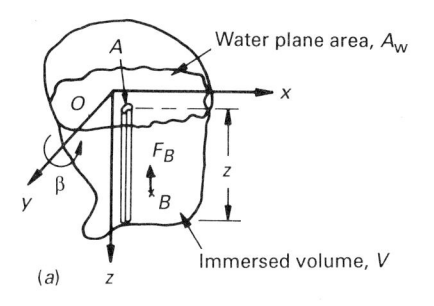

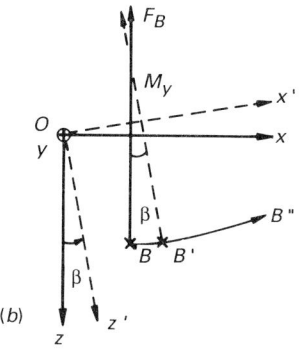

Figure 2.62 Arbitrary body floating freely at the water surface

$$\bar{y} - \bar{y}' = \frac{\beta}{V} \int_A xy \, dA \qquad (2.202)$$

This equation indicates that if the cross product of area of the water plane about Ox and Oy is zero, that is, Oxy are principal axes of the water plane area, then the centre of buoyancy will move parallel to the Oxz plane during inclination about the y axis. The condition of zero cross product of water plane area is satisfied if either Ox or Oy are lines of symmetry of the water plane. Since this condition is satisfied for the majority of floating vessels used in offshore operations, $\bar{y} = \bar{y}'$ and $\bar{x} = \bar{x}'$ are taken to be valid for inclination about the Oy and Ox axes, respectively. These conditions also imply that rotation about the Ox and Oy axes can be treated independently in hydrostatic analysis if Oxy are principal axes of the water plane area. On the other hand, if the cross product of water plane area is non-zero, it is recommended that the hydrostatic analysis is carried out from first principles without use of the relationships derived in this section.

The vertical movement of the centre of buoyancy can be written as:

$$\bar{z} - \bar{z}' = \frac{1}{2V} \beta^2 \int_A x^2 \, dA \qquad (2.203)$$

with the β^2 indicating that the movement is small or of second order. Also, the horizontal movement of the centre of buoyancy is given by:

$$\bar{x} - \bar{x}' = \frac{\beta}{V} \int_A x^2 \, dA \qquad (2.204)$$

The effect of this is illustrated by *Figure 2.62* where $BB'B''$ indicates the locus of centre of buoyancy as β increases. The upward shallow curvature of the line is due to the vertical movement in B being of second order compared to the horizontal movement. Furthermore, for small β, the vertical through the centre of buoyancy (B') goes through a fixed point M_y whose position can be defined by using triangle BM_yB' in *Figure 2.62* to obtain.

$$BM_y = \frac{BB'}{\beta} = \frac{\bar{x} - \bar{x}'}{\beta} = \frac{1}{V} \int_A x^2 \, dA = \frac{I_y}{V} \qquad (2.205)$$

with I being the second moment of water plane area about Oy. Using a similar derivation for rotation about the Ox axis, the relationship:

$$BM_x = \frac{I_x}{V} \text{ where } I_x = \int_A y^2 \, dA \qquad (2.206)$$

can be obtained. Therefore, for small rotations about the Ox and Oy axes, the centre of buoyancy will move along a surface which is concave upwards but with different curvatures in the Oy and Ox directions; the centres of curvatures being the metacentres M_x and M_y, respectively. These M_x and M_y metacentre positions are given by Equations 2.206 and 2.205 and may be considered as the effective point of application of the buoyancy force after rotations about the Ox and Oy axes, respectively. At any inclination of the body, the tangent plant to the surface of buoyancy will be parallel to the water plane. Note also that if the axes Oxy are not principal axes of the water plane (i.e. $\int_A xy \, dA \neq 0$), the metacentre positions cannot in general exist since the lines through B and B' in *Figure 2.62* will not necessarily intersect.

For stable equilibrium, the metacentre positions must both lie above the centre of gravity although the centre of buoyancy need not do so. The righting moments in the Oyz and Oxz planes, M_x and M_y can then be written as:

$$M_x = \rho g V \cdot GM_x \cdot \sin \theta = \rho g V \cdot GZ_x$$
$$M_y = \rho g V \cdot GM_y \cdot \sin \beta = \rho g V \cdot GZ_y \qquad (2.207)$$
$$GZ_x = GM_x \sin \theta, \; GZ_y = GM_y \sin \beta$$

where GM_x, GM_y are metacentric heights, θ and β are inclinations about the Ox and Oy axes, respectively, and GZ_x, GZ_y are moment arms of the restoring couples.

Calculations for initial stability of a freely floating vessel are carried out by identifying the position of five points within or around the vessel. The lowest point on the vessel vertical centre line, denoted by K (for keel level), is conventionally used as a reference. The positions of vessel centre of gravity G, centre of buoyancy B, and the longitudinal and transverse metacentres M_y and M_x are defined with respect to the keel reference K, and using the Ox axis pointing forwards and Oz vertically downwards. The longitudinal and transverse metacentric heights are given by:

$$GM_y = KB + BM_y - KG$$
and $\qquad (2.208)$
$$GM_x = KB + BM_x - KG$$

since BM_x and BM_y can be directly calculated from Equations 2.206 and 2.205, respectively.

Table 2.10 presents data on the typical positions of these points for four floating vessels used in both non-offshore and offshore applications. The relative numerical values for GM_y and GM_x reflect the large length to beam ratios of ship shape vessels compared to semisubmersibles.

2.5.2.3 Large angle stability and stability loss mechanisms

As the angle of inclination of a freely floating vessel at the free surface increases beyond small values, the analysis of the above section no longer holds true. The shape and properties of the water plane area change sufficiently that the metacentre position is no longer fixed. Furthermore, the shape of the submerged vessel

Table 2.10: Typical values of initial stability parameters

Vessel type	Distances (m)					
	KB	KG	KM_x	KM_y	GM_x	GM_y
Cargo vessel	4.70	8.81	11.93	298	3.12	289.19
Frigate	2.56	4.94	5.92	250	0.98	245.06
Destroyer	2.95	6.95	7.80	396	0.85	389.05
Semisubmersible*	6.98	14.63	20.14	21.04	5.51	6.41

*at drilling draught

volume can change substantially as keel or deck edges emerge or submerge. Large angle stability is therefore important for the design of floating offshore structures where safety or performance at larger angles of inclination need to be evaluated.

An extension of the results of the preceeding section to larger angles can be derived by the so called wall sided formula. A wall sided vessel is one in which all of the vessel's water plane intersecting sides are vertical for some distance above and below the waterline. Most semisubmersibles and conventional ship shape vessels, can be considered wall sided to a good degree of approximation since a major proportion of such vessels' sides are perpendicular to the water plane.

For a floating vessel with all of its sides perpendicular to the water plane an extension to the small angle hydrostatic stability result can be derived by considering the cross-section of the vessel as shown in *Figure 2.63* with two water planes corresponding to angles of inclination of zero and β. Since the volumes of the immersed and submerged wedges during inclination will be equal for a wall sided vessel with symmetry about the fore and aft vertical centre plane, the volume of the vessel before and after inclination can be taken as a constant V. Due to inclination through a large angle β, the centre of buoyancy will exhibit a large horizontal shift BB' and a smaller vertical shift $B'B_\beta$. The horizontal shift is given by taking moments of volume in the transverse direction about an axis through O. Thus:

$$BB' \cdot V = 2 \cdot [y \cdot y \tan \beta \cdot \tfrac{1}{2} \cdot L] \cdot \tfrac{2}{3} y \quad (2.209)$$

where L is the length of the vessel with a uniform cross-section. This yields:

$$\therefore BB' = \frac{I}{V} \tan \beta \quad \text{for} \quad I = \frac{1}{2} y^3 L \quad (2.210)$$

where I is the second moment of area of the initial (at $\beta = 0$) water plane about the axis through O. Furthermore $B'B_\beta$ is given by taking moments of volume in the vertical direction:

$$B'B_\beta \cdot V = 2 \cdot \tfrac{1}{2} y^2 L \tan \beta \cdot \tfrac{1}{3} y \tan \beta$$

$$\therefore B'B_\beta = \frac{1}{2} \frac{I}{V} \tan^2 \beta \quad (2.211)$$

Taking M and M_c as initial and large angle metacentre positions, respectively, BM_c is given by:

$$BM_c = BM + MM_c$$
$$= BB' \cot \beta + B'B_\beta$$

$$BM_c = BM \left[1 + \tfrac{1}{2} \tan^2 \beta\right] \quad (2.212)$$

The wall sided formula result can also be written in

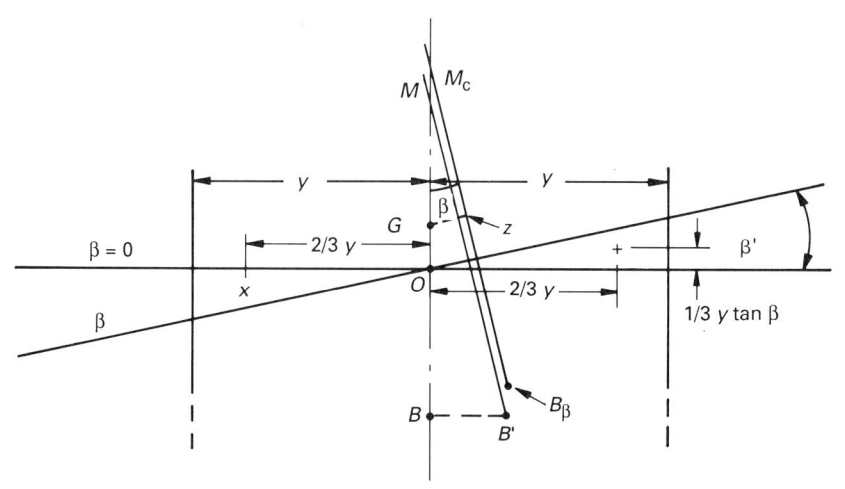

Figure 2.63 Definition diagram for wall sided formula

terms of GZ (see Equation 2.207) to yield:
$$GZ = [GM + \tfrac{1}{2} BM \tan^2 \beta] \sin \beta \qquad (2.213)$$

The wall sided formula is valid for wall sided vessels up to angles at which deck edge submergence or keel edge emergence occurs. However, as pointed out earlier, most ship shape and semisubmersible vessels are non-wall sided for a small proportion of their water plane cutting surfaces such that the wall sided formula only offers a good approximation to their stability at small to moderate angles.

Calculations of hydrostatic stability beyond the range of validity of the wall sided formula have been carried out in conventional naval architecture by using manual or computer based numerical methods to calculate the submerged volume, position of centre of buoyancy and righting moments from the hull surface definition. Rawson and Tupper (1976) present a summary of manual computations. Some of these have been converted to computer based methods which derive hydrostatics information by more extensive numerical operations on the hull form data. All of these large angle methods are not directly applicable to the non-ship shape hull forms used in offshore engineering. A more exact approach based on integration of hydrostatic pressures has been developed for use on offshore vessels and is described in more detail in Section 2.5.2.4.

The behaviour of a typical vessel floating at the free surface and inclined at large angles is illustrated by the resultant movements of the centres of buoyancy and of the metacentres. The surface defined by the movement of the centre of buoyancy is called the surface of buoyancy where as the surface defined by the centres of curvature (the metacentre) of the surface of buoyancy is defined as the evolute of the surface of buoyancy. *Figures 2.64a* and *b* show typical shapes of these surfaces. The surface of buoyancy is like an ellipsoidal shape whereas its evolute is made up of four downward or upward pointing cusps around the body shape.

In practical terms, the large angle stability characteristics of a hull form can be defined by tabulated or plotted data presented in various ways. Cross curves of stability (see *Figure 2.65*) present stability moment arms (GZ) against vessel displacement for a range of heel angles. In some cases, the variability of the centre of gravity position G, is removed from GZ by replacing it with SZ where S is an arbitrary fixed point or pole on the vessel vertical centre line. SZ can readily be related to GZ if the position of S and KG are known. Statical stability curves define the variation of GZ, SZ or righting moment against angle of heel (see *Figure 2.65*). Statical stability curves are often drawn on the same axes as wind heeling curves to check that the vessel has sufficient hydrostatic restoring moment to recover from a heeling moment due to wind. The area under either of these curves is a measure of the work done to heel the vessel over to any particular angle. A practical means of ensuring stability under wave induced motions is to require that the area under the restoring moment curves exceeds the area under the wind heeling curve by a specified fraction at least. This feature is described in more detail in Section 2.5.2.5.

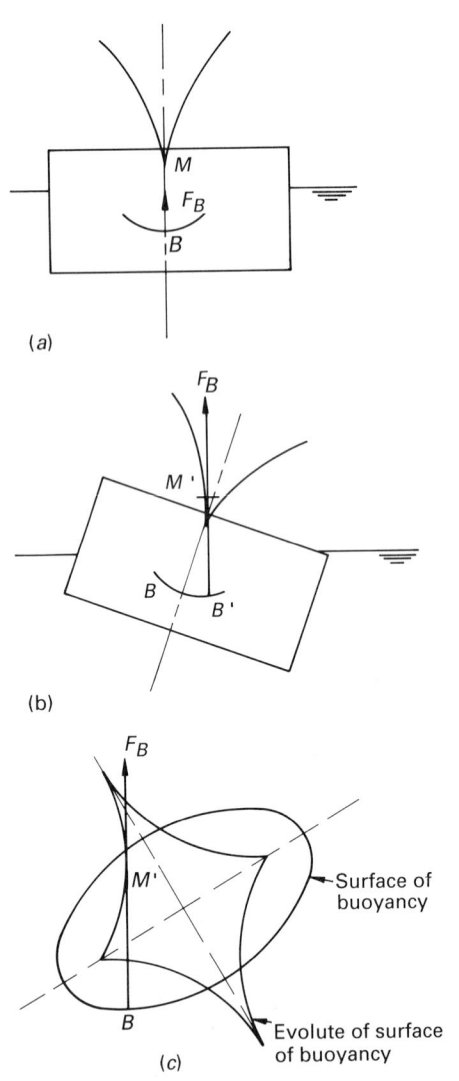

Figure 2.64 Large angle behavior of surface of buoyancy and evolute of surface of buoyancy

Curves of form can be used to define the values of displacement, KB, KM_y, KM_x, BM_y, BM_x as functions of vessel draught (see *Figure 2.65*). Safe vessel loading conditions can be defined by allowable KG curves (*Figure 2.65d*) which give the highest permissible centre of gravity heights above the keel as a function of vessel draught. All of these curves are usually presented in tabulated and graphical form in the stability manual of a floating offshore vessel.

There are a number of physical effects which can contribute to a loss in the hydrostatic stability of a vessel. These effects arise from the fact that the mass distribution of a vessel can change with the angle of inclination and the resultant centre of gravity shift is invariably such as to reduce the hydrostatic restoring moment arm GZ. The stability reductions can occur due to the effects of tanks of fluids with free surfaces inside a vessel, due to freely suspended loads from cranes and also due to loads applied by catenary moorings or riser pipe connections between the floating platform and sea bed. Calculations of these stability losses is essential during definition of the hydrostatic stability of a floating vessel.

The most common source of stability loss is the so called free surface effect. Consider a floating vessel which has partially filled tanks with free surfaces on board. Inclination of the vessel causes these free surfaces to move such that the fluid in the tanks piles up towards the sides of the tanks closest to the submerging part of the ship. This causes the centre of gravity of the tank fluid contents and therefore of the vessel to shift in the direction of movement of the vessel hull's centre of buoyancy as it inclines. There is a consequent reduction in the magnitude of the restoring couple between vessel weight and the buoyancy force. An expression for this stability loss is obtained as follows.

Consider tank A with fluid contents of weight w on board vessel B of weight W as shown in *Figure 2.66*. When the vessel inclines through a small angle β, the horizontal shift in centre of gravity of the tank contents gg', induces a horizontal shift GG', in the vessel centre of gravity such that:

$$W\,GG' = w\,gg' \qquad (2.214)$$

But gg' is identical to the centre of buoyancy shift that would occur if the shape of the tank fluid contents were to be considered as the submerged volume of a freely floating body. Then Equation 2.214 can be written as:

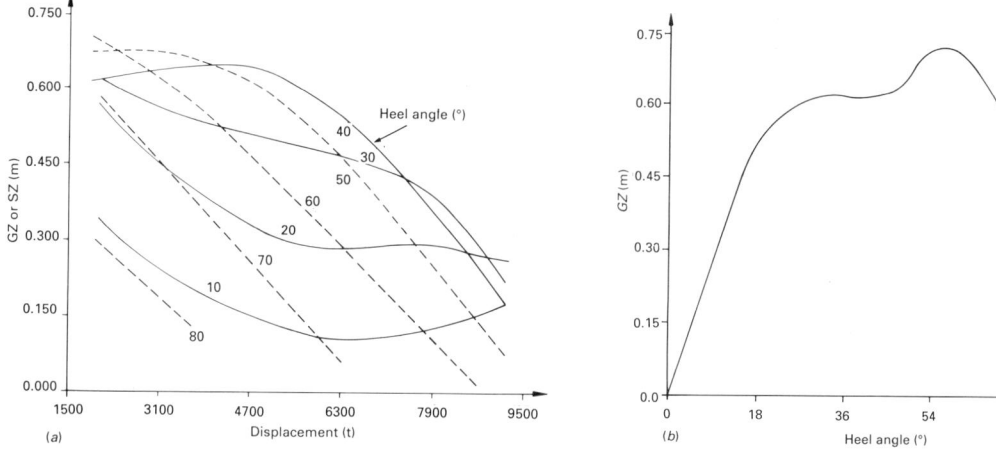

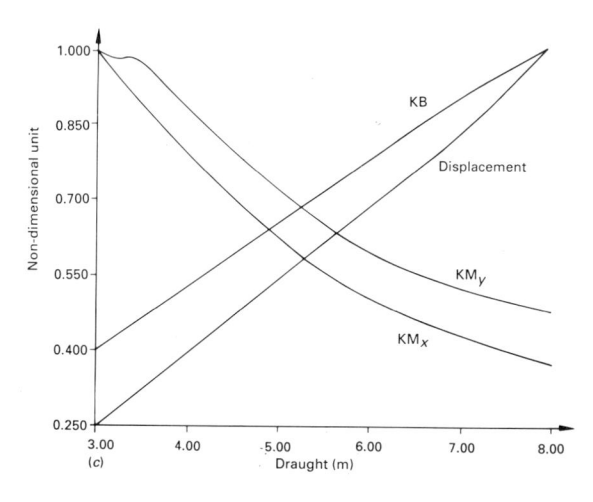

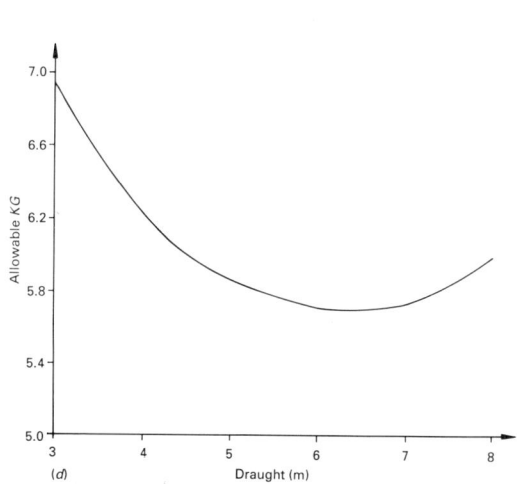

$$W\,GG' = w\,\frac{I_1}{V_1}\,\beta \qquad (2.215)$$

Figure 2.65 Hydrostatic stability curves: (a) cross curves of stability; (b) static stability curve; (c) curve of form; (d) allowance *KG* curve

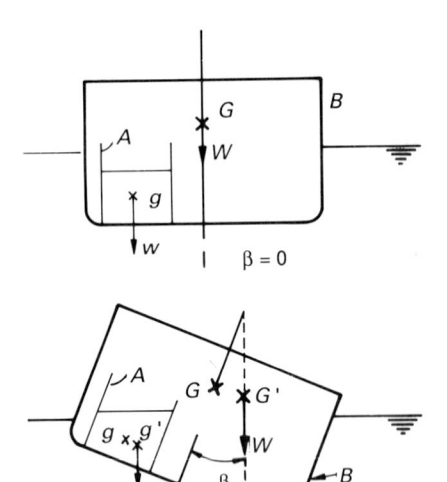

Figure 2.66 Free surface effect

where I_1 is the second moment of area of the fluid free surface about an axis through the centroid of the free surface parallel to the axis of rotation of the ship and V_1 is the volume of liquid in the tank. Converting the centre of gravity shift as a loss in metacentric height, taking $w = \rho_1 g V_1$, where ρ_1 is the density of fluid in the tank, and $w = \rho g V$ where ρ is the density of sea water and V is the submerged volume of the vessel, the metacentric height loss is obtained as:

$$\Delta(GM) = \frac{GG'}{\beta} = \frac{\rho_1 I_1}{\rho V} \quad (2.216)$$

For a ship with a large number of internal tanks (say N) with free surfaces, the total metacentric height loss is:

$$\Delta(GM) = \frac{1}{\rho V} \sum_{i=1}^{N} \rho_i I_i \quad (2.217)$$

where ρ_i is the fluid density for the ith tank and I_i is its free surface second moment of area about an axis through the free surface's own centroid and parallel to the vessel's axis of rotation.

Note that the free surface effect is independent of the volume of fluid in the tanks and would disappear if a tank was completely full; in this case the mass of fluid would act as a rigid body during vessel rotation. Free surface effects are normally reduced in the design of a floating offshore vessel by compartmentation of tanks to substantially reduce the value of I_i for each of the tanks. The loss of metacentric height or free surface correction is then subtracted from the rigid vessel metacentric height to give a net value to be considered for vessel evaluation.

Freely suspended masses on board a vessel have a similar effect to free surfaces in that the centre of gravity is shifted in the same horizontal direction as the centre of buoyancy and, therefore, tends to reduce stability. For a vessel with a N suspended masses on board each of mass m_i and with free suspension lengths l_i, the centre of gravity shift is:

$$GG' = \beta \frac{1}{W} \sum_{i=1}^{N} m_i g l_i \quad (2.218)$$

Using the same approach as before, the metacentric height loss can be written as:

$$\Delta(GM) = \frac{1}{\rho W} \sum_{i=1}^{N} m_i l_i \quad (2.219)$$

This correction to metacentric height can be applied by subtracting it from the rigid vessel metacentric height. Alternatively, the centre of gravity calculations of the vessel can be carried out by placing the masses m_i at the top end of the suspension lengths l_i. This has the same effect as the correction defined previously.

The forces applied on a vessel by a catenary mooring spread can vary with angle of inclination of the vessel to contribute to either some gain or loss in stability but such corrections are usually small and are neglected for hydrostatic calculations. The generally constant downward force applied by a marine riser to a drilling or production vessel has no effect on vessel hydrostatic stability. However, both moorings and the riser apply net downward forces on the vessel which have to be accounted for in the weight and buoyancy force balance of the vessel.

The hydrostatics of compliant vessels with rigid components that may be connected by freely articulating joints requires special treatment; Patel and Walker (1983) gives an example of one such calculation.

2.5.2.4 The pressure integration technique

The difficulty of calculating large angle hydrostatic stability using submerged volume and water plane area properties for offshore floating vessels has prompted the development of a more fundamental approach. This technique uses an integration of pressure distribution acting over the submerged body surface to yield all the necessary hydrostatic characteristics. The method is computationally more efficient when transformed into a programmed set of instructions for an arbitrary structure at any orientation. Mathematically, the surface integral of pressure can be transformed into a volume integral such that numerical quantities used in conventional naval architecture can be derived directly from the surface integrals. The computational requirement to divide all the vessel surface into panels is analogous to a requirement for surface panels in potential flow boundary integral techniques.

Consider an arbitrary body shown in *Figure 2.67* floating at a free surface with the fluid above the free surface being at a constant pressure. At any point $\mathbf{x}(x, y, z)$ in the fluid below the free surface there will be an increase in pressure given by the equation:

$$p = \rho g (d - z) \tag{2.220}$$

The incremental force $d\mathbf{F}$ acting on an element of the body surface due to this fluid pressure is:

$$d\mathbf{F} = \rho g (d - z) \, dS \tag{2.221}$$

By integrating over the submerged surface, the total force $\mathbf{F}$ is given by:

$$\mathbf{F} = \iint_S \rho g (d - z) \, \mathbf{n} \, dS \tag{2.222}$$

where $\mathbf{n}$ is the unit normal vector acting into the body and is a function of position $\mathbf{x}$. Similarly, the incremental moment $d\mathbf{M}$ is

$$d\mathbf{M} = \mathbf{x} \wedge d\mathbf{F}$$
$$= \mathbf{x} \wedge \rho g (d - z) \, \mathbf{n} \, dS \tag{2.223}$$

Integrating:

$$\mathbf{M} = \iint_S \mathbf{x} \wedge \rho g (d - z) \, \mathbf{n} \, dS \tag{2.224}$$

Now these surface pressure induced forces and moments will be in equilibrium with the weight of the body acting vertically downwards. Resolving forces and moments due to body weight, therefore, gives:

$$\mathbf{F} = W\mathbf{k} \quad \text{and} \quad \mathbf{M} = \mathbf{x}_c \wedge W\mathbf{k} \tag{2.225}$$

where W is the weight of the body and $\mathbf{x}_c$ is the centre of gravity.

For static equilibrium, Equation 2.225 must be satisfied since the fluid exerts an upward thrust on the body equal and opposite to the body's weight. This upward thrust, called the buoyancy force, must lie along the same line of action as the weight.

Now the divergence theorem states that:

$$-\iint_S \mathbf{f} \cdot \mathbf{n} \, dS = \iiint_V \nabla \cdot \mathbf{f} \, dV \tag{2.226}$$

where $\mathbf{f}$ is a vector field and ∇ is the gradient operator. The negative sign arises because the divergence theorem is written here with the unit normal vector pointing into the surface. From Equation 2.222 the buoyancy force is:

$$|\mathbf{F}| = \iint_S \rho g (d - z) \, \mathbf{k} \cdot \mathbf{n} \, dS \tag{2.227}$$

Using the divergence theorem, Equation 2.227 becomes:

$$|\mathbf{F}| = -\rho g \iiint_V \frac{\partial}{\partial z}(d - z) \, dV \tag{2.228}$$
$$= \rho g V$$

This is Archimedes' principle – the buoyancy force acting on a floating body is equal to the weight of the volume of fluid displaced by the body.

The solutions of Equations 2.225 determine the magnitude of the buoyancy force and the horizontal coordinates of the centre of buoyancy. The vertical centre of buoyancy cannot be determined from these equations since the solution for moment only gives the line of action. The solution for the vertical centre of buoyancy is given from the fact that it lies at the centroid of the submerged volume. Thus:

$$\mathbf{x}_B \cdot \mathbf{k} = d - \frac{1}{V} \iiint_V (d - z) \, dV \tag{2.229}$$

Relating the right-hand side of Equation 2.229 to the right-hand side of Equation 2.226 for the divergence theorem gives:

$$\nabla \cdot \mathbf{f} = d - z \tag{2.230}$$

The solution to Equation 2.230 is:

$$\mathbf{f} = -\frac{(d - z)^2}{2} \mathbf{k} \tag{2.231}$$

Substituting Equation 2.231 into Equation 2.229 via Equation 2.226 results in:

$$\mathbf{x}_B \cdot \mathbf{k} = d - \frac{\rho g}{|\mathbf{F}|} \iint_S \frac{(d - z)^2}{2} \mathbf{k} \cdot \mathbf{n} \, dS \tag{2.232}$$

All the hydrostatic properties can, therefore, be determined in terms of surface integrals. The advantages of these surface integrals is that they do not have to be evaluated over the water plane area since they go to zero on that surface.

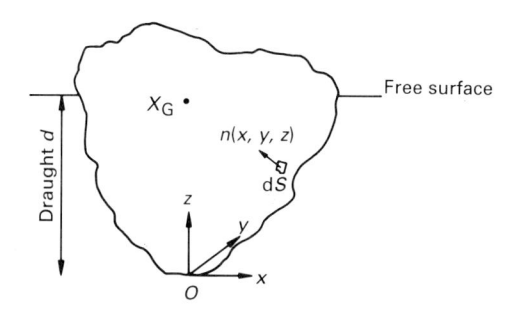

Figure 2.67 Floating body in equilibrium

The other characteristics of the water plane can also be derived from surface integrals. Taking $Ox'y'$ as the principal axes of the water plane area, these integrals can be written as follows. The second moments of the water plane area about the x and y axes are:

$$I_{x'x'} = \iint_S (x' \cdot i)^2 \, n \cdot k \, dS$$

and

$$I_{y'y'} = \iint_S (x' \cdot j)^2 \, n \cdot k \, dS \qquad (2.233)$$

The water plane area A_W is given by:

$$A_W = \iint_S n \cdot k \, dS \qquad (2.234)$$

The coordinates of the centre of flotation (x_F, y_F) are given by:

$$x_F = \frac{1}{A_W} \iint_S (x \cdot i) \, n \cdot k \, dS$$

and

$$y_F = \frac{1}{A_W} \iint_S (x \cdot j) \, n \cdot k \, dS \qquad (2.235)$$

Solutions to Equations 2.225 and 2.232 give the body's equilibrium position. However, nothing is known about the stability of the body in this position. The classical approach to stability is to apply a small linear or rotational displacement to a body and to test whether it returns to its equilibrium position of its own accord or if it goes to another. In the theory described here, a small rotation is applied to the body. The point of rotation may be anywhere in space provided that the buoyancy force remains constant. The centre of buoyancy moves to a new position x'_B and the centre of gravity moves to x'_G. Note that the centre of gravity need not more relative to the rigid body but it would move in a fixed global frame of reference.

Taking moments:

$$M_R = (-x'_G \wedge Wk) - (x'_B \wedge - Wk)$$
$$= W(x'_B - x'_G) \wedge k \qquad (2.236)$$

where M_R is the restoring moment vector. If M_R is greater than zero in the direction of rotation, then the body is stable. If M_R is less than zero, the body is unstable and will find a new equilibrium position.

The solution by pressure integration is obtained by evaluating the surface integrals described above. The surface of the body to be analysed is fully covered with a set of m surface elements. A vessel of complex geometry can, therefore, be represented by a collection of simpler surface elements. Equations 2.222 and 2.224 then become:

$$F = \sum_{i=1}^{m} F_i$$

and

$$M = \sum_{i=1}^{m} M_i \qquad (2.237)$$

where F_i and M_i are the forces and moments acting on the ith surface element. F_i and M_i are given by:

$$F_i = \iint_{S_i} \rho g \, (d - z) \, n_i \, dS_i$$

and

$$M_i = \iint_{S_i} x \wedge \rho g \, (d - z) \, n_i \, dS_i \qquad (2.238)$$

The advantage of the above is that a complex surface integral has been reduced to a summation of a set of known surface integrals. Depending on the type of surface element, Equations 2.238 may be evaluated analytically, numerically or by a combination of anlytic and numerical methods.

Two methods for the solution of these equations are presented here as examples. The first is a complete numerical summation over the element's surface. The element is treated as a region and is subdivided into a set of rectangular subelements. Equations 2.238 then become:

$$F_i = \sum_{p=1}^{M_p} \sum_{q=1}^{M_q} \rho g \, (d - x_{pq} \cdot k) \, n_{pq} \, \Delta h \Delta v$$

and

$$M_i = \sum_{p=1}^{M_p} \sum_{q=1}^{M_q} x_{pq} \wedge \rho g \, (d - x_{pq} \cdot k) \, n_{pq} \, \Delta h \Delta v \qquad (2.239)$$

where x_{pq} is the coordinate of the centroid of the rectangular subelement with dimensions Δh and Δv and a unit normal n_{pq}. There are M_p of these subelements along one submerged edge of the element and M_q along the other.

The second method is a numerical summation of a series of line integrals of finite width. The line integrals are evaluated analytically and are, therefore, exact. Equations 2.238 then become:

$$F_i = \sum_{p=1}^{M_p} \rho g \, \Delta h \int_0^{L_p} (d - x \cdot k) \, n \, dl$$

$$M_i = \sum_{p=1}^{M_p} \rho g \, \Delta h \int_0^{L_p} x \wedge (d - x \cdot k) \, n \, dl \quad (2.240)$$

where L_p is the submerged length of the pth line element. This reduction of a complete surface integral to a sum of element surface integrals is equally applicable to all the surface integral equations. This solution method is preferred because it is more accurate and computationally faster. Further details of the analytic solution for line integrals are given by Witz and Patel (1985).

The pressure integration technique can be implemented in a computer program to utilize a library of rectangular, triangular, circular and curved plate elements to discretize the surface of complex ship or semisubmersible hull forms. The surface integrals for each element are evaluated and then summed over the entire surface to yield hydrostatic properties. Two examples of the use of this technique are presented here. *Figure 2.68* offers a comparison between the pressure integration technique and the wall sided formula for a rectangular box of 10 × 10 m (33 × 33 ft) in plan view with 4 m depth and 2 m draught floating in sea water. The box offers a special case where the wall sided formula is valid over the entire 180° of rotation. The pressure integration technique is in excellent agreement with the analytic result. The two techniques are also compared in *Figure 2.70* for the more realistic case of a semisubmersible of the type illustrated in *Figure 2.69* by a line drawing of the plate geometry used to model the vessel.

The wall sided formula is often used as the first approximation in evaluating the large angle restoring moments of geometries which are not strictly wall sided. It is clear here that there are substantial changes in righting moment as the deck edge submerges with

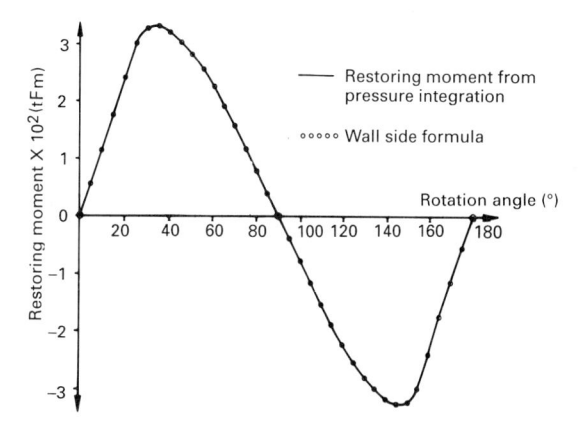

Figure 2.68 Restoring moment of a 10m × 10m × 4m box of 205 tonnes displacement in sea water

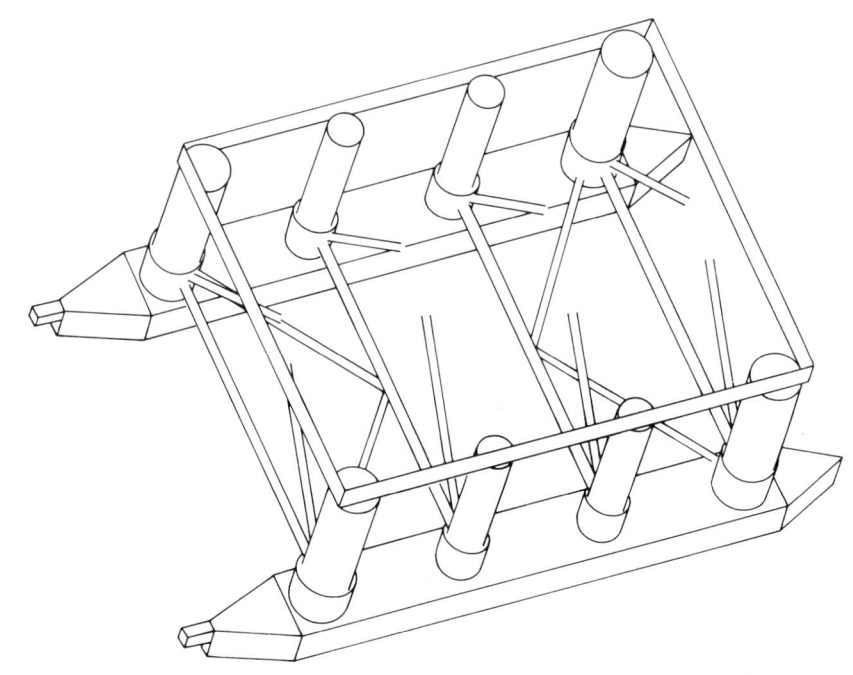

Figure 2.69 Discretization of semi-submersible for pressure integration

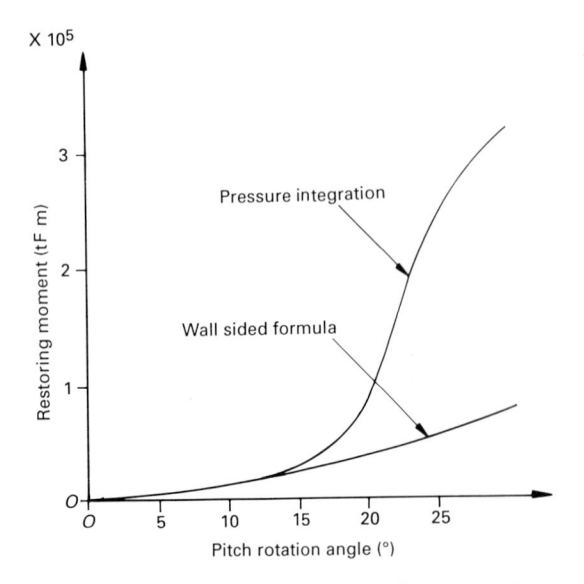

Figure 2.70 Semi-submersible restoring moments

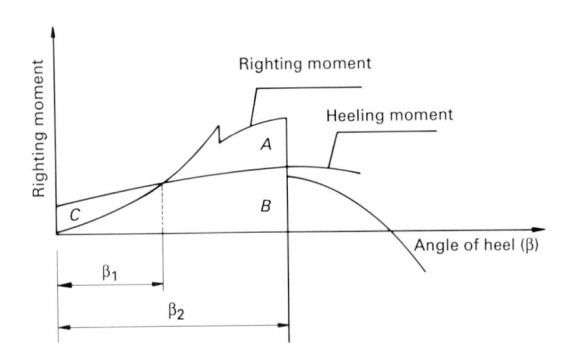

Figure 2.71 Typical righting moment and wind heeling moment against angles of heel

the wall sided formula only remaining valid up to approximately 12°. The semisubmersible case illustrates the advantages and disadvantages of both the classical approach and the pressure integration method to hydrostatic analysis. The classical approach for initial stability (that is at and around zero heel and trim angles) is easy to apply but is only valid for a small range of heel angles. Extensions of the classical approach to larger heel angles are, however, difficult to apply for complicated vessel geometries. The pressure integration method allows complex geometries such as the semisubmersible of *Figure 2.69* to be modelled accurately for large angle rotations. Its disadvantage is that it is a numerical method which necessitates the use of a computer.

2.5.2.5 Certifying authority regulations

Major ship and offshore structure classification societies such as Lloyds Register of Shipping, Det Norske Veritas and the American Bureau of Shipping act as national certifying authorities for overseeing the design, construction, installation and operation of offshore structures. The societies set out regulations for minimum standards on all aspects of the design and peration of offshore structures. They also produce guidelines for recommended practice in design procedures. The Department of Energy (1986), Lloyds Register of Shipping (1986), Det Norske Veritas (1987) and the American Bureau of Shipping (1987), publish codes of practice listing hydrostatic stability requirements that must be met for certification of a structure.

These national statutory requirements for hydrostatic stability have tended to become more stringent since the mid-1970s. A typical set of requirements for acceptable hydrostatic stability would contain the following major components:

1. Vessel light weight and centre of gravity – this part of the regulations ensure that reliable mandatory procedures are laid down to accurately determine the vessel light weight and centre of gravity position. This is done by means of an inclining test to measure metacentric height of the light ship vessel coupled with a careful weight audit of the vessel during operations.
2. Definition of operating conditions and hydrostatic particulars – the regulations also require the definition of several common loading and operating conditions for which full hydrostatic particulars need to be computed and made available in the operating manual. The stability data includes the type of information presented in *Figure 2.65*.
3. Intact stability criteria – these are designed to ensure that the righting moment against angle of heel curve for the intact, that is, undamaged vessel is of sufficient magnitude and range to withstand static overturning moments due to wind and also to tolerate wave induced overturning moments and heel angles. The latter dynamic aspect of vessel stability is difficult to quantify for design purposes but an approximate approach has been evolved by defining minimum area ratios under the hydrostatic righting moment and wind heeling overturning moment curves. These criteria are described in more detail below.
4. Damage assumptions and stability – the rules require that the vessel remains safe in terms of sinking or capsizing following a certain level of reasonable damage and flooding. The rules lay down recommended damage assumptions and the stability criteria to be met after flooding. These are normally somewhat less stringent than those for the intact vessel.
5. Survival to extreme damage – recent capsizing accidents with semisubmersible vessels have also led to

extreme damage rules which require that a specified level of vessel survivability is maintained after extreme damage such as loss of an entire vertical column.
6. Watertight integrity and freeboard – the rules for these aspects of a vessel's hydrostatic design are linked to meeting the requirements for 3, 4 and 5 above but are concerned with crew safety, vessel survival after an accident and wave induced damage in severe storms.

Most rules for mobile offshore units differentiate between ship shape, semisubmersible and jack up rigs during the transportation phase in setting slightly different criteria for each.

In order to illustrate the above requirements further, a summary of typical hydrostatic stability rules for a semisubmersible unit are presented below:

2.5.2.6 Operating conditions

1. Transit condition at low draught with pontoons piercing the water surface.
2. Operating conditions at normal drilling draught with various combinations of payload and riser pipe connections.
3. Survival conditions with a draught slightly less than that used for normal operations to improve vessel sea keeping in extreme waves.

2.5.2.7 Intact stability criteria

These are expressed in relation to the example statical stability curve presented in *Figure 2.71* where the wind heeling moment curve intercepts the hydrostatic righting moment curve at angles of heel of β_1 and β_2 with A, B and C denoting the areas between the curve boundaries shown.

1. Wind speed in operating and transit condition
   = 36.0 m/s (118 ft/s)
   Wind speed in survival condition
   = 51.5 m/s (169 ft/s)
2. The initial metacentric height after correction for free surface effects shall not be less than 1.0 m (3 ft) at transit, survival and operating draught and shall not be less than 0.3 m (0.9 ft) at temporary draught.
3. The area ratio of the righting moment curve to the wind heeling moment curve to the assumed down flooding point or second intercept should not be less than 1.3, that is:

$$\frac{A + B}{B + C} \geq 1.3 \quad (2.241)$$

The down flooding point refers to a condition or angle of heel at which a vessel opening will fall below still water level and cause progressive flooding to occur.

4. The angle of heel due to the wind heeling moment should not be greater then 15°, that is $\beta_1 \leq 15°$. This is the angle at the first crossing of the righting moment and the wind heeling moment curves.
5. The second crossing of the righting moment and heeling moment curves shall occur at an angle of heel of at least 30°, that is $\beta_2 \geq 30°$.

2.5.2.8 Damaged stability criteria

Following damage and flooding, a typical statical stability curve will look similar to that in *Figure 2.71* but the curve will be shifted to the right by a steady heel angle β_0. Then typical criteria are:

1. Wind speed = 25.8 m/s (85 ft/s)
2. Damage stability shall be investigated at transit and operating drafts.
3. Any one compartment adjacent to the sea shall be assumed to be flooded regardless of location.
4. In addition to (3), the damage is normally assumed to occur at any water level between 5 m (16 ft) above and 3 m (10 ft) below the water line in question and at least one watertight bulkhead should be assumed damaged. The vertical extent is assumed to be 3 m (10 ft) within the above zone with the horizontal extent being taken to be 3 m (10 ft) measured along the periphery of the vessel external shell. The penetration of the damage zone into the vessel is assumed to be a horizontal distance of 1.5 m (5 ft) measured radially from the shell.
5. After damage, the angle of heel should not be greater than 15°, that is $\beta_0 \leq 15°$.
6. The area under the righting moment curve up to the second crossing or any lesser angle shall be equal to or greater than the area under the wind heeling moment curve to the same limiting angle. Thus:

$$\frac{A + B}{B + C} \geq 1.0 \quad (2.242)$$

2.5.2.9 High energy extreme damage criteria

The semisubmersible will need to be provided with means of buoyancy in the deck structure sufficient to remain afloat and satisfy the criteria described below, after the loss of buoyancy equivalent to the volume of the whole, or a major part of any, one column. The loss of buoyancy is assumed to occur when the platform is at maximum operating draught and with the maximum allowable KG.

After this extreme loss of buoyancy, the semisubmersible should satisfy the following criteria:

1. The angle of heel should not exceed 35° in any direction.
2. Not taking any effect of wind and waves into account, the water line in the final condition of equilibrium should be at least 0.6 m (2 ft) below the

lower edge of any opening through which progressive flooding may take place.
3. The righting moment GZ curve in the damaged condition shall have a positive extent of at least 20° beyond equilibrium toegether with a maximum GZ value of at least 1.0 m (3 ft).

These criteria for high energy extreme damage can only be met by ensuring that the vessel deck is a watertight structure capable of providing large righting moments after major loss of buoyancy in a vertical column.

2.5.3 Dynamic response analysis

2.5.3.1 Basic equations

The dynamic response of a single freedom linear second order system is described by the equation of motion:

$$m\ddot{z} + b\dot{z} + kz = F(t) \qquad (2.243)$$

where z is the motion component, m is the total mass, b is a linear damping coefficient, k is stiffness and $F(t)$ is the wave exciting force. Some fundamental results on the free and forced dynamic response of a dynamic system described by Equation 2.243 are given here as a prelude to their application to the motion of a floating body.

The free response of the dynamic system is derived from a solution of the characteristic equation obtained by substituting:

$$z = A\, e^{\lambda t} \qquad (2.244)$$

into Equation 2.243 with the right-hand side set to zero. The two roots of the characteristic equation yield a harmonic decaying vibration if $(b/2m)^2 < (k/m)$. The vibration is described by the equation:

$$z(t) = e^{-b\omega_o t}(A_1 \cos \omega_d t + A_2 \sin \omega_d t) \qquad (2.245)$$

where the undamped natural frequency ω_o is

$$\omega_o = \sqrt{\frac{k}{m}}$$

the damping ratio β, is:

$$\beta = \frac{b}{2\sqrt{(mk)}}$$

and the damped natural frequency ω_d: (2.246)

$$\omega_d = \sqrt{\frac{k}{m} - \left(\frac{b}{2m}\right)^2}$$
$$= \omega_o \sqrt{(1 - \beta^2)}.$$

The constants A_1 and A_2 can be replaced by means of initial conditions. Taking z_o and $\dot{z}_o$ as the initial displacement and velocity, A_1 and A_2 can be found and substituted into Equation 2.245 to give:

$$z(t) = e^{-\beta \omega_o t}\left[\frac{\dot{z}_o + \beta \omega_o z_o}{\omega_d} \sin \omega_d t + z_o \cos \omega_d t\right] \qquad (2.247)$$

Figure 2.72 illustrates one possible form of $z(t)$ with z_o finite and $\dot{z}_o = 0$. It can be observed that a decaying oscillatory motion response is obtained at a natural frequency of oscillation ω_d, and an amplitude decay given by the expontential term in Equation 2.247. As the damp coefficient β, tends to zero, the oscillations of Figure 2.72 becomes an oscillation at constant amplitude and undamped natural frequency ω_o. At the other extreme, however, as β tends to 1, the damped natural frequency ω_d, tends to 0 with:

$$\left(\frac{b}{2m}\right)^2 = \frac{k}{m} \quad \text{or } b = 2\sqrt{mk}$$

In this case, Equation 2.27 becomes:

$$z(t) = e^{-\omega_o t}[\dot{z}_o t + (1 + \omega_o t)z_o] \qquad (2.248)$$

which describes a decaying motion that does not reverse direction. This motion, shown in Figure 2.73, is described as being critically damped. Thus, for a general decaying response, the damping ratio β, also defines the damping non-dimensionalized as a proportion of critical damping.

Most floating offshore structures exhibit values of damping ratio in the range 0.05–0.25. For this range of

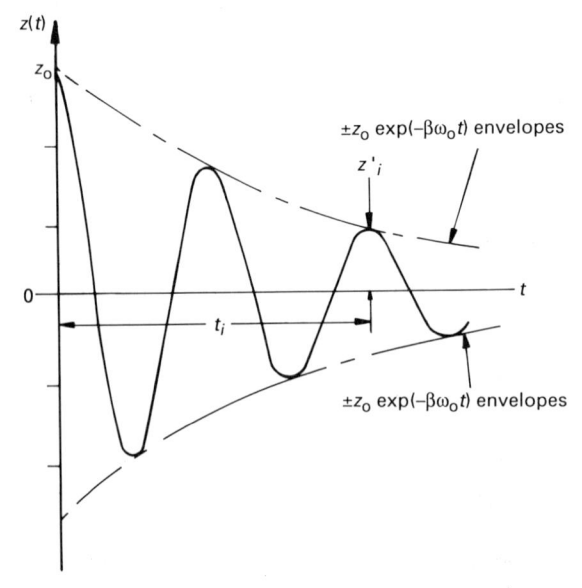

Figure 2.72 Response of single degree of freedom system with linear damping and initial displacement

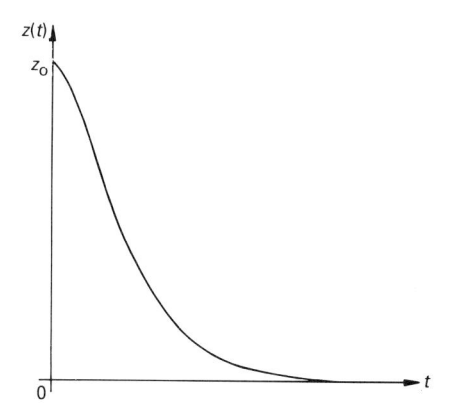

Figure 2.73 Response of single degree of freedom system with critical linear damping and initial displacement

damping ratios, the undamped and damped natural frequencies are nearly equal since for $\beta = 0.25$:

$$\omega_d = \omega_o \sqrt{(1 - 0.0625)} = 0.9375\, \omega_o$$

The damping present in a floating structure or its scaled model can be measured by recording its free vibration in still water and noting the peak values of consecutive cycles. Then the logarithmic decrement δ, is defined as:

$$\delta = \ln \left\{ \frac{z'_i}{z'_{i+1}} \right\}$$

where z'_i and z'_{i+1} are the ith and $(i+1)$th peaks of the damped response shown in *Figure 2.72*. Taking T_d as the damped natural period, the above equation can be written as:

$$\delta = \ln \left\{ \frac{e^{-\beta\omega_o t_i}}{e^{-\beta\omega_o (t_i + T_d)}} \right\} = \ln (e^{\beta\omega_o T_d})$$

$$= \beta\omega_o T_d = \frac{2\pi\beta}{\sqrt{(1 - \beta^2)}} \quad (2.249)$$

For lightly damped systems, $\sqrt{(1 - \beta^2)} = 1$ and β can be obtained by a least squares fit to the equation:

$$\ln z'_i = -\beta\omega_o t_i + \ln z_o \quad (2.250)$$

where t_i and z'_i are the coordinates of the ith peak value of the damped response (see *Figure 2.72*) and z_o is an initial displacement.

A sinusoidal force may be applied to the single degree of freedom system of Equation 2.243 by substituting:

$$F(t) = F_a\, e^{i\omega t} \quad (2.251)$$

Taking a particular integral of the form:

$$z = z_a\, e^{i\omega t} \quad (2.252)$$

and substituting this with Equation 2.251 into 2.243 gives:

$$z = \frac{F_a (k - m\omega^2 - ib\omega)}{(k - m\omega^2)^2 + b^2\omega^2} e^{i\omega t}$$

$$= H(\omega)\, F_a\, e^{i\omega t} \quad (2.253)$$

where function $H(\omega)$ is the complex transfer function between force and motion amplitude. Taking the real parts of Equations 2.251 and 2.253 also yields the governing equation and solution to be of the form:

$$m\ddot{z} + b\dot{z} + kz = F_a \cos \omega t$$

$$z = \frac{F_a [(k - m\omega^2) \cos \omega t + b\omega \sin \omega t]}{(k - m\omega^2)^2 + b^2\omega^2}$$

$$= \frac{F_a}{[(k - m\omega^2)^2 + b^2\omega^2]^{1/2}} \cos(\omega t - \epsilon) \quad (2.254)$$

where

$$\epsilon = \tan^{-1} \left[\frac{b\omega}{k - m\omega^2} \right]$$

The total solution for Equations 2.243 and 2.251 is made up of the sum of the complementary function representing the transient response of Equation 2.245 and the particular integral describing the steady state response of Equations 2.253 or 2.254.

Taking z_s to be the displacement to a static force F_o, r as the ratio of forcing frequency to natural frequency and β as the ratio of actual damping to critical damping, the magnitude of the transfer function $H(\omega)$ can be written in terms of a dynamic amplification factor as:

$$\frac{z_o}{z_s} = k|H(\omega)| = \frac{1}{[(1 - r^2)^2 + (2\beta r)^2]^{1/2}}$$

where

$$r = \omega/\omega_o, \quad \beta = b/2\sqrt{(mk)} \text{ and } z_s = F_o/k \quad (2.255)$$

and the phase lag is

$$\epsilon = \tan^{-1} \left[\frac{2\beta r}{1 - r^2} \right]$$

Figure 2.74 describes the variation of z_o/z_s and ϵ with frequency ratio r for a range of values of the damping ratio β. These figures illustrate the classical effects of frequency on a resonant system with large dynamic amplifications at a frequency ratio of 1, a quasi-static response at very low frequency ratios and an asymptotic response tending to zero at large frequency ratios. The phase lags also vary 0–180° with a value of 90° at resonance.

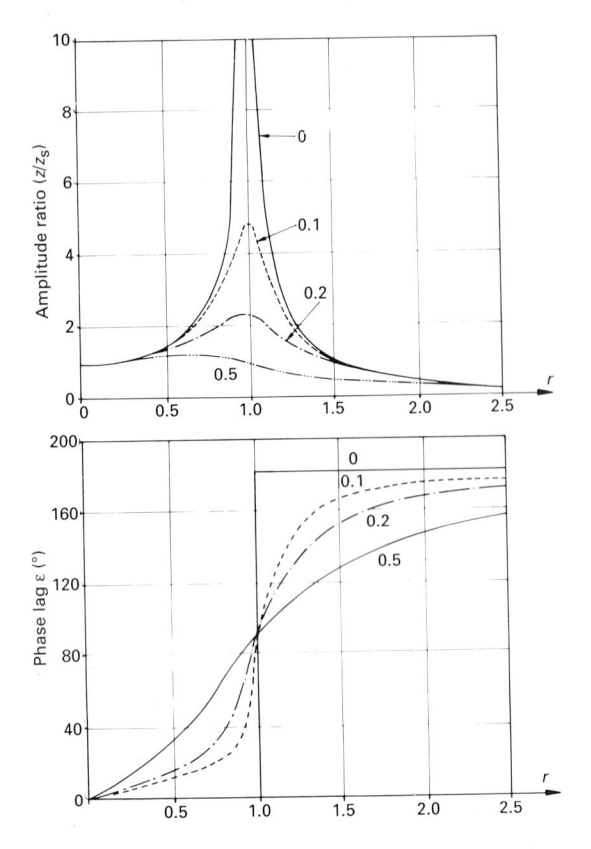

Figure 2.74 Response of linear single degree of freedom dynamic system to forced vibration. Curve labels indicate values of damping ratio β

The response of a single degree of freedom dynamic system to random excitation can be obtained by using the concept of the impulse response to represent the effect of the random exciting force as a series of infinitesimal impulses. Integration of the impulse response then yields the response of the system to the random excitation force (Warburton 1976). This leads to the result that if the dynamic system represented by Equation 2.243 is subjected to a stationary random excitation force $F(t)$ with a frequency spectrum $S_f(\omega)$, then the dynamic system's response z will have a frequency spectrum $S_z(\omega)$ given by:

$$S_z(\omega) = |H(\omega)|^2 S_f(\omega) \qquad (2.256)$$

where $H(\omega)$ is given by Equation 2.255.

Thus, if the system is excited by a force of Gaussian probability distribution and spectral density $S_f(\omega)$, the system response will also have a Gaussian probability distribution but with spectral density $S_z(\omega)$. The results of random process theory (Newland 1975) can then be used to deduce the properties of the system response. Note that the motion of an offshore structure in waves should be regarded as the product of two transfer functions: the first from wave elevation to exciting force and the second from the exciting force to wave motion response. If the wave elevation spectrum is narrow banded, then the exciting force and motion response spectra will also be narrow banded although their shapes could be substantially distorted by the transfer functions involved.

The dynamic response analysis techniques presented above can be readily extended to multiple degrees of freedom by writing the equations of motion in matrix form:

$$M \ddot{X} + B \dot{X} + K X = F(t) \qquad (2.257)$$

where the $(n \times 1)$ column vector of displacements X holds the n degrees of freedom that characterize the response of the structure and $F(t)$ denotes the $(n \times 1)$ column vector of applied forces. All of the coefficients of the n equations of motion implied in Equation 2.257 are incorporated in $(n \times n)$ matrices of mass, damping and stiffness (M, B, K). The magnitude and distribution of coefficients in these matrices reflect the dynamic characteristics of each degree of freedom and of the couplings between them.

The free and forced response analysis presented in this section can be readily extended to apply to systems with multiple degrees of freedom. A detailed description of this extension and the available methods of solution are presented by Warburton (1976) where these techniques are applied to solve for structural deformations of structures with many degrees of freedom. For floating structures that can be regarded as rigid bodies (that is structural deformations are very small compared to wave induced motions), it is usual to use only the six degrees of freedom for the surge, sway, heave, roll, pitch and yaw motions of the structure.

2.5.3.2 Application to floating bodies

1. *Heave motion of a cylindrical buoy*: the transformation of a problem concerned with fluid structure interaction of a floating body into the governing equation of a second order dynamic system is illustrated by considering heave motion of the small buoy shown in *Figure 2.75*. Coordinate y denotes vertical motion of the water surface and z denotes the resultant buoy vertical motion. Now the total force acting on the buoy will be due to added mass, damping and hydrostatic stiffness induced forces due to differential motion between the buoy and surrounding fluid. The net force due to these effects will serve to accelerate the buoy. Thus the equation of motion can be written as:

$$m \ddot{z} = A(\ddot{y} - \ddot{z}) + B(\dot{y} - \dot{z}) + C(y - z) \qquad (2.258)$$

where m is the buoy mass and A, B and C are coefficients expressing the added mass, damping and hydrostatic stiffness induced forces on the buoy. Here the damping force is taken to be linear and the Froude–Krylov force, applying through dynamic wave pressure on the buoy base, is given by the stiffness term Cy for a buoy geometry where draught h is very small compared to incident wave lengths.

Re-arranging Equation 2.258 to bring unknown buoy motion terms to the left-hand side gives:

$$(m + A)\ddot{z} + B\dot{z} + Cz = F(t) \qquad (2.259)$$

where

$$F(t) = A\ddot{y} + B\dot{y} + Cy \qquad (2.260)$$

Thus, there are two equations governing the behavior of the buoy in waves. Equation 2.257 is concerned with the mechanism of wave elevation motion y, exerting an exciting force $F(t)$ on the buoy whereas Equation 2.259 is the governing equation for a linear second order dynamic system. The form of the above equations is typical of that obtained for all floating bodies in that two transfer functions (from wave elevation to force and from wave force to motion) are involved in deriving a body motion response from wave elevation.

A typical but very simplified solution procedure for the above equations is presented here to illustrate these ideas further. Taking buoy height h, as very small compared to wave length, a wave of amplitude a with elevation:

$$y = a \cos \omega t \qquad (2.261)$$

can be used with Equation 2.260 to give:

$$F(t) = a[(C - A\omega^2) \cos \omega t - B\omega \sin \omega t] \qquad (2.262)$$

Thus the wave force is a function of A, B and C, frequency ω, and wave amplitude a, and can be expressed as an amplitude and phase transfer function of the form:

$$F(t) = F_a \cos(\omega t + \epsilon)$$

where

$$\frac{F_a}{a} = [(C - A\omega^2)^2 + (B\omega)^2]^{1/2} \qquad (2.263)$$

and

$$\epsilon = \tan^{-1}\left[\frac{-B\omega}{C - A\omega^2}\right]$$

Equation 2.263 possesses all the features of a second order dynamic system transfer function with its amplitude and phase varying with frequency of excitation. Equation 2.259 then becomes:

$$(m + a)\ddot{z} + B\dot{z} + Cz = F_a \cos(\omega t + \epsilon) \qquad (2.264)$$

and has a solution of the form:

$$z = z_a \cos(\omega t + \epsilon + \phi)$$

where

$$\frac{z_a}{F_a} = \frac{1}{\{[C - (m + A)\omega^2]^2 + (B\omega)^2\}^{1/2}}$$

and $\qquad (2.265)$

$$\phi = \tan^{-1}\left[\frac{B\omega}{C - (m + A)\omega^2}\right]$$

Note that the wave force leads the wave elevation by ϵ and the buoy motion lags the wave force by ϕ. When $m \rightarrow 0$, $\epsilon + \phi = 0$ and suggests that a massless buoy will follow the wave profile exactly without phase lag. The amplitude and phase transfer functions from wave force to buoy motion will be of the form given by *Figure 2.74* but the overall wave elevation to buoy motion amplitude and phase transfer functions will be distorted by the presence of the wave elevation to wave force transfer function. It is recommended that both of these transfer functions be evaluated independently during the motions analysis of a floating body to obtain a clearer physical understanding of its behaviour. *Figures 2.79, 2.81, 2.83* and *2.84* present these transfer functions for typical floating offshore structures.

For a cylindrical buoy of radius r, and height h, floating in water to density ρ, the terms m, C and A can be given by:

$$m = \rho \pi r^2 h$$

$$C = \rho \pi g r^2 \qquad (2.266)$$

$$A = \tfrac{1}{3} \pi \rho r^3$$

where the added mass A in vertical motions of the cylinder end is approximated by that of a hemisphere of radius r, and added mass coefficient 0.5 on the cylinder end. Evaluation of a linear damping coefficient B is examined in more detail in the next section.

2. *Heave motion of semisubmersibles*: semisubmersibles are floating platforms with a geometry that is considerably different from the conventional ship sha-

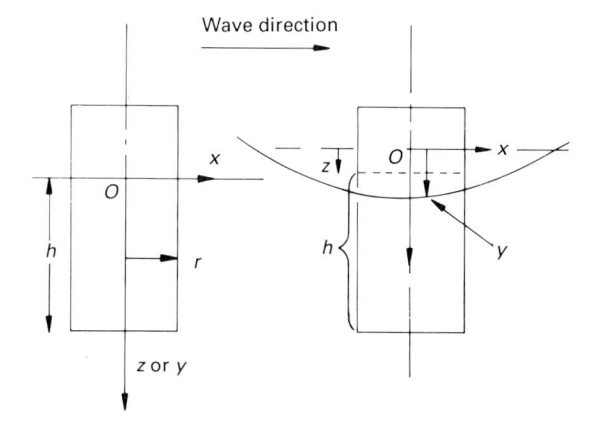

Figure 2.75 Heave motion of a buoy

pe, or monohull, form. This leads to significant differences in the hydrostatic and hydrodynamic behaviour of semisubmersibles in comparison with monohull vessels. Semisubmersibles consist of deeply submerged pontoons which are connected to the deck by several large diameter columns together with bracing members. A semisubmersible possesses low wave-induced motions due to having a large proportion of its submerged volume at a deep draught where wave pressures have decayed rapidly with depth.

The small water plane area of the vessel and the large submerged volume yield long natural periods in heave, roll and pitch. These periods are above the periods of predominant wave action, further contribution to reducing motions. However, the heave or vertical motion of a semisubmersible is still a significant factor in curtailing marine operations such as drilling or oil production. Therefore, it is desirable to refine the heave motion of a semisubmersible during design to reduce its down time due to weather.

It is relatively simple to evaluate the undamped natural frequency or period of a semisubmersible vessel. An example of how this can be done is illustrated for a twin pontooned vessel with n legs on each side as shown in *Figure 2.76*. The undamped equation of vertical motion can be written as:

$$(m + A)\ddot{z} + Cz = 0 \tag{2.267}$$

using the notation given earlier and the resultant undamped natural period is:

$$T_o = 2\pi \sqrt{\frac{m + A}{C}}$$

Using *Figure 2.76*:

$$m = \rho\,[2\pi R_2^2 L + 2n\pi R_1^2 h_1] \tag{2.268}$$

$$A = 2\pi\rho R_2^2 L$$

and

$$C = 2\pi\rho g n R_1^2$$

using an added mass coefficient of 1.0 for vertical motion of the circular pontoons. Then:

$$T_o = 2\pi \sqrt{\frac{1}{g}\left[h_1 + \frac{2R_2^2 L}{nR_1^2}\right]}$$

$$= 2\pi \sqrt{\frac{h_1}{g}[1 + 2\alpha]} \tag{2.269}$$

where α is the ratio of pontoon volume to submerged column volume. This equation demonstrates that the natural period in heave motion is strongly influenced by pontoon submergence h_1, and to a lesser extent by the volume ratio α.

Also, if the semisubmersible of *Figure 2.76* had pontoons of rectangular cross-section of width b_2 and height h_2, then terms m and A in the above equations change to:

$$m = \rho[2b_2 h_2 L + 2n\pi R_1^2 h_1] \tag{2.270}$$

and

$$A = 2\rho C_m b_2 h_2 L$$

where the added mass coefficient C_m is a function of the aspect ratio (b_2/h_2) of the pontoon cross-section. Then:

$$T_o = 2\pi \sqrt{\frac{1}{g}\left[h_1 + \frac{(1 + C_m)\,b_2 h_2 L}{\pi n R_1^2}\right]}$$

$$= 2\pi \sqrt{\frac{h_1}{g}[1 + \alpha(1 + C_m)]} \tag{2.271}$$

which is of a similar form to Equation 2.269.

The natural period in heave of a single circular column terminated at its base by a circular footing of larger radius (see *Figure 2.77*) can also be defined through the following equations using a subscript 1 to denote a single column:

$$m_1 = \rho\pi[R_1^2 h_1 + R_2^2 h_2]$$

$$C_1 = \pi\rho g R_1^2 \tag{2.272}$$

$$A_1 = \pi\rho R_2^2 C_{m1}$$

where C_{m1} is a function of the geometry of the structure through R_1, R_2 and h_2.

$$T_1 = 2\pi \sqrt{\frac{h_1}{g}\left[1 + \frac{R_2^2 h_2}{R_1^2 h_1} + \frac{R_2^3 C_{m1}}{R_1^2 h_1}\right]}$$

$$= 2\pi \sqrt{\frac{h_1}{g}\left\{1 + \alpha\left(1 + \frac{C_{m1} R_2}{h_2}\right)\right\}} \tag{2.273}$$

is again of a similar form to that found before, with α being the ratio of the volume of the column footing to the column itself. If a semisubmersible is made up of n identical column and footings of the type shown in *Figure 2.77*, then for this vessel.

$$m_n = n \cdot m_1 \quad,\quad C_n = n \cdot C_1 \quad,\quad A_n = n \cdot A_1 \tag{2.274}$$

and, therefore, the natural period remains the same as for the single column and footing arrangement. This

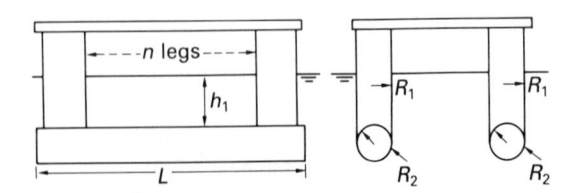

Figure 2.76 Twin pontooned semi-submersible structure

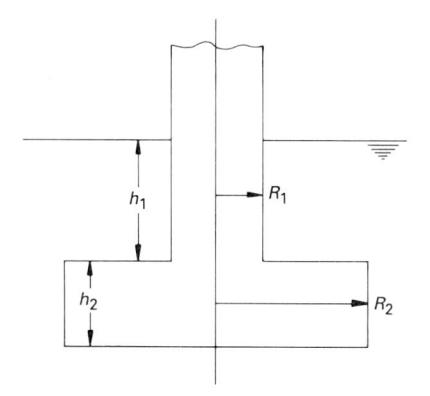

Figure 2.77 Column pontoon footing structure

result assumes that the added masses of each column and footing are unaffected by the presence of others in their vicinity.

The heave motion in waves of semisubmersibles of simple geometry can be evaluated by treating the vessel as a collection of slender members on which the loading can be calculated using the Morison Equation with deep water linear wave theory. In this particular derivation, the damping and wave induced drag forces are neglected for simplicity. Thus the equation of vertical motion for a four column, twin pontooned semisubmersible of the configuration shown in *Figure 2.78* can be written as:

$$(m + A)\ddot{z} + Cz + F(t) \quad (2.275)$$

with coefficients of the form:

$$m = 4\rho[A_v d = A_h L]$$
$$A = 4\rho C_m A_h L$$
$$C = 4\rho g A_v$$

where A_v and A_h are the areas of cross-section of the vertical columns and rectangular pontoons, respect-

ively, and C_m is the added mass coefficient for the pontoons in vertical motion.

The vertical wave exciting force on the vessel can be taken as the sum of wave forces on the pontoons and columns to yield:

$$F(t) = \int_{-L}^{L} 2\rho A_h (1 + C_m) \dot{w} \cdot dx$$
$$+ 2 A_v (p_f + p_r)$$

where $\dot{w}$ is the vertical wave particle acceleration at $z = -d$ and p_f, p_r are wave pressures at $z = -d$ and $x = +L$ and $-L$, respectively. Note that the pontoon/column intersections have been approximated by lengths d and L to simplify the algebra.

The Froude–Krylov force normal to the pontoon axis is calculated using an inertia coefficient $(1 + C_m)$ whereas the axial Froude–Krylov force on the vertical columns is calculated as a product of dynamic wave pressure and column area of cross-section. The former approach is valid for calculating normal forces on a slender member about which wave properties can be considered constant whereas the wave pressure formulation is necessary for axial forces on members whose length is comparable to incident wave length.

Using the Oxz axes of *Figure 2.78*, the wave particle vertical acceleration and pressure can be written as:

$$\dot{w} = -a\omega^2 e^{-kd} \sin(kx - \omega t)$$
and $\quad (2.277)$
$$p = + a \frac{\rho \omega^2}{k} e^{-kd} \sin(kx - \omega t)$$

and substituted into Equation 2.276 to yield the wave force relationship:

$$F(t) = 4a\rho g e^{-kd}$$
$$[-A_v \cos kL + A_h(1 + C_m)\sin kL] \sin \omega t \quad (2.278)$$

Evaluation of Equation 2.278 followed by solution of Equation 2.275 yields the wave induced force and motion response which can be written as:

$$F(t) = F_a \sin(\omega t + \epsilon)$$
and $\quad (2.279)$
$$z(t) = z_a \sin(\omega t + \epsilon + \phi)$$

Figure 2.79 presents the variations of $|F_a|/|a|, |z_a|/|a|$, ϵ and $(\epsilon + \phi)$ as functions of frequency for a semisubmersible with $d = 13.5$ m, $L = 36$ m, $A_h = 70$ m² and $A_v = 90$ m². A value of 1.2 is used for the added mass coefficient. These curves exhibit several interesting features. The wave force amplitude cancels out to zero at certain wave frequencies when instantaneous wave forces on the pontoons balance wave forces on the columns. At these cancellation frequencies phase ϵ, shifts by 180° such that over some wave frequencies, the

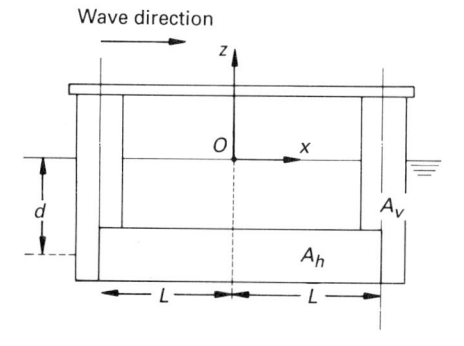

Figure 2.78 Side view of four column semi-submersible

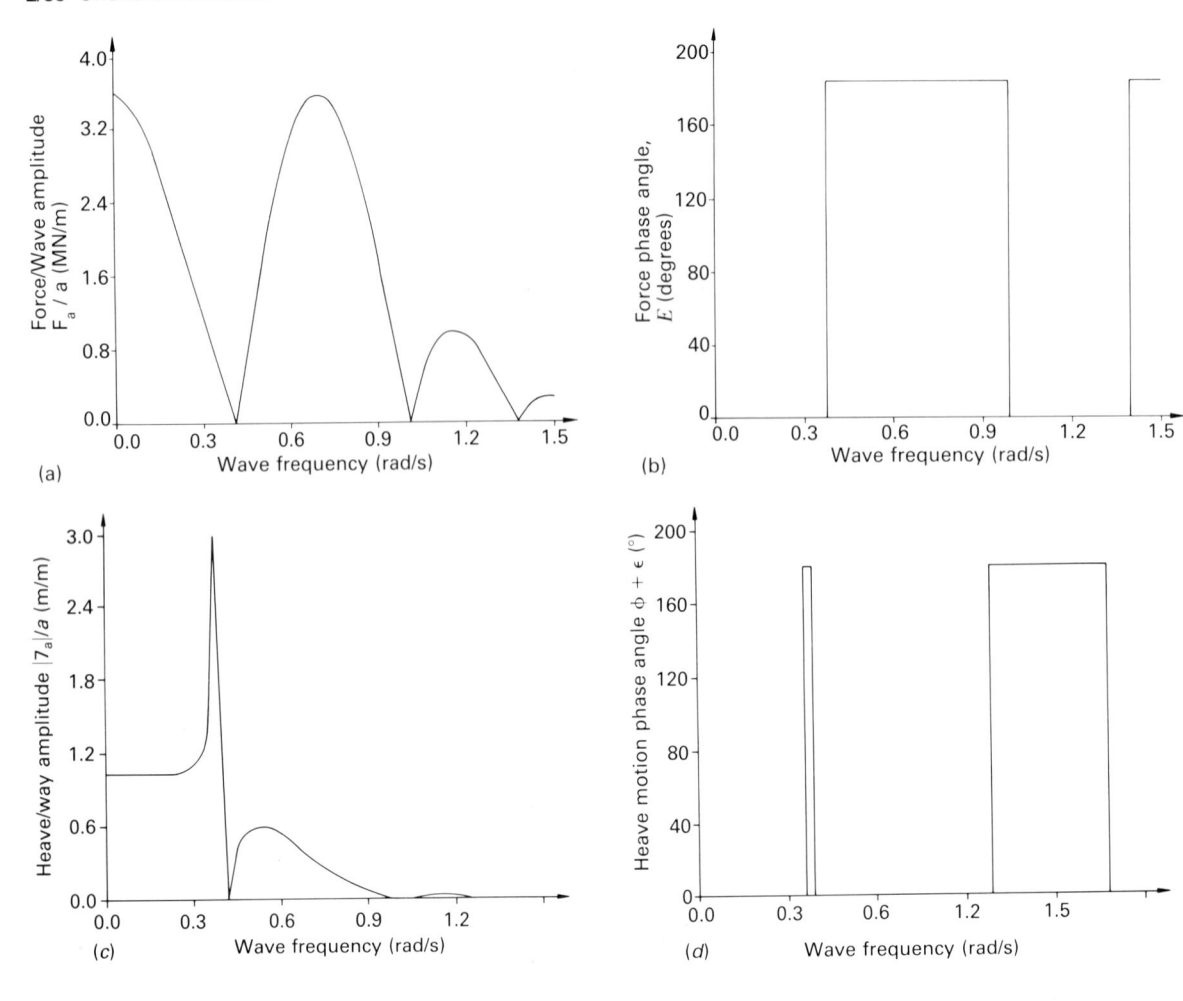

Figure 2.79 Heave force and motion response of four column semi-submersible

vertical force on the vessel is out of phase with wave elevation. The motion response transfer function shows the combined effects of the wave force cancellation and resonance with the total phase difference between vessel motion and wave elevation around the resonant frequency exhibiting some complex behaviour.

Now a major proportion of the damping of slender space frame floating structures arises from drag forces. Inclusion of this drag force damping yields an equation of motion of the form:

$$(m + A)\ddot{z} + B_v |\dot{z}|\dot{z} + C z = F(t) \quad (2.280)$$

One approach that permits the solution of Equation 2.280 is to use a linearized equivalent damping coefficient B that is selected so that the non-linear and equivalent linear damping coefficients dissipate the same energy at resonance. For a linear damping term of the form given in Equation 2.259, the energy dissipated at resonance can be written as:

$$W_L = 2 \int_{-R}^{R} B(R\,\omega_o \sin \omega_o t)\,dz \quad (2.281)$$

where $z = R \sin(\omega_o t - 90°)$ is the resonant motion at frequency ω_o. Writing the integral in terms of time gives:

$$W_1 = 2 \int_0^{\pi} B\,\omega_o\,R^2 \sin^2 \omega_o t\,d(\omega_o t)$$

$$= \pi B \omega_o R^2 \quad (2.282)$$

For non-linear drag induced damping, the work done becomes:

$$W_D = 2 \int_{-R}^{R} B_v (R^2 \omega_0^2 \sin^2 \omega_0 t)\, dz$$

$$= 2 \int_0^\pi B_v \omega_0^2 R^3 \sin^3 \omega_0 t\, d(\omega_0 t) \qquad (2.283)$$

$$= \frac{8}{3} B_v \omega_0^2 R^3$$

Equating the work done by linear and non-linear damping yields:

$$B = \frac{8}{3\pi}\left[\omega_0 R\right] B_v$$

$$= \frac{8}{3\pi}\left[\dot{z}_{max}\right] B_v \qquad (2.284)$$

Equation 2.280 can be solved by replacing $B_v |\dot{z}|$ by B from Equation 2.284. The equation has, therefore, to be solved by iteration in that a value of $\dot{z}_{max}$ has to be determined which satisfies the equation. The first solution for the iteration is usually found by assuming that the damping is linear and 10% of critical. The new value of $\dot{z}_{max}$ obtained from this initial solution is then placed in the damping term and the procedure repeated until successive values of $\dot{z}_{max}$ are within a small value of each other. This iteration converges very quickly away from resonance and within approximately ten solutions at resonance. The resultant solution for z will now also be a function of the amplitude of the applied force F_a and of wave amplitude a.

3. *Pitch motion of an articulated column*: the procedure used in the two preceeding sections can also be applied to an articulated column which is freely pinned at a fixed point below the sea surface and remains upright in still water due to an excess buoyancy in the structure. *Figure 2.80* shows a column that may typically be used as a loading or mooring terminal for an oil tanker. The equation of motion of the slender column can be written as:

$$(I + I_a)\ddot{\phi} + B_v |\dot{\phi}|\dot{\phi} + C\phi = M(t) \qquad (2.285)$$

where I and I_a are moments of inertia of physical mass and of added mass where as B_v and C are the non-linear damping coefficient and angular stiffness. All of these parameters are defined relative to pitch angle ϕ of the column about a horizontal axis through O. $M(t)$ is also the wave induced pitching moment about this axis.

These coefficients are obtained by integrating along the column length as follows:

$$I_a = \rho C_m \pi r^2 \int_0^L z^2\, dz = \frac{1}{3}\rho C_m \pi r^2 L^3$$

$$B_v = \rho C_d r^2 \int_0^L |z| z^2\, dz = \frac{1}{4}\rho C_d r L^4$$

$$I = m\,k^2$$

$$C = \rho g \pi r^2 L L_B - F_W L_G \qquad (2.286)$$

where C_m and C_d are added mass and drag coefficients. The physical inertia can be written as the product of column mass m, and the radius of gyration k, squared whereas the angular stiffness C arises from the balance of moments induced by buoyancy and weight forces acting through the centres of buoyancy and gravity (B and G) of the column.

The wave induced pitching moment is obtained by summing the contributions of the wave down the stationary upright column by the equation:

$$M(t) = \rho(C_m + 1)\pi r^2 \int_0^L \dot{u}\, z'\, dz' + \rho C_d r$$

$$\int_0^L |u| u\, z'\, dz' \qquad (2.287)$$

where $u, \dot{u}$ are horizontal components of wave particle velocities and accelerations, due to a wave of amplitude a.

The horizontal wave velocities and accelerations can be taken as:

$$u = +a\omega\, e^{k(z'-L)} \sin(kx - \omega t)$$

and $\qquad (2.288)$

$$\dot{u} = -a\omega^2 e^{k(z'-L)} \cos(kx - \omega t)$$

for deep water linear waves using an axes system of Oxz' with origin at the articulated column pivot (see *Figure 2.80*). Substituting Equations 2.288 into 2.287 yields:

$$M(t) = \rho(C_m + 1)\pi r^2 a \omega^2 \left[-\frac{L}{k} + \frac{1}{k^2}(1 - e^{-kL})\right]\cos\omega t$$

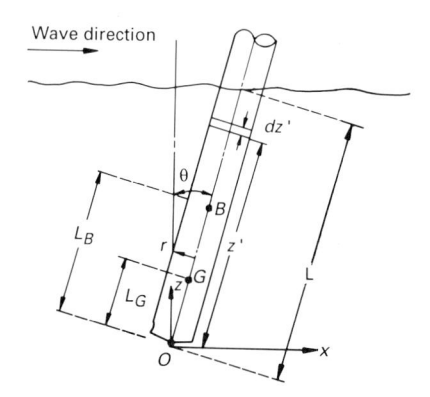

Figure 2.80 Articulated column

$$+ \frac{8}{3\pi} \rho \, C_d r a^2 \omega^2 \left[-\frac{L}{2k} + \frac{1}{4k^2} (1 - e^{-2kL}) \right] \sin \omega t \quad (2.289)$$

with application of the linearization procedure outlined earlier. Note that deep water linear wave theory is used here to simplify the algebra but, also restricts the validity of the solution to wave lengths that are less than twice the column length.

The above equations are valid for small ϕ and can be solved using the same procedure as in Section 2.5.3.1 together with linearization of the damping coefficient. *Figure 2.81* shows typical results for an articulated column of 812 t mass, 7 m (23 ft) diameter and a sea bed to water line length of 42 m (137 ft) with a moment of inertia of physical mass of 1.249×10^9 kg m^2 ($k = 41.88$ m (137.4 ft)) and L_G of 18 m (59 ft). The added mass and drag coefficients are taken to be 1.0 and 1.2, respectively. *Figure 2.81* presents the wave induced moment and motion response in terms of amplitude and phase using the equations:

$$M = M_o \sin(\omega t + \epsilon)$$
$$\phi = \phi_o \sin(\omega t + \epsilon + \phi) \quad (2.290)$$

The equations of motion for an articulated column can also be formulated for large angles and solved numerically using the time step integration routines described in Section 2.4.2. Such a solution is described in detail by Harrison (1985).

4. *Heave and surge motions of tensioned buoyant plat-*

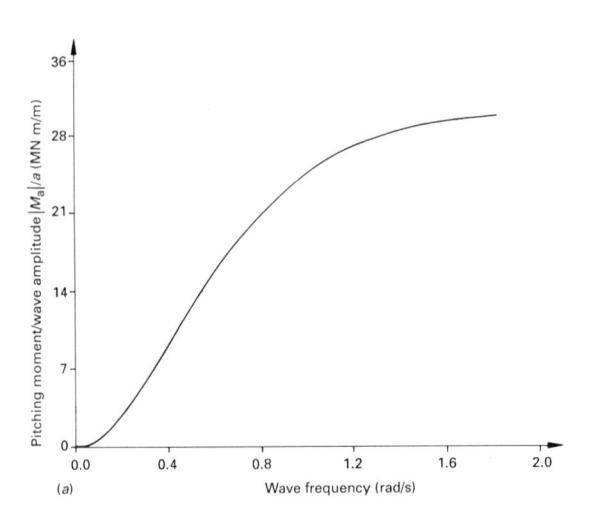

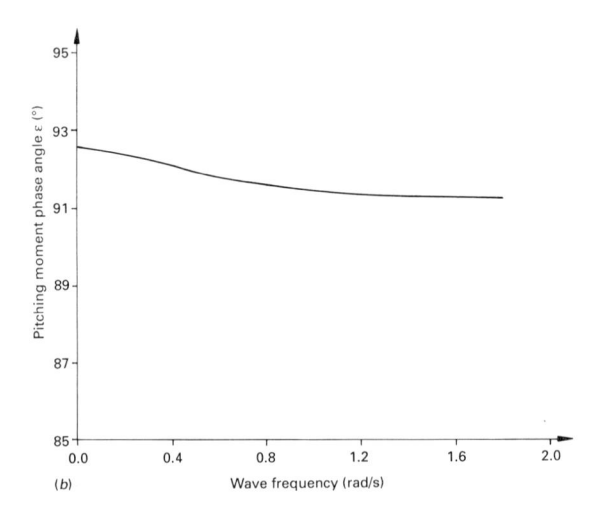

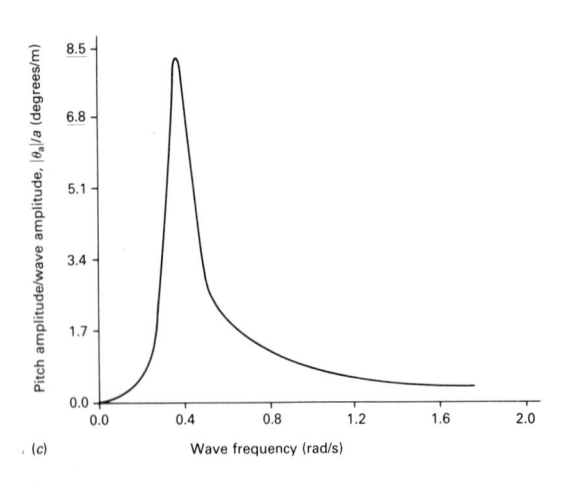

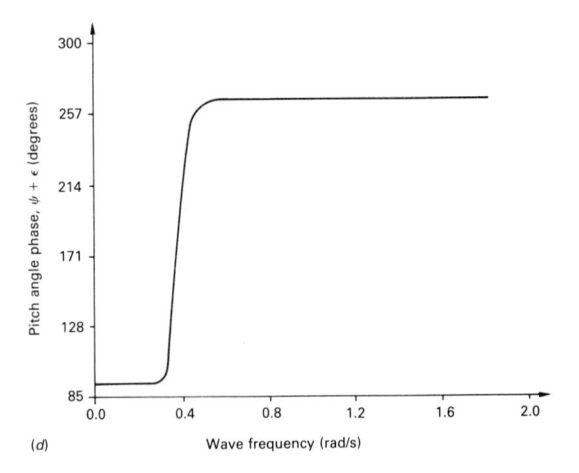

Figure 2.81 Response of an aritculated column to waves

forms: a tensioned buoyant platform (TBP) is a floating structure connected to a sea bed foundation by vertical mooring tethers which are kept in tension by excess buoyancy over weight in the platform structure. The use of TBPs as alternatives to oil production from fixed or catenary moored semisubmersibles or other floating structures is now well established. A TBP possesses a combination of desirable characteristics as an offshore work platform. The absence of a fixed rigid structure from sea surface to sea bed coupled with the excellent station keeping characteristics of tensioned vertical tethers satisfy the two important demands of economy and safe operation.

Additionally, a TBP exhibits very much smaller wave induced heave, roll and pitch motions due to the presence of the vertical mooring tethers. Large surge and sway offsets due to wind and current steady force components will, however, cause the platform to be drawn down lower in the water and exhibit a so called set down. Nevertheless, the wave frequency surge and sway motions of a TBP are desirable since horizontal wave loads are not fully reacted by the platform structure.

Conventional TBPs have structural features which are similar to those of semisubmersibles. These include the presence of deeply submerged main buoyancy chambers which give rise to a significant reduction in wave induced forces due to the exponential decay of wave particle velocities and accelerations with depth below the surface. The combination of horizontal pontoons and vertical columns supporting the working deck produces wave force cancellation at certain frequencies. This occurs because the upward wave force on vertical columns due to buoyancy changes can be almost exactly cancelled out by the downward wave force on horizontal pontoons. Additionally, the stiffness of the tethers and their axial tensions will radically influence the dynamic response of the platform to waves.

Methods of analysis for a TBP are very similar to those outlined earlier for a semisubmersible structure except for the insertion of a stiffness matrix induced by the presence of the relatively stiff platform tethers. Thus the equation of motion, the evaluation of the coefficient matrices and the solution of the equation are the same as for the semisubmersible considered previously with the tether stiffness matrix C_t included.

The governing equation for heave motion of a tensioned buoyant platform of the configuration shown in *Figure 2.82* is derived below. The platform is square in plan view with transverse pontoons (marked A in *Figure 2.82*) of length $2L$ and cross-sectional area A_h. The equation for heave motions is:

$$(m + A)\ddot{z} + (C + C_t)z = F_z(t)$$

where (2.291)

$$m = 4\rho\,[A_v d + 2 A_h L] - \frac{4T}{g}$$

with T being the tension in each of the tethers on the four corners of the platform. Also:

$$A = 8\rho\,C_{m\zeta_z} A_h L$$
$$C = 4\rho\,g\,A_v \qquad (2.292)$$

and

$$C_t = 4\lambda'$$

where C_{mz} is the added mass coefficient for the pontoon cross-section in vertical motion and λ' is the axial stiffness of each of the tethers. In this analysis, the drag damping coefficient and wave induced drag forces are neglected for simplicity. The wave force in heave is:

$$F_z(t) = \int_{-L}^{L} 2\rho\,A_h\,(1 + C_m)\,\dot{w}\,dx + 2 A_v\,(p_f + p_r)$$
$$+ 2\rho\,A_h\,L\,(1 + C_m)\,[\dot{w}_f + \dot{w}_r] \qquad (2.293)$$

where $\dot{w}_f$ and $\dot{w}_r$ are the vertical wave particle accelerations at $z = -d$ and $x = +L$ and $-L$, respectively, with the remaining notation from previous algebra in this section. Then using Equation 2.277, we get:

$$F_z(t) = 4a\,\rho\,g\,e^{-kd}\,[-A_v \cos kL + A_h\,(1 + C_m)\sin kL$$
$$+ A_h \cdot kL\,(1 + C_m)\cos kL]\sin \omega t \qquad (2.294)$$

These equations have most of the characteristics of the governing equation for semisubmersibles except for the presence of stiffness term, C_t, which is large and tends to reduce substantially the heave natural periods of such platforms in the range 3–5 s from the usual semisubmersible heave resonance period of 18–25 s. Using the definition of Equation 2.279, *Figure 2.83* presents variations of the wave force and motion

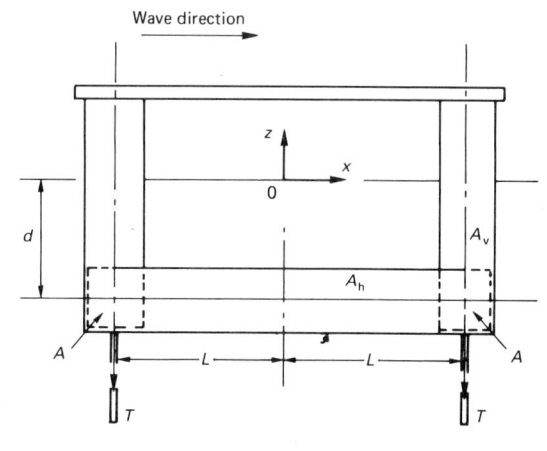

Figure 2.82 Side view of four column tensioned buoyant platform; A are transverse pontoons of length $2L$ and area of cross-section A_h

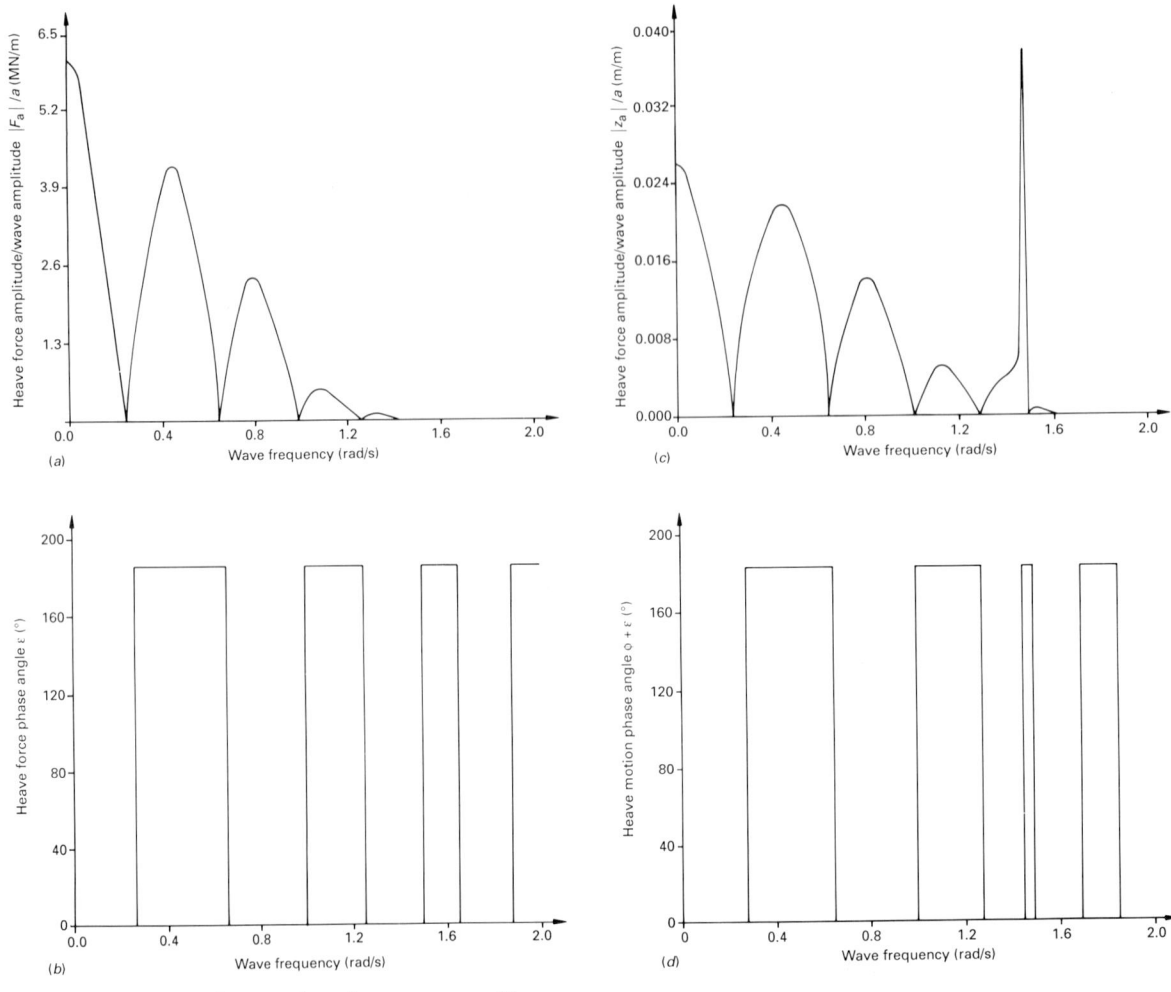

Figure 2.83 Heave force and motion response of four column tensioned buoyant platform

transfer functions with frequency for a TBP with $d = 35$ m (115 ft), $L = 50$ m (164 ft), $A_h = 100$ m^2 (1076 ft^2) and $A_v = 150$ m^2 (1615 ft). The total tether pretension and stiffness are taken to be 3.75 MN and 56.53 MN/m, respectively. The added mass coefficient for vertical motion of the rectangular pontoons is taken to be 1.2. The wave force is slightly modified due to the additional transverse pontoons but the motion response exhibits a substantial reduction in magnitude due to the shift in resonant frequency.

A similar analysis to that above can be carried out for surge motion of a TBP with a governing equation of the form:

$$(m + A') \ddot{x} + (C' + C'_t) x = F_x(t) \qquad (2.295)$$

The additional terms in the above equation are:

$$A' = 4\rho\, C_{mx}\, A_h\, L + 4\rho\, A_v d \qquad (2.296)$$

$$C' = 0 \qquad C'_t = \frac{4T}{s}$$

where C_{mx} and 1.0 are the added mass coefficients for horizontal flow normal to the pontoons and vertical columns, respectively. The hydrostatic stiffness term C' is zero and the tether stiffness is expressed in terms of tension and tether lengths. An expression for $F_x(t)$ can be derived in the same way as for heave motion to be:

$$F_x(t) = \int_{-d}^{0} 4\rho\, A_v\, [\dot{u}_f + \dot{u}_r]\, dz$$

$$+ \rho\, A_h\, L\, (1 + C_{mx})\, [\dot{u}_f + \dot{u}_r]_{z=-d} \qquad (2.297)$$

$$- 2 A_\mathrm{h} [p_\mathrm{f} - p_\mathrm{r}]_{z=-d}$$

where $\dot{u}_\mathrm{f}$, p_f and $\dot{u}_\mathrm{r}$, p_r denote horizontal wave particle accelerations and wave pressures at $x = +L$ and $x = -L$, respectively, and are given by:

$$\dot{u} = -a\,\omega^2\,e^{+kz}\cos(kx - \omega t) \quad (2.298)$$

and Equation 2.277 with C_{mx} and 1.0 taken as the added mass coefficients for horizontal flow normal to the pontoons and vertical columns, respectively. Then, the wave force $F_x(t)$ can be written as:

$$F_x(t) = -\left\{\frac{8\rho a\omega^2 A_\mathrm{v}}{k}[1 - e^{-kd}]\right.$$
$$\left. + 2\rho a\omega^2 A_\mathrm{h} L(1 + C_{mx}))e^{-kd}\right\}\cos kL \cos \omega t$$

$$- 4\rho g a\,A_\mathrm{h}\,e^{-kd}\sin kL \cos \omega t \quad (2.299)$$

Note that the only force cancellations occurring in Equation 2.299 are due to zeroes of the combined cos kL and sin kL terms.

Figure 2.84 presents the surge force and motion response transfer function calculated from these equations. The tether length s is taken to be 450 m (1476 ft) and $C_{mx} = 1.2$ with all other parameters being as in the calculation of heave response described above. Again the effect of tether stiffness dominates the motion transfer function with a low natural period and relatively large surge motions at predominant wave periods.

2.5.4 Effects of moorings

As illustrated in Section 2.5.3.2, mooring systems modify the governing equations of a floating platform by introducing stiffness terms which alter the natural pe-

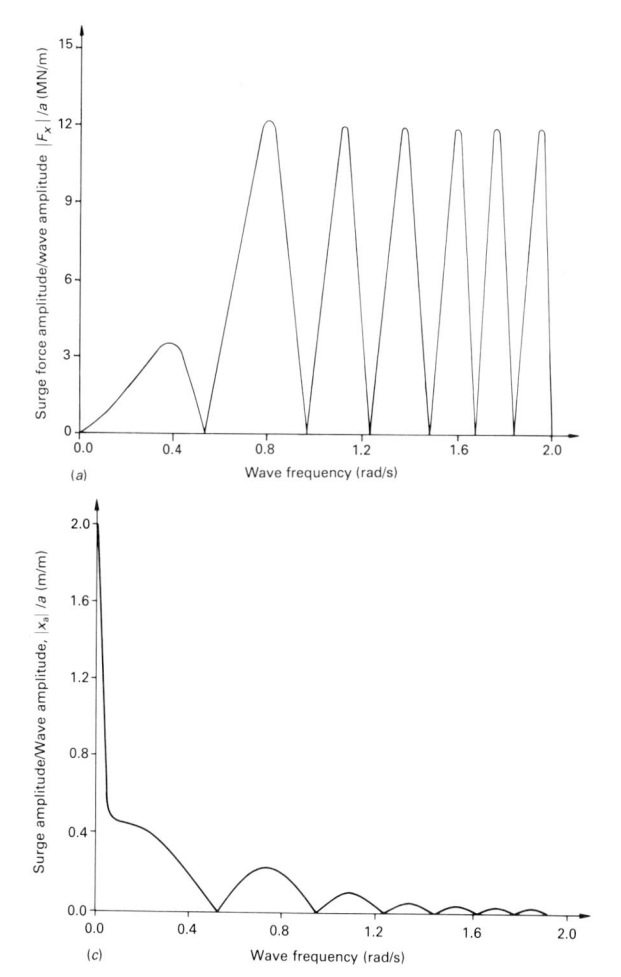

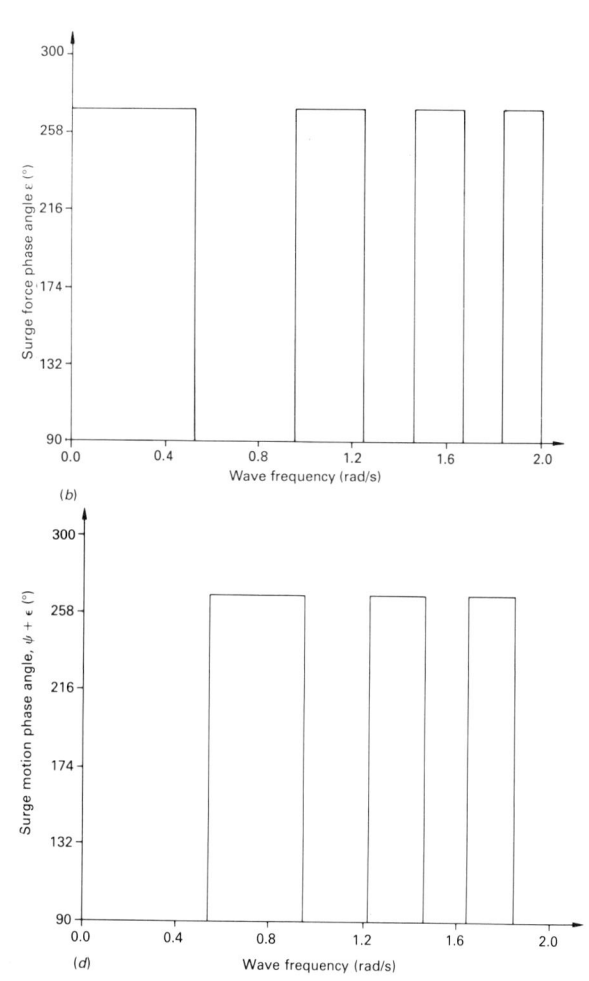

Figure 2.84 Surge force and motion response of four column tensioned buoyant platform

riods of the combined vessel and moorings to a lesser or greater extent. There are two kinds of mooring systems in common use for floating offshore structures: catenary moorings and tensioned vertical tethers. The mechanics of both of these are described in this section.

2.5.4.1 Catenary moorings

A catenary mooring system consists of freely hanging wire or chain lines connecting a surface platform to anchors on the sea bed some distance out from the platform. The mooring lines laid out, usually symmetrically in plan view, around the surface platform with a significant length laying on the sea bed and terminated by anchors. Each of the lines relies on an increase or decrease in line tension as it lifts off or settles on the sea bed to generate a restoring force on the surface platform. A spread of catenary lines thus generates a non-linear restoring force that increases with vessel horizontal offset and resists steady environmental forces.

The equivalent spring stiffness imposed on horizontal vessel motions is generally too small to significantly influence wave frequency motions although excitation by low frequency drift forces can induce dynamic magnification in platform horizontal motions and lead to high peak line tensions. The longitudinal and transverse motion dynamics of the mooring lines themselves can influence their behaviour particularly for long lengths in deep water. The role of thrusters within a thruster assisted mooring system also needs to be considered during analysis and design.

The basic theory underlying the behaviour of catenary lines is well known (O'Brien and Francis 1964). It is, however, instructive to review the basic mechanics of a catenary and to derive its governing equations.

Figure 2.85 shows a catenary mooring line deployed from points A on the submerged hull of a floating vessel to an anchor B on the sea bed. Note that part of the line between A and B is laying on the sea bed and that the horizontal dimension a is usually 10–30 times larger than the vertical dimension b. As the line mounting point on the vessel is shifted horizontally from point A_1, through A_2 A A_3 to A_4, the catenary line laying on the sea bed varies from a significant length at A_1, to none at A_4. The cable tension in the vicinity of points A is due to the total weight in sea water of the suspended line length. The progressive effect of line lift from the sea bed due to horizontal vessel movement from A_1 to A_4, increases line tension in the vicinity of points A. This feature coupled with the simultaneous decrease in line angle to the horizontal cause the horizontal restoring force on the vessel to increase with vessel offset in a non-linear manner.

This behaviour can be described by the catenary equations which are derived below. *Figure 2.86(b)* shows an element A of a line of weight per unit length w hanging freely as shown in *Figure 2.86(a)*. The element weight of $w \cdot \delta s$ is supported by a variation in tension along the line. A diferential equation for this variation is obtained by taking vertical and horizontal equilibrium of the forces acting on the line element of *Figure 2.86(b)*. Thus:

$$w\, \delta s + \frac{d}{ds}(T \sin \theta)\, \delta s = 0 \qquad (2.300)$$

$$\frac{d}{ds}(T \cos \theta) = 0 \qquad (2.301)$$

Taking $\cos \theta = dx/ds$ and $\sin \theta = dy/ds$, integration of Equation 2.301 gives:

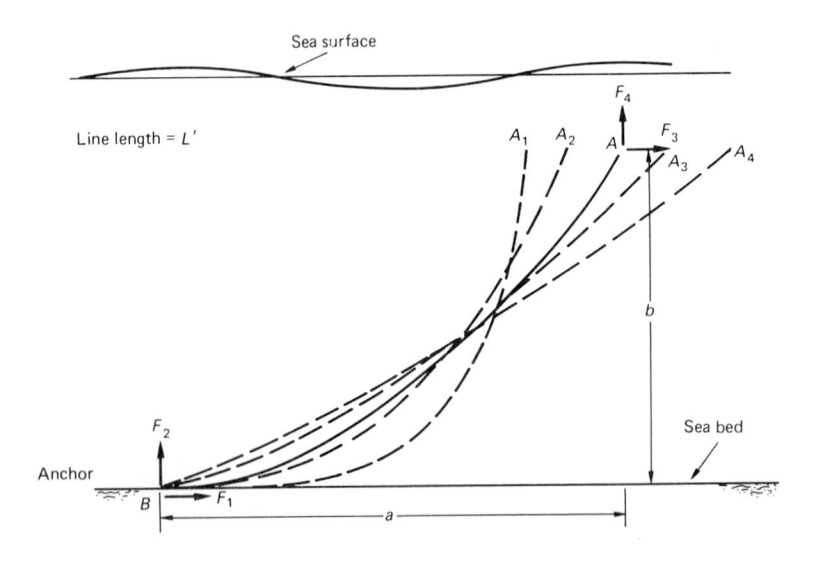

Figure 2.85 Catenary mooring line

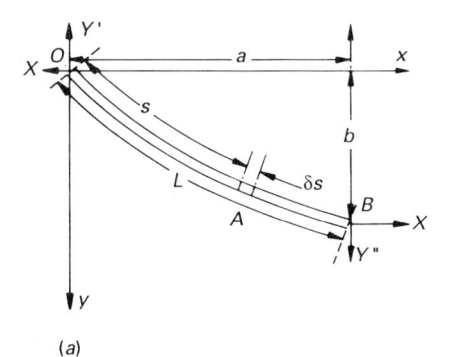

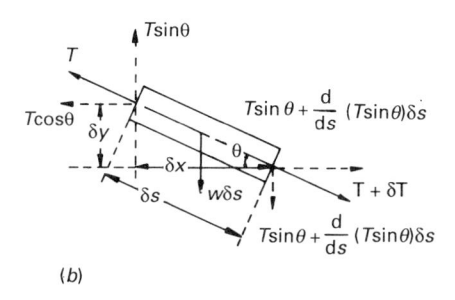

Figure 2.86 Definition diagram for a catenary

$$T\frac{dx}{ds} = X \quad (2.302)$$

where X is a horizontal force component in the line due to the fact that no external horizontal forces act on the line in still water. Substituting Equation 2.302 into Equation 2.300 then yields an equation of the form:

$$X\frac{d^2y}{dx^2} + w\frac{ds}{dx} = 0 \quad (2.303)$$

which can also be written in terms of x and y only as:

$$X\frac{d^2y}{dx^2} + w\left\{1 + \left(\frac{dy}{dx}\right)^2\right\}^{1/2} = 0 \quad (2.304)$$

These are both different forms of the differential equation for a simple catenary. The equations may be readily integrated to give, first:

$$\frac{dy}{dx} = -\sinh\left(\frac{wx}{X} - \phi\right) \quad (2.305)$$

where the constant of integration ϕ is given by using boundary conditions which constraint the line to go through points $x = 0$, $y = 0$ and $x = a$, $y = b$. With this condition:

$$\phi = \sinh^{-1}\left[\frac{\lambda\,(b/a)}{\sinh\lambda}\right] + \lambda$$

and

$$\lambda = \frac{wa}{2X} \quad (2.306)$$

Integrating Equation 2.305 again gives:

$$y = \frac{X}{w}\left\{\cosh\phi - \cosh\left(\frac{2\lambda X}{a} - \phi\right)\right\} \quad (2.307)$$

A number of worthwhile results can be derived from these equations. The length L, of segment OB can be obtained as:

$$L = \int_0^a \frac{ds}{dx}dx$$

$$= \int_0^a \cosh\left(\frac{wx}{X} - \phi\right)dx \quad (2.308)$$

$$= \frac{2X}{w}\sinh\lambda\cosh(\phi - \lambda)$$

Substituting $x = a$ and $y = b$ in Equation 2.307 and simplifying also gives:

$$b = \frac{2X}{w}\sinh\lambda\sinh(\phi - \lambda) \quad (2.309)$$

Squaring Equations 2.308 and 2.309 and subtracting yields the relationship:

$$L^2 = b^2 + \frac{a^2\sinh^2\lambda}{\lambda^2} \quad (2.310)$$

which is useful for determining L from information on variables a, b, w and x. The vertical components at each end of the line segment are obtained as follows. First, at the left-hand end:

$$y' = X\left(\frac{dy}{dx}\right)_{x=0} = X\sinh\phi \quad (2.311)$$

Now from Equation 2.306:

$$\sinh\phi = \sinh\left[\sinh^{-1}\left\{\frac{\lambda\,(b/a)}{\sinh\lambda}\right\} + \lambda\right]$$

$$= \frac{\lambda b}{a}\coth\lambda + M\sinh\lambda \quad (2.312)$$

where

$$M\cosh\left[\sinh^{-1}\left\{\frac{\lambda\,(b/a)}{\sinh\lambda}\right\}\right]$$

Taking hyperbolic cosine of Equation 2.306 gives:

$$\cosh \phi = M \cosh \lambda + \frac{\lambda b}{a} \quad (2.313)$$

Squaring Equations 2.312 and 2.315 and subtracting gives:

$$M^2 = 1 + \frac{\lambda^2 b^2}{a^2}(\coth^2 \lambda - 1) = \frac{L^2 \lambda^2}{a^2 \sinh^2 \lambda}$$

after using Equation 2.310.

$$\therefore M = \frac{wL}{2X \sinh \lambda} \quad (2.314)$$

and substituting into Equations 2.312 and 2.311 yields:

$$Y' = \frac{w}{2}[b \coth \lambda + L] \quad (2.315)$$

Also the vertical component of tension at the right-hand end is given by:

$$Y'' = X\left[\frac{dy}{dx}\right]_{x=a}$$

$$= \frac{w}{2}[b \coth \lambda - L] \quad (2.316)$$

after simplifying using Equation 2.308. Equations 2.315 and 2.316 yield the expected result that:

$$Y' - Y'' = wL \quad (2.317)$$

Similarly, the tensions at each end of the line segment can be written as:

$$T' = X\left(\frac{ds}{dx}\right)_{x=0}$$

$$= X \cosh \phi \quad (2.318)$$

$$= \frac{w}{2}[L \coth \lambda + b]$$

and

$$T'' = X\left(\frac{ds}{dx}\right)_{x=a}$$

$$= X \cosh(2\lambda - \phi) \quad (2.319)$$

$$= \frac{w}{2}[L \coth \lambda - b]$$

Equations 2.315, 2.316, 2.318 and 2.319 can be combined to yield the simple result:

$$\frac{Y' + Y''}{T' + T''} = \frac{b}{L} \quad (2.320)$$

That is, the ratio of average vertical force component to average tension at the line ends equals the ratio of the vertical projection of the line to total line length.

The above equations can be used to derive line tensions and shape for any single line of a mooring pattern. However, a typical mooring analysis requires summation of the effects of up to 12 or more mooring lines with the coordinates of the surface vessel position on the water plane introducing two further variables. The complexity of this calculation makes it suitable for implementing it within a computer program.

However, the catenary equations set out above are not entirely suitable for the basis of an efficient algorithm. The mathematical developments presented by O'Brien and Francis (1964) and Jennings (1965) give rise to an alternative set of equations describing the free hanging catenary which are much more suitable for automated calculations. Jennings (1965) has also extended the equations to apply for elastic lines and to account for temperature expansion effects.

Peyrot (1980) has converted these mathematical details into a computer algorithm to derive the end forces and tension distribution in a catenary from the knowledge of its end coordinates, line elasticity and unstretched length. The equations underlying this procedure are repeated here for completeness but utilize slightly different variables which are defined by Figure 2.87. Each cable element is described by the catenary equations:

$$L^2 = b^2 + \frac{a^2 \sinh^2 \lambda}{\lambda^2}$$

$$Y' = \frac{w}{2}\left(-b\frac{\cosh \lambda}{\sinh \lambda} + L\right) \quad (2.321)$$

where

$$\lambda = \frac{wa}{2X}$$

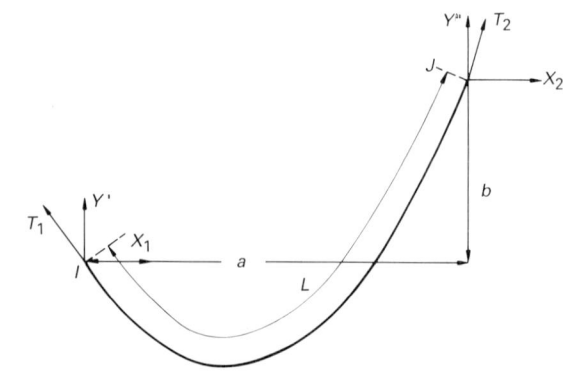

Figure 2.87 Mooring line element variables

and w is the cable weight per unit length in water and $X = |X_1| = |X_2|$ is the horizontal component of cable tension. Cable forces and tensions are related by the following equations:

$$Y'' = -Y' + w L_u \qquad X_2 = -X_1,$$
$$T_1 = \sqrt{(X_1^2 + Y'^2)} \quad \text{and} \quad T_J = \sqrt{(X_2^2 + Y''^2)} \tag{2.322}$$

where L_u is the unstressed cable length. By integrating along the cable length, the following relations are found:

$$a = -X_1 \left(\frac{L_u}{EA} + \frac{1}{w} \ln \frac{Y'' + T_J}{T_1 - Y'} \right)$$

$$b = \frac{1}{2EAW} (T_J^2 - T_1^2) + \frac{T_J - T_1}{w} \tag{2.323}$$

$$L = L_u + \frac{1}{2EAW} \left(Y_4'' T_J + Y' T_1 + X_1^2 \ln \frac{Y'' + T_J}{T_1 - Y'} \right)$$

where A is the area of cross-section of the line and E is its material modulus of elasticity.

As can be seen from this set of equations, a, b and L can be written as functions of X_1 and Y' alone. An iteration procedure is then set up in a program subroutine with the starting values of X_1 and Y' found from Equations 2.321 and 2.322 with only the first term in an expansion of $[\sinh^2 \lambda / \lambda^2]$ being used. Hence initial values of a and b are found, and linear corrections δa and δb (which depend on a and b) are then applied to X_1 and Y' until δa and δb are less than a small imposed tolerance level. Hence the forces X_1, Y' X_2 and Y'' and cable tensions are found for any catenary. The line coordinates can be determined by replacing L_u in Equation 2.323.

For mooring lines lying partially on the sea bed, the above analysis is modified using an iteration procedure so that additional increments of line are progressively laid on the bed until the suspended line is in equilibrium. Usually four or five iterations are sufficient to locate the final contact point. In many situations, multi-element lines made up of varying lengths and physical properties are used to increase the line restoring force. Such lines may be analysed by the algorithm above, where the analysis is performed on each cable element, and the imbalance in force at connecting points between element is used to establish displacements through which these points must be moved to obtain equilibrium.

A typical mooring analysis is carried out by utilizing the above algorithm to calculate the forces exerted on the vessel due to each catenary line given the line end point coordinates on the surface vessel and sea bed together with line lengths and elasticities. These forces are then summed for all lines in the mooring spread to yield the resultant horizontal restoring force and vertical forces. The restoring force and tension in the most loaded line against vessel offset curve is then calculated by displacing the vessel through prescribed horizontal distances from its initial position.

The results of a typical analysis are presented in *Figure 2.88*. The steady component of environmental force is applied to the vertical axis of this figure to obtain the resultant steady component of vessel offset from the horizontal axis. The slope of the force curve at this offset gives an equivalent linear stiffness C_t, of the mooring system in the relevant direction for use in an equation of the form:

$$(m + A) \ddot{x} + B_v |\dot{x}|\dot{x} + C_t x = F_x(t) \tag{2.324}$$

where coordinate x refers to a horizontal degree of freedom and the hydrostatic stiffness is zero.

The wave amplitude calculated from this equation together with any offsets due to low frequency drift are then applied on the horizontal axis of *Figure 2.88* on the right-hand side of the static offset component so as to read off the maximum dynamic line tension.

It is also common practice to carry out time domain numerical solutions of a hybrid equation of motion representing the vessel, environmental and mooring system forces. The equation of motion is of the form:

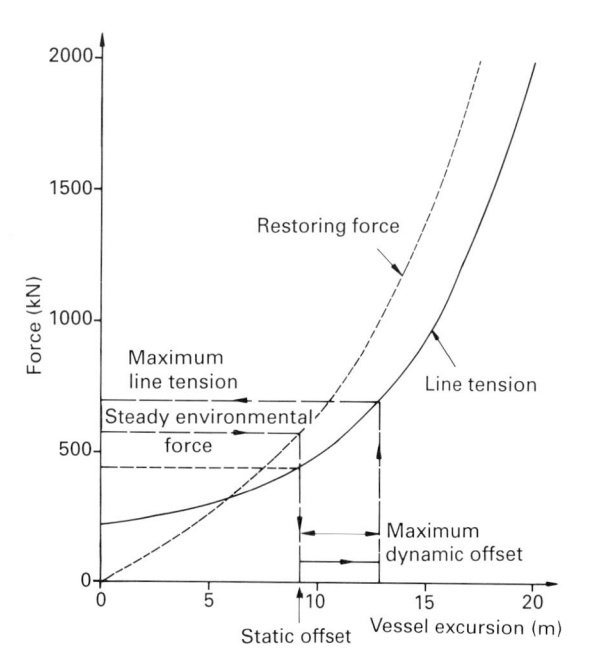

$$\ddot{M} X + B \dot{X} + B_v |\dot{X}|\dot{X} + K X = F_o(t) + F_W + F_c$$
$$+ T + F_m\{X(t)\} \tag{2.325}$$

Figure 2.88 Restoring force and most loaded line tension against vessel excursion for a typical catenary mooring system

in matrix notation for the horizontal degrees of freedom of a floating vessel described through column vector X. M, B_v and K are the total mass, drag damping and stiffness (excluding moorings) matrices of the vessel. B is a linear damping matrix contributed by damping due to waves radiating away from the vessel. $F_o(t)$ is the wave exciting force, F_W is the wind force, F_c is the current force, T is the thruster force including interaction effects, and F_m is the force exerted on the vessel by the moorings. The term $F_m\{X(t)\}$ is evaluated using the static mooring analysis calculation described above – hence this analysis is referred to as hybrid since the dynamics of the surface vessel are calculated but the mooring lines are treated quasi-statically. Equation 2.325 can be used to examine the dynamics of both intact and damaged (one line broken) mooring systems through the resultant offsets and line tensions obtained.

The wave exciting force term $F_o(t)$ in the above equation can include low frequency or second order wave drift forces. These second order wave forces arise when linearized or first order wave theory is extended to account for finite wave heights and used to calculate wave loading. Although the resultant second order wave forces are strictly only a correction to first order wave forces, these corrections have a significant effect on mooring system performance for two reasons. First, second order forces included a mean force component (wave drift force) which increases the total steady force to be reacted by the mooring system. Furthermore, second order wave forces are usually subharmonic or superharmonic in nature, that is, they have frequencies which are lower or higher than the frequencies of predominant wave action.

It is of particular interest when an irregular wave incident on a moored structure, induces second order wave forces with components at low frequencies. Due to the low stiffness in surge, sway and yaw of catenary mooring systems, their natural frequencies are also low and mooring system resonance in surge, sway and yaw can be induced by second order wave forces. The resultant large resonant motions are only limited by fluid induced damping forces and the peak mooring line tensions due to this phenomenon have to be considered in any comprehensive mooring analysis.

Mean and low frequency forces due to second order wave loading also exhibit the non-linear property that for increasing wave amplitudes, second order wave forces increase in proportion to the square of wave amplitude whereas first order wave forces increase linearly with wave amplitude. The earliest analysis of mean drift forces was made by Havelock (1940, 1942) for the mean drift force on a surface piercing body at high wave frequencies. Maruo (1960) offers an extended analysis to obtain the mean drift force on an arbitrary body from the mean rate of change of momentum in a control volume around the body.

More recently Faltinsen and Michelsen (1974) have provided a modern numerical analysis method that obtains the mean horizontal drift force and yawing moment on an arbitrary geometry using momentum conservation. Pinkster and van Oortmerssen (1976), on the other hand, compute drift forces by direct integration of fluid pressures. A systematic body surface integral equation formulation for second order forces has also been given by Standing et al. (1981) and others. Numerical solutions for low frequency drift forces for an arbitrary body in three dimensions have been given by Pinkster (1980) and by Standing et al. (1981).

However, low frequency drift forces calculated by using the above methods can only be satisfactorily applied in a design calculation through a time domain analysis which combines a drift force time history with the non-linear restoring force behaviour of a typical catenary mooring spread. The selection of floating platform added mass properties and of irregular wave spectra to be used for such design calculations pose additional problems.

One technique for integrating drift force calculations with a mooring analysis utilizes a slowly varying drift force time history generated by an inverse Fourier transform of a drift force spectrum derived from the methods outlined previously. Alternatively, more approximate methods given by Oppenheim and Wilson (1980) or by Clauss (1982) may be used. Yet another approach to the problem is given by Hsu and Blankarn (1970) and Remery and Hermans (1971) with a semi-analytical calculation of drift force time history directly from the wave elevation time history at the vessel position; this can itself be derived using an inverse Fourier transform of the wave elevation spectrum.

The solution of Equation 2.325 in the time domain with low frequency drift forces included is critically dependent on a good estimate of the damping force acting on the vessel at low frequencies. The level of damping force governs the maximum line tensions obtained and with the present state of the art, model tests are likely to yield the most accurate estimates of damping forces at low frequencies.

2.5.4.2 Tensioned moorings

The stiffness contribution of a taut mooring tether between a point on a surface platform and a sea bed foundation can be derived by idealizing the tether as weightless and perfectly elastic with a known constant tension and axial stiffness.

The stiffness contributions to all six rigid body degrees of freedom of the surface platform are derived for completeness with six elements of the restoring force vector F_m written as:

$$F_i = k_{ij} x_j$$

with k_{ij} being the (6×6) stiffness matrix with i and j denoting the row and column numbers, respectively. X_j is a (6×1) column vector where i and j denote the six rigid degrees of freedom of the surface platform in surge, sway, heave, roll, pitch and yaw, respectively.

Since the tether is assumed weightless, it is taken to lie along a straight line joining the two end coordinates. The resultant stiffness matrix is evaluated assuming that the surface platform moves through small displacements relative to the tether lengths involved.

Each tether has a tension T, an elastic stiffness λ', and is strung between coordinates (x_1, y_1, z_1) on the sea bed and (x_2, y_2, z_2) at the surface platform; both relative to the platform's principal axes system $Oxyz$ as shown in *Figure 2.89*. The direction cosines are defined as follows:

$$\cos \alpha = \frac{x_2 - x_1}{L} \quad \cos \beta = \frac{y_2 - y_1}{L} \quad \cos \gamma = \frac{z_2 - z_1}{L}$$

where the tether length is: (2.326)

$$L = \sqrt{(x_2 - x_1)^2 + (y_2 - y_1)^2 + (z_2 - z_1)^2}$$

Consider that the surface platform translates in the positive x direction through a small distance δx. Let $a = x_2 - x_1$, $b = y_2 - y_1$, and $c = z_2 - z_1$. Then to the first order, the new tether length becomes:

$$L + \delta L = L + \frac{a}{L} \delta x$$

Then the additional tension δT in the cable due to extension δx is:

$$\delta T = \lambda' \delta L = \frac{\lambda' (x_2 - x_1)}{L} \delta x \quad (2.327)$$

The resulting restoring force component along the x axis can be written as:

$$\delta T_x = (T + \delta T) \cos \alpha' - T \cos \alpha$$

where T is the cable tension and α' is given by:

$$\cos \alpha' = \frac{x_2 + \delta x - x_1}{L + \delta L}$$

$$\therefore \delta T_x = (T + \delta T) \left[\frac{x_2 - x_1}{L} \right] \frac{1 + \frac{\delta x}{x_2 - x_1}}{1 + \frac{\delta L}{L}}$$

$$- \frac{T(x_2 - x_1)}{L} = T \cos \alpha \left\{ \frac{\delta x}{x_2 - x_1} - \frac{\delta L}{L} \right.$$

$$\left. + \frac{\delta T}{T} \right\} \quad (2.329)$$

to the first order. Since $\delta L = \delta x \cos \alpha$:

$$\delta T_x = \frac{T}{L} \cos \alpha \left\{ \frac{L \delta x}{x_2 - x_1} - \cos \alpha \, \delta x + \frac{\lambda(x_2 - x_1)}{T} \delta x \right\}$$

$$= \lambda \cos^2 \alpha + \frac{T}{L} \sin^2 \alpha \, \delta x \quad (2.330)$$

In the limit as $\delta x \to 0$:

$$\frac{\delta T_x}{\delta x} = \frac{\partial T_x}{\partial x} = k_{11} = \lambda \cos^2 \alpha + \frac{T}{L} \quad (2.331)$$

which is the first term in the 6 × 6 stiffness matrix.

The remaining terms corresponding to the restoring forces due to translations can be derived in a similar fashion in terms of k_{ij} to be:

$$k_{21} = [\lambda + (T/L)] \cos \alpha \cos \beta = k_{12}$$
$$k_{31} = [\lambda + (T/L)] \cos \alpha \cos \gamma = k_{13}$$
$$k_{22} = \lambda \cos^2 \beta + (T/L) \sin^2 \beta \quad (2.332)$$
$$k_{32} = [\lambda + (T/L)] \cos \beta \cos \gamma = k_{23}$$
$$k_{33} = \lambda \cos^2 \gamma + (T/L) \sin^2 \gamma$$

The remaining 27 terms corresponding to restoring moments due to translations, restoring forces due to rotations and restoring moments due to rotations can

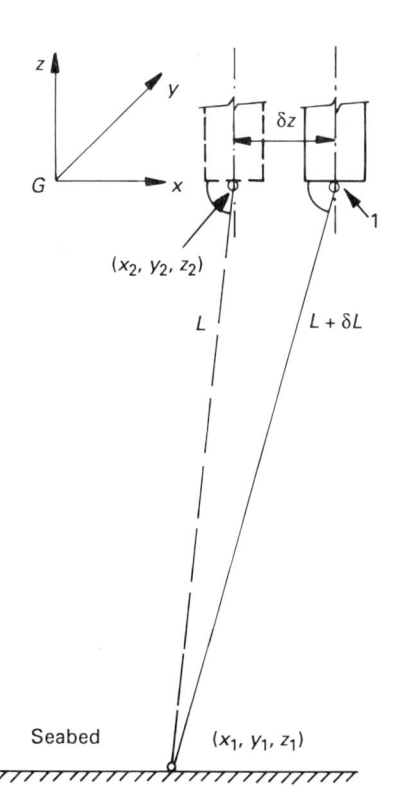

Figure 2.89 Notation for tether stiffness

all be expressed in terms of the first nine terms. These are given below for one-half of the symmetric matrix and including the leading diagonal.

$k_{41} = k_{31} y_2 - k_{21} z_2$

$k_{51} = k_{11} z_2 - k_{31} x_2$

$k_{61} = k_{21} x_2 - k_{11} y_2$

$k_{42} = k_{32} y_2 - k_{22} z_2$

$k_{52} = k_{21} z_2 - k_{32} x_2$

$k_{62} = k_{22} x_2 - k_{21} y_2$

$k_{43} = k_{33} y_2 - k_{32} z_2$

$k_{53} = k_{31} z_2 - k_{33} x_2$

$k_{63} = k_{32} x_2 - k_{31} y_2$

$k_{44} = k_{33} y_2^2 - 2 k_{32} y_2 z_2 + k_{22} z_2^2$

$k_{54} = k_{31} y_2 z_2 - k_{21} z_2^2 - k_{33} y_2 x_2 + k_{32} x_2 z_2$

$k_{64} = k_{32} x_2 y_2 - k_{22} x_2 z_2 - k_{31} y_2^2 + k_{21} y_2 z_2$

$k_{55} = k_{11} z_2^2 - 2 k_{31} x_2 z_2 + k_{33} x_2^2$

$k_{65} = k_{21} x_2 z_2 - k_{32} x_2^2 - k_{11} y_2 z_2 + k_{31} x_2 y_2$

$k_{66} = k_{22} x_2^2 - 2 k_{21} x_2 y_2 + k_{11} y_1^2$

The stiffness matrix is computed for each tether in a mooring system and summed to yield at total mooring stiffness matrix.

In some situations, such as for deep water or low displacement TBPs in shallow water, it may become necessary to calculate the dynamic interaction between the platform and its tethers with their distributed added mass and damping forces accounted for. In these cases, a mathematical model of tether vibrations needs to be coupled with the dynamics of the surface platform in order to derive more accurate estimates of platform motions, tether vibration amplitudes and tether bending moments. Patel and Lynch (1983) present an example of such work.

2.6 Analysis with multiple degrees of freedom

2.6.1 Basic equations

The general multiple degree of freedom equations of motions for a stationary floating body in gravity waves can be formulated using an approach similar to that used for single degrees of freedom in Section 2.5.3.2. It is usual to formulate these matrix equations in the six rigid body degrees of surge, sway, heave, roll, pitch and yaw described by a column vector X, although additional degrees of freedom (such as structural deformations) can be incorporated, if required.

Now the force on a general floating body of semisubmersible space frame configuration can be described by the equation:

$$M \ddot{X} = \Sigma M_A (\ddot{\eta} - \ddot{X}) + \Sigma M_{FK} \ddot{\eta} + \Sigma B_v |\dot{\eta} - \dot{X}|$$
$$(\dot{\eta} - \dot{X}) - K X - K_m X \qquad (2.334)$$

where M, M_A and M_{FK} are the (6 × 6) coefficient matrices quantifying structure physical mass, added mass and the Froude–Krylov added inertia, respectively, B_v is a (6 × 6) matrix representing the non-linear drag force contribution while K and K_m are (6 × 6) stiffness matrices contributed by the hydrostatic and mooring restoring forces acting on the vessel.

Wave particle velocities and accelerations are denoted by vectors $\dot{\eta}$ and $\ddot{\eta}$, respectively, with the summation signs in the wave force terms indicating a numerical integration over vessel submerged members to take account of spatial variations in wave properties. The contribution of linear wave radiation damping is neglected in this formulation. Furthermore, the Equation 2.334 is strictly valid only for offshore structures with slender members. Section 2.6.2.2 examines the governing equation for monolithic structures of larger cross-sectional dimensions.

Equation 3.325 can be rewritten to give:

$$(M + M_A)\ddot{X} + B_v |\dot{X}| \dot{X} + (K + K_m) X$$
$$= \Sigma(M_A + M_{FK})\ddot{\eta} + \Sigma B_v |\dot{\eta} - \dot{X}|(\dot{\eta} - \dot{X}) + B_v |\dot{X}| \dot{X} \qquad (2.335)$$

with a $B_v |\dot{X}| \dot{X}$ term added to both sides of the equation to obtain a conventional form for the left-hand side. Equation 2.335 would be difficult to solve in its present form due to the presence of $\dot{X}$ in the wave force on the right-hand side of the equation. However, two features of this wave force expression suggest a useful simplification. These are concerned with the relatively large magnitude of the inertia term in the wave force and the decay with depth of wave particle velocities which makes the $\Sigma B_v |\dot{\eta} - \dot{X}|(\dot{\eta} - \dot{X}) + B_v |\dot{X}| \dot{X}$ term small for members with significant submergence below still water level.

With these approximations in mind, Equation 2.335 can be simplified to:

$$(M + M_A)\ddot{X} + B_v |\dot{X}| \dot{X} + (K + K_m) X$$
$$= \Sigma(M_A + M_{FK})\ddot{\eta} + \Sigma B_v |\dot{\eta}| \dot{\eta} \qquad (2.336)$$

with the $\Sigma B_v |\dot{\eta} - \dot{X}|(\dot{\eta} - \dot{X}) + B_v |\dot{X}| \dot{X}$ term replaced by a $\Sigma B_v |\dot{\eta}| \dot{\eta}$ term. In order to explore the effects of this simplification, a residue wave force fraction R_i, for a particular motion (heave, say) can be defined as:

$$R_i = \frac{[\Sigma B_v |\dot{\eta} - \dot{X}_i|(\dot{\eta}_i - \dot{X}_i) + B_{vi} |\dot{X}_i| \dot{X}_i - \Sigma B_{vi} |\dot{\eta}_i| \dot{\eta}_i]}{(K_i H/2)} \qquad (2.337)$$

where the i refers to diagonal coefficients for heave motion. The residue force is non-dimensionalized with respect to the wave force amplitude at zero frequency by using the hydrostatic stiffness K_i in the mode of motion being considered and the wave height H.

Results presented later for semisubmersibles demonstrate that, after drag force linearization, the residue wave force that remains unaccounted for, due to the simplification, is negligible at wave frequencies even for large wave heights where the drag forces will be at their greatest. The drag force linearization used here is based on equal energy dissipation at resonance for the non-linear damping coefficients and their equivalent linear values. However, it should be noted that this linearization is a fundamentally different assumption from that implied in the reformulation of Equation 2.334 into Equation 2.336 which is quantified by residue factor R_i.

The formulation of the equation of motion for a monolithic ship shape hull form differs from that used for a semisubmersible since for ship shapes the diffraction of incident waves due to the large size of the hull form has to be accounted for. For these hull forms, the basic equation of motion can be written down simply as:

$$[M + M_A]\ddot{X} + B_p\dot{X} + [K + K_m] X = F\,e^{i\omega t} \quad (2.338)$$

where M, M_A, K and K_m are defined above, B_p is the matrix of potential damping coefficients and $F\,e^{i\omega t}$ is a complex vector of wave forces obtained by summing hull surface pressures from a diffraction analysis.

The above basic equations of motion are solved using different approaches for slender space framed structures and for large hull forms.

2.6.2 Coefficient matrices and wave forces

2.6.2.1 Morison Equation approach

This is applicable to slender space framed structures with characteristic cross-sectional dimensions that are less than 20% of incident wave lengths. The following additional assumptions are implied in the formulation of the equation of motion Equation 2.336 above:

1. The vessel structure is assumed to be an assembly of cylindrical elements and horizontal rectangular pontoons. The cylindrical members and rectangular pontoons are assumed to have large ratios of length to cross-sectional dimensions.
2. The motion amplitudes of the platform and wave motions are assumed to be small. As a consequence the Linear Wave Theory is used.
3. Wave forces on individual elements of the structure are computed as though other members were not present, that is, hydrodynamic interference between members is ignored.
4. The forces associated with sinusoidal wave motions are computed independently of the forces associated with absolute motions of the structure.
5. The non-linear drag damping term is linearized by assuming an equivalent linear damping which would dissipate the same energy at resonance as the non-linear damping. The contribution of wave radiation effects to the damping terms is assumed to be negligibly small.

The floating vessel is described by orthogonal coordinate axes with origin at the centre of gravity, x pointing forwards, y to port and z vertically upwards. The physical mass matrix is computed using the mass distribution of the vessel and is diagonal if the reference aces are principal axes of the vessel. The added mass matrix for the structure is obtained using a summation of the added mass matrix for each individual cylinder. These equations can readily be modified to account for rectangular cylinders.

The fluid damping matrix B_v, for the structure is evaluated in a similar manner to the added mass matrix, as a sum of the contributions from each individual members. The derivation of the generalized damping matrix for an arbitrarily orientated circular cylinder can be derived with the assumption that only drag forces normal to the cylinder axis are significant. Unlike the added mass matrix, the non-linear velocity square proportionality generates an asymmetric damping matrix. The corresponding matrices for the rectangular cylinder and non-elongated body are evaluated as special cases of the circular cylinder.

Contributions to the hydrostatic stiffness matrix K, will only arise in the heave, roll and pitch degrees of freedom due to buoyancy forces in the water plane cutting members of the hull. If, for a member number n, A_{wn} denotes the water plane area (x_w, y_w) are the coordinates of the centroid of this water plane area, then the hydrostatic stiffness matrix elements k_{ij} can be written as the summations:

$$\begin{aligned}
k_{33} &= \rho g \Sigma A_{wn} \\
k_{43} &= \rho g \Sigma y_w A_{wn} \\
k_{53} &= -\rho g \Sigma y_w A_{wn} \\
k_{54} &= -\rho g \Sigma A_{wn} x_w y_w \\
k_{44} &= \rho g V (GM)_p \\
k_{55} &= \rho g V (GM)_R
\end{aligned} \quad (2.339)$$

where V is the vessel displacement by volume and $(GM)_R$, $(GM)_p$ are metacentric heights in roll and pitch, respectively, while all other stiffness terms are zero. A linearized mooring system stiffness matrix K_m, can be obtained from the calculations described in Section 2.5.4.1 and inserted in the equation of motion to obtain motion responses of the moored rather than the free vessel.

The wave force calculation is based on the reduction of the terms on the right-hand side of Equation (2.336) into an oscillating force column vector by summing the effects of wave pressures, particle velocities and accelerations on all structural members of the semisubmersible. Conventional linear wave theory equations are presented in a compact complex number form here which is readily amenable for inclusion in an efficient computer program.

The wave surface elevation η, can be written as:

$$\eta = \eta_o\, \beta\, e^{-i\omega t} \qquad (2.340)$$

where

$$\beta = e^{i(kx\cos\alpha + ky\sin\alpha)}$$

where η_o is the wave amplitude and α is the wave direction measured clockwise positive from the x axis. The components of velocity and acceleration in the x, y and z directions are

$$u = \omega\beta v_2 \cos\alpha\, \eta_o e^{-i\omega t} \quad \dot{u} = -i\omega^2\beta v_2 \cos\alpha\, \eta_o e^{-i\omega t}$$
$$v = \omega\beta v_2 \sin\alpha\, \eta_o e^{-i\omega t} \quad \dot{v} = -i\omega^2\beta v_2 \sin\alpha\, \eta_o e^{-i\omega t}$$
$$w = -i\omega\beta v_1 \eta_o e^{-i\omega t} \quad \dot{w} = -\omega^2\beta v_1 \eta_o e^{-i\omega t} \qquad (2.341)$$

where

$$v_1 = \frac{\sinh k(d - s' + z)}{\sinh kd}, \quad v_2 = \frac{\cosh k(d - s' + z)}{\sinh kd}$$

where d is the water depth and s' is the z coordinate of still water level relative to reference axes at the centre of gravity. The wave dynamic pressure can be written as:

$$p = \rho g \beta v_2 \eta_o\, e^{-i\omega t} \qquad (2.342)$$

The three components of wave velocity and acceleration as well as wave elevation and wave pressure are complex quantities in these equations.

The above equations are used to compute wave velocities, accelerations and pressures efficiently at any point around the platform submerged members so that wave force calculations can be carried out. Normal forces on elongated cylindrical or rectangular members are computed by subdividing the members into elements, computing local wave properties and summing the consequential normal loading for all the elements. Axial forces due to wave dynamic pressures are also computed with due account being taken of surface piercing members where the dynamic pressure acts on one member end only.

The vessel response to an applied oscillating wave force vector is then computed by using an iterative technique to account for the non-linear drag damping force. A first approximation diagonal linear damping coefficient matrix is obtained by ignoring all non-diagonal terms in the total mass and stiffness matrices and assuming linear damping to be 10% of critical. The equations of motion are solved with this first approximation to the damping value. The column vector of these velocities is then substituted into the modulus sign in the damping term such that the equation:

$$(M + M_A)\ddot{X} + B_v \frac{8}{3\pi} |\dot{X}_{\text{last approx}}| \dot{X} + (K + K_m)X = F(t) \qquad (2.343)$$

is solved to obtain a better approximation for the column vector X. The equivalent damping matrix B_{eq}:

$$B_{eq} = B_v \frac{8}{3\pi} |\dot{X}| \qquad (2.344)$$

is a standard result obtained by assuming equal work done at resonance by the non-linear and the equivalent linear damping terms (see Section 2.5.3.2). The iteration is continued until a specific tolerance (such as 1%) between successive approximations is achieved.

It is of interest to explore the effects of the drag force non-linearity in Equation 2.336; both for the wave force excitation on the right-hand side and for the damping term on the left-hand side. *Figure 2.90–2.92* display the vertical wave forces and heave motion responses as functions of wave period for a typical twin pontooned semisubmersible at drilling draught. Vessel heave forces and motions are presented for two values of incident wave height and the data presented are non-dimensionalized with respect to the total heave wave force amplitude at zero frequency for heave forces and incident wave amplitude for heave motions.

Figure 2.90 separately displays the added mass and Fourde–Krylov force on the horizontal pontoons, the Froude–Krylov force on the lower ends of the vertical columns and drag force on the horizontal pontoons for 2 and 12 m (7 and 39 ft) wave heights. It needs to be pointed out that the vertical force on the horizontal

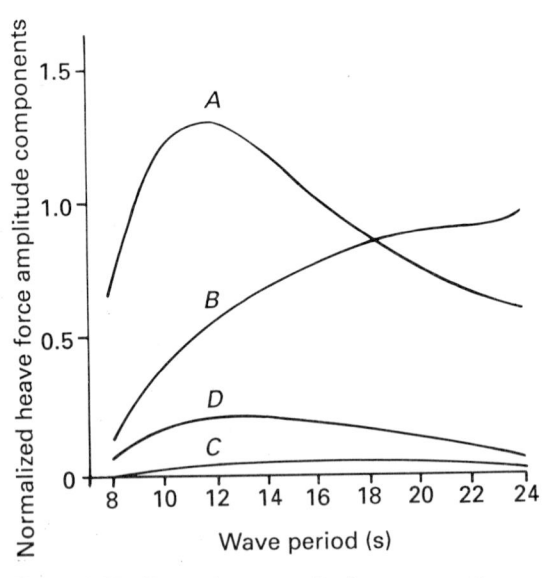

Figure 2.90 Heave force amplitude components as functions of wave period for 2 m and 12 m wave heights. Curve A denotes added mass and wave pressure force on pontoons; curve B denotes wave pressure force on vertical columns, for both 2 m and 12 m wave heights; curves C and D denote drag force for wave heights of 2 m 12 m respectively

pontoons and the vertical force on the columns are 180° out of phase so that a cancellation effect occurs at a wave period of 18.7 s for the total wave force. *Figure 2.90* illustrates how the drag force (curve C for 2 m (7 ft) wave height and curve D for 12 m (39 ft) wave height) is a relatively small fraction of the other force components on the structure.

Figure 2.91 displays the total vertical wave force for 2 and 12 m (7 and 39 ft) wave heights. The cancellation effect described above is not quite exact because of the small drag force contribution. In order to quantify the difference in wave force computed by the formulation of Equations 2.334 and 2.336 the unaccounted drag force is computed as a fractional residue term R_i as defined in Equation 2.337 and presented in *Figure 2.91*. It is apparent that the residue is negligible for wave periods up to 17 s and rises to a maximum of 0.15 for 12 m (39 ft) wave height and 20 s period. The small overall level of the residue R_i justifies the simplification made in Equation 2.336 from Equation 2.334.

The heave motion response is presented in *Figure 2.92* and shows a strong influence of the drag force non-linearity around resonance with the heave response amplitude per unit wave amplitude reducing from 4.88 m/m at 1 m (3 ft) wave amplitude to 1.26 m/m at 6 m (20 ft) wave amplitude. The vessel motion response away from resonance is not significantly affected although there is some increase in response around 16–19 s due to the corresponding increase in wave force amplitude at these periods. The large change in the unit heave response at and around resonance is to be expected since the damping force in a vibratory system is dominant at resonance.

The Morison's Equation formulation of wave force on slender space frame structure such as semisubmersibles requires the choice of applicable added mass and drag coefficients. It is common practice to use drag and added mass coefficients from recommendations made by classification societies. This procedure has been followed for these calculations with an added mass coefficient of 1.24 and a drag coefficient of 1.34 used for vertical motion of the twin horizontal rectangular pontoons.

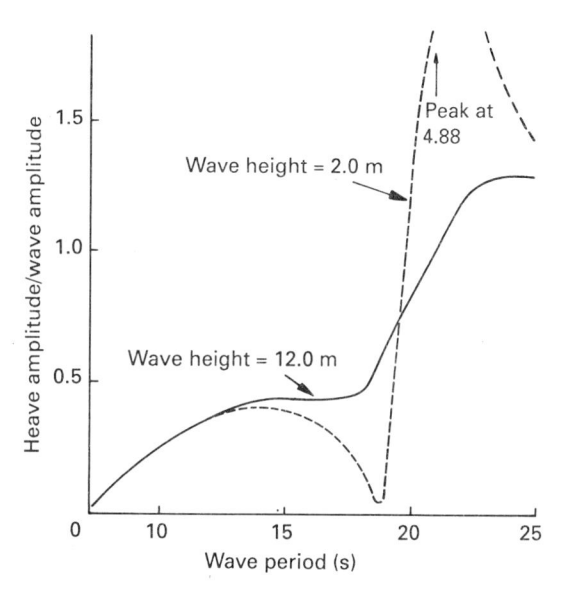

Figure 2.92 Heave motion response of semi-submersible vessel in head seas.

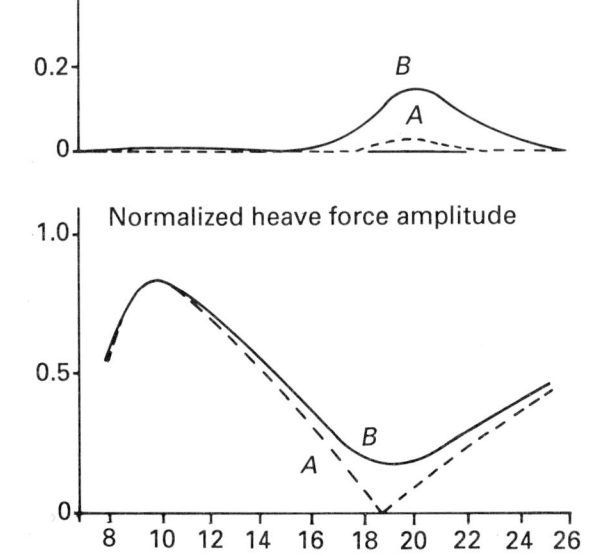

Figure 2.91 Total heave force amplitude and fractional residue of wave force as functions of wave period for 2 m (curve A) and 12 m (curve B) wave heights

In order to explore the sensitivity of the calculations to reasonable variations in these coefficients, the changes in the heave transfer function due to some of the combinations of the added mass coefficient 1.00–1.50 and of the drag coefficient 1.00–2.00 are presented in *Figure 2.93*. It is evident that the relatively large variation in the drag coefficient does not alter the heave response at wave frequencies to a significant extent. This is to be expected since variations in the drag force are small in comparison with inertia forces on the vessel. The small drag force amplitude does have some influence on the heave response amplitude at the cancellation frequency and at resonance. The difference of two large inertia related forces around the

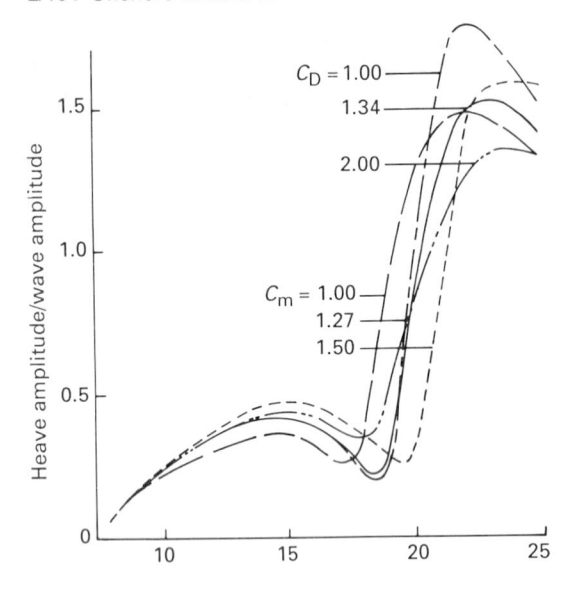

Figure 2.93 Sensitivity of heave motion response to drag and inertia coefficient variations

cancellation frequency highlights the effect of the drag coefficieitn changes. The alteration of response at resonance is due to the change in damping level induced through the drag coefficient.

It is also clear from *Figure 2.93* that the relatively large variation in pontoon added mass coefficients serves primarily to alter the heave natural period with consequential changes in the heave response at wave frequencies. These results underline the importance of selecting the added mass coefficients carefully; perhaps by reference to potential flow diffraction theory results for isolated body cross-sectional shapes.

Nevertheless, this Morison Equation calculation shows good agreement with both model test data and full scale measurements. Patel (1983) gives further information on the analyses and Lyons and Patel (1984) presents some comparison of calculations with test data.

2.6.2.2 Diffraction theory

Calculations of wave induced motions of a large non-space frame structure in gravity waves requires a solution of the wave problem with no flow boundary conditions at the moving body surface in addition to the free surface and sea bed boundary conditions. The solution can be split into two related problems: the scattering wave problem defines wave forces on a floating body when fixed in space and with waves incident on it in an identical manner to the technique for cmputing wave forces on a fixed body described in Section 2.3.8. The radiation wave problem is concerned with defining forces on the body (added mass and damping) due to its oscillation in otherwise still water. These oscillations will induce wave potentials such that the total wave potential in the fluid is the sum of the incident ϕ_w, scattered ϕ_s and forced wave potentials ϕ_f, so that:

$$\phi = \phi_w + \phi_s + \phi_f \qquad (2.345)$$

and these must satisfy the boundary conditions at the body surface given by:

$$\frac{\partial \phi_w}{\partial n} + \frac{\partial \phi_s}{\partial n} + \frac{\partial \phi_f}{\partial n} = V_n \qquad (2.346)$$

where V_n is the velocity of the body surface in the direction n normal to the surface. This boundary condition can be applied at the mean body surface since the theory is applied for small motions. ϕ, together with its three components must also satisfy the Laplace Equation and the free surface and sea bed boundary conditions. Furthermore, ϕ_s and ϕ_f must satisfy the radiation conditions.

Boundary conditions for the scattering and radiation wave problems can be split up from Equation 2.346 as:

$$\frac{\partial \phi_w}{\partial n} + \frac{\partial \phi_s}{\partial n} = 0$$

and (2.347)

$$\frac{\partial \phi_f}{\partial n} = V_n$$

respectively, both being applied on the body surface. The scattering problem is identical to the application of diffraction theory on fixed structures as described in Section 2.3.8. The radiation problem can also be solved by using either boundary integral or boundary element techniques. Only the solution using boundary integral techniques is described here for brevity. As in Section 2.3.8, the analysis assumes inviscid, irrotational flow and that wave amplitudes are small.

The unsteady flow around the floating vessel is calculated by introducing oscillating sources of unknown velocity potential on the vessel submerged surface that is discretized by a mesh of facets with an oscillating source on the surface of each facet.

A Green's Function is used to represent the velocity potential of each source which because of the form of the Green's Function, satisfies Laplace's Equation, zero flow at the horizontal sea bed, the free surface and radiation boundary conditions. Now, the solution for scattered wave potential due to the stationary floating body, subjected to incident waves of potential ϕ_w, is identical to that described in Section 2.3.8 for fixed structures. A set of linear simultaneous equations are obtained by equating the flow due to the local source plus the additional flow due to all other sources to the negative of the flow due to the undisturbed wave for each facet on the body surface. Solutions of these equations yields the unknown source strengths and, therefore, the velocity potential ϕ_s, which is used to derive pressures and wave forces by integration over the body surface. Thus, the wave force vector F, of Equation 2.338 may be obtained for an incident wave of specified frequency and direction.

The velocity potential ϕ_f is obtained in a similar way to that above except for the use of a different boundary condition which reflects the fact that ϕ_f arises from body motions in otherwise still water. Thus, at all facets, the source strengths ϕ_{fi}, are such that the flow due to the local source plus the flow due to all other sources equals the velocity component of the body along the facet normal. This velocity component will depend on the mode of motion (surge, sway, heave etc.) in which the body is moving. All of this can be represented by equating the normal velocity of the fluid and of the jth facet for the vessel moving in its kth mode of motion. This yields the equation:

$$\sum_{\text{All}\,i} \frac{\partial \phi_{fi}}{\partial n_j} \sigma_{ik} = V_{jk} \qquad (2.348)$$

where V_{jk} is the normal velocity of the jth facet with the vessel moving in its kth mode of motion. Furthermore, n_j is the normal to the jth facet, $\partial \phi_i/\partial n_j$ is the normal fluid velocity at the jth facet due to a unit source at the ith facet and σ_{ik} are the unknown source strengths required in the kth mode. Application of Equation 2.348 for all facets produces a system of complex equations to be solved for the source strengths. Once these are known, the pressures at the facets are evaluated and their effects integrated over the vessel surface to yield forces in each mode of motion to unit motion in the kth mode.

These forces may be written as a complex matrix, $G(\omega)$ which can be decomposed into its real and imaginary parts through the equation:

$$G(\omega) = \omega^2 M_A(\omega) - i\omega B_p(\omega) \qquad (2.349)$$

to yield frequency dependent added mass and damping matrices $M_A(\omega)$ and $B_p(\omega)$ which are required for Equation 2.329.

The inclusion of physical mass, hydrostatic and mooring stiffness matrices, M, K and K_m completes derivation of all of the coefficient matrices of Equation 2.338. The hydrodynamic coefficient matrices are, however, frequency dependent and require carrying out a diffraction analysis at all frequencies at which motions are required. Equation 2.338 is linear and can readily be solved to yield the displacement vector X. The exciting force vector $F(\omega)$ and the coefficient matrices $M_A(\omega)$ and $B_p(\omega)$ can also be derived using finite element methods in an analogous way to that described for the boundary integral approach described previously.

There is one further point of interest regarding the relationship between the scattered and forced wave potentials (ϕ_s and ϕ_f) for a floating vessel problem. The use of equations called Haskind Relations (see Newman 1982) enables the scattered wave potential ϕ_s, to be expressed in terms of the incident and forced wave potentiqls ϕ_w and ϕ_f. Thus, once ϕ_f is calculated, ϕ_s need not be computed by diffraction analysis but can instead be derived using the Hasking Relations.

Typical results of a boundary integral diffraction analysis for a ship shaped hull are shown in *Figure 2.95* and 2.96. The discretization of the submerged bull geometry is shown in *Figure 2.94* using 277 triangular facets on the ship half hull. The vessel is of 263.7 m (857 ft) overall length, 40.8 m (133 ft) beam and 145 937 t displacement with 14.80 m (48 ft) draught floating in deep water. *Figure 2.95* presents the variation of added mass and radiation damping coefficients with frequency for heave and pitch motions. Note that the variation in added mass is relatively small but the radiation damping shows large changes with very small values at some wave periods. Wave induced heave

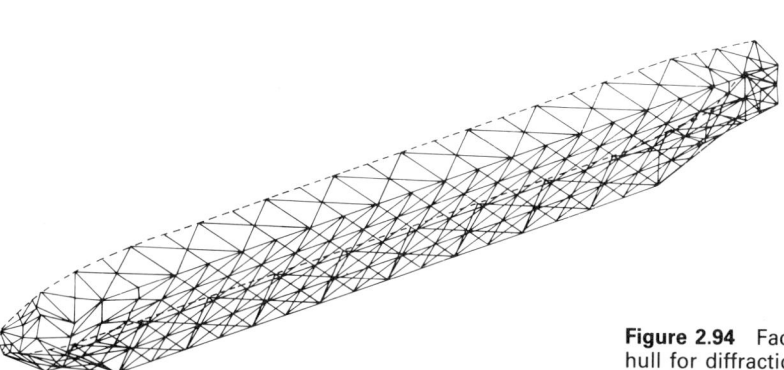

Figure 2.94 Facet discretization of a submerged ship hull for diffraction theory

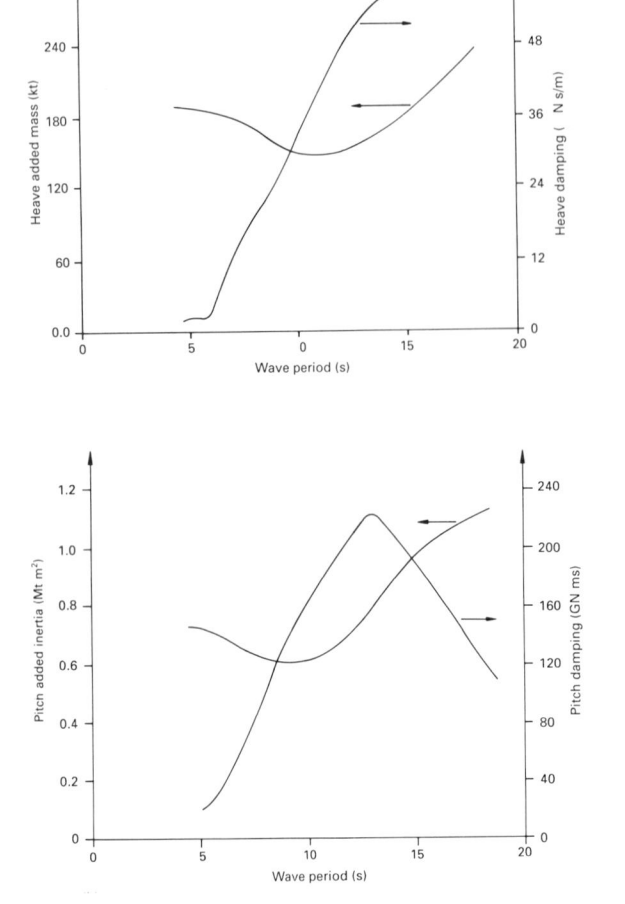

Figure 2.95 Added mass and damping coefficient variations with wave period for ship hull

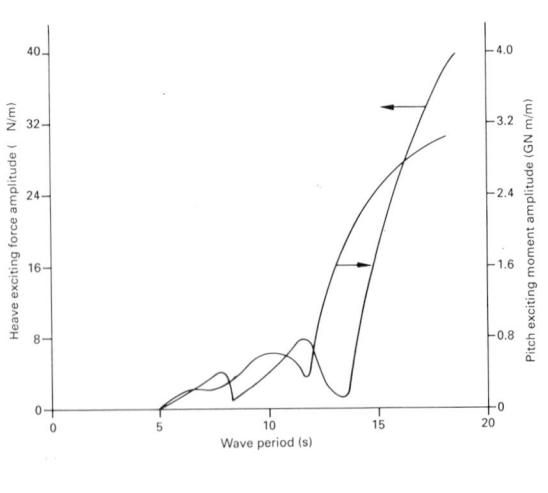

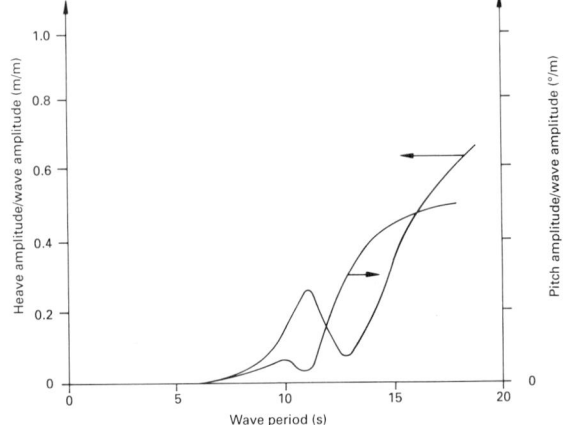

Figure 2.96 Variations of heave and pitch motion responses with wave period for a ship hull

force and pitching moments and the resultant motion responses for head seas are presented in *Figure 2.96*.

There is generally good agreement between model test data, full scale results and the results of diffraction analysis for wave induced motion response. The one exception is the roll degree of freedom near resonance where non-linear potential and viscous effects cause some discrepancy. Brown et al. (1983) presented a validation diffraction of theory predictions using model tests at two scale factors. Further work by Brown and Patel (1986) examines the reasons behind discrepancies between predictions and model tests for roll motions.

References and further reading

Airy, G. B. (1845) Tides and waves, *Encyc. Metrop.*, Art 192, 241–396

American Bureau of Shipping (1987) *Rules for building and classing mobile offshore drilling units*, ABS, Paramus, New Jersey, USA

American Petroleum Institute (1977) *Comparison of marine drilling riser analysis*, American Petroleum Institute, Bulletin 2 J, 1st edition, January 1977

American Petroleum Institute (1986b) *Basic petroleum databook*, **VI**, 3 (September 1986, API, Washington

American Petroleum Institute (1987) *Recommended practice*

for planning, designing and constructing fixed offshore platforms, Dallas, Texas, Rpt No API-RP-2A

Bathe, K. J. and Wilson, E. L. (1973) Solution methods for eigen-value problems in engineering, *International Journal for Numerical Methods in Engineering*, **6**, 213–215

Bell, K. Hansteen, O. E., Larsen, P. K. and Smith, E. K. (1976) Analysis of a wave-structural-soil system – case study of a gravity platform. In *Proceedings of the Conference on Behaviour of Offshore Structures*, **1**, BOSS 1976, 846–863

Blevins, R. D. (1977) *Flow-induced vibrations*, Van Nostrand Reinhold Company, New York

Brown, D. T., Eatock-Taylor, R. and Patel, M. H. (1983) Barge motions in random seas – a comparison of theory and experiment, *Journal of Fluid Mechanics*, (April 1983) **129**, 385–407

Brown, D. T. and Patel, M. H., (1986) On predictions of resonant roll motions for flat bottomed barges, *Transactions of the Royal Institution of Naval Architects*, **128**, 235–245

Clauss, G. F. (1982) Slow drift forces in a barge-type structure – comparing model tests with calculated results. In *Proceedings of the Offshore Technology Conference*, OTC 4435

Davenport, A. G. (1961) The application of statistical concepts to the wind loading of structures. In *Proceedings of the Institution of Civil Engineers* (August 1961) **19**, 449–472

Davenport, A. G. (1967) Gust loading factors, *Proceedings of the American Society of Civil Engineers (Structural Division)* **93**, ST3, June, 11–34

Department of Energy, (1986) *Offshore installations, guidance on design and construction*, Part II, Section 4.3, HMSO, London

Department of Energy (1986) *Development of the oil and gas resources of the United Kingdom*, Appendix 15, Department of Energy, HMSO, London

Det Norske Veritas (1982) *Rules for submarine pipeline systems*, Det Norske Veritas, Hovik, Norway

Det Norske Veritas, (1987) *Rules for classification of mobile offshore units*, Det Norske Veritas, Hovik, Norway

Dover, W. D. and Connolly, M. P. (1986) *Fatigue fracture mechanics assessment of tubular welded Y and K joints*, Paper No C141/86, Institution of Mechanical Engineers, London

Dover, W. D. and Wilson, T. J. (1986) *Corrosion fatigue of tubular welded T-joints*, Paper No C136/86, Institution of Mechanical Engineers, London

Draper, L. (1963) Derivation of a design wave from instrumental records of sea waves. In *Proceedings of the Institution of Civil Engineers*, London (October 1963), **26**, 291

Eatock-Taylor, R. and Waite, J. B. (1978) The dynamics of offshore structures evaluated by boundary integral techniques, *International Journal for Numerical Methods in Engineering*, **13**, 73–92

Eatock-Taylor, R. and Rajgopalan, A. (1982), Dynamics of offshore structures, *Journal of Sound and Vibration*, **83**, 401–431

Engineering Sciences Data Unit, (1986) *Mean forces, pressures and flow field velocities for circular cylindrical structures: single cylinder with two dimensional flow*, Engineering Sciences Data Unit, Item No. 80025 with Amendment A1 to C (June 1986), ESDU, London

Faltinsen, O. M. and Michelsen, F. C. (1974) Motions of large structures in waves at zero Froude number, *Proceedings of the International Symposium on Dynamics of marine structures in waves*, University College London, 99

Felippa, C. A. and Park, K. C. (1978) Computational aspects of time integration procedures in structural dynamics, *Journal of Applied Mechanics*, **45**, 596–611

Focht, J. A. and Kock, K. J. (1973) Rational analysis of the lateral performance of offshore pile groups. In *Proceedings of the Offshore Technology Conference*, OTC 1986

Furnes, O. & Loset, O. (1980), Shell structures in offshore platforms: design and application, *Engineering Structures*, **3** (July 1980), 140–152

Garrison, C. J. and Chow, P. Y. (1972) Wave forces on submerged bodies, *Journal of Waterways, Harbours and Coastal Division*, American Society of Civil Engineers, **98**, 375–392

Godeau, A. J. Deleuil, G. and Heas, J. Y. (1977) Statistical analysis of nonlinear dynamic response of fixed structures to random waves. In *Proceedings of the Offshore Technology Conference* OTC 3030

Grass, A. J., Simons, R. R. and Cavanagh, N. J. (1987) Vortex induced velocity enhancements in the wave flow field around cylinders. In *Proceedings of the Offshore Mechanics and Arctic Engineering Symposium*, American Society of Mechanical Engineers, **11**, 155–164

Hamilton, J. & Perrett, G. R. (1986) Deep water tension leg platform designs. In *Proceedings of the Royal Institution of Naval Architects International Symposium on Developments in Deeper Waters*, Paper 10

Harrison, J. H. (1985) *On the hydrodynamics of compliant semisubmersibles*, Ph.D. thesis Faculty of Engineering, Univerisity of London

Havelock, T. (1940a) The pressure of water waves upon a fixed obstacle. In *Proceedings of the Royal Society*, **175**, 409

Havelock, T. H. (1940b), The pressure of water waves upon a fixed obstacle. In *Proceedings of the Royal Society, London*, Series A, **963**, 175–190

Havelock, T. (1942) The drifting force on a ship among waves, *Philosophical Magazine*, **33**, 467–475

Hoerner, S. F. (1965) *Fluid dynamic drag*, Published by the author, New York

Hogben, N. and Lumb, F. E., (1967) *Ocean wave statistics*, HMSO, London

Hsu, F. H. and Blankarn, K. A. (1970) Analysis of peak mooring forces caused by slow drift oscillations in random seas. In *Proceedings of the Offshore Technology Conference*, OTC Paper 1159.

International petroleum encyclopaedia (1986) Pennwell Publications, Tulsa, Oklahoma

Jennings, A. (1965) Cable movements under two dimensional loads – a discussion, *Journal of the Structural Division*, American Society of Civil Engineers, STI (February 1965) 307–311

John, F. (1950) On the motion of floating bodies II, *Communications in Pure and Applied Mathematics*, **3**, 45–101

Kaplan, P. and Silbert, M. N. (1976) Impact forces on platform horizontal members in the splash zone. In *Proceedings of the Offshore Technology Conference*, Paper number OTC 2498

Keulegan, G. H. and Carpenter, L.H. (1958) Forces of cylinders and plates in an oscillating fluid *Journal of Research of the National Bureau of Standards* (May 1958), **60**

Lamb, H. (1975) *Hydrodynamics*, 6th edition, Cambridge University Press; Cambridge, UK

Lee, G. C. (1980) Recent advances in design and construction of deep water platforms, Part 1, *Ocean Industry* (November 1980), 71–80

Lloyds Register of Shipping (1986) *Rules and regulations for the classification of mobile offshore units*, January, Part IV, Chapter 1, Sections 2, 3, 4 and 5, Lloyds Register of Shipping,

Lyons, G. J. and Patel, M. H. (1984) Comparisons of theory with model test data for tensioned buoyant platforms, *Transactions of the ASME, Journal of Energy Resources Technology*, **106** (December 1984), 426–236

MacCamy, R. C. and Fuchs, R. A. (1954) *Wave forces on piles, a diffraction theory*, US Army Corps of Engineers, Beach Erosion Board, Tech. Memo. No. 69

Malhotra, A. K. and Penzien, J. (1970) Nondeterministic analysis of offshore tower structures, *Journal of Engineering Mechanics Division*, American Society of Civil Engineers, **96**, 985–1003

Maruo, H. (1960) The drift of a body floating in waves, *Journal of Ship Research*, **4.3**, 1

Meteorological Office (1986) *Meteorology for mariners*, 3rd edition, HMSO, London

Miller, B. L. (1980) *Wave slamming on offshore structures*, National Maritime Institute, Report No. NMI-R81, OT-R-8041

Miller, B. L. and Davies, M. E. (1982) *Wind loading of offshore structures – a summary of wind tunnel studies*, National Maritime Institute, Report No NMI R136, OT-R-8255, July 1982

Morison, J. R. O'Brien, M. B. Johnson, J. W. and Schaaf, S. A. (1950), The forces exerted by surface waves on piles, *Petroleum Transactions* **189**, TP 2846, 149

Murphy, J. E. (1978) Integral equation failure in wave calculations, *Journal of Waterways, Port Coastal and Ocean Division*, **104**, WW4, 330–334

Newland, D. E. (1984) *An introduction to random vibration and spectral analysis*. Longman, London

Newman, J. N. (1982) The exciting forces on fixed bodies in waves, *Journal of Ship Research*, **6**, 10–17

O'Brien, W. T. and Francis, A. J. (1964) Cable movements under two dimensional loads, *Journal of the Structural Division*, American Society of Civil Engineers, ST3, (June 1964) 89–123

O'Neill, M. W., Chazzaly, O. I. and Ho, B. H. (1977) Analysis of three-dimensional pile groups with nonlinear soil response and pile-soil-pile interaction. In *Proceedings of the Offshore Technology Conference*, OTC 2838

Oppenheim, B. W. and Wilson, P. A. (1980) Continuous digital simulation of the second order slowly varying wave drift force, *Journal of Ship Research*, **24**, 3, 181–189

Patel, M. H. (1983) On the wave induced motion response of semisubmersibles, *Transactions of the Royal Institution of Naval Architecs*, **125**, (July 1983) 221–228

Patel, M. H. and Lynch, E. J. (1983) The coupled dynamics of tensioned buoyant platforms and mooring tethers, *Journal of Engineering Structures* (October 1983), **5** 299–308

Patel, M. H. and Walker, S. (1983) *On the hydrostatics of floating bodies with articulated appendages*, Institution of Naval Architects (July 1983), **125** 229–236

Patel, M. H. and Seyed, F. B. (1987) Analysis and design considerations for towed pipe lines. In *Proceedings of the 6th International Symnposium on Offshore Mechanics, Offshore and Arctic Pipelines*, ASME, 17–28

Patel, M. H., Sarohia, S. and Ng, K. F. (1984) Finite element analysis of the marine riser, *Journal of Engineering Structures*, **6**, 3, 161–170

Penzien, J. and Tseng, W. S. (1978) Three dimensional dynamic analysis of fixed offshore platforms, *Numerical methods in offshore engineering* (ed O C Zienkiewicz et al.), Wiley, New York

Peyrot, A. H. (1980) Marine cable structures, *Journal of the Structural Division*, American Society of Civil Engineers, 2391–2404, (December 1980)

Pinkster, J. A. (1980) *Low frequency second order wave exciting forces on floating structures*, Publication 550, Insherlands Ship Model Basin, Wageningen, Netherlands

Pinkster, J. A. and van Oortmerssen, G. (1976) Computation of the first and second order wave forces on bodies in regular waves. In *Proceedings of the Second International Conference on Numerical Methods of Engineering*

Poulos, H. G. and Davis, E. H. (1980) *Pile foundation analysis and design*, John Wiley and Sons Chichester, UK

Rawson, K. J. and Tupper, E. C. (1976) *Basic ship theory*, Volume 1, Longman

Reese, L. C. (1977), Laterally loaded pile; program documentation. *Journal of the Geotechnical Engineering Division*, American Society of Civil Engineers, **103**, GT4, 287–305

Remery, G. F. M. and Hermans, A. J. (1971), The slow drift oscillations of a moored object in random seas. In *Proceedings of the Offshore Technology Conference*, OTC 1500

Sarpkaya, T. (1976) *In line and transverse forces on smooth and sand roughened cylinders in oscillatory flow at high Reynolds numbers*, Report No NPS-69SL76062, Naval Postgraduate School, Monterey, California

Sarpkaya, T. and Isaacson, M. (1981) *Mechanics of wave forces on offshore structures*, Van Nostrand Reinhold

Sommerfield, A. (1949) *Partial differential equations in physics*, Academic Press, New York

Standing, R. G. Da Cunha, N. M. C. and Matten, R. B. (1981) *Slowly varying second order wave forces*, theory and experiment, National Maritime Institute, Report No R138

Stoker, J. J. (1957) *Water waves*, Interscience, New York

Taudin, P. (1978) Dynamic response of flexible structures to regular waves. In *Proceedings of the Offshore Technology Conference*, OTC 3160

Thom, H. C. S. (1973) *Distribution of extreme winds over the oceans*. American Society of Civil Engineers, WW1 **99**, 1–17

Thornton D. (1979) A general review of future problems and their solution. In *Proceedings of the Second International Conference on Behaviour of Offshore Structures*, paper 88, BHRA Fluid Engineering, Cranfield, UK

Von Karman, T. L. and Wattendorf, B. F. (1929) *The impact on seaplane floats during landing*, NACA TN-321

Warburton, G. B. (1976) *The dynamical behaviour of structures*, 2nd edition, Pergamon Press Ltd

Wehausen, J. V. and Laitone, V. (1960) Surface waves, *Handbuck der Physik* (ed. S Flugge) Springer-Verlag, Berlin, **IX**, 446–778

Witz, J. A. and Patel, M. H. (1985) *A pressure integration technique for hydrostatic analysis*. Transactions of the Royal Institution of Naval Architects, **127**, 285–294

Yamamoto, T., Nath, J. H. and Slotta, L. S. (1974), Wave forces on cylinders near plane boundary *Journal of Waterways and Harbours Division*, **100**, WW4 345–359

Zienkiewicz, O. C., Bettes, P. and Kelly, D. W. (1978), The finite element method of determining fluid loading on rigid structures – two and three dimensional formulations. In *numerical methods in offshore engineering* (eds. O. C. Zienkiewicz, P. Lewis and K. G. Stass), John Wiley, Chichester, UK

3 Ships and Advanced Marine Vehicles

Charles V. Betts
(Sections 3.1–3.7)

Brian R. Clayton
(Sections 3.8–3.14)

Contents

Basic naval architecture and ship design

3.1 Introduction
   3.1.1 Basic definitions
   3.1.2 Basic ship design
   3.1.3 Types of merchant ship

3.2 Buoyancy and stability
   3.2.1 Flotation
   3.2.2 Transverse stability at small angles of heel
   3.2.3 Sinkage, heel and trim
   3.2.4 Large angle stability
   3.2.5 Flooding and damage stability
   3.2.6 Computer aided design packages

3.3 Resistance and propulsion
   3.3.1 Basic considerations
   3.3.2 Components of ship resistance
   3.3.3 Practical resistance estimation
   3.3.4 Propulsion
   3.3.5 Propeller design
   3.3.6 Full scale trials
   3.3.7 Hull form design

3.4 Manoeuvring
   3.4.1 Interaction effects
   3.4.2 Design of control surfaces

3.5 Ship motions and seakeeping
   3.5.1 Degrees of freedom
   3.5.2 Roll
   3.5.3 Heave and pitch
   3.5.4 Slamming and deck wetness
   3.5.5 Empirical methods of motion prediction
   3.5.6 Seakeeping criteria
   3.5.7 Design for seakeeping performance

3.6 Ship structures
   3.6.1 Structural materials
   3.6.2 Types of structure and modes of failure
   3.6.3 Loading
   3.6.4 Primary structure quasi-static response–hull girder bending
   3.6.5 Secondary structure
   3.6.6 Tertiary structure
   3.6.7 Structural dynamic response
   3.6.8 Ship structure design
   3.6.9 Computer aided structural design

3.7 The ship design process
   3.7.1 General considerations
   3.7.2 Initial sizing
   3.7.3 Design development
   3.7.4 Design for production
   3.7.5 CAD/CAM
   3.7.6 Classification and certification

References

High speed and multihull marine vehicles

3.8 Introduction

3.9 Planing craft
    3.9.1 Hull forms
    3.9.2 Forces on a planing craft in steady motion
    3.9.3 Estimation of forces
    3.9.4 Performance in steady motion
    3.9.5 Performance in waves
    3.9.6 Structural design
    3.9.7 Applications

3.10 Hydrofoil craft
    3.10.1 Generation of lift forces
    3.10.2 Lifting surfaces on hydrofoil craft
    3.10.3 Hydrofoils operating near the water surface
    3.10.4 Performance in steady motion
    3.10.5 Performance in waves
    3.10.6 Structural design
    3.10.7 Applications

3.11 Hovercraft
    3.11.1 Generation of supporting forces
    3.11.2 Flexible skirts
    3.11.3 Drag components
    3.11.4 Performance in steady motion
    3.11.5 Performance in waves
    3.11.6 Structural design
    3.11.7 Applications

3.12 Multihull vessels
    3.12.1 Catamarans
    3.12.2 SWATH ships

3.13 Propulsion
    3.13.1 Marine propellers
    3.13.2 Water jet systems
    3.13.3 Air propellers

3.14 Comparative performances
References

Basic naval architecture and ship design

3.1 Introduction

The traditional definition of naval architecture was the art or science of shipbuilding. Today, it is a branch of engineering of much wider scope. It encompasses research, design, construction and repair of any engineering system that moves on or under the sea. These systems include surface displacement vessels of all kinds (merchant ships, service vessels, fishing vessels, warships, yachts), high performance craft (hovercraft, hydrofoils, planing craft, multihulls, small waterplane area twin hull (SWATH) ships), underwater vehicles (submarines, submersibles, remotely operated vehicles, bathyscaphes, torpedoes, sea bed vehicles) and offshore structures (such as mobile oil rigs). An approximate breakdown of the world fleet of large marine vehicles is shown in *Figure 3.1*.

The task of the naval architect is essentially twofold. First, he or she is responsible for the overall architecture, layout and integration of the vessel. This must meet the requirements of the owner for performance, reliability and economy within constraints of safety, regulatory and other standards, resources and timescale. Second, he or she is responsible for specific aspects of the engineering of the vessel. These include the design of the hull and structure to satisfy the needs of buoyancy, stability, hydrodynamics, seaworthiness and strength and the engineering of certain systems such as pumping and cargo systems, accommodation and hotel systems and the design of the propellers, rudders and other control surfaces.

In a few countries, naval architects are required to register as licensed professional engineers. Naval architects in the UK and in many other countries become members of the Royal Institution of Naval Architects (RINA). In addition to its learned society activities, RINA, in common with other UK professional engineering institutions, sets and monitors professional standards. These cover Fellows and Members, who are usually registered with the Engineering Council as Chartered Engineers, Graduates, Associate Members and Associates, who may register with the Engineering Council as Incorporated Engineers or Engineering Technicians, respectively, Companions and Junior Members.

The basic naval architectural principles are common to all marine vehicles. However, this chapter concentrates on providing an introduction to the naval architecture of ships and advanced surface craft. Sections 3.2 to 3.6 describe the main principles and their application to monohull displacement ships. Section 3.7 describes the ship design process. The later sections describe applications particular to high speed and advanced

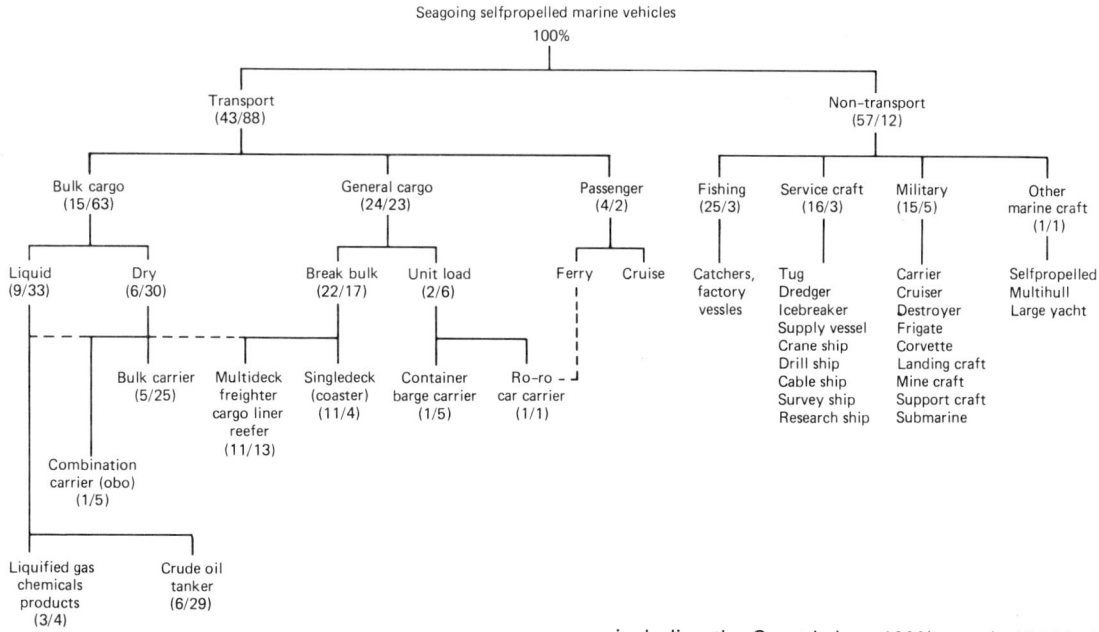

Figure 3.1 World fleet of marine vehicles, 1988. Numbers refer to percentage (by number of ships/by gross register tonnage) of the world fleet over 100 grt, including the Great Lakes; 100% equals 90 000 ships and 430 million grt (based on Lloyd's Register Statistical Tables (1988) and Buxton (1987)) (Courtesy of Lloyds' Register of Shipping)

marine vehicles such as planing craft, catamarans, hovercraft, hydrofoils, and SWATH ships. Offshore structures are covered in Chapter 2.

3.1.1 Basic definitions

It is assumed that the reader is familiar with nautical terminology and the basic parts of a ship. If required, a guide may be found in specialist glossaries such as *Glossary of Marine Technology Terms* (1980) 1, Elsevier's *Nautical Dictionary* (1978) and *The Fairplay Book of Shipping Abbreviations* (Kapoor, 1980).

3.1.2 Basic ship design

3.1.2.1 Ship size

Ship size is usually characterized by displacement (i.e. weight) or tonnage (i.e. volume). Common terms are:

Deadweight (dwt) – the weight of cargo, fuel, water, stores, crew and effects (all variable loads).

Lightweight – weight of hull and machinery and permanent fixtures (all fixed weights)

Displacement – the total weight, deadweight plus lightweight. It is equal to the weight of water displaced by the ship (Archimedes Principle). It is usually expressed in tonnes.

Gross register tonnage (grt) – a measure of the total internal volume of the ship, including the hull, the superstructure and all enclosed spaces. It is used as the basis for such things as docking, pilotage and survey fees.

Net register tonnage – essentially a function of the total volume of cargo spaces and the number of passengers, a measure of earning capacity. It is used as the basis for such things as port and harbour, light and cargo dues.

Tonnage was until recently expressed in units of 100 ft^3 (2.83 m^3). Its measurement is now governed by the International Convention on Tonnage Measurement of Ships which came into force in 1982 for all new and converted ships and from July 1994 to all other shipping. The Convention is given statutory force in the UK by the Merchant Shipping Tonnage Regulations, 1982 (see Section 3.7).

The formula for calculating gross tonnage (GT) is:

$$GT = K_1 V \quad (3.1)$$

where V is the total volume of all enclosed spaces (m^3) and K is $0.2 + 0.02 \log_{10} V$. While that for net tonnage is:

$$NT = K_2 V_c \, (4\,d/3D)^2 + K_3 \, (N_1 + N_2/10) \quad (3.2)$$

where V_c is the total volume of all cargo spaces (m^3), K_2 is $0.2 + 0.02 \log_{10} V_c$, d is the moulded draft amidships (m), D is the moulded depth amidships (m), K_3 is 1.25 $(1 + GT/10\,000)$, N_1 is the number of passengers in cabins with not more than eight berths and N_2 is the number of passengers not included in N_1. However $(4d/3D)^2$ must not be greater than 1, $K_2 V_c \, (4d/3D)^2$ must not be greater than 0.25 GT, NT must not be taken as less than 0.30 GT and where $N_1 + N_2$ is 12 or less, both are taken as zero. National regulations should be consulted for precise definitions of terms.

Ship size may also be characterized by main dimensions. Unless otherwise stated, these refer to the loaded waterline condition as designed. Length over load waterline (LWL) is length along waterline between bow and stern. This is often identical with length between perpendiculars (LBP). Length overall (LOA) is the maximum length of the ship, usually found at upper deck, whilst beam is the maximum width of the ship at the load waterline and extreme beam is the overall maximum width of the ship, usually found at upper deck. Mean draught is the draught amidships at load waterline, i.e. vertical height amidships between keel and waterline. Trim measures the difference between draughts at forward and after perpendiculars, usually designed to be by the stern, i.e. deeper draught aft. Freeboard is the vertical distance between upper deck and waterline. Finally, depth is the vertical distance between the upper deck and the outside of the keel plating, i.e. freeboard plus draught. Moulded depth, a term used in tonnage measurement and ship construction is measured between the inside of the keel plating and the underside of the upper deck at the ship's side.

3.1.2.2 Hull form definition

The three dimensional shape of a ship is traditionally represented by a lines plan (*Figure 3.2*). This shows three orthogonal sets of plane sections: the middle line plane sheer profile, the half breadth plan comprised of a number of horizontal waterplane sections and the body plan showing a number of vertical transverse sections. Due to symmetry about the middle line plane, only half sections need be shown. By convention, the body plan shows sections forward of amidships on the right hand side and sections abaft amidships on the left

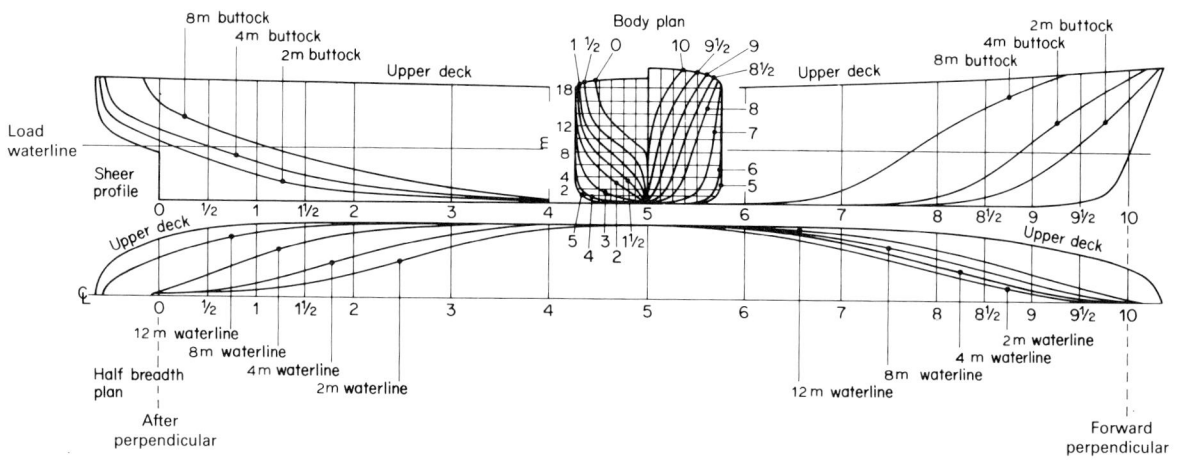

Figure 3.2 Lines plan

hand side. Intersections between the hull and vertical longitudinal sections parallel to the midline are called bow and buttock lines or buttocks.

The forward and after perpendiculars (FP and AP) are verticals drawn through fixed reference points forward and aft on the ship. They are often chosen to coincide with the intersection of the load waterplane with the stem and stern, respectively. The distance between the FP and AP is divided into a number of equal spaces, usually ten or twenty, by vertical ordinates as shown in *Figure 3.2*. The transverse horizontal distance between the centreline and a waterline at any ordinate is called an offset. A table of the full set of offsets at all the ordinates and waterlines approximately defines the hull shape in a form suitable for mathematical manipulation.

The curve of areas (*Figure 3.3(a)*) is a useful way of representing the underwater part of the hull in summary form. It is a plot of the hull cross-sectional area up to the load waterline at each ordinate. The total area under the curve is equal to the volume of displacement of the ship. The detailed shape of the curve is a useful guide to hydrodynamic performance. For this reason, choice of a suitable curve of areas usually precedes the development of the lines plan during the design of a ship.

Another useful representation is a set of Bonjean curves (*Figure 3.3(b)*). Each curve is a plot at a given

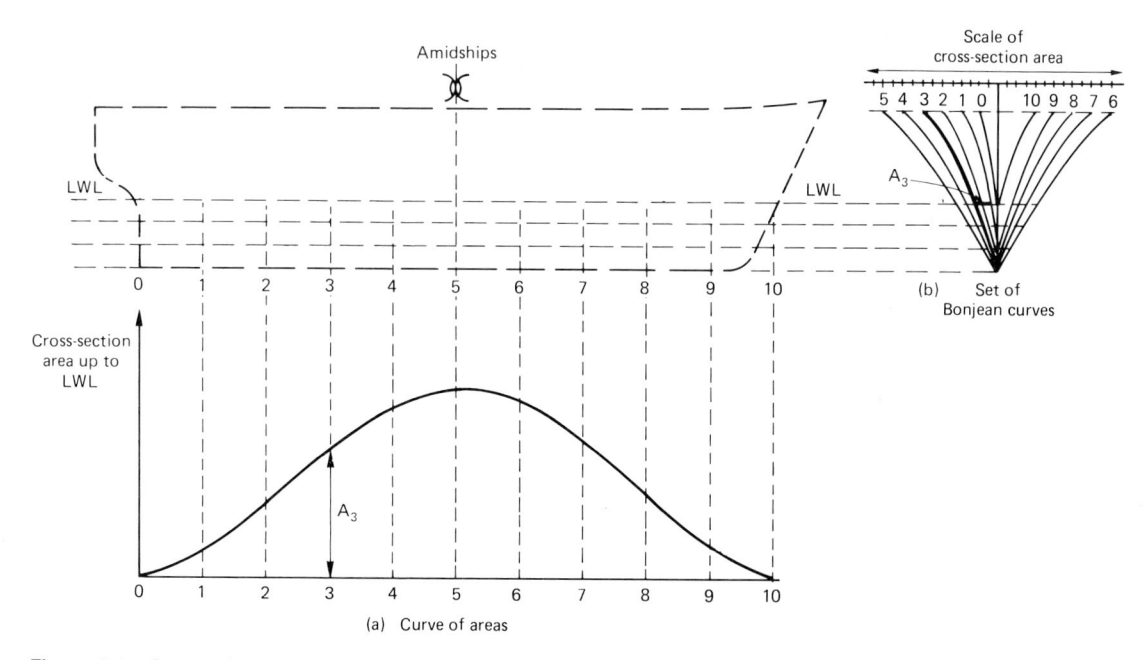

Figure 3.3 Curve of areas and Bonjean curves

ordinate of the hull cross-sectional area up to each waterline. For convenience, the cross-sectional area is plotted outwards from a vertical axis representing the draught at each waterline. A set of Bonjean curves allows the cross-sectional area at any waterline and any ordinate to be found quickly.

Form coefficients are used to characterize a hull form succinctly. They are a non-dimensional way of defining the fullness or fineness of the hull form. The most commonly used form coefficients are:

Block coefficient $C_B = \dfrac{\nabla}{LBT}$ (3.3)

Prismatic coefficient, strictly speaking, the longitudinal prismatic coefficient

$$C_P = \dfrac{\nabla}{A_M L}$$ (3.4)

Vertical prismatic coefficient

$$C_{VP} = \dfrac{\nabla}{A_W T}$$ (3.5)

Midship section coefficient

$$C_M = \dfrac{A_M}{BT}$$ (3.6)

Waterplane coefficient

$$C_W = \dfrac{A_W}{LB}$$ (3.7)

where ∇ is the volume of displacement up to the load waterline, L is the length, B is the beam, T is the draught, A_M is the hull cross-section at amidships and A_W is the waterplane area, all taken at the load (i.e. design) waterline.

3.1.3 Types of merchant ship

The main types of merchant ship in the transport category of *Figure 3.1* are described below.

3.1.3.1 General cargo ship

A general cargo ship (*Figure 3.4(a)*) is a ship with open cargo holds loaded vertically through hatches in the upper deck. The cargo is usually dry and in packaged (break bulk) form. The ship will usually have its own cranes or derricks. The holds may be divided by intermediate decks called tween decks. Some vessels have side loading arrangements and palletized cargoes. Up to 12 passengers may be carried. General cargo vessels vary in size from small single deck coasters to large multideck freighters or cargo liners of up to 10 000 grt or more. Service speeds are usually in the range 12–18 knots.

3.1.3.2 Refrigerated cargo ship (reefer)

This is a cargo ship equipped to carry all or part refrigerated cargo, such as perishable foods. The holds are insulated and a refrigeration system fitted. Higher service speeds are normal, up to around 22 knots.

3.1.3.3 Container ship

This ship (*Figure 3.4(b)*) is designed to carry large numbers of standard containers. These are stacked in holds and on the hatch covers above the upper deck. Ship cargo capacity is usually quoted in numbers of TEU, representing 20 foot equivalent units. This derives from the International Standards Organization (ISO) standard container which has nominal dimensions of 20 or 40 ft (6.09 or 12.19 m) length, 8 ft (2.44 m) width and 8 ft 6 in (2.59 m) height, although other sizes exist. Maximum permissible weights are 24 tonnes for the 20 ft container and 30 tonnes for the 40 ft container. Some containers may be refrigerated. A few container ships are designed to carry passengers. The ships ply between special terminals and container handling is normally shore based rather than by ship cranes. The ships are of open section but with double hull wing tanks and double bottoms. These provide torsional stiffness for the hull and tankage for water ballast. Ship capacities vary from a few hundred containers to over 4000 TEU, the latter requiring a displacement around 100 000 Te. The largest ships are beyond Panamax size, i.e. the largest dimensions allowed by the Panama Canal which are 32.3 m (106 foot) beam and 294 m (964.5 foot) length. Design service speeds vary from around 16 knots on short sea routes up to 24 knots or even higher on long haul routes.

3.1.3.4 Barge carrier

This ship is designed to accommodate a number of standard equal sized barges containing cargo. These are far less common than container ships. After unloading from the ship, the barges are towed or otherwise propelled to terminals which may be along inland waterways. The main designs of barge carrier are:

1. The lighter aboard ship (LASH). A ship of some 45 000 Te carrying up to 80 × 400 Te barges. These are loaded into the holds by an onboard travelling gantry crane that lifts the barges over the stern.
2. The SEABEE. A ship of some 38 000 Te carrying 38 × 1000 Te barges. These are loaded by a stern elevator and stowed on internal decks which are sealed by watertight doors.

Other designs of barge carrier include a barge carrying catamaran (BACAT).

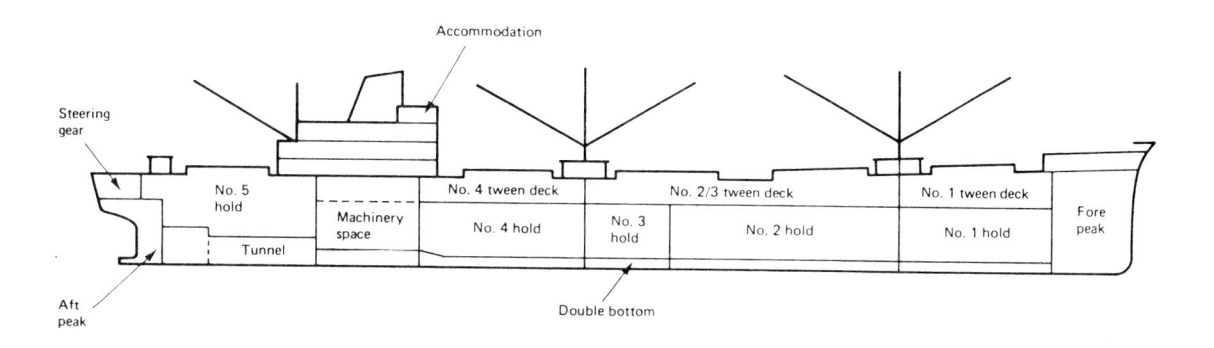

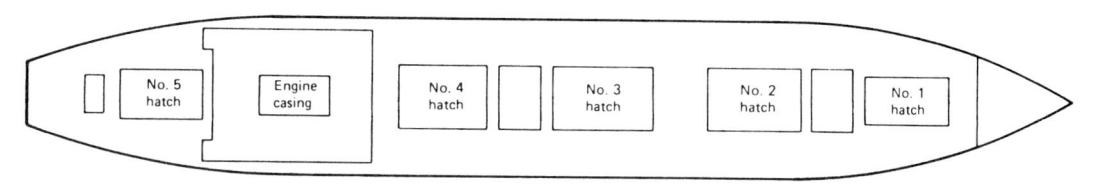

(a) General cargo ship

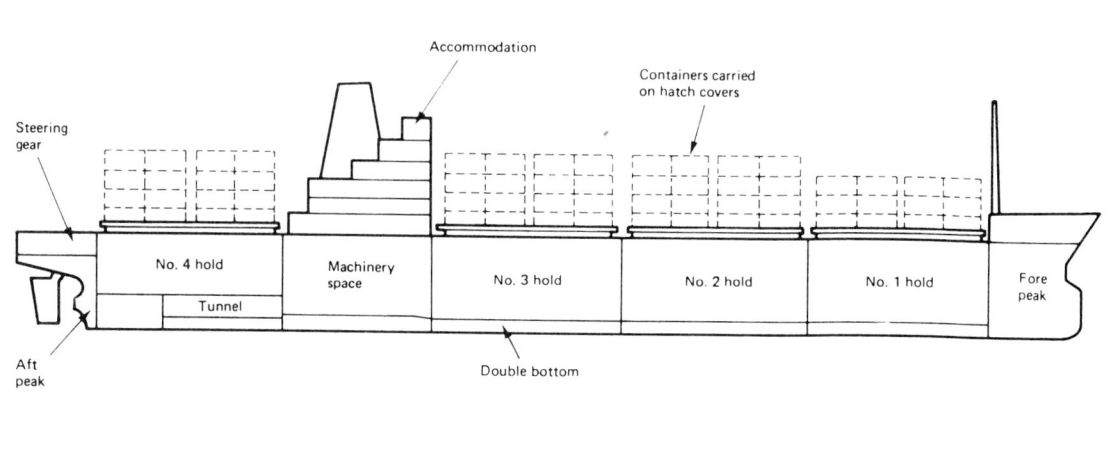

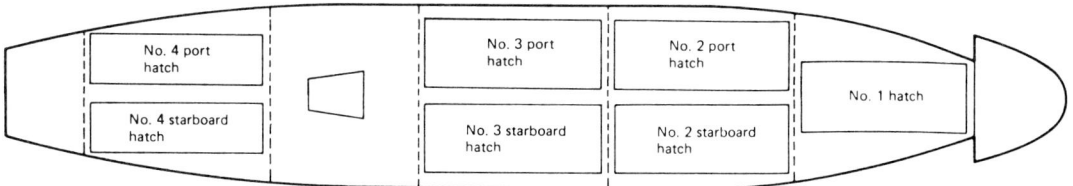

(b) Container ship (Muckle (1987), with permission)

Figure 3.4

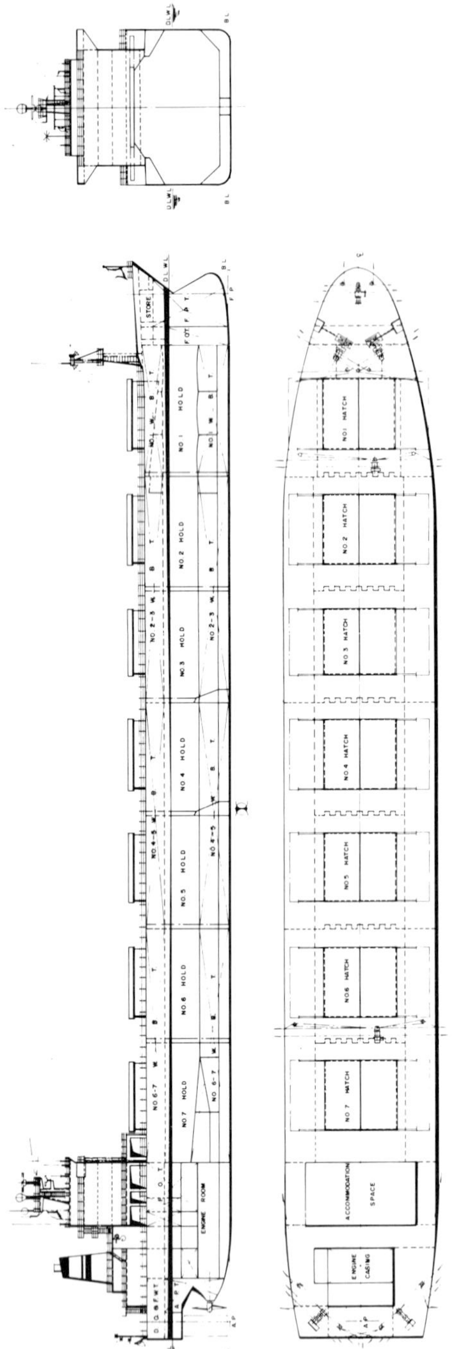

(c) Bulk carrier

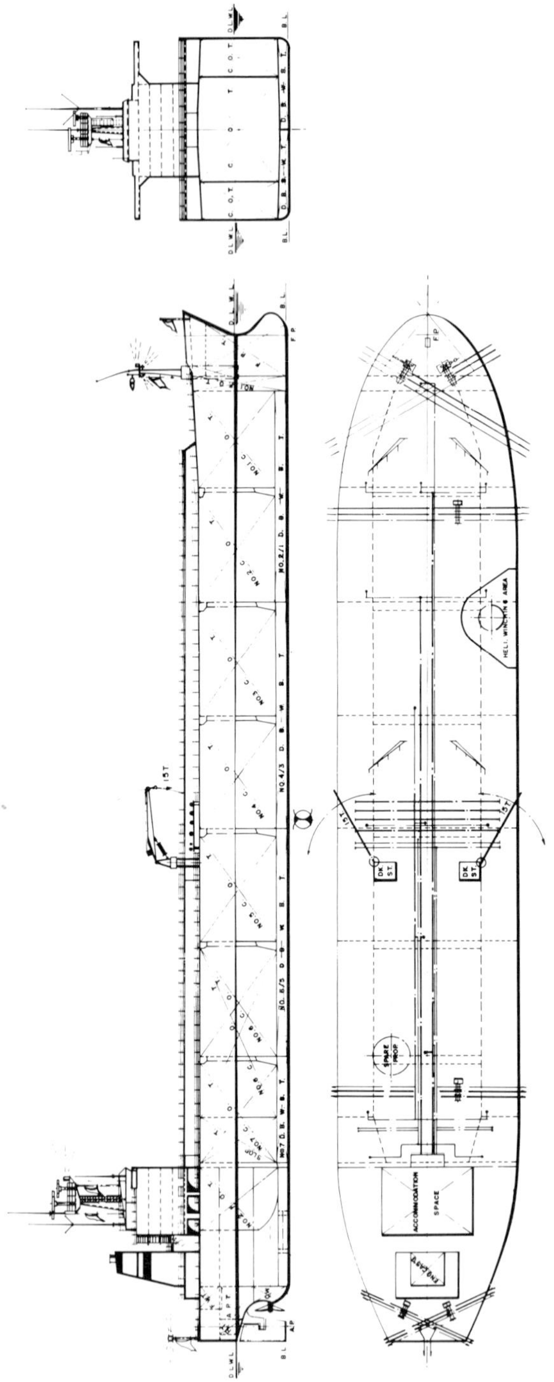

(d) Oil tanker

Figure 3.4 (cont)

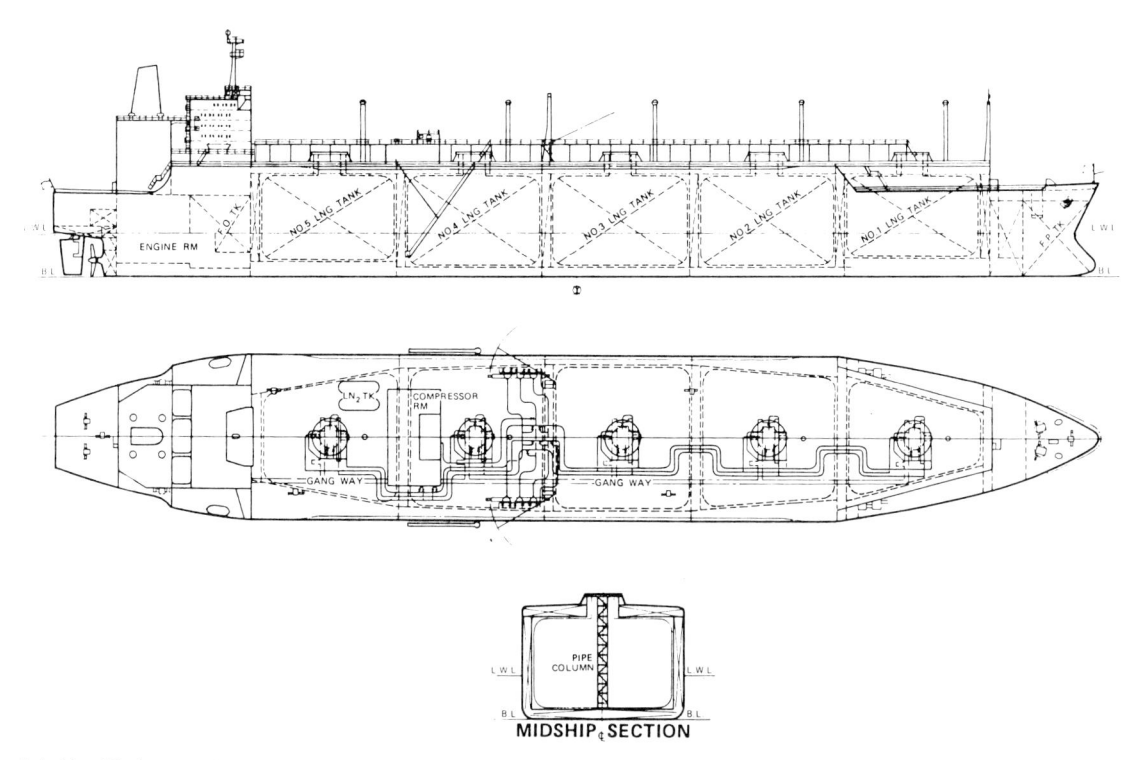

(e) Liquified natural gas carrier

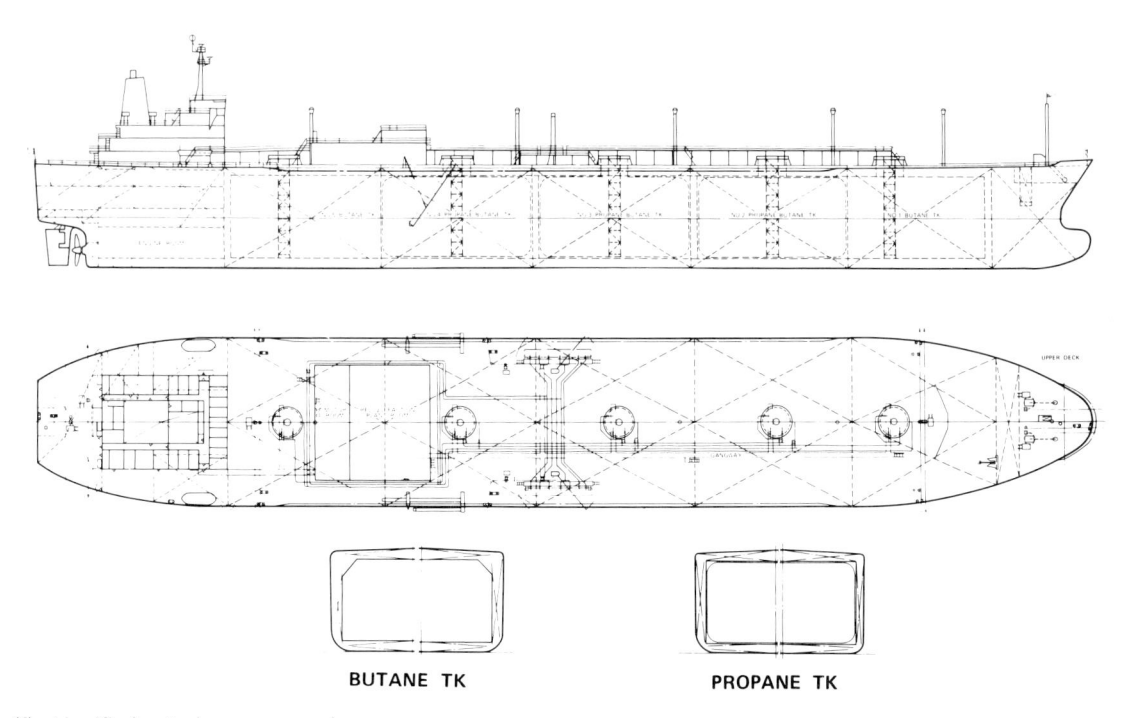

(f) Liquified petroleum gas carrier
Figure 3.4 (cont)

3.1.3.5 Bulk carrier

The bulk carrier (*Figure 3.4(c)*) is designed to carry dry cargo in bulk, i.e. grain, ore and coal, fertilizers, animal foodstuffs, sugar and various minerals and timber, often carried on deck. Loading and discharging may be by grab or suction or special built-in conveyors.

The sides of the holds are shaped to prevent cargo movement in a seaway. Saddle tanks are fitted high up at the sides of each hold. These are filled with water ballast in the loaded condition, particularly when heavy cargo such as ore is carried, to prevent the ship's centre of gravity being too low. This is undesirable as it raises the ship's roll frequency, with adverse effects on crew comfort and cargo security. A modern versatile bulk carrier may be designed to carry coal or grain in all the holds or ore cargo in alternate holds. Sizes vary from coasters to ships over 200 000 dwt. Speeds are 12–14 knots.

Many large bulk carriers are designed as combination carriers. These include the oil/ore carrier with longitudinal bulkheads either side of the main hold. The main hold is used for carrying ore and the wing tanks for oil cargo. MARPOL requirements must be met, as for tankers. Also, the oil/bulk/ore carrier (OBO) a more versatile form of carrier designed to carry combinations of dry and liquid bulk cargoes. Coal, grain or oil may be carried in all cargo holds or ore in alternate holds. Some OBO carriers can also carry petroleum products such as naptha and gasoil.

3.1.3.6 Forest products carrier

This ship is designed to carry bulk and break bulk commodities such as timber, wood pulp, board and paper rolls. These may be loaded both vertically by crane and horizontally via ramps or side loading transporters. The ship may carry bulk cargoes, such as china clay or salt and possibly containers, on the return leg.

3.1.3.7 Oil tanker

The oil tanker (*Figure 3.4(d)*) carries crude oil in a number of separate tanks in order to ease loading and unloading and to control effects on ship strength and stability. The tanks are separated by a number of transverse bulkheads and up to two longitudinal bulkheads. The size and location of the tanks is governed by the International Convention on the Prevention of Pollution From Ships (MARPOL 73/78). This requires segregated ballast tanks to provide a barrier against oil spillage in an accident or collision. Oil tanks are usually filled and discharged by fixed pumping arrangements. It can be economical to carry crude oil in very large quantities, in tankers known as very large crude carriers (VLCCs). These include the largest ships ever built, exceeding 500 000 dwt, although most are under 200 000 dwt. Service speeds depend on the market price of the oil cargo but are usually 13–16 knots.

3.1.3.8 Products carrier

This is a particular type of tanker designed to carry refined petroleum products such as naptha, gasoil, kerosene or aviation fuel. These ships are also subject to MARPOL regulations and very stringent fire prevention measures. Products carriers tend to be smaller and faster than crude oil tankers.

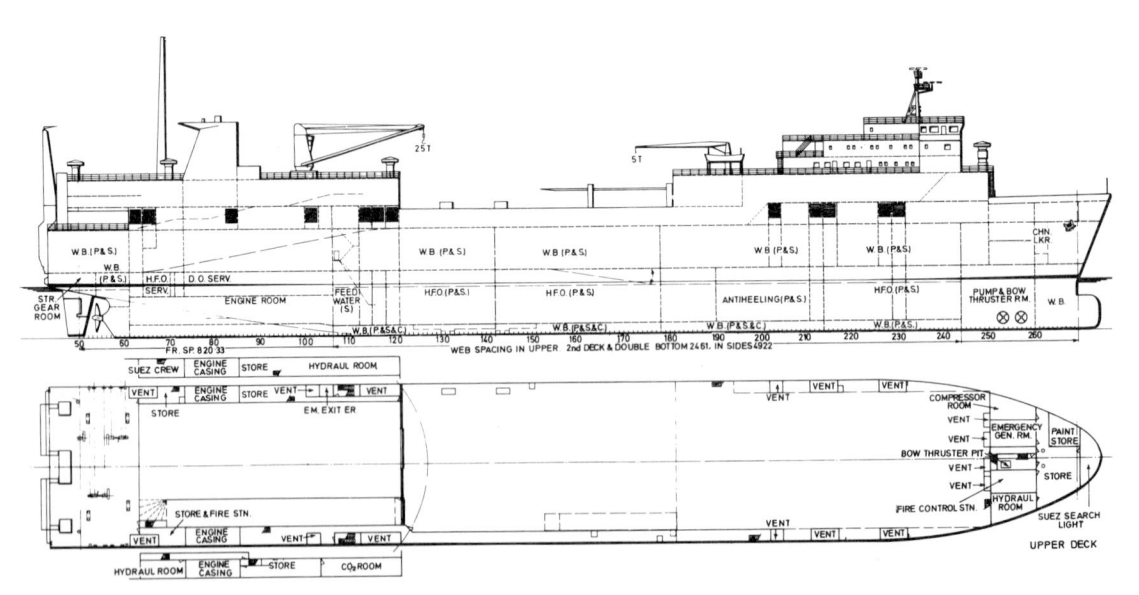

(g) Roll on–roll off ferry

Figure 3.4 (cont)

Figure 3.4 (cont)

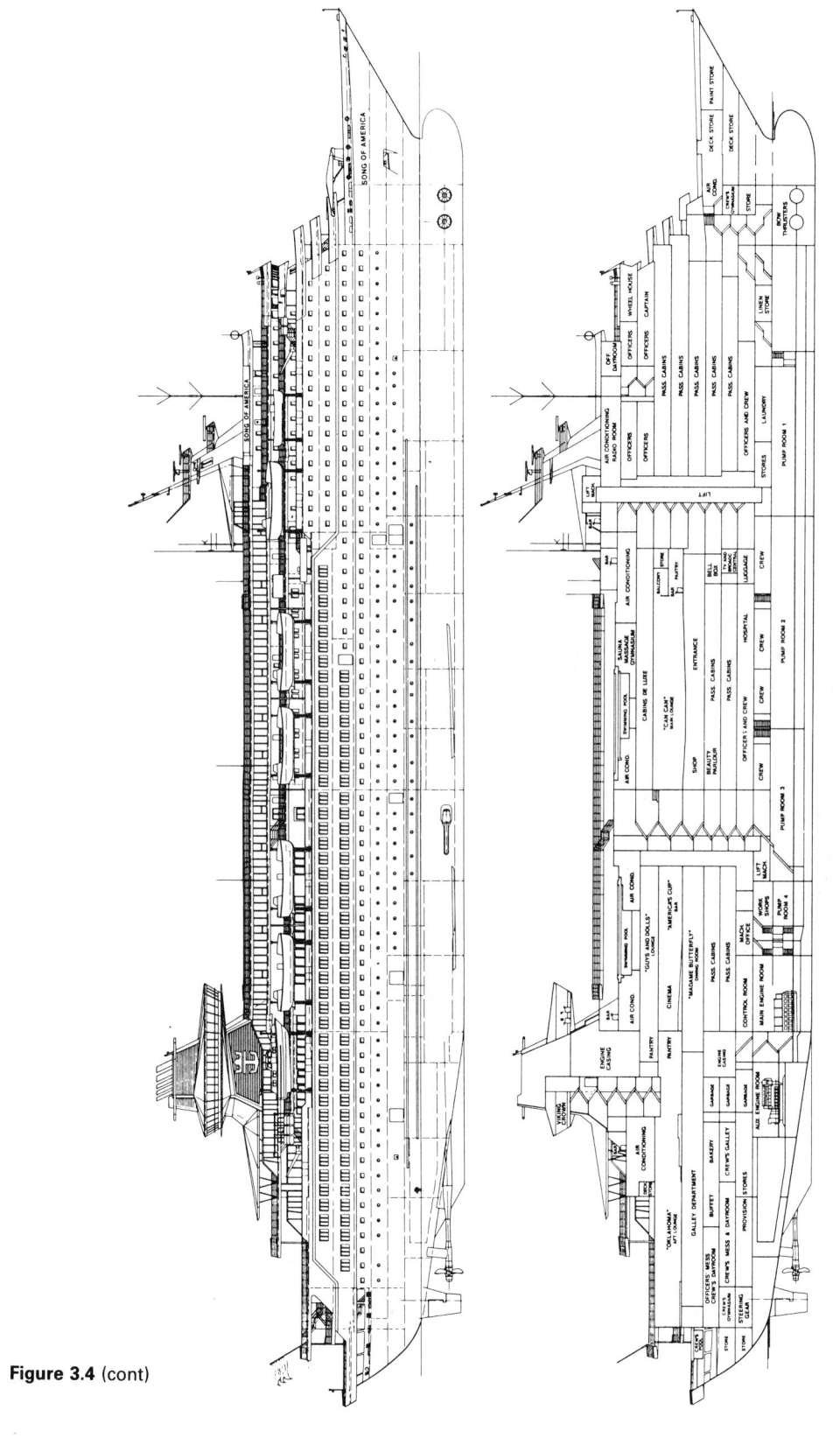

(h) Cruise ship

3.1.3.9 Chemical tanker

This specialized tanker is designed to carry chemicals in bulk. These might be heavy chemicals, e.g. acids and ammonia, molasses and alcohols, vegetable oils and animal fats, petrochemical products or coal tar products, e.g. benzene and phenol.

3.1.3.10 Parcel tanker

This is a chemical tanker designed to carry a range of chemical products.

3.1.3.11 Gas carrier

A gas carrier is a chemical tanker with tanks specially designed for the carriage of natural gas products cooled or compressed into liquid form.

3.1.3.12 Liquified natural gas (LNG) carrier

The LNG carrier (*Figures 3.4(e)*) is used for lighter products which are mostly methane. This will only liquify below −82°C, whatever the pressure and has a boiling point at atmospheric pressure of −162°C. It is carried at −165°C in insulated tanks made of special metals. There are three types of tank in service: self supporting tanks, membrane tanks supported on load bearing insulation between the tank and the hull and semimembrane tanks usually of approximately rectangular cross-section. LNG carriers are up to 130 000 m^3 in capacity, around 65 000 dwt, with service speeds up to 19 knots

3.1.3.13 Liquified petroleum gas (LPG) carrier

This ship carries the heavier products such as butane, propane, ethane and polypropylene (*Figure 3.4 (f)*). These can be liquified under pressure at ambient temperature so there is a choice of methods of transport. Three methods are used. The first is fully pressurized (about 17 bar) at ambient temperature; spherical or cylindrical tanks are required which penetrate the upper deck. The second method is semipressurized (about 8 bar) with partial refrigeration to about −10°C. The third method is at atmospheric pressure with full refrigeration to about −50°C. The three types of tanks used in LNG carriers may be utilized with the third method. LPG carriers tend to be a little smaller than LNG carriers but have similar service speeds.

3.1.3.14 Roll on–roll off (Ro-Ro) ship

This ship is designed to accept vehicular (truck and car) traffic that drives on and off the ship via ramps, as distinct from traditional Lo-Lo (lift on lift off) embarkation by crane (*Figure 3.4(g)*). To give maximum efficiency of loading and discharge, large through decks are fitted within the ship with doors at the bow and stern. This can conflict with good damage stability standards. Roll on–roll off (Ro-Ro) vessels may be designed for freight vehicles only or as combined freight, car and passenger ferries. Some are designed as roll on–roll off train ferries or even as combined rail, freight, car and passenger ferries (called combi–carriers). Ro-Ro ships range from very small car ferries to large cruise ferries of 50 000 grt that carry around 2500 passengers and 450 cars. Typical service speeds are 18–22 knots.

3.1.3.15 Passenger ships

These ships appear in two main forms: ferries and cruise ships. Most ocean going ferries are now designed as Ro-Ro ships. Cruise ships (*Figure 3.4(h)*) have replaced the old luxury liners. Some of the latter, notably the *Queen Elizabeth II* and the *France* (renamed the *Norway*), have been converted to cruise liners. Cruise ships have been a growth area of shipping and range from relatively small vessels, some even sail assisted, to ships of some 70 000 grt carrying around 2500 passengers. Service speeds are usually about 20–22 knots. Proposals have been made for giant ships of 250 000 grt carrying 5000 passengers. Cruise ships are usually designed for a specific market: the cheaper mass market, an intermediate market or the high price, luxury market with high crew to passenger ratios. All passenger ships are subject to special safety regulations (see Section 3.7).

The differences between the types of ship described above are becoming increasingly blurred. For example, some ships are designed to carry both bulk cargoes and containers (con-bulk ships) and some Ro-Ro vessels are also designed to carry lift on lift off cargoes such as containers. More detailed descriptions of the various types of merchant ship may be found in Muckle and Taylor (1987), from which the sketches in *Figure 3.4* have been taken (see also Schonknecht *et al.*, 1983; *Jane's Merchant Ships*, 1987).

3.2 Buoyancy and stability

This section is concerned with ship hydrostatics, the behaviour of a ship floating in static equilibrium in calm water. The subject is of fundamental importance to ship safety, both in harbour when loading, unloading or docking and at sea. In the latter case, the assumption of static conditions is rarely correct. However, the analytical methods presented in this section are still of value in describing the ship's stability and they give a good approximation of performance in all but extreme conditions. The requirements for adequate buoyancy and stability, particularly after damage, directly affect the design of a ship's hull form and layout. Loss of stability through poor design, maloperation, accident or extreme conditions still causes the loss of many ships and lives every year.

The principles of hydrostatics are common to all floating structures (see Chapter 2). The following is a summary of the application of the principles to ships (Clayton and Bishop, 1982; Rawson and Tupper, 1983; Lester, 1985; Muckle and Taylor, 1987; and Lewis, 1988).

3.2.1 Flotation

Figure 3.5 shows the hull of a ship in three orthogonal views: profile, plan and section. The total weight of the ship may be represented by a weight W acting vertically through the centre of gravity G. Similarly, the total buoyancy force on the ship due to the water pressure acting on the hull may be represented by an upthrust $\triangle$ acting vertically through the centre of buoyancy B.

Provided the ship is in static equilibrium, $W = \triangle$ and Archimedes' Principle tells us that $\triangle$ equals the weight of water displaced by the ship (the displacement). B and G must be vertically in line, both athwartships and longitudinally. There would otherwise be a couple, or moment, acting to rotate the ship from its equilibrium position. Note that B is at the centre of gravity of the displaced water, i.e. at the centroid of the underwater volume of the ship.

We denote the amount of the underwater volume by ∇. Then as weight equals mass × g (the acceleration due to gravity), and mass equals volume × density, we have:

$$W = \triangle = \nabla \rho_w g \tag{3.8}$$

where ∂_w is the density of the water. All these quantities must be in consistent units. Thus if ∇ is in m³ and ρ_w is in kg/m³ then g = 9.81 m/s² and:

$W = \triangle = \nabla$ (m³) × σ_w (kg/m³) × 9.81 (m/s²) in newtons

or $W = \dfrac{\nabla \rho_w}{1000}$ tonnes force (usually abbreviated to tonne) (3.9)

The value of ρ_w is 1000 kg/m³ for fresh water and approximately 1025 kg/m³ for sea water.

Also shown in *Figure 3.5* is the centre of flotation (F). This is the centroid of the waterplane shown in the plan view. It is the point about which the ship heels and trims when disturbed from equilibrium, e.g. by addition of a weight. F will normally be on the centreline, due to port and starboard symmetry.

Except in simple cases such as a rectangular barge, the underwater volume of a ship is non-prismatic (non-uniform). Neither the underwater volume nor the waterplane are then symmetric fore and aft of amidships. The calculation of volumes and areas and the position of centroids such as B and F is therefore not straightforward. Various approximate integration methods for hand calculation are available, such as the use of *Simpson's Rules* (Rawson and Tupper, 1983; Lester, 1985; Muckle and Taylor, 1987). Easier and more accurate calculation is possible using computer methods and a large number of programs are available for ship hydrostatics. Some of these use the pressure integration technique described in Chapter 2.

A good approximate formula is available for KB, the height of the centre of buoyancy above the keel. This is known as Morrish's Rule:

$$KB = \frac{5}{6}T - \frac{\nabla}{3A_w} \tag{3.10}$$

where T is the local draught, ∇ is the underwater volume of the hull and A_w is the waterplane area. This formula is quite accurate for conventional ship forms.

3.2.2 Transverse stability at small angles of heel

3.2.2.1 Basic considerations

Figure 3.6 shows a transverse section of a ship through the centres of gravity and buoyancy. The ship is upright in *Figure 3.6a*. It is inclined to a heel angle of ø in *Figure 3.6b* by an externally applied moment such as might be caused by a high wind. It is a convention to show the waterline rotated relative to the ship rather than the other way round.

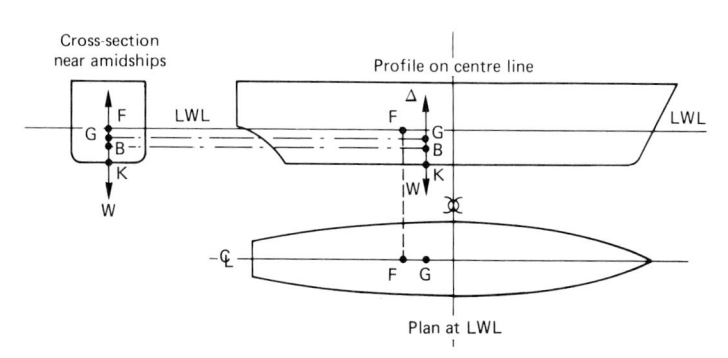

Figure 3.5 Flotation

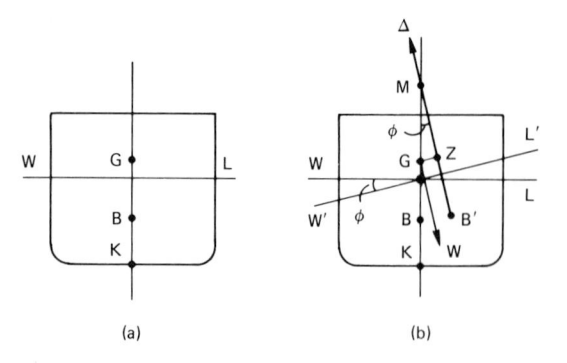

Figure 3.6 Transverse stability: (a) ship upright; (b) ship inclined

The effect of the inclination is to change the underwater shape of the hull, i.e. the displaced water volume, such that the centre of buoyancy moves from B to B'. Provided there is no movement of items within the ship, the centre of gravity remains unchanged at G. Weight and buoyancy also remain unchanged and in equilibrium, i.e. $\Delta = W$. If we now draw a vertical line (perpendicular to $W'L'$) through B', this intersects the original vertical through B at a point M. This point is called the metacentre. For ship forms, it is more or less a fixed point for small angles of heel, up to around 10°.

We can see from *Figure 3.6(b)* that there is a couple tending to bring the ship back upright (the restoring moment) of value $\Delta \cdot GZ$, where GZ is the distance between the lines of action of the weight and the buoyancy force at angle ϕ. It can be seen that:

$$GZ = GM \sin \phi. \quad (3.11)$$

As $\sin \phi \simeq \phi$ for small angles, then:

$$GZ \simeq GM \phi. \quad (3.12)$$

The restoring moment is therefore approximately $\Delta \cdot GM \cdot \phi$.

The ship is in stable equilibrium provided that the restoring moment is positive in sign, i.e. the ship is trying to return to the upright position. This implies that M must be above G (known as positive GM). If M is below G (negative GM) then the ship will be unstable and tend to heel to a larger angle (and possibly capsize). If G and M happened to be coincident, then the ship would be in neutral equilibrium.

It is evident that GM, called the metacentric height, is an important parameter in ship stability. Whilst GM should obviously be positive, it should not have too high a value. It will be seen later that GM is a measure of the ship's stiffness in roll motion and largely governs the period of roll motion. Too high a value of GM leads to a very short roll period. This can have undesirable consequences which are discussed in Section 3.5.

Typical values of GM are in the region 0.5–3 m, depending on ship size and type. The actual value of GM for a ship may be found by an inclining experiment. Alternatively, GM can be calculated. This is obviously necessary for a new design.

From *Figure 3.6b*, we have:

$$GM = KB + BM - KG \quad (3.13)$$

An approximate formula for KB was given above. KG must be calculated from the known weights and centres of gravity of the components of the ship. BM can be shown from geometric considerations (see Chapter 2) to be given by:

$$BM = I/\nabla \quad (3.14)$$

where I is the transverse second moment of area of the waterplane about the ship's centreline and ∇ is the underwater displacement volume. As an example, consider a rectangular barge of length L, beam B and draught T.

Then $I = \dfrac{LB^3}{12}$, $\nabla = LBT$, so $BM = \dfrac{B^2}{12T}$

Thus for this barge:

$$GM = \frac{T}{2} + \frac{B^2}{12T} - KG \quad (3.15)$$

The middle term illustrates the large effect that waterplane beam has on GM. A small increase in beam at the design stage can compensate for a high centre of gravity. However, while this holds for small angle stability, increased beam is not always beneficial at large angles, as will be seen later.

3.2.2.2 Wall sided formula

For small angles, we found that the righting lever $GZ \simeq GM \cdot \phi$. A more accurate formula for GZ at angles around 10° is available for wall sided ships, i.e. those having vertical sides in the region of the waterline. The derivation of the formula is shown in Chapter 2:

$$GZ = \sin \phi \left(GM + \frac{BM}{2} \tan^2 \phi \right) \quad (3.16)$$

At small angles, this degenerates to the previous approximate form. The formula is invalid at large angles, however, when the metacentre M no longer remains a fixed point, but varies in position with angle of heel. It is then more correctly called the prometacentre.

3.2.2.3 Free surface effect

We have so far assumed that the contents of the ship do not move as it heels over. In practice, a ship contains tanks of liquids such as fuel oil and water. If these tanks are pressed full then the contained liquid will behave as though it were solid. If, however, the tank is only partly full, the contained liquid will move towards the downward side as the ship heels. In *Figure 3.7* the local centre of gravity g of the liquid of density ρ_l moves to g'. This has the effect of shifting the ship's centre of gravity G to G', closer to the line of action of the buoyancy force.

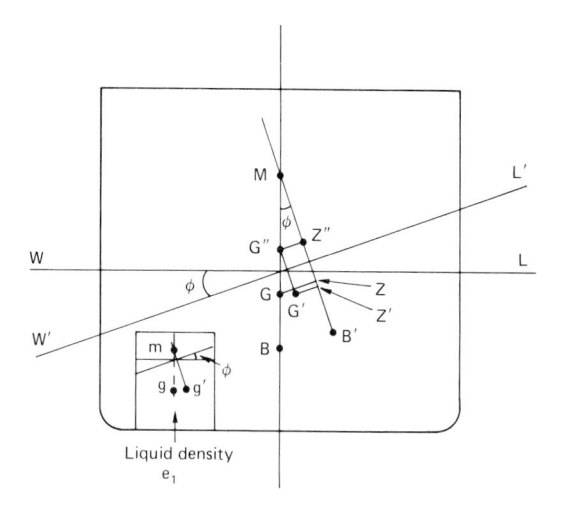

Figure 3.7 Free surface effect

For small angles, it is seen in *Figure 3.7*, that the new righting arm is given by:

$$G'Z' \simeq GZ - GG' \simeq '' GZ - GG'' \phi$$
$$(GM - GG'') \phi \quad (3.17)$$

There is an apparent loss of metacentric height, of value GG''. Taking moments about G:

$$\triangle \cdot GG' = w_1 \cdot gg' \quad (3.18)$$

where w_1 is the weight of liquid in the tank. Now we have a direct analogy between the centre of gravity of the liquid in the tank and the centre of buoyancy of the hull. Thus, from the relation $BM = I/\nabla$:

$$gg' \equiv bb' = bm \phi = \frac{i}{v} \phi \quad (3.19)$$

where i is the second moment of area of the free surface of the liquid in the tank and v is the volume of the liquid. From this we can deduce that:

$$GG' = \frac{1}{\nabla} \frac{\rho_l}{\rho_w} i \phi \quad (3.20)$$

and

$$GG'' = \frac{1}{\nabla} \frac{\rho_l}{\rho_w} i \quad (3.21)$$

This is the apparent loss in GM due to the free surface of the liquid in the tank. A fuller derivation is given in Chapter 2 in this book.

If there are a number of tanks, then the total loss of GM due to the free surface effect is the sum of the individual effects, i.e.:

$$\Sigma \, GG'' = \frac{1}{\nabla} \Sigma \frac{\rho_l}{\rho_w} i \quad (3.22)$$

where Σ denotes summation over all tanks.

The net value of GM after allowing for this reduction is called the virtual GM or GM_v. There are several points to note about the free surface effect. It is always destabilizing. It is independent of the volume of the liquids on board; even a very small amount of liquid can have a large effect if the free surface is large. It is also independent of tank position, either vertically or horizontally.

It is also very dependent on size of the free surface, particularly on width, in the same way as ship beam has a large effect on GM. The best way to reduce the effect, other than pressing tanks full, is therefore to subdivide the tanks so as to prevent the liquids moving in bulk. Note, however, that the subdivision must form completely separate tanks; subdivision of the free surface alone has no effect as the liquid as a whole can still move in bulk beneath the surface.

Finally, it can be very large. In extreme cases, the free surface effect can completely destroy the effective GM of the ship. As an example, consider the vehicle deck of a modern Ro–Ro ferry. This deck often has an area and second moment of area equal to or greater than that of the ship's load waterplane. If in an accident this is covered by even a shallow depth of sea water, then:

$$\text{Loss of } GM \, \frac{1}{\nabla} i \simeq \frac{I}{\nabla} = BM \quad (3.23)$$

This loss is more than the original value of GM (see *Figure 3.7*). The ship will therefore have negative upright stability. It will heel over rapidly and possibly capsize.

3.2.2.4 Suspended weights

A weight suspended from a ship's crane has a similar effect to a free surface. An analogy may be drawn with the liquid in the tank shown in *Figure 3.7*. When the ship heels, the liquid moves as though it were a solid body suspended from the point m (the metacentre for the liquid movement). Similarly, any weight suspended from a crane has an effect on ship stability, as though the weight's centre of gravity were fixed at the suspension point on the crane jib. People standing on the deck of a ship behave in a rather similar manner. The difference is that, in trying to remain vertical, their pivot point and therefore their effective centre of gravity is at their feet.

3.2.2.5 Inclining experiment

The purpose of the inclining experiment is to find the *GM* of a ship in a known condition. Owing to the importance of *GM* and the difficulty of estimating *KG* accurately, an inclining experiment is usually carried out on all new vessels. It is repeated after significant modifications to the ship during its service life. The experiment is based on applying a known heeling moment and then measuring the resulting angle of heel. The heel can be measured by means of a long pendulum.

Consider a known weight (usually a set of weights on wheels or slides) of amount w moved transversely across the ship a distance d. The heeling moment $w.d$ must be balanced by the restoring moment $\triangle.GM.\emptyset$, where $\emptyset$ is the small angle of heel involved.

Thus: $GM = \dfrac{wd}{\triangle \emptyset}$ (3.24)

All the quantities on the right hand side of the equation are measured as accurately as possible. The ship's displacement is found from the hydrostatic curves for the draughts measured during the experiment.

The experiment must be carried out in calm water in still conditions. Careful note is made of the ship's state, especially of the weights to go on and come off relative to some standard condition for which *GM* is required to be calculated. The *GM* as measured will include any free surface effects due to liquids on board. This virtual *GM* must be adjusted to a solid *GM* or a standard liquid condition. The information for this adjustment is obtained by dipping all tanks on board at the time of the experiment and calculating the free surface effect in each case.

3.2.3 Sinkage, heel and trim

This section is concerned with the hydrostatic behaviour of a ship under the influence of steady applied loads. We first examine what happens when an additional weight w is added anywhere on the ship's centreline. We assume throughout that w is small relative to the displacement of the ship.

Figure 3.8 shows the centreline section of a ship of displacement. The ship floats initially in equilibrium at waterline *WL*, with centre of gravity *G*, centre of buoyancy *B* and centre of flotation *F*. A weight w is added at a distance d forward of *F*. The effect of w is twofold. It makes the ship sink lower in the water so that the buoyancy will increase to match the new total weight ($\triangle + w$). It also causes the ship to trim longitudinally by an angle θ in order to bring the centre of buoyancy forward to B_1 beneath the new centre of gravity G_1. We can treat these two effects separately. This is because the addition of w is equivalent to addition of a weight w above the centre of flotation *F* plus the application of a couple $w.d$ clockwise about *F*.

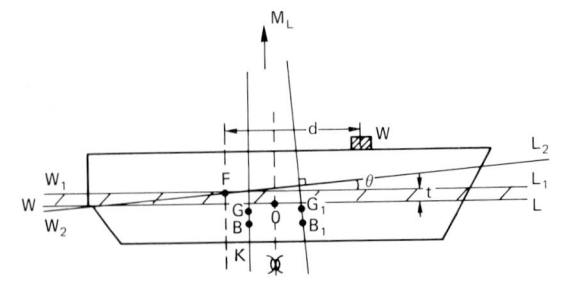

Figure 3.8 Sinkage and trim

3.2.3.1 Sinkage

Addition of a weight w at the longitudinal position of *F* has no effect on trim but causes a parallel sinkage of the hull by an amount t. Let the area of the waterplane be A; this is assumed to be approximately constant for small changes in draught. Then as w is supported by the extra displacement between waterlines *WL* and W_1L_1, we have:

$$t = \dfrac{w}{A\rho_w g}$$ (3.25)

3.2.3.2 Trim

The couple $w.d$ causes no change in displacement but trims the vessel about *F*. Now for small angles, trim may be treated in a similar manner to heel. By comparing *Figure 3.8* with *Figure 3.3* we can envisage a longitudinal metacentre M_L directly analogous to the transverse metacentre *M*. Indeed, similar formulae apply:

$$GM_L = KB + BM_L - KG$$ (3.26)

$$BM_L = I_L/\nabla$$

where I_L is the longitudinal second moment of area of the waterplane about its centroid, *F*. The practical difference from the transverse case is that I_L, being dominated by the length rather than the beam, is very large. Hence BM_L is very much greater than *BM*. As *KB* and *KG* are the same in both transverse and longitudinal senses and are much smaller than BM_L, then $GM_L \simeq BM_L$. GM_L is far larger than *GM*: typically GM_L is a few hundred metres.

Returning to the couple $w.d$, this must be balanced by the longitudinal righting moment $(\triangle + w).GM_L.\theta$.

Thus: $\theta = \dfrac{w d}{(\triangle + w)GM_L}$ (3.27)

Strictly speaking, *G* will have risen slightly due to the addition of w, but the amount is usually negligible compared to GM_L.

Knowledge of sinkage t and trim θ enables us to calculate the new draughts of the ship forward and aft. Referring to *Figure 3.3*.

Draught at the forward perpendicular increases by

$$\left(\frac{L}{2} + OF\right)\theta + t \quad (3.28)$$

Draught at the aft perpendicular decreases by

$$\left(\frac{L}{2} - OF\right)\theta - t \quad (3.29)$$

where L is the length between perpendiculars (LBP) and OF is the longitudinal distance between amidships O and centre of flotation F.

3.2.3.3 Combined heel and trim

Suppose the weight w is placed not on the centreline, but at a distance y to port. Then, provided the weight is still relatively small, the sinkage and trim are unchanged from those just calculated. The only additional effect is that a heeling couple $w.y$ appears, causing the ship to heel by an amount:

$$\emptyset = \frac{w\,y}{(\triangle + w)\,GM} \quad (3.30)$$

where GM is the transverse metacentric height.

3.2.3.4 Hydrostatic particulars

These are used to simplify the calculation of sinkage and trim. They take the form of tables or graphs, called hydrostatic curves, of the geometric properties of the ship for a wide range of mean draughts. Hydrostatic particulars are produced at the design stage for an assumed design condition which includes a standard water density and a chosen value of trim between perpendiculars. The particulars usually included are KB, KM, LCB (longitudinal position of centre of buoyancy about amidships), LCF (longitudinal position of centre of flotation abaft amidships), displacement, tonnes per unit (cm) parallel immersion (TPI) and moment to change trim 1 m ($MCT1m$).

The first four of these are purely geometric. The remainder depend on the water density. $MCT1m$ depends also on KG. $MCT1m$ may be approximated by assuming $GM_L \simeq BM_L$. A typical set of hydrostatic curves is shown in *Figure 3.9*.

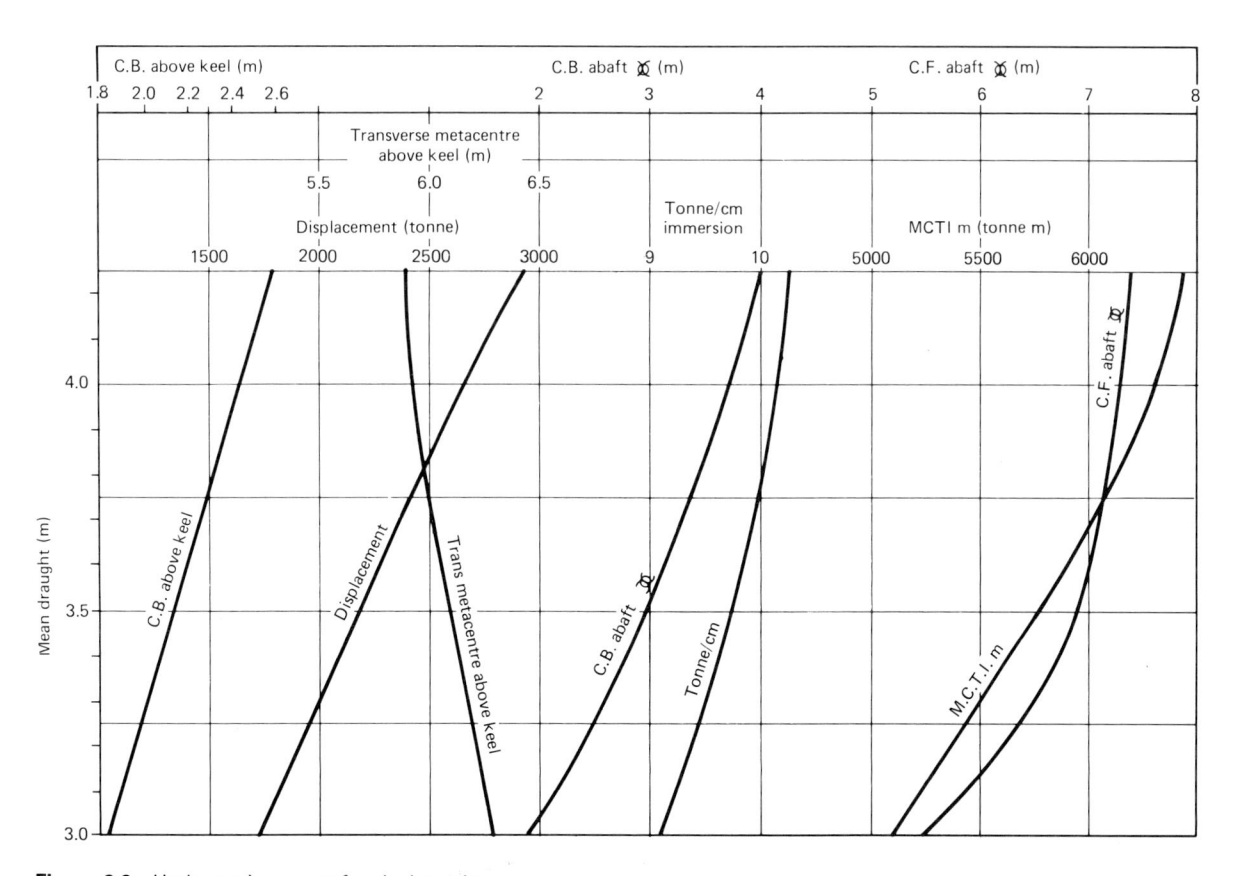

Figure 3.9 Hydrostatic curves for design trim

The tonnes per cm immersion (TPI) is the increase in displacement per cm parallel sinkage. It is related directly to waterplane area A as:

$$TPI = A \cdot \rho_w \frac{t}{100} \qquad (3.31)$$

where t is 1 cm, A is in m² and ρ_w is water density in tonnes/m³. Calculation of sinkage for addition of a weight w tonnes can now be made directly:

$$t = w/TPI \text{ cm} \qquad (3.32)$$

Similarly, the moment to change trim 1 m between perpendiculars is given in tonne–metres by:

$$MCT1m = \Delta \, GM_L \frac{1}{L} \qquad (3.33)$$

where Δ is in tonnes, GM_L is in m and L, the length between perpendiculars, is in m. Calculation of change of trim for an applied couple $w.d$ (in tonne–metres) is then given directly by:

$$\theta = \frac{wd/L}{MCT1m} \qquad (3.34)$$

where θ is in radians, or the change of trim between perpendiculars equals

$$\frac{w.d}{MCT1m} \text{ in metres}$$

3.2.3.5 Change of water density

Consider a ship of weight (displacement) Δ moving from water of density ρ_1 to water of density ρ_2. If ρ_1 is greater than ρ_2 the ship will sink a little in order to maintain equilibrium of weight and buoyancy. If ρ_1 is less than ρ_2 the ship will rise. The ship will also trim in order to keep LCB below LCG.

In may be simply shown that the parallel sinkage is given by:

$$t = \frac{\Delta}{\rho_1 gA}\left(\frac{\rho_1}{\rho_2} - 1\right) \qquad (3.35)$$

where A is the waterplane area (all in consistent units).

or

$$t = \frac{\Delta}{TPI}\left(\frac{\rho_1}{\rho_2} - 1\right) \text{ in cm} \qquad (3.36)$$

if Δ is in tonnes and the TPI (in Te/cm) applies to the initial condition of the ship in density ρ_1. If ρ_1 is less than ρ_2, t will be negative, i.e. the ship will rise

The change in trim between perpendiculars can similarly be shown to be given approximately by:

$$\delta \, Trim = \frac{a \Delta}{MCT_1}\left(\frac{\rho_1}{\rho_2} - 1\right) \qquad (3.37)$$

where a is the horizontal distance between LCB and LCF, and MCT_1 is the moment to change trim, both for the initial condition with density ρ_1. This will be in metres, if a is in metres, Δ is in the Te and MCT_1 in Te.m/m. If ρ_1 is greater than ρ_2 (and LCB is forward of LCF as is usual) the ship will trim by the bow.

3.2.3.6 Docking

A ship will normally be docked with trim by the stern. This is to ensure that the main docking load is taken at the strengthened after cut up (ACU) as shown in *Figure 3.10*. The vertical load P at the ACU will be a maximum at the instant when the ship touches (sues) all along the dock blocks.

The maximum load P can be found by treating it as a removed weight after the manner of *Figure 3.8*. In the docking calculation, P is unknown but the change of trim is known; it is the difference between the initial, free floating trim and the trim of the line of dock blocks, normally zero.

Thus change trim

$$\theta = \frac{P\,d}{L.MCT} \text{ in radians} \qquad (3.38)$$

$$\text{or } \delta \, T \simeq \frac{Pd}{MCT} \qquad (3.39)$$

where δT is change in trim (in metres) between perpendiculars

$$\text{so } P = \delta \, T \cdot \frac{MCT}{d} \qquad (3.40)$$

Note that P is proportional to δT. The initial trim should not be large or large forces will occur. Having found P, the parallel rise can be found. It equals P/TPI. Hence the ship's draughts forward and aft may also be calculated at the moment of sueing all along the dock blocks.

The upward force P will tend to destabilize the ship, as it acts below the ship's transverse metacentre. It reduces the effective transverse metacentric height GM by (P/Δ). KM, where Δ is the ship's displacement on docking and KM is the vertical distance between the ACU and the transverse metacentre, the latter calculated in the sueing condition. The reduction in GM can be dangerously large. Unless a cradle or side blocks are used, it is necessary to put steadying shores between the ship's sides and the dock walls, at the longitudinal position of the after cut up, as soon as the ACU first touches the dock blocks.

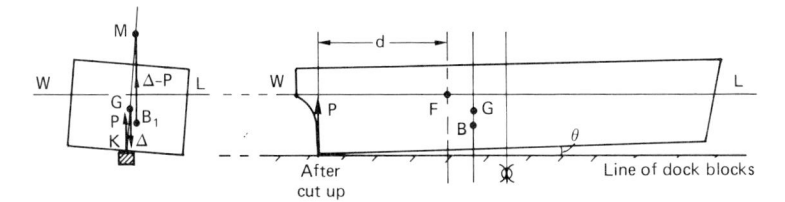

Figure 3.10 Docking

3.2.3.7 Grounding

Accidental grounding of the ship can be treated, for the calm water case, in a similar manner to docking. The vertical force at the point of grounding and the draughts of the ship as the height of tide changes can be found by equating the change in draught at the point of grounding, due to parallel rise and trim, to the known change in tide height.

3.2.4 Large angle stability

We have so far only considered transverse stability at small angles, up to about 10° of heel. We now investigate the more general case.

In *Figure 3.11*, the ship is shown inclined to an arbitrary angle of heel ø. The displacement of the ship is assumed to be constant and its centre of gravity fixed. Unlike the situation at small angles (*Figure 3.6*), it can no longer be assumed that the new waterplane is symmetrical about the ship's centreline. The prometacentre, M_1, is not a fixed point but varies with heel angle ø. Also the ship will, in general, trim fore and aft

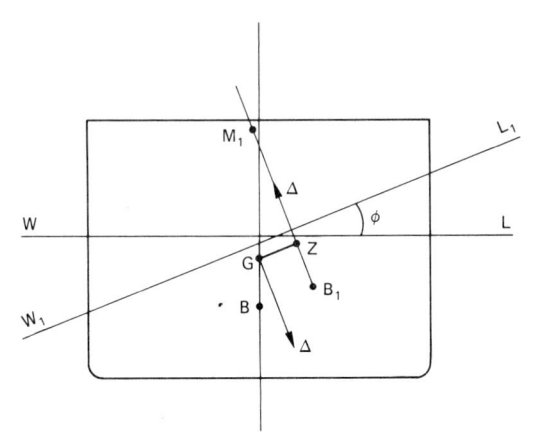

Figure 3.11 Transverse stability at large angles

as it heels, owing to the non-uniform underwater shape.

The restoring moment acting to return the ship to the upright is $\triangle . GZ$, as before. As $\triangle$ is constant, we are therefore interested in the variation of GZ with heel angle ø. Methods of calculating GZ for a range of displacements and heel angles via cross curves of stability are given in the references mentioned at the beginning of this section. Modern computer methods allow accurate calculation with full allowance for trim effects.

3.2.4.1 The GZ curve

A typical curve of GZ variation with heel angle is shown in *Figure 3.12*. This is known as a curve of statical stability or GZ curve. Important features of this curve are first, that at small angles $GZ \simeq GM$ ø (Section 3.2.2). The slope of the GZ curve near the origin is given by:

$$\frac{d\,GZ}{d\,\text{ø}} = \frac{d(GM.\text{ø})}{d\,\text{ø}} = GM \qquad (3.41)$$

Thus, if the slope of the curve at the origin is extrapolated to a value of ø = 1 radian (Point A) the ordinate has a value equal to GM, the upright metacentric height. Second, as ø increases beyond small values, the slope of the GZ curve usually increases above the initial slope. For a wall sided ship, this is predicted by the wall sided formula for GZ (equation 3.16). For a rounded bilge vessel, this increase in slope may be small or even negative.

At Point B, there is a point of inflexion in the GZ curve and the slope begins to decrease. This change is associated with immersion of the deck edge or emergence of the turn of bilge. At this point, the waterplane width (with its direct effect on the second moment of area I, and hence on BM) ceases to increase and begins to reduce. At Point C, GZ reaches its maximum value, GZ_{MAX}. A steady overturning moment applied to the ship of value greater than $\triangle . GZ_{MAX}$ would cause it to capsize.

At Point D, GZ becomes zero. This point is called the point of vanishing stability and distance OD is called the range of stability. The area under the curve is a measure of the work done in steady conditions to heel the ship to angle ø_D. It is called the *dynamical stability*, as it is related to the energy absorbing capability of the ship in roll.

GZ curves are normally produced for a range of conditions. These include deep (maximum draught), light (minimum draught) and perhaps an extreme light (harbour) condition which may be the most severe. The

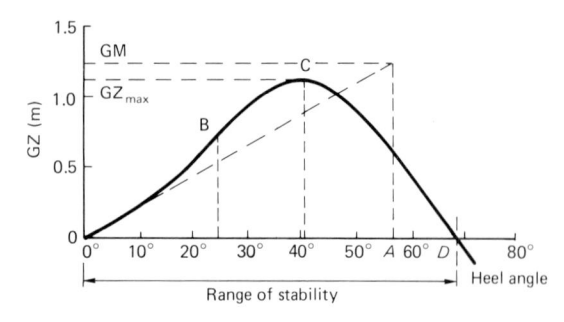

Figure 3.12 Typical GZ curve

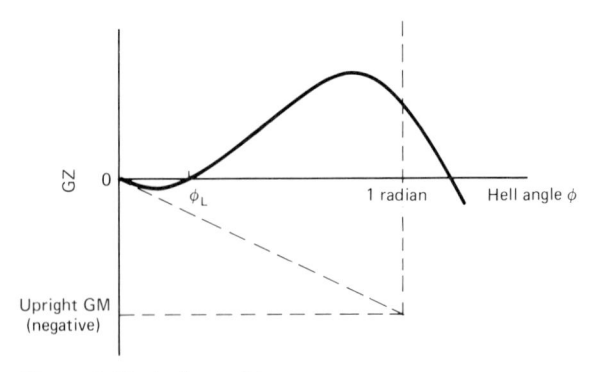

Figure 3.13 Loll condition

GZ curve is normally corrected for any expected free surface effects in the condition concerned.

Important though the GZ curve is in characterizing large angle stability, it should be recognized that it is artificial in several respects. It makes no allowance for reality at very large angles, when cargo might shift, for example. Furthermore, as drawn in *Figure 3.12* it makes no allowance for water flooding into the ship at large angles, such as through deck openings or engine intakes. This can be allowed for by equating the maximum effective heel angle of the curve and hence the range of stability to the minimum angle of downflooding. Finally, the GZ curve assumes quasi-static conditions, which are certainly not present in a heavy sea.

Nevertheless, the GZ curve continues to be the basis for international regulations governing ship stability. This will remain the case until rational alternatives are developed and accepted. Some of the research towards that goal is described in the *Proceedings of the International Conference on The Safeship Project: 'Ship stability and safety'* (RINA, 1986).

3.2.4.2 Loll

In certain conditions of loading it is possible for a ship to have negative GM when upright. The GZ curve will then look like *Figure 3.13*, where the negative GM causes a negative slope at the origin. As the ship heels to larger angles, GZ increases in value and becomes positive at angle ϕ_L. The ship is in unstable equilibrium when upright and will flop over rapidly to the angle ϕ_L, called the angle of loll. It is important to recognize the loll condition as it is potentially dangerous. It cannot, unlike heel, be corrected by applying a counter moment to the ship. This would merely cause the ship to flop over rapidly to a larger angle on the other side. A loll condition can occur when a cargo vessel unloads deep cargo holds in harbour, allowing the centre of gravity of the empty ship to rise significantly relative to that of the loaded ship. Loll can also occur when large liquid free surfaces are allowed to appear in many tanks at the same time.

3.2.4.3 Stability standards

Minimum standards for the intact stability of merchant ships and other vessels have been agreed internationally through the International Maritime Organisation (IMO) (see Section 3.7), and are kept under regular review. The standards in force in 1989 stipulate minimum values for various parameters of the GZ curve. These parameters include the initial GM, the angle of maximum GZ and the area under the GZ curve between certain angles of heel.

These standards are only intended to set minimum requirements. A ship designer should consider the operating conditions of a particular vessel before accepting its GZ curve as adequate. One approach to setting rational criteria for ship stability in terms of the GZ curve (Sarchin and Goldberg, 1962), is the basis of the stability standards of many navies. It proposes criteria for the GZ curve covering a number of situations where stability could be critical. These include beam winds and rolling in extreme weather conditions, lifting of heavy loads, crowding of passengers to one side, heeling during high speed turns, accumulation of ice on upperworks and docking. Some of these conditions can occur simultaneously and suitable criteria are proposed. The IMO has begun to recognize this approach. Its Recommendation on a Severe Wind and Rolling Criterion (Weather Criterion) for the Intact Stability of Passenger Vessels and Cargo Ships of 24 Metres Length and Over was adopted in November 1985.

In considering stability criteria, it should be realized that GM and GZ values for a given ship vary in a seaway even in a quasi-static sense. This is due to the change in shape of the waterplane and the underwater form as a wave travels past the ship. A ship in a head or following sea, particularly when the wavelength approximates to the ship length, will have its transverse stability reduced below the calm water value when the wave crest is amidships. Conversely, stability is increased above the calm water value when the wave trough is amidships. Variations in GZ of the order of ± 30% have been calculated in some cases. The possibility of related resonant rolling effects leading to capsize is discussed in Section 3.5.

3.2.4.4 Effect of flare

A major improvement to statical stability at large angles can be achieved by incorporating flare above the waterline. This is shown in *Figure 3.14*. The improvement can be substantial even with a small amount of flare. It is due to the rapid increase in waterplane area and width (hence I, hence BM) as the ship heels. For similar reasons, flare also overcomes the common problem of stability being reduced as a ship grows in displacement over its life through additions of equipment and fittings (Burcher, 1980, Aston, Rydill and Beck, 1987).

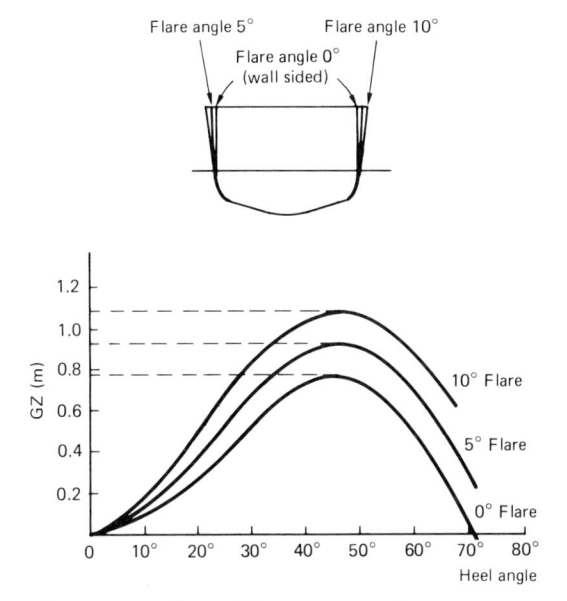

Figure 3.14 Effect of flare angle on GZ

3.2.5 Flooding and damage stability

Flooding remains a major cause of ship loss. It can occur through collision, grounding, fire, explosion, maloperation of the ship systems or enemy action. Flooding can cause loss of a ship through bodily sinkage (foundering) when all the above water reserve of buoyancy is lost, through capsize following loss of transverse stability, or through plunging by the bow or by the stern following loss of longitudinal stability.

The main design measure to prevent spread of flooding and eventual loss is provision of a watertight subdivision. This can take the form of watertight decks and double bottoms, watertight transverse bulkheads and watertight longitudinal bulkheads. The latter are rarely advisable except in very large ships or specialized ships, such as tankers. This is because longitudinal bulkheads confine flooding in such a way as to cause a heeling moment on the ship. In the damaged condition, where transverse stability is already reduced this can rapidly lead to capsize. The preferred form of watertight subdivision is therefore by transverse bulkheads.

Collision is most likely to occur at the bow. Ships are therefore required by law to have at least one collision bulkhead placed well forward, at about 5% of the ship's length from the bow. Other transverse bulkheads must be placed sufficiently close together to contain flooding arising from the assumed damage scenarios for the type of ship concerned.

3.2.5.1 Damage calculations

It is possible to treat flooding into the ship as an added weight after the manner of Section 3.2.3. If the amount of flooding is small and localized the calculation is reasonably straightforward. If, however, the amount of flooding is substantial then the assumptions of Section 3.2.3 about small additions are no longer valid and the approximations are no longer of reasonable accuracy. Furthermore, the exact amount of flooding will not normally be known at the start. The solution has therefore to be found by trial and error. In these circumstances it is much easier to treat the flooding as lost buoyancy rather than added weight.

In *Figure 3.15*, a ship is shown floating initially at waterline WL. It has been damaged such that the volume between the two watertight bulkheads has been opened to the sea. In the lost buoyancy method, the ship is assumed to retain a constant displacement. However, the buoyant contribution of the hull between the bulkheads is assumed to be lost. The ship will therefore undergo parallel sinkage until the volume of the wedges AWW_1C plus BLL_1D exactly equals the buoyancy lost in the volume $ABFE$, i.e. up to the original waterline. The ship will also trim (and heel if there is any transverse asymmetry of the flooding) until the new centre of buoyancy B_1 is vertically below the (unchanged) centre of gravity G.

The calculation procedure is to calculate lost buoyancy to waterline AB. The calculation should take account of the permeability of the ship volume $ABFE$: depending on the equipment and fittings in this part of

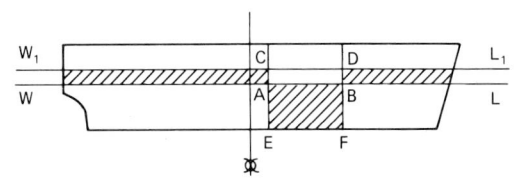

Figure 3.15 Lost buoyancy calculation

the ship, the permeability may vary anywhere between about 97% for an empty compartment to perhaps 60% in a full cargo hold. Next calculate revised hydrostatic particulars for the reduced waterplane. This is necessary because the original hydrostatic particulars are no longer valid owing to the loss of waterplane area AB and the change in buoyant underwater form. This will produce modified values of TPI, LCF, I_T (second movement of area about centreline) and I_L (second moment of area about new LCF). Hence $BM_T = I_T/\nabla$ and $BM_L = I_L/\nabla$. Note that ∇ is unchanged.

The next stage calculate parallel sinkage using modified TPI from the previous calculation and calculate movement of CB vertically and longitudinally due to transfer of buoyant volume from $ABFE$ to $AWW_1C + BLL_1D$. Hence find new KB and LCB and find new GM_T ($= KB + BM_T - KG$) and GM_L ($= KB + BM_L - KG$) noting that KG is unchanged. There is no free surface effect to be calculated due to the flooding between A and B, as that area no longer forms an effective part of the ship. Then calculate the trim required to remove longitudinal mismatch between LCB and LCG. If there is transverse asymmetry, calculate the heel required to remove transverse mismatch between B and G.

Depending on circumstances, it may be necessary to calculate the revised GZ curve, rather than just the new GM_T above. GM_L should also be studied to check for any tendency to plunge by the bow or stern. This can be a real possibility if the longitudinal extent of flooding is large and the damaged waterline intersects the upper deck. This latter possibility is prevented in theory by the floodable length regulations mentioned below.

The present international regulations for transverse damaged stability of merchant ships are minimal. They require only a residual GM of at least 0.05 m as calculated by the constant displacement (i.e. lost buoyancy) method. Passenger vessels have additional heel angle criteria applied. Some types of merchant ship have no statutory damaged stability requirements at all.

The situation for passenger ships constructed on or after 29 April 1990 is more satisfactory. Under an amendment adopted by the Maritime Safety Committee of IMO in 1988, residual stability standards of such ships after damage must take account of such factors as wind pressure, the effect of passengers on one side and the weight of survival craft being launched.

3.2.5.2 Floodable length curves

Floodable length curves are produced to check that a ship will not plunge longitudinally under specified flooding conditions. A typical curve is shown in *Figure 3.16*. The floodable length at any point along the ship is defined as that length of ship, with the point as centre, which can be flooded without immersing any point of an imaginary line called the margin line. The margin line is sited at least 76 mm below the bulkhead deck. The bulkhead deck is the uppermost watertight deck to

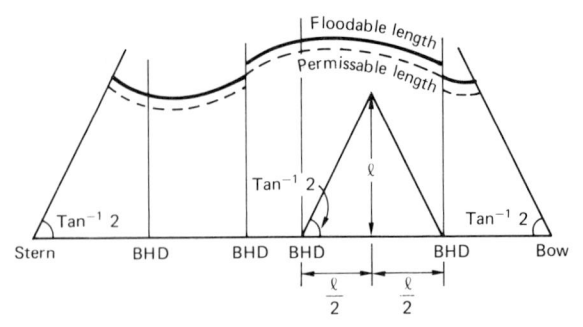

Figure 3.16 Floodable length curve

which the watertight transverse bulkheads are taken. The purpose of the margin line is to ensure a very small margin of safety before the bulkhead deck is immersed. The ship is assumed to have no list.

Calculation of the floodable length curve can be undertaken using a standard tabulated format. This is available in the UK from the Department of Transport. As seen in *Figure 3.16*, the floodable length curve is not continuous. This is due to the varying permeability of different compartments along the ship's length. Under the regulations, the floodable length is multiplied by a factor of subdivision (less than or equal to one) to yield a permissible length curve. The factor of subdivision is specified by the regulations and depends on the length of the ship and the type of service. It is more or less an inverse of the compartment standard. This represents the number of adjacent watertight compartments along the ship's length which can be flooded without immersing the margin line.

Once the permissible length curve is produced, it is possible to check that the watertight transverse bulkheads are sufficiently closely spaced to meet the required compartment standard. To carry out this check for a given pair of bulkheads, a triangle is erected with both base and height equal to the distance between the bulkheads. Provided that the vertical and horizontal scales are equal, this gives a base angle for the resulting isosceles triangle as:

$$\tan^{-1}\left(\frac{\text{height}}{\frac{1}{2}\text{base}}\right) = \tan^{-1} 2 \qquad (3.42)$$

This is shown in *Figure 3.16*. The apex of the triangle must lie below the permissible length curve. If it does not, the bulkhead spacing must be adjusted until it does.

Under current IMO agreements, cargo ships over 225 m (738 feet) overall length are required to have a one compartment standard throughout. For lengths 100–225 m (328–738 feet), this requirement is relaxed for the main machinery compartment. Under 100 m (328 feet) overall length, there is no requirement for any compartment standard; in other words, there is no floodable length requirement. An alternative, probabilistic, method for setting subdivision requirements is allowed by IMO but not yet enforced. A more general probabilistic approach to subdivision and damage stability is under review.

3.2.5.3 Freeboard and load lines

The International Load Line Convention governs minimum permissible freeboard. The main purpose is to ensure an adequate reserve of buoyancy against foundering, although freeboard also influences the *GZ* curve at larger angles.

Freeboard is measured between a still waterline and the freeboard deck. This is defined as the uppermost complete exposed deck which has permanent means of watertight closure and below which the sides of the ship are fitted with permanent means of watertight closure. Alternatively, a lower deck may be taken as the freeboard deck, provided it is a permanent deck and continuous fore and aft and athwartships.

The minimum freeboard allowed under the convention is found direct from tables. Two different tables apply depending on the type of ship. Type A ships are those designed to carry liquid cargoes in bulk and which have only small openings to the cargo tanks. Such ships usually have a high standard of subdivision. Type B ships comprise the remainder, which in general are required to have higher freeboards. Various corrections are applied in the tables depending on the ship's length: depth ratio, block coefficient, the sheer of the freeboard deck and the size of the superstructure. Different freeboards apply for different water densities, operating zones and seasons. Winter North Atlantic is considered to be the most severe environment and hence needs the largest freeboard.

The rules also cover minimum standards for watertight integrity of such things as watertight boundaries, hatch covers, freeing ports and penetrations. The various minimum allowable freeboards must be displayed on the ship's side on a load line mark, *Figure 3.17*. This used to be called the Plimsoll line.

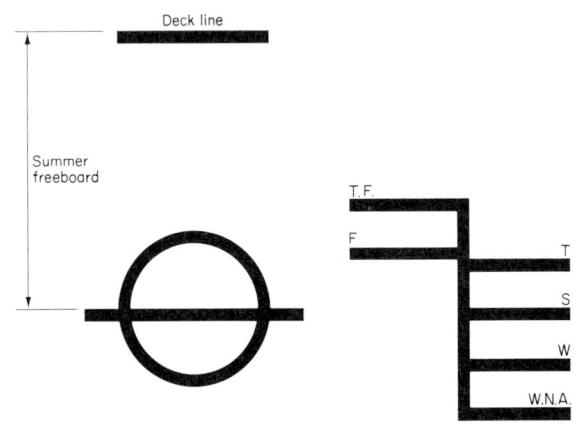

Figure 3.17 Load line mark (Plimsoll line).
TF = tropical fresh; F = fresh; T = tropical; S = summer; W = winter; WNA = Winter North Atlantic

3.2.6 Computer aided design packages

A wide range of CAD packages is available for calculating hydrostatic data, statical stability curves and damaged stability. Major design consultancy firms and most naval architecture departments in universities have packages for sale or hire. A large number of programs are available for use on microcomputers. Surveys of such software have been compiled by the RINA Small Craft Group (Roy, 1987) and by the SNAME Small Craft Committee (*Small Craft Software Catalog*, 1988). Although intended mainly for small craft applications, much of the software covered in these surveys is equally suitable for large ships (see also Section 3.7).

3.3 Resistance and propulsion

Sections 3.3, 3.4 and 3.5 are concerned with hydrodynamics, the branch of fluid dynamics that seeks to quantify the forces and moments on a body such as a ship, due to the flow of water around it. It is a complex subject that is not yet amenable to a complete theoretical treatment. Despite major theoretical advances in recent years, practically useful results can still only be obtained by a prudent mixture of theory and empiricism.

The basic theory of fluid dynamics is covered in a number of text books, such as Massey (1979). Hydrodynamics and its application to naval architecture is covered in some detail in Clayton and Bishop (1982), Rawson and Tupper (1984) and Lewis (1988). Specialist texts available include Saunders (1957) and Newman (1977) and RINA *Marine Technology Monograph No. 6*.

3.3.1 Basic considerations

A fully submerged body moving through a fluid will accelerate fluid particles from rest in its vicinity. This causes pressure fluctuations in the fluid. If the fluid is inviscid, that is to say an ideal fluid with no viscosity, then the pressure fluctuations balance out along the body and there will be no resistance to its motion. A real fluid, however, is viscous and friction will be generated between the fluid and the body.

The change in fluid velocity due to friction will be very marked close to the body. This region is called the boundary layer. A description of the development of the boundary layer is given in Chapter 2. The friction gives rise to a resistive or drag force, know as skin friction, opposing the motion of the body. Furthermore, the pressure field is itself modified and this leads to another component of viscous resistance called pressure resistance, sometimes called form drag. The viscous resistive force will, of course, be additional to the forces of gravity and buoyancy already discussed.

Consider the body approaching the interface between the fluid and another fluid of different density.

The pressure fluctuations around the body due to its motion will cause elevations and depressions of the interface. In other words, waves are formed at the interface. The pressure balance around the body is altered and energy is expended in forming these waves. This results in an increase in the total drag on the body.

We therefore see that a ship moving at the free surface between the two fluids, water and air will experience two kinds of resistance to its motion: viscous and wave making. The relationship is shown in simple form in *Figure 3.18*. By far the greater part of the viscous resistance will be due to water flow around the hull. However, there will be a small component of viscous drag, due to air flow over the above water part of the ship.

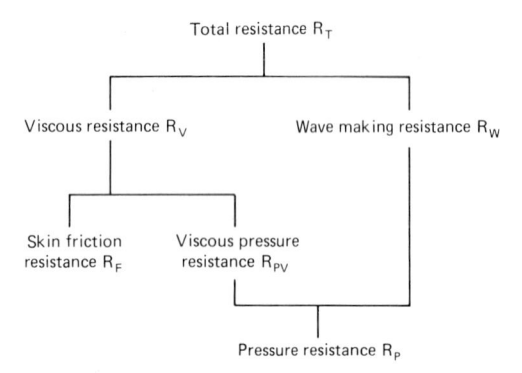

Figure 3.18 Components of ship resistance

The wave systems generated by a ship in calm water may be predicted theoretically with some accuracy. *Figure 3.19* shows the expected divergent and transverse wave systems emanating from the bow and stern regions. The waves are gravity waves as described in Chapter 1. In deep water, they move with velocity:

$$C = \sqrt{\frac{g\lambda}{2\pi}} \quad (3.43)$$

where g is gravitational acceleration and λ is the wavelength.

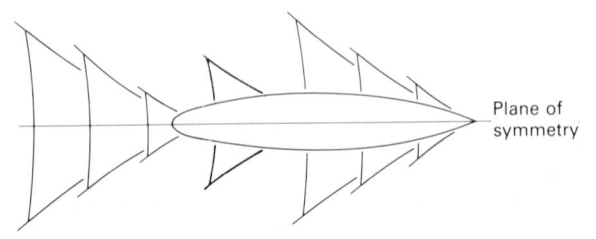

Figure 3.19 Ship wave system in calm water

3.3.1.1 Dimensional analysis

The magnitude of the total hydrodynamic force on the ship will depend on the physical attributes of the water and the ship. The basic relationship may be found by the process of dimensional analysis (Massey, 1971; Clayton and Bishop, 1982). A summary is given in Chapter 2. The following relationship is found:

$$\frac{R}{\rho V^2 L^2} = \text{function of}$$

$$\left[\frac{\rho V L}{\mu}, \frac{V}{\sqrt{gL}}, \frac{\Delta p}{\rho V^2}, \frac{V}{c}, \frac{V^2 \rho L}{\sigma} \right] \quad (3.44)$$

where R is the total hydrodynamic resistance (N), V is velocity of the ship (m/s), L is waterline length of the ship (m), ρ is water density (kg/m³), μ is water viscosity (kg/m.s), g is gravitational acceleration (m/s²), c is velocity of sound in water (m/s), Δp is difference between the static and vapour pressures in the water (N/m²) and σ is surface tension of the water (kg/s²). Consistent units must be used throughout (SI units are shown above in brackets).

The term $R/\rho V^2 L^2$ is called the resistance coefficient C.

For theoretical convenience, it is often expressed as:

$$\frac{R}{\frac{1}{2}\rho V^2 L^2}.$$

$$\frac{\rho V L}{\mu}$$

is called the Reynolds number; it is sometimes written as

$$\frac{VL}{\nu}$$

where $\nu \left(= \frac{\mu}{\rho} \right)$

is the kinematic viscosity (m²/s).

$$\sqrt{\frac{V}{gL}}$$

is called the Froude number; this is, in effect, a ratio between ship velocity (V) and the velocity

$$\left(\sqrt{\frac{gL}{2\pi}} \right)$$

of a wave of length equal to that of the ship.

$$\frac{\Delta p}{\rho V^2}$$

is the cavitation number.

$$\frac{V}{c}$$

is called the Mach number and

$$\frac{V^2 \rho L}{\sigma}$$

is called the Weber number.

The Mach number and the Weber number have no practical significance for ship hydrodynamics and are not considered further.

We then see that the resistance is a function of three quantities. One represents viscosity, expressed through the Reynolds number. The second is related to wave making expressed through the Froude number. The third term, the cavitation number, is not of direct significance for ship resistance. It is, however, important in propeller design.

3.3.2 Components of ship resistance

In equation (3.44), we found the quantities that ship resistance depends upon. In order to calculate the resistance, it is necessary to identify the function that relates the various quantities. That function is extremely complicated for a surface ship, owing to the complex interactions between the various components of resistance. For example, the ship generated waves modify the pressure distribution around the hull. This in turn affects the viscous resistance. An empirical plot of total resistance coefficient,

$$C_T = \frac{R_T}{\frac{1}{2}\rho V^2 L^2} \quad (3.45)$$

against ship speed in calm water has the typical form shown in *Figure 3.20* (curve A).

Curve B shows the plot for an equivalent deeply submerged body. The curves diverge rapidly, albeit in an irregular way as speed increases. The difference between the curves at a given speed is due to wave making at the surface. In order to simplify the analysis, it is usual to assume that wave making resistance and viscous resistance can be accounted for separately. The two components of viscous resistance (viscous pressure resistance and skin friction resistance) are also assumed to be separable, as shown in *Figure 3.18*. We can therefore write, in the terminology of *Figure 3.18*:

$$R_T = R_V + R_W = R_F + R_{pv} + R_W = R_F + R_p \quad (3.46)$$

or in coefficient form:

$$C_T = C_V + C_W = C_F + C_{pv} + C_W = C_F + C_p \quad (3.47)$$

Froude (1876) was the first to realize the essential difference between friction and wave making. He tested a series of geometrically similar scaled models and observed that the wave patterns generated were similar when the ratio $V/\sqrt{L}$ was the same for each model. He then defined a residuary resistance equal to the total resistance minus the skin friction, i.e.:

$$R_R = R_T - R_F \ (= R_{pv} + R_w = R_p). \quad (3.48)$$

He assumed R_R for each model to be a function of $V/\sqrt{L}$ only. This is, of course, a simple dimensional variant of Froude number $V/\sqrt{gL}$. His assumption is not strictly correct as the R_{pv} component of R_R is also a function of Reynolds number. However, the assumption is surprisingly accurate and ship resistance estimates are often still made using Froude's approach.

3.3.2.1 Wave making resistance

A typical plot of wave making resistance coefficient C_W against Froude number is shown in *Figure 3.21*. The curve of C_W is seen to have a series of humps and hollows. These are due to interference between the wave systems generated by the ship. This can be particularly marked for the transverse wave systems. At some speeds the waves generated at the bow reinforce the waves generated at the stern while, at other speeds, the systems will tend to cancel each other out.

It is possible to predict the positions of the humps and hollows with some accuracy (Clayton and Bishop, 1982). A major hump can be predicted theoretically when the wave length is similar to the length of the

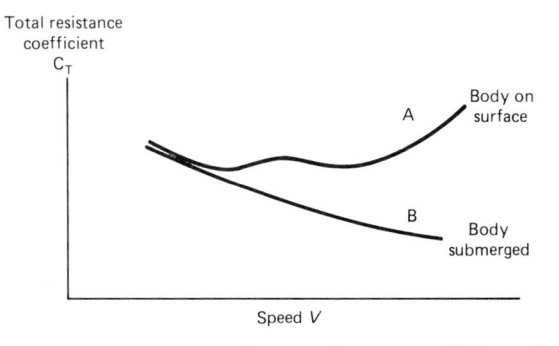

Figure 3.20 Variation of total resistance coefficient with speed V

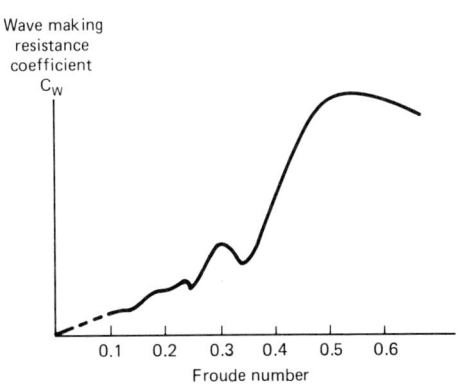

Figure 3.21 Variation of wave making resistance coefficient with Froude number

ship. This occurs at a Froude number around 0.50 and is known as the main hump. The earlier hump, occurring at a Froude number around 0.3, is known as the prismatic hump. At higher Froude numbers, the oscillations in the curve tend to die out and C_w decreases with increasing ship speed. Note that total resistance will continue to increase with speed, however, as R_w is proportional to $C_w \times V^2$ and viscous resistance increases rapidly at high speed.

Many attempts have been made at direct theoretical calculation of wave resistance (Andrew, Baar and Price, 1988). However, none has yet reached general acceptance. Estimation therefore remains semi-empirical, based on model test data.

Wave making resistance of a displacement ship can be minimized by attention to the basic hull form. It can be further reduced at higher speeds by fitting a bulbous bow. This works by modifying the pressure distribution around the hull so as to reduce the amplitude of the bow wave. However, the effectiveness of a bulbous bow varies with the ship's draught and it can actually increase resistance at low speeds. As a crude guide, a bulbous bow is worth investigation for a ship that operates for a significant part of its life at a Froude number greater than about 0.24 or for any ship having a high block coefficient (Inui, 1962; Yim, 1974; Watson, 1981; Blume and Kracht, 1985; Buxton and Logan, 1987; Lewis, 1988).

3.3.2.2 Skin friction resistance

The skin friction drag on a flat surface may be calculated by the theories of classical fluid dynamics. The skin friction resistance coefficient, C_F, is found to depend on Reynolds number Re in the form:

$$C_F = A(\text{Re})^{-n} \tag{3.49}$$

where A and n are constants.

At low Reynolds numbers, the flow in the boundary layer has a steady pattern known as laminar flow. While the flow remains laminar, A and n have fixed values. As the Reynolds number is increased, a stage is reached called the transition region. Here, the steady flow pattern within the boundary layer breaks down and the boundary layer becomes turbulent. For a flat plate this generally occurs at a Reynolds number between 3×10^5 and 10^6. At higher Reynolds numbers, the boundary layer flow remains turbulent and the constants A and n take on new fixed values. This is shown in *Figure 3.22*. For practical purposes, the flow over a ship hull is always turbulent. For a curved surface such as a ship, flat plate formulae may be used, based on the equivalent surface area. Form effect is allowed for separately via viscous pressure resistance or included in residuary resistance.

A great deal of experimental work has been done in towing tanks using flat plates and pontoons to refine the estimation of skin friction resistance. In practice,

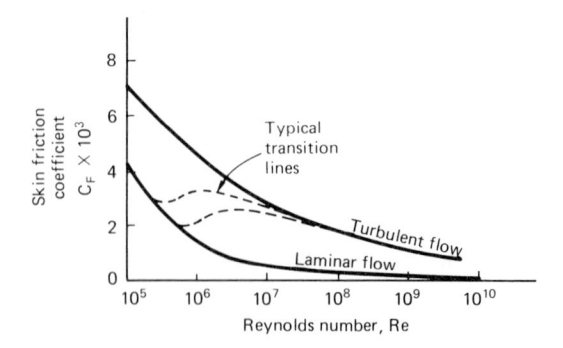

Figure 3.22 Skin friction coefficient for semi-infinite flat plate

the formula now in most common use for the hull skin friction coefficient is:

$$C_F = \frac{0.075}{[\log(\text{Re}) - 2]^2} \tag{3.50}$$

This was developed by the International Towing Tank Conference (ITTC) in 1957 as a correlation line between model resistance test results and full scale ship values. Its use in resistance estimation is discussed below.

Skin friction resistance will be increased by roughness and any fouling of the hull. The effect of hull roughness depends on the local Reynolds number. At low values it has little effect but as it increases the skin friction resistance coefficient increases above the smooth surface value. At high Re, the coefficient becomes approximately constant and hence independent of Re. The constant value is higher for greater roughness. Thus for real ships, with high values of Re over most of the hull, the effect of a given value of roughness is to add a constant increment to the skin friction resistance coefficient C_F. The formulation used in the 1978 ITTC *Performance Prediction Method* is:

$$\delta C_F = \left[105 \left(\frac{k_s}{L} \right)^{\frac{1}{3}} - 0.64 \right] \times 10^{-3} \tag{3.51}$$

where k_s is the hull roughness and L is the ship waterline length.

For a newly built hull, k_s typically has a value in the range 80–150 μm. Such a value gives $\delta C_F \simeq 0.0004$ which is commonly used in practical resistance estimation. The actual value of k_s will depend on the preparation and coating of the hull plating. Roughness tends to increase in service, the amount depending on paint type and corrosion prevention measures. Increase of roughness might typically be in the range 10–70 μm/year, leading to a power increase of perhaps 1–7% per year to sustain the same ship speed.

The fouling of hulls due to marine organisms such as weed, slime and barnacles is discussed in Chapter 1.

The rate of fouling depends on the paint systems used, areas of operation and time in port. The effect of fouling is to increase skin friction resistance. The amount has been estimated in the past by various methods. One was to assume an increase in C_F of 1/4% per day. An average operating condition was taken at six months out of dock, since the last time the hull was cleaned and new antifouling paint applied. This gave an increase in C_F of 45%. Another method was to assume an average condition in which roughness and fouling together caused a fixed increase in C_T of around 20%. With modern antifouling systems, fouling is much less and lower resistance penalties are often now assumed.

3.3.2.3 Other resistance components

So far we have considered the resistance of the bare hull in calm water. There are four other sources of resistance to be taken into account. These are appendage resistance, aerodynamic resistance, propeller–hull interaction and added resistance due to rough weather. There could also be shallow water effects to consider (Millward, 1982) and even ice resistance in Arctic waters (Peirce, 1979; Kotras, Baird and Naegle, 1983; Vaughan, 1986).

A ship hull has a number of appendages. These may include one or more propeller shafts, shaft brackets, rudders, bilge keels and perhaps fin stabilizers. These are all much shorter than the hull so that the Reynolds number of each appendage is low compared to that of the hull as a whole. Separate estimation of each appendage's resistance is therefore advisable. Empirical data is usually employed. A useful reference for calculating appendage resistance in unusual cases is Hoerner (1965). In practice, the total contribution of all appendages to the ship's resistance is rarely more than 10%, so that accurate calculation is not critical.

Air resistance is proportional to the projected area of the above water part of the ship, and the square of the relative wind velocity (Isherwood, 1973, Gould, undated; Berkelom, 1981; Rawson and Tupper, 1984). In still air, the contribution of air resistance to total ship resistance is only around 1–2%. However, in strong head winds this can increase markedly. Full scale tests on a large tanker have suggested that air resistance is 17% of the total resistance when the relative wind velocity is three times ship speed, e.g. a ship travelling at 12 knots into a 24 knot wind (Berlekom, Tragardh and Dellhag, 1975).

The presence close to the hull of a propeller developing thrust alters the pressure field in the vicinity. This causes an increase in hull resistance known as the augment of resistance. It is expressed as a fraction:

$$a = \frac{T - R_T}{R_T} \quad (3.52)$$

where T is the thrust produced by the propeller and R_T is the total ship resistance (including appendages) that would apply in the absence of the propeller. It is traditionally, though less logically, expressed as a reduction in propeller thrust rather than an augment in resistance. The thrust deduction fraction or factor is then written as:

$$t = \frac{T - R_T}{T} \quad (3.53)$$

t is invariably positive, with a value of the order of 0.1

The effect of wind on air resistance has already been mentioned. Wind generated waves will also increase resistance, partly through direct effects on the flow past the hull and partly through the ship motions caused by the waves. The efficiency of the propeller is also reduced. The overall effect can be to double the power requirement for a given speed in gale force conditions (Beaufort Force 8). Even in Beaufort Force 6 head seas, the power increase may be around 50%. In practice, the ship will usually have to reduce speed to compensate. Indeed, the captain will probably reduce speed even further in rough seas in order to avoid excessive ship motion (see Section 3.5).

Some theoretical and model work has been done on this subject. However, most of the data on power increase or speed loss in rough weather is based on analysis of full scale voyage data (Townsin and Kwon, 1983).

3.3.3 Practical resistance estimation

We have seen that there is no satisfactory theoretical method for calculating ship resistance directly. Recourse must therefore be made to the empirical techniques. We will consider these in order of increasing effectiveness.

3.3.3.1 Full scale data

This method uses the measured resistance of previous similar ships. Suitable adjustments are made to allow for any differences between the previous designs and the new design. The method can be accurate when reliable full scale trials data are available and the new design is close in hull form and type to the basis designs. Unfortunately, these two conditions are not often satisfied. Resistance estimates are therefore rarely made by this method. Even where the method is used, it is usually only for preliminary estimation at the early design stage.

3.3.3.2 Standard series data

A more generally applicable method of employing resistance data on past designs is to use standard series data. A standard series records resistance measurements on a series of model hulls whose key hull shape parameters have been varied in a systematic way, so as to cover the likely ranges of interest to a new design. A designer can then use the series both to choose a suitable low resistance hull form and to interpolate within the series to estimate the resistance of a new design. Provided series data is available for a hull type similar to that desired, the method can be fairly accurate. In practice, methods of hull design for minimum powering have improved considerably in recent years and older series data can only give an initial and probably pessimistic estimate for a new design. The main standard series that are openly available are discussed below.

3.3.3.2.1. Taylor–Gertler Series The Taylor–Gertler standard series (Taylor, 1943; re-analysed Gertler, 1954) covers a wide range of twin screw ships. Both frictional and residuary resistance are expressed in lbf/tonf of displacement ($R/\triangle$). The data are presented as continuous curves. The main parameters covered are:

Speed coefficient (range 0.30–2.0): $\dfrac{V}{\sqrt{L}}$,

where V is speed in knots and L is length in feet.

Displacement (range 20–250): length ratio

$\dfrac{\triangle}{(L/100)^3}$ where $\triangle$ is tonf, L is length in feet.

Prismatic coefficient (range 0.48–0.86):

$\dfrac{\nabla}{AL}$, where ∇ is underwater volume of displacement

(feet^3); A is underwater cross-section area amidships (feet^2).

Beam: draught ratio B/T (values 2.5, 3.0 and 3.75)

Note that the parameters are not all non-dimensional. Longitudinal centre of buoyancy is fixed amidships.

3.3.3.2.2 Series 60 Some data on single screw merchant ship forms, known as Series 60, is given by Todd (1953) and Lewis (1988).

3.3.3.2.3 BSRA series The resistance series undertaken for the British Ship Research Association (BSRA, now British Maritime Technology) is presented by Lackenby and Parker (1966). This series covers single screw merchant ship forms of varying hull proportions. The data is presented as curves of total basic hull resistance, using a Froude constant notation described in the paper. The parameters covered are:

Block coefficient ∇/LBT (range 0.65–0.80).
Longitudinal centre of buoyancy (LCB) (1% aft–3% fwd).
Breadth: draught ratio (B/T) (range 2.0–4.0).
Length/displacement ratio ($L/\nabla^{\frac{1}{3}}$) (range 4.0–6.5).

The charts cover the load draught for the ranges shown, an intermediate draught and two ballast conditions: level and trimmed (Lackenby and Parker, 1966).

3.3.3.2.4 MarAd series This comprehensive series was sponsored by the US Maritime Administration (MarAd Series, 1987). It covers full-form ships having high values of block coefficient, low length to beam ratio and high beam to draught ratio. The hull form geometry and powering estimate computer program are available in tape form (ANSI Standard D format). Further background is given by Roseman and Seibold (1985).

3.3.3.2.5 Other series Series for high speed displacement forms, covering Froude numbers up to around 1.0, are presented by Yeh (1965) and Bailey (1976). A BSRA series for trawler forms is presented by Pattullo and Thomson (1965) and Pattullo (1968).

3.3.3.2.6 Statistical analyses Various attempts have been made to produce regression equations from available resistance data (Holtrop, 1984). These are useful at the early design stage but care must be taken not to apply such equations to hull forms of different type or parameter ranges from those used to derive the equations.

3.3.3.3 Resistance model testing

The most accurate method for estimating the resistance of a new design is to measure the resistance of a scaled model of the ship. Owing to the cost involved, this is normally carried out later in the design process when the ship dimensions are known and the hull form is reasonably finalized. However, model tests should not be carried out too late to allow any modifications to the hull form suggested by the model tests.

Model testing is a skilled process. It is carried out in a long towing tank equipped with a moving carriage to tow the model hull and support the instrumentation for measuring and analysing the resistance force and other parameters of interest. A description of towing tank procedures and the characteristics of 15 of the world's major towing tanks is given in Clayton and Bishop (1982).

The method for estimating ship resistance from data obtained on a geometrically scaled model is based essentially on the theory of Froude (1876) (see Section 3.3.2). The basic procedure is as follows:

1. Measure the total bare hull resistance of the model, $R_{T(m)}$, at the same Froude number ($V/\sqrt{gL}$) as the ship, i.e. with the model speed scaled down from that of the ship in the ratio of the square root of the linear dimensions.
2. Calculate the bare skin friction resistance of the model $R_{F(m)}$ by the method of Section 3.3.2.2.
3. From (1) and (2), find the residuary resistance of the model:

$$R_{R(m)} = R_{T(m)} - R_{F(m)} \qquad (3.54)$$

4. Multiply the model residuary resistance by the ratio of ship: model displacements to obtain the ship residuary resistance:

$$R_{R(s)} = R_{R(m)} \times \triangle_{(s)}/\triangle_{(m)} = R_{R(m)} \times L^3_{(s)}/L^3_{(m)} \quad (3.55)$$

This utilizes Froude's Law of Comparison: the residuary resistance of geometrically similar forms run at corresponding speeds (i.e. same Froude number) is directly proportional to their displacements.

5. Calculate the bare skin friction resistance for the ship, $R_{F(s)}$ by the methods of Section 3.3.2.2.
6. Add the residuary and skin friction resistance for the ship to obtain the total ship bare hull resistance:

$$R_{T(s)} = R_{R(s)} + R_{F(s)} \quad (3.56)$$

It may be noted that any error in the estimation of the skin friction resistance applies to both model and ship. The error is therefore only significant in its effect on the difference between ship and model estimates.

A form factor $(1 + k)$ is usually used to correct the basic skin friction estimate for viscous pressure effects. The value of $(1 + k)$ is found from model tests at very slow speed, where wave making resistance is negligible and hence (from Equation 3.46).

$$R_T \simeq R_F + R_{pv} \equiv (1 + k) R_F \quad (3.57)$$

It is assumed that k is independent of speed and scale. Separate allowance is made for scaling appendage resistance and for roughness and fouling, as described in Section 3.3.2.

3.3.4 Propulsion

3.3.4.1 The propulsor

A ship requires a propulsion device capable of developing sufficient thrust to overcome the resistance of the ship at the maximum design speed. A great variety of propulsors is available, ranging from oars, sails and paddle wheels to pump jets, air propellers and rocket motors (Clayton and Bishop, 1982). Some of the more common types of propulsor are described in Chapter 6.

The most common and versatile propulsor is the screw propeller. This is essentially a device for converting torque, delivered along a rotating shaft, to axial thrust. The key hydrodynamic parameters of a propeller are its diameter D, speed of rotation N, surface area A, pitch P and propeller efficiency η_O. Surface area A is defined via the blade area ratio (BAR). This is the ratio of the developed area of the blades to the disc area:

$$BAR = \frac{A}{\pi D^2} \quad (3.58)$$

Pitch is defined with respect to a helicoidal surface equivalent to the propeller. Such a surface is generated by a line rotating about an axis normal to itself and advancing along that axis at constant speed. The distance along the axis that a point on the line advances during one revolution is termed the pitch. Real propellers depart from a true helicoid, so that the pitch varies with radius. A nominal pitch is therefore used, usually that at a radius of 0.7 times the maximum radius. The pitch is defined non-dimensionally as pitch ratio P/D.

Propeller efficiency is given by:

$$\eta_O = \frac{\text{useful power output by propeller (thrust} \times \text{forward speed)}}{\text{power absorbed by propeller (torque} \times \text{rotational speed)}}$$

It is defined in open water, away from the proximity of a ship hull. It can be shown that η_O has a maximum theoretical value that is related inversely to the increase in velocity of water passing through the propeller disc (Clayton and Bishop, 1982; Rawson and Tupper, 1984). Efficiency therefore tends to improve with increase of propeller diameter. There are practical limitations to the maximum diameter that can be accepted for a given ship design and values of η_O much above 70% are rarely achieved.

3.3.4.2 Propeller–Hull interaction

The proximity of the propeller to the hull causes an interaction between them. This has effects on both the resistance of the ship and the thrust and efficiency of the propeller. The increase in hull resistance has already been described in Section 3.3.2, where it was expressed as a thrust deduction factor.

The propeller is affected by the forward velocity of the water around the hull. This is different to that in open water owing to the presence of the fluid boundary layer adjacent to the hull, the shape of the streamlined flow around the hull and the orbital velocities in the waves generated by the ship. The velocity effect is denoted by the wake fraction (or factor). This is usually defined as:

$$w = \frac{V - V_A}{V} \quad (3.59)$$

where V_A is the net velocity of advance at the propeller location in the absence of the propeller, and V is the forward velocity of the ship. The wake fraction w usually has a value in the region 0.05–0.50. However, it can be (just) negative for high speed forms. The value of w can be deduced from wake survey measurements on models (Clayton and Bishop, 1982). It is usual to combine the effects of thrust deduction fraction t and wake fraction w by means of a hull efficiency:

$$\eta_H = \frac{\text{effective power required to tow appended hull at speed } V}{\text{thrust power developed by propeller at speed } V_A}$$

$$= \frac{R_T V}{T \cdot V_A} = \frac{R_T}{T} \times \frac{V}{V_A} = \frac{1-t}{1-w} \quad (3.60)$$

Another effect on the propeller is due to the flow into it being uneven and not parallel to the propeller axis. This is caused by the three-dimensional flow around the hull and the effect of appendages disturbing the flow. The overall effect is expressed as the relative rotative efficiency (RRE):

$$\eta_R = \frac{\text{efficiency of propeller behind the ship}}{\text{open-water efficiency of propeller (at speed } V_A)}$$

η_R is deduced from a comparison of model tests of the propeller in open water and fitted behind the ship model. Values of η_R are typically in the region 0.98–1.08. The three factors t, w and RRE are known as the hull efficiency elements. Data on these, related to the BSRA standard resistance series, is given by Lackenby and Parker (1966).

We can now define the overall efficiency of the propeller when propelling the ship. This is:

$$\eta_T = \frac{\text{effective power delivered by propeller}}{\text{delivered power absorbed by propeller}}$$

or

$$\eta_T = \eta_O \cdot \eta_H \cdot \eta_R \quad (3.61)$$

η_T is known as the quasi-propulsive coefficient, QPC.

Finally, we define an overall propulsive efficiency, termed the propulsive coefficient PC, as:

$$PC = \frac{\text{effective power delivered by propeller}}{\text{shaft power delivered by engine}}$$

$$= QPC \times \text{transmission efficiency}$$

Transmission efficiency is usually around 0.97–0.98 for conventional ship machinery and transmission systems.

3.3.4.3 Propulsion model testing

As seen above, two types of propulsion model test are required in addition to the bare hull resistance testing discussed in Section 3.3.3.3. The first propulsion test is to establish the open water characteristics of the chosen propeller design. These are found by running a submerged model propeller mounted on a streamlined shaft in a water tunnel, or suspended from a towing tank carriage.

Dimensional analysis similar to that in Section 3.3.1 gives the main non-dimensioned parameters as:

$$K_T = \frac{T}{\rho n^2 D^4}, \text{ the thrust coefficient}$$

$$K_Q = \frac{Q}{\rho n^2 D^5}, \text{ the torque coefficient}$$

$$J = \frac{V_A}{nD}, \text{ the advance coefficient}$$

where T and Q are thrust and torque respectively, ρ is water density, n is rotational speed, V_A is speed of advance and D is diameter. Hence torque and thrust need to be measured for a range of velocities of advance V_A and rotational speeds n. From these, the propeller open water efficiency is found as:

$$\eta_O = \frac{\text{useful power output}}{\text{input power absorbed}}$$

so $$\eta_O = \frac{TV}{Q \cdot 2\pi n} = \frac{K_T}{K_Q} \frac{J}{2\pi} \quad (3.62)$$

Testing must be undertaken with care and suitable precautions taken to avoid laminar flow and cavitation effects.

The second propulsion test is made by fitting a suitably scaled model propeller to the model hull. Any significant appendages are included on the hull. The model is then tested at a series of carriage and propeller speeds, and propeller torque and model thrust is measured.

When the model thrust, as measured on the carriage, is zero, the self propulsion point is found. From these results and the open water tests, and perhaps a wake survey, the hull efficiency elements and propeller efficiency can be calculated. A full description of the testing and analysis process, and scaling problems to be overcome, is given by Clayton and Bishop (1982).

3.3.4.4 Ship–model correlation

There are a number of difficulties involved in scaling up model test results to predict full scale resistance and powering requirements. The first problem arises from the scaling laws themselves: it is not practicable to achieve simultaneous equality of both Froude number and Reynolds number for model and ship. The scaling of appendage resistance and propeller–hull interaction is particularly uncertain. Inaccuracies of the model making and testing process must also be considered.

The second main difficulty arises from the different conditions under which model tests and full scale trials are carried out. Hull roughness and fouling, sea and wind conditions and accuracy of trials instrumentation affect the correlation achieved.

Various approaches have been made to overcoming the ship–model correlation problem for a new design.

All involve the application of some kind of fudge factor based on comparison between model tests and full scale trials on previous similar ships. Three main methods are in use:

1. The UK Ministry of Defence has traditionally used a QPC factor:

$$QPC \text{ factor} = \frac{PC \text{ from ship trial}}{QPC \text{ from model}}$$

2. The British Towing Tank Panel in 1965 recommended use of a set of three factors as follows:

$$(1+x) = \frac{\text{actual delivered power at ship } (P_s)}{\text{delivered power scaled from model tests } (P_m)}$$

$$k_1 = \frac{QPC \text{ measured at ship}}{QPC \text{ derived from model tests}}$$

$$k_2 = \frac{\text{actual ship propeller rotational speed to deliver power } P_s}{\text{predicted propeller speed from model tests to deliver power } P_m}$$

The factor $(1+x)$, often called the load factor, allows for the effects of hull roughness, fouling and trial conditions. It has a value close to 1.0 (NPL, 1966; Dawson and Bowden, 1972, 1973; Scott, 1973, 1974; Holtrop and Mennen, 1978).

3. The International Towing Tank Conference (ITTC) in 1978 proposed a method now used by most commercial towing tanks for ship performance prediction. This uses a set of correction coefficients and factors covering hull roughness (see Section 3.3.2.2), form effects, air resistance, rudder effects, and propeller scale effects. The full procedure is set out in Chapter 11 of Rawson and Tupper (1984).

3.3.5 Propeller design

The key characteristics and propulsion model testing of propellers have been described previously. The design of the propeller itself, prior to the making of a model, involves the selection of diameter, speed of rotation, number of blades and blade area and pitch. The detailed shape of the propeller must also be decided based on blade strength and cavitation requirements as well as thrust and torque.

3.3.5.1 Choice of basic characteristics

The choice of propeller type, such as fixed pitch or controllable pitch, is discussed in Chapter 6. The first estimate of basic characteristics is usually made using data from standard methodical series of propeller model tests in open water. These methodical series are comparable to the hull resistance standard model series. Methodical series in common use have been published by Troost (1950) – quotes results for two to five bladed propellers; AEW series (Gawn, 1953) – covers three-bladed propellers only, but can be used for four and five blades with acceptable accuracy for preliminary design; Wageningen series (van Manen, 1966; van Lameren et al., 1969; Oosterveld and van Oossanen, 1975) – uses charts based on two parameters (B_p and δ) that are related to K_Q and J. Data on ducted propellers is given by van Manen and Oosterveld (1966). A typical K_T–K_Q diagram from the AEW series is shown in *Figure 3.23*.

From such charts, the propeller characteristics to give maximum open water efficiency can be found. The blade area ratio is chosen to be large enough to limit the overall blade pressure to around 70–80 kPa (kN/mm^2). This is to avoid excessive cavitation. The diameter is usually chosen to be the largest possible within the geometrical constraints. These are set by the maximum acceptable draught and the minimum acceptable clearance between the blade tips and the hull plating. This clearance is necessary to avoid hull vibration set up by the pressure fluctuations from the passing blade tips. Propeller rotational speed must be compatible with the propulsion machinery arrangements; this engine-propeller matching is discussed in Chapter 6. The propeller design process is described in Clayton and Bishop (1982), Rawson and Tupper (1984) and Muckle and Taylor (1987).

3.3.5.2 Detailed propeller design

The design of the detailed blade shape of a modern propeller utilizes advanced fluid dynamics theory and large computing power. Vortex theories are used, with either two dimensional approximations, known as lifting line theory, or a full three dimensional approach, known as lifting surface theory (Clayton and Bishop, 1982; Kerwin, 1986; Muckle and Taylor, 1987; Lewis, 1988). Propeller efficiency is affected by blade roughness and this should be allowed for in the design process (Townsin et al., 1981, 1985). The resulting blade shape has to be designed for adequate strength to withstand the maximum thrust and torque forces. Allowance must be made for dynamic and fluctuating loads. A design method is given by Conolly (1961).

3.3.5.3 Cavitation

The thrust generated by a propeller arises from the hydrofoil action of the blades. High pressure on the face and low pressure on the back of each blade combine to produce lift. If the low pressure falls too far, due to increased rotational speed or too high an angle of attack of the blades, the water may boil locally. This forms bubbles of vapour, a process assisted by the presence of small particles and dissolved air in the water which act as nuclei for bubble formation. The vapour bubbles may then be swept to regions of higher

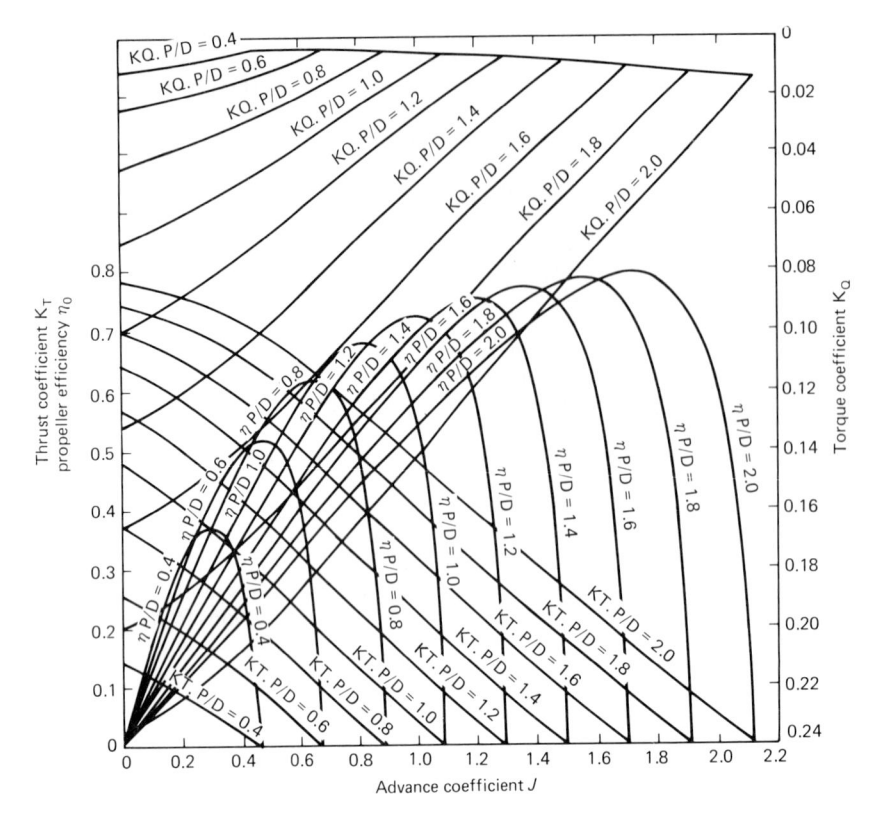

Figure 3.23 $K_T - K_Q$ propeller diagram for $BAR = 0.65$ from the AEW series

pressure whereupon they collapse. The subsequent onrush of water to fill the cavities results in high impacts loads which can cause damage to adjacent solid surfaces and generate a lot of noise.

The process of bubble formation, called cavitation, requires the local absolute water pressure to fall below the local absolute vapour pressure. At that point, the local cavitation number,

$$\frac{\Delta p}{\rho V^2}$$

introduced in Section 3.3.1.1, falls to zero. Cavitation on a propeller can take a number of forms. These include face, tip vortex and back cavitation. The latter can spread to form sheet cavitation.

As well as causing structural damage and noise, cavitation can dramatically reduce propeller efficiency. It is therefore necessary to study propeller cavitation performance using models in a special low pressure test facility called a cavitation tunnel. Prior to this testing, a guide to cavitation performance at the initial design stage is obtained from a cavitation bucket curve as shown in *Figure 3.24* (Burrill, 1955, 1963; Glover, Thorn and Hawdon, 1979; Gray and Greeley, 1982).

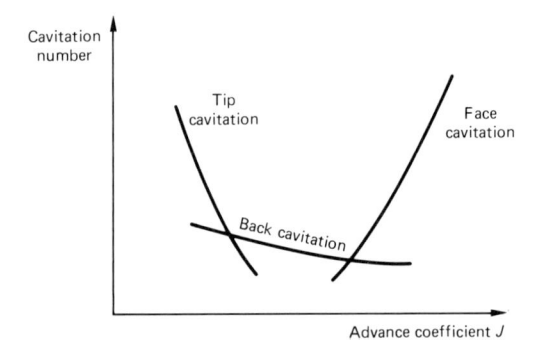

Figure 3.24 Propeller cavitation bucket curve

3.3.6 Full scale trials

Speed trials are invariably carried out on a new ship to check that the contractual speed requirement has been met. In some cases, the trials are fully instrumented in order to give accurate data on full scale propulsion performance. This gives valuable information on ship–model correlation. Propeller viewing trials may also be carried out to check cavitation performance at full scale.

The results of some well known instrumented ship trials are described in Clayton and Bishop (1982) and Lewis (1988). With modern ship–model correlation techniques, it is found that the standard deviation on predictions of full scale shaft power from model experiments is in the region of 5–10% (5% on power equates to around 1.5% on speed or, say, 1/3 knot at 20 knots).

3.3.7 Hull form design

The optimum hull form in the context of resistance and propulsion depends on the required maximum speed and operating profile of the ship. Guidance on hull form design for minimum power requirements may be found in Watson (1981), Rawson and Tupper (1984) and Lewis (1988).

3.4 Manoeuvring

Manoeuvring is concerned with the directional stability and control of a ship in the horizontal plane. This has always been a major feature of warship design but has only recently received attention in the design of merchant ships. At the same time modern design and analysis techniques have encouraged a more integrated approach to the subject of ship control as a whole. This justifies a separate chapter being devoted to developments in this field. The principles of ship control underlying ship manoeuvring are therefore included in Chapter 7. For additional information on ship manoeuvring see Inoue *et al.* (1981), Clayton and Bishop (1982), Clarke *et al.* (1983), Rawson and Tupper (1984), Mikelis (1985), Barr (1987) and Lewis (1988). An overview of the state of the art in 1987 may be found in the *Proceedings of the 1st International Conference on Ship Manoeuvrability, Prediction and Achievement* (RINA, 1987).

3.4.1 Interaction effects

When a ship hull approaches another solid underwater boundary, changes occur in the local pressure field. Such hydrodynamic interaction can cause two ships close together to collide, or a ship to hit a nearby bank lying parallel to its course (Dand, 1976, 1982; RINA, 1987).

Shallow water also changes the local hydrodynamic pressure field around a hull and can have a marked effect on ship manoeuvrability. It also leads to bodily sinkage of the ship, known as squat, and a tendency to change trim. The total effect for a ship travelling at speed in shallow water can be to increase the maximum draught by more than 10% (Dand and Ferguson, 1973; Dand, 1976; Sellmeijer and van Oortmerssen, 1984; RINA, 1987; Bernitsas and Kekerdis 1985).

3.4.2 Design of control surfaces

The forces and torques on rudders and other control surfaces are discussed in Chapter 7. (Hoerner, 1985; Hoerner and Borst, 1975; Clayton and Bishop, 1982; Rawson and Tupper, 1984; Muckle and Taylor, 1987).

3.5 Ship motions and seakeeping

3.5.1 Degrees of freedom

A ship taken as a rigid body, has six degrees of freedom. These are shown in *Figure 3.25*. As usually defined, the motions named here refer to dynamic perturbations from a mean condition and they do not include static displacements such as heel and trim. In calm water on a steady straight course, all the perturbations are zero, including surge. In a sea way that is in the presence of waves additional to the ship's own wave pattern due to its forward speed, all the motions are likely to be active. In the special case of a pure head or stern sea, the ship will heave, pitch and surge but sway, roll and yaw will be zero. In a pure beam sea, only heave, sway and roll will be excited directly.

The motions are not all independent, however. The non-uniform hull shape leads to certain couplings between motions. For example, heave and pitch are coupled, owing to the separation of the longitudinal centres of buoyancy and flotation (see Section 3.2). Similarly roll, sway and yaw tend to be coupled. There is even a coupling between roll and heave in that large amplitudes of roll can lead to heave motion.

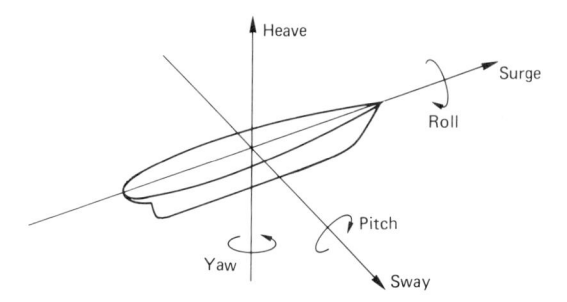

Figure 3.25 Six degrees of freedom of ship motions

3.5.1.1 Seakeeping

Seakeeping is the general term used to describe the safety, effectiveness and comfort of a ship in waves. It is mainly concerned with the motions out of the horizontal plane, i.e. with heave, pitch and roll. Interaction occurs between seakeeping and manoeuvring through roll and yaw coupling. This is of particular importance in severe following seas and requires a combined approach to manoeuvring and roll control (see Chapter 7).

The prediction and minimization of heave, pitch and roll is of great importance in ship design. Large amplitudes of these motions have a number of deleterious effects. These include discomfort and loss of efficiency of the crew, seasickness among passengers and possible injury to humans and damage to equipment from the high accelerations that can be produced. Extreme rolling can also lead to flooding, shifting of cargo and even capsize. Other direct effects on the ship include loss of ship speed, deck wetness and keel emergence.

Deck wetness occurs when bow immersion allows spray or green seas to cover the upper deck forward. The green seas condition is more severe and is taken as the definition of deck wetness in this section. Keel emergence can lead to severe impulsive loading on the hull as the keel re-enters the water at a high relative velocity such that the keel plating impacts violently on the water surface. This is known as bottom slamming. It can cause overall vibration of the hull and local damage to the plating. Another form is bow flare slamming. This occurs when a flared bow moves downward into a wave, pushing water aside rapidly. It is less violent than bottom slamming but can still cause significant loads on the hull.

The passage of waves along the ship, the accelerations due to ship motions, sea slap from green seas and slamming all cause loads on the hull structure. In fact, the shear forces and bending moments caused by these loads deform the hull structure such that the rigid body assumption made above is not strictly valid. The hull structure itself has a number of degrees of freedom or modes of distortion which interact with the rigid body modes. For the purposes of calculating ship motions, this interaction is not normally of practical significance. This is because the natural frequencies of seakeeping and hull distortion modes are very different. However, the interaction can be significant for the strength of structures having unusual geometry or greater flexibility than conventional ships (Bishop and Price, 1979). Structural strength, and the effect on it of interaction between seakeeping and hull distortion modes, is covered in Section 3.6.

The remainder of this section gives a short introduction to the theory of seakeeping. Texts covering the subject in much greater detail include Price and Bishop (1974), Newman (1977), Bishop and Price (1979), Bhattacharyya (1979), Clayton and Bishop (1982), Rawson and Tupper (1984) and Lewis (1988).

3.5.2 Roll

We have seen that with practical ship forms heave and pitch are coupled motions. For this reason, they must normally be analysed together. Roll, however, is only very weakly coupled to heave and pitch. To a reasonable approximation, it can be treated separately.

The treatment of a rolling ship-like form as a single degree of freedom dynamic system is shown in Chapters 2 and 7. For small angles of roll, the theory is simple and yields reasonably accurate estimates of roll period. The roll period for a lightly damped system is given by:

$$T_\phi = \frac{2\pi k}{\sqrt{gGM}} \qquad (3.63)$$

where k is the radius of gyration of the ship in roll and GM is the transverse metacentric height. Values of k for merchant ship and warship forms may be found from formulae such as those given in Rawson and Tupper (1984) which are based on those of Kato (1959). Typical periods of roll for most ships lie in the region 8–14 s. Larger ships have, in general, the longer periods. Roll periods of 4–6 s are preferably avoided as human susceptibility to seasickness increases significantly (see Section 3.7).

At larger roll angles, non-linear effects become significant and non-linear theories are required (Spouge, 1988). A major problem is the estimation of damping, which is low in roll. That is why a ship's roll motion is large compared to other motions and always occurs at or near the natural roll frequency (see *Figure 2.15*). The difficulty of estimating damping leads to rather imprecise calculations of roll amplitude under wave excitation. The available methods of calculation are given in the seakeeping references quoted above.

Roll amplitudes can be greatly reduced by fitting bilge keels and/or a variety of passive and active stabilizing systems. These are described in Chapter 7. An interesting development in recent years has been the use of marine aerofoils for assisting propulsion. Such devices can be optimized to reduce significantly the motions of a ship in a seaway. This also reduces ship resistance and thus the thrust required from the propeller (see Clayton and Satchwell, 1987; Clayton and Sinclair, 1988 and 1989).

3.5.2.1 Dynamic instability and capsize

The problem of estimating roll amplitudes is particularly acute at large angles. This makes prediction of capsize conditions in heavy seas very difficult. No generally satisfactory theoretical method is available in a form suitable for design use. That is why stability criteria are still based essentially on the quasistatic GZ curve (see Section 3.2). Model testing can be used instead, although even then the identification of critical conditions in a real seaway is far from straightforward.

A particular form of instability that can occur in near head or following seas of wave length equal to ship length is known as parametric resonance. This arises from the variation of GM as waves pass along the hull, as described in Section 3.2. If the wave encounter period, hence period of GM variation, is approximately half the ship's natural period in roll, a resonance is set up. Excitation of even a small roll motion by waves that are not quite dead ahead or astern leads to steadily increasing roll amplitudes with each passing wave. Severe roll angles can result. This situation is likely to be most dangerous in following seas when parametric resonance can be exacerbated by the tendency of the ship to yaw as it surges forward down a wave. If the ship surge speed approaches wave speed, broaching can occur. In this case, the yaw angle increases rapidly as the ship heels out on the turn. In extreme conditions this can leave the ship beam on to the waves and capsize can result. (RINA, 1986; Burcher, 1980; Morrall, 1981; Renilson and Driscoll, 1982; Soding, 1987; Hamamoto and Akiyoshi, 1988).

3.5.3 Heave and pitch

The damping of a hull in both heave and pitch is much greater than that in roll. Likewise the stiffness of the hull (governed by *TPI* in heave and GM_L in pitch) is far higher than that in roll (governed by GM_T). Heave and pitch amplitudes are therefore less and are unlikely to threaten ship safety directly. Nevertheless, their indirect effects can be severe as noted above.

The basic equations of motion of a simple body in pure heave and in pure pitch are given in Chapter 2 which also gives the equations of a rigid shiplike body where heave and pitch are coupled. For the axis system shown in *Figure 3.26*, the matrix equations of motion in heave and pitch in sinusoidal waves may be expressed as:

$$\begin{bmatrix} (M+M_A) & d \\ D & (I+A) \end{bmatrix} \begin{bmatrix} \ddot{z} \\ \ddot{\theta} \end{bmatrix} + \begin{bmatrix} b & e \\ E & B \end{bmatrix} \begin{bmatrix} \dot{z} \\ \dot{\theta} \end{bmatrix} + \begin{bmatrix} c & g \\ G & C \end{bmatrix} \begin{bmatrix} z \\ \theta \end{bmatrix} = \begin{bmatrix} F_o \\ M_o \end{bmatrix} e^{i\omega_e t} \quad (3.64)$$

where M is ship mass, I is moment of inertia of the ship mass in pitch about a transverse horizontal axis (Oy in *Figure 3.26*, origin O usually being taken at the ship centre of gravity), M_A is a fluid inertia effect in heave called the added mass and A is a similar added moment of inertia in pitch. The coefficients b, e, B, E are fluid damping terms and c, g, C, G are fluid stiffness terms in heave and pitch. F_o and M_o are the respective amplitudes of the fluid forces and moments acting on the ship owing to the relative motion between ship and wave. ω_e is the wave encounter frequency. At zero ship speed this equates to the wave frequency ω; at a ship speed U, the encounter frequency in deep water is given by:

$$\omega_e = \omega - \frac{U\omega^2}{g} \cos \chi \quad (3.65)$$

where χ is the angle between the direction of travel of the waves and the ship's heading (U, ω and g must be in consistent units). Thus in head seas, $\chi = 180°$, so that:

$$\omega_e = \omega + \frac{U\omega^2}{g} \quad (3.66)$$

The problem is how to calculate the fluid terms on both sides of Equation 3.64 so that the pitch and heave amplitudes can be found. There are two basic theoretical approaches. Potentially the most accurate is to treat the problem in three dimensional form. This has been intractable until quite recently but is now developing rapidly, spurred on by the needs of the offshore industry. The use of three dimensional diffraction theory and boundary integral methods is described in Chapter 2. There the method is shown for deriving heave and pitch forces and moments and hence amplitudes for a stationary ship in regular waves. Diffraction analysis for a ship with forward speed is more complicated and is not yet in everyday use.

The other basic approach is to make a two dimensional approximation to the problem. Several such methods have been developed in the past 30 years, a summary being given in Price and Bishop (1974). The most widely used method in ship design is known as strip theory.

3.5.3.1 Strip theory

The ship hull is assumed to be divided up into a number of short, vertical transverse strips as shown in *Figure 3.26*. The method is based on calculating the linear response of each strip to a sinusoidal wave of unit height passing along the hull, while satisfying compatibility conditions between each strip. The hull is assumed to be rigid, although specialized strip methods have been developed which allow for the flexibility of the hull (Bishop and Price, 1979). Linearity is assumed, such that any response is directly proportional to wave height.

Other major assumptions are that the ship's beam and draught are small compared to its length, and the flow about each strip is two dimensional. These assumptions allow great simplification by ignoring flow interaction between adjacent strips. They also imply that the wavelength is small compared to ship length so that interaction between bow and stern is negligible. These are crude assumptions but they are accepted as

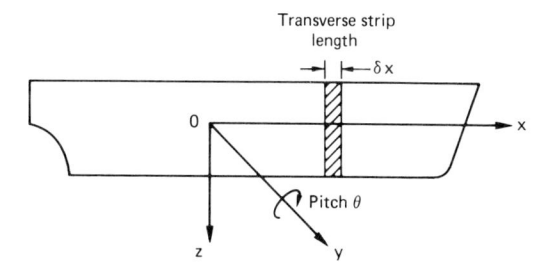

Figure 3.26 Nomenclature for heave and pitch and strip theory

the method is convenient and gives reasonable agreement with experimental results, except in extreme seas.

A number of different formulations of strip theory have been developed. The most widely used are those of Gerritsma and Beukelman (1967), Salvesen, Tuck and Faltinsen (1970), Vugts (1971) and Schmitke (1978). These differ in the details of the hydrodynamic potential flow theory used. In its most basic form, strip theory assumes that the fluid loading on a strip at a longitudinal position x is a function of the relative vertical displacement:

$$z_r(x, t) = z(x, t) - \zeta(x, t) \quad (3.67)$$

where z is the absolute vertical displacement of the strip and ζ is the wave elevation at x.

The fluid loading is taken to comprise all the fluid terms corresponding to those on both sides of Equation 3.64. These are grouped into three areas: the fluid added mass, fluid damping and fluid stiffness. The added mass and damping are functions of the shape of the underwater cross-section of the strip, the encounter frequency and the relative displacement z_r. The methods of computation vary depending on the strip theory formulation used. The fluid stiffness is simply the buoyancy term:

$$\rho g\, B(x)\, z_r(x, t) \quad (3.68)$$

where ρ is water density and $B(x)$ is the local beam at x. The ship is assumed to be wall sided over the relative vertical displacement concerned. The fluid force δF on the strip δx may be conveniently expressed by:

$$\frac{\partial F}{\partial x} = \frac{\partial J}{\partial x} + \frac{\partial K}{\partial x} \quad (3.69)$$

where J represents the fluid loading due to ship motion in still water and K represents the fluid loading due to the wave passing a stationary ship.

The total wave induced force and moment on the ship may be found by integration along the length of the hull, i.e.:

$$\begin{bmatrix} F(t) \\ M(t) \end{bmatrix} = \begin{bmatrix} K_F(t) \\ K_M(t) \end{bmatrix} + \begin{bmatrix} J_F(t) \\ J_M(t) \end{bmatrix} \quad (3.70)$$

where $F(t) = \int_L \frac{\partial F(x,t)}{\partial x} \, dx$; $M(t) = \int_L \frac{\partial F(x,t) \cdot x \cdot dx}{\partial x}$

$$K_F(t) = \int_L \frac{\partial K(x,t)}{\partial x} \, dx; \quad J_F(t) = \int_L \frac{\partial J(x,t)}{\partial x} \, dx, \text{ etc.} \quad (3.71)$$

As $z(x,t) = z_o(t) - x\theta(t)$

then the terms J_F and J_M can be expressed in terms of $z_o, \dot{z}_o, \ddot{z}_o, \theta, \dot{\theta}, \ddot{\theta}$, with coefficients now depending only on the strip cross-section shape and the wave encounter frequency ω_e.

The equations of motion of the hull can now be written as:

$$\begin{bmatrix} m\ddot{z} \\ I\ddot{\theta} \end{bmatrix} + \begin{bmatrix} J_F(t) \\ J_M(t) \end{bmatrix} = \begin{bmatrix} K_F \\ K_M \end{bmatrix} e^{i\omega_e t} \quad (3.72)$$

If the motions are assumed to be harmonic at the encounter frequency ω_e, then we can write:

$$[H] \begin{bmatrix} z(t) \\ \theta(t) \end{bmatrix} = \begin{bmatrix} K_F \\ K_M \end{bmatrix} e^{i\omega_e t} \quad (3.73)$$

where $[H]$ is a 2×2 matrix of coefficients, depending on ω_e and the strip cross-section, from the left hand side of Equation 3.72. Hence

$$\begin{bmatrix} z(t) \\ \theta(t) \end{bmatrix} = [H]^{-1} \begin{bmatrix} K_F \\ K_M \end{bmatrix} e^{i\omega_e t} \quad (3.74)$$

The ship heave and pitch amplitudes and phases are therefore found for a given wave input and encounter frequency ω_e. Velocities and accelerations at any point along the hull can also be found. The same strip theory procedure can be used for all six degrees of freedom by adding the appropriate equations of motion for roll, sway, yaw and surge.

The strip theory method is easily extended to calculate the shear forces and bending moments on the hull due to the waves. Considering symmetric motions only, the forces on an individual strip of mass metres per unit length are shown in *Figure 3.27*.

The equation of motion of the strip, relative to an equilibrium still water condition, is:

$$(m\,\delta x)\, \ddot{z}(x,t) = \frac{\partial V(x,t)}{\partial x}\, \partial x + \frac{\partial F(x,t)}{\partial x}\, \partial x \quad (3.75)$$

hence $\dfrac{\partial V(x,t)}{\partial x} = m\, \ddot{z}(x,t) - \dfrac{\partial F(x,t)}{\partial x} \quad (3.76)$

Both $z(x,t)$ and $F(x,t)$ are known functions of the harmonic quantities z, θ, ξ and ω_e.

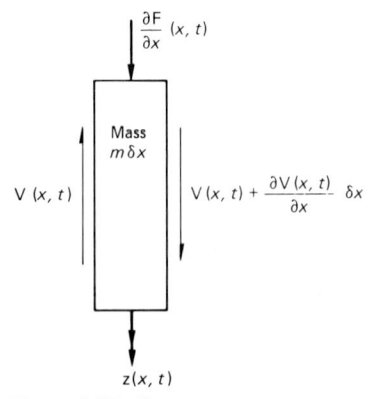

Figure 3.27 Forces on transverse strip

$\dfrac{\partial V(x,t)}{\partial x}$ is also harmonic.

The shear force and bending moment at any point x_1 can then be found by integration:

$$V(x_1) = \int_{x_1}^{L/2} \dfrac{\partial V(x_1,t)}{\partial x} dx \text{ and } M(x_1)$$

$$= \int_{x_1}^{L/2} (x - x_1) \dfrac{\partial V(x,t)}{\partial x} dx \qquad (3.77)$$

This use of strip theory is considered further in Section 3.6.

Strip theory is found to be fairly accurate for the calculation of heave and pitch motions. This is despite the large assumptions noted above. It is less accurate for roll, yaw, sway and surge. It is also less accurate for the prediction of shear force and bending moment in extreme seas, where it tends to overestimate the loadings. A major reason for this is the assumption of linearity which is invalid in large waves, where a ship is far from wall sided over the vertical displacement concerned.

A range of strip theory computer programs is available. A typical and widely used linear program covering all six degrees of freedom is SCORES (Raff, 1972). Five degrees of freedom non-linear strip theories are now appearing that overcome the assumption of linearity. These promise to give considerably more accurate estimates of motion and loads in severe sea states (Fujino and Yoon, 1986; ISSC, 1988).

3.5.3.2 Response amplitude operators

The heave and pitch response to a unit wave height can be found by one of the linear methods just described. The response to an irregular random sea is then calculated by linear superposition. The individual regular waves, modelled as sine waves, are summed to form the desired sea spectrum as shown in Chapter 1.

A response to unit regular wave amplitudes is represented by a Response Amplitude Operator (RAO), sometimes called a receptance. Typical RAOs for heave and pitch amplitudes are shown in *Figure 3.28*. This is analogous to *Figure 2.37(b)*.

For angular motions such as pitch, the motion amplitude RAO may be made non-dimensional by using wave slope rather than wave amplitude. However, wave slope is proportional to wave height for a given wave length, which compromises the linearity assumption, so wave amplitude is often used for convenience.

The RAOs illustrated apply to a given ship and ship speed only. The RAO is a function of both frequency of encounter and of the ratio of wave length λ to ship-

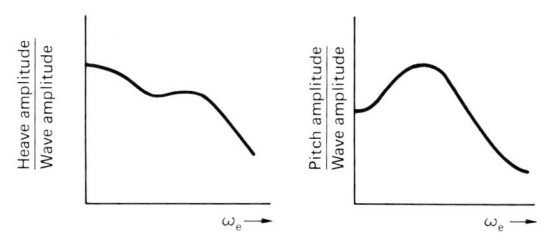

Figure 3.28 Response amplitude operators for heave and pitch, plotted against frequency of encounter ω_e

length L. RAOs are sometimes shown to a baseline of λ/L with a different curve for each ship speed.

Heave and pitch amplitudes tend to a maximum at values of λ/L in the range 0.8–1.0. This is known as ship–wave matching. As ship speed varies, the frequency of encounter with a given wave length varies, so changing the relative contribution of wave length and encounter frequency effects to the RAO. The RAOs for pitch and heave will have their overall maximum values in conditions where both ship–wave matching occurs and the encounter frequency is equal to the damped natural frequency of the ship in the motion concerned. RAOs can be derived for any response in any type of motion. Responses can include displacements, velocities and accelerations at any point on the ship and also forces and bending moments.

3.5.3.3 Motions in irregular waves

The procedure for calculating the motion at a given ship speed and heading in irregular waves can now be described. The sea spectrum in which the motion is to be calculated is identified. This may be via its characteristic wave period and significant wave height (see Chapter 1). The sea spectrum $S(\omega)$ is modified to give the encounter spectrum $S(\omega_e)$ for the ship speed and heading. This is given by the energy equivalence relation:

$$S_\zeta(\omega_e).d\omega_e = S_\zeta(\omega).d\omega \qquad (3.78)$$

$$\text{or } S_\zeta(\omega_e) = S_\zeta(\omega) \left(\dfrac{d\omega_e}{d\omega}\right)^{-1} \qquad (3.79)$$

Noting equation 3.65, $S_\zeta(\omega_e)$ is therefore obtained by multiplying the ordinate values of $S_\zeta(\omega)$ by:

$$\left(1 - 2\dfrac{U\omega}{g} \cos \chi \right)^{-1} \qquad (3.80)$$

where U is ship speed, ω is wave frequency and χ is the angle between the direction of travel of the waves and the ship's heading; U, ω and g being in consistent units (m/s, rad/s and m/s², respectively). The absissa values of $S_\zeta(\omega_e)$ are given by Equation 3.65 directly, i.e.:

$$\omega_e = \omega\left(1 - \frac{U\omega}{g}\cos\chi\right)$$

The RAO for the motion concerned is found for a range of wavelengths at the required ship speed and heading. This may be done theoretically, as described above or by model experiment described below.

The energy spectrum $S_M(\omega_e)$ of the motion is now obtained by multiplying the wave encounter spectrum ordinates by the square of the RAO:

$$S_M(\omega_e) = S_{\zeta 44444}(\omega_e)[RAO(\omega_e)]^2 \quad (3.81)$$

The process is illustrated for the case of heave in *Figure 3.29*.

Various characteristics of the motion can be obtained from its energy spectrum. It is found that the motion spectrum of a ship in an irregular long crested unidirectional sea usually approximates to a Rayleigh distribution, as does the sea spectrum itself (see Chapter 1). The probability of the motion M exceeding a value M_o in any one cycle is then given by:

$$P(M > M_o) = \exp\frac{-M_o^2}{2m_o} \quad (3.82)$$

and the probability of exceedance in n cycles is given by:

$$P(M > M_o, n) = 1 - \left[1 - \exp\left(\frac{-M_o^2}{2m_o}\right)\right]^n \quad (3.83)$$

where m_o is the area under the motion spectrum. It can be shown that average motion amplitude equals $1.25\sqrt{m_o}$ and the significant motion amplitude is $2\sqrt{m_o}$. The latter is the average amplitude of the 1/3 highest motions. The maximum motion amplitude is sometimes taken to be the average of the 10% highest motion amplitudes. This is equal to $2.55\sqrt{m_o}$.

Methods of calculating motions in a general three dimensional short crested sea are given by Price and Bishop (1974). In such head seas, heave and pitch motions are generally reduced, though other motions may increase. It is therefore common to make the pessimistic assumption that a sea is two dimensional. In other cases, a simple spreading function may be used (see Chapter 1).

The sea state, ship speed and heading for which the above calculations are made may be taken as constant only over a relatively short period, perhaps a few hours. To estimate maximum motions during a complete voyage, it is necessary to establish the range of sea states and correspondinig ship speeds and headings likely to be met over the route. The ship's motion response in each case can then be calculated and the probabilities of extreme motions identified. The likely sea states can be identified from global wave statistics (Hogben, Dacunha and Olliver, 1986).

This calculation procedure can be extended to cover the complete life of the ship. This is mainly of relevance to ship strength, where estimates of lifetime extreme loads are required for design (see Section 3.6).

3.5.4 Slamming and deck wetness

Both slamming and deck wetness are dependent on the relative motion between the ship and the sea, particularly towards the bow. A condition for bottom slamming is bow emergence, i.e. the upward motion of the ship relative to the sea surface exceeds the draught forward. Deck wetness occurs when the downward motion of the ship relative to the sea surface exceeds the local freeboard.

The relative bow motion is a function of heave and pitch and the local wave height. The relative vertical motion z_r is a function of longitudinal position x and time t and is given by:

$$z_r(x,t) = z(t) - x\theta(t) - \zeta(t) \quad (3.84)$$

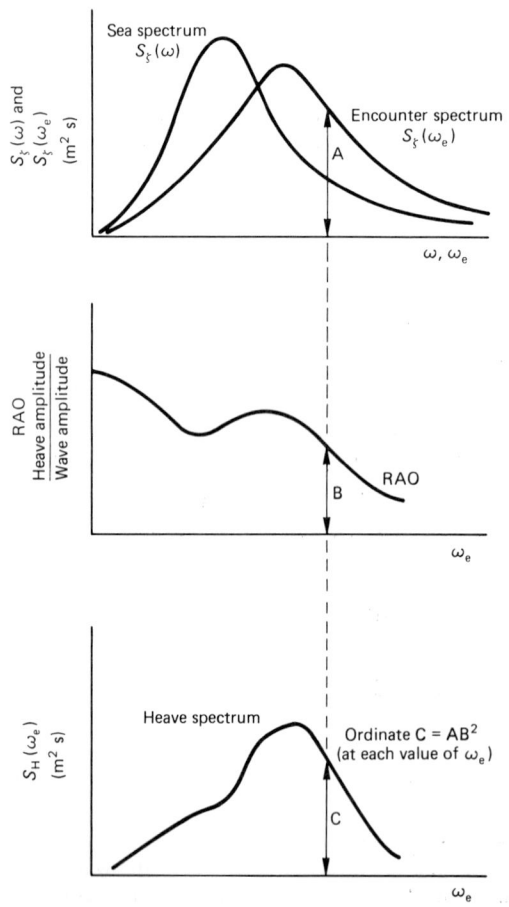

Figure 3.29 Calculation of heave spectrum from sea spectrum and RAO

where $z(t)$ is the absolute heave motion at the origin of the body axes fixed in the ship, θ is the pitch angle and ζ is the wave elevation. It is thus possible to calculate a response amplitude operator for $z_r(x,t)$ by theoretical methods such as strip theory. The energy spectrum for the relative bow motion can then be found for a given sea state, ship speed and heading in a manner identical to that shown above.

3.5.4.1 Deck wetness

It is assumed, as before, that the magnitude of relative motion between ship and waves in a given sea state approximates to a Rayleigh distribution. The probability that the relative motion z_r exceeds the freeboard F in any cycle is therefore:

$$P(z_r > F) = \exp\left[-\frac{F^2}{2m_o}\right] \quad (3.85)$$

where m_o is the area under the relative motion spectrum. This is the probability of deck wetness occurring.

Of particular interest is the number of deck wetnesses occurring in a given interval of time. This is one of the criteria used by a ship's master in deciding whether to slow down or change course in bad weather. The expected number of deck wettings per hour is given by:

$$N = \frac{3600}{2\pi}\sqrt{\frac{m_2}{m_o}} \exp\left[-\frac{F^2}{2m_o}\right] \quad (3.86)$$

where m_2 is the variance of the relative velocity. This is the second moment of the relative motion energy spectrum about its vertical axis, or the area under a curve where each ordinate of the z_r spectrum is multiplied by ω_e^2.

3.5.4.2 Bottom slamming

The probability of the relative motion forward, z_r, exceeding the local draught T in any cycle is:

$$P(-z_r > T) = \exp\left(-\frac{T^2}{2m_o}\right) \quad (3.87)$$

In the case of bottom slamming, three further conditions are necessary. A significant length of the bottom must emerge, say more than 10% of the ship's length. The bottom forward must lie roughly parallel to the water surface on re-entry; thus V-shaped bows are less likely to slam than flat U shapes. Also, the relative velocity between the ship's bottom and the water must exceed a certain critical value. That value depends on ship size and the local bottom deadrise, the angle that the bottom plating makes to the horizontal in a transverse plane (Ochi and Motter, 1973; Rawson and Tupper, 1984). For a typical cargo ship, the critical velocity is around 3.5 m/s.

The joint probability of bottom emergence and relative velocity exceeding the critical velocity V_*, assuming that these are independent events, is given by:

$$P(\text{slam}) = P[-z_r > T \cap \dot{z}_r > V_*] = P[-z_r > T].$$
$$p[\dot{z}_r > V_*] \quad (3.88)$$

For Rayleigh distributions, this gives:

$$P(\text{slam}) = \exp\left[-\frac{T^2}{2m_o} + \frac{V_*^2}{2m_2}\right] \quad (3.89)$$

where m_o and m_2 have the same meanings as in Section 3.5.4.1. The local slamming pressure is given by:

$$P = \tfrac{1}{2} k \rho_w \dot{z}_r^2 \quad (3.90)$$

where ρ_w is the density of sea water and k is a function of the deadrise angle. The expected number of slams per hour is given by:

$$N_\zeta = \frac{3600}{2\pi}\sqrt{\frac{m_2}{m_o}} \exp\left[-\left(\frac{T^2}{2m_o} + \frac{V_*^2}{2m_2}\right)\right] \quad (3.91)$$

Stern slamming is possible, though less likely than bow slamming. Its probability can be calculated by similar means to those above, using the motions and velocities at the stern. Other problems that can be analysed in a similar manner are propeller and sonar emergence.

3.5.5 Empirical methods of motion prediction

3.5.5.1 Model testing

Model tests may be used as an alternative to theoretical calculation of seakeeping performance or to confirm theoretical predictions. In testing for a new design, it is possible to use the same model as that used for calm water resistance and propulsion tests. However, a separate model is usually made, as the seakeeping model requires representation of the above water as well as the underwater hull form. It is also required to have a similar mass distribution to that of the ship, in order to give similar hydrostatic and inertial characteristics.

The basic requirements for dimensional similarity follow those described in Section 3.3. A tank with wave makers is required. The types of wave maker used and data analysis techniques employed are the same as those used in model testing of offshore structures, described in Chapter 2.

A variety of model testing techniques are available. The most straightforward is to run a captive model in a tank in regular waves of moderate amplitude to retain reasonable linearity of response. The response of the model is measured for a number of different wavelengths and forward speeds and the relevant RAOs deduced. Motions measured in head and/or following waves normally include heave and pitch and might include velocities and accelerations at several points on the model. Relative bow motions might also be measured and the incidences of deck wetness and slamming noted. This would normally be done visually or by photography. Rolling response in beam seas might be studied in a separate test. Model tests may also be used to find RAOs for shear forces and bending moments on the hull. This requires use of strain gauges and segmented models. Use of the RAOs to predict motions and forces in irregular seas is made using linear superposition as described in Section 3.5.3.3.

An alternative test method is to run the model in a known irregular wave pattern at a number of forward speeds. Several runs are necessary at each speed to gather sufficient response data, but the total testing time may be less than that required using the regular wave method. The measured responses are then analysed statistically to derive the RAOs. Special variants of this test method have been developed which reduce the number of runs required to derive reliable RAOs (Takezawa and Takekawa, 1976).

Model tests may be made in oblique waves to derive RAOs for roll, sway and yaw as well as heave and pitch. In an ordinary long and narrow test tank these tests could only be done at zero forward speed by holding the model at an angle to the tank's longitudinal axis. In practice, a free propelled model is required with its own autopilot in a large rectangular tank fitted with wave makers. Accurate measurement of response then requires sophisticated facilities. An example of free model testing to find motions and loads is given by Lloyd, Brown and Anslow (1980). Attempts have been made to produce a standard series for seakeeping behaviour, analogous to those for resistance and propulsion (Vossers, Swann and Rijken, 1960; Loukakis and Chryssostomidis, 1975).

A form of model testing that dispenses with the need for tanks is open water testing. A large instrumented, self propelled seakeeping model is run in relatively sheltered waters having small natural waves and little or no wind. The local sea state is measured by means of wave probes or wave rider buoys and the results analysed statistically. The method requires accurate measurement and very careful analysis if reliable RAOs are to be derived. The method is most suited to comparative tests of models run side by side.

3.5.5.2 Full scale trials

These are the full scale equivalent of the open water model tests described previously. A major problem is measurement of the sea state experienced by the ship at the time the response measurements are taken. This is compounded by the irregular short crested nature of real seas. Full scale trials are of use in confirming predictions from theory or model tests in a particular set of conditions (Rawson and Tupper, 1984; Aerttsen, 1969; Andrew and Lloyd, 1981; Bishop, Clarke and Price, 1984). They can also be useful in showing true response in high sea states, when these can be found, for comparison with predictions using linear superposition.

3.5.6 Seakeeping criteria

As the weather deteriorates and the sea state rises, the motions of a ship will increase. At some stage, the master will consider a particular aspect of the ship's response to be unacceptable. This judgement may be based on the comfort of passengers or crew, the ability of the crew or equipment to operate efficiently, the safety of cargo or equipment, or the well being of the ship itself.

A voluntary reduction of speed is the usual response, although a change of heading may also be made in severe conditions. This reduction of speed is additional to the involuntary reduction due to increase of resistance in waves, discussed in Section 3.3. The combined effect of involuntary and voluntary speed reduction with increasing sea state is illustrated in *Figure 3.30*. The speed reduction is usually greatest in head seas. Derivation of curves similar to *Figure 3.30* for all possible headings allows polar diagrams to be produced that show limiting speed envelopes in different sea states. These can be combined for all sea states that a ship might meet, duly weighted for their relative probabilities of occurrence. An overall measure of merit for the ship's seakeeping performance can be found by such means. The process is described by Rawson and Tupper (1984).

It is necessary during ship design to be able to estimate the likely voluntary speed reduction in heavy seas. This requires criteria to be set representing critical seakeeping motions beyond which capability is considered to be degraded. Critical values that have been considered appropriate are indicated in *Table 3.1*. The critical values for a very large ship would probably be lower than those shown in this table. Likewise, values for a passenger ship might typically be half those shown for merchant ships (Kanerva et al. 1988).

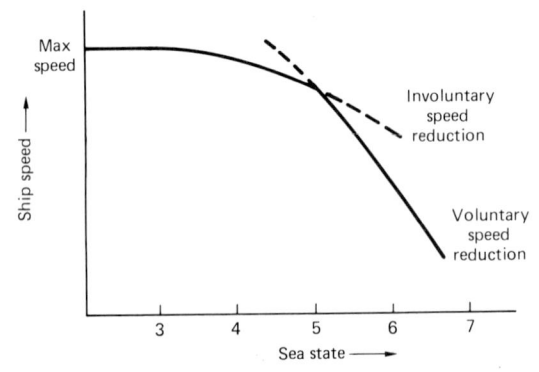

Figure 3.30 Typical speed reduction with increasing sea state

Table 3.1 Possible seakeeping criteria

Criterion	Merchant ships	Naval vessels
Roll angle	6°	6°
Pitch angle	–	1.5°
Vertical acceleration (FP)	0.2 g	0.2 g
Vertical acceleration (at bridge)	0.15 g	0.1 g
Lateral acceleration (at bridge)	0.12 g	0.1 g
Slamming	12/h	20/h
Deck wetness	30–40/h	30–90/h
Propeller emergence	120/h	
Subjective motion magnitude (SMM)	12	12

Notes: 1. Angle and acceleration values are single amplitude, root mean square (RMS).
2. Significant single amplitudes are approximately double RMS values shown.

The subjective motion magnitude (SMM) shown in the table is a measure of human response to vertical accelerations (Schoenberger, 1975). It is given by the empirical formula:

$$SMM = [30 + 13.53 (\log_e f)^2][\ddot{s}/g]^{1.43} \quad (3.92)$$

where f is the motion frequency in Hz, $\ddot{s}$ is the amplitude of the vertical acceleration (m/s²) and g is acceleration due to gravity (9.81 m/s²). If an energy spectrum is available for vertical displacement, then:

$$\ddot{s} = 2\sqrt{m_4} \quad (3.93)$$

and $f = \dfrac{1}{2\pi}\sqrt{\dfrac{m_6}{m_4}}$ \quad (3.94)

where m_4 and m_6 are the variances of the vertical acceleration and rate of change of vertical acceleration respectively, i.e. $\ddot{s}$ is the significant vertical acceleration and f is the average zero crossing frequency of the acceleration. A detailed discussion of SMM and other seakeeping criteria is given in Andrew and Lloyd (1981), Olsen (1978), Rawson and Tupper (1984), Andrew, Loader and Penn (1984) and Kanerva, Mikkonen and Nurmi (1988). Limiting criteria for helicopter operations are presented by Comstock et al. (1982) and Lloyd and Hanson (1985).

An alternative indicator of human response is the lateral force estimator (LFE). This is the apparent lateral acceleration felt by a person on a rolling ship. It is considered a better guide to limiting conditions than roll angle amplitude or absolute lateral acceleration alone (Baitis et al., 1983, 1984). A limiting value of LFE of 0.15 g has been proposed for warships (Monk, 1988).

3.5.7 Design for seakeeping performance

A number of aspects of underwater and abovewater form affect seakeeping performance. There is unfortunately no one perfect form, as certain seakeeping characteristics can only be optimized at the expense of others. For example, a large length to draught ratio is generally advantageous in reducing heave and pitch motions but is likely to increase the probability of slamming.

Nevertheless, there are some design features that are generally recognized as benefiting seakeeping performance. These include bilge keels and stabilizers to reduce roll motion, high freeboard forward to reduce deck wetness, V shaped sections forward to reduce slamming, a high waterplane area coefficient, particularly for the forward half of the ship and a low vertical prismatic coefficient, particularly forward, to reduce heave and pitch. A bulbous bow can also help by increasing damping. These features can conflict with other design requirements so compromise is necessary.

Other actions may be taken to reduce motions or accelerations locally. Equipment such as radar can be fitted on a stabilized mounting. Critical items can be sited in areas of low acceleration such as near the centre of gravity of the ship. Where seakeeping performance is vital, the adoption of a special seakeeping hull form may be justified such as SWATH (Rawson and Tupper, 1984; Lewis, 1988; Bales, 1980; Bales and Day, 1982; Blok and Beukelman, 1984; McCreight and Stahl, 1985).

3.6 Ship structures

This section is concerned with the strength of the ship's hull and its internal load bearing structures. Strength, like ship stability, directly affects ship safety, in that failure can lead directly to disaster. Even minor failures will incur an economic loss. Indeed structural repair, maintenance and preservation are a significant component of ship through life costs.

The basic physical principles underlying the strength of materials and the mechanics of engineering structures may be found in texts such as Case and Chilver (1971). Application of the principles of mechanics to the analysis of ship strength is given in Clayton and Bishop (1982). A fuller treatment of the practice of ship structural design may be found in Muckle (1967), Evans (1973, 1983), Rawson and Tupper (1983) and Lewis (1988). Modern hull structural design methods are treated in depth by Hughes (1983). The latest state of the art in all aspects of ship and offshore structures is reviewed and published every three years, with very extensive further references, in the *Proceedings of the International Ship and Offshore Structures Congress* (ISSC, 1988).

3.6.1 Structural materials

The engineering properties of the main ship structural materials are given in *Table 3.2*. Mild steel is the most commonly used material as it is readily available, cheap and easy to shape and to weld. However, it can have low toughness: this is represented in the table by the energy absorbed in a dynamic bending test known as a Charpy Test. Low toughness implies low resistance to brittle fracture, i.e. the sudden catastrophic fracture of the material by cracking, particularly at low temperatures.

Materials tougher than mild steel are preferable and are indeed a necessity for some applications. They should always be adopted for ships operating in cold regions, even if only in the form of individual crack arrester strakes. These prevent a crack that starts in brittle material from extending right around the hull.

Higher strength steels are more expensive and harder to form and weld, but can be useful in high stress areas and where reduction of structural weight is a priority. New high strength low alloy (HSLA) steels and other thermomechanically controlled processed (TMCP) and precipitation hardened steels are in use in the USA and Japan. They are cheaper and easier to weld than other high strength steels and are increasing in popularity. However, high strength steels rarely offer a weight reduction fully proportional to the increase in yield strength. This is because buckling is usually the governing failure mode for ship structures and this depends on elasticity, measured by Young's modulus, rather than on yield strength.

The disadvantages of steel are its weight and a tendency to corrode easily in the sea. Stainless steel of the 316 (austenitic) type could be used in ship structures, but it is expensive and liable to crevice corrosion. Its use is confined in practice to piping systems. Corrosion and its prevention is a major problem and is covered in Chapter 10.

Aluminium is sometimes used as a hull material. Typical properties are shown in *Table 3.2*. It is light, easily worked and is fairly resistant to corrosion. These advantages are offset by high cost, lower strength and stiffness than steel and poor fatigue resistance. Aluminium also has a low melting point, so that it loses strength rapidly if exposed to fire. In practice, aluminium is only used for the hulls of small vessels where weight is at a premium.

Non-metallic materials such as concrete, wood and glass- (and other fibre-) reinforced plastics have a place in small craft construction and in specialized ships such as mine warfare vessels. However, such composite materials are, as yet, neither suitable nor competitive for the hulls of larger ships. Further information on shipbuilding materials may be found in BS/CP 118 (1969), Boyd (1970), MIL-S-24645 (1984), BS 4360 (1986), Harris (1986), Chalmers (1988) and Sumpter *et al.* (1988). In ISSC (1988), the new high strength steels

Table 3.2 Ship structural materials

Material	Typical specification	Yield strength (MN/m^2)	Tensile strength (MN/m^2)	Young's modulus (MN/m^2)	Charpy energy (Joules at °C)	Elongation (%)	Relative cost/tonne
Mild steel	BS 4360 43 A Lloyds 'A'	245 (t > 63 mm) 275 (t < 16 mm)	430–510	201–208.5	–	22	1.00
Notch tough mild steel	BS 4360 43 D Lloyds 'D'	245 (t > 63 mm) 280 (t < 16 mm)	430–540	201–208.5	27 at $-20°C$	22	1.10
Crack arresting steel	BS 4360 50 DD Lloyds EH 32	280 (t > 32 mm) 310 (t < 22 mm)	430–590	199–207	40 at $-30°C$	20	1.15
Navy Q1N (HY 80)	NES 736 pt 1	550–655 (0.2% proof)	At least 1.4 times proof stress	197–207	70–100 at $-84°C$	20	6.00
HSLA-80	MIL-S-24645(SH)	550–690 (0.2% proof)	–	207	35 at $-84°C$	20	–
Aluminium	BS 1470 (1972) Condition 5083/0	125 (0.2% proof)	275–350	70	–	16	8.00

are reviewed in the report of Committee III.3 on material and fabrication factors; composite structures are the subject of Committee V.8 and concrete marine structures that of Committee V.2. Rules for the manufacture, testing and certification of shipbuilding materials are laid down by classification societies.

3.6.1.1 Forming and fabrication

All materials are affected by the processes used in forming and fabrication of a structure. Welding is the predominant technique in ship construction. Riveting has died out except for a very few specialized purposes. Structural adhesives are occasionally used for repair and may soon appear in shipbuilding (Hashim, Winkle and Cowling, 1989).

Welding of steel or aluminium has three main affects. One is to introduce defects into the structure which can reduce toughness and strength. These defects are checked and controlled by non-destructive testing (see Chapter 10). Another effect, arising from the rapid local heating and cooling as the weld is laid, is to introduce locked in stresses in the material around the weld. The magnitude of these residual stresses can approach the yield stress. They affect in particular the buckling strength of the structure. The third main effect of welding is a consequence of the locked in stresses. These attempt to relieve themselves by deforming the structure out of its plane. This is the cause of the unevenness, or starved horse affect, seen in ships' sides and other thin plated structures. All these effects are a function of the welding processes used (Smith *et al.*, 1988; ISSC, 1988).

3.6.2 Types of structure and modes of failure

An idealized three-dimensional view of a typical ship structure is shown in *Figure 3.31*. Apart from the basic subdivision afforded by decks and bulkheads, the plating is stiffened longitudinally and transversely by longitudinals and beams and frames, respectively. These supply the local strength and stiffness. An unstiffened structure would require far less welding and save on fabrication cost, but would need such thick plating as to be wholly impractical for a large structure.

Common types of stiffener are illustrated in *Figure 3.32*. The most efficient form is the T bar. However, angles and bulb flats are cheaper to produce and are usually chosen where weight is not an overriding consideration. Flat bar is normally used on minor structures only.

It is convenient for the purposes of structural analysis to divide the ship hull into primary, secondary and tertiary structure. This is shown in *Figure 3.33*. The primary structure is the hull taken as a whole. This behaves like a hollow beam or girder: hence the term hull girder. The overall stresses are mainly longitudinal due to bending of the hull girder and these are called the primary stresses.

The secondary structure comprises the stiffened plated structures between major boundaries such as decks and bulkheads. These orthogonally stiffened structures are called plate grillages. A grillage is subject to the in-plane primary stresses in its vicinity. In addition, the grillage deforms under local lateral loads, such as water pressure and this gives rise to bending stresses in the combined stiffener/plating elements; these are the secondary stresses.

The tertiary structure consists of the plating elements between stiffeners. These elements also bend under local lateral loading, giving rise to tertiary stresses. The total stress in a plate element is the algebraic sum of the primary, secondary and tertiary stresses.

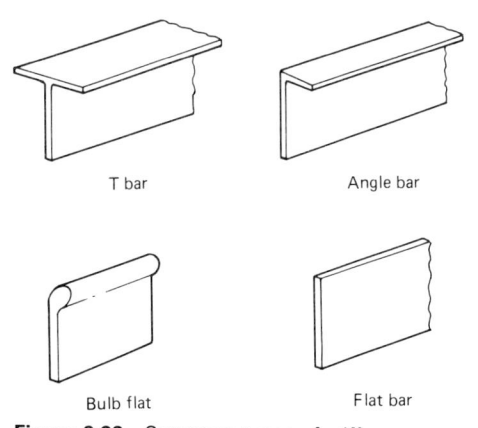

Figure 3.32 Common types of stiffener

3.6.2.1 Modes of failure

Loads on a structure, from whatever source, give rise to internal forces in that structure. The internal forces produce stresses (stress equals force per unit cross-sectional area), internal strains and external deflections. If the stresses and strains are large enough, failure will occur.

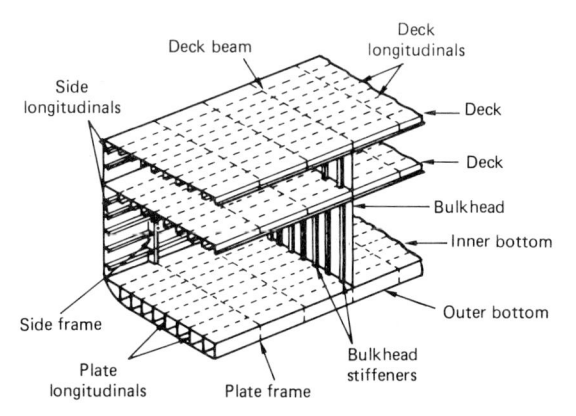

Figure 3.31 Cut away section of one side of typical ship structure

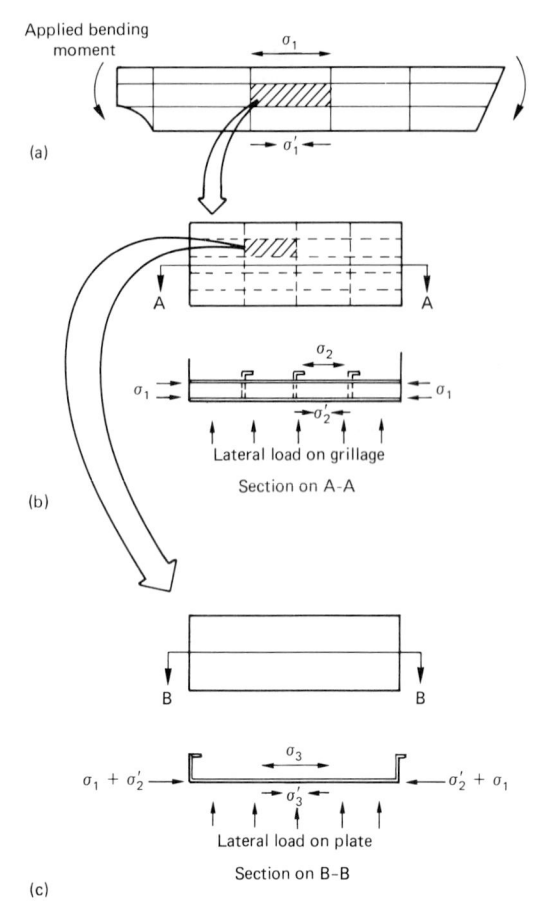

Figure 3.33 (a) primary structure (σ_1, σ'_1 are primary stresses due to bending of primary structure), (b) secondary structure (σ_2, σ'_2 are secondary stresses due to bending of secondary structure) and (c) tertiary structure (σ_3, σ_3' are tertiary stresses in plating due to bending of tertiary structures)

The modes of failure relevant to ship structures at primary, secondary and tertiary level are:

1. Plastic collapse involving ductile failure: this occurs when stresses exceed the yield stress of the material over a significant portion of the structure. Localized yielding does not necessarily lead to collapse, as it can be absorbed by redistribution of stresses in the material so as to share out the load.
2. Fatigue cracking of the material following repeated loading: the higher the number of loading cycles, the lower the stress at which fatigue failure will occur.
3. Brittle fracture: the sudden cracking of the material that can occur at low stress levels if the material has poor toughness.
4. Buckling: the instability of the structure under compression or shear loads. Pure elastic buckling is governed by stiffness rather than material strength and can occur at stresses well below yield. Practical structures tend to fail by elasto–plastic buckling, which involves a combination of buckling and yielding.
5. Excessive deflection: where the structure may deform to an unacceptable extent, causing misalignment or damage to equipment.
6. Excessive vibration or noise under dynamic loads: affects the health or safety of crew or equipment.

In specialized structures, other forms of failure may occur such as delamination of fibre reinforced plastics.

The initial design of ship structures is governed by considerations of buckling and yielding, with checks made subsequently on fatigue, deflection and vibration. Brittle fracture is avoided by correct choice of material.

3.6.3 Loading

Ship structures are subject to a great variety of static and dynamic loads. These are summarized in *Figure 3.34*. The most important load in hull structural design is the hydrodynamic loading due to waves. The worst case for normal ship forms is in head or following seas when the waves cause longitudinal bending of the hull girder.

As seen in Section 3.5, wave loading is probabilistic rather than deterministic in nature and is, strictly speaking, only definable in statistical terms. Until recently, however, theoretical means to predict loadings in this way were not available. It was therefore necessary to simplify the problem drastically. The resulting quasi-static design method was found to be adequate for the structural design of conventional ships and it is still in use along with more advanced techniques.

3.6.3.1 Quasi-static wave loading

The quasi-static standard calculation to define wave loads for hull design purposes contains simplifications. All dynamic effects are neglected. The ship is assumed to be balanced statically on a wave of length equal to that of the ship (this gives the highest bending moment). Hydrostatic buoyancy forces and ship weight are in equilibrium and the centre of buoyancy is vertically below the ships centre of gravity. The wave form is trochoidal (see Chapter 1) and its height is taken to be a function of its length.

Two worst cases are assumed in *Figure 3.35*. One is with the wave crest amidships (hogging) and the other with the wave trough amidships (sagging). In each case, the variable weights such as cargo, fuel and stores are assumed to be in the worst condition that could occur at sea. Thus, in hogging, variable weights are taken to be removed amidships but fully stored towards the ends of the ship.

The height H of the standard wave from trough to crest was for many years taken to be equal to 1/20 of its

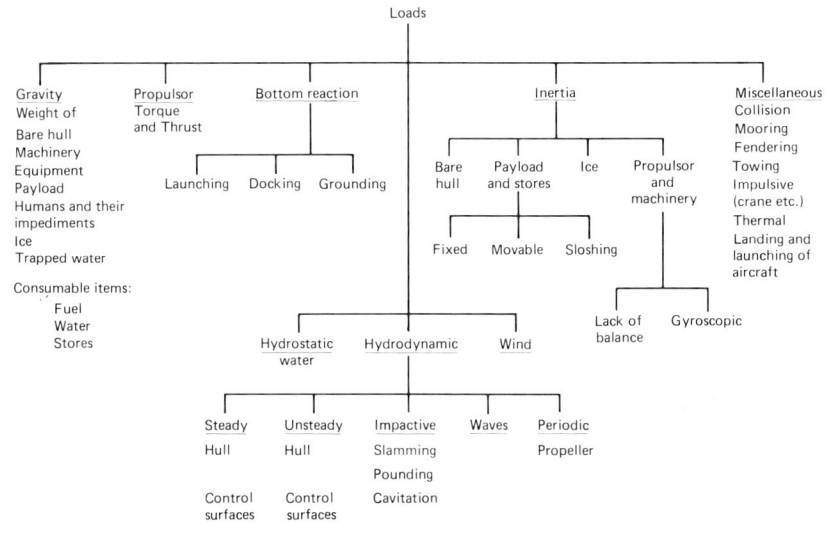

Figure 3.34 Loads on ship structures

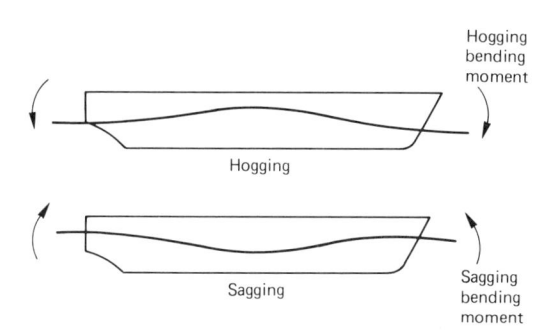

Figure 3.35 Worst cases for standard calculation

and the ship's length L. Experience suggested that this was a practical maximum for most ships although it did tend to underestimate the loads for small ships (less than 100 m length) and overestimate the loads for very large ships (more than 200 m). To counter this, another formula that became widely used was $H = 1.1\sqrt{L}$ where H and L are in feet. This translates in metric units to $H = 0.6\sqrt{L}$. For very small ships, standard waves of height $H = L/9$ have been used. Waves steeper than $L/7$ are not encountered, as they tend to break.

Having chosen a suitable worst-case condition and maximum wave height, the loads on the hull girder can be estimated by integration along the hull as shown in *Figure 3.36*. The weight per unit length is assessed at, typically, 20 positions along the length. The difference between weight and buoyancy at each position gives the net load per unit length. Total weight and buoyancy of the ship are, of course, equal. Integration of the net load along the length gives the curve of shearing force and integration of the shearing force curve gives the

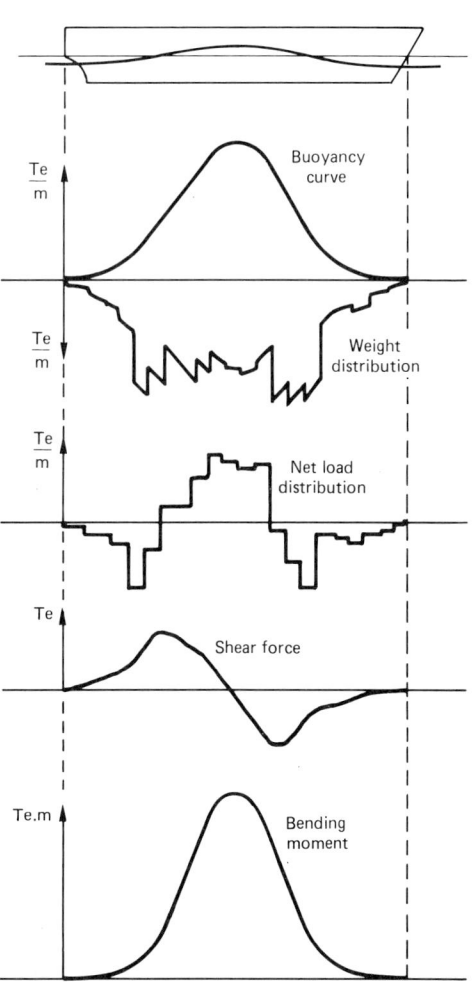

Figure 3.36 Loading, shear force and bending moment calculation in hogging condition

curve of bending moment, in accordance with classical structure beam theory.

The shear force and bending moment are calculated for the still water as well as the hogging and sagging wave conditions. This is necessary because the still water bending moment (SWBM) may be a significant proportion of the total bending moment for some ships and require specific limitation placed upon it. This can affect the permissible distribution of cargo (Rawson and Tupper, 1983; Hughes 1983; Muckle and Taylor, 1987; Lewis 1988).

The vertical bending of the hull girder is a maximum in head seas. However, in other sea conditions the hull suffers horizontal bending and torsion as well as a reduced level of vertical bending. The maximum horizontal bending moment can amount to around 40% of the maximum head sea vertical bending moment. When horizontal and vertical bending occur together, in bow or quartering seas, they are not usually in phase with one another. The net effect can be to increase maximum stresses at the deck edge by about 20% compared to vertical bending alone in head seas.

Torsion is not significant in normal closed deck vessels. However, it can be very important in vessels with wide hatch openings, such as container ships, causing large warping and shear stresses (Meek *et al.*, 1972; Hughes, 1983; Evans, 1983; Lewis, 1988).

Wave loading must also be considered for the secondary and tertiary elements of the hull structure. Quasi-static wave loads include hydrostatic pressure on grillages and plating elements. Allowing for rolling of the ship, a minimum hydrostatic pressure head of 5 m is sometimes assumed for the design of the ship's sides.

3.6.3.2 Dynamic and other loads

Dynamic wave loads include impulsive pressures from slamming (see Section 3.5.4) and sea slap; the latter occurs when waves hit exposed portions of the hull or superstructure in heavy weather. Both slamming and sea slap loads are difficult to quantify accurately. They are often modelled, once again, as equivalent static pressures. For example, a static sea slap pressure of 50 kN/m^2 is often assumed for design purposes, or 100 kN/m^2 for forward-facing exposed surfaces. Slamming loads are discussed below.

Other types of load mainly affect secondary and tertiary structure. A review and extensive further references may be found in ISSC (1988).

3.6.3.3 Extreme wave load prediction

The development of strip theory and more recent three dimensional methods was described in Section 3.5. These techniques permit structural loading to be calculated by a more rational process than the traditional standard calculation and its variants.

However, the use of strip theory still depends in its practical application on the principle of linear superposition, i.e. the assumption that wave loads are proportional to wave height. As noted in Section 3.5, this is not really valid for ship forms in large waves. The presence of non-linearity causes the wave induced sagging moment to be significantly greater than the hogging moment. The process, as currently applied, also remains quasi-static although a truly dynamic method of calculation is mentioned in Section 3.6.7. Nevertheless, strip theory methods are powerful aids to the designer and are widely used for strip structural design.

Strip theory or suitably instrumented model tests are used to derive RAOs for the wave loadings such as midship bending moment per unit wave height. From these, the probable maximum value of wave bending moment in a given sea state can be calculated assuming that the short term probability of maximum values is described by a Rayleigh distribution (see Section 3.5).

To find the maximum probable bending moment in the life of the ship, consideration must be given to the ship's operating areas and the likely speeds and headings in each case. The ISSC has agreed a two-parameter spectrum, the Bretschneider spectrum (see Chapter 2), that summarizes the long term statistics for different regions of the world). This may be used to calculate the most probable extreme bending moment in the life of the ship, usually taken to be 10^8 wave cycles. The process is illustrated in *Figure 3.37* (Faulkner and Sadden, 1979).

A margin of safety may be adopted by designing the structure to just withstand the calculated bending moment that has, for example, only a 1% probability of exceedence in the ship's life. This equates approximately to a load factor of 1.4 on the most probable

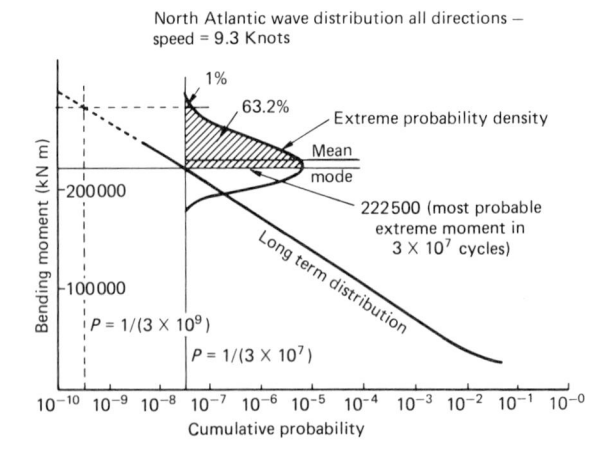

Figure 3.37 Wave bending moment probabilities for a ship hull, with life taken as 3×10^7 wave cycles (from Faulkner and Sadden, 1979) (Courtesy of The Royal Institution of Naval Architects)

extreme bending moment in the ship's life. Such statistical derivations may be compared to the standard calculation of wave loading in order to derive an equivalent effective wave height that gives the same bending moment prediction. In broad terms, for medium sized merchant ships the effective wave height does not differ greatly from the traditional $H = 0.6\sqrt{L}$ (m). For fast displacement forms, such as warships, full scale data suggest that a constant effective wave height of 8 m is most appropriate (Clarke, 1986).

3.6.3.4 Approximate formulae

Various approximate formulae for wave loading have been derived. These are useful in the early stages of design when details of the ship's hull form and weight distribution are not available (Hughes, 1983; Lersbryggen, 1978). These give the maximum wave bending moments corresponding to a probability of exceedance of 10^{-8} as:

$$M_w \text{ (sag)} = 125 \, L^2 B(C_B + 0.2) \, h_w \quad (3.95)$$

$$M_w \text{ (hog)} = 165 \, L^2 B C_B^2 h_w \quad (3.96)$$

in units of N/m where L is length between perpendiculars, B is beam, C_B is block coefficient. h_w is the characteristic value of the wave height for the vessel:

For $L < 350$ m, h_w is the smallest of

$h_w = D$, the depth of the hull girder

or $h_w = 13 - \left(\dfrac{250-L}{105}\right)^3 \quad (3.97)$

or $h_w = 13 - \left(\dfrac{L-250}{105}\right)^3 \quad (3.98)$

For $L > 350$ m, $h_w = 227/\sqrt{L} \quad (3.99)$

Wave induced shear forces are given by:

$$V_w \text{ (sag)} = M_w \text{ (sag)}/0.39L \quad (3.100)$$

$$V_w \text{ (hog)} = M_w \text{ (hog)}/[0.29 \, (C_B + 0.2)L] \quad (3.101)$$

A very approximate formula for slam induced vibratory bending moment has been deduced from full scale trials on four merchant ships (Aertssen, 1979). At a probability of exceedence of 10^{-8}, the slam bending moment is:

$$M_s = 0.000 \, 75 \rho_w g L^3 B \quad (3.102)$$

Still water bending moments can usually be calculated from first principles using suitable approximations for the weight and buoyancy distributions (Faresi, 1965; Muckle, 1967; Lewis, 1988). Alternatively, classification society rules such as Lloyds Rules (1987) can be consulted for appropriate formulae for still water and wave induced bending moments.

3.6.4 Primary structure quasi-static response–hull girder bending

It has already been mentioned that the hull girder behaves approximately like a hollow non-uniform beam. The calculated bending moment can therefore be used in conjunction with simple beam theory to derive bending and shear stresses in the hull girder.

The direct stress at a point on the beam due to the bending moment at the cross-section concerned is given by:

$$\sigma = \frac{My}{I} = \frac{M}{Z} \quad (3.103)$$

where M is the bending moment, y is the vertical distance of the point in question from the neutral axis of the beam and I is the second moment of area of the cross-section about the neutral axis. $Z = I/y$ is called the section modulus.

M has a maximum value near amidships (see *Figure 3.36*). Thus the maximum bending stress occurs amidships in either the upper deck or outer bottom, depending on the vertical position of the neutral axis. In practice, the bottom plating is invariably heavier than the deck plating, giving a neutral axis closer to the bottom than to the upper deck. Hence, the highest stresses occur in the upper deck. These are tensile stresses in the hogging condition and compressive stresses in the sagging condition (*Figure 3.38*).

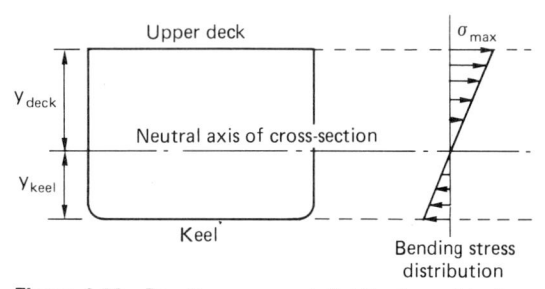

Figure 3.38 Bending moment distribution with distance from neutral axis

The maximum shear stress under wave bending occurs in the middle of the ship's side near the 1/4 points along the hull (see *Figure 3.36*). Its value is given approximately by:

$$\tau = \frac{FA\bar{y}}{It} \quad (3.104)$$

where F is the shear force at the section, $A\bar{y}$ is the first moment about the neutral axis of the cross-sectional area of material above (or below) the neutral axis, I is the second moment of area of the whole cross-section about the neutral axis and t is the total width of material at the neutral axis, i.e. twice the local material thickness.

The cross-section material used in the above calculations includes both plating and stiffeners and continuous internal decks and longitudinal bulkheads. Simple methods may be used to calculate the position of the neutral axis and the values of I and $A\bar{y}$ in order to find bending and shear stresses in a particular case (Rawson and Tupper, 1983). Note, however, that the contribution of major longitudinal bulkheads to longitudinal strength can be complex (Taylor, Payer and Swift, 1986).

3.6.4.1 Stress criteria

The acceptable level of the derived primary stresses depends on such things as the hull material, the operating conditions of the ship, the policy for corrosion control and the type of calculation made. The standard calculation is essentially comparative rather than absolute. Acceptable stresses are therefore chosen by comparison with previous successful ships. The maximum total (wave plus still water) primary stress usually considered acceptable with $L/20$ or $0.6\sqrt{L}$ waves equates to a factor of safety on yield stress of around 2 to 2.5. Maximum still water bending stress is usually limited to give a factor of safety on yield of about 4.

Under the extreme bending moments calculated using the formulae in Section 3.6.3.3, suggested stress criteria for steel ships are as shown in *Table 3.3*.

Table 3.3 Suggested stress criteria for steel ships

Stress	Criterion
Maximum bending stress under extreme combined moment (still water plus wave)	180 MN/m^2
Maximum stress range under extreme wave bending moment (a fatigue criterion)	320 MN/m^2
Maximum shear stress under extreme shear force	110 MN/m^2

3.6.4.2 Superstructures

For simple closed section hulls, simple beam theory gives a reasonably accurate estimation of primary bending stress. Difficulties arise where parts of the hull girder are not continuous along the length. A particular example is a ship's deckhouse or superstructure. This may also have major cut outs such as windows, and may be made of a different material from the main hull girder. Simple beam theory breaks down in such cases and special methods of calculation are necessary to assess the structural efficiency of the superstructure (Mitchell, 1978; Hughes, 1983; Evans, 1988; Fransman, 1988).

3.6.4.3 Fatigue

Fatigue failure is meant to be covered implicity by the overall factors of safety inherent in the standard calculation. However, it is now possible to design the hull girder explicity for a given fatigue life. This is based on the expected number of cycles of wave induced bending during the ship's life, with an allowance for higher frequency bending moments due to slam induced whipping.

The criteria applied limit the permitted stress range between the extremes of hogging and sagging. The criteria depend on the type and quality of structural joints and details where fatigue cracks might start. The process of joint classification and use of S–N curves is similar to that described for offshore structures in Chapter 2 in this book (Lersbryggen, 1978; Hughes, 1983; Lewis, 1988; ISSC, 1988).

3.6.4.4 Asymmetric loads

Methods of analysis of hull girder response under torsional, other asymmetric and thermal loads are given by Meek *et al.* (1972), Hughes (1983), Evans (1983), Pedersen (1985) and Lewis (1988).

3.6.4.5 Finite element analysis of the hull girder

The most powerful structural analysis tool for calculating primary and indeed secondary and tertiary response, is finite element analysis (FEA). In this, the structure is divided into numerous imaginary elements, usually of a rectangular or triangular shape.

The elements are assumed to be connected only at their corners, or nodes. Displacements at the nodes are related to the displacements at any point within an element by a suitable displacement function. Strains can be found from the displacements and hence stresses determined. Nodal forces are made equivalent to boundary forces on the portion of structure being analysed and displacements of the elements have to obey compatibility rules. Finally, the whole array of applied external forces and internal forces has to satisfy equilibrium conditions. The process requires the solution on a computer of a large number of simultaneous equations represented in matrix form. Finite element techniques in structural analysis are described in Zienkiewicz (1971).

FEA is now a standard tool in ship structural analysis (see Section 3.6.9) and is particularly valuable where the structural configuration is complex or contains major discontinuities. A large whole ship FEA would normally commence with a coarse mesh to obtain overall response. This could run to the order of 1000 nodes and 2500 elements. This might then be followed by a fine mesh FEA to give more detail. This could include up to 5000 nodes and 13 000 elements. A typical FEA model of a complex hull girder is shown in *Figure 3.39*.

Finite element modelling requires great care, both to achieve an efficient analysis and to avoid misinterpretation of results. A major FEA of the hull girder is a time consuming and rather unwieldy task. In many cases, simpler methods such as simple beam theory may be more appropriate for initial design. The broad structural arrangements thus produced, which should rarely need major adjustment, can then be checked in detail using FEA. Local highly stressed areas of the structure may be the subject of subsequent very detailed FEA.

3.6.4.6 Ultimate strength of the hull girder

The analysis methods described so far are based on the elastic response of the hull girder, i.e. stress is taken proportional to strain and is below the yield stress. It is also desirable to know the ultimate load bearing capacity of the hull girder. One may then ensure that this is not exceeded by the extreme wave bending load that might be experienced in the life of the ship.

An analogy may again be made with beam theory. The ultimate resistance to bending occurs when a plastic hinge forms at the cross-section concerned. The so called fully plastic moment is given by:

$$M_p = \sigma_y S \quad (3.105)$$

where σ_y is the yield stress of the material and S is the plastic modulus. S is equal to the sum of the first moments of area of the cross-section on each side of the plastic neutral axis.

For a hollow box girder, the full yield stress is unlikely to be developed in the compression flange as that will buckle elastoplastically before the stress reaches yield. The ultimate bending moment may then be represented by:

$$M_u = \phi \sigma_y S \quad (3.106)$$

where ϕ ($\phi \leq 1$) is the efficiency of the section in buckling. The value of ϕ depends on the elastoplastic collapse strength under primary loads of the individual grillages from which the hull girder is constructed. A value of $\phi > 0.8$ is unlikely in practical ship structures. The process of calculating the ultimate bending moment capacity of the hull girder is described by Smith (1977), Faulkner and Sadden (1979), Dow *et al.* (1981), Rawson and Tupper (1983), Hughes (1983), Adamchak (1984) and Lewis (1988). See also Section 3.6.5.6.

3.6.5 Secondary structure

As we saw in Section 3.6.2, the secondary structure is an assembly of grillages which are loaded by both in plane primary loads and by local lateral loads. Each grillage is therefore subject to buckling (when the in-plane loads are compressive) together with bending and shear.

An initial analysis of a grillage may be made by apportioning the loads upon it among the individual stiffeners. Each stiffener, in association with its adjacent plating, is then considered to act as a simple beam independent of the rest of the grillage (*Figure 3.40*). The length of the beam is its unsupported length. Longitudinal and transverse stiffeners in a grillage are

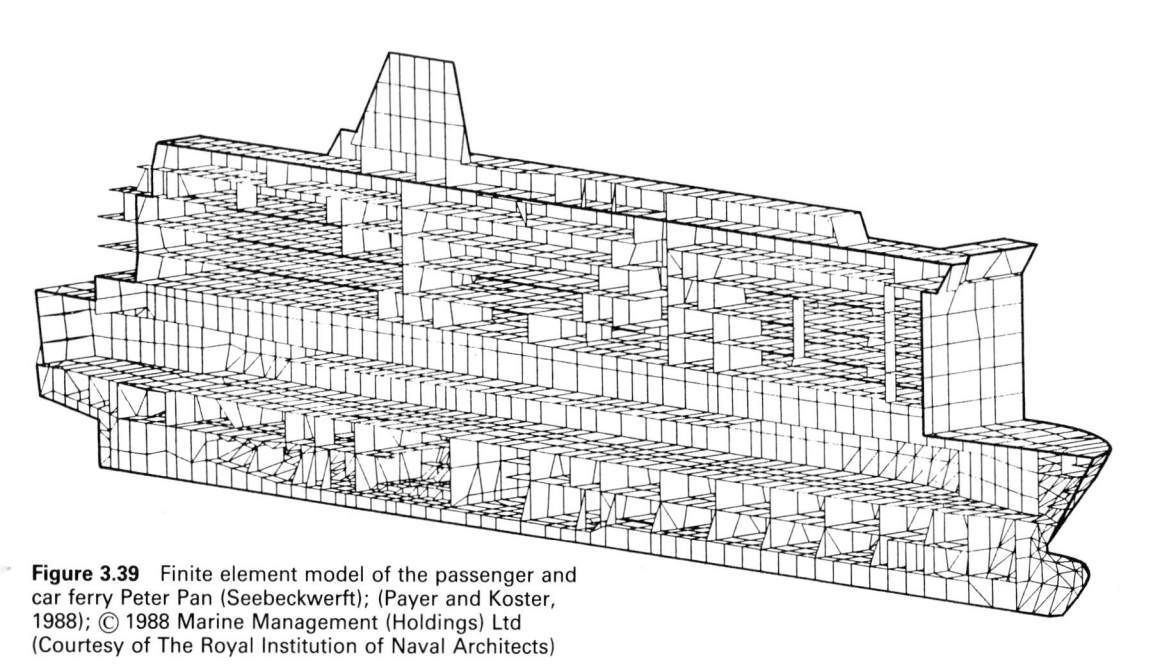

Figure 3.39 Finite element model of the passenger and car ferry Peter Pan (Seebeckwerft); (Payer and Koster, 1988); © 1988 Marine Management (Holdings) Ltd (Courtesy of The Royal Institution of Naval Architects)

invariably of different depths, for ease of manufacture. The smaller stiffeners are usually greater in number (*Figure 3.31*). Their unsupported length is that between the deeper stiffeners. The unsupported length of the deeper stiffeners is that between the boundaries of the grillage, such as decks or bulkheads orthogonal to its plane.

3.6.5.1 Effective breadth and effective width

In *Figure 3.40*, the breadth of plating b_e acting as part of the idealized beam is less than the actual breadth b between adjacent stiffeners. As the stiffeners bend under applied lateral load, the load is shared between stiffeners and plating by shear action. This is not fully effective in making the plating take its share of the load. The effect is known as shear lag. The direct stress distribution in the plating is then as shown in *Figure 3.40*. This is idealized as shown to derive b_e, called the effective breadth of the plating. It is a function of loading and geometry. Values are given in the standard text books and in Faulkner (1975) and Evans (1975). For initial design purposes, b_e may be taken very approximately as half the actual breadth b (Clarkson, 1968) or 40 × plate thickness, whichever is the less.

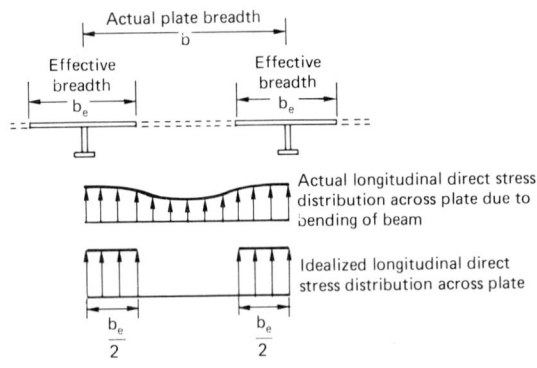

Figure 3.40 Grillage stiffener behaviour and effective breadth of plate

A somewhat similar effect to shear lag occurs when the combined stiffener/plating is loaded in-plane. The share of in-plane load taken by the central portion of the plating is reduced due to residual stresses, distortion and elastoplastic effects. The effective width, as it is now called to distinguish it from the effective breadth applicable in bending, is largely a function of fabrication processes. A value sometimes used in initial design is $40\,t$, where t is the plating thickness. A more accurate formulation by Faulkner (1975), also in Evans (1975), is based on experimental data. An iterative procedure is required to find effective width b_e but a reasonable approximation is given by:

$$\frac{b_e}{b} = \frac{2}{\beta} - \frac{1}{\beta^2} \qquad (3.107)$$

where $\beta = \dfrac{b}{t} \dfrac{\sqrt{\sigma_y}}{E}$ (3.108)

Other formulations of effective width are discussed by Hughes (1983) and Smith *et al.* (1988). The latter use non-linear finite element analysis, validated against experimental data, to give design curves for a range of plate proportions and initial deformations. Typical as-built plating deformations for merchant ships are recorded by Antoniou (1980) and Antoniou, Lavidas and Karvounis (1984). Imperfections in lighter warship-type structures are given by Somerville, Swan and Clark (1977). Having arrived at values of effective breadth and effective width, the response of the combined stiffener and plating may be analysed.

3.6.5.2 Elastic bending

Elastic analysis in bending is similar to that shown above for primary structures. The bending moment is assessed as for a beam with rigid end supports. The section modulus Z is calculated for the beam plus effective breadth of plating.

Fixed end forces and moments for a beam under typical loading conditions are given in *Table 2.7*. Bending moments for other conditions may be found in standard text books or in books of structural formulae such as Roark (1988). The ends of the beam are assumed to be clamped against rotation when the loading takes the form of uniform pressure over a large area. Simply supported (free to rotate) end conditions are appropriate for most other types of loading.

For example, under a uniform pressure p the maximum bending moment in the beam occurs at the end supports and is given by:

$$M = \frac{w\,l^2}{12} = \frac{p\,b\,l^2}{12} \qquad (3.109)$$

where w is the load per unit length on the beam ($= p \times b$), b is the full plating breadth between beams and l is the length of beam between supports. The corresponding central deflection is given by:

$$\delta_o = \frac{w\,l^4}{384\,EI} \qquad (3.110)$$

where E is the elastic Young's Modulus for the material and I is the second moment of area of the cross-section of the beam plus effective breadth of plating about their combined neutral axis. Local shear stress may also need checking under concentrated loads.

3.6.5.3 Elastic buckling

Under compressive in-plane loads, such as those arising from overall hull bending, classical Euler buckling theory is sometimes used (Bleich, 1952). The critical elastic buckling load is given by:

$$P = \frac{\pi^2 EI}{l^2} \quad (3.111)$$

where l is the effective length of the beam in buckling. This is usually taken as the actual length between supports. In practical structures, buckling is invariably inelastic rather than elastic and appropriate calculations should be made (see Section 3.6.5.6).

3.6.5.4 Combined loads

Combined lateral and in-plate loading is a common design case. The in-plane load increases the total bending moment and central deflection due to the lateral load alone. The combined response is described approximately by means of a magnification factor, $P_E/(P_E - P)$, where P_E is the critical buckling load and P is the actual in-plane load. $P = \sigma_a(A + bt)$, where σ_a is the average applied in-plane stress and A is the stiffener cross-section area. The central deflection δ_o calculated for lateral load alone is increased to:

$$\delta'_o = \frac{\delta_o \cdot P_E}{P_E - P} \quad (3.112)$$

and the bending moment M_o at the centre of the beam is increased to:

$$M'_o = M_o + \frac{P \cdot \delta'_o P_E}{P_E - P} \quad (3.113)$$

The total stress in the stiffener and plating is the algebraic sum of that due to the magnified bending moment and direct in-plane forces (*Figure 3.33*):

$$\sigma_{total} = \frac{1}{Z}\left(M_o + \frac{P\delta_o P_E}{P_E - P}\right) \pm \frac{P}{A + bt} \quad (3.114)$$

where A is the cross-sectional area of the stiffener alone and Z is the section modulus of the stiffener plus effective breadth of plating b_e. The maximum total compressive stress thus calculated must be less than the average in-plane collapse stress of the stiffener/plate panel.

3.6.5.5 Factors of safety

Elastic analysis is appropriate when studying response under working loads or cyclic loads, where large permanent deformations and fatigue failures are to be avoided. For maximum working loads, it is usual to design the structure with a factor of safety against yielding.

The factor of safety is chosen on the basis of past experience and the likely consequences of failure. Typical safety factors for secondary structure under maximum working loads are 1.2 for stiffener bending and 3 for elastic buckling (Hughes, 1983).

It is necessary also to ensure that the secondary structure does not fail completely under possible extreme loadings. In such cases, elastic analysis is inappropriate and elastoplastic or limit analysis is used.

3.6.5.6 Elasto plastic response

In bending, fully plastic analysis is appropriate. This is similar to plastic analysis of the hull girder (Section 3.6.4). The stiffener/plating combination is designed such that the fully plastic moment M_p is not less than the extreme local bending moment considered possible in the life of the ship.

Under high in-plane loads, real stiffener/plate structures will start to collapse well before the Euler buckling load is reached. This is due to initial lack of straightness, eccentricity of loading and the presence of residual stresses as well as yielding of the material as the yield stress is approached (Smith and Kirkwood, 1977).

A number of interaction formulae have been developed to account for this combination of yielding and buckling. One of the best known for practical steel plated structures is the Johnson Parabola (Faulkner *et al.*, 1973). This is shown in *Figure 3.41*. The total average collapse stress σ_{av} for plate and stiffener which should, of course, exceed the in-plane applied stress is then given by:

$$\frac{\sigma_{av}}{\sigma_y} = \frac{\sigma_u}{\sigma_y}\left(\frac{A + b_e t}{A + bt}\right) \quad (3.115)$$

where σ_y is the yield stress, A is the stiffener cross-section area, b_e is the effective width of plate, b is the actual width of plate between stiffeners, and t is the plate thickness.

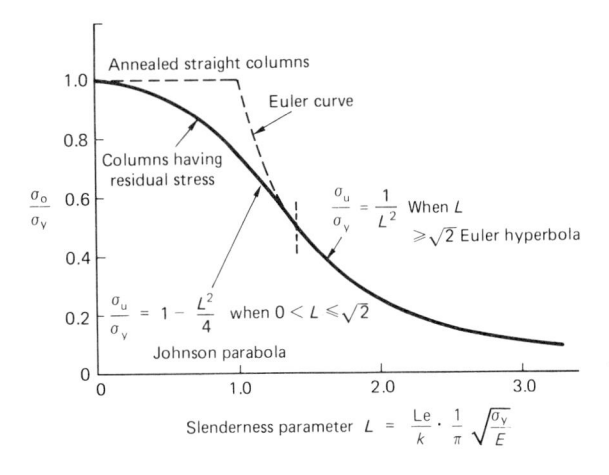

Figure 3.41 Basic column collapse interaction curve for structural steels. Notation: σ_u = Average crippling stress of column assuming it has not been annealed; σ_y = Yield stress as determined by tensile tests; Le = Effective length of column depending on end restraints; (conservatively taken as actual length); k = Least radius of gyration of column cross section $\sqrt{\frac{I}{A}}$

In reality, different types of stiffener have different collapse curves. More accurate design data is given by Evans (1975), Mansour (1976), Smith (1977) and Hughes (1983). Some approximate formulae for preliminary design against lateral and in-plane loads are given by Mansour (1986).

When the collapse strengths of the various panels that make up the full midship section are found, these can be combined to give an estimate of the ultimate bending moment of the complete section as described in Section 3.6.4.6.

3.6.5.7 Overall grillage response

Having considered the response of individual stiffener/plating sections, it is necessary to check that their combination in an orthogonal grillage structure has adequate overall strength. This is particularly necessary as a check on the adequacy of the deeper set of stiffeners, for which the methods described above are very approximate.

Analysis of a grillage in overall bending may be made by a number of methods. These are described by Clarkson (1965) who gives design curves for various types of lateral loading and edge conditions (also Chang, 1968; Evans, 1975).

Modern methods of analysis are based on matrix methods and most finite element programs have grillage analysis routines. The basic theory, given in Hughes (1983), is similar to that for the matrix analysis of frameworks outlined in Chapter 2.

A grillage can buckle in a number of different modes as shown in *Figure 3.42*. The plating can buckle between stiffeners, the smaller set of stiffeners, with plating attached, can buckle between the deep stiffeners and the grillage can buckle overall between its edge supports. Individual stiffeners can also fail by tripping, as described below. Analysis of the overall buckling of a grillage is best undertaken by a non-linear finite element method (Hughes, 1983). Hand calculation methods are available, however (Bleich, 1952; Timoshenko and Gere, 1961; Clarkson, 1965; Smith, 1975; Hughes 1983).

An approximate method useful for initial design is to model the grillage as an orthotropic plate. This is acceptable where there is a reasonably large number of stiffeners in each direction. The critical elastic buckling load for simply supported edge conditions is given by:

$$\sigma_{cr} = \frac{\pi^2 E}{12(1-v^2)} \cdot \frac{t^2}{B^2} \cdot \frac{1}{1+\delta_x} \left[2 + \frac{m^2 B^2}{L^2} \gamma_x + \frac{L^2}{m^2 B^2} \gamma_y \right]$$

(3.116)

where $\delta_x = \dfrac{A_x}{bt}$

A_x is the sectional area of a longitudinal stiffener,

$$\gamma_x = 12 \frac{(1-v^2)}{bt^3} I_x,$$

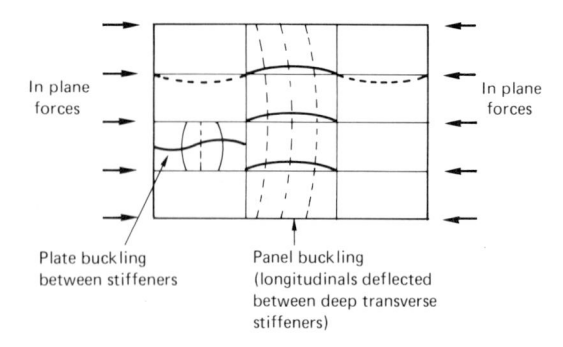

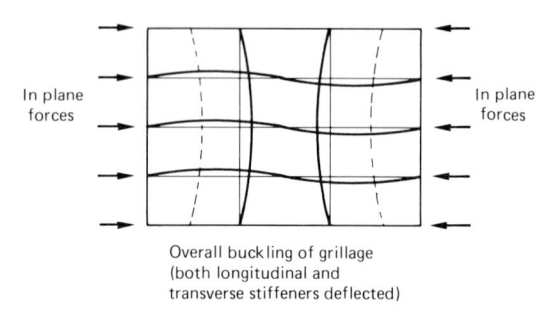

Figure 3.42 Grillage buckling modes

I_x is the second moment of area of longitudinal stiffener plus width of plating b,

$$\gamma_y = 12 \frac{(1-v^2)}{at^3} I_y$$

I_y is the second moment of area of transverse stiffener plus width a of plating, v is the material Poisson's Ratio (0.3 for steel) and m is the number of half waves in the longitudinal (x) direction into which the grillage buckles overall.

For minimum value or σ_{cr}:

$$m \simeq \frac{L}{B} \sqrt[4]{\frac{\gamma_y}{\gamma_x}}$$

This latter equation does not, in general, give an integer value for m. The integer values of m on either side of the value calculated should be investigated to find the lower value of σ_{cr}. The method may be extended to elasto–plastic analysis: Smith (1975), Evans (1975) and Hughes (1983). Application of orthotropic plate analysis to double bottom structures is given by Vilnay (1982, 1983). Shear buckling of grillage panels can be analysed using data sheets published by ESDU (1976, 1983). Methods are also given by Evans (1975) and Hughes (1983). Evans and Hughes also deal with buckling under combined longitudinal and transverse compression and give interaction formulae for combined shear and direct in-plane stress.

Tripping is the sideways or torsional buckling of the web or flange of the stiffener, causing it to rotate about its toe. This can be avoided by suitable choice of stiffener geometry. Symmetrical sections such as T bars are less prone to tripping than angle bars or bulb flats. With long unsupported stiffener lengths, fitting of small tripping brackets may be necessary as a preventive measure (Faulkner *et al.*, 1973; Hughes, 1983).

3.6.6 Tertiary structure

In the preceding section, we considered the plating only in so far as it acted as a flange to the stiffeners. We now address the behaviour of a plate element itself, bounded by longitudinal and transverse stiffeners along its edges (*Figure 3.43*).

3.6.6.1 *Plate bending*

The small deflection elastic behaviour of uniform flat plates is described by classical plate theory (Timoshenko and Woinowsky–Krieger, 1959; Evans, 1975; Hughes, 1983; Roark, 1988; Rawson and Tupper, 1983; Lewis, 1988).

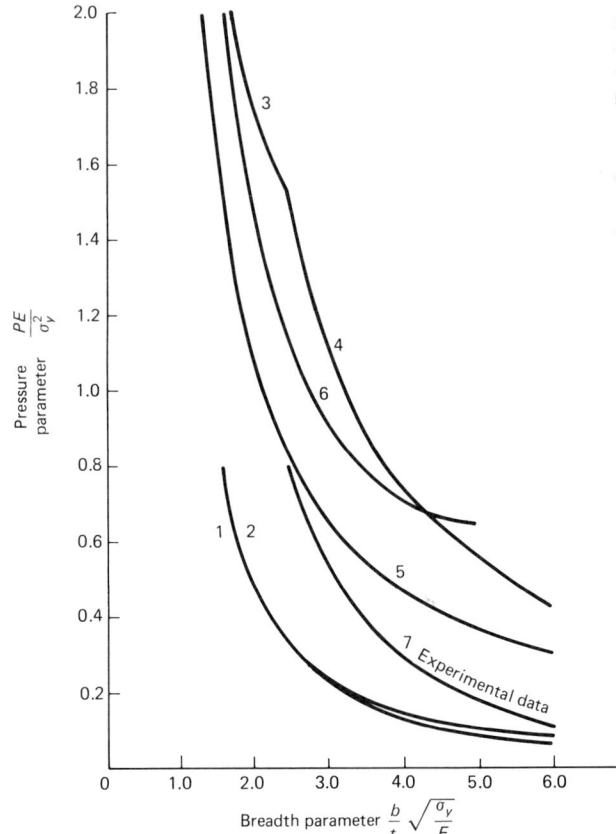

The formulae for small deflection elastic response of long plates, length a greater than about twice the width b, a condition commonly found in ship structures, are simple. The maximum bending stress under lateral pressure p, assuming fixed (clamped) edges, is:

$$\sigma_b = \frac{p}{2}\left(\frac{b}{t}\right)^2 \qquad (3.117)$$

This maximum stress occurs at the midpoint of the long edge of the plate. The corresponding central deflection of the plate is:

$$\delta = \frac{pb^4}{384D} \qquad (3.118)$$

where $D = \dfrac{Et^3}{12(1-v^2)}$

and v is the Poisson's Ratio (0.3 for steel).

Small deflection theory is valid up to a deflection of around $0.75\,t$. Above this value, membrane forces develop in the plating and large deflection theory applies (Timoshenko and Woinowsky–Krieger, 1959; Hughes, 1983).

1 Flat plate – initial yield (small deflection)
2 Flat plate – initial yield (large deflection)
3 Flat plate – plastic hinge on centre line
4 Flat plate – membrane tension = $\frac{2}{3}\sigma_y$
5 Plate with initial stress free deflection $\dfrac{\delta}{b}\sqrt{\dfrac{E}{\sigma_y}} = 0.2$ then yield
6 Plate with permanent set $\dfrac{\delta}{b}\sqrt{\dfrac{E}{\sigma_y}} = 0.2$ then yield
7 Plate with permanent set $\dfrac{\delta}{b}\sqrt{\dfrac{E}{\sigma_y}} = 0.2$ then yield

Note: For curves 1–6 the plate edges are assumed clamped against pull in as well as against rotation.
Curve 7 is from experimental data with edges free to slide

Figure 3.43 Critical pressures for long plate in bending (aspect ratio 3:1)

Under still larger lateral loads, deflection increases until regions of the plate begin to yield. Such elasto–plastic behaviour is difficult to analyse and a semi-empirical approach is utilized. When plasticity has spread throughout the plate, it will distend until the ultimate strength is reached. This occurs at loads and deflections much greater than those applying at the elastic limit. Detailed analysis of elasto–plastic and plastic behaviour is described by Evans (1975), Hughes (1983) and Lewis (1988).

Real plates have initial deformations and residual stresses that modify the theoretical behaviour. Also at larger deflections the membrane stresses cannot develop fully as the plate edges within a ship grillage are not truly fixed. Even when clamped against rotation, the edges tend to slide inwards towards the centre of the plate when high lateral loads are applied.

Figure 3.43 (Clarkson, 1962) shows critical pressures for a long plate under various elastic and elasto–plastic design conditions. The appropriate curve to use depends on the service conditions and what level of stress, deflection or permanent set (the deflection remaining when load is removed) can then be accepted. For example, a plate on the ship's outer bottom near the propellers will be subjected to a high number of load reversals and hence liable to fatigue. A low stress level would be appropriate and an elastic, small deflection theory used (Curve 1 in *Figure 3.43*). Elsewhere on the outer bottom there will be fewer load fluctuations and an elasto–plastic theory could be used. Some permanent set is probably acceptable under the design loads, bearing in mind the fact that the plate almost certainly has some initial deformation anyway. Curve 7 would then represent the most appropriate design condition. The permanent set for Curve 7 equates, for mild steel, to about 1/150 of the plate breadth b. This is around half the plate thickness for typical ship structure geometry.

Plate bending under local concentrated loads, as distinct from uniform pressure, may also be treated by classical plate theory. However, design on the basis of elastic small deflection theory is in most cases overly conservative for such loads (Jackson and Frieze, 1981; Hughes, 1983; Lewis, 1988). Some permanent set is normally acceptable and allowance for this considerably reduces the thickness of plating required.

3.6.6.2 Plate buckling

The theoretical analysis of plate buckling under in-plane loads is treated by Bleich (1952), Timoshenko and Gere (1961), Evans (1975), Dowling, Harding and Frieze (1976), Hughes (1983) and Lewis (1988). For a plate subject to in-plane compression or shear it is usual, and only slightly conservative, to assume simply supported edge conditions (edges free to rotate). The critical elastic buckling stress for a long plate of width b ($<a$) under uniaxial compression in the long direction is then given by:

$$\sigma_{cr} = \frac{4\pi^2 E}{12(1-\nu^2)} \left(\frac{t}{b}\right)^2 \quad \left(= 3.62\, E \left(\frac{t}{b}\right)^2 \text{ for steel}\right) \quad (3.119)$$

The equivalent buckling stress for a wide plate of width b ($>a$) is:

$$\sigma_{cr} = \frac{\pi^2 E}{12(1-\nu^2)} \left[1 + \left(\frac{a}{b}\right)^2\right]^2 \left(\frac{t}{a}\right)^2 \quad (3.120)$$

A wide plate loaded longitudinally buckles at a considerably lower load than a long plate. For this reason, longitudinal stiffening is preferable to transverse stiffening for resisting longitudinal compressive loads.

The critical shear buckling stress of a simply supported plate under in-plane shear is given by:

$$\tau_{cr} = E \left(\frac{t}{b}\right)^2 \left[4.8 + 3.6 \left(\frac{b}{a}\right)^2\right] \quad (3.121)$$

Buckling of plates under other boundary conditions and under biaxial compression and combined compression and shear is covered by the aforementioned references.

Under moderate levels of combined lateral and in-plane load, bending stresses can be calculated using the magnification factor approach of the Section 3.6.5.4. Lateral loads can, in certain conditions, increase critical buckling loads by suppressing the lowest buckling modes. However, this should not be relied upon in design.

The ultimate strength of plates in compression is relevant to the support that the plating gives to the stiffener when calculating grillage ultimate strength. The analysis is relatively complex, particularly when allowance is made for the effect of initial deflections, permanent set and residual stresses (Evans, 1975; Dowling, Harding and Frieze, 1976; Hughes, 1983; Cresswell and Dow, 1986; Smith *et al.*, 1987). Smith *et al.* include data curves for in plane compression and tension, suitable for design use.

3.6.6.3 Structural details

The correct design of structural details, such as connections and openings, is important if premature failure through yielding, buckling or fatigue is to be avoided (Faulkner, 1964; Munse, 1981, 1983).

Stress concentrations around openings and other discontinuities must be allowed for in design (Peterson, 1953; Rawson and Tupper, 1983; Roark, 1988; Evans, 1985). It is usual to analyse problem areas by means of detailed two-dimensional finite element analysis of the in-plane stress field. The edge conditions and loading to be applied may be found from the output of the three dimensional finite element analysis of the overall hull structure described in Section 3.6.4.

3.6.7 Structural dynamic response

The response of a ship in a seaway is a dynamic rather than a static phenomenon. Until recently, however, the mathematical techniques to define the dynamic loads and structural dynamic response were not available.

A dynamic analysis is necessary to find the response of the hull structure to loads that excite the natural distortion frequencies of the hull girder. Ordinary wave loading can do this in certain circumstances. Very long ships, such as Great Lakes ore carriers, have relatively low natural frequencies of hull girder vibration that can approach the encounter frequency of short waves. In the sea states commonly met, these waves can have sufficient energy to excite the hull, giving rise to the hull bending vibration phenomenon known as springing (Stiansen, Mansour and Chen, 1977; Troesch, 1984).

A common cause of dynamic excitation is slamming (Ochi and Motter, 1973; Stavovy and Chang, 1976; Bishop, Price and Tam, 1978; Bishop and Price, 1979; Hughes, 1983; Lewis, 1988; ISSC, 1988). This was described in Section 3.5. The structural effects of slamming are twofold. First, the severe hydrodynamic impact due to bottom slamming, sometimes called impact slamming, forward sets the whole hull girder vibrating as a free–free beam in bending. With bow flare slamming, sometimes called momentum slamming, the impact is less severe but the forces can still be sufficiently large to set the hull girder vibrating in the same way. This has significant effects on primary stresses and also on fatigue life. The stress enhancement can be as much as 50% of the normal wave bending stress amidships.

Second, the outer bottom plating at the point of slam can suffer very high pressures causing severe local deformation. Local pressure as high as 2 MPa (20 Bar) have been known to occur, with a rise time of 0.1 s.

Other dynamic loads of importance include vibration loads from propellers and machinery. Specialized vessels may have major transient dynamic loads to contend with, such as ice breakers in ramming mode and warships suffering underwater explosions.

3.6.7.1 Hull girder dynamic response

The hull girder responds to dynamic loading in a combination of vibration modes. These are similar to the modes of a free-free beam (without end constraints). The vertical vibration modes are illustrated in *Figure 3.44*. The first two modes correspond to the rigid body modes of heave and pitch. The other modes are the distortion modes that give rise to stressing and deformation of the structure. The mode of most concern to the structural designer is the first of these distortion modes, mode 2. The higher modes are more highly damped and do not usually contribute significantly to maximum hull stresses, although they may affect fatigue life. Response in horizontal and torsional modes may be of concern for open section ships such as

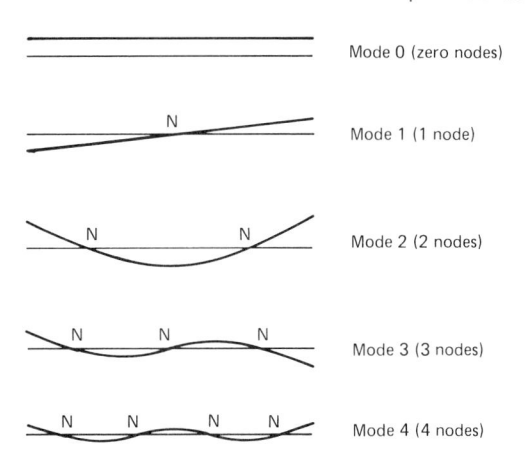

Figure 3.44 First few vertical vibration modes of hull girder

container ships. Response in these modes is complicated as there is no longer a plane of symmetry.

The unified dynamic analysis of ship motions and hull girder vibration under wave induced and slamming loads has been developed using modal techniques by Bishop and Price (Bishop, Price and Tam, 1977, 1978; Bishop and Price, 1979; Bishop, Clarke and Price, 1984; Bishop et al., 1986; Belik, Bishop and Price, 1988). The techniques are also described in Clayton and Bishop (1982), Hughes (1983), Lewis (1988) and reviewed in ISSC (1988). The modal method is based on linear superposition and yields hull girder bending moments, shear forces and stresses. A less comprehensive but non-linear approach has been proposed by Jensen and Pedersen (1979) and Hughes 1983.

The model analysis methods are available as computer programs. Dynamic FEA programs are also available for overall hull analysis. Both types of program require a large input of data on the structure and characteristics of the ship. This is not normally available at the early design stage and the methods are therefore more suitable for checking the design at a later stage. However, a suggested method for deriving useful dynamic information earlier in design is proposed by Chalmers (1988b). Guidance on control of hull vibration is also given by Chang (1979) and Chalmers (1988a).

3.6.7.2 Hull girder natural frequencies

In the early stages of design, it can be useful to have an estimate of the natural frequencies of the hull girder in advance of a comprehensive dynamic analysis. A simple approximate formula for the two node (mode 2) vertical hull vibration frequency is given by Todd (1961):

$$\omega_2 = \beta \sqrt{\frac{BD^3}{\triangle_1 L^3}} \text{ Hz (cycles per second)} \qquad (3.122)$$

where B, D, L are ship beam, depth, and length (m) and $\triangle_1$ is virtual displacement (Te). Virtual displacement includes the effect of entrained water forced to vibrate with the ship. It is given by:

$$\triangle_1 = \triangle \left(1.2 + \frac{1}{3}\frac{B}{T}\right) \qquad (3.123)$$

where $\triangle$ is actual displacement (Te) and T is draught (m).

Values of β were found to be:

Large tankers	1850
Small tankers	1375
Cargo ships at 60% load displacement	1550

Typical values for the two node vertical frequency lie around 1 Hz for large ships, and up to 2.0–2.5 Hz for small ships. The natural frequencies in the three, four and five node vertical modes are, respectively approximately twice, three times and four times the two node mode frequency. The frequency of two node horizontal vibration is invariably higher than the two node vertical mode, by around 30–40%.

3.6.7.3 Local dynamic response

It is possible to use dynamic finite element analysis to investigate the response of secondary and tertiary structures to dynamic loads such as slamming, or sloshing of liquids in tanks. The theoretical basis is the same as that described in Chapter 2. Where necessary, load inputs and edge conditions can be found from overall hull dynamic FEA analysis or directly from the modal analysis programs described above.

Once the loads are known, it is usually sufficiently accurate to undertake the analysis quasi-statically using the methods of plate and grillage described previously (Hughes, 1983; Lewis, 1988).

3.6.8 Ship structure design

So far we have considered choice of materials, estimation of loads and analysis of response. In order to design a structure, it is necessary first to decide upon the design approach to be adopted and to choose an overall structural configuration. It is also necessary to determine appropriate criteria by which to judge the success of the design.

3.6.8.1 Design approach

There are two basic approaches available for design of ship structure. The first and simplest is to design by rule. This approach uses the rules issued by one of the classification societies, such as Lloyds Rules (1987). These contain rules, formulae and tables specifying the design bending moments, materials and scantlings (form and size) of each individual plate and stiffener. The rules are based on the accumulated experience of ship structure design and performance over very many years. Rule based design is rapid and leads to quick approval by the classification society concerned.

The second, more fundamental, approach is to design the structure from first principles using the methods of load estimation and response analysis described earlier. So called rationally based design or direct calculation is more demanding than rule based design, as all possible loading mechanisms and failure modes must be identified and analysed. Furthermore, it is still necessary to obtain the approval of a classification society for the resulting design. Rational design is, however, much more likely to lead to an optimum design in terms of structural weight and cost as it is specific to the ship under consideration. For this reason, it is now in general use for all but small and very conventional ships.

Rational design is, of course, essential for a ship incorporating major novel features as the classification society rules can only be used for conventional vessels. Use of the rules outside their range of validity would be likely to lead to an unsafe or grossly inefficient structure. Once a decision has been made to undertake a rationally based design, further decisions are required on the type of analysis to be used. Are the most appropriate methods static or dynamic, deterministic or probabilistic, linear or non-linear? The answers depend on the design aims and constraints.

Initial design must at present be based on a static analysis of response and for many conventional ships this will be sufficient. However, a dynamic analysis should be made as early as possible to check the design of unconventional vessels and those having high hull girder flexibility. That flexibility is crudely measured, for vertical modes, by ship length to depth ratio. Values greater than about 12 may be considered high; values greater than 16 are not normally permitted by classification societies. Container ships and other open section vessels are flexible in torsion and should receive a dynamic analysis. A proper analysis of slamming effects on hull girder response can only be made dynamically.

Most structural design is deterministic, although load estimation is increasingly made on a probabilistic basis. A probabilistic approach to response analysis and structural reliability is discussed below. The choice of linear or non-linear analysis depends on the load conditions and the analysis tools available, as has already been discussed (Hughes, 1983).

Before finalizing the design approach to be adopted, it is advisable to consult the classification society that is to class the ship. For example, Lloyds Register of Shipping operate LR.PASS (Lloyds Register's Plan Appraisal System for Ships). The direct calculation procedure and facilities available are summarized in a report on LR.PASS current facilities.

3.6.8.2 Structural configuration

The major choice in configuration is between longitudinal framing and transverse framing. The former refers to a structure where the majority of the stiffeners runs fore and aft, with a smaller number of deeper transverse stiffeners supporting the longitudinals. This is the most common form of framing and is shown in *Figure 3.33*. It is efficient in withstanding longitudinal bending of the hull girder as the longitudinals, particularly those in deck and bottom, contribute to the hull girder section modulus and stiffen the plating against buckling under compressive primary stresses. Transverse framing involves a large number of transverse beams and frames supported by occasional deeper longitudinals. It is much less efficient under primary hull bending but is good at withstanding transverse loads such as berthing loads or water pressure from waves outside or deep tanks inside the hull. Some ships have mixed framing with longitudinal framing of the decks and bottom and transverse framing of the ship's sides.

A very important aspect of the overall structural configuration is achievement of structural continuity. This refers particularly to the avoidance of discontinuities in the longitudinal primary structure. These can lead to premature failure of the hull girder under primary bending loads. Such failure may first appear as cracking where necessary discontinuities, such as the ends of superstructures, are not carefully tied into a contiguous structure, by, for example, supporting transverse bulkheads. Such matters have to be addressed at an early design stage before the overall layout of the ship is finalized.

As the structural design is developed, it is important to ensure consistency of stiffener spacing across grillage boundaries. This requires, for example, that side frame and deck beam spacing is the same and that transverse bulkhead stiffeners end on hull or deck longitudinals. Continuity of the internal structural load path is also very important at the detailed level; too many ships are still lost through catastrophic failure of the hull girder arising from fatigue cracks or brittle fractures originating at structural details of poor design or poor quality of manufacture (see Section 3.6.6.3; Sumpter *et al.* 1988; ISSC, 1988).

3.6.8.3 Design criteria

The objective of structural design is to produce an adequately safe and serviceable structure at minimum cost to the profitability of the ship. This suggests designing the structure simultaneously for minimum production cost, minimum weight and minimum maintenance. Unfortunately, these aims are not compatible, so a compromise has to be made.

Assuming a certain quality of design and construction in order to meet safety standards and control maintenance costs, it has been suggested by Caldwell (1977) that a suitable objective function for optimization of a structure is of the form:

$$U = \frac{FW}{W_o} + (1 - F)\frac{C}{C_o} \quad (3.124)$$

where W and C are weight and cost of a proposed design, W_o and C_o are weight and cost of a baseline design, and F is a weighting factor between 0 and 1. It can be shown that if the weight saved in structure is taken up by cargo then the optimum value of the weighting factor is:

$$F = \frac{1}{1 + R_c/R_w} \quad (3.125)$$

where $R_c = \dfrac{\text{annual costs arising from initial cost}}{\text{total annual cost}}$

and $R_w = \dfrac{\text{weight of structure}}{\text{weight of cargo}}$

both R_c and R_w applying to the baseline design.

For a general cargo ship, typical values might be $R_w = R_c = 0.25$, giving $F = 1/2$. For a tanker, typical values might be $R_w = 0.2$, $R_c = 0.4$, giving $F = 1/3$.

It is relatively easy to design for reduced weight, which entails a lot of closely spaced stiffeners for structural efficiency. It is more difficult to design for reduced cost, which entails few, large stiffeners in order to minimize fabrication costs. This difficulty is due to a lack of reliable data on construction costs (Chalmers, 1986; Winkle and Baird, 1986; Section 3.7).

The choice of objectives and the process of, and techniques for, structural optimization are set out in detail by Hughes (1983). Useful reviews of the subject of structural optimisation are given by Caldwell (1986) and ISSC (1988).

3.6.8.4 Structural reliability

Significant developments in structural design methods are arising from the study of structural reliability. This extends the probabilistic approach to defining wave loads (see Section 3.6.3) to the analysis of structural response. Deterministic calculation of response uses safety factors. These implicitly cover uncertainties and likely variations in such things as material properties, accuracy of calculation methods, quality of construction and errors in operation. The probabilistic method attempts to quantify each of these uncertainties on a statistical basis. This then allows a probabilistic definition of strength. The probability of load exceeding strength may then be found. This defines the probability of structural failure. The process is shown conceptually in *Figure 3.45*. The probability of structural failure for the limit state shown in *Figure 3.45* is:

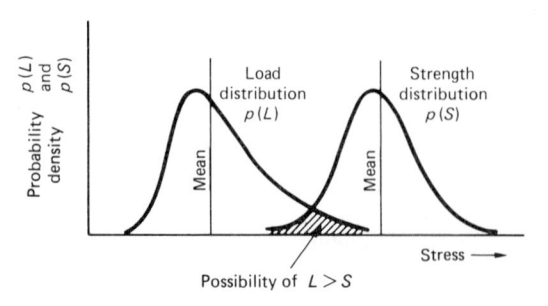

Figure 3.45 Probability distributions of load and strength

$$p_f = \text{prob}\,(L \geq S) \qquad (3.126)$$

Conversely the safety, or reliability, of the structure is:

$$R = \text{prob}\,(L < S) = 1 - p_f \qquad (3.127)$$

A fully probabilistic approach is not yet practicable. Intermediate methods are in use in the offshore industry and in civil engineering and are coming into use in ship structural design (Faulkner and Sadden, 1979; Stiansen *et al.*, 1980; Daidola and Basar, 1981; Hughes, 1983; Kaplan *et al.*, 1984; Hart, Rutherford and Wickham, 1986; Planeix, 1986; Thayamballi *et al.*, 1986; Viner, 1986; Lewis, 1988; Wirsching and Chen, 1988; ISSC, 1988). These methods use a safety index (based on means and standard deviations of the distribution) or partial safety factors.

Reliability methods require a lot of statistical data on loads, material properties, fabrication effects and response. Good quality full scale data on loads are the most difficult to obtain. This is despite many full scale ship trials and the fitting of permanently installed instrumentation. The increasing interest in real time hull strength monitoring systems (ISSC 1988), should help to supply further statistical data in due course.

3.6.9 Computer aided structural design

The synthesis of a structural design is one of trial and error. It is usual to start with the midships hull cross-section and work in terms of a mean, or smeared, thickness for each major panel. The mean thickness is the total cross-sectional area of plating and stiffeners in the panel divided by the breadth of the panel. A simplified cross-section can then be generated for analysis under the design bending moment.

Once the design criteria for primary stress levels have been met, detailed synthesis and analysis of each grillage can commence. The process is iterated until all primary, secondary and tertiary structural design criteria have been met. Clearly, much of this task can be eased by the use of computers.

A number of computer aided structural design systems (or codes) have been developed. These range from programs that merely ease the calculation of rule based structural formulae to sophisticated design and analysis codes. Some of the latter include a measure of design optimization. (Forrest and Parker, 1983; Jingen and Jensen, 1982; Byler and Walz, 1982; Pattison, Spencer and Van Griethuysen, 1982; Lersbryggen, 1983; Bailliot, Finister and Kamel, 1984; Okumoto, Takeda and Hiyoku, 1985; Garside and Das 1986; ISSC, 1988).

A comprehensive hull structural design and optimization program MAESTRO has been developed by Hughes (1983). This is based on a specially structured finite element approach. Some computer aided structural design codes are included within overall ship CAD systems. These are considered in Section 3.7.

3.6.9.1 *Finite element programs*

A wide range of FEA computer programs is available. Many have been developed for civil engineering or aerospace use. The more sophisticated programs have special preprocessors to ease data preparation for ship structural use. Some FEA programs, including the design program MAESTRO, have been developed specifically for ship structures.

A description and review of many of the FEA programs available internationally for ship structural analysis, together with full references, is in ISSC (1988). This includes coverage of non-linear FEA codes used for buckling, ultimate strength and vibration analysis. A review of finite element codes suitable for use on microcomputers is given by MacKerle (1987).

Design consultants, software houses and classification societies will undertake analysis work or hire out programs for a client's own use. In the UK, guidance on FEA programs and advice for inexperienced users may be obtained from the National Agency for Finite Element Methods and Standards (NAFEMS), part of the National Engineering Laboratory in Glasgow.

3.7 The ship design process

Ship design is the process of creating and developing a definition of a ship in sufficient detail to be able to build it, while verifying that performance and cost objectives will be met (Watson and Gilfillan, 1977; Taggart, 1980; Rawson and Tupper, 1984; Muckle and Taylor, 1987).

3.7.1 General considerations
3.7.1.1 *Design of complex systems*

A ship is a complicated engineering system made up of a number of subsystems, themselves complex. Ship design therefore follows the sequence common to all complex engineering systems.
1. The definition of the requirement includes function, performance and reliability. It also involves initial cost, through life cost in some form and timescale.

In the initial stages of a major project, the requirement is distinctly hazy or fuzzy and the designer's task is to help decide what the customer wants.
2. Major constraints can be political, regulatory, social (e.g. health and safety considerations), resource constraints, the state of research or data base limitations. A lack of past data on such things as weight, space, energy and cost elements can be a major handicap.
3. Concept studies should be broad brush in nature, with divergent thinking and creativity to the fore. They show alternative ways of meeting the need and enable broad trade offs to be made between capability, cost and risk. These assist further in refining the requirement itself.
4. Proof of feasibility, having chosen one or two of the most promising concepts, is developed to a stage where it is clear that the requirement can be met and risks have been identified and considered acceptable. Analysis is usually the key activity here.
5. Development of the design must now be strictly convergent. The purpose is to develop the design definition to a sufficient level of detail for manufacture to commence.
6. In a complex system, the detail design task invariably overlaps the build and test phase. This is both to reduce the overall timescale and to allow feedback of problems arising during production and testing.
7. In-service life is the culmination of the whole process. Although the original design task will have been completed, feedback to the designers is still required for corrective actions and for improvement of future designs.
8. Scrapping may seem to be looking a long way ahead, but the eventual need to scrap the system should be considered in design. This is now recognized in offshore engineering and could, for example, affect design of ships with nuclear power plant.

Throughout the sequence, or at least the first five stages, the design process is one of iterative problem solving. Decisions made in one area of the design impact on other areas, which then have to be re-analysed and adjusted and so on in a convergent process of ever increasing detail. Much of engineering design, particularly in the early stages, is a matter of trial and error. Compromise is usually necessary at every stage, otherwise individual subsystems will be optimized at the expense of the best overall solution.

In order to meet the agreed design requirements on time and at the right price, within the constraints prevailing, strong design management is needed. This is particularly so in the early design stages when the overall concept is being created. Decisions made at this stage, when very little money has actually been spent, have the greatest effect on the eventual capability, cost and success of the product. That is not to belittle the later design stages when the design manager may be responsible for integrating the work of a large, multi disciplined design team. Good detail design, which is done at these later stages, is essential to success. It is bad detailing that will most often aggravate the user and many engineering catastrophes can be traced to poor design of details.

3.7.1.2 The nature of ship design

Ships are among the most complex and costly engineering systems. Their design involves a wide range of technologies and the overall integration task is correspondingly very demanding. The difficulty of the task is exacerbated by design requirements and constraints that are virtually unique to ship design. These include the hostile marine environment, and the need for a ship to be self supporting in that environment for up to months at a time. In addition, no prototypes are built, except in very rare circumstances, due to the high cost and small production runs involved. Ships must work first time.

The ship design process closely follows that described above. There is, however, more architecture involved than in most engineering disciplines. The shape of the ship is important to its structural and hydrodynamic performance and to its aesthetic appeal, and the internal and external layout of the ship in both two and three dimensions is of major importance to successful design. Owing to the hostile environment, the need for mobility and the possible need to contain both fire and flood, a ship's layout is more heavily constrained by engineering and functional considerations than is the case in land based architecture. Human factors are of major concern, not only in the operational and ergonomic sense but also due to ship motion effects. The human body is particularly susceptible to certain motions and frequencies, the incidence of seasickness peaking at motion frequencies around 0.2 Hz (Schoenberger, 1975; Rawson and Tupper, 1983). The successful designer will try to minimize the incidence of seasickness by careful attention to motion frequencies and to disposition of crew and passengers to positions of minimum motion amplitude.

Conceptually, the ship design process is often illustrated as a design spiral. *Figure 3.46* shows a detailed version (Buxton, 1972). This illustrates the design of a merchant ship between the initial enquiry by a ship owner to submission of a tender for detail design and build. The spiral shows the iterative nature of design as successive aspects are checked and adjusted and the progression from broad and shallow investigations at the start to the narrow and deep nature of detail design. However, the spiral is only a rough model of design. It gives the false impression of a continuous process, whereas real design, particularly in the concept phase, involves intuitive leaps and trial solutions that sometimes fail. Furthermore, major discontinuities can occur in the sizing process, e.g. when the ship size has grown to a stage where any further growth requires a completely new fit of propulsion machinery.

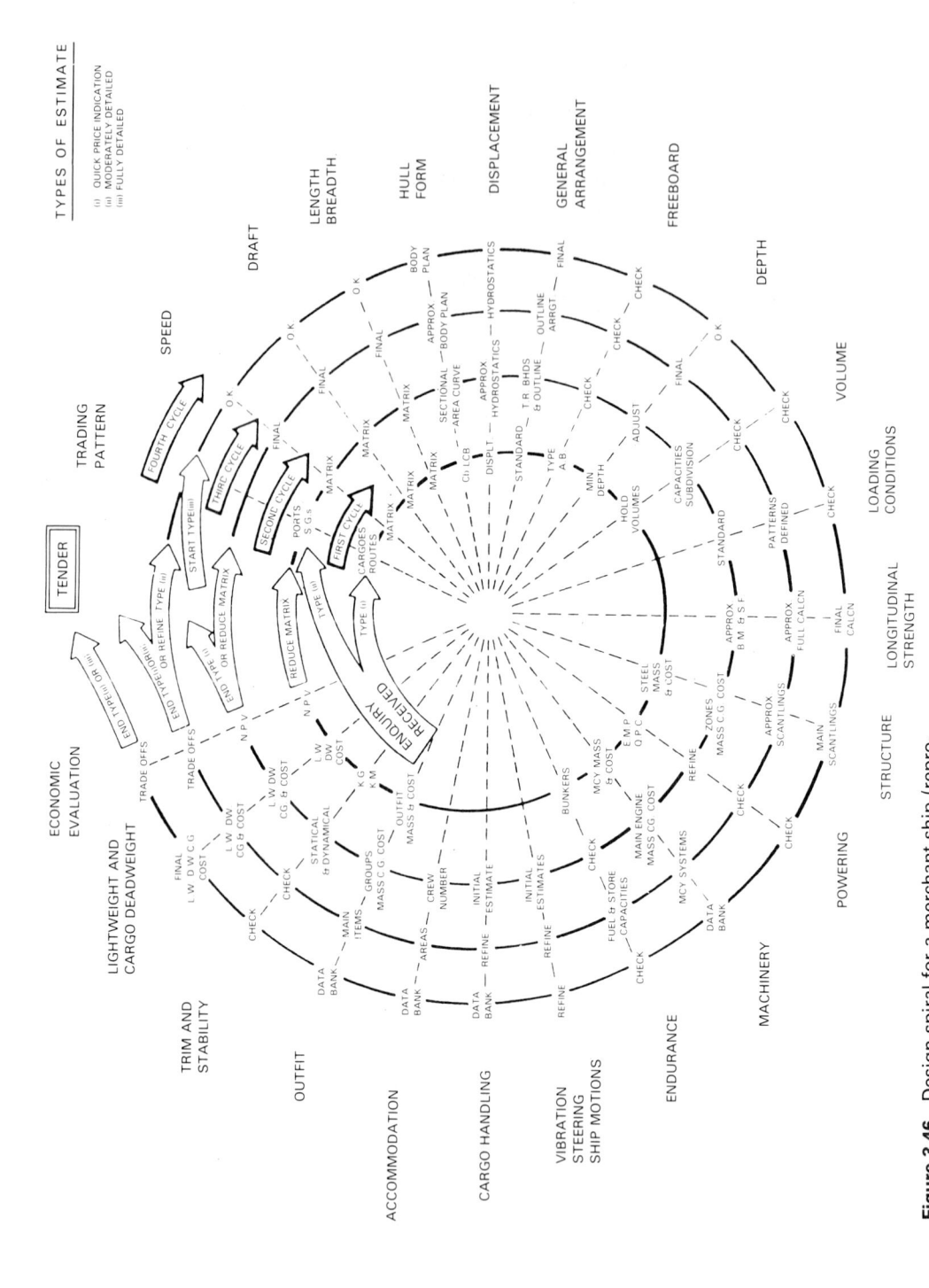

Figure 3.46 Design spiral for a merchant ship (reproduced from Buxton, 1972) (Courtesy of The Royal Institution of Naval Architects)

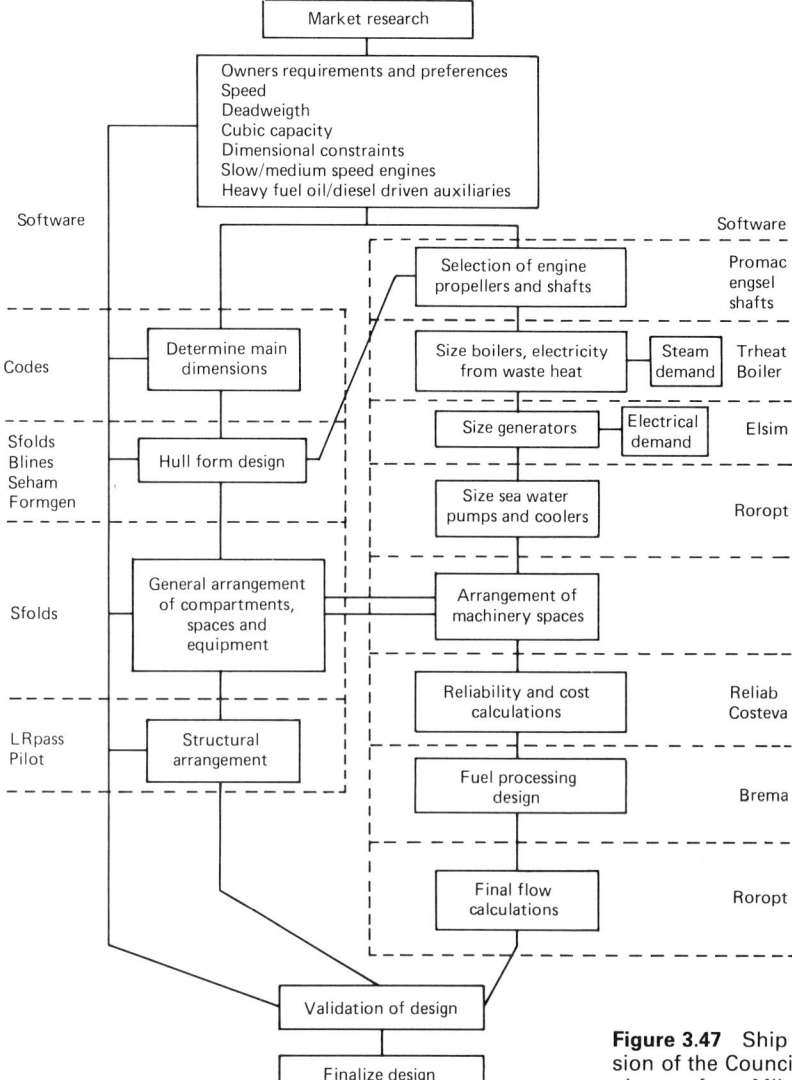

Figure 3.47 Ship design process (reprinted, by permission of the Council of the Institution of Mechanical Engineers, from Milne, 1987)

Another way of viewing the merchant ship design process is shown in *Figure 3.47* (Milne, 1987). This also shows the main computer aided design and computer aided manufacture (CAD/CAM) programs used by British Shipbuilders in the mid-1980s (see Section 3.7.5).

3.7.1.3 Economic studies

The design process for a merchant ship invariably starts with an economic appraisal, perhaps following market research by a ship owner. At this stage, the ship may be considered only as a part of a larger transportation system. The requirements for this might be met by a few large ships or many small ships. A shipowner may also wish to compare the relative merits of building a new ship, buying or converting a secondhand vessel or chartering from another owner.

A range of economic and concept design studies may therefore be carried out, from which the optimum ship type, cargo capacity and ship speed will emerge. These studies require estimates of the likely ship building and operating costs, so the process is iterative (Buxton, 1972; Goss, 1982; Buxton, 1987; Stopford, 1988). The costs are usually considered on a discounted cash flow basis over the intended life of the ship as shown in *Table 3.4* (Buxton, 1987).

Table 3.4 The costs of a ship

Cost type	Description
Capital costs	Design costs
	Building costs (down payment and installments)
	Associated loan repayments and loan interest
	Taxes
	Depreciation and profit (considered for convenience as costs)
Daily running costs	Crew expenses (wages, benefits, victualling, training, travel, etc.)
	Repair and maintenance
	Stores
	Insurance
	Administration
Voyage costs	Fuel
	Port charges and light dues
	Tugs and pilotage
	Canal dues
Cargo expenses	Cargo handling
	Cargo claims, etc.

Freight income must cover all these expenses and provide an acceptable rate of return on capital. A commonly used measure of the ship's economic viability is the required freight rate (RFR). This is the rate which the ship owner must charge per unit of cargo in order to break even.

3.7.1.4 The ship design environment

The major constraints on the ship design process are similar to those identified for engineering design generally. A discussion of the political and technological environment affecting ship design, particularly within the UK, is given by Meek (1985). This paper, together with its published discussion, also covers ship manning and crew efficiency.

As in any field of engineering, the hard learned lessons of history affect the ship design environment and are well worth study. A classic guide is Barnaby (1968). Operational aspects, ship safety and the choice of safety standards are discussed, with associated statistical data, by Meek, Brown and Fulford (1983), Brett (1988) and Spouge (1989). Risk assessment and reliability techniques for use in ship design are discussed by Clark, Lewis and Aldwinkle (1986).

3.7.2 Initial sizing

Once decisions have been made on ship type, route, required cargo capacity and ship speed, an estimate can be made of the ship size. If good data on a similar ship (a type ship) is available, sizing can be done fairly rapidly by considering differences between the type ship and the new design. Otherwise, parametric methods are required (Lamb, 1969; Watson and Gilfillan, 1977; Taggart, 1980; Kupras, 1983).

A widely used approach (Watson and Gilfillan, 1977) is to divide ships into three main categories:

1. The deadweight carrier, whose size is governed by weight of cargo, rather than its volume. This holds for a dense cargo such as ore. Such ships will have minimum permitted freeboard when fully loaded.
2. The capacity carrier, whose size is governed by volume of cargo rather than its weight. This holds for lighter cargoes, including passengers.
3. The linear dimension ship, whose principal dimensions are fixed mainly by external constraints rather than deadweight or volume requirements. Included are large Great Lakes carriers, where the beam limit of 22.8 m (74.8 ft) in the St Lawrence Seaway leads to long, thin ships and very large crude carriers whose draught is limited by the depths of shallow seas through which they must pass. Container ships, car and train ferries are also governed, in part, by the unit widths and lengths of their cargoes.

Many ships are a combination of all three categories. Both weight and volume must then be considered during the design process, together with any dimension constraints.

3.7.2.1 Initial sizing based on length

Initial sizing may commence by calculating displacement and cargo capacity for three different ship lengths. Data from past designs are used as a guide to initial choice of form parameters and dimension ratios relative to length. Provided the three calculated cargo capacities straddle the required value, then the required ship size and dimensions can be interpolated. A mass of data, estimating techniques for the necessary calculations and standard calculation sheets suitable for programming by computer are given by Watson and Gilfillan (1977).

3.7.2.2 Initial sizing based on density

An approximate but very rapid way to obtain an initial estimate of ship displacement is to use deadweight/ship displacement ratios. A variation of this is to use cargo volume fractions (ratio of volume of cargo to total enclosed volume of ship) and ship density (ratio of displacement to total enclosed volume) (Rawson and Tupper, 1984). Typical values are shown in *Table 3.5*.

Table 3.5 Typical cargo volume fractions and ship densities

Type of ship	Volume fraction	Ship density (Te/m³)
General cargo	0.50	0.50
Ro–Ro	0.55	0.35
Crude oil tanker	0.70	0.75
Container ship (including containers on deck)	0.65	0.45

With this method, a value of required cargo volume can be translated through the cargo volume fraction to an estimate of total enclosed volume. This leads, through ship density, to an estimate of ship displacement. If necessary, cargo volume can be derived from cargo weight through the cargo density. Some typical values of cargo density are shown in *Table 3.6*.

Table 3.6 Typical values of cargo density

Cargo	Density (Te/m³)
Iron ore	1.90 – 3.50
Phosphate	1.12
Coal	0.75 – 0.85
Crude oil	0.78 – 0.92
Residual oil	0.94 – 1.00
Distillates	0.84 – 0.94
Gasoline	0.74
Grain (wheat, corn, rye)	0.6 – 0.75
Containers (average)	0.34 – 0.38

As an example, a required crude oil capacity of 50 000 m³ (say 40 000 Te) leads to a tanker total enclosed volume of $(50\,000/0.7) \simeq 71\,400$ m³ and hence a ship loaded displacement of $(71\,400 \times 0.75 = 53\,600$ Te; say 55 000 Te as a first approximation.

This first estimate of ship displacement can then be used to derive initial estimates of main dimensions, i.e. length L, beam B, depth D and draught T. Typical values are taken for a set of form parameters such as L/B, B/D, D/T and C_B where C_B is the block coefficient (volume of displacement ∇/LBT). An alternative set is $\textcircled{M}$ $(=L/\nabla^{1/3})$, B/T, D/T and C_B. Using the latter set from the first estimate of displacement Δ, then $\nabla = \Delta/\rho_w$, where ρ_w is the density of water and

$$L = \textcircled{M}\nabla^{1/3}$$
$$T = \Delta^{1/3}[\textcircled{M} \cdot (B/T) \cdot C_B]^{1/2} \qquad (3.128)$$
$$B = (B/T)\,T$$
$$D = (D/T)\,T$$

A check should be made that the values of L, B and D are compatible with the total enclosed volume allowing for reasonable values of waterplane coefficient and the proportion of total enclosed volume taken up by the superstructure.

Typical values of the form parameters may be found from previous similar vessels or from regression analysis of past design data (Watson and Gilfillan, 1977; Taggart, 1980; Rawson and Tupper, 1984; Lewis, 1988). These default values are only needed for the initial design estimates. They are adjusted to more optimum values at a later design stage.

3.7.2.3 Subsystem calculations

The next stage is to refine the estimates of ship displacement and total volume by calculating weights and volumes for the individual subsystems. The usual subsystems into which a merchant ship is divided for weight purposes are cargo; machinery (propulsion systems and auxiliaries); outfit (hull and deck machinery and systems, working spaces, accommodation and fittings, paint and coverings, safety equipment, etc.); crew and effects; fuel; stores (solid and liquid); steelwork (all structure) and margin. It is convenient to use a similar categorization for volumes with the inclusion of a volume allowance for void spaces.

Methods for estimating the subsystem weights and volumes are given in the references cited. Once calculated, the subsystem weights and volumes are summed to give new estimates of ship displacement and total volume. These are compared with the initial values and the process iterated until a balance is achieved. Only a few iterations are usually required to achieve an acceptable design balance. The logic is shown in *Figure 3.48*. The procedure is repeated at intervals to check the displacement and volume as the design progresses and more detailed information on the subsystems becomes available.

3.7.2.4 Choice of dimensions and hull form

Having obtained reasonable estimates of displacement, volume and required power, it is possible to reconsider the form parameters and ship dimensions and define a suitable hull form. A good way to proceed is to undertake a parametric survey covering a range of possible forms.

This starts with systematic variation of certain key parameters to find sets or windows of feasible and consistent dimensions; main hull depth D is often a convenient starting point. These dimensions must match the overall displacement and volume and also meet requirements for minimum stability and freeboard and any overall layout constraints, such as container dimensions. External constraints must also be considered. For example, the maximum beam for the Panama Canal is 32.2 m; length may be governed by available docks and draught by port depths. Constraints set by port facilities may be checked using *Ports of the World* (1989).

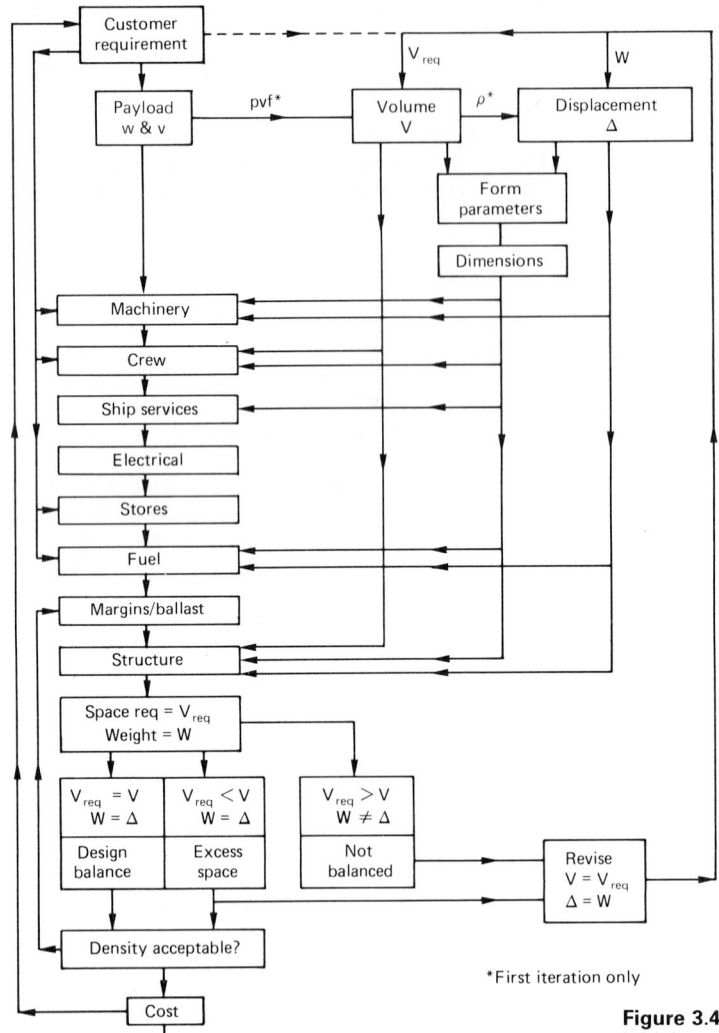

Figure 3.48 Iteration logic for design balance, after initial synthesis using payload volume fraction *pvf* and ship density ρ (after Betts, 1988)

Feasible sets of dimensions and form parameters are then checked against design requirements for powering and fuel consumption and such things as ease of cargo handling, seakeeping and manoeuvring. From these calculations, an optimum set of dimensions and form parameters will emerge. The final choice is usually made on the basis of cost. This might be minimum fuel cost (Watson, 1981) or some compromise between construction cost and operating cost. If adequate economic data are available, minimum required freight rate may be a suitable objective.

Attempts have been made to produce optimization methods giving direct calculation of optimum ship dimensions and form parameters (Lyon and Mistree, 1985). These methods have not generally been successful for real design use owing to the difficulty of defining the objective functions, the uncertain or fuzzy nature of many design constraints and requirements, and a lack of suitable data. Work on expert systems and fuzzy set approaches may overcome some of these difficulties in due course.

Development of a suitable hull form from the chosen dimensions and form parameters may be done via the curve of areas (see Section 3.3). Various methods exist for doing this. One way is to use or modify hull forms given in the various methodical series discussed in Section 3.3. In practice most design offices now have CAD programs that will synthesize a hull form from given dimensions and hull form parameters (see Section 3.7.5). Not all of these will cope with unusual geometries. If a bulbous bow is required, it may need to be faired in by trial and error.

3.7.2.5 Costing

The initial construction cost of a ship is the lightship cost. That excludes crew, stores, fuel and cargo. Detailed cost data is held by individual shipyards and design agencies, but rarely released. In the absence of such data, a method commonly used in the UK for preliminary design of merchant ships is that of Carryette (1978). This uses parametric formulae of the form:

$$\text{Ship cost} = \underbrace{\left(a\frac{W_s^{2/3}}{C_B}L^{1/3} + bW_s\right)}_{\substack{\text{labour\quad materials}\\ \text{– steelwork –}}}$$

$$+ \underbrace{(cW_o^{2/3} + dW_o^{0.95})}_{\substack{\text{labour materials}\\ \text{– outfit –}}} + \underbrace{(eP_s^{0.82})}_{\substack{\text{labour and materials}\\ \text{machinery}}} \quad (3.129)$$

where W_s is the steel weight (Te), W_o is the outfit weight (Te), L is the LBP (m), and P_s is the service power of main engines (BHP) and a, b, c, d, e are factors embracing wage rates, allowances, overall productivity levels, assumed overheads and profit, material costs, wastage allowance, delivery and handling charges and an allocation of service and miscellaneous costs. Fisher (1974) covers the effect of size on cost.

Calculation of fuel costs can be done using the fuel consumption arising from the powering calculations and choice of machinery. Other operating costs may be estimated using the references quoted in Section 3.7.1.3.

3.7.3 Design development

In parallel with the preliminary estimation of ship size and the calculation of powering requirements, the choice of propulsion system has to be considered as it impacts directly on the machinery subsystem weight and volume. Considerations underlying the choice and development of the propulsion system are given in Chapter 6.

Early attention must also be given to overall ship layout. This can determine some of the key constraints on the choice of dimensions and form. Propulsion machinery may require a minimum hull depth and the cargo holds may require certain critical dimensions, e.g. to take containers. The layout must also take account of bulkhead positioning to meet collision and damage stability requirements and structural continuity in way of deckhouses or concentrated loads. Aesthetic considerations may influence the above water profile of the ship, particularly for a ferry or cruise ship. Little has been published on aesthetics of ship shapes, although some guidance may be found in Donnelly (1985).

As the design progresses, increasing attention is given to the various ship subsystems as shown in *Figure 3.46*. Detailed development is required of layout, machinery arrangements, electrical generation, auxiliary systems, hull systems and cargo handling, propeller and control surface design, accommodation, mooring arrangements (see Chapter 8), lifesaving gear, fire fighting arrangements, pollution prevention, bridge arrangements, communication and navigation systems (see Chapter 12), ship structures and seakeeping and manoeuvring performance). Human factors need to be considered both for layout and for detailed development of subsystems (Rawson and Tupper, 1984; *Handbook of Human Factors* 1987; SNAME Bulletin, 4–22, 1988). Consideration must also be given to design for production in order to minimize construction cost.

All these aspects must be kept in balance and integrated into the total design. Furthermore, the design has to be developed to meet national and international classification and certification requirements. These are discussed in Section 3.7.6.

As well as periodic checks in weights and volumes, it is necessary to calculate and monitor vertical and longitudinal centres of gravity in order to maintain adequate stability and achieve the desired longitudinal trim. In a design that is weight or stability critical, it may be necessary to allocate weight and centre of gravity budgets to the various subsystems in order to prevent, or at least control, growth. Budgeting can be particularly useful where the design authority for a subsystem is organizationally and physically separate from the ship designer.

During the design development phase, tank tests may be commissioned to check powering predictions, supply information for propeller design and check seakeeping and manoeuvring performance. Wind tunnel tests may be arranged for smoke clearance from funnels and, if a helicopter is carried, air flow over the flight deck.

Attention must also be given to noise and vibration control. This is an important area that is subject to legal restrictions in accommodation and working areas (Morrow, 1988; Fisher and Petit, 1988; ISSC, 1988; Lloyds Register, 1988).

Reliability engineering and maintainability studies are a major consideration in warship design but have not traditionally been so in merchant ship design (Aldwinkle and Pomeroy, 1983; Rawson and Tupper, 1984; Clark, Lewis and Aldwinkle, 1986).

A fuller discussion of the detailed design phase, and the tools that can be used to manage it, is given in Rawson and Tupper (1984). A description of the design and project management of an unusual and complex commercial vessel is given by Parker and Woolveridge (1988).

3.7.4 Design for production

3.7.4.1 Through life considerations

Design for production is, by definition, concerned with minimizing building cost. However, the designer must not consider this to the exclusion of operating or through life costs. These depend on design for maintainability and ease of refit and updating. This requires some features that increase building cost. Examples include access for maintenance, good painting and preservation schemes, structural design details that are not prone to fatigue cracking and use of long life materials such as copper–nickel or stainless steel in place of mild steel for pipework. The balance to be struck will depend on the owner's requirements and careful study of cost trade offs.

3.7.4.2 Design for production

The key principles may be seen as designing for the following.

3.7.4.2.1 Compatibility This includes: optimising the design for maximum sizes of plates (lengths, widths and thicknesses) that the shipyard can handle. For example, hold lengths and heights can be made simple multiples of maximum plate dimensions, so reducing the amount of burning and welding required. Design for a hull fabrication unit breakdown of a size and weight that suits the yard facilities is also relevant.

3.7.4.2.2 Standardization This can take two forms: use of standard materials, pipe sizes, plate thicknesses and rolled stiffener sizes in accordance with British Standards or ISO ranges. Some yards develop their own standard range of parts. Second, design for a limited range of repetitive parts, that can be mass produced in the yard. This can involve rationalization of the detailed ship design to minimize the number of different materials, plate sizes, stiffener types and sizes, bracket connections and pipe sizes. Even complete fabrication blocks can be standardized. Standardization will invariably reduce building costs. The best form is the series production of ships.

3.7.4.2.3 Simplicity Work content and hence cost can be minimized by attention to the following.

1. Choice of materials. Exotic materials that are difficult to form and weld should be avoided unless essential for other reasons such as performance or maintenance.
2. Simply geometry. Flat plates are far cheaper to fabricate than curved ones. Single curvature is cheaper than double curvature. Single curvature hull forms have been built; however, fuel costs can be higher so a build cost and life cost comparison is necessary.
3. Reduction of welding. Structural design of grillages for minimum cost was considered in Section 3.6. Welding can be avoided altogether by use of pressed (swaged) forms for bulkheads instead of welded grillages.
4. Ease of welding. Unit design and butt and seam positions can be chosen at the detail design stage to maximize more efficient downhand welding and use of machine welding.
5. Simple connections. At the detail design stage, connections between stiffeners and between bulkheads and decks can be designed for ease of alignment and welding, e.g. lapped joints rather than butt joints. However, care must be taken not to compromise essential structural continuity.
6. Pipework. Attention to pipe system arrangements during design of the ship layout can lead to straightforward pipe and ventilation runs of minimum length and number of bends. Complex areas can be studied using 3-D computer graphics or physical models or mock ups.
7. Modularity. Adoption of simple, repetitive modular forms for accommodation, pumping stations, etc. is an aid to cost reduction, as are simple block forms of superstructure.
8. General layout. Uncomplicated layouts with good access for build, installation and maintenance and removal of equipment and systems should be the aim.

Guidance on design for production may be found in Taggart (1980), Kuo, MacCullum and Shenoi (1983), SNAME (1985), Lamb (1986), Milne (1987), Bruce (1987), Bong, Hills and Caldwell (1988), Hills and Buxton (1989). See also Section 3.6.8.3.

3.7.5 CAD/CAM

Until recently, computer aided design (CAD) and computer aided manufacturer (CAM) programs only interfaced late in design. Detailed information on ship structure and piece parts (brackets, joints, etc.) developed on computer aided draughting machines was passed to production planning and control of numerically controlled cutting, forming and welding machines (Fujita and Sanagana, 1979; Forrest and Parker, 1983). The available CAD programs for preliminary design work were separate and unable to communicate information directly to detailed design and manufacture (Kupras, 1983).

Recently, progress has been made on integrating the complete design and manufacture process including pipework and electrical systems as well as structure (Fijita *et al.*, 1985; Milne, 1987). This should lead eventually to computer integrated manufacture (CIM) which may include also shipyard stock control and management.

Figure 3.47 shows the main CAD programs available to British Shipbuilders. These include many developed in-house (now by Marine Design Consultants) and by the British Ship Research Association (now British Maritime Technology). The integrated system called BRITSHIPS is described in more detail by Horsham, Archer and Law (1986) and Odabasi *et al.* (1987). A concept and preliminary design suite is under development by Newcastle University as a front end to BRITSHIPS (Hills and Buxton, 1989).

The concept and preliminary design suite developed for high speed displacement ships and advanced marine vehicles at University College London is described by Andrews (1986) and Betts (1988). Developments at Strathclyde University are described by MacCallum (1982).

A review of the major international CAD/CAM systems, with extensive further references, is given in ISSC (1988). Other CAD programs for use in the preliminary design and detail design phases are listed in surveys by RINA (Roy, 1987) and SNAME (*Small Craft Software Catalog*, 1988).

3.7.6 Classification and certification

3.7.6.1 Classification

Classification societies are independent bodies concerned with the setting and maintenance of standards of design, construction and repair. They issue classification certificates to vessels which comply with their standards. Classification is not a legal requirement in the UK but is usually required for insurance purposes.

Each society's standards and rules are laid down in a range of publications covering different types of vessel. The standards apply particularly to hull materials and structure but also cover such things as main and auxiliary machinery, electrical installations, control and safety equipment. The societies may also control other aspects delegated by governments. For example, Lloyds Register of Shipping is delegated by the UK Department of Transport to issue international load line certificates, ship safety certificates and international tonnage certificates. The major classification societies each have offices and resident surveyors in the main maritime countries. These are, the American Bureau of Shipping (USA); Bureau Veritas (France); Germanische Lloyd (Federal Republic of Germany); Lloyds Register of Shipping (UK) (the oldest and largest); Nippon Kaiji Kyokai (Japan); Det Norske Veritas (Norway) and Registro Italiano (Italy). These bodies liaise through the International Association of Classification Societies (IACS).

Most classification societies have a sophisticated range of structural analysis computer programs (see Section 3.6). These are used for checking submitted designs but are also often available as part of a consultancy service to designers. To remain in class, a ship must be surveyed periodically, usually annually by classification society surveyors. Special surveys are required every four years or so when the effects of corrosion on structural scantlings and the need for replacement of wasted plating are checked.

3.7.6.2 International conventions and IMO

International regulations and codes of practice are promoted by a specialized agency of the United Nations called the International Maritime Organisation (IMO) based in London. Until 1982 it was known as the Inter-Governmental Maritime Consultative Organisation (IMCO). IMO's governing body is the Assembly consisting of 128 member states and one associate member, Hong Kong. The Assembly meets every two years, control between sessions being delegated to a Council of 32 member governments.

IMO is a technical organization and most of its work is carried out in a number of committees and subcommittees. The main ones are the Maritime Safety Committee (with ten subcommittees); the Marine Environment Protection Committee; the Legal Committee; the Technical Co-operation Committee and the Facilitation Committee. IMO has promoted the adoption of some thirty conventions and protocols and adopted over five hundred codes and recommendations concerning maritime safety, prevention of pollution and related

Table 3.7 The major IMO conventions

International convention	Entered into force
Safety of Life at Sea (SOLAS) 1974	1980
Protocol of 1978 relating to SOLAS	1981
International Regulations for Preventing Collision at Sea 1972	1977
Prevention of Pollution of the Sea by Oil 1954	1958
Prevention of Pollution from Ships 1973 and Protocol of 1978 (MARPOL 73/78)	1983
Load Lines 1966	1968
Tonnage Measurement of Ships 1969	1982
Civil Liability for Oil Pollution Damage 1969	1975
Protocols of 1976 and 1984 to Above	1981
Civil Liability in the Field of Maritime Carriage of Nuclear Material 1971	1975
Athens Convention Relating to Carriage of Passengers and their Luggage by Sea 1974, and Protocol of 1976	
Torremolinos International Convention for the Safety of Fishing Vessels 1977	

matters. Once a convention comes into force, usually after ratification by a specified number of countries, implementation is mandatory on participating countries. Codes and recommendations are not so binding but may be implemented by Governments through domestic legislation. The major IMO conventions affecting ship design are shown in *Table 3.7*.

The major IMO Codes that affect ship design include International Maritime Dangerous Goods Code (IMDG) (1965); Code of Safe Practice for Bulk Cargoes (1965); Code for the Construction and Equipment of Ships carrying Dangerous Chemicals in Bulk (1971); Code of Safe Practice for Ships Carrying Timber Deck Cargoes (1973); Code of Safety for Fishermen and Fishing Vessels (1974); Code for the Construction and Equipment of Ships Carrying Liquified Gases in Bulk (1975); Code on Noise Levels on Board Ships (1971); Code of Safety for Nuclear Merchant Ships (1981); Code of Safety for Special Purpose Ships (1983); International Gas Carrier Code (1983) and International Bulk Chemicals Code (1983). Amendments to the Conventions and Codes are agreed and published from time to time.

3.7.6.3 National regulations

International Conventions are incorporated into national law and enforced by governments. In the UK, the Department of Transport acting under the authority of the various Merchant Shipping Acts sets and enforces IMO and other standards for merchant ship safety, design and operation (see Chapter 14). The full list of current principal acts and regulations on merchant shipping is published in an annual Department of Transport Merchant Shipping Notice. Copies of Merchant Shipping Notices and further information about publications may also be obtained.

The major areas of regulation affecting ship design standards cover crew accommodation; carriage of dangerous goods; fire and lifesaving; load lines; subdivision; medical requirements; navigation and collision prevention; prevention of pollution; ship communications; ship construction and equipment for cargo ship's ship construction and equipment for passenger vessels and tonnage measurement. Special regulations cover diving and submersibles, fishing vessels and hovercraft.

The regulations covering passenger vessels are particularly extensive (Cowley, 1988). They include requirements for watertight subdivision, fire boundaries, freeboard, lifesaving appliances and transport of dangerous goods. Passenger vessels in the UK are certificated in eight different classes, depending on length and type of voyage and number of passengers. The regulations for Ro–Ro passenger ships have been enhanced since the capsize of the Ro–Ro ferry *Herald of Free Enterprise* in March 1987 and are under review by IMO at the time of publication of this book (Ro–Ro Safety and Vulnerability: the Way Ahead, 1987).

As a result of the national and international regulations, a ship needs a number of certificates before it is allowed to proceed to sea. They are an International Load Line Certificate; either Cargo Ship Safety Construction Certificate, Cargo Ship Safety Equipment Certificate and Cargo Ship Safety Radiotelegraphy Certificate; or a Passenger Ship Safety Certificate; International Oil Pollution Prevention Certificate; International Tonnage Certificate; Certificate of Registry; International Certificate of Fitness for the Carriage of Dangerous Chemicals in Bulk (Chemical Tankers); International Certificate of Fitness for the Carriage of Liquefied Gases in Bulk (Gas Carriers) and International Pollution Prevention Certificate for the Carriage of Noxious Liquid Substances in Bulk. Further International Pollution Prevention Certificates are likely to be required as the MARPOL Annexes covering sewage and garbage come into force from 1989.

References

Adamchak, J.C. An approximate method for estimating the collapse of a ship's hull in preliminary design. In *Proceedings of the Symposium on Ship Structures* (Arlington, VA, October 15–16, 1984) Society of Naval Architects and Marine Engineers, New York (1984).

Aertssen, G. Service performance and seakeeping trials on a large ore-carrier. *Transactions of the Royal Institution of Naval Architects*, **111**, 217–239 (1969).

Aertssen, G. An estimate of whipping vibration stress based on slamming data of 4 ships. *International Shipbuilding Progress*, **294** (1979).

Aldwinkle, D.S. and Pomeroy, R.V. A rational assessment of ship reliability and safety. *Transactions of the Royal Institution of Naval Architects*, **125**, 269–288 (1983).

Andrew, R.N. and Lloyd, A.R.J.M. Full-scale comparative measurements of the behaviour of two frigates in severe head seas. *Transactions of the Royal Institution of Naval Architects*, **123**, 1–31 (1981).

Andrew, R.N., Baar, J.J.M. and Price, W.G. Prediction of ship wavemaking resistance and other steady flow parameters using Newman–Kelvin Theory. *Transactions of the Royal Institution of Naval Architects*, **130**, 119–133 (1988).

Andrew, R.N., Loader, P.R. and Penn, V.E. The assessment of ship seakeeping performance in likely to be encountered wind and wave conditions. In *Proceedings of the International Symposium on Wind and Wave Climate Worldwide* (London, April 1984). Royal Institution of Naval Architects, London (1984).

Andrews, D.J. An integrated approach to ship synthesis. *Transactions of the Royal Institution of Naval Architects*, **128**, 73–102 (1986).

Antoniou, A.C., On the maximum deflection of plating in newly built ships. *Journal of Ship Research*, **24**, (March 1980).

Antoniou, A.C., Lavidas, M. and Karvounis, G. On the shape of post-welding deformations of plate panels in newly built ships. *Journal of Ship Research* **28**, (March 1984).

Aston, J.G.L. Rydill, L.J. and Beck, M.P. Improving the

safety of Ro–Ro Ships. *The Naval Architect* E137–E140 (April 1987).

Bailey, D. The NPL high speed round bilge displacement hull series. *Marine Technology Monograph* No. 4, Royal Institution of Naval Architects, London (1976).

Bailliot, J.J., Finister, D. and Kamel, H.A. *Integrating CAD–CAM Tools in the Design and Construction of Ships and Offshore Structures*. Institut de Recherches de la Construction Navale, Paris (1984).

Baitis, A.E., Applebee, T.R. and McNamara, T.M. Human factors considerations applied to operations of the FFG-8 and LAMPS Mk III. *Naval Engineers Journal* (May 1984).

Baitis, A.E., Wodaver, D.A. and Beck, T.A. Rudder-roll stabilisation for coastguard cutters and frigates. *Naval Engineers Journal* (May 1983).

Bales, N.K. Optimising the seakeeping performance of destroyer-type hulls. In *Proceedings of the 13th Symposium on Naval Hydrodynamics*, Shipbuilding Research Association of Japan, Tokyo (1980).

Bales, N.K. and Day, W.G. Experimental evaluation of a destroyer-type hull optimised for seakeeping. In *Proceedings of the STAR symposium* (New York, April 1982). Society of Naval Architects and Marine Engineers, New York (1982).

Barnaby, K.C. *Some Ship Disasters and Their Causes*. Hutchinson, London (1968).

Barr, R.A. An increased role of controllability in ship design. *Marine Technology* **24**, 4 (October 1987).

Belik, O., Bishop, R.E.D. and Price, W.G. Influence of bottom and flare slamming on structural responses, *Transactions of the Royal Institution of Naval Architects*, **130**, 261–275 (1988).

Berkelom, Van W.B. Wind forces on modern ship forms – effects on performance. *Transactions of the North East Coast Institution of Engineers and Shipbuilders*, **97** (1981).

Berkelom, W.B. van, Tragardh, P. and Dellhag, A. Large tankers – wind coefficients and speed loss due to wind and sea. *Transactions of the Royal Institution of Naval Architects*, **117**, 41–58 (1975).

Bernitsas, M.M. and Kekerdis, N.S. Simulation and stability of ship towing. *International Shipbuilding Progress*, **369** (May 1985).

Betts, C.V. Some UK developments in SWATH design research. In *Proceedings of the 1st International Conference on High Performance Vehicles* (Shanghai, 2–5 November (1988). Chinese Society of Naval Architects and Marine Engineers, Shanghai (1988).

Bhattacharyya, R. *Dynamics of Marine Vehicles*. Wiley, New York (1979).

Bishop, R.E.D., Clarke, J.D. and Price, W.G. Comparison of full scale and predicted responses of two frigates in a severe weather trial. *Transactions of the Royal Institution of Naval Architects*, **126**, 153–166 (1984).

Bishop, R.E.D. and Price, W.G. *Hydroelasticity of Ships*. Cambridge University Press (1979).

Bishop, R.E.D., Chalmers, D.W., Price, W.G. and Temarel, P. The dynamic characteristics of unsymmetric ship structures. *Transactions of the Royal Institution of Naval Architects*, **128**, 205–215 (1986).

Bishop, R.E.D., Price, W.G. and Tam, P.K.Y. A unified dynamic analysis of ship response to waves. *Transactions of the Royal Institution of Naval Architects*, **119**, 363–390 (1977).

Bishop, R.E.D., Price, W.G. and Tam, P.K.Y. On the dynamics of slamming. *Transactions of the Royal Institution of Naval Architects*, **120**, 250–280 (1978).

Bleich, F. *Buckling Strength of Metal Structures*. McGraw–Hill, New York (1952).

Blok. J.J. and Beukelman, W. The high-speed displacement ship systematic series hull forms – seakeeping characteristics. *Transactions of the Society of Naval Architects and Marine Engineers*, **92**, 125–150 (1984).

Blume, P. and Kracht, A.M. Prediction of the behaviour and propulsive performance of ships with bulbous bow in waves. *Transactions of the Society of Naval Architects and Marine Engineers*, **93**, 79–84 (1985).

Bong, H.S., Hills, W. and Caldwell, J.B. Methods of incorporating design for production considerations into concept design investigations. In *Proceedings of the 13th Symposium on Ship Technology and Research (STAR)* (Pittsburgh, June 1988). The Society of Naval Architects and Marine Engineers, New York (1988).

Boyd, G.M. *Brittle Fracture in Steel Structures*. Butterworth, London (1970).

Brett, P.O. Operational aspects in ship design: the case of the Roll on/Roll off ferry. *Society of Naval Architects and Marine Engineers Spring Meetings*, Pittsburgh (June 1988).

Brook, A.K. Improving intact stability during a vessel's design. Paper presented 30 November 1987, North East Coast Institute of Engineers and Shipbuilders, Newcastle (1987).

Bruce, G.J. Materials handling in shipbuilding – future directions. *Transactions of the Royal Institution of Naval Architects*, **130** (1988).

Bryson, Admiral Sir L. The procurement of a warship. *Transactions of the Royal Institution of Naval Architects*, **127**, 21–51 (1985).

BS 4360. *Weldable Structural Steels*. British Standards Institute, London (1986).

BS/CP 118 *The Structural Use of Aluminium*. British Standards Institute, London (1969).

Burcher, R.K. The influence of hull shape on transverse stability. *Transactions of the Royal Institution of Naval Architects*, **122**, 111–128 (1980).

Burrill, L.C. The optimum diameter of marine propellers: a new design approach. *Transactions of the North East Coast Institution of Engineers and Shipbuilders*, **72**, 57–82, D1–D22 (1955).

Burrill, L.C. and Emerson, A. Propeller cavitation: further tests on 16 inch propeller models in the King's College Cavitation Tunnel. *Transactions of the North East Coast Institution of Engineers and Shipbuilders* (1963).

Buxton I.L. Engineering economics applied to ship design. *Transactions of the Royal Institution of Naval Architects*, **114**, 409–428 (1972).

Buxton, I.L. *Engineering Economics and Ship Design* (3rd edn). British Maritime Technology, Wallsend (1987).

Buxton, I.L. and Logan, J.A. The ballast performance of ships with particular reference to bulk carriers. *Transactions of the Royal Institution of Naval Architects*, **129**, 17–42 (1987).

Byler, E. and Walz, R. US Navy CAD program for hull structure (HULSTRX). In *Proceedings of Computer Applications in the Automation of Shipyard Operation and Ship Design (ICCAS)*. North Holland Publishing (1982).

Caldwell, J.B. Theory and synthesis of thin-shell ship structures. In *Proceedings of the International Association for Shell Structures Symposium,* University of Hawaii (1971).

Caldwell, J.B. Structural optimisation – what is wrong with

it? In *Advances in Marine Structures* (eds. Smith, C.S. and Clarke, J.D.). Elsevier Applied Science, London (1986).
Carreyette, J. Preliminary ship cost estimation. *Transactions of the Royal Institution of Naval Architects*, **120**, 235–258 (1978).
Case, J. and Chilver, A.H. *Strength of Materials and Structures* (2nd Edition). Edward Arnold, London (1971).
Chalmers, D.W. Structural design for minimum cost. In *Advances in Marine Structures* (ed. Smith, C.S. and Clarke, J.D.). Elsevier Applied Science, London (1986).
Chalmers, D.W. The properties and uses of marine structural materials. *Marine Structures*, **1**, 47–70 (1988).
Chalmers, D.W. The sensitivities of the resonant frequencies of ships' hulls to changes in mass and stiffness. *Transactions of the Royal Institution of Naval Architects*, **130**, 329–336 (1988a).
Chalmers, D.W. Hull structural design using stiffness as a criterion. *Transactions of the Royal Institution of Naval Architects*, **130**, 337–351 (1988b).
Chang, P-Y. A simple method for elastic analysis of grillages. *Journal of Ship Research*, **12**(2), 153–159 (1968).
Chang, P-Y. The effect of varying hull proportions and hull materials on hull flexibility, bending and vibratory stresses. *Ship Structures Committee Report*, SSC 228, US Coast Guard, Washington D.C. (1979).
Clarke, B., Lewis, K.J. and Aldwinkle, D.S. Simulation in marine design and operation. *Transactions of the Royal Institution of Naval Architects*, **128**, 145–160 (1986).
Clarke, D., Gedling, P. and Hine, G. The application of manoeuvring criteria in hull design using linear theory. *Transactions of the Royal Institution of Naval Architects*, **125**, 45–68 (1983).
Clarke, J.D. Wave loading in warships. In *Advances in Marine Structures* (eds. Smith, C.S. and Clarke, J.D.) Elsevier Applied Science, London (1986).
Clarkson, J. Uniform pressure tests on plates with edges free to slide inwards. *Transactions of the Royal Institution of Naval Architects*, **104**, 67–80 (1962).
Clarkson, J. *The Elastic Analysis of Flat Grillages*. Cambridge University Press (1965).
Clayton, B.R. and Bishop, R.E.D. *Mechanics of Marine Vehicles*, Spon, London (1982).
Clayton, B.R. and Satchwell, C.J. An introduction to the theory, technology and economics of wind assisted ship propulsion. In *Proceedings of 9th British Wind Energy Association Conference*, Mechanical Engineering Press, Edinburgh, 299–314 (1987).
Clayton, B.R. and Sinclair, F.M. Motion damping of ships fitted with marine aerofoils. *Transactions of the Royal Institution of Naval Architects*, **130** (1988).
Clayton, B.R. and Sinclair, F.M. A comparison of the effectiveness of aerodynamic and hydrodynamic roll stabilisation methods. *Transactions of the Royal Institution of Naval Architects*, **131** (Paper W3/1989 in press).
Comstock, E.N., Bales, S.L. and Gentile, D.M. Seakeeping performance comparison of air capable ships. *Naval Engineers Journal* (April 1982).
Conolly, J.E. Strength of propellers. *Transactions of the Royal Institution of Naval Architects*, **113**, 139–160 (1961).
Cowley, J. Passenger ship legislation. In *Proceedings of IMAS 88, Transactions of the Institute of Marine Engineers* (C) **100**, Conference 1, Paper 21 (1988).
Cresswell, D.J. and Dow, R.S. The application of non-linear analysis to ship and submarine structures. In *Advances in Marine Structures* (eds. Smith, C.S. and Clarke, J.D.), Elsevier Applied Science, London (1986).
Daidola, J.C. and Basar, N.S. Probabilistic analysis of ship hull longitudinal strength. *Ship Structures Committee Report SSC- 301*, US Coast Guard, Washington DC (1981).
Dand, I.W. Hydrodynamic aspects of shallow water collisions. *Transactions of the Royal Institution of Naval Architects*, **118**, 323–346 (1976).
Dand, I.W. On ship–bank interaction. *Transactions of the Royal Institution of Naval Architects*, **124**, 25–40, 1982.
Dand, I.W. and Ferguson, A.M. The squat of full ships in shallow water. *Transactions of the Royal Institution of Naval Architects*, **115**, 237–255 (1973).
Dawson, J. and Bowden, B.S. The prediction of the performance of single screw ships on measured mile trials. *NPL Ship Report*, **168**, National Physical Laboratory, Teddington (1972).
Dawson, J. and Bowden, B.S. Performance prediction factors for twin screw ships. *NPL Ship Report*, **172**, National Physical Laboratory, Teddington (1973).
Donnelly, K. Aesthetics in warship design. *The Naval Architect*, E282-E285 (June 1985).
Dow, R.S., Hugill, R.C., Clarke, J.D. and Smith, C.S. Evaluation of ultimate ship hull strength. In *Proceedings of the Symposium on Extreme Loads and Response*, Arlington, VA (19–20 October, 1981). Society of Naval Architects and Marine Engineers, New York (1981).
Dowling, P.J. Harding, J.E. and Frieze, P.A. (eds. *Steel Plated Structures*. Crosby Lockwood Staples, London (1976).
Easton, R.W.S. Modern Warships: design and construction. *Transactions of the Institute of Marine Engineers*, **93**, Paper 53 (1983).
Elsevier's Nautical Dictionary 2nd edn (ed. Vandenburgh, J.P. and Chaballe, L.Y.) Elsevier, London (1978).
ESDU Buckling of flat plates in shear. *Data Sheet 71005, (Amendment B)*, Engineering Sciences Data Unit Ltd, London (1976).
ESDU Flat panels in shear, buckling of long panels with transverse stiffeners. *Data Sheet 02.03.02 (Amendment B)*, Engineering Sciences Data Unit Ltd (1983).
Evans, J. H. (ed.) *Ship Structural Design Concepts*. Cornell Maritime Press, Cambridge, Maryland (1975).
Evans, J. H. *Ship Structural Design Concepts*. Second Cycle. Cornell Maritime Press, Cambridge, Maryland (1983).
Faresi, R. *A Simplified Approximate Method to Calculate Shear Force and Bending Moment in Still Water at Any Point in the Ship's Length*. American Bureau of Shipping, New York (1965).
Faulkner, D. Welded connections used in warship structures. *Transactions of the Royal Institution of Naval Architects*, **106**, 39–70 (1964).
Faulkner, D. A review of effective plating for use in the analysis of stiffened plating in bending and compression. *Journal of Ship Research* (March 1975).
Faulkner, D. and Sadden, J.A. Towards a unified approach to ship structural safety. *Transactions of the Royal Institution of Naval Architects*, **121**, 1–28 (1979).
Faulkner, D., Adamchak, J.C., Snyder, G.J. and Vetter, M.F. Synthesis of welded grillages to withstand compression and normal loads. *Computers and Structures*, **3** (1973).
Fisher, K.W. The relative costs of ship design parameters. *Transactions of the Royal Institution of Naval Architects*, **116**, 129–155 (1974).

Fisher, R.W. and Petit, L.M. Computer-assisted shipboard noise predictions. *Noise Control Engineering Journal*, **31**, 111–125 (September–October 1988).

Forrest, P.D. and Parker, M.N. Steelwork design using computer graphics. *Transactions of the Royal Institution of Naval Architects*, **125**, 1–25 (1983).

Fransman, J.V. The influence of passenger ship superstructures on the response of the hull girder. *Paper No. 4, Royal Institution of Naval Architects, Spring Meetings*, London (April 1988).

Froude, W. The fundamental principles of the resistance of ships. *Proceedings of the Royal Institution of Great Britain*, **8**, 188–213 (1876).

Fujino, M. and Yoon, B.S. A practical method of estimating ship motions and wave loads in large amplitude waves. *International Shipbuilding Progress*, **33** (1986).

Fujita, Y. and Sanagana, Y. Human considerations in shipbuildings and some examples of computer aided facilities. In *Proceedings of ICCAS 79*. University of Strathclyde, Glasgow (1979).

Fujita, Y., Sanagana, Y., Mizutani, T. and Morita, Y. CIM in shipbuilding. In *Proceedings of ICCAS 85*, University of Trieste (1985).

Garside, J.F. and Das, P.K. RASDAS – rational structural design and analysis systems for ships. In *Proceedings of the International Conference on Computer-Aided Design Manufacture and Operation in the Marine and Offshore Industries (CADMO 86)*, Washington DC (1986).

Gawn, R.W.L. Effect of pitch and blade width on propeller performance. *Transactions of the Royal Institution of Naval Architects*, **95**, 157–193 (1953).

Gerritsma, J. and Beukelman, W. Analysis of the modified strip theory for the calculation of ship motions and wave bending moments. *International Shipbuilding Progress*, **14** (1967).

Gertler, M. A re-analysis of the Taylor data. *DTMB Report 806* David Taylor Model Basin, Washington DC (1954).

Gillmer, T.C. and Johnson, B. *Introduction to Naval Architecture*, Spon, London (1982).

Glossary of Marine Technology Terms. Heinemann, London (1980).

Glover, E.J., Thorn J.F. and Hawdon, L. Propeller design for minimum hull excitation. *Transactions of the Royal Institution of Naval Architects*, **121**, 267–284 (1979).

Goss, R.O. *Advances in Maritime Economics*. University of Wales Institute of Science and Technology Press, Cardiff (1982).

Gould, R.W.F. The estimation of wind loads on ship superstructures, *Marine Technology Monograph No. 8*. The Royal Institution of Naval Architects, London (undated).

Gray, L.M. and Greeley, D.S. Modelling of propeller cavitation on merchant ships. *Marine Technology* (January 1982).

Hamamoto, M. and Akiyoshi, T. Study on ship motions and capsizing in following seas (1st Report): Equations of motion for numerical simulation. *Journal of the Society of Naval Architects of Japan*, **163** (June 1988).

Handbook of Human Factors (ed. Salvendy) Wiley, New York (1987).

Harris, B. *Engineering Composite Materials*. Institute of Metals, London (1986).

Hart, D.K., Rutherford, S.E. and Wickham, A.H.S. Structural reliability analysis of stiffened panels. *Transactions of the Royal Institution of Naval Architects*, **128**, 293–310 (1986).

Hashim, S.A., Winkle, I.E. and Cowling, M.J. A structural role for adhesives in shipbuilding. *Paper No. 9, Royal Institution of Naval Architects Spring Meetings* (April 1989).

Hills, W. and Buxton, I.L. Integrating ship design and production considerations during the pre-contract phase. *Transactions of the Royal Institution of Naval Architects* **131** (1989).

Hoerner, S.F. *Fluid Dynamic Drag*. Published by the author, New York (1965).

Hoerner, S.F. and Borst, H.V. *Fluid Dynamic Lift*. Published by the author, New York (1975).

Hogben, N., Dacunha, N.M.C. Olliver, G.F *Global Wave Statistics*. British Maritime Technology, Teddington (1986).

Holtrop, J. A statistical re-analysis of resistance and propulsion data. *International Shipbuilding Progress*, **31** (November 1984).

Holtrop, J. and Mennen, G. A statistical power prediction method. *International Shipbuilding Progress* (October 1978).

Horsham, W., Archer, D.J. and Law, Q.H. Integrated preliminary design system. In *Proceedings of the International Conference on Computer-Aided Design Manufacture and Operation in the Marine and Offshore Industries (CADMO 86)*, Washington D.C., (1986).

Hughes, O.F. *Ship Structural Design: A Rationally-based Computer-Aided Optimisation Approach*. Wiley, New York (1983).

Inoue, S., Hirano, M., Kijana, K. and Takashina, J. A practical calculation method of ship manoeuvring motion. *International Shipbuilding Progress*, **28** (September 1981).

Inui, T. Wave-making resistance of ships. *Transactions of the Society of Naval Architects and Marine Engineers*, **70**, 283–353 (1962).

Isherwood, R.M. Wind resistance of merchant ships. *Transactions of the Royal Institution of Naval Architects*, **115**, 327–338 (1973).

ISSC: *Proceedings of the International Ship and Offshore Structures Congress*. Glasgow (1961), Delft (1964), Oslo (1967), Tokyo (1970), Hamburg (1973), Boston (1976), Paris (1979), Gdansk/Paris (1982), Santa Margharita, Italy (1985), Lyngby, Denmark (1988).

Jackson, R.L. and Frieze, P.A. Design of deck structures under wheel loads. *Transactions of the Royal Institution of Naval Architects*, **123**, 119–144 (1981).

Jane's Merchant Ships 3rd edn, (ed. Greenman, D.) Janes Transport Press, London (1987).

Jensen, J.J. and Pedersen, P.T. Wave-induced bending moments in ships – a quadratic theory. *Transactions of the Royal Institution of Naval Architects*, **121**, 151–165 (1979).

Jingen, G. and Jensen, J.J. A rational approach to automatic design of ship sections. In *Proceedings of Computer Applications in the Automation of Shipyard Operation and Ship Design*. North Holland Publishing, (1982).

Kanerva, M., Mikkonen, I. and Nurmi, J. Designing passenger ships for actual service conditions. In *Proceedings of IMAS 88, Transactions of the Institute of Marine Engineers* (C) **100**, Conference 1, Paper 5 (1988).

Kaplan, P., Benator, M., Bentson, J and Achtarides, T.A. Analysis and assessment of major uncertainties associated with ship hull ultimate failure. *Ship Structures Committee Report SSC-322* US Coast Guard, Washington DC (1984).

Kapoor, P. *The Fairplay Book of Shipping Abbreviations*. Fairplay, London (1980).

Kato, H. On the approximate calculation of ship's rolling

period. *Journal of the Society of Naval Architects of Japan*, **89** (1956).

Kerwin, J.E. Marine Propellers. *Annual Review of Fluid Mechanics*, **18** 367–403 (1986).

Kotras, T.V., Baird, A.V. and Naegle, J.N. Predicting ship performance in level ice. *Transactions of the Society of Naval Architects and Marine Engineers*, **91**, 329–349 (1983).

Kuo, C., MacCallum, K.J. and Shenoi, R.A. An effective approach to structural design for production. *Transactions of the Royal Institution of Naval Architects*, **126** 33–50 (1984).

Kupras, L.K. *Computer Methods in Preliminary Ship Design*. Delft University Press, Delft (1983).

Lackenby, H. and Parker, M.N. The BSRA methodical series – an overall presentation. *Transactions of the Royal Institution of Naval Architects*, **108**, 363–404 (1966).

Lamb, T. A ship design procedure. *Marine Technology* (October 1969).

Lamb, T. Design for production in basic design. In *Proceedings of the 11th Ship Technology and Research (STAR) Symposium*, Society of Naval Architects and Marine Engineers (1986).

Lersbryggen, P. (ed.) *Ship's Load and Strength Manual*. Det Norske Veritas, Hovik, Norway (1978).

Lersbryggen, P. Ship and offshore design and analysis by use of the pilot desktop computer programs. In *Proceedings of the 2nd International Symposium on Practical Design in Shipbuilding (PRADS)*, Tokyo and Seoul (1983).

Lester A.R. *Merchant Ship Stability*. Butterworth, London (1985).

Lewis, E.V. (ed.) Vol. I: *Stability and Strength*. Vol. II: *Resistance, Propulsion and Vibrations*; Vol. III: *Seakeeping and Controllability*. Society of Naval Architects and Marine Engineers, New York (1988).

Lloyd, A.R.J.M. and Hanson, P.J. The operational effectiveness of the shipborne naval helicopter. In *Proceedings of the International Symposium on The Air Threat at Sea* (London, June 1985). Royal Institution of Naval Architects, London (1985).

Lloyd, A.R.J.M., Brown, J.C. and Anslow, J.F.W. Motions and loads on ship models in regular oblique waves. *Transactions of the Royal Institution of Naval Architects*, **122**, 20–43 (1980)

Lloyds Register *Guidance Notes on Acceptable Vibration Levels and their Measurement*. Lloyds Register of Shipping, London (1988).

Lloyds Rules *Rules and Regulations for the Classification of Ships*. Lloyds Register of Shipping, London (1987).

Loukakis, T.A. and Chryssostomidis, C. Seakeeping standard series for cruiser-stern ships. *Transactions of the Society of Naval Architects and Marine Engineers*, **83** 67–127 (1975).

Lyon, T.D. and Mistree, F. A computer based method for the preliminary design of ships. *Journal of Ship Research*, **29**, 4 (December 1985).

MacCallum, K.J. Creative ship design by computer. In *Computer Applications in the Automation of Shipyard Operation and Ship Design (ICCAS IV)*, North Holland Publishing (1982).

MacKerle, J. Finite element codes for microcomputers – a review. *Computers and Structures*, **24**, 4, 657–682 (1987).

Mansour, A. Charts for the buckling and post-buckling analysis of stiffened plates under combined loading. *Technical and Research Bulletin*, **2**, 22. The Society of Naval Architects and Marine Engineers, New York (1976).

Mansour, A.E. Approximate formulae for preliminary design of stiffened plates. In *Proceedings of the 5th Offshore Mechanics and Arctic Engineering Symposium (OMAE 86)*, American Society of Mechanical Engineers, Tokyo (1986).

MarAd Series *The Maritime Administration Systematic Series of Full-Form Ship Models*. Society of Naval Architects and Marine Engineers, New York (1987).

March, J. and Trott, F. Databases for engineers. *Professional Engineering*, 29–32, (December 1988).

Massey, B.S. *Units, Dimensional Analysis and Physical Similarity*. Van Nostrand, London (1971).

Massey, B.S. *Mechanics of Fluids*, 6th edn. Van Nostrand, London (1989).

McCreight, K.K. and Stahl, R.G. Recent advances in the seakeeping assessment of ships. *Naval Engineers Journal* (May 1985).

Meek, M. Taking stock – marine technology and UK maritime performance. *Transactions of the Royal Institution of Naval Architects*, **127**, 69–94 (1985).

Meek, M., Brown, W.R. and Fulford, K. A shipbuilder's view of safety. In *Proceedings of the Centenary Conference on Marine Safety*, University of Glasgow, Glasgow (1983).

Meek, M., Adams, R., Chapman, J.C., Reibel, H. and Wieske, P. The structural design of the OCL Container ship. *Transactions of the Royal Institution of Naval Architects*, **114**, 241–292 (1972).

Mikelis, N.E. A procedure for the prediction of ship manoeuvring response for initial design. In *Computer Applications in the Automation of Shipyard Operation and Ship Design (ICCAS V)*. Elsevier Applied Science, London (1985).

MIL – S – 24645 *Steel Plate, Sheet or Coil, Age-hardening Alloy, Structural, High Strength Steel* (HSLA-80). Department of Defense, Washington DC (1984).

Millward, A. The effect of shallow water on the resistance of a ship at high sub-critical and super-critical speeds. *Transactions of the Royal Institution of Naval Architects*, **124**, 175–181 (1982).

Milne, P.A. High technology in shipbuilding, *59th Thomas Lowe Gray Lecture, Transactions of the Institution of Mechanical Engineers*, **201** (1987).

Mitchell, G.C. Analysis of structural interaction between a ship's hull and deckhouse. *Transactions of the Royal Institution of Naval Architects*, **120**, 121–136 (1978).

Monk, K. A warship roll criterion. *Transactions of the Royal Institution of Naval Architects*, **130**, 219–240 (1988).

Morrall, A. The GAUL disaster. An investigation into the loss of a large stern trawler. *Transactions of the Royal Institution of Naval Architects*, **123**, 391–440 (1981).

Morrow, R.T. Noise design methods for ships. *Transactions of the Royal Institution of Naval Architects*, **130** (1988).

Muckle, W. *Strength of Ships' Structures*. Edward Arnold, London (1967).

Muckle, W., revised Taylor, D.A. *Muckle's Naval Architecture*, 2nd edn. Butterworth, London (1987).

Munse, W.H. Fatigue criteria for ship structural details. In *Proceedings of the Extreme Loads Response Symposium*, Society of Naval Architects and Marine Engineers, New York (1981).

Munse, W.H. Fatigue characteristics of fabricated ship details for design. *Ship Structures Committee Report SSC-318*. US Coast Guard, Washington DC (1983).

Newman, J.N. *Marine Hydrodynamics*, Massachusetts Institute of Technology Press, Cambridge MA (1977).

NPL BTTP 1965 Standard Procedure for the Prediction of Ship Performance from Model Experiments. *NPL Ship Report 80*, Teddington: National Physical Laboratory, Teddington (1966).

Ochi, M.K. and Motter, L.E. Prediction of slamming characteristics and hull responses for ship design. *Transactions of the Society of Naval Architects and Marine Engineers*, **81**, 144–176 (1973).

Odabasi, A.Y., Patterson, D.R., Williams, E.A. and Davison, G.H. BRITSHIPS – application of advanced technology to shipbuilding CAD/CAM. In *Proceedings of the 6th WEMT Symposium*, June 1987, Lubeck–Travemunde (1987).

Okumoto, Y., Takeda, T. and Hiyoku, K. Modern hull structure design system "COSMOS". In *Proceedings of Computer Applications in the Automation of Shipyard Operation and Ship Design (ICCAS):* North Holland Publishing (1985).

Olsen, S.R. An evaluation of the seakeeping qualities of naval combatants. *Naval Engineers Journal* (February 1978).

Oosterveld, M.W.C. and van Oossanen, P. Further computer-analysed data of the Wageningen B-screw series. *International Shipbuilding Progress*, **22**, 251–262 (1975).

Parker, T.J. and Woolveridge, M. BP SWOPS – an operator's and shipbuilder's perspective. *Transactions of the Royal Institution of Naval Architects*, **130** (1988).

Pattison, D.R., Spencer, R.E. and van Griethuysen, W.J. The computer-aided ship design system GODDESS and its application to the structural design of Royal Navy warships. *Computer Applications in the Automation of Shipyard Operation and Ship Design (ICCAS)*. North Holland Publishing (1982).

Pattullo, R.N.M. The BSRA trawler series (Part II) – block coefficient and longitudinal centre of buoyancy variation series: resistance and propulsion tests. *Transactions of the Royal Institution of Naval Architects*, **110**, 151–183 (1968).

Pattullo, R.N.M. and Thomson, G.R. The BSRA trawler series (Part I) – beam/draft and length/displacement ratio series resistance and propulsion tests. *Transactions of the Royal Institution of Naval Architects*, **107**, 215–241 (1965).

Payer, H.G. and Koster, D. Strength and vibration investigation for passenger ships. *Proceedings of IMAS 88, Transactions of the Institute of Marine Engineers* (C) **100**, Conference 1, Paper 9 (1988).

Payne, S.M. The evolution of the modern cruise liner. *Royal Institution of Naval Architects, Spring Meetings*, London, (April 1989).

Pedersen, P.T. Torsional response of container ships. *Journal of Ship Research*, **29**, 3 (1985).

Peirce, T.H. Arctic marine technology: a review of ship resistance in ice. *Transactions of the Royal Institution of Naval Architects*, **121**, 219–235 (1979).

Peterson R.E. *Stress Concentration Design Factors*, Wiley, New York (1953).

Planeix, J.M. Recent progress in probabilistic structural design. In *Advances in Marine Technology* (eds. Smith, C.S. and Clarke, J.D.). Elsevier Applied Science, London (1986).

Ports of the World Lloyds of London Press, London (1989).

Price, W.G. and Bishop, R.E.D. *Probabilistic Theory of Ship Dynamics*. Chapman and Hall, London (1974).

Raff, A.I. Program SCORES – Ship structural response in waves. *Ship Structures Committee Report SSC-230*, US Coast Guard Washington DC (1972).

Rawson, K.J. and Tupper, E.C. *Basic Ship Theory*, 3rd edn. Longman, London (1983, 1984).

Renilson, M.R. and Driscoll, A. Broaching – an investigation into the loss of directional control in severe following seas. *Transactions of the Royal Institution of Naval Architects*, **124**, 253–273 (1982).

RINA *Proceedings of the International Conference on the Safeship Project, Ship Stability and Safety* (London, June 1986). Royal Institution of Naval Architects, London (1986).

RINA *Proceedings of the 1st International Conference on Ship Manoeuvrability, Prediction and Achievement*, London, May 1987. Royal Institution of Naval Architects, London (1987).

Ro–Ro safety and vulnerability: the way ahead. *Proceedings of International Conference*, held 14–15 December 1987. Royal Institution of Naval Architects, London (1987).

Roark, R.J. *Formulas for Stress and Strain* (6th edn) (ed. W.C. Young). McGraw-Hill, New York (1988).

Roseman, D.P. and Seibold, F. An introduction to the US Maritime Administration Systematic Series. *Marine Technology*, **22** 4 (1985).

Roy, S. Computer software survey 1986-7. *RINA Small Craft Supplement*, **19**, 137–140 (December 1987).

Salvesen, N., Tuck, E.O. and Faltinsen, O. Ship motions and sea loads. *Transactions of the Society of Naval Architects and Marine Engineers*, **78**, 250–287 (1970).

Sarchin, T.H. and Goldberg, L.L. Stability and buoyancy criteria for US naval surface ships. *Transactions of the Society of Naval Architects and Marine Engineers*, **70**, 418–458 (1962).

Saunders, H.E. *Hydrodynamics in Ship Design* (2 vols.). Society of Naval Architects and Marine Engineers, New York (1957).

Schmitke, R.T. Ship sway, roll and yaw motions in oblique seas. *Transactions of the Society of Naval Architects and Marine Engineers*, **86**, 26–46 (1978).

Schoenberger, R.W. Subjective response to very low frequency vibration. *Aviation, Space and Environmental Medicine* (June 1975).

Scott, J.R. A method of predicting trial performance of single screw merchant ships. *Transactions of the Royal Institution of Naval Architects*, **115**, 149–171 (1973).

Scott, J.R. A method of predicting trial performance of twin screw merchant ships. *Transactions of the Royal Institution of Naval Architects*, **116**, 175–186 (1974).

Sellmeijer, R. and Van Oortmerssen, G. The effect of mud on tanker manoeuvres. *Transactions of the Royal Institution of Naval Architects*, **126** 105–124 (1984).

Small Craft Survey Catalog. *Marine Technology*, **25**, 4 (October 1988).

Smith, C.S. Compression strength of welded steel ship grillages. *Transactions of the Royal Institution of Naval Architects*, **117**, 325–359 (1975).

Smith, C.S. Influence of local compression failure on ultimate longitudinal strength of a ship's hull. In *Proceedings of the International Conference on Practical Design in Shipbuilding*, Tokyo (October 1977).

Smith, C.S. and Clarke, J.D. (eds.) *Advances in Marine Structures*. Elsevier Applied Science, London (1986).

Smith, C.S. and Kirkwood, W. The influence of initial deformations and residual stresses on inelastic flexural buckling of stiffened plates and shells. In *Steel Plated Structures*,

Crosby Lockwood Staples, London (1977).
Smith, C.S., Davidson, P.C., Chapman, J.C. and Dowling, P.J. Strength and stiffness of ship's plating under in-plane compression and tension. *Transactions of the Royal Institution of Naval Architects*, **130**, 277–296 (1988).
SNAME *Design for Production Manual*. Ship Production Committee Panel SP-4. Society of Naval Architects and Marine Engineers, New York (1985).
SNAME *Bulletin 4*, 22. Use of scale models for human engineering purposes in ship design and construction. Society of Naval Architects and Marine Engineers (1988).
Soding, H. Capsizing risk analysis by means of ship motion simulation. *Schiffstecknik*, **34** (1987).
Somerville, W.L., Swan, J.W. and Clarke, J.D. Measurements of residual stresses and distortions in stiffened panels. *Journal of Strain Analysis*, **12**, 2 (1977).
Spouge, J.R. Nonlinear analysis of large-amplitude rolling experiments. *International Shipbuilding Progress*, **53** (October 1988).
Spouge, J.R. The safety of Ro–Ro passenger ferries. *Transactions of the Royal Institution of Naval Architects*, **131** (1989).
Stavovy, A.B. and Chang, S.L. Analytical determination of slamming pressures for high speed vehicles in waves. *Journal of Ship Research*, **20** (1976).
Stiansen, S.G., Mansour, A. and Chen, Y.N. Dynamic response of large Great Lakes bulk carriers to wave-excited loads. *Transactions of the Society of Naval Architects and Marine Engineers*, **85**, 174–208 (1977).
Stiansen, S.G., Mansour, A., Jan, H.Y. and Thayamballi, A. Reliability methods in ship structures. *Transactions of the Royal Institution of Naval Architects*, **122**, 381–406 (1980).
Stopford, M. *Maritime Economics*. Unwin Hyman, London (1988).
Sumpter, J.D.G., Clarke, J.D., Bird, J. and Caudrey, A.J. Fracture toughness of ship steels. *Paper No.2, The Royal Institution of Naval Architects, Spring Meetings*, London (April 1988).
Taggart, R. (ed.). *Ship Design and Construction*. Society of Naval Architects and Marine Engineers, New York (1980).
Takezawa, S. and Takekawa, M. Advanced experimental techniques for testing ship models in transient water waves. Part I, The transient test technique on ship motions in waves. In *Proceedings of the 11th Symposium on Naval Hydrodynamics*, London (1976).
Taylor, D.W. *The Speed and Power of Ships*. US Government Printing Office, Washington DC (1943).
Taylor, K.V., Payer, H.G. and Swift, P.M. Strength investigations of tanker longitudinal bulkhead plating. *Transactions of the Royal Institution of Naval Architects*, **128**, 121–141 (1986).
Thayamballi, A.K., Kutt, L.M. and Chen, Y-N. Advanced strength and structural reliability assessment of the ship's hull girder. In *Advances in Marine Technology* (eds. Smith, C.S. and Clarke, J.D.) Elsevier Applied Science, London (1986).
Timoshenko, S. and Woinowsky-Krieger *Theory of Plates and Shells*. McGraw–Hill, New York (1959).
Timoshenko, S. and Gere, J.M. *Theory of Elastic Stability*. McGraw–Hill, New York (1961).
Todd, F.H. Some further experiments on single screw merchant ship forms – Series 60. *Transactions of the Society of Naval Architects and Marine Engineers*, **61** (1953).

Todd, F.H. *Ship Hull Vibration*. Edward Arnold, London (1961).
Townsin, R.L. and Kwon, Y.J. Approximate formulae for the speed loss due to added resistance in wind and waves. *Transactions of the Royal Institution of Naval Architects*, **125**, 199–207 (1983).
Townsin, R.L., Byrne, D., Svensen, T.E. and Milne, A. Estimating the technical and economic penalties of hull and propeller roughness. *Transactions of the Society of Naval Architects and Marine Engineers*, **89**, 295–318 (1981).
Townsin, R.L., Spencer, D.S., Mosaad, M. and Patience, G. Rough propeller penalties. *Transactions of the Society of Naval Architects and Marine Engineers*, **93**, 165–187 (1985).
Troesch, A.W. Wave-induced hull vibration: an experimental and theoretical study. *Journal of Ship Research*, **28**, 2 (1984).
Troost, L. Open water tests on modern propeller forms. *Transactions of the North East Coast Institution of Engineers and Shipbuilders* **67**, 89–130 (1950).
Van Lammeren, W.P.A., van Manen, J.D. and Oosterveld, M.W.C. The Wageningen B-Screw Series. *Transactions of the Society of Naval Architects and Marine Engineers*, **77**, 269–317 (1969).
Van Manen, J.D. The choice of the propeller. *Marine Technology*, **3**, 158–171 (1966).
Van Manen, J.D. and Oosterveld, M.W.C. Analysis of ducted propeller design. *Transactions of the Society of Naval Architects and Marine Engineers*, **74**, 522–562 (1966).
Vaughan, H. Flexural response of ice-breaking cargo ships to impact loads. *Transactions of the Royal Institution of Naval Architects*, **128**, 259–267 (1986).
Vilnay, O. Flange failure of a double bottom structure. *Journal of Constructional Steel Research*, **2**, 1 (1982).
Vilnay, O. A full orthotropic plate method for double bottom structures. *Journal of Constructional Steel Research*, **3**, 1 (1983).
Viner, A.C. Development of ship strength formulations. In *Advances in Marine Technology* (eds. Smith, C.S. and Clarke, J.D.). Elsevier Applied Science, London (1986).
Vossers, G., Swaan, W.A. and Rijken, H. Experiments with Series 60 models in waves. *Transactions of the Society of Naval Architects and Marine Engineers*, **68**, 364–450 (1960).
Vugts, J.H. The coupled roll–sway–yaw performance in oblique waves. In *Proceedings of the 12th International Towing Tank Conference*, Rome (1969).
Vugts, J.H. The hydrodynamic forces and ship motions in oblique waves, *Report No. 150S*, Netherlands Ship Research Centre (1971).
Ward, G. The application of current vibration technology to routine ship design work. *Transactions of the Royal Institution of Naval Architects*, **125**, 26–44 (1983).
Watson, D.G.M. Designing ships for fuel economy. *Transactions of the Royal Institution of Naval Architects*, **123**, 505–522 (1981).
Watson, D.G.M. and Gilfillan, A.W. Some ship design methods. *Transactions of the Royal Institution of Naval Architects*, **119**, 279–324 (1977).
Winkle, I.E. and Baird, D. Towards more effective structural design through synthesis and optimisation of relative fabrication costs. *Transactions of the Royal Institution of Naval Architects*, **128**, 313–336 (1986).
Wirsching, P.H. and Chen, Y-N. Considerations of

probability-based fatigue design for marine structures. *Marine Structures*, **1**, 23–45 (1988).

Yeh, H.Y.H. Series 64 resistance experiments on high speed displacement forms. *Marine Technology*, **2**, 248–272 (1965).

Yim, B. A simple design theory and method for bulbous bows of ships. *Journal of Ship Research*, **18**, 141–152 (1974).

Zienkiewicz, I. *The Finite Element Method in Engineering Science*. McGraw–Hill, London (1971).

High speed and multihull marine vehicles

3.8 Introduction

What do we mean by high speed marine vehicles? Although this may seem an apparently straightforward question to ask, it is by no means easy to answer without considering naval architecture. For example, the SS *United States* until recently held the Blue Riband for the shortest easterly crossing of the Atlantic Ocean by covering, in 1952, the 4707 km (2942 mile) distance from Ambrose lighthouse to Bishop Rock lighthouse in three days, ten hours, forty minutes, that is, at an average speed of 35.4 knots. In 1986, Richard Branson's *Virgin Atlantic Challenger (VAC) II* took three days, eight hours, thirty-one minutes to travel between the same two points, albeit with seven hours lost because of water ingress during refueling at sea. The time difference is not all that great. But the *VAC II*, often running at 50 knots, is certainly within the common acceptance of a high speed craft whereas the SS *United States* is not. As another example, a hydrofoil craft operating as a passenger ferry in an area such as the Adriatic Sea cruises at about 32 knots, a similar speed to that available to most modern naval frigates which are not regarded specifically as high speed craft.

It is clear that size must also play a part in the concept of a high speed marine vehicle. Evidently, *VAC II* travelled at many hull lengths per unit time faster than any liner, cargo carrier or warship. In attaining the required high 'speed to size' ratios large specific fuel consumptions must be accepted, even though various techniques have been developed to minimize the rate at which fuel consumption increases with forward speed. Refuelling at sea for commercial craft is both inconvenient and uneconomic and so other elements enter the high speed vehicle concept, such as journey length, usually limited to less than 1000 km, transport efficiency and sea states, which will be examined later.

All marine vehicles that move at or near the air and water interface, create a disturbance there which, in addition to generating turbulence, shear flows and vortices, gives rise to surface water waves. The energy flux carried by these waves originates from the power required to propel the vehicle and, as discussed earlier in this chapter, the assessment of wave making resistance provides a vital ingredient in the estimation of propulsive power requirements. The variation of wave making resistance R_w, can be characterized on the basis of speed and size through the dimensionless parameter Froude number, Fr. The forward cruise or design velocity V, is usually adopted along with the length L, at the still water line to represent size, thus:

$$Fr = V/\sqrt{(gL)} \qquad (3.130)$$

in which g represents weight per unit mass and is associated with distortion of the interface.

For a marine vehicle travelling in a forward direction along its plane of symmetry, transverse waves are formed with a crest near the bow and a trough near the stern. At certain speeds, the spacing between these wave sources coincides with maximum correlation between the wave patterns, i.e. the bow generated wave system yields a trough at the stern which reinforces the stern generated wave to form a large amplitude wave system astern the hull. This condition produces a series of local maxima (humps) in the resistance versus velocity curve of a given vehicle interspersed by a series of local minima (hollows) resulting from speeds at which maximum, but always partial, cancellation of waves occurs. A typical variation of the wave making resistance coefficient C_w $(=R_w/^1/_2\rho V^2 L^2$, in which ρ represents the density of water), with Fr consisting of the superposition of both transverse and divergent wave systems, is shown in *Figure 3.49*. The largest maximum, referred to as the primary hump, corresponds theoretically to Fr = 0.52 for monohull displacement ships, a value confirmed by both model and full scale tests. For most high speed craft the primary hump may occur in the range $0.5 \leq$ Fr ≤ 0.6.

A simple formulation for the total resistance R_T is given by:

$$R_T = R_V + R_W \qquad (3.131)$$

where R_V represents the viscous resistance, which for displacement monohulls consists principally of components of skin friction and viscous pressure resistance. Division by $^1/_2\rho V^2 L^2$ through Equation 3.131 yields the dimensionless form:

$$C_T = C_V + C_W \qquad (3.132)$$

Although C_V reduces with increasing Fr, the total C_T shows a generally rising bumpy curve, as shown in *Figure 3.49*. It has already been shown earlier that C_V is largely a function of Reynolds number Re, given by:

$$Re = VL/\nu \qquad (3.133)$$

where ν represents the local kinematic viscosity of water. As a result, the maximum value of C_T shown in *Figure 3.49* may be of the order of 0.006 at one-thirtieth model scale and 0.004 at full scale. A thorough discussion of the resistance of marine vehicles is given in Lewis (1988).

Conventional displacement ships have cruise speeds related to local hollows in the C_W curve but for higher speed operation the total resistance to motion and thus power dissipation increases rapidly. *Figure 3.50* shows a typical case using as ordinate the total resistance to weight displacement ratio. The challenge facing the designers of high speed craft is to limit, by design, the rate of increase of resistance (or drag; the two words are used synonymously here according to general practice), at least up to speeds beyond which other factors prevent further progress. It is important to point out

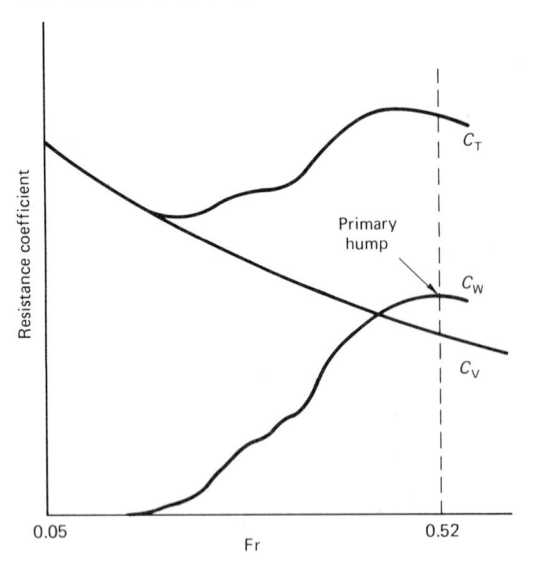

Figure 3.49 Typical variations of coefficients of wave-making resistance C_W, viscous resistance C_V and total resistance C_T with Froude number Fr

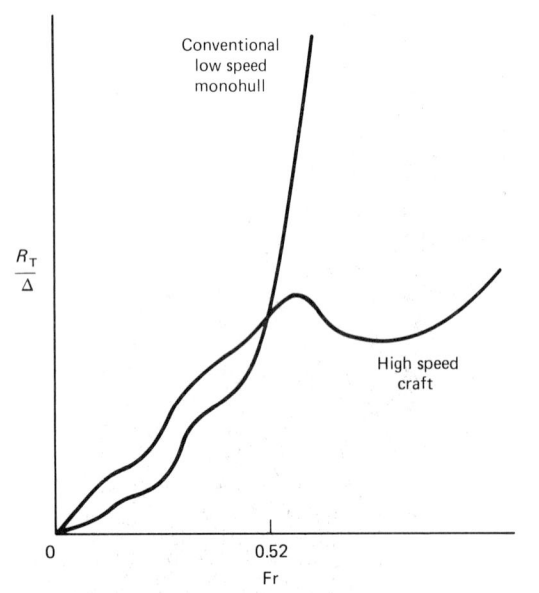

Figure 3.50 General comparison between the total resistance R_T of low speed and high speed craft, based on weight displacement Δ

that the curves in *Figure 3.50* show relative effects and are not meant to identify any particular marine vehicle that has been built, although curves typical of different high speed craft concepts are given later. Evidently, the major effectiveness of high speed craft occurs when optimum operating conditions are designed to take place at Froude numbers greater than that corresponding to the primary hump in the C_W curve. If one wished to be more precise, the condition Fr ≥ 0.7 could be adopted as that defining a high speed craft operating at its cruise condition.

Two further points are worth noting. First, the curves in *Figure 3.50* imply that a high speed craft performs rather poorly at low speeds. This is, indeed, generally true and a sufficient power plant must be installed to accelerate through the hump resistance condition as rapidly as possible. Second, the hull wetted length rather than the length at the still waterline is the significant parameter from the resistance viewpoint and for high speed craft the wetted length changes greatly with speed. Nevertheless, the length at the still waterline can be used unambiguously to compare performances of craft of the same kind.

The techniques used to limit the rate of increase of resistance to forward motion as speed increases that will be considered here are those embodied within the following craft types:

1. Planing craft and catamarans: the mono or multihull is raised in the water, by the application of an upward dynamic force resulting from forward motion which reduces the wetted surface area and length of the hulls.
2. Hydrofoil craft: the hull is raised clear of the water when forward motion takes place at high speed and the craft is wholly supported by upward dynamic forces applied to submerged appendages attached to the hull.
3. Hovercraft: the main hull is raised clear of the water and the craft supported independently or forward motion by aerostatic and aerodynamic jet reaction forces contained within the flexible walls of air cushion vehicles (ACVs), or within a combination of flexible and rigid walls (that generate additional buoyancy forces) of surface effect ships (SES).

In each of the above categories it is observed that the buoyancy force, characteristic of conventional displacement marine vehicles, plays a small, and often insignificant, part in generating the total supporting force. There is, however, some ambiguity in defining the exact proportion of the total supporting force that arises from buoyancy. This chapter also contains a discussion of SWATH ships which rely entirely on buoyancy and which, as yet, have not been developed as high speed marine vehicles.

Returning to the question at the start of this section, we shall define a high speed marine vehicle for which the design operating condition corresponds to Fr ≥ 0.7 and buoyancy accounts for no more than 60% and, for many craft, a good deal less of the total supporting force.

3.9 Planing craft

If the hull of a displacement vehicle is suitably shaped, the hydrodynamic forces developed during forward motion can be sufficient to support the craft and raise a substantial proportion of the hull out of the water. The wetted area is thus reduced and so is R_V, but the aerodynamic resistance of the above water profile is increased. The wave making resistance becomes small at high speeds and, even though R_T is then large, it is considerably less than the resistance of an equivalent conventional displacement ship moving at the same speed.

When a free jet of water impacts on a rigid flat plate, the momentum of the jet perpendicular to the plate is destroyed and the jet reaction force has a component that is perpendicular to the jet. Thus, if the jet is horizontal and the plate inclined so that its leading edge is above the centre line of the jet, an upward component of the reaction force is generated which opposes the weight of the plate. It is this principle that lies behind the effective operation of planing craft. The nearest practical equivalent to the plate and jet system, which would correspond to the mode of operation of a planing craft, is an inclined plate moving at an air and water interface. *Figure 3.51(a)* illustrates the two dimensional flow pattern relative to a flat plate which is of infinite extent normal to the plane of the paper. The plate is moving at a steady velocity V, along the surface of a stationary body of water of infinite depth. Provided that the inclination of the plate is small, the amount of water thrown forward is also small and is assumed, in theory, to travel to infinity, but in practice it degenerates to spray.

Of particular interest are the variations of local velocity q and local pressure p on the wetted surface of the plate. Results may be generalized by considering the velocity ratio q/V and the pressure coefficient $C_p = (p - p_a)/\frac{1}{2}\rho V^2$. Typical results for inviscid flow are shown in *Figure 3.51(b)*. The pressure distribution peaks near the stagnation point, where the fluid is brought to rest relative to the plate, and becomes more sharply peaked as the inclination decreases. It is also evident that because C_p is always positive the local pressure p is greater than the ambient (atmospheric pressure) p_a. The maximum pressure at the stagnation point St, can be very high. For example, at 40 knots the gauge pressure, above atmospheric, is approximately 2×10^5 Pa, or about 2 atmospheres. The integrated pressure force per unit breadth F_p', lies aft of the stagnation point and its magnitude will be modified in the flow of a real fluid by the presence of viscous shear stresses over the plate surface. Note also, that F_p' will have a rearward horizontal component that can be interpreted as a resistance opposing motion.

Either an energy or momentum analysis may be used to illustrate the relationship between F_p', the trim angle α and the amount of water thrown forward by a planing flat plate (Du Cane, 1974; Clayton and Bishop, 1982; Payne, 1988). Adopting the energy analysis, as applied by an observer outside the flow system depicted in *Figure 3.52*, we see that the only energy imparted to the water is that thrown forward from the underside of the plate. The absolute velocity of this water is $2V \cos(\alpha/2)$, neglecting the effect of the boundary layer on the wetted surface. Equating the energy flux of the water thrown forward to the rate at which work is done on the water by unit width of the plate leads to the expression:

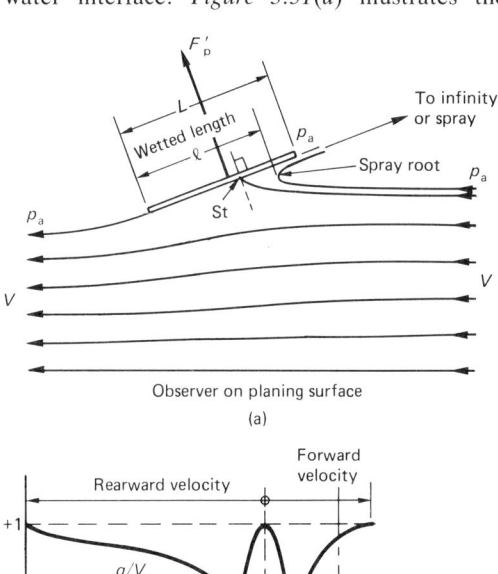

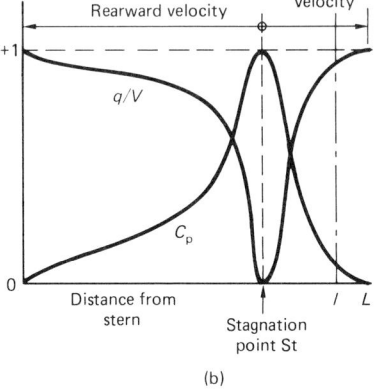

$$F_p' = \rho V^2 \delta \cot(\alpha/2) \qquad (3.134)$$

Figure 3.51 (a) Idealized two-dimensional flow past an infinite breadth plate inclined into the surface of a moving body of water; (b) Corresponding variations of non-dimensionalized local velocity and pressure on the wetted planing surface

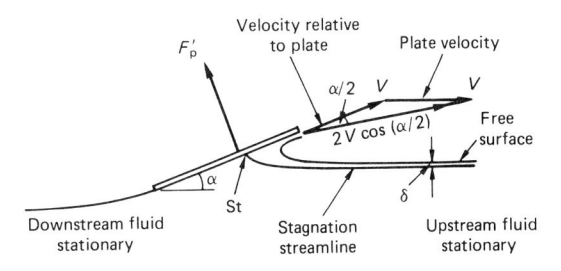

Figure 3.52 Velocity vectors for two dimensional flow past a planing flat plate

Equation 3.134 demonstrates the important point that for planing motion $\delta \neq 0$ and so spray is always present. For small α, which minimizes the rearward force, we can write the upward supporting force as:

$$F_v' = F_p' \cos \alpha \approx 2\rho V^2 \delta/\alpha \qquad (3.135)$$

whence

$$F_v'/\tfrac{1}{2}\rho V^2 L^2 = C_v' = 4\delta/\alpha L \approx (\alpha - \epsilon) \{\partial(C_v')/\partial\alpha\}_{\alpha=\epsilon} \qquad (3.136)$$

The term in braces is virtually constant for α slightly greater than zero and ϵ is a fraction of a degree which accounts for the edge effects of a finite width plate. We can therefore write

$$\delta/L \propto (\alpha - \epsilon) \alpha \text{ and } F_v' \propto (\alpha - \epsilon) V^2 \qquad (3.137)$$

In principle, therefore, an effective planing craft should operate at high speed and small trim angle in order to yield a substantial support force and a small quantity of spray by virtue of small δ.

3.9.1 Hull forms

Accepting that a flat plate is an impractical planing surface, owing to problems with directional stability and large localized hydrodynamic forces inducing high accelerations, two general hull forms have been developed. The round bottom or round bilge is shown in *Figure 3.53(a)*. Once forward motion commences the boat sinks by the stern and a flat region of the hull is required to limit the sinkage and therefore the trim. As high speeds are approached, a fine entry to the water is necessary, especially in waves and so the bows become very narrow and little use can be made of the internal space. Better high speed performance is obtained, particularly in calm water, if the under surface of the hull is transformed from the flat region near the stern into an increasingly pronounced V as the bow is approached with the leading edge sharpened, as shown in *Figure 3.53(b)*. To achieve good planing results and preserve the flat plate characteristics the sides of the craft meet the under surface V at hard angles, that is, there is no rounding off at the abrupt junction. Craft exhibiting this type of geometry are thus said to possess a hard chine, V-bottom hull and many of the present high speed boats are of this kind.

In terms of hull geometry, the overall length to beam ratio is often used to describe fast craft. For the purpose of correlating hydro and aerodynamic forces and moments the mean wetted length l, and beam b, are required. At all but the lowest speeds, water breaks away from the hard chine or from spray rails or deflectors on round bilge forms, and so b can be precisely defined in relation to the chine or spray rail shape. We define l as S/b, where S represents the wetted plan area of the hull, and so the aspect ratio AR is given by:

$$AR = b^2/S = b/l \qquad (3.138)$$

The spray velocity vector V is directed transversely (with perhaps a rearward component) relative to the hull, as illustrated in *Figure 3.54*. The more sternwards the relative velocity of the spray the lower becomes its absolute velocity and therefore its kinetic energy with the result that hydrodynamic forces on the hull are reduced to give a smoother ride. For hulls with a small deadrise angle β, such as racing power boats, the spray emerges more or less at right angles to the forward direction, which gives rise to large hydrodynamic supporting forces on the wetted hull resulting in a hard ride but lower relative resistance at very high speeds. Generally, for equivalent boat sizes and power dissipation, the round bilge form will travel somewhat slower but with a more comfortable ride. There have, however, been successful attempts to combine the best of both forms using multi-chine curved sections.

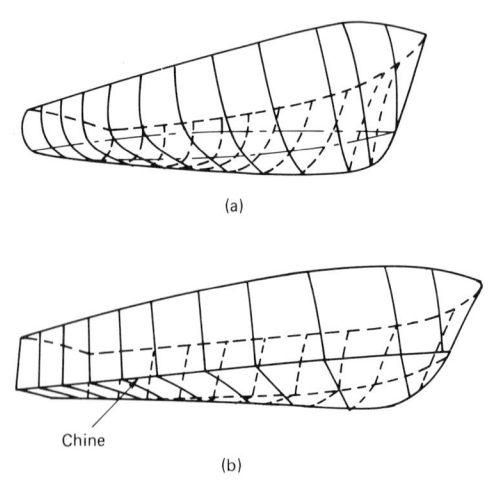

Figure 3.53 Planing craft hull forms: (a) round bilge hull form and (b) hard chine, 'V' bottom hull form

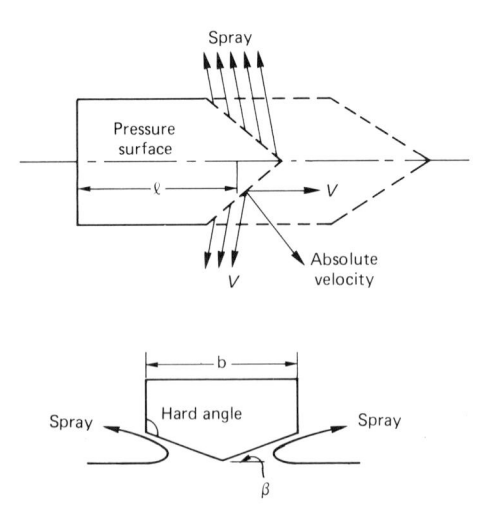

Figure 3.54 Spray direction

A figure of merit can be used in the form of the ratio of total resistance R_T to all up weight W, i.e. the total weight of the craft and contents in its cruise condition. On the basis of tests on models of simple geometry and trials on various craft, it has been found that optimum values of R_T/W lie between 0.10 and 0.14, for mean deadrise angles between 10° and 15° with aspect ratios between 0.2 and 0.3, and length to beam ratios of 4 to 6. With these values of the principal geometric parameters, longitudinal and lateral stability are maintained and the wetted surface area can be further decreased by introducing abrupt discontinuities in the hull. By riding on the resulting steps the wetted length is reduced but the vertical accelerations and loads are high. Trim can be maintained small, as it can if instead a small flap, trim tab or wedge is built into the keel at the transom (Savitsky and Brown, 1976; Wellicombe and Jahangeer, 1979).

Determination of the bottom loadings of planing craft remains a major task and finite element analyses are now being increasingly used for this purpose. Although assumptions concerning the behaviour of an inviscid fluid allow acceptable design techniques to be developed, the problem of adequately calculating viscous phenomena remains. Finally, operation in waves has resulted in the present day domination of seakeeping behaviour in design considerations. The result is often a compromise geometry with, in particular, stern deadrise angles of 15°, bow angles approaching 50° and mean values of 20°–25°, with the designer accepting a consequent increase of resistance above the optimum.

3.9.2 Forces on a planing craft in steady motion

It follows that if a resultant hydrodynamic force acts on the wetted hull of a craft at an air and water interface, some part of the hull must be submerged and water displaced. However, for a fast craft the transom remains dry since the water cannot follow the rapid change of contour at the stern, thus effectively lengthening the equivalent hull. We cannot now think in terms of a vertical buoyancy force, but a hydrostatic force acting on the surface of the hull (Clayton and Bishop, 1982).

As a result of symmetry, the net forces acting on the wetted hull of the craft are taken to lie in the fore and aft vertical plane, as shown in *Figure 3.55*, and are as follows; F_h: the hydrostatic force; F_p: the net hydrodynamic force resulting from the variation of pressure over the wetted hull; F_s: the net skin friction force on the wetted hull; T: the thrust produced by a propulsor to maintain the steady forward speed V and W: the all-up weight of the craft. The thrust is taken to act parallel to the keel, but in practice will depend on the type of propulsion unit, the inclination of shafts, etc.

In order to improve resistance at low speeds some longitudinal curvature of the hull is built into the bow region. This means that the resultant integrated hull

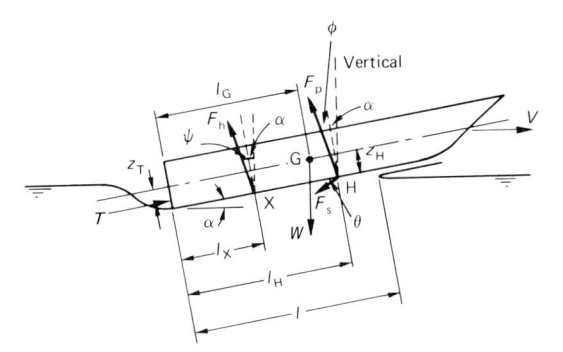

Figure 3.55 Forces on the wetted hull of a planing craft

forces shown in *Figure 3.55* are not necessarily perpendicular or parallel to the keel, as shown by the introduction of the angles ϕ, ψ and θ. These angles are always small and may usually be neglected at high forward speeds.

The equilibrium equations for steady motion are given by Clayton and Bishop (1982) and these are shown to reduce to the well known equations for a displacement ship at low speeds. When the forward speed is very high and significant planing action takes place $F_h \ll F_p$. For efficient planing α is small, $F_s \ll F_p$ and ϕ is small because β varies little over the wetted length. Furthermore, $T \sin \alpha \approx T\alpha \ll F_p$, and the force equations can be simplified to:

$$W = F_p \cos(\alpha + \phi) = F_v \quad (3.139)$$

where F_v represents the hydrodynamic vertical supporting force, and:

$$\begin{aligned} R_T &= T \cos \alpha \approx T \\ &= F_p \sin(\alpha + \phi) + F_s \cos(\alpha + \theta) \\ &= W \tan(\alpha + \phi) + F_s \cos(\alpha + \theta) \end{aligned}$$

Then, as α, θ and ϕ are small:

$$R_T = W\alpha + W\phi + F_s \quad (3.140)$$

Equation 3.140 shows that the total resistance consists of three identifiable components. The component $W\alpha = \alpha F_v$ is called the induced resistance, or induced drag, and results from the inclination of F_p from the vertical owing to the trim angle of the craft. The component $W\phi$ may be identified with wave making and viscous pressure resistance. At high speed and small immersion the wave making resistance is small. The component F_s is the skin friction resistance.

This breakdown of the total resistance illustrates a general principle associated with high speed vehicles. At the high design speed, the wave making resistance is negligible, but the craft experiences induced resistance in contrast to the reverse situation for conventional displacement vehicles at normal operating speeds.

As z_H and z_T are both generally small and nearly equal and with the small angles assumption, the equation for moments about the craft centre of mass G reduces to:

$$l_H \approx l_G \qquad (3.141)$$

when dynamic supporting forces predominate. That is, the hydrodynamic centre H is located on the hull very nearly below G and this result may be used as a design criterion for planing craft.

3.9.3 Estimation of forces

A large body of empirical and experimental data has been accumulated for hard chine hulls based initially on an analysis of the principle variables. We can therefore form the following functional relationships:

$$F_v; F_s; l_H = \text{functions } (V, \alpha, \beta, l, b, g, \rho, \mu) \qquad (3.142)$$

where μ represents the dynamic viscosity of water and g the weight per unit mass associated with the distortion of the interface. The application of dimensional analysis (Massey, 1986) leads to:

$$\frac{F_v}{\frac{1}{2}\rho V^2 b^2}; \frac{F_s}{\frac{1}{2}\rho V^2 b^2}; \frac{l_H}{b} = \text{functions}\left\{\alpha, \beta, \frac{l}{b}, \frac{V}{\sqrt{gb}}, \frac{\rho V l}{\mu}\right\} \qquad (3.143)$$

where β, l and b are considered to be mean values.

Other, mainly geometric, variables need consideration, such as forebody shape, chine shape, longitudinal curvature of the hull and so on. In principle, these are decided upon with regard to cargo, fuel, operating role, propulsion system, etc. When the hull shape has been chosen, values of F_s and l_H are required to calculate R_T and thus T for given values of $W (=F_v)$, β, b, g, ρ and μ at a steady cruise speed V.

Early work by Shoemaker (1934) and Korvin–Kroukovsky et al. (1949) was summarized by Murray (1950) to yield a series of empirical design equations for hard chine hull forms. It was found that the vertical force coefficient C_v was independent of the Reynolds number $\text{Re} = \rho V l/\mu$ and so could be written as:

$$F_v/\tfrac{1}{2}\rho V^2 b = C_v = \text{function } (\alpha, \beta, \lambda, \text{Fr}^*) \qquad (3.144)$$

where $\lambda = l/b$ and Froude number $\text{Fr}^* = V/\sqrt{(gb)}$ and is sometimes called the speed coefficient in this definition. Notice that C_v is not referred to as a lift coefficient since this term is used exclusively for hydrofoils for which, as we shall see later, the generation of supporting forces is distinctly different. Since β and Fr^* are both known at the beginning of the design stage, Equation 3.144 shows C_v as a function of two variables α and λ.

The analysis of data on planing flat plates and planar V-forms then led Murray to the following relationship for zero deadrise:

$$C_{v,0} = \alpha^{1.1} \{0.0120 \lambda^{1/2} + 0.0095 (\lambda/\text{Fr}^*)^2\} \qquad (3.145)$$

where α is measured in degrees. A modification is required to account for deadrise for which the following empirical relationship can be used:

$$C_v = C_{v,0} - 0.0065 \beta (C_{v,0})^{0.60} \qquad (3.146)$$

where β is measured in degrees.

It was also found that l_H/l was independent of both Re and Fr and so:

$$l_H/l = \text{function } (\alpha, \beta, \lambda) \qquad (3.147)$$

Data analysis indicated that the functional relationship could be written in the form:

$$l_H/l = K \lambda^{-n} \qquad (3.148)$$

where $K = (0.084 + 0.015\beta) \alpha^{-m}$, $m = 0.125 + 0.0042\beta$ and $n = 0.05 + 0.01\beta$ again with α and β in degrees. The foregoing expressions are essentially concerned with the very high speed, fully planing condition corresponding to $\text{Fr}^* > 10$. A later analysis was aimed at including predictions of semidisplacement performance in addition to that at very high speeds (Savitsky, 1964; 1985; Savitsky and Gore, 1979). Equations 3.145 and 3.148 are then modified to:

$$C_{v,0} = \alpha^{1.1} \{0.120 \lambda^{1/2} + 0.0055 \lambda^{2.5} (\text{Fr}^*)^{-2}\} \qquad (3.149)$$

$$l_H/l = 0.75 - \{5.21(\text{Fr}^*/\lambda)^2 + 2.39\}^{-1} \qquad (3.150)$$

but Equation 3.146 remains unchanged. Equation 3.149 is accurate for the ranges $0.60 \leq \text{Fr}^* \leq 13.00$; $2° \leq \alpha \leq \alpha \leq 15°$ and $\lambda \leq 4.0$. In Equation 3.149 the first term in the brackets represents the contribution of the dynamic component to the total supporting force and the second term represents the contribution from hull immersion. It is clear that for a given geometry the second term makes a rapidly decreasing contribution as speed increases. The centre of pressure equation (3.150) was derived by separate considerations of the buoyant and dynamic force components centred at one-third and three-quarters of the wetted length, respectively, forward of the transom. Excellent correlation with experimental data was found and so the implications are that α and β have only a minor effect on l_H/l which is thus a function only of λ for a given Fr^*.

The use of the preceding expressions is rather cumbersome, because λ and l are functions of forward speed and so charts have been devised, as shown by Savitsky (1964, 1985). It is, however, evident that C_v reduces with increasing deadrise and varies with the square of the wetted beam, other parameters remaining constant.

The accurate calculation of hydrodynamic frictional resistance is a vital requirement for power estimations but is difficult to assess rigorously (Clayton and Bishop, 1982). Assuming that wave making and longitudinal curvature of the hull are small, the analysis leading to Equation 3.140 gives the total resistance as:

$$R_T = W \tan \alpha + F_s \cos \alpha \qquad (3.151)$$

comprising induced drag, represented by the first term on the right hand side of Equation 3.151 and the frictional component represented by the second term. The skin friction coefficient can be written in the form:

$$C_F^* = F_s/\tfrac{1}{2}\rho V^2 b^2 = C_F(S/b^2) = C_F \lambda \sec \beta \quad (3.152)$$

where the wetted surface area of the hull $S = lb \sec \beta$ and C_F represents the skin friction resistance coefficient based on wetted area. In the past, the Schoenherr relationship for C_F has been used, but the ITTC relation given by Equation 3.50 could be equally effective, although subject to the limitations discussed by Clayton and Bishop (1982), including the use of an additional correlation allowance such as 0.0004.

Equation 3.151 can now be written in the form of a resistance to weight ratio:

$$\begin{aligned} R_T/W &= \tan \alpha + (\tfrac{1}{2}\rho V^2 b^2/W) C_F \lambda \cos \alpha \sec \beta \\ &\approx \alpha + C_F \lambda/C_v \cos \beta \\ &= R_i/W + R_F/W \end{aligned} \quad (3.153)$$

where R_i and R_F represent induced (or pressure) resistance and friction resistance, respectively. The effective power P_E required to drive the craft at its steady operating speed V is then given by $R_T V$. This power must be delivered by the propulsion system having taken account of hydrodynamic and mechanical energy losses.

Figure 3.56 shows typical variations of the resistance components as functions of trim angle and deadrise angle and these indicate that an operating trim angle between 4° and 5° provides optimum conditions.

Clement and Blount (1963) conducted tank tests on the DTMB Series 62 models in order to correlate geometry with performance. A major result of their work is shown in *Figure 3.57* which emphasises the importance of slenderness ratio, written in terms of projected chine length to maximum breadth over chines. The correlation is based on volume Froude number Fr_∇, in which the length parameter is the cube root of the displacement volume of the craft at rest. As speed, and therefore Fr_∇, increase the craft trim increases to the hump value and then decreases steadily. For all slenderness ratios the resistance increases rapidly at low speeds and again at high speeds, e.g. when $Fr_\nabla > 4.0$. The hump resistance is very pronounced for low slenderness ratios and so values of 3.0 to 5.0 are to be preferred, although modern trends have called for values in excess of 5.0, often accepting compromises imposed on or by internal arrangements.

Corresponding data for round bilge hulls have been given by Yeh (1965) for the very slender DTMB Series 64 and by Marwood and Bailey (1969) and Bailey (1974, 1976) for the lower slenderness ratios of the NPL

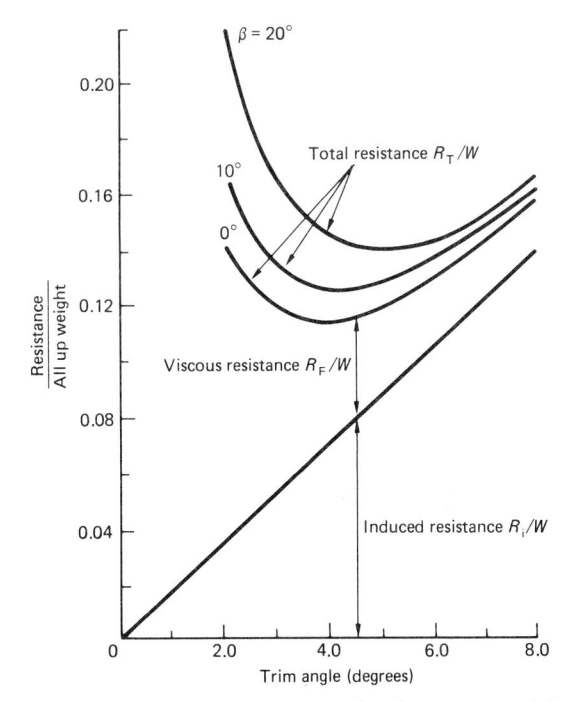

Figure 3.56 Typical variations of resistance to weight ratio for planing hulls

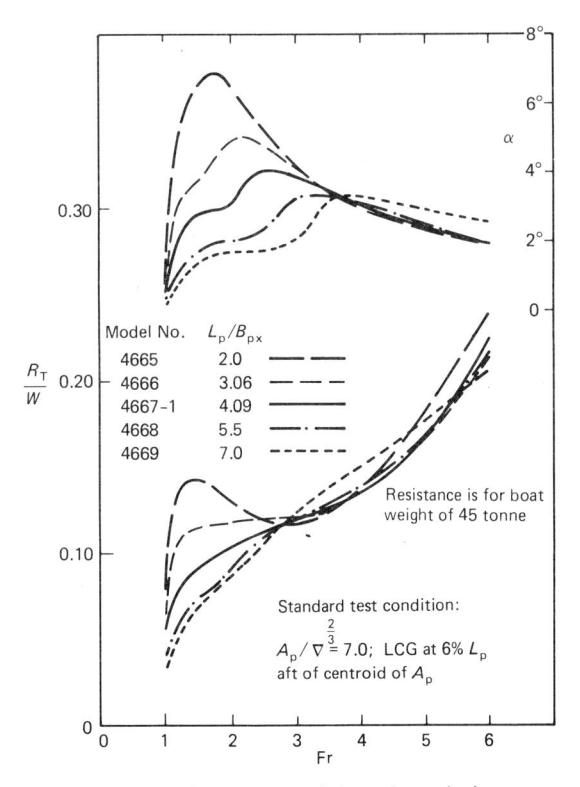

Figure 3.57 Resistance to weight ratio and trim angle α, as functions of volume Froude number Fr_∇, for the DTMB Series 62 models (after Clement and Blount, 1963). Key: L_p, project chine length; B_{px}, maximum breadth over chines; A_p, projected planing bottom area; ∇, immersed volume of hull at rest; LCG, longitudinal centre of gravity location

series. The latter were intended to cover the design of craft in the semidisplacement range of Froude number (based on wetted length), e.g. Fr < 1.2. Under these conditions the overall round bilge characteristics describing resistance and trim angle as functions of Fr_∇ are quite similar to those shown in *Figure 3.57*. It is found that trim angles are slightly less and, at lower speeds, the resistance is somewhat less for the round bilge forms. The addition of a stern wedge has been found to reduce the trim angle (by as much as 50% for a 10° wedge) and, for low slenderness ratios, to reduce the resistance at the hump speed.

Data from the NPL Series and a collection of hard chine results have been compared by Bailey (1974) on a resistance against Fr_∇ basis to deduce the optimum performance range for each type of hull for operation in calm water. *Figure 3.58* shows the recommendations derived from these data in terms of speed and ship length. It is thought that the operating regimes illustrated are likely to hold for hull shapes reasonably close to the NPL and DTMB Series. Clearly, the implication is that round bilge hulls are best for large high speed craft operating at speeds up to about 50 knots. This may go some way to explaining the often demonstrated superiority in speed of the round bilge German E boats in World War II. Nevertheless, North American designers tend to favour the hard chine hull form and contend that the differences in resistance values at low speeds are not particularly important. Indeed, it has been suggested by Hadler *et al.* (1978) that the best compromise may be the double chine, which combines the best characteristics of the two basic hull forms and yet retains the shape which offers cheaper and quicker building programmes.

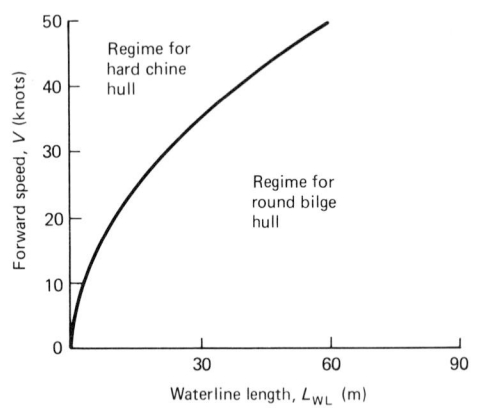

Figure 3.58 Operating regimes for hard chine and round-bilge hull forms

3.9.4 Performance in steady motion

Typical performance curves of a 21 m long craft with length to beam ratio of 4:1 operating in steady, smooth water conditions are shown in *Figure 3.59*. The earlier assumption of small trim angles is seen to hold good and at cruise speeds a value of about 4° is obtained. For this craft, and indeed for most planing craft, at low speeds the midsection falls with respect to the still water level and only begins to rise at speeds above approximately 15 knots. This settling of the hull in the water suggests that downward suction forces are generated on the wetted hull, usually near the stern, which augment the weight. For equilibrium and zero moment about the centre of gravity, the centres of buoyancy and pressure move aft and α increases. As speed increases, the dynamic supporting force increases, the centre of pressure moves forward and hydrostatic forces diminish. From 15 to 20 knots, the transom rises above its still water position with little change of trim. For the craft considered here a maximum trim angle of nearly 6° is reached at about 30 knots. At higher speeds, approaching the cruise value, the midsection rises slowly and α decreases slowly because, from Equations 3.145 and 3.149, $C_{v,0}$ reduces to provide a supporting force equal to the all up weight of the craft.

The R_T/W curve in *Figure 3.59* was based on measured values, the skin friction resistance was determined from plank data, the induced resistance was deduced from an estimate of F_v and measured angles of trim, and the residuary resistance was obtained by

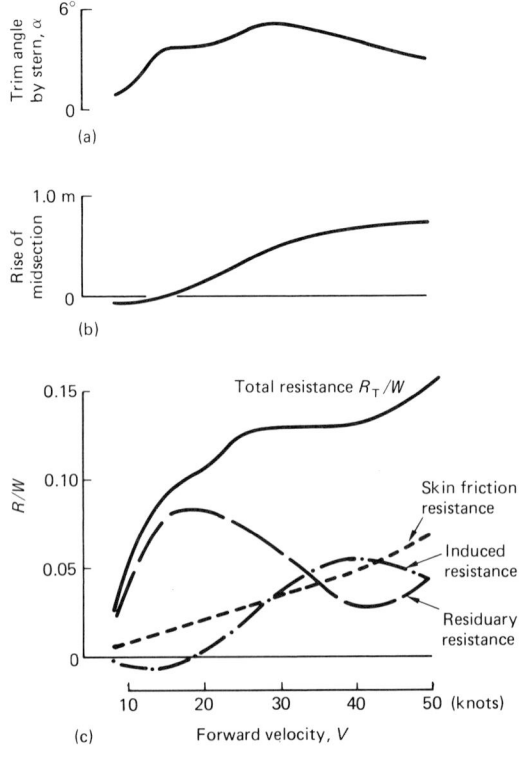

Figure 3.59 Performance of a 21 m planing craft

subtraction. The total resistance increases rapidly at low speeds, generally at a greater rate than for an equivalent size displacement ship, but then remains little changed between 25 and 40 knots before rising steadily at higher speeds. Between 10 and 20 knots, the total resistance arises primarily from surface wave generation. The residuary resistance shows a hump at about 18 knots which corresponds to a Froude number, based on waterline length of the craft at rest, of approximately 0.63. As the speed increases the wave making resistance decreases, but at high speeds viscous pressure resistance increases so rapidly that it forms the major component of the residuary resistance. As F_v is negative at low speeds, the induced resistance is then also negative but becomes increasingly positive as V rises. Above 40 knots, F_v changes little but α is seen to decrease, thus causing the product $αF_v$ to pass through a maximum and then decrease gradually. Finally, skin friction resistance increases steadily throughout the speed range since the V^2 term is always greater than the product of the wetted surface area and the skin friction resistance coefficient. For the particular craft examined here, the skin friction resistance becomes the largest component of the total resistance above 40 knots. However, the rate at which skin friction resistance increases is substantially less than that for a displacement vehicle because the wetted surface area decreases with increasing speed.

The curves in *Figure 3.59* allows an assessment of the definitions of the onset of the fully planing condition. It has been considered by, for example Newton (1962), that fully planing corresponds to the flat portion of the resistance curve when trim angle reaches a maximum, i.e. when the speed is about 28 knots. Others, such as Clement and Pope (1961) consider the fully planing condition to commence when $Fr_\nabla = 3.5$, in the present case when the speed is 40 knots. The analysis discussed in Section 3.9.3 assumed that the hydrostatic forces were essentially zero but this cannot be so, as some part of the hull is submerged and so water is displaced. Calculations on the data of the curves in *Figure 3.59* shows that $F_v = 0.5W$ at 30 knots and $F_v = 0.75W$ at 40–50 knots. Thus, even at high speeds, a significant supporting force arises from immersion of the hull in the water and similar results from Du Cane (1974) confirm this behaviour. Nevertheless, calculations based on the methods outlined earlier have been used to produce many successful series of planing craft.

3.9.5 Performance in waves

In most cases of commercial and naval exploitation of high speed planing or semiplaning craft, operation in a seaway is inevitable. Hulls must therefore be designed not only to withstand wave impact, but also to minimize the resultant acceleration (Payne, 1988). The dependence of hull acceleration at, say, the centre of mass of the craft is not simply related to wave height and so

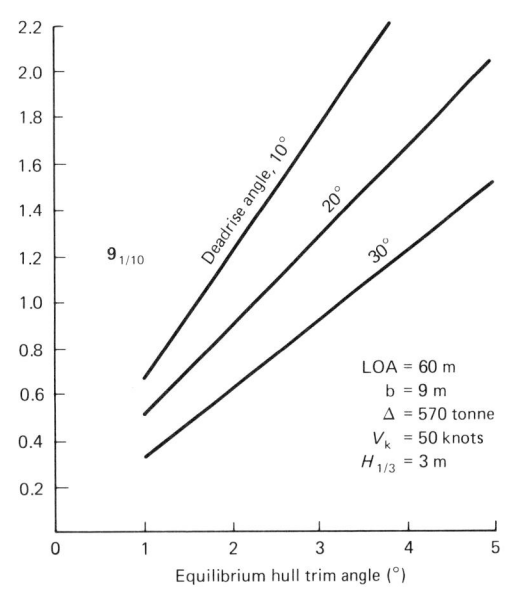

Figure 3.60 Acceleration of planing craft at centre of gravity in head seas (after Savitsky and Brown, 1976)

data from model tests are generally used. One such set has been produced and analysed by Savitsky and Brown (1976) using a Pierson–Moskowitz wave spectrum. Typical results from this work are shown in *Figure 3.60* in which the symbol H represents the significant wave height, that is, the average of the one-third highest waves. The acceleration scale represents proportions of gravitational acceleration corresponding to the average one-tenth highest acceleration which is 3.3 times the overall average acceleration. It is found that accelerations increase linearly with trim angle and may therefore be easily controlled with trim flaps or wedges, decrease linearly with deadrise angle and increase linearly with wave height and with the square of the craft speed. Accelerations are independent of length alone, are inversely proportional to craft mass and are proportional to the square of the beam. Thus narrow beam hulls, or hulls with large length to beam ratio, should yield significantly lower accelerations that broader beam craft. Substantial improvements in acceleration responses, without forsaking roll stability at lower speed operation, have been obtained with a double chine craft (Blount and Hankley, 1976).

A combination of these features leading to a long craft with a narrow beam, high deadrise and low trim angle would certainly lead to low wave impact accelerations but, unfortunately, high hydrodynamic resistance and reduced internal volume. As in all design, a compromise must be sought between the demands of numerous requirements, in this case motions, comfort, power dissipation, speed and application. The modern trend is, however, towards hulls of moderate to high deadrise and beam loadings. As a result, centre of mass acceleration levels at comparable significant wave heights, relative to the cube root of volume displacements, are at least half those of earlier designs. There is also a speed loss in waves that must generally be accepted which could be up to 20% of the design speed. Such losses tend to be more sensitive to wave height than craft speed.

Craft motions in waves are more of a problem when the hull is in a displacement mode because the wave encounter frequency may correspond to pitch and/or heave resonance. At cruise speeds, e.g. a 60 m craft running at 35 knots, the motions may well reduce by 50% resulting in a typical pitch angle amplitude of 3°. Similarly, roll motions are most severe in the preplaning speed range and the use of active roll–stabilizing systems have been investigated, as for other high speed craft. Tests have shown (Savitsky, 1985) that at low speeds roll amplitudes can be reduced by a factor of 2 and this can improve dramatically the habitability aspects for, say, crews on long duration patrols and the stability of military systems platforms.

Directional stability and control has not seen much investigation and few guide lines exist. Active rudder control can be used to retain dynamic stability, but this method is wasteful of power, or an increase in skeg area may be necessary with a quite small power increase. Cavitation presents a problem with active rudders at high speeds owing to the low pressures developed on the rudder surface as the angle of incidence is increased during fast manoeuvres. The problem can be reduced somewhat by careful selection of blade shapes and the often accompanying phenomenon of ventilation can be eased by incorporating chordwise fences on the stern mounted rudders. Care is also needed when countering the unbalanced torque of the propulsion system because the usual roll–yaw moment coupling could give dangerous heel angles. An interesting example of this type of problem is documented by Codega and Lewis (1987) related to the operation of a US Coast Guard class of high speed surf rescue boats. These 9 m long boats operate in breaking surf up to 3 m high and are capable of speeds up to 30 knots. Within the craft's normal operating envelope, it was found that a dynamic instability was manifested in which the craft trimmed by the bow, rolled to a large portside heel angle and broached violently to starboard, all within 5 s. The essential mechanism results from an unfavourable bottom pressure distribution exacerbated by the longitudinal centre of gravity being too far forward. Millward et al. (1984) have shown that transverse instability of high speed, round bilge craft with appreciable deadrise occurs when driven fast and similarly for hard chine hull shapes.

3.9.6 Structural design

Structural design of hulls depends mainly on wave impact loadings which can be large and localized. Maximum impact pressure generally occurs between 0.3 and 0.5 L aft of the bow (where L represents the length of the craft at the waterline when at rest) and this maximum value reduces by approximately 50% in the bow region and 75% in the stern region (Heller and Jasper, 1961; Allen and Jones, 1978). The materials chosen for the hull to withstand loads (both wave and internal machinery, etc.) usually comprise either aluminium, fibre reinforced plastic, steel or wood. In the main, structural design is reasonably straightforward but the limitations of the material properties must be taken into account. The most important aspect is the trade off between cost and a complex design resulting from local load analysis leading to small panels of thin plating consistent with reducing weight. Sharples (1974) has shown that from a comparison of structural weight *versus* overall length, steel hulls are significantly heavier than aluminium hulls, but that steel produces cheaper forms and greater fire resistance. *Figure 3.61* shows the results of successful craft in relation to speed, size and hull materials as given by Savitsky (1985). The prediction of local loadings using advanced analytical techniques and computer simulations will lead to the more effective use of composite materials resulting in weight reductions at reasonable cost.

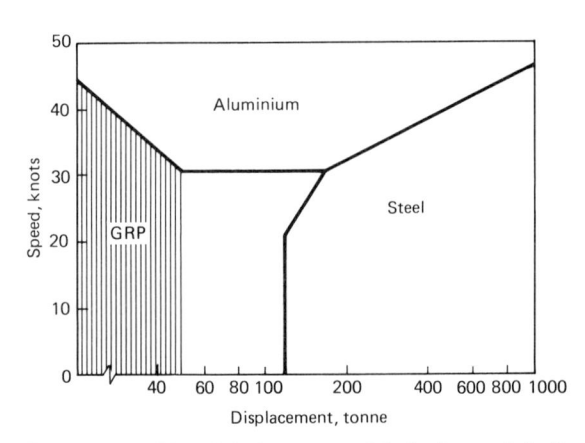

Figure 3.61 Materials for successful planing craft hulls (after Savitsky, 1985)

Survivability is a vital aspect of craft design, just as it is for conventional craft. Operational considerations such as power, speed, range and payload for small size hulls means that the endurance to remain afloat is often limited following predictable disasters such as extreme rough seas, fire damage, collision, hostile action, explosions, etc. The addition of heavy ballistic armour, watertight and fire resistant bulkheads and the like, often leads to unacceptable weight penalties. Compartmentation is generally used to meet stability criteria so that, for example, the floodable length is great enough to allow any two adjacent compartments to be flooded without the craft foundering and without hull trim or heel angles exceeding defined limits.

3.9.6.1 *Aluminium*

Aluminium is ideal for light, fabricated construction, is corrosion resistant, easily formed and joined and can be used to produce complex hull forms. However, it has a low melting point and so is susceptible to extensive fire damage. For the same reason it is not easy to weld, although improvements in pulsed arc, high frequency and electron beam techniques are leading to improvements in weld quality of thin sheets.

Traditional methods of non-destructive evaluation (NDE) consist of X-ray or dye penetrant techniques in order to detect defects such as cracks, porosity and so on. Boats constructed from aluminium generally have a longitudinal framing system which results in a strong, faired hull at an acceptable cost.

During the shell plating stage, hull construction usually takes place in the upside down position so that the plating can be wrapped round the transverse and longitudinal framing. A variety of jigs may be used to support the hull during the fabrication in which subassemblies such as deckhouses, engine foundations, transoms, tanks, etc., may be constructed in other parts of the yard and brought together at the assembly point, thus effecting substantial economies for batch production.

3.9.6.2 *Fibreglass reinforced plastic (FRP)*

For the smaller boats, fibreglass reinforced plastic (FRP) is the most popular material in use, as suggested by *Figure 3.61*. With this material there is complete resistance to corrosion, high durability and relatively low weight and cost. Care has to be taken to provide good environmental control, governed by legally enforced ventilation regulations owing to the use of noxious styrene monomers in plastic resins. The most common materials used are as follows.

3.9.6.2.1 Reinforcement Random orientated, chopped glass fibres rolled into a mat and glass fibre roving in the form of a coarse woven material using bundles of fibre strands are most commonly used. It is also possible to use non-woven reinforcement materials that require less resin fill. Sometimes, the material is unidirectional to satisfy strength requirements only in that direction. Fibres other than glass may be used; (i) aramid (Kevlar) fibres, which have much greater tensile strength than glass fibres, are lighter, have slightly greater flexural strength but lower compressive strength than FRP, (ii) carbon fibres, which are even lighter than Kevlar for the same strength. Both of these fibres and special high strength glass fibres are costly and when used are often confined to that part of the structure under high stress or where cost is not the overriding consideration such as in the design of recreational racing craft.

3.9.6.2.2 Laminating resin Isothalic and orthothalic polyester resins prepregnated with copper napthanate and catalysed with methyl ethyl ketone peroxide are most commonly used. Resins that are non-fire retardant are used for commercial applications although more expensive fire retardant resins are usually specified for naval and offshore protection/patrol craft. Vinylester laminating resins retain strength for longer in water and are attractive for use on the underwater structure, but are rather more than twice the cost of polyester resins.

3.9.6.2.3 Core material Most commonly used materials are polyurethane foam, polyvinyl chloride foam, balsa wood and plywood.

3.9.6.2.4 Techniques for FRP construction A variety of techniques are used for FRP construction of fast craft (Savitsky, 1985; Allan and Raybould, 1988; Marchant and Pinzelli, 1988). The most used method is hand layup in which skilled operators apply and saturate with resin, layers of fibreglass material (random chopped or layered woven) to a pregel coated open female mould, maintaining constant thickness throughout. Most recently, the development of a presaturated reinforcement material allows unlimited working time, but subsequent curing procedures in ovens requires costly facilities, including those for storing the material. Even for the conventional method, great care is required in the curing process, often most satisfactorily achieved in the exothermic chemical reaction of a moderately multilayered laminate, if shrinkage distortion or stress concentrations are to be avoided.

3.9.6.3 *High tensile steel*

High tensile steel has a strength to weight ratio similar to marine alloys, but is usually available to a minimum gauge thickness which may be unnecessarily large for small craft. Nevertheless, high speed planing hulls with weight displacements of at least 5 MN could prove attractive if made from high tensile steel. For smaller craft, with speeds less than about 35 knots, mild steel is used and, although heavier, steel hulls are cheaper than aluminium. The general techniques for steel hull construction follow similar procedures to those used for conventional ship forms.

3.9.7 Applications

High speed, monohull craft can be found worldwide in a large number of roles within the military, security and commercial sectors. Much of the world's naval expenditure is on marine vehicles that can cross oceans. For close shore work or in sheltered waters the small, potently armed (in both offensive and defensive senses) fast planing craft have found applications with many navies. The US Navy, for example, has used these craft for penetrating hostile coastal or river areas to deliver advance troops and to supply small calibre weapon cover. Semi-inflatable craft for use as ship's boats have also been considered but, as pointed out by Savitsky (1985), there is no clear cut future significant role laid out for planing craft in the US Navy. This is not so for many other navies, especially those of smaller nations, operating with tighter budgets which cannot cover the escalating costs of conventional shipbuilding. It is now estimated that approaching 3000 craft under 60 m overall length are now in service with world navies, most of which are in the 25–45 knot speed range.

The role of the US Coast Guard lies between the naval and commercial as it has well defined responsibilities in maritime safety, search and rescue, aids to navigation, environmental protection and law enforcement in peacetime. A substantial proportion of mission time is involved with apprehending smugglers, especially with cargoes of drugs, and illegal immigrants (Savitsky, 1985).

The commercial sector is dominated by recreational craft used as runabouts, fishing boats, cruising boats and sports racing boats which can develop speeds in excess of 80 knots. These boats vary in length from 5 to 30 m and tend to be of lightweight form as they have a less rigorous existence than work boats. It is to be expected that demands from the recreational sector will expand further, especially as techniques are developed to reduce seakeeping responses. The roles of work boats cover offshore oil rig crew transportation, commercial fishing boats, pilot and fire boats, hydrographic survey boats, etc., again within the 5–30 m length range. Patrol boats cover policing roles, especially in harbours, fisheries enforcement, customs, etc. In many cases, however, the craft used for such duties are adaptations of designs used for recreational boats and are not of the type suitable for employing the latest technology in the detect–search–apprehend scenario.

A further stimulus for a rapid growth of fast craft in modest naval programmes has been the institution of the 200 mile (320 km) territorial limit (the exclusive economic zone) imposing national jurisdiction of about 10% of the world's oceans since 1977. With the adoption of the EEZ by most maritime nations, many of whom are expanding the use of small craft as a major component of their navies, the design and building of such craft could become a major activity in naval architecture. Since estimates have shown that some 90% of living and natural resources dwell within the 200 mile limit, there could be some 500 or more additional patrol craft needed to cover operations well suited to planing craft. Conflicts in the Straits of Hormuz and the Persian Gulf during 1988 illustrated the potency of small, missile armed planing craft backed by sophisticated, high intensity, accurate gunfire. As Dorey (1977) has illustrated, the now feasible miniaturization of electronic warfare systems familiar to the weapon outfit of warships is now becoming available to the larger planing craft without compromising the advantage of high speed.

3.10 Hydrofoil craft

Unlike the main hull of a planing craft that of the hydrofoil boat is completely clear of the water at high speeds. Hydrodynamic supporting forces are developed on fully or partially submerged extensions of appendages fixed to the main hull. The supporting forces result from a suitable distribution of pressure on the wetted surfaces (the hydrofoils) once the craft is in motion. *Figure 3.62* shows two typical configurations of hydrofoil craft. The foils, attached by struts to the hull, may either pierce the water surface, or be totally submerged, or be a combination of both. Further, the main load bearing foils, which generally carry at least 65% of the total load, may either be well aft (the canard configuration), or well forward (the conventional or aeroplane configuration). In some examples, the total supporting force is divided almost equally between the foils and then the configuration is referred to as tandem. Although not shown in *Figure 3.62*, the foils may be arranged in forward and aft pairs for the fully submerged arrangement.

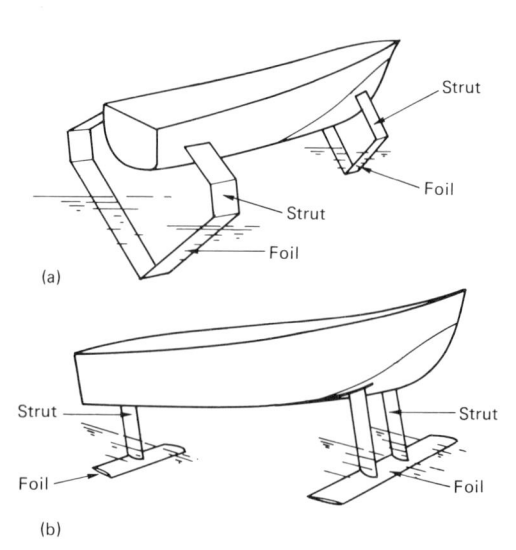

Figure 3.62 Hydrofoil craft geometry: (a) Surface piercing foils in canard configuration; (b) Fully submerged foils in conventional aeroplane configuration

The performance of a deeply submerged hydrofoil is equivalent to that of a geometrically similar, isolated aerofoil operating under dynamically similar conditions. However, at high speeds the behaviour of a hydrofoil raises problems which can only be solved with reference to specially designed lifting sections. It is worth pointing out here that although the hull itself is clear of the water the struts, foils and often the propulsor are submerged. These will, therefore, contribute a small buoyancy force to augment the hydrodynamic lift force supporting the craft.

3.10.1 Generation of lift forces

The lift force on a hydrofoil (or aerofoil) is defined as that component of the resultant hydrodynamic force which is perpendicular to the direction of the oncoming mainstream velocity vector. It may be noted that the lift force may be upwards or downwards, although for a supporting force in horizontal motion we would require an upward vertical lift force. The mechanism by which a lift force is generated on a hydrofoil is fundamentally different from that used by a planing hull to produce a supporting force F_v, as described by Clayton and Bishop (1982).

To proceed, let us consider the system in *Figure 3.63(a)* and invoke the idea of a uniform, irrotational (i.e. inviscid) flow of an incompressible fluid past a cylinder of circular cross-section and set normal to the oncoming stream. If the cylinder is of infinite span the flow is two dimensional and symmetric about both the Ox and Oy axes so that no net force exists on the cylinder.

Suppose, now, that a circulatory flow or vortex can be imposed round the cylinder in a clockwise direction, conventionally regarded as negative (see *Figure 3.63(b)*), so that the symmetry of the flow is disturbed about the x axis. The fluid flows more rapidly around the upper half of the cylinder compared with the lower half yielding low pressures above and high pressures below. The resulting pressure difference generates a transverse force (the Magnus Effect) in the Oy direction, i.e. a lift force. Also present are two locations, denoted S_1 and S_2 and called stagnation points, where the fluid is brought to rest. Mathematically, the flow pattern can be described by a stream function ψ, constant values of which represent the geometry of streamlines or, under steady conditions, flow lines. The combination of two dimensional, uniform flow past a doublet, which yields uniform flow past a circular cylinder, with a circulation (or vortex strength) Γ yields the stream function:

$$\psi = -V(r - a^2/r)\sin\theta - \Gamma\{\ln(r/r_0)\}/2\pi r \quad (3.154)$$

in which the symbols are defined in *Figure 3.63* and r_0 is a constant. It may then be shown that the lift force per unit span L' in a fluid of density ρ is given by:

$$L' = -\rho\Gamma V \quad (3.155)$$

and is referred to as the Kutta–Joukowski Law, after the two initial investigators, the latter having developed the same relationship for bodies of arbitrary cross-section. In Equation 3.155, Γ is negative and so L' is positive, i.e. an upward lift force along Oy is developed on the cylinder.

The question now to be posed is: since the flow just described has a simple mathematical form, is it possible to use this feature in developing body shapes that deform the flow to yield a lift force that does not require separate generation of circulation? The answer is, of course, yes, but this requires some explanation.

The first point to recognize is that a rotational flow cannot be induced by an inviscid fluid because no shear stresses are present, by definition. In practice, effects somewhat similar to those developed by the irrotational vortex can be obtained by rotating a circular section cylinder in a uniform viscous stream. A cross force is generated and use has been made of it in the form of vertical, rotating cylinders referred to as Flettner rotors, after Anton Flettner who used the devices attached to the deck of a ship as a means of assisting propulsion across the Atlantic Ocean (Clayton and Satchwell, 1987). There is, however, an additional viscous drag force not present in the inviscid flow model. It is clearly quite impractical to use such rotating devices as aircraft wings or as the lifting surfaces of hydrofoil craft and so alternatives must be sought.

It was Joukowski who originally showed that the flow pattern depicted in *Figures 3.63(a)* and *(b)* could be transformed to a more useful mathematically related shape which would also yield a lift force. In practice, several transformations are needed, as shown by Vallentine (1969), to obtain a hydrofoil shape which may be used to generate a lift force. The centre of the original circle is moved a small distance from the origin of coordinates, since this allows the shape to have thickness and camber (i.e. curvature about the mean line) and the oncoming flow must be rotated through α to provide an angle of incidence. The result may then look similar to that in *Figure 3.64(a)* in which the

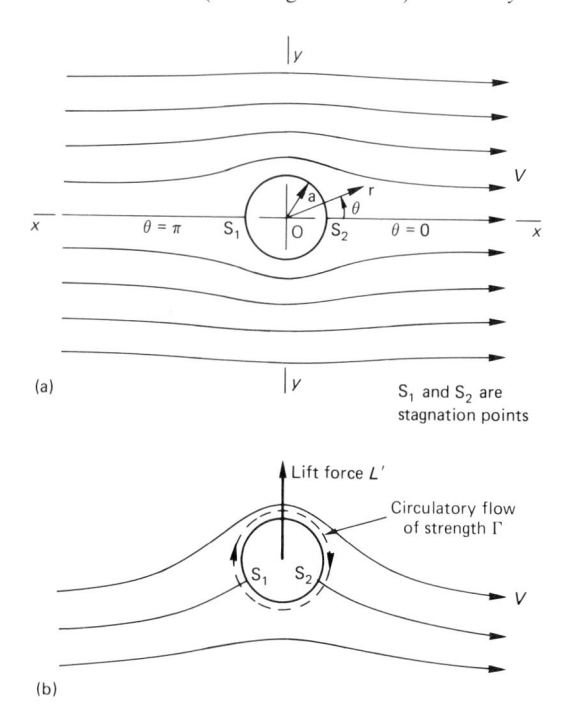

Figure 3.63 Inviscid flow about a cylinder of circular cross-section in a uniform stream: (a) no circulation and (b) with circulation

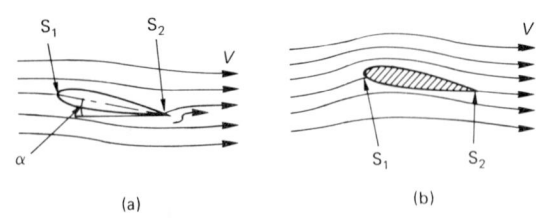

Figure 3.64 Inviscid flow about a hydrofoil

stagnation points S_1 and S_2 correspond in the transformed flow to those in *Figure 3.63(b)*. A sharp, often cusped, trailing edge is obtained from the transformation, but the required instantaneous change in flow direction round the trailing edge towards S_2 (*Figure 3.64(a)*) cannot be sustained by an inviscid fluid, even less likely by a viscous flow the energy of which is continuously reduced by shear stresses at the under surface of the hydrofoil. Consequently, the only stable position of S_2 is at the trailing edge (*Figure 3.64(b)*), and this requires another transformation. The value of the circulation necessary to fix S_2 at the trailing edge, with a corresponding movement of S_1 near the leading edge, is given by:

$$\Gamma = 4\pi V \sin(\alpha + \beta) \qquad (3.156)$$

The section lift force coefficient is then obtained from Equations 3.155 and 3.156 as

$$C_1 = L'/\tfrac{1}{2}\rho V^2 c = 2\pi \sin(\alpha + \beta) \qquad (3.157)$$

where the straight line joining the leading and trailing edges of the profile, the chord length c, is approximately $4a$ for a thin aerofoil section. The angle β is a measure of camber, that is, the departure of the generally curved mean line from the straight chord line. For some sections, such as those used for rudders, the mean line coincides with the chord line, thus $\beta = 0$; $C_1 = 0$ when $\alpha = 0$; and C_1 varies symmetrically with $\pm\alpha$. For cambered foils the lift coefficient in Equation 3.157 equals $2\pi \sin \beta$ when $\alpha = 0$. It may also be observed that the expressions for C_1 are independent of thickness of the aerofoil, provided that the section is thin.

The final point to be made in relation to the generation of a lift force on a hydrofoil requires the viscous property of a fluid for an explanation. As S_2 moves to the trailing edge (see *Figure 3.64(b)*), the presence of shear stresses between the main flow and the flow in the reverse direction towards S_2 gives rise to an eddy, called the starting vortex, which will have an anticlockwise rotation. The strength of this vortex must be equal and opposite to the circulation about the hydrofoil, called the bound vortex, to satisfy the Kelvin Theorem of constancy of net circulation, which must be zero before motion commences.

It should be noted that for real fluid flows over hydrofoils, a boundary layer of shear flow is present over the complete surface of the body and this region of retarded flow leads to the formation of a turbulent wake, which masks the identification of the rear stagnation point. Even so, predictions of lift force from the foregoing theory are remarkably accurate, in practice, for thin hydrofoils of modest camber set at angles of incidence in the range $\pm 8°$ and operating at Reynolds numbers based on chord length greater than 5×10^6. Under these conditions, corresponding to small values of both α and β the section lift coefficient is predicted to be:

$$C_1 \approx 2\pi(\alpha + \beta) \qquad (3.158)$$

and the slope of the lift curve is then

$$\begin{aligned}\partial C_1/\partial\alpha &\approx 2\pi \quad \text{(radian measure)} \\ &\approx 0.11 \quad \text{(degree measure)}\end{aligned} \qquad (3.159)$$

Variations of pressure at the surface of a typical hydrofoil section are shown in *Figure 3.65* in which St denotes the forward stagnation point and C_p a dimensionless parameter representing the difference between surface static pressure p and reference upstream static pressure p_∞. The enclosed area of the total curve is proportional to the force perpendicular to the chord line which, for small α, is very nearly equal to the lift force (perpendicular to the direction of V). Notice the important fact that a large, often the major, component of area arises from negative, or suction, pressures on one (in this case the upper) surface of the hydrofoil section. This contrasts with the planing craft hull over which the hydrodynamically generated pressure is always positive, as shown in *Figure 3.51*. For small angles of incidence and thin boundary layers at high Reynolds numbers, the effects of viscosity on the inviscid flow pressure distribution are quite small. As a result, the lift force in real flows is only slightly less than predicted

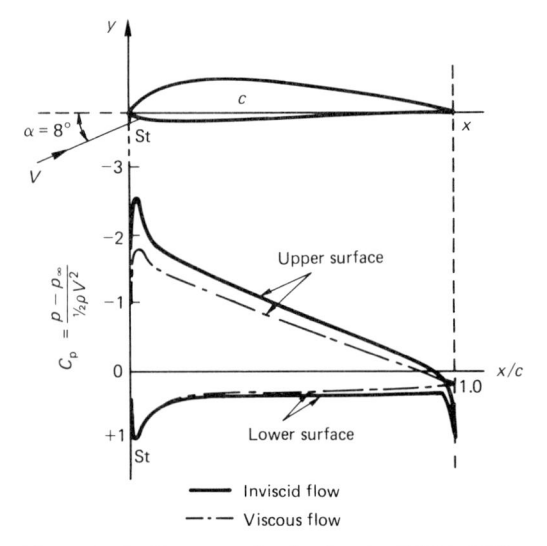

Figure 3.65 Pressure distribution for NACA 4412 hydrofoil section

by inviscid flow theory. The retarded flow in the boundary layer has insufficient energy to negotiate the adverse pressure gradient on the upper surface (i.e. gradually reducing negative C_p) and thus separates from the surface as the trailing edge is approached. A turbulent wake flow is thus developed and is shed into the flow which passes downstream. The rear stagnation point does not exist and as full pressure recovery to $C_p = +1$ does not take place a net force along the chord line remains. This force, which may incidentally be negative or positive, is supplemented by the integrated shear stress distribution over the aerofoil surface to produce a retarding (i.e. in the Ox direction of *Figure 3.65*) force nearly equal to the drag force for small α. It is clearly desirable to design hydrofoil sections that have small drag, or rather to maximize the lift to drag ratio. *Figure 3.66* shows the effects of hydrofoil shape and angle of incidence on the pressure distribution. The potentially disastrous effects of boundary layer separation can be seen, leading to complete stall and loss of lift. For thin sections, a bubble may form at the leading edge which can subsequently burst with equally dramatic effects.

A second point to note in *Figures 3.65* and *3.66* is the presence of a large negative C_p corresponding to a large negative surface pressure. If this pressure, expressed as an absolute value, should become too low relative to the local static pressure, the vapour pressure of the water may be reached and vapourization takes place. Bubbles of vapour then disrupt the flow on the upper surface, probably initiating boundary layer separation, as they flow downstream and collapse in higher pressure regions. Vibration and noise occur during this process of cavitation as well as a loss of lift force and, even worse, the violent collapse of bubbles may be so close to the hydrofoil surface that the combination of direct water impact and shock wave impact causes serious material damage. The extent of the material torn from the surfaces of propellers, rudders, stabilizers, etc., can often be seen on ships in dry dock or on a slipway. The presence of the boundary layer does, in fact, reduce the magnitude of the peak suction pressure as *Figures 3.65* and *3.66* show, but the phenomenon of cavitation is a real and ever present threat to the performance of lift generating surfaces operating in water.

The actual performance characteristics of hydrofoil sections are usually measured by some form of strain gauged force and moment balance attached to the walls of a water tunnel/channel or wind tunnel. There are a number of compendia showing data (Abbott and von Doenhoff, 1959; Riegels, 1961; Hoerner, 1965; Hoerner and Borst, 1975) and many reports on specific aero/hydrofoil sections. A typical set of data is shown in *Figure 3.67* in which the section coefficients are defined, in addition to Equation 3.157, as follows:

$$C_d = (\text{drag force/span}, D')^1/_2\rho V^2 c \qquad (3.160)$$

$$C_m = \frac{(\text{moment about quarter chord point/span})}{^1/_2\rho V^2 c^2} \qquad (3.161)$$

It may be shown (Abbott and von Doenhoff, 1959) that for thin hydrofoil theory the hydrodynamic force for small α and camber passes through the quarter chord point, and so for real flows with thin boundary layers,

——— Inviscid flow   – – – – Experiment

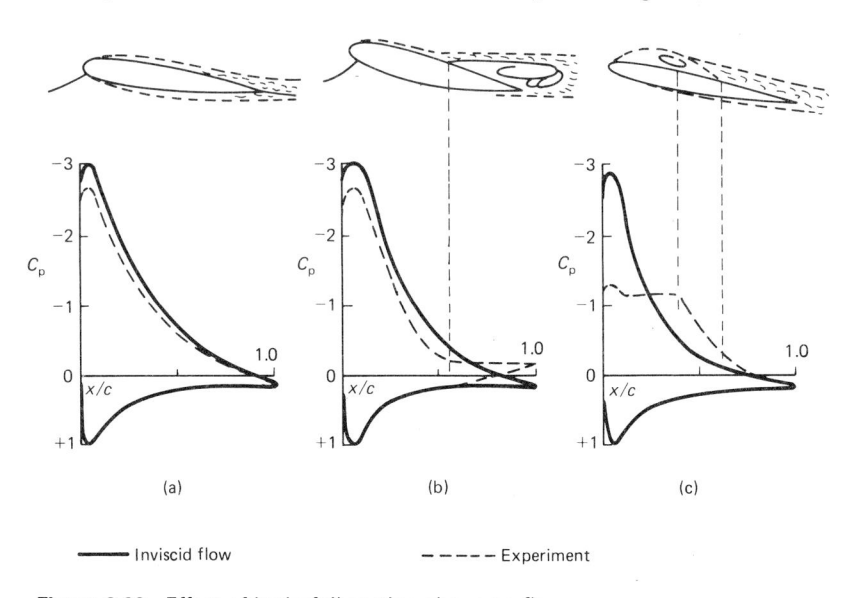

Figure 3.66 Effect of hydrofoil section shape on flow separation and pressure distribution: (a) unseparated flow; (b) rear separation; (c) leading edge separation and long bubble

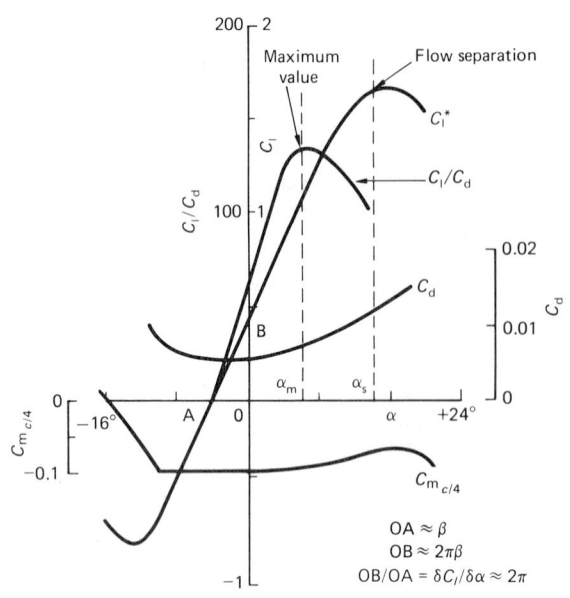

Figure 3.67 Typical data set for performance characteristics of hydrofoil sections

$OA \approx \beta$
$OB \approx 2\pi\beta$
$OB/OA = \delta C_l/\delta\alpha \approx 2\pi$

the moment measured about the quarter chord point should be small, negative (clockwise) and broadly constant.

3.10.2 Lifting surfaces on hydrofoil craft

In practice, hydrofoils cannot be of infinite span and do not usually extend much beyond the beam of the craft. The bound vortex about the hydrofoil does not end abruptly at the tips of the foil, a vortex cannot end in a fluid, only at a solid surface, but joins with vortices generated at the tips and which trail downstream. *Figures 3.65* and *3.66* show that for all angles of incidence other than that corresponding to zero lift a pressure difference occurs between the lower and upper surfaces of the hydrofoil. Flow near the ends takes place round the tips from the high to low pressure regions. Indeed, as shown in *Figure 3.68*, similar vortices are generated along each half span at the trailing edge which subsequently roll up to form a stable pair of contrarotating trailing vortices. These, in principle, join up with the starting vortex to form an interconnected set of vortices, referred to collectively as a horseshoe vortex.

At the tips, the pressure difference between the surface is zero and so at these locations the circulation and lift force must also be zero. The lift coefficient Cl therefore varies along the span and so the system can be modelled as a superposition of a series of bound vortices of different strengths, but symmetric about midspan, each of which generates a pair of trailing edge vortices. This is the classical lifting line theory. Pro-

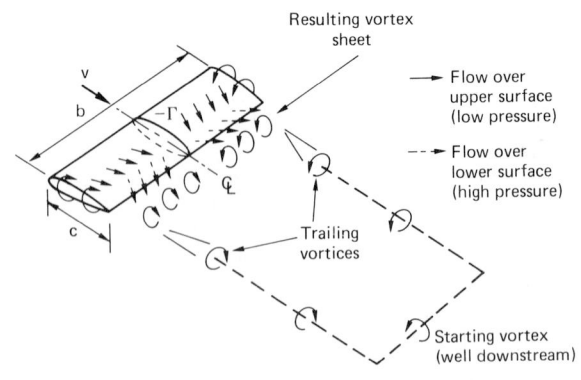

Figure 3.68 Horseshoe vortex system for finite span hydrofoil

vided that the aspect ratio $AR > 4$, it can be shown that the variation of circulation and lift force with span is semi-elliptical, with the maximum values at the midspan location.

The main effect of the tip and trailing edge vortex system is to induce downwash, i.e. a downward velocity v_i, onto the aerofoil. When the downwash is combined with the approach flow, the nominal angle of incidence α is reduced by α_i to the effective angle of incidence α_o, as indicated in *Figure 3.69*. The useful lift force L (total for the hydrofoil) corresponds to α and the main stream velocity V. The hydrofoil, however,

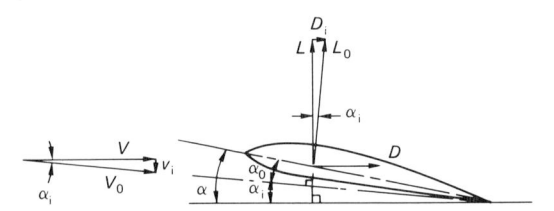

Figure 3.69 Generation of downwash velocity v_i, and induced drag D_i, on a finite-span hydrofoil

generates a lift force L_o in response to the local velocity V_o (the vector sum of V and v_i the latter constant along the span for elliptic variation of L). It may then be shown (Clayton and Bishop, 1982) that:

$$D_i/\tfrac{1}{2}\rho V^2 S = C_{D,i} = C_L^2/\pi\, AR \qquad (3.162)$$
$$\alpha_i \approx C_L/\pi\, AR \text{ (small angles; radian measure)} \qquad (3.163)$$

$$D/\tfrac{1}{2}\rho V^2 S = C_D = C_{D,0} + C_L^2/\pi\, AR = C_{D,0} + C_{D,i} \qquad (3.164)$$

where $C_{D,i}$ is the induced drag coefficient and $C_{D,0}$ is the drag coefficient of the hydrofoil when the angle of incidence corresponds to zero lift force. The slope of the lift *versus* incidence curve for small angles is given by:

$$(a_l)_{AR} = (a_l)_{AR=\infty}\{(AR)/(2 + AR)\} \qquad (3.165)$$

$$\approx 2\pi\,(AR)/(2 + AR) \quad \text{(radian measure)} \quad (3.166)$$
$$\approx 0.11\,(AR)/(2 + AR) \quad \text{(degree measure)} \quad (3.167)$$

from Equations 3.129

For the same nominal angles of incidence $C_D > C_d$ except at the angle corresponding to $C_L = 0$ and then $C_D = C_d$. Evidently, the slope of the lift curve decreases as aspect ratio decreases, as illustrated in *Figure 3.70*, and so it might be expected that foils of large aspect ratio are the more efficient lifting surfaces.

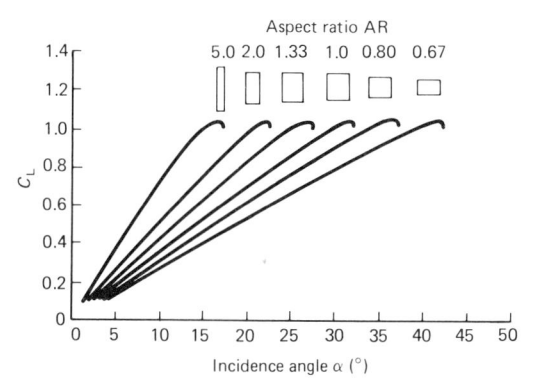

Figure 3.70 Effect of aspect ratio on the lift coefficient of a typical finite span, symmetric section hydrofoil

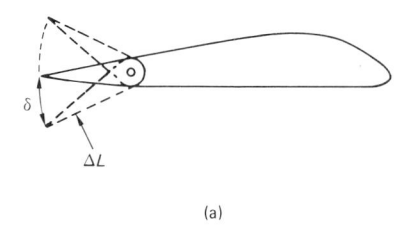

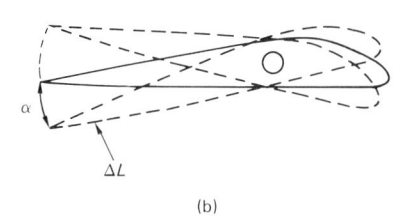

Figure 3.71 Control of hydrodynamic force: (a) trailing edge flap control and (b) variable incidence angle control

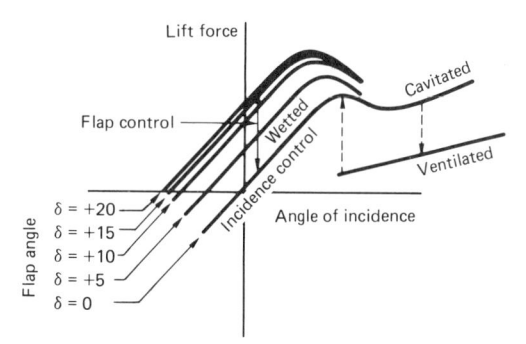

Figure 3.72 Performance of hydrofoils with lift force control

Unfortunately, there is considerable difficulty controlling and adjusting hydrofoil settings when small changes in α produce large changes in C_L. To satisfy both the limitations of strength and the susceptibility of damage to foils of large span, aspect ratios of between 3 and 4 have been adopted for practical hydrofoil craft. In order to attain a high lift force for a limited span, the plan area of the foil can be increased with sweepback accompanied by a reduction of chord along the span (taper) to reduce local bending moments. Under high loading conditions, the finite foil may suffer from substantial boundary layer separation emanating from either the junction between the strut and the foil or from the tips, depending on the distribution of spanwise lift force. Trailing edge flap adjustment or variable incidence angle adjustment of the complete foil are the usual methods of changing lift force on a given foil system for the purposes of craft control, compensation for variable payload, take off settings, etc. These techniques are illustrated in *Figure 3.71* and the effects on the lift curve are shown in *Figure 3.72*. The potentially disastrous effects of cavitation and ventilation will be discussed in the next section.

The use of flaps or incidence angle adjustment can be illustrated in terms of controlling the craft orientation and motion. *Figure 3.73* shows the forces acting during either a flat turn or a banked turn; both options are available to craft with fully submerged foil systems. In the flat turn, opposing, but equal, increments ΔL to the total lift force $L\,(= W)$ must be applied to keep the craft upright. The horizontal centrifugal force acting at the craft centre of mass must then be opposed by an equal and opposite strut force. The latter can become uncertain in waves and for tight turns the local flow may separate, cavitate or ventilate so that the surface becomes dry. The lateral hydrodynamic force on the strut can be eliminated in the banked turn, since it becomes one component of the lift force on the inclined hydrofoil.

Hydrofoil craft and monohull displacement craft react differently to wave induced roll motions. In *Figure 3.74* it is clear that a disturbance, such as a beam wave, will result in a shift of the centre of buoyancy B, of a displacement hull and the centre of lift H of a surface piercing hydrofoil system, thus inducing a roll motion. There is, in contrast, no significant additional

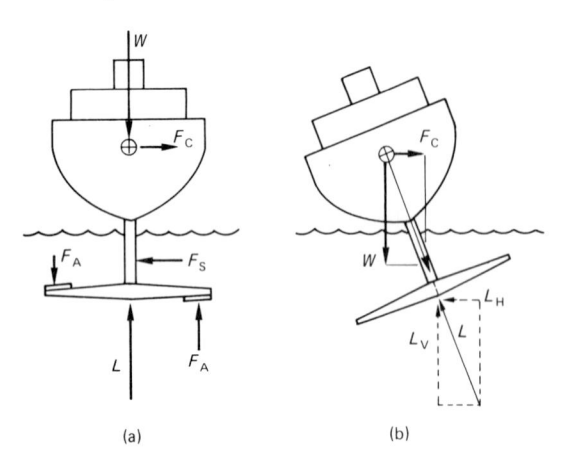

Figure 3.73 Equilibrium turning forces on a hydrofoil craft schematic: (a) flat turn: Centrifugal force F_C = Strut side force F_S; $L=W$; F_A balance moment of F_S about centre of mass; (b) banked turn: Lift force L has vertical component $L_V = W$ and horizontal component $L_H = F_C$

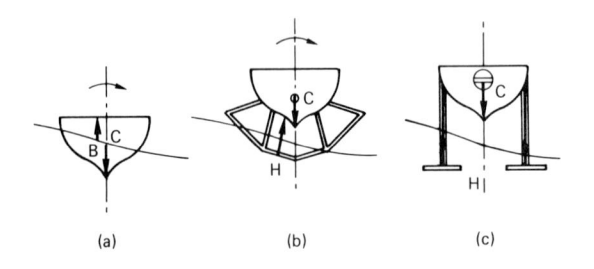

Figure 3.74 Effects of water surface disturbance: (a) displacement craft; (b) surface piercing craft; (c) submerged foil craft

force exerted on a fully submerged foil system. When a roll motion is induced, points B and H will shift to produce a roll righting moment to return each vehicle to the upright condition, as shown in *Figure 3.75*. There

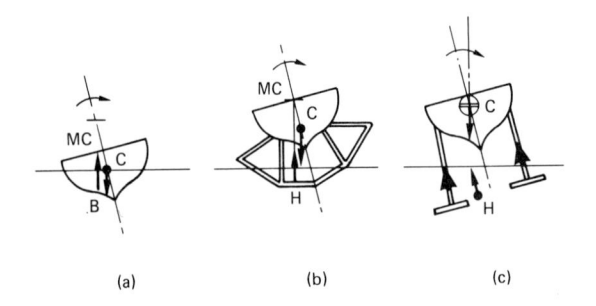

Figure 3.75 Hydrodynamic roll correction (MC refers to metacentre): (a) displacement craft; (b) surface piercing craft; (c) submerged foil craft

is, however, no such movement of the centre of lift of submerged foils and so control action, arising from differential flap control, is necessary to produce a roll righting moment. Instigation of the control action is provided by an automatic control gyroscope which senses movement from the upright condition (Johnston and O'Neill, 1974). Motion stability in terms of roll, course keeping and manoeuvring are of primary consideration for high speed hydrofoil craft.

3.10.3 Hydrofoils operating near the water surface

Hydrofoil lifting surfaces attached to high speed craft may be fully submerged but do not usually operate deeply submerged; they may actually pierce the interface. A satisfactory design must incorporate an acceptable blend of issues e.g. (a) the clearance height between the hull and the air and water interface must be great enough to avoid contact between the hull and the wave crests, (b) the size of the strut must be consistent with structural and low drag requirements and (c) the avoidance of fouling between foils and the sea bottom or jetty wall during docking which restricts the length of the strut.

The principal characteristics of low speed hydrofoils operating close to the surface are similar to those deeply submerged. Some differences in the flow pattern will occur owing to distortion of the air and water interface, whilst at higher speeds the performance of the foil may be affected by cavitation and ventilation (Eames, 1974; Du Cane, 1974).

3.10.3.1 *The dependence of lift force on depth*

The flow pattern about a hydrofoil must satisfy the condition of no flow across the interface at which the atmospheric pressure is usually constant. Now, for positive angles of incidence high velocities, and therefore low pressures (below the corresponding hydrostatic pressure), are developed adjacent to the upper surface of the foil. These low pressures are transmitted through the water, so that a depression occurs at the interface which in turn interferes with the flow over the top of the foil. As a result of this blockage effect the fluid velocities over the upper surface of the foil are reduced leading to a concomitant increase in the surface pressure there. Hence, both L and $\partial C_L/\partial \alpha$ are less than the corresponding magnitudes for the same hydrofoil deeply submerged. The extent of the interface distortion increases as the depth of immersion decreases. Thus, the lift coefficient of a foil at a given incidence angle decreases as the foil approaches the interface. This behaviour is known as the depth effect. It has been used extensively by craft operating in the calm water of rivers, lakes and inland seas in the USSR and other East European countries to ensure equilibrium conditions at various speeds. Wadlin *et al.* (1955) have given the following relationship for the lift curve slope of an infinite span hydrofoil:

$$(a_1)_{AR=\infty} = 2\pi \{1 + (4h/c)^2\}/\{2 + (4h/c)^2\} \quad (3.168)$$

It can be seen that significant changes in lift force take place in the range $0 \leq h/c \leq 1.0$, where h represents the depth below the free surface and c the mean chord length of the hydrofoil. An increase in the forward speed makes the craft rise in the water and the reduced depth of immersion, in a region already close to the interface, causes C_L to decrease. A series of equilibrium conditions is thus established, but as the range of depths is small over which the depth effect is significant the method cannot be used reliably in a seaway.

3.10.3.2 Cavitation on hydrofoils

When the absolute static pressure p of a liquid is reduced the absolute vapour pressure p_v may be reached at which the liquid boils and vapourizes. There are, of course, regions of low pressure on the foil where bubbles and vapour filled cavities may develop rapidly giving rise to the process of cavitation. In principle, cavitation commences when the local absolute pressure $p = p_v$, or in dimensionless form:

$$\sigma_L = (p - p_v)/\tfrac{1}{2}\rho V^2 \quad (3.169)$$

where σ_L represents the local cavitation number, V the velocity of the foil relative to the liquid well upstream and ρ the density of the liquid. For several reasons the actual inception of cavitation on hydrofoils starts before p is reduced to p_v. Small, solid particles in, for example, sea water act as nuclei for the vapour bubbles and so encourage premature inception. Air bubbles in the water act similarly and other important influences arise from the surface roughness of the foil, turbulence of the flow, three dimensional flows over finite span foils and non-uniform temperature distribution in the water. In other words, cavitation inception commences when $\sigma_L = \sigma_c$, a critical value obtained empirically.

A pressure coefficient can be defined as:

$$C_p = (p - p_\infty)/\tfrac{1}{2}\rho V^2 \quad (3.170)$$

where in the present case p_∞ is the upstream, absolute hydrostatic pressure at a depth corresponding to that of the hydrofoil. Thus, cavitation begins when:

$$(p_\infty - p_v)/\tfrac{1}{2}\rho V^2 + C_p = \sigma + C_p = \sigma_c \quad (3.171)$$

where σ is called the cavitation number. Rearrangement of Equation 3.171 yields the critical velocity:

$$V_c = \{(p_\infty - p_v)/\tfrac{1}{2}\rho(\sigma_c - C_p)\}^{1/2} \quad (3.172)$$

and cavitation occurs at all velocities $V \geq V_c$. The pressure coefficient is always negative at some location on the hydrofoil surface (see *Figure 3.65*) especially near the leading edge at angles of incidence other than that corresponding to zero lift. The magnitude of C_p can be large and thus V_c can be low, as Equation 3.172 indicates. In order to increase V_c it is necessary to increase p_∞, that is, to immerse the foil deeper in the water or to reduce the maximum suction pressure on the hydrofoil surface by making C_p less negative.

The following example indicates the limitations imposed by the onset of cavitation. Suppose that a foil is fixed so that the suction peak on the upper surface of the foil corresponds to $C_p = -1$, a modest value. Also assume that $\sigma_c = 0$ and the depth of immersion $h = 1.5$ m. Now, if the ambient atmospheric pressure at the water surface is constant at 100 kPa:

$$p_\infty = \rho g h + 10^5 \text{ Pa} \approx 115 \text{ kPa} \quad (3.173)$$

and so with $p_v = 2.0$ kPa for sea water of density 1025 kg/m³, we have

$$V_c = \{1.13 \times 10^5 \times 2 \div 1.025 \times 10^3\}^{1/2} \text{m/s}$$
$$= 14.85 \text{ m/s} = 30 \text{ knots} \quad (3.174)$$

When cavitation bubbles are formed near to the surface of the foil they may be swept downstream into the wake. Some remain close to the foil and then collapse violently on reaching regions of higher pressure, especially near to the trailing edge of the foil, causing widespread erosion of the material forming the struts and foils. Furthermore, extensive structural vibration and considerable noise generation accompany cavitation in addition to a sudden drop in lift and increase in drag. The preceding numerical example shows that the dependence of V_c on depth of immersion is quite small. In any case, the length of the supporting struts is limited by strength criteria and the undesirability of excessive draughts.

It should be noted that σ, in Equation 3.171, is a function of the flow conditions and the cavitation index

$$\sigma_i = -(C_p)_{min} \quad (3.175)$$

The magnitude of the suction peak pressure coefficient, is a function of the hydrofoil section shape. For the purposes of design, use can be made of a cavitation bucket diagram, of the form shown in *Figure 3.76*, which relates σ_i to the section lift coefficient C_l or, alternatively, to the angle of incidence α. Cavitation will not occur at conditions coresponding to points inside the bucket defined by the lines ABCD. The line BC corresponds to conditions at the critical σ_i near the ideal α, when minimum C_p occurs near mid-chord. Locations below BC correspond to cavitation conditions commencing at the mid-back location.

Cavitation near the leading edge on the lower surface of the foil occurs at points to the left of the critical line AB, at low C_l and α less than the ideal value, and on the foil upper surface to right of the critical line CD at high C_l. Clearly, a hydrofoil section with a flat suction pressure distribution and no large suction peak is a desirable feature, such as that which characterizes the NACA-16 series used for US craft (Johnston, 1985). By delaying cavitation, the maximum operating speed of conventional foils can be raised from 30 to about 45 knots. Some typical shapes of hydrofoil sections are

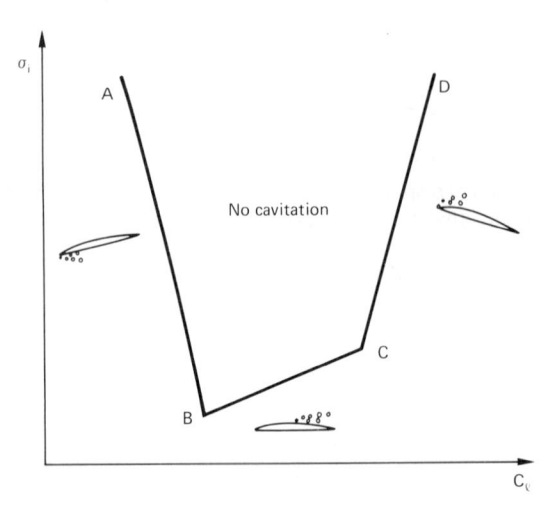

Figure 3.76 Typical form of a bucket diagram for hydrofoil cavitation

shown in *Figure 3.77*. For speeds above 50 knots, a radical change in the design of hydrofoil sections may be necessary. A curved wedge section was developed by Tulin (1956) to promote cavitation deliberately so that the flow encloses a large cavity over the upper surface, a condition known as supercavitation, as indicated in *Figure 3.78*. The sole supporting force is derived from the positive pressure distributed over the lower surface. Little useful lift component is generated by the upper surface because the cavity is at the local static pressure since the streamlines above it are all straight and horizontal. The shape of the upper surface may be chosen to satisfy manufacturing tolerances, strength requirements, etc., provided that it remains covered by the cavity. The sharp leading edge induces a large suction peak precipitating immediate and total cavitation, thus avoiding intermittent behaviour. It is possible to design the lower surface to generate an acceptable lift force without incurring excessive drag in the supercavitating range. Even so, all hydrofoil craft must operate at low speeds in the run up to the full flight cruise regime during which the supercavitating hydrofoils will have a poor performance.

At speeds above about 60 knots, neither the delayed cavitation nor the supercavitating foils are really satisfactory, especially for operating in waves. One solution to the problem has been found by operating the hydrofoil with its suction surface (i.e. the upper surface) partially vented by air, as discussed in the following section.

3.10.3.3 Ventilation on hydrofoils

Early investigators of hydrofoil craft (Hook and Kermode, 1967) observed the phenomenon of air entrainment or ventilation, as it is often called. The result was a sudden loss of lift, and although the precise nature was not understood practical techniques were developed to restore normal conditions. The low pressures on the top surface of a hydrofoil may be sufficient to draw down the air and water interface to meet the foil. Instead of the foil being fully submerged, it is now substantially enveloped by air and the contribution to lift by the suction pressures is almost totally lost. The effect is illustrated in *Figures 3.79(a)* and *(b)* which show how ventilation can occur on both fully submerged and surface piercing hydrofoils. Ventilation is not, however, confined to foils operating in the proximity of the interface. A sheet of air bubbles can form over the aft part of the suction surface of the foil if a suitable path exists (Wadlin, 1958) along which air is drawn from the atmosphere. Another flow path may be provided by the clearance between the strut and the shaft used to adjust the setting angle of the foil and therefore the incidence angle (*Figure 3.80*). Controlled flow of air from the atmosphere to the upper surface of the foil has proved successful in adjusting the lift on hydrofoils to stabilize a craft in the vertical plane (von Schertel, 1974). Further details on cavitation and ventilation, including a comprehensive bibliography, are given by Acosta (1973).

It is generally considered that the dynamic stability of a hydrofoil craft in unsteady motion is probably the most important single design problem. Owing to the significant orbital velocities of waves, the foils may be

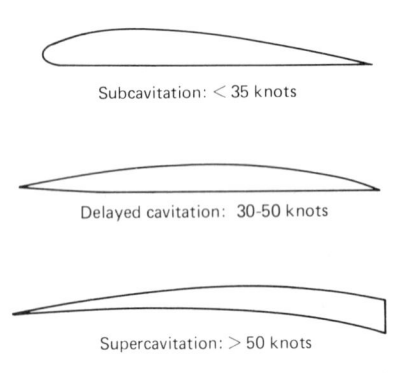

Figure 3.77 Operating speed ranges for different types of hydrofoil section

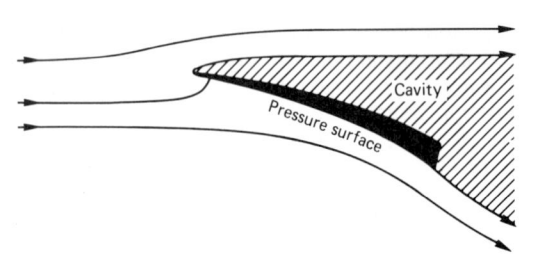

Figure 3.78 Curved wedge section for hydrofoils operating in the supercavitation regime

the upper surface at atmospheric pressure, with consequent loss of lift. Experience has shown that it is always the forward foil that is vulnerable and it is necessary that safe recovery be built into the foil configuration without loss of roll control or directional stability. Complicated combinations of cavitation and ventilation, both of which are influenced by boundary layer separation, may also result. These problems were encountered during the development of the Bras d'Or, an open ocean, antisubmarine, prototype hydrofoil ship of some 2.1 MN all up weight and 46 m overall length (Eames and Jones, 1971). Spoilers were fixed to the upper surface of the foil which encouraged boundary layer separation and prevented reattachment of the flow at low angles of incidence. Ventilation could thus be sustained over the widest possible range of incidence angle, depth of immersion and forward speed. The lift curves for the foils were then continuous with a small slope for operating incidence angles. Not surprisingly, drag was high and so the device was used only for the small front foil which operated primarily as a steerable direction controller, with the option of adjusting craft trim angle.

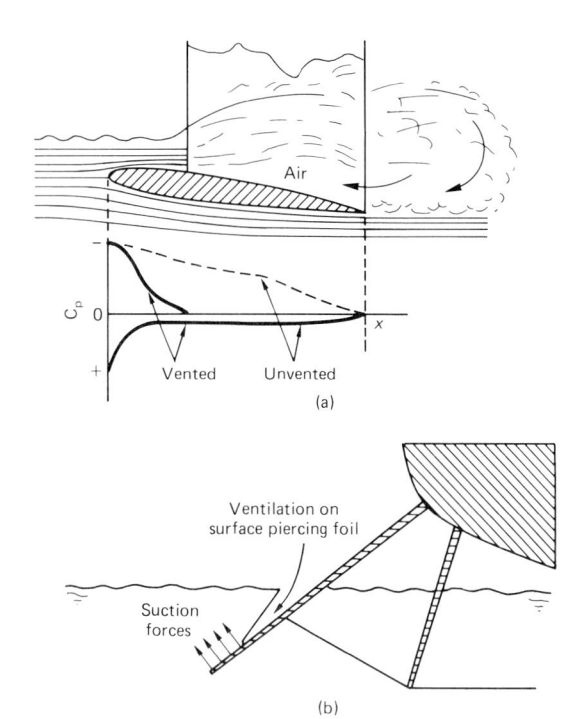

3.10.4 Performance in steady motion

The forces acting on a simple configuration, fully submerged hydrofoil craft moving steadily forward are shown in *Figure 3.81*. The craft moves parallel to its longitudinal vertical plane of symmetry and so no side forces are developed. In the case of a surface piercing hydrofoil craft, the lift force, which is perpendicular to both the direction of motion of craft and the span of the foil, must have components vertically upwards and horizontally inwards to the longitudinal plane of symmetry. Assuming that the motion is parallel to this plane and that both foils are symmetric then the horizontal components of the lift force cancel for each foil and we are left only with the vertical component.

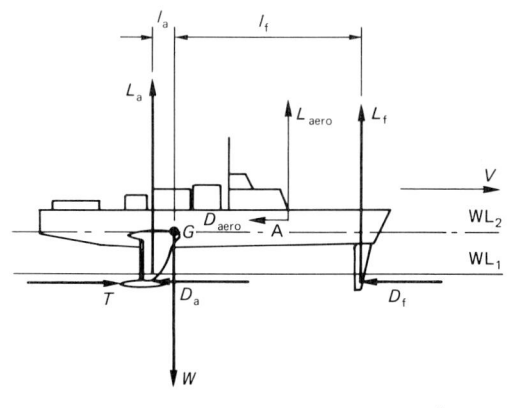

Figure 3.79 Examples of ventilation occuring on (a) a fully submerged foil system and (b) a surface piercing foil system

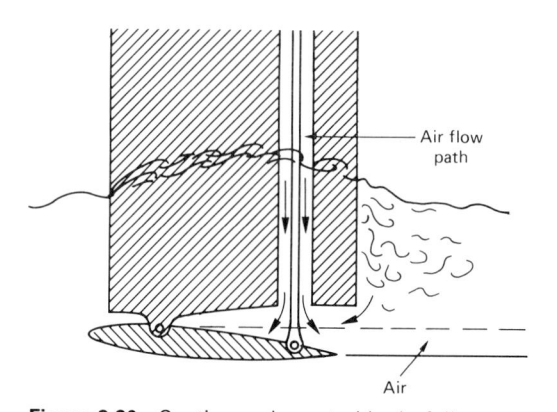

Figure 3.80 Continuously vented hydrofoil

subjected to large and rapid changes of incidence angle and therefore loading. The problem is particularly serious in a following sea for a craft with surface piercing foils when inefficient control of lift forces can result in a sea crash. Often the wave height in a seaway exceeds the length of the strut and the forward foil of a craft may fly out, or partially out, of the water, a condition known as broaching. A similar effect is obtained when the foil is sufficiently close to the water surface that subsequent ventilation leaves the whole of

WL_1: Foil borne waterline (at maximum speed) WL_2: Hull borne waterline

Figure 3.81 Forces acting on a hydrofoil craft at steady forward cruise speed

This component for each set of surface piercing foils is usually referred to as the lift force. The hull is supported clear of the water by one or more foils forward of the centre of gravity G and abaft. We may assume that in flight the submerged foils and struts displace a negligible volume of water, so buoyancy forces are far smaller than the forward and aft lift forces L_f and L_a, respectively. The corresponding drag forces are D_f and D_a. It is quite possible for significant aerodynamic lift (L_{aero}) and drag (D_{aero}) forces to act at the aerodynamic centre A of the above water profile of the craft at high speeds, say above 60 knots.

For equilibrium of forces, the all up weight W must balance the total lift force L^* and the total propulsor thrust T^*, must equal the total drag D^*, for steady motion, thus:

$$L^* = W = L_f + L_a + L_{aero} \qquad (3.176)$$

$$D^* = T^* = D_f + D_a + D_{aero} \qquad (3.177)$$

Generally, T^*, D_f and D_a are more or less in the same horizontal plane and the aerodynamic forces are small with D_{aero} acting in a plane passing close to G. Thus, equilibrium of moments about G gives:

$$l_a L_a = l_f L_f \qquad (3.178)$$

The location of G depends mainly on weight distribution, propulsion systems, operational requirements and trim at rest. The positioning of the foils depends on the ability to generate supporting forces, the control of motion at transient and cruise speeds and the adjustment of trim angle. Good manoeuvrability leads to one or more foils being positioned so that a large moment is imposed on the craft but large control forces within the craft are avoided. Consequently, one highly efficient foil system develops, close to G, a lift force which is a large proportion of W. The second foil system develops a small lift force at some distance from G and is often a relatively inefficient lifting surface. This arrangement determines the canard and aeroplane configurations, each being associated with both fully submerged and surface piercing foil systems.

Defining the total plan area of the foils as S^*, a total lift coefficient C_L^* may be defined as:

$$L^* = W = C_L^* \, (\tfrac{1}{2}\rho V^2 S^*) \qquad (3.179)$$

so that

$$S^* C_L^* = 2W/\rho V^2 \qquad (3.180)$$

assuming L_{aero} is small enough to neglect. Thus, for a craft of given weight the product $S^* C_L^*$ must be varied as forward speed is changed. Since S^* cannot be varied for fully submerged foils (retractable flaps are not yet used), C_L^* must be controlled by change of foil incidence angle, or change of camber by adjusting trailing edge flap angles. Using both forms of control allows constant trim and depth of foil immersion by adjusting independently the forward and rear foils. Many combinations of lift and control systems to achieve and maintain steady motion have been used (Trillo, 1988; Johnston, 1985; Eames, 1974) including incidence angle and flap control for, respectively, the US Navy craft Plainview and High Point; combined incidence and flap control for many Supramar and Rodriquez surface piercing craft; steerable front foil for turning the Jetfoil craft and banking with differential trailing edge flaps on the main foil (Coates, 1978); complex systems of combined control especially developed for rough water operation by the Bras d'Or (Eames and Drummond, 1973).

The product $S^* C_L^*$ can also be changed by varying S^* but keeping C_L^* constant. This is the main operating principle of the surface piercing foils shown in *Figure 3.82*. As the forward speed is increased, say, and C_L^* is constant, the craft tends to rise because $L^* > W$. However, as the craft rises less of the foil surface remains in the water and so S^* decreases to restore the equilibrium equation $L^* = W$. As submersion of the hydrofoils and struts reduces either the propellers or the intakes of hydraulic jet propulsion systems become closer to the surface, thus encouraging the possibilites of cavitation and ventilation. The presence of dihedral and/or anhedral produces sufficiently large restoring forces to counter any lateral disturbances that may arise and also reduces problems of broaching for fully submerged hydrofoil systems.

Commercial hydrofoil craft spend relatively short periods cruising hullborne, such as in harbour areas or other restricted waterways, but naval craft may spend long periods hullborne when on patrol or station keeping. The hull form should be consistent with low drag at hullborne speeds and of a planing craft type to assist

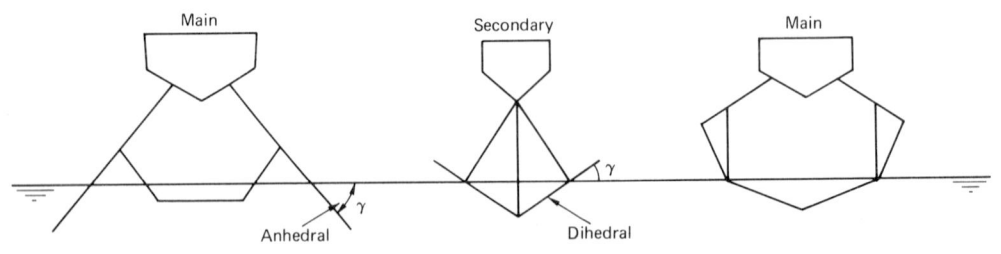

Figure 3.82 Typical arrangements for surface piercing hydrofoil systems

rapid take-off to full foilborne speeds. There is thus a balance of hull form between a narrow beam for lower speeds, a broader beam for transverse stability, and a deep V with large deadrise angle for cresting waves and take-off. The total drag just before take-off is a major factor in dictating the propulsive power requirements as it is seen from the general form of *Figure 3.83* that the total drag D^* rises steeply. The combination of the hull and all the underwater appendages produces a major hump in the drag curve which is far more pronounced than that of a planing craft. Indeed, the drag at cruise speed could be less than that required to overcome the hump, but obviously the power required is greater. Although the high speed advantage of hydrofoil craft is clear to see, a further power margin is required to overcome the additional rough water drag. Johnston (1985) suggests that as the result of a wide range of tests on seagoing craft, a power margin of some 20–25% should be adequate to allow take-off to occur in rough water in any direction.

When the hull is clear of the water it is susceptible to both spray drag and aerodynamic drag, although these are reduced by streamlining the craft. The drastic reduction of hydrodynamic induced drag, as α is decreased, and the reduction in hull drag with higher forward speeds, show as a drop in the total drag force. At speeds above take-off the power requirement rises less steeply than it would for hullborne operation, but again increases rapidly at the higher speeds until, at a steady design cruise speed V^*, a condition is met where the total available power P^* is given by:

$$P^* = TV^* = D^*V^* \quad (3.181)$$

The total drag force at the steady foilborne cruise speed is given by $D^* = D_0 + D_i$, or in coefficient form by:

$$C_D^* = D^*/\tfrac{1}{2}\rho V^2 S^* = C_{D,0} + C_{D,i}$$
$$= C_{D,0} + C_L^2/\pi\, AR \quad (3.182)$$

in analogy with Equation 3.164. The drag at zero lift D_0, includes the section profile drag and the drag effects of boundary layer separation and wake formation for both the foil and strut systems, the spray drag, and aerodynamic drag on the hull and superstructure. It can be seen that D_0 varies with V^2 and so is large at high speeds. The induced drag D_i includes the effects of downwash and wave making drag, generated by the craft in disturbing the water surface. Equation 3.182 shows that D_i varies inversely with V^2 and so is the dominant factor at low speeds. When the components are combined, it is found that the total drag will be a minimum at some particular speed, usually found to be a few knots higher than the take off speed, as shown in *Figure 3.83*.

3.10.5 Performance in waves

One of the principal advantages of hydrofoil craft over other high speed or conventional monohull ships, is the ability to operate effectively in a wide range of sea states, even though the hull size may be small in conventional terms. This attribute, particularly noticeable for craft with fully submerged foils, largely explains the US activity with this type in its naval development programmes. Johnston (1985) gives a summary of data for four craft operating in seas up to 5 m significant wave height at which the operating envelope has dropped 45 knots from 50 knots in calm water. Whilst this would be considered optimistic for a given craft, there is no doubt that such an achievement is realistic at a 4 m significant wave height, as was shown by *Tucumcari*, a 22 m, 57 tonne craft.

To be able to travel at speeds above 40 knots in rough seas, there must be available a sophisticated automatic control system (ACS). The ACS covers all the dynamics such as take-off, landing, foilborne operation, hull height above the water surface, roll and pitch stability, banking in turns and minimization of the effects of orbital velocities of water particles in wave systems. In particular, the clearance between the hull of the craft and the water must be sufficiently great so as to avoid physical contact. The vehicle may then maintain a mean clearance relative to the still-water level in a mode called platforming, as shown in *Figure 3.84*. This generally requires continuous adjustment of the foil incidence angles as the craft proceeds at a steady forward velocity and a constant trim angle. In reality, platforming occurs only in calm waters and it is reasonable to expect that in open waters contouring or some intermediate response will take place as indicated in *Figure 3.84*. Ideally, the intermediate response is sought so that the hull just misses the crests yet the foils remain immersed at the troughs to avoid broaching. A summary of the types of ACS and some indication of their complexity as developed for the US Navy seagoing craft is given by Johnston (1985), although such sophistication is not called for in commercial craft operating under less demanding sea conditions.

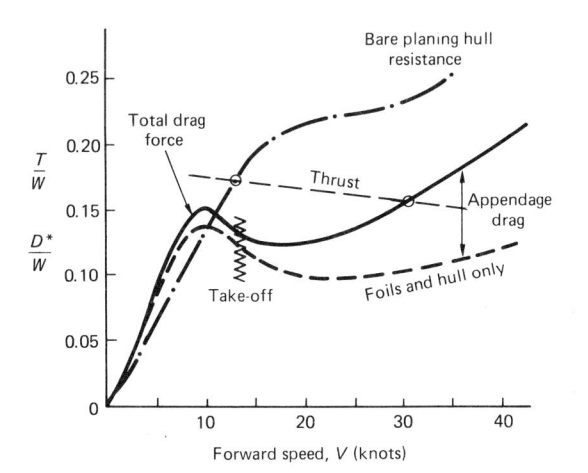

Figure 3.83 Performance characteristics of a 30 m long hydrofoil craft

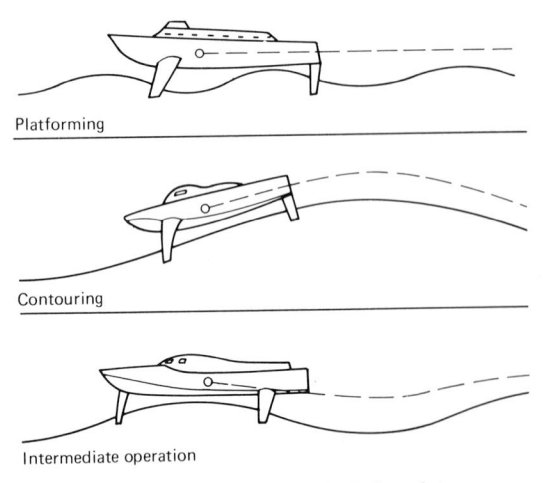

Figure 3.84 Response of a hydrofoil craft in waves

The most significant parameter governing seakeeping behaviour of a conventional ship is the length and good practice implies that no more than, say, one slam or severe deck wetting per minute occurs at a given speed in rough seas. On this basis, data shown by Johnston (1985) indicate that the hydrofoil craft may have only a modest speed advantage over conventional hulls of equal power in calm water but a speed advantage of two to four times can be maintained in rough seas of state 5 and above. However, not only must the hull and foil systems be robust enough to take the hydrodynamic loads, but also the crew and passengers must be subjected to limited motion and accelerations to maintain operational efficiency and comfort. As an example, a 200–300 tonne hydrofoil craft operating at 40 knots in rough water may display vertical accelerations of less than $0.05\ g$ at the bridge for 80% of operating day. Human beings can tolerate such an acceleration at a frequency of 1 Hz for up to 8 hours. A conventional monohull ship in the same seas at 16 knots corresponds to 60% of operating days and at 28 knots to only 45% of operating days. For larger accelerations and thus lower durations of human tolerances, the hydrofoil craft shows even greater superiority.

The preceding comments, related primarily to fully submerged foils, apply to a somewhat less dramatic extent to surface piercing foils. The 200 tonne Canadian craft *Bras d'Or*, described by Eames and Drummond (1973), had showed far better ride quality than displacement and semidisplacement craft of the same size, but was significantly inferior to a corresponding fully submerged foil craft. When not foilborne and acting in conventional displacement mode, the foil systems of *Bras d'Or*, which could not be retracted, acted as efficient motion dampers with pitching and rolling motions less than a 4000 tonne ship.

The Italian passenger craft, the *Rodriquez* RHS-200, has surface piercing foils with an automatic stability augmentation system. Rieg and King (1983) showed that the motions of this craft were considerably lower than the *Bras d'Or*.

3.10.6 Structural design

Clearly, the hull of a hydrofoil craft must be strong enough to withstand wave impact and emergency off foil crash at high speed. High and fluctuating stresses and local loadings occur at the strut attachment points and so fatigue criteria will need examination. Weight considerations are very important and so the desire for lightweight materials leads to the general adoption of welded or riveted aluminium hulls with steel reserved for the struts and foils. Cost, manufacturing techniques, maintenance and repair, and corrosion resistance are all vital factors for naval and commercial operations. Some form of cathodic protection for the hull is needed and this can take the form of sprayed zinc coating, passive zinc anodes or impressed current systems. Johnston (1985) suggests that consideration of all these factors leads to a present day hydrofoil craft hull having a length to beam ratio of 4; a sharp V forward; 20° deadrise aft; hard chine planing hull form; constructed of non-delaminating aluminium with welded frames and stringers using extended skin panels with integral stiffeners and a mass per unit enclosed volume of about 40 kg/m^3.

Further details of structural design and construction of several types of hydrofoil are given by Johnston (1985) including both fully submerged and surface piercing types. It is interesting to note that fatigue problems were met in the lead craft of the patrol hydrofoil missile (PHM) series developed in the US Navy programme. Cracks were found in the skins of the hydrofoils, which were constructed with machined leading and trailing edge sections with the skin and spanwise members connected by T welds. Subsequent analysis of operational data revealed that loadings were as expected but that high load frequencies were higher than anticipated. A revised aerofoil construction was thus devised by moving welds to low stress areas, eliminating sharp corners and machining a large proportion of the foil from solid. The camber of the original NACA-16 series foils was also increased somewhat and revised welding techniques allowed easier construction.

The current production method of the *Rodriquez* craft is devised around a build of up to ten craft per year. Sub-assemblies, such as deckhouse, bridge, etc., are constructed outside the main assembly shop and added to the hull with the shell complete. Again, aluminium plates are used for the hulls which are longitudinally framed with prefabricated web frames. The steel foils, struts and rudders are all welded with spars, ribs and a plated skin. The production techniques have been developed over 30 years and some of the original craft are still operating. On this basis it may be concluded that hydrofoil technology is well developed and future success depends on planning and exploitation of routes and applications.

3.10.7 Applications

In the US, most activity has concentrated on naval applications especially in the role of small, fast anti-submarine warfare (ASW) craft operating offshore, but not in the deep ocean. Proposals for the post-World War II development of large hydrofoil ships as trans-ocean vessels that could out run hostile nuclear powered submarines were abandoned as being unrealistic. The early 1960s saw the development of the patrol craft hydrofoil (PCH) *High Point* of length 35 m, weight 112 tonne, used as a small high speed sonar platform for local ASW work around convoys. The full sonar capability was never developed although subsequently the all up weight was increased to 126 tonne. By 1969, the largest hydrofoil *Plainview* (AGEH-1) was built and measured about 65 m in length and some 320 tonne full load displacement. The first fully operational US Navy hydrofoils were the patrol gunboat hydrofoil (PGH) craft, *Flagstaff* (PGH-1) of length 22 m and displacement 69 tonne and the slightly smaller *Tucumcari* (PGH-2). The former was propeller driven with main foil forward and the latter was powered by a waterjet system with main foil aft (canard arrangement). Both saw combat duty in Vietnam and provided valuable experience for the current hydrofoil craft, but by the mid-1980s only *High Point* remained in service.

The present fleet of hydrofoil craft consists of a number, six by 1982, of the canard configured patrol hydrofoil missile (PHM) vessels developed from the *Tucumcari* concept, with the lead craft *Pegasus* (PHM-1) delivered in 1977. All the vessels are propelled by water jets, the foils are fully submerged with the forward inverse T-foil steerable. All the struts are hinged in the hull so that the total lift system can be retracted from the water for hullborne operation. The all up weight of each PHM is about 240 tonne with an overall length of about 40 m (Johnston, 1985; Trillo, 1988).

Another potential area of activity for hydrofoil craft is with the US Coast Guard since many of its existing patrol boats are near to the end of useful life. In the past, search and rescue activities predominated, but now up to 70% of the time is involved in law enforcement, particularly in relation to the detection and apprehension of illegal drug operations. High speed is called for in the chase and apprehend mission, but then the ability to tow at low speeds is needed, requiring a mixed capability. The use of a gas turbine at high speeds and diesel engines at low speeds, using waterjet propulsion, clearly marks the PHM concept in semi-civilian form as a prime candidate.

In other countries, Italy has a group of six Nibbio class vessels (King, 1982), based on the *Tucumcari* design, in operation since the early 1980s. The former are designed as short duration (up to 12 h) fast attack day boats and possess no sleeping accommodation. Israel commenced in 1978 a series of craft based on the *Flagstaff* concept. At high speed these craft operate with the single, foil mounted propeller of 1.3 m diameter in a transcavitating mode. The two craft at present operational are powerfully armed strike vessels operating as day boats and full details have been described by Frauenberger (1982). The Soviet Union possesses a vast number of hydrofoil craft with both commercial and military variants closely related. Most are of the surface piercing type and some have the bows raised by foils with the flat keel at the stern planing on the water surface, a trail dragger, which could be accepted in calm waters. Trillo (1988) gives details of the various types, some of which have an all-up weight of 400 tonne, speeds in excess of 50 knots and combined surface piercing and full submerged foil systems. The only UK entry into the naval applications of hydrofoil craft was a series of evaluation trials on HMS *Speedy*, a variant on the *Jetfoil* design. The intended role of this 27 m long, 115 tonne vessel was in fishery protection, with rough water patrols in the North Sea. Following a very thorough appraisal of performance (Brown et al., 1984) it was decided on economic grounds to terminate the programme in favour of monohull designs.

From the naval viewpoint, one of the most interesting features under examination in the US is the concept of the extended performance hydrofoil craft. Buoyancy is incorporated in the form of a long, slender submerged body which is combined with the dynamic lift system of a fully submerged foil and strut combination. Investigations of the concept have shown (Meyer, 1981) that with a buoyancy and fuel tank providing about 50% of the total foilborne lift force an extended performance craft would have a probable range well in excess of conventional hydrofoil craft of comparable size. The benefits increase with size, say beyond 200 tonne, as a result of increased fuel to weight ratios and higher drag to weight ratios, especially at reduced foilborne speeds of 20–25 knots.

The commercial use of hydrofoil craft is found worldwide and the major users are the USSR and Eastern Bloc countries. The only such craft in the US is the Boeing *Jetfoil* (model 929) designed for 45 knots and all up weight 117 tonne with the capacity for commercial transportation over limited open ocean routes. The design was developed from the earlier naval craft and production started in the 1970s. The retractable foil system is full submerged, canard arrangement with flap control of roll, pitch and height above water. The forward strut is steered for course keeping and manoeuvring. Over 20 such craft been built and most are in current use, especially in South East Asia where there are ample opportunities for ferry services (Trillo, 1988).

Following World War II, Switzerland saw the development of surface piercing foil craft through the skills of Baron von Schertel. His efforts were developed by the Supramar company which continues with new hydrofoil concepts building on the previous experience with Rodriquez Cantiere Navale of Italy who operated under licence until 1970. Rodriquez is now the largest producer of craft having surface piercing foils with control flaps. At least 150 have been completed ranging in size from about 22 m and 30 tonne to 35 m and 120 tonne. Speeds range from 32 to 38 knots and the number of passengers from 70 to 300. Japan also has a Supramar licence and forty or so have been produced by Hitachi Zosen Company in the range 20 to 27 m and 32 to 63 tonne. Finally as stated earlier, the USSR has been producing calm water craft for many years and seagoing commercial craft since the early 1960s. The Raketa, Kometa and Kolklida are well known types which have been exported widely for coastal services. More details of these and other hydrofoil craft are given by Trillo (1986, 1988).

3.11 Hovercraft

The generic title Hovercraft is used to cover the two principal types of marine vehicle discussed in this section, namely air cushion vehicles (ACV) and surface effect ships (SES). Each type obtains a vertical supporting force by generating an overpressure above atmospheric in a region between the keel and the water surface which literally lifts the main hull clear of the water. The present day ACV has, in principle, no part of the craft in contact with the land or sea surface and is therefore completely amphibious. Air is supplied from a lift fan which provides an air flow round the periphery of the hull followed by ejection into the cushion space, as shown typically in *Figure 3.85(a)*. The SES possesses solid hulls or walls along the sides of the craft, as shown in *Figure 3.85(b)*, which are immersed in the water with

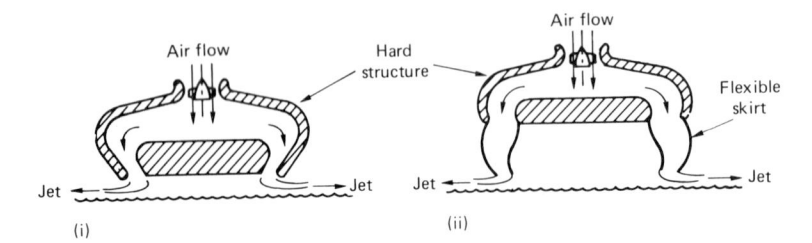

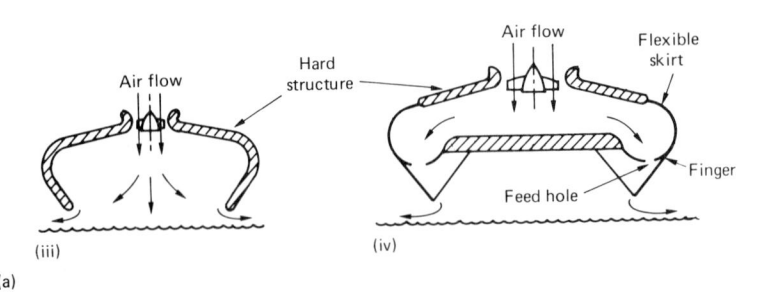

Figure 3.85 (a) Skirt forms for air cushion vehicles; (i) simple peripheral jet; (ii) skirted peripheral jet; (iii) simple plenum; (iv) skirted plenum

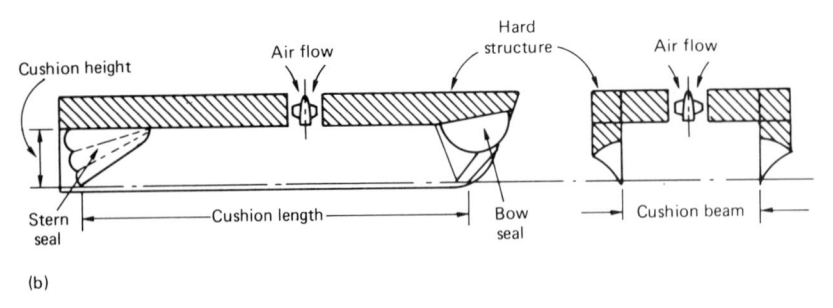

Figure 3.85 (b) Skirt forms for surface effect ships

separately inflated flexible seals (skirts) at the bow and stern. The lift air is delivered straight to a plenum, which is equivalent to the cushion, so that the supporting force comprises the sum of aerostatic and sidehull buoyancy and, possibly, planing force components.

Craft typified as ACV are the British Hovercraft Corporation (BHC) SRN4 Mk3 (*Figure 3.86(a)*; Wheeler, 1976), which provide a regular year through service across the English Channel between Dover and Calais/Boulogne and the BHC AP1-88 (*Figure 3.86(b)*; Wheeler, 1984; Mant, 1987) used for the cross Solent ferry service in Southern England and by the Scandinavian Air Systems (SAS) to transport air passengers from Kastrup Airport, Copenhagen to Malmo, Sweden. Examples of SES are provided by the Vosper Hovermarine HM218 (*Figure 3.87(a)*; Tattersall, 1983), and the larger HM5 series craft (*Figure 3.87(b)* Tattersall, 1985), and the US SES 100A which was used for prototype investigations of the larger SES 200.

The development of hovercraft can be found in many publications but those of Crewe and Eggington (1960), Stanton Jones (1968), Silverleaf (1968), Elsley and Devereux (1968), Trillo (1971), Summers *et al.* (1974), Mantle (1980), Butler (1985) and Lavis (1985) stand out as covering all the main issues, with the regular editions of *Jane's High Speed Marine Craft and Air Cushion Vehicles* providing updates of new craft and technical developments. The principal features of each of the two main types of hovercraft will now be examined together so that comparisons can be more easily made.

3.11.1 Generation of supporting forces

Typical values of over pressure (positive gauge pressure) in the cushion are between 0.015 and 0.050 atmosphere (i.e. 150–500 Pa or 0.2–0.7 lbf/in^2). Designers of British craft favour the lower values consi-

Figure 3.86 (a) SRN4 Mk3 car and passenger hovercraft, in scheduled service across the English Channel (by permission of BHC/Westland Aerospace)

Figure 3.86 (b) AP1-88 diesel-powered hovercraft, in service with Canadian Coastguard for navaid maintenance, ice breaking and search and rescue (by permission of BHC/Westland Aerospace)

Figure 3.87(a) Hovermarine HM 218 surface effect ship, of which over 100 of this series have been sold worldwide (by permission of Hovermarine International Ltd)

Figure 3.87(b) Hovermarine HM527 surface effect ship of which four of these 200 passenger craft are operating on the Hong Kong to Macao route (by permission of Hovermarine International Ltd)

stent with a softer ride and passenger comfort, whereas US craft possess higher values for more compactness of layout and very high speeds. Cushion pressure is derived from one or several fans and maintained by sealing the cushion air along the periphery using either an annular (peripheral) jet, a plenum chamber or a solid wall in combination with the first two systems. In each of these systems, shown schematically in *Figure 3.85(a)* and *(b)*, air must leak from under the cushion if the vehicle is to ride clear of the ground. Thus, the cushion pressure and a continuous leakage flow must be maintained and various types of fan or blower have been used for this purpose. It is evident, however, that if the leakage and therefore the clearance height is small contact will occur between the solid structure of the craft and rough surfaces over which the craft is moving. To avoid this problem and yet maintain a stable vehicle with small lift power, flexible skirts are now used and these will be discussed later. Essentially, the original nozzle geometry is extended downwards to provide an inflated, flexible wall or skirt and leakage air expelled through a multitude of flexible orifices into the cushion space. As a result the plenum chamber craft for both ACV and SES has been reinstated as the current design form, superseding the original ideas of Cockerell (1955).

3.11.1.1 Hovering flight

Numerous theories have been developed (Harting, 1969), especially for peripheral jet forms, and it is instructive to examine briefly some of them. Clayton and Bishop (1982) selected two for study, Elsley and Devereux (1968) examined several theories and Mantle (1980) derived a number of design parameters on the basis of the Elsley and Devereux exponential theory.

Let us first consider, for the ACV peripheral jet, a simple theory based on the force and impulse relationship, explained by Clayton and Bishop (1982). *Figure 3.88* shows a two dimensional cross-section of the nozzle, jet and base of an ACV. The flow is considered invicid and the suffixes o and c refer to the outer and cushion limits of the jet bounded by the streamlines S_o and S_c; no entrainment (a viscous phenomenon) occurs. The air velocity v_n, in the nozzle, of width a_n, is uniform, but the jet velocity v, may vary across the width a, which may be somewhat greater than a_n. In general, the jet pressure p increases from constant atmospheric pressure p_a, on S_o to the cushion pressure p_c, on S_o. For the simple theory, it is assumed that the jet pressure is constant throughout and equal to p_a so that a discontinuity in pressure occurs at S_o with the (constant) pressure in the cushion at p_c. It is also assumed that v is uniform and equal to v_n so that the total pressure in the jet is $p_t = p_a + \frac{1}{2}\rho v^2$ where ρ represents the constant air density. As with all the theories discussed here, the velocity of air in the cushion volume is assumed negligible.

Equating the force applied to the jet from left to right to the increase of momentum in the same direction over a short time interval leads to the expression:

$$(p_c - p_a) h = \rho v^2 a (1 + \cos \theta) \qquad (3.183)$$

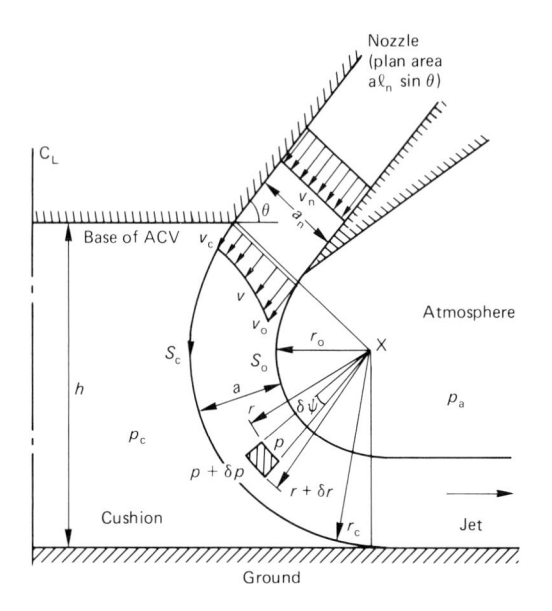

Figure 3.88 Idealized flow geometry for a peripheral jet ACV. Plan area enclosed by bottom edge of cushion wall is S

or, in a dimensionless form, the cushion pressure coefficient:

$$C_p = (p_c - p_a)/(p_t - p_a) = (p_c - p_a)/\tfrac{1}{2}\rho v^2$$
$$= 2x \qquad (3.184)$$

in which $x = a(1 + \cos \theta)/h$. Usually, pressures are measured relative to atmospheric, i.e. as gauge pressures so that $C_p = p_c/p_t$.

In the derivation of an equivalent simple theory by Elsley and Devereux (1968), it was assumed that the gauge pressure in the jet was the average of the gauge pressures at the streamlines S_o and S_c, that is, $p = \frac{1}{2}p_c$, although no logical argument is put forward for this. As a result:

$$C_p = p_c/p_t = 2x/(1 + x) \qquad (3.185)$$

A more complex, variable jet pressure, inviscid theory is illustrated by Clayton and Bishop (1982) in which the assumptions of constant velocity and constant pressure in the jet are dropped and the details of *Figure 3.88* are fully incorporated. Since the pressure and velocity of the air in the nozzle are both constant, a discontinuity occurs at the nozzle exit plane. There are, however, no discontinuities elsewhere in the jet flow. It may be shown that $v_o = v_n$ and, since p_t is constant in the jet, $v_c < v_o$ as a result of applying the radial equilibrium equation:

$$\partial p/\partial r = \rho v^2/r = 2(p_t - p)/r \qquad (3.186)$$

Integration of Equation 3.186, the application of boundary conditions at the inner and outer edges of the jet, incorporation of the jet geometry and the use of Bernoulli's Theorem between the nozzle and jet, leads to the following expression for the pressure coefficient:

$$C_p = 2x - x^2 \qquad (3.187)$$

A rather less rigorous use of the radial equilibrium equation is put forward by Elsley and Devereux (1968) in which the outer streamline is assumed to be a circular arc and its radius $r_o = h/(1 + \cos \theta)$. Then, r_o is substituted for r in Equation 3.186 on the basis that the jet is thin, i.e. the thickness a is small compared with r_o. Integration, with appropriate boundary conditions, allows development of the so-called exponential theory giving:

$$C_p = 1 - \exp(-2x) \qquad (3.188)$$

Essentially, all the derivations leading to Equations 3.184, 3.185, 3.187 and 3.188 are consistent for small x, even though their application is often extended to values of x outside the range of validity. It can be seen that when x is small all the preceding values of C_p tend to $2x$ for first order accuracy. Nevertheless, Elsley and Devereux (1968) claim predictions from the exponential theory to be close to experimental values over a wide range of x. More accurate theories sometimes assume the streamlines to be elliptical rather than circular or, alternatively, specify a particular variation of velocity across the jet. Exact solutions of the inviscid flow problem can be obtained by conformal mapping although complicated expressions are involved. Finally, we may note that several viscous flow theories, of varying complexity, have been put forward for both two and three dimensional flows. These theories allow entrainment and mixing to be accounted for and it has been found that the effects of viscosity become important when h is small, that is, for large x.

The volume flow rate of air passing through the peripheral jet area which must be supplied by the lift fan (as well as supplying other leakage sources), is given by

$$Q = l_n \int_{r_o}^{r_c} v \, dr \tag{3.189}$$

where l_n represents the peripheral length of the nozzle. The results of the integration in Equation 3.189, for the four theories represented by Equations 3.184, 3.185, 3.187 and 3.188 are, respectively:

$$Q_1 = EGx \qquad Q_2 = EGx/(1+x)^{1/2}$$
$$Q_3 = EGx \qquad Q_4 = EG\{1 - \exp(-x)\} \tag{3.190}$$

where $E = hl_n/(1 + \cos\theta)$ and $G = (2p/\rho)^{1/2}$.

In each case the nozzle power P_n, that is, the energy flux of the air in the jet, is given by:

$$P_n = p_t Q \tag{3.191}$$

and, taking the exponential theory as an example, for constant θ, the value of $\exp(-x) = 0.5$ or $x \approx 0.7$ leads to a minimum value of P_n.

Results are much simpler for plenum chamber craft. Again, assuming that the velocity of air in the cushion is negligible, the gauge cushion pressure:

$$p_c = W/S \tag{3.192}$$

where S represents the plan area enclosed by the bottom edge of the cushion wall. The cushion escape velocity v_e (see *Figure 3.89*) is thus given by:

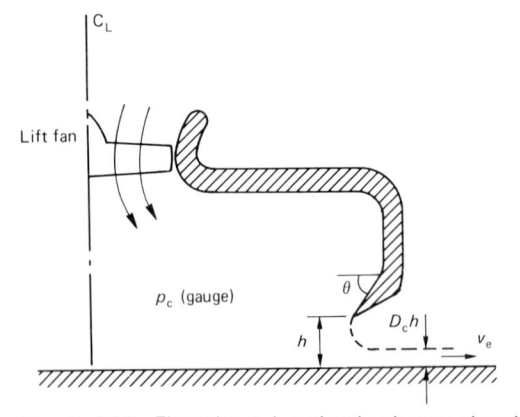

$$v_e = (2p_c/\rho)^{1/2} \tag{3.193}$$

Figure 3.89 Flow through a simple plenum chamber hovercraft

on the basis of Bernoulli's Theorem for inviscid flow. The volume flow rate of escaping air is therefore:

$$Q = v_e b l_n \tag{3.194}$$

The thickness b of the jet is usually written in the form of a discharge coefficient related to the daylight clearance h, between the wall and the ground, i.e. $b = h D_c$, whence:

$$Q = v_e h D_c l_n = h D_c l_n (2p_c/\rho)^{1/2}$$
$$= h D_c l_n (2W/\rho S)^{1/2} \tag{3.195}$$

The value of D_c depends on θ and increases from 0.5 for $\theta = 0$ to $\pi/(\pi + 2) = 0.611$ for $\theta = 90°$ but, in practice, is somewhat reduced throughout by viscous effects. For flexible sidewalls (skirts) there is considerable difficulty in obtaining accurate measures of discharge coefficient, especially if the lower portion of the skirt is comprised of many fingers. Finally, the energy flux of the escaping air, the nozzle power, is:

$$P_n = p_c Q = h D_c l_n (2p_c^3/\rho)^{1/2} = h D_c l_n (2W^3/\rho S^3)^{1/2} \tag{3.196}$$

The preceding equations for plenum craft can be used for sidewall geometries (SES) after making due allowance for buoyancy forces developed by the partially submerged twin-hull. Hydrodynamic planing forces may also be present when forward motion takes place and at higher cruise speeds, with minimum distortion of water surface, hover predictions apply with good accuracy.

The upward vertical jet reaction force for peripheral jets systems is given by:

$$F_R = \int_{r_o}^{r_c} (p + \rho v^2) l_n \sin\theta \, dr \tag{3.197}$$

Substitution for the above jet theories then yields, respectively, the following expressions for F_R:

$$F_{R_1} = Ep_c; \quad F_{R_2} = Ep_c(1 + x/2);$$
$$F_{R_3} = Ep_c(2 + x)/(2 - x);$$
$$F_{R_4} = Ep_c \left\{ \frac{1}{2x} + \frac{1}{1 - \exp(-2x)} \right\} \tag{3.198}$$

Thus, the total upward vertical supporting force F_v, which must equal the all up weight W for equilibrium, is thus given by:

$$F_v = W = p_c(S - l_n a \sin\theta) + F_R$$
$$= p_c \left\{ S - x \frac{h l_n \sin\theta}{1 + \cos\theta} \right\} + F_R \tag{3.199}$$

Taking the second simple theory example, for the purpose of illustration, we can form the peripheral to plenum nozzle power ratio, after using the second of Equations 3.198 and 3.199:

$$\frac{P_n \text{ (per)}}{P_n \text{ (ple)}} = \frac{(1+x)}{2D_c(1+\cos\theta)\sqrt{2x}}$$
$$\left[1 + \left\{\frac{hl_n \sin\theta}{S(1+\cos\theta)}\right\}\left(\frac{2-x}{2}\right)\right]^{-3/2} \quad (3.200)$$

Using the typical values $\theta = 45°$, $D_c = 0.6$, $h = 150$ mm, $l_n = 40$ m, $S = 75$ m^2 and $x = 0.5$, one finds that the nozzle power ratio is about 0.7 and about 0.8 for $\theta = 60°$, all other variables remaining unchanged. Close scrutiny of Equation 3.200 shows that for acceptable hovercraft parameters, the term in the brackets differs little from unity and the ratio depends almost wholly on $\cos\theta$. The result illustrates why the peripheral jet principle became more popular than simple plenum chamber craft, but the advent of skirts has reversed this trend.

The augmentation ratio is a parameter which gives a measure of the ground effect and is defined as:

$$A_r = F_v/al_n \rho v_n^2 \quad (3.201)$$

where the denominator is the force exerted on a nozzle of circular cross-section equal in area (al_n) to the ACV nozzle by a jet directed vertically downwards well clear of the ground. Values greater than unity are sought and, as shown by Clayton and Bishop (1982) for the simple theory, values of A_r well in excess of 10 may be achieved with acceptable geometric parameters.

The fans used to generate the air jet and hence the cushion pressure tend to be purpose built but are most often of the centrifugal or mixed flow type with backward facing blades having non-overloading power characteristics. Maximum power is therefore dissipated at the design condition of maximum efficiency. The lift fans must not only generate the total pressure in the jet at exit from the nozzle or fingers but must develop a sufficient pressure rise across the propeller to overcome the additional pressure losses in the duct system. Internal flow paths can be quite long with many bends and changes of section. Energy losses also occur at the inlet duct as air is taken in from above the craft at right angles to the external air flow. For plenum craft, with air fed directly to the cushion, losses tend to be quite low. Clearly, the aim is to provide a large proportion of static pressure rise by minimizing the flow rate and escape air velocity, thus limiting fan power. Usually one set of fans is used to provide air to the cushion and often the front seal of SES and a second set to inflate the rear bag seal. For ACV, several fans may each supply a section of the bag-finger arrangement.

Fan selection takes place on the basis of steady flow conditions but substantial variations of back pressure occur during passage over waves and the attendant cushion pumping. As a result, the instantaneous operating point traces a looped curve about the steady flow characteristics. The size and orientation of the hysteresis loop depends on frequency and amplitude of oscillation. The shape may become a figure of eight loop at low frequencies with bigger departures from the steady state curve as frequency increases, but precise predictions are not yet readily available. Fan noise increases under these unsteady conditions but, in general, tip speeds should be kept as low as possible (e.g. less than about 150 m/s for centrifugal fans or 250 m/s for axial fans). Further details are given by Butler (1985), Lavis (1985) and Mantle (1980).

Throughout the preceding development it has been assumed that the air in the cushion remains stationary. Whilst this is true for the jet shapes considered in inviscid flow, it is certainly not true for real viscous flows. Entrainment from atmosphere and deflection of flow inward, adjacent to the ground stagnation point, encourages the formation of a ring vortex for peripheral craft and general circulating flow for plenum craft. Nevertheless, such flows are at speeds sufficiently low compared with jet speeds that they may be ignored within the overall accuracy of the theory.

Finally, hovering flight over water causes a depression of the surface under the cushion, as shown in *Figure 3.90*, and this condition leads to important conclusions in respect of craft drag forces.

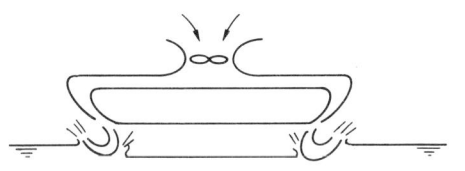

Figure 3.90 Hovering flight of ACV over water

3.11.1.2 Forward flight

As speed increases in forward flight over land, an ACV first generates a large vortex at its bow as a result of escaping cushion air being thrown forward. At high speed this vortex disappears and air flows smoothly over the forward superstructure. For a peripheral jet model, air from the forward jet is also swept underneath the cushion at high craft speeds so as to enclose the cushion in the form of a crude aerofoil. In practice, pressurized skirts, *Figure 3.85(a)*, reduce daylight clearance to such an extent that the cushion air is little affected by forward motion over flat surfaces. As air flows over the superstructure suction pressures are created there and so the craft may take on a small bow down trim angle, which can be cancelled by a trim control aileron.

Flight over water results in substantial changes from the hovering regime. The condition shown in *Figure 3.90* persists at low forward speeds but, as speed increases, water begins to pile up ahead resulting in a trim angle bows up, as indicated in *Figure 3.91*. The water surface below the cushion remains virtually parallel to the craft base. The water depression, or fluid hull, is dragged through the water, but not by physical

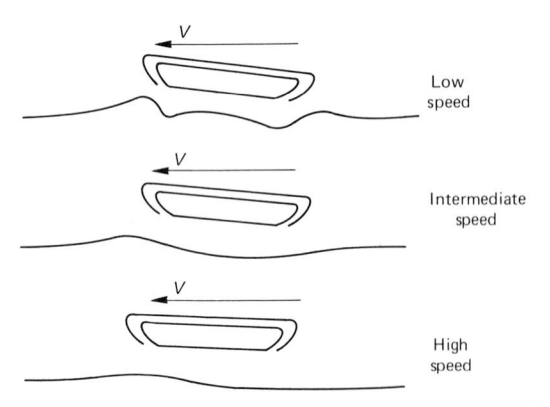

Figure 3.91 Forward flight of ACV over water

contact with the craft, and creates both a wake astern and a surface wave pattern. At intermediate speeds the craft rises in the water, trim reaches a maximum value and spray is swept aft.

At high cruise speeds trim reduces because the inertia of the water does not allow sufficiently fast reaction to the rapidly passing pressure pulse from the cushion. The water depression becomes longer and shallower and wake and wave making disturbances significantly reduce. Here again, the over water high speed motion principle is being demonstrated, i.e. the introduction of induced drag (from trim) and reduction of wave making drag, compared with conventional displacement vehicles. Similar cushion effects occur for SES but for these the sidehulls are always partially submerged in the water and so generate additional wave systems and thus further components of wave making drag. The absolute values of trim are also modified by the magnitude and location of the sidehull buoyancy forces.

3.11.2 Flexible skirts

As stated earlier, the introduction of skirts was a major influence in the development of hovercraft to a viable commercial product which demonstrated power efficiency, motion stability and acceptable operation in waves. The essential problem was to increase the structural clearance of the base of the craft above the ground to overcome obstacles and avoid contact. With the original peripheral jet concepts the required air flow through the lift fan became unacceptably excessive to maintain such a clearance. Flexible extensions to the lower edges of the walls of plenum craft and to the nozzles of the peripheral craft were therefore devised in order to maintain the cushion air with substantially increased structural clearance, but with small daylight clearance, in the manner shown in *Figure 3.92*. In the SES context the flexible skirts, often then called seals, are only installed transversely at the bow and stern,

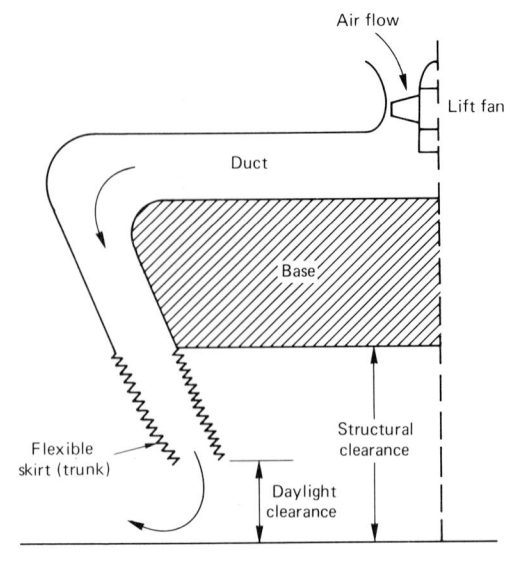

Figure 3.92 Simple flexible skirt extension

whereas ACVs have skirts around the complete periphery.

There are a number of basic forms of skirt such as the bag finger, loop finger, hinge seals, flexible bag, jupe (developed for the now abandoned SEDAM amphibious hovercraft built in France) and pericel (combining bag finger and jupe variants and used in the US Navy amphibious assault landing craft (AALC) development). Mantle (1980) and Elsley and Devereux (1968) give detailed accounts of the development of skirts with the latter concentrating on BHC designs leading to the bag finger design incorporated on, for example, the SRN4 craft, and shown schematically in *Figure 3.93*. The finger height is usually between 50% and almost 100% of the structural clearance. The loop finger skirt of Hovercraft Development Ltd has the bag as a loop with membrane *A* in *Figure 3.93* omitted. The pressure in the loop is close to cushion pressure giving rise to a softer spring and a more energy efficient lift system. It is, however, more liable to tucking under the craft when contact with waves is made which encourages plough in at the bow. The pericell skirt has effectively conical fingers mounted below the bag in-

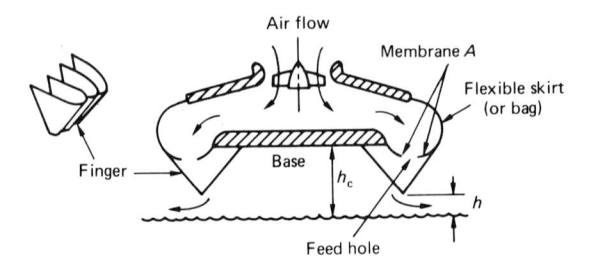

Figure 3.93 Bag finger skirt system developed by BHC

stead of the type shown in *Figure 3.93*. It is a very stiff system and so is ideal for high speeds in calm water. A hinged bow seal has been produced for use with SES in which the top edge of the sealing plate is hinged at the hard structure of the bow and swings into the cushion as forward flight commences. The extent of the backward movement is controlled by a small bag spring between the seal and the base near the hinge. Some planing forces can also be derived from this device. British SES such as the HM2 and HM5 series have a bag finger bow seal and the stern seal consists of a flexible multiple loop form as indicated in *Figure 3.85(b)*.

For ACV, the bag pressure varies with different craft but is generally 50–60% higher than cushion pressure whereas for SES the value is 10–20%. A particular problem encountered on many hovercraft is skirt bounce, a low frequency, limit cycle oscillation which appears to occur in calm water often at low speeds. Matters become even more serious if this oscillation at 1–3 Hz coincides with the natural frequency of the craft. In essence, a disturbance causes the fingers to rise, increasing the air gap which increases the escape air flow rate and that is proportional to the square root of the difference between the bag and cushion pressures. This change in pressure results in a change of pressure ratio between bag and cushion which changes the bag shape and, thus, finger depth. It is found that the bag to cushion pressure ratio can be used as a criterion for the possible onset of bounce. For ratios just above unity (a soft spring condition) bounce is prevalent with the unstable region likely to occur at ratios between 1.2 and 2.0. Values exceeding the upper limit would require unacceptable lift powers and so most hovercraft can suffer from skirt bounce.

An antibounce web has been successfully introduced by BHC into the longitudinal skirts of present day craft along with an anti-plough-in web along the bow skirt with typical pressure ratios of 1.6. Fingers are also subject to oscillations between 10 and 20 Hz which arise from fluctuating tension forces at the lower edge of the fingers in contact with the water. Other oscillations occur if the finger is required to follow waves and maintain a constant air gap (Mantle, 1980). Further details on modern skirt designs and arrangements are given by Butler (1985) and Lavis (1985), and Wheeler and Key (in Trillo, 1986) describe BHC hovercraft skirt design and manufacture.

3.11.3 Drag components

Considering the general case of an ACV moving at speed over water, we can expect the following components of drag.

3.11.3.1 Profile drag

This component of aerodynamic drag comprises skin friction drag and viscous pressure drag on the hovercraft superstructure and is related to the frontal profile area above the waterline. Hovercraft have a rather bluff geometry and so the profile drag D_p can be written in the form:

$$D_p = \tfrac{1}{2} \rho V^2 S_f C_{D_p} \qquad (3.202)$$

where ρ represents the density of air, V the craft velocity, S_f the frontal area normal to the direction of V and C_{D_p} the profile drag coefficient considered independent of V. Values of C_{D_p} vary from best values of about 0.32 for SES 100B to 0.38 for SRN6 and 0.4 for SRN4. At high cruise speeds D_p can be the highest component of drag.

3.11.3.2 Momentum drag

The horizontal velocity of the air in the atmosphere that is drawn vertically down into the lift fan is increased from zero to the craft velocity. In so doing, that mass flow of air experiences an increase in momentum and the reaction on the craft to the resulting force gives rise to the momentum drag

$$D_m = \dot{m}_f V = \rho Q_f V \qquad (3.203)$$

where $\dot{m}_f$ is the total mass flow rate of air passing through the lift fans and Q_f the corresponding volume flow rate. Engine air taken in by similar techniques is usually exhausted aft at a sufficient velocity to recover, or even exceed, the intake momentum loss.

3.11.3.3 Induced drag

Reference to *Figures 3.91* and *3.94* show that over water the craft takes on a small angle of trim so that the total upward supporting force vector F_v is tilted backwards to the vertical, although the vertical component $F_v \cos \alpha = W$ for equilibrium. A backward component, giving rising to an induced drag D_i, is therefore obtained as:

$$D_i = F_v \sin \alpha = W \tan \alpha \approx W\alpha \qquad (3.204)$$

since α is only of the order of a few degrees when the craft runs at its cruise speed.

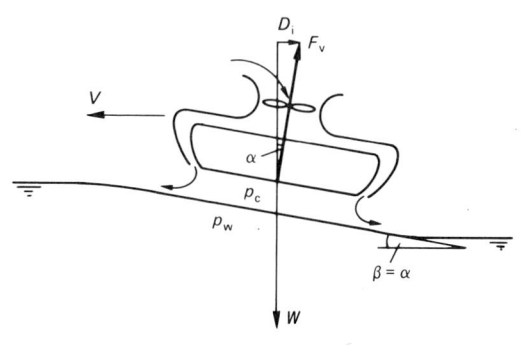

Figure 3.94 Steady forward motion of ACV over calm water

3.11.3.4 Wave making drag

Wave making drag is transmitted to the vehicle through the air cushion whether or not direct contact is made between the craft and the water. This source of drag can be interpreted as the component of pressure forces on the water depression in the direction parallel, but opposite, to V. The wave making drag D_w, is thus given by:

$$D_w = \int_{S_d} p_w \sin \beta \, dS_d = \int_{S_d} p_c \sin \beta \, dS_d$$
$$= p_c \int_{S_d} \sin \beta \, dS_d \quad (3.205)$$

where S_d and β are the surface area and inclination of the depression, respectively. For equilibrium, the water pressure at the interface $p_w = p_c$. The effect of the jet on the deformation of the interface has been neglected since we shall consider that the craft is travelling at high cruise speeds. Let us assume that the weight of the craft is supported primarily by cushion pressure forces, then:

$$W = \int_{S_d} p_c \cos \beta \, dS_d \quad (3.206)$$

If β is constant and the depression is parallel to the base of the craft (see *Figure 3.94*):

$$D_w = W \tan \beta = W \tan \alpha \approx W\alpha \quad (3.207)$$

for small trim angles.

When the craft is hovering at rest it may be shown that the volume of air in the depression, measured relative to the still water level, is equal to the volume of the water displaced by the vehicle if it were to float at the interface. It is thus reasonable to expect such a criterion to hold at low forward speeds, when D_w/W is large. The distortion of the interface reduces as V increases and so, in common with other high speed vehicles, the wave making drag of ACV is low at cruising speeds.

Lamb (1945) examined the behaviour of a two dimensional disturbance of constant pressure p_c and length L, moving steadily over an initially plane liquid surface. The theory was adapted by Crewe and Eggington (1960) specifically for hovercraft applications and they found that the ratio of the wave making drag per unit width D_w', to applied force per unit width $W' = Lp_c$, could be represented by the relationship:

$$\frac{D_w'}{W'} = \frac{2p_c}{\rho_w g L} [1 - \cos \{(Fr)^{-2}\}] \quad (3.208)$$

where the liquid is taken to be water of density ρ_w and the Froude number $Fr = V/\sqrt{(gL)}$. The maximum value of $D_w' \rho_w g L / 2 p_c W'$ occurs when $\cos (Fr)^{-2} = -1$, that is, when $(Fr)^{-2} = (2n + 1) \pi$, where $n = 0, 1, 2, \ldots$. Under these conditions the maximum value of wave making drag is:

$$(D_w')_{max} = 4W' p_c / \rho_w g L \quad (3.209)$$

and for $n = 0$ this will correspond to a primary hump at $Fr = 1/\sqrt{\pi} = 0.56$, a value confirmed in practice.

Although the preceding general trends of wavemaking drag as a function of Fr are displayed by actual hovercraft, more accurate, three dimensional theories are required to improve quantitative predictions (Newman and Poole, 1962; Barratt, 1965; Doctors, 1970). Barratt found the primary hump to be located in the range of $0.5 \leq Fr \leq 1.0$ depending on planform and his results have been supported, especially in the primary hump region in both shallow and deep water (Everest and Hogben 1967, 1969). Rather poorer agreement is found in the vicinity of the secondary and tertiary humps which appear to be significantly affected by the pressure distribution at the edge of the water depression (Newman and Poole, 1962; Doctors, 1970).

Some SES model data obtained by Wilson et al. (1978) were also well matched to theory by considering the edge effect predictions. In shallow water, the primary hump occurs at a lower Froude number than for deep water and $(D_w')_{max}$ increases as the depth of water decreases. The value of being able to accurately predict the value of $(D_w')_{max}$ is that a positive thrust margin can be accounted for in the propulsion system to allow acceleration past the local maximum of the total drag corresponding to the hump speed. Furthermore, the magnitude of $(D_w')_{max}$ reduces with increase in cushion length to beam ratio L/B, and also occurs at higher Fr as L/B increases. Data obtained by Ford (1978) for a projected 8000 tonne SES showed that for the craft to travel at 40 knots a value of $L/B > 4$ eliminated the primary hump and wave making drag was substantially lower than an equivalent low L/B craft. On the other hand, for very high speeds above 60 knots a value of $L/B = 2$ produced a craft with substantially lower wavemaking drag. Similar effects apply to other sizes of SES, and also to ACV and this topic is discussed further in the following section.

3.11.3.5 Skirt and seal drag in calm water

It is known that a substantial, often the largest, component of drag comes from direct contact between skirts or seals and the ground or water and yet there is no thoroughly satisfactory method for its prediction. Model tests and empiricism tend to be the order of the day, although reasonably reliable data are available for certain skirt systems such as the BHC bag finger geometry and the US Bell systems. The calm water skirt drag arises from skin friction at contact with ground or water surface, cushion pressure effects and spray. The magnitude of this drag component is affected by, for example, air gap clearance, craft size and shape, cushion pressure as it influences skirt stiffness and shape, and spray generation. Wheeler (1968) has examined the implications of air gap on the SRN4

Mk1 hovercraft, pointing out that calm water skirt drag can be as high as 30% of the total calm water drag, although it probably becomes progressively less significant as the craft size increases. In the context of SES and ACV over water, the calm water skirt drag, D_{scw}, is often divided into two components; wetting drag, D_{sw}, and the related wave making drag D_{swm}. In order to reduce the more obvious wetting aspect, spray inhibiting skirts have been developed to give well over 50% reductions in spray volume.

Mantle (1980) gives the following empirical relationships based on an analysis of Bell and BHC data:

$$\frac{D_{sw}}{W} = K_1 [h/l_s]^{-0.34} \left(\frac{1 + L/B}{(L/B)^{0.5}}\right) \left(\frac{\frac{1}{2}\rho_w V^2}{p_c}\right) \quad (3.210)$$

$$\frac{D_{swm}}{D_w} = K_2 \left(\frac{p_c}{L}\right)^{-0.259} - K_3 \quad (3.211)$$

where l_s represents the peripheral length of the hemline. The constants K_1, K_2 and K_3 depend on the system of units used, and these are often inconsistent. For the British units adopted by Mantle (1980), the values are 0.0058, 1.374 and 1.0, respectively. The effect of length to beam ratio on D_{sw} is fairly small; the main influences are the skirt clearance height ratio h/l_s, for skirt wetting drag and the cushion density p_c/L, for skirt wave making drag.

3.11.3.6 Skirt and seal drag in rough water

Again, satisfactory theoretical analyses are yet to be obtained but a considerable body of experimental data is available. Any techniques that are used can be considered only for sea conditions where the wave height does not exceed about 80% of the cushion height h_c (see *Figure 3.93*). For larger wave heights contact will be made between the water and the hard structure of the craft and drag is then more strongly related to craft response. As Elsley and Devereux (1968) point out, and still relevant today, the rough water drag component involves predicting the preceding drag components and attributing the rough water drag to the difference between the sum predicted and the measured total drag. They do in fact show that the rough water drag D_{srw} can be examined on the basis of the functional relation:

$$D_{srw}/W = \text{function } (p_c/h_c; h_w/h_c; V/V_H) \quad (3.212)$$

where h_w represents the significant wave height and V_H the craft speed at the primary hump.

Mantle (1980) has examined different skirt systems and shown that an analytical representation of Equation 3.212 can be obtained in the same form as that on the right-hand side of Equation 3.210 but with a different constant and with the second term in brackets replaced by $(h_w/h_c)^{5/3}$.

Skirt drag in rough water is also affected by the wearing away of material at the lower edges of the fingers (cells or segments). Results obtained from SRN 4 craft showed that, when new fingers replaced those that had worn, a speed increase of 3–4 knots took place over waves up to 2 m significant height. Another way of representing the effects of rough water drag, at wave heights less than those resulting in sea impact, is to consider speed reduction. A reasonable fit by Mantle (1980) to existing data, which also represents current trends, yields the expression.

$$V/V_c = 1 - 0.50 \, (h_w/h_c)^{0.75} \quad (3.213)$$

where V_c represents the speed in calm water obtained at the same power setting as the rough water speed V. Thus, a 30% drop in speed is to be expected when the significant wave height is 50% of the average cushion height.

From the foregoing, it can be seen that the total skirt drag can be written in the form:

$$D_s = D_{sw} + D_{swm} + D_{srw} \quad (3.214)$$

3.11.3.7 Drag over ice

Considerable opportunities, both commercial and military, are available for ACV operation over tundra, sea and ice for year round activities. In connection with oil exploration, Mantle (1980) quotes results from a US arctic research investigation using various models for operation over 1.8 m ice ridges using craft fitted with 2.7 m deep skirts, that is, $h_c = 2.7$ m. The results lend support to theory which suggests:

$$D_{ice} \propto V^{2/3} \quad (3.215)$$

A value of $D_{ice}/W = 0.02$ appears likely for 170 tonne ACV operating at 60 kt and a corresponding D_{srw}/W value could be about 0.05, albeit at a wave encounter frequency much larger than the ice ridge encounter frequency.

3.11.3.8 Sidehull drag

The total drag of SES has a major contribution from sidehull resistance and appendage resistance (from rudders, fins, propeller shafts, pods and bosses, etc). Most operating sidehull craft today (HM2, SES-100A, SES-100B) have L/B ratios approximately equal to 2 and cruise conditions consistent with $Fr > 1.5$ and so they must then exhibit high efficiency. The HM5 craft with L/B about 2.7 has an operating $Fr \approx 1.2$. Tests on the SES-100A, which possessed pods for intakes to the hydraulic jet propulsion systems, showed that at $Fr = 2$ the drag from the sidehulls and appendages amounted to about 50% of the total drag. When the speed is sufficiently high, as in the original concept for SES operations in the US, the prospects of cavitation on the sidehulls and appendages becomes important. However, it has been found that planing hull forms offer advantages in developing additional dynamic supporting force. In contrast to normal planing hull forms, discussed in Section 3.9 in this chapter, those for SES are very slender, e.g. a ratio of hull beam to length of 0.1 to 0.05. Also, the cushion side of each hull experiences a depressed water line with crossflows under the hulls so that again reliance is placed on model tests for full scale predictions.

The total sidehull drag D_h may be considered to consist of wave making drag and viscous drag components and can be written in the form:

$$D_h = 2 \{\Delta \tan \tau + {1/2}\rho_w V^2 C_f S_h \sec \tau\} \quad (3.216)$$

assuming each hull acts identically. In Equation 3.216 the vertical buoyancy force Δ is usually decided, by design, to be between 5% and 10% of the all-up weight, the skin friction force acts over the wetted surface area S_h of each side hull, C_f is the skin friction coefficient and τ the planing angle which may not necessarily equal the craft trim angle or the cushion wave drag angle implied in Equation 3.207. There is some difficulty in determining both Δ and S_h as explained by Mantle (1980).

The total drag force of a hovercraft in steady forward motion is therefore given by:

$$D = D_p + D_m + D_i + D_w + D_s \quad (3.217)$$

for ACV over water. In the case of motion over land D_w is absent and D_s is modified; for SES the sidehull component D_h must be added to the right hand side of Equation 3.217. Finally for operation over ice, D_{ice} replaces the D_{srw} component of D_s and D_w is absent.

3.11.4 Performance in steady motion

A typical variation of the total drag of a large amphibious hovercraft (SRN4), as a ratio of all up weight, is shown in *Figure 3.95*. Wave making drag at the hump speed is seen to be very important as it provides the criterion for thrust margin at low forward speeds. The induced drag is taken to be small at high cruise speeds but then profile, momentum and wetting drag components are large, as is the rough water drag. Wave making drag becomes small at high speeds.

Similar results for an SES design are shown in *Figure 3.96(a)* for $L/B = 2.0$ and in *Figure 3.96(b)* for $L/B = 6.0$. These curves are dramatically different, emphasizing the desire for high L/B ratios for operations at Fr ≈ 0.85 but low L/B ratios for Fr ≈ 1.7. Now, D_w increases as W^2 (see Equation 3.208), so with other parameters remaining unchanged and assuming p_c varies with W, it is evident from a propulsion point of view that a high L/B craft is the more suitable for variable loads and only modest speed changes. Mantle (1980) has also investigated, with some success, correlations on the basis of a drag polar, i.e. drag coefficient *versus* load or lift coefficient, which has analogy with aircraft performance.

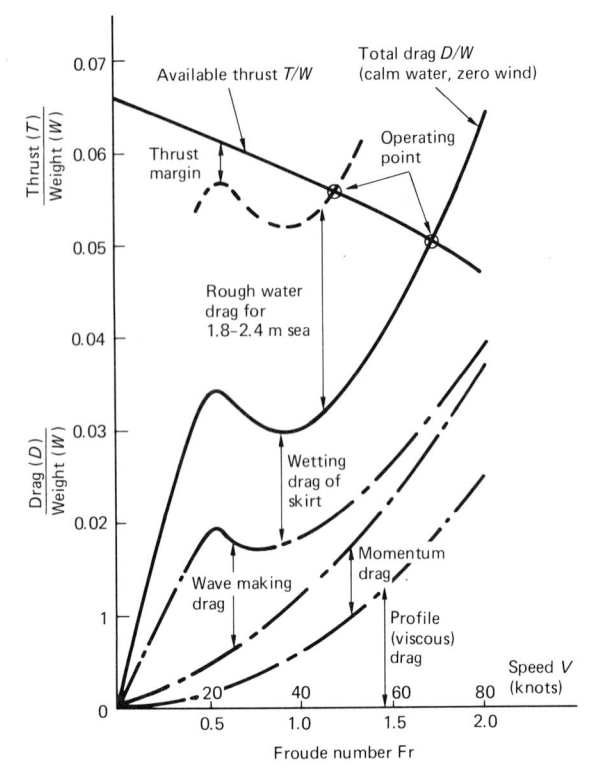

Figure 3.95 Performance characteristics of a large SRN4 ACV in calm and rough water

For steady motion the operating point is that at which the thrust curve meets the total drag curve, i.e. when $T/W = D/W$. The effective propulsive power P_e is then given by $P_e = TV_d = DV_d$ where V_d represents the design velocity of the craft. To determine the installed propulsive power requires knowledge of propulsor, transmission and overall engine efficiencies which for the sake of simplicity will be grouped here under the total propulsive efficiency η_p. Thus, the propulsive power $P_p = P_e/\eta_p$. Writing the combined efficiency of the lift fan, duct and nozzle as η_f, the installed lift fan power $P_f = P_n/\eta_f$ and so the total installed power $P = P_p + P_f$.

Typical examples of the installed power of hovercraft are:

4 × 2850 kW (BHC SRN4 Mk3), 4 × 320 kW (BHC AP1-88),
1 × 750 kW (BHC SRN6),
2 × 800 kW (Mitsui MV PP15),
4 × 660 kW (Wartsila PUC 22-2500SP),
6 × 2100 kW (Bell AALC Jeff(B)), 2 × 1350 kW (Bell LACV-30),
2 × 333 kW + 1 × 155 kW (VHL HM-218),
2 × 1050 kW + 2 × 332 kW (VHL HM-527)

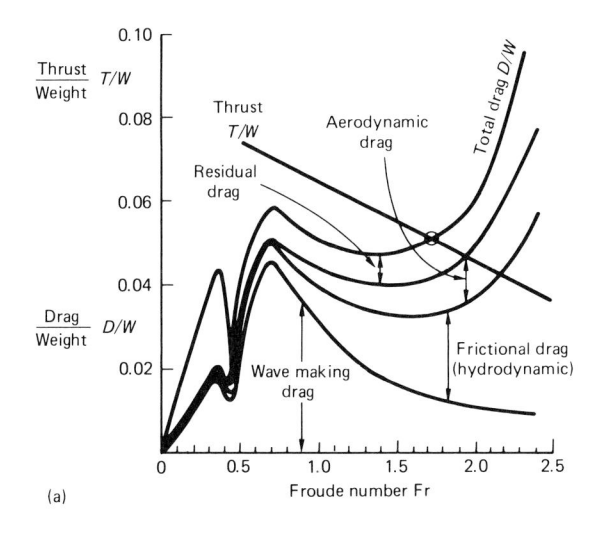

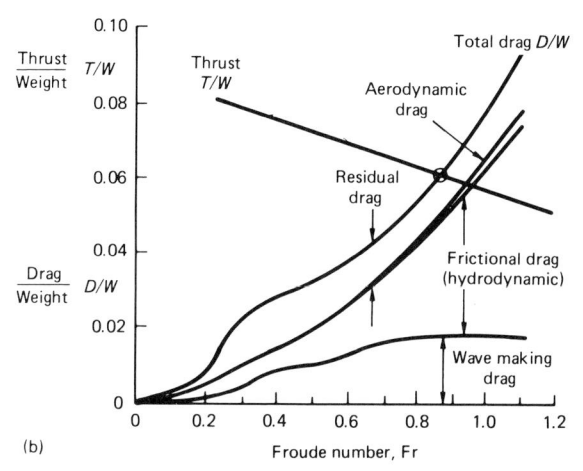

Figure 3.96 (a) Drag components of an SES with length to beam ratio of 2.0 (b) Drag components of an SES with length to beam ratio of 6.0 (after Butler, 1985)

3.11.5 Performance in waves

The response of hovercraft to surface wave excitation in terms of ride quality has, like stability analysis, been the subject of many studies. Full understanding has yet to be reached since great difficulties are met in the analysis of non-linear forces acting on the craft and the variable geometry of cushion systems as the skirts adjust to passage over random waves. Although the craft is, in principle, subjected to six degrees of freedom, we shall concentrate here on pitch and heave motions as these have a major effect on crew and passenger comfort. It may be noted that, especially for ACV, yaw and sideslip can be large. For SES, additional hydrodynamic forces arise from sidehull and appendage immersion and these forces are both exciting and damping in nature and provide lateral stability. A major complication in the scaling of model data to full size arises from cushion pressure which appears in the equations of motion as an absolute pressure. Much work has been done using simulations, but these often require many empirical coefficients which apply to a given craft and cannot be generalized. As indicated earlier, the dynamic response of the fan and its effect on the pressure flow characteristic have major influences on the response of hovercraft to motion over waves or rough ground. Indeed, one important general technique for motion control requires modification of the fan responses by means of a feedback control system sensing cushion pressure on craft motion parameters.

Generally, the worst effects of heave and pitch displacements and accelerations are felt when a hovercraft travels at cruise speed head on to short steep waves which have lengths between 2 and 0.5 times the craft length. The craft tends to contour over long waves and to platform over very short waves. In the latter case, the skirt is often excited into a high frequency oscillation, usually referred to as cobblestoning, which may occur in virtually calm water.

A first order theory of heave stability is given by Elsley and Devereux (1968) and Mantle (1980). The results of model tests tend to support the simple linear equation of motion having constant damping and stiffness coefficients. These coefficients vary from one craft to another and, in practice, it is far from straightforward to determine their magnitude with good accuracy. Mantle (1980) quotes data to show that the heave damping coefficients and craft acceleration correlations between model data and model theory are good, but dramatically poor between model data scaled to full size and full scale theory. In general, the following statements are compatible with improvements in heave static stability: increase air gap heights and therefore air flow rate; decrease skirt height; decrease cushion pressure; increase air flows and use cushion lift fans with flattish curves, i.e. modest slopes and pressure rise per unit volume flow.

Butler (1985) discussed the responses of SES in waves. The bow and stern seals should ideally follow the wave contour, although often they do not, and the keels of the sidehulls always remain submerged to provide complete sealing. Pitch and roll motions are of particular importance and it should be noted that for many present day sidehull designs (based on planing hull forms) substantial damping and added mass terms must be included in a hydrodynamic analysis of responses to waves. Hydrostatic pitch and roll restoring forces for hullborne and cushionborne operations are developed from sidehull buoyancy. Hovercraft motions depend on the extent of the tuning between wave encounter frequencies and the craft natural (damped) heave, pitch and roll frequencies. Butler (1985) shows that, taking on-cushion operation of the SES 200 as an example, heave and roll frequencies are sufficiently large to avoid tuning with large amplitude waves in high sea states. Pitching, however, can present problems if craft speeds are not very high. Present requirements perceived for SES, of up to about 1500 tonne displacement, tend to centre on cruise speeds no greater than 35–40 knots so pitching in a seaway can become severe. Fans with flattish pressure–volume ($p_f - Q$) characteristics, that is $\partial p_f / \partial Q$ modest and negative, are recommended to increase heave damping (Blyth 1986, 1988; Kaplan et al. 1981). It is clear from this work that the vertical centre of gravity location is very important in determining the stability of SES in heel. Roll stability is also reduced with increasing speed and is especially important during high speed manoeuvres performed with twin rudder control or differential thrust of propellers on each sidehull.

Pitch and roll stability can be substantially improved by using compartmentation techniques, now common on modern ACV such as those of BHC and Bell. The plenum is divided transversely by an inflated skirt, normally a bag type, and again longitudinally between the stern skirt and the transverse divider or right through to the bow skirt. This method may also be combined with techniques for ensuring a shift of the cushion centre of pressure towards the downward edge to provide a restoring moment; a scheme devised by Hovercraft Development Ltd of the UK. The essential behaviour of a compartmented cushion is illustrated in *Figure 3.97* in which the force generated by the additional cushion pressure Δp_c on the left hand side provides a restoring moment, M_ϕ in roll or M_θ in pitch, to recover equilibrium. Transverse cushion dividers have also been installed on some modern SES to

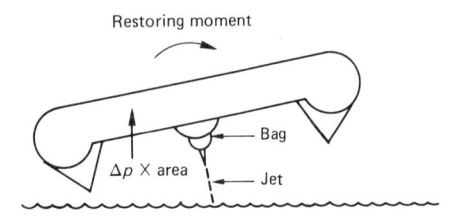

Figure 3.97 Behaviour of a compartmented cushion

improve pitch stability. The pitch and roll stiffness is then represented as a fractional (or percentage) change in the shift per degree of the centre of pressure, i.e. as M_ϕ/WB or M_θ/WL where B and L refer to cushion beam and length, respectively. Lavis (1980, 1985) has discussed stability standards of ACV and notes that increasing the height of the centre of gravity (CG) above the water line destabilizes the craft and so as this height increases the pitch and roll stiffness must also be increased. Taking cushion height as a fixed ratio of the height of the CG above the still waterline, Lavis (1980, 1985) has shown (see *Figure 3.98*) a comparison of roll stiffness values for different craft. It should be noted that the SRN5 and SRN6 craft are early versions and

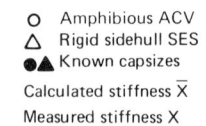

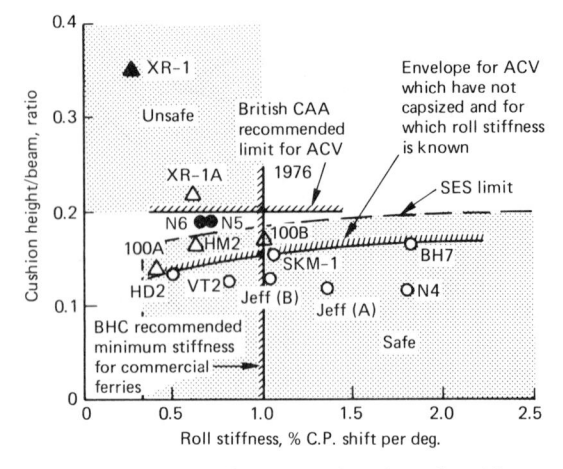

Figure 3.98 Roll stiffness as a function of cushion height to beam ratio for various hovercraft (after Lavis, 1985)

that such curves are not sufficient for a full measure of dynamic stability assessment.

A problem, particularly with the ACV, but also SES, has been plough in, which can lead the vessel to capsize. A plough in occurs when the bow skirt contacts the water because of an excessive nose down trim angle, e.g. greater than 4°. The hydrodynamic drag produced by the contact maintains the condition and decelerates the craft until trim is corrected or a wave breaks the contact. Provided contact is symmetric about the bow there is usually sufficient longitudinal stability to avoid capsize. However, if contact is asymmetric, the craft may yaw to large angles and overturn, owing to lower transverse stability and break away of the stern skirt from the water. These problems were primarily experienced by early craft and so strakes and lubrication holes were installed on the bag skirts then in use. The subsequent addition of fingers to skirts provided the required lubrication between the fingers and now the problem is large solved. Limits to speed and yaw angle are still imposed on commercial ACV so that, for example, 40 knots is the limit for a 40° yaw angle (Lavis, 1985; Mantle, 1980).

Ride control systems (RCS) have received considerable attention over the past two decades as these attempt to provide a continuous means of attenuating motions as hovercraft pass over waves. There has been more success with heave damping than with pitch damping and Butler (1985) gives an account of US Navy developments on small and large test craft. One particular RCS undergoing tests on the SES 200 regulates cushion pressure by using cushion vent valves and/or variable inlet guide vanes (VIGV) on the lift fans to modulate air flow rate. In the lower sea states, only VIGVs are used, since venting of air wastes lift fan power. However, venting is considered necessary when substantial wave pumping takes place in high sea states. Fans with VIGV also supply the bow and stern seals. The actuators on both the VIGV and vent valves are controlled by feedback signals from a cushion pressure transducer through signal conditioners and control algorithms which may be adjusted according to craft size. Adams and Beverly (1984) report that on SES 200 a 50% reduction in heave acceleration was obtained in sea states 1 and 2 and 25% reduction in sea state 4.

Clayton and Tuckey (1983) and Clayton and Webb (1986) used a wavebelt rig with large models to examine the responses to head seas. As a result, an alternative method of reducing SES motions has been developed by Clayton and Webb (1988 a,b,c) in which a variable position conical plunger is used at the centrifugal lift fan inlet region. There is no doubt that a rapid and direct response to wave excited cushion pressure changes can be obtained by adjusting the fan characteristics. The plunger system, which moves along the axis of rotation, is considered extremely energy efficient because only the axial length of the fan blade is adjusted and flow angles are unchanged. The plunger then acts as a flow controller with nominally constant pressure across the fan. For the wavebelt model tests, control was applied to several cushion fans, the bow seal fan and the rear bag seal fan. The craft was run into head sea conditions, the worst case anticipated in terms of heave acceleration, as shown in the results presented by Mantle (1974) on SES 100A tests. A model based control system was used whereby a simulation of the linearized craft dynamics is run simultaneously with the hovercraft and the parameters from the simulation used for control feedback. The simulation is corrected using the pitch and heave measurements at each time increment. There is then a significant reduction in measurement noise, because the states that are fed back are corrected by the model dynamics. All the states are also available for feedback giving greater flexibility in choosing the dynamic behaviour without the necessity for large numbers of transducers to measure the state directly. Both heave and pitch motions on the model can be reduced to between 25% and 50% depending on wave frequency and amplitude. Some pitch motion is needed to avoid large leakage flows under the skirt in wave troughs, which could cause the craft to drop partially off cushion. The system is currently under test on an SES model operating over waves and possessing sidehulls that produce substantial buoyancy forces.

3.11.6 Structural design

Essentially, the overall design concepts of ACV have tended to follow the practices for aircraft whereas SES have been based more closely on those for small boats and ships. The essential need for light weight was paramount in early ACV since skirts were relatively inefficient. Thus, to keep lift power within bounds, cushion pressure and hull loadings were low. Even later on, maximum payload requirements called for similar strong and light main structures even with substantially higher cushion pressures.

Some of the aircraft practice was relaxed when the efficient low pressure ratio responsive skirt system was installed on the SRN4 Mk3. The demand for lift power was less and drag was reduced so that high cost aircraft technology could, at least partially, be abandoned for the more competitive marine approach to structural design. High grade alloys were adopted in the fabrication, but extensive use was also made of sandwich panels for buoyancy tank and deck construction and these have proved remarkably durable. Despite many improvements during the production phase of SRN4 and BH7 craft, the structure contains many parts and innumerable mechanical fastenings requiring meticulous sealing and corrosion protection which proved to be costly and labour intensive.

The magnitude and distribution of loadings from both internal and external sources are needed for the structural design. Fatigue calculations arising from force and moment fluctuations require the specification of sea states, machinery vibrations, cargo movements and docking frequencies. Loadings can thus be grouped under the following headings: ground contact, hydrodynamic, aerodynamic, machinery and payload. Elsley and Devereux (1968) show a design approach which includes the fundamental aspects needed in analysing the hovercraft structure for size, strength and shape.

The weight and strength requirements of ACV led to the use of gas turbines, as high power to weight ratio devices, for driving both the lift and propulsion systems. However, extensive gearing is needed to transmit the lower rotational speeds required by the lift fans and the variable pitch propellers (airscrews). Manoeuvring of ACV must be done aerodynamically and on SRN4, for example, rotatable pylons support the propellers. Rudders positioned in the slipstream are available for course keeping. Hydraulic power is used to operate these systems, with the result that purchase, maintenance and running costs are at the limit of acceptance by the commercial market, especially during periods of high fuel costs.

In order to overcome these problems BHC embarked on the design of the AP1-88 as the SRN6 replacement. The aims relative to SRN6 were to halve the production cost per passenger seat place and the operational cost per passenger mile; considerably reduce external noise levels and significantly improve passenger comfort. These requirements meant extensive use of automatically welded, light alloy structures, the adoption of diesel engines and propulsion from two ducted propellers. Wheeler (1984) shows that all these aims were satisfactorily met in combination with the SRN4 Mk3 skirt system which allowed the extra weight of the structure and power plant to be propelled efficiently. To minimize the amount of welding, large extrusions were used for producing the buoyancy tank frames, top and bottom plating and roof frames. The total construction yields a weight per cushion area about twice that of SRN6. Manoeuvring and control is by use of rudders abaft the propeller and bow thrusters. A small hydraulic power assist system is included, since rudder moments are too great for manual operation. There is no doubt that the direction now being taken in the design of commercial hovercraft, epitomized by the AP1-88, is the correct one in the present competitive market.

With reference to the US military ACV, Lavis (1985) states that the basic hull structure is not the principal culprit in terms of cost. The main problem is associated with equipment systems that must be selected from existing applications in aircraft or ships. Cost and/or weight are high and incompatible with ACV operating in the marine environment. A summary of the LCAC producibility and supportability is given by Lavis (1985). The craft are essentially of aluminium alloy construction, powered by four gas turbines that drive two stern mounted, ducted air propellers and four centrifugal lift fans. The philosophy related to production follows that set in respect of the AP1-88 by making extensive use of automatic welding, modularization and extruded sections.

The principles associated with the structural design of SES are summarized by Butler (1985), specifically for US designs which use the reliability method of load determination, prediction of structural weight, design criteria and material selection. The approach is compatible with that for ships using a ship design optimization code SHIPDOC (Richardson and White, 1984). The off cushion longitudinal load due to head sea slamming in survival sea states is dominant, compared with longitudinal bending moment of ships arising from weight or buoyancy differences. The SES may be regarded as a ship in which the main payload section is hydrodynamically separate from the portion of the hull affected by the passage of water. Fatigue loading is obtained from a cumulative damage rule knowing the experimentally estimated probability density loading curves. The stress used is the nominal stress multiplied by the stress concentration factor applicable at the location under investigation. Once structural criteria specifying the combined loads and safety factors are available, there are a number of structural optimization programs to generate the familiar longitudinally framed plate and grillage arrangement.

Materials used for SES hulls are high strength marine aluminium alloys (SES 100A, 100B and 200) and GRP (VHL HM2 and HM5). High strength steels may be more attractive for larger craft. Tattersall gives a summary of the GRP production plant at VHL, covering HM2 and HM5 building, in the discussion by Butler (1985).

Skirt materials for both SES and ACV tend to be similar and based on nylon fabrics coated with Neoprene or natural rubber. This decision has been reached on the basis of an optimum of the following requirements:

1. Sufficient flexibility for good energy absorption and wave following in a seaway.
2. Sufficient stiffness to resist flutter in normal operation.
3. Good flutter damping behaviour in a seaway.
4. Sufficient tear strength and fatigue resistance at peak loads.
5. Resistance to flagellation damage at free edges.
6. Resistance to cracking and delamination in cyclic flexing and buckling.
7. Resistance to abrasion.

Bow seals for SES are either bag and finger or lobe and finger geometry similar to ACV, although semiflexible seals have been used with rigid elements incorporated to carry some of the severest hydrodynamic loads. Flexible stern seals tend to extend the full width of the cushion and to be enclosed single or multiple lobe bag form; again rigid elements have been incorporated in some US designs. Further details on skirt materials, design and behaviour are given by Mantle (1980).

3.11.7 Applications

Reviews of applications are given by Lavis (1985) for ACV and Butler (1985) for SES. The maritime future for hovercraft in the UK has been considered by Betts and Clayton (1987), and Trillo (1988) provides an overall technical review and a catalogue of craft. A brief summary of the main points is given here with concentration mainly on commercial exploitation.

Ferries have proved to make up the most widespread use of ACV starting in the UK in 1962 off North Wales and across the Solent between Ryde and Southsea using the SRN2 and now still running currently with the BHC AP1-88. The introduction of the SRN6, starting in the mid-1960s with cross-Solent ferries, allowed rapid development of applications worldwide. In terms of numbers of craft in operation, those from the British Hovercraft Corporation (BHC SRN6 and SRN4 Mks1–3) have tended to dominate the market although Japan, France and Finland have been increasingly active.

The original SRN4 Mk1 service across the English Channel between Dover and Calais commenced in 1968 and has run continuously with an additional service between Dover and Boulogne. A previous service also ran between Ramsgate and Calais but has been discontinued. The SRN4 MK3 craft (*Figure 3.86(a)*) currently used have a capacity for just over 400 passengers and 60 cars and a maximum speed of 70 knots. A similar sized French design, the Sedam N500, no longer runs because the company has been wound up. The newer routes to be inaugurated are those operating in the Finnish Archipelago since 1983, using the PVC22 – 2500 SP craft and the SAS service across the Oresund since 1984 between Kastrup Airport, Denmark and Malmo, Sweden using AP1-88 craft. It is interesting to note that these last two services operate with fuel efficient, diesel powered craft whereas previous routes have operated with gas turbine powered craft. In all, there are probably about ten regular (some seasonal) routes in operation, although others may now exist in the USSR and China.

There is little doubt that the premier commercial routes for ACV are those operated by Hoverspeed across the English Channel. Concern has, however, been expressed on the future viability of these routes now that the Eurotunnel Consortium is advancing with the construction of a railway tunnel connecting the Kent coast near Folkestone with the French coast near Calais. Several commentaries have been published on the implications (Meredith, 1986; Clayton, 1987). The general view is, that provided holiday traffic, the major revenue earner, is maintained, there will be sufficient business for hovercraft operations to continue, although new routes should also be sought. The overall infrastructure must be publicly acceptable, i.e. good road and rail access, minimum customs and immigration checks, good reliability, and improved comfort and ride. The latter points could be met by stretching further the SRN4 Mk3 to produce a 450 tonne craft (Wheeler, 1985) which would increase load capacity, allow longer sea passages and operate at higher sea states.

From a military point of view, most ACV activity has been carried out in the USA and the UK. Currently no trials are under way in the UK, although the Inter-Services Hovercraft Trials Unit did conduct a great deal of work on both ACV and SES. The reason for the subsequent closure of the ISHTU is thought to be as much a matter of financial cutbacks as to uncertainty over suitable roles for naval hovercraft. Other countries have, however, adopted BHC built ACV, such as the navies of Egypt, Iran, Iraq and Saudi Arabia. The mine countermeasure (MCM) role has proved to be particularly attractive, owing to the low underwater noise signature and high shock resistance of hovercraft.

Another US Navy programme has been to equip its Marine Corps with large numbers of amphibious landing craft air cushion (LCAC) vessels developed from the Jeff craft design commenced in 1970. These sophisticated and expensive craft deploy from docks and assault ships and enable troops, vehicles, equipment and heavy armour to be landed at great speed from an assault force positioned over the horizon up to 100 km offshore. It is claimed that the proportion of the world shorelines suitable for seaborne assault is thus raised from 17 to 70%. The LCAC operates at about twice the normal cushion pressure of the SRN4 Mk3 to give it the necessary high payload capability within the dimensional constraints of the assault ship docks. In the US Army programme the movement of equipment on and off container type ships within shoreline proximity called for a fast moving lighter with a flat platform. The 30 tonne lighter, air cushion vehicle (LACV-30) was thus developed, based largely on a military adaptation of the Bell Aerospace Textron commercial *Voyageur* (used by the Canadian Coast Guard). There are now several platoons of these craft operational. Operations over ice and in the Arctic regions have also been important with many craft seeing service over the past twenty years. Special considerations are called for which give rise to major changes in operational practice, design and construction to combat extreme weather and environmental conditions (Lavis, 1985).

The Canadian Coast Guard have used ACV in search and rescue, as well as servicing beacons, buoys and lightstations. A special role has been found in icebreaking in harbours and on rivers using ACV to run up and off the edge of the ice cover. An AP1-88 has been recently supplied by BHC as a replacement for *Voyageur* for both general CCG duties, including icebreaking. The use of a low speed, air cushion platform ahead of a ship has also been used to break up ice at a speed of about 9 knots (Clayton and Bishop, 1982).

There are many actual and possible applications of ACV including customs and law enforcement, fishery protection, evacuation and fire fighting. A summary of hovercraft applications in hydrographic and seismic surveying is given by Russell (1986). There is clearly a vital role available for hovercraft, especially ACV, as aid and rescue vehicles following disasters such as flooding, storms, fires, earthquakes, etc., as illustrated in the proceedings of a recent seminar (Clayton and Russell, 1989).

Recognizing the non-amphibious operation of SES, one can see similar applications as in the case of ACV. Vosper Hovermarine Ltd (now Hovermarine International) craft such as the HM2 series have been used as ferries, especially in south east Asia, as crewboats and workboats for offshore support and as fire fighting craft. They are also used in coast guard duties and a sprint and drift mode has been used for hydrographic tasks. The larger HM5 series has also seen ferry duties on the Kowloon to Macao route. Over 100 of the traditional, low L/B, post-hump speed SES are in operation outside the USSR (Tattersall, 1982) and probably three times as many of all types of SES inside the USSR. Operation as river ferries is best suited to high L/B ratio craft with speeds of 25 knots, owing to low drag and high efficiency. The thin sidehulls are insufficient to support the craft totally when off cushion, but, when hullborne or on partial cushion, the draught is sufficiently low to enable easy discharge of passengers with minimum jetty facilities. The latest generation of SES being developed by the US Navy, commencing with the XR-5A and BH110, have planing form sidehulls of sufficient size to provide more than 100% buoyancy, so that, off cushion, the craft floats like a catamaran. The US Navy SES programme is outlined by Butler (1985) and includes applications such as minesweepers and minehunters (MSH), but this has now been halted owing, it is believed, to problems with the GRP structure rather than the concept.

3.12 Multihull vessels

Apart from the slow speed, multihull vessels used as offshore structures (see Chapter 2), multihull ships take two main forms. These are the catamaran and the SWATH ship, shown diagramatically in *Figure 3.99*.

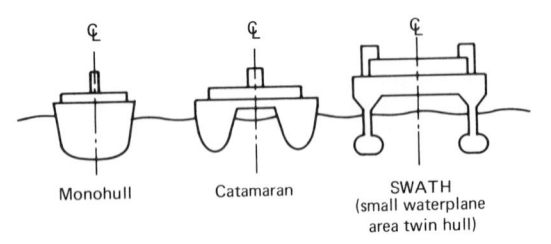

Figure 3.99 Comparison of monohull, catamaran and SWATH hull sections

3.12.1 Catamarans

A catamaran is a twin hulled vessel (see *Figure 3.99*) and its main advantages over a monohull are large deck area, some 50% greater than a monohull of the same displacement, high upright stability (GM can be ten times that of a monohull) and high speed capability if the hulls are designed in semiplaning form. Low draught and lightness of construction can also be advantages. It is usually considered that a catamaran vessel will cost more to build than a monohull of the same displacement, but less than a monohull of the same deck area.

Low speed catamarans are used where the first two of the preceding advantages are particularly sought (Michel, 1961; Corlett, 1969; Fry and Graul, 1972). These authors make proposals for very large catamarans, although few such craft have been built. The majority of powered catamarans now at sea are relatively small, high-speed vessels.

3.12.1.1 Hull forms

The individual hulls, or demihulls, of a catamaran may be symmetric or asymmetric about their centrelines. The latter can be advantageous propulsively and is usual where the hulls are relatively close together as it gives a tunnel of uniform width between the hulls. The length to breadth ratio of each demihull tends to be greater than that of monohulls, which leads to reductions in wave making. Demihull sizing is based on displacement and powering requirements and also on constraints imposed by propulsion machinery layout and draught restrictions. Hull separation and hence overall beam of the catamaran is largely a function of layout, as stability is inherently high.

High speed catamaran hulls usually have semiplaning forms with V sections and transom sterns. Novel forms of high speed wave piercing catamaran have been designed with very long, fine hulls and large crossdeck clearance in an attempt to improve seakeeping (Hercus, 1985).

3.12.1.2 Performance in steady motion

Catamarans have poor resistance characteristics at low speeds (Froude number $Fr < 0.35$) owing to a high wetted surface area and hence skin friction drag. Around the hump speed ($Fr \approx 0.5$), however, the catamaran performs better than a monohull of similar size. The hump in the resistance curve is far less pronounced as the long slender demihulls do not take up the large trim angles of a monohull at that speed. At planing speeds ($Fr \approx 1.0$) the advantage returns to the monohull as reduction in length to breadth ratio is then advantageous. The overall behaviour is illustrated in *Figure 3.100*.

Resistance and propulsion calculations for each demihull may, to a first approximation, be based on monohull methods for displacement or planing forms as appropriate to the speed regime. This will be reasonably accurate at very low speeds and, most evidence suggests, at high planing speeds. In the latter case, resistance due to spray generation and aerodynamic drag needs to be included.

At intermediate speeds, interference effects between the two demihulls can be significant. Interference arises from a combination of wave system superposition and modifications to the flow about each body. The net

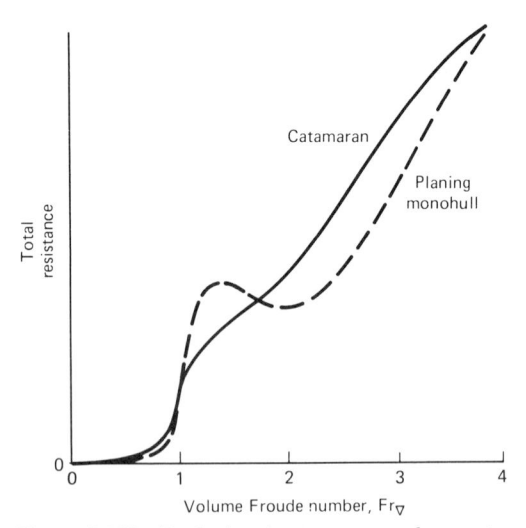

Figure 3.100 Typical resistance curves for a catamaran and a planing craft of the same displacement; volume Froude number $Fr_\nabla = V/\sqrt{g\nabla^{1/3}}$ where ∇ represents the displacement volume

effect can either reduce or increase the total resistance, depending on speed, hull form and separation of the demihulls. Adverse effects are most pronounced near hump speed and increase with reduction of demihull separation, particularly for symmetric demihulls.

Theoretical methods and experimental data for calculating catamaran resistance, with due allowance for interference effects, are given by Clement (1961), Everest (1968, 1969), Turner and Taplin (1968), Fry and Graul (1972), Yermotayev et al. (1977), Hung and Zhang (1983) and Lewis (1988). Unfortunately, these references are contradictory in some important respects and a generally agreed approach does not yet exist.

The directional stability and control of catamarans is very good compared with monohulls. There is, however, a greater tendency to broach in quartering seas, although this is offset to some extent by a lower angle of heel compared with a broaching monohull. Manoeuvrability is also good and Ozawa (1987) quotes a turning circle diameter of seven times ship length for a 41 m asymmetric hulled catamaran travelling at around 30 knots. At low speeds, the large propulsor separation may be utilized to advantage as differential devices in manoeuvring.

3.12.1.3 *Performance in waves*

The seakeeping performance of catamarans is generally inferior to that of a monohull of similar displacement. Heave and pitch motions are not very different in head seas but can be significantly worse in bow seas. Anti-pitching fins have been fitted in some catamarans, but in head or bow seas there is a tendency for slamming to occur under the cross-structure. Roll angles are limited, but the high transverse stability leads to a relatively high roll frequency and high vertical accelerations, particularly at the deck edge, both of which can be uncomfortable for passengers and crew. Little has been published on the seakeeping of catamarans but some data may be found in Lee, Jones and Curphey (1973), Hadler et al. (1974), Yamotayev et al. (1977), Rutgersson (1986) and Ozawa (1987).

3.12.1.4 *Structural design*

The primary loading on a catamaran is usually transverse rather than longitudinal wave loading. Torsional loads in bow and quartering seas and crossdeck slamming in head seas can also be significant. However, the majority of catamarans are small (less than 50 m length). For these, local secondary loads can dominate. In some areas, material thicknesses can be set by minimum corrosion and distortion requirements rather than external loadings. Lightness of structure is important in any high speed vessel. For this reason, most small fast catamarans are built in aluminium or fibre reinforced plastics rather than steel. Guidance on the structural design of catamarans is given in the rules of some classification societies (e.g. Det Norske Veritas, 1985).

3.12.1.5 *Applications*

The main applications for catamarans are as small, high speed ferries and conveyers of high value and perishable cargoes. Some modern examples are described in *Small Craft* (1989). These craft combine the advantage of speed with spacious layout and a relatively quiet environment for the accommodation area. In general, fast catamarans are also cheaper to build and operate than hovercraft or hydrofoils of the same capacity. A large number of fast catamaran ferries have been built, the leading supplier being Norway. Many are in use in the Far East, Australia, the USA and Scandinavia, with an increasing market throughout Europe. Medium speed ferries of over 2000 tonne have been built but the larger, modern high speed ferries are around 50 m overall length, 300 tonne displacement and have the capacity for some five hundred passengers at speeds up to 40 knots. Catamarans have also been built as cruise ships for use in the Far East and the USA. Examples include 40 m vessels carrying 50 passengers for cruises over two weeks at speeds up to 20 knots. Small catamarans also appear, along with trimarans, in the recreational market, as high speed planing boats and as racing and cruising sailing yachts.

Other current uses for slow to medium speed catamarans include surveying and minehunting (Robson, 1983). For both these roles, large deck area, high static and directional stability, and low underwater noise signatures are particular attractions. Proposals have also been made to use catamarans in the small, fast warship role (Rutgersson, 1986). Particulars of some high speed catamarans are given by Trillo (1988).

3.12.2 SWATH ships

The SWATH ship, or semisubmerged catamaran as it is known in Japan, came into being to give a step improvement in seakeeping performance. Unlike other advanced marine vehicles, the SWATH is not specifically intended for high speed. It is purely a displacement form.

The concept of the SWATH is shown in *Figure 3.99*. It is similar to a catamaran but with the main buoyancy structure placed well below the water surface. This improves seakeeping by decoupling the ship as far as possible from the effects of wave motion. The natural periods of heave, pitch and roll motions are greatly increased, taking them beyond the ocean wave periods that tend to excite resonant motions in conventional hulls. Exactly the same principle is invoked in semi-submersible offshore structures, the only difference being that the SWATH has to be more streamlined in order to permit forward motion at the speeds required of a ship.

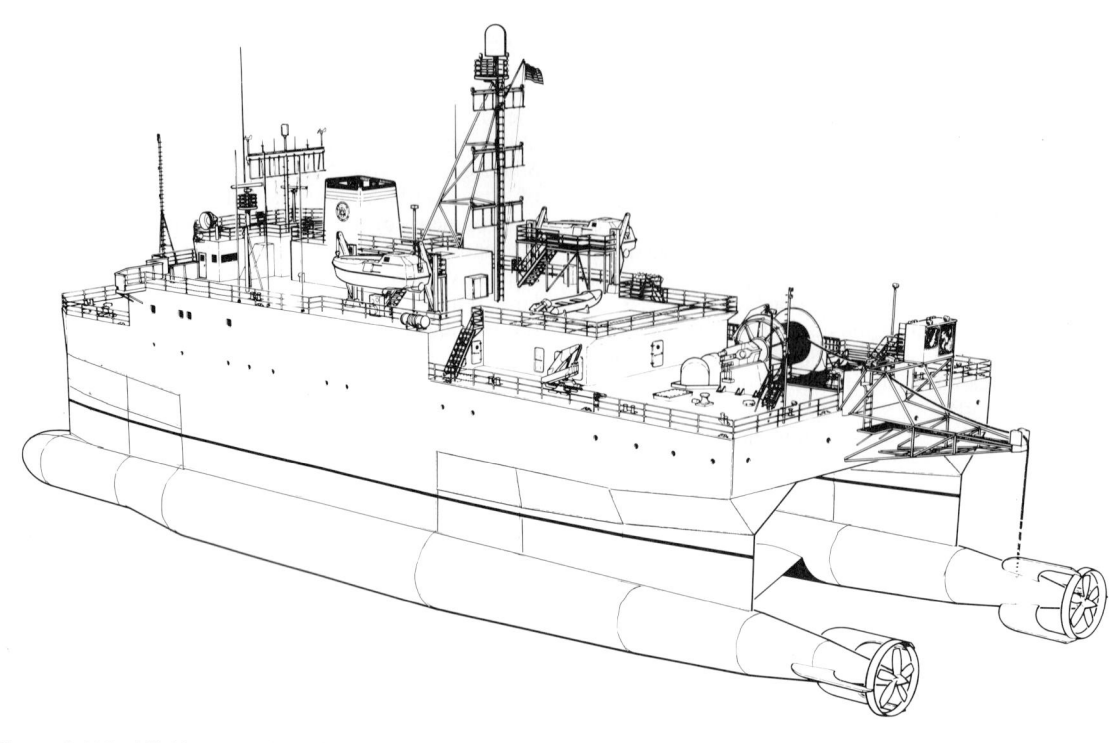

Figure 3.101 US Navy auxiliary SWATH

The SWATH has been a practical reality for less than 20 years. The first true SWATH, the *Kaimalino* of around 200 tonne, was designed in the USA and went to sea in 1973. A number of other SWATH ships followed in both the USA and Japan and the first large SWATH, the 3500 tonne Japanese diving support ship Kaiyo, commenced trials in 1984.

A major new SWATH construction programme is that of the US Navy for a class of 3500 tonne auxiliary towed array surveillance vessels outlined in *Figure 3.101* (Covich, 1987). The first of class, T-AGOS 19, was laid down in 1986 for delivery at the end of 1989. A further common hull SWATH of around 5000 tonne displacement is planned. A large SWATH research vessel is also under construction in the Federal Republic of Germany and research and design work is under way in some fifteen countries, including the UK.

A comprehensive overview of SWATH developments may be found in Gore (1985), Gupta and Schmidt (1986) and in the proceedings of the two RINA International Conferences on SWATH Ships and Advanced Multi-Hulled Vessels (RINA, 1985 and 1988). The 1988 conference included a detailed review of recent developments worldwide (Betts, 1988). Descriptions of current vessels may be found in *Jane's High Speed Marine Craft and Air Cushion Vehicles* (Trillo, 1988).

3.12.2.1 Hull forms

The nomenclature used to describe SWATH geometry is not yet fully standardized. That used here is based mainly on US and UK terminology, and is summarized in *Figure 3.102*. Elliptical rather than circular hull cross-sections are often chosen, as shown in *Figure 3.102*, typically with the width 1.5 times the height. This reduces draught and increases pitch and heave damping while giving approximately the same resistance as a circular cross-section of the same area.

Struts can be continuous on each side or intermittent. *Kaimalino* has twin struts per side but there has been a preference for single strut configurations. The reasons for this are greater static stability, better access to lower hulls, greater space utilization in the struts and improved structural arrangements. Except perhaps in very large vessels, these considerations override the potentially improved seakeeping of twin strut configurations in the platforming mode. Other advantages of single struts include reduced drag due to absence of wave interference effects between twin struts, im-

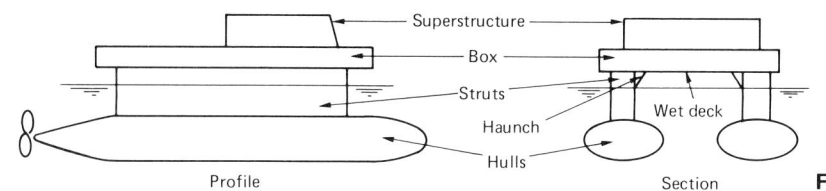

Figure 3.102 SWATH nomenclature

proved survivability after damage and better behaviour in extreme seas where twin strut configurations do not contour as well, leading to slamming on the underside of the box.

Most small SWATHs have long struts that overlap the hulls. Large SWATHs tend to adopt short struts, i.e. shorter than the hulls. Although this can complicate rudder arrangements, it reduces both resistance and longitudinal pitch excitation, and can ease the alignment of the longitudinal centres of buoyancy and flotation to minimize coupling between pitch and heave.

The lateral separation of the struts is generally chosen to suit layout and transverse stability. Resistance and propulsion are not significantly affected at practical separations. Transverse stability is generally excellent, especially at large angles, but difficulty can be found in meeting stability criteria for initial heel angles unless box clearance is kept relatively low. This conflicts with seakeeping requirements in higher sea states, where slamming can occur, so a compromise is necessary. Box clearance needs to equate approximately to the design significant wave amplitude for platforming mode operation.

Variations of the basic SWATH form have been proposed. Gupta and Schmidt (1986) discuss the relative merits of canted (angled) struts. These increase motion damping, added mass and motion periods but also increase underwater beam. Other proposals have been made by Schmidt (1986) and O'Neill (1986). The O'Neill hull form is not a SWATH but a small waterplane ship. The design consists of a central strut and underwater hull containing nearly all the buoyant volume, flanked at each side of the ship by a strut which supplies waterplane area for transverse stability but very little volume. The main hydrodynamic advantages of this form over SWATH are claimed to be a reduction in drag of about 15% and a doubling of control and low speed damping forces. Utilization of space within the central, large hull is also eased. However, some of the redundancy and survivability advantages of the twin hull form are lost.

The unconventional geometry of the SWATH leads to significant departures from monohull design characteristics in a number of respects. The advantages and disadvantages of the SWATH form relative to the monohull are summarized in *Table 3.8*. The key disadvantage of increased size and cost depends closely on the basis of comparison. The production cost per tonne is broadly comparable between SWATH and monohull

Table 3.8 Advantages and disadvantages SWATH vs monohull

Advantages	Disadvantages
1 Greatly improved seakeeping, hence operability, crew effectiveness and passenger comfort	1 Increased size and cost
2 Speed maintained in high sea states	2 Large draught
3 Flexibility of layout (rectangular box)	3 Large beam
4 Wide deck areas	4 Increased freeboard (boarding and access problems)
5 High large-angle stability	5 Sensitivity to weight change
6 Good damaged stability	6 Machinery layout problems
7 High directional stability	7 Lower calm water speed
8 Good station keeping	8 Difficult space utilization in struts
9 Good low speed manoeuvrability	9 Increased initial heel after damage
10 Low underwater noise potential	10 Possible box slamming in extreme seas
11 Good sonar platform	
12 Inherent redundancy and scope for improved reliability and maintainability	
13 Good propeller efficiency	

but a given payload requires a larger SWATH owing mainly to a significantly higher structural weight fraction. This leads to a production cost penalty of around 15–20%. Alternatively, if the basis of comparison is seakeeping, then a SWATH would be substantially smaller and cheaper than an equivalent monohull.

The design of SWATH ships (RINA, 1985, 1988) is complicated as there are many more variable design parameters than found in a monohull. On the other hand, there is a measure of greater freedom as the geometry of each part (hull, strut, box) can, to a limited extent, be optimized independently.

3.12.2.2 Performance in steady motion

The SWATH form suffers a calm water speed penalty compared with a monohull of equal displacement. This results from its larger underwater surface area and shorter waterline length. *Kaimalino* is reported to require about 20–40% more power than comparable displacement monohulls, although longitudinal shaping of the hulls can reduce the penalty to around 10%. This shaping, in the form of 'coke bottling', minimizes the tendency to a large hump in the wave making resistance curve at Fr ≈ 0.3, based on hull length. Care is needed in designing the shape as optimum forms for high speeds and intermediate speeds differ, as shown in *Figure 3.103*.

Theoretical calculation of SWATH wavemaking resistance is usually based on the approaches of Chapman (1972) or Lin and Day (1974). Resistance prediction for a new design generally utilizes either one of these theories or model tests in conjunction with the ITTC friction line for strut and hull friction, using the appropriate Reynolds number for each. A model to full scale correlation coefficient (typically 0.0005) is used with empirical appendage resistance estimates. Other approaches to wave making resistance have been proposed by Salvesen *et al.* (1985) and Chun, Ferguson and McGregor (1988) (see also RINA, 1988).

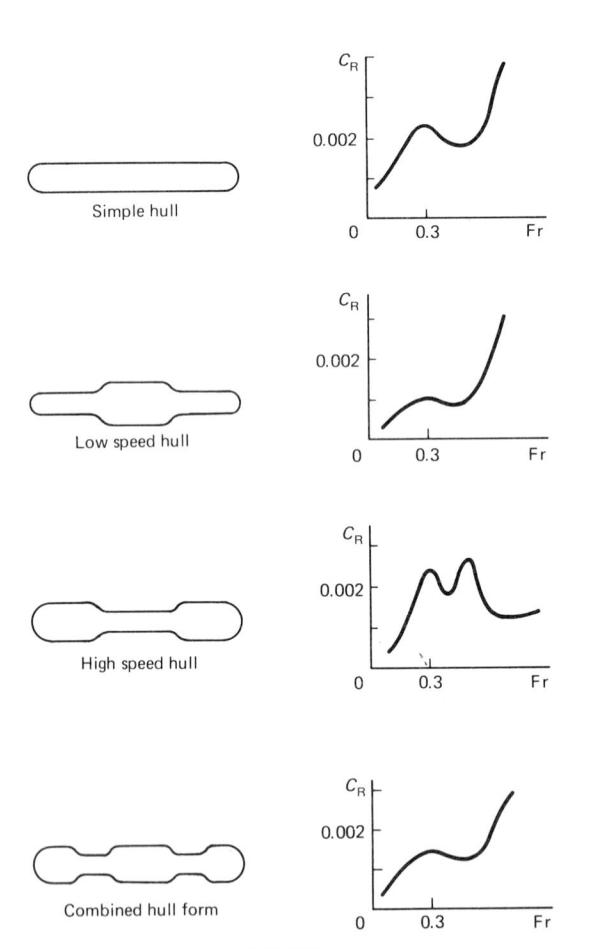

Figure 3.103 Effect of SWATH hull geometry on residuary resistance coefficient C_R

Propeller design for SWATH ships is based on monohull methods. Propulsive coefficients are slightly higher than monohull values, although data on hull efficiency elements is sparse. Contrarotating propellers are very attractive for SWATH, although it is not believed that any have been fitted. Propulsive coefficients as high as 90% are thought to be attainable with cavitation performance no worse than monohull single propeller equivalents.

The directional stability of the basic SWATH form is excellent. This is true at speed, even with one propeller trailing (Gore, 1985). The corollary is that manoeuvrability is hardly outstanding, except at low speed where a large turning moment can be applied through differential propeller thrust in the well separated hulls. Typical turning diameter to ship length ratios under full rudder for a 3000 tonne SWATH might range from about 1:1 at zero speed (with differential thrust) to 5:1 at 10 knots, and 10:1 at 25 knots. The latter ratio compares with about 7:1 for a monohull of comparable size. However, as the SWATH has a shorter length than the comparable monohull, the turning diameter is broadly similar in absolute terms.

Earlier long strut SWATH designs, including those now at sea, usually placed the rudders abaft the propellers in order to utilize the slip stream. The move towards shorter struts precludes such an arrangement and more recent designs usually show rudders sited ahead of the propeller. These can conveniently take the form of spade rudders above the after end of the hull, abaft the struts. An alternative arrangement dispenses with the normal rudders altogether. Steering is achieved by control fins between the struts cantilevered from the hulls and angled downwards by 20–30°. In this way, directional control and vertical motion control can be combined in one set of surfaces such as that fitted in T-AGOS 19.

3.12.2.3 Performance in waves

Typically, SWATH vessels are capable of operating in conditions one or two sea states higher than monohulls of comparable size. This excellent seakeeping of the SWATH form is associated with long natural periods of motion in heave, pitch and roll. Typical ranges for the natural periods of larger SWATH ships (say around 3000 tonne) are 11–13 s in heave, 14–17 s in pitch and 18–20 s in roll. Even *Kaimalino*, at only 200 tonne, has periods of 9, 10 and 16 s, respectively.

These long natural periods allow the characteristic very steady platforming motion of a SWATH in moderate to high seas where wave periods are shorter than the vessel's natural periods. Since the heave natural period is generally the lowest, SWATH design tends to be heave dominated. The motion conveniently changes to a wave following contouring motion in very high sea states, where the wave periods are longer than the vessel's natural periods, thus reducing the possibility of wet deck slamming in high waves. There is, however, a resonant region in between, which is most likely to be met in the following seas. In these conditions, behaviour of a basic SWATH form can actually be worse than a monohull of similar size. Fortunately, the provision of active stabilizer fin control can overcome this problem. It is interesting to note that the T-AGOS 19 model had no observed slamming problems when tested in the equivalent of sea state 9, significant wave height 14 m (Gupta and Schmidt, 1986).

Some motion data from sea trials of *Kaiyo* have been given by Ozawa (1987). In sea state 4, at all speeds, pitch and roll angles were within 2° significant single amplitude, as predicted by model tests, with vertical accelerations amidships below $0.1g$. Even in sea state 7 (significant wave height 9 m), where measurements were only made at low speed, pitch angle was only up to 4° and roll angle up to 6°, with vertical acceleration amidships of $0.13g$. Operating experience with the smaller Japanese SWATH vessels is also excellent with very high operating rates. Passenger motion sickness rates are reported to be of the order of 1%, far less than on much larger monohull craft. US experience has shown that the limiting conditions for helicopter operation on the 200 tonne *Kaimalino* are comparable to those on a monohull of some 20 times her displacement (Gupta and Schmidt, 1986).

A clear exposition of SWATH seakeeping, together with useful guidance and data for design, has been given by McCreight (1987) and Lamb (1988). Other work on motion prediction is reported by Wadi (1986), Hong (1986) and Dallinga, Huijsmans and Graham (1988). A number of recent developments are described in RINA (1988). Theoretical work is still proceeding on the effects of low frequency wave forces, caused by non-linear contributions to the total wave force. The effect on SWATH heave motions can lead to a significant reduction in underbox clearance, which can be aggravated by upwelling resulting from interaction between the hulls (Eatock Taylor and Hung, 1987).

It is well established that control fins can further improve the seakeeping of SWATH ships. Indeed, stabilizing fins are necessary to overcome the so called Munk moment arising from the longitudinal asymmetry of the pressure distribution on the hulls. The Munk moment, which is proportional to the added mass in heave times the square of the forward speed, can lead to pitch instability at high speeds. Fins also increase motion damping and indeed dominate it at higher speeds. Active control techniques have been reported by Wadi (1986), Caldeira–Saraiva and Clarke (1987), McGregor, Drysdale and Wu (1987) and several papers in RINA (1988).

3.12.2.4 Structural design

The large surface area of a SWATH ship leads to high structural weight. This is a major contributor to increased displacement compared to an equivalent payload monohull. The structural weight fraction tends to increase as ship size decreases, so that aluminium or fibre reinforced plastics become particularly attractive for small SWATH ships.

The dominant primary loading is the transverse wave load in beams seas. As yet, hardly any full scale data are published on SWATH structural loadings. Theory, supported by model tests, shows that the transverse load has a highly peaked maximum in beam waves of wavelength 3–4 times the underwater beam of the ship. The transverse bending moment is often expressed through an equivalent side force acting at mid-draught. In heavy seas the maximum side force to displacement ratio lies in the region 0.5–1.0, depending on geometry; the ratio does not increase further in extreme seas. A rough estimate for the expected lifetime side load F, derived from regression analysis of published data is:

$$F = K \times (\text{displacement})^{0.77} \qquad (3.218)$$

where $K = 7.94$ for single struts and 4.26 for tandem struts (Betts, 1988b). Once the major underwater dimensions are known, a better estimate can be made using the Sikora Algorithm (Sikora, Dinsenbacher and Beach, 1983).

Hull girder combined loads have to be considered. Some rules of thumb for early design purposes, suggested by Kennell (Betts, 1988) are shown in *Table 3.9*. In this table, F represents the maximum side force just discussed, M the maximum longitudinal bending moment calculated in head seas (usually less than 10% of the maximum transverse bending moment) and T the maximum torsional moment in bow or quartering seas.

Table 3.9 Hull girder combined load

Heading	Transverse load	Longitudinal load	Bending torsion
Beam	F	0.15 M	0.25 T
Bow, quartering	0.8 F	0.8 M	T
Head, following	0.15 F	M	0.1 T

The moment T is itself a function of $F \times$ strut length. A conservative estimate for initial design purposes, for single strut SWATH vessels of current proportions, is:

$$T = 0.15 \times (\text{Displacement}) \times (\text{Strut length}) \qquad (3.219)$$

The torsion and side load are assumed to be in phase.

A computational procedure for wave load prediction, based on a linear seakeeping theory with hydrodynamic interaction between the hulls, is presented by Reilly, Shin and Kotte (1988). This work also covers structural response using finite element models. Finite element analysis is essential to identify highly peaked stressed in the haunch at the junction of strut and box.

Further developments in structural design procedures are described in RINA (1988). These include prediction of slamming forces on the wet deck. Slamming pressures on the outboard faces of the struts may be as important as wet deck slamming. A comprehensive review of developments in SWATH structural design, including loading, analysis, materials, fatigue, model testing and fabrication, is given in ISSC (1988).

3.12.2.5 Applications

SWATH ships only have a clear advantage over monohulls in roles where seakeeping performance is paramount. These roles are nevertheless quite extensive, being seen as passenger and car ferrying in rough weather areas, cruising, offshore and diving support, oceanographic research, hydrographic survey, coastguard and offshore patrol, fishing, fishery protection and recreation. The SWATH form is also believed to offer significant advantages for many types of warship. A particular attraction is the provision of a stable platform for helicopter operations. SWATH ships are not suitable for carrying heavy cargoes as the small water plane area makes them very sensitive to large changes in weight.

A number of studies have been undertaken to compare SWATH and monohull solutions to the same design requirement. All such studies are faced with the question of what equivalence means. Does it mean the same size represented by, say, displacement or length? For the same displacement and design standards, the costs of a SWATH and a monohull are probably similar. For the same length, the SWATH will have a much larger displacement and cost. A more likely criterion for comparison is equivalence of payload. The latter is taken to include cargo or operational equipment (e.g. surveying gear) and the crew and stores to operate it. Similar speed and endurance are also implied. In this case, the SWATH will almost invariably be larger and more costly. However, such a comparison of ship cost per given payload hides the fact that the SWATH and the monohull design have very different characteristics. In particular, the SWATH will have far better seakeeping with consequent major benefits to operational availability, crew effectiveness and passenger comfort.

An alternative basis for comparison is seakeeping performance itself. On this basis, the SWATH scores heavily over the monohull, as indeed it should, having been conceived for just this purpose. A US Navy study quoted by Gore (1985) gave a monohull displacement 29% greater than the SWATH equivalent for equal annual operability in the North Atlantic. In another detailed US Navy study, the acquisition cost of a SWATH frigate was estimated to be some 12% greater than the payload equivalent monohull, but 12% less than a monohull giving lower but just acceptable seakeeping performance (Kennell, White and Comstock, 1985). The seakeeping advantage of the SWATH does, however, reduce at large displacements (above 15 000 tonne or so) as large monohull ships are much steadier than small ones (Cannon and McKesson, 1986).

The only truly valid method of comparison between SWATH and monohull (or any other form) is on an equivalent mission basis. This must derive from the tasks to be performed. In commercial terms, the equivalent mission basis refers to the payload, load factors, turnaround times and availability required to maximize profit or some other measure of merit. This will usually involve some aspect of seakeeping operability and will still involve judgement at the design stage on imponderables such as the attraction of SWATH vessels to potential passengers for enhanced comfort and novelty value.

3.13 Propulsion

The two main prime movers for high speed marine vehicles have been gas turbines, usually a marinized form of aircraft design, and diesel engines, now becoming more popular with the introduction of lightweight forms and a superior fuel consumption. An interesting development in propulsion, now being applied by the Japan Marine Machinery Development Association (JAMDA) to SWATH, is superconducting electromagnetic transmission for which large SWATH ships are the favoured platforms (Marine Engineers Review, 1988a).

The main forms of propulsor are marine propellers, water jet systems and aircraft type propellers which may also be surrounded by a duct. The marine propeller is subject to cavitation and ventilation problems under the high hydrodynamic loadings imposed by fast craft operation. Propellers are therefore used which possess delayed cavitation, supercavitation or superventilation characteristics. The prefix 'super' implies that the whole of the suction or back surface of the propeller blade is enveloped by a gas bubble; vapour in the case of supercavitation, air in the case of superventilation. Water vapour forms when the local pressure in the water reaches the vapour pressure. Ventilation can be introduced deliberately along controlled paths or by having the propeller pierce the free surface of the water (Kruppa, 1976). The inlets of water jet systems may either be flush mounted at the hull or have a scoop inlet or be pod mounted to face the oncoming flow, such as on the struts of hydrofoil craft. The aircraft type propeller may be on a fixed mounting or on a pylon that can swivel to allow manoeuvring at low forward speeds. In each case steady state cruise speeds are achieved when the propulsor thrust matches the craft drag.

Thorough reviews of propulsors and propulsion systems for high speed craft have been given by Barr and Etter (1975), Johnson (1968), Venturini (1980) and Mantle (1980). A discussion of propulsion generally, including systems suitable for the present applications, may be found in Lewis (1988) and Chapter 6. Some details of SWATH propulsion systems are given by Betts (1988).

3.13.1 Marine propellers

Conventional subcavitating propellers of commercial manufacture are most commonly used on planing craft up to speeds of about 30 knots. Often standard series propeller charts, such as those of Gawn (1953) and Gawn and Burrill (1957), are used to obtain the required propeller characteristics. The developed blade outlines are typically elliptical and the sections are ogive (flat face, circular arc back and sharp leading and trailing edges). There is a major breakdown of thrust on these propellers when craft speeds are increased owing to cavitation over the back of the blade compatible with high thrust loading (see Section 3.3.5.3). For higher speeds it is necessary to adopt supercavitating propellers such as those of Newton and Rader (1961) or to construct blades from the curved wedge sections devised by Tulin (1956). Blount and Hankley (1976) illustrate design procedures for selecting the optimum Newton–Rader propeller. With these propellers speeds up to 55 knots have been used for extensive periods without causing significant structural damage. The necessary hull–propeller interaction functions have been discussed by Hadler and Hubble (1971) and these correlate well with the experimental work of Blount and Fox (1976).

Both sub- and supercavitating propellers were used on the Canadian hydrofoil craft *Bras D'Or* (Eames and Jones, 1971) as well as some superventilated surfaces. Surface piercing hydrofoil craft used commercially use lightweight diesel engines driving subcavitating propellers by means of an angled shaft, power transmission system. This arrangement provides simplified construction, relative ease of maintenance and low cost. Gas turbines have been used for all major US military and commercial hydrofoil ships allowing practical design speeds greater than 40 knots. The US SES 100B was configured with a supercavitating propeller at the stern of each sidehull and three marine gas turbines. On the other hand, in the UK Vosper Hovermarine favoured two high speed diesel engines each driving a fixed pitch, subcavitating propeller (Tattersall, 1985). The HM 200 series craft performed well with three bladed propellers selected from the Gawn–Burrill Series (1957). As the result of differently arranged passenger accommodation in the HM 500 series craft, vibrations from the corresponding propeller became unacceptable. It was found that detailed modifications by Kruppa to a five bladed version led to an overall noise reduction of 9 dBA and a 6 dBA reduction in vibration level at locations above the propellers. The BH 110 and SES 200 also have Gawn–Burrill series propellers on inclined shafts.

3.13.2 Water jet systems

The water jet system comprises engine, transmission, propulsion pump, nozzle, thrust vectoring and reversing mechanisms, diffuser and ducting, inlet and appendages or fairings required for the mounting of the inlet. A further subdivision can be made according to whether the inlet is pod and strut mounted or flush with the hull. The former arrangement is mainly associated with hydrofoil craft, the latter with SES. Further specializations are listed by Barr and Etter (1975). The pumps may be axial, mixed or centrifugal flow types, single or multistaged and single or two speed running.

It is stated by Eggington and Kobitz (1975) that water jet propulsion is the preferred propulsor for SES. This statement is, however, based on the US experience with SES 100A in contrast to the propeller driven SES 100B and also with the desire to travel at very high speeds, i.e. above 50 knots with a large craft, i.e. 2000 tonne all up weight. It has also been suggested by Davison (1979) that a well designed, water jet system could be as efficient as the equivalent screw propeller for planing craft up to 20 m long.

A simple analysis of water jet propulsion is illustrated by Clayton and Bishop (1982) and is based on *Figure 3.104*. The rearward force exerted by the nozzle on the water is equal to the rate of increase of momentum of the water jet and must be equal and opposite to the forward thrust, T, that the water exerts on the craft. Thus:

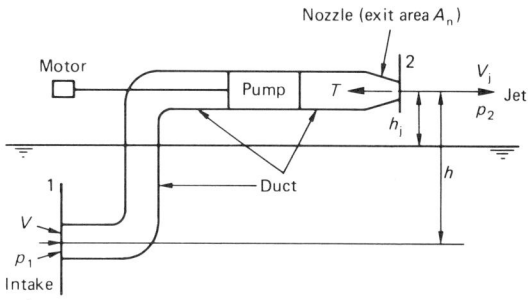

$$T = \rho V_j A_n (V_j - V) \quad (3.220)$$

Figure 3.104 Schematic of a hydraulic jet propulsion system

where V_j and V represent, respectively, the jet velocity and the intake velocity of the water relative to the craft, A_n the nozzle area and ρ the water density. The useful power obtained in propelling the craft at a steady velocity V is therefore $TV = DV$, where D represents the drag force.

Applying the steady flow energy equation between inlet and outlet and noting that pump power, P, equals the product of volume flow rate of water in the jet and the total pressure rise across the pump, one obtains:

$$P = \rho V_j A_n \{\tfrac{1}{2}(V_j^2 - V^2) + \tfrac{1}{2}kV^2 + gh_j\} \quad (3.221)$$

where k represents a duct and pump loss coefficient for the primarily turbulent flow. Denoting $\lambda_j = V_j/V$, the propulsive efficiency is then given by

$$\eta = \frac{TV}{P} = \frac{2(\lambda_j - 1)}{\lambda_j^2 - 1 + k + 2gh_j/V^2} \approx \frac{2(\lambda_j - 1)}{\lambda_j^2 - 1 + k} \quad (3.222)$$

if h_j is considered small, e.g. 1 m. The final expression in Equation 3.222 then shows that the maximum efficiency is given $\hat{\eta} = 1/(1 + \sqrt{k})$ and occurs when $\lambda_j = 1 + \sqrt{k}$. *Figure 3.105* shows the variation of η with k and λ_j and it is clear that jet velocities between 1.3 V and 2.0 V are best, according to the value of the loss coefficient.

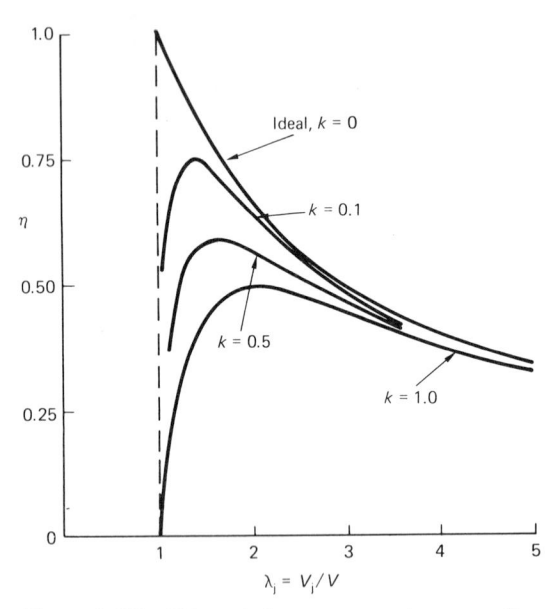

Figure 3.105 Effect of viscous energy losses on the efficiency of a hydraulic jet propulsion system

Venturini (1980) and Johnson (1968) discuss the details of pump selection. Venturini also examines the performance of water jet propulsion relative to equivalent propellers and considers losses in detail. He implies cautious optimism for the use of water jet propulsion for craft of modest size. Much depends on the inlet conditions and for the flush mounted inlets a non-uniform flow, possibly with cavitation bubbles, may well dramatically reduce the pump performance (Barr and Etter, 1975).

A well known application of the water jet propulsion system is the Boeing *Jetfoil* hydrofoil craft, and its predecessor *Tucumcari*. Both have pod struts and inlets, the former now operating successfully as a commercial ferry across the eastern section of the English Channel between Dover and Ostend. In 1979, the Royal Navy acquired a hydrofoil craft based primarily on *Jetfoil* and designated HMS *Speedy*. The main difference in the naval version was the provision of two diesel engines for hullborne cruising. The craft was evaluated in the offshore protection role with particular emphasis on fishery protection, as described by Brown *et al.* (1984). Although many of the accepted attributes of the *Jetfoil* design were confirmed the advantages were not sufficiently great to embark at great cost, in an era of cost cutting, on either running a single craft or a small fleet to replace conventional monohulls. A particularly telling factor was the low endurance (about 10 h) foilborne and this was considered unacceptable for the more remote parts of the UK EEZ. No adverse comments were made on the propulsion system. HMS *Speedy* was decommissioned in Spring 1982.

In the US SES research programme the SES 100A version was propelled by water jets, one from each sidehull, driven by an integrated lift and propulsion system of four marine gas turbines. Bliault (1989) has given details of a new series of SES now undertaking passenger ferry services between Helsingborg (Sweden) and Copenhagen. The craft hull is formed of an FRP sandwich construction with water jet inlets integral with the fabrications. The twin diesel powered, water jet systems on the so called *Jetrider* craft develop a maximum service speed of 42 knots fully loaded which is about the optimum speed for a craft 33.4 m long and 10.5 m beam. The total complement is 244 passengers and 6 crew and the total payload is 27 tonne. The initial operator did, in fact, cancel the contract. Excess weight in construction caused ingestion of air into the water jet system so that craft speed and engine settings were unacceptable. The current operator subsequently took over the craft and modified the layout so that now the two *Jetrider* craft, in service since June 1988, are providing acceptable performance but at speeds rather lower than originally conceived.

A modern catamaran, the *Oregrund*, has been built by the FBM group (formerly Fairey Marinteknik) and incorporates two diesel powered, water jet propulsion systems each using a single stage, mixed flow pump. The craft is 41.5 m long and 11.0 m beam, carries 306 passengers and 6 crew and is to operate on the Hong Kong to Macao ferry service. At a cruise speed of 38 knots it is seen as highly competitive with the Boeing *Jetfoil* craft presently operating. The *Oregrund* is fitted with a bucket deflector at each nozzle outlet which can be operated by a computer control system to allow steering and manoeuvring with sideways or rotational motion or a combination of both (Small Craft, 1989).

3.13.3 Air propellers

Air propellers are found, in the context of the present discussion, exclusively on ACV since ideal matching takes place for amphibious, high speed operations. Gas turbines also provide the best power to weight ratio as the prime mover, but considerable space is taken up by the necessarily large air intakes. As stated earlier, one particularly successful ACV, the BHC AP1-88, has broken away from tradition and employs lightweight diesel engines, as do a number of smaller ACVs. A major problem associated with air propellers is noise which is strongly influenced by propeller blade tip speed. Significant improvements have been made over the years but noise still represents a difficulty when planning new routes or terminal facilities.

A straightforward approach to ideal propeller performance can be obtained from actuator disc theory, as shown by Clayton and Bishop (1982). The flow system, considered inviscid in this model, is shown in *Figure 3.106*. The propeller is replaced by an actuator disc across which there is a sudden rise in pressure but a gradual change in velocity. The pressure p_o over the extreme edges of the stream tube surrounding the actuator disc is considered constant and V_A represents the velocity at which the propeller advances relative to the surrounding air. Conditions across each section of the stream tube are considered steady and uniform.

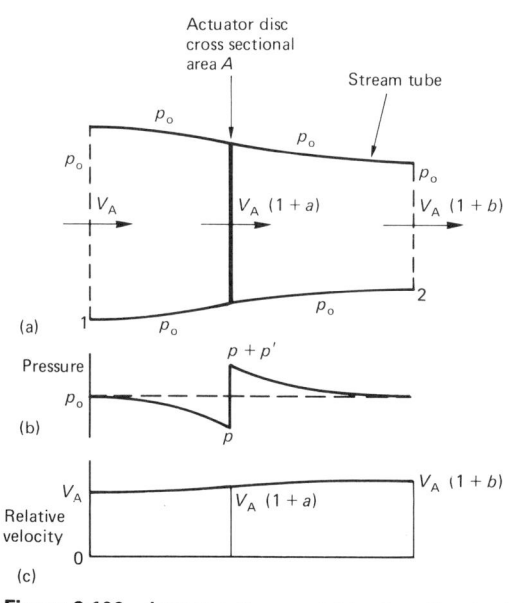

Figure 3.106 Actuator disc model to describe the ideal flow through a propeller

Application of Bernoulli's Theorem to flow ahead and behind the disc gives the increment of pressure p', as:

$$p' = \tfrac{1}{2}\rho b (2 + b) V_A^2 = T/A \tag{3.223}$$

where ρ represents the air density and T the propeller thrust.

The force exerted by the disc on the air passing through it equals the rate of increase of axial momentum of the fluid within the stream tube, if we ignore any rotation of the flow and consider the curvature of the streamlines to be small. Thus:

$$T/A = \rho b (1 + a) V_A^2 \tag{3.224}$$

and so Equations 3.223 and 3.224 show that $b = 2a$, that is, the velocity at the disc is the average of that far upstream and far downstream. The energy per unit time (power) supplied to the fluid gives rise to a rate of increase of kinetic energy and so the power absorbed by the air is:

$$P_k = 2\rho A (1 + a) a^2 V_A^3 \tag{3.225}$$

The useful work done per unit time is the output power P_o, absorbed in propelling the disc through the air, that is:

$$P_o = TV_A = 2\rho A a (1 + a) V_A^3 \tag{3.226}$$

Hence the total input power to the propeller is:

$$P = P_k + P_o = 2\rho A a (1 + a)^2 V_A^3 \tag{3.227}$$

and the ideal efficiency is given by:

$$\eta_i = \frac{P_o}{P} = \frac{1}{1 + a} = \frac{2}{\{1 + (1 + C_T)^{1/2}\}} \tag{3.228}$$

where

$$C_T = T/\tfrac{1}{2}\rho A V_A^2 = 4a(1 + a) \tag{3.229}$$

Figure 3.107 shows the variation of η_i with C_T and it is seen that efficiency drops rapidly as C_T increases. One should therefore aim for a lightly loaded propeller and so it may be inferred that the best operation occurs

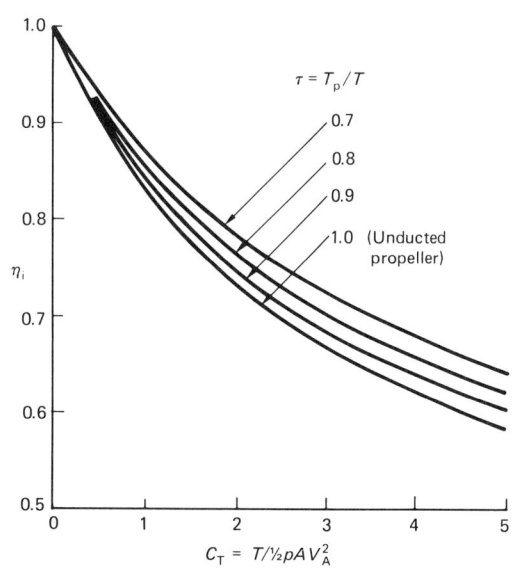

Figure 3.107 Ideal efficiency of a propeller with and without ($\tau = 1$) a surrounding duct as a function of thrust coefficient C_T

at steady, high speeds as in aircraft travelling at cruise speed and altitude.

A similar approach can be made for the ducted propeller (Clayton and Bishop, 1982) in which the duct effectively allows the capture of a greater mass of air compared with the equivalent unducted propeller. A distinction must then be drawn between the thrust from the propeller alone T_p and that from the ducted propeller T. Thus, with $\tau = T_p/T$ and C_T defined as before, the ideal efficiency of the ducted propeller is given by:

$$\eta_{id} = \frac{2}{\{1 + (1 + \tau C_T)^{1/2}\}} \quad (3.230)$$

If the duct is of sufficient length to produce a slipstream of the same cross-sectional area as the actuator disc, then it may be shown that the ideal efficiency becomes:

$$\eta_{id} = \frac{4}{\{3 + (1 + 2C_T)^{1/2}\}} \quad (3.231)$$

The effects of different values of τ are shown in *Figure 3.107* where it is seen that $\eta_{id} > \eta_i$ for a given C_T. In real viscous flows, drag forces on the duct can become a large component of the appendage drag.

Examples of the use of unducted ACV propellers can be seen on BHC SRN4 Mk3 (6.3 m diameter), BHC SRN6 variants (3 m diameter), BHC BH7 variants (6.3 m diameter), Mitsui PP5 (2.5 m diameter) and PP15; US LACV-30 and a number of Russian ACVs. In some cases, the lift air supply fans and propulsors are geared together so that with variable pitch propellers the craft operator can adjust the two power requirements as conditions change. The SRN4, for example, possesses this facility as well as having pylon mounted propellers which can rotate about the vertical to provide low speed manoeuvres. Rudders are also placed in the slipstream of the aft pair of propellers for course keeping. Having more than one propeller on a hovercraft allows reduction of thrust loading and thus higher efficiency and lower noise. It is reported by Wheeler (1976) that substantial noise reduction was made, both internally and externally, on the SRN6 Mk6 twin propeller design by reducing the propeller tip speed. There is, however, likely to be a weight penalty. Gas turbine engines themselves generate high broad band noise levels that can be reduced with silencers if space is available. The air to the engine must be clean, dry and non-corrosive otherwise compressor and turbine blades deteriorate rapidly. A filtration system is therefore required (Lavis, 1985) which can also act as an effective inlet silencer.

Ducted propellers, of the same size as equivalent open propellers, show dramatic reductions in noise level, by as much as 50% according to Lavis (1985). Examples of successful ducted propeller applications are BHC AP1-88 (2.7 m diameter); Wartsila PUC22-Larus; AALC Jeff(A) (2.3 m diameter), Jeff(B) (3.5 m diameter) and LCAC; some Russian ACV and the smaller UK Griffon and Pindair craft. Although there are benefits in terms of noise, thrust and efficiency, significant weight penalties for ducted propellers may have to be accepted (Mantle 1980).

3.14 Comparative performances

Relative performances of different craft are based on a variety of parameters such as:

1. Drag or power to weight ratio as a function of speed.
2. Rough water to calm water speed ratio as a function of significant wave height.
3. Heave, pitch and roll accelerations of equivalent craft as a function of significant wave height.
4. Various forms of transport efficiency (or effectiveness) as functions of speed or payload to weight ratio.

Leading data for some of these estimates may be found in Trillo (1988) and earlier editions of *Jane's High-Speed Marine Craft and Air Cushion Vehicles* (previously *Jane's Surface Skimmers*).

A classic and highly significant portrayal of ACV development over the years has been given by Wheeler (1976, 1984) for the BHC SRN series (*Figure 3.108*). This information represents the results of the longest production runs of any hovercraft. The results in terms of relative progress can be considered representative of most high speed craft developments. Much is owed to the gas turbine and air propeller combination for ACV to provide a commercially viable payload which overcomes the high cost of the propulsion system. Since the advent of SRN1 the installed power to weight ratio has dropped steadily from almost 90 kW/tonne to a value of 70 kW/tonne for SRN 6, 60 kW/tonne for SRN4 Mk1, 57 kW/tonne for SRN6 Mk1S and 36 kW/tonne for SRN4 Mk3. Craft developed with progressive improvements in skirts and propulsion systems and are projected to level out at values of about 27 kW/tonne. The AP1-88, which has a radically different design approach, as discussed earlier, has a value for the production craft of 33 kW/tonne as at 1984. An alternative illustration of progress is to examine the installed power/tonne–knots ratio and this is shown in *Figure 3.109* (Lavis, 1985; Chaplin, 1979). The AP1-88 value corresponds to the prototype craft and calculations from the paper by Wheeler (1984) shows the figure for the production craft to be 0.64 kW/tonne knots.

Most of the preceding improvements result from a decrease in the total installed power to weight ratio rather than speed changes. Design changes and material selection have reduced the relative weight of the SRN series but the major impact arises from power reductions because of skirts, deep cushions and low bag pressures, fingers which greatly reduce drag and the higher efficiency of lift fans and propellers designed

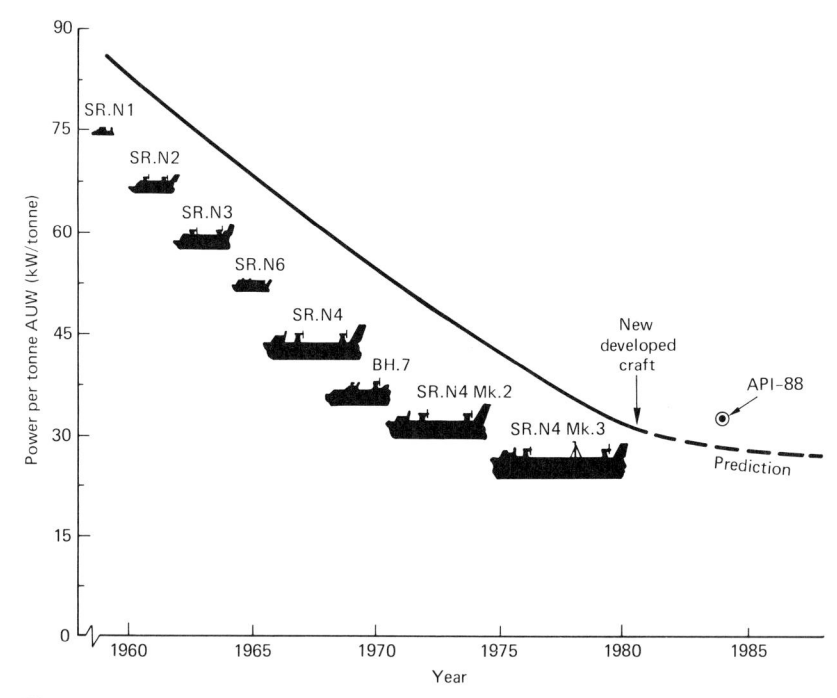

Figure 3.108 Improvement in power to weight ratio achieved by BHC hovercraft (after Wheeler, 1976, 1984)

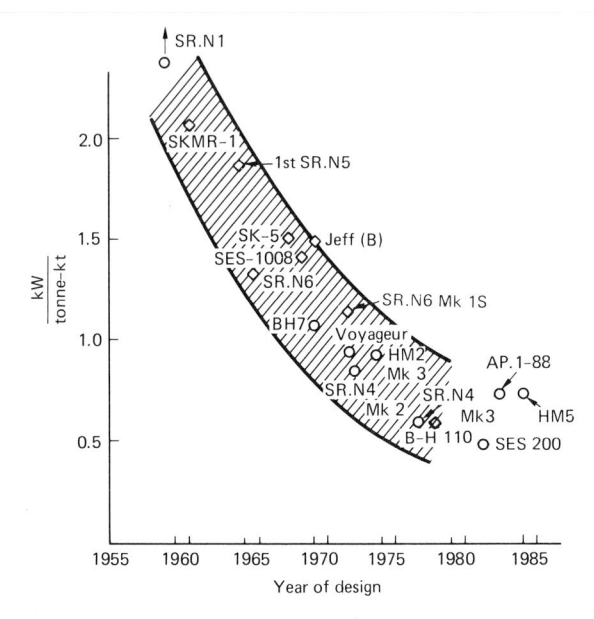

Figure 3.109 Improvements in ACV and SES performance (after Lavis, 1985)

specifically for ACV requirements. Lavis (1985) suggests that the power (kW)/tonne knots could reduce to 0.45–0.30 over the next two decades.

Examples of drag or resistance to weight ratios of different craft over the operating speed range are given in the previous sections for each craft considered and will not be discussed further here. It is, nevertheless, worth noting some values for different types of vessel. Considering the ratios of resistance to weight displacement or the equivalent drag to all-up weight for craft as appropriate, we find the following figures: 0.001 for a slow, large ship; 0.06 at 30 knots and 0.02 at 20 knots for a 1000 tonne corvette; 0.10 and 0.07 at corresponding speeds for a 100 tonne fast patrol boat of planing hull form; 0.10 at 50 knots for a 100 tonne hydrofoil craft (Eckersley–Maslin and Coates, 1978); 0.073 at 36 knots for the 106 tonne SES HM527 (Tattersall, 1985); 0.066 at 52 knots for the 38.5 tonne ACV AP1-88 (Wheeler, 1984); 0.051 at 70 knots for the 280 tonne SRN4 Mk3 (Wheeler, 1976).

Some comparative data on speed reduction of various craft in different sea states are shown in *Figure 3.110*. Seaworthiness, expressed generally in the form of *Figure 3.110*, is a fundamental attribute of any marine vehicle whether it is used for commercial or military applications. It is immediately apparent that waves affect all craft in different ways and limit the maximum achievable continuous speed. Although some effects of size are shown, within the context of

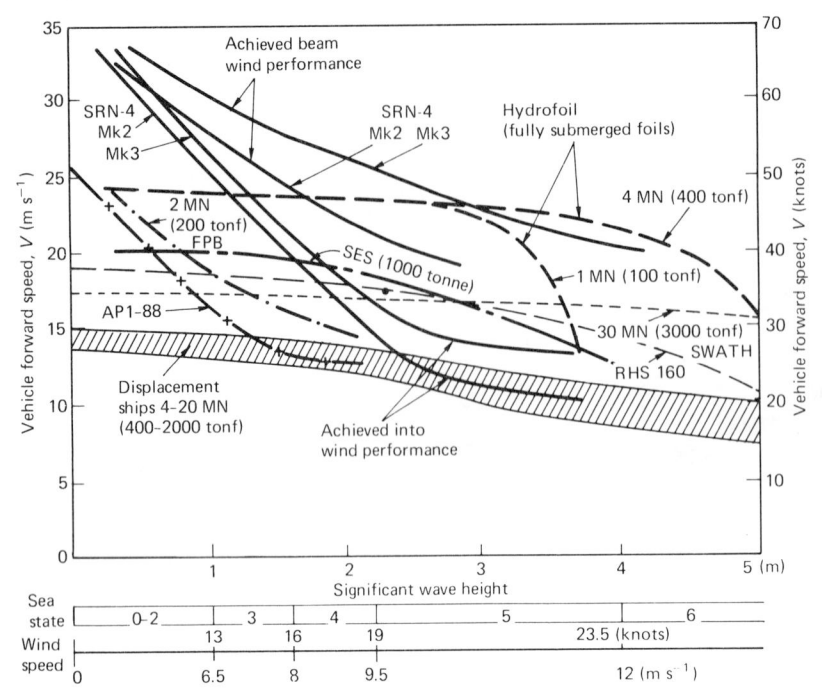

Figure 3.110 Seaworthiness of high speed craft

high speed craft and therefore small craft, the influence of size is not great enough to cause radical changes in overall trends for curves of a given type or between curves of different types.

Up to sea state 3 or 4, and depending on wind direction, it is seen that ACVs are presently the fastest practicable form of marine surface vehicle, though the performance deteriorates rapidly with the increases in wave height. Air cushion vehicles are also seriously affected by wind speed and direction. The hydrofoil craft with fully submerged foils has a lower speed initially but is able to maintain this forward speed up to sea state 5 with larger designs able to achieve sea state 6. Nevertheless, increased size and improved deeper skirt design should allow better performances from large hovercraft in the future. A hydrofoil craft with flap control, initiated by either radar or sonic height sensors at the bow, can transfer the mode of flight from platforming at low wave heights to contouring at greater wave heights in order to reduce hull contact with waves yet retain relatively low vertical accelerations. A hydrofoil craft with surface piercing foils is naturally more sensitive to surface waves especially when operating in a following sea.

Planing craft show good performance in calm water but produce very high accelerations from impact loading in waves so that drastic speed reductions take place over the whole range of operable sea states. Nevertheless, with deep V bottoms, such as those used in offshore power boats, planing craft can be driven faster up to about sea state 5, provided that sufficient power is available to compensate for the increased resistance of the deep V forms and that the crew can endure the violent motion which ensues.

In the range of sea state 6 the conventional, monohull displacement ship is capable of greater speeds than any small, high-speed surface vehicle. However, this may change in the future as, for example, larger hydrofoil craft are built; there is already an indication that the 85 tonne *Rodriquez* RHS 160 is superior in rough water. Also shown in *Figure 3.110* are predictions of a 400 tonne hydrofoil craft with fully submerged foils and a large 3000 tonne SWATH ship which can maintain a near constant speed well into waves up to a significant height of 5 m. Eggington and Kobitz (1975) considered several 1000 tonne craft designs predicted from models and sea going vessels in their support of large, fast SES. The results of *Figure 3.110* are largely borne out but some of the authors' results are vigorously questioned by proponents of other fast craft.

It has been found that operating limits on speed in a seaway are most likely to be set by increases in power requirements and by the conditions which are acceptable to the crew and passengers. The strength of the craft is usually regarded as not being at risk although many questions on structural dynamics related to marine vehicles are still largely unanswered in detail, as discussed by Bishop and Price (1980). In any given occupation and environment there are many stimuli or stressors influencing the physiological and mental processes of a crew member or passenger and it is often difficult to separate one effect from another. A general discussion of human factors, or ergonomics, related to ride motions is given by Cole (1980). Noise, vibration and low frequency motions can all become intolerable to humans in time, but quite high intensities are often sustained for short periods such as ferry crossings and even enjoyed when connected to risk events or sporting activities. Much lower limits are tolerable for a naval patrol lasting days or weeks, especially when frequently repeated. Experience and knowledge of the full facts of an event also influence human response in one way or another. Tolerable noise and vibration levels (above a frequency of 1 Hz) are fairly well defined by the International Standards Organisation (ISO) but there are problems in satisfying the specified criteria within the confines of small, high speed craft. Low frequency motions, in the range 0.1 − 0.5 Hz with exposures of 4 h duration, are particularly important for vertical accelerations above 0.75 g rms because these lead to widespread sea sickness.

It is difficult to express concisely the relative merits of different craft in response to wave excitation. For example, vertical (heave) accelerations are very sensitive to both wave steepness and frequency of encounter, which is a function of wavelength as well as craft speed. Nevertheless, the data in *Figure 3.111*, adapted from the work of Brown (1980), show the general trends, but should not be regarded as definitive for all sea and craft conditions. The excellent characteristics of hydrofoil and SWATH craft in this respect are clearly evident. An appraisal of the seakeeping capabilities of high speed craft is given by Heather *et al.*

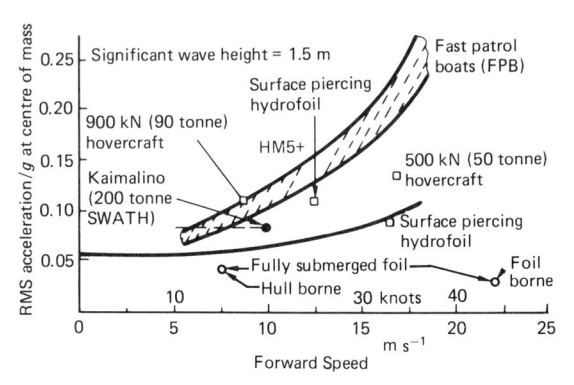

Figure 3.111 Typical vertical accelerations of high speed craft

(1978) where the discussion is, in the main, applicable to both commercial and naval craft. Additional information is given by Mantle (1980) who has shown the heave acceleration performance of ACV, SES and hydrofoil craft. The results are shown in *Figure 3.112*

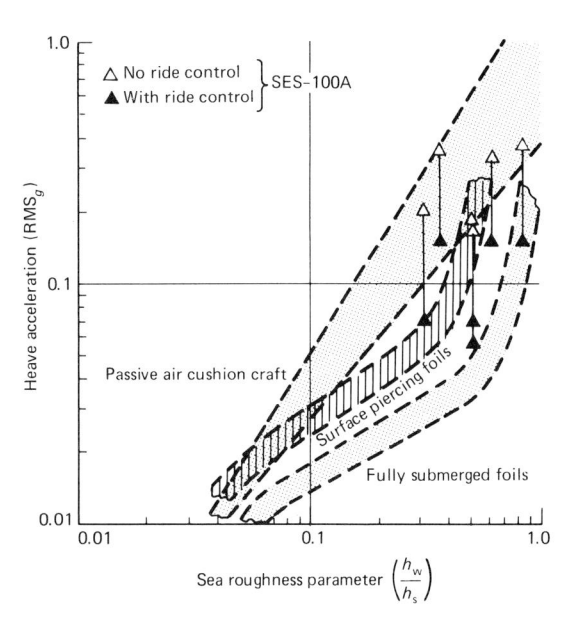

Figure 3.112 Ride quality of air cushion craft (ACV and SES) and hydrofoil craft showing effectiveness of active ride control system (RCS) on SES 100A (after Mantle, 1980)

The sea roughness parameter is defined as h_w/h_s, where h_w represents the significant wave height (average of the 1/3 highest waves in a random sea state) and h_s the clearance of the main hull above the mean waterline, which is approximately the skirt height. The effectiveness of a ride control system (RCS) on SES 100A, using an active lift fan system, is clearly seen. Again, however, care must be taken in the literal translation of these results for the reasons already mentioned.

The overall tapestry of operation for various high speed and advanced marine vehicles is summarized by Clayton and Bishop (1982), various authors in the *Naval Engineers Journal Special Edition* (1985), Betts and Clayton (1987) for ACV and SES operations in the UK, Plumb and Brown (1980) for mine countermeasures applications, Clark *et al.* (1978) for hydrofoils using the computer aided program HANDE, Davison and Fink (1980) for Jeff(A), Coles and Kidd (1980) for Jeff(B) and Wheeler (1978). The future potential for SWATH has been discussed by Eames (1980) and more recent data, including the effects of seakeeping and motion control, are given by McCreight (1987), Lamb (1988) and in RINA (1988).

An early examination of the overall performance of various modes of transport was presented by Gabrielli and von Karman (1950). They used the dimensionless term specific power, which is the inverse of transport efficiency $\Omega = WV/P$, the product of all up-weight and maximum cruise speed divided by total installed power, all in consistent units. *Figure 3.113* shows the equivalent of the Gabrielli–von Karman chart with the

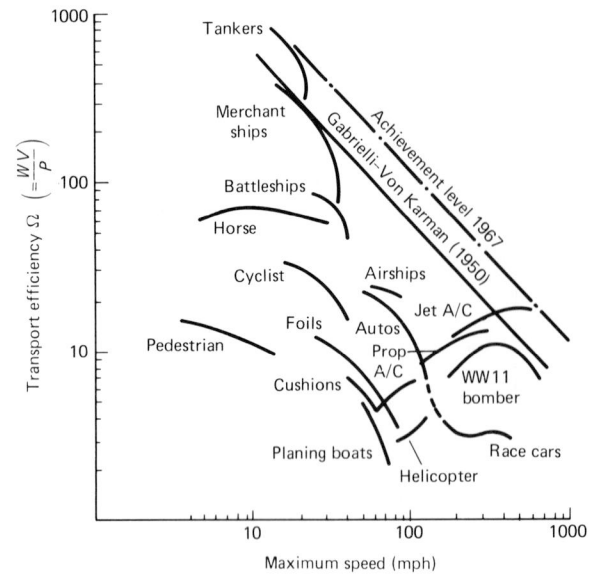

Figure 3.113 Equivalent of the Gabrielli–von Karman (1950) chart showing transport efficiency of various modes of transport as a function of maximum speed. Also shown is the limit of performance given by Mantle (1976)

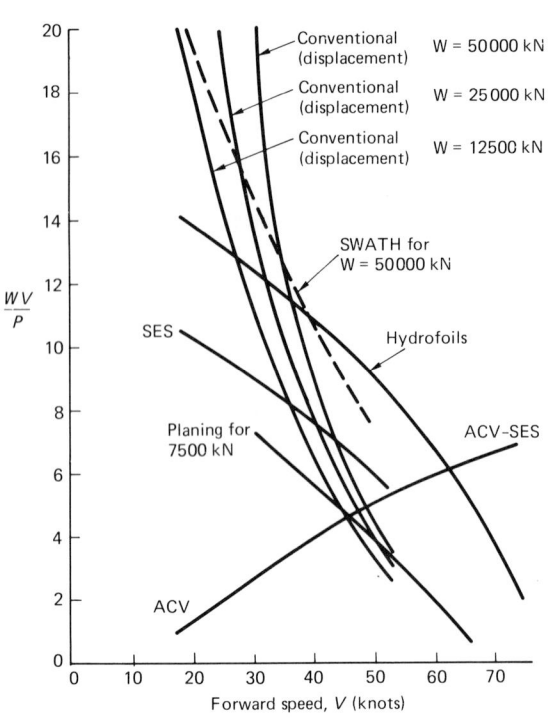

Figure 3.114 Transport efficiency of some conventional displacement ships and high speed craft as a function of forward speed (after Oossanen, 1987)

achievement line for 1950 and that for 1967 (Mantle, 1976, 1980; Oossanen, 1987). The limit for 1950 is given by $\Omega V = 5200$ knots, a value doubled to 10 400 knots by 1967. Oossanan (1987), using data from various sources including Mantle (1976) and Eames (1980), shows curves specifically for a number of advanced marine vehicles as well as selected conventional displacement ships, and these with some modern additions are shown in *Figure 3.114*.

There is clearly inferior transport efficiency within the speed range 30–200 knots which reflects the great difficulty facing designers of transport vehicles at, or near, the air and water interface. Noting that the curves in *Figure 3.114* apply for calm water conditions and that waves greatly affect performance, as discussed earlier, the following deductions can be drawn:

1. Conventional displacement monohull vessels have the highest transport efficiency (Ω) up to about 30 knots.
2. Hydrofoil craft have the highest Ω from about 35 to 60 knots.
3. ACV and SES have the highest Ω above about 60 knots
4. SWATH ships have high Ω between 30 and 40 knots bridging the performance of monohulls and hydrofoil craft.
5. Planing craft have low Ω but exceed the Ω of displacement monohulls above about 40 knots.
6. Below about 45 knots SES have values of Ω well above ACV.

Each of these conclusions is likely to be subject to quite frequent and dramatic re-evaluation as new craft, or new modifications to existing craft, are introduced. For example, a recent summary in *Marine Engineers Review* (1988b) suggests that high speed catamarans have been displacing hovercraft and hydrofoil craft as passenger ferries on a number of routes.

The effectiveness of the propulsion systems mentioned earlier can be represented in terms of propulsive efficiency *versus* craft speed. The maximum attainable open water efficiency, i.e. the value obtained with the propulsor operating in open water without the influence of the hull wake, has been used by Oossanen (1985) as a performance index. The details are shown in *Figure 3.115* from which it may be deduced that:

1. Up to about 40 knots (where conventional monohulls and SWATH are best) the subcavitating propeller is the optimum choice.
2. From about 35 to 50 knots (where hydrofoil craft are best) the transcavitating propeller is the best choice.
3. From about 50 to 75 knots (where hydrofoil craft and ACV are best) the supercavitating propeller is the best choice.
4. Above about 75 knots shrouded air propellers are best and also the best for ACV at all speeds.
5. For a wide operating speed range and compatible power and geometry requirements, the water jet

Comparative performances

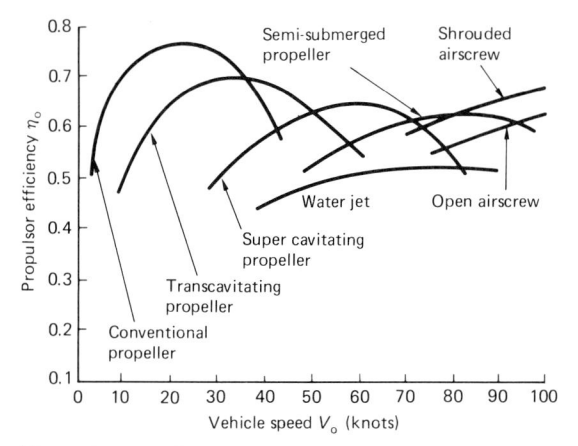

Figure 3.115 Highest achievable values of open water efficiency η_o of different types of propulsor as a function of forward speed V_o in open water (after Oossanen, 1987)

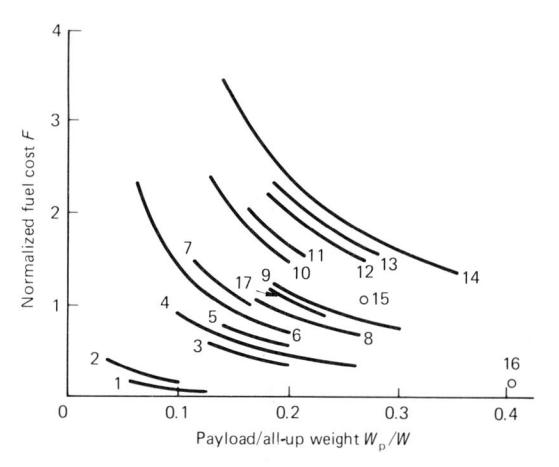

Figure 3.116 Normalized fuel cost as a function of payload ratio. Key: 1. Monohull displacement ferry; 2. High speed passenger train; 3. SWATH ship of 15 MN (1500 tonnef) weight displacement and 28 knots operating speed; 4. Modern passenger jet aircraft; 5. SWATH ship as 2, but at 33 knots; 6. Jet surveillance aircraft; 7. Surface piercing hydrofoil craft; 8. SES; 9. ACV; 10. Diesel fast patrol boat; 11. Jetfoil; 12. Slender planing boat; 13. Gas turbine fast patrol boat; 14. Helicopters; 15. Medium sized family car; 16. Motor coach; 17. Catamaran

system is favoured by many SES, catamarans and planing craft as well as some hydrofoil craft.

A detailed investigation of the economic and performance relations between a variety of marine vehicles was undertaken by Crewe (1980), with special emphasis on high-speed marine craft. Some of this work has been summarized by Clayton and Bishop (1982). Evidently, simple graphs of the type in *Figures 3.113* and *3.114* cannot contain all the influences relating costs, payloads, speeds and distances of travel. One particular correlation with further additions from Clayton (1987), is shown in *Figures 3.116*.

A normalized fuel cost F is defined as:

$$F = \left(\frac{rsfc}{\Omega}\right) \div \left(\frac{W_p}{W}\right) \qquad (3.232)$$

where *rsfc* represents the specific fuel consumption of any engine relative to a datum diesel engine and W_p the payload weight. Mean lines have been drawn corresponding to best achievements and it is evident from *Figure 3.116* that monohull displacement ferries stand out in terms of economy at low speeds and W_p/W values and that modern passenger jet aircraft are best at the other extreme. High speed marine craft are collected in the middle but have favourable values of W_p/W as a result of their light construction and propulsion systems of high power to weight ratio. The SRN4 Mk3 is positioned at the extreme right-hand end of the ACV line.

The necklace diagram of *Figure 3.117* indicates vehicles zones for different speeds V versus $V\Omega/rsfc$, and the inverse of these parameters. In popular jargon, the higher up the figure the better for economic operation. This is not necessarily a relevant measure of economic efficiency which needs to take account of earning capacity and journey distance as well as the parameters referred to earlier. When this is done, hovercraft have similar performances to passenger jet aircraft and about half that of passenger ship ferries and trains.

Nevertheless *Figure 3.117* does indicate that monohull displacement ferries are best for travel over water

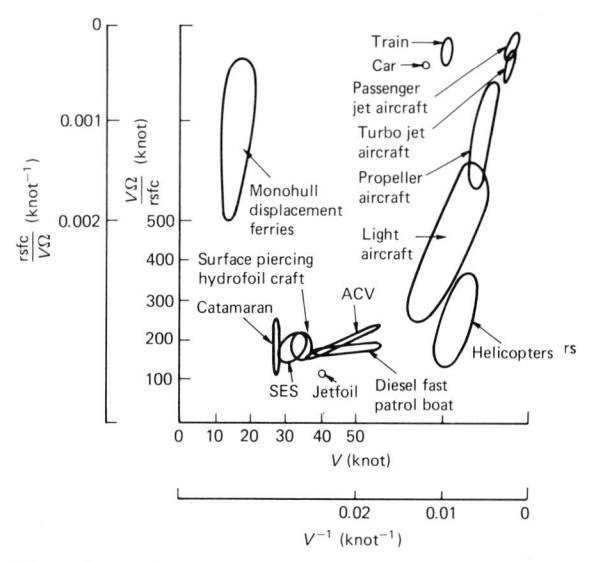

Figure 3.117 Craft zones of $V\Omega/rsfc$ as functions of forward speed

but at low speeds; the latest developments in high speed trains are best for land travel, that is, over a rigid fixed route; passenger jet aircraft are best for air travel but at medium to high speeds and long distances. Special cases such as very high speed on water, high speed on land with low passenger density, or low speed in air with low passenger density are severely penalized. These only survive when special circumstances exist, such as helicopters because of their hovering and vertical take off and landing capabilities.

It was concluded by Crewe (1980) that the escalation of fuel costs allows a simplified comparison of economic efficiency between various types of high speed marine craft and that there are fulfilling their tasks well. The generalization of the original Gabrielli–von Karman diagram to account for earnings per unit payload per unit distance related to speed allows a more valid comparison for passenger carrying craft, especially for those classified as high speed marine craft. There is, of course, still considerable room for improvement in detailed design, but this will be achieved as research and development continues.

References

Abbott, I.H. and von Doenhoff, A.E. *Theory of Wing Sections*. Dover, New York (1959).
Acosta, A.J. Hydrofoils and Hydrofoil Craft. *Annual Review of Fluid Mechanics*, **5**, 161–184 (1973).
Adams, J. D. and Beverly, W. F. Technical Evaluation of the SES–200 High Length to Beam Ratio Surface Effect Ship. *Naval Engineers Journal*, **96**(3) (1984).
Allen, H.G. and Raybould, K. High Performance Composites and Their Effectiveness in Sandwich Panels. In *Proceedings of the 6th International High-Speed Surface Craft Conference*, London, 14–15 January, Paper No. 11, (1988).
Allen, R.G. and Jones, R.R. A Simplified Method for Determining Structural Design-Limit Pressures on High Performance Marine Vehicles. AIAA/SNAME Advanced Marine Vehicles Conference, San Diego, April 1979, Paper No. 78–754 (1978).
Bailey, D. Performance Prediction-Fast Craft. *Occasional Publication 1, Paper C Royal Institute of Naval Architects* (1974).
Bailey, D. The NPL High Speed Round Bilge Displacement Hull Series. *Marine Technology Monograph* 4. Royal Institution of Naval Architects (1976).
Barr, R.A. and Etter, R.J. Selection of Propulsion Systems for High-Speed Advanced Marine Vehicles. *Marine Technology*, **12**, 33–49 (1975).
Barratt, M.J. The Wave Drag of a Hovercraft. *Journal of Fluid Mechanics*, **22**, 39–47 (1965).
Betts, C.V. A Review of Developments in SWATH Technology. *Paper No. 1 of RINA* (1988).
Betts, C.V. and Clayton, B.R. A Possible Maritime Future for Surface Effect Craft in the UK. In *Proceedings of the Symposium on Ram Wing and Ground Effect Craft*, 19 May 1987, Royal Aeronautical Society, London, 137–174 (1987).
Bishop, R.E.D. and Price, W.G. (1980) Structural Dynamics. *Naval Architect*, **6**, 247–248 (1980).
Bliault, A. Jetriders – Fast SES Passenger Craft. *Small Craft*, **24**, 17–22 (1989).
Blount, D.L. and Fox, D.L. Small Craft Power Prediction. *Marine Technology*, **13** (1976).
Blount, D.L. and Hankley, D.W. Full-Scale Trials and Analysis of High Performance Planing Craft Data. *Trans SNAME*, **84**, 251–277 (1976).
Blyth, A.G. Recent Research into the Ultimate Stability of Surface Effect Ships. In *Proceedings of the International Conference on the Safeship Project: Ship Stability and Safety*, Royal Institution of Naval Architects, London, Paper No. 4 (1986).
Blyth, A.G. Roll Stability of an SES in a Seaway. In *Proceedings of the 4th International Hovercraft Conference*, 6–7 May 1987, The Hovercraft Society, Alverstoke, Hants. Paper No. 5 (1987).
Blyth, A.G. (1988) SES Stability in Turns – The Influence of Sidewall Shape. In *Proceedings of the 1st International High-Performance Vehicle Conference*, 2–5 November 1988, Shanghai, Chinese Society of Naval Architects and Marine Engineers, Paper No. 3–1, (1980).
Brown, D.K. Hydrofoils, A Review of Their History, Capability and Potential. *Trans. Institution of Engineers and Shipbuilders*, Scotland, **123**, 49–56 (1988).
Brown, D.K., Catchpole, J.P. and Shand, A.M.J. The Evaluation of the Hydrofoil HMS *Speedy*. *Trans. RINA*, **126**, 1–16 (1984).
Butler, E.A. The Surface Effect Ship. *Naval Engineers Journal special edition*, **92**(2), 200–258 (1985).
Caldeira–Saraiva, F. and Clarke, D. Active Control of SWATH Motions. In *Proceedings of the 8th Ship Control Systems Symposium*, 6–9 October 1987, The Hague, **3**, 398–418 (1987).
Cannon, T.R. and McKesson, C.B. Large SWATHs: A Discussion of the Diminishing Returns of Increasing SWATH Ship Size. *AIAA 8th Advanced Marine Systems Conference*, September 1986, San Diego, Paper No. AIAA – 2385 (1986).
Chaplin, J.B. Air Cushion Development Yesterday, Today and Tomorrow. CASI Symposium (May 1979).
Chapman, R.B. Hydrodynamic Drag of Semi-Submerged Ships. *Journal of Basic Engineering*, ASME, **94**, 879–884 (1972).
Chun, H., Ferguson, A.M. and McGregor, R.C. New Computational Tool to Estimate the Resistance of SWATH Ships. In *Proceedings of the 1st International High-Performance Vehicle Conference*, 2–5 November 1988, Shanghai, Chinese Society of Naval Architects and Marine Engineers, Paper No. 6–1 (1988).
Clark, D.J., O'Neill, W.C. and Wight, D.C. Balancing Mission Requirements and Hydrofoil Design Characteristics. AIAA/SNAME Advanced Marine Vehicles Conference, Paper No. 78–725 (1978).
Clayton, B.R. and Bishop, R.E.D. *Mechanics of Marine Vehicles*. Spon, London (1982).
Clayton, B.R. and Russell, B.J. (ed.). *Proceedings of Seminar on Hovercraft to the Rescue*, 18 April 1989, The Hovercraft Society, Alverstoke, Hants. (1989).
Clayton, B.R. and Satchwell, C.J. An Introduction to the Theory, Technology and Economics of Wind Assisted Ship Propulsion. In *Proceedings of the 9th BWEA Wind Energy Conversion Systems Conference*, 1–3 April 1987, Edinburgh. MEP, London, 299–314 (1987).

Clayton, B.R. and Tuckey, P.R. Dynamic Response of Hovercraft to Regular Wave Excitation. In *Proceedings of the 4th High-Speed Surface Craft Conference*, 9–12 May 1983, London, 139–159 (1983).

Clayton, B.R. and Webb, R. The Dynamic Response of Surface Effect Ships in Regular Head Seas Using Model Simulations. In *Proceedings of the 5th International High-Speed Surface Craft Conference*, 7–8 May 1986, Southampton, Paper No. 2 (1986).

Clayton, B.R. and Webb, R. Evaluation of a Ride Control System for SES. In *Proceedings of the 6th International High-Speed Surface Craft Conference*, 14–15 January, London, Paper No. 8, (1988a).

Clayton, B.R. and Webb, R. Development and Analysis of SES Dynamic Ride Control Systems. *Advances in Underwater Technology, Ocean Science and Offshore Engineering*, **15**, 217–231 (1988b).

Clayton, B.R. and Webb, R. Motion Control and Analysis of SES and ACV. In *Proceedings of the 1st International High-Performance Vehicle Conference*, 2–5 November 1988, Shanghai, Chinese Society of Naval Architects and Marine Engineers, Paper No. 2–1 (1988c).

Clayton, B.R. Over or Under? The Future Role of Hovercraft as High-Speed Channel Ferries. *Endeavour*, new series, **11**(1), 29–35 (1987).

Clement, E.P. Graphs for Predicting the Ideal High Speed Resistance of Planing Catamarans. *DTNSRDC Report No. 1573* (1961).

Clement, E.P. Designing for Optimum Planing Performance. *International Shipbuilding Progress*, **26**, 61–64 (1979).

Clement, E.P. and Blount, D.L. Resistance Tests of a Systematic Series of Planing Hull Forms. *Trans SNAME*, **71**, 491–579 (1963).

Clement, E.P. and Pope, J.D. Stepless and Stepped Planing Hulls – Graphs for Performance Prediction and Design. *David Taylor Model Basin Report* No. 1490 (1961).

Coates, J.T.S. A Hydrofoil Fisheries Patrol Vessel for the United Kingdom. In *Proceedings of the Symposium on Small Fast Warships and Security Vessels*, 7–9 March 1978, Royal Institution of Naval Architects, London, 71–85 (1978).

Cockerell, C. *Improvements in or Relating to Vehicles for Travelling Over Land and/or Water*. British Patent No. 854–211 (1955).

Codega, L. and Lewis, J. A Case Study of Dynamic Instability in a Planing Hull. *Marine Technology*, **24**, 143–163 (1987).

Cole, S.H. The Human Factor Aspects of Ride Motion. In *Proceedings of the 3rd International Conference on High-Speed Surface Craft*, 24–27 June 1980, Brighton, 218–229 (1980).

Coles, A.V. and Kidd, M.A. Demonstrated Performance of the Amphibious Assault Landing Craft JEFF(B). In *Proceedings of the 3rd International Conference on High-Speed Surface Craft*, 24–27 June 1980, Brighton, 366–383 (1980).

Corlett, E.C.B. Twin Hull Ships. *Trans RINA*, **111**, 401–438 (1969).

Covich, P.M. T-AGOS 19: An Innovative Programme for an Innovative Design. *Naval Engineers Journal*, **99**(3), 99–106 (1987).

Crewe, P.R. and Eggington, W.J. The Hovercraft – A new Concept in Maritime Transport. *Trans RINA*, **102**, 315–365 (1960).

Crewe, P.R. *Hovercraft in Relation to Other Advanced Marine Vehicles, Some Particular Advantages and Problems*. Lecture on Advanced Marine Vehicles Course, October 1980, University of Southampton (1980).

Dallinga, R.P., Huysmans, R.H.M. and Graham, R. *Development of Design Tools for the Prediction of SWATH Motions*. In proceedings of the 17th Office of Naval Research Symposium on Naval Hydrodynamics, 29 August–2 September 1988, The Hague, Paper No. 2, Session VIII (1988).

Davison, E.F. and Fink, M.D. The Amphibious Assault Landing Craft JEFF(A). In *Proceedings of the 3rd International Conference on High-Speed Surface Craft*, 24–27 June 1980, Brighton, 298–314 (1980).

Davison, G.H. Modern Marine Jet Propulsion. *Naval Architect*, RINA, **6**, 234–235 (1979).

Det Norske Veritas *Rules for Classification of High Speed Light Craft*. Oslo, Norway (1985).

Doctors, L.J. *The Wave Resistance of an Air Cushion Vehicle*. University of Michigan (1970).

Dorey, A.L. The Missile-Armed Fast Patrol Boat. *International Defence Review* (1977).

Du Cane, P. (ed.) *High-Speed Small Craft*. David and Charles, Newton Abbott (1974).

Eames, M.C. Principles of Hydrofoils. Chapter 3 in *High-Speed Small Craft* (ed. Du Cane, P.). David and Charles, Newton Abbott (1974).

Eames, M.C. Advances in Naval Architecture for Future Surface Warships. *Trans. RINA*, **122**, 93–118 (1980).

Eames, M.C. and Drummond, T.G. HMCS Bras D'Or – Sea Trials and Future Prospects. *Trans. RINA*, **115**, 69–87 (1973).

Eames, M.C. and Jones, E.A. HMCS Bras D'Or – An Open-Ocean Hydrofoil Ship. *Trans. RINA*, **113**, 111–138 (1971).

Eatock Taylor, R. and Hung, S.M. Some Wave Load Effects in the Design of SWATH Ships. In *Proceedings of the 3rd International Symposium on the Practical Design of Ships and Mobile Units*, PRADS–87, Trondheim, Vol. I, 344–354 (1987).

Eckersley-Maslin, D.M. and Coates, J.F. Operational Requirements and Choice of Craft. In *Proceedings of the Symposium on Small Fast Warships and Security Vessels*, 7–9 March 1978, Royal Institution of Naval Architects, London 1–13 (1978).

Eggington, W.J. and Kobitz, N. The Domain of the Surface Effect Ship. *Trans. SNAME*, **83**, 268–298 (1975).

Elsley, G.H. and Devereux, A.J. *Hovercraft Design and Construction*. David and Charles, Newton Abbott (1968).

Everest, J.T. Some research on the Hydrodynamics of Catamarans and Multi-Hulled Vessels in Calm Water. *Trans. North-East Coast Institution of Engineers and Shipbuilders*, **84** (1968).

Everest, J.T. Some Comments on the Performance of a Single-Hull Trawler Form and Corresponding Catamaran Ships Made Up From Symmetrical and Asymmetrical Hulls. *National Physical Laboratory Report No. 129* (1969).

Everest, J.T. and Hogben, N. Research on Hovercraft in Calm Water. *Trans. RINA*, **109**, 311–326 (1967).

Everest, J.T. and Hogben, N. A Theoretical and Experimental Study of the Wavemaking of Hovercraft of Arbitrary Planform and Angle of Yaw. *Trans. RINA*, **111**, 343–365 (1969).

Ford, A.G., Wares, R.N., Bush, W.F. and Chorney, S.J. *High Length-to-Beam Ratio Surface Effect Ship*. AIAA/SNAME Advanced Marine Vehicles Conference, San Diego, Paper No. 78–745 (1978).

Frauenberger, H.C. Shimrit: Mark 2 Hydrofoil for the Israel Navy. In *Proceedings of the 1st International Hydrofoil Society Conference*, Ingonish Beach, Nova Scotia (1982).

Fry, E.D. and Graul, T. Design and Application of Modern High-Speed Catamarans. *Marine Technology*, **9**(3) (1972).

Gabrielli, G. and von Karman, T. What Price Speed? *Mechanical Engineering*, Journal of ASME, **72**, 775–781 (1950).

Gawn, R.W.L. Effect of Pitch and Blade Width on Propeller Performance. *Trans. INA*, **95**, 157–193 (1953).

Gawn, R.W.L. and Burrill, L.C. Effects of Cavitation on the Performance of a Series of 16in. Model Propellers. *Trans. INA*, **99**, 690–728 (1957).

Gore, J.L. SWATH Ships. *Naval Engineers Journal* (special edition), **97**(2), 83–112 (1985).

Gupta, S.K. and Schmidt, T.W. Developments in SWATH Technology. *Naval Engineers Journal*, **98**(5), 171–188 (1986). (See also Discussion in *Naval Engineers Journal*, **98** (7), 112–122 (1988).)

Hadler, J.B. and Hubble, E.N. Prediction of Power Performance of the Series 62 Planing Hull Forms. *Trans. SNAME*, **79**, 366–404 (1971).

Hadler, J.B., Hubble, E.N., Allen, R.G. and Blount, D.L. Planing Hull Feasibility Model – Its Role in Improving Patrol Craft Design. In *Proceedings of the Symposium on Small Fast Warships and Security Vessels*, 7–9 March 1978, Royal Institution of Naval Architects, London, 115–131 (1978).

Hadler, J.B., Lee, C.M., Birmingham, J.T. and Jones, H.D. Ocean Catamaran Seakeeping Design Based on Experiences of USNS Hayes. *Trans. SNAME*, **82**, 126–161 (1974).

Harting, A. A Literature Survey of the Aerodynamics of Air Cushion Vehicles. *AGARD Report No. 565* (1969).

Heather, R.G., Nicholson, K. and Stevens, M.J. Seakeeping and the Small Warship. In *Proceedings of the Symposium on Small Fast Warships and Security Vessels*, 7–9 March 1978, Royal Institution of Naval Architects, London, 31–46 (1978).

Heller, S.R. and Jasper, H.H. On the Structural Design of Planing Craft. *Trans. RINA*, **103**, 49–65 (1961).

Hercus, P. Design of the Wave Piercing Catamaran. Paper No. 14 of RINA (1985).

Hoerner, S.F. *Fluid-Dynamic Drag*. Hoerner Fluid Dynamics, Brick Town, NJ (1965).

Hoerner, S.F. and Borst, H.V. *Fluid-Dynamic Lift*. Hoerner Fluid Dynamics, Brick Town, NJ (1975).

Hong, Y.S. Heave and Pitch Motions of SWATH Ships. *Journal of Ship Research*, **30**(3), 12–25 (1986).

Hook, C. and Kermode, A.C. *Hydrofoils*. Pitman, London; (1967).

Hung, W.L. and Zhang, S.R. A Method of Estimating the Resistance of a Catamaran Hull. *Ship Engineering*, Chinese Society of Naval Architects and Marine Engineers, **1** (1983).

ISSC: Proceedings of the International Ship and Offshore Structures Conference. Lyngby, Denmark (1988).

Johnson, V.E. Waterjet Propulsion. In *Proceedings of the 7th Office of Naval Research Symposium on Naval Hydrodynamics*, 25–30 August 1968, Rome, 1045–1058 (1968).

Johnston, R.J. Hydrofoils. *Naval Engineers Journal* special edition, **92**(2), 142–199 (1985).

Johnston, R.J. and O'Neill, W.C. The Development of Automatic Control Systems for Hydrofoil Craft. In *Proceedings of the International Conference on Hovering Craft, Hydrofoil and Advanced Transit Systems*, 13–16 May 1974, Brighton, 265–279 (1974)

Kaplan, P., Bentson, J. and Davis, S. Dynamics and Hydrodynamics of Surface Effect Ships. *Trans. SNAME*, **89**, 211–247 (1981).

Kennell, C., White, B. and Comstock, E.N. Innovative Naval Designs for North Atlantic Operations. *Trans SNAME*, **93**, 261–281 (1985).

King, J.H. The Evolution of the Nibbio Class Hydrofoil From Tucumcari. In *Proceedings of the 1st International Hydrofoil Society Conference*, Igonish Beach, Nova Scotia (1982).

Korvin-Kroukovsky, B.V., Savitsky, D. and Lehman, W.F. Wetted Area and Center of Pressure of Planing Surfaces. Stevens Institute of Technology, *Davidson Laboratory Experiment Towing Tank Report No. 360* (1949).

Kruppa, C. Practical Aspects in the Design of High-Speed Small Propellers. In *Proceedings of the 3rd Lips Propeller Symposium*, May 1976, Drunen, Netherlands, 39–49 (1976).

Lamb, G.R. Some Guidance for Hull Form Selection for SWATH Ships. *Marine Technology*, **25**(4), 239–252 (1988).

Lavis, D.R. The Development of Stability Standards for Dynamically Supported Craft: A Progress Report. In *Proceedings of the 3rd High-Speed Surface Craft Conference*, 24–27 June 1980, Brighton, 384–403 (1980).

Lavis, D.R. Air Cushion Craft. *Naval Engineers Journal* special edition, **97**(2), 259–316 (1985).

Lee, C.M., Jones, H.D. and Curphey, R.M. Prediction of Motion and Hydrodynamic Loads of Catamarans. *Marine Technology*, **10**(10) (1973).

Lewis, E.V. (ed.) Principles of Naval Architecture, Volume II: *Resistance, Propulsion and Vibrations*. Society of Naval Architects and Marine Engineers, New York, (1988).

Lin, W.C. and Day, W.G. The Still-Water Resistance and Propulsion Characteristics of SWATH Ships. *AIAA/SNAME Advanced Marine Vehicles Conference,* February 1974, San Diego, Paper No. 74–32 (1974).

Mant, M.D. AP1–88 Operating Experience. In *Proceedings of the 4th International Hovercraft Conference*, 6–7 May 1987, Southampton, Paper No. 3, (1987).

Mantle, P.J. Large High Speed Surface Effect Ship Technology. *International Hovering Craft, Hydrofoil and Advanced Transit Systems Conference*, Brighton (1974).

Mantle, P.J. Cushions and Foils. Society of Naval Architects and Marine Engineers Spring Meeting, Philadelphia, Paper No. 2 (1976).

Mantle, P.J. Air Cushion Craft Development. *DTNSRDC Report No. 80–0212* (1980).

Marchant, A. and Pinzelli, R.F. The Role for Composite Materials in Future Marine Transportation. In *Proceedings of the 6th International High-Speed Surface Craft Conference*, 14–15 January 1988, Paper No. 9 (1988).

Marine Engineers Review From Sail to Superconducting. February 1988, 31–32 (1988a).

Marine Engineers Review Hovercraft Vs Catamarans, February 1988, 33–35 (1988b).

Marwood, W.J. and Bailey, D. Design Data for High Speed Displacement Hulls of Round Bilge Form. *National Physical Laboratory Report No. 99* (1969).

Massey, B.S. *Measures in Science and Engineering: Their Expression, Relation and Interpretation*. Ellis Horwood, Chichester (1986).

McCreight, K.K. Assessing the Seaworthiness of SWATH Ships. *Trans. SNAME*, **95** (1987).

McGregor, R.C., Drysdale, L.H. and Wu, J-Y. On the Response of SWATH Ships in the Presence of Control Fins. In *Proceedings of the 8th Ship Control Systems Symposium*, 6–9 October 1987, The Hague, **3**, 352–379 (1987).

Meredith, D. Meeting the Channel Tunnel Challenge. In *Proceedings of the 5th High-Speed Surface Craft Conference*, 7–8 May 1986, Southampton, Paper No. 1 (1986).

Meyer, J.R. Extended Performance Hydrofoils. *AIAA 6th Marine Systems Conference,* Seattle, Paper No. 81–2067 (1981).

Michel, W.H. The SeaKeeping Catamaran Ship, its Features and its Feasibility. *International Shipbuilding Progress*, **8**, 85 (1961).

Millward, A., Wakeling, B. and Sproston, J. The Transverse Stability of a Fast Round Bilge Hull. In *Proceedings of International Conference on Design Considerations for Small Craft*, Royal Institution of Naval Architects, London (1984).

Murray, A.B. The Hydrodynamics of Planing Hulls. *Trans. SNAME*, **58**, 658–692 (1950).

Naval Engineers Journal, special edition, **97** (2) (1985).

Newman, J.N. and Poole, F.A. The Wave Resistance of a Moving Pressure Distribution in a Canal. *DTNSRDC Report No. 1619* (1962).

Newton, R.N. Contribution to the Discussion on The Combination of Sail with an Alternative Source of Power, by Barnaby, K.C. *Trans. RINA,* **104**, 90–92 (1962).

Newton, R.N. and Radar, H.P. Performance Data of Propellers for High-Speed Craft. *Trans. RINA*, **103**, 93–129 (1961).

O'Neill, W.C. A New Small Waterplane Area Ship Concept. *AIAA 8th Advanced Marine Systems Conference,* September 1986, San Diego, Paper No. 86–2382 (1986).

Oossanen, P.V. Advanced Marine Vehicles: A Critical Review. *La Marina Italiana*, **85**, 51–64 (1987).

Ozawa, H. The Design and Operation of Catamaran Vehicles. *Trans. North-East Coast Institution of Engineers and Shipbuilders* (1987).

Payne, P.R. *Design of High-Speed Boats*, Volume I: *Planing*. Fishergate, Annapolis MD (1988).

Plumb, C.M. and Brown, D.K. Hovercraft in Mine Counter Measures. In *Proceedings of the 3rd International Conference on High-Speed Surface Craft*, 24–27 June 1980, Brighton, 286–297 (1980).

Reilly, E.T., Shin, Y.S. and Kotte E.H. A Prediction of Structural Load and Response of a SWATH Ship in Waves. *Naval Engineers Journal*, **100**(5), 251–264 (1988).

Richardson, W. and White, W. Extension and Application of Ship Design Optimisation Code (SHIPDOC). *Naval Engineers Journal*, **96**(3), 177–190 (1984).

Rieg, D.F. and King, J.H. Technical Evaluation of the RHS-200 for High-Speed Ferry Application and US Coast Guard Missions. *DTNSRDC Report No. SDD–83/10* (1983).

Riegels, F.W. *Aerofoil Sections: Results from Wind-Tunnel Investigations, Theoretical Foundations* (Translation by D.G. Randall). London: Butterworths, London (1961).

RINA *Proceedings of the International Conference on SWATH Ships and Advanced Multi-Hulled Vessels*, 17–19 April 1985, Royal Institution of Naval Architects, London (1985).

RINA *Proceedings of the International Conference on SWATH Ships and Advanced Multi-Hulled Vessels*, 28–30 November 1988. Royal Institution of Naval Architects, London (1988).

Robson, B.L. Development of the Royal Australian Navy GRP Minehunter Design. *Trans. RINA*, **125**, 124–142 (1983).

Russell, B.J. Hovercraft in Surveying. *Advances in Underwater Technology, Ocean Science and Offshore Engineering*, **6**, 453–460 (1986).

Rutgersson, O. Catamarans Versus Single-Hull Concepts. A Study of Stability, Powering and Seakeeping Qualities for Small Warships. In *Proceedings of the International Symposium on Coastal Defence and Assault Vessels and Systems*, 19–22 May 1986, Royal Institution of Naval Architects, London, **1**, Paper No. 4 (1986).

Salvesen, N., von Kerczek, C. H., Scragg, C.A., Cressy, C.P. and Meinhold, M.J. Hydro-Numeric Design of SWATH Ships. *Trans. SNAME*, **93**, 325–346 (1985).

Savitsky, D. Hydrodynamic Design of Planing Hulls. *Marine Technology*, **1**, 71–95 (1964).

Savitsky, D. Planing Craft. *Naval Engineers Journal*, special edition, **97**(2), 113–141 (1985).

Savitsky, D. and Brown, P.W. Procedures for Hydrodynamic Evaluation of Planing Hulls in Smooth and Rough Water. *Marine Technology*, 13 (1976).

Savitsky, D. and Gore, J.L. A Re-evaluation of the Planing Hull Form. *AIAA/SNAME Advanced Marine Vehicles Conference,* October, Baltimore MD (1979).

Schmidt, T.W. Innovation in SWATH. *AIAA Eighth Advanced Marine Systems Conference,* September 1986, San Diego, Paper No. AIAA 86–2363 (1986).

Sharples, A.K. *Small Patrol Craft Proceedings of the Symposium on Performance Predictions of Small Craft*. Royal Institution of Naval Architects, Occasional Paper No. 1 (1974).

Shoemaker, J.M. Tank Tests of Flat and V-Bottom Planing Surfaces. *National Advisory Committee on Aeronautics Technical Note No. 509* (1934).

Sikora, J.P., Dinsenbacher, A. and Beach, J.E. A Method for Estimating Lifetime Loads and Fatigue Lives for SWATH and Conventional Monohull Ships. *Naval Engineers Journal*, **95**(3) (1983).

Silverleaf, A. A Review of Hovercraft Research in Britain. *Journal of the Royal Aeronautical Society*, **72**, 1019–1028 (1968).

Small Craft Oregrund – Fast and Quiet. *RINA, No. 24*, 35–37 (1989).

Stanton Jones, R. Hovercraft. *Journal of the Royal Aeronautical Society*, **72**, 911–914 (1968).

Summers, B.J., Winter A. and Marks, J. Hovercraft. Chapter 4 in *High-Speed Small Craft* (ed. Du Cane, P.) David and Charles, Newton Abbott (1974).

Tattersall, E.G. The History and Future of the SES in the United Kingdom. *Naval Engineers Journal*, **94**, 267–276

(1982).

Tattersall, E.G. Variants of the Vosper Hovermarine SES Designs. In *Proceedings of the 4th High-Speed Surface Craft Conference*, 9–12 May 1983, London, 9–18 (1983).

Tattersall, E.G. The HM500 Series of Sidewall Hovercraft. *Trans. RINA*, **127**, 53–68 (1985).

Trillo, R.L. *Marine Hovercraft Technology*. Leonard Hill, London (1971)

Trillo, R.L. (ed.) *Jane's High-Speed Marine Craft and Air Cushion Vehicles* (19th edn). Jane's, London (1986).

Trillo, R.L. (ed.) *Jane's High Speed Marine Craft and Air Cushion Vehicles* (20th edn). Jane's, London (1988).

Tulin, M.P. Supercavitating Flows Past Foils and Struts. *National Physical Laboratory Symposium on Cavitation in Hydrodynamics*. HMSO, London (1956).

Turner, H. and Taplin, A. The Resistance of Large Powered Catamarans. *Trans. SNAME*, **76**, 180–213 (1968).

Vallentine, H.R. *Applied Hydrodynamics*. Butterworths, London (1969).

Venturini, G. Waterjet Propulsion in High-Speed Surface Craft. In *Proceedings of the 3rd International Conference on High-Speed Surface Craft*, 24–27 June 1980, Brighton, 125–142 (1980).

von Schertel, H. The Design and Application of Hydrofoils and Their Future Prospects. *Trans. Institute of Marine Engineers*, **86**, 53–64 (1974).

Wadi, Y. Motion Characteristics and Stabilisation of a SWATH-Type Passenger Ship in Waves. *Mitsubishi Juko Giho*, **23** (1986).

Wadlin, K.L. Mechanics of Ventilation Inception. In *Proceedings of the 2nd Office of Naval Research Symposium on Naval Hydrodynamics*, ONR/ACR-38, 425–445 (1958).

Wadlin, K.L. Shuford, C.L. and McGehee, J.R. A Theoretical and Experimental Investigation of the Lift and Drag Characteristics of Hydrofoils at Subcritical and Supercritical Speed. *National Advisory Committee on Aeronautics Report No. 1232* (1955).

Wellicombe, J.F. and Jahangeer, J.M. The Prediction of Pressure Loads on Planing Hulls in Calm Water. *Trans. RINA*, **121**, 53–70 (1979).

Wheeler, R.L. *The Amphibious Hovercraft*. Presented at the Diamond Jubilee International Meeting, Society of Naval Architects and Marine Engineers, New York, Paper No. 23 (1968).

Wheeler, R.L. An Appraisal of Present and Future Large Hovercraft, *Trans. RINA*, **118**, 221–244 (1976).

Wheeler, R.L. The Amphibious Hovercraft as a Warship. In *Proceedings of the Symposium on Small Fast Warships and Security Vessels*, 7–9 March 1978, London, Royal Institution of Naval Architects, 87–99 (1978).

Wheeler, R.L. Design Development and Trials of the AP1–88 Hovercraft. *Trans. RINA*, **126**, 183–294 (1984).

Wheeler, R.L. Large Amphibious Hovercraft. In *Proceedings of the Symposium 'On a Cushion of Air'*, 11 December 1985, London, Paper No. 2 (1985).

Wheeler, R.L. and Key. T. Hovercraft Skirt Design and Manufacture. In *Jane's High-Speed Marine Craft and Air Cushion Vehicles* 19th edn (ed. Trillo, R.) (1986).

Wilson, R.A. Wells, S.M. and Heber, C.E. Powering Predictions for Surface Effect Ships Based on Model Results. *AIAA/SNAME Advanced Marine Vehicles Conference*, San Diego, Paper No. 78–744 (1978).

Yeh, H.Y.H. Series 64 Resistance Experiments on High Speed Displacement Forms. *Marine Technology*, **2**, 248–272 (1965).

Yermotayev, S.G., Aframeyev, E.A., Teder, L.A. and Rabinowich, Y.S. Hydrodynamic Features of High Speed Catamarans. *Hovercraft and Hydrofoil*, **16** (1977).

4 Diving and Underwater Vehicles

Robert L. Allwood

Contents

4.1 Introduction
    4.1.1 History of underwater working
    4.1.2 Early diving technology
    4.1.3 The first signs of physiological problems
    4.1.4 Breathing systems

4.2 Diving
    4.2.1 Major developments in diving since 1900
    4.2.2 Diving physics
    4.2.3 Diving physiology
    4.2.4 Air diving
    4.2.5 Mixed gas diving
    4.2.6 Personal diving equipment
    4.2.7 Support vessels
    4.2.8 Safety of diving operations
    4.2.9 Recent developments in deep diving

4.3 Atmospheric diving suits and systems
    4.3.1 Atmospheric diving systems or suits – history
    4.3.2 Modern commercial systems
    4.3.3 Comparison of the ADS with other forms of intervention

4.4 Underwater vehicles – introduction
    4.4.1 Manned submersibles
    4.4.2 Remotely Operated Vehicles (ROVs)

4.5 Selection of underwater intervention methods for offshore oil and gas operations
    4.5.1 Operational factors affecting intervention mode selection
    4.5.2 Capabilities of the various modes of intervention
    4.5.3 The future

Further reading

4.1 Introduction

With the recent expansion of the field of marine technology, caused primarily by the rapid growth of the offshore oil and gas industry, has come an increased requirement for work to be carried out below the surface of the sea. It is perhaps ironic that man, thought to have originally evolved from the sea, now spends so much effort and ingenuity in trying to accomplish comparatively simple tasks in this alien underwater environment.

Over a period spanning many centuries man has explored a number of methods for overcoming the problems posed by this environment. Consequently, when faced with an underwater task today, there are a number of techniques available from which to choose, having to take into account suitability for the task in hand, safety of personnel and not least, its cost effectiveness.

The primary techniques available involve the use of:

1. Ambient pressure diving – where the diver enters the water and experiences the pressure changes that are a consequence of changing depth.
2. Manned submersibles – vehicles constructed to withstand external pressure changes and carry one or more divers in a near normal environment to the underwater work site.
3. Remotely operated systems – machines which carry no divers, but which are capable of carrying out work under remote control from the surface.

The development of manned submersibles and, in particular, remotely operated systems began only very recently. Modern machining techniques, new materials and micro-electronics were required before effective systems could be realized.

The aim of this chapter is to describe some of the major developments that have taken place in these fields of diving and underwater vehicles and also to present the fundamental science needed to give a basic understanding of this subject.

4.1.1 History of underwater working

It is likely that diving dates back to the very earliest of cultures when the search for food was probably the major driving force. The discovery of useful and merchantable commodities, the desire to conduct military activities and to undertake salvage operations under water ensured its development.

Although it is not known when it was discovered that divers could hold their breath and go under water, the beginnings of diving as a profession can certainly be traced back more than 5000 years. References to diving in this era are, however, scarce, consisting of no more than cave wall drawings and carvings. One of the first documented references to diving is found in the writings of the Greek historian Herodotus. He recounts the story of a diver named Scyllis who was employed by the Persian King Xerxes to recover sunken treasure in the fifth century BC.

Great technical advances have often occurred during periods of war and diving certainly appears to have benefitted in this way. From very early times, divers were active in military operations. Their missions included cutting anchor cables to set enemy ships adrift, boring holes in the bottoms of ships and the building of harbour defences at home, whilst attempting to destroy those of the enemy. Aristotle in his *Problemata* indicates that Alexander the Great not only sent divers down to destroy the Macedonian boom fortifications during the siege of Tyre, but also went under water himself to inspect the progress of work. He is reputed to have used a diving bell constructed of glass in this operation which took place in 332 BC.

Homer also refers to divers in the *Illiad*, although they are believed to be more probably swimmers operating just below the surface. Other early divers developed an active salvage industry around the major parts of the eastern Mediterranean. By the first century BC, operations were so well organized that a system of payment based on depth had been established. All these early dives relied on the technique of breath holding, a technique still used today in a few areas of the world. The breath holding skin diver usually trains from a very early age, developing the great lung capacity, stamina and confidence that are required for this activity. These dives usually last 1–2 min with a maximum depth of no more than 30 m (98.5 ft).

4.1.2 Early diving technology

The logical step required to increase the capability of the diver was to provide an air supply that would permit an extended stay under water. Numerous early works depict various attempts at designs of personal equipment to accomplish this, although almost all of them would have been at least impractical if not lethal. One exception is perhaps a system referred to by Roger Bacon in 1240 by which 'men could walk on sea or river beds without danger to themselves'. In 1532, Vegetius describes a personal ambient pressure diving dress. His drawing shows a man with his head in a leather helmet from the top of which a tube leads to the surface.

The major step forward, however, was made in the use of the diving bell to provide a local supply of air to the diver. Although there is reference to the use of a bell as early as 1531, the first well documented account of its use does not appear until well into the next century. In 1665, a bell was used for the salvage of valuable cargo from a Spanish galleon which had sunk in Tobermory Bay in Scotland seven years previously. This open one-man bell was constructed mainly of timber and suspended by chains from a vessel at the surface. This early bell was thought to have been designed by an unknown Scot.

The success of this operation is likely to have been responsible for the building of a larger bell. Given the name Spanish Bell, it was built specifically for the recovery of large sums of money in 1677 from two cargo vessels that had sunk at the Port of Cadaque. The bell, some 4 m (13 ft) high by 3 m (9.9 ft) across, was suspended by chain from a frame supported by two barges. Ballast weights of some 35 kg (77 lb) each were suspended from the rim of the bell and the diver sat on a cross-beam within the bell during ascent and descent. Excursions from the bell had to be carried out whilst breath holding in the manner of all diving operations up to this time.

A practical personal diving suit had not yet been developed, although by this time such systems were starting to receive more serious attention. By 1680, Borelli in his treatise shows a man wearing a close fitting, watertight leather suit and a brass helmet incorporating a glass window. In 1690, a further major development took place when Dr Edmund Halley operated the first diving bell with renewable air. Weighted barrels containing air were lowered to beneath his open, lead covered wooden diving bell. By removing a bung from the barrel, air was then transferred into the bell. Within this system Halley and four colleagues remained at a depth of 18 m (59 ft) in the River Thames for 30 min. Twenty-five years later, the 65-year-old Halley was able to spend over 4 h in an improved version of his bell at a depth of 21 m (69 ft).

Coinciding with Halley's achievement in 1715, a different approach to the problems of diving was being made by a Devon cooper, John Lethbridge, when he designed, built and operated what is considered to be the first atmospheric diving suit (ADS) (*see Figure 4.1*). The system consisted of a pressure resistant timber cask in which the single occupant lay horizontally. It incorporated a small window and two leather arm seals. Although his arms experienced ambient pressure, Lethbridge claims to have made dives to beyond 20 m (66 ft) for a duration of 34 min without any change of air. When air was later supplied to the system from the surface, it is claimed that dives of up to 6 h duration took place. In 1797 Klingert designed an ADS consisting of a helmet that resembled an upturned bucket and leather trousers, but there is no evidence of its construction or use.

In 1800 the American inventor, Robert Fulton, built a submersible craft which may be regarded as the forerunner to the military submarine. His two man vessel, powered by turning hand cranks, was successfully demonstrated in the River Seine in Paris. Arguably, the first self contained diving system was depicted in a work by W. Jones in 1825, which shows a diver wearing a toroidal iron tank filled with air around his waist. Again, there is no evidence that this system was constructed.

The invention of the first practical diving suit is generally accredited to Augustus Siebe (1788–1874). Actually, he was only one of several men who produced a successful system at the same time. Brothers John and Charles Deane obtained patents in 1823 on the design for a 'smoke apparatus' to enable firemen to enter burning buildings. By 1828 this had evolved into Deane's Patent Diving Dress, consisting of a heavy suit and an open bottomed helmet which rested on the diver's shoulders under its own weight. Air was pumped down to the diver through a connecting hose from the surface at a pressure equivalent to his working depth. Exhausted or surplus air simply passed out under the edge of the helmet. The system was in fact a personal diving bell and posed no problems as long as the diver remained upright. If, however, the diver should trip or bend over, then the helmet would quickly fill with water.

Siebe constructed a similar system in 1829 consisting of a copper helmet with glass ports and a heavy canvas jacket extending to the waist. At this point, he then made important modifications and sealed the helmet to the jacket, so that now the exhaust gases escaped around the bottom of the jacket, giving the diver additional safety. By 1840 he had adopted a full length waterproof suit and added an exhaust valve to it. Known as Siebe's Improved Diving Dress, it kept the diver dry, permitting the wearing of warm undergarments and allowing him to bend over without fear of flooding the helmet. It is regarded as the ancestor of today's diving suits.

By this time, several other types of diving dress, including one designed by Bethell, had also been developed and were being used in a number of operations. A major project was to provide a trial of these various systems. Under the charge of Colonel William Pasley, a unit of Royal Engineers was engaged in clearing the remains of HMS *Royal George* that had sunk and was fouling the fleet anchorage at Portsmouth. After this work, it was the Siebe suit which was formally recommended to be adopted for future operations. Diving manuals, the first of which was published by the Deane brothers in 1836 and which set out guidelines and codes of practice, were now available.

Figure 4.1 The first atmospheric diving suit; John Lethbridge, 1715

4.1.3 The first signs of physiological problems

The effectiveness of the diver by this time had been greatly enhanced by the first of the steel constructed open bells. Built by the Siebe–Gorman Company, they were of some 2.4 m (7.9 ft) diameter and had air pumped to them from the surface by sophisticated oil driven compressors. Further improvements in pump performance soon led to the construction of chambers large enough to enable several men to work at the sea or river bed. Such dry chambers became known by the French word 'caisson'. It was during this period that a new and unexplained malady began to afflict caisson workers. After completing a shift and returning to the surface, they would begin to experience pain in the joints, difficulty in breathing and dizziness. Caisson disease, as it was called, became more severe as the depth of projects increased. Fatalities were occurring with alarming frequency. Known today as decompression sickness, it is now known that it had in fact also affected all of Colonel Pasley's men who had attributed their problems to rheumatism aggravated by the cold conditions. The actual cause of decompression sickness was first identified in 1878 by the French physiologist, Paul Bert. He recommended that caisson workers should be gradually decompressed when returning to the surface. Following his advice, cases of the sickness decreased dramatically. Within a few years, specially designed 'recompression chambers' were even being situated at job sites for therapeutic purposes.

4.1.4 Breathing systems

Whilst the work of Siebe had enabled divers to work reasonably safely under water, they were still tied to the surface by the hose supplying them with air. Various inventors were looking for ways of eliminating this hose, meaning that divers would have to carry their own air or oxygen supply. A necessary component for such a system was the demand regulator. Invented in 1866 by Benoist Rouquayrol, it adjusted the flow of air from the supply tank to meet the diver's breathing and pressure requirements. However, because the high pressure tank did not yet exist, he adapted his regulator for surface supplied diving systems. Its inclusion in a self contained diving system, free of the surface, had to wait for more than 60 years.

The first self contained and closed circuit breathing system was developed by H. A. Fleuss in 1878. Using 100% oxygen, the quantity of gas required was less than what would have been needed of compressed air and thus the need for high strength gas containers was avoided. The effects of breathing pure oxygen at high pressures were not known at this time. However, despite this ignorance, in 1880 the system was used successfully by an English diver, Alexander Lambert, who walked 300 m (985 ft) in a dark flooded tunnel beneath the River Severn to operate crucial valves.

By the turn of the twentieth century, both Britain and the USA had been operating naval submarines for some years and the subject of surviving, fighting and working under water had become a complex one indeed. Research and development in the areas of diving, diving physiology and underwater vehicles was set to accelerate rapidly to this day. The most notable events associated with these various activities in this period of history will be referred to in relevant sections of this chapter.

4.2 Diving

The diver is defined as a person who enters the water, submerges and bodily experiences the ambient pressure of that water on his body. For this reason these are often referred to as 'ambient pressure divers' to distinguish them. In the commercial world, diving is regarded as simply a means to an end and divers will be judged on their ability to carry out a given task safely and efficiently at the work site.

4.2.1 Major developments in diving since 1900

At the beginning of the twentieth century an English physiologist, J. S. Haldane, turned his attentions to the problems of diving. He identified the fact that a build up of carbon dioxide was limiting diving operations to a maximum depth of 35 m (115 ft) and established a standard air supply rate which alleviated the problem. He also drew up a set of staged decompression tables which still remain the basis of today's tables. By this time divers could attain a depth of over 60 m (197 ft), limited only by the available pumps to supply the required air at this pressure. It was not long, however, before divers experienced another physiological problem, nitrogen narcosis, although it was not until the 1920s that nitrogen was linked to this.

The limit for air diving was probably established in 1915 when the USS *F-4* was salvaged from 93 m (305 ft). The US Navy divers were limited to only 10 min bottom time because of the effects of nitrogen narcosis and the restrictions due to decompression. A few years later, inventors Elihu Thomson and Dr Edgar Ende, separately theorized that helium could be used beneficially to replace the nitrogen in the diver's breathing supply. The first tests took place in 1927 and showed no adverse effects. The depth advantage of using this gas was soon established and by 1937 a diver using a 'heliox' breathing supply was compressed to a record simulated depth of 152 m (499 ft) in a pressure chamber. The first practical test of heliox came two years later when the submarine USS *Squalus* was successfully salvaged from a depth of 74 m (243 ft). During this period, the US Navy's divers were recognized as world leaders in this field. However, there were some notable exceptions. In 1937, a civilian engineer named Max Gene Nohl reached 128 m (420 ft) in Lake Michigan whilst breathing heliox and using a suit of his own design. In 1946, another civilian, Jack Brown, made a simulated heliox dive to 168 m (552 ft). In 1948, a Royal Navy diver set an open sea record of 164 m (538 ft).

In countries where helium was not easily available, workers experimented with other gases. A Swedish engineer, Arne Zetterstrom, worked with hydrogen–oxygen mixtures. Provided the oxygen concentration is kept below 4% the mixture is not explosive. However, oxygen in this concentration will not support life at the surface. Hence, a technique for changing between air and the new hydrogen–oxygen mixture was devised. In 1945, after some test dives to 111 m (364 ft), Zetterstrom reached 161 m (529 ft). However, a too rapid ascent and failure to enrich the oxygen as required resulted in his death. By 1956, a Royal Naval diver had increased the record depth to 183 m (600 ft). The decompression time required for this dive totalled 12 h.

In 1961, a young Swiss diving enthusiast, Hans Keller, stated that he could increase the existing record significantly and by changing proportions of different gases, including nitrogen, cut the decompression time to 100 min. On 3 December 1962, off the coast of Southern California, Keller and a colleague reached a depth of 305 m (1001 ft). Whilst trying to set a flag on the bottom he became entangled in it and the oxygen supply failed. Keller was rescued but his colleague and one of two standby divers died in this attempt.

During this period in which new breathing gases were tried and new records were repeatedly established, much had been happening in the design of diving equipment. Until 1928, all diving bells had been open. In that year, Sir Robert H. Davis, designed the Davis Submersible Decompression Chamber (SDC), a bell with inward opening hatches that was capable of retaining internal pressure when raised to the surface. The vessel was conceived as a method of reducing the time a diver would have to remain in the water during lengthy decompression. The increased decompression times associated with mixed gas diving prompted Davis to design a three compartment deck decompression chamber (DDC) to which his bell could be mated and permit the transfer under pressure (TUP) of the divers. This concept was a major advance in diving safety, but was not applied until the advent of saturation diving.

The saturation diving technique was first considered following the work of George F. Bond in 1957 in which he supervised a series of experiments first on animals and later on human beings in the Sealab experiments. The first practical demonstrations took place in 1962 with the Man in the Sea I programme involving the use of heliox, led by E. A. Link, and the Conshelf One programme involving the use of nitrox led by J. Y. Cousteau. Both of these groups extended both depth and duration in 1964. In that year, two men were exposed at 123 m (404 ft) for two days in Link's programme and a group of seven men at 11 m (36 ft) and 27 m (89 ft) endured 30 days in Cousteau's. In recent years, the depth capability of the diver has been continually increased in various experiments and trials too numerous to mention in this chapter. At present the deepest simulated pressure chamber dive stands at 585.5 m (1922 ft), and the deepest working dive stands at 530 m (1740 ft), in the Hydra VIII trial which re-examined the use of hydrogen in the breathing gas.

4.2.2 Diving physics

The human body has evolved to function optimally at the earth's surface where the atmosphere is rich in vital oxygen. Divers must abandon these safe surroundings and enter a totally different world consisting of a dense liquid medium devoid of gaseous oxygen. To survive they must make alternative arrangements for breathing and ensure that the body is not damaged by the higher and rapidly changing pressure that will be experienced.

In order to understand the technical and physiological problems that the divers have to face, it is necessary to consider some basic physical laws relating to matter. The most significant physical difference between air and water as far as divers are concerned is that of density. This difference alone is responsible for many of the problems.

4.2.2.1 Hydrostatic pressure

At the Earth's surface the pressure of the atmosphere (1 bar) results from a column of air some 30 km (~ 19 miles) high. By contrast in water, because of its much greater density, such a pressure results from a column just 10 m (33 ft) high. Therefore, it can be seen that as we descend in the water, the pressure we will experience will increase rapidly as shown in Equation 4.1:

$$\text{Pressure at } d \text{ m} = \left(1 + \frac{d}{10}\right) \text{ bar} \quad (4.1)$$

This hydrostatic pressure at any point acts equally in all directions and is simply a result of the 'head' or height of water present above the point of measurement.

4.2.2.2 Buoyancy

The density of water has another major effect upon an immersed body. Archimedes' principle states that an object immersed in a liquid will experience an upward or buoyant force equal to the weight of liquid it displaces. The density of the object will determine whether it floats or sinks. A body that tends to rise to the surface and float is said to be 'positively buoyant' whilst one which sinks is said to be 'negatively buoyant'. A body which experiences a buoyant force exactly equal to its weight (out of water) neither rises nor sinks and is said to be neutrally buoyant. In this case, its overall density must be identical to that of water.

Sea water, because of the various dissolved solids it contains, has a density greater than that of fresh water (1.025 kg/l) (1.29 lb/pint). Therefore objects immersed in sea water will experience a correspondingly greater buoyant force than when immersed in fresh water. The human body has an overall density very close to that of water and consequently will experience a condition close to neutral buoyancy, although there can be a significant difference between individuals.

4.2.2.3 The gas laws

Gases are subject to three inter-related factors: temperature, pressure and volume. A change in any one of these factors will result in at least a change in one of the other two. In diving, where gases are necessary for life support and large variations in pressure are encountered, the rules which predict the behaviour of gases when the above factors are changed must be well understood.

4.2.2.3.1 Boyle's law Boyle's Law states that the pressure and volume of a gas are inversely related. Expressed in Equation 4.2:

$$PV = \text{Constant} \qquad (4.2)$$

where P is the absolute pressure and V is the volume. This law enables one to calculate how a given volume of gas will change as a function of depth. *Table 4.1* shows how a volume of 1 l of gas measured at the surface changes with depth. It is important to note that the most pronounced volume changes occur close to the surface.

Table 4.1 Change in volume of a gas as depth is increased

Depth (m)	Pressure (bar)	Volume (l)
Surface	1	1.0
10	2	0.5
20	3	0.33
30	4	0.25
40	5	0.20
50	6	0.16

4.2.2.3.2 Charles' Law Charles' Law states that the amount of change in either the pressure or the volume of a gas is directly proportional to its absolute temperature. Expressed as Equation 4.2:

$$PV = RT \text{ or } \frac{PV}{T} = R \qquad (4.3)$$

where T is the absolute temperature = °C + 273 and R = a universal constant for all gases

As an example, if gas is enclosed in a rigid container (at constant V) and its absolute temperature is doubled, then its pressure will be doubled.

4.2.2.3.3 The General Gas Law It is usual to express both Boyle's and Charles' Law in a single relationship generally referred to as the General Gas Law. This law enables the behaviour of a gas to be predicted when changes are made to any or all of the factors. It is normally written:

$$\frac{P_1 V_1}{T_1} = \frac{P_2 V_2}{T_2} \qquad (4.4)$$

where the subscript 1 refers to initial values and the subscript 2 refers to final values

At this point it is perhaps worth stating the fact that liquids and solids do not obey this law. Liquids and solids are virtually incompressible and obey different laws of expansion with temperature.

4.2.2.3.4 Dalton's Law Dalton's Law, often referred to as the law of partial pressures, states that the total pressure exerted by a mixture of gases is equal to the sum of the pressures of each of the gases making up the mixture, each gas acting as if it alone was present.

The pressure of each gas in the system is called its 'partial pressure' (*PP*) and the total pressure of the system is simply the sum of the partial pressures as shown in the following equation:

$$P = PP_A + PP_B + PP_C + \ldots \qquad (4.5)$$

The partial pressure of a constituent gas in a system is simply proportional to the number of molecules of that gas present and is obtained by multiplying the total pressure by the % volume of that gas in the system as shown in Equation 4.6:

$$PP_A = P \times \frac{\% \text{ vol}_A}{100} \qquad (4.6)$$

The atmosphere consists approximately of 79% nitrogen and 21% oxygen by volume. Its total pressure is 1 bar. Therefore the partial pressure of nitrogen is 0.79 bar and the partial pressure of oxygen is 0.21 bar. At a depth of 10 m (33 ft), provided that the volume composition of this air is unchanged, then the partial pressures of nitrogen and oxygen will be doubled, *viz* 1.6 bar and 0.42 bar, respectively.

4.2.2.4 Gases in liquids

Whenever a gas is in contact with a liquid, then a portion of its molecules will enter into solution with the liquid. The same is also true to a lesser extent for solids. The amount of gas which dissolves will depend on both the gas itself and the liquid in question, the solvent. Apart from the nature of the gas and the solvent liquid, both temperature and pressure are also important factors to be considered.

4.2.2.4.1 Henry's Law Henry's Law states that at a given temperature the amount of gas which dissolves in a liquid will be almost directly proportional to the partial pressure of that gas. In diving, the gases and the solvents (the tissues of the body) are unlikely to undergo any significant temperatures changes and thus the effect of this variable need not be considered in detail. It is sufficient to say that the amount of gas in solution is reduced as temperature is increased.

As molecules of gas enter a liquid, they add to a state of 'gas tension', a term which identifies with and is analogous to the partial pressure of the gas external to the liquid. The difference between the gas tension and the partial pressure is referred to as the 'pressure gradient' and its magnitude determines the rate of flow of gas into or out of solution. After a certain time, an equilibrium will be reached when the gas tension is equal to the partial pressure and no more gas will enter the liquid. The liquid is said to be 'saturated' with the gas. A reduction of partial pressure will then result in gas leaving the liquid.

4.2.2.4.2 Graham's Law   Graham's Law states that the diffusion rate of a dissolved gas is proportional to the reciprocal of the square root of the molecular weight of the gas. It is of importance in the selection of the inert gas used in the diver's breathing mixture. Table 4.2 shows the calculated diffusion rates for various gases relative to that of nitrogen.

Table 4.2 Relative diffusion rates of gases used in diving

Gas	Calculated diffusion rate
Kr	0.58
CO_2	0.80
Ar	0.84
O_2	0.94
N_2	1.00
He	2.65
H_2	3.74

4.2.3 Diving physiology

This is the study of the effects on the body resulting from going to, staying in, and returning from, the alien underwater environment. It is a complex subject and must address the physical consequences of pressurization of the body, the effects of gases on the body at high pressures and depressurization of the body.

4.2.3.1 Direct effects of pressure

The tissues of the body, being virtually incompressible liquids and solids, can withstand extremely high pressures without any adverse effects or damage. However, the human body contains a number of natural air spaces surrounded by rigid walls which could collapse under external pressure. These spaces are the inner ears, the sinuses and, of course, the lungs. To prevent damage and pain when diving, it is necessary to ensure equalization of the pressure in these spaces with the external or ambient pressure. This should not normally be a problem as all these spaces are naturally vented to the outside of the body.

Air spaces that are formed external to the body, for example by a face mask, are also susceptible to collapse if not equalized in pressure. The general term describing this tendency to collapse, whether internal or external, is 'squeeze'. This affects only spaces which have rigid walls. Gas pockets in the intestine, for example, are not a problem as they are easily compressed.

4.2.3.1.1 Ear and sinus squeeze   The middle ear includes an air space which is separated from the external auditory canal by the eardrum; the space is vented to the throat by the eustachian tube. Even slight sudden external pressure changes cause a distortion of the ear drum which leads to discomfort. On descent into water it is necessary to equalize the inner ear pressure with the ambient pressure to prevent rupture of the eardrum. This can usually be achieved by either swallowing or blowing against a pinched nose.

The sinuses are hollow spaces located within the bones of the skull. They are lined with mucous membrane and are connected to the nasal passage. If these passages become blocked by excess mucous, then on descent blood and tissue fluids will be forced into the sinus cavities accompanied with intense pain.

To avoid ear and sinus squeeze, divers should refrain from diving if they have any signs of congestion or head cold. During descent, pain will indicate when any differential pressure develops at which point the diver may choose to abort the dive. If, however, a blockage were to occur at depth after a successful descent, then the problem is more serious. Generally, the diver's time at depth will be limited and the ascent must be made. A positive pressure in the middle ear causes the ear drum to bulge out 'reversed ear' with intense pain and probable rupture. At this point, if the diver's head is surrounded by water, for example as in scuba diving, cold water will flood into the middle ear, causing a severe attack of vertigo with accompanying nausea. A blocked sinus occurring at depth could lead to severe tissue damage on ascent.

4.2.3.1.2 Thoracic squeeze   During descent on a breath hold dive, the lung cavities will be compressed. Eventually their volume will be smaller than the residual volume of the lungs. At this point, blood and tissue will be forced into the spaces, resulting in a 'thoracic squeeze'. For this reason, an average person is not capable of breath hold diving beyond a depth of 30 m (98.5 ft), although at present the world record for such a dive stands at a remarkable 105 m (345 ft).

4.2.3.1.3 Face or body squeeze   If the air pressure in a surface supplied helmet or face mask should suddenly become lower than the surrounding ambient pressure, then the tissues of the body can be seriously squeezed. When wearing a rigid diving helmet, the consequences of such a sudden catastrophic failure in air supply pressure could prove fatal as the ambient pressure forces the diver's body into the helmet. In order to prevent such a squeeze, all surface supplied apparatus must be fitted with a reliable non-return valve which would hold the gas in the suit and helmet at pressure in the event of a gas supply failure.

4.2.3.2 Effects of gas at high pressure on the human body

Apart from the direct physical effects that pressure has on the body, it has other consequences of extreme importance to the diver. Gases which are normally harmless, and even essential to the body, become toxic when their pressures are increased.

4.2.3.2.1 Nitrogen narcosis As the partial pressure of nitrogen in the diver's breathing gas is increased and a greater amount of nitrogen (N_2) enters into solution in the body, there comes a point at which it has a narcotic effect comparable to that which arises from drinking alcohol. The effect is complex. First, it is responsible for a personality change, where the normally careful and cautious diver becomes carefree and uninhibited. Such a condition is extremely dangerous in this situation where an error of judgement is likely to have disastrous effects. Different individuals respond in different ways and whilst some may be affected at depths of less than 30 m (98.5 ft), others may reach 60 m (197 ft) before any perceptible change in personality.

The ability to reason is also impaired. This condition affects all divers without exception. It is for this reason that all tasks for deep air divers must be made as simple as possible and that they must be well briefed and supervised very closely at all times. Discussion of the detailed mechanism by which nitrogen narcosis occurs is beyond the scope of this chapter. It is sufficient to know that the abnormally high tension of nitrogen affects the nerve synapses in such a way that it impairs the transmission of messages to and from the brain.

It is known that regular diving and experience will, for an individual, lessen the effects of nitrogen narcosis, but never eliminate them. However, these effects can be intensified by high carbon dioxide levels, cold, fear, poor visibility and drinking alcohol to excess prior to the dive.

4.2.3.2.2 Carbon dioxide poisoning Carbon dioxide (CO_2) is a natural waste product of the breathing process and its presence is partly responsible for regulating the rate of breathing in the human being. A rise in the partial pressure of CO_2 and hence the level of CO_2 in the blood will cause an increase in the depth and rate of breathing. As it increases further, it may cause headaches, breathlessness, confusion and unconsciousness when it reaches 0.1 bar. Higher partial pressure will lead to convulsions and eventually death. The control of the level of CO_2 is a necessary function of any artificial breathing system and the method employed to do this is described in Section 4.4.

4.2.3.2.3 Oxygen poisoning If the partial pressure of oxygen (O_2) in the breathing mixture is increased to more than 1.6 bar, then it becomes poisonous to the human body. The first symptoms of acute oxygen poisoning are twitching of the facial muscles, dizziness, nausea, tiredness and breathing difficulties. If the partial pressure is high enough and exposure is long enough then unconsciousness will occur, followed by convulsions and fits. Such a condition is obviously extremely dangerous to a diver and anyone else working in the vicinity. However, the condition itself, provided no muscular or tendon damage has occurred, is completely reversible causing no permanent neurological injury to the diver.

Chronic oxygen poisoning is a long-term effect and can occur from breathing oxygen at partial pressures lower than those that would lead to acute oxygen poisoning (0.6–1.6 bar). The symptoms are soreness of the chest, coughing and discomfort as the lungs become irritated by the oxygen.

4.2.3.2.4 Other gases Other gases may contaminate a diver's breathing gas. The effects of poisonous gases such as carbon monoxide (CO) and hydrogen sulphide (H_2S) are increased as the breathing gas pressure is increased. Their presence in diving gas supplies must be avoided at all times.

4.2.3.3 Increased resistance to breathing air at depth

At a depth of 50 m (163 ft) the density of air has increased sixfold and consequently the breathing system experiences a significant increase to the resistance of the passage of air in and out of the lungs. Such an increase demands a corresponding increase in the effort required for breathing to take place at the necessary rate. In diving, as in any strenuous activity, such a condition results in a decrease in the effective work rate.

4.2.3.4 Depressurization on the body

During pressurization (descent), the breathing gas will flow into the spaces within the body and will also dissolve in increased quantities in the tissues of the body. At atmospheric pressure, the body tissues contain about 1 l (1.75 pints) of nitrogen. According to Henry's Law, the amount of nitrogen dissolved or released will be almost directly proportional to the change in partial pressure. Therefore at a depth of 20 m

(66 ft) (3 bar), the dissolved nitrogen content of the body could be as high as 3 l (5.25 pints).

During depressurization (ascent) the body must give up this additional gas and it is during this phase that a number of problems can occur. Depressurization of the voids within the body should be straightforward provided there are no blockages. The consequences of such blockages in relation to the middle ear and the sinuses have already been mentioned. By far the largest cavity of the body is the lung. On ascent, the gas within the lung will expand and therefore must be exhaled in order to prevent damage. If, for any reason, the gas is not exhaled, either voluntarily or due to a restriction in the respiratory system, then the diver will suffer a condition known as 'burst lung' or pulmonary baratrauma. As the greatest volume changes occur within the first 10 m (33 ft) or so of the surface, burst lung is as much as potential hazard to the sport diver in shallow water as it is to the professional.

The gases dissolved in the tissues of the body present an even greater potential hazard to the diver during depressurization, as too rapid an ascent will result in the formation of bubbles of gas within the tissues with probable dire consequences. Such a condition is prevented by adhering to strict decompression routines.

4.2.3.4.1 Pulmonary baratrauma – burst lung If the lungs should fail for any reason to exhale the expanding gases during ascent or decompression, then the excess internal pressure will eventually rupture the tissue of the lung, causing internal bleeding. Once the condition of pulmonary baratrauma has occurred, the expanding gases escape and are free to migrate within the chest cavity. Obviously a serious condition in itself, it is then compounded by the effects of the escaping gas, which are responsible for the many symptoms of this condition. Pulmonary baratrauma may present itself in several ways dependent on where the escaping gas collects; these include interstitial emphysema, pneumothorax and arterial gas embolism.

Interstitial emphysema is when gas escapes inwards into the central tissues within the chest and is free to migrate along the external surfaces of the bronchi and trachea. It may emerge under the skin around the neck (subcutaneous emphysema) or, more seriously, around the heart and major blood vessels (mediastinal emphysema) where it is likely to adversely affect the pumping of blood.

Pneumothorax results from gas being forced between the lung sac and the chest cavity wall. Such a condition leads to collapse of the lungs causing respiratory distress and eventual starvation of oxygen to the body.

Arterial gas embolism is when gas, in the form of a bubble, enters the bloodstream where it can then be transported around the body. 'Bubbles reaching the brain will restrict the flow of blood through it, starving it of oxygen and causing irreversible damage. Sizeable bubbles reaching the heart may occasionally lead to a direct cessation of the heartbeat.

4.2.3.4.2 Decompression Whereas during pressurization, breathing gases can only enter the tissues of the body through the body surfaces, alveolar surfaces and skin, in decompression gas may leave the body's tissues at any point by the formation of bubbles. The formation of such bubbles must be avoided as they exert pressure on organs and nerves resulting in 'decompression sickness', commonly referred to as the 'bends'.

The object when decompressing a diver is to ensure that all the dissolved gas within the tissues leaves the body via the same route by which it entered, usually via the lungs. Fortunately, the body can apparently hold gas in what is similar to a supersaturated solution before bubbles appear. This permits the diver to ascend progressively while allowing the excess gases to diffuse from the body. Provided the ascent rate is slow enough, any bubbles which form will be extremely small and will leave the bloodstream on their first passage through the lungs. Provided that the pressure of gases in the tissues does not exceed the ambient pressure by 1 bar, then the formation of sizeable and potentially dangerous bubbles is unlikely.

It is on such a premise that the various standard decompression tables are based. These tables have been composed to provide guidelines for controlled decompression for a wide range of diving circumstances. There are many versions of such tables and often the larger diving companies develop their own with a view to minimizing time spent by divers in decompression and hence cost. However, because of the differences in individuals and the statistical nature of gas diffusion processes, it must be stated that no table can be regarded as being infallible and cases of bends will occur even when tables have been strictly adhered to. Decompression can be speeded up by administering pure oxygen, thus decreasing the partial pressure of nitrogen in the lungs, at low pressures.

4.2.3.4.3 Decompression sickness – the bends If decompression of the body is too rapid, then gas bubbles will begin to form in the tissues of the body. The degree of formation and their location will result in numerous symptoms of varying severity referred to collectively as 'decompression sickness' or the 'bends'. Symptoms may range from a skin rash and mild discomfort, to severe pain in the joints and muscles, causing the victim to bend the limbs involuntarily (hence the term bends). More serious symptoms are numbness in the legs, paralysis, loss of hearing and vision, and unconsciousness. 'Type I' bends refer to the first set of symptoms and 'Type II' bends refer to the second, more serious, set of symptoms caused by pressure of gas bubbles on the central nervous system (*see Table 4.3*). It is important for the diver with all but the most minor symptoms to seek medical advice as soon as possible. Generally, treatment begins by immediate recompression to the working depth, which forces all offending bubbles back into solution and should relieve all symptoms.

Table 4.3 Symptoms of decompression illness

Symptom	Type	Name	Urgency/action required
Slight pain in limb	I	Niggle	Observe/possibly recompress
Rash, itching	I	Skinbend	Observe/possibly recompress
Localized soft swelling	I	Lymphatic decompression sickness	Observe/possibly recompress
Intense joint pain	I	Bend	Urgent/recompress
Swelling under skin	I	Interstitial emphysema	Observe/do not recompress
Pins and needles in feet	II	Spinal bend	Very urgent/recompress/seek medical attention
Excessive tiredness	II	Decompression sickness	Very urgent/recompress/seek medical attention
Unsteadiness, dizziness, nausea	II	Staggers/cerebral bend	Extremely urgent/recompress/seek medical attention
Headache, unconsciousness, speech defects	II	Cerebral bend	Extremely urgent/recompress/seek medical attention

Decompression must then be carefully controlled and generally follows a more conservative therapeutic table.

4.2.3.5 Aseptic bone necrosis (osteonecrosis)

Over recent years, it has been discovered by X-ray examination that many professional divers possess regions of bone where the cells have died. This condition is known as 'aseptic bone necrosis' or 'osteonecrosis' and can be of varying concern, depending on its extent and location. If it occurs at the joint of two major bones, then the bearing surfaces will be eroded, resulting in pain and the inability to move the joint. If the condition occurs in the centre section of the bone, whilst not desirable, it is less serious, provided steps are taken to stop its spread.

The exact causal mechanism of bone necrosis amongst divers is not fully understood, but the condition almost certainly arises from the restriction of the blood flow within the capillaries within the bone. Poor decompression procedures, perhaps having a cumulative effect over many dives, are a likely cause. Recent evidence, however, points to the possibility of a change in the blood flow characteristics with pressure. If confirmed, this would be worrying, as it would be unavoidable to the ambient pressure diver unless special drugs were administered.

To limit the incidence of debilitating bone necrosis amongst divers, the annual medical examination that UK commercial divers must undergo for certification by the Health and Safety Executive, includes an X-ray examination of major bones. On evidence of bone necrosis they will not be recertified and their active diving career is over.

4.2.4 Air diving

In air diving, the diver breathes natural air (21% oxygen, 79% nitrogen by volume) but at a total pressure equal to that of the surroundings. Because of the narcotic effect of nitrogen at elevated partial pressures and to some degree the increasing resistance to flow of compressed air, air diving in the UK, and in most civilized countries, is restricted to a maximum depth of 50 m (164 ft). Within this range, two modes of air diving have evolved:

1. Surface demand diving (or surface orientated). Divers are supplied by air compressed at the surface through a flexible hose or umbilical.
2. Scuba (self contained underwater breathing apparatus). As its name suggests, divers carry their own independent air supply and require no physical link with the surface.

4.2.4.1 Surface demand air diving – equipment

The basic components of a surface demand air diving system are shown in *Figure 4.2*. The compressor used must be capable of supplying clean air at a sufficient rate, which is dependent on the number of divers the system is designed to serve. The pressure it must produce must be such that after making allowances for pressure drops caused by the flow of the gas through the umbilical and its components and that caused by the demand regulator, it is at least equal to the ambient

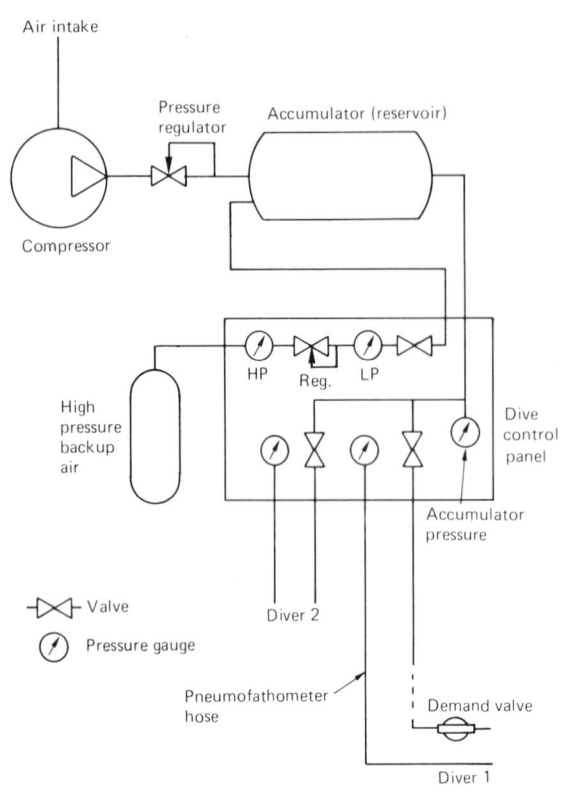

Figure 4.2 Components of a surface demand air diving system

pressure at which the diver is working. For example, if a maximum depth of 50 m (164 ft) is chosen, then the compressor must be capable of supplying gas to a pressure of greater than 8 bar. Such a compressor must also be backed up by a ballast tank or reservoir of such capacity that it is capable of supplying air to the divers for a reasonable duration in the event of compressor failure. The intake pipe of the compressor must be positioned such that it receives uncontaminated air, e.g. not next to an engine exhaust pipe. Compressors also incorporate drying stages and filters to remove moisture and other contaminants at their intakes.

As a further safety precaution, there is usually a bank of high pressure cylinders from which air can be introduced into the system through a high pressure reducing valve.

It is usual to control all operations at the dive control panel, on which the various valves and gauges are mounted with diagrammatic indication of the interconnecting pipework to aid safe operation. The main feature of this panel is the accurately calibrated pressure gauge which often reads directly in units of depth. The umbilical consists of the air hose, a separate hose with an open end at the diver which is connected to the pneumofathometer on the control panel, a communications cable, a strain member (usually rope) and sometimes a hose which carries warm water for diver heating. The demand valve is a crucial part of the equipment and its function is to control the pressure of breathing gas at the diver to be close to ambient pressure.

In commercial air diving operations, it is nearly always necessary to have a decompression chamber on site (see *Figure 4.3*). In such operations, this can be a simple single compartment pressure vessel with a single sealing hatch. This has two uses: in emergencies, if, for example, a diver suffers from the bends, it can be used to recompress the diver to relieve the symptoms and then for carrying out therapeutic decompression. It also can be used for either all or part of the diver's normal decompression. On completion of the dive, or at a certain stage during decompression whilst in the water, divers surface and enter the chamber. They are then recompressed to an appropriate level, after which they continue decompression. Such a technique is known as 'surface decompression' and requires the application of special decompression tables and often the use of increased concentrations of oxygen in the breathing mixture. By using this technique, the time which divers must spend in the water is significantly reduced and when oxygen is used, total decompression time is also reduced.

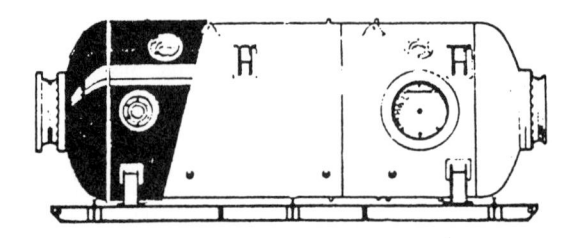

Figure 4.3 Typical simple decompression chamber (COMEX)

The advantages of surface decompression lie in the more controllable situation and a reduction of time in often very cold and chilling water. However, such advantages must be balanced against the effect of almost certain initial bubble formation in the short time at the surface before recompression takes place. This time is normally limited to 3–4 min, but could easily be inadvertently extended if, for example, the diver has difficulty in removing his gear.

4.2.4.2 Entry into the water

The means by which air divers enter the water will depend on the support vessel or structure from which they are working. When working from a small boat, divers will generally jump into the water and return by the use of a purpose built ladder attached to the side of the boat. As the surface support vessel increases in size, with a corresponding increase in free board (distance to the water surface), then alternative arrangements must be made for entry to the water. The weight of equipment that an air diver must carry is considerable and this makes climbing all but the shortest of ladders out of water extremely difficult. The usual method is to lower and retrieve the divers into and from the water in a cage (see *Figure 4.4*).

In some operations, where the support is available, then an open diving bell may be used (see *Figure 4.5*). The bell increases the protection of the divers during transit and provides a safe haven whilst at depth.

Figure 4.4 Surface demand air divers about to enter water in a cage

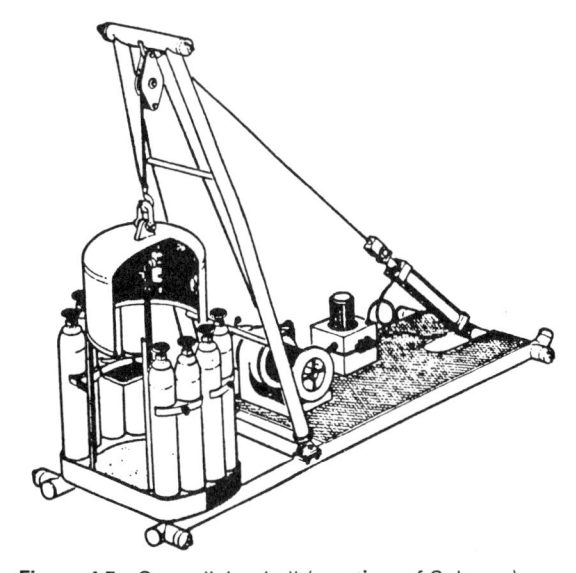

Figure 4.5 Open diving bell (courtesy of Submex)

4.2.4.3 Decompression of air divers

As stated previously, it is important to ensure that during ascent or decompression, no gas bubbles are allowed to form in the body. This is achieved by following a decompression table specific to the dive depth and the bottom time at that depth. *Figure 4.6* shows an example of such a table and indicates the depths and durations of the various stops which should be made. Decompression is usually a staged, not continuous, process.

4.2.4.4 Self contained underwater breathing apparatus – scuba

The development of scuba took place over many years and can be considered to have begun as early as 1866 with the invention of the demand regulator by Benoist

1. Move down the Maximum Depth column to find the depth – or the next greater depth.
2. Move horizontally to find the Bottom Time – or the next greater time.
3. Move vertically down to read off any decompression Stops required.
4. Note that Maximum Depths greater than 20 m and requiring decompression involve a stop at 10 m and 5 m.
5. Dives to 9 m or less require no decompression.

Max. depth (metres)	No stop time (mins.)	Bottom time in minutes				
10	232	431	–	–	–	–
12	137	140	159	179	201	229	270
14	96	98	106	116	125	134	144
16	72	73	81	88	94	99	105
18	57	59	66	71	76	80	84
20	46	49	55	60	63	67	70
Stops at 5 metres (minutes)		5	10	15	20	25	30
22	38	42	47	51	55	58
24	32	37	41	45	48	51
26	27	32	37	40	43	45
28	23	29	33	36	39	41
30	20	25	30	33	35	37
32	18	23	27	30	32	34
34	16	21	25	28	30	31
36	14	20	23	26	27	29
38	12	18	21	24	26	27
40	11	17	20	22	24	25
42	10	16	19	21	22	24
44	9	15	18	20	21	
46	8	14	17	18	20	
48	8	13	16	17		
50	7	12	15	17		
Stops at 10 metres		5	5	5	5	5
Stops at 5 metres (minutes)		5	10	15	20	25

Maximum depth is the greatest depth reached during dive.
No stop time is longest bottom time not requiring decompression stops.
Bottom time is from start of descent to start of ascent.
Descent rate is 30 metres/minute maximum.
Ascent rate is 15 metres/minute.
For more than two dives add bottom times together and decompress for greatest depth reached during dives.
For two dives only, see facing page.
No more than 8 hours in 24 spent under pressure (submerged).

Figure 4.6 Air diving decompression table (courtesy of British Sub-Aqua Club)

Rouquayrol. However, the modern open circuit scuba is generally attributed to Captain Jacques Yves Cousteau and Emile Gagnan who combined an improved demand regulator with portable high pressure air banks. Working in a small Mediterranean village in the restrictive conditions of German occupied France during World War II, they brought their system to a high state of development and successfully demonstrated it many times as they explored and photographed shipwrecks. Cousteau used his system in depths greater than 50 m (164 ft) and, after the war had ended, it rapidly became a commercial success. Today it is probably the most familiar of all diving equipments to the layperson.

When using scuba, divers must carry their own air supply in high pressure cylinders usually on their back. The capacity of the cylinders, and various sizes are available used singly or in pairs (twinsets), will determine the maximum duration of a dive at a particular depth. Typical scuba equipment is shown in *Figure 4.7*.

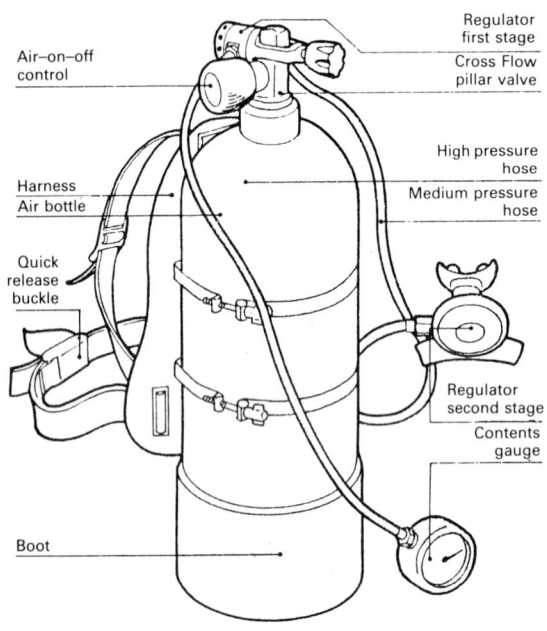

Figure 4.7 Scuba equipment (courtesy of *British Sub-Aqua Club Diving Manual*)

Although the lack of a tether might be advantageous in certain situations, one must consider the inability of the diver to communicate with the surface. This is a severe and usually unacceptable disadvantage in most commercial diving applications. Scuba diving, therefore, whilst demanding little in terms of support equipment, finds few applications in commercial diving offshore. Nevertheless, it is well suited for recreational diving.

4.2.5 Mixed gas diving

When any mixture of gas other than that identical to that of naturally occurring air is used for breathing in diving operations, the mode of diving is referred to as 'mixed gas diving'. The various gas mixtures in use commercially today are nitrox – a mixture of nitrogen and oxygen but not in the same proportions as air; heliox – a mixture of helium and oxygen; and trimix – a mixture of helium, oxygen and nitrogen. Of these mixtures, heliox is the one most widely used in deep diving operations.

4.2.5.1 Heliox diving

The problems caused by breathing air at elevated pressure prompted researchers to seek an alternative, more agreeable breathing gas mixture. It was found that by replacing the nitrogen diluant by helium the narcotic effect was eliminated and the resultant mixture offered a correspondingly lower resistance to breathing. Below 50 m (164 ft), where the use of air is prohibited by legislation, heliox has become the standard breathing gas in the diving industry.

Unlike air diving, where atmospheric air is simply compressed to the required pressure after minimal processing, heliox gas mixtures are blended specifically to meet the requirements for a given depth. In this way the oxygen content is kept within safe working limits as the total gas pressure is increased. As the diver requires more or less the same number of oxygen molecules per breath irrespective of depth, i.e. partial pressure, then it is necessary to decrease the percentage of oxygen by volume in the breathing gas mixture as depth increases. *Figure 4.8* shows the percentage by volume in oxygen in heliox as a function of depth.

In practice, it is usual to use a slightly higher partial pressure of oxygen than the 0.21 bar of the atmosphere at the surface. Partial pressures in the range 0.4–0.7 bar

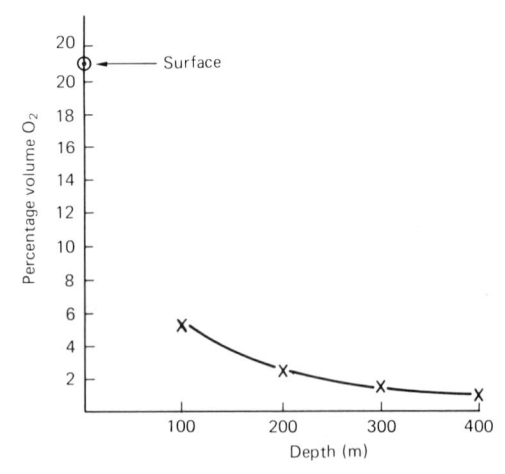

Figure 4.8 Percentage volume O_2 content in a diver's breathing mixture – 0.5 bar partial pressure

are usually used, depending on whether the diver is at rest or working hard.

Having now significantly extended the depth of diving by the use of heliox, then decompression times have also increased correspondingly. In all but the very shortest duration dives, decompression on ascent in the water is not practical. For this reason, this increase in depth now demands sophisticated diving equipment in the form of closed diving bells and deck decompression chambers (DDC).

Because of the extended decompression times required for greater depths, two different modes of heliox diving have evolved:

1. Bounce diving – the object in this mode of diving is to limit the time during which the diver experiences the elevated pressure to a minimum, hence minimizing the decompression time required.
2. Saturation diving – takes advantage of the fact that after a certain length of time at a given pressure, the body's tissues will become saturated with gas. At this point no more gas can be absorbed by the body and longer exposure to this pressure will not further increase decompression time.

4.2.5.1.1 Bounce diving The bounce diving technique is illustrated by *Figure 4.9*. Usually two divers are transported to the work site in a closed diving bell at atmospheric pressure. On reaching the work site, the bell is then pressurized with the required heliox mixture from the surface via the diving bell's umbilical. When the pressure inside the bell equals the ambient pressure, the hatch can be opened and a diver leaves the bell and sets about undertaking the task in hand. Pressurization is undertaken as rapidly as the divers can withstand.

After a set period has passed, the diver re-enters the bell, closes the hatch and decompression commences. The bell is winched to the surface and then mated with the DDC. The divers then transfer to the DDC, its pressure now corresponding to that of their first decompression stop. Decompression then takes place in the normal manner. *Figure 4.10* indicates the total decompression time as a function of depth for both 0.5 h and 1 h bottom time durations.

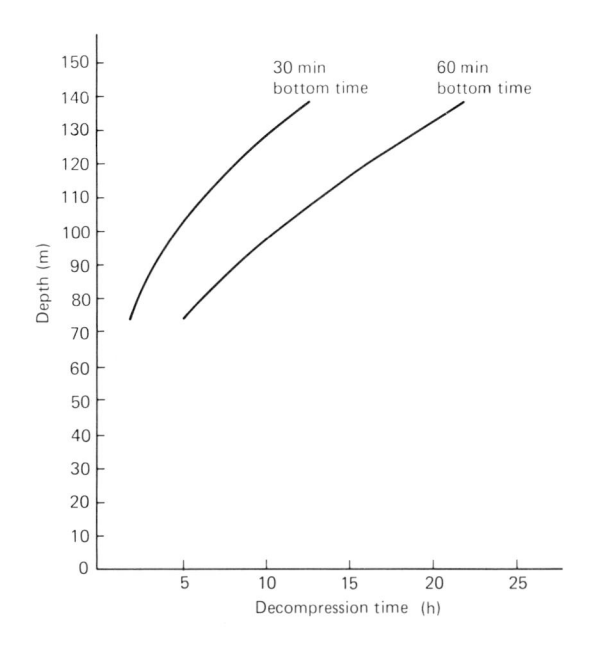

Figure 4.10 Bounce diving decompression times

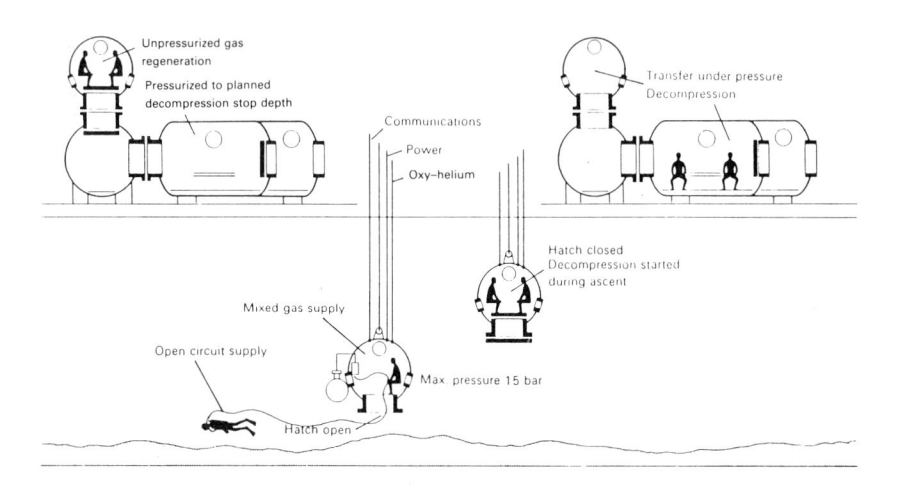

Figure 4.9 Bounce dive procedure (courtesy of Submex)

It can be seen that this technique is not efficient in terms of bottom working time to total operation time including decompression, especially when depths exceed 150 m (492.5 ft). For this reason, bounce diving is only suitable for certain short duration underwater tasks.

4.2.5.1.2 Saturation diving For tasks in deep water that cannot be completed in a short time, saturation diving is employed. Although for the given depth the total decompression time is at a maximum, far greater than that, say, for a 0.5 h bounce dive at the same depth, in this technique the divers are maintained under pressure over such a period (usually 28 days) that decompression time becomes less significant. The ability to increase the percentage of useful diving time in this way is only possible because there is a limit to the amount of gas that can be dissolved in the body's tissues at any given pressure, i.e. saturation at that pressure has occurred. As depth increases, total decompression times become lengthy (see *Figure 4.11*).

Saturation diving requires heavier, more complex and more sophisticated equipment. Unlike air diving and bounce diving, it normally requires the use of a dedicated surface support vessel. *Figure 4.12* shows a typical saturation diving system or saturation spread. Such systems vary in size and complexity and in the number of divers they can support at any one time.

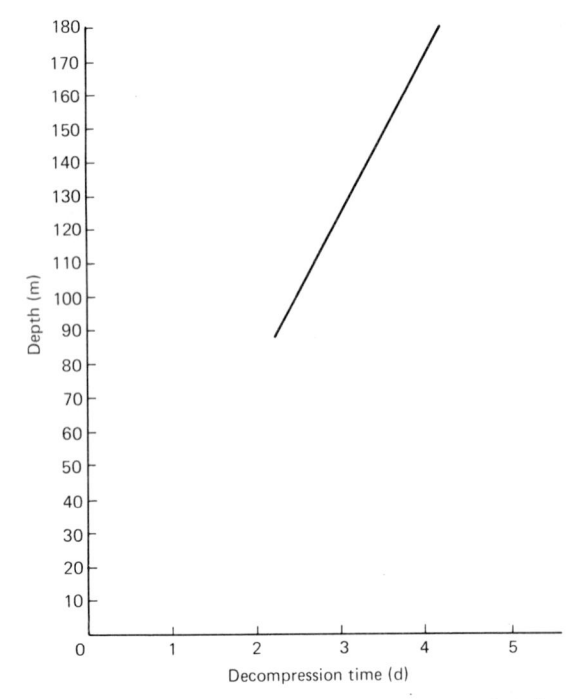

Figure 4.11 Approximate total decompression time for saturation diving as a function of depth

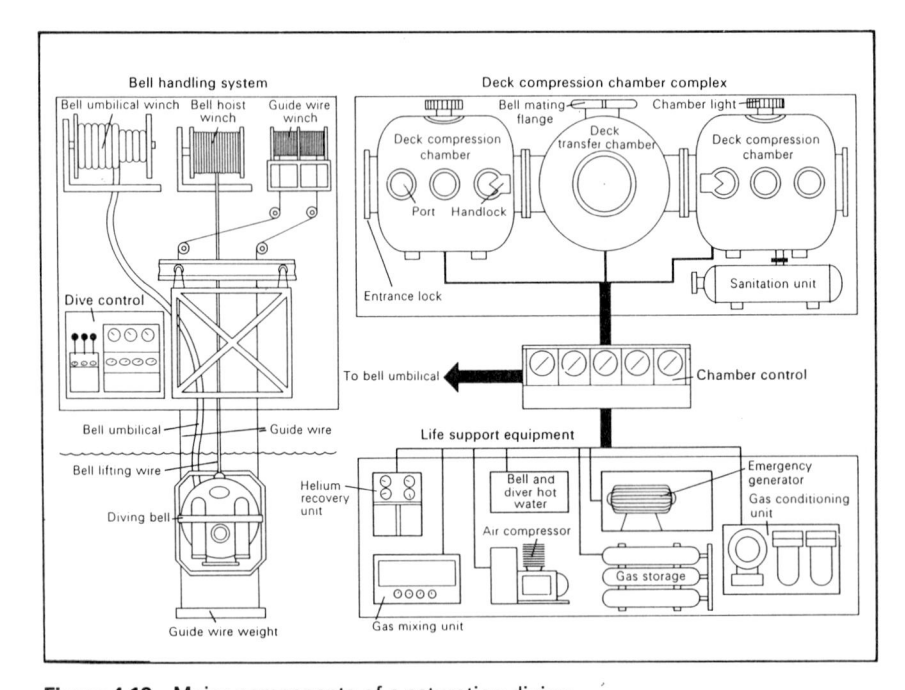

Figure 4.12 Major components of a saturation diving system (Submex)

Modern systems can often possess two bells and can support more than 12 divers under pressure. The extension in diving depth by the saturation technique has not, however, been without penalty and a number of new problems have arisen.

4.2.5.2 High pressure nervous syndrome (HPNS)

High pressure nervous syndrome (HPNS) is a condition which begins to affect divers on reaching a depth of about 200 m (657 ft). Symptoms include tremors of the hands, involuntary jerky movements of the limbs, dizziness, nausea and sleepiness. Researchers have found that changes in the electrical activity of the brain accompany these symptoms. Tremors may be followed by convulsions, which in animals, even after immediate decompression, have proved to be fatal. HPNS is therefore a serious problem and even the initial mild symptoms will dramatically reduce the work capability of the diver.

The cause of HPNS is still not known with certainty. It has been attributed to the helium at high pressure and also to pressure alone on the nervous system. It is found that individuals tend to vary in their susceptibility to HPNS. Despite the fact that the mechanism of HPNS is not fully understood, the diving industry has managed to find methods of reducing its occurrence. It has been found that a slow rate of compression can often avoid onset of HPNS and, for this reason, a compression rate of not greater than 3 m/s (10 ft/s) and preferably as low as 1 m/s (3 ft/s) is recommended for dives beyond 200 m (657 ft). Staged compression, with stops to allow the diver to assimilate to a given depth, is also helpful in avoiding this problem.

The addition of other gases into the breathing mixture, for example nitrogen (to form trimix) and more recently hydrogen, has also proved to be helpful in this respect. For the very deepest of diving operations, it is necessary to select those individuals, through trials, who are least susceptible to HPNS.

4.2.5.3 Speech distortion in a high pressure helium atmosphere

The human voice becomes seriously distorted by the presence of a helium rich atmosphere. It takes on what can only be described as a Donald Duck characteristic which becomes more marked and hence less intelligible to the listener when the partial pressure of helium is increased. For the heliox diver, such an effect can have serious consequences in that important communications from the diver to the surface support may be misunderstood. Beyond 200 m (657 ft), the distortion of speech by helium is a major problem to the diving industry.

The mechanism by which the effect occurs is not fully understood, but is thought to involve the vocal chords and the resonant passages of the chest, throat and head. The speed of sound in helium differs from that in air and from an acoustic viewpoint alters the effective size of these cavities. As the operation of the vocal chords is almost certainly linked through some feedback mechanism to these normally resonant cavities, it is perhaps not surprising that it is affected in some way.

Analysis shows that the effect that the hyperbaric helium atmosphere has on speech is complex. Although the fundamental frequency of speech remains relatively unchanged, the so-called voiced formant frequencies are shifted in both linear and non-linear fashions with respect to the change in speed of sound, dependent on their centre frequencies. There is also increased attenuation at frequencies above 5 kHz.

There are several commercial electronic helium speech unscramblers available, the better of which do, for moderate depths at least, offer a definite benefit. However, as depths increase to well beyond 200 m (657 ft) where the problem becomes very severe, it is found that systems presently available are ineffective. With practice, divers and tenders will learn to understand helium distorted speech with reasonable efficiency and in many situations they prefer not to use an electronic unscrambler. Good communications are crucial to diving operations. It is important that communications systems are continually improved as diving reaches greater depths.

4.2.5.4 Temperature control – hyperthermia and hypothermia

In normal circumstances the human body controls its temperature to a steady 37°C. The mechanisms by which it does this when surrounding and internal conditions change, are well known. Provided such changes are not too severe, the body will be maintained at this temperature at which it functions optimally.

A significant change in surrounding temperature will stimulate one of the body's mechanisms for control and at the same time, if large enough, result in a degree of discomfort. If these mechanisms fail to bring the body temperature under control, then the resulting change will cause further discomfort and other symptoms. In such a situation it is wise to assist the body's mechanisms in some way, e.g. by wearing insulating clothing in cold weather.

Hypothermia is the condition when for any reason the core temperature of the body falls abnormally low. Hyperthermia is the condition when the body temperature is abnormally high. For various reasons, divers can experience both of these conditions. For the heliox diver, surrounded by a gas which can carry heat away very rapidly, hypothermia is normally of greater concern.

The capacity of a fluid for removing heat from a body, through conductive and convective processes, is related to its thermal conductivity (K), its specific heat (C) and its density (ρ). The product $KC\rho$ provides a good indication of this capacity and *Table 4.4* gives these quantities for the various components of the

Table 4.4 Physical properties of a diver's environment

	$C(J/kg/°C)$	$K(Wm/°C)$	$\rho(kg/L)$	$KC\rho$
Air	1000	0.024	1.2×10^{-3}	0.029
Heliox				
50 m	3650	0.112	1.4×10^{-3}	0.570
200 m	4650	0.151	3.8×10^{-3}	2.670
600 m	5000	0.157	10.2×10^{-3}	8.000
Sea water	4200	0.600	1025	25850

environment encountered in heliox diving. Because the conductivity of helium is six times and its specific heat three times that of air, the heliox atmosphere is capable of removing heat from the diver's body rapidly. The problem is then further exaggerated as depth and hence gas density is increased.

From the figures in *Table 4.4* it can be seen how the hyperbaric heliox atmosphere can be a potential problem if a sufficient temperature difference between the diver's body and the surrounding atmosphere is allowed to develop.

In practice, the greatest heat loss to the heliox diver is through respiration and it is for this reason that for depths exceeding 150 m (492.5 ft), it is mandatory in the UK to provide heating for the diver's breathing gas. *Figure 4.13* shows how the minimum allowable breathing gas temperature must be increased as depth is increased. The increasing capability of heliox to extract heat from the diver's body as a function of depth, makes the diver very sensitive to changes of temperature, thus narrowing the band of temperature of the surrounding atmosphere in which he will feel comfortable (see *Figure 4.14*). At a depth of 300 m (985 ft), a change in the surrounding temperature of only 2°C will produce in the diver the complete range of temperature sensations.

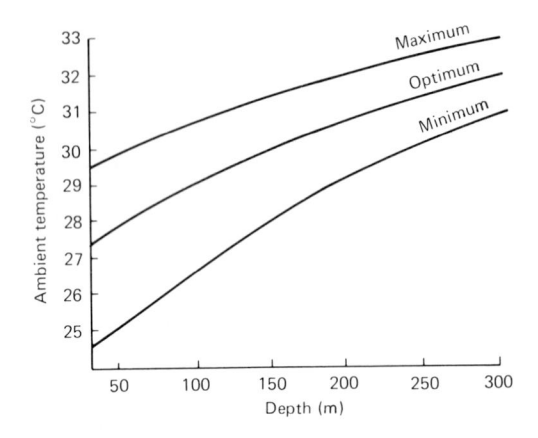

Figure 4.14 Range of comfortable ambient temperature for heliox diving

4.2.5.5 Saturation diving procedure

At the beginning of a saturation operation, divers enter the DDC. The system is then pressurized in a slow and controlled manner, especially when working at depths in excess of 200 m (657 ft). In this way the effects of HNPS are minimized and the diver has chance to become assimilated into the new conditions.

When the working pressure has been attained and the divers are comfortable, two or often nowadays three divers transfer under pressure to the bell. Hatches are closed and the bell is physically separated from the DDC. Breathing gas is pumped to the bell through its umbilical. The bell is winched down to the work site at which point its internal pressure is equal to the ambient pressure and its inward opening hatch can be opened. One diver stays in the bell to tend umbilicals and monitor various instrumentation for the working diver or divers outside the bell. The time between which the bell leaves and returns to the surface, the bell run, can last up to 8 h. Divers spend up to 4 h working in the water at any one time.

When the bell is winched to the surface, it is mated to the DDC and the divers transfer from the bell to what becomes their temporary living accommodation in which they are maintained at a constant pressure. In this way, the team of two or three divers will make a dive perhaps once each day with 16 h rest time between

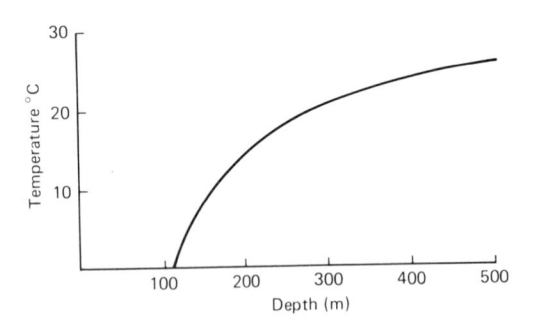

Figure 4.13 Minimum allowable breathing gas temperature for heliox diving

dives. During this time, the other divers under saturation will be utilized ensuring that work continues almost uninterrupted for 24 h/day.

At the end of the working period, the divers are then decompressed, a process for deep dives which will take several days to complete. The total time, from initial pressurization to the end of decompression, is normally restricted to 28 days. Even though decompression times are so lengthy, it can be seen that the saturation diving technique has a great advantage in terms of efficiency over bounce diving.

Divers in a saturation complex require much in terms of surface support. The composition of their gas supply must be constantly monitored and kept within strict limits. The humidity and temperature of their living quarters must also be carefully controlled in order to maintain a reasonable degree of comfort and prevent infection or injury. Meals and drinks, the taste of which cannot properly be appreciated in hyperbaric conditions, must be prepared and passed into the DDC through special purpose locks in the system. Facilities for showering and toilets must also be provided. This support requires a team of highly qualified dedicated personnel to be present.

Because pressurization and depressurization take place in a very carefully controlled manner, the saturation diver is less likely to suffer any of the adverse pressure related effects than the air diver or bounce diver. Saturation diving in the North Sea today enjoys a very good safety record which surpasses those of many of the other so-called dangerous occupations.

4.2.5.6 The saturation diving system

A typical saturation diving system is shown in *Figure 4.12*, although there are a number of designs of different size and complexity presently in use. The typical system comprises four main parts.

4.2.5.6.1 Diving bell or submersible decompression chamber (SDG) The closed diving bell may be either a two or three diver unit, the latter becoming more common in recent years. It is essentially a pressure vessel, generally spherical in shape and constructed from a high strength steel in order to minimize the necessary wall thickness required. It has an inward opening hatch at the bottom through which the divers pass into the water. It will often possess a number of small clear view ports around its side.

The bottom hatch entry point of the bell is capable of being mated in a pressure tight manner to the deck decompression chamber (DDC). Alternatively, bells will possess a side hatch to permit horizontal docking with the DDC.

Onboard breathing gas, for use usually in exceptional circumstances, is carried in a number of cylinders external to the bell. The diving bell chamber, because of its volume, is usually positively buoyant. It is therefore necessary to add a ballast weight below it. This is jettisonable to enable a buoyant ascent in an emergency. The typical modern diving bell carries a large amount of equipment essential to the well being and safety of the divers. *Figure 4.15* shows the layout of a typical bell.

4.2.5.6.2 Bell handling system Of all the major component systems, provision for bell handling and deployment is perhaps the most varied. The bell handling system should possess the following characteristics and facilities:

1. It should be designed, built and maintained to withstand the loadings under heavy weather.
2. It should be capable of passing the bell rapidly through the air–sea interface to minimize the effects of wave action.
3. It should ensure that the bell is kept well away from the surface support structure during launch and retrieval, to prevent damage.
4. It should incorporate a lifting winch of sufficient power to permit fast deployment and retrieval of the bell, whilst having precision control to permit safe approach to the sea bed and mating with the DDC.
5. It should include a translation system to move the bell from its launch position with the DDC.
6. It should incorporate separate winches for bell umbilical and main lifting line.

The diving bell may be launched over the side of the vessel or through a dedicated moonpool.

To launch over the side, an extendable A-frame gantry or davit may be used to lift the bell clear of the side of the vessel (see *Figure 4.16*). The arrangement of the DDC must be such that it allows mating of the bell to be performed with minimum translational movement.

When a moonpool is available, then the diving system is almost certainly an integral part of the vessel. A moonpool is a clear opening through the decks and hull of a vessel providing access to a comparatively sheltered area of water. Deployment through a moonpool was introduced in order to reduce downtime lost because of bad weather. Usually the bell runs down through the moonpool restrained by a vertical rail system. Where the bell is mechanically driven up and down these rails through the interface, it is known as an 'active cursor' system.

When in the water, the bell is usually restrained by a previously deployed guidewire system. The guidewires are run from a separate winch and support a heavy weight which is lowered to the sea bed. The dive control cabin, from which the winches are operated, is positioned alongside the launching point.

The configuration of the bell umbilical is dependent on the design of the bell itself and the mode of diving for which it is intended. Connecting the surface support to the bell, the modern composite umbilical will have a

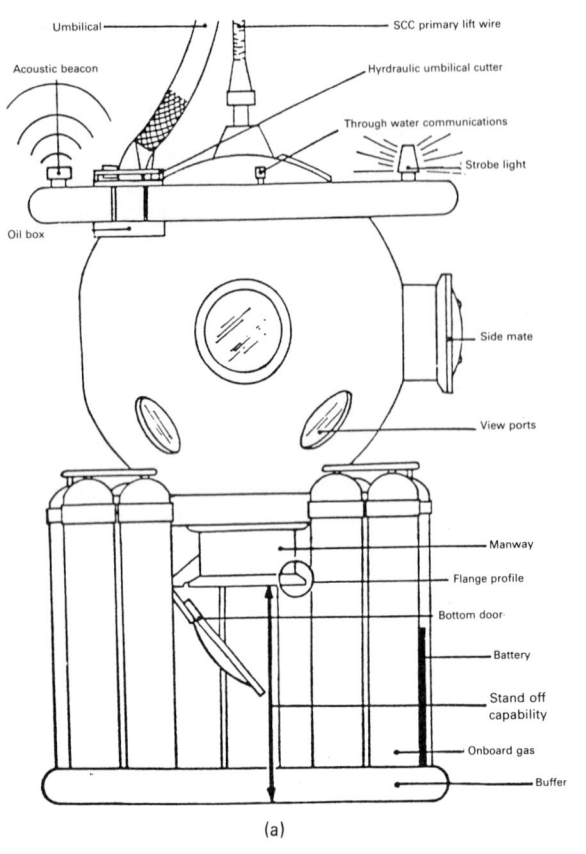

Figure 4.15 Typical closed diving bell showing (a) main external and (b) internal features (courtesy of Submex)

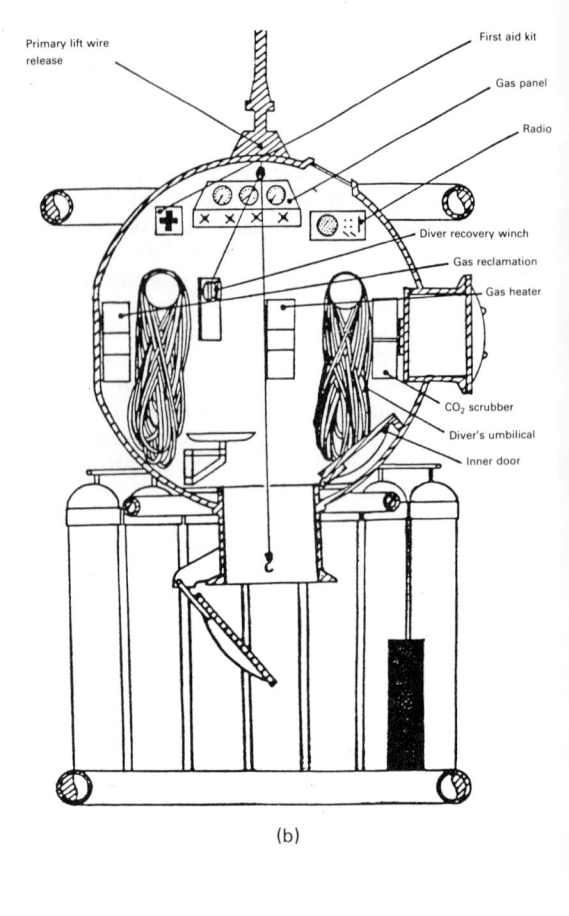

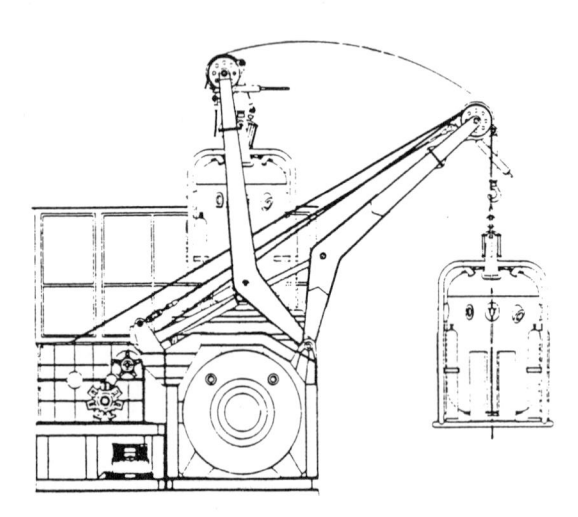

Figure 4.16 Typical A-frame bell deployment system (IUC)

diameter of up to 10 cm (4 inches). Although not normally used for lifting the bell, it should possess this capability in an emergency. The umbilical possesses the following components: hoses for the supply and often return, of breathing gas to the bell; communications lines (voice and instrumentation); a pneumofathometer hose for the accurate measurement of pressure (and hence depth) of the bell and divers; a hot water hose for the heating of divers and their breathing gas and a power cable to supply bell lighting, instrumentation and carbon dioxide scrubber.

4.2.5.6.3 The deck compression chamber (DDC) The deck decompression chamber (DDC) is a multicompartment pressure vessel mounted on the surface support platform. Generally constructed from steel, compartments tend to be cylindrical with hemispherical ends. Systems are often modular in construction to simplify their transportation and provide for system flexibility.

The modern DDC may consist of up to five or more interconnecting chambers depending on the number of occupants it is designed to house. One chamber, referred to as the entrance lock or transfer chamber is designed to receive the bell's mating flange. It is often used to store wet clothing and kit.

4.2.5.6.4 Life support system To provide a safe and comfortable environment for the divers inside the DDC, a number of properties of the atmosphere must be carefully monitored and controlled. The systems required for this are located external to the DDC, either mounted directly on its outside to minimize piping, or in a separate module. The chamber operator or life support technician is responsible for the operation of this equipment and therefore the safety of the divers. The following conditions must be closely monitored and controlled:

1. Gas pressure and composition – oxygen levels measured by monitors in the DDC and sometimes in the bell must be carefully controlled to within strict limits. Different levels may be specified for the DDC atmosphere, diver breathing gas whilst working, therapeutic purposes and during decompression. Carbon dioxide levels, again monitored in the DDC and bell, must be kept at very low levels for diver comfort and safety. Ideally the level should be kept to below 5 mbar in the DDC whilst 10 mbar would be acceptable in the bell. A level of 50 mbar would cause the divers to resort to their built in breathing systems (BIBS). Carbon dioxide is removed from the breathing gas by passing it through a scrubber, a device containing an active chemical capable of absorbing the gas, e.g. soda lime, baralime^{TM}. Nitrogen should be kept to below 0.1 bar in the heliox operations. Other contaminant gases may be extremely poisonous and their presence must be minimized if not eliminated.
2. Temperature – as previously mentioned, the diver becomes highly sensitive to temperature changes in a hyperbaric helium atmosphere. The DDC atmosphere must therefore be carefully monitored and controlled. This is often achieved by a gas recirculating heater system attached to the outside of the DDC.
3. Humidity – the DDC atmosphere should be kept preferably between 50 and 60% relative humidity. Prolonged periods of exposure outside this range can lead to respiratory problems and ear infections. The dehumidifier is often incorporated with the gas heating system external to the DDC.
4. Sanitation – the provision of the necessary showering and toilet facilities in a hyperbaric chamber creates certain technical problems solved only by elaborate external plumbing systems.

The chamber operators monitor and control the above conditions from the master control panel which is situated close to the DDC. It is from this point where they also control the compression and eventual decompression of the divers.

4.2.6 Personal diving equipment

There is a wide range of personal equipment available to the commercial diver today. The choice will be made according to the type of diving operation, its location, and to some extent by the individual diver's preference. Personal diving equipment covers helmets and masks, suits and any other equipment that the diver might normally carry.

4.2.6.1 Helmets and band masks

A helmet completely encloses the head, making a seal around the neck. In this way the diver's head is kept completely dry. A band mask, which is less massive than a helmet, makes a watertight seal around the face only.

The choice between helmet or band mask often depends on which the diver is most familiar with, as often both will be available. Normally, band masks are used for shallower work provided that the water is not polluted. The band mask is preferable if the diver is required to do a fair amount of swimming. The helmet, however, provides much greater protection when working within or near structures.

In order to overcome the problem of a build up of carbon dioxide within the volume of the helmet or mask whilst keeping the required flow rate of gas to the diver within a reasonable level, a close fitting oronasal mask is often employed within them. However, helmets without oronasal assemblies are available. Such helmets are referred to as free flow helmets, but are nowadays not in wide use.

4.2.6.1.1 The Superlite 17 diving helmet The Kirby Morgan Superlite 17 is a good example of a modern diving helmet (see *Figure 4.17*). Constructed of woven fibreglass and polyester resin, it is positively buoyant and weighs 11 kg (24 lb) in air. Weights of up to 1.5 kg (3 lb) can be added according to diver preference. The flat face plate is of 6 mm (0.25 inches) thick Lexan^{TM}. The helmet incorporates an oronasal assembly, although it can be used in a free flow mode by operating the forward facing valve on its right hand side. This facility is, however, more usually used only for clearing condensation from the inside of the faceplate. Inside the helmet is worn a head cushion for comfort. Earphones and a microphone for the necessary communication link to the surface are also incorporated within this helmet. This helmet is suitable for both air diving and heliox diving operations.

4.2.6.1.2 The band mask As with helmets, a number of band masks are available to commercial divers. *Figure 4.18* shows a typical example. When using a band mask, the diver usually wears a neoprene hood.

Figure 4.17 Superlite 17 diving helmet (S. R. Littler)

Figure 4.18 US diver's band mask (S. R. Littler)

The mask is held on to the face by a head harness or spider. The face seal consists of foam neoprene and is designed in such a way that in the unlikely event of the failure of both gas supply pressure and the return valve, it is pulled into the mask interior to prevent squeeze. The other main components of the mask, including the oronasal assembly, demand valve, free flowing facility and communications subsystems can be similar to those of the Superlite 17 helmet.

4.2.6.2 Diving suits

The function of the diving suit is to provide both physical and thermal protection to the body. *Table 4.4* has shown the capability of sea water in removing heat from the body. When diving in all but the warmest of naturally occurring water, the thermal protection offered by the suit is of extreme importance. Thermal protection can be provided either passively or actively. Various types of suit and underclothing are used in diving operations.

4.2.6.2.1 The wet suit Manufactured from a closed cell expanded foam neoprene it is a tightly fitting garment worn next to the diver's skin. Water enters the suit at its edges, cuff and neck and is trapped between suit and body. This small volume of water then reaches the surface temperature of the body, the suit providing a thermal barrier. The effectiveness of the suit would be severely reduced if it was so loose fitting so as to allow water to flow between it and the body. Wet suits are more associated with recreational water sports activities and find little use in commercial diving operations.

4.2.6.2.2 The dry suit The dry suit is a one piece garment made from a number of materials including rubberized canvas, reinforced rubber and foamed neoprene. They have seals around the cuffs and neck or face and incorporate leaktight zip fasteners. As its name suggests, the diver is meant to remain dry during diving and can therefore usefully wear thermal underwear to provide further insulation and soak up perspiration. Although there are various types of dry suit available, the standard suit consists of a loose fitting, strong thin skin with a built in inflation and deflation valve. This facility allows gas to be introduced into the suit and to provide an even balance to the external pressure. Without this the suit can be squeezed against the body in certain places causing pain and restricting movement. The gas also provides a means of adjusting buoyancy and increases the effectiveness of any thermal underwear.

One disadvantage of the dry suit is its vulnerability to damage. A puncture, depending on its extent, can lead to a loss of buoyancy and cause the dive to be aborted.

4.2.6.2.3 Hot water suit It is usually a one piece, loose fitting suit containing tubing to distribute heated water around its inside. Heated water enters a manifold at waist level and after circulation, leaves the suit at the extremities of the limbs. The manifold incorporates a valve which can be operated by the diver to alter the water flow rate and hence the rate at which heat is supplied. Fluctuations in water supply temperature are not uncommon and it is wise for the diver to wear some undergarments for protection against possible scalding.

In commercial dives below 50 m (164 ft) in the UK where diver heating is mandatory, the hot water suit is used almost universally. The heated water is usually supplied from the surface, although systems are available in which the water is heated electrically in a diver carried backpack.

In deep heliox operations, the water temperature around the body should not be allowed to fall below 32°C if hypothermia is to be avoided. To avoid scalding, the water temperature must be kept to below 45°C.

4.2.6.2.4 Electrically heated suits The electrically heated suit is a one piece, woollen garment containing heating element wires. Extending over the whole body, it is normally worn under a dry suit. Connected to a low voltage electrical supply, so that there is no risk of electrocution, it supplies heat in the same way as an electric blanket.

Whilst being energy efficient and not requiring the inconvenient hot water supply hose, it is difficult to get as much heat as the hot water suit supplies. Electrically heated suits are not commonly used in commercial operations and today find more favour in military operations.

4.2.6.3 Diver gas heater

In all operations below 150 m (492.5 ft) it is necessary to heat the diver's breathing gas. This is accomplished by mounting a heat exchanger either on the diver or on the umbilical. In either case, the same hot water supply used for body heating will be used to heat the breathing gas.

4.2.7 Support vessels

Surface support requirements vary greatly depending on the mode of diving employed and the nature and depth of the operation to be undertaken. Recreational scuba divers require no more than a small boat upon which to base their operation. At the other end of the scale, the saturation diver working on a deep sea construction project requires hundreds of tons of equipment. As commercial diving, stimulated by the offshore oil and gas industry, has advanced in terms of depth and safety, the amount and sophistication of support equipment has been steadily increasing. Few saturation spreads are portable in the true sense and the recent trend has been to build dedicated diving support vessels (DSV) to meet the requirements of the users.

Originally, most DSVs were conversions of vessels built originally for other purposes. Nowadays, when operational efficiency is perhaps more important than ever before, a number of purpose built DSVs designed to operate in poor weather conditions are available.

Most DSVs are orthodox vessels of monohull construction. However, the semisubmersible vessel, which provides excellent stability, can also form the basis of the DSV. An example of each type of vessel is described below.

4.2.7.1 The Orelia

The *Orelia* (see *Figure 4.19*) is presently one of the most advanced monohull DSVs in service. Its unconventionally shaped hull was designed for cost effectiveness in construction and to give the vessel exceptionally good stability characteristics. Some 119 m (391 ft) in length, it has 1800 m² (19 000 ft²) of deck space and a helideck. The vessel is equipped with two 100 tonne

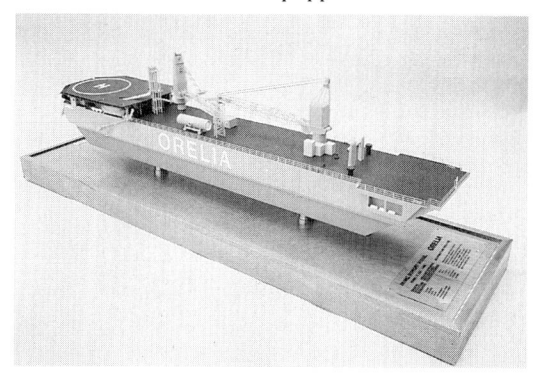

Figure 4.19 The *Orelia* diving support vessel (courtesy of Houlder Offshore)

cranes that can be used simultaneously for lifting long loads. One crane is equipped with heave compensation which is said to limit the movement of crane loads relative to the sea bed to a few centimetres.

Propulsion of the vessel is also unorthodox in that it is by six controllable pitch variable direction thrusters. The vessel possesses an advanced dynamic positioning system that can be interfaced with several surface navigation and position fixing systems.

The diving system on board the *Orelia* represents the present state of the art, consisting of a complex DDC capable of supporting 20 divers in saturation and two large bells. The bells, both possessing heave compensated handling systems, can be deployed simultaneously through the moonpools which are 25 m (82 ft) apart. The whole system, including the handling system, is rated to 450 m (1477.5 ft) water depth.

The vessel houses a permanent gas storage facility of some 18 000 m³ (618 000 ft³) capacity rated at 200 bar. There is also a gas recovery system installed on the vessel.

4.2.7.2 The Uncle John

Entering service in 1977, the *Uncle John* (see *Figure 4.20*) is a purpose built semisubmersible DSV. Supported on submerged pontoons whilst in operation (see *Figure 4.21*), the vessel is less affected by surface waves. The operators quote roll and pitch of only 2° in Beaufort 9 wind conditions as opposed to the 5–10° typical of a conventional monohull vessel. With its advanced dynamic positioning system, the vessel is able to maintain station to within a 3 m (10 ft) radius in such

Figure 4.20 The *Uncle John* multipurpose diving support vessel (courtesy of Houlder Offshore)

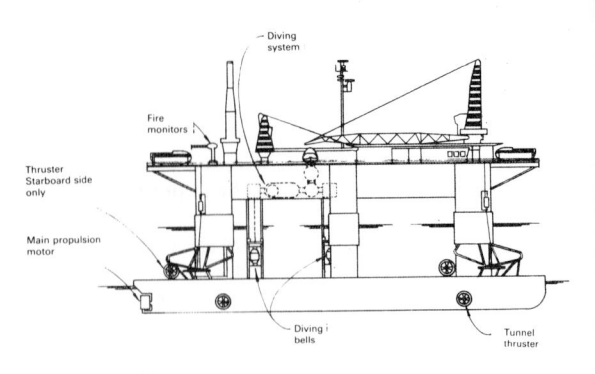

Figure 4.21 Elevation of *Uncle John* (courtesy of Houlder Offshore)

conditions, thus allowing diving operations to continue safely. The design of this vessel means that its orientation is not constrained to the wind direction, a useful capability in many operations.

Its saturation spread contains four 2.3 m (7.5 ft) diameter living chambers, each with four bunks, rated to 450 m (1477.5 ft) (see *Figure 4.22*). The system supports two bells, each rated to 350 m (1149 ft), one of which is fully heave compensated. The bells are deployed via cursors to clear the pontoons some 15 m (49 ft) below the surface. A gas reclaim system is installed and there is provision for air diving and ROV operations.

With accommodation for 102 persons, the vessel has a total clear deck space of 1800 m^2 (19 000 ft^2) and workshops on and below deck to support various activities. The vessel also has extensive craneage in the

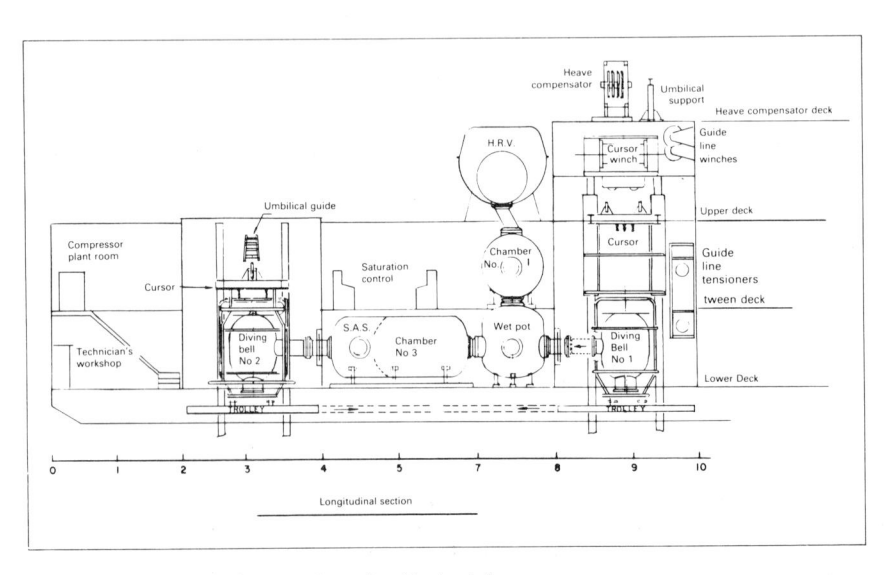

Figure 4.22 Saturation spread on the *Uncle John* (courtesy Houlder Offshore)

form of two 100 tonne cranes capable of lifting heavy loads from the sea bed, fire fighting pumps and a helideck. For these reasons, this type of vessel is often referred to as a multipurpose support vessel (MSV).

4.2.8 Safety of diving operations

When man enters an alien environment and life is totally dependent on electro-mechanical systems and the actions of others, there has to be an element of risk.

In the early days of diving when the effects of the underwater environment on the body were not understood, the number of casualties was high. We have already seen how in the mid-nineteenth century none of Colonel Pasley's men escaped the effects of decompression sickness. Although, over the years, knowledge of the effects of diving has increased significantly with a corresponding reduction in casualties, the need to achieve greater depths, under often intense commercial pressure, has occasionally caused risk to rise to an unacceptable level. Perhaps an example of such an instance corresponds to the start of the oilfield development in the northern North Sea. This location presented new difficulties to the offshore industry, both in terms of water depth and the weather conditions experienced. The diving industry was faced with new challenges for which there were lucrative rewards. The increase in risk which occurred at the time is borne out by the numbers of accidents which occurred to divers. *Table 4.5* shows figures for accidents resulting in death or serious injury that occurred to divers in the UK sector of the North Sea in the period 1974–1985.

Table 4.5 Serious and fatal accidents in UK sector of North Sea (data obtained from quarterly summaries of accidents from Department of Energy, Petroleum Engineering Department, Branch 5)

Year	Number of serious injuries	Number of fatalities
1974	1	3
1975	3	1
1976	2	6
1977	5	2
1978	5	2
1979	2	4
1980	6	0
1981	1	0
1982	7	1
1983	8	1
1984	0	0
1985	1	0

The risk to divers and hence the number of accidents, can only be reduced by making improvements to the performance and reliability of diving equipment and by applying strict procedures, possibly through legislation, in accordance with increasing knowledge and experience of this activity. It was in this way that the worrying trend in North Sea diving accidents was reversed, although figures given in *Table 4.5* do not take into account the fall off in the diver population during this time. As diving is extended to greater depths or new information arises regarding the effects on divers of existing depths, it will be necessary to revise the rules and regulations governing diving from time to time. The diver must be given the best possible protection in this way.

4.2.8.1 Regulations and guidelines

In the UK, Norway and other countries bordering the North Sea, there exist detailed legislation, codes of practice and guidelines designed to maximize the safety of diving. This is not necessarily so in other parts of the world and consequently procedures and staffing levels which are required and considered normal in the North Sea may well be considered unnecessary elsewhere.

In the UK, the relevant legislation is contained within the following Statutory Instruments:

1. Diving Operations at Work Regulations 1981 (SI1981/339) which revoked the Offshore Installation (Diving Operation) Regulation 1974 (SI1974/1229) and the Diving Operations Special Regulation 1960 (SI1974/1229 and SI1960/688).
2. Merchant Shipping (Diving Operations) Regulation 1975 (SI1976/116). (a) Merchant Shipping (Diving Operations) Regulation 1975 and amendments 1976 (SI1976/2062).
3. Submarine Pipelines (Diving Operations) Regulation 1976 (SI1976/923).
4. Merchant Shipping Craft (Construction and Survey) Regulations 1981 (SI1098), apply to operation of lock-out submersibles.

The UK legislation pertaining to commercial diving is complex in detail but the salient features are summarized below.

A diving operation must be undertaken by a team of divers plus a standby diver, under the control of a supervisor. Divers must possess an approved certificate of training, be certified medically fit and be competent to undertake the work safely. Divers are also required to keep detailed logs of all their dives.

Air diving is only permitted for depths to 50 m (164 ft) and there should be some form of communications link with the emergency services. A reserve gas supply for the divers is also required. In most situations, a communications link between diver and surface is required and a compression chamber must be available on site for all diving at depths greater than 50 m (163 ft) apart from that of the very shortest duration. At depths greater than 50 m (164 ft) a diving bell must be used and there must be a means of heating the diver. At depths greater than 150 m (492.5 ft) there must be provision for heating the breathing gas.

The legislation also governs the design of diving equipment. For example, DDCs must possess two compartments when used in operations deeper than 50 m (164 ft). Their minimum internal diameter, depending on the year the system entered service, is also laid down. All pressure vessels used must have current certification for fitness and equipment must be thoroughly maintained by a competent person.

In addition to the legislation there are also codes of practice and guidelines as follows:

1. Association of Offshore Diving Contractors Guidance Notes.
2. Each diving contractor is responsible for issuing rules for the conduct of its employees, usually in the form of a comprehensive diving procedures and safety manual.
3. The operator of the installation on which the diving contractor will work, usually an oil company, will usually impose procedural and safety requirements through the contract.

4.2.8.2 Emergencies

In an offshore emergency, divers under pressure are at a gross disadvantage compared with other workers. In the event of fire, for example, they do not have the same mobility that is required to make an escape. Rapid decompression, for reasons previously discussed, is not an option.

Emergencies fall into three categories: when the surface support platform is at risk of sinking or when the DDC is under threat locally, e.g. due to fire; when one or more divers has sustained injury for which immediate medical attention is needed; and when divers are trapped at depth either in the water or within a stricken bell. To cope with these emergencies, there are now a number of procedures that are usually available.

4.2.8.2.1 The TUP (transfer under pressure) system
This system is designed to transer the diver casualty under pressure to an onshore hyperbaric facility for the best available medical treatment. Such a facility in the UK can be found at the National Hyperbaric Centre at Aberdeen.

The patient is initially transferred from the DDC to a small portable single chamber of some 2.34 × 0.81 m (7.7 × 2.6 ft) diameter in dimensions, which is constructed of titanium for lightness. This chamber containing the patient under pressure is then carried by six people to a helicopter where it is mated with a larger chamber (2.6 × 1.14 m) (8.5 × 3.7 ft) into which the patient is transferred. At this stage, the patient will be received by one or two medics who have just been compressed prior to the transfer. The whole system including the transfer chamber and life support equipment for the larger chamber is light enough to be transported by a Sikorsky S-61 helicopter. On arrival at the hyperbaric hospital, the larger chamber is removed from the helicopter and mated to the hospital chamber.

The diver will then be attended by 'hyperbaric doctors' and treatment, which may include surgery, will begin immediately. In this way, the TUP method enables treatment to begin without delay caused by decompression.

4.2.8.2.2 Hyperbaric rescue chamber In the event of sinking of the support platform or engulfment by fire, it is necessary to evacuate all the divers as rapidly as possible. Originally, one solution was conceived as simply to detach the whole diving spread and drop it into the sea. Whilst certainly being positively buoyant, the discomfort caused to the divers would be immense and chances of survival would not be great.

A more realistic development of this dubious procedure is the hyperbaric rescue chamber. It consists of a compression chamber mated to the DDC, provided with ballast and buoyancy to ensure a stable floating configuration. It will also possess a basic self contained life support system and fenders around the outside for its protection. In the emergency, it may be lowered or dropped into the water where it can be towed or lowered on to another vessel if sea state permits. Decompression of the divers can only begin once the rescue chamber has been mated with another DDC.

4.2.8.2.3 Hyperbaric lifeboat The hyperbaric lifeboat is a logical development of the rescue chamber discussed, and consists of a compression chamber housed within a lifeboat. It is normally kept mated with the DDC system and in the emergency all divers under pressure will transfer to it. The boat, with its own crew, will then be launched into the sea and, under its own power, head away from danger. During this period the necessary life support to the divers under pressure will be provided. The lifeboat must then be recovered from the sea and mated to a DDC before decompression can commence. For all these methods to be effective, it is necessary to ensure that all chamber connecting flanges are compatible or the necessary adaptors are available.

4.2.8.2.4 Bell to bell transfer The technique of transferring divers from one bell to another whilst at depth can be used for two purposes. When a bell has become trapped on the sea bed for any reason, the greatest danger facing its occupants is that of hypothermia. In this situation, the simplest means of rescue is by the use of a second bell either associated with the same saturation system or from a totally different vessel. In practice, the rescue bell will be positioned to within 10 m (33 ft) or so of the stricken bell and its divers will assist in the transfer. The rescued divers will use umbilicals (and possibly helmets) associated with the rescue bell.

Transfer of divers from bell to bell can also be used as a means of transfer between separate DDCs in certain situations, for example between incompatible systems. It is not a fast procedure however and would not normally be recommended as an evacuation method for modern high capacity DDCs.

4.2.8.2.5 Diving bell used for evacuation Under certain conditions, the diving bell can be used as a means of evacuation from the DDC. When, for any reasons, the risk to life of remaining in the DDC becomes so great, the bell may provide the only means of escape for a limited number of divers. Used in a negatively or positively buoyant mode like a rescue chamber, the chance of survival depends on whether there is another compatible diving system close by.

4.2.9 Recent developments in deep diving

The use of helium as the inert constituent in breathing has enabled great advancements to be made in diving. Hydrogen has for many years been regarded as a possible replacement for the expensive helium gas, but early attempts to test its use ended in tragedy.

Recently however, there has been renewed interest in its use and a trial, Hydra VIII, successfully demonstrated the use of hydrogen in the deepest open water operation yet undertaken.

4.2.9.1 Hydra VIII deep diving trials

In February and March 1988 the French company Comex completed a series of deep diving trials off Marseilles in which they demonstrated the benefits of introducing hydrogen into the breathing gas.

The DSV *Orelia* took up station some 6 km (3.7 miles) south of Marseilles over the chosen work site at a depth of 530 m (1740 ft). Normally certified for a maximum depth of 450 m (1477.5 ft) the vessel and its diving system had to be specially uprated for this work. Several modifications were required which concerned the use of hydrogen. Although there is no danger of explosion in the chamber and bell atmosphere because the percentage volume of oxygen is very low, storage and handling of the gases required the utmost in added safety precautions in the form of leak detection and alarm systems. A dehydrogenation package used to remove hydrogen from the breathing gas during decompression had also to be installed.

Six divers were slowly compressed in stages over a total period of three days. At the work site they were then asked to carry out a number of observational and manipulative tasks. Their mental and physical well-being was continually monitored by psychologists and physiologists. Best results were obtained with a mix, 'hydreliox', consisting of equal parts by volume of hydrogen and helium. The trials confirmed that hydrogen used in this way was non-toxic, resulted in easier and more comfortable breathing, and considerably reduced articular and muscular pain when undertaking heavy work. It appears from the trials that the achievable work rates are considerably better than those at 400 m (1313 ft) on heliox. It was also reported that the use of hydrogen significantly reduced the occurrence of HPNS.

Decompression of the divers was achieved successfully over a period of 18 days. This trial demonstrated the ability of a carefully chosen few divers, to undertake work in depths greater than 500 m (1641 ft) of water. The company is at present planning to extend this work to a simulated depth of 700 m (2298 ft) in a hyperbaric test facility.

The major obstacle to the adoption of such deep diving commercially is the lack of suitable vessels. The cost of a purpose-built vessel with, say, 1000 m (3283 ft) capability, although technically feasible, would be prohibitively high to the oil and gas market at this present time.

4.3 Atmospheric diving suits and systems (ADS)

Even before the effects of a hyperbaric atmosphere on man were properly understood, the possibility of developing a diving suit that would protect him from the hydrostatic pressure appeared attractive. As more was learnt about the effects of pressure and diving reached ever greater depths, the advantages of such a system became more significant. If a rigid diving suit of sufficient strength with the necessary articulated joints that allowed movement of the limbs could be constructed, then it would totally eliminate many of the problems encountered during ambient pressure diving. The advantages of such a system are both physiological and economical and will be discussed together with certain disadvantages later in this section.

4.3.1 Atmospheric diving systems or suits – history

As mentioned in Section 4.1, the first atmospheric diving suit (ADS) was attributable to John Lethbridge in 1715 although the operator's arms did experience ambient pressure. At the beginning of the twentieth century there had been more than 30 ADS designs published, but it is unlikely that more than one or two might have functioned properly.

At the heart of the ADS is the articulated joint which allows movement of the operator's limbs. The joint must be leaktight and, together with the main body of the suit, capable of resisting the external ambient water pressure. Such a joint is not a trivial engineering problem, as many designers were to learn. The effect of external pressure is to cause the joint to tighten up, rendering it useless below a certain depth. The main factor delaying the development of these suits was the lack of the necessary metal working techniques at that time.

In 1913, Neufeldt and Kunke patented a design for a universal joint and various metal suits. Their joints were based on a ball and socket with ball bearings to take the thrust and fabric bellows to keep the water out. The system had articulated legs and arms with directly operated claws and included a self contained breathing system. Although the suit was designed for a depth of over 200 m (657 ft), it was found that these joints became immovable at depths greater than 100 m (328 ft). However, despite this problem, this suit aided one of the most famous salvage operations of all time, that of gold bullion from the wreck of the *Egypt* off Ushant, France in 1930. At a depth of 125 m (410 ft) the suit took on an observational role.

In 1914, Bowdoin filed an application for a patent in the USA for a novel design of rotary joint to be used in an ADS. In this design, the ambient water pressure was used to counterbalance thrust on the joint. The only source of friction was from the three sealing rings, each of which sealed against the internal and external differential pressure. However, although Bowdoin was known to have designed an ADS, there is no evidence that it was ever constructed.

In 1919, Campos filed an application for a patent for an even more advanced joint, using fluid to support the hydrostatic thrust upon the joint. Although the Campos suit is reported to have attained a depth of 200 m (650 ft), little is known as to its success.

Of the early inventors, Joseph Peress was perhaps more responsible than anyone else for the development of the present day ADS. In 1922, Peress, working in Britain, patented a new fluid supported joint. Whereas Bowdoin and Campos had patented only rotary joints, that of Peress exhibited both rotary and angular motion. The fluid in this joint was contained in a hemispherical chamber and sealed by a fabric rolling diaphragm. A suit incorporating this joint was fabricated from stainless steel but its excessive weight prevented full scale trials of the system. In 1933 Peress patented a new, similar joint which did not require the rolling diaphragm seal. A suit was built using this new joint, but unlike the previous suit, the main body and limb sections were constructed of magnesium alloy resulting in a lighter and more practical system. This suit underwent successful trials in depths greater than 150 m (492.5 ft) with the joints retaining reasonable mobility.

A few years later, the Peress suit was used in the location and survey of the wreck *Lusitania* off the south coast of Ireland. For the next 30 years, little happened in the development of the ADS: all effort seemed to be concentrated towards the development of ambient pressure diving.

It was not until 1967, when the oil and gas industry's requirements appeared to be about to exceed the capability of the ambient pressure diver, that interest in the ADS was revived. Although, at this time, an ADS design was presented by Litton Industries in the USA, the work that would determine further events began in Britain. M. Borrow and M. Humphrey of DHB Construction, together with help from Peress, after studying the original Peress system, completed a new design and produced two new prototypes. The main components were cast in magnesium alloy (RZ4). Aluminium alloy (HF15) was used for the joints and glass reinforced plastic used for the elbow and hand enclosures. The original hand was modelled on the human hand, but this design was found to be overcomplex and was later changed for a choice of simpler end-effectors or special tools.

After components of the suit had been pressure tested to 620 m (2036 ft) and certified by Lloyds, the first trial was to take place in a 20 m (66 ft) tank in November 1971. The trial was a complete success and was to be followed by numerous trials in conjunction with the Royal Navy during which the versatility of this system was demonstrated. The trials culminated in a simulated 305 m (1001 ft) dive by five different operators, a world record at this time.

During the development of this system, it became known as JIM 2 in honour of Peress' original chief diver and mechanic and was the forerunner of a series of systems which were to bear this name.

4.3.2 Modern commercial systems

4.3.2.1 JIM

JIM (see *Figure 4.23*), as already described, is a true anthropomorphic system and its development in the early 1970s marked a resurgence of interest in the ADS and stimulated the development of other systems. To date, there have been more than 20 JIMs constructed. The JIM system is primarily designed for sea bed work or work on structures which have had specially de-

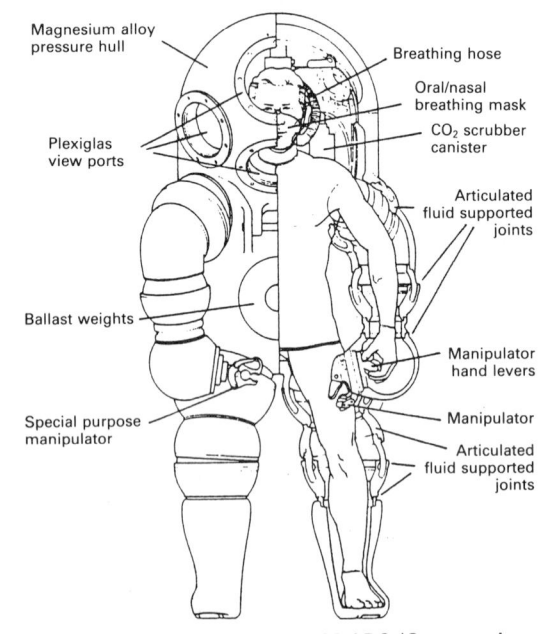

Figure 4.23 Diagram of early JIM ADS (Oceaneering International)

signed landing stages installed. For this mode of operation it is usually trimmed for approximately 300 N of negative buoyancy and in an emergency ballast weight and the umbilical can be jettisoned for free ascent.

The most recent system, type V, differs in several ways from the earlier designs (see *Figure 4.24*). The type V body is manufactured from carbon fibre reinforced plastic (CFRP), producing a very strong but light structure with potential for considerable extension of diving depth if required.

Figure 4.24 JIM (type V) (Slingsby Engineering)

A notable addition to the latest JIM system is the availability of an optional bolt on thruster pack giving the system mid-water capability. However, this concept is not entirely new, as previous systems had been similarly adapted in the past. The present system is rated for depths to 457 m (1485 ft) and according to the manufacturers is easily transportable by air.

4.3.2.2 WASP

The WASP ADS is a single submersible with mid-water work capabilities (see *Figure 4.25*). It possesses articu-

Figure 4.25 The WASP ADS (OSEL Group)

lated arms with interchangeable claws, but, unlike JIM, has no legs. Mobility is from 4 dc motor powered thrusters which are actuated by the operator's feet. The main body consists of an aluminium casting with a clear hemispherical polycarbonate dome at the top. The dome is protected by a renewable outer clear Macrilon™ cover. A built in fully automatic life support system provides 8 h endurance although this was increased to 12 h in the later series II vehicles.

In normal operation, the vehicle is attached to its umbilical which supplies power, provides for two way communication and serves as the lifting cable during deployment and retrieval.

In an emergency, the umbilical and all external equipment including the thrusters can be jettisoned providing some 450 N of buoyant force. In this situation, the operator reverts to using a backup breathing system which incorporates an oronasal mask and provides between 36 and 54 h endurance. Backup batteries housed in a pressure resistant container provide 20 min of full thrust.

Specifications
Builder – OSEL Group
Height – 2.13 m
Width (front) – 1.06 m
Width (rear) – 0.81 m
In air weight (with operator) – 470 kg

Maximum operating depth – 610 m
Operating team – 4–5 men

4.3.2.3 Hornet

The hornet is similar to functions' and based on, the WASP system. It possesses uprated thrust and control.

4.3.2.4 Mantis

Mantis (see *Figure 4.26*) is a free-swimming, tethered, ADS designed primarily for midwater operation. The operator lies in its horizontal pressure hull and operates hydraulically actuated manipulators.

The main component of the pressure hull is a filament wound glass reinforced plastic (GRP) cylinder of 0.66 m (2.1 ft) internal diameter and 1.12 m (3.7 ft) long, with a wall thickness of 50 mm (1.66 in). At one end is a hemispherical acrylic view port, at the other a cast hemispherical aluminium alloy entry hatch. On each side of the pressure hull is fitted a jettisonable variable buoyancy unit. The pods contain the buoyancy chamber, the air reservoir and the necessary control electronics and relays. Attached to each of these pods there are five thruster units enabling the system to be manoeuvred in any direction. The ten thrusters, two of which are higher powered than the others for forward and reverse motions, are controlled through a push button and slide stick unit mounted at the front of the pilot's couch.

Figure 4.26 Mantis ADS (OSEL Group)

Mounted at the front of the vehicle are two sea water hydraulic manipulators. The starboard manipulator has seven functions and incorporates a telescopic section, whilst the port manipulator has only six functions. Both manipulators have a high torque wrist rotate facility. Each manipulator is powered by its own sea water hydraulic power pack and solenoid manifold assembly.

The vehicle contains two independent life support systems which comprise oxygen bleed units and carbon dioxide scrubbers. Cabin pressure, temperature and oxygen content are constantly monitored and indications or warnings are presented to the pilot. Cameras and lights can be attached to the vehicle body for navigational purposes or to the port manipulator for close inspection. The vehicle can be fully instrumented with gyro compass, echo sounder, acoustic communications and hardwire communications.

The tether is a multicore cable enveloped in a polyethylene jacket and armoured with two layers of galvanized wire to give it a breaking strain of 20 tonnes. The conductors are co-axial cables, screened pairs and power cables insulated to suit the 660 VAC three phase power system adopted. In an emergency, the vehicle's thrusters can be powered by onboard battery packs for 4 h.

The most recent systems are rated to a depth of 700 m (2298 ft), and are delivered complete with their dedicated handling system.

Specifications of Mantis
Builder – OSEL Group
Length – 2.5 m
Width – 1.4 m
Weight in air – 1600 kg
Weight in water – Neutral (± 27 kg)
Payload – 200 kg
Life support duration – 86 h

4.3.2.5 Spider

Spider (see *Figure 4.27*) is a one-man tethered ADS, similar to WASP, possessing manually operated articulated arms but having hydraulically powered claws. The first two systems built in 1979 have now been superceded by vehicles 3 and 4 which possess a number of improvements. The vehicle is normally operated in a prone 45° position similar to that adopted by a working diver.

The lower part of the hull of the vehicle is cylindrical in shape (*Figure 4.28*). An upper backpack permanently attached to the pressure hull contains oxygen for life support, an electrical transformer and buoyancy pods. Around the lower half is fitted a jettisonable equipment frame, on which is mounted the hydraulic power pack, the hydraulic thruster units and the emergency batteries.

The upper part of the pressure hull is mounted at 45° to the lower section and possesses a hinged clear acrylic end cap through which the pilot enters the vehicle. This end cap provides good all round vision to the pilot. Also forming the top half of the pressure hull are the articulated arms which possess free movement at maximum depth.

Figure 4.27 The Spider ADS (Slingsby Engineering)

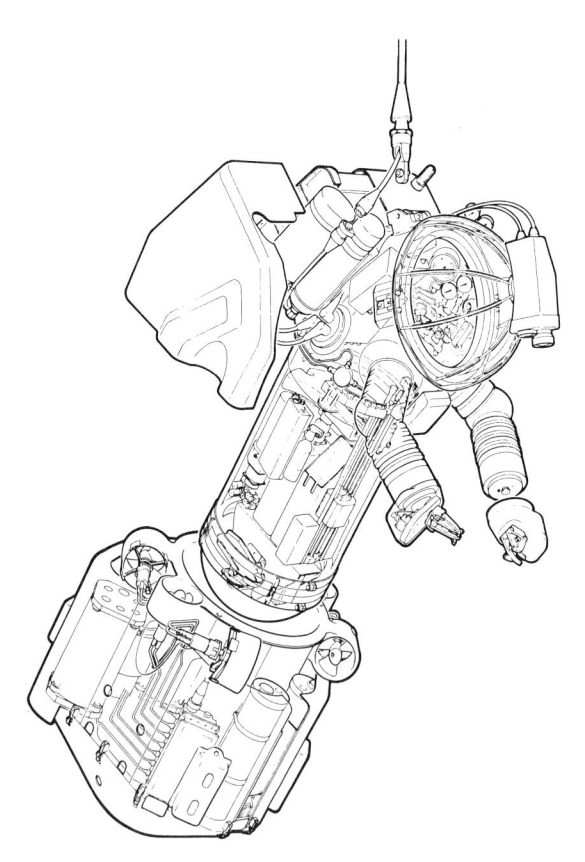

Figure 4.28 Diagrammatic view of the Spider ADS (Slingsby Engineering)

The six thrusters are operated by foot controls and the later Spider systems include an electric autopilot facility to ease the burden of the operator. The control system together with a variable trim system is said to provide for a highly manoeuvrable vehicle with a full mid-water capability.

The vehicle incorporates numerous safety features permitting 72 h of life support beyond the normal 8 h dive period. After jettisoning the umbilical, the on-board batteries allow a further 30 min of thruster operation if required.

Specifications
Builder – Slingsby Engineering
Dimensions – Height 2.2 m, width 1.5 × 1.1 m
Weight in air – 1100 kg
Depth capability – 610 m
Propulsion – 6 × 1 hp hydraulic thrusters to provide vectored thrust in any direction

4.3.3 Comparison of the ADS with other forms of intervention

Before being able to determine the safest and most efficient way of carrying out a particular underwater operation, it is necessary to fully understand the pros and cons of the various methods of underwater intervention that are available. Because the capabilities of the ADS lie between that of the ambient pressure diver and that of the ROV, it is both natural and necessary to make comparisons with both techniques.

The following points are among those which must be considered when making such an assessment:

1. The ADS operator normally experiences no increase in pressure and is therefore free of all the pressure related effects and problems that face an ambient pressure diver. This is a considerable advantage in terms of safety. Also, it is a considerable advantage economically, arising from the fact that neither the expensive equipment nor decompression time are required.
2. The ADS operator's view of the surroundings can be almost as good as that of a diver and certainly far superior than that relayed via even the best available

TV camera supported by an ROV. Vision, as in everday life, is an extremely important sense for the underwater operator.
3. Mobility and manoeuvrability of the ADS is somewhat restricted when compared with that of the diver. However, there are considerable differences between the various ADS available and much is dependent on the operator's ability.
4. In terms of manual dexterity, the ADS is generally thought to lie between that of a diver and that of an ROV supporting one or more manipulators. This is a fair assumption, although ideally the comparisons should be made for specific tasks. A good ADS operator can usually complete a most complex task but will take a much longer time than the diver. The presence of the ADS operator at the work site with the resulting direct vision and often direct mechanical control over the end effectors provides an advantage over the manipulator equipped ROV. There are many tasks for which one would not contemplate using even the most advanced of ROVs.
5. It is arguable whether it is the ADS operator or the diver who faces the higher personal risk actually during the operation. Both lives are totally dependent on mechanical and electronic systems and the competence of all the required supporting personnel. However, the ADS does give added mechanical protection to its operator and risk of a severed breathing gas carrying umbilical is not present. Certainly the ADS operator who becomes trapped has a greater chance of survival with perhaps 72 h of breathing gas and no problems of the extremely rapid heat loss resulting from a hyperbaric helium atmosphere.
6. Whilst in terms of surface support the saturation diver's requirements are great, that of a typical ADS is by no means insignificant (*see Figure 4.29*).

Launch and retrieval of an ADS is crucial and it is therefore necessary to have a launch system well capable of operating up to the desired sea state limit. This will mean a substantial hydraulic crane or A-frame. In any deep water ADS operation, it is also necessary ot have a backup system which must include a spare tether.
7. A major advantage of the ADS over the diver is its maximum depth capability. At present many ADSs are capable of operation in excess of the maximum divable depth. Thus, along with manned submersibles, they offer a solution to a category of tasks, often short duration, perhaps emergencies, in deep water that could not otherwise be economically accomplished.

4.4 Underwater vehicles

The term 'underwater vehicle', although self defining, is used to describe a multitude of systems, some of which bear little similarity. Before describing some of these systems in detail and discussing their history, it is instructive to categorize the various types of system in existence. *Figure 4.30* attempts to do this and show the relationship between the different types of vehicles.

The major distinction is usually made between the so-called manned vehicles and the remotely controlled systems. Manned vehicles are those which transport one or more persons within them, either in a standard atmospheric or hyperbaric environment. They are usually, but not always, under the control of their occupants. The remotely operated vehicle usually carries no persons and is under the control of an operator situated in a remote location, usually on a surface support vessel.

Untethered, self powered systems which operated in a near neutrally buoyant mode represent an important group within the manned vehicles. The larger systems of this type are generally known as submarines and have displacements up to 10 000 tonnes and such vessels find few if any applications outside the military sector. Their power requirements are extremely large and can only be met through the use of diesel engines and nuclear reactors. Military submarines will not be discussed in this chapter.

The smaller systems, designed and built to support scientific research or more recently the offshore oil and gas industry are generally referred to as manned submersibles. Although they share one important characteristic with the submarine, in that they incorporate a pressure resistant hull to provide a 1 atmosphere environment for their occupants, their power requirements are much smaller. Banks of lead–acid accumulators usually provide such vehicles with sufficient energy to undertake their mission.

A variation of the manned submersible is the diver lock-out submersible. This vehicle possesses a second separate pressure chamber which can be internally

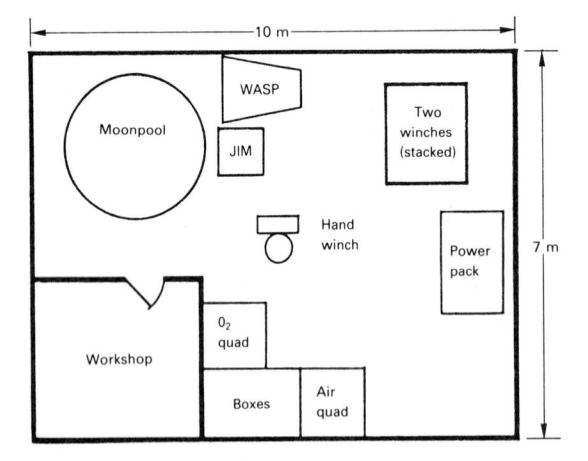

Figure 4.29 Typical deck layout for an ADS spread (note the use of a second system as backup)

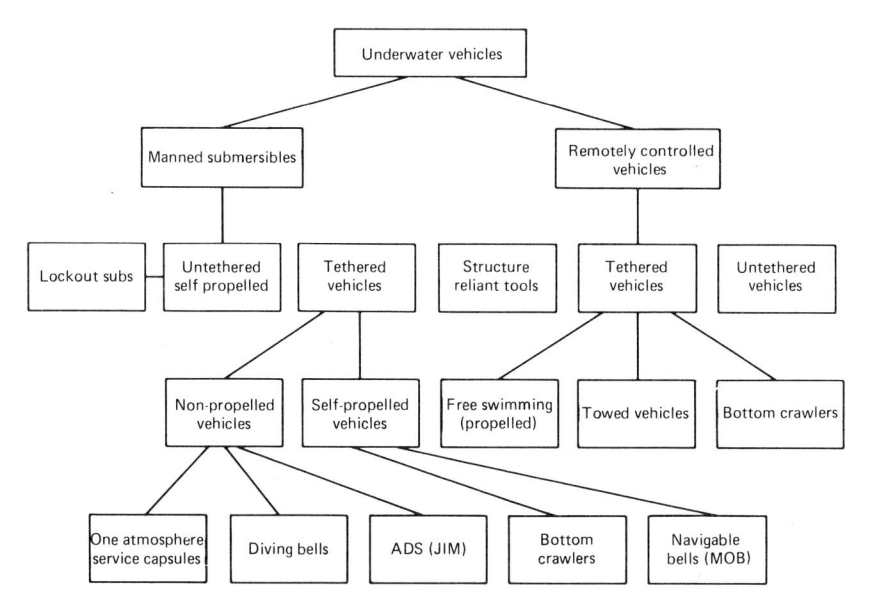

Figure 4.30 Types of underwater vehicle

pressurized to match the ambient pressure at the working depth. This chamber becomes effectively a diving bell, as it is used to transport divers to the work site. In some systems, the two pressure hulls are connected by a transfer tube to provide greater flexibility and safety.

The manned vehicles which are connected to the surface supported by a cable or umbilical, or both, can either possess their own means of propulsion or not. Those which do not possess a means of propulsion are invariably configured and operated in a negatively buoyant mode. The diving bell, previously discussed at length in Section 4.2, falls within this category. Systems which have a means of propulsion and are neutrally buoyant include some of the atmospheric diving systems which are designed to have a mid-water working capability.

The remotely controlled systems presently in existence can be classified in fewer groups. To control any system remotely requires the passage of information to and from the system. For systems operating under water, the simplest way of providing this communications link is through an electrical conductor or more recently via an optical fibre. Either method demands a physical link in the form of a tether or umbilical between system and controller. Although there are systems in existence, albeit mainly experimental in nature, that do not possess this physical link, the vast majority of these systems do have tethers.

These tethered vehicles generally possess their own means of propulsion, although there are some vehicles which are towed by the surface vessel. By far the largest group of tethered vehicles possess thrusters and operate near neutrally buoyant. Capable of being manoeuvred in three dimensions in the water column by a remote pilot, they are universally known as 'remotely operated vehicles' or ROVs. Development of this type of vehicle has been rapid over recent years as it has been seen by some as a means of replacing the diver in certain operations. Not all of these remotely controlled tethered systems are of the free swimming variety. Some are constrained to move and carry out tasks upon a structure, in which case they are known as 'structure reliant systems' or 'tools' and others move along the sea bed, *viz* the sea bed crawlers.

By now it should be apparent to the reader that there are endless possible configurations of underwater vehicle. However, by no means all of these configurations would be sensible and only a few, those mentioned above which find certain specific applications, have been developed.

4.4.1 Manned submersibles

4.4.1.1 History of manned submersibles

The history of manned submersibles is relatively short. In 1948, Auguste Piccard designed and built the first manned vehicle of this type called FNRS2. Tested unmanned, under remote control, it achieved a depth of 1220 m (4006 ft) before suffering irreparable damage. Piccard then built a modified vessel, FNRS3, which in August 1953 achieved a depth of 1080 m (3546 ft). The submersible was by no means small although it could only accommodate a crew of two persons in its 1.94 m (6.3 ft) inside diameter, 90–150 mm (3–6 inches) thick spherical steel pressure hull. It was over 18 m (59 ft) in length and had a displacement of some 150 tonnes. This vessel was named *Trieste*, from the city in which it was built. One month later it achieved a depth of 3150 m (10 342 ft). The vessel was used in several oceanographic missions in the Mediterranean until 1958, when it was sold to the US Navy.

Following the acquisition of the *Trieste* by the US Navy, a notable operation took place when Jacques Piccard (the son of the designer) and Lieutenant Don Walsh descended into the Challenger Deep in the Marianas Trench to a record depth of 10 910 m (35 821 ft) in January 1960.

In 1965–1966, the *Trieste* underwent extensive modifications which included a new buoyancy unit, a new propulsion system and a new pressure hull. Renamed the *Trieste II*, it was rerated for 6060 m (19 897 ft) and was to take part in the search for the sunken submarine *Scorpion* in 3000 m (9850 ft) off the Azores in 1969.

4.4.1.2 Manned submersible – construction

The manned submersible consists of six major components:

1. The pressure hull is the enclosure which accommodates the crew members and certain items of equipment in a 1 atmosphere environment. It must possess sufficient strength to resist the external hydrostatic pressure that will be encountered and there should be no leakage of water into it. It must obviously also incorporate an entry hatch and preferably it should possess one or more view ports to allow its crew to see outside.
2. Ascent and descent of this nearly neutrally buoyant vehicle is almost always by a variable buoyancy system.
3. The propulsion system usually consists of a number of electrically operated thrusters under the control of the pilot.
4. The power source is almost always a bank of batteries or accumulators housed within their own pressure resistance enclosures.
5. A life support system is required to maintain an atmosphere similar to that encountered at the surface, i.e. in terms of oxygen level, carbon dioxide level, temperature, humidity, pressure and contamination.
6. Instrumentation can be associated with vehicle status, navigation or execution of tasks.

4.4.1.2.1 Pressure hulls The strength of the hull of a submersible will usually determine its maximum operational depth. Strength or, more specifically, resistance to distortion or collapse under external pressure depends on shape, dimensions and the material from which it has been constructed. For a given wall thickness, material and overall diameter, the shape which is most resistant to the effects of external pressure is, of course, the sphere. The application of classical small deflection theory for the elastic buckling of a complete sphere results in the following equation for collapse pressure:

$$P_c = \frac{2E(h/R)^2}{\sqrt{3}\,(1-v^2)} \qquad (4.7)$$

where h is the wall thickness, R is the diameter of sphere to mid-wall positions, E is the modulus of elasticity and v is Poisson's ratio.

However, in practice it is found that spheres often collapse well before this pressure is reached and the following empirical relation provides a better guideline for materials that have a Poisson's ratio = 0.3:

$$P_c = 0.84E \left(\frac{h}{R}\right)^2 \qquad (4.8)$$

Because of the inherent strength of the sphere, most deep water submersibles adopt this shape for their pressure hull.

However, it is not the most convenient of shapes to occupy and, for this reason, cylindrical sections are frequently incorporated into the design. The strength of a cylinder is somewhat less than that of a sphere and its collapse pressure is given by:

$$P_c = \frac{E}{4(h-v^2)} \frac{h^3}{R^3} \qquad (4.9)$$

The collapse pressure of a cylindrical hull with hemispherical ends will lie between the two expressions given. In practice, one must consider the inclusion of entry hatches and view ports and how the method of fabrication will affect overall strength.

Submersibles possess outward opening hatches, so that there is no chance of them opening at depth. They are usually sited at the top of the hull, often on a conning tower extension, so that they can be opened if necessary while the vehicle is at the surface. Sealing of the hatch is normally through use of at least two O-rings. It is essential to maintain all mating surfaces in a scratch free state and to keep O-rings clean and lightly greased, if leakage of water into the pressure hull is to be avoided.

4.4.1.2.2 Variable buoyancy systems The buoyant force experienced by a system can be altered by changing the volume of water it displaces. This can be achieved in a number of ways, but requires the use of some form of chamber. The chamber may be flexible in the form of a leakproof bag or of rigid construction. It may also either be open at the base or completely closed. To increase the overall water displacement of the vehicle, water must be dispelled from the chamber.

When flexible lifting bags are used, water must be displaced from within them or they must displace water, with air at a pressure equal to the ambient pressure of the surrounding water. A supply of air of sufficient volume at a pressure greater than that of ambient pressure is therefore required. Normally such a supply would be carried in specially designed and constructed high pressure gas containers. The volume of air available limits the achievable buoyant force and the pressure at which it is stored limits the depth of operation. As a flexible lift bag ascends, the air inside will expand as the external pressure, and hence its pressure, decreases. In an open bottomed or parachute type lifting bag, expanding air will force the water level downwards, thus increasing buoyancy, until it escapes from the rim of the bag. In a completely closed lifting bag there must be a pressure relief valve installed, as the material will not be of sufficient strength to contain the high internal pressure when at the surface.

A rigid buoyancy chamber can be operated with an open bottom and like its flexible counterpart will not experience high pressure differentials, providing it is not too tall and therefore does not need to be made excessively strong.

When a closed buoyancy chamber is used, there is now a choice of methods for displacing the water: it can either be blown out by air as before or water can be pumped out using a high pressure pump. The water is then replaced by air at low pressure. In the former the strength of chamber required will depend on whether or not a high differential pressure is allowed to build up. In the latter the buoyancy chamber must be designed to tolerate the ambient external water pressure encountered.

Whilst systems that rely on air for displacing water are fairly simple, they are limited, as mentioned before, by the gas storage pressure and the volume of gas available. Also the gas must be constantly renewed, as when it has been used it is effectively lost. Using a high pressure pump for displacing the water places greater demands on the technology, but minimizes the consumption of air.

4.4.1.3 Operation of manned submersibles

The size and weight of manned submersibles makes their operation impossible from anything other than a dedicated support vessel fitted with a suitable deployment system. The vehicles are nearly always launched over the stern of the vessel by using a high lift capacity swinging A-frame (see *Figure 4.31*). Over the side

Figure 4.31 Launch of *Taurus* manned submersible by stern mounted A frame (courtesy of Stena Offshore, formerly BUE)

deployment of such heavy systems could cause an unacceptable amount of roll on the size of vessel likely to be used in these operations.

On launching, the submersible is attached at its main lift point by as short a line as possible to the centre of the A-frame. The A-frame is raised hydraulically and the submersible is lifted and swung clear of the stern of the vessel. Various lines are often attached to the submersible to prevent it from swinging uncontrollably. Once in the water, all lines are removed and the vessel steams ahead slowly to leave the submersible clear. On orders from the operations controller via the radio link and, provided all is well, on board the submersible, the pilot will make it dive away from the surface, so that it is no longer affected by the wave motion.

Once away from the surface, all navigation of the submersible must be done 'on instruments only', until features are visible through the view ports. In practice, when visibility is low, the pilot must rely extensively on sonar systems, acoustic tracking systems and depth sensors to find the work site. Only when very close will vision be of any use.

At the work site the crew will have a predetermined length of time during which to undertake the task in hand. It may involve only the use of an external television or still camera or it could require the skilled operation of a manipulator. During this period, the atmosphere within the pressure hull will be constantly monitored and remedial action taken manually if necessary, depending on the type of submersible in use. If provision is made, battery charge level will also be closely watched.

On completion of the task or after the predetermined dive duration, the pilot will signal the surface support if possible and ascend the submersible. It is usual for the support vessel to move a little off station after launch and it will therefore be necessary for the submersible when at the surface to make its way towards the stern of the vessel. Such a manoeuvre is generally accomplished through use of the view ports in the conning tower.

Once in position behind the vessel, the A-frame is swung over and the lift line let out. Attachment of the hook to the submersible is usually undertaken by a team of swimmers operating from a rubber boat. The submersible is then lifted from the water and lowered onto the deck of the vessel where it is then secured. Some submersible operators prefer to launch and retrieve when the vessel is slowly under way, as this has been found to improve stability.

The launch and retrieval of manned submersibles, especially those which include a diver lockout facility in which divers will be pressurized, must be undertaken with the utmost care and strictly according to any procedures laid down by the operators. Despite the obvious dangers involved, the safety record in this field, albeit small and diminishing, has been exemplary.

4.4.1.4 Commercial manned submersibles

4.4.1.4.1 LR2 LR2 is a glass reinforced plastic (GRP) submersible with a depth rating of 366 m (1190 ft) and accommodation for two crew members. The GRP construction significantly increases the payload of the vehicle, therefore allowing a greater battery capacity. Besides being non-corroding, the GRP hull provides superior insulation to the crew from the cold sea water. An acrylic front dome some 1 m (3 ft) in diameter provides the crew with an excellent view.

The propulsion system of the vehicle combines a single 7.3 kW main thruster at the rear driven by a single internal motor and four 3.5 kW thrusters, each independently operated. Together with the vehicle's dual ballast system, the thruster capability makes LR2 exceptionally manoeuvrable and well suited for underwater survey tasks. Its ability to change depth rapidly and accurately makes it particularly suited to structural surveys, where the vehicle, free of any tether, can safely enter the innermost parts of the structure.

Externally mounted sensors for acquiring the survey information include pipetracker, cathodic protection probe, echo sounder, depth sensor, television cameras, sea bed profiler, sonar array and acoustic navigation transponders. Inside the vehicle there is a gyrocompass, a video recorder and a data logger for storing the survey information. The vehicle is also fitted with a six degree of freedom manipulator which has a lift capacity of some 90 kg (200 lb) and is jettisonable in an emergency.

Oxygen content, carbon dioxide content, temperature, humidity, pressure and gas contaminants are constantly monitored within the accommodation. Carbon dioxide is removed in the usual way by a scrubber containing either lithium hydroxide or Soda Sorb™. Oxygen, stored in external cylinders, is bled into the atmosphere as required. Temperature and humidity are controlled by an air conditioning system of 2 kW capacity.

Specifications of LR2
Builder –Slingsby Engineering
Length – 7.3 m
Width – 3.0 m
Height – 2.6 m
Weight – 14 tonnes
Payload – 455 kg
Crew – 2
Maximum operating depth – 366 m
Collapse depth – 550 m
Maximum speed – 2 knots
Life support – 7 days per crew member
Pressure hull:
   Material – GRP
   Shape – cylindrical
   Dimensions – 1.3 m inside diameter

(Command module) – 50 mm wall thickness, 2.8 m long
Power source:
  Position – pods either side of pressure hull
  Battery type – 120 V Oldham OTH6 providing 438Ah
  24 V Oldham OTH8 providing 584Ah
  Total power – 67 kW
  Emergency power – Drycell, 24 V, 36Ah
Buoyancy and trim control:
  4 × GRP ballast tanks providing 7000 N lift with high pressure air available
  Two internal trim tanks providing ± 8° on pitch control
External lighting:
  Five front mounted quartz halogen lamps
  One rear mounted quartz halogen lamp
Communications:
  Underwater telephone – Subcom 2005 20B (10 or 27 kHz)
  VHF radio, range 3 miles (channels 16/33/13)
  UHF radio, range 2 miles (44.6 MHz)

4.4.1.4.2 LR3, LR5 Similarly constructed GRP to the LR2, the LR3 (see *Figure 4.32*) is a larger vehicle designed for carrying three crew members to a maximum depth of 457 m (1500 ft). It is also equipped extensively in terms of navigational, survey and communications equipment.

Specifications of LR3
Builder – Slingsby Engineering
Length – 8.75 m
Width – 3.2 m
Height – 2.75 m
Weight – 18 tonnes
Payload – 680 kg
Crew – 3 or 4
Maximum operating depth – 457 m
Maximum speed – 2 knots
Life support – 7 days per crew member
Power source – Lead acid cells providing total
  = 87.5 kWh
  Emergency dry cell battery 24 V, 36 Ah
Propulsion – Two internal 3.65 kW motors driving main thruster
  Four additional thrusters as LR2

4.4.1.4.3 PC1601 PC1601 is a deep water manned submersible capable of diving to 914 m (3000 ft). Its pressure hull consists of three 1.83 m (6 ft) inside diameter high quality carbon steel spheres (see *Figure 4.33*). The rear sphere houses survey equipment and the two forward spheres can comfortably accommodate up to four crew members. A front acrylic port 0.75 m (2.46 ft) diameter and a conning tower with seven viewports offers exceptional viewing capability. Externally, the structure of the vehicle has been streamlined by use of a fairing to reduce drag and hence optimize power consumption. Two battery pods of 0.5 m (1.64 ft) diameter and over 5 m (16.4 ft) in length form the skids on which the vehicle can rest either on the sea bed or at the surface.

Figure 4.32 LR5 manned submersible adapted for rescue from submarines (courtesy of Stena Offshore, formerly BUE)

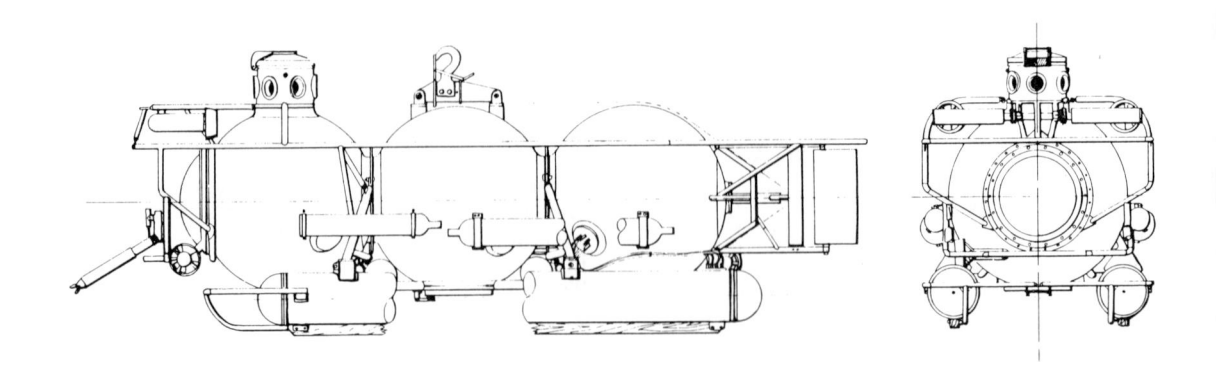

Figure 4.33 PC 1601 Manned submersible (Stena Offshore, formerly BUE)

Main propulsion is from one rear mounted thruster. The use of the rudder and four smaller thrusters gives the system full manoeuvrability. Vertical motion is accomplished through variable buoyancy. External tanks, when blown with air stored in high pressure cylinders, provide some 3400 N of lift. There is also a variable trim system that can be operated by the pilot.

The 1 atmosphere environment is monitored by carbon dioxide and oxygen analysers. The carbon dioxide is removed by an electrically powered scrubber unit and oxygen is replenished from externally mounted cylinders. The system provides for a full seven days life support for each crew member.

The vehicle is well equipped with instrumentation which includes gyrocompass, depth sounder, sector scanning sonar, depth transducers and transponders for acoustic navigation. Also included is an acoustic communications system and VHF radio for surface use. Other equipment includes ample quartz halogen lamps and video camera mounted externally and the necessary video recorder housed within. There are also two manipulators of 34 kg (75 lb) capacity mounted at the front of the vehicle which can be remotely operated from within.

Specifications of PC1601
Builder – Perry Oceanographics Inc.
Length – 7.62 m
Width – 2.44 m
Height – 2.89 m
Weight – 15 tonnes
Payload – 278 kg
Crew – Four persons maximum
Maximum operating depth – 914 m
Maximum speed – 2 knots
Normal speed – 1 knot
Duration at normal speed – 6–8 h
Power source – 120 Vdc, 52 kWh
Main thruster – 7.5 kW
Directional thrusters – 4 × 310 N thrust

4.4.1.4.4 ALVIN ALVIN was designed and built specifically for scientific operations by the Woods Hole Oceanographic Institute, Massachusetts, USA. It was originally rated to 1829 m (6005 ft). The vessel is driven by three propellers controlled from inside the vehicle. There is a small propeller on each side primarily for lift, but these can be rotated to provide thrust in horizontal directions also. The main propeller at the rear can also be turned in a similar fashion to an outboard motor to aid steering. All the propellers are fully reversible and hydraulically driven, making the vehicle extremely manoeuvrable.

A variable ballast system, as usual in such vehicles, compensates for differences in the weight of personnel and equipment, for changes in the density of water and for the weight of any samples taken from the sea bed. The system employs pressure resistant titanium spheres and a high pressure salt water pump. The overall buoyancy is altered by pumping sea water in or out of the spheres. The vessel also has a supply of high pressure air for blowing the main ballast tanks on the surface to provide added stability.

In 1973, a new pressure hull of 100 mm (4 inches) thick titanium was fitted, extending its depth capability to 3506 m (11 511 ft).

Specifications of ALVIN
Length – 7.0 m
Width – 2.4 m
Height – 3.0 m
Weight – 15 tonnes
Crew – Three men
Speed – $1\frac{1}{2}$ knots
Endurance – 72 h

4.4.1.4.5 Aluminaut Aluminaut was built in 1963–1964 as a multipurpose vehicle for exploration, research, salvage and engineering development. With its pressure hull constructed from aluminium, it was unique, requiring the largest aluminium ingots and forgings ever made at that time. Some 15.5 m (51 ft) in length and with a pressure hull of 16.5 cm (7 inches) thickness, the vehicle weighs 81 tonnes. Four 115 V, 400 A/h silver–zinc alkaline batteries provide for a range of 73 km (46 miles) and ample power for onboard systems and instrumentation.

The pressure hull gives the vehicle inherent positive buoyancy. The negative buoyancy characteristic for diving is provided by steel shot ballast contained in two saddle tanks. The shot ballast is retained by a magnetic field established in the 5.1 cm (2 inches) diameter shot discharge valve orifice. In the absence of the magnetic field, shot is discharged at a rate of 8.15 kg (17 lb)/s. In the event of failure of this system at depth there is also provision for releasing a ballast weight of some 1950 kg (4290 lb) which exceeds the maximum weight of shot which can be carried.

Aluminaut also possesses a pneumatically operated water ballast tank system which is used mainly to provide excess, stabilising, buoyancy at the surface. In 1975, *Aluminaut* underwent a major refit during which it was converted for diver lockout operations.

Original specifications of Aluminaut
Length – 15.5 m
Width – 4.9 m
Height – 5.0 m
Weight – 81 tonnes
Payload – 2680 kg
Crew – Two–three
Observers – Three–four
Maximum design depth – 4550 m
Operating depth – 3340 m
Submerged speed – 3 knots
Life support – 72 h
Thrusters – 3 × 3.65 kW dc motor driven
Equipment – Sonar, communications, television camera
Surface support – Mother ship – M/V Privateer, 41 m ex-minesweeper

4.4.1.5 The tourist submersible

Although the use of manned submersibles in support of the offshore oil and gas industry has decreased rapidly in recent years, they are now finding a new application in tourism. However, this idea is by no means a new one. At the 1964 Swiss National Exhibition, a tourist submersible, the *Auguste Piccard*, conducted 1112 dives to the bottom of Lake Geneva carrying some 32 000 paying tourists. The vehicle was some 28 m (91 ft) in length, could carry 40 passengers and dived to a depth of some 300 m (985 ft).

The commercial implications of the use of submersibles in tourism did, however, take some time to be recognized. It was not until the early 1980s that Research Submersibles Ltd acquired one of the Perry built vehicles and offered sightseeing tours off the Cayman Islands at $US 200 per dive. By 1986, the company had acquired two more vehicles which had been abandoned by the oil and gas industry and had completed some 2000 dives without incident.

In 1987, a Canadian company, Sub Aquatics Development Corporation, began the design of a 28 passenger vehicle called *Atlantis I*. Completed in 1985, it also took up service around the Cayman Islands. The following year, its sister vehicle, *Atlantis II*, took up service in Barbados.

The *Atlantis* vehicles carry 28 passengers and 2 crew members and weigh 49 tonnes. Operation of the vehicles is limited to less than 50 m (164 ft) depth. Because the vessels are operated in open water, away from any structures or other hazards and generally follow a well known route in relatively shallow water, there is no reason to suspect that there is any undue risk. All such vessels should, however, still be certified in the usual way.

Since 1986, there has been much activity in this area. A third *Atlantis* was built for operation off the US Virgin Islands and a fourth scheduled for Hawaii.

More recently, a new Canadian company, Hyco, has been established specifically for the purpose of designing and building tourist submersibles. TS-1000 *Gemini*, consisting of three acrylic spheres and capable of carrying 6 persons will dive to a depth of 1000 m (3283 ft). *Aries* will have a cylindrically shaped hull and carry up to 46 passengers to a depth of 76 m (249.5 ft).

Two Finnish and one UK company, Fluid Engineering, have also entered this market with new designs of vehicle, viz *Manea I*, *RR-Sub-49* and *Looking Glass*.

4.4.2 Remotely operated vehicles (ROVs)

In view of the hostile nature of the underwater environment, it is not surprising that systems have been developed to undertake tasks whilst leaving the operator at the surface. American patents filed as early as 1922 indicate that, even in those early days, much thought was being devoted to such a system.

Modern ROVs tend to be designed and built to accomplish a specific task or range of tasks and because of this it is possible to identify two types:

1. ROVs designed primarily for survey and inspection purposes. Such vehicles are usually comparatively small and may be of solid construction, e.g. RCV 225.
2. ROVs designed to undertake work or manipulative tasks. Such vehicles tend to be fairly large and are built around an open framework, e.g. *Scorpio*.

The ability of a remotely operated system to undertake tasks depends on the two-way transmission of information between the work site and the operator. The remote vehicle must support sensors capable of detecting information concerning the task in hand and be capable of sending that information without distortion to the operator in a manner which can be understood. To effect the task, the operator will send control information to the vehicle. In this simple description, the operator has in fact become part of a closed loop control system.

In water, as in surface based operations, vision is of major importance. It is therefore necessary that a remote viewing device forms part of the control loop which in practice means that the remotely operated vehicle should support a television camera. The television camera is therefore fundamental to ROVs and it forms part of every commercially available system. Not surprisingly, therefore, it was the availability of compact and reasonably rugged television cameras which marked the beginning of the ROV as we know it today.

4.4.2.1 Underwater television

The first widely reported use of a television camera under water was during the test of the atomic bomb at Bikini Island in 1947. The camera operating in a depth of 60 m (195 ft) was used to study movements of the sea bed during the blast.

In 1949, the Scottish Marine Biological Association used a television camera to observe fish in the sea. 1951 was an important time as it marked the introduction of the Vidicon™ type of television camera by RCA. Although television cameras had been available for many years, this was the first which could be described as reasonably compact. Together with its less demanding power and control requirements, it was therefore ideally suited for packaging into the necessary watertight pressure resistant housing.

The first commercially available systems were based on a camera built by Hydroproducts Inc. for use on the record breaking dive of the *Trieste*. Since then there have been many developments in television cameras and there are now numerous types available. Those which offer any specific advantage have almost certainly been adapted for underwater use.

4.4.2.2 The ROV – construction

The ROV is best considered as a system which comprises several components: buoyancy module, to give the vehicle the desired buoyancy, usually close to neutral; propulsion system, to give the vehicle manoeuvrability; instrumentation, for (a) vehicle status indication and (b) navigational purposes; control station, at which the pilot operates the vehicle; tether or umbilical, for supplying power to the vehicle and transmitting communications in each direction; vehicle and tether handling system, for use in vehicle deployment and retrieval; and tools and instruments, carried by the vehicle to enable tasks to be undertaken.

4.4.2.2.1 Buoyancy Net buoyancy is provided by any part of an ROV which displaces a volume of water of greater weight than itself. This will be achieved if the overall density of the item is less than that of water. Solid metal components cannot therefore provide net buoyancy, but metal walled tanks containing gas can provide a great amount of buoyancy

Most essential components of an ROV will be negatively buoyant although any pressure resistant housings for electronics or instrumentation, depending on their content, may well be buoyant. However, the essential components of an ROV are likely to be strongly negatively buoyant overall and the addition of artificial buoyancy is almost always required to bring the ROV to the desirable near neutrally buoyant state.

Unlike the manned submersibles which generally rely on variable buoyancy to provide the forces necessary for descent and ascent, the modern ROV almost always possesses fixed buoyancy with suitable thrusters to provide the vertical motion.

Overall buoyancy is not, however, the only important factor to be considered; the stability of the vehicle within the water is also of major importance. A submerged vehicle is stable only when its centre of gravity is vertically below its centre of buoyancy (see *Figure 4.34*). Whilst centre of gravity of an object is well understood, the term centre of buoyancy perhaps deserves some explanation. It is the centre of volume of an object or, put more simply, the centre of gravity of the displaced water if it could be made to retain the same shape as the object displacing it.

If forces disturb the attitude of a stable vehicle, a righting couple is set up by the buoyant and gravitational forces which tends to restore the vehicle to its original attitude. The distance between the centres of gravity and buoyancy, commonly referred to as 'metacentric height', is an index of stability; the greater this distance, the greater the couple set up in the disturbed state, and hence the greater the stability of the vehicle.

In any underwater vehicle that supports sensors, stability in the water is of major importance. For this reason, it is necessary to position any artificial buoyancy towards the top of the vehicle and keep heavy, non-buoyant components towards the bottom of the vehicle.

Artificial buoyancy can be provided in a number of ways: by purpose built closed metal tanks; by the use of spherical plastic floats of the type used in the fishing industry; by the use of special low density foams. Whichever means is chosen, the effects of increasing pressure with depth must be taken into account. A closed tank which collapses under pressure will undergo a simultaneous and unacceptable reduction in buoyancy. The same is also true of the plastic sphere which implodes and the low density foam which becomes compressed. It is apparent therefore that any buoyant system can lose its buoyancy below a certain depth. The operational depth of many ROVs is in fact determined by their buoyancy modules.

Table 4.6 compares some different systems.

Table 4.6 Typical efficiencies of common buoyant materials

Material of system	Efficiency
Syntactic foam	0.4
Aluminium cylinders	0.5
Plastic spheres	3.5

Although syntactic foam is expensive and much less efficient than, say, fishing floats, it does possess certain advantages. It can be moulded and machined to any shape thus making maximum use of available space. It can also be used at depths at which fishing floats would not survive. However, the penalty to be paid is that the provision of buoyancy through this low efficiency syntactic foam means a large increase to the weight in air of the vehicle leading to increased handling problems.

In practice, syntactic foam buoyancy units are usually covered by an outer shell of fibreglass or similar smooth material to provide protection against abrasion, good hydrodynamic characteristics and increased strength. Buoyancy units tend to be coloured yellow or bright orange so that vehicles can be easily distinguished on the surface of the sea.

4.4.2.2.2 The umbilical (tether) To an ROV the umbilical is its lifeline. It has several functions: to transmit power to the vehicle for its thrusters, instrumentation and tools; to support the transmission of information both to the vehicle, e.g. control commands and from the vehicle to the surface, e.g. television picture of task in progress; and as a physical link to aid deployment and retrieval of the ROV.

Umbilicals tend to be custom designed and manufactured to meet the requirements of a specific ROV system. In general they will comprise of at least some, if not all, of the following elements (see *Figure 4.35*):

1. Power conductors – usually of copper but today occasionally of new, lighter, highly conductive alloys in suitable numbers and of cross-section to match the power requirements of the vehicle. All conductors must, of course, be individually insulated to suit the working voltage of the system.
2. Co-axial cables – mainly for the transmission of television signals.
3. Twisted pairs – of varying types for signal transmission.
4. Optical fibres – recently introduced by some manufacturers to increase information transmission capacity and reduce interference from power sources.
5. Strain members – to provide mechanical strength especially when the umbilical is used for lifting the vehicle out of the water.

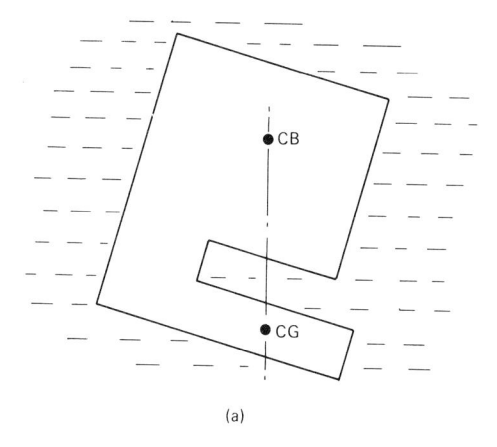

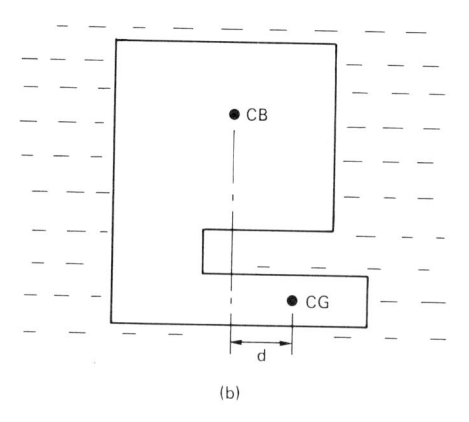

Figure 4.34 Stability in buoyancy (a) body is stable because the centre of gravity, *CG*, lies directly below the centre of buoyancy, *CB* (b) Unstable body, a righting couple proportional to the lateral separation, *d*, of *CG* and *CB* exists

Whilst some early ROVs used closed metal tanks and fishing floats to provide buoyancy, today almost every vehicle in operation uses a material specially developed for deep water buoyancy called a 'syntactic foam'. This is manufactured by blending microspheres with rigid resin systems. The resultant product is impervious to water and its mechanical properties can be tailored to suit specific depth requirements. Foams suitable for deep water, for example 6000 m (19 700 ft), need to be extremely strong and resistant to compression. Such a requirement in practice means that foams developed for deep water have to be denser and therefore give less buoyancy than foams developed for shallow water use. A useful figure of merit for buoyancy materials or systems may be defined in terms of an efficiency as follows:

$$\text{Efficiency of buoyancy system} = \frac{\text{Buoyant force provided}}{\text{Its weight in air}} \quad (4.10)$$

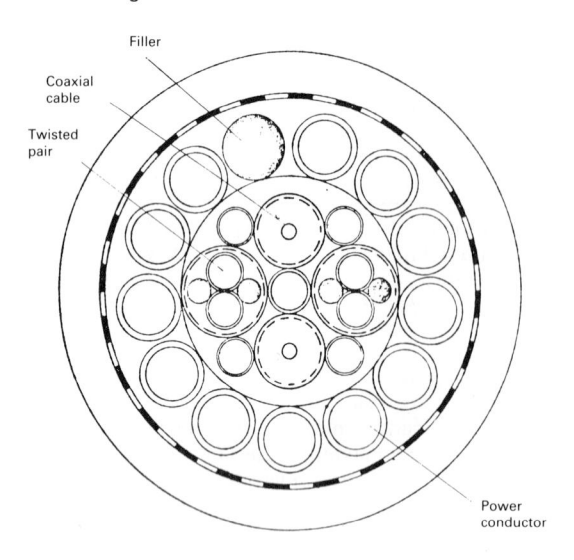

Figure 4.35 Cross-section of typical multiconductor ROV umbilical (Angus 002)

Depending on the various requirements, diameters of umbilicals can vary from one to several centimetres. It is now current practice to manufacture umbilicals that are close to neutral buoyancy, by paying close attention to the quantities of non-metallic compounds such as the protective casings of polyurethane or polyethelene that are incorporated. It was once usual to have to add buoyancy at certain intervals along the umbilical to make it neutrally buoyant, but such a practice is both inconvenient and increases the likelihood of umbilical entanglement.

It is also common practice to fill any voids that might occur between conductors with a flexible compound. In the event of damage occurring to the external protective coatings, water is prevented from travelling the entire length of the umbilical. This technique is referred to as 'water blocking'.

ROV umbilicals often experience rough handling in terms of abrasive forces and bending beyond their recommended minimum radius of curvature. It is not surprising therefore that the umbilical is regarded as the weakest link in the ROV system and from time to time will need to be replaced at great cost to the operator.

4.4.2.2.3. Propulsion systems Propulsion on the majority of free swimming underwater vehicles is effected by the use of motor driven propellers or thruster units. There is a wide variation in the designs of such propellers and their driving motors. Motors used in ROV thrusters tend to be one of three types: low voltage, dc motors with nowadays often solid state brushless commutation – such motors are typically used on the smaller, observation type of vehicle, e.g. RCV 225; high voltage, brushless ac induction type motors which have been used on the larger ROVs, and hydraulic motors which require a suitable hydraulic power pack on board the ROV. The power pack is energized usually with a high power, high voltage, ac electric motor. Hydraulically operated thrusters are generally only found on the larger ROVs. Whilst the choice of low voltage, brushless dc electric motors is the only sensible choice for small ROVs, the choice of motor on the larger ROV is not so clear cut. The various advantages and disadvantages of both electric and hydraulic motors are set out in *Table 4.7*.

At present, the vast majority of large ROVs are propelled by hydraulically driven thrusters, although it is likely that this situation has arisen as a result of a few

Table 4.7 Comparison of hydraulically driven thrusters and ac motor driven thrusters on large ROVs

ac electric motor	*Hydraulic motors*
Advantages	Advantages
Simple, cheap, easy to put together	Built from standard components
Good range of standard motors readily available	Little required for pressure compensation
	High powers available from physically small motor units
	Electrically quiet
	Present day systems given continuously variable power outputs
Disadvantages	Disadvantages
Require to be housed, sealed and usually pressure compensated	System more expensive and requires more maintenance
Variable drive through the use of variable autotransformers is cumbersome and often does not give fine enough control	Much pipework required on the ROV
Variable drive through the use of static variable frequency inverters can produce unacceptable electrical interference	Usually produces more acoustic noise

bad experiences with electric motors in the early days of development of the large ROV.

The theory of propeller design is complex and beyond the scope of this chapter. Ideally, the number of blades on the propeller, their area, their shape and their pitch are carefully chosen for optimum thrust at the available power. It is important to match the rotational velocity of the propeller with the driving motor whilst ensuring it is kept low enough to avoid efficiency reducing cavitation within the water.

In order to maximize efficiency, it is desirable to mount the propeller in a hydrodynamically shaped duct or 'Kort nozzle', as it is known (see *Figure 4.36*). Also, care must be taken to avoid restricting the flow of water past the propeller. When in-line motors are used, it is therefore necessary to keep their diameters to a minimum and extend the drive shaft so that motor and propeller are well separated.

the inlet of sea water. Such a technique is known as pressure compensation and it can be achieved through the use of oil or air.

During oil compensation all voids are filled with an inert oil and connected to a reservoir of oil which is free to experience the ambient pressure. In this way, the internal pressure is kept equal to the external pressure. The oil reservoirs must possess some kind of flexible barrier to the water to allow a flow of oil from the reservoir into the component in the event of an air pocket remaining in that component. *Figure 4.37* shows a typical arrangement where a rolling diaphragm is used for this purpose. To allow for air pockets and temperature changes, oil compensation reservoirs should have a volume of at least 10% of the volume to be compensated.

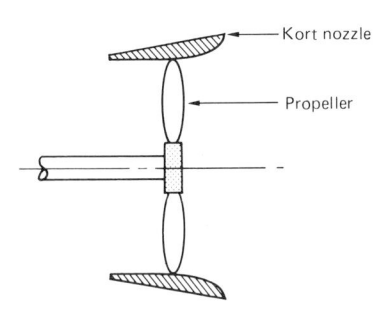

Figure 4.36 Cross-section showing propeller in a Kort nozzle

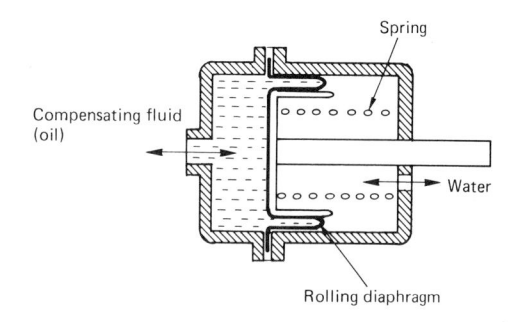

Figure 4.37 Typical oil pressure compensation reservoir; note rolling diaphragm which ensures leak free operation and spring providing constant excess pressure over ambient pressure

4.4.2.2.4 Resistance to pressure ROVs will be subject to high external pressures in operation and it is important that their various components are designed in such a way that they are not damaged or adversely affected. This can be achieved by the use of pressure resistant housings or by applying pressure compensation.

Instrumentation and electronic systems which must be kept dry and operated at atmospheric pressure are usually housed within a specially constructed pressure resistant housing. Such vessels will either be constructed from aluminium alloy and anodized or from GRP. Usually cylindrical, they will normally have removable end caps sealed by one or two O-rings. Bulkhead type connectors enable electrical connections to be made to equipment within the housing, while retaining the integrity of the vessel.

For extreme depth capability, stainless steel housings and, more recently, titanium housings are employed.

There are often hollow components and voids within the ROV structure which cannot be housed within a pressure resistant container. In such cases one must prevent the build up of differential pressures by ensuring that the pressure within the component or void is kept equal to the external pressure, whilst preventing

For small components, simply filling with oil and the connection of a short length of flexible pipe filled with oil is all that is necessary to provide adequate pressure compensation.

It is common practice to pressure compensate electrical distribution boxes and electrical motors with oil. Oil, if properly chosen, will provide good insulation whilst not affecting any materials with which it comes into contract. In mofors it will provide a path for the conductivity of heat and thus allow them to be run at higher power levels than at which they are rated. However, some energy will be lost in overcoming the extra frictional and viscous forces it creates. In dc motors which use brushes, the presence of oil may cause them to lift off the commutator and also the oil will lose its insulative characteristics as the remnants of worn brushes are dispersed within it (*Table 4.8*).

In air compensation all voids are filled with air at a pressure which balances the external pressure. Thus a reservoir of air at high pressure must be available on board the vehicle. In practice, the compensating air pressure is governed by using a valve similar to a diver's demand valve. As the vehicle dives, its internal pressure is automatically raised. For ascent, the system must

Table 4.8 Comparison of oil and air pressure compensation systems

Oil pressure compensation	*Air pressure compensation*
Must be drained prior to maintenance. Inconvenient and dirty	Maintenance straightforward. Internal components remain clean
Unlimited in depth provided no voids remain	Depth limited by reservoir pressure and volume
No refilling required provided there is no leakage	Reservoir must be regularly refilled
Provides lubrication and cooling to electric motors	Provides no lubrication and little aid to cooling
Inherently safe	Possible danger from system not depressurizing on ascent. High pressure gas cylinders require careful handling and regular certification

include a relief valve which releases the air in order to reduce the internal pressure. The amount of air used and hence the size of reservoir required will depend upon the dive depth and the volume to be compensated (*Table 4.8*).

In practice, when using any type of compensation system it is normal to arrange for the internal pressure to always exceed slightly the external pressure. In this way, any leakage that occurs will be from inside to outside, thus preventing the ingress of water.

4.4.2.2.5 Control station The control station, usually situated on board the surface support platform, is the operator's interface to the ROV and as such its design and layout are of paramount importance. *Figure 4.38* shows a typical layout. Common to all such systems is the television monitor which displays the view as seen by a vehicle mounted camera and some form of control stick by which the thrusters are operated. Other facilities will include indicators regarding the vehicle and equipment status, navigational indicators and various switches for powering up the system and its subsystems.

The current trend in modern commercial systems has been to dispense with various indicators, for example the compass repeater and depth readout, and display this information as an overlay on the television picture. If it is not detrimental to the picture or too distracting, it does have certain advantages: the pilot does not need to look away from the screen and the information can, if required, be easily recorded.

Control sticks or joysticks differ widely in design but they are generally configured in such a way that control of the ROV is simple and intuitive. Some ROV systems utilize a single joystick with all the degrees of freedom

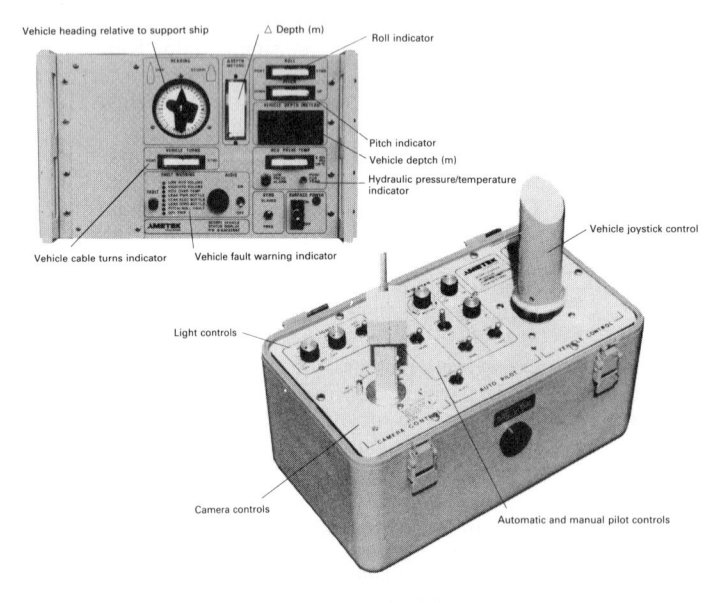

Figure 4.38 Typical ROV control unit, television monitor not shown (Ametek)

required, similar to that which controls a helicopter, others two simpler joysticks. Whilst some are integral with the control console, others are mounted on a separate smaller box or 'buddy box' which can be used some distance away from the console, e.g. on the deck of a vessel whilst viewing the ROV directly. There is no standard layout adopted and training on a specific ROV is necessary before the pilot becomes efficient.

If the ROV is fitted with a manipulator, there will be a master controller at the work station. For a modern advanced space-equivalent manipulator this would take the form of a scaled down version of the slave arm (see *Figure 4.39*).

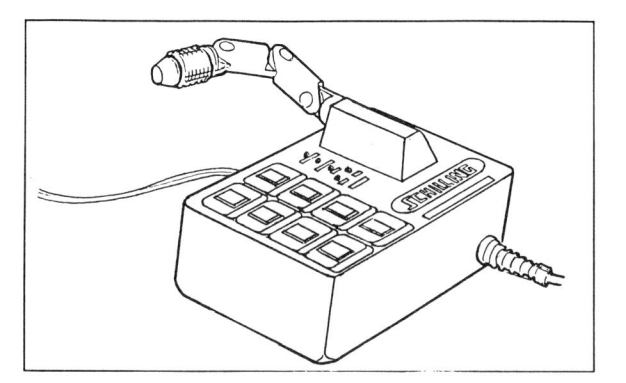

Figure 4.39 Master controller for underwater manipulator (Schilling HV5F/HV6F)

4.4.2.3 ROV deployment and retrieval

Because ROVs can vary tremendously in both design and size, their deployment and retrieval, unlike manned submersibles, can be undertaken in a number of ways. Also, because of their smaller size, dedicated support vessels for ROVs alone are not common, and in general, maximum use is made of 'vessels of opportunity'.

Motion of the support vessel and relative motion between it and the surface of the sea will increase the risk of damage to the ROV and will determine the limiting weather conditions during which the ROV can be safely launched. Any way in which these motions can be reduced will therefore increase the operability of the ROV and hence its cost-effectiveness. For this reason, operators generally launch ROVs over the side of the support vessel closest to the position which is least affected by heave motion. Usually either a hydraulically operated multijointed crane or a purpose built A-frame is used to lift the ROV in and out of the sea.

A recent trend has been to operate ROVs from a framed enclosure or garage (see *Figure 4.40*). The ROV is housed within its garage, which is negatively buoyant, and the system is lowered to the working depth on its lifting cable or umbilical. When at the required depth the ROV, after undocking from the garage, can be flown out to the work site. The garage will contain a reel system, which usually incorporates a slip ring assembly and is capable of paying out and taking in the umbilical or 'tether' as it is more commonly called when in this configuration, on command. Whereas the main lift cable must be of adequate strength and possibly armoured, the tether between vehicle and garage will normally be neutrally buoyant.

Originally, garage launching systems were only used with the smaller observational or 'eyeball' type of ROV; today they are being used increasingly with the larger work type ROV. Although garaged ROVs speed and simplify the transition through the water interface, the vehicle still has to be flown in and out of the suspended garage that will be undergoing the same heave motion as the supporting vessel. To reduce the possibility of damage during this operation, some of the larger systems now incorporate heave compensation in their lifting winch.

4.4.2.4 Dynamic response of underwater vehicles

On command, each propulsion unit will apply a certain force to the underwater vehicle along a given direction. The desired result of applying such forces is to move the vehicle from one location to another in a controlled manner. To relate the resultant motion of an ROV to the forces applied by its thrusters requires a complex analysis, which is beyond the scope of this chapter.

4.4.2.4.1 Drag A body moving within a fluid medium experiences a resistive drag force, D, given by:

$$D = \tfrac{1}{2}\rho C_d A V^2 \qquad (4.11)$$

where ρ is the density of fluid medium, A is the frontal area and V is the velocity.

The term C_d is the drag coefficient and is related to the shape and surface texture of the body. Its value for various simple shapes, for example a sphere, is well known. The value for complicated shapes, such as those of most ROVs is usually acquired by measurement. Altering the instrumentation payload of an ROV will affect its drag coefficient.

It should be noted that the resultant drag force is proportional to the square of the velocity of the vehicle. At constant velocity, the force available from the vehicle's thrusters will exactly balance the drag force. Doubling the propulsive force will not therefore double the velocity. It is this non-linear relationship which tends to limit the velocity of most ROVs to a submerged speed of between 2 and 3 knots.

4.4.2.4.2 Centre of drag The drag forces that a moving body experiences are distributed over its entire frontal area. However, this force profile can be repre-

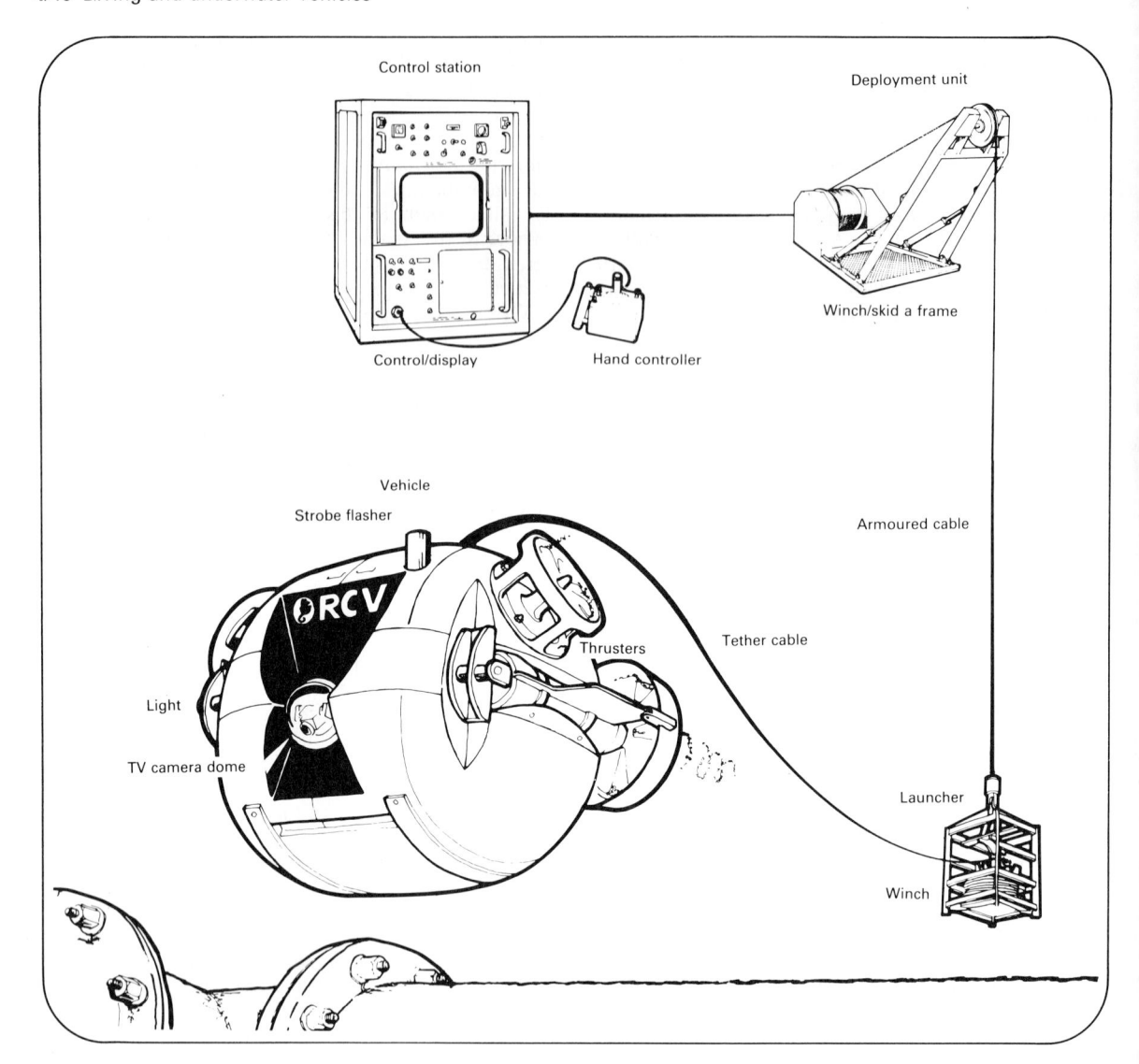

Figure 4.40 RCV 225 system, showing launcher or garage operation (Hydroproducts Inc.)

sented by a single force acting through a single point within that body, known as its 'centre of drag'. If the propulsive force is not in line with this point and opposing the drag force, then a couple will be set up which tends to turn the vehicle as it is propelled through the water. For this reason, it is important to carefully consider the siting of thrusters. For example, main forward thrusters which are positioned too low will cause the vehicle to climb when it is propelled forward.

4.4.2.4.3 Added mass Application of a force on a body which is free to move will result in its acceleration. By Newton's Second Law, the rate of acceleration is proportional to its mass. However, causing an underwater vehicle to move will also generally cause the movement of a volume of water, perhaps entrapped within a fairing. The result of this is that the vehicle under water appears to have a greater mass. This added mass that a vehicle can possess is significant and can, for certain vehicles, amount to several times that of the vehicle itself. It is therefore essential that added mass effects are considered in the design of the vehicle's control strategy.

4.4.2.4.4 Umbilical effects The effects of the umbilical on the guidance of an ROV are both complex and significant. As the ROV moves it must pull its umbilical through the water. The diameter and mass of the umbilical will generally be determined by the power requirements of the ROV. The result of drag forces on the umbilical is to create a tensional force within it. The angle the umbilical makes with the vehicle and the position of its attachment at the vehicle will determine what effect it has upon the control and guidance of the vehicle.

A full analysis of the drag generated forces on the umbilical of an ROV is extremely complex, as sea currents can vary in both magnitude and direction as a function of depth. In the simplified situation shown in *Figure 4.41* of a uniform current and stationary ROV and support vessel, the following useful relationships can be derived:

$$S = \frac{T_O}{R}(\cot \phi_1 - \cot \phi_2)$$

$$T_{1,2} = \frac{T_O(1 + f\operatorname{cosec} \phi_{1,2})}{(1 + f)} \quad (4.12)$$

$$y = \frac{T_O}{R}\left(\operatorname{Incot} \frac{\phi_1}{2} - \operatorname{Incot} \frac{\phi_2}{2}\right)$$

$$x = \frac{T_O}{R}(\operatorname{cosec} \phi_1 - \operatorname{cosec} \phi_2)$$

where R and f are constants relating to cable drag and defined as follows:

$$R = R' + F_c$$

$$f = \frac{F_c}{R'}$$

where R' is the drag on cable when normal to current, per unit length $= \frac{1}{2}\rho 0.75$ (umbilical diameter) V^2, and FF_c is the cable frictional drag when aligned with current, per unit length $\simeq 0.02R'$

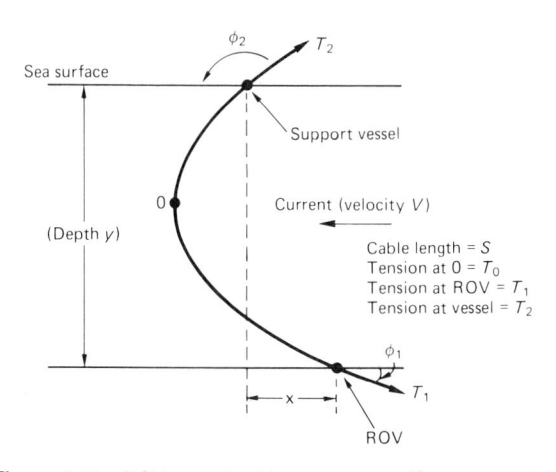

Figure 4.41 ROV umbilical in constant uniform current

4.4.2.5 Commercial ROVs in operation

From the early 1970s there have been many ROVs developed for numerous purposes. An exhaustive catalogue, there are well over 100 different types of system, is beyond the scope of this chapter. Several publications are devoted to such information (see Further Reading). The specific vehicles described here are chosen because they are of historical interest, possess special technical features or have been exceptionally successful vehicles in the commercial market.

4.4.2.5.1 RCV 225 The RCV 225 (see *Figures 4.40* and *4.42*) built by Hydroproducts was the originator of the eyeball type ROV. As RCV 225 has gained a reputation for good performance and reliability, over 80 have been produced.

The small, neutrally buoyant vehicle is roughly spherical in shape consisting of a block of syntactic foam in two halves housing a central electronics/television camera pressure housing. The six dc electric motor operated thrusters are mounted externally, as are the lights on some models.

The low light television camera incorporates a novel lens mechanism which enables the picture to be scanned vertically whilst not moving the camera body itself. Horizontal motion, 'panning', is achieved by turning this highly manoeuvrable vehicle about its own axis. Recent models possess an automatic heading facility utilizing an onboard rate gyro.

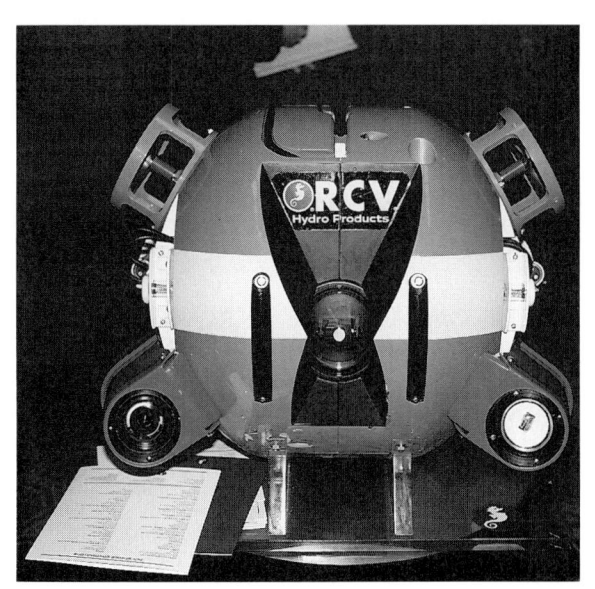

Figure 4.42 RCV 225 vehicle

The majority of RCV 225s are operated from a garage or launcher although if the application requires, the vehicle system can be configured to operate without it.

RCV 225 System specifications
Vehicle
Builder – Hydroproducts Inc.
Length – 0.51 m
Width – 0.66 m
Height – 0.51 m
Weight – 82 kg
Maximum operating depth – 400 m
Maximum operating current –
        2.5 knots limited manoeuvrability
        1.5 knots full manoeuvrability
Speed – Forward – 2.5 knots
        Reverse – 2.0 knots
        Lateral – 1.7 knots
        Up/down – 0.5 knots
        Rotation – 180°/s

Control station
Displays – Video with information overlay of vehicle depth, heading, lens pitch angle, number and direction of tether twists, tether payout length, elapsed time, date and time. Meter reading of cable continuity, voltage and insulation to sea water. Stills camera frame counter
Controls – Joystick giving proportional control for all degrees of freedom. Automatic heading and depth control with manual backup
Television camera controls – lens pitch angle (tilt), focus, 35 mm camera actuation
Deployment controls – Launcher lock, tether winch
Power required – 220 V 3-phase or 440 V 3-phase, 5 kW maximum

Deployment unit
Consists of winch, skid, A-frame and launcher
Armoured lift cable capacity – 400 m
Winch rate – Variable up to 36 m/min
Hydraulic power supply – 20 kW maximum
Launcher tether capacity – 120 m
Launcher tether winch rate – 18 m/min
Maximum pull – 220 N

4.4.2.5.2 Scorpio Amongst the ROV work systems, Scorpio and more recently its successor Super Scorpio, must rate as the most successful vehicles of this type, with 59 of these vehicles having been built.

Super Scorpio is a powerful, conventionally designed, open frame ROV. It possesses five hydraulic thrusters and can support an extensive range of instrumentation, manipulators and special tool packages. The standard system also includes the operator control unit and the umbilical on a hydraulically operated winch. Although garages are not standard equipment with this large vehicle, a number of manufacturers are able to supply such systems on request.

Super Scorpio – specifications
Builder – Ametek Inc.
Length – 2.48 m
Width – 1.48 m
Height – 1.5 m
Weight – approximately 1450 kg
Maximum operational depth – 1000 m standard, 1500 m optional, some earlier vehicles have been modified for deeper operation
Propulsion – two horizontal, two lateral, one vertical hydraulic thrusters
Speed – 2.5 knots
Control – Single joystick
Television – Three TV channels
Sonar – Ametek 25 CTFM scanning sonar range up to 610 m
Lights – 6 × 250 W variable intensity tungsten
Heading sensor – Gyro compass 1% accuracy (0.1% optional)
Depth control – 0.28 m in automatic mode
Multiplexer – 64 channels up and down
Manipulators – seven function manipulator, five function grabber
Optional features – Second seven function manipulator, navigational and tracking system interfaces, stills camera, sidescan sonar, profiler, pipe tracker, cable cutter

Figure 4.43 Example of large ROV built within an open framework, Sea Horse was operated as a trials vehicle by the Institute of Offshore Engineering, Heriot–Watt University, Edinburgh (photograph S. Hamilton)

4.4.2.6 Modularized ROVs

Experience in the offshore oil and gas industry has shown that no one work type ROV is capable of carrying out the full range of tasks that might be encountered. Rather than build many different ROVs to undertake specific tasks, some manufacturers have adapted a modular approach to the design of their systems. Such an approach allows vehicles to be optimally configured to support the tooling and instrumentation to undertake the task.

A good example of this modular approach can be found in the AOSC 2000 series. The 2000 Series is designed as a set of separate subsystems which can be used as building blocks available to be configured in the form best suited to the task for which it is required. The system uses two forms of a standard 0.5 m (1.64 ft) cube. These are either a pure buoyancy cube or a combined thruster and buoyancy cube, the latter having a conventional hydraulic thruster mounted within it. Both types of cube have identical fixing points and can be interchanged to provide a variety of different vehicle configurations.

At present, AOSC has designed two likely configurations. The smaller version, designated the 2006, has six cubes in its uppermost layer and is aimed at tasks such as drilling support. The larger vehicle, the 2008, has eight cubes in its uppermost layer, four of which are configured as vertical thrusters. This vehicle thus has a greater payload capacity and is aimed at tasks such as construction support and survey. Other examples of modular ROVs include the Dragonfly and Rigworker (*Figure 4.44*).

Figure 4.44 Rigworker R3000L shown in launcher (OSEL Group)

4.4.2.7 ROVs in diver support roles

Soon after the introduction of the eyeball type ROV, they became widely used in diving operations. At first viewed somewhat suspiciously by the divers, some of the benefits offered by the presence of an ROV were soon to be appreciated. The ability of surface personnel to see the divers at most times meant an immediate improvement in diver safety. On at least one occasion surface personnel have been aware of a diver in difficulty before the bell man. Apart from this monitoring role, it quickly became apparent that the ROV could play an active part in an operation.

The ROV provides a platform, a source of illumination, a supply of electrical or hydraulic power and a communications link or information point. It should be possible for the diver to take advantage of all such facilities with perhaps some modifications to the ROV.

The object of using ROVs in diving operations must be to improve safety for the diver and to improve operational efficiency. When an ROV is deployed alongside divers, one must take great care that it does not have the opposite effect. It is therefore vital that the vehicle is electrically and mechanically safe under water, even after failure. Also, it is preferable that its thrusters, especially the more powerful ones on larger ROVs, are adequately guarded. In such operations, there is extra responsibility on the ROV pilot to make sure that the diver and equipment are not impeded by the ROV and that its tether does not become entangled with any of the diving equipment.

In recent years it has become almost routine to deploy some type of ROV during deep diving operations. Most ROVs used in this manner are standard, although a few have undergone minor modifications. The first vehicle designed specifically for the deep water diving support role is the David.

4.4.2.7.1 David The David vehicle (see *Figure 4.45*) was designed and built specifically to support the diver. In a single package it provides the power, tools and facilities which the diver requires when working within the offshore oil and gas industry. The vehicle can be manoeuvred to the work site either under remote control from the surface or by the diver operating onboard controls. Heading, depth, pitch and roll are servo controlled for maximum stability.

Prominent on the David is its large claw shaped clamp which enables the vehicle to be fixed onto a structure at any desired angle. The vehicle's facilities include a winch, illumination, a set of hydraulically powered tools and an articulated ladder which provides a reactive force to help the diver accomplish his tasks.

For non-destructive testing and inspection David may be fitted with a full range of equipment.

David specifications
Length (with claw) – 3.8 m
Width – 2.1 m
Height – 1.8 m
Weight – 3900 kg
Thrusters – Eight hydraulic thrusters with Kort nozzles each rated at 15 kW
Diver control station – Removable to distance of 10 m for operation of hydraulic systems
Maximum depth – 1000 m
Television system –Two monochrome cameras each on pan and tilt units

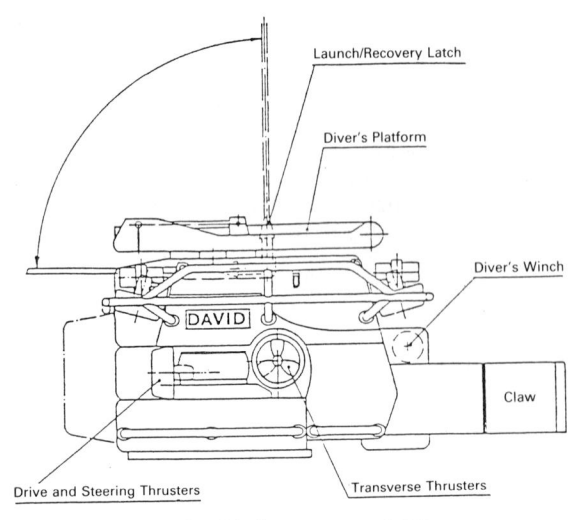

Figure 4.45 The David diver support ROV

Claw – for attachment to tubulars from 400 to 1370 mm in diameter, maximum claw force 35 kN, tilt range 15° up to 90° down, rotation ± 360°

Umbilical – 50 mm diameter, three power conductors, nine twisted pairs, thirty-six control leads, water blocked and with Kevlar™ strain member

4.4.2.8 ROVs in military operations

The use of the ROV in military operations is especially attractive. Not only have military divers to face those hazards faced by commercial counterparts, they may also be under direct or indirect threat of enemy action. Also military diving operations may take place in hostile locations outside the reach of any of the backup and emergency services which have been formed specifically for improving the safety of commercial operations. In view of such vulnerability, it is therefore sensible to replace the diver by an ROV whenever possible.

Underwater tasks within the military sector will tend to be somewhat different in nature to those encountered in civil operations. Whereas civil operations are often centred around a fixed structure at a fixed geographical location, military operations might be spread over a wider area, thus placing greater demands on the performance of an ROV to be used in such operations. One example of such an operation is that of mines countermeasures which involves their detection and eventual neutralization.

Because of the obvious danger associated with this work and because it requires a great area coverage, a properly designed ROV is the ideal tool to be used. This particular application has been recognized by a number of manufacturers who have designed and built special vehicles to meet these requirements. Of these ROVs, the PAP 104 and more recently the PAP MK5 have been by far the most successful. With more than 325 PAP vehicles sold to various navies of the world since 1970, they greatly outnumber any other type of commercially available ROV.

4.4.2.8.1 PAP 104, PAP MK5 The PAP 104 system utilizes a single wire for the transmission of data between the vehicle and the surface control. The wire is spooled off a bobbin on board the vehicle in much the same way as a wire guided torpedo. This large vehicle carries its own batteries to provide power for its two horizontal and two vertical thrusters and all instrumentation which it carries. The lack of a conventional tether enables the vehicle to work at great ranges from the support vessel and at relatively high speeds, necessary capabilities in such work.

The vehicle can be fitted with a number of different types of sonar system to work at both long and short range and various types of television camera. On detection of a suspicious underwater object by the sonar system mounted on the minesweeping support vessel, the PAP vehicle is launched. It is then piloted towards the object, its progress being monitored by the ship's sonar. When in range, the object is then located by the vehicle mounted sonar and the vehicle closes in on the object. Eventually, depending on the water visibility which prevails, the object will be detected visually. At this point the vehicle will be carefully manoeuvred towards the object until a positive visual identification can be made.

If the object is confirmed to be a mine, the action will depend upon its type. If the mine is moored on a long tether floating in mid-water high above the sea bed, the mooring may be cut by the vehicle's onboard cutter, allowing the mine to float to the surface where it can be detonated. If the mine is close to or on the sea bed, the PAP vehicle has the capability of laying a 100 kg (220 lb) high explosive charge in its immediate vicinity which can be detonated upon an acoustic command, thus destroying the mine.

After cutting the mooring or laying the charge, the PAP vehicle returns to the supply vessel either on its disposable wire or under radio control on the surface of the sea. The latest vehicle, PAP Mk5, can be used with a disposable optical fibre cable instead of a wire, providing much improved data transmission rates with corresponding improvements in vehicle capability and range. Acoustic and magnetic signatures have been deliberately minimized on these vehicles.

PAP 104 specifications
Builder – ECA
Length – 2.7 m,
Width – 1.2 m
Height – 1.3 m
Weight – 700 kg (including explosive charges)
Maximum speed – 5.5 knots

Maximum operating depth – 300 m
Battery life – up to five missions

4.4.2.8.2 CURV programme (US Navy Undersea Programme) The Cable Controlled Underwater Recovery Vehicle (CURV) Programme at the US Naval Undersea Center was organized in order to make possible remote recovery of ordnance at the Center's test ranges at Long Beach and San Clemente Island. It was to mark the beginning of the ROV as we know it today.

The first fully operational ROV developed, now referred to as CURV I, was designed for depths to 610 m (1983 ft) and began its work in 1965. Over the next two years it made many successful recoveries.

When a hydrogen bomb was lost off the coast of Palomares, Spain in April 1966, CURV I was modified for deeper operation and was instrumental in its recovery from a depth of 868 m (2821 ft). The success of this operation demonstrated the versatility of this machine and the future potential for ROVs.

It became apparent that after two years of operation, the CURV vehicle, because of wear and tear and reliability problems, could not continue to meet the requirements of the Center and it was decided to undertake two new CURV projects. CURV II, designed for 760 m (2470 ft), consisted of much of the original CURV I system but with its vehicle repaired and uprated. CURV III incorporated the successful concepts of CURV I but was a completely new system with initially a maximum operational depth of 1800 m (5910 ft) but soon changed to 2120 m (6961 ft). The vehicle was of typical open frame construction.

CURV III specifications
Size – 2 × 2 × 4.55 m
Weight in air – 2490 kg
Weight in water – Slight positive buoyancy
Propulsion – Two horizontal, one vertical three-phase electric, 10 hp motor driven thrusters, Oil-filled pressure compensated
Total power requirement – 40 kw
Frame – I section and channel section aluminium alloy
Buoyancy – Syntactic foam blocks held by rods and brackets
Instrumentation – Two video cameras
Four thallium iodide lamps, mercury vapour light, 400 exposure 35 mm stills camera, camera mounted on pan and tilt units, compass, sonar tracking systems.
Handling equipment – 5.8 m articulating hydraulic crane with winch and lift line

4.4.2.8.3 Skate The Skate ROV built by Bennico Ltd has been built specifically to locate and recover torpedoes from the sea bed. A replacement for *Cutlet*, it is operated by Butec at Kyle of Lochalsh in Scotland and is capable of locating torpedoes in depths of up to 300 m (985 ft) at a range (from the working platform) of some 200 m (657 ft). Unlike its predecessor, it has the capability of locating buried torpedoes and by use of its onboard mud pump can excavate and recover the weapon with its claw shaped grab.

4.4.2.8.4 TUMS – Towed Underwater Submersible
The TUMS system (see *Figure 4.46*) was designed and

Figure 4.46 The TUMS system (photograph courtesy of British Aerospace)

built specifically for operation from the Royal Navy's sea bed operating vessel HMS *Challenger*. Comprising a deep tethered vehicle (DTS), a garage or depressor and cable handling system, it may be operated at extremely great depths in either a free swimming or a towed mode. Its main tasks concern searching for, detection, identification, classification and retrieval of objects lying on the sea bed.

The DTS possesses thrusters which control its motion in all directions. It is equipped with sector scanning sonar, sidescan sonar, low light television, stereo stills camera and magnetometer which is trailed behind. It also possesses a six function manipulator.

The depressor isolates the DTS from the motion of the ship and provides temporary storage for items which have been retrieved from the sea bed. During launch and recovery, the depressor acts as a garage for the DTS.

4.4.2.9 LCROV – low cost ROV

As the ROV was originally developed for military applications and was then taken over by the offshore oil and gas industry, there was, originally at least, little incentive for manufacturers to build other than expensive systems. Even though some vehicles were built in fair numbers, this had very little impact on their cost to the customer. It was not until the mid-1980s that manufacturers became aware of other markets that could not support expensive vehicles and the so-called low cost ROV (LCROV) was introduced.

The forerunner in this category was undoubtedly the Minirover, a basic inspection vehicle that was lightweight, portable and priced at around $25 000, making it an order of magnitude cheaper than certain well known vehicles which possessed, on paper at least, little extra in their specifications. The Minirover was quickly followed by Searover, a similar vehicle but with uprated capability.

Since the appearance of the Minirover there have been numerous LCROVs brought on to the market. The Phantom range of vehicles with their simplicity, lightness and ruggedness was a serious attempt to introduce LCROVs into the offshore oil and gas applications market. Other vehicles have been built more specifically with onshore applications in mind, such as dam and canal lock gate inspection and work within power station cooling water intakes.

Recently a number of vehicles have been aimed at the leisure and recreation market. The ROV allows the non-diver to explore the underwater environment and the sports diver to extend frontiers, a rewarding experience to both. The use of ROVs for recreational purposes has long been recognized but until very recently their cost has prohibited their widespread use in this field. The situation has now changed with ROVs available for as little as £5000. Such developments have been made possible by the ever decreasing cost of colour ccd cameras whose performance now rivals that of the single tube camera.

An interesting example of a very low cost vehicle is the recently introduced Merlin (see *Figure 4.47*). The vehicle carries its own dry lead acid accumulators to supply power to two thrusters, lights and camera and yet weighs only 20 kg (44 lb) and is only 84 cm (2.8 ft) long. Vertical motion is achieved by a unique variable buoyancy system. Three copper tanks attached to the main pressure hull are filled with water and air as appropriate through tubes in the slender 50 m (163 ft) long tether. It is claimed that the automatic depth control system can stabilize vehicle depth to ± 15 cm (6 inches).

Figure 4.47 The Merlin low cost ROV

The Seaker Mk11 is yet smaller in size weighing only 10 kg (22 lb) and 45 cm (18 inches) in length. With two horizontal thrusters, a vertical thruster, colour video camera, compass and depth sensor, this vehicle is directly aimed at the leisure market as perhaps just another accessory to a yacht. Its cost represents only a small fraction of that of a modest vessel.

4.4.2.10 Untethered or autonomous vehicles

The tether or umbilical of the ROV is, as we have already seen, both an asset and a liability. It enables reliable, high speed data transmission between vehicle and control station and allows energy in large quantities to be transmitted to the vehicle for its propulsion system and instrumentation. However, the tether can increase operating difficulties and it is easily entangled or damaged by both underwater structures and the supporting vessel. Also its mass and drag can, under many circumstances, have a considerable affect on the performance of the vehicle. In view of this, it is not surprising therefore that several groups of researchers have sought ways in which to dispense with any physical link between control station and vehicle. In doing this, it becomes necessary for the vehicle to carry its own onboard energy supply. Also required is a means of transmitting control information to the vehicle through the water and similarly information from the vehicle's sensors back to the control station. Whilst the carrying of onboard energy is not an insurmountable problem, the reliable transmission of data through the water medium at the rates required is technically very difficult. At present, only acoustic techniques have been seriously considered and their performance falls far short of that which can be achieved by an electrical conductor.

The problem lies in the fundamental limitation in the available bandwidth of a through water acoustic link. Research effort has therefore centred on finding ways to reduce the required bandwidth. This problem has been addressed in several ways:

1. By building vehicles which possess their own artificial intelligence. Such vehicles, whilst operating under real time control, are capable of making their own decisions, thus reducing the amount of information which needs to pass between vehicle and controller.
2. By the building of preprogrammed autonomous vehicles which follow predetermined commands and thus require little if any communication between vehicle and control station. Vehicles in this category could also possess onboard intelligence which, in association with suitable sensors, would enable the programme to adapt to any changing conditions, e.g. currents, thus making the system more closely follow what was intended. Such a vehicle would be configured in such a way that it could store any information received which was relevant to the mission.
3. By employing bandwidth compression techniques on television signals. In a standard television signal which might occupy up to a 6 MHz frequency interval, much of the information can be regarded as redundant, in that a high proportion of picture elements remain unchanged between successive television fields. This is especially true in underwater scenes where relative velocities between observers and targets are generally low. There exists therefore much scope for applying various techniques which reduce the total amount of transmitted information, whilst having little effect on the subjective quality of the overall image.

Bandwidth compression techniques can also be used in conjunction with slow scan television pictures. Although slow scan television might be suitable for inspection purposes, any loss of picture continuity poses a serious problem for the pilot when manoeuvring the vehicle.

Despite the difficulties in realizing a tetherless vehicle and a reluctance on the part of the oil and gas industry to support development of such vehicles, there have been at least 17 different autonomous (untethered) vehicle projects. The most noteworthy will be discussed.

4.4.2.10.1 Epaulard The most developed of this group of autonomous vehicles is the French vehicle, Epaulard. It is a preprogrammed vehicle but with provision for changing heading, speed and depth during operation via an acoustic signal. Syntactic foam provides all positive buoyancy and weights are provided to take the vehicle to the sea bed and, on release, back to the surface.

Specification of Epaulard
Builder – ECA
Maximum depth – 6000 m
Weight – 3 tonnes
Mission duration – 8 h
Capability – Photographic, topographic profiling

4.4.2.10.2 ARCS The ARCS (see *Figure 4.48*) is an artificially intelligent autonomous remotely controlled submersible. It was designed as an under ice survey vehicle for the Canadian Arctic. It carries a digital echo sounder for bottom depth measurement, a long baseline navigation system and an obstacle avoidance sonar. The ARCS vehicle, which is some 7 m (23 ft) in length, is battery powered and preprogrammed to carry out assigned tasks. Because of bandwidth limitations, high data rate real time communication with ARCS is not attempted, although the operator is provided with regular status reports.

Figure 4.48 ARCS (courtesy of ISE)

A typical operation of ARCS has it following a preprogrammed path on an 8 km (5 mile) square and returning to its starting point upon completion of the survey. During the mission, the vehicle sends status information to the operator via an acoustic link and can receive commands sent by the operator from the surface. In this way, the operator can abort or suspend the mission or manually control the vehicle if the need arises.

4.4.2.10.3 Eave east Built primarily as a testbed for an autonomous vehicle, it is configured similarly to an ROV possessing six 0.18 kW thrusters and being slightly positively buoyant. Carrying a lead–acid battery pack of some 1260 Wh capacity, it has a mission duration of about 6 h.

Specification of eave east
Weight – 272 kg
Maximum operating depth – 91 m
Projected capability – Pipeline and platform inspection

4.4.2.10.4 ROVER ROVER (see *Figure 4.49*) was developed by the Underwater Technology Group at Heriot–Watt University, Edinburgh, as a testbed on which experiments relating to all the various problems associated with an untethered vehicle could be carried out.

Figure 4.49 ROVER (Heriot–Watt University, Edinburgh)

In order to make use of the highest possible acoustic bandwidth, the ROVER vehicle was designed to be carried to the vicinity of the work site by its mother ROV, ANGUS 003. In operation, it would remain within a 100 m (328 ft) radius of the vehicle. In this way, the acoustic path could be kept both short and horizontal to provide the best possible conditions for information transmission.

The ROVER vehicle was produced cost effectively from diver-tow units and inexpensive electric outboard motor units. Its onboard lead–acid accumulators provide for a 2 h operation duration. The vehicle possesses lights and supports any standard underwater television camera. Research work on acoustic transducer design and television signal bandwidth compression techniques suggests that several frames per second at standard resolution can be transmitted between ROVER and its mother ROV.

4.4.2.10.5 TM 308 Over the period 1983–1986, a research project jointly funded by the European Economic Community and a consortium of companies including Tecnomare, Agip and Micoperi was carried out which investigated the possible production of a completely autonomous underwater vehicle for carrying out structural inspection. The project was terminated with a complete specification of all component parts.

The vehicle is in the shape of an upright cylinder, some 2.1 m (6.9 ft) diameter and 4 m (13 ft) high and weighs some 6.6 tonnes in air. Up to 50 kW of power is generated by a closed circuit onboard diesel engine.

Piloting of the vehicle is accomplished by a supervisory control system. An unskilled operator is able to send simple high level commands to the vehicle where they are interpreted by an onboard computer. In this way, it is hoped that the pilot can concentrate efforts towards the task in hand rather than controlling the vehicle.

Television pictures are to be transmitted via a 10 kBit acoustic link. Bandwidth compression algorithms have also been developed and tested to transmit up to four television frames per second.

4.4.2.10.6 Autonomous vehicles in marine physics and marine geology The expected advent of several environmental remote sensing satellites in the early 1990s will bring about an enormous increase in available surface or near surface oceanographic data. To fully complement this data with subsurface oceanographic data and permit accurate ocean modelling for climate studies and weather forecasting is a near impossible task. At present the necessary oceanographic data in terms of conductivity and temperature as a function of depth, are recorded by instruments lowered from ships. The high cost of ship time and technical support, and the slowness of the profiling weighs against the prospect of ever gaining sufficient data by this means.

The solution proposed by the Institute of Oceanographic Sciences is to mount the required instrumentation on an untethered autonomous underwater vehicle. The system, which is capable of vertical and horizontal movement over very long ranges, has been given the name DOLPHIN (Deep Ocean Long Path Hydrographic Instrument). Its primary objectives are: to obtain data more cheaply and thereby obtain more data than other techniques; to reduce the support vessel time normally required for such work; to undertake preprogrammed tasks, taking profiles along predetermined courses; to communicate data via satellites which will mean surfacing from time to time; and to navigate through the use of satellites.

In contrast to the oceanographer's desire to measure over vast areas of the ocean, is the marine geologist's desire to investigate relatively small areas of the sea bed often at very great depth.

Although manned submersibles and to some extent ROVs, have been used for this purpose, there are serious drawbacks. The support required for manned submersible operations and the minute coverage of the sea bed obtainable, makes this technique prohibitively expensive in all but the most exceptional circumstances. The use of an ROV or towed vehicle with a tether poses operational difficulties in deep ocean situations, with drag forces making vehicle manoeuvring and control extremely difficult.

The Institute of Oceanographic Sciences has also proposed a solution to this particular problem in the form of an untethered autonomous underwater vehicle. DOGGIE (Deep Ocean Geological and Geophysical Instrumented Explorer) would contribute towards the discovery, assessment and understanding of sea bed mineral resources. Designed primarily to make a range of acoustic and magnetic measurements over a predetermined area, it can be launched from a support vessel which is then free to go about other duties. In this way, DOGGIE offers improved ship utilization and furthermore, one ship could operate several DOGGIEs simultaneously if the need arose.

4.4.2.11 Other systems

4.4.2.11.1 ROV Mantis and DUPLUS II Although the Mantis ADS had been very successful for several years in the role of offshore drilling support, the early 1980s saw a trend by the oil companies to turn towards ROVs. To avoid losing their market share, the vehicle's operators and manufacturers converted Mantis for unmanned remote operation. Retaining all of the original systems and with improvements to thrusters, including the addition of vertical thrusters, the vehicle could either be operated in an unmanned or a manned mode.

All controls were duplicated at the surface station and inside the vehicle, as far as possible keeping the controls identical to avoid confusion when pilots switched between the two modes. This configuration also allowed the vehicle to be remotely controlled whilst manned, thus allowing the pilot to concentrate on his task which might involve the use of the manipulators. Such a facility was also useful from a training point of view, where a pilot under instruction could be actively supervised by experienced personnel at the surface.

The success of this concept prompted the development of a completely new vehicle which has this dual capability, DUPLUS II. With a completely revised and again updated thruster system, the vehicle is seen to fill a niche between the diver and the ROV.

DUPLUS II specifications
Manufacturer – OSEL Group
Length – 2.45 m
Width – 1.8 m
Height – 1.4 m
Weight – 1700 kg
Maximum operating depth (manned) – 800 m
Thrusters – 8 × 600 V three-phase ac
  2 × 80 V dc
Manipulators – 1 × sea water hydraulic six function arm
  1 × sea water hydraulic seven function arm
  1 × sea water hydraulic three function arm
  Telescopic grabber
Controls – Auto depth and altitude ± 0.3 m
  Auto heading ± 2°
Video – Two independent channels for monochrome or colour TV with a switchable third camera
  Built in monitor for onboard pilot
Buoyancy – 50 kg air reservoirs
Instrumentation – Depth sensor, gyro compass, echo sounder – all with onboard and surface readouts

4.4.2.11.2 SID The Structure Inspection Device (SID) is a structure reliant vehicle. It is based on an astable ROV which can roll through 360° and pitch to ± 90°. This capability is claimed to allow it to approach a structure in the most convenient orientation thus emulating a diver's flexibility. The system comprises a vehicle, tether management system and surface control station. SID is attached to the structure using suction from a pressure drop created within a three leaf flexible foot; adjusting the pressure drop allows the vehicle to glide over the surface of the structure. It supports three five-function manipulators and a water jet cleaning system. There is also capacity for carrying various NDT systems.

4.5 Selection of underwater intervention methods for offshore oil and gas operations

The techniques and systems described so far in this chapter are simply a means to an end, their purpose being to enable the many necessary tasks that must be performed under water to be accomplished. The selection of the particular mode of intervention, whether it be a diver or some type of vehicle, involves a complex decision process.

In the offshore oil and gas industry, the cost of an operation is of great importance and will play a major part in determining the operational method. Safety to personnel, to installations and the environment are also important factors which must be taken into account.

Any underwater operation must be planned meticulously if its success is to be assured. The optimum mode of intervention in terms of cost and safety can only be determined after a number of aspects concerning the operation have been carefully considered and the limitations of each intervention techniques are well known and understood.

4.5.1 Operational factors affecting intervention mode selection

The main technical factors which will affect the mode of intervention selected for an underwater operation are described as follows.

4.5.1.1 Task difficulty

Part of the operation planning involves its breakdown into a series of tasks. Tasks encountered in offshore oil and gas operations are wide ranging (see *Table 4.9*) and can be classified as to their difficulty. At one end of the scale is the task of general observation, which in clear water is regarded as perhaps the simplest of all tasks. At the other end of the scale there are those tasks which require a high degree of precision manipulation, perhaps also involving the application of large forces and torques. Such work requires an adaptability, only available through fast and accurate decisions.

It is also useful to categorize tasks into those which are observational only and those which also involve a degree of work or manipulation before selecting the mode of underwater intervention.

4.5.1.2 Maximum depth of operation

The depth of the operation can, if great enough, be the sole determining factor in the selection of the mode of intervention. It will, however, always be important in the selection process.

At present, commercial diving operations are limited through available equipment and procedures to about 400 m (1313 ft) depth, although it has been shown that diving to beyond 500 m (1642 ft) is a feasible albeit costly proposition. However, it is worth pointing out that this depth surpasses the capability of many ADS, manned submersibles and even ROVs.

When determining the mode of intervention, depth can be used initially to exclude either techniques or systems from consideration. For example, one would not plan to use an air diving team to work at a depth of 60 m (197 ft) on a salvage operation, or employ a particular ROV in waters of greater depth than its maximum operating depth.

4.5.1.3 Operation (or task) duration

The duration of the underwater task, or of the operation if it comprises tasks of a repetitive nature, will through economic and technical reasons, have a direct bearing on the selection of the intervention technique.

For example, it is not cost effective to employ heliox bounce divers in most constructional operations. Such

Table 4.9 Typical underwater tasks in the offshore oil and gas industry

Observation/measurement

Pipeline route survey	Geological observations
Inspection of installed pipelines	Site surveys
Cathodic protection surveys	Cathodic protection surveys
Inspection of SBM hoses and manifolds	Pipeline inspection SBM inspections
Scour surveys	Assessment of fouling
Mooring lines, chains inspection	Anchor dragging assessment
Location of pipelines	Wreck identification
	Bottom profiling

Observation/manipulative

Debris removal prior to pipelaying and installation	Small object retrieval by attachment of lines
Loosen and tighten bolts	Rock collection
Collect hard rock core samples	Hard rock drilling
Stud insertion	Drill bit recovery
Operating pipeline valves	Cathodic protection voltage readings
Wire brushing	Water jetting prior to inspection
Concrete chipping	
Cable burial	Non-destructive testing (NDT)
Assisting trenching operations	Flowline tie in
Guidewire changing	AX ring change out
Preparation of abandoned well for re-entry	
Maintenance of moorings and platforms	
Welding, cutting operations	

operations, whilst requiring the manipulative dexterity of the diver, would normally be undertaken using saturation divers.

On the other hand, if a long duration operation involved only observational tasks, for example a pipeline inspection, it is more likely that a suitable ROV would be employed.

4.5.1.4 Available support

The available support at the site of an operation is often a factor in determining the mode of intervention. On many occasions saturation divers have undertaken the simplest of tasks that could have been carried out by an ROV simply because the divers were on site and the ROV was not. The cost of mobilizing an ROV could easily outweigh the cost of the dive, especially if divers were already in saturation. The available support, or for that matter the lack of it, does on some occasions override the logical decisions.

4.5.1.5 Special considerations

Occasionally there are special circumstances which must be considered. There is often some element of danger which precludes the use of any manned form of intervention. An unsafe structure, a blowout or contaminated water are examples where the expendible ROV can be put to use.

4.5.2 Capabilities of the various modes of intervention

The assessment of the capability of a technique or system described in this chapter is not a simple task. Although the capability of, say, a surface demand air diver might be well recognized, there will obviously be some variation between individuals. When trying to assess the capability of an ROV, the problem is much greater as there is an extremely wide variation in type and specification. Also it is well known that the manufacturer's specifications can at times border on optimism. There is no substitute for experience of a particular vehicle or system in a previous similar operation.

Table 4.10 attempts to present the capabilities of the various types of intervention described. The absolute present day costs of the various services are not given. There tends to be great variations in such costs dependent on the state of the market. Instead, only a comparative cost is given here.

Only when one has a thorough knowledge of the various techniques listed is it possible to choose the most suitable mode of intervention for the underwater operation. Even then it may be that the choice is not clear cut. For example, although a particular ROV might on paper be capable of performing a task in deep water, one must carefully consider the cost implications if it is found that the ROV takes longer than expected to complete the task. For such a task it may be less costly overall to employ a mixed gas bounce diver, with perhaps higher confidence in his ability, than the ROV. It can be seen that cost rates themselves are not of absolute importance. The relative efficiency of the technique must also be taken into account. Studies comparing, for example, the efficiency of divers to that of the ROV have been undertaken and not surprisingly the results are very dependent upon the task.

4.5.3 The future

Although developments in the fields of diving and underwater vehicles have been rapid in recent years, the simplest of underwater operation is still regarded as a major undertaking. There remains considerable scope for improvements to be made in all aspects of underwater intervention. The following serves to highlight some which require most urgent attention.

4.5.3.1 Communications

Communications between divers and diver and surface personnel even today are unsatisfactory, with a considerable proportion of actual dive time wasted because of misunderstood messages. As depths increase, the problem of speech distortion due to high pressure helium and hydrogen will become even more pronounced. Improvements in diver communications systems are therefore of prime importance, if optimum efficiency is sought.

4.5.3.2 Diver assistance

Diver efficiency could be improved dramatically by the more widespread use of active diver support ROVs. The ROV can provide power, support tooling and transport materials, thus relieving the diver of much exhaustive work.

4.5.3.3 Enhanced ROV control

The capability of present day ROVs will not be extended until significant improvements in ROV control and positioning have been made. Such an advance is dependent on the introduction of improved motion and position sensors. It is likely that such sensors could be optically based, rather than a further refinement of acoustic technology.

4.5.3.4 ROV pilot training

The effectiveness of the ROV, however advanced its control system, tooling and instrumentation, is very much dependent on its operating team. There is therefore a requirement for more formalized training for ROV operators. The more widespread use of simulators could be helpful in this respect.

Table 4.10 Comparison of underwater intervention techniques

Intervention mode	Maximum depth (m)	Maximum duration	Capability	Disadvantages	Relative cost
SCUBA diving	50 (mandatory in UK)	Bottom time severely limited by decompression routines and available breathing gas, especially near maximum depth	Extremely flexible and capable of performing tasks requiring high degree of dexterity, although this capability is somewhat reduced when at maximum depth	Limited communications with surface and other divers. Possible nitrogen narcosis at maximum depth	Low
Surface demand air diving	50 (mandatory in UK)	Dependent on depth; bottom time limited by decompression. Greater than scuba	As for Scuba, but with advantage of communications with surface	Somewhat restricted by umbilical, care required not to entangle or damage it. Possible nitrogen narcosis at maximum depth	Low to moderate
Mixed gas bounce diving	Although there is no fixed limit, dives are normally restricted to maximum depth of 150 m because of physiological problems especially during compression	Usually restricted to $\frac{1}{2}$h or 1 h bottom time to prevent decompression time becoming too lengthy	As for surface demand air diving. Work capability decreases with increasing depth	Rapid compression not ideal for deeper depths. Helium distorts speech communications. High cost of helium	High
Mixed gas saturation diving	Commercial dives have been undertaken in depths approaching 400 m. Experimental dives have reached 530 m. 700 m seems feasible depending on available support	In UK, time limited in saturation to 28 bottom time therefore depends on depth. Individual divers will spend a maximum of 8h/d actually working	As for bounce diving, but work rate greater because diver is acclimatized to the ambient pressure. Work rate decreases with increasing depth	Speech distortion by helium. High cost of helium even when reclaimed. Usually dedicated support required	Very high
ADS	700	Dependent on task and conditions; 3–4 h dives are preferred, 8 h dives are possible	Manipulative work, if well planned, can be undertaken; largely dependent on the skill of the operator	Access to worksite limited	Moderate to high
Manned submersible	Dependent on vehicle typically 300–1000 m for oil and gas support systems	Dependent on vehicle, limited by battery life and crew. Typical 6–8 h per mission. Spare battery pack and relief crew enables repetitive missions	Capable of direct observation and very basic manipulative work. Can support much instrumentation: sidescan sonar, sub-bottom profiler etc.	Access to worksite extremely limited. Dedicated support vessel required	Very high
ROV	Typically 400–1000 m but no fundamental limit apart from umbilical problems.	Continuous operation is possible, provided reliability is good and relief crews are available	Indirect observation. Work ROVs have performed numerous tasks, but few tasks are yet routine. Success depends on the skill and sometimes ingenuity of the operating team. Can support much instrumentation	Limited viewing capability, limited control capability and poor reliability in some cases	Moderate

4.5.3.5 Task specific tools

A solution often adopted for undertaking a complicated task is to design and build a special tool which is dedicated to that task. By this means ROVs have for example replaced the AX ring gasket in a subsea wellhead. There is much scope for extending the capabilities of ROVs and for that matter autonomous systems in this way, but such a course demands a greater interaction with the subsea system designers at an early stage.

4.5.3.6 Underwater viewing

The importance of visual information in the operation of ROVs cannot be undervalued. Ideally a remote viewing system should fool operators into believing they are actually at the work site and only then will they work efficiently. Unfortunately, most television systems have the opposite effect. There exists much scope for improvement in present day systems especially in the form of the image display.

Whereas, in the late 1970s, there was much talk of the ROV replacing the diver, it is now generally accepted that there will always be a requirement for the ambient pressure diver. The ROV must be regarded as complementary to diving, although there will obviously be areas where it is strongly competitive.

The extension of man's capabilities under water by any means provides a great challenge, comparable to that of the exploration of outer space.

Further reading

Diving

A Guide to Diving Operations at Work, 1981 Health and Safety Executive, HMSO (1981).
Aseptic bone necrosis in commercial divers Underwater Engineering Group, Technical Note UTN/25 (1981).
Cross, M., The selection of divers for deep operations, *Petroleum Times*, Sept. 16 pp 27–29 (1977).
Diving Manual, Royal Navy, HMSO, London
Hollien, H., Shearer, W. and Hicks (Jr) J.W., Voice fundamental frequency levels of divers in helium–oxygen speaking environments, *Undersea Biomedical Research*, **4** 199–207 (1977).
Hydrogen helps divers to 531 m in comfort, *Offshore Engineering*, March (1988).
Rey, L. (ed.) *Arctic Underwater Operations*, Graham and Trotman (1988).
Rules and Regulations for the Construction and Classification of Submersibles and Diving Systems, Lloyd's Register of Shipping, London (1980).
Sisman, D. (ed.) *The Professional Diver's Handbook*, Submex, London (1982).
Sports Diving Manual, British Sub-Aqua Club (1986).
United States Navy Diving Manual, Best Publishing Co., California (1987).
Walker, P.A.; (ed.) *Safety of Diving Operations*, Graham and Trotman (1986).
Werts, M.F. and Shilling, C.W. (ed.) *Underwater Medicine and Related Sciences*, Volumes 2–5, Plenum Publishing (1971–1981).

Vehicles

Allwood, R.L. and Fyffe, A. *Pilot Study on Diver Replacement*, Institute of Offshore Engineering, Edinburgh (1979).
Allwood, R.L., Holmes, R.T. and Virr, L.E. *ROVs in diver support roles: military applications*, ROV '85 Conference Proceedings San Diego, Marine Technology Society (1985).
Boulton, S.B. Dual ROV-manned vehicles *Advances in Underwater Technology, Ocean Science and Offshore Engineering*, **5** Submersible Technology. pp 7–13, Society for Underwater Technology, Graham and Trotman (1985).
Busby, R.F. *Manned Submersibles*, US Office of the Oceanographer of the Navy, US National Technical Information Service (1976).
Busby, R.F. *Remotely Operated Vehicles*, US Office of the Oceanographer of the Navy, US National Technical Information Service (1979).
Dunbar, R.M., Roberts, S.J. and Wells, S.C. *Communications, bandwidth reduction and system studies for a tetherless unmanned submersible*. Oceans '81 Conference, Boston, Mass. Marine Technology Society, Washington DC.
Holmes, R.T. and Dunbar, R.M. Fourth generation ANGUS may launch free-swimming ROVER, *Offshore Engineer* (November), pp 19–20 (1979).
Myers, J.J. et al. *Handbook of Ocean and Underwater Engineering*, McGraw-Hill (1969).
Nicoll, G.R. and Dunbar, R.M. *Feasibility Study of a Free-Swimming, Unmanned, Remote Controlled Submersible*, Heriot-Watt University, Edinburgh, (Department of Industry Ref. EMRA RD 1195/08) (1979).
Operational Guidelines for Remotely Operated Vehicles, Marine Technology Society, Washington, DC (1984).
Remotely Operated Vehicle/Diver Involvement, Association of Offshore Diving Contractors (1985).
ROV Review 1986, Windate Enterprises (1986).
Society for Underwater Technology (ed.) *Submersible Technology*, Graham and Trotman (1986).
Talkington, H.R. *Undersea Work Systems*, Marcel Dekker (1981).
Trillo, R.L. (ed.) *Jane's Ocean Technology*, MacDonald and Jane's, London (1978).

Selection of underwater intervention method in oil and gas operations

Bachrach, A.J. and Egstrom, G.H. *Stress and Performance in Diving*, Best Publishing Co. (1987).
Chew, M.J. *Effective use of divers and remotely operated vehicles offshore*, Proceedings of 2nd International Conference on the maintenance of maritime and offshore structures, Inst. of Civil Engineers, London (1986).
Inspection costs at least £250,000 per anum per structure, *Offshore Engineer* (November), p 35 (1977).
Simpson, J. *The cost-effectiveness of ROV versus the diver in the role of inspection*, IRM '86 Conference Proceedings Offshore Conferences and Exhibitions Ltd, Kingston-upon-Thames.
Wardle H. *An assessment of divers' capabilities: a sample survey of industrial opinion 1977/78*, Underwater Engineering Group (UR15). London (1979).

General

Bevan, J. and Hunter, T. *The Underwater Industry Bibliography*, Submex, London (1988).
Myers, A. *Current Bibliography of Offshore Technology and Offshore Literature Classification*, ASR Marketing.

5 Marine Risers and Pipelines

Minoo H. Patel and Geoffrey J. Lyons

Contents

5.1 Introduction
    5.1.1 Marine risers
    5.1.2 Vertical and flexible risers
    5.1.3 Horizontal pipelines
    5.1.4 Analysis methods and design techniques
5.2 Marine risers
    5.2.1 Governing equations
    5.2.2 Methods of analysis
    5.2.3 Vortex shedding induced effects
    5.2.4 Design considerations
5.3 Pipelines
    5.3.1 Pipeline design procedures
    5.3.2 Sea bed stability and burial
    5.3.3 Pipe laying techniques
    5.3.4 PIGS and PIGging
    5.3.5 Types of PIG

References

5.1 Introduction

The operation of fixed and floating offshore structures requires the use of pipe connections between surface facilities and the sea bed as well as pipes laying on or below the sea bed for the transportation of oil and gas.

5.1.1 Marine risers

Pipes bridging the vertical separation between surface vessel and sea bed are called marine risers and are of two fundamental types. Since the 1950s drilling operations from fixed and floating offshore structures have been carried out by using jointed steel pipes of 0.204 m (8 in)–0.762 m (30 in) external diameter to act as a conduit for the drill pipe penetrating the sea bed. Such drilling risers connect the surface platform to the subsea well head. Drilling mud at high pressure is transported to the drill face through the hollow drill pipe and returns up to the surface vessel through the annulus between the drill pipe and drilling riser.

Marine risers are also used to transport oil and gas from producing fields for processing up to a surface platform and back down for export through a subsea pipeline or a tanker loading system. Vertical steel marine risers used for drilling or production split into two categories. Fixed offshore structures tend to use risers which are clamped at intervals to structural members of the platform along their vertical run up to the surface. On the other hand, floating or compliant offshore structures tend to use freely strung risers which are only connected at the surface vessel and sea bed. Such risers have to be held up with a sufficiently high tension at their top to prevent buckling due to self weight of their very slender geometry. Such risers also need to have heave compensating slip joints at their top end to take up the relative motion between the moving surface vessel and stationary sea bed.

5.1.2 Vertical and flexible risers

In recent years, the offshore industry has pioneered the use of pipes of composite steel and elastomer construction for use as marine risers. These so-called flexible risers are strung in non-vertical catenary shapes from the surface platform and are often supported by an intermediate buoy. Flexible risers are able to operate with much larger surface platform offset (from above the subsea well head) than is permissible with more rigid vertical steel risers which are generally limited to offsets of 7–10% of water depth.

Both vertical steel and flexible risers need to be carefully analysed during design to ensure that the pipes have acceptable levels of deformations, stresses and fatigue lives due to forces induced by currents, waves and surface vessel motions. The presence of internal hydrostatic pressure and external sea water pressure has a fundamental effect on the governing equations for these tubular structures as does the influence of the momentum of fluid flow within the pipe. The first part of this chapter is concerned with presenting a brief overview of analysis methods and design considerations for both vertical steel and flexible catenary marine risers.

5.1.3 Horizontal pipelines

Horizontal pipelines placed on the sea bed are an essential part of the production and transportation of offshore oil and gas. Such pipelines connect step out satellite wells to single centrally placed surface processing facilities as well as providing an export means either to offshore tanker loading terminals or to a shore base. Land pipelines have been extensively used for the transportation of fluids and today such pipelines can carry products as diverse as oil, gas, water, chemicals, slurries and even dry powders. The extension of land based pipeline technology to the offshore environment has required the development of remote pipe laying maintenance, repair and operating techniques.

The design of subsea pipelines has to take account of more rigorous external pressure induced loading and the thermal effects of being immersed in sea water. The line has to be designed to have acceptably low internal stress levels during laying and when subjected to functional and environmental loads. The line must also be provided with sufficient protection from internal and external corrosion. The second half of this chapter presents an overview of theoretical and practical aspects of the engineering of offshore pipelines.

5.1.4 Analysis methods and design techniques

A brief literature survey of analysis methods and design techniques used in marine risers is presented as a prelude to the remaining sections of this chapter. Morgan (1974–76) presents an interesting description of the historical development of marine riser technology and describes all the fundamental features that govern vertical steel riser design and selection.

Most analysis work on steel risers has been carried out to determine lateral motions and corresponding stresses due to forces induced by ocean currents, waves and surface vessel motions. Gardner and Kotch (1976), Sparks (1979), Patel *et al.* (1984), McIver and Lunn (1983), McNamara *et al.* (1986) and Wang (1983) present some typical analysis methods. The role of internal and external hydrostatic pressure in modifying the governing equations of motions was first identified by Young and Fowler (1978). The resultant concept of effective tension has now been incorporated generally into design methods for vertical steel risers. Vertical steel risers at high water depths are also susceptible to axial vibrations. These have been investigated by Sparks *et al.* (1982) and Miller and Young (1985).

In recent years, attention has been focused on design and analysis methods for flexible catenary risers. The work of De Oliveira *et al.* (1985), Ractliffe (1985),

Hansen and Bergan (1986), Owen and Qin (1986) and Narzul and Marion (1986) addresses the non-linear behaviour of catenary risers and the special considerations that have to be applied to their design. Many of the analysis techniques apply equally to pipeline analysis but Cowan and Andris (1977) and Kirk and Etok (1979) present two examples of such analyses applied to pipeline laying. Malahy (1986) gives a more modern analysis technique for pipe and cable laying conditions.

5.2 Marine risers

5.2.1 Governing equations

A vertical marine riser may be regarded as a hollow beam column. The difference between a column subjected to lateral loading and a marine riser is that the riser is subjected to both internal and external hydrostatic pressure as well as axial and lateral loadings.

If the riser is simply considered as a beam column then the governing differential equation used for lateral static deflection is:

$$\frac{d^2}{dy^2}\left(EI\frac{d^2x}{dy^2}\right) - T(x)\frac{d^2x}{dy^2} - w\frac{dx}{dy} = f \quad (5.1)$$

where EI is the riser stiffness, T is axial tension in the riser pipe wall, w is the weight per unit length of riser and contents and f is the lateral force per unit length. The coordinate system used is shown in *Figure 5.1* with x measured from the bottom of the riser and positive upwards while y denotes horizontal riser deflection from a vertical through the riser base.

If, however, the hydrostatic pressure is included in the analysis, a slightly different form of Equation 5.1 is arrived at. The force due to the hydrostatic external pressure distribution which exists around the riser and also the force due to internal pressure (which is related to well head pressure) are resolved into horizontal and vertical force components and incorporated into the governing equation for static deflection to give:

$$\frac{d^2}{dy^2}\left(EI\frac{d^2x}{dy^2}\right) - [T(y) + A_o p_o - A_i p_i]\frac{d^2x}{dy^2} \quad (5.2)$$

$$- (\gamma_s A_s - \gamma_o A_o + \gamma_i A)\frac{dx}{dy} = f$$

where the additional terms are the external hydrostatic pressure around the riser, p_o, the internal hydrostatic pressure, p_i, with A_o being the cross-sectional area of riser bore and wall, A_i, the cross-sectional area of riser bore only and A_s, the cross-sectional area of riser wall. γ_i is the specific weight of fluid in the riser bore, γ_o is the specific weight of fluid surrounding the riser tube (sea water) and γ_s is the specific weight of riser pipe wall material.

Equation 5.2 will be derived later but is valid for small deflections only, that is, for offset angles less than 10° from the vertical and, therefore, the error in applying this equation to a vertical steel riser is usually negligible. Some interesting points concerning the effects of pressure on the riser may be deduced by further consideration of the second term in Equation 5.2. The $(A_o p_o - A_i p_i)$ term comes from the lateral effect of external and internal hydrostatic pressure. Its effect is similar to that of the actual tension in the riser wall since this term also multiplies the second derivative of displacement, x. The pressure term does not modify the actual riser axial tension or the resultant direct stress in the riser wall. For this reason, the collection of parameters that multiply the second derivative is sometimes called effective tension, T_e, given by:

$$T_e = T_o + A_o p_o - A_i p_i \quad (5.3)$$

The concept of effective tension is a convenient mathematical grouping of parameters that have a similar effect. Equation 5.3 demonstrates that the effect of external hydrostatic pressure is similar to that of a tensile axial force, while the internal pressure influences riser behaviour as would a compressive force. $(\gamma_s A_s + \gamma_i A_i - \gamma_o A_o)$ is equivalent to the corresponding term w in Equation 5.1.

Now, the differential equation describing the static behaviour of a marine riser of arbitrary geometry is derived using the notation of *Figure 5.2* and the element of *Figure 5.3*. The analysis is restricted to two dimensions for simplicity. The static forces acting on the pipe element of *Figure 5.3* can be listed as follows:

1. An axial tension and shear force within the pipe wall material.
2. A horizontal force due to the resultant of external

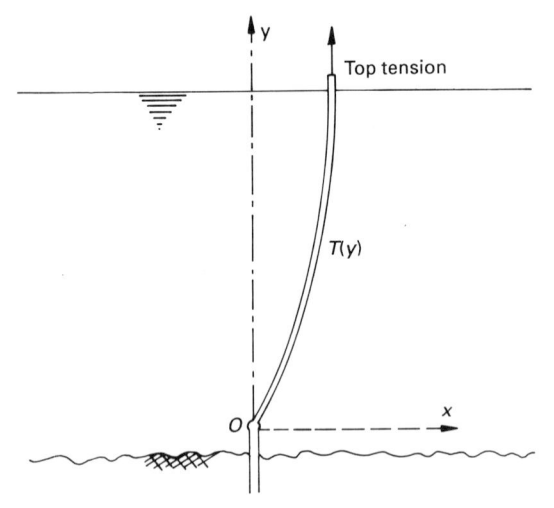

Figure 5.1 The conventional vertical riser coordinate system notation, with x measured from the bottom of the riser

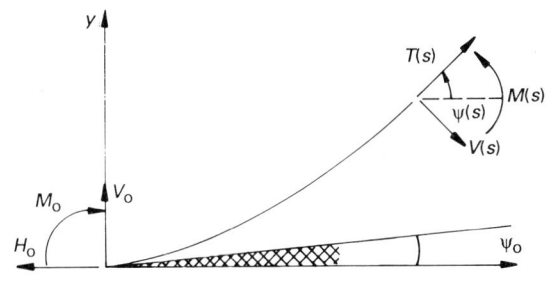

Figure 5.2 Notation for the static behaviour of a riser of arbitrary geometry, with the analysis restricted to two dimensions

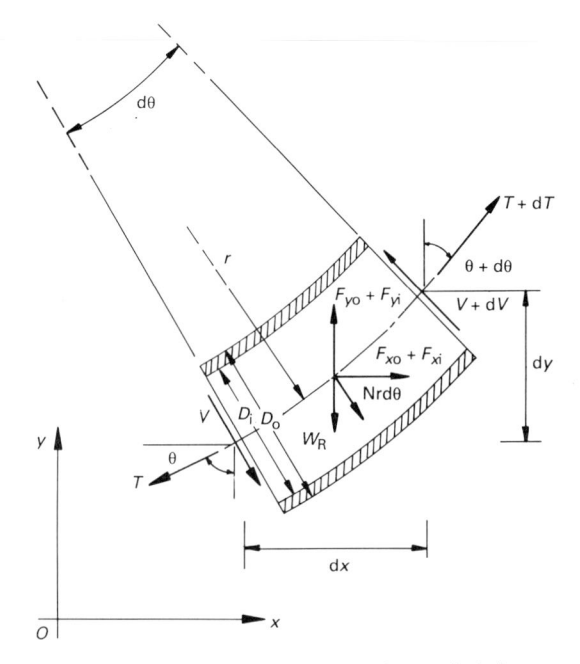

Figure 5.3 Riser element describing the static behaviour of a marine riser of arbitrary geometry

and internal hydrostatic pressures, called $(F_{xo} + F_{xi})$.
3. A vertical force due to the resultant of external and internal hydrostatic pressures $(F_{yo} + F_{yi})$.
4. A drag force due to steady current. The velocity vector is resolved into components normal and tangential to the element, with only the normal component assumed to exert a distributed force of N per unit length.
5. The weight of the element (W_R) acting vertically downwards.

Summing components of force in the y direction for the element in *Figure 5.3* yields the equation:

$(T + dT) \sin(\theta + d\theta)$
$- T \sin \theta - (V + dV) \cos(\theta + d\theta) + V \cos \theta$
$+ (F_{yo} + F_{yi}) - W_R - N \cos \theta \, r \, d\theta = 0$ (5.4)

where W_R is the element weight. Similarly, summing forces in the x direction yields:

$(T + dT) \cos(\theta + d\theta) - T \cos \theta$
$+ (V + dV) \sin(\theta + d\theta) - V \sin \theta$
$+ (F_{xo} + F_{xi}) + N \sin \theta \, r \, d\theta = 0$ (5.5)

These equations can be simplified for small $d\theta$ to:

$(T \cos \theta + V \sin \theta) d\theta + dT \sin \theta - dV \cos \theta + (F_{yo} + F_{yi}) - W_R - N \cos \theta \, r \, d\theta = 0$ (5.6)

and

$-(T \sin \theta - V \cos \theta) d\theta + dT \cos \theta + dV \sin \theta$
$+ (F_{xo} + F_{xi}) + N \sin \theta \, r \, d\theta = 0$ (5.7)

Combining these expressions gives:

$T \, d\theta - dV + (F_{yi} + F_{yo} - W_R) \cos \theta - (F_{xo} + F_{xi}) \sin \theta - N r \, d\theta = 0$ (5.8)

Continuing with the above analysis requires that the forces on a cylindrical element due to internal and external hydrostatic pressure $(F_{xo}, F_{xi}, F_{yo}, F_{yi})$ be defined. This is done using the derivation presented below.

A hollow cylindrical member submerged in a fluid and containing a fluid within itself will experience a force due to the external and internal hydrostatic pressures of both fluids acting on the surfaces of the cylinder. An element of the cylinder is shown in *Figure 5.4*. The resultant force is obtained by finding the force on an arbitrary section of the element (shaded portion of *Figure 5.4*), and resolving it into components before integrating to obtain the total force on the element. Note that only the force on the curved walls of the cylinder due to hydrostatic pressure is evaluated. The force on the end cross-sections is not considered here since the cylinder is taken to be very long and the end cross-section will usually terminate to a coupling such that hydrostatic pressure will not act on the cylinder cross-sections. Furthermore, the axes system used in *Figure 5.4* is such that the hydrostatic pressure is taken to increase linearly along the vertical axis.

Angle ϕ is used to describe position on the circumference of the element to be analysed. Initially, only external pressure is considered in the derivation. Forces due to internal pressure can be readily deduced from those due to external pressure by a simple reversal of signs and change of diameter.

As shown in *Figure 5.4*, the length, ds, of any strip on the cylinder circumference parallel to its axis is given by:

$$ds = \left(r + \frac{1}{2} D \cos \phi\right) d\theta$$ (5.9)

where r is the element radius of curvature and D is the diameter of the pressure bearing surface. θ and ϕ are

5/6 Marine risers and pipelines

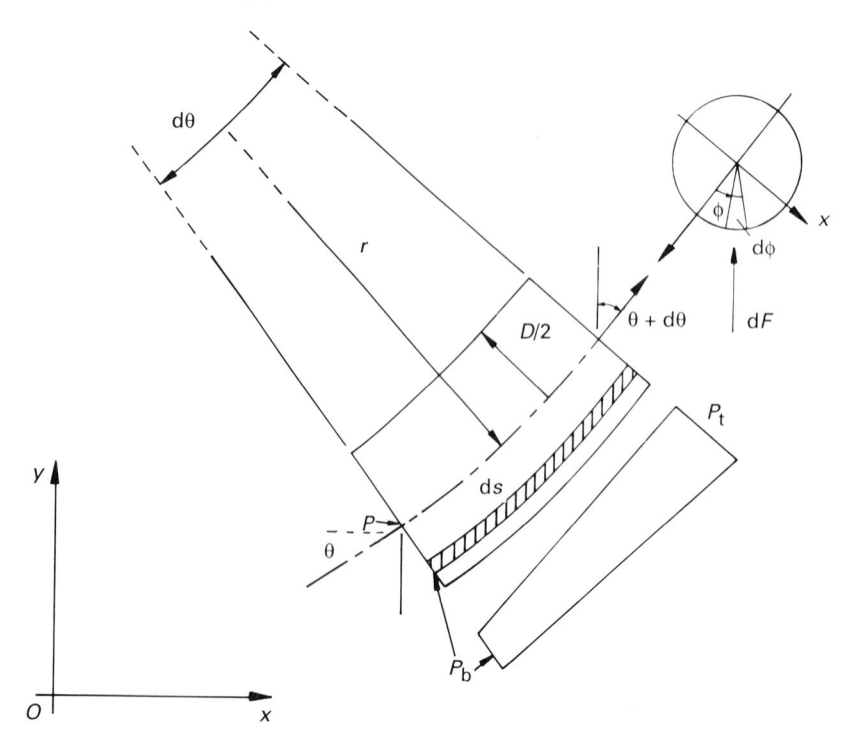

Figure 5.4 An element of a hollow cylindrical member, submerged in a fluid and containing fluid within itself

defined in *Figure 5.4*. If the hydrostatic pressure on the centre line of the element at its lower end is p, then the pressure, p_b, at various levels along the bottom surface is given by:

$$p_b = p - \frac{1}{2}\gamma D \cos\theta \cdot \sin\phi \qquad (5.10)$$

where γ is the weight per unit volume of the fluid medium. Also, the corresponding pressure p_t at the top of the element is hydrostatic, and given by the equation:

$$p_t = p_b - \gamma \cdot \sin\theta \qquad (5.11)$$

since

$$\sin\theta = \frac{dy}{ds}$$

The area of section of element described by arc $d\phi$ is given by:

$$dA = \frac{1}{2}D \cdot ds \cdot d\phi \qquad (5.12)$$

The force which acts on this section of the element is then:

$$dF = \frac{1}{2}(p_b + p_t)\,dA \qquad (5.13)$$

Substituting the expressions derived for dA, p_b, p_t into Equation 5.13 gives:

$$dF = \frac{1}{2}\left(p_b + p_b - \gamma\,ds\,\sin\theta\right)\left(\frac{1}{2}D \cdot ds \cdot d\phi\right)$$

$$dF = \left(p - \frac{1}{2}\gamma D \cos\theta \cdot \sin\phi - \frac{1}{2}\gamma\,ds\,\sin\theta\right)$$

$$\cdot \left(\frac{1}{2}D\,ds\,d\phi\right) \qquad (5.14)$$

Replacing Equation 5.9 for ds leads to:

$$dF = \left\{p - \frac{1}{2}\gamma D \cos\theta \cdot \sin\phi - \frac{1}{2}\gamma\left(r + \frac{1}{2}D\cos\phi\right)\right.$$

$$\left. \cdot d\theta\,\sin\theta\right\}\left(\frac{1}{2}r + \frac{1}{4}D\cos\phi\right)D\,d\theta\,d\phi \qquad (5.15)$$

Expanding the individual terms gives:

$$dF = \left[\frac{1}{2}p\,D\,r\,d\theta - \frac{1}{4}\gamma D\,r^2 \sin\theta\,(d\theta)^2\right]d\phi$$

$$+ \left[\frac{1}{4}p\,D^2\,d\theta + \frac{1}{4}\gamma D^2\,r\,\cos\theta\,d\theta\right.$$

$$-\frac{1}{4}\gamma D^2 r\sin\theta \,(d\theta)^2\bigg]\sin\phi \,s\phi + \bigg[\frac{1}{8}\gamma D^3 \cos\theta \,d\theta$$

$$-\frac{1}{8}\gamma D^3 \sin\theta \,(d\theta)^2\bigg]\sin^2\phi \,d\phi \quad (5.16)$$

The differential force may be resolved into it's three directional components: F_x, F_y and F_z along the x, y and z axes, respectively. In this case, the analysis is restricted to two dimensions and since there is no deformation out of the x–y plane, the resultant force in the z direction is taken as zero:

$$F_z = 0.0$$

$$dF_x = -dF \sin\theta \sin\phi \quad (5.17)$$

$$dF_y = dF \cos\theta \quad (5.18)$$

$$F_x = \int_{\phi=0}^{\phi=2\pi} -dF \sin\theta \sin\phi \quad (5.19)$$

$$F_y = \int_{\phi=0}^{\phi=2\pi} dF \cos\theta \sin\phi \quad (5.20)$$

Therefore:

$$F_x = -\bigg[\frac{1}{2}p\,D\,r\sin\theta \,d\theta - \frac{1}{4}\gamma r^2 D \sin^2\theta \,(d\theta)^2\bigg]$$

$$\cdot \int_{\phi=0}^{2\pi} \sin\phi \,d\phi - \bigg[\frac{1}{4}p\,D^2 \sin\theta \,d\theta$$

$$+ \frac{1}{4}\gamma D^2 r \sin\theta \cos\theta \,d\theta - \frac{1}{4}\gamma D^2 r \sin^2\theta (d\theta)^2\bigg]$$

$$\cdot \int_{\phi=0}^{2\pi} \sin^2\phi \,d\phi - \bigg[\frac{1}{8}\gamma D^3 \cos\theta\sin\theta \,d\theta$$

$$- \frac{1}{16}\gamma D^3 \sin\theta (d\theta)^2\bigg] \cdot \int_{\phi=0}^{2\pi} \sin^3\phi \,d\phi \quad (5.21)$$

Using:

$$\int_0^{2\pi} \sin\phi \cdot d\phi = 2, \quad \int_0^{2\pi} \sin^2\phi \cdot d\phi = \pi,$$

$$\int_0^{2\pi} \sin^3\phi \cdot d\phi = \frac{4}{3} \quad \text{and} \quad A = \pi D^2/4$$

gives the force in the x direction on a curved element as:

$$F_x = -[pA + r\gamma A(\cos\theta - \sin\theta \,d\theta)] \sin\theta \,d\theta \quad (5.22)$$

When the force due to internal pressure is considered, its form will be the same but of opposite sign. Combining the effects of internal and external pressure for the most general case gives:

$$(F_{xo} + F_{xi}) = [(p_i A_i - p_o A_o) + r(\gamma_i A_i - \gamma_o A_o)]$$
$$(\cos\theta - \sin\theta \,d\theta z0] \sin\theta \,d\theta \quad (5.23)$$

where p_o, p_i are the external and internal pressures, respectively, at the level of the bottom of the element centre line.

The vertical force F_y is obtained in a similar way. Before the integration is performed, the expression for the force in the vertical direction appears as:

$$F_y = \bigg[\frac{1}{2}p\,D\,r\,d\theta - \frac{1}{4}\gamma D\,r^2 \sin\theta \,(d\theta)^2\bigg]\cos\theta \int_0^{2\pi}$$

$$\sin\phi \,d\phi + \bigg[\frac{1}{4}p\,D^2\,d\theta + \frac{1}{4}\gamma D^2 r \cos\theta \,d\theta$$

$$-\frac{1}{4}\gamma D^2 r \sin\theta \,(d\theta)^2\bigg]\cos\theta \int_0^{2\pi} \sin^2\phi \,d\phi$$

$$+ \bigg[\frac{1}{8}\gamma D^3 \cos\theta \,d\theta - \frac{1}{8}\gamma D^3 \sin\theta \,(d\theta)^2\bigg]\cos\theta$$

$$\int_0^{2\pi} \sin^3\phi \,d\phi \quad (5.24)$$

In evaluating this integral the following are used:

$$\int_0^{2\pi} \sin\phi \,d\phi = 0$$

$$\int_0^{2\pi} \sin^2\phi \cdot d\phi = \pi \quad (5.25)$$

and

$$\int_0^{2\pi} \sin^3\phi \,d\phi = 0$$

Thus Equation 5.24 becomes:

$$F_y = [p\,A + r\,\gamma\,A\,(\cos\theta - \sin\theta \,d\theta)] \cos\theta \,d\theta \quad (5.26)$$

As before the effect of including the internal pressure acting on the element can be done quite easily. The final expression for the vertical force on a curved inclined element due to both internal and external pressure is given by:

$$(F_{yo} + F_{yi}) = [(p_o A_o - p_i A_i) + r(\gamma_o A_o - \gamma_i A_i)]$$
$$(\cos\theta - \sin\theta \,d\theta)] \cos\theta \,d\theta \quad (5.27)$$

Substituting Equations 5.23 and 5.22 for the resultant horizontal $(F_{xo} + F_{xi})$ and vertical forces, $(F_{yo} + F_{yi})$ due to internal and external hydrostatic pressure together with equations for the element weight and drag force into Equation 5.6 yields:

$$T \cdot d\theta - dV - [(p_i A_i - p_o A_o) + r(\gamma_i A_i - \gamma_o A_o) \\ (\cos\theta - \sin\theta \, d\theta)] \cos^2\theta \, d\theta - W_R \cos\theta - [(p_i A_i \\ - p_o A_o) + r(\gamma_i A_i - \gamma_o A_o)(\cos\theta - \sin\theta \, d\theta)] \sin^2\theta \, d\theta \\ - N r \, d\theta = 0 \quad (5.28)$$

and after simplication this becomes:

$$[T + p_o A_o - p_i A_i] \, d\theta - dV + \{(\cos\theta - \sin\theta \, d\theta)(\gamma_o A_o \\ - \gamma_i A_i) - \gamma_s A_s \cos\theta - N\} \, r \, d\theta = 0 \quad (5.29)$$

with $W_R = \gamma_s A_s \, r d\theta$ where γ_s is the weight per unit volume of the pipe material and A_s is the pipe wall area of cross-section. It is of interest at this stage to rewrite Equation 5.9 for a nearly vertical pipe. This can be done by using ϕ as the angle between the pipe element and the vertical such that $\phi = \pi/2 - \theta$ and $d\phi = -d\theta$.

Then Equation 5.9 can be rewritten in terms of ϕ as:

$$-[(T + p_o A_o - p_i A_i] \, d\phi - dV - \{(\sin\phi + \cos\phi \, d\phi) \\ (\gamma_o A_o - \gamma_i A_i) - \gamma_s A_s \sin\phi - N\} \, r \, d\phi = 0 \quad (5.30)$$

Now for small ϕ, the expressions:

$$\cos\phi = 1, \sin\phi = -\frac{dx}{dy}, r \, d\phi = -dy, \frac{d\phi}{dy} = -\frac{d^2x}{dy^2}$$

(5.31)

are substituted into Equation 5.30. After neglecting products of differentials, dividing by dy and using the small deflection equation:

$$\frac{dV}{dy} = \frac{d^2}{dy^2}\left(EI \frac{d^2 x}{dy^2}\right)$$

the equation becomes:

$$\underbrace{\frac{d^2}{dy^2}\left\{EI \frac{d^2 x}{dy^2}\right\}}_{A} - (T + p_o A_o - p_i A_i) \underbrace{\left\{1 + \left(\frac{dx}{dy}\right)^2\right\}^{-1}}_{B}$$

$$\cdot \frac{d^2 x}{dy^2} - (\gamma_s A_s - \gamma_o A_o + \gamma_i A_i) \underbrace{\frac{dx}{dy}}_{D} = N \underbrace{\left\{1 + \left(\frac{dx}{dy}\right)^2\right\}^{1/2}}_{C}$$

(5.32)

Note that term A in the above equation arises from the lateral effects of internal and external pressure and is the source of the concept of effective tension outlined earlier. Terms B and C, on the other hand, are due to the effects of riser orientation. Now because small deflections are assumed in vertical riser analysis, terms B and C in Equation 5.32 may be equated to unity to give:

$$\frac{d^2}{dy^2}\left(EI \frac{d^2 x}{dy^2}\right) - (T + p_o A_o - p_i A_i) \frac{d^2 x}{dy^2} \\ - (\gamma_s A_s - \gamma_o A_o + \gamma_i A_i) \frac{dx}{dy} = N$$

which is of similar form to Equation 5.2.

For a flexible riser, however, the approximation leading to elimination of terms B and C is not applicable. A more general approach is thus required. Now using the expression:

$$\frac{d\theta}{ds} = \frac{1}{r} = \frac{\dfrac{d^2 y}{dx^2}}{\left[1 + \left(\dfrac{dy}{dx}\right)^2\right]^{3/2}} \quad (5.33)$$

in Equation 5.29, dividing by (ds) and using:

$$\cos\theta = \frac{dx}{ds} = \frac{1}{\left[1 + \left(\dfrac{dy}{dx}\right)^2\right]^{1/2}}$$

$$\sin\theta = \frac{dy}{ds} = \frac{\dfrac{dy}{dx}}{\left[1 + \left(\dfrac{dy}{dx}\right)^2\right]^{1/2}}$$

$$\frac{dV}{ds} = \frac{dV}{dx} \cdot \frac{dx}{ds} = \frac{dV}{dx}\left[1 + \left(\dfrac{dy}{dx}\right)^2\right]^{-1/2}$$

reduces term of zeroth order to:

$$(T + p_o A_o - p_i A_i) \frac{d^2 y}{dx^2}\left[1 + \left(\frac{dy}{dx}\right)^2\right]^{-1} - \frac{dV}{dx}$$

$$+ (\gamma_o A_o - \gamma_i A_i - \gamma_s A_s) - N\left[1 + \left(\frac{dy}{dx}\right)^2\right]^{1/2} = 0 \quad (5.34)$$

Taking

$$T_e = T + p_o A_o - p_i A_i$$
$$w_e = \gamma_o A_o - \gamma_i A_i - \gamma_s A_s \quad (5.35)$$

gives

$$T_e \frac{d^2 y}{dx^2}\left\{1 + \left(\frac{dy}{dx}\right)^2\right\}^{-1} + w_e - \frac{dV}{dx}$$

$$- N\left\{1 + \left(\frac{dy}{dx}\right)^2\right\}^{-1/2} = 0 \quad (5.36)$$

Equation 5.36 is the final form of the governing equation for a flexible riser. A comparison of Equations 5.36 and 5.2 shows that some of the terms are common to both equations. For example, the effective tension, T_e and effective weight w_e occur in both equations. The additional terms T_v and w_v together with T_e and w_e now account for both the horizontal and vertical effects of hydrostatic pressure.

Equation 5.36 may also be rewritten in terms of ψ and s (see *Figure 5.2*) by using the following expressions:

$$\frac{dy}{dx} = \tan \psi, \quad \left[1 + \left(\frac{dy}{dx}\right)^2\right] = [1 + \tan^2\psi] = \sec^2\psi$$

and

$$\frac{d^2y}{dx^2} = \sec^3\psi \frac{d\psi}{ds}$$

to yield

$$T_e \frac{d\psi}{ds} + w_e \cos\psi - \left(\frac{dV}{ds} + N\right) = 0 \quad (5.37)$$

It is informative at this stage to transform Equation 5.37 into that for a simple catenary *in vacuo* so as to check its validity. Thus if:

$$p_o = p_i = 0, \; \gamma_o = \gamma_i = 0, \; T_e = T = H_o \sec\psi \quad (5.38)$$

and

$$w_e = -\gamma_s A_s = -w$$

then

$$T \frac{d\psi}{ds} - w \cos\psi - \left(N + \frac{dV}{ds}\right) = 0 \quad (5.39)$$

For zero drag force and zero shear variation, we get:

$$H_o \sec^2\psi \frac{d\psi}{ds} = w \quad (5.40)$$

which is the equation of the simple catenary. The complexity of the governing equations for a flexible riser requires that the solutions be obtained through a numerical finite element or finite difference computation.

5.2.2 Methods of analysis

A typical finite element analysis method used for vertical marine risers is described here. There are, however, some special features of such vertical marine risers that have to be accounted for in any analysis. These are briefly discussed first.

It is important to distinguish between the tension and non-tension contributing riser internal contents. Since the marine riser is a long, slender structure with relatively small bending stiffness, it needs to be kept in tension to prevent buckling collapse. Thus a tension is applied to the riser at its top, and it is the weight in air of the riser pipe, associated choke and kill lines and the vertical force due to internal and external hydrostatic pressure on a non-vertical pipe segment or buoyancy module which cause a variation in tension along the riser's length. The weight of the separately tensioned drill pipe and the riser fluid contents do not directly affect the tension variation. However, the non-tension contributing elements in a riser cross-section must be accounted for when computing an effective lateral force component (coefficient A in Equation 5.32).

For deep water risers, the top tension requirement to prevent buckling collapse can become excessive. In order to reduce top tension, buoyancy modules can be attached along the length of the riser. The distribution of buoyancy modules influences the tension variation in the riser, thus altering its structural response and internal stresses. However, the increase in diameter of a riser cross-section due to buoyancy modules also increases current and wave forces. This introduces considerable scope for optimizing the intensity and distribution of buoyancy modules in deep water applications.

5.2.2.1 Static analysis

The finite element analysis presented here is based on a governing equation of the form given by Equation 5.32 which is restricted for the moment to vertical risers by neglecting coefficients B and C in Equation 5.32. The description of the analysis is also restricted to two dimensions for simplicity. The vertical riser pipe is idealized as an assembly of beam elements as shown in *Figure 5.5*. Each element possesses six degrees of freedom, two translations and one rotation at each end. Consequently, the numerical computation is two dimensional with all external forces on the riser, including forces due to current and waves, acting on one plane.

The current loading q per unit length along the riser due to a lateral drag force is:

$$q = \tfrac{1}{2} \rho_o C_D d |U_c| U_c \quad (5.41)$$

where ρ_o is the density of sea water, C_D is a drag coefficient, d is an effective riser external diameter, and U_c is the vector of current velocity. The variation of current velocity with depth needs to be known.

A static analysis can also be used to relate riser deflections and stresses to current and wave loadings in a quasi-static manner. For a known regular wave height and period, the current velocity U_c can be superimposed on to the horizontal component of the wave particle velocity U_w at any instant. The quasi-static hydrodynamic loading can then be written as:

$$q = \tfrac{1}{2} \rho_o C_D d |U_c + U_w|(U_c + U_w) + \tfrac{1}{4} \rho_o \pi d^2 C_m \dot{U}_w \quad (5.42)$$

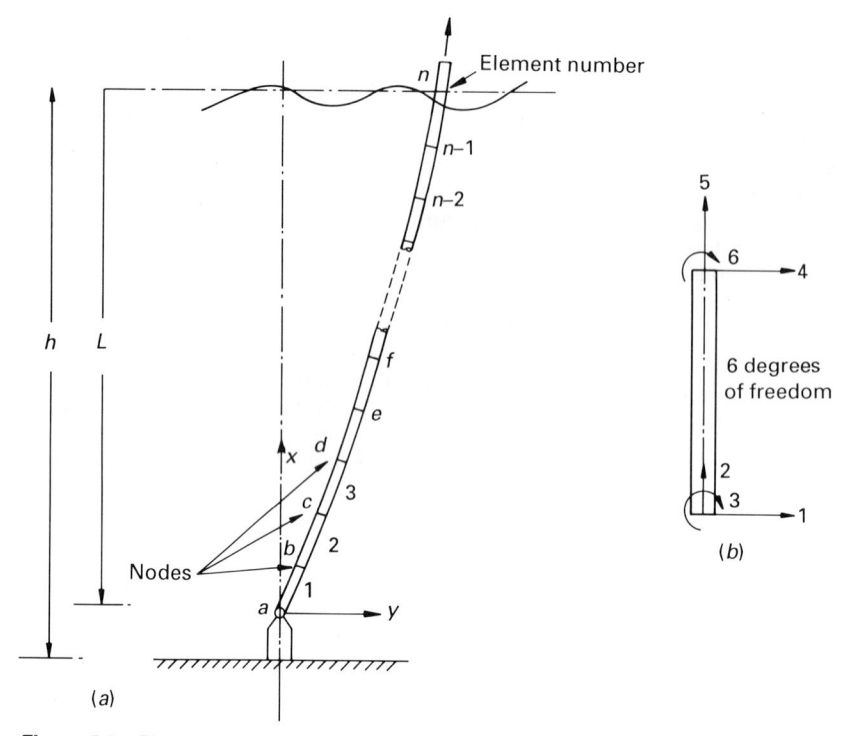

Figure 5.5 Element and global mode description for a finite element idealization of a vertical riser. (a) Riser element nodes; (b) Single beam element

$\dot{U}_w$ is the horizontal local component of wave particle acceleration and C_m is an applicable inertia coefficient.

The total stiffness matrix $\mathbf{S}_m$ for each beam element is derived as the sum of the standard elastic stiffness matrix $\mathbf{S}_e$ and a geometric stiffness matrix $\mathbf{S}_g$, which is a function of deflected element geometry and axial force on the element. Thus:

$$\mathbf{S}_m = \mathbf{S}_e + \mathbf{S}_g \tag{5.43}$$

For an element of length L and an effective axial tension T', $\mathbf{S}_m$ in member axes is given by the sum of:

$$\mathbf{S}_e = \frac{EI}{L^3}\begin{bmatrix} AL^2/I & & & & & \text{symmetric} \\ 0 & 12 & & & & \\ 0 & 6L & 4L^2 & & & \\ -AL^2/I & 0 & 0 & AL^2/I & & \\ 0 & -12 & -6L & 0 & 12 & \\ 0 & 6L & 2L^2 & 0 & -6L & 4L^2 \end{bmatrix} \tag{5.44}$$

and

$$\mathbf{S}_g = \frac{T'}{L}\begin{bmatrix} 0 & & & & & \text{symmetric} \\ 0 & 6/5 & & & & \\ 0 & L/10 & 2L^2/15 & & & \\ 0 & 0 & 0 & 0 & & \\ 0 & -6/5 & -L/10 & 0 & 6/5 & \\ 0 & L/10 & -L^2/30 & 0 & -L/10 & 2L^2/15 \end{bmatrix} \tag{5.45}$$

where A is the area of steel cross-section, E is Young's modulus and I is an appropriate second moment of area. The local effective axial tension T' is calculated by accounting for the modification due to hydrostatic pressures in the surrounding fluid, as described earlier.

The fixed end action vectors $\mathbf{A}_{mL}$ are obtained by using an assumed shape function $N(x)$ in conjunction with a total lateral load distribution $w(x)$. This load is due to both the hydrodynamic loading q and an effective lateral load derived from term D in governing Equation 5.32 with dx/dy obtained from an initially assumed undeflected riser configuration. Thus:

$$\mathbf{A}_{mL} = -\int_0^1 w(x)\, N(x)\, dx \tag{5.46}$$

where x is the vertical distance from the bottom ball joint and l is the total riser length. The final static member and actions $\mathbf{A}_m$ are then obtained from:

$$\mathbf{A}_m = \mathbf{A}_m = \mathbf{A}_{mL} + \mathbf{S}_m \mathbf{D}_m \qquad (5.47)$$

where $\mathbf{D}_m$ is the nodal displacement matrix. These combined end actions are applied incrementally in order to account for the changes in term D and the non-linear behaviour caused by large deflections of the riser pipe. Thus $\mathbf{A}_m$ is divided into a specified number of equal increments $\Delta \mathbf{A}_m$ which are applied progressively to obtain the incremental displacements $\Delta \mathbf{D}$ through the equation:

$$\Delta \mathbf{D} = \mathbf{S}^{-1} \Delta \mathbf{A}_m \qquad (5.48)$$

where $\mathbf{D}$ and $\mathbf{S}$ are the overall displacement and overall stiffness matrices in global coordinates. The overall stiffness matrix is re-evaluated after each load increment to account for the change in geometry due to large deflections.

The static analysis is executed in different ways depending on the type of dynamic analysis which is to follow. For frequency domain dynamic analysis, the effects of current are considered in the preceding static analysis. However, the non-linear time domain method requires that the steady current, unsteady wave velocities and riser velocity be summed before applying the non-linear square law drag force. For this reason, the static analysis preceding the time domain dynamic analysis only accounts for self weight, buoyancy and pressure forces on the riser and excludes the current velocity. The loading due to current is then accounted for in the time domain dynamic analysis.

5.2.2.2 Dynamic analysis

The differential equation of motion for a system with many degrees of freedom and having a mass matrix $\mathbf{M}_T$ can be written as:

$$\mathbf{M}_T \ddot{\mathbf{D}} + \mathbf{C}\dot{\mathbf{D}} + \mathbf{S}\mathbf{D} = \mathbf{F} \qquad (5.49)$$

where $\mathbf{D}$ is the matrix of nodal displacements, and $\mathbf{C}$ and $\mathbf{S}$ are the structural damping matrix and the overall stiffness matrix, respectively; all are defined in global riser axes.

The external force matrix $\mathbf{F}$ due to wave action on the system is obtained from a modified form of Morison's equation:

$$\mathbf{F} = \rho_o \mathbf{V}\dot{\mathbf{U}} + \rho_o \mathbf{V}(C_m - 1)(\dot{\mathbf{U}} - \ddot{\mathbf{D}})$$
$$+ \mathbf{B}|\mathbf{U} - \dot{\mathbf{D}}|(\mathbf{U} - \dot{\mathbf{D}}) \qquad (5.50)$$

where $\mathbf{V}$ is the vector of elemental volumes, $\mathbf{B}$ is the matrix of hydrodynamic drag coefficients and $\mathbf{U}$ and $\dot{\mathbf{U}}$ are the horizontal components of wave particle velocities and accelerations. It is assumed here that the fluid induced forces on a structure are given by the linear superposition of a drag force and an inertia force. The first two terms of Equation 5.50 signify Froude–Krylov and added mass forces, respectively, while the last term describes the drag force.

By substituting Equation 5.50 into Equation 5.49 and replacing $[\mathbf{M} + \rho_o(C_m - 1)\mathbf{V}]$ by $\mathbf{M}_T$, $\rho_o C_m \mathbf{V}$ by $\mathbf{M}_H$ and re-arranging we get:

$$\mathbf{M}_T \ddot{\mathbf{D}} + \mathbf{C}\dot{\mathbf{D}} + \mathbf{S}\mathbf{D} = \mathbf{M}_H \dot{\mathbf{U}} + \mathbf{B}|\mathbf{U} - \dot{\mathbf{D}}|(\mathbf{U} - \dot{\mathbf{D}}) \qquad (5.51)$$

The above matrix equation cannot be used directly for incorporating the boundary condition at the surface vessel which requires that the riser top end must follow the surge motion of the surface platform. This known horizontal riser nodal translation at the surface (denoted by suffix B) can be separated from all other unknown degrees of freedom (denoted by suffix A), through the following matrix partitioning:

$$\begin{bmatrix} \mathbf{M}_{T\,AA} & \mathbf{M}_{T\,AB} \\ \mathbf{M}_{T\,BA} & \mathbf{M}_{T\,BB} \end{bmatrix} \begin{bmatrix} \ddot{\mathbf{D}}_A \\ \ddot{\mathbf{D}}_B \end{bmatrix} + \begin{bmatrix} \mathbf{C}_{AA} & \mathbf{C}_{AB} \\ \mathbf{C}_{BA} & \mathbf{C}_{BB} \end{bmatrix} \begin{bmatrix} \dot{\mathbf{D}}_A \\ \dot{\mathbf{D}}_B \end{bmatrix}$$
$$+ \begin{bmatrix} \mathbf{S}_{AA} & \mathbf{S}_{AB} \\ \mathbf{S}_{BA} & \mathbf{S}_{BB} \end{bmatrix} \begin{bmatrix} \mathbf{D}_A \\ \mathbf{D}_B \end{bmatrix} = \begin{bmatrix} \mathbf{M}_{H\,AA} & \mathbf{M}_{H\,AB} \\ \mathbf{M}_{H\,BA} & \mathbf{M}_{H\,BB} \end{bmatrix} \begin{bmatrix} \dot{\mathbf{U}}_A \\ \dot{\mathbf{U}}_B \end{bmatrix}$$
$$+ \begin{bmatrix} \mathbf{B}_{AA} & \mathbf{B}_{AB} \\ \mathbf{B}_{BA} & \mathbf{B}_{BB} \end{bmatrix} \times \begin{bmatrix} |\mathbf{U}_A - \dot{\mathbf{D}}_A| \cdot (\mathbf{U}_A - \dot{\mathbf{D}}_A) \\ |\mathbf{U}_B - \dot{\mathbf{D}}_B| \cdot (\mathbf{U}_B - \dot{\mathbf{D}}_B) \end{bmatrix} + \begin{bmatrix} 0 \\ \mathbf{F}_B \end{bmatrix}$$
$$(5.52)$$

Here, $\mathbf{F}_B$ is a force required to cause the specified surge motion at the surface. The dynamic response of the riser structure in terms of the remaining degrees of freedom can be obtained solely from the upper set of equations in Equation 5.52, which do not contain $\mathbf{F}_B$.

5.2.2.3 Element property formulation

In the formulation of the beam element mass matrix, the lumped mass or the consistent mass approach may be used. In the former, the entire mass is assumed to be concentrated at nodes where the translational degrees of freedom are defined. For such a system, the mass matrix has a diagonal form. Off diagonal terms disappear since the acceleration of any nodal point mass would only produce an inertia force at that point. The consistent mass formulation, however, makes use of the finite element concept and requires that the mass matrix be computed from the same shape functions that are used in deriving the stiffness matrix. Coupling due to off diagonal terms exists, and rotational as well as translational degrees of freedom need to be considered.

In theory, this consistent mass approach can lead to greater accuracy, although this improvement is believed to be small. On the other hand, the lumped mass formulation is easier to apply because fewer degrees of freedom are involved, leading to a simpler definition of element properties. The lumped mass formulation is chosen for this analysis because the advantages of a small improvment in accuracy for the consistent mass approach are outweighed by the additional computational effort entailed in its implementation.

It having been noted that off diagonal terms of $\mathbf{M}_H$, $\mathbf{M}_T$ and $\mathbf{B}$ are zero for the lumped mass formulation, the following equations are obtained from Equation 5.52:

$$\mathbf{M}_{T\,AA}\ddot{\mathbf{D}}_{2A} + \mathbf{C}_{AA}\dot{\mathbf{D}}_A + \mathbf{S}_{AA}\mathbf{D}_A = \mathbf{M}_{H\,AA}\dot{\mathbf{U}}_A$$
$$+ \mathbf{B}_{AA}|(\mathbf{U}_A - \dot{\mathbf{D}}_A)|(\mathbf{U}_A - \dot{\mathbf{D}}_A) - \mathbf{C}_{AB}\dot{\mathbf{D}}_B - \mathbf{S}_{AB}\mathbf{D}_B \qquad (5.53)$$

At the end of the static analysis, the stiffness matrix of the structure in its deformed position is available. In modelling the dynamic response about this mean statically deflected shape, the stiffness matrix is assumed to remain constant throughout the dynamic analysis.

In the lumped mass approach, all the rotational degrees of freedom need to be substructured out. Since vertical wave forces are not significant for the riser system, the vertical translation degrees of freedom can also be eliminated. This feature can lead to a substantial reduction in computer time and storage in the dynamic analysis. The horizontal degrees of freedom having been segregated, the force deflection equations can be written in partitioned form as:

$$\begin{bmatrix} \mathbf{S}_{HH} & \mathbf{S}_{HN} \\ \mathbf{S}_{NH} & \mathbf{S}_{NN} \end{bmatrix} \begin{bmatrix} \mathbf{D}_H \\ \mathbf{D}_N \end{bmatrix} = \begin{bmatrix} \mathbf{F}_H \\ 0 \end{bmatrix} \qquad (5.54)$$

where subscripts H and N denote the horizontal and the other group of vertical and rotational degrees of freedom, respectively.

From Equation 5.54:

$$\mathbf{D}_N = -\mathbf{S}_{NN}^{-1}\mathbf{S}_{NH}\mathbf{D}_H \qquad (5.55)$$

The condensed stiffness matrix suitable for use in the equations of motion is then:

$$\mathbf{S}_{HH}^* = \mathbf{S}_{HH} - \mathbf{S}_{HN}\mathbf{S}_{NN}^{-1}\mathbf{S}_{NH} \qquad (5.56)$$

The matrix is further partitioned to separate the top surge degree of freedom:

$$\mathbf{S}_{HH}^* = \begin{vmatrix} \mathbf{S}_{AA} & \mathbf{S}_{AB} \\ \mathbf{S}_{BA} & \mathbf{S}_{BB} \end{vmatrix} \qquad (5.57)$$

where subscript B denotes the vessel motion as before.

The mass matrix for each element is built up by concentrating half of the total mass of mud, pipes and buoyancy material at each end of the element. For a fully submerged vertical element, the added mass associated with unit horizontal body acceleration is $\rho_o(C_m - 1)V$, where C_m is an inertia coefficient which includes the Froude–Krylov effect. Taking half the added mass to be lumped at each node, the added mass submatrix for each element is:

$$\begin{bmatrix} \tfrac{1}{2}\rho_o(C_m - 1)V & 0 \\ 0 & \tfrac{1}{2}\rho_o(C_m - 1)V \end{bmatrix} \qquad (5.58)$$

This added mass matrix and the real mass matrix are summed together to give the total mass matrix $\mathbf{M}_{T\,AA}$.

The manner in which the partially submerged element at the water surface is idealized depends on the amount by which the element is wetted at the mean sea level. If the wetted length L_s is less than half the element length, all the added mass is lumped at the lower node and the element submatrix becomes:

$$\begin{bmatrix} \rho_o(C_m - 1)A_x L_s & 0 \\ 0 & 0 \end{bmatrix} \qquad (5.59)$$

where A_x is the total cross-sectional area of the riser element, including buoyancy elements when present. Should L_s be greater than half the element length L, the added mass associated with the lower half of the element is concentrated at the lower node, while the rest of the hydrodynamic effects are taken to act on the top node. The element submatrix for such a situation is:

$$\begin{bmatrix} \tfrac{1}{2}\rho_o(C_m - 1)A_x L & 0 \\ 0 & \rho_o(C_m - 1)A_x(L_s - L/2) \end{bmatrix} \qquad (5.60)$$

For the riser structure, this appears to be a simple and logical way to treat the element at the water surface in the lumped mass formulation. The hydrodynamic mass matrix $\mathbf{M}_{H\,AA}$, which includes Froude–Krylov forces, is built up from element submatrices in a similar manner. The submatrices corresponding to Equations 5.59 and 5.60, respectively, are:

$$\begin{bmatrix} \rho_o C_m A_x L_s & 0 \\ 0 & 0 \end{bmatrix} \qquad (5.61)$$

$$\begin{bmatrix} \rho_o C_m A_x L_s & 0 \\ 0 & \rho_o C_m A_x(L_s - L/2) \end{bmatrix} \qquad (5.62)$$

Due to the unit relative horizontal velocity $(U - \dot{D})$, the horizontal drag force on a fully submerged element is $(1/2)\rho_o C_D L d$. The hydrodynamic damping submatrix for such an element is:

$$\begin{bmatrix} \tfrac{1}{4}\rho_o C_D L d & 0 \\ 0 & \tfrac{1}{4}\rho_o C_D L d \end{bmatrix} \qquad (5.63)$$

$$\begin{bmatrix} \tfrac{1}{2}\rho_o C_D L_s d & 0 \\ 0 & 0 \end{bmatrix}, \text{ for } L_s \leq L/2 \qquad (5.64)$$

and

$$\begin{bmatrix} \tfrac{1}{4}\rho_o C_D L d & 0 \\ 0 & \tfrac{1}{2}\rho_o C_D(L_s - L/2)d \end{bmatrix}, \text{ for } L_s > L/2 \qquad (5.65)$$

The structural damping matrix may be explicitly defined as:

$$\mathbf{C} = \alpha_0 \mathbf{M}_T + \alpha_1 \mathbf{S} \qquad (5.66)$$

To obtain the coefficients α_0 and α_1, the damping ratios ζ_1 and ζ_2 in any two modes need to be specified.

An eigen value analysis is carried out to find the natural frequencies corresponding to the two modes chosen.

For Rayleigh damping:

$$\begin{bmatrix} \zeta_1 \\ \zeta_2 \end{bmatrix} = \frac{1}{2} \begin{bmatrix} \frac{1}{\omega_1} & \omega_1 \\ \frac{1}{\omega_2} & \omega_2 \end{bmatrix} \begin{bmatrix} \alpha_0 \\ \alpha_1 \end{bmatrix} \quad (5.67)$$

From Equation 5.67:

$$\alpha_0 = \frac{2(\zeta_1 \omega_2 - \zeta_2 \omega_1)}{\frac{\omega_2}{\omega_1} - \frac{\omega_1}{\omega_2}} \quad (5.68)$$

$$\alpha_1 = \frac{2(\zeta_1 \omega_2 - \zeta_2 \omega_1)}{\frac{\omega_2}{\omega_1} - \frac{\omega_1}{\omega_2}} \quad (5.69)$$

A damping ratio of 5% in the first two modes is usually chosen for all the analyses carried out in this work. The actual level of structural damping that should be specified is rather unclear in current literature.

5.2.2.4 Solution in the frequency domain

A linearized form of the equation of motion may be obtained by replacing the drag term in Equation 5.53 with a suitable equivalent linear damping term which is proportional to the relative velocity ($U_w - D_A$). For such a linear system:

$$M_{T\,AA}\ddot{D}_A + (C_{AA} + B_{eq\,AA})\dot{D}_A + S_{AA}D_A = M_{H\,AA}\dot{U}$$
$$+ B_{eq\,AA}U - C_{AB}\dot{D}_B - S_{AB}D_B \quad (5.70)$$

Since the current velocity imposed is not sinusoidal, only the wave particle velocity U_w and the structure velocity $\dot{D}_A$ can be included in the fluid interaction term. The stiffness matrix in the frequency analysis will therefore be obtained from the final statically deformed shape caused by current and riser internal forces. From linear wave theory, the elevation ξ of a single wave train may be represented by:

$$\xi = r\cos(ky - \omega t) \quad (5.71)$$

The corresponding horizontal wave particle velocities and accelerations are given by:

$$U_w = \omega r \frac{\cos(x - 1 + h)k}{\sinh kh} \cos(ky - \omega t) \quad (5.72)$$

$$\dot{U}_w = \frac{\omega^2 r \cos(x - 1 + h)k}{\sinh kh} \sin(ky - \omega t) \quad (5.73)$$

Rewriting Equation 5.72 in complex form gives:

$$U_w = \mathrm{Re}\left[\omega r \frac{\cos(x - 1 + h)}{\sinh kh} e^{iky} e^{-i\omega t}\right] \quad (5.74)$$

or

$$\mathrm{Re}(U_w')\,e^{-i\omega t} \quad (5.75)$$

where U_w' is a complex amplitude. Similarly:

$$\dot{U}_w = \mathrm{Re}(-i\omega U_w' e^{-\omega t}) \quad (5.76)$$

In the steady state, the response of the system represented by Equation 5.70 to a sinusoidal wave will also be proportional to $e^{-i\omega t}$. Thus:

$$D_A = \mathrm{Re}(D_A' e^{-i\omega t}) \quad (5.77)$$

where D_A' is complex. Differentiating Equation 5.77 and substituting Equations 5.75 and 5.76 into Equation 5.70 gives:

$$[S_{AA} - M_{T\,AA}\omega^2 - i\omega(C_{AA} + B_{eq\,AA})]D_A'$$
$$= M_{H\,AA}(-i\omega U_w) + B_{eq\,AA}U_w' = F' \quad (5.78)$$

where F' is a complex forcing function and B_{eq} is an equivalent damping matrix described in the following section.

Since the matrix B_{eq} contains a term in $\dot{D}$, available only from the final solution, an iterative calculation scheme needs to be derived. Starting from a trial solution for the velocities $\dot{D}$, B_{eq} is estimated and the simultaneous complex Equation 5.78 are solved for a new set of displacements and velocities D_A and $\dot{D}_A$. These velocities are compared with the original values ($\dot{D}$) and the whole calculation is repeated with a better estimate of B_{eq} until the real and imaginary parts of $\dot{D}$ and $\dot{D}_A$ differ by a small specified tolerance.

5.2.2.5 Linearization of the damping term

Since damping forces are responsible for the dissipation of energy in a vibratory system, the obvious and most common way of obtaining B_{eq} is to equate the work done by the linearized and the non-linear forces such that:

$$B_{eq}(U - \dot{D}) \equiv B|(U - \dot{D})|(U - \dot{D}) \quad (5.79)$$

For the purpose of illustration, a convenient node where y is assumed to be zero is chosen. From Equation 5.72, the wave particle velocity is:

$$U = R\cos\omega t \quad (5.80)$$

where

$$R = \frac{\omega r \cosh k(x - 1 + h)}{\sinh kh}$$

Let the corresponding riser nodal velocity be defined by:

$$\dot{D} = Q\cos(\omega t - \phi) \quad (5.81)$$

where Q is the amplitude of vibration velocity and ϕ is an arbitrary phase difference. The relative velocity is:

$$(U - \dot{D}) = R \cos \omega t - Q \cos(\omega t - \phi)$$
$$= R_T \cos(\omega t - \Psi)$$

where

$$R_T = (R^2 - 2RQ \cos\phi + Q^2)^{1/2} \quad (5.82)$$

$$\tan \Psi = \frac{-Q \sin \phi}{R - Q \cos\phi}$$

The work done by the damping force $B|(U - \dot{D})|(U - \dot{D})$ over an elemental displacement dD may be written as:

$$dW = B|R_T \cos(\omega t - \Psi)|R_T \cos(\omega t - \Psi) Q \cos(\omega t - \phi) d(\omega t) \quad (5.83)$$

On substituting β for $(\omega t - \Psi)$, we can express the work done over a complete wave period by this non-linear term as:

$$W = \int_{-\Psi}^{2\pi-\Psi} B|R_T \cos\beta|R_T \cos\beta\, Q \cos(\beta + \Psi - \phi)\, d\beta$$

$$= \int_{-\Psi}^{2\pi-\Psi} B|R_T \cos\beta|R_T \cos\beta\, Q \cos(\beta + \gamma)\, d\beta$$

$$= \frac{8}{3} Q B R_T^2 \cos\gamma \quad (5.84)$$

by splitting up the limits of integration to account for the modulus sign, and assuming that $\gamma = (\Psi - \phi)$ is time independent.

The work done by an equivalent linearized damping force $B_{eq}(U - \dot{D})$ over a wave cycle is readily obtained from:

$$W = \int_0^{2\pi} B_{eq} R_T \cos(\omega t - \Psi) Q \cos(\omega t - \phi)\, d(\omega t)$$

$$= \pi Q B_{eq} R_T \cos\gamma \quad (5.85)$$

Finally, equating the work done by the two damping terms gives:

$$B_{eq} = \frac{8}{3\pi} B R_T$$

Hence, the damping coefficient can be used in Equation 5.78:

$$B_{eq} = \frac{8}{3\pi} B|(U - \dot{D})_{max}| \quad (5.86)$$

5.2.2.6 Initial trial solution

To ensure that Equation 5.78 converges rapidly to the final solution, a reasonably accurate initial estimate of the displacements $\mathbf{D}$ and thus the velocities $\dot{\mathbf{D}}$ is required for evaluating the equivalent damping matrix from the total mass matrix and the diagonal terms of the stiffness matrix. Then, assuming a damping ratio of 10%, the initial estimate of $\mathbf{B}_{eq}$ is taken to be:

$$\mathbf{B}_{eq} = $$

$$0.2 \begin{bmatrix} (M_{TAA}^{11} S_{AA}^{11})^{1/2} & 0 & 0 \\ 0 & (M_{TAA}^{22} S_{AA}^{22})^{1/2} & 0 \\ 0 & 0 & (M_{TAA}^{NN} S_{AA}^{NN})^{1/2} \end{bmatrix}$$

(5.87)

This matrix is substituted into Equation 5.78, which is subsequently solved for the initial trial solution. This method leads to rapid convergence with only two or three iterations required for forcing frequencies away from the structure resonant frequencies. Up to ten iterations may be necessary at and around resonance frequencies.

5.2.2.7 Time series analysis

The basic method of analysis here involves integrating Equation 5.49 through discrete steps in time and accounting for the non-linear drag loading without a linearization approximation.

In the equation of motion (Equation 5.51) the generalized fluid velocity can be decomposed into the static current velocity $\mathbf{U}_c$ and a wave particle velocity $\mathbf{U}_w$. Thus Equation 5.51 becomes:

$$\mathbf{M}_T \ddot{\mathbf{D}} + \mathbf{C} \dot{\mathbf{D}} + \mathbf{S} \mathbf{D} = \mathbf{M}_H \dot{\mathbf{U}}_w + \mathbf{B}|(\mathbf{U}_w + \mathbf{U}_c - \dot{\mathbf{D}})|$$
$$\cdot (\mathbf{U}_w + \mathbf{U}_c - \dot{\mathbf{D}}) \quad (5.88)$$

where $\dot{\mathbf{U}}_c$ is taken to be zero for the current velocity. The requirement to sum the current and wave velocities before applying the resultant loading through the square law relationship requires that the current velocity be ignored in the static analysis that precedes this time domain calculation.

The time step integration of the equation of motion also allows irregular wave sequences (and the corresponding surface vessel surge responses) to generate dynamic excitation forces on the riser. This wave sequence can be specified in two ways. A wave elevation spectrum of the incident irregular wave can be used to compute the corresponding spectra of the subsurface wave velocities and acceleration as well as the spectrum of surface vessel surge motions. These spectra can be Fourier transformed to generate corresponding time series of these quantities for use in the dynamic analysis. However, this procedure is cumbersome and computationally time consuming. Therefore, a simple alternative method is usually employed. The incident wave elevation is specified as a 'frequency comb' sum of individual sinusoidal components with randomly distributed phase angles. The subsurface wave kinematics and surface vessel surge response are then readily computed by summing the effects of all the sinusoidal components in the wave spectrum.

The numerical time step integration technique proposed by Newmark is used with the following relations:

$$\dot{\mathbf{D}}_{t+\Delta t} = \dot{\mathbf{D}}_t + [(1-\delta)\ddot{\mathbf{D}}_t + \delta\ddot{\mathbf{D}}_{t+\Delta t}]\Delta t$$
$$\mathbf{D}_{t+\Delta t} = \mathbf{D}_t + \dot{\mathbf{D}}_t\Delta t + [(\tfrac{1}{2}-\beta')\ddot{\mathbf{D}}_t + \beta'\ddot{\mathbf{D}}_{t+\Delta t}]\Delta t^2 \quad (5.89)$$

where β' and δ are parameters which can be varied to achieve an acceptable integration accuracy and stability. Subscript t denotes the variable at the begining of the time interval Δt.

The direct integration analysis does rely on selection of an appropriate time step which must be small enough to obtain sufficient accuracy, although a time step smaller than necessary would reflect on the cost of the solution. Bathe and Wilson (1976) have analysed the stability and accuracy of various numerical integration schemes and suggested that, for reasonable accuracy, the time step to period ratio be not more than 1/6 for the highest significant mode. In its standard form, the Newmark technique is unconditionally stable.

The two parameters δ and β' introduced in Equation 5.89 indicate how the acceleration is modelled over the time interval. $\delta = 1/2$ and $\beta' = 1/6$ correspond to a linearly varying acceleration. Newmark's original scheme which is pursued here uses $\delta = 1/2$ and $\beta' = 1/4$ and gives a constant average acceleration based integration scheme. Using these latter values in Equation 5.89 and re-arranging gives:

$$\ddot{\mathbf{D}}_{t+\Delta t} = \frac{4}{(\Delta t)^2}[\mathbf{D}_{t+\Delta t} - \mathbf{D}_t - (\Delta t)\dot{\mathbf{D}}_t] - \ddot{\mathbf{D}}_t$$

$$\dot{\mathbf{D}}_{t+\Delta t} = \frac{2}{\Delta t}[\ddot{\mathbf{D}}_{t+\Delta t} - \ddot{\mathbf{D}}_t] - \dot{\mathbf{D}}_t \quad (5.90)$$

Then expressing Equation 5.88 explicity at instant $(t + \Delta t)$ and using the lumped mass approach with the top vessel surge motion duly separated as in Equation 5.53, we get:

$$\mathbf{M}_{T\,AA}\ddot{\mathbf{D}}_{A,\,t+\Delta t} + \mathbf{C}_{AA}\dot{\mathbf{D}}_{A,\,t+\Delta t} + \mathbf{S}_{AA}\mathbf{D}_{A,\,t+\Delta t}$$
$$= \mathbf{M}_{H\,AA}\dot{\mathbf{U}}_{WA,\,t+\Delta t} + \mathbf{B}_{AA}|(\mathbf{U}_{WA} + \mathbf{U}_{CA}$$
$$- \dot{\mathbf{D}}_A)|_{t+\Delta t}\,(\mathbf{U}_{WA} + \mathbf{U}_{CA} - \dot{\mathbf{D}}_A)_{t+\Delta t} - \mathbf{C}_{AB}\dot{\mathbf{D}}_{B,\,t+\Delta t}$$
$$- \mathbf{S}_{AB}\mathbf{D}_{B,\,t+\Delta t} \quad (5.91)$$

Substituting Equation 5.90 into 5.91 and re-arranging gives:

$$\left[\frac{4}{(\Delta t)^2}\mathbf{M}_{T\,AA} + \frac{2}{(\Delta t)}\mathbf{C}_{AA} + \mathbf{S}_{AA}\right]\mathbf{D}_{A,\,t+\Delta t}$$
$$= \mathbf{M}_{H\,AA}\dot{\mathbf{U}}_{WA,\,t+\delta t} + \mathbf{B}_{AA}|(\mathbf{U}_{WA} + \mathbf{U}_{CA} - \dot{\mathbf{D}}_A)|_{t+\Delta t}$$
$$\cdot (\mathbf{U}_{WA} + \mathbf{U}_{CA} - \dot{\mathbf{D}}_A)_{t+\Delta t} - \mathbf{C}_{AB}\dot{\mathbf{D}}_{B,\,t+\Delta t}$$
$$- \mathbf{S}_{AB}\mathbf{D}_{B,\,t+\Delta t}\left[\frac{4}{\Delta t}\mathbf{M}_{T\,AA} + \mathbf{D}_{AA}\right]\dot{\mathbf{D}}_{A,\,t}$$
$$+ \left[\frac{4}{(\Delta t)^2}\mathbf{M}_{T\,AA} + \frac{2}{(\Delta t)}\mathbf{C}_{AA}\right]\mathbf{D}_{A,\,t} + \mathbf{M}_{T\,AA}\ddot{\mathbf{D}}_{A,\,t}$$
$$= \mathbf{F}_{t+\Delta t} \quad (5.92)$$

This is the basic equation used in the time step integration scheme.

The solution scheme assumes that displacement, velocity and acceleration vectors at time zero, denoted by subscript 0, are known and the solution is required from time zero to time τ. The given time span τ is subdivided into equal time intervals Δt (where $\Delta t + \tau$ divided by the number of time intervals). The algorithm calculates the solution at the next required time from known information at the previous time steps. The process is repeated until the solution at all discrete time points is known.

To initialize the numerical solution, the acceleration corresponding to zero time is derived from the reduced form of Equation 5.92 giving:

$$\ddot{\mathbf{D}}_A = \mathbf{M}_{T\,AA}^{-1}[\mathbf{M}_{H\,AA}\dot{\mathbf{U}}_{WA0} + \mathbf{B}_{AA}|(\mathbf{U}_{WA} + \mathbf{U}_{CA})_0|$$
$$\cdot (\mathbf{U}_{WA} + \mathbf{U}_{CA})_0 - \mathbf{C}_{AB}\dot{\mathbf{D}}_{B0} - \mathbf{S}_{AB}\mathbf{D}_{B0}] \quad (5.93)$$

In arriving at Equation 5.93 the unknown value of velocity $\dot{\mathbf{D}}_{A,\,t+\Delta t}$ of the forcing vector of Equation 5.92 has been approximated to $\dot{\mathbf{D}}_{A,\,t}$. The approximation gives an acceptable degree of accuracy provided the time step chosen is sufficiently small. An alternative approach to this would require an elaborate iterative scheme with a significantly greater computation effort.

From the set of simultaneous Equations 5.92 the displacements are simply obtained from:

$$\mathbf{D}_{A,\,t+\Delta t} = \mathbf{J}^{-1}\mathbf{F}_{t+\Delta t} \quad (5.94)$$

where

$$\mathbf{J} = \frac{4}{\Delta t^2}\mathbf{M}_{T\,AA} + \frac{2}{\Delta t}\mathbf{C}_{AA} + \mathbf{S}_{AA} \quad (5.95)$$

The inversion of matrix $\mathbf{J}$ in the above equation can be made more efficient by the use of banded equation solvers as suggested by Bathe and Wilson (1976). However, $\mathbf{J}$ is independent of time and needs to be inverted once only.

When $\mathbf{D}_{A,\,t+\Delta t}$ is known, the accelerations and velocities at $(t + \Delta t)$ are derived from Equations 5.90.

5.2.2.8 Typical results

Finite element calculations of the type presented here can be validated by a number of methods.

Table 5.1 Ten element idealization of weightless tensioned 152 m (500 ft) beam

Parameters:		
Total length	152.400 m	
Applied tension	54.422 tf	
Uniform load intensity*	6.035×10^{-3} t/m	
Results:	Analytical solution	FE idealization
Slope at ends (rad)	0.007 1	0.007 1
Maximum moments (tf–m)	0.863 1	0.868 9
Lateral displacements (m)		
Node		
1,11	0.000 0	0.000 0
2,10	0.104 4	0.104 4
3,9	0.191 4	0.191 3
4,8	0.254 9	0.254 8
5,7	0.293 3	0.293 1
6	0.306 1	0.306 0

*Equivalent to load caused by 0.5 m/s current

Table 5.2 Input parameters for API cases

Distance from mean sea level to riser support ring	15.24 m
Distance from sea floor to bottom ball joint	9.144 m
Water depth	152.4 m
Riser pipe outer diameter	0.406 4 m
Riser pipe wall thickness	0.015 87 m
Choke line outer diameter	0.101 6 m
Choke line wall thickness	0.016 51 m
Kill line outer diameter	0.101 6 m
Kill line wall thickness	0.016 51 m
Buoyancy material outer diameter	0.609 6 m
Modulus of elasticity of riser pipe	2.1×10^7 t/m^2
Density of sea water	1.025 t/m^3
Density of mud	1.438 t/m^3
Drag coefficient	0.7
Added mass coefficient	1.5
Effective diameter for wave–current load	0.660 4 m
Density of buoyancy material	0.160 2 t/m^3
Current at surface	0.257 4 m/s
Surface vessel static offset	4.572 m
Weight per unit length of riser joint in air	0.2565 tf/m
Wave height	6.096 m
Wave period	9.0 s
Vessel surge amplitude	0.609 6 m
Vessel surge phase angle	15°

For the static analysis, the finite element formulation can be checked by comparison with the analytic result for an idealized weightless tensioned beam. A typical such comparison is shown in *Table 5.1*. Such comparisons can confirm the validity of the computational procedure as well as indicating the number of finite elements required for an acceptable level of accuracy.

The American Petroleum Institute committee on the standardization of offshore structures defined a set of

Table 5.3 Results of static analyses of API cases

Case	Max. bending stress		Max. total stress (ksi)	Angles from vertical (deg)	
	Value (ksi)	Location* (ft)		Lower BJ	Top
500–0–1	3.99 (2.53)	104 (111)	5.46 (4.34)	3.64 (2.94)	0.44 (0.82)
500–0–2	1.62 (0.94)	104 (115)	6.98 (6.80)	2.58 (2.20)	0.96 (1.21)
500–20–1S	5.92 (5.86)	442 (461)	9.43 (9.51)	4.35 (3.66)	−1.17 (−0.79)
500–20–2S	3.90 (4.27)	442 (463)	10.381 (10.54)	2.79 (2.51)	0.04 (0.24)

*Above lower ball joint
†Mean figures from the API bulletin 2J are quoted below each solution in parentheses
‡ksi is kilo pounds force per square inch BJ is ball joint

test risers as a basis for comparing the performance of riser analysis methods for both static and dynamic loadings. Nine anonymous participants to this study submitted solutions for the various test cases and API *Bulletin* 2J (1977) gives the overall comparisons. *Table 5.2* gives the input data for one of the API test cases, and *Table 5.3* displays the corresponding static analyses results. These are displayed in terms of maximum bending stress value and position, maximum total stress (axial plus peak bending), as well as upper and lower riser angles from the vertical. Results of the analysis presented here are given in *Table 5.3* with the mean values from the nine API test cases displayed in parentheses.. It is clear from these comparisons that results from the API *Bulletin* and the present method agree reasonably well.

The frequency domain and time domain dynamic analyses presented here have also been compared with the dynamic analyses in the API *Bulletin*. *Figures 5.6* and *5.7* show typical results for one of the API test risers; the plotted API values are the maximum and minimum of the combined results from the nine calculations compiled in the bulletin. The frequency domain analysis is computed conventionally using a regular wave period of 9 s and wave height of 6.096 m (20.3 ft). The time domain analysis uses a single frequency comb to produce equivalent data but with the non-linear drag force due to current and wave velocities included in the calculations. It should be emphasized that none of the results published in the API *Bulletin* has to our knowledge been directly validated by measurements on full scale risers. Nevertheless this comparison gives an indication of agreement between the other methods and the analysis presented here.

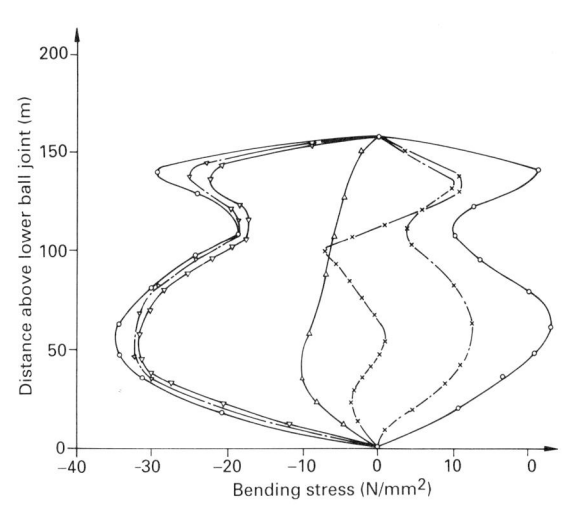

Figure 5.7 The effects of non-linear fluid loading on riser response; displacement and stresses for a marine riser, compared with API case 500-20-1D; bending stresses. Key: Δ = static value; 0 = API results; ∇ = static and dynamic minimum; + = static and dynamic maximum; — = frequency domain analysis; - - - = time domain analysis

5.2.2.9 Influence of non-linearity on structural response

A comparison of the time domain and frequency domain analyses presented in *Figures 5.6* and *5.7* gives an indication of the effects of non-linear fluid loading on the riser response. A static current profile is included, and so time domain and frequency domain results differ markedly owing to the effect of the square law drag force with and without linearization. However, the frequency domain results are at lower values for the induced stresses.

The finite element analysis and the frequency domain and time domain solutions outlined in this section attempt to balance the small computing cost advantages of linearization against the additional accuracy available from the non-linear time domain calculation. The frequency domain analysis uses the linearization approximation of equal energy dissipation between non-linear damping and equivalent linear damping in the solution. An alternative linearization technique for frequency domain analysis has been tested by Krolikowski and Gray (1980). It is based on a statistical minimization of mean squared error between the non-linear damping force and its linear representation used in the analysis. The statistical approach uses linearization at the discrete frequency components of a wave spectrum to arrive at a global linearized damping force with a least squares minimized error. This technique allows a frequency domain method to be applied over a wider frequency range, in contrast to the linearization method used in the analysis presented here which is used for regular waves only.

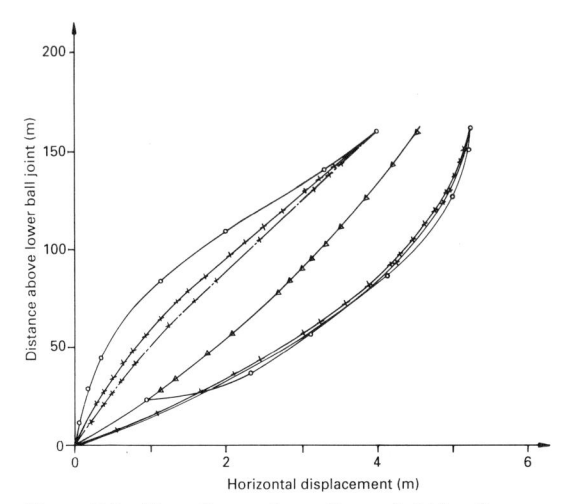

Figure 5.6 The effects of non-linear fluid loading on riser response; displacement and stresses for a marine riser, compared with API case 500-20-1D; horizontal displacement

The technique of linearization by least squares minimization is not followed up in the frequency domain analysis presented here. This is because both riser methods developed here have been aimed at computing riser motions and stresses, the latter for feeding into fatigue calculations based on linear elastic theory or fracture mechanics. The fracture mechanics approach demands that representative stress time histories for a marine riser in waves be known in detail, particularly in terms of the sequences of stress cycles that are likely to occur. A computationally efficient time domain analysis is capable of producing this information, whereas frequency domain analyses, whatever their level of sophistication in linearization, operate in the frequency domain where the phase information which governs wave sequencing is lost.

A further feature which has promoted the use of an efficient time domain analysis for riser calculations is based on the comparative performance of the frequency domain and time domain analyses which shows that there are substantial differences in peak stresses between the two analyses. These discrepancies may be reduced by a more sophisticated linearization technique in the frequency domain analyses, but the discrepancies do highlight the importance of modelling the non-linear fluid loading on the riser cross-section in a physically representative manner.

An additional problem associated with marine risers occurs in the analysis of multitube production risers of complex cross-sectional geometries. These may be made up of a central structural riser with a number of large diameter satellite flow lines or as a bundle or array of flow lines. The beam finite element analysis techniques described in this paper need to be extended to these production risers. Patel and Sarohia (1982) suggest one solution by equivalencing a production riser of complex cross-section to a simpler single-tube marine riser, which is then used for the finite element analysis. This approach is sufficient for a global riser analysis, but it needs to be used with care when localized riser fluid forces or member stresses are required. Krolikowsky (1981) presents an alternative frequency domain approach.

The analysis of flexible catenary risers is also carried out by finite element methods which are extended to take account of the catenary goemetry of a flexible riser and also of non-linear behaviour with large deformations. It is usual in analysis methods for flexible risers to use an elastic catenary solution to obtain an initial riser profile for the static analysis prior to applying current loads. The resultant deformed riser profile is then used as the initial configuration for a frequency domain or time domain dynamic analysis. The tendency is to use the time domain method since both geometric and fluid drag induced non-linearities are believed to have a greater effect on the behaviour of flexible risers than they do on vertical steel risers.

Figure 5.8 presents a typical configuration of a flexible riser with an intermediate mid water depth buoy. *Figure 5.9* presents the corresponding displacement envelope of the riser as it is excited by current, waves and surface vessel motions.

There are also examples of essentially horizontal flexible pipelines that arise when subsea pipelines are fabricated at an onshore location and towed out to an offshore site for installation. Such lines can be 1–5 km in length and 0.3–1.0 m (1 ft–3 ft 3 in) in diameter. These extremely slender and flexible structures have to be kept in tension during the towing process and their weight per unit length, wave induced motions and

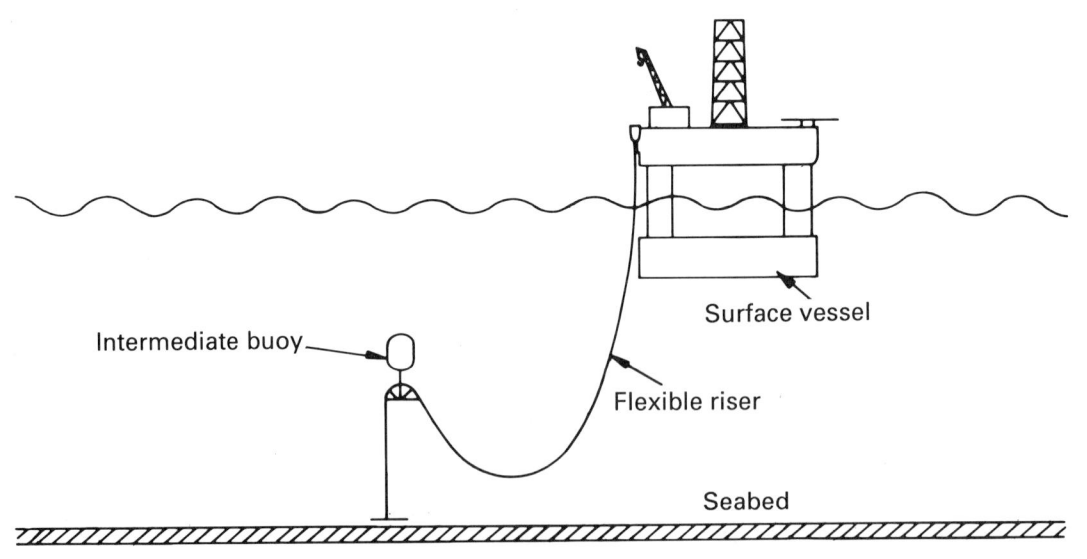

Figure 5.8 Typical configuration of a flexible riser with an intermediate mid water depth buoy

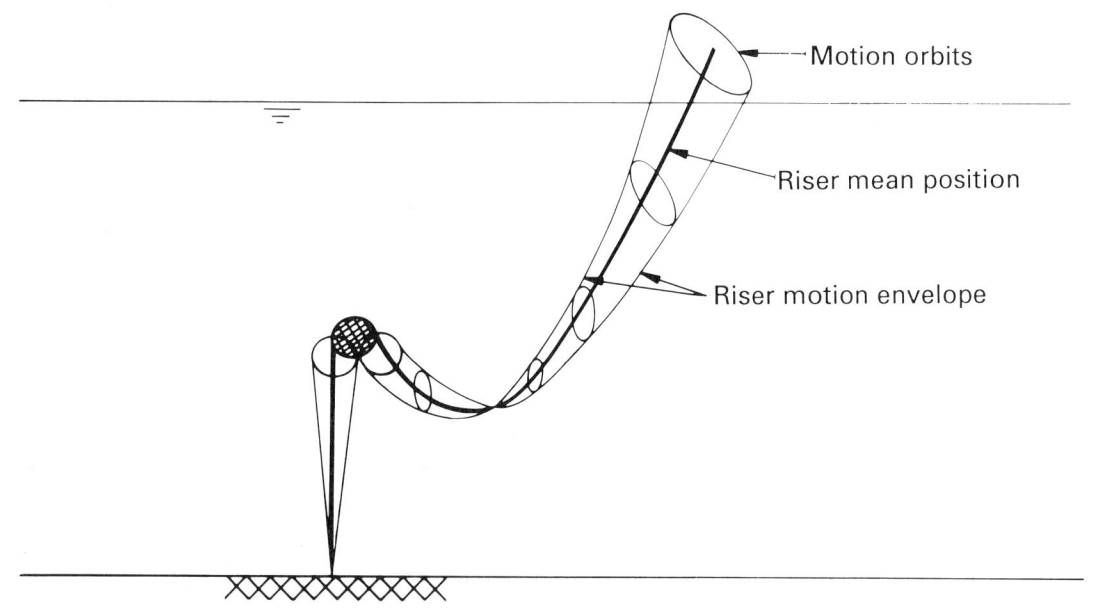

Figure 5.9 Deformation behaviour of a flexible riser motion envelope excited by current, waves and surface vessel motions

stresses as well as installation procedures have to be carefully designed to prevent catastrophic loss of the line during installation. Patel and Seyed (1987) present typical calculations that can be carried out for towed pipelines.

5.2.3 Vortex shedding induced effects

It was Strouhal (1878) who found that Aeolian tones generated by a wire were proportional to the wind speed V divided by the wire thickness D. Later Rayleigh (1986) proved that the vortex shedding frequency was not only a function of V/D but also of Reynolds number ($Re = VD/\nu$) where ν is kinematic viscosity. *Figure 5.10* shows the variation of Strouhal number, S, with Reynolds number for a smooth stationary circular cylinder as determined by more recent researchers. Absolute values of S have been shown to depend also on cylinder surface roughness, length to diameter ratios and turbulence levels, Sarpkaya and Isaacson (1981).

The physical mechanism of vortex shedding from bluff cylinders is as follows. A particle flows towards the leading edge of the cylinder, the pressure in the fluid particle rises from the free stream pressure to the stagnation pressure. The high pressure near the leading edge impels the developing boundary layers around both sides of the cylinder. However, the pressure forces are not sufficient to force the boundary layers around the back side of bluff cylinders at high Reynolds numbers. Near the widest section of the cylinder, the boundary layers separate from each side of the cylinder

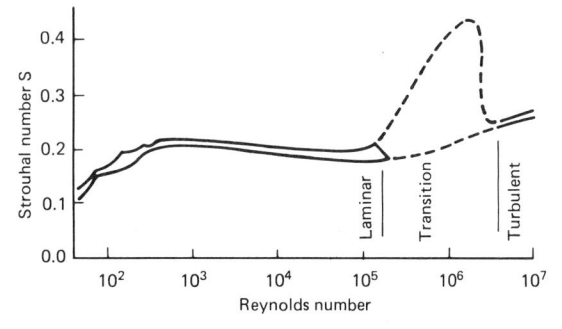

Figure 5.10 Strouhal number variation with Reynolds number for a smooth stationary circular cylinder

surface and form two free shear layers that trail aft in the flow. These two free shear layers bound the wake. Since the innermost portion of the free shear layers moves much more slowly than the outermost portion of the layers which are in contact with the free stream, the free shear layers tend to roll up into discrete, swirling vortices. A regular pattern of vortices is formed in the wake which can interact with the cylinder motion and is the source of vortex induced vibrations.

The major regimes of vortex shedding from a circular are given in *Figure 5.11*, adapted from Leinhard (1966). The vortex sheet evolves constantly as it flows downstream of the cylinder with the lateral to streamwise spacing necking down to a minimum a short distance downstream of the cylinder, before increasing,

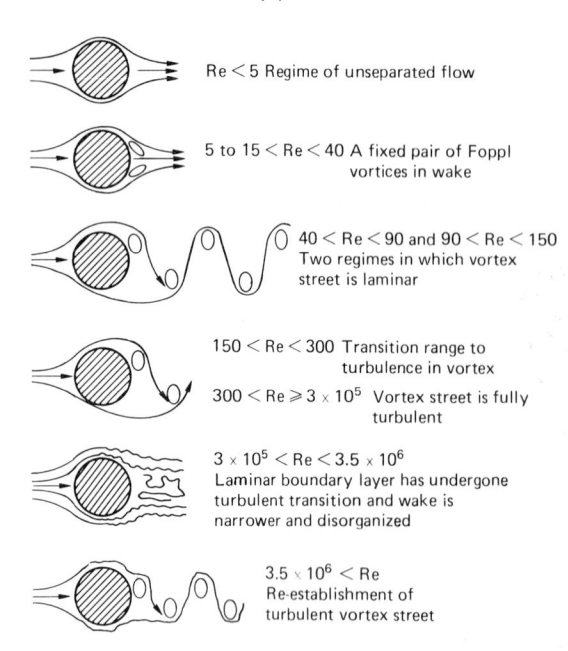

Figure 5.11 The major regimes of vortex shedding in fluid flow across circular cylinders

Re < 5	Regime of unseparated flow
5 to 15 < Re < 40	A fixed pair of Foppl vortices in wake
40 < Re < 90 and 90 < Re < 150	Two regimes in which vortex street is laminar
150 < Re < 300	Transition range to turbulence in vortex
300 < Re ⩾ 3 × 10^5	Vortex street is fully turbulent
3 × 10^5 < Re < 3.5 × 10^6	Laminar boundary layer has undergone turbulent transition and wake is narrower and disorganized
3.5 × 10^6 < Re	Re-establishment of turbulent vortex street

Scraeffer and Eskanazi (1959). It has been shown that the wake can be strongly three dimensional, Humphreys (1960) and Roshko (1953).

5.2.3.1 Vortex induced vibrations

Vortex induced forces that act along the in line and transverse direction to the excitation flow give rise to additional riser response. Considerable work has been done to investigate these effects in steady flow, Chryssostomidis and Patrikalakis (1984), Every et al. (1981), Griffin et al. (1973), (1980), (1984), Hall (1981), Jacobsen et al. (1984), King et al. (1973), Pelzer and Rooney (1984), Schafer (1984), Tsahalis (1984). However, the flow around a riser will generally vary with time and axial location owing to the oscillatory and depth decaying nature of waves, possibly complicated by surface vessel motions. Griffin and Ramberg (1982), Blevins (1977), King (1977), Sarpkaya and Isaacson (1981), Shaw (1979) and Simpson (1978) give comprehensive reviews of the state of the art in respect of vortex shedding and associated vibrations. The CIRIA Report, Hallam et al (1978), presents a background and some example solutions for simplified vortex induced vibration problems.

Many of the theories that have been developed to predict the vortex induced oscillations of bluff cylindrical members attempt to include physical phenomena underlying the fluid mechanics of vortex behaviour and the structural member response to the consequent loading.

The first of these is lock on of vortex shedding frequencies (determined by the Strouhal relationship) to a natural frequency of the cylinder. Thus:

$f_v = f_n$ at lock on

and

$f_v = f_s \, SV/D_s$ otherwise

The following parameters are of major importance in determining the amplitude of vibrations and the range of lock on or synchronization for a given body:

$$\text{reduced damping, } k_s = \frac{2m\delta}{\rho D_s^2} \quad (5.97)$$

$$\text{reduced velocity, } V_r = V/(f_n \, D_s) \quad (5.98)$$

The reduced damping is the product of the logarithmic decrement of structural damping (it does not include fluid damping) and the mass density of the structure relative to the fluid.

The reduced velocity may be used to determine the existence and degree of vortex induced vibration. For transverse vibrations of a cylindrical bluff body, it has been shown from experiments in water by many researchers that excitation begins when the reduced velocity reaches a value of between 3.5 and 5.0. A peak occurs around 6.0 and decay to no vibration at around 8.0–12.0 (see *Figure 5.12* and *Figure 5.13*).

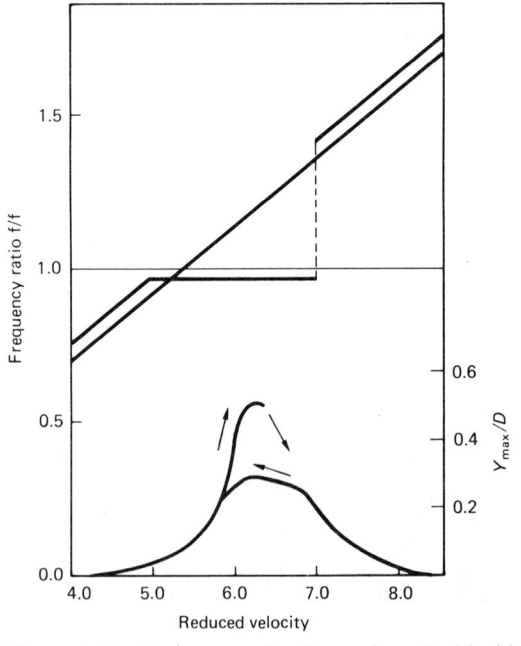

Figure 5.12 Transverse vibrations of a cylindrical bluff body; resonance of a rigid right cylinder

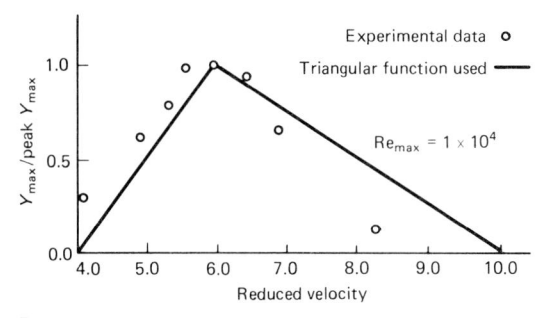

Figure 5.13 Transverse vibrations of a cylindrical bluff body; oscillations of a circular cylinder

For in line oscillations, the onset of vibration occurs at reduced velocities around 1.0 –1.5 for the case where two vortices of opposing sign are shed symmetrically and continues to a reduced velocity of about 2.5, when a stream of alternating vortices is formed. Vibration ceases for the in line case at a reduced velocity of 3.0–3.5 (see *Figure 5.14* adapted from Dean *et al.* (1977)).

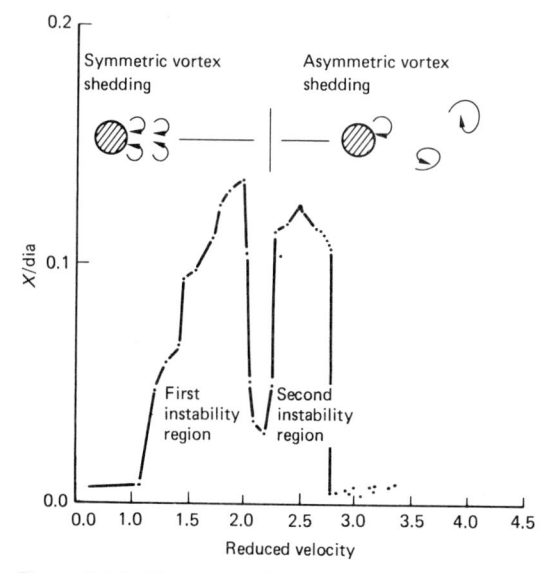

Figure 5.14 The onset of vibration for in-line oscillations of a circular cylinder

It has been shown for right circular cylinders in uniform flow, that there is a relationship between the maximum amplitude of transverse vibrations and the reduced damping which applies for flexible and flexibly mounted rigid cylinders. Various workers give this relationship as follows and as compared in *Figure 5.15*: Griffin *et al.* (1973) give:

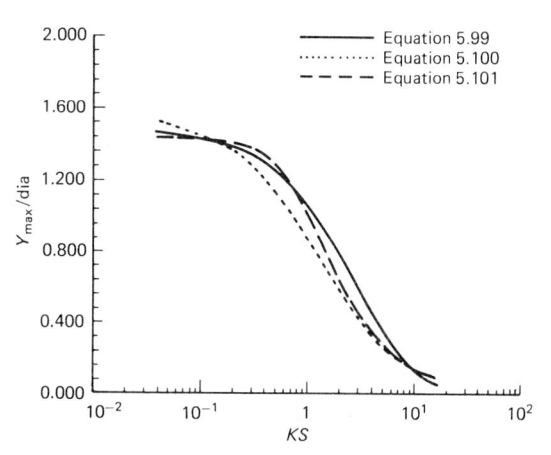

Figure 5.15 The relationship between maximum transverse vibration amplitude and reduced damping for mounted cylinders, using Equations (5.99)–(5.101) for circular cables

$$\frac{Y_{max}}{D} = \frac{1.29\,\gamma}{[1 + 0.43\,(2\pi S^2\,k_s)]^{3.35}} \qquad (5.99)$$

Blevins (1977) gives:

$$\frac{Y_{max}}{D} = \frac{0.07\,\gamma}{(1.9 + k_s)\,S^2}\left[0.3 + \frac{0.72}{(1.9 + k_s)\,S}\right]^{1/2} \qquad (5.100)$$

Iwan (1975) gives:

$$\frac{Y_{max}}{D} = \frac{\gamma}{[1 + 9.6\,(k_s/\pi)^{1.8}]} \qquad (5.101)$$

where a geometric function of mode shape:

$$\gamma = \xi_{max}\,(x/L)\left[\frac{\int_0^L \xi^2(x)\,dx}{\int_0^L \xi^4(x)\,dx}\right] \qquad (5.102)$$

is used to collapse the data for the different modes of response for the systems shown in *Figure 5.16*. ξ_{max} is the maximum value of the modal shape ξ over the span extending from $x = 0$ to $x = L$.

5.2.3.2 Analysis models

The non-linear wake oscillator model initially proposed by Bishop and Hassan (1964) and pursued by others, including Blevins (1977) and Hartlen and Currie (1970), is based on a modified Van der Pol Equation. This has been developed because it exhibits many of the features of interaction between the structure and its wake at resonance. Model parameters must be determined from curve fitting of experimental data. Some success has been achieved using this method for steady flow. Nordgren (1982) applied Iwan and Blevins' (1974) version of this model, with riser equations

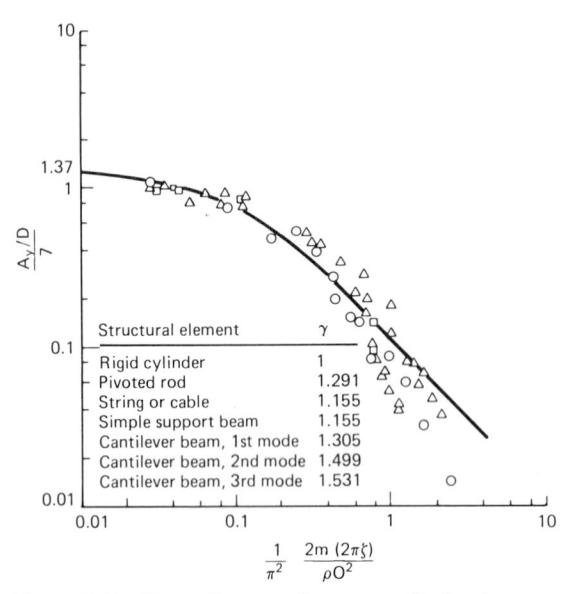

Figure 5.16 Normalized maximum amplitude of response *versus* mass ratio damping parameter (Iwan, 1975 by permission)

derived from the theory of elastic rods. He applied a strip theory approach with the vortex model acting only on the portion of the riser exposed to a current which varied with depth. It is not apparent that this analysis accounted for the effects of the limited spatial extent of lock on and the fluid damping of inactive elements.

The correlation model developed by Blevins and Burton (1976) and Kennedy and Vandiver (1979) is a specialized dynamic analysis using random vibration theory based on a representative spanwise correlation and cylinder amplitude dependence of vortex induced forces. Experimental data on correlation lengths and lift functions or resonant cylinder vibration amplitudes are used to determine model parameters. This approach is useful in making predictions of non-resonant response condition which may occur at low amplitudes of vibration where it is probably superior to the wake oscillator model. However, existing data is limited to steady flow conditions for stationary cylinders. The validity for straightforward extensions to non-steady flow conditions is questionable, especially in respect of correlation length parameters. Whitney and Nikkel (1983) applied this method to uniform and sheared flows. Their results for uniform flow compare well with laboratory and field tests. However, they were not able to validate conclusively their predictions for sheared flow.

Empirical models based on measured fluid dynamic force coefficients have been used to predict resonant transverse vibrations in steady and harmonic flow, Sarpkaya and Isaacson (1981), Rajabi (1979) and Zedan and Rajabi (1981). Rajabi *et al.* (1981) applied empirical correlations for lift coefficients and shedding frequencies to an analytic frequency domain model for vortex induced vibration of risers. It assumes lock on with one mode and perfect vortex correlation along the length. It makes use of the relationship between a lift amplification parameter C_L/C_{Lo} and KC/KC^*, where KC^* is Keulegan–Carpenter number KC at perfect synchronization, with KC defined as $V_{max}T/D$. KC/KC^* is equal to the corresponding ratio of reduced velocities V_r/V_r^*, C_{Lo} is the lift coefficient of a stationary cylinder and is a function of Reynolds number This relationship, shown in *Figure 5.17*, is analogous to that of Y_{max}/D versus V_r shown in *Figure 5.13*. Apparently this model takes no account of the influence of one mode upon another nor of the consequences of the limited spatial extent of lock on.

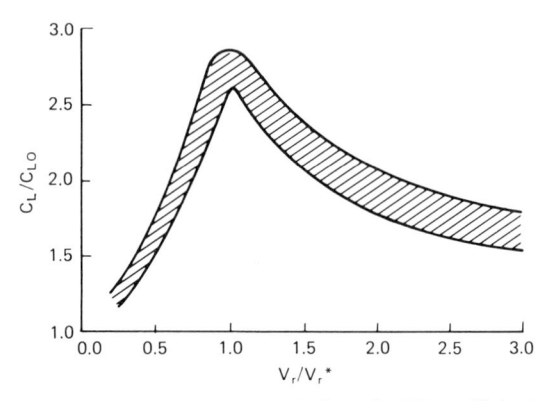

Figure 5.17 Empirical correlations for lift coefficients and shedding frequencies for an analytic frequency domain model for vortex induced vibration of risers

A statistical vortex shedding linear model based upon flow oscillator governing equations has been developed by Benaroya and Lepore (1983). This uses a variation of the Hartlen and Currie (1970) uniform flow model developed by Landl (1975) which introduces a fifth order fluid dynamic damping term to account for the hysteresis effect and the cases of soft and hard excitation. Hard excitation refers to a reduced velocity range for which two stable states are possible for one value of reduced velocity; the position of rest and a vibration of finite amplitude. To get an oscillation from rest in this case, it is necessary for an external disturbance to exceed a certain displacement threshold. In the case of soft excitation, the rest position is unstable so that an oscillation is always generated. The statistical model assumes perfect spanwise correlation of the flow and, therefore, is not in this form fully applicable to the varying flow cases to which risers are generally subjected.

Other methods such as discrete vortex models and numerical solutions of the time dependent Navier–Stokes equations in the presence of an oscillating cylinder are computationally expensive for the results obtainable. Sarpkaya and Shoaff (1979) developed a comprehensive discrete vortex model based on potential flow and boundary layer interaction, rediscretization of the shear layers and circulation dissipation to determine the characteristics of an impulsively started flow. The evolution of the flow from start to large times, lift and drag forces, Strouhal number, oscillations of the stagnation and separation points and the vortex street characteristics, were all calculated and found to be in good agreement with experiment. This numerical model was then applied to flow about a transversely oscillating cylinder. It produced many of the experimentally observed features of the lock on phenomenon. Apparently it took about 3 h of computer time on a CDC-6600 to reach a steady state equivalent to a simulated time of 400 s. Whilst such a model provides a useful tool for numerical experiments to investigate the underlying physics of vortex shedding and associated vibration, it does not in this form provide a method of simulating the vortex induced vibration response of tethers or risers for engineering design purposes.

The method favoured by Lyons and Patel (1986) for application to the dynamics of marine risers and tethers invokes the following assumptions:

1. The vortex shedding phenomenon is dependent on instantaneous relative flow velocity
2. Transverse vibration is approximated to begin at a reduced velocity of 4, reach a maximum of 6 and cease beyond 10.
3. The amplitudes of vibration for each mode may be calculated with a scheme devised by Iwan (1981) described below where the regions of excitation are those defined in (2).
4. Regions exciting higher modes do not excite lower modes, i.e. modal priority of higher modes occurs.
5. The drag coefficient, which will vary with time and along the length, is fixed at 2.0 for computational simplicity.
6. The added mass coefficient, which is also likely to vary with time and along the length, is fixed at 1.0.
7. Similarly the Strouhal number is fixed at 0.2.
8. For pinned end members, all higher natural frequencies are integer multiples of the fundamental natural frequency. Although it is likely that buoyancy dependent effective tension, which varies along the length, and added mass will have some effect on natural frequencies and mode shapes.
9. Lock on frequencies do not vary from the natural frequencies of the member.
10. The mode shapes are given by $\xi = \sin \dfrac{n\pi x}{L}$
11. For all such modes, the mode shape factor I_n, has a value of 1.155.

Iwan's scheme presents a simple analytical model for the vortex induced transverse oscillation of

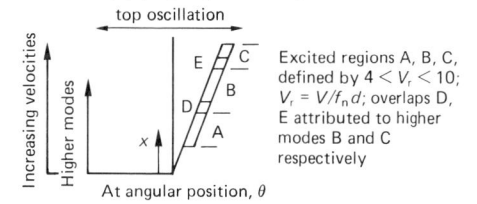

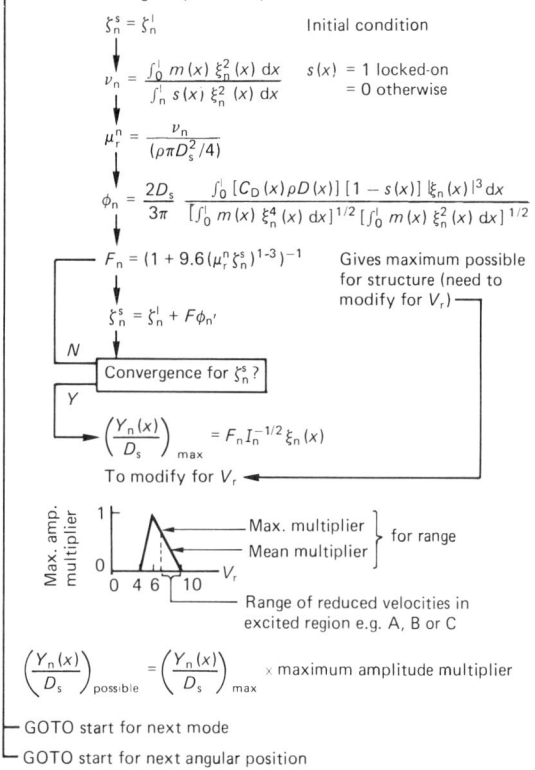

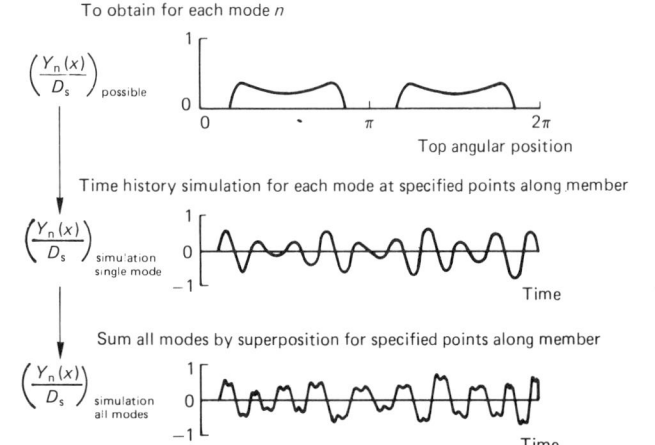

Figure 5.18 A simple analytical model for the vortex induced transverse oscillation of non-uniform structures, based on a modal decomposition approach

non-uniform structures in which the effects of limited spatial extent of lock on and fluid damping of inactive elements are accounted for. The theory is based on a modal decomposition approach.

The appropriate equations used are given below. *Figure 5.18* shows the solution scheme graphically.

The amplitude of locked on oscillation of the structure is given by:

$$Y_n(x) = S_s F_n I_n^{-1/2} \xi_n(x) \qquad (5.103)$$

where D_s is the cylinder diameter and the modal shape factor is:

$$I_n = \frac{\int_0^L m(x) \xi_n^4(x) \, dx}{\int_0^L m(x) \xi_n^2(x) \, dx} \qquad (5.104)$$

The amplification factor is taken to be:

$$F_n = (1 + 9.6 \, (\mu_r^n \, \zeta_n^s)^{1.8})^{-1} \qquad (5.105)$$

where ζ_n^s is the effective damping, although expressions by other authors may be used as in *Figure 5.15*. A particularly important parameter is the effective mass ratio:

$$\mu_r^n = \frac{v_n}{(\rho \pi \, D_s^2/4)} \qquad (5.106)$$

in which the effective mass is given by:

$$v_n = \frac{\int_0^L m(x) \xi_n^4(x) \, dx}{\int_0^L m(x) \xi_n^2(x) \, dx} \qquad (5.107)$$

where $s(x) = \begin{bmatrix} 1 & \text{for those portions of the structure where vortex shedding is locked on to the structural motion} \\ 0 & \text{otherwise} \end{bmatrix}$

The effect of the position of locked on regions determined by this parameter on the amplitude of vibration is demonstrated in *Figure 5.19* for the first mode of vibration. It is clearly seen that the amplitude of vibration is greater when the region of excitation is near the centre (antinode) and increases with the extent of the excited region. Similar effects result for all other modes.

Figure 5.18 presents a flow chart of the implementation of the time domain theoretical model. Relative velocities along the length are calculated and the extent of regions of corrected vortex shedding excitation are identified for incremental time steps. Iwan's method is implemented for the length of the member in each mode which is excited to obtain the modal amplitude. Since this amplitude is the peak resonant amplitude, it is necessary to modify it to determine the amplitude of vibration at the reduced velocities in the region of excitation. Use is made of assumption (2) above. From this an amplitude multiplier is determined, by which the peak resonant amplitude is modified dependent on the range of reduced velocities in the excited region.

The method utilizes the maximum reduced velocity in each region of excitation. The amplitude values for each mode are constructed into time histories which are then superimposed to obtain the overall member vibration time history. During vortex induced vibration of a particular mode, the member amplitude is set at the value given by the above procedure. When this mode is inactive, however, its vibration is taken to be due to its damped motion in still water from the vortex induced vibration during its last active condition. This time history procedure thus accounts for the following features: decay of vibration using the member structural and viscous damping in still water, ζ_n^T; phase of vibration changes randomly if a mode has a period of

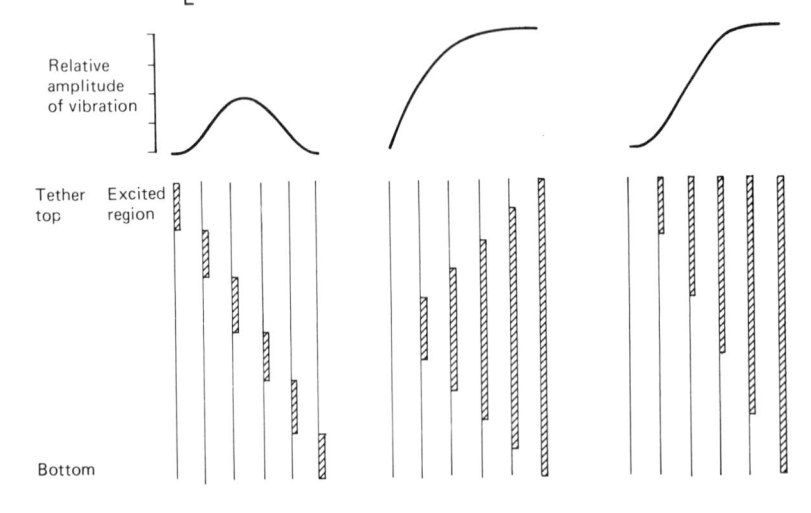

Figure 5.19 The effect of position of lock on; variation of amplitude of vibration with extent of excited region for the first mode of pinned-end sinusoidal vibrations

inactivity, and vibration amplitude for any mode not being lower than that due to decay from a previous event.

The value of drag coefficient has been fixed within the computation to permit a simplified implementation with good agreement with measurements. In reality, the drag coefficient is a function of Reynolds number and of vibration amplitude. Griffin and Ramberg (1982) give this function from the results of full scale measurements in current flow as:

$$C_D/C_{Do} = 1 + 1.16/(W_r - 1)^{0.65} \quad \text{for } W_r > 1$$

and

$$C_D/C_{Do} = 1 \quad \text{for } W_r < 1 \quad 5.108$$

where the wake stability parameter, $W_r = (1 + 2Y/D)/(V_r S)$.

King (1977) compares the relationships between steady drag coefficient and Strouhal number with Reynolds number variation as shown in *Figure 5.20*.

The definition of drag coefficient to be used remains an unclear area but the above equations may be used within this method to incorporate a more refined variation of drag coefficient. Values of C_D/C_{Do} of up to 4.5 have been demonstrated. The situation is complicated for multiriser bundles. Depending upon the configuration, the entrained fluid may often result in riser behaviour as if it were a single body. Patel and Sarohia (1982) have demonstrated multitube vortex flow visually shown as in *Figure 5.21*. Drag coefficients for bundled risers have been obtained by Demirbilek and Halvorsen (1985).

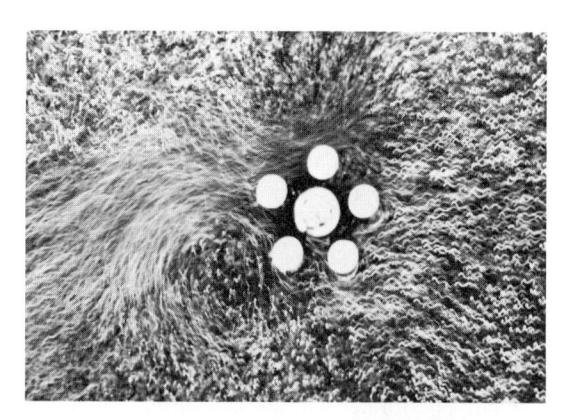

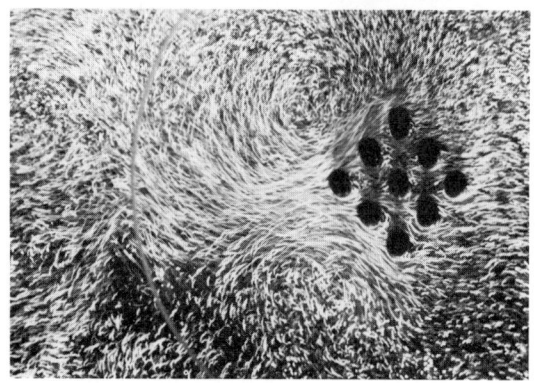

Figure 5.21 Multiple vortex flow; depending on the configuration, the entrained fluid may result in the riser behaving like a single body

5.2.3.3 Suppression

In order to reduce or avoid troublesome vortex induced vibrations, there are two approaches which may be adopted. One may consider altering of the riser physical properties to increase k_s or changing the natural frequencies so as to avoid vortex shedding frequencies. This approach is often not possible because of other design constraints. The second approach is to attach some form of flow spoiling or damping device along the riser length. For practical purposes whilst they may reduce the vibrations their attachment and riser deployment is often unsatisfactory and their service

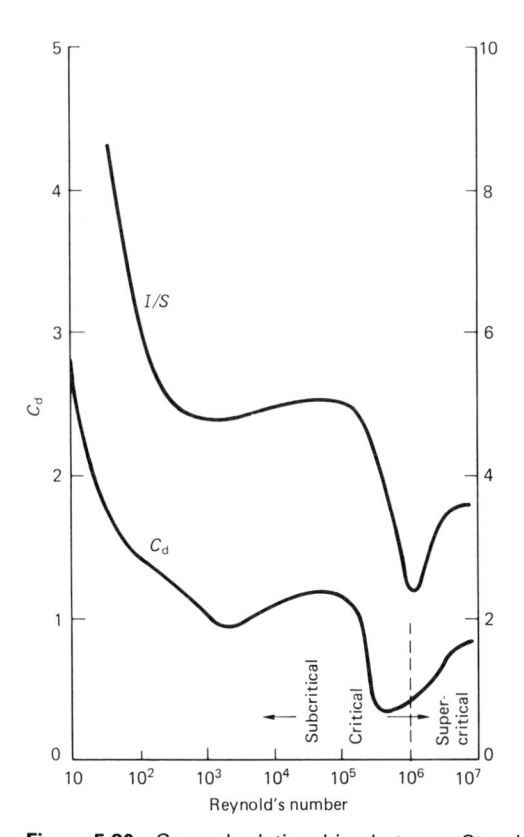

Figure 5.20 General relationships between Strouhal number, steady drag coefficient and Reynolds number

mance variable due to marine fouling and mechanical failure.

Flow spoiling devices can be categorized into three main groups: those which affect the separation lines or separated shear layers, those which affect the entrainment layers and those which act as near wake stabilizers and inhibit the switching of the confluence point of the entrainment layers.

These devices are discussed in detail by Zdravkovich (1981) and Every et al. (1982). They include splitter plates, fairings, guide plates and vanes. These have the disadvantage that they are unidirectional in action and can cause large lateral forces when the flow is at an angle to the design direction. Since this is often the case in the offshore environment it is benficial to allow them to weathervane. In its simplest form this can comprise flags along the riser length. However, these may wrap themselves around the riser. A more sophisticated design is the use of a series of rotatable aerofoil shaped sections. These may also provide the advantage of reduced steady drag force. Omnidirectional performance may also be obtained by the use of helical strakes as commonly used in chimneys. However, they have the disadvantage of increasing the steady drag force.

Similarly, perforated shrouds and shrouds of vertical slats offer suppression with somewhat less of a drag penalty. Investigations have also been made into the vortex suppression capabilities of plumes of rising air bubbles disrupting the flow around the riser with success, but this is probably only viable as a temporary measure.

An interesting proposal for riser vibration suppression is the nutation damper. This device has been used successfully in spacecraft. It is essentially a torus which is part filled with a sloshing liquid. Its damping characteristics are shown by Modi et al. (1986) to be sensitive to the physical properties of the liquid used, its height in the torus, damper geometry and dynamical parameters representing amplitude and frequency. Advantages include requirement for only a few dampers along the riser length with minimal increase in steady drag, with optimal reduction in vibration when placed at antinodes of maximum vibration.

5.2.4 Design considerations

Whether the riser is for drilling or production duty, the fitness for purpose of the design is determined by the estimates of its likely loading conditions. These conditions include environmental forces and if applicable surface support motions. Confident estimates of these are essential. They are not limited to static behaviour, but should include dynamic response. For rigid risers the likely causes of failure are local material yielding and Euler column buckling.

The design of all types of tensioned risers is affected by motions of the surface facility, tensioner stroke limits and response rates, bottom connection angle limits and distribution of buoyancy modules.

Additionally, drilling risers are particularly affected by mud weight, drill string tension and possible abnormal gas pressure.

Whilst production risers are particularly affected by buoyancy modules for the free standing mode, drag of multiple piping, rigidity of multiple piping and installation, repair and maintenance procedures.

5.2.4.1 Sources of failure

It is important to understand the likely causes of riser failure when designing a riser system. Almost inevitably this understanding comes from past experience. Morgan (1974–6) indicates the following for tensioned risers:

Response	Cause
1. Buckling	Failure to predict multiple curvature
	Failure to predict high curvature
	Inadequate top tension available
	Inadequate tensioner rate
	Excessive bending in free handing condition
	Failure of buoyancy modules
2. Ball joint damage Drill string fatigue BOP fatigue damage Blowout risk	Drill bit, collars, casing causing mechanical damage as a result of excessive ball joint angle
3. Riser conductor failure	Excessive bending moment due to vessel excursion and BOP weight
4. Emergency disconnect failure	Excessive bending causing binding
5. Riser to supplementary buoy overstressing	Out of phase dynamics of system elements
6. Conductor pipe failure and BOP stack collapse	Resonant excitation of BOP

5.2.4.2 Riser top tension and supplementary buoyancy

A truly vertical riser connected at the sea bed has no buoyancy force. This is because buoyancy is the resultant net force acting vertically on a body and if there is no horizontal surface on which the hydrostatic pressures may act the resultant force is zero. However, disconnect the riser from the sea bed or incline it and it will exhibit a buoyancy force. Generally for risers the combined effects of self weight and buoyancy yield a net negative force which is destabilizing in that the riser will continue to move away from the vertical unless restrained. This restraint is provided by means of top tensioning which may be aided by the use of supplementary buoyancy modules along the riser length.

Near optimum choice of top tension can be arrived at by calculating the sum of the reduction in bending stress and the increase in axial stress with increase in top tension (see *Figure 5.22*). Care must be taken to ensure that the lateral component of top tension does not result in excessive horizontal deflection of the bottom BOP stack, *Figure 5.23*. The moment due to the BOP weight and its eccentricity may lead to bending failure of the sea bed conductor column. Reduction in top tension requirement is particularly advantageous in very deep water. Care must be taken to ensure that such reductions do not lead to local compression which is more likely to occur near the sea bed.

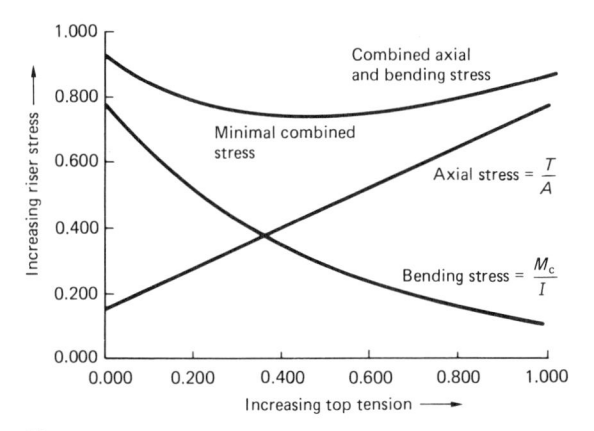

Figure 5.22 Top tension of risers, calculated as the sum of the reduction in bending stress and increase in top tension

Buoyancy modules in use include air filled cans in which the volume of air may be controlled from the surface and so alter the buoyancy available. Other forms do not offer this control but have cost advantages in certain cases. Materials for these include cellular polystyrene, cellular vinyls, cellular silicones, cellular acetate, synthetic foams which may contain spheres of various materials and foamed aluminium. Some of these materials can deteriorate with time resulting in a change in buoyancy. Manufacturers should be consultated for suitability of depth ratings.

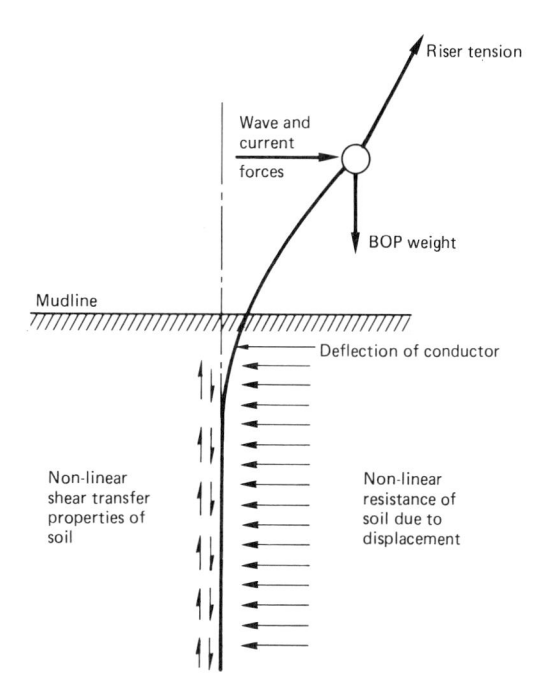

Figure 5.23 Optimization of top tension; BOP eccentricity, resulting in bending of the sea bed conductor column

5.2.4.3 Production riser configurations

Whilst drilling risers include choke and kill lines attached to the main drill–mud tube, production risers generally consist of more exotic configurations. For an in depth exposé of production risers the reader is directed to a series of articles by G W Morgan (1974–6). Special care is needed in their design when used with floating production systems. *Figures 5.24* and *5.25* summarize the fitness for purpose of production risers which are currently considered for use with floating production systems, Shotbolt (1983).

They are tensioned compact or enclosed bundle riser, tensioned riser with stand off arms, array or tensioned single well risers, ribbon riser with weighted seabed flexing arm or boom, catenary flexible single well risers, and catenary flexible riser with comingled flow.

The particular design considerations are choke manifold location, operation and servicing, number of lines in the riser, rig–riser interface, moonpool area operations, wireline and major workover of the rig well cluster, weather sensitivity, emergency disconnect, number and spacing of wells in the rig well cluster, risk of damage to subsea equipment by dropping objects, operational versatility of the rig (drilling–production) and water depth sensitivity.

Flexible risers are being increasingly considered for production duty. Claimed advantages include

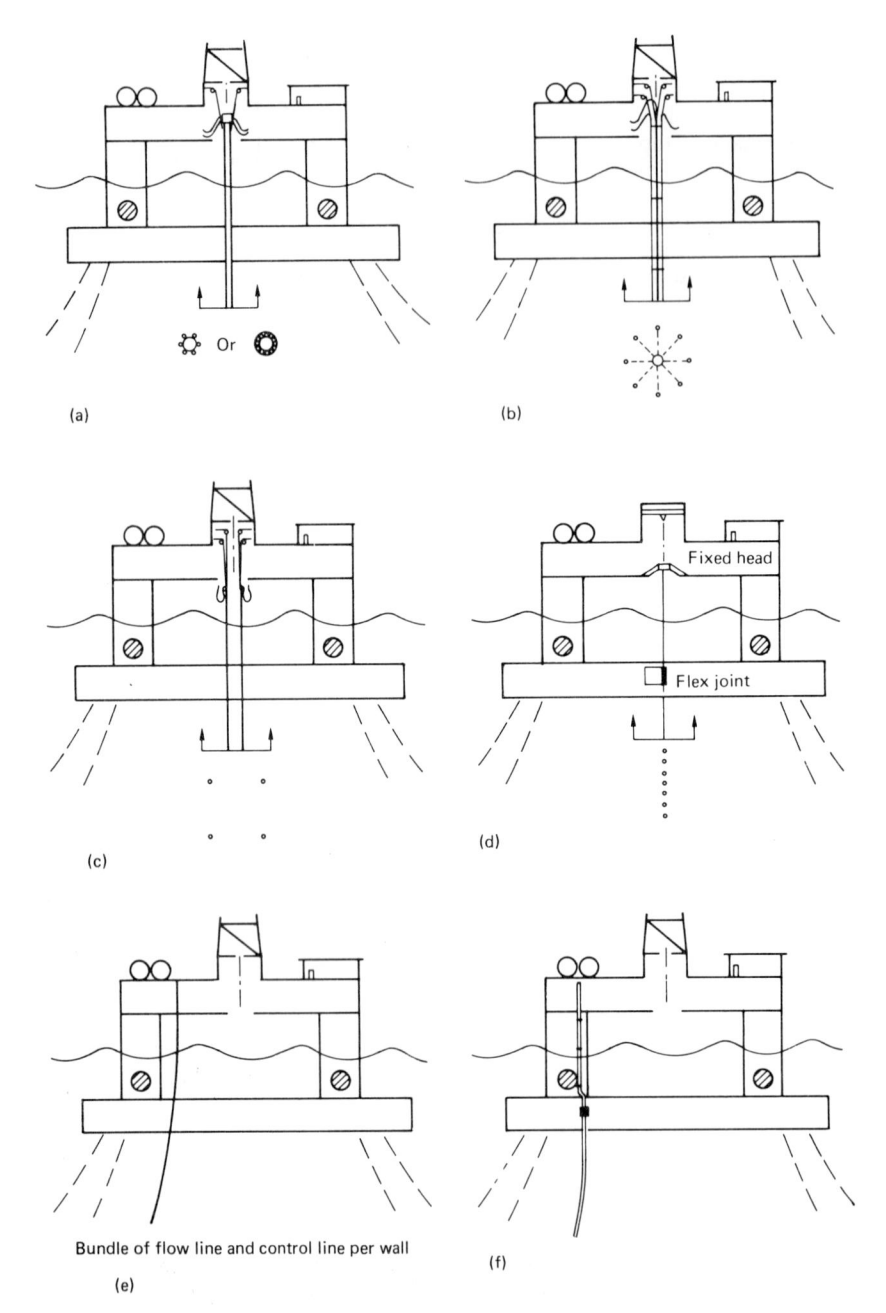

Figure 5.24 Risers for use with floating production systems; an illustration of alternative systems: (a) tensioned compact or enclosed bundle; (b) tensioned riser with stand-off arms to support smaller lines; (c) array of tensioned single well risers; (d) ribbon riser tensioned at sea bed by weighted flexing arm or boom; (e) catenary flexible single well risers, bundle of flow line and control line per well; (f) catenary flexible riser with commin flow, steel line through wave zone

Aspects of configuration and operation	Riser system A	B	C	D	E	F
Choke manifold location	●	●	●	●	●	·
Number of lines in riser	●	·	·	·	·	●
Riser/rig interface	●	·	·	●	●	●
Moonpool area operations	●	·	●	●	●	●
Workover of rig cluster	·	·	●	·	●	●
Weather sensitivity	●	·	·	●	●	●
Emergency disconnect	●	·	●	·	●	●
Number and spacing of wells in rig cluster	●	●	·	·	·	●
Risk of damage to subsea/equipment from dropped objects	·	·	●	·	●	●
Operational versatility of rig	●	●	·	·	●	●
Water depth sensitivity	●	·	·	●	●	●

Figure 5.25 Risers for use with floating production systems; a table comparing relative benefits, with dot size proportional to the benefit of each system illustrated in Figure 5.24

accommodation of large relative motions, good thermal insulation, good sound and vibration attenuation, will not melt in severe fire, corrosion resistant, re-usuable, resistant to dynamic fatigue, variable weight to suit application, easily routed over or around obstructions, compensation for misalignment, accommodation of uneven seabed, easy to lay with small vessel, low cost installation, can be reel wound and easy connection to manifolds.

Examples of possible configurations are shown in *Figure 5.26*. The construction and applications of some flexible risers are shown in *Figure 5.27*.

Whilst the traditional tensioned rigid riser system offers the following advantages: smaller line physical size particularly for high pressure lines, smaller bundle size, lower weight, proven service life, no gas permeability problems, less uncertain fatigue life, buoy systems not required and single rather than multiple upper and lower connections.

5.3 Pipelines

This subsection covers various issues that arise in the design of a subsea pipeline. The technical complexity of the necessary analysis is reduced from that required for marine risers although the number of interacting design constraints tend to be greater for pipelines. The behaviour of the sea bed soil for bottom standing or trenched

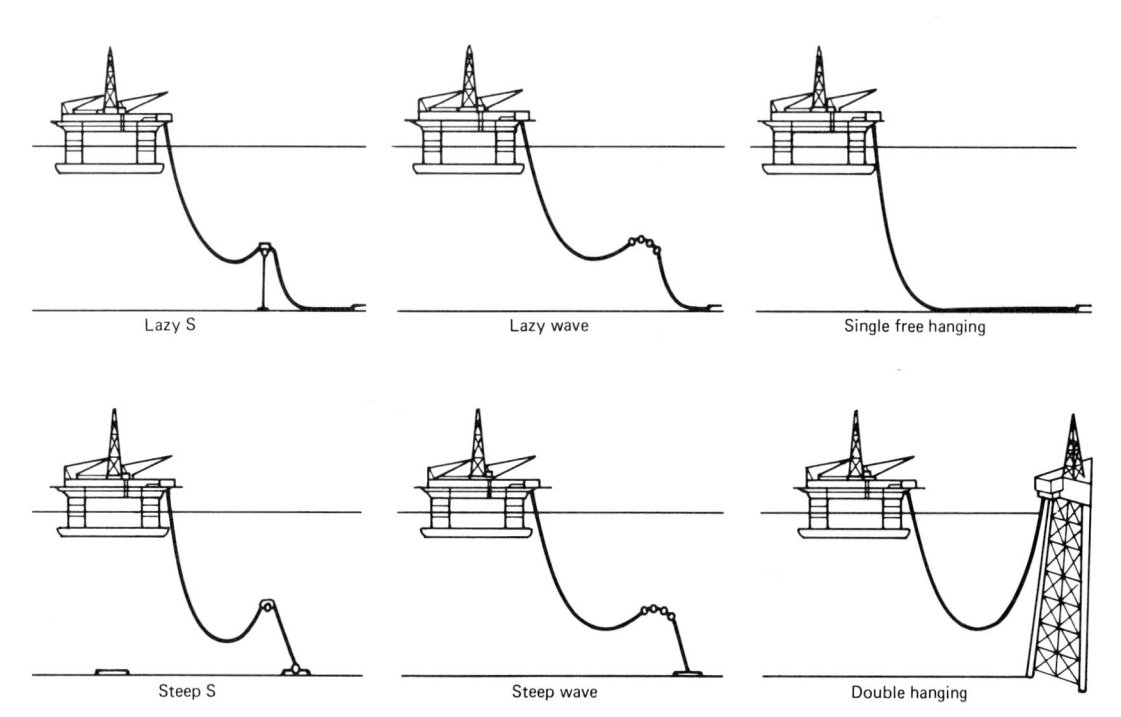

Figure 5.26 Flexible risers; an illustration of current configurations

(Configurations shown: Lazy S, Lazy wave, Single free hanging, Steep S, Steep wave, Double hanging)

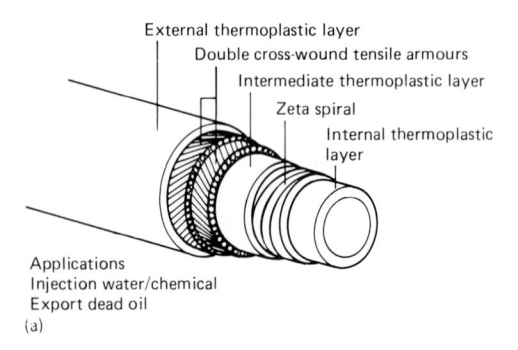

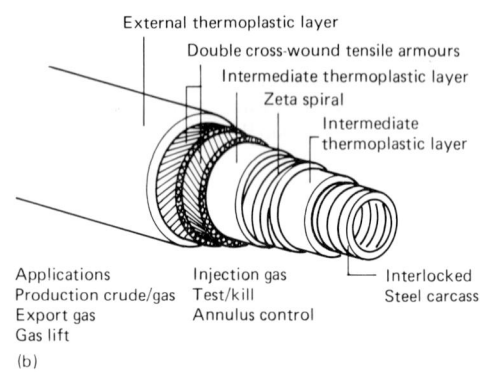

Figure 5.27 Flexible risers; two typical structures; (a) smooth bore (b) rough bore (courtesy of Coflexip (UK) Ltd)

pipelines also introduces factors that are difficult to quantify and require a degree of empiricism during design.

5.3.1 Pipeline design procedures

There are five principal requirements that govern the design of a subsea pipeline:

1. The line should be of sufficient diameter to ensure that the required volume or mass flow rate is achieved. The selection of pipe size is a compromise between pipeline diameter and hence its initial capital cost against the power requirement to maintain flow and the resultant cost of operation which increases with decreasing pipe diameter.
2. In many cases of subsea pipelines carrying hydrocarbons the line must have sufficient insulation to maintain the temperature of the flowing medium above a reasonable limit. This requirement is particularly important for liquid hydrocarbons since their viscosities increase with drop in temperature.
3. The structural integrity of the pipeline must be ensured during initial pipe laying and trenching as well as during operation. Pipe stresses can arise due to external and internal hydrostatic pressures, pipe curvature on an undulating sea bed, temperature induced stresses and due to external impacts. All of these have to be analysed for design.
4. Subsea pipelines laid directly on the sea bed without burial have to possess sufficient stability, that is, submerged weight to ensure that lateral current and wave forces do not cause the pipe to move. The required submerged weight is achieved either with concrete coating or by the use of steel pipe that is thicker than necessary for the applied loads. The question of pipeline stability is considered further in Section 5.3.2.
5. The internal and external surfaces of the pipelines must be capable of resisting corrosion due to the internal product and to external sea water. Such corrosion protection is achieved in a variety of ways as described elsewhere in this book.

Each of the above requirements are considered in more detail below. However, most detailed subsea pipeline designs are carried out to one of the four generally accepted codes of practice described below:

1. ANSI B 31.4: American National Standards Institute
   Liquid petroleum transportation piping system.
2. ANSI B 31.8: American National Standards Institute
   Guide for gas transmission and distribution piping system.
3. IP-6 1973 : Institute of Petroleum
   Petroleum pipelines safety code.
4. DnV 1981 : Det Norske Veritas
   Rules for the design construction and inspection of submarine pipelines and pipeline risers.

There are four primary load conditions that are considered during the design of a subsea pipeline. These conditions are pipe laying and burial, if specified, installed but non-operational, testing prior to operation and normal operation.

The installation and testing condition is normally subject to being able to withstand specified environmental criteria (current, wave height, period) which are unlikely to be exceeded during these phases. When in the other conditions, however, the pipeline is expected to survive the maximum likely environmental conditions (say a 100 year return period current speed and wave height) with an acceptable margin of safety.

5.3.1.1 Pipeline sizing

The requirement to attain a particular product flow rate through a pipeline tends to be a trade off between pipe diameter and pressure drop across its ends. Larger pipe diameters reduce the flow velocity and hence the pressure required to maintain flow but at increased capital cost. Decreasing pipe diameter increases the required driving pressure and hence the operating cost.

The following equations can be used to calculate and balance the requirements of pipe diameter, product flow rate and pressure loss across the pipe ends.

The mass flow rate Q in a pipeline is given by:

$$Q = \rho A \overline{U} \tag{5.109}$$

where ρ is the product density, A is the pipe internal area of cross-section and $\overline{U}$ is an average velocity within the pipe. Fully developed flow within a pipe exhibits a variation of velocity across the pipe section; such typical velocity profile shapes are shown in *Figure 5.28*. The laminar profile is observed when flow within the pipe is at low velocity and adjacent layers of fluid in the pipeline slip smoothly relative to each other without much mixing transverse to the flow direction. For larger flow velocities or pipe wall roughnesses, the internal pipe flow becomes turbulent, that is, the flow is characterized by large velocity fluctuations both along and transverse to the flow. This flow turbulence induces much more mixing in the flow and leads to a velocity profile with a flatter central section as shown in *Figure 5.28*.

(a) Laminar flow

(b) Turbulent flow

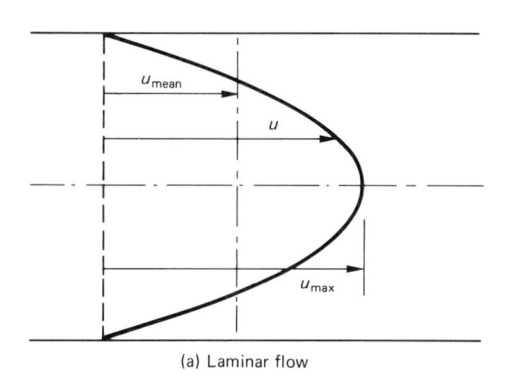

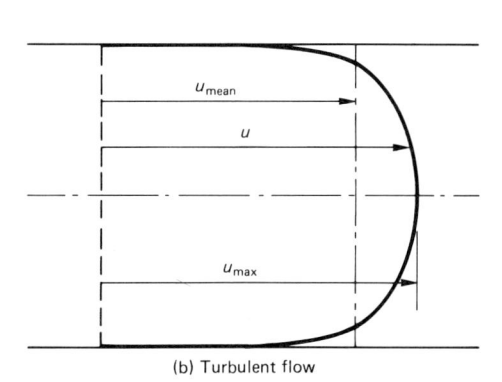

Figure 5.28 Pipe flow velocity profiles for (a), Laminar flow and (b), Turbulent flow; laminar flow is associated with low velocity and low transverse mixing, turbulent flow with the reverse

The Reynolds number is a useful non-dimensional parameter that quantifies the ratio of inertia to viscous forces in the flow. The Reynolds number Re is defined as:

$$\text{Re} = \frac{\overline{U} D}{\nu} \tag{5.110}$$

where $\overline{U}$ is the average flow velocity, D is the pipe internal diameter and ν is the kinematic viscosity of the flowing medium. It is known from experimental work that the flow regime in a pipe will remain laminar for Re < 2000 and will become turbulent for Re > 4000. A transitional region exists for 2000 > Re > 4000 where the existence of a laminar or turbulent velocity profile will also depend on other factors such as pipe wall internal roughness and pressure gradient along the pipe axis.

The static pressure p in a fluid at a depth h below its free surface can be written as:

$$p = p_a + \rho g h \tag{5.111}$$

where p_a is the pressure above the free surface, ρ is the fluid density and g is the acceleration due to gravity. Clearly, Equation 5.111 will apply as much to external sea water as to the product inside the pipe.

The Bernoulli Equation describes the static pressure and velocity behaviour of a flowing fluid. If subscripts i and o are taken to be conditions at the inlet and outlet of the pipe, respectively, then the total head, H_i, at the inlet can be written though Bernoulli's Equation as:

$$H_i = z_i + \frac{p_i}{\rho g} + \frac{\overline{U}_i^2}{2g} \tag{5.112}$$

where z_i is the vertical elevation of the inlet about a datum point, p_i is the static pressure at inlet, $\overline{U}_i$ is the average velocity at the inlet, ρ is the product density and g is the acceleration due to gravity.

Similarly, at the outlet one can write:

$$H_o = z_o + \frac{p_o}{\rho g} + \frac{\overline{U}_o^2}{2g} \tag{5.113}$$

where all the terms have the same definition except referring to the outlet. Note that parameter H is a measure of pressure expressed as the height of a column of product liquid in the earth's gravitational field. The pressure head loss ΔH due to frictional effects of flow from the inlet to the outlet can be written in terms of the heads as:

$$\Delta H = H_i - H_o \tag{5.114}$$

For a pipeline of uniform diameter, $\overline{U}_i = \overline{U}_o$ and the head loss is governed by frictional forces and changes of pipe elevation only. Once the head or pressure loss is known, the power P required to drive fluid flow through the pipe can be written as:

$$P = \rho g \Delta H \cdot Q_v \tag{5.115}$$

where Q_v is the volume flow rate through the pipe.

The head loss due to viscous shear stresses in the fluid at the pipe wall (ΔH_f) is commonly obtained by use of the formula:

$$\Delta H_f = \frac{fL\overline{U}^2}{2Dg} \quad (5.116)$$

where f is a friction factor, $\overline{U}$ is the average velocity of fluid flow in the pipe, L is the pipe length, D is the pipe internal diameter and g is the acceleration due to gravity.

Experimental observation has shown that the friction factor is a function of Reynolds number for laminar flow and of Reynolds number and pipe roughness in turbulent flow. This variation of f with Re is presented in *Figure 5.29*. It demonstrates that internal roughness plays an important part in the head loss for turbulent flow.

The above equations apply to the transportation of essentially incompressible liquids through pipelines. For gas flows, the compressibility of the gas must be taken into account. Pure gases without significant liquid content can be treated as perfect gases obeying the gas law equation:

$$pV = mRT \quad (5.117)$$

where p is pressure, V is the gas volume, T is the absolute temperature, R is the gas constant and m is the mass of the gas.

Well insulated gas pipelines can be treated as exhibiting isothermal (that is constant temperature) internal flow. The mass flow rate of a gas through such pipelines requires careful consideration of the compressibility of the gas. Shapiro (1953) presents the constitutive relations for such flows.

5.3.1.2 Temperature drop along a pipeline

In most cases of subsea pipelines carrying hydrocarbons, there is a need to reduce heat loss from the products to the surrounding sea water or soil in order to maintain product temperature. This is necessary because a drop in product temperature leads to an increase in its viscosity, and therefore, increased pumping pressure loss. In some cases of heavy or waxy crude oils, the flow may actually cease due to high pressure losses caused by an increase of viscosity due to cooling.

The heat transfer through the wall of a subsea pipeline will be governed by Fourier's law which can be stated as:

$$\dot{q} = -kA\frac{dT}{dr} \quad (5.118)$$

where $\dot{q}$ is the rate of flow of heat through a material of thermal conductivity k, area of cross-section A and temperature gradient dT/dr. Fourier's law is based on the assumptions that the conducting material is isotropic, that the direction of heat flow is in steady state and parallel to dimension r only.

For a thick cylinder of internal radius r_i and external radius r_o, Equation 5.118 can readily be integrated to give:

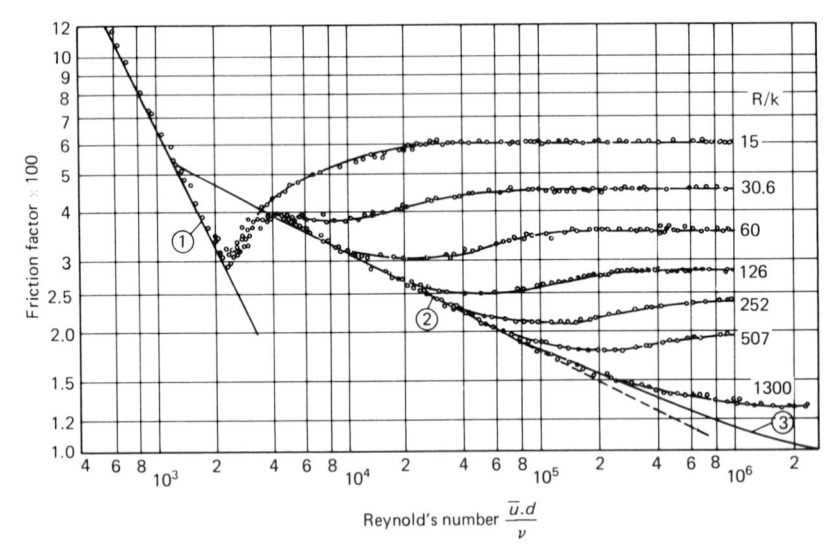

Figure 5.29 The friction factor as a function of Reynolds number for smooth and rough pipes; curve 1 is for laminar flow; curves 2 and 3 for turbulent flow in a smooth pipe. The remaining curves are for ratios of R/k, where k is the height of roughness elements, with R the pipe radius

$$\dot{q}_c = \frac{2\pi k(T_i - T_o)}{\ln\left(\dfrac{r_o}{r_i}\right)} \tag{5.119}$$

where $\dot{q}_c$ is the rate of flow of heat per unit length of the cylinder.

However, a typical pipeline wall will have a number of layers which can be listed as internal fluid film, steel cylinder, corrosion protection coat, concrete weight coating and sea bed soil or surrounding water.

The composite heat flow rate through these layers is described by the following equation:

$$\dot{q}_c = \frac{2\pi L(T_i - T_n)}{\dfrac{1}{r_i h_i} + \dfrac{\ln(r_2/r_1)}{k_{12}} + \dfrac{\ln(r_3/r_2)}{k_{23}} + \ldots + \dfrac{\ln(r_n/r_{n+1})}{k_{n-1,n}} + \dfrac{1}{2\pi r_n h_o}}$$

where subscripts 1, 2,...,n refer to the interface between layers and subscripts 12 and 23, etc., refer to the properties of the layer in between the interfaces. h_i and h_o are the unit surface conductances on the inner and outer walls. $\dot{q}_c$ is, as before, the rate of flow of heat per unit length of the cylinder.

Note that the determination of unit surface conductances h_i and h_o due to the fluid film inside and the sea water outside the pipe needs to take account of convection effects. This is not described here due to lack of space but readers are referred to Krieth (1965) which sets out the basic theory and calculation procedures for convection induced heat transfer around the pipe. Krieth (1965) also illustrates the derivation of an expression for the exponential drop in temperature along a pipe length once the heat transfer rate across the cross-section of the pipe is established.

5.3.1.3 Pipe stressing

A subsea pipeline will experience internal material stresses due to the effects of internal and external hydrostatic pressure, bending stresses due to pipe curvature during laying, residual tensile and bending stresses left after laying is complete and temperature induced stresses.

The tangential stress σ_t due to the differential between internal pressure p_i and external pressure p_o can be simply written as

$$\sigma_t = \frac{(p_i - p_o)D_m}{2t} \tag{5.121}$$

where D_m is the pipe mean diameter and t is wall thickness.

Equation 5.121 shows that the wall thickness, t, is a function of both design pressure and allowable stress level. The allowable stress level is given by the quality of the steel from which the pipeline is to be manufactured and is usually defined as a percentage of the yield strength (stress).

For subsea pipelines, the commercially available quality of line pipe steel ranges between Grade B and Grade X-70. The Grade B material has a specified minimum yield stress of 36 000 psi (2530 kgf/cm^2) and pipe of Grade X-70 material a specified minimum yield stress of 70 000 psi (4920 kgf/cm^2). Higher grade steel up to X-100 (100 000 psi or 7030 kgf/cm^2 yield stress) are under development.

The tangential stress described above will gives rise to an axial strain, ϵ_a, of magnitude:

$$\epsilon_a = -\frac{\nu \sigma_t}{E} \tag{5.122}$$

where ν is Poisson's ratio and E is Young's modulus. Since σ_t would normally be tensile, ϵ_a is a compressive strain. Now for an axially restrained pipe, ϵ_a would lead to a compressive axial stress of magnitude $\nu \sigma_t$. For an axially unrestrained pipe, the strain ϵ_a would lead to a change in length and no compressive stress. Note that sea bed friction is the normal mechanism restraining axial movement of a subsea pipeline.

Temperature induced stresses in a pipeline also depend on the degree of axial restraint. Stresses are developed whenever thermal expansion or contraction of the pipeline is prevented. The best method of calculating temperature induced stresses is to determine the expansion or contraction that would occur assuming the pipe was unrestrained and then calculate the stresses produced by the force required to return it to its original dimension.

As an example, if a pipeline's temperature is increased from T_1 to T_2, the compressive stress due to increase in temperature of the axially restrained line is given by:

$$\sigma_a = E \alpha (T_2 - T_1) \tag{5.123}$$

where α is the coefficient of linear thermal expansion and E is Young's modulus.

When a pipeline is laid around a horizontal radius, is resting on an irregular sea bed or is spanning, bending stresses proportional to the curvature of the pipeline are induced. The magnitude of the bending stress σ_b is:

$$\sigma_b = E Y \frac{d^2 y}{dx^2} \tag{5.124}$$

where E is Young's modulus for the pipe material, d^2y/dx^2 is the pipe curvature, and y is the distance from an axis through the cross-section centroid to any part of the cross-section. σ_b can also be written as:

$$\sigma_b = \frac{M y}{I} \tag{5.125}$$

in terms of applied bending moment M and the second moment of area of the cross-section about an axis through the centroid I. Note that the bending stress varies linearly with y to give an equal compressive and tensile component on opposite parts of a pipe cross-section with zero axial tension.

When a pipeline is being laid, a tensile force is applied to the pipe to minimize bending and hence the chance of buckling (see Section 5.3.3). The level of tensile force depends upon the method of laying but once installed on the sea bed, a percentage of this tension can remain locked in the pipe wall. To allow for this in the calculation of likely stresses, a percentage of the applied pipe laying tension, say 25%, is sometimes assumed.

All the tangential axial stresses (and radial stresses in a thick walled pipe) need to be combined using von Mises or Tresca criteria to determine a combined stress against which likely pipe failure can be judged.

During both its installation and operation, a pipeline is prone to collapse and buckling due to either external hydrostatic pressure, to bending at a high curvature, or to a combination of both. When designing a pipeline, it is therefore important to ensure that the line is strong enough to resist collapse and buckling.

The parameters affecting collapse and buckle of a pipeline are its diameter to wall thickness ratio, steel yield stress and its out of roundness. There are in fact two distinct failure mechanisms which are circumferential elastic buckling of the cylinder due to external pressure or buckling due to longitudinal bending. Only the former is considered here.

For a submerged pipeline, if the external hydrostatic pressure exceeds the pressure of internal contents, then a compressive hoop stress, proportional to the pressure difference, is induced in the wall of the pipeline. For a pipeline that is perfectly circular, this pressure difference at which the pipeline will collapse is termed the critical pressure.

An equation for predicting a critical pressure difference ΔP_{cr} for collapse is given by:

$$\Delta P_{cr} = \frac{2E}{(1-\nu^2)} \left[\frac{t}{D}\right]^3 \qquad (5.126)$$

where D is the pipe outside diameter, E is Young's modulus, t is wall thickness and ν is Poisson's ratio. The above formula applies for collapse with the pipe material remaining elastic and requires some modification for elasto-plastic material behaviour during collapse.

Where a pipeline is operating or under hydrostatic test, failure occurs when the difference between the external and internal pressure, is equal to the critical pressure. In practice, pipelines are not perfectly round, and have out of roundness from manufacturing imperfections or from deformation (ovalling) due to self weight when resting on the sea bed. To take account of this out of roundness, the formula for critical pressure is modified as described by Tam and Croll (1986).

Two other parameters need to be examined in the analysis of buckle and collapse. First, the pressure difference which is required to initiate a buckle. This initiation pressure is primarily a function of the D/t ratio and pipe material properties. From experimental work, a lower bound for initiation pressure, Δp_i, has been established and is given by:

$$\Delta p_i = \frac{6.055 \times 10^5}{(D/t)^{2.06}} \text{ psi} \qquad (5.127)$$

The formula is valid for the D/t ratio of 20 through to 50 and may be used up to 70 with caution.

The second parameter is the propagation pressure difference and is defined as the lowest pressure difference (between external and internal fluid) at which the pipeline will continue to buckle after buckling has been initiated. The propagation pressure difference is the threshold pressure difference experienced by the pipeline, below which buckle cannot occur. An empirical expression for the propogation pressure difference, ΔP_p, is given by Kyriakides and Babcock (1981) as:

$$\Delta P_p = \sigma_y \left[10.7 + 0.54 \frac{E}{\sigma_y}\right] \left[\frac{t}{D}\right]^{2.25} \qquad (5.128)$$

where σ_y is the yield stress and E is Young's modulus.

5.3.2 Sea bed stability and burial

A pipeline resting on the sea bed is subjected to a number of static and dynamic forces. The forces acting in a plane perpendicular to the pipe axis consist of the static forces of self weight, buoyancy and soil friction. In addition, current flow can induce static drag and lift forces. Dynamic forces arise from wave induced drag and inertia forces which act along the flow direction parallel to the sea bed as well as lift forces which act perpendicular to the sea bed plane.

The pipeline needs to remain stable under the action of all these forces which requires that the largest instantaneous lateral force on the pipe is lower than the limiting pipe friction with the sea bed. Wave induced forces on the pipe are time dependent but so is the limiting friction force since the pipe normal reaction on the sea bed will depend on the oscillatory lift force and will, therefore, itself be oscillatory.

The governing equations for pipeline stability are summarized below for a pipe of external diameter D resting on the sea bed with a normal current velocity U incident on the pipe (see *Figure 5.30*).

Then the steady forces per unit pipe length acting normal to the pipe axis are: buoyancy force per unit length

$$F_B = \tfrac{1}{4} \rho g \pi D^2 \qquad (5.129)$$

where ρ is sea water density.
Weight per unit length

$$F_W = \sum_{i=1}^{N} \rho_i g A_i \qquad (5.130)$$

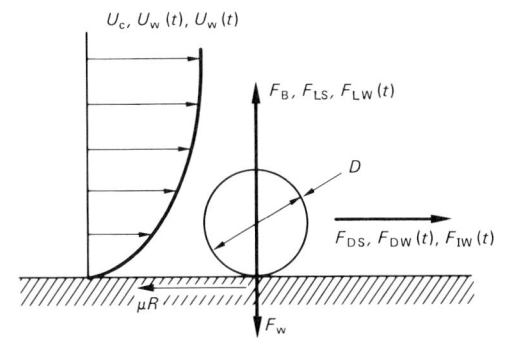

Figure 5.30 Stability of a pipeline resting on the sea bed; the pipeline of external diameter D, has a normal current velocity U incident upon it

where A_i are the internal cross-sections of pipe components such as concrete coating, steel and internal fluid, each of density ρ_i. An external carrier pipe containing internally bundled pipes can also be accounted for using Equation 5.130 with N internal pipe cross-section elements that must include all internal fluid and solid elements.

The forces per unit lengths induced by current U_c are given by:

steady drag force per unit length

$$F_{DS} = \tfrac{1}{2} \rho\, D\, C_D |U_c| U_c \qquad (5.131)$$

steady lift force per unit length

$$F_{LS} = \tfrac{1}{2} \rho\, D\, C_L |U_c| U_c \qquad (5.132)$$

where C_L and C_D are appropriate lift and drag coefficients.

Then for steady conditions, the reaction force per unit length R between pipe and sea bed is:

$$R = F_W - F_B - F_{LS} \qquad (5.133)$$

and the available limiting lateral friction force per unit length is:

$$\mu(F_W - F_B - F_{LS}) \qquad (5.134)$$

where μ is the limiting coefficient of friction. Then for pipeline stability, the condition below must be satisfied:

$$F_{DS} < \mu(F_W - F_B - F_{LS}) \qquad (5.135)$$

In Equations 5.134 and 5.135, note how buoyancy and lift forces serve to reduce the magnitude of reaction force and thus the available friction force. Jones (1976) gives further information.

In unsteady flow induced by wave action, additional unsteady forces due to wave fluid velocity and acceleration have to be accounted for. Due to the non-linear behaviour with velocity of the lift and drag forces, these equations need to be reformulated to allow current and wave velocities to be added before they are inserted into the above equations.

Thus, for unsteady flow, Equations 5.129 and 5.130 still apply but Equations 5.131 and 5.132 need to be replaced by:

$$F_{DW}(t) = \tfrac{1}{2} \rho\, D\, C_D |U_c + U_w(t)|[U_c + U_w(t)] \qquad (5.136)$$

and

$$F_{LW}(t) = \tfrac{1}{2} \rho\, D\, C_L |U_c + U_w(t)|[U_c + U_w(t)] \qquad (5.137)$$

where $U_w(t)$ is the horizontal wave velocity which is an oscillatory function of time. The forces F_{DW} and F_{LW} are then also functions of time.

The wave induced inertia force per unit pipe length is:

$$F_{IW}(t) = \tfrac{1}{4} \rho\, \pi\, D^2\, C_m\, \dot{U}_w(t) \qquad (5.138)$$

where C_m is an added inertia coefficient and $\dot{U}(t)$ is wave fluid acceleration parallel to the sea bed.

Then, as in Equations 5.133 and 5.134, the instantaneous normal reaction force and instantaneous limiting friction force (both per unit length) are:

$$R(t) = [F_W - F_B - F_{LW}(t)] \qquad (5.139)$$

and

$$\mu[F_W - F_B - F_{LW}(t)] \qquad (5.140)$$

For pipeline stability in waves, the condition:

$$[F_{DW}(t) = F_{IW}(t)] < \mu[F_W - F_B - F_{LW}(t)] \qquad (5.141)$$

must apply at all values of time t within a cycle. Note that Equation 5.141 has absorbed the steady flow stability condition of Equation 5.135 such that only Equation 5.141 needs to be satisfied for pipeline static and dynamic stability.

Wave velocities and accelerations are calculated at the pipe centre using either shallow water linear theory or Stokes' fifth order wave theory. In very shallow water, other special purpose wave theories may need to be used.

There are difficulties associated with applying the above equations in that both the current velocity U_c and the wave velocity $U_w(t)$ may decay close to the sea bed due to boundary layer effects and exhibit a typical velocity profile of the type shown in *Figure 5.30*. It is common practice to use either the velocity at the pipe centre or an average velocity across the pipe diameter for stability calculations.

The added inertia, lift and drag coefficients will be dependent on the Reynolds number:

$$\mathrm{Re} = \frac{U_m D}{\nu} \qquad (5.142)$$

and the Keulegan–Carpenter number:

$$K = \frac{U_m T}{D} \qquad (5.143)$$

where ν is the kinematic viscosity of sea water, U_m is the maximum wave velocity and T is the wave period.

Values for these coefficients have to be selected specifically for a pipe in contact with the seabed since this proximity has a substantial influence on the coefficients. Det Norske Veritas presents recommended coefficient values which are reproduced in the chapter on Offshore Structures. Sarpkaya (1976) presents experimental data for hydrodynamic lift, drag and added mass coefficients as functions of sea bed clearance.

A stability analysis is generally carried out by computations to ensure that Equation 5.141, with a known value of μ, is satisfied at all times during a wave cycle. The analysis can also be carried out by replacing the inequality in Equation 5.141 with an equals sign and calculating the limiting coefficient of friction required for stability at a large number of time instants within a wave cycle. The minimum value of such a coefficient of friction is then the design value that must be available for pipe stability. Typical coefficients of friction may vary between 0.5 and 0.8 depending on sea bed soil properties, pipe embedment in the sea bed and pipe surface roughness.

The above analysis can be readily applied for regular long crested waves propagating in a direction perpendicular to the pipe axis. In survival conditions, this approach may be too conservative since short crested seas which are generally approaching at angles other than perpendicular to the pipe axis will induce non-coherent lateral forces along the pipe length and thus generally tend to ensure that the pipe is more stable than suggested by calculations based on long crested seas propagating in a direction normal to the pipe axis. The simplest way to deal with a regular long crested sea at an angle, α, to the direction perpendicular to the pipe axis is to reduce both wave velocity and acceleration by multiplication with $\cos \alpha$ and then apply the above equations.

A number of subsea pipeline design problems described in Sections 5.3.1 and 5.3.2 above can be solved by using a trenching method to bury the pipeline some distance below the sea bed. The burial of a pipeline has several advantages – chief among these are:

1. The line is protected from vessel anchor, mooring chain and fishing trawl board impact.
2. Dynamic excitation of the pipeline due to vortex shedding or spanning, i.e. unsupported sections, is eliminated.
3. The pipeline's stability may be designed to be just sufficient to remain in place after laying and prior to burial.
4. Pipeline movement due to current and wave action is eliminated.

The disadvantage of pipe burial lies in the fact that it becomes more difficult to survey the pipeline for maintenance purposes, detect leaks and make repairs.

There are four main techniques and associated equipment used for the burial of subsea pipes. These are described briefly below:

5.3.2.1 Low pressure jetting

This technique uses low pressure water jets mounted on a jetting frame to remove sea bed material, create a trench and bury the pipeline. The water pumps are usually mounted on the submerged machine with an umbilical cable to supply electrical power and control signals. Low pressure jetting carried out from a frame of the type shown in *Figure 5.31(a)* is very suitable for loose unconsolidated soils and onshore approaches.

5.3.2.2 High pressure jetting

This method of pipe burial is similar to that of low pressure jetting except that high pressure jets, powered by pumps on the surface vessel, are used to obtain a greater rate of trenching and to work on a bigger range of soil types.

5.3.2.3 Mechanical cutters

Mechanical cutters such as the type shown in *Figure 5.31(b)* offer a positive cutting action and spoil removal and can usually operate on difficult stiff soils. However, the positive cutting action is obtained at the price of more difficulty in removing the trenched soil since this is cut away in larger particle size without the fluidizing effect of the jetting technique.

5.3.2.4 Mechanical ploughs

The mechanical plough offers one of the best methods of pipe burial available since its action combines the processes of cutting into the soil and removing spoils from the trench. Ploughs are also capable of working in a large variety of soil types. *Figure 5.31(c)* shows a typical design of a pretrenching plough, which prepares a trench prior to pipe laying. A post trenching plough (not shown here) digs a trench beneath the pipeline and is used after the pipe laying to dig a trench and drop the pipe into it.

5.3.3 Pipe laying techniques

The most commonly favoured method to date for submarine pipe laying is the use of a lay barge, whilst the tow method is gaining acceptance for tows of lines of up to 15 km in length. Additionally the J-lay method is being developed which has benefits for deep water pipelines, as has the reeled pipeline.

5.3.3.1 Lay barge method

This has the advantage of laying the pipe directly on to its final location ready for burial. The pipeline is of steel tube construction and requires a purpose designed pipeline fabrication and laying vessel.

In order to provide as stable a working platform as possible for extreme environments, semi submersible

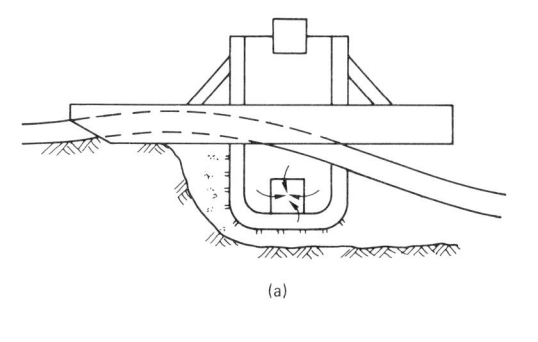

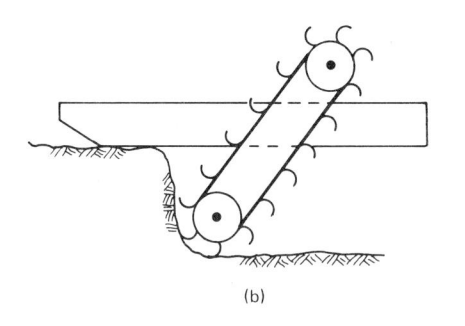

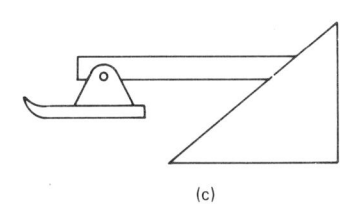

Figure 5.31 Machines for pipe burial (a), High or low pressure jetting sled, (b), Mechanical cutter, (c); Mechanical plough

designs are used with a centre laying stinger as shown in *Figure 5.32*. More favourable conditions otherwise permit the use of monohulls which may be flat bottomed or ship-like, Sriskandarajah (1988), with a side lay of pipe if more convenient (*Figure 5.33*). Whichever type of vessel is used, the critical requirement is to avoid overstressing the pipeline whilst lowering it to the sea bed.

Typically a lay barge has a large deck area for pipe joint storage. Mechanical handling equipment transports the joints in turn into the enclosed fabrication chain. The protective end caps are removed and if necessary any remedial weld preparation performed. The joint is then positioned for welding by hand or more recently by automatic equipment. The lay barge advances one pipe joint length in the direction of pipe laying by pulling in on its fore anchor lines and slackening those aft. A new joint is added at the beginning of the fabrication chain.

The number of welding stations in this chain is dependent upon the weld specification for the particular pipeline. It is a function of bead thickness and hence number of passes required. Non-destructive examination is carried out using an internal X-ray or radioactive source. When the weld is complete the weld area is blasted before protective coatings are applied.

The most critical stress regions in the S-lay method are the over bend and the sag bend, *Figure 5.34*. Control of the stresses in these regions is provided by the ramp and stinger angles along with pipe tension using tensioners. To this end, pipe laying vessels are available with a range of stingers of different lengths, fixed or hinged at the stern. The roller heights on the ramp and stinger may be altered to ensure correct curvature in the pipe. Additional variation in water entry angle may be obtained by appropriate ballasting of the vessel.

The choice of the pipe laying method is dependent upon the water depth and size of pipe to be layed. Analysis of the stresses induced during pipe laying is

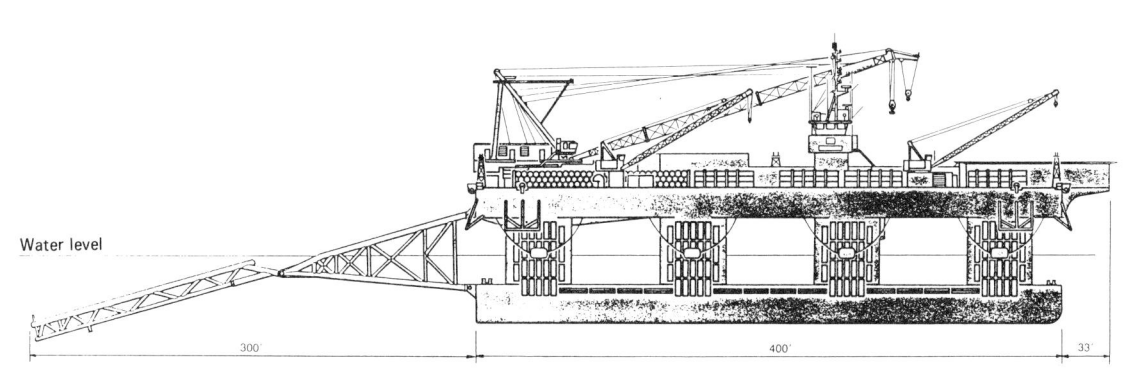

Figure 5.32 A semi submersible working platform with a centre laying stinger, designed to be as stable as possible in extreme environments

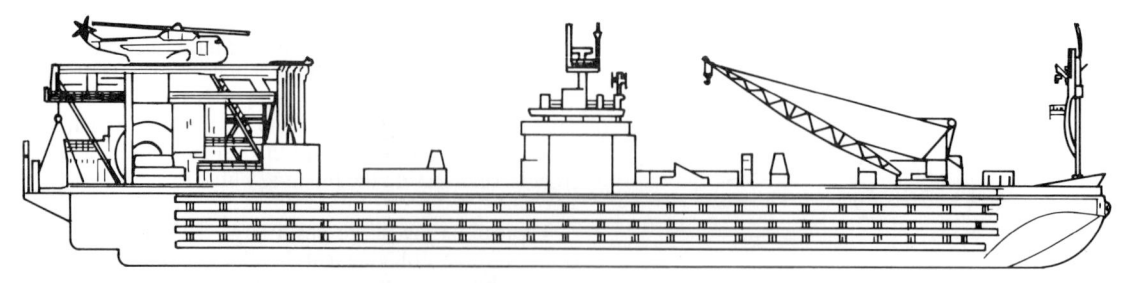

Figure 5.33 A ship shape monohull with a side lay of pipe, used in more favourable conditions than in Figure 5.32

essentially a problem of determining the geometry of the deflected shape. The pipeline may be considered as a continuous beam of varying loads and stiffnesses as shown in *Figure 5.34*. The large angular changes involved do not generally permit small deflection theory to be used. The bending moment must be calculated using the formula:

$$M = EI \cdot \frac{\dfrac{d^2y}{dx^2}}{\left[1 + \left(\dfrac{dy}{dx}\right)^2\right]^{3/2}}$$

Finite difference equations may be used to approximate the differential equations to effect a numerical solution as shown be Wilkins (1970). Clearly, when the pipeline has significant stiffness, a simple catenary analysis cannot be used. However, a limited analysis form has been developed (Dixon and Rutledge, 1967) known as the stiffened catenary which has application with low stiffness pipelines or deep water where stiffness of the pipe is not a major factor.

A typical bending moment distribution is shown in *Figure 5.34*. As the pipe approaches the lay barge it emerges from the water and may undergo a change in apparent weight per unit length of as much as 10:1. In order to reduce tensioning equipment requirements, and stinger length which is susceptible to wave damage; the stresses in the overbend are often designed to be high. On the barge the pipe may undergo an angular change in excess of 10°. At the touchdown point the pipe is subjected to external hydrostatic pressure in addition to bending and tensile stresses.

5.3.3.2 Towed method

The installation of subsea pipelines or pipe bundles by fabricating them at an onshore site and towing them offshore is being increasingly considered as an economic means of laying pipe particularly for infield applications. The fabricated pipe assembly can be transported offshore using bottom tow, off bottom tow, by towing at the surface or by towing at a controlled depth, Rooduyn (1985). The selection of which of these to use is dependent on the total length and submerged weight of the pipeline as well as water depth and sea bed topography including man-made obstructions along the tow route. This technique is also being considered for installation of tethers for deep water tension leg platforms. Transportation through distances up to 450 miles has been acheived, R J Brown (1988). A typical off bottom towing spread is shown in *Figure 5.35*.

Although the onshore fabrication of a towed pipe holds considerable economic advantages, these are to some extent counterbalanced by the constraints placed on the line design in having to ensure an optimum combination of the conflicting demands of pipe submerged weight per unit length, on bottom stability, towing tension and wall thickness.

The use of bundled lines within a carrier pipe has proved an effective design. This feature, however, does not remove all of the constraints described above. The

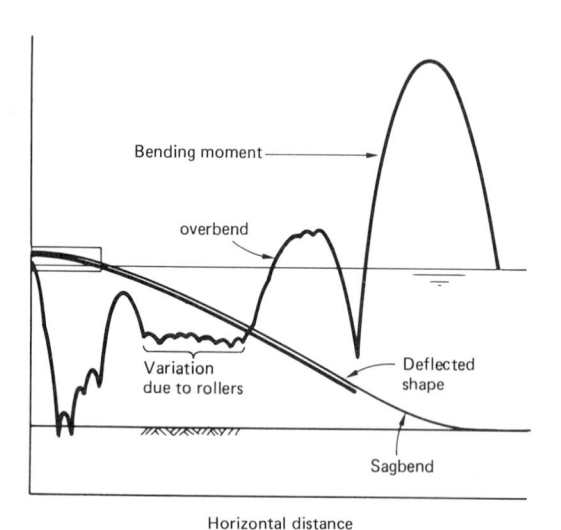

Figure 5.34 A typical bending moment distribution, using the S-lay method

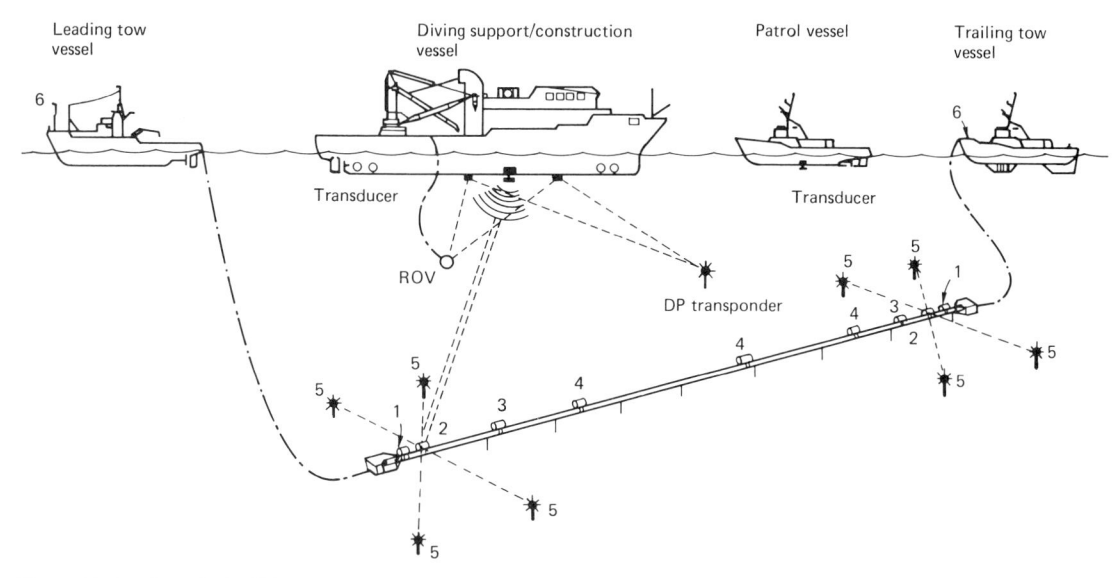

Figure 5.35 A typical off bottom towing spread for the installation of subsea pipelines: (1) Internal pressure/depth package; (2) Towhead relay transponder; (3) Relay transponder; (4) Depth sensors ΔT; (5) Sea bed transponder; (6) Tensometers

advantages include simplification of field layout, small bundle diameter not requiring burial and reduction in thermal gradient. The disadvantages include multiple risk of damage due to a single incident, difficulty of repairs and external inspection of individual flowlines.

For tows off the sea bed, the preceding statical considerations are further complicated by dynamic excitation of the towed pipeline by wave action. This requires a solution for the displacements and hence the stress distribution of a very long and flexible horizontal cylinder loaded by wave action in the drag dominated regime in or near the wave zone. The effects of sled masses (at the pipe ends), towing hawser elasticities, towing speed and tension on the dynamic stresses within the pipe have all to be accounted for in the design process.

The calculation of the static and dynamic behaviour of the towed pipeline or bundle is conventionally done by a finite element or equivalent numerical analysis (Rooduyn and Boonstra, 1985, and Redshaw and Stalker, 1979). Similar calculation work has also been done for the hydrodynamics of towed instrument arrays used in offshore surveying (Kato et al., 1986, and McDearman et al., 1986). However, the typical geometries of towed pipeline bundles are such that it is possible to introduce simplifications which permit analytical solutions to be used in the preliminary static and dynamic analysis stages of the design process (Patel and Seyed, 1987).

For pipelines towed along the sea bed sufficient towing force must be available to overcome friction. These effects may to some extent be reduced with the aid of buoyancy modules temporarily attached at intervals along the pipeline length. The limitation is the available capacity of towing vessels or land based winches in the case of land pulls from an offshore lay barge.

5.3.3.3 J-lay technique

Conventional S-laying of pipelines is probably limited by cost effective positioning capabilities and stinger dimensions to a maximum depth of around 800 m (440 fathoms). The J-lay method is being developed to enable laying of small and large diameter pipes in water depths of 300–3000 m (165–1650 fathoms). This method essentially avoids an overbend and its associated stresses (*Figure 5.36*).

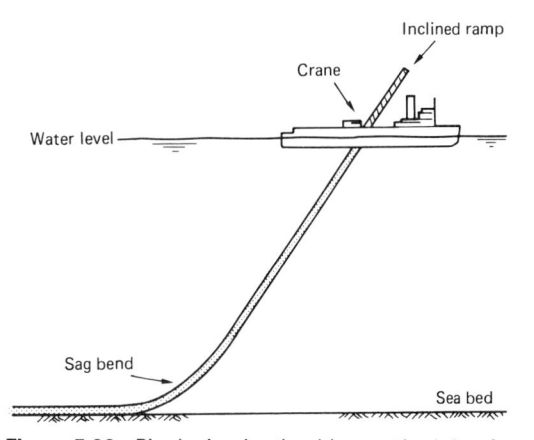

Figure 5.36 Pipelaying by the J-lay method developed to lay small and large diameter pipes in depths of 300–3000m

Also horizontal restoring forces are lower than for an S-lay pipeline, so positioning requirements are easier to meet. It is relatively insensitive to variations in water depth, and requires only one set of equipment (supports and ramps) for the full range of pipe diameters. However, since the pipe must be joined in a near vertical position, physically the opportunity for multiple welding, coating and inspecting stations is reduced. In view of this, novel joining processes have been developed. These include: electron beam welding (de Sivry *et al.*, 1981), flash butt welding (Langer and Switaiski, 1981), and screwed joints (Kaluszynski *et al.*, 1981). Each of these can be accomplished in a much smaller span than the horizontal S-lay method.

5.3.3.4 Reel pipe laying

This has now been developed to the extent that 9.12 km (5.7 miles) of 400 mm (16 in) pipe can be spooled and 320 mm (12.75 in) steel pipe can be laid in depths of up to 900 m (500 fathoms). Additionally, flexible lines can also be laid by this method. Steel pipe is welded and field coated on land in controlled conditions to form stalks which are spooled on in series, each welded to the preceding stalk to form a continuous pipe. The spool on one current vessel has a horizontal axis of rotation with 2000 tonne capacity, Dickson (1981), *Figure 5.37*. In order to remove curvature induced by spooling, a pipe straightener is required. This, along with the pipe tensioner and pipe clamp, is located in the adjustable stern ramp of a ship shape vessel. Sophisticated load and position computer control ensures that the laying operation is relatively straightforward with laying rates of 60 m/min (200 ft/min) being possible. This method also permits the simultaneous laying of umbilicals.

5.3.3.5 Landfall

Two methods are commonly used for the construction of shore approaches – the options being to pull the pipeline from the shore or to pull the pipeline to the shore from a lay barge. When pulling from the shore a pull barge or tow vessel may be used. When pulling from a lay barge on which the pipe is being fabricated, heavy duty drum or linear winches are employed on the beach. In both cases a specially designed pulling head is attached to the lead end of the pipeline. Large diameter pipes of several kilometres length are not uncommon. The force required to pull a pipeline is roughly equal to its total weight. It is often advantageous to add buoyancy modules along the length to reduce the required pulling force and thus the necessary winch or tow capacity. These may be of metal can construction, or for economy, synthetic foam.

The zone where the pipeline crosses the shoreline is often subjected to forces from breaking waves which can cause instability of the sea bed and pipeline. For this reason it is preferable to construct the line into a prepared trench providing protection. Coffer dams are sometimes constructed in the surf zone. The profile of the pipeline path may be smoothed to reduced excessive bending in the pipe and might require removal of sand dunes, for example. Where a gently sloping shoreline is not available, the approach may be to bore through the land so that the pipeline can be pulled from a lay barge through a tunnel.

5.3.3.6 Survey

It is necessary to carry out surveys before and after laying the pipeline. Initially this is to ensure the fitness for purpose of the proposed route and finally to check for correctness of laying. There are many hazards, both natural or man-made, which may exist along the pipeline route. Amongst the natural hazards which must be considered are obstructions which are part of the sea bed topography, but one must also consider the environmental forces due to waves and current, the likelihood of soil movements and possibly earthquakes. Man-made hazards include dragging anchors, sunken vessels and other pipelines.

These hazards must be identified and a suitable route devised by a marine survey before the pipe can be layed. A major part of this survey is the continuous profiling of the sea bottom along the pipeline route.

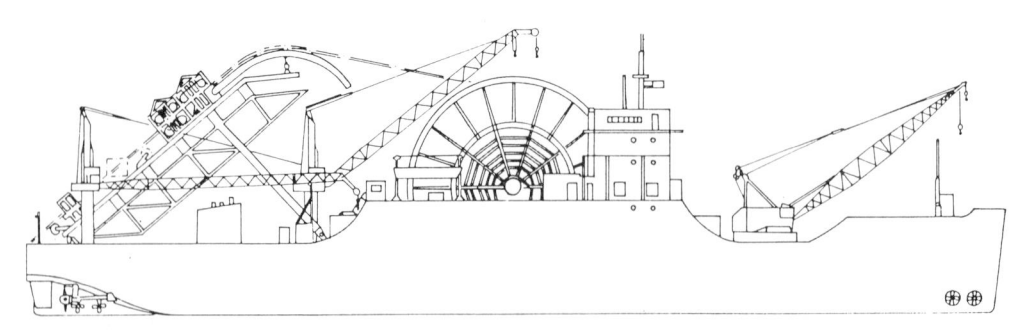

Figure 5.37 A real pipelaying vessel; up to 2000 tons of 12.75 in. Steel pipe can be laid by this method

This is achieved by towing fish of various kinds to measure the depth of the sea floor, for location of objects, and sub bottom geological structure. These devices are mostly acoustic, such as echo sounders and side scan sonar. Magnetometers are used for the detection of metal objects such as shipwrecks and pipelines. The width of the survey route is dependent on the water depth and topography. Brown (1988) indicates that 600–3000 m (333–1667 fathoms) has been necessary for a very deep water pipeline bottom tow route. It is of course important to accurately track the pipeline. This is done with acoustic transponders coupled with satellite navigation.

Once the pipeline has been layed, it is usual to survey it initially and then at regular, possibly annual, intervals. The objective of these surveys is primarily to detect any condition of the line which may lead to a pipe failure either immediately or later in its service life. The as-laid survey requirement of Det Norske Veritas (1982) is for provision of at least the following information:

1. Detailed plot of the pipeline position.
2. Thickness of cover or depth of trench (if applicable) and description of the pipe's condition of laying on the sea bed due to scour, spanning or naturally induced burial.
3. Verification that the condition of weight coating or the anchoring systems which provide for on bottom stability are in accordance with the approved specifications.
4. Description of wreckage, debris or other objects which may affect the cathodic protection system or otherwise impair the pipeline.
5. Description and location of damage to the pipeline, its coating or cathodic protection system.

Depending upon the nature of inspection required and the local conditions of the area of inspection, alternative methods need to be adopted. These include the use of divers, acoustic survey and submersible survey with manned or unmanned vehicles. For routine line surveys Bares et al. (1980) indicates specific survey requirements for submersibles: sonar for location of the pipe when the submersible first reaches the seabed or if the submersible loses the pipe, pipe tracking system for following the pipeline which may be intermittently buried, good observational capabilities, video recording with voice track, colour still photography at points of interest such as damage, debris, eroded anodes, depth of burial measurement, transverse trench profiles and corrosion protection measurements such as electric potential measurements.

5.3.4 PIGs and PIGging

From construction, through commissioning and for the operational life of a pipeline the major tool used for maintenance and monitoring is the PIG (Pipeline Inspection Gauge). PIGs are devices which are passed through pipelines to remove obstructions, provide interface separation between fluids, enable maintenance and enable monitoring.

The life of a pipeline can be divided into five phases: construction, commissioning, operation, maintenance and inspection. Each of these require different types of PIGging operations to be performed.

5.3.4.1 Construction

Debris introduced into the pipeline during construction is removed using a simple PIG. The debris may be accidently introduced, e.g. a welding sealing bladder or welding rods in submarine pipelines. For such lines, the PIGs are often run during the installation operation as the pipe sections are joined together. These may include a buckle detector or gauging PIG towed inside the pipeline beyond the sea bed touchdown point, and possibly non-destructive testing equipment using X-ray imaging. Preinstallation of dewatering PIGs in the pipeline initiation head can allow emptying of the pipeline should it become damaged during construction.

On completion of pipeline construction, it is usual to run a PIG train to remove loose mill scale and corrosion. This is important because internal surface roughness increases resistance to fluid flow, thus reducing pipeline efficiency. The pipeline must then be tested for dimension and integrity. The pipeline diameter is measured with either a gauging plate PIG or a caliper PIG. Integrity is checked by pressure testing. Generally this involves filling the pipeline with test water by using batching PIGs to ensure adequate control and efficient removal of air. The filtered test sea water may be dosed with corrosion inhibitors, biocides and oxygen scavengers to neutralize it.

5.3.4.2 Commissioning

Once pressure testing is complete, the test water must be removed. For oil pipelines it may be sufficient to fill the pipeline with oil behind a short PIG train to displace the water directly. For gas pipelines it is important that the pipeline is effectively dried to avoid the formation of hydrates and ice. The choice of method of commissioning is complex, being dependent on pipeline size and available resources. Options include vacuum drying, nitrogen drying, use of dry air or natural gas, in each case with foam swabbing PIGs, or methanol glycol swabbing with multiple cup PIGs. Such drying agents may be gelled enabling a reduction in the number of mechanical PIGs in the drying train. A newly developed technique using high expansion aqueous foams is capable of dewatering and drying tortuous pipelines in which mechanical pigs alone or in conjunction with gel PIGs would become damaged.

5.3.4.3 Operation

During pipeline operation, PIGs have the two major roles of product separation and removal of liquid drop out.

Pipeline efficiency may be improved by permitting different products to be transported in separate batches. To this end PIGs are used to minimize mixing between the batches.

Variations in pressure and temperature along the pipeline can lead to liquid drop out in gas pipelines. This reduces the flow efficiency and can lead to the formation of liquid slugs which may overload or damage downstream processing plant. Regular removal of this liquid using PIGs can prevent these problems.

5.3.4.4 Maintenance

The internal pipe wall will deteriorate during operation and reduce the flow efficiency. This deterioration can be due to corrosion or deposition of products such as wax or barium scale. Control of these may be achieved by PIGging using scraper PIGs and chemical inhibition. Batch inhibition using a slug of inhibitor between two PIGs may be used.

5.3.4.5 Inspection

A range of intelligent PIGs has been developed to enable internal inspection of pipelines. The primary mode of operation is by magnetic flux induction techniques for the measurement of metal loss and cracks. Hence the level and location of corrosion and defects may be determined.

5.3.5 Types of PIG

The types of PIG that are available can be separated into two groups: non-intelligent passive devices and intelligent active devices which provide data on the condition of the pipeline. The former group includes cleaning, gauging, batching and swabbing PIGs.

5.3.5.1 Instrument PIG

An instrument PIG offers an example of an intelligent device that can travel along the pipeline bore, sometimes under its own power, to detect and record any irregularities, such as dents or obstructions. Instrument PIGs are used to survey newly laid pipelines or to inspect old lines for internal damage and are capable of measuring deviations of a few millimetres in large diameter pipes. This type of PIG usually incorporates an acoustic transponder by means of which its location can be determined, should it meet an obstruction and become lodged in the pipe.

Intelligent PIGs are used for diameter and bend measurement, corrosion detection and measurement, crack detection, leak detection and visual inspection. Some PIGs in an effort at optimization are capable of combining the above tasks.

5.3.5.2 Diameter and bend measurement

The simplest form of PIG which provides intelligence to the user is the gauging PIG. This is not a very intelligent PIG. However, it informs the user whether the pipeline is free from obstruction throughout its length or not. As shown in *Figures 5.38* and *5.39* in its simplest form it comprises two cups with an intermediate body. The steel gauging plate at the front is of slightly smaller diameter than the nominal bore of the pipeline. Any obstructions in the pipeline will cause damage to this plate. However, passing of the PIG over an obstruction may also damage the pipeline. It is not enough to know that there is an obstruction, it is important to know where it is. Recent designs permit position of defects in the pipeline to be measured and recorded using miniature onboard computers which can be interrogated on arrival at the receiving trap. The steel gauging plate is replaced with a strain gauged polyurethane disc which will not cause damage to the pipeline. Bend measurements may be obtained using similar PIGs which have an additional pipewall roller to provide data on curvature.

5.3.5.3 Corrosion detection and measurement

Although ultrasonic and other methods are being developed the method generally used for corrosion detection and measurement is electromagnetic. These PIGs act by inducing a magnetic field and measuring the disturbance in the magnetic flux due to metal loss using search coils. This data is either stored for later analysis, or processed on board the PIG. These PIGs are relatively large, often in three sections, i.e. batteries, data storage and sensors.

5.3.5.4 Crack detection

The above detection method works well for circumferential cracks. However, in order to detect longitudinal cracks it is necessary to rotate either the PIG or its sensors as it travels along the pipeline so that lines of flux may be induced at an angle.

5.3.5.5 Leak detection

PIGs capable of measuring pressure decay and flow are used for leak detection and location. Flow detection methods using ultrasonic emissions and radioactive tracing have been developed.

5.3.5.6 Visual inspection

Systems are being developed using colour stereoscopic cameras for damage detection. More detailed information on the history of PIG development and operator experience with PIGs is provided in Tiratsoo (ed) (1987).

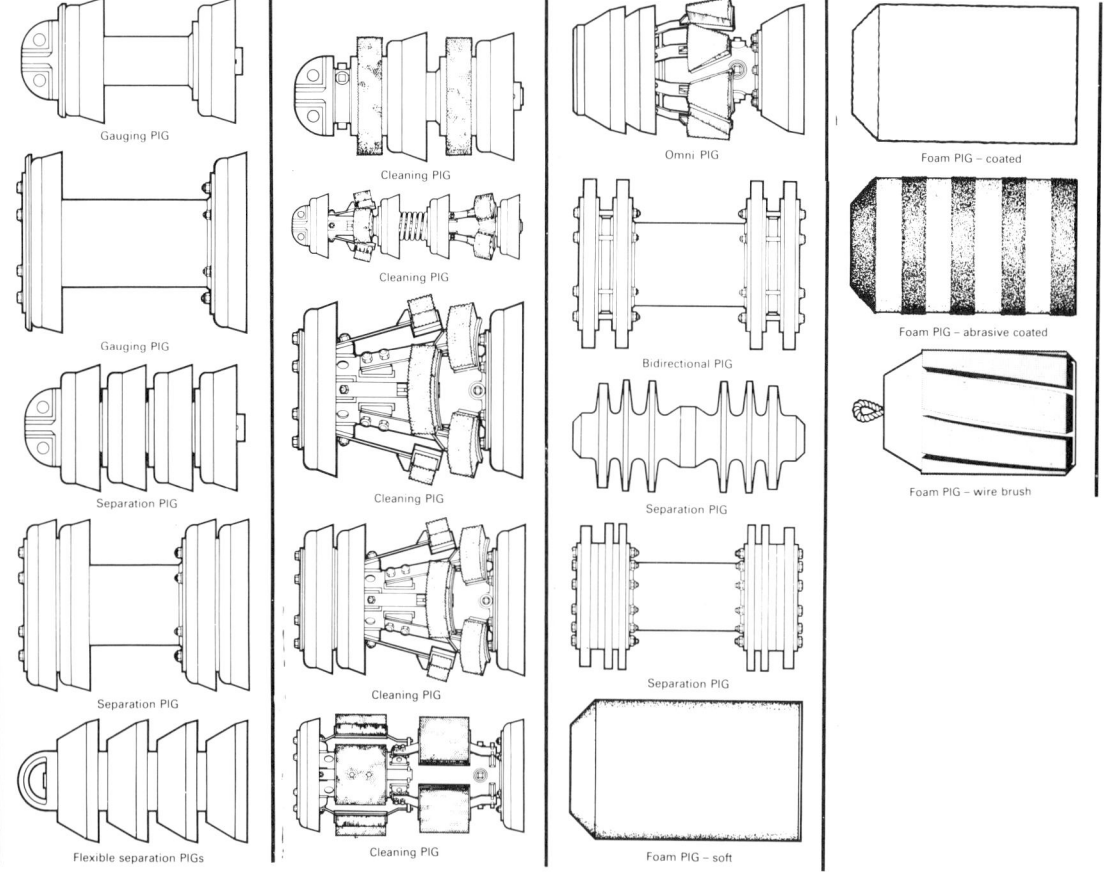

Figure 5.38 PIGs for separation, gauging and cleaning (courtesy of Pipeline Engineering and Supply Co Ltd)

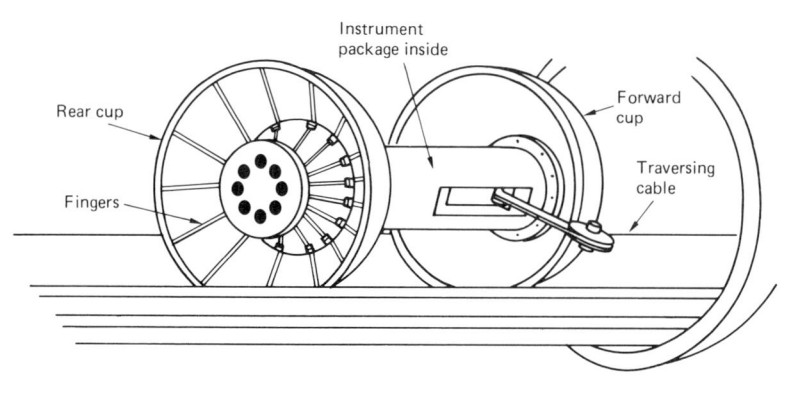

Figure 5.39 A simple instrument pig

References

American Petroleum Institute. Comparison of marine drilling riser analysis. *API Bulletin 2J*, January (1977).
Bares, A. Harve, J. L. and Ling, M. Inspection and maintenance of frigg pipeline transpotation system. *Proceedings of European Offshore Petroleum Conference and Exhibition*, EUR158, London, 105–117 (1980).
Bathe, K. J. and Wilson, E. L. *Numerical Methods in Finite Element Analysis*. Prentice Hall (1976).
Benaroya, H. and Lepore, J. A. Statistical flow-oscillator modelling of vortex shedding. *Journal of Sound and Vibration*, **86**(2), 159–179 (1983).
Bishop, R. E. D. and Hassan, A. Y. The lift and drag forces on a circular cylinder in a flowing field, *Proceedings of The Royal Society* London, Ser A, **277**, 51–75 (1964).

Blevins, R. D. and Burton, T. E. Fluid forces induced by vortex shedding. *Journal of Fluid Engineering*, 19–26 (1976).
Blevins, R. D. *Flow Induced Vibration*. Van Nostrand Reinhold (1977).
Brown, R. J. Deepwater bundle pipeline and connections, Gulf of Mexico case study. *Proceedings of the 1988 Offshore Mechanics and Arctic Engineering Conference*, ASME, Houston, Texas, Volume V, pp. 187–192 (1988).
Chyssostomidis, C. and Patrikalakis, N. M. A comparison of theoretical and experimental prediction of the vortex induced response of marine risers. *Proceedings of the 3rd International Offshore Mechanics and Arctic Engineering Symposium*, ASME, Volume 1, pp. 318–327 (1984).
Cowan, R. and Andris, R. P. Total pipe laying system dynamics. *Proceedings of the 1977 Offshore Technology Conference*, Houston, Texas, Paper NO OTC 2914 (1977).
Dean, R. B. Milligan, R. W. and Wooton, L. R. An experimental study of flow induced vibration. *EEC Report 4*, Atkins Research and Development, Epsom (1977).
Demirbilek, Z. and Holvorsen, T. Hydrodynamic forces on multitube production risers exposed to currents and waves. *Proceedings of the 4th Offshore Mechanics and Arctic Engineering Symposium*, pp. 363–370 (1985).
De Oliveira, J. G., Goto, Y. and Okamoto, T. Theoretical and methodological approaches to flexible pipe design and application. *Proceedings of the 1985 Offshore Technology Conference*, Houston, Texas, Paper No OTC 5021 (1985).
de Sivry, B., Kaluszynski, M., Haman, J. P. and Jegouse, M. Electron beam welding for J-curve pipelines. *Proceedings of Deep Ocean Technology Conference*, Palma de Mallorca, Spain, **3**, 7–26, October (1981).
Det Norske Veritas. *Rules for Submarine Pipe Line Systems*, Det Norske Veritas, PO Box 300, Hovik, Norway (1982).
Dickson, L. J. 'Apache' – The worlds first pipe laying reel ship. *Proceedings of Deep Ocean Technology Conference*, Palma de Mallorca, Spain, **3**, 74–82, October (1981).
Dixon, D. A. and Rutledge, D. R. Stiffened catenary calculations in pipeline laying problems. *Transactions of American Society of Mechanical Engineers*, Petroleum Division, Paper 67-Pet-6, September (1967).
Every, M. J., King, R. and Griffin, O. M. Hydrodynamic loads on flexible marine structures due to vortex shedding. *ASME Journal of Energy Resources Technology*, **104**, 330–336 December (1981).
Gardner, T. N. and Kotch, M. A. Dynamic Analysis of riser and caissons by the finite element method. *Proceedings of the 1976 Offshore Technology Conference*, Houston, Texas, Paper No OTC 2651 (1976).
Griffin, O. M., Pattison, J. H., Skop, R. A., Ramberg, S. E. and Meggitt, D. J. Vortex excited vibrations of marine cables. *Journal of Waterways: Post Coastal and Ocean Division*, **106**, 183–205 (1980).
Griffin, O. M. and Ramberg, S. E. Some recent studies of vortex shedding with application to marine tubulars and risers. *ASME*, **104**, 2–13 March (1982).
Griffin, O. M., Skop, R. A. and Koopman, G. H. The vortex excited resonant vibrations of circular cylinders. *Journal of Sound and Vibration*, **31**(2), 235–249 (1973).
Griffin, O. M. and Vandiver, J. K. Vortex induced strumming vibrations of marine cables with attached masses. *Proceedings of the 3rd International Offshore Mechanics and Arctic Engineering Symposium*, **1**, 300–309 (1984).
Hall, S. A. Vortex induced vibrations of structures, PhD Thesis, California Institute of Technology (1981).
Hallam, M. G., Heaf, N. J. and Wooton, R. L. Dynamics of marine structures. *Construction Industry Research and Information Association (CIRIA) Report UR8*, London, pp. 175–206 (1978).
Hansen, H. T. and Bergan, P. G. Non-linear dynamic analysis of flexible risers during environmental loading. *Proceedings of the 1986 Offshore Mechanics and Arctic Engineering Symposium*, ASME, Tokyo, Japan (1986).
Hartlen, R. T. and Currie, I. G. Lift oscillation model for vortex induced vibration. *Journal of Engineering, Mechanical Division, ASCE*, **96**, 577–591 (1970).
Humphreys, J. S. On a circular cylinder in a steady wind at transition Reynolds numbers. *Journal of Fluid Mechanics*, **9**, 603–612 (1960).
Iwan, W. D. The vortex induced oscillation of elastic structural elements. *ASME Journal of Engineering for Industry*, **91**, 1378–1382 November (1975).
Iwan, W. D. The vortex induced oscillation of non-uniform structural systems. *Journal of Sound and Vibration*, **79**(2), 291–301 (1981).
Iwan, W. D. and Blevins, R. A model of vortex induced oscillation of structures. *Journal of Applied Mechanics*, **41**, 581–586, (1974).
Jacobsen, V., Nielsen, R. and Fines, S. Vibration of offshore pipelines exposed to current and wave action. *Proceedings of the 3rd International Offshore Mechanics and Arctic Engineerging Symposium*, **1**, 291–299 (1984).
Jones, W. T. On bottom pipeline stability in steady water currents, *Proceedings of the 1976 Offshore Technology Conference*, Houston, Texas, Paper No OTC 2598 (1976).
Kaluszynski, M., Magloire, C., and Mandry, P. The use of threaded pipe for the laying of small diameter pipelines in deep water. *Proceedings of Deep Ocean Technology Conference*, **3**, 35–42 (1981).
Kato, N., Koda, S. and Takahashi, T. Motions of underwater towed system. *Proceedings of the 1986 Offshore Mechanics and Arctic Engineering Symposium*, ASME, Tokyo, Japan, 426–33 (1986).
Kennedy, M. and Vandiver, J. K. A random vibration model for cable strumming predictions, *Civil Engineering in the Oceans*, IV, ASCE, pp. 273–292 (1979).
King, R. A review of vortex shedding research and its application. *Ocean Engineering*, **4**, 141–171 (1977).
King, R., Prosser, M. H. and Johns, D. J. On vortex excitation of model piles in water. *Journal of Sound and Vibration*, **29**(2), 169–188 (1973).
Kirk, C. L. and Etok, E. U. Wave induced random oscillations of pipe lines during laying. *Applied Ocean Research*, **1**, No 1 (1979).
Krieth, F. *Principals of Heat Transfer*, International Textbook Co., 2nd edn (1965).
Krolikowski, L. P. Modern production risers, Part 9 – The frequency domain method for production riser analysis. *Petroleum Engineering International*, p. 88 July (1981).
Krolikowski, L. P. and Gray, T. A. An improved linearisation technique for frequency domain riser analysis. *Proceedings of the 1980 Offshore Technology Conference*. Houston, Texas, Paper No OTC 3777 (1980).
Kyriakides, S. and Babcock, C. D. Experimental determination of the propagating pressure of circular pipes. *Transactions of American Society of Mechanical Engineers*, **103**, 328–336 (1981).
Landl, R. A mathematical model for vortex excited vibration of bluff bodies. *Journal of Sound and Vibration*, **42**(2), 219–234 (1975).
Langer, J. and Switaiski, B. Pipe laying in deep sea by

J-method with flash butt welding technique. *Proceedings of Deep Ocean Technology Conference*, **3**, 62–69 (1981).

Leinhard, J. H. *Synopsis of Lift, Drag and Vortex Frequency Data for Rigid Circular Cylinders*. Washington State University, College of Engineering, Research Division Bulletin 300 (1966).

Lyons, G. J. and Patel, M. H. A prediction technique for vortex induced transverse response of marine risers and tethers. *Journal of Sound and Vibration*, December (1986).

Malahy, R. C. A non-linear finite element method for the analysis of offshore pipelines, risers and cable structures. *Proceedings of the 1986 Offshore Mechanics and Arctic Engineering Symposium*, ASME, Tokyo, Japan (1986).

McDearman, G. F., Scherch, R. P. and Dunne, A. L. Arrays of arbitrary geometry towed underwater in three dimensions, *Proceedings of the 1986 Offshore Mechanics and Arctic Engineering Symposium*, ASME, Tokyo, Japan, 371–6 (1986).

McIver, D. B. and Lunn, T. S. Improvements to frequency domain riser programs. *Proceedings of the 1983 Offshore Technology Conference*, Houston, Texas, Paper No OTC 4550 (1983).

McNamara, J. F. O'Brien, P. J. and Gilroy, S. G. Non-linear analysis of flexible risers using hybrid finite elements. *Proceedings of the 1986 Offshore Mechanics and Arctic Engineering Symposium*, ASME, Tokyo, Japan (1986).

Miller, J. E. and Young, R. D. Influence of mud column dynamics on top tension of suspended deepwater drilling risers. *Proceedings of the Offshore Technology Conference*, Houston, Texas, Paper No OTC 5015 (1985).

Modi, V. J., Welt, F. and Irani, M. B. On the nutation damping of fluid structure interaction instabilities and its application to marine riser design. *Proceedings of the 1986 Offshore Mechanics and Ocean Engineering Symposium*, Tokyo, **3** (1986).

Morgan, G. W. Applied mechanics of marine riser systems, *Petroleum Engineer*, Parts 1–3, October 1974 – May 1976.

Narzul, P. and Marion, A. Static and dynamic behaviour of flexible catenary risers. *Proceedings of the 1986 Offshore Mechanics and Arctic Engineering Symposium*, ASME, Tokyo, Japan (1986).

Nordgren, R. P. Dynamic analysis of marine risers with vortex excitation. *Journal of Energy Resources Technology*, **104**, 14–19, March 1982.

Owen, D. G. and Qin, K. Model tests and analysis of flexible riser systems. *Proceedings Of the 1986 Offshore Mechanics and Arctic Engineering Symposium*, ASME, Tokyo, Japan (1986).

Patel, M. H. and Sarohia, S. On the dynamics of production risers. *Proceedings of the Third International Conference on Behaviour of Offshore Structures (BOSS 82)*, **1**, 599 (1982).

Patel, M. H. Sarohia, S. and Ng, K. F. Finite element analysis of the marine riser. *Engineering Structures*, **6**, p 00 (1984).

Patel, M. H. and Seyed, F. B. Analysis and design considerations for towed pipe lines. *Proceedings of the 6th International Symposium on Offshore Mechanics – Offshore and Arctic Pipe Lines*, ASME, 17–28 (1987).

Pelzer, R. D. and Rooney, D. M. Near wake properties of a strumming marine cable: an experimental study. *Proceedings of the 3rd International Offshore Mechanics and Arctic Engineering Symposium*, **1**, 310–317 (1984).

Ratcliffe, A. T. The validity of quasi-static and approximate formulae in the context of cable and flexible riser dynamics. *Proceedings of the Fourth International Conference on Behaviour of Offshore Structures (BOSS 85)*, Amsterdam, Netherlands (1985).

Rajabi, F. Hydroelastric oscillation of smooth and rough cylinders in harmonic flow, PhD Thesis, Naval Postgraduate School, Monterey, California, December (1979).

Rajabi, F. Zedan, M. F. and Mangiavacchi, A. Vortex shedding induced dynamic response of marine risers. *Journal of Energy Resources Technology*, **106**, 214–221, June (1984).

Rayleigh, Lord. *Theory of Sound*, London Macmillan, 2nd edn (1896).

Redshaw, P. R. and Stalker, A. W. Explosive welding combines with bottom tow for new subsea pipeline construction technique. *Proceedings of the 1979 Offshore Technology Conference*, Houston, Texas, 30 April – 3 May, OTC 3523, 1431–8 (1979).

Rooduyn, E. G. Submarine flowlines transportation of pre-fabricated pipelines with the controlled depths tow method, *Advances in Underwater Technology and Offshore Engineering, volume 2, The Design and Installation of Subsea Systems*, SUT, pp. 217–237 (1985).

Rooduyn, E. J. and Boonstra, H. Design aspects of the controlled depth tow method for pipeline bundles. *Proceedings of the Third Deep Offshore Technology Conference*, Italy, **1**, Paper 6, 1–21 (21–23 October (1985).

Rosho, A. On the development of turbulent wakes from vortex streets, National Advisory Committee for Aeronautics, Report NACA-TN-2913 (1953).

Sarpkaya, T. Vortex shedding and resistance in harmonic flow about smooth and rough circular cylinders. *Proceedings of Behaviour of Offshore Structures Conference*, Trondheim, **1**, 220–235 (1976).

Sarpkaya, T. Vortex induced oscillations, a selective review. *Journal of Applied Mechanics*, **46**, 241–258, June (1979).

Sarpkaya, T. and Isaacson, M. *Mechanics of Wave Forces on Offshore Structures*, Van Nostrand Reinhold (1981).

Sarpkaya, T. and Shoaff, R. L. A discrete vortex analysis of flow about stationary and transversely oscillating circular cylinders *Technology Report No. NPS-69LS79011*, Naval Postgraduate School, Monterey, California (1979).

Schafer, B. Dynamical modelling of wind induced vibrations of overhead lines. *International Journal of Nonlinear Mechanics*, **19**(5), 455–467 (1984).

Schlichting, H. *Boundary Layer Theory*, 6th edn, McGraw Hill Book Co. (1968).

Scraeffer, J. W. and Eskanzi, S. An analysis of the vortex street generated in a viscous fluid. *Journal of Fluid Mechanics*, **6**, 241–260 (1959).

Shapiro, A. H. *The Dynamics and Thermodynamics of Compressible Fluid Flow*. Volume 1, Ronald Press Co., New York (1953).

Shaw, T. L. (ed.) *Mechanics of Wave Induced Forces on Cylinders*, Pitman (1979).

Shotbolt, JK. The influence of production-riser design on the configuration and operation of semisubmersible floating production systems. *Proceedings Offshore Europe Symnposium*, Aberdeen, 6–9 September (1983).

Simpson, A. S. Cables: dynamic stability aspects – a review, Department of Aeronautics Engineering, University of Bristol. *Symposium on Mechanics of Wave-induced Forces on Cylinders*, 3–6 September (1978).

Sparks, C. P. Mechanical behaviour of marine risers mode of influence of principal parameters. *Proceedings of the Winter Annual Meeting of the American Society of Mechanical Engineers*, **7**, New York, 2–7 December (1979).

Sparks, C. P., Cabillic, J. P. and Schawann, J. C. Longitudinal resonant behaviour of very deep water risers. *Proceedings of the 1983 Offshore Technology Conference*, Houston, Texas, Paper No OTC 4317 (1982).

Sriskandarajah, T. *Installation of Steel Flowlines From a Divind Support Vessel*, Institute of Marine Engineers, London, February (1988).

Strouhal, V. Uber eine besondere Art der Tonerregung, *Ann. Physik* (Leipzig) (1878).

Tam, C. K. W. and Croll, J. G. A. Buckle initiation in damaged subsea pipelines. *Proceedings of the 8th Annual Energy Sources Technology and Exhibition*, Dallas, Texas, Pipeline Engineering Symposium, 71–78 (17–21 February (1986).

Tiratsoo, J. N. H. *Pipeline Pigging Technology*, Pipes and Pipelines International Scientific Surveys Ltd (1987).

Tsahalis, D. T. Vortex induced vibrations of a flexible cylinder near a plane boundary exposed to steady and wave induced currents. *Journal of Energy Resources Technology*, **106**, 206–213, June (1984).

Wang, E. Analysis of two 13 200-ft riser systems using a three dimensional riser program. *Proceedings of the 1983 Offshore Technology Conference*, Houston, Texas, Paper No OTC 4563 (1983).

Whitney, A. K. and Nikkel, K. G. Effects of shear flow on vortex shedding induced vibration of marine risers. *15th Offshore Technology Conference*, OTC 4595, 127–137, May (1983).

Wilkins, J. R. Offshore pipeline stress analysis. *Proceedings of the 1970 Offshore Technology Conference*, Houston, Texas, paper No OTC 1227 (1970).

Young, R. D. and Fowler, Mathematics of marine risers. *The Energy Technology Conference and Exhibition*, Houston, Texas, November (1978).

Zdravkovich, M. M. Review and classification of various aerodynamic and hydrodynamic means for suppressing vortex shedding. *Journal of Wind Engineering and Industrial Aerodynamics*, **7**, 145–189 (1981).

Zedan, M. F. and Rajabi, F. Lift forces on cylinders undergoing hydroelastic oscillations in waves and two dimensional harmonic flow, *Proceedings NHL, International Symposium on Hydrodynamics in Ocean Engineering*, **1**, 239–262 (1981).

6 Marine Engines and Auxiliary Machinery

Chris Wilbur

Contents

6.1 Machinery arrangements
    6.1.1 Choice of main engine
    6.1.2 Factors influencing machinery selection
    6.1.3 Typical applications of machinery
    6.1.4 Single screw plants
    6.1.5 Multiple screw plants
    6.1.6 Plants with mixed machinery
    6.1.7 Loiter drives
    6.1.8 Auxiliary propulsion
    6.1.9 Uniform machinery
    6.1.10 Father and son arrangement
    6.1.11 Electric drive
    6.1.12 Power station concept
    6.1.13 Mechanical and electrical drive
    6.1.14 TNT system

6.2 Propulsion diesel engines
    6.2.1 Diesel engine principle
    6.2.2 Four stroke cycle
    6.2.3 Two stroke cycle
    6.2.4 Slow speed diesel engine
    6.2.5 Uniflow scavenged two stroke engine
    6.2.6 Crosshead engine applications
    6.2.7 Advantages of direct drive
    6.2.8 Slow speed engine data
    6.2.9 Medium speed diesel engines
    6.2.10 Medium speed engine applications
    6.2.11 Medium speed advantages
    6.2.12 Medium speed engine data
    6.2.13 High speed diesel engine
    6.2.14 Typical high speed design
    6.2.15 Automotive type engines
    6.2.16 Reasons for marinization
    6.2.17 Truck engine advantages
    6.2.18 Multiple engine plants
    6.2.19 Air cooled engines

6.3 Steam plant
    6.3.1 Steam propulsion machinery
    6.3.2 Steam plant principle
    6.3.3 Steam turbine features
    6.3.4 Most advanced steam plant
    6.3.5 Future outlook

6.4 Gas turbines
    6.4.1 Gas turbine plant
    6.4.2 Operating principle
    6.4.3 Gas turbine applications
    6.4.4 COGOG
    6.4.5 COGAG
    6.4.6 CODLAG
    6.4.7 Gas turbine advantages
    6.4.8 A typical modern design
    6.4.9 Performance improvements
    6.4.10 ICR plant
    6.4.11 RACER systems

6.5 Nuclear power
    6.5.1 Nuclear history
    6.5.2 Basic principle
    6.5.3 Nuclear advantages
    6.5.4 Applications

6.6 Propeller shaft drives
    6.6.1 Power transmission
    6.6.2 Shaft bearings and seals
    6.6.3 Propulsion gears
    6.6.4 Clutches and couplings
    6.6.5 Couplings
    6.6.6 Clutches

6.7 Electrical power generation
    6.7.1 Auxiliary requirements
    6.7.2 Auxiliary power
    6.7.3 Uni-fuel concept
    6.7.4 Shaft generators
    6.7.5 Modes of operation
    6.7.6 Power turbines
    6.7.7 CODAG generators

6.8　Steam generation
　　6.8.1　Current marine boilers
　　6.8.2　Modern coal fired boilers
　　6.8.3　Coal fired boiler design
　　6.8.4　Stoker firing
　　6.8.5　Coal handling
　　6.8.6　Auxiliary steam
　　6.8.7　Waste heat recovery

6.9　Ancillary machinery
　　6.9.1　Marine pumps
　　6.9.2　Positive displacement pumps
　　6.9.3　Gear pump
　　6.9.4　Screw pump
　　6.9.5　Non-positive displacement pumps
　　6.9.6　Centrifugal pump
　　6.9.7　Propeller pump
　　6.9.8　Jet pump
　　6.9.9　Cooling systems
　　6.9.10　Centralized cooling system
　　6.9.11　Lubrication systems
　　6.9.12　Fuel treatment
　　6.9.13　Fuel problems
　　6.9.14　Fuel system design
　　6.9.15　Bilge and ballast systems
　　6.9.16　Compressed air
　　6.9.17　Refrigeration
　　6.9.18　Air conditioning
　　6.9.19　Minor ancillaries
　　6.9.20　Oil fuel system
　　6.9.21　Cargo pumping system
　　6.9.22　Sanitary system
　　6.9.23　Steering gear
　　6.9.24　Electro–hydraulic steering gear
　　6.9.25　Rotary type
　　6.9.26　Deck machinery
　　6.9.27　General cargo ships
　　6.9.28　RoRo ships
　　6.9.29　Container ships
　　6.9.30　Bulk carriers
　　6.9.31　Tankers

6.10　Propulsion
　　6.10.1　Propulsion concept
　　6.10.2　Propeller application
　　6.10.3　Multiple screws
　　6.10.4　Overlapping propellers
　　6.10.5　Ducted propellers
　　6.10.6　Screw propeller features
　　6.10.7　Propeller efficiency
　　6.10.8　Controllable pitch propellers
　　6.10.9　Advantages of CPP
　　6.10.10　Economic loading
　　6.10.11　Typical CPP design
　　6.10.12　Voith Schneider propeller
　　6.10.13　Self pitching propeller
　　6.10.14　Contra rotating propellers
　　6.10.15　Grim wheel
　　6.10.16　Water jets
　　6.10.17　Pump jet
　　6.10.18　Azimuthing thrusters
　　6.10.19　Z-drive advantages
　　6.10.20　Contra rotating thruster
　　6.10.21　CR propeller–thruster system

Further reading

6.1 Machinery arrangements

Marine propulsion is one of the earliest of applied sciences put to good use by man. It became useful when the first prehistoric man straddled a convenient floating log to cross a river and found that with a piece of wood in his hands, he could control the direction in which the floating vessel could be made to travel. The tree branch was soon shaped to form a paddle of greater propulsive efficiency and eventually oars were developed to obtain more effective application of effort through the mechanics of a simple lever. Man later learned to harness the natural energy of the wind through sails to assist oars and, except for the smallest of craft, man relied on the wind alone.

The invention of the steam engine was soon seized upon as a useful power source to propel marine craft, particularly as various propulsors had already been invented and only needed a suitable machine to drive them. The steam revolution did not take place overnight, however, as for decades steamships were still fitted with sails until confidence in steam grew to the point where ships could rely on engines alone.

The steam age was also short lived, because even the earliest of internal combustion engines were more efficient than steam plants and today the huge efficiency gap has already made the steamship virtually obsolete. Within the last decade very few steamships have been built as the only application of steam turbines now considered worthwhile is for the propulsion of Liquified Natural Gas carriers, as boil-off from the LNG cargo, which would otherwise be lost to the atmosphere, can be burned under a boiler and hence contribute to the propulsion of the ship. Steam is still used in naval submarines fired by nuclear reactors but developments with anaerobic diesel engines and public antipathy to anything nuclear, could soon make steam obsolete for naval vessels as well.

Ships and boats of all sizes are thus powered in most cases by diesel or petrol engines of one sort or another driving propellers or water jets, though there are still some special applications for gas turbines, electric motors and indeed some steam turbines, while propulsors such as paddles, water pumps, oars, etc., still survive.

Today, more than ever, the science of marine propulsion is dominated by economics and the different types of prime mover and their methods of installation are governed by economic criteria such as initial cost, fuel cost, size, weight and cost of operation and maintenance.

Unlike other forms of transport such as land vehicles and aircraft, where the prime mover types are basically similar, albeit with make variations, ships can be powered by some completely different forms of heat engine, with variants within each type, not to mention many similar competing engine makers. Broadly speaking, ships today are either fitted with diesel engines, steam turbines, or gas turbines and there are a number of ways of connecting the main engine to the propeller or other main propulsion device. Among the three commonly used main propulsion power sources, there are, however, many different makes and types to choose from. There are even some ships still in service powered by steam reciprocating engines.

Furthermore, there is also a significant number of vessels relying on wind power though practically all are also fitted with auxiliary engines. The latest applications of machinery for propulsion and auxiliary duties aboard ships forms the basis of this chapter.

6.1.1 Choice of main engine

The choice of main propulsion engine for a ship is by no means an easy one. It was not too long ago that most ship owners chose either a direct coupled slow speed engine driving a fixed pitch propeller, or a four stroke medium speed engine geared to either a fixed pitch or controllable pitch propeller. Today, however, the quest for fuel efficiency has encouraged owners to select new or revived alternatives such as electric drive systems, small crosshead engines where trunk piston types would hitherto have been expected and in some cases slow speed engines have been geared to the propeller.

One normally expects a direct drive engine to be used in a liner type cargo vessel, tanker or bulk carrier, etc., and geared trunk piston engines to be used in smaller craft and those requiring good manoeuvrability such as tugs, ferries, offshore support vessels, etc. In recent years many interesting arrangements have been applied, usually to gain some economic advantages or increased flexibility of operation. The merits of typical current machinery arrangement are discussed elsewhere in this chapter.

6.1.2 Factors influencing machinery selection

When it comes to selecting the propulsion engines for a ship and its installation arrangement, many factors influence the final choice. There are ship owners who prefer to remain with a particular make or type of engine for reasons such as experience of reliable operation in the past, crew familiarity, spare parts control, good service backup from the manufacturer and so on, but new engine designs are introduced with alarming frequency these days because of the intense competition between engine builders. In many cases an owner who has not built a vessel for some years will have to choose between a number of machinery plants with which he is unfamiliar.

Decisions are nowadays based on factors such as the type and cost of the fuel to be burned, maintenance costs, manning levels, availability of spare parts and, not least, the initial purchase price of the engine. Life cycle costs are also taken into account as the initial choice of main engine can have a considerable influence on the operating costs, over, for example, 20 years operation.

The primary factors which influence the choice of engine can be summarized as:

1. The ability to burn the heaviest and cheapest fuel without detrimental effect on the engine components and maintenance costs.
2. The maintenance work requiring periodic attention, workload for the crew to be carried, cost and availability of spare parts.
3. Suitability of the plant to be operated unattended for long periods, thus saving on manpower through elimination of watch keeping.
4. Propulsive efficiency of the plant, i.e. the ability to turn the shaft at low enough speed to drive the largest diameter and hence most efficient propeller.
5. The size and weight of the propelling machinery as this can affect the revenue earning capacity of the vessel.
6. The purchase and installation costs of the whole machinery plant.
7. The reputation of the particular engine type for reliability and operation for long periods without attention.

The selection of the main engine and its ancillary machinery are decisions of major importance but of equal importance are the methods used to drive the propeller and indeed, the type of propellers used, as these can considerably alter the economic operational profile for the vessel. Current practice is to design the whole machinery installation to suit the vessel's planned operational programme, as a vessel which is frequently manoeuvred such as a ferry, and one with widely varying loadings such as a tug or trawler, will have far different requirements of the power plant than a ship spending most of its time at maximum speed, for example a container ship.

6.1.3 Typical applications of machinery

As mentioned earlier, ships today are normally powered by slow speed, medium speed or high speed diesel engines, other types of internal combustion engine burning petrol, paraffin or gas, steam turbines with boilers fired by coal or oil, gas turbines burning distillate fuels, or nuclear, with a reactor used to generate steam for turbines. Vessels are also fitted with sails as an auxiliary propulsion device or an engine can be the auxiliary to sails. There are no fixed rules as to the type of machinery needed for a given ship type, e.g. identical-looking cruise liners could be equipped with different types of diesel engine, steam turbines or even gas turbines. However, each ship type usually has a machinery arrangement preferred by the majority of ship owners. Machinery installations designed specifically to achieve certain fuel saving advantages, or for greater operational flexibility, are described in the remainder of this section.

6.1.4 Single screw plants

Most ships are provided with single screw propulsion as this arrangement of propulsion engine and shafting gives simplicity and the best distribution of machinery weight inside the vessel. However, the one propeller may be driven by one or more engines with, in the latter case, most commonly two engines connected to a twin input/single output gearbox. Single engines are usually direct coupled to the propeller shaft, particularly in the case of slow speed two stroke engines, (*Figure 6.1*) while medium speed engines are normally geared as the engine speeds are mostly too high for the propeller, with the engine set off the central line to facilitate the lateral offset of the gear wheels, or if an epicyclic type gear is used, the engine can be in line with the shafting.

The argument for single engine–single screw is one of simplicity but with a fixed pitch propeller the engine must be direct reversing to give astern operation. Many modern geared medium speed plants, however, are provided with Controllable Pitch (CP) propellers (*Figure 6.2*), which greatly enhance manoeuvrability as the engine runs at more or less constant speed in one direction only and the thrust is varied ahead or astern with the propeller.

The indirect or geared drive of one propeller is considered to be best in terms of flexibility, economy and reliability. Two engines geared to one propeller through one plain reduction gear fitted with suitable clutches and couplings give positive benefits as follows:

1. For vessels with more than one engine, reliability is improved considerably as the vessel can still operate with one engine if a breakdown occurs at sea.
2. When vessels are running light, partially loaded or slow steaming, one engine can be run at its normal rating while the other is shut down, avoiding the undesirable condition of running engines for long periods at part load which leads to fouling and deterioration in performance.
3. For a given power requirement two smaller engines have smaller more manageable components leading to easier maintenance and cheaper spare parts.
4. Engines can be overhauled at sea by steaming on the other engine which suits the trend today for ships to spend the minimum time in ports.
5. By fitting the obligatory reduction gear between engine and propeller the optimum shaft revolutions to suit the propeller design can be precisely selected, leading to highest propulsive efficiency.
6. The weight of the machinery and space required (particularly headroom) is much reduced, an important feature for vehicle deck vessels such as car ferries.
7. By modifying the number of engines per ship and cylinder numbers per engine to suit individual horsepower requirements, the propelling machinery for a fleet can be standardized on a single bore size, with subsequent savings in spares costs, etc. For vessels

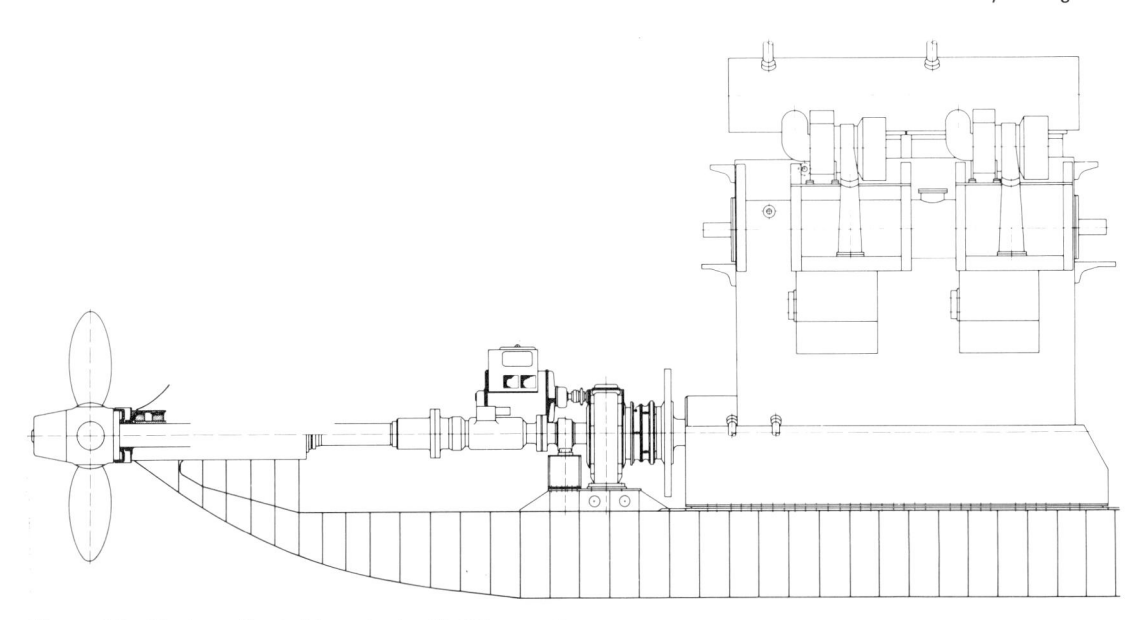

Figure 6.1 Modern direct drive plant with CP propeller and shaft generator

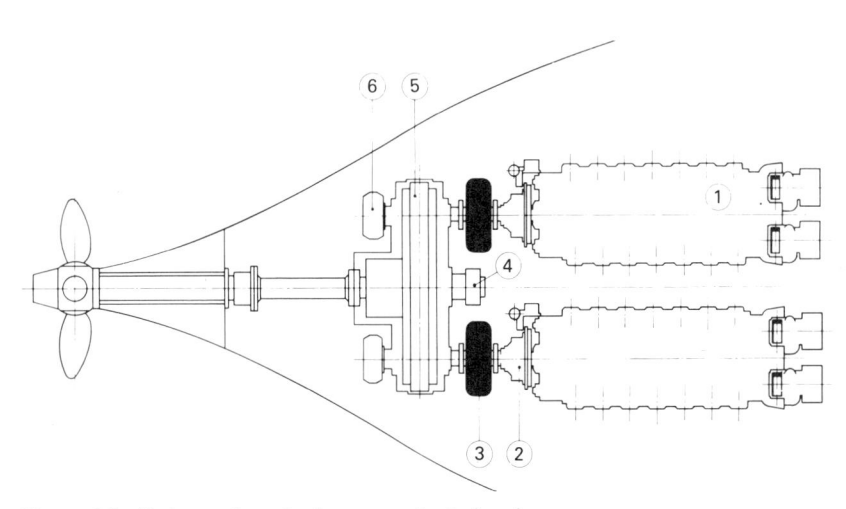

Figure 6.2 Twin engine–single screw plant showing: (1) Main engine, (2) Flexible coupling, (3) Hydraulic coupling, (4) CP propeller controls, (5) Reduction gear, (6) Clutch/coupling

with high auxiliary loads the same engine types can also be used as generator sets.

The above arguments refer mainly to four stroke medium speed engines but smaller two stroke crosshead type engines can equally be geared in pairs to drive one propeller, though such an arrangement is by no means common. When comparing the two engines types, however, other factors such as the differences in specific fuel consumption, lubricating oil consumption, time between overhauls of individual components and numbers of cylinders required for equivalent outputs are also important considerations.

6.1.5 Multiple screw plants

Ships may be fitted with one, two, three or even four propellers. The latter type is rarely encountered these days as the quadruple screw transatlantic liners of yesterday required such high installed horsepowers which could only be met with multiple power units.

Today modern engines are easily built to deliver sufficient power and the twin screw arrangement offers the best advantages. Twin propellers are fitted to most passenger liners, ferries and offshore supply vessels, or any ships which need the propulsion power to be spread over two propellers.

Two main engines, each direct coupled to its own propeller, offer similar advantages for twin engines geared to one screw (*Figure 6.3*). The reduction gear and clutches, etc., are not necessary so transmission losses are reduced.

By fitting two propellers, the diameter of each can be reduced allowing a shallower draught, which is indeed a major reason why twin screws may be specified in the first place. Also, two engines will be smaller and lighter than the single screw alternative. For vessels requiring high installed horsepower, such as ferries and cruise liners, a commonplace arrangement to meet loadings from very slow to high top speeds is to have four stroke medium speed engines coupled in pairs of two engines per propeller shaft in a twin screw arrangement. Clutches allow any of the four engines to be brought into service at will and loading of individual engines can be precisely controlled and optimized for the operating mode at any time.

Triple screws are not so popular for merchant vessels but for ferries one arrangement is to fit fixed pitch propellers at the wings and to have the centre engine driving a CP propeller which is thus put to good use when manoeuvring in and out of port. A similar arrangement has also been fitted in very large and fast container ships. Triple screws are also popular for naval patrol vessels to give a good power range from the three propulsion units to better match loiter, cruise and boost speed and power characteristics.

6.1.6 Plants with mixed machinery

The earliest examples of what could be termed mixed machinery are from the mid-nineteenth century when completely separate steam propelling machinery plants were fitted on a ship for separate powering of a pair of paddle wheels and a screw propeller, as with the famous Brunel designed liner *Great Eastern*. However, a more apt description of a mixed plant is where the two prime movers are of completely different type; a very well known example from early this century was the liner *Titanic* where massive four cylinder triple expansion steam engines were directly coupled to wing propeller shafts and these both exhausted into one low pressure steam turbine directly coupled to the centre propeller. The steam engines were manoeuvred in port and once full away, the exhaust steam turbine was brought into operation, providing more power at no extra cost as the steam for the turbine would otherwise be exhausted direct to the condenser. Numerous smaller ships were also provided with the Bauer–Wach system in which an exhaust steam turbine was mechanically geared to a steam reciprocating engine.

Modern mixed plants usually comprise combinations of diesel engines, steam turbines, gas turbines or elec-

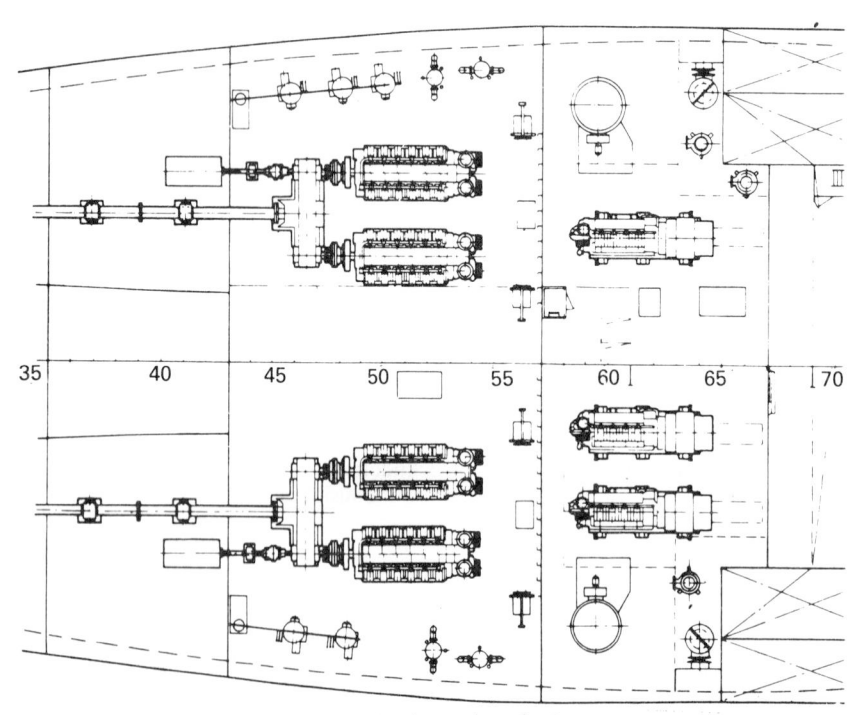

Figure 6.3 Typical twin screw–quadruple engine plant

tric motors. Such plants are more commonly found aboard naval vessels, though one significant merchant ship example is the ferry *Finnjet* which has both a main gas turbine drive and a diesel–electric system of drive for the main propellers, to suit winter and summer sailing schedules and to reduce fuel costs.

Other examples are ferries which have been retrofitted with electric motor or diesel engine power azimuthing main propulsion thruster units to both increase service speed and improve manoeuvrability.

Typical modern mixed propulsion machinery systems (*Figure 6.4*) are thus:

1. COSAG: Up until the early 1960s most warships of frigate size were powered by steam turbines with both cruise and boost turbines geared to the propeller to meet top speed and normal cruising power levels. With the advent of the aero type gas turbine, which in the marinized version became available for marine propulsion, the use of a gas turbine of high power to replace the boost steam turbine became popular. The advantage of the COmbined Steam And Gas plant is that the vessel can be made ready for sea very quickly whilst steam can be raised when the vessel is already under way under gas turbine power, while the acceleration characteristics of the gas turbine are much better. Also, very high outputs can be obtained from a compact power unit, sufficient for a frigate required to have a top speed of 28–32 knots. The general decline in popularity of steam machinery for both merchant and naval vessels because of poor economy and performance, means that there is now little interest in COSAG plants.
2. CODOG: The COmbined Diesel Or Gas plant simply uses diesel engines for the cruise mode, as modern high speed and lightweight diesel engines offer the most economical and reliable form of propulsion machinery for most of the operating time of a warship. For boost speeds, however, the much more powerful gas turbine is put into operation. With this system only the gas turbine or the diesel engines can be in use at the same time.
3. CODAG: The COmbined Diesel And Gas system differs from the above in that the gearing and control systems are designed such that the gas turbines and diesels can be operated together with suitable load sharing systems in operation. Most large warships are twin screw so it would be possible to have one shaft driven by diesel engine and the other by gas turbine, though this is unlikely to be so in practice other than in the case of complete breakdown of one of the four prime movers.
4. CODLAG: This system still uses a gas turbine for maximum speed operation but has an electric motor incorporated into the propeller shaft line to give slow speed operation on motor drive alone. The power for this propulsion motor is obtained from the ship's auxiliary generators and the result is a

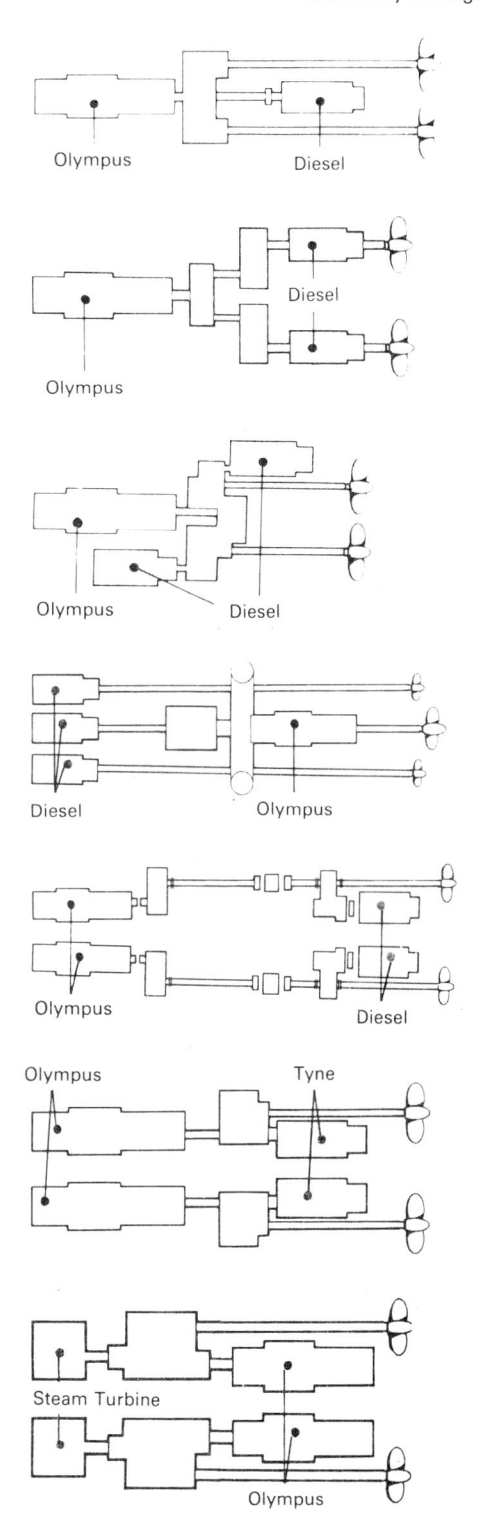

Figure 6:4 Mixed machinery with Olympus gas turbines, diesel engines and steam turbines

considerable reduction in the weight and space occupied by the machinery. A particular advantage for an antisubmarine frigate is the silent running feature of the plant at slow speed, as the reduction gearing and gas turbines are stopped when on motor alone and the diesel generator sets are mounted in heavily noise insulated containers. For this system it is normal for the motor to be electrically excited at all times so at full power the motor will still contribute to propulsion of the vessel.

5. CODAL: This system uses high speed diesel engines as the primary propulsion system but also coupled to each main reduction gear is an electric propulsion motor. Again the power for the propulsion motor is obtained from the auxiliary generator sets and by use of diesel engines or electric motors a wide range of speed and powers is obtainable from the plant at the lowest specific fuel consumption rates.

6.1.7 Loiter drives

The systems mentioned above are all used for warships of medium to large size but there are many smaller craft of both naval and commercial types that can use a mixed system to advantage. For patrol boats, as one example, it is now common to have electric motors connected to the propeller through the main gearbox for very slow speed loiter drives with the diesel main engines shut down.

Similar arrangements are to be found in research ships and indeed offshore support vessels, as described elsewhere in this chapter. In such craft the auxiliary engines provide a convenient power source but in some vessels the generators may have insufficient reserve power to drive both the loiter motors and the ship's service alternators, while to increase the size of the auxiliaries may mean low load operation of these at sprint speeds and in port when the electric motors are shut down.

A solution is to use a small diesel engine in place of the electric motor, to drive into the gearbox through a suitable clutch. Alternatively, an hydraulic motor may be used for loiter speed propulsion, in place of the electric motor or diesel engine.

6.1.8 Auxiliary propulsion

The aforementioned propulsion systems each feature alternative power plants to drive a ship's propellers, to suit variable roles and operating modes. A completely different approach is to employ auxiliary propulsion systems with the thrust units and their prime movers comprising completely separate plants.

An azimuthing propeller driven by small diesel engine, electric motor or hydraulic motor, gives both slow speed and manouvrability and has the added advantage that it can be sited at any convenient point in the hull. A disadvantage is the considerable drag caused by the unit when the vessel is at speed on the main propellers, unless the thrusters fitted are made to retract into the hull. Retracting thruster units for auxiliary propulsion are fitted particularly to tugs and offshore service craft.

Another alternative is to use a water jet driven by a small diesel engine with the jet unit mounted on the transom. There is no drag on the hull, manoeuvring qualities are good, but the efficiency of the jet at slow speeds is not as good as a small propeller, which is the other obvious choice for auxiliary propulsion. There are also examples of very powerful pleasure craft fitted with wing propellers driven by high speed diesel engines for normal cruising, while at the centre a water jet powered by a gas turbine gives a phenomenal top speed to the craft.

There are other common multiple machinery installations in ships but in these, respectively, CODAD, CODOD, COGAG and CODOG arrangements, the same prime mover type but with different output is used. These are described elsewhere in this chapter.

6.1.9 Uniform machinery

Multiple medium speed diesel engines geared to one or more propellers offer the flexibility that the prime movers can be loaded more effectively at ship speeds below the normal continuous service speed. For power generation purposes a number of smaller engines, most probably of different type from the main engines, are similarly loaded to meet the auxiliary power demand.

With some vessels which have high auxiliary power requirements, notably ferries, cruise liners and offshore service craft, uniform machinery can be employed. Here both main and separate auxiliary engines are of the same bore and stroke and type, but do not necessarily have the same number of cylinders. The advantage is that spare parts are common for all engines installed in the vessel with subsequent savings in the need for spare parts, maintenance workload, etc.

6.1.10 Father and son arrangement

An approach to the problem of matching power requirements to suit a ship's operating schedule is application of the so-called father and son system. This is normally a geared medium speed drive of one or more propellers but the engines simply are of different size, either with fewer cylinders in the case of the son or this engine may be of completely different type or indeed of different make. To operate successfully, however, the father and son engines must be coupled to gears in such a way that they can run together or either unit can operate alone, thus providing three separate power levels to match the differing speeds needed by, say, a cruise liner.

A good example of a very flexible father and son plant is seen in the installation from the cruise ship *Crown Odyssey* (*Figure 6.5*). Four main engines are

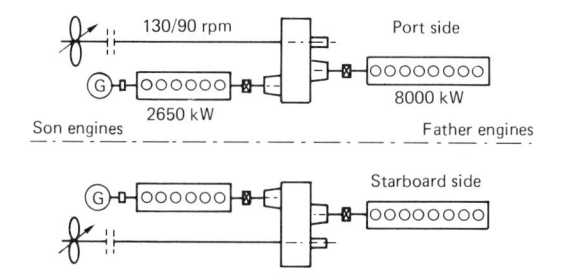

Figure 6.5 Father and son machinery plant

installed: at each side of the centre line is located an eight cylinder unit of 8000 kW and a six cylinder engine of different type and smaller bore developing 2650 kW. The two large father engines also operate at a lower nominal speed than the two smaller engines. A reduction gear with suitable clutches connects the father and son at that side to the corresponding propeller and also driven from each gearbox is a large shaft alternator. Two additional main engines identical to the son engines are fitted as dedicated auxiliary engines with no connection to the propelling plant.

The special feature of this installation is that a two speed stage is incorporated into each reduction gear which allows a slower propeller speed to be attained with the father engines disengaged and with the drive from the son engines to the propeller through double reduction stages in the gear. By virtue of the fact that both the father and son engines can be connected and disconnected at will by hydraulically operated clutches incorporated in the gearboxes, the following modes of operation are possible:

1. For maximum speed, all four engines are connected to the system together with the shaft generators and minor speed changes are effected with the CP propeller.
2. The two father engines only are connected and the propellers can be on combinator control permitting changes to both engine speed and propeller pitch. Auxiliary power is from the son engines or the auxiliary engines as the shaft generators cannot be driven by the father units.
3. The two son engines alone are connected and through the two speed gear cause the propeller to operate at reduced speed. Combinator control is possible and electrical power is obtained from the auxiliary engines.

The three possible operating modes and the choice of engines for auxiliary power generation, not to mention the option to operate the ship on one propeller alone, leads to reduced fuel consumption, in particular at lower ship speed. This is mainly attained by displacing the curves of the favourable specific fuel consumption into the particular operating range of the ship as the engines operate at higher loads for given speeds. This arrangement means that 65% of maximum speed can be obtained from only 25% of installed power.

6.1.11 Electric drive

The concept of using a number of prime movers such as diesel engines or steam turbines to drive one or more propellers relies on an effective power transmission, usually mechanical gears with suitable clutches and couplings to control the loading of engines. However, in the early part of this century the manufacture of acceptable gears was a problem so an electric transmission system was adopted.

A good example of this was during World War II when the lack of gear manufacturing facilities caused the large scale construction of turbo–electric plants for destroyers and the well known T2 turbo–electric tankers. Turbo–electric propulsion, that is the drive of the propeller by direct coupled or geared electric motors with the power supplied by one or more generators coupled to steam turbines, was also applied to a number of notable passenger liners, such as the French *Normandie*, and numerous ferries, particularly of the double ended type were given electric drive mainly because manoeuvring was made relatively simple through bridge control of reversible electric motors.

After World War II, however, gear making technology improved significantly and the electric drive system fell from favour on account of the higher cost of these plants, the greater transmission losses in converting from mechanical to electrical and back to mechanical power, and the greater complexity of the machinery installation.

In recent years, though, electric drive has made something of a comeback as some of the inherent advantages are becoming more recognized. Whereas electric drive was until fairly recently applied mainly to very specialized ships such as research vessels, buoy tenders, ice breakers, etc., it is now being used in many large passenger liners (the liner *Queen Elizabeth* 2 has been re-engined and now has two 42 MW main propulsion motors, the largest electric motors afloat), a wide range of offshore craft such as supply ships, survey ships pipelayers, standby ships and semisubmersibles for drilling or work vessels. Electric drive is now also popular for ferries, ice breakers, and even fishing vessels.

Notwithstanding the 8–10% loss in efficiency through power conversion, and the higher cost of the installation in machinery and control systems, electric drive is becoming more popular because of the following advantages:

1. There is a considerable increase in reliability because there are several prime movers instead of just one.
2. Maintenance is easier because the main engines are smaller in order to avoid large electrical machines and the plant is of lower weight and there is much more interchangeability of spare parts.
3. One or more engines can be shut down at sea and the service speed of the ship can be varied from the number of generator sets connected to the circuit.

This is of importance to a ferry operating to a timetable and any loss of speed through adverse weather can be made up by using more prime movers than normal.
4. Conversely, when a ship is running light or at slow speed, particularly in the case of cruise liners, the main engines can be more effectively and efficiently loaded by using fewer engines to match the reduced propulsion requirement.
5. It is only necessary to employ engines of one bore and stroke size throughout the ship and there is the option of using the main generator sets for auxiliary power generation as well or at least using electric motor–generator sets if auxiliary power supplies need to be at a different voltage or frequency.
6. The main engines can be located at any convenient position in the ship while the populsion motors can be right aft eliminating propeller shafting. A recently constructed series of RoRo ferries have ten generator sets housed in containers on the main deck, allowing complete elimination of a below decks engine room.
7. Control of the propeller shaft speed and direction from the bridge is simplified with electrical controls compared with the electrical and mechanical control devices needed for direct coupled and geared diesel propulsion engines.
8. Port turnaround times are much quicker as all maintenance work can be carried out at sea.
9. The weight of the machinery is reduced as is the headroom requirement for the engine room.
10. There is much flexibility concerning power distribution between propellers and power requirements can be matched to suit the vessel's operating profile.

The major disadvantage of the electric system is the higher fuel consumption resulting from the loss of transmission efficiency and the higher initial cost of the machinery. Nevertheless, there is the benefit that engines need not be operated at part load, which is detrimental to their performance, leading to savings in maintenance costs.

6.1.12 Power station concept

Some types of ships require both a high propulsion power and a high auxiliary load. A prime example is a large cruise liner where the hotel load created by more than 2500 persons aboard and air conditioning, ventilation, etc., creates a huge demand for electricity. A popular solution is the power station concept where as many as nine large generator sets are installed, each coupled to a similar alternator to generate power for all consumers as in a shore power station. In such a plant engines may be started and stopped automatically to suit any load situation and different voltages throughout the ship are produced by rotary converters.

Another type of craft using this system is a work vessel for the offshore oil industry. Such ships need power for production machinery, drill rigs, cranes, pumps, accommodation, etc., but they are also quite often equipped with a dynamic positioning system. A number of electrically driven thruster units of either fixed or CP type are fitted around the vessel and its position relative to the seabed is maintained by computer control of the thrust of each propeller. The power plant system is of particular advantage here in meeting the irregular surges in load from the thrusters and the large number of electrical consumers in the ship.

Early electric drive systems invariably used direct current motors and generators, usually with a Ward–Leonard control system. This couples the armatures of the generators and the motors in a loop in which the motor field is kept constant and the generator field varied to control the voltage in the loop and therefore the speed of the propulsion motor. Most common in marine installations, however, is to control the generator voltage to keep a constant current in the loop while the excitation of the motor(s) is altered to give the required shaft speed.

The modern trend is to use alternating current systems for ships on the power generation side, but either ac or dc propulsion motors (*Figure 6.6*). For the latter it is normal to use thyristors to control the dc motor but the most recent trend is to use ac propulsion motors. Recently introduced cycloconvertors and frequency controllers now allow very large and powerful synchronous motors to be used for propeller drive. The weight of such motors is about half that of a corresponding dc motor and the electrical efficiency is higher.

6.1.13 Mechanical and electrical drive

A system of propeller drive that has been installed in a number of offshore support vessels is a combination of geared diesel and diesel-electric drive (*Figure 6.7*). A typical arrangement has two propellers, each connected through a reduction gear to two medium speed diesel engines.

Also connected to each reduction gear is an electric generator which will also operate as an electric motor. By using one of the electrical machines as a generator and the other as a motor it is possible to have a low power diesel electric propulsion system to satisfy the low power required for the propeller when the vessel is in a standby role. Such a twin screw plant can have one main engine driving the alternator and the propeller, while the other main engine would be shut down and the second propeller would be driven by the shaft generator acting as a propulsion motor.

There is also the considerable flexibility of four engines installed in the vessel and the added advantage of electric drive for very low speed and power demand.

A further option is to fit dedicated propulsion motors as well as shaft alternators which, though much more complicated and costly, give a greater degree of flexi-

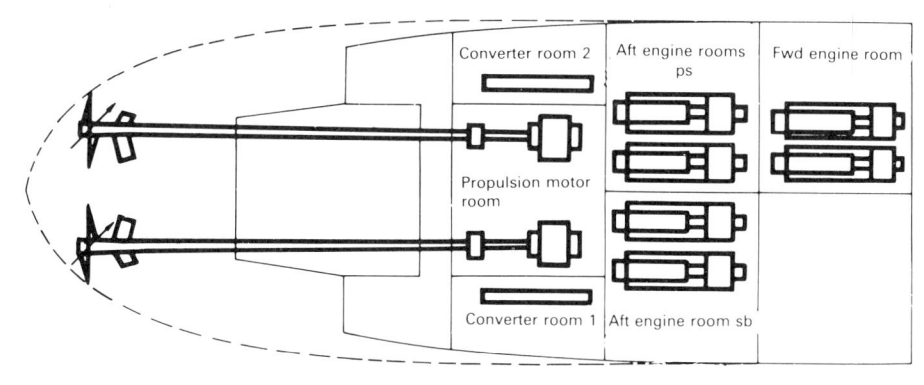

Figure 6.6 Diesel–electric plant for a cruise liner

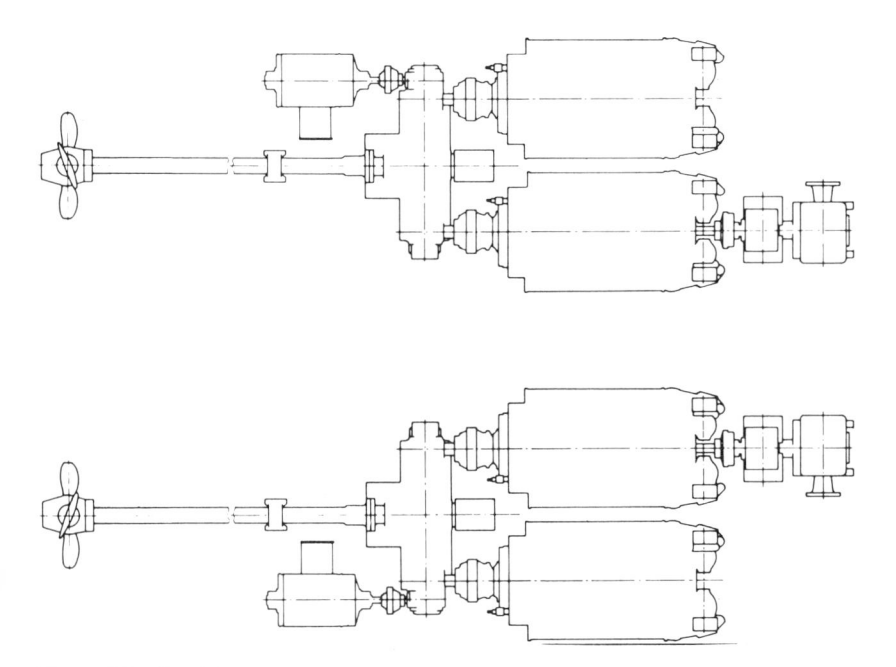

Figure 6.7 Combined diesel and electric drive

bility and potential for fuel savings. For such power plants to operate successfully, controllable pitch main and side thruster propellers are essential.

6.1.14 TNT system

An innovative machinery equipment arrangement also developed for an offshore supply vessel is the so-called TNT system. This is for a twin engine twin screw plant but by means of a transverse power shaft between the two main reduction gears which allows either or both of the propellers to be driven by either of the propulsion engines. The drive between the gearboxes is facilitated by bevel gears and Cardan shafts and multidisc clutches connect one side of the system with the other. A shaft generator is also coupled to each reduction gear unit.

The purpose of this system is to increase safety should either engine fail, to reduce engine operating times, reduce fuel consumption, to have uniform load distribution between engines and to facilitate operation of either or both shaft generators irrespective of which main engine is operating.

In the first mode of operation each engine drives its propeller on that side and the shaft generator, with the cross-connecting drive system disengaged. This would be normal for a passage to and from a rig at normal speed.

In the second mode either engine is used to drive both propellers and the intermediate shaft between gears is engaged. In this mode, 73% of the ship's maximum speed can be obtained with only 25% of the engine power and it is claimed that the TNT system leads to fuel savings of 14%, compared with a conventional geared propeller arrangement.

6.2 Propulsion diesel engines

6.2.1 Diesel engine principle

The compression ignition or diesel internal combustion engine is today by far the dominant prime mover for ship propulsion, it being selected for practically all new ships, with the decline in present interest in steam and gas turbines. Diesel engines of slow, medium and high speed type can be of either two or four stroke cycle, but nowadays, two stroke working cycle engines of interest for ship propulsion are slow speed (50–250 r.p.m.) while medium speed (250–1000 r.p.m.) and high speed (above 1000 r.p.m.) engines are practically all of the four stroke cycle.

In the original patent of the famous German engineer, Rudolf Diesel, the engine operated on the diesel cycle in which heat was added at constant pressure, using blast injection. Nowadays the term refers to any internal combustion reciprocating engine in which heat generated by compressing air in the cylinder ignites a finely atomized spray of fuel. In the theoretical diesel cycle air is compressed adiabatically to a point where fuel injection begins and heat is added partly at constant volume and partly at constant pressure. Expansion of the gas in the cylinder follows adiabatically until the heat is rejected from the cylinder to the exhaust at constant volume. Typical cylinder pressures are shown at full load in *Figure 6.8*.

A diesel engine works on the two or four stroke cycle and all currently available models are single acting, i.e. the working fluid acts on the top side of the piston only.

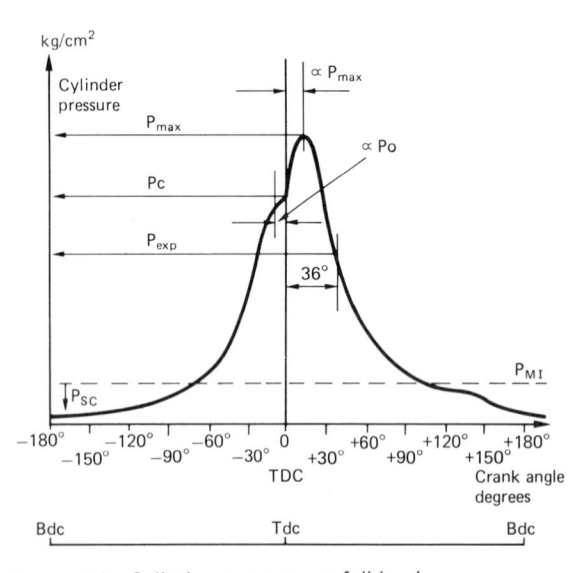

Figure 6.8 Cylinder pressures at full load

6.2.2 Four stroke cycle

The four stroke cycle is so named because four piston strokes or two complete shaft revolutions take place between charge air renewals. The events are controlled by poppet type valves in the cylinder head which are timed to open and close at set points in the cycle to create:

1. An induction stroke in which air is drawn or forced into the cylinder.
2. A compression stroke in which the air trapped in the cylinder with inlet and exhaust valves both shut, is compressed to the temperature required for fuel ignition.
3. An expansion stroke or power stroke, after injection and combustion of the fuel just before TDC.
4. An exhaust stroke in which the exhaust valve opens for gases to be expelled from the cylinder, ready for a new charge of fresh air and another cycle.

A typical four stroke timing diagram is shown in *Figure 6.9*.

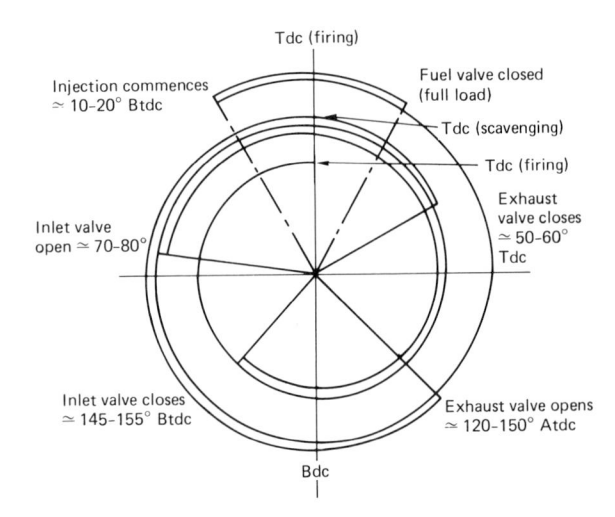

Figure 6.9 Four stroke timing diagram

6.2.3 Two stroke cycle

As the name implies, for the two stroke cycle the sequence of events above are accomplished in two piston strokes or one revolution of the crankshaft. Modern designs only operate on the uniflow principle, in which air is admitted through ports into the cylinder near the bottom, and exhaust gases pass out through a poppet valve in the cylinder head. Two strokes are possible because the exhaust valve opens at around 120° after TDC and closes at up to 150° before TDC, the air admission having begun just before the exhaust closes to facilitate scavenging of gas from the cylinder. With both air and exhaust shut off, compression takes place until fuel injection at 10–20° before TDC. Combustion after TDC normally lasts for 30–50°, as with the four stroke engine.

A typically two stroke engine timing diagram is shown in *Figure 6.10*.

The actual timing points of cycle events vary within engine types and makes, but the basic principle of operation is the same for all. The applications of the different diesel engine types, their advantages and main points of design are discussed in the following section.

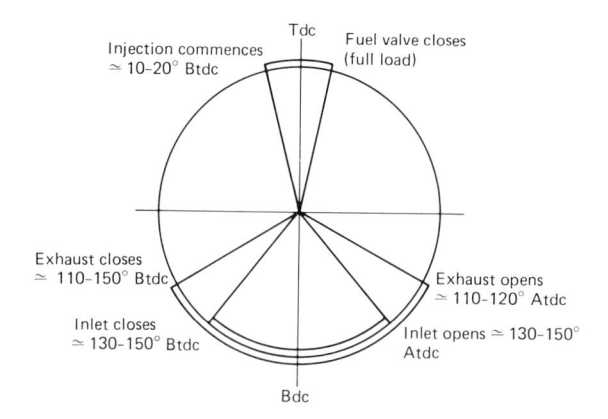

Figure 6.10 Two stroke timing diagram

6.2.4 Slow speed diesel engine

The prime mover type which is used to power the great majority of ocean going ships is the slow speed marine diesel engine. This is a power unit which is unique to ship propulsion as its horsepower, crankshaft speed, fuel consumption, maintenance procedures, etc., have been evolved over the years to especially suit marine conditions. The engine is used ashore for power generation in large stationary plants but here it is nowhere near as popular as the higher speed four stroke engine.

The marine slow speed diesel engine evolved from the steam reciprocating engine, as in the earliest days the steam cylinders on top were replaced by ones suitable for internal combustion of liquid hydrocarbon fuels, whilst the open crank chamber of the steam engine was also a feature of early diesel engines.

Since the first ocean going motor ship in 1912, slow speed engines have taken on many forms. Typical types were the single acting four stroke and the rather troublesome double acting four stroke types. Two stroke engines were to become most popular, built in both single acting and double acting versions while another type is the opposed piston two stroke as in the case of the British Doxford engine. Fuel systems initially used blast injection, soon to be followed by solid injection as used today. Scavenging of the cylinders was by means of engine driven piston type scavenge air pump or by the underpiston arrangement using the undersides of the engine pistons as air pumps

to deliver pressurized air to the cylinders for combustion. Such systems have now been completely superceded by exhaust gas turbocharging in which the energy in the exhaust gases is used to power a rotary air compressor which delivers air to the cylinders. Engines with different turbocharging and scavenging systems have been in use for some time but market forces have affected engine development to the point that for the last few years only one type of engine is now produced, the three surviving competitive engine models being similar in general outline and operating principle.

6.2.5 Uniflow scavenged two stroke engine

The slow speed marine diesel engine now produced for ship propulsion ranges in output from about 2000 bhp from an engine with cylinders of 260 mm bore up to around 65 000 bhp from a single 12 cylinder engine of 900 mm bore. Shaft speeds range from 250 r.p.m. for the smallest type down to 60 r.p.m. with the largest most powerful engines. Herein lies the major advantage of this engine type in that its slow shaft speed at full power means that the engine can be directly coupled to the propeller which results in a simple arrangement requiring no clutches, couplings or reduction gears.

Most new ships have their machinery located aft so only a short length of shafting is needed to connect the engine crankshaft with the tail shaft which carries the propeller. The average shaft speed of around 100 r.p.m. is adequate for most propellers for optimum propulsive efficiency, and astern operation of the propeller is normally obtained by direct reversing of the main engine. This is accomplished by mechanical alteration of the fuel injection and exhaust valve timing, to suit operation of the engine in the opposite direction.

The common design of modern slow speed engines comprises three major construction items connected by long tie rods. The cast iron or fabricated steel bedplate carries the crankshaft and houses the thrust bearing at its aft end. It is at this point that the thrust of the propeller is transferred to the ship structure. The entablature section is fitted on top of the bedplate and this forms a closed chamber for the engine running gear and as a mounting for the crosshead guides. Mounted above the frame box are the cylinder jackets which house the cylinder liners with individual cylinder heads fitted on top of each. Because the engines are of crosshead types – that is with a separate piston rod terminating in a crosshead to which is coupled the connecting rod attached to the crankpin through the large and bearing – there is complete separation of the crankcase from the power cylinders by means of a stuffing box around the piston rod. This is a major advantage for this engine type as any products of combustion leaking past the piston cannot contaminate the oil in the crankcase.

The engines are single acting only and operate on the uniflow scavenging principle (*Figure 6.11*). Towards the end of the power stroke, the piston uncovers ports

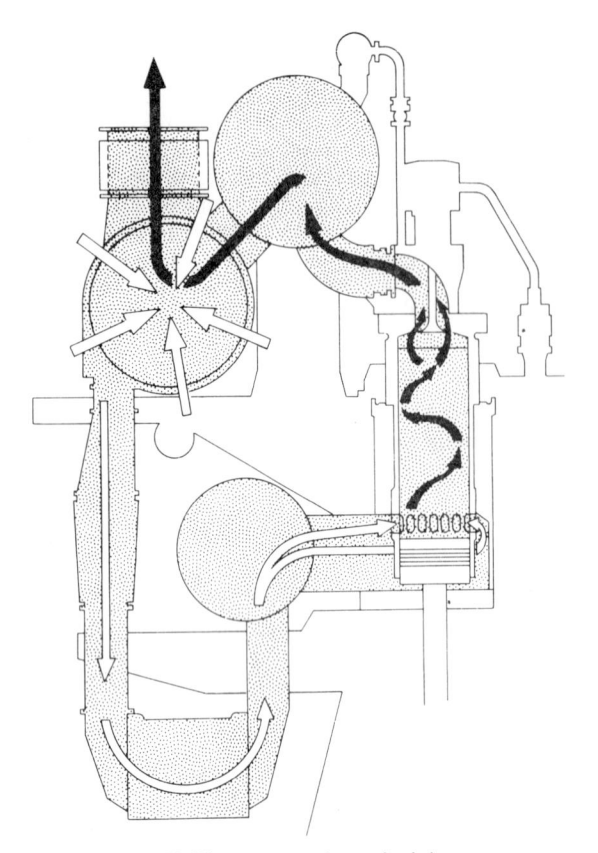

Figure 6.11 Uniflow scavenging principle

in the lower part of the cylinder to admit pressurized charge air which drives the exhaust gas from the cylinder through the open exhaust valve mounted in the cylinder cover. The piston passes its bottom dead centre and in rising in the cylinder covers the ports to shut off the scavenge air supply. The exhaust valve closes and thus the air trapped in the cylinder is compressed to a suitable pressure and hence temperature to cause fuel injected into the cylinder just before top dead centre to ignite and burn, and in so doing, the gas expands and drives the piston downwards for the next power stroke. The engine fires once per revolution and the term 'uniflow' relates to the flow of scavenge air and exhaust gas from the bottom of the cylinder upwards to the exhaust outlets.

Two stroke engines are all fitted with exhaust gas turbochargers to supply the charge air, and fuel is injected into the cylinders at very high pressure by means of plunger type fuel pumps and atomizing fuel valves. Such engines are very well suited to operation on heavy residual fuels of poor quality such as are widely available today.

6.2.6 Crosshead engine applications

The simplicity of the direct drive concept has naturally resulted in this machinery arrangement becoming the most popular for medium to large size ships, it being the natural progression from direct drive of propellers by steam reciprocating engine. Before the adoption of geared medium speed diesel engines in the early 1960s, the steam turbine was the alternative propulsion plant to the direct drive diesel. Nowadays this latter type is installed in more than 70% of ocean going ships, from small freighters up to supertankers of around 250 000 tdw.

The ideal application for this engine is a large vessel which operates over long distances at maximum speed, such as a supertanker or container ship. As it is not widespread practice to fit vessels powered by slow speed engines with controllable pitch propellers, such engines are slower to manoeuvre and take some time to develop their full output from a standstill and are thus not popular for ferries and other ships that are frequently in and out of port. Nevertheless, quite a number of passenger vessels have been fitted with the engine, the owners preferring the simplicity and reliability of direct drive.

6.2.7 Advantages of direct drive

Direct coupling to the propeller offers the simplest mechanical drive from the crankshaft. There is, however, the disadvantage that the propeller has to operate at the design speed of the engine, which may not necessarily be the optimum to suit the particular design of propeller. At one time a small loss in propulsive efficiency was accepted for the sake of simplicity but development of engines has gone most of the way in eliminating this drawback.

Engine builders now offer standard models at a selection of speed and power outputs within a narrow band, so the latest tendency is to design the optimum propeller in terms of its diameter, pitch and blade area to suit the required thrust, and try as closely as possible to match this propeller's design revolutions and power requirement with the engine. The introduction of long stroke and super long stroke engines with reduced shaft revolutions coupled with derating, has done much to reduce the propulsion losses with direct drive.

As direct drive engines develop high outputs per cylinder, particularly the larger bore models, it is relatively simple to meet most power requirements with an engine having a small number of cylinders, usually six on average. It is only the largest of vessels, particularly high speed container ships, that would normally be fitted with a 12 cylinder engine. Fewer cylinders means less maintenance work on a periodic basis and less spare parts need to be carried.

In most vessels the machinery space height is less of a problem than length, so a larger bore engine with fewer cylinders will inevitably result in a shorter engine room space and more space in the ship for cargo. It is also proven by practice that larger bore engines have a lower specific fuel consumption than smaller engines and the bigger engines have a good tolerance to burning heavy fuels of poor quality.

6.2.8 Slow speed engine data

Whilst at one time many very different designs of slow speed two stroke marine diesel engines were available, market forces led by the need to offer the lowest possible specific fuel consumption rates have, in recent years, led to a consolidation of design by all manufacturers. The result is that the three engine designing companies, Sulzer (Switzerland) MAN–B&W (Denmark) and Mitsubishi (Japan) each offer only a single acting, uniflow scavenged two stroke engine for direct drive ship propulsion.

The engines differ only in minor design details but all use single acting pistons coupled to the crankshaft through crosshead and connecting rod and most are offered from 4 to 12 cylinders in a range of bore sizes 260–900 mm. The engines are all turbocharged and intercooled and the common feature to all is the use of a single hydraulically actuated exhaust valve mounted in the individual cylinder covers and air intake ports arranged low down in the cylinder liner, to facilitate operation on the uniflow scavenging principle.

A detailed account of engine design is beyond the scope of this book but the cross-sectional sketches of the Sulzer RTA series (*Figure 6.12*), MAN–B&W, K–MC (*Figures 6.13*) and Mitsubishi UEC–LS type (*Figure 6.14*), show many of the engine design features. Similarly, technical parameters for the complete range of two stroke engines currently available are given in *Table 6.1*.

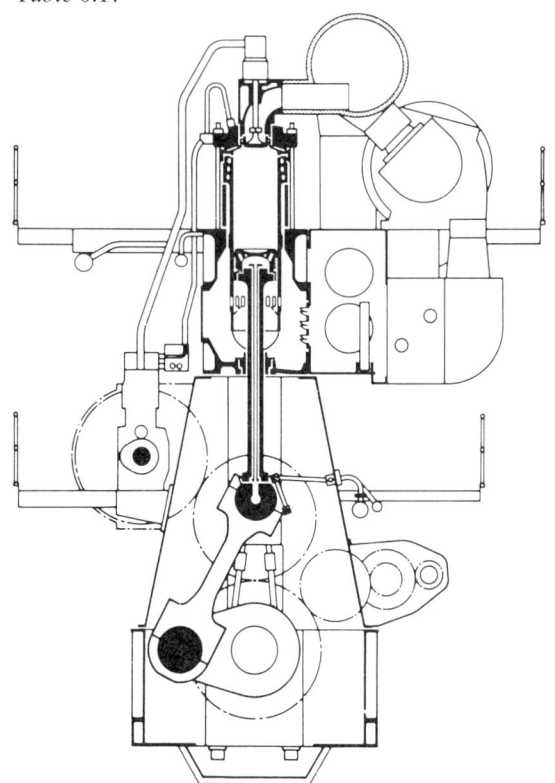

Figure 6.12 Cross-section of Sulzer RTA type

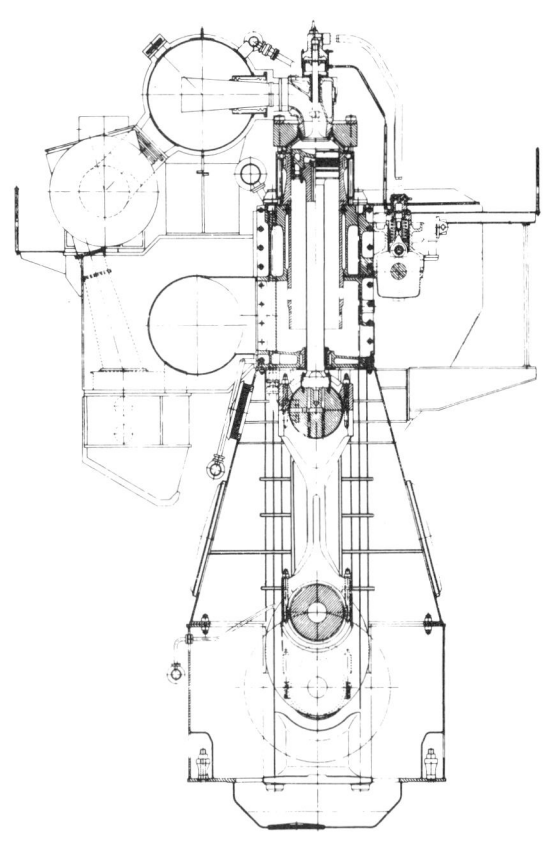

Figure 6.13 Cross-section of MAN–B&W K80MC

6.2.9 Medium speed diesel engines

The trunk piston four stroke medium speed diesel engine (*Figure 6.15*) first made its appearance as a submarine prime mover, but it was not until the 1950s that the higher powered models, of a size that could be considered as a direct replacement for two stroke crosshead engines, began to be developed for merchant ships.

The French firm, SEMT, was a leading pioneer in initiating a completely new philosophy in propulsion engines, that of using engines designed to run faster than a normal propeller and the introduction of geared drive to larger ships. The geared medium speed engine had already been used for some time in small craft such as tugs, coasters and ferries, as they continue to do so in practically all cases today. By the mid-1960s medium speed engines were running at speeds between 450 and 600 r.p.m. and developing around 500 bhp/cylinder.

By comparison, today there are engines developing nearly 2000 bhp/cylinder at 425 r.p.m., or a theoretical maximum output of 36 000 bhp from one 18 cylinder model. Normal practice is to build engines up to nine cylinders in line and from 10 to 20 cylinders in vee form where cylinders on opposite banks share a common crankpin through side by side connecting rod bearings. Engines for marine propulsion duty are rarely ordered

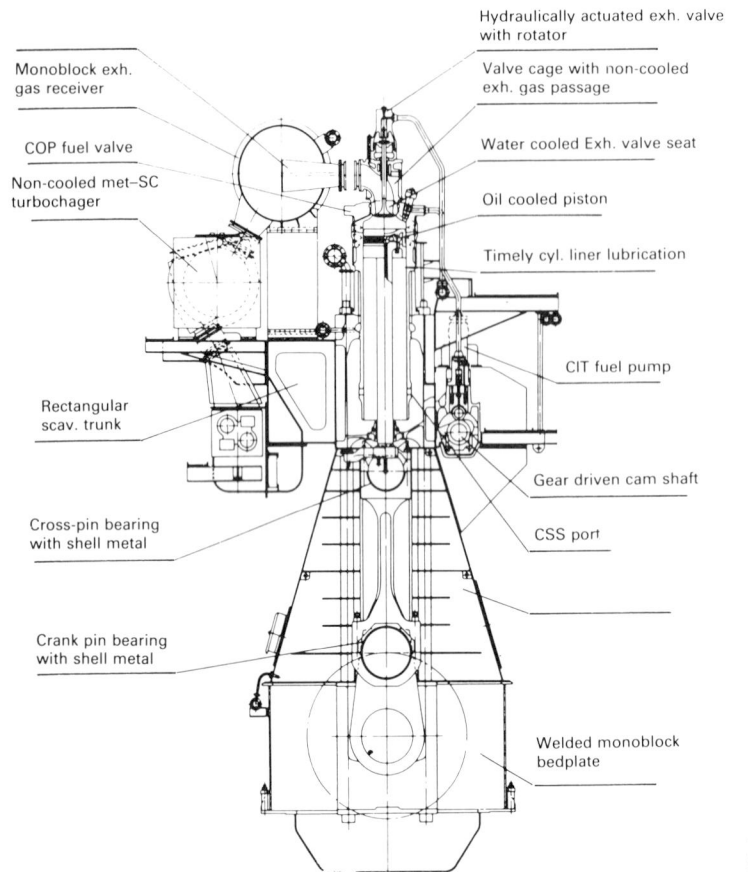

Figure 6.14 Cross-section of Mitsubishi UEC-LS type

Figure 6.15 Mirrlees–Blackstone medium speed engine

Table 6.1 Slow speed engine data (dimensions for 6 cylinder engines)

Type	Output (Bhp)	Bhp/cyl.	Bore (mm)	Stroke (mm)	RPM	Bmep (bar)	L (mm)	B (mm)	H (mm)	Wt (tons)
Sulzer										
RTA38	2000–8370	930	380	1100	196	16.7	5610	2300	6130	100
RTA48	3240–13 320	1480	480	1400	154	16.8	7295	2700	7945	210
RTA52	4240–15 440	1930	520	1800	130	17.1	6960	3030	9075	240
RTA58	4720–19 440	2160	580	1700	127	16.7	8475	3280	9910	320
RTA62	6080–22 080	2760	620	2150	109	17.2	8300	3560	10 630	380
RTA68	6480–23 600	2950	680	2000	108	16.6	9810	3780	11 465	500
RTA72	8200–29 840	3730	720	2500	94	17.2	9680	4090	12 260	580
RTA76	8080–44 160	3680	760	2200	98	16.6	10 930	4100	12 510	670
RTA84M	8200–60 840	5070	840	2400	81	17.2	12 000	4680	14 530	945
RTA84C	11 440–62 400	5200	840	2400	100	17.2	12 000	4320	13 520	850
MAN–B&W										
S26MC	950–3970	495	260	980	250	17.1	3990	1880	910	40
L35MC	1600–6080	760	350	1050	200	16.9	5333	1980	5360	66
L42MC	2240–9280	1160	420	1360	168	16.5	6184	2460	6905	130
L50MC	3160–13 200	1650	500	1620	141	16.5	7450	2710	8199	210
S50MC	3400–14 240	1780	500	1910	123	17.4	7612	2914	9145	235
L60MC	4520–18 880	2360	600	1944	117	16.5	8488	3228	9664	340
S60MC	4880–20 400	2550	600	2292	102	17.4	8736	3478	10 950	370
L70MC	6160–25 600	3200	700	2268	100	16.5	9946	3842	11303	530
S70MC	6720–27 920	3490	700	2675	88	17.4	10 218	4250	12753	570
L80MC	8120–50 520	4210	800	2592	88	16.5	11 329	4388	13015	765
K80MC	8160–50 880	4240	800	2300	100	16.5	11 329	4388	13015	760
K80MC-C	855–52 300	4360	800	2300	104	16.2	11 329	4388	13015	760
S60MC	8640–54 720	4560	800	3056	77	17.4	11 729	4824	14666	825
L90MC	10 160–63 720	5310	900	2916	78	16.5	12 811	4936	14603	1025
K90MC-2	12 800–64 320	5360	900	2300	100	16.5	12 811	4936	14603	1010
K90MC	10 240–64 320	5360	900	2550	90	16.5	12 811	4936	14603	1015
K90MC-C	22 600–66 100	5500	900	2300	104	16.2	12 811	4936	14603	1015
Mitsubishi										
UEC37LA	2 800–5 600	700	370	880	210	15.9	5610	1900	4275	75
UEC45LA	4 800–9 600	1200	450	1350	158	15.9	6265	2560	5660	155
UEC52LA	6 400–1 280	1600	520	1600	133	15.9	7270	3000	6720	239
UEC60LA	8 400–16 800	2100	600	1900	110	15.9	8320	3450	7845	370
UEC52LS	7 200–14 400	1800	520	1850	120	17.2	7310	3220	7380	256
UEC60LS	9 600–21 600	2400	600	2200	100	17.4	8385	3720	8650	402
UEC75LS2	16 000–32 000	4000	750	2800	84	17.3	9700	–	–	610

with more than 16 cylinders and there is in fact a marked preference for in line types.

The medium speed four stroke engine is in essence an enlarged version of the high speed types used for small ships or generator drive. The running gear comprises only a piston and connecting rod and, as the cylinders are open to the common crankcase, system oil is also used for cylinder liner lubrication, rather than through separate systems as with crosshead engines. Modern engines are turbocharged and intercooled and lately have been adapted for exhaust gas power turbines. Considerable development has resulted in specific fuel consumption rates to practically equal crosshead engines and persistent problems such as high exhaust valve failure rates and bearing problems have now been virtually eliminated. Furthermore, the engines are now well proven to operate on very poor quality heavy residual fuel without severe maintenance cost penalties and provided routine maintenance is attended to they operate as reliably as the alternatives.

Notwithstanding the above and despite the considerable technical progress during the last 20 years, the medium speed engine still faces stiff market resistance and so far the goal of displacing of the direct drive engine as natural choice for bulk ships, tankers and cargo vessels, etc., seems as far from reach as ever. The most significant applications still remain in ferries, cruise liners and smaller vessels, etc.

6.2.10 Medium speed engine applications

By virtue of their dimensions, power range and wide availability, there is in theory no type of ship that cannot be equipped with four stroke trunk piston medium speed engines. In fact, in the past they have been fitted to most ship types, except the very biggest of tankers, and their only limits are in power output. In reality, the medium speed engine presence is most marked in certain ship types.

The medium speed engine cannot drive a propeller efficiently at crankshaft speed so a reduction gear is necessary. This engine type is therefore very suitable for ships where more than one engine can be used to advantage and the gear is used to conveniently connect more than one engine to a propeller shaft. The height of the engines is much less than the alternative so they are much favoured for ships with restricted headroom in the machinery space, such as car ferries. The speed of the engines is also suitable for direct drive of ac generators, so medium speed engines are particularly popular for ships with electric propulsion, such as ferries, cruise liners, icebreakers, research ships, etc. Where small outputs, say as little as 2000 bhp is required, comparatively small engines can be supplied. Hence, medium speed engines are used in an enormous number of small ships such as coasters, ferries, offshore craft, etc.

To summarize, medium speed engines are popular for particular ship types, e.g. small ships of all types, restricted headroom ships such as ferries, RoRo ships, multi-engine plant vessels such as passenger ships, ice breakers, oil rigs, offshore support vessels, etc., and warships where the use of heavy direct drive engines is completely out of the question.

The concepts of multi-engine plants, mixed machinery and other practices involving medium speed engine installations are fully described elsewhere in this chapter.

6.2.11 Medium speed advantages

For any type of ship that is fitted with medium speed engines the major advantages are generally those relating to geared drive of the propeller as the shaft speed can be selected precisely through the gear ratio, to give the most efficient combinations of propeller dimensions and design revolutions. Such points are raised elsewhere in this chapter. In addition, there are other key advantages, many of which can sway the choice of machinery to the four stroke engine.

The main advantages are:

1. Completely free choice of the propeller speed.
2. Lower cost per horsepower of plant including the cost of gear, compared with direct drive.
3. Flexibility afforded by use of more than one engine, greater redundancy, more economical fuel consumption and better loading of engines at reduced ship speeds.
4. Easier installation as engines are transported and fitted in one piece. Much simpler flexible mounting arrangements.
5. Considerable saving in weight of machinery and saving in engine room length and ability to fit in low headroom engine rooms.
6. Easier handling of components of more manageable weight and size.
7. Suitability for use in diesel–electric plants.
8. Ability to operate on heavy fuels.

The above are acknowledged advantages for medium speed engines but some limiting factors persist, namely the requirement for high cylinder pressures to achieve high thermal efficiency, with consequent higher stress of components, component temperatures, etc. Bore sizes and hence outputs, are limited by maximum acceptable cylinder centre distances.

The main disadvantages are high lubricating oil consumption, shorter experience of burning some heavier fuels, and more items requiring maintenance and a generally higher standard of maintenance, fuel treatment, cleanliness, etc., needed to ensure consistent reliable operation.

The disadvantages above may be significant for some ships but the recent tendency has been to prefer medium speed engines to have fewer cylinders, large bore and bore/stroke ratios, moderate piston speeds and very precise control of combustion chamber components through cooled cylinder heads, cylinder liners and piston crowns.

6.2.12 Medium speed engine data

There remains today many different makes of medium speed diesel engines available for main propulsion and nearly all (except some US built two stroke types) operate on the four stroke principle. All use trunk pistons, cylinder heads with either two or four valves and some have special design features such as armoured exhaust valves and oil or water cooled piston crowns, cylinder liners and cylinder heads to facilitate operation on heavy fuels. A selection of engine types are shown in cross-section in *Figures 6.16* to *6.20* and technical data for some leading engine makes is given in Table 6.2.

6.2.13 High speed diesel engine

Given the large number of existing fishing and work boats, small ferries and indeed leisure craft, as well as ocean going ships equipped with auxiliary engines, the high speed diesel is the most prolific of all marine power units. Accordingly, there are more manufacturers of high speed diesels than any other type.

The high speed diesel engine is not to be confused with the marinized automotive type as the latter is a truck or bus engine adapted for operation aboard a vessel, as discussed later in this section, whereas the

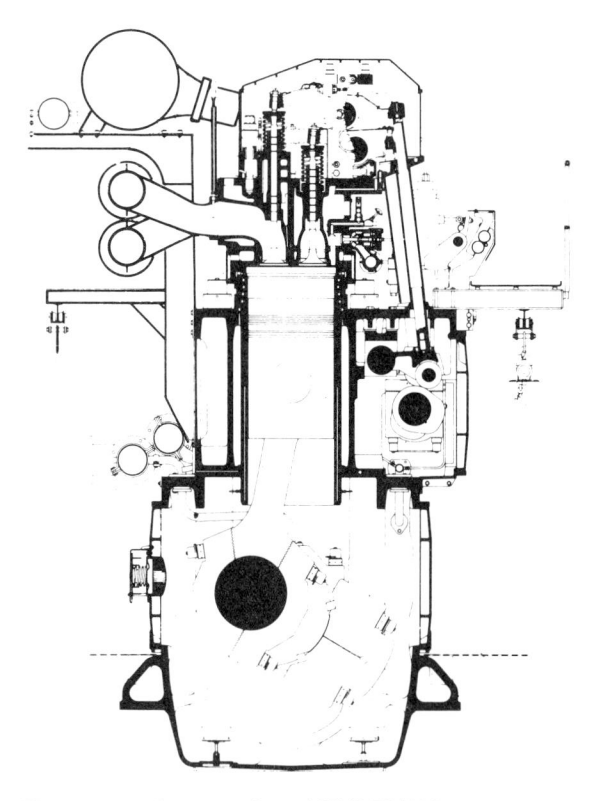

Figure 6.16 Cross-section of SWD TM620

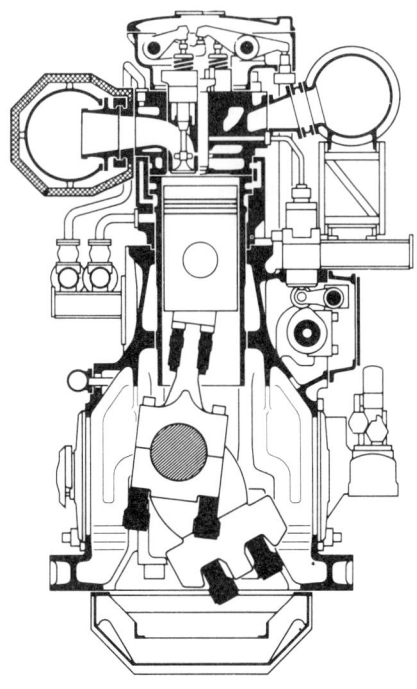

Figure 6.17 Cross-section of MAN–B&W L58/64

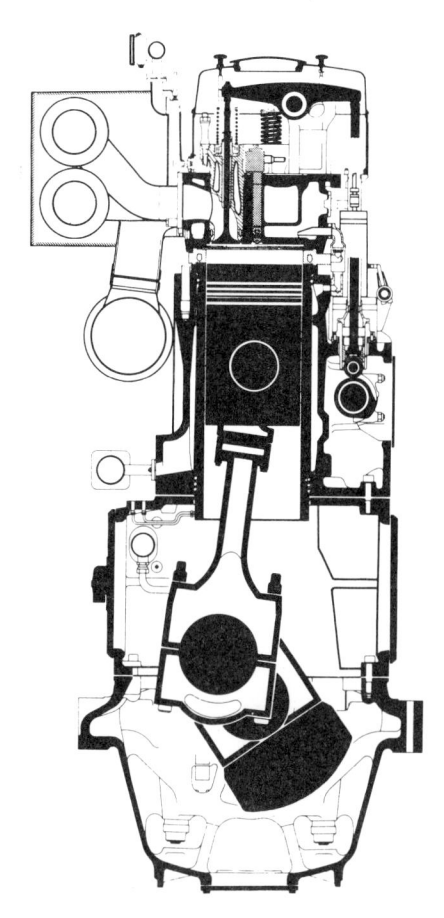

Figure 6.18 Cross-section of MaK M552C

high speed engine is usually of heavier duty construction, has a much longer service life and periods between overhauls and is despatched from the manufacturer's works ready for installation.

The high speed diesel engine is usually regarded as a smaller, lighter, less powerful and higher speed version of the four stroke trunk piston medium speed diesel engine. The combination of high operating speed (usually in excess of 1000 r.p.m.) and multiple cylinder units up to as much as 20 cylinders, also means that many small units develop more power than much larger medium speed models, making them very suitable for propulsion applications where low weight and high power are both required features. High speed engines are therefore widely used in high speed military craft, fast catamarans, luxury yachts and other pleasure craft and as auxiliary engines in a very wide range of ships and boats.

The advantages of a typical high speed engine are thus:

1. Very high power/weight ratio and compact overall dimensions.

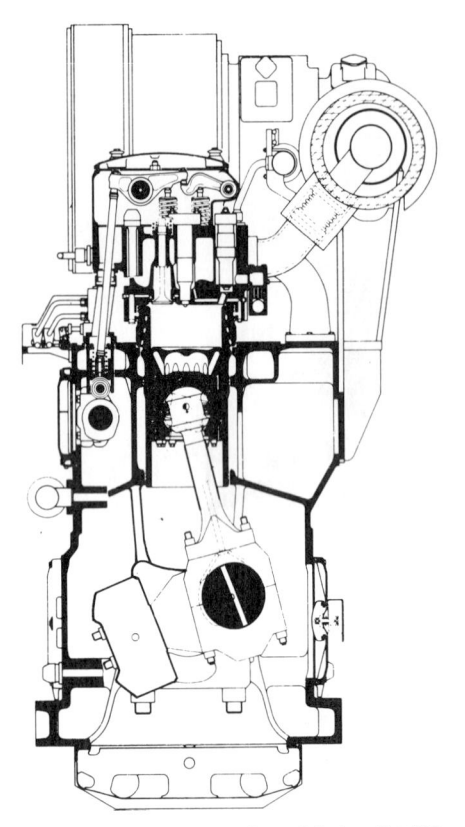

Figure 6.19 Cross-section of Sulzer ZA40S

2. Wide range of engines to choose from with over 40 engine makes on the market.
3. Engines designed from the outset for marine application and supplied ready to fit.
4. Small components making maintenance easier and lower cost of spare parts.
5. Engines suitable for unattended operation, with self contained auxiliary equipment such as pumps, filters and coolers, etc.
6. Lower specific fuel consumption compared with automotive type engines and some types now have proven experience of operation on residual fuels without operating problems.

All high speed engines are of four stroke type, with separate cylinder heads housing two or four valves and having either bedplate mounted or underslung crankshafts. They are built typically 4–8 cylinders in line and 8–20 cylinders in V form. A typical design is discussed below.

6.2.14 Typical high speed design

One of the most successful of high speed diesel engines is the range produced by the West German manufacturer, MTU. One model of this range used in military, commercial and leisure craft installations is the Series

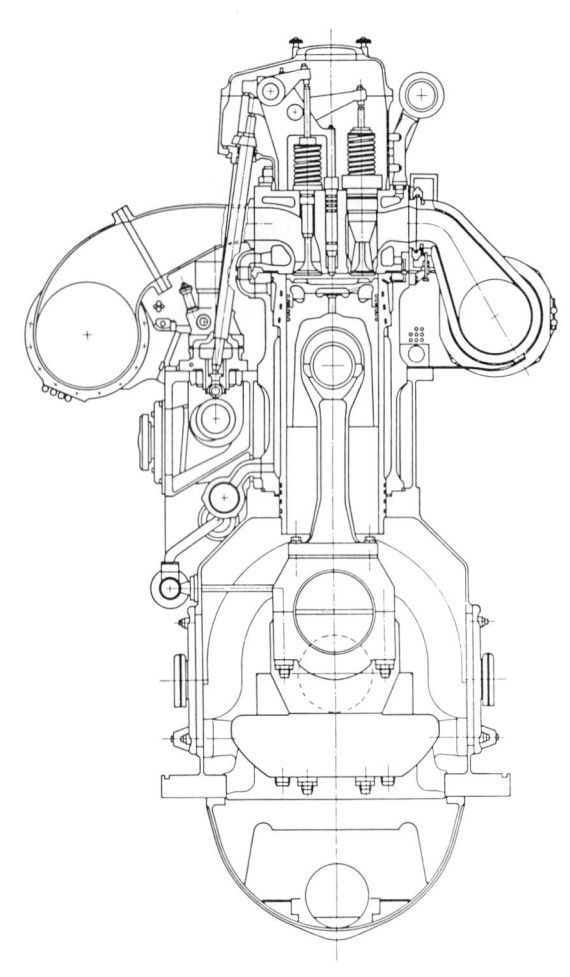

Figure 6.20 Cross-section of SEMT–Pielstick PC40L

396 built in 6, 8, 12 and 16 cylinders, developing powers 380–2560 kW at speeds up to 1800 r.p.m. (*Figure 6.21*).

The engine is built with a one piece nodular cast iron crankcase with an integrated gearcase at the timing end and with a light alloy oil pan and separate flywheel housing bolted on. Individual cast iron cylinder heads with replaceable valve seat inserts are fitted and each head contains two inlet and two exhaust valves together with a centrally placed fuel injector.

The forged crankshaft is machined all over and supported in sleeve bearings and has bolted on counterweights. The connecting rods are forged, machined all over and, with an angular split and serrated big end bearing, are arranged side by side on the crankpins to serve two opposing cylinders. The pistons have a light alloy skirt and a bolted on steel crown and are given internal cooling from a fixed oil spray located in the crankcase structure. Two compression rings are fitted in the crown and an oil control ring is fitted between the crown and skirt. Valve actuation is by two camshafts

Table 6.2 Medium speed engine data (dimensions for 6 cylinder engines)

Type	Output (Bhp)	Bhp/cyl.	Bore (mm)	Stroke (mm)	RPM	Bmep (bar)	L (mm)	B (mm)	H (mm)	Wt (tons)
Allen										
S37	2640–3960	440	325	370	750	17.5	5085	1710	2825	29
Bergen										
KRM(B)6	1500–1650	275	250	300	825	20	3875	1285	1860	13
BRM6	3300	550	320	360	750	22	5425	1738	2470	27
Deutz–MWM										
628	1025–2570	295	240	280	1000	21	3345	1370	2570	8
510B	4406	550	330	360	750	21	5195	2110	3545	24
640	2570–4930	616	370	400	650	21	6348	2249	4075	29
Grandi Motori Trieste										
A420	4200–11 200	700	420	500	500	17	5580	2575	3550	48
B420	4620–12 320	770	420	500	500	20	5650	2575	3550	48
B550	8820–29 400	1470	550	590	450	21	7396	3005	5550	108
Krupp MAK										
M35	3000–5135	640	350	450	600	22	4842	1900	2840	34
M551	4000–6250	780	450	550	450	18	5415	2134	3325	49
M552	5000–6660	835	450	520	500	18	5565	2340	3325	53
M601	8000–12 250	1500	580	600	425	20	7500	3327	3854	110
MAN–B&W										
L+V28/32	925 –5 400	300	280	320	750	18	4700	1500	3375	21
L+V32/36	3300–9900	550	320	360	750	22	5900	1950	3900	36
L+V40/45	4500–14 850	825	400	450	600	22	7100	2550	4550	59
L40/54	4500–7425	825	400	540	514	21	7400	2600	4350	75
L52/55B	8040–12 050	1340	520	550	500	20	8250	3100	4500	102
L58/64	10 800–16 200	1800	580	640	428	22	9700	3450	5150	149
Mirrlees Blackstone										
K Major	3500–13 140	730	400	457	600	19	7250	2438	4264	52
RUSTON										
RKC	1250–4520	283	254	305	1000	17	4800	1830	2310	20
RK270	1500–6160	385	270	305	1000	20	4785	1830	2396	21
AT350	2000–4500	500	350	368	600	21	5240	1500	2930	26
Semt–Pielstick										
PC2-5	3900–11 700	650	400	460	520	–	7076	1840	3063	35
PC20	4480–6725	750	400	550	450	–	7308	2512	2586	53
PC30	5000–9000	1000	425	600	450	–	7460	2345	3980	82
PC4-2	8250–29 700	1650	570	620	400	–	9030	3880	6095	122
PC40	8250–14 850	1650	570	750	350	–	9465	3200	2626	149
Suller										
Z40	4080–12 000	750	400	480	560	–	6570	1540	3730	50
ZA40	4500–15 660	870	400	480	580	–	7040	1750	4150	58
ZA40S	4500–16 200	900	400	560	510	–	7040	1750	4150	60
Stork Werkspoor										
SW280	1970–7920	440	280	300	1000	19	3820	1775	2355	16
TM410	5050–15 200	840	410	470	600	20	5906	2720	4510	61
TM620	11 500–17 300	1925	620	660	425	20	8255	3300	6685	173
Wartsila										
Vasa 32E	2200–10 026	557	320	350	800	22	4955	1960	3480	26
Vasa 46	4920–22 130	1240	460	580	??4	25	7820	2435	5150	86

Figure 6.21 MTU type 396 engine

arranged in a 90° V-form between the cylinder banks, via roller tappets, push rods and rocker arms.

Two features distinguish the MTU engine: sequential turbocharging employs a number of turbochargers which are automatically cut in/out as a function of engine speed and power requirements. In addition to increased torque, this system gives reduced fuel consumption and lower exhaust temperatures. Additionally, cylinder cut out improves the combustion of fuel at idling speed by shutting the fuel off one bank of cylinders to increase the loading of the opposite bank at low engine loadings. On models fitted with charge transfer, the cylinders with fuel cut off act as air compressors to increase the amount of combustion air available at reduced loading.

There are minor design differences among the numerous engine makes and models but the concept of producing the maximum output from a lightweight yet robust engine, is a feature common to all high speed marine diesel engines.

6.2.15 Automotive type engines

There exist today an enormous number of vessels which are too small to take, or their owners are unwilling to pay for, a traditional medium speed marine diesel engine. For these vessels, mainly pleasure craft, work boats, small ferries and fishing boats, the automotive type diesel engine (*Figure 6.22*) offers the most economical and convenient propulsion machinery choice.

This engine gets its name from the fact that it is exactly the same motor as is used to power lorries and industrial equipment but, to make such engines which are available up to 12 litre capacity and about 500 bhp, suitable for marine use, they must first be marinized.

Marinization is essentially the process of converting the basic factory delivered automotive diesel to make it suitable for operation in a boat. The basic engine is built by firms with household names such as Ford, General Motors, Volvo, etc., and is purchased by specialist converters, who attach the necessary components and in some cases make some internal modifications, for resale as ready to use marine engines. Generally, the marinizer aims to achieve an engine of highest output at greatest reliability from the basic block turned out by the manufacturers. From the basic factory models the marinizers are able to offer a wide power range of engines, some turbocharged, others naturally aspirated, intercooled, etc., and at different ratings for leisure, work and high speed craft applications.

6.2.16 Reasons for marinization

The reason why so many automotive type diesel engines are supplied for marine use is purely commercial with the main argument being one of cost. As the engines themselves are mass produced the initial cost is less and despite the amount of work necessary to make the engines ready for a boat, the automotive engine may be half the price of an engine built purposefully as a marine engine. The cost differential can be most significant in the upper power range, but smaller marine engines tend to be more competitive with marinized automotive units.

The major manufacturers need outlets as well as the vehicle market to maintain high levels of production. They therefore see the marine market as a worthwhile

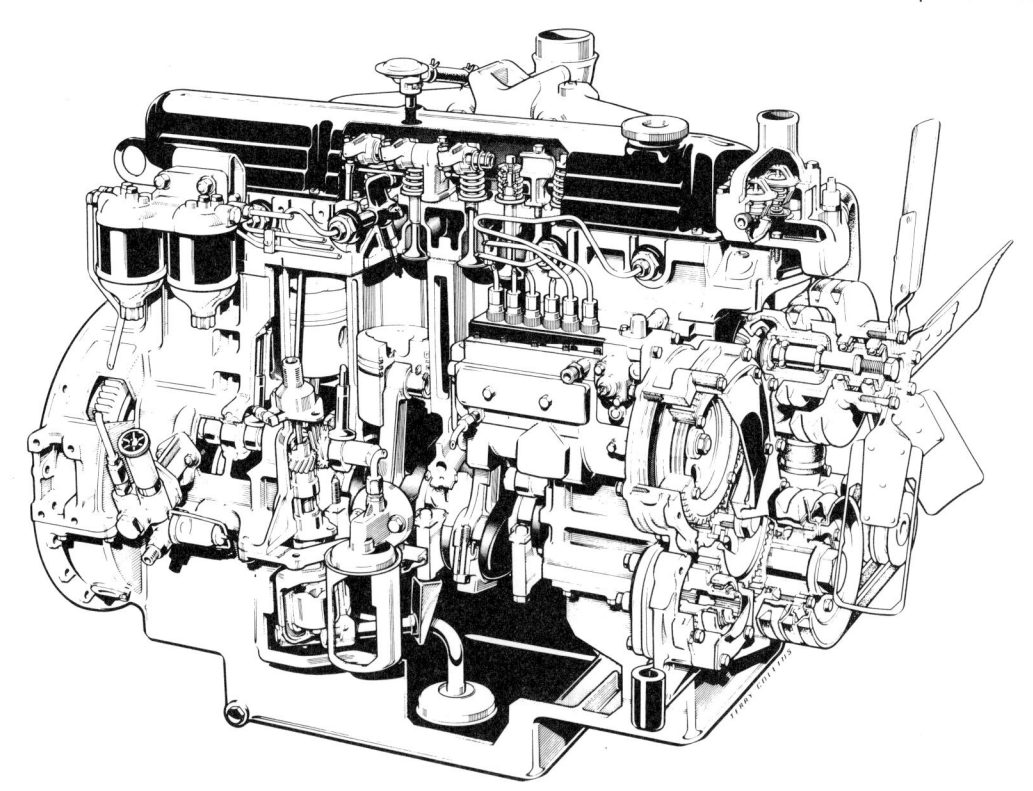

Figure 6.22 Cut-away sketch of Ford diesel engine

one and engine sales here make a major contribution to recovering the initial development costs.

Having chosen the basic engine to be offered the task of the marinizer is to ensure that it will withstand marine corrosion and any parts that are unsuitable have to be replaced or treated to inhibit corrosion. By virtue of the high running temperatures of these small diesel engines they must be fresh water cooled as the engines cannot sustain wastage by corrosion if sea water cooling is used. Fresh water cooling requires the fitting of a heat exchanger and sea water circulating pump while normally added is a water cooled exhaust manifold to reduce the engine space temperature and that of the exhaust pipe. A further heat exchanger normally fitted is an oil cooler to control the temperature of the engine and gearbox lubricating oil. Such a cooler is normally sea water cooled but some units use the engine jacket cooling water instead.

A further function of the marinizer is to fit a suitable gearbox (*Figure 6.23*) and these small engines operating between 2000 and 4000 r.p.m. require a reverse/reduction gear of between 2 and 5:1 reduction ratio depending upon whether the engine is in a high speed pleasure boat or a work boat where a slow turning propeller of coarse pitch is desirable. To fit the gearbox it is normally necessary to replace the flywheel supplied with the engine to one that will match the desired type of reduction gearbox.

Also normally supplied by the marinizer is a remote control unit for the engine together with an instrument panel, and the engine may also be delivered to the boat yard complete with stern gear, propeller shaft and propeller.

6.2.17 Truck engine advantages

The small work boat or ferry is operated in a very different manner to a ship. Operation is usually on a day basis during which time the craft will be constantly in use and, therefore, reliability of the engine is paramount. Small workboats do not carry an engineer or maintenance man, so any defects arising during the day must be attended to after the boat's working period, either by a mechanic employed by the operator, or by a service engineer supplied by the engine manufacturer or a dealer or repair establishment. Maintenance work aboard the vessels tends to be minimal and when the engine is in operation it is rarely attended. Most work to the engine will be undertaken during a seasonal lay up or, on breakdown, when an urgent repair becomes imperative. The craft is operated similar to driving a car or truck and when working, load variations on the engine from idle to full r.p.m. in seconds is both normal and frequent.

This mode of operation of the typical work boat is very punishing for any propulsion engine but it is the

Figure 6.23 Marinized engine with marine gearbox

automotive type engine that stands up to it well. The owner of a small work boat buys this type of engine at a rating specified by the manufacturer with the object of using the engine at any power and speed up to this rating. The engine has an expected life on which it is sold and a complete overhaul or replacement by a new or rebuilt engine will usually be carried out with the unit lifted out of the vessel for overhaul ashore or replaced. A fleet owner may prefer to standardize on one particular type of engine and may indeed have spare engines which can be used as replacements at short notice.

Automotive type engines are well suited for work boat propulsion duty and their main advantages can be summarized as follows:

1. The engines are much cheaper to purchase and install than a heavier duty longer lasting and most probably more economical marine diesel engine.
2. They are lighter and take up less space, which is important for small work boats.
3. Installation is simple with few control connections so replacement of an engine or removal from the boat for repair or replacement is easily accomplished.
4. As the engines are basically similar to vehicle units, spare parts are easily available from a large number of dealers and such firms can also provide a service back up.
5. The engine comprises a complete propulsion plant and no other auxiliary equipment other than a fuel tank and remote controls are needed. The engines are equipped with cooling pumps, starter, generator, etc., and power take off drives can be provided for bilge pumps, hydraulics, larger generators, etc.

Engines for work boats are delivered to a boat yard with an attached gearbox and in most installations drive of a fixed pitch propeller is the most common. However, water jet propulsion is also popular for fast vessels and craft designed to operate in shallow waters. The jet pump unit may be directly coupled to the engine without a gearbox which gives a simple total installation.

6.2.18 Multiple engine plants

Many types of small craft require more horsepower than is obtainable from one truck engine and, because of weight and space limitations, an owner may choose to install more than one automotive unit, rather than a traditional high speed marine engine of greater power. Ferries and fishing boats particularly, often have more than one engine and the method of installation may be to have two or more shaft lines or, as is quite common, to fit a custom built twin input–single output reverse–reduction gear so that two engines can drive to one propeller. Also popular, particularly in Scandinavia, is to connect two or more engines to one propeller by multiple V-belts and with suitable clutches provided to enable disconnection of individual engines. V-belts are reliable and well proven and result in a very cheap installation but the absence of a reverse gear usually requires the use of a controllable pitch propeller or azimuthing type thruster or cycloidal propeller, as the propelling unit.

Multiple automotive engines in a work vessel bring a number of major advantages:

1. More than one engine allows the power to be matched to different operating modes, such as free running, towing or trawling, which bring fuel savings as operation of small diesel engines at part load is uneconomical.
2. Lower engine wear rates and longer periods between overhauls because the engines work most of the time at their design performance level.
3. In the event of an engine failure the vessel can still proceed on the remaining units without too great a loss of speed because of the small horsepower involved.
4. Considerable reduction in machinery space size and weight of the plant compared with one heavy duty marine engine of the same total power.
5. The maintenance of a fleet of vessels can be planned ahead for scheduled overhaul of particular engines by removal at night or lay up periods or with the vessel underway on the other engines.

6.2.19 Air cooled engines

Many types of small craft with unattended machinery spaces, such as fishing and work boats, require main engines up to around 500 bhp and most use water cooled automotive or other types of diesel engines. However, a further alternative is the lightweight and self contained air cooled diesel engine, of the same type as used in contractors' plant and earthmoving equipment ashore. A new marine application for the air cooled engine is drive of the lift fans and airscrews of hovercraft, replacing the gas turbine on account of the diesel engine's light weight, cheapness, lower fuel consumption and easier and cheaper maintenance.

The air cooled engine resembles the automotive type in general outline but makes greater use of aluminium alloy castings in the design and the major difference is the provision of an engine driven fan which discharges air externally over the cylinder block and heads through suitable ducting. One drawback of this engine type is that adequate ducting for air inlet and exhaust must be provided (*Figure 6.24*), but at the same time there are major advantages:

1. Reliable operation due to absence of water carrying components such as pumps, filters, valves, manifolds and piping, which are all subject to corrosion.
2. Blockages of cooling system by mud, sand, etc., leading to engine overheating is avoided.
3. Less thermal stress on the engine in cold climates.
4. No preheating is required for starting.
5. Hotter running means less tendency to cold corrosion with fuels of high sulphur content.
6. Engine can still be operated with vessels aground or in dry dock.
7. Simpler engine installation with less auxiliary equipment.
8. Light weight making the engines also suitable for auxiliary and bow thruster drive duties.

6.3 Steam plant

6.3.1 Steam propulsion machinery

By the turn of this century practically all ships were powered by steam reciprocating engines and by that time the steam engine had reached the peak of its development with very large quadruple expansion engines in common use. The next major step, in the early 1900s, was the introduction of the steam turbine, a rotary machine in which the dynamic force of velocity energy in steam acting on blades fixed to a wheel attached to a shaft, causes the shaft to rotate. Steam turbines were initially installed as large units directly

Figure 6.24 Installation of aircooled engines

coupled to the propeller shaft but the advent of reduction gearing brought many advantages. Compared with the reciprocating engine of the time the weight of the main engine was considerably reduced and as steam in a turbine is expanded right down to the condenser pressure, more work is obtained from the steam, giving a higher overall efficiency and reduced fuel consumption.

Steam turbine machinery was popular for a wide range of merchant ships and naval vessels up until the early 1950s, by which time the diesel engine had moved so far ahead of the turbine in terms of efficiency, manning requirements and running costs, thus the use of steam machinery rapidly declined. A comeback was experienced in the 1970s when the very large crude oil tankers were built in large numbers, but eventually very powerful diesel engines were specified for these ships as well, leading to the total collapse of the turbine market. Steam plants hung on for a time in warships but here as well the advantages of the gas turbine eventually forced steam out of the reckoning for new naval machinery installations.

The current situation remains very grim for steam turbine builders as for the last decade very few ships indeed have been ordered with steam machinery. The one remaining application for merchant ships is the propulsion of LNG tankers in which boil off from the gas cargo is fed to the ship's boilers instead of being vented to the atmosphere. The amount of gas burnt in this way for propulsion depends upon the quantity of boil off gas available and such ships are of necessity built with a dual fuel gas and/or fuel oil combustion system. The other application is in nuclear submarines where the heat generated by nuclear fission has to be used to produce steam which is then best used to drive steam turbines for main propulsion and auxiliary power. This application is very specialized, tends to be classified and the number of plants built each year is limited.

The outlook for the steam machinery is not at all encouraging, unless in the future there is widespread use of coal again at sea – a boiler being the only plant that can burn solid fuel without problems, as earlier experiments with diesel engines and gas turbines to use various forms of coal, have not been too successful. In view of the poor future prospects for steam turbine machinery in future ships, the following section deals only with basic concepts.

6.3.2 Steam plant principle

The propulsion of ships by steam was the earliest major form of mechanical power, giving rise to the generic term steam ship. The beginning was the use of very simple reciprocating engines supplied with steam at near atmospheric pressure from iron boilers fed with sea water. The principle of steam machinery is very simple.

Water is heated in a pressure vessel until it reaches the temperature required for evaporation into steam to take place. Further heating raises the temperature and pressure of the steam until it reaches the plant's designed working pressure, while superheating of the steam at its working pressure increases its volume, giving a greater weight of steam produced for each unit volume of water boiled and is thus an efficiency improving measure. Steam is generated in the boiler at the same rate as it is consumed in the engines, to maintain a constant working pressure. Steam is fed at the desired working pressure to a reciprocating engine where expansion of the steam in an enclosed cylinder causes displacement of a piston and useful work done through the transfer of reciprocating motion of the piston to rotary motion of the shaft through the connecting rod and crankshaft. The higher the steam pressure and speed of the shaft, the more power is developed, so engines and steam conditions are designed to suit the required output. A more efficient steam prime mover is the turbine which uses the velocity of a steam jet to drive a bladed wheel rather than static pressure acting on a piston.

After leaving the engine or turbine the steam enters a condenser for cooling by sea water back to fresh distilled water which is again pumped back to the boilers. The feed water is filtered and preheated before returning to the boiler and all the main machinery of a steam ship is designed to maintain this closed circuit. Steam losses to the atmosphere are made up with stored distilled water or freshwater is made using on board sea water distillers. The fuel for the boilers of steam ships may be coal, heavy fuel oil or gas, while timber has been used in the past in numerous river vessels.

Coal firing had practically disappeared by the 1950s but it re-emerged in the early 1980s when a number of automated coal fired turbine ships were built and these are all still trading successfully. Also at the time, there was hope for a revival of the steam reciprocating engine in modern form, but sadly none of the much heralded projects came to fruition.

6.3.3 Steam turbine features

There are two types of steam turbine used aboard ships. The impulse turbine consists of a ring of nozzles followed by a row of blades mounted on a wheel and facing the nozzles. Steam is expanded in the nozzles and passes out as a high velocity jet which impinges on the specially shaped rotor blades, resulting in an applied force causing rotation of the wheel. All of the pressure drop takes place in the fixed nozzles and there is no pressure drop across the blades.

The reaction turbine consists of a ring of fixed blades acting as nozzles followed by a row of similar blades mounted on the rotor. One half of the pressure drop takes place in the fixed blades and the rotor blades themselves act as moving nozzles and expand the steam over the second half of the pressure drop.

Normal practice is for turbine plants to have rotors built up with many stages with typical impulse blading used on high pressure turbines (HP) and reaction blading for the low pressure turbine (LP). The turbines are not reversible so a separate astern turbine is incorporated into the casing of one of the units with the steam inlet and blading designed for the opposite direction of operation: steam is admitted to the astern turbine after the main ahead steam supply is shut off.

Turbines may be of simple, compound or triple cylinder compounded arrangement, comprising many expansion stages in each turbine casing. The usual arrangement is for the HP turbine to comprise one separate cylinder and the LP turbine the second.

After expansion in the HP unit, steam exhaust from the HP casing and by cross-over pipe is fed to the LP turbine casing for further expansion before finally exhausting to the condenser. The turbines are normally mounted parallel to one another and connected to separate pinion shafts in the common double reduction gear. An intermediate pressure turbine (IP) may be mechanically joined to the HP turbine rotor and the astern turbine enclosed in the LP turbine casing. In a double or triple turbine arrangement, controlled steam nozzles are fitted only to the HP turbine. Various combinations have been built but most common for marine propulsion is the two turbine cross compounded arrangement.

The main components of a turbine plant (*Figure 6.25*) are the steam turbines themselves, reduction gear incorporating a thrust block, a condenser to cool exhaust steam and return it to feed water, feed pumps to supply the boilers and, of course, the boilers themselves.

For many years marine boilers have been of the water tube type. In addition, the turbines and gears require a lubrication system and feedwater heating and various other smaller items such as drains coolers, feed water tanks and filters, etc. A more advanced plant is shown in *Figure 6.26*.

The collapse of the market for steam turbine plants has practically halted turbine development. However, the last significant new system, which incidentally failed to find a customer, was called the 'Very Advanced Plant' from the Swedish company Stal Laval. This is described below.

6.3.4 Most advanced steam plant

The Very Advanced Propulsion (VAP) system designed after the oil crisis of 1973 when orders for steam ships were cancelled and the steam plant market dried up virtually overnight. By the early 1980s the manufacturer Stal Laval had designed a completely new plant employing a number of novel features not least of which is the use of fluidized bed combustion (FBC) in the boiler plant where it is designed to act as a reheater–superheater in conjunction with a conventional water tube boiler.

The advantage of fluidized bed combustion is that it allows steam temperatures of 600°C and above without the risk of corrosion in the superheater, while a wide variety of solid and liquid fuels can be used. The other innovative feature is the speed of the turbines at nearly double that of the previously most efficient design, allowing smaller more efficient turbines.

The VAP arrangement (*Figure 6.27*) has the HP and IP turbine units in separate casings and a separate LP turbine. The HP turbine has a barrel type casing and seven stages while the IP turbine is of similar construction with six stages. These two turbines are mounted parallel to each other and each is rigidly coupled to a separate pinion of the single stage twin input helical

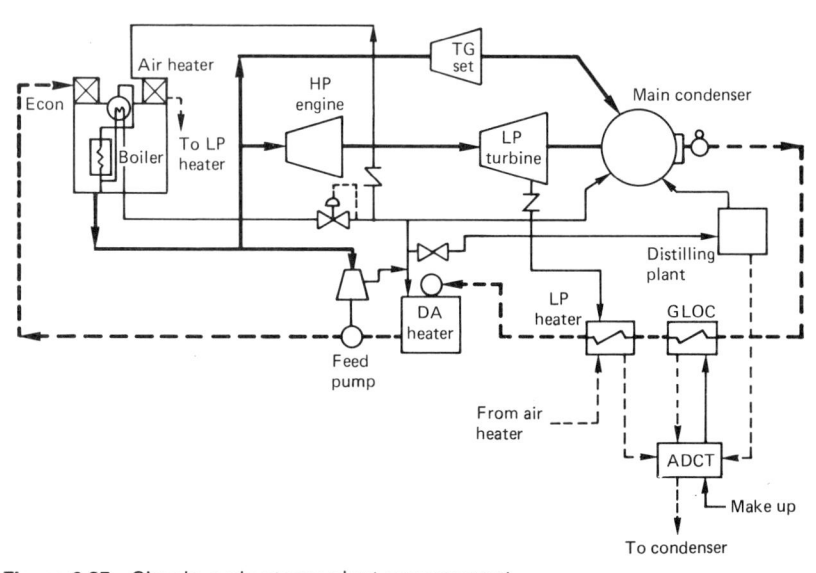

Figure 6.25 Simple cycle steam plant arrangement

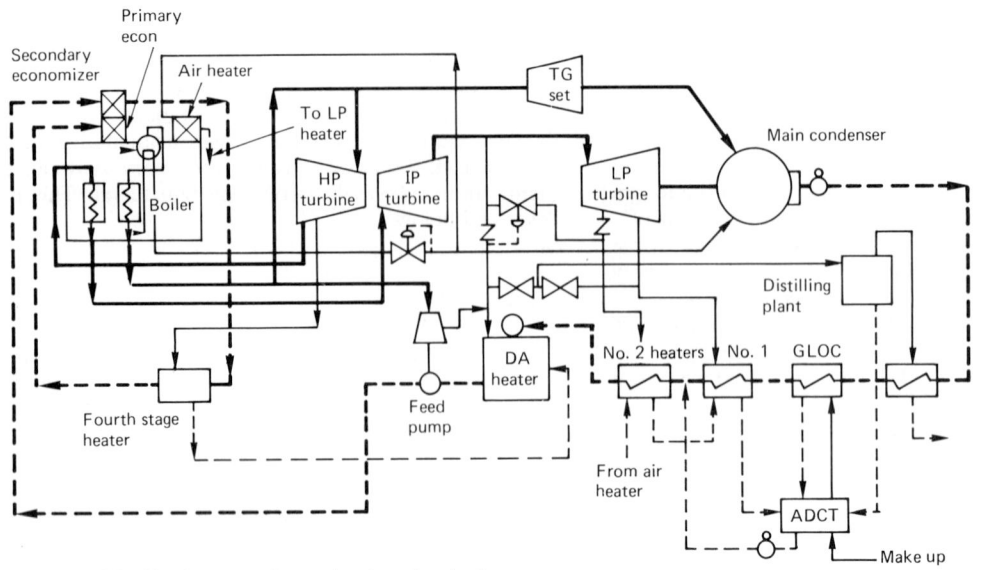

Figure 6.26 Dual economizer reheat cycle plant

primary reduction gear unit. The LP turbine has a horizontal split casing and is connected to the primary reduction gear through a separate epicyclic reduction gear and it also contains the separate astern turbine.

The final component of the power train is a two stage epicyclic gear coupled to the primary gear output and with connection to a separate thrust block and the propeller shaft turning at 70 r.p.m. The condenser has titanium tubes and is mounted forward of the LP turbine on the same axis.

The main boiler is of Babcock MR radiant water tube type producing steam at 500°C at the superheater outlet. The separate topping superheater is separately oil fired in a fluidized bed system giving a final superheat and reheat temperature of 602°C. After passing through the HP turbine, the steam is reheated to about 410–602°C in the FBC and then enters the IP turbine and thence to the LP turbine and finally the condenser.

The VAP turbine is designed for outputs of 10–50 MW and has a best specific fuel consumption rate of around 223 g/kWh, which is still somewhat higher than a diesel engine but still represents a major leap forward for steam machinery. Such a plant could have a major place in future marine propulsion installations if there is a significant switch to coal firing.

6.3.5 Future outlook

As already mentioned, the steam turbine displaced the steam reciprocating engine because of the former's much better power–weight ratio and the better Rankine cycle efficiency and hence considerable reduction of fuel consumption. It is, however, the disadvantages of the turbine in comparison with the diesel engine which have caused the demise of steam main propulsion machinery.

The disadvantages are thus:

1. Fuel consumption is considerably higher than the diesel engine.
2. Turbines must run at high speed for best efficiency so a high ratio reduction gear is essential to obtain an economical propeller speed.
3. A high pressure boiler system with strict attention to feed water quality is essential.
4. A steam plant is much more complicated and requires very careful operation and maintenance, particularly with regard to boilers.
5. Greater efficiency is obtained through higher steam temperatures and pressures, needing special heat resistant materials for HP turbine construction.
6. A separate astern turbine has to be provided and astern power is much less than that of ahead, resulting in very slow response to manoeuvres, compared with a diesel engine.

These major disadvantages preclude the steam turbine from new projects in the immediate future but, in the longer term view, present day consumption of fuel oil can only lead to a rapid depletion of known oil resources resulting in rapid and ongoing rises in the price of oil. Steam has the advantage that it can consume alternative fuels such as coal and gas and may well return if the cost of fuel oil needed for diesel engines becomes unacceptably high.

Given that future steam turbine plants will have to burn coal, gas or other solid or liquid fuel other than hydrocarbon oils, there are development possibilities for marine turbine plants, summarized below:

1. Reheat cycles using heat in the exhaust gases to raise the temperature of the steam between stages in the turbines, to raise the plant thermal efficiency.

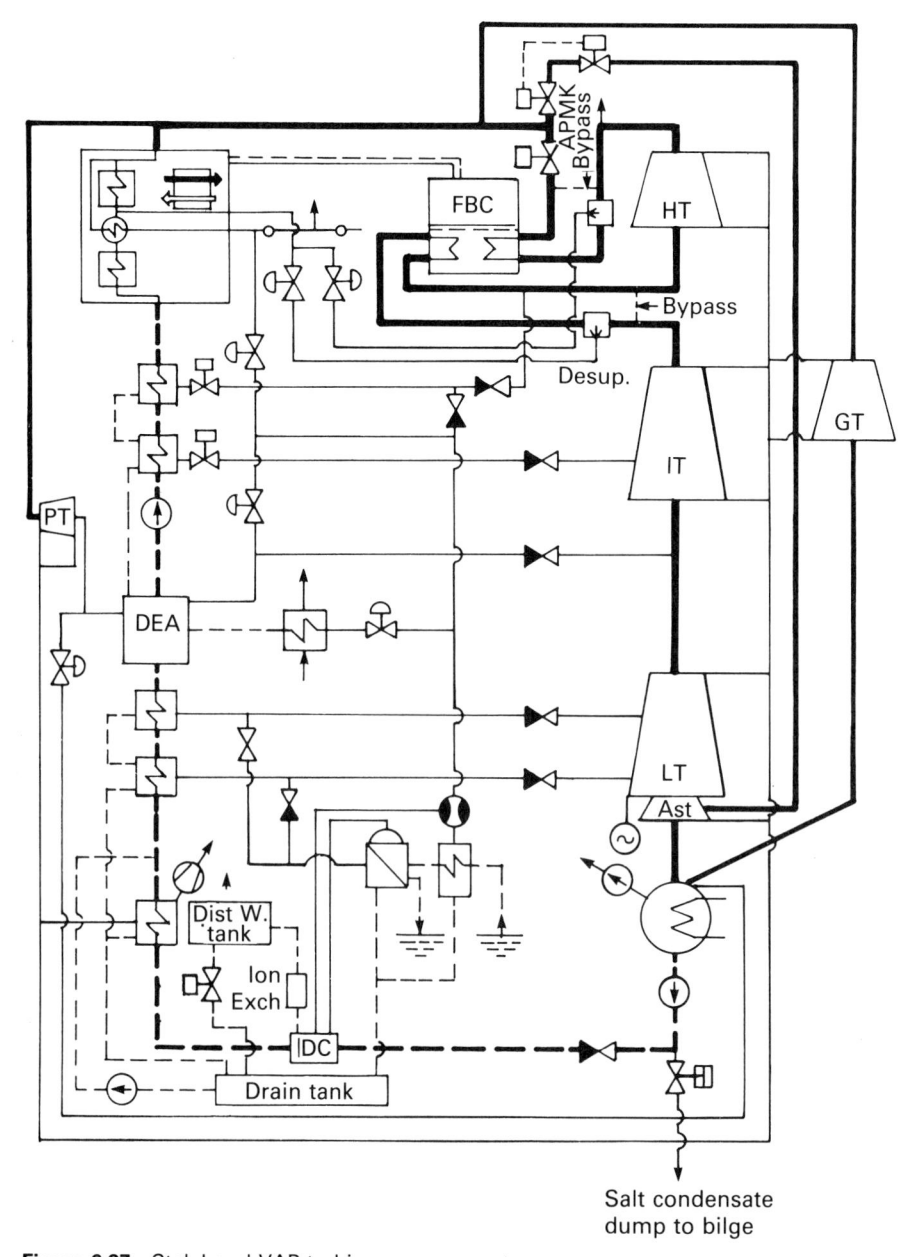

Figure 6.27 Stal–Laval VAP turbine arrangement

2. Use of very high pressure plants with flash type boilers.
3. Application of nuclear power to merchant ships.
4. Use of the supercritical steam cycle.
5. Full scale use of current technology of fluidized bed combustion taking advantage of the wide range of fuels that can be burned without boiler corrosion problems.
6. Use of the binary vapour cycle.

The future may see the application of any of the above for ship propulsion and even combinations of the above. There is also much potential for application of the steam turbine in conjunction with other prime movers such as gas turbines whereby waste heat in the exhaust from the gas turbine is used to generate steam for a turbine which is also coupled to the propulsion gear train. This concept, albeit with diesel engines providing the waste heat, has already been used in numerous ships with the turbine driving a ship's service generator. Waste heat recovery steam turbines have also been coupled to the main propulsion system.

6.4 Gas turbines

6.4.1 Gas turbine plant

Soon after the end of World War II, it was recognized that the gas turbine being developed at that time for aircraft could make a promising prime mover for marine propulsion. The earliest shipboard experiments were carried out by the US Maritime Administration with the re-engined Liberty ship *John Sergeant*, while in Britain the pioneering work was mainly done by Rolls Royce in conjunction with the Royal Navy and the first successful marine installation was in the naval patrol vessel *Brave Border*. Following these early experiments, work was done by Shell with various types of turbines aboard its tanker *Auris*.

In the 1970s a number of gas turbine powered merchant vessels were built and were technically successful. Most of these have since been re-engined with diesel engines because the gas turbine's lower efficiency and low tolerance to poor fuel quality made these ships very expensive to operate after the fourfold increase in fuel prices in the mid-1970s.

The higher operating costs of gas turbines mean that at the present time there is very little interest in such plant for commercial ships but this prime mover's advantages are widely recognized for naval vessels ranging from fast patrol boats to aircraft carriers.

The gas turbine or jet engine is very similar to the propulsion engine for the majority of civil and military jet aircraft, it being referred to as the aeroderivative gas turbine, though it is specially designed and adapted for the marine environment. Whereas in an aircraft of jet type the exhaust from the gas generator is expelled to the atmosphere to provide thrust, in a ship this exhaust gas is passed through a power turbine causing it to rotate thus providing a mechanical power drive through gearing to the propeller. Gas turbines in use today are mainly of aerotype but there have been a number of examples of the heavy duty type at sea. This latter type is of heavier construction and similar to the type used in industrial applications ashore.

6.4.2 Operating principle

The marine aeroderivative gas turbine (*Figure 6.28*) is a fully rotating machine which operates on a steady flow cycle and works as an open cycle heat engine using atmospheric air as the largest mass fluid in the working cycle. Air is drawn into the compressor section of the unit, compressed and heated by combustion of liquid or gas fuel in special combustion chamber cans where heat is added at constant pressure before expansion of the exhaust gas through the power turbine which is mechanically coupled to the propeller. In modern turbines the compressor section is normally in two stages, each stage being driven by its own power turbine, leading to the so-called two spool turbine type.

There are different arrangements but all gas turbines are made up of three basic units: compressor, combustor and power turbine.

The function of the compressor is to provide the required air mass at sufficient pressure to burn the required amount of fuel to generate the acceptable output from the power turbine. Axial flow type compressors are normally used in the current generation of turbine, in which air is compressed as it passes from stage to stage axially along the shaft. In a centrifugal type compressor the impeller draws air in near its centre of rotation and accelerates it radially outwards by centrifugal force. Axial types are more efficient and

Figure 6.28 Marine Spey gas turbines

are easier to build with pressure ratios above 20:1, so this type is most favoured for the latest machines.

The combustion chamber is the component in which the compressed air and fuel are mixed, ignited and burned and comprises a casing with a perforated inner shell with a fuel nozzle and an initial ignition device such as a spark plug. Large turbines may have as many as 20 such combustion cans which are normally configured as an annulus shape to deliver the hot products of combustion to the turbine first stage inlet guide vanes and first row of rotating blades. Part of the compressed air from the compressor, called primary air and usually about 30%, passes directly to the combustion chambers where it is mixed with atomized fuel so that the mixture can be ignited and burned. The remaining 70% secondary air passes into the combustion chamber through the perforated inner shells and mixes with the gases of combustion. The secondary air cools the combustors and reduces the temperature of the combustion gases to a level acceptable for the turbine section without damaging the blading.

The turbine section is located immediately downstream of the combustion chamber and comprises a fixed stator and moving rotor. The turbine is nearly always of axial flow type and consists of one or more stages which produce rotational power from the kinetic energy of the gases after passage through the nozzle blades. This rotational energy is to power turbines for first and second stage compressors and to drive accessories. After extracting power for the compressors and accessories, the remaining gas exhaust from this section is used to drive the power turbine.

Modern marine gas turbines have the power turbine as a completely separate unit through the air inlet elbow, gas generator, power turbine and exhaust elbow are built into one module with the components enclosed in the housing to make it suitable for naval vessels from the point of view of noise emission, resistance to shock, damage control, fire fighting, etc. The output of the power turbine is passed to a separate reduction gear which may be of reversing type for use with a fixed pitch propeller or unidirectional incorporating only clutches and usually a double reduction double helical gear pinions and wheels, with output to the propeller shaft. Gearboxes for naval gas turbine plants tend to be of complicated design because of the many different machinery arrangements employed.

6.4.3 Gas turbine applications

The gas turbine can be used to power any type of ship as it has in fact been used in diverse types such as trawlers, container ships, bulk carriers, ferries, gas tankers and, of course, warships from patrol boats up to large aircraft carriers. Nowadays, its use is mainly confined to craft where both a high output and light weight are essential, a good example of this being hovercraft where high machinery weight makes more difficult the lifting of the craft onto a cushion of air and for success a good balance between weight and power has to be achieved. Gas turbines can also be built small enough for auxiliary duties such as driving generators or pumps. They have also been used to power fast yachts and power boats.

Gas turbine engines are mainly used in naval vessels for the operational reasons outlined elsewhere. In frigates they are used in combination with other machinery, such as diesel engines or electric motors, with the turbines fulfilling their function of providing sufficient power to meet the high top speeds required. Modern warships are tending to also use all gas turbine plants, as the manufacturers improve the performance and efficiency of the current generation of engines. Typical arrangements of all gas turbine plant are described below.

6.4.4 COGOG

The earliest all gas turbine plants used two sizes of engine, each coupled to a gearbox and propeller in a twin screw arrangement (*Figure 6.29*). The Royal Navy has many frigates with Rolls Royce Olympus turbines for boost power and the much smaller Tyne also installed for the cruise function. These two types of turbine are only operated independently with the changeover effected by self synchronizing clutches built into the gearbox. A CP propeller is needed for astern operation and the smaller cruise turbine avoids the considerable loss of efficiency from operating the boost units at part load.

6.4.5 COGAG

The improvement in efficiency of current gas turbines and development of new intermediate output engines has given rise to COGAG plants where a number of similar units are installed and can be operated together to meet speed requirements from cruise to boost. The smaller units have better efficiency at lower powers and in the case of the Royal Navy *Invincible* class aircraft carriers, four Olympus turbines provide sufficient power for maximum speed operation, while the vessel can cruise quite well on one alone.

6.4.6 CODLAG

The latest generation of frigates uses a CODLAG system (*Figure 6.30*) in which two gas turbines provide sufficient power for the top speed, and cruising is undertaken using an electric motor incorporated into each propeller shaft. The motor obtains its power source from the ship's auxiliary generators and as well as providing a very easily controllable slow speed drive, it allows silent running which is important for antisubmarine operations and a propeller reversing facility, so fixed propellers can be used.

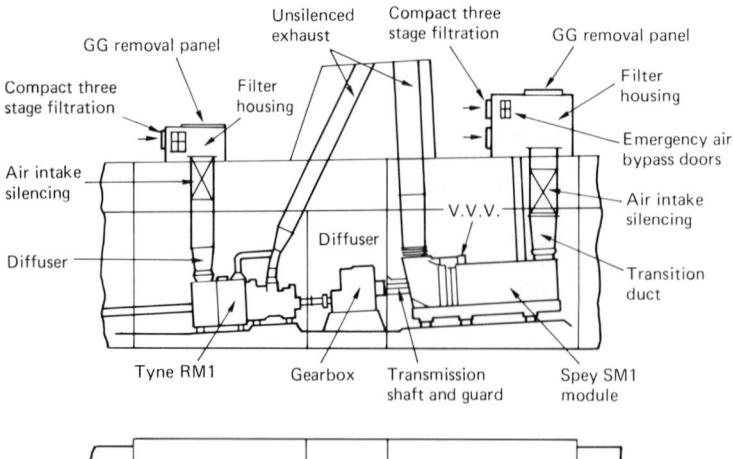

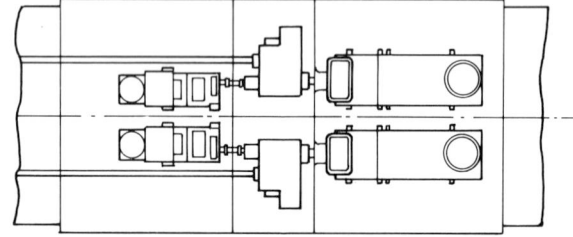

Figure 6.29 COGOG machinery arrangement for a frigate

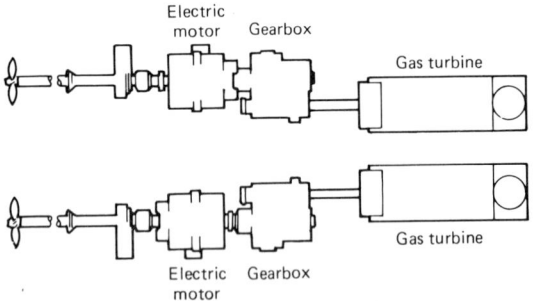

Figure 6.30 CODLAG arrangement with electric motor drive

6.4.7 Gas turbine advantages

The gas turbine is rapidly becoming the preferred primary marine propulsion unit for a range of craft up to small aircraft carriers and in practically all warships of frigate and destroyer types. The reason why the gas turbine is now preferred is on account of its major advantages which are summarized as follows:

1. The gas turbine has a very low specific weight in terms of lbs/shp which suits naval designers requiring ever more power from a smaller machinery space, as more space is needed for weapons systems. Turbine units developing as much as 36 000 shp per unit are installed in ships taking up a fraction of the space needed by an equivalent output steam or diesel engine plant.
2. Turbines cause much less vibration than other machinery types and are thus easier to quieten by suitable mounting and enclosure in the engine room.
3. The plant is much simpler as the cooling requirements are negligible, with main cooling of the power unit effected by the incoming air, and with vital auxiliaries such as lubrication, fuel, etc., normally attached, making gas turbines self contained power units.
4. A gas turbine can provide its full power in less than 5 min from cold, an important feature for warships needing to put to sea rapidly.
5. The gas turbine is an extremely reliable prime mover as a result of its extensive development for aircraft propulsion where reliability is paramount.
6. Maintenance work is simpler as the tendency is to remove complete gas generator units ashore for complete overhaul at regular intervals, with replacement by freshly overhauled units. However, maintenance by replacement requires an adequate shore repair facility or contract arrangement with the manufacturer.

The above major advantages make gas turbines well suited for naval applications but, despite their vast potential, the disadvantages and their limitations are enough to make them increasingly unattractive for commercial applications.

The major disadvantages which adversely effect significant use of gas turbines in large merchant ships are:

1. The gas turbine is a high speed unidirectional machine making both a reduction gear and a reversing mechanism essential, if it used to drive a fixed pitch propeller. However, the greater use and acceptance of controllable pitch propellers is making unidirectional operation far less of a problem.
2. The gas turbine is notoriously inefficient at part load and therefore it is only suited for applications where most operation is at full power only, a fact recognized even in naval vessels where lower powered separate turbines or alternative prime movers are used for slow speed operation.
3. The aerotype gas turbine operates only on distillate fuel or gas because of likely corrosion damage to components caused by metals such as sodium and vanadium in residual or blended fuels and the combustion system requires clean fuel and very careful fuel treatment. Distillate fuels burnt in quantity will considerably add to the fuel bill of a merchant ship not only because of the higher cost of distillate but also because of the much increased consumption resulting from the thermal efficiency which is much lower than a diesel engine.
4. Gas turbines require a considerable quantity of inlet air causing large areas of the ship to be taken up by ducting and measures have to be taken to prevent ingress of salt spray to the engine.
5. The very specialized nature of the gas turbine means that normally only simple routine maintenance and emergency repairs are undertaken *in situ*. Specialist shore maintenance of gas turbines is expensive.

The above advantages and drawbacks of the marine gas turbine clearly show why this prime mover is growing in popularity with navies and rapidly losing favour with commercial ship operators. However, this situation could easily change if there is to be a future for high speed commercial ships requiring exceptionally high horsepowers at an acceptable weight burden and where the cost of fuel is of less importance in shipping economics than it is today.

Further technical developments to improve the thermal efficiency of the gas turbine using ceramic components, reheat cycles, etc., should make this prime mover much more attractive for most ship types. For very high powered ships of the future, given a plentiful supply of light fuel or gas, the turbine would automatically be preferred to a nuclear plant for merchant ships.

6.4.8 A typical modern design

Gas turbines in current use aboard warships are of simple cycle without reheat but are termed second generation units and all recent applications use turbines developed from aircraft engines. An example of the latest technology is the Marine Spey SM1A produced by Rolls Royce (*Figure 6.31*).

The SM1A turbine is supplied as a module which forms a complete propulsion package comprising the Spey gas generator, acoustic enclosure, cascade air inlet bend, ancillary systems, electrical power points and fire protection systems, and the propulsion power turbine, all mounted on a common baseplate. This package is a standard unit with its own local controls and is provided complete with heat and noise insulation and self contained ventilation system.

The gas generator is of twin spool axial flow type having a full power pressure ratio of 21.8:1.. The low pressure compressor has five stages and is driven by a two stage turbine, while the high pressure compressor has eleven stages and is also driven by its own two stage turbine.

The combustion system has been designed for low emissions with a reflex airspray burner to premix air and fuel which is then burnt in a twin vortex circulation resulting in near full combustion efficiency.

The power turbine is of two stage type with shrouded rotor blades and an unsplit casing. The rotor is carried in hydrodynamic bearings housed in a rigid centre body which is supported by suitably insulated vanes running

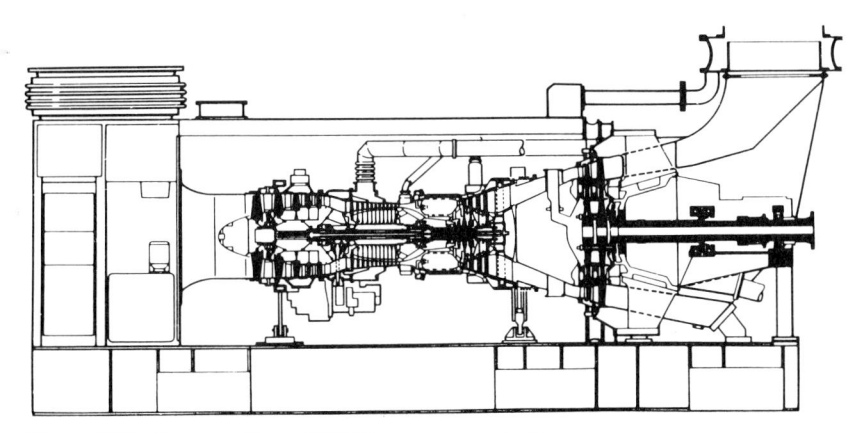

Figure 6.31 Cross-section of SM1A marine gas turbine

across the gas stream to a cold mounting ring. This in turn is supported by two struts from the baseplate. The exhaust volute has an unswept bend and the inner wall of the exit features a connection point to sense volute static pressure depression and provide automatic ventilation of the module enclosure.

The acoustic enclosure for the turbine is a rectangular structure made of steel covered on the outside with thick mineral wool insulation held in place by Zintec sheeting. It is open ended for attachment of the power turbine and intake, allowing the gas generator to be removed through the air intake end. The module is provided with an access door, viewing port, internal lighting, ventilation and remote controls and monitoring equipment.

The gas generator is started by an air motor together with an automatic clutch and reduction gearing. This motor spins the HP rotor until ignition and gas production causes the unit to run unaided. Lubricating oil is contained in a tank mounted on a baseplate at the side of the air intake cascade bend. Oil is taken from the tank to a gas generator driven pump and thence the engine bearings. A scavenge section of the lubricating oil pump returns oil to the tank. The power turbine is lubricated from the ship's main gearbox lubrication system.

The latest version of the SM1 is the Marine Spey SM1C which is presently rated at 18 MW (26 150 bhp) and has been specified for Dutch M class frigates and the Royal Navy Type 23. The unit is to be further developed for higher output with a predicted fuel consumption rate of 0.227 kg/kWh. The expected maximum power rating is 19.5 MW.

6.4.9 Performance improvements

The major reason for the gas turbine's fall from favour is its lower thermal efficiency (36% for simple open cycle types) compared with modern diesel machinery (around 50%) which results in a much higher fuel consumption and cost for the same power. The largest single energy loss in the simple cycle machine is that in the exhaust gas heat rejected to the atmosphere. Current projects being supported by navies aim to improve the thermal efficiency of marine gas turbine plants by recovering part of this lost heat energy, while at the same time manufacturers are trying new materials such as ceramics which allow higher gas temperatures through the turbine. Two projects are actively being pursued.

6.4.10 ICR plant

A portion of the heat lost in the exhaust can be recovered by placing a regenerator between the compressor and combustor sections of the plant, the extra heat given to the air provided by power turbine exhaust passing through the regenerator which is mounted atop the power turbine. Further efficiency improvement can be obtained by fitting an intercooler between the low and high pressure compressor stages. Both these measures are incorporated in the intercooled regenerated plant (ICR) such as that being developed by Rolls Royce for the US Navy.

With the ICR system it is possible to increase the thermal efficiency to about 43% and maximum output by 20%, while regeneration of the combustion feed air allows a good efficiency to be obtained at part load and an enhanced full power efficiency for much of the engine power range. It is hoped to result in an overall efficiency higher than some diesel engines while retaining a significantly higher power to weight ratio. This development will allow the one turbine type to perform both boost and cruise functions. It will also make the engine more attractive for merchant ship applications.

6.4.11 RACER systems

A second approach uses the heat in the exhaust gas to generate steam in a system known as Rankine Cycle Energy Recover (RACER). An economizer and a superheater are placed in the exhaust uptake and steam produced is fed to a steam turbine which is coupled to the same gearbox as the gas turbine main engine. By proper design the pressure drop of the exhaust gas through the steam generator can be controlled so that the combined output of both cycles will yield a thermal efficiency higher than either cycle operating independently. To achieve the high output, improved efficiency and greater range required by naval designers, the plant must be capable of sustained operation as a combined system.

The systems discussed above are two avenues of research to improve the efficiency and attractiveness of the marine gas turbine. This work is ongoing in parallel with research and development by engine designers into new materials to improve the performance and efficiency of the turbine itself.

6.5 Nuclear power

6.5.1 Nuclear history

There is a modern form of ship propulsion which is surrounded by mystery, is viewed with considerable mistrust and faces massive public antipathy. Nuclear power has been used as the primary heat source for steam generation in both commercial and military vessels and, with the exception of the latter, the experiment has not been wholly successful. No truly commercial nuclear powered ships are still in service despite the absence of any reported accidents with the nuclear reactors of the ships previously operated and the obvious advantages of nuclear renergy for steam raising. The celebrated ships *Savannah*, *Mutsu* and *Otto Hahn*, have either been re-engined or withdrawn from service, but nuclear powered ice breakers are still used by the Soviet Union and indeed a new vessel was built in Finland as late as the end of 1988.

Problems were experienced with manning the pioneer vessels but the most insurmountable obstacle was the refusal of many port authorities to allow these nuclear powered vessels to enter ports, severely restricting their sphere of operations. The current attitude to nuclear ships is that the supply of fossil fuels would have to be terminated, or severely restricted, before interest among ship owners in nuclear power could be aroused.

The most forthright advocate of nuclear power is the US Navy which now uses it to power almost all of its submarines, large aircraft carriers and several cruisers. This navy's experiments with nuclear power began in the mid-1950s when the submarine USS *Nautilus* was built and proved successful.

Nuclear power is popular for naval vessels because of the tremendous amount of energy that is released when certain atoms are split apart in a process known as fission. Also, the process takes place without needing a supply of air, so this method of generation of steam to drive turbines is ideal for submarines, giving them limitations related only to the crew requirements, rather than the machinery as in conventional submarines relying on batteries or other methods for submerged operations.

6.5.2 Basic principle

The theory of nuclear fission is beyond the scope of this book so only a brief account of the process is given.

In a nuclear powered vessel the heat energy results from fission of the uranium setting up a reaction, with energy production the final result. This energy is in the form of heat generatd by friction caused by the interactions of neutrons with atomic nuclei. A nuclear reactor is a device to maintain and control the nuclear fission chain reaction. The central section is the reactor core in which uranium dioxide fuel elements are used for the fission process, with water passed through the reactor as coolant. The fuel elements heat up and this heat is transferred to the reactor coolant as it circulates through passages between the fuel elements. The heat produced by fission must be constantly removed from the fuel elements to prevent them from melting and the reactor vessel must also be kept under high pressure to prevent the coolant from boiling and forming steam which has a lower heat capacity than water; if steam entered the reactor coolant pump cavitation would cause loss of circulation.

Since the reactor cooling water is prevented from boiling in the reactor core, the steam needed for the propulsion turbines is produced externally in a heat exchanger of shell and tube type, with the coolant passing through the tubes while the outside surfaces of these tubes are in contact with the feed water circulated for what is basically a conventional steam plant. The feed water boils in the heat exchanger producing high quality wet steam.

Control rods in the reactor form the dual function of keeping the thermal–neutron flux at the desired level in the reactor core and providing a means of stopping the fission process. When the rods are inserted into the vicinity of the fuel they capture neutrons, diverting them from the process and slowing it down. Control is effected by withdrawing the rods until the reactor becomes critical and, typically over the lifetime of the core, the control rods have to be further and further withdrawn to reach the critical state as the fuel is depleted. When the reactor is at its critical state any movement of the control rods will change its operating temperature by varying the thermal–neutron flux level. The power level of the reactor is controlled by altering the steam demand on the system as it is in fact a self balancing system.

6.5.3 Nuclear advantages

Despite the political and other factors thwarting the significant use of nuclear power in ships, the system has some key advantages, but also some major disadvantages. For warships the chief advantage is the far greater endurance possible than with a fossil fuelled ship, e.g. an aircraft carrier steaming for something like 200 000 miles before refuelling, compared with around 1500 miles for an oil fuelled ship at maximum speed.

The major advantages are:

1. Long periods between refuelling operations and considerable endurance range for the vessel after each fuelling operation.
2. Huge quantities of fuel need not be transported with resultant weight savings and space needed for fuel and a huge reduction in manpower requirement for fuelling operations in naval applications.
3. As nuclear power is not dependent on air for combustion it is a very useful choice for submarine propulsion. For surface ships no air inlet is required and similarly there is no exhaust to give the ship a heat signature and no pollution of the atmosphere.
4. Because large quantities of fuel are not carried nor consumed, there are no changes in ship draft and trim as the fuel is used.
5. Nuclear plant is very simple to control, it responds instantly to load demand changes and can supply huge quantities of high pressure steam.

Not all of the advantages mentioned above are significant enough to make nuclear plant more attractive for merchant ships. The disadvantages now appear formidable:

1. Reactor plants are heavy and require very dense shielding to contain radiation, which eliminates many of the advantages of not carrying fuel oil. The power to weight ratio of nuclear plants is only of advantage in large vessels as plants small enough for

for, say, a frigate or medium sized container ship are still very heavy.
2. The costs of building and maintaining a nuclear plant are very high because of the very stringent quality control necessary to ensure reliability and, extremely important, the safety of the plant and hence the ship and its crew.
3. The high costs of purchase and operation is a major deterrent to commercial operators who will be concerned with profitable operation and return on investment. It is not certain that life cycle costs for a nuclear plant will be less than the most efficient diesel engine alternative, particularly as experience of full life operation of a nuclear vessel under commercial trading conditions is non-existent.
4. The training of crews competent enough to operate nuclear plants is both time consuming and expensive. Experience has shown that there is great difficulty in attracting suitably qualified scientists to serve aboard ship. Training for nuclear plant operation is best undertaken in a military environment.

6.5.4 Applications

In view of the aforementioned disadvantages of nuclear power for ships, its applications are somewhat limited and it is concluded that this propulsion system is best suited either for government funded and operated vessels and, of course, large warships where the life cycle costs of very powerful ships, such as the huge aircraft carriers of the US Navy, are in fact much lower than a fossil fuelled plant of equivalent output. The application in large submarines is now also very well established in many navies, notably for Britain's Polaris type and forthcoming Trident submarines.

For merchant or quasi-military vessels which could accommodate the heavy machinery and at the same time require very high machinery output, these are considered suitable for nuclear power. Ice breaking ships are probably the best example; ice breakers to operate in far northern latitudes and possibly large gas or oil tankers to transport fuel reserves exploited from Arctic regions. For general shipping, the most likely future application is in very large and fast container type vessels and probably huge submarine tankers.

Whereas economics normally affect the choice of propulsion machinery in a ship, in the case of nuclear propulsion, this is more likely to be influenced by political developments and whether nuclear power really needs to be adopted as the energy replacement for oil or gas.

6.6 Propeller shaft drives

6.6.1 Power transmission

The main propulsion power transmission system of a vessel has three functions. These are to transmit the motion and power from the engine crankshaft to the propeller and to turn the propeller at as near as possible to the optimum r.p.m. for which the particular propeller is designed on the basis of its diameter and pitch ratio. This power is transmitted by suitably dimensioned hollow or solid steel shafting, which also fulfils the function of transferring the thrust of the propeller to the hull of the vessel.

The type of transmission employed depends on the speed of prime mover and desired propeller speed and whether some form of control is built into the system, i.e. clutches, and the number of prime movers to be coupled to the propellers. Additionally, modern ships make good use of the main transmission system to drive auxiliary machinery also, as outlined elsewhere in this chapter.

The simplest form of marine transmission is direct drive and for ocean going ships, numerically this is most common with the slow speed crosshead type diesel engine which is connected by lengths of shafting to the tailshaft supported in the sterntube and which carries the propeller. With the tendency to place the machinery as far aft as possible, the shafting tends to be very short, within the minimum length laid down by the classification societies and the usual coupling method is flange to flange with the power transmitted by fitted bolts in the flange bolt holes.

Other methods use sleeve type connectors which work on an interference fit and are assembled and disassembled using hydraulic equipment. Ships' tailshafts need to be periodically surveyed so provision has to be made for a length of the propeller shafting which can be removed to provide room for the propeller shaft to be drawn inboard, though with the modern tendency to balanced space rudders and no stern post, many vessels have removable flanges on the tailshaft to allow it to be taken out through the stern bush in a dry dock.

An essential part of the drive system is the thrust block, which forms a separate thrust shaft section and which consists of a collar faced on both sides by tilting pad segments to withstand the thrust load of the propeller. The thrust bearing may be separate and flanged coupled into the shaft line between the engine and tailshaft or, as is more common nowadays, be built into the aft end of the engine itself. The direct drive system thus simply consists of a length of shafting suitably supported in bearings and with couplings designed to transmit the full torque of the engine. More complicated arrangements are in respect to ships powered by medium and high speed diesel engines and steam and gas turbines.

All prime movers other than slow speed diesel engines are designed to be operated at speeds higher than optimum propeller speeds so a speed reduction system is essential. The economic advantages, relative simplicity and long history of successful service experience make toothed gearing the first choice for power transmission from diesel engines and turbines. The gearbox is located near to the output side of the engine with perhaps a flexible coupling between the two and may be of plain reduction type for ships with reversible main engines or ships equipped with controllable pitch propellers, or of the commonly used reverse–reduction type for unidirectional engines and fixed pitch propellers, which is practically universal for small craft.

6.6.2 Shaft bearings and seals

In older types of ships with the main engines located amidships, the line shafting extends from the engine room to the aft peak bulkhead through a tunnel built in the holds, of sufficient size to allow access and maintenance work. Modern ships with engines aft have done away with the shaft tunnel and, in so doing, the long line of tunnel bearings to support the propeller shaft.

A shaft line will comprise three separate sections beginning with the thrust shaft closest to the engine output or gearbox and provided with a thrust bearing (*Figure 6.32*) of Michell type (segment pads) or of self aligning spherical roller bearing type as used in smaller vessels. Thrust bearings are normally lubricated from the main engine oil system. One or more sections of intermediate shafting are normally fitted aft of the thrust shaft, these are usually provided with fixed flanges or may have couplings of SKF oil sleeve type which are removable. A typical main shaft bearing or plummer block (*Figure 6.33*) has its lower half surface white metalled and has a lubricating ring resting on the shaft which dips into an oil well located beneath the bottom bearing shell, to spread oil to the bearing. The alternative today is the grease lubricated self aligning spherical roller bearing.

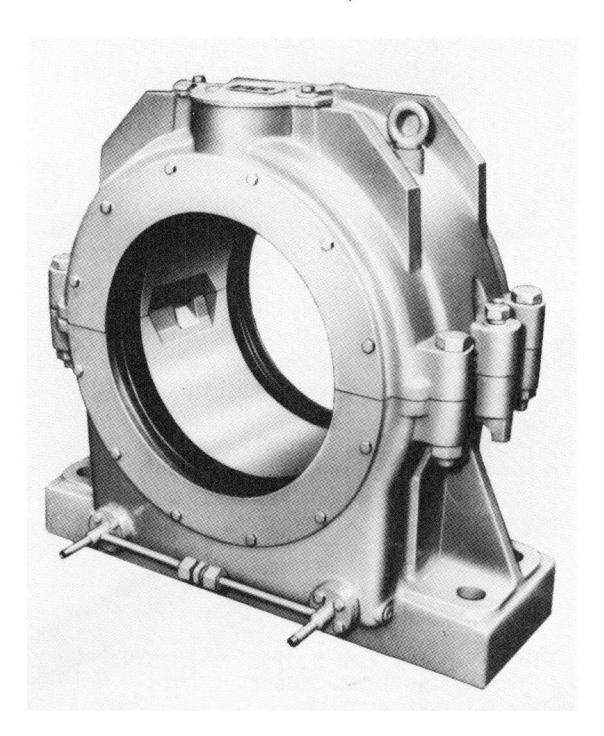

Figure 6.33 Typical propeller shaft bearing

The final aftermost length of shaft is the tailshaft, which runs in a tube known as the stern tube (*Figure 6.34*) and has bearings and seals. This steel tube is usually welded between the aft peak bulkhead and the stern post. It is suitably bored for alignment with the shafting and engine and is equipped with bearings and seals to prevent water ingress to the vessel. Older stern tubes used lignum vitae for the bearing surface with these wood strips pressed into bronze bushes which are themselves pressed into the fore and aft ends of the

Figure 6.32 Michell type thrust bearing

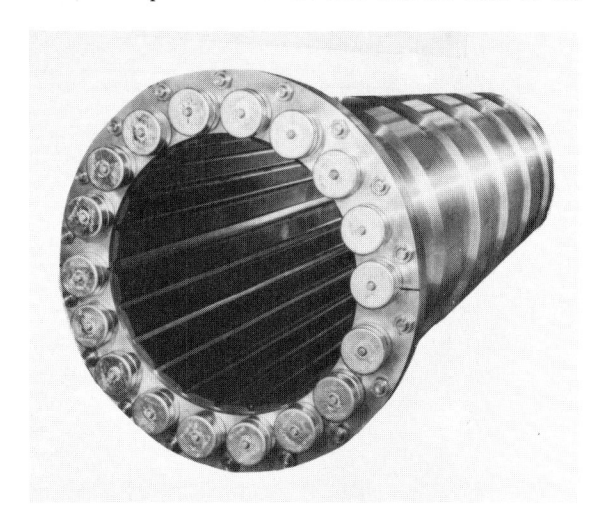

Figure 6.34 Complete tailshaft stern tube

stern tube. Lubrication is by water and a bronze sleeze is shrunk onto the propeller shaft to give a running surface and prevent rusting and galvanic corrosion of the shaft. Sealing of the shaft is by a stuffing box with soft packing at the fore end of the stern tube.

Another type of bearing uses natural or artificial rubber or nylon, but most common nowadays is white metal lined with lubrication by oil, necessitating a seal system at both the fore and aft ends of the tube. Such a special stern tube seal system is the Simplex type.

The Simplex seal system has a stuffing box consisting of a housing secured to the stern post containing sealing rings and a chrome–steel bush attached to the propeller and providing a running surface for the seals. Bellows type lip seals which are kept in place by either spring, oil or water pressure, prevent oil flow outside the vessel and sea water ingress. A number of types of stern tube sealing systems are available using radial or lip type seals.

6.6.3 Propulsion gears

Marine gear units for ships tend to be custom designed for each application while for small ships, work boats and pleasure craft, the gearboxes are normally standard types offered with a range of reduction ratios within a given size carcase and at ratings to suit light through to heavy duty applications.

Small marine gears invariably comprise a clutch or clutches to engage the propeller in one direction or both ahead and astern directions of operation (*Figure 6.35*).

Gearboxes can be single or double reduction, can accept from one to four power inputs and may or may not incorporate reverse operation. They may incorporate epicyclic and/or spur gearing, usually of helical type for most applications, or double helical (herringbone) type (*Figure 6.36*) for large turbine reduction gears, and they may have the output co-axial with the input, or offset in any direction. For small craft it is not uncommon to have gear units with vertical or horizontal offset and a down angled output to align with raked shaft lines and facilitate engine and gear installation.

A V-type gear installation has the gear unit located forward of the engine with an angled gear drive to a propeller shaft angled down below the engine. This space saving installation allows the engine to be located right aft next to the transom giving a larger clear deck area.

A typical plain reverse–reduction gear will have its input shaft coupled to the outer member of a clutch to engage a pinion meshed with the main gearwheel fixed to the output shaft and to a second clutch coupled to a second pinion meshed to an immediate pinion also meshed with the main wheel. Neutral is obtained with both clutches disengaged, and ahead and astern through the appropriate clutch to engage the double or triple gear train between the input and output shafts.

Usually the mode involving fewer gears is designated for ahead operation resulting in engine and propeller turning in opposite directions. If this is not possible then it is usually acceptable for astern to be used so that engine and propeller turn in the same direction. If a co-axial configuration, or extra reduction is required, this is normally obtained from an additional gear train before the clutch input shaft. Designs of gears vary considerably but the principle of separate power paths engaged by clutches is common to most marine gears.

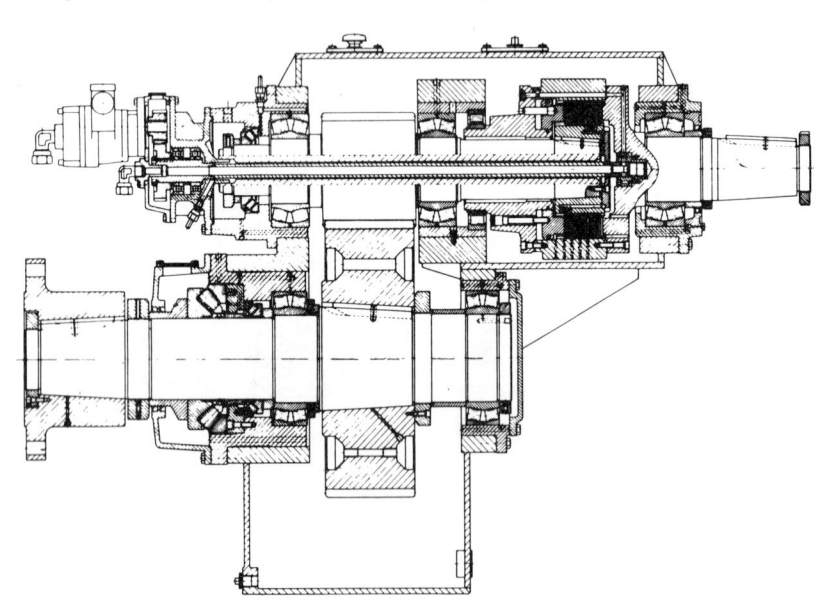

Figure 6.35 Small marine gear with built in clutch

Figure 6.36 Large double helical marine gear

Gear casings may be of suitable cast iron or steel fabrication and the gears are usually cut to a shallow helix with the teeth cut to as large a modulus as possible and are hardened and ground to extremely precise limits to ensure quiet operation. Pinions are often made integral with the input shafts, while wheels are usually keyed or press fitted on an appropriately tapered seat on the output shaft.

Internal clutches are invariably oil operated and of multiplate design. The plates are of dissimilar metals, such as sintered bronze and steel, and to permit prompt disengagement one set of plates is produced in concave form so that releasing the engaging pressure enables them to push away from the flat plates. In the event of hydraulic oil failure there is normally provision to lock the plates together. Hydraulic pumps are normally integrally driven from the input side of the clutches to ensure an oil supply before engagement.

Bearings of large gearboxes are usually plain and the built in thrust bearing is of the classical Michell tilting pad type. Smaller units increasingly use roller bearings because of their lower friction, greater precision, and higher load capacity combined with much experience of operation. Forced lubrication of the gears through internal sprayers is normal, with the system comprising gear driven oil pump, filter, cooler, pressure relief valve to provide for clutch operating pressure, etc.

Very large gears for diesel engines and turbines usually have separate external gear oil pumps driven by electric motors and external oil cooling and treatment equipment with filters and centrifugal separators for oil purification.

Some gears are designed to also include the actuating mechanism for a CP propeller, perhaps using the same oil circuit, in which case the oil control box may be mounted at the forward end of the output shaft or in the hub of the main gearwheel itself. Additionally, the gearbox may be designed with a secondary output with its own clutch to drive a generator, usually through a step up drive from the input to allow operation of the Power Take Off (PTO) with both ahead and astern clutches disengaged. Small gears, say for a fishing boat or yacht, may also have a trolling valve facility to allow the gears to be turned by the propeller when underway with the main engine stopped, without fear of damage to the gears with the oil pump stationary. Slipping clutches are another feature to provide a variable speed output from a constant input, such as when a shaft generator is engaged and it is required to vary the speed of the propeller.

Unless they are of planetary or epicyclic type (*Figure 6.37*), which seem to have limited appeal for propulsion drives, marine gears are usually of parallel shaft type using precision cut spur gears. A non-reversing twin

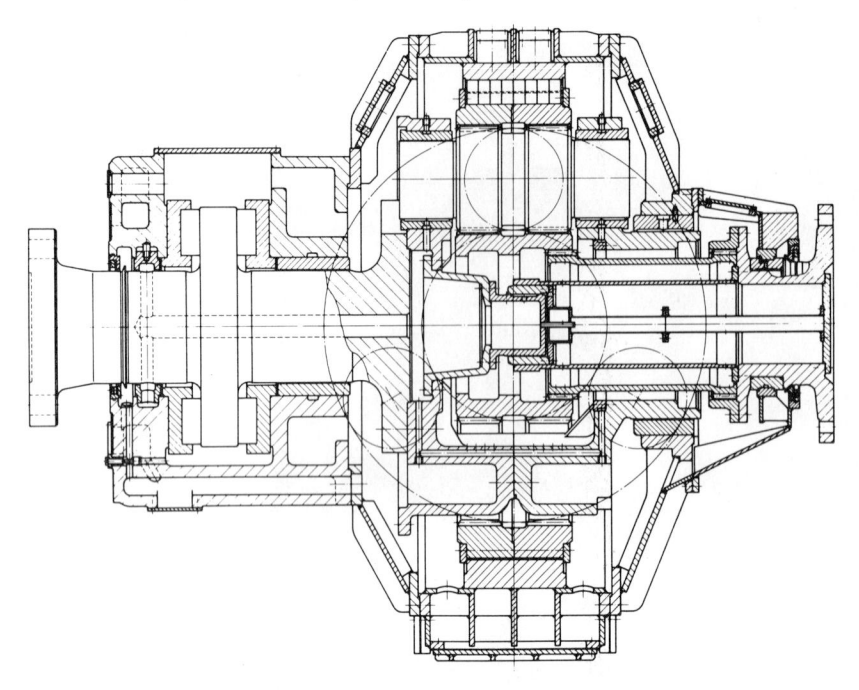

Figure 6.37 Planetary type reduction gear

input–single output plain reduction gear would normally have externally mounted clutches to engage either engine at will and it is becoming increasingly normal practice to fit flexible couplings between the engines and gear to control torsional vibrations. The engine may be flexibly mounted whereas the gears are rigidly bolted to the hull. Naval installations, in which extremely low noise levels are essential, particularly in antisubmarine frigates, invariably have both the prime movers and gears flexibly mounted on a raft structure and with suitable aligning couplings fitted where necessary.

Modern marine gears are now proven to be extremely reliable, with failures rare and by providing a simple method of matching a fixed engine speed to a required propeller speed through selection of the appropriate gear ratio, they make a major contribution to the economical operation of the vessel. Further advantages are obtained from special gear installations that, for example, allow different powered engines to drive a propeller, one engine to drive two propellers and gears for mixed machinery such as CODOG systems for warships, as well as auxiliary machinery PTO drives. Such economy arrangements are mentioned elsewhere in this chapter.

6.6.4 Clutches and couplings

Modern geared propulsion systems require reliable clutches and couplings for economical operation and to enable the plant to be used to best advantage. A clutch is a mechanical device to allow a powered shaft to be disengaged or brought into operation. A coupling is a device to solidly or flexibly join two axial shafts.

In a marine propulsion system, clutches are used to disengage an engine or other prime mover from a gearbox and in multi-engine plants one or more engines can be connected to a single gearbox by individual clutches between each engine and the gear. The clutches may be fitted into the gear itself, in which case a solid coupling would be used to connect the engines to the reduction gear.

Couplings are also used to connect the gear output with the propeller shaft and sections of propeller shafts to each other. Shaft generators driven from an engine or gearbox may be fitted with a disengaging clutch to enable the generator to be stopped, or only a coupling may be fitted, so disconnection is only electrical.

The desire to reduce noise and vibration on ships means that an increasing number of vessels are using flexible couplings between machinery items to eliminate torsional vibrations and to allow radial and axial misalignment or deflection in the case of flexibly mounted machinery. Some clutches are automatic in operation, usually relying on centrifugal force for engagement at a preset input shaft speed.

6.6.5 Couplings

Some of the coupling types typically in use aboard ships are as follows.

6.6.5.1 Flange coupling

This is the simplest of all couplings and consists of a pair of identical flanges fitted to axial shafts and rigidly coupled face to face by equi-spaced bolts. Common uses are joining sections of propeller shafting, coupling motors to pumps, etc.

6.6.5.2 Muff coupling

This is simply a bush, usually split at one side or into two half sections split along the centre plane and held together by bolts, to firmly grip two axial shafts. Uses are propeller shafts of small craft, pump drives, etc.

6.6.5.3 Gear coupling

This consists of a spur gear wheel attached to one shaft and an internal toothed rim in engagement with the gear and connected to the second driven axial shaft. It is used in steam turbine rotor connection to the reduction gear because axial movement due to heat expansion is possible.

6.6.5.4 Hydraulic coupling

This is a sleeve type coupling (*Figure 6.38*) in which hydraulic pressure is used to force two slightly tapered sleeves against one another to provide a strong friction grip between the sections. It is used in propeller shaft lines and rudder drive mechanisms.

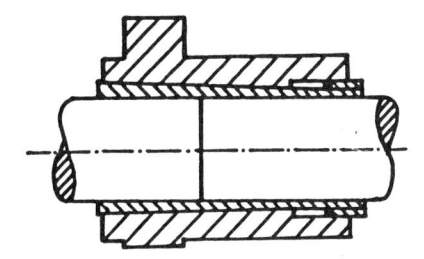

Figure 6.38 Sleeve type shaft coupling

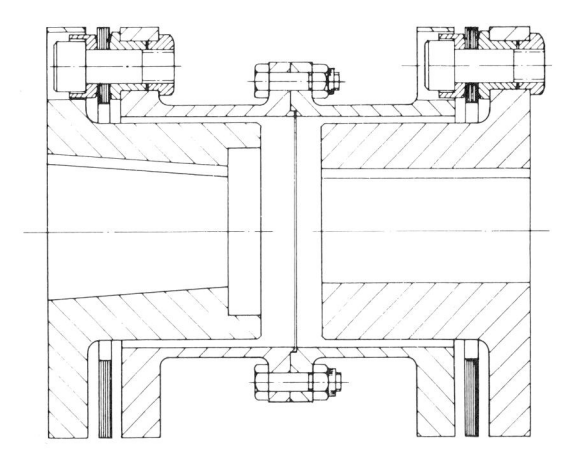

Figure 6.39 Membrane type flexible coupling

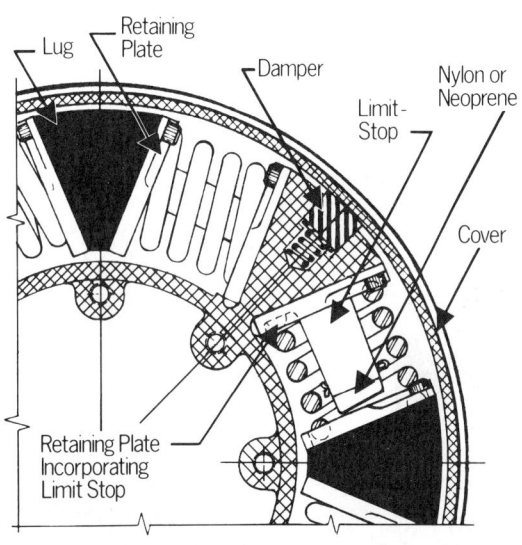

Figure 6.40 Helical spring type coupling

6.6.5.5 Flexible couplings

Relative movements between driving and driven machines, such as when a main engine is mounted on spring type or rubber antivibration mounts, necessitates a flexible coupling to joint the two components. A number of flexible coupling types are in regular use. The simplest is of flange type but instead of the coupling bolts being a neat fit in the flange bolt holes, they are fitted into rubber bushes. This is common for pump drives. Metallic membrane couplings of metal disc or ring type are used in naval transmission systems (*Figure 6.39*). A flexible coupling very often used in diesel engine drives uses a series of helical springs (*Figure 6.40*) arranged around a pitch circle with the ends of each spring connected to an inner and outer ring, respectively. Another type uses radially arranged metal springs, their ends joined to inner and outer half members to provide a flexible drive.

6.6.6 Clutches

A clutch allows a drive to be interrupted. Some of the types used on ships are as follows.

6.6.6.1 Friction clutch

The simplest type of clutch comprises two half sections mounting driving and driven plates of different materials. They are brought into contact by manual or hydraulic pressure and held together to give a friction grip drive. Alternatively, the plates may be held together by spring pressure and forced apart for disengagement. These are very common in small craft reduction gearboxes.

6.6.6.2 Centrifugal clutch

This is an automatic engaging coupling in which flyweights fitted with friction linings are mounted to the input drive member (*Figure 6.41*). As the shaft rotates the pivoted weights move radically outward to engage with an internal rim or drum fitted to the output member. The point of engagement depends upon the shaft input speed causing sufficient centrifugal force to create the friction grip. Another type of centrifugal clutch is the powder coupling in which a fine powder material is used as the friction creating media between the two halves.

Figure 6.41 Elements of a centrifugal clutch

6.6.6.3 Pneumatic clutch

This device has an actuating tube of neoprene rubber located behind friction shoes under a steel rim assembly (*Figure 6.42*). Application of compressed air inflates the tube causing the shoes to engage the rim, locking the inner and outer drive members. It is used in main propulsion drive systems.

Figure 6.42 Pneumatic clutch mounted on reduction gear

6.6.6.4 Fluid coupling

The fluid coupling (*Figure 6.43*) is basically two facing members, one forming an hydraulic impeller and the driven member forming a turbine. Circulation of oil between the two halves results in a positive drive between the two.

There are many benefits with this type: chiefly no wear as parts are never in contact; torsional vibrations are isolated in the coupling; there are no problems from overheating; there is no inertia load on the engine during start up and control is simple and effective. A controlled slip function allows drive and driven shafts to operate at different speeds, depending on whether they are fixed oil fill types which run at nearly the same input and output speeds, or scoop control types where slip can be adjusted by the amount of fluid inside the coupling. Partially filled couplings engage at a certain speed and with scoop filling types engagement can be controlled by the amount of fluid introduced into the coupling. Hydraulic couplings of various types are popular for propulsion systems with heavy fixed pitch propellers that are frequently started and stopped, such as on tugs. They are also used for warship transmission systems and lately are being used in shaft generator drives.

Other types of clutch–coupling, such as the electromagnetic type, cone type friction clutches, etc., are rarely used in ships today.

6.7 Electrical power generation

6.7.1 Auxiliary requirements

Each type of ship's propulsion engine, with the exception of the smallest diesel powered vessels, requires ancillary machinery and equipment to enable it to operate. Referred to as auxiliary machinery, the ancillary plant of all ships has a number of dedicated functions, all of which are necessary for the safe and efficient operation of the vessel.

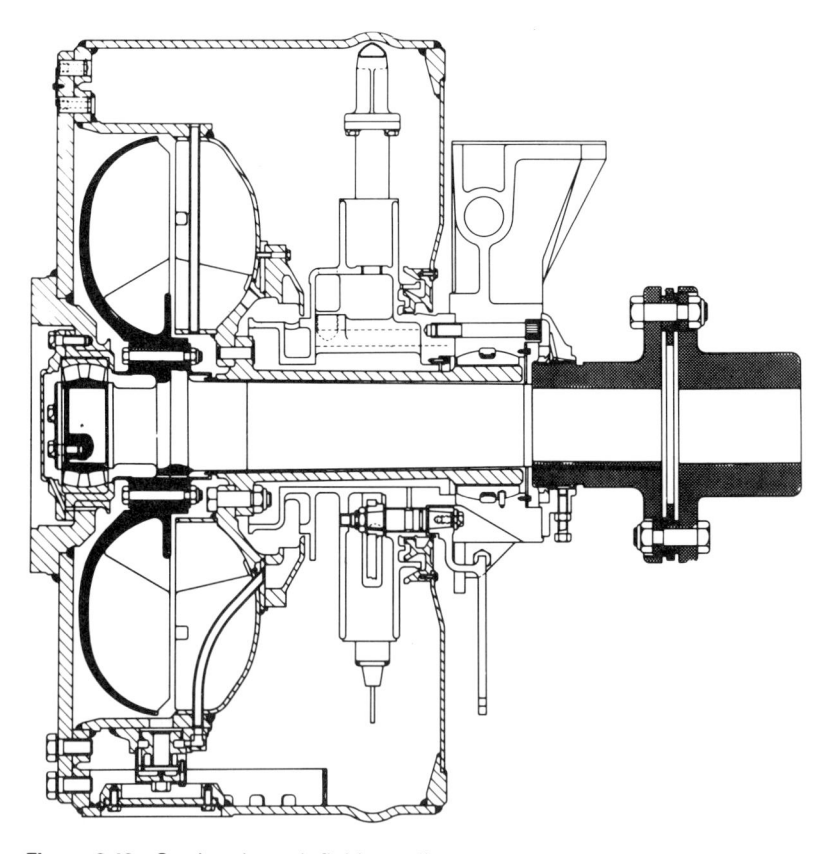

Figure 6.43 Section through fluid coupling

These functions broadly speaking are: generation of electric power, ancillary machinery needed for the main engine to operate, systems to preserve the safety of the ship and its crew, hotel services for the crew and passenger living requirements and equipment and systems for cargo and water ballast operations.

The earliest of steam ships used very little in the way of auxiliary machinery as electric power had not yet been used aboard ships and any pumps or other devices needed for the propulsion plant, or safety, etc., were either driven from the main engine or by independent steam engines. The advent of electricity first brought electric light to replace oil lamps, eventually to be followed by dc electric motors to drive pumps, fans, refrigeration machinery, etc.

A period of mixed steam and electrically driven auxiliary machinery followed, but by the 1950s practically every diesel powered ship was equipped with auxiliary diesel generator sets to provide prime electrical power for lighting and all power supplies to motor driven auxiliaries. Steam usage became confined to heating of fuel, cargo on tankers and accommodation and water heating, with the steam supplied from small auxiliary boilers and/or exhaust gas heated economizers.

The aforementioned type of auxiliary machinery arrangement is more or less the standard for modern vessels, with the exception of current steam turbine ships, but there is a popular and increasing tendency to drive auxiliary generators, at least, from the main engine and to exploit the waste heat in engine exhaust gases to improve the overall plant efficiency.

Diesel engines require forced oil lubrication, fuel treatment, handling and supply equipment, cooling systems, and air supply arrangements. Large engines have these functions supplied from separate ancillary systems whereas most medium sized and small engines are supplied with some auxiliaries built on, particularly oil and water pumps, coolers, etc., so installation of these smaller four stroke engines is simpler.

Ships must be provided with means to remove water from inside the vessel, so bilge pumping systems are important, as are separate ballast water handling systems. Fuel is stored in double bottom and/or deep tanks so means must be provided to enable fuel transfer. Further subsystems provide compressed air for engine starting, operation of control equipment and for ship maintenance tools, while hotel services include potable and sewage water supply, sewage treatment, cabin and water heating and cargo and domestic refrigeration, air conditioning and ventilation. Auxiliary machinery relating to cargo operations will include cargo pumps for tankers and hydraulic or electric supply and motor units for deck cranes, winches, steering gear, etc.

Auxiliary machinery arrangements may vary considerably for different types and sizes of vessels, but general shipboard systems are discussed in the following section.

6.7.2 Auxiliary power

Second in importance to the main propulsion plant in a ship are the auxiliary engines installed to generate electrical power, without which nothing including the main engine can operate. Today, as much as possible is electrically operated as electric machines are easy to install and simple to control and lead to a neater, less cluttered arrangement of the engine room machinery; pump drives are a good example as vertically mounted electric motors take up little space.

Since the early 1950s when ship owners began a large move away from steam propulsion and auxiliary machinery, there has been a steady rise in electrical demand aboard ships. Not only were steam driven generators and pumps replaced with diesel driven ac generators (*Figure 6.44*) and ac electric motors for pumps, compressors fans, etc., but the cargo handling gear and mooring equipment also became increasingly electrified. Whereas a typical sea load of around 30 kW would have sufficed for a steam ship, the electric loading of motor ships quickly shot up to around 500 kW at sea and as much if not more in port when working cargo. Today average sea loads can be much higher, particularly on refrigerated vessels, on passenger ships, of course, where a high demand is caused by the hotel requirements, and on offshore work and supply vessels with high loads for cranes, anchor winches, drilling equipment, thrusters, etc.

Ships vary considerably but a normal cargo ship would be provided with three diesel generator sets using typically in each case a six or eight cylinder turbocharged four stroke engine operating at a constant speed of 720 or 750 r.p.m. or indeed 1000 r.p.m. directly coupled to a three phase ac generator developing around 600 kW at 415 or 440 V. Higher voltages such as 3.3 or 6.6 kV are also in use but these tend to be for ships such as dredgers, offshore craft, etc., where high output electric motors are used for work or cargo operations and advantage is taken of higher voltage control and transmission equipment. For most ships the practice is to generate three phase 50 or 60 Hz power, which is used mainly for electric motors and to transform down the main power supply to single phase 220 or 240 V power supplies for lighting and appliances or other voltages to suit, say, navigation equipment, computers, etc.

The installed electrical power is regulated by the classification societies but the three auxiliary engine ship would have one to carry the normal sea or harbour load, one to act as a standby to the running unit and to be used in addition to the running unit when the vessel is manoeuvring to provide a margin of safety, and the third unit available for maintenance work to be carried out on it while the vessel is on passage. A ship such as a refrigerated cargo liner may need as many as four large generator sets to cope with the huge additional burden of the refrigeration plant once the vessel is loaded. Generator sets are normally operated on a rota basis to ensure that the total running hours for each is kept

Figure 6.44 Diesel generator set for a small ship

roughly similar and to ensure that full maintenance and periodic survey work is carried out on each machine within the survey time cycle.

All generators in a ship are connected to an electric switchboard where they can be connected and disconnected at will either manually or by automatic means. The advent of the unmanned engine room has brought rules to ensure that auxiliary engines are kept on standby at all times so that should there be a power failure, or increase in demand, the automation systems will start up to the standby set and parallel it with the running units with the load subsequently shared equally. Similarly, when the load is reduced, the control system will automatically stop a predesignated running set. Safeguards built into every ship's system also provide for preferential load shedding should the running auxiliary be unable to meet the demand.

The greater sophistication of auxiliary power plant automation and controls has also been brought to prominence by the popular use of main engine driven shaft generator systems. In the event of failure of power supplies from the shaft generator or its inability to meet an increase in demand, standby auxiliary sets must be brought into service immediately.

Unmanned engine rooms have also called for a greater degree of reliability from the auxiliary engines, so it has long been held that auxiliaries should burn clean distillate fuels to enable part load operation without danger of excessive internal fouling and separate auxiliary cooling and fuel systems, etc.

However, in recent years the types of engines available for auxiliary duty have improved considerably and, coupled with the massive oil price rises in the mid-1970s, there has also been a tendency to use heavy or blended fuels in auxiliary engines. The greater acceptance of heavy fuel and increased reliability has also led to the adoption of so-called one fuel ships and greater integration of ancillary systems, such as centralized cooling systems and common fuel treatment plants, etc., which are mentioned elsewhere in this chapter.

6.7.3 Uni-fuel concept

Heavy fuel has been used in ship main engines for more than 30 years but this has not been the case with the smaller auxiliary engines. It has taken a lot longer for such small four stroke medium and high speed diesel engines to be developed for reliable operation on residual fuels. The introduction of heavy fuel to auxiliaries has been in stages, moving from distillate to marine diesel and blended fuels of varying viscosity.

Today, however, there are a number of engines which are marketed to burn residual fuels of the same viscosity as the main engines. Such engines recently introduced have been designed from the outset for the heaviest residuals and their availability and proven service experience has enhanced the adoption of the so-called uni-fuel concept.

Prior to introduction of the latest engines, efforts to reduce fuel costs have led to the use of intermediate fuel or blended fuel for auxiliary engines. This created the complication of three separate fuel systems, needed to handle residual fuel, marine diesel oil and the blended product of these two for the auxiliaries, as well as tankage, separators and the blending equipment, etc.

With the uni-fuel concept (*Figure 6.45*) only two fuels and one fuel system are required. These are heavy residual fuel fed to both the main engines and the

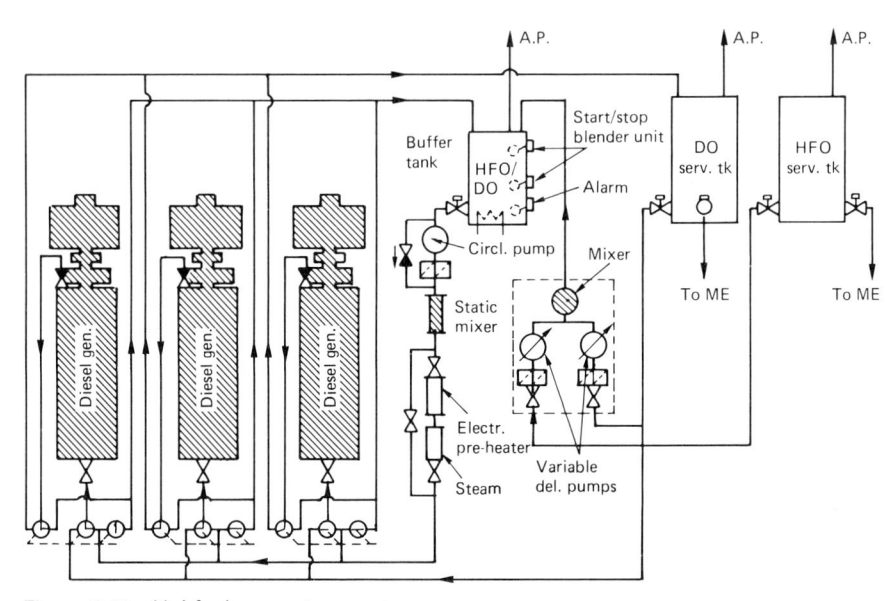

Figure 6.45 Uni-fuel generators system

auxiliaries and marine diesel oil which is needed for flushing through the fuel systems before engine overhaul or dry docking, starting up with a cold ship when fuel preheating is not possible and for emergency operation.

The advantages of one fuel operation are as follows:

1. All engines on board, both main and auxiliary, run on the same bunker fuel oil.
2. Residual fuels in the range 380–700 cSt at 50°C are used only.
3. Direct financial benefits in reduced capital costs and lower fuel costs plus less space required for the much simpler fuel system.
4. Blending equipment is not required.
5. Potential problems with incompatible fuel blends are avoided.
6. The engines are started and stopped on bunker fuel oil and this fuel is used at all auxiliary engine loadings.

In a typical uni-fuel system the heavy fuel for the auxiliaries is taken direct from the pressurized supply circuit of the main engine. A pressurized fuel oil system is popular today because it avoids the problem of any water in the system evaporating at the high supply temperatures required by high viscosity fuels to achieve the low viscosity for injection into the engine. A low pressure supply pump delivers fuel from the service tank to the high pressure circuit. A booster pump then feeds fuel to the engines through the final heater, viscosity meter and indicator filter. A small buffer tank serves to equalize the temperature between the hot surplus fuel returned to the booster pump and fuel coming from the service tanks. Such an arrangement allows hot fuel to be circulated through the auxiliary engines' fuel systems when stationary, enabling the auxiliaries to be started at short notice.

6.7.4 Shaft generators

An old idea which has been revived in the last few years to a considerable extent is to use the power of the main engine to produce electrical energy by means of a shaft generator. Earlier shaft generators were usually a small dc generator driven via V belts from a pulley mounted on the propeller shaft to provide electricity for mainly lighting, whilst the vessel was underway. Voltage control was by simple rheostat and small variations in shaft speed and hence voltage were acceptable.

Today, with ships using ac electricity, the problems of voltage control are more complex and shaft generators are much larger being designed to carry the normal full sea electrical load including auxiliary machinery. Generators are not very often now driven by belts but a number of arrangements (*Figure 6.46*) are in regular use aboard a wide range of ships.

There are several methods of driving a ship's service generator from the main propulsion plant. The simplest is to incorporate a large diameter slow speed alternator

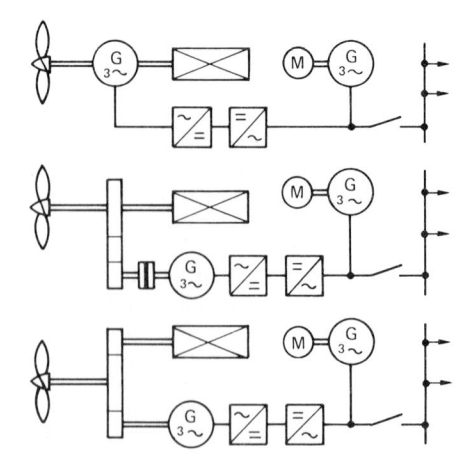

Figure 6.46 Shaft generator drives arrangement

directly into the shaft line with the short shaft carrying the alternator rotor flange coupled to the propeller shafting.

A second method is to directly mount a slow speed alternator to the free end of the main engine, while on the Sulzer two stroke engine, a special gear train is taken off the camshaft timing gears for the drive of an externally mounted smaller and higher speed alternator. The so-called tunnel gear drive (*Figure 6.47*) has a step up gear drive taken off the propeller shaft to drive a high speed alternator located adjacent to the shafting; however, this system is restricted by the space available around the shaft in the narrow confines of the aft end of the engine room. For vessels equipped with medium speed engines, the most convenient system is to incorporate a step up gear drive into the main gearbox to which a shaft alternator is attached by flexible coupling and/or a clutch.

6.7.4.1 Shaft generator advantages

Consideering that the normal method of generating electric power aboard is by high speed auxiliary diesel engines burning distillate or perhaps heavy fuel, all main engine driven generators offer the follow advantages:

1. The specific fuel consumption of the main engine is lower than that of auxiliary engines, so the fuel cost per kW generated is less.
2. The fuel used in the main engine is normally cheaper than fuel for auxiliaries.
3. The auxiliary engines can be stopped at sea leading to reduced maintenance costs from fewer running hours.
4. With auxiliary engines stopped at sea there is much less noise in the engine room.
5. Maintenance work can be carried out to the auxiliary sets at sea.

move the auxiliary engines from the network as demand drops.

6.7.5 Modes of operation

Irrespective of where a shaft generator is positioned, or how it is driven from the propulsion plant, there are definite modes of operation of a generator.

Operation of the generator will depend on whether the engine drives a fixed or controllable pitch propeller. For ac generators it is important to maintain a constant rotor speed for acceptable constant frequency, so if the engine drives a fixed pitch propeller, a frequency converter system is essential and the generator can only be used within a speed band of about 80–105% of the nominal engine speed. A constant frequency can be obtained by governing the engine for constant speed operation but this precludes the ability to operate at reduced speeds, while frequency variations will increase in bad weather. Normally a main engine designed to operate at constant speed would be coupled to a CP propeller to allow independent manoeuvring of the vessel with the shaft generator in use, but the combination of high revolutions and fine propeller pitch, at reduced ship speeds, may be detrimental to the propeller.

For vessels with fixed pitch propellers a solution is to fit a constant speed drive (*Figure 6.48*) (CSD) between the engine PTO and the generator input. This is a gearbox of epicyclic type in which the ratio can be altered to maintain a constant output irrespective of the input speed. This mechanism is a derivative of gearboxes used for tracked vehicles where relative speed outputs of the gearbox is used to steer the vehicle.

For ships the CSD principle (*Figure 6.49*) is the reverse. The gear ratio changes are effected by hydraulic motor–pump units connected to the differential planetary gears. This system is designed to be effective within the 70–100% nominal main engine speed range and experience has shown that the system is sensitive enough for successful parallel operation at sea.

Other methods of shaft generator drive are feasible but so far have had little appeal in practice. One is the so-called variable frequency system in which auxiliary equipment that is not so sensitive to changes in frequency, such as lighting and large motors, can be connected to the shaft generator, but weather conditions at sea would cause unavoidable frequency changes in the absence of electrical frequency controllers. Another alternative is to use a hydraulic motor to drive the alternator with the oil supply from a shaft driven hydraulic pump. This is similar in principle to a rotary converter system comprising a dc motor driving an alterator with the electric power supplied from a shaft driven dc generator.

The above briefly describes the main methods of producing electric power from the main engine and service experience is showing substantial fuel cost sav-

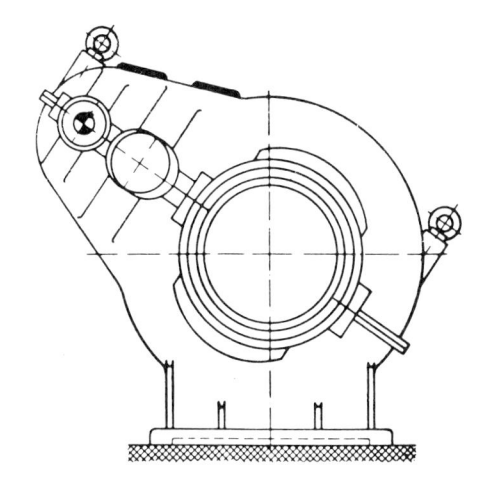

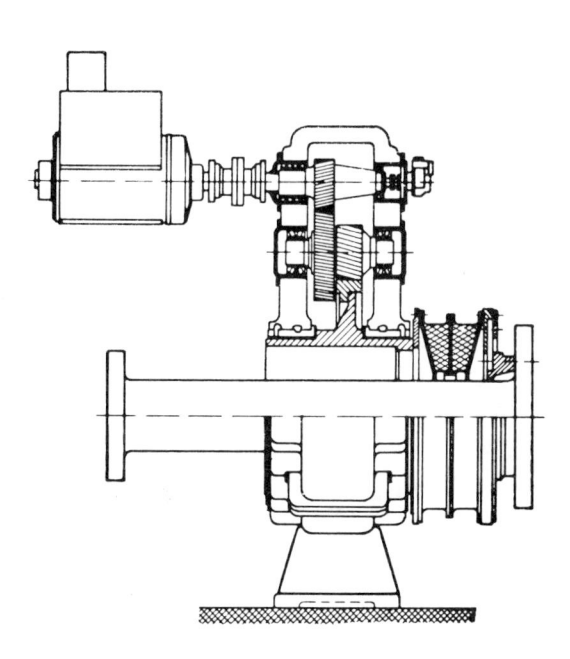

Figure 6.47 Tunnel gear shaft generator drive

6. Potential fouling and increased maintenance of auxiliaries are avoided as they are less likely to be operated at sea under part load conditions.

Shaft generators are designed to provide the normal sea load but in modern vessels the shaft generator is designed for parallel operation with the auxiliary generators; automatic control systems are used to start up a standby generator should the shaft unit be unable to cope with the load demand alone and, similarly, re-

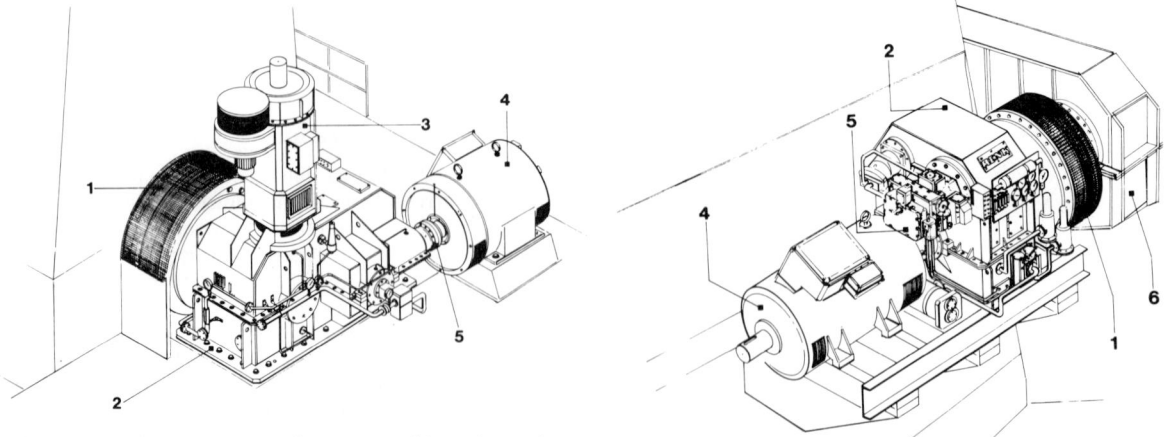

Figure 6.48 Arrangements of generator drives through CSD units: (1) Input coupling; (2) CSD casing; (3) Dc motor; (4) Ac generator; (5) Coupling; (6) Step up gear

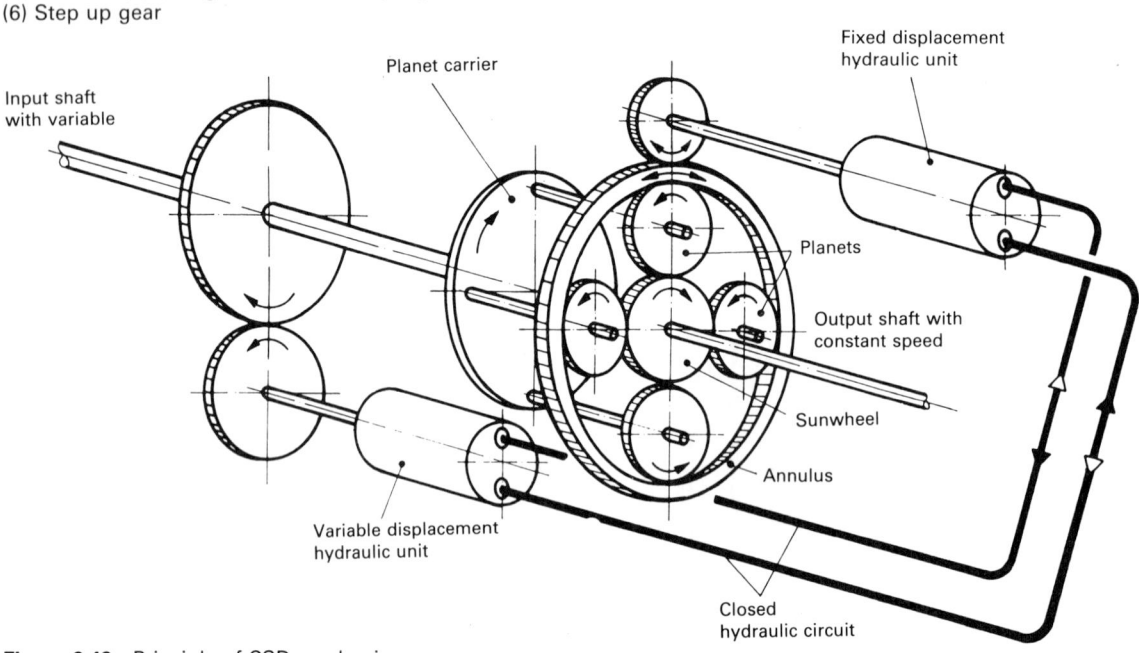

Figure 6.49 Principle of CSD mechanism

ings. The current trend is to drive an alternator from the main gearbox in medium speed engined ships and for direct drive installations to drive the alternator from the engine via a special CSD, a number of which are available.

6.7.6 Power turbines

One of the benefits of recent developments with exhaust gas turbochargers is that their high efficiency means that they can supply more than adequate charge air for the diesel engine combustion system, resulting in an excess of exhaust gas energy available for operation of the engine. This excess exhaust gas is put to good use in so-called power turbines, which operate on diesel engine exhaust gases bled from the exhaust manifold while still leaving a sufficient mass of gas to power the main turbocharger whose compressor delivers charge air to the engine cylinders. The small power turbine produces mechanical power which is fed into the main propulsion system by means of suitable reduction gearing.

This turbo compound system (*Figure 6.50*) has the benefit that while it provides useful additional power which would otherwise be lost in the engine exhaust gases. It also reduces the specific fuel consumption of

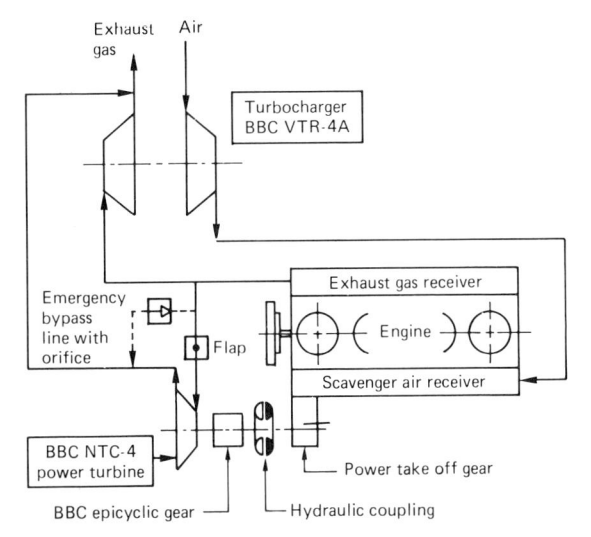

Figure 6.50 Sulzer Efficiency Booster power turbine system

the main engine and improves performance at part load when the gas supply to the turbine can be shut off causing the charge air pressure to rise in the cylinders resulting in cleaner and more efficient combustion of the fuel. At normal full load, with both turbines in operation, the additional power delivered to the engine crankshaft from the power turbine is now proven to reduce the fuel rate by as much as 4%. This improvement contributes to a reduction in fuel cost for a given horsepower delivered to the propeller.

The output of the power turbine can be used in a number of ways. Most popular is connection to the engine through a power take in (PTI) (*Figure 6.51*)

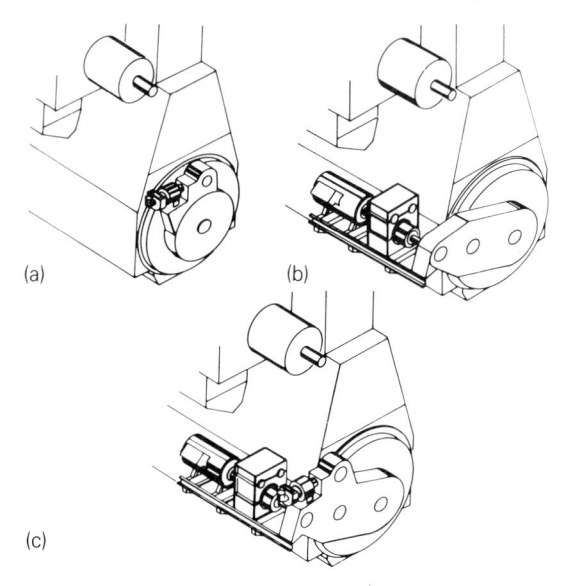

Figure 6.51 Power take in (a) and CSD system (b) combined for complete system (c)

built onto the engine (similar to the power take off (PTO) used to drive a shaft generator). An epicyclic reduction gear is part of the power turbine assembly and between this and the PTI on the engine it is normal to fit a fluid coupling, which may also be of the oil filling type, to enable disconnection of the turbine drive.

Another option is to have the power turbine coupled to an electric generator which feeds power to the electric supply network. However, this is not proving as popular as the system developed by Renk and MAN–B&W where the power turbine is combined with a shaft generator such that the turbine contributes useful power to both propulsion and electricity generation. Should the shaft generator be disconnected from the network, the full output of the power turbine is consequently delivered to the engine crankshaft.

6.7.7 CODAG generators

A further application of a power turbine to improve overall performance is its use in a CODAG GenSet (*Figure 6.52*). With this arrangement the power turbine is not geared to the main engine crankshaft but is coupled to one of the ship's auxiliary generator sets in a system patented by MAN–B&W. This plant also uses an integrated charge air system (ICS) which, by taking some of the charge air from the scavenge manifold of the main engine and feeding it the turbocharger of the auxiliary, allows the latter to operate continuously on heavy fuel at all loads from full power down to and including idling.

By coupling the power turbine to the free end of the auxiliary engine through a suitable epicyclic reduction gear and taking the auxiliary combustion air supply from the main engine, the generator set will normally run at near idle conditions and act as a speed governor controlling the frequency of the electrical network.

At the same time, this auxiliary engine acts as a backup machine ready to supply additional power if the available amount of exhaust gas from the main engine becomes too small to yield the necessary driving force for the generator load. Should the power turbine be cut out, the running auxiliary engine becomes an instant standby machine to take over the full electrical load.

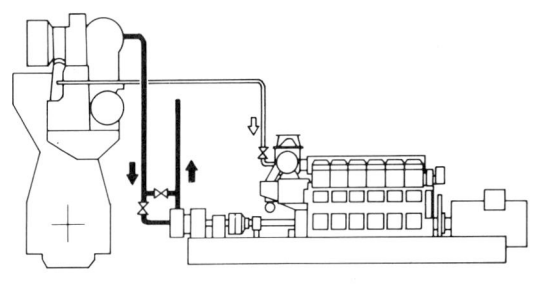

Figure 6.52 Arrangement of CODAG generator system: the power turbine of a main engine turbo compound system coupled to ICS diesel-generator

This system is considered to be ideal for ships where the electrical power demand at sea is normally higher than the free energy available from a turbocompound system, such as on a typical container ship.

For such a system to be economically feasible the following must apply:

1. The maximum output of the power turbine must be no more than 90% of the base load of the electrical network, leaving 10% for the auxiliary engine so that it can control the speed and frequency of the set as a whole.
2. The maximum output of the generator set should be at least 10% higher than the baseload in order to cope with peak loads.
3. The maximum output of the diesel generator should be at least 20% higher than the maximum output of the power turbine.
4. The auxiliary engine must operate under all load conditions on the same heavy fuel as the main engine.
5. The CODAG GenSet should be of the same size and power as other auxiliaries fitted and be capable of parallel operation.

The major benefit claimed for this system is a considerable saving in the cost of electrical power produced, together with a simpler and cheaper utilization of the free excess energy available in the main engine exhaust gases. Other benefits are compact installation of the power turbine, lower installation costs, and increased flexibility and security of supply with no risk of blackouts.

6.8 Steam generation

6.8.1 Current marine boilers

The efficiency drawbacks of using steam as a power source have meant that in order to compete with other forms of energy, the design of marine boilers for main propulsion has had to be improved continually through better materials, improved heat transfer and higher steam pressures and temperatures. This evolution has led to the superheated and reheated water tube boiler which is fitted to many of the steam ships still at sea.

The design of modern water tube boilers is essentially simple with an upper steam drum which acts as a reservoir connected to one or more lower water drums by a series of tubes passing through a large furnace fired by oil, coal or gas. A water level is maintained in the upper steam drum to ensure that at all times the tubes are constantly filled with water and the direction of flow in these tubes or circulation of the boiling water is natural, as a result of the difference in specific weights of steam and water. Steam under pressure is drawn from the upper steam drum, superheated in special tubes passing through the furnace and supplied to the propulsion and auxiliary machinery as a power source. There are many types of water tube boiler still used in ships but as renewed interest in steam appears unlikely, oil fired boilers are not described here.

6.8.2 Modern coal fired boilers

In the early 1980s at a time of very high oil fuel prices there was a revival of interest in again using coal to fire marine boilers. Orders were placed for a number of new bulk carriers to be equipped with stoker fired boilers and steam turbines and in Spain a tanker was converted from oil to coal as were some gas carriers in the US. These ships are still trading but the expected upsurge in interest at that time failed to materialize. It now appears unlikely that there will be any further interest in coal in the short term.

Coal firing of ships discontinued after World War II but continued ashore, where there were many developments in coal handling and combustion systems, which were put to good use in the reborn marine coal fired plants (*Figure 6.53*) of the early 1980s. As coal burns more slowly than oil a greater volume of combustion chamber is needed for a coal fired boiler and there are the added problems of coal storage on the ship, handling of the coal from the bunkers to the furnace, the combustion system and disposal of the ash left after combustion. The relatively new features of marine installations are discussed below.

6.8.3 Coal fired boiler design

A type of boiler used in the majority of recently built ships is the Combustion Engineering V2M9-S which has a dropped furnace supported at the mid point by the side wall header and the water drum saddles. The furnace comprises welded wall panels in the roof, front, rear and side walls and this construction continues on the side walls to a point where the generating bank starts. The superheater consists of a primary and a secondary stage, each of the vertical in line multipass fully drainable type. The spacing of the water tubes is much wider than found on oil fired units, to prevent ash build up, and the superheater floor is sloped to direct cinders to a hopper for re-injection into the furnace. A balanced draught system is applied using a forced draught fan and an induced draught fan, which results in a slightly negative pressure in the furnace and allows casing requirements to be minimized and prevents gas and soot leakage from the boiler. A small overfire fan is also fitted to inject air above the fuel bed through overfire air nozzles in the front and rear walls of the furnace, to provide additional turbulence in the furnace, giving increased time for combustion, more complete burning of carbon particles and reduced ash carry over.

6.8.4 Stoker firing

The latest boilers are fitted with mechanical stokers (*Figure 6.54*) to burn solid coal. The continuous dis-

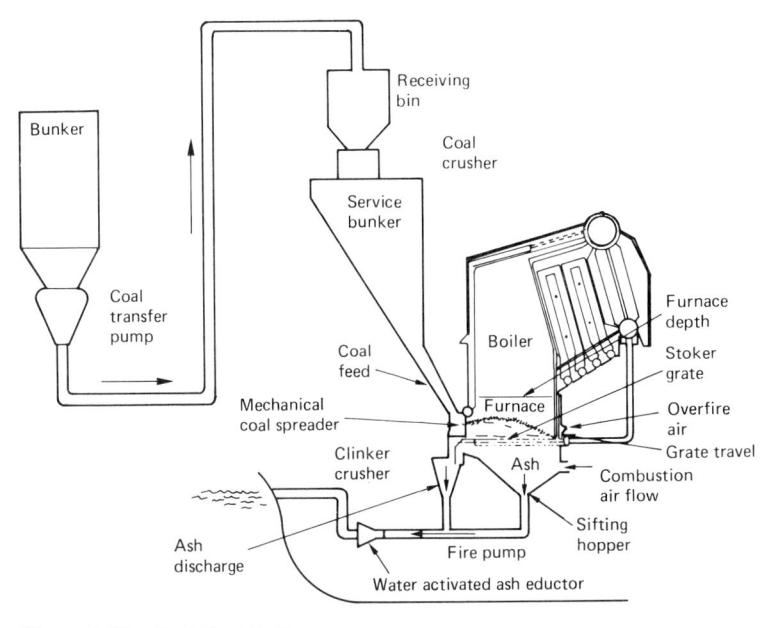

Figure 6.53 Coal fired boiler system

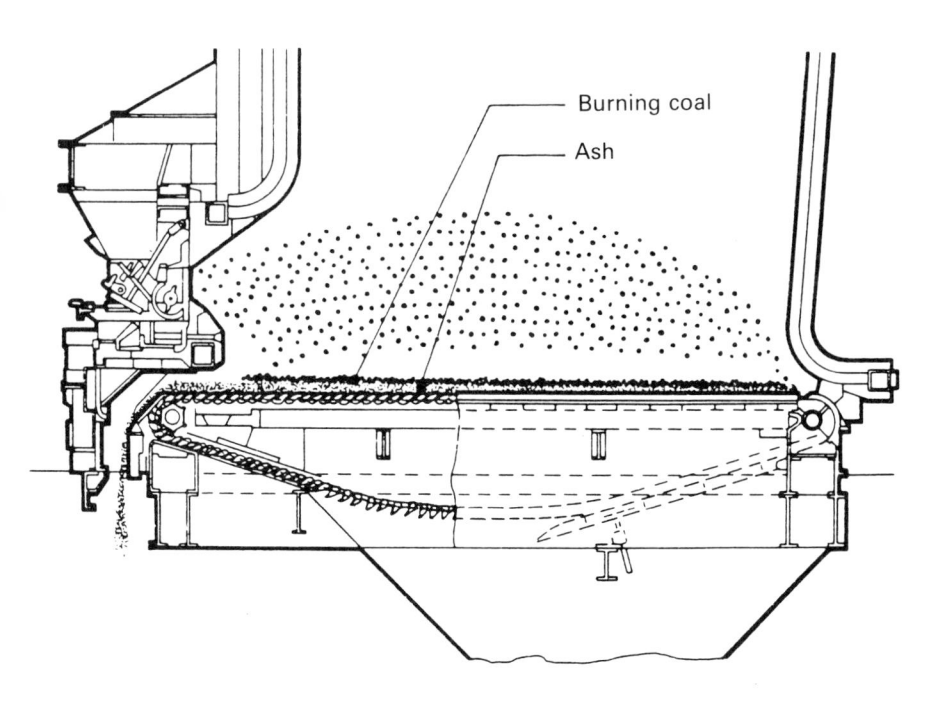

Figure 6.54 Mechanical stoker for coal fired boilers

charge spreader stoker has a travelling grate portion comprised of carrier bars and keys, driven by geared electric motor arrangement, with the speed of the grate adjusted to maintain a constant 3–4 in ash thickness at the discharge end where the ash falls from the grate to a collection tank. The normal direction of operation is towards the front of the boiler with the spreader system physically throwing the coal towards the back, to be consumed as it travels forward on the grate. *Figure 6.54* shows a typical stoker arrangement.

6.8.5 Coal handling

Coal must be stored in suitable bunker holds and a usable ship plant depends on the automatic mechanical transfer of the coal to the boiler and disposal of the ashes. The possible handling methods are bucket elevators, dry or wet belt or chain type conveyors or pneumatic systems. The latter have proven most effective in the current generation of coal fired ships.

The system recently applied has been adapted from shore boiler plants. Known as the Denseveyor system, it operates on the dense phase principle in which coal is fed into pipes as a slug of solid material pushed by a controlled supply of compressed air. A Denseveyor hopper is automatically filled with a fixed amount of coal and once the coal inlet valve to the unit is shut off, compressed air is introduced to this chamber to force the coal to the daily service hopper in the boiler room. A number of Denseveyors are fitted to the bottom of the bunker hoppers and are filled by gravity feed.

After combustion the ashes collect in a tank from where they are transferred to a storage tank or pumped overboard using a high pressure water jet or by vacuum creating devices to suck ashes from the collecting system.

6.8.6 Auxiliary steam

Whereas large quantities of steam are not required on modern ships for main propulsion there is, nevertheless, still a demand on motor ships for low pressure steam for heating purposes, particularly on oil tankers where cargo may have to be heated for discharge and, with the widespread use of heavy fuel, steam is used for preheating the oil before combustion. Steam is also used for accommodation heating, in the galley, providing hot water, etc.

On modern motor ships steam is generated thus: in oil fired auxiliary boilers usually of vertical water tube type, in composite boilers which can be fired by oil or use the waste heat in the main engine exhaust gases as the heating medium and in waste heat economizers relying only on exhaust gases for heating.

Which system is fitted will depend upon the likely steam demand as a tanker, for example, needs much larger separately fired auxiliary boilers than a small cargo ship which can probably do away with auxiliary boilers altogether, using the waste heat in the cooling water system of diesel engines as well as exhaust gases and perhaps relying on, say, electric heating in port when no steam is available.

6.8.7 Waste heat recovery

To effectively use some of the heat loss in the exhaust gases of the main engine, and perhaps even the auxiliaries, waste heat recovery systems are now in widespread use. These vary in complexity ranging from low pressure small capacity economizers fitted in the exhaust uptakes to raise steam for heating when the ship is at sea only, to plants capable of providing steam for heating and a turbogenerator to supply the ship's electrical power at sea (*Figure 6.55*). The latter presupposes a main engine of sufficient output to supply enough heat in the exhaust. A number of systems are commonly in use as described below.

6.8.7.1 Composite boiler

This system is suitable for a small motorship and uses a composite boiler which may be oil fired together with, or separately from, the main engine exhaust gas flow. The boiler has two heating sections which can be separately adjusted for steam output and they are usually designed for automatic operation with the burner cutting in when the output from the exhaust system is insufficient. Such boilers are popular because of their simple design and relatively low cost.

6.8.7.2 Separate systems

A waste heat arrangement suitable for larger ships uses two separate steam raising components, one an oil fired boiler and the other an exhaust gas boiler or an economizer. While normally at sea the economizer only would be in operation and the oil fired unit shut down, the steam space of the latter is used as a pressure vessel for the economizer. Both units are thus held at the same pressure and are able to produce steam when required. An alternative arrangement has the two boiler units operating as independent units each with their own feed water pumps and both capable of supplying a common steam range.

6.8.7.3 Double evaporation units

A steam system which is popular on large diesel engined tankers uses large water tube boilers of double evaporation type in conjunction with an exhaust gas economizer (*Figure 6.56*). This system is expensive and only really justified for ships with very powerful main engines and must be capable of supplying sufficient steam for turbogenerators, cargo heating, fuel heating and domestic services. The double evaporation boiler comprises an oil fired unit producing steam in separate drums at two pressures.

Figure 6.55 Packages turbogenerator used in waste heat recovery plants

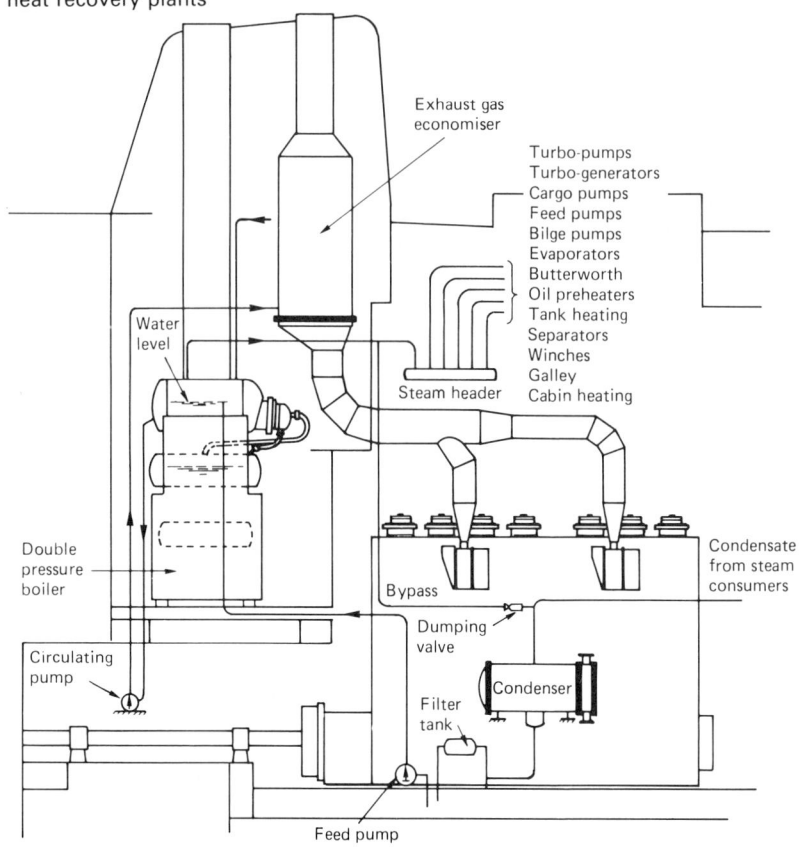

Figure 6.56 Typical auxiliary steam system with double evaporation boiler and exhaust gas economizers for a motorship

A quite common arrangement to supply large quantities of steam for cargo tank heating is a steam to steam generator section in which high pressure steam is used to generate low pressure steam for heating. In this way possible contamination of the main boiler feed water system in the event of cargo oil leaks into the heating coil system is avoided. In such a plant the exhaust gas waste heat unit operates as an economizer supplying a steam–water mixture to the secondary steam drum (lower pressure) of the double evaporation boiler, which acts as a steam reservoir when the primary system is shut down but still provides a supply of steam to the cargo tanks.

6.9 Ancillary machinery

6.9.1 Marine pumps

The pump is one of the earliest of mechanical devices to be used aboard ships, as a means of removing water from the hull of a vessel and thus forming a major item of safety equipment. Marine pumps, however, are used for many more tasks than pumping water overboard from ships' bilges and a wide range of fluids are handled. Pumps are used to circulate sea water for cooling, fresh water for cooling and drinking, lubricating oils, fuel oils, refrigerating media such as brine, many types of liquid cargoes from heavy bitumen to white spirit, chemical, liquified petroleum gases and many other substances.

Different fluids also call for different types of pumps such that a modern ship will have installed many pumps of centrifugal type, others of positive displacement, screw type, vane type, diaphragm type, gear type, venturi type, and so on. The uses of the various types are briefly discussed below.

A pump is a device that uses an external power source to apply a force to a fluid in order to move the fluid from one place to another. Pumps develop no energy of their own but simply use the work or force provided by human hand, electric motor, turbine, steam pressure, diesel engine, etc., to move the fluid by overcoming the static pressure differential between pump suction and delivery and overcoming the friction forces encountered by the fluid in the system. The greater the applied force the higher the pump pressure and/or volume of fluid moved.

Although designs and types of pumps are legion, there are basically three types: positive displacement; non-positive displacement and the jet pump.

6.9.2 Positive displacement pumps

In the positive displacement pump a fixed volume of the fluid is displaced or pumped during each pumping cycle. It works by drawing the fluid into a cavity and by reducing the cavity volume after opening up a flow patch (discharge) into the system. The most common mechanism is a piston that reciprocates in a cylinder with inlet or suction valves and discharge valves fitted to the pump to ensure one way flow of the fluid.

The most common uses for reciprocating piston positive displacement pumps are in emergency boiler feed, fire pumps, bilge pumps and cargo stripping pumps on oil tankers. Because the piston operates in a closed cylinder, a vacuum or suction is created so the pumps are self priming and are thus considered to be more reliable. The vertical pump was much more common in the days of steam as the direct coupling of a steam piston rod to the pump piston rod comprised a simple and effective mechanical solution. Pumps may be single or double acting with suction and delivery valves fitted at both ends of the pump cylinder, giving a constant pressure delivery.

6.9.3 Gear pump

A positive displacement type used for fuel and lubricating oil systems is the gear pump, which uses spur gears that are meshed together with one gear driven by an electric motor. As the gears rotate (in opposite directions) the fluid is trapped in the spaces between the gear teeth and the pump casing and is carried to the discharge side of the pump where a pressure build up forces the oil into the discharge line. A modified form of the simple gear pump is the herringbone gear type which gives a more even pressure discharge while another type uses a pair of helical gears and is used for low viscosity fluids at high speeds and heavy fluids at low speeds.

6.9.4 Screw pump

A rotary screw pump (*Figure 6.57*) normally uses a pair of intermeshing screws which are mounted on parallel shafts connected by timing gears to keep the screws in the correct relation to one another to transfer the fluid. The timing gears maintain close clearances between the screws as they rotate. The most common type has opposite hand threads so that liquid introduced at the ends of the shafts is carried to a centre discharge space, resulting in two streams discharged at the centre. The pitch of the screws has a significant effect on the handling characteristics of these pumps, which are very common in engine lubricating oil systems.

6.9.5 Non-positive displacement pumps

A non-positive displacement pump does not discharge a fixed volume fluid in a given period of time because the principle of operation is by imparting a velocity to the fluid and converting this velocity into pressure causing the fluid to move by virtue of the pressure difference in the system.

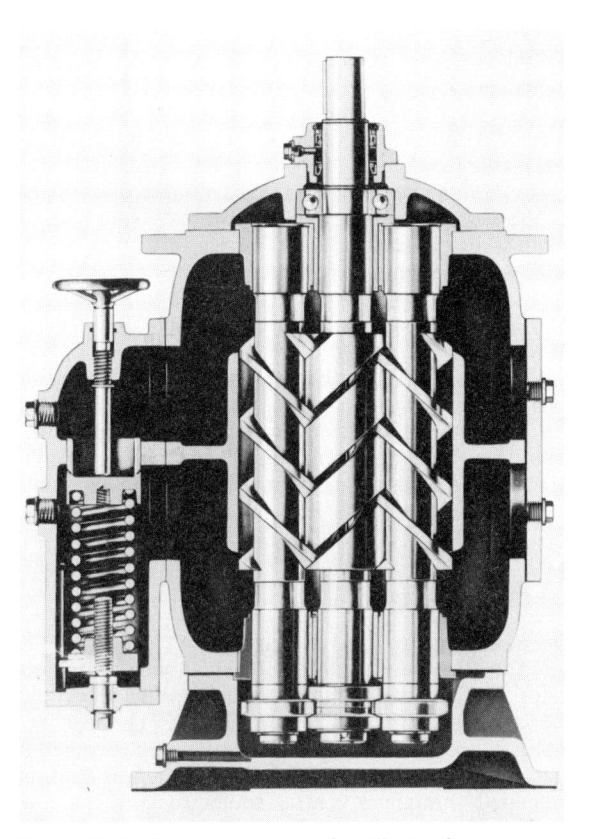

Figure 6.57 Screw type pump for oil transfer

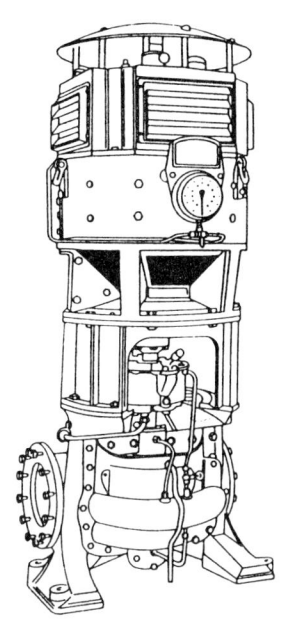

Figure 6.58 Sketch of marine centrifugal pump

6.9.6 Centrifugal pump

The most common type of non-positive displacement pump is the centrifugal pump (*Figure 6.58*). This is basically an impeller (paddle wheel) operating inside an eccentric casing (volute) with centrifugal force generated by the high speed rotation of the impeller. This latter consists of a disc on which is radially mounted a series of curved vanes with provision for entry of the liquid at the centre of the impeller. Fluid enters the impeller at the centre and the high speed rotation throws the liquid radially outwards, imparting to it a high velocity. After leaving the vanes of the impeller the high velocity liquid enters the stationary pump casing which encloses the impeller, the clearance between impeller and casing increasing progressively towards the outlet. The purpose of this volute casing is to convert the kinetic energy of the mass of fluid into pressure, by allowing the fluid mass to slow down as it passes through the widening channel, creating build up of pressure.

Centrifugal pumps may be single or multistage, the latter having two or more impellers housed together in the one casing. It is normal to have impellers operating in series, that is discharging into the suction of the next impeller or stage. Centrifugal pumps are widely used aboard ships, mainly for water pumping and as such are installed as sea and fresh water cooling pumps; ballast pumps, cargo pumps and horizontal multistage types are used as boiler water feed pumps.

6.9.7 Propeller pump

Another type is the propeller pump which uses the action of a screw propeller in a casing to push the fluid in an axial direction, leading the pump to be called an axial flow type. It has very little suction or lifting power and is therefore only used for low head pressure applications, typically as a main sea water circulating pump in steam vessels.

6.9.8 Jet pump

The jet type pump has no moving parts but uses a high pressure steam or water jet which is allowed to expand into a region of lower pressure, thus increasing in velocity. When this jet impinges upon the fluid to be pumped at the low pressure end of the nozzle, kinetic energy is transferred from the jet to the fluid causing motion to the fluid in the direction of flow of the jet. The mixture passes to a diverging chamber (diffuser) where the decreasing volume causes the mixture to slow down and pressure to increase near the outlet.

Two types of jet pumps are eductors and ejectors. Eductors are used for pumping water from bilges and ballast tanks using a water jet from, say, a fire pump as the driving force. Ejectors are used to pump gases, the most common use being air ejectors fitted to initiate and sustain vacuum in condensers, with steam used to activate the fluid flow.

All three latter categories of pumps are used in shipboard systems but by far the most prevalent are centrifugal pumps, particularly for fresh and sea water systems. Some of these systems are described further below.

6.9.9 Cooling systems

A very important engine room ancillary function is that of cooling of main and auxiliary engines, steam in condensers, lubricating oil, and minor items such as shaft tunnel bearings, steam drains, etc. The traditional cooling media on ships has always been sea water but as this causes corrosion and galvanic action when in contact with certain metals in a cooling system, there is a tendency these days to limit the use of sea water as much as possible. Many small craft still use sea water directly for engine cooling and all that is required is a suction filter and an engine driven pump to move the raw water through the engine jackets and thence overboard. Ships, though, generally use fresh water as the cooling medium, this water being cooled in a heat exchanger of shell and tube type or plate type, with sea water as the low temperature media component of the system.

With the tendency to use fresh water for as many cooling functions as possible in order to contain corrosion, the one or more sea water circulating systems will use centrifugal type main pumps and heat exchangers to extract heat from either combined or independent engine fresh water cooling circulation systems. Temperatures in the fresh water system are regulated by controlling the speed of sea water pumps, throttling suction valves or through bypass valves at the heat exchangers.

These days such temperature control functions are carried out automatically. Instead of heat exchangers, a keel cooling system can be used in which fresh water is passed through heat exchange grids mounted externally on the vessel's hull, but this is normally confined to smaller vessels such as tugs, ferries, etc., and small leisure craft. Such a system prevents the ingress of sea water into the vessel's piping but temperature control is poor, so keel cooling is hardly suitable for vessels trading from the Arctic to the Tropics.

Whereas a considerable number of ships use the conventional sea water–fresh water heat exchangers system, the popular trend for ocean going ships is the centralized cooling system which can also use microelectronic controls to promote efficiency and economy of operation. Such a system is described briefly below.

6.9.10 Centralized cooling system

The modern centralized cooling system (*Figure 6.59*) is based on the simple concept of employing two fresh water circuits, one high temperature and the other low and a very simple sea water circuit. The heat exchangers used are of plate type employing corrugated titanium plates to carry the cooling water. Sea water is circulated through a central cooler to cool the low temperature fresh water circuit which is used for the cooling media feed of, for example, main engine charge

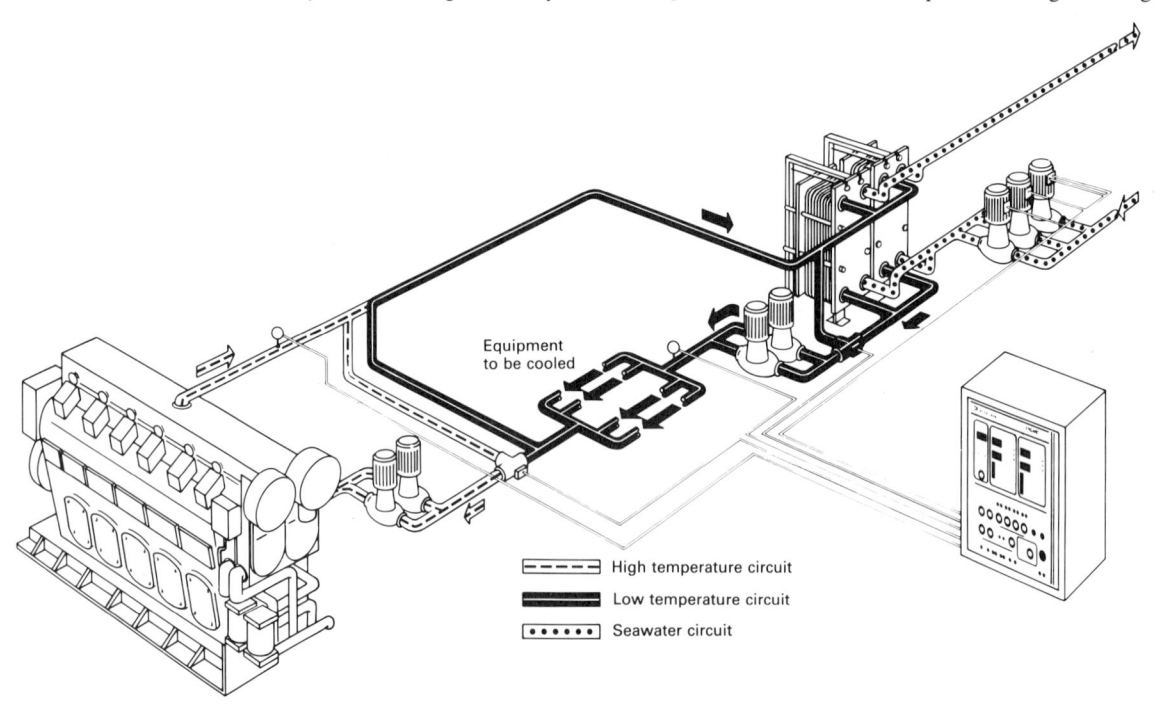

Figure 6.59 Centralized cooling system

air coolers, lubricating oil coolers, auxiliary engine coolers, etc., while the high temperature fresh water circuit is used exclusively for the main engines.

Temperatures in the two systems are controlled by regulating valves to direct the hotter main engine jacket water through its own cooler connected to the low temperature fresh water circuit with regulation through bypass valves. An alternative arrangement is to allow the two fresh water systems to mix together in controlled proportions with temperatures controlled by valves, and the number and speed of the main sea water circulating pumps in operation.

Microprocessors control all variables in the system to bring the following advantages: minimized energy consumption, automatic start and stop of sea water pumps, exact control of temperatures in both high and low temperature fresh water circuits and minimized sea water piping within the ship.

The fully automatic centralized cooling system may be considered too complex for many small ships, but for ocean going vessels it gives the best overall economy and fits in well with the latest trend to operate heavy fuel burning diesel engines at elevated cooling water temperatures.

6.9.11 Lubrication systems

The function of a lubricating oil system (*Figure 6.60*) is to circulate oil to the bearings of a diesel engine, turbine, gearbox, gear-wheels, or metal parts in contact and to carry away the heat generated by friction. With smaller engines and gearboxes, etc., it is common to have the lubrication system as an integral part of the machine with gear type oil pumps driven from a rotating part such as the crankshaft. Larger ships' main engines, such as direct drive two stroke engines, however, are not equipped with integral oil pumps and thus lubrication of the main engine (or bearings and reduction gear of a steam turbine plant) is through a separate anciliary system.

A typical main engine lubricating oil system will have two or more independent electrically driven lubricating oil pumps which draw oil from a lub oil drain tank, usually located beneath the main engine. A coarse mesh suction filter is fitted between the pumps and tank. The pump discharges into a fine mesh filter and to an oil cooler and thence to the oil inlet manifold on the engine for distribution to individual engine bearings, oil sprayers, etc.

Some engines use oil for cooling the piston crowns and here separate lubricating oil and cooling oil branches are made at the engine inlet, with suitable control valves to regulate flow in the two systems. After passing through the engine bearings the oil drops to the engine oil pan and through a sieve back to the lub oil drain tank.

Control of oil temperature is effected at the sea or fresh water cooled lub oil heat exchanger, of shell and

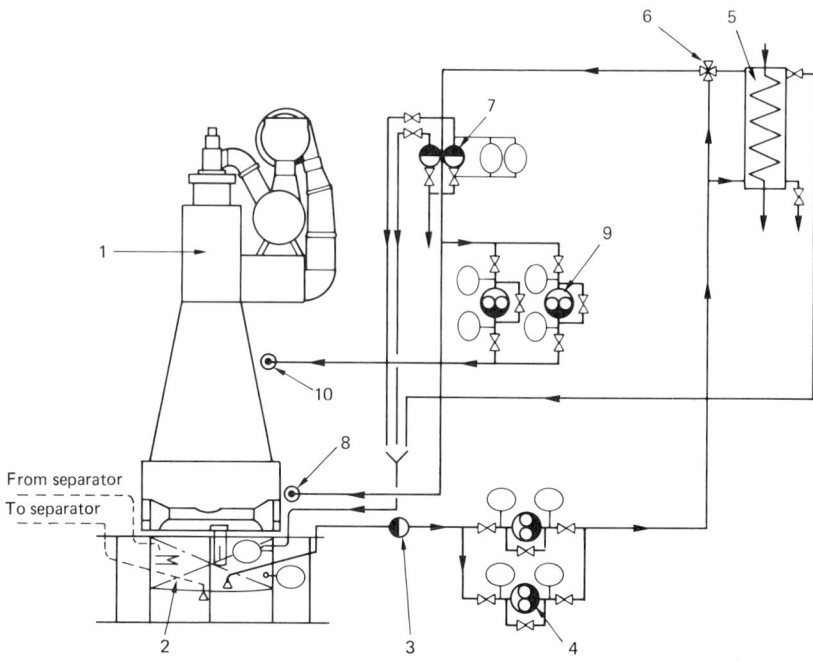

Figure 6.60 Engine lubrication system: (1) Main engine; (2) Oil drain tank; (3) Suction oil filter; (4) Oil pump; (5) Oil cooler; (6) Temperature control valve; (7) Oil filter; (8) Bearing oil inlet; (9) Crosshead oil pump; (10) Crosshead oil inlet

tube or plate type, using a thermostatically controlled bypass valve. The oil filters are usually of duplex type which means that they can be switched over for cleaning during operation. Pressure control of the system is usually by means of a spring loaded bypass valve on the pump discharge, with connection back to the suction side of the pump. It is normal practice for the oil system to be provided with at least one centrifugal separator to remove impurities from the oil on a regular basis. It is of great importance to remove metal particles that have found their way into the oil system as a result of wear in components and to remove sludge that may form in the oil, from, for example, microbiological degradation and any water that may contaminate the system. The oil must also be regularly analysed by a laboratory to assess its lubricating ability and Total Base Number (TBN) levels, water content, etc., to prove the oil fit for further use.

Lubricating oil pumps are as a rule of gear or screw type which have the advantage of being self priming and in most cases are electrically driven by ac motors. As these run at constant speed, pressure is regulated by the system control and bypass valves. On four stroke engines with their own pumps, two pumps are normally fitted, one drawing from a storage tank to the bearings, etc., and a second scavenger pump is used to remove oil from the engine sump back to the storage tank. The requirement is for more than one main engine oil pump to be fitted and with independent systems, as indeed engine driven pumps, the standby pump is brought into operation on main pump failure by the automatic control system normally provided.

6.9.12 Fuel treatment

The subject of fuel quality, chemistry and treatment is a very complex one beyond the scope of this book but it suffices to say that with the wide variations in quality of fuels supplied to ships, some form of fuel treatment is necessary. This is to remove impurities from the fuel which may cause damage to the engines or clog up pipelines, filters, etc., which can lead to an interruption of the fuel supply and power failure with potentially disastrous effects.

Liquid fuels supplied to ships are generally categorized by their viscosity, which is a measure of thickness, specific gravity or density in relation to water, and ignition quality usually referred to as Cetane number, while the pour point indicates the degree of pumpability of the fuel and asphaltenes indicate solid contaminants.

The normal fuels supplied to ships are:

1. Gas oil or distillate which is similar to that used for diesel powered road vehicles and is a refined product derived from crude oil.
2. Marine diesel which is also a distillate fuel but is a little heavier than pure gas oil.
3. Intermediate fuel which is a mixture of distillate and residual or heavy fuel, in proportions to suit the customer.
4. Heavy fuel which is practically all residual fuel but is lightened with distillate to make it acceptable for many medium and slow speed diesel engines. Typical grades are 1500 and 3500 Seconds Redwood 1 viscosity.
5. Residual fuel is in theory the absolute remains of the refining process and has a viscosity of around 7000 Seconds Redwood 1. In practice, very little straight residual is supplied to ships but it is used as the base stock to produce lighter blends by mixing with distillate.

6.9.13 Fuel problems

Few problems are nowadays expected with distillate fuels but all vessels, no matter how small, require a storage tank to enable water in the fuel to be drawn off and a filtration system to remove sludge or metal particles, sand, etc., before delivery to the engine. Ships burning heavier fuels need additional centrifugal separators to remove water and solids and for heavy fuels heating and viscosity control systems are also required. Specific problems encountered with fuels, in particular heavy fuels, are as follows.

6.9.13.1 Storage problems

The storage problems is a build up of sludge in the bunker leading to problems in handling. The sludge which settles in the bunker tanks, or finds its way to the fuel lines, tends to overload the fuel separators with the loss of usable fuel and can cause problems with the fuel injectors and wear of the engines through abrasive particles. Sludge is commonly caused by instability of fuels of different grades or bunkered at different locations. A detergent type chemical additive can be used to reduce sludge formation in the tanks and careful operation of the fuel treatment plant is necessary.

6.9.13.2 Water in the fuel

Water enters fuel during transport and storage and free water can seriously damage fuel injection equipment, cause poor combustion and lead to excessive cylinder liner wear. Sea water contains sodium which will contribute to corrosion of the engine when combined with vanadium and sulphur during combustion. Water is removed by draining of the settling tanks and operation of centrifugal separators.

6.9.13.3 Burnability

The fuel qualities used to indicate a fuel's burnability are conradson carbon residue, asphaltene content, cetane number and carbon to hydrogen ratio. These parameters have a strong influence on combustion, degree of fouling, etc. Effective use of separators, settling tanks and filters and the maintenance of correct viscosity is essential to ensure burnability. Chemical additives which employ a reactive combustion catalyst can also be used to reduce the products of incomplete combustion.

6.9.13.4 High temperature corrosion

Vanadium is the major fuel constituent influencing high temperature corrosion by combining with sodium and sulphur during the combustion process to form eutectic componds with melting points as low as 530°C. These molten compounds are very corrosive with piston crowns and exhaust vales being very susceptible. The main defence against high temperature corrosion has been to reduce the running temperature of engine components through design, to levels below that at which vanadium components are melted.

6.9.13.5 Low temperature corrosion

Low temperature corrosion is generally caused by sulphur which in the combustion process combines with oxygen to form sulphur dioxide (SO_2) while some of the sulphur dioxide further combines to form sulphur trioxide (SO_3) which reacts with moisture to form sulphuric acid (H_2SO_4). When the engine metal temperatures are below the acid dew point (160°C) the vapours condense as sulphuric acid, resulting in corrosion. The solution is to maintain temperatures in the engine above the dew point through good distribution and control of the cooling water. The running temperatures of the engine have thus to be maintained to avoid both high and low temperature corrosion.

6.9.13.6 Abrasives

The normal abrasive impurities in fuel are ash and sediment compounds while solid metals such as sodium, nickel, vanadium, calcium and silica can cause significant wear of fuel injection equipment, cylinder liners, piston rings and ring grooves. An even more dangerous contaminant is metallic catalyst fines composed of very hard and abrasive alumina and silica particles which are residues from the catalytic cracking refinery process. The only effective treatment is by centrifugal separator with manufacturers' recommending series operation (a purifier followed by a clarifier) at throughputs as low as 20% of the rated value.

6.9.14 Fuel system design

A fuel system (*Figure 6.61*) suitable for a ship burning heavy fuel must be designed to perform the following functions:

1. A handling system to enable fuel bunkered at a suitable point on deck to be delivered to the various double bottom, wing, side or deep tanks in the vessel used for bunkers. A transfer system is needed to pump fuel from any bunker tank to the fuel settling tanks.
2. Settling tanks are required to store fuel, usually for at least 24 h, to allow water and sludge to settle out of the fuel by gravity difference, with drain points provided to allow water to be drawn off.
3. Centrifugal separators are needed to remove water and solid impurities with the separator taking its supply from the settling tank and discharging cleaned fuel to a daily service tank which contains the treated fuel for supply to engines. Normal practice is to fit two settling tanks and two daily service tanks which are changed over as required.
4. Heating, usually by steam, thermal fluid or electricity, is required for the fuel before it is passed through the separators and before injection to ensure that the fuel is supplied to the engine at the correct viscosity for the fuel injection system. Automatic devices are normally fitted for viscosity control.
5. Filters are necessary to finally remove any contaminants before delivery to the engine and are normally

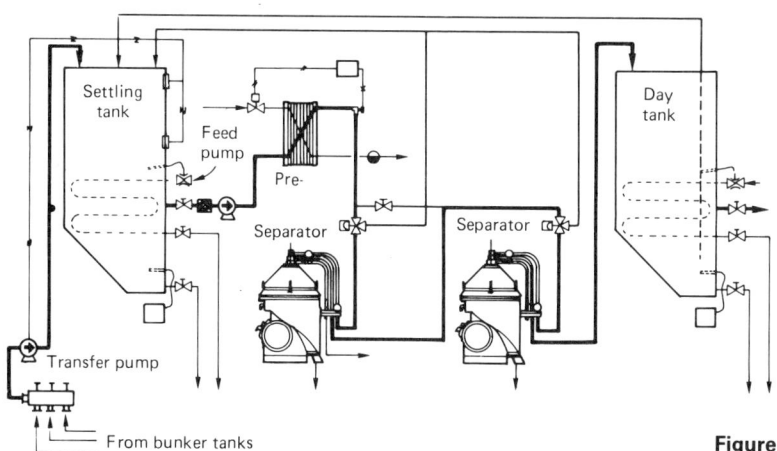

Figure 6.61 Simplified fuel treatment system

placed between the fuel booster pump, sucking from the service tanks, and the fuel pumps on the engines.

In addition a fuel system may include fuel blenders which allow on board mixing of residual fuel and distillate and homogenizers which serve to compact the fuel with any impurities so that the latter can be burned in the engine.

The diagrammatic sketch of the fuel system in *Figure 6.62* shows the very latest recommendations of the technical organization, CIMAC, with regard to treatment of so-called poor quality heavy residual fuels.

6.9.15 Bilge and ballast systems

The bottom of a ship is the bilge, which by virtue of the transverse and longitudinal bulkheads in the hull is made up of many separate bilge areas and water ballast is carried in dedicated double bottom, deep or wing tanks which form part of the ship's structure. These tanks may also be used as fuel bunker tanks with adequate devices to prevent oil entering the ballast discharge system.

Bilge and ballast systems are separate but together they must make it possible to remove water from every bilge and ballast space in the vessel. The ballast system is connected by piping to all spaces in which ballast water can be supplied whereas the bilge system is connected to gutters or wells in the cargo holds, engine room, etc. Centrifugal type pumps are normally used for these operations but they must have adequate suction and by suitable piping cross connections a number of pumps, say bilge, ballast or general service may perform both functions.

Classification rules on bilge systems are strict with, for example, suction lines to be provided on both sides of a bilge space and more than one pump must be capable of connection to the bilge piping system, non-return valves to be used and so on. Other rules require that all bilge water must be passed through an oily water separator to prevent pollution of the sea, particularly in harbour.

6.9.16 Compressed air

Compressed air is stored energy and as such is used aboard ships for tasks such as operating pneumatic tools, control devices and actuators and the most important function, that of starting diesel engines. Air

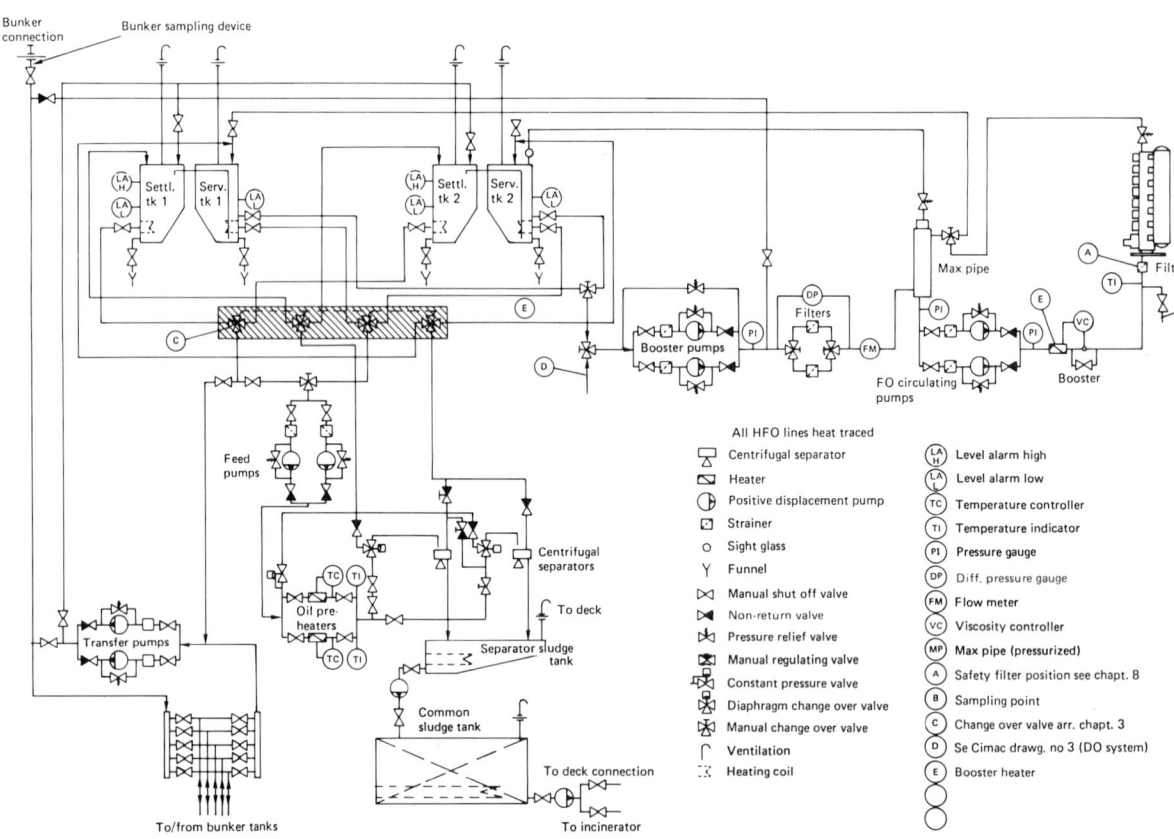

Figure 6.62 Latest CIMAC recommendations for a fuel treatment system

is pressurized by a compressor and stored under pressure in air receivers for use at any time.

A compressor is a machine which operates on the principle that the pressure of air is increased if its volume is decreased. This is achieved in a displacement type of compressor using a reciprocating piston in a cylinder equipped with suitable valves to prevent discharge of the air from the cylinder before compression is complete. A compressor may be single or multistage with a two stage machine having one low pressure cylinder and a smaller high pressure one with the LP discharge leading to the HP suction. To increase the compressor efficiency an air intercooler would normally be fitted to the LP discharge and an aftercooler to the HP or final discharge from the machine.

Compressors are totally enclosed machines cooled by water and lubricated by a forced feed system and/or splash from the crankcase. For shipboard use this type of compressor is invariably used for main air supplies, though smaller air cooled units may be used for separate control air systems and completely separate refrigeration systems which use special gases such as Freon instead of air.

A rotary type compressor usually comprises a rotor fitted with radially sliding vanes, mounted off centre in a cylindrical housing. As the unit rotates the blades are forced out by centrifugal force to seal against the cylinder walls and compressing air trapped between the rotor and casing. Compression takes place in the second half of each revolution as the volume between rotor and casing progressively reduces. Such compressors are not normally used for high pressures and are thus mainly fitted in ships for cargo refrigeration plant. A further type of rotary compressor is a fan, used extensively for ventilating cargo, living and machinery spaces on ships and for forced draught of boilers. Rotary compressors and fans may be of radial or axial flow type.

The function of the air receiver is to provide air storage capacity and it acts as a damper to the pressure pulsations caused by a piston type compressor. These are normally cylindrical with hemispherical ends and more than one may be fitted to a ship. Complete air systems comprising multiple compressor units, air receivers, water traps and drains and piping systems are normally designed for automatic operation to keep a constant reserve of air at working pressure and to satisfy all demands, particularly at peak load times during engine manoeuvring.

6.9.17 Refrigeration

Refrigeration is used on all ships to maintain food provisions in good order during a voyage and on a large scale for temperature control of cargo holds on refrigerated cargo liners, fishing vessels and on gas carriers to prevent cargo loss through boil off.

Refrigeration is the cooling of a space or object and maintaining its temperature lower than that of the surrounding atmosphere. The most common system is the so-called vapour compression cycle. In this cycle, the working fluid or refrigerant undergoes a repeated conversion between liquid and gaseous states through the processes of compression, condensing at high pressure and expansion into an evaporator.

The main components of the system are thus the refrigerant gas, expansion device, evaporator, compressor and receiver. Halogenated hydrocarbon compounds are used as the refrigerant, Freon R22 being very common. The expansion device or reducing valve controls the reduction in refrigerant pressure as it moves from the high to the low pressure side of the system and it controls the amount of refrigerant entering the evaporator, thus regulating the capacity of the system. The evaporator is an indirect heat exchanger in which the refrigerant absorbs heat from the substance being cooled; this would be air in domestic systems or chilled water and liquid brine which is often used in cargo refrigeration plants.

As the liquid refrigerant leaves the expansion device and enters the evaporator at low pressure and thus low saturation temperature, heat from the substance being cooled is absorbed by the refrigerant, which thereby vapourizes as its temperature rises. Superheating of the vapour by a small amount in the evaporator, ensures complete vapourization and prevents liquid carry over to the next stage, that of compression. The compressor transfers the refrigerant from the low to the high pressure side of the cycle and the refrigerant leaves the compressor as a high pressure, high temperature superheated vapour. The vapour passes at high pressure to the condenser which is an indirect heat exchanger in which the refrigerant rejects heat to a cooling medium, usually sea water on ships. When the temperature has dropped to the saturation temperature, the vapour condenses and rejects the latent heat of evaporation absorbed in the evaporator. A receiver is used as storage and a surge tank for the subcooled refrigerant flowing from the condenser and prevents vapour from entering the expansion valve.

The above cycle, simply described, operates continuously for each cycle and control of temperatures in the cooled spaces is usually by a combination of air control through fans and regulation of the cooling media piped into each space and spread around the area through grids. Cargo spaces are extensively insulated and remote reading temperature gauges and controls for fans, brine flow and temperature, etc., are nowadays under computerized control and monitoring.

6.9.18 Air conditioning

Air conditioning is the process of maintaining the atmosphere of an enclosed space at any required temperature and degree of humidity and purity. To achieve this, an air conditioning plant cools, humidifies or dehumidifies, ventilates and purifies the air that it circulates to the spaces. A major part of air conditioning plant is a refrigeration system which cools and dehumidifies the air.

For heating of the air a steam or electric coil may be used. For ships, air conditioning plants may use a refrigerant circulation system, chilled water circulation system, or self contained units.

A refrigerant system is a vapour compression refrigeration system in which several evaporators are used to control the temperatures of various spaces, with hot air from the spaces drawn through ducting and passed over the evaporators. The refrigerant in the evaporator absorbs heat from the air and excess moisture condenses on the outside of the evaporator. In a large ship's system, such as on a passenger vessel, each evaporator has its own expansion device to control the temperature of the refrigerant and hence the spaces being cooled.

Air purity is maintained by filter units and circulation by motor driven fans. In a chilled water system cold water is used as a secondary refrigerant passed to cooling coils located in the ventilation ducts. This is the most common system used aboard ships and has the advantage that all refrigeration machinery can be centrally located, primary refrigerant leaks are more contained and easier to locate and good temperature control is obtained from multiple chilling coil units.

Self contained air conditioning units are small complete plants fitted into metal cabinets and may be located in individual spaces, such as control rooms, cabins, etc., on ships not initially provided with air conditioning.

6.9.19 Minor ancillaries

Major main engine service systems and fuel treatment plants are discussed elsewhere but a ship has many other smaller ancillary systems, not all essential for ship operation but mainly provided to improve living conditions for crew and passengers. These are briefly described below.

6.9.20 Oil fuel system

A vessel on a long voyage will consume a large amount of oil fuel which must be carried. Fuel is normally stored in the double bottom tanks of the ship which are provided with filling and discharge pipes and are normally filled from the bunkering point on deck by gravity through valve boxes and manifolds. The tanks are provided with overflow pipes and sounding equipment so that their contents can be measured.

Before oil is treated, for burning in the machinery, it will be stored in settling tanks in the engine room and the fuel oil transfer pump is used to move fuel from the bottom or side bunker tanks to the settling tanks. This pump sucks through a course filter and as well as filling the settling tanks it can be used to transfer oil between any of the bunker tanks as is necessary to regulate the ship's list and trim as fuel is consumed. Screw type pumps are normally used for fuel oil transfer.

6.9.21 Cargo pumping system

An oil tanker carries liquid cargo which must be removed by a pumping system. The hull is divided into tanks, each with a suction and filling line and the normal arrangement is to have a pump room between the aftmost tanks and the engine room. From two to four main cargo pumps are driven through the engine room bulkhead by steam turbines, or electric motors, diesel engines or gas turbines, these pumps being of centrifugal type of high capacity. The pumps are controlled from the deck or special cargo control room, and valves in the tanks, remotely operated from the deck, are used to direct the movement of cargo from the tanks to shore loading point. For final emptying of the tanks during discharge, positive displacement type stripping pumps are used.

6.9.22 Sanitary system

A fresh water and a sea water sanitation system is provided to supply washing water and flushing water for toilets in cabins etc. For these two separate systems, the most common is the hydrophore system in which an electrically driven centrifugal or mono pump draws from water tanks, or the sea, to a pressure storage vessel which is filled partly with water and partly with air. As water is drawn off this tank its pressure drops and a pressurestat controls the operation of the supply pump, to pump in more water. The pressure is maintained at around 4 bar, sufficient to reach taps, showers and baths in the accommodation and, for the sea water system, individual WC units. Fresh water is normally drawn from the system to a heater which may operate on steam, electricity or main engine cooling water.

Some further systems fitted to ships are oil–water separators, evaporators to produce fresh water from the sea, ventilation equipment, fire fighting pumps and ring mains, hydraulic equipment for powering mooring equipment and cargo winches and cranes, separate cooling and lubricating systems for auxiliary engines and minor equipment too numerous to mention here.

6.9.23 Steering gear

For any boat or ship, the most vital function in terms of manoeuvrability is that of steering. Also, loss of steering control can result in loss of the vessel. Except for vessels equipped with steerable thruster units, all powered craft are fitted with a rudder to deflect the propeller slip stream and cause changes in the heading of the vessel. Some form of mechanical device is needed to move the rudder, such as the simple hand operated tiller on small craft to large electro–hydraulic powered steering gear on seagoing ships.

Remote hand operation of the tiller is common on all small craft where the rudder forces are low enough for manual movement. The mechanisms in use are typically: rod and chain, the chain passing around the steering wheel and movement to the rudder transferred to the tiller arm by rods or cables; sheathed cable connecting the tiller to a steering head in which the cable is connected to a drum that is turned by the steering wheel; hydraulic systems with a wheel operated plunger in a pressure cylinder hydraulically connected to an actuating cylinder connected to the tiller. Seagoing ships, however, practically all use powered steering gear, either using electric motor geared to the rudder stock, or electro–hydraulic ram type or rotary hydraulic types linked to the rudder.

6.9.24 Electro–hydraulic steering gear

The most common type of steering gear fitted to ships is the electro–hydraulic type. A gear for large ships would comprise four hydraulic rams connected at opposite sides to the tiller in so-called Rapson slide arrangement. These four rams are supplied by hydraulic oil from Hele Shaw pumps which are of variable stroke multiple radial piston type and run at constant speed driven by an electric motor. Normally two pumping units are provided, the second unit acting as standby in case of failure and also run during manoeuvring.

The function of the gear is as follows. Control of the stroke of the pump pistons and hence delivery of oil to the power rams is by telemotor control from the steering head on the bridge. By increasing the stroke, oil is pumped through valve blocks and to two of the hydraulic rams (diagonally opposite units), the oil being drawn from the other two cylinders. Two pipes are led from each hydraulic pump to the valve block and from there to two of the cylinders with cross connections to the valve block of the other pump and to the other two cylinders. Alternate suction and delivery ports in the pumps allow discharge and suction in either direction and hence movement of the rudder in response to telemotor control from the wheelhouse.

6.9.25 Rotary type

The rotary type of hydraulic steering gear is very popular on small to medium sized vessels. This ingenious device comprises a cylinder of endless type forming a complete circle and containing a hydraulic piston. The cylindrical casing forms an oil tank and the pump unit is integrated into the housing. Movement of the piston is transferred directly to the rudder stock. Hydraulic manoeuvring valves are remote controlled electronically. The advantage of this type is that it takes up very little space and torque is constant over the full $2 \times 45°$ of rudder movement.

6.9.26 Deck machinery

Most machinery provided for a ship is fitted below decks in the engine room but deck mounted machinery is also required to achieve mooring and anchoring, materials and stores handling and cargo handling. These operations usually involve the lifting of objects so machinery required is mainly hauling winches or, as is more common nowadays, pedestal mounted jib cranes.

At one time all deck machinery was steam operated, with each cargo winch or windlass comprising a self contained unit with twin steam cylinders geared to a winch drum around which was wrapped the hauling cable. Steam winches were replaced by electric motor powered units usually connected to the winch drum by worm and wheel reduction gearing and for cargo handling the union purchase system, whereby a pair of winches located at each corner of a hatch end having their hauling wires terminated at a common block, was most common. A pair of derricks, one with its head positioned above the hatch centre and the other out over the loading quay, provided the means of transfer of objects between shore and ship, with movement of the load controlled by paying in or out of the respective winch hauling wires.

Hydraulic motors are also popular for cargo winch drive using motors of either radial piston or vane type, operating on either low pressure or high pressure systems. The hydraulic pump power packs are usually located in the engine room or in deck houses to feed pressurized hydraulic fluid to the winches. This system is very popular for fishing vessels to power the fishing gear, such as net winches and the mooring equipment.

At one time most general cargo ships and bulk carriers were fitted with numerous cargo handling winches, usually four per hatch, but the modern preference is to employ jib cranes so positioned that they can be spotted over any point in the open hatch area. These may be electric, electro–hydraulic or hydraulic and are equipped with an enclosed cab for the driver. On bulk carriers of medium to large size often no cargo handling machinery is fitted with shore cranes being relied upon for cargo operations. Winches or hydraulic systems are usually fitted only for opening and closing of the hatch covers.

Whereas all ships are provided with a windlass mounted on the foredeck for raising the anchor, and winches or a capstan at the stern to handle mooring ropes, the cargo handling equipment fitted depends upon the ship type. Typical arrangements are briefly discussed below.

6.9.27 General cargo ships

As mentioned above dry cargo ships may be equipped with winches and derricks or, more commonly, jib cranes (*Figure 6.63*) mounted on pedestals between the holds.

Figure 6.63 Dry cargo ships may be equipped with winches and derricks or, more commonly, jib cranes

6.9.28 RoRo ships

The Roll-on/Roll-off ship was developed to allow integration of the ship into a complete transport system by enabling cargo on wheels to be loaded into the vessels. Ramps fitted at the stern, quarter, side or bow allow vehicles such as trailers, lorries, cars, coaches, etc., to be driven directly aboard, resulting in fast port turnarounds. Ferries make use of the RoRo concept and in addition to the shore access equipment some vessels have internal ramps for access between decks or special hydraulically or wire operated lifts may be fitted. A typical RoRo cargo vessel is depicted in *Figure 6.64*.

6.9.29 Container ships

The container ship was developed to provide a more efficient means of transport by loading cargo into standard sized metal containers. The ships are equipped with cellular guides in which the containers are stacked and normally shore mounted gantry cranes are used for the lifting operation.

6.9.30 Bulk carriers

The gearless bulk carrier is most common but on smaller vessels it is usual to fit jib cranes, though gantry

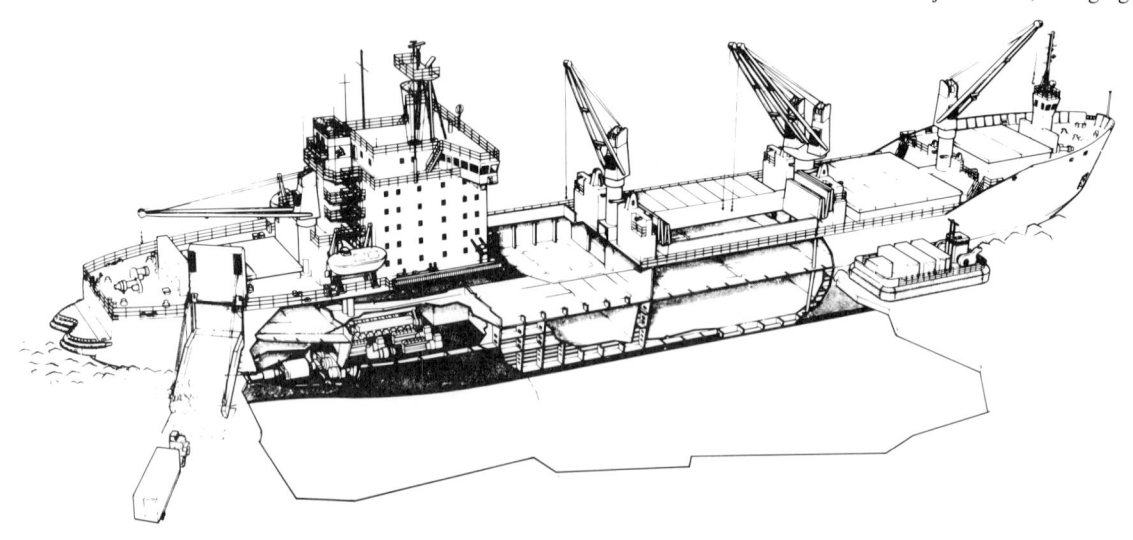

Figure 6.64 The RoRo concept applied to a multipurpose cargo ship

type cranes fitted on rails mounted on the deck are sometimes used.

6.9.31 Tankers

Oil and chemical tankers carry only liquid cargoes and a normal tanker has its hull subdivided into many small tanks to allow parcels of liquid cargo to be carried. A piping system is connected to cargo pumps and through the complex of valves and pipes any tank on the ship can be filled or emptied.

6.10 Propulsion

6.10.1 Propulsion concept

Propulsion is the term used to describe mechanical means of enabling a floating vessel such as a ship or boat to propel itself through the water. Practically all craft are propelled by marine screws; some are still propelled by paddle wheels for their novelty value; others use water jets because of restricted water depth or high speed requirements; and other concepts such as using an electro–magnetic force to induce thrust are being studied.

For any self propelled vessel the following functions are essential:

1. A propulsion system which provides the necessary thrust to give the vessel in question a specified speed and in some cases the static thrust is of importance. The normal solution is a conventional marine propeller with shaftline, stern tube with seals, line shafting bearings, reduction gear in the case of medium speed engines and, of course, the propulsion engine itself. Remote control of the machinery from the bridge is nowadays normally provided.
2. A steering system which provides control of the ship's direction and course keeping at speeds at which a rudder is effective. The standard solution is a rudder to deflect the water stream aft of the propeller, with rudder post and bearings and a steering machine with auxiliaries. Rudder control from the wheelhouse is by remote control of the steering gear from a wheel or lever. An interface to an autopilot system is normally provided.
3. A manoeuvring system for safe and rapid ship control in confined waters where speed and direction control of the propellers in combination with rudder movement is still inadequate. To achieve better results it is normal to fit transverse thrusters at the bow and in same cases also at the stern. Such a side thruster is normally in a tunnel, with a suitable drive system, and remote controls from the wheelhouse.

Not all vessels are identically equipped, for among the above three main functions, many options exist, particularly with regard to marine propellers, with merits for each particular application. The main concepts of ship propulsion are discussed in the following section.

6.10.2 Propeller application

For reasons of simplicity, efficiency and the need to minimize initial costs, the preferred choice for ships is to fit a single fixed pitch propeller of optimum diameter, pitch, etc., in open water arrangement with a balanced type spade rudder aft of the propeller. For some vessels, however, there are advantages from using more than the one propeller, with two being the most common. Propellers may also be ducted to improve thrust performance or CP propeller types may also be selected for vessels required to manoeuvre frequently.

6.10.3 Multiple screws

The use of twin or triple screw propulsion is an adaptation of the concept of utilizing the largest possible propeller diameter but, in cases where the diameter is limited by the hull or draught, multiple propellers allow a greater mass of water to be utilized than the single screw alternative. In twin screw installations the benefit from the energy in the wake is lost due to the off centre position of the propellers but this can be obtained with triple screws where the centre screw absorbs most of the power. Though multiple propellers may not provide the most efficient form of propulsion, there are, nevertheless, other benefits. Some advantages are: the power required is distributed over a number of smaller engines; enhanced safety in case of failure of one propeller or prime mover; enhanced manoeuvrability as each propeller can be operated ahead and astern independently.

6.10.4 Overlapping propellers

Interlocking or overlapping propellers are in effect twin screws arranged in such a way that the blade discs overlap each other, with a view to utilizing some of the wake energy. This system is rarely used in practice because if the units are fitted in the same longitudinal position some form of gear arrangement is necessary to prevent the blades from interfering with each other.

Another solution is to have one propeller displaced longitudinally. The reduced shaft centres resulting from this arrangement could make the positioning of main engines more difficult.

6.10.5 Ducted propellers

A ducted propeller (*Figure 6.65*) is one which is encircled by a duct of special shape such as the Kort nozzle. The propeller operates with a small clearance between the blade tips and the internal wall of the nozzle, while the nozzle itself extends forward of the propeller and has a hydrofoil cross section.

Figure 6.65 Ducted azimuthing type propulsion units on a tug

The shape of the nozzle and its installation results in the following positive advantages: At high thrust levels the efficiency is higher, with, for example, tugs fitted with a Kort nozzle achieving efficiency improvements of as much as 20% and increases in bollard pull of 30% or more, compared with a conventional propeller absorbing the same power; reduction in propeller efficiency in a seaway is less with a nozzle; course stability is substantially improved by the nozzle.

Kort nozzles may be of fixed or steerable type with the nozzle replacing the rudder and supported by a rudder stock and possible lower pintle at the keel. The steerable type has a somewhat lower efficiency than the fixed type because the gap between the propeller and internal wall must be greater to provide clearance to turn the unit for steering.

Ducted propellers are now very well established for tugs and trawlers where a high thrust is required at a low ship speed. Their use on larger seagoing vessels, though, is not so well established, mainly due to negative features of the system.

Major disadvantages are:

1. Manoeuvrability and directional stability when going astern is impaired.
2. In shallow water the nozzle tends to draw in sea bed shingle, etc., which can cause damage, as can operation in ice.
3. Because of the pressure drop that takes place in the nozzle due to its foil shape, there is a risk of cavitation and erosion damage.

6.10.6 Screw propeller features

The most common form of propulsor for ships is the fixed pitch or conventional screw propeller. It is a simple device, as it is mostly cast in one piece from metals such as aluminium, cast iron, stainless steel and bronze alloys. For merchant ships most propellers are made from bronze alloys containing copper, aluminium and nickel, combining strength with resistance to corrosion.

The complete marine propeller (*Figure 6.66*) comprises the boss or centre hub which is bored, usually tapered, to suit the propeller shaft which may be keyed and the blades which are shaped on their back or pressure face to resemble a screw thread in some ways. A propeller operates by causing an increase in velocity to the mass of water passing through it. The number of blades varies considerably: small leisure craft propellers have two or three blades, merchant ships three to five and high performance small craft can have as many as eight. The pitch of the propeller is the theoretical distance that it will advance through the water in one revolution, though the primary purpose is to provide thrust behind the vessel. The actual propeller advance is affected by slip but this can be both positive and negative, depending upon sea conditions.

Regarding the propeller blades, the edge that cuts into the water is known as the leading edge while the opposite side is the trailing edge. When viewed from astern a right handed propeller rotates clockwise and a left handed unit turns anticlockwise. Both types are extensively used.

Figure 6.66 Large diameter keyless propeller for a VLCC

6.10.7 Propeller efficiency

The performance and efficiency of a propeller is greatly affected by the main dimensional parameters, some of which are discussed below.

6.10.7.1 Propeller diameter

The most important factor affecting propeller efficiency is the propeller diameter and associated pitch ratio. Normally for a given propeller diameter, the pitch is fixed from power absorption requirements. Pitch ratios outside the norm for a given power absorption can lead to heavy and light propellers.

For a given speed, power and revolutions there is an optimum propeller diameter and pitch or, conversely, optimum revolutions for a given diameter, speed and pitch. As it is more efficient to impart a small increase in velocity to a large mass of water, the propeller diameter should be as large as possible for a given installation. This is affected by the draught of the vessel and minimum clearance available between the blade tips and hull. As the propeller diameter is increased, the optimum revolutions reduce, so the propeller shaft speed should be as low as possible to allow the largest diameter propeller that can be fitted. The optimum diameter is affected by the wake factor, usually around 5% for a large vessel.

6.10.7.2 Number of blades

The number of blades fitted to a particular propeller is mainly chosen with consideration to vibrations but for given revolutions the lower the number of blades the greater the propeller efficiency. Over the normal range of blade numbers from four to six, a four bladed propeller will be about 2% more efficient while with three and even two bladed propellers, the greater efficiency can be more significant. However, the vibration characteristics of low blade numbers represent a high risk and more blades of smaller diameter may be the optimum choice for a passenger ship, for example where vibrations are unacceptable.

6.10.7.3 Blade surface areas

The surface area of the blades has a direct effect upon efficiency as it is associated with friction caused by the propeller in operation. However, the surface area is greatly affected by the need to control cavitation under all conditions. Cavitation is the production of bubbles or cavities in the water from pressure reduction at the blade edges, causing a reduction in thrust and efficiency and erosion of the propeller material.

6.10.7.4 Blade thickness

The thickness of the blades has little effect on efficiency but has a direct effect on the strength of the propeller blades needed to resist bending and fracture.

6.10.7.5 Boss size

The size of the propeller boss effects efficiency as it is known that the larger boss needed for CP propeller to house the actuating gear leads to a reduction in propeller efficiency.

6.10.7.6 Rake

Rake has no effect on propeller efficiency but the rake angle of the blades will affect the position of the blade tips relative to the propeller aperture which can significantly affect efficiency through hull–propeller interaction.

6.10.7.7 Skew

The sweep back shape or skew of propeller blades reduces the pressure impulses on the hull adjacent to the propeller. This results in reduced vibrations so it may be possible to use a larger diameter and more efficient propeller than one without skew back.

6.10.7.8 Blade clearance

Propeller blade clearances are important as they affect the propeller–hull interaction and, as clearances reduce, the thrust deduction increases, therefore blade clearances should be as large as possible.

6.10.7.9 Propeller roughness

The frictional drag of a propeller in operation is dependent on the surface condition of the blades and the absorbed torque is a function of the drag. Therefore, a propeller will become less efficient with increasing blade roughness associated with fouling, damage or erosion.

The extensive subject of propeller design theory is beyond the scope of this book but propeller design does indeed have much influence on the efficiency of the ship propulsion plant as a whole. The design factors as outlined above each have an effect on propeller efficiency. For conventional ships' propellers the following guidelines are important for achieving optimum efficiency: the diameter should be as large as possible, the number of blades should be as low as practical, the blade area should be as low as possible, the boss diameter should be as small as possible and propeller clearance should be as large as possible.

6.10.8 Controllable pitch propellers (CPP)

A normal ship's propeller has its pitch designed for the application but the desire to have a propeller where the pitch can be altered is as old as the screw propeller itself. The simplest method of pitch adjustment is where propeller blades are flanged to the boss and elongated bolt holes in the flanges allow the blades to be swivelled on their axis to the desired pitch setting. However, this method has little appeal as the ship must be docked for pitch adjustment but it can be useful to obtain the optimum pitch setting following sea trials.

The more widespread and practical type of CP propeller (*Figure 6.67*) has the blades adjusted in unison from inside the vessel at will, using a mechanical device for blade angle settings. The fully adjustable CP propeller is most versatile and because of the many advantages of using this propulsion device, the last few years have seen a rapid increase in its use on many different vessels types of all sizes.

Figure 6.67 CP propeller of highly skewed type

6.10.9 Advantages of CPP

The advantages of a CP propeller apply to all types of vessels but are of special importance for ships which operate under varying conditions of resistance to forward motion, such as tugs and trawlers and even cargo vessels, which are not always running fully loaded and thus have a different power requirement at different stages of a voyage.

Another important aspect is manoeuvrability which affects ferries, supply vessel, tugs, etc., where gains are to be had in terms of speed, accuracy and safety. Greater manoeuvring speed is obtained by the immediate and direct remote control of the propeller thrust from the bridge without the time lag required to reverse the propellers as on fixed pitch installations. This faster reversal gives the vessel a shorter stopping time and length while greater accuracy is obtained because the engine power can be used at any required ship speed with continuous control of propeller thrust ahead and astern irrespective of the power developed by the main engine. The greater safety of the vessel is a result of the enhanced speed of operation and accuracy of control. Also, the duration of manoeuvring is not dependent on the availability of compressed air for engine starting, while in the case of rudder damage on multiscrew ships, the CP propeller gives an emergency mode for manoeuvring and steering.

There has long been a dispute about the reduction in fuel consumption resulting from a CP propeller but a significant point is that the CP propeller does allow the operator to make better use of the available engine power which brings a higher performance. An example is on a trawler where a higher trawling speed results in a bigger catch, while a tug will take less time for a long tow.

The improvements may not necessarily be measured as a reduction in fuel burnt per mile but will result in improved overall economy when the increase in income is compared with the total operating cost of the vessel. For example, three fishing boats with CP propellers will often bring home the same catch in a season as four conventional boats.

Propulsion efficiency is dependent on factors such as propeller speed, diameter and pitch, etc., which is matched for a given loading but with a CP propeller the propeller pitch can be adjusted for different loadings, avoiding operation with a propeller that is too light or too heavy.

When the load on the engine is reduced, the relationship between horsepower and r.p.m. can be altered by changing the propeller pitch so that the optimum efficiency is obtained for the power being developed. This is of particular importance in multiple engine plants where one or more prime movers can be cut out for shorter or longer periods.

For vessels powered by diesel engines the use of a CP propeller will result in a number of advantages. For a start the engine does not require to be made direct reversing, meaning a cheaper and simpler engine and frequent starts and stops during manoeuvring is avoided. Thus, starting systems can be of smaller capacity and there is less wear and tear on the machinery.

There is also the benefit that engine overloading is avoided by using an automatic pitch control system to reduce pitch and hence engine loading if the maximum permitted cylinder pressures are likely to be exceeded. Constant speed operation of the engine means that operation on heavy fuel for most of the time is possible, leading to fuel cost savings; and barred speed ranges are avoided as the engine is not run from stop to full speed and vice versa during manoeuvring, as with direct drive systems. Another very important advantage is that constant speed operation of the engine allows a simple form of power take off for the drive of alternators, pumps, etc.

With vessels powered by steam or gas turbines, an astern turbine or reversing gearbox is avoided and for steam turbines the windage losses from the astern turbine during normal operation is also avoided. The turbines can be run at relatively high speed at all times, improving their thermal efficiency, and the entire engine power can be used for astern thrust instead of the reduced percentage normally only available with separate astern turbines.

Vessels with electric propulsion also benefit as smaller and less expensive ac motors can be used for constant speed operation, rather than dc motors with speed control devices. Dynamically positioned vessels for the offshore industry are more easily controlled when constant speed motors driving CP propellers and thrusters are used. Full thrust variation is possible without affecting the motor speed and frequency.

6.10.10 Economic loading

The loading of a propulsion engine driving a fixed pitch propeller is governed by the propeller law whereby the power varies according to the cubed power of speed. The maximum power installed is that required to achieve the maximum ship speed and the tendency is to fit engines with some 10–30% more power than required by the hull to give the so-called sea margin to cover for weather and hull fouling.

It is claimed that a vessel with a CP propeller will achieve the same results in terms of free running speed, and on trawlers and tugs during towing, as a conventional propellered vessel. The most economical operation of a diesel engine is between 80 and 90% of maximum speed and power. A CP propeller helps to achieve the best specific fuel consumption by operation of the engine at maximum speed and using the propeller pitch adjustment to keep the engine at maximum load and, thereafter, the engine r.p.m. are dropped by about 5% to set the engine loading to optimum performance. There is also the option to lower the engine speed and increase the propeller pitch to find the optimum settings for each vessel loading condition.

On the question of costs it is indisputable that CP propellers are more expensive, as complicated mechanical design and controls are necessary. However, the additional costs of the CP system should be looked at against a background of possible operating cost benefits, such as:

1. Possible improvement in propulsive efficiency, simpler engine plant, greater manoeuvrability, less wear and tear, etc., as mentioned above.
2. Spare propeller blades are cheaper than a complete spare solid fixed pitch propeller.
3. In a fleet standardized on one type of CP propeller, spare part costs are reduced.
4. Reduction in the cost of main engines and some auxiliary machinery.
5. The ability to drive shaft generators and pumps, etc., from the main engine results in a direct saving in fuel costs, particularly if auxiliary engines operating on light distillate fuels is the alternative.
6. Greater manoeuvrability could lead to savings in harbour costs if tugs are not needed.

The above savings and others relating to crew costs and maintenance are said to easily ensure the payback of the additional capital cost burden.

6.10.11 Typical CPP design

A controllable pitch propeller is an ingenious piece of machinery (*Figure 6.68*) which must be designed and

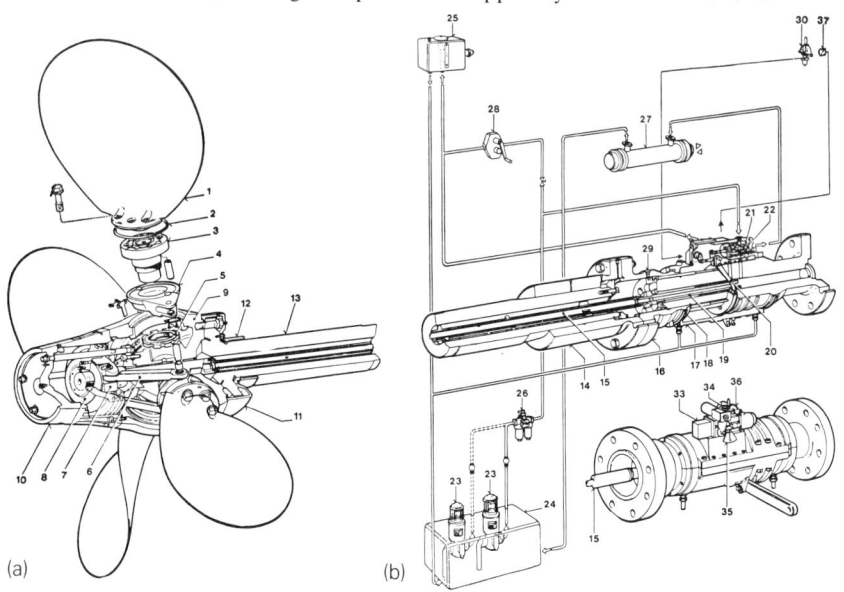

Figure 6.68 Components of a CP propeller system. (a) Blade parts: (1) Blade; (2) Blade seal; (3) Blade trunnion; (4) Adjusting crank; (5) Trunnion nut; (6) Link; (7) Crosshead; (8) Servomotor piston; (9) Hub; (10) Servomotor cylinder; (11) Shaft flange; (12) Shaft seal; (13) Propeller shaft

built to be completely reliable under all operating conditions. It must also be efficient as a thrust device, must withstand very substantial cyclic loads and at the same time the pitch must be accurately set and maintained, while rapid pitch changing is essential. The heavy demands made on the mechanism of a CP propeller have meant that practically all modern propellers use a hydraulic servo system, with the main difference between competitive types being whether the actuating piston mechanism is located within the propeller hub or in the shaft line inboard, within the latter case movement of the servo piston being transferred to the swivelling propeller blades by push–pull rods inside the shafting. A typical design is that produced by Messrs Escher Wyss.

In the Escher Wyss type CA propeller hub collar bearings and a sliding block arrangement are used to alter the blade angle. The unit consists of a monoblock hub body, blade carriers and the servomotor piston with characteristic L-shaped slots accommodating the sliding blocks for changing pitch. Other variants use trunnion bearings and a crank ring arrangement. To avoid potential problems which could require the vessel to be docked for repairs, all hydraulic control valves are located inside the vessel and two concentric oil tubes are accommodated in the bore of the propeller shafting. The actuating unit is located inboard and the oil tubes follow every movement of the piston and hence every pitch change movement of the propeller blades, with control of the oil flow from inside the ship.

The function of the inboard actuating unit is to receive command signals from a control unit and to route the oil flow through the respective path in the oil tubes to the ahead or astern side of the servo piston, to ensure that the blades change to the precise pitch required. As a safety measure the oil tubes can be mechanically locked from inside the vessel to allow emergency operation in full ahead pitch.

6.10.12 Voith Schneider propeller

A type of controllable pitch propeller which is more akin to a paddle wheel, is the Voith Schneider propeller (*Figure 6.69*). This cycloidal type of propeller manufactured by Messrs Voith typically uses four or five aerofoil shaped blades which are mounted in the vertical plane with their axes of rotation in the horizontal plane.

The Voith propulsion unit operates in one direction only at a constant speed. Each of the blades is pivoted and connected to a central star shaped arm by levers, such that as the unit rotates the angle of the blades changes, similar to the mechanism of a feathering paddle wheel. However, the essential feature of the V/S propeller is that the central connection points of the linkages can be moved away from the centre of rotation. The crank type kinematics of the unit mean that the thrust of the unit can be in any direction, according to the position of the steering centre relative to the propeller centre. The normal arrangement has a bevel gear drive of the vertically mounted propeller shaft and the only connection between the engine and the propeller is a torque converter to absorb shock loads of rapid pitch changes and provide no load starting of the engine.

The ability of the unit to have its thrust directed at any angle to the ship's centreline makes it highly suitable for tugs and ferries requiring good manoeuvrability. Some vessels have a unit at the stern and one at the bow, which is popular for ferries, while for tugs it is normal to have the propulsion units side by side at the stern or in so-called tractor arrangement with the units forward of amidships.

6.10.13 Self pitching propeller

A propeller system of neither fixed nor controllable pitch type, which is deceptively simple but has been pursued for more than a century and only recently perfected, is the self pitching propeller. Small units have now been tested on leisure craft with quite impressive results.

The innovative feature of this unit is that the blades are free to pivot through 360° on their axes and thus adopt any attitude to the water stream, depending upon load, speed, etc. The blades themselves are offset from the pivot point in their base and while centrifugal force from the mass of the blades themselves causes the blades to swivel in one direction, the water motion tends to force them in the opposite direction against the centrifugal force. The result is a point of equilibrium which sets the propeller pitch, which will change automatically under differing conditions.

A normal propeller of fixed pitch is designed for optimum ahead running speed; astern performance is unavoidably inferior and when stopped the propeller causes drag on a vessel under way, such as on a cruising yacht. The self pitching propeller, by virtue of its design, has the following advantages:

1. When the shaft is stopped the propeller blades are automatically feathered.
2. On a fast boat the propeller pitch becomes coarser as the boat accelerates.
3. On a motor sailer the propeller pitch adjusts automatically to take maximum advantage of wind assistance.
4. The astern performance is remarkable as the blades assume the same configuration as in ahead but facing in the opposite direction. Ahead and astern thrust is identical.

6.10.14 Contra rotating propellers (CRP)

Co-axial contra rotating propellers (CRP) are now beginning to be fitted to ships though they have been around for some time, the best known example being on torpedoes.

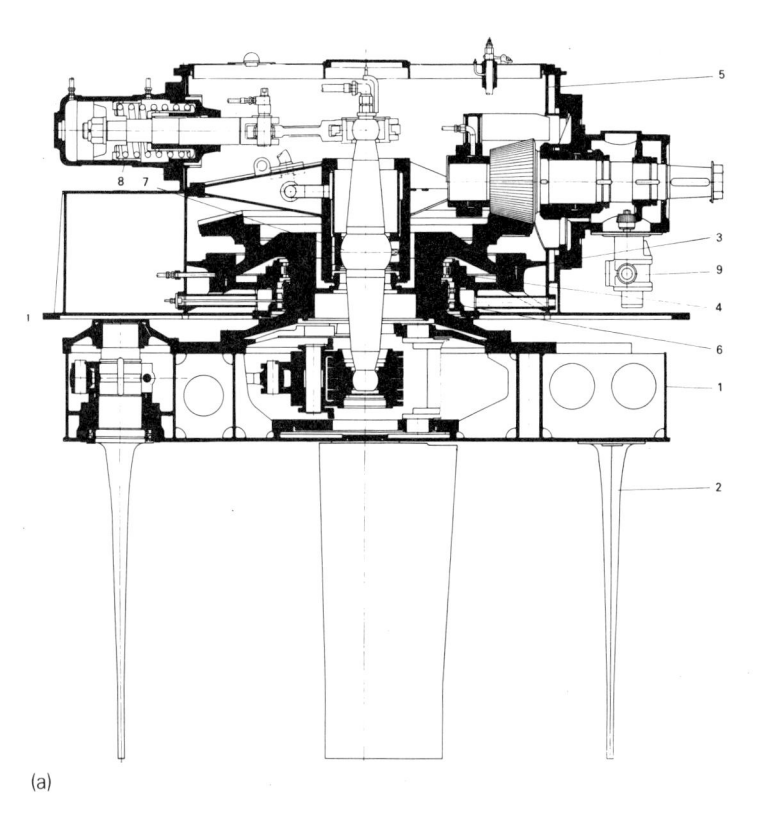

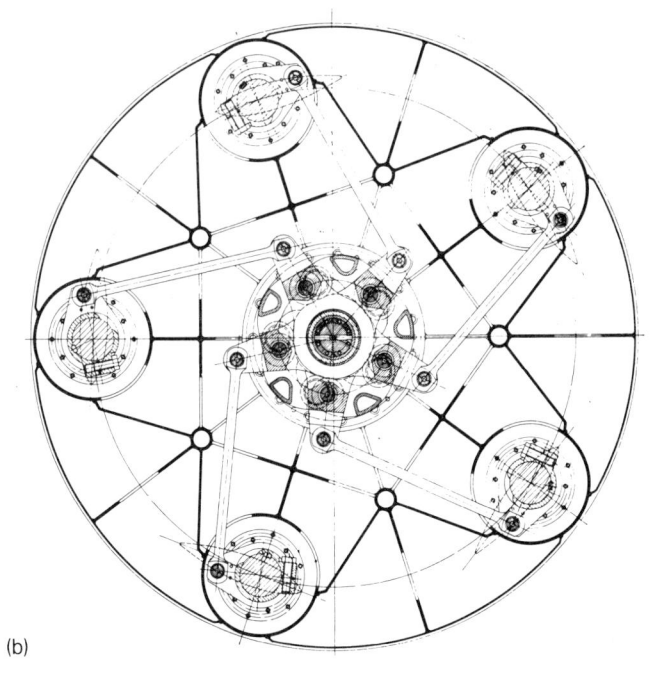

Figure 6.69 Sketch of the Voith Schneider propeller. (a): (1) Rotor casing; (2) Propeller blade; (3) Bearing lantern piece; (4) Roller bearing; (5) Propeller casing; (6) Driving sleeve; (7) Control rod; (8) Servomotors; (9) Oil pump. (b): Plan view of unit showing connection of five propeller blades to the centre control mechanism

By fitting two propellers on the same shaft operating in opposite directions, the normal rotational losses of a propeller are eliminated as the swirl energy is absorbed by the aft propeller. The usual rotational losses are typically 8–10%. The result is an improvement in propulsion efficiency though the major advantage of CRPs is that opposite propeller rotation brings natural torque compensation giving much better directional stability to the vessel, hence the torpedo application where veering off track, once aimed, is highly undesirable. Additionally, the input power is absorbed by two propellers instead of one, resulting in reduced loading on the propellers and higher propulsive efficiency.

To drive two propellers in opposite directions brings much mechanical complication which has hitherto inhibited the use of CRPs. The shaft has to be bored to take an inner shaft for the second propeller and in the engine room special gear such as a star type planetary gear is necessary to cause the two driven shafts to turn in opposite directions, particularly if one main engine is used.

Another problem is the mounting of the inner shaft bearings and the outer sealing arrangements. So far the mechanical complexities, with additional costs, have not been considered justified by the increase in efficiency. There is also the danger of cavitation at the aft propeller and to prevent this the aft propeller is normally made with smaller diameter and may be of different pitch or blade number.

The CRP system has only very recently been applied to large merchant ships with the first application being to the car carrier *Toyofuji 5*, in 1988, with a CRP built by Mitsubishi, Japan. It is claimed that the fuel saving on this vessel amounts to 16% and large vessels such as VLCCs with large diameter slow turning CRPs, could save 16–20% compared with conventional propellers.

At the present time, however, a much wider application is on outdrive units for small craft using the Volvo DuoProp system. Contra rotating propellers are also being fitted to azimuthing propulsion units where reverse rotation for the second propeller is obtained by simply adding a second bevel gear in the lower gearbox and inner and outer propeller shafts.

6.10.15 Grim wheel

A large diameter slow turning propeller is most efficient but worthwhile gains in propulsive efficiency can be achieved by use of the so-called Grim wheel. This device, named after its inventor, is a turbine like free spinning vane wheel which is mounted on an extension to the propeller shaft aft of the main propeller.

The Grim wheel (*Figure 6.70*) has more blades than the usual propeller and is some 20% larger in diameter. The vanes of the Grim wheel are so designed that the inner section acts like a water turbine to extract from the propeller slipstream a large amount of energy which would otherwise be lost. This recovered energy is converted directly into additional and useful thrust in the aft part of the vanes which act as a propeller. This function of making use of the active propeller spin and jet energy results in a saving either in the form of an increase in thrust and ship speed or a reduction in the horsepower input required for a given ship's speed. The Grim wheel has now been fitted to a number of ships and recorded improvements in propulsive efficiency range from 5 to 15%, depending upon the vessel type and propeller loading.

Figure 6.70 Grim wheel fitted aft of a ship's propeller

A typical Grim wheel has seven vanes with the boss and vanes cast as a single unit. The unit rotates on low friction roller bearings on a stub shaft flanged to the propeller boss and radial lip seals prevent water ingress to the bearings. It can be used with either fixed pitch or CP propellers and has been used with both.

Compared with a conventional propeller system the addition of the Grim wheel brings the following advantages:

1. Propulsion efficiency increase by up to 15%.
2. The main propeller can be operated at higher r.p.m. favourably affecting the weight and cost of the propulsion machinery.
3. Grim wheels can be as large in diameter as possible since the large number of blades and the low speed of the wheel allow small vertical clearances with the hull to be acceptable.
4. There is less resistance from a rudder fitted behind a Grim wheel.
5. Ships fitted with the device have a better stopping capability.

It is normal to fit Grim wheels to single screw ships but twin wheels can be employed. Most applications so far have been to vessels provided with around 12 000 kW, though the largest units ever made were of 6.7 m diameter and were fitted on the passenger liner *Queen Elizabeth 2*.

6.10.16 Water jets

A water jet (*Figure 6.71*) is basically a marine propulsor in which water is fed through an inlet duct at the bottom of the vessel, to an axial flow water turbine or pump which adds energy before expelling the water through a nozzle at a much higher velocity than the incoming stream. The resultant change in momentum provides thrust to the vessel and the jet stream can be deflected either side of centre by a moveable nozzle and/or scoop to provide steering and reverse control. Water jets are typically directly driven by a diesel engine or gas turbine and are popular for high speed craft and vessels required to operate in shallow waters.

A major advantage of the water jet is that for a given power loading a higher static thrust can be provided than could be obtained from an equivalent conventional or shrouded propeller. This is because for low speed applications the pump can be operated on or close to the point of optimum efficiency. Another feature is the ability to absorb the full driving engine power at all water speeds without cavitating. Compared with a propeller this results in superior acceleration characteristics as well as excellent directional control as the maximum thrust is immediately available.

A typical design of water jet has axial flow impeller blades with a very wide chord, running close to the pump casing to minimize tip losses and stator blades to straighten the water flow downstream of the impellers to avoid rotation losses. To overcome a problem with debris a grille is normally fitted over the intake and for steering the outlet nozzle pivots about an axis set at 45° to the vertical. When the nozzle is swung either side of the straight ahead position, the whole jet stream is deflected, with minimum loss for positive steering action. Selection of ahead through neutral to astern is obtained by the progressive lowering of a scoop or bucket which rotates about horizontal pivots mounted on the outlet nozzle. Jet units range in power from a few horsepower up to as much as 15 000 hp absorbed by a single unit.

Typical applications for water jets range from small sports boats, to work boats, luxury yachts, military amphibious vehicles and work boats, safety boats where protruding propellers pose a danger to swimmers and any craft required to operate in very shallow water. An important application these days is for propulsion of high speed catamaran ferries.

6.10.17 Pump Jet

A variation on the water jet is the pump jet (*Figure 6.72*) developed by Schottel. This is basically a centrifugal type pump in which the impeller axis is mounted vertically and it works in a volute type casing. The suction port is mounted against the flat bottom of the boat and the water is energized in the pump and expelled through a nozzle back into the water at a down angle of 15°, at the bottom of the craft. The complete volute casing can be rotated through 360°, thus allowing

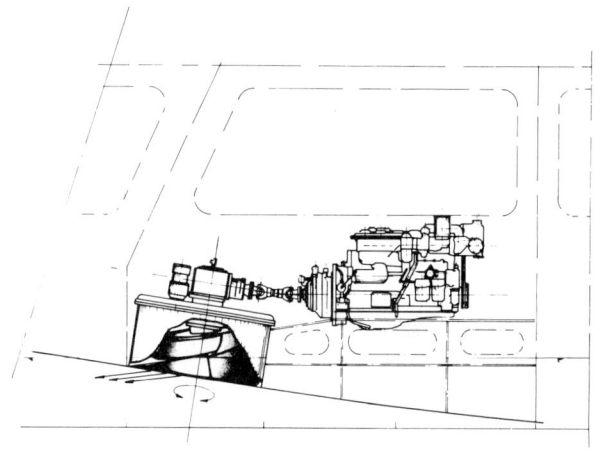

Figure 6.72 Installation of a pump jet system

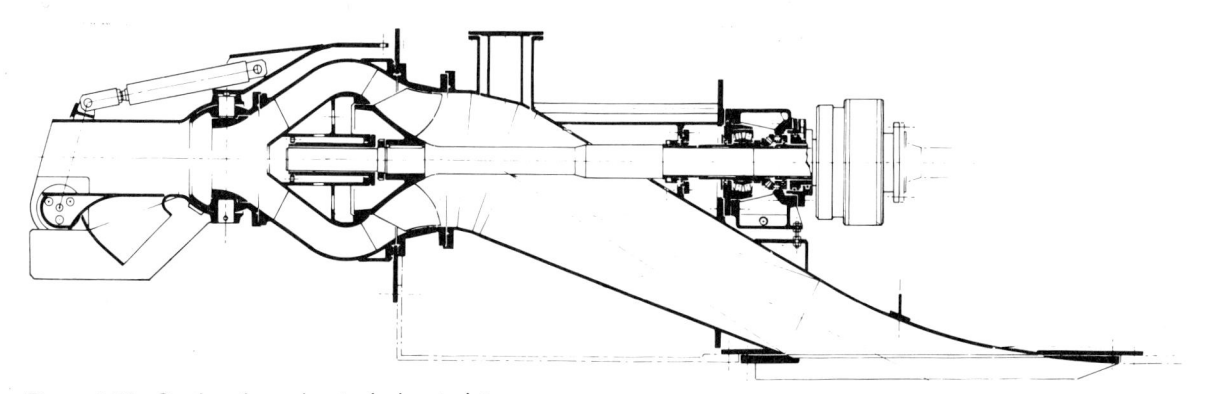

Figure 6.71 Section through a typical waterjet

the water jet stream to be directed at any angle to the boat centreline. This leads to a highly manoeuvrable craft, particularly if twin units are installed.

The top of the pump casing is fitted with a bevel gearbox to facilitate coupling of any suitable prime mover through a clutch and Cardan shaft. Typical driving engines are diesels 30–550 hp. The unit develops considerable thrust, as much as 80–93 N kW.

The pump jet can be used for main propulsion or as a bow thruster unit and is particularly suitable for work boats which require a high degree of manoeuvrability in adverse shallow water and high current conditions. The unit will operate with a bottom clearance of only 10 cm.

6.10.18 Azimuthing thrusters

A normal propulsion installation requires a rudder for vessel steering, but a simple rudderless system which has become very popular for a wide variety of vessels, is the Z-drive or azimuthing thruster unit (*Figure 6.73*). Rather than a propeller mounted to a shaft supported in the ship's structure, a steerable leg allows change in direction of the propeller thrust in the manner of the outboard motor used on a vast number of small boats. The application to ships is thus far limited in size to around 6000 bhp, but as yet such a large unit has not been used for normal ship propulsion, rather as a thrust unit on offshore work vessels.

Figure 6.73 Schottel azimuthing propeller unit

The Z-drive is a mechanically simple device, basically comprising a leg construction to which is attached a lower gearbox housing a bevel gear and the propeller shaft and an upper gearbox for another bevel gear drive and the azimuthing or slewing mechanism. The upper input shaft and the propeller shaft are parallel to each other and are connected by a vertical drive shaft and the pairs of bevels. The normal azimuthing mechanism is a worm and wheel drive mounted at the thruster head to allow the complete unit to be turned through a full 360° so that the propeller thrust can be directed anywhere in the horizontal plane. The unit is lubricated internally and a separate electric or hydraulic motor is used for the steering mechanism. Propellers used are either of fixed pitch, in which case the engine speed is varied and astern thrust is obtained by turning the thruster so that the propeller faces forward, or it may be of controllable pitch type operating at more or less constant speed and using pitch changing for speed control and manoeuvring.

Installation of Z-drives is relatively simple as the complete unit is bolted to a mounting fabricated in the ship. The unit is dropped in place and the connection to the propulsion engine is normally through a Cardan shaft coupling the thruster upper gearbox input shaft and the engine output and there is usually a clutch fitted in the drive line either at the engine or at the thruster input. The azimuthing drive motor is controlled through a system in which normally a single lever is used for engine speed and thrust direction; the lever assembly is simply turned to face the direction of desired thrust and the steering control thus turns the thruster unit to follow suit. Twin thruster units fitted side by side at the stern or further forward in tractor arrangement, usually have independent controls.

6.10.19 Z-drive advantages

The azimuthing thruster unit has for some time been used to propel small low powered vessels such as barges, ferries, etc., while it has made a very significant impact in the tug propulsion field with the considerable popularity of the tractor tug concept. In small vessels the Z-drive comprises a very convenient to fit propulsion unit, while the outstanding manoeuvring qualities are the main reason why the system is used in the first place. The combination of steering and thrust executed by the one mechanical device gives major advantages such as:

1. Most effective application of thrust as the whole unit is turned to direct propeller thrust where it is wanted rather than relying on deflection by rudder.
2. Very precise and effective steering of the vessel through thrust direction.
3. Ease of vessel control through single lever system or microprocessor controllers linking more than one thruster unit.
4. Much simpler machinery installation with no shaft bearing housing to be bored, giving more flexibility

in stern section design of the vessel.
5. Simpler maintenance of the whole propulsion machinery plant.
6. Great flexibility in the choice of propulsion machinery as electric drive motors, for example, can be flange coupled direct to the vertical drive shaft, or any suitable prime mover can be coupled to an upper bevel gear mechanism if this conception is employed.
7. If a constant speed thruster with a CP propeller is used, full use can be made of the engine or transmission system to also drive generators, pumps, etc.
8. The system is simpler, making it very suitable for remote operation and surveillance.

Benefits as outlined above, and others, have secured the Z-drive a strong position in the small vessel population market. However, further benefits should also make the system an attractive choice for the propulsion of larger vessels. These benefits are:

1. A simple stern, for instance the so-called barge stern can easily be adopted. This reduces shipbuilding costs and allows the construction of a broader stern, gaining displacement and improving stability.
2. The number of individual engine bedplates and seatings and engine systems is reduced and machinery installation is quicker and cheaper.
3. The engine room length can be remarkedly shortened meaning the payload section length of the vessel can be increased by around 5–10%
4. The thruster unit with upper gearbox can be mounted in a container which is flanged to a corresponding well made in the vessel's stern. The unit can thus be installed late in construction and even after launching.

6.10.20 Contra rotating thruster

The contra rotating propeller, which is also mentioned elsewhere in this chapter, was patented more than 150 years ago but is only now being applied to the propulsion of cargo type vessels. A version has also been developed for an azimuthing drive unit with the unit now on the market known as the CRP–Aquamaster, built by Messrs Hollming, Finland.

Steerable propulsion units have favourable features for the adaption to CR propellers. The right angle bevel gear in the lower body allows the use of two driven wheels so CRP is simply accomplished through an extra bevel wheel and concentric shaft to carry the two propellers. The bearing, shafts, seals and gears are extra but otherwise the system is conventional Z-drive.

The CRP–Aquamaster units that have been evaluated in service on cargo vessels, have four blades on the forward propeller and five on the aft propeller, these being optimized to efficiency whilst the diameter of the aft propeller is 10–20% less than that forward, to guarantee that the tip vortex of the fore propeller will not hit the aft propeller, and for optimum efficiency. The thrust load is divided to two propellers and nine blades, resulting in low loads per blade and fewer vibrations and cavitation. Also, the side force of the CRP set is nil, allowing identical units to be used on twin screw vessels, resulting in fewer spare parts required.

The manufacturer of the CRP–Aquamaster unit claims efficiency improvements of around 8% as well as the benefits of extremely good manoeuvrability and compactness of the machinery package.

6.10.21 CR propeller–thruster system

A variant of the above which has also been the subject of recent investigation is the concept of a conventional ship's CP propeller fitted in its usual position and with a rotatable Z-drive thruster working as a contra rotating propeller aft of the larger conventional propeller (*Figure 6.74*). The two propellers are arranged as close as possible to each other, hence the rudder blade section of the thruster is behind the propeller, which acts as a pulling thruster unit.

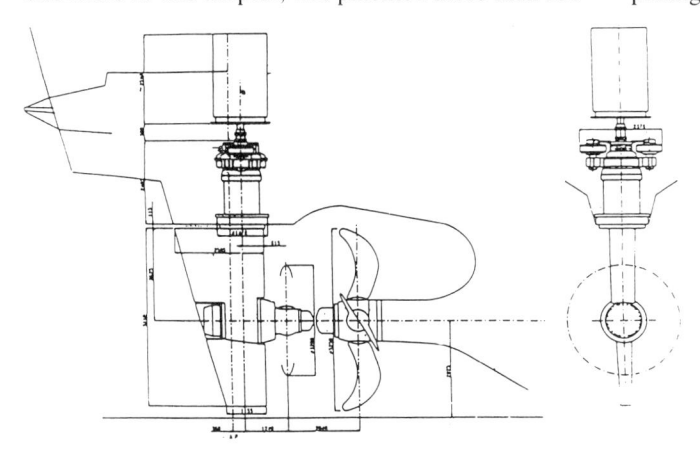

Figure 6.74 Concept of CP ship's propeller and contra rotating thruster system

The main advantages are improved manoeuvrability and flexibility and claimed efficiency gains of 5–10%. One disadvantage, however, is that the CR system requires rudder angles which are 2.5–3 times greater than a conventional rudder at a certain desired steering force, as the rudder area of an azimuthing thruster is limited.

Thrusters of types mentioned above are all used to increase propulsive efficiency and/or improve manoeuvrability. Such units are, however, of complicated mechanical design and expensive to manufacture. Consequently, the most popular arrangement for commercial ships is a large diameter as slow as possible turning single fixed pitch main propeller at the stern and a fixed side thruster unit fitted at the bow.

Further reading

Blank, D.A. and Richardson, D.J. (1985) *Introduction to Naval Engineering*, second edition, Naval Institute Press, Anapolis, USA.
Knak, C. (1979) *Diesel Motor Ships Engines and Machinery*. GEC GAD Publishers, Copenhagen, Denmark.
McBirnie, S.C. (1980) *Marine Steam Engines and Turbines*, fourth edition, Butterworths, London.
Milton, J.H. and Leach, R.M. (1980) *Marine Steam Boilers*, fourth edition, Butterworths, London.
Schneekluth, H. (1987) *Ship Design for Efficiency and Economy*, Butterworths, London.
Smith, D.W. (1983) *Marine Auxiliary Machinery*, sixth edition, Butterworths, London.
Wilbur, C. and Wight, D. (1984) *Pounders Marine Diesel Engines*, sixth edition, Butterworths, London.

7 Marine Control Systems

Ian McCallum

Contents

7.1 The requirement for control
   7.1.1 Steering control – requirement
   7.1.2 The effect of disturbances on course-keeping requirements
   7.1.3 Position control – the track-keeping problem
   7.1.4 The dynamic positioning problem
   7.1.5 Speed control and fuel economy requirements
   7.1.6 Whole ship control
   7.1.7 Information requirements
   7.1.8 The requirement for ship controllability
7.2 Control methodology
   7.2.1 Systems, inputs and outputs
   7.2.2 Disturbances or noise
   7.2.3 System description, open and closed loop performance
   7.2.4 System parameters
   7.2.5 Performance analysis, cost functions
   7.2.6 System analysis
7.3 The ship as a dynamic system
   7.3.1 Axis systems
   7.3.2 Ship equations of motion
   7.3.3 Effectors and control surfaces
   7.3.4 The effects of wind and waves on ship motions
7.4 Ship controllability
7.5 The role of the classification societies
7.6 Heading control – autopilots
   7.6.1 Conventional autopilots – controls
   7.6.2 Adaptive autopilots – methodology
   7.6.3 Autopilot control in a seaway
   7.6.4 Achieving optimal autopilot response

7.7 Track keeping and position control
   7.7.1 Requirements
   7.7.2 Information requirements – sensors
   7.7.3 Effector requirements
   7.7.4 Control methodology
7.8 Roll control – ship stabilizers
   7.8.1 User requirements
   7.8.2 Roll motions of a ship in waves
   7.8.3 Minimizing roll motion – hull design
   7.8.4 Minimizing roll motion – imposed devices
   7.8.5 Minimizing roll motion – operational methods
   7.8.6 Rudder roll stabilization
7.9 Machinery Control
   7.9.1 Requirements for a digital machinery control system
   7.9.2 Microprocessor based controllers
   7.9.3 Fault detection and diagnosis
7.10 Integrated ship control
   7.10.1 Requirements for integrated ship control
   7.10.2 Ship integration methodology
   7.10.3 Implications of integrated control on personnel
7.11 Assessing controllability – ship trials
   7.11.1 Assessing controllability
   7.11.2 Ship trials – purpose and conduct
   7.11.3 Improving controllability
   7.11.4 Predicting controllability
   7.11.5 Controllability as a port–ship problem

References

7.1 The requirement for control

The financial and operational constraints which most shipping companies have faced over the past decade have had a pronounced effect on the design of ships of all types. The trends have been towards larger ships, with simpler powering arrangements, based largely on heavy fuel, oilburning slow speed diesel engines, often with a high freeboard. Such ships frequently use ports where the operational clearances, in terms of available width in the lock, available depth of water under the keel, or room to turn, are severely curtailed (*Figure 7.1*). For large container ships, operational schedules demand that the time spent in port is minimized, so that there is a trend towards estuarial or riverside berths for these ships, which therefore have to be able to operate in tidal conditions. The effect of these trends towards operating ships at minimum clearance is effectively to reduce the available safety margins. There is therefore an increasing need for ships to be able to be controlled in a more efficient manner, so that the effects of these physical limitations may be minimized.

Figure 7.1 Ship in confined water (Courtesy of British Associated Ports)

The need for more precise control is apparent also in those ships which have to operate relative to the sea bed, as the available tolerances for positional errors are small. Offshore supply vessels have to be able to maintain station to a few metres in heavy seas close to a rig. Survey vessels have to be able to carry out soundings along a grid relative the sea bed, and dredgers have to be able to position themselves so that they are dredging new spoil effectively. These ships therefore have to be controlled so that they can move along a fixed track relative to the sea bed, as opposed to a fixed course relative to the water. Minesweeping operations also require the ship to move relative to a track, to ensure effective sweeping operations with no gaps.

A consequence of the trend towards large ships moving in more restricted areas is that the likelihood of collision or grounding is increased, while the results of such groundings are more severe. A ship displacing several tens of thousands of tonnes is not going to be able to withstand any collision with dock wall or other obstruction at speeds above half a knot without considerable damage occurring to both ship and obstruction. Similarly, the use of anchors on very large ships is confined to keeping the ship stopped; an anchor cannot be used to slow a large ship from speeds over a knot (Royal Institution of Naval Architects, 1978), as the results of such an exercise will be simply to damage the anchoring facilities.

The requirement to be able to produce a desired state or condition of a ship has resulted in the definition of ship controllability used by the 14th International Towing Tank Conference (ITTC), 1975:

> Controllability is 'that quality of a ship which determines the effectiveness of the controls in producing any desired change at a specific rate, in the attitude or position of the moving ship'.

The above definition accentuates the effectiveness of controls. The fact that environmental conditions are not mentioned implies that control should be effective throughout a reasonable range of environmental conditions. Thus a ship which cannot be adequately controlled in high winds, or will not manoeuvre astern, cannot be said to be adequately controllable, even if its behaviour in calm conditions going ahead is impeccable.

Controllability is related to stability, in that many ships which are dynamically unstable in yaw are very controllable, and indeed a measure of instability can assist in controllability. The relationship between controllability and stability, and the limitations placed on an adequate definition of controllability, are discussed in Section 7.3.

7.1.1 Steering control – requirement

The two most basic controls required in a ship are those associated with control of the direction or heading of the ship, and of its speed. For many ships, these are the only navigational controls available.

Single screw, single rudder ships predominate in the deep sea trades, for obvious economical reasons, and the control arrangements on the bridge consist of a single helm control, assisted by an autopilot, and a single engine control lever, which may operate in either bridge control, where the desired shaft speed is set by the position of the lever, or engine room control, where the lever position is transferred to the engine control room for action to be taken there.

The basic requirements for heading control is that the ship's head is maintained to within a given band of the desired value. The size of the band, or steering error, will depend in turn upon the dynamic properties of the ship, the effectiveness of the steering arrangements, the disturbances present (wind and waves), and on the perceived requirements of the Officer of the Watch, which will in turn vary with his assessment of the navigational situation.

The process of steering a ship is shown diagrammatically in *Figure 7.2*. The difference between the desired course and the actual course is assessed. This may be done in a number of ways, depending on how the ship is being steered. If under helmsman control, the helmsman may be given instruction to steer a compass course (i.e. a course relative to the earth's north–south lines of longitude), or may be given instruction to head towards a fixed object, or simply to keep in the middle of the channel. If the ship is in autopilot control, a desired course will have been set on the autopilot by the Officer of the Watch, usually by turning a control on the autopilot. In either case, the difference between the desired and actual course of the ship is defined as the heading error.

The helmsman or autopilot will act on this error, and will alter the demand to the rudder control mechanism. This signal is shown in *Figure 7.2* as the desired rudder angle. In most large ships, this action will be in the form of a signal to an amplifier or to a servo valve in a hydraulic control mechanism. In smaller or older ships, the control action may be directly passed to the rudder. The result of the control action is that the rudder will, after a time lag, assume the value of the desired rudder angle, within the bounds of error of the control system. The rudder will then (usually) act on the slipstream of the propeller, and create a turning moment on the ship. The effectiveness of the rudder depends on a number of factors, which will be analysed in Section 7.2, but its effect will generally be to turn the ship in the appropriate direction. As the ship turns, the error will reduce and eventually the ship will assume a heading approximately equal to the desired heading, at which time the error is zero. As the link between the rudder angle and the ship's behaviour is complex, the ship will generally not stay at the desired heading, unless a continuous control action is applied. The nature of the control action will in turn depend on the required steering performance and on the ship's design and operating condition.

The rudder or rudders are usually placed at the stern of the ship, immediately aft of the propellers. The reasons for siting the rudders at that position are concerned with the effectiveness of the controls. The ship will turn, under the combined influences of rudder, inertial and hydrodynamic forces, about a point which will usually be some distance forward of the mid-point of the ship. In some ships this pivot point, which may be defined as the point at which there is no component of sway velocity, is situated some distance forward of the ship (*Figure 7.3*).

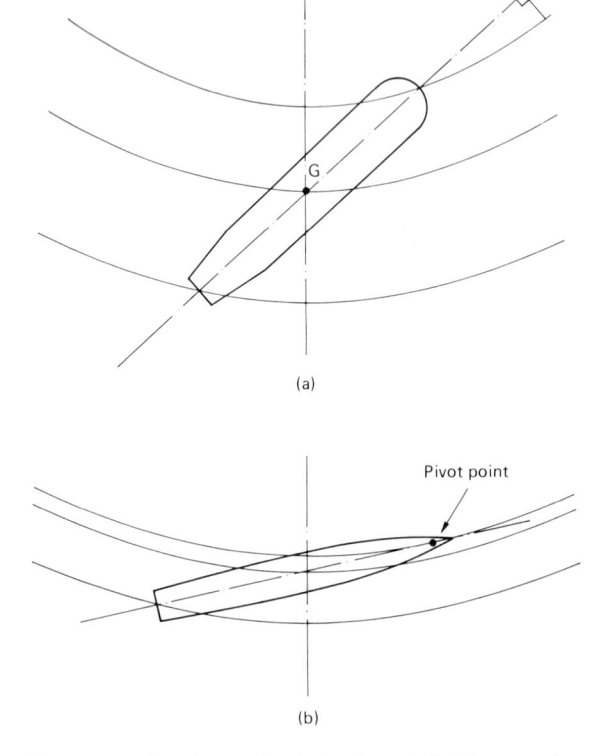

Figure 7.3 Steady turning behaviour of full form and slender ships: (a) VLCC; (b) Frigate

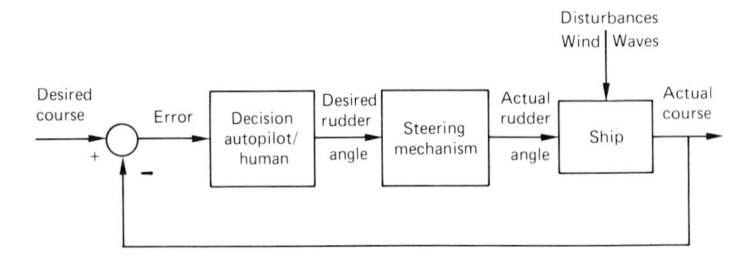

Figure 7.2 Block diagram of ship turning behaviour

A rudder situated at the stern of the ship has two main effects: it is able to exert a large lever arm about the pivot point and is able to position the ship such that the hydrodynamic forces assist in the turn. Some ships are able to turn effectively going astern, when the turning effect is largely caused by the hydrodynamic forces on the rudder alone, but the turning ability of a ship is very much enhanced if the hull hydrodynamic forces augment the rudder forces. A further reason for placing the rudder at the stern of the ship is that it can be positioned so that the propeller slipstream augments the flow of water over the rudder. As rudder forces are heavily dependent on the velocity of flow across them, the effectiveness of the rudder is enhanced by this positioning. When the ship's engines are put astern however, the propeller slipstream can be confused, sometimes with a total loss of rudder effectiveness. Ships which have to manoeuvre astern frequently and reliably, such as cross channel ferries, are occasionally provided with stern rudders and propellers (*Figure 7.4*) or have a completely symmetrical configuration so that they may travel in either direction with equal facility.

The traditional form of rudder control has remained largely unchanged for many centuries. The Greek Trireme, shown diagrammatically in *Figure 7.5*, used trailing rudders to cause the ship to turn. Despite the large length–breadth ratio of the ship, it turns effectively under rudder control as shown in *Figure 7.6* (Lowry and Squire, 1988).

The required turning ability of a ship will depend to some degree on its role. An offshore supply vessel, tug or ferry will usually have to have a better turning ability than will a bulk carrier or container ship. However, there is a discernable trend towards a reduced degree of tug assistance for ships berthing, and many conventional cargo vessels are required to have a greatly enhanced turning ability at low speed. Special rudder arrangements can be used to augment the traditional configuration, such as flapped rudders or thrusters. These arrangements are discussed further in later sections. The turning requirements of offshore supply vessels are particularly severe and conventional methods have largely given way to a variety of special effectors, such as fully azimuthing thrusters and a range of transverse thrust arrangements.

Submarines and submersibles have similar attitude or directional control requirements, with the additional requirement that control is required in the vertical as well as in the horizontal plane. A submarine will normally control its heading and attitude by means of two sets of control surfaces at the rear of the hull, and additional control planes forward for vertical motion control (*Figure 7.7*). The ways in which these planes are altered will depend on the speed of the submarine and on the type of manoeuvre required.

A small research or reconnaissance submersible may have totally different control methodologies to effect a change in heading. The motion of the spherical submersible (*Figure 7.8*) is controlled by means of four small

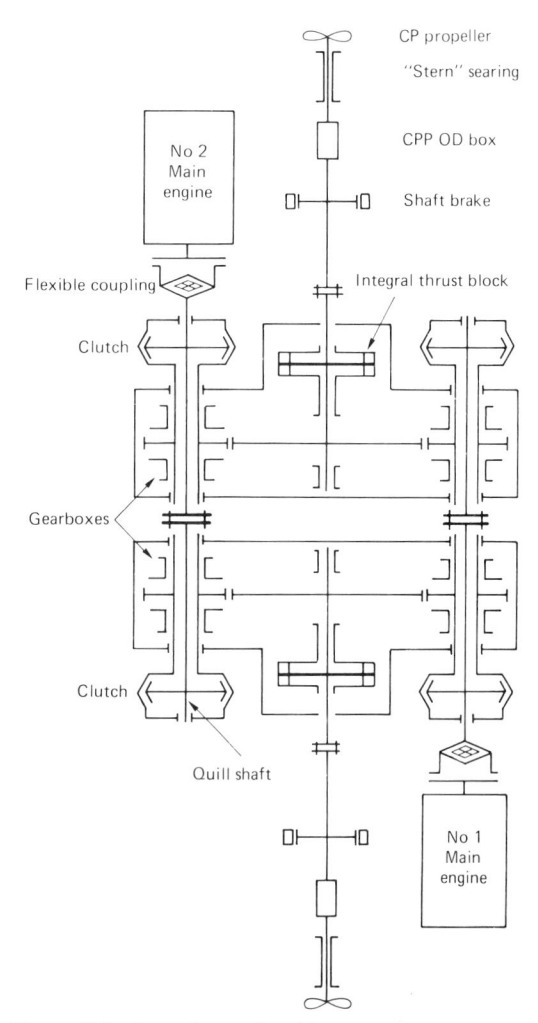

Figure 7.4 Propeller and rudder arrangements on a P & O ferry (courtesy of Urban Transit Authority, New South Wales, Australia)

Figure 7.5 Greek Trireme Olympias (Courtesy of The Trireme Trust)

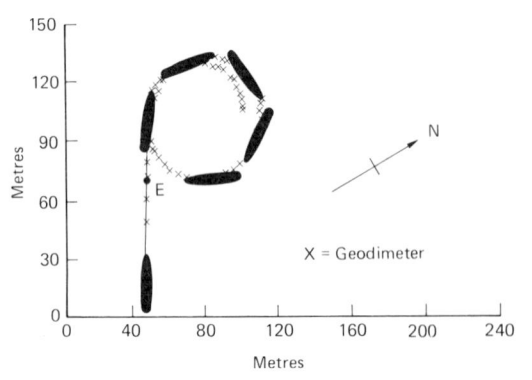

Figure 7.6 Trireme turning circles (Courtesy of I J Lowry, T Squire)

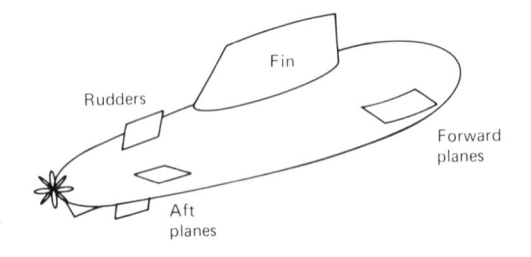

Figure 7.7 Submarine control surfaces

Figure 7.8 Reconnaisance submersible (Courtesy of Hydrovision Ltd)

thrusters, the effect of which is to enable motion in any desired direction to be produced.

7.1.2 The effect of disturbances on course-keeping requirements

Disturbances will act on the ship as shown in *Figure 7.2*. The most common form of disturbance will be the presence of wind and waves, but other effects which will affect the course control behaviour will include:

- The presence of the sea bed, in that ships behave differently in shallow water to the way they do in deep water. Turning ability is reduced and the diameter of turn increased.
- The presence of banks and other ships. Ships tend to turn away from banks and there are complex interactions between two ships passing close to each other. In certain circumstances, a small ship can be uncontrollably drawn under the bow of a larger ship as a result of interaction forces.

The way in which the steering ability of the ship is affected by the presence of the wind depends on the shape of both the above water and underwater hull, and on the strength and direction of the wind. The effect of waves will generally be to reduce the effectiveness of the control mechanisms and to make steering a course within a given margin more difficult. These effects are discussed in greater detail in Section 7.2.

7.1.3 Position control – the track-keeping problem

As the aim of most voyages is to get the ship and its cargo from one point on the earth's surface to another, the track-keeping problem is central to the operation of most ships most of the time. However, for most practical purposes, the track-keeping problem can be reduced to one of course-keeping, in that the effects of the disturbances which will tend to prevent the correct track being steered can be allowed for the accuracy required to ensure that the ship arrives at the correct destination.

For more exacting applications, however, there is a requirement for more precise track-keeping. Examples of this enhanced requirement can be found in survey ships and boats, in minesweepers, in dredgers and in offshore supply vessels, mobile drilling rigs, etc. All these vessels need to be able to keep to a fixed track rather than a fixed course, within clearly defined limits. A ship at the end of a voyage will also have to assume a track-keeping role, as it must navigate an approach channel and enter a lock or dock, often with under 1 m (3.3 ft) clearance. This normal approach phase is a type of track-keeping.

The principle of track-keeping is similar to that of course-keeping, in that an error is defined, and the ship's controls operated in a manner so as to reduce the error ideally to zero, but more usually to within a satisfactory limit. The principle is shown in *Figure 7.9*. A reference track is defined, which is fixed relative to the sea bed or in relation to shore features. The

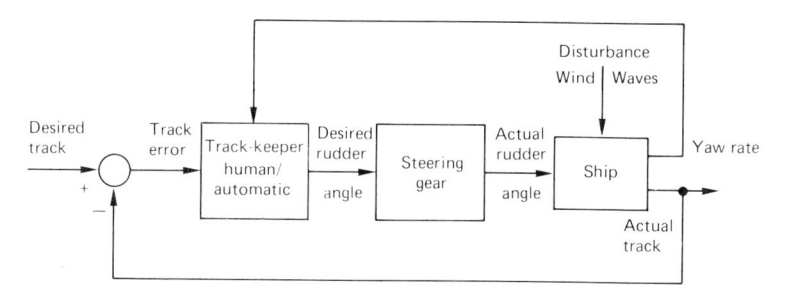

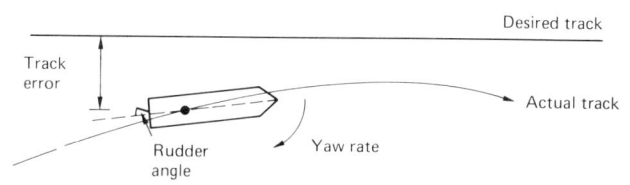

Figure 7.9 Track-keeping problems: block and schematic diagrams

distance from the reference track is determined, either by eye or mechanically and this distance is used as part of an error signal. Other components of the error signal might be the yaw rate of the ship and its heading. Decisions on the control of the ship are made as a function of the track error, as modified by the other inputs, and signals sent to the rudder control and engine control mechanisms.

With the notable exception of dynamic positioning systems (see Section 7.1.4), there is usually a human controller in the chain in a decision making role. Thus, if we take the example of a ship making a landfall along a shore based leading mark with a cross tide and wind, the mariner will assess his lateral distance off his reference track by observing the leading marks, will assess the effects of the wind and current on his ship and will order a course and speed so as to transit the channel effectively, within a perceived distance from the reference track. *Figure 7.10* shows the track of a simulated ship transiting the site of a channel in the River Severn, with a 6 knot current towards the southwest and a strong easterly wind. The pilot observes his position relative to a transit line consisting of two leading marks (shown in the plot) and orders a course which in his judgement will keep the ship on a satisfactory track. In this case, the ship has to be misaligned to the reference track by some 17° in order to keep parallel to the track. If the pilot had wished to stay exactly on the reference track he would have had to alter his desired, and hence actual, course by more than 17°.

Some research has been carried out (Parsons and Hua, 1981) on the possibilities of a fully automatic track-keeper for a Great Lakes bulker required to transit tight bends in Great Lakes channels. The methodology used positional information from a Decca Hi-Fix or radar systems. The ship control signals were evaluated using

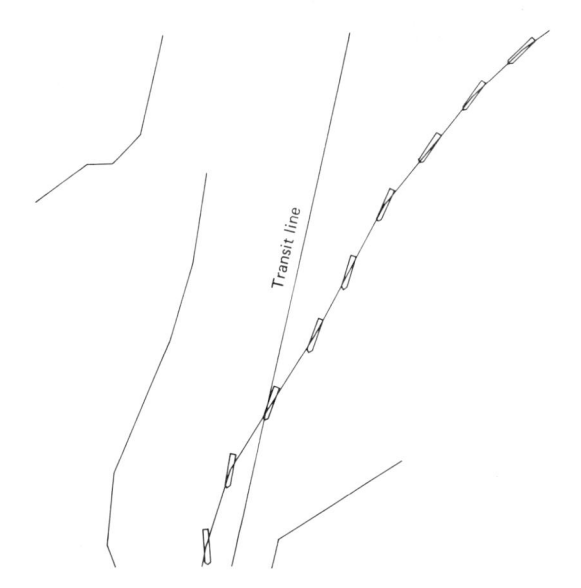

Figure 7.10 Track-keeping in the presence of wind and currents

Kalman filter techniques to minimize the effects of system noise, and it was able to be shown in simulation that it is feasible to control a large ship adequately in confined waterways, in the presence of system noise and uncertainties about the system behaviour.

Sophisticated methods are available to enable the mariner to know his track error for particular operations. In the case of a dredger, it is necessary to know within a very few metres where the ship is relative to the reference track in order to ensure that the dredging operation is satisfactorily carried out. If a shore based hyperbolic navigation system is used to determine the instantaneous position of the ship, it is possible to display the ship position electronically on an electronic chart. *Figure 7.11* shows such a system, in use on an Associated British Ports dredger.

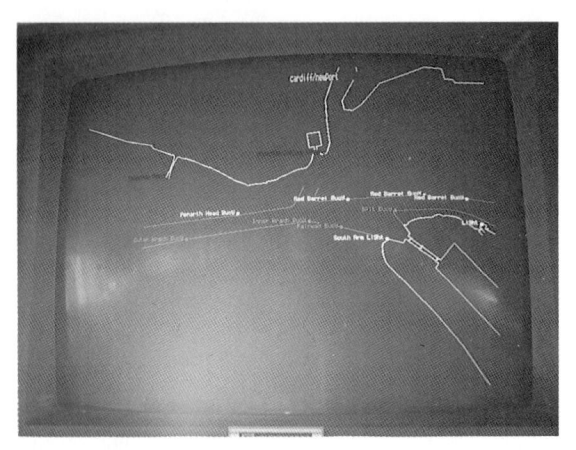

Figure 7.11 Electronic position display equipment on dredgers, courtesy Associated British Ports

7.1.4 The dynamic positioning problem

In a number of cases associated with the offshore industry, there is a requirement for a ship or platform to maintain its position relative to a fixed datum on the sea bed. This is a special case of track-keeping, where the reference track is a single point. Because of the accuracy requirements for dynamic positioning (DP), a number of specialist devices are in use for both identifying the positional error and controlling the ship motion.

Figure 7.12 shows a typical DP controlled vessel, with its position sensors and controllers. Position information is obtained from a range of sources. A taut wire may be suspended from the ship to the sea bed, and its angle from the vertical used to convey information about the position of the vessel relative to its datum. Acoustic information may be obtained from reference transponders attached to the sea bed, and Doppler sonar information can be used to give velocities in both fore and aft and athwartships directions relative to the sea bed.

The error signal for position is then known, and a set of signals to the effectors can be calculated to produce an optimal vessel response to get the ship into the correct position. A wider range of effectors is employed in drill ships and offshore supply vessels, including fixed pitch propellers in nozzles, controllable pitch propellers, rotatable thrusters and fixed thrusters.

In severe weather conditions, it is sometimes impossible to control the ship using the manoeuvring thrusters alone. In these cases, an alternative control methodology can be employed whereby the ship is manoeuvred to allow the main engines to be used to stem the weather. This system is further described in Section 7.7.

7.1.5 Speed control and fuel economy requirements

The cost of fuel is one of the two most significant operating costs of a ship (the other being crew costs), and much effort is expended in ensuring that a ship is designed to operate as economically as possible over most of its operating range. Most merchant ships are designed to operate effectively at a single speed, which may be the maximum speed of the ship. Other ships, such as tugs, trawlers, offshore supply vessels, and all warships, have to be capable of operating effectively over a range of speeds, and may have very complex engine control arrangements to enable this to be achieved. An antisubmarine frigate, for example, will need to be able to operate for prolonged periods at a speed slow enough for its sonars to be effective, which may be at comparatively slow speeds. Once a target is detected, there may be need to have available a far greater range of speeds to enable the ship to prosecute a contact, or simply to shift operations to a new site. This need for a range of speeds is frequently achieved by equipping the ship with more than one engine per shaft, so that different powers, and hence speeds, can be achieved by using either, or sometimes both, of the available engines on each shaft.

For offshore supply vessel operations, an even greater range of operating speeds is required. The ship may be required to achieve effectively a zero speed, in the presence of wind and wave disturbances, when it is operating in the support role, where for instance it will need to maintain precise station under a crane hook. As soon as the crane transfer is complete, the vessel is likely to need to transit at a reasonable speed to a new station or to return to shore. This dual role may be achieved in a number of ways. The normal transit role is often achieved by fairly conventional means, with the positioning capability achieved by directional thrusters. The requirement for precise control of engine speed for positioning is severe, however.

The majority of merchant ships have far simpler engine control requirements. With most cargo vessels

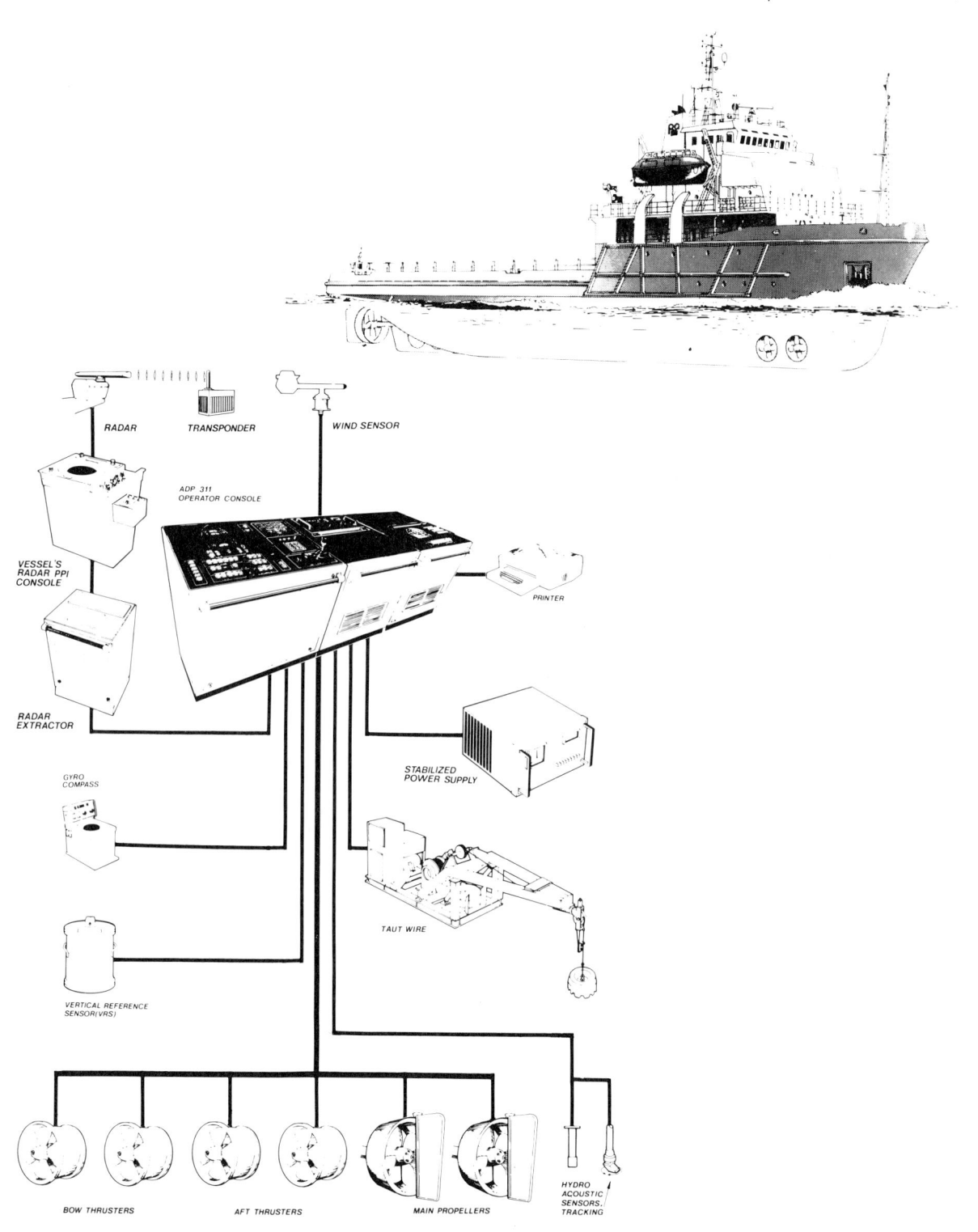

Figure 7.12 Dynamically controlled offshore vessel: sensors and effectors (Courtesy A/S Kongsberg Vapenfabrikk)

being fitted with a single slow speed diesel engine, and designed for operation at a single speed, the main control requirement is simply that of maintaining a desired engine speed, in the presence of disturbances. The control of engine speed is shown diagrammatically in *Figure 7.13*. A desired engine speed is set by the operator (either directly from the bridge, or from the engine control room under command from the bridge). The difference between the actual engine speed and the desired value is called the error. A controller, or governor, will act on this error and regulate the supply of fuel to the engine in such a way as to tend to reduce the error to zero. The controller will, however, need to take account of the operating condition of the ship in a number of ways, to avoid damage to the machinery, or to stay within a number of ship design parameters. For example:

- Diesel engines cannot operate below a certain speed, so that the engine may stop if too low a speed is ordered.
- It may be impossible to achieve the desired engine speed, because of adverse weather, or because the ship is manoeuvring.
- During the first few hours of operation from cold, the maximum allowable speed is limited by thermodynamic considerations, and the controller may be designed to allow for only a gradual build up of engine speed for several hours.
- If any of a number of engine operating conditions varies outside acceptable limits, the controller may be designed to slow the engine down or to stop it altogether. For example, if the supply of lubricating oil fails, the engine can suffer irreversible damage in a few seconds of operation. The controller will therefore cut off the supply of fuel to the engine in these conditions, it being reckoned that the overall safety of the ship will be better served with the engine stopped than with it continuing until damaged. In some ships, particularly warships, an override on this facility is likely to be provided.

An alternative form of ship speed control is for a controllable pitch propeller (CPP) to be fitted to each shaft. The effective thrust is then determined by both the shaft speed and the angle of the controllable pitch propeller blades. A range of control strategies is found to control power as a function of these two variables, with the controller sometimes being referred to as a combinator in these circumstances. The control of these devices is discussed further in Section 7.9.

The two largest items of expenditure associated with most shipping operations remain the cost of crew and the cost of fuel. There is therefore a continuing requirement to develop more efficient ways of using fuel, and towards the reduction of crew numbers. The most significant developments towards greater fuel economy have in recent years been the almost universal adoption by the deep sea trades of the slow speed diesel engine as a propulsion plant, with greater use of automated engine controls to achieve an optimal fuel performance. The use of heavier grades of fuel has enabled significant cost savings to be made, with most engines now being able to both start and run on heavy distillates.

Cost savings can be made by simplifying the overall propulsion system design, principally by using a single large engine, directly coupled to a single fixed pitch propeller. *Figure 7.14* shows a typical modern engine installation in a container ship of 17,600 tonnes deadweight, where it will be seen that the main propulsion unit consists of a single five cylinder slow speed diesel engine. The advantages of such a power plant over the more traditional installation of a steam power plant lie almost entirely in reduced fuel consumption. The slow

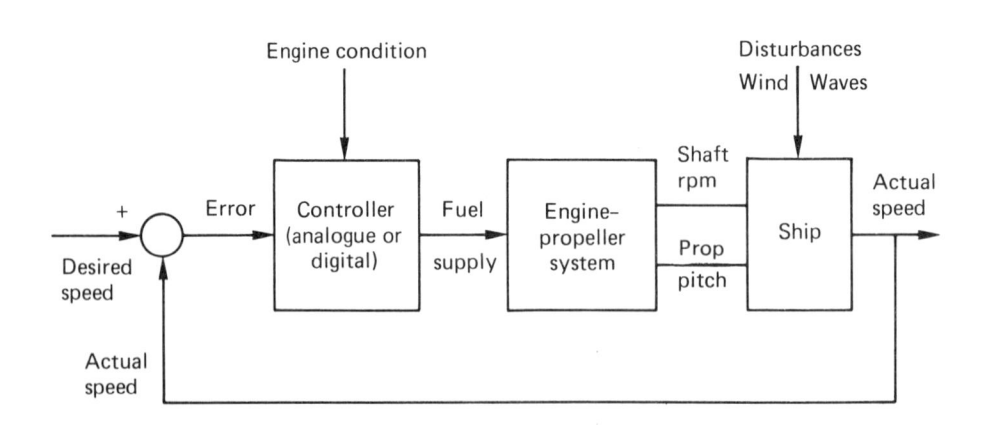

Figure 7.13 Control of engine speeds: block diagram

speed diesel installation will be more expensive in its first cost, and will have a greater maintenance load, because it has more moving parts and is larger in size overall than an equivalent steam plant of the same power rating.

The disadvantages of the common single engined, single fixed pitch propeller installation, from a ship controllability point of view, are concerned with the difficulties of slow speed operations, the problems of close quarters manoeuvring and the inherent unreliability of having only one main power source.

A slow speed diesel engine will rotate, at its slowest, at about 22 rpm without stalling. This will correspond, in some ships, to a speed in excess of 5 knots, which is likely to be far too high for many operations in rivers or when approaching a berth. In order to go slower, the engine will have to be stopped, which will result in a loss of directional control, because of the reduced water flow across the rudder. Once the engine is stopped, it has to be started again using the ship's air supply, and there are a finite number of starts available to the master, depending on the charge level in the air bottles.

In order to slow the ship, or reverse it, the engine has to be stopped and restarted in the reverse direction. This will also deplete the available air supply and occasionally it will prove impossible to restart the engine at all, particularly if the plant is still below its normal operating temperature. The available air supply may also not be able to start the engine in a reverse direction at all if the ship is travelling in a forwards direction above a certain speed, so that the ship cannot always be relied upon to achieve its full stopping performance.

For a large ship, there is an inherent weakness in a design which relies on the continued functioning of a single power source for ship propulsion, without any sensible level of auxiliary provision. Most single shaft installations may be thought of as being deficient in this respect, as any significant defect in the main engine will render the ship totally powerless, requiring tug assistance for it to be taken into port for repair. The

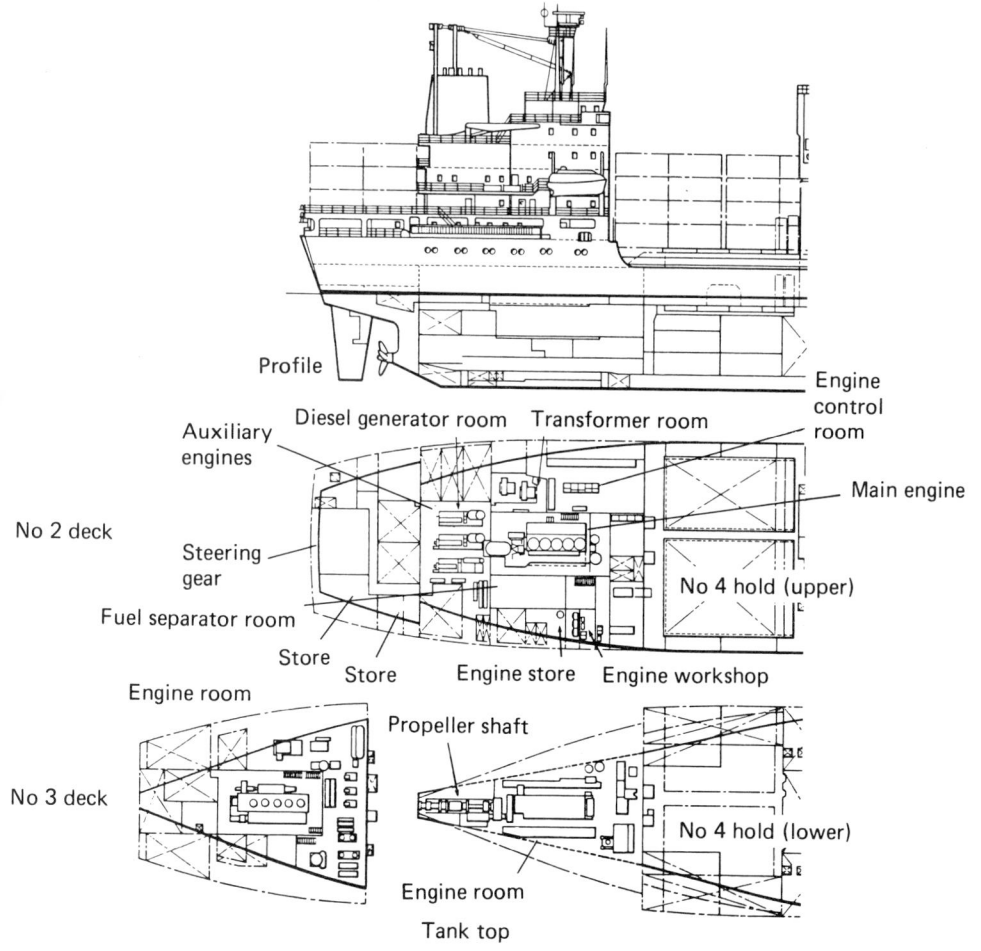

Figure 7.14 Bulker engine installation: general layout
(Courtesy The Motor Ship)

largest ships, in excess of 250 000 deadweight tonnes, may have only a very few repair berths which can accommodate them, necessitating long journeys under tow for repair. It is often now beyond the capacity of a ship's crew to carry out many repairs on board, because of the size and weight of the replacement units which have to be manoeuvred into position.

A twin screw installation presents a greater level of fundamental reliability, in that a significant level of propulsive ability remains with only one power plant serviceable, but there are penalties to be paid, in terms of cost and a more complex stern design, to accommodate the two shafts. For controllability purposes, it is unsatisfactory to have a twin screwed installation with only one rudder, as the rudder is largely out of the slipstream of the propellers in normal operations, and so additional cost needs to be incurred to install a twin rudder configuration also.

7.1.6 Whole ship control

This term is used to define the situation where all ship operations, navigation, machinery control and cargo operations are controlled from a central position in the ship. Because of the overriding requirement in all ships to keep a good lookout at all times, this control station is likely to be situated at or adjacent to the ship's bridge. The logic for such a control methodology for a ship is clear when one considers that most control functions today consist of monitoring a number of computer based control systems, most of which will give an indication of malfunction. It makes little sense to have a watchkeeper on the bridge, one in the engine room and one in the cargo control space, all watching an automatic watchstander.

Because of the inherent infallibility of machinery, especially that exposed to the marine environment, there is little support from regulatory bodies for the totally unmanned ship, although the technical problems associated with such a ship have been available for at least a decade. It is almost universal in merchant ships now to have an unmanned engine space, with control and monitoring being left to automatic systems. Thus, the technical problems associated with whole ship control are largely those of siting the appropriate control consoles with their monitoring and alarm systems and arranging for appropriate actions to be taken in the event of malfunction. There are some design problems associated with the man–machine interface, as it is inappropriate for a single watchkeeper to be presented with a large amount of redundant information, and some human related problems associated with obtaining and training ship's personnel who are capable of responding with equal skill to a problem associated with a malfunctioning engine and a navigational situation. Many countries are now, however, addressing this last problem, by introducing dual qualifications for selected officers.

Figure 7.15 shows a proposed liquid natural gas carrier designed in the early 1980s for Arctic opera-

Figure 7.15 Proposed liquid natural gas carrier, courtesy Arctic Pilot Project

tions, which had some of the features of whole ship control built into it. Control was to be exercised from the bridge located forward at the top of the superstructure. The ship was never built, but incorporated a number of design features to increase its controllability, including:

- Twin screws, with controllable pitch propellers, the propellers being ducted to enhance the thrust from them;
- the use of gas turbine main propulsion (using cargo boil-off to provide the gas) which would give a very good transient response for use in the ramming mode of icebreaking, in which repeated backings and fillings are required;
- the use of a spoonbow, in which the bow shape resembles that of a teaspoon, rather than that of the more traditional wedge shape, and a reamer section, in which the widest part of the ship is at the icebreaking bow shape. This design allows easier breaking of ice in the continuous mode, by the ship carving out a channel a little wider than the beam of the vessel.

There has been little progress towards full implementation of whole ship control in the past decade, although increasing automation of bridge, engine control and cargo handling equipment separately has been achieved. On the bridge, the most significant developments have been towards automating the navigational function. Automatic watch can be kept by an automatic radar plotting aid (ARPA), which can detect and plot a number (typically 50) of radar targets and give an alarm when any comes into a guard zone set up by the Officer of the Watch. A number of electronic navigation systems, principally satellite navigation systems (SAT-NAV), Loran and Decca, can be integrated into a single system, in which an optimal estimate of the ship's position is given, based on the likely accuracy of each of the individual navigation systems. A typical system is shown in *Figure 7.16*.

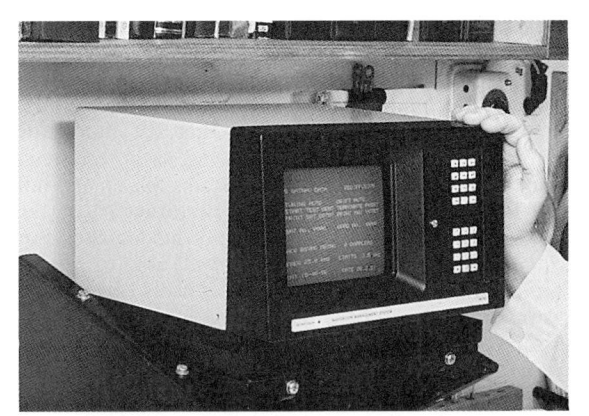

Figure 7.16 Typical integrated navigation system (Courtesy of Rediffusion Radio Systems Ltd)

A further development towards greater navigational automation is the potential use of the electronic chart. This is a possible method of augmenting or possibly replacing the traditional paper charts by information stored on disk and rescaled onto a VDU on the bridge as required. Advantages of such a system are considered to be a considerable saving in space as fewer traditional charts could be required, and the ability to select what information is displayed. For example, a ship drawing 5 m (16.4 ft) is not going to be interested in bottom contours deeper than, say, 15 m (49.2 ft). Charted information can be displayed wherever it is best used, e.g. at the front of the bridge, as daylight view VDUs can be used, and the scale of the chart is able to be varied at will.

Difficulties inherent in such a system include the problems of updating information, although this can theoretically be done automatically by satellite communication from shore. The problem of responsibility for the accuracy of the information supplied, and for ensuring that it is complete when the user has a high level of selectivity in deciding what to see, have also exercised those responsible for the introduction of these systems. Of continuing concern is the position when the ship suffers a complete blackout, in which case there will be no chart information available exactly when it is most likely to be required. Even ships fitted with good standby generation capabilities have been known to be without power for some considerable time. If it is going to be necessary to have a full outfit of charts available for backup purposes, some of the arguments for having the electronic system in the first place disappear.

In the engine room, the most significant trends have been towards increasingly complex monitoring arrangements. It is common today to have a ship's power plant monitored with several hundred sensors for temperature, pressure, flow and viscosity. The sensed variables are either displayed directly as gauges, or go into a computer system, where they are compared with the maximum allowable values, and an alarm signal produced if any variable is outside the safe band. These alarms are displayed on a simple alarm panel and logged in the computer, so that the sequence and identity of each alarm can be determined. This information is used by the engineer to assist in the diagnosis of an engine fault.

The automatic diagnosis of engine faults from this information is a logical development from simply displaying information and the use of expert systems techniques to achieve this is beginning to be achievable. Work at Newcastle University has been directed towards producing a prototype system for this task.

7.1.7 Information requirements

For the control of a ship to be carried out automatically, there is a need for the control system to be informed about the current state of the ship's systems. *Figure 7.2* shows a block diagram of the simplest control system, with only one controlled variable. In this case, the input to the controller consists of the error between the desired value of the controlled variable and its actual value. There must therefore be some form of measurement to produce a signal which is compatible with the requirements of the controller. In nearly every case with modern control mechanisms, the controller will consist largely of some form of microprocessor, so that the error signal needs to be in the form of an electric signal, either in analogue form or as digital information. Similarly, the input signal usually needs to be converted into the same form as the output so that they can be compared.

In the example of a ship steering system shown in *Figure 7.2*, the desired signal will be obtained either from the position of the helmsman's wheel or from the autopilot signal. In either case, this will be in the form of an electrical signal by the time it gets to the controller or autopilot. The actual value of rudder angle will be converted, usually very close to the rudder, by some form of rotary converter such as a synchro, to give the required feedback signal.

An engine installation in a merchant ship will typically have several hundred sensors. Each will be converted into an appropriate voltage signal by a transducer and used as an input to a monitoring system. This requires a considerable investment in transducers, system writing and computer power, but is essential if a responsibly engineered unmanned machinery system is to be incorporated into the ship. Once the necessary signals are in the computer, it is possible to carry out whatever signal processing is necessary to produce the alarm signals. These then need to be displayed, either on a VDU, or as lights in an alarm panel. The requirements of a computer based monitoring and alarm system are discussed more fully in Section 7.10.

7.1.8 The requirement for ship controllability

Working from the basic definition of controllability in Section 7.1, it will be seen that there is a need for a ship's ability to be controlled in all aspects of its behaviour to be more clearly defined. At present the requirement for adequate controllability is not at all well defined.

Ship operators have at the most a vague idea of what a good standard of controllability means, and none of the regulatory bodies have clear standards or definitions which can be said to define the property adequately. Because the concept is not well defined, a ship owner cannot specify in a meaningful manner what aspects of ship controllability he requires in his new ship and certainly the ship builder is not in a position to guarantee any specific level or degree of controllability, as such a quality is not readily measurable.

Similarly, insurers, who might be thought to be in a position to benefit from more controllable ships, as such ships could be thought of as having a smaller likelihood of collision or grounding, have had no input at all in the definition of controllability.

There are two international bodies which do, however, have some interest in the matter. The International Maritime Organization (IMO) has a standing committee on ship manoeuvrability, which is producing a set of guidelines, and the manoeuvrability panel H10 of the US Society of Naval Architects and Marine Engineers (SNAME) have worked in this area.

The last named panel surveyed a large number of pilots in an attempt to obtain an opinion on those aspects of a ship's manoeuvres which were most desirable from a controllability point of view. The most significant factors listed were:

- slow speed manoeuvrability (86% of pilots);
- adequate backing power and straight line stopping ability (79%);
- short response time following rudder or engine commands (77%);
- adequate swing control with moderate rudder angles (66%).

Any quantitative measure of controllability must therefore contain elements to assess each of the above quantities if an adequately controllable ship is to be achieved as far as the users are concerned. The methods whereby controllability may be assessed and measured are discussed further in Section 7.3.

7.2 Control methodology

This section describes the fundamental principles of control, from a user viewpoint. The aim of a control system is to produce a desired state of a system through the action of relevant controls. The control actions required to achieve the desired state are defined by the design of the control system. Thus, to achieve a desired course in a ship in a seaway, a control action is applied to the rudder and some form of strategy is defined in either the autopilot or asked of the helsman to achieve this state. Control may either be automatic, or may have a crew member in the loop. Generally, for well defined systems, the desired system state can be achieved much more readily and reliably using automatic control, with the advantage that the automatic controller will not tire, need relieving, or be erratic in its performance. However, in systems which are not well defined or understood, the performance of the automatic controller may be totally inadequate. For example, most autopilots are not well able to steer a ship down a heavy quartering sea. This is because the disturbances to the system are large and can be unpredictable. A good helmsman in these circumstances can, by predicting when the waves are likely to hit the ship, often steer a better course in these special circumstances.

7.2.1 Systems, inputs and outputs

A typical control system for steering a ship is shown in *Figure 7.2*. For the purpose of ship control methodology, we may define a system thus as 'a collection of interrelated physical objects'. The system may consist of the whole ship, the steering mechanism or a small part of a mechanism, such a servo control valve. The person carrying out the study has complete freedom to define his system in the manner he deems of most use to that particular study. Clearly, the equations he obtains will be different for each system defined.

It is common to draw a system boundary around the system, to show clearly what is included in the system. *Figure 7.17* represents the ship shown in *Figure 7.2* as a

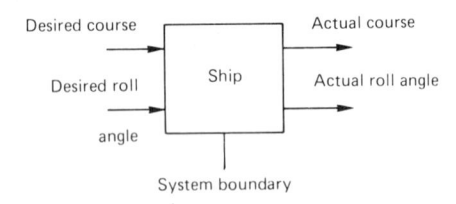

Figure 7.17 The ship as a dynamic system

system by drawing a system boundary round it. A system will in general have one or more inputs, and one or more outputs. We may define system inputs and outputs as 'a system input is an independent or controlling variable' and 'a system output is a dependent or controlled variable'. Thus, for the ship's steering system *Figure 7.2*, we may define the system input as the desired course and the system output as the actual course. If we wished to consider the ship in a seaway, with roll motions, we may wish to include the stabilizers as part of the defined system, in which case the system inputs would consist of: the desired course and the desired roll angle (usually zero); and the outputs would consist of the actual course and the actual roll angle.

The equations of motion for the ship will necessarily differ if we have to include considerations of roll, and so we would only include this motion if the problem under consideration required it. We may also include the actual rudder angle as an output and this variable will normally be calculated in the system analysis, but it is not necessary to define any particular variable as an output if we do not need it.

When defining the system and its inputs and outputs, it is important to distinguish between a physical quantity leaving the system (which may or may not be an output as defined) and a system output. For example, if we wished to examine the relationship between the level of water in a cooling header tank and the input and output flow rates of water, we could define a system as shown in *Figure 7.18* consisting of the tank and its inflow and outflow valves. Our system boundary

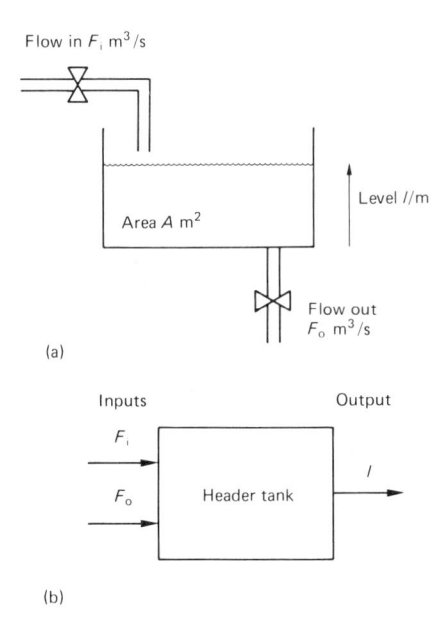

Figure 7.18 A simple dynamic system: (a) physical and (b) system diagrams of the header tank

will thus be drawn around the tank as shown. Water may be let into the header tank by controlling the inflow valve. An input to the system (a controlling variable) is thus the inflow to the tank, measured in m³/s. The outflow from the tank may be controlled by the outlet valve. The flow out of the tank is therefore another input (or controlling variable). The level in the tank, assuming no leakages or overflow, is defined by the relationship between the inflow and outflow. The output from the system is the tank level (a controlled variable).

We have control over the system inputs, whereas system outputs are controlled solely by the effects of the inputs on the system (as affected by disturbances). In this simple example, the flow out of the tank is a system input, although water is in fact flowing out of the tank. The distinction between a physical outflow and a system output is clear once the definitions are recalled.

For the simple system as defined, we may relate the output (tank level) to the input (inflow and outflow) by the relationship:

$$\text{Area} \times \text{change of level} = (\text{inflow} - \text{outflow}) \quad (7.1)$$

or, expressed quantitatively:

$$A \cdot dl/dt = (F_i - F_o)$$

or $dl/dt = (F_i - F_o)/A$ \quad (7.2)

7.2.2 Disturbances or noise

In the simple system of the header tank in *Figure 7.18*, it is possible that the relationship described will not effectively describe the change of level in the tank. If the tank has a leak in it, or the water is close to boiling point and evaporates, the relationship will not be precise. We can say that there are disturbances which affect the output. For the purpose of system analysis, we may define a disturbance as 'a variable which will affect the output of a system, but is not controllable'. Usually a disturbance will affect the system in an undesirable manner. For most ship control problems, disturbances are caused by wind and waves. Disturbances are shown on the system diagram as entering the system boundary, usually at the top.

7.2.3 System description, open and closed loop performance

Two fundamental ways in which a control system may operate are in the open loop or the closed loop mode: 'In an open loop control system, the control action of the system is unaffected by the value of the output'; 'In a closed loop control system, the control action of the system is affected by the value of the output'.

The difference between these two forms of control may be illustrated by considering a simple speed control system for an internal combustion engine (*Figure 7.19*). In the open loop configuration. (*Figure 7.19a*) the amount of fuel allowed to go to the engine is controlled by a valve or controller. The engine responds to the amount of fuel it is given and an output

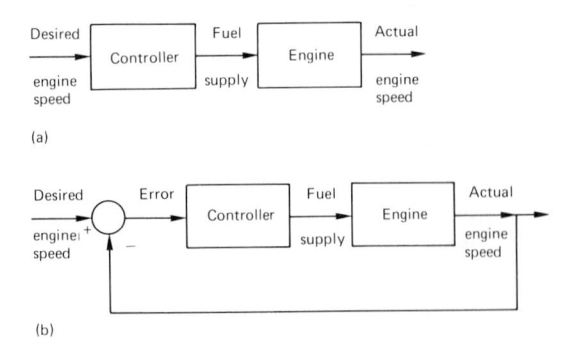

Figure 7.19 (a) Open and (b) closed loop control

speed results. This form of control is that adopted in a car, where the accelerator pedal controls the amount of fuel and the engine responds, in general, by speeding up in response to increased fuel provided. However, this mode of control suffers from a number of disadvantages. The way in which the engine responds to fuel supplied is a function not only of the amount of fuel, but also of the conditions obtained at the time. Thus, if the engine is operating outside its effective range (such as would occur going up a hill in too high a gear), increasing fuel supply may actually cause the engine to slow down. The relationship between speed demand position and the output speed is not at all simple to predict, so that the resulting output conditions are not necessarily well defined in systems with a range of disturbances or operating conditions.

In many systems, however, open loop control provides a cheap, simple method of achieving the required results. A domestic pop-up toaster is a form of open loop controller, in that the degree of brownness of the toast depends only loosely on the settings. However, provided the operating conditions are kept the same for successive operations (such as thick or frozen bread), the results will be largely predictable.

If a tighter control is required of the output, which is required to adopt the value largely independently of conditions, some knowledge is needed on the way the output differs from its desired value. This difference (or error) can then be used to drive the system, and the controls can be made to operate in such a manner as to attempt to reduce the error to zero. This feedback characterizes the closed loop form of control.

In *Figure 7.19b*, the desired speed is compared with the actual engine speed, and an error produced equal to the difference between them. This error then controls the fuel supply in such a way as to increase the speed of the engine in response to an increased demand. Should the conditions be such that a larger amount of fuel than usual is required (such as might be the case if a car were going uphill, or a ship were operating in waves), the control system will compensate by allowing even more fuel to be used. The resulting engine speed will be kept close to the desired or input value, within the design limits of the control system. Should the engine be operating well outside its design conditions, even the closed loop system may not be able to supply the correct amount of fuel to achieve the desired value. For instance, if a car is going downhill, the engine speed may be high despite there being almost no fuel at all going to the engine. Similarly, if a ship is turning under maximum rudder at full speed, the shaft speed may fall, even with maximum fuel being fed to the engine.

Often, a human controller will be used to provide the necessary feedback to enable closed loop operation to be achieved by what is essentially an open loop system. In the example of the speed control of an internal combustion engine, it is very common for the operator to be provided with instrumentation stating what the output (actual speed) is. The operator will then take some form of control action, and change the input so as to achieve the desired speed. This is the normal method of operation by a car driver.

7.2.4 System parameters

Any dynamic system may be described by giving values to the constants of the system, such as its size, power, operating voltage, etc. Additionally, there will be a number of system characteristics which are constant for any particular run or sequence of operations, but which may be varied to alter the performance of the system. Such characteristics are called parameters: 'A parameter is a quantity or system variable, which is constant in the case considered, but which is variable in other cases.'

Much of marine control engineering is concerned with optimizing the performance of a system by adjusting its parameters. An autopilot will have a number of parameters, such as maximum rudder angle and amplifier gain, which are adjustable by the user to achieve changes in the system performance. Such adjustment may be carried out automatically, according to some simple or complex control law, so as to achieve an optimum performance without user intervention. This type of operation is known as adaptive control.

An increasingly familiar example of a user controlled parameter adjustment is the mode control on a large car automatic gearbox. A lever allows two or three settings to be selected, giving different control programmes to the gearbox selection ratios. In an economy condition, the speeds at which the gearbox changes into a higher gear will be raised, while in a power setting they will be lowered. This gives the user a measure of control over the gearbox performance, and it will be up to the user to decide what is an optimum type of performance for the application. If he is late for a business meeting, the user may select the power setting, while if he is going home along a highway, he may be more concerned with the fuel saving benefits associated with the economy settings.

7.2.5 Performance analysis, cost functions

An aim of most control systems is to achieve the designed performance from the plant. By suitable adjustment of the system parameters, a range of different levels of performance can be achieved. The performance of the plant may also be affected by disturbances, environmental conditions or failures of the system or the controller. For example, the shaft speed achieved for a given fuel flow rate may depend on: the hull resistance; the state of the engine (in particular the cooling temperatures and the conditions of the fuel injectors); the presence of failures in the engine. The speed achieved by a loaded ship for a given shaft speed may depend on: the wind; the presence of waves or swell; whether the ship is manoeuvring; the state of the hull.

If the control is to be effective, it is necessary to be able to measure the behaviour of the system, to analyse it and to assess the system performance against some criterion. Ideally, the performance criterion should be quantitative, to enable meaningful comparisons to be made with other systems, or with the same system at other times.

The performance of a given system may be assessed by obtaining, from a number of measurements, a set of relevant data, and then preparing a combined figure, which presents a meaningful measure of that performance. Such a process is carried out, in a subjective and imprecise way, by every shopper faced with an array of similar items on a supermarket shelf. Relevant measures will include the price of the item, its perceived quality, the packaging, the brand name and perhaps the colour. All these quantities, some of which may not be able to be measured precisely, will be assimilated by the potential buyer and some form of assessment made. The assessment criteria will vary widely from shopper to shopper. For some the price will be predominant, for others the quality or brand name. Assessments by the same shopper on different occasions will also differ, so that the choice of item after pay day may be different from that just before pay day.

The process of assessment in all cases is the same, however. An attempt is made to combine a number of desirable criteria into a single function, the maximization or minimization of which will give a clear indication of the best solution.

A similar process is used to assess the performance of a system under the action of a controller. A number of relevant measures of the performance of the system is made and a combination of these measures used to define a single quantity, which may then be used to produce the best (or optimum) performance of the system. The measure is known as the cost function or performance criterion: 'A cost function is a single quantity which is used to assess the performance of a system.'

A cost function is typically built up of a number of factors, each of which has to be given a quantitative measure. In some cases, each of the factors can be expressed in common units (which may or may not be expressed in money terms), while in other systems some form of weighting criterion has to be prepared for each factor.

The cost function for an autopilot may be expressed in terms of its two most important performance requirements, the course error and the rudder activity. The cost function will be made up of both of these component parts, so that it may be written:

$$CF = f(\text{course error}) + W \times f(\text{rudder angle}) \qquad (7.3)$$

where the designation $f(\ )$ denotes a function of the variable. To ensure that the performance of the autopilot is measured for long enough to give a meaningful reading, the components are usually integrated over a period of time. The total cost function is made up of the sum of the two components. Because the components will not necessarily be of equal importance, one of them is weighted, by the factor W. The cost function will thus indicate by its value how well the autopilot is performing. If both the course error and the rudder angle were zero for the required period of time, the value of the cost function would be zero, indicating perfect performance. In practice this will not be feasible, because of the influence of disturbances and perhaps the ship's directional instability. The aim, however, is to adjust the autopilot in such a way as to produce a minimum cost function. The performance of the autopilot is then said to be optimized.

Changing the value of the weighting factor W will give a different emphasis to each of the two components of the cost function, so that a minimum value of the cost function, and hence an optimum performance, will be obtained for a different type of ship behaviour. Increasing W will give a greater emphasis to the size of the rudder angle, and less to the amount of course error. Optimum performance will therefore be achieved with smaller rudder movements, by allowing the ship to yaw slightly off course. This type of performance would be suited to operations away from the confines of a port area, where it is more important to minimize rudder activity than to steer a very precise course. For close quarters operations, it is more important to steer accurately on course and large rudder angles are more acceptable, so that a small value of W would be appropriate. Thus, varying W will change the type of performance which is considered optimal. The autopilot problem is discussed in more detail in Section 7.5.

The cost function has a number of important characteristics:

- It may have nothing to do with actual cost in financial terms. A cost function associated with the performance of a racing car engine will generally have little to do with financial considerations, except in a a loose way, while one associated with the performance of a marine engine plant will generally be concerned to a high degree with costs in financial

terms. Either engine can, however, have some form of cost function developed for it which will enable its performance to be assessed in a way which is meaningful to its operators or owners.
- There is no unique cost function associated with any given system to define its performance. A cost function is made up of a number of individual elements, each of which is weighted in accordance with the relative importance given to it by the system designer. There is thus a scope for qualitative judgement both in deciding what factors are to be included and in deciding the appropriate weighting factors. The optimal performance of the system will change according to the relative sizes of the weighting factors and on the terms included in the cost function.
- As it is likely that each of the components of the cost function will be in different units, the weighting factors must take the units into account so as to produce a meaningful total. As a cost function is used essentially in a comparative manner, it is of little importance what its units are.
- For most systems, as the cost function is based on the outputs of the system, it is necessary to be able to measure the outputs in order to evaluate the cost function.

7.2.6 System analysis

As part of the design process of a control system, it is necessary to be able to analyse and assess its performance in response to changes in its parameters. This process is simpler to carry out if it is performed in a logical manner. Most control systems are sufficiently similar in their operation that the stages in their analysis are the same.

The analysis of a system may conveniently be divided into a number of consecutive stages. These are: (1) system definition; (2) analysis; (3) synthesis; (4) measurement; (5) performance assessment; and (6) optimization.

7.2.6.1 Define the system, with its inputs and outputs

This involves drawing the system in a diagrammatic manner so as to be able to define what is being analysed and therefore what the independent variables (inputs) and the corresponding dependent variables (outputs) are. A system diagram can then be drawn, which may be either a simple box, with the inputs and outputs identified, with a system boundary drawn, or it may define the intermediate subsystems in more detail. *Figure 7.20* shows a system consisting of the rudder control mechanism of a ship. This system can be shown, as in *Figure 7.20a*, as a simple single input and single output system, or it may be represented in rather more detail as a set of interconnected subsystems, as shown in *Figure 7.20b*. Each of these is a system in its own right and could be analysed as such. This representation of a complex system by a series of interconnected subsystems is known as the block diagram of the system. The use of a block diagram can greatly simplify the analysis of a system, as the relationship between each subsystem is clear.

7.2.6.2 Analysis

This process consists of writing a number of equations to describe the behaviour of the system in a quantitative manner. This involves identifying the physical laws which apply to the system and producing a set of algebraic or differential equations to describe its behaviour. In order to write these equations, it is necessary to make a number of assumptions about the behaviour of the system. These will generally take the form of simplifying assumptions. Thus, in the system of *Figure 7.20b* we may wish to assume that the servo valve has a linear performance, such that the spool valve input is always proportional to the desired rudder. In fact there will be some form of dynamic delay or lag associated with the operation of the servo valve. For the purpose of many analyses, this lag will not be significant, but for others it will be most important. Clearly, if the analysis were to be concerned with the performance of the servo valve itself, it would be totally unsuitable to use the assumption of linearity.

The equations of the system will tend to be statements of one or more physical laws. For mechanical and hydraulic systems, such as those used in many marine control systems, the laws of conservation of mass and energy, Bernoulli's Theorem and Newton's Laws of Motion are likely to be of use. Familiarity with the basic laws relating to a range of engineering disciplines is required of the systems designer. Examples of the ways in which system equations are formed will be given in later sections, dealing with the various types of control required in a modern ship.

The results of the analysis will usually be in the form of a number of algebraic and differential equations. For many systems, these equations will be non-linear and may be somewhat complex. For example, the analysis of a slow speed diesel engine system and its auxiliary cooling and lubricating systems will contain several hundreds or thousands of equations, depending on the complexity of the analysis required.

7.2.6.3 System synthesis

Very significant changes have taken place over the past two decades in the methods used to analyse and solve complex system equations. From the 1950s to the end of the 1960s, the science of control engineering was being developed rapidly, the fundamental concepts of stability were well understood, and the ability to write defining equations for quite complex systems was well established. It was not, however, possible to solve the equations using the techniques available at the time in any but a simplified way. With computers in their infancy, it was feasible in the better equipped laborato-

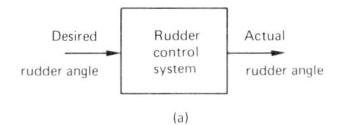

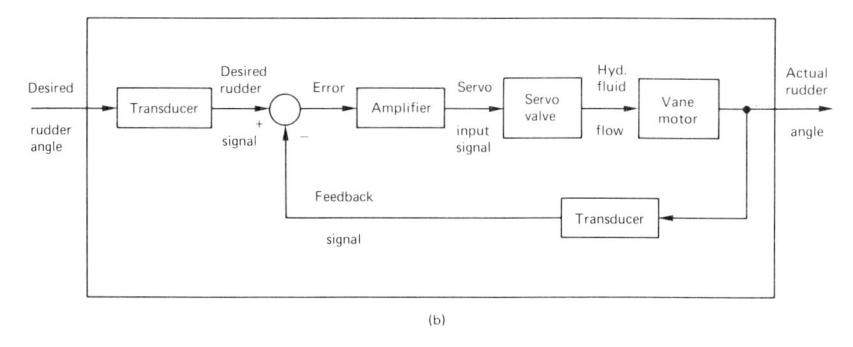

Figure 7.20 System control of a rudder control system: (a) simple schematic and (b) block diagrams

ries to represent small systems on analogue computers, so as to examine their behaviour. However, the ability of even the largest analogue computers to handle substantial non-linearities was severely limited.

Methods of analysis were therefore developed which relied on reducing systems to relatively simple, linear systems of differential equations, the response of which could be solved, for simple inputs, by using Laplace Transforms. A range of design criteria were developed, usually relating to the open loop frequency response, which enabled some predictions to be made about the closed loop, non-linear behaviour of the system.

With the advent of very cheap, powerful computers, the ability to solve large systems of non-linear differential and algebraic equations is within the capability of all systems engineers, and the problems of synthesis have largely given way to those of simulation, in which the behaviour of the system under investigation is represented on one or more digital computers, so that the effect of changes to the system parameters can be examined in a direct manner. The stability of the system can be examined, and in many cases instability can be used to enhance the dynamic performance of a system. For many purposes a system simulation, or mathematical model, is used either as a reference for the behaviour of the actual plant, or instead of the actual plant in training or research simulators.

A range of tests must be devised to ensure that the synthesis of the system is correct, within acceptable bounds, and the results of the synthesis used to assess the viability of each of the assumptions made in drawing up the system equations. For example, if it is found, in the system of *Figure 7.20b*, that the response of the system output, the actual rudder, does not correspond in simulation to the response given in reality, it may be necessary to examine the validity of the assumption that there is a linear response in the servo valve. It may, for example, be that the time taken for the servo valve to respond to signals is a significant proportion of the time taken for the hydraulic valve or ram to respond. In that case, the assumption of linearity would be unjustified and would have to be relaxed by writing a more representative set of equations for the servo valve.

An example of the importance of time delays in system behaviour is found in the response of ships of different sizes to rudder demands. For large ships, such as bulkers and VLCCs, the speed of response of the rudder to helm orders does not significantly affect the turning behaviour of the ship, as the inertia of the vessel is such that its yaw rate builds up far slower than the rudder angle changes. For a small ship of, say, 100 tons displacement, the turning behaviour of the ship can be at least as fast as the response of the rudder, so that, in effect, the overall turning behaviour of the ship is largely governed by the rate at which the rudder can be moved. It would seem sensible, therefore, for smaller ships to have a far faster rudder performance specified than large ones, but this is rarely done.

7.2.6.4 Measurement

Once the system is analysed and synthesized, it is necessary to measure its outputs. In digital simulations, this is a relatively simple task, as all the output variables are known at all times and only need extracting from the computer program. In actual systems, measuring the system variables can be difficult and expensive. In current military ship systems, in excess of 2500 variables are continuously monitored and stored on a data highway, covering both machinery control and surveillance and damage control requirements. In merchant ships, where the impact of first cost is much stronger, a typical installation may have some 200 data points measured and displayed. An assessment of the system variables is needed so as to be able to predict or analyse the performance of the system.

7.2.6.5 Performance assessment

The performance of the system is determined by noting the values of one or more system outputs for given system inputs. For more complex systems, combinations of system output variables are used to produce a suitable cost function, either continuously or on demand. For simpler requirements, all that may be required is to compare the value of the output variable with a limiting or maximum value. Thus, in an engine monitoring system, the exhaust temperatures may simply be compared with the maximum allowable values and an alarm signal generated if the alarm value is exceeded.

7.2.6.6 Optimization

The objective of most control systems is to arrange events so as to produce some form of optimum performance (i.e. fastest, least fuel for given speed, minimum rudder angle excursions, temperature within $\pm 1°$, etc.). To achieve this requires some form of tuning to be carried out on the system parameters for an optimal result. With suitable measurements, this is a feasible procedure, using either manual or automatic methods. The implementation of this technology in marine systems is most frequently found in adaptive autopilots. These will be described in Section 7.5.

7.3 The ship as a dynamic system

An overall system diagram for a ship can be represented as in *Figure 7.21*. The ship is shown as a single block, with a number of inputs and outputs. Disturbances are represented by wind and wave inputs.

The choice of which inputs and outputs to use is entirely a matter for the person carrying out the investigation or analysis. The particular set of variables shown is suitable for an investigation into the motion of the ship in a seaway. It is of no interest to an engineer seeking to analyse the behaviour of the engine controller or the behaviour of the refrigeration system in the ship. For each investigation, a different system, with different inputs and outputs, will be more suitable. For many investigations into ship motions, we are interested only in motion in three degrees of freedom, by assuming that the ship is able to proceed only in a horizontal plane. The outputs associated with the remaining motions, in roll, pitch and heave, are not then needed, as we will assume these motions are zero. Throughout any investigation, it will be necessary to ensure that this assumption is not violated by the uses to which a system synthesis or mathematical model is put. This section is concerned only with ship motions, and so the inputs and outputs shown in *Figure 7.21* are adequate.

7.3.1 Axis systems

The study of ship motions and control has two main components: steering control, in which the main motions are in yaw and seakeeping, or stabilization, where motion at least in roll is of concern, but motion in all six degrees of freedom may need to be studied. The axis system most commonly used for studying ship motion is shown in *Figure 7.22a*. These body-centred axes are fixed in the ship, meaning that they move and turn with the ship. It is also feasible to define a set of axes fixed relative to the earth (*Figure 7.22b*). This system is

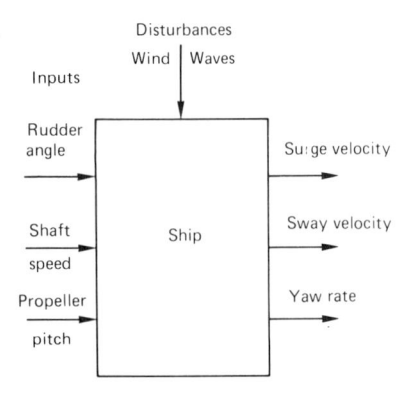

Figure 7.21 Ship motion, illustrating the ship as a dynamic system

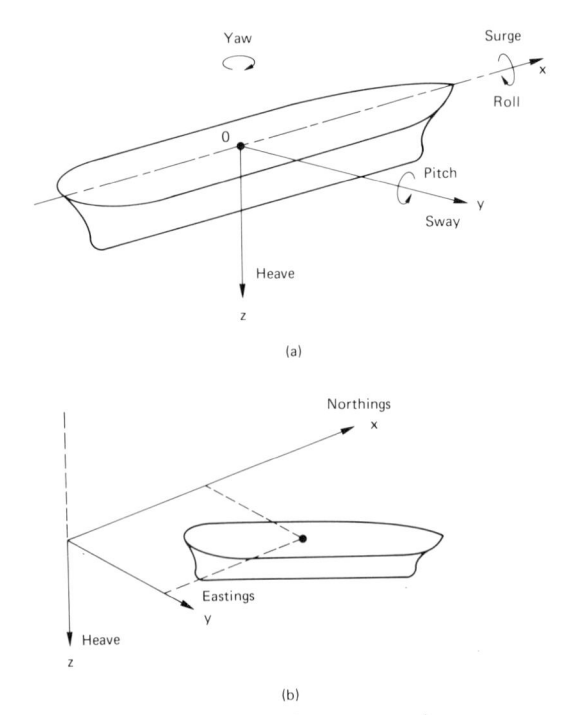

Figure 7.22 Axis systems for manoeuvring

frequently used for seakeeping studies and the results of most analyses are usually transposed to earth-centred axes, as it is the results of motion relative to the earth that is usually of most interest.

In the ship-centred system, the origin of the axes are conveniently taken as being at either the centre of gravity of the ship or at its mid-point. Additional terms in the motion equations will be necessary if the axes are not taken as being at the CG, to allow for the motion of the CG. The motion equations of the ship will also differ with changes in the axis system.

As the ship is totally free to move in response to forces acting on it, it will have a total of six degrees of freedom, three translational and three rotational (*Figure 7.22a*). The three translational motions are called surge, or motion in a forward direction Ox. Positive surge motion is in the ahead direction; sway, or motion in an athwartships or sideways direction Oy. Positive sway motion is to starboard; heave, or motion vertically, in direction Oy. Positive heave is downwards. These three motion directions form an orthogonal or right angled axis system.

The three rotational motions are called:

1. Roll, or motion about the Ox axis. Positive roll is taken as being clockwise looking from aft, i.e. starboard side down.
2. Pitch, or rotation about the Oy axis. Positive pitch is taken as being in a clockwise direction looking out from the origin, i.e. with the bow rising.
3. Yaw, or rotation about the Oz axis. Positive yaw is taken as being in a clockwise direction looking out from the origin, i.e. with the ship turning to starboard.

In keeping with the convention that positive angles are in a clockwise direction, a positive rudder angle is one which will tend to turn the ship to port.

7.3.2 Ship equations of motion

Ship equations of motion are required for a number of purposes, such as:

- to enable the motion of the ship to be studied for primary research purposes;
- as an aid to ship hull design;
- to assist in the design of thrust and control surfaces;
- to represent the ship in a range of simulators, for research, design and training purposes.

The accuracy and comprehensiveness of the equations of motion of the ship will differ greatly for different uses of the model. To represent a ship accurately when it is acted upon by wind and waves and manoeuvring in a shallow Channel close to a berth requires a high level of complexity, both to represent the basic hydrodynamic equations and to model the effects of all the environmental forces. For a model in a radar training simulator, all that is required is that the ship model behaves approximately as a real ship would in response to helm and rudder commands. The first example requires a quantitatively correct representation of the ship, the second only an approximation.

The hydrodynamics of a ship's hull are not well understood. Design of hulls has for many centuries been a process of gradual change and development from established designs. In recent years, however, there have been rapid developments in ship design, leading to wide divergences from established forms. While in most cases ships leaving the yard have been able to be steered, stopped and controlled, this has not always been the case, and occasionally ships have been built which simply do not respond to controlling forces in a predictable manner, or only do so in a totally unacceptable manner. The ability to represent the dynamic behaviour of a ship mathematically is therefore becoming of greater importance.

The basic equations of motion of a ship can best be studied by considering the case of a ship moving with three degrees of freedom, surge, sway and yaw. In this case we assume that the ship is moving on a flat sea, with no rolling or pitching motion. For many purposes this is an entirely reasonable assumption. The basic structure of the equations is found by considering Newton's Second Law of Motion in each of the three degrees of freedom. Referring to the axis system in *Figure 7.22a*:

(Mass × acceleration in Ox) = sum of forces in Ox (7.4)

(Mass × acceleration in Oy) = sum of forces in Oy (7.5)

(Inertia × angular acceleration about Oz) = sum of moments about Oz (7.6)

Because the axes are fixed in the ship, there will be coupling terms in the equations in surge and sway. It is found also that ships manoeuvring behave as though their mass were significantly greater than their displacement, so that the total mass to be considered is equal to the actual mass, or displacement, and an additional 'added mass'. As the ship moves, it will be necessary to accelerate an amount of water into the vicinity of the ship. Some of this water may move with the ship and some may be displaced away from it. The amount of water being accelerated will depend on the type of motion of the ship. It will be least when the ship is moving straight ahead and greatest when the ship is moving sideways. This added mass may be taken as being only of the order of 10% of the displacement in the surge direction, but can be about equal to the displacement in the sway direction. Similarly, there will be an added inertia about the yaw axis of the ship, which may be about half the calculated moment of inertia. The equations of motion therefore become:

$$m_1 \dot{u} - m_2 v r = X \quad (7.7)$$

$$m_2 \dot{v} + m_1 u r = Y \quad (7.8)$$

$$I \dot{r} = N \quad (7.9)$$

where m_1 and m_2 are the total masses in Ox and Oy,

and I the total inertia; u, v and r are the accelerations in surge, sway and yaw, and X, Y and N the sums of the forces and moments about the respective axes.

The forces acting on the hull consist of hydrodynamic, effector and environmental forces, and each must be evaluated and summed to give the total forces and moments acting on the hull. *Figure 7.23* from Pourzanjani *et al.* (1987) shows some of the forces and moments which must be considered when representing a ship in a harbour situation. This modular method of representing the forces and moments has considerable advantages in that is is possible to consider each of the components separately, and then produce the overall model by simply summing them as shown in the diagram. For different applications, different degrees of sophistication may be required. If a range of models for each module is developed, the overall model can be assembled to any required degree of complexity.

There are several methods of representing the hull hydrodynamic forces. Three of the most common methods are:

1. To consider the forces and moments as polynominal Taylor expansions about an operating point, expressed in terms of the linear and angular velocities u, v and r and the rudder angle. Typical representations of this type of model are by Gill, (1979) and Systems Control Inc. (1981). The latter paper presents the mathematical model in one of the largest ship simulators, which is used for a range of research applications, including port design, human factors and controllability studies. This method is widely used and produces models which can be made to represent a range of ship types in a variety of operational conditions. Because there is no readily identifiable relationship between the structure of the model and physical behaviour of the ship, it is not easy to produce the model. A large number of experimental tank tests are usually required to obtain the coefficients for the model and changes are not easy to accomplish.
2. To represent the forces and moments in a direct manner, using lumped parameters. Forces represented are typically those shown in *Figure 7.24*. Hull lift and drag are considered as single forces acting about the centre of pressure of the hull. Similar lift and drag forces are associated with the rudder, and the propeller contributes thrust and side forces. There is additionally a viscous torque resisting yaw, Nv. The representation of the hull forces and moments in this way is economical in computing power and there is a clear and identifiable relationship between the hull forces and the structure of the equations, as shown in McCallum (1980).

Additionally it is possible to represent the fluid flow around the hull using finite element analysis techniques and using the laws of conservation of mass and momentum, as shown in *Figure 7.25*, described by Rizzi (1987). This type of analysis has the great advantage that there is theoretically a good relationship between the model and reality. The

Basic structure of a modular ship model

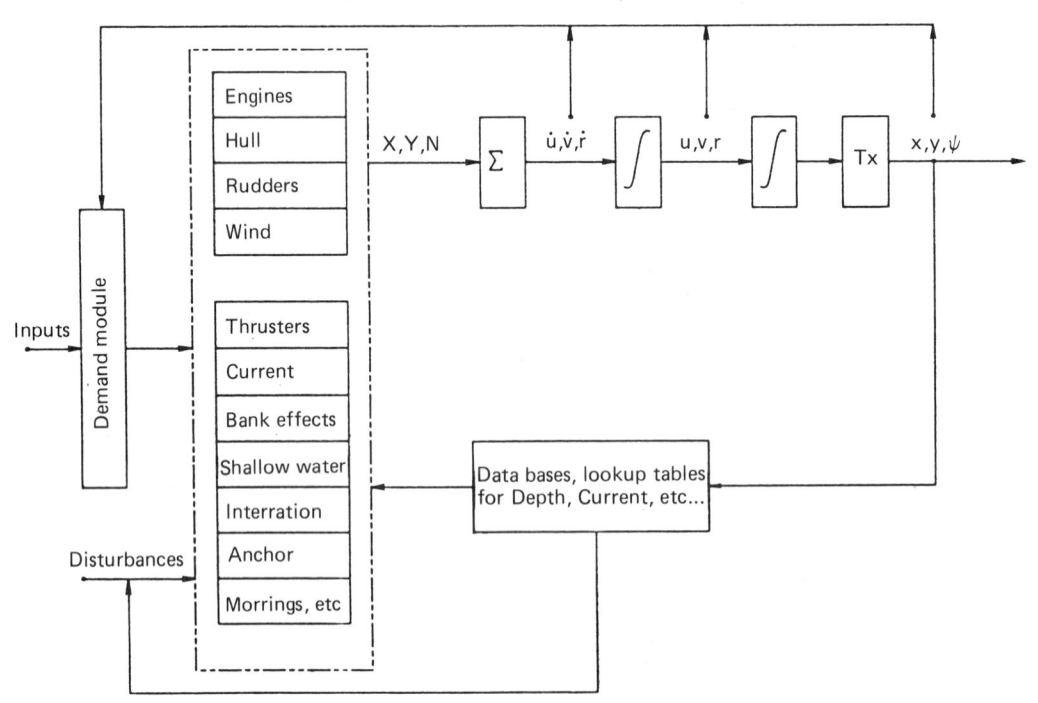

Figure 7.23 Basic structure of a modular ship model

Figure 7.24 Forces and moments on a manoeuvring ship

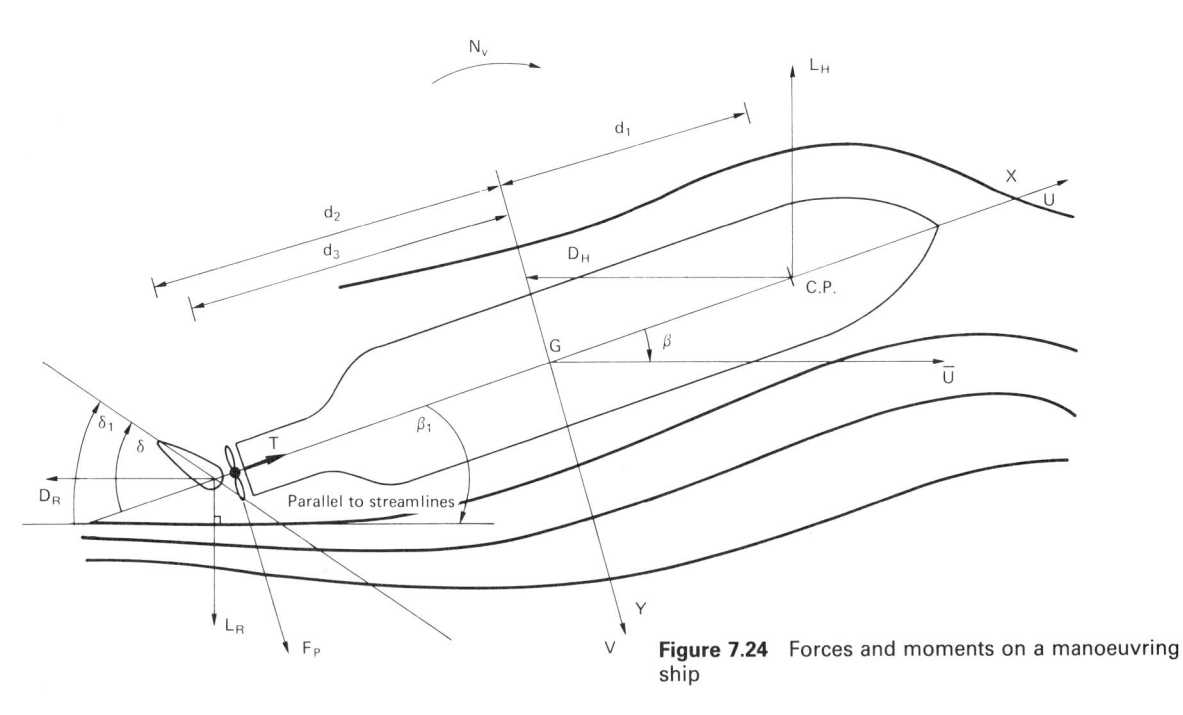

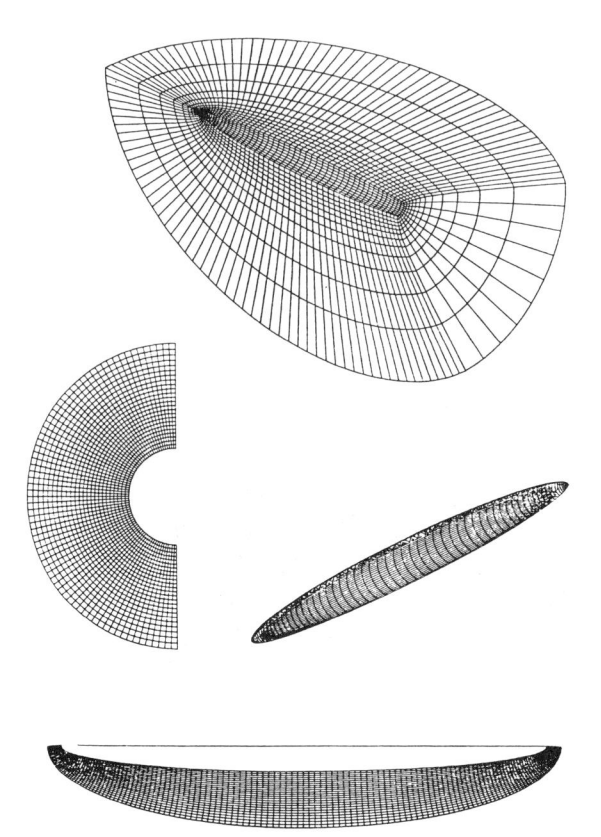

Figure 7.25 Streamline flow around a hull form (Courtesy of A Rizzi)

major problem inhibiting the greater growth of this type of representation is that the method requires the use of a supercomputer if reasonable update rates are to be achieved. As the real cost of computing power is falling very quickly at present, it is likely that there will be a greater use made of this type of model as more computers with adequate power become available.

The remainder of the forces and moments acting on the ship are calculated system by system, and the resultant forces and moments summed. Some influences on the model, such as currents and shallow water effect, depend on the geographical position of the ship, and so there is a feedback path from the ship's position to a database of depths and currents, so that the correct current and shallow water forces can be calculated.

Referring to *Figure 7.23* the process of calculating the ship motion in earth coordinates is clear. The forces and moments from all sources are summed. From Newton's Second Law, the accelerations in the three body-centred directions of surge, sway and heave are evaluated, knowing the total mass in each direction. For most simple ship motions, the total masses can be evaluated using constant figures for added mass and inertia (usually taken as 10% in surge, 100% in sway and 50% in yaw).

Integration of the accelerations will produce the velocities, and further integration yields the distances moved and the ship rotation, still in body-centred coordinates. Simple transformation will then give the displacements of the hull in earth-centred axes, as shown in *Figure 7.26*.

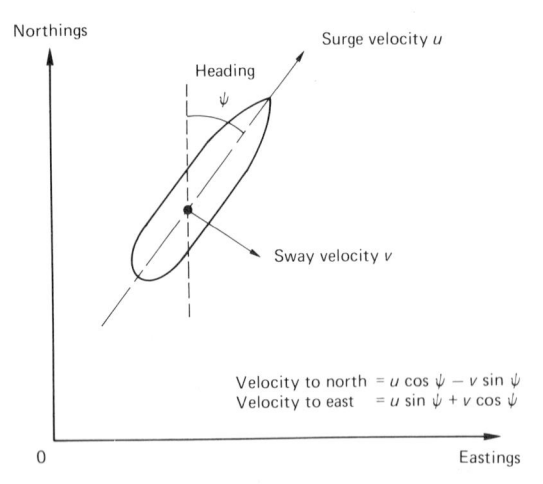

Figure 7.26 Transformation from body-centred motion to earth-centred motion

Velocity to north = $u \cos \psi - v \sin \psi$
Velocity to east = $u \sin \psi + v \cos \psi$

The feedback paths shown in *Figure 7.23* indicate the route by which the error signals are generated from which the demand signals to control surfaces or effectors may be generated. Acceleration inputs may be generated from vertical reference sensors, velocity inputs from Doppler logs and position sensors from a range of systems, including taut wire systems for offshore installation, hyperbolic electronic measures such as Decca, or from radar or sonar information.

The above analysis is adequate for most studies, where it is sufficient for the ship to be considered as moving on a flat plane, in three degrees of freedom. This method of analysis will clearly not do for studies involving wave action or roll control; neither will it do for submersibles. For these studies, it is necessary to consider motion in all six degrees of freedom or in those motions for which the study is relevant. For example, a study of roll control on manoeuving ships will need the addition of an equation in roll such as that of Kallstrom and Ottosson (1982). The study of a submarine depth control system can be carried out to a large extent by considering motion in the vertical plane only, i.e. in surge, heave and pitch.

Very few adequate models have been produced for motion in all six degrees of freedom, for three reasons:

1. The work needed to produce adequate validation data, either at full scale or in a ship model is very large and few civil authorities are prepared to pay for this work to be done. Those models produced for full motion study of naval submarines rarely find their way into the open literature.
2. The demand for such expensive models is low. Most models are required to design one particular aspect of a ship's control system and so a limited model of three or four degrees of freedom is normally adequate. Even commercial ship simulators with nominally full pitch roll and heave motions rarely have fully coupled, validated mathematical models.
3. In the absence of ship or model data, there is little point in producing a model of the necessary complexity.

The basic structure of a fully coupled, six degree of freedom model can be found in Hayes (1971). This remarkable work develops the full six degree of freedom models for a deep submergence rescue vehicle. Hayes points out that for an asymmetric representation of a freely manoeuvring model there are 1296 coefficients to be evaluated. The general equations of motion are also presented in Mandel (1967).

7.3.3 Effectors and Control Surfaces

The process of control involves identifying an error and producing a control force to counteract that error. Control of ship motions is exercised in two main ways: by imposing an inclined surface to the streamline flow of water around the ship's hull to provide mainly a force perpendicular to the motion of the ship, or by causing a surface to move in such a way as to provide a reaction force driving the ship in the required direction. All such surfaces are known as effectors.

A simple bulk carrier may have only the simplest set of effectors, consisting of a single fixed pitch propeller for moving the ship in the surge or fore and aft directions and a single rudder designed to provide a transverse force at the stern of the ship, to place the ship in such a condition that the hydrodynamic forces can turn it. It should be noted that if the ship is of a high block coefficient, the rudder contributes little to the actual turning of the ship, this being achieved by the changed momentum of the water near the bow of the ship. In ships with finer lines, the rudder contributes more to the turn, and the hydrodynamic forces less, with the result that the ship turns rather less readily than does the bluffer shaped ship.

Twin screwed ships will manoeuvre more readily if their rudders are sited directly abaft the screws, so that they operate in the propeller slipstream. Fortunately, few ships are now defined with twin screws but a single rudder. This configuration tended to give poor controllability, as for most of the rudder's travel it was outside the propeller slipstream, but for large rudder angles there was an enhancement of rudder performance. This could yield unpredictable performance. In ships where rudder performance is of particular importance, multiple rudders might be provided. Large tow boats, for example, such as are used in the Mississippi, are frequently provided with a total of six rudders. Two are sited immediately aft of the ducted propellers and are used mainly for going ahead. Four others, known as flanking rudders, are situated immediately forward of the propellers and are used when going astern (*Figure 7.27*). The two steering rudders are coupled together, and the four flanking rudders are also coupled together.

Some cross-channel ferries, which have to manoeuvre well going both ahead and astern, have quite so-

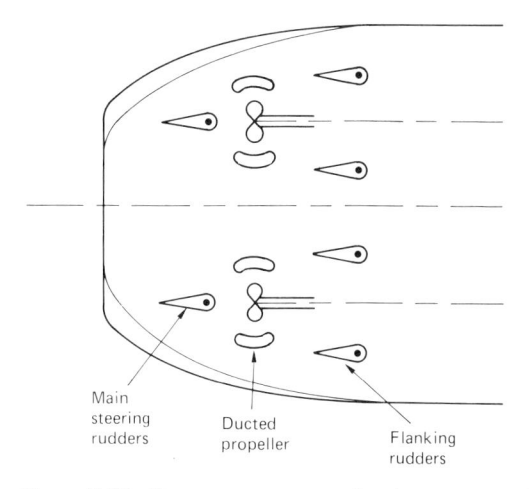

Figure 7.27 Stern arrangement of twin screw tow boat

phisticated steering arrangements. For going ahead, triple screws are fitted, with a single rudder abaft the centre screw. This enables the ship to manoeuvre well by putting the two outer engines astern and the centre one ahead. This provides a net zero thrust, but allows for the deflection of the centre propeller slipstream, giving a strong directional thrust. For going astern, a single low powered propeller and rudder are fitted at the bow of the ship, giving adequate directional control with the engines astern.

Various devices have been used at times to increase the effective lift of a rudder. The lift may be assumed to be proportional to the angle of incidence of the rudder to the local streamlines and to the square of the fluid velocity across the rudder. Increased lift may therefore be obtained by ensuring that the rudder is of a reasonable size and is placed in the slipstream of the propeller. Control surfaces will, however, stall at large angles of incidence (above, say, 35°), and so a number of methods have been used to delay the onset of stall. These can take the form of flaps or rotating cylinders, which will impede separation, or specially shaped rudders with top and bottom plates, which are designed to operate at angles of inclination of up to 70° without separation.

Control surfaces similar to rudders in shape may be used projecting from the side or the turn of the bilge of a ship to control roll. These active stabilizers operate in a very similar manner to rudders, and have similar control mechanisms. These are discussed in Section 7.7. As they are not placed in the slipstream of a propeller, they rely on the forward speed of the ship to produce any roll effect and so cannot operate at zero or slow speeds.

The other method of producing or controlling ship motions is to cause an inclined surface to move relative to the water, thus producing thrust in the required direction. The most usual example of this is the ordinary fixed pitch propeller. The design of these devices is extremely complex, as they must operate over a range of conditions without cavitating or causing excessive vibrations, and a large variety of shapes may be seen, designed to achieve these desirable criteria. A propeller can be at its most efficient only in one operating condition. For most merchant ships, this condition will correspond to the maximum speed, or 'full away' condition, as it is for this condition that the ship is designed to operate for most of its time. For other ship types, this criterion may not be the most satisfactory. For example, an antisubmarine frigate may need to operate at its most efficient and quiet condition at a relatively slow speed, so that its sonar can operate effectively. Maximum speed operations can afford then to be less than optimally efficient.

Greater propeller thrust can be achieved by placing the propeller in a duct, which may be fixed or designed to swivel to afford some measure of directional control. Some offshore vessels use directionally controllable nozzles as the only form of steering. Transverse propellers in ducts placed across the hull are used to provide transverse thrust, as bow and stern thrusters. Stern thrusters are less commonly found, as a large degree of stern transverse thrust is obtained when the rudder is applied.

The blades of a propeller may be made to swivel under control, giving a controllable pitch propeller. This is less efficient than a fixed pitch propeller at its design condition, but gives a wide measure of control, including the ability to provide stern power very quickly. The controllable pitch propeller has a large boss into which the blades are fitted and requires a hydraulic system and suitable controls to operate it. A disadvantage of this type of propeller is that with the engine set to Stop, there is in effect a disc of metal spinning in the water just in front of the rudder, so that there is very little control available. Also, the blades can catch lines during the berthing manoeuvre. These disadvantages are, however, small compared with the very real advantages of smooth, fast thrust control.

Thrust may be obtained by imparting a velocity to a large volume of water and propelling it out of the stern of the vessel. This form of water jet is used to propel a range of vessels from small boats needing to operate in areas where a propeller presents a hazard, such as in swamps, to the jetfoil.

Vertical blades operating in a cyclical motion are used to impart horizontal thrust in the most flexible way, as there are no limits to the direction in which the thrust is applied. They are therefore able to be used for tugs. A further advantage of this form of provision is that the thrust is centred about the midships part of the tug and so there are fewer out of balance forces to contend with.

7.3.4 The effects of wind and waves on ship motions

The necessity for a ship to be able to cope with the effects of wind and waves presents the ship designer with some of his greatest challenges. The forces encountered can be very large (the steady wind force on a large container ship in a force 7 wind of about 35 knots can be in excess of 100 tonnes; and they are not entirely predictable. Many ships have been lost through simply not being able to cope with the wave forces imposed on them. In some areas of the world, notable the Agulhas Current off South Africa, waves can build up to the extent that a large well found ship can simply fall into a hole in the sea, and fail to come up from it.

Waves are entirely wind formed, in that, in the long term absence of wind, there will be no waves. As waves store energy, they will, however, take some time to die down once the wind has ceased. Waves can be characterized by their wavelength and frequency and a number of wave spectra have been published giving the characteristics of waves in different parts of the world (Moskowitz, 1963).

The main responses of a ship to waves will be influenced to a large degree by the direction in which the waves are approaching the ship. For waves head on, or nearly so, the main effect is to increase the pitch of the ship, so that the bow slams into the waves. Careful bow design can assist in keeping the majority of the water from coming onboard, but the main solution to excessive ship motion in heavy bow seas is to slow down, to alter the apparent incoming frequency. Large waves will cause high bending moments in the hull.

For waves nearly beam on, the main effect is an increase in the rolling moment. If the wave frequency is nearly the same as the natural roll period of the ship, very large roll angles can build up, as there will be little to dampen out these induced roll motions. Bilge keels and active stabilizers will assist in dampening this type of roll, but again the most effective countermeasure is likely to be to alter the effective frequency, in this case by altering course.

Large waves coming from the quarter can cause very large yawing motions, with ships changing course by several tens of degrees in response to a single wave. This type of yaw motion will also cause heavy rolling, for which stabilizers will be largely ineffective. A good helmsman can counter this type of motion to some extent by predicting when the wave will strike. Autopilots are in general not very effective at countering single large waves, as they are essential reactive devices, since they have to have an error before they will act, whereas for a large wave, it is essential to get some corrective action applied before the wave has been experienced.

7.4 Ship controllability

While the definition of controllability used by the International Towing Tank Conference (1975), referred to in Section 7.1, is clear from a user's viewpoint, it achieves little in terms of enabling the ship owner or designer to be able to specify what he means by controllability in precise, or quantitative, terms. It is possible to be precise about what is meant by a 'fast' ship, as the speed may be specified in knots and this feature normally forms part of the contractual arrangement between the ship builder and the owner. It is not possible at present, however, to be at all precise about the ship's controllability, as there are few terms which can be used to define this apparently desirable quality. In this section, we examine what features of a ship's structure and equipment contribute to controllability, and indicate what is meant by controllability in rather more precise terms. We examine the requirements of those responsible for defining and enforcing standards of ship construction and operation in this area.

In Section 7.2, the concept of optimum performance was examined and it was concluded that the performance of a system is optimal if a predefined cost function achieves a minimum or maximum value in the given operating circumstances. The cost function, if it is to be useful, should be defined in user terms, so that the optimal performance of the system is recognizable to the users as being close to what they might define in qualitative terms as being a 'good' performance. In attempting, therefore, to define ship controllability in quantitative terms, we need to find out what the users of ships require of them.

In an attempt to define more rigorously what was meant by adequate controllability, Landsburg et al. (1983) carried out a survey, in which a number of pilots were asked what features of a ship's operation were considered to be important, as far as the ease of operation of the ship was concerned. In order of importance, the factors were: slow speed manoeuvrability; adequate backing power and straight line stopping ability; short response time by the ship following rudder and/or engine commands; adequate swing control with moderate rudder angles. The design features which, in the pilots' opinion, contributed to manoeuvrability included: a large rudder; bow and/or stern thruster; adequate ahead and astern power; trim by stern; reliable bridge equipment; a rate-of-return indicator; good bridge layout; good visibility from the bridge.

It will be seen that there is no single factor which uniquely contributes to controllability. Both straight line and turning performance is seen as being of importance. The requirements for controllability must be met at the design stage and tested at sea, using quantitative trial measurements. The assessment of controllability, and the trials necessary to enable the assessment to be carried out, will be discussed in Section 7.10.

The correlation between a ship's hull form and its manoeuvring behaviour is not clearly defined. As the turning ability of the ship is governed by the balance between a number of large hydrodynamic forces, as shown in *Figure 7.24*, it should not be surprising that quite small changes in the hull shape will cause large differences in behaviour. In particular, the degree of bluffness of the bow will have a marked effect on the turning ability, and hence the stability in turn, as a bluffer bow will cause the hull hydrodynamic turning forces to be both larger (because of the greater change of momentum of the water flowing past the hull) and situated further forward. Some bulk carriers with high block coefficients enter the turn quickly and have a relatively small turning circle diameter of under two ship lengths. Because of the high hydrodynamic forces assisting the turn, however, such ships may be slow to come out of the turn. Cases have been reported of ships with large block coefficients having an unacceptably low ability to come out of a turn. In one example, a products carrier had a 70° overshoot on a 10/10 zig–zag trial; in other words, after a turn had been initiated using 10° of rudder, the ship took a further 70° or so to come out of the turn.

Conversely, many ships with finer lines are stable directionally, but cannot be said to be adequately controllable, as they do not turn readily enough in response to the helm. Typically, large Penamax container ships, which have a beam restriction of around 32 m (104.98 ft), are built with a length to beam ratio of 9. Because of the necessity for such ships to have a high maximum speed, they are built with fine lines. This type of hull shape will have smaller hull hydrodynamic turning forces, acting further aft, and so will not turn well. Turning circle diameters of around six ship lengths, or nearly 1 nautical mile, are not exceptional. Such ships, although being much more stable dynamically in turn than the bluffer bulk carriers, still cannot necessarily be considered to be adequately controllable, as far as their directional behaviour is concerned.

There is in general an inverse relationship between stability and controllability in turn, in that ships which are dynamically unstable will tend to turn more quickly in the steady state and to initiate a turn quicker. The degree of controllability of a ship in turn may be enhanced by a number of extraneous measures. These involve the selection and siting of a number of effectors and the design of skegs of other forms of stabilizing influences, generally sited towards the stern of the ship. A range of rudder types has been developed which will enhance the turning ability of ships, and the siting of rudders in the slipstream of the propellers will in general assist the turning response of the ship, as the rudder forces will be increased by the faster wake they are in. Devices to enhance the lift force of a rudder include:

- flapped rudders, which operate by increasing the stall angle of the control surface, thus allowing effective rudder sideways force at high angles;
- rudders with rotating cylinders in front of them, which operate by enabling the incoming stream to adhere to the control surface at high rudder angles, similarly increasing the effective lift of the rudder;
- rudders with special shapes, again designed to operate at large angles; an example of this type is the Schlling Rudder (Bingham and Mackey, 1987), which operates by deflecting the propeller slipstream to an angle of up to 90°. The rudder is shaped so as to minimize stalling effects at rudder angles of up to 75°.

Figure 7.28 shows a comparison of the lifting capability of one of these devices.

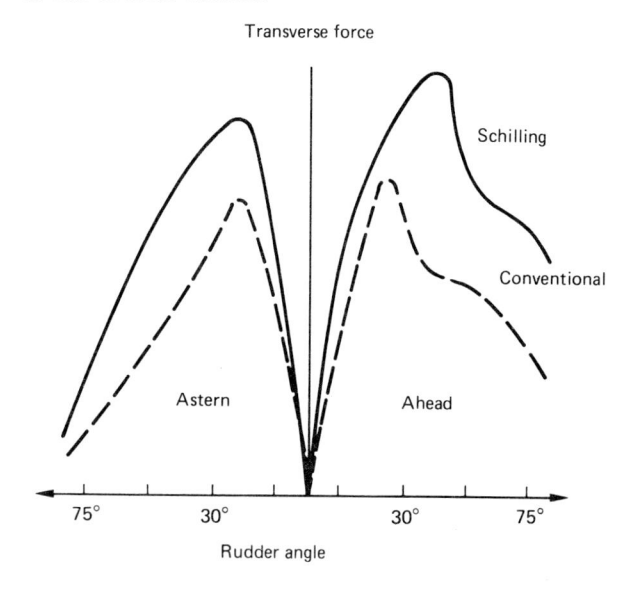

Figure 7.28 Enhanced lift rudders, courtesy Schilling Rudders

Alternative methods of enhancing the directional controllability of a ship include the use of auxiliary turning devices such as bow and stern thrusters, which operate by providing a direct thrust athwartships. These are usually situated in tunnels, which limit their effectiveness to ship speeds of below about 4 knots. On ships which have a high need for good manoeuvrability, such as tugs, offshore supply vessels and drill ships, a range of auxiliary devices is used, including:

1. Azimuthing thrusters, where an auxiliary propeller is designed so as to be capable of rotating through 360°. These devices are limited in power, and are frequently retractable.
2. Steerable nozzles surrounding the main propeller, which operate by vectoring the streamline of the propeller. These devices are limited by geometrical considerations to angles of about 45°.
3. Vertical axis propellers, such as the Voith–Schneider system, in which vertical blades rotate in such a way as to produce a thrust which is variable in direction throughout 360°.
4. A twin Schilling rudder system, in which twin, independently operated rudders are sited immediately abaft the propeller. Operating these rudders enables thrust to be produced in any direction from a single fixed pitch propeller.

The role of the rudder in a turning manoeuvre with the ship going ahead may be considered as being largely that of getting the ship into the situation in which the hydrodynamic hull forces can turn it. In ships with large block coefficients (such as tankers and bulk carriers), the forces turning the ship are largely the hydrodynamic forces acting near the bow. The function of the rudder is to deflect the stern of the ship so that a drift angle is set up and the hydrodynamic forces can act. In *Figure 7.29a* a dynamically unstable ship is shown going straight ahead, when the rudder is placed to port. The effect of the rudder is to deflect the streamlines in its vicinity to move the stern of the ship to starboard. As there are no sideways forces acting on the bow of the ship initially, the bow will tend to continue on in the same direction, so that the ship is inclined at an angle to its original direction. Although the ship may have turned through several degrees, the path of the CG of the vessel will not have moved significantly; in other words, the ship is still going in its original direction,

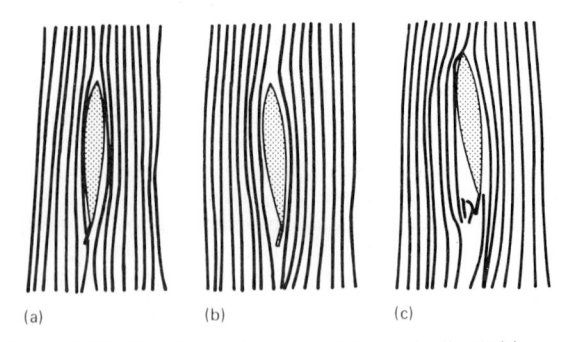

Figure 7.30 Turning behaviour of dynamically stable ships: (a) initial, (b) transitional and (c) steady state conditions

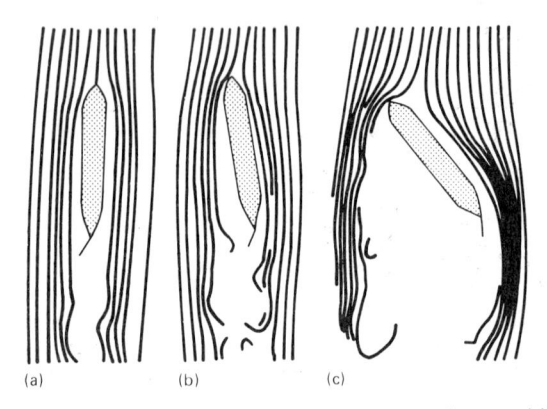

Figure 7.29 Turning behaviour of dynamically unstable ships: (a) initial, (b) transitional, and (c) steady state condition.

although it has started to turn. This situation produces a large hydrodynamic force, caused by the deflection of the incoming water stream by the hull. The hydrodynamic forces act in direction so as to increase the angle of the ship to the streamlines, and hence to further increase the hull forces. It is these hydrodynamic forces which largely turn the ship, the rudder providing little or no turning effect once the turn has been established.

In ships with a smaller block coefficient (*Figure 7.30*), the initial action of the rudder is still to move the stern of the ship to starboard. Because of the shape of the hull, the hydrodynamic forces are both smaller and centred further aft, so that the turning moment is much less and the rudder will continue to assist in the turn.

The turn will be opposed by the drag of the hull through the water, so that when the rudder is placed over to the opposite side (applying counter rudder), the rudder forces will in general be sufficient to overcome the hydrodynamic forces. Rarely, and fortunately unusually, some ship designs are such that this is not the case and a ship might continue to turn in the original direction for a long time, until the speed eventually falls so that the hull forces are insufficient to oppose the rudder, or power is removed to allow the ship to slow.

A ship's stopping ability will to a large extent be affected by the type of power plant and propellers fitted. Controllable pitch propellers, in which each blade is able to rotate about an axis perpendicular to the shaft, so that the effective pitch of the blade can be altered from the bridge or engine control room, will give a fast response to engine command, and the engine will be able to deliver a substantial part of its power in the astern mode. This type of propeller is found most frequently in ships which need frequent reversals of thrust, such as ferries, warships and offshore supply vessels. Because the blades have to be designed to operate effectively over a wide range of pitch angles, this type of propeller will not be as efficient as a fixed pitch propeller operating at its design condition, and it will be slightly less fuel efficient. A disadvantage of the controllable pitch propeller is that, at minimum pitch, with the engines set to Stop, the propeller is effectively a spinning disc of metal situated just in front of the rudder. This will blank off most of the fluid flow to the rudder, which will greatly reduce the directional controllability.

For ships with a reduced need for manoeuvring, such as bulk carriers and tankers, it is very common to fit a slow speed diesel engine coupled to a fixed pitch propeller. This arrangement can produce a substantial amount of stern thrust, but as the engine has to be physically stopped and started again in the opposite direction in order to get any stern power, there will be a significant delay in applying that thrust. This inability to provide reverse power quickly is a limiting factor in the controllability of this type of power plant.

Steam turbines do not provide a large stern thrust unless coupled to a CPP and, because of the vacuum in the condenser of the steam plant, a steam turbine set to Stop will provide a very small retarding thrust. Because of the large powers involved, the use of auxiliary devices to enhance stopping is very limited. In a study by the US Coast Guard (Card *et al.*, 1979), undertaken at a time when there was great public concern about the ability of new large VLCCs to stop effectively, a number of auxiliary devices were examined with a view to their feasibility in enhancing a tanker's stopping ability. Consideration was given to a range of thrusting devices (including jet engines and rockets), and to a number of drag enhancing devices, including flaps and parachutes. In no case did any commercially feasible enhancement reduce a vessel's stopping ability by more than 20%, and the most effective ways of enhancing a ship's slowing ability included using a slower approach speed and turning, where there was enough sea room. The use of a turning manoeuvre to stop the ship is extremely effective with full form ships, as the ship will adopt a high drift angle, presenting a large area to the flow of the water. The report also concluded that 'there are no national or international standards which require manoeuvring or stopping ability of tank vessels to be considered in the design process'.

It will be seen that assessing controllability is a matter of achieving a balance between conflicting requirements. The problem facing the ship designer is that of ensuring an adequate balance between these conflicting requirements.

7.5 The role of the classification societies

Shipping is regulated by a number of national and international bodies. If a ship is to be able to trade, it has to be 'in class' with one of the classification societies such as Lloyds, Det Norske Veritas or Bureau Veritas. Each of these societies provides detailed guidance on the way the ship is to be constructed, and the appropriate standards need to be achieved if the ship is to be classed. On the subject of manoeuvrability and controllability, the societies are largely silent, however. In a study of the relevant guidance given by a number of classification societies, Lowry (1988, personal communication) found that, while most societies have extensive rules about the construction of rudders, the guidance as to their effectiveness is minimal. Typical is the statement of Germanischer Lloyd (1986) that, 'Each ship is to be provided with a manoeuvring arrangement which will guarantee sufficient manoeuvring qualities.'

It will be noted that, while the need for adequate manoeuvrability is stressed, no attempt is made to define what adequate manoeuvrability might constitute, nor to give any qualitative expressions to this quality. This is in marked contrast to the detailed specification used for all structural aspects of the ship's design. It should be stressed that the example quoted is typical of the approach currently adopted by the classification societies.

Regulation of shipping activities is also exercised by governmental authorities, with the aim of increasing ship safety. It is found, however, that there is similarly no guidance given on the manoeuvring qualities expected of ships, nor on the aspects of ship design which might be expected to yield good controllability. Typical of the approaches adopted by governmental authorities are the remarks of the Marine Directorate of the UK Department of Transport (Lowry, 1988) that, 'There is no legislation on ship controllability or manoeuvrability for the Department of Transport to enforce and there are no plans to make such legislation.'

Ship owners might be expected to be interested in the controllability of their ships, in the expectation that more controllable ships could be safer to operate and hence be less frequently involved in collisions or groundings. Ship owners are mainly interested in transporting goods at minimum cost. Fuel saving measures will frequently be incorporated into a design, as the cost of these measures can readily be set against the expected fuel savings. The cost of additional measures to enhance the controllability of a ship may similarly be readily identified at the design stage, but the consequent savings are not readily apparent to the ship owner. Also, as there are no widely accepted quantitative cost functions which can be used to measure the gain in controllability, it is difficult for the ship owner to be well advised on the potential improvements to be gained from enhanced controllability.

Insurers might similarly be expected to have a direct interest in the controllability of ships, on the ground that a more controllable vessel could be expected to have fewer losses than one with poorer controllability. The response most frequently found from insurers, however, is that the design of ships is a matter for the ship owner and classification society, not the insurers.

Finally, it could be supposed that port authorities would be interested in the ability of ships to manoeuvre safely in their port and it is here that in recent years there has been most direct interest shown from some authorities. In some parts of the world, notably in the United Kingdom, and increasingly in Australia, it is becoming the norm for port authorities, when designing new or changed ports, to investigate the ability of particular classes of ships to use the port. By using quantitative design tools, such as ship simulators, port authorities are able to ensure that a port is safe to accept the types of ships which are likely to call there (McCallum, 1987). The desirable feedback path from the port authority, through the ship owner to the naval architect, which could lead to safer and more controllable ship designs, is largely absent, however.

The current situation on ship controllability thus appears to be that:

- There are no agreed standards of ship controllability.
- Without such standards, no classification societies or government departments are in a position to issue codes or regulation governing the controllability of ships.
- Without enforceable codes, ship owners cannot quantify to naval architects the degree of controllability which they require in their ships.

- As enhanced controllability is normally achievable only at an additional cost to a ship design, it is unlikely that such measures will be incorporated into a design, as it could be argued that to incorporate enhanced controllability features will render a ship uncompetitive.
- Insurers have no incentive to load premiums in favour of more controllable ships, as they have no measures with which to assess controllability.
- While port authorities can recognize poor controllability and design their ports to take account of it, there is no apparent feedback path to the ship designer.

The result of this situation is that ships continue to be designed and operated with unnecessarily poor controllability. As an example, a modern high speed container ship will be designed with the following features, all of which will contribute to its economical operations:

- A high freeboard, with up to five layers of deck-mounted containers. This will give the ship a very large windage, with the result that its slow speed performance cannot be guaranteed.
- A single slow speed, directly coupled diesel engine. For a large container vessel with a maximum operating speed of 26 knots, the lowest speed at which the ship is capable of operating continuously could be as high as 6 knots (slow speed diesel engines cannot operate at speeds below about 22 rpm, which will correspond to about a quarter of the ship's full speed). The ship can go at a slower speed than 6 knots only by putting its engines to Stop, which in a slow speed engine means stopping the engines completely. With a very much reduced flow, the rudder will be less effective, giving poorer directional control. It will not be feasible to control the direction of the ship at 6 knots by use of auxiliary devices such as thrusters or tugs, as these do not operate effectively above about 4 knots.
- In order to achieve a high operating speed, the ship is likely to have relatively fine lines, particularly around the bow. While this achieves a greater stability directionally, it decreases the ship's turning ability, so that the typical turning circle of such a ship is of the order of 1500 m or five ship lengths. This imposes severe limitations on the ship's unaided manoeuvrability.

The design of such a ship will be governed to a very large extent by consideration of financial cost, both of initial cost and running costs.

If the ship were to be designed more with adequate safety and controllability in mind, the following design changes might be considered.

- The bow of the ship might be made rather fuller. This would tend to decrease the directional stability of the ship, but increase its manoeuvrability and hence its controllability. There would be a price to pay, however, in terms of the power required to propel the ship at its design speed.
- The windage may be reduced by limiting the number of layers of containers to four or even three on deck. This would have a clear adverse implication in revenue terms.
- The ship might be powered by twin engines and twin rudders. This would enhance the turning ability and give a greater margin of safety, since the failure of one engine would not totally incapacitate the ship.
- The propellers could be replaced by controllable pitch propellers (CPPs), which would enable the ship to go slowly. CPPs are, however, more costly and slightly less efficient than fixed pitch propellers at their design speed.

It will be seen that the sum of measures proposed, while likely to yield a ship with significantly enhanced capability to respond to controls, is unlikely to be an attractive proposition to a ship owner, as each of the enhancements to the ship design is going to cost him money, thus reducing his competitiveness.

7.6 Heading control – autopilots

As was indicated in Section 7.1, the main function of an autopilot is to attempt to keep the ship on a constant heading. The autopilot will usually also be able to be used to change the ship's course. It will be appreciated that the problems associated with course-keeping are rather different from those associated with course-changing.

The basic functions of an autopilot are shown in *Figure 7.31a*. The desired course is set into the autopilot by the user and this course is compared in the autopilot with the actual course achieved. The difference between the two, the course error, is then acted on by the autopilot and an output signal is sent to the rudder control mechanism. This demanded rudder command corresponds to the helm or wheel order with the rudder control system in hand control.

The rudder control mechanism will then operate, until actual rudder angle equals desired rudder angle, within the accuracy limits of the mechanism. The rudder will then control the ship's course and the actual course will be fed into the autopilot signal. The degree of success with which the autopilot will be able to control the ship's course depends on a number of factors.

- The speed of the ship. At zero ship speed, clearly any movement of the rudder will have no effect on the ship's course at all. As the speed increases, the amount of rudder to be used for a given ship response will be less until, at maximum speed, a very small alteration will suffice. For this reason, the ship's speed is usually fed into the autopilot.
- The environmental conditions. Clearly, the presence of wind and waves will affect the ship response. In heavy quartering seas, it is highly unlikely that the autopilot will be able to control the ship satisfactorily at all. The presence of a strong beam wind will

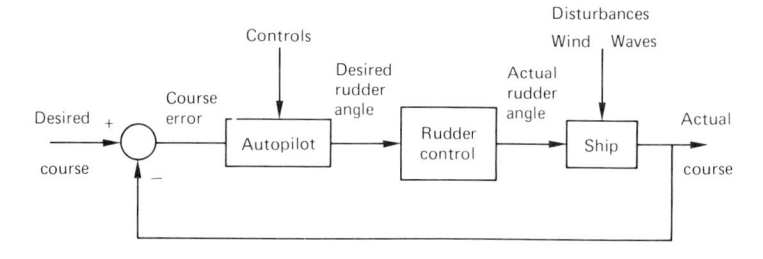

Figure 7.31 Basic functions of an autopilot

require a continuous helm signal and the presence of swell will induce cyclical yaw motions which may be unsatisfactory.
- The condition of the autopilot. Conventional autopilots will have a number of manual settings with which the user can obtain what he deems are optimum settings. Adaptive autopilots will attempt to produce an optimal performance automatically.
- The number of pumps running in the steering gear. With two pumps running, the rudder will move at approximately twice the speed at which it will operate with only one. In smaller ships, this will have a marked effect on the turning ability. This is because the rate at which a smaller ship will turn is commensurate with the speed of operation of the rudder. For large ships, the time constants associated with yaw behaviour of the ship are large compared with the speed of movement of the rudder, and so the effect on ship movement of the number of pumps operating is smaller.

The system diagram of an autopilot may be drawn as in *Figure 7.32*. The inputs are the desired course, the actual course, and the ship's speed. The output is the desired rudder angle. The disturbances consist of wind and waves and there will be some system parameters.

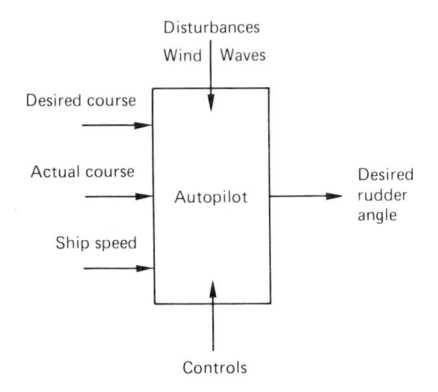

Figure 7.32 System diagram of an autopilot

7.6.1 Conventional autopilots – controls

The essential function of an autopilot can be seen as being the task of changing the course error signal into a desired helm command. The way in which nearly all conventional autopilots operate is similar. If we imagine a situation (*Figure 7.33a*) in which the ship is on a course of 028° and the required course is 030°, there will be an error signal equal to the difference between the desired and actual courses, or 2°. The autopilot will then calculate a rudder demand of a size and direction such that the ship will come round to starboard. The size of the rudder setting will depend on the settings of the autopilot and on the ship's speed, but could typically be in the range 0.5–5.0°. For larger course errors, the calculated rudder angle will be correspondingly larger. In other words, there is an element of proportionality in the calculation of the desired rudder angle.

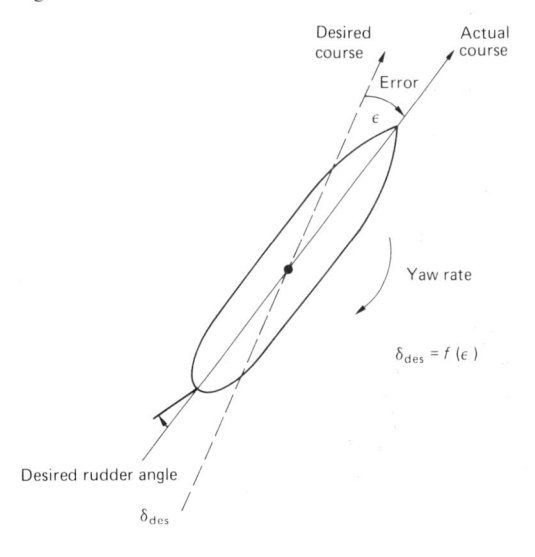

$\delta_{des} = f(\epsilon)$

Figure 7.33 Conventional autopilots: controls

If the ship is turning towards the desired course, it will be necessary to apply a correspondingly smaller rudder angle. Similarly, if the ship is turning away from the desired course, it will be necessary to apply a greater rudder angle. There is thus an element in the calculation which depends on the rate of turn of the ship. In practice, as is shown in the system diagram of the autopilot, the inputs do not include a signal for rate of turn, so this has to be calculated internally. With an autopilot containing a microcomputer, estimating the rate of turn is reasonably simple, as an estimate can be obtained from the difference in heading angles over a short period of time. As this type of differentiation is likely to be somewhat erratic, a smoothing circuit or filter will be necessary. For those autopilots built up from analogue circuitry, the necessary differentiation will need to be carried out electronically, by, for example, a phase advance circuit.

If the ship were subjected to an asymmetrical force for a long time, such as side wind, it would be necessary to keep a consistent rudder angle to counter the disturbance. To achieve this under automatic control would need a constant course error, which would be unsatisfactory. If, however, the error signal is integrated over a period of time, and a demanded rudder angle generated dependent on this integral signal, a zero mean course error will be produced. In practice, the implementation of this integral signal can also be achieved by putting in a component of demanded rudder proportional to the mean rudder angle experienced over a period of time.

This three-part type of automatic control, known as proportional, derivative and integral (PDI) control, forms the basis of most commercial and naval conventional autopilots. Devices with this type of control will enable a ship to be steered effectively in most conditions. However, there will frequently be occasions when the operator will need to exercise more control over the behaviour of the ship and so a number of additional controls will be fitted. The names used for these controls vary widely from one type of controller to another, but the functions are similar. *Figure 7.34* shows the front panels of the autopilot currently in use with the Royal Navy. The design of this autopilot is simple and reliable and its behaviour may readily be understood, as all the operator controls are on the front panel. The desired course is set on the centre dial of the set course unit (*Figure 7.34a*) by rotating a knob to move the ship shape on the course dial. The actual course of the ship is shown in the same unit by a smaller ship shape behind the desired course. The course error is modified by proportional and derivative action, to give a desired rudder angle to be passed to the rudder control mechanisms. The value of the desired rudder will be modified internally, according to ship type, to achieve an optimal performance for that particular class of ship, and by the user, according to the settings on the front panel of the auto steering unit. The desired rudder settings will in general be reduced at higher speeds and so there is a facility to input the ship's speed. This is done either automatically from the ship's log, or by the user on the auto steering unit (*Figure 7.34b*). It may be desired to limit the rudder angle applied, for example, to reduce the amount of roll in turn, and so a facility for setting the rudder limits is provided.

In non-critical manoeuvres, there are advantages in allowing the ship to move off course for a few degrees without rudder action, as this gives a smoother passage of the ship through the water, with a smaller drift angle and hence less resistance. The rudder movements are also reduced. In critical manoeuvres, it is more important to stay on course accurately and the allowable yaw must be reduced. A yaw control on the front panel of

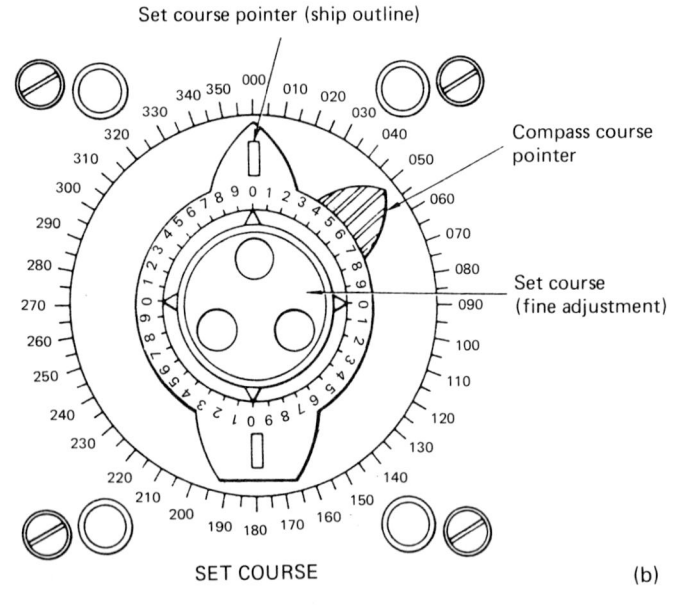

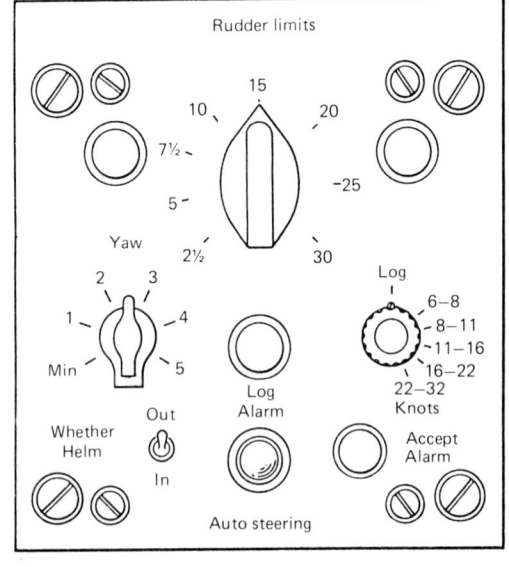

Figure 7.34 Royal Navy Set Unit Course (a) and autopilots (b); front panels (Courtesy of Royal Navy)

the auto steering unit allows the amount of course alteration before corrective action is taken to be varied.

To allow for weather effects, the weather helm circuits evaluate the amount of rudder necessary to keep the ship on the correct course, and a signal proportional to this rudder angle is applied when the weather helm control is On. An autopilot will need additional facilities to allow for large alterations in the set course. For example, if the desired course control is turned through more than 180° some form of logic needs to be built into the autopilot to enable the ship to be turned either the shortest way round or in the direction in which the control knob was turned. Usually the latter is the more logical choice of action, as it can be assumed that the user has a good reason for turning the control knob in a particular direction.

7.6.2 Adaptive autopilots – methodology

The ship's officer needs to set the controls of a traditional autopilot in such a way as to provide the optimum performance of the ship for the conditions prevailing and the ship's task. As he is able to observe only the rudder angle and ship's heading, it can be difficult to achieve a good or optimal set of control values. Additionally, to continue to achieve optimum performance, he will need to change the settings from time to time. This will rarely be achieved in practice, both because of the tedious and difficult nature of the adjustments and also because it is difficult to know when the optimum performance has been achieved. The process of changing the autopilot settings manually to cater for changes in the ship's circumstances is shown diagramatically in *Figure 7.35a*.

To attempt to solve these problems, a number of adaptive autopilots have been designed, with the aim of producing better course-keeping and course-changing by automatically adjusting the autopilot parameters. There are several ways in which the necessary adaption may be achieved. Some or all of the parameters of the autopilot can be changed as simple functions of the external conditions. This process is essentially automating what the conscientious operator will do. Examples of quantities which are used in different autopilots include the speed of the ship (as in the log input to the Royal Navy autopilot), the state of loading of the ship and the depth of water. The behaviour of the ship will vary, sometimes to a marked degree, as each of these conditions changes, and so it is logical to vary the autopilot parameters to preserve close to optimum performance over a wider range of operating conditions. Such adaptive autopilots are described by Sugimoto and Kojima (1978). The methodology is shown diagramatically in *Figure 7.36b*.

It is to be expected that a noticeable improvement to ship steering performance will be achieved. As the adaption methodology is essentially open loop, it relies on the accuracy of the control algorithms, which may not have had to have been drawn up without too much knowledge of the ways in which the ship could be expected to respond to the changes in conditions. Such a form of adaption will also not be able to cater for other extraneous conditions, which have not been preplanned.

To extend the range of optimal behaviour of the autopilot further, it is possible to alter the autopilot setting as a result of a series of evaluations of a relevant cost function. With appropriate design, the autopilot settings can then be continually tuned so as to minimize the cost function on a continuous basis. The methodology is shown diagramatically in *Figure 7.35c*. Relevant ship measurements are taken or derived and input to a computation module. The module calculates the cost function and, depending on its value relative to a minimum value, the autopilot coefficients or settings are changed. Such autopilots are described by Kallstrom (1982), Tosi (1980) and Reid (1978). Typical cost functions will be based on weighted sums of the course error, the rudder activity and the yaw rate. It would be expected that only the last few values of the cost function will be used, as the behaviour of the system a long time ago is not relevant to the present behaviour. The algorithms used for altering the autopilot constants will vary from design to design and the choice of cost function and adaption algorithms will determine to a large extent the success of the autopilot.

As most of the measurements will be noisy, it is necessary to filter the measured signals to obtain the best estimate of their values. Kalman filtering techniques are now usually employed for this task. A further method whereby the parameters of the autopilot may be adaptively tuned is to use a model reference technique. In this method, shown diagramatically in *Figure 7.35d*, a mathematical model of the ship is subjected to the same inputs as the actual ship, and to the same disturbances. The ship model is tuned so as to give an optimal performance, so that, if the actual ship's autopilot is giving its best performance in the prevailing circumstances, the performance measures of the ship will correlate well with those of the model. The differences in output between the model and the ship are compared and the differences used to tune the ship's autopilot so that the differences are minimized. It may then be assumed that the ship's performance is optimal. Difficulties may be experienced with this form of autopilot in defining the mathematical model of the ship, with sufficient scope for it to be able to perform well in an adequately large range of environmental conditions.

7.6.3 Autopilot control in a seaway

A well tuned autopilot will be able to cope with a range of environmental conditions, and this range can be extended if the built in adjustments are used intelligently or if a good adaptive autopilot is used. In many cases, the steering behaviour of an autopilot will be better than most helmsmen, not least because the autopilot does not tire or become distracted.

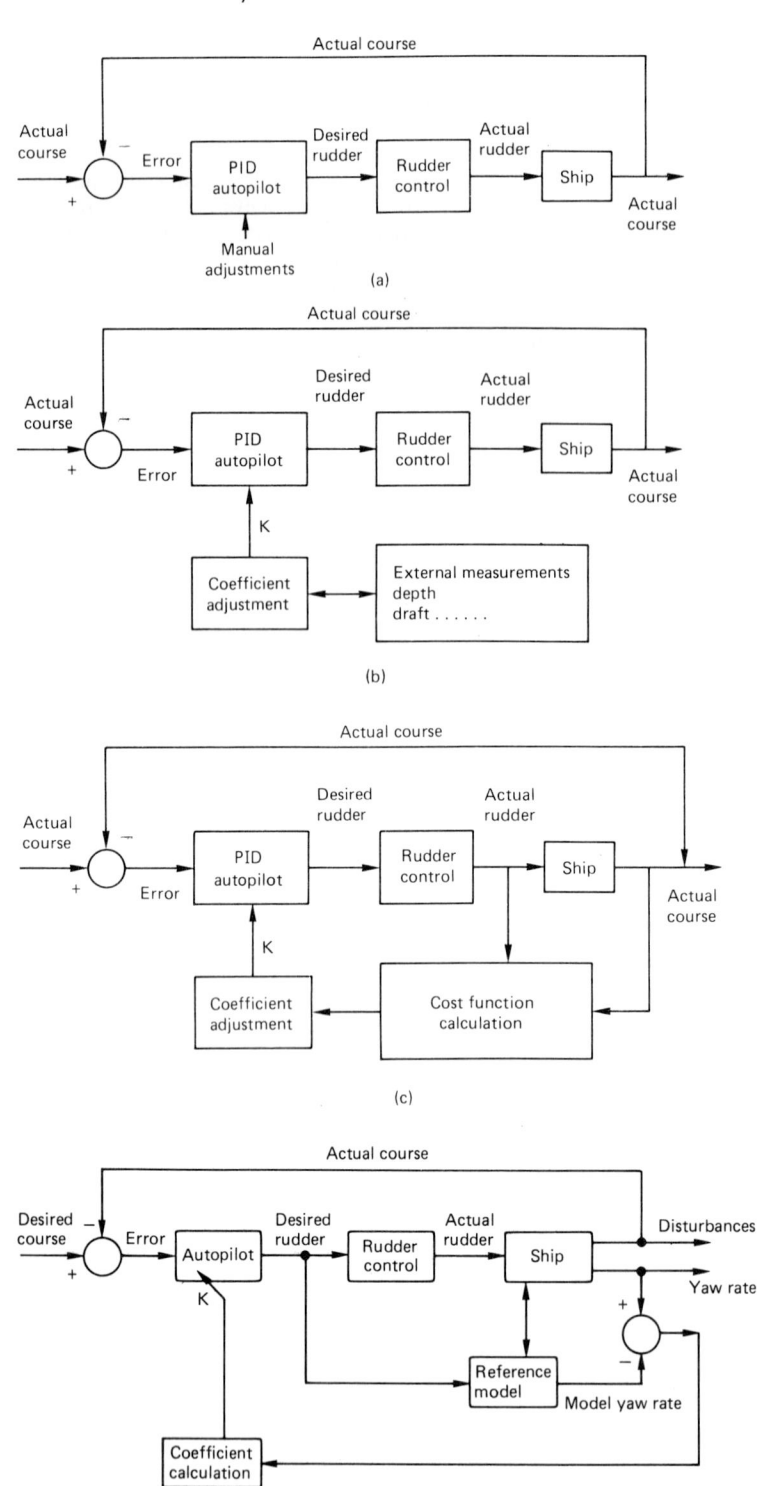

Figure 7.35 Adaptive autopilot; (a) manual adjustment; (b) adaption with reference to external conditions; (c) adaption by cost function minimization; (d) adaption of model reference

In a heavy quartering sea, however, many autopilots cannot function effectively because they do not contain an adequate level of information about the oncoming sea. A helmsman in such conditions can look at an approaching large wave, or can sense its onset by the vertical movements of the deck and take appropriate steering action sufficiently quickly so as to anticipate the wave to some degree. In conditions where the effective wavelength is an appreciable proportion of the ship's length, very large yawing motions can occur if this anticipatory type of action is not able to be used. One solution proposed is to use some form of shipboard sensor to determine the presence of a large wave. This could be achieved by some form of laser based device which would be aligned up weather, and used to sense the presence of a wave some seconds before it strikes the ship.

For smaller ships, the critical factor limiting steering behaviour is the rate at which the helm can be applied in response to a manual or autopilot signal. The rudder rate required by most classification societies is around 2.8°/s, so that to apply a 10° counter helm from an initial helm order of 20° will take about 11 s. In a large ship, say over 5000 dwt, the time taken for the ship's yaw to build up is large compared with the time taken to apply the rudder. For a small ship of a few hundred tonnes deadweight, however, the ship will respond in yaw much faster than the rudder can act, so that effectively the dominant influence on turning behaviour is the performance of the rudder engine rather than the inherent dynamics of the hull. It is felt that a higher steering performance could be achieved for smaller ships if the rudders were designed to move very much faster in small ships.

As the rudders are placed low down in the ship, below the centre of gravity and below the centre of lateral hull resistance, the rudder action will also tend to cause roll in the ship. In large hulls, this is not particularly significant, but in long, thin, small hulls, such as those of warships, this effect is very noticeable. Initially, as the rudder is applied, and before the ship starts the turn, the roll will tend to be inwards. Later in the turn, as the ship starts to yaw, this inward roll will be countered by the centrifugal force opposing the lateral hull resistance and so there will be a tendency for the hull to roll outwards. The overall roll behaviour can therefore be rather complex and will vary from ship to ship. The fact that a rudder will cause roll can be used beneficially in some ships, by using the rudder in an active manner to assist in roll stabilization. This is discussed further in Section 7.9.

7.6.4 Achieving optimal autopilot response

An autopilot is designed to achieve a desired course and does this by using the rudder. A balance has therefore to be struck between the opposing requirements of keeping a tightly controlled course and using too much rudder activity. If many large rudder commands are used, there will be an increase in rudder drag, the ship will be set at larger drift angles, increasing the hull drag, and the rudder movements themselves will increase wear in the rudder control mechanism. A correct balance has therefore to be struck between the requirements of the autopilot.

What the precise balance is will depend on the task of the ship. In open seas, there is a strong preference for reducing the drag of the ship and small excursions are generally acceptable. A seemingly large mean course excursion will add only an imperceptible amount onto the distance travelled. For example, a mean course error of 1° (equivalent to a course variation of 2° overall) will add less than 0.02% to the total distance travelled. Conversely, when a ship is in close quarters, such as when transiting a narrow channel, the course error is of paramount importance, as small course errors will develop into off-track errors. In these situations, considerations of rudder wear or hull and rudder drag are of much less importance than good course keeping.

The controls of a conventional autopilot can be adjusted to give either type of behaviour. For the close quarters situation, the gain of the autopilot will need to be increased to give greater emphasis on course-keeping, while for the open sea situation the gain will need to be reduced. The rate control would be adjusted to give a response with no overshoots. A steering quality index, or cost function based on the course error and rudder activity, can similarly be altered to give either response by changing the weighting of the two terms.

7.7 Track-keeping and position control

7.7.1 Requirements

The autopilot is a relatively simple device, which relies entirely on shipborne sensors to function. It has, however, a major disadvantage in that its output, the ship's course, is not actually required in many cases. The task of most ships is to travel in a predictable manner along a path or track fixed relative to the earth's surface, rather than on a fixed course through the water. For most purposes, it is acceptable for the ship operator to make the necessary calculations with respect to the external conditions to provide the desired course, which is then used as the input to the autopilot.

For certain critical conditions, however, it is necessary for the path over the sea bed to be kept with greater precision than is feasible with an autopilot alone. These applications include:

- Surveying. Clearly, if the water depth is to be recorded at a given point on the sea bed, it is necessary to know with precision where the ship is at the moment the depth reading is taken. Also, it is helpful if a known track can be steered so that readings are in a reasonable line, so that a uniform coverage can be obtained.
- Minesweeping and minehunting. It is vital for mine-

sweeping operations for the lanes to be known to be clear of mines. Therefore, a swept path must be maintained relative to the sea bed for this operation to be effective.
- Dredging. Similarly, it is necessary to know with some degree of precision that the channel has been dredged adequately. This requirement has assumed rather more importance in recent years, with large ships regularly operating with very small design clearances.
- Port approach. In some specialist operations, particularly those where poor visibility is frequently encountered in conditions where the ship has a very small clearance for the approach, additional assistance must be given to the operator. This is a track-keeping operation, as the position of the ship relative to the sea bed is of interest.
- Offshore operations. It is in this area where many of the developments in the field of track-keeping have arisen in recent years. As many of the offshore installations are fixed on the sea bed, their support requires ships to keep station relative to a rig for often prolonged periods while, for examle, crane operations transfer goods from the deck of an offshore supply vessel to the rig. The ability to maintain a track consisting of a single point is a specialist application of track-keeping, known a dynamic positioning, (DP) and is the subject of Chapter 8.
- Underway replenishment. This is a particular form of track-keeping, where the requirement is to keep the ship, usually a warship, a predetermined (small) distance away from another ship, so that stores or fuel may be transferred. This problem is exacerbated by the hydrodynamic forces between the ships and the fact that the store ship may not be keeping a perfectly steady course (Whitesel and Wavle, 1981).

7.7.2 Information requirements – sensors

In these applications, the need is the same, to control the position of the ship relative to a known fixed position relative to the sea bed or, in the case of underway replenishment, relative to another moving ship. For this, it is clearly necessary to know the present position of the ship. The basic block diagram of a track-keeping system is very similar to that of an autopilot (*Figure 7.36*), except that the primary comparison is between the ship's desired position and its actual position. Additionally, many track-keepers will also control the engine performance.

All systems require as inputs information on the present position of the ship, so that this may be compared with the desired position to obtain a system error. In some systems, a speed error will also develop. The methods for obtaining this information vary extensively. One of the most promising methods is the global positioning system, based on satellites, which can now give a position accurate to a few metres at worst. This may be sufficiently accurate for many track-keeping tasks and has the advantage of being contained within the ship.

Close into shore, a number of high precision systems may be used, based on infrared or radio waves. These all require one or more shore stations to be set up, and so are most useful for tasks where a ship is repeatedly carrying out the same role. Figure 7.11 shows a high accuracy positioning system used in a dredger owned by Associated British Ports in South Wales. The system produces a high definition display, showing the position of the dredger in relation to the shore, with the dredged channel shown. The display is used only to give information to the ship's staff, with the conning of the ship being carried out in a conventional manner, using helm and engine orders. However, the overall operational methodology is the same as that shown in *Figure 7.36*, with a human in each of the control loops.

For surveying operations, it may be feasible to set up position lines ashore (leading marks) and to con the ship down the line thus defined, using normal conning commands. This is a preferred method of harbour approach for many ships, with the advantage that it is cheap and reliable. Leading marks cannot be used, however, in conditions of poor visibility. Radar can be used to fix position in relation to shore features, but can suffer from very bad distortions close to bridges across a river.

Offshore vessels have particularly stringent requirements for position fixing in relation to the sea bed or to a fixed structure such as a rig, and a range of systems are used, based on sonar, radar, inertial navigation and on a taut wire attached to the sea bed. Many ships employ more than one of these systems (see Chapter 8).

One factor in common with most position measuring devices is that the data is noisy. Almost all trackkeepers using measured data rely on some form of smoothing or estimation to obtain the most reliable data. The technique of Kalman filtering is commonly used for this task (Quick, Scott and Morley, 1987).

7.7.3 Effector requirements

For most track-keeping applications, the ship's normal complement of propellers and thrusters is adequate, as the track-keeping controller may be thought of as simply a development of the autopilot. For more severe requirements, special effectors are required to ensure the vessel can maintain its designed station in a range of weather conditions. *Figure 7.37* shows the types of thruster commonly fitted aboard many offshore vessels. Typically controllable pitch propellers are used, often sited inside fixed or azimuthing nozzles to increase their effective thrust. Both now and stern thrusters are used, frequently in multiple units. Transverse thrusters cannot be used effectively in multiple units. Transverse thrusters cannot be used effectively at ship speeds of over about 4 knots, as the hydrodynamic flow past the nozzles destroys their effectiveness, and so rotatable thrusters are sometimes fitted. Specialist vessels such as tugs can use thrusters consisting of a series of vertical vanes, which move in a cyclical manner, such as the Voith – Schneider system (Todd, 1967).

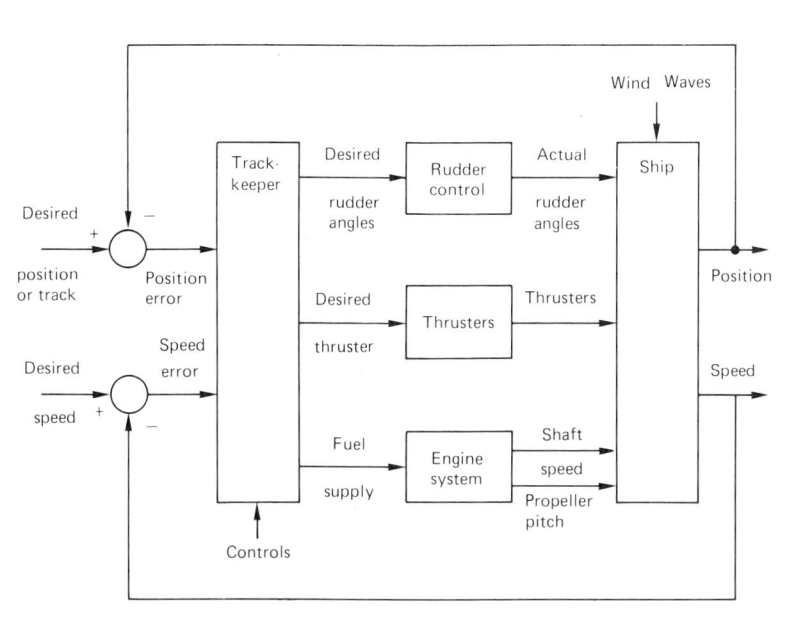

Figure 7.36 Track-keeping: operational principles

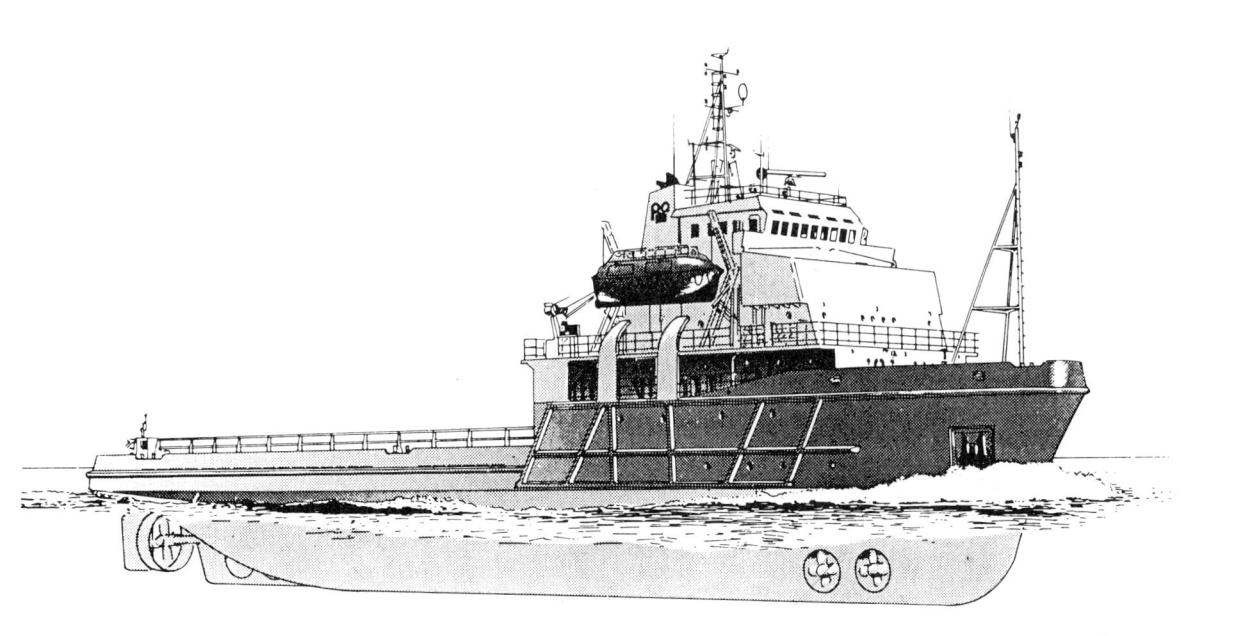

Figure 7.37 Thrusters for track-keeping and position control (Courtesy of A/S Kongsberg Vapenfabrikk)

7.7.4 Control methodology

The methods employed in track-keeping devices are similar to those used in autopilots, although, as there is a greater complexity in the control problem, there is greater variety in the control methods used. In general, some form of stategy is used to control the rudders which incorporates the position error information as well as error information based on course and rate of turn information. For example, the control function used to determine the desired rudder angle for the prototype replenishment system developed by the US Navy (Whitesel and Wavle, 1981) involved elements of the following quantities:

- course error between the two ships;
- rate of turn;
- yaw acceleration;
- the distance off the replenishing ship;
- the lateral separation rate;
- the integral of separation distance.

Each of these quantities was given an appropriate weighting, and the rudder angle calculated as the weighted sum of these quantities.

It is useful, with a track keeping device used for assisting in navigating in a channel, for there to be some measure of prediction in the control algorithm, so that the rudder is applied in plenty of time before a bend in the channel. In a device used for designing channel layouts, developed by Cheng, Williams and McCallum (1982), using simulation techniques, the track-keeping algorithm uses the concept of a 'look ahead' distance to evaluate the required rudder angle (*Figure 7.38*).

A point is defined some distance 'a' along a reference track, which may for example be along the centreline of a desired channel. The aim of the track-keeper is to reduce to zero the angle θ_a between the ship's head and the heading which would bring the ship on to the designated reference track at the point distance 'a' along the track. Later modification of this sytem defined the angle θ_a as that between the ship's current trackline and the heading which would bring the ship on to the designated reference track at the point distance 'a' along the track. Control is achieved by a simple proportional and differential controller acting on the rudder. The system also incorporates a controller to achieve the desired speed along the ship's track. The use of the look ahead concept is analagous to the ship's pilot anticipating the bend and getting the helm on in good time. Varying the look ahead distance will represent different pilot behaviour. The system is used to evaluate the feasibility of ships being able to transit particular channel layouts with the aim of optimizing a port layout for a range of ship types in differing environmental conditions.

The minehunting role requires precise positioning of a ship, while a mine is found and destroyed. For such a ship, with limited thruster capability, it is not always possible in severe weather conditions for position control to be achieved by the use of thrusters alone (*Figure 7.39*). In severe conditions, a different form of control may be used, known as Position Control through Manoeuvring (PCM), as indicated by Quick, Scott and Morley (1987).

In light weather conditions, the thrusters are sufficient to move the ship into its desired position without change of heading, so that the ship can face into the prevailing conditions. In heavier weather, the thrusters are insufficient to be able to control the ship against the weather conditions and so control is lost. If, however, the thrusters are used to position the ship so that its main propulsion unit can propel the ship into its new position, the system will continue to be operable in very heavy weather. The choice of which model to use is left to the discretion of the minehunter commanding officer.

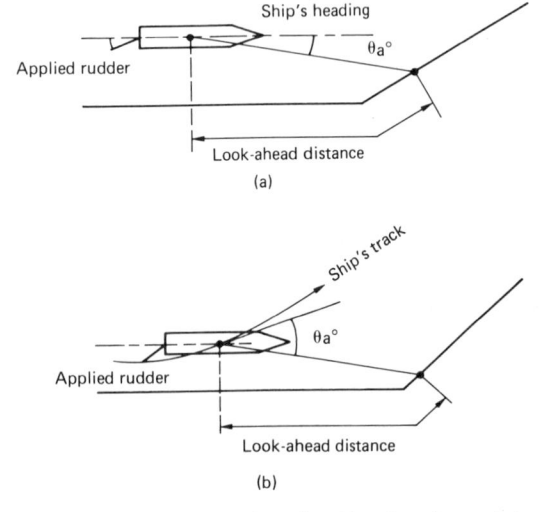

Figure 7.38 Track-keeping algorithm for channel design; (a) original and (b) revised concepts

7.8 Roll control – ship stabilizers

7.8.1 User requirements

Because of the underwater shape of a ship's hull, which is designed to cause minimum resistance to motions consistent with adequate cargo or weapon carrying capacity, all ships will roll to a greater or lesser extent, under the action of waves. This rolling effect is entirely deleterious to the effective performance of the ship's role. Among the problems which occur due to excessive roll action are:

1. Loss of speed for a given power output.
2. Loss of efficiency of ship staff, either directly because of the motion and the necessity to hang on, or because of actual seasickness.
3. Loss of commercial effectiveness because of the necessity to secure cargo or ship equipment firmly. An obvious example is the requirement to secure

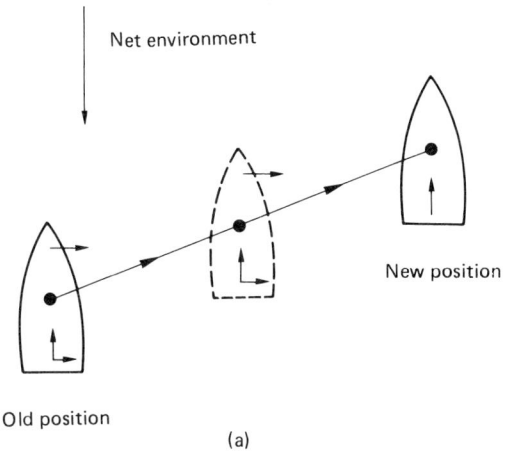

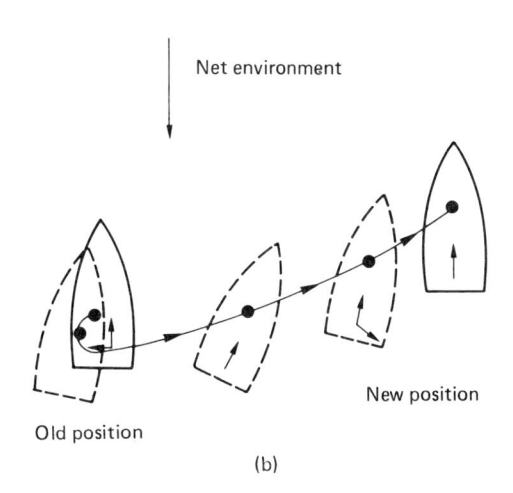

Figure 7.39 Position control of minehunter in light and heavy weather

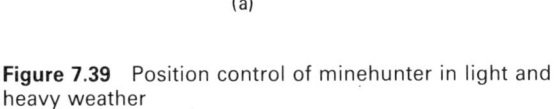

each of perhaps 1000 vehicles for a short sea crossing on a Ro–Ro ferry.
4. Loss of commercial attractiveness of a ship caused by the unpleasant motion. Customers will not wish to pay for a cruise if they are to be sick all the time.
5. Damage to ship equipment or cargo caused solely by rolling.
6. Complete loss of the ship if rolling becomes too severe, or if cargo shifts.
7. Loss of accuracy and effectiveness of warship weapons systems caused by excessive roll motion.
8. Reduction in operational capability of a warship because it is unable to operate its helicopters due to excessive roll.

It is not surprising therefore that considerable attention has been paid over the past century to reducing the extent of roll motion in a ship. Because a ship's hull is between five and ten times as long as it is wide, the equivalent problems associated with pitch are much less severe.

The basic methods of achieving roll reduction are concerned with three main methods of approach:

1. altering the basic hull design so that it does not roll excessively,
2. adding devices to control the roll motion, and
3. operating the ship to reduce its propensity to roll.

All three methods are in common use in various ships.

7.8.2 Roll motions of a ship in waves

A ship will roll in waves under the combined influence of asymmetrical effects of the buoyancy effects of the wave, and the ship's dynamic inertia. A ship upright in the water (*Figure 7.40a*) will be acted on by its weight, passing through the CG of the ship, and by a buoyancy force which, as stated by Froude (1861), acts perpendicular to the water at the hull. These two forces will produce a rolling moment, which will vary cyclically according to the frequency of the waves, and will force the ship into some form of oscillatory motion. The type of motion produced will depend on the relative frequencies of the exciting wave and the undamped roll frequency of the ship. It is unfortunate that the undamped roll frequency of many ships is in the same band of frequencies as many waves, and it requires very substantial alterations to the ship's geometry to alter this situation. In *Figure 7.40b*, where the roll period of the ship is greater than that of the wave (i.e. the ship takes longer to complete one roll cycle than does the wave, or the ship's natural frequency is lower than that of the wave) the ship will tend to react more slowly to the incoming wave than does the surface of the water. Thus the ship will tend to roll into the wave. Conversely, if the roll period of the ship is less than that of the wave (*Figure 7.40c*) the ship will tend to roll with the wave.

In either case, the maximum roll angle will be out of phase with the wave by approximately 90°. In order to

7.8.3 Minimizing roll motion – hull design

It is seen above that a ship will roll more if the buoyancy forces are asymmetrical and if the natural frequency of the hull is similar to that of the incoming wave. If either of these aspects of the waves can be altered significantly, roll will be reduced. Altering the design of a hull radically is excessively costly and can obviously be achieved only at the design stage. Ships designed primarily to reduce ship motion are therefore to be found only where the effects of such motion will totally preclude effective operations, such as drilling ships or some passenger ferries. The two most usual types of design are both based on a multihull concept. A catamaran, or twin hulled vessel, will have a very short roll period, because any asymmetrical buoyancy forces will have a large righting moment, because of the large separation of the hulls. Thus although such a hull will tend to follow the incoming wave scope (*Figure 7.41a*) for a large wave, the resulting roll motion will tend not to be accentuated to a marked degree by the resonance effect of the wave excitation. In small waves with a high frequency, such as would be close to the natural resonance of the hull, the wave height must be relatively small (*Figure 7.41b*) and so the hull can still ride with a small roll angle.

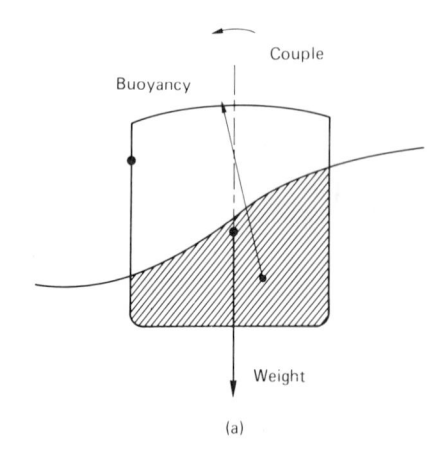

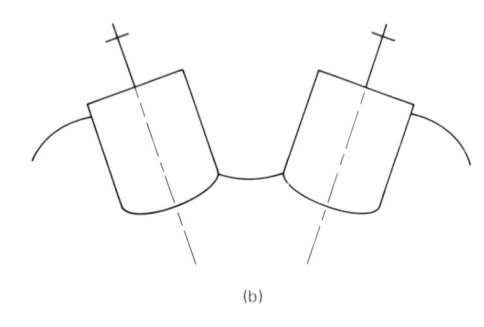

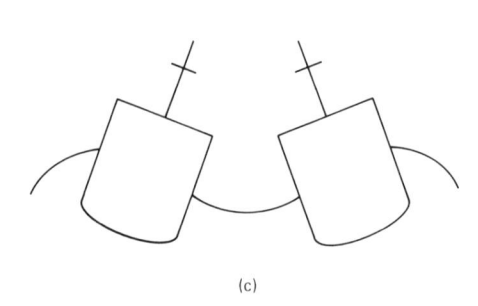

Figure 7.40 The onset of roll motion in waves: (a) rolling moment of ship in beam sea; (b) ship rolling into wave; (c) ship rolling with wave

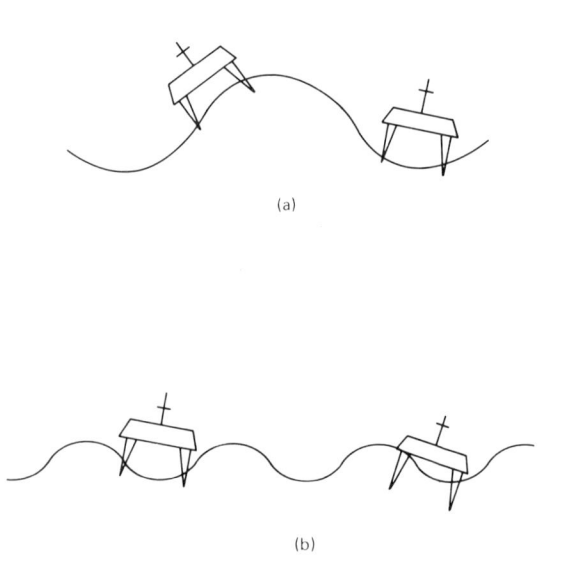

Figure 7.41 Catamaran hull form in waves: (a) wave following in large waves; (b) small roll angles in small waves

reduce the roll amplitude, it is therefore necessary to induce a countering roll moment which will also be out of phase with the incoming wave train by about 90°. In practice, waves are not true sinusoids and so the resulting ship motion is itself neither regular nor sinusoidal. Devices to reduce the roll motion will therefore need to be capable of being tuned so as to produce the optimal effect on ship motion.

A catamaram hull will still, however, respond to the asymmetrical buoyancy effects of the wave. If a hull is designed so that the wave surface does not affect its buoyancy, there will be little reason for roll to be induced in the first place and so roll damping will tend to be of less importance. Such a hull is in regular use in offshore operations, and is known as the Small Waterplane Area Twin Hull or SWATH (*Figure 7.42*)

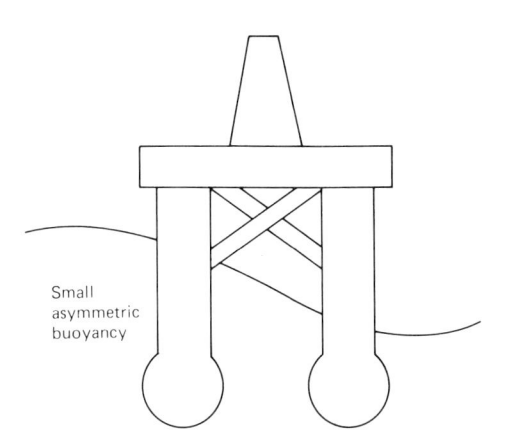

Figure 7.42 SWATH vessel in waves

In this type of vessel, the large majority of the hull form is situated significantly below the waterline, at a depth such that the two hulls will be submerged in all expected waves conditions. The hulls are joined to the upperworks by a structure of as small a cross-sectional area as is structurally feasible. In this way, the changes in buoyancy forces in waves are almost totally absent and the vessel will therefore remain relatively static in most wave conditions. Should waves occur of a size such that the lower hull becomes exposed, the buoyancy forces will become asymmetrical, and so some rolling may be expected, but this will occur only in very severe conditions. It will be appreciated that this type of hull is very expensive to produce and is of very special applicability. For most ships, other methods have to be found to reduce roll.

7.8.4 Minimizing roll motion – imposed devices

For most ships, the basic hull design is effectively fixed by considerations other than those of effective roll damping, so that other methods have to be found to control roll. The three most usual devices are bilge keels, passive tanks and active fins.

Bilge keels (*Figure 7.43*) are very widely fitted to all types of ship, as they are cheap to install, cost very little to run as they have no moving parts, and are reasonably effective, reducing roll amplitudes by up to 35%, compared with ships not so fitted. They also operate well at zero speed. A bilge keel needs to protrude beyond the boundary layer, so that it is operating in relatively undisturbed water. In a large ship, the boundary layer may extend to 0.5 m (1.64 ft) away from the hull. The bilge keels will need to be confined

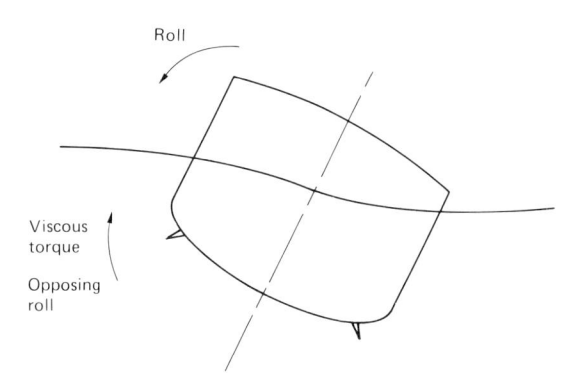

Figure 7.43 Bilge keels

to that part of the hull where they will not protrude beyond either the maximum beam of the ship or the maximum depth, or damage will occur.

Bilge keels operate by imposing a resistance to the rolling motion, so that they become more effective at high roll velocities. They are aligned with the undisturbed streamlines of the hull, to reduce their resistance to forward motion, but when underway in a seaway, they will no longer be aligned with the prevailing streamlines and so some added resistance is found.

Passive tanks have been in use for over 100 years, the earliest example of their use being found around 1874. They consist (*Figure 7.44*) usually of a tank at either beam of the ship, connected by a passageway across the ship. Variations exist where the passageway consists of the sea, and the two beam tanks are open at the bottom. Other examples have been produced where the side tanks are high in the ship, which contributes also to reducing the rolling by raising the centre of gravity and so increasing the roll period away from resonance with wave conditions.

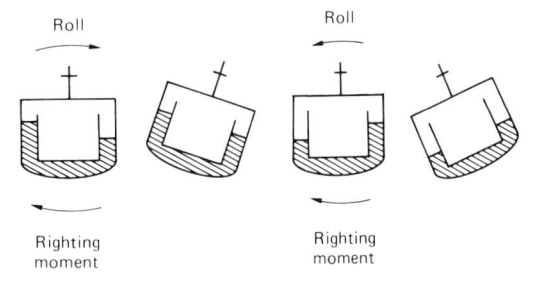

Figure 7.44 Passive roll tanks; operation to control roll

The operation of passive tanks may be best understood by considering their ability to damp out an oscillation in flat water. The sequence of diagrams shows how gravity effects will cause the water in the higher tank to flow towards the lower. If the system is tuned correctly, by the time the water has flowed into the lower tank, the ship will have rolled and the water will again be in the higher tank. Thus, by careful design, the necessary 90° of phase difference between the restoring couple and the wave action can be obtained. Control of the water flowing through the cross-passage may be achieved by fitting some form of restriction in the water path, which may be varied to allow for different wave frequencies. The frequency of greatest concern, however, is that of the natural rolling frequency of the ship and so if a fixed system is to be used, this will be tuned to that frequency. Most cargo ships will have a range of frequencies of natural roll, depending on their loading condition.

Passive tanks have major advantages in that they require almost no maintenance, are simple to fit and operate at very slow and zero speeds. To obtain a significant reduction in roll requires a weight and space of water which may reduce the cargo carrying capacity by up to 4%. Passive tanks can reduce the roll effect by up to 70%.

For ships where a stable platform is necessary, but whose hull shape is constrained to be conventional, such as passenger liners and warships, an active system is usually employed, consisting of a number of retractable fins, typically two or three sets on either side of the ship, which are actively controlled so as to produce a restoring torque to counteract the roll. The operational principles are similar to those of an autopilot (*Figure 7.45*).

The ship's roll is detected in a sensor unit. This will typically be a roll gyro. A computation unit, which may be expected to be a digital processor, calculates the desired fin angle, which will then be applied to the fin control unit. The control algorithm used is typically the weighted sum of the roll angle, its velocity and its acceleration (Wolford, 1978).

This unit controls power to the fin rotation mechanism and there will be some form of feedback, so that the fin stops at the correct position. The dynamics of the entire unit will be adjusted so that the required roll reduction is adequate. The dynamics of the fin power unit will need to be taken into account, as the fins will take a finite time to reach the desired value, by which time the desired value itself will have changed. Typically, the response of the fins will need to be significantly faster than the roll period of the ship. A typical fin installation is shown in *Figure 7.46*. It is necessary for the fins to be retractable to avoid damage during berthing. Although a set of fins could be constructed to lie totally within the confines of the bilge keels, because they have to be free to move, they will not be capable of taking any form of impact.

7.8.5 Minimizing roll motion – operational methods

Roll motions will be worst if the period of the hull rolling is close to that of the incoming waves, because of the resonance effect of the oscillatory motion. This will be worse in ships with small damping, such as those with inadequate bilge keels, or a very rounded hull, such as some minesweepers or fishing boats. Should rolling be particularly bad, it is possible to alter the period of the incoming waves by altering the course or speed of the ship. Often the worst rolling will occur in a quartering sea, when the incoming wave period is increased because the ship speed is close to that of the wave train, and the effective period is thereby lengthened, so that even a relatively small sea can cause severe rolling motions. All that is then required is to alter either the ship speed or heading by a significant amount to change the incoming wave period so that it is different from the natural rolling period of the ship, but that is a matter for the judgement of the command.

For ships fitted with active roll stabilizers, increasing speed may be effective, as these devices do not operate at slow speeds. For ships fitted with passive tanks, a

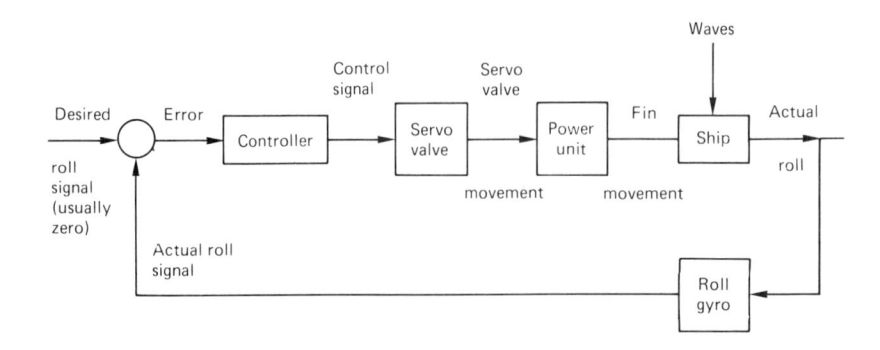

Figure 7.45 Active fin stabilizers: operational principles

Roll control – ship stabilizers

been counteracted by the autopilot. Similarly, the rudder, because it is placed below the CG, will cause some rolling motion in addition to its yawing effect. The situation can therefore exist where the autopilot and the fin control servos are acting in opposition (Savill, Waugh and Britten, 1980). Certainly in most ships there is no attempt made to combine the actions of the fin controllers and the autopilot into one unit. This problem is discussed further in Section 7.9.

One area in which significant amounts of research have been carried out is to recognize the rolling effect of the rudder and to attempt to use it as a stabilizer instead of the fins. If successful, this would enable one set of large control surfaces, the fins, to be dispensed with, as well as preventing the two systems working against each other. The essence of the problem, as discussed by van Amerongen and van Nauta Lemke (1982), is whether the two motions can be satisfactorily decoupled, to prevent the roll stabilization causing undesired yawing motions. Essentially, the roll motion requires small, fast corrections, while the control of yaw requires the application of a rudder angle for a longer period. Provided that there is a sufficiently large difference in the frequencies associated with the two motions, it is likely to be a feasible process. As yet, the work is not significantly past the experimental stage, although full scale trial results have indicated that the process is feasible (van Amerongen, van der Klugt and Pieffers, 1984). A prerequisite for the system to be successful is for the speed of operation of the rudder to be increased by a factor of about six times, to approximately 15°/s.

7.9 Machinery Control

The fundamental changes which have taken place in ship and control system design in the past decade have had a fundamental effect on the way marine propulsion machinery is controlled. The main trends in ship design have been caused indirectly by the large increases in fuel prices and crew costs. The increase in fuel prices has led to fundamental changes in the way ships are propelled and controlled, with a current, almost total, predominance in large merchant ships of slow speed, single acting diesel engine propulsion systems. Smaller vessels show an equivalent predominance on single or multiple medium speed diesel engine installations. In warships, because of the need for small, light, efficient and flexible power plants, there has been a significant growth in the use of gas turbine propulsion units, with medium speed diesel installations as a cruising speed engine in many navies. This predominance towards diesel engine propulsion has enabled far more energy efficient installations to be fitted, with a number of energy saving auxiliary systems, such as shaft generators and waste heat boilers. Diesel engines for merchant naval applications have been adapted to burn

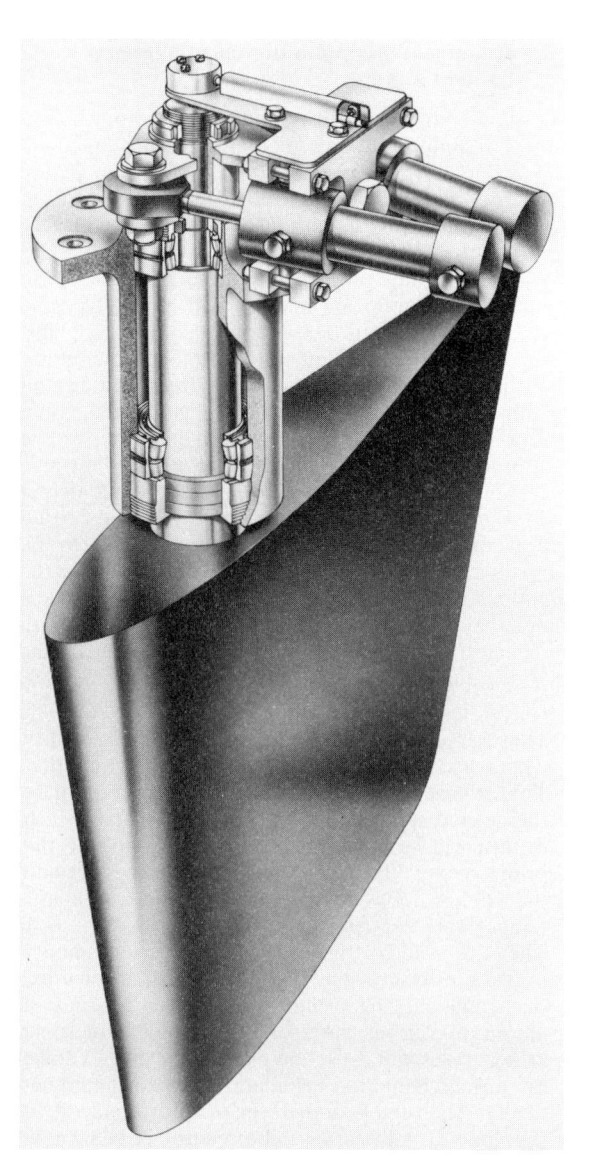

Figure 7.46 Fin system, courtesy Vosper Thorneycroft PLC

better response will be obtained if they are tuned, so that the period of oscillation of the water in them is that of the incoming waves.

7.8.6 Rudder roll stabilization

Active stabilization fins are situated at the turn of the bilge. One of the effects of moving the fins is that, in addition to affecting the rolling motion of the ship, the fins also produce a turning motion, caused by that component of the fin force which acts horizontally about the yaw axis. This yaw motion will have

high viscosity distillate fuels, requiring a substantial amount of fuel treatment and heating before they can be used.

The trend towards reducing crew numbers in engine rooms has led to the development and universal application of unmanned engine rooms, with a consequent reliance on automatic engine performance monitoring arrangements. Coupled with these trends has been the requirement to reduce maintenance levels. This has led to the development and implementation of a range of condition monitoring systems, which are able to predict an incipient fault before it becomes sufficiently serious to jeopardize operational integrity. Complex, usually computer based, systems of alarms are in service, which are able to log engine system faults as they occur and cause alarm indications to sound in strategic places around the ship, such as on the bridge and in the engineers' quarters.

Because the engine control room is unmanned for most of the time, it becomes necessary for sophisticated display equipment to be fitted, to enable an engineer to be able to tell the state of his engine plant very quickly and to diagnose any abnormal operational situations. The man–machine interface (MMI) has assumed a far higher importance with the development of unmanned machinery spaces than was the case when a watchkeeping team was always present. As the watchkeeper is present only intermittently until a fault condition is detected, there can be considerable difficulty in diagnosing the cause of a fault, particularly if there is any of the information presented on the gauges. Developments are currently in hand towards automatically diagnosing a range of fault conditions, using artificial intelligence techniques, based on the information presented on the performance monitoring systems.

Control of a ship's machinery installation is now routinely carried out from the bridge. An engine plant will typically be started and brought to a condition of readiness from the control room and control then passed to the bridge. In warship installations, there is a need for control to be exercised from the bridge, from the machinery control room, or from local consoles situated close to each item of machinery.

These changes in the way marine plant is controlled have been brought about by the ready availability of powerful, cheap computing facilities. Almost all machinery control operations are now digitally controlled, enabling far greater precision to be achieved in the control functions. The ability of digital systems to handle large amounts of data quickly enables a far greater amount of information on the state of the propulsion system to be processed, displayed and recorded. It is also possible to transmit the current status of the machinery plant, via a satellite communication link, to a shore based office where maintenance routines can be planned to fit in best with the operational requirement of the ship.

7.9.1 Requirements for a digital machinery control system

The main components of a modern digital machinery control system are shown diagrammatically in *Figure 7.47*, while a typical control room is shown in *Figure 7.48*. The principles of operation are the same for all types of prime mover, although the details will vary from type to type.

The present state of the plant is determined from the performance monitoring system, which will typically have several hundred sensors throughout the plant, measuring among other quantities a range of temperatures, pressures, viscosities, flows, torque, speed and the amount of black smoke being generated. In a typical warship installation, over 1000 quantities will be measured. Each quantity will be measured by a transducer and converted into a form suitable for digital analysis and presentation. This process will be carried out at intervals. Some quantities will need to be sampled more frequently than others. In particular the quantities associated with combustion need to be sampled more frequently than do those associated with the heat transfer processes in cooling systems. The recorded quantities are used for a number of purposes.

1. They serve as a measure of the achieved quantity where a control loop is controlling that quantity. Thus, the actual shaft speed will be used to compare the desired speed to provide an error signal, which in turn will be used to alter the fuel supply to the prime mover. Temperature measurements around the cooling loops will be used to alter automatic temperature control valves.
2. They are used to trigger an alarm when a quantity exceeds a preset value. This alarm will be recorded on a data logging printer, will trigger a series of alarms in various parts of the ship, and in some critical cases will cause the plant to be slowed down or, in merchant ship installations, to stop automatically. An example of this type of autostop facility is the loss of lubricating oil pressure. Classication societies' rules specify that, if the lubricating oil pressure falls below a certain value, the engine is to be immediately stopped automatically. The reasoning behind this action is that the engine will cease to be serviceable in a few tens of seconds if it is allowed to run in this condition and so it is preferable to stop it before critical damage is done.
3. They provide data for the performance monitoring system, to be used to check that the plant is running within its design limit and is tuned adequately. For example, if excessive black smoke is detected other than transiently, the engine cannot be running correctly and investigation needs to be instigated.
4. They provide data for the health monitoring system, to be used for planning maintenance and keeping the plant in good condition in the long term.

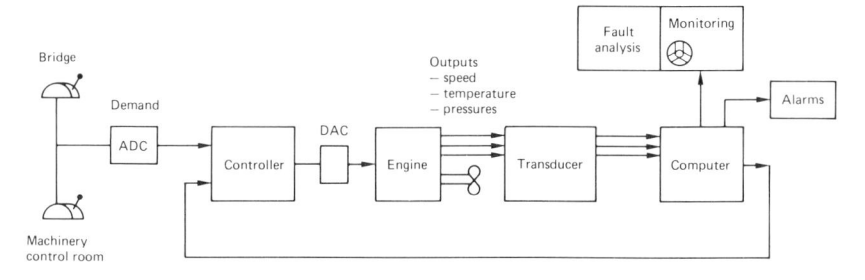

Figure 7.47 Modern digital machinery control system

Figure 7.48 Typical control room (Courtesy The Motor Ship)

5. They display to the plant operator the current state of the plant.

Control may be exercised from the bridge, the control room or, if fitted, a number of local control stations, fitted adjacent to the individual items of plant. Control consists in most cases of providing a set point for the automatic control mechanism. This frees the user from all requirements to keep the plant within its operating envelope. The demand lever can be put from Full Ahead straight to Full Astern if desired and the plant controllers will take account of critical stresses and speeds, providing the desired shaft speed and direction as soon as operationally feasible. With digital controls, the establishment and altering of operational envelopes are relatively straightforward at the design stage.

Typically, a single layer will be used to control each shaft set. If a controllable pitch propeller is fitted (*Figure 7.49*), the single lever will control a programmed set of pitch and speeds, designed to enable the ship to achieve its desired role in an optimal manner. A number of programs for this type of device, often called a combinator, are available. Commonly fitted to small ships, one option is for the engine to keep a constant speed, while the lever alters only the pitch of the propeller blades. This has some major disadvantages. If the engine speed is to be constant, it must be constant at the maximum available shaft speed. This is uneconomical at low ship speeds, and presents a danger at manoeuvring speeds, as a fast moving set of

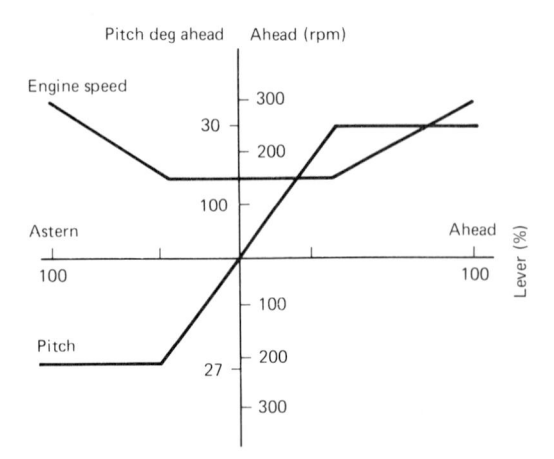

Figure 7.49 Controller pitch propeller: combinator diagram

blades is liable to catch mooring lines. An alternative strategy is shown in *Figure 7.49* where both the pitch and engine speed are altered as a function of the lever position. At manoeuvring speeds, the engine is turning slowly and only pitch is altered. Once the pitch is set to full, either ahead or astern, further movement of the lever accelerates the engine. This combinator program enables better and more economical control to be achieved.

Control of each item of plant is achieved at individual control units, under the overall control strategy determined by the machinery control unit. Each of these items of equipment will be microprocessor controlled. The output of the local control units will consist typically of a movement of, for example, a fuel rack or an electrical impulse, such as to a solenoid operated hydraulic control valve. Many items of plant will have their own internal servo loops, which may or may not be incorporated into the machinery monitoring system.

7.9.2 Microprocessor based controllers

The main requirements for a digitally based machinery control system are that it should be reliable, available, maintainable by shipboard personnel, have a low first cost, have a low throughlife cost, be modular and require little manning. Those for a naval environment should be able to withstand action damage, mechanical shock, electromagnetic shock and interference and be nuclear hardened.

The difference between reliability and availability is important in this context. An item of equipment may be very reliable. However, if it is difficult to maintain, it may be that it is out of action for a long time every time it does go defective. The normally accepted definition of availability is:

$$\text{Availability} = \frac{\text{(Mean time between failures)}}{\text{(Mean time between failures + mean time to repair)}}$$

(7.10)

High availability figures can thus be achieved by either increasing the time between failures or making the equipment easier and hence quicker to repair. In practice, attention is paid to both these areas. Computing boards for marine use need to be ruggedly held in an enclosure which is vibration resistant, and attention needs to be paid to the mounting arrangements, to isolate the equipment mechanically from shipborne vibrations. Cooling and air filtration arrangements may need special attention. If the computing equipment is housed in the control room, it is likely that the ambient conditions in the room will be adequate for the computers.

For digital equipment, repair of many defects will consist mainly of diagnosing a defective computer board using built in test equipment (BITE), which can identify a defect to a particular board, and replacing the board. Often, individual boards are cheap enough to throw them away when defective. Otherwise, a board is returned to shore for repair at the manufacturers. This philosophy places a high reliance on the ship having a comprehensive spares holding of boards. This problem can be mitigated by designing control equipment in such a way as to use a few standard types of board, with the specific control requirements being undertaken in the software.

The essential features of a digital engine controller are described as follows:

1. Demand levers may be expected to be on the bridge and in the machinery control room (MCR). A method of deciding which lever is in control will be fitted. The controlling lever will produce a signal which is a function of the desired speed or engine setting. This may be a DC voltage, supplied from a potentiometer fitted to the lever, or may be synchro, which gives an AC voltage corresponding to the position of the lever.
2. An analogue to digital converter (ADC), which will convert the desired setting into a digital format readable by the computer.
3. A microcomputer, which may be programmed with the control logic pertaining to the particular installation in which the controller resides. The software may simply contain a program to compare the desired and actual speeds and alter the fuel rack accordingly, or may be more complex, containing for example a program to take the speed of the engine up very slowly between Full Ahead and Full Away in a merchant ship, so as not to cause excessive thermal stresses in the engine. The computer

may also contain starting logic, so that the engines can be remotely started from the bridge.

4. A digital to analogue, converter (DAC) which changes the output signals from the computer into a form suitable for operating the fuel rack. The main requirement of the output stage is to provide an increase in power to enable a large mechanical fuel rack shaft to be move. This may take several forms, including a DC servo with its own feedback system, or a stepper motor. The latter form of output has the advantage that it does not require any local form of feedback, as its position is uniquely determined by the number of steps sent to it by the computer.

   For control applications requiring a larger output force, required to move the blades of a controllable pitch propeller system, the output from the computer may be sent to the servo valve of a hydraulic system. Power magnification of many thousand times is possible in this way, allowing large mechanical control surfaces to be moved in a controlled manner.

5. Speed sensing device. It is necessary to know the actual speed of the engine, so that an error or difference signal can be calculated in the computer. This can consist of a tachogenerator, which produces a signal proportional to engine speed. It is more usual, and a simpler solution with digital control, to fit a small magnetic or electromagnetic pickup on the flywheel of the engine, which sends a pulse every time it passes a sensing coil.

Although the constructional details of an individual control system may change, the principles are the same. They involve defining a desired value of a quantity (engine speed, propeller pitch, etc.), measuring the actual quantity, operating on this error signal in a microcomputer, and calculating a control signal. This control signal is then amplified and used to operate a controller, to effect the desired condition in the controlled plant.

7.9.3 Fault detection and diagnosis

A problem associated with unmanned engine rooms is that an incipient fault may develop and continue unnoticed for an appreciable time, particularly if the engine is operating at less than full power, so that the temperatures and flow rates in the engine are likely to be below their maximum values. Before the advent of unmanned engine rooms, it was necessary for an engineer or mechanic to go around the plant at intervals to take a number of readings of local gauges and enter them in a log. The very act of going round the engine room enabled the engineer to detect an incipient fault by small changes in perceived or actual temperatures or even by sound and smell. The presence of the engineer on watch thus assisted the diagnostic process. With unmanned engine rooms, this direct contact with the plant is largely missing. If any of a number of selected variables goes outside its preset tolerance, an alarm will sound and usually a computer printout will be made, to establish the order in which the alarms occurred. The engineer will typically be faced with a number of alarms showing, and will have to use his skill to establish the nature of the fault and the correct remedial actions. One effect of the very high level of reliability of modern engines, coupled with the decrease in time spent in the engine room or control room, is that engineers have much less operational experience of fault conditions, and so may be unable to detect the nature of a fault readily from the information presented on the alarm panels or system screens. They may be aided in the diagnostic procedure by being able to call up a series of system diagrams on VDUs in the control room, shown in the centre of the control panel in *Figure 7.48*.

This present method of fault diagnosis is shown diagrammatically in *Figure 7.50*. The importance of the Man-Machine Interface MMI in the design of the control panel needs to be emphasized, as the engineer is trying to understand the behaviour of a number of complex, interrelated systems, using only the informa-

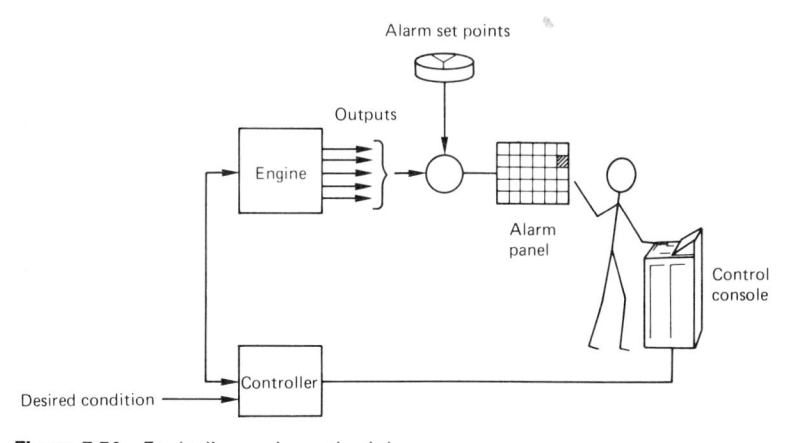

Figure 7.50 Fault diagnosis methodology

tion at the control panel. Older control panels, consisting of a row of up to fifty identical gauges, are examples of what is perhaps the worst possible way of presenting this information. Present systems make extensive use of mimics, with live gauges and indicators forming part of the mimic, and large VDUs showing the current state of the plant in a diagrammatic way.

It is now possible to use artificial intelligence (AI) technology to assist in the diagnosis of engine faults (Stansfield, Katsoulakos, Ruxton et al., 1987; Penny, MacIsaac and Saravanamuttoo, 1982; and Kawamura, Kurosaki, Inagaki et al., 1987). The basic concept of an AI or expert system based fault diagnosis system is shown in *Figure 7.51*. The engine plant is operated in parallel with a mathematical model of the plant, with both the actual engine system and the mathematical model being subjected to the same inputs and control variables, speed demand, pump states, etc. If the mathematical model were correct in every respect, there would be no difference between the outputs of the model and the actual engine plant in normal operating circumstances. In fault conditions, there will be a difference between the outputs of the model and those of the engine, which can be detected by comparing the two sets of outputs, using the performance monitoring system for the engine, and the simulation outputs. A combination of the differences detected and the actual engine outputs is then fed to an expert system, which is programmed with a knowledge base of information obtained from one or more human experts. The knowledge base is designed to relate the observed symptoms to a specific fault. It may be that there is insufficient knowledge immediately available within the knowledge base for the system to be precise. The system may then request more information to be obtained, either automatically or by operator input. In this way, the probable nature of a fault is expected to be displayed on the system output screen.

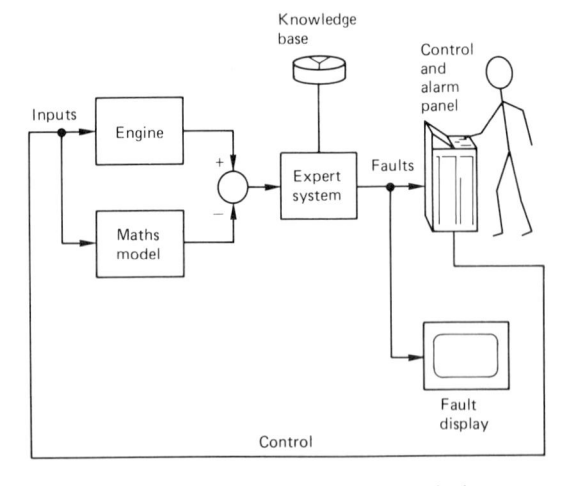

Figure 7.51 Expert system based fault analysis

It should be stressed that this system is still being developed and is unlikely to be seen at sea for a number of years. It is of interest, however, in showing the ways in which the techniques of engine control are progressing, aided by the extensive use of onboard microcomputers.

7.10 Integrated ship control

7.10.1 Requirements for integrated ship control

The control methodology of ships has grown over the past few decades in a rather haphazard and uncoordinated manner. Ship builders are accustomed to building ships singly, so that there is a tendency to design ships piecemeal, using whatever control equipment is commercially available. Different types of company are traditionally associated with the different features of a ship control system, from the heavy hydraulic equipment of a steering gear or fin stabilizer, to the electronic disciplines of a collision avoidance radar.

As has been referred to in Section 7.8, it is possible to have a ship designed at present with many interacting, but unrelated, control systems operating in the ship. The rudder will be controlled by an autopilot and its movement will affect the speed of the ship and its rolling motion. The roll of the ship will be controlled by stabilizers, which will affect the speed and yaw. The speed of the ship is controlled by the engine governor and the thrust of the ship will affect the yaw performance. Additionally, each of these control systems is likely at present to contain one or more microcomputers and possibly hydraulic control equipment. There is frequently no attempt for any commonality of supply for the computing and control equipment, so that there is a need for ship's staff to be familiar with a wide range of diverse equipment, and for the ship to hold documentation and spares for many different types of computer.

Two main influences have come to bear on the problem of controlling a ship in a more logical and cost effective manner. The large increase in fuel prices in 1973 caused ship owners to examine critically the economics of operating their ships in a fuel efficient manner, while the decreasing availability and increasing costs of employing large numbers of crew have resulted in policies to reduce their numbers.

A major effect of these two influences has been for ship owners to examine the methods whereby ships are controlled and operated. If the ship can be operated as a single system, there may be expected to be savings in overall efficiency, both as regards fuel consumption and manning levels. There may be expected to be penalties to be paid in the level of complexity in the ship and in the levels of competence in the remaining crew personnel, so that there is likely to be a level of control integration beyond which it is uneconomical to go.

7.10.2 Ship integration methodology

At present, most ships are controlled as a number of largely separate systems. *Figure 7.52* shows the control paths in the four main areas of navigation, engine and hull, stabilization, and cargo handling. For a warship, the concepts of sensor control and weapon allocation need to be considered in place of the cargo handling concept of a merchant ship. Ironside (1981) has pointed out that some current warship weapon systems are inflexible in their operation, and that much ship integration consists of a collection of independent systems.

If substantial savings are to be made in the most sensitive areas of crew and fuel efficiency, a sensible first step is to examine each of the four subdivisions singly, as there is little point in attempting to integrate a set of inefficient subsystems, with the aim of producing an efficient whole. It is necessary to start with subsystems which are in themselves at the highest level of efficiency which it is economically feasible to achieve. For example, 1 or 2% of fuel may be saved by making a hull smooth and painting it with self-polishing paint. It needs to be carefully worked out whether the additional cost of the smoothing work and the paint can realistically be saved by the fuel savings in a reasonable time. In the case of an antisubmarine frigate, however, the hull may need to be smoothed so as to obtain an adequate level of sonar detection. In this case, a totally different set of economic calculations would need to be performed, which would balance the increased effectiveness of the ship against the increased cost.

The main areas in which individual improvements in efficiency may be sought may be thought to include:

1. Navigation: autopilot performance (modified cost functions); collision avoidance programs; electronic charts (reducing the amount of correction work); integrated position finding systems (SATNAV, Decca, Omega); advanced communication systems.
2. Engine and hull: selection of fuel efficient engine arrangements; use of cheaper fuel distillates; unmanned machinery control; condition monitoring equipment; variable injection timing; improved maintenance arrangements, leading to greater availability; automatic fault detection and diagnosis; improved hull shape for reducing drag; slower speed propeller; wake recovery systems; improved high lift rudder arrangement.
3. Stabilization: improved cost functions in control mechanism.
4. Cargo handling: automatic loading and unloading control; condition and loading calculations; self discharge facilities (allowing the use of cheaper ports).

Some of the above systems, while contributing to overall efficiency in one area, will degrade the efficiency of the ship in another area. Thus, fitting a single slow speed diesel with a large fixed pitch propeller, while resulting in a fuel efficient power plant, will degrade the controllability of the ship overall, as it will be difficult for the ship to go slowly and there are a finite number of manoeuvres available before the starting air is exhausted.

While the overall gains of fitting each improvement in efficiency may be difficult to quantify, it is certain that most will add to the first cost and in some cases the running costs of the ship. It is therefore necessary to determine an optimal balance of efficiency to obtain the best overall solution.

Once the individual areas of concern in a ship design have been attended to, so that the ship will perform as well as is feasible in each department, it is then possible to address the problem of integrating all the functions

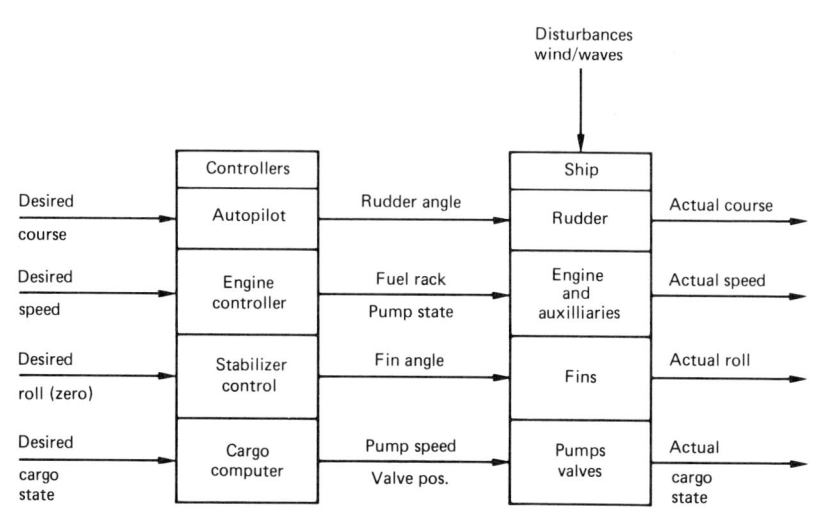

Figure 7.52 Ship control methodology; non-integrated

of the ship so that it performs well as a single system. This will introduce problems of overall control philosophy, data gathering and distribution, and manning and training.

The overall control philosophy for operating a ship as a single system must imply a single control position. Because of the requirement for the ship controller to keep a lookout, this control position must be situated where a clear view of the surrounding sea may be obtained, such as the bridge. The present practice of siting the bridge aft in a large ship may need to be examined, as in many ships there is a large area of sea which is invisible to the Officer of the Watch. For ships with substantial deck gear or a large number of containers on deck, this view can be seriously deficient. It is not adequate either to rely on electronic detection of navigational hazards, as many small targets, such as small wooden fishing boat trailing nets, may be undetectable by radar and yet will require to be given a large berth. The tendency in many ships, such as the proposed Arctic Pilot Project ship of *Figure 7.15*, and most ferries, is to control the ship from a control position sited high in the ship and placed forward.

If a single watchkeeping officer is to control the entire ship, adequate information must be provided. This implies a data gathering system, which needs to be shipwide, with sensors at perhaps 300 key points for a merchant ship and up to 10 000 for a warship. This information needs to be transmitted to the control position and to a number of local control consoles, so that alternative control of machinery is possible in the event of breakdown of the central control equipment. The requirements of this data transmission highway may be considered to be:

1. It should be comprehensive, taking in data from all key areas and making it available where required.
2. It should have a low first cost and low through-life costs.
3. It must have a built in redundancy, to allow for breakdown and, for warships, to be able to sustain action damage.
4. It must be capable of modification and expansion as the role of the ship or the equipment fitted changes.
5. It must not, of itself, require excessive maintenance or impose heavy training requirements on staff.

Typical of a warship distributed data system is the Canadian ship integrated processing and display system (SHINPADS) (Ironside, 1981), which is shown diagrammatically in *Figure 7.53*. The data bus carries serial data all round the ship at a rate of 10 megabits per second.

It will be seen that data concerning all aspects of ship control are contained on the bus and that access to all control data is therefore available to each user. Flexibility is thereby ensured, both for a present configuration of machinery, sensors and weapons and also for further upgrades. To cope with the requirements of action survivability, the bus has a total of six cables, distributed around the ship.

For a merchant ship, the same principles apply, except that the requirements are an order of magnitude less stringent. The number of sensors is far fewer and the update rate may be slower. Also, there will be a need for two or three data lines in order to allow for breakdown. The other requirements, commonality of data and standardization are, however, similar. A typical merchant ship data transmission layout is shown in *Figure 7.54*.

A major problem of selectivity of information arises with a shipwide distribution system. While it is perfectly feasible to provide the ship controller with continuously updated information on all aspects of the ship

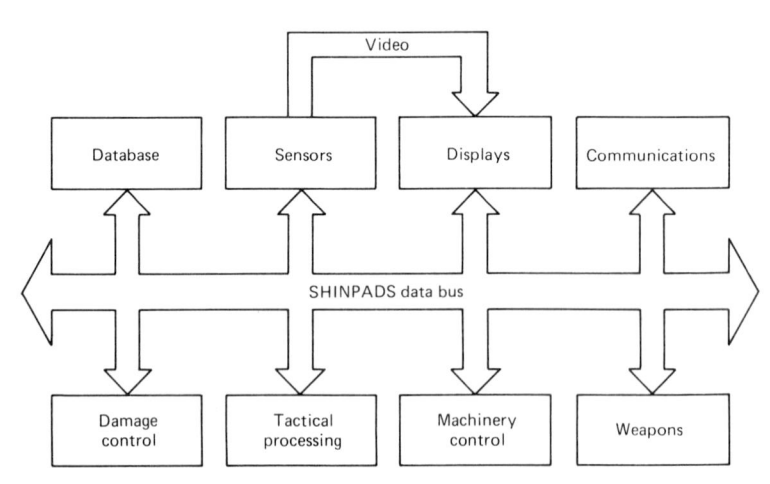

Figure 7.53 Integrated ship control: the SHINPADS system, courtesy Royal Canadian Navy

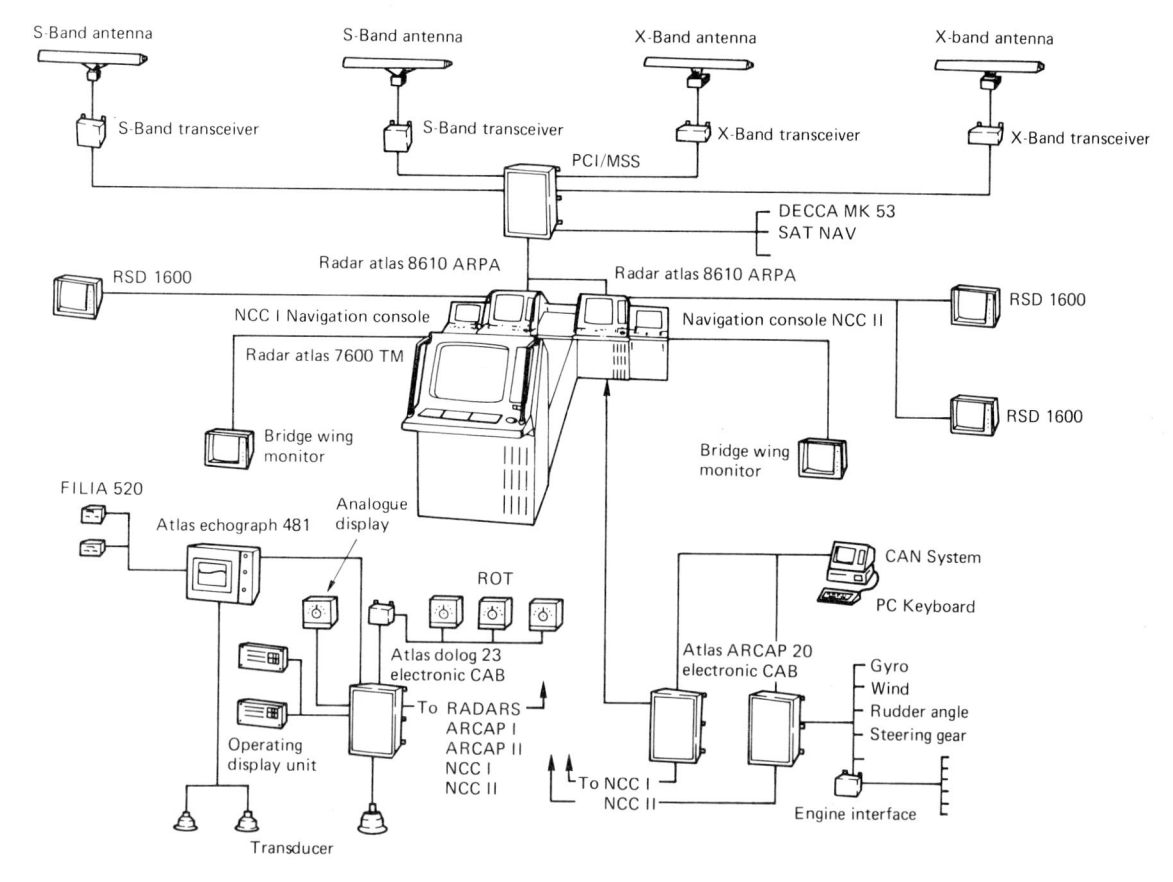

Figure 7.54 Integrated ship control: typical merchant ship installation (Courtesy of Krupp-Altas Elektronik)

systems, this will result in a large amount of information, in which the essential information will be hidden. In order for the ship operator to have effective control, he must have sufficient but not excessive information. For example, operators keeping watch on the bridge will need to know little about the engine systems until there is an incipient or actual fault condition, in which case they will need a large amount of detail about the particular system which is faulty. Similarly, a Principal Warfare Officer controlling his ship from the operations room in a frigate has available to him all the details about perhaps a hundred friendly, neutral and potentially hostile targets. It is impossible to assimilate this information at a rate commensurate with the requirements of a modern engagement and so the data system has to be capable in itself of choosing what information will be needed, and of altering that information continually, as new threats appear.

With an integrated control system there is much scope for standardization if the system can be designed from scratch. While this may be possible for a new class of warship, it is rarely possible in a merchant ship installation, which has to have considerations of first cost high on the priority list. Integration in this case is often more an exercise in making a number of relatively disparate systems operate together, with at the most a centralized control system to prevent the rudder and stabilizers acting against each other. If it is possible to design an integrated system from the first concept of a merchant ship, there should be scope for reductions in first and through-life costs by using a standard range of computing systems throughout the ship, which will have an impact on spares holdings, documentation and training.

7.10.3 Implications of integrated control on personnel

One of the main aims of implementing integrated control and automation concepts is that it is feasible to reduce crew numbers significantly. In the 1960s, it was common to have a crew of over 40 in an ocean going cargo ship. Early actions resulting in fewer staff aboard included operating a centralized messing system and adopting the general purpose rating concept, whereby ratings were able to be transferred to either deck or engineering skills depending on the daily requirements of the ship. These small changes were able to be accomplished with little additional ship costs. Further reductions in crew numbers needed maintenance to be conducted or planned from shore, an unmanned engine room and a smaller watch on the bridge, routinely using the autopilot. The consequences of these changes are felt most in the engine room, where a comprehensive monitoring and control arrangement is required. The cost of this has to be balanced against the reduction in crew costs over the life of the ship. At this stage the crew numbers are likely to be in the high teens. Further reduction to single numbers will require a full data link implementation, enabling most tasks to be controlled from the bridge, a large degree of redundancy to be built into the engine room equipment and sophisticated emergency power arrangements. Crew at all levels will need to be fully integrated, with the three officers apart from the master being trained to be able to cope with both bridge and engine room duties. There will be a need at this stage for an officer to cope with the extensive computer provision in the ship. This officer will need computer qualifications, specialist training and a backup of spares and tools to keep the computer controlled equipment fully operational. Because a crew of perhaps six or eight will be physically unable to change major engine components at sea, additional reliability may need to be built into the ship, such as using a twin engine configuration, with additional standby generator power. Communications with shore will be required to a greater extent, with telex, FAX and satellite communications being required.

It will be appreciated that there is likely to be a point at which the additional complexity of imposing more and more automation to reduce crew numbers becomes uneconomic. At the end of the 1980s, this point appears to have been reached at between six and ten staff for cargo vessels with high cost crews, with perhaps twelve to eighteen for ships with lower cost crews. As crew numbers are reduced, the cost of reducing by one further person increases, since more and more automation is required. The incremental cost of additional crew decreases as crew numbers increase, as the highest paid personnel are needed even with very small crew numbers (a ship with a single crew member will need a very high salary indeed) but the costs of automation are also very high, so that this concept is likely to be totally uneconomic. It is also impossible to abide by regulations requiring a continuous lookout to be kept with fewer than about four crew, and difficult with numbers considerably under ten.

There are additional implications on training for integrated control. There is a requirement to train officers to be competent at both the deck and engine room aspects of their profession and also to be able to use a much more complex electronic system involving all aspects of the ship's machinery. There is a need for deep specialism in the electronics officer, who needs to be trained to operate and maintain complex electronic and computer systems. Similar cross-training is required for ratings, so that they can operate complex machinery and carry out maintenance tasks.

In navies, the concept of the user maintainer has been seen for several decades, on the principle that if a system is being used, it cannot be maintained, and *vice versa*. Although frigates typically have crews of almost 200, the problems of adequate training are still severe, as the level of complexity of warship equipment is very high and shipwide integration needs specialists in computer engineering to be provided.

7.11 Assessing controllability – ship trials

In Section 7.1, the concept of controllability was defined in terms of the effectiveness of the controls in affecting a desired ship performance. The desirable requirements of a ship in achieving an adequate degree of controllability were indicated in Section 7.4 to be mainly those of manoeuvrability at slow speeds and the effectiveness of the controls. In this section, possible methods of assessing ship controllability in a quantitative manner are described and the details of relevant ship trials outlined. As it would be most desirable if controllability were able to be assessed before a ship is built, some indications of methods of predicting it from the ship's lines plans are described, although at present these are very much at the research stage.

7.11.1 Assessing controllability

In 1983, Landsburg *et al.* defined the four most important aspects of a ship's behaviour, which were thought by a large number of pilots to contribute towards controllability, as:

- slow speed manoeuvrability,
- adequate backing power and straight line stopping ability,
- short response time to commands; and
- adequate swing control with moderate rudder angles.

It will be noted that these desirable features are all expressed in qualitative terms, so that it is not simple to transform these requirements into a quantitative measure of performance.

The usual approach is to devise trials which are pertinent to each of the perceived desirable qualities, to measure the results and to devise a formula to relate the trial results to presumed performance needs. As each step in this chain of events requires a decision to be made on what is suitable or relevant, it is perhaps not surprising that there is currently little uniformity in evaluating suitable measures of controllability. Some agreement is, however, possible on what might constitute a suitable range of trials to carry out, to at least determine some of the performance characteristics of a ship.

7.11.2 Ship trials – purpose and conduct

Ship trials can be inconvenient, expensive and time consuming to carry out, as they usually require the full availability of the ship, in good condition, under predetermined environmental conditions, often with extensive trials equipment and personnel embarked. Three of the best conducted sets of trials involved two tankers *Esso Bernicia* (Clarke, Patterson and Wooderson, 1970), and *Esso Osaka* (Crane, 1979), and a fast cargo ship USS *Compass Island* (Morse and Price, 1961).

Ship trials are normally carried out for one of several purposes:

1. As a contractual check at the end of a ship building contract. Such trials are normally short in duration, and limited to a small number of contractual points such as maximum straight line speed.
2. First of class trials. Where a fleet of several ships is being built, such as will commonly occur with a warship class, it is worthwhile testing the first ship to be built more extensively, as a check on the overall effectiveness of design on the hull and machinery.
3. As a check on manoeuvrability. It is these tests with which we will be most concerned in this section.
4. To provide data for ship manoeuvring mathematical models. These trials, as they need to examine the ship behaviour in all manoeuvring regimes, will be very extensive in scope. Unfortunately, ships are very rarely made available for such trials to be performed.

The desirable features of a particular ship trial are:

1. That the starting conditions and test inputs are easily reproducable.
2. That it should be able to be conducted and recorded with a minimum of added ship equipment.
3. That the trial should bear a close relationship to actual ship operating conditions.

For any trials it is necessary for the ship's position to be fixed relative to either the shore or to the water, at intervals of every few seconds. This may be done using shore based tracking equipment, such as the geodimeter shown in *Figure 7.55*, which is able to give a reading directly to a microcomputer about every 2 s to an accuracy of well under 10 cm. It is possible also to measure ship position using a shipborne set of equipment incorporating an inertial navigation package, which will give the accelerations of the ship, and hence its displacements.

For accurate trials results, additional recording equipment needs to be installed in the ship, to measure variables such as shaft speed, rate of turn and heading angle. The fitting of this equipment onto a trial vessel can greatly increase the total planning and execution

Figure 7.55 Geodimeter in use for ship trials, courtesy Geotronics Ltd

time of a trial, as some dockyard work is likely to be required to develop or install interfaces. Less accurate trials, which may nevertheless be adequate for a preliminary assessment of controllability, can be achieved using only ship's equipment, with quantities being recorded manually. Position can be recorded either using a ship's own navigation radar on a fixed object, or using the radar from an adjacent ship.

To assess controllability, one or more trials will be necessary for each of the desirable qualities referred to above. The trials most commonly associated with measuring controllability are the crash stop, the turning test or circle manoeuvre, and the zig-zag, or Z manoeuvre. In many menus of ship trials, the manoeuvres are carried out using full speed and full rudder. However, for assessing controllability, it may be argued that slow speed manoeuvres will yield more information, as they will more closely represent the conditions pertaining when controllability is at a premium, in harbour manoeuvres.

The crash stop is a manoeuvre designed to test the vessel's stopping ability. The ship proceeds on a straight course at a meaningful speed for the manoeuvre concerned, and the engine is put to Full Astern. The distance travelled before the ship stops, and the lateral deviation, are measured along with the time taken to stop (*Figure 7.56a*).

The circle manoeuvre is of particular interest, in that it gives information about the transient and steady state manoeuvring ability of the ship. Again, the ship starts on a steady course and speed. At the execute time, the rudder is put over to an appropriate predetermined angle and held there. The ship then starts to turn, and will eventually reach a steady turning condition (*Figure 7.56b*). Measures of interest will include the advance, transfer and turning diameters, which give an indication of the ship's turning ability. If the helm is put to midships after the ship has reached a steady turn, a simple assessment may be made of its course stability. This manoeuvre is known as the pullout manoeuvre. Unstable ships will continue turning, while stable ships will return to an approximately straight path. (For single screwed ships, the propeller will cause even a stable ship to turn slightly with the rudder midships.) As far as controllability is concerned, a degree of course instability can be advantageous, as the turning and checking ability may be faster than with a similar size of ship with a directionally stable hull, as was indicated in Section 7.4.

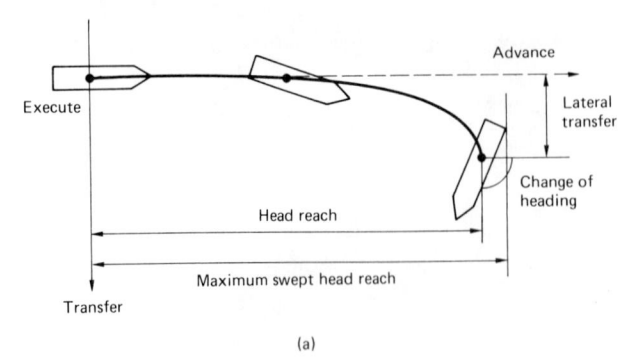

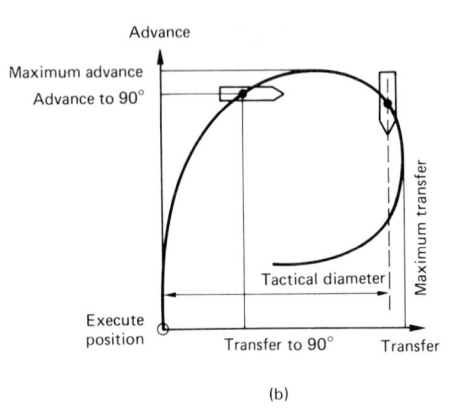

The zig–zag manoeuvre is perhaps the closest to actual ship operations. The ship again starts in a straight line at constant speed and the rudder is placed over by a specified amount (the rudder angle). This is typically 5, 10 or 20°, depending on the size and manoeuvrability of the ship. After the ship has turned through a determined number of degrees (the check angle), the helm is put over the other way by an amount equal to the rudder angle, and the ship will then turn in the opposite direction, until the check angle is reached on the other side of the initial course, when the rudder is again reversed (*Figure 7.56c*). Measures of interest are the overshoot angles after the helm has been reversed, and the swept path of the ship. The additional swept path after the helm has been reversed may also be of interest.

For controllability assessment, slow speed manoeuvres with moderate helm angles are more likely to be relevant than full speed manoeuvres with large helm angles. There is to date no widely accepted method of combining the figures obtained from these or any other

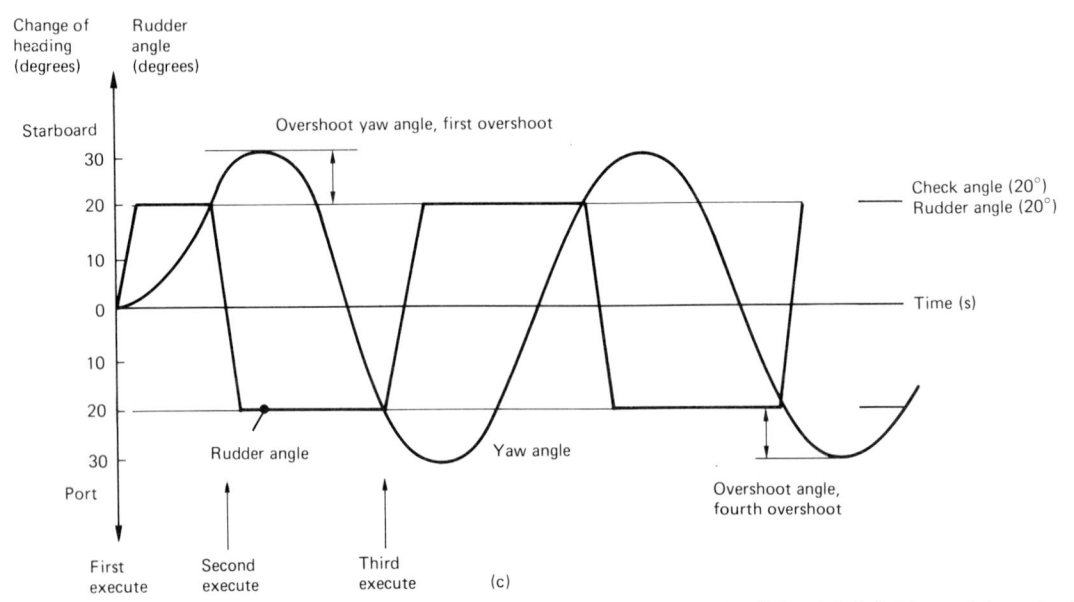

Figure 7.56 Ship trial definitions: (a) crash step; (b) turning trial; (c) zig–zag behaviour

sets of trials into one 'controllability index'. A discussion paper leading towards the establishment of controllability standards by the US Coast Guard was presented by Cojeen, Landsburg and MacFarlane (1987), which proposed tentative recommendations towards considering manoeuvring performance in the design of new vessels, incorporating the results of the above trials, with some others. Clarke (1987), reported on deliberations of IMO on the same topic, but considered the use of indices based on manoeuvring standards of a linear mathematical model (the Nomoto K–T indices), for assessment of controllability. Both papers recommend the use of a pilot card, containing details of the ship's manoeuvring ability.

It would be feasible to give a single quantitative grade for controllability of a ship if a suitable number of individual indices of performance were combined into a single grade. *Table 7.1* shows one possible method. For each of four trials, a number of measures is taken, and a grade awarded. The grade would need to be defined according to the size and type of the ship. Grades may be on a simple scale of five, ranging from excellent, through good, acceptable and marginal, to unacceptable. The total grade achieved is then the lowest of any individual trial, on the assumption that a ship has to be satisfactory at all aspects of manoeuvring for it to have an adequate level of controllability. The concept is illustrated in *Table 7.1*.

Table 7.1 Controllability assessment: possible implementation

Trial	Measure	Units	Grade
zig–zag	Max overshoot	deg	
Circle	Max advance/Lpp	–	
	Max transfer/Lpp	–	
Stopping	Max advance/Lpp	–	
	Max transfer/Lpp	–	
	Time to stop	sec	
Reduced speed manoeuvring	Max course deviation	deg	
	Achieved grade		

The reduced speed trial is designed to assess the ship's ability steer at speeds of between 4 and 6 knots. The concept of being able to give a ship a quantitative measure of controllability is attractive, but significant work is necessary before there is a widely accepted methodology for this to be achieved.

7.11.3 Improving controllability

If a ship is found to have an inadequate controllability, a number of measures can be taken to improve the situation in some cases. Where the rudder performance is inadequate, as could be shown by excessive first overshoot angle in a zig–zag trial, a rudder with greater lift can be retrofitted. Similarly the stopping ability and the slow speed manoeuvring ability may be able to be improved by fitting a controllable pitch propeller. In some cases, however, the basic hull form is at fault, leading to excessive overshoots and virtually uncontrollable behaviour. In these cases, some degree of palliative action may be taken by increasing the size of the skeg aft, increasing the size of the rudder or by fitting additional control surfaces. These attempts to impose controllability on an otherwise uncontrollable hull are not in themselves a satisfactory solution, but merely a way of making the best of a poor design. A hull should be inherently controllable, without relying too much on imposed devices to make it so. Of particular interest is the precise shape of the forward part of the hull, as the hydrodynamic lift and drag forces are to a large extent generated at the forebody of the hull. If this forepart is too bluff, generating an excessive amount of lift at the forward extremity of the hull, it could be that the hull will become totally uncontrollable in yaw with normal control surfaces.

7.11.4 Predicting controllability

It would be preferable to be able to predict the controllability of a hull–effector combination at a very early stage in the design of a ship. To find out at trials time, or even in service, that the ship has undesirable control characteristics, cannot be considered as sound design practice. It will be seen, though, that the controllability of a ship is made up of a number of aspects, some of which, such as the swing control, depend on the interrelationship of both hull and effectors. Captive model trials are unlikely to give more than an indication of the controllability, as they rarely incorporate realistic representation of the hull and rudder interaction in manoeuvring conditions. Some progress is being made with the ability of researchers to model the manoeuvring behaviour of ships before they are built from a combination of free running model tests and mathematical models (Martinussen and Linnerud, 1987).

7.11.5 Controllability as a port–ship problem

The controllability of a ship does not only affect the ship itself. If a ship has a poor level of controllability, it is likely to require greater sea room to manoeuvre or will be operating closer to the margins of safety in a port. For example, if a high sided container ship, with a single slow speed diesel and fixed pitch propeller, is leaving an up- river berth shortly after high water, with a long trip down river, it may be that it will be travelling at some time with a 4 knot current. If that ship can go at 6 knots at the minimum, because of its engine design, it will be travelling at 10 knots relative to the ground. The ship may then have to traverse a set of bridge piers with minimal clearance and with a side wind blowing. In these, very typical, conditions, the safety margins in a port may be minimal. Techniques to assess the suitability of a ship to transit a port area safely include the use of a range of ship simulation techniques, where the ship and the port are modelled using a number of computers, and the ability to transit the port safely in a range of environmental conditions. The simulation requires there to be a mathematical model of the ship under consideration, with its controllability adequately modelled. This can be achieved using a range of techniques, including carrying out special ship trials, performing free running model trials (with their disadvantages) or by modelling the ship directly from a knowledge of its lines, plans and effector design (McCallum, 1987). A number of such models have now been produced which have been tested against actual ship results, and a measure of confidence is able to be built up in the techniques. The use of ship simulators for developing the suitability of a port for particular ship operations is now becoming a routine part of the port design process.

References

Bingham, V. P. and Mackey, T. M. High Performance Rudders – with Particular Reference to the Schiling Rudder. *Marine Technology*, 24, 4, SNAME, New York (1987).

Card, C. C., Cojeen, H. P., Spencer, J. S. and Harmon, J. P. Report to the President on an Evaluation of Devices and Techniques to Improve Manoeuvring and Stopping Abilities of Large Tank Vessels. *US Department of Transportation Report* No. CG-M-4-79, NTIS Springfield, VA, (1979)

Cheng, R.C. H., Williams, J.E., McCallum, I. R. Flexible Automatic Ship Controllers for Track-keeping in Restricted Waterways. Proc. 4th International Symposium on Ship Operation Automation, (ISSOA '82), Instituto Internazionale Della Communicazione, Genoa, Sept 20–22, 1982, pp. 185–194

Clarke, D. Assessment of Manoeuvring Performance. In *Proceedings of the International Conference on Ship Manoeuvrability*, 29 April–1 May 1987, Paper 2, RINA, London (1987)

Clarke, D., Patterson, D. R. and Wooderson, R. K. Manoeuvring Trials with the 193 000 tonne d.s. Tanker *Esso Bernicia. British Ship Research Association Report* NS295 (1970)

Cojeen, H. P., Landsburg, A. C. and MacFarlane, A. A. One Approach to the Development and Achievement of Manoeuvring Standards. In *Proceedings of the International Conference on Ship Manoeuvrability*, 29 April–1 May 1987, Paper No. 3, RINA, London (1987)

Cooling, J. E. Propulsion Control System for the 1980s. In *Proceedings of the Ship Control Systems Symposium*, 30 October–3 November 1978. Annapolis, pp. R4-1–18. Dept of Defense, Bethesda, MD (1978)

Crane, C. L. Manoeuvring Trials of a 278 000 dwt Tanker in Shallow and Deep Waters. *Trans SNAME*, 87, pp. 251–283 (1979)

Froude, W. On the Rolling of Ships *Transactions of the Institute of Naval Architects* (1861). Reprinted in *The Papers of William Froude 1810–1879*, The Institution of Naval Architects, London pp. 40–85 (1955)

Gill, A. D. Mathematical Modelling of Ship Manoeuvring. In *Proceedings of the Institute of Mathematics and its Applications Conference on Mathematical Aspects of Marine Traffic*, London, September 1987, pp. 193–227 (1987)

Hayes, M. N. Parametric Identification of Nonlinear Stochastic Systems Applied to Ocean Vehicle Dynamics. DSc thesis, Department of Ocean Engineering, Massachusetts Institute of Technology, Cambridge, MA (1971)

Iijima, Y. Ship Automation – Past, Present and Future. In *Proceedings of the 4th International Symposium on Ship Operation Automation*, 20–22 September 1982. pp. 215–223. Instituto Internationale della Communicazione, Genoa (1982)

Ironside, J. E. SHINPADS – An Integrated Philosophy for the 21st Century. In *Proceedings of the 6th Ship Control Systems Symposium*, Ottawa 26–30 October 1981, National Defence Headquarters, Ottawa (1981) pp. D2 2–1–D2 2–8.

Kallstrom, C. G. Identification and Adaptive Control Applied to Ship Steering. *Swedish Maritime Research Centre*, SSPA, 93 (1982)

Kallstrom, C. L. and Ottoson, P. The Generation and Control of Roll Motion of Ships in Close Turns. In *Proceedings of the 4th International Symposium on Ship Operation Automation*, Genoa, 20–22 September 1982, pp. 43–50. International Federation for Information Processing (1982)

Kawamura, Y., Kurosaki, Y., Inagaki, M. and Miyamoto, Y. *Proceedings of the 8th Ship Control Systems Symposium*, 6–9 October, 1987. pp. 3.277–287. Ministry of Defence, The Hague (1987)

Landsburg, A. C., Card, J. C., Crane, C. L., Alman, P. R. et al. Design and Verification for Adequate Ship Manoeuvrability, *Trans. SNAME*, 91, pp. 351–401 (1983)

Lowry, I. J. and Squire, T. M. *Trireme Olympias Extended Sea Trials, Poros 1988*. Department of Maritime Studies, University of Wales College of Cardiff, (1988)

MacGillivray, P. J. and Marshall, D. J. A Modern Machinery Control Console for Warships. In *Proceedings of the 7th Ship Control Systems Symposium* 24–27 September 1984. Bath, pp. 233–243. Ministry of Defence, Bath (1984)

Mandel, P. Ship Manoevring and Control. In *Principles of Naval Architecture*, J.P. Comstock (ed.), pp. 463–574. The Society of Naval Architects and Marine Engineers, New York (1967)

Martinussen, K. and Linnerud, I. Techniques for Predicting Manoeuvring Characteristics of Ships at the Design Stage. In *Proceedings of the International Conference on Ship Manoeuvrability*, 29 April–1 May 1987, Paper no. 18, RINA, London (1987)

Marwood, C. T. Experience in Developing a Digital Distributed Control and Surveillance System. In *Proceedings of the 6th Ship Control Systems Symposium*, 26–30 October, 1981. Ottawa, pp. 03-1–20. National Defence Headquarters, Ottawa (1981)

McCallum, I. R. A Ship Steering Mathematical Model for All Manoeuvring Regimes. In *Proceedings of the IIC Conference on Ship Steering Automatic Control*, Genoa, 25–27 June 1980, pp. 119–139, Instituto Internazionale della Communicazione (1981)

McCallum, I. R. Dynamic Quantitative Methods for Solving Actual Ship Manoeuvrability Problems. In *Proceedings of the International Conference on Ship Manoeuvrability*, 29

April–1 May 1987, Paper 26, RINA, London (1987)

Morse, R. V. and Price, D. *Manoeuvring Characteristics of the Mariner Type Ship* (USS *Compass Island*), *in Calm Seas*. Sperry Polaris Management, Sperry Gyroscope Company, New York, (1961)

Moskowitz, L. Estimates for the Power Spectra for Fully Developed Seas for Wind Speeds of 20 to 40 knots. *Technical Report for US Naval Oceanographic Office*, New York University (1963)

Parsons, M.G. and Hua, T.C. Surface Ship Path Control Using Multivariabe Integral Control . Dept of Naval Arch. & Marine Engineering Publication No. 233, University of Michigan, Ann Arbor, Michigan USA, Jan. 1981

Penny, P. V., MacIsaac, B. D. and Saravanamuttoo H. I. H. Development of a Diagnostic Model for Naval Gas Turbines. In *Proceedings of the 4th International Symposium on Ship Operation Automation*, 20–22 September 1982. Istituto Internationale della Communicazione, Genoa, pp. 95–104 (1982)

Pourzanjani, M. M., McCallum, I. R., Flower, J. O. and Zienkiewicz, H. K. Hydrodynamic Lift and Drag Simulation For Ship Manoeuvring Models. In *Proceedings of the 4th International Conference on Marine Simulation*, Trondheim, Statens Treningssenter for Skipsmanoevring (1987)

Reid, R. E. and Williams, V. E. A New Ship Control Design Criterion for Improving Heavy Weather Steering. In *Proceedings of the 5th Ship Control Systems Symposium*, Vol. 1, Annapolis, MD (1978)

Rizzi, A. Large Scale CYBER–205 Simulation of Vortex Flowfields Around Submarines . Proc. 1st Intercontinental Symposium on Maritime Simulation, Munich June 1985, pp. 114–124. Springer-Verlag

Roberts, G. N. and Towill, D. R. Integrated Control of Warship Manoeuvring. In *Proceedings of the 8th Ship Control Systems Symposium*, 6–9 October 1987. pp. 2.56–68. Ministry of Defence, the Hague (1987)

Royal Institution of Naval Architects. *Dictionary of Ship Hydrodynamics*. Marine Technology Monograph **6**, (1978)

Stansfield, J. T., Katsoulakos, P. S., Ruxton, T. and Isaias, L. The Application of Simulation Techniques in Advanced Engine Monitoring and Control. In *Proceedings of the 8th Ship Control Systems Symposium*, 6–9 October, 1987. Ministry of Defence, The Hague. pp. 2.385–396 (1987)

Sugimoto, A. and Kojima, T. A New Autopilot System with Condition Adaptivity. In *Proceedings of the 5th Ship Control Systems Symposium*, Vol. 4, Annapolis, MD (1978)

Systems Control Inc. A System for Estimation of Hydrodynamic Coefficients for Full Scale Ship Tests. *US Department of Commerce, Maritime Administration Report* NMRC–207 (1981)

Tosi, F. and Verde, E. Microprocessor Based Adaptive Autopilot: System Development and Sea Trials. In *Proceedings of the Symposium on Ship Steering Automatic Control*, Genoa, 25–27 June 1980. Istituto della Communicazione, Genoa (1980)

van Amerongen, J., van der Klugt, P. G. M. and Pieffers, J. B. M. Model Tests and Full Scale Trials with a Rudder Roll Stabilization System. In *Proceedings of the 7th Ship Control Systems Symposium*, 24–27 September 1984, Ministry of Defence, Bath (1984)

van Amerongen, J. and Udink ten Cate, A. J. Model Reference Adaptive Autopilots for Ships. *Automatica*, **11**, 445, (1975)

Wolford, J. C. A Microprocessor Based Stabiliser Fin Control System . Proc. Fifth Ship Control Symposium, David W. Taylor Naval Ship Research and Development Center, Annapolis Maryland USA, Oct 30–Nov 3, 1978, paper R1

8 Mooring and Dynamic Positioning

David Bray

Contents

Introduction
8.1 Mooring
    8.1.1 Alongside moorings for vessels
    8.1.2 Mooring of vessels using anchors
    8.1.3 Mooring of vessels using buoy systems
    8.1.4 Multi-anchor spread mooring

8.2 Dynamic positioning
    8.2.1 DP principles and application
    8.2.2 Historical background to DP development
    8.2.3 Comparison of position using DP and anchoring techniques
    8.2.4 DP system elements and equipment
    8.2.5 Redundancy and its provision
    8.2.6 DP control modelling
    8.2.7 Position reference systems
    8.2.8 DP operations, planning and watch keeping procedures
    8.2.9 Thruster assisted mooring
    8.2.10 DP vessels and their function

References

Introduction

This chapter concerns itself with the business of maintaining the position of any floating object, be it a vessel or other structure. The maintenance of position may be achieved by means of ropes, wires or chain cable to the shore, other fixed structure, or to the sea bed; this is referred to as mooring. Otherwise the maintenance of position may be achieved by use of propellers and thrusters, with no connections to fixed points. This latter is dynamic positioning, and is dealt with in the second section of this chapter.

8.1 Mooring

The first section of this chapter deals with mooring in its various forms: alongside mooring of vessels, mooring of vessels using anchors, mooring of vessels using buoy systems and multi-anchor spread mooring of vessels and rigs. Thruster assisted mooring systems must also be considered in which the vessel position is actively maintained by a combination of anchor spread mooring and thruster control. These systems are akin to dynamic positioning systems and are described in Section 8.2.9.

8.1.1 Alongside moorings for vessels

8.1.1.1 Conventional mooring techniques

This section deals with the more conventional aspects of mooring vessels to quaysides and other locations. Mooring arrangements must allow not only for the mechanics of keeping the vessel in position, but also for the manoeuvring of that vessel during docking and undocking operations. This will include such operations as warping the vessel from one spot to another, working through locks, working with tugs, using moorings in various shiphandling evolutions, towing another vessel and being towed.

In general, the above requirements, together with those connected with anchoring, are fulfilled by a range of simple and robust equipment fitted to the vessel, able to handle rope, wire and anchor chain cable (*Figure 8.1*). For a medium sized merchant vessel the requirements might be as follows: facilities for running, heaving and securing mooring ropes and wires at both ends of the vessel, configured to be able to work on both sides of the vessel. Also facilities for handling two anchors with chain cable from forward. A vessel moored in conventional manner alongside will use moorings ropes at bow and stern, breastropes and springs; these latter are usually of wire.

8.1.1.2 Winches

Winches for the handling of ropes and wires may be powered by electric motor, or may be of hydraulic or steam power (*Figure 8.2*). The commonest warping arrangements consist of a powered drum on to which the mooring rope is turned and tailed off by hand. Once tensioned the rope is stoppered off and then transferred to mooring bitts. A more modern arrangement has the rope of wire spooled onto the winch drum; the winch can haul and slack as necessary, be braked to hold the

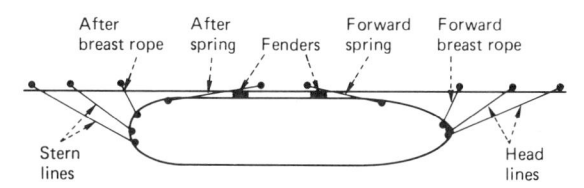

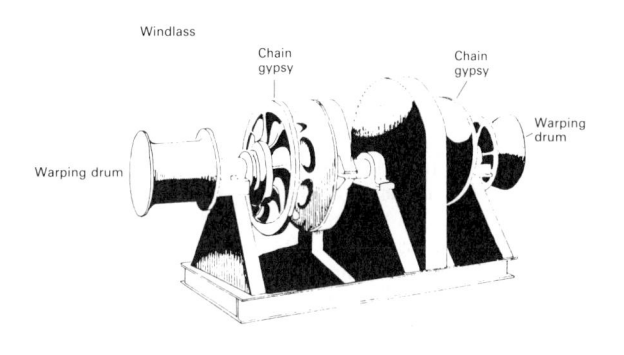

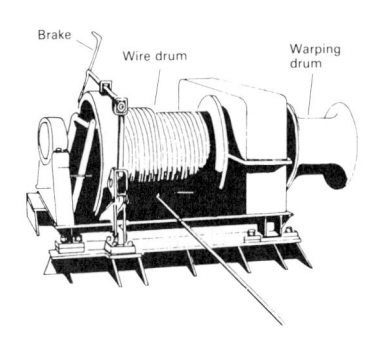

Figure 8.1 Alongside moorings

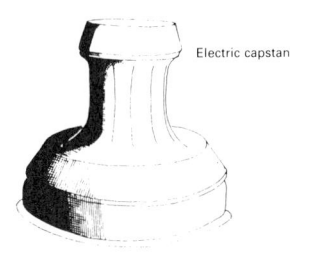

Figure 8.2 Mooring equipment: windlass, mooring winch and electric capstan

rope fast, or be placed into a tension mode. This latter is able to accommodate vessel movements against the quayside and is thus able to cope with tidal movements and rapid changes in freeboard. The rope or wire is permanently anchored and stored on to the winch drum.

An alternative to the winch drum is the capstan. This is a powered winch drum turning on a vertical axis, usually used as a mooring warping drum but not for permanent stowage of rope or wire. Control of mooring machinery may be local, from a control box on the winch, or adjacent; otherwise a remote control station may be provided nearby or on the bridge. Alternatively a portable control box may be carried by the mooring officer. The mooring arrangements must be adequately provided with rollers and fairleads such that a direct lead can be obtained for any rope or wire without suffering a sharp nip or bend.

8.1.1.3 Mooring ropes and wires

Mooring ropes and wires will be sized in proportion to the size of the ship. Mooring ropes are commonly of man-made fibre, these have almost totally superseded natural fibre (manila, sisal, coir, etc.) ropes. Usually mooring ropes are of eight strand plaited construction, either of nylon or polypropylene. As an example, a merchant vessel of around 10 000 tonnes would carry mooring ropes of 64 mm diameter, each of around 200 m in length. Wire springs would be 24 mm diameter and about 100 m long. Up to eight ropes and four springs could be carried.

8.1.2 Mooring of vessels using anchors

8.1.2.1 Bower anchors

A merchant vessel is required to carry two bower anchors. These will be located forward, permanently shackled on to chain cables and handled by means of windlass or anchor capstans. As an example a cargo vessel of 160 m length will carry two anchors of 5.4 tonnes, each connected to 11 shackles of chain cable. (The shackle is the unit of length for vessel anchor cables. Being 15 fathoms (90 ft) it does not lend easily to metrication, but is about 27 m.) A spare bower anchor will be carried on deck and may be somewhat smaller than the working anchors. Modern ships' anchors are of the stockless type, a variety of designs being available. Two are shown in *Figure 8.3*.

The vessel will be provided with windlass and other equipment for anchor handling. Chain cable is stowed in the chain lockers, a purpose built space adjacent to the forepeak. The windlass may be independent or combined with the mooring winch units. The windlass is fitted to allow the cable to be hove or slacked under power, braked or free dropped under brake control. The cable hauling element is a drum wheel known as the gypsy or cable lifter. This wheel is grooved and

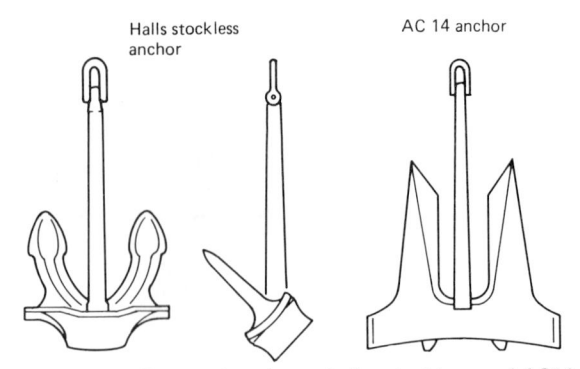

Figure 8.3 Types of anchors: halls, stockless and AC14 anchor

fitted to engage the chain links. The gypsy can be clutched in to its drive shaft, or put out of gear (declutched) for free dropping under control of the brake. Other deck fitments include a bow stopper which positively engages the cable, taking the load when the vessel is laying to the anchor (*Figure 8.4*).

8.1.2.2 Chain cable

Chain cable may be manufactured from forged or cast steel. Cable is of stud link chain; the size is measured from the diameter of the bar from which the cable is made. Each shackle of chain is joined to adjacent links by means of a D joining shackle or a Kenter lugless joining shackle (*Figure 8.5*). Connection to the anchor may be by means of a D-shackle or by a shackle swivel. The inboard end of the cable is secured in the chain locker by means of a pin passed through lugs welded to a reinforced section of bulkhead. This allows the cable to be slipped if necessary.

All anchors, cables and other associated gear (joining shackles, swivels, etc.) must be tested. Cast anchors are subjected to a drop test and a bending test and all anchors are subjected to a proof load. Chain cable is subjected to a proof load, while a sample section (three links) is tested to a tensile breaking stress. Anchors and cables so tested are marked and a test certificate is provided.

8.1.2.3 Anchoring techniques

Whilst it is not the function of this chapter to detail all of the practical seamanship involved in the use of anchors, a few notes upon anchoring techniques are not out of place.

The commonest use of an anchor is to allow a vessel to temporarily secure itself in shallow water whilst waiting to enter port or for some other reason. Here a single anchor will be used. Before dropping, the anchor will be walked back out of the hawsepipe under power, until it is hanging just above the water. Then, with the brake on, the windlass can be put out of gear and the

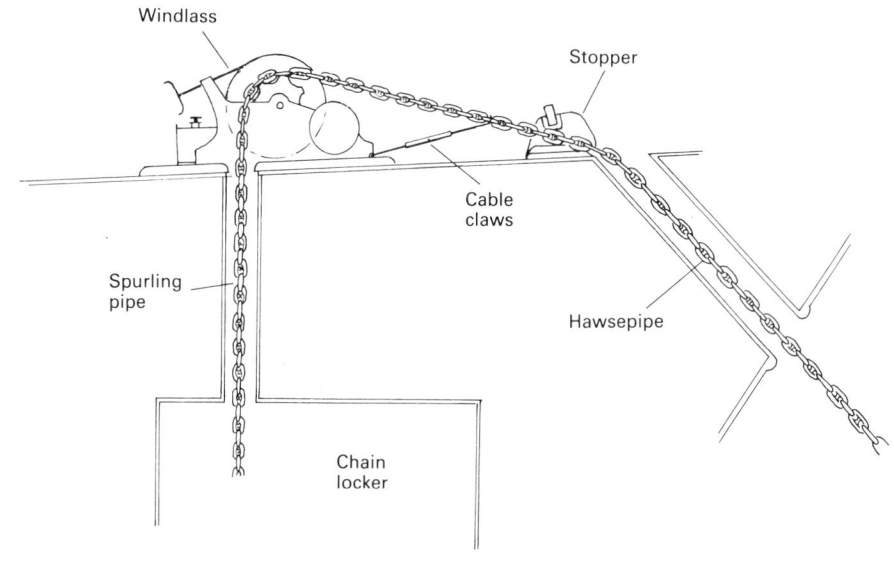

Figure 8.4 Anchor handling arrangements in a conventional merchant ship

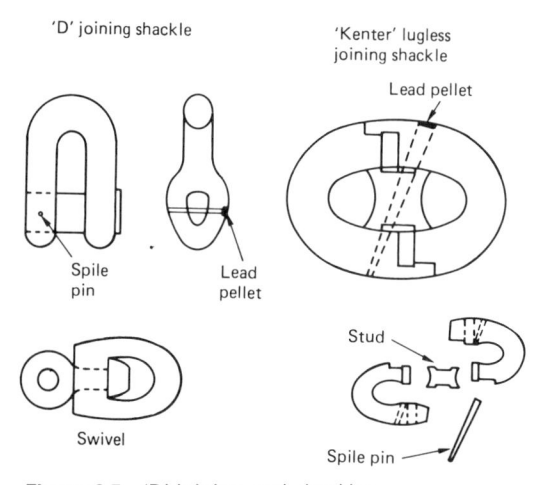

Figure 8.5 'D' joining and shackles

brake released when the order is given. The amount of cable to be veered or paid out will depend upon a number of factors, mainly the depth of water, the quality of the bottom (holding ground), the strength of tide, the expected weather conditions and the amount of swinging room available. Four times the depth of water may be taken as a rule of thumb minimum amount of cable to use. When dropping the anchor on the brake, the cable should be kept under control so that it does not pile up on the anchor on the sea bed. Joining shackles are marked with white paint such that the amount of cable veered may be monitored.

In order to reduce the amount of swinging room required by the vessel when at anchor, it is possible to use two anchors. When two anchors are used the vessel is said to be moored (as opposed to anchored using a single anchor). The normal method of mooring in this way is known as the standing moor. It is assumed there is a tidal stream running and the final configuration will be, in this example, an upstream anchor and a downstream anchor, each veered to four shackles. The vessel stems the tide and drops the upstream anchor. It veers cable to a total of eight shackles, ranging this cable out on the sea bed. It then drops the other anchor underfoot. It can now heave in four shackles on the uptide anchor while veering cable on the second, downtide anchor (*Figure 8.6*).

The same anchor configuration may be achieved by means of a running moor, where the vessel steams slowly into the tide, dropping the downtide anchor first.

8.1.3 Mooring of vessels using buoy systems

8.1.3.1 Circumstances

A number of circumstances may arise involving mooring a vessel to buoys (*Figure 8.7*). The vessel may need to secure to a harbour mooring buoy as an alternative to anchoring or mooring alongside. In this case it may use ordinary mooring arrangements with one or more ropes or wires secured to the buoy.

Alternatively it may use its anchor cable for securing to the buoy. This makes for a more secure moor, especially if the vessel is to occupy the buoy for a long period of time, or if there are strong tides in that location. The disadvantage here is that the anchor must be disconnected from the cable beforehand. This process is known as hanging off and consists of walking back cable until the anchor is just clear of the hawse-

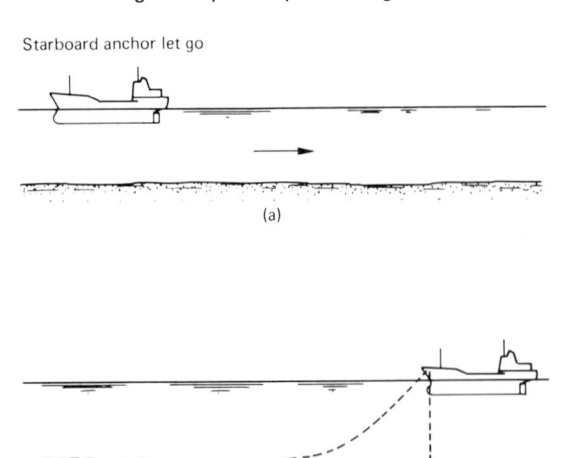

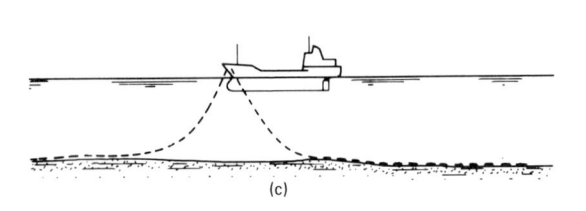

Figure 8.6 Standing moor: (a) starboard anchor uptide, port anchor downtide, four shackles on each; (b) starboard anchor veered to eight shackles, port anchor dropped underfoot; (c) starboard anchor hove, port anchor veered to four shackes on both

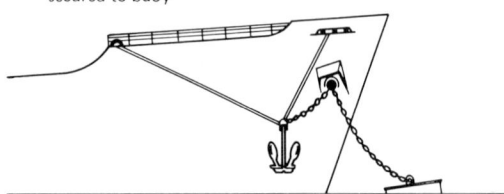

Figure 8.7 Vessel moored to buoy

pipe, hanging it from wires secured on the foredeck, unshackling from the cable, then passing the cable down the hawsepipe to the buoy and shackling on there.

In many ports throughout the world, river moorings are provided using buoys. A vessel may moor to buoys forward and aft, or if strong tide or bore conditions prevail then a four buoy mooring is used, all using anchor cable for connection to the buoys. If a cable mooring is required to buoys forward and aft, then the procedure is to hang off an anchor forward and unshackle a length of cable. The remaining cable on the windlass may be shackled to the forward buoy, while the spare uncoupled length is taken aft and used for connecting to the after buoy.

Usually the mooring facilities aft are designed only for handling rope and wire, so the heaving and securing of chain cable presents difficulties. Vessels regularly visiting ports where this mooring method is used are often designed and constructed with hawsepipes and cable handling facilities (windlass, chain lockers, etc.) aft.

8.1.3.2 Mooring large tankers

A different aspect of buoy mooring is presented in the requirement to moor large tankers bow-on to a buoy or similar structure. Operations of this type are usually associated with the export of crude oil from the producing field, or with the discharge of deep draught tankers at a roadstead where deep water alongside mooring facilities are not available.

The problems arising here are mainly due to the large size of the vessels concerned coupled with the exposed environment obtaining, especially at the offshore oilfield. Conditions for securing in position a large tanker could hardly be less favourable than at a number of locations in the northern North Sea in winter.

Generally, a tanker of the type concerned will be fitted with a complete system package designed for use at one or more specific loading and discharge terminals. The package will include the design of the forward mooring arrangements, the mooring handling arrangements at the buoy structure, together with the loading hose handling arrangements.

A variety of offshore loading buoy structures is at present in use ranging from tethered floating buoys to semirigid articulated towers. Some of these structures are very large, often containing a large reservoir of export oil themselves. A further variation on the theme is the floating storage unit (FSU). In this arrangement a converted supertanker is permanently moored by means of a yoke to an articulated tower. The export tanker couples up by the bow to the stern of the FSU. Export cargo is then transferred through loading lines to the shuttle tanker for transport to the shore refinery. When loading the FSU is allowed to adopt a weathervane heading. Often the shuttle tanker will not lie to this heading, there being a danger of the tanker swinging around to lie athwart or alongside the FSU. This problem is often forestalled by employing a tug made fast at the aft end of the shuttle tanker physically to control the heading.

8.1.3.3 Mooring recovery and breakaway

This chapter is not concerned with the mechanics of providing loading hose connections and their handling and systems. Of more interest are the arrangements for mooring, including mooring recovery and breakaway.

At the design stage it is necessary to understand the response of both vessel and buoy structure to the effects of wind and waves on the location concerned during the most severe environment expected. A computer study of such factors will indicate the maximum hawser pull likely to be encountered and will set the design parameters for the strength of the elements used.

Likewise, maximum environmental conditions can be arrived at for the continuing of loading operations; conditions exceeding these will necessitate breakaway and the vessel waiting on weather for the resumption of the loading operation.

Irrespective of the type of structure to which the vessel is to be moored, the bow mooring arrangements, equipment and mooring recovery methods are similar for all ships (*Figure 8.8*). Equipment fitted in the vessel will include a conventional winch for inhaul of light messenger lines, a heavier mooring or traction winch for inhaul of the main messenger, a quick disconnect stopper unit for securing the hawser and providing a continuous read out of hawser tension and suitable fairleads.

8.1.3.4 Mooring procedures

For recovery of mooring by the vessel, the buoy crew will stream a messenger on the tide. This is a light rope of around 100 m in length, of 30 mm polypropylene. This is buoyed at intervals for visibility and flotation. Connected to this is the main hauling messenger of a further 200 m or so, typically 80 mm diameter polypropylene. To this is connected the main hawser, of chain or chain and wire combination, of about 50 m in length (*Figure 8.9*).

Once the messengers have been streamed and are floating down wind or tide of the buoy (in light weather conditions a work boat may be used to deploy the messengers) the tanker is brought slowly up toward the buoy. The light messenger is picked up either by a grapnel on a heaving line or with work boat assistance. Inhaul is provided by a twin drum traction winch, each drum containing six sheaves, the messenger passing drum to drum to maintain adhesion without manual assistance. The messenger passes through the hawser stopper unit during inhaul. The end of the messenger is spliced into the stopper plug, which is on the end of the

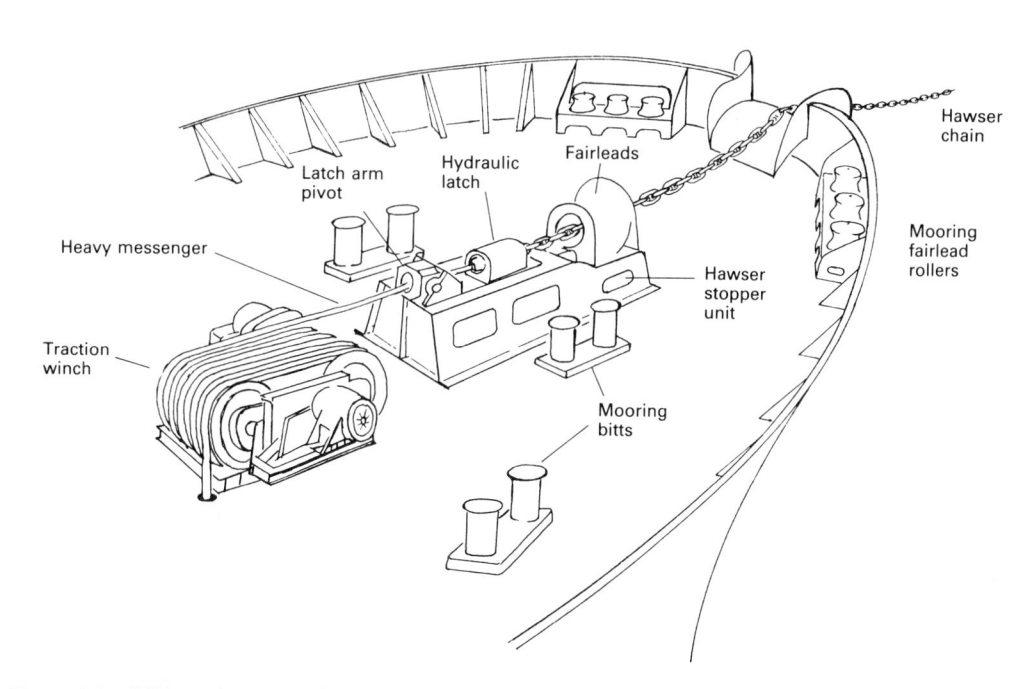

Figure 8.8 Offshore bow mooring arrangement

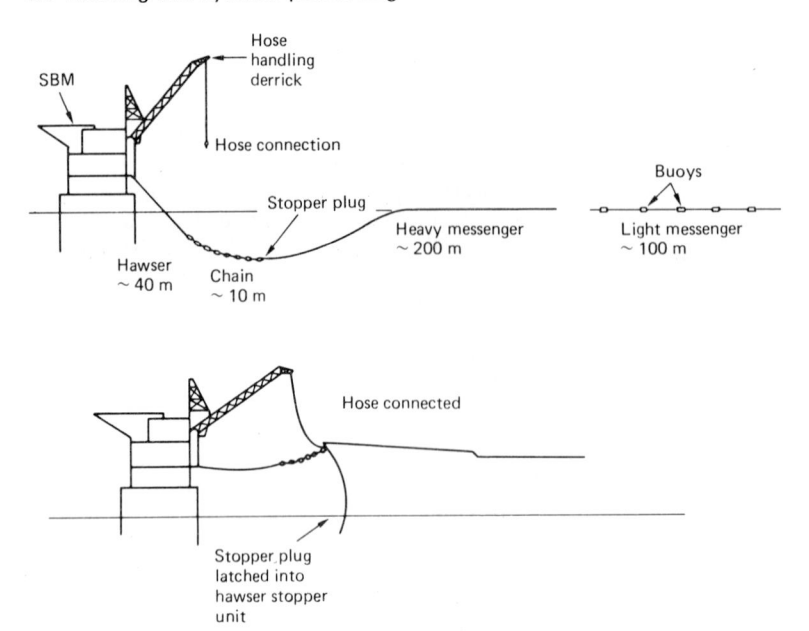

Figure 8.9 SBM moorings

hawser. The stopper plug is hove into the stopper housing where it is positively latched by hydraulic actuator.

All winch and stopper controls are placed in a local control cab also containing load monitoring and recording instrumentation. Hawser load is measured by a load cell incorporated into the stopper unit. Winch drives and all other actuating equipment are usually driven hydraulically; most tanker operators specify no electrical equipment to be fitted on deck.

The mooring arrangements will incorporate means to break off from the mooring in an emergency. Usually, breakaway can be achieved remotely from the control cabin. Upon breakaway the buoy crew will recover the mooring hawser back to the buoy by means of a recovery line.

8.1.3.5 Loading hoses

Not mentioned in the above description is the handling arrangements for the loading hoses. These may be handled by derrick or crane on the buoy or by the ship's own gear. In some cases the hoses are suspended in air while in others they are floating. In all cases connection is made to a receiver on deck with positive hydraulic sealing to avoid spills and pollution. Hose handling control is provided in the cab together with the mooring controls. Emergency breakaway arrangements must allow hose disconnection before hawser disconnection; also emergency stop of loading must incorporate means to dampen out water hammer effects that might occur if the shutdown is too rapid.

Typical values for environmental maximum conditions for buoy loading are as follows: current 1.5 knots, wind 40 knots and wave height 4.5 m significant, 8 m maximum. Breakaway will be initiated when the monitored hawser load reaches around 200 tonnes. The above figures are representative only; individual installations will have their own criteria.

Vessels using a thruster assisted mooring system or dynamic positioning system can usually operate in more severe environmental conditions.

8.1.4 Multi-anchor spread mooring

8.1.4.1 Semisubmersible drilling rigs

A number of vessel types and functions require an offshore fixed location; typical of these vessels is the semisubmersible drilling rig. Other vessels that may require such a fixed location are crane barges, floating accommodation platforms and production platforms. An offshore location may be in deep water and a harsh environment regarding wind, current and waves. Three methods may be used to provide the positioning; a multi-anchor spread mooring, a thruster assisted spread mooring and dynamic positioning. This section deals with spread moors.

In water depths of up to 500 m, a vessel (e.g. a semisubmersible drilling rig) may be positioned by a spread mooring. The mooring pattern may consist of eight, nine or ten anchors placed away from the vessel in different directions (*Figure 8.10*). A drilling rig may require to be on station for a number of months to complete one or more wells. During that period the vessel must stay on location to within a distance equal to about 3% of water depth. Drilling should be able to proceed in environmental conditions of 50 knot winds,

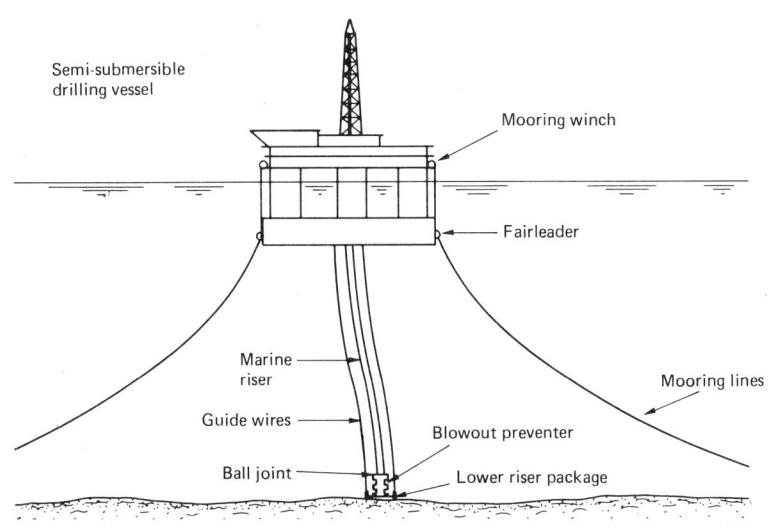

Figure 8.10 Drilling vessel moorings, semisubmersible drilling

11 m wave height and a current of 1 or 2 knots. Under more severe conditions drilling can be suspended, the riser disconnected from the wellhead and the vessel can ride out the storm.

8.1.4.2 Catenary curve

When deployed, the mooring line (either chain or wire) will take up a curve known as a catenary. The direction of pull on the anchor is horizontal and the amount of line deployed will be between five and seven times water depth.

In *Figure 8.11* the tension in the mooring lines is shown as T, while H is the horizontal component of T. The environmental force F is that representing the total effect of wind, wave and current forces. When in equilibrium $F = H_1 - H_2$. In rough weather the tension in the windward lines may approach the maximum allowable load. An improvement may be achieved by slacking one or more leeward or downwind lines, thus reducing the value of H_1.

A variation of the conventional spread mooring system is the turret moored drillship (*Figure 8.12*). The vessel is a monohull type, fitted with tunnel or azimuth thrusters in addition to main propeller and rudder arrangements. The drilling rig is mounted upon a central turret which is able to rotate about the drilling axis. Drilling takes place through a central open well, over which the turret is located. Anchors, mooring lines, winches and other mooring gear is located on the turret, while the vessel is able to rotate to the most favourable weather heading, under the control of the thrusters.

8.1.4.3 Wire or chain mooring lines

Mooring may be achieved by wire or chain mooring lines, or a combination of both. In this latter case, the

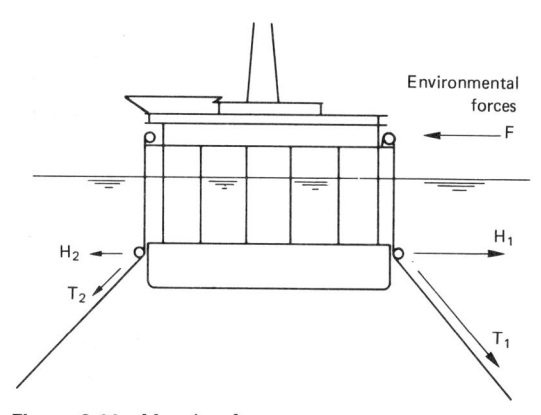

Figure 8.11 Mooring forces

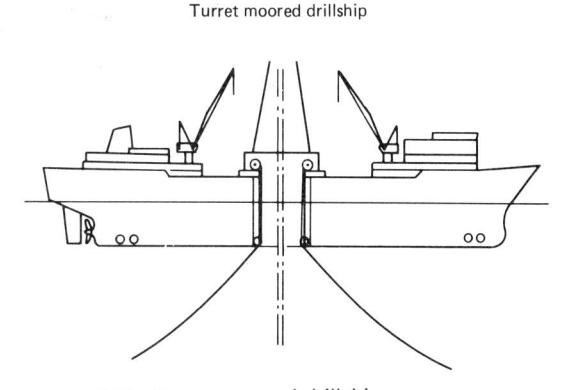

Figure 8.12 Turret moored drillship

anchor will be connected to a length of chain cable, with the remainder of the mooring line of wire. This arrangements gives rise to special difficulties in handling. For handling wire, all the wire spools onto the hauling drum, but for chain the winch must be fitted with cable lifters (gypsy wheels or wildcats) and the chain stows in cable lockers in the rig columns.

Anchors may be to a variety of designs, mostly variations upon the stockless Danforth theme (*Figure 8.13*). Types include the Flipper delta, the Moorfast and the Stato. A single fluke fixed anchor, the Bruce type, is occasionally used.

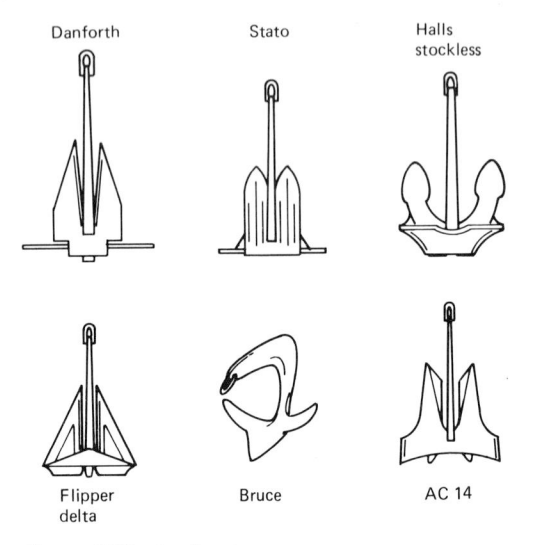

Figure 8.13 Anchor types

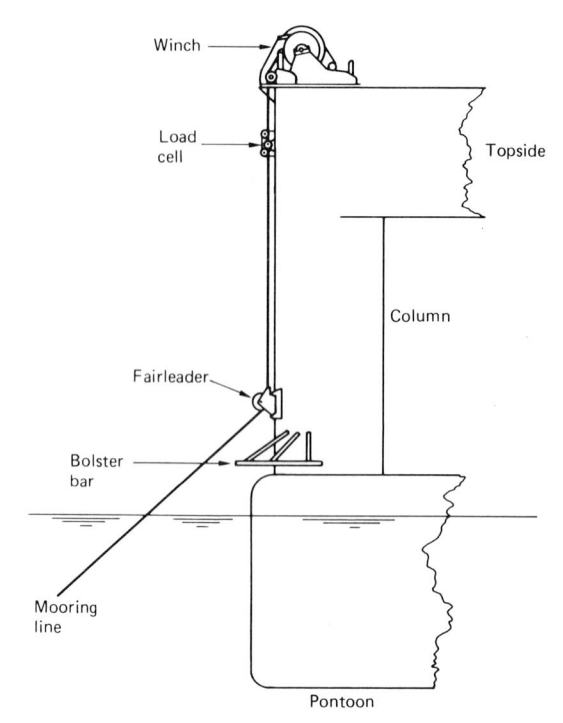

Figure 8.14 Rig mooring arrangements

Rig winches are usually located at the upper deck level with suitable fairleaders installed (*Figure 8.14*). The anchors themselves are racked at pontoon level, being stowed by being hove up close into a bolster bar. The winches may be driven electrically or hydraulically and will be fitted with facilities to heave, veer and brake as required with local control and central remote control upon the bridge or control room. Mooring line tension will be continually monitored and recorded by load cells fitted to the winch and fairleaders.

8.1.4.4 Anchor handling vessel

An essential part of the spread mooring facility is the anchor handling vessel (*Figure 8.15*). A drilling rig upon a spread mooring will always have one of these vessels in attendance and when laying the spread may use two or three. Multipurpose vessels, they have towing facilities as well as anchor handling, and they are also able to act in a supply boat mode and often are fitted for firefighting and rescue roles. A typical anchor handling tug supply vessel will be 60 m in length by 15 m beam, be fitted with twin screw CP propellers and a bow tunnel thruster and have about 9000 HP. A multipurpose anchor handling and towing winch of the waterfall type will be fitted, able to handle mooring wire, mooring chain and towing wire. The winch will be capable of heaving up to 200 tonnes and holding to 300 tonnes. The vessel's bollard pull will be in the region of 100 tonnes.

8.1.4.5 Anchor spread configurations

The establishment of a spread mooring for a drilling rig upon a work site requires a considerable amount of planning. The site will be surveyed and sea bed conditions checked. The work site will be marked with acoustic beacons for position reference. Liaison must be established between all interested parties such as the oil company, the drilling contractor and the anchor handler crews. The anchor spread configuration will be planned; the number of anchors, the length of mooring line and the sequence for setting and retrieving anchors. This location plan must take into account the prevailing winds, tidal flow directions, water depth and sea bed quality. The positions for the anchors will be marked with buoys and the final location for the rig will also be marked. When under way the rig will be under the control of one or more tugs, assisted by the rig's own thrusters. The following routine is given as a typical example of the procedure adopted in setting up a mooring spread on eight anchors (*Figure 8.16*).

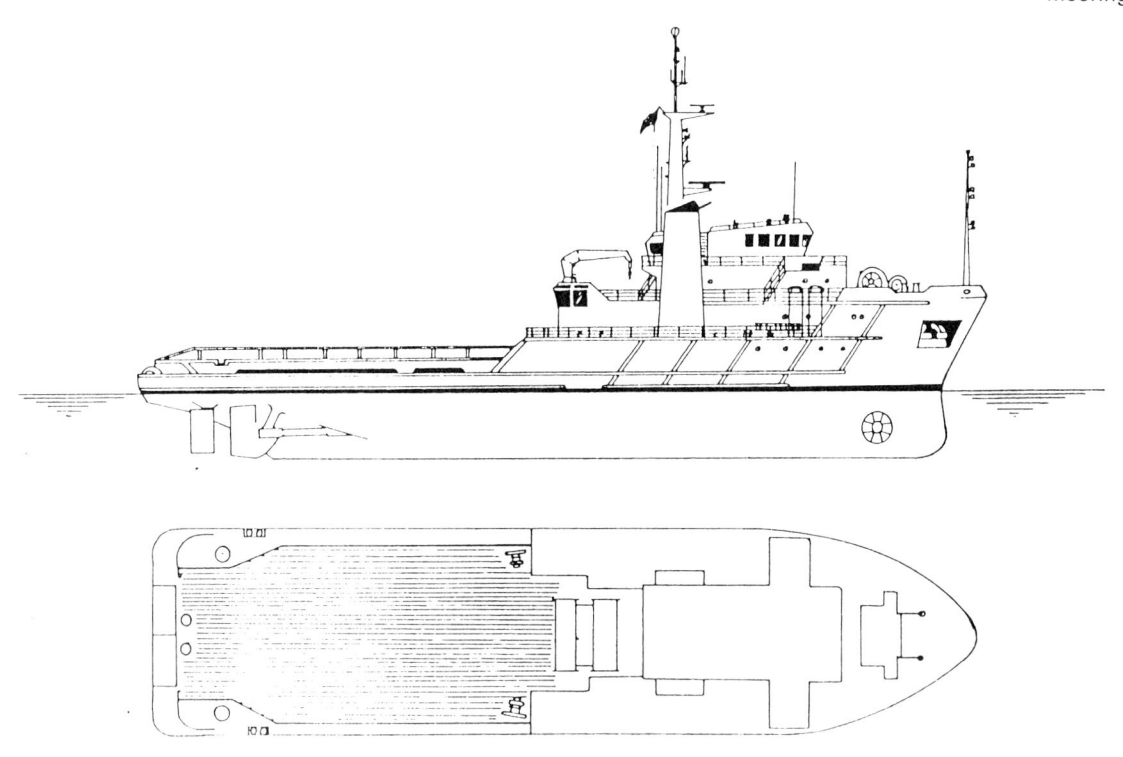

Figure 8.15 Anchor handling/supply tug

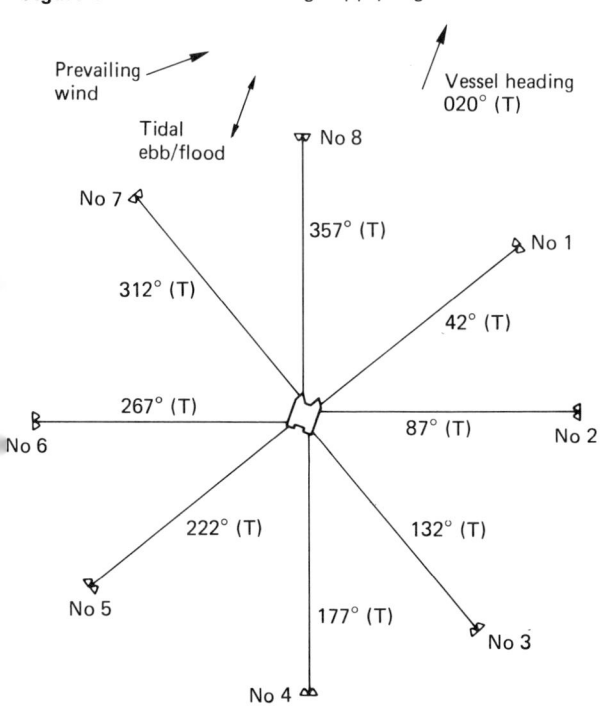

8.1.4.6 Procedure

Upon completion of the transit passage, the rig will be stopped at a staging area just outside the anchor spread pattern. Anchor 5 will be backed off the bolster until it is hung about 20 m clear of the pontoon. Anchors 1 and 8 are then transferred to separate anchor handling tugs. These two vessels now maintain position clear of the rig, while the final approach is made under tow from the staging area to the drilling location. Anchor 5 is dropped as the rig passes marker buoy 5 and cable is paid out. When the rig arrives on location, cable 5 is held on the winch brake and the rig's position and heading maintained by the tugs. The anchor handling vessels can now run anchors 8 and 1, after which anchor 4 will be run. With these four anchors run and set the rig can tension cable to approximately 45 tonnes and set the rig on to its final location by heaving or slacking anchor cables (*Figure 8.17*). This completed, the towing vessels can be released, and the final four anchors run and set. It may now be necessary to return to anchor 5, recover and reset it, as it may not have been dropped in its correct position at the outset.

Once all anchors in the mooring spread have been run and set, all cables can be preloaded by tensioning cables to approximately 100 tonnes. This should be done by each opposing pair of cables in turn, such that the rig does not move off station. At this stage, anchors can be checked for drag; a dragging anchor will be indicated by a cable tension oscillating between 20 and

Anchor setting order: 51842637 (reset 5)

Figure 8.16 Spread mooring location plan (eight-anchor)

variation to the described method is the use of chasers. A chaser is a ring encircling the anchor stock. The tug's pendant and work wire are shackled to the chaser. The anchor is run and set normally, but instead of buoying off the pendant, the pendant is chased back to the rig, the chaser running along the cable. Once back at the rig the chaser pendant is transferred by crane to the rig (*Figure 8.18*).

If an anchor will not hold when pretensioned, it will be necessary to piggyback or install a second anchor on the same cable (*Figure 8.19*). Assuming a buoyed pendant has been used, the anchor handling tug is able to pick up the buoy, remove it from the pendant, and attach a second anchor. A pendant is attached to this second or piggyback anchor for setting; the second anchor is then set in the same manner as the first one and the pendant buoyed off. It may be necessary to deploy a second piggyback if the anchors already set still show a tendency to drag.

The length of buoy pendant must be enough to cope with any tidal range at the location. If the pendant is made too long, the excess may pile up on the sea bed and foul the anchor. Common practice is to install a spring buoy into the pendant (*Figure 8.20*).

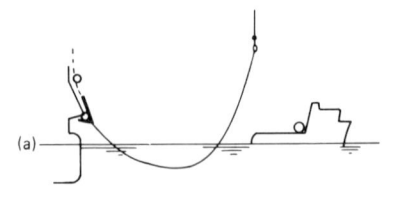

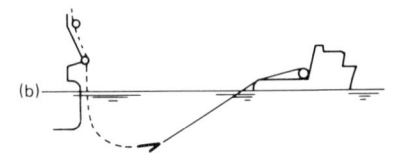

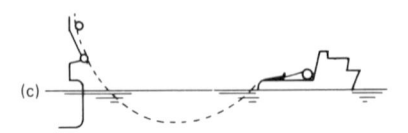

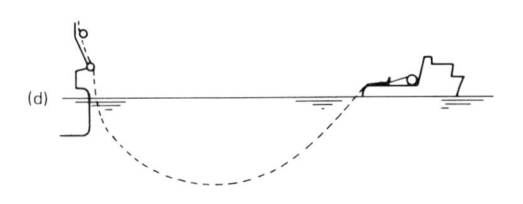

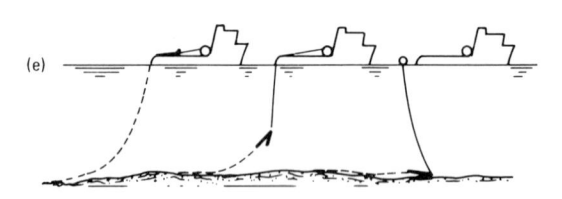

Figure 8.17 Anchor transfer, haul out and setting: (a) anchor pendant passed to tug by rig crane; (b) pendant connected to tug's work wire; rig winch walked back. Tug moves away from rig; (c) anchor hove to tug stern roller or onto afterdeck; (d) tug hauls mooring line out upon bearing required. Rig winch pays out to predetermined distance; (e) anchor lowered to sea bed on buoy pendant, buoy connected and dropped

80 tonnes while tensioning. Once it has been determined that all anchors are holding, then final tensions can be set at 100–125 tonnes.

8.1.4.7 Variations

The above procedure is given as an example. There will be variations dependent upon the number of anchors used, the number of anchor handling tugs available, expected weather conditions, depth of water, etc. One

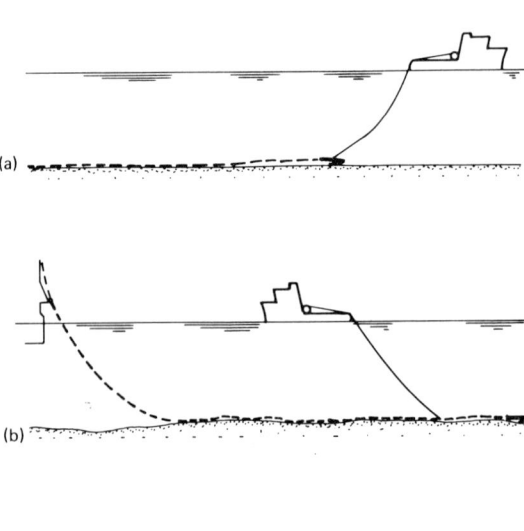

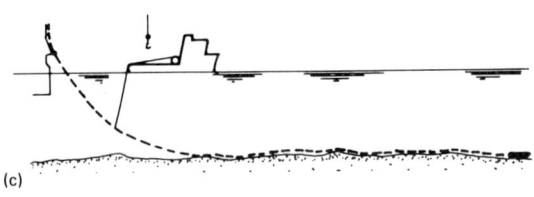

Figure 8.18 Chaser strip off. If a chaser is used to haul out and set anchors, then this must be stripped off. (a) anchor set using chaser pendant on winch; (b) chasing back pendant length approximately three times water depth; (c) on approach to rig, pendant shortened up and transferred to rig crane

Setting a piggyback anchor

(a)

(b)

(c)

(d)

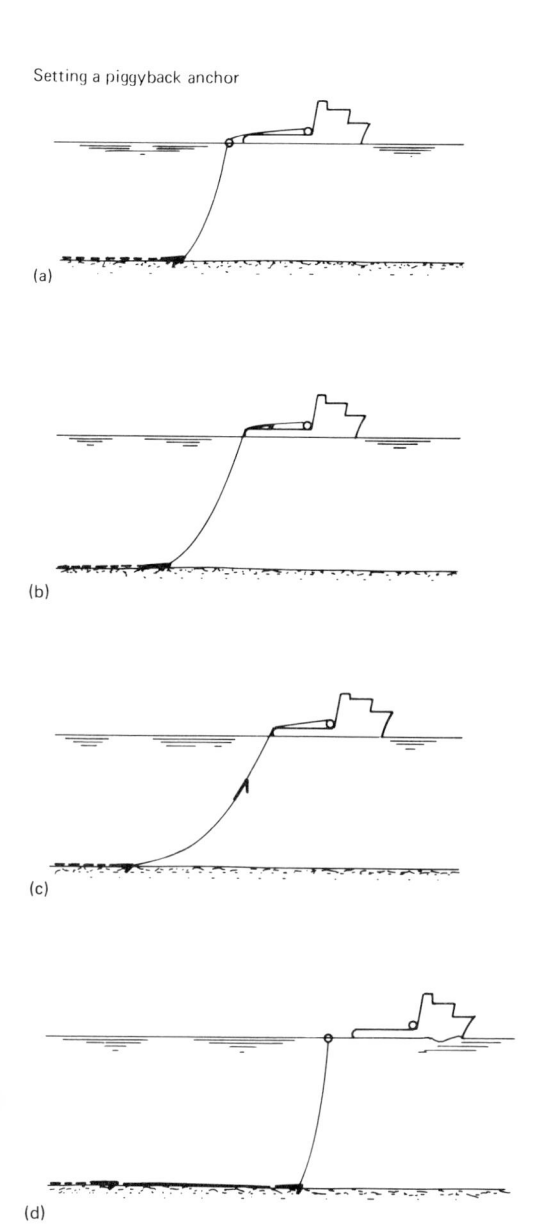

Figure 8.19 Setting a piggyback anchor: (a) tug picks up buoy with work wire, buoy removed and pendant hauled; (b) pendant shackled to piggyback anchor; fresh pendant on winch; (c) piggyback anchor lowered on pendant, tug manoeuvred to lay out anchor; (d) piggyback anchor set, pendant buoyed off

8.1.4.8 Anchor retrieval

When it becomes necessary to retrieve an anchor, either to reposition it, or upon the rig preparing to move off location, then the anchor handling tug is used in much the same way as before (*Figure 8.21*). The rig will slacken off tension on the cable, and the tug manoeuvres on to the buoy, which is lassoed and

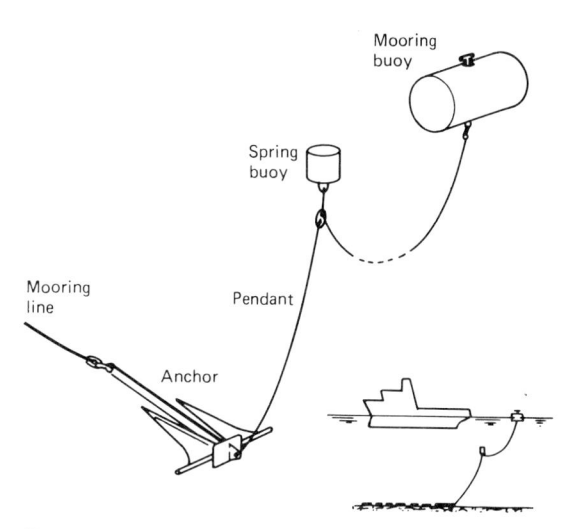

Figure 8.20 Arrangement of spring buoy

hauled on board by the tug's work wire. The buoy removed, the work wire is shackled to the pendant and the anchor broken out. Using the pendant the anchor is hove up to the stern roller and on to the afterdeck. The rig will then winch the anchor and tug toward the rig. When the vessel is close in to the rig the anchor can be veered away on the pendant as the rig winches in. When the anchor is racked onto the bolster bar of the rig the pendant can be transferred to the rig crane.

The order of anchor retrieval is normally the reverse of the setting order. All parts of the mooring spread should be inspected upon recovery. Wires should be checked for corrosion and chafe. If signs of wear, broken wires or kinks are evident the wire should be condemned. Shackles, rings, swivels, etc., must be checked for cracks and distortion; chain must be checked for missing studs.

In storm conditions, adjustments will need to be made to the anchor spread of the rig in order to avoid excessive cable tensions. Leeward lines can be slackened to reduce the loading on the windward ones, although at least one leeward line should remain tensioned in order to prevent loss of position. In extreme weather conditions, the riser will be disconnected from the wellhead and all lines slackened. Those on the windward side should be run out to their full extent. The rig should be evacuated until weather conditions improve.

8.2 Dynamic positioning

8.2.1 DP principles and application

8.2.1.1 Definition

Dynamic positioning (DP) can be defined as a system which automatically controls a vessel to maintain its position and heading exclusively by means of active

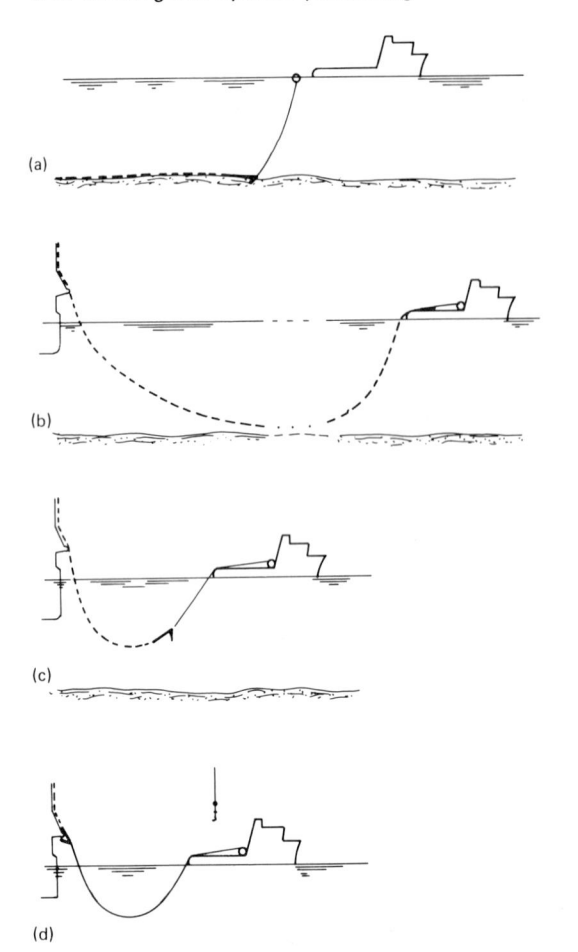

Figure 8.21 Anchor retrieval: (a) rig untensions line, tug picks up buoy, buoy lassoed and hauled on board; (b) buoy removed from pendant, anchor hove on board, rig winches in; (c) when close in to rig, tug slackens away on pendant, rig winches in; (d) anchor racked, rig picks up pendant with crane hook

thrust. By using the word 'automatically' we exclude systems which are reliant upon a joystick or other operator input in order to maintain control over the vessel, although manual control is one of the functions of a DP system. In addition to controlling the vessel to maintain a given or set point position and heading, the system also caters for changes to position and/or heading being implemented by a variety of means.

In the past, most ships of whatever type spent most of their time travelling from A to B by which means money was being earned. Complex manoeuvring was limited to port arrivals and departures when tug assistance, pilotage, etc., was available. Ships of these types were not fitted with any high precision manoeuvring or navigation capability until recently, it being deemed uneconomic. More recently, the fitting of bow tunnel thrusters to trading vessels proved of value in reducing utilization of harbour tugs. Further manoeuvrability is provided in specialist vessels such as ferries in order to expedite port arrival and departure evolutions.

In recent years a further requirement has arisen; that of maintaining the vessel upon a fixed location so as to facilitate operations such as firefighting, diver support, drilling, surveying or dredging. Further, a requirement exists for precision manoeuvring in such functions as pipe and cable laying. Traditionally, the maintenance of a fixed position has been dependent upon an anchor spread, but there are many situations where this method of operation would incur severe penalties in terms of time and cost. In many areas the presence of sea bed hardware precludes the use of anchors for positioning. Also, the cost of an anchor spread increases with water depth. In very deep water DP may be the only rational method of maintaining position.

8.2.1.2 How DP works

In simple terms, a DP system consists of a central processor linked to a number of position reference and environment reference systems. The ship is provided with sufficient power and manoeuvrability by means of a variety of thrusters and propellers. The measured position of the vessel is compared to the desired or set point position; the computers then generate appropriate thruster commands to maintain or restore vessel position. Effects of wind forces and other environmental forces are taken into account. A bridge control console allows the operator to communicate with the system and *vice versa* and vessel control to be effected (*Figure 8.22*).

8.2.2 Historical background to DP development
8.2.2.1 The 1960s

Dynamic positioning as a technique first appeared in the early 1960s. The earliest vessels to which DP was applied were designed for drilling, coring, diver support or cable laying. It is therefore only 25 years or so since the introduction of this technique. The early DP vessels were often conversions of conventional ships, being fitted with extra thruster capability, analogue computers to process information, with the ship's position measured from a single reference system. The commonest position reference then was a taut wire system in which a boom projects over the ships side carrying a wire over a sheave. A weight anchors the wire to the sea bed while a tension winch keeps the wire taut against ship movements. Ship position is then defined by the angle of the wire, the depth of water and the position of the weight. Other position reference systems were utilized, such as hydro–acoustic system communicating with sea bed located beacons or transponders. By the early 1970s all DP computers were digital. At this time the major manufacturers involved were American and French.

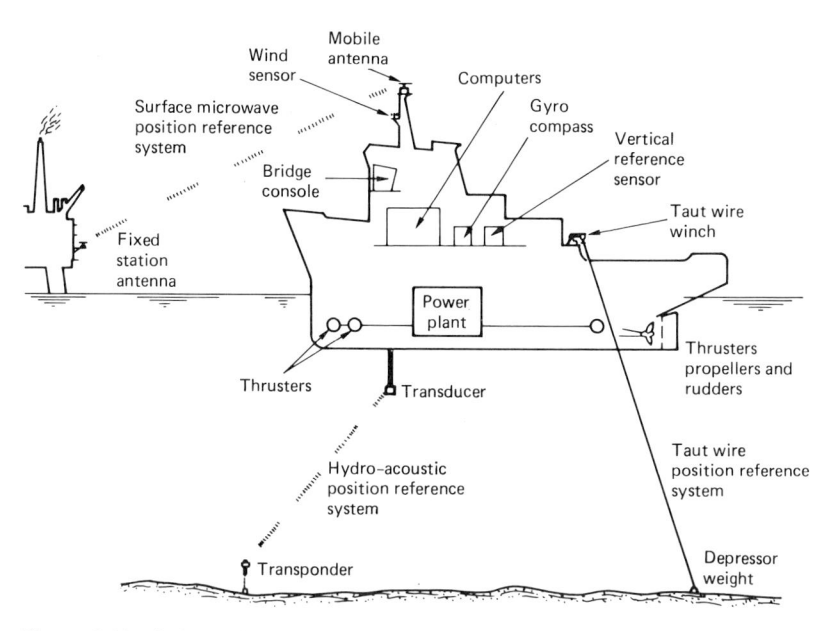

Figure 8.22 A dynamic positioning installation

One of the more well-known of DP vessels at this time was the *Glomar Challenger*, a coring vessel of around 10 000 tons which spent many years engaged in geological investigations worldwide, often in very deep water.

During the 1970s a large amount of development was under way in the middle and northern sectors of the North Sea. The water depths of 80–150 m raised new problems for the developers and their operators. Elsewhere in the world where oil and gas related exploration was taking place, shallow water conditions allowed operations to take place from vessels and barges positioned by means of a four or six point anchor spread. The deeper water and more severe weather conditions in the North Sea precluded this method of operation and so DP was adopted as a technique. Unfortunately, the reliability of the earlier DP systems was not good, particularly as regards diving support vessels (DSVs). A number of incidents occurred which led to DP being regarded with a great deal of suspicion by divers, their operators and charterers.

8.2.2.2 The 1970s and 1980s

During the 1970s and early 1980s, great strides were taken in improving the reliability of DP facilities, particularly in DSVs. Full system redundancy has become a requirement and many operators and charterers are demanding fully trained, qualified and competent operators to run the system. Reliability is currently vastly improved and this has accompanied a diversification of the types of vessel and operation utilizing DP techniques. Some of the vessel types using dynamic positioning are: diving support vessels, semisubmersible multirole support vessels, heavy lift crane barges, platform supply vessels, survey/ROV support vessels, drilling vessels, coring vessels, cable/pipe laying vessels, floating oil production vessels, well stimulation and workover vessels, firefighting vessels, dredgers and hopper barges, trenching and stone dumping vessels, offshore loading tankers and flotel accommodation units.

8.2.3 Comparison of positioning using DP and anchoring techniques

8.2.3.1 Alternative methods

Any consideration of the application of DP techniques must make a comparison with alternative methods of positioning. Often the only alternative positioning technique is the use of an anchor spread, but this is limited by depth of water. If this exceeds around 100 m then cost, time and space considerations preclude it. Often the amount of hardware on the sea bed renders it impossible to run anchors safely, if at all. The presence of anchor cables may prevent the vessel from positioning close to platform structures and may also provide obstructions to other vessels.

The establishment of an anchor spread can take many hours and require the attendance of an anchor handling tug; likewise recovery and shifting. Acquisition of a precise position and heading within an anchor spread is also problematical. Using DP, however, a desired position can acquired with precision inside of 1 hour and usually much faster. Position may be adjusted at will, a shift of position or heading takes only a few minutes. Position can be selected with minimum reference to sea bed or other obstructions. This allows divers to locate close to the work site and use short umbilicals.

Also, a DP vessel is able to react very quickly to changes of weather. Often, if the task permits, the vessel may be weathervaned or kept head to wind and/or sea to reduce power requirements and vessel movement. These features enable DP vessel downtime (waiting on weather) to be minimized.

8.2.3.2 Drawbacks of DP

Some vessel functions require precise track following or manoeuvring characteristics; e.g. cable laying or ROV support. These are easily achieved using DP but anchoring is not appropriate. To balance the picture, some of the drawbacks of DP must be mentioned. Any DP vessel is complex and dependent upon sophisticated technology to maintain positioning capability. Any system is subject to failure, despite all efforts to avoid this. A diver, for instance, may prefer to work from a vessel secured by anchors and cables than one kept in place by some form of black magic. Another drawback from the diver's viewpoint is the disturbance of the water caused by constantly running thrusters and propellers; indeed it is possible for divers and umbilicals to become drawn into thrusters. Another DP related diver hazard is the presence of a taut wire and weight upon the sea bed; another down line for the diver to tangle with.

Nevertheless, DP is now a well established technique in use worldwide, and it is probably true to say that without DP much of the oil and gas related development now taking place would not have been possible.

8.2.4 DP system elements and equipment

8.2.4.1 Main elements

The DP system at its simplest consists of six separate groups or elements. It must be realized that the DP is dependent upon many shipboard services (such as power supply and management) and as such these must be considered to be part and parcel of the DP system. Also part of the system is the human element; the DP watchkeeping officers on the bridge, the engineers and electronics operators who are required to provide running and maintenance functions, also repair.

A schematic diagram of a DP installation is shown in *Figure 8.23*. Here, the six elements consists of: control, position reference systems, heading reference, environment reference systems, power supply elements and manoeuvring elements.

8.2.4.2 Control element

Central to the DP system configuration is the control element. This consists of the computers or processors, the bridge control console and not forgetting, of course, the DP operator.

The computer installation will be configured according to the level of reliability required for the DP function. The consideration here is one of redundancy. For ship types and operational modes where DP failure would cause serious risk to life or severe economic penalty it is necessary that the whole system be protected from catastrophic failure. Discussion upon these aspects follows in a later section of this chapter. It is

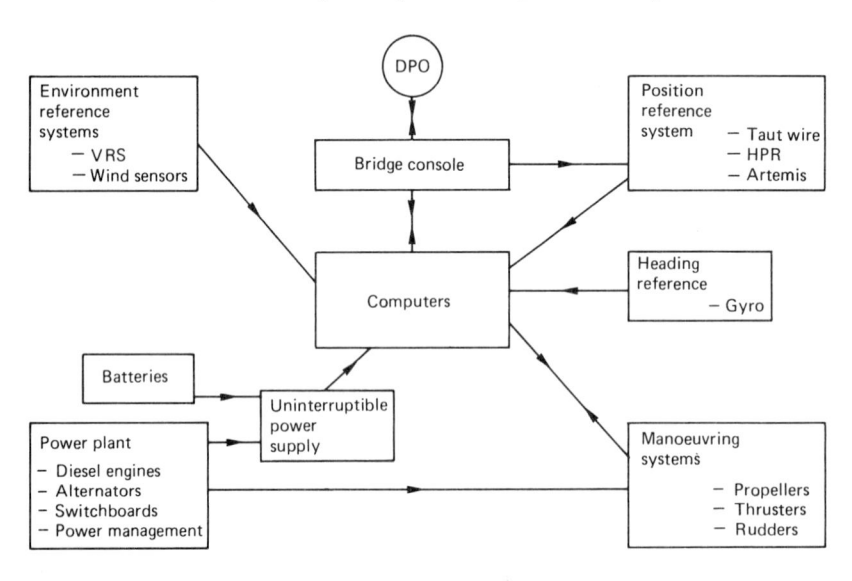

Figure 8.23 Elements of a DP system

obvious that for DSVs, drill ships and multirole support vessels, a fully redundant installation is required. For other vessel types, e.g. ROV support, stone dumping, dredging or offshore loading redundancy is not essential and since the provision of redundancy as a whole is a very expensive exercise it is usually uneconomic to provide such a fit in these vessels.

In any DP system, the basic processor principles are the same, as are the system requirements regarding ship positioning (*Figure 8.24*). Data is fed into the system from sensors such as position reference sensors, gyro compass, wind sensors, vertical reference sensors together with feedback data from the power plant, thrusters, etc. A number of program subroutines continually check the validity of the data against preset limits and values predicted from mathematical modelling. Multiple sensor inputs are compared, yielding data upon accuracy for each individual sensor input. Modern DP processors utilize mathematical modelling to provide data upon the vessel's dynamic behaviour. Sensor measurements update the vessel estimator, the output from which is an estimate of vessel position, heading and velocity. From this the controller can determine the thrust allocation required to maintain or restore set point position and heading.

The processor's program routines are initially determined at the vessel design stage, usually being the program from previous similar vessel installations. After completion the vessel will undergo extensive DP trials in a variety of weather conditions during which the systems can be tuned to provide optimum DP performance, not only for position keeping characteristics but also regarding power requirements and hence fuel consumption. Subsequent alterations to the vessel may require a software update to accommodate, e.g. structural alterations to the vessel (a helideck added, a new crane installed, a deckhouse removed) will alter the sail area of the vessel and its handling characteristics. Likewise provision of extra power generating capacity or alterations to the thruster configuration will necessitate software modifications.

8.2.4.3 Bridge control console

The physical control of the vessel is in the hands of the DP operator at the bridge control console. This is the point from which the operator–system–ship communication is provided. When in DP mode the operator has no need to leave this console and all manoeuvring functions are carried out from here. The system provides comprehensive feedback data upon one or more VDU screens, together with all control, function, display and alarm facilities. The location of this console will depend upon the vessel function. Usually on the bridge, affording the DPO an adequate view of the operation from the console position, the console is often placed (in DSV installations) in an aft-facing configuration in the after wheelhouse.

Other vessels have the DP console forward (or even sideways) facing, while a few vessels have the DP installation away from the bridge with no outside view whatever. This latter arrangement cannot be recommended since, despite all technology, the DPO's prime reference is his/her eyes. If a DSV is on DP only 20 m from a platform structure then it is essential that the DPO can see the platform at all times. When things go wrong it is often detected by eye before the DP raises the alarm.

8.2.4.4 Position reference

Position reference is an essential input to any DP system. Ideally the position accuracy of systems used should be 0.1–1.0 m. A variety of position reference systems are in use, with varying levels of effectiveness. Each system employed will have its own characteristics of range, accuracy and operational limitations; these must be taken into account by the operator when establishing position reference for DP.

Although there have been in the past around 40 different position reference systems used in conjunction with DP, three types predominate. These are the taut wire system, the hydro–acoustic position reference (HPR) and the surface microwave position reference. Other PRSs in common use include the Syledis hyperbolic radio survey system, a Radactor or radar extraction system, TV tracker system using TV cameras mounted on board and others such as Pulse/8, Argo, Hyper-Fix and Miniranger.

A more detailed discussion of the three predominant types follows in later sections of this chapter. The simplest DP installations are only able to use one single position reference input with no redundancy available. However, most DP systems have the facility to work with two or more PRSs. A typical redundant DP installation is able to interface with up to six PRSs. A modern DSV may have: a Simrad hydro–acoustic position reference system, an Artemis microwave position reference system, a taut wire port, a taut wire starboard and a Syledis.

The DP system is able to process position data from any or all of these systems simultaneously. The mathematical modelling technique enables all data to be pooled to provide a best estimate of the vessel's position. This ensures that the fix obtained takes into account all available incoming data. Also each input is checked for credibility against the estimated position of the vessel allowing a weighting factor to be applied to each PRS input commensurate with its observed performance. If two or more PRSs have been acquired and the input from one of them fails then the information from that source can be discounted by the system and the remaining PRSs used. A warning will be initiated for the operator but DP performance should not be seriously degraded.

For redundant operation it is necessary that more than one PRS be in use. For diving and drilling

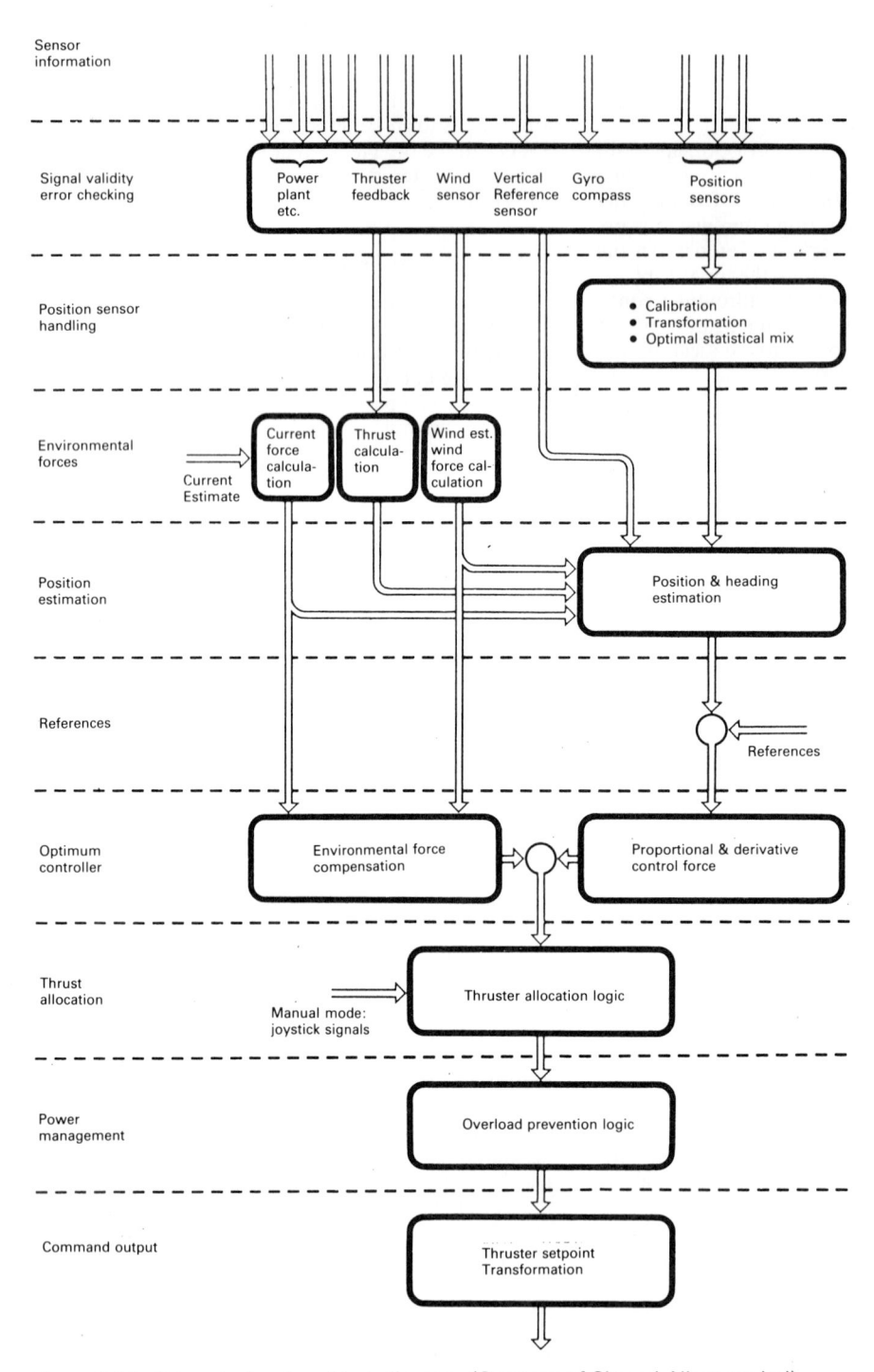

Figure 8.24 DP control system block diagram. (Courtesy of Simrad Albatross Ltd)

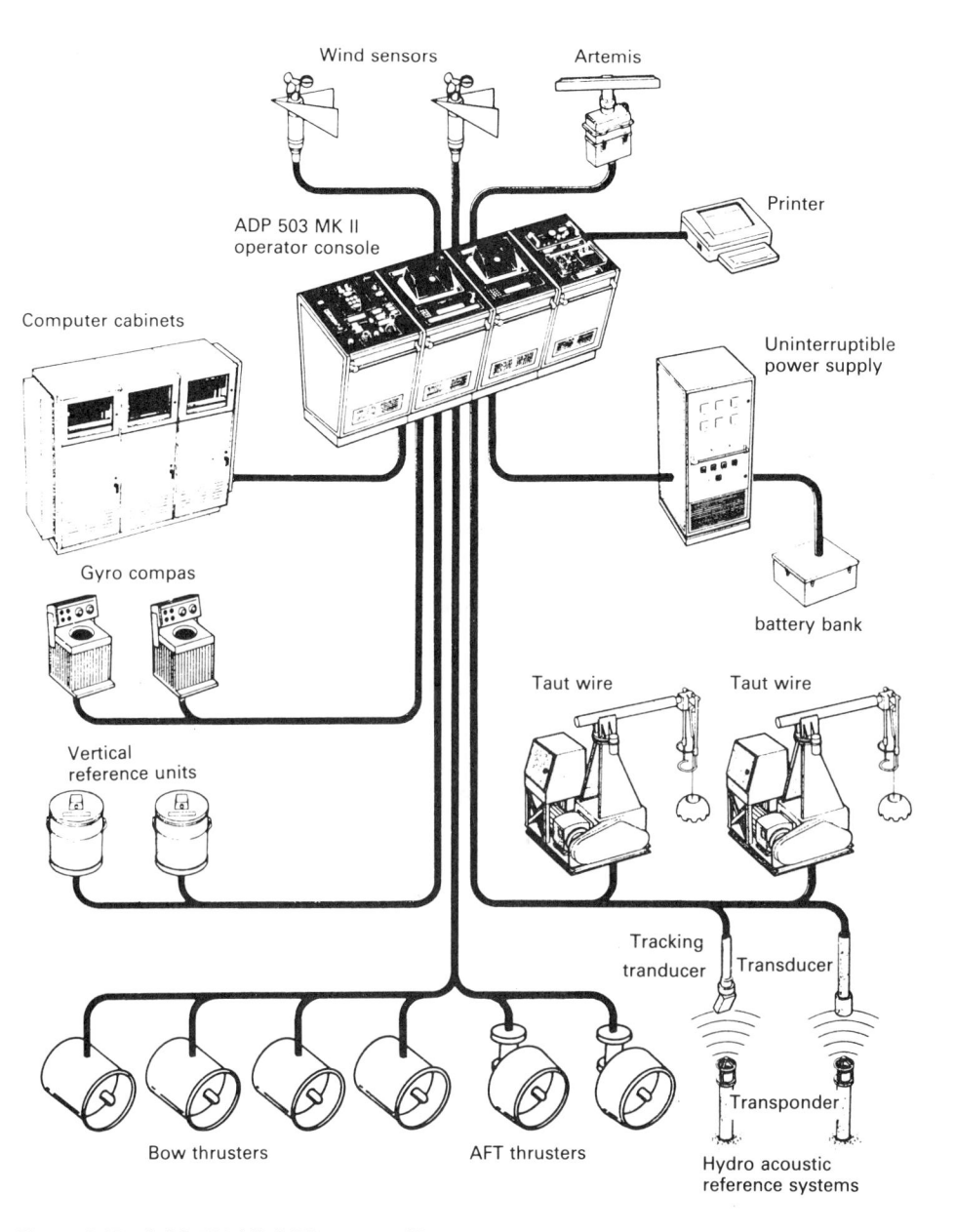

Figure 8.25 A DP 503 Mk II DP system. (Courtesy of Simrad Albatross Ltd)

functions the recommendation given in the *Guidelines for the Specification and Operation of Dynamically Positioned Diving Support Vessels (Guidelines)* is for two PRSs in use with a third on standby mode. The PRS pooling system mentioned above does not attach any significance to a PRS on standby so the common interpretation here is simply three PRSs in the system before commencing diving or drilling. A loss of one of these three would necessitate suspension of the operation. In selecting the PRS the operator will keep in mind the possibility of common mode failure. To this end, his/her three PRSs deployed should all operate upon different principles to avoid simultaneous failure of two or more PRSs. If two taut wire systems were deployed it is possible that both could fail at the same time due to violent vessel rolling. Under some circumstances it is not possible to acquire the most desirable configuration of position reference systems.

The operator must then deploy the best alternatives and together with all interested parties, make an assessment of the reliability and accuracy of those available and make a decision as to whether the operation can proceed.

8.2.4.5 Heading reference

Heading reference is provided from one or more gyro compasses. The gyro compasses used are normally identical units to those provided in conventional vessels; indeed, one of the gyros provided for DP purposes also doubles as the master gyro for navigational purposes, driving repeaters for such functions as autopilot, radar stabilization and azimuth repeaters.

8.2.4.6 Wind sensors

Wind sensors are provided to give the system a constantly updated value for wind direction and strength. Short term variations in both must be compensated for if precise positioning is to be achieved. Wind sensors or transmitting anemometers are fitted, usually in duplicate, to provide feed-forward computer signals direct to the thrusters to compensate for wind induced movement of the vessel from its set point position and heading.

Problems arise, associated with wind sensors due to wind shadow of the sensor element by the ship's structure or from adjacent platform structure. This will be particularly pronounced if the vessel is downwind of a structure; it may happen that the wind sensor at masthead height is wind shadowed by the platform topside, while the bulk of the vessel hull and superstructure is less wind shadowed by the lattice structure of the platform jacket. Other wind sensor problems arise due to helicopter disturbance, particularly if the wind sensor installation is close to the helideck. Helicopter downdraft can induce false momentary wind sensor inputs, which are not representative of the wind forces acting upon the ship as a whole. Since wind sensor input has an immediate feed-forward effect, the result may be a rapid drive-off as the DP attempts to compensate for apparent gusts (*Figure 8.26*).

Often, two wind sensors are fitted, allowing the operator to select which input is likely to be the most representative. The two sensors may be fitted at different heights on the mast, or at opposite ends of an athwartships yard. In the latter case the operator would probably select the windward sensor in order to avoid

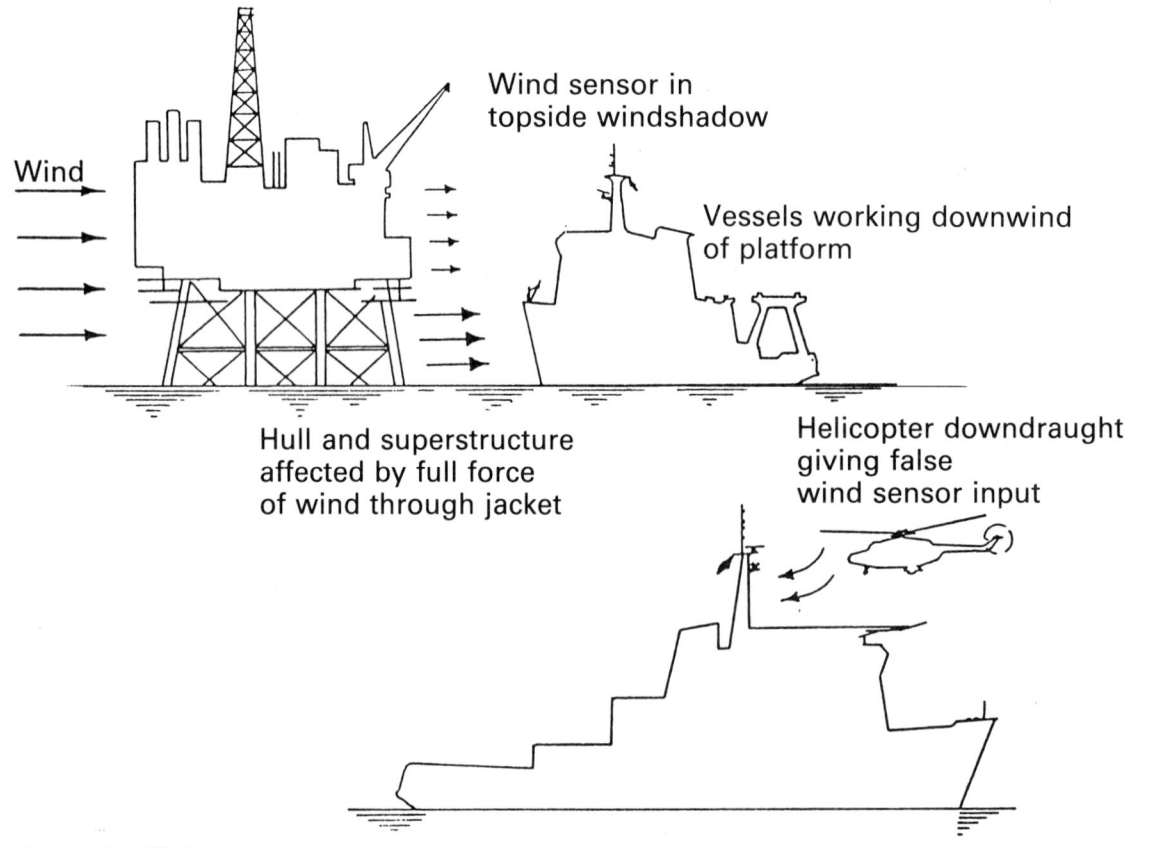

Figure 8.26 Windsensor errors

as much as possible disturbance caused by the ship's structure. One remedy suggested for the problem of wind shadow from adjacent upwind platform structure is to deselect the wind sensors from the DP altogether. However, caution must be exercised here since there will be no wind feed-forward available. Problems will arise when changes in wind speed and/or direction occur, resulting in a deterioration in positioning of the vessel. It must also be remembered that while all wind sensors are deselected the processor will continue to use the model values for the wind, i.e. the values last recorded. If, when a wind sensor is reselected into the system, there are different wind values (which is likely) then a temporary loss of position control may be expected.

In vessels where the helideck is adjacent to the wind sensor location, or where helicopter wind sensor disturbance has been observed, it is common to deselect wind sensor input during helicopter movements. In some cases this will necessitate temporary suspension of the operation in progress.

8.2.4.7 Vessel attitude

It is necessary to provide an input to the processor regarding vessel attitude, i.e. angles of roll and pitch, on a continuous basis. Several of the position reference systems function by measuring angles relative to the shipboard sensor element. Roll and pitch of the vessel will introduce errors into these angle inputs which will translate into position errors. By providing the system with constant roll and pitch angle data, the position reference input data may be corrected to the true vertical. Roll and pitch information is provided from a vertical reference unit or sensor (VRU or VRS). There are three types of VRSs: those using a vertical gyro, those using a pendulous mass and those using accelerometers for angle measurement. The latter type are commonly used in current DP installations.

From one manufacturer, two types of vertical reference sensor are available. The simplest type yields angular information regarding roll and pitch. A more complex type utilizes more accelerometers to yield values for vessel heave (vertical bodily movement) also. Although a consideration of heave is not essential to the DP function, this output can be utilized elsewhere, in heave compensation for diving bells or crane hooks, or in fire monitor stabilization.

8.2.4.8 Power supply and distribution system

Central to the operation of any DP vessel is the power supply and distribution system. Power needs to be supplied to the thrusters and other manoeuvring systems fitted and also to the DP control elements and reference systems. A more detailed consideration of power supply and distribution is given in a later section; a brief overview is given here.

The thrusters will generally absorb more power than any other consumer on board. Dynamic positioning requires more power than other ship functions so it is often found that DP vessels have a very high level of installed power. The DP function often requires large unpredictable changes of power load such as occurs when a vessel is on DP headed into the wind and the wind rapidly freshens and changes direction. The result may be the vessel maintaining station beam-on to a gale force wind, requiring much more power than hitherto. It can be seen that the power generation system needs to be flexible in order to avoid unnecessary fuel consumption. Many DP vessels are fitted with a diesel-electric power plant with all thrusters and consumers electtrically powered. Power is generated by a number of diesel alternators. Another configuration consists of part diesel direct drive and part diesel-electric. A vessel may have twin screws as main propulsion driven direct from diesel engines. Bow and stern thrusters are electrically driven taking power from shaft alternators coupled to the main diesels or from separate diesel alternators. There are many alternatives.

8.2.4.9 Thrusters

The manoeuvring capability of the vessel is provided by the thrusters. In general, three main types of thruster are fitted in DP vessels: main propellers, tunnel thrusters and azimuth thrusters (*Figure 8.27*). Main propellers, either single or twin screw are provided in a similar fashion to conventional vessels. In DP vessels where such main propulsion forms part of the DP function propellers are usually controllable pitch running at constant r.p.m. This facilitates the use of shaft driven alternators as these could not be used if the shaft drive is not at constant r.p.m.; the DP function is not best served by fixed pitch propellers continually starting, stopping and reversing, particularly if the power source is direct drive diesel. Main propellers are usually accompanied by conventional rudders and steering gear. Generally (though not exclusively) the DP system does not include rudder control; the autopilot being disconnected and the rudder set amidships when in DP mode.

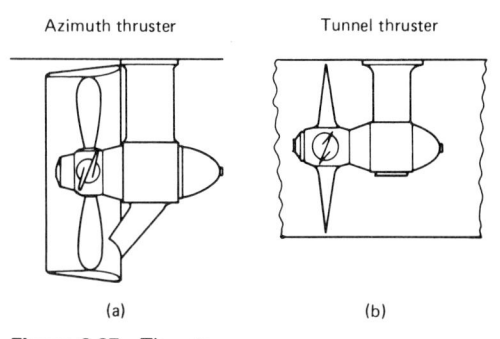

Figure 8.27 Thrusters

In addition to main propellers, any DP vessel will employ a variety of thrusters for manoeuvrability. Typically a DP vessel will use six thrusters, three at the bow and three aft. Forward thrusters tend to be tunnel mounted, operating athwartships. Here a CP propeller (or impeller) is mounted in a tunnel and runs at constant r.p.m. Drive is from above using a bevel gearing from an electric motor or diesel engine. Thrust is produced by pitching the blades port or starboard. Usually two or three tunnel thrusters are fitted in the bow, with control applied identically to all. The resultant turning moment applied to the vessel is most marked if the vessel does not have appreciable head or sternway. Once the vessel is making way the effect of tunnel thrusters drops off considerably.

Tunnel thrusters may be used at the stern also; alternatively it may employ azimuth or compass thrusters. These units consist of a CP propeller mounted in a short tunnel. The unit projects beneath the bottom of the vessel and can be rotated to provide thrust in any direction. Propeller drive is by bevel gearing from above. The whole unit may in some cases be retracted into the hull. Azimuth thrusters have the advantage that they can provide thrust in any direction (compared with tunnel thrusters) and are often used as main propulsion. However, they are more troublesome to locate satisfactorily. If fitted below the bottom of the hull they increase the draught considerably and need to be retractable and, to state the obvious, if they are retracted in shallow water, their power is not available. A wide range of azimuth thrusters is available from a number of manufacturers, ranging from 600 kW to 6000 kW (800 HP–8000 HP) with propeller diameters ranging from around 2.0 m to over 4.0 m.

8.2.5 Redundancy and its provision

8.2.5.1 Redundancy in DP systems

For many DP applications, the provision of redundant operation is not required and is not cost effective. The function of redundancy is to provide greater system reliability in the face of component failures by means of providing backup and standby systems. This provision is obviously expensive and is not fitted in vessels where DP failure would not cause risk to life, serious damage or expensive delays. Examples of non-redundant DP function include: ROV support, stone dumping, off-shore loading of shuttle tankers and platform supply vessels. In these operations the failure of the DP system would constitute an annoying nuisance, perhaps time consuming but not vital.

Often, the operation can continue with the vessel in manual control; otherwise, the operation can be safely abandoned in the absence of DP function. The classification societies recognize the various capabilities of DP vessels and apply notations to a vessel's class accordingly. Lloyds Register of Shipping is able to assign the notation DP(AM) to a vessel fitted with automatic controls for position keeping with manual standby. This equates to non-redundant DP capability. Det Norske Veritas assigns the notation AUT. For fully redundant DP systems complying with the classification societies' rules for provision of redundancy the notations are DP(AA) (Lloyds) and AUTR (Det Norkse Veritas).

Redundancy may be achieved in several ways. Equipment may be duplicated or there may be various means of providing a function. It must be realized that simple provision of a backup system does not automatically provide adequate redundancy. It is worth examining the function of redundancy; what do we need it for? In a non-redundant vessel, e.g. an ROV support vessel, it is possible for a single component failure to result in loss of DP function. An example might be computer failure. The loss of DP function might manifest itself in a drift-off, a gradual loss of station due to lack of controlling thrust being applied, or it may suffer a drive-off with thrusters running away out of control. Obviously the latter is more serious than the former but in either case the DP operator on the bridge is able to take manual control, perhaps stop the errant thruster and regain control of the vessel quickly enough so that the ROV is not lost.

8.2.5.2 Dive support vessels

If that ship were a dive support vessel, however, with divers deployed on a work site with numerous downlines and umbilicals in use and work in progress, then a drive-off or even a drift-off would be serious or even fatal, especially if the ship were stationed only 20 m away from a fixed platform structure. Under these circumstances it is essential that the vessel remains on station irrespective of the status of the DP and other systems. If any part of the DP system (including thrusters and power supply) fails then the standby element should automatically and immediately function so that the operator on the bridge retains full positioning and manoeuvring control. The DP system should then provide an alarm warning the operator of the failure, indicating a degraded status so that he can take appropriate steps to secure the job safely.

The important feature is that the back up element has functioned without the operator having to initiate it. Section 8.2 of the *Guidelines* states that 'a fundamental principle of all DP diving vessel design and operation is that no single fault should cause a catastrophic failure'. The term 'catastrophic failure' is intended to mean the vessel moving away from its set point heading or position.

8.2.5.3 Forms of redundancy

Provision of redundancy arrangements may take a number of forms. The simplest level is to merely duplicate an element, with a manual selection of one or other components. This may be the case with wind sensors where the operator may choose port or starboard wind sensor selected into the system. This is acceptable since there is no possibility of a catastrophic failure pending the loss of wind sensor input. The DP is able to function quite happily without wind sensor input for a short period of time providing there are no large variations in wind direction and strength.

The above arrangement is not suitable for more vital elements of equipment. When considering the control system computer a more comprehensive arrangement is required. In many installations, two complete identical computers are fitted, both working in an operating mode but only one of which is online and the other on standby. Upon failure of the online computer, the operator is able to switch immediately to the other unit and since this is running and programmed the change-over should not result in operational discontinuities. This configuration requires operator intervention to switch computers.

A more satisfactory arrangement is to provide three computers: a pair to provide facility as above (online and standby) and a third unit which monitors the two. This is the configuration used in the Albatross ADP 503 system in which either A or B computers may be online, the other on standby while the C Computer continually monitors the A and B units. The C computer is able to detect online computer malfunction and switch the standby unit online. This allows for a true bumpless transfer as far as the operator is concerned.

Another approach to the problem is to provide triple redundancy or triplex operation. This it the arrangement seen in the Albatross ADP 703 system where three computers are provided, all operational and online. Three computers in a triad perform exactly the same job operating on exactly the same data from sensors, etc. If one of the computers fails this is automatically detected and isolated. The voting logic of the system allows the malfunctioning unit to be identified. The advantage of this configuration is that the operation can continue subsequent to a failure since there are two operational units remaining giving a measure of redundancy. This also gives rise to the facility of hot repair to the malfunctioning unit. The triplex philosophy is carried through the whole DP system so that all major sensors are triplicated (VRS, Gyro, etc.). Whilst increasing the level of reliability of the system it must be pointed out that triplicating systems is more expensive than duplicating them so there is usually a cost penalty.

8.2.5.4 Redundancy of position reference systems

Redundancy of position reference systems is broadly achieved by using two or three PRSs together. As mentioned previously, a modern DP system is able to pool position reference data and achieve a best fix from several different sources. If three PRSs are used together then the loss of one of them is not catastrophic. The operator should not subject himself/herself to common mode failure by deploying two PRSs of the same type since it is then possible for both units to fail simultaneously due to the same cause. For instance, two taut wire PRSs can fail together if the vessel rolls violently such that the winches cannot match the accelerations. In all cases the PRSs should be safeguarded against loss of power supply.

8.2.5.5 Redundancy of power supply

Redundancy of power supply arrangements is a more complex subject. It is necessary to provide sufficient power, whether diesel–electrical plant is installed or direct drive diesel, such that the vessel's operational capability can be maintained subsequent to the failure of any single power unit. Power management arrangements must be provided so that when consumption of power reaches the level of power available then non-essential loads are shed in reverse order of their importance.

In a diesel–electric installation with a number of diesel alternators providing power, then a spinning reserve must be maintained of the equivalent of at least one alternator capacity. Power distribution arrangements must be made such that no single fault within the switchboard, cabling and distribution network can prevent sufficient thruster supply to maintain vessel position and heading. The electrical arrangement of main and auxiliary busbars is normally sufficiently versatile to allow power to be maintained despite a considerable amount of failure within the system. The DP system itself is supplied from an uninterruptible power supply (UPS) which is redundant within itself, takes power from two separate busbars and also has a 30 min battery back up. The UPS, it must be stressed, only supplies the DP system (console, computers, reference systems) and not thrusters.

8.2.5.6 Redundancy in thrusters and propellers

The provision of thrusters and propellers for the vessel must also take on board the need for redundancy. In simple terms this means that the vessel must have sufficient thruster capability so as to be able to remain on station and heading subsequent to losing thrust from any one thruster. Fully redundant DP vessels generally are fitted with a minimum of six thrusters. The system has alarm functions set at 80% of thruster output. This warns the DP operator that he is reaching the point at which he would be short of thrust if for any reason one thruster were lost. The situation becomes more complicated again when considering vessels having a direct drive diesel element to the manoeuvring arrangements.

Consider the case where twin diesels are coupled to CP propellers, with electrical power generated from shaft alternators, additional power being provided from separate diesel alternators. A single fault could immobilize one main diesel, stopping that propeller and leaving the vessel deficient in electrical power since that shaft alternator capacity has also been lost. This is a worst case and must be allowed for in the documented capability levels for the ship.

It must be realized that the best arrangements possible regarding redundancy cannot achieve total reliability. Redundancy arrangements can be negated by the physical location of equipment. In multicomputer systems the two, three or four computers are usually located in the same compartment. It is common to see both gyro compasses and VRSs side by side; likewise the two elements of the UPS are usually located adjacent to each other. System cabling is often grouped together in a common trunking. This arrangement is prone to fire, explosion or flood damage. Likewise, a fire on the bridge which destroys the bridge console is likely to knock out the DP in its entirety. Similarly, the best redundancy arrangements cannot allow for multiple simultaneous failures. It is well known that things always go wrong in threes and this is especially true shipboard, with a harsh environment. Severe weather conditions with water flying about can result in many failures in a short time.

8.2.6 DP control modelling

8.2.6.1 Feedback control

The DP system is an example of closed loop or feedback control using mathematical modelling. A separate model is acquired for the dynamics of the vessel hull, the thrusters, each position reference system, and also the various other sensors used. The mathematical model takes time to build up to an optimum level, acquiring data upon all of the variables. Since the majority of these variables change their characteristics only slowly if at all then the model can be based upon a number of minutes sampling of values. It must be remembered that a system model is an approximation which characterizes the behaviour of that physical system. The accuracy of the model will vary; the more accurate the model the more complex it is and the more costly to produce. The ultimate function of the control system is to compute thruster commands that will, when applied to the thrusters, maintain the vessel on station or return it to the set point position and heading. This task must be carried out in the face of a variety of external forces such as wind and tide, other ship-induced forces such as thrust from firefighting monitors, or the drag induced when pipe laying due to the maintained pipe tension.

One of the most important forces that must be compensated for is that resulting from the wind. Since the wind speed and direction are subject to very rapid changes and since the vessel is rapidly influenced by wind forces, it is necessary to provide direct thruster compensation for measured wind variations. This is referred to as feed-forward and requires an accurate wind-sensor input in order to function. Without wind feed-forward, changes in wind speed and direction would not be compensated for until the model was fully updated.

The use of a sophisticated mathematical modelling technique allows a number of benefits to the realized. An effective dead reckoning mode is available such that DP control is not lost if all position reference input is lost. The system continues to position the vessel using the information in the model. The model will deteriorate in time and accuracy will degrade but meantime the vessel can be kept in automatic DP mode with often better results than by switching to manual control. The use of modelling techniques also allows for combination of several position reference systems in a pooling arrangement; observed position spreads allow a bias to be allocated to each system so that the optimum value for the ship's position is continuously calculated.

8.2.7 Position reference systems

8.2.7.1 Repeatability

Every DP system needs a position reference and this reference must possess sufficient repeatability to give the vessel the required positioning accuracy. Many different types of position reference have been used in conjunction with DP systems over the years, with greater or lesser accuracy. We will confine our remarks here to a detailed consideration of the three most extensively used PR systems, with a brief mention of some other frequently used PRSs.

Ideally, we would like a repeatability value for a PR system of about 0.1 m and currently used systems approach this, but failing this, a PR system with accuracy values of around 1–2 m are of value. Lower accuracies than this are of limited use. As a consequence, normal run of the mill navigation systems such as Decca Navigator, RDF and Loran-C are not normally used as DP position reference. Specialist radio survey systems are often used, however; these usually provide greater accuracy. Such systems as Syledis, Pulse/8 and Argo are occasionally found interfaced with DP installations. The three main PR systems found in DP vessels are hydro–acoustic position reference (HPR), surface microwave position reference and taut wire position reference. Many DP vessels carry all three, often with one or two systems duplicated. It is rare to find a DP vessel not carrying one of these three systems.

The centre of rotation (CR) of a vessel is the designated spot within the vessel that we are positioning. In a conventional monohull DSV the CR will be located in the centre of the moonpool, such that if the vessel's heading is changed whilst the diving bell is deployed, the bell simply rotates rather than translates laterally. The CR may thus be approximately amidships or there may be other arrangements. A cable laying vessel with a stern mounted lay-sheave may have the CR positioned upon this spot. A vessel may have more than one CR, selected by the operator; a DSV may have two diving bell positions with a CR selectable on each one. More complex vessels may have a larger number of CRs (e.g. 12) corresponding to diving locations, workpool location and crane hook positions, all operator selectable. Another possibility is that the operator can select the position of his CR by entering longitudinal and athwartships values for the point reference to a datum position. When a CR is selected, the offet values (longitudinal and athwartships) between individual PR sensors and the selected CR are allowed for. If the CR is changed then these offset values are automatically changed.

8.2.7.2 Hydro–acoustic position reference (HPR)

8.2.7.2.1 Simrad system A number of manufacturers are involved in the production of HPR equipment, a good example of which is the Simrad system (*Figure 8.28*). The principle of position measurement involves communication at hydro–acoustic frequencies between a hull mounted transducer and a sea bed located transponder.

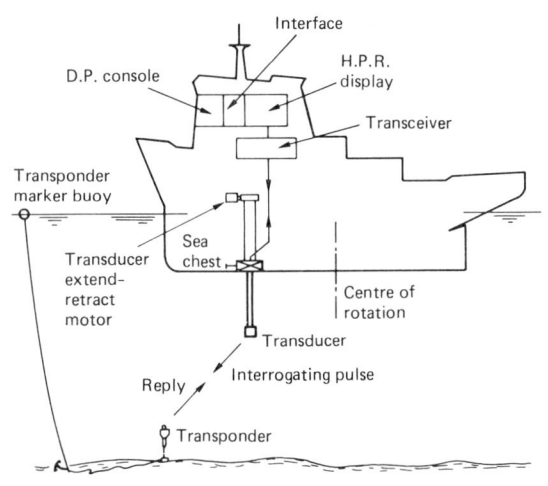

Figure 8.28 Elements of a HPR system

The system is based upon the supershort baseline principle with all acoustic transmit–receive elements mounted in a single transducer unit. An interrogating pulse is transmitted from the transducer. This pulse is received by the transponder on the sea bed, which is triggered to reply. The transmitted reply is received at the transducer. The transmit–receive time delay is proportional to slant range. The hull mounted transducer is able, by means of its supershort baseline configuration, to determine the angles of the incoming reply with respect to the vertical, both longitudinal and athwartships. These angles and range define the position of the ship relative to that of the transponder. A number of corrections are made automatically. The angles must be compensated for values of roll and pitch from the VRS. The determined position will be that of the transducer and offsets will be allowed for to give the position of the CR of the vessel.

8.2.7.2.2 Basic elements The basic elements of an HPR system are shown in *Figure 8.28*. Frequencies used are in the 20–40 KHz range. The Simrad system uses a discrete range of channels of transponder communication each channel has a designated interrogation frequency and reply frequency (*Figure 8.29*). Up to 16 channels are available. Position reference may take place from a single transponder laid on the sea bed. Greater reliability and accuracy can be obtained from using more transponders. Any number of transponders can be used provided they operate upon different channels, up to the number of channels installed.

Transponder channels

TP CH. no.	TP symbol	Interrogation frequency	Reply frequency
11	⊡	21552	27173
22	⊙	22779	28409
33	△	23981	29762
44	✕	25188	31250
55	⋰	26455	32468
1	1●	20492	29762
2	2●	21552	30488
3	3●	22124	31250
4	4●	22727	31847
5	5●	23364	32468
6	6●	24038	27173
7	7●	24510	27778
8	8●	25000	28409
9	9●	26042	29070
88	A●	38500	37500
99	B●	39500	

Figure 8.29 Transponder channels. (Courtesy of Simrad Albatross Ltd)

The HPR system is very versatile; it is not simply used for position reference for DP vessels. It is also used for position monitoring of sea bed or subsea vehicles, and also for marking for relocation of subsea features, e.g. wellheads, pipelines, etc. It may also be

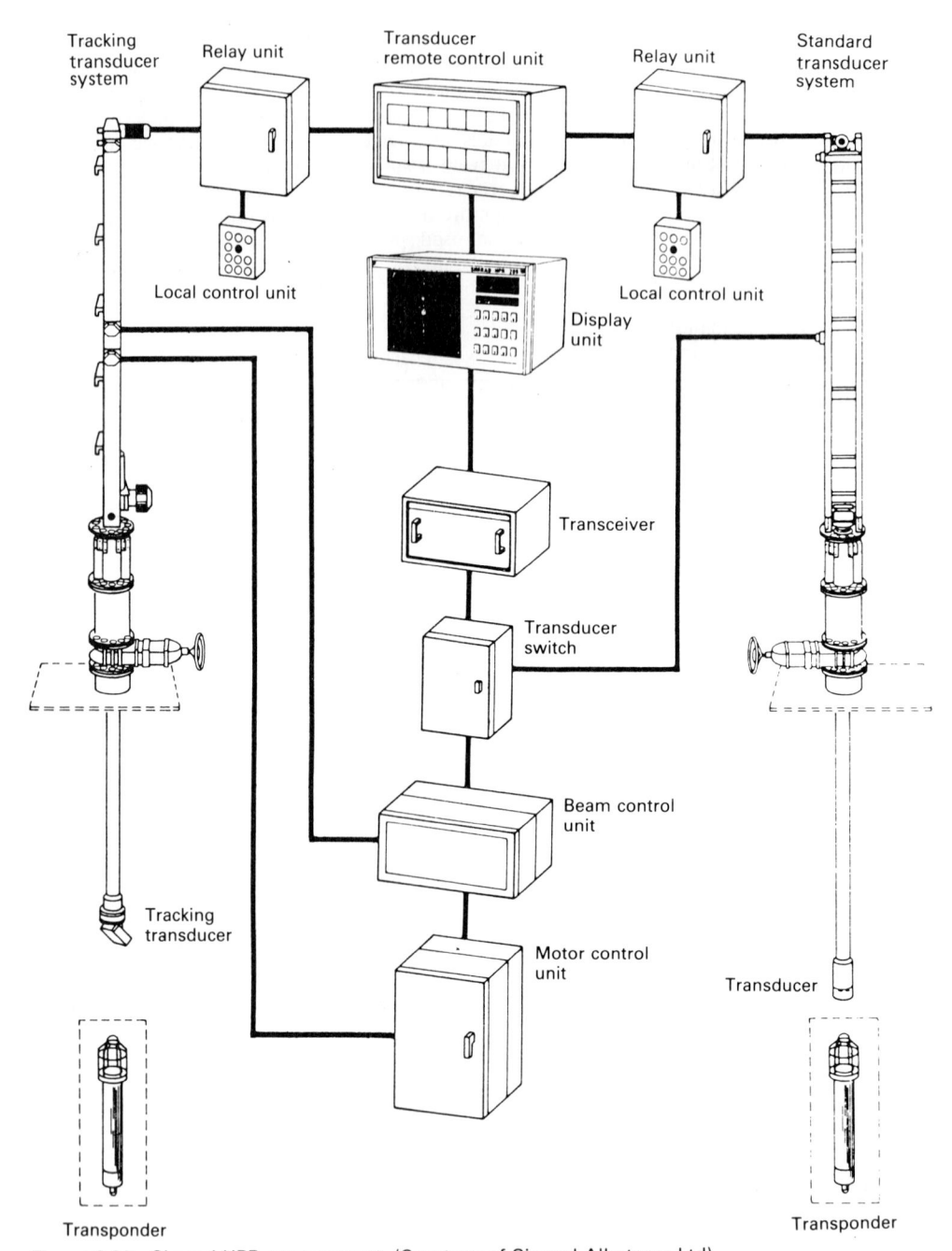

Figure 8.30 Simrad HPR arrangement. (Courtesy of Simrad Albatross Ltd)

used for location and monitoring of the position of a diver or diving bell.

Transducers are fitted to the bottom of the vessel in the form of a probe, able to extend approximately 4 m below the shell plating. Accordingly they are provided with an electric motor to enable them to extend and retract with both local and remote controls at the HPR display unit. A ship's bottom sea chest or gate valve is provided to ensure the watertight integrity of the vessel when the probe is withdrawn into the hull. The probe is normally raised when the HPR is not in use, particularly when the vessel is under way.

Two types of transducer are available; the standard or fixed unit and tracking type. Often a vessel is fitted with one of each type, often located under the forebody of the vessel. Each unit is able to operate in a wide or narrow beam mode. A narrow beam has a scope of about 30° whilst a wide beam has about 160°. The tracking transducer has the facility of being able to direct its narrow beam in azimuth and elevation (or depression) angles so as to follow a selected transponder. This gives greater ranges on a particular transponder and also helps prevent interference from other sources.

The operator is able to select the interrogation rate of the system. A high rate of interrogation is used where the vessel or transponder may be moving, or where up to three transponders are active and ranges are relatively short (up to about 250 m). Typical high interrogation rates may be 1.0, 2.0 or 4.0 s. Low interrogation rates with intervals of 8.0–40.0 s may be used where the position situation is fairly static or where it is necessary to save transponder battery capacity. Under normal circumstances, the operator would select wide beam operations until transponder communication is established; he/she may transfer to narrow. If transponder communication is lost then the system automatically reverts to the wide mode.

8.2.7.2.3 Control and display unit The control and display unit contains pushbutton controls through which the system is operated, together with a CRT display operating in plan position indicator (PPI) mode, similar to a radar display.

A typical CRT display has a number of facilities. It is able to portray the ship in the display centre as either relative or true. Relative may display ships head up, or in the case of the HPR forming part of a DP console fitted into the after bridge of a DSV, hence aft-facing, relative mode would then display sternup. True display, when selected, will put north at the top of the screen. The display range may be selected by the operator; typical ranges available are: 10, 25, 50, 100, 500, 1000, 2500 and 5000 m.

When in use, the display shows the vessel at screen centre and deployed transponders in their appropriate locations. Each transponder has a symbol, either a number (1 to 9) or a symbol square ($\square$), delta ($\triangle$), circle ($\bigcirc$), X or Y and the transponder symbol appears on the CRT accompanied by positional coordinates. The coordinate mode may be selected by the operator as polar, in which case the coordinates displayed would be range, bearing and depth (horizontal range and depth in metres, relative bearing in degrees). The operator may alternatively select Cartesian coordinates such that the three values displayed represent X, Y and Z offsets (distance starboard, distance forward, depth, all in metres). The operator also has the facility of placing a selected transponder at the display centre.

Once communication with one or more fixed transponders has been established, the position data may be fed through to the DP system by means of a suitable interface, for position reference. Again, compensation is made to allow for the offset distances between the transducer sensor and the CR of the vessel. If more than one transponder is in use, then the DP pooling logic is able to resolve the best positional data from each. It may happen that one (or more) of the transponders selected is not in a fixed position, e.g. one is mounted upon an ROV. Since the DP must ignore position reference indications from this transponder the operator is able to 'mobile select' that transponder. Thus, it will appear on HPR and DP displays such that the operator can monitor its position, but its input will be ignored by the DP as regards position measurement.

8.2.7.2.4 Transponders A number of different types of transponders are available for different functions and situations. A standard transponder may be secured by bracket to subsea structure, or it may be moored to the sea bed (*Figure 8.31*). The recommended method of mooring is to fit the transponder with a divingcell float, and attach it to a 150 kg sinker by means of 1–2 m of mooring chain. The unit is then deployed onto the sea bed by means of a light wire line, either from the deck of the ship or from a small boat, with the wire buoyed off.

Another method of deploying a transponder is to secure it into a purpose built tripod which is then lowered to the sea bed on a wire rope. Once located, sufficient slack is paid out to accommodate subsequent vessel movement. Deployment and recovery is by a winch and davit arrangement.

Other types of transponder are available for use in special circumstances. A Release Transponder may be laid upon the sea bed without a downline. Operating normally the transponder provides position reference. After use the transponder can be recovered by acoustic command, releasing it from the sea bed. The transponder floats to the surface and can be recovered by boat or from the ship.

An Inclinometer Transponder provides information not only for position reference but also regarding angular attitude. A unit of this type may be used in subsea installation, or in drilling. In the latter case the transponder would be secured to the lower ball joint of the marine riser. In a DP drill ship information upon ball joint angle is used to position the vessel, the object being to maintain the riser vertical at the sea bed in the face of constantly varying tides.

A standard transponder may be approximately 1 m in height by 120 mm in diameter. For small vehicles, such as RCVs, miniature transponders are available. Transponders such as these can also be carried by a diver.

More reliable communications may be obtained by using a responder instead of a transponder. A responder is used upon such things as ROVs or sea bed crawler vehicles, where a hardline umbilical is available. A spare umbilical channel is used to trigger the

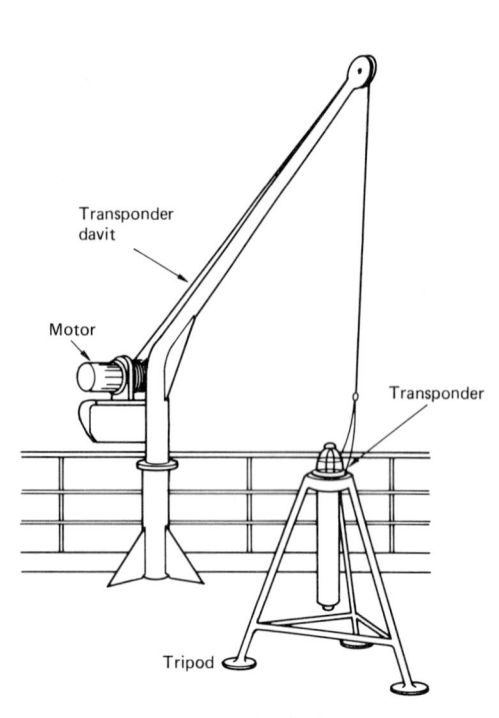

Figure 8.31 Transponder deployment

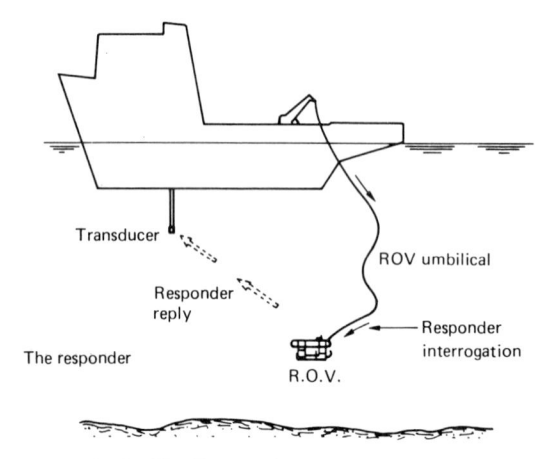

Figure 8.32 The Responder

responder such that only one-way acoustic signals are used (*Figure 8.32*).

8.2.7.2.5 Other systems The system described above is only one method of utilizing hydro–acoustic principles for position reference. Although it is the commonest it must be realized there are others.

A short baseline system uses a single acoustic pinger placed on the sea bed, transmitting continuously. Onboard ship the acoustic pulses are received by usually four passive hydrophones placed at different locations. Since the physical locations of the hull mounted hydrophones are known, then the times of arrival of individual pulses at each can be compared and the time differences will define the direction and distance of the pinger from the vessel.

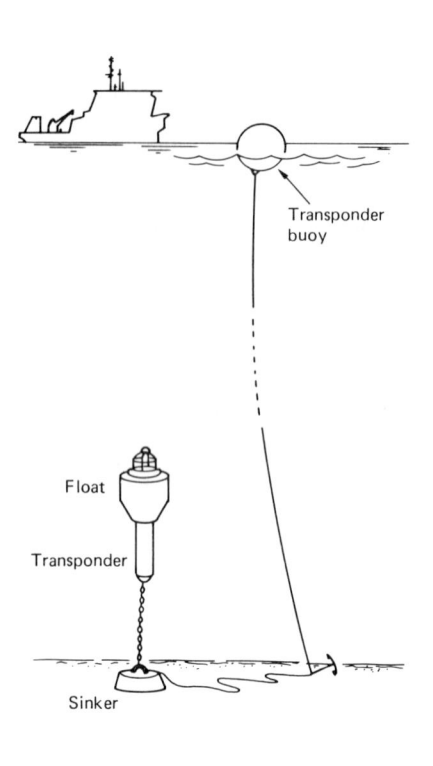

The long baseline system uses an array of three or more transponders laid on the sea bed in the vicinity of the work site. One transducer upon the vessel interrogates the transponder array, but measures only ranges. Position reference is obtained from range–range geometry from the transponder locations. This principle provides greater accuracy than the short baseline system since the baseline length is not limited by ship dimensions, but it is reliant upon two-way acoustic communication.

HPR is a versatile and widely used position reference for DP and other purposes. It does, however, suffer from some limitations. Its main problems are acoustic noise and aeration. Transponder communication may suffer noise interference from the ship, from machinery or thruster noise, or from external noise from nearby installations. Acoustic interference may originate from rough seas or rainfall. There may also be acoustic reflection from any structure. Aeration from any cause will result in loss of signal. Common causes of aeration are thrusters (vessel or ROV), mud or stone dumping, diver breathing gases and rough seas.

Careful positioning of transponders by the operators can obviate many of these anticipated problems. Other problems result from acoustic refraction, particularly at long horizontal ranges from the transponder, caused by temperature or density layers in the water.

System accuracy is dependent upon a number of factors, some of which, such as water temperature, cannot be allowed for. In general, accuracy can be taken to be 1–3% of slant range. For best accuracy, horizontal range should not exceed water depth; at horizontal ranges larger than this there is an increased risk of loss of signal due to refraction.

8.2.7.3 Surface microwave position reference system

8.2.7.3.1 Principles In this system, position reference is obtained by means of communication at 9 GHz (X-band, 3 cm wavelength) radio waves or microwaves. The system described is the Artemis system manufactured in the Netherlands, and involves two stations, one located on board the DP vessel itself and the other at some fixed location ashore or upon a platform installations (*Figure 8.33*).

The position reference is in the form of range and bearing. The station on board ship is referred to as the mobile station, while that ashore is the fixed station. Position data is thus obtained as the range and bearing of the mobile station with reference to the fixed station. Each station consists of a control data unit and an antenna unit.

The principle is simple. The two antennae automatically train so as to face each other at all times when a microwave link is established. The mobile station transmits a signal, which is received by the fixed station and retransmitted as a reply. The time lapse between mobile station transmission and reply reception is proportional to range between antennae. The azimuth or bearing is measured at the fixed station and is transmitted, encoded, as part of the reply. The signal used is a very brief interruption in a continuous wave transmission; this interruption being detected at the fixed sation and a similar reply interruption initiated. For accuracy, successive displayed ranges are obtained by averaging a great number of observed time lapses (1000 or 10 000, depending upon function selected).

Position reference is thus obtained in a continuous, reliable, accurate manner utilizing only one external fixed station. Once a microwave link is established voice communications are possible using handsets. Position reference can be obtained whenever there is a line of sight between the fixed and mobile stations which give the system greater operating ranges than HPR or taut wire. Typical maximum range is about 30 km but for DP purposes 10 km is a more realistic figure. At ranges greater than this the bearing accuracy of about 0.03° leads to a deterioration in positional accuracy inadequate for DP purposes. Range accuracy of around 0.5 m is obtained. Range and bearing obtained are then passed into the DP system by means of a suitable interface and corrections are applied for vessel roll and pitch values which cause the antenna to move and for antenna offsets relative to the CR.

8.2.7.3.2 Artemis fixed station The Artemis fixed station is often set up either permanently or as a temporary installation upon a fixed production platform or similar. Before use the fixed station must be calibrated for bearing. To do this it is necessary to obtain a visual reference direction to identify some fixed object nearby, another platform perhaps, to which the bearing can be obtained by reference to terrestial coordinates. A small telescope is shipped into a mounting on the fixed station antenna and the antenna trained by hand until the reference object lies on the crosswire in the tele-

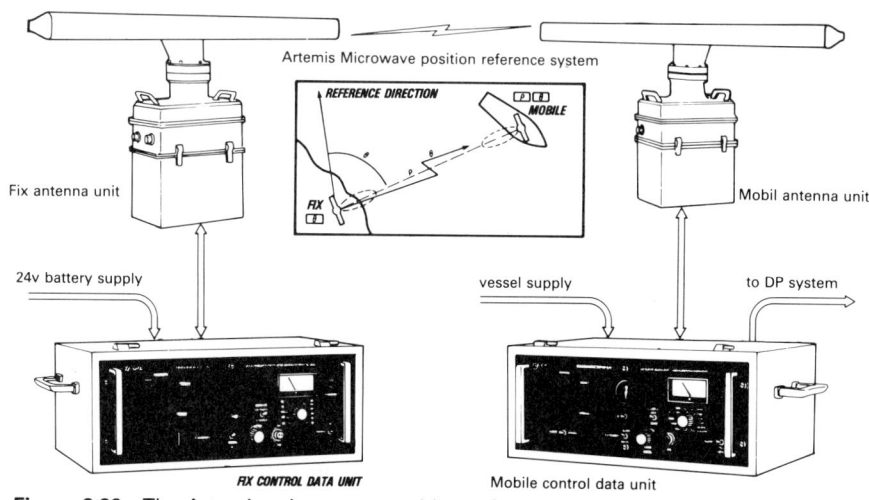

Figure 8.33 The Artemis microwave position reference system. (Courtesy of Simrad Albatross Ltd)

scope. The antenna is thus trained on to the reference bearing, which is entered into the unit display. The reference bearing may be referred to true north or any local grid north. Subsequent to such calibration the telescope can be unshipped and bearing readout for any antenna direction should follow correctly.

The fixed station may be a temporary installation, placed upon a platform for one particular job of limited duration. In this case it is necessary for the equipment to be installed, taking care that the antenna will cover the sea area of the DP vessel's work site, also that the antenna can rotate without fouling the platform structure. It is also necessary to ensure that the antenna position will not interfere with other platform functions. A power supply must be obtained, preferably with a redundant battery backup supply and platform staff will need to be instructed in the operation of the system. If the vessel is to work all around the platform then two or more fixed stations will need to be installed to cover 360°. The elevations of the fixed station antennae must be compatible with that of the mobile station, since, with only a 22° vertical beam width it is possible, at close ranges, for the signals to be lost if fixed and mobile antennae are at different heights (*Figure 8.34*).

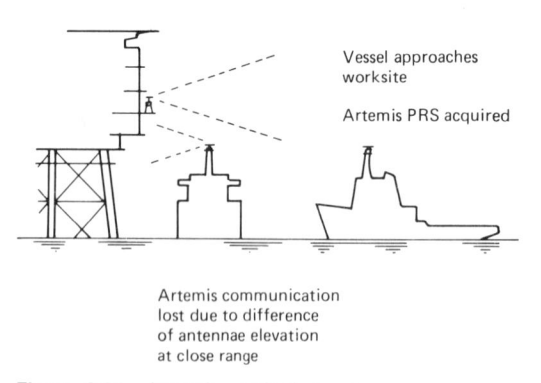

Figure 8.34 Artemis vertical range

8.2.7.3.3 Range From the viewpoint of the DP operator, the major advantage of Artemis compared with many other PRSs is its range. When approaching the work site the Artemis link can be established when a mile or more away and provided line of sight is maintained then position reference is available immediately DP is engaged.

However, there are drawbacks. Interference can be experienced from radar transmissions from other vessels or from one's own at 3 cm frequency. Also, loss of signal will occur due to any line of sight break, such as a vessel passing through the beam, or personnel on the platform working in the vicinity of the fixed antenna. Precipitation may also cause loss of signal. An operational disadvantage is the lack of control over the fixed station. Being sited upon a remote platform it is not immediately accessible to ship's staff and is subject to interference from unauthorized platform staff. Occasionally, for example, the batteries are borrowed for another purpose and not replaced.

Under some circumstances it may not be possible to establish an Artemis fixed station. This may be where the fixed station location is not positively fixed in position. An example of this situation is an offshore spar buoy used for shuttle tanker loading.

Increasingly such tankers are fitted with DP for positioning during the loading operation, and Artemis is used for position reference. The spar buoy is, however, mobile to a certain extent and may be able to rotate. In this case the fixed station is replaced by an Artemis beacon. The beacon is a transponder with a broad beam antenna. The antenna itself is fixed. The system is used in exactly the same manner as the standard Artemis. Range data is obtained in the same way with identical accuracy. Bearing is obtained at the mobile station end, using the mobile antenna direction and gyro heading reference. Bearing accuracy is thus reduced, being only as good as the gyro compass perhaps 0.25–0.5°. For this reason it is suitable for shorter range operations.

8.2.7.4 Taut wire position reference

8.2.7.4.1 System description Taut wire differs from previously discussed position reference systems by being chiefly mechanical in principle, not reliant upon radio or acoustic transmissions.

A taut wire system consists of a constant tension winch unit fitted on deck, with a boom or derrick projecting over the side of the vessel. Wire from the winch drum passes over a sheave at the end of the boom, through a sensor head and terminates in a depressor weight on the sea bed. Position reference is obtained from measurements of wire angle and water depth, the position of the vessel being defined relative to the location of the stationary depressor weight.

A typical taut wire system is the Simrad Albatross lightweight taut wire mark IV, illustrated in *Figure 8.35*. In this system the depressor weight has a mass of 350 kg, while 500 m of 5 mm wire is provided. Maximum wire angle is 35° to the vertical in any direction, and the maximum water depth for use is 350 m. The motor and drive unit are mounted on deck at the side of the ship, with the system control panel adjacent. The boom stows in a horizontal position lying fore and aft. To deploy the boom it is first traversed through 90° to the outboard position, then extended to full reach hydraulically. The weight is then lowered to the sea bed. Once at the sea bed the system automatically puts the winch into tension or 'mooring' mode. At this moment the wire length deployed is read by the system. Subsequent movements of the vessel are accommodated by the spooling of the winch, while wire angles to the vertical, both in the longitudinal (fore and aft) and in the athwartships plane are continuously monitored by the sensor units at the end of the boom.

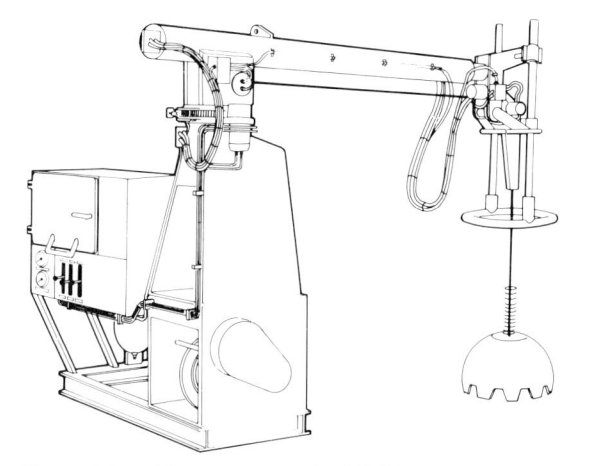

Figure 8.35 Albatross taut wire Mk IV system

Taut wire signals are fed back to the DP through a suitable interface and can be accepted by the operator at the DP console. The system continually corrects input data for values of roll and pitch such that wire angles are relative to the true vertical instead of the local ship vertical. Corrections are also applied to allow for the offset distance of the position of the sensor head relative to the CR of the vessel. The geometry of the position reference is very simple, as shown in *Figure 8.36*.

Frequently, the taut wire system is fitted with remote controls such that the DP operator may deploy the boom and weight from the bridge. Often, however, this facility is not used unless the operator can actually see the system he/she is deploying. It is not a particularly safe practice to remotely operate the taut wire without being able to see what is happening.

8.2.7.4.2 Advantages and drawbacks A taut wire system is particularly robust and reliable, together with being particularly accurate. Maintenance is necessary but at least this is under the control of the vessel operators; there is no remote and inaccessible equipment involved. Spare wires and depressor weights can be carried and the system may be regularly oiled and serviced to schedule. A worn, stranded or frayed wire can be replaced. In constant use the wire may wear at the same spot all the time. To overcome this problem the wire can be unshipped from the weight and cropped back approximately 10 m every couple of weeks or so. This brings a new portion of wire onto the sheaves when working and also freshens the connection to the depressor weight, another point of possible failure.

One drawback of the system is its limited operating range, due to the 35° wire angle limit (*Figure 8.37*). This limit is imposed due to the increasing risk of dragging the weight at larger angles. A dragging weight could, of course, lead to immediate position errors from the system. The operating range is thus dependent upon water depth; the deeper the water the greater the

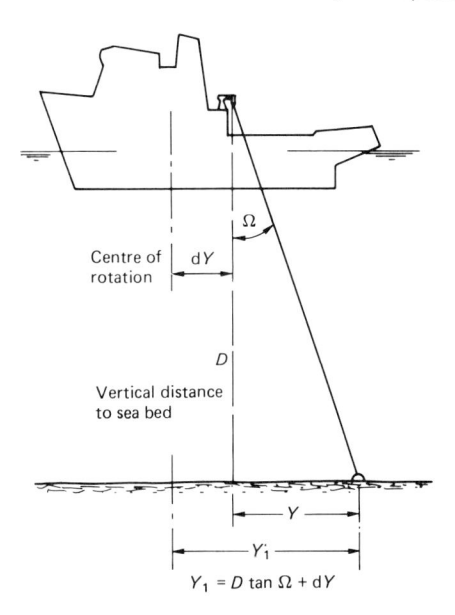

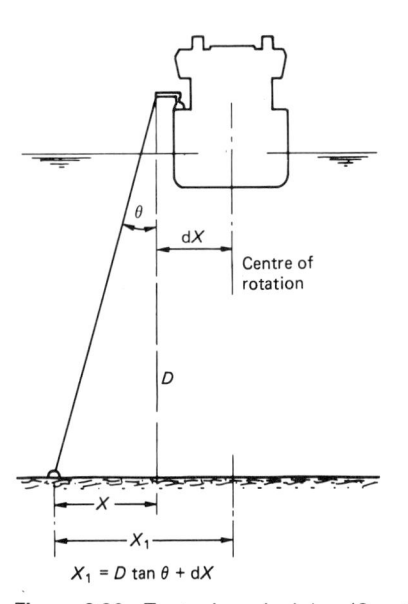

$Y_1 = D \tan \Omega + dY$

$X_1 = D \tan \theta + dX$

Figure 8.36 Taut wire principles. (Courtesy of Simrad Albatross Ltd)

range of cover. Another limit is imposed by the bilge keel of the vessel impinging upon the wire before the 35° angle is reached.

8.2.7.4.3 Use of two taut wire systems Often a vessel is fitted with two taut wire systems, one on each side, allowing the operator to select the most advantageous system for the circumstances. If the vessel is working close alongside a platform, with divers deployed into the platform, then the taut wire on the side away from

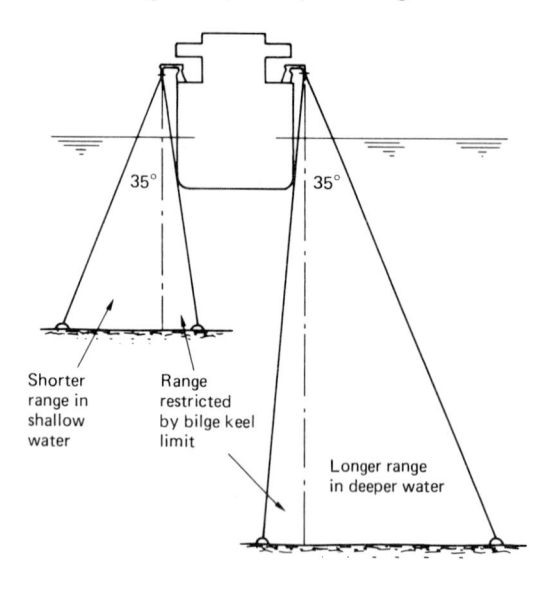

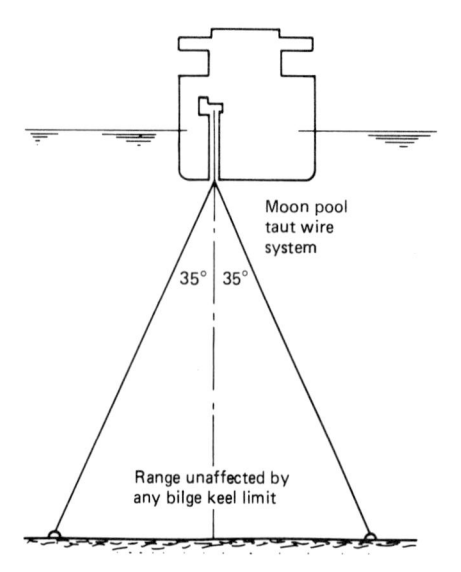

Figure 8.37 Taut wire range

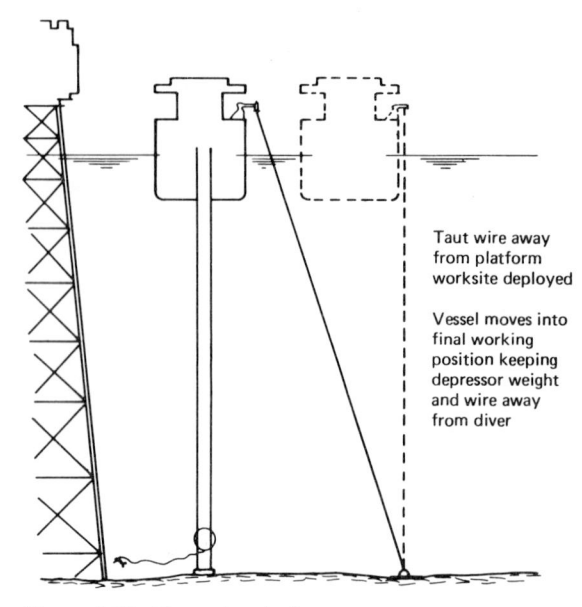

Figure 8.38 Taut wire deployment

the platform will be used. This keeps the wire away from the divers and also allows the vessel to deploy the taut wire before making its final move onto the work site without the wire coming onto the bilge keel (*Figure 8.38*).

In some vessels the taut wire system is fitted at the forward extremity, right on the bow of the vessel. One problem associated with this location is the large accelerations experienced in moderate to rough seas due to vessel pitching. Occasionally system dropout will result from the winch spooling rate being unable to cope with the pitching.

When using taut wire as a position reference, it is necessary to plan with care the position for deployment of the depressor weight. The sea bed can be a cluttered place and often a field operator will stipulate that there must be nothing placed on the sea bed within a certain distance, perhaps 10 or 20 m of pipelines, control lines or other sea bed installations. In addition to this constraint the operator should aim to place the weight in a position so that the final working position of the vessel does not result in the wire angle being almost at its 35° limit. This would prevent the vessel moving in one direction.

A further drawback of the taut wire system is that there is no geographical reference for position. It is never known with certainty exactly where, in terms of geographical coordinates, the depressor weight has landed. Since position reference is from the weight then exact coordinates for the ship are not available as they would be if using Artemis from a precisely known location of a fixed station.

The taut wire requires a continuous power supply and it uses a fairly large amount of power when compared with other position references. As such it is not connected to the UPS system powering the remainder of the DP.

8.2.7.4.4 Types of taut wire systems Several other types of taut wire system are in use. For some types of vessel a platform type of taut wire can be used. This may be fitted in a bridge or cellar deck of a semisubmersible vessel where access to the sea is available directly beneath the taut wire location platform. While similar to the type already described, this unit dispenses with the extending telescopic boom. Instead the weight simply lowers directly away from its housing structure.

For some vessels a moonpool taut wire is suited (*Figure 8.39*). This unit is mounted inboard with a depressor weight deployed through the bottom of the vessel through a small moonpool or wet well. The sensor head is incorporated into an elevator unit which is lowered from the tweendeck stowage level down to the keel level. The depressor weight and wire are then lowered onwards from there. Movement compensation is provided by hydraulic accumulator and positional data is obtained and processed in exactly the same way as in the types previously described. The function of a moonpool taut wire is to enable a DP vessel to operate in surface ice conditions. It also has the advantage of obviating the bilge keel angular limit.

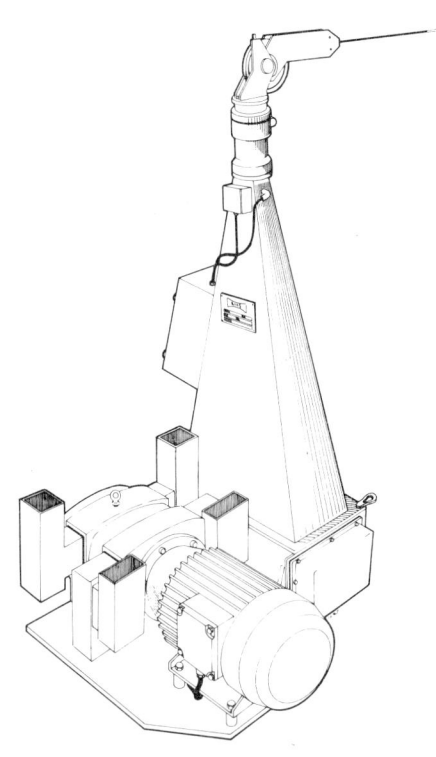

Figure 8.40 Surface taut wire system. (Courtesy of Simrad Albatross Ltd)

8.2.7.5 Other position reference systems

The three PR systems already discussed are those most commonly found in DP vessels. Other systems are used, however, and over 40 different position references have been used in conjunction with DP systems over the years. Several of these are described in the chapter on navigation and positioning systems.

8.2.8 DP operations, planning and watch keeping procedures

8.2.8.1 Planning

It is necessary, before commencing operations with a DP vessel, to undertake detailed planning of the operation proposed. Prior to this stage, an assessment must be made of the capability of the vessel to maintain station under the expected conditions. Station keeping capability is related to many factors such as available power, thruster configuration, hull size and shape, superstructure windage or sail area, expected environmental conditions and factors relating to the task in hand. It is also necessary that the vessel has the required level of system redundancy, reserve power and general reliability such that the task can be completed safely and on schedule.

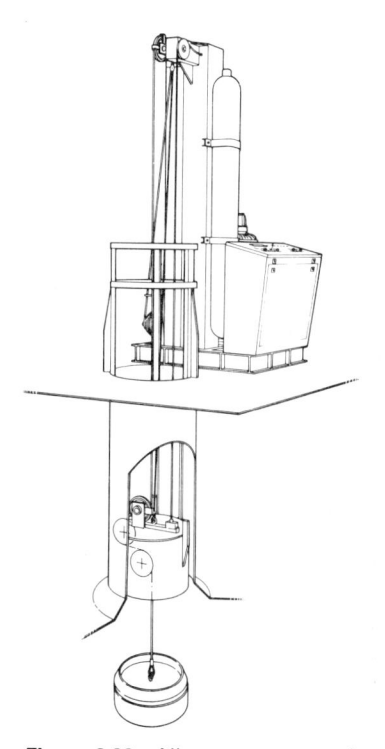

Figure 8.39 Albatross moonpool taut wire. (Courtesy of Simrad Albatross Ltd)

Another style of taut wire is the horizontal or surface taut wire system. This unit gives position reference relative to a fixed structure. The wire is passed across to the platform adjacent and secured. The geometry is different to the vertical taut wire but principles are the same. No boom is needed; instead the sensor is located atop a short vertical tower. Range is limited to about 100 m wire length but the wire is wholly in view unlike the vertical taut wire system (*Figure 8.40*).

To assess the station keeping ability of the vessel, a potential charterer may refer to a set of statistics known as the environmental regularity numbers (or ERNs). These are percentages, quoted for a particular set area, of the amount of time that a vessel will be able to maintain station. Three figures are given, respectively all thrusters working, the least significant thruster stopped and the most significant thruster stopped. Typical ERNs for a DSV may be 99, 96 and 81.

ERNs are only able to give an outline of the capability of the vessel. A more detailed assessment may be made by inspecting a DP operational capability graph, or footprint diagram (*Figure 8.41*). A footprint diagram is a polar graph showing weather heading against wind speed. A number of curves may be plotted representing (for example) maximum capability, operational capability with 20% power in reserve, capability subsequent to worst case switchboard failure. Environmental conditions assumed for the presentation of data will be stated on the diagram, e.g. 1 knot wind induced current.

——— Maximum DP capability
—·— DP diving capability
- - - - Capability under worst case switchboard failure

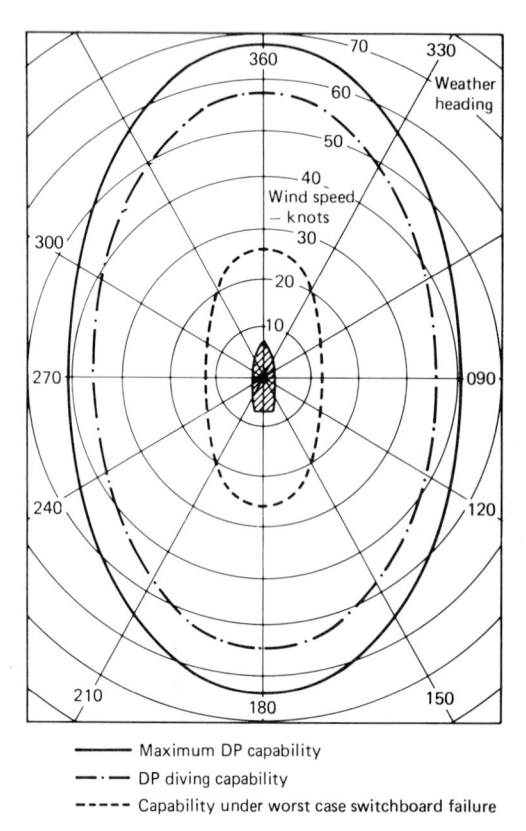

Figure 8.41 DP diving capability (footprint) diagram

8.2.8.2 Safety audit

Prospective charterers will require a detailed audit of the vessel under consideration to be carried out, in order to satisfy themselves that the vessel is suitable for the intended task.

Barber (1984) states that the purpose of a safety audit can be set out under five different headings: assessment of vessel design, maintenance, reliability, vessel performance and operational capability. The audit will include the following elements, carried out by a team of systems engineers: inspection of drawings, completion of checklist, inspection of ship's equipment, review of ship's company, inspection of operations manuals and sea trials.

The checklist is an essential part of the audit. The audit engineer may have his/her own or his/her clients checklist to hand, the UKOOA has drawn up a standardized checklist for use where no other is available or suitable. Inspection of equipment will include maintenance schedules and records, spares carried on board, general state of the machinery. The ship's company must be adequate in skills, experience, qualifications and numbers. This is particularly important regarding engineers, electricians and electronics technicians, and watch keeping DP operators. Operations manuals should be available detailing all operational procedures including pre-operations checks and operational failure actions. Following the above, the vessel will be taken on a short sea trial to prove the DP system and its redundancy arrangements.

8.2.8.3 Detailed planning

The planning of DP operations should be undertaken in great detail. Depending upon the task the client will have laid down the operational manoeuvres of the vessel in order to carry out the work. The ship's staff will need to interpret this planning in the light of the practicalities of conducting the vessel under DP. Factors which must be taken into account include the availability of position references upon the work site, environmental conditions expected, presence of surface or sea bed obstructions and own vessel capability. The planning of the job may involve use of large scale work site plans showing detailed information of all sea bed and surface features. A useful tool is a vessel template to the same scale, showing all relevant information, which may be positioned and moved on the work site plan to obtain and verify positional information.

Part of the planning procedure will include contingency planning, with escape manoeuvres designated at each stage of the operation. Prior to commencing the operation, a conference may be convened of all interested parties to ensure that all concerned are aware of the planned operation, sequence of events, etc. Communication, both on board and external, must be maintained. Where communications are vital to the safety of operations, i.e. bridge/DPO/dive control, then several independent lines of communication will be provided.

Pre-operational checklists should be provided, suitable for the operation concerned, i.e. pre-dive checklist, pre-DP checklist. Logbook records should be maintained containing time reference data. In many vessels, voice communications and conversations are tape recorded, either on the bridge or in dive control.

Bridge or DP control room manning must be adequate. When on DP it is usual for two qualified DP operators to be on duty. One of the DPOs will attend exclusively to the DP console while the other will carry out all other watch keeping and communications tasks. It is important to realize the boredom factor associated with DP watch keeping. When an operation is in progress long periods elapse with no movements required and otherwise nothing happening. He/she must, however, remain fully alert and ready in all respects to cope instantly with any emergency or failure. To this end the DPO should not spend more than 2 h on the console, a common arrangement being that the two duty DPOs switch roles at hourly intervals.

8.2.9 Thruster assisted mooring

8.2.9.1 Function

Closely related to the technique of dynamic positioning is the subject of thruster assisted mooring. This finds application in a number of vessel types utilizing spread mooring, e.g. semisubmersible drilling rigs, crane barges, production vessels.

The function of a thruster assisted mooring (TAM) or position mooring (PM) system (*Figure 8.42*) is automatically to monitor and control the mooring spread wire tensions and lengths so as to maintain the required vessel position and heading. Additionally, active control is applied to the vessel's thrusters to reduce vessel oscillations and mooring line tensions.

A TAM system will utilize many of the elements previously discussed in connection with dynamic positioning: position reference is supplied from hydro-acoustic systems and taut wire systems, heading reference from gyro compasses, vessel attitude from vertical reference sensors. Suitable interface units are provided between the central processors and the various elements such as power supply plant, thrusters, mooring winches, sensors, etc.

At its simplest, a TAM system may be used to monitor and control mooring line tensions and lengths in order to maintain vessel position. VDU displays will give information upon anchor spread configuration, line tension history, position reference status. The main function of the system, however, is to allow the vessel's thrusters to share some of the load of vessel positioning, such that mooring line tensions remain at the values obtaining during calm weather conditions. Using thrusters in this way improves the quality of position keeping by removing low frequency oscillations about the set point position. Displayed data will include history of position and heading deviations, thruster configuration (force and azimuth of applied thrust) thruster history, anchor line catenary and fairlead angle data, power generation and distribution data, vessel motion data (roll, pitch and heave).

A range of warnings and alarms are provided, covering such variables as line tensions, vessels position and heading, power available. Simulation facilities allow the operator to predict the consequences of a hypothetical line break or thruster failure or severe weather conditions. A shift in vessel position may be achieved by entering data for the new set point position and/or heading; the new mooring winch configuration is calculated and displayed. The shift may then take place, assistance being provided by the thrusters.

Other facilities may include monitoring of various environmental elements such as wave height (from a waverider buoy) and current direction and strength. A drilling rig may have a floating set point mode in which the position of the rig may be automatically varied in order to maintain a 0° riser angle.

8.2.10 DP vessels and their function

DP vessels are built to fulfil a variety of functions and tasks. Some vessels are conversions or retrofits, usually associated with a change in vessel function. Most DP installations, however, are designed with that application in mind. In this section we examine typical DP applications, with examples of vessels of each type detailed.

8.2.10.1 Dive support – DSV Stena Seaspread

Dive support is perhaps the most prevalent utilization of DP. A large number of dive support vessels are in service worldwide and many of them are DP capable. Most are capable of being positioned by a four-anchor spread mooring, but their DP capability usually renders the anchor facility redundant. A DP DSV frequently carries out subsea tasks close in to fixed platforms. Saturation diving on a shift system allows underwater operations to proceed continuously for weeks at a time.

Stena Seaspread is an example of this type of vessel (*Figure 8.43*). Built in 1980 it is one of a class of four of a common design. In recent years this vessel has been modified to fulfil the requirements of Ministry of Defence charter, the following description refers to the vessel prior to these modifications.

Leading dimensions are as follows: length OA, 111 m, moulded breadth 20.5 m, maximum draught 6.70 m, gross tonnage approximately 7000 tonnes, maximum deadweight approximately 4670 tonnes and service speed 16 knots.

The vessel is designed as a highly capable, versatile offshore support unit. The diving spread consists of a 12 man saturation complex rated to 450 m depth. A diving

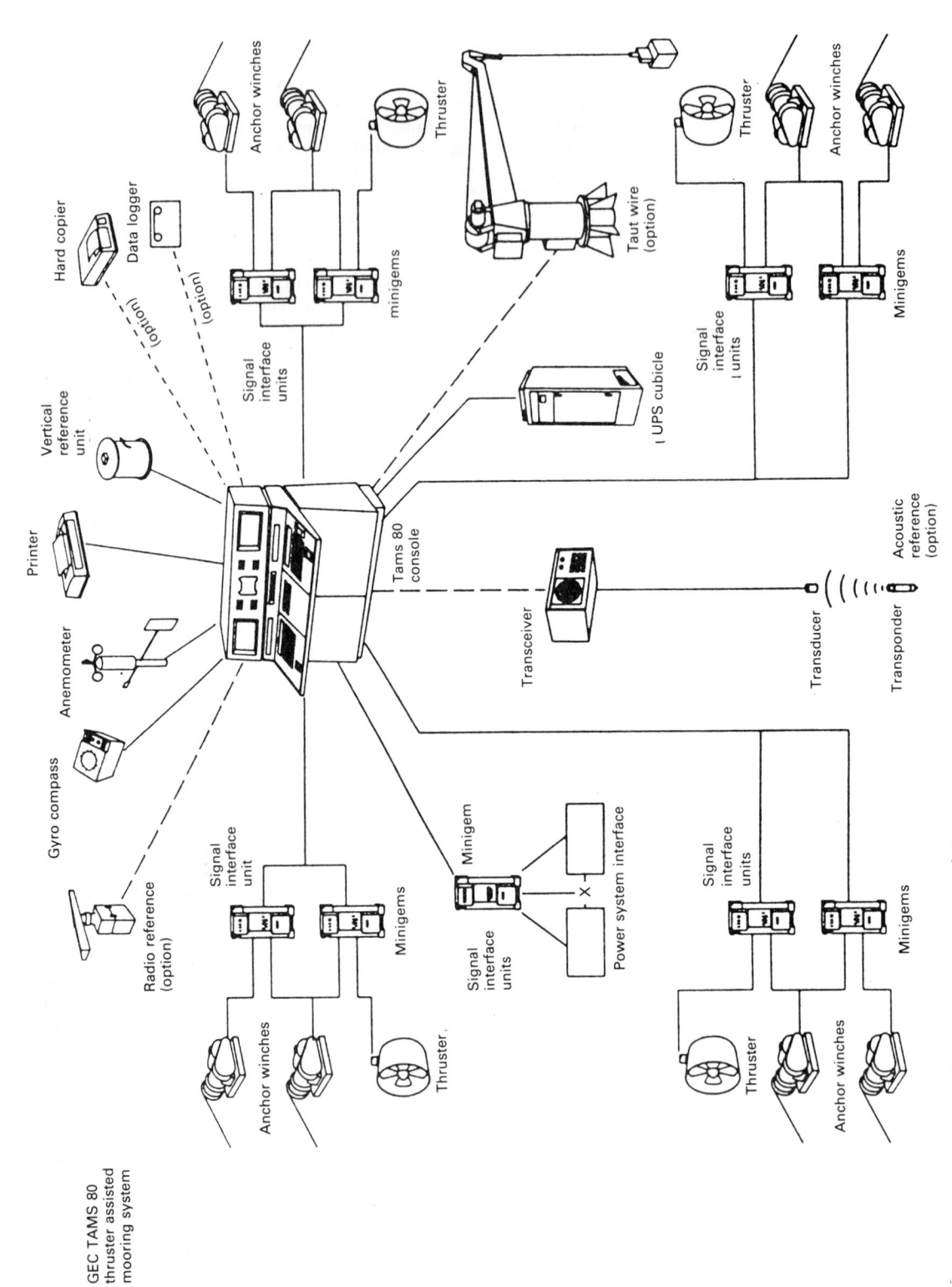

Figure 8.42 GEC TAMS 80 thruster-assisted mooring system. (Courtesy of GEC Projects Ltd)

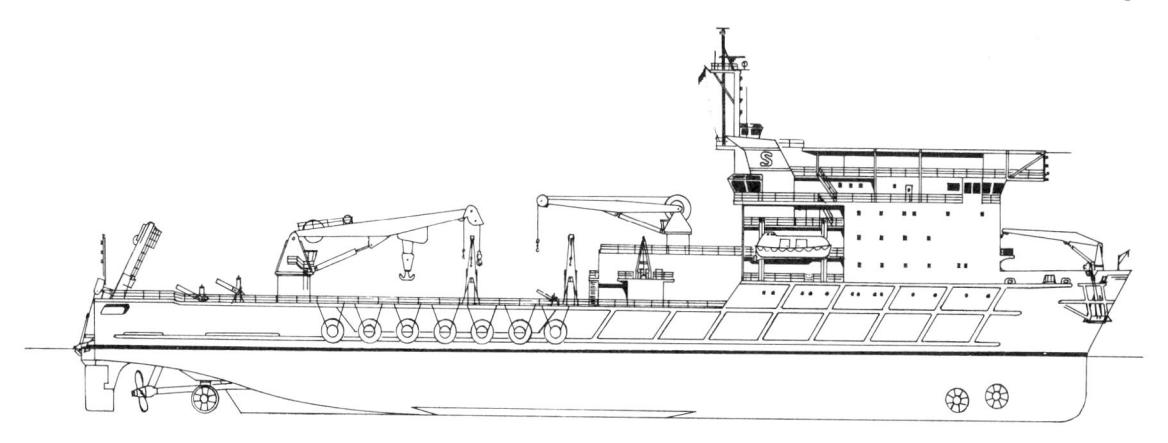

Figure 8.43 DSV Stena Seaspread

bell works through a central moonpool and is fully heave compensated; the bell can accommodate three divers. Bell crosshaul facilities are provided. The diving operations are controlled from the dive control room adjacent to the moonpool. Here, the dive superintendent takes charge of operations with full communications with the divers and other parts of the vessel. A separate saturation chamber control room is provided where life support for the divers in the saturation chambers is maintained. The saturation chambers can communicate directly with a hyperbaric lifeboat to provide escape facilities to the divers under pressure in the chambers. Gas bottles totalling about 13 000 m^3 provide the diver's helium–oxygen breathing mix. For diving at depths up to 50 m an air diving station is provided on deck with a three man wet bell deployed over the side of the ship. The vessel is equipped to operate a variety of remotely operated submersible vehicles or sea bed crawler vehicles.

Other features of the design of the vessel include deck crane capacity of 100 tonnes, a service crane of 15 tonnes capacity, a 100 tonne stern A-frame and pipe davits of 120 tonnes capacity. Open deck space of about 58 m length is available with workshop space adjacent. Firefighting capability is provided by four monitors, each of 2300 m^3/hr. A fixed water spray system protects exposed areas of the vessel against heat. A four point mooring system is installed, each consisting of a 5 tonne Delta Flipper anchor with 1500 m wire cable.

A helideck is fitted atop the accommodation block, capable of taking helicopters up to S61 type. Accommodation is provided for a crew of 24 and up to 73 supernumeries. Domestic provision includes a hospital, conference room, cafeteria, cinema and gymnasium

The dynamic positioning system consists of an Albatross ADP 503 fully redundant system. Three computers and dual uninterruptible power supplies are provided. Position reference is obtained from an Artemis microwave system, a Simrad hydro–acoustic system and two vertical taut wires. The DP system is fitted in the after bridge together with ballast control, communications and other operational systems. Manoeuvrability is provided from a main propeller, controllable pitch, powered by electric motors totalling about 6000 HP. A single balanced rudder is installed abaft the propeller but is kept amidships when in DP. Two 1500 HP azimuth thrusters are fitted aft and two 1500 HP tunnel thrusters are fitted forward.

The vessel is diesel–electric, with power generated by five NOHAB diesel alternators, each of 3600 HP, 6000 V/60 Hz on three phase. Main consumers use 6 kV while smaller equipment is supplied at 440 or 220 V. High and low voltage switchboards are placed in separate watertight compartments and a power management system maintains load sharing between generators, while autostart facilities ensure an optimum number of generators online to meet demand.

Since construction, the vessel has spent most of its time in the North Sea, mainly servicing the Thistle field. During the 1982 Falklands conflict it was chartered by the Ministry of Defence and refitted as a forward repair ship providing repair and maintenance services for about 40 vessels.

Initially working in South Georgia it later moved to the total exclusion zone east of the Falklands and, after the Argentinian surrender, to San Carlos Bay. During the conflict it was joined by its sister ship, the *Stena Inspector*. This latter vessel was subsequently sold to the Ministry of Defence to become the RFA *Diligence* continuing its forward repair ship role in the Falkland Islands. More recently, *Diligence* has been deployed to the Arabian Gulf to provide support facilities to the RN mine countermeasures force, and *Stena Seaspread* was again chartered by the Ministry of Defence to replace *Diligence* in the Falklands. Other vessels to this design are the *Bar Protector* and the *Stena Constructor*.

8.2.10.2 Semisubmersible drilling vessel – Ocean Alliance

Offshore drilling rigs can take a number of forms. In shallow water a jack up rig is used. In deeper water, up to about 500 m, a drilling vessel may be positioned using a spread mooring arrangement. In deeper water still, up to about 1000 m, a thruster assisted mooring system may be used. Dynamic positioning must be used in deeper water than this. Drill ships may be monohull or semisubmersible in configuration.

Ocean Alliance (*Figure 8.44*) was finally delivered to Ben–Odeco in 1988 after a long period of construction at the troubled Scott–Lithgow yard. It is a semisubmersible drilling unit of some 46 000 tonnes operating displacement. Its overall length is approximately 120 m, and its breadth 70 m. It is of twin hull, eight column semisubmersible configuration. Self propelled, it has a transit speed of 12 knots on a draught of 8.8 m. Operating draught is approximately 25 m.

Designed for deep water drilling in severe environments, it can operate in water depths of up to 1800 m. Positioning and propulsion is provided by eight azimuth thrusters of 4000 HP each. The dynamic positioning is provided by an Albatross ADP 503 Mk II system, fully redundant with three computers, with position reference provided by two separate hydro–acoustic systems. A separate system, an ADP 311, provides further DP backup against system failure. With any drilling system, the vessel is connected to the wellhead by the marine riser, containing the drill string. It is essential that the riser be maintained in a vertical position when drilling. If a tide is running, the riser will bow downtide, such that the riser makes an angle with the wellhead at the wellhead ball joint. In deep water a drill ship may position herself to maintain the ball joint angle within set limits. An inclinometer transponder may be located upon the lower riser such that this angle can be monitored. The DP system uses this and other transponders to provide position reference, also the ball joint angle input is used to calculate the optimum position for the vessel in order to maintain a vertical lower riser. Thus the positioning of the vessel is based upon this floating set point. A watch circle is shown upon the DP display, usually set at a radius equal to 3% of the water depth and centred upon the floating set point.

In water depth of up to 900 m the positioning may be achieved by DP alone, or by thruster-assisted mooring. The vessel is provided with eight 20 tonne Moorfast anchors, each provided with 1600 m of 83 mm stud link chain. In water of less than 500 m the positioning may be achieved by the anchor spread alone. The DP system is intended to achieve station keeping with one generator and one thruster in reserve in wind speeds of up to 65 knots, significant wave heights of up to 8 m (maximum 15 m) and a 2 knot current. Maximum capability for station keeping is intended to be a wind speed of 70 knots, gusting 87, with a current of 2 knots and a maximum wave height of 30 m.

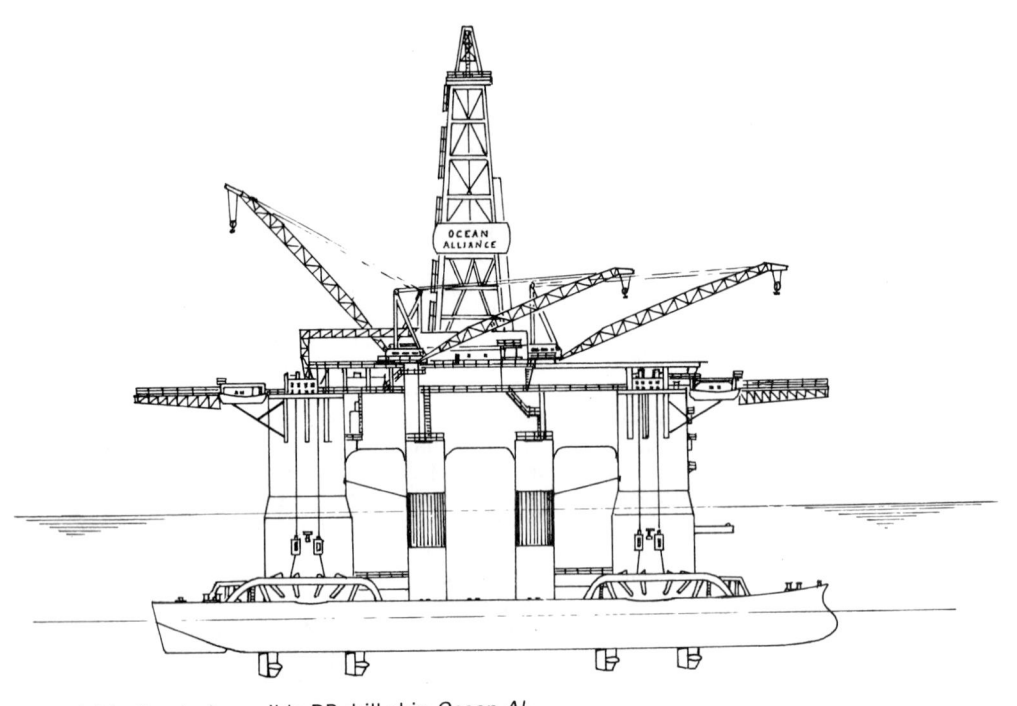

Figure 8.44 Semisubmersible DP drill ship *Ocean Alliance*

Commensurate with its function as a deep water drilling unit, *Ocean Alliance* has a large storage capacity for drill pipe, liquid mud, drill water and other consumables. Its total capacity is a 6000 tonne variable load. Power is produced by six Pielstick diesel alternators producing about 40 000 HP. Three 59 tonne cranes are located on deck for materials handling on deck and working supply boats. Sophisticated BOP control is provided with test facilities. A data acquisition system constantly monitors the condition of all major equipment, operation of systems and drilling parameters of the vessel. Accommodation is provided for 110 persons.

8.2.10.3 *Heavy lift crane vessels – McDermott DB50 and DB102*

Many crane vessels and barges are at work throughout the world. Most are engaged in oil and gas exploration related work. Some are conversions, others purpose built. Some are monohull types, others semisubmersible. Most are self propelled with positioning obtained by spread mooring, while a few are DP capable. In this latter category are the derrick barges DB50 and DB102, both operated by McDermott.

Derrick Barge 50 (*Figure 8.45*) entered service in 1988 after a troubled construction period. Ordered by ITM (Offshore) Ltd the vessel was originally named the ITM *Challenger*. Construction took place at the North Sands yard of North East Shipbuilders in Sunderland. ITM went into receivership in 1986, and the vessel was sold by the builders to Lombard Initial Leasing, and leased by McDermott on a long term basis.

DB50 is a monohull crane vessel of some 151 m length overall, 46 m breadth and a gross tonnage of some 30 000. It is equipped with a crane of lifting capacity 4000 tonnes. Its intended operations include offshore construction, jacket and topside installations, subsea installations, pile driving and pipe laying.

Positioning of the vessel may be achieved by an eight anchor spread mooring arrangement; alternatively the anchor spread may be complemented by the thrusters, or positioning can be by DP alone. The system is a GEC combined DP and TAMS 80. Four azimuth thrusters provide manoeuvrability, two located forward and two aft. Mooring arrangements consist of 8–12 tonne flipper Delta anchors each provided with 2350 m of 74 mm wire rope. The eight single drum winches may be controlled locally or from a console in the wheelhouse. Control is from two GEM80 microprocessors. Position reference for the DP/TAMS is provided by two taut wires, an Artemis system and a Simrad hydroacoustic system.

Power is provided from five Allen S37 diesel alternators each developing 3800 HP at 6.6 kV. Thrusters, crane machinery, firefighting pumps and motion com-

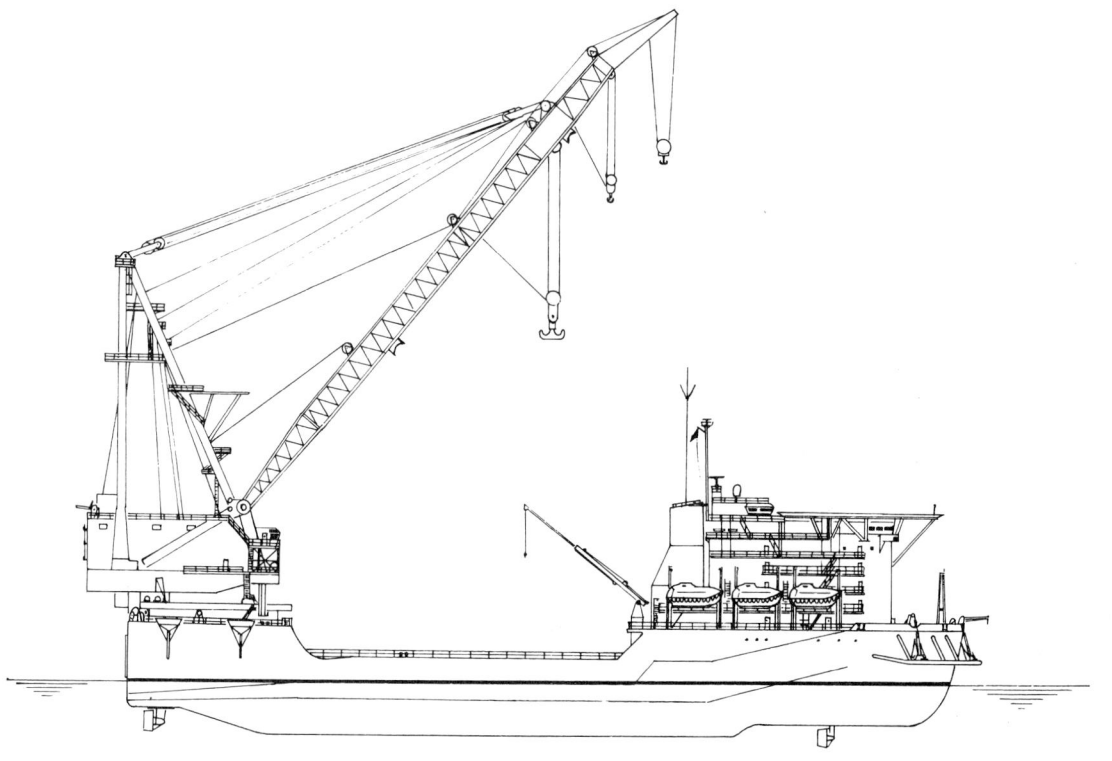

Figure 8.45 *Derrick Barge 50*

pensation compressors are all supplied at 6.6 kV, while auxiliary machinery is supplied at 440 V through 6.6 kV/440 V transformers.

The crane is able to fully revolve with a load of 3200 tonnes on the hook, or fixed and tied back can lift 4000 tonnes at an outreach of 37 m. A maximum test lift of 4400 tonnes was carried out on trials.

The vessel is provided with a motion suppression and heel compensation system which is able to reduce both wave induced motions and crane induced heeling. Eight motion suppression tanks (four on each side) are fitted, open to the sea at the bottom. Water levels in the tanks can be regulated by air supply and exhaust, allowing the GM of the vessel to be altered and rolling to be minimized. Heeling moments caused by the crane are compensated for by pressurizing the tanks on the side toward which the vessel is heeled, depressing water levels on that side and providing a compensating moment. A comprehensive computer system provides information upon stability, weight distribution, vessel motions and environmental data during crane operations.

Derrick Barge 102 (*Figure 8.46*) is a semisubmersible crane vessel, self propelled, carrying two cranes of maximum capacity 6000 tonnes each. Entering service in 1986, its first tasks included the Ekofisk subsidence project. To compensate for a sinking sea bed a number of platform topsides and bridges had to be raised by about 6 m; DB102 was utilized to assist in this task.

The vessel is 202 m in length by 97 m broad. Construction is of twin hull, eight column type with a hull depth of 49.5 m to main deck level. Transit speed is eight knots on a draught of 13.2 m, while operational draught is between 27 and 32 m. Six 3000 HP azimuth thrusters provide main propulsion, mooring assist and dynamic positioning. Two DP systems are installed, an Albatross ADP 503 Mk II, also an ADP311. Position reference includes taut wire, HPR and Artemis. Also installed is a position mooring system providing thruster assist for her twelve 20 tonne anchors. The anchors are deployed upon wire cables but provision is made for handling chain cable if required.

The two cranes are each able to fully revolve with a 6000 tonne load at 43 m radius; in tandem a lift of 12 000 tonnes is possible. Auxiliary hooks can lift 900 tonnes at 79 m radius. A recent task for DB102 was the recovery of the Piper Alpha accommodation modules.

Deck capacity load is 12 000 tonnes and a crawler service crane of 158 tonnes capacity is provided. Power is provided from six diesel alternators producing a total of about 37 000 HP. Accommodation is provided to a high standard for up to 750 persons and the helideck can take the largest types of commercial helicopters currently in use.

Everything about DB102 is larger than life. A short description of this nature does not do it justice. However its lifting capacity is exceeded by the Italian vessel *Microperi 7000*. Of similar dimensions and design and provided with similar DP and PM systems, this vessel is fitted with twin cranes of 7000 tonnes capacity each, providing a tandem lift capacity of 14 000 tonnes.

8.2.10.4 Pipe laying – Lorelay

Subsea pipelaying is a common function for DP fitted vessels. In many cases the vessel used is a DSV specially mobilized for a pipe laying task. A different approach to the pipe laying problem is provided by

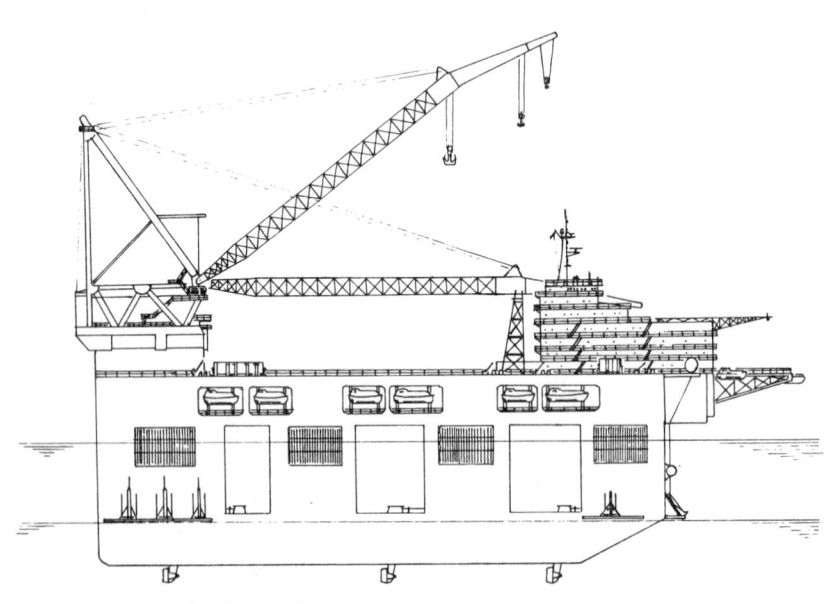

Figure 8.46 *Derrick Barge 102*

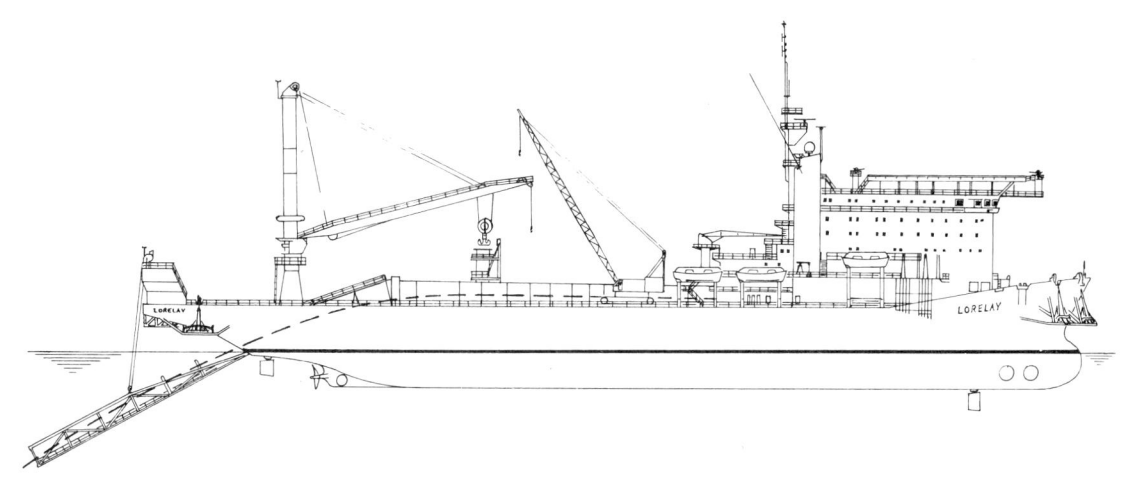

Figure 8.47 Pipe lay vessel *Lorelay*

Lorelay (*Figure 8.47*). The vessel has the ability to load barges through an open stern into a floodable hold. This provides particular advantages regarding the speed at which pipe loading operations can be conducted. Pipe may be preloaded into pipe carrying barges, which may be floated into the vessel's hold and docked. Its large cargo capacity allows about 4000 tonnes of pipe to be carried.

The vessel is of a conventional monohull configuration, 182 m overall length by 26 m beam. Its operating draught is 6.8 m and is able to cruise at 16 knots. Manoeuvrability is provided from a main propeller, two azimuth thrusters aft, one tunnel thruster after and two tunnel thrusters forward. Power is provided from four main diesels providing a total of about 19 200 kW. The vessel was built in Germany in 1973 and was converted to its present form in 1986 in Holland. The conversion to a pipe layer required considerable changes. DP and mooring facilities were fitted together with cranage and helideck. Accommodation is provided for a total of 186 men.

The mooring system comprises four bow and two stern Delta Flipper anchors of 10 tons each. Dynamic positioning provision consists of an Albatross ADP 503 Mk II system, fully redundant with three computers. A second system, an ADP 311 is provided in a supervisory capacity. Cranage is provided, consisting of a motion compensated 300 tonne crane able to work to 400 m depth, and a 120 ton crawler crane. Pipe laying capability is from 100 mm to 600 mm diameter pipe, to a water depth of 360 m. Six welding and testing stations are provided on deck. The pipe lay and trenching crawler vehicle Digging Donald operates upon the sea bed, remotely controlled from the vessel. This unit is able to trench pipe into the sea bed to a depth of 2 m.

The vessel is operated by Allseas Marine Contractors SA of Switzerland and is registered in Panama.

8.2.10.5 Multirole field support – Iolair

Semisubmersible vessels of this type provide a range of support facilities to an oil and gas field. Vessels such as Iolair are able to provide a range of construction, maintenance and inspection facilities, with attendant diving, cranage and workshop capability. Another vital function of the vessel is emergency support, and to this end it is capable of carrying out firefighting, platform evacuation and well-kill operations.

Iolair (*Figure 8.48*) was built at the Scott–Lithgow yard on the Clyde and delivered in 1982. Operated by BP, it is intended to provide services in the Forties field. It is a six column semisubmersible of 102 m length, 51 m breadth with a maximum displacement of around 20 000 tonnes. It flies the British flag, being registered in Dundee.

Four point anchor mooring is provided but its main positioning capability derives from DP. It is fitted with an Albatross 503 Mk 1 system, with position reference provided from HPR, taut wire, Syledis, Artemis and Miniranger. Propulsion is provided by two CP propellers, one at the aft end of each pontoon, together with four tunnel thrusters. A transit speed of 12 knots is possible on a draught of 6.9 m, while operational draught is 15.2 m. Power is provided from two segregated engine rooms, each housing three diesel generators totalling 20 400 kW of power. All propulsion motors are electric. Normal operation is possible in wind speeds of up to 70 knots and wave heights of up to 15 m.

Diving capability is provided by a six man saturation system rated to 300 m with dual chambers for split level diving. A three man diving bell is deployed through one of the pontoon columns, with a 15 m crosshaul capability. Surface air diving takes place from a wet bell deployed from the upper deck. An ROV can be deployed when required. Other facilities on board include a comprehensive hospital, workshops, cranes;

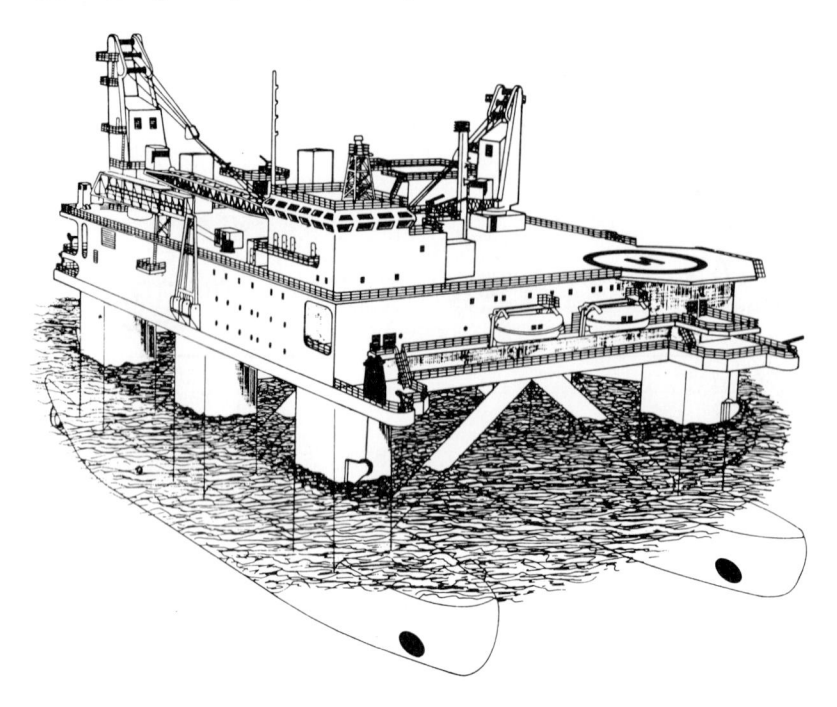

Figure 8.48 Multirole support vessel *Iolair*

40 and 100 tonne capacity, accommodation for 62 crew and 162 supernumaries. It carries its own small helicopter and can handle helicopters up to S61 type. A motion compensated active gangway allows personnel evacuation from adjacent platforms in emergency. Firefighting capability is provided from 14 fire monitors and two water cannon. Total firefighting capacity is 50 000 gal/min from four pumps. Self drenching can also be applied to keep vessel steelwork cool when engaged in firefighting. Recent activities include assistance in rescue operations during the Piper Alpha platform emergency and shortly after, the Ocean Odyssey blowout.

8.2.10.6 Further applications

The foregoing examples give a cross-section of DP applications. Described here briefly are a few further DP vessels and applications.

A number of hydrographic and geophysical survey vessels are provided with a DP capability. The vessels are usually small (40–80 m) and are often conversions from freezer trawlers or other vessels types. DP systems are usually non-redundant, often fitted with a track follow capability. Examples of this type are *Master Surveyor* and *Southern Surveyor* (*Figure 8.49*).

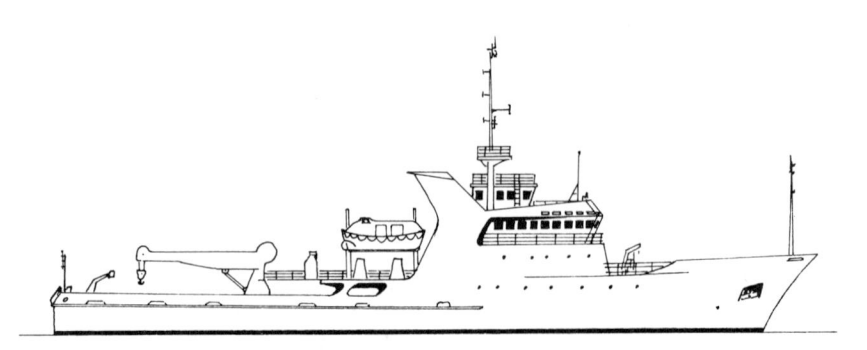

Figure 8.49 DP survey vessel *Southern Surveyor*. (Courtesy of Simrad Albatross Ltd)

Well stimulation is a task that may require DP station keeping characteristics. Several vessels have been converted to this role, such as *Skandi Fjord* and *Vestfonn* while others, such as *Stena Seawell* and *Big Orange XVIII* are purpose built.

Stone dumping is another task suited to DP techniques. Vessels of this type are able to carry out pipe or cable bury operations. Of bulk carrier design, a controlled fall pipe is supplied with stone from a conveyor discharge system. DP with track follow capability is fitted. Vessels of this type include *Rocky Giant*, *Trollness* (*Figure 8.50*) and *Seaway Sandpiper*.

A DP capability is often provided in offshore-loading shuttle tankers, used for bulk crude oil transport from deep water oilfields. The shuttle tanker is provided with tunnel thrusters fore and aft, in addition to her main propeller. A DP system is fitted in a forward control room, with an Artemis reference system linked to a beacon on the mooring buoy. Once the vessel is coupled to the buoy the DP system maintains a weathervane heading with the vessel kept a fixed distance from the buoy. The hawser may be left slack, allowing offtake operations to take place in more severe weather conditions then possible without DP assistance. Examples of this type of vessel are *Norissia* and *Esso Fife* (*Figure 8.51*) both used for offshore loading on the Brent spar buoy.

References

Guidelines for the Specification and Operation of Dynamically Positioned Diving Support Vessels. Norwegian Petroleum Directorate/Department of Energy (1983)
The Safety Auditing of DP Vessels. S. W. Barber, Institute of Marine Engineers (1984)
DP Guidance Notes and Checklist. UKOOA (1985)

Acknowledgements

The author thanks the following for their assistance:

Mr I Buchanan of Simrad Albatross Limited
Mr J Bannigan of GEC Electrical Projects Ltd
Mr A Marriott
Mr N Nail
Mr I Cardno
Mr S Woodward
Mr J Gilburt
Mrs A Cleverly
Miss L Andrews

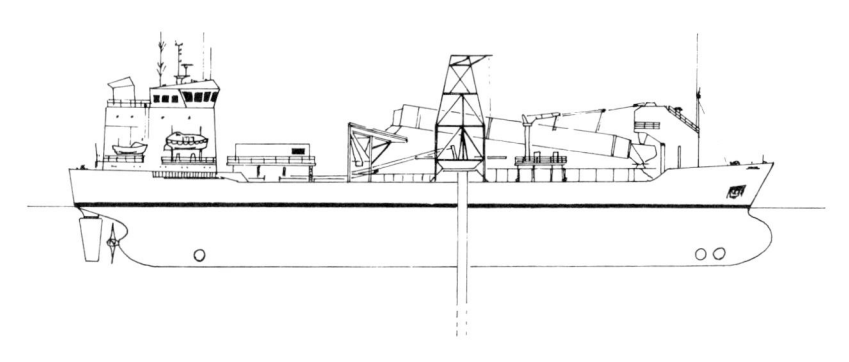

Figure 8.50 Stone dumping vessel *Trollnes*

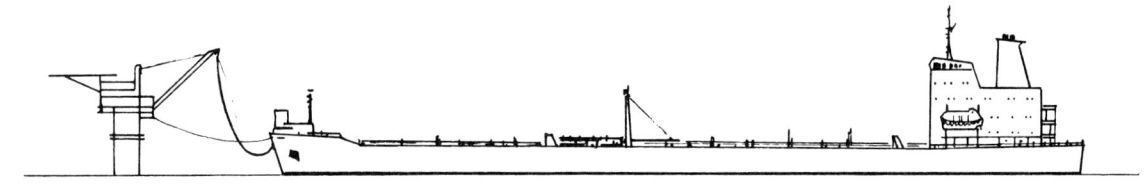

Figure 8.51 Offshore loading tanker *Esso Fife*

9 Marine Salvage

James Kearon

Contents

9.1 Introduction
    9.1.1 Examples of salvage
    9.1.2 Salvage remuneration
    9.1.3 Insurance
    9.1.4 Freight
    9.1.5 Out of pocket expenses
    9.1.6 Salvage award

9.2 Contractual salvage
    9.2.1 Types of contract
    9.2.2 Standard salvage contracts
    9.2.3 Arbitrator
    9.2.4 Examples of other standard forms of salvage agreement

9.3 Salvage related organizations
    9.3.1 The Salvage Association
    9.3.2 International Salvage Union (ISU)
    9.3.3 International Maritime Organisation (IMO)
    9.3.4 Protection and indemnity associations

9.4 Types of salvage
    9.4.1 Parties to a salvage operation

9.5 Salvage operations
    9.5.1 Cargo discharge
    9.5.2 Free surface
    9.5.3 Flooding

9.6 Salvage methods and risks
    9.6.1 Cofferdams
    9.6.2 Fire
    9.6.3 Pumps
    9.6.4 Inert gas
    9.6.5 Hot tap machines
    9.6.6 Side scan sonar
    9.6.7 Underwater excavation
    9.6.8 Air lift
    9.6.9 Beach gear
    9.6.10 Carpenter stopper
    9.6.11 Air bags
    9.6.12 Water damage protection
    9.6.13 Salvage tugs
    9.6.14 Ground reaction
    9.6.15 Bollard pull

9.7 Towing
    9.7.1 Towing points
    9.7.2 Bimco Towcon
    9.7.3 United Kingdom Standard Towing Conditions (revised 1986)

9.8 Pollution control
    9.8.1 Oil skimmers
    9.8.2 Dispersant spraying equipment

References

9.1 Introduction

Kennedy's *Civil Salvage*, an authorative publication on salvage, defines a salvage service for practical purposes as:

> A service which saves or helps to save a recognized subject of salvage when in danger if the rendering of such service is voluntary in the sense of being solely attributable neither to pre-existing contractual or official duty owed to the owner of the salved property, nor to the interest of self preservation.

This is a very legalistic definition and the International Maritime Organisation has since defined a salvage operation as any act or activity undertaken to assist a vessel or any other property in danger in navigable waters or any waters whatsoever.

The salvor's right to reward for salving maritime property has long been enforced in English Law in the Admiralty Court as part of Maritime Law, but has more recently been extended by statute to cover the salvage of lives and of aircraft.

In order to advance a claim for marine salvage a number of conditions must be satisfied and are exhaustively discussed elsewhere, they include the following:

1. The salvage services must have been rendered on the high seas, or at a place within statutory limits which can vary from jurisdiction to jurisdiction.
2. The property saved must be a recognized subject of salvage.
3. The claimant must fall within one of the recognized categories of salvors.
4. The services must have been rendered voluntarily.
5. There must have been a danger from which the salved property or lives were saved.
6. The services must have achieved some success.
7. The claimant must not have been at fault.
8. With the exception of life salvage, the services must not have been rendered against the owner's will.

Central to the salvor's claim for salvage is the danger to which the property is subject. The degree of danger is clearly in the eye of the beholder but must be real; it is considered that the moment of danger need not be imminent or absolute. The vessel may be in a position where it could be exposed to danger at some future time. A suitable test would appear to be that defined by Kennedy:

> so much a just cause of apprehension that, in order to escape out of it or to avoid it no reasonably prudent and skilful seaman in charge of the venture would refuse the salvor's help if it were offered to him upon the condition of his paying for it the salvor's reward.

9.1.1 Examples of salvage

Examples of successful salvage services are numerous but include towing an immobilized vessel to a suitable place of safety, extinguishing fires, recovering cargo from sunken ships or those otherwise in peril and refloating a stranded vessel. It is also noteworthy that the salvage services are rendered to the salved property and not, except incidentally, to its owners. Indeed, it does not matter in principle whether they appear or not, since the property can subsequently be sold without reference to its owners in the event that the salvage services are not paid for by the contracting parties.

9.1.2 Salvage remuneration

In English Law there is no absolute rule or fixed scale of remuneration unless fixed by pre-agreement. The amount of the salvor's reward is at the discretion of the court or arbitration as the case may be and must come from the fund available from the values of ship, cargo, bunkers and freight at risk.

Before arriving at suitable remuneration, the court must assess the material circumstances in the salvage services. These are classified by Kennedy as follows:

1. As regards the salved property: the degree of danger, if any, to human life, the degree of danger, if any, to property and the value of the property as salved.
2. As regards the salvors: the degree of danger, if any, to human life, the salvor's classification, skill and conduct and the degree of danger, if any, to the property employed in the salvage service, its value and the cost of its transportation to the site.
3. Also, the time occupied, the work done in the performance of the salvage service, the responsibilities incurred in the performance of the salvage service such as risk to the salvors insurance or third parties, and loss or expense incurred in the performance of the salvage services, such as detention, loss of profits, repair of damage and fuel consumed.

Where all or many of these elements are found to exist or some of them are found to exist to a high degree, a large award is usually given; where a few of them are found or they are present only in a low degree the salvage remuneration award is comparatively small.

As a general rule, the Courts or arbitrator will not normally award the salvor more than half the value of the salved property, whether it is derelict or not. In those cases where the award has exceeded half then it is likely that either the salved property was derelict or abandoned, the salved values were low, the salvors' property incurred expenses were high in proportion to the salved values or the services were especially meritorious.

The value of the salved property remains a contentious issue. It can be seen that the salved fund is central to the salvor's remuneration and clearly it is in the interest of owners of the salved property to put forward a low value and conversely the salvor will argue for higher values.

In English Law, the value of the salved property is considered to be the value at the termination of services, taken at the nearest safe port or redelivery. It is the responsibility of the owner of the salved property to provide evidence of the value, usually in the form of the opinion of sale and purchase brokers, or if sold then the sale agreement. The salved value is generally taken as the sound value minus certain deductions.

9.1.2.1 Sound value

The sound value of the ship is, unless sold following termination of salvage services, arrived at by estimate and may be subjective to the owner. For example, the sound value of a ship can vary alarmingly and its market value could be collapsing as the salvage work proceeds. On the other hand, the value may be enhanced by the owner's contractual arrangements at the time. The converse would prevail if an unfavourable charter were taken into account. There has been much debate within the legal profession as to whether the salvor should benefit or suffer from this uplift or reduction in ship valuation. However, In English Law, the courts and arbitrators normally take into account the market value of the ship prevailing at the time when the salvage operation is completed.

9.1.2.2 Deductions

From the sound value deductions are made, based upon the cost of bringing the casualty back to its precasualty condition. Typical deductions, supported by appropriate vouchers, may be as follows:

1. Crew's wages during the salvage operation and the repair period; crew's bonus, overtime, maintenance and victualling; disbursements at port of refuge, including agents' disbursements.
2. Cost of temporary repairs, including any cargo handling costs to facilitate these to enable the vessel to proceed on passage or to a repair port; cost of permanent repairs, including any cargo handling costs to facilitate these; survey fees; superintendence; owners' supplied items.
3. Engine room consumption during the salvage operation and repairs period; telephone, telex, cable and wireless message expenses, arranging salvage assistance, during salvage operation and arranging for repairs; owners' miscellaneous expenses.
4. Special hull insurance premiums taken out on owners' behalf during the period of salvage operation and while under tow to repair port; pilotage and towage costs on leaving the port of refuge and while in the repair port; cost of towage to repair port; cost of items of ship's equipment used during the salvage/repair operations and replaced by the owners.

9.1.2.3 Scrap value

If the cost of repairs is excessive then the salved value (if a repaired vessel is considered) may be less than the scrap value of the ship, in which case the scrap value at the nearest scrapping port or 'as is, where is' should be obtained and from this the cost of getting the casualty to the scrapping port (i.e. towage, ship husbandry for the voyage, etc.) is deducted. Providing the shipowners can show that the scrap value exceeds the salved value, for example by producing Salvage Association or other acceptable reports or producing vouchers, then the courts may decide that the salved value is the scrap value. For the purpose of assessing the award, the salvors are entitled to have taken into account the higher of the scrap value or salved values. Bunkers, stores and freight (if at risk onboard ship) should be added to the salved ship value.

9.1.2.4 Cargo values

Cargo values, for the purposes of salvage litigation, can be more complicated than the ship value. However, in arbitrations held in England the cost-insurance and freight (c.i.f.) values are generally accepted but may be open to argument during litigation.

If the casualty is salved and able to continue the voyage then the value of the consignment is taken as evidenced by the property invoice. If, on the other hand, the property is salved but the contract of carriage is frustrated then the cargo value is taken 'as is, where is'. Alternatively, the invoice value may be agreed, less the cost of transhipment to the port of destination.

Deductions are made, as in the case of ship values, from the sound value to obtain the salved cargo value. These deductions may include damage to the cargo and the expenses involved in unloading, storage, sale, brokerage, etc.

9.1.3 Insurance

The insurance is proved by the premium paid on the consignment.

9.1.4 Freight

The term freight is used to describe monies paid or to be paid for the transportation of cargo between prescribed destinations. The freight is the responsibility of the person who does not hold the freight, for instance if the freight is prepaid then it is at risk of the cargo interests.

If after the salvage operation the vessel is able to continue the voyage then all freight is salved and all the freight contributes to the salved fund. However, if the cargo is not carried to the destination under the contract of carriage no freight is earned unless the ship owner is prevented by the cargo owner from transhipping the consignment or the cargo owner prefers to take delivery of this consignment at a port of refuge.

9.1.5 Out of pocket expenses

Under English Law, provided the salved fund is sufficient, a salvor may expect to be compensated for out of pocket expenses, provided they are property hired by the salvor in the furthering of the salvage services and before the casualty has been placed in a position of safety and directly occasioned by the performance of the salvage services. Examples would include charter of aircraft to airlift the salvor's equipment to the casualty site, charter of local tugs and the cost of repairing damage to a salving vessel during the services. As with the salved property interests, the burden of proof of expenses lies with the salvor.

The salvor should not, however, unnecessarily hazard his personnel, vessel or equipment. Similarly, if the damage occurs while the salving vessel is undertaking a task necessary for the success of the operation but the damage is in fact due to the negligence of the salvor's personnel, the salvor is unlikely to be fully compensated for the cost of repairs. When putting forward these expenses the salved property interests are entitled to ask the salvors to submit strict proof as to the expenses. Contrary to what had been considered the case for many years, salvors can be sued for damage caused by the negligence of their personnel. In the event that the negligent act or acts take place inside the confines of the salvage vessels then these vessels' tonnages could be used for the limitation of liabilities.

9.1.6 Salvage award

The salvage award is generally made against salved properties in gross, each salved interest paying in proportion to its salved value against the total salved fund. The remuneration to owners, masters and crew of the salving vessels will include agreed losses and/or expenses made by the salvor during the services. If the court is requested to apportion the award between owners and crew any losses and expenses may be taken into account since clearly the salvage crew should not benefit from their employers' losses which are added into the gross award. The owners, masters and crew of all salving vessels may claim salvage and salved property interests often ask to be indemnified by the salvor from claims of masters and crews of vessels not on salvage articles.

To circumvent the foregoing problem, salvage clauses are incorporated into the crew articles of agreement of salvage tugs. The following is a typical example:

> The master and the members of the crew agree that it is a condition of this agreement that they shall not be entitled to claim Salvage Remuneration in respect of any work or services rendered during the period of Agreement and that the Owners of the vessel reserve to themselves the right to make such additional payments either to the Master or any member of the crew as they (the Owners) in their sole discretion consider the Master or any member of the crew merit.

9.2 Contractual salvage

A salvor may render services without a contract as a volunteer salvor (however unless the vessel is abandoned he must have the owner's consent to start and continue the salvage operation) and his remedy for remuneration would be through the courts after issuing a writ against the salved property. In certain areas of the world the judiciary may not have a history of salvage law and tend to be biased towards one side or the other. The expense of suing through the courts can be costly and slow with the outcome uncertain. In the modern situation, it is normal for an owner to be fully aware of the circumstances of the casualty and probably be in contact with potential salvors. Almost all salvage, therefore, is now carried out under contract between the parties concerned. Notwithstanding, there are jurisdictions where the courts may be particularly generous to salvors and hence in those areas it could be difficult for an owner to make a commercial contract.

Where salvors have doubts about the potential salved fund or indeed the sound value they sometimes ask for these aspects to be agreed in advance. If the services are performed under a salvage contract the parties may agree upon a sound valuation of the casualty at the time of making the agreement in an attempt to circumvent possible disputes in litigation. This valuation clause will then be inserted into the salvage contract and might, where the insured value is higher than the sound value, refer to the insured value as being the sound value for the purposes of arbitration. It is of course a matter of negotiation between the parties whether such a clause can be inserted and upon what basis the valuation will be made.

Whilst the ship owners or the authorized agents may contract upon whatever terms they wish, it does not follow that cargo interests will necessarily be bound by the ship value unless they also sign the subject salvage contract or have indicated their agreement separately. Clearly, they would only agree to an insured ship value provision which placed them in an advantageous position on their proportions of the award payable and would be unlikely to do so if the resulting total fund would be likely to make an appreciable upward difference to their part of award.

9.2.1 Types of contract

The salvors may, depending upon the circumstances, offer their services on a variety of terms such as daily hire, possibly with a bonus upon successful completion; a lump sum, either on the basis no cure–no pay or with stipulations; a standard form of salvage agreement or a combination of the above such as a lump sum based on a specific time table with a daily rate above this period. The degree of risk and the possibility of competitive tenders from other salvors are factors that the salvor will consider in arriving at the contractual financial offer.

It is unusual for tug masters at the casualty site to negotiate contractual terms as they invariably have only their principal's authority to offer their preferred form of salvage agreement.

In circumstances where the casualty is found to be old and of dubious value, or where the salvage operation is likely to be difficult or may result in the salvor incurring considerable expenses, the salvor may estimate that the residual salved fund will be insufficient to provide adequate salvage remuneration. In these circumstances, the salvor will likely only enter into a contract on the basis of a lump sum, daily hire (time and equipment) or insist that a minimum salved fund be agreed in the contract before mobilizing the equipment.

Drawing up a contract invariably involves the services of lawyers and protracted negotiations on the wording of non-standard clauses. Those closely involved in marine salvage will be aware that once a vessel suffers a casualty or is placed in a position of danger and requires salvage services its condition and the prospect of a successful salvage will rarely improve with time. It is not unknown for a ship to be lost or its condition to deteriorate seriously whilst contractual terms are negotiated.

Figures 9.1 and *9.2* taken on the same day, illustrate the possible rate of deterioration.

It follows that under such circumstances a ship or cargo owner whose property is in immediate peril will be forced to negotiate under duress. However, courts or arbitrators will be likely to throw out any unjust contractual clauses made under duress. On the other hand, it is often the case that in the hours (or days) following a casualty, information surrounding the plight of the ship may be limited and it may be difficult to assess just how much danger the ship or cargo is in, or indeed what form the salvage operation should take.

Since in the vast majority of salvage operations the salvor may only be paid out of the salved fund available, then it is in the salvor's interests to agree a suitable contract as soon as possible, mobilize all equipment and start work before the situation deteriorates. For these reasons, a large proportion of salvage operations are undertaken on a standard form of salvage agreement, a number of which have their origins in the last century.

Figure 9.1 Vessel stranded on rocks in heavy seas (courtesy of The Salvage Association)

Figure 9.2 Same vessel as in Figure 9.1, now broken up. Note that the bow and bridge are upside down (courtesy of The Salvage Association)

9.2.2 Standard salvage contracts

These are preprinted contract forms in which the salvors essentially undertake to make their best endeavours to salve the ship and cargo and make themselves responsible for the safety of both the ship and cargo in respect of their actions. The more widely used contracts are, in general, founded on the principle no cure–no pay and the remuneration for services rendered, if not fixed in advance, is by arbitration, court awards or mutual agreement.

The standard salvage contract can be readily agreed between the masters of the casualty and the salving vessel. It is not imperative that the form be signed by both parties prior to commencement of services, a witnessed radio signal or telex message will suffice. It is not uncommon for the contract to be agreed and signed during or following the services. Modern communication systems and particularly the widespread use of satellite telephone and telex connections enable the owner to be advised immediately following the casualty so that in many salvage operations the contract is agreed and signed directly between the salvage company and owner. The complexity of the casualty and the particular nature of the situation will however dictate how contractual terms are handled.

9.2.2.1 Lloyd's Standard Form of salvage agreement

This contract is widely known throughout the maritime world as Lloyd's Open Form and was first introduced in 1892. The contractual terms have been modified over the years and was most recently revised in 1980. The contract and the ensuing arbitration are governed by English Law.

As an indication of the extent of its use in salvage operations, a 10 year analysis of salvage operations carried out on standard salvage contracts by members of the International Salvage Union showed that some 80% were carried out under Lloyd's Standard Form conditions.

Lloyd's Standard Form is on the basis no cure–no pay and the salvors undertake to make their best endeavours to salve the ship and/or cargo, bunkers and stores and to deliver the salved property to a safe or agreed location for redelivery to owners. The payment to the salvors for the services is made from the value of the property salved. The Lloyd's form can be signed on behalf of the salvors/contractors on the one hand by any properly authorized agent or representative of the salvors. It is not essential that the form be signed on their behalf by their tug master, though it frequently is. It is however preferable for the form to be signed by the master of the casualty on behalf of salved property interests, rather than owners or agents.

Although Clause 17 of the form provides that cargo and freight and the respective owners are bound by the signature to Lloyd's form, cargo interests have on occasion sought to establish that, where the form was signed without concurrence of cargo interests, they were not bound. The International Marine Organisation draft salvage convention addresses this problem and gives the master in times of peril the right to act on behalf of cargo interests.

It is generally considered that the master of a vessel in real danger (or with a reasonable apprehension of real danger) would bind cargo interests as an agent of necessity if it would have been impossible, or at any rate impractical, for the master to have communicated with the cargo interests or their representatives, the action taken by the master of the casualty was necessary for the benefit of the cargo interests and the master of the casualty acted in a bona fide manner in the interests of the parties concerned.

It may obviously be time consuming and impractical for the salvors to confirm that the Lloyd's form was being signed with the concurrence of cargo interests, especially in the case of a general cargo or container ship with numerous consignees. However, in cases where the casualty is carrying a homogeneous cargo for a single consignee then the owner, master or salvor would be wise if possible (and provided the delays caused did not further imperil the cargo) to confirm, at any early stage, that cargo interests and/or charterers agree to the Lloyd's form.

9.2.2.2 Best endeavours

Whilst the salvor contracts to make the best endeavours, the salvor does not contract that the services will be successful; however, having entered into the agreement the salvor should not hold back or withdraw from the contract because it becomes apparent that the likely cost of the operation may exceed the anticipated award. Salvors are not expected to bankrupt themselves by the operation and if continuing efforts would entail costs which would greatly exceed the salved fund, then it may be reasonable for them to withdraw.

9.2.2.3 Safety net

In an endeavour to encourage salvors to render assistance to vessels which could cause pollution and whilst the Lloyd's form is still on the basis no cure–no pay, the 1980 form provides for a payment in the case of services to a laden, or partly laden tanker with a cargo of fuel oil, heavy oil, crude oil or lubricating oil where, without negligence on the part of the salvor the services are unsuccessful, the services are partially successful or the contractor is prevented from completing the services.

In such circumstances the salvor may be awarded, solely against the tanker owner, reasonably incurred expenses together with an increment of up to 15% of his expenses. The expenses are in addition to actual out of pocket expenses such as that incurred in hiring outside equipment and includes a fair rate for the salvor's own tugs, craft, personnel and other equipment used on the project. Negotiations are taking place to extend the scope of this clause, both in respect of the type of ship (so that vessels carrying toxic cargoes are included) and the size of the payment.

9.2.2.4 Redelivery

On completion of the salvage services and when the ship is safely afloat at the agreed location, it is customary for the master of the casualty to sign the salvor's release note, accepting redelivery and confirming satisfactory completion of the services.

Unfortunately, redelivery is often the most contentious aspect of the contract and requires good will on the part of both the owner and salvor. In recent years there has been a tendency for owners to demand extended services from salvors before accepting redelivery and in some instances ships have been delivered into dry dock before the owner could sign a release note. This is certainly not in the spirit of the contract and the owner ought to take redelivery when in a position to make normal commercial contracts for the care of the ship. Unfortunately when the salved value is low it is often in the best financial interests of the owner to delay redelivery or try to avoid it altogether. As an illustration of the problems that can sometimes occur an actual case is related.

A loaded vessel entered into a strange port to land an injured seaman. It anchored in a position such that when the tide went out it sat on rocks breaching almost all double bottom tanks and pipe tunnel. The engine room flooded by water leaking in from the tunnel.

A Lloyd's Standard Form of Salvage was agreed between the owner and a salvage company and work started right away. The vessel was refloated within days by using compressed air into the double bottom, expelling the water contained therein and by pumping out the engine room.

The water was found to have leaked into the engine room via a small hole which, being located behind some sea water pipes, was found to be inaccessible. Despite weeks of effort, it was not possible to stem the leak completely, but it was by the building of cement boxes reduced to almost a trickle which could be pumped out using a small pump for a few minutes daily. The owner refused to accept redelivery on the grounds that as the vessel needed constant pumping it was not safely afloat.

The salvor subsequent towed the vessel some 3000 miles where it was discharged and dry docked some 6 months after the casualty. It was only then that the leak could be stopped completely.

9.2.2.5 *Port of refuge difficulties*

Difficulties can arise where the authorities of a port of refuge named in the form do not permit the casualty to enter the port, or even into the area, particularly in the case of badly damaged ships or where the authority believes there to be a threat to the environment from pollution.

The Lloyd's Form 1980 attempts to address redelivery problems in Clause 2, where salved property interests undertake to cooperate fully with the salvors during the services, including obtaining entry to ports of refuge and to accept prompt redelivery. However, despite the provision of this clause the foregoing difficulties invariably remain.

9.2.2.6 *Security*

Following safe redelivery, the salvors notify the ship owners who notify the other salved property interests and Lloyd's of the amount of security, including costs, expenses and anticipated interest which is required. Prior to completion of the services, therefore, salvors must make an approximate assessment of the value of and danger to the salved property: such details as are available of the salvage services and the salvor's knowledge of the level of awards will pitch security requirements at what is considered to be the correct level.

The security is generally lodged at Lloyd's, who will only accept guarantees on their preprinted forms and from UK guarantors. Alternative arrangements acceptable to the salvor can be made and given direct. If the security is too low and is exceeded in the ultimate award, obvious problems result, not least in obtaining recovery from the owners of cargo which has already been released. On the other hand, if security is considered too high by the arbitrator and provided a formal protest is made by the salved property before security is lodged the salvors may be ordered to pay the cost incurred in providing security in excess of the amount considered reasonable. The arbitrator when making the award is not normally aware of the amount of security that has been lodged.

The ship owners are not normally obliged to provide security for the salvor's claim against cargo; however Clause 5 of the 1980 Lloyd's Form provides for the ship owner to use the best endeavours to assist the salvor to ensure cargo interests put up their security. Frequently, ship owners ask salvors to split their security requirements amongst the various items of salved property.

The assessment of the initial security figure and of split security figures and the closing of acceptable guarantees require particular care in order to cover all interests. In this respect, salvors usually seek the advice of firms of solicitors specializing in maritime law.

Clause 5 of Lloyd's Form 1980 also provides for the salvor's maritime lien on the salved property pending provision of adequate security to the satisfaction of the Committee of Lloyd's and the contractor. The salved property has a period of grace of up to 14 working days to provide the security. During this period, the salvor agrees not to arrest or detain the salved property. The exception of this is when the salvor has reason to believe that the removal of the property is contemplated contrary to the agreement.

9.2.3 Arbitrator

Once requested under Clause 6, an arbitrator is appointed by the Committee of Lloyd's from a panel comprising barristers with experience in the field of Admiralty Law.

Any of the parties to the salvage agreement may make a claim for arbitration, namely the ship owners, cargo or part cargo owners, owners of freight, or any part thereof, separately at risk, the salvor/contractor, the owners of bunkers and/or stores and any other person who may be party to the agreement.

If any of the parties to the arbitration wish to be heard or put forward evidence they should advise the Committee of Lloyd's accordingly and nominate a solicitor to represent their interests.

Even after an arbitrator has been appointed, there is no reason why an amicable settlement cannot take place. Many cases are settled either by the parties themselves or more usually by their solicitors, subject to the prior agreement of their underwriters to be bound to any such agreement.

9.2.3.1 *Evidence of salvage services*

Prior to the arbitration, the evidence of the respective

parties has to be carefully prepared. It is frequently the case that one of the firms of lawyers practising in this field will have taken statements depending on which side is being represented, from the casualty master, salvage master and/or tug master shortly after the salvage operation, while the matters are still fresh in their minds. Before statements are taken, it is useful if as many as possible of the following, some of which are disclosable documents in any event, are made available to the respective solicitors:

1. General requirements: a photocopy of deck log covering salvage services for each of the vessels involved; a master's report for same period for each of the above vessels; a radio log extract where relevant and cables for same period for each of the above vessels; a report of the salvor's salvage officer or underwriter's surveyor; any existing printed particulars of the above vessels and value of each at time of services; photographs of the casualty, salvage craft employed and principal events in salvage operation. Other general requirements are official weather reports where applicable – preferred to forecasts; details of out of pocket expenses and vouchers in support; particulars of special equipment used, hired or purchased for the service; certificate of redelivery; details of any alternative assistance available; details of work carried out by the ship's crew during the salvage and details of any damage caused by the salvors.
2. Additional requirements in cases of: stranding, which includes tide gauge readings, HW/LW soundings in and around casualty, divers' reports if any, tidal predictions, a large scale chart/plan of grounding site; towing, including track chart where applicable if damage has been sustained by the vessel, a tug or its gear, details of cost and survey evidence in support; fire, which includes information on hazardous cargo, and fire equipment on tugs (foam portable pumps, inert gas systems, etc.).

9.2.3.2 Criteria considered by the arbitrator

As mentioned elsewhere, the arbitrator will consider, among other things, the risks and dangers faced by the property salved, those faced by the salvors and the salving vessels, the value of what was salved, the value of the salving vessels and the merits of the salvage services. In recent years any award has been enhanced where pollution has been averted. In making the award the arbitrator should have in mind the need to encourage salvors. Arbitrators look closely at the status of the salvor and are more often generous to professional salvors who maintain fully equipped salvage tugs and shore depots.

The arbitrators often obtain guidance from earlier awards but, this apart, the awards are confidential to the parties. Subject to the time limits provided in the form any of the parties represented at the arbitation may appeal or cross appeal against the award of the arbitrator to a sole appeal arbitrator. Any award on appeal is final and binding on all the parties concerned.

On a point of law only the parties may go to the High Court and/or in appropriate cases, all the way to the House of Lords. If any of the parties fail to pay their proportion of the salvage award then under Clause 15 the Committee of Lloyd's may realize or enforce the security and pay from it to the contractor. The award is, under Clause 1, made in the currency preferred by the salvor and inserted into the contract where provided. If the space provided is left blank then any award will be made in pounds sterling. The value is taken as that prevailing on termination of services.

9.2.4 Examples of other standard forms of salvage agreement

9.2.4.1 Japan shipping exchange salvage agreement

This contract is used most often when assistance is rendered to a Japanese vessel or by Japanese salvors and in the main is a simplified version of Lloyd's Open Form. The contract is naturally governed by Japanese Law and any action under the contract must be brought before the Tokyo District Court.

In this no cure–no pay agreement, the salvor agrees to use the best endeavours to salve the vessel and/or cargo and to take them to the nearest place of safety or other place to be agreed for delivery to the salved party. The salvor also agrees to use the best endeavours to prevent the escape of oil from the vessel whilst performing the services.

An interesting variation to Lloyd's Open Form is that under this agreement the salvor must make daily reports to the master and the owner concerning the condition of the vessel and the progress and status of the salvage. The amount of salvage remuneration will depend on the circumstances of the case taking into account the following: the costs and expenses of the salvor and the difficulties met, skill of the salvor, value of the salved property and any other relevant factors.

As with the Lloyd's Open Form special arrangements apply where the vessel is a tanker and where without negligence on the part of the salvor the operation is not successful, only partly successful, or the salvor is prevented from completing the services.

9.2.4.2 Chinese salvage contract

This is the standard form of the Maritime Arbitration Commission of the China Council for the Promotion of International Trade. The contract is based on the principle of no cure–no pay and is subject to arbitration.

In this contract the salvor undertakes to perform operations to salve the vessel or cargo and any other property and take them into a pre-ordained port. The contract may stipulate a fixed remuneration for the job or a compromise settlement if the salvage operations are only partially successful. If it is not possible to reach an agreement on the fixed amount then this amount can be left to arbitration. The contract also allows for the amount of security that will be required following the salvage to be inserted. If the parties fail to reach an agreement in respect of the security or one of the parties fails to perform, the salvor shall be entitled to notify the Maritime Arbitration Commission of the amount required. The Chairman of the Commission shall make decisions regarding measures of security.

Clause 6 of the agreement stipulates that, pending the completion of the measures of security the vessel and property salved shall not, without consent in writing of the salvor or the chairman of the Maritime Arbitration Commission of the China Council, be removed.

The salvor has an absolute option from the beginning and until the end of the service to determine whether or not the salvage or assistance shall give a satisfactory result and whether or not the vessel and the values on board are valuable enough to meet the expenses of the salvage and assistance. The salvor is entitled to cancel the agreement and to abandon the business even if salvage activities have already started. In the event of cancellation of the agreement or abandonment of the contract, the salvor cannot claim for any loss; however, if the vessel or cargo is partly salved from the salvor shall be awarded from the salved values.

9.2.4.3 Contrat d'assistance maritime

This is a French version of the Lloyd's Open Form and based on no cure–no pay terms where the contractor agrees to use the best endeavours to salve the vessel and the cargo, providing at own risk the proper machinery and other systems and labour and take ship and cargo into a designated place. As with other similar types of salvage contract the remuneration shall not exceed the value of the salved property.

The amount of the award for salvage will be fixed by artibration and the contract leaves open the place of arbitration. If the parties concerned are unable to agree on arbitrators, the contract allows them to be appointed by the president of the Tribunal of Commerce. The arbitrator's decision is final and binding and both parties must declare that they accept such a decision in advance and give up any appeal or recourse against the same.

9.2.4.4 Turkish salvage contract

The Turkish Maritime Organisation Salvage and Assistance Agreement is normally only presented to vessels grounding in Turkish waters. The contract is on the basis of no cure–no pay and as would be expected arbitration is held in Istanbul.

One arbitrator is appointed by the salvor and one by the master or ship owner embracing the ship cargo and freight. In the event that the master or ship owner does not appoint the arbitrator and notifies the fact to the salvor within one week after the salvor notifies them of the arbitrator, the arbitrator can be selected by the Istanbul Commercial Court.

One interesting paragraph in the contract is in relation to the fee to be paid to the arbitrators and translation is as follows:

> The fee to be paid to the arbitrators shall be computed over a sum adjudged. In the event of a dispute being resolved by two arbitrators this fee shall be 10% of the sum adjudged and in the event of the dispute being settled by three arbitrators 12%. The fee due over this rate shall be paid by the defendant and it shall be equally divided amongst the arbitrators. The arbitrators shall reckon a 10% interest on the sum adjudged as from the day following the date of the completion of the salvage assistance operation.

Clause 7 also states that should, by the time the salvor arrives at the site of the accident, the casualty salvage itself by its own means the salvor shall have the right to receive an appropriate fee and expenses that are incurred and any loss suffered as a result of such a situation. If the vessel is salvaged by a third party, the vessel agrees to pay a full salvage assistance award.

9.2.4.5 International Salvage Union Subcontract Agreement

This contract is used by all members of the International Salvage Union where one company has agreed to salvage a vessel under Lloyd's Standard Form of Salvage Agreement and wishes to engage the services of another salvor. Under the contract it is agreed that the contractor engages the subcontractor on no cure–no pay terms to assist in the performance of obligations under this form.

The subcontractor agrees to use the best endeavours including the provision of such personnel, equipment and services as set out in a schedule or are reasonably requested by the contractor during the performance of the services. In consideration of this, the contractor agrees to share with the subcontractor the remuneration as finally awarded or agreed by the parties concerned. The subcontractor in turn expressly agrees not to claim salvage against the owners of the salved property or make any claim for remuneration in respect of the services rendered.

9.2.4.6 Other types of salvor's contract

In addition to the foregoing standard forms various salvage companies have their own salvage contracts. Most are based on the principle of no cure–no pay and allow the salvor, if it becomes clear that the work involves a certain loss, the right to withdraw from the contract.

9.3 Salvage related organizations
9.3.1 The Salvage Association

In the mid-nineteenth century it became apparent to shipping interests in London that an organization was required to provide a central, trusted and technically proficient directorate. This organization could then, amongst other duties, advise them on how to deal with specific casualties and the prospects for salvage of hulls and cargoes. Accordingly, a general meeting of the Committee of Lloyd's on 2 July 1856 approved the formation of The Association for the Protection of Commercial Interests as Respects Wrecked and Damaged Property. The original finance was provided by a number of Lloyd's underwriters, the then five marine insurance companies and The Royal Mail Steam Packet Company. It is interesting to note that this was the first cooperative venture between Lloyd's and the London marine insurance companies. The Association was granted a royal charter in 1867.

In 1895, a hull surveying office was opened at Cardiff and the advantage of having surveyors trained exclusively for the Association's work soon became apparent. It would obviously be uneconomical and impractical for any single section of the maritime industry to maintain a worldwide organization to protect its interests or to arrange for the recovery of property. The Salvage Association, backed by the Lloyd's agency network, soon became the largest surveying organization of its kind. Today, there are 29 branch offices of the Association located strategically around the world in the principal ship repairing areas and some 404 Lloyd's agencies.

Over the years the Association became known as The Salvage Association (often called the London Salvage Association) and when in 1971 a new charter was granted (widening the scope of the Association's area of activities to meet the changing needs of the shipping and insurance industry) this new name was incorporated. Although formed for and by London marine underwriters, The Association also now acts for underwriters, banks, governments, lawyers, or owners of ships or cargo worldwide.

Damage occurs to ships and cargoes unexpectedly, often in remote and inaccessible places, and the circumstances in which a vessel may need salvage assistance are many and varied. The Salvage Association is often asked to assess the most appropriate method and advise on the type of equipment available and the likely cost thereof. Unlike the Liverpool and Glasgow Salvage Association of 1881, which until 1968 maintained its own salvage equipment, the Salvage Association has never owned any salvage plant. Its surveyors, however, can if required coordinate operations and take decisions on behalf of instructing principals throughout the world.

9.3.1.1 Salvage officers

The Association has its own staff salvage officers available to give immediate guidance in cases where expert knowledge is required and, in particular cases, with the approval and authority of the owner, take charge of the entire salvage operation.

In many casualties, assistance is provided by commercial salvors under a Standard Form of Salvage Contract. While such a contract allows the salvor to make all decisions regarding the salvage operations, in practice the presence of the Association's surveyor not only provides the salvor with advice and cooperation on the spot, but also enables an independent report of the salvage operations to be provided for use when the salvage award is subsequently assessed at arbitration. Once the salvage operations are completed and the vessel delivered to a safe port, any damage sustained in consequence of the casualty must be assessed, appropriate repairs agreed and the costs agreed if considered reasonable.

9.3.1.2 Underwriter's surveyor

For most ship owners serious casualties are a one off occurrence, that is the ship owner or operator normally has little or no experience regarding salvage or other actions that might be necessary to preserve the property. Underwriters on the other hand deal daily with casualties and most underwriters either have their own inhouse surveyors or have access to surveyors who are familiar wth the type of casualty in question.

In salvage, an underwriter's surveyor usually serves three purposes. The first is to advise underwriters and owners to take such action as will best protect the property. The second is salvage expertise gained over numerous salvage operations can be passed on to the owner of the vessel or cargo in trouble. It may be the first time that the owner has been involved in a salvage operation. Finally, when an owner enters a contract with a salvor the surveyor attends its execution and finally presents an unbiased report on the operation.

Casualties normally divide into two basic types. The first is where the vessel is in a position of peril, e.g. drifting on to rocks, on fire or sinking and needs immediate assistance. The second is where the casualty has already run its course (e.g. where the vessel has capsized, is stranded in a sheltered position or has even sunk and where time allows, the situation can be evaluated and then competitive tenders sought). When surveyors first attend a salvage case such as a stranding, it is likely that no claim has yet been made upon underwriters and they can only advise the owner on the most prudent actions to take in the circumstances.

The first responsibility of the surveyors is to determine if salvage is technically feasible and, if so, whether the cost of the salvage and repair can be carried out within the insured value. If the surveyors decide that salvage should go ahead then they can advise the owner on the type of contract that may be obtained and the subsequent relative merits of salvors' proposals.

Where vessels are in a position of peril it is not at all unusual for professional salvors to be unavailable; professional salvors to be unwilling to quote or who will only quote on terms which are unacceptable; countries to refuse to allow foreign salvors to operate in their waters; it to be more practical for the owners to engage or hire equipment locally and to undertake the operation themselves; or the only salvor available or allowed is inexperienced in the type of operation.

Following a report of a casualty, owners often seek advice. The advice needed by the owner might include, for example, what action needs to be taken and what facilities are available locally, including names and contacts. It must be remembered that it is not only necessary to obtain assistance but that assistance must be adequate to cope with the situation. For example, as the condition of the vessel is normally deteriorating it serves no purpose to engage a contractor who does not have equipment available to deal with the problem in an acceptable time frame.

Other advice needed might be what type of contract is available and should be entered into and assistance with the evaluation of any salvage offers including terms or exclusions. In any of the above scenarios, the underwriter's surveyor can, with the authority of the owner, arrange and supervise the entire salvage operation.

9.3.2 International Salvage Union (ISU)

The International Salvage Union (ISU) is an association which represents the interests of marine salvage contractors operating throughout the world. The aim of the ISU as set out in their various publications is to seek to generate a better understanding, both within and outside the maritime community, of the responsibilities, activities and remuneration of marine salvors. The ISU also represents the interests of its members in international forums concerned with maritime law and practice, insurance and safety at sea. Membership of the ISU is open to marine salvage contractors. There are currently 34 members of the ISU, based in 22 different countries.

9.3.3 International Maritime Organisation (IMO)

Following some very large tanker casualties and in particular those of the *Torrey Canyon* and *Amoco Cadiz*, both of which spilled large quantities of crude oil on to beaches in Europe, it was recognized that action must be taken to encourage salvors to go to the aid of such tankers. The prime objective was to try and prevent massive damage to the environment. In 1980 the Lloyd's Standard Form of Salvage Agreement was revised to include a clause that provided that the salvor should also use the best endeavours to prevent the escape of oil from the vessel while performing salvage services. In the event of failure to salve the vessel, the clause provided for payment to the salvor of reasonably incurred expenses plus an increment of up to 15% of these expenses. This was an exception to the no cure–no pay principle and is commonly known as the safety net. The safety net applied only to loaded or partly laden tankers and did not apply to vessels carrying toxic or hazardous waste. In addition to the changes in the Lloyd's Form the International Maritime Organisation (IMO) asked the International Maritime Committee (CMI) to prepare a new draft salvage convention. The purpose of the new convention was to update the 1910 Brussels Salvage Convention. It was also intended that the new convention would encourage salvors to render assistance to vessels loaded not only with oil but also with toxic cargoes.

The completed draft was presented to the Diplomatic Conference in 1989. The articles in the draft include many interesting items which are covered here briefly. A full draft is available from the IMO, 4 Albert Embankment, London SE1 7SR, UK.

Article 1 tries to define the various aspects of a salvage operation which is itself defined as meaning any act or activity undertaken to assist a vessel or any other property in danger in navigable waters or any other waters whatsoever. Vessel is defined as any ship, craft or structure capable of navigation and property as any property not permanently and intentionally attached to the shoreline and includes freight for the carriage of the cargo whether such freight be at risk of the owner of the goods, the ship owner or the charterer. One of the most important definitions is that which constitutes damage to the environment as meaning substantial physical damage to human health or to marine life or resources in coastal or inland waters or areas adjacent thereto caused by pollution, contamination, fire, explosion or similar major incidents. Payment is defined as any reward remuneration, compensation or reimbursement due under this convention.

Under the provisions of Article 4 the master of a ship has the authority to conclude a salvage contract on behalf of the owners of the vessel and the master or the owner of the vessel shall have the authority to conclude a salvage contract on behalf of the property on board that vessel. This provision has become necessary as in recent years many cargo owners have refused to give the master the authority to enter into contracts on their behalf.

Article 6 lays out the duties of all parties to the salvage operation and imposes on the salvor a duty to exercise due care to prevent damage to the environment and to seek assistance when required. The article also defines the duty of the owner and masters.

Article 10 defines the criteria for assessing the award and states that the award is to be fixed with a view to encouraging salvage operations taking into account the following considerations without regard to the order in which presented:

1. The value of the property salved.
2. The skill and effort of the salvor in preventing or minimizing damage to the environment.
3. The measures of success obtained by the salvors.
4. The nature and degree of the danger.
5. The efforts of the salvor in salving the vessel property and life, including the time used and expenses and losses incurred with the salvor.
6. The risk of liability and other risks run by the salvor's equipment.
7. The promptness of the service rendered.
8. The availability in use of the vessel's rudder and equipment intended for salvage operations.
9. The state of readiness and efficiency of the salvor's equipment and the value thereof.

It is intended that any reward shall be borne by the property interests in proportion to their value. It also states that the awards that are recoverable shall not exceed the value of the salved property.

In Article 11, Special Compensation, if the salvor has carried out salvage operations on a vessel which itself or its cargo threatened damage to the environment and failed to earn a reward under Article 10, then the salvor is entitled to compensation from the owner of the vessel of expenses plus an increment of up to 100%. This figure was proposed by the CMI.

Article 15 states that the salvor may be deprived of the whole or part of the payment due under the convention if the salvage operations have become necessary or more difficult because of the salvor's fault or neglect or if the salvor has been guilty of fraud or other dishonest conduct.

Article 21 on jurisdiction allows that unless the parties have agreed to the jurisdiction of another court or to arbitration, an action for payment under the convention may, at the option of the salvor, be brought in a court which is competent according to the law of the state where the court is situated and within the jurisdiction of which is situated in one of the following places: the principal place of business of the defendant; the port or place to which the salved property has been brought; the place where the property salved has been arrested; the place where security for the payment has been given and the place where the salvage operation took place.

9.3.4 Protection and indemnity associations

These associations, the Protection and Indemnity (P and I) Clubs, provide ship owners with protection against claims to third parties and other liabilities which are not usually covered under standard marine insurance policies. They are run on a mutual basis and are non-profit making. The ship owners are the members of the club and each contributes on the basis of the number, type and age of the ships, where they trade and the claims record.

With respect to salvage operations these liabilities may result from wreck removal, where a sunken, stranded or derelict vessel is either causing an obstruction or damage to the environment, pollution, whether by oil or other toxic substances or damage caused to third party property by the casualty. Whilst the P and I associations will monitor a salvage operation, they do not normally become involved with this aspect of the casualty until after the vessel has either been declared a constructive total loss or total loss.

They are however involved in the prevention of pollution and if the casualty has caused pollution they will be involved from the outset. If there is serious risk of pollution, the associations often arrange for specialist equipment and personnel to be assembled or dispatched to the casualty and thus minimize the time that would be needed to mobilize and start any clean up operation.

It is said that approximately 85% of all ocean going ships are entered into P and I associations and some 98% of tanker owners participate in the Tanker Owners' Pollution Federation who manage the TOVALOP (tanker owners' voluntary agreement concerning liability for oil pollution) scheme.

Depending on the circumstance of the case and the size of the threat, the owner, club and federation might organize the clean up operation. Where the oil pollution is on a large scale, it is more often the case that governments assume overall control.

9.4 Types of salvage

For practical purposes salvage falls into five main categories:

1. Conventional salvage, the most publicized and frequent type of salvage, is a service provided where a vessel has suffered a casualty such as stranding, immobilization, fire, etc., and is unable to proceed entirely with its own means. The object of this type of service is normally to enable it to complete its voyage or where applicable to return to service after necessary repairs have been completed. *Figures 9.3 and 9.4.*
2. Wreck removal or port clearance is the next most common type. This is a service performed for the benefit of safe navigation, port operation, site clearance, such as removal of debris following a major offshore accident, or prevention of pollution. This type of operation is normally carried out within port limits, oilfields or near an environmentally sensitive area. Apart from oilfield clearance, an operation of

Figure 9.3 Conventional salvage. Vessel stranded on an exposed shore; sand buttress built to stop vessel moving further inshore (courtesy of The Salvage Association)

Figure 9.4 Conventional salvage. Large tanker stranded on an exposed shore (courtesy of The Salvage Association)

this nature normally takes place where some statutory authority, for example the government or port authority, has the necessary powers to order the removal of the debris, vessel or its cargo. Some authorities insist on total removal whilst others only ask for the wreck to be cleared to a specific level such as the sea bed (*Figure 9.5*). Removal of redundant offshore gas or oil platforms also comes into this category.

3. Cargo salvage is when on many occasions a vessel may be so badly damaged that it cannot be salved or alternatively the hull cannot be salved economically but the cargo or part of it can (*Figure 9.6*).
4. In underwater cargo recovery salvage of vessels sunk in deep water is seldom economically feasible; however, in some instances, where vessels are carrying highly valued cargo, usually non-ferrous metals or oil, it may be economical to salve the cargo and leave the ship. It is difficult for an owner to repudiate ownership and hence the associated liability and it is to be noted that the ownership of wrecked and sunken ships and cargo does not become the property of anyone who seeks to take possession. In the case of a cargo recovery operation, where the ship's structure is to be damaged for the purpose of gaining access to the cargo, permission of the ship's owner must be obtained before commencement.
5. Since scuba diving has become popular there has been great interest in archaeological salvage by persons of good and bad intent. This has caused many governments to take action to preserve these wrecks or to involve themselves in the arranging of contracts for the recovery and/or sale of either cargo or artefacts.

Figure 9.5 Wreck removal. Sheer legs using a wrecking grab to disperse a wreck (courtesy of The Salvage Association)

Figure 9.6 Cargo salvage. Vessel broken in two, but some cargo can be saved (courtesy of The Salvage Association)

9.4.1 Parties to a salvage operation

Although wreck removal operations may be undertaken on behalf of or at the instigation of a conservancy authority (normally on a specifically drawn contract) almost regardless of expense, the salvage of ships and cargoes which are to be put to further use must provide an economic benefit for the parties involved.

There are normally three main parties involved in a salvage operation, each with a different financial interest. A ship owner must, before entering into a contract, decide whether the cost of salvage, repairs and disbursements will exceed the value of the vessel. In addition due regard must be taken of the interests of mortgagees, cargo owners, charterers, third parties and underwriters.

The underwriter or representative must consider whether the cost of salvage and repairs will exceed the insured value written into the policy of insurance before agreeing it, or advising that the owner should enter into a salvage contract.

The salvor before offering services on a Standard Form of Salvage Agreement or any no cure–no pay agreement, must consider whether the sound market value of the property in distress, less the eventual cost of repairs, leaves sufficient to cover expenses and anticipated profit.

9.5 Salvage operations

9.5.1 Cargo discharge

A frequent aspect of salvage operations is that cargo must be discharged or transferred. The problems associated with this are numerous and are often more difficult to solve than many conventional ship salve operations. The reasons for the discharge could be to lighten the vessel to facilitate refloating; to facilitate the onward carriage of the cargo to its destination; to gain access to a damaged internal structure; to prevent pollution or other damage to the environment; or to avoid damage or further damage to the cargo.

It follows that the damaged vessel could be either aground, afloat or even sunk and in any event it is rare that the damaged ship is able to manoeuvre and proceed to a safe port for discharge. If afloat, the vessel will in all probability be disabled and possibly at anchor.

With the exception of underwater salvage, which will be considered later, the easiest way to salvage or save the cargo would normally be first to salve the ship together with the cargo on board. In the majority of operations, this is not only the easiest course of action, but also has the big advantage in that it reduces handling of the cargo in what are usually far from ideal conditions. It is well known that this invariably reduces its value, by damage, loss, delayed delivery or cost of transport to destination, so that its value after salvage is often less than its sound value. This applies even if no damage was caused during the casualty itself and a forced sale of cargo at a port of refuse regularly only realizes about 25% of the true value. Considering low valued bulk cargoes such as iron ore, rock phosphate or rock chippings, it is often cheaper for the receiver to buy a new cargo from source and arrange another vessel to ship it than to arrange for its transhipment. This is because freight and handling costs represent a large proportion of the delivered value of many bulk cargoes. With such cargoes the problems include the possibility of finding a suitable dumping place.

9.5.1.1 Cargo salvage from damaged vessels

There are many instances where a ship sustains a casualty, such as engine failure, fire, collision or even arrest by some authority where it cannot continue to its destination. If the ship owner then abandons the voyage, the cargo owner has the option to either remove the cargo for transhipment, sell it as is where is or abandon it. In this connection, it is interesting to note that particularly in times of shipping recession, but not invariably so, the value of the cargo regularly exceeds the value of the ship (*Figure 9.7*).

If the ship is in relatively sound condition, it may be towed to destination. Provided the ship owner agrees the cargo owner can arrange this. If the total cargo is owned by a single party, arrangements can be fairly straightforward. Where several consignments of general cargo are involved, there can be many problems, not least that of getting the owners to agree on a concerted course of action. Indeed, it may be difficult to contact them at all. Delays of up to a year and sometimes more are not uncommon in trying to trace the respective owners and agree on a course of action.

It is clearly better if one organization can act on behalf of all cargo owners as this often reduces many of the complications. For the owner of one consignment

Figure 9.7 Vessel disabled following a collision and cargo being transferred for onward transportation (courtesy of The Salvage Association)

of cargo to proceed alone involves many problems, for example:

1. If the parcel of cargo is overstowed then it may be necessary to handle cargo belonging to other parties, who might want payment or guarantees against damage.
2. It may be necessary during the operation for the owner of one package of cargo to arrange storage for other people's cargo.
3. The ship owner might want indemnities from the cargo owner in the event that the ship is damaged or endangered.
4. If not prepaid, the freight may have to be paid and indeed sometimes the ship owner's port and other expenses must also be settled before the port authorities will allow the cargo to be removed.

It is often preferable for both the cargo owners and underwriters to sell the cargo as is where is and so avoid the costs and risks involved with its discharge and transportation. This said, there are occasions where this action might not be desirable and their points of view often differ. For example, the cargo owner might want the consignment delivered to the destination for a specific purpose and non-delivery could cause losses in excess of its actual value. Indeed, the very survival of its owner's business might depend on its delivery. In addition to this, cargo values, like all commodities, can rise and fall. On occasion the value can rise above its precasualty or insured value. Conversely, the cargo owner might not want the consignment, e.g. a replacement cargo may have been purchased or the market conditions could have changed since the ship's departure.

Where the voyage is abandoned due to the insolvency of the ship owner, then it is worth noting that the cargo owner will probably not be indemnified by the underwriters because this is excluded from most cargo insurance policies.

If it is considered that the cargo will be a constructive total loss under the insurance policy, then it is essential that notice of abandonment is given to the underwriters. They, however, do not normally wish to assume the role of a trader, nor to accept the liabilities which may be attached to the cargo and rarely accept abandonment of the cargo or take over its ownership. When declining to accept abandonment underwriters normally agree to place the assured in the same position as if on the date of the refusal the assured had obtained a writ to legalize the abandonment.

9.5.1.2 Underwater salvage (commonly known as wet salvage)

The origins of underwater salvage are obscure, but mention was made by Aristotle of a diving bell used by Alexander the Great in 332 BC. Diving bells and suits have been developed and refined over the years, but to all intents and purposes only those developed after about the middle of the nineteenth century are of any real consequence.

The method of removing cargo from sunken steel vessels was pioneered by a Commander Quaglia who recovered gold valued at over £1 000 000 from the liner *Egypt* in the late 1930s. The methods used are still valid today and only the tools have changed. The salvage ship was secured over the wreck on a multipoint mooring and then, using a man in an observation chamber giving directions, explosives were laid on the structure to break open the wreck. Cargo was then removed by grabs. This method is colloquially known as smash and grab and has stood the test of time and technology. It has been used to remove numerous cargoes of non-ferrous metals and coal in depths of up to about 300 m (975 ft).

Cargoes recovered range from gold bars to coal and their salvage may involve high technology equipment. It is always possible that the cargo involved may be toxic or pose a threat to the environment and as such its removal is more like wreck removal and economic factors play a less than normal part.

The vast majority of easily accessible wrecks in Europe with valuable cargo aboard have already received salvor's attention and focus is shifting to deeper and more exposed waters. Recent development and advances in technology have meant that cargo salvage can now be contemplated from depths beyond consideration only a few years ago. It is now one of the few growth sectors of salvage, although it is a high cost operation and requires large investment and hence high value cargoes (*Figure 9.8*).

Recent developments and advances include dynamically positioned and diving support vessels which are able to keep on location directly over a sunken wreck; manned and remote controlled underwater vehicles which can undertake operations at great depths; sophisticated searching devices, e.g. high resolution side scan sonar that can pinpoint wrecks; saturation diving techniques and the use of mixed gases;

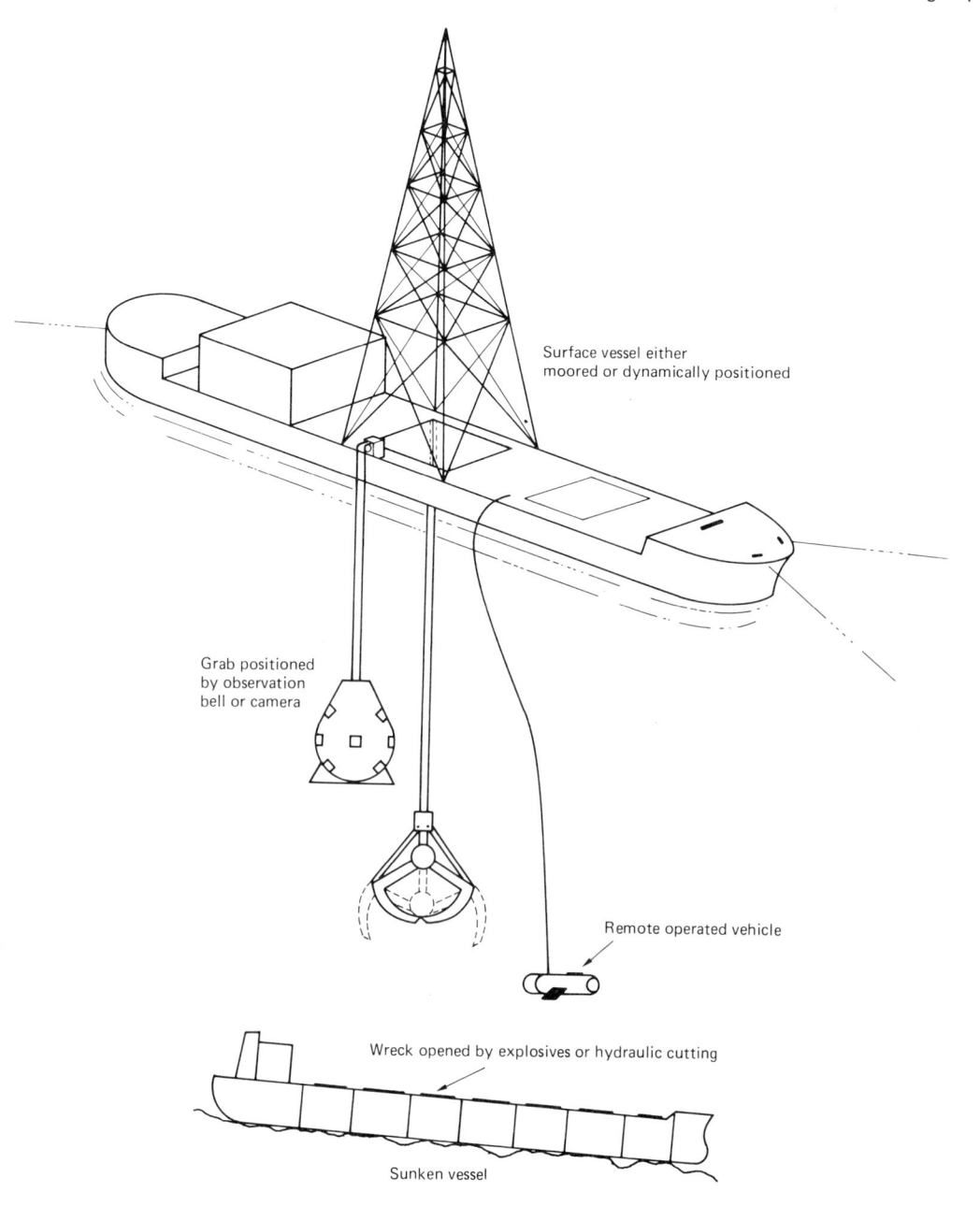

Figure 9.8 Smash and grab underwater cargo recovery

extremely accurate satellite and hyperbolic positioning equipment for the searching phase, together with advanced acoustic devices for returning to the precise position; techniques such as friction welding and hydraulic cutting for gaining access and putting attachments on the hull.

With these advances in technology goes the exponential rise in the cost of the operation and the salvor must ensure that the cost of the recovery and restoration or sale does not exceed any likely return.

9.5.1.2.1 Deep water cargo recovery A cargo recovery operation can be divided into three phases: research, search and recovery. During the operation of research, the probable location and cargo onboard a vessel must first be ascertained. Thereafter the person having title to the goods should be traced if possible, in order to agree contractual terms and to obtain stowage and general arrangement plans. Only when proof is obtained that a vessel was carrying a certain cargo and did sink in a specific location can progress be made. In this respect, the Salvage Association maintains a database listing sunken vessels, probable positions and other details. Whenever a ship sinks containing a valuable cargo, and is not salvaged and the Salvage Association is involved, then the file is transferred to a special department where it is kept indefinitely. The likely weather conditions and operational windows should be investigated using probabilistic methods.

In the operation of search, the area around the probable location of the wreck must be systematically surveyed using side scan sonar or other appropriate equipment. If a wreck is detected, it must be identified using a remotely operated vehicle, observation chamber or diver and its aspect and condition verified. The problem is not usually finding a wreck but finding the right one. On one occasion, a salvor had to eliminate more than 80 before deciding on the right one. The problem of identification is also difficult and is only very rarely a matter of simply reading the name. A vessel which has lain on the sea bed for many years may no longer even look like a ship, let alone like any particular one.

It should be remembered that accurate navigational positioning equipment has only recently been available and the recorded sinking position, which may have been erroneous in the first instance, can be many miles out. One salvor considered a position within ten miles as good and a search area of hundreds of square miles as not uncommon (*Figure 9.9*).

And finally during recovery, either a deep water multipoint mooring or a dynamically positioned vessel must be located over the wreck to facilitate the recovery, which may be carried out using either manned submersibles, deep diving or smash and grab techniques. Whatever the attitude of the ship, the structure over the cargo must first be removed. Once access to the cargo space has been achieved, then valueless cargo might have to be removed as the commodity being searched for can be under thousands of tonnes of unwanted material. It must be remembered that valuable cargoes are normally heavy and placed low in a ship.

9.5.1.3 *Lightening of dry cargo vessels*

Where dry cargo vessels are involved, the main problem is whether the ship is equipped with derricks or cranes and whether these can be made operational. Placing mobile cranes on the deck of a casualty is a difficult operation under any conditions and at the location of a stranding, usually requires the use of a floating crane. Alternatively, the cranes can be carried on the lightening vessel or on a handling barge that can be moored between the casualty and the lightening vessel.

Container vessels are an extreme example of the problems that can be encountered in the lightening of dry cargo vessels, e.g. most container vessels are not fitted with derricks or cranes; the top tier can be very

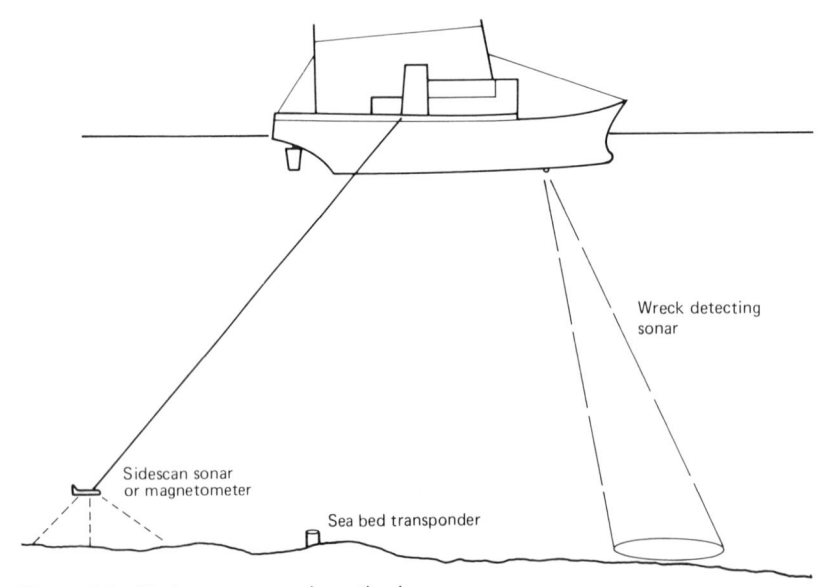

Figure 9.9 Underwater search method

high and out of reach of most mobile cranes; the containers are usually stowed tight together and the weight is such that they are not easily handled without a crane; or on occasion it has been necessary to utilize a helicopter to remove cargo before the vessel could be refloated. *Figure 9.10* shows such an operation.

Figure 9.10 Helicopter lightening a stranded container vessel (courtesy of The Salvage Association)

9.5.1.4 Ship to ship transfer in salvage operations

Whenever large tankers are under way and loaded, their momentum is such that, even if they ground at slow speed, they can very rarely be refloated using tug power alone and almost always require lightening. Large tankers have been known to ground so lightly that the crew did not notice, and the casualty still resulted with a ground reaction of several thousand tonnes. The ground reaction is indicative of the amount of pull needed to free the casualty and, as such, it is clear that tugs alone will rarely be powerful enough to refloat this type of vessel.

Transfer of crude oil, gas or chemicals from a stranded vessel is an operation that must be planned carefully and not rushed into; hasty decisions can easily make the situation worse and it serves no purpose to have two stranded vessels where there was only one before. It is often the case, however, that the situation does not become clear until after the work has started and in practice many of the initial assumptions will be proved wrong.

On rare occasions, a vessel may strand near the time of low water so that as the tide rises it would float free. In these circumstances, all that is needed are tugs to hold it in place until that happens. Assuming this is not the case, then before any transfer or lightening operation can be planned the condition of the vessel, damage, remaining strength, cargo distribution, where and how it is aground, depths, currents, the nature of the sea bed, etc., must be determined. Before planning for and chartering a lightening tanker or before attempting to berth any lightening ship alongside a stranded casualty, a careful assessment must also be made of the hydrographic situation by way of a bathymetric survey of the area.

Before lightening, the vessel should also be secured in position by way of ground tackle and ballasting so that it does not move further inshore or refloat before it is intended. It may or may not be possible to bring the lightening vessel of a similar size alongside and the length and type of hose needed can vary enormously. Ground tackle is not normally effective for holding very large tankers in position. In these instances, the ship must be ballasted until the situation is suitable for refloating. A large proportion of the pumping capacity must be reserved for these operations (*Figure 9.11*).

The casualty may be stranded in a position where a small tanker can be brought alongside which can then act as a shuttle to either another larger vessel moored nearby or alternatively to an oil terminal onshore (*Figure 9.12*). On the other hand, it may not be possible to berth any vessel alongside in which case the lightening vessel must be moored close by and hoses floated to it (*Figure 9.13*).

In the last two decades, manpower, economies of scale and other costs have led to larger tankers being used to carry crude oil. Whilst many deep water terminals were developed to handle these vessels, at some locations this was not possible and lightening operations, where the cargo is offloaded from the larger crude carriers to smaller vessels for onward carriage, have become routine. Safety procedures have been developed to reduce the risks associated with these operations and guidelines have been developed and published by the International Chamber of Shipping and the Oil Companies International Marine Forum. These guidelines for both the transfer of petroleum and liquefied gases are readily available.

In salvage operations, it is not always possible to carry out cargo operations as safely as if both vessels were fully operational and undamaged, but the guidelines are normally followed as closely as possible with additional recommendations to cover the special problems of discharging or lightening a stranded vessel.

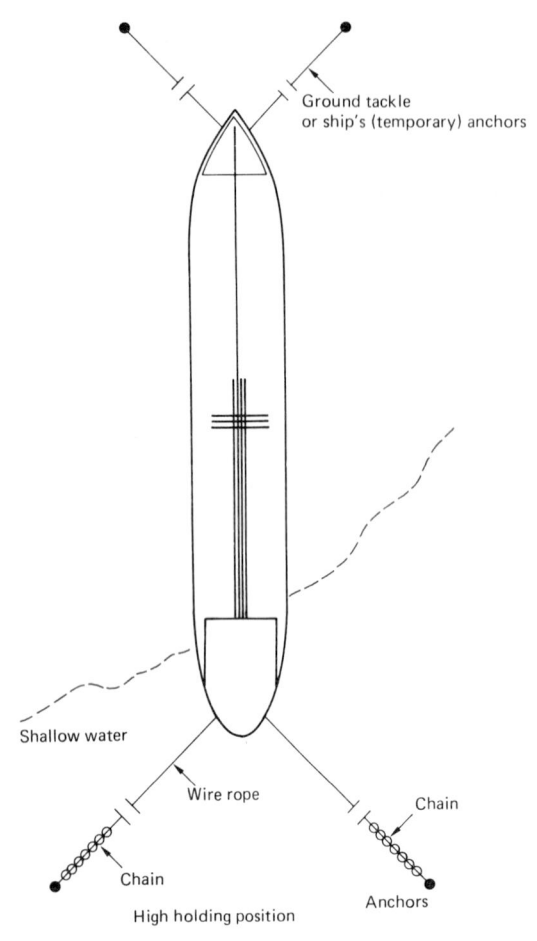

Figure 9.11 Ground tackle holding a tanker in position

9.5.1.5 Tanker discharge

As mentioned elsewhere, on oil and chemical carriers it is preferable to make use of the vessel's own piping and pumping systems and efforts should be made to achieve this. On occasion, this is not possible and it is necessary to discharge the cargo by placing pumps into the tanks and pumping the cargo 'over the top'. While the discharge of oil cargoes over the top, where pumps and their discharge hoses are lowered into the tanks from the deck, has been carried out on numerous vessels, it is still an operation that should be treated with extreme caution. The fact that the tanks are open during the discharge makes it necessary to be more than usually careful in the discharge. It is normal to place polythene covers with sand or sawdust bags around openings to reduce the amount of gas escape to the deck. This is inevitable and, as the pumps must be handled through the openings, this makes for an operation that should not be entered into lightly.

The pumps have been improved, in both size and

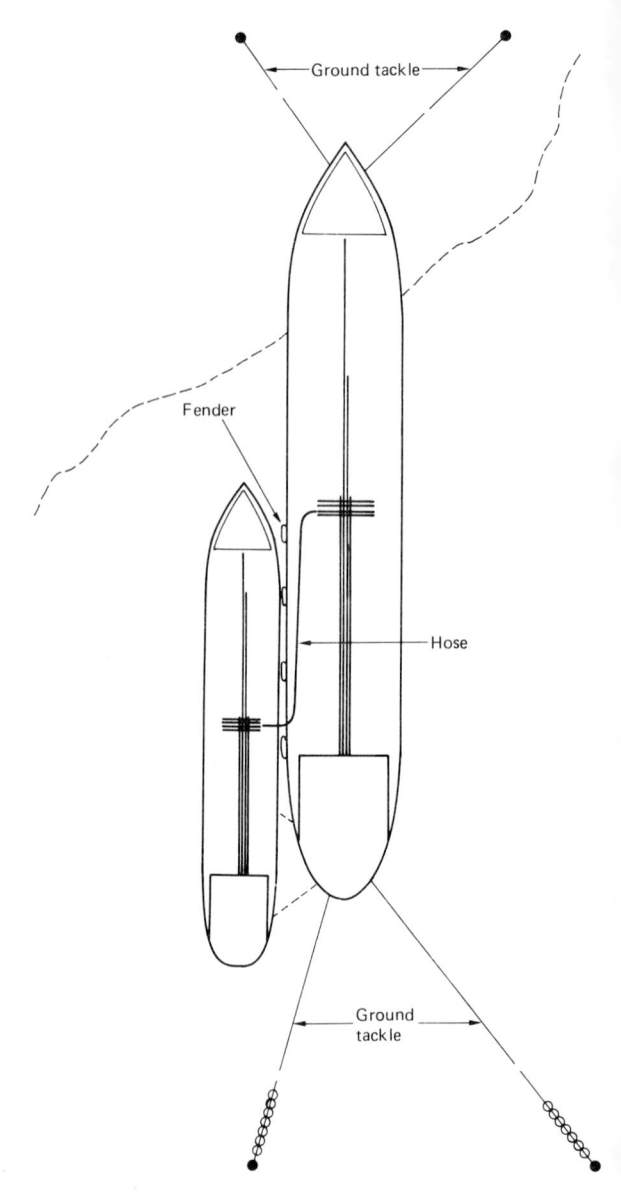

Figure 9.12 Small tanker alongside a casualty

performance, so that it is now possible to use ones specifically designed for over the top discharge. They are designed to be lowered through a Butterworth Opening (openings in the deck of a tanker used for tank cleaning and normally obstruction free between the deck and the tank bottom (*Figure 9.14*)). Although it is possible to carry out the operation with intrinsically safe electric pumps, the method most commonly adopted is by submersible hydraulic pumps, some of which were specifically designed for this particular work. These have a diesel prime mover which is spark free and can be located in the least gaseous area remote from the operation. Larger capacity pumps are also

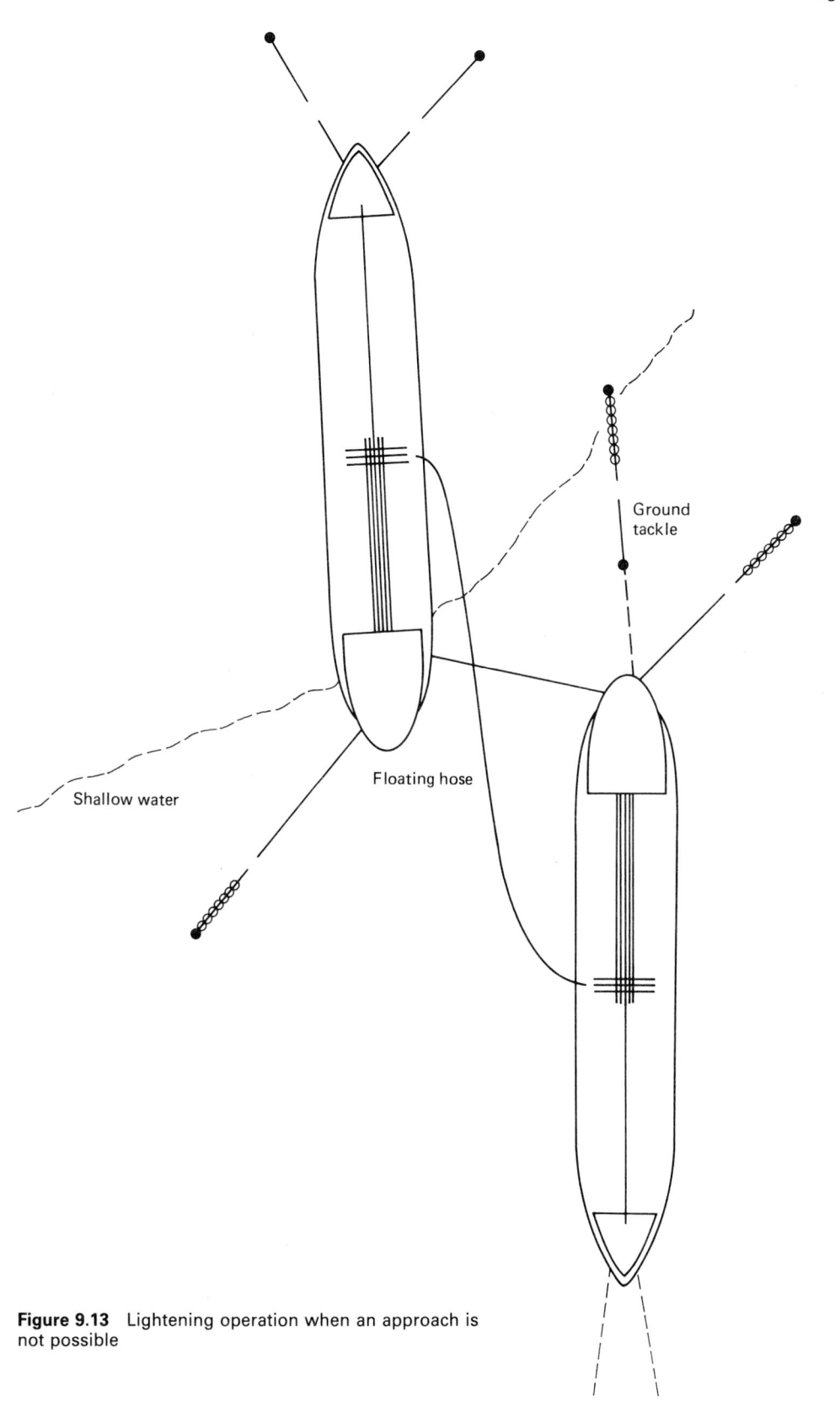

Figure 9.13 Lightening operation when an approach is not possible

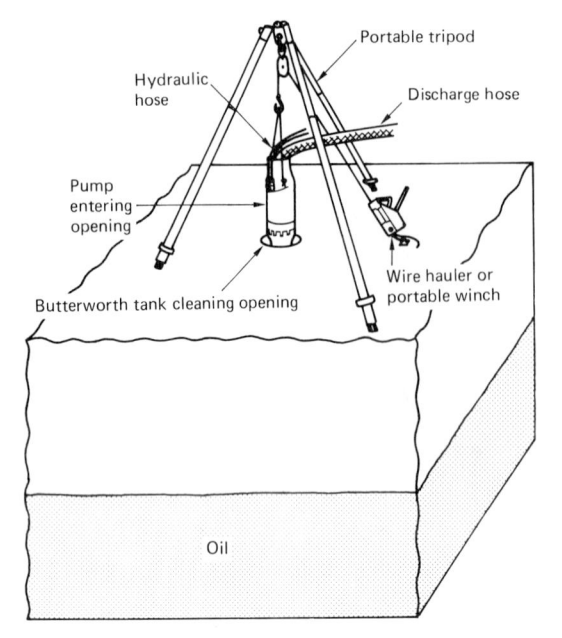

Figure 9.14 Hydraulic pump being inserted into a Butterworth Opening

available, but are too large to fit through Butterworth Openings and must be lowered down via the larger tank access openings, which are not normally free of obstructions. This often means putting a diver into the tank wearing breathing apparatus to manoeuvre the pump past obstructions. If the casualty is carrying dangerous chemicals then toxicity could be a problem and the manufacturers should be consulted.

9.5.1.6 Fendering

The preferable method of fendering, which is desirable for all ship to ship transfer operations and is absolutely essential for large ships, is the use of inflatable fenders of the Dunlop, Seacushion or Yokohama type. Whilst these are the types more often used, other fenders are available and have been used successfully in salvage operations. *Figure 9.15* shows some typical fenders used in salvage, including one of the Yokohama or Sea cushion type with tyres and chain netting for handling and greater stand off together with one of the Dunlop type.

As time is usually critical in an emergency, availability is normally the major criterion, but if, after giving consideration to the outreach of the cranes or derricks, a choice still exists then the largest fenders available should be used. These are normally secured to the lightening vessel so that maximum protection is provided during the berthing and transfer operation.

The fender manufacturers supply tables that indicate the size of fenders required for particular situations and are based on energy absorption during berthing opera-

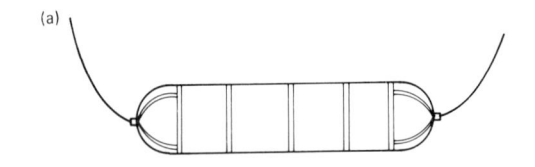

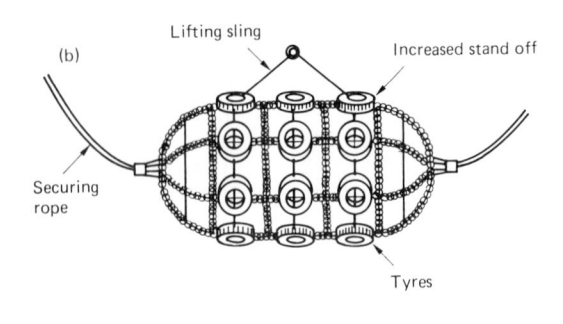

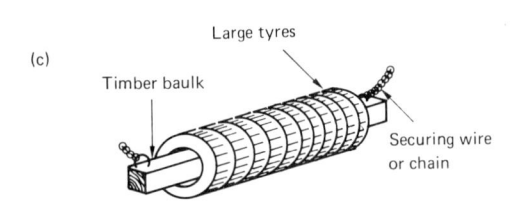

Figure 9.15 Types of fender; (a) Dunlop type inflatable fender; (b) Sea cushion or Yokohama fender, fitted with chain and wire net for handling; (c) Emergency fender of timber baulk and large tyres, usually built on site

tions. The main factors which must be considered include the size of the vessels, exposure of the site, rate of approach during berthing and whether it is possible to keep the vessels upright during the operation.

For fendering two different size vessels of deadweight A and B, the equivalent deadweight C can be obtained by the following equation:

$$C = 2 \times A \times B/(A + B) \tag{9.1}$$

Table 9.1 is taken from the International Chamber of Shipping Oil Companies' International Marine Forum and is only a guide. Where fendering is required for a particular operation, the manufacturer's specification must be consulted. Sometimes good fenders are not available and have to be constructed from timber and large tyres (*Figure 9.15*). These fenders are, however, only suitable for use in sheltered waters.

9.5.1.7 Stability and strength during discharge

Discharging operations on damaged vessels can sometimes make a critical stability or longitudinal strength

Table 9.1 International Chamber of Shipping Ship to Ship Guide: for fendering two ships of deadweight A and B tonnes the equivalent value C can be obtained from $C = 2AB/(A + B)$ for two ships of equal deadweight $C = A = B$

Size of ship (or equivalent value C for two ships)	Berthing velocity	Effective berthing energy at $^1/_4$ point contact	Tender details (high pressure pneumatics or foam)		Tender details (low pressure pneumatics)	
DWT	m/sec	tonnes	Diam. × length (m)	Minimum quantity	Diam. × length (m)	Minimum quantity
1 000	0.30	3	1.0 × 2.0	3	1.5 × 4.0	3
3 000	0.30	8	1.5 × 3.0	3	1.8 × 6.0	3
6 000	0.30	14	2.0 × 3.5	3	2.3 × 8.0	3
10 000	0.25	15	2.0 × 3.5	3	2.3 × 8.0	3
25 000	0.25	36	2.5 × 5.5	3	2.75 × 12.0	3
50 000	0.20	45	2.5 × 5.5	4	2.75 × 12.0	3
100 000	0.15	48	3.3 × 4.5	4	3.2 × 12.0	3
200 000	0.15	91	3.3 × 6.5	4	2.75 × 16.0	3

This table is intended purely as a guide and in deciding the fendering rquired for a particular transfer, reference should be made to the supplying manufacturer's specifications and guidance on the use of their fenders in order to determine if the fenders are in fact suitable in terms of energy absorption and stand-off capability.

situation worse and careful checks should be carried out prior to and during any operation.

In salvage, it is generally accepted that problems associated with structural strength must be expected when working with very large vessels. Local and overall stresses may be increased to unacceptable limits during the operation to remove the cargo. Incorrect ballasting, loading or lightening procedures could easily break a weakened or even a sound vessel in two and has done so on occasion. Where small vessels are involved, whilst strength may also be a concern, the problem is normally one of stability.

When built, large vessels are usually fitted with computers for calculating the bending moments and shear stresses for different stages of loading. If the vessel is in an undamaged condition and the sequence of loading or discharging is controllable, then acceptable stresses can be obtained and maintained throughout using these computers. Where, however, a vessel is aground or afloat with structural damage to the hull girder, then much more care must be taken. Early indications of hull deflections can be monitored by the use of theodolites or lines set up on the deck but, nevertheless, it is preferable if correct pumping sequences can be determined beforehand.

As an example of the problems that might be encountered, *Figure 9.16* shows a large vessel aground amidships with a damaged bottom and breached tanks. In such an operation the vessel could be refloated by transferring the cargo from the forward and aft tanks, but such action could involve stress problems in the damaged bottom. Extreme care must be taken.

It should be borne in mind that it is not unknown for the loading computer to have been damaged, lost, or removed by the time the salvage operation gets under way and in this respect those attending should be prepared to make good estimates from approximate formulae or first principles. For one off situations, the calculations can be carried out by hand, but where different scenarios need to be tested computers are

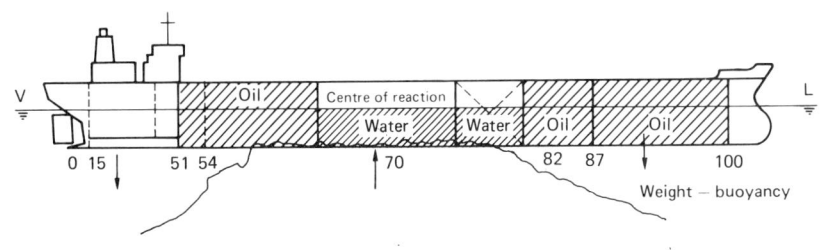

Figure 9.16 Large vessel in weakened condition; there is structural damage to the bottom and flooding

essential. In recent years, computer programs have been developed for calculating bending moments and other loads on damaged vessels. Advances in computer technology now mean that computations can be carried out on board the damaged vessel using portable computers with obvious advantages.

The vessel's classification society and the ship's builders and designers usually have enough information at hand to enable them to carry out what can sometimes be very involved stress calculations. The main disadvantage of this is the difficulty associated with changes in plan and often the remoteness of the casualty. In complex salvage operations, where telex or fax terminals are available, it is not uncommon for the salvor's or underwriter's representative to be in regular contact with the builders or classification society, enabling information transfer.

Irrespective of how the strength analysis is carried out, as with all studies the accuracy of the results depends on the reliability of the input data. Information required includes line plan, weight distribution and section modulus. The section modulus must be that prevailing and must take into account any damaged and distorted structure; this involves many assumptions. On occasion, accurate data is either not available or would take too long to obtain and in such an event it is sometimes possible to estimate a fairly accurate line plan, sound section modulus and weight distribution from information normally available onboard. Whilst this may not be very accurate, it will often indicate the areas of concern and facilitate first approximation manipulations.

As an example of the approximation methods that are available and based on the theory that all tankers are relatively similar and built to standard classification society rules, it is possible to estimate the minimum sound section modulus knowing only the length, breadth, depth and draft of the vessel (*Table 9.2*).

9.5.1.8 Stability

It has long been the general rule that, in salvage operations, whilst longitudinal strength is the major problem with very large vessels, stability is normally a problem with smaller ships. Stability of stranded or damaged vessels must be carefully analysed to determine whether it is a problem and if so whether or not it will endanger the vessel when it refloats.

Stability during a salvage operation may be adversely affected in a number of ways including breaching of compartments, flooding as a result of fire fighting operations, discharge of cargo, discharge of ballast and ground reaction. When vessels strand, it is not at all unusual that cargo, fuel or ballast must be discharged or moved so that the ground reaction is reduced to facilitate the refloating. Where there is no cargo or ballast or where the removal or shifting of these would not be advantageous, it might be necessary to reduce weight by removing equipment and fittings from the vessel. This is usually a last resort as it normally means much effort and cost for little effect.

The lightening or transfer sequence on a stranded cargo vessel is normally removal or transfer of ballast; removal of fuel; discharge of that cargo which is easiest to handle, readily accessible and gives the most advantage. This is usually the large heavy items that are easy to sling and lift such that the operation does not require much labour input, permitting removal of equipment.

From the above, it can be seen that the easiest options usually involve items low down in the vessel, that is, those that add to the stability, whilst the cargo and equipment are usually high up and reducing the stability (*Figure 9.17*). From this, it follows that during salvage operations all pressure from this aspect, both as regards time and cost, is towards methods that reduce the vessel's stability. It is often necessary to operate with stability less than would be acceptable under normal conditions, but this also means that even more care should be taken.

9.5.2 Free surface

The effect of free surface is another major factor in salvage operations, as it is not at all uncommon to have compartments partially flooded or breached. The sub-

Table 9.2 Minimum section modulus requirements for ocean going ships

$Z = f T B \text{ cm}^2 \times \text{m}$

where T = draft (m)
B = breadth (m)
f = factor from table
L = length (m)
Z = minimum section modulus
D = depth up to strength deck

L	f	L	f
30	3.540	109	18.480
36	3.930	115	20.260
42	4.620	122	22.019
48	5.309	128	23.887
54	6.194	134	26.755
60	7.078	140	27.819
67	8.257	146	29.785
73	9.437	152	31.948
79	10.715	158	34.110
85	12.189	164	36.273
91	13.664	170	38.534
97	15.138	176	40.893
103	16.809	183	43.252

Limiting ship dimensions for ocean going ships.

$$\frac{L}{33} + 1.5 \leq B \leq \frac{L}{33} + 6$$

$$10 \leq L/D \leq 13.5$$

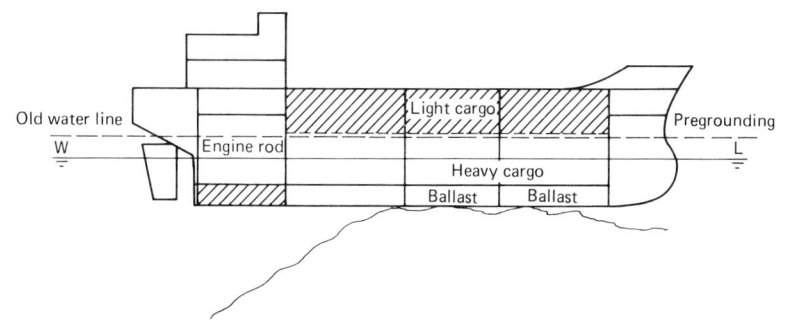

Figure 9.17 Vessel aground with only light cargo at the top

ject is dealt with more fully elsewhere but for convenience we recap.

When a tank or space is completely filled with water, the liquid cannot move and, as far as stability is concerned, the liquid behaves as a solid load acting at the centre of gravity of the space. If the space is only partially filled, the situation changes completely and as the vessel lists, the water flows over to the low side and in so doing effectively reduces the vessel's stability. *Figure 9.18* shows a vessel with a flooded hold, together with a diagram of its effect.

In general, the reduction in stability from free surface effect depends on both the length and breadth of the flooded compartment and the volume of displacement of the vessel. The virtual loss of stability can be calculated by the equation:

$$I/V \times d_1/d_2 \times 1/n^2 \tag{9.2}$$

where V is the vessel's volume of displacement, d_1 is the density of the liquid in the tank, d_2 is the density of the liquid in which the vessel is floating, n is the number of longitudinal compartments into which the space is subdivided and I is the second moment of area of the free surface. For a rectangle I can be calculated by:

$$I = L \times B^3/12 \tag{9.3}$$

where L is the length of free surface and B is the breadth of free surface. It can be seen that the breadth of the compartment is the most dominant factor, as the reduction in transverse stability varies as the cube of this term.

For calculation purposes, the quantity of liquid in the compartment is usually ignored; however, if there is only a small amount of liquid, it will pocket out once the vessel lists, leaving the hull side of the tank dry. Similarly if a tank is nearly full, the liquid will pocket out on the low side under the deck. Dry cargo in the hold and machinery will also reduce the free surface breadth and thus the effect. For calculation purposes, as the benefit from cargo is difficult to determine, on most operations it is to be ignored.

When a vessel is breached and a space fills with water, the effects may be analysed either by consider-

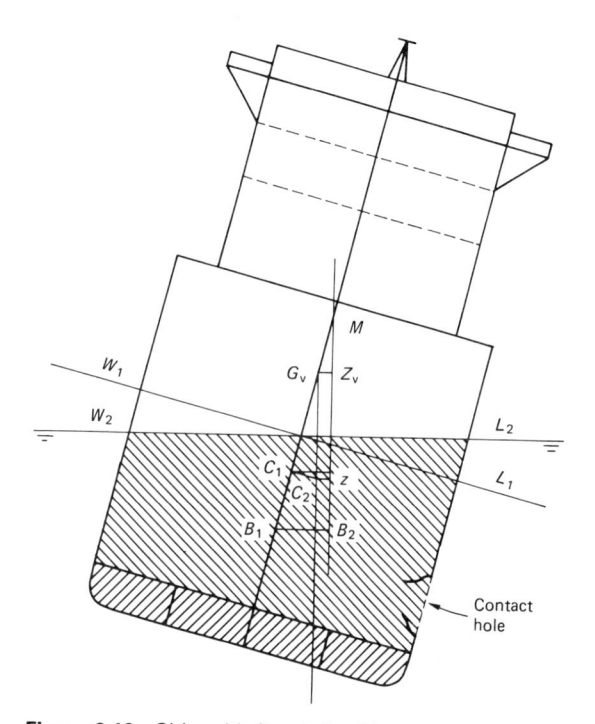

Figure 9.18 Ship with flooded hold

ing the space as lost buoyancy (i.e. the space is no longer supporting the vessel) or by considering the water that entered the vessel as an added weight. For convenience, the latter method is usually adopted. On undamaged and operating vessels, the reduction in stability caused by slack tanks is normally kept within acceptable or manageable limits by good practice and compliance with conditions and recommendations set out in the vessel's stability booklet.

It is not uncommon for spaces to become flooded either during an accident, such as a grounding or stranding or later as the result of actions taken by others. For example, there have been many instances where vessels have capsized as a result of water being pumped in to extinguish a fire. Where a space or large compartment is flooded, either intentionally or accidentally, as for example in a flooded hold in a cargo ship, the reduction in stability can be quite large and could easily lead to the capsize of the vessel.

A vessel sitting upright on the sea bed, even when flooded, is normally perfectly stable; even though potentially in the floating condition it may not be so. The vessel is being held upright by the ground reaction and the fact that the centre of gravity is inside the rotation point at the bilge. If the vessel is unstable then, once the vessel floats, the situation changes and, unless stability is ensured, the ship will capsize (*Figure 9.19*).

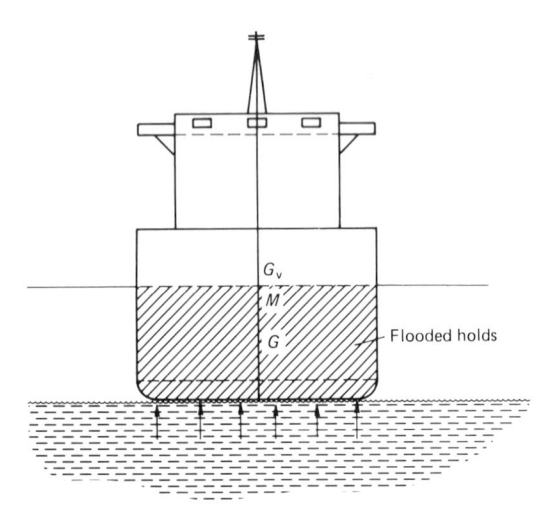

Figure 9.19 Vessel unstable, but held upright by the sea bed

On sunken or flooded vessels, extreme care must be taken to ensure either that the dewatering is carried out in such a manner that the free surface effect is kept down to an acceptable amount or that the vessel is held upright by other means until enough spaces are pumped out to restore stability. This is often one of the main problems in refloating sunken or flooded vessels. *Figure 9.20* show floating cranes being used to provide stability whilst the vessel is being pumped out.

9.5.2.1 Stability (effect of ground reaction)

When a ship is stranded, part of its weight is being borne by the ground and this causes an upthrust at the ship's bottom. This has the same effect as if a weight had been removed from a position at the keel and adversely affects the vessel's stability.

As with the problem of free surface, provided the vessel is fully aground it is prevented from capsizing by the ground itself supporting the bilges. If, however, the vessel is not flat bottomed, only aground at one end or passing through this stage when refloating, then the loss of stability caused by this upthrust must be kept in mind during any refloating operation. In practice, most vessels have a sufficient reserve of stability and this aspect does not cause problems, but it should always be considered. *Figure 9.21* illustrates the upward forces on a grounded vessel. The weight of the vessel acts downwards through the centre of gravity G. The force P acts upwards through the keel and is equal to the ground reaction. For equilibrium, buoyancy (B) provides the balance, the upthrust and this also acts upwards but through the metacentre M. When these two parallel forces are analysed, it can be shown that the vessels statical stability is virtually reduced by the amount $P \times KM/W$, where P is the upthrust from the sea bed, KM is the distance between the keel and the transverse metacentre and W is the vessels displacement.

9.5.3 Flooding

When flooding is carried out intentionally, e.g. the flooding of a ship's hold to extinguish a fire, it should always be remembered that this is a highly dangerous process and should not be undertaken lightly and preferably not before the stability and strength implications have been assessed.

Passenger ships are accepted as being particularly vulnerable to capsize when fire fighting is taking place in port. Some of the world's better known salvage operations can be directly traced back to incidents where the consequences were not considered before water was pumped in to extinguish a fire. The list includes such names as *Normandie*, *Empress of Canada*, and the *Seawise University* (ex *Queen Elizabeth*).

If the flooding is intentional and controlled then once the immediate danger is passed, removing the water is not normally difficult. It is usually only a matter of reversing the operation. If the vessel has not been damaged and the flooding was not accidental then most dewatering operations are carried out using the vessel's own or small portable pumps.

If the flooding was accidental, the result of some casualty, such as striking a rock, explosion or collision then before pumping the vessel must be sealed. Patching or sealing a hole before pumping is often very problematic and indeed very often the damage cannot be reached because the vessel is lying or sitting on the opening.

The most common method of fitting a patch is to use timber or steel attached to the shell by either hook bolts or welding (*Figure 9.22*) or where large holes are involved using a cox gun. Where patching is necessary the conditions for underwater welding are seldom ideal. For example changes in the hydrostatic head, which are caused by wave action or ship movement which occurs even when conditions appear to be flat calm and perfect, invariably cause water to flow into and out of the flooded compartment. When the flow rate is high, it is often necessary to restrict the work done by divers to short periods when no tidal rise or fall is occurring.

Figure 9.20 Sheer legs providing stability during pumping out (courtesy of Smit International)

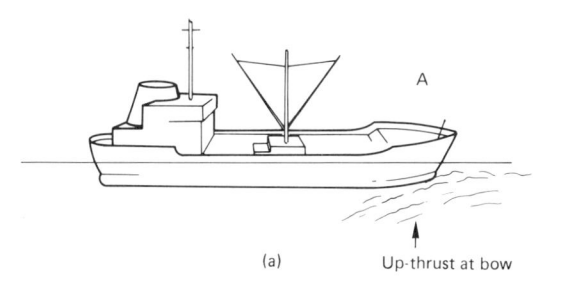

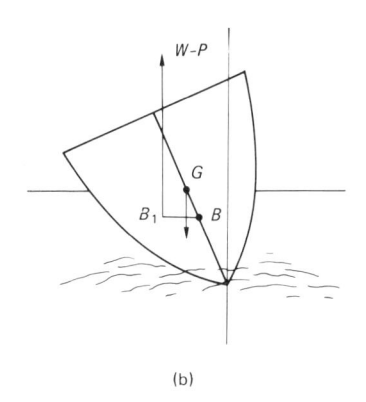

Figure 9.21 Buoyancy after grounding equals weight of vessel minus upthrust; (a) small vessel grounded forward (b) cross-section showing equilibrium

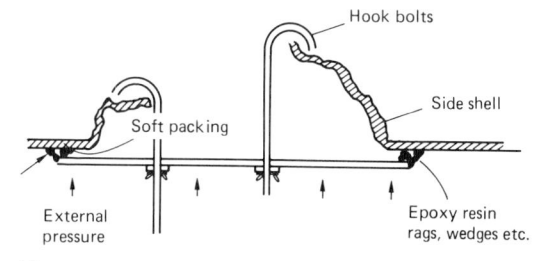

Figure 9.22 Patch using hook bolts

Soft packing is usually placed between the patch and the shell to try to stop the major flow of water and polythene sheeting or epoxy resin is often used to cover a patch or to seal the lesser leakages around the patch. When sealing patches, it is essential that the space is also pumped out concurrently to provide a positive head of pressure (*Figure 9.23*).

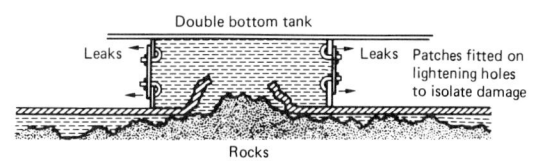

Figure 9.23 Water pressure distorting an internally fitted patch, causing a leak

9.5.3.1 Inside patching

As a stranded ship is often lying on the damage or hole, it is sometimes necessary to try and seal a breach in the hull from the inside. Where patching is being done from the inside, it is often extremely difficult to stop a leak. Many internal patches will start to leak as the pressure builds up during the pumping out stage (*Figure 9.22*).

In addition to the problems associated with the patch itself internal patching work is also highly dangerous as:

1. The diver is working in bad conditions in that the only light is that taken down.
2. It is extremely easy for a diver, even with guide ropes, to get lost and or disorientated.
3. The divers are often working in conditions where the visibility can be measured in centimetres, which makes any work slow and laborious.
4. Divers should not work in the vicinity of an operational pump.

9.5.3.2 Cement boxes

Cement boxes are simply constructed by building a framework of shuttering around the leak and filling it with concrete. As cement will not set where water is flowing, drain pipes must first be fitted to divert any flow. Once the cement has set, the pipes can be capped (*Figure 9.24(a)*). Cement boxes can also be built underwater and the concrete delivered by hose or canvas shute. On occasion, large quantities of concrete have been pumped into a vessel's hold or tank to try to seal leaks that proved inaccessible or difficult to stem. Whilst they are still frequently used, large cement boxes often prove very expensive to remove and indeed the final repair can be increased severalfold because of the removal problems associated therewith.

If circumstances permit, it is often better to construct a steel box around the leak possibly using the ship's frames or floors to form some of the sides and to isolate the damage by welding whilst keeping the water down using pumps. *Figure 9.24(b)* shows such a repair.

9.5.3.3 Cox's bolt gun

It often transpires that, before a flooded vessel can be pumped out, it is necessary to carry out patching operations. The damage may have resulted from a stranding, collision, mines, explosion or other type of casualty and the hole can be any shape or size. Where vessels are holed significantly below the waterline, such as may be the case following a collision or explosion, the patching can be a major project.

Where the hole is small, patching or plugging is relatively easy and can involve materials as common as wooden wedges and rags or small plates and packing held in place by hook bolts or welded brackets. Conversely, if the hole is large, then the plate must be stiffened to withstand hydrostatic pressure and the

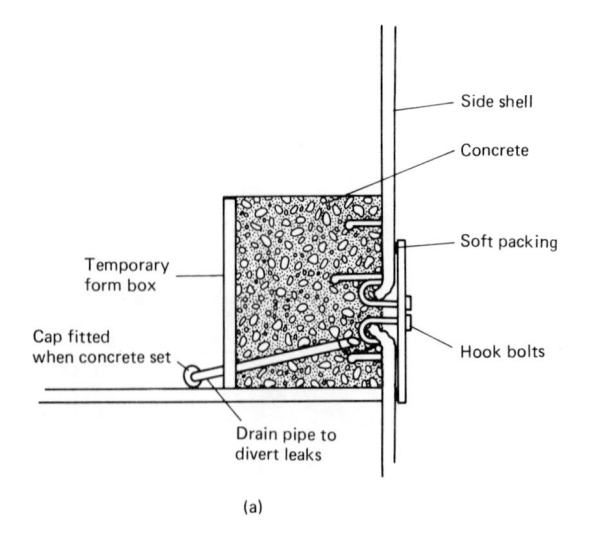

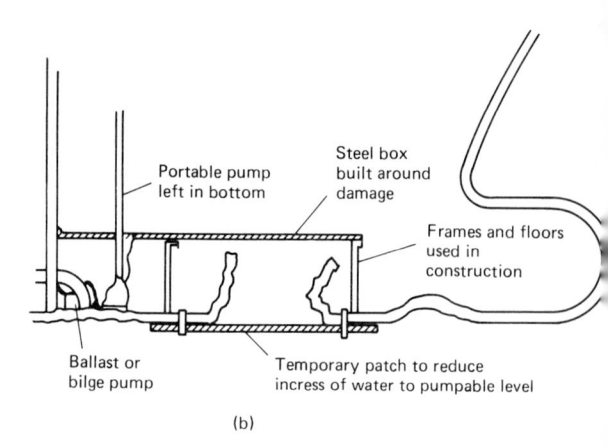

Figure 9.24 (a) Cement box; (b) Steel box built around floors and frames

weight of the patch can be substantial. The plating can still be held in place using hook bolts or indeed by underwater welding, but it becomes much more cumbersome and time consuming.

One method of attaching a patch is by the use of a gun which can fire threaded bolts into the shell of the damaged vessel. These were widely used during World War II for injecting air into submerged submarines. The gun fired hollow bolts through steel shell plates. The potential for this and the securing function soon became apparent (*Figure 9.25*). The bolts are driven by an explosively activated gun designed to drive solid or hollow bolts into steel plate either above or below the surface. The bolts can penetrate up to 25 mm (1 inch) high strength steel plate.

As high temperatures are generated, extreme care must be taken to ensure that the bolts are not driven into spaces containing an explosive mixture. The bolts can be placed around the hole and then used to hold the

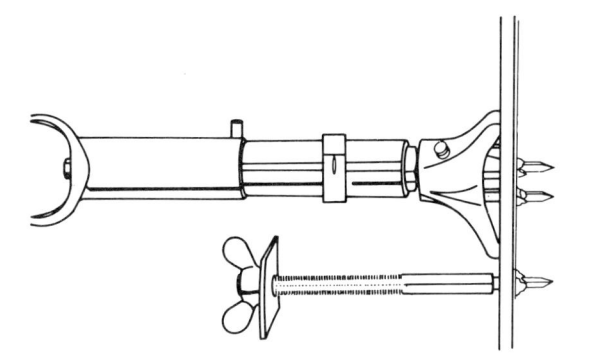

Figure 9.25 Cox submarine bolt driving and punching gun for ship's plate

patch in place, extra bolts should be used, because on occasions the quality of adherence may be suspect.

In addition to the securing of steel patches the bolts can be very useful in the construction of cofferdams, where a structure must be held in position. For cofferdam construction and wooden patches where longer bolts are required, extension bolts are available that screw on to the fired bolt. If timber baulks of up to 12 inches are being used holes can be drilled in the timber beforehand and then the bolts driven through these holes into the steel hull.

This device fires an explosively powered projectile and extreme care must be taken in handling the gun and ammunition. The gun, whilst simple and easy to operate, could with careless handling cause injury or death. Thus the same precautions must be taken as when handling any other type of firearm.

Manufacturers stress that the following precautions must always be observed:

1. Never pass a loaded barrel with the muzzle aimed at any part of your own body, or in line with any other person.
2. Never place the hand over the muzzle when inserting a loaded barrel into the gun holder.
3. Never depress the firing catch until ready to fire.
4. Never attempt to use the gun with either the firing catch or retaining catch removed.
5. Never use a charge which is obviously much too strong for the plate being operated upon; the projectile may pass right through.
6. When punching, always ensure that no damage will be caused by the punching of metal or the punch itself. Bolts often pass completely through the plate and may still possess considerable velocity.
7. When shooting into plate, the thickness of which is doubtful, always ensure that no persons will be endangered on the other side, should the charge prove too strong and pass through the plate at high velocity.
8. Always ensure that the firing catch is free and not jammed down through particles of sand or other foreign matter in the mechanism. The firing catch is a safety device and a firing stroke is quite impossible unless the catch is depressed.

9.6 Salvage methods and risks

9.6.1 Cofferdams

It is generally easier to dewater a space by pumping than by air pressure, in that water flow can be stemmed much more readily than stopping air escaping. Where a vessel is only partially submerged and where the hull opening is below the surface it is often possible to extend the opening upwards by building a cofferdam to such a height that any rise in tide or waves will not allow water to enter or flow back into the space. The cofferdam will allow a pump to be placed inside the flooded compartment whilst preventing water from flowing back into the space while it is being pumped out (*Figure 9.26*).

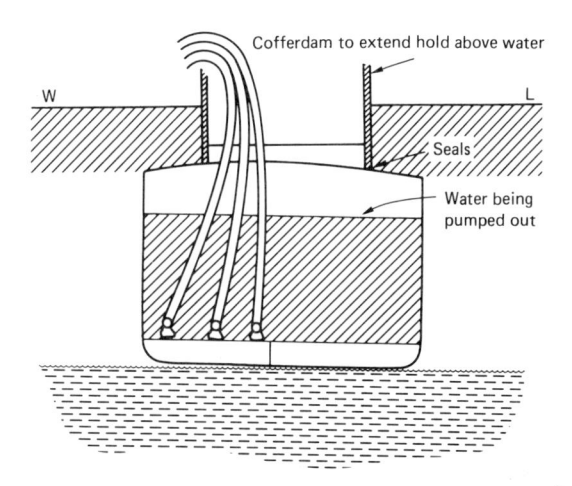

Figure 9.26 Cofferdam built around a vessel's hold

A cofferdam could be a large and complex structure or something as small as pipes fitted to hatch covers or tank lids and extending above the water line. In modern salvage, however, it is unusual for salvors to construct the large complicated cofferdams that were common when labour was cheap and other facilities such as cranes and sheerlegs were not readily available. A great disadvantage with a complex cofferdam is that it is very weather sensitive and weeks of work can be wiped out in a few minutes.

9.6.1.1 Underwater epoxy resin

Most people will be aware of the many brands of epoxy resin that are available for the repair of corroded and damaged car bodies. This resin is similar in many ways to the type used by salvors to seal leaks around patches.

The resin is produced by some of the major oil companies and adapted for specific purposes by specialist formulators. Additives have been developed so that the resin can be applied under water and adheres to almost all clean surfaces.

Whilst it was originally produced for the civil construction industry for sealing leaks and cracks in road or concrete structures epoxy resin has proved the ideal solution for sealing small leaks under water. It is now widely used in the marine and offshore oil industry.

A hardening agent is added to the resin and the mixture sets rock hard within hours. The specially adapted resin and hardener are mixed above water to a putty-like consistency and can then be applied by divers. Epoxy resin suitable for underwater application has been available since the 1960s and is now frequently used. Several brands are available and some of the smaller salvors simply use the car repair variety with good results. One manufacturer gives the life, during and setting times of its products (*Table 9.3*).

Figure 9.27 Fire fighting operation on a large tanker, hit by a rocket (courtesy of The Salvage Association)

Table 9.3 Pot life and cure times (courtesy of Sealocrete Products (UK) Ltd)

Temperature of Wataseal substrate and environment °C	Usable life 1 litre mix	Setting time	Cure schedule
5	5 1/2 h	24 h	1 wk
10	3 1/2 h	12 h	72 h
15	2 1/2 h	8 h	48 h
20	1 h 10 min	4 h	38 h
25	45 min	2 1/2 h	22 h
30	25 min	2 h	18 h

For a good seal it is important that the patch and shell plate is free of grease, marine growth or loose rust scale, as otherwise the epoxy resin when hard will be adhering to an unsound base which might come away. It must be applied in the direction of flow so as to push the epoxy into the opening. To achieve this a positive head of pressure must be maintained by pumping the compartment. If applied correctly, the seal that can be achieved is often near perfect and infinitely better than that achieved using wedges, rags, polythene or cement.

9.6.2 Fire

This is probably the most horrific of casualties that occur to ships. Its type and ferocity will depend on its location on board, i.e. cargo spaces, accommodation, or machinery space. On vessels where a severe fire has taken place it is not uncommon to find the actual steel and superstructure melted and this is indicative of the very high temperatures generated and the difficulty in extinguishing it. *Figure 9.27* shows a large tanker on fire after being hit by a rocket.

Frequently, fire is followed by flooding which can be either carried out intentionally to help extinguish the fire or accidentally where gaskets are burned out of pipes in the engine room or the plating is damaged or distorted by the heat.

When severe fires take place the fire fighting is often more containment than actual extinguishing. The complexity of alleyways, staircases and bulkhead openings usually preclude the possibility of fire fighters entering the smoke filled and hot accommodation module and on many occasions the fire must burn itself out or nearly so. If large capacity monitors are used carelessly considerable extra damage can be incurred to the burning structure; however, if a tug must fight a fire with monitors, then it is probable that the area involved will be destroyed within the boundaries being sprayed.

9.6.2.1 Fire monitors

Most deep sea salvage tugs are fitted with fixed water and foam monitors, and in addition transportable fire pump sets have been developed by some of the major salvors. These portable fire pump sets are designed so that they can be flown to a casualty anywhere in the world at a moment's notice and are capable of delivering 10 000 l/m with a range 85–100 m. Though vessels have been constructed with monitors capable of moving 4000 tons/h, in most instances this would be about 1000 tons/h. *Figure 9.28* shows a typical fire monitor and performance details.

A number of vessels have been built for fighting fires on offshore structures and carry classification society notation to this effect. The society takes into account fire fighting capability, stability and position keeping capability when using the monitors and self protection against external fires. If satisfied the classification

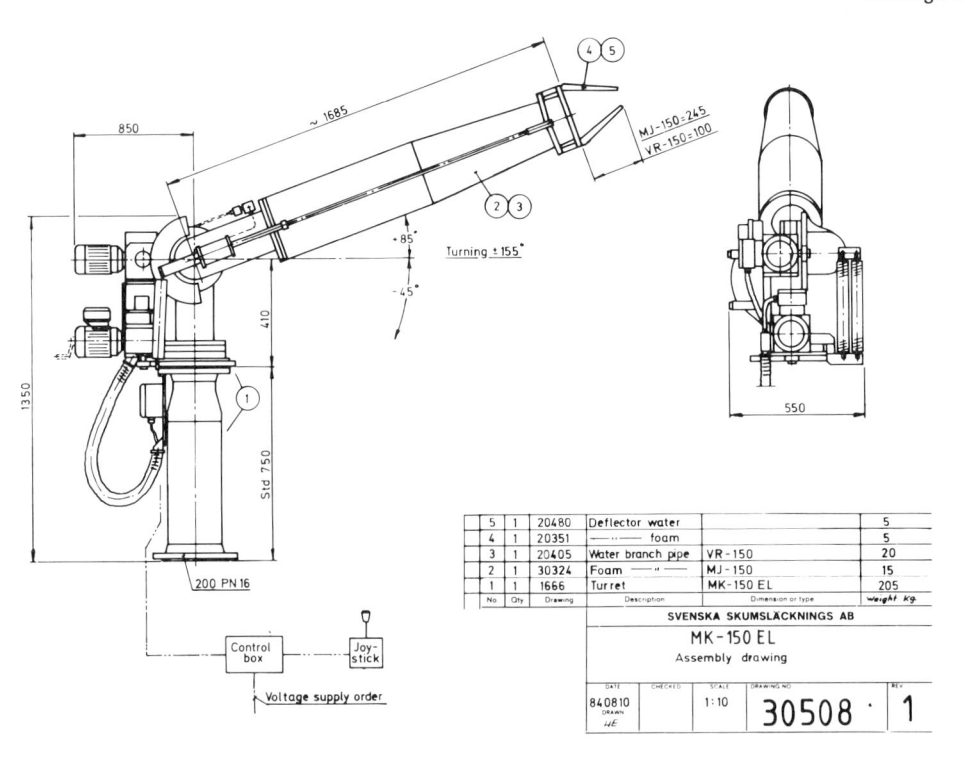

Electric Remote-Controlled Fire Monitor MK-150 EL equipped with
either MJ-150 combined foam/water branchpipe
or   VR-150 water branchpipe

Drawing 30508

Inlet pressure, MPa	0.5	0.8	1.1	0.5	0.8	1.1	0.5	0.8	1.1
Water capacity, l/min	5000			9000			13000		
Range of jet, MJ-150, m (approx)	55	70	77	60	77	90	65	82	96
Foam expansion ratio (approx)	1:10 (depending on type of foam conc.)								
Range of jet, VR-150, m (approx)	65	80	90	70	85	100	75	90	110
Reaction force, N	2600	3300	3900	4425	5775	6850	5525	7675	9350

NOTE: Both the MJ-150 and the VR-150 branchpipes are individually calibrated prior to delivery for any water capacity/pressure relation within 5000–13000 l/min and 0.45–1.40 MPa.

Figure 9.28 SKUM fire fighting monitor

society may give the vessel the notation: Fire Fighter (FI FI) I, II or III. Notation I implies that the vessel has been built for early stage fire fighting and rescue operations close to the structure on fire. Notation II or III implies that the vessel has been built for continuous fighting of large fires and for cooling of structures on fire.

The water monitor system capacities vary with the FI-FI notation as shown in *Table 9.4*.

Vessels carrying the notation III should also be equipped with two foam monitors, each of capacity not less than 5000 l/min with a foam expansion ratio of maximum 15 to 1. In addition the foam concentrate tank is to have capacity for at least 30 min of maximum foam generation from both foam monitors. It is worth noting that the reaction force from these monitors is considerable, such that it is said that four monitors when operating if pointed astern can drive a supply

Table 9.4 Water monitor system capacity

Notation	I	II	III
Number of monitors	2	3 or 4	4
Capacity of each Monitor in m³/h	1200	2400 or 1800	2400
Number of pumps	1–2	2–4	2–4
Total pump capacity (m³/h)	2400	7200	9600
Length of throw (m)	120	150	150
Height of throw (m)	45	70	70
Fuel capacity (h)	24	96	96

vessel at several knots; thus on occasion position keeping can pose problems.

In order to save some vessels the salvage team must board the vessel to finalize the extinguishing operations. This is a hazardous business and often the salvage crew or fire fighters are unaware of what the ship contains or what gases are being produced.

A modern method of extinguishing fires is to use high expansion foam, sometimes made with nitrogen. It is not usual to find high expansion foam or inert gas plants onboard a tug and if they are required, then in all probability they will have to be flown from home base or obtained at the nearest port. Portable fire monitors are readily available which can be located onboard a tug or supply vessel. Most cargo vessels are fitted with extinguishing systems using CO_2 or halogen but these installations are normally small and only of use for extinguishing fires if caught at once.

9.6.3 Pumps

Pumps are by far the most important item of mechanical equipment used by the salvage industry and stocks of pumps are kept by all salvors of consequence. They can also be obtained quite readily from the many pump manufacturers and plant hire firms. The types of pump generally available can be powered by electric, hydraulic, air, diesel, petrol or steam prime movers. The type used depends to a large extent on the application, though during many salvage operations the governing factor is simply the type available.

9.6.3.1 Diesel pumps

Diesel pumps are used extensively in the offshore oil industry during loadout operations where a barge must be kept both level with the quay and also without trim. For such an operation, where it is often necessary to have pumping capacity to match a rise or fall of tide together with trimming moments caused by large moving weights, they are ideal. These large pumps are however very bulky and difficult to handle and can weigh up to 2 tonnes each. Diesel pumps are sometimes used in salvage operations but because of their bulk and difficulty in handling by a small crew, they are only used when they can be lifted by crane or derrick. They do however have the advantage that once in place with the hoses rigged they can work for very long periods with only minimal maintenance, moving large volumes of water at low pressure. Like all non-submersibles pumps they are limited to a suction head of about 8 m.

9.6.3.2 Electric submersible

These pumps are the most versatile and are found on almost all salvage tugs (*Figure 9.29*). They are also generally available onshore for hire and are easily transportable. The advantages include their ease of handling, the fact that no preparatory installation work is required and that they can be lowered directly into the flooded space with the discharge hose led overboard (*Figure 9.30*).

Depending on the pump specification, they may or may not be intrinsically safe for use in explosive atmospheres. The designs cover a variety of head pressures and volumes. The pumps also come in various capacities and designs and suitable for most uses, the large 25 cm type can pump up to 1000 tonnes/h depending on the type of pump and head of water. Smaller diameter pumps are available which are designed for use in confined spaces or placing through small openings and it is these that find favour with most salvors. *Figure 9.31* shows a selection of electric pumps together with performance curves.

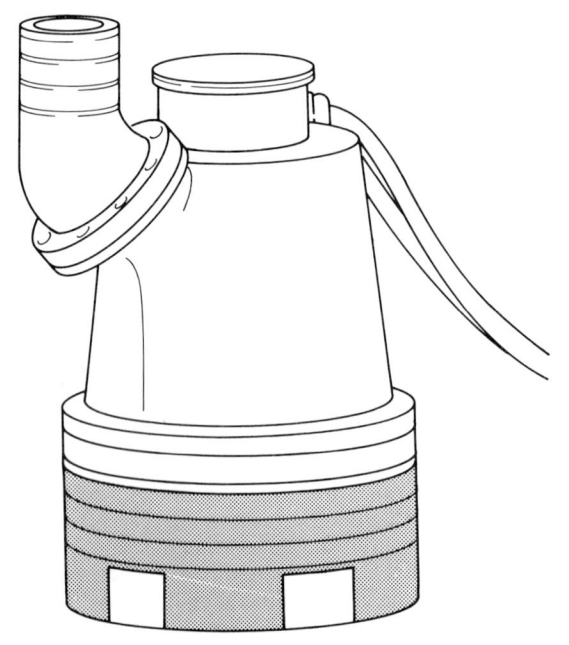

Figure 9.29 Typical electric submersible pump (courtesy of Flygt Pumps)

Figure 9.30 Selection of submersible pumps and hoses ready for placing in a sunken vessel (courtesy of The Salvage Association)

9.6.3.3 Air pumps (diaphragm pump)

These pumps are not normally high capacity but they are usually very dependable and safe for use in hazardous areas. They will operate for long periods unattended and even run dry for long periods without incurring damage and in addition are very light and easy to handle (*Figure 9.32*).

Another advantage of this type of pump is that they can be connected directly on to bilge valve chests or pipes, which can be very useful where access to the flooded compartment is obstructed.

They generally cannot operate against high pressure and some diaphragm materials may not be resistant to oil or chemicals. The pumps also pulsate when being operated and need to be lashed securely during operation. In a space which is over-rich in hydrocarbon gas consideration must be given to the effect of the exhaust air/gas mixture and the possibility of producing an explosive atmosphere.

9.6.3.4 Hydraulic pumps

Hydraulic pumps were in the main only introduced into the marine salvage industry after some major tanker casualties where the ship's pumping systems were rendered inoperable. It became clear that pumps were required which were intrinsically safe, able to operate in a gaseous or explosive environment and independent of the vessel's fixed installation.

Most of the available hydraulic pump packages have been designed so that the pump can be lowered through the Butterworth Openings in the tanker's deck. Butterworth Openings are designed for tank cleaning and are normally obstruction free (*Figure 9.33*). Larger capacity units are available for use where large obstruction free openings exist (*Figure 9.34*).

Hydraulic power packs can be either electric or diesel driven and are designed to be readily transportable by air or vessel to a casualty. Many of the major salvors and pump manufacturers maintain emergency pumping systems incorporating power pack, pump, hoses, fuel, etc., ready for immediate use.

9.6.4 Inert gas

9.6.4.1 Explosion

An explosion can take place when a source of ignition is applied to a flammable mixture which is contained in a closed space or one with a restricted opening and the resulting combustion and expansion causes a build up of pressures that the space cannot contain. For a gas to be flammable, it is generally accepted that it must contain at least 11% by volume of oxygen (*Figures 9.35 and 9.36*).

It is fairly normal for a salvor to be working with unknown products and with little reliable information. The risk of explosion during an operation is often the greatest fear of contractors. Many salvage operations take place on vessels where for various reasons the tanks or spaces contain an explosive mixture. Many deaths have been caused when salvors or other contractors have cut into a steel tank which was assumed to be safe but which in fact contained an explosive mixture.

During a salvage operation there are many sources of ignition, some of which are obvious and some less so. Examples of these are cutting into or welding attachments to tanks containing an explosive mixture; explosive driven bolts; open flames (which could be an actual fire), including smoking, lighted matches and cigarette lighters; hot flying particles such as burning embers or incandescent soot from a funnel, electric welding or gas cutting; sparks from tools and friction sources; sparks or arcs from electrical equipment such as portable generators or non-intrinsically unsafe pumps lights or radios and hot surfaces, such as hot machinery, and high temperature steam lines. Other sources may include faults or earths on switchboards or electric wiring; electric storms and lightening; static electricity generation which could result from using CO_2 direct from pressurized cylinders to inert a pumproom and tank spaces; different electric potential between vessels involved in the operation (sparking could occur if an intermittent and low resistance path developed between the vessels). Some materials when damp or soaked in oil are liable to ignite without the application of flame through a build up of heat produced during oxidation, and pyrophoric combustion is another source of ignition.

Three conditions must be fulfilled to cause an explosion or fire and form what is generally known as the fire triangle (*Figure 9.37*). The removal or elimination of any one of the three conditions means that an explosion or fire can no longer occur. All possible sources of

9/34 Marine salvage

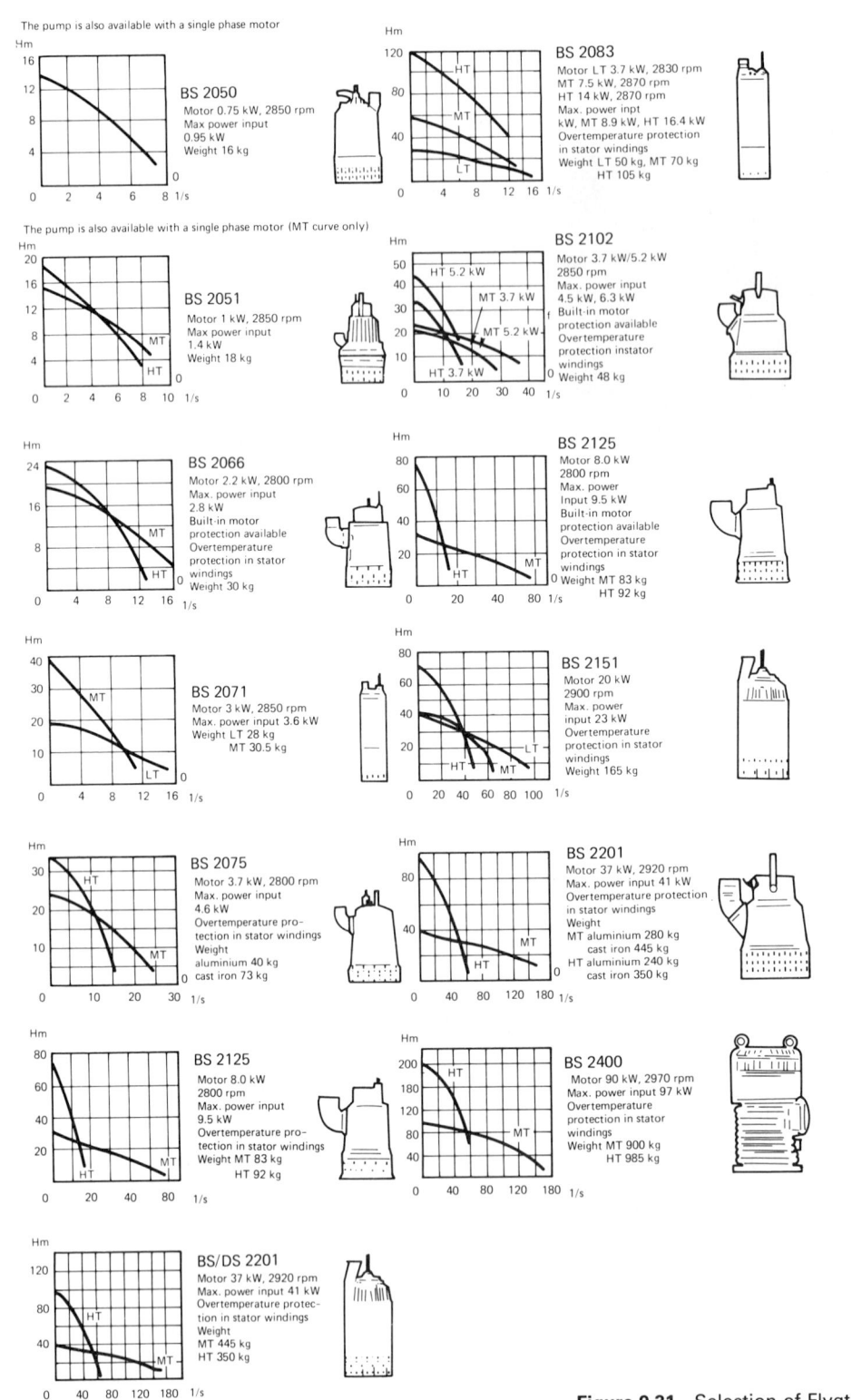

BS 2050 — Motor 0.75 kW, 2850 rpm; Max power input 0.95 kW; Weight 16 kg. The pump is also available with a single phase motor.

BS 2051 — Motor 1 kW, 2850 rpm; Max power input 1.4 kW; Weight 18 kg. The pump is also available with a single phase motor (MT curve only).

BS 2066 — Motor 2.2 kW, 2800 rpm; Max. power input 2.8 kW; Built-in motor protection available; Overtemperature protection in stator windings; Weight 30 kg.

BS 2071 — Motor 3 kW, 2850 rpm; Max. power input 3.6 kW; Weight LT 28 kg, MT 30.5 kg.

BS 2075 — Motor 3.7 kW, 2800 rpm; Max. power input 4.6 kW; Overtemperature protection in stator windings; Weight aluminium 40 kg, cast iron 73 kg.

BS 2125 — Motor 8.0 kW, 2800 rpm; Max. power input 9.5 kW; Overtemperature protection in stator windings; Weight MT 83 kg, HT 92 kg.

BS/DS 2201 — Motor 37 kW, 2920 rpm; Max. power input 41 kW; Overtemperature protection in stator windings; Weight MT 445 kg, HT 350 kg.

BS 2083 — Motor LT 3.7 kW, 2830 rpm; MT 7.5 kW, 2870 rpm; HT 14 kW, 2870 rpm; Max. power input kW, MT 8.9 kW, HT 16.4 kW; Overtemperature protection in stator windings; Weight LT 50 kg, MT 70 kg, HT 105 kg.

BS 2102 — Motor 3.7 kW/5.2 kW, 2850 rpm; Max. power input 4.5 kW, 6.3 kW; Built-in motor protection available; Overtemperature protection in stator windings; Weight 48 kg.

BS 2125 — Motor 8.0 kW, 2800 rpm; Max. power Input 9.5 kW; Built-in motor protection available; Overtemperature protection in stator windings; Weight MT 83 kg, HT 92 kg.

BS 2151 — Motor 20 kW, 2900 rpm; Max. power input 23 kW; Overtemperature protection in stator windings; Weight 165 kg.

BS 2201 — Motor 37 kW, 2920 rpm; Max. power input 41 kW; Overtemperature protection in stator windings; Weight MT aluminium 280 kg, cast iron 445 kg; HT aluminium 240 kg, cast iron 350 kg.

BS 2400 — Motor 90 kW, 2970 rpm; Max. power input 97 kW; Overtemperature protection in stator windings; Weight MT 900 kg, HT 985 kg.

Figure 9.31 Selection of Flygt pumps with performance curves

Salvage methods and risks **9**/35

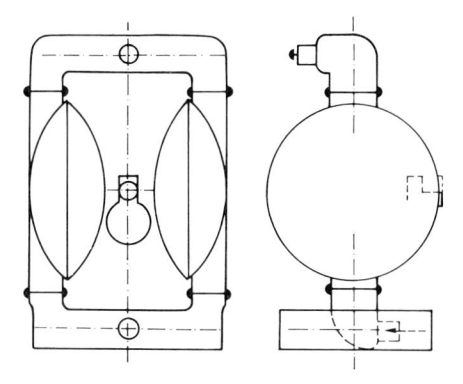

Figure 9.32 Wilden diaphragm pump

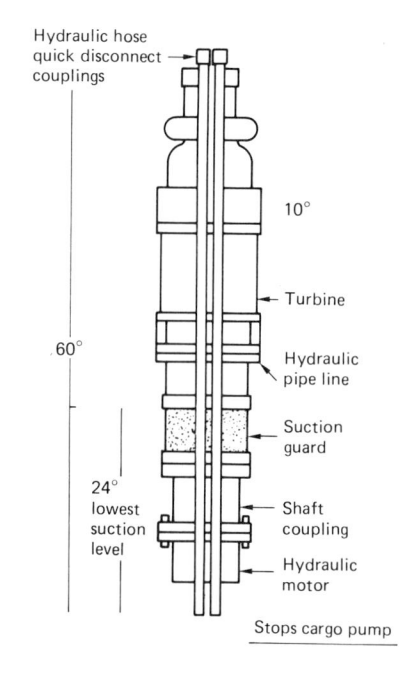

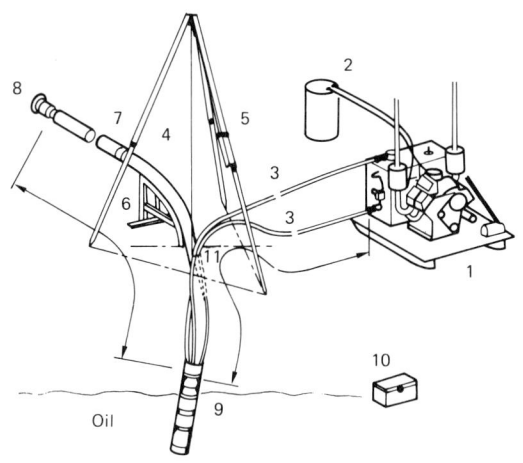

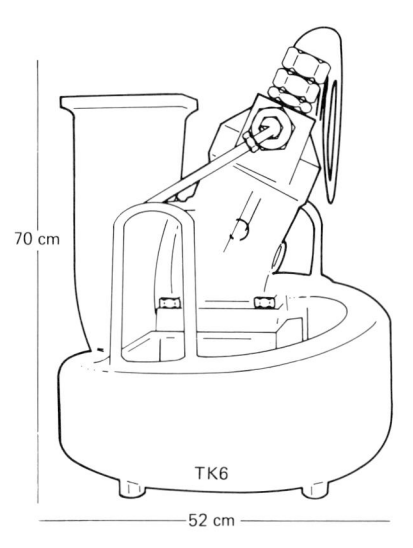

Figure 9.34 Frank Mohn hydraulic pump; maximum capacity 500 m³/h at 35 m head.

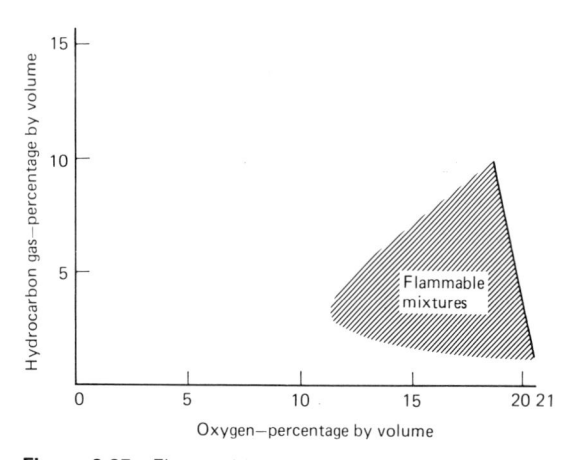

Figure 9.35 Flammable mixture range shown as a percentage of oxygen and hydrocarbon gas

Figure 9.33 General arrangement when using a Stops pump (1) prime mover unit – Bernard diesel engine coupled to high pressure hydraulic pump; (2) fuel supply; (3) hydraulic hoses – max. length 000 ft; (4) tripod; (5) :griphoist; with lifting wire; (6) hose guide; (7) collapsible rubber hoses – 68 ft lengths all connections; (8) to API adaptor flange; (9) STOPS pump unit; (10) tool box; (11) Butterworths/manhole opening in deck.

Figure 9.36 Large tanker following an explosion (courtesy of The Salvage Association)

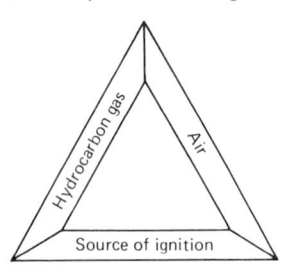

Figure 9.37 Conditions for explosion of fire for hydrocarbon gas mixed with air and ignited (the fire triangle)

ignition should be subjected to rigid inspection and eliminated whenever possible. Pools of oil and other potentially combustible material should be removed when possible and generators, spare equipment and tools should be kept away from the working or gaseous areas.

In 1969, explosions on board three very large crude carriers set in train investigations into the causes with a view to eliminating them. These led to much safer practices and to a situation where most large tankers are now required to be fitted with inert gas producing installations designed to ensure that the gas in cargo tanks can be kept out of the explosive range at all times. Records must be kept regarding the operation of this equipment and are subject to inspection by state officials.

9.6.4.2 Inert gas

Inert gas is normally defined as any gas or mixture of gases which is not capable of supporting a fire and include nitrogen, carbon dioxide and gases containing less than 11% oxygen. It is generally accepted that a fire should not start or be sustained if the amount of oxygen in a confined and gaseous space is less than 11% by volume; but to give a margin of safety an oxygen content of not more than 5% by volume is usually aimed for.

9.6.4.3 Portable inert gas generators

The lessons learned by the oil companies were not lost on the salvage industry and portable inert gas generators were developed. Many inert gas plants have been made air transportable in two modules one of about 1.7 tonnes and the other of about 1.9 tonnes and able to produce up to 1300 m^3 of gas per hour. They are normally used more for the prevention or containment of a fire or explosion rather than extinguishing it. Many of the major salvage companies own portable inert gas plants and they are also available for hire from a few plant hire firms.

Portable inert gas generators were primarily intended for use in tankers but other uses have since been found, such as providing a flow of gas to a ship's hold (ships usually only carry a limited amount of carbon dioxide) which is on fire or likely to catch fire as a result of heating or spontaneous combustion. In addition, inert gases can be compressed and used as a substitute for compressed air as part of a salvage operation where a tank or space contains an explosive mixture.

Many portable inert gas generators come in two parts which can be coupled up on site (*Figure 9.38*). Typical dimensions of each part:
Length: 213 m
Width: 1.30 m
Height: 2.30 m
Total weight of generator: 3500 kg
Capacity: 1000 m^3/h
Composition of inert gas
Nitrogen: 86%
Carbon dioxide: 14%
Oxygen: max 0.7%
Fuel: light marine diesel oil
Consumption: 90–110 kg/h

Other problems associated with the risk of explosion have been encountered in the oil industry and equipment developed to solve them can often be adapted for use in the marine salvage industry.

9.6.5 Hot tap machines

Hot tap machines were developed in the oil industry for cutting holes into tanks which were either full of, or contained, oil under pressure. The tool is used for fitting discharge pipes and valves to a tank for the purpose of pumping out or otherwise removing the oil whilst maintaining the tank's integrity. Before the hot tap tool can be used, a spool piece and gate valve must first be connected to the tank. This is usually achieved using explosive driven bolts. An alternative means of fixing such is friction welding (*Figure 9.39*).

Figure 9.38 Portable CPS generator, owned by Smit International

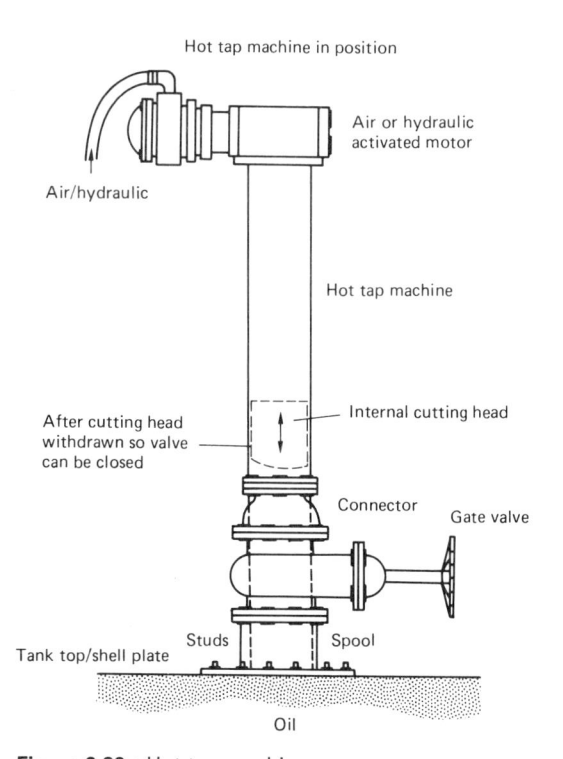

Figure 9.39 Hot tap machine

The tapping tool itself may be air or hydraulically driven and is fitted with a special bit and cutting head. The hole is cut through an open gate valve which is then closed after removal of the cutting tool. The machine is normally operated remotely from its power source, that is the power source is often located on the vessel whilst the machine is working under the water. The tool is most frequently used in the removal of bunkers or fuel from sunken vessels.

Proposals were recently examined for the removal of a cargo of crude oil from a vessel sunk in a depth of some 500 m. To facilitate this removal, it was proposed that the tool would be operated and fitted by remote controlled vehicles and the spool piece would be connected in place using friction welding techniques. The cutting tools most frequently used weigh approximately 300 kg.

9.6.6 Side scan sonar

Sonar is widely used by the military, fishing and offshore oil industry and is often used to support salvage services by way of sea bed profiling, locating lost items, aircraft or indeed wrecks for charting or salvage purposes. The system has been developed from acoustic origins and works on a similar principle to the echo sounder, whereby an acoustic pulse–pressure wave is generated by a sound source and propagated in a given direction. The pulse is reflected back off the sea bed to a hydrophone which converts it back to electrical energy. The depth may be derived from the equation:

$$d = V_p \cdot t_1 \qquad (9.4)$$

where d is the depth and V_p is the velocity of propagation of the sound pulse and depends on the modulus of elasticity of the medium, the temperature of the medium and the density of the medium.

Side scan sonar incorporates two identical systems, port and starboard so that pulses are transmitted and received on either side simultaneously to give a complete stabbed profile of the track.

The resolution is improved when the transmitter is closed to the reflecting medium and as it also helps to eliminate noise and motions, the transmitter is towed beneath the surface and as deep as possible. The unit is located on a fish which emits high frequency acoustic pulses to either side at right angles to its track and angled downwards. The tow depth, or tow fish height above the sea bed, is controlled by the length of cable deployed and also the speed of the towing vessel and angle of the fish fins (*Figure 9.40*).

The acoustic pulses transmitted from the fish are reflected off objects and features on the sea bed and return to the receiver unit which is also mounted on the fish. The signals received pass to the graphic recorder on board the survey vessel via an umbilical. The recorder typically has two channels (port and starboard) which give a continuous permanent strip chart recording. The chart will show the sea bed directly beneath the fish as well as the terrain and obstructions on either side. Modern processing systems can additionally record the data on video cassette for subsequent analysis.

The operator should be aware that features will be relative to the towed fish; however, navigational data on board the survey vessel can also be fed into the processors in order to give absolute positions of features. The range covered on either side of the fish may be up to 600 m.

Targets give a stronger reflected signal accompanied by a shadow on the recording. A shadow behind the object indicates a projection, whilst one in front of the object indicates a hole. An experienced operator can interpret most records at a glance, recognizing not only significant features and objects, but also more subtle data such as the composition and relative hardness of the sea bed, and the shape and condition of submerged objects.

To prevent fishtailing, considerable research has been carried out to ensure a hydrodynamically stable platform. These take various forms varying from a twin fish catamaran, to torpedo shapes and ray fish type shapes.

It is not necessary to use a dedicated survey support ship and in this respect harbour tugs can be readily modified to stream, tow and recover the fish and process the signals. Most tugs can be fitted out in one day.

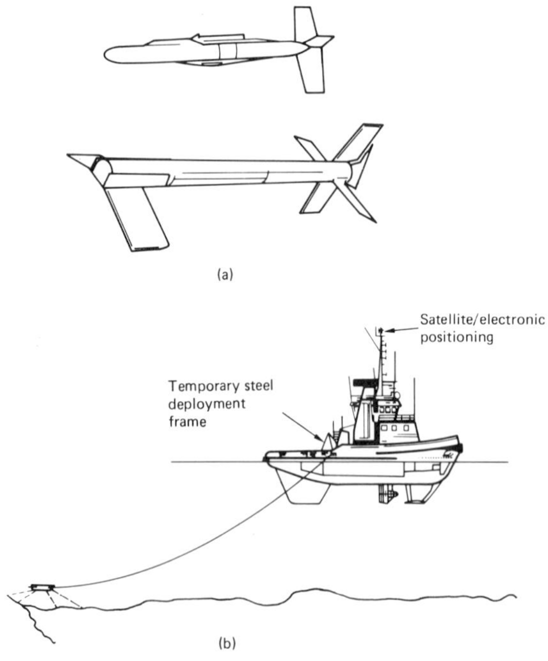

9.6.7 Underwater excavation

9.6.7.1 Dredging systems

Electric submersible gravel and sand pumps have been used industrially for some time in quarries and mines. They have proved useful in salvage operations for the removal of iron ore coal, stone, etc., from flooded spaces. More recently, the offshore oil industry has developed advanced hydraulic systems for removing sand, silt, mud and other debris from around the legs and cell tops of oil rigs. These systems have also been used in the dredging of sand from land quarries and the moving of coal spoils. The installations can be quite large for long term, large scale operations or semi-portable for short term operations.

The portable systems are powered by a skid mounted diesel pump set which supplies water through a flexible hose to a manifold on a submersible frame. A central jet pump and a suction head jet pump are powered from this manifold. The material is drawn in at the suction head and sent back to the central pump where it is further boosted into a discharge line. An additional outlet on the manifold can be used to power a disintegrator for breaking up hard packed or cohesive materials such as clay. As the units are designed for use offshore, they are readily transportable and the total unit can fit inside one (20 ft) container (*Figure 9.41*).

Figure 9.40 (a) Two of the many types of Sonar fish; (b) Side scan Sonar as fitted on a small tug

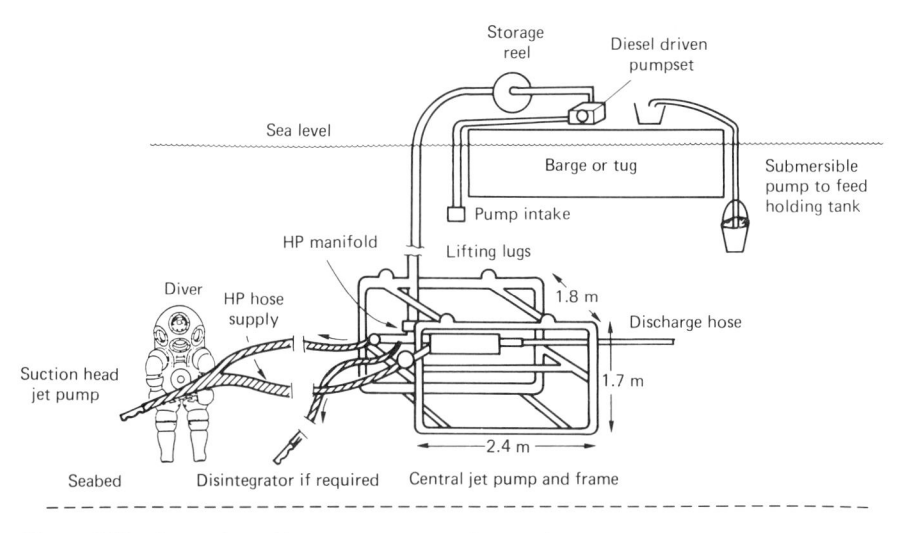

Figure 9.41 General working arrangement of an underwater excavation system (courtesy of Genflo Subsea Ltd)

As the removal of debris from around oil rigs is carried out on a regular basis, it is likely that other developments such as the above will eventually find their way into conventional salvage operations but meanwhile they are rarely available and where excavation is needed a simple air lift is used.

9.6.8 Air lift

9.6.8.1 Principle of air lift

The introduction of compressed air into the base of a submerged pipe reduces the hydrostatic head over the water inside the pipe as opposed to the water outside the pipe. This difference in pressure creates an upward flow of water which will carry with it particles of mud, sand, stones, etc.

The efficiency of the air lift depends on the construction, but in salvage operations the differences in efficiency are not normally a major factor. Almost any type and size of pipe can be used and the design will depend not only on the requirements of the salvage operation but also on the available material and the compressed air capacity on site (*Figure 9.42*).

The pipe should be durable and capable of taking the air connection and withstanding the pressure differential. If time and material allows and to reduce the risk of damage, the bottom section should be made of steel section; however, during one salvage operation a coal cargo was removed using a 30 cm rigid plastic drainage tube. Water is sometimes pumped down by external pipes to a point slightly below the airlift orifice, in order to both agitate the sea bed or cargo and prevent material clogging up the pipe. If cargo is being dis-

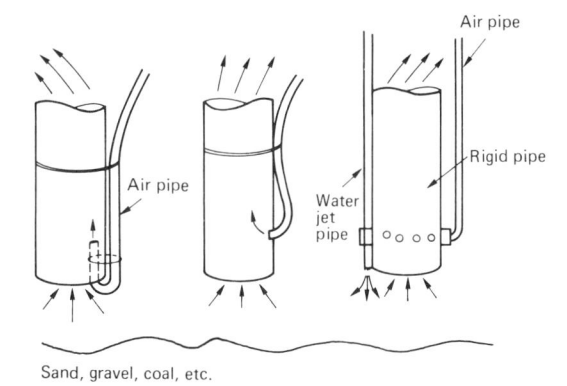

Figure 9.42 Air lift pipes

charged from an enclosed hold then a continuous supply of water will be required as the air lift also removes water from the space.

9.6.9 Beach gear

Beach gear or ground tackle (*Figure 9.43*) has been used in refloating or uprighting ships from the earliest time when a rope secured to an anchor or other weight was used to pull a stranded vessel free or parbuckle a vessel upright.

On sailing vessels, the use of tackles was commonplace and these were used whenever extra pulling power was required. The principle of beach gear has not changed much over the years but now instead of hand power, winches are used and the anchors are

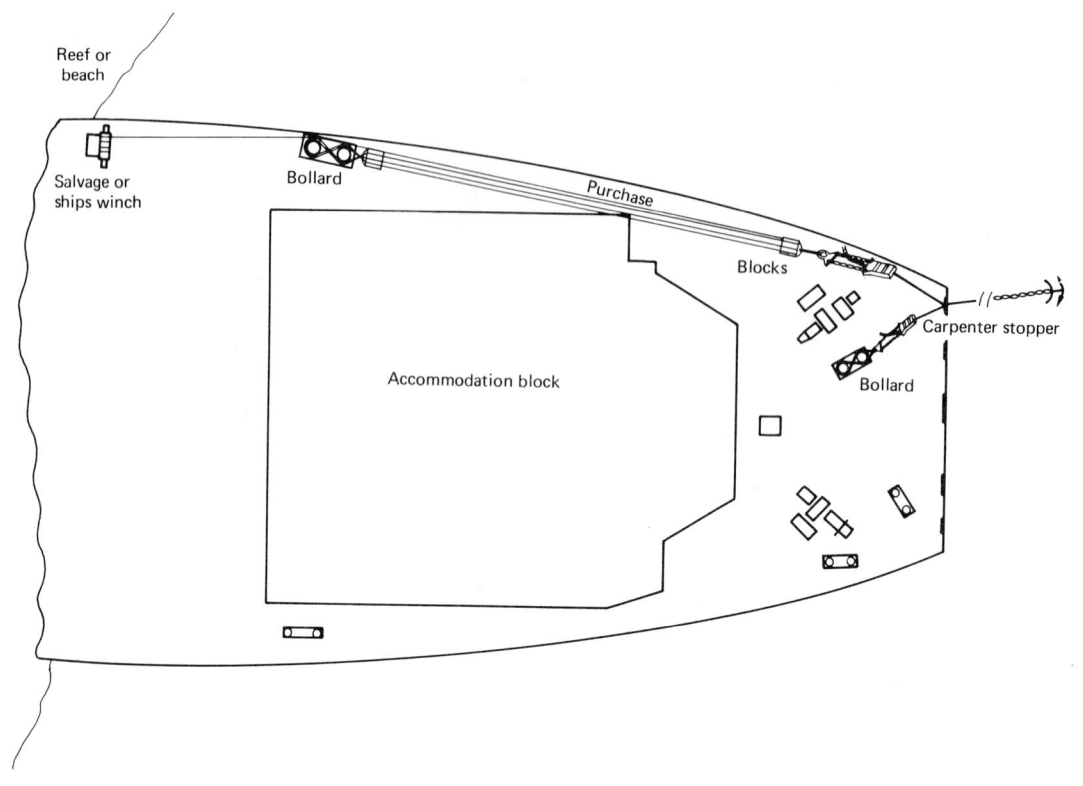

Figure 9.43 Beach gear or ground tackle rigged on the stern of a stranded vessel.

more efficient. *Figure 9.44* shows a selection of anchors. In addition to its weight, the ability of an anchor to hold depends on a number of factors, such as the shape of the anchor flukes; the fluke angle in relation to the shank; the sea bed; the angle of pull exerted on the shank in relation to the bottom; the distance the anchor is dragged over the bottom to permit it to dig in; and the scope of and weight of the anchor chain.

In some recent salvage operations where the holding capability of the sea bed was in question, the anchors were replaced by piles driven into the sea bed.

The purpose of beach gear is normally twofold, to hold the vessel in position while preparations are made to salvage it and to help the salvage operation by adding pulling power when it is required.

Whenever a vessel is stranded on an exposed shore and whether the exposure is to weather, sea or currents, it is outside of its natural environment and the objective must be to refloat it as soon as possible. Most salvage operations cannot be carried out instantaneously and they sometimes take considerable time to arrange; e.g. repairs might have to be carried out to enable the vessel to float or some cargo might have to be discharged. During this time the vessel must be protected if possible and prevented from moving further inshore. *Figure 9.45* shows ground tackle rigged on board the SS *Manhattan* aground off Korea.

It is also very important that the vessel does not refloat itself and move into deep water before all arrangements have been made to ensure the vessel will float safely. To illustrate this danger of not securing a vessel against moving inshore, we relate an actual case.

A loaded vessel was grounded in the mouth of a river in Africa at neap tides. The position was relatively exposed and the ship was subjected to continuous but not severe wave action. The ship was virtually undamaged and a local tug was engaged to refloat it on the basis of Lloyd's standard salvage agreement. The salvors were not able to free the vessel at the first attempt and they decided to wait until the spring tide when the water would be deep enough to float the vessel. As the tide's wave action pushed the vessel daily further inshore so that even at the top of spring tides the tug was still unable to refloat it, concern was felt and eventually a salvage officer was dispatched to the scene. Unfortunately by this time the vessel had moved so far inshore that whilst it was still relatively undamaged and it was still technically possible to refloat it, the cost of doing so was very high. The vessel was subsequently abandoned when it became apparent that the cost of refloating the vessel would exceed the insured value.

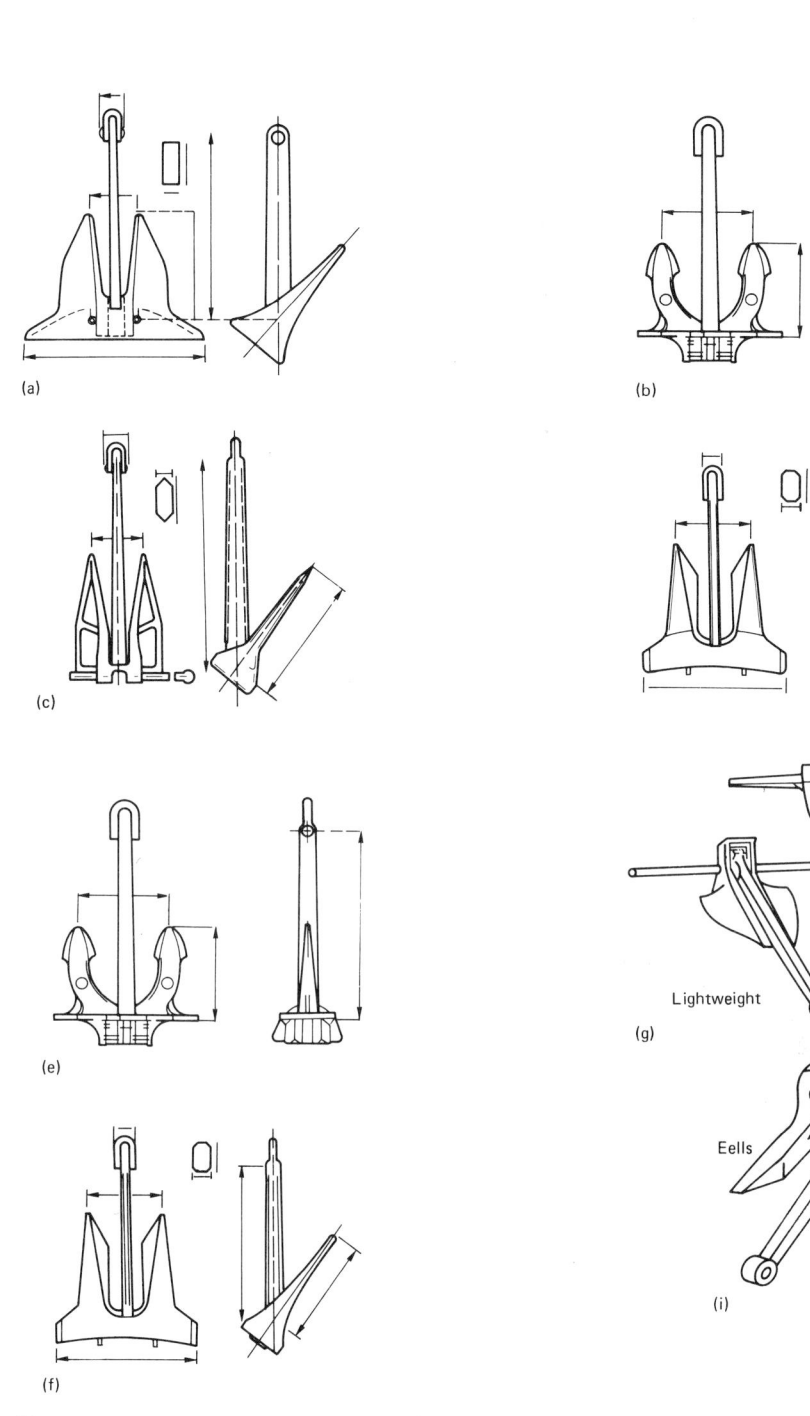

Figure 9.44 Selection of anchors (a) Stokes high holding stockless anchor; (b) union stockless anchor; (c) meon mark 3 high holding anchor; (d) AC14 high holding stockless anchor; (e) union stockless anchor; (f) AC14 high holding stockless anchor; (g) lightweight anchor; (h) stato anchor; (i) eells anchor

It is clear from the above that had the vessel been secured in position the rising spring tide would have refloated it without much effort or cost and the equipment was available locally to do so.

In addition to ground tackle, tugs can also be used for the purpose of holding a vessel in position and are frequently used to do so. The disadvantages are that they are not always able to maintain station in bad

9.6.10 Carpenter stopper

Carpenter stoppers form an integral part of any ground tackle where large wires must be handled. The stopper grips and holds the steel wire without causing kinks or damaging the wire rope. It works on a wedge principle. The stopper is closed around the steel wire and the hinge kept closed by a locking pin. The wedge after being lightly tapped into position will, when the load is applied, automatically tighten. The stopper can only be loosened by removing the locking pin and is designed so that it can be opened and closed by one man using a heavy hammer. *Figure 9.46* shows a carpenter stopper with the cover opened to show the wedging principle.

Figure 9.45 Ground tackle rigged on the bow of a stranded tanker (courtesy of The Salvage Association)

weather or over a long period. Whilst large tackles can be rigged, the most commonly used and easily manageable ground tackles develop a pull in the order of 40–50 tons and when this is compared with the pull developed by the average tug, the advantages can be seen. For example, the bollard pull of even the largest tug is only in the order of 200 tons.

Standard beach gear for refloating, i.e. anchors and tackles rigged onboard the casualty, is still routinely used in salvage operations in the Far East, but is becoming a rarity in Europe where high powered tugs or salvage craft are available equipped with deck tackles or hydraulic rams.

9.6.10.1 Lifting

When vessels sink in anything other than shallow water, the salvage cost is normally so high and the flooding damage so extensive that the cost of salvage and repair exceeds the value of the vessel. Notwithstanding this, small vessels such as tugs are frequently lifted from water of depths of up to 50 m deep.

9.6.10.2 Sheer legs

Whilst not originally designed or intended for salvage work, they have become an integral part of it. Sheer legs are simply advanced A frame type lifting structures fitted on a suitably strengthened pontoon (*Figure 9.47*).

Figure 9.46 Carpenter stopper

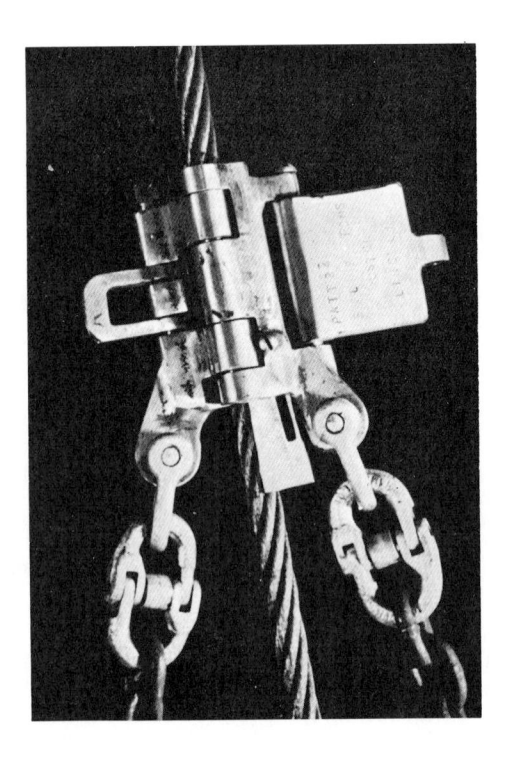

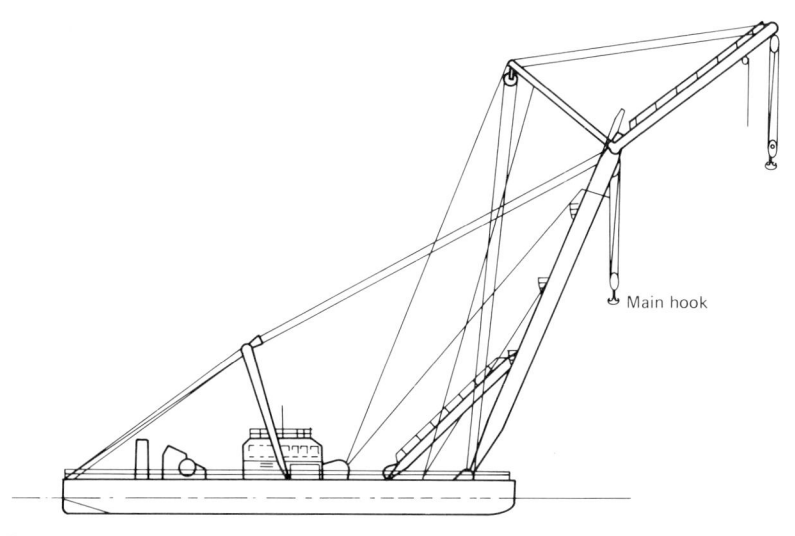

Figure 9.47 Small sheer leg

They are usually purpose built together with the pontoon and are often fitted with accommodation and propulsion units for manoeuvring purposes. These propulsion units are not normally intended for use on long distance moves and the sheer legs are towed from one location to another. Many of the sheer legs are designed whereby the A frame can be lowered and stowed during exposed sea passages.

Modern sheer legs can lift between 50 and 3000 tons on the main hoist tackles. In addition to the main hoist many of the larger sheer legs are fitted with what are generally termed gin tackles and bow rollers. These are essentially deck tackles and can be used to lift objects up to water level. When making a combined lift using main hoists and deck tackles even some of the smaller sheer legs can lift fairly large vessels up to a level where pumps can be used.

Whilst these units are more often engaged in marine civil engineering and wreck removal type projects they are also regularly used on major salvage operations, not only for lifting but also for providing stability to flooded vessels whilst they are being dewatered. Because of their pontoon shape, sheer legs have a very short rolling period which makes them unsuitable for operation in exposed sea areas in other than reasonable weather. In Europe and the Far East there is a plentiful supply of floating sheer legs but they are less common elsewhere.

The use of these craft have made the salvage of many vessels which would have been very difficult and time consuming relatively easy. Sheer legs are frequently used for removing wrecks, where it is often necessary to break a wreck into manageable pieces using cutting chains (where large chains are led under a wreck and then by using two hooks the chain is used in a sawing action cutting the hull into sections) to facilitate lifting. Wrecks are also sometimes demolished with explosives, or parts of the structure can be torn free and lifted using special wrecking grabs. *Figure 9.48* shows a typical salvage grab.

9.6.10.3 Heavy lift vehicles

During the last 15 years or so lifting craft have evolved from sheer legs to crane barges to crane ships to semisubmersible crane barges. The development was needed not only to increase the lifting capabilities but also to improve the sea keeping characteristics so that cranes could operate for longer in exposed areas (*Figure 9.49*).

Owing to the increase in oil exploration there are now a great number of derrick barges available some having two slewing cranes of up to 7000 tonnes capacity each on the main hooks. The highly damped motions of the large semisubmersible lifting cranes is such that even during severe winter storms the dynamic pitch and

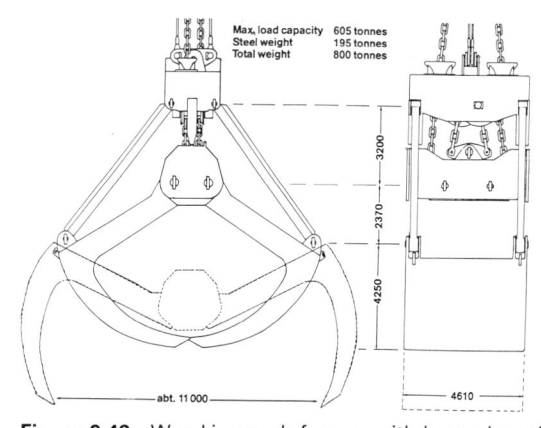

Figure 9.48 Wrecking grab for use with large sheer leg

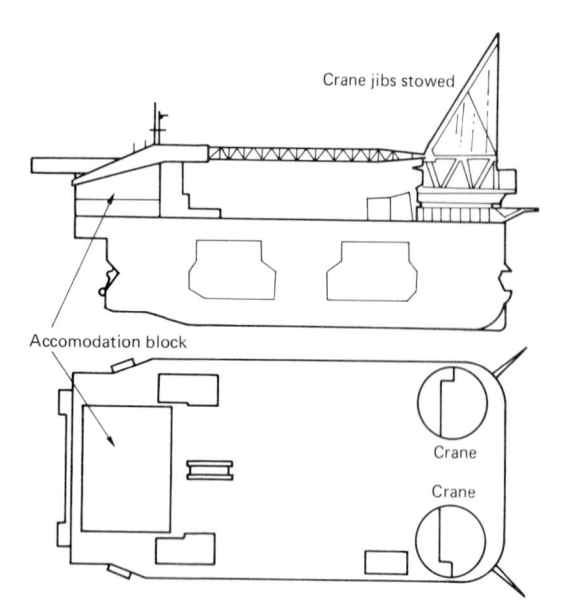

Figure 9.49 Parbuckling operation of a large vessel

roll angles do not normally exceed a few degrees. They do have a disadvantage that as with all types of cranes the wires are generally too short for lifts from all but esturial waters. For special projects longer wires can be fitted or fleeting arrangements where the item is lifted in stages can be adopted.

When a lifting or removal operation is being undertaken on a commercial basis the fund of money available often precludes the use of large and costly pieces of equipment. However, it has often transpired that a salvage or removal that could have been completed in a relatively short time using a large crane or sheer legs, has taken several weeks using less costly methods and on other occasions vessels have had to be abandoned.

9.6.10.4 Compressed air

Compressed air may be used to dewater a flooded compartment in order to restore buoyancy when the space is either open to the sea or it is not possible to use pumps. The method basically entails blowing compressed air into the compartment to displace water through fractures in the shell plating or alternatively via sounding pipes or air pipes (*Figure 9.50*).

There are two separate uses of compressed air in dewatering a compartment, each with its own problems. The first is to dewater a compartment during refloating. The second is to provide buoyancy to a compartment of a sunken vessel as part of an operation to bring the vessel to the surface.

In preparing the compartment for dewatering by this method difficulty is invariably experienced in making the compartments air tight (particularly engine rooms

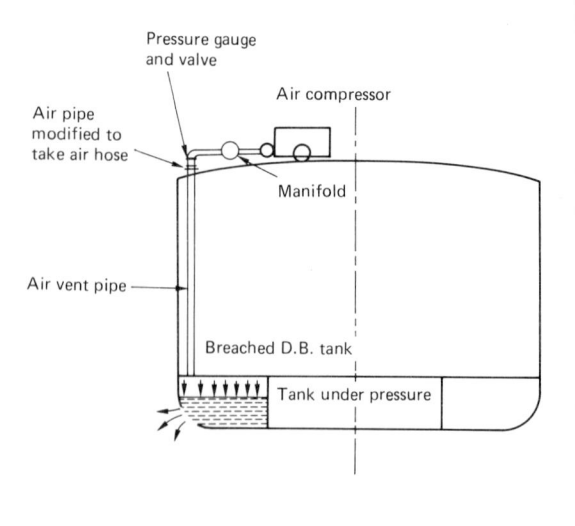

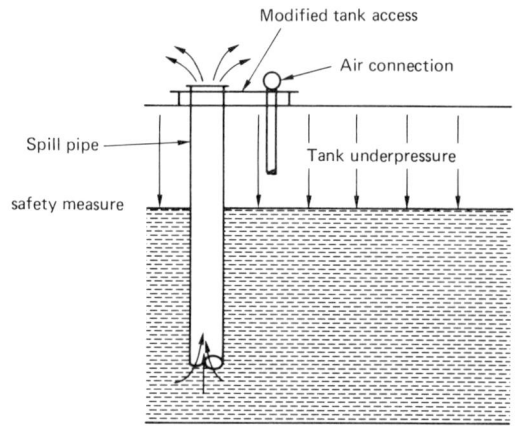

Figure 9.50 Use of compressed air to dewater a flooded compartment

and accommodation) to prevent loss of air to adjacent spaces and ultimately to the surface. Air will escape through the smallest of apertures and the preparation time by divers in sealing bulkheads, electrical conduits, etc., is often protracted. Refloating operations are sometimes delayed while contractors try to locate and seal leaks.

When using compressed air to dewater a damaged compartment where the ingress has been through a fracture or hole then the fractures in the shell plating must be big enough to ensure that not only the water quantity being displaced equals the quantity of air being injected, but that they are also able to cope with the rapidly expanding air whilst the vessel is being raised. On occasion it is necessary to arrange spill holes or spill pipes to take the excess quantity of air.

It should be remembered that tanks, shell and bottom plating are designed to withstand pressure from the outside and care must be taken not to overpressurize the compartment. In addition, hatches or other boundaries may require strengthening to withstand the pressure in the compartment. In practice, the use of compressed air is generally confined to tanks as the work necessary to make large spaces airtight often outweighs the advantages.

The air compressor may be connected to the flooded compartment through convenient air or sounding pipes, adapted to take the air hose and pressure gauges. When a number of compartments are to be dewatered by pressurizing, then it is advantageous to connect the air compressors to a central manifold, often called a pig, from which the supply hoses to the compartments can be led. Air injection is thus more easily controlled (*Figure 9.51*).

Once the compartment is dewatered to the level of the uppermost fracture (or source of leakage) the air valve can be closed. Any escape of air can be replaced periodically by further air injection, thus maintaining the pressure.

Compressed air, although cheap and extensively used in salvage operations to dewater compartments, has the disadvantages that whilst this method is often convenient it is more difficult to control than conventional pumping. As it is not possible to take any soundings, the salvor can only estimate the water quantity that has been expelled by comparing the air pressure of the compartment with the depth of the outside water. In addition, as it is not always possible to examine all boundaries for their strength or indeed integrity, care must be exercised to prevent structural damage. Extra care must be taken when a vessel is being raised because as the water pressure reduces, the air expands and indeed many salvage attempts have failed or damage has been caused by the failure to handle rapidly expanding air.

9.6.10.5 Foam in salvage

The use of foam or foam spheres is very attractive to salvors, as the compartment into which the buoyancy is required need not be air or watertight. The foam or foam spheres is formed by mixing two fluids and is often pumped down in the form of a slurry. Whilst the space need not be air or water tight, it must be strong enough to withstand the upward pressure and if not it must be stiffened. Foam has a great advantage in that its expansion is extremely small with reduction in water pressure. Foam is affected by pressure, in that its density increases with depth so that its effectiveness is limited to depths in the order of 50 m.

On occasion various forms of polyurethane foam have been used to impart buoyancy to sunken vessels (*Figure 9.52*) where adequate cranage was not available

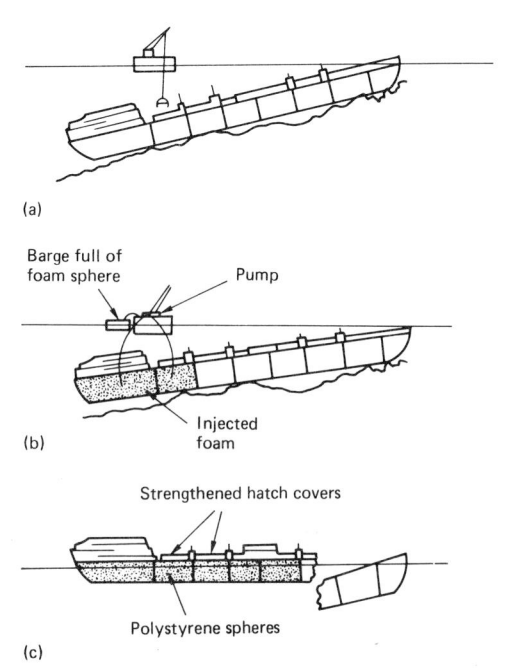

Figure 9.52 Wreck removal of the London Valour using polystyrene spheres: (a) Stage 1 – carco removed and hatch covers stiffened to withstand upward pressure; (b) stage 2 – polystyrene spheres produced on shore, then transported out to the vessel in barges and pumped down; (c) stage 3 – all except bow raised using foam assisted by compressed air, bow section later broken up and removed.

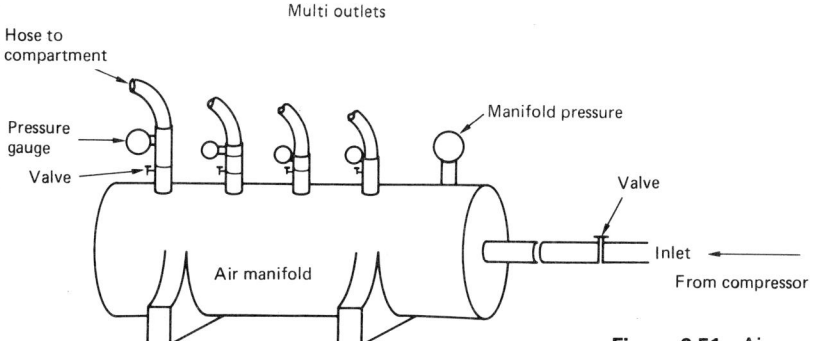

Figure 9.51 Air manifold for use where more than one compartment has been breached

or could not be used, but its use has not gained general acceptance owing to the high cost of the material. In addition, the actual insertion and subsequent removal and disposal are expensive operations. Fast acting biodegradable foam is being developed but is not yet available.

9.6.11 Air bags

Ships are kept afloat by buoyancy and when enough of this is removed they sink. It follows therefore that in addition to the usual method of lifting and pumping, if buoyancy is added or restored, the vessel will come to the surface. Air bags can be used to accomplish this or to reduce weight so that available cranage can achieve the task. *Figure 9.53* shows a salvage company raising a large fishing vessel using air bags.

Figure 9.53 Large fishing vessel being raised using air bags (courtesy of PB Eurosalve)

Air bags have been both successfully and unsuccessfully used in salvage operations for many years. The problems normally found are:

1. Lack of connecting points.
2. Buoyancy loss when the bags reach the surface.
3. Difficulty in submerging the bags prior to connecting owing to residual air.
4. Chafe cutting of the bag material.
5. Loss of the bags if bad weather is encountered during the operation.
6. The operation usually taking much longer than if cranes were used and the outcome becoming uncertain.
7. The air tending to leak from the bags; where parachute type bags are being used it is necessary to keep topping up the air.
8. Difficulty of using in strong currents.
9. Large number of bags required.

The bags are made of modern material such as polyester with nylon or polyester straps and the bags readily available vary in size from 100 kg to 35 tonnes lift; larger sizes can be manufactured if time permits and in practice the bags are often made to suit the project. *Figure 9.54* shows the approximate dimensions of the bags together with their weight and folded dimensions.

When raising sunken vessels, the use of air bags on their own, i.e. without crane assistance, is not widespread, as more convenient and efficient methods of lifting are generally available. In many salvage operations the values involved are so low that if a large crane or other expensive plant were used the salvage operation would become uneconomical. It is in this type of situation that air bags are most frequently used.

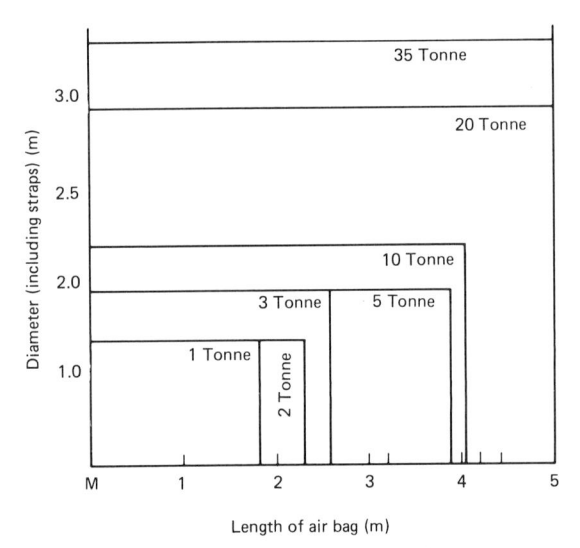

Figure 9.54 Approximate dimensions of enclosed air bags, based on information supplied by Auto Marine

Where vessels are raised using air bags (unless the tide range is such that the vessel can be pumped out at low water) alternative methods must be used to provide stability during the pumping out stage. Air bags have on occasion been used as added buoyancy where they have reduced the draught of structures so that they could float or pass over an obstruction.

9.6.11.1 Parbuckling

Certain types of ships are notorious for capsizing. The most spectacular are passenger ships which have been on fire and the extinguishing efforts have capsized the vessel. The most regular are roll on–roll of ships.

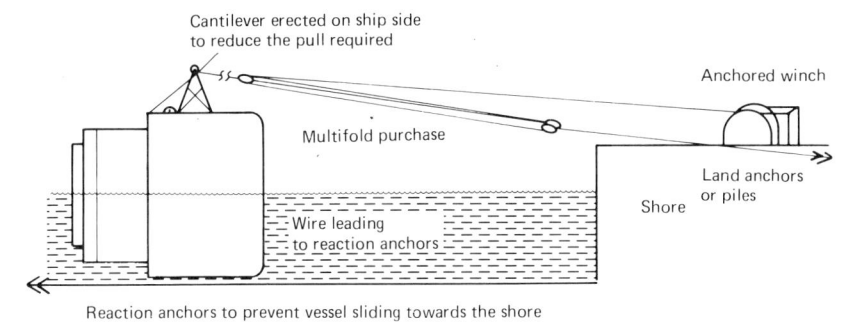

Figure 9.55 Parbuckling where the vessel is near the shore

The most commonly used procedure for uprighting a ship is parbuckling. The classical method for carrying out this operation is to erect cantilevers on the side of the ship. On a convenient shore, winches are anchored down and between the winches and the arms multifold tackles are fitted (*Figure 9.55*). The system has the simplicity of being self correcting because the righting arm gives the maximum leverage at the beginning of the operation. As the ship approaches the upright position the lever diminishes. This reduces the risk of pulling the vessel over in the opposite direction. If the capsized vessel is lying offshore then winches are placed on a barge, with anchors or other securing points such as steel piles driven into the sea bed. In both these methods some opposite restraint is normally required to present the ship sliding towards the pulling force.

Parbuckling for smaller vessels is often carried out using floating sheer legs where wire strops are placed around the vessel so that when the tackles are heaved the vessel rotates (*Figure 9.56*). Care must be taken to spread the load over the ship's side and to avoid turning the casualty over in the opposite direction. Another method sometimes used when parbuckling small vessels with sheer legs is to use both deck and jib tackles in opposite direction so that the casualty is rotated in the slings. *Figure 9.57* shows the method used to parbuckle the *Herald of Free Enterprise*.

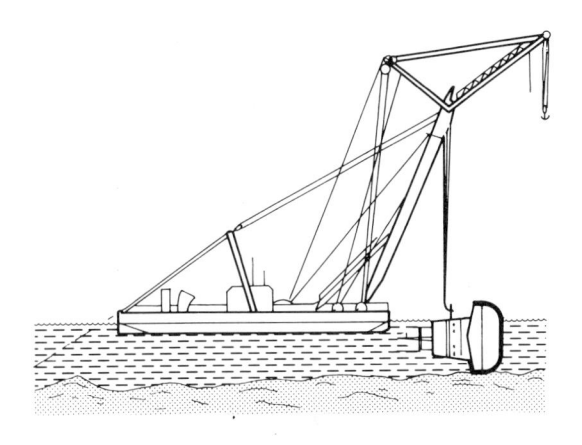

Figure 9.56 Using sheer legs to parbuckle small vessel

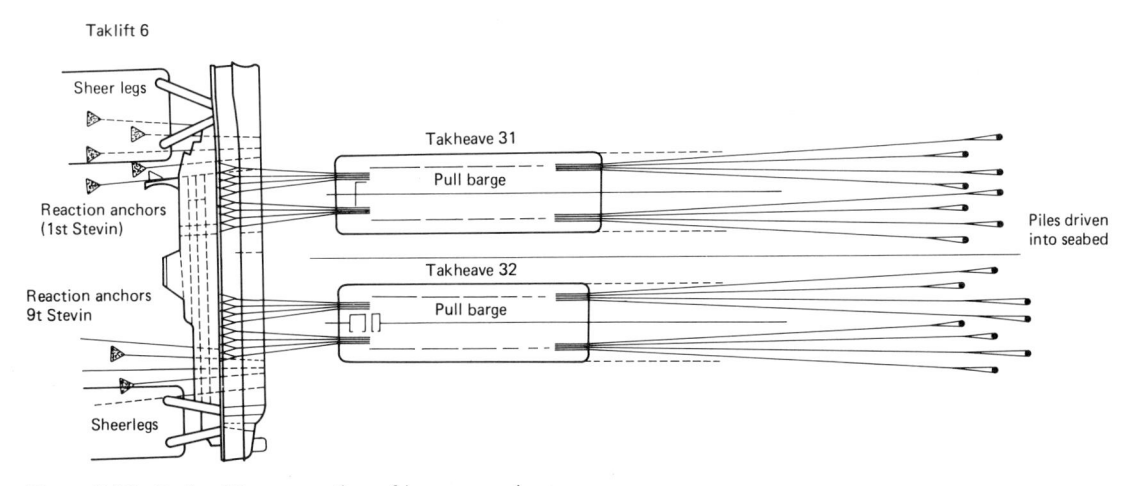

Figure 9.57 Parbuckling operation of large vessel

9.6.12 Water damage protection

The object of most salvage operations is to reduce the financial loss and to save as much as possible. In other words, it is about values rather than material. There are of course exceptions to this but this is the general case. If the salvage contract is based on the principle that the salvor will receive an award out of the salved fund then it will in all probability be the view.

Whenever a ship sinks or is flooded, the machinery and electronics are probably immersed in a very corrosive medium, sea water. If the water has a low oxygen content then most metals will not corrode at any significant rate so whilst the machinery is submerged the corrosive action is curtailed. Once the space has been pumped out or the vessel is raised, the machinery is exposed to the atmospheric oxygen and the components and surfaces if not protected are subjected to both chemical and corrosive attack. The rate of such deterioration depends on the temperature and other environmental factors but it is very fast and within days corrosion will have started and be accelerating.

Around 40% of the value of any vessel is in its machinery and equipment. If after flooding the plant is to be saved and losses reduced then the equipment must be protected as soon as possible after dewatering. In bygone days steam engines were of very simple construction (with the exception of turbines) with most parts exposed and easily cleaned. In addition to this, it was not unusual for the engines and boilers to be operated in what would now be considered less than ideal conditions. After the flooding of an engine room, it was still possible (after checking for damage) to light charcoal fires or flash up the boiler slowly to dry the brickwork and insulation and then to run the engine on steam whilst oiling and greasing the parts. Conversely, the construction of modern machinery is such that the majority of the moving parts are enclosed lightly loaded and not easily accessible for dewatering, cleaning and oiling. As a consequence, if the losses are to be kept to a minimum then much more care must be taken in protecting the components from corrosion.

Chemicals have been developed that have a preferential affinity for metals and which form an anti-corrosive/protective coating of around 2μm thick on the surface. Low surface tension also means that the liquid can spread into remote nooks and crannies. The chemical displaces the water and also protects the components from the condensation that inevitably occurs in an engine room that has just been pumped out. One manufacturer of this type of chemical describes the firm's product as, 'a first aid treatment for situations when electrical or mechanical engines/machinery are exposed to air after accidental contact with water'. Various chemicals are available to achieve the desired result and in addition to the usual well known brand names there are many products from specialist companies.

As it is not unknown for machinery to be immersed in sea water for six months or more without significant corrosion, it is preferable if the machinery space is left flooded until the chemicals have been obtained and the preparation for the preservation work is complete. It is not unusual for the application work to be carried out as the space is being dewatered and usually by a team familiar with the work.

9.6.12.1 Preparation

For best results it is preferable before applying the chemicals that the machinery, electric motors, etc., are washed down with high pressure fresh water to wash away the salt. In practice, however, this is not always possible and the chemicals are sometimes applied immediately after dewatering.

The high capillary action of the chemical helps in the application as it penetrates many areas that are not directly accessible but nevertheless the more thorough the application the better the result. The chemical can be either sprayed on or pumped through the oil ducts. Small electric motors, components or easily forgotten items such as spare parts can be dipped into a bath filled with the chemical. Where plant is being treated the effectiveness is considerably enhanced if the machinery can be turned during the application. To avoid the need for a costly re-application, once treated the machinery space should not be allowed to flood up again and before dewatering all precautions should therefore be taken to ensure that this will not occur.

Experience has shown that in almost all cases where the chemical is correctly applied the surface is protected from corrosion for up to three months. If repairs are not well in hand by that time it is recommended that chemicals are re-applied. The protection depends on the chemical being active at the time of application and on occasion this has not been the case. It is prudent to keep the treated components under constant review, so that any deterioration is detected and treated before it becomes a problem. As an indication of the quantities involved to correctly treat the average engine room, in excess of 1000 l of chemical would be needed and about three crew for four days.

9.6.13 Salvage tugs

It could be said that the first tugs were powered by oars but the most accepted concept is that of powered vessels with the first real tug having steam as the source of power. During the early days of sailing vessels most admirals knew that to win a naval battle, the fleet had to be in the correct position at the right time and in the 1700s the first efforts were made to design steam powered vessels.

In 1801 a tug capable of operating under its own power was built to tow barges loaded with farm produce and whilst it was not very successful people very quickly realized that this invention was the perfect solution to a time honoured problem, that of berthing and unberthing sailing ships. It was only a small step from there to towing sailing ships from and out to the sea, which naturally progressed to the carrying out of rescue towing services. By 1830 Lloyd's List records show that tows by steam powered tugs was so commonplace that they were no longer specifically mentioned.

Some ports and sea areas were more dangerous than others and vessels repeatedly needed emergency assistance. The term 'salvage tug' was born when some owners found it was more lucrative to keep their tugs on standby to render emergency type assistance than to have them engaged in commercial towing.

In 1924, one of the first tugs intended for ocean rescue was placed on station at the Azores, the stationing of these tugs had been made possible by the introduction of radio transmitter's onboard vessels. Prior to the widespread use of radio, tugs could only be stationed near ports and known danger points. As the number of salvage tugs grew, they were placed on continuous station at locations where it was most lucrative, e.g. positions where strandings, collisions or storms were commonplace.

Some of the favourite salvage stations were the Dardenell's; the Straits of Singapore; the Cape of Good Hope; the North Sea; the south west approaches to English Channel; the north west coast of Spain and Gibraltar. Salvage tugs were also usually found near the coaling stations as they themselves needed frequent refuelling.

As the vast majority of salvage operations were carried out on the basis of no cure–no pay and everything depended on success, the tugs became increasingly better equipped and carried everything that was considered necessary for a successful salvage operation. Tug owners also invested in building large purpose built and dedicated salvage vessels. A typical dedicated salvage tug carried a mountain of equipment with which to tackle most salvage operations. Equipment included items such as diving equipment; assorted electric and diesel pumps; portable generators and cables; portable air compressors, pressure valves and hoses; steel plate welding and cutting gear; wooden and inflatable work boats and beach gear. *Figure 9.58* shows one of the last salvage tugs still in operation.

By the late 1970s with the advent of radar and sophisticated navigational equipment, together with the introduction of traffic separation schemes in congested waters, the number of vessels colliding and stranding started to decrease. This coupled with the increase in size, reliability and consequent reduction of the number of ships meant that keeping tugs on salvage station was no longer economically viable.

In recent times, the practice has to all intents and purposes disappeared. One exception was in the Persian Gulf where the war between Iran and Iraq and attacks on shipping had kept the number of vessels needing assistance relatively high. The fall in the number of salvage operations and the high cost of keeping expensive tugs on station meant that tug owners found that they had no option but to seek other work for their salvage vessels. This change in use and the need to have multipurpose as opposed to dedicated vessels has resulted in no pure salvage tugs being built in recent years. The number of such vessels is falling continuously and they are, unless some solution to their loss of profitability is found, likely to disappear altogether within the next decade.

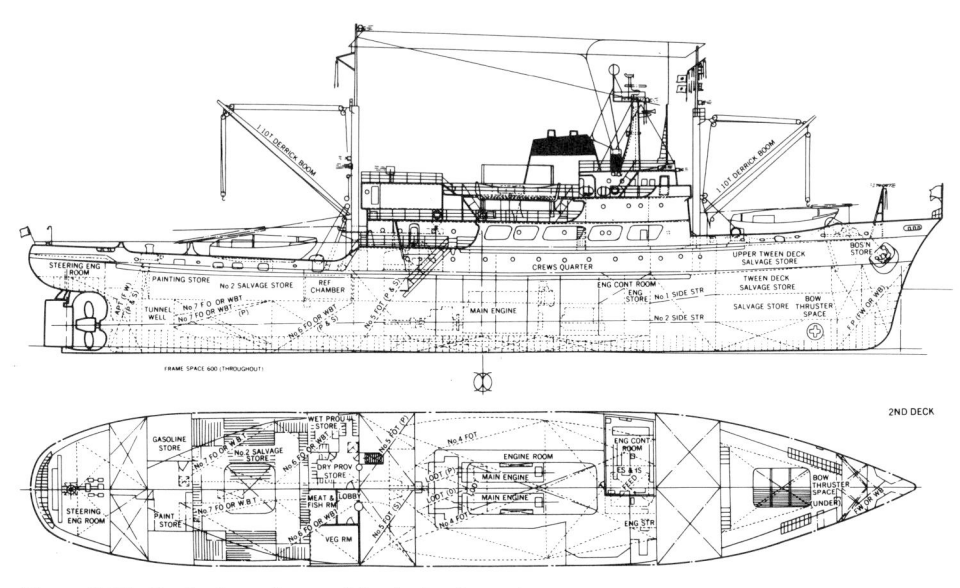

Figure 9.58 Particulars of one of the last salvage tugs

9.6.13.1 Supply vessels

The development of the offshore oil fields has meant that a large number of supply vessels with high horse power were built for laying anchors, towing rigs and servicing platforms. When the downturn in the offshore oil industry came the supply boat owners entered the tug market and consequently these vessels are now being used in salvage operations on a regular basis. Most offshore supply and anchor handling vessels were designed to be highly manoeuvrable with a large clear deck for anchor handling and cargo storage. These advantages have meant that in many ways they are ideal for use as work bases in salvage and indeed many of the old salvage tugs were constructed so that they needed their own assisting tug when working in restricted areas. *Figure 9.59* shows a typical offshore supply vessel.

One of the few disadvantages in using offshore supply vessels is that they do not normally carry any salvage equipment or indeed much towing equipment and if this is needed it must be either obtained locally or flow out to the casualty. This places a limitation on their usefulness in rescue towing.

Another problem is that of crews in that they generally have no experience of salvage or general towing and often baulk at the work that is often called for not only in salvage but general towage.

9.6.13.2 Salvage equipment availability

The major salvors still keep a substantial amount of equipment on standby for use in salvage operations but again the economics of keeping high capital cost items on standby has meant that this stock is also running down rapidly.

Fortunately as one stock is run down other sources emerge and most but not all equipment can be obtained in a relatively short time from shore based plant hire firms and flown to the casualty. If one piece of equipment is not readily available others usually are and improvization has played and will always play a large part in salvage.

When operating far from their home base the major salvage companies, in order to keep costs down, relied on hiring equipment and labour locally. In recent years this practice has become even more pronounced and salvage companies are emerging who do not own any equipment and who rely on hiring plant when it is needed. In addition many items of equipment developed for the oil industry can also be used or modified for use in salvage operations and are often available.

9.6.14 Ground reaction

Whenever a vessel is stranded or aground and not afloat, the sea bed is exerting an upward force on the hull, i.e. some of the weight of the vessel is resting on the sea bed. This is termed the ground reaction and must be overcome before the vessel can be refloated.

The ground reaction is a measure of the change in displacements between those corresponding to the prestranded and the stranded mean draughts. Allowance must be made to the floating draught for the loss of buoyancy from flooded spaces (*Figure 9.60*). The change in trim can be used to determine the position of the uptrust from the ground reaction.

The ground reaction, by exerting a force upwards at the vessel's keel, has virtually the same affect as removing a weight from the bottom and thus also adversely affects the vessel's stability. This aspect is, however, considered elsewhere.

A vessel proceeding at full speed contains a large amount of kinetic energy and if it then strands the change in draught and likely damage is a function of this energy. As an indication, vessels have been known to rise several metres forward when they grounded at speed. The ground reaction can be reduced by discharging or shifting cargo or ballast and on occasions it is even necessary to either lower the ship by dredging or raise the water level by building dykes around the vessel. After grounding and discharging the pull needed to free the vessel is a function of many items including the remaining ground reaction, coefficient of friction of the sea bed, wave action, and current pressure, but generally:

$$F = g \times u \quad (9.5)$$

where F is the force required to free the vessel, g is the ground reaction and u is the coefficient of friction of the sea bed.

The approximate coefficients of friction of sea bed are as follows: coral between 0.5 and 0.8; sand between 0.3 and 0.4; rock (if not impaled) between 0.8 and 1.5; shingle between 0.3 and 0.5.

Notwithstanding the above, the type of bottom is only one of the parameters affecting the coefficient of friction and others include suction developed by the bottom, e.g. mud; possible damage to the ship's bottom; shape of ship's bottom in contact with the sea bed and impalement of the ship's bottom by rocks or other debris.

9.6.15 Bollard pull

Bollard pull is defined as the amount of force expressed in tonnes that a tug can exert under given conditions. The static bollard pull is normally found by attaching a dynamometer with a scale higher than the expected maximum pull of the tug to a fixed point on the dock and connected to the tug which is run at full power. The bollard pull of a tug normally falls off with speed. The tug owners' objective when quoting the bollard pull of their tug is to give an indication of the tug's towing power under normal operating conditions. However, the figure is in effect only a guide to the power in calm water and under the test conditions during certification.

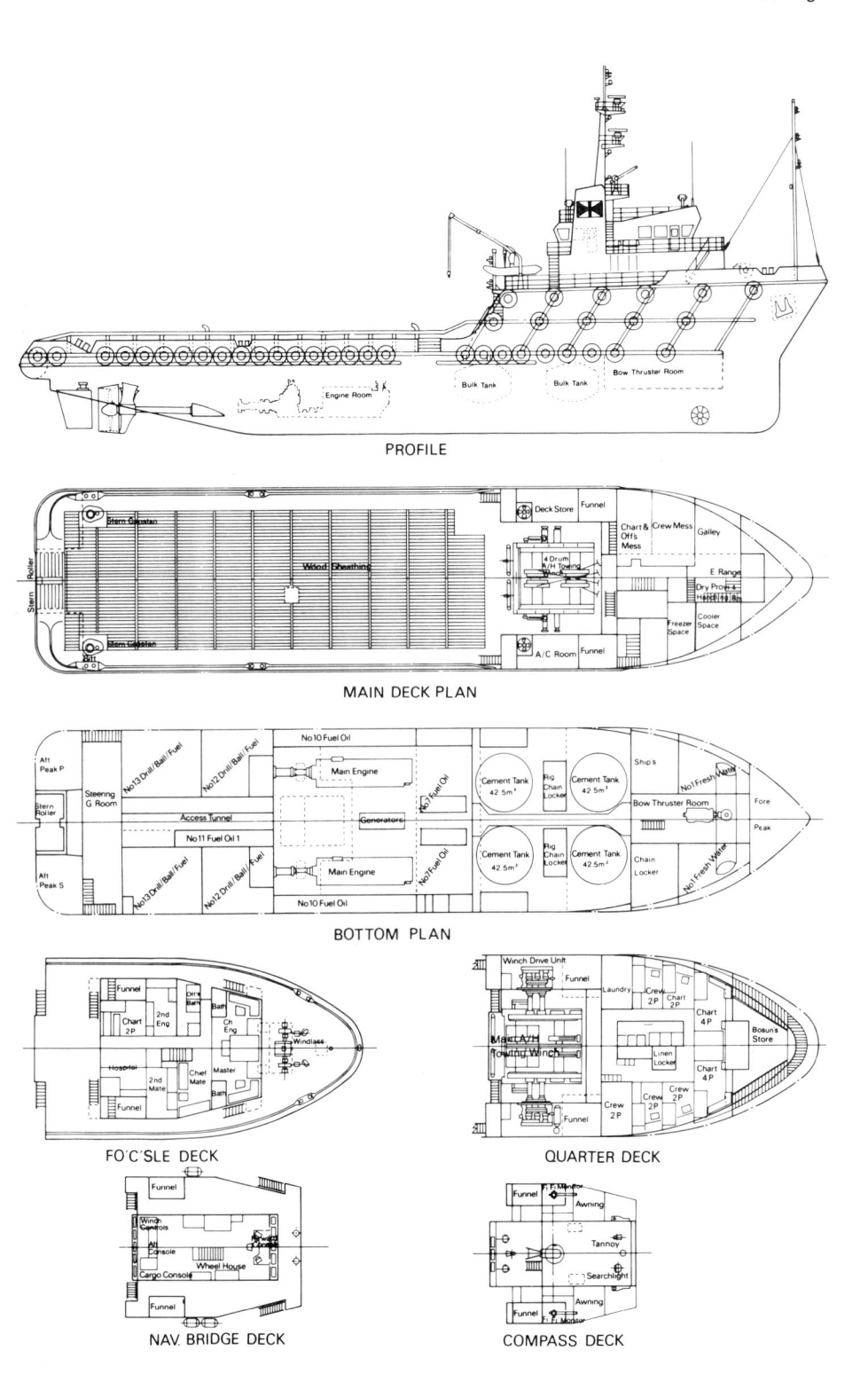

Figure 9.59 Anchor handling tug supply vessel

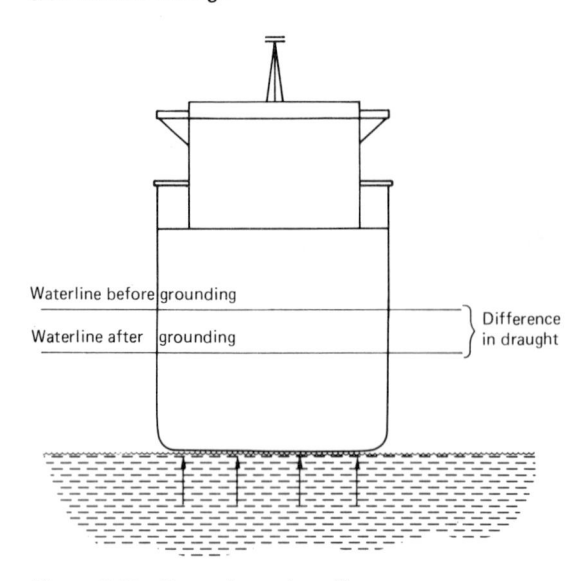

Figure 9.60 Ground reaction; if no compartments are breached, the difference in displacements at the respective waterlines is a measure of this

Bollard pull may be expressed in a number of ways. Maximum bollard pull is equal to the maximum average of recorded tension in the tow wire over a period of 1 min during the test and is normally associated with maximum engine output and optimum propeller pitch. Continuous bollard pull represents the continuously maintained tension in the tow wire over a minimum period of 5 min although 10–15 min is considered preferable. Effective bollard pull gives an indication of the bollard pull which the tug can develop in an open sea way and represents a proportion (usually 0.75%) of the continuous bollard pull after making due allowance for weather.

The power of the tug is open to intepretation and misrepresentation and accordingly Lloyd's Register of Shipping and Det Norske Veritas in particular have developed recommendations in order to arrive at a uniform set of criteria for bollard pull certification. These briefly cover four major areas.

9.6.15.1 Testing environment

The location should provide a shore anchorage point of adequate strength; distance between tug and shore of minimum two ship lengths or 300 m; clear water around the tug of minimum one ship length; water depth in a 100 m radius of the tug should be a minimum of 10 m or twice the deepest draft; a maximum current speed 1 knot; the sea surface should be calm and winds should be less than 10 knots.

9.6.15.2 Machinery

The main engine should be run at the manufacturer's recommended maximum continuous output. If the test is performed in main engine overload conditions then the certificate should be endorsed accordingly. Main engine or propeller shaft driven auxiliaries should be connected during the test, otherwise the certificate should be endorsed accordingly.

It should be noted that rudder angle will adversely affect the bollard pull and shaft revolutions, torsion-meter readings and all other engine data coupled with recordings of rudder angle during the test should be accurately synchronized with bollard pull readings.

9.6.15.3 Vessel

The test should be carried out with the displacement corresponding to full ballast and fuel capacities. The tug should be trimmed even keel or by the stern less than 1% of the vessel's length. The towing equipment should be part of the tug's normal outfit.

9.6.15.4 Dynamometer

The test instrument can be of the form of a mechanical load gauge or an electric load cell and should produce a continuous readout in numerical and graphical form. The dynamometer should form part of the towing wire connection and may be located either on the shore or on the deck of the tug.

As can be seen above, the bollard pull depends on numerous factors such as depth of water, size of tug, type and size of propeller, etc. Where no tests have been carried out, a good estimation can be found as follows: for open propellers the bollard pull in tonnes is usually in the order of BHP/100. For kort nozzles, the bollard pull in tonnes is usually in the order of BHP/70.

When deciding on the suitability of a tug the bollard pull should not be considered in isolation as the tug's displacement is also important. For example, when towing and in adverse weather a large displacement tug will be able to maintain headway where a smaller but more powerful tug cannot. When assisting a stranded vessel where the tug can be allowed to sheer from side to side a large displacement can be a distinct advantage and the effective pull can be 50% greater than that measured on a bollard pull test.

9.7 Towing

Before any vessel can be towed or even moved its condition must be evaluated and preparations made. In some extreme circumstances a vessel must be moved to a location where preparation work can be carried out in safety. For example, the vessel may have stranded in an exposed area breaching several compartments and where it would not survive. Alternatively, the vessel may have to be taken to a sheltered or shallow area where it cannot sink or for beaching to facilitate the salvage repairs.

On occasions, because of the circumstances and urgency of the situation, it is not possible to make any preparation for the tow and risks must be taken to save the vessel. When these risks are known, accepted and appreciated by all concerned then the tow can proceed. The only action that can be taken will be to place a salvage crew on board together with portable pumps, air compressors and life saving appliances. Fortunately these occasions are rare and normally the vessel can be readied for the tow.

The amount of preparation necessary for a tow will depend to a large degree on where the vessel is being towed from and to, the likely weather that will be encountered during the tow and whether the vessel will be manned or unmanned.

Preparation of the vast majority of vessels for towing whether in an emergency situation or for a normal towage operation requires careful attention to the following:

1. All openings are to be closed and made weather and water tight.
2. The rudder is to be put amidships and secured.
3. The propeller shaft is to be locked to prevent rotation (*Figure 9.61* shows some methods of securing).
4. All sea and overboard discharge vales are to be closed and the stern gland tightened.
5. If sea valves are leaking and pipes suspect, lines are to be blanked.
6. All tanks are to be sounded and contents recorded.
7. Stability condition is to be calculated and ensured to be adequate.
8. All derricks, cranes, cargo and other items to be secured.
9. At least one anchor to be secured whereby it can be released in an emergency.
10. Bollards and underdeck structure are to be checked to ensure they are strong enough (*Figure 9.62* and *Table 9.5*). The table is based on a standard bollard design and is often used to determine the probable strength of a ship's bollard. If bollards are not adequate alternative brackets should be fitted.
11. Fairleads are to be checked and ensured as large and strong enough.
12. If possible the vessel is to be trimmed by the stern (a good approximation would be in the order of 1 m for each 100 m in length) and deep enough forward to prevent pounding.
13. Chafe chains are to be used from connection points to at least outside the fairleads and to be of sufficient strength. *Table 9.6* gives the Lloyd's Register of Shipping test loads for stud link chain cables and is based on new chain.
14. Emergency towline is to be rigged and to include a trailing floating line.
15. White lines are to be painted each side of the bow and stern as close to the waterline as possible and large enough to be seen from the tug.
16. Portable pumps are to be placed on board the vessel if its own pumps are not working.
17. If damaged or suspect the strength of the hull structure is to be checked for bending, sheer and torsional forces.
18. Navigation lights and day towing signal are to be fitted.

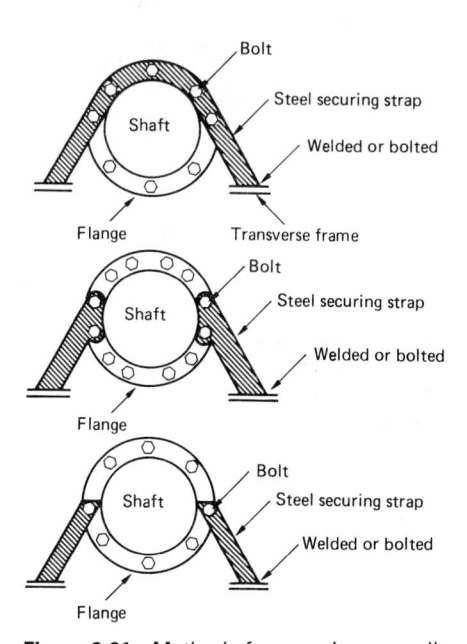

Figure 9.61 Methods for securing propeller shafts

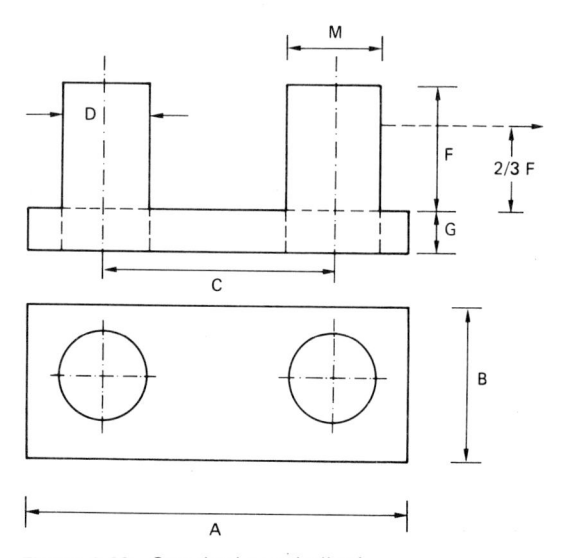

Figure 9.62 Standard type bollard

Table 9.5 Bollard strength

P tonnes	A mm	B mm	C mm	D mm	F mm	G mm	M mm
8	580	230	350	110	195	75	130
12	705	280	425	168	235	90	155
20	910	360	550	219	300	100	200
30	1135	450	685	273	380	130	255
44	1340	530	810	324	450	150	305
53	1470	580	890	356	490	150	335
70	1675	660	1015	406	560	185	385
83	1895	750	1145	457	630	185	140
107	2080	830	1250	508	700	215	490
127	2310	910	1400	558	770	235	550
153	2525	1000	1525	610	840	235	585

P in tonnes is the breaking strength of the wire and the bollard is assumed to lie in the line of pull and to be in as new condition. The foregoing is for guidance only and based on standard design bollards.

9.7.1 Towing points

Mobile offshore oil rigs and barges are generally fitted with proper towing pads and chain bridles. Large oil tankers which are designed for mooring to offshore terminals are often fitted with mooring points which have also proved ideal as a towing point (*Figure 9.63*).

On dry cargo vessels, the strength and position of the towing point and fairleads is often problematic as most conventional vessels were not designed for towing. The only towing points available are normally the bollards which are often inadequate in size, condition or indeed are badly sited.

Where the bollards are inadequate on occasion it is still possible to still use them by leading the chains around a second set and behind the windlass. *Figure 9.64* shows a typical connection where the bollards are considered inadequate.

Figure 9.63 Mooring points fitted to a very large crude oil carrier (courtesy of The Salvage Association)

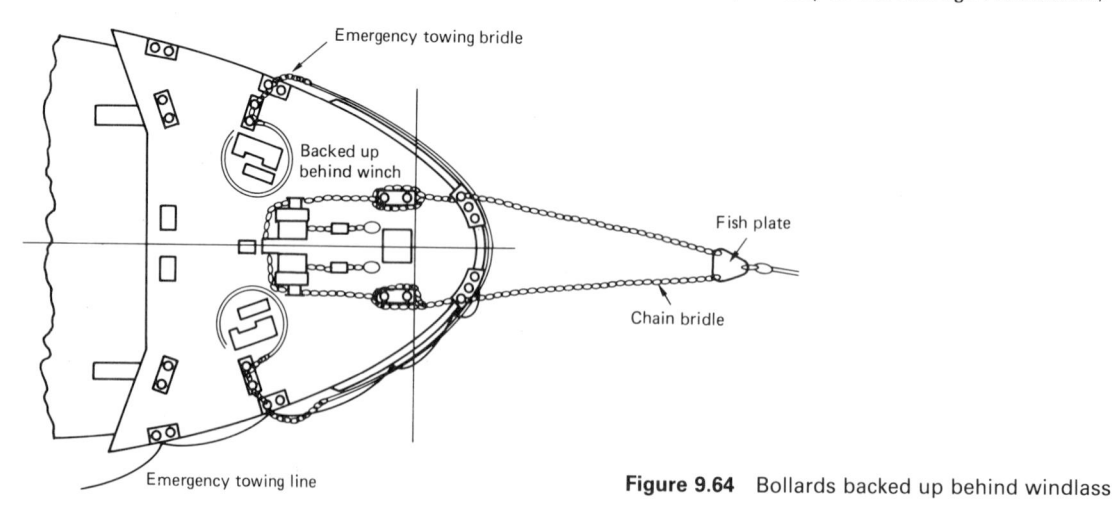

Figure 9.64 Bollards backed up behind windlass

Table 9.6 Lloyd's Register test loads for stud link chain cables (courtesy of Lloyd's Register of Shipping)

Chain diameter mm	Grade U1		Grade U2		Grade U3		Grade OS4	
	Proof load kN	Breaking load kN	Proof load kN	Breaking load kN	Proof load kN	Breaking load kN	Proof load kN	Breaking load kN
12.5	46	66	66	92				
14	58	82	82	116				
16	76	107	107	150				
17.5	89	127	127	179				
19	105	150	150	211				
20.5	123	175	175	244	244	349		
22	140	200	200	280	280	401		
24	167	237	237	332	332	476		
26	194	278	278	389	389	556		
28	225	321	321	449	449	642		
30	257	368	368	514	514	735		
32	291	417	417	583	583	833		
34	328	468	468	655	655	937		
36	366	523	523	732	732	1050		
38	406	581	581	812	812	1160		
40	448	640	640	896	896	1280		
42	492	703	703	981	981	1400		
44	538	769	769	1080	1080	1540		
46	585	837	837	1170	1170	1680		
48	635	908	908	1270	1270	1810		
50	686	981	981	1370	1370	1960	2160	2740
52	739	1060	1060	1480	1480	2110	2330	2960
54	794	1140	1140	1590	1590	2270	2500	3170
56	851	1220	1220	1710	1710	2430	2680	3400
58	909	1290	1290	1810	1810	2600	2860	3630
60	969	1380	1380	1940	1940	2770	3050	3870
62	1030	1470	1470	2060	2060	2940	3240	4120
64	1100	1560	1560	2190	2190	3130	3440	4370
66	1160	1660	1660	2310	2310	3300	3640	4630
68	1230	1750	1750	2450	2450	3500	3850	4890
70	1290	1840	1840	2580	2580	3690	4060	5160
73	1390	1990	1990	2790	2790	3990	4390	5580
76	1500	2150	2150	3010	3010	4300	4730	6010
78	1580	2260	2260	3160	3160	4500	4960	6300
81	1690	2410	2410	3380	3380	4820	5320	6750
84	1800	2580	2580	3610	3610	5160	5680	7220
87	1920	2750	2750	3850	3850	5500	6060	7690
90	2050	2920	2920	4090	4090	5840	6440	8180
92	2130	3040	3040	4260	4260	6080	6700	8510
95	2260	3230	3230	4510	4610	6440	7100	9010
97	2340	3340	3340	4680	4680	6690	7370	9360
100	2470	3530	3530	4940	4940	7060	7780	9880
102	2560	3660	3660	5120	5120	7320	8050	10230
105	2700	3850	3850	5390	5390	7700	8480	10770
107	2790	3980	3980	5570	5570	7960	8760	11130
111	2970	4250	4250	5940	5940	8480	9350	11870
114	3110	4440	4440	6230	6230	8890	9790	12440
117	3260	4650	4650	6510	6510	9300	10240	13010
120	3400	4850	4850	6810	6810	9720	10700	13590
122	3500	5000	5000	7000	7000	9990	11010	13980
124	3600	5140	5140	7200	7200	10280	11320	14380
127	3760	5350	5350	7490	7490	10710	11790	14980
130	3900	5670	5570	7800	7800	11140	12270	15580
132	4000	5720	5720	8000	8000	11420	12590	15990
137	4260	6080	6080	8510	8510	12160	13400	17020
142	4520	6450	6450	9030	9030	12910	14220	18060
147	4790	6840	6840	9560	9560	13660	15050	19120
152	5050	7220	7220	10100	10100	14430	15890	20190
157	5320	7600	7600	10640	10640	15200	16740	21270
162	5590	7990	7990	11180	11180	15980	17600	22350

In an endeavour to make the connecting operation easier one of the towing companies designed and developed a towing bracket which can be fitted directly on to the deck of a ship or barge over a suitable strong point. This connection is now widely used and is known as a Smit Towing Bracket (*Figure 9.65*). *Figure 9.63* shows mooring points fitted to the world's largest vessel.

9.7.2. Bimco Towcon

Whilst still relatively new, the use of this contract has become widespread and is now the one most commonly used in sea towing.

It was introduced in 1985 following an investigation by the International Salvage Union which showed that virtually every towing company was using its own contract. Some of the contracts incorporated the UK standard towing conditions or the Netherlands towing conditions and it was said that whilst some of the agreements were detailed and explicit others were not. The new contracts differ from others still in use by making a number of areas that were previously for the benefit of the tug owner, to be for the mutual benefit of both parties in the venture.

The Bimco Towcon was drawn up by the Baltic and International Maritime Council, the European Tug Owners' Association (ETA) and the International Salvage Union (ISU).

It is the recommendation of the ISU and ETA that their members should use the agreements for all future ocean towage work and as mentioned above the contract, whilst still relatively new, is already widely used on towages and salvage. The contract is usually used where a lump sum agreement has been made, i.e. the hirer agrees to pay the tug owner a lump sum for a specified tow.

The contract lump sum is often broken into milestone payments and Box 32 in the agreement details the following: lump sum towage price; amount due and payable on signing the agreement; amount due and payable on sailing of tug and tow from place of departure; amount due and payable on passing of tug and tow off a certain point; and amount due and payable on arrival of tug and tow at destination.

The contract also takes into account the price of bunkers and any difference in cost will be payable by the tug owner or hirer to the other.

9.7.3. United Kingdom Standard Towing Conditions (revised 1986)

Until very recently these conditions were used extensively in the towage industry and were normally attached to and formed part of towage contracts. In recent years, at least as far as general towage is concerned, the UK standard towing conditions have been superseded by the Bimco tug or tow hire conditions. The UK conditions are now in the main only included as part of harbour towage agreements in Britain and by some British operators for local tows.

The UK Standard Towing Conditions were often considered unfair by ship owners in that they attempted to remove any liability from the tug owner; e.g. Clause 4 stated

'Whilst towing, or whilst at the request either expressed or implied of the hire rendering any services of whatsoever nature other than towing:

A) The tug owner shall not (except as provided in 4(c) and 4(e)) be responsible or liable for

   1. Damage of any description done by or to the tug or tender or done by or to the hirer's vessel or done by or to any cargo or other thing on board or being loaded on board or intended to be loaded on board the hirer's vessel or the tug or tender or to any other object or property
   or
   2. Loss of the tug or tender or the hirer's vessel or of any cargo or other things on board or being loaded on board or intended to be loaded on board the hirer's vessel or the tug or tender or any other object of property.

   Any claim by a person not a party to this agreement for loss of damage of any description whatsoever.

arising from any course whatsoever including (without prejudice to the generality of the foregoing) negligence at any time of the tug owner his servants or agents, un-seaworthiness, un-fitness or breakdown of the tug or tender, its machinery, boiler, towing gear, equipment, lines, ropes or wires, lack of fuel, stores, speed or otherwise.

B) The hirer shall (except as provided in clause 4c and e) be responsible for, pay for and indemnify the tug owner and in respect of any loss or damage in any claims of whatsoever nature or howsoever arising or caused, whether covered by the provisions of clause 4(a) hereof or not suffered by or made against the tugowner and which shall include without prejudice to the generality of the foregoing any loss of or damage to the tug or tender or any property of the tug owner even if the same arises from or is caused by the negligence of the tug owner or agents'.

9.8 Pollution control

In recent years pollution and damage to the environment have become a matter of topical concern. In salvage operations the new draft IMO convention places emphasis on this aspect together with obligations on the salvor to prevent pollution if possible *(Figure 9.66)*.

It is unclear whether or not the obligation to prevent or minimize damage to the environment also means that oil slicks already existing close to the casualty when

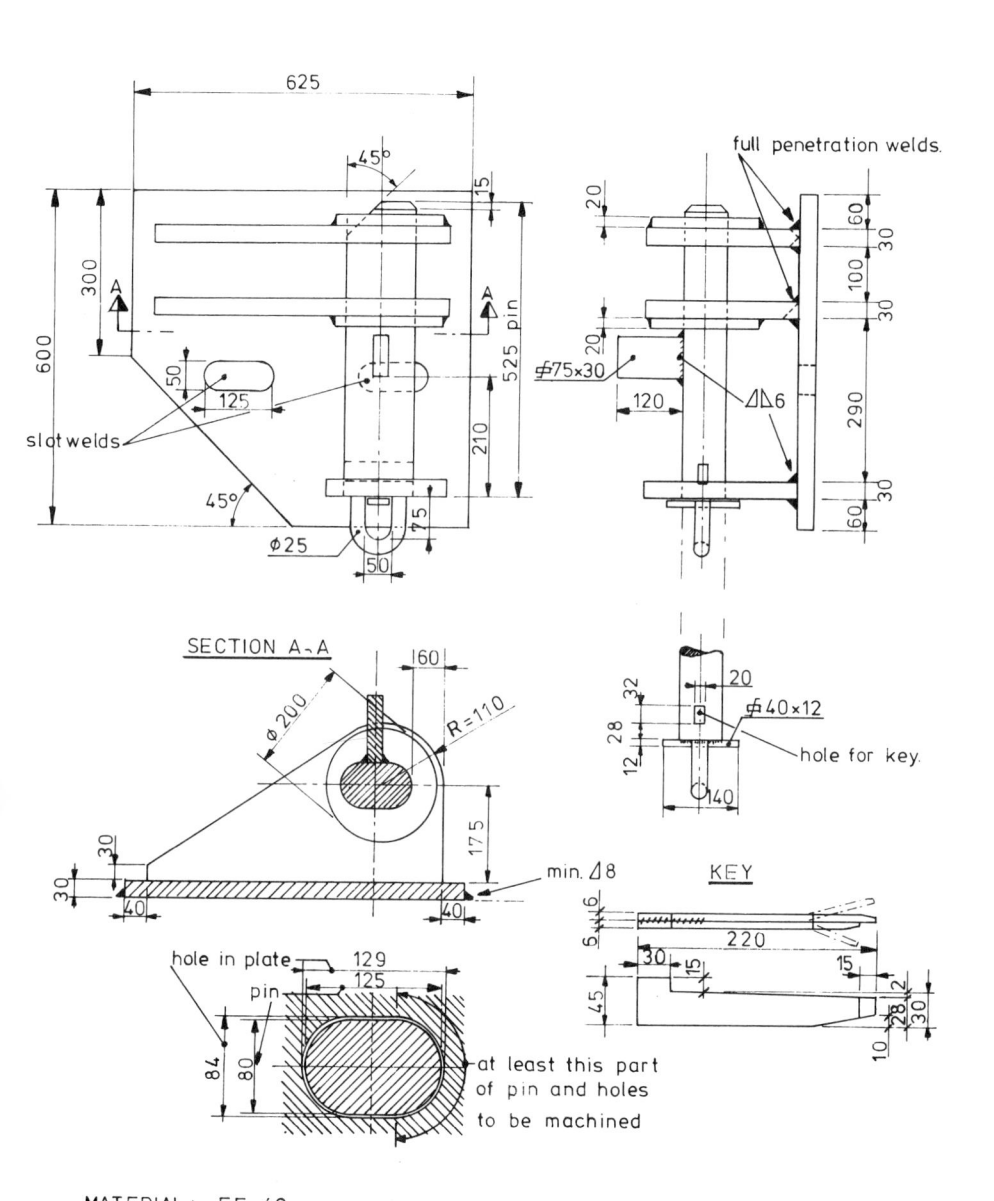

Figure 9.65 Smit Touring Bracket connection for $2^{3}/_{4} - 3$ inch towing chain (designed by Smit International)

Figure 9.66 *Amoco Cadiz* broken in two and spilling oil

the salvor arrives must be cleaned up. This will likely depend on the circumstances of each particular case. Oil slicks are particularly damaging to both environment and amenities when the slicks reach the coast line and are also difficult and more expensive to clear up than if attacked whilst still in open waters.

Where a vessel has sunk inside a port oil booms can be used to prevent any oil that has escaped from spreading (*Figure 9.67*). When related to pollution, oil booms are simply floating barriers which prevent oil from escaping. They are basically divided into two types, internal foam flotation or air inflated. *Figure 9.69* shows a few of the many types available (*Figure 9.57*).

High buoyancy foam flotation booms have the advantage that puncture of any cylinder does not result in the boom being unable to perform its function. In addition, this type of barrier is easy to handle and store. The main disadvantage is that of poor performance in waves of any significant size.

Inflated booms are generally more expensive than the internal foam type and in addition they are difficult and expensive to clean and repair. They are, however, excellent for handling, have a high buoyancy to weight ratio and are ideal for transportation by air to remote areas. The main disadvantage is that if puncture resistant material is not used it is relatively easy to puncture an air compartment and in addition if a valve malfunctions the section concerned can lose buoyancy. These booms may be either self or pressure inflated. Pressure inflatable booms are easy to deploy and recover; however, the deployment is slow as compartments have to be inflated separately.

Most manufacturers design their booms for different applications, e.g. open seas, sheltered bay, harbour or river. The dimensions of the boom, fin and skirt will depend on the application and will range from say 20 cm height above water on a type suitable for a river to

Figure 9.67 Sunken vessels; note the oil booms

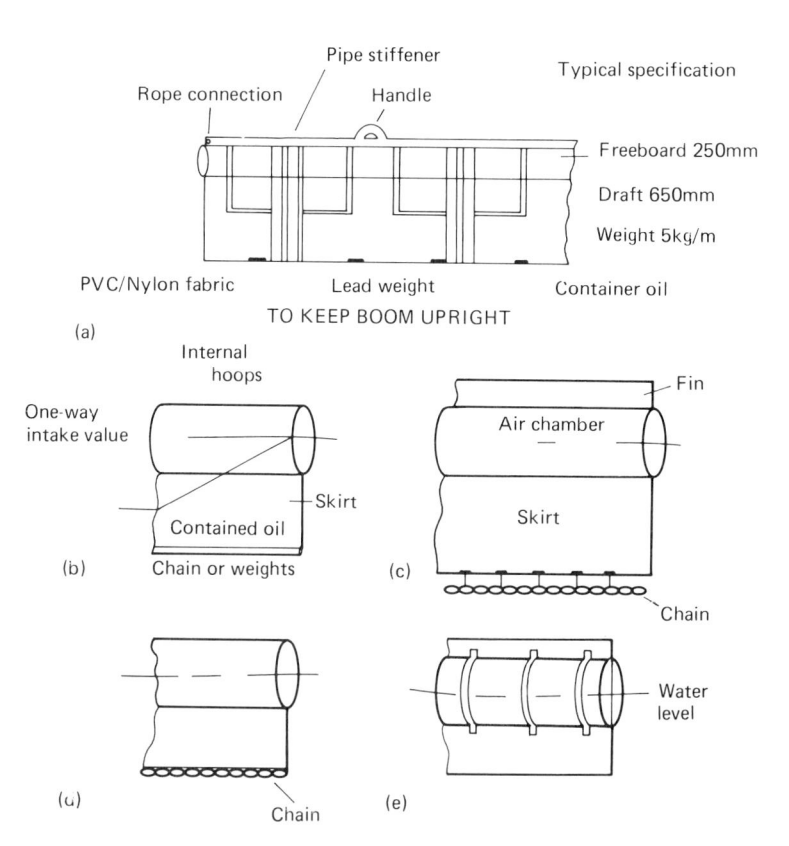

Figure 9.68 Selection of oil booms; (a) Internal form flotation oil boom; (b) self inflated boom (c) pressurized boom; (d) High buoyancy internal foam; (e) Externally fitted foam

50 cm height above water on an open sea model. The depth of skirt could range from say 30 cm on a river model up to say 110 cm on a open sea model. Many of the booms are fitted with a chain which serves as ballast in order to hold the skirt vertical. These chains can also be utilized as strong points when the boom is used as a sweep to draw the oil to a collecting point.

Pollution prevention booms work satisfactorily in harbours or sheltered waters. However, in exposed sea areas the wave action often prevents them from operating without splashover occurring. The inflatable boom appears to work best in exposed water but no boom is really satisfactory in open seas.

The boom can only stop the oil from spreading. The oil must then be removed by mechanical means or dispersed using chemicals. The salvor and/or contractor should always ensure that the appropriate boom is utilized to match the conditions, otherwise the laying of a boom around a vessel may be no more than a poor cosmetic exercise.

9.8.1. Oil skimmers

Skimmers work by removing the oil and water mix from the surface of the sea, separating, then storing the oil for stowage and disposal. It is only possible to cover a representative sample as there are various types of skimmers available.

Designs range from conveyer belt and paddle wheel to oleophilic rope type. In the former rotating paddles pick up the oil and water mixture and deposit it on a conveyor belt which carries it to a separator. After removing the water the oil is deposited in a holding tank. The tank is then either continuously unloaded or is large enough for this to be done periodically.

The oleophilic rope type skimmer works by pulling an endless rope mop through the oil. After passing through the oil the rope mop is wrung out by hydraulically driven rollers. The oleophilic mops selectively attract a wide range of oil and this type of skimmer whilst basically only a calm water model has the advantage of a high pick up rate. The catamaran type also have integral oil storage facilities and mobility for the tracking of individual slicks. A wide choice of sizes are available and some are hand portable, whereby the unit can be placed on the deck of a vessel with the rope inside the ship's hold (*Figure 9.69*).

In the combination boom skimmer a boom is towed by one or two vessels to collect the oil which is then skimmed (*Figure 9.70*).

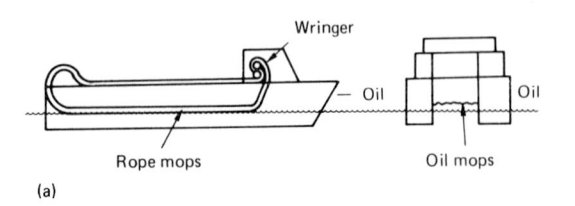

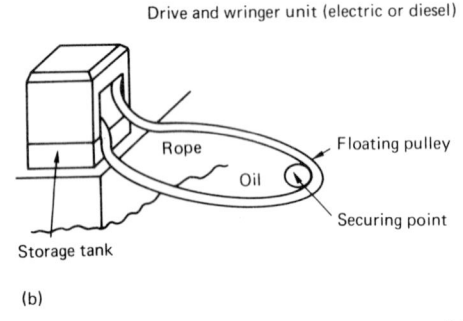

Figure 9.69 Skimmers; (a) Catamaran type; (b) Small portable rope mop type

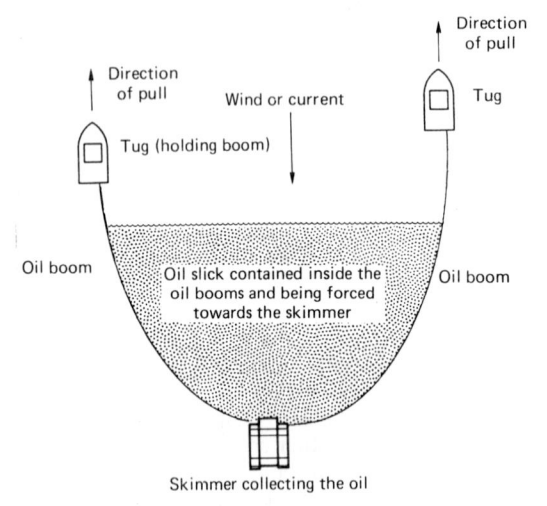

Figure 9.70 Oil slick recovery, using a combination of oil boom and skimmer

9.8.2 Dispersant spraying equipment

When oil is spilled in an open sea way its recovery is difficult. Most of the oil skimmers and booms available can only be used in relatively calm waters and often therefore the only remedy is by spraying oil dispersants on the slicks. Oil dispersants have been available for many years and a great deal of research has been undertaken on their effect on the environment.

Most spraying booms are designed to fit on virtually any vessel (*Figure 9.71*) and are generally supported by a guy rope attached to a mast or the superstructure so that the boom can be raised or lowered. A pump unit containing a dispersant pump and a sea water pump controls the mixture and output. The dispersant is fed via the hollow boom to the nozzles which are designed to give the correct flow without clogging. Hand spraying equipment may be used for smaller oil slicks or at locations otherwise inaccessible to craft mounted equipment.

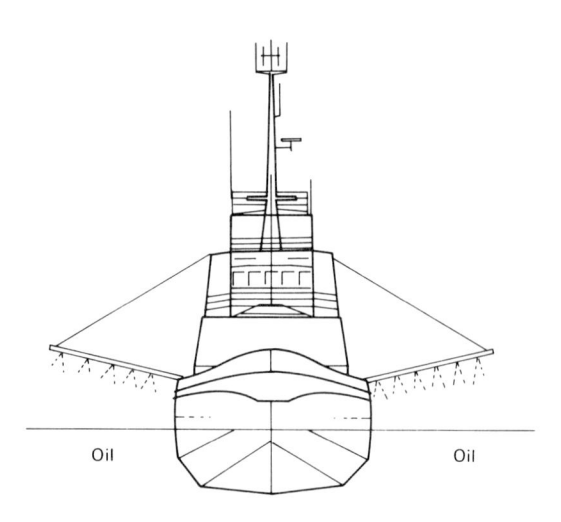

Figure 9.71 Dispersant spraying equipment fitted on a tug

Further reading

Blank, T. S. (1989) *Modern Towing*. Cornhill and Maritime Press Inc., Cambridge, Maryland
Brady, E. M. (1960) *Marine Salvage Operations*. Cornhill Maritime Press Inc., Cambridge, Maryland
Brady, E. M. (1967) *Tugs, Towboats and Towing*. Cornhill and Maritime Press Inc., Cambridge, Maryland
Brice, G. (1983) *Maritime Law of Salvage*. Stevens, London
Derrett, D. R. (1984) *Ships' Stability for Masters and Mates*. Stanford Maritime Ltd, London
Det Norske Veritas (1981) *Rules for Classification: Mobile Offshore Units*
Donar, M. F. (1958) *The Salvager*. Rose and Hanns Inc., Minneapolis
Gores, N. (1972) *Marine Salvage*. David and Charles, Newton Abbot
Hancocks, D. (1987) *Reeds Commercial Salvage Practice*,

Vols. 1 and 2. Thomas Reed Publications Ltd, Sunderland
ICS (1978) *Tanker Safety Guide (liquified gas)* Witherby, London
ICS/CCIMF (1988) *Ship to Ship Transfer Guide (petroleum)*. Witherby, London
ICS/CCIMF (1986) *International Safety Guide (oil tankers and terminals) (ISGOTT)*. Witherby, London
ICS/CCIMF (1980) *Ship to Ship Transfer Guide (liquified gases)*. Witherby, London
Lloyd's of London Press Ltd (1984) *Lloyd's List, 250th Anniversary Supplement*. Lloyd's of London Press Ltd
Office of the Supervisor of Salvage, US Navy (1970) *U.S. Salvage Manuals*, Vol. 1 and 2
Office of the Supervisor of Salvage, US Navy *U.S. Salvage Petroleum Products Emergency Transfer Handbook*
Steel, D. W. and Rose, F. D. (1985) *Kennedy's Law of Salvage*. Stevens, London
Williams, M. (1978) *No Cure No Pay*. Hutchinson Benham, London

10 Corrosion and Defect Evaluation

Bob Reuben

Contents

10.1 Introduction

10.2 The chemistry of corrosion
    10.2.1 The electro-chemical series
    10.2.2 Thermodynamics and potential for corrosion
    10.2.3 Kinetics of corrosion and polarization
    10.2.4 Evans Diagrams and mixed potential theory
    10.2.5 Surface corrosion products and passivity

10.3 The mechanisms of corrosion
    10.3.1 Purely electro-chemical effects
    10.3.2 Combined electro-chemical/mechanical effects
    10.3.3 Specific effects of interest in marine environments

10.4 Corrosion resistance of marine alloys
    10.4.1 Carbon–manganese and low alloy steels
    10.4.2 Stainless steels
    10.4.3 Cast irons
    10.4.4 Copper and nickel alloys
    10.4.5 Aluminium and titanium alloys
    10.4.6 Other marine alloys
    10.4.7 Summary of marine uses and corrosion properties

10.5 Corrosion protection
    10.5.1 Coatings
    10.5.2 Cathodic protection
    10.5.3 Inhibition and modification of the environment
    10.5.4 Anodic protection
    10.5.5 Materials selection and design

10.6 Defects and non-destructive testing
    10.6.1 The genesis of defects
    10.6.2 The significance of defects
    10.6.3 The growth of defects
    10.6.4 The detection of defects – NDT methods
    10.6.5 Monitoring methods
    10.6.6 Defects and reliability

10.7 Codes, test methods and standards
    10.7.1 Standards relating to materials and their properties
    10.7.2 Codes and standards relating to design
    10.7.3 Standards relating to manufacturing processes
    10.7.4 Standards relating to inspection and testing

References
Codes and standards

10.1 Introduction

The intention of this chapter is to give a concise outline of the principal aspects of corrosion as applied to marine and related environments and to introduce some of the important considerations in defect evaluation. These two subject areas are probably the most important for the materials engineer interested in marine and offshore design. The level is a non-specialist one and is aimed at designers, plant engineers, scientists and others with an interest in marine materials engineering.

The sections deal with the basic scientific principles of corrosion, the mechanistic aspects of how corrosion manifests itself, an overview of useful marine alloys and their corrosion resistance, the ways of protecting against corrosion and applications in marine engineering, a review of the significance and the detection of defects including corrosion and sources of standard procedures for testing, specification and design. It is hoped that each section can be approached independently of the others.

10.2 The chemistry of corrosion

Corrosion can be defined as chemical or electro-chemical degradation of metal. Although it is often enhanced by purely physical effects, the corrosive element can always be described in terms of chemical thermodynamics and kinetics. This section is concerned with the extent to which the tendency and rate of corrosion of metals can be understood using chemical principles and, although a detailed treatment of this area is not generally required by a corrosion engineer, it is hoped that the outline provided here will serve to aid in understanding the more applied aspects of the subject covered in the rest of the chapter.

10.2.1 The electro-chemical series

The electro-chemical activity of metals is usually defined in terms of their electrode potential, which is a measure of the ability of the metal to oxidize or form positive ions through the half cell reaction:

$$M = M^{z+} + z\,e^- \tag{10.1}$$

where M is the metal of interest, z is the valency of M and e^- represents an electron.

It can be shown thermodynamically (Moore, 1972) that the free energy change (ΔG) for an electro-chemical reaction is related to its potential (E) through the simple equation:

$$\Delta G = -zFE \tag{10.2}$$

where F is the Faraday constant (96 500 C/g mole).

This means that an electro-chemical reaction is subject to the same equilibrium principles as other chemical reactions, including the Law of Mass Action which states that 'a reaction will move to oppose any perturbation from its equilibrium position'. In simple terms, this means that in the reaction above the removal of product electrons would lead to a continuation of the corrosion reaction in an attempt to re-establish equilibrium.

Conversely, the provision of electrons would inhibit, or even reverse, any tendency to corrosion. The half cell equilibrium is also influenced by the ion concentration so that it is necessary to use a standard concentration for comparison of half cell potentials. This gives rise to the electro-chemical series (*Table 10.1*) in which the most noble (inactive) metals are shown at the top of the series and the base (active) metals are shown at the bottom. A number of other half cell reactions of relevance are also shown in *Table 10.1*. Since potential is not a quantity which can be measured absolutely, the standard hydrogen electrode has been adopted as standard in *Table 10.1*.

Table 10.1 Standard electrochemical potentials of relevance to marine corrosion

Half cell reaction	Potential (mV)
$Au^{3+} + 3\,e^- = Au$	1500
$Fe^{3+} + e^- = Fe^{2+}$	771
$O_2 + 2H_2O + 4e^- = 4OH^-$	401 (pH = 14)
$Cu^{2+} + 2e^- = Cu$	337
$AgCl + e^- = Ag + Cl^-$	222
$H^+ + e^- = 1/2 H_2$	0
$Ni^{2+} + 2e^- = Ni$	−250
$Fe^{2+} + 2e^- = Fe$	−440
$Zn^{2+} + 2e^- = Zn$	−763
$Ti^{2+} + 2e^- = Ti$	−1630
$Al^{3+} + 3e^- = Al$	−1660
$Mg^{3+} + 3e^- = Mg$	−2370
$Li^+ + e^- = Li$	−3045

Data taken from Parsons (1959) at 25°C referred to the standard hydrogen electrode

10.2.2 Thermodynamics and potential for corrosion

The electro-chemical series can be used in a quantitative fashion to assess the thermodynamic tendency in combinations of half cells. First, the half cell potentials can be modified to account for non-standard ion concentrations using the Nernst Equation:

$$E = E_o + \frac{RT}{zF} \ln C^{z+} \tag{10.3}$$

where E is the standard electrode potential, C_n is the ion concentration, E_o is the potential at concentration $C = 1$ and F is the Faraday constant.

Most corrosion reactions can be regarded as the sum of two half cell reactions where the oxidation reaction is designated anodic and the reduction reaction is designated cathodic. The anodic reaction in any pair of half cells is that which has a baser potential (i.e. lower in the electro-chemical series). The anodic and cathodic reactions may both take place on the same surface and, for corrosion to take place, at least one of these must involve metal dissolution. The cathodic reaction can be regarded as a net consumer of electrons and the anodic reaction as a producer of electrons. A good example of the interdependence of the anodic and cathodic reactions is given by the corrosion of steel in neutral electrolytes:

Anodic reaction:
$$Fe = Fe^{2+} + 2e^- \quad (10.4)$$
$$(E_{0.1} = +440 \text{ mV})$$

Cathodic reaction
$$O_2 + 2H_2O + 4e^- = 4OH^- \quad (10.5)$$
$$(E_{0.2} = +401 \text{ mV})$$

From earlier considerations it can be seen that, for a corrosion reaction to proceed, the overall cell potential must be positive, corresponding to a negative free energy change and combining the above half cell reactions in the ratio 2:1, the overall standards cell potential can be obtained:

$$2Fe + O_2 + 2H_2O = 2Fe^{2+} + 4OH^- \quad (10.6)$$
$$(E_{cell} = E_{0.1} + E_{0.2} = +841 \text{ mV})$$

For details on combining electro-chemical half cell reactions see Uhlig and Revie (1985).

Since E_{cell} is positive this means that, when the ionic concentrations are at the standard level, the aqueous corrosion of iron is thermodynamically feasible. However, in neutral electrolytes saturated with atmospheric oxygen, the hydroxyl ion concentration is about 10^{-7} molar and the partial pressure of oxygen is about 0.2 atmospheres.

Furthermore, when a substantial bulk of water is available, it is unlikely that the iron ionic concentration at the corroding surface will exceed about 10^{-6} molar. Thus, using the Nernst Equation:

$$E_1 = E_{0.1} + \frac{0.059}{2} \log(10^6) \quad (10.7)$$

and

$$E_2 = E_{0.2} - \frac{0.059}{4} \log\left[\frac{(10^7)^4}{0.2}\right]$$

$$= E_{0.2} - 0.059 \text{ pH} + 0.015 \log(p_{O_2}) \quad (10.8)$$

so that increasing alkalinity retards the corrosion of iron, as does reduced oxygen concentration or a build up of ferrous ions at the corroding surface. Note that the appropriate value for $E_{0.2}$ is 1.23 V since that in Table 10.1 only holds at pH = 14 (i.e. $-14 \times 0.059 = 826$).

10.2.3 Kinetics of corrosion and polarization

Thermodynamic considerations will indicate the tendency and ultimate equilibrium point of reactions but give no indication as to the rate at which these will proceed. The application of chemical kinetics to corrosion reactions centres around a knowledge of the corrosion current as opposed to the cell potential. In most cases, it is the corrosion current density which is of importance since this is related to the rate of loss of metal (by weight) per unit area which has implications for the rate of loss of metal thickness (see Section 10.3.1):

$$C = \frac{10^{-3} i M}{F} \quad (10.9)$$

where C is the rate of metal loss, expressed as kg/m^2 s, F is the Faraday Constant (C/g mole), i is the corrosion current density (A/m^2) and M is the molecular weight (g).

It is of course very difficult to measure the current flowing between the anodic and cathodic half cells, especially since these may be taking place on the same surface, but the concept of corrosion current is very important in understanding the kinetics of corrosion. If current is flowing in a corrosion cell, it follows that the cell cannot be at equilibrium since the potential of the electrode will shift so as to oppose the corrosion current.

In an active cell the difference between the equilibrium potential of the anode or cathode (E_{eq}) and the corrosion potential (E_{corr}) is known as the overpotential. Since decreased potential tends to stifle anodic reactions and assist cathodic reactions, care must be taken in sign conventions in order that the correct potential difference can be calculated when combining half cell reactions. The shift in potential of an anode (η_a) when a current flows is generally in a positive direction, whereas that at the corresponding cathode (η_c) will be in a negative direction, the overall tendency in the cell being for the potential difference to become smaller.

Even at equilibrium, a half cell reaction is still going on except that the rates are equal in the forward and reverse directions. This means that a half cell reaction like that for dissolution of a metal could be either anodic, going in the direction $M \rightarrow M^{z+}$ or cathodic, going in the direction $M^{z+} \rightarrow M$. At equilibrium, the current densities in the anodic and cathodic directions are equal and this is known as the exchange current density (i_o). The development of an overpotential by virtue of current flow is known as polarization and this can occur through two basic mechanisms. These are dealt with in detail by Fontana and Greene (1982) and are known as concentration polarization which is due to a diffusional limit on the current density and activation polarization which results from control of the current density by some step in the surface reaction. Activation polarization causes the overpotential to increase with the logarithm of corrosion current whereas concentration polarization gives rise to a limiting current density (i_L) which cannot be exceeded no matter what the overpotential may be.

Generally, activation polarization can affect both anodic and cathodic reactions whereas concentration polarization tends only to affect some cathodic reactions. The polarization curve for a typical half cell would therefore look like *Figure 10.1* depending upon whether the cathodic reaction shows concentration polarization (curve A) or not (curve B).

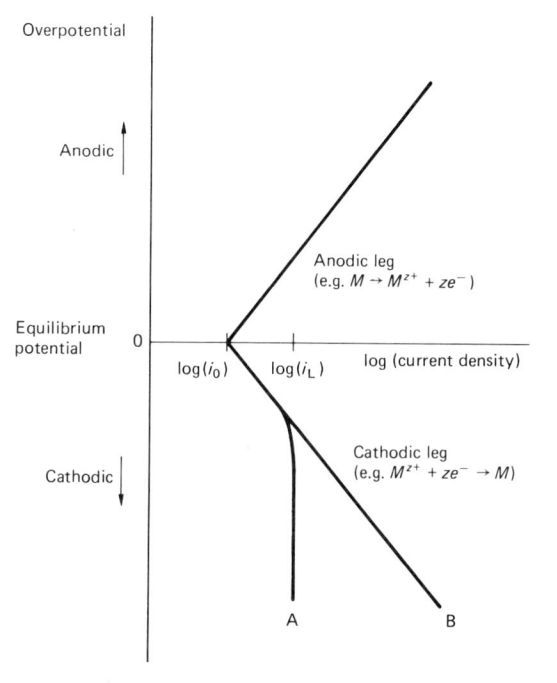

Figure 10.1 Polarization of a hypothetical half cell in its anodic direction, showing activation polarization and in its cathodic direction, showing activation polarization, B and activation/concentration polarization, A

10.2.4 Evans Diagrams and mixed potential theory

One important way in which the theory of polarization outlined in Section 10.2.3 may be used is in the Evans Diagrams. These diagrams are plots of potential *versus* current or current density in various half cells and can be used to understand a wide range of corrosion phenomena. A number of useful examples are given by Fontana and Greene (1982) although the term Evans Diagram is not used by these authors.

Figure 10.2 shows a schematic diagram of the interplay between two half cell reactions involving two species C and A and their respective ions. Both species are shown as being activation polarized and the effect of possible concentration polarization of C is shown by

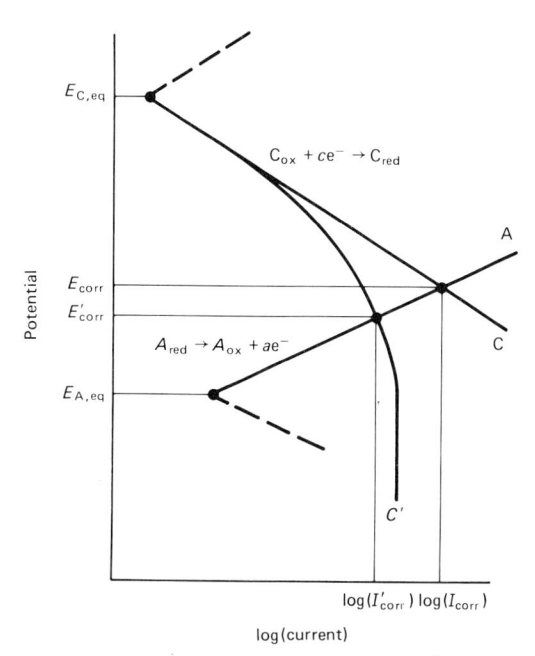

Figure 10.2 Evans type diagram illustrating the application of mixed potential theory to two half cell reactions

curve C'. Since the equilibrium potential of C ($E_{C,eq}$) lies above that of A ($E_{A,eq}$), this means that reaction C will behave cathodically and reaction A will behave anodically. If the cell is short circuited, the anodic and cathodic reactions must proceed at the same rate to prevent charge build up so that the corrosion current (I_{corr}) and potential (E_{corr}) are given by the intersection of the anodic and cathodic polarization curves. The effect of concentration polarization on the cathodic reaction can also be seen in *Figure 10.2*. Specific examples of this type of theory will be dealt with later.

10.2.5 Surface corrosion products and passivity

Many corrosion reactions give rise to the formation of insoluble products on the corroding surface and these often have a strong influence on the kinetics of the process. Furthermore, some metals and alloys will form a surface film or scale prior to service, for example in the atmosphere, and this may deliberately be enhanced or encouraged, as in anodizing of aluminium and its alloys. Surface films or layers show a variable degree of corrosion protection depending mainly upon their permeability to the corrodant and their ability to reform in the environment (self healing). As far as dry oxidation is concerned, the subject is extensive and outwith the scope of this chapter. The reader is referred to Kubaschewski and Hopkins (1962) for a detailed treatment but most interest in oxidation refers to high temperatures whereas the marine corrosion engineer is more interested in oxides which may form under normal atmospheric or natural submerged conditions.

The films which form during aqueous or atmospheric corrosion can be thin and relatively protective, in which case the alloy is said to be in a passive state, or bulky and loosely adherent in which case the retardation of the corrosion rate is likely to be slight. An example of the former type of layer is that which forms on titanium and an example of the latter is that which forms on carbon–manganese steel (mild steel). Although non-protective layers like that which forms on steel result in a reduction of corrosion rate with time (Schumacher, 1979), it is normal for corrosion engineers not to rely on this and to consider the corrosion rate to be constant with time, usually measured over a period of about ten years. Similarly, a layer which is considered to be protective usually permits an often negligibly small corrosion rate.

The formation of a passive layer can be demonstrated by both thermodynamic and kinetic theory. The thermodynamic theory was established largely by Pourbaix (1974) and is summarized in a volume of potential pH diagrams. These diagrams use chemical and electrochemical thermodynamic data to establish a type of phase diagram for a metal or alloy in aqueous media. A simple example is given by the behaviour of iron in water of variable pH (*Figure 10.3*). It should be noted that the water is supposed to contain only hydrogen and hydroxyl ions and the presence of other species, especially chloride, can have a substantial effect on the zones of passivity. The development of part of the diagram can be illustrated by a continuation of the argument outlined in Section 10.2.2. From the Nernst Equation, it has already been seen that the potential of the iron dissolution half cell can be given by:

$$E = -0.44 + 0.0295 \log C_{Fe}^{2+} \text{ (at 25°C)} \quad (10.10)$$

However, the chemical equilibrium which describes the formation of ferrous hydroxide must also be considered:

$$Fe(OH)_2 + 2H^+ = Fe^{2+} + 2H_2O \quad (10.11)$$

and this has the equilibrium constant:

$$K = 10^{13.29} \text{ (at 25°C)} \quad (10.12)$$

The water and hydroxide concentrations can be taken as unity so that the ferrous ion concentration can be expressed in terms of the pH;

$$\log C_{Fe}^{2+} = 13.29 - 2\text{ pH} \quad (10.13)$$

If, as before, it is considered that the ferrous ion concentration at the surface is about 10^{-6}, then the above two equilibria can be described by the equations:

$$E = -0.62 \text{ for Fe/Fe}^{2+} \quad (10.14)$$

and

$$\text{pH} = 9.6 \text{ for Fe/Fe(OH)}_2 \quad (10.15)$$

these two lines are shown in *Figure 10.3* and represent two of the boundaries of the Fe^{2+} phase field. The remaining lines can be deduced from similar thermodynamic principles. Note that when an equilibrium

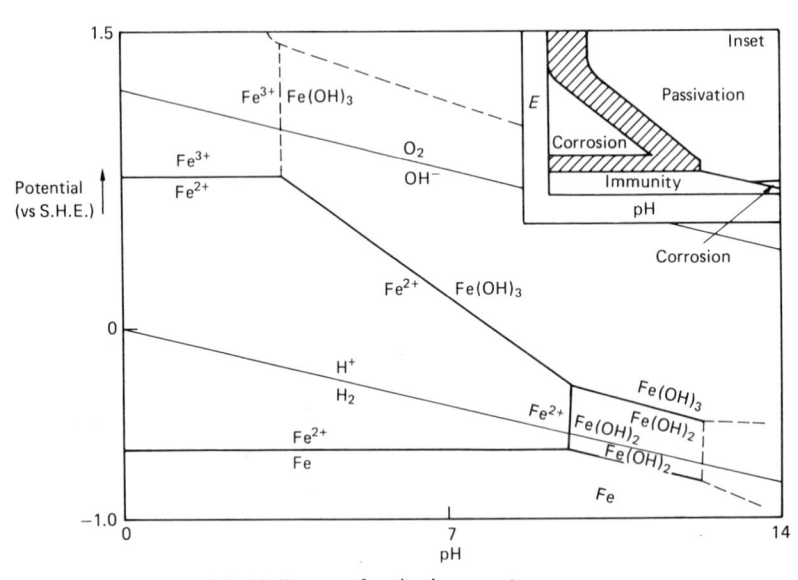

Figure 10.3 Potential-pH diagram for the iron–water system at 25°C also showing oxygen reduction and hydrogen evolution lines for normal atmospheric conditions. Inset shows regions of corrosion, immunity and passivity (adapted from Pourbaix 1974)

involves electron transfer, it is potential-sensitive and when proton transfer is involved, it is pH-sensitive.

To complete the diagram, it is helpful to include the oxygen reduction and hydrogen evolution equilibria since these are likely cathodic reactions in the aqueous corrosion of steel:

$$O_2 + 2 H_2O + 4e^- = 4 OH^- \quad (10.16)$$

giving

$E_{O_2} = 1.22 - 0.059$ pH (for $p_{O_2} = 0.2$ Ats)

and

$$H^+ + e^- = 1/2 H_2 \quad (10.17)$$

giving

$E_{H_2} = -0.059$ pH (for atmospheric pressure).

The diagram may now be interpreted in terms of the changes which are likely to occur over the range of pH and potential displayed. The phase fields which contain Fe^{2+} or Fe^{3+} are those in which corrosion is likely. Those containing the hydroxides, which eventually age to oxides, represent areas of stable corrosion product formation although, as pointed out by Uhlig and Revie, (1985) this does not strictly imply passivity. The area of stable iron represents immunity from corrosion since the dissolution of iron is thermodynamically unfavourable.

The inset in *Figure 10.3* shows these areas with the boundaries thickened to show the relatively small effect of several orders of magnitude change in the surface iron ion concentration. The area below the oxygen reduction line is that in which this reaction may provide the cathodic half of the corrosion cell in water saturated with atmospheric oxygen and the area below the hydrogen evolution line is that in which this reaction may produce hydrogen at 1 atmosphere pressure. The ways in which these changes might be exploited in corrosion protection will be considered later.

The kinetic aspects of passivity can be best illustrated on an Evans Diagram. A schematic diagram illustrating passivation is shown in *Figure 10.4* (curve A). The cathodic leg of the half cell is of no interest here and is shown dotted. If the alloy is initially at equilibrium in its active state and the potential is raised it will at first show an increased corrosion current in accordance with activation polarization. However, a potential will eventually be reached (E_p) where the corrosion current drops off dramatically to a relatively low value (i_p). This primary passive potential is a function of the alloy involved, although some alloys show no tendency to passivate, and also the environment, chloride and hydrogen ions being particularly known to affect passivity phenomena. As the potential is increased further, no substantial increase in the corrosion current is observed until a second critical potential, shown as E_T, is reached where the current again increases with potential, often more rapidly than before.

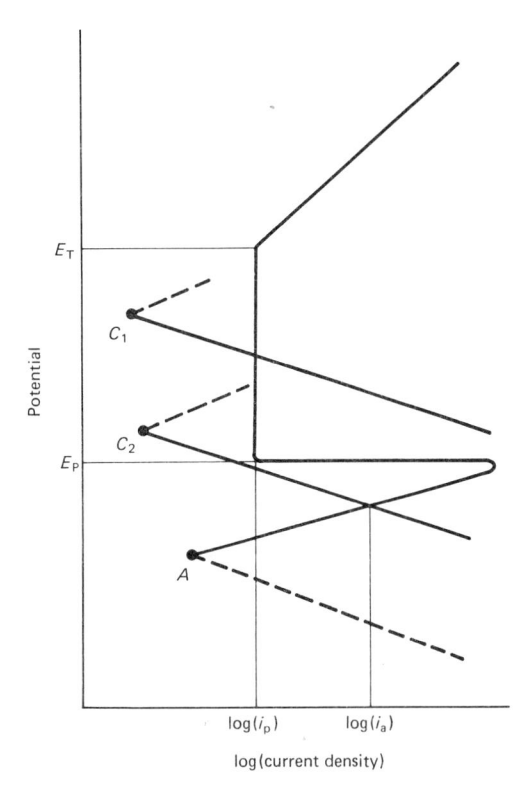

Figure 10.4 Typical potential-current density, showing passivation of the anodic half cell, curve A and the effect of two possible cathodic half cells, curves C_1 and C_L (from Shreir, 1976)

Figure 10.4 also illustrates the effect of a second, cathodic half cell on passivation behaviour. Two different possible, although not simultaneous, cathodic half cells are shown. In the case of the upper of the two cathodic half cells (C_1), the anodic corrosion would be at the low, passive level (i_p). However, if the cathodic reaction were to be as shown for the lower of the two possibilities, then the anodic metal would be in the active state and would corrode relatively rapidly (at i_a). It can be seen that multiple intersections of the anodic curve are possible and these, more subtle, effects are dealt with in Fontana and Greene (1982).

Passivation is very important in the corrosion behaviour of a number of engineering alloys and some of the more practical aspects will be treated later. The sensitivity of the passivation phenomenon to electrolyte chemistry is a very important factor which can limit the performance of passivating alloys.

10.3 The mechanisms of corrosion

The popular conception of corrosion is coloured by the everyday observation of atmospheric corrosion of steels. In fact, such broad-fronted corrosion is only one of a number of mechanisms or morphologies by which corrosion can proceed. Some of these mechanisms only occur under special conditions unrelated to the marine environment and will be considered no further here. This section is therefore concerned with corrosion mechanisms and other specific phenomena which may take place in marine and related environments. An appreciation of these mechanisms is essential for the marine engineering designer in order to know what type of corrosion resistant properties are required and to avoid problems caused by design or manufacture.

The corrosion mechanisms can be divided into those which are of purely electro-chemical origin and those which have some mechanical element. Some aspects of these mechanisms which are of particular interest in marine environments are also discussed.

10.3.1 Purely electro-chemical effects

Figure 10.5 illustrates the different ways in which a purely electro-chemical phenomenon can give rise to corrosion. In such mechanisms, the source of primary damage is always by dissolution of anode material, the differences being in the way such dissolution is distributed over the corroding surface.

10.3.1.1 General corrosion

In general or broad fronted corrosion, the anodic sites are evenly distributed over the metal surface, resulting in a broadly even loss of section thickness. Such corrosion is readily designed against, since losses over a period of time can be predicted. It is usual to express general corrosion rates in terms of the rate of loss of metal thickness (e.g. μm/year) and this can be related to the rate of weight loss per unit area described in Section 10.2.3:

$$r = 3.15 \times 10^{13} \frac{C}{D} \qquad (10.18)$$

where r is the thickness loss (μm/year), C is the weight loss (kg/m^2 s) and D is the metal density (kg/m^3).

In a normal design, a penetration rate of less than about 25 μm/year would be considered to be acceptably small and rates above 200 μm/year would require some protection. These figures are, however, approximate since an easily replaceable thick walled component in a non-critical area may permit corrosion rates of a few thousand μm/year.

The principal uncertainty in the application of corrosion allowances is that the corroded profile will not, in general, be flat, so that the designer usually requires some idea of the maximum pit depth. Also, corrosion rates are not usually constant with time, but reported values often contain a degree of conservatism to allow for this.

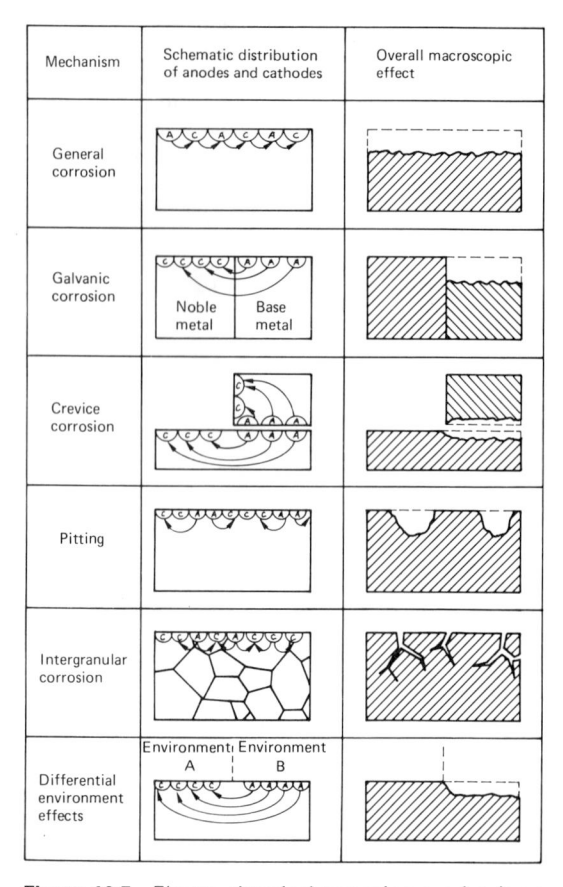

Figure 10.5 Electro-chemical corrosion mechanisms. Areas marked A are anodic where the corroding metal dissolves to produce metal ions in solution. In areas marked C, the oxidized form of the cathodic species combines with electrons to its reduced form, e.g. oxygen reduction reaction. (a) General corrosion; (b) galvanic corrosion; (c) crevice corrosion; (d) pitting; (e) intergranular corrosion; (f) differential environment effects

10.3.1.2 Galvanic corrosion

Galvanic or bimetallic corrosion occurs when a second, more noble metal is connected to the corroding metal. Depending on the noble to base metal area ratio and the polarization characteristics of the two metals, the corrosion rate may be increased substantially.

The electro-chemical series shown in *Table 10.1* is only of limited use in deciding the galvanic activity of an alloy, since it does not allow for passivity effects. For this purpose, a more practical galvanic series for engineering alloys in sea water has been devised and this is shown in *Table 10.2*. The effect of galvanic coupling on corrosion can be understood in terms of an Evans Diagram such as *Figure 10.6* using the example of iron as the base metal connected to copper as the noble metal in sea water. In the absence of copper, the

Table 10.2 A practical galvanic series for engineering alloys in sea water (from LaQue 1975)

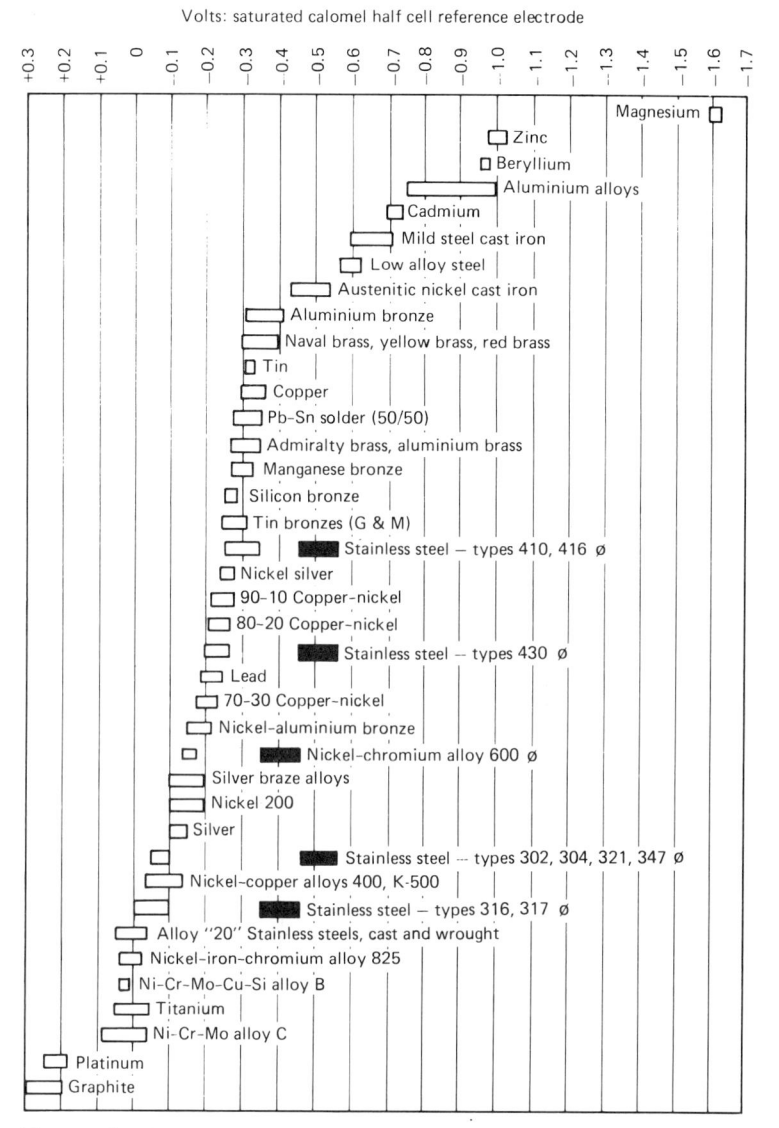

Volts: saturated calomel half cell reference electrode

(Potentials range from +0.3 to −1.7 V)

- Magnesium
- Zinc
- Beryllium
- Aluminium alloys
- Cadmium
- Mild steel cast iron
- Low alloy steel
- Austenitic nickel cast iron
- Aluminium bronze
- Naval brass, yellow brass, red brass
- Tin
- Copper
- Pb-Sn solder (50/50)
- Admiralty brass, aluminium brass
- Manganese bronze
- Silicon bronze
- Tin bronzes (G & M)
- Stainless steel — types 410, 416 ø
- Nickel silver
- 90-10 Copper-nickel
- 80-20 Copper-nickel
- Stainless steel — types 430 ø
- Lead
- 70-30 Copper-nickel
- Nickel-aluminium bronze
- Nickel-chromium alloy 600 ø
- Silver braze alloys
- Nickel 200
- Silver
- Stainless steel — types 302, 304, 321, 347 ø
- Nickel-copper alloys 400, K-500
- Stainless steel — types 316, 317 ø
- Alloy "20" Stainless steels, cast and wrought
- Nickel-iron-chromium alloy 825
- Ni-Cr-Mo-Cu-Si alloy B
- Titanium
- Ni-Cr-Mo alloy C
- Platinum
- Graphite

Alloys are listed in the order of the potential they exhibit in flowing sea water. Certain alloys indicated by the symbol ø in low velocity or poorly aerated water and at shielded areas may become active and exhibit a potential near − 0.5 volts

corrosion current of the iron will be I_a due to the cathodic polarization curve for oxygen reduction on the base metal, similar to *Figure 10.2*.

The effect of connecting a piece of copper will be to provide an additional cathodic surface which will in general have different values for exchange current density (i_o) and diffusion limited current density (i_L). Also, as the noble metal area is increased, the exchange current, as opposed to current density, will increase, as will the diffusion limited current for oxygen reduction. The net effect is for the corrosion current to increase from I_c to I_c' to I_c'' as the noble metal area is increased.

In the case of copper and steel, a copper to steel area ratio of 3:1 will approximately double the corrosion rate of the steel and a ratio of 20:1 will result in about a tenfold increase in the corrosion rate of the steel (LaQue, 1975). Similar data for other galvanic couples are available (Astley and Rowlands, 1985) noting that

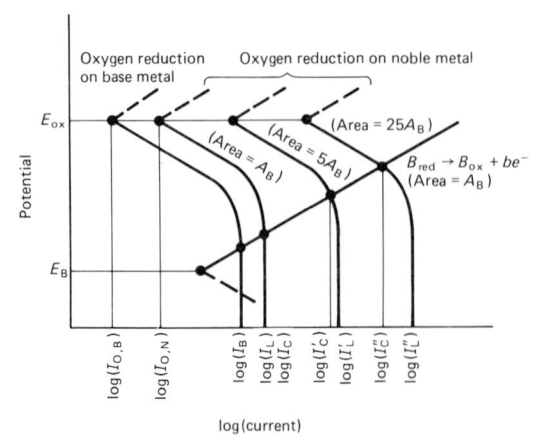

Figure 10.6 Effect of galvanic coupling on the corrosion rate of a base metal, B, with a concentration polarized cathodic reaction such as oxygen reduction. The cathodic reaction is shown as taking place on one of three sizes of a noble metal or on the base metal

cathode to anode area ratios of less than about 0.1 do not cause significant corrosion acceleration. Alloys which are made up of more than one phase often show galvanic corrosion between the different phases.

Furthermore, the galvanic effects set up between the substrate and adhering mill scale, damaged passive layers and even damaged paint layers can be sufficient to give rise to localized corrosion. An interesting example also occurs between corroded and new cast iron where the corroded iron may have a layer of graphite and behave cathodically. Other subtler galvanic effects involving more than one corroding metal are described by Fontana and Greene (1982).

10.3.1.3 Crevice corrosion

Crevice corrosion is caused by local composition differences produced in the electrolyte by the corrosion reaction. An important part of the mechanism is the prevention of mixing at the corrosion site so that it tends to be aggravated by stagnant conditions. Crevice corrosion is normally associated with passivating metals (e.g. stainless steels) in electrolytes which contain an aggressive anion (e.g. chloride), and the mechanism has been explained by separation of the cathodic oxygen reduction reaction and the anodic dissolution of iron (Shreir, 1976).

Taking stainless steel as an example, the access of dissolved oxygen to the sites within the crevice is restricted so that the balance between anodic and cathodic sites is disturbed resulting in a build up of iron ions in the crevice. This positively-charged area then attracts negatively-charged ions, notably chloride, which tend to depassivate the stainless steel resulting in further local dissolution of iron and so on.

10.3.1.4 Pitting

Pitting is the name given to the highly localized penetration shown in *Figure 10.5(d)* and is geometrically distinct from the slight variations across a surface which shows general corrosion, although the term is often used to describe local penetration in a generally corroding surface.

There are a number of different submechanisms which can give rise to pitting but these generally involve local breakdown of a passive or other protective layer. In this sense, pitting resembles crevice corrosion although the latter requires a macroscopic change of environment, whereas pitting tends to initiate at microscopic heterogeneities. Once initiation has been achieved, the sheltered environment can be produced by the shape and orientation of the pit itself or by a cap of corrosion products. Because of the requirement for shelter and, in some cases, for poorly oxygenated conditions, pitting tends to be found more often in stagnant conditions. Shreir (1976) has reviewed the mechanisms for pitting in a number of different alloy systems.

10.3.1.5 Intergranular corrosion

In some circumstances certain parts of a metal surface are more susceptible to corrosion than others. This can lead to a variety of localized corrosion effects amongst which is intergranular corrosion. The best known example of intergranular corrosion takes place in some stainless steels which have been poorly heat treated. If the stainless steel is of an austenitic type, contains sufficient carbon and is heated to between 500 and 800°C for a sufficiently long time, the carbon precipitates at the grain boundaries as a chromium carbide. This leaves areas close to the grain boundaries deficient in chromium and affects the passivity at these sites. Subsequent exposure to a corrosive environment can then lead to rapid penetration along the grain boundaries.

One way in which stainless steels can become sensitized is by welding, in which case the localized corrosion adjacent to the weld line is known as weld decay. Sensitization is now such a well-documented phenomenon that it now rarely occurs since sensitization-resistant stainless steels are now readily available.

10.3.1.6 Differential environment effects

The Nernst Equation, Section 10.2.2, states that the electrode potential of a metal or a half cell reaction can be influenced by the composition of the electrolyte. This can give rise to a phenomenon, not unlike galvanic corrosion where parts of the same metal surface are exposed to varying environmental conditions. The best known of such differential environment effects is probably the differential aeration cell, where parts of the same surface exposed to a higher oxygen level will tend to behave cathodically to parts with a lower supply of oxygen and small areas of metal exposed by local coating failure can show very rapid corrosion. Both crevice corrosion and pitting are special cases of this general effect, where the difference in environment is that produced by splitting the anode and cathode reactions.

10.3.2 Combined electrochemical/mechanical effects

Corrosion and mechanical damage are often found in association. An obvious example is corrosion at sites of mechanical damage to coatings and it is well established that metal which has been plastically deformed behaves anodically to unstrained material.

There are, however, a number of other perhaps more subtle mechanisms which require consideration. One important feature of these conjoint phenomena is that the proportion of the damage which is due to electrochemical effects as opposed to mechanical effects can vary quite substantially. Also, the mechanical effect is often confined to damage of a passive surface film or corrosion product layer rather than direct damage to the underlying metal. The mechanisms under discussion are illustrated in *Figure 10.7*

10.3.2.1 Fretting corrosion

Fretting corrosion is the name given to a set of phenomena which involve corrosion and abrasion. Depending upon which is the dominant component, this can be seen as corrosion aggravated by wear or vice versa. The mechanism can operate by a wear–corrosion or corrosion–wear route depending upon which component actually removes metal from the surface (Fontana and Greene, 1982). The latter of these two submechanisms is illustrated in *Figure 10.7*.

Fretting itself is normally distinguished from other types of wear by its low amplitude of relative reciprocating motion between the wearing surfaces, as might be experienced in a wire rope or between two relatively

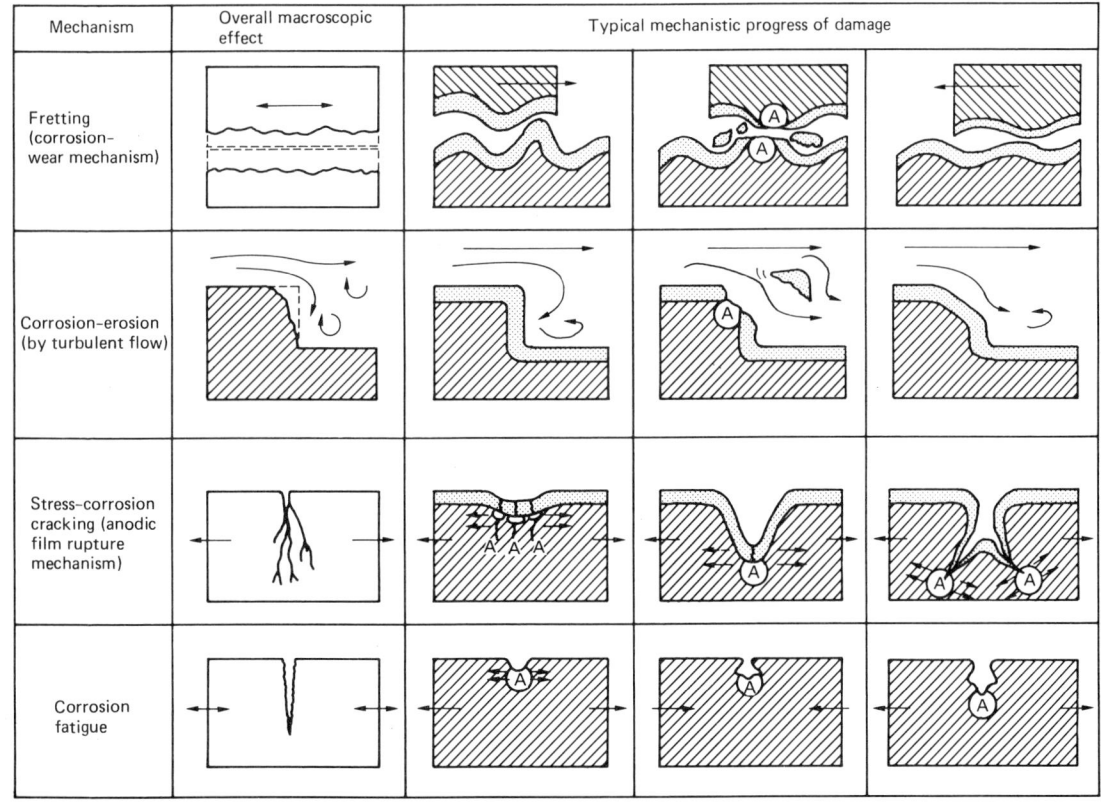

Figure 10.7 Combined electro-chemical–mechanical corrosion mechanisms. Cross-hatched areas denote metal section and shaded areas denote films or corrosion product layers. Areas marked A show sites of anodic corrosion

lightly bolted surfaces. The wear–corrosion route to damage could be more strictly regarded as fretting rather than fretting corrosion, since the corrosion component only acts after metal has been removed from the surface by mechanical wear. The oscillatory nature of the loading can also result in contact fatigue damage which may be further aggravated by corrosion. Waterhouse (1972) deals with both the mechanical and electro-chemical aspects of the phenomenon.

10.3.2.2 Corrosion–erosion

The term corrosion–erosion is normally used to describe those phenomena where the mechanical component is associated with fluid wear. Such mechanisms normally involve extremes of fluid flow although some types of corrosion (e.g. corrosion by dissolved carbon dioxide) are very susceptible to flow effects.

The fluid usually acts to remove corrosion product forming on the surface, thus defeating the beneficial effect of surface films and continually exposing new metal to the corrosive action of the fluid. The erosion can come about through impingement, cavitation or by turbulent flow conditions, such as at bends in pipes. Cavitation, in particular, can remove metal without there being a significant corrosion component, so that in some cases the corrosion resistance of the metal becomes irrelevant in favour of more physical properties. Suspended solids in the fluid stream also have a substantial effect upon the erosional component, as do the corrosion products themselves.

A certain amount of work has been carried out on hydrodynamic limits for corrosion–erosion in order to give designers some figures for its avoidance. This design guidance is often given as a velocity limit although more detailed analyses have been carried out taking account of flow conditions. For a review of the physical and electro-chemical aspects of corrosion–erosion, the reader is referred to Preece (1979).

10.3.2.3 Stress corrosion cracking (SCC)

Stress corrosion is one of the most destructive types of corrosion and must be avoided at all costs. Cracks form at regions of stress concentration on the metal surface and these cracks propagate under the combined action of a crack opening stress and a corrosion reaction at the crack tip. The reaction thus attacks a very small proportion of the total exposed area and progresses rapidly through the section with consequent loss of load bearing capacity.

Detailed mechanisms vary with the specific alloy/environment/stress combination and these are dealt with in detail by Logan (1966). Only one of these is shown in *Figure 10.7*, but the macroscopic effect is general. One fortunate aspect is that stress corrosion requires both the existence of a crack-opening stress, usually a surface tensile stress and a specific combination of alloy and environment and that most of these combinations of practical importance are well documented. It is rare for the engineer to risk any likelihood of stress corrosion and the normal design philosophy is to avoid the known susceptible combinations since the avoidance of stress is not often a practicable proposition. It is important to realize that residual as well as applied stresses can provide the mechanical element of this phenomenon so that even lightly loaded components may suffer cracking.

Crack propagation rates are dependent upon a number of factors within a particular alloy–environmental system. For example, austenitic stainless steels may crack in warm chloride-containing solutions. At a given concentrated stress level the time to failure is dependent upon such factors as the temperature, composition of the steel, chloride content of the solution, condition of the steel, potential and oxygen content of the solution (Sedriks, 1979). Other systems will have different sets of critical factors but all share a pronounced sensitivity to stress, normally displaying a threshold level below which cracking will not take place.

10.3.2.4 Corrosion fatigue

Corrosion fatigue is more commonly regarded as fatigue aggravated by corrosion rather than the converse. Fatigue is the propagation of surface cracks under an alternating stress and can proceed without a corrosive element. One way of describing the phenomenon of fatigue is by the relationship between endurance (number of stress cycles to failure, N) and the stress range (difference between maximum and minimum stress, S). A typical relationship is shown in *Figure 10.8* along with the general effect of a corrosive component. Since the corrosive component is time-dependent and the mechanical component is dependent upon the number of cycles, it follows that the rate of crack progress through the section depends upon the ampli-

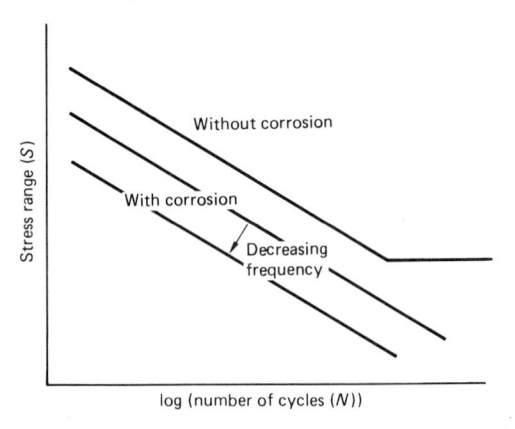

Figure 10.8 General effect of corrosion on the *S–N* curve for fatigue

tude of the stress fluctuations and the loading frequency.

The mechanistic effect of a corrosive environment is preferentially to dissolve the anodic areas at the crack tip and increase the rate at which cracks propagate. Fatigue and corrosion fatigue are of particular importance in the marine environment where alternating loads due to waves, currents and wind are likely to occur. Mechanistic details are covered in Parkins and Kolotyrkin (1983) and specific relevant aspects are dealt with in Section 10.3.3.

10.3.3 Specific effects of interest in marine environments

As can be seen from Chapter 1, the marine environment is a complex one from the point of view of biological, physical and chemical factors. For example, temperature and oxygen levels vary quite considerably with both depth and location (LaQue, 1975). Dexter (1987) has reviewed global variations in salinity, dissolved oxygen, temperature and other effects likely to affect the corrosion of metals in sea water.

As well as the natural environment, it is also important to consider the synthetic environments which are created in association with the activities carried out on and under the sea (e.g. sea water cooling, oil and gas recovery). When designing against marine corrosion, it is therefore important to be aware of those aspects of the environment which affects corrosivity and then to be aware of the corrosion mechanisms which may arise in specific areas.

10.3.3.1 Environmental factors

For the corrosion of steel, the marine environment can be divided into five zones, shown in *Figure 10.9*, each of which has its own set of corrosivity factors.

The marine atmosphere is more corrosive than that inland because of wind borne salt and spray which can be deposited on metal surfaces. The factors which affect corrosivity are therefore wind, temperature, vertical and horizontal distance from the sea surface, existence of solids (e.g. sand) which may be entrained in wind, attitude of metal surface, rainfall and air and water pollution. All these factors cause an immense variability in marine atmosphere corrosion rates throughout the world. Many exposure tests have been carried out over the years so that a large body of data is available on many common alloys in a variety of locations (Griffin, 1987). A map of atmospheric corrosivities for the British Isles which includes coastal areas was published by the UK Ministry of Agriculture, Fisheries and Food in 1986. Accelerated test methods are also used to evaluate coatings for marine atmospheres and these are considered further in Section 10.7.4

The splash zone is the region where alternate wetting and drying takes place and is recognized as the most

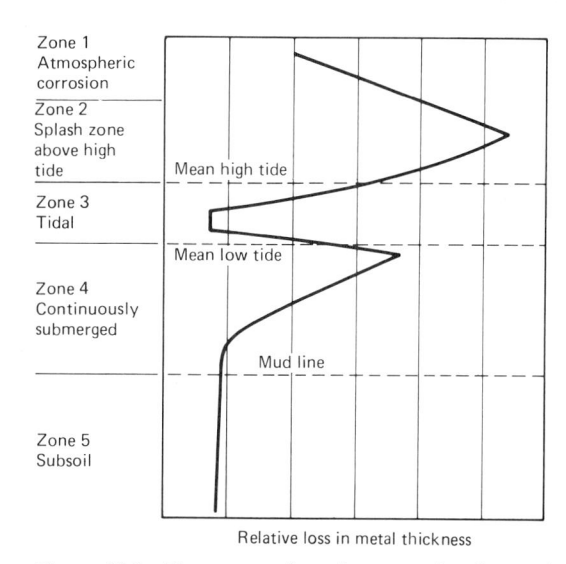

Figure 10.9 Five zones of marine corrosion for steels (from LaQue 1975)

susceptible in the marine environment. The main reason for this is that a plentiful supply of both electrolyte and oxygen is available and this is the principal factor in the splash zone where total immersion does not occur with the consequent mitigations outlined below. Temperature and sea state will also affect splash zone corrosion rates.

The intertidal zone is also alternately wet and dry but is totally immersed for at least part of the time. This results in a number of effects, the sum of which normally results in a rather low corrosion rate compared with the submerged zone and the splash zone. The first of these is that biofouling layers will form on the surface and these can restrict the supply of oxygen and provide a physical barrier to corrosion and the possible erosional effects of high currents.

However, fouling layers may also give rise to adverse chemical environments and, in particular, the trapping of silt can produce conditions in which sulphate reducing bacteria (SRB) may thrive. Hard fouling layers can, by their shell growth, shave coatings from the surface, resulting in reduced protection. The intertidal zone can also be protected to a certain extent by differential aeration effects driven by the relatively rapid corrosion in the splash zone.

The main factors affecting corrosion in the submerged zone are temperature and dissolved oxygen content. Since both these factors depend on depth, corrosion rates in very deep water tend to be rather low. In shallower waters, circulation can be such that oxygen content shows little variation with depth. Other factors which may influence corrosivity in the submerged zone include velocity of the sea water, salinity, electrical conductivity of the sea water and pollution. Data on corrosion rates for submerged and deeply submerged steel and other alloys are collected in Schumacher (1979).

Below the mud-line in the subsediment zone oxygen supply is generally limited and corrosivity is often dictated by biological processes. In particular, SRBs (see Section 10.3.3) may find suitable conditions in which to produce sulphide which will enhance corrosivity. The biological activity of the sediment is dependent upon a large number of different factors including depth, temperature and water column biology (King, 1979).

Corrosion of steels in the above zones has been studied for a number of geographical locations and relative corrosion profiles of the form of *Figure 10.9* have been published to aid the design of sheet piles and offshore structures (Morley and Bruce, 1983).

10.3.3.2 Selective leaching

A number of alloys contain elements which are widely separated in the galvanic series (e.g. copper–zinc alloys or brasses). When such alloys are immersed in sea water there is the possibility that microgalvanic cells can be set up where the baser metal will dissolve into the electrolyte, leaving behind a porous sponge of the more noble metal. This can be difficult to detect since the volume of the component may remain constant although its strength will be radically reduced. Not all alloys with components of different nobility will show selective leaching and those which do can often be inhibited by, for example, the addition of a small amount of a cathodic poison, such as arsenic in Admiralty brasses.

The principal alloy systems which may show microgalvanic effects are the brasses (copper–zinc), aluminium bronzes (copper–aluminium), and the grey cast irons (iron–carbon), the effects being referred to as dezincification, de-aluminification and graphitization, respectively. The detailed mechanism of dezincification and de-aluminification is different to that for graphitization.

In the first two cases, the element which is leached out is present initially as a solute in the matrix metal (in both cases copper) so that for any significant depth of de-alloying both the copper and the base metal must become dissolved in the electrolyte, the copper plating back on to the surface as a spongey layer. In cast irons, the graphite is present as a separate phase and the iron simply dissolves leaving a skeleton of graphite with some rust residue.

10.3.3.3 Bacterial corrosion and hydrogen sulphide

Bacteria do not cause corrosion although a number of their metabolic products are known to have an adverse effect. For sea water, the most important group comprises the sulphate reducing bacteria (SRB). Under suitable conditions, amongst which are the presence of sulphate ion and organic nutrient and the relative absence of oxygen, these bacteria produce sulphide ions which can have a detrimental effect on steels and a number of other alloys.

Although the metabolic pathway is quite complex (Tiller, 1983) the sulphate reduction step can be simplified to:

$$SO_4^{2-} + 4 H_2 = S^{2-} + 4 H_2O \qquad (10.19)$$

Hydrogen sulphide may arise from direct bacterial production on the metal surface or elsewhere by bacterial or other means. Whatever the cause, dissolved hydrogen sulphide, as well as being directly corrosive producing deep pits and rapid loss of section (Williams, 1985), also produces hydrogen at a very high partial pressure. The detailed mechanism of corrosion, especially when the bacteria colonize the metal surface, is still a matter of some discussion (Tiller, 1983) but there is little doubt as to the phenomenology, in terms of loss of section and evolution of hydrogen.

Some of this hydrogen dissolves in the metal being corroded, usually steel, and can produce various forms of hydrogen embrittlement and internal cavity formation (Terasaki *et al.*, 1984). This hydrogen damage is usually categorized as blistering, where hydrogen re-evolves inside the steel forming cavities, stepwise or hydrogen induced cracking (HIC), where small cracks form between the cavities and, in the worst cases and where a surface stress is present in a susceptible material, large scale cracking can occur and this is often referred to as sulphide stress corrosion (SSC). The materials most susceptible to SSC are high strength steels and there exist recommendations for steels and other alloys which are suitable for use in stagnant conditions or where hydrogen sulphide is known to be present (NACE Standard MR–01–75).

10.3.3.4 Sea water handling

There are a large number of areas in which the containment or movement of sea water is important and the range of temperatures, pressures and chemical conditions which may arise in such situations deserves separate attention. One design advantage that sea water systems have over immersed components is that a certain amount of inhibition and/or de-aeration may be possible in some circumstances.

One major group of sea water systems is used to cool machinery or to dump waste heat in condensers. Such cooling systems often require large flow rates and it is very uncommon to recirculate the coolant. This means that chemical treatments are limited to the addition of small amounts of relatively harmless biocides or film forming agents. Temperature, flow conditions and chemical condition of the water vary considerably with design and location and it is often economical to use more expensive and resistant alloys. The commonest corrosion problems encountered in sea water cooling are corrosion–erosion, crevice corrosion under deposits, sulphide corrosion, stress corrosion, selective leaching, galvanic effects between components, hot-spot pitting and corrosion–fatigue (Scanlon and Emms, 1979). Tubes, tube plates and water boxes are often made of different materials and the different selection criteria are discussed by LaQue (1975).

Sea water is also used extensively for ballasting and elsewhere in storage systems to displace less dense fluids. The absence of flow usually means that coatings can be used in such systems to far greater effect. Cathodic protection, de-aeration, and inhibition are also available and the use of mild steels in such systems means that many of the problems associated with the more resistant alloys (e.g. stress corrosion and crevice corrosion) can be avoided.

However, the stagnant conditions and possible presence of nutrients could lead to a build up of SRB activity although cathodic protection can be used to mitigate this somewhat.

In offshore oilfields, it is often the practice to inject sea water into reservoirs to enhance recovery. The amount and accessibility of the tubulation used for such purposes is such that resistant alloys are often prohibitively expensive but a long and predictable lifetime is required. A great deal of care is therefore taken over the condition of injection water with the use of inhibitors, biocides and other chemicals being almost universal (Ostroff, 1979). Flow rates are generally so high in such systems that even small asperities on the internal bore, such as those at the butt coupling between two tubes, are often sufficient to precipitate local wall penetration due to corrosion–erosion over time.

Pumps, pipes, valves, impellers and other fittings for sea water use also require special design consideration. Here again, the main problems are associated with corrosion–erosion. Galvanic effects may also be encountered as often more than one alloy is used in a piping system (Robson, 1984). Systems employing raw sea water are of course the most suceptible, for example de-aeration units or offshore fire pumps. These pumps require to be reliable even after long periods of ideleness so that corrosion resistance under stagnant conditions is also of importance. For impellers and diffusers, high velocity effects such as cavitation often limit designs, so that materials resistant to such effects may be required. Pipes are likely to encounter a range of velocities from stagnation to conditions where turbulent erosion may occur. At intakes, fouling resistance can be a helpful property obviating the need for special coatings and all tubulation requires to be readily weldable.

Desalination units have similar operating conditions to heat exchangers at the reject heat areas, but in heat recovery sections conditions are usually somewhat less severe due to lower oxygen levels (LaQue, 1975). Schumacher (1979) has reviewed desalination experience for a number of alloys including carbon and low alloy steels, stainless steels, and copper-, aluminium- and titanium-based alloys.

10.3.3.5 Corrosion of steel in association with concrete

The large proportion of limestone in most cements means that they are alkaline in nature. The pH is usually around 12.5, so that a passive film is formed on mild steels and, under normal atmospheric conditions, reinforcing steel in concrete is considered to be free from corrosion (see *Figure 10.3* and Section 10.2.5).

However, as with most passive films, the one which forms under the circumstances described above is suceptible to disruption by chloride ions. Since concrete is permeable to water, chloride ions in sea water will eventually find their way to reinforcing steel embedded in concrete submerged in sea water. Once the steel has been depassivated, there is the possibility that localized corrosion may occur by differential aeration provided there is sufficient oxygen access to other parts of the reinforcement (*Figure 10.10*). Such differential aeration cells are much larger than the microcells which normally form on steel in sea water with the cathodic areas being separated by as much as a few metres.

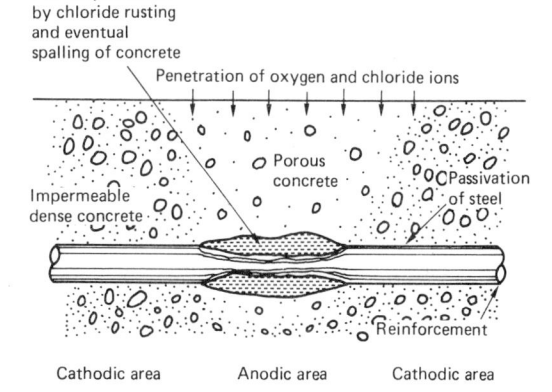

Figure 10.10 Schematic representation of a macrocell for the corrosion of steel embedded in concrete (from Browne and Domone 1975)

This has led to the use of the term macrocell (Browne and Domone, 1975) to describe such cells and there is some evidence that such a mechanism leads to concrete spalling in marine structures, although totally submerged concrete is likely to be far more resistant, due to the limited access of oxygen to the surface of the steel. The kinetics and hence the corrosion rate of the steel are rather complex since, as well as the penetration of chloride to anodic areas and oxygen to cathodic areas, the relative areas of anodes and cathodes are also important. Some detailed analyses of permeation rates (Browne and Domone, 1975) and polarization studies

(Wilkins and Lawrence, 1980) have suggested that, although chloride penetration to the reinforcement can take place within a few years, the depassivation effect does not seem to be as severe as one might expect, unless the concrete quality is locally poor (i.e. unsound).

Furthermore the cathodic reaction shows a diffusion limited current density which is much lower than that in sea water, due to the presence of the concrete cover. The overall effect is that steel embedded in sound concrete corrodes only slowly with little damage, but that local unsoundness can produce problems. The amount of deterioration in a given time has been shown to be a strong function of the amount of concrete cover (Slater, 1983).

10.3.3.6 Corrosion fatigue of structural steel

The fatigue of structural steel has received a large amount of attention because of the use of steel jacket structures for the offshore production of hydrocarbons. Although these structures are cathodically protected, the influence of sea water on fatigue, with or without cathodic protection, has been extensively studied. Both the effects of electro–chemical factors and mechanical factors have been reviewed in Jaske et al. (1981). The effects of some of the electro-chemical factors are summarized in *Figure 10.11*, but temperature, water pressure, velocity and pH are all recognized as having some effect.

Although loading frequency has a significant effect in corrosion fatigue (see *Figure 10.8*), the range of wave loading frequencies is small enough that this factor is often ignored, a conservative approach being taken. Fatigue design curves and design guidance for structural steel in sea water are extensively available in a number of different standards and these can be readily adapted for a wide range of structural configurations (see Sections 10.6.3 and 10.7.2).

In general, most of such design work is concentrated on the mechanical element of corrosion fatigue with the electro-chemical element playing a relatively minor role and most codes contain guidance on both the S–N and the fracture mechanics approaches.

10.3.3.7 Ships' propulsion and steering systems

The high fluid velocities and large pressure drops associated with marine propellers and, to some extent, rudders, make these components particularly prone to corrosion–erosion effects of which cavitation is probably the most damaging. Good hydrodynamic design goes a long way towards eliminating such problems, but it is normal to use specially developed cavitation resistant alloys for large propellers and, even then, lifetimes can be somewhat limited. For smaller craft, unless these are of particularly high performance, a few standard materials tend to be used, but the conditions under which modern heavier shipping operates require careful materials selection. LaQue (1975) has reported the main stern gear failure types as cavitation erosion of propellers and rudders, wear of bushings and liners and fatigue at propeller roots, rudders and critical areas of the aft hull and screw shaft.

Shafts are normally designed for strength and torsional stiffness, but care must be taken over corrosion resistance. It is sometimes practice to have a sleeve over the outboard portion of a steel shaft made from a material more galvanically compatible with the propeller. Other solutions include weld overlaying and bonding the shaft and propeller to the ship's cathodic protection system, usually with brushes. Other possible dangers include fatigue, fretting and galvanic corrosion. For a more detailed discussion of propeller shafts, the reader is referred to LaQue (1975).

10.3.3.8 Corrosion in oil and gas production

Hydrocarbons produced from below the sea bed have first to be carried to a primary processing area (usually offshore) where dewatering and oil and gas separation are carried out. They then have to be transported to the shore and this is often done using pipelines.

In the latter case, the processing stage allows a substantial reduction in the corrosivity of the product. However, if the carbon dioxide, hydrogen sulphide and oxygen levels exceed certain critical values, coupled with a sufficiently high humidity, some corrosion avoidance strategy is advisable (e.g. Det Norske Veritas Standard, Rules of Submarine Pipeline Systems). Such a strategy may involve inhibition, internal coating, materials selection or the adoption of an extra thickness of material (corrosion allowance). Some of these aspects are dealt with by Hill and Trout (1986).

External protection of submarine and buried pipelines is normally achieved by a combination of anti-

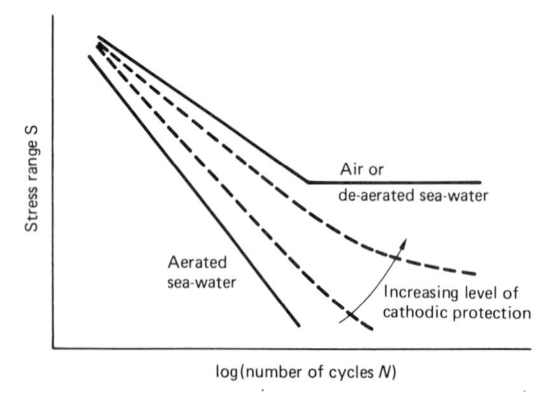

Figure 10.11 Electro-chemical parameters affecting the corrosion fatigue of steels in sea water

corrosion coating backed up with cathodic protection (Mollan and Andersen, 1986). The tubulars carrying the raw product to the primary processing system (production tubulars) can contain sand, oxygen, water, hydrogen sulphide and carbon dioxide. The corrosivity of the oxygen–sea water mixture has already been dealt with and the effects of hydrogen sulphide and carbon dioxide are a continuing subject of study (Williams, 1985).

Sand entrained in the production stream can aggravate all of the corrosive effects by adding a further erosional component and carbon dioxide corrosion is known to be particularly affected by erosion even in the absence of sand particles (deWaard and Milliams, 1975).

Inhibitors are often injected into the production stream in order to improve lifetimes and hydraulic studies are occasionally used to identify areas of particular turbulence in order that resistant alloys may be used locally (Tischuk and Huber, 1984). Corrosion phenomena and their control in petroleum production are summarized in a National Association of Corrosion Engineers' publication (NACE, 1979)

10.3.3.9 Marine fouling and corrosion

Marine fouling can in itself be a nuisance, for example through its hydrodynamic effect on ships and structures or in the obstruction of sea water intakes or even the obliteration of underwater markers. The pattern of fouling at a particular location is dependent not only upon depth, but the degree and range of species will vary with time. Fouling patterns are also dependent upon geographical location, globally and within a relatively small area such as the North Sea (Forteath *et al.*, 1984).

The effect of fouling organisms on corrosion varies with the type of organism, the type of protection used and the nature of the metal substrate. For instance, barnacles may initiate pitting in alloys which are prone to crevice corrosion, but their action on a painted surface is to grow into the painted layer, shaving it off and thus reducing protection. Dense fouling mats may also interfere with cathodic protection systems or act as collectors of silt which can subsequently be colonized by sulphate reducing bacteria.

The influence of fouling on corrosion has received relatively little attention (Terry and Edyvean, 1984) being perhaps confined to studies on microalgae, especially on ships. The reason for this is perhaps that marine exposure tests, if conducted for a long enough period, include any effects of the local fouling organisms and, for engineering purposes, there is no need to separate this out, unless the balance is disrupted in some way. Fouling can be reduced by the application of antifouling paints and coatings and also by the use of fouling resistant (copper base) alloys (Miller *et al.* 1984).

10.4 Corrosion resistance of marine alloys

With a few exceptions, alloys used in marine environments do not differ from those in general engineering use. Also, the selection of an alloy for a particular marine application rarely depends on its corrosion resistance alone. For this reason, the following section will include some general observations on the properties of each group of alloys before considering their corrosion properties. With the large number of alloys in common use it is not possible to consider each separately. Instead, the alloys have been split into groups, each being considered with reference to literature where more detail can be found. A summary of useful corrosion properties is also provided at the end of this section.

10.4.1 Carbon–manganese and low alloy steels

This class of materials comprises the major proportion of alloys used in general as well as marine engineering. The properties which make this so are generally mechanical, notably the high strength and elastic modulus. The enormous variability in properties, particularly strength, of iron–carbon based alloys has led to a high availability and hence low cost compared with other alloys.

There is an array of carbon–manganese and low alloy steels but, for the purposes of this section, these can be broken into the low carbon, 1.5% manganese steels which are widely used for structural and pressure vessel purposes, the low alloy steels with medium carbon and a few percent of other alloying elements such as chromium and manganese and others with higher carbon and alloying additions. These last are normally only used in special areas such as bearings where corrosion protection consists simply of exclusion of the marine environment and will not be considered further here.

The remaining two groups are metallurgically quite different with the carbon–manganese steels having a ferrite–pearlite structure (mixture of iron and iron–iron carbide laminate). The low alloy steels are usually heat treated to a tempered martensite structure (a solid solution of carbon in iron with some fine spheroids of iron carbide), making them generally stronger than the first group.

However, this metallurgical distinction is not a sharp one and certain structural steels which contain small additions of niobium, vanadium, aluminium, titanium and other grain refining elements often are designated high strength low alloy (HSLA) steels although they have more in common metallurgically with carbon–manganese steels. Furthermore, a number of the higher strength pipeline and structural steels occupy a middle ground between the two metallurgical types.

As mentioned earlier, the main factor affecting the corrosion of immersed steel is the presence of oxygen (Schumacher, 1979).

Figure 10.12 shows a typical variation of corrosion rate for some ferrous alloys with dissolved oxygen content. However, corrosion rates will be dependent also on time, temperature, salinity, sea water velocity and other factors already identified in Section 10.3.3, and Dexter (1987) has dealt with the global variability and interdependence of a number of these factors.

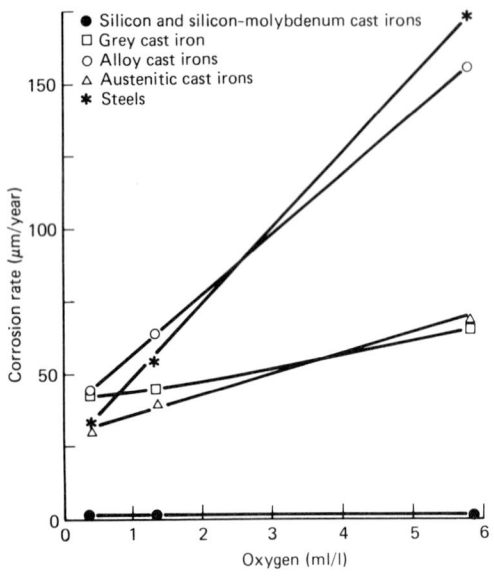

Figure 10.12 Effect of oxygen on 1 year corrosion rates of some ferrous alloys (from Schumacher 1979)

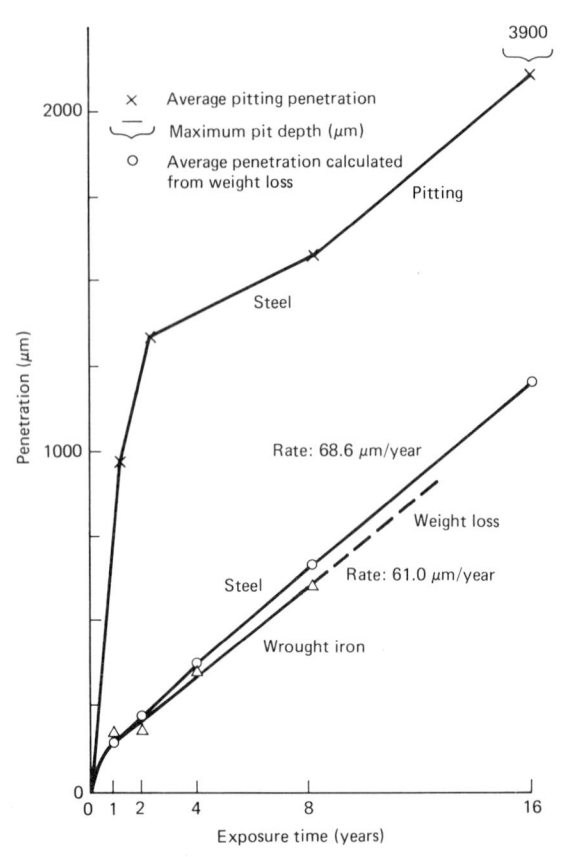

Figure 10.13 Variability of corrosion penetration due to localized pitting effects (from Schumacher 1979)

The comparison of immersion test results is complicated by the effects of time and non-uniformity of attack as shown in *Figure 10.13*. As can be seen, the average weight loss when expressed as loss in wall thickness can give a misleading value for the expected penetration, since local areas may show as much as twice the average penetration. It might also be noticed that the tendency is for the average and maximum penetration rate curves to become parallel and linear at longer times so that corrosion penetration with time can be expressed as:

$$D = D_o + kt \tag{10.20}$$

where D_o is depth of penetration after time t, D is depth of penetration after 1 year and k is the average corrosion rate.

Schumacher (1979) has suggested that k should be around 60μm/year and D around 60μm for normal temperate sea water.

There is no evidence to suggest any significant difference between immersed corrosion rates in low alloy and carbon–manganese steels. The major element likely to cause any difference is chromium, but this is not normally present in sufficient large quantities (Chandler, 1985).

The effect of water temperature is complex since the solubility of oxygen decreases and the amount of marine growth increases with increasing temperature. Both of these factors tend to reduce the increase in corrosion rate expected with increasing water temperature. *Figure 10.14* shows the effect of sea water temperature in the absence of marine growth protection reported by LaQue (1975)

In deeper water, substantial reductions in oxygen level and temperature occur so that corrosion rates for steel can be reduced substantially as shown in *Figure 10.15* (Reinhart and Jenkins, 1972; LaQue, 1975). The corrosion rate curves can be seen to follow almost exactly the oxygen concentration curve.

In the splash and intertidal zones, corrosion can be as much as ten times, although more commonly about five times, that in the submerged zone. The intertidal region can, however, be cathodically protected by more rapidly corroding steel if there is a large enough expanse of this. This, coupled with the possible protective effects of fouling layers, usually means that the intertidal region of a partially immersed structure can show a lower corrosion rate than the splash or the submerged zone.

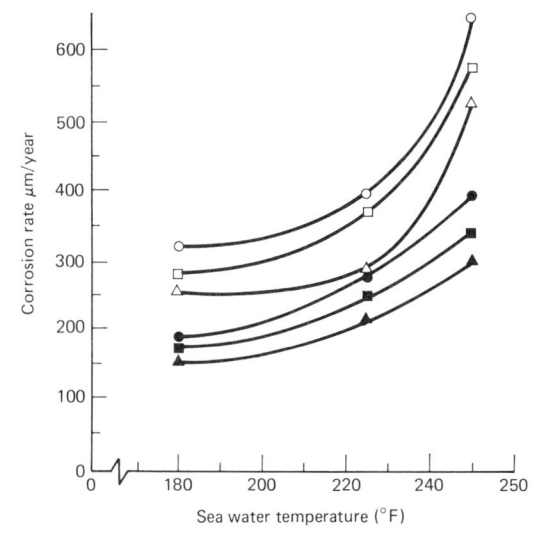

Figure 10.14 Effect of temperature and velocity on the corrosion of steel in sea water (from LaQue 1975)

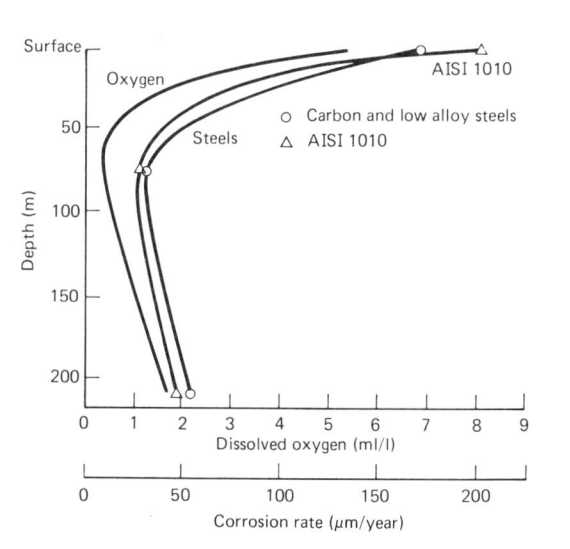

Figure 10.15 Effect of depth on dissolved oxygen level and on the corrosion rate of steels (from LaQue 1975)

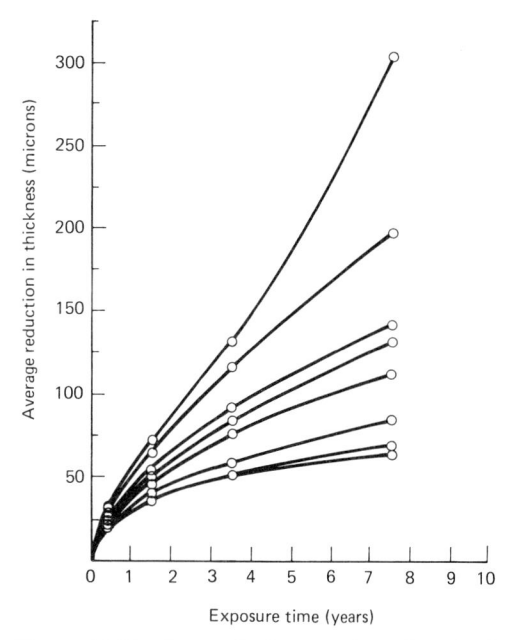

Figure 10.16 Atmospheric corrosion of steels showing the effects of alloying additions (from Schumacher 1979)

In the marine atmosphere, the corrosion rate of steel is rather more sensitive to alloy content. The single most effective element at low concentrations has been shown to be copper (Larrabee, 1953) and this is deliberately added to some steels at a level of about 0.5%. Such steels have a number of proprietary names but are often referred to as weathering grades and *Figure 10.16* shows typical weathering curves for steels with and without a deliberate copper addition.

It can also be seen that a combination of about 1% chromium and 0.5% copper is most effective in resisting the marine atmosphere. Chandler and Bayliss (1985) have reviewed more recent experience on weathering steels and have concluded that some care should be taken before adopting these steels, especially in areas where chlorides are present.

One important difference between carbon–manganese and low alloy steels is their tolerance of hydrogen. Generally, steels containing dissolved carbon are embrittled by hydrogen and are therefore susceptible to SSC as described in Section 10.3.3.3. This susceptibility is normally quantified by the hardness of the steel and a NACE Standard (MR–01–75) provides guidance for acceptable hardness levels in areas where hydrogen sulphide might be expected.

10.4.2 Stainless steels

When the chromium content of steel exceeds about 11%, a highly impermeable mixed iron–chromium oxide is formed on the surface on exposure to air. This means that such alloys can be passivated so that resistance to corrosion in a wide range of media is enhanced. The metallurgical effect of such a large alloying addition is such that it is not generally possible to exploit the effects of carbon on strength as is possible in the steels discussed in Section 10.4.1. This has broadly meant that the corrosion resistance of stainless steels has been bought at the cost of reduced strength and versatility.

Also, in the commonest group, the austenitic stainless steels, an addition of around 8% nickel is necessary for phase stability and increases the cost of such alloys. The metallurgy of stainless steels is based around the stability of two of the allotropic forms of iron, ferrite and austenite. Ferrite is the form normally extant at room temperature in carbon–manganese steels and it is this which responds to strengthening by carbon producing either ferrite–pearlite or tempered martensite structures after appropriate heat treatment. Austenite is normally only stable at elevated temperatures but suitable additions of nickel and/or manganese will allow austenite to exist at room temperature.

The different types of stainless steels are classified in terms of whether the predominant structure is ferritic, austenitic or martensitic. Austenitic stainless steels are the commonest group with compositions based on about 18% chromium: 8% nickel or 25% chromium: 20% nickel. The strength of these steels is generally quite low compared with carbon–manganese steels but some higher strength grades containing nitrogen are available.

The martensitic stainless steels contain around 11–18% chromium with little or no nickel. These alloys can contain anything from 0.1 to 0.75% carbon giving a range of strong, wear resistant, but brittle, steels after appropriate heat treatment.

Ferritic stainless steels are not heat treatable since the carbon is always present as carbides. They are generally stronger than austenitic grades but weaker than martensitic grades. Compositions are similar to the martensitic steels with rather lower carbon levels.

Duplex stainless steels combine the strength of ferrite with the toughness of austenite. Compositions are such that not quite enough nickel is present to stabilize austenite fully, resulting in a mixed ferritic–austenitic structure which combines some of the advantages of both groups.

The last group, the precipitation or age hardening stainless steels, includes small additions of titanium, aluminium, copper or molybdenum to form small precipitates within the alloy during a special heat treatment. The matrix material may be any of those mentioned above and the effect of the precipitate is to increase the strength of the alloy.

Although very resistant to general corrosion both in the atmospheric and submerged conditions, stainless steels suffer from the corrosion problems often associated with passivating alloys. This means that they are susceptible to pitting and crevice corrosion, particularly under conditions where the oxygen supply is relatively poor and chloride ion is present. Such environments clearly include stagnant sea water so that, for such service, more advanced pitting resistant stainless steels are required (Shone et al., 1988).

Where oxygen supply is plentiful, the austenitic grades normally perform quite well in submerged service, provided that crevices are avoided, but some grades will rust in the marine atmosphere. The non-austenitic grades do not perform particularly well in either atmospheric or submerged service.

Austenitic steels and some other grades suffer from stress corrosion in solutions containing chloride ions, even at very low concentrations. However, a temperature of above about 60°C is required so that this problem is not likely in normal submerged conditions, but care should be taken if the steel is used to contain a warm fluid and has external contact with sea water. Ferritic stainless steels are much more resistant to this form of stress corrosion (Sedriks, 1979).

The duplex stainless steels are finding increasing use in environments where there is a danger of sulphide and chloride stress corrosion. These steels appear to combine the sulphide and hydrogen resistance of austenitic stainless steels with the chloride resistance of ferritic stainless steels (Wilhelm and Kane, 1984).

The atmospheric and splash zone resistance of the stainless steels is generally good, although some superficial rusting may mar the appearance of the steel. In such applications, the avoidance of crevices and water traps is very important. There is no marked evidence to suggest any effect of depth on the corrosion performance of stainless steels (Schumacher, 1979).

10.4.3 Cast irons

Cast irons have a somewhat old fashioned reputation mainly because they have been in use for so long. In fact, cast irons are metallurgically quite sophisticated alloys, particularly some of the modern grades for specific applications.

The alloys are all typified by a high carbon content, usually around 4%, which in many cases produces bulk (about 50μm in size) graphite within the structure. This graphite occupies typically around 15% of the volume of the iron and the remainder can be ferritic, ferritic–pearlitic, austenitic or martensitic. Some irons do not contain free graphite but these find few, if any, marine applications. The commonest and cheapest are the grey irons which contain typically 2–4% carbon and 1–3% silicon. The strength can be improved and the brittleness alleviated somewhat by control of the matrix structure by heat treatment or alloying of the graphite shape by inoculating with a small amount of magnesium.

A number of more elaborate alloy irons exist for special applications. The most important of these for marine service are the range of austenitic irons based on alloying additions of 14–32% nickel and 1.75–5.5% chromium.

Under immersed conditions, unalloyed irons corrode at approximately the same rate as carbon–manganese steels (Chandler, 1985), but this has not prevented their widespread use as valve bodies, water boxes and impellers for sea water use, probably due to the relative cheapness of large section thicknesses. Grey irons, because of their high volume fraction of (noble) graphite show a form of selective leaching known as graphic corrosion. In marine atmospheres, the corrosion rate of unalloyed irons is around half that of carbon–manganese steels. The high nickel irons show corrosion rates of one-third to one-tenth those of their unalloyed counterparts in immersed and atmospheric conditions (Chandler, 1985).

10.4.4 Copper and nickel alloys

Alloys based on copper and nickel have a high inherent nobility (i.e. they are high in the electro-chemical series). This means that corrosion rates are generally low in these alloys without some of the difficulties associated with the passivating alloys, although this is more so for copper than nickel. Copper alloys are the traditional metals for marine use and still occupy an important place. Nickel alloys are relatively expensive, although very resistant and have replaced copper alloys in some of the more extreme conditions.

The copper alloys have the unique property amongst metals in that they have a natural resistance to marine fouling. This resistance is related almost entirely to the copper content (Fulmer Research Institute, 1980) and, although the physical and microbiological mechanism is quite complex (Kirk et al., 1985) this can most easily be understood in terms of the toxicity of copper. It would appear that a certain amount of corrosion of the alloy must take place in order that a sufficiently high copper ion concentration is built up near the metal surface. This corrosion rate has been estimated at about $25\mu m$/year although a number of other compositional and environmental factors have an effect (LaQue, 1975).

Many of the copper alloys have been developed as casting alloys and this is a reflection of the typical applications of such alloys which in turn determines their shape. It is convenient to divide the copper alloys into the following groups: high coppers (mostly copper), brasses (copper–zinc), bronzes (copper–tin), aluminium bronzes (copper–aluminium), cupro–nickels (copper–nickel) and complex alloys.

These groupings are somewhat traditional and many of the more modern alloys cross the boundaries between them so that the terms brass and bronze are not always applied to Cu–Zn and Cu–Sn alloys alone. The corrosion resistance of these alloys is good, but there is a tendency to be affected by pollution, so that some alloys may show limited performance in inshore areas such as harbours (Francis, 1985).

The copper based alloys also are sensitive to differential environment effects and in particular to local differences in the copper ion concentration produced by the slow dissolution of the alloy. For example, immersed crevices can show a build up of copper ions which can become cathodic to the area immediately outside the crevice. Furthermore, some copper alloys can be susceptible to velocity effects and this can limit applications for sea water handling. The multiphase nature of some copper alloys can produce microgalvanic effects and others, such as brasses, can be affected by selective leaching (dezincification).

The high coppers find some applications in ships for piping but are quite susceptible to velocity effects and to polluted water (Chandler, 1985). The brasses find a number of marine applications provided they have had additions to improve resistance to dezincification (Admiralty brass with 0.1% arsenic or antimony). Even then, both the 70% copper: 30% zinc and the 60% copper: 40% zinc alloys can be affected by high flow rates and the 60/40 alloys are difficult to inhibit against dezincification; this problem tends to affect alloys in both submerged and atmospheric service (Schumacher, 1979).

Bronzes begin to lose their corrosion resistance when the tin content becomes high enough to form a separate phase, but phosphor bronzes with 0.4% phosphorus and up to 8% tin find some marine uses. The aluminium bronzes are heat treatable and both cast and wrought forms are amongst the strongest of the copper alloys. Provided that care is taken in the heat treatments, corrosion resistance is good and, although selective leaching is possible under some circumstances, this can be eliminated by compositional and heat treatment control (Copper Development Association, 1981). One of the principal uses of aluminium bronzes has been in marine propellers due to their relatively high resistance to cavitation, although for heavy duty applications, rather more resistant alloys are available.

The cupro–nickels are perhaps the most corrosion resistant of the copper alloys. Two grades based on 70% copper: 30% nickel and 90% copper: 10% nickel have become standard in marine condensers, due to their resistance to pitting, resistance to velocity effects and ease of fabrication. These properties are usually enough to justify higher first cost, particularly in areas where maintainability is difficult.

The more complex alloys include those developed directly from other groups for specific purposes such as high strength cupro–nickels and aluminium brass. The former use precipitation strengthening by addition of aluminium and titanium whereas aluminium brass contains 76% Cu, 22% Zn, 2% Al, 0.05% As, and was developed specifically for resistance to velocity effects. A number of other more complex brasses and bronzes are available for specific applications but special mention should be made of the gunmetals or G-bronzes (copper–tin–zinc alloys, occasionally with lead additions) which are widely used for the cast parts of sea water piping systems, such as pump and valve bodies.

The main characteristic of nickel alloys is the exceptional resistance of some to marine corrosion. This is balanced by the very high cost of nickel, so that uses tend to be restricted to areas where this cost can be justified in terms of, for example, increased reliability. Some nickel alloys can, however, suffer from pitting in stationary sea water and some from crevice corrosion. Resistance to velocity effects is generally good. There are three main groups of nickel alloys which are used in marine applications: nickel–copper alloys, nickel–chromium (–iron) alloys and nickel–chromium–molybdenum (–iron) alloys.

The nickel–copper alloys are based on about 30% copper. The two principal alloys are known as 400 and the higher strength K500 which contains precipitation hardening additions. These alloys, particularly K500, find application as fasteners and shafts for atmospheric and submerged conditions but are very resistant in the marine atmosphere. They are often used as sheet liners or weld overlays on carbon–manganese steels to reduce cost over that which would be required to construct the entire component from a nickel-based alloy. The Ni–Cr–Mo–Fe alloys are very resistant to sea water, being virtually inert under quiet conditions (Schumacher, 1979).

10.4.5 Aluminium and titanium alloys

Aluminium and titanium alloys are both typified by high elastic modulus and high strength when expressed per unit density. This means that both have a high structural efficiency, a property related to strength or elastic modulus and density (Crane and Charles, 1984) rendering them particularly useful in designs where rapid transportation is important. Both aluminium and titanium are relatively base metals, relying on the formation of a passive film for protection against corrosion.

The relative cheapness of the aluminium alloys means that they have found applications in ships, particularly in superstructures. The corrosion resistance in submerged and atmospheric conditions is generally good but care should be taken over possible localized effects due to lack of oxygen supply (i.e. crevice corrosion and pitting). This can limit corrosion resistance under conditions of deep submergence (Beccaria and Poggi, 1986).

A wide range of aluminium alloys are available, mostly with small additions of one or more elements. Some of these alloys are not suitable for marine use, particularly those containing copper. Also, some of the higher strength alloys can suffer from stress corrosion in sea water (Schumacher, 1979). Aluminium's position in the galvanic series means that it is a useful metal for use as a cathodic protection anode and special alloys have been developed for this purpose. This also means, however, that great care should be taken to ensure that galvanic corrosion of structural aluminium does not take place and special attention should be paid to avoiding combinations of aluminium alloys and copper alloys. The corrosion resistance of all aluminium alloys can be improved by an artificial thickening of the oxide layer known as anodizing.

Aluminium alloys are grouped into series according to their principal alloying element, these alloying additions normally being quite small, usually no more than a few percent.

The main alloying elements are: copper (2000 series), manganese (3000 series), silicon (4000 series), magnesium (5000 series), magnesium plus silicon (6000 series) and zinc (7000 series). Despite the relatively small compositional differences, quite marked differences in corrosion resistance, particularly under marine conditions, are experienced. The main potential problems are pitting, crevice corrosion, exfoliation and stress corrosion, and Schumacher (1979) has summarized experience on submerged, atmospheric and inter-tidal corrosion performance. In general, the 1000, 3000, 5000 and 6000 series alloys provide the majority of those in marine use with the 7000, 2000 and 3000 alloys performing less well (Fulmer Research Institute, 1980).

Titanium is expensive and its use is limited to small scale, low maintenance applications or to those where cost is not a primary concern. Although the electrode potential of titanium is rather low, since it is an inherently base metal, its very high affinity for oxygen means that it is passive over a very wide range of chemical conditions. Titanium alloys are therefore resistant to both pitting and crevice corrosion in sea water and their general corrosion rate is negligible, this resistance extending even to deep water.

Since titanium is so expensive, it is normally used in association with other metals and it normally behaves cathodically to them. This means that care has to be taken when designing with titanium that it does not cause unacceptable galvanic corrosion in metals to which it is connected. However, provided that cathode–anode size ratios are small, as is likely with such an expensive element as titanium, relatively few problems are encountered. Under certain severe conditions titanium has been observed to suffer from crevice corrosion and also stress corrosion cracking although the former normally requires high temperature oxygen starved conditions and the latter a high stress concentration (Logan, 1966).

Shreir (1976) has observed that failure experience of titanium is rather rarer than might be expected from the results of laboratory studies. Titanium has been considered to show advantage over copper alloys and stainless steels for cooling systems in power station condensers using coastal, estuarine or polluted water (Heaton et al. 1979).

10.4.6 Other marine alloys

A number of other metals and alloys are used from time to time in marine environments. The commonest of these are the zinc and, to a lesser extent, magnesium alloys used as sacrificial anodes in cathodic protection systems. It is often necessary that these alloys have a low self corrosion rate and special additions need to be made to prevent such problems as intergranular corrosion of zinc alloys in mud. It is therefore important that anodes are made from alloys specially prepared for the purpose of cathodic protection (Todd and Perkins, 1976).

Cobalt and cobalt based alloys occasionally appear as overlays and these are very resistant to marine corrosion. The precious metals are used as coatings in electrical applications and graphite, platinized niobium or titanium and silicon iron are all used as anodes in impressed current cathodic protection systems.

A series of ultrahigh strength steels containing substantial additions of nickel (about 20%), cobalt (about 10%) and molybdenum (about 5%) were developed with particular applications for very deep water pressure hulls. These maraging steels have been found to be susceptible to stress corrosion cracking in sea water when heat treated to their highest strength levels (LaQue, 1975). However, the resistance to SCC is markedly improved if the full strength is not developed and these alloys show improved performance over heat treated low alloy steels for the same strength level.

10.4.7 Summary of marine uses and corrosion properties

Table 10.3 has been compiled to give an overview of the uses and limitations of the types of alloys which have been discussed in this section. The table is not intended to be exhaustive but to point to the main considerations and range of measured data on marine alloys. The alloys have been split into broad groups to give a general feeling for the important properties, although designers will normally require rather more detailed information. *Table 10.3* has been compiled largely using data from LaQue (1975), Shreir (1976), Schumacher (1979), Fulmer Research Institute (1980) Chandler (1985), Dexter (1987) and Griffin (1987).

10.5 Corrosion protection

It is not common for the engineering designer to have the luxury of selecting a material for its corrosion resistance alone, although this will be a factor in the decision, along with other properties such as strength, toughness, fabricability and others. For this reason, it is often necessary to institute some strategy for corrosion protection.

There are four main methods of corrosion protection: coatings, cathodic protection, anodic protection and inhibition. These involve either alteration or exclusion of the environment or modification of the thermodynamics or kinetics of the metal surface. A fifth course involves avoidance of the systems in which corrosion can occur and this can be achieved by careful materials selection and design. Some aspects of this strategy have already been dealt with in Section 10.4 in the identification of susceptible and resistant alloys for particular conditions.

10.5.1 Coatings

Coatings have been employed to protect steel from corrosion for many years and have developed from single coat paints and washes to the modern multilayer highly resistant coatings used to protect offshore structures and ships. Since most large items are built from either steel, concrete or masonry it is not surprising that large structures or vessels for marine use are constructed from steel or concrete.

Since concrete, amongst its other functions, acts as a coating which protects the underlying steel reinforcement, almost all the coating technology developed for the marine environment relates to the protection of carbon–manganese steel although coatings are occasionally used to prevent permeation into concrete. Coatings are most often applied to steelwork exposed to the marine atmosphere, splash and intertidal zones. They are less commonly applied in the submerged zone since here cathodic protection is generally more effective and reliable, although coatings can be used to reduce the cathodic protection requirement for submerged structures (Thomason *et al.* 1987).

Anticorrosion coatings are almost invariably used in combination with cathodic protection for submerged and buried pipelines (Roche and Samaran, 1987) where long maintenance free periods are essential. Again, in vessels containing non-flowing sea water cathodic protection is often applied, although coatings may be used more especially if there is likely to be flow or a change of liquid level. In sea water handling, there is also the possibility of modifying the corrosivity of the fluid by de-aeration or stripping of other harmful dissolved material.

The materials used for coating can be metallic, inorganic or organic in nature and examples of each of these might be, respectively, galvanized layers (zinc), anodized layers (aluminium oxide) and epoxy resin. Important properties of coating materials include self corrosion resistance and galvanic activity (for metallic coatings), impact and abrasion resistance and resistance to water and ionic absorption and penetration. Coatings may include more than one type of material. For example, it is not uncommon for an inorganic material to be bound with an organic resin and to contain a metallic dust.

There are three basic classifications of functions which coatings perform in the prevention of corrosion of the underlying substrate; these are shown in *Figure 10.17*. These functions can be summarized as exclusion of the environment, sacrificial action (cathodic protection) and self healing or inhibitive action. All coating systems will show one or more of these effects.

It is immediately apparent that any coating will exclude the corrosive from the metal surface to a certain degree and this can be quantified in terms of its permeability and electrical conductivity since the existence of a moisture film at the metallic surface can lead to rapid degradation by blistering and rusting.

Protection by exclusion of the environment is potentially unstable since the possible effects of defects (or holidays) in the coating or permeation by the corrosive must be considered. Small holes in the coating can provide a very small anodic area which, if the cathodic reaction can proceed on the coating surface or if

Table 10.2

Alloy system		General† metal loss μm/y	Pitting Rating 1: poor – 4: excellent	Pitting* penetration μm/year	Crevice resistance 1: poor 5: inert	Submerged Depth effect	Velocity ϕ effect	Stress-corrosion in sea-water	Marine atmosphere metal loss μm/year	Mud (compared with overlying sea-water)	Remarks and general marine uses
Carbon–manganese steels		100–200 (▵ 400 in splash)	2 (1 if scaled)	380–760 (510–1020)	3	Approx. 75% reduction in rate over 2000m	Very susceptible	Resistant	30–100	Generally lower but susceptible to SRB	General and structural use for submerged and atmospheric service. Requires protective measures
Low alloy steels		As above	As above	As above	3	As above	As above	Resistant	10–40 (varies with alloy content)	Generally lower; danger of SSC	High strength applications like pressure vessels, hulls, pipelines, fasteners, tethers, chains, petroleum engineering use
Stainless steels	Austenitic	<25	1–2	500–1800	1–2	Reduction in general corrosion rate	Higher velocities reduce susceptibility to pitting	Austenitics susceptible above about 60°C especially with crevices. Some p.h. and martensitic grades also susceptible	Negligible	Tendency to lose passivity and pit	Some alloys/treatments susceptible to intergranular corrosion. Submerged uses limited by tendency to localized attack although special sea-water grades available. Some petroleum engineering applications.
	Martensitic	<25	1	>1250	1						
	sea-water grades	<25	3	180 (approx)	1–2						
Cast irons	Grey/S.G. irons	60–500	3	100–300	3	Approx. 60% reduction in rate over 2000m	Susceptible	Resistant	10		Water boxes, casings, impellers for sea-water handling. Graphitize
	Austenitic Ni-irons	20–70	3	50–100	3	Approx 50% reduction in rate over 2000m	Moderately susceptible	Resistant	<3	–	As above but improved resistance for more critical applications
Copper alloys	Cupro-nickels	2.5–40	3	25–130	4	No substantial effect	Generally susceptible Cupro-nickels. G-bronzes and aluminium bronzes least susceptible	Some bronzes may be susceptible – can be overcome by appropriate alloying and/or heat treatment	0.6–2.5	Susceptible to pollution	Sea water piping, propellers, shafts, condensers, fasteners, cladding, deck fittings, desalination plant, sea-cages. All alloys have some anti-fouling properties dependent on copper content. Some alloys show selective leaching
	Copper	12.5–70	3	150–300	2						
	Admiralty brass	12.5–50	3	150–300	4						
	Aluminium brass	12.5–50	3	180	4						
	G-bronze	25–50	3	130–350	4						
	Aluminium bronze	50–60	3	75	4						
Aluminium alloys		Mostly pitting	2–3	Varies with alloy 7.5–1300	1–2	Greater tendency to pitting with depth	Susceptible –very susceptible	Some high strength alloys may be susceptible	0.5	Tendency to pit	Almost all attack is by pitting. Deckhouses, superstructures, coatings, cathodic protection anodes
Nickel alloys	Ni-Cr-Mo	0	4	0	2–5	50–80% reduction in corrosion rate up to 2000m	Little effect at moderate velocities	Resistant	Negligible	Alloys which show crevice or pitting susceptibility will also pit in mud	Fasteners, shafts, cladding, impellers. Areas where high reliability is required
	Ni-Cr	0	1–3	▵>1530	1–2						
	Ni-Cu	<25	2	130–380	2–3						
	Ni	<25	1	250–1250							
Titanium		0	4	0	5	Maintains resistance with depth	Resistant	Immune except for some specific severe conditions	Negligible	Resistant	Small scale use in severe conditions. Some applications for condenser tubes for polluted waters

ϕ In absence of direct mechanical damage such as cavitation
† Average loss in quiet sea water
* Pitting penetration expressed as depth of penetration is quite unreliable for design purposes and is provided here as an indication only

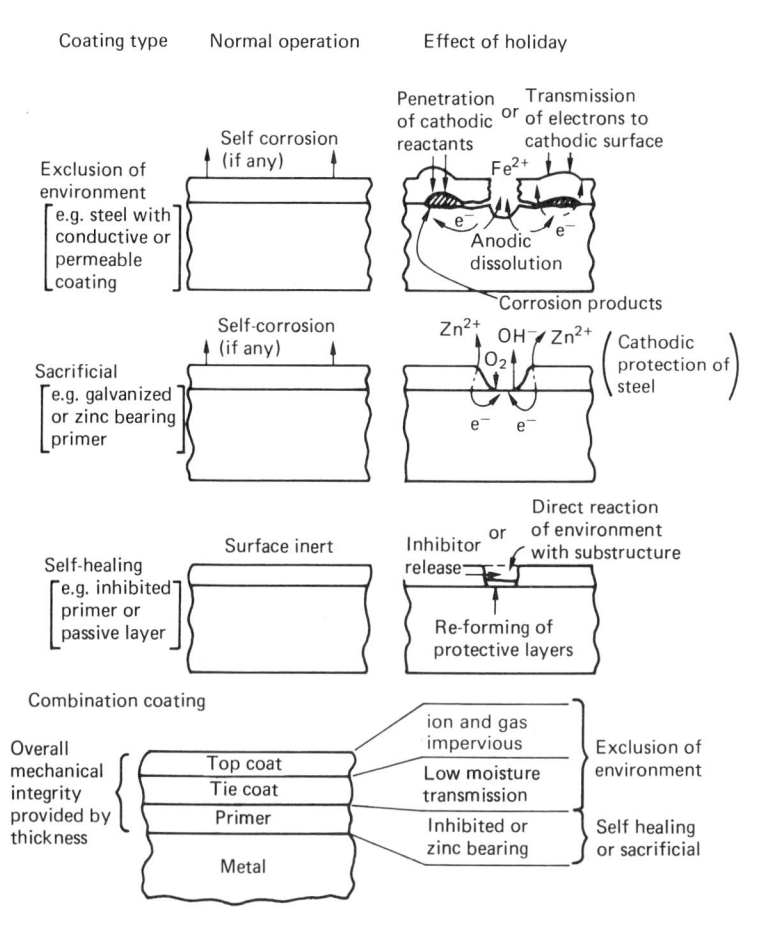

Figure 10.17 Classification of the essential functions of anticorrosion coatings

cathodic reactants can penetrate to the surface, may result in very rapid corrosion. Most organic coatings eventually fail by penetration of water to the surface at holidays or into areas of poor adhesion and eventual spreading of the corrosive effect under the coating, followed by blistering and further penetration. For this reason, the preparation of steel prior to coating and the ability of the coating to resist undercutting are extremely important (Gabe, 1983).

Metallic coatings which are cathodic to the underlying substrate are particularly susceptible to holidays since here the substrate will cathodically protect the coating with a large anode to cathode area ratio. It is therefore inadvisable to use noble metal coated steel in sea water unless this coating is mechanically very resistant.

Conversely, excellent resistance to holidays can be provided by coatings with a sacrificial content. Once corrosive has penetrated to the substrate, a galvanic cell is set up where the sacrificial component of the coating acts as the anode, thus cathodically protecting the damaged region.

Since the area ratio of anode to cathode is relatively large, protection can be effective for quite some time. A similar benefit can be derived from coatings which contain an inhibitive component. When the coating is originally applied, the inhibitor reacts with the steel surface to produce a protective film. Should the coating and film subsequently become ruptured, the inhibitor once again should come into contact with the steel, so that the film becomes healed. A similar effect takes place in passivating alloys where the self healing of the passive layer takes place by direct reaction of the substrate with the environment.

Most coating systems show a combination of the above effects and a typical example is shown in *Figure 10.17*. The overall thickness of the coating provides an additional level of mechanical integrity.

The selection of a coating system is subject to many factors of which service conditions, required life and economic considerations are probably the most important. For marine applications, the greatest challenge is probably offered by the splash zone, although walkways have the additional difficulty of abrasion. To give an idea of the type of protection required in the splash zone, a typical coating system for UK waters would consist of the following (Whitehouse, 1983):

1. Blast clean the surface to a white metal finish to achieve good adhesion.
2. Apply a primer such as zinc silicate or zinc rich epoxy.
3. Apply a top coat, such as coal tar epoxy or vinyl tar or glass flake reinforced polyester.

A typical dry film thickness would be about 400μ and the expected time to first maintenance about 8 years (Carruthers, 1985) and Saville (1982) has listed the main causes of coating failure as:

1. Poor preparation of surface (practice, profile and application environment).
2. Low thickness and poor hiding at asperities.
3. Painting on to moist surfaces.
4. Contamination between coats and insufficient curing time.
5. Poor handling of coated components.
6. Poor specification.

As well as the obvious wind, weather and sea water resistance, splash and intertidal zone coatings must also be resistant to disbondment which might be caused by the alkali produced by the cathodic protection system for the submerged zone. Alternatives to the traditional splash zone protection are sometimes applied with the justification of longer maintenance free periods and, in some cases, weight saving. Such systems usually involve cladding with a noble metal such as Ni–Cu alloy 400 or a cupro–nickel, or alternatively a thick layer of rubber, such as neoprene.

Coatings to be used in submerged service along with cathodic protection are often much thinner than those described above and maintenance would not be carried out on such coatings.

For service in the marine atmosphere, a similar system to that used in the splash zone might be used, but with a dry film thickness of around 250 μ. Aggregates such as carborundum can be incorporated into the topcoat while it is still tacky to provide antiskid and wear resistant properties for helidecks and walkways (Whitehouse, 1983).

10.5.2 Cathodic protection

Cathodic protection of steel immersed in water consists of bringing the steel towards the area of immunity in the Pourbaix Diagram where the dissolution of iron becomes a thermodynamically unfavourable process (see Section 10.2.5). This means that the normal tendency of steel to corrode can be made more difficult by depressing its potential from the natural equilibrium value. This can be achieved in practice by connecting it to a baser metal (such as zinc, aluminium or magnesium for steel) in which case the process is known as cathodic protection by galvanic anodes.

Alternatively, the metal can be connected to the negative pole of a DC source with the positive pole connected to a suitable anode material such as silicon iron. This method is known as cathodic protection by impressed current. In both cases, both the anode and cathode must be immersed in the same electrolyte. These two methods are shown in *Figure 10.18*. Galvanic anodes tend to be preferred when the danger of alkali or hydrogen damage is to be avoided, but are of limited use in high resistivity media (e.g. soil). Although a smaller number of anodes is required for impressed current protection, the requirement for a separate power supply can be inconvenient and care must be taken not to overprotect. Because electrons are supplied from the power source rather than by dissolution of the anode, impressed current anodes are not consumed at anything like the rate of galvanic anodes.

The scientific basis of cathodic protection is best understood in terms of an Evans Diagram. As pointed out in Section 10.2.4, when an anodic and a cathodic reaction take place, the corrosion current and corrosion potential are fixed by the intersection of the polarization curves for both half cell reactions. If a third half cell reaction of a generally more anodic nature is

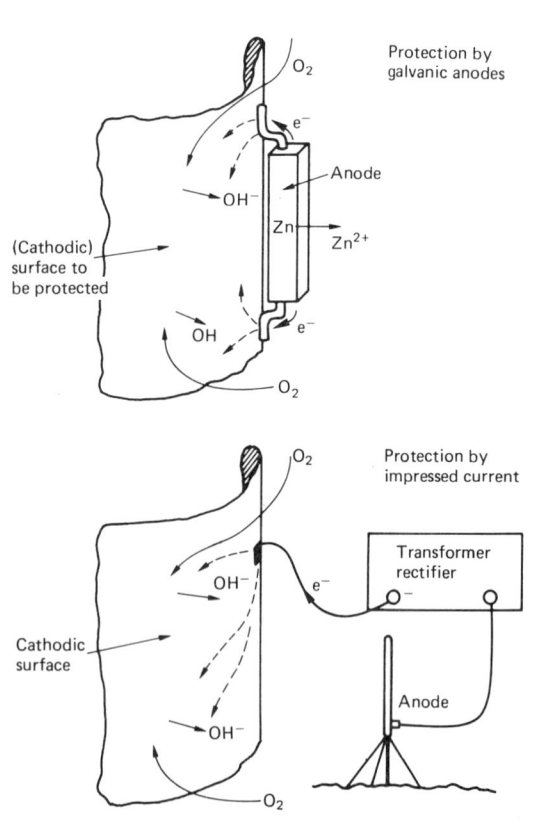

Figure 10.18 The impressed current and galvanic anode systems for cathodic protection

introduced, it will take place preferentially, shifting the corrosion potential and hence reducing the corrosion current for the original anodic half cell reaction. This is illustrated for the steel/zinc/sea water system in *Figure 10.19(a)* where the potential at the steel surface is reduced from E_{corr} to E_{prot}.

In general, there will, however, be a certain resistance between the steel and the zinc giving rise to the potential differences shown in *Figure 10.19(b)* (Leach, 1979). Since the corrosion current is on a logarithmic scale, the reduction in corrosion of the steel is quite considerable (from $i_{corr\ m}$ to $i_{corr\ m'}$). It is normal to depress the potential of steel in sea water by around 0.2 V from its natural value (Ashworth, 1986). When the Fe/Fe^{2+} half cell reaction is suppressed, it follows from *Figure 10.19* that the oxygen reduction reaction is driven, giving rise to the production of hydroxyl ions at the cathodic surface. This process can be seen as the consumption of dissolved oxygen to balance the sacrificial anodic corrosion rate, hence repressing any dissolution of ferrous ions.

Paints and some other coatings can be damaged by alkaline conditions and, before a coating can be used in conjunction with cathodic protection, it is necessary to ensure that it is resistant to such cathodic disbondment.

The hydroxyl ion concentration can also affect uncoated structures by exceeding the solubility product of calcium hydroxide, so that a calcareous layer may be precipitated on to the cathodic surface. This has the beneficial effect of reducing the cathodic protection load for the structure, thus increasing anode life. If the potential of the steel is depressed too far, there is a danger that some contribution to the cathodic reaction may be made by hydrogen evolution, leading to the possibility of embrittlement of steels. It is normal, therefore, to have a minimum potential below which the steel should not be taken by cathodic protection. This is not normally a problem with galvanic anodes, since the natural potentials of both zinc and aluminium are above the value required for hydrogen evolution, but care has to be exercised with impressed current systems where the potential is not limited by natural factors.

As well as reducing general corrosion, cathodic protection has been used successfully to prevent pitting, crevice corrosion, stress corrosion, galvanic effects and to alleviate the chemical components of corrosion–erosion and corrosion fatigue (Sedriks, 1979).

When making field measurements of potential, it is quite impracticable to use a hydrogen electrode as a reference, so that a number of other, more practical, standards have evolved. For sea water use, the silver–silver chloride–sea water reference cell is the commonest, with a potential of 250 mV more positive than the hydrogen electrode. The saturated calomel electrode (SCE) happens to have the same potential as the Ag–AgCl–sea water electrode and it is common practice to refer potentials to SCE, probably because of the convenience of the abbreviation. For soils and potential measurement of steel embedded in concrete, the copper–copper sulphate electrode is more commonly used, with a potential of 320 mV more positive than the hydrogen electrode.

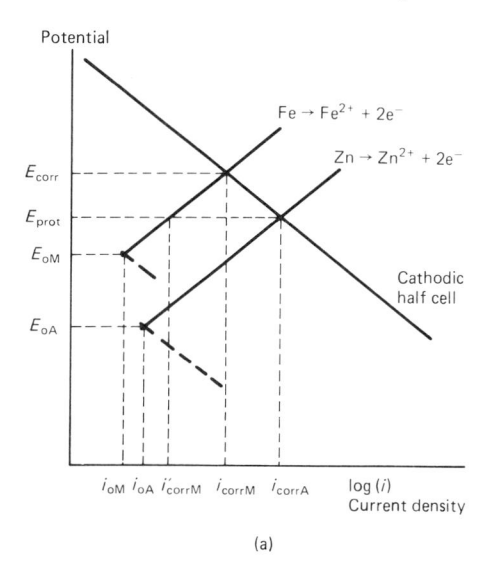

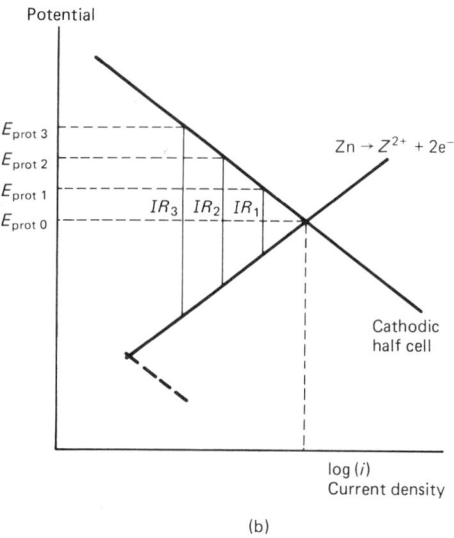

Figure 10.19 Potential–current diagrams illustrating the principles of cathodic protection (reproduced by permission of the Council of the Institution of Mechanical Engineers from Leach 1979)

Cathodic protection design can be quite a complex matter, but consists of ensuring that the entire area to be protected is within a potential band which will ensure a sufficiently low corrosion rate without the dangers of overprotection. Since conditions can change with time, for example, the formation of calcareous deposits, changes in water temperature and consumption of anode material, it is also necessary to ensure that an acceptable potential profile is maintained for the lifetime of the structure. Ultimately, this would require solution of Laplace's Equation for time varying boundary conditions but, fortunately, this is rarely necessary and simplified procedures can be followed for all but the most complex geometries.

For offshore installations, including pipelines, the design process consists of calculating the mass of sacrificial material required to supply a known protective current density with a known potential difference between anode and cathode. The number of anodes required is dependent upon their current output, which is a function of driving potential and anode resistance which in turn depends on the electrolyte resistivity and anode shape. The effects of coatings can also be taken into account to reduce the anode requirement (Mollan and Andersen, 1986).

For impressed current systems, the use of an active control system introduces a greater degree of versatility. An interesting example of impressed current for the protection of warship hulls is given by Tighe–Ford et al. (1988).

10.5.3 Inhibition and modification of the environment

This group of protective measures involves modification of the corrosive fluid and is therefore confined to systems where the amount of this is limited. Also, if the corrosive is to be discharged to the general environment, there are limits to the types and quantities of additives which may be used. The simplest method of environmental modification is to remove the corrosive element from the fluid. One example of this is in mild steel systems where a substantial reduction in corrosion rate can be achieved by de-aeration of sea water. As far as additives are concerned, it is commonplace to use biocides, mostly sodium hypochlorite, to prevent micro- and macrofouling and other biocides can be used to limit the activity of SRBs (NACE, 1976).

The use of film forming inhibitors has been briefly dealt with in the context of coatings in Section 10.5.1 and inhibitors can also, of course, be more generally distributed within the stream of closed systems. The chemistry of corrosion inhibition is a complex area and, for a detailed treatment, the reader is referred to Robinson (1979).

There are a number of different mechanisms by which the broad class of materials we call corrosion inhibitors can act. First, some materials can act as anticatalysts, interfering with the anodic or cathodic reactions. These include substances such as the organic amines which retard the metal dissolution process and hydrogen evolution poisons, such as arsenic or antimony ions, which can be effective in conditions where the cathodic reaction is hydrogen evolution (e.g. in acid solutions or in the presence of hydrogen sulphide).

A second group act by scavenging for specific corrosive components, such as sodium sulphite which removes oxygen by the reaction:

$$2\,Na_2\,SO_3 + O_2 = 2Na_2\,SO_4 \quad (10.21)$$

Inhibition can also be achieved by the formation of a film on the metal surface. This can be a stabilization of the natural oxide film (e.g. on iron by chromates or nitrates) or by the formation of an insoluble salt layer such as zinc, calcium or manganese salts.

Inhibitors are normally sold for a specific metal–environment combination and detailed compositions are rarely disclosed by manufacturers, so that the engineer is often ignorant of the scientific details of the inhibition process. Some typical examples include the vast array of inhibitors for use in association with oilfields and additives used to retard the corrosion of reinforcement in concrete.

In addition to the aqueous phase inhibitors there is also a range of vapour phase inhibitors which might be used to prevent corrosion in critical airtight areas which might otherwise be subject to the marine atmosphere. An example of this might be a junction box containing critical electrical connections in an exposed location.

In summary, the principal uses of inhibitors in marine environments are in the areas of petroleum production and transport with some applications in closed water systems (Bregman, 1963). For once through systems, such as sea water cooling, chemical treatment is rarely used for reasons of effluent control (Scanlon and Emms, 1979).

10.5.4 Anodic protection

In contrast to cathodic protection, anodic protection uses the application of an anodic current to the surface to be protected. This corresponds to the abstraction of electrons from the dissolution half cell reaction:

$$M = M^{z+} + z\,e^- \quad (10.22)$$

and this would appear, by the LeChatelier principle, to encourage corrosion. However, as pointed out in Section 10.2.5 certain metals show an active–passive transition where the corrosion current rapidly drops off at a certain potential (primary passivation potential). Anodic protection is therefore only applicable to metals showing such a transition and depends on raising the metal surface above this potential, also ensuring that it does not depassivate in service.

The breakdown potential, when going from the passive to the active state, is known as the Flade potential and is usually a few millivolts more negative than the primary passivation potential (Shreir, 1976). The passive films formed during anodic protection are also subject to the limitations already described with regard to chloride ions and this means that anodic protection is not likely to be used for steel in sea water.

Current uses of anodic protection are mostly in the area of sulphuric acid handling, but developments into a number of other areas are expected (Riggs et al. 1981).

10.5.5 Materials selection and design

Perhaps the simplest method of corrosion control is to avoid the combinations of metal and environment which are known to give rise to difficulties. This approach is, of course, severely limited by the economics and availability of the more resistant materials and any proposed use of these must be carefully balanced against the protection and maintenance costs which might be incurred by the use of a cheaper but less durable material. The enormous cost of anticipated corrosion is undeniable, but this is not generally sufficient to justify the uncritical adoption of materials about which limited design experience exists.

Formalized methods for economic calculations on coatings (Chandler and Bayliss, 1985) and generalized corrosion control methods (Institution of Metallurgists, 1974) are available and these normally involve an analysis of first cost along with possible future maintenance or repair costs in a discounted cash flow model. An interesting example of such an approach is given by Tischuk and Huber (1984) where the use of stainless steel tubing for the primary production of offshore hydrocarbons is justified over the more traditional carbon–manganese steel tubing for a specified life.

To adopt an alternative material, or to select the material for a new design, the engineer must not only be aware of the general and corrosion properties of that material but must also be aware of any design limitations involved in its adoption. This requires a thorough knowledge of the corrosion mechanisms as outlined in Section 10.3 (e.g. the need to avoid crevices when using stainless steel) as well as access to reliable data on corrosion performance. A number of authors have listed design rules to be adopted in the avoidance of corrosion (Shreir, 1976; Chandler and Bayliss, 1985; Fontana and Greene, 1982; Pludek, 1977). These rules usually contain a statement of the less obvious circumstances under which corrosion might take place, such as bimetallic contacts, stray cathodic protection current, crevices, differential environment effects, corrosion fatigue, fretting corrosion, effects aggravated by fluid flow, stress concentration and stress corrosion cracking, hot-spot effects and effects aggravated by welding.

Other ways in which designers can reduce corrosion problems are to avoid water or mud traps, which may provide stagnant conditions or allow the development of harmful bacteria and in the general reduction of heterogeneity, whether of stress, deformation, microstructure or environment. When surfaces or assemblies are to be coated, care must be taken to ensure that accessibility both for initial coating and also for any maintenance is adequate and that the surface is smooth enough to allow total coverage, but not so smooth that adhesion is affected. This normally involves the adoption of a surface standard fo blast cleaning. Care must also be taken in joint design to ensure that an even, uninterrupted coating is provided (Pludek, 1977) and the hiding qualities of the coating are very important here.

The designer should always be aware of the option of allowing corrosion to take place at its natural rate provided that a rational maintenance or changeout strategy is available and that this is justifiable in the face of the economics and logistics of all foreseeable alternatives.

10.6 Defects and non-destructive testing

Although not related solely to corrosion, the subject of non-destructive testing (NDT) has implications for corrosion monitoring as well as in the detection of other defects. Because of the importance of defects in marine engineering this section will deal with the broader aspect of the subject, as well as the detection and monitoring of corrosion. Since no manufactured product is dimensionally perfect and since defects can initiate and propagate whilst a structure is in service, it is necessary to have a conception of the ability of a structure or component to tolerate defects. The way in which defects grow under stress is dependent upon geometry, environment and loading conditions as well as properties of the material and the formalism by which this is treated is known as fracture mechanics. Also of importance are the ways in which defects arise and propagate and the significance of these in terms of reliability.

This section is therefore concerned first with the ways in which defects arise and goes on to consider the significance of defects, the ways in which they might grow, the means of detecting and monitoring defects and, finally, the effect that they might have on the final reliability of a structure or component.

10.6.1 The genesis of defects

Defects are most conveniently described as areas of reduction in section thickness from a prescribed initial value. One of the most important properties of such defects is their radius of curvature which can range from the infinitely small (cracks) to the infinitely large (general loss of thickness) as shown in *Figure 10.20*. A defect may also be defined in terms of the number of dimensions required to characterize its shape. *Figure 10.21* illustrates the difference between a one-dimensional (through thickness) crack, a two-dimensional crack and a cavity (three dimensions).

In general, the most important of these dimensional parameters is the crack depth (a), although the length (c) may also be critical depending upon the geometry of the component and of loading. The dimension in the direction of loading (b) is usually of little importance, although it may indirectly affect the radius of curvature of the crack tip.

10/30 Corrosion and defect evaluation

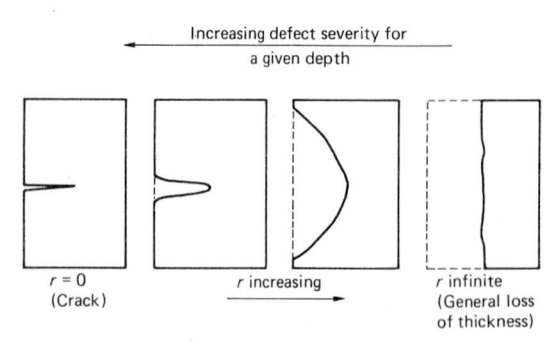

Figure 10.20 Effect of root radius of curvature on defect severity

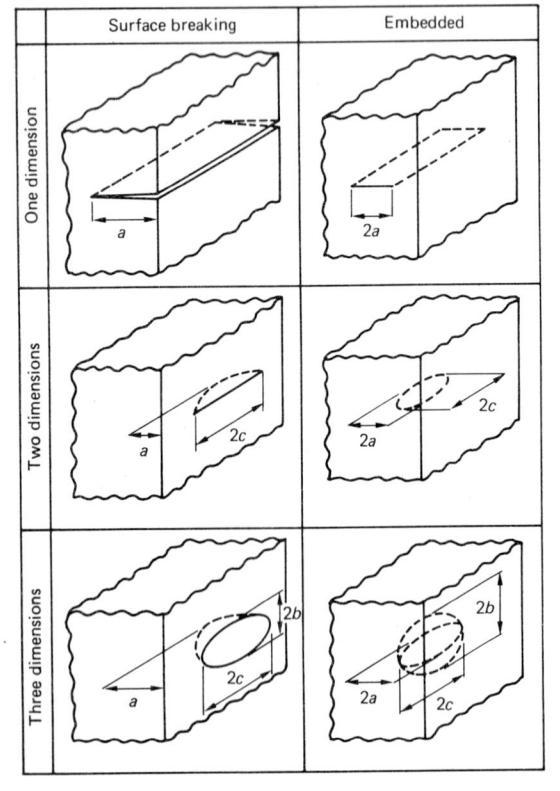

Figure 10.21 Dimensionality of embedded and surface breaking defects

For mathematical treatment, it is normal to idealize defect outlines to the nearest enveloping simple geometric shape and the commonest of these are the rectangle, ellipse and circle for embedded defects and the corresponding semishapes for surface defects (*Figure 10.22*). The convention of using dimensions measured from the centre of symmetry of the defect (e.g. *a* and 2*a* for surface and embedded defect depths) will be adhered to in this section, although some authors referred to do not adopt this.

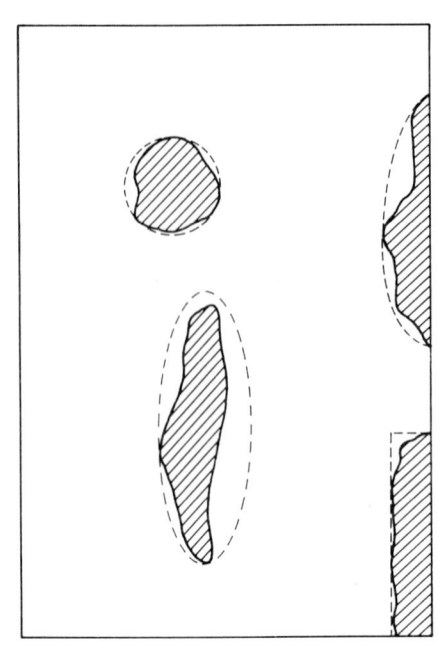

Figure 10.22 Idealization of planar defect shapes

Defects can appear at almost any stage in the life of a manufactured product. For instance, a steel plate may contain embedded laminations as a result of rolling of non-metallic inclusions, may become cracked during welding or flame cutting and/or may corrode or suffer from fatigue during service. Defects which appear during primary processing like casting, forging and rolling are usually detected before use and, if significant, result in scrapping or reprocessing. Secondary processing such as adding the final component or subassembly shape is normally a more expensive process so that repair of significant defects is often undertaken.

Fabrication processes, such as welding, may introduce additional defects which, again, will normally be repaired if significant. *Table 10.4* shows typical defects which may arise from the various manufacturing processes and *Table 10.5* indicates methods which can be used to detect these (Alexander *et al.*, 1985; Birchon, 1975). Although residual stress and undesirable metallurgical changes are technically material defects, this section will only address the detection and significance of discontinuities in the metallic matrix (e.g. cracks, cavities, slag pockets, etc.).

Testing of structures and components in service for defects is normally referred to as monitoring. Although many non-destructive testing methods can be used for monitoring, some methods are specific to the latter. Defects which arise during service are normally caused by the effects of corrosion, wear or fatigue, although other physical or chemical mechanisms are possible. Wear and general corrosion both produce a broadly uniform loss of wall thickness, whereas some forms of

Table 10.4 Genesis of typical defect types

Typical manufacturing processes

Primary processing	Casting of ingots, rolling of plates, sheet or sections
Secondary processing	Near net castings, drawing tubes or wire, forging, welding tubulars, machining, heat treatment
Fabrication/assembly	Welding, bolting, adhesives

Typical defects by process type

Working processes	Laps, cold shuts, edge cracks, centre burst, residual stress, non-uniform deformation
Casting processes	Gas porosity, shrinkage cavities, non-metallic inclusions, segregation, hot cracking
Welding processes	Gas porosity, slag inclusions, hot and cold cracking, residual stresses, metallurgical changes (e.g. sensitization) undercut and overlap
Heat treatment	Distortion, grain growth, sensitization (stainless steels), quench cracking (C–Mn and low alloy steels)
Machining or surface removal	Grinding cracks, chatter marks, surface working and smearing.
Service	Fatigue, corrosion (general, pitting, SCC, etc), wear (fluid and/or solid)

Table 10.5 Capabilities of NDT methods

Method	Materials	Defect types	Minimum detectable and measurable defect size
Liquid penetrant	Most	Clean, surface breaking	8 mm min; depth not measurable
Magnetic particle	Must be ferro-magnetic	Clean, surface breaking or just sub-surface	0.25 mm min; depth not measurable
Ultrasonics	Most	Mostly internal but can be used for surface	5 mm long × 2 mm deep at surface. 2.5 mm long × estimate of depth sub-surface
Eddy current	Metals and coating thickness measurement	Mostly surface but some sub-surface	Approx. 8 mm × 8 mm, but detectable depth varies
Radiography	Most	Mostly internal but will show surface defects	Varies with orientation relative to beam

corrosion and fatigue produce much more localized loss of section and require to be treated as cracks.

10.6.2 The significance of defects

The significance of uniform loss of section thickness is relatively simple to assess and this has led to the traditional concept of corrosion allowance in pressure vessel design, where sections are made thicker than required by an amount which should reflect the envisaged corrosion loss over the life of the vessel. For more localized corrosion or other local defects, the radius of curvature is an important factor.

In the worst case of cracks, a fracture mechanics approach is required where the tolerance of the material to defects (its toughness) is compared with the severity of the defects, related to its size, shape and the applied loading. Although the concepts are a little more subtle, this approach can be compared to strength analysis where the tolerance of the material to stress (its yield strength) is compared with the severity of the loading (the applied stress). There are several ways of assessing defect severity and toughness and some of these have been incorporated into standards for setting maximum allowable defect sizes in (e.g. weldments) (BSI Standard PD6493). The point of commonality between all fracture mechanics approaches is that they seek to establish the point at which a defect becomes unstable under a given loading regime and fracture takes place. This allows the designer to set either maximum allowable defect sizes for a given load (design case) or a maximum permissible load which a given defect can tolerate (maintenance case). This process is often referred to as Engineering Critical Assessment (ECA) (Smith *et al.*, 1985).

The way in which toughness is measured for fracture mechanics analysis depends largely upon whether the material is in a compliant enough condition to permit significant plastic deformation at the tip of a very sharp crack. If the material, thickness and loading conditions are such that little plastic deformation can take place, a linear elastic analysis is sufficient (Knott, 1981) and toughness can be described by the plane–strain fracture toughness K_{IC}. The object of fracture safe design then becomes simply to ensure that the severity of any defect, measured by the stress intensity factory K_I, is always less than K_{IC}.

$$K_I < K_{IC} \quad (10.23)$$

Thus K_{IC} is the material property (like yield strength) and K_I is a function of the loading conditions (like applied stress). The value of K_I in terms of geometry can be calculated by linear elastic finite element analysis and this has already been done for a wide range of possible geometries (Rooke and Cartwright, 1976). In general, K_I can be expressed as:

$$K_I = Ys\sqrt{\pi a} \quad (10.24)$$

where Y is a function of geometry, s is the applied stress and a is the crack depth. Thus, for a given geometry, a crack is more likely to propagate if it is longer or if the applied stress is higher.

Similar considerations hold for the cases where there is substantial plasticity at the crack tip, although the toughness is usually measured by the amount by which the crack faces can be opened in the direction of applied stress before unstable fracture takes place. This quantity is normally referred to as the crack tip opening displacement (CTOD) and given the symbol δ_c.

For less sharp defects and for more ductile materials there is the possibility that the fracture criterion will be governed by plastic instability, rather like a standard tensile test. Such a limit load analysis is often applicable to pressure vessels and pipelines (Jones *et al.*, 1988) and limit loads for a variety of defect and structural geometries have been reviewed by Miller (1987). The particular case of maximum allowable localized corrosion in pipelines has been studied by Kiefner and Duffy (1973).

10.6.3 The growth of defects

If defects simply affected the integrity of structures in proportion to the loss of cross-sectional area which they occupy, it would be a relatively simple matter to set design requirements. This is, of course, the case for defects of very large radius of curvature, but as the sharpness of the defect becomes more pronounced, the intensification of any applied stress at its tip becomes greater and a point may be reached with increasing stress, defect dimensions or sharpness where the defect becomes unstable and propagates through the section and this is the condition identified in Section 10.6.2. The main object of fracture mechanics design is to ensure that instabilities do not take place for any envisaged defects under the envisaged loading conditions. Since the defect dimensions affect the point of instability, it is important to ensure that critical defects do not exist prior to service and that non-critical defects do not grow to a critical size during service. For this reason, it is essential to have an idea of how defects might grow in service.

10.6.3.1 Unstable or fast fracture

In order for a defect to propagate in an unstable manner more energy must be released than is consumed in the process. This fact was first quantified by Griffith (1921) and validated on experiments with glass.

Early observations on the fracture behaviour of metals, particularly steels, led to a distinction between two types of fracture, originally referred to as ductile and brittle. In modern fracture mechanics, the terms plastic and elastic are rather more likely to be used. In qualitative terms, a brittle fracture can be seen as one in which the strain around the crack tip is entirely elastic (*Figure 10.23*). The energy consumed in the

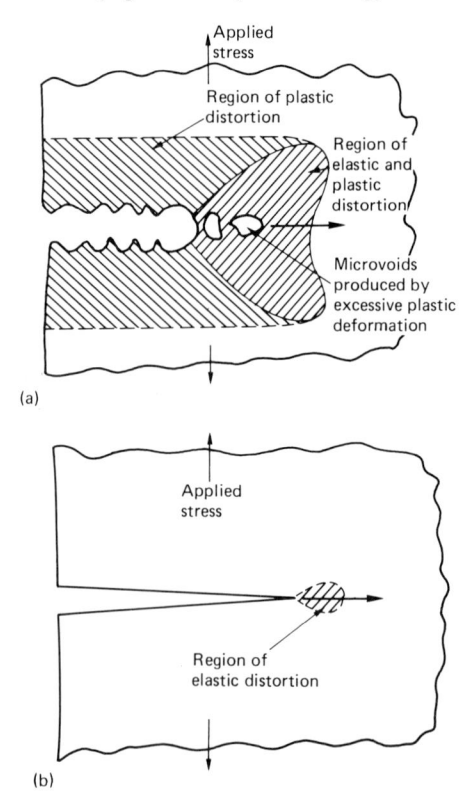

Figure 10.23 Qualitative representation of (a) an elastic–plastic (ductile) and (b) a perfectly elastic fracture (brittle)

process of crack propagation is only that required to create the two new surfaces, since after fracture all the elastic strain energy is recovered.

In the ductile mode (*Figure 10.23*) some plastic deformation occurs around the crack tip, so that, after fracture only the elastic component of the strain energy is recovered. Since the plastic strain energy is consumed in the process, it follows that materials which undergo plastic deformation relatively easily are more difficult to fracture. Tough materials can be designated as those which have large critical crack sizes for a given stress and geometry and brittle materials as those with relatively small critical crack sizes. In terms of linear elastic fracture mechanics, the critical crack size can be expressed in terms of the fracture toughness:

$$a_{crit} = K_{IC}^2 / \pi S^2 Y^2 \qquad (10.25)$$

10.6.3.2 Fatigue

As mentioned previously, it is possible for defects to propagate without becoming unstable. The principal mechanical method by which this can take place is fatigue, where a defect advances by a process of reversed plastic deformation at the crack tip (*Figure 10.24*). The mechanism depends upon repeated plastic deformation on planes inclined at about 70° to the crack plane (Knott 1981) and, as such, requires unsteady loading, although reversing of the sense of loading is not necessary. Clearly, as the fatigue crack grows, the fracture stress will fall and a point will eventually be reached where the crack becomes unstable at the maximum value of the unsteady stress and fracture occurs.

Although stress concentrators are normally present on all manufactured surfaces which will act as initiation points for fatigue cracks, it is possible for cracks to self initiate on highly polished surfaces by the formation of intrusions and extrusions (Engel and Klingele, 1981).

As well as the S–N approach to fatigue outlined in Section 10.3.3.6, a fracture mechanics approach may be taken based on the Paris Law (Griffiths and Richards, 1973):

$$\frac{da}{dN} = C \triangle K^n \qquad (10.26)$$

where da/dN is the rate of crack propagation with number of cycles, C and n are constants of the material and $\triangle K$ is the range of stress intensity factor.

Using

$$K = Y \sqrt{\pi a} \triangle s \qquad (10.27)$$

the Paris Law may be integrated to give the endurance:

$$N = \frac{\triangle s^{-n}}{C} \int_{a_{initial}}^{a_{critical}} (\pi a)^{-n/2} Y^{-n} \, da \qquad (10.28)$$

which can then in principle be calculated from a knowledge of the material constants C and n for any loading condition and initial defect geometry. Although this approach is mathematically equivalent to the S–N approach it requires an estimate of initial defect size and is therefore more often used in defect assessment situations than in initial design.

10.6.3.3 Stress corrosion, corrosion fatigue and hydrogen embrittlement

The mechanical propagation of a crack can be aided by a corrosion component, as is the case in corrosion fatigue, or the corrosion mechanism itself may be responsible for the initiation and propagation of cracks under a steady applied stress, as is the case in stress corrosion or hydrogen embrittlement associated with sulphide stress corrosion. In any case, the electrochemical conditions at the crack tip are such that preferred dissolution or hydrogen evolution occurs at these points and the crack is able to propagate into the section. Critical stress intensity factors for hydrogen embrittlement and stress corrosion cracking have been measured for a number of systems (Opoku and Clark, 1980).

10.6.4 The detection of defects – NDT methods

Once the maximum allowable defect size is known, it is necessary to institute a programme of testing which will ensure, to a reasonable level of certainty, that a defect of critical proportions does not enter service. Furthermore, a monitoring or inspection system is required to ensure that critical defects do not develop during service. Such testing needs to be carried out without damaging the component so that an array of non-destructive testing (NDT) methods has evolved.

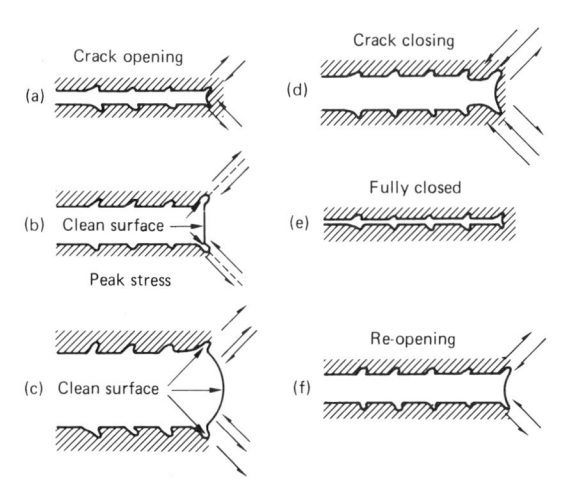

Figure 10.24 Fatigue crack advance by the plastic blunting process (from Knott 1981)

These methods can broadly be divided into those developed to search for surface breaking defects and those developed for embedded defects, although some techniques are capable of both functions. Since surface breaking defects are easier to find, the techniques are normally simpler and cheaper. Some NDT methods can be used to measure section thickness and these are used in the monitoring of corrosion and wear. A special set of techniques has been developed for corrosion monitoring and these will be discussed. The principal NDT techniques discussed in this section are summarized in *Table 10.5*.

10.6.4.1 Highlighting techniques

Two principal techniques are used to enhance the visibility of surface breaking defects. Since these techniques depend upon direct observation they are subject to its limitations and also depend heavily on good surface preparation for their correct operation.

Magnetic particle inspection (MPI) employs a special ink containing magnetic particles which is sprayed on to the metal surface. A magnetic field is applied to the metal surface and the ink particles align themselves along the flux lines bridging any surface discontinuities (*Figure 10.25*). A number of variants exist employing background paints, fluorescent links and different ways of applying the magnetic field. The major limitation is that the technique is only applicable to ferromagnetic materials. Also, defects which are aligned along the lines of magnetic flux will not be detected. Dye (or liquid) penetrant inspection involves the use of a liquid with good wetting characteristics containing a strong dye. The liquid is applied to the surface to be inspected and penetrates into any cracks by capillary action. After a suitable length of time, excess penetrant is removed and a developer is applied to the surface. The developer draws penetrant from any surface defects showing as stains approximating the shape of the defect (*Figure 10.25*). Dye penetrant inspection is rather more versatile than MPI, being applicable to a wider range of materials.

10.6.4.2 Ultrasonic techniques

These techniques depend upon differences between the acoustic impedance of the test material and that of air or some other medium. The acoustic impedance of a material is the product of its density and the speed of sound in that medium and this determines the degree of reflection as sound waves cross an interface between two media. This means that a pulse of ultrasound transmitted through a solid will be reflected in part when it reaches the far wall. If the transducer is able to detect as well as transmit sound, then the time taken for the pulse to return to the transducer will be related to the wall thickness. Furthermore, if a defect of sufficiently different acoustic impedance is present in the solid, there will be a subsidiary reflection from this defect which can be used to determine its approximate size and depth (*Figure 10.25*).

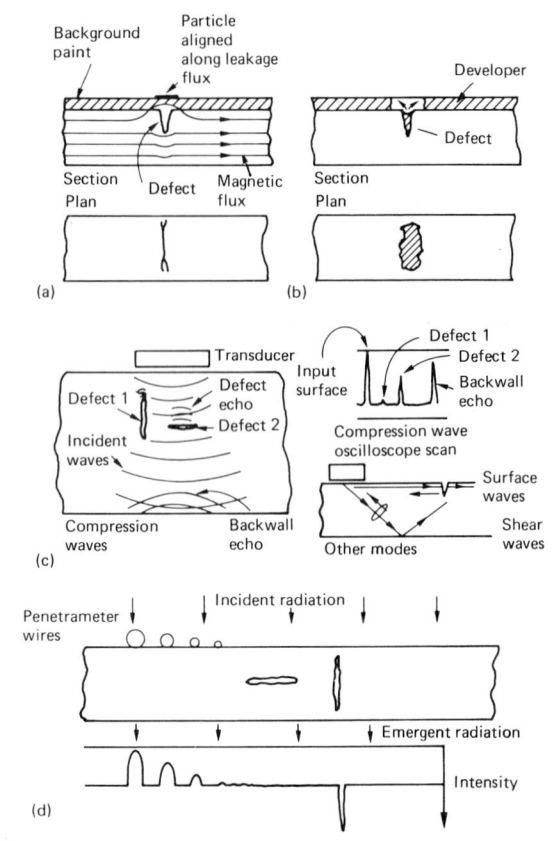

Figure 10.25 Some major non-destructive test methods: (a) magnetic particle, (b) liquid penetrant, (c) ultrasonics, (d) radiography

The use of a standard block containing manufactured defects allows a calibration of the system. A number of variations exist with different methods of display and scanning and ultrasound pulses are often injected into the solid at an angle or even along the surface under which conditions they can be used to detect surface defects. Because the display does not generally give a pictorial representation of the defect distribution, some skill and experience is required on the part of the operator. Difficulties may also arise if defects are distributed (e.g. porosity) or have no part of their surface aligned perpendicular the beam. Ultrasonic non-destructive testing is suitable for the detection of embedded defects in a wide range of materials provided that a suitable frequency is used.

10.6.4.3 Radiographic techniques

Radiographic non-destructive testing exploits the differences in absorption coefficient for high energy electromagnetic radiation between the material being tested and air or some other medium. When X-ray or gamma radiation passes through a solid, it is absorbed according to:

$$I(x) = I(o) \exp(-kx) \qquad (10.29)$$

where $I(x)$ is the intensity a depth x into the solid, $I(o)$ is the incident intensity and k is the absorption coefficient.

Clearly, the intensity of the beam when it has passed through the solid can be used to determine the thickness of the solid and, by irradiating an extensive area, defects can be detected. Most radiographic techniques employ a photographic film to display the emergent intensity and this gives an internal pictorial view of the solid much like the traditional medical X-ray. One disadvantage of such a view is that double imaging has to be used to obtain a conception of depth and this is only normally done if such information is specifically required. Choice of film and incident intensity is normally made by reference to standard tables and, for detailed work, standard blocks of known thickness can be tested and emergent intensities measured by microdensitometer readings on the carefully developed films.

For most purposes, however, it is sufficient to use a penetrameter, normally consisting of wires of known diameter, along with a single film exposure to obtain an idea of the position and size of defects (*Figure 10.25*). In contrast to ultrasonic methods which work by reflection, radiographic NDT is relatively insensitive to thin defects because these have a small effect on the overall metal thickness.

10.6.4.4 *Electrical methods*

A number of methods are available which employ electric current as the probe. The most important of these is eddy current testing which employs the sensitivity of induced currents to surface and near surface electrical conductivity. The back emf, produced in a coil carrying alternating current by virtue of the test surface alters its impedance and this change is displayed for the purpose of flaw detection.

Disruption of the normal eddy current pattern by a flaw can be detected by a characteristic disruption in the voltage across the coil. This is normally displayed on an oscilloscope through a resonant circuit tuned for the normal inductance of the coil. The display can take a number of formats, but perhaps the simplest is to plot a reference impedance (reactance *versus* resistance) on an oscilloscope screen along with the measured impedance at the probe. Use of a storage oscilloscope allows a view of the change in impedance with time. Eddy current testing is still under active development and for a recent review the reader is referred to Hull and John, 1988.

10.6.4.5 *Other NDT methods*

The methods described in Sections 10.5.3.1 to 10.5.3.4 comprise the majority of those in engineering use. However, new methods are being developed and a number of others gain importance from time to time for special materials and/or applications. Among these are fibre optic methods, such as borescopes; thermography, employing variations in thermal conductivity; acoustic emission, employing the sound waves which accompany deformation or cracking; and laser stimulated ultrasonics, using a high energy laser to produce an ultrasonic pulse.

10.6.5 Monitoring methods

As pointed out in Section 10.6.3 monitoring involves the examination of components or structures in service and is therefore in a general sense related to the propagation of defects which were either not present, insignificant or undetected at the manufacture or fabrication stage.

Many of the NDT methods described in Section 10.6.3 can be used for monitoring purposes and it is not intended that these be reiterated here. Instead this section will be devoted to corrosion monitoring and monitoring for fatigue, these being the two principal mechanisms by which defects may be created or propagated in service. Monitoring for mechanical damage will not be dealt with except in the special case of underwater inspection.

10.6.5.1 *Corrosion monitoring*

The corrosion monitoring methods can be summarized under three headings: use of NDT methods, measurement of corrosion current or potential and comparison methods.

Most of the NDT methods which are used in corrosion monitoring give wall thickness measurements and are therefore most useful for monitoring general corrosion. The method in common use for this purpose is ultrasonics although eddy currents, radiography and thermography may all be used. A number of very sophisticated NDT systems have evolved for corrosion monitoring in difficult environments typified by the magnetic inspection vehicles developed recently for internal pipeline condition monitoring (Shannon and Knott, 1985). Localized corrosion such as cracking or pitting may also be detected by the NDT methods and the use of acoustic emission for the detection of stress corrosion continues to develop.

The second category of methods relies on the fact that, as indicated in Section 10.2, the corrosion current and corrosion potential are related by the Tafel Equation (which gives rise to the logarithmic relationship observed in Evans-type Diagrams). This means that the so-called polarization resistance can be used to determine the corrosion current in an electrode of the metal of interest. This is done by connecting a sample of the corroding material to a circuit which allows either the potential or the corrosion current to be perturbed by a very small amount from the natural condition. The polarization resistance to such perturbations (dE/dI) is directly proportional to the corrosion current (Mansfield, 1976). More detailed information on the surface kinetics can be obtained from the polarization impedance but this is more likely to be used as a research tool than as a monitoring technique.

Both of the above techniques require the insertion of probes into the electrolyte and rates are measured on the probe rather than on the surface of interest. It is sometimes possible to use less sophisticated electrical measurements for monitoring purposes. For example, measurements of the potential of a cathodically protected surface can be carried out by simply immersing a reference electrode near the surface and using a high impedance voltmeter; this is normally sufficient to establish whether the protection is adequate. Under conditions where the cathode and anode of a corrosion cell can be separated, for example in bimetallic corrosion, a zero resistance ammeter can be used to measure the corrosion current directly although the number of cases where this can be used is small.

The third category of methods involves the insertion of samples or coupons into the corrosive medium and the periodic removal of these samples for laboratory examination. Such examination normally involves the measurement of weight loss and, if appropriate, assessment of pit depth and special coupons can be prepared for monitoring of crevice corrosion or stress corrosion.

10.6.5.2 Fatigue monitoring and underwater inspection

Because of the unsteady nature of wave, current and wind loading, structures which are subject to these forces must not only be designed against fatigue, but a programme of monitoring is required to ensure that fatigue cracks do not develop and grow during the life of the structure. In the case of ships and mobile offshore units such monitoring can be undertaken in a dry, or at least sheltered area whilst the unit is out of service and this would usually consist of some combination of MPI, visual inspection and ultrasonics or radiography for areas of concern.

Fixed offshore structures must, however, be examined *in situ* and at least part of this examination must be carried out under water. The problem of monitoring the fixed steel structures associated with the recovery of offshore hydrocarbons has prompted the development of a number of underwater inspection techniques (Moncaster, 1982). Due to the very large areas involved, it is necessary to assign priorities to different areas and this can be quite easily done since the structural nodes are known to be the most susceptible to fatigue. Non-critical areas will normally only be subjected to visual inspection whereas the nodal welds would normally be examined using MPI with special fluorescent inks. Ultrasonics and, less commonly, radiography and eddy current testing may be used for areas which are particularly crucial or where other methods have indicated a requirement for further inspection.

Other methods of guarding against fatigue include the use of witness devices which are bonded to the surface and break with the development of a surface crack and condition monitoring where the dynamic response of the structure is examined for any changes which may indicate the failure of a member.

10.6.6 Defects and reliability

Although not by any means the only factor, the size distribution of defects has a strong influence on the reliability of a structure or pressure vessel. Fitness for purpose analysis is a branch of engineering which accepts that no manufactured product is perfect and is aimed at keeping the risk of failure acceptably low by ensuring that the probability of the existence of a defect of size greater than the critical size is known and controlled. Such an analysis starts with an assessment of the defect size distribution in the manufactured product and an assessment of the maximum allowable defect size, based on toughness measurements of the material.

The probability of failure if the component was put into service in this condition would be given by the interaction of the two distributions (i.e. the probability that a_{act} exceeds a_{all} in *Figure 26(a)*). If an NDT method capable of detecting defects as small as a_c is applied to the component, the size distribution of defects will be modified as shown in *Figure 26(b)*. As can be seen, there is a small but finite probability that defects of size greater than the inspection threshold will go undetected. The application of a service test, such as a pressure test, would further truncate the defect size distribution. When in service, defects may grow and new defects may be initiated and furthermore the maximum allowable defect size may become smaller (due, for example, to wall thinning by corrosion). The net effect of this is to cause the two distributions to move together as shown in *Figure 26(c)*. The probability of failure is therefore calculable at any time in the life of the product. As well as giving an assessment of the reliability of a structure containing defects such an analysis can be used, for example, to optimize inspection intervals (Wolfram, 1983) and to optimize the NDT sensitivity (Thomas, 1981).

10.7 Codes, test methods and standards

In a publication of this length, it is not possible to outline all available codes and standards in the area of marine corrosion and defect analysis. Instead, the purpose of this section is to refer the reader to areas where data, details of test methods and specifications may be obtained. When selecting a material for a marine application, the engineer will require data on available materials which are normally obtained from handbooks or manufacturer's literature, occasionally from scientific and engineering journals. For some

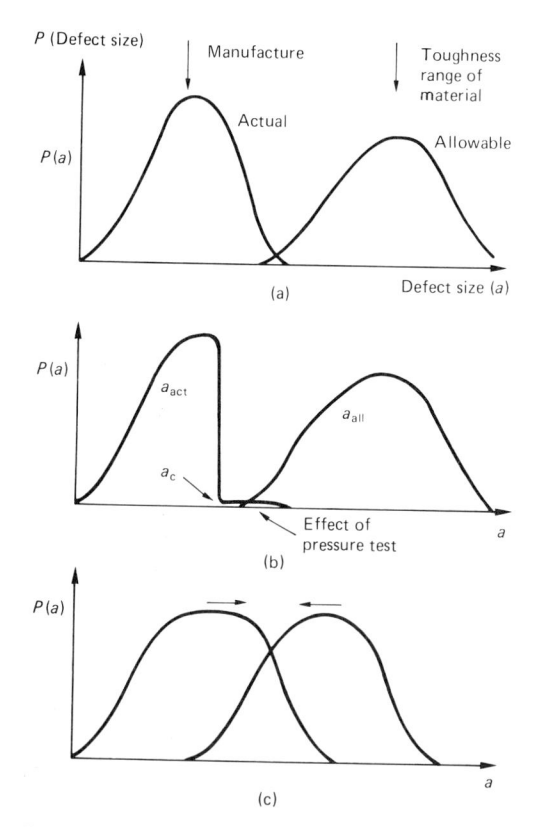

Figure 10.26 Probability density functions illustrating fitness for purpose analysis: (a) initial, (b) after NDT pressure test, (c) after service period

specific problems, for example fatigue or SSC, design guides in the form of standards or handbooks are available and these outline the appropriate criteria and considerations.

When specifying a material to a supplier, it is necessary to refer to a standard so that the material is supplied in the required condition with the required properties, although the purchaser is often free to alter specific clauses of the standard. Manufacturing processes such as welding or casting are often governed by standards which are written to ensure a reasonable level of quality and design codes for critical equipment such as pressure vessels and structures are widely available.

Finally, test methods such as those required to assess susceptibility to corrosion or fatigue are often outlined in standard documents as are inspection methods such as pressure testing and NDT. A number of Institutions referred to frequently in this section are therefore used in abbreviated form: British Standard (BS), American Society for Testing and Materials (ASTM), National Association of Corrosion Engineers (NACE), American Iron and Steel Institute (AISI), American Petroleum Institute (API), American Society of Mechanical Engineers (ASME) and Det norske Veritas (Norwegian Certifying Authority) (DnV).

10.7.1 Standards relating to materials and their properties

Engineering designers often use standards to define the form, chemical composition and other properties of engineering alloys. Such standards are published by a number of Institutions but only some examples will be referred to here. For an exhaustive list, the reader is referred to the catalogue of the appropriate Institution.

A typical standard relates to a group of materials often for a specific purpose (e.g. structural steels) and/or in a specific product form (e.g. hollow sections). It will also usually refer to a number of different grades of alloys, the gradation often being carried out in terms of mechanical properties, such as yield strength. The chemical composition of each grade is also normally specified. Two examples should suffice to illustrate this.

Example 1: ASTM A694-84: Forgings, carbon and alloy steel, for pipe flanges, fittings, valves and parts for high pressure transmission service. This standard identifies eight grades of steel (F42–F65) suitable for use in the applications and condition described in the title. The gradation is carried out in terms of mechanical properties, grade F42 having a minimum yield strength of 290 MPa (42 ksi) and F65 having a minimum yield strength of 448 MPa. The chemical composition of the steel is only loosely defined leaving some leeway for agreement between manufacturer and purchaser. Certain other specifications are made regarding manufacturing, inspection and ordering information so that no misunderstanding need arise between purchaser and supplier over the quality of the product.

Example 2: BS1400: Copper alloy ingots and copper alloy and high conductivity castings. This standard relates to cast copper alloy products of which a ship's propeller might be an example. The standard specifies such items as strength, chemical composition, soundness and microstructure and covers required inspection procedures and test methods. An appendix includes a brief guide to alloy selection including considerations of corrosion resistance. This standard is more wide ranging than the previous example in that most copper alloy types from G-bronze to high conductivity coppers are covered.

10.7.2 Codes and standards relating to design

A number of guides exist for general design in marine engineering. This section will only consider those aspects relevant to materials engineering. The most likely cases in which an engineer is likely to require to consult a design guide are offshore structures (fixed or mobile), ships, pressure vessels and pipelines because of the potentially disastrous consequences should a failure occur.

The UK Department of Energy issues guidelines on the design and construction of offshore installations with the primary objective of explaining how such installations may achieve fitness for purpose certification, which is a legal requirement in UK waters. Similar guidelines are issued by DnV, API and others. Besides more general design rules, these codes contain guidance on materials of construction, fabrication and inspection procedures. For example, the Department of Energy document refers to BS4360 for specification of structural steels and BS5135 amongst others for welding procedures. The two main pressure vessel codes are ASME (Boiler and Pressure Vessel Code) and BS5500, both of which have detailed sections on manufacture, inspection and materials selection. Again, other standards are often called from these.

Although technically pressure vessels, pipelines have other significant design limitations and are therefore dealt with separately and the Institute of Petroleum, ASME and DnV pipeline codes all have provision for subsea pipelines. Steels for pipelines are normally specified from API 5LX and API also publish standards for a number of other petroleum related components. Separate welding codes are normally used for pipelines and the commonest of these for offshore use are probably API 1104 and BS4515.

Some specific aspects of detail design are also dealt with by standard documents. The design of cathodic protection systems is outlined in documents by NACE (RP-01-76), BS (CP 1021) and DnV (TNA 02) amongst others and in some cases these include specific guidance on the calculation of required anode weights and expected anode lifetimes. Coating systems for protection of iron and steel against corrosion are described in a BS Code of Practice BS5493 and also in NACE RP-01-76 and others. A BS policy document BS PD6484 gives qualitative guidance on expected corrosion rate acceleration factors due to galvanic effects for various anode: cathode area ratios for a wide range of metals and alloys.

Fracture safe design is becoming an important area of engineering and a number of codes are now available to assist the engineer. Both ASME in the Boiler and Pressure Vessel Code and BS PD6493 provide procedures for the assessment of maximum tolerable defect sizes in welded construction and both also allow for fatigue design based on fracture mechanics principles.

Fatigue design can also be carried out using S–N curves which are plots of stress range (difference between maximum and minimum values) *versus* number of cycles which can be endured at that stress range. ASME, DnV, BS and others have published S–N curves for a variety of weld geometries in constructional steels both with and without a sea water corrosion component and these are discussed in a (British) Welding Institute publication (1983).

10.7.3. Standards relating to manufacturing processes

A number of the materials standards referred to in Section 10.7.1 contain specifications for some aspects of manufacturing processes, particularly those relating to quality control (e.g. soundness of castings). In this chapter the most important manufacture related standards are those which cover welding since this is normally responsible for the majority of any defects present in a component as it goes into service and weldments are often the principal sites for fatigue because of the geometrical stress concentrations produced.

For this reason, welding of or on any structural or containment part is normally carried out to a set of codes which govern details of the welding process, inspection procedures, allowable defect sizes, consumables, qualification of welders and qualification of procedures. Typical examples of such Codes are the American Welding Society's Structural Welding Code, AWS D1.1 and BS4515, covering the process of welding of steel pipelines on land and offshore. As far as qualification is concerned, BS4871 is written specifically for approval testing of welders working to approved procedures and approval testing of such procedures is covered in BS4870.

10.7.4 Standards relating to inspection and testing

These standards are frequently called from others referred to above and normally outline the methods to be used for inspection and testing rather than the results which are required.

10.7.4.1 Test methods

One important group of standards relating to test methods concerns the mechanical testing of materials and includes those for commonplace tests such as tension testing for strength (e.g. ASTM E8 and BS18) to the more specialized procedures required for the assessment of toughness or fatigue resistance. A number of designers still use the Charpy or Izod impact tests (e.g. BS 131) for the assessment of toughness but these have the disadvantage of being qualitative in their application and at best provide a gradation of toughness for alloys of a similar type. Fracture mechanics tests such as the COD or K_{IC} (fracture toughness) tests are more directly applicable to a programme of fracture control (e.g. BS5447 and BS5762). Testing for fatigue resistance can be carried out using the traditional S–N approach or by measuring rate of crack growth with range of stress intensity factor, ΔK. Both approaches are covered by standard specifications as to the loading range, number and size of specimens required as well as other parameters such as ASTM E466.

Corrosion testing is subject to a wide range of standards depending on the type of corrosion which is to be examined. For atmospheric exposure a range of *in situ* tests exist but these can be very lengthy. This illustrates a basic problem with corrosion testing in that the expected service lives often exceed 20 years but test results are needed in a much shorter time. Furthermore, the weight loss is rarely linear with time and is often spatially heterogeneous. Most test methods are aimed at overcoming such difficulties but a thorough knowledge of corrosion properties only comes with a combination of field experience and testing of a particular alloy.

For atmospheric corrosion, a high degree of acceleration can be obtained by use of the hot salt fog test (e.g. ASTM B117) but some debate exists as to the exact degree of acceleration achieved so that the test is largely comparative (Lee and Money, 1984). For stress corrosion, SSC, HIC, exfoliation, dezincification and other specific types of corrosion, the method is generally again comparative using a very severe exposure and measuring the degree of damage produced. For example, the ASTM method for chloride stress corrosion in stainless steels (ASTM G36) employs a solution of boiling magnesium chloride although the results of such a test will not, in general, allow a prediction of the time to failure in another chloride containing solution at another temperature.

For localized corrosion, such as pitting, standard methods of exposure and quantification of damage are specified, such as ASTM G48. Methods also exist for the measurement of cavitation and corrosion erosion (e.g. ASTM G73). Perhaps the most quantifiable type of corrosion is that which occurs uniformly under immersed conditions. Sensitive measurements can be made in the laboratory using linear polarization resistance (e.g. ASTM G5) or linear polarization impedance and, provided that the surface and solution properties have been correctly reproduced, these are directly applicable to field exposure with the proviso that some cognisance must be taken of possible changes in corrosion rate with time. Test methods for pipeline and petroleum engineering components subject to sulphide environments are given by NACE (TM 01 and TM 02).

10.7.4.2 Inspection and monitoring

Non-destructive examination standards are called from most of the manufacturing standards, especially those involving welding. The structural and pressure vessel codes also have a particular emphasis on non-destructive examination. Furthermore, the law in some countries, including Department of Energy guidance for offshore installations, calls for periodic inspection of important marine plant such as offshore structures, ships, pipelines, pressure vessels and lifting gear. Such inspection often uses the NDT methods outlined in Section 10.6.4. The radiographic, ultrasonic, MPI and dye penetrant methods are widely covered by standards and others are covered to a lesser extent. For example, ASTM Specification E138 deals with the practice of magnetic particle inspection whereas BS PD6513 covers the magnetic principles involved in such testing as well as the practice. DNV Classification Note No. 7 details procedures and calibration for quantitative ultrasonic NDT including defect sizing.

Methods for corrosion monitoring are given by NACE, ASTM and BS. These include methods for installation and examination of corrosion coupons, (e.g. NACE-RP-07) and use of other techniques such as linear polarization probes and hydrogen probes. In general, the NDT methods used for corrosion monitoring are dealt with in the NDT standards for more general use.

References

Alexander, W. O. *et al.* (1985) *Essential Metallurgy for Engineers*, Van Nostrand Reinhold, Wokingham, UK
Ashworth, V. (1986) The theory of cathodic protection and its relation to the electrochemical theory of corrosion. In *Cathodic Protection*, eds. Ashworth, V. and Booker, C. J. L., Ellis Horwood, Chichester, UK
Astley, D. J. and Rowlands, J. C. (1985) *British Corrosion Journal*, **20 (2)**, 90–94
Beccaria, A. M. and Poggi, G. (1986) *British Corrosion Journal*, **21 (1)**, 19–22
Birchon, D. (1975) *Non-Destructive Testing*, Design Council Engineering Design Guide 9, Oxford University Press, Oxford
Bregman, J. I. (1963) *Corrosion Inhibitors*, Macmillan, New York
Browne, R. D. and Domone, P. L. J. (1975) The long term performance of concrete in the marine environment. In *Proceedings of Institution of Civil Engineers Conference on Offshore Structures* (London, 1974), ICE, London
Carruthers, R. (1985) The use of 90/10 copper–nickel as a splash zone cladding. In *Proceedings of a Conference on Copper Alloys in Marine Environments*, (Birmingham, 1985), Copper Development Association, Potters Bar, UK
Chandler, K. A. (1985) *Marine and Offshore Corrosion*, Butterworths, London
Chandler, K. A. and Bayliss, D. A. (1985) *Corrosion Protection of Steel Structures*, Elsevier, London
Copper Development Association (1981) *Aluminium Bronze Alloys Corrosion Resistance Guide*, Copper Development Association, Potters Bar, UK
Cotterell, A. H. (1958) *Transactions of the American Institute of Mining Metallurgical and Petroleum Engineers*, **212**, 192–200
Crane F. A. A. and Charles, J. A. (1984) *Selection and Use of Engineering Materials*, Butterworths, London
deWaard, C. and Milliams, D. E. (1975) Prediction of carbonic acid corrosion in natural gas pipelines. In *First International Conference on Internal and External Protection of Pipes* (Durham, 1975), BHRA Fluid Engineering, London
Dexter, S. C. (1987) Seawater. In *Metals Handbook Vol. 13, Corrosion*, American Society for Metals, Ohio
Engel, L. and Klingele, H. (1981) *An Atlas of Metal Damage*, Wolfe, London
Fontana, M. G. and Greene, N. D. (1982) *Corrosion Engineering*, McGraw-Hill, Tokyo
Forteath, G. N. R, Picken, G. B. and Ralph, R. (1984) Patterns of macrofouling on steel platforms in the Central and Northern North Sea. In *Corrosion and Marine Growth on Offshore Structures*. (eds. J. R. Lewis and A. D. Mercer) Ellis Horwood, Chichester, UK
Francis, R. (1985) *British Corrosion Journal*, **20 (4)**, 167–182
Fulmer Research Institute (1980) *Fulmer Materials Optimiser*, Fulmer Research Institute, London

Gabe, D. L. (1983) *Coatings for Protection*, Institute of Production Engineers, London
Griffin, R. B. (1987) Marine Atmospheres. In *Metals Handbook Vol. 13, Corrosion*, American Society for Metals, Ohio
Griffith, A. A. (1921) *Philosophical Transactions of the Royal Society*, **A221**, 163–178
Griffiths, J. R. and Richards, C. E. (1973) *Materials Science and Engineering*, **11 (6)**, 305–318
Heaton, W. E. Edgeley, J., Andrews, E. F. C. and Patient, B. (1979) A review of the factors leading to the use of titanium as a steam condenser tube material within the CEGB. In *Proceedings of the Institution of Mechanical Engineers Conference on Cooling with Seawater* (London, 1979), Mechanical Engineering Publications, London
Hill, R. T. and Trout, S. G. (1986) *Pipes and Pipelines International*, July-August 1986, 15–22
Hull, B. and John, V. (1988) *Non-Destructive Testing*, Macmillan, London
Institution of Metallurgists (1974) *Economics of Corrosion Control*, Autumn Review Course Publication, (York, 1974), Institution of Metallurgists, London
Jaske, C. E. Payer, J. H. and Balint, V. S. (1981) *Corrosion Fatigue of Metals in Marine Environments*, Springer Verlag/Battelle Press, Columbus, Ohio
Jones, D. G. Hopkins, P. and Clyne, A. J. (1988): Assessment of weld defects in offshore pipelines. In *Proceedings of a Conference on Offshore Pipeline Technology* (Stavanger, 1988), IBC Technical Services/Norwegian Petroleum Directorate, Stavanger
Kiefner, J. F. and Duffy, A. R. (1973) Criteria for determining the strength of corroded areas of gas transmission lines. In *American Gas Association Operating Section Proceedings*, 1–86
King, R. A. (1979) Prediction of corrosiveness of seabed sediments. In *Proceedings of a Conference Corrosion 1979*, (Atlanta, 1979), National Association of Corrosion Engineers, Houston.
Kirk, W. W., Lee, T. S. and Lewis, R. O. (1985) Corrosion and marine fouling characteristics of copper–nickel alloys. In *Proceedings of a Conference on Copper Alloys in Marine Environments* (Birmingham, (1985), Copper Development Association, Potters Bar, UK
Knott, J. F. (1981) *Fundamentals of Fracture Mechanics*, Butterworths, London
Kubaschewski, O. and Hopkins, B. E. (1962) *Oxidation of Metals and Alloys*, Butterworths, London
Leach, J. S. L. (1979) The fundamentals of cathodic protection. In *Proceedings of Institution of Mechanical Engineers Conference on Cathodic Protection in Steam and Process Plant* (London, 1979), Mechanical Engineering Publications, London.
LaQue, F. L. (1975) *Marine Corrosion – Causes and Prevention*, John Wiley, New York
Larrabee, C. P. (1953) *Corrosion*, **9 (8)**, 259–271
Lee, T. S. and Money, K. L. (1984) *Materials Performance*, **23 (7)**, 28–33
Logan, H. L. (1966) *The Stress Corrosion of Metals*, John Wiley, New York
Mansfield, F. (1976) *Advances in Corrosion Science and Technology*, **6**, 163–267
Miller, A. G. (1987) Review of limit loads of structures containing defects, *CEGB Report TPRD/B/0093/N82*
Miller, D., Cameron, A. M. and Shone, E. B. (1984) Application of novel anti-fouling coatings on offshore structures. In *Corrosion and Marine Growth on Offshore Structures* (eds. J. R. Lewis and A. D. Mercer), Ellis Horwood, Chichester, UK
Ministry of Agriculture, Fisheries and Food (1986) *United Kingdom Atmospheric Corrosivity Values*, MAFF, London
Mollan, R. and Andersen, T. R. (1986) Design of cathodic protection systems. In *Proceedings of Conference Corrosion '86*, (Houston, 1986), National Association of Corrosion Engineers, Houston
Moncaster, M. B. (1982) *Journal of the Society of Underwater Technology*, August 1982, 7–16
Moore, W. J. (1972) *Physical Chemistry*, Longman, London
Morley, J. and Bruce, B. W. (1983) Survey of steel piling performance in marine environments, *British Steel Corporation Technical Report T/RS/1195/21/82/C*, Sheffield, UK
National Association of Corrosion Engineers (1976) *The Role of Bacteria in the Corrosion of Oilfield Equipment*, TPC Publication No. 3, NACE, Houston
National Association of Corrosion Engineers (1979) *Corrosion Control in Petroleum Production*, TPC Publication No. 5, NACE, Houston
Opoku, J. and Clark, W. G. (1980) *Corrosion – NACE*, **36 (5)**, 251–259
Ostroff, A. G. (1979) *Introduction to Oilfield Water Technology*, National Association of Corrosion Engineers, Houston
Parkins, R. N. and Kolotyrkin, Y. M. (1983) (eds.) *Proceedings of UK/USSR Seminar on Corrosion Fatigue* (USSR, 1980), The Metals Society, London
Parsons, R. (1959) *Handbook of Electrochemical Constants*, Butterworths, London
Pludek, V. R. (1977) *Design and Corrosion Control*, Macmillan, London
Pourbaix, M. (1974) *Atlas of Electrochemical Equilibria in Aqueous Solutions*, NACE, Houston
Preece, C. M. (1979) *Erosion – Vol. 16 of Treatise on Materials Science and Technology*, Academic Press, New York
Reinhart, F. M. and Jenkins, J. F. (1972) *Technical note N-1213*, US Naval Civil Engineering Laboratory, Port Hueneme, California
Riggs, O. Locke, C. E. and Hamner, N. E. (1981) *Anodic Protection – Theory and Practice in Prevention of Corrosion*, Plenum, New York
Robinson, J. (1979) Corrosion inhibitors – recent developments, *Chemical Technology Review* **132**, Noyes Data Corporation, Park Ridge, New Jersey
Robson, D. N. C. (1984) Mixed metal galvanic problems associated with seawater pumping equipment. In *Corrosion and Marine Growth on Offshore Structures* (eds. J. R. Lewis and A. D. Mercer) Ellis Horwood, Chichester, UK
Roche, M. and Samaran, J. P. (1987) *Materials Performance*, **26 (11)**, 28–34
Rooke, D. P. and Cartwright, D. J. (1976) *Compendium of Stress Intensity Factors*, HMSO, London
Saville, R. (1982) *Anti-corrosion*, **29 (8)**, 13–14
Scanlon, A. J. and Emms, D. (1979) Minimising corrosion of auxiliary tubular heat exchangers used in seawater cooling systems. In *Institution of Mechanical Engineers Conference on Cooling with Seawater* (London, 1979), Mechanical Engineering Publications, London
Schumacher, M. (1979) *Seawater Corrosion Handbook*, Noyes Data Corporation, Park Ridges New Jersey
Sedriks, A. J. (1979) *Corrosion of Stainless Steels*, John Wiley, New York

Shannon, R. W. and Knott, R. N. (1985) On-line inspection – Development and operating experience. In *Proceedings 17th Annual Offshore Technology Conference* (Houston, 1985), OTC, Dallas, Texas
Shone, E. B. Malpas, R. E. and Gallagher, P. (1988) Stainless Steels as replacement materials for copper alloys in seawater handling systems. Paper produced for and presented at the Institution of Marine Engineers, (London, 1988), Institution of Marine Engineers, London
Shreir, L. L. (1976) *Corrosion*, Newnes–Butterworths, London
Slater, J. E. (1983) *Corrosion of Metals in Association with Concrete*, American Society for Testing and Materials STP 818, ASTM, Philadelphia
Smith, I. J. Pisarski, H. G. and Smith, H. G. (1985) ECA procedures for reliability of offshore welded construction. In *17th Annual Offshore Technology Conference Proceedings*, (Houston, 1985), OTC, Dallas, Texas
Terasaki, F., Ohtani, Y., Ikeda, A. and Nakanashi, M. (1984) Steel Plates for Pressure Vessels in Sour Environment Applications. In *Proceedings of an International Conference on Fusion Welded Steel Pressure Vessels* (London, 1984), Mechanical Engineering Publications, London
Terry, L. A. and Edyvean, R. G. J. (1984) Influences of microalgae on the corrosion of structural steel. In *Corrosion and Marine Growth on Offshore Structures* (eds. J. R. Lewis and A. D. Mercer) Ellis Horwood, Chichester, UK
Thomas, J. M. (1981) Quantitative risk analysis of offshore welded structures. In *Proceedings of an International Conference on Fitness for Purpose Validation of Welded Construction* (London, 1981), Welding Institute, London
Thomason, W. H., Pape, S. E. and Evans, A. (1987) *Materials Performance*, **26 (11)**, 22–27
Tighe-Ford, D. J. McGrath, J. N. and Wareham, M. P. (1988) Evaluation of warship impressed current cathodic protection systems, Paper presented at the Institution of Marine Engineers, (London, 1988), Institution of Marine Engineers, London
Tiller, A. K. (1983) Electrochemical aspects of microbial corrosion. In *Proceedings of a Conference on Microbial Corrosion* (London, 1983), The Metals Society, London
Tischuk, J. L. and Huber, D. S. (1984) Use of economic analysis to select the most cost-effective method of downhole corrosion control. In *Proceedings of a Conference U.K. Corrosion 1984*, (London, 1984), Institution of Corrosion Science and Technology, London
Todd, J. M. and Perkins, J. (1976) *Naval Engineers' Journal*, August 1976, 55–72
Uhlig, H. H. and Revie, R. W (1985) *Corrosion and Corrosion Control*, John Wiley, Toronto
Waterhouse, R. B. (1972) *Fretting Corrosion*, Pergamon, Oxford
Welding Institute (1983) *Proceedings of 2nd International Conference on Offshore Welded Structures* (London, 1982), Welding Institute, London
Whitehouse, N. R. (1983) Survey of Painting Practice for Protection of Offshore Structures, *UK Paint Research Association Report*, Teddington, UK
Wilhelm, S. M. and Kane, R. D. (1984) *Corrosion-NACE*, **40 (8)**, 431–439
Wilkins, N. J. M. and Lawrence, P. F. (1980) Fundamental mechanisms of corrosion of steel reinforcement in concrete immersed in seawater. *Technical Report Number 6 for Concrete in the Oceans Management Committee*, CIRIA/UEG/Cement and Concrete Association/Department of Energy, London
Williams, R. J. (1985) Corrosion in oil and gas flowlines. In *Proceedings of a Conference on Offshore Oil and Gas Pipeline Technology* (Amsterdam, 1985), Oyez, London
Wolfram, J. (1983) *Journal of the Society of Underwater Technology*, Summer 1983, 27–32

Codes and standards

This list gives details of Codes and Standards referred to in the text. It is not intended to be an exhaustive list of Standards on the subjects treated.

American Society or Testing and Materials (ASTM)

ASTM A694: Forgings, carbon and alloy steel, for pipe flanges, fittings, valves and parts for high pressure transmission service
ASTM B117: Method of salt spray (fog) testing
ASTM E8: Tension testing of metallic materials
ASTM E138: Magnetic particle inspection
ASTM E466: Recommended practice for constant amplitude axial fatigue tests for metallic materials
ASTM G36: Practice for performing stress corrosion cracking tests in boiling magnesium chloride solution
ASTM G5: Practice for standard reference method for making potentiostatic and potentiodynamic anodic polarisation measurements
ASTM G48: Test method for pitting and crevice corrosion resistance of stainless steels and related alloys by use of ferric chloride solution
ASTM G73: Practice for liquid impingement corrosion testing

British Standards (BS)

BS 5447: Methods of test for plane strain fracture toughness (K_{IC}) of metallic materials
BS 5762: Methods for crack opening displacement (COD) testing.
BS 1400: Copper alloy ingots and copper alloy and high conductivity castings
BS 4360: Weldable structural steels
BS 5135: Metal arc welding of carbon and carbon--manganese steels
BS 5500: Unfired fusion welded pressure vessels
BS 4515: Process of welding of steel pipelines on land and offshore
BS 5493: Code of practice for protective coatings of iron and steel structures against corrosion
BS 4870: Approval testing of welding procedures
BS 4871: Approval testing of welders working to approved welding procedures
BS 18: Methods for tensile testing of metals
BS 131: Methods for notched bar tests
BS PD6513: Magnetic particle flaw detection
BS PD6484: Commentary on corrosion at bimetallic contacts and its alleviation

BS PD6493: Guidance on some methods for derivation of acceptance levels for defects in fusion welded joints
BS CP1021: Code of practice for cathodic protection

National Association of Corrosion Engineers (NACE)

NACE RP–01–76: Control of corrosion on steel fixed offshore platforms associated with production
NACE RP–07–75: Preparation and installation of corrosion coupons and interpretation of test data in oil production practice
NACE MR–01–75: Sulphide stress cracking resistance of metallic materials for oil field equipment
NACE TM–01–77: Testing of metals for resistance to sulphide stress cracking at ambient temperatures
NACE TM–02–84: Test method for evaluation of pipeline steels for resistance to stepwise cracking

Det norske Veritas (DnV)

DnV Rules for the design, construction and inspection of offshore structures
DnV Rules for submarine pipeline systems
DnV TNA 02: Fabrication and installation of sacrificial anodes
DnV Classification Note No. 7: Ultrasonic inspection of welded connections

Other codes and standards

API 1104: Standards for welding pipelines and related facilities
API 5LX: High test line pipe
ASME/ANSI B31.4 and B31.8: Code for pressure piping
ASME: Boiler and pressure vessel code
Department of Energy: Offshore installations – Guidance on design and construction
Institute of Petroleum (IP6): Model code of safe practice – Petroleum pipelines
AWS.DI.1: American Welding Society, Structural Welding Code

11

Marine Safety

David J. House

Contents

11.1 Introduction

11.2 Onboard safety information
    11.2.1 Organization of safety procedures
    11.2.2 Shipboard safety communications
    11.2.3 Watch keeping duties
    11.2.4 Vessels lying alongside one another
    11.2.5 Navigation safety – passage planning
    11.2.6 Navigational safety – prior to sailing
    11.2.7 Keeping a safe navigational watch
    11.2.8 Conducting a safe anchor watch
    11.2.9 Mooring the vessel by use of anchors

11.3 Operational deck safety
    11.3.1 Safe access to and from the vessel
    11.3.2 Safe practice with derricks and cranes
    11.3.3 General advice on working aloft and overside

11.4 Safety in cargo operations
    11.4.1 Safe shipment and handling – general cargo
    11.4.2 Hold cleaning requirements
    11.4.3 Safe stowage of deck cargoes
    11.4.4 Safe working practice with deep tanks
    11.4.5 Ventilation of ships cargoes
    11.4.6 Associated hazards with the shipment of bulk cargoes
    11.4.7 Recommended practice for the stowage of bulk cargoes
    11.4.8 International Maritime Dangerous Goods (IMDG) Code
    11.4.9 The recommendations
    11.4.10 Safe handling of dangerous goods

11.5 Tanker work safety
    11.5.1 Safe practice in the carriage of oil cargoes
    11.5.2 Oil cargoes, tanker work
    11.5.3 Tanker work
    11.5.4 Tanker operations
    11.5.5 General preparation of tanks – prior to loading a tanker vessel
    11.5.6 Associated equipment
    11.5.7 Static electricity – causes and dangers
    11.5.8 Prevention of static electricity to avoid ignition
    11.5.9 Example use of the inert gas system

11.6 Safety in offshore working practice
    11.6.1 Introduction to offshore safety
    11.6.2 Offshore safety information
    11.6.3 General practice affecting clothing regulations offshore

11.7 Marine emergencies
    11.7.1 Collision
    11.7.2 Man overboard
    11.7.3 Man overboard – associated actions
    11.7.4 Safe operation in marine helicopter activity
    11.7.5 Fire fighting
    11.7.6 Fire in port – suggested line of action
    11.7.7 Fire at sea – suggested line of action
    11.7.8 The safe use of breathing apparatus
    11.7.9 Breathing apparatus – operational checks

11.8 Marine survival
    11.8.1 Survival procedure prior to abandonment
    11.8.2 The abandonment phase
    11.8.3 The survival phase
    11.8.4 The rescue phase
    11.8.5 Life saving appliances and arrangements
    11.8.6 Class VII foreign going non-passenger, non-tanker vessels
    11.8.7 Class VII(T) foreign going tankers
    11.8.8 General rates applicable to these applicances
    11.8.9 The 1983 amendments to the International Convention for Safety of Life at Sea 1974

11.9 Search and rescue operations
    11.9.1 Introduction
    11.9.2 International distress signals
    11.9.3 General arrangements for search and rescue procedures
    11.9.4 Guide to the duties of the on scene commander
    11.9.5 Current rescue methods
    11.9.6 Alternative helicopter use
    11.9.7 Communications
    11.9.8 Amphibian (rotor wing) aircraft (wet surface landing capability)
    11.9.9 SAR aircraft equipment
    11.9.10 Droppable equipment from aircraft

Acknowledgements
References
Further reading
Merchant shipping (M) notices

11.1 Introduction

This chapter is concerned with responsibility of the individual in the marine environment. This responsibility being towards one's fellow man and directly towards the mariner's prime objective, which is the safety of life at sea.

The text will refer to safe working practices employed aboard ships and installations. The use of marine equipment and publications are discussed, together with accepted procedures for emergency and survival operations. However, with continued developments, no text can hope to cover every aspect, for every occasion. To this end, marine safety should become a subject to be read with an eye to the practical aspect of shipboard operations and with commonsense prevailing.

The subject is essential in any business, but perhaps more so when aboard a vessel which may find itself many miles from the nearest assistance. The seagoing professions will continue to operate with hazardous cargoes and negotiate navigational difficulties because it is all part of the commercial shipping business. Current information regarding safe procedures, special cargoes, emergency practices, and routine operations must be readily available to the mariner.

11.2 Onboard safety information

11.2.1 Organization of safety procedures

The Master is ultimately responsible for the safety of the ship and crew. Crew members have a duty to the Master to ensure safety is maintained in those matters within their control. To monitor the overall safety situation, a safety officer or committee will be elected. This can be any one of the ship's officers. Some vessels will have a safety committee which will perform a similar function, but, whatever the system in use, the aim is to maintain the highest possible standard of safety on board the vessel.

11.2.1.1 The safety officer

The main aim of the safety officer is in accident prevention. A positive approach is needed so that safety measures are properly carried out. The safety officer should:

1. Foster among the ship's complement a high degree of safety consciousness and an active interest in accident prevention.
2. Provide a channel by which suggestions for improving safety may be transmitted to appropriate personnel on board and ashore.
3. Ensure the compliance by all personnel to the safety rules and instructions in force at the time.
4. Investigate any accidents and unsafe occurrences, working practices or conditions on the ship.

11.2.1.2 Safety committee

This performs a similar function to that of the safety officer and is chaired by the Master or deputy. Representatives of all departments on the vessel are generally present. With a committee, it is recommended that proper minutes are kept and distributed to other vessels and the company office when deemed necessary. Every shipboard crew member has a duty to inform, through channels, the company representatives of potential hazards. The 'I'm alright Jack' attitude does not work where safety is concerned.

A general reference for onboard safety information is the *Code of Safe Working Practice*, published by the UK Department of Transport and distributed by Her Majesty's Stationery Office. It contains information about numerous aspects of safety including fire precautions, emergency procedures, safety officers and safety committees, protective clothing and equipment, safety signs, notices and colour codes, permit to work systems, accommodation ladders, gangways and other means of access, movement about the ship, entering enclosed or confined spaces, manual lifting and carrying, tools and materials, welding and flame cutting operations, painting, working aloft and outboard, anchoring, mooring and casting off, lifting and mechanical handling appliances, hatches, work in cargo spaces, work in machinery spaces, hydraulic and pneumatic equipment, overhaul of machinery, servicing radio and associated electronic equipment, storage batteries, work in the galley, pantry and other food handling areas, work in the ship's laundry, general cargo ships including dangerous goods, carriage of containers, tankers and bulk product carriers, ferries, Ro-Ro carriers and car carriers, ships serving offshore gas and oil installations and a list of British Standard Specifications relevant to recommended safe practices.

11.2.1.3 Muster lists

Clear instructions shall be provided for every person to follow in the event of an emergency. Muster lists, which specify the requirements laid down, must be exhibited in conspicuous places throughout the ship, including the navigation bridge, engine room and crew accommodation spaces. Illustrations and instructions in the appropriate language, are posted in passenger cabins and displayed at muster stations and passenger spaces to inform passengers of their muster station, the essential actions to take in an emergency and the method of donning lifejackets.

11.2.1.4 Emergency parties

The function of most emergency parties is to be able to respond quickly and effectively to tackle or act as backup against whatever incident has arisen (see Section 11.7). The types of emergency which may be encountered aboard a ship at sea are too numerous to attempt to list and detail. Consequently, selected ones have been used to highlight the most common, and so familiarize the mariner with a guide to an expected line of action.

11.2.2 Shipboard safety communications

11.2.2.1 General emergency alarm signal

The general emergency alarm system should be capable of sounding the general alarm signal which consists of seven or more short blasts followed by one long blast on the ship's whistle or siren and additionally on an electrically operated bell or klaxon or other equivalent warning system. These are powered from the ship's mains supply and the emergency source of electrical power required by the regulations.

The system should be capable of operation from the navigation bridge and, except for the ship's whistle, also from other strategic points. The system must be audible throughout all accommodation and normal crew working spaces and supplemented by a public address or other suitable communication system. An emergency system comprising of either fixed or portable equipment or both, shall be provided for two way communications between emergency control stations, muster and embarkation stations and strategic positions on board.

11.2.3 Watch keeping duties

11.2.3.1 Keeping a watch in port

It is essential that a level of watch keeping duties is maintained while the vessel is in port. That level will be dependent upon the amount of activity taking place on board. For example there may be ongoing cargo working operations or the taking in of stores, bunkers or water in progress. Alternatively the vessel may be lying alongside with no work in progress.

11.2.3.2 Cargo work in operation

It is normal practice during cargo working operations to have at least one deck officer on deck watch to conduct the loading or discharging of the cargo in a proper manner and be aware of all safety precautions relating to it. If the operation is particularly complex or hazardous, then additional officers may be used.

During the cargo operations, the officer in charge of the deck must consider the moorings of the vessel, the position of gangways, fire precautions, access by unauthorized personnel and the keeping of the chief officer's log book of all events taking place. The officer of the deck is responsible for all operations being performed by crew and shore personnel and must ensure that everything is carried out in a safe manner.

11.2.3.3 Taking on stores, bunkers or water

During these operations the duty officer must supervise activities which are usually carried out by other people in relevant departments. Records must be kept of start and finish times, quantities loaded and/or discharged together with any problems or accidents. Particular care must be taken during bunkering operations to avoid any spillage. The officer must ensure that appropriate signals are displayed and safety procedures employed.

11.2.3.4 Lying alongside with no work in progress

During this period an officer of the watch (OOW) is appointed. The OOW is responsible for ensuring that a 24 h gangway watch by crew or watchman is maintained; making frequent fire patrol checks; ensuring security at all times by frequent checks and rounds of the vessel; and maintaining adjustments to moorings and gangways.

11.2.4 Vessels lying alongside one another

11.2.4.1 Safe access

Gangway access is usually provided between ships lying alongside one another. The higher ship would normally provide the gangway. Whilst it is in position, the following precautions should be taken: adequate lighting to be provided by both vessels during the hours of darkness; a gangway net should be in position; a lifebuoy and line should be readily available at the foot of the gangway, which must be tended if either vessel is working cargo, taking bunkers or water or working ballast.

11.2.4.2 Fenders

It is customary for the vessel which is coming alongside another to provide the fenders. These fenders should be positioned before drawing alongside the other vessel.

11.2.4.3 Etiquette when crossing an inside vessel

Crew members crossing over other vessels should respect the privacy and quiet of the host ship. This is particularly so when coming back from ashore late at night. It is important to observe safety notices and regulations.

11.2.4.4 Moorings

When putting out moorings from the outside vessel, it is an advantage if some of the lines can be run ashore. Springs must be run to the other vessel. It is always advisable for the inside vessel to put out additional moorings before the arrival of the second vessel.

11.2.5 Navigation safety – passage planning

11.2.5.1 Principles of passage planning

In the interest of safety at sea, every vessel should establish a passage plan prior to departure from a port. The passage plan should provide detailed information reflecting the forthcoming voyage from berth to berth. The completed passage plan should be constructed in four sections: appraisal, planning, execution, and monitoring.

11.2.5.2 Appraisal

Appraisal is the gathering of all information from all available sources including up to date charts, pilot books and ocean passage books. Additional information concerning local weather, tidal data and any restrictions due to the vessels draught, engines manning levels and equipment should also be considered.

11.2.5.3 Planning

Having completed the fullest possible appraisal using all of the available information on board the vessel, the navigating officer should now act upon the Master's instructions to prepare the detailed plan. This should include the plotting of the intended passage on the chart with due regard to navigational hazards. Particular attention should be made to radar conspicuous objects, ramarks or racons which may assist with position fixing. Key elements of the plan should be highlighted, such as course alteration points, safe speed and speed alterations to achieve desired estimated times of arrival (ETAs), minimum under keel clearances in critical areas, primary and secondary methods of position fixing in critical areas, positions where a change of machinery status is required and where allowances should be made for manoeuvring characteristics, heel effect and squat, need to be taken into account. Additional relevant information should also be noted, i.e. times of sunset and sunrise, and times and heights of high and low waters.

The planning stage will probably not cover every eventuality of the voyage. Much of what is planned may have to be changed after embarking pilots or due to weather restrictions. In any event the real value of the plan should not be lost, the fact that it marks in advance where the ship should not go and highlights the precautions to be taken en route, to provide initial warning of the ship standing into danger.

11.2.5.4 Execution

Once the planning of the intended passage has been finalized, the method and tactics to execute the plan should be decided upon. The following factors should be taken into account when entering the execution stage of the passage plan:

1. The condition and reliability of ship's equipment.
2. The ETAs at critical points for high and low water levels.
3. The meteorological conditions known to frequently exist in particular areas, especially those renowned for poor visibility.
4. The relevance of daytime or night time passing of danger points and the effect this might have on position fixing.
5. Anticipated traffic conditions expected at navigational focal points.

The Master may after due consideration of the plan include special features, such as increasing personnel at specific points to ensure that possible hazards and prevailing conditions can be adequately negotiated.

11.2.5.5 Monitoring

Once the execution of the passage plan is in operation, the close and continuous monitoring of the ship's progress must be established to effect a safe passage. To this end, full use of all ship's equipment should be made and carefully checked prior to sailing and at regular intervals throughout the passage. Every fix should if possible be based on three position lines and estimated positions should be projected and plotted at a convenient interval of time after a known position is established.

11.2.6 Navigational safety – prior to sailing

11.2.6.1 Navigational equipment check list

One of the most important check lists employed on board a vessel is the navigational check list. It is customary to perform the check one hour before sailing and the fact that the checks have been made should be reported to the Master and recorded in the deck log book. Modern ships have increased levels of equipment and the use of the check list is essential to ensure that all systems are seen to function correctly and that nothing is omitted. A typical list should contain the following details together with any special requirements for any particular ship:

Navigation lights (main and emergency) and spares
Radio equipment (VHF and mains set)
Telephone systems
Portable radio communicators
Whistles, sirens (main and emergency and automatic fog)
Clear view screens and window wipers
Steering gear
Automatic pilot system
Rudder indicators
Tachometer
Clocks and chronometers
Telegraph or bridge control operations
Smoke detector system

Master gyro compass together with all repeater alignments
Azimuths rings fitted as required
Magnetic compass inspected and comparisons made
Gyro error calculated
Radar equipment operational and performance checked
Echo sounder
Panel and indicator displays inspected for illumination
Relevant charts and publications available to meet requirements of the passage plan
Direction finder inspected
All other navigational equipment observed and noted as operational
Flag locker inspected

11.2.7 Keeping a safe navigational watch

Attention is drawn to the relevant Merchant Shipping Notices M1040, M1158, M1102, M854 and the IMO Recommendations on Navigational Watchkeeping.

Standing and maintaining an efficient safe navigational watch is one of the most important duties that the deck officer will be asked to perform. Diligence in these duties should never be relaxed and the Master's standing orders are to be observed at all times. Different levels of activity will be required, depending upon the prevailing circumstances including: in pilotage waters; in busy and/or confined waters, during deep sea passages; in heavy weather and in restricted visibility.

11.2.7.1 Pilotage

It is important that bridge personnel should work as a team, particularly during a pilotage situation. A combination of Pilot, Master, officer of the watch (OOW), helmsman and lookouts, if all are working together provide maximum efficiency and the safest watch keeping method. Preplanning (M 854) should be carried out and a full and frank exchange of information passed between the team. Final responsibility of course rests with the Master and delegated representative, the officer of the watch.

11.2.7.2 Busy and confined waters

The vessel may frequently be navigating in confined or busy waters without a pilot and the team is subsequently reduced. The same principles apply. Preplanning, with full use being made of all navigational equipment and freedom of information to all personnel on the bridge is considered essential.

11.2.7.3 During deep sea passage

Bridge manning during a long deep sea passage is often reduced to one or two bridge watch keepers. Great care must be taken that a proper and effective lookout is maintained at all times. The officer of the watch, should also have the additional facility of calling on extra personnel for any reason which could affect the safe navigation of the vessel.

11.2.7.4 In heavy weather

Normal watch keeping duties should be performed with special attention to any movement of cargo, motion in the seaway that may cause damage to the ships hull and the safety of ship's personnel in their movements about the vessel.

11.2.7.5 In restricted visibility

Restricted visibility means any condition in which visibility is restricted by fog, mist, falling snow, heavy rainstorms, sandstorms or any other similar cause.

Rule 3, (1) of the Regulations for the Preventing Collisions at Sea 1972 (as amended by resolution A464 XII) states

> 'Company regulations and the Master's standing orders will dictate the manning levels for the keeping of a safe navigational watch during these periods of restricted visibility. The doubling of watches would not be unusual to ensure that an effective lookout was maintained by all available means including by sight, hearing and radar observation.'

11.2.8 Conducting a safe anchor watch

The anchored position of the vessel will need to be frequently ascertained to ensure that external influences are not causing the ship to drag its anchor and so move from its designated position to one of uncertainty. The anchored position of the vessel should be checked by a combination of visual anchor or transit bearings, radar distances or radar bearings or combined Decca readings or ARPA automatic radar plotting aid.

11.2.8.1 Visual anchor bearings

Suitable shoreside objects should be selected for taking bearings during the day and night. The ship's position should be plotted on the chart and a note made of the bearing of each object. Objects which provide a large acute angle should be selected to give a distinctive position fix.

11.2.8.2 Transit bearings

Objects in transit often provide a rapid visual check for the OOW. However, the use of transits alone should not be relied upon, but only used as an additional indication of the ship's position.

11.2.8.3 Radar distances

The use of three radar ranges from well separated and identified marks can be usefully employed to define the ship's position. The use of radar should be employed with an alternative system of position fixing.

11.2.8.4 Radar bearings

Radar bearings can be used to fix the vessel's position when at anchor, but due regard to the inaccuracies in radar bearings should be borne in mind and they should be used with caution.

11.2.8.5 Radar range and bearing

The combined use of range and bearing from an identified object can provide indication of the ship's position. However, the observer should be aware of the limitations of the equipment being used and employ an alternative method of comparison when position fixing.

11.2.8.6 Decca readings

Observations of Decca readings should be maintained. These can be expected to change slightly over a 24 h period due to variable errors. Some Decca receivers have an anchor watch facility and, provided that the variable error throughout the period is reasonable, a good indication of the vessel dragging its anchor would be observed.

11.2.86 ARPA watch

Vessels fitted with an ARPA can use the anchor watch facility to great advantage. Attention must be paid to the information which will be set into the equipment to monitor the amount of movement of the vessel.

11.2.8.7 Procedure in restricted visibility

If a vessel is anchored in restricted visibility, the following points should be borne in mind:

1. The international regulations in respect of fog signals should be strictly observed.
2. Extra lookouts should be employed.
3. Greater than usual attention to radar information for anti-collision data and for position monitoring should be expected procedure.

11.2.9 Mooring the vessel by use of anchors

11.2.9.1 The principle of anchoring

It is the Master who will decide to anchor the vessel. This may require it to wait for any period of time from 20 minutes to several weeks. The vessel should always be anchored safely. Bearing this in mind, certain factors must always be considered prior to letting go an anchor: the depth of water beneath the vessel; the type of holding ground that the anchor will have to dig into; the expected weather conditions; the total amount of cable to use; the type of mooring (single or two anchors); the sea room available and the length of time that the vessel will be anchored.

11.2.9.2 Use of a single anchor

A single anchor would be used where the time at the anchorage will be small. This assumes that the weather, holding conditions, and so on, do not make it necessary to use a second anchor. The minumum length of cable normally used for anchoring will be four times the depth of water. This is to ensure that up to 50% of the cable lies along the sea bed so that the pull on the anchor will be horizontal. A non-horizontal lead to an anchor will cause loss of hold power (see *Figure 11.1*). If strong wind or tidal conditions exist, more cable may be needed to provide the horizontal lead in conditions of increased tension on the cable.

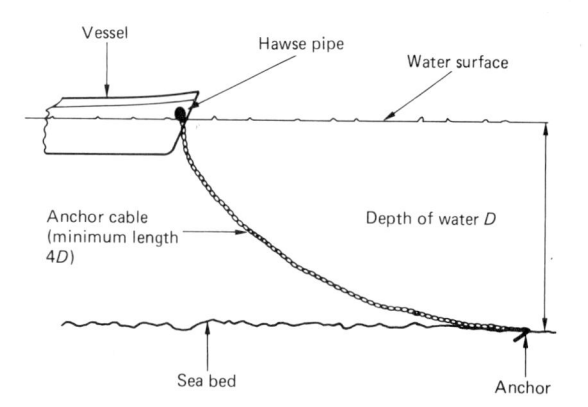

Figure 11.1 Vessel riding to a single anchor

11.2.9.3 Preparing an anchor for letting go

Once power has been obtained on deck and the windlass has been oiled and checked, the anchors must be made ready to let go. This operation must be carried out carefully and systematically to ensure that the operation will run smoothly. This is very important because if a proper routine is established when time is not limited, when an emergency occurs, the anchoring procedure is more likely to go smoothly and quickly.

Once deck power is obtained, the following operations are carried out:

1. Put the windlass into gear.
2. Check that the windlass brake is on and holding.
3. Remove the hawse pipe cover.
4. Remove the devil's claw.
5. Remove any additional lashings.
6. Remove guillotine/compressor bar.

7. Take off the brake and walk back the cable to break the cement pudding inside the spurling pipe entrance.
8. Clear away the old cement.
9. Walk the anchor out of the hawse pipe down to the water.
10. Screw the brake on hard and check that it is holding.
11. Take the windlass out of gear and report to the bridge that the anchor is ready to let go.

11.2.9.4 Safe anchorage

While the anchors are being prepared for use, the Master will be identifying the position where the anchor is to be dropped. The vessel will then be manoeuvred into such a position as to be able to let go the anchor. Once the anchor is ready for dropping, the Master will be informed. The vessel will then be allowed to gather sternway by use of the engines and tide, if any. If the vessel is heading into a strong tide sternway may mean that the vessel steams slowly ahead into the tide yet moves backwards over the sea bed. It is this moment which is the critical factor. In conditions of no tide, the wake of the propeller passing forward along the vessel's side is a good indication of sternway through the water.

Motion of the vessel through the water will be needed to maintain effective steering. On command, the brake is removed from the windlass and the anchor allowed to go. As the vessel drops astern of the anchor, the required amount of cable is paid out in a straight line and the brake will then be firmly applied. The backwards motion of the vessel will be slowed by forward applications of engines if necessary and the cable will draw taut, then bend again as the ship draws to the anchor. The officer in charge of the anchor party will then know that the vessel is brought up and that the anchor is holding. If the anchor fails to hold, three courses of action are: more cable will be paid out; a second anchor is let go; or a better location is used.

11.2.9.5 Heaving up the anchor and securing for sea

The actual decision to use the engines will again depend upon the tidal conditions. In slack water, the windlass alone will probably be used to heave in the anchor cable. The recovery procedure is simply the reverse process to letting go, the securing arrangements being applied and the spurling pipes blanked off. A vessel should never put to sea with open spurling pipes otherwise flooding of the fore end of the vessel may occur. This will adversely affect the trim of the ship and cause loss of stability, a potentially fatal situation.

11.2.9.6 Emergency anchoring

In confined waters, it may be necessary to let an anchor go in an emergency. In most cases, the anchors will already be in a state of readiness. If this is the case, then there may be time to walk out the anchor clear of the hawse, but at other times it could be necessary simply to let it go from the stowed position. Much will depend on the prevailing conditions as to the correct sequence of events to effect an emergency anchoring. If it is a tide ahead, it may possibly be a question of stopping engines, then going astern in order to stop the vessel. As the tide moves the vessel astern, then let go the anchor and veer (pay out) cable to the appropriate amount.

If there is a tide astern, then the engines should be stopped, put astern if possible and cant the vessel by putting the helm hard over. Let go the appropriate anchor on a short stay, i.e. the anchor should be on the bottom with a short length of cable. This will hold the bow, while the tide swings the stern. As the ship is nearing completion of the turn, it may be necessary to put engines at slow ahead to ease the tension and weight on the cable. Once the turn is complete, veer the cable to the required amount, stopping engines if appropriate (see *Figure 11.2*).

Due to strong wind or current, there could be occasion to let both anchors go to avoid the vessel being set down onto a lee shore, appropriate if the ship has lost power or suffered a steering gear failure. Two anchors would provide greater holding power against the elements and effectively increase or eliminate the approach time towards the danger.

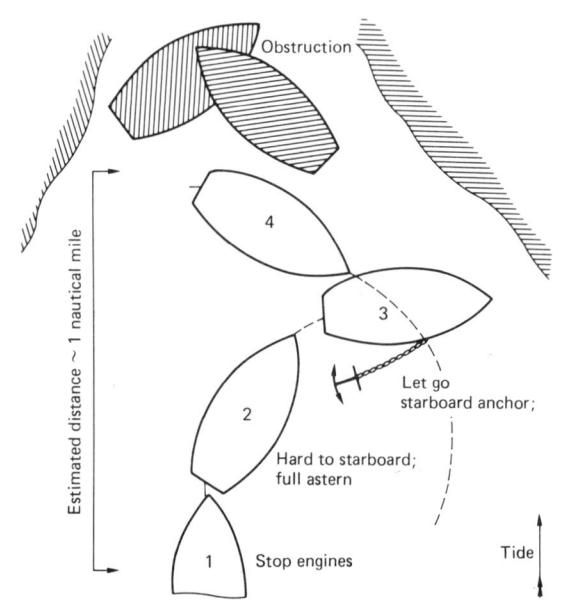

Figure 11.2 Emergency anchoring: tide astern

11.3 Operational deck safety

11.3.1 Safe access to and from the vessel

It is the responsibility of the owners of the vessel to provide safe and suitable means of access to the ship. However, it is the Master's responsibility to ensure that the rigging of gangways and access ladders is carried out safely, in order to provide safe methods of authorized boarding or disembarking from the ships.

The gangway or accommodation ladder, as it has more commonly become known, is manufactured to meet the requirements of both the Factory and Workshops Act and the Docks Regulations (see *Figure 11.3*). When rigged, they should be properly secured and fenced on either side by means of upper and lower rails, ropes or chains.

Figure 11.3 Safe access: Department of Trade style of gangway; stanchions and manropes rigged and gangway net in position

All persons using an accommodation ladder should exercise caution when boarding or disembarking from the ship by this means. A lifebuoy, with self activating light and line attached should be rigged and kept ready for use close to the access point. Should the access ladder and the approaches to the ladder be in use during the hours of darkness, the operational area of the gangway should be well illuminated.

Gangways should be kept within specified limits and should not be used at an angle of inclination greater than 30° from the horizontal unless it is specifically designed for such. Accommodation ladders would not normally be operated at a greater angle of 55° below the horizontal. Suitable adjustments should be carried out to the gangway to take account of tidal movement, to ensure that the gangway remains within the specified limits and continues to provide a safe means of access.

Prior to rigging, gangways and their associated fittings should be inspected for visible defects. Any inadequacies should be reported to a responsible officer to allow corrective measures to be taken. A continual assessment of the rig should be made while in use and supporting equipment such as guard ropes or chains should be kept taut. A safety net should be rigged between the ship's side and the quayside to arrest the fall of a person slipping off the accommodation ladder.

11.3.1.1 Pilot ladder – construction

This construction is applicable to a Class VII vessel (see *Figure 11.4*). The side rope of the pilot ladder will be of manilla rope with a diameter of 1.8 cm (0.6 inch). Each rope must be continuous and be left uncovered. The steps of the ladder should be of ash, oak, elm, teak or other equivalent hard wood. Each step must be made from a piece of wood which is free of knots. A non-slip surface must also be applied. Each step must be not less than 48 cm (1.6 ft) long, 11.5 cm (0.38 ft) wide and 2.5 cm (1 inch) in depth. Steps must be placed not less than 30 cm (1 ft) or more than 38 cm (1.26 ft) apart. They must be secured in such a manner that they remain horizontal. The four lower steps may be made of rubber of sufficient strength and stiffness or of other suitable material of the same character.

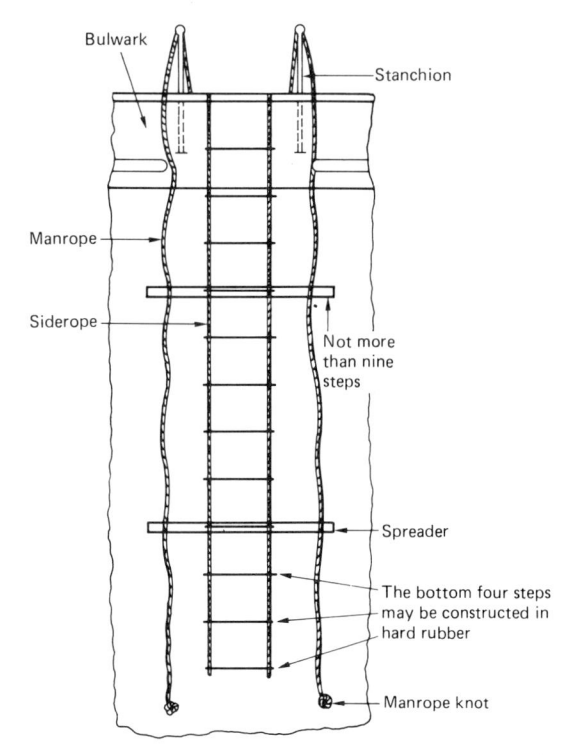

Figure 11.4 Rigging of pilot ladders (courtesy of Blackpool and the Fylde College, Nautical Campus Fleetwood)

The man ropes must be of not less than 20 cm (0.8 inch) diameter and have to be secured to the ship. The hard wood battens have to be between 180 and 200 cm (6 and 6.67 ft) long and need to be placed at such intervals as will prevent the ladder from twisting. This means that they are usually fixed so that the lower batten is not lower than the fifth step from the bottom and that the interval between battens is not more than nine steps. Each batten should be constructed of ash, oak or similar hard wood and be free from knots like the steps. No pilot ladder is allowed to have more than two replacement steps secured in a different manner from the original method of securing. These must be secured in place by the original method as soon as possible.

11.3.1.2 Additional points

Once the pilot ladder has been rigged securely, a light should be provided during the hours of darkness. At all times, a lifebelt and rescue line should be readily available. The lifebelt should be equipped with a self igniting light. A responsible officer should inspect and check the rigging of the pilot ladder prior to its use. The pilot ladder is normally secured forward of the bridge from the uppermost continuous deck. Many modern vessels will have the bulwark or rails cut away to form an access gate. This gateway will normally be closed by a portable rail or hinged door. Should a purpose-made access not be available to the pilot, the bulwark ladder should be employed.

11.3.1.3 Mechanical pilot hoists

Many modern ships, especially large tankers, are equipped with a powered appliance for hoisting pilots on board. The source of power used may be either electrical, hydraulic or pneumatic.

The hoists will normally comprise:

1. Two separate wire rope falls made of flexible steel wire rope of adequate strength and treated so as to be resistant to corrosion.
2. The design of the winch will include a brake system together with a drive system, such as a worm drive, capable of supporting the working load. Winch controls should be clearly marked Hoist, Stop or Lower. An Emergency Stop to cut off power is needed. A drum is incorporated to accommodate the wire falls.
3. Crank handles must be provided for manual operation. They are fitted in such a manner that once in position and interlocked, the source of power is cut off.
4. Automatic safety cut outs must be built into the system so that, should the hoist come against any obstruction, the power is cut to avoid overstress on the falls.
5. The ladder will be raised by the hoist when the pilot is ready. It will consist of a rigid section for the transport of the pilot and a boarding ladder to allow the pilot to reach the rigid section.

11.3.1.4 Operation aspects

The rigging and testing of the hoist should be supervised by a responsible officer. Personnel engaged in the actual operation of embarking or disembarking pilots should be correctly instructed in correct hoist operaprior to its use. The hoist must be tested prior to each use and a pilot ladder should be rigged adjacent to the hoist so that it is available for immediate use. A stowage position away from the weather and the dangers of ice formation should be available when the hoist is not in use. At night, a light should be provided to illuminate the hoist and the point of boarding. The position of the hoist control should also be adequately lit. On boarding the hoist, there should be facilities for adequate communication between pilot, the hoist operator and the officer in charge of the embarkation. The pilot should have means to activate an emergency stop. This control should be within easy reach of the pilot when he is on the hoist. See also M Notice 898 for further details.

11.3.2 Safe practice with derricks and cranes

11.3.2.1 Safe handling practice of derricks

All derrick rigging should be regularly maintained under a planned maintenance programme and, in any event, prior to use of the rig all rigging should be visually inspected for any defect (see *Figure 11.5*). Before a derrick is to be raised, lowered or adjusted with a topping lift span tackle, the hauling part of the topping lift should be flaked down the deck clear of the operational area, all persons being warned of the operation and to stand clear of bights in the wire. When topping lifts are secured to cleats, bitts or stag horns, three complete turns should be taken before the additional four cross turns on top. A light lashing should be placed about the top two or three turns to prevent the wire from jumping adrift due to the natural springiness of the wire.

When the rig of a derrick is to be changed or altered in any way, as with doubling up, then the derrick head should be lowered to the crutch or to deck level in order to safely carry out alterations. Winch drivers should take instructions from a single controller who would pass orders from a place of safety from which a clear and complete view of the operation must be available. When derricks are being raised or lowered, winch drivers should operate winches at a speed consistent with the safe handling of the guys.

Cargo runners should be secured to winch barrels by use of a U Bolt or proper clamp and when fully extended a minimum of three turns should remain on

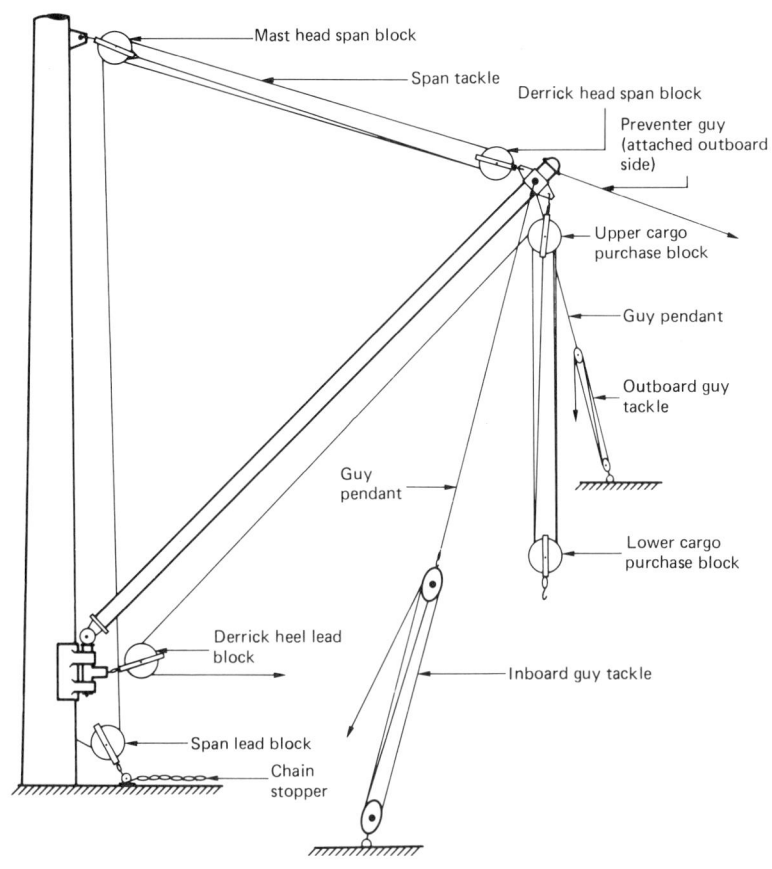

Figure 11.5 Typical purchase span derrick (courtesy of Blackpool and the Fylde College, Nautical Campus Fleetwood)

the barrel of the winch. Should it be necessary to drag heavy cargo from under tween decks, the runner should be used direct from the heel block via snatch blocks to avoid placing undue overload on the derrick boom.

11.3.2.2 Union purchase

This system is used for the rapid handling of light loads, up to 2 tonnes. Pairs of derricks are rigged to move parcels of cargo. One derrick is fixed in position over the hatch and the second over the quayside. Neither derrick will be moved at all during cargowork, being fixed by guys and a preventer (see *Figure 11.6*).

Unloading by union purchase: sequence of operations:

1. The load is lifted out of the hold using the first derrick; the second runner of derrick two is slack.
2. The second derrick is then used to haul the suspended load across the deck while the operator of derrick one allows that runner to slacken as required.
3. When all the weight has been transferred to derrick two, the load will be over the quayside and can be lowered.

Loading the vessel follows the reverse procedure.

11.3.2.3 The use of cranes

The crane, although being part of standard port or harbour equipment, has been incorporated on board the modern cargo vessel with extremely successful results. Not only is the crane a labour saving device, needing only one driver, but the manoeuvrability of the cargo hoist is much more versatile than derrick handling. Most shipboard cranes may be fitted to swing through 360° but for the purpose of safe handling, limit switches often act as cut outs to avoid the jib of the crane fouling obstructions. Limit switches are also fitted to the luffing operation of the jib, as well as the cargo hoist wire, to prevent off setting of the jib boom, and the cargo hook fouling the upper sheaves of the hoist.

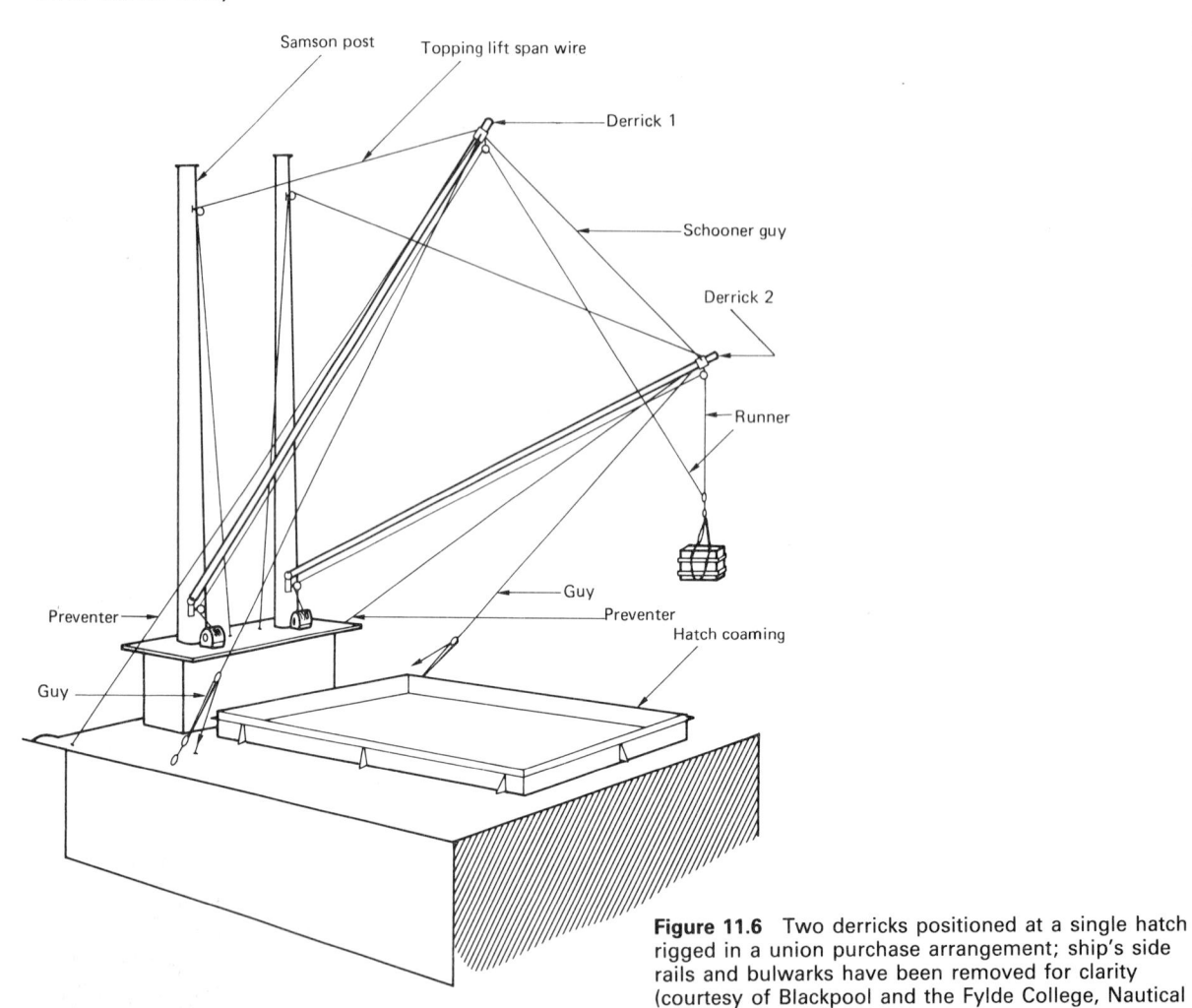

Figure 11.6 Two derricks positioned at a single hatch rigged in a union purchase arrangement; ship's side rails and bulwarks have been removed for clarity (courtesy of Blackpool and the Fylde College, Nautical Campus Fleetwood)

All cranes are provided with individual motors to permit luffing, slewing and cargo hoist operation. They can operate against an adverse list of approximately 5° together with a trim of 2°. When twin cranes are fitted, these may operate independently or may be synchronized to work under one driver from a master cabin. *Figure 11.7* illustrates the use of hand signals for deck crew working with cranes.

The seafarer should be aware that there are many types of cranes on the commercial market and their designs vary with customer requirements. As a general rule, the majority of cranes confirm to the following design:

Machinery platform – accommodates the dc generator, gear boxes for luff, slew and hoist operations. Also houses the slewing ring, together with the jib foot pins.

Driver's cabin – integral with the crane structure. Welded steel construction with Perspex windows. Front window to open. Internal lighting.

Jib – of a welded steel construction. Supporting upper sheaves for topping lift and cargo hoist.

Sheaves – mounted in friction resistant bearings.

Topping lift and hoist – galvanized steel wire ropes having a minimum breaking strain of 180 kgf/m^2. Depending on the Safer Working Load (SWL) of the crane this BS could be greatly increased.

11.3.2.4 Testing

All equipment will be tested to obtain the SWL value. Usually test lifts will be performed using water tanks or

Figure 11.7 Visual communication for deck operations; a code of hand signals when working winches, cranes or derricks (courtesy of Blackpool and the Fylde College, Nautical Campus Fleetwood)

spring balances. The testing will be in excess of the SWL figure to be assigned. This test will be to a special proof load according to the table below.

Table 11.1 Calculation of SWL

Designated SWL	Proof load
up to 20 tonnes	SWL plus 25%
20–50 tonnes	SWL plus 5 tonnes
over 50 tonnes	SWL plus 10%

Because of the possibility of equipment failure under proof conditions, extra care will be taken.

With cranes which have a variable radius, loads will be hoisted and swung out to the maximum radius. Proof value for maximum and minimum radius will be used during the test. If a crane is hydraulic, the test will also involve the lifting of the maximum load in excess of the SWL that can be achieved.

In any work with cranes or derricks you must follow all safety procedures. These can be found in various publications, including the Code of Safe Working Practice, Dock Regulations and M Notices.

11.3.3 General advice on working aloft and overside

In this section reference should be made to the Code of Safe Working Practices for Merchant Seaman. Proper precautions should be taken for the safety of personnel who are detailed to work at heights where the operator cannot or may not be able to give full attention to guard against falling. This is especially important aboard a vessel at sea, where movement of the ship cannot be anticipated when in a seaway. In any event where a risk of falling more than 2 m is present, a safety harness and lifeline should be worn by the operator.

Young seafarers under the age of 18 or persons with less than 12 months' sea time, should not be detailed to work aloft unless adequately supervised by an experienced seaman. No person should be expected to work overside while the vessel is under way. Neither should work on stages or bosun's chairs take place in the vicinity of cargo operations, unless that work is of an essential nature.

Where work is scheduled to take place on the radar scanner, ship's whistle, radio aerials or funnel, the appropriate officer should be informed, as well as the duty watch officer, the appropriate precautions being taken to ensure the safety of the operator prior to commencing the work. Once completed, the duty officer should be informed, to return the site to its normal working condition.

When overheard work is being carried out, extreme care should be taken to avoid risk to persons below. Tools should be taken aloft by means of a tool belt or by use of a line and suitable container. These tools should not be left exposed so as to be liable to be accidentally dislodged on to persons below. The handling of such tools should not be carried out with cold or greasy hands, or where the tools themselves are greasy.

Whenever such work is being carried out, suitable warning and display notices should be set up in the immediate vicinity. These notices should cover all appropriate access and exit points and highlight any area of potential risk.

11.3.3.1 Safe working aloft or overside

11.3.3.1.1 Precautions when rigging working stages
Wooden stages should be stowed in a dry, well ventilated space and should be load tested prior to use to four times the weight for which they are intended. Plank stages should be carefully examined before use. They should be free from defect and be clean and clear of grease. Additional equipment, such as lizards, gantlines and blocks should be inspected prior to use and be seen to have no visible defect. The anchor points for the securing of stages should be of adequate strength and where practical, be permanent fixtures of the ship's structure. Stages should be rigged over water.

Gantlines which secure stages should be of such a length as to provide an additional lifeline to the operator, should he or she fall into the water. The rigging of gantlines should be clear of sharp leads and edges. Portable rails and stanchions should not be employed as anchoring points for the rigging of stages. Persons working from stages should be equipped with a lifeline and safety harness. A lifebuoy and line should be kept ready for immediate use in the event of an operator falling into the water. A side ladder should be rigged close to stages to provide safe access to and from the operating level. A stand-by person should tend the staging to raise the alarm in the event of operational problems. Hanging stages should be restricted in there movement by means of bowsing lines when practical.

Correct lowering hitches should be secured, which will allow the movement of the stage to be carried out in a controlled manner. The bowlines, set into gantlines, should be of an adequate height so as to provide stability to the stage. Stages should not be rigged overside while the vessel is under way at sea. Stages should not be worked in or around a cargo working area. If stages are fitted with guard rails or fencing, then this protection should extend to 1 m from the floor. Where operators are required to carry out movement to the stage, any such move should be small and made in a controlled manner.

11.3.3.1.2 Precautions for the use of the bosun's chair
Prior to use, the bosun's chair should be inspected for any defect. The condition of the rope bridle should be carefully examined and seen to be in good condition. Again prior to use, the chair should be load tested to four times the weight which is intended.

The operator of the chair should wear a lifeline and safety harness. A winch should never be used to hoist the occupied chair. Motive power to the chair should be provided by hand only.

When the chair is to be used in conjunction with a gantline, the latter should be inspected and seen to be in good condition, it should be secured to the chair by means of a double sheetbend, and the associated equipment lizard or tail block should be inspected and load tested prior to use.

Where a lowering hitch is to be used, both parts of the gantline should be securely strapped together in order to secure the chair, prior to completing the lowering hitch. It is considered unsafe practice to hold on with one hand while at the same time securing the lowering hitch.

In general, hooks should not be employed with bosun's chairs, unless, because of their construction, they cannot be accidentally dislodged.

11.3.3.1.3 When riding a stay
A bow type shackle should be used to secure the bridle of the chair to the stay. The bow of the shackle should be over the stay, not the bolt. The bolt of the shackle should be moused with seizing wire.

11.4 Safety in cargo operations

11.4.1 Safe shipment and handling – general cargo

11.4.1.1 Some cargo considerations

The chief officer will usually have to calculate all the information relevant to the loading of the vessel. The amount of bunkers, fresh water and stores carried will have to be included, as these also affect the departure draught, stability and trim. Some considerations are:

1. Cargo plan – the production of a cargo plan showing the correct stowage of the cargo to be loaded.
2. Dock labour – the ordering of cargo gangs who will handle the cargo.
3. Working method to determine whether working will be overside to lighters or onto the quay.
4. Special equipment – the determination of whether special gear will be needed for cargo handling.
5. Power requirements – the briefing of engineers so that power for winches and other equipment will be available when required.
6. Preparations – the preparation of hatches; the removal of lashings from deck cargo prior to discharge; the preparation of heavy lift derricks, if required; the rigging of other cargo handling equipment; the efficient lighting of holds and all safety procedures must be observed.
7. Crew briefing – the familiarization of crew and officers with all aspects of correct loading and discharge procedures.
8. Special provisions – the provision of special requirements, such as tally clerks and lock-up stow for special cargo.

Figures 11.8 and *11.9* illustrate the securing of preslung cargoes, while *Figures 11.10* and *11.11* deal with roll on–roll off unit loads.

Figure 11.9 Preslung cargoes on board an offshore supply vessel as it manoeuvres towards an operation position for the cranes of the installation

Figure 11.10 Roll on–roll off unit loads being secured by chains and loadbinders; club feet being anchored into flush star insert or star domes

Figure 11.8 Preslung pipe loads secured by chain lashings aboard an offshore supply vessel

Figure 11.11 Roll on–roll off unit load being manoeuvred over the stern door/ramp of an Irish sea ferry vessel. Once the unit is positioned inside the vehicle deck, the powered tug unit is detached from the container unit

11.4.2 Hold cleaning requirements

Sweep down the compartment and remove all traces of previous cargo. The hold may also require a salt water wash, followed by a fresh water water wash, to remove excessive dust or residues of previous cargo.

Clean out bilges; this will include bilge suctions, tween deck scuppers and drain wells. It is necessary to check that all the holes in rose boxes are clear and that the complete drainage system and pumps to bilges are working satisfactorily. If the previous cargo was a bulk cargo, ensure that the plugs at the bilge to deck angle have been removed to allow drainage. All the limber boards should be checked to ensure they are in good condition. If the bilges are contaminated by an odourous cargo residue, then on completion of cleaning, the hold should be sweetened with a chloride of lime wash. This coating will help prevent corrosion as well as acting as a disinfectant.

Examine the spar ceiling and sections should be replaced where necessary. Check the cement chocks in the tween decks are in place and test them if the next cargo is to be loaded in bulk. Remove them if the next cargo is of a general nature. The tween deck hatch boards should be in good condition and should fit well. If the vessel has wooden main hatch covers then these (and the tarpaulin cover) should be examined.

An examination of dunnage should take place and any soiled dunnage should be removed and new dunnage laid in the direction of the drainage. The fire smothering system and the fire detecting system in the hold should be checked. Ensure that guard rails are in position around any hatch openings. Ensure that the hold lighting is working and, if necessary, ensure that additional light is fitted. Make a final inspection of the hold to ensure that all is correct and ready for the new cargo.

11.4.3 Safe stowage of deck cargoes

Special attention must be given to the securing of all deck cargo. Due regard must be given to adequate timber under each parcel of cargo to prevent deck contact. Cargo on the fore deck will usually encounter more weather than that stowed on the after deck. Also check the height of the cargo. If the cargo has a high centre of gravity, then more lashings will be required. Protection for those areas of the cargo which come into direct contact with the securing wires is required and all bottle screws must be checked shortly after sailing and the tension adjusted accordingly. Reasonable access must be provided around the deck cargo to enable the crew to perform their normal duties. Any steps provided must have correct hand rails fitted. Fire hydrants and sounding pipes must be accessible.

11.4.4 Safe working practice with deep tanks

Deep tank preparation when a cargo of vegetable oil is to be carried should be as follows.

Sweep and collect any residue cargo or dunnage from the previous stow. If general cargo has been carried in the tank during the previous voyage, sweeping and clearing up the debris will be necessary. If the previous cargo was a liquid any residue will similarly have to be removed. This involves puddling the cargo towards the bilge well and collecting the residue liquid into drums, which can then be lifted clear of the tank.

The internal tank surfaces and steelwork must be cleaned. This task obviously varies with whatever cargo was carried previously and the standard of cleanliness required. For vegetable oils, this would be high, due to the substantial value and foodstuff nature of the cargo.

Protective clothing, eye shields or goggles may be necessary and the tank must be well ventilated before entry and throughout the working period. Take note of appropriate measures for entry to enclosed spaces. Solvents may be necessary for cleaning, particularly where tank coatings are employed. Bare steelwork may need to be wire brushed and scraped to remove loose corrosion.

Rig and test the heating coils. Most vegetable oils will need heating to maintain the oil in liquid form. Most heating coils are portable and rigged when they are required. Heating is achieved by passing low pressure steam through the steel tubes which form the coils. The steam valves are cracked open and the coils checked to ensure there is no leakage of steam from the coils into the tank which would damage the vegetable oil during the voyage.

Check the bilges. The bilge strums will need to be thoroughly cleaned and tested. Fill and pressure test the tank. A cargo surveyor may wish to see this routine procedure followed. Part of the cleaning procedure may be incorporated in this routine. For example, as the tank is nearly empty, chemical solvents may be added to the last few tonnes of water and with the aid of a submersible pump or tank washing machine the internal structure washed. The tank is emptied, drained and dried. Seal the bilges. The bilge line will not be required for the loaded voyage. The oil cargo is loaded and discharged through the tank top hatch. Seal other line systems. Fire detection, smotherline lines and ventilator trunkways are battened down when carrying vegetable oil cargoes. The oil vapours would clog all the apertures, causing considerable maintenance problems.

Rig thermometer tubes. The temperature of the cargo will need to be monitored throughout and the tube will allow a thermometer to be lowered or temperature reading devices installed at different levels over the depth of the tank. Rig ullage pipes. The normal practice is to measure the ullage, the measurement from the surface of the oil to the tank top. Rig tank lid relief valves. During the voyage, changes in temperature, together with the motion of the ship, can create liquid expansion and or vapour pressure levels which need to be relieved to avoid damage to the tank structure. On occasions, expansion pipes or trunks are incorporated with the relief value.

11.4.5 Ventilation of ship's cargoes

To stop ship's sweat, we have to ventilate efficiently in order to change all the air in the hold and replace the warm moist air with cooler drier air. This includes air between parcels of cargo and not just that near the hold sides.

11.4.5.1 Cargo sweat

This occurs when a cargo contains moisture which subsequently evaporates into the air of the hold. Cargo sweat can only be avoided by proper ventilation to carry the moisture away. It may also occur if the hold is ventilated at the wrong time, so that warm moist air enters a comparatively cool compartment. In this instance, the originally warm air from outside may be cooled below its dewpoint when it comes into contact with the cooler cargo and so condensation results. This type of sweat is a particular hazard in a vessel which had loaded in the UK in cool weather and proceeds to the Far East. If weather conditions prevent ventilation until warmer temperatures are experienced, the hold temperature will be much the same as when it was loaded. However, the outside temperature will be higher, so if the hold is ventilated, sweat could be caused due to the cooling of externally introduced air. To avoid this, we have to compare the hold temperature with the dewpoint of the outside air. If the dewpoint of the outside air is lower than the hold temperature then we can ventilate. If the dewpoint is higher, then it is not safe to ventilate as this will cause sweat.

11.4.6 Associated hazards with the shipment of bulk cargoes

The hazards associated with the carriage of bulk cargoes may be considered under three headings.

11.4.6.1 Structural damage

Many materials carried in bulk are of a low stowage factor (SF), therefore the cargo spaces of a ship are only partially filled when the ship is fully laden. If care were not taken, the cargo weight would be concentrated over limited sections of the ship's steel structure. This may promote structural failure, resulting in severe corrosion, or more serious fatigue cracks or brittle fractures.

Further consideration is that many ships carrying bulk cargoes are of considerable length and the longitudinal distribution of cargo must allow for, and maintain within acceptable limits, the overall stress levels in the ship's girder length.

11.4.6.2 Loss or reduction of stability

During the course of a voyage, some loss of stability may occur due to the consumption of fuel and water from double bottom (DB) tanks. Allowance should be made for this when planning the voyage. However, if stability is critical, exceptional circumstances could create a serious situation. This could include an unexpected shift in the cargo stow causing change of trim or, more significantly, a list. A large list or angle of heel reduces the stability on the listed side; the range of positive stability is less and the resistance to further heeling of the ship when working in a seaway is less.

A shift of cargo usually occurs during heavy weather. A shift in the stow may be influenced by improper distribution of the cargo or inadequate trimming. Other cargoes partially liquify as a result of the motion and vibration of the ship and this is related to the level of moisture (water) content in the stow.

11.4.6.3 Chemical properties

Some materials can cause harm with skin contact or exude toxic fumes or gases. Other materials may be flammable or possess corrosive properties or reactive characteristics.

11.4.6.4 Terminology related to the carriage of bulk cargoes

Concentrates are materials that have been derived from a natural ore by physical or chemical refinement or purification processes. They are usually in small granular or powder form. Moisture content is that percentage proportion of the total mass which is water, ice or other liquid. Moist migration is the movement of moisture contained in the bulk stow when, as a result of settling and consolidation in conjunction with the ship's motion, water is progressively displaced. Part or all of the bulk may develop a flow state, which occurs when a mass of granular material is saturated with liquid to such an extent that it loses its internal shear strength and behaves in a liquid form. Incompatible materials are those which may react dangerously when mixed and are subject to recommendation for segregation. Coal includes sized grades, small coal, coal duff, coal slurry and anthracite.

Trimming is a manual or mechanically achieved adjustment to the surface form or shape of a bulk stow in a cargo space. The angle of repose is the natural angle between the cone slope and the horizontal plane when bulk cargo is emptied onto this plane in ideal conditions. A nominal value, arrived at from experiments carried out with a tilting box, can be given for a given type of cargo. The angle of repose value is used as a means of registering the likelihood of a cargo shift during a voyage.

The transportable moisture limit (TML) of a cargo that may liquify is the maximum moisture content which is considered safe for carriage in ships other than those which, because of specialized fittings, may carry cargo with a moisture content above this limit. Cargoes having an angle of repose greater than 35° (see *Figure 11.12*) have a low stowage factor; they also have considerable mass or weight per unit volume. Generally they will only partially fill the cargo space in which they are loaded. By trimming, the peak pile height will be reduced, and the mass will be distributed over a larger area of the tank top.

Although these cargoes are not as likely to shift as readily as those with an angle less than 35°, the possibility still exists, particularly in small ships, with the vigorous rolling motion associated with the inevitable stiff condition of stability. In these circumstances the trimming should be taken further to reduce the cone slopes away from the hatch square.

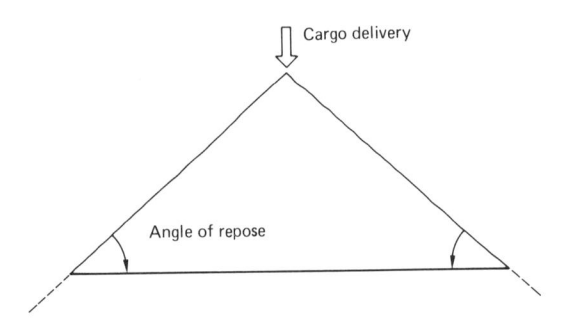

Figure 11.12 The 35° angle of repose, is used as an arbitrary line between bulk cargoes which have considerable weight, where care must be taken to avoid structural damage and, alternatively, those bulk cargoes where major concern is the lack of stability in the stow and a possible shift of cargo may occur during the voyage.

11.4.6.5 Bulk cargoes which may liquify

These include concentrates, certain coals and other materials generally consisting of small particles and of similar physical properties. Examples of such materials are shown in the copy from Appendix A of the IMO Code of Safe Working Practice for Bulk Cargoes.

The moisture content of such cargoes is a critical factor. Any bulk material with a moisture content above the TML must not be carried in an ordinary general cargo ship, since it can go into a state of partial liquification, aided by the compaction, the motion of the ship and the vibration during the voyage. The cargo may flow from one side of the ship to the other with the rolling motion of the vessel. This could cause a progressive roll motion where the ship may reach a dangerous angle of heel and be caused to capsize. It is essential to know the moisture content at the time of loading. Test certification should be sought and particular checks made if the circumstances show any reason to believe that the moisture content may be rising, e.g. if heavy rain has been experienced prior to or during loading.

Once loading has been completed or during the voyage, any moisture entering the cargo space would have the effect of raising the moisture content and taking the bulk materials towards the flow state. Naturally, the cargo hold would be secured against the ingress of water and the ventilation system should not allow access of moisture. Bilge lines would also be pumped at regular intervals.

11.4.6.6 IMO grain rules

Grain here means wheat, maize, barley, oats, rye, pulses or seeds whether processed or not which, carried in bulk, behaves like grain in that it is liable to shift transversely across a cargo space of a ship, subject to normal seagoing motion. The term 'filled', when applied to a cargo space means the space is filled and trimmed to feed as much grain into the space as possible, trimming under the decks and hatchcovers, etc., whilst the term 'partly filled' means the level of bulk cargo is less than this. The cargo should always be trimmed level with the ship upright. There may be a limit to the number of partly filled spaces that a vessel is permitted.

In order to carry grain, a ship must have a document of authorization or an appropriate exemption certificate. This means that the ship has been surveyed and proper grain loading information supplied to the ship for use by the deck officer responsible. The principles of the Grain Rules should be understood by all deck officers.

11.4.7 Recommended practice for the stowage of bulk cargoes

Modern ships usually carry comprehensive loading information and often mechanical or electronic calculator aids to assist the ship's officer towards safe load distributions when handling these low SF bulk cargoes. The IMO rules give advice which should be followed in the absence of loading information referred to above, this when dealing with bulk cargoes with a SF in the region of 0.56. The recommended practice is the fore and aft distribution of weight should not differ radically from that employed for a general cargo and there is a maximum number of tonnes which should not be exceeded when loading any one cargo space. This can be calculated from the empirical formula

$$0.9LBD* \qquad (11.1)$$

where L is the length of the hold (m), B is the average breadth of the hold (m) and D is the summer load draught (m)

Where cargo is untrimmed or only partially so, a maximum height for the pile peak above the cargo space floor can be determined from the equation:

* $1.1 \times D \times SF$ * SF is the stowage factors (m³/tonne) (11.2)

If the bulk material is trimmed entirely level the total mass or weight in a space can be 20% above the quantity found from Equation 11.1. Where a cargo space exists aft of the machinery space a 10% increase over Equation 11.1 is acceptable, because of the extra stiffening effect of the shaft tunnel.

When it is necessary to carry some of the high density cargo in a tween deck or other higher cargo space, care must be taken to ensure deck area is not overstressed or the stability of the ship as a whole not reduced to an unsatisfactory level, making proper allowance for the intended voyage. An empirical formula can be used to assess the maximum height of stow:

Height of stow = $\dfrac{SF}{1.39} \times$ (height of tween deck)(11.3)

For example, if a tween deck were 2.6 m in height and the bulk cargo had a SF of 0.6:

Recommended maximum height of stow –

$\dfrac{0.6}{1.39} \times 2.6$ (11.4)

= 1.12 m

11.4.8 International Maritime Dangerous Goods (IMDG) code

The IMDG code and its provisions are designed to assist in compliance with legal requirements of the International Convention for the Safety of Life at Sea currently in force, regarding the carriage of dangerous goods by sea. The code is also intended as the basis for individual requirements in order to provide harmonization between member countries engaged in such trade.

The IMDG code contains recommendations relevant to the carriage of dangerous substances. It specifies, in detail, the stowage, packing and labelling of dangerous cargoes, the precautions and the documentation which is required when the cargo is presented for shipment; the correct technical name, its UN number and the class of substance to which it belongs are also included.

The code was approved by the Maritime Safety Committee in 1965. Since this time, amendment and updating have taken place and the code now consists of five volumes containing the relevant information and the agreed recommendations.

11.4.8.1 The Blue Book

The Blue Book is a report of the Standing Committee in 1978 on the carriage of dangerous goods in ships, carried out on behalf of the Department of Trade. It contains relevant recommendations and reiterates Statutory Instrument No. 1747 of 1981, The Merchant Shipping (Dangerous Goods) Regulations.

The basic recommendations contained in the Blue Book are those of the IMDG code, but it does differ slightly from the IMDG code, and where this occurs, then the recommendations of the Blue Book for individual schedules take precedence over the corresponding entries in the IMDG code for British ships.

11.4.8.2 Conclusion

The carriage of dangerous goods involves the use of the IMDG Code, the Blue Book, the IMO Code of Safe Practices for bulk cargoes, the Code of Safe Working Practice for merchant seaman and the relevant M notices, plus other publications available at the time. These, together with information from the shippers and their agents should enable dangerous goods to be carried in a safe manner on board ship. In the event of dangerous goods being presented for shipment without the necessary documentation or relevant information as to its correct carriage, then it should be refused for shipment. This will very quickly achieve a response either to get the information or cause questions to be asked before it is allowed to be loaded.

11.4.9 The recommendations

11.4.9.1 Classification

Dangerous goods must be classified according to their properties and in the event of a particular substance having properties fitting it for inclusion in more than one class, then it is placed in the class appropriate to the most dangerous property when carried in ships.

Class 1: explosives are divided into five divisions: substances and articles which have a mass explosion hazard and which have a projection hazard but not a mass explosion hazard; substances and articles which present no significant hazard; and very insensitive substances which have a mass explosion hazard.

Class 2: gases are compressed, liquified or dissolved under pressure, and subdivided for segregation and stowage purposes into inflammable gases, non-inflammable gases and poisonous gases.

Class 3: inflammable liquids are subdivided further: a low flashpoint group of liquids having flashpoints below −18°C, closed cup test; an intermediate flashpoint group of liquids having a flashpoint of −18°C up to, but not including 23°C, closed cup test; and a high flashpoint group of liquids having a flashpoint of 23°C up to and including 61°C, closed cup test.

Class 4: inflammable solids or substances are subdivided further. Inflammable solids are those possessing the properties of being easily ignited by external sources, such as sparks and flames and of being readily combustible or of being liable to cause or contribute to fire through friction. Substances liable to spontaneous combustion are either solids or liquids possessing the common property of being liable spontaneously to heat and to ignite. Substances emitting flammable gases when wet are either solids or liquids possessing the common property, when in contact with water, of evolving inflammable gases. In some cases gases are liable to spontaneous ignition.

Class 5: oxidizing substances (agents) and organic peroxides are subdivided further into oxidizing substances (agents). These are substances which, although in themselves are not necessarily combustible, may, either by yielding oxygen or similar processes, increase the risk and intensity of fire in other materials with which they come into contact. Organic peroxides are organic substances which contain the bivalent $-0-0$ structure and may be considered derivatives of hydrogen peroxide, where one or both of the hydrogen atoms have been replaced by organic radicals. Organic peroxides are thermally unstable substances, which may undergo exothermic self accelerating decomposition. In addition, they may be liable to explosive decomposition, burn rapidly, be sensitive to impact or friction, react dangerously with other substances and/or cause damage to the eyes.

Class 6: poisonous (toxic) and infectious substances are subdivided into poisonous (toxic) substances. These are substances liable either to cause death or serious injury or to harm human health if swallowed or inhaled or by skin contact, and infectious substances containing viable micro-organisms or their toxins which are known or suspected to cause disease in animals or humans.

Class 7: radioactive substances which spontaneously emit a significant radiation and of which the specific activity is greater than 0.002μ curie/g.

Class 8: corrosives are solids or liquids possessing, in their original state, the common property of being able, more or less severely, to damage living tissue. The escape of such a substance from its packaging may also cause damage to other cargo or to the ship.

Class 9: miscellaneous dangerous substances are those which present a danger not covered by other classes.

11.4.9.2 Shipping name

Any items under a particular class must also carry the correct technical name of the produce. This may be identified by a proper shipping name. The purpose of indicating the proper shipping name of the substance or article being transported on documentation accompanying a consignment, and of marking this name on the package containing the goods, is to ensure that the substance or article can be readily identified during transport. This ready identification is particularly important in the case of a spill or leak of dangerous goods, in order to determine what response actions, emergency equipment or, in the case of poisons, antidotes, are necessary to deal effectively with the situation.

11.4.9.3 Index

An index or table providing a list of the most dangerous substances and a reference to where they can be found in the publications, together with a table of UN numbers for the substances and corresponding IMDG code numbers.

11.4.9.4 Packaging

Indicates the recommended method of packing and illustrations of construction of packages for individual substances such that no danger will arise from the substance being wrongly packed. Before the goods are taken on board the ship, the Master is furnished with a declaration by the shipper that the goods are packaged in accordance with the requirements of the regulations. An example of such recommendations for packing could be:

PACKING
Class 3.1 – Flammable liquids – Carbon Disulphide. Hermetically sealed contents would include glass bottles packed with inert cushioning and absorbent material together in a wooden box; metal cans packed together in a wooden box; cylinders and/or metal drums.

11.4.9.5 Stowage

The recommended method of stowage for individual substances is that dangerous goods are stowed safely and appropriately according to the nature of the goods. Incompatible goods must be segregated from one another, such that any possibility of interaction between dangerous cargoes cannot occur. This stipulates the degree of separation between any two classes to avoid interaction occurring. The term 'separated from' means in different holds. An example of such recommendations could be:

STOWAGE
Class 3.1 – Flammable liquids – Carbon Disulphide. Unless approved by the competent authority, the maximum quantity in packaging on any ship: 500 kg, equivalent to 450 litres (1100lb) (98 gallons).

Prohibited on any ship carrying explosives (except explosives in division 1.4. Compatibility Group S; see also paragraph 6.1.8 of the Introduction to Class 1).

Keep as cool as reasonably practicable.

Cargo ships, or passenger ships which are carrying not more than 25 passengers or 1 per 3 metres (10 feet) of length } ON DECK ONLY

Other passenger ships: PROHIBITED

11.4.9.6 Labelling

Labelling shows illustrations of recommended labels covering each type of dangerous goods under the class to which it is allocated. No package shall be taken on board a ship unless it is clearly marked with the correct technical name and an indication (label) of the nature of the danger to which the goods may give rise (*Figure 11.13*). If dangerous goods are carried in a freight container then the container must be clearly labelled, indicating the nature of the danger to which the goods give rise as well as the package itself. Such labels and markings would follow the recommendations of the IMDG code.

Figure 11.13 Dangerous cargo example labels (courtesy of Blackpool and the Fylde College, Nautical Campus Fleetwood)

11.4.10 Safe handling of dangerous goods

11.4.10.1 Loading

Ensure correct declaration and certification. Check for correct packaging and labelling as per IMDG Code/ Blue Book. Ensure correct segregation and stowage. Make sure firefighting equipment is available as appropriate and sources of ignition eliminated as required, supervised by a responsible officer. Ensure there is adequate information on handling precautions and personnel wear protective gear as appropriate. Check for adequate ventilation and anti-spill measures. If shipped in bulk, the dangerous goods booking list should be consulted to establish the identity and hazards of the cargo. Ensure rescue/resuscitation and first aid equipment is available, together with relevant information and publications available.

11.4.10.2 On passage

Location of dangerous goods must be clearly shown on cargo plan. There should be regular inspection of cargo segregation and monitoring of compartments for fire outbreak indications. All personnel should be alert to hazards, particularly if the cargo is on deck.

11.4.10.3 Discharging

Discharging should be supervised by a responsible officer. Thorough ventilation is essential before entry. Check gas levels as necessary and check for damaged cargo and act accordingly. Safety precautions will be as for loading.

11.5 Tanker work safety

11.5.1 Safe practice in the carriage of oil cargoes

11.5.1.1 Oil cargoes (petroleum)

Petroleum covers crude oil and all its products. Crude oil contains all of the various products which are refined from it at different stages of the refining process. Petroleum can then be broken down into two categories of clean oils and black oils. The former are refined products which leave little or no residue in tanks and are usually the more volatile type of oil. The latter are so called because they leave a residue in tanks, which may contaminate clean oils. Although usually less volatile, black oils may include crude oil, which can be volatile.

This type of classification does not, of course, give an indication of the danger from explosion or fire; it is purely an indication of carriage to avoid the contamination problem. Cargoes are generally classed according to their degree of volatility and flammability. Volatile petroleum has a flashpoint below 60°C by closed cup test or 66°C by open cup method. Flashpoint is the lowest temperature at which a liquid gives off sufficient vapour to form a flammable mixture. Taking note of the volatility and flammability of the products, they can be split into group A, grades having a flashpoint below 23°C including motor spirit, aviation spirit, benzole, benzine and crude oil; group B, with a flashpoint 23–60°C, including kerosine, burning oil, white spirit and some crude oils; and group C, with a flashpoint above 65°C including gas oil, diesel oil, furnace oil and lubricating oil.

11.5.2 Oil cargoes, tanker work

Owing to the special nature of oil cargoes, consideration must be given to the hazards associated with the loading, carriage and discharge from tanker vessels. It is not within the scope of this chapter to cover every aspect of tanker work. The major risk areas are fire or explosion causing pollution and health and toxicity causing a reaction.

11.5.2.1 Fire and explosion

Petroleum liquid and its products do not ignite but the vapour given off from them does. The flammability of the liquid is the ability to burn if the conditions are favourable. For most inflammable vapours there is a high limit above which the mixture of air and hydrocarbons is too rich and a low limit below which the mixture is too weak and it will not burn. The limits of flammability are usually upper fire limit: 10% gas to 90% air; lower fire limit 1% gas to 99% air. An explosimeter will not give readings outside these limits.

A graph known as a flammable envelope can be constructed for all products and will show the area over which combustion can occur. Once the flammable envelope has been produced, a range of limits can be established. Provided the atmosphere in the tank can be maintained outside these limits, no fire or explosion will take place. There are of course, several factors which can effect the flammable envelope, the obvious ones being the percentage mixture of oxygen to hydrocarbons. There are dangers with using the too rich method, as when gas freeing the tank. The atmosphere may pass through the flammable envelope in order to get a 0% hydrocarbon reading. When using the too lean method it is possible that residues in the tank may become stirred up and give off hydrocarbon gases causing the tank atmosphere to pass up into the flammable envelope. One distinct way to solve the problem is to replace the oxygen in the tank with an inert gas such that the mixture in the tank cannot possibly lead to combustion.

11.5.2.2 Health and toxicity

Personnel coming into skin contact with oil cargoes through spillage, leakage or any other reason are in danger from corrosive elements of the cargo affecting the skin direct and by absorption of poisons, the most vulnerable organ being the eyes, risking blindness. Other skin ailments which result from contact with oil cargoes are dermatitis, burnt and dry skin, acne and in the long term skin cancer.

11.5.2.2.1 Hydrogen sulphide Many crude oils contain trace quantities of hydrogen sulphide, but some, known as sour crudes contain this in sufficient quantities as to create a hazard in their handling. Hydrogen sulphide is highly volatile and as a result will be present in large quantities in the vapour from the oil. This means that gas vented during the loading period could be extremely hazardous if inhaled. The presence of hydrogen sulphide as a vapour can be quickly identified by its most offensive and pungent odour, somewhat similar to rotten eggs. Its toxic effect, however, is paralysis of the nervous system and one of the first senses to be rendered ineffective is smell. Depending on the concentration and time of exposure unconsciousness would follow, respiratory failure would occur and unless quickly restored, a fatal result can be expected.

11.5.2.2.2 Benzene This is of the aromatic family of hydrocarbons which have a greater toxic property than conventional oils. Effect of liquid contact on the eyes causes great irritation. If contact is made with the skin a slight irritation occurs, but with continuous contact, defatting with possible secondary infection. Systemic poisoning can result from sustained contact.

The effect of the vapour will cause irritation to the eyes, especially in high concentrations, but there is no effect of vapour on the skin. When inhaled, it can increase irritation, dizziness, headache and drowsiness. It has a cumulative effect which leads to fatigue, loss of appetite, nervousness and can lead to blood disorders.

11.5.2.2.3 Toxicity The toxic effects of petroleum vapour will prevent the supply of air to a person breathing in a confined space, but in addition, some petroleum vapours are toxic and may have both chronic and acute effects:

0.1% gas in air (1000 ppm) leads to irritation to eyes within 1h.
0.1% gas in air (7000 ppm) can cause symptoms of drunkenness.
2.0% gas in air (20 000 ppm) can cause unconsciousness, paralysis and death.

Some heavier hydrocarbon vapours are toxic well below the lower fire limit of 1% and this necessitates showing the threshold limit value or toxicity of an inhaled gas in parts per million (ppm).

11.5.3 Tanker work

11.5.3.1 Ship and shore safety check list (additional reference should be made to the tanker safety guide)

A responsible officer, who is familiar with the arrangement of lines, valves and venting system in the tanker, should supervise and control all cargo handling, the responsible officer should be satisfied, as appropriate that: appropriate ship and shore personnel have been notified that cargo handling is about to begin; warning notices are displayed as required; no unauthorized persons are on board the tanker or unauthorized craft alongside.

Additionally, the officer should confirm that craft alongside are advised that cargo handling or ballasting operations are to begin and that necessary safety measures are to be observed; that no unauthorized work is being carried out; that fire appliances are in good order and available for immediate use; emergency towing-off wires are positioned; the agreed ship and shore communication system is working; the ship's portable R/T sets are of approved design; when necessary, adequate safe lighting is available; no naked lights are being used; there is no smoking on board except in places permitted by the Master; that galley precautions are observed, and that all doors and ports required to be closed are, in fact, closed and ventilators are suitably trimmed.

Other precautions include all cargo tank lids, tank washing openings, ullage openings, sighting ports and similar fittings (except those to be used initially) being closed; scuppers to be properly plugged; that all sea and overboard discharge valves connected to the cargo system are closed when not in use and preferably lashed; that all cargo lines, including stern discharge lines, which are not in use, are isolated, if possible, and appropriate valves closed; any necessary shore connections are properly made and supported; all valves to cargo lines and bunker lines, which are required for use, are properly set; cargo tanks to be used are open to atmosphere via designed venting system; and finally, the safety check list for the terminal is completed and signed.

11.5.3.2 Ship and shore communications

The maximum loading or discharging rate for the bulk transfer of cargo should be agreed between the responsible officer and the shore personnel, having regard to the nature of the cargo to be loaded or discharged; the arrangement and capacity of the tanker's pumps, cargo lines and gas venting system; the maximum allowable pressure during discharge in the ship and shore hose connections or shore pipelines; and any other limitations.

11.5.4 Tanker operations

Loading plan: tankers

Prior to loading, a cargo plan will be drawn up so that the disposition of cargo and loading sequence is known. Everyone involved with loading can then be briefed about the loading particulars and sequence. This task involves the calculation of how much cargo the vessel can load, the determination of where it is to go and why. The problems of deciding how much to load will not be discussed here, but we will cover the considerations when making up the loading plan.

The vessel's loaded deadweight must be strictly controlled in conjunction with the freeboard and draught. These should coincide with tabulated values but a vessel may hog or sag; the freeboard is always the crucial factor and must never be reduced. The cubic capacity of the tanks must be matched to the volume, type and weight of cargo to be loaded. The relative density of the cargo controls the weight of cargo for a fixed volume and so must also be considered. The lower the relative density, the lower the weight of cargo for a particular volume. Therefore, if a tank is filled to capacity, the weight of cargo will change with the relative density.

The amount of bunkers, stores, fresh water and ballast must be calculated beforehand and subtracted from the loaded deadweight. This will then give the amount of cargo that can be loaded, weightwise but not necessarily volumewise. The seasonal zone that the vessel is in upon loading will determine the maximum loaded deadweight. This must be strictly applied, although allowances can be made for fuel and water consumption before reaching the open sea. If loading in a summer zone and during passage the ship passes into a winter zone, one must ensure the vessel will not be overloaded on reaching that zone. This implies that the intended route must be known.

Due consideration must be given to temperature changes, loading to the expansion or contraction of the cargo when going from cold to warm climates or *vice versa*. The cargo when heated will expand and this may cause an overflow from a tank. Careful assessment of a vessel's trim on loading, during the voyage and at destination, must be made. Adequate stability must be maintained throughout and the consumption of bunkers, fresh water, etc., must be taken into considertion. The loading sequence of different products must be considered so that contamination of highly pure cargoes does not result. The same is true on discharge, particularly if discharge is at more than one port. Also, when loading and discharging, adequate stability must be maintained. This can make loading and discharging highly complex. Loading should avoid free surface effects occurring in too many tanks at the same time. The number of partly filled tanks should be kept to a minimum at all times during loading. Loading so as to avoid undue stresses and moments must be done. A good, even stowage pattern is also essential.

Cargoes stowed next to each other must be compatible chemically. If adjacent cargoes inadvertently get mixed, no hazards, such as explosion or the production of harmful vapours, must result. It must be ensured that separate loading and discharging lines are used if more than one grade of cargo is being loaded or discharged. You must make sure that the loading rate suitable for the ship is fully understood and not exceeded by the shore station which may have a higher rate. Finally, you must ensure that all personnel are aware of the overall situation. If necessary, notices should be posted to inform all crew members.

11.5.5 General preparation of tanks – prior to loading a tanker vessel

Having completed the loading plan, we can then proceed with the preparatory work so the vessel is ready to load on time. In the preparation for loading, we are assuming there will be readied tanks available, tanks which have been prepared for the cargo to be loaded.

During tank preparation complete a loading plan and ensure that personnel are aware of it. Check and test heating coils for any leaks. Heating coils may be needed as some cargoes are carried hot. Check that all cargo tanks are clean and ready to load. This will normally also be done by a surveyor representing the shippers, who will either accept or reject the tanks on behalf of the shipper. Check that all valves in the system are free and set them to the required position for loading the tanks in question. Plug the main deck scuppers and any other openings that may allow passage of spilt cargo overside. Generally, wood plugs covered by cement will be used. Place drip trays at the manifolds to catch any spillage. Test the manifold system with air pressure for leaks if required.

Ensure that all tank openings are battened down. If there are any that are required to be left open, these must be covered with fire gauze to prevent an explosion. If the vessel uses a Whessoe gauge readout, then test this device and set it for an empty tank reading. All ports and doors must be closed. If the vessel has air conditioning, it must be put on recirculation and not an external fan. This will exclude flammable vapours. If the vessel has pressure vacuum valves, ensure they are set to correct pressure and vacuum readings. A PV valve is a valve which will open due to pressure above or below 1 atmosphere being applied. They prevent large changes in tank pressures during cargo operations.

Ensure that towing wires are in position, ready for tugs to tow the vessel in the event of fire. If bonding is required, ensure that it is done correctly. Most countries have now realized that the bonding wire does not suffice and the pipelines themselves require bonding. This is the earthing of the vessel to prevent a build up of electrical charge that could cause a spark and so cause an explosion. Finally, dry powder extinguishers should be placed at manifolds and the fire main system put under pressure with hoses run out ready for use.

Notices must be posted in prominent and correct positions for

No unauthorized personnel
No smoking
No lights, matches, etc.
No electrical appliances that are not intrinsically safe
No metal tools or spark producing footwear
No welding or hot work

A communication system will be set up with shoreside personnel. Ensure that the shore cargo line is suitably supported to avoid kinking and that it is connected to the manifold correctly: a bolt in every hole and not every other one. If a chick san is used for loading, ensure that its weight is supported by a jack and not by the manifold.

Run up the B flag (see *Figure 11.14*) on the mast. This means: I am taking in or discharging or carrying dangerous goods.

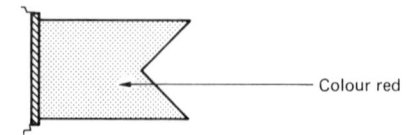

Figure 11.14 The B flag, coloured red

To assist cargo flow, you will need to station a person at the manifold valves ready to open them if necessary, also place someone to confirm that there is cargo entering the correct tank. Confirm that the shore checklist is completed. Inform the shore personnel that the vessel is now ready to load, with only the manifold valve to open when cargo pumping commences.

11.5.6 Associated equipment

11.5.6.1 Explosimeter

This is an instrument designed to test and measure the degree of flammability of the atmosphere in a compartment, but it may also be used to indicate whether a compartment is sufficiently clear of hydrocarbons to permit entry of personnel. Any reading of the meter when sampling is taken to indicate that further venting is required before entry can be permitted. It should be clearly understood, however, that although the instrument measures hydrocarbons, it does not indicate oxygen content. Hence, a compartment may be clear of hydrocarbons and declared gas free, but may be deficient in oxygen, such that it will not sustain life.

Prior to use of the instrument, it must be flushed through with clean fresh air several times, as it is quite possible that residues of hydrocarbons from previous readings may have remained in the instrument and this could provide incorrect readings. It should also be tested and calibrated at regular intervals to ensure its performance is correct.

11.5.6.2 Oxygen analyser

This is an instrument designed to test and measure the percentage oxygen content of a compartment or gas flow. It is similar in principle to the explosimeter, except that the oxygen content is being measured. A sample is drawn through the instrument and, as long as the reading shows the full 21% oxygen required to support life, the compartment could be entered once the hydrocarbon content is correct.

11.5.6.3 Draager

This is a trade name for a chemical absorption gas detector. When working with product carriers or chemical carriers, it is not sufficient to use just an explosimeter and oxygen analyser, as the products in question are not detectable by either of these instruments. The Draager is designed strictly for this.

11.5.7 Static electricity – causes and dangers

Static electricity can cause sparks that ignite flammable gas, although not all sparks have sufficient energy to do this. All materials normally contain equal numbers of positive and negative charges; these are usually distributed evenly throughout the material. Electrostatic charging occurs whenever this equality is disturbed and charges from one sign are separated from those of the opposite sign, causing a voltage build up. When the voltage build up is substantial, it may be sufficient to break down the atmosphere in the vicinity and so cause a rapid discharge in the form of a spark.

11.5.7.1 Oil flow

For practical purposes, all flow in petroleum pipelines is turbulent; this scoops up an inner layer of charged molecules and distributes it throughout the bulk of the petroleum in the pipe, leaving the outer layer at the interface with the pipe wall. The charged layers are subsequently separated and the petroleum in and emerging from the pipeline becomes charged. The faster the flow in the pipe, the greater the turbulence and generally the greater the charging. This means that whenever turbulence is present, there is the possibility of a build up of static, which could result in a spark being generated.

11.5.7.2 Water flow

When water droplets settle through a depth of petroleum in a tank, it will cause separation of the two charged layers. Such charging will continue throughout the period of water settlement and therefore may persist long after pumping into the tank has ceased.

11.5.7.3 Air flow

The passage of air through petroleum will not in itself generate static, but if air is passed into the bottom of a tank and then releases water to the surface, this action could result in a static build up. Subsequent settling of the water back through a low conductivity distillate could result in a powerful electrostatic charge, which will persist until the disturbed water has settled.

11.5.7.4 Steam

Water droplets issuing at high velocity with a jet of wet steam can become charged by contact with the nozzle through which the jet is issuing. This may give rise to a charged mist in the tank and any unbonded conductor in the tank may accumulate electrostatic charge by the settlement of the mist upon it.

11.5.7.5 Pipelines

When charged petroleum flows through a rubber flexible pipeline, part of the charge may be picked up by the hose flanges which are normally in contact with the liquid in the hose. A dangerous voltage may accumulate on the exterior of the flange.

11.5.8 Prevention of static electricity to avoid ignition

Cargo hoses should be thoroughly bonded so that there is adequate and continuous earthing of all metal parts through the tanker and its manifold to the sea. In the initial stages of loading the rate should be low in order to reduce the generation of static electricity with admixtures of water and oil in the pipeline. It will also reduce the amount of turbulence in the tank and so allow the water in the cargo to settle more quickly; this will allow it to lay relatively undisturbed when the loading rate is increased. Volatile or non-volatile petroleum at a temperature above its flash point should not be loaded or transferred overall. Non-volatile petroleum may be loaded overall only if the tanks are gas free and provided no contamination with volatile petroleum can occur. Water should not be loaded overall into a tank which has contained volatile petroleum distillates until the tank has been stripped.

During the loading of any cargo which may give rise to flammable vapours, no earthed conducting probe and no unearthed conductor should be introduced into the tank. An example of this would be a metal weight on the end of a tape or a tin can sampling receptacle. The two examples refer to ullaging and sampling of the tank. Great care must be taken that no spark is allowed to occur either at the ullage port or at the surface of the liquid. The possibility of ignition by static tanks following discharge, but which are not gas free, arises when a conductor is suspended into a tank and steam is present. No tank washing machine or other conductor, whether earthed or unearthed, should be permitted in a tank which has steam injection, unless of the direct fixed type.

Compressed air or inert gas should not be used to clear pipelines back to the tank after loading of a distillate oil which may produce a flammable vapour, unless adequate precautions are taken to prevent air or gas entering the tank and causing turbulence with water in the bottom. To minimize the accumulation of static electricity in shore tanks at the beginning of discharging of distillate oils, which may produce a flammable vapour, pumping speeds should be kept low until shore personnel advise that they can be increased.

11.5.9 Example use of the inert gas system
11.5.9.1 Sequence of operations

Prior to loading operations, the inert gas system will be switched on and the tank brought under positive pressure; the system is then switched off during actual loading operations and the gas is then exhausted through the pressure relief valve riser as the cargo enters the tank. On completion of loading, the system is switched on again to ensure that the tank is under positive pressure and inerted still. The system is then closed down. During the passage the system will be operated at regular intervals, e.g. every morning, as the gas may suffer from seepage and be lost to the atmosphere. Hence the tank needs to be kept topped up with inert gas.

During discharging operations, the system is put into operation such that, as the cargo is being discharged, the inert gas is replacing it. As the last of the cargo is removed from the tank it is now full of inert gas and under positive pressure. On completion of cargo, the tank is left inert. During tank cleaning operations the inert gas system is left running to maintain the tank in the inerted condition. It has to be left running as during tank cleaning, gas escapes through the Butterworth Openings, so pressure has to be maintained to ensure that the tank remains inert. The operation of the inert gas system for ballasting operations follows that of loading and discharging of cargo.

Initially the system will have been on, to ensure that the tank is inert during the washing cycle. Once washing is completed then the tank is inerted. The system is now shut down and forced venting with air now takes place, pushing the inert gas out through the risers or vent pipes. The atmosphere in the tank should be sampled regularly to establish that the tank is being maintained in its correct condition, i.e. inert correctly or gas free. If the tank is allowed to stray outside the strict limits, it may cause the atmosphere in the tank to enter into the flammable envelope range and a danger of combustion will develop. Alternatively, gas may be allowed to seep into a gas free tank, putting lives at risk if the tank is to be entered.

11.6 Safety in offshore working practice
11.6.1 Introduction to offshore safety

In the interest of operational safety, health and welfare on or near an offshore installation, the attention of all personnel is always drawn to the legal requirements of Statutory Instrument 1976 No. 1019.

The reasons for a positive attitude towards safety offshore are numerous and the following should be considered. Any accident or hazardous situation offshore could easily escalate beyond controlled dimensions. A considerable number of people are often concentrated in a small area, remote from the shore. Over 500 is not an unusual figure during the construction stage. An extensive mixture of work skills is present – this can lead to lack of familiarity with other people's working practice. Adverse weather and open sea conditions would influence an evacuation situation and make it difficult. Ignorance of safety procedures and the correct usage of equipment, including personal safety gear, affects even the most experienced of operators. It is essential that in a 24 hour offshore environment safety attitudes are maintained, off shift being equally as important as on shift working.

11.6.1.1 Definitions

A cold platform is defined as an installation which has not commenced production or drilling. A hot platform is defined as an installation which has commenced drilling or commenced supplying hydrocarbons to the platform's systems (see *Figures 11.15* and *11.16*).

Figure 11.15 Jack up mobile drilling rig designed to operate in depths up to 100 m (325 ft)

Figure 11.16 Fixed platform constructed by means of a steel jacket which is secured to the sea bed by steel piles. The helideck and the position of the survival craft are easily identified

11.6.2 Offshore safety information

The safety procedures amongst offshore operators differ considerably from company to company. However, the principle of maintaining a safe working environment remains common to all interests. To this end, those intending to engage in employment within the perimeter of the Offshore Industries should familiarize themselves with the expected levels of knowledge concerning the following.

11.6.2.1 Helicopter boarding pass

The boarding pass will include identity details of the individual, which may be obtainable through a computerized system. It will also probably contain a declaration statement, which will refer to the state of health of the person travelling offshore and to the fact that no unlawful act is being carried out, e.g. no drugs, alcohol or firearms, etc., are being transported. The boarding pass will briefly detail aircraft safety and could specify installation detail, such as allocated accommodation or survival craft space. They are often double passes to meet the requirements for both inward and outward journeys.

11.6.2.2 Station identify card

This is an information and instructional card found in every cabin on the installation. It will provide details of all the alarms for the installation and the significance of each alarm. It will identify the individual's name, cabin or bunk number and the respective lifeboat or muster station. A map or escape route plan of the main deck layout together with the position of lifesaving and fire fighting appliances is often included.

11.6.2.3 Safety handbooks

Most operators currently issue all persons with details of safety procedures affecting the installation, within a pocket sized handbook. The contents are usually reinforced by a lecture and question session conducted by the safety officer, for all persons who are visiting the installation for the first time. Specific areas are covered by the handbook, such as helicopter procedures or clothing regulations, permit to work systems, accommodation regulations, escape routes, etc. It is the duty of every person working offshore to become aware of hazards and the relevant safety procedures affecting themselves and pertaining to their particular installation. The alarm systems are detailed, together with fire and accident prevention methods. It should be read and understood by individuals prior to the emergency.

11.6.2.4 Safety information display posters

These are usually laminated instructional posters which are set up in restaurants, games rooms, lounges, etc. They display the types of alarms which operate on the installation and provide information regarding the immediate actions people should take in the event of each specific alarm or alert occurring.

11.6.3 General practice affecting clothing regulations offshore

When working offshore most companies require a minimum requirement from personnel regarding the wearing of clothes suitable to the task being performed.

When working outside, a non-metal safety helmet, suitable protective clothing and reinforced toecap workboots should be worn. When working in exposed situations thermal underwear, a waterproof coverall, work gloves suitable for the task, a safety helmet liner and ear defenders when appropriate are standard requirements. In addition, suitable eye protection should be required, in the form of goggles, when working in the vicinity of chipping, grinding, lathe work, heavy hammer work or near oxyacetylene cutting or welding equipment.

Safety belts and harness must be worn in accordance with the procedures as specified by work permits or whenever someone is required to work in an exposed position where the risk of falling is present. Should risk of falling into the sea also exist, then a buoyancy aid or lifejacket should also be worn. Those working in enclosed spaces must be fitted with a lifeline and safety belt, in conjunction with breathing apparatus wherever the space might contain a deficiency of oxygen, toxic or other noxious gases.

11.6.3.1 The permit to work system

The purpose of a permit to work system either aboard an installation or on board a ship is to ensure the personal safety of those engaged in the work, and all others on the platform and to ensure that the overall safety of the installation is not put at risk.

The permit to work is a legal requirement which is effective for all work on platforms, with the exception of routine operations and catering. Safe working practice can subsequently operate through forward planning of the work required in cooperation with all parties concerned: communication of work plans to those who will become directly or indirectly involved in those plans; as well as the observation and implementation of all precautions and monitoring systems guarding the work being carried out, including ensuring the operation of all safety devices on equipment and the supply of protective gear as and when appropriate; control, observation and monitoring of the actual work whilst being carried out; and the requirement for the operational site to be left in a clean and safe state so that nothing is left that may endanger anyone or the overall safety of the installation.

11.6.3.2 Operation detail – affecting permit to work schemes

Example types of permit:

1. Hot work permit – e.g. Flame cutting and welding
2. Cold work permit – e.g. Crane operations
3. Electrical permit – e.g. All electrical work
4. Preparation/reinstate permit – Required for gas tests, isolation of safety or emergency systems, electrical isolation or de-isolation
5. General work permit – All relevant work not covered by specific permits
6. Entry into confined space – permit – e.g. Tank entry

All permits are immediately suspended on the sounding of any platform alarm signal.

It is normal practice that every permit to work is issued in quadruplicate and copies colour coded:

Original	being retained by the supervisor or control room
1st Copy	displayed at the working site so that operators know of the hazards and what safe guards are required
2nd Copy	To the offshore installation manager (OIM)
3rd Copy	Usually issued to the section head who has authorized the work

The issue of a permit to work will not in itself make any job safe, but it will provide a guide to operatives and to supervisors of the safe procedures required for whatever work is to be carried out.

11.7 Marine emergencies

11.7.1 Collision

In the event of a collision occurring to a vessel at sea, the watertight integrity of the ship is most likely to suffer. The collision may involve another vessel, land mass, or ice flow; in any event shipboard personnel will experience a degree of shock. No precise guidelines can be specified for every situation but a general format of probable actions by the Master and crew can be anticipated:-

1. Raise the general alarm
2. Depending on circumstances, stop main engines. It may be considered prudent action to maintain engines at dead slow to prevent the rapid withdrawal from a stricken vessel. A gashed hull form could sink quickly once the separation takes place between collision parties.
3. Close all watertight doors and fire doors immediately.
4. Activate bilge pumps and/or ballast pumps to the damaged area of the vessel if there is obvious ingress of water.
5. Carry out a damage assessment as soon as possible after the event.
6. Establish communications and damage control party at incident scene.
7. The Master should evaluate the risk to life and property. When practical the Master should exchange with the other party, name and port of registry of the ship, ports of departure and destination; render all possible assistance to the other vessel and cause a statement of the incident to be entered into the log book.
8. The damage sustained and whether the vessel is to be abandoned would dictate the instructions to the radio officer. In either case an urgency signal would probably be despatched, which may be followed by a distress signal.
9. The Master, or officer in charge, should report the incident as soon as practicable or in any event within 24 h after the ship's arrival in port. The report being made to the controlling authority: for the UK, the Department of Transport.
10. Following any collision, consideration should be given to flooding and loss of buoyancy; risk of fire, toxic chemicals or gas and possible explosion and adverse weather effecting the vessel. Taking into account the associated hazards, alternative solutions should be explored such as beach the vessel in shallows to prevent sinking; make a collision patch to reduce ingress of water; provide a list to the vessel to clear the damaged area over the water line; and delay tactics to prepare survival craft, prior to abandonment.

11.7.2 Man overboard

If a person falls overboard, it is essential that he or she is recovered as quickly as possible. If the person is observed when falling overside, then the recovery operation can be executed quite quickly. There are four methods of turning the ship in such a situation: the Williamson turn; the single delayed turn; the elliptical turn; and turning short round in confined waters. Each method has its merits, depending on whether the person overboard is visible, the state of actual visibility at the time, night or daylight conditions, etc. In all cases, however, the objective is to recover the casualty as soon as practical without exposing other crew members to unnecessary hazards.

11.7.3 Man overboard – associated actions

In the four methods of recovery additional actions by the OOW will be required, in addition to ship handling operations. On all occasions, the vessel should be positioned to the weather side of the person in the water, as this will allow the leeward side rescue boat to be launched to effect recovery. If the person is seen to fall overboard then the lifebuoy on the bridge wing should be released. It may be prudent to release the second bridge buoy to provide an indication of line of approach, once the ship has turned. The light or smoke signal on these buoys will provide a distinct search area for lookouts, who should be detailed off as soon as practicable after the incident is noted.

Additionally, main engines of the vessel should be placed on standby and a reduction of speed initiated during the turn. The emergency signal should be sounded and the Master informed as soon as possible. When appropriate, the vessel should be placed on manual steering, the emergency boat's crew mustered and the hospital area made ready to treat for hypothermia and/or shock. Other shipping in the area should be informed and Morse code, O, should be sounded on the whistle if other shipping is in close proximity. During the hours of daylight, the international flag O should be exhibited.

Lookouts should maintain observation throughout any operation, preferably from a high vantage point, especially important when knowing that a single person makes a difficult target to see in any but smooth waters.

11.7.4 Safe operation in marine helicopter activity

11.7.4.1 *Helicopter and ship operations*

In all combined ship and helicopter operations marine personnel are advised to carry out operations to the safety standards published by the international Civil Aviation Authority (ICAC) and in accord with the International Chamber of Shipping's guide to helicopter and ship operations.

11.7.4.2 *Responsibilities*

The Master is responsible for the overall safety of the ship. The aircraft pilot is responsible for the safety of the helicopter at all times. The helicopter winchman and/or cabin attendant is responsible to the pilot for passengers entering or leaving the aircraft, supervision of passengers in emergencies, observation of pilot's blind spots, supervision of loading and unloading the helicopter, completing relevant documentation, e.g. manifests, customs declaration, etc., and assisting the officer of the deck to ensure safe conduct in the vicinity of the helicopter. The employer or user is responsible for selecting a suitable helicopter to complete a safe operation, bearing in mind the constraints of the ship and the limitations of the aircraft.

11.7.4.3 *Landing or deck officer's check list*

All rigging stretched aloft, all stays and aerials, etc., should be secured, lowered or removed to prevent interference with the aircraft. All loose objects inside the operational area and adjacent to this area should be removed or secured against the downdraft from the helicopter's rotors. A rescue party should be detailed with at least two members wearing fireman's protective suits. The ship's fire pump should be operational and a good pressure must be observed on branch lines. Fire hose branch lines should be near to but clear of the operational area, preferably upwind of the operation. Foam extinguishing facilities should be standing by close to the operating area. Foam nozzles should be pointing away from the aircraft. The ship's rescue boat should be in a state of readiness for immediate launch in the case of such a situation developing. All deck crew involved in and around the area should wear highlighting coloured work vests. Protective helmets should not be worn unless a substantial chin restraining strap arrangement is attached to the helmet. Emergency equipment should be readily available including a large axe, portable fire extinguishers, a crowbar, a set of wire cutters, first aid equipment, red emergency signalling torch and marshalling batons at night.

Correct navigation lights and signals should be exhibited by the ship to indicate that it is restricted in its ability to manoeuvre, because it is engaged in the launching or recovery of aircraft. All non-essential crew and/or passengers are clear of the operational area. The hook handler should be adequately equipped with strong rubber gloves and rubber-soled shoes to avoid the danger of static. The OOW should be informed when all preparations are complete and when the ship is ready to receive the helicopter.

11.7.5 Fire fighting

Fire at sea is probably the most feared type of emergency that the Master and crew of any vessel can expect. Just as threatening to life is the fire on board a vessel in port. In every case, large fires are initially generated from a smaller fire and prevention is much preferred to the cure. Fire prevention can be established within crews by correct attitudes with the safety of life and the ship being the top priority. Practical and effective training for personnel at all levels with recall and refresher training at periodic intervals has proved beneficial in the past. Positive participation in drills and exercises while on board creates confidence in personnel and familiarization with fire fighting facilities. The following suggested lines of action are presented for both a fire at sea and in port. They are not meant as rigid guidelines and persons involved should allow the necessary flexibility in their interpretation, depending on the circumstances of the case.

11.7.6 Fire in port – suggested line of action

Raise the alarm. This should be carried out immediately on board the vessel once the fire has been discovered. Furthermore, the alarm should be passed to the shore port authority as soon as practicable after discovery. Probably the easiest and most practical way of contacting the port or harbour authority would be by means of the ship's VHF radio or ship to shore telephone, if established, or by means of a messenger.

Local fire brigade contact should be relayed direct or via the harbour authority. All non-essential personnel should be removed from the vessel as soon as possible after the alarm has been sounded. A fire party will remain on board, together with the necessary back up manpower to restrict and tackle the fire. All fires should be tackled as soon as practicable after discovery.

The immediate action at the scene of the fire should be to close down all ventilation to the compartment on fire and restrict the oxygen flow. Obtain water pressure on the ship's fire main and commence running branch lines to the fire area. Restrict heat transfer by commencing boundary cooling as soon as possible after water pressure is available. Every fire has six sides, and boundary cooling should take place on all or as many sides as practical. An officer or competent crew member should be detailed to meet the arrival of the fire brigade. The international shore connection should be ready to hand over together with the fire envelope and damage stability information.

The actions of the fire party will vary with the type of fire, its location and the possibility of casualties caught inside the fire area. Depending on circumstances and reasonable access, casualties should be removed from the immediate area of danger as soon after the fire has been discovered as is practical. A direct attack on the fire must be considered a priority if it is accessible. This action may not extinguish the fire, but will act as a holding operation, to restrict the spread of the blaze until the local fire brigade arrive. Care should be taken to assess the nature of the fire before exposing fire fighters. An assessment is doubly important in order to determine by which means and by what medium the fire should be tackled.

Efficient means of communications should be established between key personnel and respective stations. These include the Master of the vessel; the chief officer in charge of the fire party; the gangway watch officer; the port or harbour authority; the fire brigade chief officer and the duty engineer. The main stations will probably be the engine room; the navigation bridge control; the scene of the fire; the gangway and quayside; and the harbour control office.

11.7.7 Fire at sea – suggested line of action

When raising the alarm, the Master of the vessel should be given a preliminary assessment as soon as information becomes available, specifically the scene of the fire, number of casualties, if any, and the nature of the fire. The immediate response actions by the OOW will be to place the vessel on standby; reduce the vessel's speed and manoeuvre the vessel stern to the wind to effectively reduce the draught within the vessel; inform the engine room of the situation and obtain water on the deck fire main; order the closing of all watertight doors; and obtain a position on the chart.

The probable actions of the Master will be to assume command of the navigation and control bridge; obtain an updated situation report from the fire party at the scene of the fire. Communicate with the chief officer and other heads of department regarding the situation and the method of approach, following assessment; order the radio officer to standby to transmit an urgency signal. Depending on circumstances, this may later be transmitted as a distress message.

Meanwhile the probable actions of the fire party will be to close off all ventilation to the fire area and at the same time check the area for possible casualties; depending on the type of fire, rig and activate hose branch lines to attack the fire as soon as practical. This action should take place on as many of the six sides of the fire as possible. Fire fighters should not be committed until a full assessment of the fire scene has been made and then only when fully equipped with protective clothing, breathing apparatus, necessary fire fighting equipment and back up personnel in attendance.

Ancillary fire fighters will commence boundary cooling to surrounding areas of the fire, carry out a roll call of the ship's company when practical and act as relief and back up to the main fire party. They will ensure that an adequate supply of foam compound, air bottles, etc., is kept continually supplied to the scene of the fire.

The engine room actions will be to respond to all bridge orders regarding the movement of main engines, continue to operate water pumps to ensure a desired level of pressure on the fire main and stand by to use bilge pump system and ballast systems, as required. This assumes that the fire is sited outside the engine room.

The Master should then order communications to be established between the scene of the fire and the navigation control bridge. At an appropriate time, the Master should be advised of the situation to allow correct decisions to be made regarding the sending of an urgency or distress signal. A safe port option should be investigated, if this is viable. A statement should be entered in the log book once the situation is brought under control and persons in general should be kept informed of the situation to avoid unnecessary panic.

Associated hazards include the possibility of dangerous or hazardous goods being in the proximity of a fire at sea. Investigation of the cargo plan and the general area surrounding the fire should be thoroughly examined. Additional stability problems may also arise with continual and concentrated amounts of water being employed to extinguish the blaze. The damage or flooding stability information should be consulted as appropriate.

11.7.8 The safe use of breathing apparatus

There are several different manufacturers of self contained breathing apparatus (BAs) and the following guidelines are based on the Siebe Gorman International Mark II model. Operators should exercise extreme caution and follow the manufacturer's instructions relating to whatever model type is in use in the practical environment.

The International Mark II apparatus consists of two compressed air cylinders mounted on a framework, which is worn by the operator. The cylinders are of lightweight design and when fully charged the equipment should not exceed 17.3 kg (38 lbs) weight. Cylinder volume provides 4 litres of air, which is sufficient for the wearer to be engaged for approximately 20 min hard work. The level of work being carried out by the operator will affect the amount of air being consumed and subsequently the duration that the operator can remain engaged with the situation. Guidelines supplied by the manufacturers suggest hard work: 40 min (two cylinders); moderate work: 62 min and at rest: 83 min.

11.7.9 Breathing apparatus – operational checks

There are several preparations for use. Demist the mask visor with antidim solution. Don the apparatus and adjust the harness for a comfortable fit. Open the cylinder valves. Put on the mask and adjust to fit by pulling the two side straps before the lower ones. Inhale deeply two or three times, to ensure that the air is flowing freely from the demand valve and that the exhalation valve is functioning correctly. Hold your breath and make certain that the demand valve is shutting off on exhalation or that leakage, if any, is slight. Close the cylinder valves and inhale until air in the apparatus is exhausted; inhale deeply. The mask should crush on to the face indicating an airtight fit of both the mask and the exhale valve. Finally, reopen the cylinder valves.

Pre-operational checks must be carried out monthly unless otherwise used. Ensure that the bypass control is fully closed. Open the cylinder valves. The whistle, if fitted, will momentarily be heard as pressure rises in the set. Check the cylinders are fully charged. Any leaks in the apparatus will be audible and should be rectified by tightening the appropriate connections. Do not overtighten. Close the cylinder valves and observe the pressure gauge and, provided it does not fall to zero in less than 30 s, the set is leak tight. Depress the demand valve diaphragm to clear the circuit of compressed air. Close pressure gauge shut-off valve and reopen cylinder valves. The pressure should remain at zero. Reopen valve. Gently open the emergency bypass control; air should then be heard to escape from the demand valve. Close the control. Close the cylinder valves. Gently depress the demand valve diaphragm and observe the pressure gauge. When it falls to approximately 43 atmospheres (44.5 kg/cm^2) the whistle should sound.

11.8 Marine survival

Following any emergency at sea where the risk of evacuation from the parent vessel becomes a possibility, marine personnel should be prepared to become involved in survival procedures. In every case, the individual will require full control of all facilities in order to participate effectively as a member of a team or as an independent survivor. An awareness of the potential hazards, together with alternative methods for prolonging survival time, must be considered essential knowledge for any person employed at sea.

An awareness and confidence in lifesaving equipment on board can only be created by active participation in survival techniques. These experiences can be acquired by training courses in shore based establishments, or by positive involvement in emergency drills on board. Additional information can be obtained from publications, review bodies, accident reports, official notices and manufacturer's literature.

Once the basic knowledge is known, the survivor, in order to be a survivor, will pass through three phases: the abandonment from the ship or installation; survival either aboard survival craft or in the water; followed by rescue probably to another vessel, aircraft or similar support vehicles. Each of these phases will vary in the length of time, depending on the conditions of the situation. Each will also be associated with variable circumstances which the would-be survivor would be expected to cope with.

11.8.1 Survival procedure prior to abandonment

From the outbreak of an emergency on board, the actions that are taken by the individual could well dictate their survival.

11.8.1.1 Initial actions

Put on warm clothing. Don an immersion or survival suit or waterproof coverall. Secure a lifejacket in the proper manner. Proceed swiftly to the emergency or muster point. A roll call will be taken by the person in authority and individuals may be designated to carry out various actions. If you are not given a designated task, await further instructions in a calm manner; listen and watch developments in the situation. Do not hinder those persons engaged with life saving appliances.

11.8.1.2 Subsequent actions

You may be instructed to move towards a helideck for evacuation by air or to enter a survival craft. Depending on circumstances, time may permit the collection of additional survival equipment, such as blankets, food and water, medical supplies, communications equipment, etc. Follow the instructions of those persons who are trained and experienced to handle this type of situation, e.g. lifeboat, coxwains, the helicopter landing officer or despatcher, the offshore installation manager, ship's Master or the officer in charge.

11.8.2 The abandonment phase

Time permitting, you will have prepared yourself for survival and are adequately clothed and protected. You should endeavour to remain dry through out this period if at all possible, bearing in mind that the human body will probably lose its heat 26 times faster once immersed in water, in comparison to the limited heat loss from an individual who remains in the dry condition.

It will be the responsibility of the officer in charge to remain with the parent vessel or installation up to that time that the position is no longer retainable. Once it is clearly shown that the situation can only deteriorate then the order to abandon or evacuate would be given by word of mouth. At this moment the life support facilities, such as heat, nourishment and protection from the parent vessel will be lost to the individual.

Abandonment may take considerable time, especially if the incident involved a large passenger liner, or accommodation or installation platform. To this end, panic should be avoided, and a helpful attitude by all concerned could assist the situation. Large numbers of people will be difficult to control, but segregation into viable batches of approximately 20–25 can be manipulated towards helidecks or survival craft. This number also allows a head count to be made of evacuees and is a suitable figure to accommodate survival craft, i.e. 40 seat boats, 20–25 person life rafts, 10–20 capacity helicopters.

11.8.2.1 Order of priority evacuation (once the abandonment phase has been reached)

By helicopter
By standby vessel or other support craft
By rigid totally enclosed survival craft
By marine escape slides and rescue craft
By davit launched life rafts
By inflatable life rafts
By entering the water close to a survival craft

11.8.3 The survival phase

The actions of personnel during this phase will obviously depend largely on the prevailing circumstances, but also on the type of survival craft in use, assuming a survival craft is being used. The immediate dangers to personnel during and following the abandonment will be during the launching period from the parent vessel. The dangers of taking a boat or life raft away from a ship's side or installation structure are considerable. This is especially the case if the incident and evacuation are taking place during the hours of darkness, possibly with expected bad weather and sea conditions.

Statistics indicate that if mariners can survive the first hour of this phase, their chances of winning through are high, provided they carry out what must be considered normal survival practice. Once at surface level, the type of survival craft in use must be compatible with conditions. Life rafts which cannot be made airtight would not sustain life in a toxic atmosphere; neither could they expect to last as long as a totally enclosed boat in a fire.

If a satisfactory launch occurs, the survival craft should endeavour to clear the immediate vicinity of danger about the mother vessel. If practical, the craft should be manoeuvred away and upwind of the incident area, clear of gas or other toxic chemicals. A sharp lookout should be maintained for other possible survivors while at the same time attempting to establish a position within approximately two miles of the scene. Circumstances may well dictate that an attempt should be made to make a landfall, but failing this, rescue forces could normally expect to respond to the last known position. This assumes again that a situation report and position were passed prior to evacuation.

The survival phase is very often the longest. Periods of depression can set in amongst personnel, which tend to destroy the will to survive. A sense of loss becomes strong and may cause change in character. The phase can continue for many hours or even days, being interrupted with periodic hopes of being located. The geography will reflect the type of search and rescue facilities available and the efficiency and speed of location which takes place. The unknown factor throughout the abandonment and survival phases is always the weather. Location of survivors might be quickly achieved; with poor visibility the actual rescue of those survivors could take considerably longer.

Persons who find themselves in this situation can aid their location by prudent use of distress signals and following the recommended survival practice. All survival craft are equipped with basic life support facilities including provisions and water, distress and communication equipment, first aid kit, and protection from the elements. More recently the use of Emergency Position Indicating Radio Beacons (EPIRBs) has played a major role in the location of survivors. Not all survival craft are fitted with this aid, but many ships and installations now have them, provided seafarers have the forethought to place them aboard during evacuation.

11.8.4 The rescue phase

This is generally a period of anticipation, which unfortunately can often go badly wrong as was seen with the recovery of personnel from the mobile offshore drilling unit, *Ocean Ranger*. In this incident, the stability of the survival craft was destroyed during an uncontrolled approach, in which potential survivors were attempting to transfer to the deck of a standby support vessel. People left the protective canopy of the totally enclosed boat and came on deck. This action alone would have raised the centre of gravity of the boat, causing a loss of stability. Combined with sea conditions and the wake current of the standby rescue vessel, the survival craft was seen to capsize and the occupants subsequently perished.

This incident is one of many which highlights the need for caution throughout all of the three phases, abandonment, survival and rescue. No one is a survivor until they have passed successfully through the final phase of rescue and have both feet firmly set in a safe haven. The need for continued discipline, forethought and awareness in this last and final phase are essential, as this is one of the most hazardous periods that survivors will pass through.

Personnel in this situation should realize that many rescue activities require a degree of self help, e.g. ascending scrambling nets, climbing into rescue baskets, etc. The authorities involved with rescue operations are normally well trained and prepared for most eventualities. Having said this, many rescues in the past have been carried out quite successfully by the first available craft, very often involving persons without specialist training or equipment.

The value of training marine personnel is especially relevant when disaster strikes. That training in each of the three phases discussed should provide each individual with the basic knowledge to overcome the sea.

11.8.5 Life saving appliances and arrangements

The following information is taken from:

Life-Saving Appliances and Arrangements
CLASS VII (F.G.) NON PASSENGER
CLASS VII (T) TANKER VESSEL

and takes into account the 1983 amendments to the International Convention for the Safety of Life at Sea, 1974. The following information relates to the regulations designated by the UK Department of Transport. Slight differences exist between these regulations and the regulations of the IMO and other authorities. Specific confirmation should be obtained from the regulations of the country of origin of the vessel.

11.8.6 Class VII foreign going non-passenger, non-tanker vessels

11.8.6.1 Life rafts

1. For ships of 500 tons or over, there should be sufficient lifeboats on each side of the vessel to accommodate all the persons on board. If the vessel is 1600 tons or over, the boats should be not less than 7.3 m in length.
2. For a vessel of 500 tons but less than 1600 tons there should be sufficient life rafts to accommodate all persons on board. If there are 16 or more persons on board, then there must be at least two rafts of approximately equal capacity. For a vessel of 1600 tons or more there should be sufficient life rafts for at least half the persons on board.
3. For ships of 500 tons but less than 1600 tons the provision of life boats and life rafts may be varied as follows. The provision may be as described above or, instead of the lifeboats, life rafts may be provided on each side of the vessel of sufficient capacity to accommodate all persons on board. In addition, there must be other life rafts of sufficient capacity to accommodate all persons on board unless the rafts which are mounted on each side are able to be launched easily from either side of the vessel. There must also be a life boat fitted with a motor, a Class C boat fitted with motor or an inflatable boat with motor, all of the davit-launched type.

   The life rafts, in this case, must be of the launching appliance type if the distance from the embarkation position to the water exceeds 4.5 m with the ship in the lightest sea going condition.
4. For vessels of less than 500 tons the lifeboat requirements are as for vessels of 500 tons or over as described above plus life rafts as described above or a life boat or Class C boat or inflatable boat (all of which are capable of being launched on one side only) plus at least two life rafts of total capacity to accommodate twice the number of persons on board.
5. Ships of 150 m length or more which have no midships structure are required to carry an additional life raft of six persons' capacity or more which must be sited either as far forward or as far aft as appropriate, i.e. at the opposite end of the ship from the accommodation.
6. Unless stated previously, it is assumed that all life rafts are able to be launched from either side of the vessel.
7. The life boats as described above must be attached to gravity davits unless the boats weigh not more than 2300 kg in the turning-out condition, i.e. boat plus equipment and two persons. In this case, luffing davits are permitted. If the vessel is 1600 tons or more, one of these boats must be a motor boat.
8. Each ship of this class must carry portable radio emergency equipment.

11.8.6.2 Lifebuoys

1. Ships under 500 tons are required to carry at least four buoys. Two will be fitted with light and smoke signals and will be capable of quick release from the bridge. Two will be fitted with a buoyant line of a suitable length but not with a light.
2. Ships of 500 tons or more are required to carry at least eight buoys, two of which will be fitted with light and smoke signals and be capable of quick release from the bridge. At least two more buoys will have lights fitted and at least two more (one on each side) will be fitted with a line of a suitable length but no light.

   Lifebuoys must have an outside diameter 76 cm and an inside diameter of 45.5 cm.

The weight of the buoy is to be not more than 6.15 kg.

If the release of a self-igniting light depends upon the weight of the buoy, then that buoy must be not less than 4.3 kg.

A lifebuoy must support 14.5 kg of iron in fresh water for 24 h.

11.8.6.3 Lifejackets

Lifejackets (*Figures 11.17, 11.18* and *11.19*) must be provided as follows: one for each person on board who weighs 32 kg or more plus one for each person on board who weighs less than 32 kg. In addition: if there are over 16 persons on board, 25% extra jackets must be carried; if 4–16 persons are on board, four extra jackets must be carried; if less than 4 persons are on board, two extra jackets must be carried.

There are two sizes of jackets, the larger to support persons of 32 kg or more, the smaller to support persons of less than that weight. Each type has its own buoyancy specification which is tested over a 24 h period. The securing tapes on lifejackets must be able to withstand a load of 140 kg.

Figure 11.17 A once-only suit, as used by the Royal Navy. The general purpose lifejacket is also shown (courtesy of Avon Inflatables Ltd)

Figure 11.18 Royal Naval lifejacket. The jacket is designed to meet BS 3595 (1981) and is available as an air (oral inflation) lifejacket which provides 15.8 kg (35 lb) minimum buoyancy when fully inflated. Alternatives are an air/CO_2 which is inflated by a pull toggle or a CO_2 automatic inflated by water activation, manual or oral inflation (courtesy of Avon Inflatables Ltd)

(a)   (b)

Figure 11.19 Lifejacket (a) prior to and (b) after inflation (courtesy of Avon Inflatables Ltd)

Figure 11.20 Expect maximum deflection limits, either side of the target when engaged in line throwing operations (plan view)

18.6.4 Line throwing appliances

A line throwing appliance must be carried on all ships of this class. This appliance will consist of a pistol and four individual rockets and four lines or four separate self contained units, each with line and rocket. The lines are approximately 250 m in length and at least 4 mm in diameter. The rockets must be capable of taking the line to a minimum distance of 230 m in calm weather. The lateral deflection of the rocket must not exceed 10% of the length of flight.

11.8.6.5 Parachute distress rockets

At least 12 parachute distress rockets are to be carried on the bridge of each ship of this class. Electric lighting, operated by main and emergency power, must be provided for alleyways, stairways, exits and launching areas on these vessels.

11.8.7 Class VII(T) foreign going tankers

The majority of requirements for these vessels are identical to those for Class VII vessels and in these cases we will simply refer to the Class VII explanation previously mentioned. There are, however, certain adjustments in the provision of boats and rafts which will be covered in detail.

1. For ships of 500 tons or over there should be sufficient life boats on each side of the vessel to accommodate all persons on board. If the vessel is 1600 tons or over, the boats should be not less than 7.3 m in length.

   In addition, for a vessel of 500 tons but less than 1600 tons, there should be sufficient life rafts to accommodate all persons on board. If there are more than 16 persons on board, then there must be at least two rafts of approximately equal capacity. For a vessel of 1600 tons or more, there should be life rafts for at least half the persons on board.
2. For ships of 3000 tons or more, there should be at least two life boats on each side, of sufficient aggregate capacity to accommodate all persons on board. Two of these boats will be carried amidships and two carried aft. If the ship has no midships structure, then all four boats will be carried aft. If it is not practicable to carry four boats aft, then it may be permitted to carry only two boats aft (one each side) provided that: the boats are not more than 8.5 m in length; the boats are stowed as near sea level as possible; and the boats are stowed as far forward on the after structure as possible and in such a position that the after end of the boat is at least 1.5 times the length of the boat forward of the ship's propeller. In addition, there should be life rafts for at least half the persons on board.
3. For ships under 500 tons the lifeboat requirements are the same as for vessels of 500 tons or over, as described in Section 11.8.6.1 plus life rafts as described in Section 11.8.6.2, or a life boat or class C boat or an inflatable boat (all capable of launch on one side only) plus at least two life rafts of total capacity to accommodate twice the number of persons on board.
4. For ships of 150 m or more in length with no midships structure, additional raft requirements are as for Class VII above.
5. All life rafts be capable of launch on each side.
6. At 500 tons and over and 3000 tons and over, lifeboats are to be attached to gravity davits. Luffing davits are permitted if the ship is under 1600 tons and the boat weighs not more than 2300 kg in the turning out condition. At least one life boat on each side must be a motor boat if the ship is 1600 tons or over.
7. Portable radio requirements are as for Class VII.
8. Lifeboat requirements are as for Class VII.
9. Line throwing appliances are as for Class VII.
10. Lifebuoy requirements are as for Class VII.
11. Lifejacket requirements are as for Class VII.
12. Parachute distress rocket requirements are as for Class VII.
13. Electric lighting requirements are as for Class VII.

11.8.8 General notes applicable to these appliances

11.8.8.1 Lifeboats

The maximum weight of a lifeboat in a fully laden condition must not exceed 20 300 kg; it must not exceed 150-person capacity; unless it is a motor boat, it must not exceed 100-person capacity; and it must not exceed 60-person capacity unless it is a mechanically propelled boat.

11.8.8.2 Life boat davits

Luffing davits must be able to launch the boat if the ship has a 15° adverse list. Gravity davits must launch the boat if the ship has a 25° adverse list. Both types of davit must be fitted with a winch and have wire falls. The falls must be end for end every two years and renewed every four years. Stainless steel wire falls will last considerably longer than the galvanised steel wire falls and do not have to conform to this requirement (see *Figure 11.21*).

11.8.8.3 Lifebuoy signals

Lights are to be self igniting, tankers to have the electric battery type. They must be of a type which will

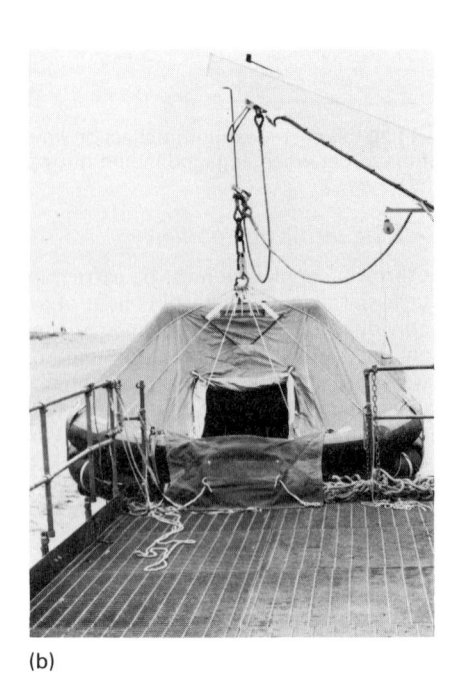

Figure 11.21 (a) Single arm davit turned out, showing the inflated life raft at the embarkation deck ready for boarding; (b) Single fall, suspension securing and the triung of the davit launched life raft. The access boarding flap is clearly seen in the foreground; (c) Container retaining lines following inflation

not be extinguished in water and must burn for not less than 45 min at a power of not less than 2 candelas. Smoke signals must have a self contained means of ignition and must produce dense orange smoke for at least 15 min. They are to be replaced every three years or sooner as necessary.

11.8.8.4 Parachute distress rockets (for boat, raft or bridge)

These must produce a single red flare and go to a height of at least 300 m. The flare must fall at not more than 5 m/s. They must burn for not less than 40 s at 30 000 candelas, and must burn out at a height of not lower than 50 m above sea level. These also are renewed every three years or earlier.

11.8.8.5 Hand flares (for boats and rafts)

These must produce a red light with a minimum of 15 000 candelas for at least 1 min. They are renewed every three years or sooner.

11.8.8.6 Life boat smoke signals

These must produce dense orange smoke for not less

Figure 11.22 Lifeboat smoke signals (courtesy of Viking A/S Nordisk Gummibadsfabrik)

than 2 min and not more than 4 min and must also be renewed every three years or sooner (see *Figure 11.22*). All life rafts other than those prescribed in Classes VII and VII(T) are to be stowed so that they will float free if the ship sinks. If lashings are used to secure the rafts, then a hydrostatic release must be incorporated into the lashing to automatically release those lashings when submerged not more than 4 m.

11.8.9 The 1983 amendments to the international convention for safety of life at sea 1974

These regulations came into force on 1 July 1986. Ships built after that date will conform to the new regulations completely. For ships which were built before that date, certain new regulations will have to be complied with by 1 July 1991. These will be indicated in the following paragraphs which detail some of the principal changes.

11.8.9.1 Cargo ships

These vessels will carry totally enclosed life boats on each side to accommodate all persons on board. Cargo ships other than oil tankers, chemical tankers and gas carriers which, in the opinion of the authorities, operate under favourable climatic conditions, may carry partially enclosed life boats instead. Chemical tankers and gas carriers will carry totally enclosed boats with self contained air support systems. Oil tankers, chemical tankers and gas carriers which carry cargoes with flash point not exceeding 60°C will carry fire protected totally enclosed boats.

All the above will also carry life rafts of total capacity to accommodate all persons on board. If these rafts cannot be easily transferred to each side, then there must be life rafts on each side to accommodate all on board. Existing vessels do not need to conform to the new rules for life boats unless the ship is replacing the life boats and launching gear. If the lifeboats are being replaced but not the launching appliance (or *vice versa*) then they may be replaced by a similar item to that which is being replaced. Existing vessels must, however, provide life rafts for all on board by 1991.

If the vessel is less than 85 m in length (other than oil or chemical tankers or gas carriers), they may carry, instead of these items, life rafts on each side for all on board. If these cannot be easily transferred to either side, then they will carry additional rafts so as to produce a total capacity on each side to accommodate 150% of the total number of persons on board. For vessels which choose this option, if any one survival craft is lost or is rendered unserviceable, there shall be sufficient survival craft available for use on each side to accommodate all persons on board.

On cargo ships, where the survival craft are stowed more than 100 m from the stem or from the stern, an additional life raft will be carried and stowed as far forward or aft as appropriate. In some cases, the vessel will require one forward and one aft. The rule will also apply to existing vessels by 1991.

All cargo vessels built from 1 July 1986 will carry at least one rescue boat. This may, in fact, also be a life boat, provided that it conforms to the requirements for a rescue boat. This may be either of rigid or inflated construction or a combination of both. It must be not less than 3.8 m and not more than 8.5 m in length and must be capable of carrying at least five seated persons and a person lying down. It must also be capable of towing the largest, fully loaded survival craft carried on the ship at a speed of 2 knots.

All the life boats on vessels built after 1 July 1986 will be motor boats and be fitted with a searchlight and have quick release mechanism. The launching appliances for survival craft on vessels built after 1 July 1986 will all be operated on a gravity system. The single arm davit for launching life rafts does not require the turning out operation to be by gravity. All lifesaving appliances are to be fitted with retroreflective material. This also applies to existing vessels by 1 July 1991.

Each ship will carry at least one manually operated EPIRB on each side of the ship and stowed so as to be rapidly placed in any survival craft (except those mentioned above). This also applies to existing vessels by 1 July 1991. Each ship must carry two way radio telephone apparatus for use between survival craft, between ship and survival craft and between ship and rescue boat. There must be at least three such units on each ship. These may also be used for normal onboard communications provided they are compatible with the specifications for the survival craft units. This rule will apply to existing vessels by 1 July 1991.

Each ship will carry immersion suits for each member of the rescue boat. The ship will also carry at least three such suits for each boat on board plus thermal protective aids (TPAs) for all persons on board who do not have an immersion suit. These three suits plus TPAs are not required if the boats are totally enclosed and can accommodate all persons on board on each side. Vessels less than 85 m having survival craft detailed above, will carry immersion suits for all on board unless the rafts are davit launched or are served by launching appliances which do not require persons to enter the water. All survival craft will be provided with thermal protective aids for 10% of the number of persons the craft is permitted to accommodate, or two, whichever is greater. Existing ships will comply with this rule by 1 July 1991.

Lifejackets. The rules are basically the same as previously stated except that all lifejackets are to have lights fitted. This also applies to existing ships by 1 July 1991.

Lifebuoys. New vessels will have a different criteria for lifebuoys; vessels under 100 m in length will carry at least eight; those between 100 and 150 m will carry at least ten. Vessels 150 to 200 m will carry at least twelve and over this limit will carry at least fourteen. Not less than half will have lights of which not less than two (equally distributed each side) will also have a smoke signal and be quick release from the bridge. At least one buoy on each side will have a buoyant line of length not less than twice the height from stowed position to waterline with the ship at its lightest sea going condition, or 30 m, whichever is greater. These buoys with lines will not have lights. At least one buoy is to be placed in the vicinity of the stern. The external diameter of the lifebuoys is to be not more than 800 mm and the internal diameter is to be not less than 400 mm. The weight is to be not less than 2.5 kg. If the buoy is to release the light or smoke signal, then the buoy must be of sufficient weight to effect the release, or 4 kg, whichever is greater.

Life rafts will be provided with two buoyant smoke signals as are the boats. These signals will produce smoke for at least 3 min. Radar reflectors will be provided in all boats and rafts. Falls used in launching will be turned end for end at intervals of not more than 30 months and be renewed at intervals of not more than five years or earlier if necessary.

11.9 Search and rescue operations

11.9.1 Introduction

With any marine emergency various organizations and authorities tend to become immediately involved, depending on the type and extent of the emergency incident. The most common participants are the coastguard, the military, Royal National Lifeboat Institute, air sea rescue stations, hospital and medical back up services and local shipping.

Where a major disaster occurs, such as the offshore oil installation *Piper Alpha* in the North Sea in July 1988, the combined efforts of all rescue services are employed by a coordinating centre. It is clear that no text could detail every eventuality but a knowledge of the resources, their availability, location and potential, could well provide on scene commanders with the means of establishing a successful outcome and a limitation to the numbers of casualties.

Rescue vehicles are varied. They may be land based or sea based, surface craft or aircraft. Whatever the type of rescue vehicle, it should be realized that it has its limitations: fuel, range, night flying capability, weather restrictions, self protection, lack of speed potential, payload capacity, etc. The survivor tends not to tolerate the limitations and quite rightly anticipates a professional solution to the immediate problem by the rescue services. However, in any incident, the human factor cannot be ignored and time is often the major problem for the survivor.

11.9.2 International distress signals

Most marine emergencies and subsequent rescue operations are triggered by the use of one or more of the recognized marine distress signals. These are as follows:

1. A gun or other explosive signal fired at intervals of about a minute.
2. A continuous sounding with any fog signalling apparatus.
3. Rockets or shells, throwing red stars fired one at a time at short intervals.
4. A signal made by radiotelegraphy or by any other signalling method consisting of the group • • • — — — • • • (SOS) in the morse code.
5. A signal sent by radiotelephone consisting of the spoken word 'Mayday'.
6. The international code signal of distress, indicated by NC.
7. A signal consisting of a square flag having above or below it a ball or anything resembling a ball.
8. Flames on the vessel as from a burning tar or oil barrel.
9. A rocket parachute flare or a hand flare showing a red light.
10. A smoke signal giving off orange smoke.
11. Slowly and repeatedly raising and lowering arms outstretched to each side.
12. The radio telegraph alarm signal.
13. The radio telephone alarm signal.
14. Signals transmitted by emergency position indicating radio beacons (EPIRBs).
15. Approved signals transmitted by radiotelecommunication systems.

The use of these signals for any other purpose other than distress is strictly prohibited.

11.9.3 General arrangements for search and rescue procedures

11.9.3.1 Marine casualties

Following the transmission of a distress signal by a vessel in distress, and the subsequent reception of this signal the coast radio station will rebroadcast the message across all distress frequencies, and will inform the relevant organizations. For the UK, these are Coastguard Radio Liaison Station (CRLS) and Lloyd's.

The CRLS, depending on the circumstances of the case, will ascertain which Maritime Rescue Co-ordination Centre (MRCC) or which Maritime Rescue Sub-Centre (MRSC) will coordinate the Search and Rescue (SAR) activities. This will continue until a successful conclusion is achieved or until that moment in time when the search is called off. In the event that the casualty is an aircraft it would be expected that the air traffic control centre would inform the appropriate rescue coordination centre.

11.9.3.2 Guidance for the Master or officer in charge of a distress

Information relevant to assistance and rescue methods can be found in the MERSAR manual, obtainable from the International Maritime Organisation. The contents of the manual will include communication methods, search pattern techniques and instructions on equipment use.

11.9.4 Guide to the duties of the on scene commander

In general, an experienced party with SAR training will be designated as the on scene commander (OSC). However circumstances may not make this possible, especially in the initial stages of an incident becoming apparent. The OSC could well be the captain of the first ship on the scene, who could effectively operate as the coordinator surface search (CSS). It would be the responsibility of the OSC to conduct the rescue and recovery operation in accordance with the plan and instructions of the SAR mission coordinator (SMC). The OSC will be expected to modify the plan depending on circumstances and the available facilities. Such modifications would be relayed to the SMC.

Effective communications between the OSC and the SMC should be maintained together with the continual monitoring of the weather and sea conditions prevailing. Other search and rescue units would maintain contact with the OSC and frequent situation reports would be made by him to the mission coordinator. Any results of search units or further recommendations would be communicated to the SMC. Search units would be released from their duties by the SMC, probably following the advice of the OSC.

Figure 11.23 (a) Inflatable boats 395 and 455, designed to meet the requirements of the Department of Transport for all vessels under 500 gross registered tonnes (courtesy of Lifeguard Equipment Ltd); (b) The Avon 7.4 m *Sea Rider*, also available in 6, 7 and 8 m sizes (courtesy of Avon Inflatables Ltd)

11.9.5 Current rescue methods

The mariner should always remember that location is not the same as recovery and rescue. Once a distress situation has been located, many factors could influence the time before actual rescue operations can be achieved, least of all the local weather conditions. However, assuming that location has been achieved the method of recovery could involve one or more alternatives.

Surface craft recovery will vary for different types of vessel. High freeboard ships would probably employ additional means to aid the rescue such as scrambling nets, knotted ropes, a crane or derrick basket, etc. It should be noted that with the use of such equipment a degree of self help by the casualty would be required and this may not always be possible if the party is injured.

Fast rescue craft (FRC) vary in design and include rigid, semirigid, and inflated types of hull. They are all usually of a low freeboard to assist the recovery of persons from the water. They are normally operated by two or three persons, a coxwain and swimmer/observer/medic. The advantage of the FRC is in its speed and versatility. Being a small fast craft, it can often get into confined areas, like the surrounding waters of an offshore installation. Examples of the current designs of fast rescue craft can be seen from *Figure 11.23(a)* and *(b)*.

Helicopter recovery, without doubt, has become the most widely used and accepted method of rescue for today's marine industry. It has all the advantages of the perfect rescue vehicle with few of the disadvantages. Its vantage point from the air is supreme and, when coupled with its ability to hover, can provide the ideal observation platform to aid location. Once location is achieved, it can provide the necessary recovery and ambulance service, wihtout the need, in most cases, for calling in back up surface craft (see *Figures 11.24* and *11.25*).

11.9.6 Alternative helicopter use

Land on evacuation is where a helideck, or ship's open deck space is available, when this method of recovery is preferred. A speedy rescue is achieved of several persons at once, up to the maximum payload capacity of the aircraft. Difficulties arise with heavy sea conditions, causing excessive motion on the ship's deck, or where the structure and hence the helideck are listed over due to damage or external influence. Potential hazards to the rescue aircraft could also exist with the proximity ot toxic chemicals, fire and/or local explosions in the deck landing area.

Figure 11.24 Designated Westland Wessex helicopter of the Royal Air Force Search and Rescue Wing

Figure 11.25 A Westland Sea King helicopter of the Royal Air Force engaging in a stretcher casualty recovery operations (courtesy of Westland Helicopters)

Search and rescue operations 11/41

Figure 11.26 Use of rescue and transfer recovery baskets. The photograph shows a training exercise in the use of a rescue basket. A semi rigid fast rescue craft stands by in the background (courtesy of Viking A/S Nordisk Gummibadsfabrik)

Figure 11.27 Use of the helicopter lifting strop (a) Arms up through. (b) The strap up under the arms. (c) The clamp is tightened. (d) Arms down along the body, or grip the clamp (courtesy of Viking A/S Nordisk Gummibadsfabrik)

Figure 11.28 A Westland Sea King helicopter operated by the Royal Air force engages in a strop recovery operation (courtesy of Westland Helicopters)

Single strop use is extensively employed for use from a deck or from the water. Whenever possible, an aircrew member will be lowered with a strop for use by the casualty (see *Figure 11.28*). However, with some operators, the survivor may find it necessary to position the strop personally (see *Figure 11.27*) and in this eventuality the following procedures is recommended:

1. Grasp the strop and place both arms and head through it.
2. Ensure the padded back strap is positioned as high as possible across the back, with the straps passing under the armpits and upwards in front of the face.
3. Pull down the toggle as far as possible.
4. When ready to be lifted, look upwards towards the helicopter, and give the thumbs up sign with the arm fully extended.
5. Place both arms back at the sides of the body although some operators expect the survivor to grip the strop.

During evacuation from small surface craft, the helicopter is often called upon to evacuate one or more persons from a small surface craft like a life raft or fast rescue boat. In every situation the circumstances will dictate the method and mode of operation. Where several persons are to be evacuated, the Hi-Line method may be adopted.

The proximity of the aircraft to the surface craft could have a direct bearing on the rate of drift, and/or steerage control of the vessel on the surface, due to the effects of downwash from the aircraft's rotors. Daylight or nighttime would also influence the position of the aircraft in relation to the surface craft. Also the type of craft which is engaged with the aircraft has a direct influence on the performance of the operation. Life rafts have no engine power. Lifeboats have limited power or speed ratio. Fast rescue craft have a comparative high level of speed and manoeuvrability, but are often adversely affected by noise and spray, because of the exposed hull form.

11.9.7 Communications

Wherever small vessels are engaged with aircraft, it is not unusual to encounter difficulties with radio communications. All SAR aircraft carry 2182 kHz. Most designated SAR aircraft also have the capability to receive and transmit, on 156.8 MHz, a VHF maritime distress frequency, which is generally carried by most vessels. Both SAR and civil aircraft can operate with 3023 and 5680 kHz when carrying HF communication equipment, which enables coordination in SAR operations and with the coast radio stations. In addition, the aeronautical distress frequencies of 121.5 MHz and 243 MHz are carried by civil and SAR aircraft.

11.9.8 Amphibian (rotor wing) aircraft (wet surface landing capability)

The design of certain helicopters, like the Sikorsky S61N and the HH-3F Pelicans, as operated by the US Coast Guard, can carry out surface landings on the sea. However, wet touch downs must be considered extremely hazardous and should not be considered a routine operation. Although the aircraft is designed with a sealed watertight hull it would generally only put down on the sea surface in an emergency and if weather and sea conditions allowed.

The mariner should bear in mind that few aircraft have this facility and if alternative methods of recovery are appropriate these should be adopted in preference to landing on the surface. In the event that an amphibious aircraft was so engaged, the main rotor would be kept operable to provide semihover capability and stability to the aircraft at surface level. It would always be carried out at the discretion of the pilot and the aircraft would remain for a minimum time at surface level.

11.9.9 SAR aircraft equipment

Designated SAR helicopters will carry specified equipment, but this will differ from country to country, and also between different types of aircraft having alternative payload capacity or alternative configuration. Items carried include life raft, strops, double lift harness, Aldis lamp, navigation equipment bag, set of translations, smoke of floats, marine markers, first aid box, Verey pistol, cartridges, breathing apparatus mask or fins, set of weight and knife, weighted message bag, long lead and headset, despatcher harnesses, lifejacket with personnel locator beacon, stretcher, blankets, Neil Robertson stretcher strops, ear defenders, wet winching helmet, thermal blanket, hi-line valise 90 and 150 ft rope, nylon cord (breaking strength 400 lbs), chemical night sticks, weighted bags, grapple, harness cutting tool, Karabina clips and, finally, lost diver marker buoy.

11.9.10 Droppable equipment from aircraft

Most air and maritime SAR facilities include stocks of basic survival equipment held ready for immediate use. These emergency packs are often in the form of several different coloured parcels and in the case of maritime operations these should also be buoyant. The contents of a marine pack consist of rations, medical supplies, blankets and protective clothing and miscellaneous items such as a fishing kit, compass, axes and lifejackets.

A combination pack of mixed contents are multicolour coded. Additionally life rafts could also be dropped when survival craft have become damaged, existing survival craft are overcrowded, where survivors are located in the water or when survival craft have become unserviceable.

Acknowledgements

In compiling this work, I would like to acknowledge the assistance of the following: Blackpool and the Fylde College, Nautical Campus; Viking A/S Nordisk Gummibadsfabrik; Lifeguard Equipment Limited; Avon Inflatables Limited; and Westland Helicopters Limited.

References

A Report of the Standing Advisory Committee on the carriage of Dangerous Goods in Ships, Department of Transport (1978).
Specification for Lifejackets, British Standard No. 3595 (1981).
Code of Safe Working Practice, Department of Trade, HMSO, London (1978).
Code of Safe Working Practice for Solid Bulk Cargoes, Intergovernmental Maritime Consultative Organisation (1987).
Docks Regulations, HM Government (1934).
Factories Act, HM Government (1961).
Grain Rules and Regulations, Intergovernmental Maritime Consultative Organisation (1982).
International Convention for the Safety of Life at Sea, Intergovernmental Maritime Consultative Organisation (1974 and amendments 1983).
International Maritime Dangerous Goods Code, Intergovernmental Maritime Consultative Organisation (1988).
International Regulations for Preventing Collision at Sea, Intergovernmental Maritime Consultative (1985).
The Merchant Shipping (Fire Protection) Regulations (Statutory Instrument No. 1218), HM Government (1985).
Merchant Ship Search and Rescue Manual, Intergovernmental Maritime Consultative Organisation (1986).
Offshore Installation, Guidance on Lifesaving Appliances, Department of Energy, HMSO, London (1978).
Offshore Installations (Operational Safety Health and Welfare) Regulations (Statutory Instrument No 1019), HM Government (1976).

Further reading

Annual Summary of Admiralty Notices to Mariners **4** HM Admiralty (1980)
Guide to Helicopter/Ship Operations (2nd edition) International Chamber of Shipping (1982)
Helicopter Landing Officers' Handbook (3rd edition) Offshore Petroleum Industrial Training Board, Montrose, Scotland (1983).
House, D.J. *Seamanship Techniques* vols. I and II, Heinemann, London (1987).
House, D.J. *Marine Survival and Rescue Systems*, E & F N S, London (1988).
International Manual of Maritime Safety Polytech International, London (1979).
Lavery, H. I. *Shipboard Operations* Heinemann, London (1984).
S.O.S. Manual Polytech International, London (1977)
Tanker Safety Guide (2nd edition) International Chamber of Shipping (1984).

Merchant shipping (M) notices

Navigation Safety (M854) Department of Transport (1978).
Operational Guidance for Officers in charge of a Navigation Watch (M1102) Department of Transport (1984).
Pilot Ladders and Mechanical Hoists (M898) Department of Transport (1979).
Radio Telephone. Distress Procedure (M743) Department of Transport (1976)
Use of Automatic Pilot (M1040) Department of Transport (1982).
Use of Radar and Electronic Aids to Navigation (M1158) Department of Transport (1984).

12 Radar and Electronic Navigation

Sean Waddingham
(Sections 12.1–12.6)
Kenneth MacCallum
(Sections 12.7–12.13)
Victor J. Abbott
(Sections 12.14–12.19)
Stewart Grimes
(Sections 12.20–12.26)
Brian Whiting
(Sections 12.27–12.33)
Colin M. Brown
(Sections 12.34–12.43)

Contents

Navigation systems

12.1 Introduction
    12.1.1 The electronic revolution
    12.1.2 Change of emphasis
    12.1.3 Basis of navigation

12.2 The model of the Earth
    12.2.1 Geodesy

12.3 Hydrography
    12.3.1 Data acquisition
    12.3.2 Marine charts
    12.3.3 Physical features
    12.3.4 Survey methods
    12.3.5 Tidal datum
    12.3.6 Chart production
    12.3.7 Chart corrections
    12.3.8 New developments in hydrography

12.4 Location of observer
    12.4.1 Coastal and ocean navigation
    12.4.2 Dead reckoning
    12.4.3 Speed and heading measurement
    12.4.4 Steering and autopilot
    12.4.5 Tidal or current set
    12.4.6 Position fixing
    12.4.7 Ocean navigation
    12.4.8 Astronomical navigation

12.5 The beginnings of electronic navigation
    12.5.1 Shipboard factors affecting direction finding
    12.5.2 Environmental factors affecting accuracy
    12.5.3 Shore radio stations
    12.5.4 Usage of direction finding

12.6 Navigation in the modern world

Acoustic sensing and positioning systems

12.7 Introduction
    12.7.1 Why use acoustics?
    12.7.2 Historical development of underwater acoustics

12.8 Acoustic theory
    12.8.1 The sound field
    12.8.2 Transmission
    12.8.3 Propagation
    12.8.4 Reflection

12.8.5 Noise
12.8.6 Prediction

12.9 Acoustic system components
12.9.1 Introduction
12.9.2 Transducers
12.9.3 Signal processing
12.9.4 Measurement and display

12.10 Sensor systems
12.10.1 Echo sounders
12.10.2 Side scan sonars
12.10.3 Swathe sounding systems

12.11 Acoustic navigation and positioning systems
12.11.1 Introduction
12.11.2 Long baseline and intelligent systems
12.11.3 Phase comparison or very short baseline systems
12.11.4 Short baseline systems

12.12 Other acoustic systems
12.12.1 Doppler log and navigation systems
12.12.2 Oceanographic systems
12.12.3 Telemetry

12.13 Conclusion

Earth-based electronic navigation systems

12.14 Introduction

12.15 Propagation of low and very low frequency waves

12.16 Racal–Decca navigator
12.16.1 System description
12.16.2 Transmitters
12.16.3 Receivers
12.16.4 Propagation anomalies
12.16.5 Accuracy

12.17 Loran C
12.17.1 System description
12.17.2 Transmitters
12.17.3 Receivers
12.17.4 Propagation anomalies
12.17.5 Operational details
12.17.6 Accuracy
12.17.7 Conclusion

12.18 Omega
12.18.1 System description
12.18.2 Transmitters
12.18.3 Receivers
12.18.4 Propagation anomalies
12.18.5 Operational details
12.18.6 Accuracies

12.19 Conclusions

Earth-based electronic precise positioning systems

12.20 Introduction

12.21 Categorization of systems
12.21.1 System modes
12.21.2 Phase comparison systems
12.21.3 Pulse systems

12.22 Laser systems
12.22.1 Atlas Polarfix
12.22.2 GOLF 2

12.23 Microwave systems
12.23.1 Artemis
12.23.2 Axyle
12.23.3 Micro-Fix
12.23.4 Trisponder

12.24 UHF systems
12.24.1 Trisponder
12.24.2 Syledis

12.25 HF systems
12.25.1 Argo
12.25.2 Geoloc
12.25.3 Hi-Fix 6
12.25.4 Hyper-Fix

12.26 LF systems
12.26.1 The Datatrak system
12.26.2 Pulse 8

Satellite navigation

12.27 Introduction

12.28 Satellite orbits

12.29 The Navy Navigation Satellite System (NNSS)

12.30 Navstar Global Positioning System

12.31 Starfix

12.32 Geostar

12.33 Navsat

Radar

12.34 The history of radar
12.34.1 Marconi
12.34.2 The impetus of war
12.34.3 The demand for centimetric wavelengths
12.34.4 Component development

12.34.5 The breakthrough
12.34.6 The completed system

12.35 Parts of radar equipment
    12.35.1 Basic radar
    12.35.2 Transmitter
    12.35.3 Receiver

12.36 Cathode ray tube
    12.36.1 Control settings
    12.36.2 Time base coils and unit

12.37 Antenna
    12.37.1 Slotted waveguides
    12.37.2 Scanner rotation

12.38 Time synchronized components
    12.38.1 Anti-sea clutter
    12.38.2 Brightening pulse
    12.38.3 Calibration rings
    12.38.4 Variable range marker

12.39 General aspects of radar performance
    12.39.1 Radar horizon
    12.39.2 Pulse length
    12.39.3 Pulse repetition frequency (PRF)
    12.39.4 Target characteristics
    12.39.5 Frequency bands
    12.39.6 Radar beacons – racons
    12.39.7 Identification of ships by transponder

12.40 Collision avoidance interpretation of radar data
    12.40.1 Relative motion
    12.40.2 True motion

12.41 Computer aided systems
    12.41.1 Collision avoidance data
    12.41.2 Automatic radar plotting aid (ARPA)

12.42 Radar in vessel traffic systems
    12.42.1 VTS research

12.43 Electronic chart display systems (ECDIS)

Navigation systems

12.1 Introduction

The pace of development in seafaring over the centuries has generally been slow, with a gradual improvement in technique and refinement in design of vessels and equipment. The sailing vessels of the early nineteenth century were not so very different to those of four hundred years before. Ships were still limited in size by their construction in wood; speeds had increased little through improved hull designs and vessels were more able to withstand bad weather. However, none of these refinements to the basic design could overcome the dependence of the sailing ship upon the wind. Speed was related directly to wind strength and direction. Adverse winds could make progress in a given direction impossible, with major diversions from the direct route being common to take advantage of prevailing winds. As a result, transit times port to port could vary widely. Severe weather conditions often led to damage or loss of vessels.

The second half of the nineteenth century saw changes in seafaring which irrevocably altered its very nature. Engineers and naval architects produced first, steam powered ships independent of the wind and later, iron and steel hulled vessels. Steam powered ships made consistent speeds and transit times between ports possible, leading to more organized trading, coinciding with improved communications due to the laying of international telegraph lines. Larger vessels were possible with the adoption of iron and, later, steel in hull construction. Higher cargo capacities resulted and these vessels were able to withstand wind and weather far better.

Advances in the technique of navigation tended to lag behind those in engineering. The new, faster ships were little better informed about their position and the relative location of hazards and other environmental features than were the vessels at the beginning of the century. Safety margins for the avoidance of hazards still had to be large, since poor visibility could rob the vessel of positional information from astronomical observations.

12.1.1 The electronic revolution

The changes brought about by electronics, including radio, in this century have been as far reaching as the engineering advances of the last. Radio and, later, satellite communications have made contacting a vessel anywhere on the globe as easy as making an international trunk call. In the last 40 years, radar has largely overcome the difficulties created by darkness and fog, and huge advances have been made in the area of navigation, using dedicated transmissions from shore stations.

The earliest use of radio signals for navigation purposes was by direction finding of shore based transmissions and was basically an extension of the technique of visual bearings from land masses. The military positioning needs of World War II created the basis for ranging (measured time delay) navigation systems, which include the Decca Mainchain in common use today.

By the 1960s, the use of artificial satellites in space to transmit radio navigational data to vessels anywhere in the world allowed position fixing at intervals of a few hours. At the start of the 1990s, new satellite systems and computers will give the navigator the position to within a few metres at any time in good or bad weather, anywhere.

12.1.2 Change of emphasis

This precision will remove the uncertainty of position of the vessel, leaving only uncertainties about the location of hazards. In the near future, the challenge of navigation will be the mapping of all hazards and obstacles to enable the accurately located vessel to avoid them with the minimum clearance and therefore loss of time. The art and science of the navigator using astronomical observations, relative locations from nearby land masses and navigation aids, including buoys and lighthouses, to derive an uncertain position will have been replaced. Continuous positioning using systems, including the global positioning system (GPS), will have a worldwide usage and be applicable to ships, aircraft and land vehicles. It may even be offered as a dashboard display on the automobile of the future and be accurate enough to decide on which side of the highway the automobile is located.

12.1.3 Basis of navigation

Even in an era of rapid change, the principles of navigation hold true. The navigator's task remains to guide the vessel between two points via the shortest route, maintaining adequate safety margins away from known or possible dangers. It is necessary to have a database of information on the environment and that the observer can determine the vessel's position accurately.

12.2 The model of the Earth

12.2.1 Geodesy

Geodesy is the study of the shape of the Earth and by implication the creation of coordinated systems to define the locations of points on its surface. This has been a continual process of refinement over the centuries as scientific techniques have improved, with recent rapid advances associated with the advent of satellite mapping techniques.

One of the earliest discoveries must have been the east to west movement of stars and planets and therefore the location of the pivot point about which the sky appears to rotate. This knowledge of direction independent of terrestrial landmarks was the first real awareness of navigational skills.

12.2.1.1 Directions on the Earth

In general, marine navigation considers the Earth to be a perfect sphere rotating around a polar axis which passes through the North and South geographic Poles. The Equator is a circle running around the Earth half-way between the Poles and at 90° to the polar axis. The Earth's rotation moves a point on the surface eastwards in a direction parallel with the Equator, with west being its opposite. These four directions, termed the cardinal points of the compass, North, South, East and West are at right angles to one another at any one point on the surface. Modern practice subdivides the compass into 360°, starting at 000° for north and continuing clockwise through 90° at east and so on, until returning via 359° to 000° again. The use of a 360° system for direction can be seen in *Figure 12.1* where the relationship of two vessels is shown.

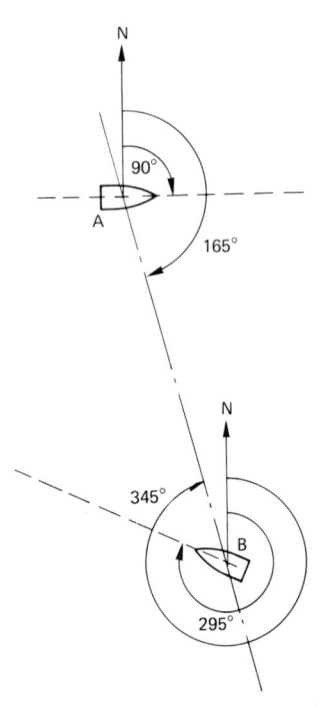

Figure 12.1 Headings and bearings: ship A is on a heading (or course) of 090° (referenced to north); ship B is on a heading of 295°; ship B is on a bearing of 165° from ship A; and ship A is on a bearing of 345° from ship B

The four cardinal directions based on 90° intersections tend to impose a familiar two-axis coordinate grid, but it cannot be forgotten that the Earth's surface is spherical and therefore two axes are really insufficient. Points on the surface due east or west of each other are equidistant from the Equator. This is not true of the north–south axis, since from any two points on the Earth's surface the north directions must converge at the North Pole.

12.2.1.2 Coordinates, latitude and longitude

The spherical nature of the Earth has resulted in a coordinate system based on angular measure in degrees. Since the plane of the Equator is at 90° to the polar axis, the distance between is divided into 90° of latitude, the Poles being 90° north or south and the Equator 0°. Lines joining places of equal latitude are termed parallels of latitude, since they are truly parallel and equidistant, with 1° of latitude being the same distance at the Pole and the Equator. The same is not true of the coordinate running through the Poles, termed longitude.

The meridians, lines of equal longitude, all pass through the North and South Poles and therefore converge as they approach those points. Therefore the measured length of 1° of longitude reduces from a maximum at the Equator towards zero at the Pole. The convention of longitude is that the circumference of the Earth is divided into two 180° sectors with a common 0° meridian, internationally agreed to pass through Greenwich, just south-east of London. Longitudes are expressed as degrees east or west from the Greenwich Meridian. The 180° meridian is on the opposite side of the Earth from the 0° meridian and is used for most of its length for the International Date Line.

12.2.1.3 Shortest distances between points

The meridians and the Equator are all lines travelling the full circumference of the Earth and as such are termed great circles. A great circle is a true circumference of the Earth, that is, a line on the surface whose plane passes through the centre of the Earth. Since it cuts the surface at 90°, it represents the shortest distance between any two points along its length. Of the parallels, only the Equator is a great circle, because none of the others have planes which pass through the centre.

Straight lines which maintain a constant heading (compared to north) while joining two points on the Earth's surface are termed rhumb lines. However, if the line is extended around the Earth, the plane surface generated will probably not pass through the centre of the globe. Therefore, the line between the two points is not a great circle course and is not the shortest distance. This effect is most noticeable on east–west routes in the higher latitudes, i.e. transatlantic or transpacific where the difference between great circle and rhumb line

The model of the Earth 12/7

courses may amount to hundreds of kilometres. The most extreme cases are transpolar air routes where two points on the same latitude are closer across the Pole than by following a constant course around the parallel. When travelling between two points on the Equator or on the same meridian, the great circle and rhumb lines coincide.

Figure 12.2a shows the shortest distance between A and B through the Earth as a chord which lies on the plane passing through the centre at C. The plane meets the surface to give the great circle course which crosses the meridians at differing angles. The practical difficulty with using a great circle course is that theoretically it is a continually changing heading. In practice, a series of short rhumb line (and therefore constant heading) courses are steered to approximate to the great circle.

Figure 12.2b attempts to explain the apparent contradiction of a line which curves in direction and yet represents the shortest distance between its ends. It shows a segment of the Earth's surface with lines joining diagonal corners. In the upper diagram, the observer's eye is on the plane of the great circle through X and X'. The line joining Y and Y' is a rhumb line maintaining a constant heading across all the local versions of north which are converging on the Pole. However, when this convergence is removed, as on the normal marine chart, the rhumb line is rendered as a straight line due to its constant heading. The great circle is then displayed as an arc curving towards the Pole, as shown in the second diagram.

12.2.1.4 Ellipsoids and geoids

As we have seen, standard marine navigation considers the Earth to be a perfect sphere, but over the centuries and especially since satellite remote sensing, it has become obvious that the planet is far from a regular shape. Even with topographical features like mountain ranges and ocean trenches averaged out there are regional 'highs' and 'lows' which, in an age of increasingly accurate navigation systems, means that more realistic mathematical models of the Earth are necessary. This smoothed out shape is termed the geoid and is based on average sea level. Therefore a number of standard ellipsoids representing ever closer approximations to the actual measured shape of the Earth have evolved. Some have a general application and are used as standards worldwide.

The ellipsoid called WGS-72 is used by the Transit satellite navigation system. Positions derived from it are slightly different from those plotted on a chart which represents a spherical planet. In some cases, an ellipsoid has been generated which more closely approximates to the geoid within a limited area of the Earth's surface; for instance, ED 50 represents the actual geoidal shape within Europe better than other worldwide standards. These higher accuracy ellipsoids are primarily of interest in specialist fields like undersea oil exploration.

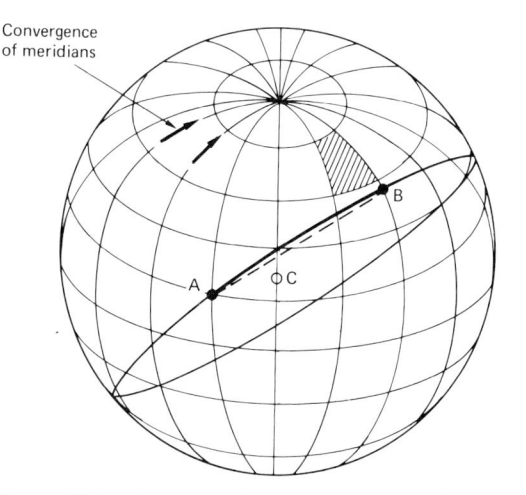

Figure 12.2a Shortest distances between points: the shortest route between A and B travels through the earth (----); the line projected on the Earth's surface represents part of a circumference based on the Earth's centre and passing through A and B

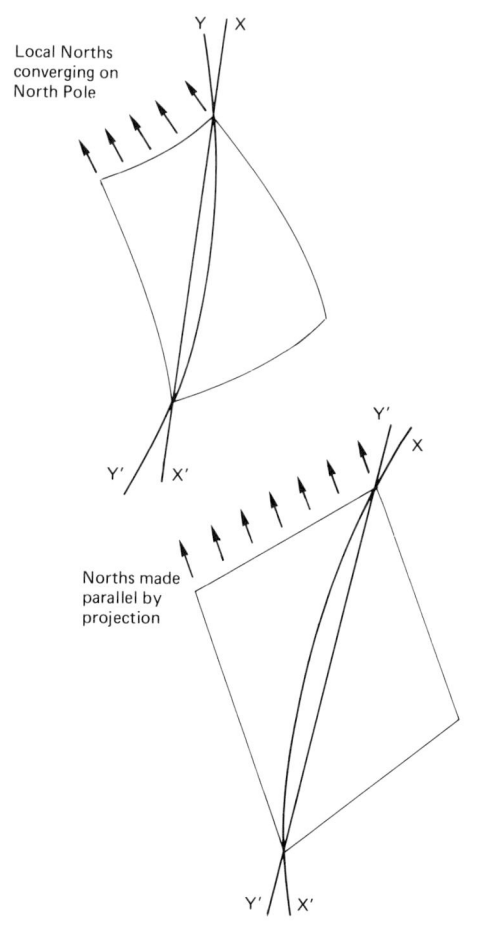

Figure 12.2b Shortest distances between points

12.2.1.5 Marine chart projections

The representation of part of a spherical surface on a flat two-dimensional chart means that an element of distortion is inevitable. Maps and charts of the globe are made using the principle of projection. Locations on the chart are produced as if by projection from a point or points of origin through the corresponding locations on a transparent globe and on to a surface which may be curved or flat.

Figures 12.3 and *12.4* show visualizations of the two most important projections for marine charts, Mercator and Gnomonic. In both cases, the projection point is at the centre of the globe. In the case of the Mercator, *Figure 12.3* (right), the projection surface representing the chart is in the form of a cylinder contacting the globe at the equator. It can be seen in the cross-section (left) that projection rays from high latitudes and the Poles would not touch the cylinder unless it was of infinite length. In the Gnomonic example (*Figure 12.4*), the projection surface is a plane touching the globe at one point of contact only. This may be anywhere on the globe, with the Pole frequently being chosen to overcome the lack of coverage of the Mercator in this area.

The Mercator projection is the most common form used for marine charts and when applied to the complete surface of the globe produces the traditional school atlas view of the Earth's surface. *Figure 12.5* is a Mercator world map showing the deviations from the geoidal reference. All parallels and meridians are rendered as straight lines, making the derivation of co-ordinates for physical features easy. Since there is no representation of the convergence of meridians toward the Poles, the direct bearing between points A and B is the rhumb line track. However, there is distortion in both horizontal and vertical directions and from this results the familiar expansion of land masses in the more extreme Northern and Southern Hemispheres. A degree of longitude on, for example, the 50° parallel is physically shown the same size as one on the Equator, and therefore land masses and oceans are stretched laterally.

Due to the system of projection, the physical size on the chart of a degree of latitude becomes larger as one moves away from the Equator, therefore again there is expansion toward the Poles. This expansion is geometrical in nature and precludes use of the projection in polar regions. Since the rhumb line course between two points is a straight line, then the great circle route plots as an arc, curving toward the nearer Pole. If the comparison between rhumb and great circle is made at points further towards the Pole then the degree of curvature in the latter becomes more pronounced due to increased distortion.

The gnomonic projection has different characteristics in that great circles and meridians appear as straight lines while parallels are curved. The most familiar usage of this projection is of Antarctica centred on the South Pole. Here again there is expansion of scale, this time in all directions away from the tangent point of the projection, in this case the Pole itself. Gnomonic charts used for ocean passages allow the plotting of the great circle course between two points as a straight line. On this type of chart, the rhumb line course appears as a curve. Since it is difficult to steer a great circle, it is from a chart on this projection that coordinates along the route are derived. These are then transferred to a Mercator to produce a series of short segments, each with its own compass heading which would together approximate to the great circle course.

12.2.1.6 Chart projections and distance

Since distortion is produced inevitably when a projection is used, it is obvious that scale will vary within the

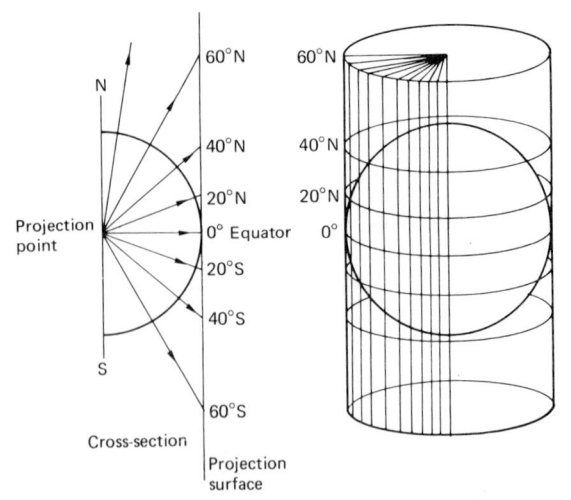

Figure 12.3 Mercator chart projection: the poleward increase in vertical distortion restricts its usefulness; a parallel is rendered the same diameter as the Equator, hence lateral distortion is introduced

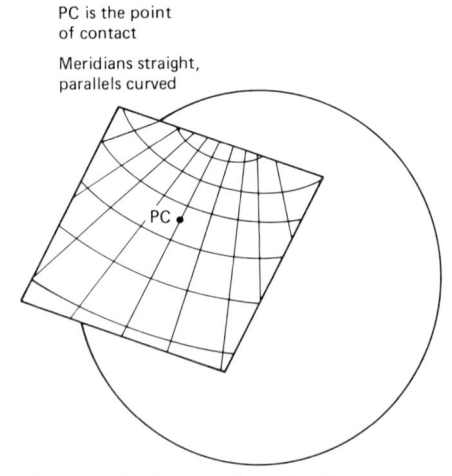

Figure 12.4 Gnomonic projection

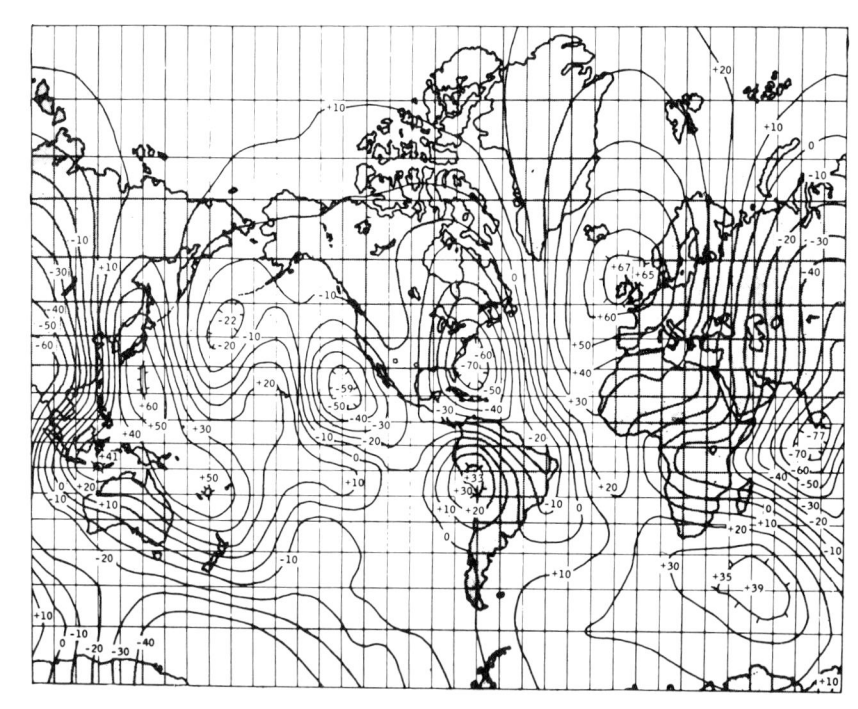

Figure 12.5 Geoidal chart showing deviations (m) of the Earth's shape, represented by mean sea level, from the shape of a reference ellipsoid; the level in the North Sea is about 65 m (213 feet) higher and near Sri Lanka about 77 m (252 feet) lower than reference (from Sonnenberg, 1987, courtesy of the author and publishers)

area covered by the chart. Since the angles used to produce latitude and longitude are constant through this distortion, they are used as a guide. The distance on the surface corresponding to 1 min of arc at the centre of the earth is 1 sea mile. The ellipsoidal nature of the planet means this may vary from 1844 to 1862 m (6050 to 6109 ft) and an internationally agreed standard for the nautical mile of 1852 m (6076 ft) has been established. A British Standard value of 1853.18 m (6080 ft) also exists. The nautical mile may be divided decimally, a tenth of a mile archaically being known as a cable.

It should be remembered that the degree of longitude covers a reducing distance as one moves polewards due to the convergence of meridians, so it cannot be used as a distance scale on the chart. The scale of minutes on the latitude border of the chart can be used, although on small scale charts the distance in nautical miles must be read on the same latitude as it is to be used. This is because the scale varies so much within the chart that it would be incorrect to transfer a measured value from one part to another.

A speed of 1 nautical mile/h is 1 knot and this unit is used for vessel, wind and water current speeds. Therefore, for example, a distance of 12 nautical miles would be covered in 2 h at a speed of 6 knots. In specialized circumstances, speeds may be expressed to two places of decimals of a knot, but vagaries of water movement make further subdivision of limited usefulness.

12.3. Hydrography

12.3.1 Data acquisition

The acquisition of data and the subsequent production of marine charts is the task of the hydrographer. Since they have a responsibility for the safety of ships and lives, the bodies which gather hydrographic data are normally government departments, very often part of the country's navy. Most maritime countries have their own hydrographic service and produce charts of their own home waters as well as making this information available to other hydrographers. Some, like the (British) Admiralty Hydrographic Office, offer surveying and charting services to other countries. The National Ocean Survey has hydrographic responsibility for US home waters and possessions while the (US) Naval Oceanographic Office produces charts covering the rest of the world.

The increasing pace and diversity of marine activities have heightened the pressure on the world's charting community to an unprecedented degree. Larger vessels, competitive pressures on routing and the special requirements of the offshore oil industry have created a demand for modern survey information on a scale that cannot be satisfied with the limited resources available. In addition, the rapid changes in navigational technology have themselves conspired to render old data obsolete. Despite the pooling of information and resources under the auspices of bodies such as the International Hydrographic Organization, large areas of the world's coastal waters are inadequately surveyed by modern standards.

In an age of satellite sensing and aerial survey, it is possible that techniques from these fields will assist further in hydrographic surveying. At the moment they are largely restricted to the delineation of coastlines and other surface features. To the mariner, these are more than matched in importance by the need to look under the water surface and sense depth and to a certain extent the nature (rock, sand, etc.) of the sea bed. Complicating this third dimension in most parts of the world is the tidal factor, which may change the usable water depth by a number of metres. Many channels and approaches to harbours are only navigable when high tide gives additional clearance beneath the ship's keel. Modern vessels may draw over 25 m (82 ft), whilst as recently as the late 1950s, 15 m (49 ft) was considered deep draughted. As a result, large areas of the continental shelf, formerly considered to be deep enough not to warrant detailed charting, have been added to the list of areas where vessels might be in danger of grounding.

In view of the disparity between the need for modern charting around the globe and the resources available, it is necessary to concentrate the efforts of hydrographers in areas of greatest need. Therefore, the more heavily travelled sea routes receive the most attention, whilst the more remote corners of the globe are unlikely to be surveyed so often or in as great detail. Data on the survey date of a chart and of subsequent revisions are usually displayed upon it and the limitations of contemporary techniques in addition to any likely physical changes since the time of the survey should be taken into account upon use.

12.3.2 Marine charts

Marine charts are produced in a wide variety of scales, depending on their intended usage. High detail surveys are necessary for port and harbour areas to produce charts with a scale ratio of 1:5000 to 1:12 500. These are termed large scale charts or plans. The approaches to harbours and offshore anchorages may be surveyed in sufficient detail to produce charts to a scale of 1:25 000. Coastal areas passed through on passage may only be covered from 1:50 000 to 1:150 000 and are referred to as small scale charts.

Since these charts at different scales will almost certainly overlap each other, it is obvious that data gathered at a density suitable for a harbour plan could not all be incorporated on a coastal chart covering the adjacent coastline as well. Therefore, as scale is reduced, detail has to be suppressed in order to avoid visual clutter without losing vital elements. The principal elements must be retained to represent the general characteristics of the area, but at the same time critical features must be emphasized. For instance, a sandbank with a complex outline may be considerably simplified for a smaller scale chart whereas an isolated rock pinnacle must be reproduced however small the scale. On a chart covering the whole of the English Channel, the principal lighthouses and buoys are shown for the benefit of vessels in transit through the area. Larger scale charts would be referred to for details of buoy patterns for the major port approaches.

Water depths are usually represented with isolated numbers and also contour lines joining points on the sea bed of equal depth. Depending on the age of the chart edition, depth may be given in fathoms (obsolete measures equivalent to 1.829 m) or metres. It should be noted that in some cases new charts have been issued using old data converted into metric units. Most physical features, including rock outcrops, wrecks, buoys and lightships, etc., are represented by symbols. In most chart series there is a key chart for identification of these representations, the British Admiralty example being number 5011.

12.3.3 Physical features

Changes in the terrestrial landscape are generally slow processes, with hills and valleys being eroded over hundreds of thousands, if not millions, of years. However, the high energy environment of the sea means that waves and currents can radically reshape their surroundings in far shorter periods. Single storms may erode cliffs, throw up beaches or shift millions of tons of sea bed deposits. More often, changes are gradual but their rate of change due to local conditions will dictate the frequency at which new surveys are required. A harbour mouth prone to silting would therefore require resurveying more often than one with a rocky and therefore more unchanging bottom.

In addition to charting the natural environment, the influence of the human race must be recorded. Structures including piers, dredged channels, oil platforms, undersea pipelines and cables, not to mention buoys, lighthouses and other aids to the navigator, must also be printed on the chart. Notional features like the traffic separation zones used to segregate in and outbound vessels in congested waters must also be faithfully recorded. Special editions of charts are produced which are overprinted with the lattice scales necessary to use radio navigation systems including Decca Navigator, Loran and Omega. In many cases, it is alterations in these manmade features that necessitate the revision of the chart.

12.3.4 Survey methods

A marine survey consists, in essence, of water depths recorded along a line representing the path of the recording vessel. Formerly, these were taken at discrete intervals, with a weighted line with marks at a regular spacing. This use of leadlines continued as late as the 1930s, when the echo sounder came into use. This equipment uses the principle of acoustic two-way travel time together with a known through water velocity to produce a continuous graphic display of water depth directly below the recording vessel. However, it gives little information about depth on either side of the vessel's path, so the spacing of survey lines is critical to the overall resolution of the survey (*Figure 12.6*). The water depth measured must be corrected for the depth below the surface of the echo sounding head and also for the state of the tide to a reference height known as chart datum. The measuring of water depths and application of corrections is termed bathymetry. Echo sounders are now standard fitments aboard most ordinary tracing vessels for ordinary navigation.

In the 1970s, side scan sonar was introduced, using the same principle of sound ranging, but transmitting sideways, from a towed body behind the vessel at an oblique angle to the sea bed, to provide a sound picture on either side of the vessel's track. This allowed the identification of shoals, wrecks or rock outcrops away from the line of survey. However, the lack of precise knowledge about the angle at which the sound waves travelled meant that distance from the line of survey and therefore depth could not be computed for features identified by this method.

The vessel's position along the survey line needs to be located very accurately, to position the depth data precisely. In coastal areas, this was formerly done by visually measured angles from the coastline, where reference points would have been set up. Increasingly, modern surveys are utilizing radio navigation systems to provide all weather surveying with no deterioration in accuracy with distance offshore. For short ranges, laser ranging may be used for very high accuracy.

The linear nature of this bathymetric survey means that the resolution it will measure in terms of size of features on the bottom depends on the spacing of the lines. Therefore, a detailed survey in a harbour entrance is less likely to miss an isolated rock pinnacle than a general coastal study where the lines are more widely spaced.

12.3.5 Tidal datum

As has been mentioned, the depths recorded by the surveyor must be referenced to a datum level to compensate for the state of the tide at the time of survey. This enables a subsequent user of the chart to apply a correction to the depth to datum, based on tidal conditions at the time of use. This gives the chart user the actual depth to be expected, given average tidal conditions. In order to provide a margin of safety, the chart datum has been chosen to represent the lowest low water situation likely to be encountered. The standard chosen has, in the past, varied considerably between countries and indeed between surveys. It is now the policy on most new surveys to adopt the lowest astronomical tide (LAT). This is the lowest predictable tide under average meteorological conditions. Formerly, mean low water springs (MLWS) was used, but had the disadvantage that water depths were frequently lower than those shown on the chart.

Reference works, including the Admiralty Tide Tables, give the predicted heights and times of low and high water at larger ports. Tables of differences are provided to allow derivation of values for smaller ports nearby. Cotidal and corange charts allow extrapolation to estimate tidal height and timing offshore. Tidal atlases show the direction and strength of tidal currents at different states between high and low water in a graphical form.

All these forecasts are based on the astronomical component of the final tidal height and are calculated months in advance. Local meteorological conditions make considerable differences to the actual limits of the tidal range, especially in confined waters like those of the southern North Sea. The timing of high and low water is also affected under these circumstances, both in when they occur and their duration.

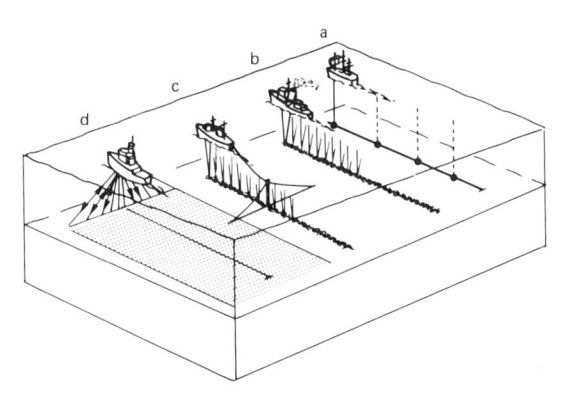

Figure 12.6 Progress in bathymetric coverage: lead lining a shows the discrete nature of the measurements, while the echo sounding vessel b secures continuous coverage along the survey line; with side scan sonar c the vessel can sight features and obstructions off the line of survey; d shows a vessel using an array of hull mounted transducers to provide true depth coverage on a wide swathe; the first vessels using this technique are now coming into service

12.3.6 Chart production

As with most other aspects of navigation, electronic techniques have made possible vast steps forward in surveying and subsequent chart production. Data from electronic navigation systems, echo sounders, side scan sonars and other sensors are processed by computer, digitally recorded on tape and used to form a database. This data is further processed before being used to drive the draughting plotters which draw the master charts.

Although attempts are being made to automate still further the layout of the final chart, the decisions necessary to display data graphically in the most helpful way must still be made by the cartographer.

After publication, charts may be purchased from various sources, but, to ensure that they have been subsequently kept up to date, they should be supplied by an Admiralty chart depot or agent. In order to cover large areas, a series of geographically linked charts are frequently made available as a standard folio. Alternatively, the abridged folio is available for vessels only making passage through an area. The catalogue of admiralty charts lists the full range of charts, folios and other publications by the British Hydrographic Department.

12.3.7 Chart corrections

For reasons already considered, charts become out of date quite rapidly, so after initial publication it is essential that they are modified on the basis of the latest available information. The sources used may include the Admiralty *Notices to Mariners*, which are published on a weekly schedule and for the most urgent information, radio navigational warnings.

Radio navigational warnings are transmitted at fixed times from coastal radio stations and give details of new wrecks or obstructions, damage or alteration to lights or buoys, cable laying operations or other information of use in coastal navigation within the local area covered. These details are normally only added in pencil to a chart, since they may be only temporary, whereas if more permanent in nature, they will be covered in the next *Notices to Mariners*.

The *Notices to Mariners* are booklets available from chart agents and depots and contain the information needed to keep the full range of Admiralty charts and other publications up to date. Descriptions of the corrections to be drawn on the charts are included and where these are too complicated, a redrawn section of the chart is reproduced at the correct scale. This new section is termed a block and may be cut out and pasted on the chart in the correct location. After corrections have been made to a chart, the relevant reference number should be inserted in the margin, so that a subsequent user can see at a glance how up to date the information is. Eventually the number of corrections becomes such that a new chart edition becomes necessary.

12.3.8 New developments in hydrography

The recent rapid advances in technology which are revolutionizing position finding are also giving help to the hydrographer in coping with the demand for improved charting. The application of improved data processing and new sensing techniques will help at the survey stage and electronic charts and real time corrections utilizing better telecommunications at the user stage.

In order to overcome the limitations of the traditional bathymetric survey lines, new vessels are coming into service incorporating multibeam sonars. These use an array of echo sounder transducers mounted in the hull so as to give a fan shaped output extending up to 52° either side of the vertical. Highly sophisticated pitch and roll sensors allow compensation for vessel movement so that the foreshortened oblique angle view familiar from side scan sonar can be turned into a series of accurate slant ranges. These in turn can be related to vessel track and water surface so as to give the vessel bathymetric coverage over a swathe across the sea bed. The width of the swathe varies with water depth, but the vessel can now create complete sea bed coverage by sailing traditional survey lines.

Whilst the survey vessel and its launches offer a high capability system, there is scope in many areas for sensing outfits with more specialized characteristics. The LIDAR system uses an array of downward facing laser range finders with water penetrating capability, mounted in an aircraft. At optimum operating height a swathe 270 m (886 ft) wide is covered at aircraft speeds over 100 mph with water depth sensing accuracies not too far from those achieved by traditional echo sounding. The technique has already been used for the production of charts in northern Canada and offers considerable potential in areas with good water clarity and depths of less than 30 m (98 ft). Within these limitations, the primary advantage is a capability to cover up to 35 km^2 per aircraft hour and consequently offer very low operating costs in comparison with a vessel based survey.

In parallel with the advances in sensor technology, great strides are being made in the development of systems to display the data for the eventual user, the vessel navigator. Perhaps the most far reaching is the development of electronic chart display and information systems (ECDIS). Work is going on via a number of bodies to overcome the limitations of the paper chart as a data storage medium and to provide a computer controlled database outputting to high resolution video monitors. This would be capable of displaying all the information available on a traditional chart at a variety of scales with the capability to selectively include categories of information. These categories of information would include coastlines, depth information, wrecks and pipelines, separation zones and anchoring/mooring prohibitions, etc. These could be displayed in combinations and at scales to suit the requirements of the user.

The data necessary to generate the display is held in memory and can be electronically updated. The display would eventually become part of the integrated system, taking input from navigation sensors and also from radar to truly produce an electronic model of the environment around the ship and with a capability to suggest courses of action to the navigator to avoid obstacles and other vessels. Automatic adjustment of depths displayed could be incorporated on the basis of published tide tables and compared with the vessel's own echo sounder to assess differences from the published values. There are systems already in use utilizing some of these capabilities, but the main problems may be to do with the need for international standardization rather than technological limitations.

In the US, data gathering and processing improvements are being applied to the Notices to Mariners system. The Defense Mapping Agency, National Ocean Service and US Coastguard's *Notices*, service a total of over 5000 charts worldwide. The establishment of a database enables the notification, verification and dissembling of information with a far more rapid turnaround than was previously the case. Use of improved telecommunications means that hard copy versions of the *Notices* can be bypassed and corrections sent direct to ships at sea via radio or satellite communications (INMARSAT). This method, known as INFONET could provide the basis for automatic updating of ECDIS so as to provide real time chart corrections under the control of the chart issuing body.

12.4 Location of observer

12.4.1 Coastal and ocean navigation

Having established a model of the surrounding environment in the form of a chart, it is necessary for the navigator to establish the position within that model. From this can be derived the necessary courses and speeds to reach the destination whilst staying clear of charted hazards. This normally requires observation of external features to the vessel to derive its relative position. Within a suitable range of coast this might take the form of visual bearings of landmarks identified from the chart, referenced to North. This could be extended by radar ranges and bearings in poor visibility or darkness. Radio navigation aids, including Loran and Decca, can provide accurate positions well away from the transmitting stations on land.

Further offshore, astronavigation is used to derive a position based on visual observations of the sun, stars and planets from the observer's location on the spinning Earth. Modern satellite based navigation systems use radio transmissions to calculate the position of the vessel relative to the precisely located spacecraft in their orbits.

12.4.2 Dead reckoning

All the methods mentioned so far use observations of external objects to derive a position relative to those objects whose own location is known from the chart. It is possible, however, to calculate a position based on distance and direction of movement from a known point of origin. This is termed dead reckoning (DR), possibly being derived from deduced reckoning. The speed of the vessel is determined from measurement of water movement past the hull using a device called a log or the number and relationship of propeller revolutions for the ship. The distance moved is a function of speed over a given time period. The direction of movement is determined from the heading of the vessel's centreline relative to North observed from magnetic of gyro compass.

This technique seems to have much to offer in reducing the number and complexity of observations to two (ie speed and direction), and if used upon a solid surface, would be a feasible system. However, because of the variables intrinsic in the vessel's movement through the water both with regard to heading and speed and the shifting of that fluid relative to the solid Earth, errors over time degrade the value of the method.

Only in the case of the highly complicated inertial navigation systems produced for military purposes can these errors be negated. This equipment uses a combination of accelerometers and gyroscopes to derive true motion of the vessel, without reliance on external information or on active sensors which might give its position away. The technology necessary to achieve these results, on the basis of which nuclear missiles may be launched, is extremely complex and expensive. It is therefore unlikely to become available for general navigation for many years, if at all.

In the past, DR was used extensively, being the only source of continuous positioning information allowing the derivation of a position at any given moment. So, a vessel might leave a known location, say a particular buoy, sailing on a heading of 090° at a speed of 10 knots at 9.00 am. Since a knot is a speed of 1 nautical mile/h, by 10.00 am it will be 10 miles east of the buoy, by 10.30 am 15 miles east and so on. These positions may be plotted on the chart and turned into a latitude and longitude. However, any error in the speed or direction values used results in a derived position which is incorrect. In addition, since the updated position is already in error, the errors are cumulative as the vessel travels further on. Since the direction and value of the errors are not known, a circle or ellipse of error exists around the derived position, the diameter of which must be estimated.

12.4.3 Speed and heading measurement

The direction of travel of the vessel is determined relative to North using a compass. This might be the magnetic type which aligns itself broadly to the local magnetic field of the Earth, or the gyrocompass which uses a combination of the local rotational motion of the Earth and gravity, to align itself to the rotational axis of the Earth. With the compass pointing to North, the heading of the centreline of the ship should represent the direction of travel.

The vessel's speed through the water can be measured in a number of ways and any device used for this purpose is called a log. Water flow past the hull may be used to drive a small propeller or may cause a difference in hydrostatic pressure. Alternatively, ultrasonic signals may be reflected from particles in the sea water and the Doppler principle used to derive their movement relative to the hull.

12.4.3.1 Magnetic compasses

In the case of the magnetic compass the alignment is towards the magnetic axis of the Earth, which is different to the rotational axis. The magnetic North Pole is currently in northern Canada and moving slowly in a direction that can be forecast. Consequently, for any point on the Earth's surface there is a difference between the direction of the magnetic North Pole and True North. This is termed variation. In a location from which the two poles are in line, i.e. due south of both, the variation is zero. For other locations it will be an angular value east or west. Marine charts have the local value of variation, its direction east or west and its annual rate of change printed on them, so as to allow correction of magnetic bearings.

Local magnetic anomalies in the sea bed caused by particular geological outcrops may distort the magnetic field and major upsets to the field can be caused by solar flares. These factors may cause inaccurate derivation of True North from a compass.

Another source of error is the distortion in the Earth's magnetic field caused by the vessel itself. Normally the hull and/or the engines are made of steel and therefore the orientation of the vessel with regard to the magnetic field results in heading dependent errors. These errors are termed deviation and must be determined empirically for the vessel throughout the 360° of true heading. Plus or minus angular values for a given magnetic heading will then exist, which can be applied to magnetic bearings taken on that heading.

12.4.3.2 Gyrocompasses

The gyrocompass has an internal rotor spinning at very high constant speed. The rotation of the Earth, together with its gravitational pull, results in the gyro aligning itself with the Earth's axis of spin. The gyrocompass therefore indicates True North, although it must be set up for approximately the right latitude and speed of vessel. It is unaffected by the heading of the vessel or environmental features of the operating area but will, like any mechanical device, have a possibility of error. This will probably be independent of vessel heading and will have been identified by external measurements. Such an error would normally be small, typically 0.5° and easily compensated for. To ensure correct performance, the speed of rotation must be maintained accurately, so the electrical supply to the driving motors must be precisely controlled. The gyrocompass requires time to settle down and achieve equilibrium after being switched on and large and rapid changes of ship's heading may also cause it to meander either side of the true reading for a while.

Today, all vessels of any size use a gyrocompass for ship's heading but in case of failure would have a magnetic compass in reserve. Small vessels including fishing boats and yachts normally rely solely on a magnetic compass.

12.4.4 Steering and autopilot

Most vessels today maintain their course via an autopilot, which uses output from the compass to apply corrections to the rudder. The course determined by the navigator is entered into the autopilot which then adjusts the rudder to compensate for changes in the ship's heading caused by wave action or other external forces. It is therefore only the orientation of the vessel that is maintained and this is unlikely to be the direction of actual movement across the globe for a number of reasons.

The vessel should travel straight through the water due to a balance of hydrodynamic pressures on either side of its hull. In practice, because of wind pressure exerting an unequal force on different parts of the above water structure, the vessel may well crab through the water at an angle to its centreline axis. This angle is termed leeway and the differential force exerted is called windage. Obviously, the orientation and strength of the wind, as well as the above water configuration of the vessel, will affect the value of the leeway angle. This angle is also a product of vectors, so the speed of the vessel through the water is also a factor, with a higher speed through the water minimizing the effect of the wind. This effect of windage is a force attempting to push the vessel off the desired heading and so the autopilot may apply a fairly continuous rudder adjustment to compensate for it. Depending on whether this is towards or away from the prevailing wind, this compensation is termed weather or lee helm.

In addition there may be a tendency for wave action to move the ship bodily in the direction the waves are travelling, offsetting its track through the main water body.

12.4.5 Tidal or current set

By far the most important factor degrading DR estimates of position is the movement of the water itself relative to the sea bed as a result of tidal or ocean currents. In parts of the deep ocean, this factor may be of minimal importance, but in some areas the tidal speed may be as high as, or higher than, the vessel's speed through the water. It follows that the actual movement of the ship across the Earth's surface is the result of combining the vectors for its movement through the water with those of the water across the sea bed.

Whilst the vessel's heading and speed through the water may be reasonably well known, the current speed and heading can only be an estimate. This might be based on published tide tables or on measurement of ocean currents but will certainly be in error to a greater or lesser degree. This introduces a factor of uncertainty into a DR position which grows larger with the passage of time. The amount which the derived position is offset in the direction of water movement is termed the set.

Figure 12.7 shows how water movement affects the distance and direction travelled over the Earth's surface. The actual distance and direction travelled are termed the distance and course made good and may differ considerably from the course steered and speed through the water. In order to achieve a desired course made good, the navigator may use an estimate of current set to derive a compensated course to steer, but the limitations of the estimate will determine how close the final result is.

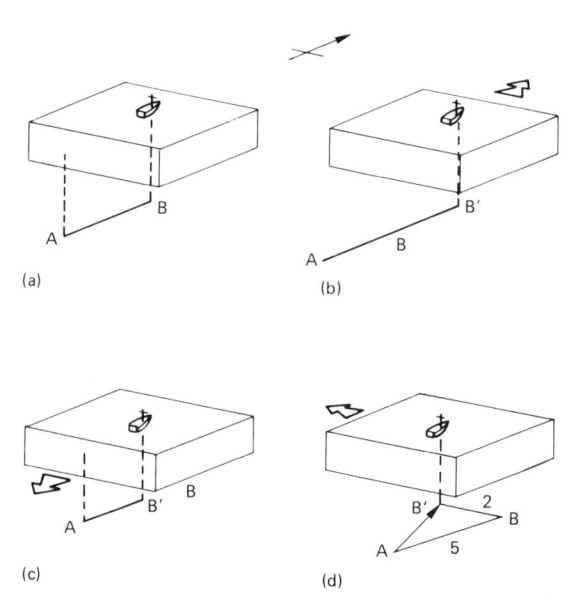

Figure 12.7 Relative water movement: the diagrams illustrate the tidal effects on a vessel's movement over the sea bottom; in all cases the ship is steaming due north at 5 knots through the water. In example (a), there is no tide and the vessel covers 5 nautical miles over the ground in an hour. In example (b), a 2 knot tide from behind moves the water mass with the vessel to give it 7 miles over the ground. In example (c), the tidal direction is reversed and only 3 miles are covered. In the final case (d), the tide is from the side moving the water mass, and hence the ship. The result is that the vessel covers 5.38 miles on a course of 338°; dead reckoning ignores these effects and plots 5 miles on a heading of 000° in all cases; in each instance, the vessel actually moves from A to B'

12.4.6 Position fixing

The limitations of DR have meant that, whenever possible, observations of external phenomena should be used to derive positions or fixes which can be used to update the DR position. Before the advent of electronic navigation, fixes were made periodically, by taking the compass bearings of landmarks or by astronomical sights and DR was used to derive position between these upates. Systems like the Transit satellites still provide only discrete fixes and DR must be used to fill in the gaps between satellite passes. Continuous fixing, made possible by radio positioning systems like Decca Mainchain and computers to turn their readings into continuously updated latitude and longitude removes the need for any DR. The GPS satellite system goes one better, since by having a number of spacecraft in the visible sky at any one time will provide continuous fixing worldwide.

12.4.6.1 Three-way fix

The simplest kind of fix is a single measurement of the horizontal angle between North and the line of sight bearing to a known landmark. The landmark might be a physical feature like a headland, a conspicuous building or possibly a buoy or structure offshore. This bearing can be drawn on the chart from the landmark's position and the observer can be said to be somewhere along this line. By taking two other bearings as well and bringing the three together at an intersection, a more accurate fix has been created. These lines are termed 'lines of position' (LOPs) and, due to observational error, will probably show a misclosure. The resulting triangular shape is frequently known as a cocked hat and it is expected that the actual position will lie within it (*Figure 12.8*). Two bearings only would give a fix of sorts, but error in either measurement contributes maximum displacement to the final position. Shallow angles of intersection will contribute a larger possibility of error, as will sightings to more distant targets.

Another example of an LOP is a situation where two objects are aligned from the observer's point of view. By identifying the objects on the chart, a line can be constructed upon which the observer's position must lie. This situation is called a transit and again may involve natural or manmade features. The safe approach route through a channel is often marked by having conspicuous white posts or lights at night, called leading marks and located on the shore. By lining up the marks and steering a course towards them to maintain their alignment, the vessel passes along the safe channel.

The LOPs may be range related rather than bearings and radio navigation systems like Decca Mainchain and Loran use this principle. The intersecting LOPs represent the distances to shore stations expressed as a difference in time delays, given the known value of radio wave propagation and are consequently hyper-

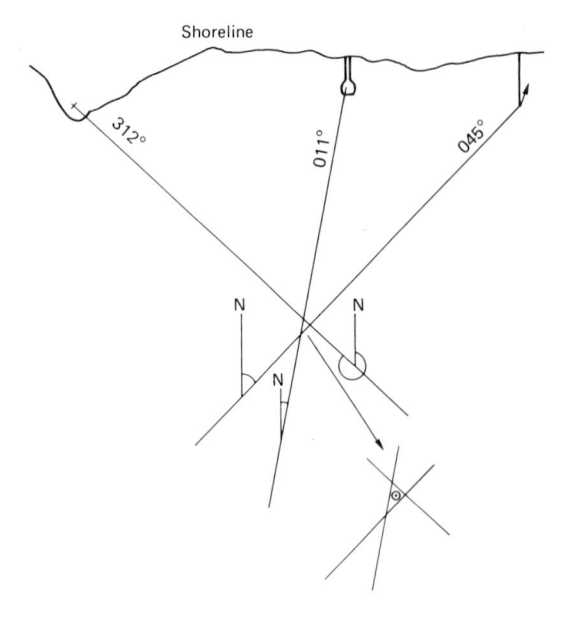

Figure 12.8 Three way fixing: shore landmarks; in this case three bearings are taken from shore landmarks simultaneously and the angles referenced to North; the bearings are all LOPs, as the observer must be located somewhere along their length; when plotted on the chart, there will often be a misclosure as shown in the inset; by plotting the intersection of the bisection of the angles, the most probable position is located

bolic in form. Here again, the geometry of the LOPs determines the quality of the fix. Ranges to landmarks may also be derived by radar giving arc shaped LOPs and these are used to plot a fix, although for general navigation a single radar range and bearing are usually considered sufficiently accurate.

12.4.6.2 Running fix

The running fix combines the features of DR and fixing by transferring an LOP in a direction determined by DR. This is useful when there are a limited number of landmarks, as a bearing from one feature may be carried along a DR course until a second bearing is available to intersect with it and thus provide a limited accuracy fix. The limitation in accuracy is contributed by the DR element, but again is useful in the absence of other information.

12.4.7 Ocean navigation

Most of the fixing techniques already discussed are restricted to that relatively small part of the world's seas within visual sighting range of land. Buoys and lightships extend this coverage further offshore but, once the shallower parts of the continental shelf have been left behind, other methods are brought into play.

The electronic navigation revolution has wrought many changes. Systems like Decca Mainchain and Loran are restricted in range and coverage although the latter can be used over a very large proportion of the Northern Hemisphere.

At the moment, only comparatively low accuracy radio transmission systems like Omega are available on a truly worldwide basis. This system uses very low frequency transmissions and gives fix accuracies which can be as low as 6–7 nautical miles.

The Transit satellite system gives worldwide coverage using a number of spacecraft in orbits passing over the Poles. These allow position fixing to a high degree of accuracy using a special radio receiver and computer which measures the Doppler shift of their transmissions. However, there are not enough of these units for there to be one above the observer's horizon at all times and reception is only on a 'line of sight' basis. It takes data gathered from a complete pass across the sky arc to provide a single fix, and DR must be used between fixes which may be some hours apart. The system has been in use since the 1960s but is due to be supplanted in the mid-1990s by the GPS.

GPS also uses satellites, but in a far higher orbit pattern, and will deploy sufficient units that a number will be within the sky arc at any time from any point on Earth. Their mode of operation will use very accurate timing and a special radio transmission to determine ranges from the satellite to the observer. This will result in a circular LOP on the Earth's surface that will intersect with others from other satellites to form fixes. The lowest accuracies that these should offer will be within 100 m (328 ft) and for specialist usage will be far higher. This will be continuous fixing, since a number of satellites will be in the sky at any one time and will be available worldwide. As such, the system may well supplant nearly all other forms of navigation for general maritime purposes and at least relegate them to backup roles.

Before the advent of the various radio navigation methods, offshore work was the preserve of DR and positioning using observations of the stars and planets. Astronomical navigation still provides a very useful source of corroberative information today.

12.4.8 Astronomical navigation

This method uses the principle that heavenly bodies, including the sun, moon, some planets and certain stars can be used by visual observation and knowledge of exact time to determine the observer's position on the Earth.

It is commonly known that Polaris, the pole star, gives an approximate idea of True North, as its position in the sky is close to an extension of the rotational axis of the Earth. However, since it appears to be directly above the North Pole, it follows that its altitude above the horizon at that point will be 90°. Equally, on the

Equator, 90° of latitude away, the star will appear to be on the local horizon and to the north of the observer.

Most people also understand that at noon the sun is as high in the sky as it is going to be and many realize that it may be directly overhead under some circumstances in the tropics. The understanding of these two facts provide at least a foundation for a comprehension of the principles of celestial navigation.

12.4.8.1 Longitude and the passage of time

Having established Greenwich as an origin for longitude it is obviously possible to use it as a standard for time also. As the Earth rotates, the zero meridian will pass through a line joining the centre of the planet and the sun. This will be noon on the Greenwich meridian and the sun will have achieved its highest altitude above the horizon at this time. Since 24 h later this must occur again, it follows that the meridian 15° to the west of Greenwich must pass through the line 1 h later, that is, 1/24 of 360°. By the same token, 180° west of Greenwich represents 12 h difference on Greenwich and can also be considered as 12 h ahead of Greenwich, hence its use as the International Date Line. In order to avoid a system based on incremental differences in local time, these 15° meridians have been used as the basis for the twenty four time zones used around the world.

For navigators, these wide bands of standardized time are of little interest and they require a system of angular distance west of the Greenwich meridian which can be related directly to the difference in time. This is termed the Greenwich hour angle (GHA) and is really the continuation of west longitude around the world and back to Greenwich again.

It must be pointed out that due to various astronomical forces acting on the Earth, its orbital speed around the sun is not absolutely regular. Therefore the time at which the sun achieves a GHA of 0°, that is, directly over the Greenwich meridian, will hardly ever be at exactly 12.00 GMT. That perfect relationship is only achieved by the theoretical mean sun.

12.4.8.2 Geographical position and declination

As has already been considered, there is a point on the Earth's surface which lies directly between the centre of the planet and the sun and this is termed the geographical position (GP) of the sun. Due to the rotation of the Earth, this point is moving quite fast across the surface, to complete a circuit of the Earth in 24 h. Just as GHA can be used to define its position in the east–west direction, the term declination delineates it in the latitudinal sense. As with latitude, when on the Equator the point's declination is 0° and it is expressed as degrees north or south.

Given sufficient astronomical observations and subsequent calculations, the GHA and declination of the sun can be worked out for any GMT on any date throughout the year. This data, together with that for a number of other significant navigational heavenly bodies including the major planets, moon and principal stars, are published as a series of tables in a nautical almanac.

12.4.8.3 Celestial fixing

The geographical positions of the various astronomical bodies is considered as a number of landmarks which can be used to produce lines of position and therefore fixes, much like terrestrial landmarks. The significant difference is that these landmarks may be hundreds or thousands of miles distant from the navigator and that they will be moving westward at high speed.

Whilst the GPs themselves are too far away to see and are theoretical points anyway, the bodies themselves can be seen if they are above the observer's horizon under the correct conditions. Ideally, it would be best to measure the angular distance between the body and a position directly above the observer, that is, normal to the Earth's surface at that point. The incoming light rays are normal to the surface at the GP and those arriving at the observer's position come in at an angle to the direction normal to the Earth's surface at that point. That angle is unique for points lying on a large diameter circle on the Earth's surface with the GP being at its centre.

The situation can be envisaged as a tube embedded in a sphere, which has an infinite number of spines on its surface, all pointing away from its centre, and where the parallel edges of the tube represent beams of light from the celestial body (*Figure 12.9*). At the point where the central axis of the tube strikes the surface, there will be a single spine which points directly up that axis. This can be considered as the geographical point (GP). Where the tube joins the sphere, there is a circle of spines where the angle θ between the base of the

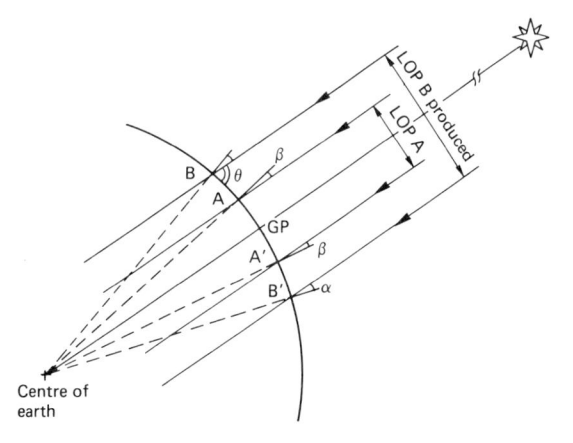

Figure 12.9 Celestial navigation: since A and A′ are on an LOP closer to the GP than B and B′, angle β is smaller than angle α; angle θ is the altitude above the horizon (90° − α)

spine and the tube wall is the same. If we choose any individual spine on the surface and adjust the tube diameter so that it intersects the sphere at the spine's base, we can measure the spine and wall angle β and locate a new circle of spines with the same relationship. This is an LOP on the sphere and we can say that, for a given angle, the spine must lie on the circle.

Having established a single LOP by this method, the technique could be used for other bodies and would thus create a fix at the intersection point. However, it should be remembered that, unless the observer is very close to the GP, the circle to be drawn could be thousands of miles in diameter and thus would be impossible to plot on the scale of chart used for navigation. On a normal chart, the LOP would appear as a straight line, although it is in reality an arc.

There are some other difficulties to be borne in mind. Whilst in terrestrial surveying a plumb bob can be used to determine the vertical, such a system would be impractical on the heaving deck of a vessel. In addition, as we have already established, the GP is in rapid motion across the face of the Earth and therefore so is the LOP. The angle between the spine and the tube will only be valid for a moment as the sphere turns and an instant later a different spine will have the same relationship with the tube.

12.4.8.4 Sextant and chronometer

In order to overcome these difficulties, two instruments have been developed over the centuries, gradually evolving from more primitive forebears. The sextant provides the angular measurement, whilst the chronometer provides a temporal datum by which the Earth's position in its rotational cycle can be fixed.

The sextant, instead of attempting to measure the angle between the vertical and incoming light beam (termed the zenith distance), instead measures between the horizontal and the beam (angle θ in *Figure 12.9*). A well-delineated horizon at sea is a good measure of the horizontal, as the planet's surface curves away from the observer in all directions. The uninterrupted boundary between sea and sky can be considered as being level with the observer. As the observer's height above the surface increases, so he or she can see further around the curvature of the Earth and the visual boundary between sea and sky no longer represents the horizontal. For theoretical purposes, measurements should be taken from sea level, but in practice a correction factor is applied, based on the height of the observer above sea level.

The chronometer is a very accurate clock and allows the observations made in a remote location to be tied to astronomical predictions of celestial bodies' positions relative to the Earth. In former times, the chronometer would be set to an official source at the beginning of a voyage and no form of subsequent correction for error in its time keeping could be applied. With the advent of radio, time signals could be transmitted to ships at sea allowing them to monitor their chronometers' performance. This establishment of a time standard made it possible to utilize the astronomical data gathered by observatories at known locations around the world to make predictions about the apparent positions of the sun, moon, planets and stars at a given time.

By comparing the navigator's observations of celestial bodies with the predictions of the GPs of those bodies, published in a nautical almanac, and by using a common time standard, LOPs can be determined by a number of methods. Some use direct calculation, whilst others use precomputed tables, but the former approach has been made considerably easier by the widespread use of electronic calculators.

12.4.8.5 Azimuths and intercepts

In most cases, the system used is to compare the observed altitude, once all the necessary corrections have been applied to it, with the figures for an assumed position which is close to the actual position. An azimuth or direction for the GP of the body from the assumed position is found first and the position and a line of bearing plotted on the chart.

Next, the zenith distance for the assumed position is found and the difference between it and the observed value noted. If the observed value is larger than that of the assumed position, then the LOP intercepts the azimuth closer to the GP. If the value is smaller, LOP and intercept will be further away from the GP than the assumed position. The LOP is plotted crossing the azimuth at right angles and within the scale of the chart appears as a straight line. As with other LOPs, a number can be used simultaneously to produce a fix or the running fix principle can be used to combine sights taken at different times.

12.4.8.6 Limitations of celestial navigation

A number of corrections have to be applied to sextant readings to correct for refraction of light in the atmosphere, the apparent size of the sun and moon and various other characteristics of the celestial body or the instrument itself. In addition, the most obvious limitation is the dependence on suitable visibility. Meteorological conditions can obscure the skies for weeks and even a hazy horizon can make accurate fixing impossible.

Whilst modern electronic systems have largely relegated celestial navigation to a secondary role, it does offer a valuable source of information should major equipment failure affect these systems. As a result, knowledge of its techniques are likely to remain a requirement for seagoing officers for many years to come.

12.5 The beginnings of electronic navigation

Virtually all of the navigational techniques until this century relied on visual information to fix the position of the vessel. Bearings from landmarks and, at night, the identification of the coded light sequences of buoys and lighthouses, all depended on being able to see the objects over distances of many kilometres. Fog, snow or heavy rain could easily prevent this, leaving the navigator with only DR and estimates of tidal and other factors to derive updated positions.

For deep ocean navigation, cloud cover obscuring the skies could make reliance on DR for long periods a reality, resulting in highly inaccurate positions, often with fatal results.

In the early part of the twentieth century, the first system using an alternative to the visible light part of the electromagnetic spectrum came into use: radio direction finding. This extended the principle of taking bearings from shore stations, but used a directional aerial which was aligned to give a bearing based on the vessel's compass. By using radio signals, the transmission could be received in darkness and bad weather conditions and is still in use today. The transmissions can also be received over the horizon and so therefore are not limited by the need for a line of sight. Early systems were based on a rotating loop aerial about 1 m (3.28 ft) in diameter, which had a coil winding inside it. The aerial receives the maximum signal and therefore has the maximum induced voltage in it when it is at 90° to the electromagnetic lines of force of the incoming transmission.

As the aerial is rotated toward the position where the loop is parallel to the lines of force, the signal level drops rapidly until it reaches the null point where the signal has dropped to zero. Passing through this point, signal strength rises rapidly, reaching a maximum when once again at 90° to the lines of force (*Figure 12.10*).

The mechanical limitations of the rotating aerial, together with the requirement for more flexible location, resulted in the Bellini–Tosi system, where the single rotating loop is replaced by two loops set at 90° to one another. One loop is parallel to the fore and aft axes of the vessel and the second is aligned across the ship. Field coils are attached to the two aerial frames and a further search coil can be adjusted within them. The two aerials resolve the incoming signals into resultant voltages at 90° and the search coil senses the field from the two coils (*Figure 12.11*). The entire assemblage is called a radio goniometer. In both systems an auxiliary antenna may be used to overcome the problem of ambiguity, since the main antenna will indicate two direction 180° apart.

Although direction finding has become increasingly sophisticated over the years, it has a number of limitations which have meant that, in comparison with modern radio navigation systems, its restricted accuracy relegates it to an auxiliary role.

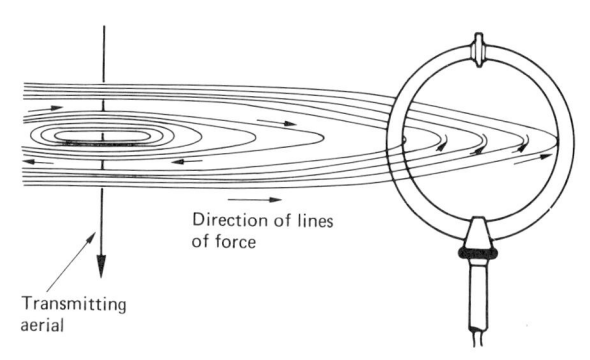

Figure 12.10 Propagation of magnetic lines of force generated in the transmitting aerial: because the number of lines of force inside the loop aerial changes continuously, an alternating voltage is generated in the loop windings (from Sonnenberg, 1987, courtesy of the author and publisher)

12.5.1 Shipboard factors affecting direction finding

The aerial system used in direction finding is fairly limited in its null sensitivity and as a result the overall accuracy of the equipment is in the region of ± 1° but other factors may degrade this performance. The hull of the vessel, including the superstructure, masts and rigging, will tend to reradiate radio energy, which will arrive at the aerial in different phases and strengths. These will distort the incoming signals received by the aerial but, because of alignment, will show little or no effect in the longitudinal axis of the vessel or at 90° to this. Maximum distortion will be experienced at 45° to these directions and will tend to slew the perceived direction of the signals towards bow and stern. This is termed quadrantal error.

Another aerial system on board may contribute its own semicircular error and to determine the combined error curves it may be necessary to carry out a series of empirical tests. In many ways, this is similar to determining the error curves for deviation of a magnetic compass. The radio bearings derived from stations are compared with optical readings to the same sites and a polar diagram of corrections relative to the ship's longitudinal axis is drawn up. These must then be applied to subsequent radio bearings taken aboard the ship. Since other aerials and structures contribute to this fingerprint, it follows that conditions must be the same during subsequent usage as in the calibration tests.

12.5.2 Environmental factors affecting accuracy

These are largely concerned with the propagation of radio waves through the Earth's atmosphere and may result in major inaccuracies of the measured bearings. The radio waves received by a vessel may be distorted by reflection from the ionosphere or refracted by passing over a coastline and it must also be borne in

sunset. This night effect may cause signals to fade and flutter, as well as spreading the null point.

The velocity of propagation of radio waves is different over land and sea and this results in bending of the direction of travel, which is most marked at low angles of incidence. This may be as much as 6 or 7° where the required bearing runs parallel to the coast but reduces to zero where the bearing is at 90° to the shoreline. Marine radio beacons are usually positioned close the shore to avoid this effect, but those located primarily for air navigation, and which may also be used at sea, may be subject to it.

Both these factors are difficult to quantify and therefore bearings which are subject to them must be used with caution. An error of 1° at a range of 10 miles moves the fix 0.17 miles, so large errors can develop when fixing at ranges of over 100 miles. The difference between the great circle bearing and the rhumb line needed to plot on a chart is only important at these longer ranges and tables of half convergency or a calculator can be used to derive the correction.

12.5.3 Shore radio stations

The location, frequencies and other characteristics of the beacon transmitting stations are published in the Admiralty *List of Radio Stations* and its equivalents from other bodies. Some shore stations operate their own direction finding equipment, which by virtue of its fixed location and superior calibration can derive more accurate bearings on transmitting vessels, then radio this information to the vessel concerned. This feature is of particular use when a distress signal is received and a number of shore stations can operate together to produce a fix on the victim for the rescue services.

Some specialized radio beacons exist which use particular transmission techniques to indicate the receiving vessel's position. These include radio lighthouses which transmit a VHF signal on channel 88 with a null signal sweeping round like the beam of a searchlight. On receiving the null on an ordinary marine VHF and noting the time of reception, the navigator can read from a table the bearing from the transmitting station.

12.5.4 Usage of direction finding

The development of more modern methods of electronic position fixing, including Loran and Decca, has eclipsed radio direction finding and relegated it to a supporting role in many parts of the world. Vessels are still equipped with it, but it is primarily considered to be for locating vessels in distress.

12.6 Navigation in the modern world

The pace of development in navigation in the last twenty years has been extremely rapid, and with the

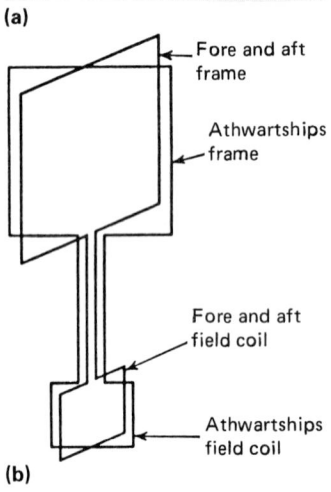

Figure 12.11 (a) Bellini–Tosi aerial; (b) the two circuits of the fixed frame aerials and the two field coils (from Sonnenberg, 1987, courtesy of the author and publisher)

mind that they follow a great circle route around the Earth's surface.

The signal received by a station is made up of two elements, a ground wave which travels around the curvature of the Earth and a sky wave which is reflected back from the ionosphere. This second element is stronger at night than by day and its distorting effect on the signals is most pronounced close to sunrise and

imminent promise of GPS it would seem that the navigator's skill is just about redundant. Continuous bridge readout of position to an accuracy of less than 100 m (328 ft) in all weather conditions from the GPS should make the positioning problem a thing of the past.

However, as has already been pointed out, the improvement in positioning technology is by no means the whole answer, since there is now a change in direction in prospect. Economic pressures and the highly integrated nature of international trade are extending further the demands put upon navigational techniques. The safety margins in terms of clearance of known hazards which were applicable in the age of chronometer and sextant navigation are not really tenable in an age of satellite navigation. There must be duplication and cross-checking of navigation systems and other safety measures in order that these reduced spatial margins do not compromise overall safety by simple equipment failure or operator error.

Already there is pressure in some areas to reduce the number of traditional navigational aids in terms of lighthouses and buoys, since the modern navigational systems have supplanted them for the majority of users. However, it should be pointed out that there is a very wide variation in the navigational standards of vessels using the world's ocean routes and until it is possible to ensure that all users are equipped with the latest systems, a range of technologies will have to be allowed for.

To avoid vessels wasting time waiting to enter harbour, there is a requirement to make the best use of a harbour's tidal range and experiments are already underway with automatic real time water depth monitoring systems to avoid dependence on predetermined tables. Such systems are taking advantage of modern technology to reduce safely the generous margins formerly necessary due to imprecise knowledge.

Integrated navigation, radar and charting systems will offer the display facilities to make best use of this information, but its performance will only be as good as the data input. The new navigation systems will allow very accurate positioning and modern radars incorporating computers give accurate information about relative course and speed. At all times such systems will have to be monitored carefully, possibly by traditional navigational techniques, in order to ensure that they accurately reflect the real world.

Acoustic sensing and positioning systems

12.7 Introduction

12.7.1 Why use acoustics?

The energy sources and sensors used for measurement on land cannot be used underwater. The sea is almost opaque to visible light and radio waves, although there has been limited success using lasers for the collection of bathymetric data. Scattering and the attenuation of light means that, at depths greater than 100 m (328 ft), there is little or no natural light and an observer at that depth will be experiencing a pressure of about 11 atmospheres.

An energy source and system are required which will propagate to the required range, have sufficient resolution to detect and identify the target data and operate at a known speed. To achieve this, the transmitter and receiver must have good coupling with the medium and there must be high conductivity within the medium (*Tables 12.1 and 12.2*).

Such equipment requires a high energy transmitter with matched receiver and a complex signal processor. In nature, there are examples of such equipment in bats and dolphins. However, it is only relatively recently that man has successfully used acoustics for underwater work, but there are limitations.

The speed of sound in sea water is of the order of 1500 m s^{-1}. This makes for a low rate of data collection. If a ship using a device to mechanically scan a 60° sector with a 5° beam to a range of 1 km, the time taken would be 16 s. At 10 knots this would mean a movement of 48 m (157.5 ft). A consequence of this is that most echo sounders and sonars use a sonar beam on a discrete bearing or have to resort to electronic beam steering or scanning.

The attenuation of high frequencies means that the practical upper limit for sonar systems is 500 kHz, whilst the lower limit is 5 kHz. Following from the relationship wavelength equals speed/frequency, this produces a wavelength of 0.0075 and 0.3 m (0.025, 0.98 ft) respectively, which limits the resolution ability of acoustic systems.

The speed of sound in the sea varies with temperature, salinity and pressure. This can result in substantial refraction producing sound channels, shadow zones

Table 12.1 Acoustic quantities and units

Quantity	Name	Symbol	SI unit	Equivalent
Speed of sound	–	c	m s^{-1}	–
Density	–	ρ	kg m^{-3}	–
Particle velocity	–	u	m s^{-1}	–
Acoustic impedance	rayl	Z	kg m^{-2} s^{-1}	ρc
Force	newton	N	kg m s^{-2}	J m^{-1}
Energy (E)	joule	J	kg m^2 s^{-2}	N m
Pressure (p)	pascal	Pa	kg m^{-1} s^{-2}	ρc u
Power (P)	watt	W	kg m^2 s^{-3}	J s^{-1}
Intensity (I)	–	I	kg s^{-3}	p^2 (ρc)$^{-1}$

I = flow of energy across a unit area per unit of time; $I = P/A$, where A is the area through which acoustic energy will flow; P = intensity over an area; $P = I/A$; E = power over an interval of time; $E = Pt$, where t is the interval of time.

Table 12.2 Typical acoustic values

Material	Density kg m^{-3}	Speed m s^{-1}	Impedance kg m^{-2} s^{-1}	Coupling	Conductivity
Space	nil	nil	nil	nil	nil
Air	1.293	340	439.6	poor	poor
F.W.	1000	1450	1 450 000	good	good
S.W.	1025	1500	1 537 500	good	good
Sediment	2300	2000	4 600 000	good	variable to good
Concrete	2600	3000	7 800 000	excellent	excellent to good
Granite	2700	5500	14 850 000	excellent	excellent

and poor bearing data. Noise is an unwanted signal which reduces target detection, identification and reliability of communications. Noise can be classed as ambient, self-generated or reverberation. Increased transmitted power can help overcome the first two but not the latter. Also there is a limit on transmitted power caused by cavitation bubbles forming and restricting transmission.

From the above, it is apparent that the use of sound underwater has severe limitations, which requires the user to have a good working knowledge of underwater acoustics. Otherwise when sound is slowed, bent, reflected, attenuated, scattered, distorted, lost, found again or speeded up in the course of work, there will be little appreciation of what is going on and the operator could recourse to increasing the smoothing.

All of these effects depend upon or are controlled by the physical character and biology of the ocean volume and boundaries. They must be understood.

12.7.2 Historical development of underwater acoustics

Nearly 2500 years ago, Aristotle noted that sound could be heard underwater. This was confirmed by Leonardo da Vinci in 1490 when he stated that ships could be detected using an underwater listening device. The step from curiosity to science was made by Colladon and Sturm; measuring the speed of sound on Lake Geneva in 1827 they obtained a remarkably accurate value of 1435 m s^{-1} (4664 ft s^{-1}) in fresh water at a temperature of 8°C.

Work in Australia by Threlfall and Adair in 1889 showed an apparent correlation between transmitted power and speed of sound. The sound was taking the least time path and was being refracted below the sea bed.

On the night of 15 April 1912, there was an event which had a major impact on the development of underwater acoustics. The RMS *Titanic* struck an iceberg and sank with a loss of five hundred lives. On 10 May 1912, L F Richardson filed patent number 11,125; 'for detecting the presence of large objects underwater by means of the echo of compressional waves having a wavelength of 30 centimetres, directed in a beam by a projector'.

It was not proceeded with, but at the same time, in the USA, Fessenden developed his highly efficient reversible electroacoustic transducer. By 1914, he was able to detect an iceberg at a range of 2 miles. During the experiments, a secondary echo was observed. It proved to be the sea bed and was the start of echo sounder development.

Between 1914 and 1918, there was unrestricted submarine warefare. This caused considerable cooperation between the UK and France, culminating in the development of piezo-electric transducers and the ability to detect submarines at a range of 8 km. The paper analogue recorder was invented by Marti in 1919 and was used for a cable route survey in 1922. The main parts of underwater acoustic systems were now in place.

Expansion has been massive since 1939, in response to both military and commercial pressures. This has meant a wide range of equipment allowing the investigation of the water body, sea bed and sub bottom, plus the development of acoustic navigation and positioning systems.

12.8 Acoustic theory

12.8.1 The sound field

Underwater sound, no matter how produced, is governed by the laws of physics. For the majority of sonar systems, sound is transmitted from a transducer, which has a vibrating surface in contact with the sea, producing alternate zones of compression and rarefaction propagating as a wave through this elastic medium. It is a longitudinal wave, with motion parallel to the direction of propagation.

The propagation can be described completely using the wave equation. This is a fundamental equation of physics; it combines the equations of state, force, motion and continuity and it is a description of the variance of pressure in a sound wave with time and space.

Below is a form of this equation where the acoustic pressure p is related to position x, y, z, time t and the square of the velocity c.

$$\frac{\partial^2 p}{\partial t^2} = c^2 \left(\frac{\partial^2 p}{\partial x^2} + \frac{\partial^2 p}{\partial y^2} + \frac{\partial^2 p}{\partial z^2} \right) \tag{12.1}$$

To solve the wave equation, normal mode theory is used and while the solution is complete, it is difficult to satisfy the particular conditions of source, propagation and boundary, when using the real surfaces of sea and sea bed.

The simplest way that the propagation of this wave can be described is to look at it in one dimension, x. For this, the solution can be shown to be that of a sinusoidal plane wave as set out below:

$$P(xt) = P \text{ max. } \cos(kx - \omega t + \phi) \tag{12.2}$$
$$D(xt) = D \text{ max. } \sin(kx - \omega t + \phi) \tag{12.3}$$

where P is pressure, D is displacement, f is frequency, λ is wavelength, t is time, ϕ is phase shift, k is $2\pi/\lambda$ = angular wave number and ω is $2\pi f$ angular frequency.

This sound wave propagates as a mechanical disturbance causing local variations in the pressure and density of the medium. The pressure variations are at a maximum when the displacement of the transducer face is zero. The variation in density sets up zones of compression and rarefaction for the period of transmission.

The speed of propagation of the wave through sea water is approximately 1500 m s^{-1} (4921 ft s^{-1}) when comparing air and underwater acoustics, similar frequencies produce much shorter wavelengths in air, due to the much lower speed of sound. This, when allied to the much lower density of air, means that devices such as loudspeakers and microphones employed in the transmission and reception of airborne sound, cannot be used underwater.

12.8.2 Transmission

The units, relationships and nomenclature commonly used in underwater acoustics are given in Tables 12.1 and 12.2. The power of an acoustic wave depends upon its amplitude of oscillation. Amplitude is reduced by attenuation; this can result in large variations in the power, pressure and intensity of an acoustic transmission. By expressing these variations as a logarithmic scale of ratios, they become easier to handle, calculations are simpler and there is the additional benefit that the human audio response is approximately logarithmic. This is the decibel scale and is the measure of a change or a ratio expressed as a logarithm base 10. It allows sound to be described as having a certain level, the number of decibels (dB) by which its intensity differs from a reference intensity. It is also used to describe attenuation of intensity with range from an acoustic source.

Sonar transmissions are almost always pulsed. The main characteristics of an acoustic pulse are shown in *Figure 12.12*. If one assumes a square pulse with no distortion on its journey to and from the target, then the theoretical range discrimination is pulse length/2. In the non-ideal world of underwater acoustics, a realistic value is double this.

The maximum power of an acoustic pulse is limited by the cavitation threshold where the negative pressure of the pulse is less than the vapour pressure of the water and causes bubbles to form at the transducer face, distorting and reducing the output. It is by increasing the pulse length that more energy is transmitted.

The wavelength of transmission is a function of frequency and speed of propagation ($\lambda = c/f$). Pulse length and frequency of a system are bound together by system bandwidth. The minimum pulse length is reached when it is the reciprocal of the bandwidth. The bandwidth of a sonar system is normally controlled by the transducer; this is generally narrow, approximately 0.2 of the frequency. Using these relationships, a sonar system, with a frequency of 50 kHz, would have a bandwidth of 10 kHz and a minimum pulse length of 0.1 ms. This indicates the beginning of a conflict. Long range requires a long pulse, discrimination requires a short pulse. This conflict of range *versus* discrimination is a feature of underwater acoustics. It is resolved by priority or compromise.

Sonar transducers, in addition to being resonant, have a sensitivity which will vary with direction. This is termed the beam pattern or, more correctly, the directional power response curve (DPR). The axis of the main lobe of the beam is the direction of maximum sensitivity; the DPR curve also exhibits unwanted sidelobes and back radiation *Figure 12.13*.

In an echo sounder, once an acoustic beam diverges from the transducer face, there will be an error in depth measurement, as, unless on a flat sea bed, the true depth will not be recorded. The additional sidelobes and back radiation shown in *Figure 12.13*, can produce false returns. Echo sounders are used on boats and ships and, unless stabilized and heave compensated,

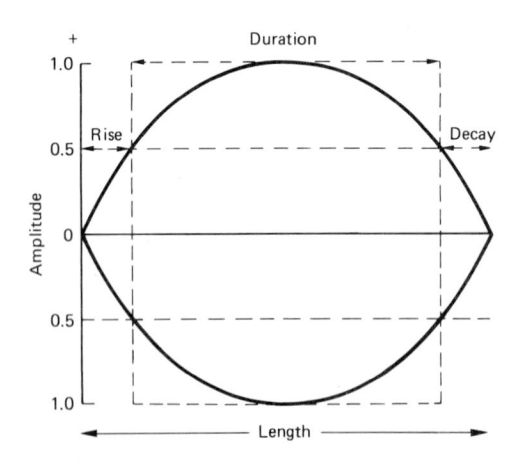

Figure 12.12 Acoustic pulse

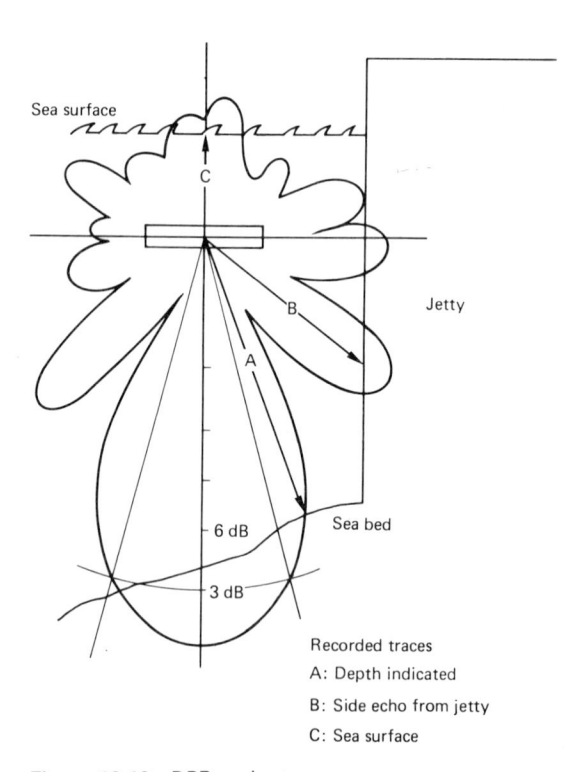

Figure 12.13 DPR and returns

Recorded traces
A: Depth indicated
B: Side echo from jetty
C: Sea surface

the measured depth will again be in error. Increasing the beamwidth can compensate for rolling, but at the expense of discrimination.

The beamwidth of a sonar system is normally defined as the beam angle contained between the 3 dB points. This is approximated by:

$$\text{B.W. (rads)} \approx \lambda/L \qquad (12.4)$$

provided that $L \gg \lambda$ and is in the far field where the transducer can be considered as being a point source. These effects can be summarized:

> frequency = < range, > resolution
> bandwidth = < range, > resolution
> pulselength = > range, < resolution
> beamwidth = < range, < resolution, > coverage
$$(12.5)$$

The beam pattern of the transducer can also be expressed in dB as the directivity index (DI). The DI equals $10 \log_{10} I_a/I_o$, where I_a is intensity at the beam axis and I_o is intensity of an omnidirectional source, both measured at the same point. It can be approximated by:

$$\text{DI} \approx 10 \log_{10} 4\pi A/\lambda^2 \qquad (12.6)$$

For a 50 kHz system with a transducer 0.2×0.2 m (0.65×0.65 ft)

DI $\approx$ 27.5 dB and beamwidth to 3 dB points $\approx 9°$.

In terms of output, the transmission can be expressed in dB as the source level (SL). This is convenient shorthand quantifying the transmission. It is the ratio in dB of the radiated sound pressure to a reference pressure of 1μ Pa at 1 m (3.28 ft) from the source. This can be shown to be:

$$\text{SL} = (171 + 10 \log W + 10 \log E + \text{DI}) \text{ dB, ref. } 1\mu \text{ Pa} \qquad (12.7)$$

where W is the transmitter power in watts and E is the transducer conversion efficiency. For the 50 kHz system with $W = 100$ W and $E = 0.5$, then

$$\text{SL} = 215 \text{ dB, ref } 1 \mu \text{ Pa} \qquad (12.8)$$

12.8.3 Propagation

Sound radiated from a transducer spreads with range; the advancing wave front covers an increasing area. Consequently, the sound intensity decreases with the square of the range. This spherical spreading becomes cylindrical upon reaching boundaries and is time stretched by multipath propagation. Even so, the approximation of spherical spreading can be considered realistic.

This element of the transmission loss can be expressed in dB. If I_1 and I_2 are intensities at range$_1$ and range$_2$, where range$_1$ is 1 m, then

$$\text{HL } 10 \log_{10} I_1/I_2 = 20 \log_{10} r_2 \qquad (12.9)$$

In addition, there is the true energy loss, termed absorption. This is due to the acoustic energy being dissipated by the thermal conductivity, viscosity and molecular readjustments taking place in the sea and preventing lossless propagation. Absorption has been extensively studied and can be shown to vary directly with frequency for sea water of a given temperature, salinity and pressure. It is proportional to range and the coefficient of absorption a can be approximated by:

$$a = \left(\frac{0.17}{T+18}\right) f^2 \cdot 10^{-3} \qquad (12.10)$$

where T is the temperature °C and f is the frequency in kHz.

There is one more element in transmission loss. This is the transmission anomaly A, an estimated value in dB caused by the refraction, diffraction and scattering of the transmission.

The total transmission loss is shown in the following equation:

$$\text{HL} = 20 \log_{10} r + ar + A \qquad (12.11)$$

Figure 12.14 illustrates transmission loss for a frequency of 50 kHz at a temperature of 10°C.

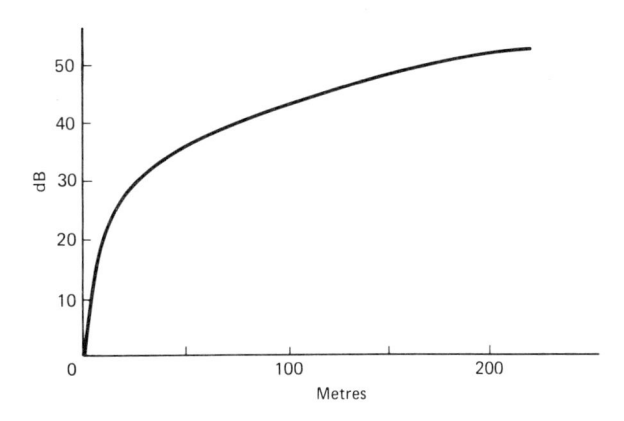

Figure 12.14 Transmission loss at 50 kHz and sea temperature 10°C

Variations in temperature, salinity and pressure, in addition to effecting absorption will cause variations in the speed of sound in sea water, which can be expressed as:

$$C_{sw} = (K/\rho)^{1/2}$$

where K is the bulk modulus and ρ is the density of sea water. Temperature has the greatest effect with an increase of 1 °C giving an increase of speed of 2 m/s (6.5 ft/s).

A wavefront in an homogeneous sea will propagate according to Huyghens' Principle. Variations in C_{sw} will mean that the wavefront will be distorted and the rays normal to the advancing wavefront will be bent by an amount and in a direction dependent upon the velocity gradient. Snell's Law applies here as it does in optics (*Figure 12.15*).

Knowledge of the velocity profile allows the path of the ray to be traced and zones of shadow and convergency to be identified. This is important for sonar systems looking in directions other than the vertical. For all sonar systems, the accuracy of range measurement depends upon a knowledge of the speed of sound.

Snell's law

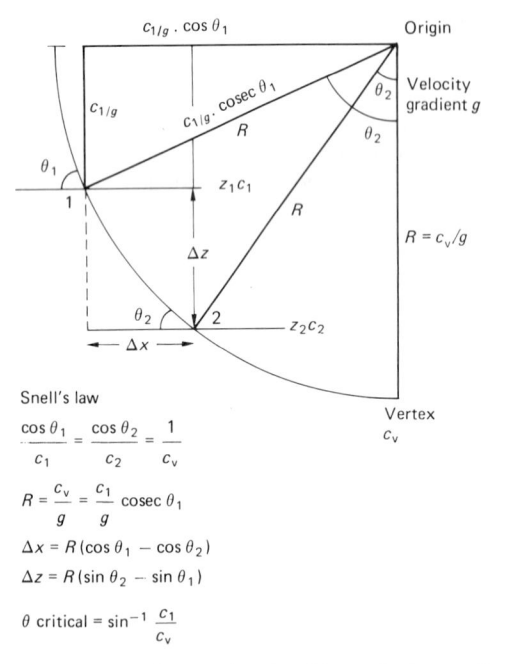

$$\frac{\cos \theta_1}{c_1} = \frac{\cos \theta_2}{c_2} = \frac{1}{c_v}$$

$$R = \frac{c_v}{g} = \frac{c_1}{g} \operatorname{cosec} \theta_1$$

$$\Delta x = R(\cos \theta_1 - \cos \theta_2)$$

$$\Delta z = R(\sin \theta_2 - \sin \theta_1)$$

$$\theta \text{ critical} = \sin^{-1} \frac{c_1}{c_v}$$

Figure 12.15 Ray tracing (positive velocity gradient)

12.8.4 Reflection

Reflection of acoustic energy will take place if there is a change of acoustic impedance (Z) within the medium. The more abrupt the change, the greater the reflection. Reflections can come from targets where size $\gg \lambda$ or from scatterers where $\lambda \gg$ size. The reflection can be wholly contained within the beam, but if larger than the beam, the reflection is termed reverberation from sea bed, sea surface or volume.

Assuming a target is an anchor on the sea bed, the target strength TS is the intensity of reflected sound received at the transducer. The anchor is acting as a sound source and the TS is a function of the shape, size, material and aspect of the anchor. This can be summarized as the effective cross-sectional area σ and can be estimated as:

$$\text{TS} = 10 \log_{10} \sigma/4\pi \text{ dB} \qquad (12.13)$$

Returns from small targets, where $\sigma \ll \lambda$ have their value for σ reduced by $\pi d/4\lambda$. The sound is now scattered rather than reflected. This was first demonstrated by Lord Rayleigh, who also showed that the value for σ was a function of the density and elasticity of the particles and that small gas and air bubbles resonated. This is a useful feature, as the resonant frequency is controlled by the bubble size and pressure. This has been used to identify plankton and to specify the optimum frequency to counter aeration around a transducer.

If present in sufficient numbers, small particles such as these cause volume reverberation. If caused by fish, then it is of value to the fisherman, but is an unwanted signal if measuring depth. Reverberation from the sea surface is unwanted, unless wave or tide height are being measured. However, the back scattering from the sea bed is a form of reverberation which is often the main target for the hydrographic surveyor.

The level of reverberation received is directly related to source level, pulse length and beamwidth. The rate of decrease is less than the target return of a discrete signal; also the active area of the beam increases with range while a discrete target becomes a progressively smaller proportion of this area and harder to detect. Target detection in an area of diffuse reverberation improves with a system having a wide band, short pulse and narrow beam.

Scattering from the sea bed is the strongest form of reverberation. It is a function of the grazing angle of the ray θ and the coefficient of reflectivity R, where R is equal to reflected pressure/incident pressure between two zones, 1 and 2. It can be shown that:

$$R = \frac{Z_2/Z_1 - \left[1 - \left(\frac{C_2^2}{C_1^2} - 1\right)\tan^2 \theta\right]^{\frac{1}{2}}}{Z_2/Z_1 + \left[1 - \left(\frac{C_2^2}{C_1^2} - 1\right)\tan^2 \theta\right]^{\frac{1}{2}}} \qquad (12.14)$$

For values of $\theta < 10$, R is virtually independent of θ. Then

$$R \approx \frac{Z_2 - Z_1}{Z_2 + Z_1} \qquad (12.15)$$

when $Z_2 \gg Z_1$ $R \to 1$ is a rigid boundary, i.e. the sea bed; when $Z_1 \gg Z_2$ $R \to -1$ is a soft boundary, i.e. the sea surface, and when $Z_1 = Z_2$, $R \to 0$ equals no boundary.

This is a partial indicator of why fluid mud is difficult and rock is easy to detect with an echo sounder. However, reverberation is a complex subject and while it can allow recognition of patterns on a side scan sonar record, it can restrict the range and accuracy of acoustic navigation systems. Whatever the effect, it is important to remember that it is a direct result of acoustic transmission.

12.8.5 Noise

Noise is any unwanted signal. Its source can be natural or manmade, remote or self generated; it can be directional or isotropic and can be a limiting factor of sonar performance at all ranges. The frequency band of interest is 5–500 kHz. In this band, the dominant noise is from surface waves and molecular agitation in the sea.

Wave noise depends upon wind speed and wave action and the Knudsen spectra were obtained from observations relating noise to sea state. In shallow water it is isotropic, while in deep water it tends to decrease away from the vertical. In this frequency range, the spectrum level can range from 20 to 70 dB ref. 1 µ Pa. It is proportional to frequency and falls off at a rate of 6 dB per octave.

Thermal noise differs from other noise sources in that it has a positive slope of 6 dB per octave. Above 50 kHz it tends to be the dominant noise source, ranging from 15 to 35 dB. It is the lower limit of hydrophone sensitivity at these frequencies. The source of this noise is the molecular agitation in the sea striking the transducer face.

There are other noise sources, such as those generated by fish, mammals, rain and industrial noise. They tend to be intermittent and anomalous and consequently are difficult to predict.

12.8.6 Prediction

In order to predict the range performance of a sonar system, it is necessary to ascertain the strength and path of the acoustic signal. This may be achieved by using the sonar equation and the technique of ray tracing.

The sonar equation uses the logical grouping of a range of parameters linking medium, target and equipment. These relationships allow simple calculations to be made which can determine maximum range of particular equipment or determine the equipment for specific target and medium. It uses the equality between the signal level E and noise level N to determine the recognition differential M. The parameters used are:

1. Equipment: source level (SL), self noise (NS), directivity index (DI) and figure of merit (FM).
2. Medium: transmission loss (HL), reverberation (NR) and ambient noise (NA).
3. Target: target strength (TS).

These parameters are arbitrary; the degree of sophistication of the sonar equation controls the amount by which they are expanded. The grouping of the parameters into equations is limited in that they are not general but specific.

The application of the sonar equation is best illustrated by the following example:

Sonar: frequency 108 kHz, output 100 W, DI 27.5 dB, M −6 dB.
Medium: sea temperature 10 °C, noise 35 dB, reverberation 50 dB.
Target: Target strength −3 dB.

What is the maximum range over which the target will be detected (FM)?

$M = (SL + TS - 2HL) - (NA + NR - DI)$
$FM = 2HL$ (12.16)
$FM = 2HL = (SL + TS - M) - (NA + NR - DI)$ (12.17)

By entering in the appropriate values and carrying out the iteration, the reader should obtain a range of approximately 245 m (803.8 ft).

Ray tracing depends upon Snell's Law. This describes the refraction of rays in the medium and, as long as the velocity profile is known, the ray path can be predicted. From *Figure 12.15* it can be seen that in a zone of constant velocity gradient the ray follows the arc of a circle. Each ray will have a different radius of curvature and will always curve towards the minimum velocity.

Following *Figure 12.15* the method is: determine the velocity gradient; then the radius of curvature of the rays of interest; next determine the limiting rays and the vertical and horizontal displacement of the ray. If the ray meets a boundary between two zones of differing velocity, the launch angle into the new zone will be the same as the arrival angle. Lastly, it will be necessary to determine the new velocity gradient and repeat these steps. These computations for sonar prediction are repetitive and ideal for a computer.

12.9 Acoustic system components

12.9.1 Introduction

While it is possible to design and construct a purely mechanical sonar system, today, electronics are a major feature of such systems. The result is complex equipment designed for particular tasks, but which have common features.

Figure 12.16 illustrates the major components of an active sonar system. A pulse of electrical energy is formed and converted into mechanical energy or sound and transmitted by the projector. The reflected or retransmitted acoustic signal is received by the hydrophone and converted to electrical energy. This operation can occur at the same point provided that the two are separated in time, when a single transducer is used for both functions. The received signal is then subjected to a degree of signal processing. This will include amplification, filtering and possibly some form of rectification, before being fed to the display, storage or recording part of the sonar. This completes the cycle.

12.9.2 Transducers

These are electro–mechanical devices. The quality of the data transmitted and received are controlled by the conversion of events from electrical to mechanical and vice versa, i.e. transduction. The ability to carry out this process depends upon the materials used having

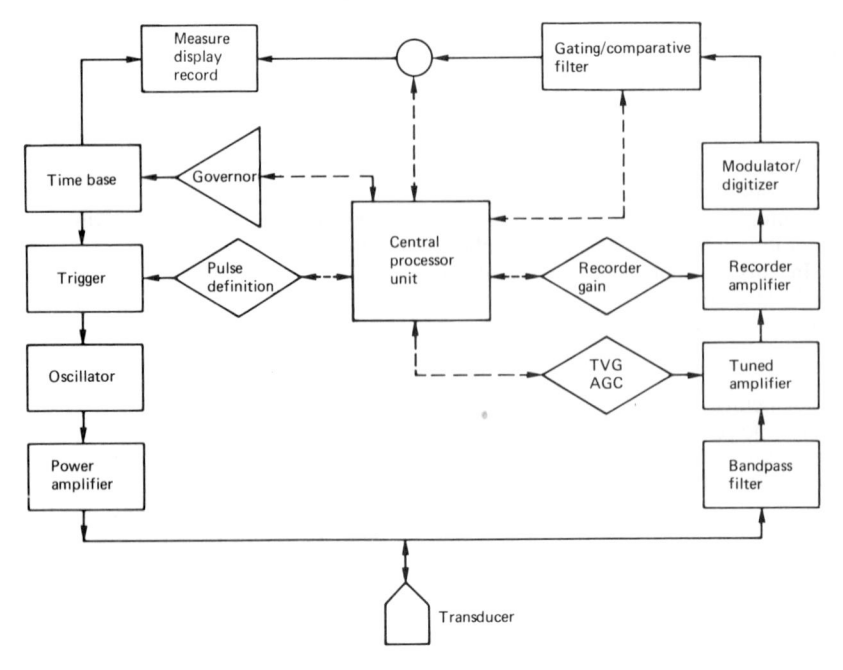

Figure 12.16 Block diagram of sonar system components

the properties of either magnetostriction, piezo-electricity or electrostriction. Today all are available, although electrostrictive transducers are the commonest.

These are ferroelectric materials which show a dimensional change in response to an applied electric field. The transducer is polarized and it has a linear response in both directions. It has a transducing efficiency of about 70% and moderate impedance. It is a ceramic and can be moulded to the exact size and shape required. It will operate over a wide frequency range, although, at low frequencies, problems of driving are overcome by sandwich construction with the ceramic element being bonded between a metal head and tail mass.

Magnetostrictive materials are ferromagnetic; they are of low efficiency when used as transducers and are generally falling into disuse. Piezoelectric quartz crystals used in the early development of transducers are little used today except as hydrophones.

The acoustic beam width is inversely proportional to transducer size. But the beam of an extended face transducer can be modified for specific purposes. If the transducer is composed of an array of individual elements then, by adjusting the element responses, the pattern of the beam can be tailored to a required shape. However this is modification of a single beam on a discrete bearing. If the transducer is composed of an array of elements, either individually or in staves, the array may be steered electronically by applying time delays or phase shifts to the elements or staves. The effect of this is to rotate or steer the main lobe of the beam in a predetermined direction by delaying and summing the output of the array elements into one receiving channel. By this means transducers can be designed to receive signals from any chosen direction, a sector can be swept or a fan of preformed beams can be generated. This technique has found wide applications in swathe sounding and acoustic navigation systems.

A low frequency, narrow beam sonar with no side lobes is large, expensive and difficult to construct using conventional techniques. The principles of non-linear acoustics have been known for some time, but it is only recently that they have been applied to marketable hardware to produce such a system. This is achieved by using two collimated, high power, high frequency sources. The difference frequency is modulated by the near field in front of the transducer; a parametric signal is produced which has a narrow beam and no side lobes. This principle has been successfully applied to the range of deep water parametric echo sounders which is currently available.

12.9.3 Signal processing

A relatively weak signal must be enhanced by some form of processing prior to transmission and subsequent to reception if performance of a system is to be optimized.

The transmitter is relatively straightforward. It controls the interval, length, sequence, pattern and number of pulses which the gated oscillator feeds to the power amplifier for the subsequent output to the transducer. These features are either system or operator controlled and will vary with the system type, range and resolution required. It will take into consideration interference by or to other systems.

The response of the transducer to the applied power is a major factor is maximizing the range and signal to noise ratio of the sonar system. The limit is reached by the onset of cavitation or the maximum power level that the array elements can handle. However, short range performance can be degraded by too high a source level and some systems feature a power control which balances output power with range.

The time between the transmission and the return of the reflected or retransmitted signal is a measure of the range. The accuracy of this range depends upon the knowledge of the speed of sound and the timing circuits of the sonar. The received signal at the transducer needs to be filtered and amplified in a controlled manner before further processing and measurement. This is the most complex unit in the sonar system.

The purpose of this unit is to carry out a precisely controlled amplification of the received signal and to pass it on to the remaining stages of filtering, decoding, etc., at the required level. The transducer and amplifier are matched in terms of frequency and bandwidth and at this stage often include a bandpass filter to reject wideband interference.

The gain systems will either be automatic (AGC) or time varied (TVG). The purpose of a TVG system is to compensate for the range terms contained in the transmission loss element of the sonar equation. Survey echo sounders often have a TVG function of $20 \log r$, although it is more usual to have controls which allow the operator to control the amount and time by which the gain can be varied. This is particularly important when using side scan sonar which builds a picture of the sea bed normal to the vessel's movement by printing the returns from the isonified strips as adjacent lines on the recorder display. On most modern side scan sonars, this is automated and the TVG will vary according to the signal strength received.

An AGC system will adjust the gain according to the received noise and signal level, so that there is a constant signal output, despite variations in the input. For systems where the requirement is a range or ranges to discrete points, the AGC ensures constant signal strength under all conditions. This is often used as a regulator with a TVG ensuring that with increasing range and high background noise the signal is detected and is within the dynamic range of the receiver.

The analogue signal is now further processed to allow it to be approved and possibly decoded before being used. It is then demodulated to produce a digital output. This can be further filtered in terms of a simple range gate or a more sophisticated statistical filter before being fed to the measuring, recording and display elements of the system.

12.9.4 Measurement and display

All systems require a time base in order to measure the range and control the rate at which repetitive transmissions are made. The time base must move at a constant speed which is determined by the speed of the acoustic wave, which will vary. It initiates the trigger pulse, which is the zero point of the scale; it marks the point where the acoustic transmission takes place. Transmission is normally a repetitive operation and the pulse repetition frequency (PRF) is a function of the expected maximum range of the system.

The output of the time base can be in digital or analogue form on a variety of media; it will display the range as factored time or in units of time as detected by the system. For storage and further processing, these data, together with all the operational settings, are made available by suitable interface to a computer based data processing system. On the large integrated systems currently found offshore, a computer is used to control the operational functions of systems according to the criteria established by the program.

The simplest form of display is alpha numeric and consists of a block of LEDs or LCDs. These are the least informative and are normally used as a means of verifying system operation. They are sometimes used in conjunction with an analogue output as a remote display providing information, such as depth, for the helmsman.

Analogue displays were first used on echo sounders. They were important in that they provided a data store in the form of a profile. The storage medium is paper; today this is normally dry and sometimes metallized. The time base can be a single or multipen system driven across the paper at constant speed by mechanical means, although on some side scan sonars this is replaced by a drum mounted rotating helix.

Recent developments include the thermal printer where the time base is an electronic multi-element comb with no moving parts. This has the particular advantage to the operator of producing no carbon dust or smell, having greater dynamic range and increased resolution. For systems where amplitude information is a requirement, colour printers are an advantage; these have recently been introduced on survey echo sounders.

The VDU was first used with acoustic navigation and positioning systems. Position was either shown relative to a grid system or as a PPI with the ship as the centre point. The VDU can be monochrome or colour, this latter feature having made them attractive for use with side scan sonar, enabling relative position, shape and amplitude information to be shown on a single display. This type of display is transient and data storage has to be carried out separately.

The display of information is the interface with the operator. While it is important, the use of computer generated imagery could either obscure or enhance the true system performance. An understanding of the total system is necessary.

12.10 Sensor systems

12.10.1 Echo sounders

The basic principles of the echo sounder are well known. An acoustic pulse is transmitted, reflected by the sea bed and returned to the receiver. The calculation, depth is $C_{sw} \times t/2$, this is performed taking into account variations in C_{sw} and t caused by environmental and instrumental conditions. Providing that it is not a false return from fish, bubbles, etc., the result is the depth to the sea bed.

This defines the task of an echo sounder which is to measure depth which is subsequently related to position. But how accurate is this measurement? The IHO Standards for Hydrographic Surveys (1982) states that errors should not exceed 0.3 m (0.98 ft) from 0–30 m (98 ft), 1 m (3.28 ft) from 30–100 m (98–328 ft), 1% of depths > 100 m (328 ft). The sources of error in depth measurement are temperature, salinity and tidal effects of the sea, sea bed material and topography, the echo sounder itself, including the position of the transducer, vessel movement and rotation. From this, it would appear that the only accurate sounding would be obtained from a securely moored platform floating on a flat homogenous non-tidal sea over a flat homogeneous sea bed using a perfectly adjusted echo sounder. This is patently not the working environment.

To date, these problems have not received the attention they merit. This is possibly because there is not a lot, statistically, that can be done with a single depth measurement. This is unlike multiposition line fixing where the attendant redundancy and adjustment is used to confirm positioning system accuracy. There have been few attempts to carry out simultaneous, comparative trials of echo sounders. In the main, such work has been done to prove the validity of swathe sounding systems for the collection of bathymetric data.

The operating environment of the echo sounder can be broadly divided into three regions. The most specialized is that of the port and its approaches. Here one finds the greatest variety of sea bed type and morphology. Tidal effects and temperature salinity variations can be large. To this add industrial noise, the effect of wakes and suspended sediment and it becomes a most testing acoustic environment.

In spite of this, one type of echo sounder commonly used in port surveys is portable and has a single frequency of 200 kHz beamwidth of 8° and a pulse length of 0.2 ms with analogue output, manual phase control and annotation, depth range 120 m (394 ft). This is a cost effective approach as long as there is minimal automation of the survey operation, and the sea bed is well defined. While this type of echo sounder is capable of being automated, the single frequency can be a limitation.

A large number of ports suffer from the phenomenon of fluid mud flows. These are loose mud deposits with a very low shear strength. The high frequency echo sounder will only show the top surface of the layer. A lower frequency echo sounder will show this surface and penetrate the layer; if there is a compacted sediment 1 m (3.28 ft) or so below this then this may be recorded. Surveys for navigation and dredging require a minimum of a dual frequency echo sounder. As ships can manoeuvre in fluid mud of a density of up to 1200 kg m^{-3}, ideally this level needs to be identified. This requires a measurement system able to define the density variation with depth.

Echo sounders which have dual frequency capability are the most flexible, sophisticated and expensive. While in common use in major ports, they are universal on the continental shelf and beyond. Here water depths are much greater, but sea bed morphology and type tend to be more regular. A typical specification for such an echo sounder would be simultaneous operation at 33 and 210 kHz, beamwidth 9°/18°, pulse length 0.1/2.0 ms., analogue and digital output, automatic phasing and annotation, fitted with TVG, AGC and APC, depth range 200/2000 m (656/6561 ft). An echo sounder with such a specification has a wide range of applications and has become the standard echo sounder for offshore work.

Working in the deep ocean requires low frequency for range, but, if sea bed irregularities are to be detected, a narrow beam is required. An echo sounder operating at 3.5 kHz with a 2° beam requires, however, a transducer of 12.5 m (41 ft) diameter. A parametric system can achieve this with a transducer array of less than 1 m with a depth range of up to 12 000 m. Whilst the transducer is of manageable size, the system requires a much higher transmitter power; consequently this type of system is suitable for ship fit only.

A major error source in depth measurement is the movement of the survey vessel in a seaway. In removing this error, the surveyor's eye and filtering are only effective if ship motion and sea bed variations are widely separated in frequency, otherwise detail of the sea bed will be lost. The best method is to measure heave, the vertical movement of the vessel and have some form of electronic or mechanical stabilization of the transducer beam. However, such systems are not an overall panacea. They are sensitive to course alterations and to the effect of a following sea when the survey vessel is scending. It is worth realizing that even when the sounding is correctly compensated for vessel movement, it could be located in the wrong position. With a vessel rolling through 40°, a vertical spacing of 25 m (82 ft) between the aerial of the positioning system and the transducer could mean an error in position of 8.5 m (27.88 ft).

The dual frequency, heave and roll compensated, digitized echo sounder will be of no use unless properly calibrated. The traditional method is that of the bar check, where an acoustic target is deployed on a calibrated line beneath the transducer; this is effective to no more than 15 m (49 ft). While this is possibly the best method to determine the depth of the transducer, the calibration is best carried out by determining the correct speed of sound from a temperature and salinity profile, or directly using a velocimeter. This value is then input to the echo sounder or the associated computer. Once calibrated, the sounding operation can continue, but it is important that, where data are being collected and logged digitally, verification is carried out using the analogue record. This is an essential aspect of quality control.

Future trends in echo sounders are becoming apparent. The same basic echo sounder is being used for survey, oceanography and fisheries with modifications for each specialist purpose. Thermal printers and colour displays are available. All systems have digital outputs; some are designed to be linked to microcomputers. Analogue mechanical controls are being replaced by menu driven keyboards. Whilst this will make the interface between the machine and the operator easier, it does not reduce the need for the operator to understand the acoustic processes involved.

12.10.2 Side scan sonars

An echo sounder measures the depth beneath the survey vessel. There is no indication of what lies between the lines of soundings. If one was searching for an obstruction in 50 m (164 ft) depth with an echo sounder with a beamwidth of 20°, line spacing would need to be 17.6 m (57.7 ft) to give 100% cover. The demand equipment to improve this situation was met by the development of side scan sonar. This originated as a naval system used for the detection of bottomed submarines. During the early 1960s geologists realized the potential of side scan sonar for sea bed mapping. As more systems became available, surveyors adopted them to carry out searches for wrecks, debris and obstructions and to provide a qualitative infill between lines of soundings.

A typical mode of operation is shown in *Figure 12.17*. An acoustic pulse from the transducers in the fish isonify the sea bed on either side of the track of the fish. Signals which are reflected and back scattered to the transducers are received, processed and displayed on the recorder as a single trace. This cycle is repeated as the fish is towed along the track and the recorder builds up an acoustic picture of the sea bed.

The beam pattern of the transducer is designed to give maximum resolution along track with a narrow horizontal beam and maximum coverage across track, with a wide vertical beam. *Figure 12.18* illustrates this together with the problem of representing on the recorder that part of the reflection and back scattering

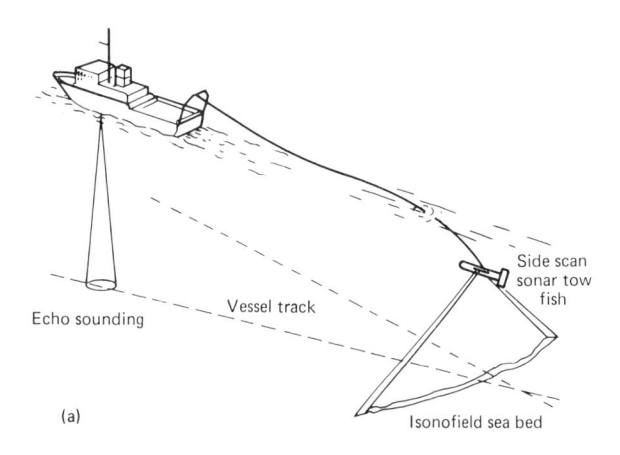

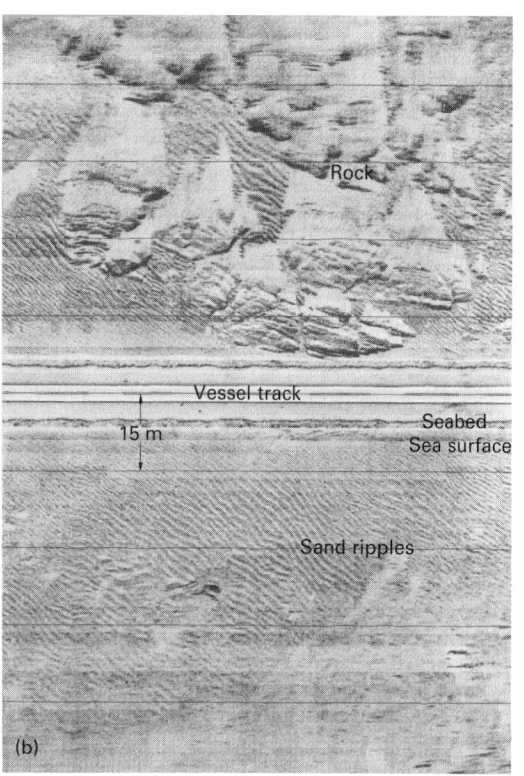

Figure 12.17 (a) Side scan sonar (b) side scan sonar record

due solely to the target or reverberation surface. The level of the returned signal can have a dynamic range in excess of 90 dB. The dynamic range of recorders varies, but rarely exceeds 25 dB.

As can be seen, variations in the range and the height of the fish above the sea bed will cause variations in the two-way transmission loss, the strike angle and transducer response. If a usable acoustic picture of the target area is to be displayed by the recorder, then the amplification stages of the side scan sonar must have a

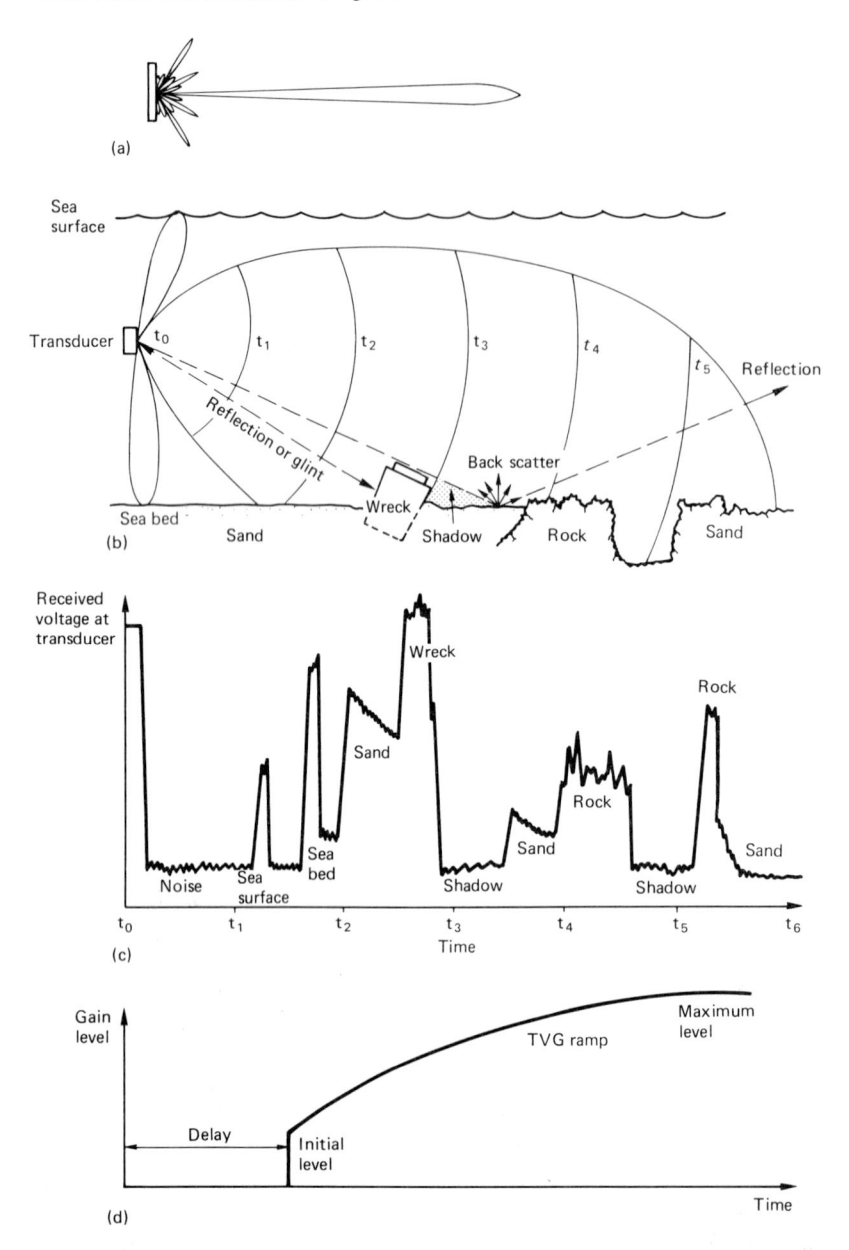

Figure 12.18 Diagrammatic representation of side scan sonar and its measurements: (a) horizontal beam pattern of side scan sonar; (b) vertical beam pattern of side scan sonar over sample sea bed; (c) signal received by side scan sonar transducer in (b) above; (d) example of TVG system minimizing effect of transmission loss and strike angle

TVG/AGC system which will normalize the signal output to the recorder.

On early side scan sonars, manual controls were fitted which consisted of a delay and a ramped TVG. These allowed the operator to adjust record quality in real time to produce a good sonar record. The quality of this record was always marred by being subjective and by the TVG values used being a rough approximation.

A range of automatic or hands off tuning systems was soon developed and has continually increased in its effectiveness. Today such systems have totally replaced manual tuning. Automatic tuning adopted two main approaches. the earlier was basically an automatic TVG, where the variations to the gain were calculated using the sonar equation and the height or the fish with enough range to counter the effects of transmission losses, strike angle and transducer response. A later development designed to overcome problems caused by the incorrect assumptions made during these calculations was a secondary contrast control. It is a negative feedback system, where the output to the recorder is sampled at two time terms, 100 μs and 10 s; should the signal level exceed a defined threshold, the gain is reduced and vice versa. Variations on either or both of these methods are used in the majority of side scan sonar signal processors.

Side scan sonars can be divided into two classes, conventional, in the frequency range 50–500 kHz and GLORIA II, which operates at a frequency of 6.5 kHz using a 4 s chirp pulse from an array which is electronically steered to remove the effects of yaw; the range is in excess of 30 km. The fish is 7.75 m (25.42 ft) long and the weight of the specialist handling gantry and fish is in excess of 12 tonnes. This, together with the necessary support facilities and personnel, means that it cannot be used on a ship of opportunity. It is currently being used to help chart the EEZ of the USA.

With the exception of those sonars fitted as part of a package in deep tow survey systems, the majority of side scan sonars are portable and can be used in survey craft as small as 6 m (19.6 ft). *Table 12.3* gives examples of the range of equipment available. The most recent development is the higher frequency sonars which, when used on submersibles and controllable tow vehicles, allow detailed pipeline inspection.

During operations the aim is to achieve the highest resolution and 100% cover to maximum range. Within the constraints of beam pattern and pulse length this can be optimized by the length of tow cable streamed and the speed of the survey vessel. From *Figure 12.18* it can be seen that the nearer the fish is to the sea bed, the shorter the range but the greater the across track cover; also small targets will be easier to detect by virtue of the greater shadow length. Depth of tow is increased either by reducing speed or lengthening the tow. Slow speed has the additional advantage of increasing coverage and resolution along the vessel's track. Add to this the ability to alter the shape and angle of the vertical beam of the side scan and it can be seen that side scan sonar operations need to be carefully planned and executed.

The side scan record will show, in addition to the target information, sea surface and volume reverberation in addition to any ambient, self and industrial noise. The towed fish, while decoupled from the survey vessel, is still affected by vessel movement. The position of the fish, unless tracked acoustically, is in doubt in areas where the tidal stream is strong and the heading of the fish is not known. In any sort of a seaway, it is likely that the fish will be yawing.

The result of this is that interpretation of the record is difficult and subjective and errors can be made. Although side scan geometry would appear to lend itself to simple quantification of the record, the variables indicate that any quantification should be treated with extreme caution.

Nevertheless, side scan sonar is one of the most valuable acoustic systems available to the surveyor. As long as the limitations are appreciated, it is a relatively cheap and effective search system.

12.10.3 Swathe sounding systems

The echo sounder was a tremendous advance over the lead line as a means of collecting data. It allowed the sea bed to be portrayed numerically in a geographical reference system. The side scan sonar added to the picture, but did not produce quantitative data. Swathe sounding systems combine and extend these two systems in that a swathe of referenced soundings is obtained.

The considerable number of systems currently available or under development can be grouped under three headings: boom mounted transducer arrays, beam forming arrays and interferometers. Boom mounted transducer systems are a means of parallel sounding applicable to calm, shallow water. They consist of up to fifty transducers at a frequency of 210 kHz on separate channels, pulse length 100 μs effective range up to 60 m (197 ft). These systems are particularly suited to survey

Table 12.3 Characteristics of side scan sonar systems

Type	Frequency (kHz)	Beamwidth	Pulse length (ms)	Source level (dB)	Maximum effective range (m)
Long range	50	2° horizontal 60° vertical tilt 0°, 20°, 40°	0.3	230 ref 1μPa	600
General purpose	100	1° horizontal 20°/40° vertical tilt 0°, 10°, 20°	0.1	220 ref 1μPa	300
High resolution	500	0.2° horizontal 40° vertical 10° tilt	0.02	120 ref 1μPa	100

operations in inland waterways and sheltered harbours. Data are output on line in the form of charts and profiles. The whole system is computer controlled for data acquisition, processing and storage. The application for such a system is specific, but within this narrow field it is an efficient method of reducing the spacing of depth data.

Multibeam systems, using beam forming techniques to transmit a fan of discrete beams from a single transducer array, have been commercially available since the mid-1970s. Although such a system had been patented some ten years previously, it was the development of computers with enhanced processing power and data storage which allowed their commercial development. To date, over twenty vessels have been fitted with these systems and since the mid 1980s a number of comparable and competing systems have been developed.

The acoustic element of the system can consist of a crossed fan of separate projector and hydrophone arrays or dual purpose transducer arrays, mounted in an L or T configuration on the ship's hull. This has caused problems of masking by hull generated air bubbles; the solution appears to be to have a fine entry to the vessel's hull. The number and size of the beams generated controls the swathe width. The number of beams varies from twenty to sixty four and the beam-widths from 2° to 5°. Extensive use is made of beam forming and steering, which, together with shading, attempts to produce a uniform footprint on the sea bed and maintain the quality of the depth measurement.

The echo returns, once detected, are passed to the signal processor. Here the depth is digitized and corrections for transducer movement and refraction of the beams are applied. The system accepts inputs from the vessel's position and heading sensors which allows it to compute and display profiles and contour plots, in addition to storing these data.

These systems were initially designed for deep water work and the width of the swathe was 75% of the water depth. Later systems have increased the swathe width to as much as 340% for shallow water work; the frequency has also increased from 15 to 95 kHz. However, the application of these systems to shallow water bathymetry is limited by the width of the swathe and cost, as they are large, expensive ship mounted systems.

Interferometers originated with systems depending upon the acoustic equivalent of the Lloyd mirror effect of optics. They were not successful as sounding systems because of the problems identifying the interference fringes and obtaining an adequate number of depth samples. Development is currently concentrated on phase measurement interferometers, with phase ambiguity being resolved either by calibration of the system over a flat sea bed or, more elegantly, by using two interferometers of unequal spacing and using the difference as a vernier.

These systems are generally of much higher frequency, but can generate a swathe width of up to 700% of transducer altitude. The transducers are fitted in towfish and are far simpler and cheaper to construct than the multibeam systems. However, for accurate depth determination, the fish requires attitude sensing equipment or some form of stabilizing, in addition to an acoustic tracking system to establish fish position. The systems have a very high data rate, which results in the majority of the data processing having to be done offline, as there is a high computation load in reducing the depths from the output of two interferometers and checking for errors caused by volume reverberation.

These systems produce high quality side scan records in addition to the swathe of soundings. They are portable and, to a degree, complement the multibeam systems, although the long range, low frequency interferometers do not appear to have the same accuracy.

12.11 Acoustic navigation and positioning systems

12.11.1 Introduction

These systems are a product of military necessary, occasioned by the loss of the USS Thresher in 1963 in over 2000 m (6561 ft) of water, outside of the cover of accurate surface positioning systems. As a consequence, the US Navy initiated development of acoustic systems to establish the position of surface and subsurface vessels in deep water. In 1966, an early system was successfully used by the US Navy to locate nuclear weapons which had been lost off Palomares in Spain.

During the 1970s, a large variety of systems were developed and were being used by the offshore industry in the North Sea. However, it took further developments in the range and type of equipment available, with the application of microprocessors, purpose engineered software, increased reliability and reduced operator complexity, before systems overcame their initially bad reputation. They are now used for a wide range of operations offshore.

There are tremendous advantages in using acoustic systems for offshore work. The beacons or transponders are portable and the establishment of a system is straightforward. As yet, no licence is required and the operator has an independant system available for 24 h operation. They can be used for surface and subsurface navigation and positioning and can be integrated with other surface positioning systems. The two main types used, whilst overlapping in their applications, complement rather than compete.

12.11.2 Long baseline and intelligent systems

The principle of operation is that an array of transponders is deployed on the sea bed. They are interrogated simultaneously using a common interrogation frequency (CIF). The transponders reply on their individual reply frequencies (IRF). With ranges from three or more transponders a 3D trilateration can be carried out. The position solution is the intersection of spheres which have radii equal to the received slant ranges. If the interrogation vessel is static, then a simultaneous solution is obtained. Errors, which will show by the spheres not intersecting at a point, will be instrumental and environmental. But if the vessel is underway, then the volume of uncertainty of the intersection will, in addition, reflect the movement of the vessel. This requires the position line generated by each range to be deskewed or referred to a common point.

This is termed the simultaneous or reply diversity mode of operation and is the most time efficient method for single vessel positioning or navigation within an array. If a submersible is being used, then it can be positioned using the sing around mode. This is achieved acoustically by the surface vessel transmitting on the CIF to the array and the submersible. They reply on their IRF and the IRF of the submersible causes the array to reply on their IRF a second time. From this collection of ranges, the surface vessel and the submersible can establish their position. The submersible with its thruster units can often be too noisy for acoustic interrogation; then a hard wired responder is fitted and is triggered or interrogated electronically.

All the above implies that the relative and absolute positions of the transponders forming the array are known. It is not practical to deploy the transponders on precisely located references on the sea bed and so their positions must be established. This is commonly termed calibration, an unusual and ambiguous term, as calibration is analogous to the establishment of a control network in trigonometrical survey by trilateration. For relative positioning, calibration is carried out by slowly steaming through the array and measuring the slant baselines. The horizontal distance can be defined, provided the depths of the transponders are known; this is a lengthy operation. Should the array position be required in an absolute reference system, then either a grid on grid calibration of the complete array or a box in calibration of selected transponders must be carried out. These operations are illustrated in *Figure 12.19(a), (b)* and *(c)*.

If surface or satellite positioning is available, then it is common practice to carry out an absolute calibration initially. A sufficiently large number of simultaneous surface position and acoustic range measurements is required to allow the best fit transponder coordinates to be derived using the method of least squares. A similar result could be achieved by boxing in all the transponders, but this would take the longest time.

The initial systems were developed for long range, deep water work; as a result low frequencies of 8–15 kHz were used. When used in the North Sea, they never realized their range potential, the frequency band was noisy, the pulse length was long and resulted in accuracies of 2–5 m (6.5–16.4 ft). In addition accurate velocity measurements were required to compensate for the effects of ray bending and velocity variations. Despite these factors, the systems were widely used for a range of operations including navigation, tracking submersibles and towed devices, remote control and telemetry and the positioning of subsurface structures.

Development was aimed at improving performance and increasing the range of applications for acoustic systems. An early development was the dual mode transponder (DMT). For static operations in a noisy environment, the use of reply diversity means that for each ranging cycle there is increased reverberation with the possibility of signal loss or reduced accuracy. In a static situation DMT transponders can be interrogated in turn, sequentially; this allows all transponders to reply on a common reply frequency (CRF). A transponder fitted with CIF and CRF plus an individual channel frequency (ICF) can be used in either mode. There are the additional advantages that individual transponders can be commanded to enable, disable and release using the ICF; similar advantages are available with the sing around mode. It was a major increase in operational flexibility. This was further improved by using the quieter medium frequency band, 20–30 kHz, which allowed a shorter pulse.

The year 1979 saw another advance. A DMT fitted with a microprocessor came into use; this had the addition of an individual command frequency for telemetry (ITF). Commands could now be received, decoded and executed. Data could be formatted and transmitted in digital form. Instead of just reacting to simple commands, the transponder was now intelligent. In addition to action as a standard DMT, the transponders in the array now had the ability to carry out a relative calibration by direct baseline measurements. Each transponder could be commanded to act as a master and interrogate the remainder of the array *Figure 12.19(d)*. These ranges plus the measured temperature, salinity and depth at each transponder produced a far more accurate calibration.

Prior to these developments, the application of acoustic positioning to structure positioning and pipelay was limited by having to use the sing around mode and a fixed transducer position. However, the ability to command a transponder to act as a remote master and measure the ranges to a calibrated array meant that, when fitted to a pipeline being laid, the point of touchdown could be determined accurately. As the transducer on the vessel is now used as a telemetry link to the remote master transponder, it could be sited in the optimum position to maintain a reliable acoustic path (*Figure 12.19(e)*). This increased operational diversity has been applied to a wide range of operations offshore including ROV navigation, integrated telemetry links, structure positioning and to direct navigation.

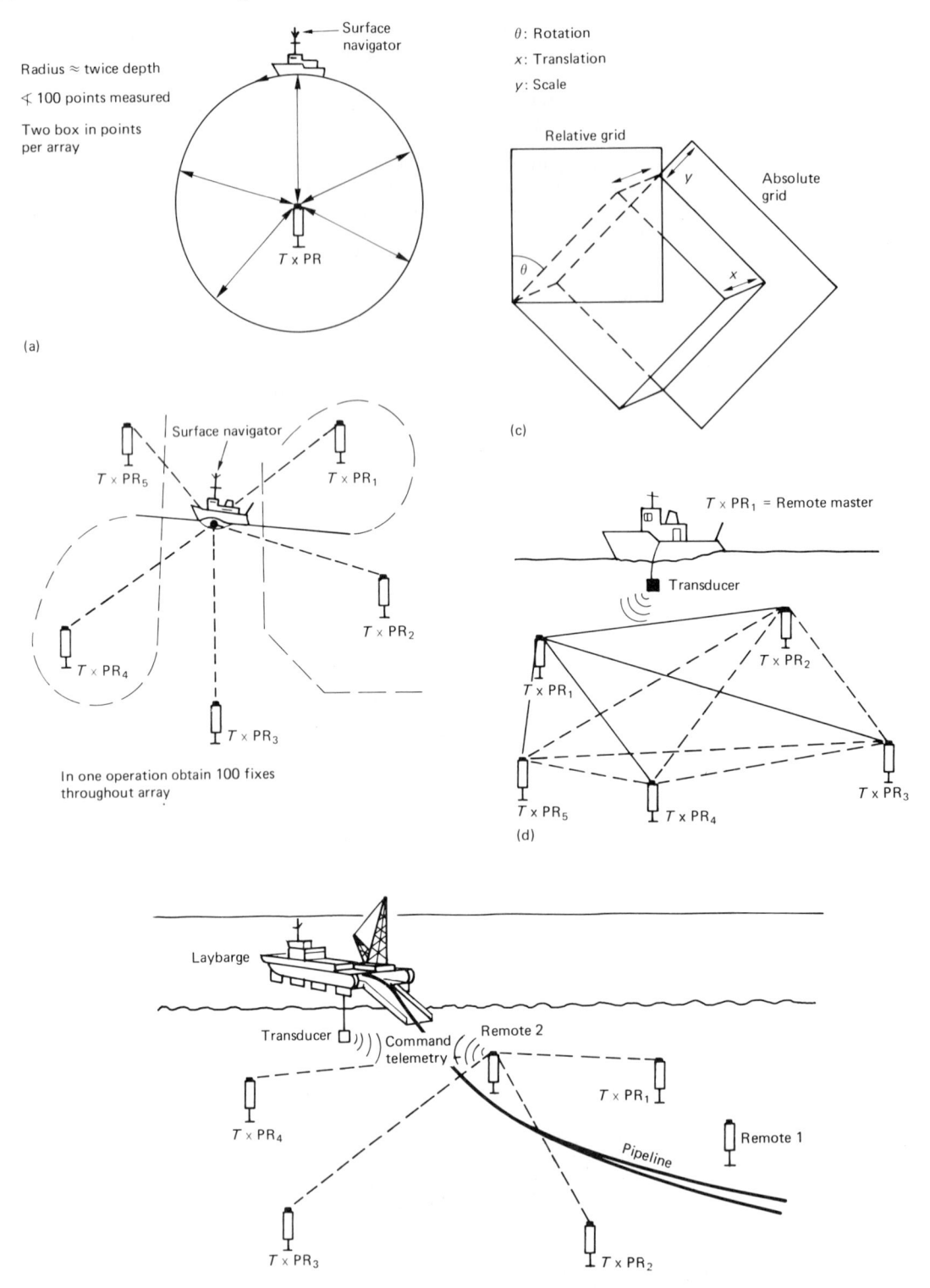

Figure 12.19 (a) Box in calibration; (b) Grid on grid calibration; (c) Grid on grid adjustment; (d) Relative calibration; (e) Pipe laying using an intelligent system

12.11.3 Phase comparison or very short baseline systems

The long baseline systems determine position in relation to a fixed array of sea bed transponders. In complete contrast to this, the phase comparison systems only require a single sea bed transponder or pinger with a three element transducer array mounted on the surface vessel. This would appear to be a much simpler system in terms of hardware. But it is countered, at least partially, by the fact that the transducer must be precisely fitted on board the vessel in terms of position and attitude and a vertical reference sensor (VRS) is required to measure pitch and roll.

It was during the mid-1970s that these systems first made their appearance. The simplest and cheapest system is a free running pinger deployed on the sea bed which transmitts a short acoustic pulse at spaced intervals. The pulses are received by a three element hydrophone array mounted in an L shape, one element being the reference, the other two being mounted on the x and y axes of the vessel. The spacing of the elements is of the order of a half-wavelength; as long as this distance is much less than the distance to the pinger, then the acoustic wavefront can be considered planar and the mechanical angle (θ) can be seen to be:

$$\sin x = C_{sw}(t_1 - t_0)/D \tag{12.18}$$

With angles in x and y and a known depth of the pinger, the position relative to the pinger can be established (*Figure 12.13(b)*).

The accuracy of such a system is better than 1% of range when nearly or above the pinger, this decreases rapidly as the pinger moves away from the vertical. By using several frequencies, a number of pingers may be used; even so there are limits to the application of the system to operations such as station keeping, monitoring fallpipe position on stone dump vessels or riser re-entry.

This system has a mechanical equivalent that is standard fit to all dynamically positioned (DP) vessels. It is the taut wire equipment (TWE). It can only be used for static positioning but is fundamental equipment on fire fighting and safety vessels used in support of the offshore industry. They could be working, in DP, in an area of high intensity noise which would effectively disable any acoustic system. TWE consists of several hundred metres of 12–15 mm (0.4–0.5 inches) steel wire on the end of which is a depressor weight. This is deployed on the sea bed from a self tensioning winch, via a davit fitted with a sheave able to measure the angle and amount of cable out. These data are fed to the positioning system where the displacement in x and y from the depressor is determined, which is then fed to the DP system. This is a simple, elegant solution.

When a phase comparison system is used in conjunction with a transponder or transponders, the additional cost is compensated by the increased range, accuracy and flexibility available. The systems work in the medium to high frequency band of 20–70 kHz; multiple transponders can be positioned or tracked at the same time, the number depending upon the frequencies available to the system. The transducer, now consisting of hydrophone array and projector, can be fitted to a vessel of opportunity, but the installation of the transducer and the VRS needs to be attended to with great care.

In operation, once the transmitted pulse has been verified and accepted by the transponder, it transmits a reply. The phase of the received reply pulse is detected, the mechanical angles computed, range is determined from the elapsed time and the result is displayed on a computer controlled CRT or VDU relative to the screen centre and the vessel's heading. If a gyro compass is fitted and interfaced to the system and the position of the transponder is known in some geographical reference system, then the absolute position of the vessel can be determined and displayed. The display need not be centred on the vessel; any of the transponders may be put at the origin of the display. The displays are keyboard controlled and menu driven. All data are available to be output to a computer based data logging system.

Although claiming an accuracy of 0.5% of range, the cover is limited by the beam pattern; here there is a conflict. The wider the vertical beam, the greater the cover, but the angular resolution is decreased. The two graphs shown in *Figure 12.20(d)* illustrates this effect. Graph (i) shows the full curve where $D = \lambda/2$. While it has a mechanical angle (θ) of 180°, the rate of change slows as θ reaches 90° so that 1° of phase measurement will give a bearing resolution of 6°. Graph (ii) in *Figure 12.20(d)* shows a dramatic improvement if D is increased to $3.\lambda$, there is the bonus of the narrow beam having increased gain and reduced susceptibility to noise, because of the lower side lobe level but now there is ambiguity in the measurement of θ.

This is resolved by having a transducer switchable from wide to narrow beam and instead of one narrow beam on a discrete bearing, to steer it electronically. This allows the system to search in wide beam, locate and identify the transponder, switch to narrow beam for measurement. With the transducer being able to rotate in azimuth, accurate measurements out to 2 km can be made. Used in a tracking mode the system requires memory and prediction, which is achieved using a Kalman filter in the software which automatically incorporates a procedure for dealing with a lost signal.

Such systems have wide applications, particularly for tracking ROVs and towed devices such as side scan sonar and seismic streamers and for DP operations. They do have the capability of operating in pseudo-long baseline mode, but do not have the all round efficiency of intelligent systems.

By deploying a single transponder, there is no requirement for relative calibration, and absolute calibration would consist of the box in of one tran-

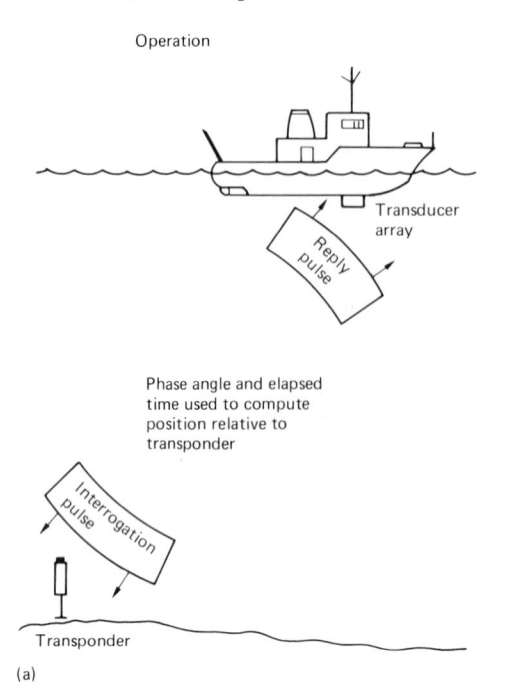

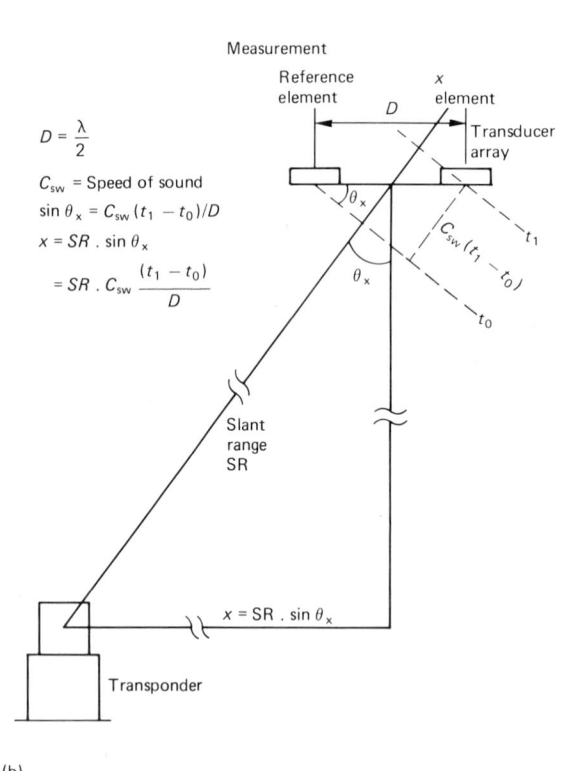

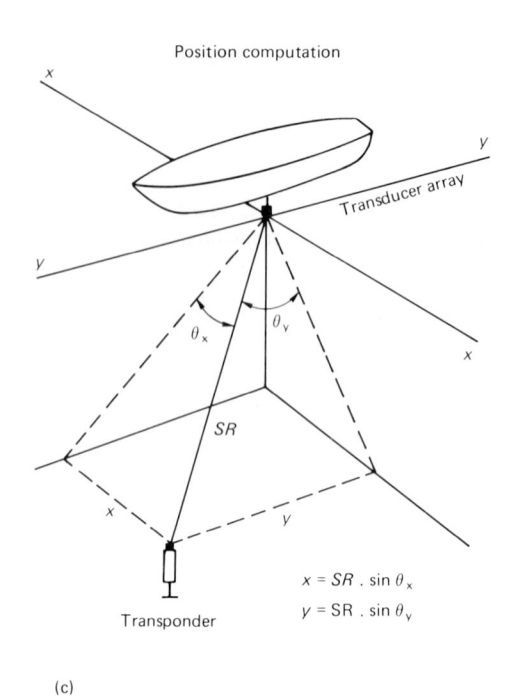

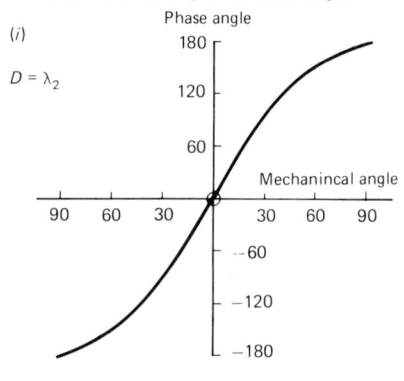

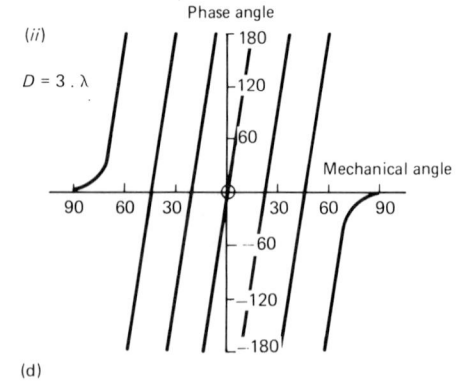

Figure 12.20 Phase comparison of an acoustic navigation system (a) operation: (b) measurement; (c) position computation; (d) graphs of phase angle/mechanical angle D

sponder. However, there has been substantial discussion over the calibration of these systems. To use the trigonometrical survey analogy again, this calibration is the equivalent of field checks on a theodolite. The process is essential and can be divided into two: determination of fixed errors which should be carried out on installation of the equipment and checking for variable errors which the prudent surveyor would carry out prior to each operation. Detail of these operations are well documented and are beyond the scope of this text.

12.11.4 Short baseline systems

These systems consist of an array of a minimum of three transducers or hydrophones mounted on or below the hull of the vessel, which determine the slant ranges to a single transponder or responder. With the array mounted on the vessel, a VRS must be used to compensate for their movement relative to the transponder. The dimensions of the array require extremely accurate time measurements to be made in order to maintain system accuracy.

Problems associated with the accuracy of short baseline systems and the successful development of phase comparison systems have prevented their widespread adoption. However, they are to be found on some semisubmersible drilling rigs and stone dump vessels.

A more recent development has been for streamer tracking on vessels carrying out three-dimensional seismic operations. Here transducers are mounted on the vessel and on the seismic sources; this can give a baseline of up to 150 m normal to the vessels track from the transducers mounted on the source arrays, whose position is found by interrogation from the hull mounted transducers. The array is now working in the horizontal and there is no requirement for a VRS. The outboard transducers interrogate transponders on the streamer and the ranges allow their positions to be trilaterated Although not the complete answer, when used in conjunction with streamer compasses and tail buoy data, it does provide improved position data for the streamer.

12.12 Other acoustic systems

12.12.1 Doppler log and navigation systems

The effect of Doppler shift of frequency is named after Christian Johann Doppler who in 1853 demonstrated the link between frequency shift and source speed. For acoustic transmission in the sea, this relationship can be shown to be:

$$\triangle f = f_C \, 2V \cos \theta / C_{sw} \qquad (12.18)$$

From this, it can be seen that:

$$V = \triangle f \, c_{sw}/2 \, f_C \, \cos\theta \qquad (12.19)$$

where V is the velocity, $\triangle f$ is the shift in frequency, C_{sw} is the speed of sound in sea water, f_C is the frequency of transmission and θ is the transmission angle.

In a Doppler sonar, the frequency shift of the backscattered signal received from the sea bed is measured, the transmission angle and frequency are known and the speed of sound is derived.

If the vessel is pitching in a sea way or the trim is changed, the value for the transmission angle will be varied and put error into the system. This is greatly reduced by fitting two transducers in the Janus configuration, looking forward and aft. However, for large angles of pitch or trim, input from VRS is required to maintain accuracy.

The intensity of the received signal needs only to be sufficient for detection. Variations in this intensity are controlled by limiting and AGC circuits. Consequently, the type of sea bed has no effect on the measurement of Doppler shift.

Doppler sonars with computer control either measure the sea temperature via a thermistor mounted on or near the transducer and compute the value, or estimated values are entered for computation. This can be a major source of error; one manufacturer has removed this by controlling the geometry and element spacing of the transducer array, so that the Doppler shift in frequency is a function of vessel speed and element spacing.

A single axis system will only give speed over the ground; direction still has to be estimated. However, if two transducer systems are mounted in the x and y directions on the vessel's hull, then movement over the ground relative to the vessel's head will be determined. When a gyrocompass is integrated, it then becomes a Doppler navigation system.

Such systems are in common use, particularly on vessels collecting geophysical data, where it is part of the satellite navigation system. Operating frequencies are around 100 kHz and sea bed lock can be maintained to 600 m (1968 ft). Once this is lost, the accuracy of the system is degraded as the Doppler shift is now measured from scatterers in the water body. Other applications are increasing and include fitting to subsurface vessels and on vessel docking systems.

12.12.2 Oceanographic systems

There have been various applications of acoustics to oceanographic sensors. One of the earliest and least successful was the application of pulsed transmissions to measure tide and wave height. In effect, it was an inverted echo sounder on the sea bed. Today this type of measurement is usually carried out by pressure sensors and wave rider buoys.

There are two types of acoustic current meters available. One measures the speed of an acoustic pulse in two directions along the same path. The travel time is dependant upon the speed and direction of the water flow. This means that the time difference for the two pulses is a direct function of the water speed along the transducer axis. If this is measured along two orthogonal axes, the speed and relative direction may be determined. To obtain true direction, the orientation of the axes must be established.

Doppler current meters require sufficient scatterers in the water body to give a measurable return. If this exists, then the Doppler systems have the protection to measure the current profile over the complete depth. They can be mounted on the sea bed, a buoy or a surface vessel. Frequencies, beam patterns and pulse rates are a function of the range and resolution required of the measurements. It is an area of measurement which has considerable potential for development.

Measurements fundamental to all aspects of offshore work are continuous seismic reflection profiling. These systems cope with all types of sea bed in terms of range of penetration and resolution. This is reflected in the vast range of seismic sources and detectors available. It is a separate and large subject.

12.12.3 Telemetry

There has been increasing use of acoustics for control and data retrieval since the mid-1960s. The alternative is an inflexible hard wire link. Acoustic telemetry allows data to be observed, collected and logged in real time. Initial systems were single channel and could only be used for data with a slow rate of change such as temperature and salinity, where multiples could be restricted by simple gating. Similar methods were also used to control the basic functions of equipment such as the first of the acoustic positioning systems.

There has been rapid development in acoustic telemetry in response to the requirement for the handling of large amounts of data at faster rates. These developments all relied upon some form of modulation. Three types have been developed and used: pulse pattern modulation (PPM), frequency modulation (FM) and frequency shift keying (FSK).

PPM is the simplest and was used to extend control for acoustic position systems. A series of pulses was transmitted at varying time intervals, with the pattern defining the message. While adequate in noisy conditions, it was susceptible to the effects of multipath propagation. FM was found to be resistant to the effects of noise, interference, fading and multipath effects because of the characteristic of locking in to the first return and ignoring secondary returns. However, it tended to be limited in the range of data carried.

FSK overcame this limitation and in the broadest sense is a combination of PPM and FM. It is in common use today. Within each pulse the frequency is switched or shifted to represent the 0 and 1 of a binary message. This means fewer pulses for a given message and rejects reflections because of the long time interval between pulses at the same frequency.

Acoustic telemetry is used for a wide range of operations offshore, including control of BOP stacks through monitoring the control functions of wellheads, template placement and data telemetry. These operations all take place in an acoustically hostile environment; all systems are duplicated and designed to be secure and failsafe.

12.13 Conclusion

Any offshore operation working on or beneath the sea has a need for some sort of acoustic system and a knowledge of acoustics on the part of the operator. This has been a broad introduction to the subject. There are omissions, such as the range of sonars specifically designed for ROV operations in support of construction and maintenance of offshore structures.

There are some competing systems, such as synthetic aperture radar mounted in satellites, airborne laser bathymetric systems and the humble hand lead. But none of these invalidate the echo sounder or swathe sounding systems.

New technology is being applied, an example of which is the improvement of the signal to noise power ratio by the use of a swept signal or chirp, a long pulse having a wide bandwidth with consequent increases in range and resolution. This has already been applied to an acoustic positioning system. Others will follow; acoustic sensing and positioning is a developing field.

Earth-based electronic navigation systems

12.14 Introduction

This section describes three long range navigation systems. Racal–Decca Navigator covers out to 450 km from the master transmitter, Loran C to 2000 km by direct transmission or 4500 km using a reflected wave, and Omega provides worldwide coverage, with an effective 15 000 km coverage from a station.

In their simplest form, each operates in the hyperbolic mode, allowing passive reception by any number of vessels. The hyperbolic mode is a result of a receiver measuring the difference in arrival time or phase angle from two transmitters. No connection with the transmitters is necessary and the time of the transmission is irrelevant. The value of the time or phase difference is a result of the positioning chain characteristics and the receiver's position in relation to the chain.

Racal–Decca Navigator and an early form of Loran were developed during World War II. First trials of the classified system, designated QM (later Decca Navigator), took place in the Irish Sea in 1942. Following experiments with lower frequencies, those in use at present were chosen and the system helped in the success of the D-Day landings.

Loran A, operating at just below 2 MHz, had grown to seventy transmitting stations and 75 000 sets of receivers by the end of the war. The operating system for Loran A is similar to Loran C, but the lower frequency of the latter enables greater coverage. The number of Loran C sets is in the order of 450 000 worldwide.

Omega development came later out of 1947 proposals for a very low frequency system (VLF). Trials started in 1961 with three transmitters, then in 1967 with four and in 1968 the decision to go ahead with full implementation was taken. Eight stations were available by 1982. Omega provides 24 h, all weather, worldwide coverage. It will not be superseded in that capability until there is a full constellation of GPS satellites. However, the accuracies obtainable will show a dramatic improvement when the change is made from Omega to GPS.

12.15 Propagation of low and very low frequency waves

Radio waves cover the lower part of the electromagnetic (Em) spectrum, with wavelengths from around 1 m (3.28 ft) to around 1000 km. This is equivalent to the bands VHF to ELF. Long range radio navigation systems cover the LF and VLF bands, that is with wavelengths from 1 to 100 km, and frequencies from 300 to 3 kHz.

Racal–Decca Navigator operates between 70 and 130 kHz, Loran C at around 100 kHz, and Omega between 10 and 14 kHz (*Figure 12.21*). The lower the frequency the greater the range possible from the transmitter to the receiver, but the lower the accuracy in position, all other things being equal. Propagation takes place in two forms: groundwave and skywave.

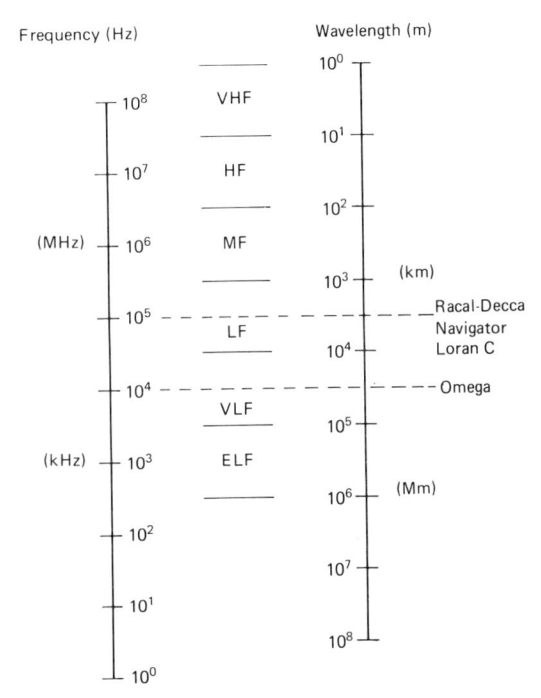

Figure 12.21 Frequency and wavelength ranges of the Racal–Decca Navigator, Loran C and Omega

Low frequency groundwaves follow the curvature of the Earth, giving rise to terminology of spheroidal or ellipsoidal distances. When calculating travel time or phase angle, it is necessary to allow for these true paths, for instance when producing lattices for a navigation chart. For the mariner using the charts, the corrections have already been included and direct plotting on the lattices is possible. Similarly, receivers supplying position in latitude and longitude have calculated their results upon a particular shape of the Earth.

The precise values may not match each navigation chart exactly, as a different mathematical definition of the shape of the Earth may have been used in each. However, the differences may be no greater than the normal inaccuracies expected with these systems. For long range navigation, the chart scales in use are small and small inaccuracies will not show.

Skywave is that part of the signal from the transmitting antenna which travels towards the ionosphere and returns to Earth after reflection from one or more of the ionospheric layers. These layers vary in height daily and seasonally. The effect also varies with sunspot activity, which has an eleven year cycle. The sunspot activity increases the strength of the ionospheric layers and decreases their height above the Earth. These layers are zones of highly conductive free electrons caused by the sun's ultraviolet radiation splitting molecules into their component parts. The layers can be shown approximately as in *Figure 12.22*.

The path taken by the skywave signal means that it covers a longer distance than is covered by the groundwave. Hence there is a longer travel time. The difference in arrival time at the receiver can allow it to separate the direct (groundwave) signal, and the reflected (skywave) signal. At longer ranges, the signal paths become similar in length, until the signals mix. There is still a difference in signal strength, though this will gradually bias to the skywave and there will be a difference in phase.

12.16 Racal–Decca Navigator

12.16.1 System description

Decca Navigator was used in the landing operations on the Normandy beaches in June 1944, following first sea trials in September 1942. Those first trials had been with frequencies of 305 and 610 kHz. The operating system for the beach landings was in the band used today, with frequencies between 70 and 130 kHz. By 1946, the system was available commercially and today consists of about fifty chains around the world (*Figure 12.23*, reproduced from a Racal–Decca brochure).

The Decca positioning chains were financed by hiring out receivers. This provided a regular income to allow for system development, maintenance, servicing and the construction of new Decca chains. However, the inability to make a one-off payment caused annoyance to the users and there was a ready market for receivers sold by rival companies in the 1980s. These sales would have left Decca, now owned by Racal, with all the operating costs and none of the income. A number of attempts were made in the courts to halt sales of these sets. Racal could not secure a sound commercial basis for the system and host countries of the transmitting chains were obliged to take over responsibility for their operation In the UK, the General Lighthouse Authorities took responsibility with the day to day maintenance subcontracted to Racal–Decca Marine Navigation Ltd. This also enabled Racal to sell receivers.

The method of position determination is by phase comparison. The system is passive, in that a receiver is carried on board the vessel, the transmitters are ashore and there is no direct link between them. The receiver notes the phase of the incoming signal. If the phase of the signal at the transmitters was known at any instant, then a computation could be performed using the wavelength to calculate the distance from the stations. However, as there is no direct contact, this is not possible. Instead, the phase of the signal from a second transmitter is compared with the first and the difference measured. Two phase difference measurements (using three stations) can be used to find the receiver location.

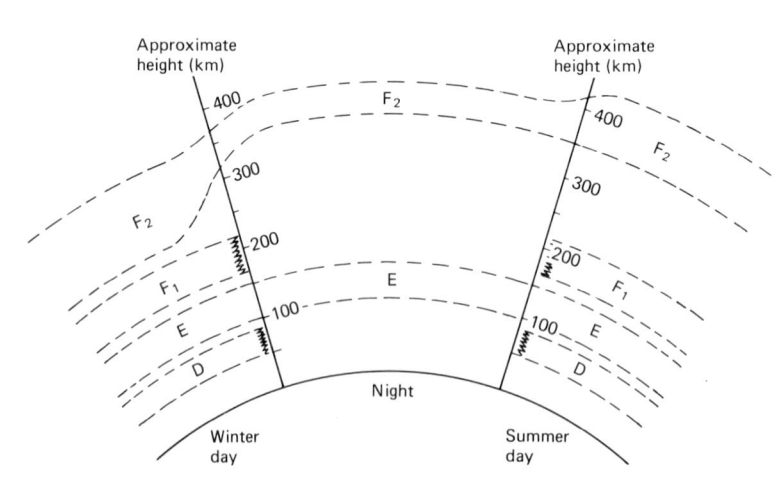

Figure 12.22 Ionospheric layers: the D layer is absorbing, changes of height alter reflated signal paths

12.16.2 Transmitters

Most chains consist of a master and three slaves, although some have just two slaves. The slaves are distinguished by calling them Red, Green and Purple, and the phase difference patterns produced are named after the respective slave station. The baseline length between master and slave varies, but for the operating range of the system, best results are obtained with distances between 130 and 200 km (81 and 124 miles). The angles between the master and slaves also varies and chains are not usually symmetrical.

All transmitters work on a fundamental frequency of about 14 kHz, from which the transmitting frequencies are built up by multiples. The master transmission frequency is six times the basic frequency and lies at or between 84 and 86 kHz. As each chain operates with its own set of frequencies, the receiver automatically selects the correct group when chain selection is made. The multiples of the fundamental frequency for the slaves are Purple – five times, Red – eight times, and Green – nine times.

The number of lanes on a baseline can be found from the baseline length divided by the lane width. However, these are not designated by a simple numerical sequence, but by letters denoting zones, and a number of lanes within each zone. The Red slave has twenty four lanes in each zone, the Green eighteen and the Purple, thirty. Lane numbers start at zero for Red, thirty for Green and fifty for Purple, thereby avoiding a source of ambiguity. Indeed naming the colour of the pattern is not strictly necessary as lane 56.27 must be a Purple pattern. Zones are lettered A to J. Should the baseline contain more than 10 zones, the lettering starts again at A. Since the second zone A will be at least 100 km (62 miles) away from the first, this ambiguity gives no practical difficulty.

12.16.3 Receivers

The Decca Navigator receiver has undergone a transformation in the last few years. The design of even the straightforward Racal–Decca Mark 53 incorporates press button controls, digital displays and automatic conversion to latitude and longitude. The circular dials still remain with the Mark 21. Decca coordinates in zones, lanes and hundredths of a lane are still available, but the computing power within the receiver gives a host of extra facilities.

The calculation sequence commences with the reception of the master signal and up to three slaves, all on different frequencies. The receiver multiplies these up to common comparison frequencies for master/red, master/green and master/purple. The difference in the phase of the signal is output as hundredths of a Decca lane.

Obtaining the number of lanes could be achieved by reference to a known point and setting in the correct values. However, with the lane identification (LI) feature, the receiver can resolve the position to within a zone. This is achieved by using the larger lane width associated with the lower, fundamental frequency of about 14 kHz. This frequency is not actually transmitted, but is resolved from the supplementary transmission of all four standard frequencies from each of the shore stations. This occurs over 2 s in every 20 s of the normal transmission cycle. This technique can produce a more stable result than the standard transmission at extreme range. Navigation can then continue using these LI signals. The feature is known as multipulse.

In the Mark 21 receivers, it is necessary to go through the process of referencing the output. This is to allow for any inequality in the signal paths for the different frequencies and their relevant multipliers and discriminators. When checking the reference, the counters need to be set to zero so operations can continue.

12.16.4 Propagation anomalies

Lattices, and hence latitude and longitude, are computed using a fixed theoretical speed of propagation for the transmitted waves. The value used is 299 700 km/s (186 233 miles/s). This makes no allowance for variations in the atmospheric conditions of temperature or water vapour. Ground conductivity is another variable. The value for the speed of propagation over sea water is fairly constant, but with the signals passing overland between the transmitters, or overland on the way to the receivers, the average speed will be less than that allowed. Ground conductivity is low over bare rock and will vary over soil that is dry and wet with sunshine and rain. Tidal beaches will also have a variable affect.

Changes due to ground conductivity have been quantified by comparison to more accurate systems. This has occurred mostly in coastal areas. These fixed errors are supplied in the manufacturer's data sheets to enable the navigator to plot the vessel's position more accurately. Values, which can be positive or negative, are applied to the lattices before they are drawn on some nautical charts. A further application would result in an error in the plotting.

Decca Navigator has approved coverage from the Department of Transport to 450 km (280 miles) from the master transmitting station. During daylight, coverage can extend to 700 km (435 miles), but at night and with bad reception, it can be as low as 370 km (230 miles). Near the magnetic equator, there is a reduction in coverage to the east and west of a chain.

12.16.15 Accuracy

At the outset of this section, a distinction must be made between absolute and relative accuracy. The former gives the derived position a level of quality in comparison to the true position in the world. Knowledge of

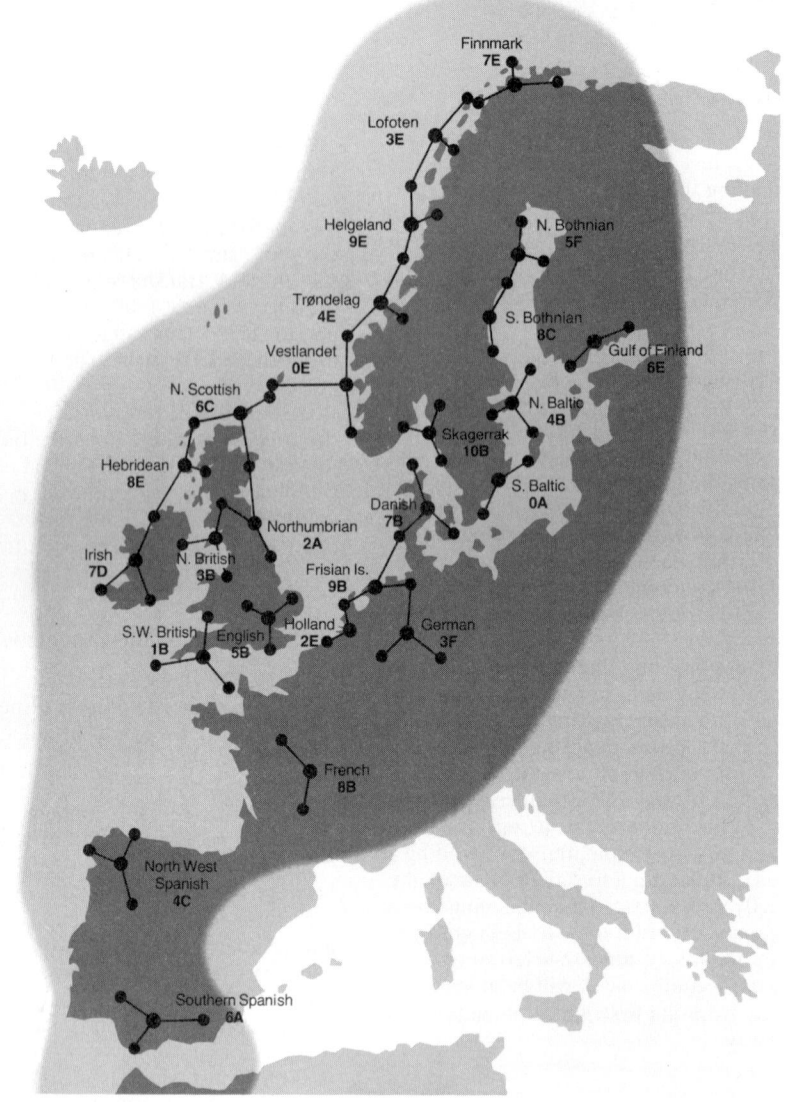

Chain	Code
Finnmark	7E
Lofoten	3E
Helgeland	9E
Trøndelag	4E
Vestlandet	0E
N. Bothnian	5F
S. Bothnian	8C
Gulf of Finland	6E
N. Baltic	4B
Skagerrak	10B
S. Baltic	0A
N. Scottish	6C
Hebridean	8E
Northumbrian	2A
Danish	7B
Irish	7D
N. British	3B
Frisian Is.	9B
S.W. British	1B
English	5B
Holland	2E
German	3F
French	8B
North West Spanish	4C
Southern Spanish	6A

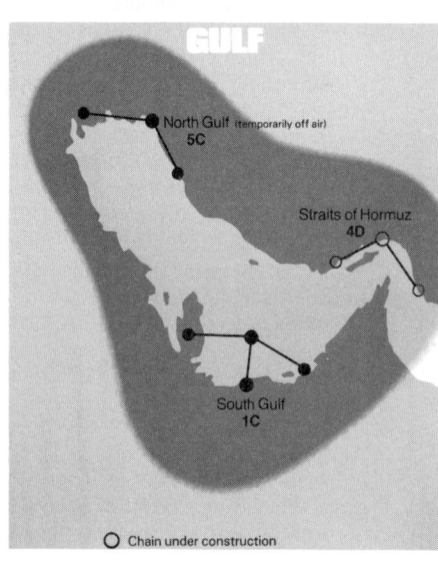

GULF

Chain	Code
North Gulf (temporarily off air)	5C
Straits of Hormuz	4D
South Gulf	1C

○ Chain under construction

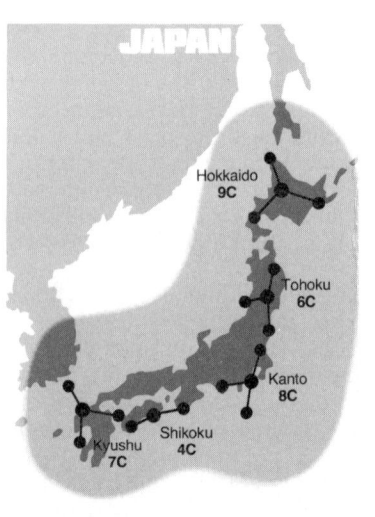

JAPAN

Chain	Code
Hokkaido	9C
Tohoku	6C
Kanto	8C
Shikoku	4C
Kyushu	7C

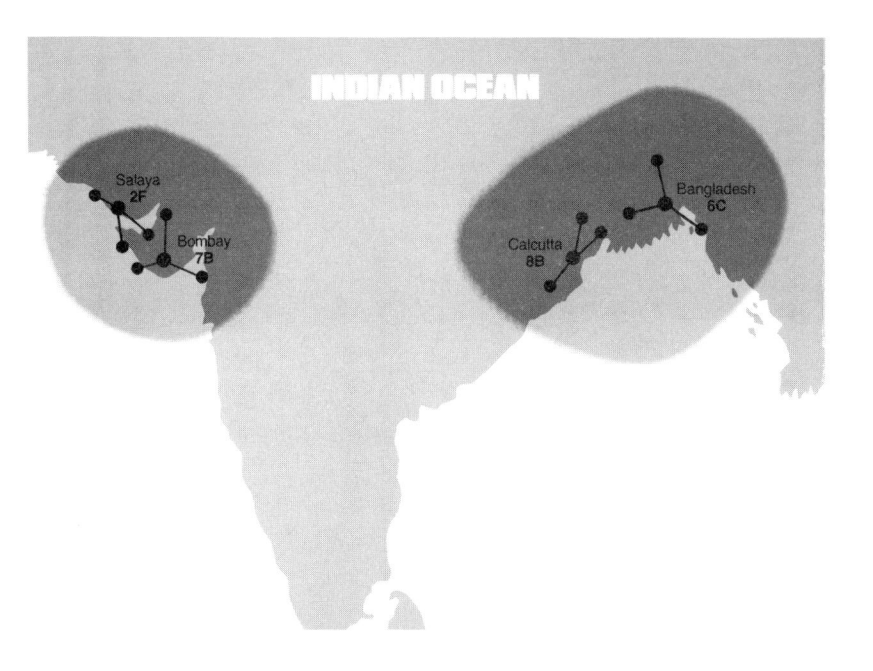

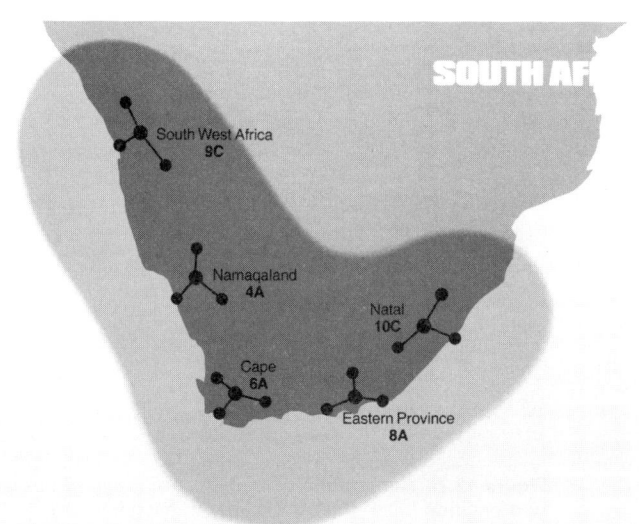

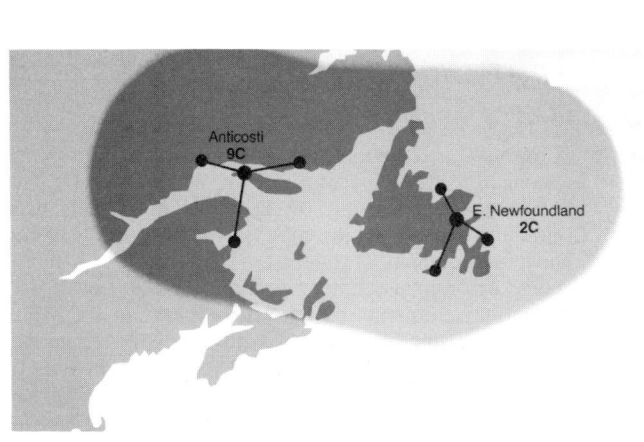

Figure 12.23 Racal–Decca Navigator coverage diagrams

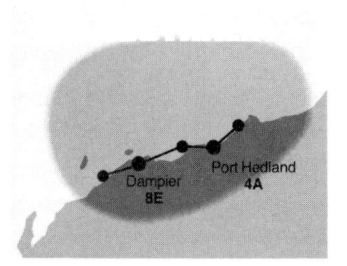

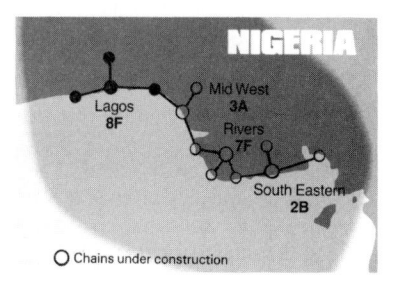

absolute accuracy is only possible by comparison to a more accurate system. Relative accuracy indicates the ability to return to a site designated in the system coordinates, yet which may be unknown in true latitude and longitude. As an illustration, when moored alongside a jetty, the observed Decca coordinates may plot somewhere further along the river, showing that the absolute accuracy is low. However, if the Decca coordinates are the same on each return to the jetty, then the relative or repeatable accuracy is high.

Fishermen may take their trawl alongside a wreck night after night. They might never know the exact location of the wreck, but do not lose their nets.

Accuracy is dependant upon three fundamentals: frequency, lane expansion and angle of cut of the position lines. Then other factors have an influence: receiver inaccuracies or their ability to resolve the signal from noise, the interference of skywave or the varying speed of propagation with meteorological changes.

The frequency propagated is a multiple of the fundamental frequency of the chain. This frequency is further multiplied to obtain a comparison frequency. The lattice lines are based upon this comparison frequency. As an example, using round numbers, a fundamental frequency of 14 kHz is multiplied by eight to give a transmission frequency of the Red slave of 112 kHz. This is multiplied by three within the receiver to give a comparison frequency of 336 kHz. The wavelength of this is approximately 890 m (2920 ft). One lane is 445 m (1460 ft).

This lane width applies only to the baseline between the master and Red slave. Away from the baseline, the lanes diverge. This is a function of the hyperbolic lattice generated by any passive differencing positioning system. The lane expansion can be computed, being a function of the angle at the receiver subtended between the two stations. Lane expansion increases away from the interstation baseline. The pattern for a particular slave is unusable on the baseline extensions, in line with the master and slave.

The angle of cut of two position lines generated by two pairs of stations is the intersection angle between the two lattice lines. Taking the acute angle, the figures accepted in hydrographic surveying are between 90 and 30°. An illustration of the effect of angle of cut coupled with the uncertainty in the position line is given in *Figure 12.24*.

Accuracy is not a simple figure that can be stated for the whole coverage of the chain. Reference needs to be made to charts of predicted coverage and the time of day and year. Using these, a reference table will indicate the accuracy likely for a majority of the time with figures varying from under a couple of hundred metres to a few kilometres. A possibility of lane slip can make some parts of the coverage unreliable. All the tables are available in the Admiralty List of Radio Signals, 1988.

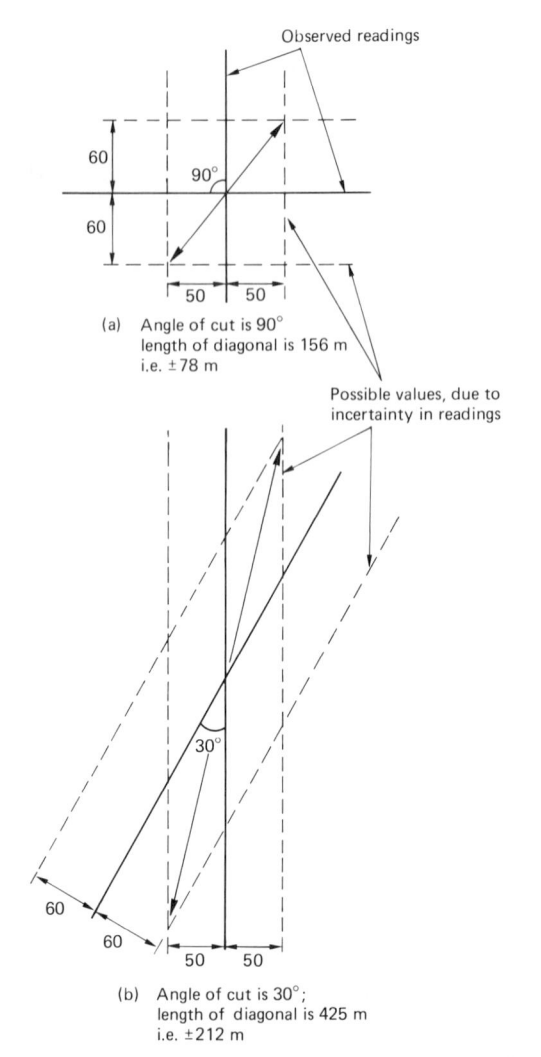

(a) Angle of cut is 90°
length of diagonal is 156 m
i.e. ±78 m

(b) Angle of cut is 30°;
length of diagonal is 425 m
i.e. ±212 m

Figure 12.24 Illustration of the effect of angle of out for two position lines: the uncertainty in each line of position is ± 50 m (164 feet) and ± 60 m (197 feet)

12.17 Loran C

12.17.1 System description

Loran A, a forerunner to Loran C, was, like Decca Navigator, developed for use in World War II. The principle of operation of both Loran types is very similar. The two major changes with Loran C are the lower frequency of operation and the extra phase comparison measurement within the pulse. Loran A is a medium frequency (MF) system operating at 2 MHz. Its range using the groundwave is 700 km (435 miles). Loran C is a low frequency (LF) system operating at 100 kHz, with a range using groundwave of 2000 km (1243 miles).

Loran C obtained US Government approval in 1970 following development of the first chain in 1957. It was named as the preferred navigation system in the coastal area in the National Plan for Navigation. It is managed by the US Coast Guard (USCG), with civil use transmitting stations overseas operated by nationals of the country. A tradition of mutual cooperation has been established and this is leading to the establishment of a chain in the Bering Sea, jointly operated by the US and the USSR. The Russians already have coverage of their land territories using three chains, Tchaika, similar to Loran C, the signals of which are picked up by the Loran C monitor in Turkey. Cooperation has also been established with Racal–Decca who designed and operate a Pulse 8 hydrographic survey positioning chain. This is a very similar system to Loran C, but of lower power and with shorter baselines between transmitting stations.

In addition, there are two chains with six stations operated by Saudi Arabia. Japan is considering expanding Loran C coverage for civil users. India has interests in establishing Loran C facilities.

A position line is obtained by measuring the difference in arrival time of the master signal and those of the slaves. This passive receiving system means no knowledge of the transmitting times of the signals is necessary. The lines of constant time difference follow hyperbolae around the foci of the transmitting stations. Three stations are needed for two lines of position and hence a position fix.

Coverage is good in the Northern Hemisphere, providing information in the Mediterranean, for crossing the North Atlantic, the eastern seaboard of North America and large parts of the northern Pacific. Fourteen chains were operational in early 1989 (*Table 12.4*).

Table 12.4 Loran C coverage chains operational, early 1989

Chain	Code
Mediterranean sea	7990
Norwegian sea	7970
Icelandic	9980
Labrador sea	7930
Canada East Coast	5930
USA Northeast	9960
USA Southeast	7980
Great Lakes	8970
USA West Coast	9940
Canada West Coast	5990
Gulf of Alaska	7960
Central Pacific	4990
North Pacific	9990
Northwest Pacific	9970

12.17.2 Transmitters

A Loran C chain consists of a master and either two, three or four slaves. There is at least one monitor associated with the chain, which ensures that the signals at that place remain within prescribed operational limits.

Chains are distinguished from each other by the time between repetitions of the group of signals. This is quoted to four digits, to which adding a zero gives the group repetition interval (GRI) in microseconds. Thus the Norwegian chain, designated 7970, means that the full cycle of transmissions repeats every 79 700 μs.

The signal structure is such that the master is distinguishable from the slaves as it transmits nine pulses, each lasting about 250 μs, with a gap of 1000 μs between consecutive pulses (2000 μs) between the eighth and ninth.

The slaves transmit just eight pulses, but with the same length and interval. These eight pulses give the name to Racal's Pulse 8 system.

Slaves are distinguished one from another by their order of reception following the master transmission. Therefore, it is necessary for the order to remain the same wherever the receiver is in the chain. Coding delays are established such that each slave's transmission is delayed long enough that the order is maintained. The timing for the Norwegian chain is illustrated in *Figure 12.25*.

It is not possible to produce a square wave pulse with a straight rise at the face without an infinite bandwidth of transmission frequency. To use the allowed bandwidth of 20 kHz, the pulse is built up as quickly as possible and dies away more slowly. It contains 20–27 oscillations of the 100 kHz carrier and is of the form shown in *Figure 12.26*.

The pulse can be made to start rising on the positive side or the negative. This facility is used to avoid confusion with skywave signals. It is called phase coding.

12.17.3 Receivers

Modern Loran C receivers are capable of many functions like interchain fixing, obtaining a fix from different positioning systems and output via a microprocessor in latitude and longitude or distance to sail. The problems for a Loran receiver to solve are chain identification, station identification, discrimination from and of skywaves and time difference measurement on pulses with coarse and fine measurements.

Chain identification comes from the correct GRI. Station identification comes from establishing the master (nine pulses) and knowing the order of slave reception, which is always the same due to the coding delays.

A groundwave signal reaches a receiver 35–1000 μs before the skywave signal. The receiver has to lock on

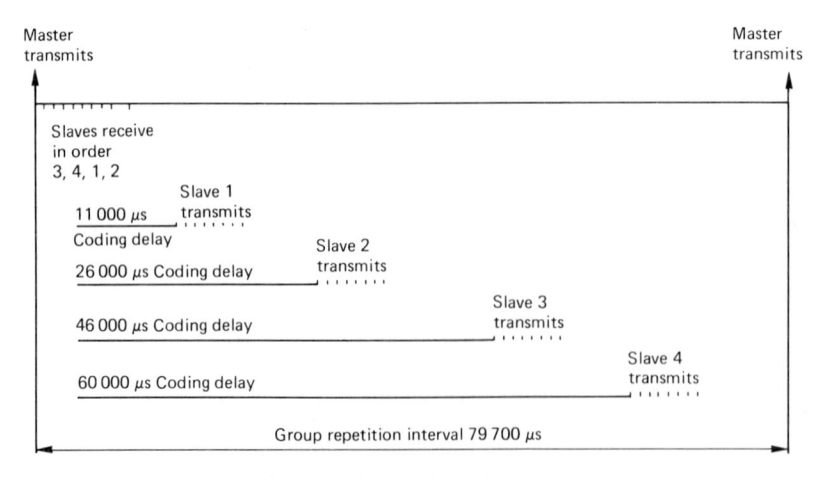

Figure 12.25 Diagram of Norwegian chain timing

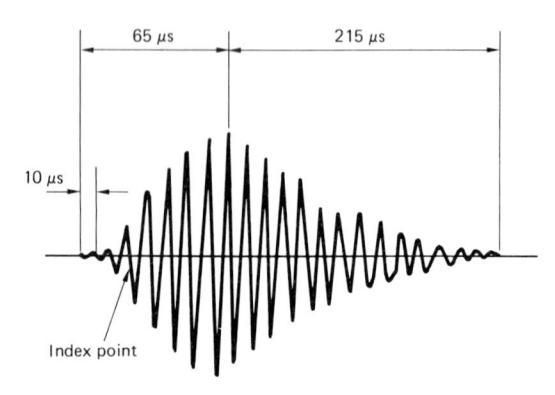

Figure 12.26 Loran C pulse shape

to the groundwave signal and perform its fine measurement before it is made unusable by mixing with the skywave pulse. The groundwave can be used up to 2000 km (1243 miles) from the transmitting station; then there is a section where the skywave interferes and then the skywave signal can be used, albeit at lower accuracies.

The receiver detects a pulse by a signal of the correct frequency reaching the required amplitude. A coarse time measurement can be taken by matching the pulse envelopes of the master and slave signals. The precision is to tens of microseconds. A fine measurement is made by phase comparison of the 100 kHz carrier wave within the pulse, to a precision of tenths of a microsecond. A reference point is needed, being the third crossing point of a full cycle within the pulse.

The ninth pulse of the master transmission can be switched on and off (blinking) to indicate faults in the master transmission and for transmitting messages to slave stations. Slave station blink is achieved by switching the first and second pulses of that slave on and off and indicates faults in that transmitted data. Phase coding of the transmissions assists in automatic lock on. It also prevents contamination of the second pulse travelling by groundwave being contaminated by the delayed first pulse travelling by skywave. Location of the third cycle reference point can be achieved through the amplification of the pulse and adding it to the same pulse amplified 1.35 times and phase shifted through 180°. A minimum is obtained at exactly 30 μs and the receiver can be locked to the signals.

12.17.4 Propagation anomalies

Groundwave propagation is useful to about 2000 km (1243 miles) from the transmitting stations. Ground conductivity influences the speed of propagation. The higher velocities are over sea water at 0.9997 times the speed of light, to low velocities over low conducting ground at 0.9977 times the speed of light. However, charts are usually produced using the velocity over sea water. If there is land path, then a correction may be available and can be applied. Some coastal charts have lattices allowing for land path printed on to the charts.

Skywave propagation is that signal reflected from the ionosphere. It travels further than the groundwave and can be reflected from different ionospheric layers as well as to the ground and back. Heights and reflectivity of the layers vary with the amount of ultraviolet light produced by the sun and the density of the gases in the atmosphere. Loran C is monitored by at least one dedicated monitor station within every chain. Each receives and records data from one or more master and slave pairs and transmits corrections as necessary to be implemented at the transmitting stations. Where there is more than one monitor in a chain, one of them is designated as control. Many monitors are fully automated and data can be sent over a dedicated land line or a microwave link.

The coverage of a chain is designated as that within which the signals remain within tolerance for 95% of the time. This is affected initially by chain geometry and angle of intersection of the LOPs and also by outside parameters like noise.

The four variables affecting the quality of signal reception are the output power of the transmitting stations, signal attenuation, the atmospheric noise level and pulse distortion. Coverage diagrams are available in Loran C handbooks.

12.17.5 Operational details

Due to land path in certain areas, variations in the speed of propagation affect the signals and to achieve a high absolute accuracy there must be a calibration against a more accurate system. Application of the corrections enables the position to be improved. This is standard practice in the hydrographic surveying version of Loran C – Pulse 8. Seasonal and meteorological changes can also affect the speed of propagation, especially for that section of the signal over land. Fixed corrections can be to the charts or applied within the receiver.

The bandwidth of Loran C allows interference from signals near the 100 kHz central frequency. Signal processing and notch filtering give some protection. The continuous wave signals of Decca Navigator come within this band.

12.17.6 Accuracy

Loran C does not have the absolute accuracy of the proposed full coverage GPS. However, the repeatable accuracy is better, so for returning to Loran coordinates or for a rendezvous at stated Loran coordinates, the accuracy value is better than GPS.

A value for accuracy depends upon the location of the receiver with respect to the distance from the shore stations, the signal patch, the angle of cut of the lines of position, receiver noise, other signal interference and the presence of skywave. Accuracy charts are available from receiver manufacturers showing, e.g. the coverage expected for 95% of the time with an expected accuracy of 400 m (1312 ft).

12.17.7 Conclusion

The expansion of Loran C is to continue. Debate is underway on the establishment of a north west European navigation system. Most coastal states are participating and the present facilities will likely be incorporated into an overall system. Some new transmitters would be needed, but, if adopted, the future of Loran C in coastal waters of the Northern Hemisphere would be assured.

12.18 Omega

12.18.1 System description

Omega is the only worldwide positioning system giving continuous position fixing. There have been eight stations giving full coverage since 1982. The US Department of Defense intends to keep using Omega until at least the end of 1994, and a full or partial service is likely to be available for some time after that.

Omega development started in 1957. Initial development came under the US Navy, but it is now the responsibility of the USCG. It had extensive use by 1977. This was without the full complement of stations and hence without worldwide coverage. The coverage figure now stands at 92% of the Earth's surface. Despite its lower positional accuracy compared to other systems, it is widely used because it is available where there are no other systems. It has been certified by the Federal Aviation Authority for North Atlantic navigation since 1977 and as a supplemental means of navigation for high altitude navigation across the continental US. In the Southern Hemisphere there is no Loran C and the Omega system is indispensable for continuous position fixes as Transit satellites will only provide positions every hour.

The number of users is put at about twenty thousand and is expected to remain fairly constant up until 1992 at least. Differential Omega is offered by a commercial firm in the equatorial Atlantic and the Far East. A great improvement in accuracy is available within 1000 km (621 miles) of the local differential transmitters.

12.18.2 Transmitters

During the experimental stages, there were transmitters at New York and Trinidad. The eight permanent stations are now in Norway, Liberia, Hawaii, North Dakota, La Reunion, Argentina, Australia and Japan. They are designated A to H, respectively. Apart from the two in the US, the others are run under bilateral agreements between the US and the host nations. Support is provided by the USCG in varying amounts to the host nations. Once the US need for Omega ends, this funding arrangement may change.

The transmitters come in two forms: a valley span as in Norway and Hawaii, and a tower at the other stations. Australia is illustrated in *Figure 12.27*.

Each station transmits on five frequencies, four common to all in the chain and one peculiar to each station. Each transmits its five frequencies in a particular order for precise periods of time which vary from 0.9 to 1.2 s. There is a quiet time of 0.2 s between each transmission. The exact format is shown in *Figure 12.28*.

The whole signal is repeated every 10 s, within which there are eight transmissions from each station. Only one station is transmitting on any one frequency at any one time. A receiver can identify a particular station and produce a phase comparison by retaining the phase of the first until the second arrives.

All the transmitters keep to the same time base, which began at 0000 (GMT) on 1 January 1972. Omega was then set coincident with Coordinated Universal Time (UTC), but due to the occasional input of 1 s into UTC, there is now some difference. This is not important for most purposes.

To achieve the same time base, an atomic clock used to be flown from station to station on a monthly basis. Now, time transfer is achieved using GPS satellites.

12.18.3 Receivers

As for the previous systems in this section, the modern Omega receiver can put out the basic lane readings of each pattern or through its integral microprocessor, positional and navigational information directly. The effect of this is seen in a change of navigational chart sales away from the latticed variety as the information is plotted directly as Latitude and Longitude from the receiver. There are a number of points for the receiver designer to take into account.

12.18.3.1 Station identification

It is possible for the receiver to identify which signal is coming from which station without reference to clock time. For example, with the 10.2 kHz frequency, a signal reception length of 0.9 s followed by 1.0 s identifies the stations A followed by B. With the signal format known within the receiver, many combinations exist with different stations and different frequencies to achieve correct synchronization.

Reference could be made to Omega standard time, to which all transmissions are referenced. With an atomic clock on UTC and the difference between the time systems known, the time of each station transmission would be known.

There has to be an accurate internal timing and phase measurement over the 10 s of the signal transmission format. This allows accurate phase comparison measurements of the signals from any two stations.

12.18.3.2 Lane identification

As in all hyperbolic phase comparison systems, measurement is undertaken within one lane of zero phase difference to zero phase difference. Computation of which lane it is, anywhere in the world, is not possible without some form of external input. On the baseline of two stations, using the 10.2 kHz frequency, the lane width is 15 km (9.3 miles). If the navigator can determine the position to better than ± 7.5 km (4.7 miles), the lane number can be determined and continuous navigation without external help is possible. By use of the frequencies in differing combinations, the ambiguity can be extended to 530 km (329 miles), and a positional requirement of ± 265 km (165 miles). A

Figure 12.27 The Australian transmitter

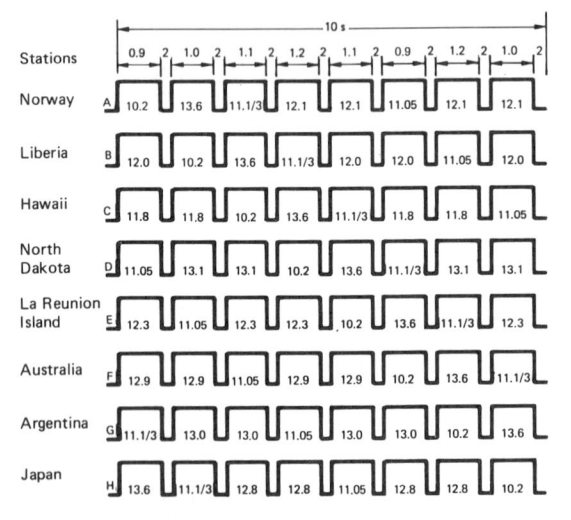

Figure 12.28 Omega signal transmission format (from *Omega Global Radionavigation: a Guide for Users*, U.S. Department of Transportation)

single Transit satellite pass is good enough to provide the ± 7.5 km (4.7 miles). The precision in position necessary decreases away from the baseline. Once working, the receiver will keep count of changes in lane number unless there is a break in signal reception.

12.18.4 Propagation anomalies

Omega signals, being of VLF, propagate between the Earth's surface and the base of the ionosphere. There is little or no penetration into the ionosphere. The phase delay on a transmission path can be predicted with considerable accuracy under normal conditions. With reference to the diagram of ionospheric layers, *Figure 12.22*, it can be seen that the path will alter with the time of day and with the season. The variations can be predicted with a greater accuracy by day than by night. For a transmission path crossing the line of dawn and dusk, the prediction is of lower accuracy still. This applies more to a north–south path than an east–west one.

The reflections within the Earth and ionosphere wave guide produce many modes of propagation, which, especially within 500 km (311 miles) of the transmitting station, can interfere with each other. Attenuation of the signal reduces the effect of all but the first and second modes beyond this. Only the first-order mode is significant beyond 1000 km (621 miles). As the modes propagate with different phase velocities, erroneous values may be received where the amplitudes are equal. Signals can be unusable. This interference affects the transmitters on the geomagnetic Equator more than the others. It is unwise to use signals in areas of possible modal interference.

One unpredictable form of interference is the sudden ionospheric disturbance (SID). Just as the ionosphere is a result of the Sun's radiation striking the atmosphere, any increase in radiation changes the state of the ionosphere. A solar flare will increase the state of ionization on the sunlit side of the Earth, causing a phase shift at the receiver for any part, or whole, daytime transmission path. The maximum effect occurs after 5 or 6 min, and lasts from 45 min to 3 h. It is not possible to broadcast corrections for this phenomenon.

A second unpredictable interference is also due to solar flares. With a SID, more abundant X-rays were the reason for the change. With polar cap anomalies (PCA), the emission of protons by the sun produces an effect when they are concentrated by the Earth's magnetic field in the polar regions, and produce a lowering of the base of the ionosphere. This effect occurs a few hours after a solar flare and lasts from one to three days. Navigational warnings are issued when a PCA occurs.

Two other points are worthy of mention: signal travel over permafrost and long path interference. In the former, Omega signals can be attenuated so much, due to the low conductivity of permafrost, that signal ability can be reduced. In long path interference, the point on the other side of the Earth's surface from an Omega transmitting station (the antipodes) suffers from unstable signals due to undeterminate phase.

Propagation correction tables exist for improving the fix given by Omega lines of position. These have been obtained as a result of an extensive monitoring program. There were fifty four monitor sites worldwide in 1983. Each site is occupied for two years.

12.18.5 Operational details

An Omega receiver antenna is best mounted away from other ship's installations to avoid ship's noise, or re-radiation into the antenna. Height is not required in itself to improve Omega reception.

Information on Omega status is available from a number of sources, including recorded telephone messages, radio broadcasts (HYDROLANT/HYDROPAC and NAVAREA IV/XII), and in *Notices to Mariners*. In addition, station maintenance takes place from a few days to a few weeks in the following months (*Table 12.5*).

Table 12.5 Omega station annual maintenance

Month	Station	Code
February	Liberia	B
March	Argentina	F
June	Hawaii	C
July	North Dakota	D
August	Norway	A
September	La Reúnion	E
October	Japan	H
November	Australia	G

12.18.6 Accuracies

Omega gives positional accuracies of 4 km (2.5 miles) during the day and 8 km (5 miles) at night at best. Many factors affect this, some within operator control and others not. Monitoring, and the validating program for defining system capabilities and limitations, continues (*Table 12.6*).

Table 12.6 Omega data available

Area	Date
North Atlantic	1980
North Pacific	1980
South Atlantic	1983
Indian Ocean	1987
South Pacific	1988
West Pacific	1989
Mediterranean	1990

Differential Omega provides a way of improving the fix accuracy, especially for coastal navigation. By knowing the position of a local monitor and comparing the received signals with those computed for that point, the difference can be transmitted as a correction to all special receivers within the locality.

Positional accuracy is improved to 1 km (0.62 miles) at 200 km (124 miles) from the differential station and 2 km (1.24 miles) at 900 km (559 miles) from the station. Corrections are broadcast for the relevant Omega signals. Variations due to SID, PCA and other variables can be allowed for.

Sixteen differential stations, manufactured by Sercel, are available in the North Atlantic, with another one planned (*Figure 12.29*). There are five in Indonesia with another fifteen planned in the Far East altogether.

12.19 Conclusions

The longevity of all these systems must be in doubt with the approaching full constellation of the GPS. An argument can be put forward for the continued use of Omega as against GPS on cost grounds. For navigation purposes, that is, with the accuracy required for shipping, differential Omega could be made available at a more reasonable price. The reality is that GPS will come into all our lives in the way that the Transit satellite system has. With a decreasing receiver price, 24 h availability, worldwide navigation and higher order of navigational accuracy, the writing is on the wall for these land based systems. There is likely to be at least a 15 yr handover period. Used within their capabilities, these systems will continue to provide dependable navigation information.

Acknowledgements

The US Coast Guard staff in London and the US were most helpful about Loran C and Omega. Racal Marine/Racal Group Services Limited provided information about the Racal–Decca Navigator. Ormston Technology Ltd provided information about Sercel products and Differential Omega coverage diagram.

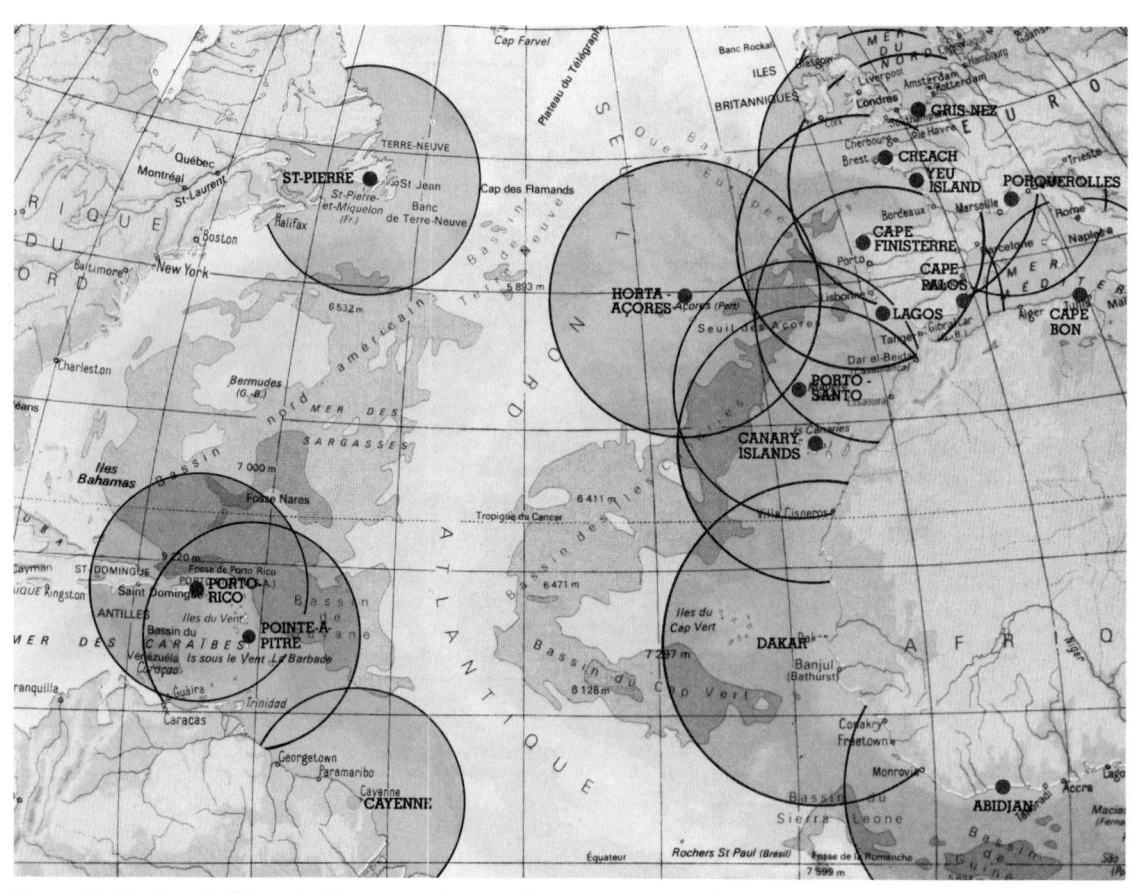

Figure 12.29 Sercel differential Omega stations and coverage

Earth-based electronic precise positioning systems

12.20 Introduction

Over the past twenty years, the demands of the offshore hydrocarbon industry have revolutionized the requirements of Earth-based electronic position fixing systems. In the early years of exploration in the North Sea, a considerable amount of the work was carried out using the Decca Survey receiver type 80309A and DL-21. Receiver type 80309A was a derivative of the Decca Navigator Main Chain receiver but was designed and built to obtain the maximum accuracy from the Decca Main Chain. The DL-21 receiver was designed to utilize both the Loran A medium frequency and Loran C low frequency chain transmissions. As the industry has grown, the standards of positioning accuracy have had to improve to cope with the needs of the end user of more precise repeatable positioning.

Position fixing control for the initial stages of offshore exploration can be provided by a single shore based system with long baselines between the stations. For North Sea exploration, a considerable amount of work was undertaken using the Pulse 8 system which provides the user with an accuracy of better than ± 50 m (± 164 ft) from the coast of Holland to beyond 62°N. Loran C can still be used for the initial stages or satellite systems are used in conjunction with a shore based system.

As development progresses beyond the initial exploration stages, the position fixing requirement becomes more and more critical. Using only one position fixing system does not provide sufficient data or cross-reference of the position fix to confirm the readings. What is required from any positioning system is repeatability of the position fix to allow the relocation of well heads, positioning of bury barges over pipelines and setting production platforms over existing templates. Conversion to true geodetic position is carried out using the customer's own computer equipment to allow the appropriate weighting of the position fix to be an end user selectable option.

Field developments in the North Sea and other offshore locations around the world are extending into ever deeper water as existing oil fields are expended. This increases both the costs and the risks to the offshore operators, who in turn are demanding a higher standard of accuracy from the survey contractors. The requirement to position a 20 000 ton steel jacket over a template 150 m (492 ft) deep to an accuracy of ± 0.5 m (1.64 ft) and to be able to monitor its attitude and orientation to an accuracy of $\pm 0.3°$ demands standards that are the equal of docking a spacecraft onto Skylab.

Whilst very much stressing the needs of the offshore hydrocarbon industry for position fixing, Earth-based positioning systems have their uses both in commercial and defence applications. Commercially, the increased accuracy of Earth-based electronic precise positioning systems has refined dredging operations of ports and waterways. Derivatives of systems that were originally designed for work offshore are now used over land for tasks such as crop spraying where the improved accuracy ensures the maximum coverage with the minimum wastage of materials.

The surveying of coastal areas and the production of hydrographic charts has been revolutionized by companies such as Qubit UK Ltd (Lynchborough Rd, Passfield, Hampshire GU30 7SB UK). Both Qubit and other companies have produced survey processing systems which use the latest advances in position fixing aids and computer technology to provide a complete onboard hydrographic survey package. With the Qubit system, data are collected, collated and processed to produce charts on board the survey vessel that are the equal of shore based facilities.

In defence applications, Earth-based electronic precise positioning systems are used to fulfil the mine countermeasures role. The ability to sweep channels accurately is essential to allow the free flow of vessels in and out of restricted waterways in wartime.

The latest application where Earth-based electronic precise positioning systems are being used is in the field of vehicle location and reporting systems. Datatrak Ltd, a subsidiary of Securicor plc, have developed a vehicle location system that uses an Earth-based electronic precise positioning system. The system coverage is presently being deployed to cover Great Britain and has an accuracy of at least 50 m (164 ft) for 24 h a day. Allied to the reporting system, Datatrak have produced a system that is unique in its potential, particularly in the field of security.

12.21 Categorization of systems

Earth-based electronic positioning systems can be conveniently categorized into distinct groups by their frequency and mode of operation. The higher the frequency, the greater the potential accuracy but the shorter the potential maximum range of the system. Conversely, the lower the frequency the less the accuracy but the greater the potential range. Frequency of operation of Earth-based systems and the actual systems covered are as follows:

1. Laser: Atlas Polarfix; Golf 2.
2. Microwave: Artemis; Axyle; Micro-Fix; Trisponder.
3. UHF: Trisponder; Syledis.
4. HF: Argo; Geoloc; Hi-Fix 6; Hyper-Fix.
5. LF: Datatrak; Pulse 8.

12.21.1 System modes

Earth-based precise positioning systems are generally based upon the measurement of distance (range–range mode) or upon the difference of distances (hyperbolic mode) between a mobile and fixed base or shore stations. The mobile is at an unknown location and the shore stations are at known fixed positions.

12.21.2 Phase comparison systems

Precise, accurate and repeatable results are obtained using phase comparison techniques with relatively simple instrumentation and using a fairly narrow bandwidth. In general, the higher the transmission frequency, the greater the resolution of the system. There are three major factors that limit the frequency of operation: range, ambiguity resolution and size of equipment.

To produce an accurate measurement, only systems that follow the Earth's curvature (groundwave systems) can be used as skywave is totally unpredictable. The maximum range of a system is determined by frequency as range varies as the inverse of the frequency. Phase measurements are determined within a 360° phase rotation which is referred to as a lane and the lane width is equal to half the wavelength of the transmitted frequency. As frequency increases, the width of a lane becomes so small that the ambiguity resolution becomes harder to achieve. Systems are used where a second frequency is radiated and ambiguity of position fix is extended to 1 km^2 (0.40 square miles) which is regarded as minimum acceptable level.

Equipment size must be taken into account, as the lower the frequency, the larger the antennae required. Logistically, systems in the LF band are fixed permanent sites which cannot be quickly moved. Taking these points into consideration, a good compromise for phase comparison systems is using the HF or 2 MHz band of frequencies. Range is limited by skywave, particularly at night, and lane width is in the order of 100 m (328 ft) providing an accurate system. At frequencies below this, accuracies of 50 m (164 ft) are to be expected and are suitable for navigation systems.

Phase comparison systems have limitations over range–range systems in that the angle of cut and lane expansion factors are significant. In general, the increase in the lane width, coupled with a reduction of the angle of cut as the receiver moves away from the shore stations, results in a loss of positional accuracy.

With hyperbolic systems, the fix should be ignored if the angle of cut is less than 30° and/or lane expansion is greater than three.

12.21.3 Pulse systems

Precise, accurate and repeatable results are obtained with pulse systems when sharp pulses are used, which in turn require a broad frequency spectrum. The bandwidth and frequency spectrum allocation are defined by CCIR rules and controlled and limited by government regulations. These are presently under review, but pulse systems operate in the VHF/UHF/SHF bands. At these frequencies, to extend the range beyond line of sight would require extreme high power transmission. This is expensive, making equipment cumbersome, and it is impractical.

Pulse systems have inherent problems and limitations. With a pulse or range–range system, the mobile needs to transmit and get responses from the shore stations. When used on warships, there is a requirement not to radiate on any frequency and hence range–range systems are a major disadvantage. At high frequencies, the size of the onboard antenna is reasonably compact but, at 2 MHz, radiating antennae are quite significant.

12.22 Laser systems

12.22.1 Atlas Polarfix

12.22.1.1 System description

Atlas Polarfix is a microprocessor controlled dynamic position fixing system based on laser ranging technology. It is designed for various applications in which precise tracking and position fixing of moving objects are required. Its use is particularly applicable for survey work of inland and coastal waters, for tracking sensing buoys and position fixing of offshore units such as tugs, support vessels, etc.

Polarfix provides a range and bearing to the target from the fixed shore station. The accuracy of the position fix from Polarfix is in the order of ±0.1 m (0.3 ft) for each kilometre of the measured range, and the bearing accuracy is in the order of ±0.001°. Polarfix requires only one shore station, which uses a pencil sharp laser beam to measure the position of a target, the target being equipped with a ring of prism reflectors. This means that a single tracking station can provide coverage of a complete circle around itself with a radius of the laser range. As there are no geometrical requirements for the angle of cut, the shore station can be placed close to the water line, thus minimizing the usual logistical efforts needed to survey in and maintain equipment sites. Because of the laser principle, there will be no signal reflections or phase cancellations that will degrade the accuracy of the system. Even in foggy weather, the operational range of the system is still 1.5 times the visibility.

After the initial set-up and lock on to the target, the Polarfix shore station tracks the vehicle automatically and no operator control is required to maintain or to reacquire target contact. Where the position information is required on board the target vessel, a UHF telemetry link is used between the ship and shore equipment. The shore station can also be controlled through the telemetry link or directly by an operator.

12.22.1.2 Shore station equipment

The shore based transmitting equipment consists of a sensor control unit and a sensing head mounted onto a tripod with a telescope mounted on top of the sensing head and in line with the laser beam. The transmitter can either be a Class 1 or 111a laser giving maximum ranges of 3000 and 5000 m (9842 and 16 404 ft), respectively. Selection of the class of laser is by means of an attenuation filter. The transmission wavelength of the laser signal is 904 mm with a pulse width of 20 ns, pulse interval of 2.3 ms and a pulse power of 7 W maximum.

The sensing head can be rotated at a maximum of 10°/s and tilted at a maximum of 2.5°/s. Shore station supply is from an internal battery and the unit's power consumption is 45 W maximum. This gives between 4 and 6 h running from the battery. A battery with charger option is available which will extend the length of operation. The total weight of a Polarfix shore station is in the region of 40 kg (88 lb), which allows ease of deployment in inaccessible sites. The equipment is robust and is capable of working in temperatures between -10 and $+55$ °C. The system has a limited range by its frequency of operation and can only be used by one user at a time.

12.22.1.3 Ship or sensor equipment

Fitted on board the ship or sensor is the reflector ring which weighs in the order of 10 kg. When control of the shore station is required from the ship, a processor and UHF link is required. The supplier is Krupp Atlas Elektronik GMHB, Selbaldbrucker Heerstrasse 235, D-2800 Bremen 44, Postbox 44 85 45 West Germany. Telephone (421) 4 57-0. Telex 24 5746-0 ka d.

12.22.2 GOLF 2

12.22.2.1 System description

GOLF 2 is an acronym for Gyro Orientated Laser Field-Ranger Mk 2. GOLF 2 is similar in design to Polarfix, in that it again uses laser technology for its operation. GOLF 2 sends a narrow pencil laser beam to a target and measures the time taken for the pulse to return. This time, divided by two and multiplied by the speed of light, provides the distance between the instrument and the target. GOLF 2 has a range accuracy of ± 2.5 m (8.2 ft) and a bearing accuracy of 0.1°.

GOLF 2 is designed for one-person operation from a single shore site. It is designed to be highly portable for ease and speed of deployment and use. Unlike the Polarfix system, there is no requirement to fit the target vessel with a responder or reflector ring. As there are no geometrical requirements for the angle of cut, the shore station can be placed close to the water line, minimizing the usual logistical efforts to survey in and maintain equipment sites. Because of the laser principle, no signal reflections or phase cancellations will degrade the accuracy of the system. By means of an integrated digital gyro, GOLF 2 provides an instant range and bearing to a target with a digital indication relative to a fixed reference azimuth or True North. Where the position information is required on board a vessel, a UHF telemetry link is used to transfer the data. In a similar manner to Polarfix, the GOLF 2 system is limited to one user at a time. It does have the advantage of not having to fit a reflector ring on the target, allowing rapid change from one target to another.

12.22.2.2 Shore station data summary

The shore based transmitting equipment consists of an encoder, control unit and telescope mounted on a purpose built tripod. The transmitter is a laser giving ranges between 150 m (492 ft) (minimum) and 9000 m (29 527 ft) (maximum). The transmission wavelength is 1064 mm with a pulse width of 10 ns and output power of 4 mJ nominal. Shore station supply is from a rechargeable NiCad battery pack which provides approximately 1800 shots between charging. Total weight of a shore station is in the region of 36 kg (79 lb), allowing ease of deployment, and the equipment is capable of working in temperatures between -30 °C and $+55$ °C.

12.22.2.3 Ship or sensor equipment

In the standard mode of operation, there is no requirement to fit any equipment on to the ship or target. When data is required on the ship, an additional remote display unit and UHF link is required both on the ship and at the shore station. This remote display unit is approximately 500 × 500 × 500 mm (1.64 ft) and weighs in the order of 20 kg (44 lb). The supplier is Hydroquip Ltd., Unit 11, Murcar Commercial Park, Denmore Road, Bridge Of Don, Aberdeen, Scotland, AB2 8JW. Telephone (0224) 824141. Telex 739506 (MDLAB).

12.23 Microwave systems

12.23.1 Artemis

12.23.1.1 System description

Artemis is a microwave hydrographic and geodetic position fixing system that provides range and bearing to a mobile station from a fixed shore station. The accuracy of the position fix is in the order of 1.5 m (4.92 ft) and bearing accuracy is in the order of 0.04°.

Artemis uses two radar type antennae, one on the mobile station and the other on the fixed reference station ashore. These are automatically self-tracking regardless of the manoeuvring speed of the mobile, allowing the system to be used by ships, vehicles, helicopters, etc. The azimuth reading can be aligned at the shore station by reference to a known reference point. This ensures that the mobile station is provided with a correct azimuth value.

Once aligned, range and bearing is displayed at the mobile station on LED displays. A wideband data channel is available that allows the vehicle's azimuth, as measured at the fixed station, to be relayed to the mobile. Alternatively, TV and voice can be transmitted over this link. With Artemis, one fixed station is sufficient for position fixing of a mobile within line of sight. No calibration is required prior to each range measurement and the system uses a very low radiated power. The system is insensitive to both active and positive interference due to its antenna tracking characteristics.

12.23.1.2 Shore station equipment

The shore station equipment consists of an antenna unit, a control unit, data unit and a telescope. The shore station is powered from a 24 V dc battery. The antenna unit is in the region of 400 × 300 × 500 mm (1.3 × 0.98 × 1.64 ft) and weighs some 26 kg (57 lb). The control unit is 500 × 220 × 350 mm (1.64 × 0.72 × 1.14 ft) and weighs 14 kg (31 lb), the data unit is 520 × 220 × 270 mm (1.70 × 0.72 × 0.88 ft) and weighs 8 kg (18 lb). The frequency of transmission is in the range 9.2–9.3 GHz and the radiated power is approximately 30 mW. If range extension to 30 km (18.64 miles) is required, power output can be increased to 200 mW by the use of a microwave amplifier.

12.23.1.3 Mobile station equipment

The mobile station equipment consists of an antenna unit, a control unit and a data unit. The station is powered from a 24 V dc battery. Dimensions of the antenna and the control unit are the same as the shore station equipment. Equipment is capable of working in the temperature range −20 to +55 °C and has a BCD parallel output for peripheral equipment. The supplier is Christiaan Huygenslaboratorium B.Y., Koningin Astrid Boulevard 56, Noordwijk 2460, The Netherlands. Telephone 01719-193 02.

12.23.2 Axyle

12.23.2.1 System description

Axyle is a high precision radio positioning system that uses two spot centimetric frequencies. The system provides an XY position fix at a high output rate and is designed for terrestrial, airborne and seaborne usage at ranges of up to approximately 5 km (3.11 miles).

Axyle consists of a mobile station which radiates a pulse at 2.442 GHz to up to eight remote beacons. The eight beacons reply to the mobiles interrogation at a fixed frequency of 5.7875 GHz and the mobile uses the four best beacons for its positional fix. The system is therefore working in the range–range or circular mode of operation. The frequencies used and the spectrum width comply with the ISM and FCC International standards for which a frequency allocation is not required. Axyle is a multirole system which allows up to four mobiles to operate at the same time. The system is designed for 24 h operation in all weather conditions. In typical operating conditions, the Axyle system will provide an accuracy of approximately 300 mm (1 ft). Output is in XY coordinates that are user selectable at an output rate of better than 200 ms between updates. The Axyle network has been designed so that it can be configured through a special software package installed on any PC compatible computer. This package will also select the different displays and can be used to set up the serial input and output port to end user requirements. Due to its short range operation, the system does not suffer from the effect of nulls or secondary reflections that are associated with equipment providing longer ranges.

12.23.2.2 Shore station equipment

The shore station equipment consists of a lightweight waterproof beacon mounted on a tripod. The beacon is approximately 215 × 185 × 185 mm (0.70 × 0.60 × 0.60 ft) and weighs in the region of 6 kg (13 lb). Power requirements are for a dc supply in the range 9–16 V and consumption is 10 W maximum. The system is therefore very quick and easy to deploy and maintain, a standard 12 V vehicle battery providing greater than 24 h coverage.

12.23.2.3 Mobile station equipment

The mobile station equipment consists of an antenna module and a processing module. The antenna module is the same size as the beacon but is powered from the processing module, which is approximately 465 × 236 × 308 mm (1.52 × 0.77 × 1 ft) and weighs 10 kg (22 lb). Power requirements are for a dc supply in the range 9–36 V and consumption is 80 W maximum which includes the antenna module. The supplier is Sercel-France, B.P.64.44471 Carquefou Cedex, France. Telephone (3) 40 30 11 81.

12.23.3 Micro-fix

12.23.3.1 System description

Micro-Fix is a short range position fixing system combining microwave interrogation techniques to achieve a repeatable accuracy of, typically, ±1 m (3.28 ft). The system utilizes a low power solid state transmitter operating in the 5 GHz band providing a working range of 80 km (50 miles) with line of sight. A basic Micro-Fix system consists of two principal units, these being a microwave transponder (T/R) and a control measurement unit (CMU). T/R units are sited at known shore locations around the survey area and are referred to as remote stations. A master station can interrogate up to eight remote stations from a possible total of thirty one with each remote T/R unit preset to recognize its own unique station code. By the inclusion of a chain code, up to four separate chains can be deployed in the same survey area without interstation interference. Multiuser capability allows a maximum of sixteen users for each deployed chain depending on the number of remotes and interrogation demands.

The T/R unit incorporates automatic adjustment and calibration to compensate for the errors due to the turnaround delays that are normally associated with microwave ranging systems. This eliminates the need for predeployment calibration and, since it is a continuous self-adjustment, the accuracy is maintained over the whole operating temperature range of the system. The deployed T/R operate on the same single frequency of 5.48 GHz (other frequencies are available to special order) and can be interchanged from site to site without affecting the performance of the system. This common unit concept presents logistical advantages to the user both for ease of deployment and for spares holdings.

Each of the deployed T/R units can be fitted with either an omnidirectional or sector antenna. In normal conditions, a T/R designated as the master T/R is fitted with an omnidirectional antenna and the remote T/Rs are fitted with a sector antenna. Circular polarization is used by the system to reduce the null effect phenomenon and to ensure the rejection of any reflections. Interference suppression is inherent in the operation of the T/R unit.

12.23.3.2 Remote stations

The equipment at a remote station consists of a T/R mounted on a surveyor's tripod, a sector antenna fitted to the T/R and a suitable 24 V battery supply. The current drawn by each T/R is 12 W when operating and 3 W in stand by. The units are fully waterproof and will float if dropped in the sea.

12.23.3.3 Master station

The equipment at a master station consists of a T/R mounted on the masthead, an omnidirectional antenna and a CMU. The CMU is housed in a strong weatherproof case which can be free standing or mounted in a 480 mm (19 inch) equipment rack. The CMU has an ac or dc power supply which is specified by the user. The ac PSU is for 110 or 230 V 50 to 400 Hz inputs whilst the dc PSU is for supplies in the range 16–36 v dc. Power consumption from either the ac or dc supply is approximately 40 W.

A wide range of peripheral equipment can be used with the CMU by means of the sophisticated data communications packages available. Options available are simultaneous incremental plotter drive and serial range and guidance data outputs; simultaneous HP-1B plotter drive and serial range and guidance data outputs; serial range and guidance data outputs; serial track and guidance for helmsman's display and composite video output of front panel and track guidance data. The supplier is Racal Marine Systems Ltd, Burlington House, 118 Burlington Road, New Malden, Surrey, KT3 4NW. Telephone 01 942 2464.

12.23.4 Trisponder

12.23.4.1 System description

Trisponder is a family of short and medium range microwave and UHF positioning systems which use the time difference techniques to measure ranges between user vessels and up to eight shore stations. In addition to displaying measured ranges, the system also computes vessel position by trilateration and provides guidance information to the vessel's helmsman.

The vessel equipment comprises a control unit and an interrogator, while the shore stations comprise a transponder together with a suitable power supply. The control unit initiates range measurements by instructing the interrogator to transmit a set of coded pulses but only the one which recognizes its own code replies to the interrogation. The reply is received back at the interrogator and the control unit then computes the range from the time difference measurement. Each shore station is interrogated in turn until all ranges have been measured. Up to eight vessels can use the same set of deployed shore stations simultaneously.

Trisponder systems are available using either microwave frequencies (8800–9500 MHz) for line of sight applications or using UHF frequencies (427.5 or 435 MHz) for over the horizon applications. There is also a range and bearing option available for the microwave system which enables a vessel to obtain its position using just one shore station. Microwave Trisponder must have radio line of sight but otherwise there are no significant limitations. UHF Trisponders' maximum range and accuracy beyond the radar horizon are dependent upon the prevailing propagation conditions. Rhotheta must have radio line of sight and also good visibility as an operator is required at the shore station to use the optical azimuth tracker.

Trisponder signals received at the master station may sometimes consist of a direct component and a component which is reflected from the surface of the sea. Multipath signals received in this way may be 180° out of phase and so combine with each other to produce a complete signal loss, referred to as a null. This null effect is dependent upon the antenna heights and it is most pronounced when the sea surface is smooth. An antinull option is available for Trisponder which comprises a dual antenna system with automatic changeover. With this option fitted the digital distance measuring unit (DDMU) automatically switches between antennae if a null is detected and so provides the system user with uninterrupted positioning. A similar effect can result from reflections from large fixed objects. These will give large and obviously incorrect range readings which are rejected by the DDMU range prediction routines. Trisponder is easily deployed and requires minimal field support, and is therefore ideal for inshore and near offshore survey applications.

12.23.4.2 Transponders

A standard transponder unit is used as the master station transmitter and receiver unit and the remote station transmitter and receiver unit. The identification plate on the handle indicates whether a unit is a remote or a master. All remote transponders are identical except for the pulse repetition interval code which is individually assigned by setting a series of code switches within the unit. The code setting should be indicated on the unit and all remotes deployed must be set to different PRI codes so that the master can distinguish between stations.

Transponders are lightweight waterproof units which are designed to withstand a considerable amount of rough handling. They can be mounted on a standard surveyor's tripod (the normal method of installation for remote stations) or on a specially designed mounting bracket (the normal method of installation aboard ship).

The antenna is attached to the transponder using a coupling ring which provides a waterproof waveguide joint and enables the antenna to be fitted quickly and easily. The transponder power supply is provided by a 24 V battery pack in the case of remote stations or via the DDMU power and signal cable in the case of master stations.

12.23.4.3 Antenna

A range of slotted waveguide antennae is available for use with transponders. Most are horizontally polarized, though vertical versions are available for special applications. Omnidirectional antennae are used with master transponders where all-round cover is required. These have a 360° horizontal beamwidth and a 19° vertical beamwidth. Directional antennae are normally used at the remote stations and these have a horizontal beamwidth of 110° and a vertical beamwidth of 7°. Both types of antennae are interchangeable so it is possible to use omnidirectional antennae at remote stations when all-round coverage is required.

12.24 UHF systems

12.24.1 Trisponder

Refer to Section 12.23.4 for details.

12.23.4.4 Digital Distance Measuring Unit

The DDMU is used at the master station to control system operation and to display and output positioning data. It is connected to the master transponder by a combined power and signal cable, and requires a 24 V dc power supply. The front panel comprises a CRT display, together with a membrane keyboard and a power switch. All connectors are situated on the rear panel. The unit is splashproof and is capable of withstanding a considerable amount of rough handling. A wide range of data inputs and outputs are provided for interfacing with external equipment including Serial RS 232C (baud rate selectable); Parallel BCD; ARINC; IEEE 488 (Using optional adapter); and incremental plotter (using optional plotter drive). The supplier worldwide is Del Norte Technology Inc., 1100 Pamela Drive, Euless, Texas 76040, Telephone 817–267–3541; the European centre being Del Norte Technology Ltd, Unit D, The Dorcan Complex, Faraday Road, Swindon, Wilts. SN3 5HQ, Telephone (0793) 511700.

12.24.2 Syledis

12.24.2.1 System description

Syledis is a medium range system that operates on one spot frequency in the 420–450 MHz UHF band and has a bandwidth of ±2 MHz at 60 dBs. This means that only a single frequency assignment is required for the system. The Syledis system uses pulse correlation techniques to keep the peak power low but still achieve ranges well beyond line of sight.

In most systems, a very short RF pulse is radiated to achieve good accuracy and to provide reasonable separation between multiple path effects. To achieve this, the radiated power must be very high. The Syledis system avoids this problem by using PSK modulation. The modulating code is a repetitive 127 element pseudorandom sequence that is 66.6 µs long, each element being 0.52 µs. Over the 66.66 µs period, the 127 pulses are the equivalent of a single pulse that is 0.52 µs long with a peak power 127 times higher than the output power. The basic waveform is repeated a total of forty times to further increase the effective peak power of the system. Thus Syledis radiates during a 2.66 µs period a 20 W peak power signal whose properties are the same as a 0.52 µs pulse with a peak power of 100 kW. This method allows low power transistorized circuitry to be used and thus improve the MTBF over other more conventional high power pulse systems.

With the Syledis system, the transmissions are time shared between the various units of the deployed network. The format can be between a five and sixty time slot format and each time slot may be assigned as a transmitting slot from one unit as well as a receiving slot for another unit in the system. The Syledis ship units are known as mobiles and the shore stations as beacons. Syledis can be operated in the range, hyperbolic or compound mode. In the range–range mode the interrogator on the mobile transmits a signal to the deployed beacons which each respond to the interrogation. In the range mode, up to four mobiles can work in conjunction with three beacons. Leapfrogging is allowed where each beacon is selected from eight with three mobiles or from six with four mobiles. In the hyperbolic mode, the reference signal is transmitted by one beacon and received by the other deployed beacons which radiate in phase signals with the master beacon. Any combination of beacons can be selected to construct two or three patterns of hyperbolic lines of position. In the compound mode, a combination of range–range and hyperbolic is used. The number of mobiles using range–range is restricted to three.

12.24.2.2 Shore stations

The minimum requirements for each shore station are a beacon, a power amplifier, a filter and an antenna. The standard beacon is a fully waterproof unit whose dimensions are 177 × 448 × 384 mm (7 × 17 × 15 in) and weighs 16 kg (35 lb). The unit accepts an input supply in the range 10–35 V dc and draws between 10 and 50 W from the supply dependent upon the selected operating mode and the configuration used. This falls to 7.5 W when the unit is in the standby mode. The power amplifier is the same physical size as the beacon and weighs 17 kg (37 lb). Power requirement for the power amplifier is in the range 22–35 V dc and current drain is 5 W during reception, 85 W during transmission. Output RF power from the station can be set to a maximum of 25 W for normal use or to 100 mW for short range harbour work. The filter circuit is connected between the amplifier and the antenna where it ensures that 95% of the radiated energy is within the bandwidth of F ± 1.25 MHz. There is a choice of antennae that can be used with a beacon and the appropriate antenna is selected dependent upon required range and beamwidth required. Antennae are available with gain from 3 to 11.5 dB and with beamwidths between 80 and 180°. Weights of antennae are between 7 and 11 kg (15 and 24 lb) and have the ability to be simply installed against a wall or hand rail. When long ranges are required, an optional booster power amplifier is deployed at each shore station and on the mobile if range–range is being used. This unit increases the radiated power from 20 to 320 W.

12.24.2.3 Mobiles

Each mobile is fitted with a Syledis transceiver (capable of working in any of the three modes of operation), a power amplifier and an antenna. The transceiver is 225 × 440 × 470 mm (9 × 17 × 19 in) and weighs 18 kg (40 lb). The unit requires a supply in the range 19–35 V dc and draws between 55 and 72 W depending upon the operating mode and configuration. The power amplifier is identical to that used at each shore station.

The mobile antenna is an omnidirectional whip type antenna, three types of which are available. Antenna selection is dependent upon required beam width, gain and output power. The same booster power amplifier as is used on shore stations can be used on the mobiles to increase range of the systems. The supplier is Sercel-France, B.P.64.44471 Carquefou Cedex, France. Telephone (3) 40 30 11 81.

12.25 HF systems

12.25.1 Argo

12.25.1.1 System description

The Argo HF position fixing system is a multi-user, multi range system designed for use in ships and helicopters. Argo used a single frequency in the range 1605–2000 kHz that is time shared around the shore and mobile stations of the chain. One station is designated as the master and transmits the systems synchronization pulse to which all stations are tied. The expected accuracy of the Argo system is 5 m (16.4 ft). Argo uses phase comparison techniques for its range computation. Each station phase compares its transmitter output voltage against the internal clock, the antenna current against the internal clock, and received signals against the internal clock. The shore stations process this data and transmit in phase with received signals. Processing of phase measurement in the mobile provide the phase difference between mobile and shore station, which equate to range.

As Argo is primarily a range–range system and the transmission cycle is in the order of 2 s, the number of potential users or number of ranges available is as follows: two range – twelve users plus lane ident; three range – nine users plus lane ident and four range – seven users plus lane ident. In addition to these modes, three software packages are available to support the following: six user circular and unlimited hyperbolic, high speed (200 knots) unlimited user hyperbolic and seven user circular with unlimited passive ranging.

As with all HF systems, Argo is affected by skywave interference. Daytime range is in the order of 750 km (466 miles) across seawater but is reduced between 250 and 420 km (155 and 261 miles) for night time effects due to multipath and noise effects. Argos is virtually unaffected by rain, fog or light snow but is affected by atmospheric noise associated with electrical storms. The best performance from Argo is in higher temperate latitudes and during the winter where 24 h operation can be achieved at ranges in excess of 400 km (249 miles).

Argo is type certified by the US Government Federal Communications Commission. The signal bandwidth of the system's transmissions is less than 100 Hz and the average of mean radiated power is less than 10 W. The stability of the transmission is within ± 1 Hz and up to 16 frequencies can be programmed to the nearest Hz. When using Argo, it is the end user's responsibility to obtain the necessary frequency allocation approval from the relevant authority. Positioning with Argo requires the ranges on the mobile station to be entered or calibrated to the correct values that correspond to true distance from the mobile station to the shore station. This is normally carried out at a calibration point where the true ranges are known and quantified.

12.25.1.2 Shore stations

Each shore station requires a range processing unit, an antenna loading unit and an antenna. Antennae between 9 and 30 m (29 and 98 ft) are available, the higher the antenna the greater the output power and hence greater potential range. The small portable 9 m (29 ft) antenna is designed for use where short ranges are required and where the 30 m (98 ft) antenna is either impractical to deploy or is unnecessary for the task to be carried out.

All units are supplied in knock down containers and can be deployed fairly rapidly. Extra time is required when the larger antenna is used. The range processing unit is not fully waterproof and needs to be sheltered from rain, seawater and excessive solar radiation in a suitable ventilated container. Stations supply requirement is for a dc supply between 22 and 32 V dc with a mean current drain of 3.5 and 20 A peak. Operation of a shore station is automatic after initialization, and stations can be left unattended if at a secure site.

12.25.1.3 Mobile stations

Each mobile station requires a range processing unit, an antenna loading unit and control display unit and an antenna. Antennae between 7 and 12 m (23 and 39 ft) are available, size used being determined by the job requirement and the physical restraints of the equipment. Range processing unit and control display unit are not waterproof and must be protected from rain, etc. Mobile stations require a dc supply between 22 and 32 V dc with a mean current of 6 and 24 amps peak. The control display unit has outputs for RS 232, HP-IB and a four channel analogue output to drive a chart recorder. The manufacturer is Cubic Western Data, 5650 Kearny Mesa Road, San Diego, California. Telephone (714) 268-1550.

12.25.2 Geoloc

12.25.2.1 System description

Geoloc is an HF radiopositioning system that radiates an energy spread in the 2 MHz band. The energy spread is kept automatically under the measured atmospheric noise level. This means that there is no requirement to obtain frequency allocation for the system. Ranges from the system of approximately 1000 km (621 miles) can be obtained round the clock with no effects from any ionospheric reflected waves.

Geoloc has four possible working modes, two satellite assisted and two autonomous modes. In the satellite assisted modes there is the H mode and TSS mode. In the H mode all transmitters and receivers are equipped with a cesium or rubidium oscillator. Updating of the Geoloc fix is given by a Transit or NAVSTAR receiver helped by the Geoloc speed vector. In the TSS mode, the transmitters only, or transmitters and receivers, are synchronized through associated satellite receivers.

In the autonomous modes there is the AS mode and the GEOSYL mode. In the AS mode all transmitters are synchronized through a monitoring station and two-way HF link. In the GEOSYL model all transmitters are synchronized through a two-way time transfer Syledis chain. In all working modes the maximum mobile speed is limited to 20 knots.

12.25.2.2 Transmitting stations

Each transmitting station consists of a transmitter, a time keeper and a backup time keeper. The transmitter has automatic adjustment of the radiated power in relation to the atmospheric noise level and is in the range 10 mW to 10 W maximum in 1 dB steps. The backup time keeper is only used at start up for the synchronization of all components of a network. Each transmitting station uses a 24 m (78 ft) antenna with its own matching and discharge box.

12.25.2.3 Mobile stations

Each mobile station consists of a receiver, time keeper, backup time keeper and central processor. The receiver has four independent channels, each one dedicated to signals from one transmitting station. Microprocessor control notch filters provide protection against jamming. Central processor computer a fix with real time corrections for both propagation speed and the effect of land paths. Each mobile station uses a 3 m (10 ft) whip antenna. Both the mobile and transmitting stations are designed to be run from a 24 V dc power supply. The supplier is Sercel-France, B.P.64. 44471 Carquefou Cedex, France. Telephone 40 30 11 81.

12.25.3 Hi-Fix 6

The Hi-Fix 6 position location system is designed to provide an optimum repeatable accuracy of better than ± 1 m (3.28 ft) and uses 2 MHz phase comparison techniques. Hi-Fix 6 employs a maximum of six transmitting stations, each radiating two spot frequencies in the 1.6–5.0 MHz band on a time shared basis and has a typical operating range in temperate latitudes of 300 km (186 miles). The measured phase relationship of these frequencies, received from selected pairs of stations, define both fine and coarse lines of position within patterns focused on the stations; three fine and coarse pattern values being presented on the receiver to a resolution of 0.01 lane and at a data update rate of 260 ms. The intersection of fine pattern lines of position yield the fix and a continuous lane identification facility, provided by the coarse pattern, defines this fix uniquely within an area of about 1 km^2.

The accuracy of the system depends on pattern geometry, propagation effects and instrumentation factors. This corresponds to 0.01 of a lane under the best operating conditions. On the base line, 0.01 of a lane represents 0.3 m (0.98 ft) at 5 MHz and 0.94 m (3.1 ft) at 1.6 MHz.

Hi-Fix 6 has three modes of operation, these being hyperbolic, circular and compound modes. Compound is a mixture of hyperbolic and circular modes. In the standard hyperbolic mode, the system can be used by an unlimited amount of users. In the circular mode only two vessels can operate when both fine and coarse patterns are required, but four vessels can operate when only fine patterns are required. In the compound mode of operation, an unlimited number of users can use the hyperbolic patterns with one vessel using the circular pattern.

Operational flexibility is a feature of Hi-Fix 6, but the system is not designed for instant usage. Each transmitting site will require approximately one man day to install, and a further two days to ensure that the chain is stable and has been corrected for random and systematic error before being fully usable. The system is then highly stable and will provide repeatable data to the user.

12.25.3.1 Shore station equipment

Shore station equipment consists of a control unit, power amplifier, antenna matching unit and an antenna. The control unit is the signal source for the station transmissions and, at secondary stations, performs phase lock functions. By using time shared transmissions the antenna at each site serves for both reception and transmission. Two types of antenna are available, these being 10 and 30 m (32.8 and 98.4 ft) variants. Where maximum radiated power is required, the 30 m (98.4 ft) antenna improves power from 30 to 90 W. At full transmitter output a transmitting station consumes approximately 12 amps of power at 24 V.

12.25.3.2 Ship equipment

In normal hyperbolic mode of operation only a receiver and antenna are required on the ship. The latest variant of receiver can be either rack mounted or supplied in a custom designed robust package. The receiver can be powered from a 115/230 V ac supply or 16/36 V dc supply. The unit has standard serial and parallel outputs and options are available to suit client requirements.

When circular or compound mode of operation is required, a transmitting antenna and an amplifier must also be fitted on the ship. This greatly extends the time and logistics required to install a ship fit. The supplier is Racal Marine Systems Ltd, Burlington House, 118 Burlington Road, New Malden, Surrey, KT3 4NW. Telephone 01 942 2464.

12.25.4 Hyper-Fix

12.25.4.1 System description

Hyper-Fix is a third generation medium frequency phase comparison positioning system and uses advanced microprocessor techniques. The range, accuracy and reliability of Hyper-Fix shows an improvement over the earlier Hi-Fix 6 equipment. The system radiates frequencies in the band 1600–3400 kHz and depends on the groundwave mode of propagation.

Hyper-Fix suffers the same effects as all other HF positioning systems in night time effect. Being microprocessor controlled and utilizing software phase lock loops as opposed to hardware loops, the time constant of the loop can be readily adjusted to overcome some of the undesirable affects. Long time constants are used at night to overcome short term changes affecting the pattern stability.

A chain consists of from three to eighteen transmitting stations. One station transmits the trigger signal which controls the timing of the chain. At each station, the phase of the transmitted signal can be referenced to the signal from any other station. In previous systems, the phase reference was allied to the trigger or prime station which limited the chain configuration. Hyper-Fix is thus able to provide the optimum performance and maximum chain coverage. Because the system is microprocessor based, most of the operational functions and system modes of operations are software controlled. Hyper-Fix has two modes of operation, these being hyperbolic and circular modes.

In the hyperbolic mode the transmitting stations are installed at known locations and hyperbolic patterns are generated between pairs of stations. The intersection of hyperbolic patterns produces a hyperbolic lattice from which the ship's position can be found. Up to six stations can be used to achieve the required area of coverage. There is no limit to the number of ships using the chain, as all that is required on board is a receiver. In the circular mode of operation, the principals are the same but patterns are generated between the ship and selected shore stations. The position fixing lattice is then formed by two or more circular patterns focused on selected shore stations. Only two ships can operate in the circular mode of operation. As Hyper-Fix is so versatile and because timing and phase memory is software controlled, up to three sequences can be transmitted within a single timing cycle. Each sequence can be individually configured to be either hyperbolic or circular.

For long duration deployment or when attempting to obtain the best results from the system, the transmitting sites are carefully surveyed and the chain is calibrated to compensate for random and system errors in the chain. The time involved for this is quite considerable. A rapid deployment package has been designed to use standard Hyper-Fix equipment, but keep installation time to a minimum. Two experienced personnel can install a transmitting station within 40 min. The transmitting antenna kit supplied consists of a 10 m (32.8 ft) top dressed carbon fibre mast complete with a four radial earth mat system. A custom built equipment shelter houses the equipment stack containing receiver and controller, power amplifier and battery charger and antenna tuning unit. Power is supplied by a thermo–electric generator running off bottled propane and butane. Fuel consumption is approximately 0.25 kg/h at a power output of 120 W.

12.25.4.2 Shore stations

There are three separate equipment configurations which may be employed at a transmitter station. SRT2 is the simplest arrangement and comprises the three basic units of the systems: receiver and controller, power amplifier and antenna tuning unit. It is also used with the rapid deployment package. SRT1 is a four unit arrangement which uses a supply protection unit which provides protection for the equipment against lightning strikes. LRT1 is a five unit arrangement which is used whenever long baselines are necessary. The secure supply unit incorporates a rubidium frequency standard. This stable frequency source enables long time constants to be employed in the control units. The result of these long time constants is to minimize the short term effects of skywave and atmospheric noise.

12.25.4.3 Ship equipment

There are two receiving configurations for hyperbolic operations. HMR1 is the simplest system employing a receiver and receiving antenna assembly. The receiver may be powered using a dc or ac power supply unit as specified by the user. In the HMR2 system the secure supply unit is used to provide a no break supply in the event of mains failure. It is recommended for all offshore applications. The supplier is Racal Marine Systems Ltd, Burlington House, 118 Burlington Road, New Malden, Surrey, KT3 4NW. Telephone 01 942 2464.

12.26 LF systems

12.26.1 The Datatrak system

12.26.1.1 System information

The Datatrak vehicle location and reporting system provides vehicle fleet operators with real time information on the position and status of all vehicles under their control. Each vehicle is fitted with a unit called a locator unit, which is linked to a transmit–receive antenna mounted on the roof of the vehicle. The antenna receives low frequency (LF) radio signals transmitted by a network of navigation transmitting stations and phase compares these signals to determine the position of the vehicle relative to the transmitting station network.

This position is software converted to provide a map grid position for the vehicle in terms of eastings and northings. Once the vehicle position is established, it is then transmitted, together with status information, to a network of base stations on a specially allocated UHF radio data channel. Each base station is linked to a regional computer centre via a telephone line. The regional computer centre distributes data to each user's display and logging centre from the regional exchange again over telephone lines. A modem and a suitably configured computer system for displaying and logging the incoming data is all that is required at the user display and logging centre.

12.26.1.2 Positioning subsystem

Positioning coverage over the required area is provided by a network of time synchronized navigation transmitting stations that radiate low frequency phase coherent signals (130–150 kHz) on a time shared basis in a recurring timing cycle. The locator unit in the vehicle measures the phase difference between the signals it receives from each transmitter and stores this information in its memory. The information stored by the locator unit is used to generate a hyperbolic line of position (line of equal phase difference) between a pair of stations. When a second hyperbolic line of position is used from another pair of stations, the intersection of the two hyperbolae uniquely locates the vehicle. As the transmitting stations are located at approximately 160 km (99 miles) intervals, a vehicle will always be in range of signals from at least three stations which will provide three hypebolae. There will normally be at least four transmitting stations in range which will provide six hyperbolae giving an accurate positional fix (within 50 m (164 ft) in most circumstances) and allows for a degree of redundancy.

A unique feature of the Datatrak system is its ambiguity resolution. In most hyperbolic systems, the user has to enter the mobile's position into the receiver to an accuracy of 1 km^2 (0.39 square miles). The ambiguity resolution of the Datatrak system is so large that no operator intervention is required and, once the receiver is powered, it will locate its position within the network. Thus Datatrak is truly a hands off system. Provided that the receiver has not been moved since it was last switched off, as would be the case of a normal vehicle, acquisition of vehicle's position at switch on takes a matter of a few seconds.

12.26.1.3 Transmitting stations

The transmitting stations are remotely controlled and are run by Datatrak. The chain transmissions are constantly monitored and corrected for changes in velocity of propagation and ground conductivity. The transmitting stations are remotely controlled over a modem line allowing changes of settings to be made. The system is very versatile and transmitting stations can also be monitored and controlled from other transmitting stations using on line LF transmissions.

12.26.1.4 Receiver

The Datatrak receiver is a small microprocessor controlled unit fitted in a vehicle that acts as the LF receiver and UHF transmitter. The unit runs from the vehicle's battery supply and has a power save feature where power is removed from the unit when the vehicle is left for any length of time. The supplier is Datatrak Limited, Groundwell Industrial Estate, Swindon, SN2 5AZ. Telephone (0793) 722549. Fax (0793) 728302.

12.26.2 Pulse 8

12.26.2.1 System description

Pulse 8 is an LF position fixing system which operates at a fixed frequency of 100 kHz. The system is designed for 24 h operation at ranges up to 800 km (500 miles) where a repeatable accuracy of better than ± 50 m (164 ft) is required. A transmitting chain consists of a prime station and up to five secondary stations. Each station radiates synchronized groups of pulses at a controlled repetition rate. The time differences between groups of pulses from selected station pairs is measured by the onboard receiver to a resolution of 10 ns.

Particular problems with the system are the proximity of other strong signals near to the Pulse 8 frequency. In particular is a transmission at 101.1 kHz. These unwanted frequencies need to be rejected by the receiver using notch filters.

12.26.2.2 Transmitting stations

The transmitting stations consist of a container housing the electronic equipment and a 100 m (328 ft) guyed lattice antenna. The stations are permanent installations, run and operated by Racal Survey. The station's transmissions are continuously monitored and adjusted as necessary to correct for propagation errors.

The five Pulse 8 service chains presently deployed and in use are shown in *Table 12.7*.

Table 12.7 Pulse 8 service chains in use

Name	Approximate area of coverage	
North Sea north	65°N/8°W	56°N/8°E
North Sea south	60°N/2°W	52°N/10°E
Porcupine Bank	54°N/13°W	51°30′N/10°W
Fastnet/Biscay	53°N/18°W	40°N/01°W
Yellow Sea	38°N/118°E	32°N/126°E

12.26.2.3 Mobile station

The only requirement on the ship or mobile installation is for a receiver and an antenna. The Pulse 8 Mk 4 receiver is 177 × 483 × 387 mm (7 × 19 × 15 in) and weighs approximately 17 kg (37 lb). The unit in this form can be rack mounted in a standard 480 mm (19 inch) equipment rack or be fitted in a heavy duty case for protection. The receiver features filter circuits that provide > 40 dBs of rejection of the 101.1 kHz frequency and > 60 dBs at the 90 and 110 kHz frequencies. The units power requirements are for 115 or 230 V 50/60 Hz supply at 85 W.

With the addition of extra equipment, the unit is capable of synthesizing a local transmitting station signal which allows hyperbolic or three range operating modes to be used. The ranging mode provides greater area of coverage and, unlike most conventional ranging systems, is available to an unlimited number of users and does not require transmissions from the ship.

Satellite navigation

12.27 Introduction

The use of satellites for navigation had its origin in the 1950s with the beginning of the space age. The first man made satellite, Sputnik 1, was launched on 4 October 1957 by the USSR. By modern standards this was very small, but included a radio transmitter on its payload. The signals from this were monitored on Earth and Drs Gruier and Weiffenbach of the Applied Physics Laboratory of the Johns Hopkins University, USA, were particularly interested in the substantial Doppler frequency shift that was observed. This led them to develop algorithms for determining the orbits of the satellite from measurements of the change in frequency received at a single ground tracking station.

Drs McClure and Kershner, also of the Applied Physics Laboratory, suggested that the process could be inverted, i.e. the position of a point on Earth could be determined from Doppler measurements on the signals from a satellite where position at a given moment of time was available. This resulted in funding being given to the development of the Transit navigation satellite system, the first truly worldwide all weather navigation system in December 1958.

There are at present five satellite based radio navigation systems actually transmitting signals: Transit (US Navy Navigation Satellite System), NAVSTAR GPS (US Department of Defense), Tsicada (USSR), GLONASS (USSR), Starfix (John Chance Inc.). Of these Transit, Tsicada and Starfix are fully operational. Unfortunately, little is known about Tsicada and GLONASS from the USSR, but they are very similar in operation to Transit and NAVSTAR GPS, respectively. Many other systems have been designed. Many have not progressed further, whilst limited testing has been carried out on others.

12.28 Satellite orbits

The use of satellites for the determination of position on Earth requires a knowledge of the position of the satellite at the time that measurements are taken. This may be achieved by monitoring the satellite from known points on Earth, but the information will not be available to a user at the time of measurement, i.e. real time. For navigation purposes, it is therefore necessary to calculate the satellite position from a prediction of the satellite orbit.

The orbit which a satellite occupies is determined from its launch conditions. For example, the angle that the orbit makes with the Equator (inclination) is mainly determined from the location of the launch site. Geosynchronous satellites, which appear stationary with respect to points on the Earth's surface, lie in the plane of the equator (zero inclination) and take 24 h to orbit the Earth (period), must be launched from a point of low latitude.

The theory with regard to orbits was developed by Kepler in the sixteenth century and can be summarized as:

1. An orbit is an ellipse with the Earth at one of the foci.
2. The area swept by a vector drawn from the satellite to Earth is proportional to time.
3. The square of the period of the orbit is proportional to the cube of its mean altitude above the Earth's surface.

These laws assume that only gravitational forces act on the satellite, which is the ideal, and it is usual to refer to the orbit so described as a normal orbit.

An orbit may be determined by monitoring the position of the satellite from known ground stations and determining a set of parameters which describe the position of the orbit with respect to the Earth, e.g. inclination, size, shape, altitude, and the position of the satellite within the orbit, by speed, period, time at given location. Using this information, it is possible to predict the location of the satellite at a particular instant. In order to do this, it is necessary to assume a model for the gravity field of the Earth, Sun and Moon and the solar radiation pressure.

Although gravity models have improved greatly with the studies of the behaviour of satellites, it is only possible to use these predicted orbits over a maximum of two days. Clearly, the longer the time period, the less accurate they become. It is therefore necessary to predict the orbit of each satellite regularly and to pass on these data to the user with great speed and at frequent intervals.

2.29 The Navy Navigation Satellite System (NNSS)

The development of the Navy Navigation Satellite System (NNSS), or Transit, started in December 1958. The aim was to provide a means by which Polaris submarines could obtain an accurate position update for their inertial navigation equipment. It was therefore necessary to provide a worldwide, all weather, passive (no transmissions) navigation system. It became operational in January 1964, but was not released to the civilian community until July 1967.

The satellites themselves are small, measuring less than 40 cm (15 inches) in size with four solar panels 1.7 m (5.57 ft) long to charge the internal batteries. The signals are transmitted by a lampshade antenna

Figure 12.30 The Transit satellite

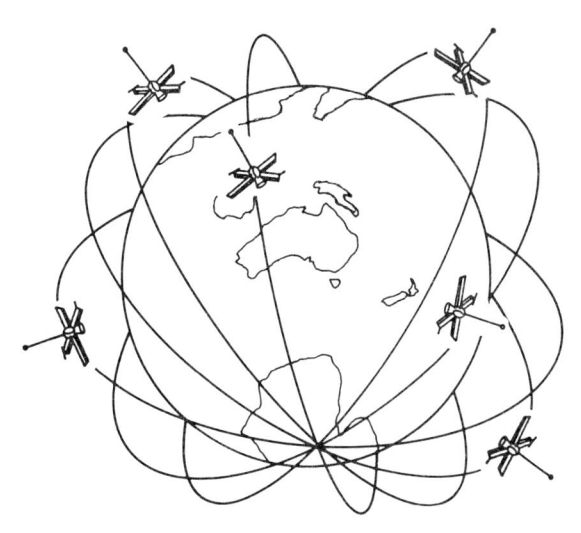

Figure 12.31 Transit satellite birdcage

which is constrained to point down towards the Earth by a gravity gradient stabilization boom (*Figure 12.30*). Along the solar panels, there are magnetic hysteresis rods to damp out the tendency to sway back and forth by interaction with the Earth's magnetic field.

The satellites used are in near circular, polar orbits approximately 1075 km (668 miles) high with a period of 107 min. This constellation has been described as forming a birdcage around the rotating Earth such that each satellite is viewed in turn (*Figure 12.31*). The original constellation consisted of five orbiting satellites (with 12 reserves stored on Earth). This gave an average time interval between fixes of 35 and 100 min, depending on latitude.

The daily operation is controlled by the US Navy Astronautics Group which has its headquarters at Point Mugo, California with other tracking stations at Prospect Harbour, Maine, Rosemount, Minnesota and Wahiawa, Hawaii. These four stations are therefore spread across the mainland USA and out into the Pacific, providing a large longitudinal spread. They form the operations network (OPNET). Observations made at these stations are sent to the Point Mugo computing centre and used to determine the orbit parameters for each satellite and to predict each orbit several hours into the future. This information forms part of the navigation message which is transmitted to the appropriate satellite from the injection stations at Point Mugo and Rosemount at the next opportunity. Each satellite receives a new message approximately every 12 h, but the capacity of each satellite would allow operation for 16 h. This spare capacity ensures operation in the event of erroneously missing an opportunity to inject data.

In addition to OPNET, there are about 15 stations spread around the world which continuously track one satellite. These stations are part of the Transit Network (TRANET). The results are sent to the US Defense Mapping Agency, which computes the observed, rather than predicted, positions of the satellites or precise ephemeris. This information is available to authorized users about a week after observation to allow a more precise computation of position to take place if required.

Unlike surface radio navigation systems which require simultaneous measurements on signals from several locations, Transit measurements are with respect to successive positions of the one satellite as it passes across the sky. The movement of the satellite with respect to the user is monitored by determining the Doppler frequency shift of the 150 and 400 MHz signals it transmits. It is therefore necessary to track the satellite for 10–16 min, during which time the satellite travels 4400–7000 km (2734–4349 miles), providing an excellent base in order to obtain a position. This does, however, create a problem for vessel navigation, since the movement of the ship must be considered in the calculation.

Position is determined from measurements of the change in the frequency of the signals received from the satellite due to its motion with respect to the receiver. This is obtained by comparing the received signals with those generated locally by a stable reference oscillator in the receiver (*Figure 12.32*). The number of frequency cycles received in a given time period from the satellite is subtracted from the number generated locally to give the Doppler count, N. The transmitted

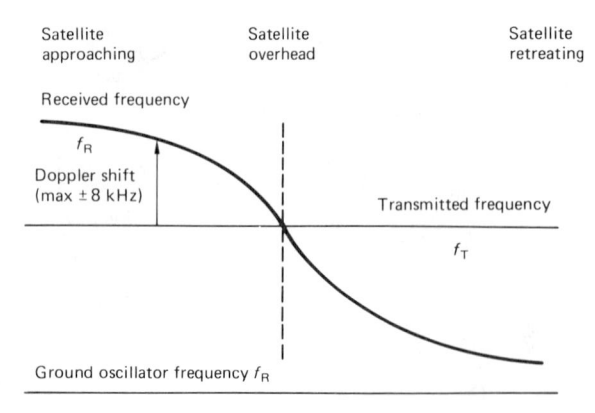

Figure 12.32 Doppler shift with relation to satellite orbit

message contains a number of timing marks and these are used to signal the start and end of a time period. The Doppler count can be represented for the period from time T_1 to time T_2 (*Figure 12.33*) by:

$$N = \int_{T_1 + R_1/C}^{T_2 + R_2/C} (f_G - f_R)\, dT \quad (12.20)$$

where R_1 and R_2 are the ranges of the satellite at times T_1 and T_2, respectively, C is the velocity of light and f_G is the locally generated frequency and f_R is the received frequency. Note that the time mark transmitted by the satellite at time T_1 is received at time $(T_1 + R_1/C)$ and similarly for time T_2.

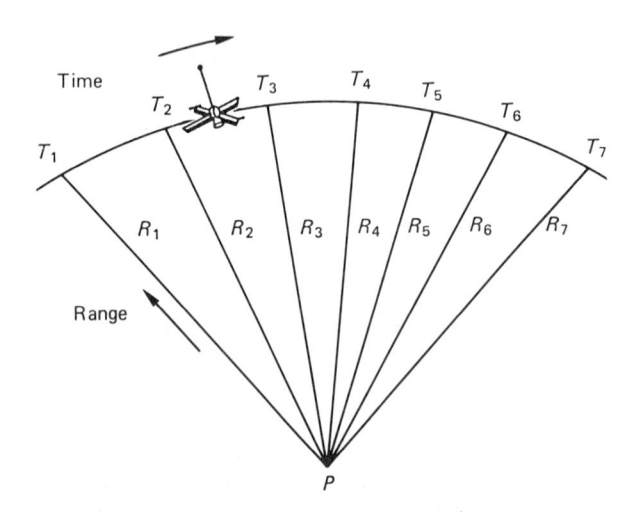

Figure 12.33 Determining position by Doppler count

Equation 12.20 can be expanded to give:

$$N = \int_{T_1 + R_1/C}^{T_2 + R_2/C} f_G\, dT - \int_{T_1 + R_1/C}^{T_1 + R_2/C} f_R\, dT \quad (12.21)$$

The second part of Equation 12.21 represents the number of cycles received between the two timing marks which must be equal to the number of cycles transmitted. Hence it becomes:

$$N = \int_{T_1 + R_1/C}^{T_2 + R_2/C} f_G\, dT - \int_{T_1}^{T_2} f_T\, dT \quad (12.22)$$

Integrating Equation 12.22 gives:

$$N = f_G\,[(T_2 - T_1) + (R_2 - R_1)/C] \\ - f_T\,(T_2 - T_1) \quad (12.23)$$

Rearranging Equation 12.23 gives:

$$N = (f_G - f_T)\,(T_2 - T_1) \\ + f_G\,(R_2 - R_1)/C \quad (12.24)$$

Equation 12.24 shows the two parts of the Doppler count. The first part is the constant frequency difference, both oscillators considered stable, multiplied by the time interval. The second is a measure of the change in range of the satellite, measured in wavelengths of the ground reference frequency, from the beginning to the end of the time period. This change in range is analogous to the range difference obtained in surface (hyperbolic) navigation systems. However, since Transit is a three dimensional position fixing system, a surface, rather than a line, of position can be generated from each measurement, i.e. a hyperboloid of position. The intersection of a number of such surfaces enables a three dimensional position to be determined.

A usable satellite pass for a two dimensional position may be considered as one which is above the horizon for 10–18 min. During this time, twenty to forty counts will be collected to give a redundancy of measurements together with good fix geometry. Three dimensional positions require longer high elevation passes.

In order to determine the position of the observer, it is necessary to know where the satellite was at each timing mark. The satellite message, modulated on both frequencies, contains the broadcast ephemeris: the orbit information computed and injected from OPNET. Using this information, the positions of the satellite can be predicted and the position of the observer computed. The position is generally given as latitude, and longitude as defined in the geocentric datum used to describe the satellite orbits. Until January 1989, this was the World Geodetic System 1972 (WGS-72), but was changed to WGS-84 at this time for consistency with NAVSTAR GPS.

If the receiver is moving during the satellite pass, as in the case of ship navigation, the motion must be included in the position fix calculation. The north–south component is particularly important since the satellite essentially travels in this direction. Automatic speed and heading inputs are often used for this purpose. Each knot of error in the speed results in an error of about 450 m (1476 ft) in position. It will have been noted that the satellites broadcast on two frequencies, 150 and 400 MHz. This is to enable the effects of atmospheric refraction to be modelled, since each will be affected by a differing amount.

The Transit satellites have performed well. The main problems experienced are in the ability of the satellites to maintain their orbit separation. Since the satellites have no means to adjust the orbit alignment, some have moved close together (bunched) and left large gaps in the birdcage. To overcome some of these problems, a new generation of NOVA satellites was developed (*Figure 12.34*). These were fully compatible with the existing satellites, but had a disturbance compensation system (DISCOS) to give increased stability of orbit. Three such satellites have been put into orbit.

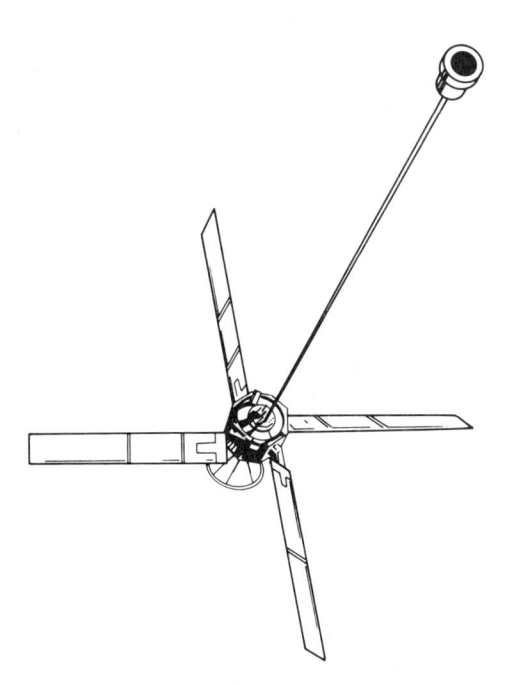

Figure 12.34 The NOVA Transit satellite

The majority of the satellites which were held in reserve have been launched, with some being used to fill gaps, some replacing failed satellites and the others stored in orbit awaiting the command from the ground to transmit.

The accuracy of a position fix is dependent on a number of error sources. The total fix error of a stationary receiver is, however, unlikely to exceed 50 m (163 ft) using the broadcast ephemeris. For a static system the main error sources are:

1. Uncorrected propagation effects (tropospheric and ionospheric) – error 1–5 m (3.28–16.40 ft).
2. Equipment noise (e.g. oscillator errors) – error 3–6 m (9.8–19.60 ft).
3. Uncertainties in gravity model used in orbit generation – error 10–20 m (32.8–65.6 ft).
4. Uncertainties in receiver altitude – error 10 m (32.8 ft).
5. Incorrect modelling of forces on satellites – error 10–25 m (32.8–82 ft).
6. Rounding off errors in ephemeris – error 5 m (16.40 ft).

Some of these can be reduced significantly by using special techniques for static positioning. Translocation involves the use of a second receiver at a known point. Simultaneous measurements are made at both points (known and unknown) and are post-processed together. In this way, it is possible to evaluate some of the errors at the known point and apply them to the unknown point. This can give a potential accuracy of about 1 m (3.28 ft).

Transit receivers vary in their size, complexity and price. Relatively inexpensive, single frequency, navigation receivers are available. Some are combined with receivers for surface radio navigation systems, such as Loran C and Omega. These allow continuous navigation between Transit fixes. There are also a number of Geodetic receivers for static point positioning.

With the advent of GPS, the use of Transit will decline. However, now that the majority of the spare satellites have been launched, the system is at its best. Its future has been guaranteed until 1996 and it will remain the most accurate satellite system for static point positioning, unless some of the restrictions on GPS are removed.

12.30 NAVSTAR Global Positioning System

The Navigation Satellite Timing And Ranging (NAVSTAR) Global Positioning System (GPS) project started in April 1973. The aim was to develop a system which would provide an all weather real time positioning system anywhere in the world, 24 h a day to an accuracy of about 10 m (32.8 ft) on a common datum. It was the combination of two projects; Timation of the US Navy and System 621B of the US Air Force. The original plan was to complete testing of the system by 1984 for full scale operation in 1988 and for the satellites to be launched by the US Space Shuttle. The Shuttle launch disaster in 1986 caused a delay in the programme and plans have since had to be changed. It is now expected that most of the satellites will be launched by an expendable rocket launcher, the Delta II, with only a few put into orbit by the Shuttle. The expectation is that there will be sufficient coverage for continuous two dimensional positioning by the end of 1990 and the full constellation for continuous three dimensional positioning will be operational by the end of 1991.

The control segment consists of four monitoring stations at Hawaii, Ascension Island, Diego Garcia and Kwajalein, with the master control station at the Falcon Air Force base in Colorado Springs. These stations are therefore spread around the world. They track the satellites and send their data to the master control station, where the orbit parameters, clock corrections, etc, are computed. Three of the monitoring stations, Ascension Island, Diego Garcia and Kwajalein, are able to upload this information to the satellites. Uploading takes place approximately every 8 h.

The space segment was originally planned to have eighteen active satellites. This has now been increased to twenty one with three in orbit as spares. The satellites occupy near circular orbits with a period of about 12 h and an altitude of some 20 000 km. The satellites are to be in six orbit planes inclined at an angle of 55° to the Equator. The constellation has been designed such that at least four satellites will be in view at any point worldwide at any time.

The high altitude means that each satellite is in view of a point on Earth for about 5 h per orbit. It also means that there is little atmospheric drag and that the gravitational fluctuations are kept to a minimum, but there is more effect from solar radiation pressure.

The satellites themselves are larger and much more complex than Transit satellites. The early test satellites (Block I) weighed about 400 kg (88 lb) and the production satellites (Block II) are larger at over 845 kg (1859 lb). Each satellite contains a number of time standards, including a high accuracy caesium clock which is accurate to 1 s in 300 000 years. The replacements to the Block II production satellites are being designed for launch around 1995 and it is expected that these will be able to communicate with each other.

The transmissions from the satellites take place at two frequencies; 1575.4 MHz (L_1) and 1227.6 MHz (L_2) with wavelengths of 19 cm (0.62 ft) and 24 cm (0.8 ft), respectively. These frequencies are multiples (154 and 120) of a base frequency of 10.23 MHz (f_o). The signals are encoded with two pseudorandom (PRN) codes; the C/A code or P code.

The C/A, coarse/acquisition or clear/access, code is the basis of the standard positioning service (SPS) and is available to everyone but is only broadcast on the L_1 frequency. It consists of a sequence of 1023 pseudo-random binary biphase modulations with a rate of 1.023 MHz and therefore repeats every millisecond.

The P, precise or protected, code is much longer, 267 days at a rate of 10.23 MHz and is the basis of the precise positioning service (PPS). It is transmitted on both the L_1 and L_2 frequencies, and while all satellites are equipped with equal P code generators, each satellite is only allotted a seven day section of the code. It therefore repeats itself each week. The individual satellite codes are mutually uncorrelated, so that there is no interference between satellite signals. Therefore, there are thirty seven code sections available and hence a maximum of thirty seven satellites in the constellation. It is, however, only available to authorized users.

The two frequencies are also modulated by the navigation message. This contains detail orbit data for that particular satellite (ephemeris), information on all satellites in the constellation (almanac), GPS time and ionospheric and tropospheric delays.

The basic measurement of GPS is one of range from the receiver to the satellite. This is obtained by determining the travel time for signals to reach the receiver from the satellite. Using the velocity of propagation this can be converted to a distance. The measurement is made via the pseudorandom C/A or P code. The code is generated in the receiver and correlated with that received from the satellite (*Figure 12.35*). The locally generated code is shifted until maximum correlation is achieved (i.e. in step) with that received. The shift in the code is a measure of the travel time of the signals if the satellite transmissions are synchronized with the locally generated code. It is, however, unlikely that these are in perfect unison and hence there is an unknown clock offset to be accounted for. Measurement is therefore not a pure range, but a pseudorange. The accuracy of the travel time measurement is approximately 1% of one bit of the code used, which equates to 1 µs for the C/A code and 0.1 µs for the P code giving a range accuracy of 3 m (9.8 ft) and 30 cm (1 ft), respectively.

In order to determine the position of the receiver from pure ranges, it is necessary to make three measurements. For each range, the receiver can be considered to lie on the surface of a sphere, centred at the satellite, computed from the ephemeris and radius equal to the range. The intersection of three such spheres will give a position (three unknowns: X, Y and Z). With pseudoranges, however, this is not possible since each range is contaminated by a single unknown clock offset. This fourth unknown can, however, be

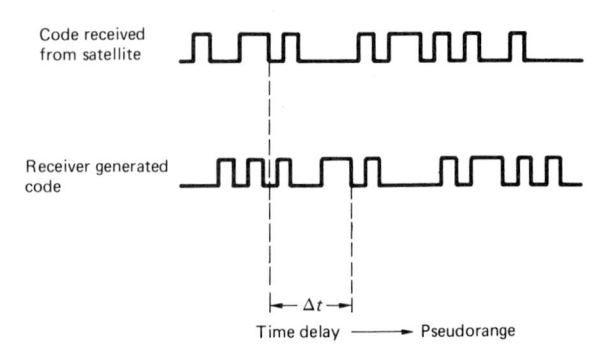

Figure 12.35 Pseudorange measurements

solved for by the use of another satellite, i.e. four in total. Hence the requirement of the constellation that four satellites are always in view.

It is possible to determine position from just three satellites by constraining the height. That is, by reducing the number of unknowns to three (latitude, longitude and clock offset). Alternatively, by introducing a highly stable oscillator into the receiver, it is possible to remove the clock offset. Use of both can reduce the number of satellites necessary to two. At present, with limited satellite (Block I) coverage, this can greatly increase the usable time period or satellite window.

The accuracy of the position fix depends on a number of factors. Since the P code has a higher frequency, this will give a higher order of accuracy in the range determination and hence a better fix accuracy, about 10 10 m (32.8 ft). Unfortunately, it is unlikely that this will be available to the majority of users and an accuracy of around 20–30 m (65.6–98.4 ft) can be expected using the C/A code. This far exceeds the accuracy expected and it has been stated that for defence reasons, the production (Phase II) satellites will be subjected to a downgrade in the accuracy obtainable using the C/A code (selective availability). It is thought that this will be achieved by introducing errors into the ephemeris and timing information and will limit the accuracy of the position to about 100 m (328 ft). It has also been stated that a precise ephemeris will be available about two weeks later to allow a more accurate post-processing of position.

The accuracy of a GPS position can be improved using a differential technique similar to that adopted with the Omega system (*Figure 12.36*). Hence monitoring stations are sited at points of known positions. This allows a correction to be determined, which can take one of two forms.

Firstly, the monitor can determine the pseudorange to each satellite and compare this to that calculated from the knowledge of the satellite's position (broadcast ephemeris) and that of the monitor (known point).

The difference, pseudorange correction, for each satellite forms a correction, which can be applied at an unknown point in the vicinity (*Figure 12.37*).

Alternatively, the monitor can determine its position from the pseudoranges and compare this to the known position of the station. The difference (position correction) can then be applied at an unknown point in the vicinity.

This technique corrects, to some extent, any timing or ephemeris errors in the position fix computation. Clearly, the pseudorange correction is a measure of these errors in the direction of the monitor station from each satellite. The position correction is the effect of these errors on the final computed position.

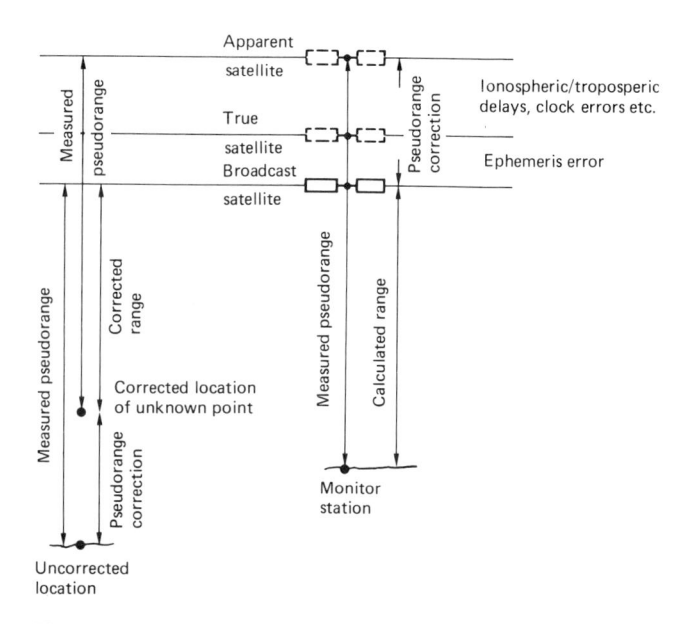

Figure 12.37 Pseudorange correction

To use this method of navigation, it is necessary to obtain the correction information on the vessel in real time. This can be achieved by transmitting it from the monitor to receivers in the vicinity, perhaps via communication satellites. No internationally agreed frequency exists but the Radio Technical Commission for Maritime Services has made recommendations on a format that the correction messages should take, RCTM SC-104. This allows for a number of parameters which include the differential corrections and their rate of change.

The effectiveness of differential GPS depends on maintaining the same satellite directions at both points. In addition, the position correction depends on the constellation and configuration of satellites used to obtain the position. The high altitude of the satellites means that this remains constant over a relatively large area. Tests carried out in the North Sea have shown

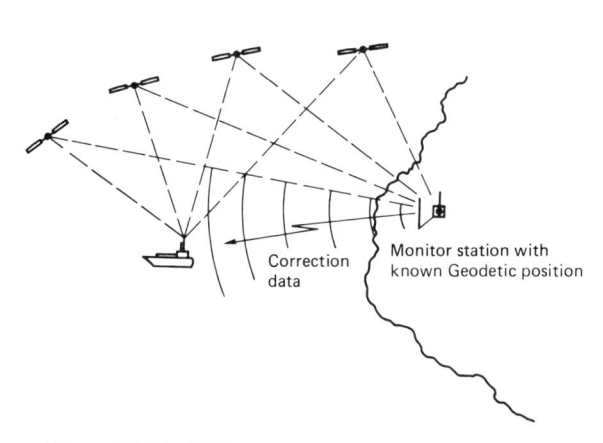

Figure 12.36 Differential GPS

both types of correction to be effective some 1000 km (621 miles) from the monitor with an accuracy of 5–10 m (16.4–32.8 ft) in position. It should be noted that the use of differential GPS could well remove much of the selective availability applied to the production (Block II) satellites.

Another improvement in single point positioning is the use of a pseudolite, a special form of GPS. It consists of a transmitter at a known point which sends signals at the L_1 frequency which resemble satellite transmissions and hence act as a pseudosatellite. These signals can include differential type correction information. This technique has the advantage that it is not necessary for the GPS receiver to have extra circuitry for the differential correction and gives another pseudorange for the position fix. It does, however, require software to choose and identify the pseudolites' signals and there must be a clear line of sight between the user and monitor.

There are techniques which make measurements on the carrier wave of the satellite signals. These techniques are limited to static and low dynamic situations since they require continuous monitoring of the satellite signals over a minimum of 30 min. These methods, often referred to as interferometric and kinematic GPS, do not determine absolute position but give millimetric accuracy in relative position, i.e. difference in latitude, longitude and height or chord distance and direction from a known point.

Regardless of the technique employed, there is a limit to the accuracy with which position can be determined. The uncertainties in the measurements can be translated into the accuracy obtainable in the computed position. This accuracy is highly dependent on the intersection angles of the lines of position or the geometry of satellites with respect to the point. The best accuracy is obtained when the satellites are at right angles to one another. The worst is when they are bunched in approximately the same direction. This can be quantified by the dilution of precision (DOP). (*Figure 12.38*), which is the ratio of the positioning accuracy to the measurement accuracy or:

$$\sigma = \text{DOP}(\sigma_o) \quad (12.25)$$

where σ is the measurement accuracy and σ is the positioning accuracy.

There are many varieties of DOP, depending on which particular coordinate or combination of coordinates is of interest, for example, VDOP: height (vertical); HDOP: two dimensional (horizontal) position; PDOP: three dimensional position; TDOP: time and GDOP: three dimensional position and time (geometrical).

The accuracy of a position fix is dependent on a number of error sources. The satellite clock is highly accurate, but still requires correction from the control segment. The measurements are totally dependent on it. In practice, any clock error is indistinguishable from ephemeris errors, but the error after correction may amount to some 3 m (9.8 ft). The positions of the satellites are given by the ephemeris in the World Geodetic System 1984 reference datum (WGS 84) and hence the final position will be given in the same system. The positional accuracy of the satellite obtained from the broadcast ephemeris is about 6 m (19.7 ft). However, the biggest problem area is the refraction of the signals in the ionosphere and troposphere. Since most users only have access to the C/A code (L_1 frequency), it is not possible to model the effects as in Transit. It is therefore necessary to model the effects using an empirical formula (several exist) to make a correction, using information from the satellite message. The resulting error may be of the order of 20 m (65.6 ft). Another source of error is from reflected signals or multipath interference. This may be reduced by careful antenna design.

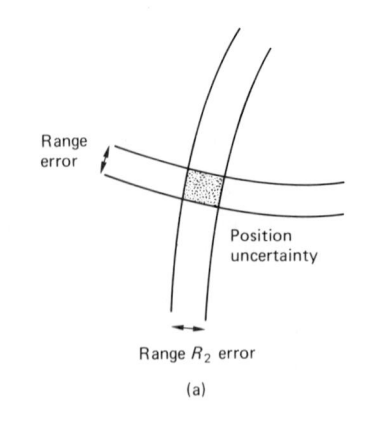

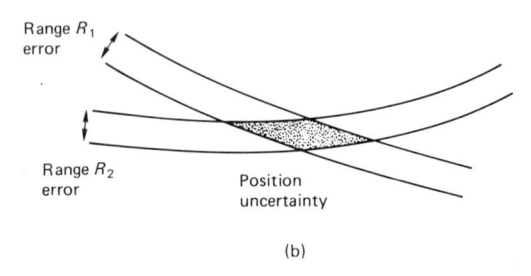

Figure 12.38 Dilution of precision (DOP)

The number of GPS receivers on the market is increasing but they can all be categorized by the manner in which they track satellites. A continuous tracking receiver has a receiving channel dedicated to each satellite and maintains lock on the signal. This type of receiver, therefore, has circuitry dedicated to each satellite it is tracking, but it is important that an interchannel calibration is carried out to ensure there are no timing delays. A switching receiver is one in which each channel samples more than one satellite signal. In this way, the amount of circuitry is reduced, but more demands are put on the microprocessor and the software is more complex. Such a receiver can take longer to regain lock onto a satellite if lost for any reason. A number of manufacturers combine continuous and switching channels in a receiver which is often referred to as a hybrid.

Switching channels can be subdivided by the time required to sequence through the signals. A multiplexing channel is one that sequences through all its assigned satellites in 20 ms, the time taken to transmit one bit of the satellite message. In this way, the message from each satellite can be read simultaneously. A sequencing channel switches between signals at a rate which is not related to the message bit rate. They can be fast or slow, taking seconds or minutes, respectively, to sequence through the signals. Receivers with sequencing channels generally have an extra channel to obtain the satellite message, as data will be lost during switching.

There have been great advances in receiver technology, which is increasing their power and reducing their size. Some receivers have a number of modes of operation: navigation, static, kinematic and differential, while others have specific applications. Some use a single frequency (L_1) but there are those that make use of both (L_1 and L_2), mainly for static positioning. It is unlikely that there will ever be any commercial P code receivers. At present, the smallest receiver would fit into a pocket, but it is expected in the future that a receiver could be included in a wrist watch and become part of everyday life.

12.31 Starfix

Starfix is the only commercial satellite based positioning system available at this time. It became operational in 1986 after some three years' development. The system offers a 24 h per day operational service to an unlimited number of users, but its coverage is limited to the continental United States and up to 700 km (434 miles) offshore into the Gulf of Mexico. It does not use dedicated satellites, unlike GPS and Transit, but channels on four communication satellites. These satellites are almost motionless with respect to a point on Earth (geostationary) at an altitude of about 36 000 km (22 370 miles) (*Figure 12.39*).

A network of tracking stations has been established at known locations across the United States with the master station in Austin, Texas. In the event of a failure at the master station, control will be switched to a backup at Houston, Texas. All the tracking stations send their data by wire to the master station, where the position of each satellite is computed together with correction data. This information is then transmitted to the satellite at a frequency of 6 GHz.

Unlike other systems, the satellite does not need to store data or perform calculations, since it is in view of the master station at all times. Instead it acts as a relay by retransmitting the data at around 4 GHz, using one

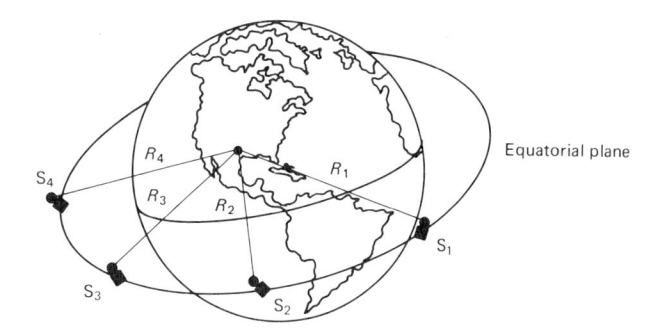

Figure 12.39 Starfix constellation

of the standard communication transducers, which have a reasonably constant delay between receive and transmit. These signals are modulated by a pseudo-random code with a rate of 2.4576 MHz. This code is used to determine the travel time of the signals by comparison with a locally generated version of the code in the receiver and is converted into a pseudorange, and finally the point position is determined as in GPS (*Figure 12.40*).

The tracking stations also act as differential monitor stations and the master station also includes a pseudorange correction in the satellite message, which can be applied to the observed pseudoranges to give a better position.

The techniques used are basically the same as for GPS and hence there are four unknowns to be determined; three for position and a receiver clock offset. It is common for the height of the observer to be considered known, hence only three satellites are required instead of four for the full solution. The quoted accuracy of position is 3–7 m (9.84–23 ft).

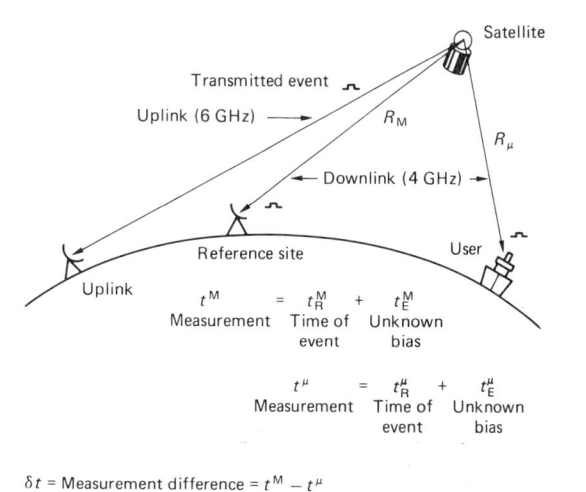

$t^M = t_R^M + t_E^M$
Measurement Time of Unknown
event bias

$t^\mu = t_R^\mu + t_E^\mu$
Measurement Time of Unknown
event bias

δt = Measurement difference = $t^M - t^\mu$

$$= (t_R^M - t_R^\mu) + (t_E^M - t_E^\mu) = \frac{(R_M - R_\mu)}{C_V} + \Delta t$$

Figure 12.40 Starfix measurement principle

Because of the frequencies used, the receiver requires a four antenna array, with each directed towards its satellite. This necessitates the use of an azimuth controller to maintain lock on the satellites and results in a relatively large, powered antenna installation. This type of system suffers a little from the orbits of the satellites. Since all are in equatorial orbits with only longitudinal separation, they are all in one general direction, south. This limits the accuracy of height determination and it is generally considered necessary to input an accurate height into the solution. It is possible to integrate another system such as GPS to improve the solution. The use of geostationary satellites does, however, mean that the constellation does not change and hence there is no variation in the DOP.

12.32 Geostar

The American Geostar system was first publicized in 1982 as a two way digital communication system with the ability to determine the position of the user. It is navigationally the same as the European system LOCSTAR from the French Centre Nationale d'Etudes Spatiale (CNES). These systems are intended for use by an organization to determine the location of its mobiles rather than as a navigational aid.

The systems use geostationary satellites with special dedicated packages aboard. The system determines position from satellite ranges to each user by transmitting signals from a central control to the user via one satellite. The user then retransmits the signal, together with identification information. This is relayed by two or three satellites to the control which then calculates delay times, ranges and the position of the user which can then be relayed back if required.

The system does not have an unlimited multi-user capability, since the mobile users are required to transmit. The maximum number of users on the system will depend on the time required between fixes. For example, the number of users has been estimated at 457 for a 1 s update and more than 100 000 for a 10 min update. The accuracy of position is not known, since it has not yet been demonstrated. It is however expected to be coarse, of the order of several hundred metres. The system suffers from all the problems of Starfix in terms of geostationary satellites and its accuracy would be improved by the input of height into the solution.

12.33 Navsat

Navsat is a system designed by the European Space Agency (ESA). Its present stage of design is for a mixture of geostationary communication satellites with up to twelve more in highly elliptical orbits, providing signals for long periods (for about 5 h at high latitudes. This would provide a continuous worldwide navigation system. The system is similar to Starfix, in that all timing and message formatting is done on the ground with a continuous uplink to the satellites. The satellites relay these signals at a different frequency using standard communication transponders. The system is still very much at the design stage, with no tests having been carried out, even using any of the existing geostationary satellites.

Radar

12.34 The history of radar

The word radar is obtained from radio detection and ranging, and the original principles were derived from acoustic methods of ranging, i.e. by timing the elapsed period between transmission and reception of an acoustic signal, multiplying by the speed of propagation and dividing the resultant range by two, as is done by echo sounders or, with a greater degree of sophistication, by Asdic. Contrary to some opinion, it can be claimed that radar itself was not a product of war, but that the development of radar was. As early as 1885, the great American inventor and entrepreneur Thomas Edison had filed a patent aimed at the prevention of collisions at sea, based on the reflection of electromagnetic waves and in the following year, 1886, Heinrich Hertz had demonstrated the basic principles of electromagnetic reflection in the proof of Maxwell's equations.

12.34.1 Marconi

However, the precise definition of the problem to be solved was the contribution of Guglielmo Marconi. In 1922, in a paper given at the American Institutes of Electrical Engineers and of Radio Engineers, Marconi said,

'As was first shown by Hertz, electrical waves can be completely reflected by conducting bodies. In some of my tests, I have noticed the effects of reflection and detection of these waves by metallic objects miles away. It seems to me that it should be possible to design apparatus by means of which a ship could radiate or project a divergent beam of these rays in any desired direction; should these rays come across a metallic object such as another steamer or ship, they would be reflected back to a receiver screened from the local transmitter on the sending ship, and thereby immediately reveal the presence and bearing of the other ship in fog or thick weather'
(quoted in Palandri and Calamia, 1985).

Problem definition was followed by research and practical demonstrations were given by Marconi in 1933.

12.34.2 The impetus of war

It is stated by Swords (1985) that '. . . even if military influences had been absent, radar was destined to be born in the 1930s' and, to support this, the anticollision system fitted to the French Liner *Normandie* in 1935 and the perfection of the radio altimeter by Bell Telephone laboratories in 1937 are cited.

Nevertheless, in the 1930s war was never far away and, commencing with the Daventry experiment in 1935, by the end of 1939, at the commencement of hostilities, a system of twenty early warning stations was in operation. These stations, known as the chain home system, extended from the Isle of Wight to the Firth of Tay. Intended to detect incoming bombers, the system operated at high frequency, in the band 22–27 MHz, with wavelengths around 13 m and utilized arrays of comparatively large antennae. Early identification friend or foe (IFF) transponders discriminated between enemy aircraft and *some* of our own (Swords, 1986).

12.34.3 The demand for centimetric wavelengths

Components used in early radars could not produce, or receive, high power centimetric, rather than metric, wavelengths. There were good reasons for wanting to do this.

Quoting the radar pioneer Dr Henri Gutton (Molyneux–Berry, 1985):

'The highest frequencies of the Hertzian spectrum have the great advantage of permitting highly directive antennas and, as a consequence, a substantial gain on the beam axis.

Thus for two wavelengths λ_1 and λ_2, using the same area of antenna, the effective power transmitted in the optimum direction is related by:

$$\frac{G_1}{G_2} = \left(\frac{\lambda_2}{\lambda_1}\right)^2$$

so, for $\lambda_1 = 10\lambda_2$, $G_2 = 100\,G_1$

For a communications link using the same antenna area for transmission and for reception, the power received at wavelengths 10 times shorter is 10 000 times greater. This means that, for equal transmission powers and reception quality, the range using the shorter wavelength is 100 times greater. This is true for a telephony link, and also for radar.

This argument, despite the difficulties of the work, made us choose to use the shortest possible wavelengths.'

Since transmitted and received power vary with antenna size, centimetric wavelengths would allow efficient antennae to be installed within the confines of an aircraft fuselage. Also, since the directivity of an antenna of a given size varies directly with frequency, or inversely with wavelength, shorter wavelengths allow narrower beam widths. These would reduce side lobe effects, minimize ground return signals in aircraft and produce increased target bearing accuracy in addition to greater target discrimination.

12.34.4 Component development

All those working with radar at that time were aware, along with Henri Gutton, of the advantages which shorter wavelengths would bring and they were equally aware of the problems. These were largely related to the components with which they had to work, particularly those of oscillatory circuits and detectors which utilized thermionic valves and had made great strides since the early crystal detectors of communications systems. But perhaps one of the limitations of advancement is a reluctance on the part of those advancing to look backwards for inspiration.

In the summer of 1939, a number of university research workers had been invited to spend some time at operational chain home stations. A group from the University of Birmingham was stationed at Ventnor, Isle of Wight and was made aware of the limitations of the metric wavelength system, particularly with respect to its poor resolution of individual aircraft within a group. Discussions led to the establishment of the Admiralty Laboratory within the Physics Department at Birmingham University, with the task of obtaining sufficient transmitter power and receiver sensitivity at centimetric wavelengths (Shearman and Land, 1985).

Prominent amongst this group were J T (later Sir John) Randall and H A H Boot. Shearman and Land (1985) note that the original team were all nuclear physicists with only 'a passing knowledge of radio', stating that this was deemed to be 'an advantage, not a handicap, for it meant that they were free from prejudices and preconceptions. They would be willing to work from first principles.'

12.34.5 The breakthrough

First principles led Randall back to the work of Heinrich Hertz, in *Electric Waves* translated by D E Jones, in which his innovative imagination was fired by the illustrations of the circular wire resonator with a spark gap, which, with a little artistic licence, can be equated to *Figures 12.41(a)* and *(b)*.

The first trial of the cavity magnetron on 23 February 1940 is described by Shearman and Land as follows.

'Randall and Boot were astonished to find ample power emerging from the wire attached to the coupling loop, as evidenced by a sizzling blue glow. They had no ready means of measuring the power, but made attempts by decapping car headlamps of increasing wattage and hanging the filament leads on the coupling loop. "In all cases the lamps were burnt out. This was a C.W. system as no modulators were as yet in existence and we later established that our first valve produced about 400 Watt at a wavelength of 9.8 cm".'

This trial led rapidly to the development of operational pulsed cavity magnetrons.

12.34.6 The completed system

With respect to receivers able to detect microwave signals, a reversion from thermionic diodes back to point contact crystal detectors, not dissimilar to the old fashioned cat's whisker systems, solved part of the

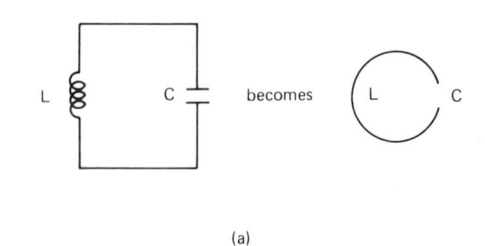

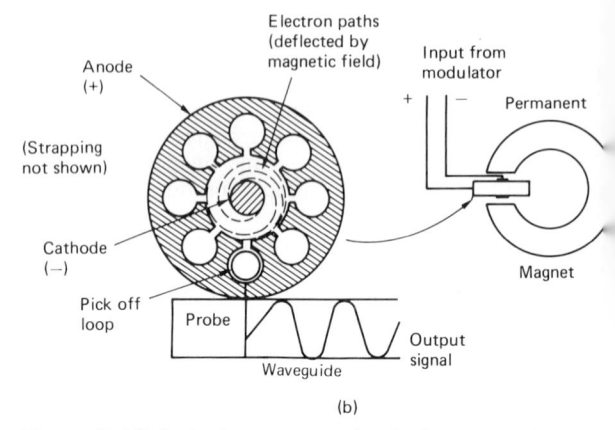

Figure 12.41(a) In the resonant circuit, the resonant frequency, f_R, is equal to

$$\frac{1}{2\pi\sqrt{LC}}$$

where L is inductance (Henrys) and C is capacitance (Farads). To increase the resonant frequency, f_R, both L and C should be reduced to minimum values; effectively a single turn for L, and the ends of the wire for C; (b) to reduce resistance, multiple cavities are cut from a solid block of copper, which are strapped together to maintain tune intensity.

problem, whilst the development of the resonant cavity klystron as the local oscillator for superheterodyne operation finally provided the solution (*Figure 12.42*). Trials of radar systems very similar to those now in use began in November 1941 and operational systems were introduced shortly afterwards.

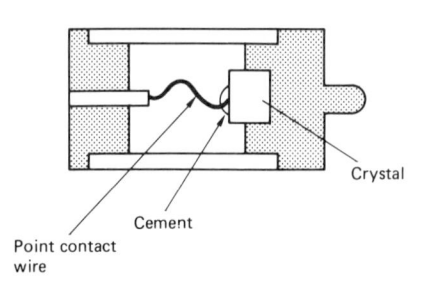

Figure 12.42 Point contact crystal detector

12.35 Parts of radar equipment

12.35.1 Basic radar

Figure 12.43 (*a*) and (*b*) shows a block diagram of a simplified basic radar and a typical manufacturer's specification, which may be broken down into five sections:

1. Transmitter (T_x) – units 1 to 4.
2. Receiver (R_x) – units 5 to 11.
3. Cathode ray tube – units 12 to 14.
4. Antenna and synchronized components – units 15 to 17.
5. Time synchronized components – units 18 to 22.

These sections can now be examined unit by unit.

12.35.2 Transmitter

12.35.2.1 Trigger unit (Unit 1)

This provides short pulses at intervals, varying from about 500 up to about 3500 pulses per second, depending on the radar range in use (note connection with unit 22). Timing precision is not important, since the same pulses trigger both transmitter and receiver (note – the reasons for variations in pulse intervals are discussed later.

12.35.2.2 Delay line (Unit 2)

More correctly, this should be termed the artificial delay line, since they are now built using banks of inductors, capacitors and resistors, the chosen length of which determines the length of the transmitted pulse. The delay line is essentially a power reservoir which has a long charging period and a short high powered discharge. For example, if the pulse interval between trigger unit pulses is 1000 μs, and the required pulse length to be transmitted is, say, 0.5 μs, then the ratio of delay line charge to discharge periods is 2000:1. Delay lines discharge in the form of a square dc pulse.

12.35.2.3 Modulator (Unit 3)

This is a very low resistance valve which is closed during delay line charging and opened by the trigger unit pulse, allowing all the power in the chosen delay line length to discharge via a pulse transformer to the magnetron. A simple analogy to make with units 1, 2 and 3 preceding is that of a bathroom cistern, where the header tank trickle charges until full, then operation of the flush control triggers the discharge valve, allowing a short high powered release of the accumulated water.

12.35.2.4 Magnetron (Unit 4)

The principle of the magnetron has been discussed in Section 12.34.5 with *Figures 12.41 a* and *b*. Its function is to convert the dc pulse to an ac pulse of the same of the same length and at the required frequency which for 3 cm radars, will be in the band 9300–9500 MHz.

Quoted commercial magnetron frequencies are not precise. Micron variations in manufacture will produce differences in what appear to be identical magnetrons and temperature variations produced during the operation of the radar and by ambient environmental conditions will cause the output frequency to vary. Commercial marine radar magnetrons are not tuneable and an important operational factor to note is that the receiver has to be tuned at fairly frequent intervals, during use of the equipment, to the transmitted magnetron frequency. Manufacturers usually specify transmitting frequencies as, for example, 9410 ± 30 MHz (Racal–Decca RD 80 series).

12.35.2.5 Waveguides

Coaxial cable may be used, particularly in shorter range, lower power radars, to carry the signal pulse from the magnetron to the antenna. With the majority of high power systems, however, waveguides are used.

Waveguide theory is based upon transmission line principles and essentially states that a rectangular conducting box section, with internal dimensions which are a quarter wavelength $\lambda/4$ across the narrowest gap and half a wavelength $\lambda/2$ across the widest, will contain an electromagnetic field within the box; that is, with minimum resistance loss within the conducting metal itself. The signal will travel along the waveguide as a series of reflections from wall to wall.

A waveguide system is more efficient than coaxial cable in terms of power losses, but losses will occur, particularly where the waveguide suffers from bending, where the cross-section changes from the rectangular which is necessary at the point of rotation of the scanner and when dirt or moisture enter the waveguide. This occurs because of changes in the reflections taking place within the guide.

12.35.2.6 Magnetron probe

The majority of communications and electronic position fixing system antennae are quarter wavelength grounded dipoles. Taking, as a simplification, the speed of propagation of a radio signal to be 3×10^8 m/s, then:

wavelength $(\lambda) = 3 \times 10^8 / f_T$

and

optimum antenna size $= \dfrac{\lambda}{4}$

For example:

1. Omega $f_T \simeq 10$ kHz, where f_T is transmitted frequency

$$\lambda \simeq \dfrac{3 \times 10^8}{10^4} \simeq 3 \times 10^4 \text{ m}$$

Optimum antenna length $\simeq 7500$ m

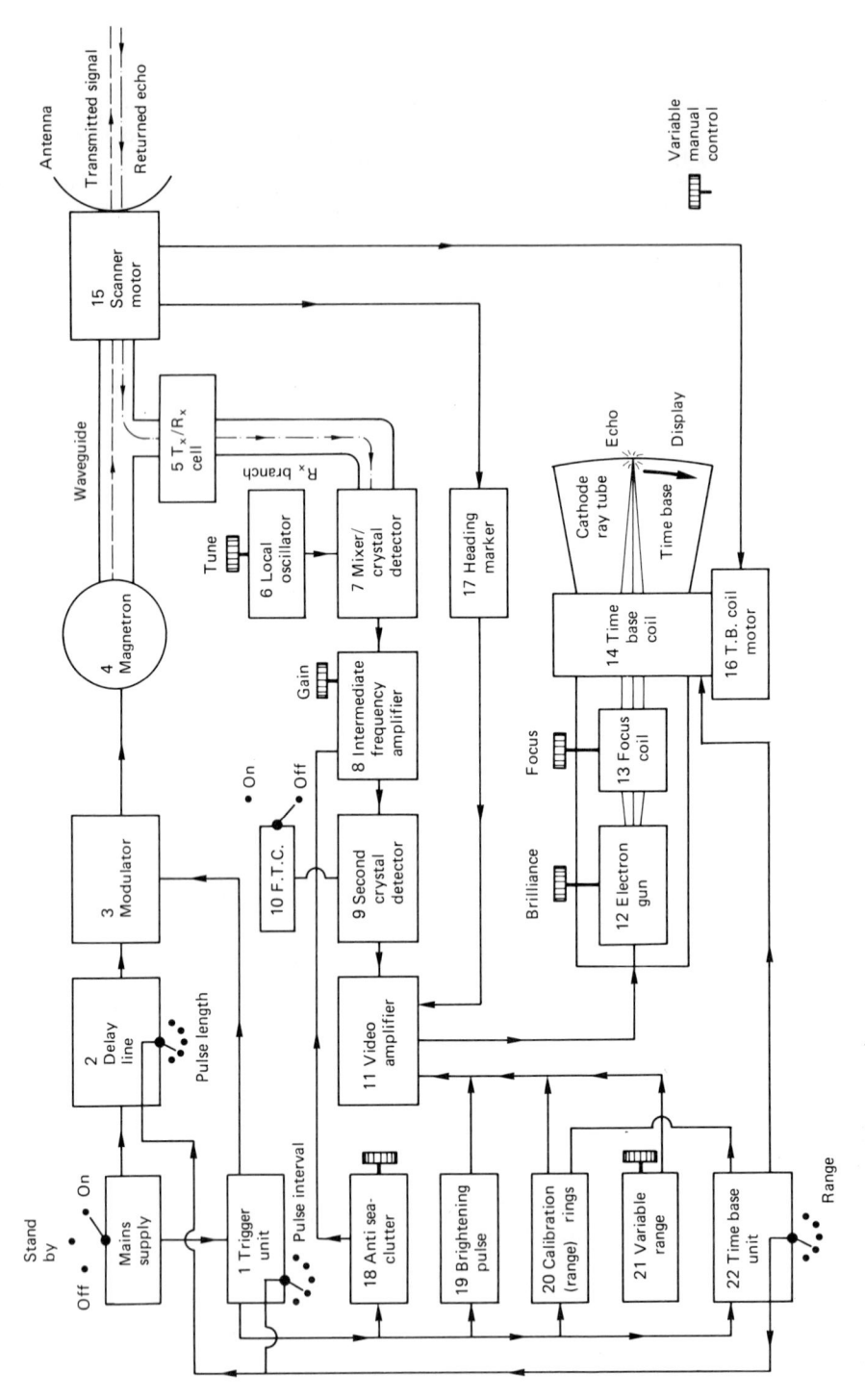

Figure 12.43 Simplified basic radar

2. Hyperfix $f_T \simeq 2$ MHz

$$\lambda \simeq \frac{3 \times 10^8}{2 \times 10^6} \simeq 150 \text{ m}$$

Optimum antenna length $\simeq 37.5$ m.

If we apply the same criteria to radar, then:

Marine radar band $f_T \simeq 10$ GHz

$$\lambda \simeq \frac{3 \times 10^8}{10^{10}} 3 \text{ cm}$$

Optimum antenna length = 7.5 mm

Surprising though it may be, the radar antenna, in radio terms, may be considered to be the probe, from the magnetron pick off loop, which is inserted in the short dimension of the waveguide and which produces an electromagnetic field similar to the fields produced by any other type, or size, of antenna, and which are then channelled by the waveguide to the scanner (*Figure 12.44*).

12.35.3 Receiver

An overall discussion of radar ranges and powers is detailed later; suffice to say, at the moment, that the amount of energy returned to the receiver is very small compared to that which is transmitted. Additionally, a rule of thumb would state that the returned energy of a target varies inversely as the fourth power of the ratio of its range variation.

For example if, say on the 12 mile displayed range, we have two identical targets A and B, A being at a range of a quarter mile and B at a range of 10 miles:

$$\text{Range ratio/variation} = \frac{\text{Range B}}{\text{Range A}} = \frac{10}{0.25} = 40$$

Roughly Echo strength of B = $\dfrac{\text{Echo strength of A}}{40^4}$

= $\dfrac{\text{Echo strength of A}}{2\,560\,000}$

The receiver therefore has to process signals very much weaker than those transmitted.

12.35.3.1 Transmit and receive cell (Unit 5)

The waveguide and scanner unit is a duplex system; that is, the same waveguide and scanner are used for both transmission and reception, since each is required at a different time. In the transceiver, the waveguide branches from transmitter and to receiver.

It is necessary to protect the receiver detector crystals while the transmitter is operating, since they would be damaged by the high powered transmission signal. The T_x/R_x cell ionizes during transmission and effectively blocks the receiver branch. After transmission ceases, the cell de-ionizes and the branch opens to received signals.

12.35.3.2 Tuning (Units 6 and 7)

As stated in Section 12.35.2.4, the transmitted frequency may vary and it is necessary to tune the receiver to the transmitted frequency. The transmitted frequency itself is too high to be efficiently amplified and therefore radar receivers, in common with many other radio receivers, operate on the superheterodyne principle. This method converts the incoming signal to a fixed frequency which can be amplified by a fixed tuned amplifier. This fixed frequency is known as the intermediate frequency (IF).

In radar, this is achieved by mixing the signal derived from a local oscillator (LO, Unit 6) with the incoming magnetron frequency. The local oscillator is tuned such that the difference between its frequency, f_{LO}, and the

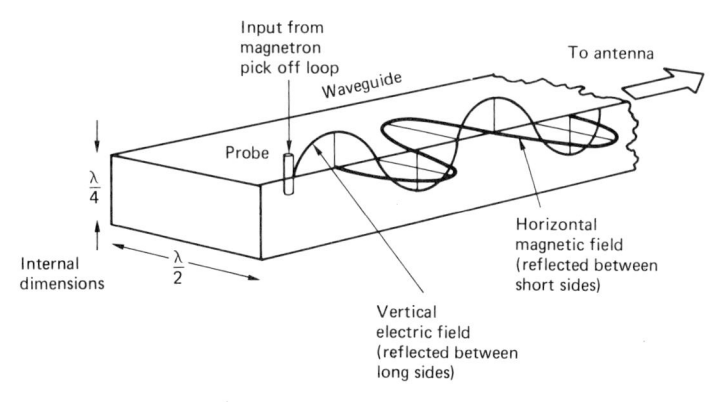

Figure 12.44 Waveguide propagation (see also Figure 12.41(b))

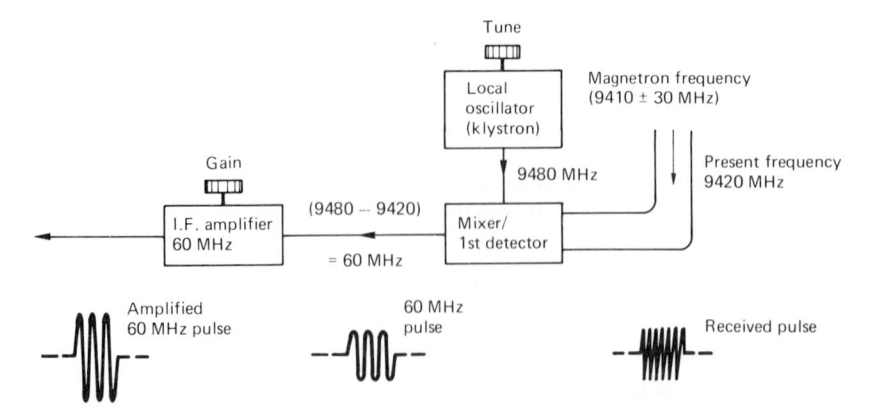

Figure 12.45 Tuning and IF amplification

received frequency, f_m, is equal to the IF. The mixer and detector (unit 7) detects this as the beat frequency of f_{LO} and f_M. The IF used in radar is usually about 60 MHz, i.e. the incoming frequency drops, in one stage, from around 10 GHz to 60 MHz (*Figure 12.45*).

In *Figure 12.45*, the incoming frequency, f_M, is 9420 MHz (9.42 GHz) and the required, IF, is 60 MHz.

Detector output = $f_M \sim f_{LO}$

therefore if IF = 60 MHz, then $f_M \sim f_{LO}$ = 60 MHz

Required f_{LO} turning is to 9480 MHz (or 9360)

The local oscillator (Klystron) itself is usually a tuneable frequency resonant cavity, which will produce super high frequencies, but at relatively low power.

12.35.3.3 Intermediate frequency (IF) amplifier (Unit 8)

The frequency of this amplifier is chosen on the basis that, firstly, it is a frequency at which the amplifier will be efficient and, secondly, minimal interference should occur, both from external noise and from internal feedback. However, some interference is always present at all frequencies and therefore the bandwidth over which the unit will amplify is made purposely very narrow, to limit interference which might mask weak radar echoes. This bandwidth required, however, varies inversely with the transmitted pulse length, since the pulses themselves occupy, as deeply modulated signals, wide bandwidths.

For example, the IF bandwidth required for a short pulse of 0.01 μs may be as high as 20 MHz, whereas that for a long pulse of 1.0 μs may be as little as 4 MHz.

12.35.3.4 The importance of frequent tuning

Taking the latter case and the example in the previous section, we can illustrate what happens if the radar tuning is not frequently checked, i.e. at 1 h or 2 h intervals during normal use and more frequently when the equipment is first warming up. In that example, we had the magnetron producing 9420 MHz and the local oscillator (Klystron) producing 9480 MHz, the beat frequency of which is the 60 MHz required for optimum amplification by the IF amplifier. If, however, over a period of time, the magnetron frequency alters due, say, to ambient temperature changes, to 9410 MHz, then the beat frequency now becomes 9480 − 9410 MHz = 70 MHz.

The IF bandwidth at this pulse length is 4 MHz, which means that the amplifier will operate efficiently between 58 and 62 MHz. The signal from the first detector, at 70 MHz, is, however, outside this range, therefore incoming echoes will not be amplified and will not show on the display (*Figure 12.46*). This can obviously be a dangerous situation, both with respect to other vessels, if the radar is in use for collision avoidance and with respect to other obstructions, if being used for navigation.

Tuning is best performed on land echoes. If out of range of land, tuning can be made on other vessels, or on sea clutter. In the absence of and in addition to, all these sources, tuning can be performed on tuning indicators and on performance monitors which com-

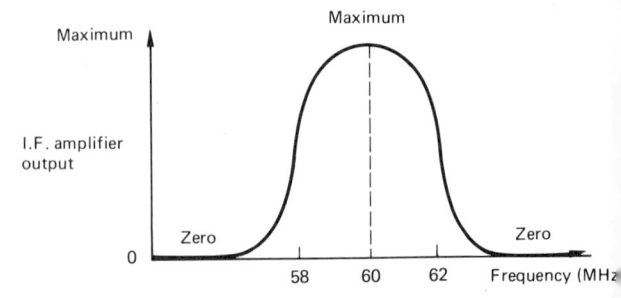

Figure 12.46 IF bandwidth

pare previously determined optimum transmitter and receiver performance with current levels.

12.35.3.5 Gain control

Amplification should be adjusted by means of the gain control such that weak targets are detected, but not to the extent that interference becomes obstructive. When adjusting the gain control manually, this is done by increasing gain until the display has a light background speckling of interference (thermal noise or thermal grass).

Referring back to a previous section which notes that relative echo strengths within a single range scale may vary by a factor of millions, it should be obvious that amplification factors cannot be the same for all echoes within that range and that there must be an amplification limit on higher powered echoes. This is the reason why excessive amplification of interference may mask target echoes and, perhaps more importantly, is the reason for modification of the amplification factor within the sea clutter area close to the ship using radar (input to the IF amplifier from unit 18).

12.35.3.6 Second crystal detector (Unit 9)

The output from the IF amplifier is now an amplified pulse, still the same length as that transmitted, but at a frequency of around 60 MHz. The required input signal to the cathode ray tube is a dc pulse and the purpose of the second detector is this conversion from ac to dc (*Figure 12.47*).

12.35.3.7 Fast time constant (Unit 10)

This unit, which may also be termed the 'rain differentiator', sharpens the outlines of large echo blocks and operates, when switched in, on the second detector. Its objective is to break down areas of saturation on the display and it may be used to discriminate specific required targets within land mass echoes, or, more frequently, within the clutter which is produced by rain and to some extent other precipitation such as hail or snow.

Briefly, it operates as follows. Rainfall can be considered to reflect over large, fairly constant echo strength areas. As noted previously, amplification limits may cause higher echo strength objects, such as other ships, to be displayed at the same intensity on the screen as the surrounding rain and thus to be masked. Unlike sea clutter, rain clutter does not necessarily occur in the region around the transmitting ship itself, but can occur anywhere on the display and therefore cannot be reduced by the same method as sea clutter.

However, rain clutter can be considered as a fairly homogeneous echo block and, using a differentiating circuit within the second detector, only those echoes which increase in amplitude will be output. This has the effect that the edge of rain will be displayed, along with any rising amplitude echoes within the rain, such as other ships, as illustrated in *Figure 12.48*. The effect on land mass echoes is similar.

12.35.3.8 Video amplifier (Unit 11)

The dc output of the second detector is further increased by the video amplifier and this unit also collates the dc outputs of the various time synchronized units 19, 20 and 21. The output of this unit is, therefore, the information data stream required for interpretation by the cathode ray tube, in the form of dc pulses (*Figure 12.49*). Note that the signal started as a dc pulse from the delay line, was then converted to SHF by the magnetron, then, after reception, to an intermediate frequency by the first detector, then finally back to a dc pulse by the second detector.

12.36 Cathode ray tube

Positive dc pulses from the video amplifier are applied to the grid of the electron gun (unit 12) which produces

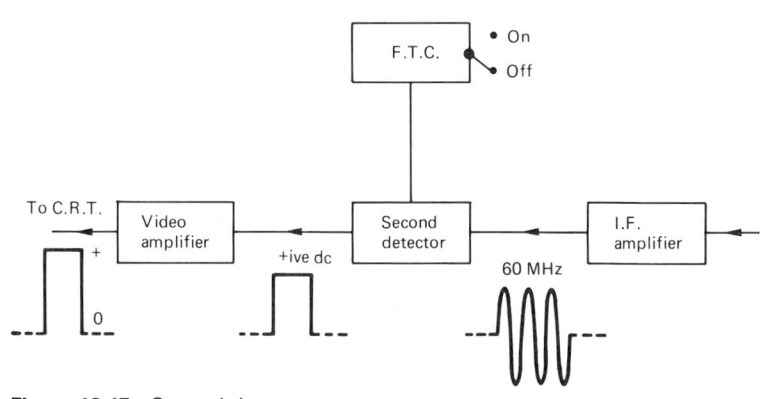

Figure 12.47 Second detector

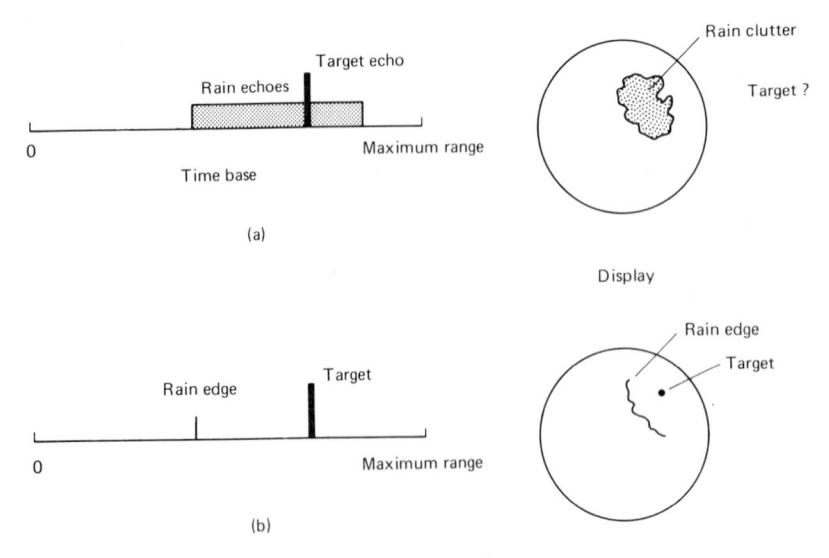

Figure 12.48 Fast time constant (differentiator)

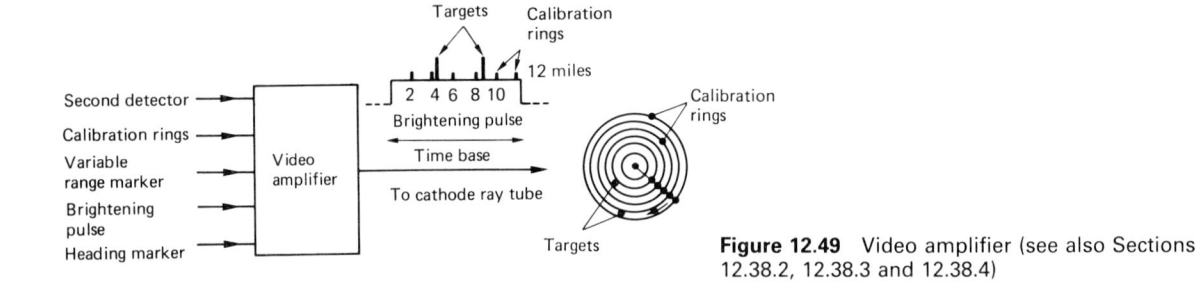

Figure 12.49 Video amplifier (see also Sections 12.38.2, 12.38.3 and 12.38.4)

a stream of electrons for the duration of the pulse, this duration being determined during transmission by the delay line. Electron stream density is governed by the grid bias of the electron gun and is varied by the brilliance control. This setting is critical and is achieved in operation by increasing the brilliance until the rotating time base trace becomes visible, then turning it down until the time base just disappears, i.e. just at grid bias cutoff. This ensures, firstly, that weak echoes are not swamped by excessive, unnecessary brilliance and, secondly, that such echoes will raise the grid above cutoff.

12.36.1 Control settings

These should be performed in the order of brilliance gain and tune, all three controls initially being turned down, i.e. adjust brilliance as above, then turn up the gain to give a light speckling of interference, then tune as described in Section 12.35.3.4. A further manual control may be provided to focus the electron beam (unit 13) or this may be automatically controlled by a feedback loop.

12.36.2 Time base coils and unit (units 14 and 22)

These units determine the maximum range to be displayed on each scale and present the range of target echoes in analogue form. In *Figure 12.50*, the coils are set across the axis of the cathode ray tube (CRT).

If current is passed through the coils, the solenoid effect will produce a magnetic field across the CRT and, in line with Flemings' left hand rule will deflect the electron stream at right angles to the magnetic field, i.e. across the face of the display. Increasing the current through the coils will increase the deflection and a specific value of current will produce full scale deflection, that is, to the edge of the display.

If, therefore, current through the coils is increased linearly over a specific time period, then the trace on the display can be made to move from the centre of the display as in relative motion radar, but from any other point, as may be required in true motion radar, to the extreme edge of the display, in that same time period.

12.36.2.1 Displayed range

Assuming a simple radio wave velocity of propagation of 3×10^8 m/s and a nautical mile of 1852 m (6076 ft),

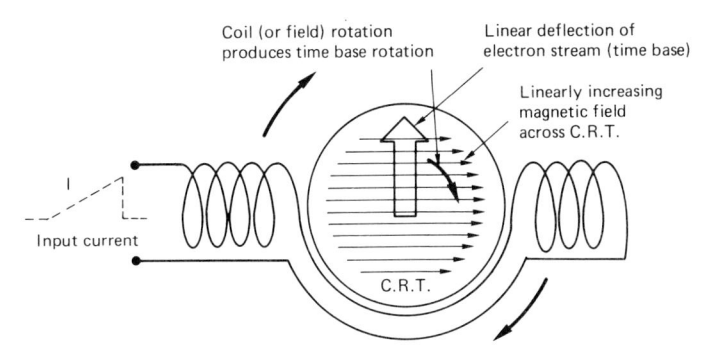

Figure 12.50 Time base coils

then the time taken by a radio signal in travelling one mile is approximately 6.17 µs. The time taken for a signal to travel to a target 1 mile away and to return to the receiver, is, therefore, approximately 12.34 µs.

If, therefore, a range of, say, 12 miles is required to be displayed, then the current through the time base coils must rise linearly from zero, to full scale deflection in $12 \times 2 \times 6.17$ µs = 148.08 µs. If the range scale is required to be changed to, say, 24 miles, then the rise time must be increased to 296.16 µs (*Figure 12.51*).

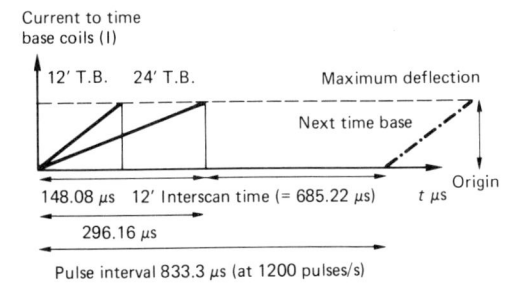

Figure 12.51 Time base unit output current (see also Table 12.8)

This time base current is provided by the time base unit (unit 22), for a particular range selected by the operator, this control also selecting the pulse length to be transmitted (via unit 2). Note that the trigger unit (1) synchronizes the transmission of the pulse to the start of the time base.

In the absence of signals to the CRT electron gun grid, the beam is suppressed, but when it is illuminated by an echo or by signals from the calibration or variable range marker units 20, 21, or by interference, the beam will indicate the relative range of signal by its deflection from the time base origin.

12.36.2.2 Second trace echoes

After full scale deflection the beam returns to the origin to await transmission of the next pulse (flyback). There then follows a resting period, or interscan time, the period of which must be sufficient to allow echoes from ranges greater than the displayed range to return, before the next time base commences. In simple terms, if there were no interscan period and the next time base commenced immediately after flyback, then a target at 14 miles range would show at its correct range, if the 24 miles display was selected, but at 2 miles if the 12 miles display was then chosen.

This effect, known as a second trace echo can still occur in non-standard propagation conditions of super refraction, when detectable echoes can be received from targets at ranges greater than the equivalent range of the pulse interval. For example, if the pulse interval in the previous illustration was 500 µs, this equates to an equivalent range of

$$\frac{500}{12.34} \text{ miles} = 40.5 \text{ miles}$$

and a target at a true range of 45 miles would therefore show at a second trace range of 4.5 miles.

12.37 Antenna

The function of the antenna or scanner is to produce a transmitted beam width which is narrow in azimuth, i.e. in horizontal width, to produce good target discrimination, but with a wide vertical width, to allow the vessel to roll and pitch whilst maintaining echo contact. This is necessary, since the scanner is simply bolted to the ship's structure and is not stabilized, as it might be in military installations.

No scanners have yet been designed which are perfectly directional and therefore energy will be transmitted in directions other than required. Such sidelobes are a nuisance and can produce echoes, particularly at short ranges, which are confusing.

Minimum performance specifications for approved radar equipment are set by the Department of Transport (1982) and are quoted in decibel (dB) units of power relative to that of the maximum power of the main beam in the horizontal plane. These are that:

1. inside ±1° from the maximum, power must fall by −3 dB (0.5 maximum),
2. inside ±2.5° from the maximum, power must fall by −20 dB (0.01 maximum),
3. inside ±10° from the maximum, power must fall by −23 dB (0.005 maximum),
4. outside ±10° from the maximum, power must fall by −30 dB (0.001 maximum),

$$\left(\text{where dB} = 10 \log_{10} \frac{\text{Power A}}{\text{Power B}}\right)$$

The first of these requirements determines the main beam width, essentially stating that, outside a beam width of a maximum of 2°, power radiated must be less than half the maximum radiated.

The last requirement is intended to reduce the all round side lobe echo nuisance, in confining the maximum power radiated to a value less than one-thousandth of the maximum. Remember, however, that in Section 12.35.3 it was noted that the receiver is capable of detecting echo powers varying by factors of millions and therefore sidelobe echoes, particularly from short range targets, can cause nuisance echoes on the display (*Figure 12.52*).

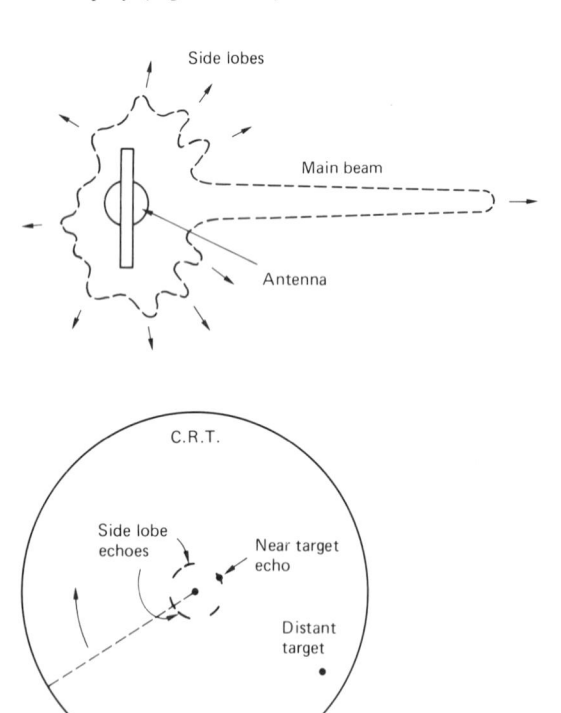

Figure 12.52 Side lobe echoes

Side lobe echoes can be removed from the display by reducing the receiver gain, but at the possible penalty of obliterating weaker, long range echoes. Modern equipment is invariably better than the minimum standards laid down by the Department of Transport. For example, Racal–Decca's 2990 BT series give horizontal beam widths (−3 dB) of 1.3° for a 1.8 m (6 ft) antenna and 0.8° for a 2.7 m (9 ft) antenna.

12.37.1 Slotted waveguides

As noticeable above, the larger the antenna, the narrower the beam width. Earlier parabolic reflector type antenna are now largely obsolete, superceded by slotted waveguide systems. The earliest HF radar antennae consisted of arrays of dipoles from which phase integration gave directivity. Slotted waveguides are not dissimilar, although centimetric wavelengths require different manipulation. As the name suggests, the antenna consists of the end of the waveguide in which the signal is being conducted from the magnetron, slots being cut in the narrow side of the waveguide rectangular box section. Inside the waveguide, standing waves are set up, from the wave travelling to the waveguide end, and those reflected back from it.

The spacing of the slots is such that in phase energy is radiated from each of the slots and the system acts in a similar manner to an array of small dipoles. As illustrated previously, the equivalent size of quarter wavelength grounded dipole would be around 7 mm, so the analogy is not too outlandish.

As with all arrays, the greater the number of elements, in this case, the greater the slotted waveguide length, the greater the directivity (*Figure 12.53*). Vertical beamwidth is restricted by a simple reflector, to about 20°. This is to the −3 dB vertical power points.

The beam transmitted therefore, between −3 dB points, is effectively a fan shape, under 2° wide horizontally and about 20° wide vertically. 'Beam' in this sense is perhaps confusing, since the transmitted signal is very short. For example, 0.5 μs pulse will occupy a physical length of about 150 m (492 ft) (*Figure 12.54*).

The electromagnetic field radiated by such a system as described would be horizontally polarized. There are means however of varying the field polarization through 360° (circular polarization) and of causing signals from rain clutter to be rejected, whilst accepting signals from irregularly shaped objects, i.e. the targets required. The gain of a directional antenna can be thought of as the ratio between the energy received from its main beam at a given range and that which would have been received from a theoretical non-directional antenna (isotropic radiator) fed with the same power, and at the same range. Again this ratio will be quoted in decibels, a typical value being 30 dB (1000:1).

echo is required to remain visible on the display, therefore the screen coating is required to have a fairly long period of afterglow unlike, for example, a television screen where the afterglow period is negligible. Note that this requirement is not necessary with raster scan displays.

The number of returns would be increased, in the formula above, by:

1. increasing the beamwidth which would give poor bearing discrimination;
2. increasing the pulse repetition frequency which would reduce interscan time and give rise to the increased possibility of second trace echoes;
3. reducing the scanner speed, which would increase the afterglow period required.

In practice, a compromise combination of the above factors is made.

12.37.2.1 Heading marker (unit 17)

All bearings are made relative to the ship's head, not relative to the course which the vessel may be making good. The heading reference is obtained mechanically, by the scanner tripping a microswitch in its mounting, which illuminates the full time base for two or three scans via the video amplifier, and this produces a solid line on the display. This heading line is referenced to the 360° relative heading, or in a gyro compass stabilized display, to the course steered.

12.38 Time synchronized components

In addition to the time base unit 22 already discussed, the following are also synchronized to the pulse transmission timing, by the trigger unit.

12.38.1 Anti-sea clutter (unit 18)

If the gain is set to amplify distant targets, then returns from the sea surface close to the operating ship, particularly if waves are pronounced, may saturate the display in that region and obliterate the echoes of larger targets, such as ships.

If gain was simply reduced, to highlight near echoes, then the more distant, weaker targets may be lost. The answer is to employ a swept gain circuit which reduces the gain level when the pulse is first transmitted, but allows it to return to the operator set level in a chosen period of time, this period being equal to the range at which the sea clutter return becomes unobtrusive.

12.38.2 Brightening pulse (unit 19)

After full scale deflection the trace returns to its origin, where it remains during the interscan time. Echoes returning during this time will be processed by the receiver and could produce brightening of the electron

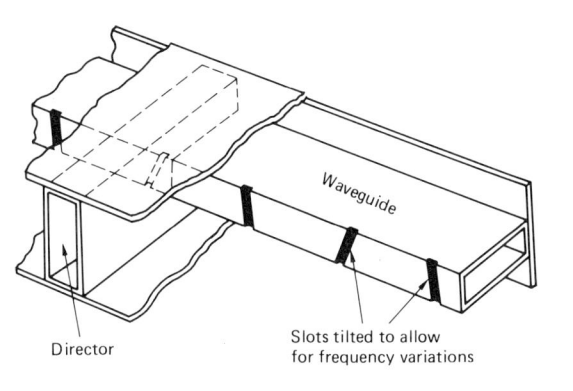

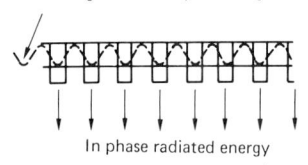

Figure 12.53 Slotted waveguide antenna. Slotted waveguide is encased in an aerodynamic glassfibre case to eliminate water and dirt effects

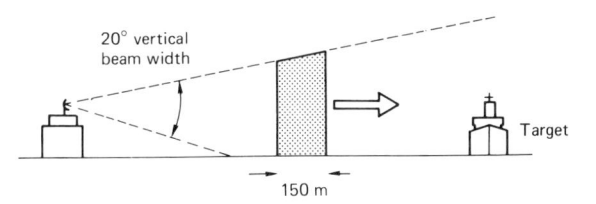

Figure 12.54 Transmitted pulse

12.37.2 Scanner rotation

To be detected, a target has to return a number of echoes, which are then integrated. If this number equals N, then for a target to be detected:

$$N < \frac{\text{beamwidth to } -3 \text{ dB points}°}{\text{scanner rotation speed} \times 360°} \times \text{pulse repetition frequency}$$

where scanner rotation speed is expressed in degrees per second and PRF is expressed in pulses per second.

The scanner motor (unit 15) is synchronized with the time base coil motor (unit 16) or with a three phase rotating magnetic field system, such that the time base drawn on the display rotates in synchronism with the beam direction of the antenna. To achieve the number of echo returns N as described, the scanner speed is usually low, about 25 rpm, which means that the time base is drawn, at a particular point on the display, at just under 3 s intervals. Between time base passes, the

beam. This would lead to false echoes being displayed during flyback, which is not instantaneous, and general deterioration of the screen coating in the region of the trace origin.

To prevent this, the CRT grid bias is dropped during the flyback and interscan periods, to a level at which, even with the addition of an echo signal from the video amplifier, the grid bias will not rise above extinction and the electron beam will not be illuminated.

12.38.3 Calibration rings (unit 20)

This unit produces short pulses at specific intervals, which, as the time base rotates, draws concentric circles on the display. If 1 mile rings are required, then such pulses are produced at 6.17 × 2 μs = 12.34 μs intervals. The 2 mile rings would be produced by pulses at 24.68 μs intervals and so on.

12.38.4 Variable range marker (unit 21)

This produces a pulse at a variable time, the control being operated such that the circle produced passes through the point of which the range is required and the time is then calibrated as range.

12.39 General aspects of radar performance

Various factors affect radar performance and these can be discussed under the headings below.

12.39.1 Radar horizon

Super high frequencies propagate by the direct wave, which is unaffected by ground conductivity and does not suffer from contamination by skywaves. This mode is often referred to as line of sight, which is not quite true as the radar horizon is approximately 6% further than the visible horizon. Since radar frequencies are lower than those of visible light, refraction is greater.

A rough rule of thumb is that the radar horizon range = $\sqrt{h_T}$ miles (4.06 $\sqrt{h_T}$ km), where h_T is the height of the scanner above the sea surface in metres. The detectable range of an object, therefore, with respect solely to the radar horizons involved would be detectable range = 2.2 ($\sqrt{h_T}$ + $\sqrt{h_o}$) miles, or 4.06 ($\sqrt{h_T}$ + $\sqrt{h_o}$) km where h_o is the height of the object, also in metres (*Figure 12.55*).

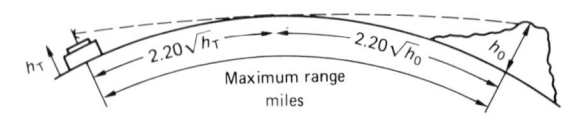

Figure 12.55 Detectable range

12.39.1.1 Sub and Superrefraction effects

These figures are based on standard meteorological conditions, however and non-standard conditions may produce changes in range. Subrefraction occurs when atmospheric density increases with height at more than the standard rate and will reduce the radar range. This effect mainly occurs in the Polar regions, when cold continental air crosses a relatively warm sea surface.

Superrefraction is effectively the reverse of this and is more frequently encountered, particularly in tropical regions, when warm continental air passes over a relatively cold sea surface. Radar ranges may be greatly increased, to the extent that multiple trace echoes may occur.

In mid-ocean these effects are less likely, but superrefraction can occur in cyclonic systems due to rising air masses and, conversely, subrefraction may occur in anticyclones due to subsiding, cold air. In extreme conditions, superrefraction may be termed ducting.

12.39.2 Pulse length

The length of the transmitted pulse largely determines the transmitted energy, but since this length can be expressed not only in time, but also in distance, it also determines range discrimination, picture definition and the minimum range at which targets can be detected, which is important in confined waters such as rivers and estuaries.

If a pulse is 1 μs in length, this corresponds (at 3 × 10^8 m/s (9.84 × 10^8 ft/s) to a distance of 300 m (984 ft). Bearing in mind that radar measures total response time, but that displayed ranges are a half of this two-way distance, the above pulse would not be able to discriminate between two targets which were less than 150 m (4926 ft) apart and on the same bearing. Additionally, since the T/R cell does not open up the receiver until transmission has ceased, with a pulse of this length, targets less than 150 m (492 ft) away could not be detected, i.e. the minimum range.

At long ranges, however, discrimination and indeed minimum range, are of minor importance in comparison with, perhaps, the detection of land. Therefore long, high power pulses are used. At short ranges, power is less important, but in operational terms, discrimination is a higher priority, therefore pulses as short as 0.05 μs may be utilized, i.e. with 7.5 m (24.6 ft) discrimination and minimum range.

12.39.3 Pulse repetition frequency (PRF)

High PRFs give high sensitivity and definition. However, high PRFs mean short pulse intervals and the possibility of multiple trace echoes. Therefore, on long ranges, where high power pulses are transmitted, PRFs are low, to give long interscan times, whereas, on short ranges, where the pulses are short and low powered, PRFs may be increased. *Table 12.8* shows the Racal–Decca 360/500 series (3 cm) as an example.

Table 12.8 Range, pulse length and pulse repetition frequency, Racal–Decca 360/500 series

Range scales (km)	Pulse length (μs)	PRF (pulses/s)
0.2–2.4	0.08	2400
4.8–38.4	0.3	1200
76.8–153.6	1.0	600

12.39.4 Target characteristics

The characteristics of a target which affect its reflecting properties are as follows.

12.39.4.1 Material

Since a reflected signal is in reality a reradiated signal, the conductivity of the target material is critical. Non-conducting materials, such as glass reinforced plastics, do not return echoes.

12.39.4.2 Shape

The aspect of any reflecting surface must be such as to direct energy towards the receiving antenna, adverse aspects producing divergent reflection. Plane surfaces at right angles to the transmitted signal patch will provide good echoes, but at any other angle are poor reflectors; large, slab sided, welded hulls of ships being a prime example. Sloping coastlines, sandbanks, hills and some icebergs may also be poor radar targets. Buoys and particularly spherical buoys, are also poor.

Corner reflectors, consisting of two or three plane surfaces mutually at right angles, occur naturally in topographic features on land and within ships' structures and return echoes in an advantageous direction. The movement of a target ship relative to the operating ship can cause aspect changes which result in the centre of reflection wandering along the length of the hull. This effect is known as glint or scintillation and can cause tracking problems, particularly if the target ship is at close range.

Radar reflectors, used on small targets such as buoys, and particularly on glass reinforced plastic pleasure boats, are built of an array of corner reflectors, designed to reflect from any angle of incidence (*Figure 12.56*).

12.39.4.3 Size

It does not necessarily follow that the larger the target the greater the echo strength. It is so horizontally, up to a size where the target is larger than the beamwidth. Since each echo is returned individually, only that part of the target which is illuminated by the beam can instantaneously reflect. As the beam rotates, however, other areas of the target will later return echoes.

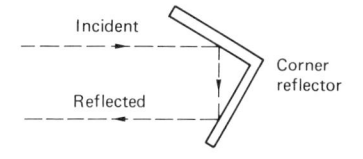

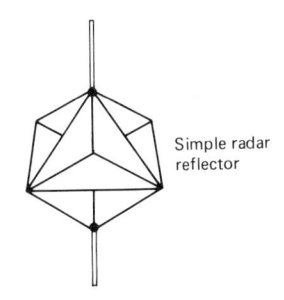

Figure 12.56 Corner reflectors

With respect to height of a target, if the reflecting surface is perpendicular, larger echoes may be returned with height. In the case of large objects such as hills or mountains, the reflecting surface is generally sloping and therefore only that area on which the pulse is incident can instantaneously return an echo. The slope itself may also reflect adversely.

In effect, large objects such as hills, mountains and coastline features return a *series* of echoes, separated in time and limited in extent to a few metres horizontally and vertically (*Figure 12.57*).

12.39.5 Frequency bands

Frequencies available for commercial marine radars are shown in *Table 12.9*.

Table 12.9 Frequencies for commercial marine radars

Band	Wavelength (cm)	Frequency (GHz)
L	30 – 15	1 – 2
S	15 – 3.50	2 – 4
C	7.50 – 3.75	4 – 8
X	3.75 – 2.50	8 – 12
Ku	2.50 – 1.67	12 – 18
K	1.67 – 1.11	18 – 27

A 3 cm radar operates in the X band between 9300 and 9500 MHz. This frequency is a compromise, taking into account range required, attenuation in atmospheric gases, penetration through precipitation, the ability to detect objects close to the sea surface, and good bearing and range discrimination.

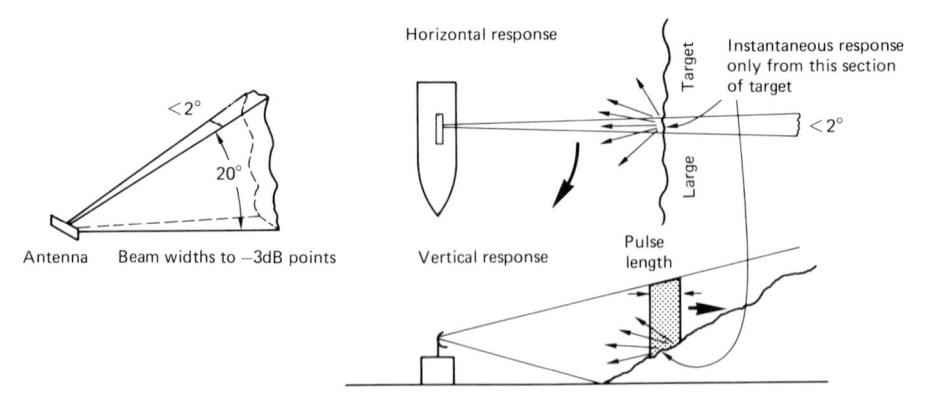

Figure 12.57 Large target response

A 10 cm radar operates in the S band. Using this frequency greater range can be obtained (5–10% more than X band). Penetration through precipitation, is better and less sea clutter is returned.

However, bearing discrimination is less and much larger antennae are required. For example, in the Racal–Decca 2690 BT series quoted in Section 12.37, using 3 cm a 2.7 m (9 ft) antenna will give a beam width of 0.8° whereas using 10 cm a 3.6 m (12 ft) antenna is only just within the minimum specification at 2° width (to −3 db points.) Therefore 10 cm picture definition and discrimination tends to be poorer than 3 cm.

However, with respect to collision avoidance, search and rescue and the use of radar beacons (racons) there are advantages in using 10 cm. The wider beam width and greater power (10 cm magnetron peak powers tend to be 20% higher than 3 cm) give larger and more powerful echoes on the display, more easily detected both visually and automatically and the slight reduction in definition can be tolerated.

The other penalty in choosing a 10 cm system is cost, currently around £8000 more than the comparative 3 cm radar. For a small ship radar, where an X band system may cost around £9000 and a comparative S band set £17 000, this differential may be prohibitive. For larger ships, using an integrated automatic radar plotting aid (ARPA), X and S band costs, respectively, may be £25 000 and £33 000.

For an initially more expensive vessel, this differential is more acceptable. Indeed, a number of ship owners will now specify twin interswitched 3 and 10 cm radars in the fitting out of new large vessels. Magnetron peak powers for such systems are usually around 25 kW for 3 cm systems and 30 kW for 10 cm.

12.39.6 Radar beacons – racons

Whereas radar reflectors are passive enhancers, racons can be termed active enhancers, since they effectively reflect an amplified and time extended return. The devices are transponders which, on reception of a ship's transmitted signal, respond on the same frequency or on a fixed outband frequency. This discrete, individual response depends upon no two ships within the racon area having both the same transmission frequency and the same pulse repetition frequency.

The general purpose of racons is to identify specific landmarks, or installations such as buoys, light vessels and rigs. Since the response occurs after the reception of the ship's signal, the display occurs on the far side of the true position of the structure carrying the racon, and may be Morse coded to increase identification (*Figure 12.58*).

12.39.6.1 Types of racon in use or under development

1. Swept frequency. The racon receiver sweeps over the radar band and responds, when it is tuned to a

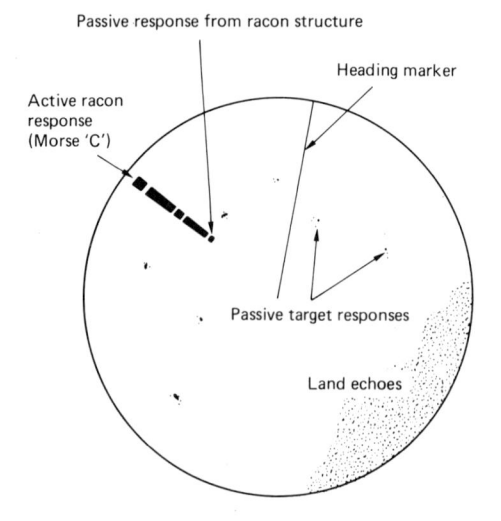

Figure 12.58 Racon response

ship's radar signal, at the same frequency as that received.
2. Frequency agile (interrogation frequency response). The racon receiver here computes the incoming interrogation signal frequency and responds on that frequency.
3. Fixed frequency. Reception by the racon is as the swept frequency type, but transmission is on a fixed frequency of 9310 ±10 MHz for 3 cm radar and 2910 ±10 MHz for 10 cm. Ship's radar must be modified to receive these frequencies.
   Since reception of these racon responses is at a different frequency to passive echoes, they can be displayed independently from them, i.e. racon response can be switched on or off as required, useful in areas where unwanted racon responses may clutter the display.
4. Offset frequency agile (offset interrogation frequency response).
   Similar to the frequency agile system above, but responds on a frequency which is different from that of the interrogating signal by a fixed amount. This again allows independent display of the racon response.
5. Interrogation Time Offset Frequency Agile Racon. The ITOFAR is based on the frequency agile systems, but with a fixed time delay in the response of the racon, which is allowed for in the ship's radar processing. This is currently under development and is likely to be adopted by the International Association of Lighthouse Authorities (IALA) and therefore by the International Maritime Organisation (IMO), as an international standard.

12.39.7 Identification of ships by transponder

Carry-aboard transponders have been used by pilots in ports around the world for a number of years to identify ships on harbour radar systems, in a similar manner to the military IFF systems. There is no reason why individual ships at sea could not identify each other by transponder signals, which could also be coded, quite simply, to carry additional data such as ship type, length, course, speed and ship's name.

However, there would be considerable commercial, not to mention military, opposition to this and such a system would not be efficient if all ships were not so fitted. International agreement to such an innovation would also take many years to come into force.

12.40 Collision avoidance interpretation of radar data

12.40.1 Relative motion

Using relative motion radar, either in head up or north up modes, the displayed motion of a target is relative to own ship. Analysis of this relative motion vector gives an easy indication of whether the relative motion track is dangerous, the miss distance, if not actually a collision situation, and the times at which these will occur.

By applying own ship's course and speed to the relative motion of the target, simple trigonometry reveals, firstly, the true course and speed of the target and, by manipulation, the alteration of course or speed of own ship which may be required to resolve the situation (*Figure 12.59*).

Manual plotting, as above, can be performed on paper plotting sheets or, more commonly by a reflection plotter, fitted over the screen itself and by using chinograph pencils. Note that the accuracy of relative plotting is not affected by currents or tidal streams, since both ships are assumed to be similarly influenced.

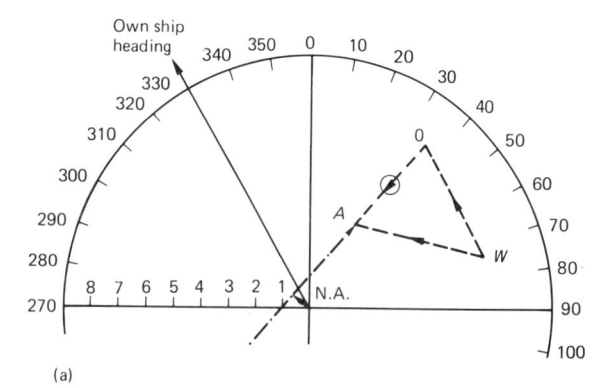

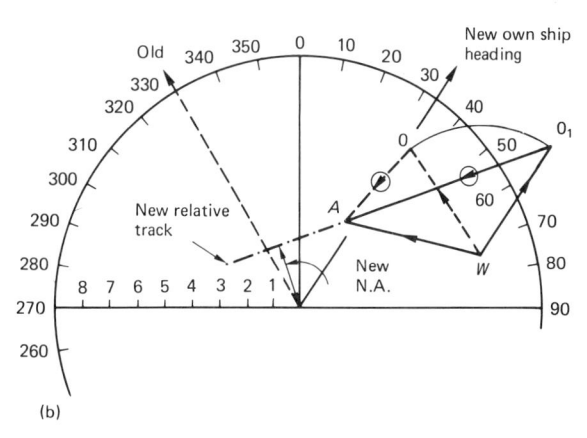

Figure 12.59 Relative motion plotting. In (a), from the plotted target ranges and bearings: 0A is the target's relative track over time 't'; NA is the projected nearest approach; WO is own ship's heading and distance run over time 't' (330°T). Therefore, WA is the target's true motion over time 't'. In (b), if own ship's projected alteration of course is to 033°T, then WO_1, with target's true track WA, gives a new relative track O, A, and an increased nearest approach, NA_1

12.40.2 True motion

Relative motion can appear confusing to an observer without experience in using it, and for navigational purposes in confined waters such as harbours, rivers and estuaries, has a limited value. True motion displays attempt to display the true motion of both own ship and targets. This is achieved by causing the origin of the time base to be slowly deflected across the display at a rate, and a direction, similar to the course and speed of the ship. This is done by additional time base coils deflecting north/south and east/west at very slow rates. For example, the time taken to traverse the time base origin of a ship travelling at 15 knots, across the display of 24 miles range diameter display, would be 1 h 36 min. At the instantaneous position of the origin at each transmission, the time base is drawn in a few tens of microseconds and is revolved around the origin, in synchronism with the scanner, in a period of under 3s.

In these circumstances motion of the origin, with respect to the instantaneous display and, effectively, to the rotation of the scanner, can be considered to be negligible (*Figure 12.60*). The display is said to be ground stabilized if, in addition to own ship inputs of course and speed through the water, allowance is also made for the set and rate of any current and tide. If no such allowances are made, the display is sea stabilized. These allowances are difficult to calculate exactly, being always based on past data or approximations. If the display is not correctly ground stabilized, courses and speeds obtained from it will not be true courses and speeds made good. Additionally, land and fixed object echoes will move on the display instead of remaining stationary.

Danger of collision, using true motion, is not as immediately obvious as in relative motion and the tracks of both own and target ships have to be extrapolated to obtain an indication of risk. True motion data can be plotted in the reverse manner of a relative motion plot, to give the targets' relative motion. Generally, relative motion would be used for open water collision avoidance, whilst true motion may be useful for close inshore work. All radars which have true motion mode are capable of being used in relative motion.

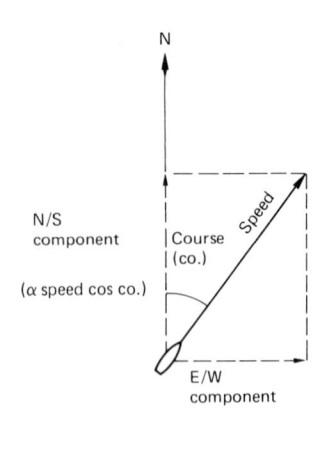

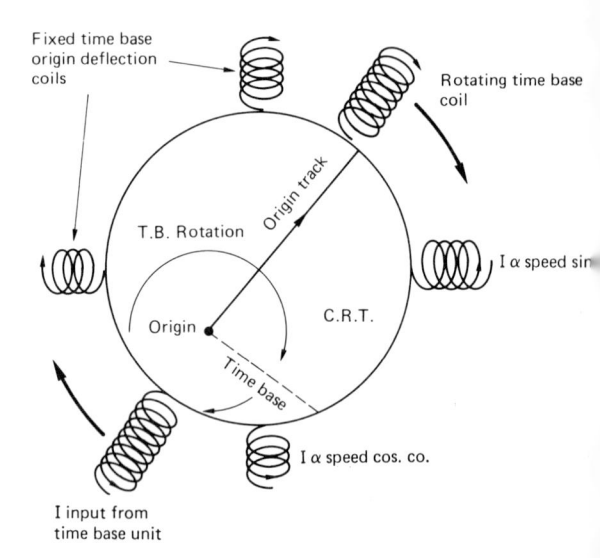

Figure 12.60 True motion display

12.41 Computer aided systems

As has been seen in previous sections, one rotation of the scanner and thus one rotation of the time base takes about 3 s. If the average pulse repetition frequency is 1000 pulses per second, then around 3000 time bases will be drawn during this rotation.

The position of an echo or of interference and clutter, is therefore expressed in polar form, that is, as a bearing and a range along about 3000 time bases. Bearings and ranges can be stored in a digital computer, after conversion from polar to Cartesian coordinates. Ranges are stored by means of banks of switch registers which are effectively a series of binary digit switches the length of which, in time, is made equal to the length of the time base. Each switch therefore corresponds to a discrete section of the time base range and any signal received within that particular range section will change the switch to the on position (*Figure 12.61*).

Initial clutter and interference elimination can be effected by having signal amplitude threshold levels, below which the switch register will not accept signals, but there are dangers that weak echoes from small targets such as fishing vessels and pleasure craft may be

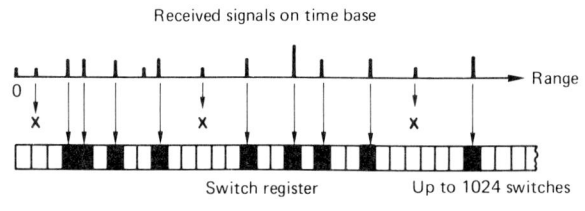

Figure 12.61 Switch registers. Signal amplitude below threshold level (x); switch on switch off (■); Signals above may be sea clutter, interference or targets, random signals disgarded by analysis of 'N' switch registers

discarded, particularly if such threshold levels are fixed values. Statistical elimination of interference, and clutter to a lesser extent, may be performed by comparing a series of shift registers, each one of which corresponds to a full, discrete, time base.

Interference and, to a lesser extent again, clutter will appear randomly on each time base. Therefore, by testing the number of occasions at which a signal is recorded at the same range on each of a series of registers and allowing for the fact that small targets will not echo on every time base, random signals may be discarded. Note that the length of the register is a fixed number of switches and that the range discretion of each switch depends upon the radar range required to display (e.g. with a switch register of, say, 256 switches: on the 3 miles radar range each switch corresponds to a range block of

$$\frac{3 \times 1852}{256} = 21.7 \text{ m } (71.2 \text{ ft})$$

Whereas on the 24 miles radar range, each block represents

$$\frac{24 \times 1852}{256} = 173.6 \text{ m } (569.6 \text{ ft})$$

Thus the accuracy of stored range data degrades with increasing range required to be displayed.

Bearings are stored by means of shaft encoders, which are similar to complex segmented commutators, built in a series of concentric rings, from which each discrete section corresponds to a particular arc of 360° rotation. Shaft encoders giving bearing resolution of less than one-tenth of a degree are in use and are synchronized with the rotation of the scanner. Further filtering of registered data can be performed, particularly in the case of clutter, by an analysis of successive antenna sweeps.

However, even with stored bearing of this accuracy, it should be noted that positional accuracy again degrades with range, since the length of the arc subtended at 3 miles range is 9 m (29.6 ft). But the length of the arc subtended at 24 miles range equals 70 m (229.6 ft) (*Figure 12.62*).

Due to this storage format, the actual smooth motion of a target at sea tends to be extracted from the computer store in step functions, range changing from switch to switch of the register and bearing from sector to sector of the shaft encoder.

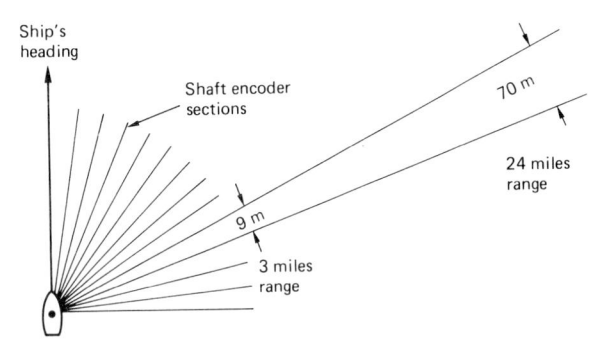

Figure 12.62 Shaft encounter sectors

12.41.1 Collision avoidance data

Stored range and bearing data, converted from Polar to Cartesian coordinates, can now be output to raster scan displays, i.e. using horizontally drawn lines. This has the considerable advantage over simple radar displays in that, since the picture is drawn from the computer store at short refresh intervals, afterglow problems do not exist, and daylight viewing displays can be used, with the inclusion of various colours to differentiate aspects of the displayed picture.

12.41.2 Automatic Radar Plotting Aid (ARPA)

More importantly, the stored target data can be processed by the system to give the same type of information as could be obtained by manual plotting, such as target course, speed, nearest approach, time of nearest approach and aspect. Trial manoeuvre facilities can allow the navigator to see what effect a specific alteration of course or speed would have in a particular situation. Such navigational data are provided in both analogue form, as relative or true motion displayed vectors and in alphanumerics listing the above parameters. In all ARPA systems, the data is extracted from the radar system at the amplifier stage, such that, should the computer fail, basic radar data is available for the navigator to fall back on.

12.41.2.1 Additional information

So far, we have discussed ARPA as an instrument which will perform the plotting functions of a human operator, but more rapidly and with the ability to handle multiple targets. The minimum number of manual selected targets which the ARPA must be able to track, process and display is ten (U.K.D.Tp. Marine Automatic Radar Plotting Aid (ARPA) Performance Specification (1981)).

Targets may, however, having been identified as such by the system, also be automatically processed; in which case, lacking the prioritization of the human operation, a minimum of 20 targets is required.

12.41.2.2 Operational warnings

These warnings, visual and audible, must warn the operator when a target is within a specific range from where it might become a threat, when it is forecast to pass within a chosen danger range with time as a variable and also if a tracked target is lost, this latter being particularly important in heavy sea clutter. Undue reliance should not be placed on such warnings, since no ARPA is 100% efficient at detecting or tracking targets, particularly small echoes at or near threshold levels.

12.41.2.3 Predicted areas of danger

Output navigational information so far discussed has been in a form similar to that which would be obtained by manual plotting. The computer system, however, has the ability to perform calculations not available to the human observer in the time available. For example, in the Sperry predicted areas of danger (PAD) system, those courses which it would be dangerous for the ship to steer are displayed.

They are derived by choosing a minimum miss distance for all traffic and calculating, for a particular target, those two courses of own ship which would pass ahead and astern of the target at this miss distance, respectively.

The two points on the targets' track at which own ship's ahead and astern courses cross form the two ends of a hexagon, the semimajor axis of which is the miss distance (*Figure 12.63*). This is an oversimplification, the PAD structure becoming complex with varying speed ratios and aspect angles, but it does illustrate the basic concept.

The main value of such a representation is that simply by looking at the display, the navigator can note whether the ship's heading line intersects any of the PADs. If it does not, then the ship will clear all targets by more than the selected minimum miss distance. If the heading line does intersect one or more target PADs, then a new course can be selected where intersection does not occur. This data is provided in addition to that alphanumeric information which is a

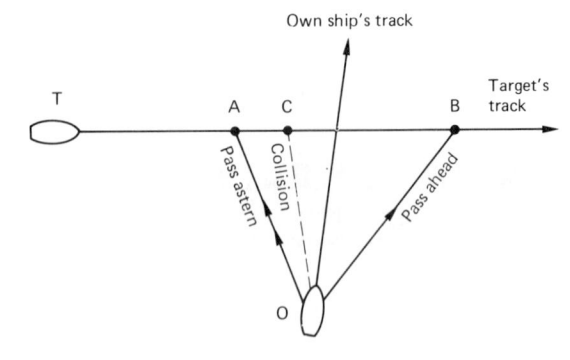

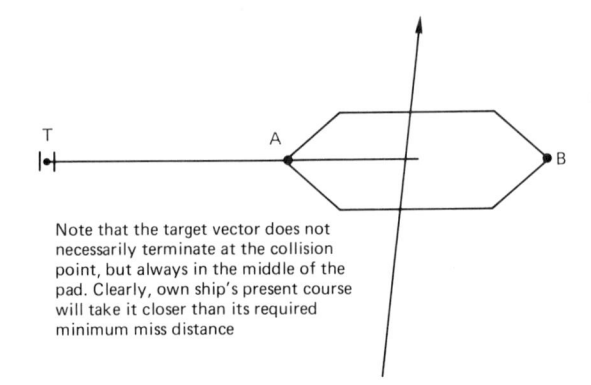

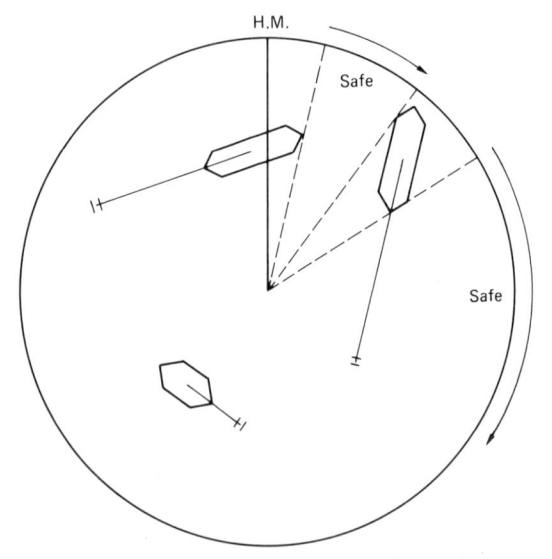

Figure 12.63 Predicted area of danger concept. Course to steer to pass astern of target at minimum required miss distance 'M' (OA); course to pass ahead of target at distance 'M' (OB); collision course (OC)

requirement for each target and is an aid to the navigator's decision making (Bole and Jones, 1981).

With respect to ergonomic considerations, Sperry Marine's latest system, the rascar (rasterscan collision avoidance radar) which incorporates the PAD concept, has no mechanical controls, all such knobs and switches being replaced by touch sensitive screen controls (*Figure 12.64*).

What is therefore to be the future automation of ship control? It is, of course, *technically* feasible to remove humans from the decision-making loop completely, but this is not *economically* viable at present.

12.42 Radar in vessel traffic systems (VTS)

Radar has been used ashore in the control of marine traffic since the first installation in Liverpool in the 1940s, and has developed from single, standalone units, to multiple site vessel traffic systems (VTS), in which the various radar sites are linked to the VTS control centre site by a variety of data links, including the use of telephone land lines, microwaves and VHF. In addition to permanent landmarks such as shorelines, buoyage and fixed hazards may be displayed.

Targets, automatically detected and tracked, are superimposed as symbols which represent their position course, speed and track history. Risk of collision, or other danger such as grounding, is continuously assessed by the system. Using the Sperry Marine system, which is based on the Rascar type unit, as an example, the operator has immediate access to all relevant data on targets by simply touching an area of the screen. Similarly, he or she may zoom the display into a specific area of interest, or obtain a split screen view of several areas of the port, simultaneously, from the different radar units (*Figure 12.65*).

Definition and accuracy of short based radars are inherently better than those of ship mounted systems. Firstly, they tend to be more expensive; secondly, because of the rigid mounting and available space they tend to have narrow vertical beamwidths (no rolling and pitching) and, with comparatively large antennae, narrow horizontal beamwidths.

Fixed site calibration is also good with known reference bearings, which do not have to rely on gyro compass outputs, as do ships' radar. Harbour radars also may be sited such that sea clutter does not pose a problem. Secondary displays will relate targets to other vessel data, including name, call sign, vessel type, country of registry, cargo, etc.

Figure 12.64 Sperry Marine Rascar. Reproduced by kind permission of Sperry Marine Inc.

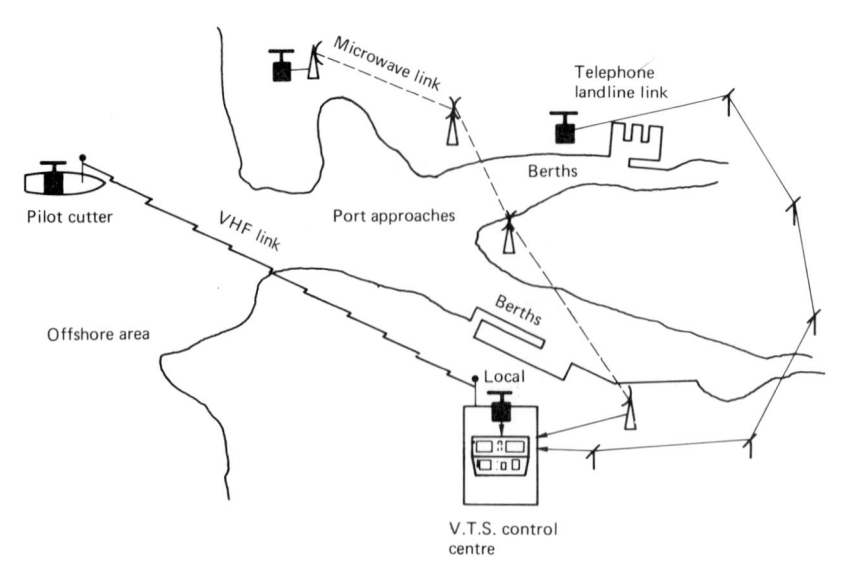

Figure 12.65 Vessel traffic system, control radars. Remote site radar picture converted to digital form for transmission to VTS control

12.42.1 VTS research

VTS systems of many different types exist in various ports throughout the world and are installed on a number of offshore structures. The Commission of the European Communities has just completed an investigation into shorebased marine navigation aid systems (COST 301, 1988) in which the requirements of VTS are examined, particularly with respect to the integration of the various European systems.

One requirement is a long range radar, described as over the horizon (OTH) radar. Latest research shows that the original high frequency radar (< 30 MHz), as used in the wartime chain home system, may provide the answer, since the surface wave (ground wave) component of the propagated signal extends well beyond the normal radar horizon. Such systems would be of great value in the general control and surveillance of traffic in large sea areas such as the North Sea, operating in a similar manner to the Channel navigation system, which uses shore radar at St Margarets Bay and Cap Gris Nez.

12.43 Electronic Chart Display Systems (ECDIS)

Returning to ships' radar once again, one ergonomic problem is that radar is only part of the navigation equipment fit. EPF, echo sounding, helm indicators, compass and radar all tend to be sited in various parts of the wheelhouse, while charts, books and other static data, such as tide tables, tend to be in a separate area.

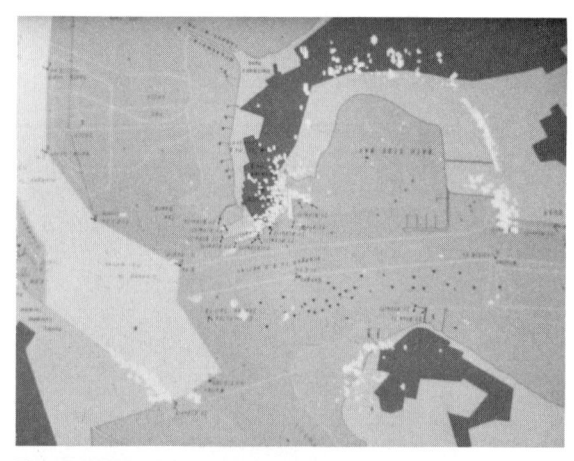

Figure 12.66 Leaving Harwich – using the Canadian hydrographic service Electronic Chart Testbed. Ship positioned by electron position fixing systems (Decca navigator in this case). Depth contours colour banded. Radar response overlaid (photograph by Author)

Above surface dangers are avoided by reference to radar, while below surface dangers are avoided by reference to paper charts.

On numerous occasions, ships have run aground while attempting to avoid collision, and *vice versa*. The problem is particularly acute at night, where lighting conditions in the wheelhouse and chart room will be different (Brown, 1989). VDU displays of digitized chart data, customized for a particular user and integrated with the radar display, may increase the overall efficiency of navigation, and a considerable amount of work has already been accomplished in this direction particularly by the Canadian Hydrographic Service. Most north west European countries are now involved in development projects (The North Sea Project, 1988).

Figure 12.66 shows the Norwegian Hydrographic Service vessel M/V Lance leaving Harwich, chart and radar information being shown on the Canadian electronic chart testbed.

Acknowledgements

Examples of typical manufacturers' specifications reproduced by kind permission of Racal Marine Electronics Ltd.

References and further reading

Admiralty List of Radio Signals, 5 (1988)
Appleyard, S. F. Marine *Electronic Navigation*, Routledge and Kegan Paul, London (1980)
Blanchard, W. F. Civil Satellite Navigation and Location systems, *Journal of Navigation*, 42, 2 (1989)
Bole, A. G. and Jones, K. D. *Automatic Radar Plotting Aids Manual*, Heinemann, London (1981)
Bomford, G. *Geodesy*, Clarendon Press, Oxford (1980)
Brown, C. C. *Developments in Short-Sea Navigation Systems*, Short Sea Europe Conference, 14–15 March 1989, London (1989)
Burnside, C. D. *Electromagnetic Distance Measurement*, 2nd edn, Crosby Lockwood Staples (1982)
Clay, C. S. and Medwin, H. *Acoustical Oceanography*, Wiley Interscience (1977)
COST 301 *Shore-based Marine Navigation Aid System*, Main Report Eur 11304 EN, Commission of the European Communities, Brussels (1988)
Cross, P. A., Hollwey, J. R. and Small, L. G. *Geodetic Appreciation*, Polytechnic of East London Working Paper No. 2 (1981)
Department of Transport *Marine Radar Performance Specification*, HMSO, London (1982)
Department of Transport *Marine Automatic Radar Plotting Aid (ARPA) Performance Specification*, HMSO, London (1981)
Dutton's Navigation and Plotting, 4th edn, Naval Institute Press
Hydrographic Society *Symposium on Depth Measurement and Sonar Sweeping*, S.P.5., Hydrographic Society (1978)
Hydrographic Society *Symposium on Position Fixing at Sea*, S.P.7., Hydrographic Society (1980)
Hydrographic Society *Evaluation of Echo Sounders*, BPA (1988)
Ingham, A. E. *Sea Surveying*, John Wiley (1975)
Laurila, S. H. *Electronic Surveying and Navigation*, John Wiley (1976)
Milne, P. H. *Under Water Acoustic Positioning Systems*, Spon (1983)
Mitson. Sonar in Fisheries, *Fishing News* (1985)
Molyneux-Berry, R. B. *Dr Henri Gutton, 1905–1984*. Seminar paper, Institution of Electrical and Electronic Engineers 10–12 June 1985, London (1985)
Ott, L. Starfix: Commercial Satellite Positioning, Advances in Underwater Technology, *Ocean Science and Offshore Engineering* 16 (1988)
Palandri, R. and Calamia, M. *The History of the Italian Radio Detector Telemetro*, seminar paper, Institution of Electrical and Electronic Engineers, 10–12 June 1985, London (1985)
Russell–Cargill, W. J. (ed.) *Recent Developments in Side Scan Sonar Techniques*, University of Capetown (1982)
Shearman, E. D. R. and Land, D. V. *The Beginnings of Centimetric Radar in the United Kingdom*, Seminar paper, Institution of Electrical and Electronic Engineers, 10–12 June 1985, London (1985)
Sonnenberg, G. J. *Radar and Electronic Navigation (6th edn)*, Butterworth, London (1988)
Stansell, T. A. The Transit Navigation Satellite System, Magnavox (1978)
Swords, S. S. *The Beginnings of Radar*, seminar paper, Institution of Electrical and Electronic Engineers, 10–12 June 1985, London (1985)
Swords, S. S. *Technical History of the Beginnings of Radar*, Peter Peregrinus, London (1986)
The North Sea Project; a test project for the electronic navigation chart, Norges Sjokartverk, Norway
Toft, H. *GPS Satellite Navigation*, Shipmate Marine Electronics (1985)
Urick, R. J. *Principles of Underwater Sound*, McGraw Hill (1983)
Urick, R. J. *Sound Underwater*, McGraw Hill (1984)
Wells, D. *et al. Guide to GPS Positioning*, Canadian GPS Associates (1986)
Wylie, F. J. (ed.) *The Use of Radar at Sea*, Royal Institute of Navigation, Hollis and Carter, London (1978)

13

Maritime Law

Robin R. Churchill

Contents

13.1 Introduction
    13.1.1 International law
    13.1.2 Sources of the international law of the sea
    13.1.3 Role of international organizations

13.2 Maritime zones
    13.2.1 Baselines
    13.2.2 Internal waters
    13.2.3 Territorial sea
    13.2.4 Archipelagic waters
    13.2.5 Contiguous zone
    13.2.6 Continental shelf
    13.2.7 Exclusive economic zone (EEZ)
    13.2.8 High seas
    13.2.9 International Seabed Area
    13.2.10 Boundaries between maritime zones

13.3 Offshore oil and gas exploitation
    13.3.1 Evolution of the continental shelf regime
    13.3.2 Legal definition of the continental shelf
    13.3.3 Coastal State's rights and duties on its continental shelf

13.4 Mining of manganese nodules in the international sea bed area

    13.4.1 Mining under the 1982 Convention
    13.4.2 Resolution II and the regime of preparatory investment protection
    13.4.3 Reciprocating States Regime
    13.4.4 Pollution from nodule mining

13.5 Shipping
    13.5.1 Nationality of ships
    13.5.2 Safety standards
    13.5.3 Traffic management
    13.5.4 Pollution

13.6 Fishing
    13.6.1 International fisheries law before the mid-1970s
    13.6.2 Fisheries under the 1982 Convention

13.7 Military uses of the sea
    13.7.1 Use of force generally
    13.7.2 Deployment of particular weapons and military equipment

13.8 Conclusions

References and further reading

13.1 Introduction

How far off its coast may a State license oil companies to explore for oil and gas beneath the sea bed? Can a State refuse a ship carrying a dangerous cargo, such as toxic waste or radioactive matter, admittance to its ports, or require it to keep a certain distance offshore? Are the two superpowers allowed to place submarine listening devices off each other's coasts?

These are the kinds of questions with which the international law of the sea, which forms the main subject of this chapter, is concerned.

Marine activities are regulated by a mixture of the international law of the sea and national law, i.e. the law of a particular State. The international law of the sea sets a general framework, but often, though not invariably, leaves matters of detail to be regulated by national laws, which obviously can vary widely from one State to another. An example may make this point clearer. In relation to offshore oil and gas exploitation, the international law of the sea prescribes the area within which any particular State can regulate such activities and allows it to permit drilling platforms to be erected, provided that such platforms do not cause excessive interference with other users of the sea. However, matters of detail, such as safety standards for such platforms or the amounts of tax to be levied on the oil companies, are governed by the national law of the State concerned, not the international law of the sea.

The totality of rules found in the international law of the sea and national laws governing marine activities is often referred to as marine law or maritime law. Although this chapter is entitled Maritime Law, because national laws vary so widely and because this book is not directed to readers in any one country, it concentrates on the international law of the sea component of maritime law. In particular, the chapter focuses on how the international law of the sea regulates those activities of particular interest to marine technology – exploitation of marine resources, shipping and military activities.

The international law of the sea is, as its name suggests, a branch of international law. Because laypeople usually have a very hazy idea of what international law is, it is best to begin by saying a few words about this.

13.1.1 International law

The layperson often thinks international law is something to do with the law of foreign countries or the law dealing with multinational companies. International law is neither of these things; simply stated, it is the legal system which governs the relations between different countries or States, to use the more correct legal terminology. Thus, while a national legal system, say that of France, regulates people and artificial legal persons such as companies in France, such people and companies being referred to as the *subjects* of the French legal system, so international law regulates States and international organizations composed of States (such as the United Nations (UN)), such States and organizations being referred to as the *subjects* of international law.

It will perhaps help to make these rather abstract statements more concrete by giving some examples of the kinds of matters with which international law is concerned. They include, outside the marine context: the drawing of frontiers between States, diplomatic relations, the framework of trade, economic and other forms of cooperation, air travel between different States, the settlement of disputes, the use of force and the functioning and establishment of international organizations, such as the UN and its specialized agencies, the European Community (EC) and other regional bodies.

Although a broad analogy has been drawn between international law and a national legal system, it is important to emphasize that there are radical differences between national legal systems, such as those of France and the UK, and international law. This is particularly the case when it comes to the making and application of legal rules. In a national legal system we are accustomed to finding a central legislative body, such as Parliament in the UK, Congress in the USA and the Assemblé Nationale in France. In international law, however, there is no such centralized legislature; even the UN does not fulfil this role. Instead legal rules are made either through custom or by agreements between two or more States, known as treaties or conventions.

Looking at the first of these sources of legal rules, a rule of custom is formed when there has been a period of fairly uniform and consistent practice by States in regard to a certain matter and States have acted thus because they felt legally obliged or permitted to do so. Custom used at one time to be the major source of international law, but nowadays is being increasingly replaced by treaties which provide the certainty, detail and speed which custom lacks. Treaties or conventions, the second source of international law, are agreements between States containing rules which are intended to be legally binding. They may be concluded by two States only (bilateral), or by States of one particular region, or by States on a general worldwide basis (multilateral).

Normally a treaty is drawn up at an international conference (or, in the case of a bilateral treaty, at a meeting between the two ministers or governments concerned). The treaty is then opened for signature. Signature does not normally imply being bound by the treaty; it usually only tokens broad agreement with its contents. To become bound by the treaty normally requires a further act, ratification (i.e. formal confirmation by the government of the State concerned, often after having obtained the approval of the legislature, that it will be bound by the treaty). Most multilateral treaties usually stipulate a minimum number of ratifications before they enter into force at all and then, of course, they are only in force for, and binding on, those States which have ratified them.

A second major difference between international law and a national legal system, following on from the difference in law-making, is when it comes to legal disputes. In a national legal system we take it for granted that two people who cannot settle a legal dispute such as an alleged breach of contract or a claim for compensation following an accident between themselves, can refer the matter to a court for a judge to decide. In international law the position is quite different. Although there is an international court (which sits in The Hague in the Netherlands), disputes which cannot be settled amicably between the States concerned cannot be referred to the international court unless all the States involved agree.

Another important difference relates to law enforcement. In a national legal system the criminal law is enforced by the police and offenders are dealt with by the courts. In international law there is no international policeman to deal with States which break the law. It was originally intended that the UN, particularly through the Security Council, should be able to take on such a role, but because of disagreements among the major powers, particularly the USA and the USSR, this has only been done in isolated incidences, e.g. in the Middle East after the Suez Crisis in 1956 and in the Congo in 1960. In the absence of an international law enforcement agency, States are compelled to rely, as best they can, on self-help.

The absence of a central legislature, compulsory recourse to courts and a law enforcement agency often leads non-lawyers to say that international law is not law at all. This is wrong. In fact, international law is a system of law, although a rather underdeveloped one. With its emphasis on consent and self-help, it has many of the features of the simple legal systems found in primitive communities. That international law is law is attested by the fact that the governments of all States regard themselves as bound by legal rules and seek to justify their actions, however apparently illegal, by reference to international law. Thus, for example, both the USA and USSR sought to justify their respective interventions in Grenada in 1983 and Afghanistan in 1979 as being in accordance with international law.

Finally, before leaving the topic of international law, we must consider the relationship between international law and national law, a matter which also bears on how international law relates to individuals. This relationship is determined, not by international law, but by each State's national law, which tends to adopt one or other of two basic approaches. In some States, such as France and the Netherlands, international law is automatically part of the law of the land, and may even override national law which is in conflict with it. Other States, notably the UK and the Scandinavian countries, take the opposite approach. International law is not automatically part of the law of the land; for international law rules to become part of the national legal system they must be transformed by adopting national legislation which incorporates such rules into the national legal system.

Having looked briefly at the nature of international law generally, we can now turn to examine in more detail that part of international law concerned with marine matters, the international law of the sea. We will begin by looking at the sources of this body of law.

13.1.2 Sources of the international law of the sea

As a branch of international law, the sources of the international law of the sea are obviously the same as the sources of international law generally, which were discussed above, namely custom and treaties. Before 1945 the law of the sea was nearly all found in custom. The basic features of this customary law were a narrow territorial sea (a term which will be explained in more detail shortly) and common, almost wholly unregulated, use of the high seas beyond (the doctrine of the freedom of the seas). At the time this very simple form of legal regulation was perfectly adequate, since the uses that were then made of the sea were few, chiefly navigation and fishing, there was no real conflict between them and they did not present any danger of over-utilizing the sea's resources.

Soon after World War II the UN decided that it would be useful to codify this customary law in the form of one or more treaties and at the same time to take into account some post-war technological developments, particularly the possibility of obtaining oil and gas from the sea bed and improved fishing techniques. Accordingly, therefore, the first United Nations Conference on the Law of the Sea was held in Geneva in 1958. The Conference, on the basis of drafts prepared by the International Law Commission (a UN body of legal experts), succeeded in drawing up four conventions, dealing, respectively, with the Territorial Sea and the Contiguous Zone, the High Seas, Fishing and the Conservation of Living Resources and the Continental Shelf. Each of these conventions has been in force since the early 1960s and has been ratified by between 36 and 57 of the world's approximately 170 States.

The Conference was not, however, completely successful. It failed to agree on the breadth of the territorial sea or on special fishing rights for coastal States beyond the territorial sea. A second conference was held two years later, also in Geneva, to reconsider these questions, and again, this time by the slimmest of margins, failed to reach agreement.

In spite of the major codification of the law which took place in 1958, strong feelings of dissatisfaction were being voiced by many States only a decade later. There were three basic grounds of complaint. First, there were differences over the limits of coastal States' rights. The failure of the Geneva Conferences to agree on a maximum limit for the territorial sea and on special fishing rights for coastal States had led to a proliferation of claims to all kinds of zones. Claims to the territorial sea varied in breadth from 3 miles (mainly the West European States, Japan and the USA), through 12 miles (some 55 States in different parts of the world) to 200 miles (mainly South American States). In addition to a territorial sea, many States claimed an exclusive fishing zone, ranging from 12 to 200 miles. Some States also claimed special zones for other purposes, including the prevention of smuggling and pollution control.

Second, the law was felt to be out of date or inadequate for dealing with technological developments. The latter had led to new uses of the sea, including the possibility of obtaining manganese nodules from the deep sea bed, fish farming, the erection of artificial islands and the use of the sea as a dustbin for polluting wastes. Existing uses of the sea had also changed in degree: ships had become larger and less manoeuvrable, while at the same time traffic in the busy waterways of the world had increased and more dangerous cargoes (oil, chemicals, liquid natural gas) were being carried. Fishing techniques had been developed, particularly by the Japanese and the Russians, for vacuuming whole areas of the sea. Increased and diversifying use of the sea meant, too, that there was likely to be increasing conflict between different uses, e.g. offshore drilling rigs might interfere with shipping or fishing nets catch on pipelines. In short, the laissez-faire concept which had underlain the development of the law right up to the Geneva Conferences of 1958 and 1960 needed to be replaced by planned marine management.

The third ground of complaint with the existing law of the sea came from the developing countries. Many of them had not been independent at the time of the Geneva Conferences and so had not been represented there. They felt that the existing law did not adequate meet either their needs or their aspirations.

These factors led the UN to decide in 1970 to hold a third conference on the law of the sea. The conference took place, over 11 sessions, between 1973 and 1982. It ended by adopting a treaty, the United Nations Convention on the Law of the Sea. The length of time taken by the conference is explained by the large number of participants (over 150 States, nearly twice as many as at the 1958 Conference); the scale and diversity of the subject matter, relating to almost all aspects of the sea; the widely differing interests of the participating States; the absence of a draft text at the beginning of the Conference (unlike in 1958) and the decision to proceed by way of consensus, searching for areas of agreement without formal votes.

The Convention has been signed by 159 States and other entities (including the EC). The major non-signatories are the USA, the UK and West Germany, all of which are opposed essentially only to those parts of the Convention dealing with deep sea mining. The Convention requires 60 ratifications to enter into force. As of May 1988 it had received 35 ratifications, nearly all from developing countries. It is thus likely to be several years before the Convention comes into force. When the Convention does come into force, it will supersede, for States parties to it, the 1958 Conventions.

The Convention is an enormous and complex document, running to nearly 200 pages and regulates virtually every use of the sea. In many cases, however, the Convention does no more than establish a framework of basic principles, leaving the elaboration of precise rules to other bodies, such as national governments and international organizations. Although the Convention is not in force, some parts reflect pre-existing customary law and repeat provisions from the 1958 Conventions. Other parts, notably the provisions dealing with the 200 mile exclusive economic zone, have already passed into customary law because of recent State practice. These parts of the Convention, therefore, are binding on all States now, even though the Convention has not come into force.

Although the only treaties so far mentioned are the four Geneva Conventions of 1958 and the United Nations Convention of 1982, these are by no means the only treaties concerned with the law of the sea. There is, in fact, a whole host of treaties dealing with detailed matters such as fisheries, pollution and maritime boundaries which complement the basic framework treaties of 1958 and 1982. Many of these treaties are referred to in this chapter.

From the above discussion it can be seen that the question of what law applies to any particular maritime activity is often a tricky one. The matter may be regulated by customary law whose content may be uncertain or disputed, by one of the 1958 Conventions, or by the 1982 Convention where it comes into force or before that, if the matter in question relates to a provision of the Convention which has already become part of customary law, or there may be some other, more specialized, treaty which is applicable.

Before turning to the substantive rules of the international law of the sea, a few words should be said about the role of international organizations in the formulation and application of those rules.

13.1.3 Role of international organizations

Although as mentioned earlier, there is no centralized legislature in international law, a number of international organizations have and do play a part in law making. The role of the UN in convening the three conferences on the law of the sea has already been mentioned. Of the UNs' specialized agencies, the International Maritime Organization (IMO) plays a particularly important role in this regard. The IMO, which is the specialized agency for shipping, has convened a number of conferences which have adopted over two dozen conventions on the safety of shipping, the regulation of navigation and pollution from ships (some of which are discussed in Section 13.5). Furthermore, the IMO Assembly, comprising the 130 or so State members of the organization, plays a quasi-legislative role in relation to these conventions, in that it can amend them and such amendments are binding unless a sufficient number of State parties to the convention in question object.

The other UN bodies which are particularly involved in marine matters are the Food and Agriculture Organization (FAO) and the United Nations Educational, Scientific and Cultural Organization (UNESCO), though neither of these bodies has a law making role. FAO reviews the state of the world's fisheries, establishes, where necessary, regional fisheries bodies to advise on fisheries management and provides assistance to many poorer countries in developing their fishing industries. UNESCO promotes and coordinates marine scientific research.

At the regional level there are various regional fisheries and pollution commissions which can adopt regulatory measures binding on their members and generally oversee the implementation of regional fisheries and pollution treaties. In a special category, because of the degree to which it can legislate on behalf of its members, is the EC, which has adopted its own body of fisheries law, a number of measures aimed at preventing pollution, and various measures concerning shipping. In addition it has become party to a number of treaties in its own right.

13.2 Maritime zones

Legally, the sea is divided up into various zones. The nearer a zone is to land, the greater are the rights of the adjacent State (known as the coastal State) to exploit the resources of the zone and regulate activities in the zone, and the fewer are the rights of other States. Correspondingly, the further from land the zone, the fewer are the rights of the coastal State and the greater the rights of other States. Beyond the zones subject to the varying jurisdiction of the coastal State are the high seas, which are in principle open to equal use by all States. Historically, the zones subject to the jurisdiction of the coastal State were limited in number and narrow in breadth. With the greater use of the sea and the increasing exploitation of its resources, most of which are found relatively close to land, has come, over the past 50 years or so, a steady trend towards coastal States claiming zones for an increasing variety of functions and of increasing width. Most of these claims have now been sanctioned by the United Nations Convention on the Law of the Sea.

In this section, the various zones into which the sea is divided will be examined. In reading what follows, it may be found useful to refer to Figure 13.1.

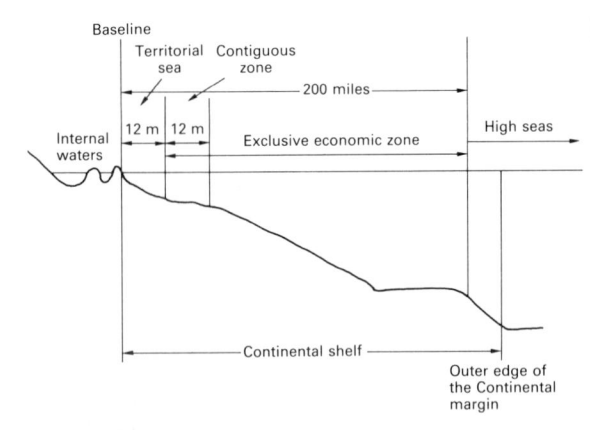

Figure 13.1 Maritime zones

13.2.1 Baselines

The breadth of maritime zones is usually defined as being so many nautical miles, measured from the baseline. Thus, before considering any of the various zones it is useful to know what is meant by a baseline.

Normally the baseline is the low-water line along a stretch of coast. But there are many exceptions to this rule. The two most important exceptions concern straight baselines and bays. Where a coast is deeply indented and/or fringed with islands, as for example much of the coast of Norway, Chile or the west of Scotland, then instead of using the low-water mark as the baseline, a State may draw straight baselines connecting the outermost points on the coast and/or islands, provided such lines follow the general direction of the coast.

The second important exception concerns bays. Where the mouth of a bay is less than 24 nautical miles in width, a line may be drawn across it, and this line then serves as the baseline instead of the low-water mark around the shore of the bay.

There are a number of other less important exceptions, but it is beyond the scope of this chapter to give details of them.

13.2.2 Internal waters

Internal waters are those waters lying on the landward side of the baseline. Thus, they include bays, fjords, ports, rivers and canals. Internal waters are part of the coastal State's territory and so other States have no general right of access to such waters, let alone to exploit their resources. Nevertheless, States usually give the ships of other States access to their ports by entering into bilateral treaties on the matter. Likewise, there are treaties giving foreign ships the right to use the more important navigable rivers (e.g. the Rhine and the Danube) and the principal interoceanic canals (the Kiel, Panama and Suez Canals).

Where a foreign ship is in a coastal State's internal waters, it is in principle subject to the full jurisdiction of that State, i.e. that State can extend its laws, e.g. on pollution control, to a foreign ship and enforce them. By convention, however, coastal States refrain from exercising their jurisdiction in relation to internal matters on board the ship (e.g. petty criminal offences committed by one crew member against another).

13.2.3 Territorial sea

The territorial sea is a belt of waters extending seawards from the baseline which is subject to the sovereignty of the coastal State. Until recently the breadth of the territorial sea was a controversial matter. As we have seen, neither the first (1958) nor the second (1960) United Nations Conferences on the Law of the Sea were able to reach agreement on this question. The third conference, however, was more successful. The 1982 Convention now provides that the maximum breadth of the territorial sea is to be 12 nautical miles.

Although the Convention is, as noted above, not yet in force, this provision now almost certainly represents customary law. Of the world's 139 coastal States, 99 now claim a 12 mile territorial sea and a further 17 some lesser breadth. This, it is true, still leaves 23 States (mainly in Latin America and Africa) which claim a territorial sea of more than 12 miles. A number of these claims, however, resemble more an exclusive economic zone than a territorial sea as traditionally understood and defined in the Convention. Furthermore, it is likely that a number of these States will abandon their claims as and when they ratify the Convention and/or when it comes into force.

Because the territorial sea is subject to the coastal State's sovereignty, other States' rights in this zone are fairly restricted. However, unlike internal waters, foreign ships do enjoy a general right of access to this zone. Both custom and the 1958 and 1982 Conventions confer what is known as a right of innocent passage on foreign ships. This is the right to pass through the territorial sea provided the ship does not act in a manner which is prejudicial to the peace, good order or security of the coastal State, and the 1982 Convention lists certain activities (e.g. weapons practice, serious pollution, fishing, research or survey activities) which are deemed to be prejudicial in this way.

A ship which is exercising the right of innocent passage remains subject to the jurisdiction of the coastal State. Thus, it must comply with the coastal State's regulations relating to such matters as navigation routes and pollution prevention, although the coastal State's laws (according to the 1982 Convention) must not affect the design, construction, manning or equipment of foreign ships.

The right of innocent passage is the only right enjoyed by foreign States in the territorial sea. Thus the exploitation of sea bed resources, fishing and the carrying out of scientific research remain within the exclusive province of the coastal State.

The adoption of a 12 mile territorial sea means that some 116 straits less than 24 miles in width now consist entirely of territorial sea. These include many straits of great strategic and commercial significance, such as the Straits of Dover, Gibraltar, Bab-el-Mandeb, Hormuz, Malacca and the Baltic. Under the pre-existing law there would be only a right of innocent passage for foreign ships through such straits. This gave rise to anxieties among the main military and shipping powers at the third UN Conference. They were worried in particular by three factors: the possibility of States bordering such straits controlling traffic through them, the fact that submarines exercising a right of innocent passage are required to navigate on the surface and that the right of innocent passages does not extend to aircraft flying over the territorial sea.

To accommodate these concerns the 1982 Convention provides for a right of transit passage through straits of the type described above. Transit passage is the exercise of the freedom of navigation solely for the continuous and expeditious transit of a strait. This right allows submarines to transit while submerged and extends to aircraft. The coastal State's jurisdictional competence is also more restricted than in the territorial sea generally.

Whether the Convention's provisions on transit passage have already passed into customary law is a controversial matter. Not surprisingly, those who stand to gain most from such a development, such as the USA, have argued most vehemently that the right of transit passage is now a rule of customary law, but the evidence to support such an assertion is fairly limited. What should be noted, however, is that some straits are subject to special treaties, predating the 1982 Convention, which give foreign ships a right akin to that of transit passage. The most notable of these treaties is the Montreux Convention of 1936, which regulates passage through the Dardanelles and the Bosphorus.

13.2.4 Archipelagic waters

Some States, particularly in the Caribbean, Pacific and Indian Ocean, such as Indonesia, the Philippines and Fiji, consist of one or more archipelagos. Before the third UN Conference there was no special provision in international law for such archipelagic States. The 1982 Convention, however, contains a special regime for archipelagos.

The main features of this regime are as follows: an archipelagic State may now draw archipelagic baselines joining the outermost points of the archipelago. The territorial sea and other maritime zones are measured seawards from such zones. The waters enclosed within archipelagic baselines are known as archipelagic waters. These waters are subject to the sovereignty of the archipelagic State, but foreign vessels enjoy a right of innocent passage through them.

Furthermore, the archipelagic State must designate sea lanes and air corridors through the archipelago within which foreign ships and aircraft enjoy a right of archipelagic sea lanes passage, a right which is to all intents and purposes the same as the right of transit passage through straits.

Notwithstanding the fact that the 1982 Convention is not yet in force, about 14 archipelagic States have adopted legislation embodying the Convention's provisions. Given the acquiescence of other States in this legislation, the convention's regime for archipelagos would seem well on the way to becoming customary law.

13.2.5 Contiguous zone

The contiguous zone was originally the creation of custom, but is now governed by the 1958 and 1982 Conventions. The zone, which is adjacent to and beyond the territorial sea, is basically a law-enforcement one, where the coastal State may exercise the control over foreign ships necessary to prevent infringement of its customs, fiscal, immigration or sanitary regulations within its territory or territorial sea, or to punish such infringements. The 1958 Convention set a maximum breadth to the zone of 12 miles, but with agreement reached on a 12 mile territorial sea, the 1982 Convention has extended the maximum breadth of the contiguous zone to 24 miles

The contiguous zone is an optional one. States are not obliged to claim such a zone and in practice only about a quarter of all coastal States have exercised this option. Most States that do claim such a zone, however, claim the 24 miles provided for by the 1982 Convention.

13.2.6 Continental shelf

The development of the legal concept of the continental shelf and its legal nature are considered in detail in Section 13.3. The continental shelf is the area of sea bed beyond the territorial sea (its precise outer limit being left for discussion in Section 13.3), on which the coastal state enjoys sovereign rights for the purpose of exploring and exploiting its natural resources.

13.2.7 Exclusive Economic Zone (EEZ)

The EEZ is a relatively recent concept in the law of the sea, being the creation of the third UN Conference. The zone is a reflection of the aspiration of developing countries to gain more control over the economic resources off their coasts, particularly fish stocks, which in the past have largely been exploited by the distant water fleets of developed States. At the same time the zone is something of a compromise between those States that claimed territorial seas extending beyond 12 miles (some Latin American and African States) and those developed States, such as the USA, Japan and most western European States, which favoured narrow coastal State limits in order to protect their distant water fishing and navigation interests.

The 1982 Convention provides that the EEZ may extend up to a maximum limit of 200 nautical miles measured from the baselines. Within the EEZ the coastal State enjoys various rights, all essentially connected with the exploitation of the zone's natural resources, while other States have rights related to transit and communications. The easiest way to understand the nature of the EEZ is to elaborate the various rights accorded by the Convention to the coastal and other States.

The Convention confers essentially six rights on the coastal State. First, the coastal State has the right to exploit the non-living resources of the EEZ, such as oil and gas, sand and gravel. Second, the coastal State has the right to exploit the living resources of the EEZ. This right is, however, subject to various management duties which are explained in more detail in Section 13.6. Third, the coastal State has the right to exploit the EEZ for other economic purposes, such as the production of energy from the waves, wind or currents. Fourth, the coastal State has the right to authorize and regulate the construction and operation of artificial islands, installations and structures, such as drilling rigs for oil and gas exploitation or wave barrages. More is said about this right in Section 13.3.3.1. Fifth, the coastal State has the right to regulate scientific research in its EEZ. It must normally give its consent to pure research by other States in its EEZ, but it may withhold its consent to resource orientated research.

Finally, the coastal State has various rights to control pollution in its EEZ. In particular, it can regulate pollution from offshore installations, the deliberate dumping of waste from ships and aircraft and, within certain limits, other forms of pollution from ships.

Turning now to the rights of States other than the coastal State, such States enjoy the freedoms of navigation, overflight by aircraft and the laying of submarine cables and pipelines within the EEZ.

Although the 1982 Convention is not yet in force, such was the level of agreement on the EEZ at the UN Conference that since 1977 there has been a wave of unilateral claims to an EEZ. At the present time some 75 States claim an EEZ. A further 20 or so States, acting on the same inspiration but not yet wishing for various reasons to claim a full EEZ, have claimed a 200 mile fishing zone. The main States that have not claimed a 200 mile zone of some kind are those that border semi-enclosed seas such as the Baltic, Mediterranean and Persian Gulf, where for obvious geographical reasons it is not possible for zones of any great width to be claimed. Such has been the level of practice and the lack of protest to it that the International Court of Justice in 1985 had no hesitation in declaring that the EEZ was now part of customary international law.

The significance of the EEZ is that, together with the territorial sea, it covers most of the world's commercial fisheries and known submarine hydrocarbon deposits, those areas where most marine scientific research takes place and many of the world's major shipping routes.

Thus, the way in which coastal States exercise their new found rights and responsibilities in the EEZ will largely determine how well the marine environment and marine resources will be managed in the future.

At the same time a related point must also be made. In spite of the rhetoric of the developing countries at the third UN Conference, few of them are among the principal beneficiaries of the zone. Those States which have gained most from the EEZ are those States that front the great oceans of the world and, as a quick glance at an atlas will show, many such States are developed. Of the 15 leading beneficiaries, in order of magnitude of area gained, only seven are developing countries: Indonesia, Brazil, Mexico, Kiribati, Papua New Guinea, Chile and India.

13.2.8 High seas

The high seas comprise the waters lying beyond the coastal State's various zones. Before 1945 this meant that the high seas were the waters beyond the territorial sea. Even after the development of the legal concept of the continental shelf in the years following 1945, this remained the position, because the continental shelf does not affect the status of the waters lying over it (see Section 13.3.3). However, with the emergence of the 200 mile EEZ as a customary rule of international law and the widespread claiming of such zones, the high seas now begin at the 200 mile limit, at least in those areas where a 200 mile EEZ is claimed by the local coastal States.

The basic principle governing the high seas is that they are open to use by all States. This principle is usually referred to as the freedom of the high seas. The freedom includes, but is not limited to, the freedoms of navigation, overflight, fishing, research and the laying of submarine cables and pipelines. In exercising its freedom of the high seas, each State is required to have reasonable regard to the interests of other States in their exercise of the freedom of the high seas. It follows from the principle of freedom of the high seas that no State may validly purport to subject any part of those seas to its sovereignty. The broad rules described in this paragraph are the creation of customary law and are codified in both the 1958 Convention on the High Seas and the 1982 Convention.

The open access, non-territorial nature of the high seas has important consequences as far as the regulation of high seas activities is concerned. Under both custom and the conventions a State may prescribe and enforce legal rules on the high seas only in respect of ships flying its flag. This is the principle of flag State jurisdiction. To this principle there are a number of exceptions, the most important of which are hot pursuit, piracy and, under the 1982 Convention, pirate broadcasting. Further exceptions are also sometimes created by treaty, e.g. in 1981 the UK and USA concluded an agreement which allows US ships to board, search and seize British ships suspected of drug trafficking in certain areas in the Caribbean, Gulf of Mexico and off the eastern coast of the USA.

It follows from the principle of flag State jurisdiction that activities on the high seas will be regulated primarily by individual States, applying such laws as they deem appropriate to their ships. Thus, if it is desired to take action in respect of some common interest on the high seas, such as the management of certain fish stocks or the prevention of pollution, interested States must get together and agree on the necessary measures. Each State will then have to apply such measures to its ships on the high seas and ensure its ships observe those measures. Other States will not normally, unless special provision is made for this in the agreement, be able to take action in this regard.

13.2.9 International Seabed Area

Traditionally, the high seas and its associated freedoms have been regarded as encompassing not only the water column but also the air space above and the sea bed below. As far as the sea bed is concerned, some modification must now be made to this traditional position.

First, in areas where the legal continental shelf extends beyond 200 miles, or in areas where no 200 mile EEZ is claimed, the sea bed comprising (in legal terms) the continental shelf, even though lying beneath the high seas, is subject not to the regime of the high seas, but to that of the continental shelf. The latter regime is discussed in detail in Section 13.3.

Second, the sea bed beyond the continental shelf, even though lying beneath the high seas, will, when the 1982 Convention enters into force, become the International Seabed Area, and subject to the regime governing that area. This regime is described in detail in Section 13.4.

13.2.10 Boundaries between maritime zones

Where States adjoin one another facing the open sea, such as Peru and Chile on the Pacific coast of South America or the Netherlands and West Germany in the North Sea, it will be necessary to draw a boundary between their various maritime zones – the territorial sea, contiguous zone if claimed, continental shelf and 200 mile EEZ or fishing zone if claimed. Similarly, it will be necessary to draw a territorial sea boundary between States whose coasts are opposite one another and less than 24 miles apart (e.g. Malaysia and Indonesia in the Straits of Malacca) and a continental shelf and EEZ–fishing zone boundary between States whose coasts are more than 24 but less than 400 miles apart. The international law governing the drawing of such boundaries is currently one of the least clear areas of the law of the sea.

In trying to explain this area of law, it is perhaps best to start with the Conventions. As far as the boundary between neighbouring States' territorial seas is concerned, the 1958 Convention on the Territorial Sea provides that the boundary is to be the median line (i.e.

a line equidistant between the coasts of the States concerned), unless historic title or other special circumstances dictate a different line. A similar provision is found in the 1982 Convention.

Continental shelf boundaries are dealt with by Article 6 of the 1958 Continental Shelf Convention. This provides that the boundary between neighbouring States' continental shelves is to be the median line unless another boundary line is justified by special circumstances. These are not defined in the Convention, but are generally taken to include such factors as small offshore islands or particular configurations of the coast which distort the median line to such an extent that it no longer appears a reasonable or fair boundary.

The customary rules on maritime boundary delimitation have been elaborated in some half dozen cases heard before the International Court of Justice or arbitration tribunals since 1969. These courts have rejected the median line as being the starting point or being the predominant method in maritime boundary drawing. Instead, these bodies have said that delimitation is to be effected in accordance with equitable principles and taking account of all the relevant circumstances. It has not been made clear, however, what exactly equitable principles comprise or what circumstances are relevant.

The more recent cases have laid stress on the delimitation producing an equitable result and whereas the earlier cases' doctrine of equitable principles applied only to continental shelf delimitation, the later cases suggest that the same principles apply to all forms of boundary delimitation. The idea also seems to be emerging in the later cases that the only circumstances relevant to delimitation are geographical ones, such as the configuration of the coast and the presence of islands.

A number of principles also appear to be becoming recognized as essential to producing an equitable result. These include a reasonable degree of proportionality between the lengths of the respective coastlines of the States concerned and the areas of maritime zone awarded to each as a result of the delimitation; giving offshore islands less than full effect in the generation of maritime zones, and use of the median line as the starting point in delimitations between opposite, but not adjacent, States. It must be stressed, however, that international courts and tribunals emphasize that the circumstances of each delimitation are unique, so that the solution found in one case cannot serve as a precedent for other disputed boundaries, let alone a general rule on delimitation.

When the third UN Conference came to discuss continental shelf and EEZ boundary delimitation, a strong division emerged between supporters of the median line, basing themselves on the 1958 Convention, and proponents of equitable principles (inspired by the earlier cases referred to above). A compromise between these points of view was eventually found and put into the Convention, but, more than most compromises, seems to amount in practice to very little of substance. The 1982 Convention provides that the delimitation of the continental shelf and EEZ between neighbouring States is to 'be effected on the basis of international law . . . in order to achieve an equitable solution'.

Of the approximately 400 potential maritime boundaries in the world, about one-quarter have so far been agreed, eight as the result of a decision of an international court or arbitral tribunal, the remainder by treaty. This leaves about 300 boundaries still to be settled, a number of which are currently the subject of serious dispute. Most of these will no doubt in the course of time be settled by agreement between the States concerned, while a few will probably be referred to international courts for adjudication.

Having looked at the different zones into which the sea is divided, we will now turn to examine the law relating to those uses of the sea to which maritime technology is most relevant, beginning with offshore oil and gas exploitation.

13.3 Offshore oil and gas exploitation

It follows from what was said in Section 13.2.3 that any oil or gas found in the bed of the territorial sea belongs to the coastal State. This State has the exclusive right to engage in and regulate the exploitation of such resources. The main obligation imposed by international law on the coastal State is that the latter must ensure that oil and gas exploitation does not hamper the innocent passage of foreign ships through its territorial sea. The other international law obligations relate to pollution and are discussed in Section 13.3.3.4.

Beyond the territorial sea the regulation of offshore oil and gas exploitation falls under the legal regime of the continental shelf. It is this regime that is the main topic of this section. We will begin by looking at the evolution of the continental shelf regime, then examine the legal definition of the continental shelf and finally consider the rights and duties of the coastal State in respect of its continental shelf.

13.3.1 Evolution of the continental shelf regime

The first drilling for oil and gas from the sea bed began in the USA shortly before World War II. Although this was only in shallow water and close to shore, its potential led President Truman in 1945 to issue a proclamation in which he claimed all the hydrocarbon resources of the continental shelf adjoining the US coast and beyond the territorial sea as the exclusive property of the USA. This proclamation was followed by similar claims made by many other States. The authors of these claims justified them on the grounds that the close relationship of the continental shelf to the land made the coastal State the obvious entity to exploit them; furthermore considerations of military security required exclusive coastal State control.

These claims received formal international recognition with the adoption at the first UN Conference in 1958 of the Convention on the Continental Shelf. This Convention contains a fairly comprehensive framework of rules governing the nature and extent of States' rights on the continental shelf. Apart from the definition of the continental shelf, these rules are reproduced largely unchanged in the 1982 Convention. In addition, many of these rules, particularly those concerning the entitlement of the coastal State to the natural resources of its continental shelf, are found in customary law.

13.3.2 Legal definition of the continental shelf

The first and most important point to realize is that international law defines the term continental shelf in a radically different way from geologists and geomorphologists, whose use of the term was explained in Chapter 1.

In seeking to define the continental shelf for legal purposes, the drafters of the 1958 Convention began, naturally enough, with its geomorphological concept. However, since the continental shelf varies in different parts of the world in both depth and distance from the shore, being for example almost non-existent off the Pacific coasts of South America while extending for several hundred miles off Australia and Argentina, it proved impossible to arrive at a general definition based on the continental shelf's natural features.

Instead, after much discussion a compromise was reached, and the continental shelf was defined for legal purposes as being the sea bed and subsoil of the submarine area adjacent to the coast but outside the area of the territorial sea, to a depth of 200 m or, beyond that limit, to where the depth of the superjacent waters admits of the exploitation of the natural resources of the said areas.

In 1958 it was not technically possible to drill in more than 200 m of water, but since then drilling has taken place in much greater depths. The developing countries feared, therefore, that by defining the continental shelf not only in terms of the depth of superjacent waters but also in terms of exploitability, the 1958 Convention had introduced a seemingly open ended definition to the continental shelf, which would allow the technologically advanced countries to extend their continental shelves seaward until they met in the middle of the ocean. The desire to prevent this happening, by establishing a precise and unambiguous outer limit to the continental shelf, was one of the reasons for convening the third United Nations Conference on the Law of the Sea.

When the proposal for a 200 mile EEZ became generally acceptable at the conference, it seemed that this would solve the problem, by establishing 200 miles as the outer limit for both the EEZ and continental shelf. However, a sufficiently large and important group of States, including Argentina, Australia, Canada, India, Norway, the UK and the USA, had continental margins stretching beyond 200 miles. (The continental margin includes not only the continental shelf, but also the continental slope and rise which connect the continental shelf to the deep sea bed (see Chapter 1) and which are also thought to contain extensive hydrocarbon resources). These States insisted that they should also be entitled to the resources of this outer part of the continental margin. This was eventually agreed to.

As a result the 1982 Convention defines the outer limit of the continental shelf as being either 200 miles from the baseline or the outer edge of the continental margin, whichever is the further (Article 76). Where, under this definition, the outer limit of the shelf is the outer edge of the continental margin (rather than 200 miles), the Convention seeks further to define and confine this limit. Thus the Convention provides that for legal purposes the outer edge of the continental margin is either a line connecting points not more than 60 miles apart, at each of which points the thickness of sedimentary rocks is at least 1% of the shortest distance from such point to the foot of the continental slope, or a line connecting points not more than 60 miles apart, which points are not more than 60 miles from the foot of the slope. In each case the points referred to are subject to a maximum seaward extent; they must be either within 350 miles of the baseline or within 100 miles of the 2500 m isobath. Clearly, this is a complex formula which will be difficult to apply in practice.

It is therefore to be welcomed that the Convention establishes a Commission on the Limits of the Continental Shelf, consisting of 21 experts in geology, geophysics and hydrography, to assist States in delimiting the outer limit of their continental shelves where these extend beyond 200 miles.

In return for meeting the demands of States with broad continental shelves/margins, the UN Conference inserted a provision in the 1982 Convention requiring States whose continental shelves extend beyond 200 miles to pay to the International Seabed Authority, discussed in Section 13.4.1., a proportion of the proceeds from any sea bed resources exploited beyond the 200 mile limit. Such proceeds would then be distributed among developing countries. Currently, little if any exploitation of sea bed resources takes place beyond the 200 mile limit, so it remains to be seen whether any sizable aid will be generated for developing countries in this way.

Although the 1982 Convention is not yet in force, its provisions on the outer limit of the continental shelf have a growing importance in State practice. Currently, some 22 States define the outer limit of their continental shelf as being either 200 miles or the outer edge of the continental margin (though without specifying which); while a further three States define the outer limit as being the edge of the continental margin, 350 miles and 100 miles beyond the 2500 m isobath, respectively.

Under the Continental Shelf Convention of 1958 all islands, regardless of size, are entitled to a continental shelf (in the legal sense). The 1982 Convention modifies this position, by providing in Article 121(3) that islands in the shape of rocks which cannot sustain human habitation or economic life of their own are not entitled to a continental shelf (or EEZ).

The aim of this provision is to overcome some of the inequities which might be thought to result from the ability of small, isolated islands to generate possibly extensive continental shelves. The provision, however, is not well drafted. It does not define what a rock is, nor suggest how one may distinguish between rocks and other islands. In addition, the question of whether any particular rock can sustain human habitation or economic life is one that may admit of more than one answer because of the vagueness of the phrases used. In practice, most uninhabitable rocks lie immediately offshore and thus have little effect on the claiming of continental shelves, whether Article 121(3) applies or not.

Thus, the practical application of the provision, once the 1982 Convention comes into force (it seems unlikely that Article 121(3) has become part of customary law because of the dearth of State practice), will be largely confined to the few very small isolated islands that exist, such as Rockall in the North Atlantic. The UK, which owns this islet of 624 m^2, lying 290 miles west of the Scottish mainland, has claimed both a continental shelf and a 200 mile fishing zone around it, a claim which is currently disputed by Iceland, Ireland and Denmark (on behalf of the Faroes).

Before leaving the question of the definition and extent of the continental shelf, it should be pointed out that most States will have to determine the boundary between their continental shelf and the continental shelves of neighbouring States. In Section 13.2.10 some indication was given of the rules of international law relating to the drawing of such boundaries.

13.3.3 Coastal State's rights and duties on its continental shelf

The rights a coastal State has on its continental shelf are defined, in both the 1958 and 1982 Conventions, as being sovereign rights for the purpose of exploring it and exploiting its natural resources. In other words, the coastal State's rights are limited to the exploitation of natural resources; the continental shelf is not part of its territory. These rights of the coastal State are exclusive in the sense that if it does not explore the continental shelf or exploit its natural resources, no one else may undertake such activities without the express consent of the coastal State.

The Conventions define the natural resources over which the coastal State has sovereign rights as being the mineral and other non-living resources of the sea bed and subsoil, together with living organisms belonging to sedentary species, that is to say, organisms which at the havestable stage, either are immobile on or under the sea bed or are unable to move except in constant physical contact with the sea bed or subsoil. Thus, included in this definition are obviously oil and gas, as well as sand and gravel, tin and other minerals mined from the sea bed. Sedentary species include oysters, clams and abalone. More doubtful are crabs and lobsters. Several disputes have arisen in the past as to whether such creatures are sedentary species, e.g. between Japan and the USA in respect of king crabs in the Behring Sea in the 1960s. Now that the waters overlying the continental shelf are largely EEZ, where the coastal State has the right to regulate fishing (see Section 13.2.7), such disputes are unlikely to recur.

The relatively limited rights which the coastal State has on its continental shelf and the non-territorial character of the latter, are stressed by a provision common to both the 1958 and 1982 Conventions that the coastal State's rights do not affect the legal status of the superjacent waters. Before the third UN Conference the legal status of these waters was as high seas. Since the Conference and the emergence of the 200 mile EEZ in customary law, the legal status of these waters is now as a EEZ for those States claiming such a zone, except in those areas where the continental shelf extends beyond 200 miles; here the status of the superjacent waters remains high seas.

Where the status of the superjacent waters is as high seas, there is an obligation on the coastal State, in exercising its sovereign rights over the resources of the continental shelf, not to interfere unjustifiably with other users of the high seas. Where the superjacent waters are EEZ, the obligation is one of not interfering unjustifiably with the navigation and other rights of other States in the EEZ.

Having looked at the coastal States' rights and duties on its continental shelf in general terms, we will now turn to look at the specific rights and duties related to offshore oil and gas exploitation.

13.3.3.1 Erection and operation of installations

The exploitation of oil and gas from the continental shelf obviously requires the use of mobile drilling rigs during its exploratory phase and the use of large, permanent platforms during its production phase. Both the 1958 and 1982 Conventions allow the coastal State to authorize and regulate the emplacement and erection of such installations. However, such installations must not be established where interference may be caused to the use of recognized sealanes essential to international navigation (1958 Convention, Article 5(6); 1982 Convention, Article 60(7)). To minimize the possibility of further interference with shipping (and other uses of the sea), the Conventions provide that due notice must be given of the construction of installations and permanent means for giving warning of their presence must be maintained, such as lights and foghorns.

The Conventions allow the coastal State to establish safety zones around drilling rigs and production platforms. Unauthorized ships may not enter such safety zones. The purpose of these zones is to reduce the possibility of a collision between a ship and an installation. Under the 1958 Convention the maximum extent of safety zones is 500 m, measured from the outer edge of the installation. The 1982 Convention retains the same figure, but provides that it may be exceeded where authorized by generally accepted international standards or as recommended by the competent international organization. No such standards or recommendations currently exist. Thus, the maximum extent of safety zones remains 500 m for the time being.

Both the 1958 and 1982 Conventions give the coastal State jurisdiction over the installations on its continental shelf. Thus, the coastal State can extend its criminal and civil law to such installations and enforce its law there, regardless of the nationality of the oil company operating the installation or the nationality of those working on the installation. So, for example, if a Dutchman murdered a Norwegian on a production platform, owned and operated by an American oil company in the British sector of the North Sea, a British policeman could fly out to the platform and arrest the Dutchman, who could subsequently be tried before a British court.

Likewise, if a French worker were injured on the same platform because of the operator's negligence, he or she could sue the operator for compensation in a British court, which would apply the relevant British law.

It is by virtue of the jurisdiction that it has over oil and gas installations that the coastal State can lay down construction, operating and safety standards for such installations, require the payment of taxes on the revenues earned from oil and gas exploitation, and at a more basic level establish the licensing system under which offshore oil and gas operations are usually conducted. All these matters are regulated by national law, rather than international law. Not surprisingly, the details of such things as safety standards, tax rates and licence fees and conditions vary quite considerably from one coastal State to another.

13.3.3.2 Decommissioning of installations

Offshore oil and gas are obviously finite resources, and eventually their fields will become exhausted. It is calculated that some North Sea fields will be exhausted by the early 1990s. When this happens, what is to become of the production platforms and other installations used to exploit such fields which have thus become redundant?

The 1958 Continental Shelf Convention provides, in quite categoric terms, that any installations which are abandoned or disused must be entirely removed (Article 5(5)). This Convention was drawn up, it must be remembered, at a time when experience of offshore oil and gas activities was very limited. By the time of the third UN Conference it was realized that it would be very costly to remove completely all obsolete installations. Further, in the case of the larger and heavier installations, such as those used in the northern North Sea, complete removal would be technically very difficult.

The 1982 Convention therefore takes a different approach to the 1958 Convention. Article 60(3) provides as follows:

"Any installations or structures which are abandoned or disused shall be removed to ensure safety of navigation, taking into account any generally accepted international standards established in this regard by the competent international organization. Such removal shall also have due regard to fishing, the protection of the marine environment and the rights and duties of other States. Appropriate publicity shall be given to the depth, position and dimensions of any installations or structures not entirely removed."

The competent international organization referred to is IMO, on which see Section 13.1.3. In 1988 the IMO drew up the international standards referred to in Article 60(3). These standards, which are in the form of guidelines, provide that in general abandoned installations should be removed, except in certain circumstances. A decision to allow an installation to remain, in whole or in part, should take into account such factors as the potential effect on the safety of navigation, the rate of deterioration of the material from which the installation is constructed, the potential effect on the marine environment, the risk that the installation will shift its position, the costs, technical feasibility and dangers involved in removal and the determination of a new use or other reasonable justification for allowing the structure to remain.

When a decision regarding removal is made, the IMO guidelines state that the following standards should be taken into account:

1. All abandoned or disused installations standing in less than 75 m of water and weighing less than 4000 tonnes excluding deck and superstructure should be entirely removed.
2. All abandoned or disused installations emplaced after the beginning of 1998, standing in less than 100 m of water and weighing less than 4000 tonnes should be entirely removed.
3. Where an installation is only partially removed, at least 55 m of water should be left above that part of the installation remaining.
4. After the beginning of 1998 no installation should be erected unless the design is such that complete removal is feasible when the installation becomes obsolete.

As their description suggests, these IMO guidelines are not as such legally binding. Nevertheless, the 1982 Convention, which is not in force, does require States to take account of them and it may be expected that most, if not all, States will act in accordance with the guidelines.

In spite of the fact that the 1982 Convention is not yet in force, the trend of State practice, reflected both in national legislation and in the work of the IMO and inspired by the economic and technical factors referred to earlier, is moving towards the formulation of a new rule of customary law that would require no more than the partial removal of some obsolete installations. Such a rule would supplant the seemingly unequivocal obligation of the 1958 Convention to remove all obsolete installations entirely.

13.3.3.3 Laying of submarine pipelines

In most cases of offshore oil and gas production, the oil and gas is brought ashore through pipelines laid between the production platform and the coast. It follows from the general rights of a coastal State on its continental shelf that it can authorize and regulate the construction of such pipelines, including prescribing such matters as the route of the pipeline, whether it should be entrenched in the sea bed, etc.

In some cases a coastal State might want a pipeline to cross the continental shelf of a second State, either to land the oil or gas in the latter State or in a third State. In such cases the first State has the right under international law to lay a pipeline on the continental shelf of another State. Under the 1958 Convention the latter State may not intervene in the first State's action, except where necessary to take reasonable measures in connection with the exploration and exploitation of its own continental shelf. Under the 1982 Convention the scope for intervention is greater; the State across whose continental shelf the pipeline is being laid must give its consent to the route of the pipeline, and where the pipeline enters its territorial sea, can lay down conditions relating to the pipeline.

In practice, States which lay oil or gas pipelines across the continental shelves of other States usually conclude agreements with the latter States to regulate the construction and operation of such pipelines. Thus, for example, Norway has concluded agreements with the UK and West Germany concerning pipelines carrying oil and gas from the Norwegian continental shelf to those two States.

13.3.3.4 Pollution from offshore oil and gas activities

Neither exploratory drilling for, nor production of, oil and gas will cause any significant pollution of the sea if properly conducted. Furthermore, any escape of oil or gas is contrary to the commercial interests of the operator, for such an escape represents lost production and may give rise to claims for compensation. It is thus not surprising that little effort has been spent on negotiating international agreements dealing with deliberate pollution resulting directly from offshore oil and gas exploitation, although both the 1958 and the 1982 Conventions address a general exhortation to States to avoid pollution, requiring them to adopt regulations to prevent pollution from pipelines and oil and gas installations. The main efforts of international law in this area have been directed at seeking to avoid accidental pollution and dealing with the question of liability for pollution.

As regards the avoidance of accidental pollution, the law in this area is rather sketchy. There has been some bilateral cooperation, notably between Norway and the UK, over safety standards for pipelines and installations. In addition, the IMO has drawn up a recommended code of the construction and equipment of mobile offshore drilling units. Accidental pollution resulting from a collision between an installation and a ship is sought to be avoided by the requirement that installations must be sited away from the main shipping lanes and their presence adequately indicated.

In addition, the IMO has recommended that fairways or routing systems for shipping should be established through offshore exploitation areas where the proliferation of oil installation or traffic patterns warrant it. Such fairways have, for example, been established in the Gulf of Mexico. Furthermore, the IMO has also recommended that ships not involved in the offshore oil and gas industry should avoid certain areas, such as the Gulf of Campeche, because of the degree of offshore oil and gas activity in such areas.

Finally, where accidental pollution does occur, particularly on a large scale, there are various regional and bilateral agreements that provide for the States' parties to cooperate in tackling such pollution, e.g. assisting each other with equipment and know-how for dealing with pollution emergencies.

Turning to the question of liability for pollution, one of the problems often faced by the victims of pollution of whatever kind in seeking compensation is to prove the necessary fault on the part of the polluter. In order to try to overcome this problem for the victims of pollution from offshore oil and gas activities in northwest European waters, the States bordering those waters got together in 1977 and concluded a treaty on the matter – the Convention on Civil Liability for Oil Pollution Damage resulting from Exploration for and Exploitation of Sea bed Mineral Resources.

The Convention provides that the operator of an installation causing pollution is, subject to one or two very limited exceptions, automatically liable for all damage caused by the pollution. As a *quid pro quo* for this strict liability, the operator's liability is limited to 40 million Special Drawing Rights (about £32 million or $50 million), although it is open to a State party, if it so wishes, to provide that the liability of the operator of installations on its continental shelf shall be higher or even unlimited in respect of pollution damage caused in that State.

In any case, an operator's liability is unlimited if the pollution damage occurred as a result of an act or omission by the operator himself/herself, done deliberately with actual knowledge that pollution damage would result.

Although the Convention was concluded as long ago as 1977, it has not been ratified by any State and, for reasons which are not entirely clear, appears unlikely ever to enter into force. In practice, this is not as unfortunate as might at first appear, because the oil companies operating in north-west European waters have concluded a private liability scheme, the Offshore Pollution Liability Agreement (OPOL), which is substantially similar to the 1977 Convention, although the upper limit of liability is $60 million. Outside north-west Europe there are no treaties or industry schemes dealing with liability.

Finally, apart from large scale pollution resulting directly from offshore oil and gas activities, miscellaneous deliberate, but relatively minor, pollution from oil and gas installations may occur in a number of ways; the discharge of drilling muds and cuttings, production water and displacement water, which contain oil and chemicals, the dumping of debris and the discharge of sewage and garbage. Such pollution is dealt with by a number of treaties. The 1973 International Convention for the Prevention of Pollution from Ships regulates the amount of oil which can be discharged in offshore operations, as well as regulating the discharge of sewage and garbage. Various treaties concerned with the dumping of wastes at sea regulate the deliberate dumping of wastes from sea bed installations, although they do not cover the disposal of wastes incidental to or derived from the normal operation of installations. Lastly, in north-west European waters the 1974 Convention for the Prevention of Marine Pollution from Land-based Sources applies to continental shelf installations and the commission set up by the convention has adopted various discharge standards for oily waste water and drilling muds.

13.4 Mining of manganese nodules in the international sea bed area

Lying on the deep sea bed, mainly beyond the geological continental shelf, are large quantities of manganese nodules. These are potato-sized lumps, composed of high grade metal ores, of which the most important are manganese, iron, nickel, copper and cobalt. Although the existence of manganese nodules has been known for over a century, it was not until the 1960s that they began to attract widespread interest. This was largely fuelled by a book by an American geologist, John Mero, which suggested that enormous riches could be had from exploiting these nodules. With the benefits of hindsight, it is now possible to see that Mero exaggerated the commercial possibilities of manganese nodules by underestimating both the cost and complexity of developing the necessary technology to gather and process the nodules.

Nevertheless, Mero's book was enormously influential when published and a number of companies, mainly in the USA, began research into how manganese nodules might be exploited. There was considerable concern, particularly among developing countries, that if nodule mining became a commercial possibility, all the economic benefits would go to the developed countries. This was either because such mining would be done on the basis of freedom of the high seas (see Section 13.2.8) and only the developed countries would have the necessary technology in practice to exercise this freedom, or because the flexible definition of the continental shelf (see Section 13.3.2) would allow States to extend their sea bed jurisdiction indefinitely seawards. Because of their geographical position and possession of various colonial territories, developed States would benefit most from this extension of the continental shelf.

These concerns, that developed countries would be the main, if not sole, beneficiaries of nodule mining, were articulated in a speech to the UN General Assembly made by the Maltese ambassador to the UN, Dr Arvid Pardo, in 1967. He proposed that the sea bed lying beyond the limits of national jurisdiction, i.e. beyond the legal continental shelf, should be reserved exclusively for peaceful purposes and that the resources of this area should be used in the interests of all mankind.

The General Assembly responded by setting up a committee to study Dr Pardo's proposal. After three years' work the committee produced a Declaration of Principles on the legal regime for the deep sea bed. This Declaration, which was approved by the General Assembly by a very large majority in 1970, stated that beyond the limits of national jurisdiction which were not defined was an area which should not be subject to appropriation by any State. Instead, this area, which, with its resources, was described as being the common heritage of mankind, was to be regulated by an international regime to be drawn up and, in particular, its resources were to be exploited for the benefit of mankind as a whole.

The elaboration of the international regime referred to in the Declaration became subsumed in preparations for, and eventually the work of, the third UN Conference on the Law of the Sea. The regime for deep sea mining was one of the most contentious issues at the Conference. The developing countries proposed that an international sea bed authority should be set up having the power both to engage in nodule mining itself and to control mining by private and State-owned companies. Royalties from the latter and profits from the authority's own mining would be distributed among all States as the common heritage of mankind. In contrast, the developed States wanted the authority to be limited to a licensing body, with few regulatory powers of substance.

As the Conference progressed, the developed States moved considerably towards the position of developing States. This was facilitated because an attempt was made to assuage their fears that the authority would interfere for purely political reasons in mining operations by confining the authority's discretion within closely defined limits and by establishing an elaborate system of decision making. The resulting provisions of the 1982 Convention form an extraordinarily detailed and complex legal regime for regulating manganese nodule mining.

This complexity has been further added to, because in the last days of the UN Conference, in an attempt to meet objections from some Western countries (notably the USA), a radical modification was made to the Convention provisions by introducing in Conference Resolution II a special supplementary regime to provide preferential treatment for pioneer investors in nodule mining. In spite of these last minute concessions, the USA, together with the UK and West Germany, did not and have not signed the Convention. These States, together with some other Western States, believe that it is possible to mine manganese nodules on the basis of the freedom of the high seas and have concluded a number of treaties to regulate such mining as between themselves. The legal regime developed by the latter is known as the Reciprocating States Regime.

Thus, there are three legal regimes which relate to manganese nodule mining: the 1982 Convention, Resolution II and the Reciprocating States Regime. Each of these will be discussed in turn. In the case of the first two regimes, it will also be necessary to have regard to the work of the Preparatory Commission (known as PrepCom). PrepCom, which comprises representatives of States which have signed the Convention, was set up by the UN Conference to prepare for the establishment of the International Sea bed Authority, drafting its rules and procedures, and to administer Resolution II's regime of preparatory investment protection.

It is also important to realize that each of these three regimes is essentially concerned with what is currently a future activity. Although much research and development on nodule mining has been carried out, mining on a commercial scale has not yet begun and it is generally thought such mining will not begin before the end of the present century.

13.4.1 Mining under the 1982 Convention

As mentioned above, the regime for nodule mining laid down by the 1982 Convention is an extremely detailed and complex one and only a bare outline of its provisions can be attempted here. All mining in the international sea bed area (i.e. the sea bed beyond the legal continental shelf) is to take place only under the authorization of the International Seabed Authority (ISA).

The ISA is composed of three main organs: the Assembly (on which all States' parties to the 1982 Convention are represented), the Council, comprising 36 States elected by the Assembly, and the Secretariat. The Council is served by two specialized bodies, the Economic Planning Commission and the Legal and Technical Commission. In addition, the ISA has a mining arm, the Enterprise, which has its own Governing Board, Director General and staff. The ISA will be financed initially by contributions from its member States and later, as nodule mining gets under way, from fees and royalties from nodule mining companies and the Enterprise.

The system of mining in the international sea bed area is the so-called parallel system, under which mining is to be undertaken by both the Enterprise and by commercial operators. The latter must be companies, either privately or State-owned, possessing the nationality of States party to the Convention and effectively controlled by them or their nationals. Such companies must also meet the financial and technical standards laid down by the ISA.

A company, meeting the above conditions, wishing to engage in nodule mining must submit an application to the ISA. The application must specify two nodule sites of equal estimated commercial value and each large enough to support a mining operation. The ISA, acting through the Council and the Legal and Technical Commission, may then approve a plan of work relating to one of the two sites. Approval may only be refused on certain limited grounds, e.g. to prevent the risk of damage to the marine environment, or to prevent one State, through its companies, having too great a share of nodule mining. If the proposed plan of work is approved, the ISA must then designate the second site specified in the company's application as a reserved site. Such sites are reserved for the Enterprise or for developing States.

To engage in commercial production of its site, the company requires further, separate authorization from the ISA. It must be given if all applications in each four-monthly round can be approved without hitting the production ceiling. This ceiling, designed to protect the land based producers of the minerals found in manganese nodules from adverse competition from nodule mining, is based on the growth of world nickel consumption. If the production ceiling is not reached and thus the authorization is granted, it must specify the amounts of nickel and other minerals that may be produced from the nodules each year.

If, on the other hand, the production ceiling is reached, the ISA must choose between applicants, giving priority to those applicants who were refused in previous rounds, who offer better assurances of performance or earlier prospective benefits to the ISA, or who have invested most in research.

A company authorized to engage in nodule mining by the ISA must pay fees and royalties. There is an initial fee of $500 000 and then a fee of $1 million a year. The system of royalties is complicated and in fact there are two systems. The first, designed mainly for State-owned companies, involves only a production charge, which is 5% of the market value of the processed metals extracted from the nodules for each of the first ten years of production, and 12% of that value each year thereafter.

The second system, designed for privately owned companies, provides that during the period when the company is still recovering its development costs a production charge of 2% of recovered metal market value is payable, together with a share of attributable net proceeds, which varies between 35 and 50%, depending on the profitability of the mining venture. Thereafter, the production charge rises to 4% and the share of the attributable net proceeds to between 40 and 70%.

Apart from the payment of fees and royalties, companies granted an authorization to mine must comply with the Convention's provisions concerning the transfer of technology. These require the company to make available to the Enterprise, on fair and reasonable terms and conditions, any technology which it uses in nodule mining which it is legally entitled to transfer, when the Enterprise cannot obtain such technology on reasonable commercial terms on the open market. If the company is not entitled to transfer the technology, it is to get an undertaking from the owner of the technology to transfer on the same terms. If such an undertaking cannot be obtained, the company must not use the technology in question.

Once the company has received the necessary authorization and begun mining, its operations will be supervised by the ISA to ensure that it complies with the plan of work that was authorized and the ISA's regulations on such matters as protection of the environment. Non-compliance can lead to the imposition of monetary penalties by the ISA or, in extreme cases, to an order to suspend or terminate mining. Disputes over such matters may be referred to the Seabed Disputes Chamber of the International Tribunal for the Law of the Sea, which the Convention establishes.

Turning from nodule mining by companies to mining by the Enterprise, the latter may mine the sites reserved to it either on its own or by means of joint ventures with companies. Like companies, though with some significant differences, the Enterprise requires authorization from the ISA. It, too, is governed by the production ceiling and it must pay fees and production charges.

The fees and royalties from companies and the Enterprise are to be distributed by the ISA, once it has deducted its own operating costs, equitably among States. Although the Convention does not say so directly, this must mean distribution among developing countries.

Although it is impossible at this stage to calculate how much money will be available for distribution in this way, it seems unlikely that the sums will, relatively speaking, be very large. Apart from these probably moderate gains, developing countries will also obtain some benefit from the know-how and technology which will be passed on to the nationals of developing States through the training programmes which the ISA is to promote. It is in these ways that the 1982 Convention seeks to ensure that the international sea bed area, the common heritage of mankind, is exploited for the benefit of mankind as a whole.

13.4.2 Resolution II and the regime of preparatory investment protection

As mentioned earlier, the third UN Conference on the Law of the Sea made radical additions to the 1982 Convention scheme at the eleventh hour in order to meet the concerns of the USA and some other western States, in particular the concern that there was a lack of protection for investments which had already taken place in nodule mining. These additions in favour of what are known as pioneer investors are to be found in Resolution II adopted by the Conference.

The Resolution identifies as pioneer investors four States (France, India, Japan and the USSR) and four multinational consortia, largely dominated by US companies. To obtain the benefits of Resolution II pioneer investors must register their claims to pioneer areas with the PrepCom. Only States which have signed the Convention may register as pioneer investors. Since the USA has not signed the Convention and cannot do so (since the time limit for signature has expired), it follows that the four multinational consortia cannot register as pioneer investors.

On the other hand, the four States mentioned (France, India, Japan and the USSR) were all registered as pioneer investors during 1987, having beforehand resolved (as required by the Resolution) their overlapping claims to areas of sea bed. It is also possible for East European and developing States to become pioneer investors if such States have each invested at least $30 million before the Convention comes into force.

Pioneer investors must detail two sites, each of up to 150 000 km^2. One site is to be allocated by PrepCom to the pioneer investor and the other banked for eventual use by the ISA. In their sites, pioneer investors are to have immediate and exclusive rights to conduct exploratory activities short of commercial production. Such production cannot begin until authorized by the ISA, which means, of course, that it cannot begin until after the 1982 Convention has come into force. When this happens, pioneer investors will have priority, within the production ceiling described above, over all other applicants, apart from the Enterprise.

In return for this preferential treatment, pioneer investors must pay an initial fee of $250 000 and an annual fixed fee of $1 million plus the production levy when they begin production and maintain investment in their site at levels to be prescribed by PrepCom. They must also undertake, prior to the entry into force of the Convention, to perform its obligations concerning transfer of technology. Furthermore, in order to advance the progress of the ISA's involvement in nodule mining, each pioneer investor must, at the ISA's request, explore the site reserved to the ISA on a cost plus 10% basis and also train designated personnel.

As mentioned, Resolution II was not sufficient to overcome the objections of the USA, West Germany and the UK to the 1982 Convention regime. These three States, together with some other western States which are less than completely satisfied with the Convention regime, have established what is known as the Reciprocating States Regime, which we must now examine.

13.4.3 Reciprocating States Regime

Under this regime a number of western States have adopted similar national laws which interlock so as to provide for comprehensive regulation of sea bed mining. The USA led the way by enacting the first legislation of this kind in 1980, followed by West Germany later in the same year, the UK and France (even though a pioneer investor) in 1981, Japan (again, even though a pioneer investor) in 1982 and Italy in 1985. Belgium and the Netherlands are also believed to be planning to enact such legislation.

All these laws follow the same basic pattern. Each provides that companies and nationals of the State concerned may not engage in nodule mining unless licensed by that State or by one of the other Reciprocating States. Companies which have been licensed are obliged to pay a levy equivalent to 3.75% of the value of the nodules mined. Such levies will be handed over by the Reciprocating States to the ISA if and when the 1982 Convention enters into force for them; otherwise the levies will be distributed as the State concerned sees fit.

By means of treaties of 1982 and 1984 the Reciprocating States have agreed to procedures for avoiding the overlapping of areas of sea bed which they license. More remarkably, by a series of agreements concluded in 1987, the Reciprocating States, including two pioneer investors, France and Japan, resolved overlapping claims with the third pioneer investor, the USSR. With the fourth pioneer, India, there is no problem as India's desired area of sea bed lies in the Indian Ocean, whereas all the other pioneer investors and the Reciprocating States are interested only in the sea bed of the Pacific.

As can be seen, the laws of the Reciprocating States differ from the Convention regime in important respects. The levy is only about half that envisaged by the Convention and no provision is made for the banking of reserved sites for the Enterprise. Nor is there any requirement concerning the transfer of technology. The Reciprocating States emphasize, however, that their legislation is interim and will apply only pending the entry into force for them of the 1982 Convention.

However, many of the Reciprocating States may not ratify the Convention until some time after it has come into force, if they ratify it at all. There is a fear, therefore, that these States' laws may become permanent, thus constituting a regime for nodule mining outside that of the Convention and possibly effectively in practice supplanting it.

The question arises, therefore, whether it is permissible under international law for the Reciprocating States to engage in nodule mining on the basis of this regime, both before and after the 1982 Convention enters into force. The Reciprocating States argue that such mining is an exercise of the freedom of the high seas (see Section 13.2.8) and therefore perfectly lawful. This argument is rejected by the developing States. They argue that the 1970 Declaration of Principles and other UN resolutions have established a new rule of customary law whereby the deep sea bed is not subject to the legal regime of the high seas and may only be exploited in accordance with the regime established by the 1982 Convention. Although UN General Assembly resolutions are not as such legally binding, it is generally recognized that they may constitute evidence of State practice and thus encapsulate rules of customary law.

There is, therefore, a good deal of force in the developing States' argument that a new rule of customary law of the nature described has been developed. However, it is also generally recognized that although rules of customary law are normally binding on all States, a State which persistently objects to a rule of customary law during the period of its formation is not bound by that rule. This, the Reciprocating States say, is the position here. They have persistently objected to the rule alleged by the developing States and so are not bound by it.

While in terms of strict law the position of the Reciprocating States is defensible, it is unfortunate, for the prospects of international cooperation between States, that a fundamental split over manganese nodule mining has arisen between a few rich western nations and the rest of the international community.

13.4.4 Pollution from nodule mining

Before leaving the topic of manganese nodule mining, a few words should be said about the question of pollution. Until mining on a commercial scale becomes much more imminent, it is difficult to assess whether nodule mining poses a serious threat of pollution. Some legal provision has already been made, however, to deal with such a threat should it materialize.

In respect of mining which takes place under the Convention regime, the 1982 Convention provides that the ISA is to adopt rules to prevent pollution from nodule mining. Such rules will be enorced both by the ISA and the States whose companies are engaged in nodule mining.

In the case of mining under the Reciprocating States Regime, little international law is at present applicable. The 1973 International Convention for the Prevention of Pollution from Ships and the various conventions on the dumping of waste at sea, which are discussed in more detail in Section 13.5.4, would regulate pollution to some degree. In time the Reciprocating States may feel it necessary to conclude a special treaty on the matter, or at least harmonize their national laws which are currently very diverse in this respect.

13.5 Shipping

The marine activities considered in the previous two sections, offshore oil and gas exploitation and manganese nodule mining, are among the newest uses of the sea. Shipping, on the other hand, is, along with fishing, the oldest of marine activities and still remains one of the most important. It is a marine activity to which technology has considerable application in the design and construction of ships, in navigational aids, including satellites, and in antipollution equipment.

This section concentrates on the public law, regulatory side of the international law relating to shipping, which is of most relevance to marine technology and omits discussion of the private law, commercial side of shipping. The section will consider in turn safety standards, traffic management and pollution.

First, however, it is necessary to begin by saying a few words about the nationality of ships. Ships, like people, have a nationality. When international rules on such matters as safety standards and pollution are adopted, they do not apply directly to ships as such. Because they are international law rules, they apply to, and are binding on, States only.

Such rules can be applied to ships in one of two ways. A State can apply the rules to all ships having its nationality, a form of jurisdiction known as flag State jurisdiction. Or a State can apply the rules to all vessels within its maritime zones (coastal State jurisdiction), although international law may restrict the ability of a coastal State to apply and enforce such rules to foreign vessels and restrict even more its jurisdiction to apply national laws which in content go beyond the international rules. Broadly speaking, the further the maritime zone is from the coast, the more the coastal State's jurisdiction is restricted.

The jurisdiction of flag States and coastal States to prescribe rules is often concurrent. However, on the high seas, it follows from what was said earlier (see Section 13.2.8) that flag State jurisdiction is the only basis for applying and enforcing rules to ships. Furthermore, because in the EEZ ships enjoy freedom of navigation (see Section 13.2.7), flag State jurisdiction is the main way in which rules are applied to ships in that zone, the chief exception being a very limited competence which coastal States are given to prescribe and enforce pollution rules.

Because flag State jurisdiction is such an important basis for the prescription and enforcement of rules to ships, it is desirable to know in what circumstances States may grant their nationality to ships and thus obtain such jurisidction.

13.5.1 Nationality of ships

Originally international law gave States a complete discretion as to the conditions on which they granted their nationality to ships. Not surprisingly, therefore, national laws concerning the nationality of ships have varied enormously. Some States lay down strict conditions for the grant of nationality, e.g. requiring the ship to be built in their territory, or the shareholders of the company owning the ship to have their nationality.

At the other extreme are States which are prepared to grant their nationality to virtually any ship. Such States are known as flag of convenience or open registry States. The principal States are Liberia, Panama, Cyprus and Singapore, whose ships account for nearly 30% of the world fleet. Foreign ship owners are encouraged by the low fees, low taxation and lower crew costs (which result from low wages and manning levels) to register their ships in flag of convenience States. By having their operating costs reduced as a result of these economic incentives, ship owners operating under flags of convenience usually enjoy a significant competitive advantage over ship owners whose ships are registered under other flags.

The complaint against flags of convenience, which is not surprisingly often motivated by economic considerations, is that these States have no proper control over their ships and therefore do not adequately enforce safety standards and antipollution measures. Past practice certainly supports this argument; the statistics show that flag of convenience States have more than their fair share of ships lost at sea and some of the most serious pollution incidents (e.g. the *Torrey Canyon* and *Amoco Cadiz*) have involved flag of convenience ships. It must be emphasized, however, that substandard and polluting ships are not confined to flags of convenience.

A number of attempts have been made by international law to deal with flags of convenience. The Convention on the High Seas of 1958 provides that there must exist a genuine link between a State and a ship to which it grants its nationality and that the flag State must effectively exercise its jurisdiction over its ships. This provision has not been very successful for a variety of reasons. First, no flag of convenience State has ratified the Convention. Second, the term 'genuine link' is not defined, nor is there any indication of what kind of links between a ship and its flag State would amount to genuine ones. Third, the provision about effective exercise of flag State jurisdiction is vaguely phrased and lacks obvious sanction in cases of non-compliance.

In spite of the lack of success of the 1958 Convention, similar provisions are found in the 1982 Convention. More recently, a further effort to deal with flags of convenience has been made with the adoption in 1986 of the UN Convention on Conditions for Registration of Ships. First, the Convention aims to strengthen the link between a ship and its flag State by requiring a clear link with the State in the form of an appropriate level of participation, a matter on which each State is left with considerable discretion, by its nationals in the ownership or crewing of the ship. Second, the Convention seeks to ensure that States effectively exercise jurisdiction and control over their ships by requiring States to maintain a competent and effective maritime administration to secure compliance with national and international shipping standards.

It is likely to be a number of years before the 1986 Convention comes into force, so it remains to be seen whether it will be any more effective than the 1958 Convention at dealing with the problems of flags of convenience.

Having looked at the way in which international law deals with the question of how a ship acquires the nationality of a particular State, which then applies its laws to that ship, we must now turn to consider how international law prescribes what the content of such laws shall be, beginning first with safety standards.

13.5.2 Safety standards

Since World War II the need for safety standards and regulations has become increasingly apparent. This is due to a number of factors. First, there has been an enormous increase in the number of ships; total world tonnage increased fivefold between 1948 and 1978, creating serious traffic problems in the world's busiest waterways. Second, the size of ships has increased enormously, with consequent reductions in manoeuvrability, e.g. a supertanker travelling at full speed takes several miles to stop. Third, ships are now carrying more dangerous cargoes, such as oil, liquified natural gas, toxic chemicals and radioactive matter, thus making the consequences of any accident more serious.

In theory, each State could adopt its own laws, applying such safety standards as it felt appropriate to its ships and to foreign ships entering its ports or territorial sea. In practice, this would be a recipe for chaos, because standards would vary widely and might well be incompatible, thus making it impossible for a ship travelling between the ports of two States with incompatible laws to comply with the laws of both these States. Furthermore, because improved safety standards usually involve extra costs for ship owners and because shipping is a very competitive industry, most States are reluctant to impose stricter safety standards on their ship owners unless other States do the same.

For these reasons, therefore, the international community has developed a set of uniform international standards to promote the safety of shipping. These standards are contained in a number of conventions adopted under the auspices of the IMO.

The main such convention is the 1974 International Convention for the Safety of Life at Sea (SOLAS Convention). This is the latest in a succession of SOLAS Conventions, the first of which was inspired by the sinking of the Titanic. The 1974 Convention is a bulky document running to nearly 300 pages and contains a large number of complex regulations laying down standards relating to the construction of ships, fire safety measures, lifesaving appliances, the carriage of navigational equipment and other aspects of the safety of navigation, the carriage of dangerous goods and special rules for nuclear ships. These standards are binding on about 98% of the world fleet.

Another important convention is the International Convention on Load Lines of 1966. This Convention, which is binding on about 99% of the world fleet, deals with the problem of overloading, often a cause of casualties to ships, by prescribing the minimum freeboard or the minimum draught to which a ship is permitted to be loaded.

Inadequately trained or qualified crews are a major factor in the cause of shipping accidents. To deal with this problem IMO drew up an International Convention on Standards of Training, Certification and Watchkeeping for Seafarers in 1978. The Convention, which is binding on about 73% of the world fleet, lays down mandatory minimum requirements for the certification of masters and other officers, and prescribes basic principles for keeping navigational and engineering watches.

In addition to these conventions, the IMO has adopted numerous recommendations, guidelines and codes relating to the safety of ships. Such measures, being usually in the form of resolutions of the IMO Assembly, are not as such legally binding. Nevertheless, some of these measures do subsequently become incorporated in conventions and are thus legally binding, e.g. the Code for the Construction and Equipment of Ships Carrying Dangerous Chemicals in Bulk, originally adopted as an IMO Assembly resolution in 1971, was incorporated into the SOLAS Convention by means of amendments adopted to the Convention in 1983.

The various measures in the three conventions discussed above are to be prescribed and enforced for their ships by flag States' parties to these conventions. There is a danger, particularly as regards flag of convenience States, that flag States will be lax in enforcing the conventions' standards. To deal with this problem a number of steps have been taken.

First, the three conventions provide that a port State, i.e. a State in one of whose ports a particular ship happens to be, can see that ships and their crews in its ports have valid certificates of the kind required by the conventions. Where valid certificates are lacking or where the condition of a ship no longer corresponds to the particulars of the certificate, the port State shall ensure that the ship does not sail until it can proceed to sea without danger to the ship or those on board.

Second, the International Labour Organisation (ILO) in 1976 adopted a Convention concerning Minimum Standards in Merchant Ships. Under this convention a State which believes that a foreign vessel in one of its ports does not conform to certain specified safety standards may inform the flag State and may take measures necessary to rectify any conditions on board which are clearly hazardous to safety or health, provided that it does not unnecessarily detain or delay the ship.

Lastly, 14 western European States signed a Memorandum of Understanding (i.e. not a formal treaty) on Port State Control in 1982. Under this memorandum the maritime authority of each State undertakes to maintain an effective system of port State control to ensure that ships visiting its ports comply with the three IMO Conventions, the ILO Convention and with the 1973 International Convention for the Prevention of Pollution from Ships, discussed in Section 13.5.4. Each authority must inspect a minimum of 25% of the ships using its ports. Where an inspection reveals deficiencies that are clearly hazardous to safety, health or the environment, the hazard must be removed before the ship is allowed to proceed to sea.

In spite of all the measures referred to above, the most recent annual reports of the Memorandum of Understanding warn that there is some way to go before the problem of substandard ships is eliminated.

13.5.3 Traffic management

A number of international measures have been taken to improve the safety of navigation at sea, in particular to seek to reduce the risk of collisions. The most important such measure is perhaps the 1972 Convention on the International Regulations for Preventing Collisions at Sea, the latest of a series of such regulations. The current Regulations, which are binding on about 96% of the world fleet, are principally concerned with a ship's conduct and movements in relation to other ships, particularly when visibility is poor, for the purpose of avoiding collisions, and with the establishment of common standards for light and sound signals. Under the 1982 Convention, ships exercising their right of innocent passage through the territorial sea or their right of transit passage through straits must observe the Regulations, regardless of whether their flag State or the coastal State is a party to the 1972 Convention.

An important means of reducing the risk of collisions is the use of traffic separation schemes to separate shipping in congested areas into one-way lanes. While coastal States may prescribe such schemes for their territorial sea, most such schemes in fact are prescribed by the IMO, which began doing so in 1967 and has to date prescribed over 90 schemes. Observance of IMO prescribed schemes is mandatory for ships whose flag States are parties to the 1972 Collisions Regulations Convention. Non-compliance may be punished by the flag State or, where the infringement occurs in the territorial sea, by the coastal State.

As well as traffic separation schemes, the IMO also recommends deep water routes, areas to be avoided, which are areas in which navigation is particularly hazardous or where it is exceptionally important to avoid casualties, e.g. for environmental reasons, and other routing measures.

In future it seems likely that the management of marine traffic will go beyond traffic separation and other routing schemes and become more comprehensive, though it seems unlikely that it will ever reach the precision or sophistication of air traffic control. Already the IMO has taken some action in this regard, e.g. in both the Baltic Straits and the Straits of Malacca it has recommended such measures as speed restrictions, the reporting by vessels of their position and the use of pilots.

In addition to the international measures for traffic management described above, there are a number of national schemes, mainly in the form of Vessel Traffic Services in the approaches to ports and harbours. The IMO has recommended guidelines for Vessel Traffic Services, which national authorities are urged to follow in the interests of international harmonization and improving maritime safety.

Before leaving the topic of traffic management, two further international measures should be noted. First, the SOLAS Convention requires States' parties to establish and maintain such aids to navigation, including radio beacons and electronic aids, as, in their opinion, the volume of traffic justifies and the degree of risk requires. Second, a worldwide maritime communications satellite system, known as INMARSAT, has been established under the IMO's auspices. This system was described in Chapter 2.

13.5.4 Pollution

Ships may cause pollution of the marine environment in a variety of ways. The form of pollution which first attracted the attention of international legislators was oil pollution. When an oil tanker has unloaded its cargo, a certain amount of oil remains clinging to the tanks. This oil must be removed before a new cargo can be taken in board. One way of doing this is for tankers to wash out their empty tanks at sea. In addition, an empty tanker returning to pick up its next cargo often uses sea water as ballast and this water, containing residues of oil, has of course to be pumped out before a new cargo can be taken on board. Other ships also use sea water as ballast in their empty fuel tanks and this too is eventually pumped out.

To deal with these forms of pollution the International Convention for the Prevention of Pollution of the Sea by Oil was drawn up in 1954. The Convention prohibited the discharge of oil and any oily mixture having an oil content of more than 100 ppm within 50 miles of land and in certain other areas. These discharge standards were tightened up by amendments to the Convention adopted in 1962, but they remained relatively ineffective, partly because of the difficulty of detecting violations of the standards and partly because of the reluctance of many flag States to prosecute the masters of their vessels where violations were detected.

The failure of the 1954 Convention and its 1962 amendments to prevent oil discharges at sea prompted the search for alternative ways of disposing of unwanted oil. The result was the development in the 1960s of the load on top system. This system separates the oil from the oily water on board; the almost oil-free water is returned to the sea and the oil residues are retained on board. The next cargo can, in most cases, then be loaded on top of the residues.

In 1969 further amendments to the 1954 Convention were adopted, the effect of which is that tankers must either operate the load on top system or retain all oily residues on board for eventual disposal ashore. Ships other than tankers must discharge oily water through a suitable separator and retain sludges from heavy fuel oils on board.

The 1954 Convention with its 1969 amendments has now been superseded, for the States' parties to it, by the International Convention for the Prevention of Pollution from Ships (MARPOL), adopted under IMO auspices in 1973. Annex I of the Convention is similar to the 1954 Convention as amended, but includes a number of significant additional provisions. The most important of these are: the introduction of special areas such as the Mediterranean and Baltic where no discharges at all are permitted, even by tankers operating the load on top system; the requirement for ships other than tankers to be fitted with oily water separating or filtering equipment and adequate sludge tanks; and the requirement for most tankers to be fitted with segregated ballast tanks and for crude oil washing and for new non-tankers over 4000 GRT to be fitted with segregated ballast tanks.

The remaining annexes of MARPOL, which are both lengthy and complex, deal with all other forms of deliberate pollution from ships, except dumping, which is discussed below. Briefly, they provide as follows. Under Annex II the discharge of residues containing noxious liquid substances must be made to a reception facility, unless they are adequately diluted, in which case they may be discharged into the sea in accordance with detailed regulations.

The Annex also contains provisions for minimizing pollution in the event of an accident. This last is the chief concern of Annex III, which seeks to prevent or minimize pollution from harmful substances carried in packaged forms by laying down regulations concerning packaging, marking, labelling, documentation, stowage and quantity limitations. Annex IV prohibits the discharge of sewage within 4 miles of land unless a ship has in operation an approved treatment plant. Between 4 and 12 miles from land, sewage must be comminuted and disinfected before discharge. Finally, Annex V sets specified minimum distances from land for the disposal of all the principal kinds of garbage and prohibits the disposal of all plastics.

As mentioned above, one of the weaknesses of the 1954 Convention was the reluctance of some flag States to enforce the Convention adequately in respect of their ships. The MARPOL Convention attempts to remedy this problem by giving increased powers to port States. Under the Convention the authorities of a port State may inspect a foreign ship and where the condition of the ship warrants it, they may detain the vessel until it can proceed to sea without presenting an unreasonable threat of harm to the marine environment.

Furthermore, where the inspection indicates a violation of the Convention, the authorities of the flag State are to be informed and must take legal proceedings if there is sufficient evidence. The 1982 UN Convention takes this a step further, by giving port States the right to arrest and prosecute foreign ships alleged to have violated the international rules either in the port State's territorial sea or EEZ or on the high seas.

Enforcement action can also be taken by coastal States in respect of violations of the rules occurring in their territorial sea and under the 1982 Convention in the EEZ in the case of pollution causing or threatening major damage. However, because of the difficulties of arresting an unwilling ship in passage, coastal State enforcement action is likely in practice to be less effective and less used than port State action.

Apart from international pollution rules, it is also possible for States to enact their own rules. Under the 1958 Convention and customary law, States have an unfettered discretion to legislate in respect of pollution for foreign ships passing through their territorial sea.

However, because of the possibility of incompatible national rules resulting from the exercise of this discretion (cf. the discussion on safety standards earlier), the 1982 Convention limits the scope for national rules by providing that coastal States must not legislate for the design, construction, manning or equipment of foreign ships unless they are giving effect to generally accepted international rules or standards (Article 21(2)).

Within the EEZ the 1982 Convention provides that a coastal State may only enact pollution legislation which conforms and gives effect to generally accepted international rules and standards (although an exception is made for ice covered areas, where the coastal State can adopt non-discriminatory pollution regulations).

Apart from the development of coastal and port State jurisdiction, a further incentive for ship owners to comply with pollution rules is the fact that international rules have been developed to facilitate the bringing of claims for compensation by the victims of pollution. As pointed out in Section 13.3.3.4, one problem for the victims of pollution in seeking compensation is to prove the necessary fault on the part of the polluter. A further problem in the case of pollution from ships is that where the pollution is caused by a foreign ship, the victim may find it difficult to sue the ship owner before the national courts because the latter may be reluctant to assume jurisdiction. Even if the victim wins the case, it may be difficult to enforce the judgement.

To deal with these problems the International Convention on Civil Liability for Oil Pollution Damage was adopted under IMO auspices in 1969. The Convention provides that where a ship causes oil pollution damage, the ship owner is, subject to three very limited exceptions, strictly liable for the damage and for the cost of any remedial measures taken. In return for this strict liability, the ship owner's liability is limited to a maximum of 14 million Special Drawing Rights (about £11 million or $17 million). The ship owner must have insurance cover for this amount. Claims must be brought by the victims of pollution damage in the courts of the State where the damage occurred. The judgements of such courts are enforceable in all State parties to the Convention. Where the oil pollution damage exceeds the ship owner's limits of liability under the 1969 Convention, the International Oil Pollution Compensation Fund, set up by a convention of 1971 and financed by a levy on oil imports, will reimburse victims for the excess, up to a limit of 45 million Special Drawing Rights (about £36 million or $56 million).

The world's leading tanker and oil companies have established two industry schemes, known as TOVALOP and CRISTAL, which broadly mirror the provisions of the 1969 and 1971 Conventions. These schemes apply in the many States which are not parties to the two Conventions.

So far we have considered all forms of deliberate pollution from ships, except the dumping of waste, which has been treated separately by legislators. Since the 1960s dumping at sea has become an increasingly popular way of disposing of waste resulting from land based activities, partly because of its relative cheapness and ease and partly because of the tightening of pollution controls on land.

International measures to control such dumping were introduced in 1972 with the adoption of the Convention on the Prevention of Marine Pollution by Dumping of Wastes and Other Matter (commonly known as the London Dumping Convention). The Convention prohibits dumping of the most toxic substances and lays down strict guidelines as to how and where other wastes may be dumped. The London Convention is worldwide in scope; it has been followed by regional conventions, substantially similar in content, for the north-east Atlantic and North Sea (1972), the Baltic (1974), the Mediterranean (1976) and the South Pacific (1986).

Since the 1970s the incineration of toxic wastes at sea in specially designed ships has become an increasingly popular way of dealing with the waste problem on land. Legally, incineration has been dealt with under the dumping conventions. In 1978 the States' parties to the London Convention adopted guidelines governing the incineration of wastes at sea. This has been followed, a decade later, by a decision taken in October 1988 to ban all incineration of wastes at sea from 1994.

13.6 Fishing

Prior to 1900, fishing, although along with shipping the oldest use of the sea, was technologically very primitive. During the present century, however, there has been an enormous application of technology to the fishing industry in such ways as the development of more sophisticated nets and gear, the introduction of acoustic equipment for locating shoals of fish and the design and construction of larger and stronger fishing vessels, many of which have elaborate fish processing and freezing facilities on board. It is these technological developments that are largely responsible for the enormous increase in the world fish catch in recent decades, from about 20 million tonnes a year in 1950 to about 66 million tonnes a year in the mid-1980s.

International fisheries law relates relatively little to the use of technology in the fishing industry, although, as will be seen, it is predicated upon a constant scientific monitoring of the size of, and recruitment to, fish stocks. For this reason, therefore, the discussion of the law in this area will be briefer than in the preceding sections relating to other uses of the sea, where there is a much more obvious correlation between the law and the relevant technology.

The development of the international law relating to fishing falls into two broad phases. The first phase is the period before the mid-1970s. This period was characterized by generally narrow coastal State zones and a fair amount of international cooperation in fisheries management through the medium of international fishery commissions. The second, current, phase is characterized by 200 mile coastal State zones and reduced international cooperation.

We will begin by looking briefly at the system of regulation during the first phase in order to understand the present system.

13.6.1 International fisheries law before the mid-1970s

During this phase coastal State limits were generally narrow. Within the territorial sea, which, apart from off the coasts of some Latin American and a few African States, did not extend to more than 12 miles, the coastal State had the exclusive right to regulate and exploit all fish found there.

In the case of States whose territorial sea was less than 12 miles in breadth, many such States from 1960 onwards began claiming a 12 mile fishing zone. Within this zone, which rapidly attained the status of a rule of customary international law, the coastal State had the exclusive right to regulate fishing, though engaging in fishing was not always reserved to the fishing vessels of the coastal State; other States' vessels which had traditionally fished in the waters concerned were often given the right to fish in the outer six miles of the zone.

Beyond the 12 mile territorial sea or fishing zone was the high seas. It follows from what was said about the nature of the high seas earlier (see Section 13.2.8) that in this area the fishing vessels of all States had the right to unrestricted fishing under the doctrine of the freedom of the high seas. A number of consequences followed from this freedom to fish on the high seas.

First, although all States in theory enjoyed this freedom, in practice the freedom was exercised most by those countries with the necessary technology and capital to take advantage of it. These were, of course, the developed countries, particularly those countries with large distant water fishing fleets such as Japan, the Soviet Union, Poland and West Germany.

The second consequence of the freedom of high seas fishing was for more fishing vessels to take part in a fishery than a fish stock could stand, i.e. overfishing occurred. In order to try to prevent overfishing and because no one State could act on its own, a score or so of international fisheries commissions were set up to regulate fishing in particular regions or for particular species.

Although in theory such bodies would seem to be the answer to the problem of overfishing, in practice they were not very successful. There were a number of reasons for this. First, the commissions were often unable, because of the competing demands of their member States, to agree on sufficiently strict regulatory measures. Second, even where measures were agreed, the constitutions of most commissions allowed dissatisfied member States to opt out of being bound by them. Finally, where the measures were binding, they were often poorly observed and enforced.

By the early 1970s there was much dissatisfaction with the international law of fisheries. Most developing coastal States were resentful of the fact that the vessels of distant developed States, equipped with the latest technology, were catching fish a comparatively short distance off their coasts. Even if the former States did not have adequate vessels of their own to fish their offshore waters, they wished at least to be able to control the activities of foreign operators and to be able to obtain some revenue through licence fees and to gain access to technological know-how.

At the same time some developed coastal States were not happy with the existing legal regime, either because they wanted greater access to, or more control over, their offshore fishery resources, or because they were sceptical of the ability of international fishery commissions effectively to regulate fishing in the face of the increasing pressure on stocks resulting from ever more intensive methods of fishing.

These dissatisfactions meant that when preparations began for the third UN Conference on the Law of the Sea, many coastal States seized the opportunity to press for a fundamental revision of the international law of fisheries in their favour. These aspirations have been met, and the 1982 Convention whose provisions we now turn to examine contains a very different regulatory framework from the previous law.

13.6.2 Fisheries under the 1982 Convention

The core of the provisions of the Convention dealing with fisheries is the 200 mile EEZ, which we have already looked at in general terms earlier (see Section 13.2.7). Since the universal establishment of 200 mile EEZs would cover an area where over 90% of the world's commercial fishing currently takes place, the regime of the EEZ is the dominant framework of regulation for fisheries under the Convention.

Within the EEZ the coastal State has sovereign rights for the purpose of exploring and exploiting, conserving and managing the fish stocks of the zone. These rights are subject to a number of duties. The coastal State must take such conservation and management measures as will ensure that fish stocks in its EEZ are not endangered by over-exploitation and that such stocks are maintained at or restored to 'levels which can produce the maximum sustainable yield, as qualified by relevant environmental and economic factors, . . . and taking into account fishing patterns, the interdependence of stocks and any generally recommended' subregional, regional or global minimum standards (Article 61(3)). Subject to this, the coastal State is required to promote the objective of optimum utilization of the fish stocks of its EEZ. Finally, the coastal State is to establish the total allowable catch (TAC) for each fish stock within its EEZ.

These duties are obviously formulated in very broad and general terms and the coastal State is given extensive discretion in their application. This is particularly the case in relation to setting the TAC, where the management objective of maximum sustainable yield is so heavily qualified that a coastal State could legitimately set practically any size of TAC, provided that it did not lead to over-exploitation which endangered fish stocks.

Once the coastal State has set the TAC for each stock, it must give foreign vessels the right to fish for that part of the TAC which its own fishermen are not capable of catching. The coastal State has again considerable discretion in deciding which foreign fishermen are to be allowed to fish, although fishermen from the landlocked and geographically disadvantaged States of the region must be allowed to fish for part of the catch allocated to foreign fishermen.

In respect of those foreign fishing vessels which are allowed to fish in its EEZ, the coastal State may prescribe and enforce conditions to govern their fishing. These conditions may include requiring foreign fishermen to have licences, to observe the coastal State's conservation measures, to carry out research programmes, to land part or all of their catches in the coastal State, to train coastal State personnel and to transfer fisheries technology to the coastal State.

The 1982 Convention gives the impression that most fish stocks confine themselves to the EEZ of a single coastal State. In fact in some areas, such as the north-east and east central Atlantic, this is very far from being the case; many stocks migrate between the EEZs of two or more States, usually known as shared or joint stocks, and/or the waters beyond (straddling stocks).

The Convention contains one brief provision dealing with this situation. Article 63 exhorts the States concerned (which in the case of straddling stocks include States fishing on the high seas for such stocks) to agree upon the measures necessary for the conservation of such stocks. Nothing further is said, for example, about management objectives or allocation of the catch among interested States.

The above rules are considerably modified in the case of anadromous, catadromous, sedentary and highly migratory species and marine mammals. In the case of anadromous species, such as salmon, the State in whose rivers such fish spawn is primarily responsible for managing such species; it is not, however, under any obligation to establish total allowable catches or to admit foreign States to its economic zone to fish for any surplus there may be. Furthermore, fishing for anadromous species beyond 200 mile limits is normally forbidden. In the case of catadromous species (e.g. eels), such fishing is always forbidden.

In other respects, however, the general rules apply, but are supplemented by an obligation on coastal States through whose waters catadromous species migrate to cooperate over management with the State in whose waters the species spend the greater part of their life cycle. In relation to sedentary species, such as shellfish and some crustaceans, the coastal State is under no obligation to set total allowable catches or accommodate foreign fishermen.

In the case of highly migratory species such as tuna, the coastal State's normal management functions are supplemented by an obligation to cooperate with other interested States, either directly or through regional fisheries commissions, which are either already in existence (such as the Atlantic and Inter-American Tuna Commissions), or are to be specially created. Finally, in the case of marine mammals, the coastal State is entitled to limit or prohibit the exploitation of such species, rather than establishing a total allowable catch.

In the area beyond the 200 mile EEZ, i.e. on the high seas, fishing continues to be open to all States. Nevertheless, the Convention imposes a duty on interested States to cooperate in the management and conservation of high seas fish stocks, making use, where appropriate, of international fishery commissions.

Although the 1982 Convention is not yet in force, the concept of the EEZ, and particularly its fisheries provisions, had commanded such a degree of consensus by a relatively early stage of the UN Conference that, as noted (in Section 13.2.7), from about 1977 onwards a large number of States have claimed a 200 mile EEZ or fishing zone without waiting for the Conference to end or the Convention to enter into force.

While the right to an EEZ or fishing zone has now become part of customary international law, it is much less certain whether the conservation and management duties which the Convention imposes on the coastal State in its EEZ have also passed into customary law.

This is because, first, few States' legislation refers to these duties, though the explanation for this may be not because such duties are not accepted, but because they are thought an inappropriate matter for legislation, relating as they do to administrative practices and second, because the duties may be too vague to become rules of custom.

Many other aspects of the 1982 Convention regime relating to fisheries are reflected, to a greater or lesser extent, in the practice of States. Thus, there exists a large number of bilateral agreements providing for the access of foreign fishermen to the EEZ of coastal States; a number of agreements have been concluded relating to the management of joint stocks; several new international fisheries commissions have been created, and a number of existing commissions adapted, for the regulation of fishing beyond 200 miles. To give any details of these matters is, however, beyond the scope of this chapter.

13.7 Military uses of the sea

Finally, we come to the last, but certainly not the least, important use of the sea of which we have to consider the legal regulation; the use of the sea for military purposes. This is obviously not only a very important use of the sea but one to which technology has a major application. We will consider the law relating to this use of the sea under two main headings: the use of force generally, and the deployment of particular weapons and military equipment.

13.7.1 Use of force generally

The UN Charter and customary international law prohibit States from using or threatening force, whether by land, sea or air, against the territorial integrity or political independence of any other State (Article 2(4) of the UN Charter). To this fundamental principle there are three lawful exceptions.

First, a State is allowed to use force in self-defence. A recent marine example is the naval task force despatched by the UK to retake the Falkland Islands following their invasion by Argentina in 1982. A second, though infrequent, exception is where force is used under the authorization of the UN. An example from the marine area is the blockade by the UK in 1966 of the port of Beira in Mozambique, then a Portuguese colony, to enforce economic sanctions against the Smith regime in Rhodesia.

The final, and even more infrequent, exception is enforcement action taken by regional organizations under the authorization of the UN Security Council. Thus, unless it falls under one of these three exceptions the use of military force at sea which amounts to an act or threat against another State's political independence or territorial integrity will be illegal.

While the deployment of naval force in ways which do not amount to acts or threats of the kind just described is in principle permissible, there are nevertheless restrictions on the ways in which naval force may be deployed in various of a coastal State's maritime zones. Warships, like merchant ships, enjoy a right of innocent passage through the territorial sea. For passage to qualify as innocent, it will be recalled (see Section 13.2.3), it must be in a manner which is not prejudicial to the peace, good order or security of the coastal State.

Among the activities which the 1982 Convention lists as deemed to be so prejudicial are weapons practice, the collection of information to the prejudice of the defence or security of the coastal State and the launching, landing or taking on board of any aircraft or military device.

Two further limitations on the right of innocent passage of warships should be noted. First, submarines are required to navigate on the surface (in practice virtually all submarines are military vessels). Second, some States require warships intending to exercise a right of innocent passage through their territorial sea to give them advance notification or in some cases to request prior authorization. There are doubts as to the legality of this practice.

In territorial seas comprising straits, the 1982 Convention, it will be recalled (see Section 13.2.3.), gives foreign ships a right of transit (rather than innocent) passage. This rather greater right requires all ships to 'refrain from any activities other than those incident to their normal modes of continuous and expeditious transit' (Article 39(1)).

As far as warships are concerned, this prohibition would seem to cover most, if not all, of the acts mentioned above which are proscribed in the territorial sea generally. On the other hand, unlike the territorial sea, submarines may navigate through straits in a submerged position.

Moving outwards from the territorial sea to the EEZ, in the latter zone, it will be recalled (see Section 13.2.7), foreign ships enjoy the freedom of navigation 'and other internationally lawful uses of the sea related to [this freedom], such as those associated with the operation of ships' (Article 58(1)). Whether this freedom extends to foreign warships engaging in naval manoeuvres or weapons exercises in a State's EEZ is something of a moot point. While some developing States claim such manoeuvres and exercises can be carried out in the EEZ only with the consent of the coastal State, most developed States take the view that the coastal State's consent is not required.

On the high seas naval manoeuvres and weapons testing are clearly permitted as an exercise of the freedom of the high seas, provided they are carried out in such a way that they have reasonable regard to the interests of other States in their exercise of the freedom of the high seas. During the 1950s the USA, and to a lesser extent the UK, carried out nuclear tests in the Pacific which involved temporarily designating large areas of sea which ships were strongly advised not to enter. This testing was claimed to be an exercise of the freedom of the high seas, though whether it can really be said to have had reasonable regard to other States' use of the high seas must be doubtful. In any case such testing is now prohibited under the 1963 Nuclear Test Ban Treaty.

Having looked at the way in which the law regulates the use of force at sea in general terms, we turn now to consider the law relating to the deployment of particular weapons and military equipment.

13.7.2 Deployment of particular weapons and military equipment

The weapons with whose deployment the international law of the sea has been most concerned are, not surprisingly perhaps, nuclear weapons. On the global level the 1971 Treaty on the Prohibition of the Emplacement of Nuclear Weapons and Other Weapons of Mass Destruction on the Seabed prohibits the emplacement of nuclear weapons and other weapons of mass destruction, and of installations designed to store, test or use them, on the sea bed beyond 12 miles from the coast. At the regional level the Latin American States, through the Treaty of Tlatelolco (1967), and the South Pacific States, through the Treaty of Raratonga (1985), prohibit the testing and stationing of nuclear weapons on the bed and in the waters of their 12 mile territorial seas. Finally, at the bilateral level the Strategic Arms Limitation Agreement (SALT) of 1972 between the USA and USSR limits, among other things, the number and armament of submarines carrying nuclear missiles.

The above mentioned treaties are the only measures that at the present time limit the deployment of particular weapons. Apart from the limitations discussed, naval vessels are free to carry such weapons as they wish, although the use of such weapons will be governed by the general rules relating to the use of force, discussed above. The same is true, too, of other forms of military equipment carried on ships. However, the position is different as regards military equipment used outside ships, such as submarine surveillance systems and other monitoring devices.

While a State could position such devices in its various maritime zones as a result of its general jurisdiction in those zones (see Section 13.2) and on the high seas as an exercise of the freedom of the high seas, it could obviously not place such devices in the territorial sea or internal waters of other States without their consent, since the latter zones are subject to the coastal State's sovereignty. Whether a State could place devices of the kind discussed on the EEZs or continental shelves of other States is more problematical. It could be argued that such devices are structures. If so, and if the devices interfered with the exercise of the coastal State's rights in its EEZ or continental shelf, their emplacement by other States would only be permitted under the 1982 Convention if the coastal State gave its consent.

On the other hand, some monitoring devices could be regarded as cables, in which case a State would be able to place such devices in the EEZ and on the continental shelf of other States by virtue of its freedom to lay cables in those zones. If such monitoring and similar devices were neither structures nor cables, the 1982 Convention does not give a clear answer as to the permissibility of their emplacement in the EEZ. Such emplacement would fall neither within the rights of the coastal State nor within the rights of other States. In the case of uses of the EEZ which fall within neither category of rights, the 1982 Convention provides that the permissibility of such uses is to be determined 'on the basis of equity and in the light of all relevant circumstances' (Article 59).

13.8 Conclusions

It will be clear by now that the international law of the sea is a complex subject and that the legal position on many questions is not very clear, due in part, but not entirely, to the uncertain status of the 1982 Convention on the Law of the Sea. This chapter has done no more than give a rather rough sketch of those areas of the law most relevant to marine technology, and the discussion of many problematical matters has had to be brief and over-simplified. Readers who wish to know more and who want fuller answers to particular questions should consult the reference list.

References and further reading

Bardonnet, D. and Virally, M. (eds.). *Le nouveau droit international de la mer*. Editions A. Pedone, Paris (1983).
Churchill, R.R. and Lowe, A.V. *The Law of the Sea*. 2nd edition, Manchester University Press, Manchester (1988).
Dupuy, R-J and Vignes, D. (eds.). *Traité du Nouveau Droit de la Mer*. Economica, Paris and Bruylant, Brussels (1985).
Mangone, G.J. *Law for the World Ocean*. Stevens, London (1981).
O'Connell, D.P. *The International Law of the Sea*. 2 Vols., Clarendon Press, Oxford (1982 and 1984).
Sohn, L.B. and Gustafson, K. *The Law of the Sea*. West Publishing Company, St Paul, Minnesota (1984).

Index

Abyssal plains, 1/5
Accommodation ladders, safety aspects, 11/9
Acoustic systems, 12/22–40
  components, 12/27–30
    measurement/display, 12/29–30
    signal processing, 12/27–28
    transducers, 12/27–28
  development, 12/23
  Doppler, 12/39
  limitations, 12/22–23
  for navigation/positioning, 12/34–39
    development, 12/34
    intelligent, 12/35, 12/36
    long baseline, 12/34–35, 12/36
    phase comparison, 12/37–39
    short baseline, 12/39
  for non-destructive testing, 10/35
  oceanographic, 12/39–40
  quantities/units, 12/22
  sensor systems
    echo sounders, 1/4, 1/8, 12/11, 12/24–25, 12/29, 12/30–31
    side scan sonar, 1/8, 9/37–38, 12/11, 12/31–33
    swathe sounding systems, 12/33–34
  telemetry, 12/40
  theory, 12/23–27
    noise, 12/26–27
    prediction, 12/27
    propagation, 12/23, 12/25–26
    reflection, 12/26
    sound field, 12/23–24
    transmission, 12/24–25
  see also Hydro-acoustic position reference (HPR)
Acoustics, underwater, 1/12
Admiralty Hydrographic Office, 12/9
Aerofoils, marine, 3/34
AEW series for propeller design, 3/31
Air bags in salvage, 9/46
Air conditioning, 6/61
  see also Compressed air
Air cushioned vehicles (ACV), *See* Hovercraft
Aircraft, search and rescue, 11/42
  see also Helicopters
Airlift in salvage, 9/39
Albatross, *see* Taut wire position reference

Alloys
  corrosion resistance, 10/17–23
    summary data, 10/22, 10/23
  selective leaching, 10/14
  stress corrosion cracking, 10/12
Aluminaut submersible, 4/39
Aluminium
  alloys
    corrosion resistance, 10/22
    selective leaching, 10/14
  for planing craft, 3/86, 3/87
  for ships, 3/42
ALVIN submersible, 4/38
Anchor watch, 11/6–7
Anchoring, 8/4–5
  versus dynamic positioning, 8/15–16
  multi-anchor spread, 8/8–13
    anchor handling vessel, 8/10
    anchor types, 8/10
    catenary curve, 8/9
    configurations, 8/10
    procedure, 8/11–12
    retrieval, 8/13
    semisubmersible drilling rigs, 8/8–9
    wire/chain lines, 8/9–10
  procedures for safety, 11/7–18
Anchors
  Bower, 8/14
  for multi-anchor spread, 8/10
  in salvage, 9/40, 9/41
  slowing ability, 7/3
Ancillary machinery, 6/54–65
  air conditioning, 6/61
  bilge/ballast systems, 6/60
  on bulk carriers, 6/64–65
  compressed air, 6/60–61
  on container ships, 6/64
  cooling systems, 6/56–57
  deck machinery, 6/63
  fuel systems, 6/59–60
  lubrication systems, 6/57–58
  minor systems, 6/62
  pumps, 6/54–56
  refrigeration, 6/61
  on Ro–Ro ships, 6/64
  steering gear, 6/62–63
  on tankers, 6/65
  see also Machinery arrangements
Anodes in cathodic protection, 10/22
Anodic protection, 10/28

Anodic protection (*cont'd*)
  see also Cathodic protection
Aramid, 3/87
Archimedes principle, 2/70, 2/77
Archipelagic waters, 13/7
ARCS autonomous ROV, 4/53–54
Argo positioning system, 12/59–60
Aries submersible, 4/39
Artemis positioning system, *see* Surface microwave position reference system
Arterial gas embolism, 4/10
Artificial intelligence in fault diagnosis, 7/48
Aseptic bone necrosis from diving, 4/11
Astronomical navigation, 12/16–18
Atlantic Ocean, 1/4
  Mid Atlantic Ridge, 1/5
Atlantis submersibles, 4/39
Atlas Polarfix, 12/54–55
Atmosphere, 1/17–19
  ionosphere, 1/19
  weather systems, 1/18–19
  winds, 1/18
Auguste Picard submersible, 4/39
Automatic Radar Plotting Aid (ARPA), 7/12, 12/89–91
Automation
  digital control systems, 7/44–48
    fault detection/diagnosis, 7/47–48
    microprocessor based, 7/46–47
    requirements for, 7/44–46
  trend towards, 7/43–44
  in whole ship control, 7/12–13
  see also Integration
Automotive engines, 6/22–25
  advantages, 6/23–24
  marinization, 6/22–23
  multiple plants, 6/24–25
Autopilots, 7/4, 7/30–35, 12/14
  adaptive, 7/33, 7/34
  control in a seaway, 7/33, 7/35
  conventional, 7/31–33
  cost function, 7/17
  versus helmsman, 7/14
  and navigation, 12/14
  response, 7/30–31
    optimal, 7/35
Auxiliary power
  requirements, 6/42–44

2/Index

Auxiliary power (cont'd)
  steam, 6/52
  see also Generators
Axyle positioning system, 12/56
Aximuthing propulsion, 6/8
  thrusters, 6/74–75, 8/21, 8/22
Azimuths, 12/18

B flag, 11/24
Ballast systems, 6/60
  sea water, and corrosion, 10/15
Bandwidth compression in ROV communication, 4/53
Barges
  carriers, 3/6
  derrick, 8/39–40
    in salvage, 9/43–44
  lay, 5/36–37
Baselines for maritime zones, 13/6
  archipelagic waters, 13/7
Basins
  marginal, 1/4, 1/26
  tides in, 1/16
Bathymetry, 12/11
Bauer–Wach system, 6/6
Bays and maritime zone baselines, 13/6
Beach gear in salvage, 9/39–42
Bearings, radar storage, 12/89
  see Compasses; Heading; Navigation; Track keeping/position control
Bearings/seals
  for gearboxes, 6/39
  for propeller shafts, 6/37–38
Beaufort Scale, 1/20
Bellini–Tosi aerials, 12/19, 12/20
Bells, diving, 4/19
  early use
    pre-1900, 4/3–4
    post-1900, 4/6
  in emergencies
    bell-to-bell transfer, 4/26
    for evacuation, 4/27
  handling, 4/19–20
  see also Decompression
Bends, 4/10–11
Benthonic organisms, 1/21
Benzene hazard from crude oil, 11/22
Bernoulli's equation
  in linear wave theory, 2/27
  for pipe flow, 5/31
Bilge keels, 7/41
Bilge systems, 6/60
Bimco Towcon, 9/56
Biocides in corrosion protection, 10/28
Biological oceanography, see Oceanography, biological
Black Sea, 1/31
Blistering from bacterial hydrogen, 10/14
Blue Book for dangerous goods, 11/19
Body squeeze, 4/8–9
Boilers
  coal fired, 6/50–52
    coal handling, 6/52
    design, 6/50

Boilers (cont'd)
    stoker firing, 6/50–52
  principle, 6/50
Bollard pull, 9/50, 9/52
  certification, 9/52
Bolt gun, 9/28–29
Bone necrosis from diving, 4/11
Bonjean curves, 3/5–6
Booms, oil, 9/58–59
Bottom slamming, 3/24, 3/38, 3/39
Bounce diving, 4/15–16
Boundary layer theory, 2/23–24
  separation, 2/24
Bow flare slamming, 3/24
Bower anchors, 8/4
Boyle's Law, 4/7
Brass
  corrosion resistance, 10/21
  selective leaching, 10/14
Breathing apparatus, 11/30–31
Bretschneider spectrum, 3/46
British Ship Research Association, resistance series, 3/27
British Standards, see Standards/codes of practice
BRITSHIPS, 3/67
Brittle fracture in ship structure, 3/44
Bronze, corrosion resistance, 10/21
Buckling in ship structure, 3/44
Bulk cargoes
  shipment hazards, 11/17–18
  stowage, 11/18–19
Bulk carriers, 3/10
  equipment, 6/64–65
Bunkering, and safety, 11/4
Buoyancy, 4/6–7
  of risers, control, 5/26–27
  of ROVs, 4/40–41
  salvage systems, 9/44–46
  submersible systems, 4/34–35

Buoyancy factor, 2/70
Buoys
  heave motion analysis, 2/84–85
  for mooring, 8/5–8
    circumstances, 8/5–6
    large tankers, 8/6
    loading hose handling, 8/8
    recovery/breakaway, 8/7–8
  navigational, 12/19
    pressure to reduce, 12/21
    radar reflection, 12/85
  see also Lifebuoys
Bureau Veritas, on ship controllability, 7/29
Burnability of fuel, 6/58–59

Cable (distance), 12/9
Cables, see Chain cables
Canyons, marine, 1/5, 1/6
Capsize, 3/19, 3/34–35
  of hovercraft, by plough-in, 3/114
  and longitudinal bulkheads, 3/21

Capsize (cont'd)
  stability, and free surface effect, 9/25–26
Capstan, 8/4
Carbon dioxide poisoning, 4/9
Carbon monoxide in diving gas, 4/9
Carbonate sediments, 1/7
  transportation, 1/8
Cargo
  operations
    safety, 11/15–21
    watch-keeping during, 11/4
  pumping system, 6/62
  salvage, 9/15–24
    from damaged vessels, 9/15–16
    lightening, 9/18–19
    from tankers, 9/19–24, 9/20, 9/22
    underwater, 9/16–18
    value, 9/4, 9/15
  ships, 3/6–12
    safety equipment, 11/36–38
    see also particular type
  sweat, 11/17
  tankers, 11/21–22
    handling checklist, 11/22–23
  volume fraction for ship sizing, 3/62–63
Carpenter stoppers, 9/42
Cast irons, corrosion resistance, 10/20–21
Catamarans, 3/118–119
  oil skimmer, 9/59–60
  propulsion, water jet, 3/126
  roll, 7/40
  see also High speed vehicles; SWATH ships
Catenary moorings, 8/9
  dynamic response analysis, 2/92–97
  and hydrostatic stability, 2/76
Catenary risers, analysis, 5/18
Cathode ray tube for radar, 12/79–81
Cathodic protection, 10/23, 10/26–28
  anodes, 10/22
  coatings, 10/25
  design, 10/27–28
  field measurement, 10/27
  principle, 10/26–27
  see also Anodic protection
Cavitation, 10/12
  on hydrofoils, 3/91
    near-surface, 3/95–96
  number, 3/24
  on propellers, 3/31–32, 6/67, 10/16
  on rudders, 10/16
    on planing craft, 3/86
  threshold in acoustic transmission, 12/24
Cement boxes in salvage, 9/27
Centrifugal pumps, 6/55
Chain cables, 8/4
  cutting, in salvage, 9/43
  test loads, 9/55
  see also Towing
Channel tunnel, implications for hovercraft, 3/117

Charles' Law, 4/7
Charts
　corrections, 12/12
　data representation, 12/10
　electronic, 7/13
　electronic position fixing for, 12/53
　production, 12/11–12
　projections, 12/7–8
　　and distance, 12/8–9
　tidal, 12/11
　see also Electronic chart display and information systems (ECDIS); Hydrography; Surveying
Chronometers, 12/18
Circulation
　in estuaries, 1/27
　oceanic, 1/12–17
　　deep sea, 1/14–15
　　remote sensing, 1/12–13
　　surface currents, 1/13–14
　　tides/tidal currents, 1/15–17
　see also Currents; Gyres
Classification societies
　and control systems, 7/29–30
　for ships, 3.67
　see also Det Norske Veritas; Lloyds Register of Shipping
Clutches, transmission, 6/40, 6/41–42
　see also Couplings; Gears
Coast Guards, US, 3/88
　hydrofoil craft for, 3/101
Coastguard Radio Liaison Station (CRLS), 11/38
Coatings, protective, 10/23, 10/25–26
　cathodic, 10/25
　for concrete, 10/23
　failure, 10/26
　splash zone system, 10/26
Cobalt alloys, 10/22
CODAG generators, 6/49–50
Codes of practice, see Standards/codes
Cofferdams, in salvage, 9/29
Collisions
　avoidance with radar
　　computer system data, 12/89
　　relative motion, 12/87
　　true motion, 12/88
　procedure in, 11/27
　regulations for preventing, 13/21
　and watch keeping, 11/6
Combinators, in CP propeller control, 7/45–46
Compasses, 12/14
　gyro, 12/13, 12/14
　for dynamic positioning, 8/20
　points of, 12/6
Compressed air, 6/60–61
　in salvage, 9/44–45
　see also Air conditioning
Computers
　CAD/CAM in naval architecture, 3/23, 3/58, 3/66–67
　in dynamic positioning, 8/16
　radar systems, computer aided, 12/88–91

Computers (cont'd)
　bearing storage, 12/89
　clutter/interference elimination, 12/88–89
　collision avoidance data, 12/89
　for strength/stability in tanker discharge, 9/23–24
　see also Processors in dynamic positioning
Concrete
　platforms, 2/10–12, 2/17, 2/18, 2/19
　protective coating, 10/23
　reinforcement corrosion, 10/15–16
　see also Cement boxes in salvage
Container ships, 3/6, 3/7
　controllability, 7/27
　design and, 7/30
　direct drive engines, 6/14
　equipment, 6/64
　salvage of cargo, 9/18–19
Continental margins, 1/6
Continental shelf, 13/8
　coastal State's rights/duties, 13/12–15
　　decommissioning installations, 13/13–14
　　erection/operation of installations, 13/12–13
　　pipeline laying, 13/14
　legal definition, 13/11–12
　legal evolution, 13/10–11
　see also Territorial sea
Contra rotating propellers, 6/70, 6/72
Contra rotating thrusters, 5/75
　propeller/thruster system, 6/75–76
Contracts
　for towing, 9/56
　for salvage, 9/5–11
Control systems, 7/1–57
　automation trend, and costs, 7/43–44
　classification societies and, 7/29–30
　digital, 7/44–48
　　fault detection/diagnosis, 7/47–48
　　microprocessor based, 7/46–47
　　requirements for, 7/44–46
　integrated, 7/48–52
　　implications on personnel, 7/51–52
　　methodology, 7/48–51
　　need for, 7/48
　methodology, 7/14–20
　　analysis, 7/18–20
　　cost function, 7/17–18
　　definition of system, 7/14–15
　　description, open/closed loop, 7/15–16
　　disturbances, 7/15
　　parameters, 7/16
　　performance analysis, 7/16–17
　requirement, 7/3–14
　　controllability, 7/14
　　course keeping, and disturbances, 7/6
　　dynamic positioning, 7/8, 7/9
　　information, 7/13–14
　　safety margins, pressure to reduce, 7/3

Control systems (cont'd)
　　speed control, 7/8, 7/10–12
　　steering, 7/3–5
　　track keeping, 7/6–8
　　whole ship control, 7/12–13
　roll control, 7/40–44
　　and hull design, 7/40–41
　　imposed devices, 7/41–42
　　operational methods, 7/42
　　rudder in, 7/43
　and ship motions, 7/20–26
　　analysis, and equations of motion, 7/20–24
　　axis systems, 7/20–21
　　effectors/control surface, 7/24–25
　　wind/wave effects, 7/25–26
　track keeping/position control, 7/35–38
　　effector requirements, 7/36–38
　　information sensors, 7/36
　　methodology, 7/38
　　operations requiring, 7/35–36
　see also Autopilots; Dynamic positioning; Track keeping/position control
Controllability, 7/14, 7/26–29
　definition, 7/3
　and hull form, 7/26–27
　improving, 7/55
　and port problems, 7/55–56
　predicting, 7/55
　regulatory standards, lack of, 7/29–30
　requirement, 7/14
　ship trials for, 7/52–55
　　conduct, 7/53–55
　　purpose, 7/53
　stopping ability, 7/27–28
　in turning, 7/27–28
Controllable pitch propellers, 6/4, 6/67–70, 7/25, 7/27
　advantages, 6/68–69
　combinator control, 7/45–46
　cost benefits, 6/69
　design, 6/69–70
　engine loading, 6/69
　Voith Schneider, 6/70, 6/71
Cooling systems, 6/56–67
　sea water, corrosion, 10/14–15
Copper alloys
　corrosion resistance, 10/21–22
　selective leaching, 10/14
Coriolis effect, 1/14
Corrosion, 10/3–29
　alloys, resistance, 10/17–23
　biological, 1/23
　bacterial, 10/14
　chemistry, 10/3–7
　　electrochemical series, 10/3
　　Evans Diagrams, 10/5
　　kinetics, and polarization, 10/4–5
　　passivation, 10/5–7
　　thermodynamics, 10/3–4
　codes/standards, 10/36–38
　　for design, 10/37–38
　　materials, 10/37

Corrosion (*cont'd*)
    non-destructive testing, 10/38–39
    defect propagation and, 10/33
    by fuel, 6/59
    marine fouling and, 10/17
    and marine zones, 10/13–14
    mechanisms, 10/7–17
        electrochemical effects, 10/8–11
        electrochemical/mechanical, 10/11–13
    monitoring/detection, 10/35–36
        in pipelines with PIGs, 5/42
    in oil/gas production, 10/16–17
    of propulsion/steering systems, 10/16
    protection, 10/23, 10/25–29
        anodic, 10/28
        cathodic, 10/26–28
        chemical, after immersion, 9/48
        coatings, 10/23, 10/25–26
        inhibition/modification of fluid, 10/28
        materials selection/design, 10/29
    and sea water handling, 10/14–15
    and selective leaching, 10/14
    of steel
        in concrete, 10/15–16
        structural, fatigue, 10/16
    see also Cavitation; Defects; Non-destructive testing
Cost function, 7/17–18
Couplings, 6/40–41
    fluid, 6/42, 6/43
    see also Clutches; Gears
Course keeping
    autopilot *versus* helmsman, 7/14
    and disturbances, 7/6
    see also Autopilots; Compasses; Steering; Track keeping
Cox-bolt gun, 9/28–29
Cracking
    detection in pipelines, 5/42
    stress corrosion, 10/12
    see also Corrosion; Defects
Crane vessels, 2/21
    Derrick Barges, 8/39–40
    for platform installation, 2/17
    for salvage, 9/43–44
    semi-submersible, 2/9
    see also Sheer legs
Cranes
    safe use, 11/11–12
        hand signals, 11/13
    on ships, and stability, 3/15
Crash stop, 7/53
Crevice corrosion, 10/10
Crews, and integrated/automated control, 7/12, 7/51–2
Cruise liners, power station concept, 6/10
    see also Passenger ships
Cupro-nickel alloys, corrosion resistance, 10/21
Currents
    coastal, 1/26–27
    deep sea, 1/14–15

Currents (*cont'd*)
    in equatorial/tropical oceans, 1/28, 1/29
    longshore, 1/27–28
    and navigation, 12/14–15
    offshore structure loading, 2/19, 2/27
    and sediment transport, 1/8
    surface, 1/13–14
    tidal, 1/17
    see also Circulation; Gyres
CURV programme, 4/51
Custom, rule of, in international law, 13/3, 13/4
Cyclones, 1/19
Cylinders, collapse pressure, 4/34

Dalton's Law, 4/7
Dangerous goods
    code of practice, 11/19–21
    handling, 11/21
    International Maritime Dangerous Goods (IMDG) code, 11/19–21
Datatrak, 12/53, 12/62–63
David ROV for diver support, 4/49–50
Davidson Current, 1/26
De-aluminification, 10/14
Dead reckoning, 12/13
Decca Navigator
    accuracy, 12/43, 12/46
    commercial background, 12/42
    coverage, 12/44–45
    development, 12/41
    in integrated system, 7/13
    Mainchain, 12/5, 12/15
    operation frequency, 12/41
    principle, 12/41, 12/42
    and propagation anomalies, 12/43
    readings, for anchored position, 11/7
    receivers, 12/43
    transmitters, 12/43
Decibels
    requirements of radar antennae, 12/81–82
    scale, 12/24
Decks
    access, safety aspects, 11/9–10
    cargo stowage, 11/16
    machinery, 6/63
    wetness, and ship motions, 3/34, 3/38–39
Declination, 12/17
Decompression, 4/10
    of air divers, 4/13
    chambers, 4/12
        legislation, 4/26
        life support system, 4/21
        for saturation diving, 4/19, 4/20–21
        see also Bells, diving
    after heliox diving, 4/15, 4/19
        in bounce diving, 4/15
    history, 4/5
    illness, 4/10–11

Defects
    assessment, 10/31–32
    characterization, 10/29–30
    codes/standards, for welding, 10/38
    cracking
        detection in pipelines, 5/42
        stress corrosion, 10/12
    genesis, 10/30, 10/31
    growth, 10/32–33
        and corrosion, 10.33
        fatigue, 10/33
        plastic/elastic, 10/32–33
    monitoring/detection, 10/35–36
        corrosion, 10/35–36
        fatigue, 10/36
    non-destructive testing for, 10/33–35
    and reliability, 10/36
    see also Corrosion
Denseveyor coal handling, 6/52
Derrick Barges, 8/39–40
    in salvage, 9/43–44
Derricks, safe use, 11/10–11
Desalination units, corrosion, 10/15
    see also Salinity
Design
    for corrosion control, 10/29
    of pipelines, 5/30–34
    for roll control, 7/40–41
    standards
        design process, 10/37–38
        materials, 10/37
    see also Naval architecture
Det Norske Veritas
    redundancy classification, 8/22
    on ship controllability, 7/29
Dezincification, 10/14
Diaphragm pumps in salvage, 9/33
Diesel generators, 6/44–45
Diesel engines for propulsion, 6/12–25
    air cooled, 6/25
    automotive type, 6/22–25
        advantages, 6/23–24
        marinization, 6/22–23
        multiple plants, 6/24–25
    for dynamic positioning vessels, 8/21
    high speed, 6/18–20, 6/22
        advantages, 6/19–20
        applications, 6/19
        design, 6/20, 6/22
    medium speed, 6/15–18
        advantages/disadvantages, 6/18
        applications, 6/17, 6/18
        data, 6/18, 6/21
    principle, 6/12
        four stroke cycle, 6/12
        two stroke cycle, 6/12–13
    slow speed, 6/13–15
        applications, 6/14
        data, 6/15, 6/17
        direct drive advantages, 6/14
        uniflow scavenged two stroke, 6/13–14
    see also Gas turbines
Diesel pumps in salvage, 9/32
RFA Diligence, 8/37

Direct stiffness analysis, 2/49–56
Dispersant spraying for oil pollution, 9/60
Distress signals, 11/35–36, 11/38
Diving, 4/5–32
  accident statistics, North Sea, 4/25
  air diving, 4/11–14
    scuba, 4/13–14, 9/14
    surface demand, 4/11–13
  atmospheric suits/systems, 4/27–32
    commercial, 4/27–31
    history, 4/27–28
    ROV dual systems, 4/55
  comparisons, ADS/ROV/saturation, 4/31–32
  developments
    needs, 4/57, 4/59
    recent, 4/27
    since 1900, 4/5–6
  emergencies, 4/26–27
  history
    air supply, 4/5
    decompression, 4/5
    suit, 4/3–4
  mixed gas, 4/14–21
    heliox, 4/14–17
  personal equipment, 4/21–23
    band masks, 4/21–22
    gas heater, 4/23
    helmets, 4/21
    suits, 4/22–23
  physics, 4/6–8
  physiology, 4/8–11
    air breathing at depth, 4/9
    bone necrosis, 4/11
    depressurization, 4/9–11
    gas effects, 4/9
    pressure effects, 4/8–9
  regulations/guidelines, 4/25–26
  support
    ROV, 4/49–50
    vessels, 4/23–25, 8/16, 8/22, 8/35, 8/37
  see also Bells, diving/ Manned submersibles; Remotely operated vehicles (ROV); Underwater working
Docking of ships, and trim, 3/18
  see also Ports
DOGGIE geological ROV, 4/55
DOLPHIN oceanographic ROV, 4/54
Doppler systems
  current meters, 12/40
  log/navigation, 12/39
  see also Satellite navigation
Doxford engine, 6/13
Draager, 11/25
Drag
  on hovercraft, 3/109–112
  on offshore structures, 2/20, 2/37–40
  on ROVs, 4/45–46
  on umbilicals, 4/47
Dredging
  hydraulic systems in salvage, 9/38–39
  track keeping, 7/8, 7/35

Drilling mud, 2/3–4
Drilling rigs
  jack-up, 2/3
  semi-submersible, 2/3–6
    operation from, 2/3–4
  see also Offshore structures, fixed; Offshore structures, floating
Ducted propellers, 6/65–66
Dumping of waste, 13/23
Duplex stainless steels, 10/20
DUPLUS II, 4/55
Dye penetrant inspection, 10/34
Dynamic positioning, 2/4–6, 7/8, 7/9, 8/13–43
  versus anchoring, 8/15–16
  control modelling, 8/24
  control system, 7/8, 7/9
  definition, 8/13–14
  drawbacks, 8/16
  elements/equipment, 8/16–22
    attitude, 8/21
    control, 8/16–17, 8/18
    control console, 8/17
    heading reference, 8/20
    position reference, 8/17, 8/19–20
    power system, 8/21
    thrusters, 8/21–22
    wind sensors, 8/20–21
  history, 8/14–15
  planning, 8/34–35
  position reference systems, 8/24–33
    hydro-acoustic, 8/25–29
    redundancy, 8/23
    repeatability, 8/24
    surface microwave, 8/29–30, 12/55–56
    taut wire, 8/30–33, 12/37
  principle, 8/14
  redundancy, 8/22–24
  requirement for, 8/14
  station keeping assessment, 8/33–34
  vessels, 8/35, 8/37–43
    applications, 8/42–43
    crane vessels, 8/39–40
    DSV Stena Seaspread, 8/35, 8/37
    field support, 8/41–42
    pipe laying, 8/40–41
    safety audit, 8/34
    semisubmersible drilling, 8.38–39
  see also Control systems; Mooring; Thrusters; Track keeping/position control
Dynamometer for bollard pull, 9/52

Ear squeeze, 4/8
Earthquakes, see Tsunamis
Eave East ROB, 4/54
ECDIS, 12/12–13, 12/92–93
Echo sounders, 1/4, 1/8, 12/11, 12/30–31
  displays, 12/29
  errors, 12/24–25
  see also Sonar; Swathe sounding systems
Eddies, oceanic, 1/14
Edison, Thomas, 'radar' patent, 12/73

Ekman Spiral, 1/14
Electric drive, 6/9–10
  combination mechanical drive, 6/10–11
Electric submersible pumps for salvage, 9/32, 9/34
Electrical non-destructive testing, 10/35
Electricity, static, hazard for tankers, 11/25
  see also Auxiliary power; Generators
Electrochemical series, 10/3
Electronic chart display systems (ECDIS), 12/12–13, 12/92–93
Ellipsoids, 12/7
Embrittlement of steel, 10/33
  and cathodic protection, 10/27
Emergencies, 11/27–31
  anchoring, 11/8
  breathing apparatus, 11/30–31
  distress signals, 11/35–36, 11/38
  fire, 11/29–30
  man overboard, 11/27–28
  parties, 11/3–4
  position indicating beacons (EPIRBs), 11/32, 11/37
  procedures for diving, 4/26–27
  see also Collisions; Survival
Engineering Critical (ECA) of defects, 10/32
Engines
  automatic monitoring, 7/13
  automotive, 6/22–25
  main, choice/selection, 6/3–4
  single screw plants, 6/4–5
  in speed control, 7/8, 7/10–11
  see also Diesel generators; Diesel engines for propulsion; Gas turbines; Machinery arrangements; Propulsion; Steam plant
Environmental regulatory numbers (ERNs), 8/34
Epaulard autonomous ROV, 4/53
Epoxy resins, underwater, 9/29–30
Erosion–corrosion, 10/12
Escher Wyss CP propellers, 6/70
Estuaries, 1/27
  tides in, 1/16
European Community, legislative role, 13/6
Evans Diagrams in corrosion chemistry, 10/5
Exclusive Economic Zone (EEZ), 13/8–9
  fast craft for policing, 3/88
  and fishing, 13/24–25
Expert systems in fault diagnosis, 7/48
Explosimeters, 11/24
Explosion and salvage, 9/33, 9/36
  see also Fire

Face squeeze, 4/8–9
Factors of safety
  for ship hull girder, 3/48
  for ship secondary structures, 3/51

Failure
  of pipelines, 5/34
  of risers, 5/26
  structural, bulk cargoes and, 11/17
  see also Fatigue
Fans, sedimentary, deep sea, 1/8
Fans for hovercraft, 3/107
Fast rescue craft (FRC), 11/39
Father and son machinery
    arrangements, 6/8–9
Fatigue
  analysis of offshore structures, 2/66–67
  corrosion, 10/12–13
    of structural steel, 10/16
  cracking
    propagation, 10/33
    in ship structure, 3/44
  of hull girder, 3/48
  monitoring, 10/36
  steel behaviour, 10/22
  see also Failure
Fault detection/diagnosis in control
    systems, 7/13, 7/44, 7/47–48
  see also Defects
Fendering
  in ship to ship transfer, 9/22
  in tanker cargo salvage,
    vessels alongside, and safety, 11/4
Ferries, 3/10, 3/12
  catamaran, 3/119
  medium sized engines for, 6/18
  multiple screw plants, 6/6
  steering arrangements, 7/24–25
  see also Herald of Free Enterprise;
    Ro–Ro vessels
Fibre optic NDT, 10/35
Fibreglass reinforced plastic (FRP), for
    planing craft, 3/87
Finite element analysis
  of flow around ship, 7/22–23
  of hull girder, 3/48–49
  in hydrostatics, 2/78
  of offshore fixed structures
    dynamic, 2/59–65
    static, 2/48–56
  for ship structural design, programs,
    3/58
  of wave diffraction, 2/47–48
  see also Risers: analysis
Finnjet mixed power plant, 6/7
Fins, active, 7/42
Fire
  procedures, 11/29–30
  and salvage, 9/30–32
  see also Breathing apparatus;
    Explosion and salvage
Fishing
  in coastal water, 1/27
  international law on, 13/23–25
    and Exclusive Economic Zones,
      13/24–25
    and freedom of the high seas,
      13/23–24
Fishing boats, controllable pitch
    propeller for, 6/68

Fixing, 12/15–16
  celestial, 12/17–18
Fjords, 1/27
Flags of convenience, 13/19–20
Flettner rotators, 3/89
Floating storage units, (FSU), 8/6
Flow
  finite element analysis of, 7/22–23
  static charge hazard, 11/25
Fluid mechanics
  concepts, 2/20, 2/22
  Reynolds number, viscous forces and,
    2/22–26
    bluff body flows, 2/24–26
    boundary conditions, 2/23–24
    Froude number, 2/23
Foam
  for ROV buoyancy, 4/41
  in salvage, 9/45–46
Food and Agriculture Organization
    (FAO), 13/6
Forest products carriers, 3/10
Fouling, 1/22–23
  copper alloy resistance, 10/21
  and corrosion, 10/17
    in intertidal zone, 10/13
  roughness, 2/37, 2/39–40
  and skin friction, 3/26–27
Fourier's Law for heat loss, 5/32
Fracture zones, 1/5
Free surface effects, 2/75–76, 3/14–15
  and salvage, 9/24–26
Freeboard, 3/23
Fretting corrosion, 10/11–12
Froude number, 2/23, 3/77
  and ship resistance, 3/24
  and wave making resistance, 3/25–26
Froude–Krylov force, 2/87
Fuel
  bunkering, 11/4
  costing in design, 3/65
  economy, 7/10
  grades, 6/58
  problems, 6/58–59
  quality, 6/58
  system, 6/62
    design, 6/59–60
  uni-fuel concept, 6/45–46
  see also Diesel generators; Diesel
    engines for propulsion

Gabrielli–von Karman chart, 3/132,
    3/134
Galvanic corrosion, 10/8–10
Gangways, safety aspects, 11/9
Gas carriers, 3/9, 3/12
  boil-off for steam raising, 6/3, 6/26
  whole ship control, design for, 7/12
Gas, *see* Offshore operations
Gas laws, 4/7
Gas turbines, 6/30–34
  advantages, 6/32
  applications, 6/31
  arrangements, 6/31–32
  design, 6/33–34

Gas turbines (*cont'd*)
  development,
    current, 6/34
    historical, 6/30
  disadvantages, 6/33
  principle, 6/30–31
  see also Diesel propulsion engines;
    Machinery arrangements;
    Propulsion; Steam plant
Gases
  atmospheric, 1/17
  in diving
    early developments, 4/5–6
    mixed, 4/14–21
    at pressure, physiology of, 4/9
  flow in pipes, 5/32
  inert
    in salvage, 9/36
    system for tankers, 11/22, 11/26
  laws, 4/7
  measuring instruments for tankers,
    11/24–25
  physics of, in liquids, 4/7–8
  in sea water, 1/10
  see also Compressed air
Gear pump, 6/54
Gears, 6/38–40
  see also Clutches; Couplings
Gemini submersible, 4/39
Generators
  diesel, 6/44–45
  for dynamic positioning vessels, 8/21
  power turbines, 6/48–49
    in CODAG generators, 6/49–50
  shaft, 6/46–48
    advantages, 6/46–47
    methods, 6/46
    modes of operation, 6/47–48
  uni-fuel concept, 6/45–46
  see also Auxiliary power
Geneva Conventions on the Law of the
    Sea, 13/4–5
Geodesy, 12/5–9
  chart projections, 12/7–8
  and distance, 12/8–9
  coordinates, 12/6
  directions, 12/6
  ellipsoids/geoids, 12/7
  shortest distance between points,
    12/6–7
Geodimeter, 7/53
Geoids, 12/7
Geoloc positioning system, 12/60
Geology, ROVs for, 4/54–55
Geostar, 12/72
Geostrophic currents, 1/14
Glint in radar tracking, 12/85
Global Positioning System (GPS), *see*
    NAVSTAR GPS
Glomar Challenger, 8/15
GLORIA, 1/4, 1/12, 12/33
Gnomonic projection, 12/8
Golf 2, 12/55
Graham's Law, 4/8
Grain rules, 11/18

Graphite corrosion of grey irons, 10/20
Graphitization, 10/14
Gravity waves, 2/27–28
Great circles, 12/6, 12/8
Great Eastern power plant, 6/6
Green's Function, in wave analysis, 2/103
Greenwich hour angle, 12/17
Grillage analysis, 3/52–53
Grim wheel, 6/72–73
Ground reaction, 9/26, 9/50
   see also Sea bed; Soil/pile interactions
Ground tackle in salvage, 9/39–42
Grounding of ships, and trim, 3/18–19
Groundwaves, 12/41–42
   propagation, and Loran C, 12/48
Gulf Stream, 1/14, 1/26
Guyed tower production platform, 2/15, 2/16, 2/17
Guyots, 1/5
Gypsy, 8/4
Gyres, 1/13–14
   and organism distribution, 1/22
Gyrocompass, 12/13, 12/14
   for dynamic positioning, 8/20

Hand signals for lifting gear, 11/13
Handbooks for offshore safety, 11/27
Hanging off, 8/5–6
Haskin Relations, in wave force analysis, 2/103
Heading
   measurement, 12/13–14
   references for dynamic positioning, 8/20
Health, and crude oil toxicity, 11/22
   see also Safety
Heat storage by water, 1/9
Heave/pitch, 3/35–38
   and dynamic positioning, 8/21
   and irregular waves, 3/37–38
   motion analysis
      of articulated column, 2/88–89
      of cylindrical buoy, 2/84–85
      of semisubmersible, 2/85–88
      of tensioned buoyant platforms, 2/90
   response amplitude operators, 3/37
   stability of hovercraft, 3/113–114
   strip theory, 3/35–37
Helicopters
   boarding pass, 11/27
   safety, 11/29
   search and rescue, 11/39–42
   and wind sensors, 8/20, 8/21
Heliox diving, 4/14–17
   bounce diving, 4/15–16
   first use, 4/5–6
   oxygen percentage, and depth, 4/14
   saturation diving, 4/15, 4/16–21
      bell/bell handling, 4/19–20
      deck decompression chamber, 4/20–21
      high pressure nervous syndrome (HPNS), 4/17

Heliox diving (cont'd)
   life support system, 4/21
   procedure, 4/18–419
   speech distortion, 4/17
   temperature control, 4/17–18
Helmets, diving, 4/21
Henry's Law, 4/7–8
Herald of Free Enterprise, 3/68
   parbuckling, 9/47
Hertz, Heinrich, 12/73
Hi-Fix 6 positioning system, 12/60–61
High pressure nervous syndrome (HPNS) in heliox diving, 4/17
High seas, 13/9
   freedom of, and fishing, 13/23–24
High speed vehicles
   definition difficulties, 3/77
   performances, 3/128–134
   propulsion, 3/124–128
      air propellers, 3/126–128
      marine propellers, 3/125
      water jet, 3/125–126
   resistance, 3/77
      limitation, 3/78
   see also Catamarans; Hovercraft; Hydrofoil craft; Planing craft; SWATH ships
HMS Challenger, 4/52
Hogging in ship loading analysis, 3/44
Hold cleaning, 11/16
Horizon radar, 12/84
Hornet diving system, 4/30
Hot tap machines, 9/36–37
Hovercraft, 3/102–117
   applications, 3/116–117
   drag components, 3/109–112
   example types, 3/103
   flexible skirts, 3/108–109
   performance
      in steady motion, 3/112
      structural design, 3/115–116
      in waves, 3/113–115
   development, 3/128–129
   principles, 3/102–103
   propulsion, air propellers, 3/126–128
   support generation, 3/103–108
      cushion pressures, 3/103–104
      forward flight, 3/107–108
      hovering flight, 3/104–107
      leakage air, 3/104
   see also High speed vehicles; Hydrofoil craft; Planing craft
Hull–propeller interaction, 3/29–30
Hurricanes, 1/19
Huyghen's principle in wave propagation, 12/25–26
Hydraulic pumps in salvage, 9/33, 9/35
Hydro-acoustic position reference (HPR), 8.25–29
   limitations, 8/28–29
   Simrad systems, 8/25–28
Hydrofoil craft, 3/88–102
   applications, 3/101–102
   design, 3/100–101
   geometry, 3/88

Hydrofoil craft (cont'd)
   lift generation, 3/89–92
   lifting surfaces, 3/92–94
   near-surface hydrofoils, 3/93–97
      cavitation, 3/95–96
      lift and depth, 3/93–95
      ventilation, 3/96–97
   performance
      in steady motion, 3/97–99
      in waves, 3/99–100
   propulsion
      marine propellers, 3/125
      water jet, 3/126
   see also High speed vehicles; Hovercraft; Planing craft
Hydrogen
   from bacterial corrosion, 10/14
   as diving gas
      early use, 4/6
      recent development, 4/27
   see also Embrittlement of steel
Hydrogen sulphide
   in diving gas, 4/9
   hazard from crude oil, 11/22
Hydrography
   data acquisition, 12/9–10
   developments, 12/12–13
   physical features, 12/10
   survey methods, 12/10–11
   tidal datum, 12/11
   see also Charts; Oceanography, regional; Surveying
Hydrostatics of floating structures, 2/69–81
   concepts, 2/69–70
   curves for ship trim, 3/17
   pressure, 4/6
      integration, 2/76–79
      regulations, 2/79–81
   stability, 2/70–74
      large angle, 2/72–74
      loss, 2/74–276
Hyper-Fix positioning system, 12/61–62
Hyperthermia in heliox diving, 4/17
Hypothermia in heliox diving, 4/17–18

Ice
   breakers, whole ship control design for, 7/12
   formation, 1/11
   hovercraft drag, 3/111
   and salinity, 1/28
Icebergs, 1/11
Incineration of waste at sea, 13/23
Inclining experiment for ship stability, 3/16
Indian Ocean, 1/4
   Indian Ocean Ridge, 1/5
Inert gas
   in salvage, 9/36
   system for tankers, 11/22, 11/26
Inertial navigation systems, 12/13
INFONET, 12/13
Infrared imagery, 1/12–13
INMARSAT, 12/13, 13/21

Insurers, on ship controllability, 7/29
Integration
　control, 7/48–52
　　implications for personnel, 7/51–52
　　methodology, 7/48–51
　　need for, 7/48
　navigation systems, 12/12–13, 12/21
　processing and display system (SHINPADS), 7/50
　see also Automation
Intergranular corrosion, 10/10
International Hydrographic Organization, 12/10
International Labour Organization (ILO), on minimum standards, 13/20
International Load Line Convention, 3/23
International Maritime Dangerous Goods (IMDG) code, 11/19–21
International Maritime Organisation (IMO), 3/67–68
　grain rules, 11/18
　legislative role, 13/5–6
　on manoeuvrability, 7/14
　on offshore activities
　　decommissioning installations, 13/13
　　pollution, 13/14
　safety standards, 13/20
　salvage convention, 9/12–13
　on ship stability, 3/20
　Weather Criterion, 3/20
　see also United Nations
International Salvage Union, 9/12
International Seabed Authority, and nodule mining, 13/16
Iolair field support vessel, 8/41–42
Ionosphere, 1/19
Iron, grey cast, selective leaching, 10/14
　see Cast irons, corrosive resistance; Steels
Islands
　and maritime zone baselines, 13/6
　Pacific, 1/4
　volcanic, 1/5
　see also Archipelagic waters

J-lay pipelaying, 5/39–40
Jack-up rigs, 2/3
Jacket offshore structures, 2/6–10
Jet pump, 6/55–56
　see also Water jet propulsion
Jetting for pipeline burial, 5/36
JIM atmospheric diving systems, 4/27–28
Johnson Parabola, 3/51

Karman vortex street, 2/26
Kepler's laws, 12/64
Keulegan–Carpenter number
　in pipeline stability, 5/35
　in wave loading, 2/34
　and vortices, 2/40–41
Kevlar, 3/87
Knot (speed), 12/9

Kort nozzle, 6/65–66
　of ROVs, 4/43
Kutta–Joukowski Law, 3/89

Labelling of dangerous goods, 11/20–21
Ladders, safety aspects, 11/9–10
Lagrange's equation, 2/59–60
Lasers
　LIDAR, 12/12
　in non-destructive testing, 10/35
　positioning systems, 12/54–55
　　Atlas Polarfix, 12/54–55
　　Golf 2, 12/55
　for wave sensing, 7/35
Law, maritime, 13/3–27
　fishing, 13/23–25
　　and freedom of the high seas, 13/23–24
　　and United Nations Convention 1982, 13/24–25
　international law
　　custom/treaties, 13/3–4
　　international organizations in, 13/5–6
　　principles, 13/3–4
　　United Nations Conventions, 13/4–5
　manganese mining, 13/15–18
　　and pollution, 13/18
　　Reciprocating States Regime, 13/18
　　and United Nations Convention 1982, 13/15–18
　maritime zones, 13/6–10
　　archipelagic waters, 13/7
　　baselines, 13/6
　　boundaries, 13/9–10
　　continental shelf, 13/8
　　continuous, 13/8
　　Exclusive Economic Zone (EEZ), 13/8–9
　　high seas, 13/9
　　internal waters, 13/6
　　international seabed area, 13/9
　　territorial sea, 13/6–7
　and military use, 13/25–27
　　deployment restrictions, 13/26
　　force, use of, 13/25
　　weapons/equipment deployment, 13/26–27
　and national law, 13/3
　offshore exploitation, 13/10–15
　　coastal State's rights/duties, 13/12–15
　　continental shelf evolution, 13/10–11
　　continental shelf legal definition, 13/11–12
　shipping, 13/19–23
　　nationality of ships, 13/19–20
　　pollution, 13/21–23
　　safety standards, 13/20–21
　　traffic management, 13/21
　see also Legislation, UK, for diving safety
Lay barges, 5/36–37

LCROV, 4/52
Leading marks, 12/15
Leak detection in pipelines, 5/42
Legislation, UK, for diving safety, 4/25–26
　see also Law, maritime
Leisure craft
　auxiliary propulsion, 6/8
　high speed engines, 6/19
LIDAR, 12/12
Life saving regulations, 11/33–38
　1983 amendments, 11/36–38
　Class VII, 11/33–35
　Class VII(T), 11/35
　general requirements, 11/35–36
　see also Search and rescue; Survival
Lifeboats/rafts, 11/35, 11/36
　hyperbaric, 4/26
　requirements
　　for cargo ships, 11/36–37
　　for Class VII ships, 11/33
　　for Class VII(T) ships, 11/35
　in survival, 11/32
Lifejackets, 11/34
　for cargo ships, 11/37
Light transmission by sea water, 1/11–12
　and organism distribution, 1/21
Lighthouses, 12/19
　pressure to reduce, 12/21
Linear Wave Theory, 2/28–31
Liners, see Passenger ships
Lloyds Register of Shipping, 3/67
　redundancy classification, 8/22
　rules for ship structure design, 3/56
　on ship controllability, 7/29
　standard form of salvage agreement, 9/7–9
　　arbitration, 9/8–9
　see also Classification societies; Salvage Association
Load lines, International Convention on, 13/20
Loading of cargo, see Cargo
Loading on ship structures, 3/44–47
　approximate formulae, 3/47
　dynamic, 3/46
　extreme prediction, 3/46–47
　hull girder bending, 3/47–49
　quasi-static wave, 3/44–46
　SWATH ships, 3/123–124
Loading on offshore structures, 2/17, 2/19–20, 2/22–48
　currents, 2/19, 2/27
　dynamic response, 2/59–65
　fluid mechanics concepts, 2/20, 2/22
　lift/drag coefficients, 2/20
　quasi-static analysis, 2/48–59
　　foundation effects, 2/56–59
　　methods, 2/48–56
　and sea bed proximity, 2/41, 2/42–43
　viscous forces and Reynolds number, 2/22–26
　　bluff body flows, 2/24–26
　　boundary conditions, 2/23–24
　　Froude number, 2/23

Loading on offshore structures (*cont'd*)
  vortex shedding, 2/40–41
  wave diffraction, and large bodies, 2/43–48
  waves, 2/27–40
    design, wave theories in, 2/32
    drag/inertia coefficients, 2/37–40
    flow regimes, 2/33–34
    gravity waves, 2/27–28
    linear wave theory, 2/28–31
    Morison's equation, 2/33–34, 2/34–37
  winds, 2/17, 2/19, 2/26–27
LOCSTAR, 12/72
Loiter drives, 6/8
Loll, 3/20
London Dumping Convention, 13/23
Loran A, 12/41, 12/46
Loran C, 12/46–49
  accuracy, 12/49
  coverage, 12/47
  development, 12/41
  expansion, 12/49
  in integrated system, 7/13
  operation frequency, 12/41
  operational details, 12/49
  and propagation anomalies, 12/48–49
  receivers, 12/47–48
  system, 12/46–47
  transmitters, 12/47, 12/48
  see also Pulse 8 positioning system
Lorelay pipe laying vessel, 8/40–41
LR2 submersible, 4/36–37
LR3 submersible, 4/37
Lubrication systems, 6/47–58

Mach number, 3/25
Machinery arrangements
  azimuthing propulsion, 6/8
  electric drive, 6/9–10
  father and son, 6/8–9
  loiter drives, 6/8
  mechanical/electrical drive, 6/10–11
  mixed plants, 6/6–8
  multiple screw plants, 6/5–6
  power station concept, 6/10
  single screw plants, 6/4–5
  TNT system, 6/11–12
  uniform machinery, 6/8
  see also Ancillary machinery; Diesel propulsion engines; Engines; Gas turbines; Propeller shaft drives; Propulsion; Steam plant
Macrocell, 10/15
MAESTRO for ship structural design, 3/58
Magnesium alloys for sacrificial anodes, 10/22
Magnetic particle inspection (MPI), 10/34
Magnetron for radar, 12/74, 12/75
Magnetron probe for radar, 12/75, 12/77
Magnus effect, 3/89
Man overboard, 11/27–28

Manganese mining, 1/10
  and pollution, 13/18
  Reciprocating States Regime, 13/8
  UN Convention 1982, 13/15–18
    Declaration of Principles, 13/15
    International Seabed Authority, 13/16–17
    Resolution II and pioneer investors, 13/17–18
Manganese nodules, 1/10
Manned submersibles, 4/33–39
  commercial, 4/36–39
  construction, 4/34–35
    buoyancy systems, 4/34–35
    pressure hull, 4/34
  history, 4/33–34
  operation, 4/35–36
  steering requirements, 7/5–6
  types, 4/32–33
  see also Remotely operated vehicles (ROV); Semisubmersible vessels; Underwater working
Manoeuvring of ships, 3/33
Mantis atmospheric diving system, 4/30
  ROV dual system, 4/55
Mapping, *see* Surveying
MarAd resistance series, 3/27
Maraging steels, 10/23
Marconi, Guglielmo, on early radar, 12.73
Marine Directorate (UK), on ship controllability, 7/29
MARPOL (prevention of pollution convention), 13/22
Masks, diving, 4/21–22
Mass Action, Law of, 10/3
Mediterranean, 1/30
  pollution, 1/31
Mercator projection, 12/8
Merchant ships, 3/6–12
  see also particular type
Merlin LCROV, 4/52
Mesosphere, 1/17
Micro-Fix positioning system, 12/56–57
Microperi 7000, 8/40
Microrelief, 1/5–6
Microwave positioning systems, 12/55–58
  Artemis, 8/29–30, 12/55–56
  Axyle, 12/56
  Micro-Fix, 12/56–57
  Trisponder, 12/57–58
Military use of the sea, law on, 13/25–27
  force deployment restrictions, 13/26
  force, use of, 13/25
  see also Warships
Mineralization in sea water, 1/10
Minesweeping/minehunting
  electronic position fixing for, 12/53
  by ROV, 4/50
  track keeping for, 7/35, 7/38, 7/39
Modelling (mathematical)
  of dynamic positioning control, 8/24
  of vortex shedding effects on risers, 5/21–25

Modelling (of vessels)
  for controllability, 7/55, 7/56
  correlation, ship/model, 3/30–31
  for motion prediction, 3/39–40
  for propulsion, 3/30
  for resistance, 3/27–28
Montreux Convention (1936), 13/7
Mooring, 8/3–13
  alongside, 8/3–4
  and safety, 11/4
  to buoys, 8/5–8
    circumstances, 8/5–6
    large tankers, 8/6
    loading hose handling, 8/8
    recovery/breakaway, 8/7–8
  thruster assisted, 8/35, 8/36
  see also Anchoring; Anchors; Dynamic positioning
Moorings
  catenary, 8/9
  dynamic analysis, 2/92–97
  and hydrostatic stability, 2/76
  tensioned, dynamic analysis, 2/97–98
  see also Chain cables
Morison's equation
  in heave analysis of semisubmersible, 2/86
  for wave loading, 2/33–34, 2/34–37
    for arbitrarily inclined cylinders, 2/36–37
    and drag/inertia coefficients, 2/37, 2/38–39
    on floating structures, 2/99–102
    for vertical/horizontal cylinders, 2/35–36
Morrish's rule, 3/13
Multihull vessels, *see* Catamarans; SWATH ships
  see also High speed vehicles
Munk moment of SWATH ships, 3/123
Muster lists, 11/3

NASTRAN, 2/49
National Ocean Survey (US), 12/9
Nationality of ships, 13/19–20
Nautical mile, standards, 12/9
Naval architecture, 3/3–75
  CAD/CAM in, 3/23, 3/58, 3/66–67
  classification/certification, 3/67–68
  complex system design, 3/58–59
  conceptual illustrations, 3/59–61
  costing, 3/65
  definitions, 3/4
  design process, peculiarities of, 3/59
  development of design, 3/65
  economic studies, 3/61–62
  environment for design, 3/62
  flotation, 3/13
  full scale trials, 3/32–33
    for motion prediction, 3/40
  hull form design, 3/4–6, 3/33
  merchant ships, 3/6–12
  motions/seakeeping, 3/33–34
    criteria, 3/40–41
    deck wetness, 3/34, 3/38–39

10/Index

Naval architecture (cont'd)
    degrees of freedom, 3/33
    heave/pitch, 3/35–38
    manoeuvring, 3/33
    performance, design for, 3/41
    prediction, 3/39–40
    roll, 3/34–35
    slamming, 3/34, 3/38, 3/39
  production, design for, 3/66
  professional registration, 3/3
  propeller design, 3/31–32
  propulsion, 3/28–31
    model testing, 3/30
    propeller–hull interaction, 3/29–30
    propulsors, 3/29
    ship–model correlation, 3/30–31
  resistance, 3/23–28
    dimensional analysis, 3/24–25
    estimation, 3/27–29
    friction effects, 3/23
    skin friction, 3/26–27
    wave making, 3/24, 3/25–26
  sinkage/heel/trim, 3/16–19
    and docking, 3/18
    and grounding, 3/18–19
    hydrostatics, 3/17–18
    and water density change, 3/18
  sizing, 3/4, 3/62–64
  stability
    flooding/damage, 3/21–23
    large angle, 3/19–21
    small angle, 3/13–16
  structural dynamic response, 3/55–56
  structure design, 3/56–58
    approach, rule versus rational, 3/56
    computer aided, 3/58
    configuration, 3/57
    criteria for structural optimization, 3/57
    reliability, 3/57–58
  structures, primary, 3/41–58
    failure modes, 3/43–44
    hull girder bending, 3/47–49
    loading, 3/44–47
    materials, 3/42–43
    types, 3/43
  structures, secondary, 3/49–53
    combined loads, 3/51
    effective breadths/widths, 3/50
    elastic bending/buckling, 3/50–51
    elasto-plastic response, 3/51–52
    factors of safety, 3/51
    grillage response, 3/52–53
  structures, tertiary, 3/53–54
    details, 3/54
    plate bending/buckling, 3/53–54
Naval Oceanographic Office (US), 12/9
Naval vessels, see Warships
  see also Military use of the sea, law on
Navigation
  automation, 7/12–13
  electronic systems, see Loran C; Omega; Decca Navigator
  equipment check list, 11/5–6

Navigation (cont'd)
  historical perspective, 12/5
  location of observer
    astronomical navigation, 12/16–18
    dead reckoning, 12/13
    ocean navigation, 12/16
    position fixing, 12/15–16
    speed/heading measurement, 12/13–14
    steering/autopilot, 12/14
    tidal/current set, 12/14–15
    visual fixing, 12/13
  passage planning, 11/5
  radio direction finding, 12/19–20
  system developments, 12/20–21
    cautionary note, 12/21
  watch keeping, 11/6
  see also Acoustic systems; Geodesy; Hydrography; Radar; Satellite navigation
Navsat, 12/72
NAVSTAR GPS, 12/15, 12/16, 12/52, 12/67–71
  accuracy, 12/70
  corrections, 12/69–70
  position fixing, 12/68–69
  receivers, 12/70–71
  space segment, 12/68
Navy Navigation Satellite System (NNSS), 12/15, 12/16, 12/64–67
  accuracy, 12/67
  control, 12/65
  NOVA satellites, 12/67
  position determination, 12/65–67
Nekton, 1/21
Nernst equation, 10/4
Newmark-β method in structure analysis, 2/65
Newton–Rader propeller, 3/125
Nickel alloys, corrosion resistance, 10/21–22
Nickel, see Manganese mining
Nitrogen narcosis, 4/9
  first experienced, 4/5
  see also Bends
Nitrox, 4/14
Nodules, mineral, 1/10
  see also Manganese mining
Non-destructive testing, 10/33–35
  codes/standards, 10/38–39
    for inspection/monitoring, 10/39
    methods, 10/38–39
  for corrosion monitoring, 10/35
  for defect monitoring, 10/30, 10/31
  electrical, 10/35
  highlighting, 10/34
  radiographic, 10/34–35
  ultrasonic, 10/34
  see also Defects
Notices to Mariners, 1/12
  US improvements, 12/13
Nuclear power, 6/34–36
  advantages/disadvantages, 6/35
  applications, 6/36
  marine use, 6/34–35

Nuclear power (cont'd)
  principle, 6/35
Nuclear testing at sea, 13/26
Nuclear weapons deployment, 13/26

Ocean Alliance semisubmersible, 8/38–39
Ocean Ranger rescue incident, 11/32–33
Oceanic provinces, 1/3–8
  continental margins, 1/6
  marginal basins, 1/4
  ocean depths, 1/4
  oceans, 1/3–4
  sea bed stability, 1/8
  sediments, 1/6–8
    distribution, 1/7–8
    transportation, 1/8
    types, 1/6–7
  topography, 1/4–6
  see also Circulation; Sea water; Waves
Oceanography, biological, 1/19–23
  distribution of organisms, 1/21–22
  marine lifestyle, 1/19–21
  marine technology, organisms and, 1/22–23
  types of organism, 1/21
Oceanography, regional, 1/25–31
  acoustic systems in, 12/39–40
  coastal ocean, 1/26–27
  deep ocean, 1/26
  equatorial/tropical, 1/28
  estuaries, 1/27
  marginal basins, 1/26
  polar, 1/28
  ROVs for, 4/54
  surf zone, 1/27–28
  temperature oceans, 1/28, 1/30–31
  see also Hydrography
Offshore operations
  corrosion problems, 10/16–17
  and maritime law, 13/10–15
    coastal State's rights/duties, 13/12–15
    continental shelf evolution, 13/10–11
    continental shelf legal definition, 13/11–12
  pollution, rights/duties, 13/14–15
  position reference needs, 12/53
  safety, 11/26–27
    clothing, 11/27
    definitions, 11/26
    permit to work system, 11/27–28
    personnel information, 11/27
  support vessels, power station concept, 6/10
  see also Dynamic positioning
Offshore Pollution Liability Agreement (OPOL), 13/15
Offshore structures, fixed, 2/3–68
  ancillary equipment, 2/17
  decommissioning, rights/duties, 13/13–14

Offshore structures (cont'd)
  erection/operation, rights/duties, 13/12–13 design
    concepts, 2/17, 2/18
    problems, 2/3
    requirements, 2/67–68
  certification requirements, 2/67–68
  cost, and need for compliant structures, 2/12–17
  drilling depth capacity, 2/6
  failure/fatigue
    life, 2/66–67
    monitoring, 10/36
  loading, 2/17, 2/19–20, 2/22–48
    currents, 2/19, 2/27
    dynamic response, 2/59–65
    fluid mechanics concepts, 2/20, 2/22
    lift/drag coefficients, 2/20
    and sea bed proximity, 2/41, 2/42–43
    and slamming, 2/41, 2/43
    vortex shedding, 2/40–41
    wave diffraction, and large bodies, 2/43–48
    waves, 2/27–40
    winds, 2/17, 2/19, 2/26–27
  platform selection, 2/12
  quasi-static analysis, 2/48–59
    foundation effects, 2/56–59
    methods, 2/48–56
  rigs, 2/3–6
  structures, 2/6–12
    concrete platforms, 2/10–12
    jacket structures, 2/6–10
  subsea engineering/maintenance, 2/17
Offshore structures, floating, 2/69–105
  analysis, 2/98–105
    equations, 2/98–99
    wave forces, 2/99–105
  characteristics, 2/69
  dynamic response analysis, 2/81–92
    catenary moorings, 2/92–97
    equations, 2/81–84
    heave motion, 2/84–88
    pitch motion of articulated column, 2/88–89
    tensioned buoyant platforms, 2/89–92
    tensioned moorings, 2/97–98
  hydrostatics, 2/69–81
    concepts, 2/69–70
    pressure integration, 2/76–79
    regulations, 2/79–81
    stability, 2/70–72
    stability loss, 2/74–76
Oil, see Lubrication systems
Oil cargoes
  classification, 11/21–22
  tanker work with, 11/22
  vegetable, tank safe practices, 11/16
Oil pollution, see Pollution
Oil tankers, see Tankers
Omega, 12/16, 12/49–52
  accuracies, 12/51–52
  coverage/usage, 12/49, 12/52

Omega (cont'd)
  development, 12/41
  operation frequency, 12/41
  operational details, 12/51
  and propagation anomalies, 12/51
  receivers, 12/50–51
  transmitters, 12/49–50
Ore carriers, spring, 3/55
Orelia DSV, 4/23
  in deep diving trials, 4/27
Osteonecrosis from diving, 4/11
Oxygen analysers, 11/24–25
Oxygen poisoning, 4/9

Pacific Ocean, 1/3–4
  East Pacific Rise, 1/5
Paints, see Coatings, protective
Palmgren–Miner fatigue damage rule, 2/66
PAPnaval ROVs, 4/50–51
Parametric resonance, 3/34–35
Parbuckling, 9/46–47
Paris Law, in crack analysis, 10/33
Passage planning, 11/5
Passenger ships, 3/11, 3/12
  abandoning ship, 11/32
  cruise liners, power station concept, 6/10
  Queen Elizabeth 2, electric drive, 6/9
  regulations, 3/68
  see also Ferries; Titanic
Passivation, 10/5–7
  potential pH diagrams, 10/6–7
Passive tanks, 7/41–42
PC1601 submersible, 4/37–38
Pelagic clays, 1/7–8
Permit to work system, 11/27–28
pH of sea water, 1/10
PIGs, 5/41–43
Piles for offshore structures
  jacket structures, 2/7
  loading analysis, 2/56–59
Pilot access, 11/9–10
Pilotage, team work during, 11/6
Pipe laying vessel Lorelay, 8/40–41
Pipelines, 5/3, 5/29–43
  burial, 5/36
  corrosion protection, 10/16–17
  design procedures, 5/30–34
    codes of practice, 5/30
    heat losses, 5/32–33
    requirements, 5/30
    sizing, 5/30–32
    stressing, 5/33–34
  laying, 5/36–41
    coastal State's rights and duties, 13/14
    J-lay, 5/39–40
    landfall, 5/40
    lay barge, 5/36–38
    reel, 5/40
    towed method, 5/38–39
  PIGs, 5/41–43
  stability, 5/34–36

Pipelines (cont'd)
  static charge hazard, 11/25
  surveying, 5/40–41
  towed to installation site, 5/18–19
  see also Risers
Pipes, see water, and corrosion, 10/15
Pitch, see Heave/pitch
Pitting, 10/10
Planing craft, 3/78–88
  applications, 3/87–88
  energy analysis, 3/79–80
  forces
    estimation, 3/82–84
    in steady motion, 3/81–82
  hull forms, 3/80–81
  performance
    in steady motion, 3/84–85
    in waves, 3/85–86
  principle, 3/79
  stability, 3/86
  structural design, 3/86–87
  see also High speed vehicles; Hovercraft; Hydrofoil craft
Plankton, 1/21
  distribution, 1/22
Plastic collapse in ship structure, 3/44
Platforming of hydrofoil craft, 3/99
Platforms, see Concrete platforms; Offshore operations; Offshore structures, fixed; Offshore structures, floating
Pleasure craft, see Leisure craft
Plimsoll line, 3/23
Pneumothorax, 4/10
Polar oceans, 1/28
Pollution
  control, 9/56, 9/58–60
  booms, 9/58–59
  dispersant spraying, 9/60
  skimmers, 9/59–60
  and IMO salvage convention, 9/12–13
  international law on, 13/21–23
  of the Mediterranean, 1/31
  from nodule mining, 13/18
  oceanic transport of, 1/12
  from offshore activities, rights/duties, 13/14–15
  and Protection and Indemnity Associations, 9/13
  see also Waste dumping incineration
Port State Control Memorandum, 13/21
Ports
  and controllability, 7/55–56
  fire in, 11/29–30
  Vessel Traffic Services guidelines, 13/21
  watch keeping in, 11/4
  see also Docking of ships, and trim; Fendering
Position reference systems, 8/17, 8/19–20, 8/24–33
  electronic, 12/53–63
    categorization, 12/53–54
    HF, 12/59–62
    laser, 12/54–55

Position reference systems (*cont'd*)
 LF, 12/62–63
 microwave, 12/55–58
 offshore industry needs, 12/53
 UHF, 12/58–59
 hydro-acoustic, 8/25–29
  baseline systems, 8/28
  limitations, 8/28–29
  Simrad system, 8/25–28
 redundancy, 8/23
 repeatability, 8/24
 surface microwave, 8/29–30
 taut wire, 8/30–33, 12/37
 see also Dynamic positioning; Track keeping/position control
Power turbines, 6/48–49
 see also Auxiliary power; Generators; Nuclear power; Steam power
Pressure, and depth, 1/11
 compensation in ROVs, 4/43–44
 and diving
  direct effects, 4/8–9
  gas effects, 4/9
Probabilistic approach to design, 3/57–58
Processors in dynamic positioning, 8/17
Products carrier, 3/10
Propeller pump, 6/55
Propeller shaft drives, 6/36–42
 bearings/seals, 6/37–38
 clutches
  gearing, 6/38, 6/39, 6/40
  transmission, 6/40, 6/41–42
 couplings, 6/40–41
 gears, 6/38–40
 power transmission, 6/36
Propellers, 3/29, 7/25
 arrangements, 6/65
 azimuthing, 6/8
 cavitation, 3/31–32, 6/67, 10/16
 contra rotating, 6/70, 6/72
  combined propeller/thruster system 6/75–76
 controllable pitch, 6/4, 6/67–70, 7/25, 7/27
  advantages, 6/68–69
  combinator control, 7/45–46
  cost benefits, 6/69
  design, 6/69–70
  engine loading, 6/69
  Voith Schneider, 6/70, 6/71
 design, 3/31–32
 for dynamic positioning vessels, 8/21–22
 efficiency, and dimensions, 6/66–67
 fixed pitch characteristics, 6/66
 grim wheel, 6/72–73
 for high speed vehicles
  air, 3/124, 3/126–128
  marine, 3/124, 3/125
 for ROVs, 4/43
 self pitching, 6/70
 for SWATH ships, 3/122
 variable pitch, 7/10
 see also Fans for hovercraft

Propulsion, 3/28–31, 6/65–76
 azimuthing, 6/8
 concept, 6/65
 of high speed vehicles, 3/124–128
  air propellers, 3/126–128
  marine propellers, 3/125
  water jet, 3/125–126
 of hovercraft, 3/126–128
 history, 6/3
 of hydrofoil craft
  marine propellers, 3/125
  water jet, 3/126
 modelling, 3/30
 ROV systems, 4/42–43
 system corrosion–erosion, 10/16
 turbo-electric, 6/9
 water jet, 3/125–126, 6/73–74, 7/25
 see also Diesel propulsion engines; Engines; Machinery arrangements; Gas turbines; Nuclear power; Propeller shaft drives; Propellers; Steam power; Thrusters
Protection and Indemnity Associationns, 9/13
Provinces, *see* Oceanic provinces
Pulmonary baratrauma, 4/10
Pulse 8 positioning system, 12/47, 12/63
 calibration, 12/49
 see also Loran C
Pump jet, 6/73–74
Pumps, 6/54–56
 for cargo discharge, 6/62
 for oil, 6/58
 in salvage, 9/32–33, 9/34–35
  tanker discharge, 9/20, 9/22
 sea water, and corrosion, 10/15
Pycnocline, 1/11, 1/15

QPC factor in model correlation, 3/31
Qubit survey system, 12/53
Queen Elizabeth 2, electric drive, 6/9

Racal, *see* Decca Navigator; Pulse 8 positioning system
Racons, 12/86–87
Radar, 12/73–93
 for anchored position, 11/7
 antennae, 12/81–83
  scanner rotation, 12/83
  slotted waveguides, 12/82–83
 beacons, 12/86–87
 block diagram, 12/76
 cathode ray tube, 12/79–81
 in collision avoidance
  computer system data, 12/89
  relative motion, 12/87
  true motion, 12/88
 computer aided systems, 12/88–91
  automatic plotting aid (ARPA), 7/12, 12/89–91
 bearing storage, 12/89

Radar (*cont'd*)
 clutter/interference elimination, 12/88–89
 collision avoidance data, 12/89
 electronic chart display systems (ECDIS), 12/92–93
 frequency bands, 12/85–86
  and cost, 12/86
 history, 12/73–74
 pulse length/repetition frequency, 12/84
 radar horizon, 12/84
 receiver, 12/77–79
  and target characteristics, 12/85
 time synchronized components, 12/83–84
 transmitter, 12/75, 12/77
 transponders for identification, 12/87
 in vessel traffic systems (VTS), 12/91–92
Radio direction finding, 12/19–20
Radio transmission
 and the ionosphere, 1/19
 propagation, 12/41–42
Radiographic NDT, 10/34–35
Rankine Cycle Energy Recover (RACER) systems, 6/34
Rascar, 12/91
RCV 225 ROV, 4/47–48
Receptance, 3/37
Recreation
 leisure craft
  high speed engines, 6/19
  auxiliary propulsion, 6/8
  planing craft, 3/88
 ROVs, 4/52
 scuba diving, 4/13–14
 submersibles, 4/39
 see also Cruise liners, power station concept
Refraction, radar sub-/super-refraction, 12/84
Refrigerated cargo ships, 3/6
Refrigeration, 6/61
 for air conditioning, 6/62
Regulations
 collision prevention, and watch keeping, 11/6
 for diving safety, 4/25–26
 for merchant ships, 3/68
 ship floodable length, 3/22–23
 ship transverse damage stability, 3/22
 see also Standards/codes of practice
Remote sensing, 1/12–13
Remotely operated vehicles (ROV), 4/32, 4/33, 4/39–55
 and atmospheric diving suits
  dual systems, 4/55
  systems compared, 4/31–32
 autonomous, 4/52–55
 buoyancy, 4/40–41
 commercial vehicles, 4/47–48
  RCV 225, 4/47–48
  Scorpio, 4/48
 control station, 4/44–45

Remotely operated vehicles (contd)
  deployment/retrieval, 4/45
  development needs, 4/57, 4/59
  diver support, 4/49–50
  dynamic response, 4/45–47
    added mass, 4/46
    drag, 4/45–46
    umbilical effects, 4/46–47
  low cost (LCROV), 4/52
  military, 4/50–52
  modularized, 4/48–49
  propulsion, 4/42–43
  resistance to pressure, 4/43–44
  Structure Inspection Device, 4/55
  television, 4/40
  umbilical, 4/41–42
  see also Manned submersibles;
    Underwater working
Replenishment systems, 7/35, 7/38
Rescue phase in survival, 11/32–33
  see also Search and rescue
Resins in planing craft construction,
  3/87
Responders in hydro-acoustic
  positioning, 8/27–28
Response amplitude operators (RAO),
  3/37
Reynolds number
  and drag/inertia coefficients, 2/37,
    2/38–39
  in pipe flow, 5/31
  in pipeline stability, 5/35
  and skin friction resistance, 3/24, 3/26
  viscous forces and, 2/22–26
    bluff body flows, 2/24–26
    boundary conditions, 2/23–24
    Froude number, 2/23
Rhumb lines, 12/6, 12/8
Right of innocent passage, 13/7
  for warships, 13/26
Right of transit for warships, 13/26
Rigs, see Drilling rigs
Risers, 2/17, 5/3, 5/4–29
  analysis, 5/9–19
    damping term linearization, 5/13–14
    dynamic, 5/11
    element property formulation,
      5/11–13
    frequency domain solution, 5/13
    non-linearity and structural
      response, 5/17–18
    static, 5/9–11
    tension factors, 5/9
    time series, 5/14–15
    trial solution, 5/14
    validation, 5/15–17
  design conditions, 5/26–29
  buoyancy control, 5/26–27
  configurations for production, 5/27–29
  sources of failure, 5/26
  equations, 5/4–9
  types, 5/3
  vortex shedding effects, 5/19–26
    analysis models, 5/21–25
    suppression, 5/25–26

Risers (cont'd)
  vibrations, 5/20–21
  see also Pipelines
Ro-Ro vessels, 3/10, 3/12
  equipment, 6/64
  regulation review, 3/68
  see also Ferries
Rockall controversy, 13/12
Rodriquez RHS-200, 3/100
Roll, 3/33–35
  control, 7/40–44
    and hull design, 7/40–41
    imposed devices, 7/41–42
    operational methods, 7/42
    rudder in, 7/43
  and dynamic positioning, 8/21
  hovercraft stability, 3/113–114
  of hydrofoil craft, 3/93–94
  motions in waves, 7/39–40
  stabilization, rudder in, 7/43
  and operation efficiency, 7/38–39
  see also Heave/pitch; Stabilizers
Ropes, mooring, 8/4
  see also Anchors; Anchoring; Chain
    cables
Roughness of hulls
  from marine growth, 2/37, 2/39–40
  and skin friction, 3/26
  and scale testing, 3/30, 3/31
ROVER autonomous ROV, 4/54
Rudders, 7/4–5, 6/65
  cavitation, 10/16
  on planing craft, 3/86
  classification societies' rules, 7/29
  cost function control, 7/17
  design for enhanced performance,
    7/27
  mechanism of action, 7/27–28
  rate, and ship size, 7/35
  in roll stabilization, 7/43
  siting, 7/24–25
  for SWATH ships, 3/122
  see also Autopilots; Control systems
Running fix, 12/16

Sacrificial protection, see Anodic
  protection; Cathodic protection
Safety, 11/1–43
  aloft/overside working, 11/14
  anchor watch, 11/6–7
  in anchoring procedures, 11/7–18
  audit of DP vessel, 8/34
  in cargo operations, 11/15–21
  deck access, 11/9–10
  derricks/cranes, 11/10–13
    cranes, 11/11–12
    derricks, 11/10–11
    hand signals, 11/13
    testing, 11/12–13
    union purchase, 11/11
  diving, 4/25–27
    accident statistics, North Sea, 4/25
    emergencies, 4/26–27
    regulations/guidelines, 4/25–26

Safety (cont'd)
  in emergencies, 11/27–31
    alarm signal, 11/4
    breathing apparatus, 11/30–31
    collision, 11/27
    diving, 4/26–27
    fire, 11/29–30
    man overboard, 11/27–28
  in helicopter activity, 11/29
  margins, and control efficiency, 7/3
  navigation
    equipment check list, 11/5–6
    passage planning, 11/5
    watch keeping, 11/6
  offshore, 11/26–27
  organization onboard, 11/3–4
  standards for ships, international,
    13/20–21
  survival, 11/31–38
    abandonment phase, 11/32
    appliances/arrangements
      regulations, 11/33–38
    pre-abandonment procedure, 11/31
    rescue phase, 11/32–33
    survival phase, 11/32
  tankers, 11/21–26
    cargo handling check list, 11/22–23
    communications, ship/shore, 11/23
    gas measuring instruments,
      11/24–25
    inert gas system, 11/26
    loading plan, 11/23
    oil cargoes, 11/21–22
    static electricity, 11/25
    tank preparation, 11/24
  vessels alongside, 11/4
  watch keeping
    anchor, 11/6–7
    navigation, 11/6
    in port, 11/4
  see also Health, and crude oil
    toxicity; Search and rescue
Sagging in loading analysis, 3/44
Salinity, 1/9–10
  in coastal ocean, 1/27
  in deep ocean, 1/26
  in equatorial/tropical oceans, 1/28
  and freezing/thawing, 1/11, 1/28
  in the Mediterranean, 1/30
  see also Desalination units, corrosion
Salmon conservation, 13/25
Salvage, 9/1–60
  airlift, 9/39
  award, 9/5
  beach gear, 9/39–42
    versus tugs, 9/41–42
  buoyancy systems, 9/44–46
    air bags, 9/46
    compressed air, 9/44–45
    foam, 9/45–46
  cargo discharge, 9/15–24
    from damaged vessels, 9/15–16
    lightening, 9/18–19
    from tankers, 9/19–24
    underwater salvage, 9/16–18

Salvage (cont'd)
  value reduction, 9/15
  Carpenter stoppers, 9/42
  cofferdams, 9/29
  conditions for, 9/3
  contracts, 9/5–11
  definitions, 9/3
    freight, 9/4
  equipment availability, 9/50
  examples, 9/3
  excavation, underwater, 9/38–39
  expenses, 9/5
  explosion risk, 9/33, 9/36
  fire monitors, 9/30–32
  and flooding, 9/26
  and free surfaces, 9/24–26
  ground reaction, 9/26, 9/50
    stability and, 9/26
  hot tap machines, 9/36–37
  inert gas, 9/36
  lifting, 9/42–44
    heavy vehicles, 9/43–44
    sheer legs, 9/42–43
  organizations, 9/11–13
    International Maritime
      Organisation (IMO), 9/12–13
    International Salvage Union, 9/12
    Protection and Indemnity
      Associations, 9/13
    Salvage Association, 9/11–12
  parbuckling, 9/46–47
  patching, 9/26–29
    epoxy resins in, 9/29–30
  pollution control, 9/56, 9/58–60
    booms, 9/58–59
    dispersant spraying, 9/60
    skimmers, 9/59–60
  pumps, 9/32–33
  remuneration, 9/3–4
  side scan sonar in, 9/37–38
  towing, 9/52–56
    points, 9/54, 9/56
    preparation, 9/52–53
  tugs, 9/48–50
    bollard pull, 9/50, 9/52
    supply vessels as, 9/50, 9/51
  types, 9/13–15
  water damage protection, 9/48
Salvage Association, 9/11–12
Sanitary systems, 6/62
  see also Sewage disposal legislation
Satellite navigation, 12/5, 12/64–72
  development, 12/64
  Geostar, 12/72
  Navsat, 12/72
  NAVSTAR GPS, 12/15, 12/16, 12/52, 12/67–71
    accuracy, 12/70
    corrections, 12/69–70
    position fixing, 12/68–69
    receivers, 12/70–71
    space segment, 12/68
  Navy Navigation Satellite Systems
    (NNSS), 12/15, 12/16, 12/64–67
    orbits, 12/64

Satellite navigation (cont'd)
  SATNAV, 7/13
  Starfix, 12/71–72
  see also Remote sensing
SATNAV, 7/13
Saturation diving, 4/15, 4/16–21
  development, 4/6
  see also Heliox diving
Schilling rudder, 7/27
Scintillation in radar tracking, 12/85
SCORES for strip theory, 3/37
Scorpio ROVs, 4/48
Screw pump, 6/54
Scuba diving, 4/13–14
  and archaeological salvage, 9/14
Sea Beam sounder systems, 1/4, 1/12
Sea bed
  proximity, and offshore structure
    loading, 2/41, 2/42–43
  stability, 1/8
  surveying, 1/4, 1/8
  topography, 1/4–8
  see also Ground reaction; Soil/pile
    interactions
Sea slap, 3/46
Sea water, 1/8–12
  acoustics, 1/12
  chemistry, 1/8–9
  density
    distribution, 1/11
    and sinkage/heel/trim, 3/18
  dissolved gases, 1/10
  light transmission, 1/11–12
  mineralization, 1/10
  pH, 1/10
  salinity, 1/9–10
    in coastal ocean, 1/27
    in deep ocean, 1/26
    in equatorial/tropical oceans, 1/28
    and freezing/thawing, 1/11, 1/28
    in the Mediterranean, 1/30
  temperature distribution, 1/10–11
Seakeeping, 3/33–34
  criteria, 3/40–41
  deck wetness, 3/34
  degrees of freedom, 3/33
  manoeuvring, 3/33
  performance, design for, 3/41
  prediction, 3/39–40
  see also Heave/pitch; Roll; Slamming
Seaker LCROV, 4/52
Seals, see Bearings/seals
Search and rescue, 11/38–42
  aircraft
    amphibian, 11/42
    equipment, 11/42
    helicopters, 11/39–42
  communications, 11/42
  coordination arrangements, 11/38
  on-scene commander's duties, 11/39
  surface craft, 11/39
  see also Distress signals; Life saving
    regulations; Lifeboats/rafts;
    Lifejackets
Seasickness, 3/34

Sediments, 1/6–8
  distribution, 1/7–8
  transportation, 1/8
  types, 1/6–7
Seiches, 1/25
Seismic sea bed survey, 1/8
  see also Acoustics systems
Self pitching propellers, 6/70
Semisubmersible rigs, 2/3–6
  multi-anchor spread, 8/8–9
Semisubmersible vessels
  derrick barge, 8/40, 10/2
  diving support, 4/23–25
  drill ship Ocean Alliance, 8/38–39
  heave motion analysis, 2/85–88
  support vessel Iolair, 8/41–42
  see also Manned submersibles
Sewage disposal legislation, 13/22
  see also Sanitary systems
Sextant, 12/18
Shackles, 8/4
Shaft drives, see Propeller shaft drives
Shear lag, in ship structure, 3/50
Sheer legs, 9/42–43
Side scan sonar, 1/8, 12/11, 12/31–33
  in salvage, 9/37–38
Siliceous sediments, 1/7
Simplex seal system, 6/38
Simrad hydro-acoustic positioning
  system, 8/25–28
  see also Taut wire position reference
Sinkage, 3/16
  hydrostatics, 3/17–18
  squat, 3/33
Sinus squeeze, 4/8
Skate ROV, 4/51
Skimmers, oil, 9/59–60
Skywave, 12/42
  propagation, and Loran C, 12/48
Slamming, 3/24
  bending moment, 3/47
  bottom, 3/24, 3/38, 3/39
  structural effects, 3/55
  of SWATH ships, 3/124
  on offshore structures, 2/41, 2/43
Slip forming of concrete platforms, 2/12
SM1 Marine Spey gas turbines, 6/33–34
Small waterplane area twin hull ships,
  see SWATH ships
Smit towing bracket, 9/56, 9/57
Snell's Law, 12/26, 12/27
Soil/pile interactions, 2/56–59
Solar flares, Omega interference, 12/51
Solar radiation, 1/17–18
Sonar
  beamwidth, 12/25
  Doppler, 12/39
  equation, 12/27
  multibeam, 12/12
  sensitivity, 12/24
  side scan, 1/8, 12/11, 12/31–33
    in salvage, 9/37–38
  see also Echo sounders; Swathe
    sounding systems
SOS, 11/38

Sounders, *see* Acoustic systems; Echo sounders; Sonar
Speech distortion in heliox diving, 4/17
Speed measurement, 12/13–14
Spey SM1 gas turbines, 6/33–34
Spheres, collapse pressure, 4/34
Spider atmospheric diving system, 4/30–31
Splash zone, 10/13
  protective coating for, 10/26
Springing, 3/55
Sputnik I, 12/64
Squat, 3/33
Squeeze in diving, 4/8–9
Stability
  of hovercraft, 3/113–114
  of pipelines, 5/34–36
  of planing craft, 3/86
  of ships
    and cargo stowage, 11/17
    and flooding/damage, 3/21–23
    large angle, 3/19–21
    small angle, 3/13–16
    standards, 3/20–21
  see also Controllability
Stabilizer systems, 7/25, 7/41–42
  for planing craft, 3/86
  see also Roll: control
Stainless steels, corrosion resistance, 10/19–20
Standardization in ship design, 3/66
Standards/codes of practice
  for corrosion, 10/36–38
    for design, 10/37–38
    materials, 10/37
    non-destructive testing, 10/38–39
  for dangerous goods, 11/19–21
  defect sizes, allowable, 10/31
  design, 10/37–38
  fatigue behaviour of steels, 2/66
  materials properties, 10/37
  nautical mile, 12/9
  non-destructive testing, 10/38–39
  for pipeline design, 5/30
  safety, 13/20–21
    working practices, 11/3
  on ship stability, 3/20–21
  for survival appliances/arrangements, 1/33–38
  on welding, 10/38
Starfix, 12/71–72
Static electricity, hazard for tankers, 11/25
Station identity card, 11/27
Steam power
  declining use, 6/25–26
  development potential, 6/27–28
  disadvantages, 6/27
  features, 6/26–27
  generation, 6/50–54
    auxiliary steam, 6/52
    coal fired boilers, 6/50–52
    from waste heat, 6/52–54
    water tube boiler principle, 6/50
  obsolescence, 6/3

Steam power (*cont'd*)
  principle, 6/26
  static charge hazard, 11/25
  turbines, stopping ability, 7/28
  Very Advanced Propulsion (VAP) system, 6/27–28
  see also Gas turbines; Nuclear power
Steels, 3/42
  corrosion
    of carbon/manganese low alloy, 10/17–19
    in concrete, 10/15–16
    of stainless steel, 10/19–20
    structural, fatigue, 10/16
  embrittlement, 10/33
    and cathodic protection, 10/27
  for pipelines, 5/33
  for planing craft, 3/86, 3/87
  ultra-high strength, 10/23
Steering
  control, 7/3–5
    heading control, 7/3–4
    requirement, 7/5
  gear, 6/62–63
  systems, 6/65
    corrosion-erosion, 10/16
  see also Autopilots; Compasses; Course keeping; Rudders; Track keeping
Stena Inspector DSV, 8/37
Stena Seaspread DSV, 8/35, 8/37
Stone dumping vessels, 8/43
Stopping
  ability, 7/27–28
  crash stops, 7/53
Storm surges, 1/25
Straits, right of transit through, 13/7
Stratification, 1/15
Stratosphere, 1/17
Stress corrosion cracking (SCC), 10/12
Strip theory, 3/35–37
  for wave load prediction, 3/46
Structure Inspection Device (SID), 4/55
STRUDL, 2/49
Submarines, 4/32
  acoustic detection, early development, 12/23
  nuclear, steam power, 6/3, 6/26
  right of transit, 13/7
  steering requirements, 7/5
  see also Manned submersibles; Nuclear power; Remotely operated vehicles (ROV); Submersibles;
Submersibles, steering requirements, 7/5–6
  see also Manned submersibles; Remotely operated vehicles (ROV); Semisubmersible vessels
Sulphate reducing bacteria, 10/13, 10/14
Sulphide stress corrosion, 10/14
Supercavitating hydrofoils, 3/96
Supercavitating propellers, 3/125
Supply vessels
  effectors for positioning, 7/37

Supply vessels (*cont'd*)
  for salvage work, 9/50, 9/51
  speed control requirement, 7/8
  steering requirements, 7/5
Support vessels
  power station concept, 6/10
  diving, 4/23
    dynamic positioning hazard, 8/16
    dynamic positioning redundancy, 8/22
    Orelia, 4/23, 4/27
    semisubmersible, 4/23–25
    Stena DSVs, 8/35, 8/37
  field support semisubmersibles, 8/41–42
Surf zone, 1/27–28
Surface microwave position reference system, 8/29–30, 12/55–56
  fixed station, 8/29–30
  principles, 8/29
  range, 8/30
Surge analysis of tensioned platforms, 2/90–92
Surveillance devices, submarine, law on, 13/26–27
Surveying
  hydrographic methods, 12/10–11
  for pipelines, 5/40–41
  Qubit system, 12/53
  of sea bed, 1/4, 1/8
  track keeping for, 7/35
  vessels, 8/42
  see also Hydrography; Pulse 8
Survival, 11/31–38
  abandonment phase, 11/32
  appliances/arrangements regulations, 11/33–38
    1983 amendments, 11/36–38
    Class VII, II/33–35
    Class VII(T), 11/35
    general requirements, 11/35–36
  pre-abandonment procedure, 11/31
  rescue phase, 11/32–33
  survival phase, 11/32
  see also Emergencies
SWATH ships, 3/119–124, 7/40–41
  advantages/disadvantages, 3/121
  applications, 3/124
  comparisons with monohull, 3/124
  concept, 3/119
  developments, 3/119–120
  hull forms, 3/120–121
  performance
    in steady motion, 3/121–122
    in waves, 3/122–123
  structural design, 3/123–124
  see also High speed vehicles
Swathe sounding systems, 12/33–34
Sweat of cargo, 11/17
Syledis positioning system, 12/58–59
Syntactic foam for buoyancy, 4/41

Tafel equation, 10/35
Tanker Owners' Pollution Federation, 9/13

Tankers, 3/8, 3/10, 3/12
  anti-pollution legislation, 13/21–23
  cargo salvage from, 9/19–24
    discharge, 9/20, 9/22
    fendering, 9/22
    lightening vessels, 9/19, 9/20, 9/21
  equipment, 6/65
    life saving, 11/35
  mooring to buoys, 8/6
  safety, 11/21–26
    cargo handling check list, 11/22–23
    gas measuring instruments, 11/24–25
    inert gas system, 11/22, 11/26
    loading plan, 11/23
    oil cargoes, 11/16, 11/21–22
    ship/shore communications, 11/23
    static electricity, 11/25
    tank preparation, 11/24
  shuttle, dynamic positioning, 8/43
  stopping ability, 7/28
  supertankers, direct drive engines, 6/14
  see also Passive tanks
Taut wire position reference, 8/30–33, 12/37
  advantages/disadvantages, 8/31, 8/32
  deployment, 8/32
  principles, 8/30–31
  two-wire systems, 8/31–32
  types, 8/32–33
Taylor Gertler resistance series, 3/27
Tchaika, 12/47
Telemetry, acoustic system, 12/40
Televisions
  for ROVs, 4/47
  underwater, 4/40
Temperature control in heliox diving, 4/17–18
Temperature distribution in oceans, 1/10–11
  ice formation, 1/11
  from infrared imagery, 1/12–13
Territorial sea, 13/6–7
  see also Continental shelf
Thermal protective aids (TPAs), 11/37
Thermoclines, 1/11
  in coastal waters, 1/26
  in equatorial/tropical oceans, 1/28
Thermographic NDT, 10/35
Thoracic squeeze, 4/8
  see also Pulmonary baratrauma
USS Thresher, 12/34
Thrusters, 6/74–76, 7/25, 7/27, 7/37–38
  azimuthing, 6/74–75, 8/21, 8/22
  contra rotating, 5/75
  propeller/thruster system, 6/75–76
  in dynamic positioning, 8/21–22
    redundancy, 8/23
  for minehunters, 7/38
  in mooring, 8/35, 8/36
Tides, 1/15–16
  currents, 1/17
    and navigation, 12/14–15
  levels, datum/tables, 12/11

Titanic
  and acoustic system development, 12/23
  power plant, 6/6
Titanium alloys, corrosion resistance, 10/22
TM 308 autonomous ROV, 4/54
TNT machinery arrangement, 6/11–12
Topography
  of ocean surface, by remote sensing, 1/13
  of sea bed, 1/4–6
Tornadoes, 1/19
Torpedoes
  contra rotating propellers, 6/72
  ROV recovery, 4/51
Towing, 9/52–56
  chain cable test loads, 9/55
  contracts, 9/56
  of pipes, 5/18–19
    for laying, 5/38–39
  points, 9/54, 9/56
    Smit Bracket, 9/56, 9/57
  preparation, 9/52–53
  see also Tugs
Toyofuji 5, contra rotating propellers, 6/72
Track keeping/position control, 7/6–8, 7/35–38
  effector requirements, 7/36–38
  information sensors, 7/36
  methodology, 7/38
  operations requiring, 7/35–36
  see also Dynamic positioning; Position reference systems
Traffic management, international conventions, 13/21
Training
  standards, international convention, 13/20
  for whole ship control, 7/12
Transducers, 12/27–28
  in hydro-acoustic positioning, 8/26–27
  response, 12/29
Transfer Under Pressure (TUP) system, 4/26
Transit, see Navy Navigation Satellite System (NNSS)
Transit passage through straits, 13/7
Transponders
  in hydro-acoustic positioning, 8/27–28
  for identification, 12/87
Trawlers, controllable pitch propellers, 6/68
Treaties in international law, 13/3–4
Trenches, 1/5
Trieste submersible, 4/33–34
Trim of ships, 3/16
  and docking, 3/18
  and grounding, 3/18–19
  hydrostatics, 3/17–18
  and water density change, 3/18
Trimix, 4/14
Trireme, 7/5
Trisponder positioning system, 12/57–58

Troposphere, 1/17
Tsunamis, 1/24
Tugs
  bollard pull, 9/50, 9/52
  certification, 9/52
  in salvage, 9/41–42, 9/48–50
  see also Towing
TUMS, 4/51–52
Tuna management, 13/25
Turbidity currents, 1/8
Turbines, see Gas turbines; Power turbines; Steam power
Turbo-electric propulsion, 6/9
Turning trial, 7/53–54
Turrent moored drillship, 8/9

Ultrasonic NDT, 10/34
Umbilicals
  for diving bells, 4/19–20
  for ROVs, 4/41–42
    dynamic effects, 4/46–47
Uncle John, 4/23–25
Underwater working, 4/1–60
  history, 4/3–5
    air supply, 4/5
    decompression, 4/5
    diving suit, 4/3–4
  method selection, 4/55–57, 4/58
    capabilities compared, 4/57, 4/58
    operational factors, 4/56–57
  see also Diving; Manned submersibles; Remotely operated vehicles (ROV)
Uni-fuel concept, 6/45–46
Union purchase derrick operation, 11/11, 11/12
United Nations, 13/3
  in law enforcement, 13/4
  see also International Maritime Organisation (IMO); Law, maritime
SS United States, 3/77

Vegetable oils, tank safe practices, 11/16
Ventilation of cargoes, 11/17
Vessel traffic systems (VTS), radar in, 12/91–92
Virgin Atlantic Challenger II, 3/77
Voith Schneider CP propeller, 6/70, 6/71
Volcanoes/volcanic islands, 1/5
  see also Tsunamis
Vortex shedding effects on risers, 5/19–26
  analysis models, 5/21–25
  suppression, 5/25–26
  vibrations, 5/20–21

Wageningen series for propeller design, 3/31
Wake fraction, 3/29
Ward–Leonard control system, 6/10

Warships
　catamarans as, 3/119
　constant effective wave height, 3/47
　gas turbine arrangements, 6/31–32
　hydrofoil craft as, 3/101
　patrol vessels, triple screw plants, 6/6
　planing craft as, 3/87–88
　power plants, 6/7–8
　right of innocent passage/transit, 13/26
　speed control requirement, 7/8
　see also Military use of the sea, law on; Remotely operated vehicles (ROV): military
WASP atmospheric diving system, 4/29–30
Waste dumping/incineration, 13/23
Watch keeping
　anchor, 11/6–7
　navigational, 11/6
　in port, 11/4
Water, molecular structure, 1/8–9
　see also Sea water
Water jet propulsion, 3/125–126, 6/73–74, 7/25
Wave equation, 12/23
Waveguides for radar, 12/75
　slotted, 12/82–83
Waves (radio), propagation, 12/41–42
　and Loran C, 12/48

Waves (water), 1/23–25
　characteristics, 1/23–24
　and course keeping, 7/6
　generation by vessels, 3/24
　　high speed vehicles, 3/77
　offshore structure loading, 2/27–40
　　fixed, 2/27–40, 2/43–48
　　floating, 2/99–102
　performance of vessels in
　　catamarans, 3/119
　　hovercraft, 3/113–115
　　hydrofoils, 3/99–100
　　planing craft, 3/85–86
　　SWATH ships, 3/122–123
　and pipeline stability, 5/34–36
　radar sensing, 7/35
　and ship motions, 7/25–26
　　roll, 7/39–40
　and ship stability, 3/20
　types, 1/24–25
　wavemaking drag/resistance
　　of hovercraft, 3/110
　　of ships, 3/25–26
　see also Heave/pitch; Loading of ship structures; Seakeeping; Surf zone
Weather Criterion of the IMO, 3/20
Weather systems, 1/18–19
Weber number, 3/25
Welding

Welding (cont'd)
　codes/standards, 10/38
　problems in construction, 3/43
Well stimulation vessels, 8/43
Whole ship control, 7/12–13
Winches
　for mooring, 8/3–4
　for rig anchors, 8/10
Wind, 1/18
　Beaufort Scale, 1/20
　and course keeping, 7/6
　offshore structure loading, 2/17, 2/19, 2/26–27
　sensors for dynamic positioning, 8/20–21
　and ship motions, 7/25–26
　speeds, 1/19
Windage, 12/14
Wires, mooring, 8/4
　see also Taut wire position reference
World War II, and radar development, 12/73

Z-drive, 6/74–75
Zig zag manoeuvre, 7/54
Zinc alloys as sacrificial anodes, 10/22
　see also Dezincification